Wheat Evolution and Domestication

Moshe Feldman • Avraham A. Levy

Wheat Evolution
and Domestication

 Springer

Moshe Feldman
Department of Plant and Environmental Sciences
Weizmann Institute of Science
Rehovot, Israel

Avraham A. Levy
Department of Plant and Environmental Sciences
Weizmann Institute of Science
Rehovot, Israel

ISBN 978-3-031-30177-3   ISBN 978-3-031-30175-9 (eBook)
https://doi.org/10.1007/978-3-031-30175-9

This Springer imprint is published by the registered company Springer Nature Switzerland AG
The registered company address is: Gewerbestrasse 11, 6330 Cham, Switzerland

To Adriana, my best friend for over 65 years, the mother of my children, and the love of my life, who was a pillar of strength and endless support in helping me to fulfill my dreams.

Moshe Feldman

To my family for their constant support and love.

Avraham A. Levy

Preface I

This book could have been entitled "For the love of wheat". It is the brainchild of my mentor, Prof. Moshe Feldman, whose intimate knowledge of wheat is the result of a lifetime dedication. Moshe spent more than 60 years with wheat and its relatives, in the field, in the natural habitat, in the laboratory, or under the microscope. He obtained a Ph.D. on wheat evolution, a postdoc on wheat cytogenetics, and in his own laboratory he did research on wheat genetics and evolution and remained active as Emeritus Professor. I started my Ph.D. with Moshe, in 1982, working on genetic diversity in wild emmer wheat. After that, I studied molecular mechanisms of genome evolution in several plant species. I established my own laboratory in 1992, and since then, I have been involved in collaboration with Moshe, on projects related to wheat evolution, in particular on the genome's response to polyploidization. The contents of this book cover a century of research on wheat taxonomy, genetics, and evolution, accompanied by most recent genomic studies.

Rehovot, Israel   Avraham A. Levy

Preface II

The pioneering discovery of the accurate chromosomal number of wheats in 1918 by Tetsu Sakamura in Japan and Karl Sax in the USA paved the way to comprehensive research on the biology, cytogenetics, genetics, genomics, and evolution of wheats and their wild relatives. Already, in the first steps of this endeavor it became apparent that the wheat species comprise a polyploid series containing diploids (einkorn wheat), tetraploids (emmer and durum wheats), and hexaploids (spelt and bread wheats). The fact that the economically most important wheats, bread and durum wheats, are polyploids triggered wheat scientists to identify the parental diploid species that donated their genome to the polyploid wheats. In addition, it promoted studies on important genomic changes, involving the cytological and genetic diploidization processes that contributed to the successful establishment and great adaptability of the polyploids.

During the last hundred years, a large number of researchers from various parts of the globe gathered a vast amount of information on the various wheats and their wild relatives. Much of the gathered material centered on cytogenetics, evolution patterns, and phylogenetic relationships between domesticated wheat as well as between domesticated wheats and their wild relatives. A critical study that took place in the wheat-wild relatives has revealed that they are a rich reservoir of useful genes that, upon transfer to wheat, may improve yield, quality, resistance to diseases, tolerance to abiotic stresses, and adaptation to a variety of new marginal environments. The last several decades have witnessed dramatic advances in the field genomics providing invaluable information on genome sequences, structure, gene identification, and organization in various species of the group, as well as development of efficient cytogenetic and molecular techniques to exploit the vast wild genetic resources for wheat improvement. We feel that the time is ripe to present an overview linking up the accumulated classical information with molecular know-how. The purpose of the book is, therefore, to review the major discoveries that have led to the current understanding of the genetic and genomic structure as well as evolution of the wild relatives in nature and of the domesticated wheats under cultivation.

The book deals with the Triticeae tribe of the grass family. Its chapters describe the taxonomy and evolution of the Triticeae genera, the dating and location of their origin, genome analyses, phylogenetic relationships, and evolutionary trends, as well as genome structure of the economically important species. The various species of the sub-tribe Triticineae, that include wheat and rye, are described in greater details, and special emphasis is given to the wheat group (the genera *Amblyopyrum*, *Aegilops*, and *Triticum*). Evolutionary aspects of the diploid and polyploid species, and the evolution of domesticated wheats under cultivation, are reviewed, and a reasonable anticipation concerning future prospects of wheat improvement is presented.

It is our great pleasure to thank Prof. Lydia Avivi for helpful discussions and assistance in writing and organization of the book; Dr. Eitan Millet for interesting discussions on the evolution of the group, and for donating a few of the plant figures. Thanks are also due to Dr. Cathy Melamed-Bessudo for contributing to our studies, to Sagie Brodsky for collecting data on crops productivity and valuable discussions on gene regulation, and to Naomi Avivi-Ragolsky for taking excellent care of the seed stocks and the plants and photographing part of them. We also thank Ms. Yehudit Rosen for reading and commenting on the text, Ms. Noa Ilan for drawing some of the figures, and the photographic division of the Weizmann Institute of Science for drawing several figures.

We wish to express our gratitude to the Weizmann Institute of Science which provided us with a fertile environment and the freedom to follow our passion. Finally, we are grateful to our students, postdocs, and many colleagues in Israel and abroad whose work contributed so much to our understanding of the genetic structure and evolutionary relationships of the wheats and their wild relatives—a contribution that inspired our imagination.

Rehovot, Israel   Moshe Feldman
2022   Avraham A. Levy

Contents

List of Figures

List of Tables

Introduction

1.1 The Importance of Wheat as a Staple Food

As one of the first cereals to be domesticated, the history of domesticated wheat and that of human civilization have been interwoven since the dawn of agriculture. In the course of its domestication, the wheat plant lost its ability to disseminate its seeds effectively and became completely dependent on human for seed dispersal. Man, in return, fostered this cereal to such an extent that it is now one of the world's foremost crops. The domestication of wheat, and that of other edible plants, have provided humankind with the ability to produce sufficient food, leading to population increase and spread to almost all parts of the globe. It also enabled the colossal development of human civilization. As a result, humans became completely dependent on wheat and other domesticated plants for their survival.

Since the beginning of the cultivation of diploid and tetraploid wheat (first cultivation of wild forms, about 12,000 C^{14}—calibrated years ago, and later, appearance of domesticated einkorn and emmer 10,800 calibrated years ago, and naked tetraploid wheat 10,200 years ago, and of hexaploid wheat about 10,000 years ago (Table 1.1), wheat has become one of the most important staple food nourishing mankind. Domestication of wheat has led to the formation of new wheat species, such as hexaploid wheat, *Triticum aestivum*, many subspecies and numerous cultivars. Most modern cultivars belong to bread wheat, since it is high-yielding, grows well in a variety of climates and soils, and very suitable for bread making because of the high gluten content of its endosperm. (The elastic gluten protein entraps the carbon dioxide formed during yeast fermentation enabling the leavened dough to rise). Most of the remaining modern cultivars belong to durum wheat, which is mainly grown in relatively dry regions, particularly in the Mediterranean basin, Australia, India, Russia and in the low-rainfall areas of the great plains of the USA and Canada. Its relatively large grains yield a low-gluten flour, suitable for pasta, flat bread (pita) and semolina products.

Throughout 10,000 years of cultivation, bread wheat and, to a lesser extent, durum wheat, have been of supreme importance in facilitating and sustaining the development of human civilization in southwestern and central Asia, Europe, North and South Africa, North and South America, and Australia. From a plant that grew in a relatively small region in the Fertile Crescent, South-West Asia, wheat is now grown on more area than any other crop (219 Mha in 2021) and with global production of 766 million metric tons (FAOSTAT 2021). This high production makes wheat the second most-most produced cereal after maize (1018 Mt), that is extensively used for animal feed, and more than rice (745 Mt), the main human food crop in Eastern Asia (FAOSTAT 2021).

Bread wheat is high yielding in a wide range of environments, ranging from 67°N in northern Europe to 45°S in Argentina; however, in the subtropics and tropics, its cultivation is restricted to higher elevations (Feldman et al. 1995). It provides food to one-third of the global human population, about 20% of the global caloric requirements for human consumption and 20% of the protein consumed (Feldman et al. 1995; Shewry and Hey 2015). The world's highest wheat-producing regions are the EU (160 Mt), China (125 Mt), India (100 Mt), Russia (60 Mt), the USA (60 Mt), Canada (34 Mt), Pakistan (24 Mt), Ukraine (24 Mt), Australia (23 Mt), and Turkey (20 Mt). Wheat makes up a significant portion of the calories consumed by mankind, where its grain contains most of the nutrients essential to man. These are carbohydrates (70–80%), proteins (8–15%), fats (1.5–2.0%), minerals (1.5–2.0%), and vitamins, such as the B complex and vitamin E. Globally, wheat is the leading source of vegetable protein in human food, bearing a higher protein content than maize and rice. In addition to the relatively high yield and good nutritive value of the wheat grains, their low water content, ease of processing and transport and good storage qualities have made wheat an important food staple for more than 35% of the world's population. During the last 70 years, the global wheat area has increased by more than 50% and average yields have

© The Author(s) 2023
M. Feldman and A. A. Levy, *Wheat Evolution and Domestication*,
https://doi.org/10.1007/978-3-031-30175-9_1

Table 1.1 Chronology in uncalibrated and ^{14}C calibrated radiocarbon years BP (before present) for the Late Epipalaeolithic and the Neolithic periods in the Levant, the western flank of the Fertile Crescent

Uncalibrated date range[a]	[14]C Calibrated range (Approximate)[b]	Period	Major events in wheat cultivation[c]
13,000–10,300	15,500–12,000	Late Epipalaeolithic (Natufian)	Harvesting from wild emmer and einkorn stands—agrotechnical development
10,300–9500	12,000–10,800	Pre-Pottery Neolithic A (PPNA)	Cultivation of brittle forms of emmer and einkorn—the first phase of cultivation
9500–9000	10,800–10,200	Early pre-Pottery Neolithic B (E-PPNB)	Appearance of non-brittle emmer and einkorn; cultivation of brittle and non-brittle types in mixture—the second phase of cultivation
9000–7500	10,200–8300	Late pre-pottery Neolithic B (L-PPNB)	Appearance of free-threshing, naked tetraploid wheat; cultivation of wild and domesticated emmer and naked tetraploid wheat in mixture; expansion of wheat culture to all regions of the fertile crescent; appearance of non-brittle hexaploid wheat; significant increase in human population and site size—the third phase of cultivation
7500–6200	8300–7100	Pottery Neolithic	Spread of wheat culture to central Asia, southern Europe, and Egypt—expansion of agriculture

[a] Uncalibrated dates and periods after Harris (1998)
[b] Calibrated dates for the start and end of each period were calculated using the calibration software OxCal v.4,2Bronk Ramsey ©2020 and the new dataset of the IntCal20 in Reimer et al. (2020)
[c] After Kislev (1984)

increased from 1.0 to 2.5 t/ha, reaching levels as high as 12 t/ha, mainly due to improved cultivars, wider use of fertilizers and improved-agronomic practices.

1.2 Interest in the Origin and Evolution of Domesticated Wheat

It is a reasonable assumption that in a pre-agricultural society, each gender had distinct roles relating to food provision, with men hunting, and women collecting and gathering seeds, fruits, and other plant materials. While collecting seeds of wild wheat and barley, they noticed that fallen seeds were later responsible for the growth of new plants. Consequently, women became aware of the profitability of sowing their surplus seeds for next year's food (Kislev et al. 2004), and deliberately decided to plant seeds of these cereals in more desirable fields close to their dwelling (Kislev 1984). It is not clear when this brilliant discovery was made. One of the most ancient sites bearing signs of collecting and processing wild emmer wheat, *Triticum turgidum* ssp. *dicoccoides*, was found in Ohalo II, a 23,000-year-old hunter-gatherers camp on the south-western shore of the Sea of Galilee, Israel (Snir et al. 2015). About 10,500 years ago, wild wheat begun to be cultivated, and their spike remnants appeared in the archeological records discovered in several sites of the Levant. Few hundred years

later on, archaeological remnants indicate replacement of wild wheat by domesticated emmer which, characterized by non-fragile spikes, and consequently, were not capable of dispersing their seeds and were dependent on human for their propagation. The beginning of cultivation of wheat, as well as other cereals and pulses, marked the Neolithic (or Agricultural) Revolution, was one the most important revolution in human history, laying the foundation for the development of human civilization.

The view that women were the first to cultivate plants is reflected in the biblical story: "And when the woman saw that the tree was good for food, and that it was pleasant to the eyes, and a tree to be desired to make one wise, she took of the fruit thereof, and did eat, and gave also unto her husband with her; and he did eat" (Genesis 3:6). This event describes the transition of man from a hunter and gatherer to a farmer; "… And unto Adam he said: Because thou hast hearkened unto the voice of thy wife, and hast eaten of the tree, of which I commanded thee, saying: Thou shalt not eat of it; cursed is the ground for thy sake; in sorrow shalt thou eat of it all the days of thy life. Thorns also and thistles shall it bring forth to thee; and thou shalt eat the herb of the field. In the sweat of thy face shalt thou eat bread … Therefore, the LORD God sent him forth from the garden of Eden, to till the ground…" (Genesis 3:17–23). This biblical myth, describing the expulsion of Adam and Eve from the Garden of Eden, may be considered a reflection of the Neolithic

Revolution, during which humankind assumed control over its own food production. Prior to this period, the pre-agricultural hunter-gatherers gradually became acquainted with nature's periodicity and with the life cycle of plants that produced edible seeds or fruits in their environment. During the Neolithic Revolution, man put his observations to practice and succeeded to domesticate a number of the local edible plants. Several early Jewish scholars of the second and third centuries AD assumed that the "tree of knowledge" was the wheat plant (Talmud Babylonian, Berakhot, 40: A; Bereshit Rabba, 15:8) and interpreted the ancient biblical story as an expression of man's wish to fulfill his creation "in the image of God" by mastering his own food production.

The rather traumatic narrative of the biblical version of the beginning of agriculture, may indicate that man started to cultivate plants against the God's command. By contrast to it, in many ancient nations, where mythology prevailed, cultivated plants were considered a generous gift given to man by gods. For this reason, the ancient Egyptians bestowed gratitude to Isis and Osiris for introducing wheat and barley into Egypt from Mt. Tabor in Israel, and for teaching people the secrets of their cultivation. Similarly, the ancient Greeks ascribed the gift of these important cereals to Demeter, and the Romans to the goddess Ceres. The Mayans in Mexico considered maize as the gift of God, and in Shaanxi, China, one of the cradles of Chinese civilization, there is a statue of a godlike personage that brought plants to people and taught them how to cultivate them.

Human interest in the origin of domesticated plants, the geographical sites of their origin, identification of their wild progenitors, and their evolution under cultivation, dates back to the beginning of historic time. These subjects have always attracted great interest, and have excited and stimulated man's curiosity and imagination. Botanists, geneticists, agronomists, breeders, ethno-botanists and students of agricultural history have grappled with these mysteries by conducting extensive botanical, cytogenetic, molecular, and evolutionary studies on the genetic and genomic structure of the Triticeae tribe, in general, and of various species of the wheat group (the genera *Aegilops* and *Triticum*), in particular. The scientific study of the Triticeae tribe began with Linnaeus (1753), who classified the species of wheat in a separate genus *Triticum*, and has since been subjected to many taxonomic, morphological, eco-geographical, cytogenetic, molecular, and evolutionary studies.

One of the first attempts to identify the progenitor of domesticated wheat, specifically of common wheat, was made in the middle of the nineteenth century, by several botanists who concluded that the natural inter-generic hybrid and hybrid-derivatives of *Aegilops geniculata* Roth (=*Ae. ovata* L.) x common wheat, were the ancestral forms of this wheat [the nomenclature of the species of *Triticum* and

Aegilops in this book is as defined by van Slageren (1994). In 1821, Requien discovered such a hybrid in southern France, that grew from a spike of *Ae. geniculata*. Later, similar hybrids were also collected in northern Italy and in North Africa. Because of its resemblance to common wheat, Requien (see Fabre 1855) named it *Aegilops triticoides*. Since backcrossed progeny of these hybrids to common wheat exhibited an intermediate morphology between *Ae. triticoides* and some lines of common wheat, Fabre concluded, in 1855, that bread wheat had originated from *Ae. geniculata*. He assumed that under cultivated conditions, *Ae. geniculata* gradually transforms into bread wheat. This hypothesis was disproved by Godron (1876), who confirmed that, *Ae. triticoides* and all other intermediate forms between *Ae. geniculata* and bread wheat, are hybrids and hybrid derivatives. Godron produced similar forms by crossing bread wheat with *Ae. geniculata* and backcrossing the hybrids to the wheat parent. Of note, because *Ae. triticoides* is a natural inter-generic hybrid, where *Ae. geniculata* is the female parent, and not a true species, *Ae. triticoides* should be referred to as x *Aegilotriticum triticoides* (Req. ex Bertal.) van Slageren and not *Ae. triticoides* (van Slageren 1994).

Revelation of the origins of x *Aegilotriticum triticoides* led de Candolle (1886) to understand that historical, linguistic, and folkloristic types of evidence, alone, are insufficient to reveal the origins of domesticated plants; botanical, genetic, and archaeological studies are essential. The origin and evolution of a domesticated plant can be best studied following the identification and analysis of the current and past distribution of its wild progenitor. This may indicate the changes that led to domestication, as well as the site of the initial cultivation. However, when such a wild progenitor is not found, or is extinct, understanding the complete history of the domesticated plant is greatly impaired.

Consequently, the x *Aegilotriticum triticoides* saga emphasized the need for a clear definition of the key features characterizing the wild progenitors of domesticated wheat. These progenitors must have been valid species and therefore self-propagating. They are assumed to have had a spike similar in its basic features to that of domesticated wheat, but with a brittle rachis that upon maturation, disarticulates into single spikelets (dispersal units), thereby facilitating self-dispersal. In addition, the wild progenitors, like all other wild grasses, seemingly had tightly closed glumes, resulting in "hulled" grains protected against extreme climatic conditions and herbivores. Eco-geographically, they should have occupied specific geographic regions and well-defined primary habitats. However, most of today's domesticated wheat has undergone considerable morphological changes and only a few have retained ancestral morphological features, such as hulled grains, as in domesticated einkorn, emmer, and spelt wheat (see Chap. 10). Also, most domesticated wheat grow nowadays much larger areas as their progenitors. This renders the identification of the wild pro-genitors and the site of domestication more complex.

Domesticated wheats are classified into three main groups: diploids (einkorn), tetraploids (emmer, durum, rivet, Polish and Persian wheat), and hexaploids (spelt, bread, club and Indian shot wheat) (see Chap. 10). At the end of the nineteenth century, most botanists assumed that the domesticated wheat taxa had a polyphyletic origin and that at least two species of wild wheat progenitors, namely single-grained (einkorn) and double-grained (emmer) wheat, were taken into cultivation. The wild progenitor of domes-ticated einkorn was discovered in the middle of the nine-teenth century, and that of domesticated emmer and durum at the beginning of the twentieth century. These discoveries enabled the use of the gene pool of both the progenitors and of other related species for wheat improvement. Only in the second half of the twentieth century, it became clear that there are no wild hexaploid progenitors and that all types of domesticated hexaploid wheat were formed in farmers' fields by hybridization between domesticated tetraploid wheat and a wild diploid species of *Aegilops*.

1.3 The Need to Exploit Wild Wheat Relatives for Wheat Improvement

Given that only a small number of wild genotypes were selected for domestication, the genetic basis of domesticated wheat in the early stages of agriculture was relatively nar-row, representing only a fraction of the large variation that existed in the wild progenitors. Yet, during the 10,000 years of wheat cultivation, the genetic basis of domesticated wheat has been broadened, to some extent due to mutations and sporadic hybridizations with their wild progenitors and other closely related species in southwest Asia. Moreover, the tendency of traditional farmers in many parts of the world, to grow a mixture of genotypes in one field (polymorphic fields) or even a mixture of species of different ploidy levels (Zeven 1980), enabled hybridization and introgression of genes among the various genotypes. This, coupled with wheat's ability to self-pollinate, greatly facilitated the for-mation and selection of many distinct genotypes. Traditional farmers selected and planted grains of the lines most desir-able for their specific needs and consequently, selection pressures were thus consistently exerted, albeit, in different directions, by farmers in different localities. These efforts resulted in numerous landraces that demonstrated better adaptation to a wider range of climatic and edaphic condi-tions and to diverse farming regimes. But, under modern plant breeding practices, which began towards the end of the nineteenth century, the wheat fields have become genetically uniform (one elite cultivar is grown not only in one field but in a whole region), so that spontaneous gene exchange

between different cultivars in the farmer's fields has become less likely. On the other hand, gene migration has been greatly increased by worldwide introduction and exchange of cultivars. Crosses between these cultivars have been restricted to the breeder's experimental fields. In the breeder experimental stations, hybridizations have been mainly confined to intraspecific crosses and little use has been made of neither the gene pools of other wheat species nor of those of wild relatives, toward improvement of bread wheat. Such breeding practices, particularly the replacement in many countries of traditional varieties (landraces), suitable for local climatic and agronomic conditions, by a small number of elite, high-yielding cultivars (mega varieties), have greatly eroded the genetic basis of bread wheat. The Green Revo-lution, which started in the 60s of the previous century is responsible for the replacement of numerous landraces in India, Pakistan Turkey and North Africa, by a relatively small number of high-yielding cultivars and failure to con-serve the replaced traditional landraces (Feldman and Sears 1981). Such erroneous practice still continues and con-tributes to the loss of genetic diversity in bread wheat, consequently reducing its adaptability to abiotic stresses, increasing its susceptibility to biotic pressures, and consid-erably limiting the ability of breeders to improve further its yield and quality. This outcome of the erosion in the genetic diversity of wheat has become a more serious problem in light of the current climate changes, combined with contin-uous growth of the world human population, which requires elevated wheat yields, to improve its resistance to biotic stresses and tolerance to abiotic ones, and to produce genotypes that can provide reasonable yields in new habitats.

Attempts to increase desirable genetic variation via irra-diation or chemical treatment, yielded poor results. More-over, transgenesis, which could broaden the gene pool for wheat improvement, is not yet commonly practiced in wheat. Therefore, utilizing the germplasm of the various Triticeae species remains one of the best options for addressing the challenges of genetic erosion and of yield demands (Feldman and Sears 1981). Consequently, advanced efforts have been made to support more efficient exploitation of the wild gene resources for wheat improve-ment. The vast genetic resource of the Triticeae contains numerous economically important genes that can be exploited to create a potentially new variation of domesti-cated wheat. Many of the species of the two sub-tribes of the Triticeae, the Triticineae and the Hordeineae, can be crossed with bread and durum wheat, and economically important genes can be transferred to the domesticated background, via the use of various cytogenetic manipulations.

Although this option is not new, during recent decades, only a small number of genes were transferred from wild species to domesticated wheat. Breeders preferred to use domesticated sources instead of wild ones because they did not know how to overcome the challenges in using them. Nevertheless, a number of inter-generic and inter-specific hybrids have been produced between several Triticeae species and bread or durum wheat. These hybrids, most of which were viable, have been used for a variety of purposes, including genomic analysis, studies of speciation, phylogeny and evolution, and as the starting point in efforts to introduce alien variation into domesticated wheat.

Already in the beginning of the previous century, Aaronsohn (1910), while discovering wild emmer (*Triticum turgidum* subsp. *dicoccoides*) in nature, was impressed by its wide range of adaptation. He noticed that certain forms of this wild taxon possess several valuable traits, namely, large grains, ability to grow in relatively dry habitats, and resistance to rust. Consequently, he recommended utilizing wild emmer in breeding programs, especially to improve the resistance of domesticated wheat to drought, extreme climatic and soil conditions, and rusts, and to increase grain size and yield. Aaronsohn believed that "the cultivation of wheat might be revolutionized by the utilization of wild wheat. Such utilization might facilitate the formation of many new varieties, some of which will be hardy and able to grow in dry and warm habitats or in areas with poor soil and can thus expand the wheat growing area" (Aaronsohn 1910, p. 52).

Aaronsohn's belief that wild emmer can be utilized in the improvement of domesticated wheat was shared by Schweinfurth (1908), von Tschermak (1914), and several other early wheat geneticists. The fertile or partially fertile F_1 hybrids between bread wheat and wild emmer, produced by von Tschermak (1914), indicated that gene transfer from wild tetraploid wheat into hexaploid domesticated types is possible by simple breeding procedures. It was hoped that wild emmer could be used for production of domesticated varieties that would be adapted to arid regions (von Tschermak 1914).

Vavilov (his work and ideas reviewed in Vavilov 1951) further advanced the notion of exploitation of wild species to improve wheat and other crops. He identified regions where the world's major crops were first domesticated but still contained the greatest diversity of their wild relatives. McFadden (1930) was one of the first breeders that transferred a gene for stem rust resistance, later designated Sr2, to a variety of bread wheat from a domesticated emmer (a tetraploid wheat), thus producing the cultivars Hope and H-44, that was resistant to rust that severely infected wheat fields in the USA. Production of synthetic allopolyploids with new genomic combinations was one approach to evaluate and utilize wild germplasms. Broad hybrids in the Triticeae tribe have been attempted and studied for over 100 years. The first such hybrid was between wheat and rye (Wilson 1876). Rimpau (1891) described 12 plants recovered from seed of a wheat-rye hybrid that represented the first triticale. The idea was that such an allopolyploid would combine the cold tolerance of rye and the grain quality of wheat. But Triticale does not possess the grain quality of bread wheat and it is used mainly as an animal feed (Oettler 2005). In 1947, McFadden and Sears (1947) produced synthetic allopolyploids between different species of *Triticum* and *Aegilops* as a starting material for evaluation and transfer of desirable genes to a domesticated background.

Over the years, cytogenetic methods for proper genetic analysis and interspecific transfer of desirable characters have been developed, making the use of the wild gene resources more efficient. Sears (1972) described these methods, including induction of homoeologous pairing and recombination between wild and domesticated chromosomes by genetic means and use of ionizing radiation to translocate alien chromosome segments into a domesticated one. The induction of homoeologous pairing is, by far, the simplest gene transfer method (Sears, 1972), but induced translocations have also produced some favorable results (e.g., Sears 1956; Knott 1971).

The production of aneuploid lines of bread wheat (Sears 1954), as well as several alien addition, substitution, and translocation lines (Riley 1965; Sears, 1969, 1975; Feldman and Sears 1981; Feldman, 1988; Millet et al. 2013, 2014), enabled the genetic analysis of individual alien chromosomes or even chromosome arm on the genetic background of domesticated wheat and facilitated the transfer of selected chromosomal segments without affecting the rest of the domesticated genome. Using these techniques, several key genes, primarily those affecting qualitative traits, namely, genes improving disease and virus resistance, have been transferred from species of *Triticum*, *Agropyron*, *Aegilops*, *Amblyopyrum*, *Secale*, and other Triticeae, into bread and durum wheat (Sharma and Gill 1983; Feuillet et al. 2008; Millet et al. 2014). Quantitative traits that are controlled by several or many genes that are widely distributed throughout the genome, are much more difficult to manipulate.

Harlan and de Wet (1971) classified primary, secondary, and tertiary gene pools of wild relatives of a crop, as determined by their genetic distance from the crop. A primary gene pool consists of species with genome(s) homologous to that of the crop, including land races of the crop as well as other related domesticated crops and wild species. Among members of this gene pool crossing is easy, hybrids are generally fertile, and exhibit good chromosome pairing, approximately normal gene segregation and generally simple gene transfer by conventional breeding methods. Transfer of genes from species with different ploidy levels required some manipulations, such as production of synthetic allopolyploids. A secondary gene pool consists of species with closely related homoeologous (partially homologous) genome(s), that are crossed relatively easily with the crop, but whose chromosomes do not pair regularly with those of the crop and therefore, transfer of genes from these species required cytogenetic manipulations such as the use of genes inducing homoeologous pairing or the use of genotypes lacking homoeologous pairing suppressors. The tertiary gene pool consists of more distantly related taxa, and exploitation of this gene pool required special cytogenetic and molecular manipulations (e.g., embryo rescue in the F_1 hybrids, induction of chromosome pairing and recombination at meiosis of the hybrid, or induction of translocation with ionizing irradiation, transformation with selected genes affecting desired characteristics) in order to overcome inter-generic genetic barriers.

In the Triticeae, the primary gene pool of bread wheat, consists of hexaploid landraces and domesticated and wild tetraploid wheats (*T. turgidum* and *T. timopheevii*), diploid wheat (*T. urartu* and *T. monococcum*) and *Ae. tauschii*, all of which have a homologous genome(s) with that of bread wheat. The gene pool of tetraploid wheat has been exploited, to some extent, for wheat improvement via direct crosses (Millet et al. 2013), whereas that of *Ae. tauschii* has been subjected to allopolyploid bridging (crosses between bread wheat and synthetic allopolyploid of tetraploid wheat x *Ae. tauschii*). The secondary gene pool contains species of *Aegilops* and several species of *Agropyron*. Their genome(s) is homoeologous to that of common wheat and consequently, there is little pairing and recombination at meiosis of the F_1 hybrids. However, pairing can be induced relatively easily by the use of mutants of genes that prevent homoeologous pairing (e.g., *ph1b*, *ph1c*, and *10/13*) and genes from the diploid species *Aegilops speltoides* or *Amblyopyrum muticum*, that promote homoeologous pairing in hybrids with wheat. The tertiary gene pool of bread wheat contains all other Triticeae species. The wide morphological and ecological variation of the various Triticeae species may indicate that this tribe contains a very rich gene pool that can be exploited to widen the genetic basis of wheat. Most, if not all, of the Triticeae species, can be crossed with wheat and produce viable hybrids. However, in most cases, the hybrids are sterile, due to the lack of chromosomal pairing at meiosis, resulting in the production of imbalanced gametes; this limitation can be overcome by more radical cytogenetic manipulations (Feldman and Sears 1981; Feldman 1988). Thus, the gene pool of the entire tribe may serve as an important source of useful traits for the improvement of wheat.

During the twentieth century, studies in areas of taxonomy, eco-geography, cytogenetics, and evolution, have mainly focused on the genera *Triticum*, *Hordeum*, and *Secale*, as well as on the closely related genera *Aegilops* and several species of *Agropyron* (now *Elymus*). These studies have provided important information on the genetic structure of members of the primary and secondary gene pools, as well as on phylogenetic relationships between these gene pools and bread wheat. Consequently, several genes, mainly those conferring resistance to biotic stresses, that were not found in the domesticated gene pool of wheat, were successfully transferred to bread wheat (Fedak 2015; Zhang et al. 2015). It is estimated that genes were transferred from at least 52 species belonging to 11 genera (*Aegilops*, *Agropyron*, *Amblyopyrum*, *Dasypyrum*, *Elymus* (=*Thinopyrum*, *Lophopyrum*, *Pseudoroegneria*), *Agropyron*, *Hordeum*, *Leymus*, *Psathyrostachys*, *Secale*, and *Triticum*) (Wulff and Moscou 2014). However, currently, information on many of the wild relatives, mainly on the tertiary gene pool, is scanty and fragmentary. Although molecular studies in recent decades have improved, to some extent, the ability to identify, allocate and isolate useful genes in several Triticeae genomes (Paux and Sourdille 2009; Stein 2009; Krattinger et al. 2009; Hein et al. 2009; Eversole et al. 2009), their transfer to the domesticated wheat background has still encountered many obstacles. Further studies of these gene pools, in which information from the fields of cytogenetics and genomics should be combined, are essential.

References

Aaronsohn A (1910) Agricultural and botanical explorations in Palestine. Bull. Plant Industry, U.S. Dept. of Agriculture, Washington, No. 180, pp 1–63

de Candolle A (1886) Origin of Cultivated plants. Kegan, Paul, Trench and CO Second Edition du ble et de l'orge, London. Ann Sci Nat, Ser I (Paris) IX:61–82

Eversole K, Graner A, Stein N (2009) Wheat and barley genome sequencing. In: Feuillet C, Muehlbayer GJ (eds) Genetics and genomics of the Triticeae. Springer, pp 713–742

Fabre E (1855) On the species of *Aegilops* of the south of France and their translation into cultivated wheat. J Roy Agric Soc England 15:160–180

FAOSTAT (2021) Food and Agriculture Organization of the United Nation (FAO), Statistical Database, Statistical Division. Rome

Fedak G (2015) Alien introgression from wild T*riticum* species, *T. monococcum*, *T. urartu*, *T. turgidum*, *T. dicoccum*, *T. dicoccoides*, *T. carthlicum*, *T. araraticum*, *T. timopheevii*, and *T. miguschovae*. In: Molnár-Láng M, Ceoloni C, Dolezel J (eds) Alien introgression in wheat. Springer, pp 191–220

Feldman M (1988) Cytogenetic and molecular approaches to alien gene transfer in wheat. In: Proceedings of 7th international wheat genetics symposium, vol 1, Cambridge, England, pp 23–32

Feldman M, Sears ER (1981) The wild gene resources of wheat. Sci Am 244:102–112

Feldman M, Lupton FGH, Miller TE (1995) Wheats. In: Smart J, Simmonds NW (eds) Evolution of crop plants, 2nd edn. Longman Scientific & Technical, pp 184–192

Feuillet C, Langridge P, Waugh R (2008) Cereal breeding takes a walk on the wild side. Trend Genet 24:24–32

Godron DA (1876) Un Nouveau chapitre ajouté a l'histoire des Aegilops hybrids. Mém De L'acad De L'acad De Stanislas (nancy) 9:250–281

Harlan JR, de Wet JMJ (1971) Towards a rational classification of cultivated plants. Taxon 20:509–517

Harris DR (1998) The origins of agriculture in southwest Asia. Rev Archaeol 19:5–11

Hein I, Kumlehn J, Waugh R (2009) Functional validation in the Triticeae. In: Feuillet C, Muehlbayer GJ (eds) Genetics and genomics of the Triticeae. Springer, pp 359–385

Kislev ME (1984) Emergence of wheat agriculture. Palaeorient 10:61–70

Kislev ME, Weiss E, Hartman A (2004) Impetus for sowing and the beginning of agriculture: ground collecting of wild cereals. Proc Natl Acad Sci, USA 101:2692–2695

Knott DR (1971) The transfer of genes for disease resistance from alien species to wheat by induced translocations. Mutation breeding for disease resistance. International Atomic Energy Agency, Vienna, pp 67–77

Krattinger S, Wicker T, Keller B (2009) Map-based cloning of genes in Triticeae (wheat and barley). In: Feuillet C, Muehlbayer GJ (eds) Genetics and genomics of the Triticeae. Springer, pp 337–357

Linnaeus C (1753) Species Plantarum, Tomus I, May 1753: i–xii, 1–560; Tomus II, Aug 1753:561–1200, plus indexes and addenda, 1201–1231

McFadden ES (1930) A successful transfer of emmer characters to *vulgare* wheat. J Am Soc Agr 22:1020–1034

McFadden ES, Sears ER (1947) The Genome approach in radical wheat breeding. Agron J 39:1011–1026

Millet E, Rong JK, Qualset CO, McGuire PE, Bernard M, Sourdille P, Feldman M (2013) Production of chromosome-arm substitution lines of wild emmer in common wheat. Euphytica 190:1–17

Millet E, Rong JK, Qualset CO, McGuire PE, Bernard M, Sourdille P, Feldman M (2014) Grain yield and grain protein percentage of common wheat lines with wild emmer chromosome-arm substitutions. Euphytica 195:69–81

Oettler G (2005) The fortune of a botanical curiosity—Triticale: past, present and future. J Agric Sci 143:329–346

Paux E, Sourdille P (2009) A toolbox for Triticeae genomics. In: Feuillet C, Muehlbayer GJ (eds) Genetics and genomics of the Triticeae. Springer, pp 255–283

Reimer PJ, Austin WEN, Bard E, Bayliss A et al (2020) The IntCal20 northern hemisphere radiocarbon age calibration curve (0–55 CAL kBP). Radiocarbon 62:725–757

Riley R (1965) Cytogenetics and evolution of wheat. In: Hutchinson J (ed) Essays in crop plant evolution. Cambridge University Press, London, pp 103–122

Rimpau W (1891) Kreuzungsprodukte Landwirtschaftlicher Kulturpflanzen. Land–wirtschaftl Jahrb 20:335–371

Schweinfurth G (1908) Über die von A. Aaronsohn ausgeführten Nachforschungen nach dem wilden Emmer (Triticum dicoccoides Kcke.). Ber Deutsch Bot Ges 26:309–324

Sears ER (1954) The aneuploids of common wheat. Missouri Agric Exp Stn Res Bull 572:1–58

Sears ER (1956) The transfer of leaf-rust resistance from *Aegilops umbellulata* to wheat. In: Genetics in plant breeding Brook-haven symposia in biology, pp 1–22

Sears ER (1969) Wheat cytogenetics. Annu Rev Genet 3:451–468

Sears ER (1972) Chromosome engineering in wheat. In: Fourth Stadler Symposium, Columbia Missouri, pp 23–38

Sears ER (1975) The wheats and their relatives. In: King RC (ed) Handbook of genetics, vol 2, pp 59–91

Sharma HC, Gill BS (1983) Current status of wide hybridization in wheat. Euphytica 32:17–31

Shewry PR, Hey SJ (2015) The contribution of wheat to human diet and health. Food Energy Secur 4:178–202

Snir A, Nadel D, Groman-Yaroslavski I, Melamed Y, Sternberg M, Bar-Yosef O, Weiss E (2015) The origin of cultivation and proto-weeds, long before Neolithic farming. PLoS ONE 10: e0131422

Stein N (2009) Physical mapping in the Triticeae. In: Feuillet C, Muehlbayer GJ (eds) Genetics and genomics of the Triticeae. Springer, pp 317–357

van Slageren MW (1994) Wild wheats: a monograph of *Aegilops* L. and *Amblyopyrum* (Jaub. and Spach) Eig (Poaceae). Agricultural University, Wageningen, The Netherlands

Vavilov NI (1951) The origin, variation, immunity and breeding of cultivated species. Selected writings, translated from the Russian by Chester KS. Chronica Botanica 13: Chronica Botanica Co. Waltham, Mass. And Wm. Dawson and Sons, Ltd., London, pp 1–366

von Tschermak E (1914) Die Verwertung der Bastardierung für phylogenetische Fragen in der Getreidegruppe. Zeitschr. Pflanzenzucht 2:291–312

Wilson AS (1876) On wheat and rye hybrids. Trans Proc Bot Soc, Edinburgh 12:286–288

Wulff BBH, Moscou MJ (2014) Strategies for transferring resistance into wheat: from wide crosses to GM cassettes. Front Plant Sci 5: article 692

Zeven AC (1980) Polyploidy and domestication: the origin and survival of polyploids in cytotype mixtures. In: Lewis WH (ed) Polyploidy—biological relevance. Plenum Press, New york, pp 385–407

Zhang P, Dundas IS, Robert A. McIntosh RA et al (2015) Wheat–*Aegilops* introgressions. In: Molnár-Láng M, Ceoloni C, Doležel J (eds) Alien introgression in wheat; cytogenetics, molecular biology, and genomics. Springer, Heidelberg, New York, London, pp 221–243

2.1 General Description

The wheat group (the genera *Amblyopyrum, Aegilops* and *Triticum*) is classified in the tribe Triticeae Dumort of the grass family Poaceae (Gramineae). The relatively young tribe diverged from other tribes of the subfamily Pooideae about 25 million years ago (MYA) (Huang et al. 2002b; Gaut 2002) and constitutes a distinct natural and probably most advanced group in this subfamily (Renvoize and Clayton 1992). It is the most economically important tribe of the family, giving rise to the domesticated cereals wheat, rye and barley, and to several important, mostly perennial, fodder grasses such as *Elymus, Leymus, Psathyrostachys* and others. It has a characteristic spiked morphology that distinguishes it from other tribes in the Pooideae and its inflorescence is solitary bilateral raceme, which is an advancement from the simple panicle that predominate in the Pooideae (Renvoize and Clayton 1992).

The progenitors of the Triticeae were probably all diploids with large chromosomes and a symmetric karyotype, perennials and allogamous, and bore several multi-floret spikelets on each rachis node. From an originally paniculate inflorescence, several genera developed a spike with three or two and other with one spikelet at each rachis node (Runemark and Heneen 1968; Sakamoto 1973). Clayton and Renvoize (1986) suggested that the distinctive bilateral spike inflorescence of the Triticeae, derived from a condensed panicle. This is well shown by *Leymus* species, particularly *L. condensatus*. Sakamoto (1973) assumed that the primitive Triticeae resembled current *Psathyrostchys* species.

Traditional taxonomy assigns 18 genera to this tribe, (*Brachypodium* P. Beauv. is not included in the tribe), 17 described by Clayton and Renvoize (1986), and one, *Amblyopyrum*, separated from the genus *Aegilops* by Eig (1929b) and van Slageren (1994) (Tables 2.1 and 2.2). The genera are classified in two sub-tribes, Hordeineae and Triticineae, the first contains 7 genera and the second 11 genera (Table 2.1). Based on the presence of two spikelets

on each rachis node, the two genera *Taeniatherum* and *Crithopsis* were placed in the Hordeineae (Tzvelev 1976; Clayton and Renvoize 1986). However, because molecular phylogenetic studies show that these two genera are closer to the Triticineae than to the Hordeineae (Hsiao et al. 1995a; Mason-Gamer and Kellogg 1996b: Petersen and Seberg 1997; Seberg and Petersen 2007; Mason-Gamer et al. 2002; Escobar et al. 2011), we decided to include them in the Triticineae. The Triticeae includes about 330 species (Clayton and Renvoize 1986), which grow in temperate and arctic zones, principally in the northern hemisphere, and in southwest and central Asia. It includes genera with only perennial, perennial and annual, or only annual species (Table 2.3). About 250 species are perennials that are distributed mainly in temperate-arctic regions, while the annuals, including the progenitors of wheat, rye and barley, are mainly distributed in the east Mediterranean and central Asiatic regions (Table 2.3). Some of the species are obligatory allogamous (self-incompatible), others are facultative allogamous and the rest are facultative autogamous.

Based on their geographical distributions, Sakamoto (1973, 1986, 1991) classified the Triticeae genera into two major groups, Arctic-Temperate and Mediterranean-Central Asiatic (Table 2.3). This classification was later supported by the analyses reported by Hsiao et al. (1995a) and Fan et al. (2013). The Arctic-Temperate group, distributes across the arctic-temperate regions of the world, and has evolved into many endemic species in each area. Six of the genera are perennial, with two or three spikelets at each rachis node [*Psathyrostachys, Elymus* (several species have solitary spikelets at each rachis node)*, Hordelymus, Lemus, Histrix,* and *Sitanion*]. Only *Hordeum* includes perennial and annual species. One noteworthy characteristic of this group is the extensive inter-generic and inter-specific hybridization across their entire area of distribution. The Mediterranean group, distributed across the Mediterranean-central Asiatic region, consists mainly of eight annual genera, six of them (*Heteranthelium, Eremopyrum, Amblyopyrum, Aegilops, Triticum,* and *Henrardia*) have solitary spikelet at each

M. Feldman and A. A. Levy, *Wheat Evolution and Domestication*,
https://doi.org/10.1007/978-3-031-30175-9_2

Table 2.1 The genera of the Triticeae[a]

Sub-tribe	Genus	Number of species[b]	Ploidy level	Type of polyploidy[f]	Growth habit[g]	Pollination mode[h]	Distribution[i]	Number of spikelets on node[j]
Hordeineae	*Elymus* L.	~150	2x–12x	Au-Al-Au/al	P	C, S	TA, M	G, S
	Hystrix *Moench*	9	4x	Al (?)	P	C	TA	G
	Sitanion Raf.	4	4x (?)	Al	P	C	TA	G
	Leymus Hochst.	~40	4x–12x	AL	P	C	TA	G
	Psathyrostachys Nevski	7	2x	—	P	C	TA	G
	Hordelymus (Jessen) Harz	1	4x	Au	P	C	M	G
	Hordeum L.	~40	2x–6x	Au	P, A	C, S	TA, M	G
Triticineae	*Agropyron* Gaertn.	~15	2x–6x	Au	P	C	M	S
	Eremopyrum (Ledeb.) Jaub. & Spach.	5	2x–4x	Al	A	S	M	S
	Heteranthelium Jaub. & Spach.	1	2x	—	A	S	M	S
	Secale L.	3[c]	2x	—	P, A	C	M	S
	Dasypyrum (Coss. & Dur.) Dur.	2	2x–4x	Au	P, A	C	M	S
	Triticum L.	6[d]	2x–6x	Al	A	S	M	S
	Amblyopyrum (Jaub. & Spach.) Eig	1[d]	2x	—	A	C	M	S
	Aegilops L.	24[d]	2x–6x	Al	A	C, S	M	S
	Henrardia C.E. Hubbard	2	2x	—	A	S (?)	M	S
	Taeniatherum Nevski[e]	1	2x	—	A	S	M	G
	Crithopsis Jaub. & Spach[e]	1	2x	—	A	S	M	G

[a] According to Clayton and Renvoize (1986)
[b] The number of species is underestimated
[c] According to Frederiksen and Petersen (1998)
[d] According to van Slageren (1994)
[e] Based on molecular phylogenetic studies, *Taeniatherum* and *Crithopsis* were placed in the subtribe Triticineae (see Chap. 2)
[f] Au = Autopolyploidy; Al = Allopolyploidy; Au/Al = Auto-allopolyploidy
[g] P = Perennials; A = Annuals
[h] C = Cross pollination; S = Self pollination
[i] TA = Temperate-Arctic; M = Mediterranean-Central Asiatic
[j] G = Spikelets in group; S = Solitary spikelets

rachis node whereas two genera (*Crithopsis* and *Taeniatherum*) have two spikelets at each rachis node. The Mediterranean species of *Hordeum* have three spikelets at each rachis node. Muramatsu (2009) suggested that six genes located on chromosomes of homoeologous group 2 of bread wheat, determine spikelet per rachis node solitariness, while the two–three spikelets per rachis node is a recessive trait. The genera *Hordeum*, *Dasypyrum* and *Secale* have perennial and annual species. Two genera (*Elymus* and *Agropyron*) are perennial. Each genus of the Mediterranean group is morphologically distinct. *Aegilops* is the largest genus (24 species) and *Taeniatherum, Crithopsis, Heteranthelium* and *Amblyopyrum* are monotypic. Natural inter-generic hybridization is more restricted in this group than in the Arctic-Temperate group, perhaps due to their

larger genomic diversification that creates stronger inter-generic barriers.

The degree to which polyploidy has occurred in the Triticeae varies greatly among genera. In several genera all species are diploids (*Psathyrostachys, Secale; Amblyopyrum, Heteranthelium, Henrardia, Taniatherum,* and *Crithopsis*) while in others all species are polyploids (*Hystrix, Sitanion, Leymus,* and *Hordelymus*) (Table 2.1). About 102 (31%) of the species are diploids, 3 (1%) triploids, 148 (45%) tetraploids, 56 (17%) hexaploids, 17 (5%) octoploids, 1 (0.3%) decaploid and 1 (0.3%) dodecaploid. The genus *Elymus* displays the larger series and highest level of polyploidy, ranging from 2× to 12× (Table 2.1). Polyploidy occurs either by interspecific or intergeneric hybridization (allopolyploidy), or within a

Table 2.2 The genera of the Triticeae and their synonyms

Genus	Synonyms
Elymus L.	*Anthosachne* Steud.; *Australopyrum* (Tzvelev) Á. Löve; *Braconotia* Godr.; *Brachypodium* sect. *Festucopsis* C. E. Hubbard; *Campeiostachys* Drobov; *Clinelymus* (Griseb.) Nevski; *Cryptopyrum* Heynh.; *Crithopyrum* Steud.; *Elymus* sect. *Clinelymus* Griseb.; *Elytrigia* Desv.; *Elytrigia* sect. *Pseudoroegneria* Nevski; *Festucopsis* (C.E. Hubbard) Melderis; *Goulardia* Husn.; *Hystrix* Moench; *Kengyilia* C. Yen et J. L. Yang.; *Lophopyrum* Löve; *Pascopyrum* Löve; *Pseudoroegneria* (Nevski) Löve; *Roegneria* C. Koch; *Semeiostachys* Drobov; *Sitanion* Raf.; *Sitospelos* Adans.; *Terrellia* Lunell; *Thinopyrum* Löve;
Hystrix Moench	*Asperella* Humboldt; *Gymnostichum* Schreb.; *Stenostachys* Turcz.; *Cockaynea* Zotov
Sitanion Raf	*Polyanthterix* Nees
Leymus Hochst	*Triticum* sect. *Anisopyrum* Griseb.; Anisopyrum (Griseb.) Gren. & Duval.; *Aneurolepidium* Nevski; *Malacurus* Nevski
Psathyrostachys Nevski	–
H ordelymus (Jessen) Harz	*Cuviera* Koeler; *Elymus* sect. *Leptothrix* Dumort.; *Orostachys* Steud.; *Elymus* sect. *Medusather* Griseb.; Hordeum subgen. Hordelymus Jessn; *Hordeum* subgen. *Hordelymus* Jessen; *Leptothrix* (Dumort.) Dumort.; *Medusather* Candargy
Hordeum L.	*Zeocriton* Wolf; *Critesion* Raf.; *Critho* E. Mayer
Agropyron Gaertn	*Kratzmannia* Opiz; *Costia* Willkomm; *Agropyron* sect. Australopyrum Tzvelev; *Australopyrum* (Tzvelev) Löve
Eremopyrum (Ledeb.) Jaub. & Spach	*Triticum* sect. *Eremopyron* Ledeb
Heteranthelium Jaub. & Spach	–
Secale L.	–
Dasypyrum (Coss. & Dur.) Dur	*Secalidium* Schur; *Triticum* sect. *Dasypyrum* Coss. & Dur.; *Triticum* sect. *Pseudosecale* Godr.; *Haynaldia* Schur; *Pseudosecale* (Godr.) Degen
Triticum L.	*Crithodium* Link; *Gigachilon* Seidl; Frumentum Krause;
Amblyopyrum (Jaub. & Spach) Eig	*Aegilops* L. subgen. *Amblyopyrum* Jaub. & Spach; *Aegilops* sect. *Amblyopyrum* (Jaub. & Spach) Zhuk.; *Aegilops* sect. *Anathera* Eig;
Aegilops L.	–
Henrardia C. E. Hubbard	–
Taeniatherum Nevski	*Triticum crinitum* (Schreb.) Nevski (=*T. caput-medusae*)
Crithopsis Jaub. & Spach	–

According to Melderis et al. (1980), Clayton and Renvoize (1986), Renvoize and Clayton (1992), and Van Slageren (1994)

species when genetically differentiated sub-populations of that species come back into contact and hybridize (autopolyploidy; Stebbins 1950).

The diploid Triticeae species, as all the diploids of the family Poaceae, are considered paleopolyploids. This consideration assumes that all genera which have a basic chromosome number of x = 12 or more are derivatives of lines that underwent genome duplication at some time during their evolutionary history (Stebbins 1971). Because all primitive grass sub-families have a basic chromosome number of x = 12, it implies that the ancestor of the grasses was itself a polyploid. Accordingly, all the diploid Poaceae species would be paleopolyploids that underwent cytological and genetic diploidization, whereas the large number of present-day polyploid Poaceae (>60%; Goldblatt 1980), distributed in all the clades, are species that underwent an additional cycle(s) of chromosome doubling during their formation. In these species, the duplicated genomes did not diverge much from those of their diploid (=paleopolyploid) progenitors and chromosome number and cytological behavior are still indicative of genome duplication. Most of these neopolyploids (hereinafter polyploids) were derived from distant inter-specific or inter-generic hybridizations, giving rise to new allopolyploid species and the remaining derived from intra-specific hybridizations giving rise to new autopolyploid cytotypes (Stebbins 1971).

All polyploid species in the genera *Hordelymus*, *Hordeum* and *Dasypyrum* are autopolyploids, whereas those in *Eremopyrum*, *Triticum* and *Aegilops* are allopolyploids. The genus *Agropyron* contains auto and allopolyploid species and *Elymus* contains autopolyploid, allopolyploid and auto-allopolyploid species (Table 2.1).

Table 2.3 Classification of the genera of the Triticeae according to their distribution group and growth habit (after Sakamoto 1973)

Distribution group	Growth habit			Rachis node With:
	Perennial	Perennial + Annual	Annual	
Arctic-temperate Group (Primitive group)	*Elymus* *Hystrix* *Sitanion* *Leymus* *Psathyrostachys* *Hordelymus*	*Hordeum*	—	Spikelets in groups
Mediterranean-Central Asiatic group (Advanced group)	—	*Hordeum*	*Taeniatherum* *Crithopsis*	
	Elymus *Agropyron*	*Dasypyrum* *Secale*	*Heteranthelium* *Eremopyrum* *Triticum* *Amblyopyrum* *Aegilops* *Henrardia*	Solitary spikelets

The autopolyploids can be divided further into two types: typical autopolyploids, e.g., *Hordeum bulbosum,* characterized by multivalent pairing at meiosis (Morrison and Rajhathy 1960a, b; Jørgensen 1982). The formation of multivalents during meiosis is often associated with partial sterility and multisomic inheritance. One advantage of typical autopolyploids is the capacity to maintain high levels of heterozygosity, with multiple alleles per locus, or more rarely, to reach homozygosity with multiple dosages of a given allele. Hence, in spite of the partial sterility, selection in such autopolyploids may favor multivalent formation, and therefore, will act against genomic changes that may lead to cytological and genetic diploidization. It is not surprising therefore, that typical autopolyploids are prevalent among perennial allogamous species that bear the capacity for vegetative propagation, in addition to the sexual reproduction, that compensates in many species for the partial sterility. Generally, typical autopolyploidy does not form new species but rather, increases intra-specific genetic variability and eco-geographical flexibility. Typical autopolyploidy can tolerate and consequently, accumulate more mutations than its diploid cytotype.

Bivalent-forming autopolyploids, e.g., tetraploid *Elymus elongatus,* characterized by exclusive bivalent pairing at meiosis in spite of the fact that they contained four homologous chromosomal sets (Heneen and Runemark 1972; Charpentier et al. 1986, 1988). The genus *Hordeum* contains several tetraploid cytotypes exhibiting exclusive bivalent pairing at meiosis (Gupta and Fedak 1985). These species are either bivalent-forming autopolyploids or segmental allopolyploids (von Bothmer et al. 1995; Blattner 2004; Jakob et al. 2004). This type of autopolyploids are more fertile than the typical autopolyploids, and are prevalent among annual species. Many of them, especially those that

underwent some degree of genetic diploidization, are characterized by disomic inheritance.

Eilam et al. (2009) determined the amount of nuclear DNA in diploid and tetraploid cytotypes of several species containing either typical or bivalent-forming autopolyploids. While the typical autotetraploids had close to the expected sum of their diploid cytotype, most of the bivalent-forming autopolyploids had considerably less nuclear DNA (10–23%) than the expected sum value. A newly-synthesized autotetraploid line of *Elymus elongatus,* that had significantly smaller amount of nuclear DNA than the expected additive value of its diploid parental plant, had similar amount to that in the natural autotetraploid cytotypes of *E. elongatus* (Eilam et al. 2009). This indicates that genome downsizing in this autopolyploid was reproducible and occurred immediately after autopolyploidization and there were no further changes in genome size during its life history.

The shift from potential multivalent pairing towards a bivalent type of pairing in the bivalent-forming autotetraploids was presumably brought about by instantaneous elimination of DNA sequences that are involved in homology recognition and in initiation of meiotic pairing. It is assumed that elimination of sequences occurred in two out of the four homologous chromosomal sets. It is important to refer in this regard to Dvorak (1981b) who noticed that two out of the four subgenomes of the tetraploid cytotype of *E. elongatus* appeared to be a modified version of the diploid genome. He proposed that this divergence facilitates pairing between fully homologous chromosomes, thus, leading to bivalents at meiosis and to disomic inheritance (Dvorak 1981b). This group of autopolyploids benefit from full fertility and permanent fixing of heterozygosity between alleles of the partly diverged subgenome pairs.

Allopolyploid species exhibit exclusive intra-subgenomic bivalent pairing of homologous chromosomes at meiosis and consequently, regular segregation of chromosomes at first meiotic anaphase, full fertility and disomic inheritance. Thus, allopolyploids are characterized by homozygosity within subgenomes and permanent heterozygosity between subgenomes. Allopolyploidy is more prevalent in annual, autogamous species. Allopolyploids are new species, largely isolated from their diploid progenitors, but not finally isolated from other allopolyploids sharing a common genome that can hybridize and exchange genes (Zohary and Feldman 1962).

Sha et al. (2010) found that several *Leymus* species (genome NsNsXmXm) have the Ns and others have the Xm Plasmon, indicating multiple origins of these allopolyploids. Evidence for multiple origins of *Aegilops triuncialis* (genome UUCC or CCUU) came from the findings that several lines of this species contain the U and others the C plasmon (Kimber and Tsunewaki 1988). It is reasonable to assume, therefore, that other Triticeae allopolyploids, and presumably also autopolyploids, have multiple origins.

In accordance with the conclusion of Leitch and Bennett (2004) stating that genome downsizing following polyploid formation is a widespread phenomenon in angiosperms, the DNA content in most allopolyploid Triticeae species is significantly smaller than the expected value calculated from the sum of the DNA content in their two parental species (Table 2.4). These data align with findings demonstrating that allopolyploidization leads to instantaneous elimination of DNA sequences in the wheat group (Feldman et al. 1997; Liu et al. 1998a, b; Ozkan et al. 2001; Shaked et al. 2001; Han et al. 2003) and in triticale (Ma and Gustafson 2005, 2006; Ma et al. 2004). A decrease in genome size already occurred in the first generation of the allopolyploids, indicating that the change was a rapid event (Ozkan et al. 2001). The low variation in nuclear DNA content at the intra-specific level (Eilam et al. 2008) also suggests that the changes in genome size occurred soon after the formation of the allopolyploids and no further significant changes occurred during the life history of the allopolyploids. The similarity in DNA content of natural and synthetic allopolyploids having the same genomic combinations (Eilam et al. 2008), also shows that changes in genome size are reproducible and occur during or soon after allopolyploidization.

The Triticeae tribe must certainly be a comparatively young group with much of the critical differentiation perhaps started first during the middle of the Tertiary in the Oligocene (Table 2.5). The wheat lineage has diverged from the barley lineage about 8 to 15 MYA during the Miocene (Table 2.6). The ancestors of wheat and *Aegilops* have diverged from rye about 7 MYA. Divergence of the diploid *Aegilops* and *Triticum* species begun about 7 MYA but most species diverged 2 to 4 MYA. The allopolyploid wheats and perhaps also those of *Aegilops* were formed between 1.0 to 0.01 MYA (Table 2.6).

It is generally accepted that all the Triticeae are derived from the same common ancestor. Therefore, their genomes still maintain considerable homology, i.e., they are homoeologues, and their chromosomes maintain similar gene order. However, gross structural rearrangements have occurred during the evolution of some of the taxa. On the other hand, there are several species whose genome apparently do not possess a substantially restructured karyotype and so, it is assumed that their genome does not differ in chromosome structure from the chromosomal structure of the progenitor of the tribe.

2.2 Triticeae Taxonomy

The Triticeae presents a prime example of the complexity of multiple taxonomy systems (Renvoize and Clayton 1992), introducing much confusion into the scientific literature, as different names have been used for the same taxa. The taxonomy of the tribe is complicated by several special factors, such as ancient and recent inter-generic and inter-specific hybridizations and allopolyploidy, which are largely responsible for the blurred boundaries between genera. Being a relatively young tribe, the Triticeae shows an exceptional capacity for such hybridizations, that is unparalleled by any other Poaceae group. Studies indicated that most Triticeae genera hybridize with each other (Sakamoto 1973; Cauderon 1986). These inter-generic hybridizations imply a more close-knit reticulate pattern of relationships between the various genera and thus, create problems both in the theoretical concept of generic rank and lineage, and in the practical construction of taxonomy keys (Clayton and Renvoize 1986). Consequently, many obstacles in the identification and classification of the various genera are encountered in taxonomic research and many disagreements still remain. The fact that natural hybridization occurs among many of the genera, indicates that the tribe represents a genetic category called comparium (Clausen et al. 1945; Stebbins 1950). Comparium, which is characteristic of young groups, includes all the genera between which hybridization is possible, either directly or through intermediates (Clausen et al. 1945). The exchange genes between a number of taxonomic Triticeae genera makes difficult the traditional taxonomic classification.

Table 2.4 Species of the Triticeae, their synonyms, genome symbol, and 1C nuclear DNA content

Genus	Species	Synonym	Genome[a]	1C DNA (pg)[b]
Elymus	*Reflexiaristatus* (Nevski) Melderis ssp. *strigosus* (M. Bieb.) Melderis	Elymus *strigosus* Rydb.; *Pseudoroegneria strigosa* (M. Bieb.) Á. Löve; *Elytrigia strigosa* (M. Bieb.) Nevski; *Agropyron strigosum* (M. Bieb.) Boiss.	StSt	4.9[c], 4.9[d]
	reflexiaristatus (Nevski) Melderis subsp. *reflexiaristatus*	*Agropyron reflexiaristatum* Nevski; *Agropyron strigosum* subsp. *reflexiaristatum* (Nevski) Tzvelev; *Elytrigia strigosa* subsp. *reflexiaristata* (Nevski) Tzvelev	StSt	—
	libanoticus (Hackel) Melderis	*Agropyron libanoticum* Hack ex Kneuck; *Pseudoroegneria libanotica* (hack.) D.R. Dewey	StSt	4.0[c], 4.0[d]
	spicatus (Pursh) Gould	*Agropyron spicatum* Pursh; *Pseudoroegneria* spicata (Pursh) Á. Löve; *Elytrigia spicata* (Pursh) D.R. Dewey	StSt	4.7[c], 4.7[d]
	stipifollus (Czern. ex Nevski) Melderis	*Pseudoroegneria stipifolia* (Czern. ex Nevski) Á. Löve	StSt	4.0[c], 4.0[d]
	tauri (Boiss. & Bal.) Melderis	*Pseudoroegneria tauri* (Boiss, & Bal.) Á. Löve; *Agropyron tauri* Boiss & Bal.; *Elytrigia pertenius* (C. A. Mey.) Nevski; *Elymus pertenuis* Assadi	StStPP	—
	panormiitanus (Parl.) Tzvelev	*Agropyron panormitanum* Parl.	StStPP	—
	deweyi	*Pseudoroegneria deweyi* K.B. Jensen, S. L. Hatch & Wipff; *Elytrigia deweyi* (K. Jensen & al.) Valdés & H. Scholz	StStPP	9.3[c]
	abolinii (Drobow) Tzvelev	*Roegneria abolini* (Drobow) Nevski	StStYY	8.7[c], 9.3[d]
	ciliaris (Trin) Tzvelev	*Agropyron ciliare* (Trin) Franch; *Roegneria ciliaris* (Trin) Nevski	StStYY	8.7[c], 8.7[d]
	caucasicus (K. Koch) Tzvelev	*Roegneria caucasica* K. Kock	StStYY	—
	canadensis L.	*Elymus philadelphicus* L.; *Elymus brachystachys* Scribn. & Ball.	StStHH	10.8[c],10.5[d]
	caninus (L.) L..	*Agropyron caninum* (l.) Beauv.; *Roegneria canina* (l.) Nevski; *Roegneria behmii* Melderis	StStHH	9.6[c], 8.5[d]
	glaucus Buckl.	*Elymus mackenzie* Bush; *Elymus parishii* Burt Davy & Merr.; *Elymus virescens* Piper	StStHH	9.3[c], 8.6[d]
Elymus	*lanceolatus* (Scribn & J.G. Sm.) Gould	*Agropyron dasystachyum* (HOOK.) Vasey; *Elytrigia dasystachya* (Hook.) Á. & D. Löve	StStHH	8.4[c], 8.3[d]
	mutabilis (Drobov) Tzvelev	*Agropyron mutabile* Drobov; *Roegneria mutabilis* (Drobov) Hyl.	StStHH	8.7[c], 8.7[d]
	sibiricus L.	*Clinelymus sibiricus* (L.) Nevski	StStHH	8.3[c], 8.3[d]
	trachycaulus (Link) Gould ex Shinners	*Roegneria pauciflora* (Schweinitz) Hyl.	StStHH	9.6[c], 9.6[d]
	dahuricus Turcz. Ex Griseb.	*Clinelymus dahuricus* (Turcz. Ex Griseb.) Nevski	StStHHYY	13.2[c], 13.2[d]
	drobovii	*Roegneria drobovi* (Nevski) Nevski; *Agropyron drobovii* Nevski	StStHHYY	—
	alatavicus (Drobow) Á. Löve	*Agropyron alatavicum* Drobow; *Elytrigia alatavica* (Drobow) Nevski; *Kengyilia alatavica* (Drobow) J. L. Yang, C. Yen & B. R. Baum; *Kengyilia longiglumis* (Keng) S. L. Chen; *Roegneria longiglumis* Keng & S. L. Chen	StStPPYY	15.1[c], 15.2[d]
	batalinii (Krasn.) Á. Löve	*Triticum batalinii* Krasn.; *Agropyron batalinii* (krasn.) Roshevitz ex Fedtsch.; *Elytrigia batalinii* (Krasn.) Nevski; *Kengyilia batalinii* (Krasn.) S. L. Chen	StStPPYY	—
	thoroldianus (Oliv.) Singh	*Kengyilia thoroldiana* (0liv.) S. L. Chen; *Agropyron thoroldianum* Oliv.; *Roegneria thoroldiana* (Oliv.) Keng	StStPPYY	—
	transhyrcanus (Nevski) Tzvelev	*Roegneria transhyrcana* Nevski; *Agropyron lepyourum* (Nevski) Grossh.	StStStStYY	—
	scabrous (R. Br.) Á. Löve	*Agropyron scabrum* (R. Br.) Beauv.; *Anthosachne australica* C. Yen & J. L. Yang; *Anthosachne scabra* (R. Br.) Nevski	StStWWYY	—

(continued)

Table 2.4 (continued)

Genus	Species	Synonym	Genome[a]	1C DNA (pg)[b]
	repens (L,) Gould	Agropyron repens (L.) Beauv. Elytrigia repens (L.) Desv. Ex Nevski	StStStStHH	13.0[c]
	smithii (Rydb.) Gould	Pascopyrum smithii (Rydb.) Barkworth & D. R. Dewey; Agropyron smithii Rydb.: Elytrigia smithii (Rydb.) Á. Löve	StStHHNsNsXmXm	17.7[c], 17.7[d]
Elymus	pungens (Pers.) Melderis	Psammopyrum pungens (Pers.) Á. Löve; Triticum pungens Pers.; Agropyron pungens (Pers.) Roem. & Schultes;	EEStStStStPP	—
Elymus	elongatus (Host) Runemark ssp. elongatus	Agropyron elongatum (Host) Beauv.; Lophopyrum elongatum (Host) Á. Löve; Thinopyrum elongatum (Host) D.R. Dewey	EeEe	5.6[c], 6.1[d], 5.8[e]
	elongatus (Host) Runemark ssp. flaccidifolius (Boiss. & Heldr.) Runemark	Agropyron scirpeum C. Presl; Agropyron scirpeum var. flaccidifolium Boiss. & Heldr.; Agropyron elongatum var. flaccidifolium (Boiss. & Heldr.) Boiss. & H eldr.; Agropyron flaccidifolium (Boiss. & Heldr.) Candargy; Agropyron elongatum Host ssp. scirpeum (C. Presl.) Ciferri & Giacom.; Elymus elongatus ssp. flaccidifolius (Boiss. & Heldr.) Runemark; Elytrigia scirpea (C. Presl) Holub; Lophopyrum scirpeum (C. Presl) Á. Löve; Thinopyrum scirpeum (C. Presl) D. R. Dewey	Ee1Ee1Ee2Ee2	10.4[e]
	elongatus (Host) Runemark subsp. turcicus (P.E. McGuire) Melderis	Elytrigia turcica P. E. Maguire; Elytrigia elongata subsp. turcica (P. E. McGuire) Valdés & H. Scholz; Elytrigia pontica subsp. turcica (P. E. McGuire) Jarvie & Barkworth; Lophopyrum turcicum (P. E. McGuire) McGuire ex Löve; Thinopyrum turcicum (P. E. McGuire) Cabi & Dogan	EeEeEbEbStStStSt	—
	elongatus (Host) Runemark ssp. ponticus (Podp) Melderis	Triticum ponticum Podp; Thinopyrum ponticum (Podp.) Barkworth & D. R. Dewey	EeEeEbEbExExStStStSt	22.6[c], 22.6[d]
	nodosus (Nevski) Melderis ssp. caespitosus (K. Koch) Melderis	Lophopyrum caespitosum (K. Koch) Á. Löve; Agropyron caespitosum K. Koch; Thinopyrum caespitosum; Elytrigia caespitosa (K. Koch) Nevski	EeEeStSt	—
	nodosus (Nevski) Melderis ssp. nodosus	Lophopyrum nodosum (Nevki) Á. Löve; Agropyron nodosum Nevski; Thinopyrum nodosum	EeEeStSt	—
Elymus	bungeanus (Trin.) Melderis	Pseudoroegneria geniculata (Trin.) Á. Löve; Agropyron geniculatum Trin. Ex Ledeb.; Agropyron scythicum Nevski; Elytrigia geniculata (Trin.) Nevski; Triticum geniculatum Trin.	EeEeStSt	9.9[c]
	hispidus (Opiz) Melderis	Trichopyrum intermedium Á. Löve; Triticum intermedium Host; Agropyron intermedim (Host) Beauv.; Elytrigia intermedia (host) Nevski; Thinopyrum intermedium (Host) Barkworth & D.R. dewey; Elymus intermedius (Host) Beauv; Agropyron trichophorum (Link) Richt; Triticum trichophorum Link	EeEeEeEeStSt	12.9[c]
	farctus (viv.) Runemark ex Melderis ssp. bessaribicus (Savul. & Rayss) Melderis	Agropyron bessarabicum Savul. & Rayss; Elytrigia juncea ssp. bessarabicum (Savul. & Rayss) Tzvelev]	EbEb	7.4[c], 7.4[d]
	curvifolius (Lange) Melderis	Lophopyrum curvifolium (Lange) Á. Löve; Agropyron curvifolium Lange; Elytrigia curvifolia (Lange) Holub; Thinopurum curvifolium (Lange) D. R. Dewey	Eb1Eb1Eb2Eb2	—
	farctus ssp. boreali-atlanticus (Simonet & Guinochet) Melderis	Agropyron junceum ssp. boreali-atlanticum Simonet & Guinochet; A. junceiforme (Á. & D. Löve) Á. & D. Löve; Elytrigia juncea ssp. boreoatlantica Hyl.]	EbEbEeEe	11.8[c], 13.0[d]
	farctus subsp. rechingeri (Runemark) Melderis	Agropyron rechingeri Runemark; Elymus rechingeri Runemark in Runemark & Heneen	Eb1Eb1Eb2Eb2	—
	farctus var. sartorii (Boiss. & Heldr.) Melderis	Thinipyrum sartorii (Boiss. & Heldr.) Á. Löve; Agropyron junceum ssp. sartorii (Boiss. & Heldr.) Maire; Agropyron sartorii (Boiss. & Heldr.) Grecescu; Elytrigia sartorii (Boiss. & Heldr.) Holub; Triticum sartorii (Boiss. & Heldr.) Boiss. & Heldr.	EbEbEeEe	—

(continued)

Table 2.4 (continued)

Genus	Species	Synonym	Genome[a]	1C DNA (pg)[b]
	farctus ssp. *farctus* (L.) Á. Löve;	*Agropyron junceum* (L.) Beauv.; *Thinopyrum junceum* (L.) Á. Löve; *Elytrigia juncea* (L.) Nevski	Eb1Eb1Eb2Eb2EeEe	16.3[c], 16.4[d]
	distichus (Thunb.) Melderis	*Thinopyrum distichum* (Thunb.) Á. Löve; *Agropyron distichum* (Thunb.) Beauv.; *Elytrigia disticha* (Thunb.) Prokudin ex Á. Löve; *Triticum distichum* Thunb.	EbEbEeEe	—
Elymus	*serpentinus* (?)	*Festucopsis serpentini* (C.E. Hubb.) Melderis; *Brachypodium serpentini* C.F. Hubb.	LL	—
	sanctus (?)	*Peridictyon sanctum* (Janka) Seberg, Fred. & Baden; *Festucopsis sanctum* (Janka) Melderis; *Brachypodium sanctum* (Janka) Janka; *Festuca sancta* Janka	XpXp	—
	athericus (link) Kerguélen	*Psammopyrum athericum* (Link) Á. Löve; *Elytrigia atherica* (Link) Kerguélen ex Carreras Martnez; *Triticum athericum* Link; *Agropyron athericum* (Link) Samp.;	LLEE	—
Leymus	*akmolinensis* Drobow	*Elymus akmolinensis* Drobow; *Elymus dasystachys* Glaber Korsh.; *Aneurolepidium akmolinensis* (Drobow) Nevski; *Leymus paboanus* Tzvelev	NsNsXmXm	11.3[c], 11.3[d]
	chinensis (Trin.) Tzvelev	*Elymus chinensis* (trin.) Keng; *Agropyron chinensis* (Trin.) Ohwi; *Aneurolepidium chinensis* (Trin.) Kitg.	NsNsXmXm	9.8[c]
	flavens (Scribn. & J. G. Sm.) Pilg.	*Elymus arenicola* Scrib. & J. G. Sm.; *Elymus flavescens* Scribn. & J. G. Sm;. *Leymus arenicola (Scribn. & J. G. Sm.) Pilg.*	NsNsXmXm	12.2[c], 12.2[d]
	racemosus (Lam.) Tzvelev	*Elymus racemosus* Lam.; *Elymus sabulosus* (M. Bieb.) Tzvelev: *Leymus sabulosus* (M. Bieb.) Tzvelev.; *Elymus giganteus* Vahl	NsNsXmXm	11.4[c], 10.6[d]
	ramosus (Trin.) Tzvelev	*Elymus ramosus* (Trin.) Filatova; *Elymus trini* Melderis; *Agropyron ramosum* (Trin.) K. Richt	NsNsXmXm	10.1[c], 10.2[d]
	secalinus (Georgi) Tzvelev	Elymus secalinus (Georgi) Bobrov; *Elymus dasystachys* Trin.; *Triticum secalinum* Georgi	NsNsXmXm	10.8[c], 10.7[d]
	triticoides (Buckley) Pilg.	*Elymus triticoides* Buckley; *Elymus orcuttianus* Vasey; *Elymus condensatus* J. Presl. Var. *triticoides* (Buckley) Thurb.	NsNsXmXm	11.2[c], 11.2[d]
Psathyrostachys	*fragilis* (Boiss.) Nevski	*Hordeum fragile* Boiss.; *Elymus fragilis* (Boiss.) Griseb.	NsNs	8.4[c], 8.4[d]
	juncea (Fischer) Nevski	*Elymus junceus* Fischer	NsNs	7.8[c], 7.8[d]
	stoloniformis C. Baden	-	NsNs	8.9[c], 8.9[d]
Sitanion	*hystrix* (Nutt.) J.G. Smith	*Elymus elymoides* (Raf.) Swezey; *Sitanion elymoides* Raf.;	StStHH (?)	—
Hystrix	*californica* (Bol. ex Thurb.) Kuntze	*Lemus californicus* (Bol. ex Thurb.) Barkworth; *Asperella californica* (Bol. ex Thurb.) Beal	—	—
	patula Moench	*Elymus hystrix* L.	—	—
	japonica (Hack.) Ohwi	*Asperella japonicas* Hack.; *Elymus japonicas* (Hack.) Á. Löve; *Hystrix hackelii* Honda; *Hystrix duthiei* subsp. *japonica* (Hack.) C. Baden, Fred. & Seberg	—	—
	longearistata (Hack.) Honda	*Asperella longearistata* (Hack.) Ohwi; *Asperlla sibirica* var. *longearistata* Hack.; *Elymus asiaticus* ssp. *longearistatus* (Hack.) Á. Löve	—	—
Hordelymus	e *uropaeus* (L.) Harz	*Elymus europaeus* L.; *Hordeum europaeum* (L.) All.; *Hordeum sylvaticum* Huds.; *Cuviera europaea* (L.) Koeler	XoXoXrXr	—
Hordeum	*vulgare* L.	*Hordeum zeocriton* L.; *Critesion vulgare* Raf.; *Hordeum vulgare* ssp. *agriocrithon* A. E. Åberg; *Hordeum distichon* L.	HH	5.5[c]
	spontaneum C.Koch	*Hordeum ithaburense* Boiss.; *Hordeum vulgare* ssp. *spontaeum* K. Koch	HH	5.5[c], 5.5[e]
	bulbosum L.	*Critesion bulbosum* (L.) Á. Löve	HH	5.5[c], 4.6[d], 4.6[e]

(continued)

Table 2.4 (continued)

Genus	Species	Synonym	Genome[a]	1C DNA (pg)[b]
	bulbosum L.	*Critesion bulbosum* (l.) Á. Löve	HHHH	11.0[c], 8.9[d], 8.8[e]
	chilense Roem et Schult.	*Critesion chilense* (Roem et Schult.) Á. Löve	II	5.4[c], 4.9[d]
	geniculatum All.	—	II	5.4[c]
	geniculatum All.	—	IIII	10.8[c]
	glaucum Steud.	—	II	5.5[c]
	jubatum L.	*Critesion adscendens* (Kunth) Á. Löve; *Critesion jubatum* (L.) Nevski; *Elymus jubatus* link; *Hordeum adscendens* Kunth	IIII	10.9[c]
	leporinum Link	*Hordeum murinum* ssp. *leporinum* (Link) Arcangeil	XuXuXuXu	10.9[c]
	leporinum Link	*Hordeum murinum* ssp. *leporinum* (Link) Arcangeil	XuXuXuXuXuXu	16.4[c]
Hordeum	*marinum* Hudson	*Critesion marinum* (Huds.) Á. Löve; *Hordeum gussoneanum* Parl.; *Hordeum maritimum* With.	XaXa	5.5[c], 5.2[e]
			XaXaXaXa	10.3[e]
	glaucum Steud.	—	XuXu	—
	murinum L.	*Critesion murinum* (L.) Á. Löve	XuXu	5.5[c], 6.4[e]
			XuXuXuXu	11.1[c], 10.4[e]
	pusillum Nutt.	*Critesion pusillum* (Nutt.) Á. Löve	II	5.5[c]
	roshevitsii Bowden	—	II	5.5[c], 5.6[d]
	secalinum Schreber	*Zeocriton secalinum* (Schreb.) Beauv.; *Critesion secalinum* (Schreb) Á. Löve	IIXaXa	11.2[c]
	violaceum Boiss. & Hohen.	*Critesion violaceum* (Boiss. & Hohen.) Á. Löve; *Hordeum brevisubulatum* (Trin.) Link ssp. *violaceum* (Boiss. & Hohen.) Tzvelev	II	5.5[c]
	bogdani Wilensky	*Critesion bogdani* (Will.) Á. Löve	II	5.0[c], 4.7[d]
	brachyantherum Nevski	*Critesion brachyantherum* (Nevski) Barkworth & D. R. Dewey; *Critesion brachyantherum* (Nevski) Weber; *Hordeum jubatum* ssp. *brachyantherum* (Nevski) Bondar	II	4.8[c], 4.8[d]
	californicum Covas & Stebbins	*Hordeum bracyantherum* Nevski ssp. *californicum* (Covas & Stebbins) von Bothmer et al.; *Critesion californicum* (Covas & Stebbins) Á. Löve	II	4.7[c], 4.8[d]
	comosum J. Presl	*Hordeum andinum* Trin.; *Hordeum divergens* Nees & Meyen ex Nicota; *Critesion comosum* (J. Presl) Á. Löve	II	4.6[c], 4.6[d]
	flexuosum Nees ex Steud.	*Critesion flexuosum* (Nees ex Steud.) Á. Löve	II	4.4[c], 4.4[d]
	aplophilum Griseb.	*Critesion aplophilum* (Griseb.) Barkworth & D. R. Dewey	II	4.4[c], 4.4[d]
	stenostachys Godr.	*Critesion stenostachys* (Godr.) Á. Löve	II	5.0[c], 5.0[d]
Agropyron	*cristatum* (L.) Gaertn.	*Eremopyrum cristatum* (L.) Willk.	PP	7.1[c], 7.1[d],
			PPPP	13.5[f]
			PPPPPP	21.7[f]
	mongolicum Keng	—	PP	7.8[c]
	desertorum (Fisch. ex Link) Schult.	*Agropyron cristatum* ssp. *desertorum* (Fisch. ex Link) A. Löve; *Agropyron cristatum* var. *desertorum* (Fisch. ex Link) Dorn	PPPP	13.2[c]
Agropyron	fragile (Roth) P. Candargy	—	PP	—
			PPPP	—
	pectinatum (Labill) Beauv.	*Australopyrum pectinum* (Labill.) Á. Löve; *Agropyron bowne* (Kunth) Tzvelev	WW	—

(continued)

Table 2.4 (continued)

Genus	Species	Synonym	Genome[a]	1C DNA (pg)[b]
Eremopyrum	*triticeum* (Gaertn.) Nevski	*Agropyron triticeum* Gaertn.; *Agropyron prostractum* (L. f.) Beauv.	FF	5.5[c]
	bonaepartis (Spreng.) Nevski	—	XbXb	—
	bonaepartis	—	XbXbXdXd	—
	distans (K. Koch) Nevski	—	XdXd	—
	orientale (L.) Jaub. & Spach	—	XdXdFF	—
Dasypyrum	*villosum* (L.) Coss. & Durieu ex P. Candargy	*Haynaldia villosa* (L.) Schur; *Secale villosum* L.	VvVv	5.3[c], 5.2[e]
	breviaristatum (H. Lindb.) Fred.	*Haynaldia breviaristata* H. Lindb.; *Dasypyrum hordeaceum* (H. Lindb.) Maire &Weiller	VbVb	—
	breviaristatum (H. Lindb.) Fred.	—	VbVbVbVb	—
Secale	*strictum* (C. Presl.) C. Presl. ssp. *strictum*	*S. montanum* Guss.	RR	8.2[c], 9.4[e]
	strictum (C. Presl.) C. Presl. ssp. *africanum* (stapf) K. Hammer	*S. montanum africanum*	RR	7.4[c]
	Sylvestre Host	*S. fragile* M. Bieb.; *S. campestre* Kit.; *S. glaucum* d'Urv.	RR	7.21[c]
	Cereale L. ssp. *cereale*	*S. cereale* ssp. *indo-europaeum* Antropov & Antropova in Roshev; *S. trijlorum* P. Beauv.;	RR	8.3[c], 8.6[e]
	Cereale L. ssp. ancestrale Zhuk.	*S. ancestrale* (Zhuk.) Zhuk.:	RR	—
Heteranthelium	*piliferum* (Banks & Sol.) Hochst	*Elymus pilifer* Banks & Sol.; *Agropyron piliferum* (Hochst) Benth. Ex Aitcl	QQ	4.5[c], 4.2[e]
Triticum	*monococcum* L.	*Crithodium aegilopoides* Link; *T. aegilopoides* (Link) Balansa; *T. boeoticum* Boiss.; *T. thaoudar* Reut.; *T. spontaneum* Flaksb.; *Crithodium monococcum* (L.) Á. Löve	A^{m}A^m	6.7[c], 6.5[e]
	urartu Tumanian ex Gandilyan	—	AA	4.9[c], 6.0[e]
	turgidum L.	—	BBAA	12.2[c], 12.9[e]
	timopheevii (Zhuk.) Zhuk.	*T. dicoccoides* (Körn. ex Asch. & Graebn.) Schweinf. ssp. *armeniacum* Jakubz.; *T. armeniacum* (Jakubz.) Makush.*T. araraticum* Jakubz.; *T. turgidum* L. ssp. *armeniacum* (Jakubz.) Á. Löve; Gigachilon timopheevii (Zhuk.) Á. Löve	GGAA	11.3[c], 11.9[e]
	aestivum L.	*T. vulgare* L.	BBAADD	17.3[c], 18.0[e]
	zhukovskyi Menabde & Ericz.	—	GGAAA^{m}A^m	18.1[c], 17.7[e]
Amblyopyrum	*muticum* (Boiss.) Eig	Aegilops mutica Boiss.; Triticum muticum (Boiss.) Hack.; Aegilops tripsacoides Jaub. & Spach.; T. tripsacoides Bowden;	TT	6.3[c], 5.8[e]
Aegilops	*speltoides* Tausch	Ae. aucheri Boiss; Ae. ligustica (Savign.) Coss.; T. speltoides (Tausch) Gren; Sitopsis speltoides (Tausch) Á. Löve	SS	5.1[c], 5.8[e]
	bicornis (Forssk.) Jaub. & Spach	*T. bicorne* Forssk.; *Sitopsis bicornis* (Forssk.) Á. Löve	S^{b}S^b	7.1[c], 6.8[e]
	longissima Schweinf. & Muschl.	*T. longissimum* (Schweinf. & Muschl.) Bowden; *Sitopsis longissima* (Schweinf. & Muschl.) Á. Löve	S^{l}S^l	6.0[c], 7.5[e]
	sharonensis Eig	*T. sharonense* (Eig) Feldman & Sears; *Sitopsis sharonensis* (Eig) Á. Löve	S^{sh}S^{sh}	7.1[c], 7.5[e]
	searsii Feldman & Kislev ex Hammer	*T. searsii* (Feldman & Kislev) Feldman & Kislev; *Sitopsis searsii* (Feldman & Kislev ex Hammer) Á. Löve	S^{s}S^s	5.9[c], 6.6[e]
	tauschii Coss.	*Ae. squarrosa* L.; *T. tauschii* (Coss.) Schmahlh.; *T. aegilops* Beauv. ex Roem. & Schult.; *Patropyrum tauschii* (Coss.) Á. Löve	DD	5.2[c], 5.2[e]

(continued)

Table 2.4 (continued)

Genus	Species	Synonym	Genome[a]	1C DNA (pg)[b]
Aegilops	caudata L.	Triticum caudatum (L.) Godr. & Gren.; Triticum dichasians Bowden; Ae. dichasians (Bowden) Humphries; Ae. markgrafii (Greuter) Hammer; Orrhopygium caudatum (L.) Á. Löve	CC	4.6[c], 4.8[e]
	comosa Sm. In Sibth. & Sm.	T. comosum (Sm. In Sibth. & Sm.) K.Richt.; ; Comopyrum comosum (Sm. In Sibth. & Sm.) Á. Löve	MM	6.2[c], 5.5[e]
	uniaristata Vis.	T. uniaristatum (Vis.) K. Richt.; Chennapyrum uniaristatum (Vis.) Á. Löve	NN	6.3[c], 5.8[e]
	umbellulata Zhuk.	T. umbellulatum (Zhuk.) Bowden; Kiharapyrum umbellulatum (Zhuk.) Á. Löve	UU	5.0[c], 5.4[e]
	peregrina (Hack. In J. Fraser) Maire & Weiller	Ae. variabilis Eig; T. variabilis; T. preginum Hack. In J. Fraser; Aegilemma peregrina (Hack. In J. Fraser) Á. Löve	S^vS^vUU	13.8[c], 12.5[e]
	kotschyi Boiss.	T. kotschyi (Boiss.) Bowden; Aegilemma kotschyi (Boiss.) Á. Löve	S^vS^vUU	12.3[c], 12.6[e]
	triuncialis L.	T. triunciale (L.) Rasp.; Aegilopodes triuncialis (L.) Á. Löve	UUCC, CCUU	10.5[c], 10.9[e]
	geniculata Roth	Ae. ovata L.; T. ovatum (L.) Gren. & Godr.	M^oM^oUU	9.4[c]
	biuncialis Vis.	Ae. lorentii Hochst.; Ae. macrochaeta Shuttlew & A. Huet ex Duval-Jouve; T. macrochaetum (Shuttlew & A. Huet ex Duval-Jouve) K. Richt.	UUM^bM^b	11.3[c]
	columnaris Zhuk.	Triticum columnare (Zhuk.) Morris & Sears	UUX^nX^n	10.5[c], 10.9[e]
	neglecta Reg. ex Bertol.	Ae. triaristata Willd.; T. negletum (Reg. ex Bertol.) Greuter; T. triaristatum (Willd.) Godr. & Gren.	UUX^nX^n	15.5[c], 10.6[e]
	recta (Zhuk.) Chennav.	Ae. triaristata ssp. recta Zhuk.; T. rectum (Zhuk.) Bowden	UUX^nX^nNN	21.6[c], 16.2[e]
	cylindrical Host	T. cylindricum (Host) Ces. Pass & Gibelli; Cylindropyrum cylindricum (Host) Á. Löve	DDCC; CCDD	9.59[e]
	Ventricosa Tausch	T. ventricosum (Tausch) Ces. Pass & Gibelli; Gastropyrum ventricosum (Tausch) Á. Löve	DDNN	9.6[c], 10.6[e]
	crassa Boiss.	T. crassum (Boiss.) Aitch. & Hemsl.; Gastropyrum crissum (BOiss.) Á. Löve	$D^cD^cX^cX^c$	10.5[c], 10.9[e]
	crassa Boiss.	T. crassum (6x); Gastropyrum crassum	$D^cD^cX^cX^c$DD	15.7[c]
	juvenalis (Thell.) Eig	T. juvenale Thell.; Aegilonearum juvenile (Thell.) Á. Löve	$D^cD^cX^cX^c$UU	18.8[c]
Aegilops	vavilovii (Zhuk.) Chennav.	Ae. crassa Boiss.; Ae. crassa Boiss. Var. palaestina Eig; T. syriacum Bowden; Gastropyrum vavilovii (Zhuk.) Á. Löve	$D^cD^cX^cX^cS^sS^s$	18.3[c]
Henrardia	persica (Boiss.) C. E. Hubb.	—	OO	—
	pubescens (Bertol.) C. E. Hubb.	—	OO	—
Taeniatherum	caput-medusae (L.) Nevski	Taeniatherum crinitum (Schreb.) Nevski	TaTa	4.4[c]
Crithopsis	delileana (Schult. & Schult. F.) Roshev.	Elymus delileanus Schult	KK	4.7[c]

[a] Genome symbols after Wang et al. (1995), Wang and Lu (2014), Symbols of Hordeum after Blattner (2018), Symbols of Eremopyrum modified from Sakamoto (1991), Symbols of Dasypyrum after Ohta and Morishita (2001), Symbols of Aegilops after Dvorak (1998), Unknown or unverified genomes are designated with the letter **X** followed by a lowercase letter for the species; Modified versions of a basic genome are designated by superscripts in small letters indicative of the species carrying such modified genomes. The genomes of the diploid or the tetraploid parents that donated the cytoplasm to the polyploid species of Triticum and Aegilops, according to Kimber and Tsunewaki (1988), are on the left side of the genome symbols. The genome symbol of Elymus elongatus ssp. ponticus is after Li and Zhang (2002) and Liu et al. (2007)

[b] 1C-value in picograms

[c] The Angiosperm C-value database (http://www.rbgkew.org.uk/cval/) at Kew Botanic Gardens

[d] Vogel et al. (1999)

[e] Eilam et al. (2007, 2008, 2009)

[f] Yousofi and Aryavand (2004)

Table 2.5 Geological epochs in the Cenozoic era (65 million years ago to the present) and major climatic, ecological and Triticeae evolutionary events

Period	Epoch	Million years ago	Major climate & ecological events	Events in the evolution of the Triticeae
Quaternary	Holocene	0.01–present	Warmer climates; conversion of many grasslands and forests into cultivated areas; increase in human population	Domestication of wheat, barley and rye; formation of hexaploid (bread) wheat and domesticated forms of tetraploid wheat
	Pleistocene	1.8–0.01	Global cooling; four major ice ages; most temperate zones were covered by glaciers during the cool periods and uncovered during the warmer interglacial periods	Formation of the allotetraploid species of *Triticum* and the allotetraploid and allohexaploid species of *Aegilops*
Tertiary	Pliocene	5.3–1.8	Cooler and drier global climates; accumulation of ice at the poles; development seasonal climate (cold and humid winters and hot and dry summers) in the east Mediterranean and south-west Asia; development of today's landscapes; further spread of grasslands	Evolvement of the diploid species of *Aegilops* and *Triticum*
	Miocene	23.8–5.3	Warmer global climates; disappearance of the Tethys Sea and the climate cooled off towards the end; diversification of temperate ecosystems and new ecological niches opened; expansion of grasslands;	Diversification of grasses; divergence of the Triticineae (wheat lineage) from the Hordeineae (barley lineage)
	Oligocene	33.7–23.8	Cold and dry climates; transformation of vegetation to something similar to that of today	Appearance of early Triticeae
	Eocene	54.8–33.7	Warm and humid climates, became cooler towards the end; forests got smaller while grasslands and savannas increased	Further development of grasses

Table 2.6 Time of beginning divergence of the Triticineae lineages in million years ago

Lineages	Beginning divergence time	Method of study		References
		Sequencing of	RFLP of	
Barley and rye-wheat	14.0–10.0	chloroplast genome	—	Wolfe et al. (1989)
	15.0–11.0	—	BAC libraries	Ramakrishna et al. (2002)
	11.6	nuclear genes	—	Chalupska et al. (2008)
	11.0	two nuclear genes	—	Huang et al. (2002a, b)
	10.6	chloroplast genome	—	Gornicki et al. (2014)
	10.1	four nuclear genes	—	Dvorak and Akhunov (2005)
	8.90–8.13	chloroplast genome	—	Middleton et al. (2014)
Wheat and rye	7.0	275 nuclear genes	—	Marcussen et al. (2014)
	4.0–3.7	chloroplast genome	—	Middleton et al. (2014)
Wheat and *Aegilops*	6.5	three chloroplast genes	—	Marcussen et al. (2014)
	2.7 (4.1–3.4)	four nuclear genes	—	Dvorak and Akhunov (2005)
	4.5–2.5	two nuclear genes	—	Huang et al. (2002b)
	2.9–2.1	chloroplast genome	—	Middleton et al. (2014)

Several classifications, attempting to reflect the evolutionary history of the tribe, have been proposed for the various taxa of the Triticeae over the past 100 years (reviewed by Barkworth 1992). Since the great diversity between the various taxa of the Triticeae lies in the structure of the inflorescence, including the glumes and the lemmas which envelope the flower and seed (Stebbins 1950), many classifications have been based on morphological traits, particularly of the spike. Although morphological similarities may indicate a close phylogenetic relationship, closely related species can also display morphologically difference when grown in different environments (Yen et al. 2005). Thus, a taxonomic approach solely relying on morphological analysis will inevitably include misclassifications. Indeed, classifications on a morphological basis have always been a subject of taxonomic disagreements, as is obvious from several recent discussions (Tzvelev 1976; Melderis et al. 1980; Dewey 1982, 1984; Löve 1982, 1984; Gupta and Baum 1986, 1989; Baum et al. 1987; Barkworth 1992). Significant disagreement exists between classical taxonomists and those suggesting a new pattern of classification founded on cytogenetic relationships between genera rather than on morphological features.

Because of their economic importance, various taxa of the Triticeae have been recurrently used for basic and applied studies, where intensive cytogenetic studies (reviewed in Löve 1982, 1984; Dewey 1982, 1984; summarized in Wang et al. 1995) have uncovered the wide variety in the generic makeup of the Triticeae. Krause (1898) combined *Elymus*, *Hordeum*, *Agropyron*, *Secale*, and *Triticum* to form a new genus, *Frumentum* Krause. Since there are essentially weak genetic barriers between the Triticeae genera, Stebbins (1956) recommended merging all these genera into one genus. However, this suggestion has been rejected by Runemark and Heneen (1968) and Mac Key (1966, 1981) as being artificial, single-character subdivision of the tribe and as such creates more problems than it solves.

The taxonomic philosophy advocating classification that reflects phylogeny and biological relationship (Löve 1982, 1984), and classifications that are based on the assumption that a genus is a group of species with a common genome, led to the rise of genomically-based genera (Löve 1982, 1984). According to this concept, taxa bearing similar genomes are treated as congeneric (Dewey 1984). The similarity between genomes is determined by genome analysis, where complete chromosomal pairing at the first meiotic metaphase in the F_1 hybrid of two taxa indicates similarity between their genomes. The genome classification of Love (1982, 1984) assumes that (1) a genome type equals a genus, and (2) a genome type should be the single most important determiner for the designation of terminal taxa in any forthcoming phylogenetic analysis. Accordingly, the classification of most genera was drastically reorganized (Löve 1982, 1984; Dewey 1984; Wang et al. 1995), dividing the tribe into 38 different mono-generic genera. For instance, *Triticum* species with the A genome constitute the genus *Crithodium*; *Triticum* and *Aegilops* species with the B genome constitutes *Sitopsis*; the D genome constitutes *Patropyrum*; the AB genome complex constitutes *Gigachilon*; and the ABD genome complex constitutes *Triticum* (Löve 1984). For *Aegilops* this has led to its split-up into 12 genera (without *Amblyopyrum*). Some of these 'split genera' are mono-genomic, can therefore considered truly monophyletic and should consequently appear as terminal taxa in a cladistics analysis. This is the case for *Amblyopyrum* and the diploid *Aegilops* 'genera', which appear as terminal clades in a parsimonious tree of the mono-genomic groups in the Triticeae (Kellogg 1989). The genome-based classification was principally accepted by Dewey (1984) for perennial Triticeae, and is being followed by others (Anonymous 1986).

The genomic classification was criticized by Baum et al. (1987) who presented a number of arguments against the formation of genomic genera. Although the genome may be more important than certain other trivial plant traits, Baum et al. (1987) opined that classification should be based on a profile of as many characteristics as possible, rather than on a single feature. They claimed that the genomic classification system gives rise to far too many monotypic genera and that genomic genera are not recognizable morphological units. Moreover, Baum et al. (1987) argue that the genomic system is unstable, necessitating changes with every new genome combination recognized; and that genomes are not good characters anyway. The genome classification, though attractive in theory, is sometimes difficult to translate into practical morphological diagnoses, i.e., there is incongruence between the genomic and morphological data.

The complexities of genome-based classifications are clearly evident upon phylogenetic study of the wheat group. The diploid donors of two out of the three subgenomes of hexaploid wheat, A and D, were identified on the basis of genome analysis and morphological characteristics [Kihara (1944) and McFadden and Sears (1944, 1946) identified the D genome donor; Dvorak (1976) and Chapman et al. (1976) identified the A genome donor]. Extensive work has shown that the phylogenetic relationships of the third subgenome of hexaploid wheat, the B subgenome, could not be ascertained on the basis of genome analysis (Riley 1965; Feldman et al. 1995). More recently, even for the A subgenome, it has been shown that chromosome 4 of diploid wheat is absent in hexaploid wheat (Wazuddin and Driscoll 1986). Baum et al. (1987) predict that new genera will have to be erected and species will need to be transferred from one genus to another when new genome(s) or genome combination(s) are discovered. These authors provide several examples: For *Pseudoroegneria tauri* (Boiss. & Bal.) A. Löve [currently *Elymus tauri* (Boiss. & Bal.) Melderis], a completely new genome combination (PPStSt) has been discovered, necessitating the erection of a new genus (Wang et al. 1986). Genome constitution of *Pseudoroegneria spicata* A. Löve [currently *Elymus spicatus* (Pursh) Gould] has been shown to be StStHH, on the basis of karyotype, instead of StStStSt, which required the transfer of this species to the genus *Elymus* (Wang 1985a). Another case relates to *Thinopyrum bessarabicum* (Savul. & Rayss) A. Löve [currently *Elymus farctus* (Viv.) Runemark ex Melderis] and *Lophopyrum* (Host) A. Löve] [currently *Elymus elongatus* (Host) Runemark] with the J and E genomes, respectively, which have been shown to be closely related. These findings suggest that either genome J should be designated as Eb (Dvorak 1981a, b; McGuire 1984) or E as Je (Wang 1985b). Later, these two closely related genomes were designated Eb and Ee, respectively (Wang et al. 1995). On this basis, it was recommended that *Lophopyrum elongatum* be transferred to the genus *Thinopyrum* (Wang 1985b). Several other genera erected by Löve (1984), based on the assumption that they are autopolyploids, were found to be allopolyploids and, as such, will require new generic names. Many of these classification transfers result in taxonomic realignments of already well-defined and well-marked taxa. In view of the above, Baum et al. (1987) believe that phylogenetic relationships based on the genome alone are nonoperational due to instability of the genomic system; every newly characterized genome combination necessitates classification changes. They concur with Kimber's (1984) suggestion that the sum of morphological characteristics of diploid taxa might also be a good indicator of evolutionary relationships and can be used to supplement the cytogenetic data.

Moreover, since evolution in the Triticeae is highly reticulated, cladistics algorithms (Baum 1982, 1983), only relevant when analyzing groups where reticulation has not occurred, is not suitable for the Triticeae. While Jauhar and Crane (1989) believed that a multidisciplinary approach to taxonomic classification is certainly advisable, such an approach still considers genome analysis the most useful in revealing phyletic relatedness and deserving of consideration for biosystematic delimitation of taxa. The genome analysis approach is generally a good measure of homology in spite of the following limitations: intimate chromosome pairing as seen at the first meiotic metaphase of F_1 hybrids may not represent identity of nucleotide sequence at every locus. Second, the amount of pairing in hybrids may be influenced by genotypes of the parents. Third, genome analysis is possible only if two taxa can be hybridized and viable

hybrids obtained; genome analysis is more useful at the species level, but its usefulness decreases in broader hybrids. Chromosome pairing in diploid hybrids does not reliably indicate chromosome relationships (Jauhar 1975; Kimber and Feldman 1987), because chromosomes have an inherent tendency to pair even with less related chromosomes when their own homologous partners are not available. The precise nature of specificity of pairing is not yet fully understood, however. Other techniques of genome analysis, such as karyotype analysis based on chromosome measurements, chromosome banding patterns, in situ hybridization with specific probes, and restriction enzyme analysis of the chloroplast, mitochondrial, or whole nuclear genome, have also been successfully employed. Each of these techniques brings certain inherent advantages and disadvantages. Whole-genome sequencing of related taxa may generate the ultimate information for assessing phylogenetic relatedness. However, it will not be sufficient to define speciation events that prevent cross-fertilization or affect hybrid viability and fertility. A combination of criteria will be needed to reach a coherent classification of genera and species.

Further arguments against genomic genera include the difficulty to reach agreement regarding the definition of genomic similarity. Alonso and Kimber (1981) claimed that the ability of chromosomes to pair roughly estimates similarity of total DNA. Chromosome pairing as an indicator of phylogenetic relationships has been discussed in general terms by Jackson (1982, 1984, 1985). He pointed out that classical genome analysis can misinterpret phylogenetic relationships in a multiple allelic system of pairing control mutations. Moreover, while pairing between chromosomes may indicate similarity, failure of pairing does not necessarily indicate dissimilarity, since individual genes or groups of genes may suppress pairing between very closely related genomes (Okamoto 1957; Riley and Chapman 1958; Sears 1976; Jackson 1982). Pairing may also fail between homologous chromosomes due to desynaptic genes (for a review see Kaul and Murthy 1985) or B-chromosomes (Mochizuki 1964; Dover and Riley 1972a; Vardi and Dover 1972). Likewise, high pairing in hybrids does not necessarily indicate full homology, as certain genes promote pairing between partially homologous chromosomes, like those existing in *Amblyopyrum muticum* and *Aegilops speltoides* (Dover and Riley 1972b; Dvorak 1972; Chen and Dvorak 1984). Genetic induction of meiotic pairing in hybrids between common wheat and rye, was also shown to be present in *Secale cereale* (Gupta and Fedak 1986). Furthermore, the genomic classification of genera (Dewey 1984) assumes that events leading to the origin of species are different from those that lead to the origin of genera, the former involving changes without leading to a divergent genome, and the latter involving addition, substitution, deletion, or complete transformation of a genome. However, such a difference between the origin of species and genera does not consistently exist. Moreover, as was pointed up by Kellogg (1989), genomes are not arbitrary divisions, and therefore, genome-based classifications exhibit the same weakness as the biological species classification approach.

The generic classification of the Triticeae is currently in a state of flux, and the eye of critical disputes, whose outcomes are difficult to predict. The genome-based classification, although attractive in theory, is often incongruent with morphological observations and cannot serve, by itself, a basis for classification. Consequently, in this book, the genera were classified according to Melderis (1978), Melderis et al. (1980), Clayton and Renvoize (1986), and Renvoize and Clayton (1992) who based their classification on morphological and eco-geographical characteristics (Table 2.2). Likewise, the nomenclature used in this book is as per Melderis et al. (1980); Clayton and Renvoize (1986); Van Slageren (1994); and (Table 2.2).

There are three complexes of genera that are especially difficult to classify: the complex *Triticum-Aegilops*, the complex *Elymus-Agropyron-Sitanion-Hystrix*, and the complex *Hordeum–Hordelymus*.

Triticum-Aegilops—The classification of these genera is still under debate. Bowden (1959), Morris and Sears (1967) and Kimber and Feldman (1987) included the two genera in one genus. *Triticum* was included because tetraploid wheat contains one subgenome that derived from an *Aegilops* species and hexaploid wheat contains two such subgenomes. This classification was criticized by Gupta and Baum (1986) and was not accepted by most taxonomists who dealt with this group. Taxonomists, such as Zhukovsky (1928), Eig (1929a), Bor (1968), Melderis (1980), Hammer (1980) and van Slageren (1994), objected to the unification of these two genera, since the morphological separation between them is distinct, in sharp contrast to the genetic relationships uniting most of the genera in the tribe.

Elymus-Agropyron-Sitanion-Hystrix—Nevski (1934) Melderis (1953), Bor (1968) and Sakamoto (1974) regarded these taxa as separate genera, while Runemark and Heneen (1968) united the four genera into one genus *Elymus*. In Melderis' (1980) definition of the genus *Elymus*, only the genus *Sitanion*, the sections *Roegneria* and *Elytrigia* from *Agropyron* and the species of *Clinelymus* were included. While they did not mention the genus *Asperella*, Runemark and Heneen (1968) regarded this genus close to *Elymus*. *Elymus* and *Leymus* are contiguous genera, whose separation has been contentious (Clayton and Renvoize 1986). Melderis separated species from

Elymus to *Leymus* and left the species of *Agropyron* in the genus *Agropyron*. *Hystrix* (including *Asperella*) and *Sitanion* share a genome with *Elymus* and can be considered related to the latter (Clayton and Renvoize 1986).

The number of spikelets on each rachis node served as the marker that distinguished between *Agropyron* and *Elymus* (in *Agropyron* there is one spikelet and in *Elymus* there are several), but this marker is valueless, since in several species of *Agropyron*, there are pairs of spikelets in some of the rachis nodes, while several species of *Elymus* sometimes display solitary spikelets on each rachis node. In the new classification, Melderis (1980) left only species with solitary spikelet on each rachis node (species containing the P genome) in the genus *Agropyron* and transferred the remaining species (i.e., *elongatum* and *junceum* and related species with the Ee or Eb genomes) to *Elymus*.

Hordeum-Hordelymus—the classification of these genera is less complex and based mainly on morphological traits. Runemark and Heneen (1968) separated the two genera, as well as *Critesion* from *Hordeum*. In contrast, Melderis (1953, 1980) included *Critesion* in the genus *Hordeum* and left *Hordelymus* as a separate genus.

The various genera of the Triticeae tribe are classified into two sub-tribes: Hordeineae, the barley lineage, (seven genera) and the Triticineae, the wheat lineage, (eleven genera) (Table 2.1 and Fig. 2.1). The genus *Brachypodium* P. Beauv diverged approximately 32–39 MYA from the pre-Triticeae lineage (Bossolini et al. 2007; International Brachypodium Initiative 2010). In spite of its morphological intergradation with *Elymus serpentini* (Renvoize and Clayton 1992), this genus is not included in the Triticeae (Catalan et al. 1995; Hasterox et al. 2004). Despite the early proposal to combine *Brachypodium* (e.g., Krause 1913) or *Bromus* (e.g., Avdulov 1931) with Triticeae, Hubbard (1948) noted the morphological similarity between all three, without actually merging them. With regard to morphology as well as other aspects, however, it seems best to keep the three separate (Macfarlane and Watson 1982). *Bromus* is the most obviously distinct, differing from the other two in several important traits. *Brachypodium* scarcely differs from Triticeae in morphology, with the exception of the racemose (rarely paniculate) inflorescence.

2.3 Time of Origin of the Genera

The critical diversification of the tribe, mainly at the diploid level, started during the Oligocene (33.7–23.8 MYA), in the middle of the Tertiary (Fan et al. 2013; Table 2.5). During the Oligocene geological epoch, cold and dry climate prevailed in the temperate zone of the northern hemisphere of the old

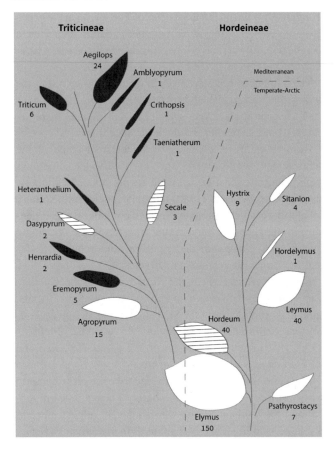

Fig. 2.1 Relationships of the genera of the Triticeae (modified from Clayton and Renvoize 1986 and from Feldman and Levy 2015. Genera with perennial species = white; genera with annual species = black; genera with perennial and annual species = patterened. The number of species is written below the genus name. The genera *Taeniatherum* and *Crithopsis* that were placed in the Hordeineae for the presence of two spikelets on each rachis node, were transferred to the Triticineae because of molecular data showing greater similarity to *Triticum* than to *Hordeum*. (For the divergent time of the main groups see Table 2.6)

facilitated the expansion of grasslands and brought about the development of the annual, autogamous species.

It is therefore assumed that the radiation of the Triticeae might have been triggered by the late Miocene climate. This is in accordance with the findings of Fan et al. (2013 and reference therein), which demonstrated that a major radiation of the Triticeae occurred during a relatively narrow period of time in the late Miocene (9.2–6.1 MYA). Diversification in the Mediterranean lineage of Triticeae not only stimulated the formation of many new genera, but also provided the opportunity for the production of many allopolyploids (Fan et al. 2013).

The wheat and barley lineages diverged from one another during the Miocene, about 15–8 MYA (Wolfe et al. 1989; Ramakrishna et al. 2002; Huang et al. 2002a, 2002b; Dvorak and Akhunov 2005; Chalupska et al. 2008; Fan et al. 2013; Middleton et al. 2013, 2014; Marcussen et al. 2014; Gornicki et al. 2014) (Table 2.6). Based on analysis of 275 single nuclear gene copies of hexaploid wheat, Marcussen et al. (2014) concluded that the ancestors of the wheat group diverged from rye during the Miocene, about 7 MYA, while on the basis of chloroplast DNA analysis, Middleton et al. (2014) suggested that this divergence occurred in the Pliocene, 3–4 MYA. The diploid *Triticum* and *Aegilops* species began to diverge from one another during the late Miocene (7.0–5.3 MYA) and continue to diverge in the Pliocene, about 4–2 MYA (Huang et al. 2002b; Dvorak and Akhunov 2005; Middleton et al. 2014), and possibly also in the Pleistocene (1.8–0.01 MYA). The allopolyploids of this group were formed 1.0–0.01 MYA, during the Pleistocene and Holocene epochs (Huang et al. 2002b; Dvorak and Akhunov 2005; Marcussen et al. 2014; Gornicki et al. 2014; Tables 2.5 and 2.6).

2.4　Phylogenetic Studies in Perennial Diploid Species

There is a general agreement that the Triticeae tribe is monophyletic (Watson et al. 1985; Kellogg 1989; Soreng et al. 1990; Hsiao et al. 1995a). Over the years, chromosome pairing was analyzed in several inter-specific and inter-generic hybrids between perennial diploid Triticeae, but yielded, in several cases, ambiguous results (Wang 1989). By large, studies assessing the relationships among basic genomes of perennial diploid Triticeae species have been insufficient and lag far behind those of annual species (Wang and Lu 2014). Hence, phylogenetic studies in these taxa were mainly based on a variety of morphological and molecular analyses (Seberg and Petersen 2007; Escobar et al. 2011). Early studies on the basis of morphological characteristics led to the construction of several

world, triggering the transformation of vegetation to something similar to that of today. The cold climate channeled the evolution of the tribe toward the development of genera that are adapted to mesophyllic habitats, i.e., the early Triticeae that grew in the temperate-arctic zones. Later on, during the Miocene era (23.8–5.3 MYA), warmer climates prevailed, but towards the end of this geological epoch the Tethys Sea, that covered large area of southwest Asia, disappeared and the east Mediterranean region rose. The climate in the east Mediterranean and central Asia became seasonal (cold and humid in the winter and hot and dry in the summer). This climate change, characterized by a relatively short growth period in the winter and long drought in the summer, led to the development of diversified ecosystems and consequently, the opening of new ecological niches. Such changes phylogenetic trees (Baum 1982, 1983; Baum et al. 1987; Kellogg 1989; Frederiksen and Seberg 1992) that were reviewed and discussed by Seberg and Frederiksen (2001). Phylogenetic trees have also been constructed based on several types of molecular analyses, namely, isozymes (McIntyre, 1988), restriction site data (Monte et al.1993; Mason-Gamer and Kellogg 1996a), sequence data from a number of different coding and/or non-coding regions, viz. 5S DNA (Kellogg and Appels 1995), internal transcript sequences (ITS) (Hsiao et al. 1995a, b), *rpoA* a plastid gene (Petersen and Seberg 1997), and *waxy,* a low-copy nuclear gene (Mason-Gamer et al. 1998). However, incongruence exists between the phylogenetic trees constructed based on morphology and on molecular data, as well as between trees constructed using various kinds of molecular data (Seberg and Petersen 2007; Escobar et al. 2011). Trees constructed using a combination of morphological information and several types of molecular data did not yield satisfactory results either. For example, Seberg and Petersen (2007) combined morphological data with nucleotide sequence data from five different genes, two chloroplast genes, a mitochondrial gene, and two single-copy nuclear genes. With the exception of the mitochondrial and nuclear sequences, the data of the morphological and molecular analysis were incongruent.

Most morphological and molecular trees shared the *Aegilops* clade, while the morphological trees also included *Amblyopyrum* and *Henrardia* in the *Aegilops* clade (Seberg and Petersen 2007). The most striking difference between the morphological and molecular trees was the position of *Hordeum senso lato* and *Psathyrostachys*. *Triticum monococcum* is included in the *Secale* clade according to several morphological trees, but is a sister clade to the *Aegilops* clade in some molecular studies (Kellogg and Appels 1995; Mason-Gamer and Kellogg 1996a). Hsiao et al. (1995a, b) and Kellogg et al. (1996) considered *T. monococcum* to be the sister group to *Elymus elongatus* (formerly *Agropyron elongatum*). In an attempt to shed more light on the Triticeae phylogeny, Mason-Gamer (2001, 2005), Petersen and Seberg (2002), and Helfgott and Mason-Gamer (2004) studied single-copy nuclear genes, and Petersen and Seberg (1997) and (Yamane and Kawahara (2005) studied chloroplastic genes; these studies failed to lead to any consensual definition of clades.

The fact that no consensus was reached concerning the phylogenetic relationships between the various diploid taxa of the tribe, was either due to a limited number of samples (Kellogg and Appels 1995; Kellogg et al. 1996; Mason-Gamer and Kellogg 1996a; Escobar et al. 2011) or to the small number of genes that were analyzed (Hsiao et al. 1995a, b; Kellogg and Appels 1995; Petersen and Seberg

1997; Helfgott and Mason-Gamer 2004; Mason-Gamer 2005). This ambiguity of the phylogenetic relationships in the Triticeae is also result from extensive inter-generic hybridizations and introgression events, as well as to incomplete lineage sorting of ancestral polymorphisms, indicating an intricate, reticulate pattern of evolution in this tribe (Kellogg 1996; Komatsuda et al. 1999; Nishikawa et al. 2002; Mason-Gamer 2005; Kawahara 2009; Escobar et al. 2011). Such a reticulate pattern of evolution presents a considerable challenge in phylogenetic analyses since different genes may exhibit conflicting genealogical histories (Escobar et al. 2011).

A recent attempt to overcome this obstacle in the Triticeae was made by Escobar et al. (2011), who used a molecular dataset, including one chloroplastic and 26 nuclear genes to (i) test whether it is possible to infer phylogenetic relationships in the face of large-scale introgressive events and/or incomplete lineage sorting, (ii) identify parts of the evolutionary history that have not evolved in a tree-like manner, and (iii) decipher the biological causes of gene-tree conflicts in this tribe, including the frequency of recombination, chromosomal location of analyzed sequences and evolution rate. They showed that combining information from several loci located on different chromosomes and in different cellular compartments (nucleus and chloroplast), enabled the identification of major clades and resolved most parts of the Triticeae phylogeny. Escobar et al. (2011) succeeded to construct a comprehensive, multigenic phylogeny of the diploid taxa of Triticeae, and to identify the biological groups that most likely underwent reticulate evolution. Their phylogenetic hypotheses suggested the existence of 5 major clades within Triticeae, with *Psathyrostachys* and *Hordeum* being the deepest genera, and *Psathyrostachys* branches being sister to the remaining Triticeae which followed by the sequential branching of *Hordeum* (Fig. 2.2). The divergence of these two taxa occurs in a tree-like manner. The suggestion that *Psathyrostachys* is the deepest taxon in the tribe is in accordance with the assumption that this species resembles the primitive Triticeae (Sakamoto 1973). Several previously published phylogenies, including some derived from nuclear (Mason-Gamer 2001; Kellogg and Appels 1995; Petersen et al. 2006) and chloroplastic DNA analyses (Petersen and Seberg 1997; Mason-Gamer et al. 2002), also recognized the early divergence of *Psathyrostachys* and *Hordeum*, although several other studies disagreed (Mason-Gamer 2005; Helfgott and Mason-Gamer 2004; Kellogg and Appels 1995; Hsiao et al. 1995a, b). Nevertheless, the data of Escobar et al. (2011) strongly reinforce the suggestion that *Psathyrostachys* is the sister group of all other Triticeae.

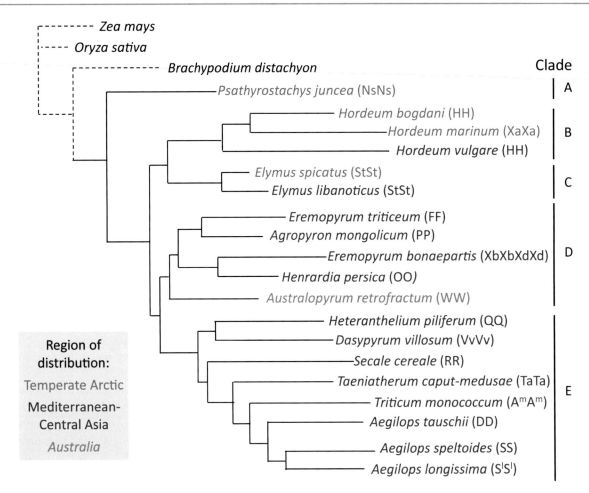

Fig. 2.2 Phylogenetic tree of the Triticeae genera adapted from Escobar et al. (2011)

These major clades were also recently defined using the nuclear *phosphoglycerate kinase* (*PGK*) gene that codes for plastid *PGK* isozyme (Adderley and Sun 2014). Bieniek et al. (2015), who studied phylogenetic relationships among the Triticeae diploid species through DNA barcoding of three chloroplastic genes, provided support of the classification into the above clades. They also revealed a close relationship between the *Elymus* (=*Pseudoroegneria*) and the *Taeniatherum* clades, as well as a clear distinction between the *Psathyrostachys* and the *Hordeum* clades, which is consistent with previous molecular studies (Petersen and Seberg 1997; Escobar et al. 2011; Fan et al. 2013). The results of Bieniek et al. (2015) showed the *Hordeum* clade to be an independent group. Moreover, their analysis separated the three studied genomes of the *Hordeum* clade, namely, I (in diploid *H. bogdani*), H (in diploid *H. bulbosum*) and Xu (in diploid *H. murinum*). These results are consistent with previous studies of chloroplast and nuclear DNA sequences (e.g., Blattner 2009; Naghavi et al. 2013).

The *Elymus* clade is the largest clade, consisting of representatives of the St and E genomes (Bieniek et al. 2015). It contains the diploid species of the genus *Elymus* senso lato,

(Melderis et al. 1980), namely, species of *Pseudorogneria* (genome St), *Agropyron elongatum* (=*Lophopyrum elongatum*) (genome Ee) and *Agropyron junceum* [=*Tinopyrum bessarabicum* (genome Eb).

Löve (1984) used the genome symbols J for the genome of *Elymus farctus* (=*Thinopyrum bessarabicum; Agropyron junceum*), and E for that of *Elymus elongatus* (=*Lophopyrum elongatum; Agropyron elongatum*), whereas Dewey (1984) considered the J and E genomes as the same basic genome. Studies of chromosome pairing in hybrids between these species supported Dewey's conclusion (Wang 1985b; Wang and Hsiao 1989), whereas Jauhar (1988) reached different conclusion by studying chromosome pairing in the same plant materials. Since a literature review indicated that most studies regard J and E genomes as members of the same cluster (see Table 1 in Wang and Lu 2014), it is now generally accepted to regard them as very closely related genomes, supporting the use of a common basic genome symbol, E (Seberg and Frederiksen 2001; Yen et al. 2005; Fan et al. 2007; Liu et al. 2008; Sha et al. 2010; Yan et al. 2011; Wang and Lu 2014). Thus, the genomes of *Elymus farctus* was designated Eb and that of *E. elongatus* Ee

(Wang et al. 1995; Table 2.4). Both Eb and Ee are close to subgenomes A, B, and D of the wheat group (reviewed in Wang and Lu 2014).

The St genome of diploid *Elymus* species and P of *Agropyron* are moderately related to Ee genome of *Elymus elongatus* and Eb genome of *E. farctus*, respectively (Wang 1989). Bieniek et al. (2015) found that the nucleotide sequences of the diploid Ee, Eb and St taxa are almost identical, with only one substitution within the *matK* gene differentiating genome Eb from the Ee and St genomes. Petersen and Seberg (1997) and Wang and Lu (2014) confirmed the very close relationship among the Ee, Eb and St genomes.

In addition, the multigenic network structure (Escobar et al. 2011) highlights parts of the Triticeae history that did not evolve in a tree-like manner; *Dasypyrum, Heteranthelium, Secale, Taeniatherum, Triticum* and *Aegilops* have evolved in a reticulated manner. Moreover, the results of Escobar et al. (2011) provided strong evidence of incongruence among single-gene trees, with different portions of the genome exhibiting different histories. They determined the role of recombination and gene location in the incongruence, and demonstrated that loci in close physical proximity are more likely to share a common history than distant ones, due to a low incidence of recombination in proximal chromosomal regions of Triticeae (Akhunov et al. 2003a, b; Luo et al. 2000, 2005; Lukaszewski and Curtis 1993).

In conclusion, the study of Escobar et al. (2011) showed that in spite of strong tree conflicts, not all clades of Triticeae are affected by introgression and/or incomplete lineage sorting. Notably, *Psathyrostachys, Hordeum* and *Agropyron, Eremopyrum* and *Henrardia* diverge in a tree-like manner, whereas the evolution of *Elymus, Dasypyrum, Heteranthelium, Secale, Taeniatherum, Triticum* and *Aegilops* is reticulated. There is no straightforward way to determine whether incongruence in Triticeae results from introgression or incomplete lineage sorting. In order to determine whether incongruence in Triticeae results from introgression or incomplete lineage sorting, whole genome sequence of most species of this tribe will be needed. Recombination could be an important evolutionary force in exacerbating the level of incongruence among gene trees.

2.5 Genome Analysis of Polyploid Species

The genomes of the diploid Triticeae are presented in Table 2.7 and those of the polyploid taxa in Table 2.8. At the diploid level, the genus *Aegilops* contains the largest number

Table 2.7 The genomes of the diploid Triticeae

Region of distribution	Clades					
	Psathyrostachys[a]	Hordeum[a]	Elymus[a]	Agropyron-Eremopyrum-Henrardia[a]	Dasypyrum-Secale-Heteranthelium	Aegilops-Triticum-Taeniatherum-Crithopsis
Temperate-arctic	NsNs	II	StSt			
		XaXa	LL			
		XuXu	XpXp			
Mediterranean-central Asiatic		HH	StSt	PP	VvVv	A^mA^m
			EeEe	WW	VbVb	AA
			EbEb	FF	RR	TT
				XbXb	QQ	SS
				XdXd		S^bS^b
				OO		S^lS^l
						S^hS^h
						S^sS^s
						DD
						CC
						MM
						NN
						UU
						TaTa
						KK

[a] According to Escobar et al. (2011)

Table 2.8 The genomes of the polyploid Triticeae

Region of distribu-tion	Ploidy type	Clades[a] containing polyploid species				
		Hordeum	Elymus	Agropyron-Eremopyrum	Dasypyrum	Aegilops-Triticum[b]
Temperate-arctic	Auto-polyploids	HHHH				
		XaXaXaXa				
		XuXuXuXu				
		HHHHHH				
	Allo-polyploids		StStYY			
			StStHH			
			LLEE			
			NsNsXmXm			
			NsNsXrXr		'	
			StSt HHYY			
			StStPPYY			
			StStWWYY			
	Auto-allo-polyploids		StStStStYY			
			StStStStHH			
			StStHHNsNsXmXm			
Mediterranean-central asiatic	Auto-polyploids	HHHH	StStStSt	PPPP	VbVbVbVb	
			EeEeEeEe	PPPPPP		
			EbEbEbEb			
			EeEeEeEeEeEeEeEeEeEe			
	Allo-polyploids		StStPP	XbXbXdXd		BBAA
			EeEeStSt	XdXdFF		GGAA
			EbEbEeEe			S^{l}S^{l}UU
						UUCC
						M^{o}M^{o}UU
						UUM^{o}M^o
						UUYY
						DDCC
						DDNN
						D^{c}D^{c}XX
						BBAADD
						GGAAA^{m}A^m
						D^{c}D^{c}XXDD
						D^{c}D^{c}XXUU
						D^{c}D^{c}XXS^{s}S^s
	Auto-Allo-polyploids		EeEeEeEeStSt			
			EbEbEbEbEeEe			

[a] According to Escobar et al. (2011)
[b] Genome symbol of allopolyploids of *Aegilops* and *Triticum* according to Dvorak (1998)

(10) of different genomes, *Elymus* contains five and *Hordeum* four different genomes. Several genera (*Taniatherum, Crithopsis, Secale. Heteranthelium, Amblyopyrum* and *Henrardia*) contain only one genome. The large number of genomes within *Aegilops* and *Elymus* indicates considerable genomic divergence within these two genera at the diploid level. Genus's age did not correlate with the extent of genomic divergence, as exemplified by the *Aegilops* species, which most of them are relatively young, while several species of *Elymus* and *Hordeum* are old. Similarly, no association was observed between the extent of divergence at the diploid level and the pattern of evolvement of the various species, since the monotypic genera (*Heteranthelium, Taeniatherum, Amblyopyrum,* and *Secale*) and those with a small number of genomes (*Triticum* and *Dasypyrum*) evolved in a reticulate pattern, just like the multi-genomic genus *Aegilops* (Escobar et al. 2011). The existence of four monotypic genera (*Taeniatherum, Crithopsis, Heteranthelium* and *Amblyopyrum*), three genera with a small number of species (*Triticum, Dasypyrum* and *Secale*) and the multi-genomic genus *Aegilops* (Table 2.7) in the Mediterranean-Central Asiatic region, indicates that the conditions that prevailed at the Pliocene era in this region (Table 2.5) drove this vast divergence. *Aegilops,* that diverged into at least ten different species, two species of *Elymus* containing the Ee and Eb genomes, and *Hordeum* species containing genome H, all evolved in the Mediterranean region. However, most of the genomes of *Hordeum* and *Elymus* clades evolved in the temperate-arctic region (Table 2.7).

Autopolyploids developed in four genera, whereas allopolyploids exist in seven genera. Three genera (*Hordeum, Agropyron* and *Dasypyrum*) comprise only autopolyploids, while *Elymus* contains autopolyploids, allopolyploids and auto-allohexaploids. *Elymus* and *Aegilops* have the largest number of genomes at both the diploid and allopolyploid levels. In *Elymus,* allopolyploids were formed via hybridization between species belonging to the same clade as well as between species of different clades (StStPP and StStHH), whereas, *Aegilops,* allopolyploids were formed solely between hybridization of species of the same clade. Most of the polyploids of the *Hordeum* and *Elymus* clades were formed in the temperate-arctic region, whereas those of the *Aegilops-Triticum* and *Agropyron–Eremopyrum* clades were formed in the Mediterranean–central Asiatic region (Table 2.8).

The genus *Elymus* sensu lato (Melderis et al. 1980; Clayton and Renvoize 1986; Renvoize and Clayton 1992) contains about 150 perennial species, including, among others, the traditional species of *Elymus* L., *Australopyrum* (Tzvelev) Löve, *Kengylia* Yen and Yang, *Lophopyrum* Löve, *Pseudoroegneria* (Nevski) Löve, and *Tinopyrum* Löve (Table 2.3). The genus consists of diploid, allo-, auto-, and auto-allopolyploid species. The diploid species contain the St, Ee, Eb, L, and Xp genomes, most of the allopolyploids share a common St subgenome in different combinations with subgenomes H, Y, P, W, and Ee, others have the genomic combination of EbEbEeEe, whereas the autotetraploids contain the Ee or Eb genomes (Table 2.8). In their search for the diploid contributors of genomes to allopolyploid *Elymus,* Sun et al. (2009), Hodge et al. (2010), Fan et al. (2012), and Bieniek et al. (2015) concluded that the diploid species of *Elymus* (=*Pseudoroegneria*) (genome StSt) is the maternal genome donor of allopolyploid *Elymus* (also including *Kengylia*). Study of cpDNA of species in the genus *Elymus* (Redinbaugh et al. 2000; Mahelka et al. 2011; Yan and Sun 2012; Mason-Gamer 2013; Adderley and Sun 2014; Dong et al. 2015) supports the finding that the chloroplast genome of the allopolyploids carrying the St subgenome was inherited from the diploid parent with the St genome. Hence, the allopolyploids of *Elymus* that contain the St subgenome presumably share very closely related nuclear and chloroplast St genomes. The minor differences among the St subgenomes of different allopolyploid *Elymus* species may have resulted from the process of allopolyploidization and inter-generic hybridization (Dong et al. 2015). Furthermore, Dong et al. (2015) data also suggest that diploid *Elymus* species (=Pseudoroegneria species) from central Asia and Europe are more ancient than those from North America. Consequently, they hypothesized that the *Elymus senso lato* species originated in central Asia and Europe and later spread to North America.

The H subgenome of *Elymus* is believed to derive from an unknown diploid species of *Hordeum,* the P subgenome from a diploid species of *Agropyron* (Petersen et al. 2011; Sun et al. 2009; Fan et al 2013; Dong et al. 2015) and the W subgenome from *Elymus* (=*Australopyrum*) (Petersen et al. 2011). The origin of the Y subgenome that is found in several allotetraploid and allohexaploid species of *Elymus* is still puzzling (Wang and Lu 2014). No putative Y genome diploids have been identified and several researchers assumed that the St and Y genomes of *Elymus* are related and may have originated from the same ancestral genome (Liu et al. 2006; Okito et al. 2009; Zhang et al. 2009; Dou et al. 2012). Alternatively, Sun et al. (2008), Sun and Komatsuda (2010), Yan et al. (2011), and Fan et al. (2012)

concluded that the Y subgenome is not close to the St genome and evolved from a different diploid ancestor. Moreover, Sun et al. (2008) found that the Y genome was close to other genomes in the genus, namely, W, P, and Eb/Ee genomes, whereas Sun and Komatsuda (2010) showed that the Y genome shared a common lineage with W and Eb genomes. In conclusion, the donor of the Y genome has not been confirmed yet (Wang and Lu 2014).

Elymus hispidus (=*Thinopyrum intermedium*; *Agropyron intermedium*) is an auto-allohexaploid with two closely related subgenomes and one more distant subgenome (Stebbins and Pun 1953; Dewey 1962). Liu and Wang (1993) designated EbEbEeEeSS as the genome constitution of this species (the S was later changed to St). The two closely related subgenomes are derived from diploid *Elymus farctus* (=*Th. bessarabicum*) (genome Eb) and diploid *E. elongatus* (=*Th. elongatum*) (genome Ee), respectively. The presence of the St genome of diploid *Elymus* (=*Pseudoroegneria*) in *Elymus hispidus* has been verified by all subsequent studies (Zhang et al. 1996, 1997; Chen et al. 1998; Tang et al. 2000; Kishii et al. 2005; Mahelka et al. 2011). The Xp genome of diploid *Elymus* sanctus (=*Peridictyon sanctum*; *Festucopsis*) was found to be close to the Y subgenome of the allohexaploid species *Elymus alatavicus* and *E. batalinii* (=species of *Kengyilia*) (genome StStPPYY), suggesting that the Y genome is the progenitor of the Xp genome (Fan et al. 2012).

The fact that most *Elymus* allopolyploids share the St genome (Table 2.8) eases inter-specific hybridizations between these allopolyploids. It thereby serves as the pivotal genome, assuring some fertility in the inter-specific hybrids and thus, enables the differential genome to undergo some recombination leading to the formation of recombinant, modified genomes (Zohary and Feldman 1962).

All the tetraploid species of the genus *Leymus* are allotetraploids (Zhang et al. 2006; Liu et al. 2008; Sha et al. 2008, 2010; Fan et al. 2009), whose genomes were designated NsNsXmXm (Wang et al. 1995). The Ns subgenome of *Leymus* derived from *Psathyrostachys* (Zhang and Dvorak 1991; Mizianty et al. 1999; Anamthawat-Jónsson and Bödvarsdóttir 2001; Culumber et al. 2011; Adderley and Sun 2014; Bieniek et al. 2015), but the identity of the donor of the Xm subgenome remains unclear (Guo et al. 2014). The studies of Fan et al. (2009) and Sha et al. (2010) indicated that the Xm subgenome may have originated from an ancestral lineage of *Agropyron* (genome P) and *Eremopyrum triticeum* (genome F). Further studies are required to determine

whether the Xm subgenome originated from the P- or the F-genomes (Wang and Lu 2014). Evidence for multiple origins of allotetraploid *Leymus* came from the work of Sha et al. (2010), who found that some *Leymus* species have the Xm and others have the Ns plasmon. Thus, *Leymus* species can be divided into two groups based on the maternal donors of the Ns cytoplasm versus the Xm cytoplasm.

The allotetraploid *Hystrix species* have the NsNsXmXm genome (Svitashev et al. 1998; Sha et al. 2010). Löve (1984) considered *Hordelymus* an allotetraploid species containing the H genome of *Hordeum* and the Ta genome of *Taeniatherum*. von Bothmer et al. (1994) ruled-out the presence of the H genome in *Hordelymus* and demonstrated the presence of the Ns genome instead, an indication that was supported by Petersen and Seberg (2008). The non-coding chloroplast *trnS-psbC* sequences of *Hordelymus* are similar to those of *Taeniatherum caput-medusae*, *Psathyrostachys juncea* and *Hordeum bogdani* (Ni et al. 2011). The study of Bieniek et al. (2015) revealed a close affinity between the chloroplast genomes of *Hordelymus* and the Ns-carrying taxa *Psathyrostachys* and *Leymus*, but not with *Taeniatherum* and *Hordeum*, supporting the presence of the Ns genome in *Hordelymus europaeus*. Their data indicated that *Psathyrostachys juncea* (genome NsNs) contributed its genome as a maternal parent to the allotetraploid *Hordelymus europaeus*. Wang et al. (1995) proposed to temporarily denote the *Hordelymus* genomes as XoXoXrXr, but recent findings (Ni et al. 2011; Bieniek et al. 2015) suggest a genomic formula of NsNsXrXr.

2.6 Evolutionary Trends

2.6.1 Steps in the Development of the Tribe

The Triticeae is a relatively young tribe in the family Poaceae, but very evolutionarily dynamic. Little is known about processes that led to the development of the various genera and speciation in the tribe. Runemark and Heneen (1968) and Sakamoto (1973) delineated five steps instrumental in the development of the various Triticeae genera: (1) evolution of diploid, allogamous, perennial genera in the temperate-arctic zones; (2) intensified autopolyploidization and allopolyploidization and geographical spread; (3) evolvement of diploid annual genera in the hot and dry summer area of the east Mediterranean and central Asiatic regions; (4) evolution of autogamous genera;

(5) intensified allopolyploidization of the autogamous annuals in the genera *Aegilops*, *Triticum*, and *Eremopyrum*. Since all the primitive genera distribute in the arctic-Temperate zone, this is probably the center of origin of the tribe, while the east Mediterranean-central Asia region is the center of variation.

It can be assumed that speciation processes advanced at different rates in different localities, in accordance with environmental pressures. In those cases where large environmental changes occurred, namely, in the Mediterranean and central Asia, evolution of the genera and species was more rapid than that in those cases where environmental changes were mild, i.e., in the temperate–arctic zone.

The Mediterranean and central-Asiatic climates are unstable, introducing continuously sizeable changes in many habitats. These incessant changes in the environmental conditions, resulting in the destruction of old habitats and formation of new ones, produced a pressure that accelerated speciation processes, leading to the evolution of new taxa. In reaction to these environmental unstable conditions, annualism and autogamy evolved, enabling rapid colonization of new habitats by new, more adapted, genotypes that were partially ecologically isolated from the progenitor genotypes. The great variation in the inflorescence traits and in seed dispersal techniques in the Mediterranean and central-Asiatic genera (*Aegilops*, *Heteranthelium*, *Eremopyrum Crithopsis*, *Taniatherum*, and *Henrardia*) reflect rapid adaptation to the wide radiation that occurred in the habitats of this region (Sakamoto 1973). The initial steps of such differentiation occurred at the diploid level. Afterward, this divergent evolution was accompanied by a convergent evolution, resulting from allopolyploidization of inter-generic and inter-specific hybrids, a process which has become an important factor in the evolution of the tribe. Allopolyploidization was followed by divergence at the polyploid level, and, on the other hand, it has facilitated considerable gene transfer between species and genera, further enhancing convergent evolution. Thus, the Triticeae species, mainly the Mediterranean and central Asiatic ones, developed in cycles of divergence and convergence.

It is important to note that in the diploid species of the Mediterranean-central Asiatic group, the ecological, physiological and genetic inter-specific barriers are relatively stronger than in the arctic-Temperate group. Hybrids between species that their genome is partially diverged, exhibit reduced pairing at meiosis and complete or partial sterility.

The evolutionary trends in the tribe are reflected in spike structure (erect or nodding, number of spikelets per node, fragile versus tough rachis, pedicel absence or presence, fragile versus tough rachilla, unawned or awned glume tip, back-rounded or keeled glume, caryopsis free or adherent to lemma), longevity (perennialism or annualism), pollination mode (cross or self), polyploidy (autopolyploidy or allopolyploidy), geographical distribution (temperate-arctic or Mediterranean-central Asiatic), and habitat (closed and humid or open and dry). The main evolutionary trends in the tribe are detailed below.

2.6.2 From Tall to Short Plants

A number of species are tall, with few tillers, while others are short, with many culms. In most species, the short plants have short spikes and the tall plants have long spikes. Tall plants with few culms are better adapted to wet and closed habitats, while short plants with many tillers are better adapted to open and dry habitats. The short stature with many tillers sustains moisture among the tillers and protects the plant from drying up by the winds, by preventing over-evaporation in dry habitats. Tall plants are considered more primitive than short plants. This trait is not uniform within genera and usually characterizes only single species. Several species of *Agropyron* and *Elymus* exhibit intra-specific variation in plant height. It is possible to consider the general tendency of transition from perennial, tall genera to annual, short ones (*Eremopyrum*, *Heteranthelium*, *Crithopsis*, specie of *Aegilops*).

2.6.3 From Perennialism to Annualism

Perennialism can be considered advantageous in stable habitats, where the water regime is not a limiting factor. In such habitats, it is more beneficial to the plant to invest less in seed dispersal mechanisms and in germination assurance and more in niche occupation for number of years and to propagate through vegetative reproduction. Thus, perennial species are usually less evolutionarily dynamic and occupy more well-defined and specialized habitats. This type of life cycle is prevalent in the temperate-arctic regions.

Annualism is considered a more efficacious means of addressing unstable and sometimes disturbed habitats. Thus, annual species are relatively uncommon in cool temperate regions and predominant in warm, dry ones, with seasonal rainfall (Raunkiaer 1934), like the Mediterranean and central Asiatic climate. Annualism is a life form more suitable to climate that is characterized by a long dry and hot season and unpredictable rainy winter. Annuals are efficient colonizer species that form dense populations. They prefer to invest efforts in the production of seeds, in their dispersal and in germination assurance. They do not proliferate vegetatively. Phylogenetic analyses indicate that annualism has evolved independently several times in the tribe (Kellogg 1989; Frederiksen and Seberg 1992).

Most Triticeae genera are uniform in their life form, with the exception of *Hordeum*, *Dasypyrum* and *Secale* (Table 2.3), that have both perennial and annual species. While the perennial species may be either cross- or self-fertilized, depending on the species, the annuals are almost exclusively facultative self-pollinated. The perennial species with rhizomes are, almost in all cases, self-incompatible and cross-fertilized.

2.6.4 From Allogamy to Autogamy

There are two types of pollination systems in the tribe, cross-pollination either obligatory or facultative, and self-pollination, with occasional cross-pollination. Rye (*Secale cereale*) is normally a cross-pollinated species and produces weak and abnormal offspring when forcibly inbred, while wheat is typically self-pollinated, with occasional cross-pollination. The two types of pollination systems do not randomly distribute among species. Most of the cross-pollinated plants are perennials, while most of the self-pollinated ones are annuals.

Most of the perennial species in the tribe reproduce through complete (self-incompatible) or facultative allogamy and many of them also by vegetative reproduction via rhizomes or bulbs. They usually have large anthers, with a considerable number of pollen grains that remain viable for a relatively long time, and ramified stigma, with a large surface to absorb the pollen spread by neighboring plants. Facultative autogamy mainly developed in the annual species in order to enable rapid fixation of adapted genotypes. These species have relatively small anthers, with small amounts of pollen grains, that lose their viability relatively rapid, and a short duration of floret opening.

2.6.5 From Simple to Improved Dispersal Units

In the Triticeae, as in many other tribes of the grass family, there is an inverse correlation between the persistence of and capacity for vegetative reproduction and the degree of sophistication of its seed dispersal mechanism (Stebbins 1950). Increased seed dispersal efficiency is through the development of specialized structures that occurred independently in a number of different evolutionary lines in this tribe.

In the Triticeae, some of the most conspicuous reproductive characteristics used for the separation of species and genera, consist of a series of various devices that enable more efficient seed dispersal. A principal change driving more efficient seed dispersal in the tribe was the development of awns on the end of the glumes and lemmas, that either burry the grain in the soil, through movements of the awns due to changes in humidity (Elbaum et al. 2007) or help in dispersing the seeds by clinging to various parts of animals. The second change was the in development of spike tendency to disarticulate into spikelets at each node of the rachis.

The primitive genera of this tribe (*Elymus* and *Agropyron*) lack these specializations. *Hordeum* developed accessory awns for more efficient seed dispersal, through the sterilization of two of the three spikelets at each node of the spike and reduction of their glumes and lemmas to prolonged awns. Another line that developed a similar type of specialization is represented by the genus *Aegilops*. In this genus, the accessory awns develop through prolongation of the nerves of the glumes or lemmas.

Several perennial species from the temperate region have a tough, non-fragile rachis and a dispersal unit consisting of a single grain. Such dispersal units that fall on the ground without any protection or burying mechanism, exist in species of *Elymus*, *Leymus* and *Agropyron*. In other perennial species, the dispersal unit is somewhat more advanced, where inflorescences disarticulate to single spikelets that can bury themselves in the soil (species of *Elymus*, *Agropyron*, *Hordeum*, *Psathyrostachys* and *Dasypyrum*). Development of awns on the lemmas have assisted in burying and protecting grains from drought, winds, burning and various herbivores. The Mediterranean-central Asiatic annual species have more advanced dispersal units, where exhibit an arrowhead shape or barrel shape, with a segment of the rachis in its base or side and awns in its upper part. Such

Fig. 2.3 Spikes and dispersal units in the Triticineae: **a** Wedge; **b** Barrel; **c** Umbrella. Details in the text

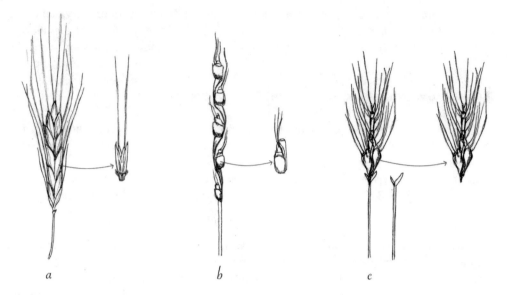

<table><tr><td>*a*</td><td>*b*</td><td>*c*</td></tr></table>

dispersal units have several variations that presumably developed independently in the different annual genera. In some of the genera, the rachis disarticulates above the spikelets (Wedge type; *Triticum*, species of *Aegilops*, *Amblyopyrum*, *Secale*, *Eremopyron*, *Henrardia*). In several species of *Aegilops*, the rachis disarticulates below the spikelets (barrel-type), and in other species of *Aegilops*, the rachis disarticulates below the spike and the entire spike serves as the dispersal unit, with several awns in its head (umbrella-type) (Fig. 2.3). The Umbrella type may consist of a long spike with many spikelets or short and a compact spike with one to two small spikelets at the top. In several genera, the wedge-type dispersal unit carries several spikelets on one rachis segment (e.g., *Hordeum*), while in others, it carries both fertile and sterile spikelets (*Heteranthelium*).

The type of the dispersal unit is usually fixed in the genus and in each species (except for *Aegilops speltoides* that has both wedge type and umbrella type dispersal units). The genus *Aegilops* is unique in that its species exhibit a variety of dispersal units. Another trend occurring in the dispersal units of the tribe is the transition from a grain joined with the palea, to a free grain.

2.6.6 Changes in Spikes Make-Up

Changes in spike structure occurred as follows:

(a) From multi-florets spikelets to spikelets with only a few florets—improvement of pollination modes and seed-dispersal apparatuses was followed by a reduction in the number of flowers in the spikelets. This reduction occurred in the annual genera and in perennial genera featuring spikes that disarticulate into spikelets at maturity. This is manifested by the transition from multi-florets spikelets to spikelets with one, two or three florets, or by the degeneration of florets or spikelets in the spike. In the self-pollinating annual genera, the upper florets are androgenic and in some (in *Heteranthelium* and the *Aegilops* species with the umbrella-type dispersal unit), the upper spikelets also degenerate and bear more awns were added to the dispersal unit.

(b) From two to several spikelets on each rachis node to solitary spikelet on each node—this trait is not significant in the evolution of the tribe since the number of seeds in the seed dispersal unit is not determined but the number of spikelets on each node but by the number of spikelets in dispersal unit. However, reduction in the number of spikelets per node increased the efficiency of the dispersal unit in burrowing the grain in the soil. In a primitive genus like *Agropyron* there is a reduction in the number of spikelets per node even though it is meaningless from evolutionary point of view since their dispersal unit is a single floret.

(c) Increase in grain size—independent of the reduction in the number of seeds per spike, spikelets or dispersal unit, the more developed species grain size tended to increase on the account of grain count. Large grains have an advantage in assuring rapid and successful germination as well as successful competition with other seedlings in the Mediterranean or sub-Mediterranean dwarf-shrub formations. Large seeds may be advantageous in dispersal unit burying in the soil.

2.6.7 From Symmetric to Asymmetric Karyotype

The prototype of the tribe had large chromosomes and a symmetric karyotype, i.e., all centromeres were median or sub-median. All the genera in the sub-tribe Hordeineae have a symmetric karyotype, while the advanced genera in the sub-tribe Triticineae have an asymmetric karyotype either with small chromosomes (*Eremopyrum* and *Henrardia*) or with large chromosomes (the advanced species of *Aegilops*, namely, *caudata*, *comosa*, *uniaristata* and *umbellulata*). Increasing asymmetry may result from pericentric inversion (the inverted segment includes the centromere), from extra-radial intra-chromosomal translocation, where a chromosomal segment changes its position from one chromosomal arm to the other, and from inter-chromosomal translocation, either by transposition or unequal reciprocal translocation of a portion of the chromosome arm. The evolutionary advantage of an asymmetric karyotype consists in successful maintenance of linked-gene combinations (Stebbins 1971) and to some extent, genetic isolation of the species from species with a symmetric karyotype.

2.6.8 From Diploidy to Polyploidy

As most genera contain diploid species that presumably derived from ancestral taxa of the genus (Sakamoto 1973), the first steps of differentiation in the tribe are assumed to have occurred, in most genera, on the diploid level. The diploids of the genera that distribute in the temperate-arctic regions contain the basic genomes of the tribe (e.g., genomes St, E, I, and others).

Polyploidy enables to overcome the isolating barriers between species, to fix heterotic interactions and to create new intraspecific-polyploid cytotypes or allopolyploid species. Polyploidy can be considered as superimposed on the main evolutionary routes of the tribe that relate to life forms, reproductive systems and dispersal system. Polyploidy can be of the autopolyploidy type, which originates through chromosome doubling of an intra-specific hybrid or a diploid taxon, and consequently, contains multiples of

the same or very similar genomes. Alternatively, polyploidy can be of the allopolyploidy type, which originates from chromosome doubling of an inter-specific or inter-generic hybrid and thus, contains two or more dissimilar subgenomes.

2.6.9 Differences in Genome Size Between Diploid Triticeae Species

There are considerable differences in genome size between the diploid species of the Triticeae (Table 2.4; Fig. 2.4). Whereas diploid species of *Hordeum* display no change in genome size, several other groups show an increase in DNA amount in the more advanced taxa. Thus, whereas *Elymus* diploids containing the St genome have a relatively low amount of DNA (1C DNA ranges from 4.0 to 4.9 pg) the diploids of *E. elongatus* (genome Ee) and *E. farctus* (genome Eb) have higher DNA content (5.6–6.1 and 7.4 pg, respectively). Moreover, the P genome of diploid

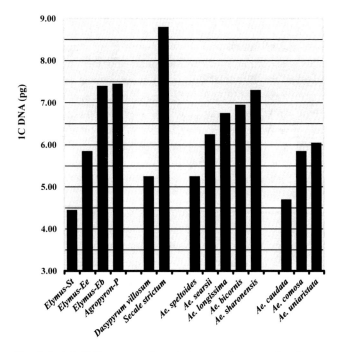

Fig. 2.4 1C DNA content (in pg) in several ancestral and more recently evolved diploid species of the Triticineae, showing increase in DNA content in younger species See Table 2.4 for the source of the data

Agropyron species, which is moderately related to the genomes of diploid *Elymus* (Wang 1989; Bieniek et al. 2015), has a relatively large DNA content (1C DNA ranges from 7.1 to 7.8 pg). Likewise, the genome of *Dasypyrum villosum* is relatively small (5.2–5.3 pg) while those of *Secale* species are large (1C DNA is more than 8 pg). Similar trend exists in *Aegilops* section Sitopsis; the genome of *Ae. speltoides*, the basal species of the section, is relatively small (1C DNA is 5.1–5.4 pg) and those of the other Sitopsis species, namely, *Ae. bicornis*, *Ae. searsii*, *Ae. sharonensis* and *Ae. longissima*, are relatively large (1C DNA content ranges from 5.9 to 7.5 pg). Also, the genome of *Ae. caudata* is small (1C DNA = 4.6–4.8 pg) while those of the more advanced species, *Ae. comosa* and *Ae. uniaristata* are larger (1C DNA is 5.8–6.2 in *Ae. comosa* and 5.8–6.3 in *Ae. uniaristata*).

The DNA fraction responsible for the increase in DNA content is the repetitive DNA, which, in the Triticeae species, consists primarily of transposable elements (TEs) (Kumar and Bennetzen 1999; Vicient et al. 2001; Sabot et al. 2005; Senerchia et al. 2013). TEs, whose activation may be induced by genetic and environmental stresses, have the potential to affect genome size, structure and function through transposition, ectopic recombination and epigenetic re-patterning (Fedoroff 2012). As such, TE are major contributors to genome plasticity and divergence, to genetic diversity and speciation (Bariah et al. 2020). The large differences in genome size between the various diploid species of the Triticeae (Table 2.4 and Fig. 2.4), indicate that encountering environmental challenges such as the unstable climate of the Mediterranean region, may have led to bursts of TEs and consequently, to gain of nuclear DNA that may have played a significant role in the speciation of several Triticeae species, mainly those belonging to the sub-tribe Triticineae.

References

Adderley S, Sun G (2014) Molecular evolution and nucleotide diversity of nuclear plastid phosphoglycerate kinase (PGK) gene in Triticeae (Poaceae). Gene 533:142–148

Akhunov ED, Akhunova AR, Linkiewicz AM, Dubcovsky J, Hummel D, Lazo G, Chao SM, Anderson OD, David J, Qi LL, Gerard G et al (2003a) Synteny perturbations between wheat homoeologous chromosomes caused by locus duplications and deletions correlate with recombination rates. Proc Natl Acad Sci USA 100:10836–10841

Akhunov ED, Goodyear AW, Geng S, Qi LL, Echalier B, Gill BS, Miftahudin J, Gustafson JP, Lazo G, Chao SM et al (2003b) The organization and rate of evolution of wheat genomes are correlated with recombination rates along chromosome arms. Genome Res 13:753–763

Alonso LC, Kimber G (1981) The analysis of meiosis in hybrids. ii. triploid hybrids. Can J Genet Cytol 23:221–234

Anamthawat-Jónsson K, Bödvarsdóttir SK (2001) Genomic and genetic relationships among species of Leymus (Poaceae: Triticeae) inferred from 18S–26S ribosomal genes. Amer J Bot 88:553–559

Anonymous (1986) Genetic resources in the Triticeae. FAO/IBPGR Plant Genet Resour Newslett 64:2–4

Avdulov N (1931) Karyo-systematisce Untersuchung der Familie Gramineen. Bull Appl Bot Genet PL Breed, Leningrad, Suppl 43:1–428

Bariah I, Keidar-Friedman D, Kashkush K (2020) Where the wild things are: transposable elements as drivers of structural and functional variations in the wheat genome. Front Plant Sci 18:585515

Barkworth ME (1992) Taxonomy of the Triticeae: a historical perspective. Hereditas 116:1–14

Baum BR (1982) Cladistic analysis of Triticeae by means of Farris's distance Wagner procedure. Can J Bot 60:1194–1199

Baum BR (1983) A phylogenetic analysis of the tribe Triticeae (Poaceae) based on morphological characters of the genera. Can J Bot 61:518–535

Baum BR, Estes JR, Gupta PK (1987) Assessment of the genomic system of classification in the Triticeae. Amer J Bot 74:1388–1395

Bieniek W, Mizianty M, Szklarczyk M (2015) Sequence variation at the three chloroplast loci (*matK, rbcL, trnH-psbA*) in the Triticeae tribe (Poaceae): comments on the relationships and utility in DNA barcoding of selected species. Pl Syst Evol 301:1275–1286

Blattner ER (2004) Phylogenetic analysis of *Hordeum* (Poaceae) as inferred by nuclear rDNA ITS sequences. Mol Phylogenet Evol 33:289–299

Blattner FR (2009) Progress in phylogenetic analysis and a new infrageneric classification of the barley genus *Hordeum* (Poaceae: Triticeae). Breed Sci 59:471–480

Blattner FR (2018) Taxonomy of the genus *Hordeum* and Barley (*Hordeum vulgare*). In: Stein N, Muehlbauer GJ (eds) The barley genome, compendium of plant genomes, https://doi.org/10.1007/978-3-319-92528-8_2

Bor NL (1968) Gramineae. In: Towsend CC, Guest E, El-Rawi A (eds) Flora of Iraq, vol 9, pp 210–263

Bossolini E, Wicker T, Knobel PA, Keller B (2007) Comparison of orthologous loci from small grass genome *Brachypodium* and rice: implications for wheat genomics and grass genome annotation. Plant J 49:704–717

Bowden MW (1959) The taxonomy and nomenclature of the wheats, barleys and ryes and their wild relatives. Can J Bot 37:657–684

Catalán P, Shi Y, Amstrong L, Draper J, Stace CA (1995) Molecular phylogeny of the grass genus *Brachypodium* P. Beauv. based on RFLP and RAPD analysis. Bot J Linnean Soc 117:263–280

Cauderon Y (1986) Cytogenetics in breeding programmes dealing with polyploidy, interspecific hybridizationand introgression. In: Horn W, Jensen CJ, Odenbach W, Schieder O (eds) Genetic manipulation in plant breeding, proceedings international symposium organized by EUCARPIA, 8–13 Sept 1985, Berlin, Germany, Walter de Gruyter, Berlin, pp 83–104

Chalupska D, Lee HY, Faris JD, Evrard A, Chalhoub B, Haselkorn R, Gornicki P (2008) ACC homoeoloci and the evolution of wheat genomes. Proc Natl Acad Sci USA 105:9691–9696

Chapman V, Miller TE, Riley R (1976) Equivalence of the A genome of bread wheat and that of *Triticum urartu*. Genet Res 27:69–76

Charpentier A, Feldman M, Cauderon Y (1986) Genetic control of meiotic chromosome pairing in tetraploid *Agropyron elongatum*. I. Pattern of pairing in natural and induced tetraploids and in F$_1$ triploid hybrid. Can J Genet Cytol 28:783–788

Charpentier A, Cauderon Y, Feldman M (1988) The effect of different doses of *Ph1* on chromosome pairing in hybrids between tetraploid *Agropyron elongatum* and common wheat. Genome 30:974–977

Chen KC, Dvorak J (1984) The inheritance of genetic variation in *Triticum speltoides* affecting heterogenetic chromosome pairing in hybrids with *Triticum aestivum*. Can J Genet Cytol 26:279–287

Chen Q, Conner RL, Laroche A, Thomas JB (1998) Genome analysis of *Thinopyrum intermedium* and *Thinopyrum ponticum* using genomic in situ hybridization. Genome 41:580–586

Clausen J, Keck DD, Hiesey WM (1945) Experimental studies on the nature of species. II. Plant evolution through amphiploidy and autoploidy with examples from the Madiinae. Publ Carnegie Inst Wash 564, Washington, DC

Clayton SD, Renvoize SA (1986) Genera graminum. In: grasses of the world. Distributed for Royal Botanic Gardens, Kew bulletin. Additional series; 13, Kew, London, pp 1–389

Culumber CM, Larson SR, Jensen KB, Jones TA (2011) Genetic structure of Eurasian and North American *Leymus* (Triticeae) wildryes assessed by chloroplast DNA sequences and AFLP profiles. Pl Syst Evol 294:207–225

Dewey DR (1962) The genome structure of intermediate wheatgrass. J Hered 53:282–290

Dewey DR (1982) Genomic and phylogenetic relationships among North American perennial Triticeae. In: Estes JR, Tyrl RJ, Brunken JN (eds) Grasses and grasslands: systematics and ecology. University of Oklahoma Press, Norman, pp 51–88

Dewey DR (1984) The genomic system of classification as a guide to inter-generic hybridization with the perennial Triticeae. In: Gustafson JP (ed) Gene manipulation in plant improvement. Plenum Press, New York, pp 209–279

Dong ZZ, Fan X, Sha LN, Wang Y, Zeng J, Kang HY, Zhang HQ, Wang XL, Zhang L, Ding CB, YangRW, Zhou YH (2015) Phylogeny and differentiation of the St genome in *Elymus* L. sensu lato (Triticeae; Poaceae) based on one nuclear DNA and two chloroplast genes. BMC Plant Biol 15:179

Dou QW, Lei YT, Li XM, Mott IW, Wang RRC (2012) Characterization of alien grass chromosomes in backcross derivatives of *Triticum aestivum*, *Elymus rectisetus* hybrids by using molecular markers and multi-color FISH/GISH. Genome 55:337–347

Dover GA, Riley R (1972a) Variation at two loci affecting homoeologous meiotic chromosome pairing in *Triticum aestivum* × *Aegilops mutica* hybrids. Nature New Biol 235:61–62

Dover GA, Riley R (1972b) Prevention of pairing of homoeologous meiotic chromosomes of wheat by an activity of supernumerary chromosomes of Aegilops. Nature 240:159–161

Dvorak J (1972) Genetic variability in *Aegilops speltoides* affecting homoeologous pairing in wheat. Can J Genet Cytol 14:371–380

Dvorak J (1976) The relationship between the genome of *Triticum urartu* and the A and B genomes of *Triticum aestivum*. Can J Genet Cytol 18:371–377

Dvorak J (1981a) Genome relationships among *Elytrigia* (= *Agropyron*) *elongata*, *E. stipifolia*, "*E. elongata* 4x," *E. caespitosa*, *E. intermedia* and "*E. elongata* 10x". Can J Genet Cytol 23:481–492

Dvorak J (1981b) Chromosome differentiation in polyploid species of *Elytrigia*, with special reference to the evolution of diploid-like chromosome pairing in polyploid species. Can J Genet Cytol 23:287–303

Dvorak J (1998) Genome analysis in the *Triticum-Aegilops* alliance. In: Slinkard AE (ed) Proceedings of 9th international wheat genetics symposium. University Extension Press, University of Saskatoon, Saskatoon, Saskatchewan, Canada, pp 8–11

Dvorak J, Akhunov ED (2005) Tempos of gene locus deletions and duplications and their relationship to recombination rate during diploid and polyploid evolution in the *Aegilops-Triticum* alliance. Genetics 171:323–332

Eilam T, Anikster Y, Millet E, Manisterski J, Feldman M (2007) Genome size and genome evolution in diploid Triticeae species. Genome 50:1029–1037

Eig A (1929a) Monographisch-Kritische Ubersicht der Gattung Aegilops. Reprium nov. Spec Regni Veg 55:1–288

Eig A (1929b) *Amblyopyrum* Eig. A new genus separated from the genus Aegilops. PZE Inst Agric Natural Hist Agric Res 2:199–204

Eilam T, Anikster Y, Millet E, Manisterski J, Feldman M (2008) Nuclear DNA amount and genome downsizing in natural and synthetic allopolyploids of the genera *Aegilops* and *Triticum*. Genome 51(8):616–627

Eilam T, Anikster Y, Millet E, Manisterski J, Feldman M (2009) Genome size in natural and synthetic autopolyploids and in a natural segmental allopolyploid of several *Triticeae* species. Genome 52:275–285

Elbaum R, Zaltzman L, Burgert I, Fratzl P (2007) The role of wheat awns in the seed dispersal unit. Science 316:884–886

Escobar JS, Scornavacca C, Cenci A, Guilhaumon C, Santoni S, Douzery EJ, Ranwez V, Glémin S, David J (2011) Multigenic phylogeny and analysis of tree incongruences in Triticeae (Poaceae). BMC Evol Biol 11:181–198

Fan X, Zhang HQ, Sha LN, Zhang L, Yang RW, Ding CB, Zhou YH (2007) Phylogenetic analysis among *Hystrix*, *Leymus* and its affinitive genera (Poaceae: Triticeae) based on the sequences of a gene encoding plastid acetyl-CoA carboxylase. Plant Sci 172:701–707

Fan X, Sha LN, Yang RW, Zhang HQ, Kang HY, Zhang L, Ding CB, Zheng YL, Zhou YH (2009) Phylogeny and evolutionary history of *Leymus* (Triticeae; Poaceae) based on a single-copy nuclear gene encoding plastid acetyl-CoA carboxylase. BMC Evol Biol 9:247

Fan X, Sha LN, Zeng J, Kang HY, Zhang HQ, Wang XL, Zhang L, Yang RW, Ding CB, Zheng YL, Zhou YH (2012) Evolutionary dynamics of the Pgk1 gene in the polyploid genus *Kengyilia* (Triticeae: Poaceae) and its diploid relative. PLoS ONE 7:e31122

Fan X, Sha LN, Yu SB, Wu DD, Chen XH, Zhuo XF, Zhang HQ, Kang HY, Wang Y, Zheng YL, Zhou YH (2013) Phylogenetic reconstruction and diversification of the Triticeae (Poaceae) based on single-copy nuclear *Acc1* and *Pgk1* gene data. Biochem Syst and Ecol 50:346–360

Fedoroff NV (2012) Transposable elements, epigenetics, and genome evolution. Science 338:758–767

Feldman M, Levy AA (2015) Origin and evolution of wheat and related Triticeae species. In: Molnar-Lang M, Ceoloni C, Dolezel J (eds) Alien introgression in wheat, cytogenetics, molecular biology, and genomics. Springer, Cham, Switzerland, pp 21–76

Feldman M, Lupton FGH, Miller TE (1995) Wheats. In: Smart J, Simmonds NW (eds) Evolution of crop plants, 2nd edn. Longman Scientific & Technical, pp 184–192

Feldman M, Liu B, Segal G, Abbo S, Levy AA, Vega JM (1997) Rapid elimination of low-copy DNA sequences in polyploid wheat: a possible mechanism for differentiation of homoeologous chromosomes. Genetics 147:1381–1387

Frederiksen S, Petersen G (1998) A taxonomic revision of *Secale* (Triticeae, Poaceae). Nord J Bot 18:399–420

Frederiksen S, Seberg O (1992) Phylogenetic analysis of the Triticeae (Poaceae). Hereditas 116:15–19

Gaut BS (2002) Evolutionary dynamics of grass genomes. New Phytol 154:15–28

Goldblatt P (1980) Polyploidy in Angiosperms: Monocotyledons. In: Lewis WH (ed) Polyploidy. Basic Life Sciences, vol. 13, Springer, Boston, MA

Gornicki P, Zhu H, Wang J, Challa GS, Zhang Z, Gill BS, Li W (2014) The chloroplast view of the evolution of polyploid wheat. New Phytol 204:704–714

Guo GY, Yang RW, Ding CB, Fan X, Zhang L, Zhou YH (2014) Phylogenetic relationships among *Leymus* and related diploid genera (Triticeae: Poaceae) based on chloroplast trnQ–rps16 sequences. Nordic J Bot 32:658–666

Gupta PK, Baum BR (1986) Nomenclature and related taxonomic issues in wheats, triticales and some of their wild relatives. Taxon 35:144–149

Gupta PK, Baum BR (1989) Stable classification and nomenclature in the Triticeae: desirability, limitations and prospects. Euphytica 41:191–197

Gupta PK, Fedak G (1985) Genetic control of meiotic chromosome pairing in the genus *Hordeum*. Can J Genet Cytol 27:515–530

Gupta PK, Fedak G (1986) The inheritance of genetic variation in rye (*Secale cereale*) affecting homoeologous chromosome pairing in hybrids with bread wheat (*Triticum aestivum*). Can J Genet Cytol 28:844–851

Hammer K (1980) Vorarbeiten zur monographischen Darstellung von Wildpflanzensortimenten: *Aegilops* L. Kulturpflanze 28:33–180

Han FP, Fedak G, Ouellet T, Liu B (2003) Rapid genomic changes in interspecific and intergeneric hybrids and allopolyploids of Triticeae. Genome 46:716–723

Hasterox R, Draper J, Jenkins G (2004) Laying the cytotaxonomic foundations of a new model grass, *Brachypodium distachyon* (L.) Beauv. Chromosome Res 12:397–403

Helfgott M, Mason-Gamer RJ (2004) The evolution of North American *Elymus* (Triticeae, Poaceae) allotetraploids: evidence from phosphoenolpyruvate carboxylase gene sequences. Syst Bot 29:850–861

Heneen WK, Runemark H (1972) Cytology of the *Elymus* (*Agropyron*) *elongatus* complex. Hereditas 70:155–164

Hodge CD, Wang H, Sun G (2010) Phylogenetic analysis of the maternal genome of tetraploid StStYY Elymus (Triticeae: Poaceae) species and the monogenomic Triticeae based on rps16 sequence data. Plant Sci 178:463–468

Hsiao C, Chatterton NJ, Asay KH, Jensen KB (1995a) Phylogenetic relationships of the mono genomic species of the wheat tribe, Triticeae (Poaceae), inferred from nuclear rDNA (internal transcribed spacer) sequences. Genome 38(2):11–223

Hsiao C, Chatterton CJ, Asay KH, Jensen KB (1995b) Molecular phylogeny of the Pooideae (Poaceae) based on nuclear rDNA (ITS) sequences. Theor Appl Genets 90:389–398

Huang S, Sirikhachornkit A, Su X, Faris JD, Gill BS et al (2002a) Genes encoding plastid acetyl-CoA carboxylase and 3-phosphoglycerate kinase of the *Triticum/Aegilops* complex and the evolutionary history of polyploid wheat. Proc Natl Acad Sci, USA 99:8133–8138

Huang S, Sirikhachornkit A, Faris JD, Su X, Gill BS et al (2002b) Phylogenetic analysis of the acetyl-CoA carboxylase and 3-phosphoglycerate kinase loci in wheat and other grasses. Plant Mol Biol 48:805–820

Hubbard CE (1948) Gramineae. In: Hutchinson J (ed) British Flowering Plants, pp. 248–348

International Brachypodium Initiative (2010) Genome sequencing and analysis of the model grass *Brachypodium distachyon*. Nature 463:763–768

Jackson RC (1982) Polyploidy and diploidy: new perspective on chromosome pairing and its evolutionary implications. Amer J Bot 69:1512–1523

Jackson RC (1984) Chromosome pairing in species and hybrids. In: Grant WF (ed) Plant biosystematics. Academic Press, Toronto

Jackson RC (1985) Genomic differentiation and its effect on gene flow. Syst Bot 10:391–404

Jakob SS, Meister A, Blattner R (2004) The considerable genome size variation of *Hordeum* species (Poaceae) is linked to phylogeny, life form, ecology, and speciation rate. Mol Biol Evol 21:860–869

Jauhar PP (1975) chromosome relationships between *Lolium* anf *Festuca* (Gramineae). Chromosoma 52:103–121

Jauhar PP (1988) A reassessment of genome relationships between Thinopyrum bessarabicum and T. elongatum of the Triticeae. Genome 30:903–914

Jauhar PP, Crane CF (1989) An evaluation of Baum et al'.s assessment of the genomic system of classification in the Triticeae. Amer J Bot 76:571–576

Jørgensen RB (1982) Biosystematics of *Hordeum bulbosum* L. Nordic J Bot 2:421–434

Kaul MLH, Murthy TGK (1985) Mutant genes affecting higher plant meiosis. Theor Appl Genet 70:449–466

Kawahara T (2009) Molecular phylogeny among *Triticum-Aegilops* species and of the tribe Triticeae. Breed Sci 59:499–504

Kellogg EA (1989) Comments on genomic genera in the Triticeae (Poaceae). Amer J Bot 76:796–805

Kellogg EA (1996) When the genes tell different stories: the diploid genera of Triticeae (Gramineae). Syst Bot 21:321–347

Kellogg EA, Appels R (1995) Intraspecific and interspecific variation in 5S RNA genes are decoupled in diploid wheat relatives. Genetics 140:325–343

Kellogg EA, Appels R, Mason-Gamer RJ (1996) When genes tell different stories: the diploid genera of Triticeae (Gramineae). Syst Bot 21:321–347

Kihara H (1944) Discovery of the DD-analyser, one of the ancestors of *Triticum vulgare*. Agric Hortic 19:13–14

Kimber G (1984) Evolutionary relationships and their influence on plant breeding. In: Gustafson JP (ed) Gene manipulation in plant improvement. Plenum, New York, pp 281–293

Kimber G, Feldman M (1987) Wild wheats: an introduction. Special Report 353. College of Agriculture, Columbia. Missouri, USA, pp 1–142

Kimber G, Tsunewaki K (1988) Genome symbols and plasma types in the wheat group. In: Miller TE, Koebner RMD (eds) Proceedings of 7th international wheat genetics symposium, Cambridge, pp 1209–1210

Kishii M, Wang RRC, Tsujimoto H (2005) GISH analysis revealed new aspect of genomic constitution of *Thinopyrum intermedium*. Czechoslovakia J Genet Plant Breeding 41:92–95

Komatsuda T, Tanno K, Salomon B, Bryngelsson T, von Bothmer R (1999) Phylogeny in the genus *Hordeum* based on nucleotide sequences closely linked to the *vrs1* locus (row number of spikelets). Genome 42:973–981

Krause EHL (1898) Floristische Notizen II. Gräser. 1. Zur Systematik und Synonymik Botanisches Centralblatt (Jena) 73:337–343

Krause EHL (1913) Beiträge zur Gramineen-Systematik (Fortsetzung). Beih Bot Centralbl 30:111–123

Kumar A, Bennetzen JL (1999) Plant retrotransposons. Annu Rev Genet 33:479–532

Leitch IJ, Bennett MD (2004) Genome downsizing in polyploid plants. Biol J Linnean Soc 82:651–663

Li DY, Zhang XY (2002) Physical localization of the 18S–5.8S-26S rDNA and sequence analysis of ITS regions in *Thinopyrum ponticum* (Poaceae: Triticeae): implications for concerted evolution. Ann Bot 90:445–452

Liu ZW, Wang RRC (1993) Genome analysis of *Elytrigia caespitosa, Lophopyrum nodosum, Pseudoroegneria geniculate* ssp. *scythic*a, and *Thinopyrum intermedium*. Genome 36:102–111

Liu B, Vega JM, Segal G, Abbo S, Rodova H, Feldman M (1998a) Rapid genomic changes in newly synthesized amphiploids of *Triticum* and *Aegilops*. I. Changes in low-copy noncoding DNA sequences. Genome 41:272–277

Liu B, Vega JM, Feldman M (1998b) Rapid genomic changes in newly synthesized amphiploids of *Triticum* and *Aegilops*. II. Changes in low-copy coding DNA sequences. Genome 41:535–542

Liu QL, Ge S, Tang HB, Zhang XL, Zhu GF, Lu BR (2006) Phylogenetic relationships in *Elymus* (Poaceae: Triticeae) based on the nuclear ribosomal internal transcribed spacer and chloroplast trnL-F sequences. New Phytol 170:411–420

Liu Z, Li D, Zhang X (2007) Genetic relationships among five basic genomes St, E, A, B and D in Triticeae revealed by genomic Southern and *in situ* hybridization. J Integr Plant Biol 49:1080–1086

Liu ZP, Chen ZY, Pan J, Li XF, Su M, Wang LJ, Li HJ, Liu GS (2008) Phylogenetic relationships in *Leymus* (Poaceae: Triticeae) revealed by the nuclear ribosomal internal transcribed space and chloroplast trnL-F sequences. Mol Phylogenet Evol 46:278–289

Lukaszewski AJ, Curtis CA (1993) Physical distribution of recombination in B-genome chromosomes of tetraploid wheat. Theor Appl Genet 84:121–127

Luo M-C, Yang Z-L, Kota RS, Dvorak J (2000) Recombination of chromosomes 3A (m) and 5A (m) of *Triticum monococcum* with homoeologous chromosomes 3A and 5A of wheat: the distribution of recombination across chromosomes. Genetics 154:1301–1308

Luo M-C, Deal KR, Yang Z-L, Dvorak J (2005) Comparative genetic maps reveal extreme crossover localization in the *Aegilops speltoides* chromosomes. Theor Appl Genet 111:1098–1106

Löve Á (1982) Generic evolution of the wheatgrasses. BioI Zentralbl 101:199–212

Löve Á (1984) Conspectus of the Triticeae. Feddes Repert 95:425–521

Ma X-F, Gustafson JP (2005) Genome evolution of allopolyploids: a process of cytological and genetic diploidization. Cytogenet Genome Res 109:236–249

Ma X-F, Gustafson JP (2006) Timing and rate of genome variation in triticale following allopolyploidization. Genome 49:950–958

Ma X-F, Fang P, Gustafson JP (2004) Polyploidization-induced genome variation in triticale. Genome 47:839–848

Mac Key J (1966) Species relationship in *Triticum*. In: Proceedings of 2nd international wheat genetics symposium, Lund, Sweden, Hereditas, Suppl 2:237–276

Mac Key J (1981) Comments on the basic principles of crop taxonomy. Kulturpflanze 29(S):199–207

Macfarlane TD, Watson L (1982) The classification of Poaceae subfamily Pooideae. Taxon 31:178–203

Mahelka V, Kopecky D, Paštová L (2011) On the genome constitution and evolution of intermediate wheatgrass (*Thinopyrum intermedium*: Poaceae, Triticeae). BMC Evol Biol 11:127

Marcussen T, Sandve SR, Heier L, Spannagl M, Pfeifer M (2014) The international wheat genome sequencing consortium. Jakobsen KS, Wulff BBH, Steuernagel B, Klaus FX, Mayer KFX, Olsen OA (eds) Ancient hybridizations among the ancestral genomes of bread wheat. Science 345(6194):288–291

Mason-Gamer RJ (2001) Origin of North American *Elymus* (Poaceae: Triticeae) allotetraploids based on granule-bound starch synthase gene sequences. Syst Bot 26:757–768

Mason-Gamer RJ (2005) The B-amylase genes of grasses and a phylogenetic analysis of the Triticeae (Poaceae). Am J Bot 92:1045–1058

Mason-Gamer RJ (2013) Phylogeny of a genomically diverse group of *Elymus* (Poaceae) allopolyploids reveals multiple levels of reticulation. PLoS ONE 8:e78449

Mason-Gamer RJ, Kellogg EA (1996a) Chloroplast DNA analysis of the monogenomic Triticeae: phylogenetic implications and genome-specific markers. In: Jauhar PP (ed) Methods of genome analysis in plants. CRC Press, Boca Raton, Florida, pp 301–325

Mason-Gamer RJ, Kellogg EA (1996b) Testing for phylogenetic conflict among molecular data sets in the tribe Triticeae (Gramineae). Syst Bio l 45:524–545

Mason-Gamer RJ, Weil CF, Kellogg EA (1998) Granule-bound starch synthase: structure, function, and phylogenetic utility. Mol Biol Evol 15:1658–1673

Mason-Gamer RJ, Orme NL, Anderson CM (2002) Phylogenetic analysis of North American *Elymus* and the monogenomic Triticeae (Poaceae) using three chloroplast DNA data sets. Genome 45:991–1002

McFadden ES, Sears ER (1944) The artificial synthesis of *Triticum spelta*. Records Genet Soc Amer 13:26–27

McFadden ES, Sears ER (1946) The origin of *Triticum spelta* and its free-threshing hexaploid relatives. J Hered 37(81–89):107–116

McGuire PE (1984) Chromosome pairing in triploid and tetraploid hybrids in *Elytrigia* (Triticeae; Poaceae). Can J Genet Cytol 26:519–522

McIntyre CL (1988) Variation at isozyme loci in Triticeae. Pl Syst Evol 160:123–142

Melderis A (1953) Generic problems within the tribe *Hordeae*. In Osvald H, Aberg E (eds) Proceedings of 7th international bot congress, Stockholm 1950, Uppsala, Almquist & Wiksells, pp 1450–1485

Melderis A (1978) Taxonomic notes on the tribe Triticeae (Gramineae), with special reference to the genera *Elymus* L. *senso lato*, and *Agropyron* Gaertbner *senso lato*. In: Tutin TG, Heywood VA, Burges NA, Moore DM, Valentine DH, Walters SM (eds) Flora Europea, vol 5, pp 369–384

Melderis A (1980) Taxonomic notes on the tribe Triticeae (Graminae), with special reference to the genera *Elymus* and *Agropyron*. Fl Europ Not sys, No 20

Melderis A, Humphries CJ, Tutin TG, Heathcote SA (1980) Tribe Triticeae Dumort. In: Tutin TG et al (eds) Flora Europaea, vol 5. Cambridge University Press, Cambridge, pp 190–206

Middleton C, Stein N, Keller B, Kilian B, Wicker T (2013) Comparative analysis of genome composition in Triticeae reveals strong variation in transposable element dynamics and nucleotide diversity. Plant J 73:347–356

Middleton CP, Senerchia N, Stein N, Akhunov ED, Keller B, Wicker T, Kilian B (2014) Sequencing of chloroplast genomes from wheat, barley, rye and their relatives provides a detailed insight into the evolution of the Triticeae tribe. PLoS ONE 9:e85761

Mizianty M, Frey L, Szczepaniak M (1999) The *Agropyron-Elymus* complex (Poaceae) in Poland: nomenclatural problems. Fragm Flor Geobot 44:3–33

Mochizuki A (1964) Further studies on the effect of accessory chromosomes on chromosome pairing (in Japanese). Jap J Genet 39:356–362

Monte JV, McIntyre CL, Gustafson JP (1993) Analysis of phylogenetic relationships in the Triticeae tribe using RFLPs. Theor Appl Genet 86:649–655

Morris R, Sears ER (1967) The cytogenetics of wheat and its relatives. In: Quisenberry KS, Reitz LP (eds) Wheat and wheat improvement. American Society of Agronomy, Madison, Wisconsin, USA, pp 19–87

Morrison JW, Rajhathy T (1960a) Chromosome behaviour in autotetraploid cereals and grasses. Chromosoma (Berl.) 11:297–309

Morrison JW, Rajhathy T (1960b) Frequency of quadrivalents in autotetraploid plamts. Nature 187:528–530

Muramatsu M (2009) A presumed genetic system determining the number of spikelets per rachis nodein the tribe Triticeae. Breed Sci 59:617–620

Naghavi MR, Rad MB, Riahi M, Taleie A (2013) Phylogenetic analysis in some *Hordeum* species (Triticeae; Poaceae) based on two single-copy nuclear genes encoding acetyl-CoA carboxylase. Biochem Syst Ecol 47:148–155

Nevski SA (1934) Hordeae Bentb. In: Komarov VL (ed) Flora of the USSR II: 590

Ni Y, Asamoah-Odei N, Sun G (2011) Maternal origin, genome constitution and evolutionary relationships of polyploid *Elymus* species and *Hordelymus europaeus*. Biol Plant 55:68–74

Nishikawa T, Salomon B, Komatsuda T, von Bothmer R, Kadowaki K (2002) Molecular phylogeny of the genus Hordeum using three chloroplast DNA sequences. Genome 45:1157–1166

Ohta S, Morishita M (2001) Relationships in the Genus *Dasypyrum* (Gramineae). Hereditas 135:101–110

Okamoto M (1957) Asynaptic effect of chromosome V. Wheat Inf Serv 5:6

Okito P, Mott IW, Wu Y, Wang RRC (2009) A Y genome specific STS marker in *Pseudoroegneria* and *Elymus* species (Triticeae: Gramineae). Genome 52:391–400

Ozkan H, Levy AA, Feldman M (2001) Allopolyploidy—induced rapid genome evolution in the wheat (*Aegilops-Triticum*) group. Plant Cell 13:1735–1747

Petersen G, Seberg O (1997) Phylogenetic analysis of the Triticeae (Poaceae) based on *rpo*A sequence data. Mol Phylogenet Evol 7:217–230

Petersen G, Seberg O (2002) Molecular evolution and phylogenetic application of *DMC1*. Mol Phylogent Evol 22:43–50

Petersen G, Seberg O (2008) Phylogenetic relationships of allotetraploid *Hordelymus europaeus* (L.) Harz (Poaceae: Triticeae). Pl Syst Evol 273:87–95

Petersen G, Seberg O, Yde M, Berthelsen K (2006) Phylogenetic relationships of *Triticum* and *Aegilops* and evidence for the origin of the A, B, and D genomes of common wheat (*Triticum aestivum*). Mol Phylogenet Evol 39:70–82

Petersen G, Seberg O, Salomon B (2011) The origin of the H, St, W, and Y genomes in allotetraploid species of *Elymus* L. and *Stenostachys* Turcz. (Poaceae: Triticeae). Plant Syst Evol 291:197–210

Ramakrishna W, Dubcovsky J, Park Y-J, Busso C, Emberton J, SanMiguel P, Bennetzen J (2002) Different types and rates of genome evolution detected by comparative sequence analysis of orthologous segments from four cereal genomes. Genetics 162:1389–1400

Raunkiaer C (1934) The life form of plants and statical plant geography. Clarendon Press, Oxford, UK

Redinbaugh MG, Jones T, Zhang Y (2000) Ubiquity of the St chloroplast genome in St-containing Triticeae polyploids. Genome 43:846–852

Renvoize SA, Clayton WD (1992) Classification and evolution of the grasses. In: Chapman GP (ed) Grass evolution and domestication. Cambridge University Press, Cambridge, UK, pp 3–37

Riley R (1965) Cytogenetics and evolution of wheat. In: Hutchinson J (ed) Essays in crop plant evolution. Cambridge University Press, London, pp 103–122

Riley R, Chapman V (1958) Genetic control of the cytologically diploid behaviour of hexaploid wheat. Nature 182:713–715

Runemark H, Heneen WK (1968) *Elymus* and *Agropyron*, a problem of generic delimitation. Bot Notiser 121:51–79

Sabot F, Guyot R, Wicker T, Chantret N, Laubin B, Chalhoub B, Leroy P, Sourdille P, Bernard M (2005) Updating of transposable element annotations from large wheat genomic sequences reveals diverse activities and gene associations. Mol Genet Gen 274:119–130

Sakamoto S (1973) Patterns of phylogenetic differentiation in the tribe Triticeae. Seiken Zihô 24:11–31

Sakamoto S (1974) Intergeneric hybridization amomg three species of *Heteranthelium*, *Eremopyrum* and *Hordeum* and its significance for the genetic relationships within the tribe Triticeae. New Phytol 73:341–350

Sakamoto S (1986) Genome analysis of a polyploid form of *Haynaldia hordeacea* in the tribe Triticeae, Gramineae. In: Li Z, Swaminathan MS (eds) Proceedinsg of 1st international symposium on chromosome engineering in plants. Xian, China, pp 52–53

Sakamoto S (1991) The cytogenetic evolution of Triticeae grasses. In: Tsuchiya T, Gupta PK (eds) Chromosome engineering in plants: genetics, breeding, evolution. Elsvier Science Publications B.V, Amsterdam, The Netherland, pp 469–482

Sears ER (1976) Genetic control of chromosome pairing in wheat. Ann Rev Genet 10:31–51

Seberg O, Frederiksen S (2001) A phylogenetic analysis of the monogenomic Triticeae (Poaceae) based on morphology. Bot J Linn Soc 136:75–97

Seberg O, Petersen G (2007) Phylogeny of Triticeae (Poaceae) based on three organelle genes, two single-copy nuclear genes, and morphology. Aliso J Syst Evol Bot 23:362–371

Senerchia N, Wicker T, Felber F, Parisod C (2013) Evolutionary dynamics of retrotransposons assessed by high throughput sequencing in wild relatives of wheat. Genome Biol Evol 5:1010–1020

Sha LN, Yang RW, Fan X, Wang XL, Zhou YH (2008) Phylogenetic analysis of *Leymus* (Poaceae: Triticeae) inferred from nuclear rDNA ITS sequences. Biochem Genet 46:605–619

Sha LN, Fan X, Yang RW, Kang HY, Ding CB, Zhang L, Zheng YL, Zhou YH (2010) Phylogenetic relationships between *Hystrix* and its closely related genera (Triticeae; Poaceae) based on nuclear *Acc1*, *DMC1* and chloroplast *trnL-F* sequences. Mol Phylogenets and Evol 54:327–335

Shaked H, Kashkush K, Ozkan H, Feldman M, Levy AA (2001) Sequence elimination and cytosine methylation are rapid and reproducible responses of the genome to wide hybridization and allopolyploidy in wheat. Plant Cell 13:1749–1759

Soreng RJ, Davis JI, Doyle JJ (1990) A phylogenetic analysis of chloroplast DNA restriction site variation in Poaceae subfam Pooideae. Pl Syst Evol 172:83–97

Stebbins GL (1950) Variation and Evolution in plants. Columbia University Press, New York

Stebbins GL (1956) Taxonomy and evolution of genera. With special reference to the family Gramineae. Evolution 10:235–245

Stebbins GL (1971) Chromosomal evolution in higher plants. Edward Arnold Ltd., London

Stebbins GL, Pun FT (1953) Artificial and natural hybrids in the Gramineae, tribe Hordeae. V. Diploid hybrids of *Agropyron*. Amer J Bot 40:444–449

Sun GL, Komatsuda T (2010) Origin of the Y genome in *Elymus* and its relationship to other genomes in Triticeae based on evidence from elongation factor G (*EF-G*) gene sequences. MolPhylogenet Evol 56:727–733

Sun GL, Ni Y, Daley T (2008) Molecular phylogeny of *RPB2* gene reveals multiple origin, geographic differentiation of H genome, and the relationship of the Y genome to other genomes in *Elymus* species. Mol Phylogenet Evol 46:897–907

Sun GL, Pourkheirandish M, Komatsuda T (2009) Molecular evolution and phylogeny of the *RPB2* gene in the genus *Hordeum*. Ann Bot 103:975–983

Svitashev S, Byrngelsson T, Li X, Wang RRC (1998) Genome specific repetitive DNA and RAPD markers for genome identification in *Elymus* and *Hordelymus*. Genome 41:120–128

Tang S, Li Z, Jia X, Larkin PJ (2000) Genomic *in situ* hybridization (GISH) analyses of *Thinopyrum intermedium*, its partial amphiploid Zhong 5, and disease-resistant derivatives in wheat. Theor Appl Genet 100:344–352

Tzvelev NN (1976) Poaceae URSS. Tribe III. Triticeae Dum. USSR. Academy of Sciences Press, Leningrad, pp 105–206

van Slageren MW (1994) Wild wheats: a monograph of *Aegilops* L. and *Amblyopyrum* (Jaub. and Spach) Eig (Poaceae). Agricultural University, Wageningen, The Netherlands

von Bothmer R, Lu B-R, Linde-Laursen I (1994) Intergeneric hybridization and C-banding patterns in *Hordelymus* (Triticeae, Poaceae). Pl Syst Evol 189:259–266

von Bothmer R, Jacobsen N, Baden C, Jørgensen RB, Linde-Larsen I (1995) An ecogeographical study of the genus *Hordeum*, 2nd edn. In: Systematic and ecogeographic studies on crop genepools, vol 7. International Plant Genetic Resources Institute, Rome

Vardi A, Dover GA (1972) The effect of B chromosomes on meiotic and pre-meiotic spindles and chromosome pairing in *Triticum/Aegilops* hybrids. Chromosoma (Berl.) 38:367–385

Vicient CM, Jääskeläinen J, Kalendar R, Schulman AH (2001) Active retrotransposons are a common feature of grass genomes. Plant Physiol 125:1283–1292

Vogel KP, Arumuganathan K, Jensen KB (1999) Nuclear DNA content of perennial grasses of Triticeae. Crop Sci 39:661–667

Wang RR-C (1985a) Identification of intergeneric hybrids in the tribe Triticeae by karyotype analysis. Agronomy Abstracts American Society of Agronomy, Madison, WI, p 74

Wang RR-C (1985b) Genome analysis of *Thinopyrum bessarabicum* and *T. elongatum*. Can J Genet Cytol 27:722–728

Wang RR-C (1989) An assessment of genome analysis based on chromosome pairing in hybrids of perennial Triticeae. Genome 32:179–189

Wang RR-C, Hsiao CT (1989) Genome relationship between *Thinopyrum bessarabicum* and *T. elongatum*: Revisited. Genome 32:802–809

Wang RR-C, Lu BR (2014) Biosystematics and evolutionary relationships of perennial Triticeae species revealed by genomic analyses. J Syst Evol 52:697–705

Wang RRC, Dewey DR, Hsiao C (1986) Genome analysis of the tetraploid *Pseudoroegneria tauri*. Crop Sci 26:723–727

Wang RR-C, von Bothmer R, Dvorak J, Fedak G, Linde-Laursen I, Muramatsu M (1995) Genome symbols in the Triticeae (Poaceae). In: Wang RR-C, Jensen KB, Jaussi C (eds) Proceedings of 2nd international Triticeae sympsoium, 20–24 June 1994. Logan, UT, pp 29–34

Watson L, Clifford HT, Dallwitz MJ (1985) The classification of the Poaceae: subfamilies and supertribes. Austral J Bot 33:433–484

Wazuddin M, Driscoll CJ (1986) Chromosome constitution of polyploid wheats: introduction of diploid wheat chromosome 4. Proc Natl Acad Sci, USA 83:3870–3874

Wolfe KH, Gouy M, Yang Y-W, Sharp PM, Li W-H (1989) Date of the monocot-dicot divergence estimated from chloroplast DNA sequence data. Proc Natl Acad Sci USA 86:6201–6205

Yamane K, Kawahara T (2005) Intra- and interspecific phylogenetic relationships among diploid *Triticum-Aegilops* species (Poaceae) based on base-pair substitutions, indels, and microsatellites in chloroplast noncoding sequences. Am J Bot 92:1887–1898

Yan C, Sun GL (2012) Multiple origins of allopolyploid wheatgrass Elymus caninus revealed by *RPB2*, PepC and *TrnD/T* genes. Mol Phylogenet Evol 64:441–451

Yan C, Sun GL, Sun DF (2011) Distinct origin of the Y and St genome in *Elymus* species: Evidence from the analysis of a large sample of St genome species using two nuclear genes. PLoS ONE 6:e26853

Yen C, Yang JL, Yen Y (2005) Hitoshi Kihara, Áskell Löve and the modern genetic concept of the genera in the tribe Triticeae (Poaceae). Acta Phytotaxonomica Sinica 43:82–93

Yousofi M, Aryavand A (2004) Determination of ploidy levels of some populations of *Agropyron cristatum* (Poaceae) in Iran by flow cytometry. Iranian J Sci Techn Trans A 28:137–144

Zhang H-B, Dvorak J (1991) The genome origin of tetraploid species of *Lemus* (Poaceae: Triticeae) inferred from variation in repeated nucleotide sequences. Amer J Bot 78:871–884

Zhang X-Y, Dong Y-S, Wang RR-C (1996) Characterization of genomes and chromosomes in partial amphiploids of the hybrid *Triticum aestivum* x *Thinopyrum ponticum* by in situ hybridization, isozyme analysis and RAPD. Genome 39:1062–1071

Zhang X-Y, Wang RR-C, Fedak G, Dong Y-S (1997) Determination of genome and chromosome composition of *Thinopyrum intermedium* and partial amphiploids of *Triticum aestivum*—*Th. intermedium* by GISH and genome specific RAPD markers. Chin Agric Sci 1997:71–80

Zhang HQ, Yang RW, Dou QW, Tsujimoto H, Zhou YH (2006) Genome constitutions of *Hystrix patula*, *H. duthiei* ssp. *duthiei* and *H. duthiei* ssp. *longearistata* (Poaceae: Triticeae) revealed by meiotic pairing behavior and genomic in-situ hybridization. Chromosome Res 14:595–604

Zhang C, Fan X, Yu HQ, Zhang HQ, Wang XL, Zhou YH (2009) Phylogenetic analysis of questionable tetraploid species in *Roegneria* and *Pseudoroegneria* (Poaceae: Triticeae) inferred from a gene encoding plastid acetyl-CoA carboxylase. Biochemic Syst Ecol 37:412–420

Zhukovsky PM (1928) A critical systematical survey of the species of the genus *Aegilops* L. Bull Appl Bot Genet and Pl Breed 18:417–609

Zohary D, Feldman M (1962) Hybridization between amphidiploids and the evolution of polyploids in the wheat (*Aegilops-Triticum*) group. Evolution 16:44–61

3.1 Chromosome Karyotypic Features

3.1.1 Introduction

Early studies on genome structure commenced by determining karyotypic features that were visible with a light microscope, such as chromosome number, size, centromere position and arm ratio. The Triticeae species contain the basic chromosome set of x = 7. Chromosome morphology in most genera of the tribe is characterized by a symmetric karyotype, with large metacentric or sub-metacentric chromosomes. Several genera (e.g., *Eremopyrum* and *Dasypyrum* (= *Haynaldia*) have asymmetric karyotypes, with small sub-metacentric and acrocentric chromosomes, whereas several species of *Aegilops* have asymmetric karyotypes with large sub-telocentric chromosomes. More advanced cytogenetic tools revealed the structure of important chromosomal regions such as centromeres, telomers, and nucleolar organizers, as well as types of chromatin (hetero and eu) and repetitive DNA and their distribution along the chromosomes. Finally, whole genome sequences provided a more accurate description of the chromosome. Nevertheless, understanding the connection between sequences and chromosome behavior, dynamics and expression remains a challenge. In this section, we describe those morphological features of chromosomes that are visible by cytological approaches, as well as the more recent insight on genome and chromosome structures that have emerged from whole genome sequences.

3.1.2 Centromeres

The centromere is a chromosomal region that is not condensed at mitosis and thus, appears as a constriction on each condensed chromosome during the mitotic metaphase. This constriction, referred to as the primary constriction, is the site where the chromosomal-spindle fibers, now known to be microtubules, are attached during cell divisions (Flemming 1882). The chromosomes of the Triticeae species possess a permanently localized centromere region.

The centromere is a nucleoprotein complex that is an essential part of the chromosome, due to its indispensable role in chromosome segregation during cell divisions. Acentric chromosomal segments do not move, nor segregate during cell divisions and thus get lost. The centromere is the site of assembly of the kinetochore, a group of proteins with microtubule binding activity. The kinetochore functions in tying sister chromatids together, and in generating attachments to spindle microtubules in a bipolar fashion, leading to regular chromosome segregation during mitosis and meiosis (Valente et al. 2012; de Rop et al. 2012; Birchler and Han 2013). The kinetochore-forming domain of the centromere is differentiated in plants from the rest of the chromosome by the presence of CENH3 (CENP-A in vertebrates), a histone H3 variant that replaces the canonical histone H3 in nucleosomes of active centromeres (Palmer et al. 1991). The sequence of the core domain of CENH3 is conserved among species. Several studies revealed that CENH3 nucleosomes and histone H3 nucleosomes are interspersed in the centromere region (Zhang et al. 2002).

Functional centromeres are formed by the binding of the histone variant, CENH3, at a specific location on the chromosome that generally coincides with the presence of centromere-specific repeats. Chromatin immuno-precipitation with a CENH3 antibody enabled to define a single region in bread wheat, ranging between $\sim$ 5 and 9 Mb, in each chromosome (IWGSC 2018). In some of the bread wheat lines studied, the centromere position was shifted due to pericentric inversions, for example on chromosomes 4B and 5B (Walkowiak et al. 2020).

In plants, the centromeric sequences are mainly composed of arrays of tandem repeat satellite DNA and interspersed retrotransposons, whose sizes can range into megabases. Specific centromeric DNA sequences are not necessary for centromere formation (Feng et al. 2015), suggesting that the role of the DNA sequences at the centromere is structural rather than functional. Discoveries in

M. Feldman and A. A. Levy, *Wheat Evolution and Domestication*,
https://doi.org/10.1007/978-3-031-30175-9_3

the past decades, including "neocentromeres" and "centromere inactivation", indicated that centromere identity, in both plants and other organisms, is determined by epigenetic mechanisms and not by the centromeric-DNA sequences (Feng et al. 2010; Ekwall 2007; Henikoff and Furuyama 2010; Valente et al. 2012; Birchler and Han 2013). Hence, CENH3 and other proteins are critical to centromere function, whereas the DNA sequence is not necessarily a determining factor (Birchler and Han 2013). In rice, the centromere was shown to be composed of a 155-bp satellite repeat sequence and CRR (centromere retrotransposon of rice) (Cheng et al. 2002). The maize centromere sequence composition is similar to that of rice; it is composed of two types of sequences—centromere repeat C and CRM (centromeric retrotransposon of maize) (Birchler and Han 2009). The number of repeats of these sequences varies among different chromosomes (Jin et al. 2004). The Triticeae centromeres are no exception to other plants, forming through the assembly of the CENH3 proteins with arrays of satellite repeats and retroelements (Cheng and Murata 2003). The primary constrictions of barley, wheat, *Aegilops*, and rye chromosomes, i.e., the physical locations of centromeres, harbor retroelement-like sequences (Presting et al. 1998; Fukui et al. 2001; Cheng and Murata 2003). The retroelement *cereba*, was isolated from a barley primary constriction (Presting et al. 1998), and satellite sequences consisting of an AGGGAG motif, were found at the core of the barley centromeres (Hudakova et al. 2001).

In wheat, a detailed chromatin immunoprecipitation analysis, using a CENH3 antibody followed by sequencing of the precipitated DNA fraction, enabled to identify a distinct dynamic structure for wheat centromeres, possibly due to evolution through frequent hybridization and allopolyploidization (Su et al. 2019). Unlike typical plant promoters which carry a satellite of tandem repeats ranging from 150 to 180 bp, the wheat CENH3 nucleosomes were associated with two different types of repeats much larger than other plant centromeric repeats, namely 550 and 566 bp long respectively (Su et al. 2019). Moreover, different subgenomes tended to contain different repeats types and some chromosomes lacked satellite repeats altogether. Phylogenetic analyses indicated that the repeat signals were stronger in diploids than polyploids and that the centromere structure had rapidly evolved at the polyploid level (Su et al. 2019). Two types of retroelements are localized in centromeric regions of wheat, namely the *cerebra*-like retroelement (Cheng and Murata 2003), also known as *crew* or *crw* (centromere-retroelement of wheat) and the less abundant *Quinta* element (Li et al. 2013). Finally, two CENH3 genes and protein variants are present in wheat that have evolved at the diploid levels. Altogether, the different repeats, their types and abundance, the retroelements and the CENH3 variants (Yuan et al. 2015) all seem to have contributed to

the dynamic nature of wheat centromeres (Li et al. 2013; Su et al. 2019), which in turn might have contributed to genome stability in a allopolyploid background.

The centromere typically divides longitudinally at mitosis and at second meiotic division, thus ensuring regular segregation of daughter chromosomes to opposite poles. Accidental transverse (misdivision) instead of longitudinal division of the centromere at meiosis or mitosis of bread wheat, may yield two functional halves of the centromere that may give rise to stable telocentric chromosomes and/or isochromosomes, demonstrating that the centromere structure is a reverse repeat (Steinitz-Sears 1966). By analyzing misdivision derivatives, Kaszás and Birchler (1996) provided molecular evidence of repeat DNA units in maize centromeres, and demonstrated that a change in copy number does not impact centromere function. In common wheat, the frequency of misdivision, as well as the relative frequency of one-chromatid and two-chromatid misdivisions, is chromosome-specific and is affected by the genetic background (Sears 1952; Steinitz-Sears 1973; Makino et al. 1977; Morris et al. 1977; Vega and Feldman 1988).

Wagenaar and Bray (1973) noticed that at first meiotic metaphase of wheat hybrids, the two sister kinetochores of a univalent chromosome are located adjacent to each other, as in normal bivalents, and then move to take on a typical mitotic configuration, with sister kinetochores on opposite faces of the chromosome, interacting with microtubules of opposite poles. During this shift in orientation, one of the sister kinetochores is simultaneously attached to microtubules originating from both polar regions, suggesting not only the presence of several microtubule-binding sites within a given sister kinetochore, but also, the independent activities of these sites. The possibility of multiple microtubule-binding sites on each sister kinetochore was supported by Vega and Feldman (1988), who observed a pair of parallel fibers perpendicularly protruding from each sister kinetochore in dividing univalent chromosomes of bread wheat. This is in agreement with the observations of Zinkowski et al. (1991), who reported that multiple fragments resulting from detached mammalian kinetochores still progress through mitosis.

3.1.3 Telomeres

Telomeres are DNA sequences located at chromosome ends that stabilize chromosomes and protect them from deterioration or from fusion with neighboring chromosomes. Their absence leads to abnormal chromosome behavior, e.g., the induction of breakage-fusion-bridge cycle due to chromosome break (McClintock 1942). Telomeres cannot normally be transposed to intercalary positions in the chromosomes. They have a compound structure, a special cycle of division and a tendency for non-homologous association at the beginning of first meiotic prophase.

Telomeres are complex nucleoprotein structures consisting of several proteins and non-coding DNA repetitive nucleotide sequences at each end of a chromatid. Most plant telomeres have TTTAGGG repeats (Riha and Shippen 2003; Fajkus et al. 2005; Watson and Riha 2010). The telomeres of barley, rye and wheat chromosomes contain an array of repeats that hybridize with the *Arabidopsis thaliana* TTTAGGG telomeric sequence (Schwarzacher and Heslop-Harrison 1991; Werner et al. 1992; Röder et al. 1993; Cheung et al. 1994; Mao et al. 1997). Telomeric sequences can be added de novo to broken wheat chromosome ends, presumably during gametogenesis (Werner et al. 1992). Triticeae species contain telomere-associated sequences next to the telomere. These sequences have been subjected to divergence, amplification, and deletion processes in different lineages of Triticeae, so that different variants of the telomere-associated sequences and different quantities of these sequences are present in different lineages (Bedbrook et al. 1980; Appels et al. 1989). The wheat sub-telomeric regions contain several features, such as genes, transposable elements, repeats, GC content, recombination hotspots and sequence motifs for relevant DNA-binding proteins. Comparison of these features among wheat chromosomes shows a high polymorphism between homoeologous chromosomes (Aguilar and Prieto 2020). This polymorphism might provide a physical basis to differentiate homologs from homoeologous chromosome during meiotic pairing initiation.

Telomeres are complex nucleoprotein structures consisting of non-coding DNA and several proteins. They have a unique mode of replication that involves telomerase, a specialized and specific ribonucleoprotein enzyme complex (Greider and Blackburn 1985). This enzyme adds repetitive nucleotide sequences to the ends of the DNA, thereby elongating the telomere. In plants, telomerase remains active in organs and tissues containing dividing cells (Fitzgerald et al. 1996; Heller et al. 1996; Riha et al. 1998) and is regulated in a cell cycle-dependent manner, with a peak level in S-phase. Telomere dynamics are coupled to meristem activity and continuous growth, disclosing a critical association between telomere length, stem cell function, and the extended lifespan of plants (González-García et al. 2015). Correspondingly, the lengths of telomeres are maintained during plant development. Telomere shortening to a critical level is a signal to stop the cell cycle and start the processes of cellular senescence (Dvorackova et al. 2015).

The very distal end of the telomere is comprised of a 300-bp single G-rich strand, which forms a specific structure, termed the t-loop (Griffith et al. 1999). The t-loop is believed to be essential for telomere capping (reviewed in de Lange 2004), stabilizes the telomere and prevents the telomere ends from being recognized as break points by the DNA repair machinery. In plants, the G-rich strand at the 3′ end of the chromosome is longer than the C strand. The G-rich strand forms duplex telomeric DNA, effectively hiding the end of the chromosome. The t-loop is held together by several telomere-specific binding proteins, referred to as the shelterin complex (Martínez and Blasco 2010). Shelterin provides protection against double-strand break, repair by homologous recombination and non-homologous end joining (Lundblad 2000; Martínez and Blasco 2010). T-loops have been observed in humans, protozoans and plants (Griffith et al. 1999; Cesare et al. 2003).

The function of telomeric repeats in protecting the ends of chromosomes is well established (Blackburn 1986; Harper et al. 2004; Scherthan 2007), but the function of the telomere-associated sequences is not known. Their conserved occurrence in the Triticeae genomes implies that they are important, presumably to discriminate between telomeres of different species.

3.1.4 Nucleolar Organizers

Satellites are chromosome segments that are separated from the rest of the chromosome by a constriction, called a secondary constriction. In each Triticeae species, there is at least one pair of chromosomes with satellites, referred to as SAT-chromosomes. Most species contain two, and in few cases, even three such pairs. Most satellites exist on the short arm of the chromosome, in a sub-telomeric position, but in *Eremopyrum distans*, the satellites are on the long arm. The secondary constrictions are associated with the formation of the nucleolus and thus, referred to as the nucleolar organizing region (NOR) (McClintock 1934). The NOR is active in nucleolus formation and contains ribosomal (rDNA) genes that code for the ribosomal RNA (rRNA), the precursor particles of ribosomes that are assembled in the nucleoli and are active inside the nucleolus. The extent of the activity of the rDNA genes is proportional to the size of the nucleolus (Birnstiel et al. 1971; Appels et al. 1980). Consequently, nucleolus formation is considered evidence for the expression of the ribosomal genes, and the lack of nucleolus indicates the absence of rRNA transcription (Flavell et al. 1986). Moreover, the relative size of nucleoli within the same nucleus has been taken as a measure of the differential activity of ribosomal genes of one versus another nucleolar organizer (Flavell et al. 1986).

The ribosomal genes are highly redundant, with several hundreds or even thousands of tandemly arranged copies per secondary constriction. Each copy is composed of genes that code for the 18S, 26S, and 5.8S rRNA, separated by a non-coding spacer DNA (Flavell and O'Dell 1979; Appels and Honeycutt 1986). Only transcriptionally active NOR loci give rise to a nucleolus and there is a relationship between the number of rDNA genes, nucleolus organizer activity, and nucleolus size (Flavell and O'Dell 1979). Since these genes are also active in prophase, the region is still uncondensed at metaphase and therefore, appears constricted.

The number of rRNA-encoding genes in each of the four NOR sites of allohexaploid wheat was first determined by Flavell and coworkers. In the standard laboratory Chinese Spring cultivar of bread wheat, chromosomes 1A and 5D contain a very small proportion of the rRNA-encoding genes (10%), while chromosomes 1B and 6B possess 30% and 60% of these genes, respectively (2700 and 5500 copies, respectively) (Mohan and Flavell 1974; Flavell and O'Dell 1976). In full accord with these findings, chromosomes 1A and 5D produced very small nucleoli or none at all in the Chinese Spring cultivar (Crosby 1957; Crosby-Longwell and Svihla 1960). These findings align with the small proportion of total rDNA gene complement in chromosomes 1A and 5D. Similar patterns were found in allotetraploid wheat (Frankel et al. 1987). The total number of rDNA units in the fully sequenced genome of Chinese Spring was estimated at 11,160 copies corresponding to 100 Mb (Handa et al. 2018), 30.5% of which are on the Nor-B1 locus (Chr. 1B), 60.9% on Nor-B2 (Chr. 6B) and 8.6% in other NORs (Handa et al. 2018). These numbers based on genome sequence analysis are consistent with earlier estimates (Flavell and O'Dell 1976). Four main subtypes of rDNA units were identified in Nor-B1 and Nor-B2 with one particular subtype, S1, more strongly expressed than the other three, even though it was not the most abundant (Handa et al. 2018).

Dubcovsky and Dvorak (1995) summarized data showing that the major NOR loci are located in homoeologous groups 1, 5, and 6 across the Triticeae. A single major rDNA locus per genome is present in rye and Ae. tauschii on chromosomes 1 and 5, respectively (Appels 1982; Lawrence and Appels 1986; Lassner et al. 1987). When two loci are present per genome, all possible pairwise combinations have been recorded; they are on chromosomes 1 and 5 in the T. monococcum, T. urartu and Ae. umbellulata genomes (Gerlach et al. 1980; Miller et al. 1983), on chromosomes 1 and 6 in the Ae. speltoides and wheat B subgenomes (Crosby 1957; Crosby-Longwell and Svihla 1960; Dvorak et al. 1984), and on chromosomes 5 and 6 in the Elymus elongatus (= Lophopyrum elongatum) and barley genomes (Dvorak et al. 1984; Saghai-Maroof et al. 1984). Three major NOR loci were identified in the Ae. longissima genome, on chromosomes 1, 5, and 6 (Friebe et al. 1993), and in the Psathyrostachys fragilis genome, but in this species the chromosomes harboring these loci were not identified (Linde-Laursen and Baden 1994). The variation in number and position of the NOR loci indicate that they are mobile in the Triticeae genomes (Dubcovsky and Dvorak 1995).

In addition, several minor loci can be present per genome (Mukai et al. 1991; Leitch and Heslop-Harrison 1992; Jiang and Gill 1994; Dubcovsky and Dvorak 1995). rDNA evolved in concert and loci on different chromosomes share the same sequence variants (Appels and Dvorak 1982; Dvorak and Appels 1982).

Triticeae 5S DNA loci harboring tandem arrays of the 5S rRNA genes, evolve like rDNA (Dubcovsky et al. 1996). In Triticum, Aegilops, and Elymus (= Lophopyrum), 5S DNA loci are on the short arms of the chromosomes of homoeologous groups 1 and 5 (Dvorak et al. 1989). However, they are on the long arms of chromosomes 2H and 3H in barley (Kanazin et al. 1993). Like rDNA, the 5S DNA loci are mobile due to translocation into new sites.

In hybrids and allopolyploids, the rDNA genes of one parental set are transcribed, while most or all rDNA genes inherited from the other parent remain silent. This phenomenon is known as nucleolar dominance (Navashin 1928, 1934; Pikaard 1999, 2000) and is the general phenomenon in the allopolyploid species of the genera Aegilops and Triticum (Feldman et al. 2012). The diploid species of wheat, T. monococcum and T. urartu, contain two nucleolar organizer regions, one on chromosome arm 1AS and the second on 5AS (Gerlach et al. 1980; Miller et al. 1983). In the allopolyploid wheat species, the NOR of 1AS is inactive, while that of 5AS was lost (Miller et al. 1983; Jiang and Gill 1994). Thus, allohexaploid wheat (genome BBAADD) possesses four pairs of NORs on the short arm of chromosomes 1A, 1B, 6B, and 5D (Crosby 1957; Crosby-Longwell and Svihla 1960; Bhowal 1972; Darvey and Driscoll 1972). In this species, the nucleolar organizers of the B subgenome suppress the nucleolar organizers of the A and D subgenomes (Crosby 1957; Crosby-Longwell and Svihla 1960; Darvey and Driscoll 1972; Flavell and O'Dell 1979). Similarly, the nucleolar organizers of the B subgenome suppress those of the A subgenome in allotetraploid wheat (genome BBAA) (Frankel et al. 1987) and those of the R subgenome in 6x and 8x triticale (genome BBAARR and BBAADDRR, respectively) (Darvey and Driscoll 1972; Cermeño et al. 1984a; Martini and Flavell 1985; Appels et al. 1986; Brettell et al. 1986). Nucleolar dominance was also observed in all allopolyploid species of Aegilops (Cermeño and Lacadena 1985; Cermeño et al. 1984b). In these species, the U genome from Aegilops umbellulata completely suppresses the NOR activity of the M, S and D subgenomes (Cermeño et al. 1984b). The nucleolar organizers of the U genome also suppress the activity of the NOR loci of the rye R genome in hybrids between allopolyploid species of the U genome-bearing Aegilops and Secale cereale or S. vavilovii (Cermeño and Lacadena 1985).

Nucleolar dominance in the allopolyploid species of the wheat group is achieved either by elimination of rRNA-encoding genes, as is the case of 5AS, or by suppression of their activity. Silencing of NOR loci is brought about by increased cytosine methylation at their CCGG sites (Gustafson and Flavell 1996; Houchins et al. 1997). Similarly, Chen and Pikaard (1997) found that silencing of rRNA-encoding genes in *Brassica* allotetraploids is achieved by DNA methylation and histone acetylation. Reversal of the suppression of the NOR loci of genome R in wheat x rye hybrids and in hexaploid triticale by treatment with the demethylating agent 5-Azcytidine, are in keeping with the role of cytosine methylation in nucleolar suppression (Vieira et al. 1990; Neves et al. 1995; Amado et al. 1997).

Newly synthesized allopolyploids exhibit genetic and epigenetic changes in their rRNA-encoding genes similar to those occurring in natural allopolyploids, i.e., the same nucleolar organizers were affected in natural and synthetic allopolyploids having the same genomic combinations, indicating that these changes are reproducible. Moreover, these changes in the newly synthesized allopolyploids show that they were generated during allopolyploid formation (Shcherban et al. 2008; Baum and Feldman 2010). Wheat 5S DNA also undergoes immediate elimination of unit classes in response to allopolyploidization (Baum and Feldman 2010). This elimination was reproducible, indicating that no further elimination occurred in the unit classes of the 5S DNA during the life of the allopolyploids.

A detailed molecular analysis of the fate of wheat NORs was done in several allopolyploids, including the synthetic and natural BBAADD genomes (Guo and Han 2014). It shows that the NORs from the B subgenome are dominant in several genomic combinations, and that the elimination of the other NORs proceeds in two steps—first through silencing and hypermethylation in the first generations of the nascent allopolyploids—interestingly, this silencing is not reversible when ploidy level is reduced—then, elimination of the non-B rDNA copies takes place progressively, starting in the fourth and ending by the 7th generation since polyploidization (Guo and Han 2014).

3.1.5 Use of C- and N-Banding for Chromosome Identification

The distribution of eu- and hetero-chromatin along chromosomes has been determined through the position, size and intensity of Giemsa stain (C-banding). Giemsa stain interacts specifically with constitutive heterochromatin and thus, the C-bands expose the position of this type of chromatin in the chromosomes (Gill 1987). The dark (stained) bands and light (unstained) bands represent heterochromatic and euchromatic regions, respectively. The C-banding technique stains all classes of constitutive heterochromatin and identifies each of the 21 chromosomes *in T. aestivum* (Endo and Gill

1984), and each of the 14 chromosomes of *T. turgidum* (Seal 1982; Bebeli and Kaltsikes 1985; Badaeva et al. 2015).

Chromosomes 1B, 3B, 5B, 6B and 7B of the B subgenome and chromosome 4A of the A subgenome are most heavily C-banded in the standard laboratory cultivar Chinese Spring of allohexaploid wheat (Endo and Gill 1984). As in allohexaploid wheat, subgenome B in allotetraploid *T. turgidum* ssp. *durum* is also more heavily C-banded than subgenome A; the same chromosomes that are heavily banded in allohexaploid wheat also heavily banded in ssp. *durum* (Seal 1982; Bebeli and Kaltsikes 1985; Badaeva et al. 2015). Yet, there is widespread banding polymorphism among different lines of wild and domesticated forms of *T. turgidum* (Badaeva et al. 2015). For example, Badaeva et al. (2015) found that karyotypes of wild and domesticated emmer showed an extremely high diversity of C-banding patterns. B subgenome chromosomes were more polymorphic than A subgenome chromosomes. The lowest diversity of C-banding patterns was found for chromosome 3A, while chromosomes 2A and 4A proved to be most variable among the A subgenome chromosomes. On the B subgenome, the lowest polymorphism was observed for chromosome 4B and the highest, for chromosomes 3B and 7B, respectively.

The C-banding method has been used in studies of several aspects of cytogenetics and evolution of *T. turgidum*. For example, Bebeli and Kaltsikes (1985) constructed the karyotypes of cultivars Capeiti and Mexicali, two durum wheat cultivars, on the basis of C-banding of their chromosomes. Using C-bands, Seal (1982) identified all 14 *T. turgidum* chromosome pairs in hexaploid triticale (genome BBAARR). While little variation was found between genotypes in the distribution of C-bands, considerable variation was found in their size, total number and total length. The bands in both A and B subgenomes were concentrated in the centromeric, distal and terminal regions.

Natarajan and Sharma (1974), using Giemsa C-banding, examined the distribution of heterochromatic regions in the chromosomes of diploid, allotetraploid and allohexaploid wheats and found that the distribution pattern of heterochromatin of *Ae. speltoides* was more similar to that of the B subgenome chromosomes than the patterns of other Sitopsis species, but was not identical to it. C-banding results regarding the phylogeny of *T. turgidum*, and particularly the origin of its B subgenome, were not conclusive, which may be due, in part, to the wide occurrence of intraspecific polymorphism in C-banding patterns, especially differences in the distribution of intercalary and telomeric C-bands (Badaeva et al. 2015).

C-banding has been used quite extensively in the identification of intraspecific chromosomal rearrangements in the different subspecies of *T. turgidum* (e.g., Badaeva et al. 2007, 2015, 2019). During a C-banding survey of a large collection of cultivars of domesticated emmer, ssp. *dicoccon*,

Rodríguez et al. (2000) and Badaeva et al. (2015) identified different types of chromosomal rearrangements, some of which were novel to *T. turgidum*. Chromosomal rearrangements were represented by single translocations and or multiple translocations, as well as by paracentric and pericentric inversions. The use of C-banding enabled the detection of the position of translocation breakpoints, which was either at or near the centromere, or interstitial. Centromeric translocations significantly prevailed in tetraploid wheat (Badaeva et al. 2015).

Use of the C-banding technique in the identification of individual chromosomes and chromosome arms at first meiotic metaphase of F_1 hybrids involving *T. turgidum* and related species, enabled determination of the type of pairing of each *T. turgidum* arm, i.e., homologous versus homoeologous pairing. Thus, for example, Naranjo (1990) analyzed meiotic pairing in the hybrid tetraploid triticale (genome BARR × rye) and identified the arm homoeology of A-B chromosomes of *T. turgidum* using the C-banding technique. Results confirmed that the homoeologous relationships between chromosome arms of the A and B subgenomes in *T. turgidum* are the same as in *T. aestivum*, and that a double translocation involving 4AL, 5AL, and 7BS, and a pericentric inversion involving a substantial portion of chromosome 4A, are present in *T. turgidum* as well as in *T. aestivum*. C-banding studies also successfully identified alien chromosomes, which were added to the hexaploid wheat complement or which substituted one of its chromosomes (Gill 1987).

C-banding was also applied in *T. turgidum* to determine the approximate location of the *Ph1* gene in the 5BL arm. Dvorak et al. (1984) C-banded the long arm of chromosome 5B of a mutant line of the Italian cultivar Cappelli of *T. turgidum* ssp. *durum* deficient for the *Ph1* gene and of another Cappelli line bearing a duplication of part of the long arm of 5B that carries *Ph1*. Compared with arm 5BL of the parental cultivar, the 5B long arm of the *Ph1* mutant was shorter, owing to a deletion of one of two inter-band regions in the middle of the arm. In the line suspected to have a duplication, the 5BL arm was longer than in 'Cappelli' and the interband region that was absent in the *Ph1* mutant was twice as long.

Genetic mapping of polymorphic C-bands also enables direct comparisons between genetic and physical maps (Curtis and Lukaszewski 1991). More specifically, Curtis and Lukaszewski used eleven C-bands and two seed storage protein genes on chromosome 1B, polymorphic between cultivar Langdon of ssp. *durum* and four accessions of ssp. *dicoccoides*, to study the distribution of recombination along the entire length of the chromosome. The genetic maps obtained from the four individual ssp. *dicoccoides* chromosomes were combined to yield a consensus map of 14 markers (including the centromere) for the chromosome.

In contrast to the C-banding technique that stains all types of constitutive heterochromatin (Gill 1987), the N-banding technique specifically reveals heterochromatin containing polypyrimidine DNA sequences (Dennis et al. 1980). Using this technique, Gerlach (1977) was able to identify nine of the twenty-one chromosome pairs of hexaploid wheat cultivar Chinese Spring. These nine chromosomes, 4A, 7A and all of the B subgenome chromosomes, showed distinctive N-banding patterns. The remaining chromosomes show either faint bands or no bands at all. Wild allotetraploid wheat, *T. turgidum* ssp. *dicoccoides* showed banded chromosomes similar to those observed in hexaploid wheat. Of the diploid species, wild and domesticated *T. monococcum*, *T. urartu* and *Aegilops tauschii* showed little or no banding as would be expected of donors of the A and D subgenomes. *Ae. speltoides* had a number of N-banded chromosomes as would be expected of a species closely related to the B subgenome donor. Endo and Gill (1984) using an improved N-banding technique, succeeded to identify 16 of the 21 chromosomes of *T. aestivum*.

3.2 Main Components of the Triticeae Genomes

In this section we describe the general features of the genome's main components shared among the Triticeae species. We discuss both the relative stability of genome size within species and genera, as well as the dynamic processes that can cause genome expansion or reduction in size.

3.2.1 Genome Size

Genome size, determined via analysis of nuclear DNA amount, is known for a large number of Triticeae species (Table 2.4). These studies showed that Triticeae species are characterized by a relatively large genome, compared with that of *Oryza sativa* and *Brachypodium distachyon*, in which 1C nuclear DNA size is 0.51 and 0.30 pg, respectively (Bennett and Leitch 2005), while in Triticeae diploids, it ranged from 4.0 pg in *Elymus libanoticus* and in *E. stipifolius* to 8.9 pg in *Psathyrostachys stoloniformis* or to 9.4 pg in *Secale strictum* (Table 2.4). This, more than two-fold difference in genome size between diploid Triticeae species, indicates that, even though the tribe evolved in a monophyletic manner (Fig. 2.4), genome size changed rapidly already in the primordial species. These large differences in genome size among the diploid species of the tribe imply that genome size expansion during evolution is not a general characteristic of the tribe.

No significant intra-specific variation in 1C nuclear DNA size was found among diploid, allopolyploid, and autopolyploid Triticeae species (Eilam et al. 2007, 2008, 2009). The genome-size stability at the intra-specific level is striking in view of the fact that retrotransposons comprise a significant fraction of the genomes of many Triticeae species (Kumar and Bennetzen 1999; Vicient et al. 2001; Sabot et al. 2005) and as such, these genomes have a considerable potential to undergo rapid changes in nuclear DNA amount.

Relatively little variation in 1C nuclear DNA size is seen within genera, the exception are the genera *Elymus* and *Aegilops*. Diploid *Elymus* species with St genome have 4.0–4.9 pg whereas *Elymus* species with Ee and Eb genomes have 5.85 and 7.4 pg, respectively (Table 2.4). *Aegilops caudata* has 4.84 pg and *Ae. sharonensis* 7.3 pg (Table 2.4). In contrast to the intra-specific and intra-generic levels, there are large differences in nuclear DNA size at the inter-generic level, ranging from 4.0 pg per 1C nucleus in *Elymus libanoticus* to 8.9 pg in *Psathyrostachys stioloniformis* (Table 2.4). At the diploid level, self-pollinating species and cross-pollinating species present similar nuclear DNA sizes. Likewise, nuclear DNA sizes of the perennial species were within the same range as those measured for the annual species [only *Secale strictum* (= *S. montanum*)] has a larger genome than its annual relative). Diploid *Aegilops* species that grow in the southern part of the distribution area of the genus, i.e., in hotter and drier habitats (*Ae. bicornis*, *Ae. searsii*, *Ae. longissima* and *Ae. sharonensis*) have significantly more DNA than diploid species growing in other parts of the species distribution region.

As found for the grass family (Caetano-Anollés 2005), the Triticeae tribe also shows no clear trend of genome size evolution on the inter-generic level; in some species, genome size increased while in others, it decreased (Table 2.4). An increase in genome size can be brought about by the activation of transposons, especially retrotransposons (Bennetzen and Kellogg 1997; Bennetzen 2000, 2002; Wendel et al. 2002), whereas a decrease in DNA amount can result from a variety of recombinational mechanisms, such as unequal homologous recombination between homologous chromosomes, sister chromatids, or intra-chromatids, where the latter, can take place between long terminal repeats (LTRs) of retrotransposons (Vicient et al. 1999; Shirasu et al. 2000; Bennetzen 2002; Devos et al. 2002), or via deletions resulting from non-homologous end joining upon double-strand break (DSB) repair (Gorbunova and Levy 1997). The relative extent of these two counteracting mechanisms determines the direction of genome size change.

The inter-generic differences in genome size mostly reflect variation in the non-coding DNA sequences, since grass genomes contain, more or less, a similar number of genes (Bennetzen et al. 2005; Table 3.1). Differences in genome size are mainly related to differences in transposable element, primarily retrotransposons, content, (Fedoroff 2000,

2012; Bennetzen et al. 2005). The regulation of transposable elements proliferation and the evidence for their role in genome size evolution are discussed in Sect. 3.2.2.2. The contrast between the low variation in DNA size at the intra-specific and intra-generic levels and the high variation at the inter-generic level (Table 2.4), suggests that genome size differences occurred mostly during genera divergence.

3.2.2 Repetitive DNA

Repetitive DNA is the largest component of the genome in species from the Triticeae. It can reach a percent > 90% as in the genome of the Sitopsis species of the genus *Aegilops* (Table 3.1). It is a major contributor to plant chromosome structure and genome evolution (Flavell 1986; Fedoroff 2000, 2012; Bennetzen 2005; Wessler 2006). It is generally packaged as heterochromatin, giving rise to the banding patterns described in Sect. 3.1.5. Repetitive DNA sequences are present either in arrays of tandemly repeated sequences (satellite DNA), or in repeats dispersed throughout the genome, and are classified into two major classes, depending on their structure, position on the chromosomes, and/or the mode of multiplication (transposable elements). The main components of the repetitive DNA described in this section are the Transposable elements and Satellite DNA. Large repeats arrays such as in the nucleolar organizer (Sect. 3.1.4) are also significant components of the genome.

3.2.2.1 Satellite DNA

The satellite DNAs of plants are organized as tandemly-arrayed, highly-repetitive and highly-conserved monomer sequences that are predominantly organized in the genome in uninterrupted tracts (Mehrotra and Goyal 2014). The monomer unit of satellite DNA may range from a few nucleotides to 400 base pairs in length. The tandemly repeated satellite DNAs are found preferentially in the constitutive heterochromatin at specific positions of the chromosomes, such as the pericentromeric, sub-telomeric, telomeric or intercalary regions, e.g., the rDNA in the nucleolar organizing regions. These arrays can be visualized along chromosomes with the C-banding technique that has been widely used in wheat to stain specifically the constitutive heterochromatin in cytological preparations of somatic and meiotic metaphase chromosomes (Gill 1987). Dennis et al. (1980) isolated satellite sequence from bread wheat and found it to be a repeat of $(GAA)_m(GAG)_n$, where m and n may have different values in different arrays. In situ hybridization showed that this satellite sequence is located on all seven chromosomes of the B subgenome and chromosomes 4A and 7A of the A subgenome in hexaploid wheat, equivalent to the location of the N-bands (Gerlach et al. 1979).

Table 3.1 Genome size and features in fully sequenced species from the Triticeae and other grasses

Species	Genome formula	Genome size in Gbp	1C DNA amount (pg)	Transposable elements as percent of genome	Number of high-confidence genes	References
Oryza sativa	OsOs	0.372	0.51	37	41,046	International Rice Genome Sequencing Project (2005)
Brachypodium distachyon	BdBd	0.272	0.30	26.1	25,532	International *Brachypodium* Initiative (2010)
Triticum urartu	AA	4.94	6.02	81.4	37,516	Ling et al. (2013, 2018)
Aegilops tauschii	DD	4.3–4.5	5.17	84–85.9	42,828–39,622	Jia et al. (2013), Luo et al. (2017), Zhao et al. (2017), Zimin et al. (2017)
Ae. speltoides	SS	4.60–5.13	5.81	75.2	36,928–37,607	Li et al. (2022), Avni et al. (2022)
Ae. searsii	S^SS^S	5.55	6.65	82.5	37,995	Li et al. (2022)
Ae. sharonensis	$S^{sh}S^{sh}$	6.07–6.7	7.52	84.3	31,198–38,440	Li et al. (2022), Avni et al. (2022)
Ae. bicornis	S^bS^b	5.73	6.84	83	40,222	Li et al. (2022)
Ae. longissima	S^lS^l	6.22–6.7	7.48	81	31,183–37,201	Li et al. (2022), Avni et al. (2022)
T. aestivum cv. CS Subgenome A	AA	5.95	–	85.9	35,345	International Wheat Genome Sequencing Consortium (2018)
T. aestivum cv. CS Subgenome B	BB	6.29	–	84.7	35,643	International Wheat Genome Sequencing Consortium (2018)
T. aestivum cv. CS Subgenome D	DD	4.79	–	83.1	34,212	International Wheat Genome Sequencing Consortium (2018)
T. aestivum ssp. *aestivum* cv. CS	BBAADD	17	17.67	84.7	107,891	Murat et al. (2014b), International Wheat Genome Sequencing Consortium (2018)
Triticum turgidum ssp. *dicoccoides*	BBAA	10.5 SubA: 4.9 SubB: 5.2	12.91	82.5	65,012	Avni et al. (2017)
Triticum turgidum ssp. *durum*	BBAA	12.31	12.8	82.2	66,559	Maccaferri et al. (2019)
Secale cereale	RR	7.91	8.65	> 80	–	Martis et al. (2013), Bauer et al. (2017)
Hordeum vulgare	HH	5.10	5.50	84	26,159	The International Barley Genome Sequencing Consortium (2012)

Simple sequence repeats (SSRs) is a group of microsatellites occurring universally in plant genomes as tandem repetitions of short sequence motifs. SSRs that are located in non-coding DNA may not affect the organism fitness. This, together with their structure that promotes slippage by DNA polymerase, allows them to accumulate mutations that have been used as highly-polymorphic genetic markers in mapping, identification, characterisation and management of wild and domesticated genetic resources, as well as in tagging genes controlling traits that are essential for wheat breeding (Röder et al. 1998; Pestsova et al. 2000; Gupta and Varshney 2000; Gupta et al. 2002; Somers et al. 2004). On the other hand, SSRs that are located in coding regions, can lead to genetic changes. Analysis of more than 15,000 *Arabidopsis* and more than 16,000 rice SSRs indicated that they may affect the expression of a large number of genes (Sharopova 2008). Data on DNA methylation, histone acetylation, and transcript turnover suggest that SSRs may affect gene expression at transcriptional and posttranscriptional levels.

3.2.2.2 Transposable Elements

The second group of repetitive DNA sequences comprises a vast array of DNA sequences, with a dispersed organization, which are scattered throughout the genome. This group includes transposable elements (TEs), that range in size from a few hundred base pairs (bp) to 15 kb. TEs have the ability to move to new sites in genomes either directly by a cut-and-paste mechanism involving DNA intermediates (transposons; Class 2) or indirectly through a copy-and-paste mechanism involving RNA intermediates (retrotransposons; Class 1) (Fedoroff 2012). Other DNA elements include smaller families such as *Mutator*, *Harbinger*, *Mariner*, *Miniature inverted repeats elements* (MITEs) or *hAT* elements that are less abundant. Helitrons are an interesting but small class of DNA elements that transpose via DNA replication rather than excision-insertion and that are thought to contribute to gene duplication (Morgante et al. 2005).

Class 2 DNA, cut-and-paste transposons include several types of families, the most abundant of which is the *CACTA* family. These TEs have inverted repeats at their termini that contain the conserved *CACTA* motif. In addition, they code for a transposase protein.

Class 1 DNA, retrotransposons or retroelements, comprise two main types: (1) long terminal repeat (LTR) retrotransposons, flanked by LTRs, and (2) non-LTR elements [such as long interspersed nuclear elements (LINEs) and short interspersed nuclear elements (SINEs)]. LTR retrotransposons are the most abundant mobile elements in plant genomes (Feschotte et al. 2002). Indeed, in some grasses, LTR retrotransposons represent up to 90% of the genome (Bennetzen and Kellogg 1997; Kumar and Bennetzen 1999; Vicient et al. 2001; Feschotte et al. 2002; Sabot et al. 2005; Senerchia et al. 2013; Table 3.1). Due to their mechanism of reverse-transcription, LTR-elements can be classified as young, when the LTR are identical, or old, when LTRs are different. Sequence divergence between the LTRs of an element provides thus a useful tool to deduce the age of transposition events. This tool has enabled to date different waves of transposition during wheat evolution and speciation (Wicker et al. 2018).

In bread wheat TEs constitute ∼ 80–85% of the total genome, depending on the lines, the overall composition is ∼ 70% long terminal-repeat retrotransposons (LTR) and ∼ 12% DNA transposons (Walkowiak et al. 2020), altogether wheat TEs were divided into 505 families (IWGSC 2018). Among DNA elements, *CACTA* elements represent the largest class accounting for ∼ 15% of all genome while other elements constitute less than 0.5% of the genome. By contrast retroelements constitute ∼ 70% of the wheat genome (∼ 69% LTR retroelements and ∼ 1% LINEs and SINEs). LTR elements that are "young", i.e., that have transposed recently, tend to be located in the gene-rich recombinogenic distal part of chromosomal arms, while "old" LTR elements tend to be conserved among wheat lines and to be located in the pericentric heterochromatic regions of the chromosomes (Walkowiak et al. 2020). Overall TEs amount is relatively similar in the three subgenomes of allohexaploid wheat, however, they do account for part of the subgenome size difference between B and D and their distribution is highly variable between subgenomes.

TEs have the potential to affect genome structure and function through transposition, ectopic recombination and epigenetic re-patterning (Shalev and Levy 1997; Bennetzen 2005; Slotkin and Martienssen 2007; Fedoroff 2012). They have served as building blocks for epigenetic phenomena, both at the level of single genes and across larger chromosomal regions (Slotkin and Martienssen 2007). Since TE activity is governed by epigenetic regulation (Slotkin and Martienssen 2007; Fedoroff 2012), their activation might be induced by genetic and environmental stresses (Fedoroff 2012). Hence, TEs may mutate genes, alter gene regulation, and generate new genes, in response to environmental challenges, thus providing fuel for evolution (Kidwell and Lisch 2000).

Remarkably, while the intergenic TE composition is not conserved between the subgenomes, the overall TEs composition, the spacing between genes and the enrichment of TEs near genes is highly conserved. TEs associated with genes are particularly interesting as they can affect transcription of neighboring genes as shown for retroelements (Kashkush et al. 2002, 2003) or they can affect splicing as shown for wheat SINEs that are enriched in introns and may affect intron retention (Keidar et al. 2018). MITEs also show a strong association with wheat genes, being near genes, or even within the transcriptome and are thus able to affect both expression and protein composition (Keidar-Friedman et al. 2018). All this suggests that TEs are drivers of dynamic genomic changes and modulation of gene expression and contribute massively into shaping genome structure, expression and evolution. However, these dynamic changes, which can lead to deleterious genome instability are mitigated by epigenetic genome stabilizing factors that suppress TEs activities (Fedoroff 2012). Repressive protein complexes, histone methylation, RNA interference (RNAi) and RNA-directed DNA methylation, as well as recombinational regulatory complexes may cause epigenetic silencing of TEs (Law and Jacobsen 2010; Feng et al. 2010; Zhang and Zhu 2011; Simon and Meyers 2011). Plants have a more complex and redundant array of epigenetic silencing mechanisms than animals, making use of multiple DNA methylation mechanisms, chromatin protein modification, and feedback mechanisms involving small noncoding RNAs (Zaratiegui et al. 2007; Law and Jacobsen 2010; Simon and Meyers 2011). Plants methylate C residues in nucleotides within all sequence contexts (C^mG; C^mNG, or C^mNN), thus stabilizing the silencing and inactivation of genes and other genetic elements in plants (Bird 2002; Law and Jacobsen 2010).

Despite these multiple silencing mechanisms, TEs have retained some extent of mobility throughout wheat evolution as suggested from their high diversity between the subgenomes of bread wheat (Wicker et al. 2018) and between varieties in the same subgenome. While there is no evidence for a TEs mobility burst upon allopolyploidization (Choulet et al. 2014; Wicker et al. 2018), transcriptional activation (but not transposition) of wheat retroelements was shown in a wheat synthetic allopolyploid (Kashkush et al. 2002, 2003) and smoking gun evidence for mobility of MITEs in a newly synthesized wheat allohexaploid was reported (Yaakov and Kashkush 2012). Allopolyploidization might be to some extent a trigger for TEs mobility and mutation tolerance in allopolyploids might have facilitated mobility in allopolyploid backgrounds during wheat evolution. Nevertheless, TE mobility is highly controlled through epigenetic modifications, such as cytosine methylation and small RNAs (Kenan-Eichler et al. 2011) and through histone modifications. The H3K27me2 modification has been shown for example to contribute to TEs stability in wheat euchromatin (Liu et al. 2020).

Middleton et al. (2013) found that the abundance of several TE families varies considerably between the Triticeae species, indicating that TE families can thrive extremely successfully in one species, but go virtually extinct in another. In this regard, Senerchia et al. (2013) suggested that ancestral TE families followed independent evolutionary trajectories among related species, highlighting the evolution of TE populations as a key factor of genome differentiation. The balance between genome expansion through TE proliferation and contraction through deletion of TE sequences, drives variation in genome size and organization (Bennetzen and Kellogg 1997). Indeed, TEs were found to be one of the main drivers of genome divergence and evolution in the Triticeae (Yaakov et al. 2012; Wicker et al. 2018). Remarkably TEs, which are probably under low selection can rapidly "decay" and be eliminated from genomes, or when a transposition burst occurs they rapidly expand in copy number (Wicker et al. 2018). The genome shrinking and expansion cause a great diversity in the size of the genomes in the Triticeae and its close relatives (Table 2.4). Two extreme cases are Brachypodium, with a small ∼ 272 Mb genome and ∼ 25% TEs (International Brachypodium Initiative 2010) versus Rye, which has the largest genome in the Triticeae (7.9 Gb) and also the highest percentage of TEs (∼ 90%) (Rabanus-Wallace et al. 2021 and Li et al. 2021).

The genome size stability at the intra-specific level of Triticeae (Eilam et al. 2007, 2008, 2009) is striking in view of the fact that retrotransposons comprise a significant fraction of the genome (Kumar and Bennetzen 1999; Vicient et al. 2001; Sabot et al. 2005; Senerchia et al. 2013). This is likely due to epigenetic silencing (Fedoroff 2012). On the other hand, the expansion and diversification of TEs in different lineages of the Triticeae, point to an interesting correlation between speciation and transposition. Silent inactive transposons can be activated by a variety of environmental and genetic stresses that cause a "genomic shock" (McClintock 1984). A variety of DNA-damaging agents, biotic and abiotic stresses, as well as pathogen infection and the passage of plant cells through tissue culture were indeed shown to activate TEs, supporting McClintock's genomic shock hypothesis (Hirochika et al. 1996; Grandbastien et al. 1997; Kim et al. 2002; Ito et al. 2011). Transposon activation is also triggered by interspecific hybridization and allopolyploidization (Kashkush et al. 2002, 2003; Madlung et al. 2005; Kenan-Eichler et al. 2011). Such activation can cause bursts of transposition that result in genome expansion for a long period before it is silenced or silenced the triggered activity quickly within few generations (Ito et al. 2011). The activation also contributes to speciation due to the mutagenic effect of transposition bursts that potentially reduce the viability and fitness of the interspecific hybrid, thereby contributing to formation of a genetic barrier between species (Levy 2013).

3.3   Whole Genome Sequencing in Species of the Triticeae

3.3.1   *Triticum aestivum* L. ssp. *aestivum* (Bread Wheat)

3.3.1.1   Assembly of the Bread Wheat Genome

Common wheat, *Triticum aestivum* (genome BBAADD), is an allohexaploid species with a huge genome (17,000 Mbp), about five times larger than the human genome, ∼ 40 times the genome of *Oryza sativa* and ∼ 60 times that of *Brachypodium distachyon* (Table 3.1). Sequencing the genome of allohexaploid wheat, *T. aestivum*, has been a challenging task, due to its large size (a haploid genome, 1C of ∼ 16 pg), the high proportion (∼ 85%) of repetitive sequences, and its allohexaploid nature harboring three related subgenomes that share partial homology (homoeology). An important milestone for the wheat community has been the publication of a first version of an annotated high-quality sequence of the standard laboratory cultivar, Chinese Spring (CS) of bread wheat, ssp. *aestivum*, (International Wheat Genome Sequencing Consortium (IWGSC) 2018). This version presented the assembly of 97% of CS genome into large contigs and scaffolds mapped along the 21 chromosomes. This accomplishment has been achieved through the integration of data from various sources derived from several years of community efforts, including BAC sequencing, molecular markers, and more recent Illumina sequencing data together with NRGene's deNovoMagic2 assembly algorithm and Hi-C data. Another later milestone has been the publication of sequencing and de novo assembly of 15 additional wheat lines that enabled to get insight into the genomic variation between varieties that accumulated following hexaploid wheat formation and to start defining a wheat pangenome (Walkowiak et al. 2020). While cv. Chinese Spring has been a standard for the wheat community, owing to its excellent combining ability that lead to a broad use in early studies on wheat evolution and genome composition (Sears 1954; Sears and Miller 1985), it turns out that it is a landrace that is little bit of an outgroup from the pool of modern varieties (Walkowiak et al. 2020). Nevertheless, we keep referring to it as the standard and mention other varieties only when CS is a clear outlier.

New updates are published by the IWGSC and are available at the IWGSC web site for both genome and transcriptome data. The sequence data has provided new insights into the hexaploid wheat genome structure that are summarized here.

3.3.1.2 Protein Coding Genes

Genes are obviously the most important component of the genome, even-though they represent only a small fraction of the total wheat genome (roughly 5–10%, depending on inclusion of coding regions, introns, UTRs, promoters). There has been contradicting values published on the number of genes in the bread wheat genome. This was due to low quality of the genomic sequences in early estimates and also due to the difficulty to define a gene, or a genuine open-reading frame, and because many genes have decayed and became truncated or mutated pseudo-genes. Some standard emerged in the field, sorting genes as "high confidence" (HC) genes, or as "low confidence" (LC) genes or pseudo-genes. Sequence analysis, searching for protein-coding open reading frames in CS (IWGSC 2018), has identified a total of 107,891 "high confidence" (HC) genes, which are distributed almost equally between the A, B and D subgenomes (Table 3.1). Presence of a transcript and a good prediction of a function (based on homology with known genes from other species) was attributed to 85 and 82% of HC genes, respectively. Another group of 161,537 putative genes, referred to as "low confidence" (LC) genes, i.e. that had weaker features expected for genes, was identified. Only 49% of these putative genes had a corresponding transcript. Finally, 303,818 pseudogenes, i.e. DNA segments that share homology to genes but do not code for proteins, due to truncation or stop codons, were detected. Interestingly, pseudogenes were less abundant in subgenome D (27%) than in A (33%) and B (36%). Overall, if we consider only HC genes, the number of genes per

subgenome is similar to other plant species. However, if we consider the allohexaploid nature of wheat and the fact that ~ half of the LC genes have some coding potential, and maybe some of the pseudogenes too, wheat seems to have much more genes than other cereals and than most other plant species.

3.3.1.3 RNA Genes

RNA genes, that code only for an RNA but not a protein, include as main groups Ribosomal RNAs (rRNAs), the long non-coding RNAs (lncRNAs), micro-RNAs (miRs), small interfering RNAs (siRNAs) and transfer RNAs (t-RNAs).

Ribosomal RNAs (rRNAs) are transcribed from rDNA genes, and do not code for proteins but rather are a major component of the ribosomes. Their structure, expression, and their rapid changes in expression and copy number as a result of allopolyploidy have been described in detail in Sect. 3.1.4.

lncRNAs are typically transcripts > 200 nt that are in general 5′ capped and polyadenylated (Budak et al. 2020). They have not been fully curated in bread wheat but their number is probably > 100,000 considering that in tetraploid wheat they were estimated at 89,623 for wild emmer wheat (Akpinar et al. 2018), 115,437 for durum cultivar Svevo (Maccaferri et al. 2019) and 20,338 in *Ae. tauschii* (Luo et al. 2017). They regulate a broad range of functions such as various stress responses such as cold stress (Lu et al. 2020) or developmental processes such as germination (Budak et al. 2020).

MicroRNAs are ~ 21 nt-long single strand RNA molecules, processed from a hairpin precursor. They function as post-transcriptional regulators of gene expression (Axtell and Meyers 2018). Their function is typically of suppressors of stress-related or developmental processes. They were shown to be associated with response to nitrogen levels, grain development (Hou et al. 2020), water deficit, heat stress and germination (Liu et al. 2020). With the availability of the wheat genome sequence, it has been possible to predict their abundance, based on a combination of structural features and homology to known microRNAs as well as with support from their expression, in particular from libraries of transcribed small RNAs (Jaiswal et al. 2019). With such partial validation, their total number ranges from ~ 2500 to ~ 4500 depending on the prediction and validation tools (Jaiswal et al. 2019). Their distribution is very variable, between homoeologues and between groups from the same subgenome in a way that is not fully related to chromosome size—for example chromosome 3B counts 614 predicted microRNAs while 3D only 45 and 7B counts 52.

Small interfering RNAs (siRNAs) are an abundant class of small non-coding RNAs produced from a double-stranded RNA precursor and processed through a variety of Dicer-like endonucleases with different properties which can generate small RNAs of 21, 22, 23, or more commonly 24 nt in size (Pikaard and Mittlestein-Scheid 2014). They provide pathways for RNA-directed DNA methylation, transcriptional silencing of transposons and repeats, viruses, and transgenes. Considering the abundance of transposons in the wheat genome (discussed below), the importance of the thousands of 24 nt siRNAs species in guiding the methylation machinery to transposons and in mediating their silencing is critical for the maintenance of the wheat genome integrity (Kenan-Eichler et al. 2011). The control of transposons by siRNAs can in turn affect neighboring genes and have consequences on gene expression. For example, gametocidal action in wheat was found to be related to the activity of both miRNAs and siRNAs (Wang et al. 2018). Another example, from the study of synthetic tetraploid wheats is the correlation between genome-wide changes in siRNAs and in the expression of associated genes or transposons and of chromatin modifications (Jiao et al. 2018).

3.3.1.4 The Wheat Transcriptome

In parallel to the publication of the wheat genome (IWGSC 2018), a thorough analysis of the wheat transcriptome was reported (Ramírez-González et al. 2018) based on data from 850 RNA-Seq experiments, derived from 32 tissues and different growth stages or stress treatments. This work provides a comprehensive description of gene expression in wheat as summarized below. This analysis showed that ∼ 85% of the HC genes were expressed versus ∼ 50% of LC genes. Non-expressed genes might have a very low expression, or are expressed under very specific conditions not tested yet or might be defective, e.g., in promoter region, as might be the case for most LC genes. Differential expression between genes was mostly related to differences in tissue origin rather than developmental stage or stress. Only ∼ 10% of the expressed HC genes were tissue-exclusive. These tissue-specific genes were enriched for reproductive functions, transcripts were shorter in length and had a weaker expression that ubiquitous house-keeping genes. Interestingly, genes located in distal regions of the chromosome tended to have a weaker expression level than genes in proximal regions. Expression levels correlated with typical activating or repressive chromatin modifications. Inspection of the expression of homoeologs for which all three subgenomes were represented (triads) showed that for ∼ 70% of the triads, the level of expression was balanced, i.e., homoeoalleles were generally expressed at the same level. For 30% of the triads, expression was unbalanced with one of the three subgenomes being suppressed (most common) or one of the three being active (less common). The D-subgenome was slightly less frequently suppressed than the A or B subgenomes. When looking at gene expression regulation, modules of co-expressed genes were defined. In 83.6% of the cases, the members of the triads were co-regulated, i.e., were in similar modules of expression. The remaining 16.4% of triads showed differential expression patterns in the three homoeoalleles, suggesting sub- or neo-functionalization. Syntenic triplets where more coregulated than non-syntenic ones. Likewise, in cases where only two homoeologs are present, conservation in expression patterns was more prominent than divergence and synteny was a good predictor of conserved expression. Differential expression was often associated with the diversity of TEs in proximity to the promoter region.

A comparison of the grain development transcriptome of bread wheat as well as of its diploid and tetraploid progenitors/relatives showed that the factors affecting divergence in gene expression were (by order of importance): the tissue, the developmental stage, the subgenomes and the species (Xiang et al. 2019). This work constitutes a transcriptome atlas for grain development, starting from the embryonic phase (two cells) through endosperm and embryo development as well as for the pericarp, in bread wheat, tetraploid *Triticum turgidum* (ssp. *durum*), and in diploid ancestors *Triticum urartu*, *Aegilops speltoides*, and *Aegilops tauschii*. Patterns of gene expression seen in bread wheat might be the result of an evolutionary process caused among other factors by selection under domestication or alternatively by intergenomic interactions established upon merging of the sub-genomes.

To understand the interactions between sub-genomes, an insightful experimental system has been to compare gene expression in parents and derived hybrids and allopolyploids at the early stages of interspecific hybridization and allopolyploidization that gave rise to bread wheat. Indeed, bread wheat can be re-synthesized from its tetraploid and diploid progenitors as shown by McFadden and Sears (1944). Since then, several studies on synthetic allopolyploids have enabled us to learn on interactions between the subgenomes and the effect of dosage on overall gene expression (see review in Li et al. 2015). When comparing the expression of homoeoalleles in the diploid (Genome D) and tetraploid (Genome BBAA) progenitors, to that of the derived synthetic allohexaploid wheat (Genome BBAADD), mid-parent values is the prominent pattern of gene expression, nevertheless significant levels of non-additivity were reported, the extent of which was variable between studies, ranging from ∼ 3 to 20% (Pumphrey et al. 2009; Akhunov et al. 2010; Chague et al. 2010; Li et al. 2014). In these studies, biased expression dominance towards the AB sub-genome progenitor was observed using either microarrays or RNA-seq analysis. Moreover, while non-additive expression of homoeoalleles established upon allopolyploidization was not always transmitted to the next generations, there is strong evidence showing that these new expression patterns are often maintained in modern bread wheat.

Gene expression is not only affected by the interactions between different subgenomes in allohexaploid wheat, but also by the ploidy level. Aneuploidy, often encountered in newly formed allopolyploids, can affect gene expression due to imbalance between homoeologous chromosomes. However, even when chromosome dosages are balanced, ploidy can affect both phenotype and gene expression. Comparison of gene expression was done between the polyhaploid, genome BAD, to the allohexaploid, genome BBAADD (Wang et al. 2011) and between the allohexaploid and a nonaploid, genome BBBAAADDD (Guo et al. 2020). Only a small fraction ($\sim$ 0.2%) of the genes were affected (silenced or activated) in the polyhaploid (Wang et al. 2011). By contrast, in the nonaploid, there was a significant dysregulation of gene expression ($\sim$ 25% of the genes) (Guo et al. 2020) and nevertheless the relative ratio between triplicated homoeologs remained relatively unchanged.

Overall these experiments show that there is both a high potential for plasticity as a result of allohexaploidy and at the same time, the system is robust with additivity and mid-parent values remaining the most prominent patterns of gene expression. Understanding the mechanisms that contribute to plasticity and robustness of gene expression and how this contributes to fitness is a major challenge for future research. Genome wide studies on intergenomic interactions in advanced model systems, such as budding yeast (Tirosh et al. 2009) have shown how divergent gene expression in parental species can be caused by cis or trans factors. The same kind of interactions must be taking place in allohexaploid wheat with a higher degree of complexity due to the higher ploidy level and the epigenetic regulators absent in yeast (e.g. cytosine methylation and small RNAs). For example, mutations in *cis*-acting factors, e.g. promoter regions or coding regions, are maintained in an additive manner upon genome merging of the two related species. However, divergences due to mutations in *trans*-acting factors in parental species, e.g. transcription factors, small RNAs, chromatin remodelers, can abolish differences between homoeologs in the allopolyploid and can trigger a rewiring of gene expression by activating or suppressing genes that were otherwise silent. Certain types of methylation can be stably inherited in *cis* or modified due to *trans* effects as with RNA-dependent DNA methylation.

Small RNAs (siRNAs or MicroRNAs) expression can be affected by cis and trans factors and small RNAs can act as trans-acting factors, being synthesized by one subgenome and suppressing genes or transposons in another subgenome. This makes for a high potential for small RNAs-mediated regulation and rewiring of target genes expression during allopolyploidization. Kenan-Eichler et al. (2011) and Li et al. (2014) have shown non-additive and dosage-dependent expression of small RNAs and of their targets in allohexaploid wheat. Typical targets of siRNAs are transposons,

and a correlation between the presence of certain siRNAs whose expression was non-additive in allohexaploid wheat together with methylation and transcriptional activity of their transposons targets has been found (Kenan-Eichler et al. 2011). Specific genes are also affected by non-additive expression of microRNAs: for example, miR9863 was hypothesized to control the enhanced powdery mildew resistance in synthetic hexaploid wheat (Li et al. 2014). Finding more evidence for small RNAs-Targets expression-phenotypic function has been limited so far but should become more effective using current genomics tools and high-quality reference genomes.

On top of these interactions, dosage effects related to ploidy levels also seem to contribute to fitness in a not yet fully understood manner. For example, it seems that the ability to establish in new niches is affected by the ploidy level: hexaploid wheat is more widespread than tetraploid wheat which in turn is more widespread than diploid wheat. However, higher ploidy levels (> 6) seem to have a reduced fitness (Guo et al. 2020) suggesting an optimal ploidy level for wheat fitness and maybe for productivity. The reasons for this success are largely unknown.

A search for the genes that have contributed to wheat success as a crop, in relation to polyploidy, has started in recent years, establishing a causal relationship between such genes and fitness is an important and challenging task. An interesting example is the immediate and increased salt resistance that was obtained in nascent allohexaploid wheat compared to its diploid progenitors due to an altered upregulation of a High Affinity K transporter (Yang et al. 2014). Remarkably this salt tolerance which was obtained already upon allopolyploidization has persisted in natural hexaploid wheat and might have enabled wheat to settle in a broader range of soils.

3.3.2 *Triticum turgidum* L.

3.3.2.1 Ssp. *dicoccoides* (Körn.exAsch. and Graebn.) Thell. (Wild Emmer)

Wild emmer wheat, *Triticum turgidum* ssp. *dicoccoides*, (2n = 4X = 28, genome BBAA) is the direct progenitor of the A and B subgenomes of both tetraploid macaroni wheat and hexaploid bread wheat. Its domestication, which gave rise to emmer wheat (*Triticum turgidum* ssp. *dicoccon*), was described in Chap. 10, Sect. 10.3.2. Its chromosomes show full pairing and recombination with the A and B subgenomes of bread and macaroni wheat, therefore it is an important source of diversity for breeders. The whole genome sequence of emmer wheat was therefore an important milestone (Avni et al. 2017). Moreover, it was the first wheat genome where a high quality of assembly was achieved, with very few gaps. The whole genome size was estimated at

10.5 Gb, out of which 98.4% of the whole genome could be assembled in very large scaffolds from small Illumina Reads (Avni et al. 2017). This achievement was at large due to a new assembly algorithm developed by NRgene combined with a fine genetic mapping of markers that served as anchors to the DNA scaffolds from the Distelfeld lab together with Hi-C data. In total, 10.1 Gb (out of 10.5 Gb whole genome size) was assigned to the 14 chromosome pairs and 0.4 Gb was unassigned. 65,012 High-Confidence genes were identified; 30,730 in sub-genome A and 32,083 in sub-genome B and 82% of the genome was annotated as transposable elements. The wild emmer wheat genome enabled to identify key gene(s) that contributed to domestication, such as the *Brittle-Rachis* loci (*BTR1-A* and *BTR1-B*), which when mutated give rise to the non-fragile rachis (Avni et al. 2017; Nave et al. 2019). It also enabled to evaluate the extent of introgression of wild emmer wheat DNA into the hexaploid genome (see details in Sect. 13.2).

Gene expression studies, under 20 different conditions, indicated that 30.4% of the genes are expressed in all conditions, 48% in at least one (but not all) condition, and 21% were expressed at a low level or not at all (Avni et al. 2017). The number of genes expressed in each sub-genome was very similar, and there was a slight (5%) higher expression on average for genes of the A sub-genome.

3.3.2.2 Ssp. *durum* (Desf.) Husn. (Macaroni Wheat)

Durum wheat, *Triticum turgidum* ssp. *durum* (2n = 4X = 28, genome BBAA), is an allotetraploid species whose genome is genetically very close to that of its domestic progenitor, domesticated emmer wheat *Triticum turgidum* ssp. *dicoccon* (2n = 4X = 28, genome BBAA), and of its wild progenitor, wild emmer wheat, *Triticum turgidum* ssp. *dicoccoides* (2n = 4X = 28, genome BBAA) (Feldman 2001). The F$_1$ hybrid between these species is fully fertile. Durum wheat is a leading crop used for pasta, bulghur, couscous, frikeh and certain types of bread. It represents ∼ 7% of the global wheat production. Sequencing its genome has therefore been an important goal. Early studies consisted in high-resolution mapping of the durum genome using ∼ 30,000 genetic markers, mostly SNPs, spread throughout the genome, yielding a genetic size of 2631 cM for a physical size estimated at 12 Gb (Maccaferri et al. 2015). Whole genome sequencing was first performed on the durum cultivar Svevo (Maccaferri et al. 2019). Genome assembly was performed using the method described for wild emmer (Sect. 3.3.2.1). It gave rise to 10.45 Gb sequences, out of which 9.96 Gb were assigned to the 14 chromosome pairs and 449 Mb were unassigned scaffolds. Overlapping the SNP markers of the genetic map (Maccaferri et al. 2015) with the physical map (Maccaferri et al. 2019) showed that a large pericentric region representing

44% of the genome has a very low recombination rate (107 Mb/cM on average) while distal regions representing ∼ 22% of the genome, had high recombination rates (1.8 Mb/cM on average). Genome annotation revealed 66,559 HC genes and expression studies, from a set of 21 different RNA-Seq samples, showed that 90.5% of these genes were expressed in at least one of the 21 conditions tested.

3.3.3 *Triticum urartu* Tum. ex Gand. (Donor of Bread Wheat A Subgenome)

The genome of *Triticum urartu*, wild einkorn wheat, was sequenced by Ling et al. (2013), first as a draft, and later as a high-quality genome (Ling et al. 2018). Being a diploid (2n = 2x = 14; genome AA) it was of interest due to its reduced complexity. Moreover, it is the diploid donor of the A subgenome of tetraploid *T. turgidum* species (genome BBAA), *T. timopheevii* (genome GGAA) and of hexaploid wheat (genome BBAADD). The high-quality genome was sequenced combining bacterial artificial chromosome (BAC)-by-BAC sequencing, whole-genome shotgun sequencing, and optical mapping (Ling et al. 2018). Sequences assembly generated 4.86 Gb of scaffold sequences, which is very close to the estimated 4.94 Gb genome size. Genome annotation identified 37,516 high-confidence and 3991 low confidence genes. Average gene size is of 1453 bp transcripts, 332 amino acids proteins and 4.5 exons per transcript, which is similar to that of other grasses. In addition, 31,269 microRNAs (miRNAs), 5810 long non-coding RNAs (lncRNAs), 3620 transfer RNAs (tRNAs), 80 ribosomal RNAs (rRNAs) and 2519 small nuclear RNAs (snRNAs) were identified (Ling et al. 2018). Repetitive DNA represented 3.90 Gb (81.42%) of the genome, out of which 3.44 Gb (71.83%) were retrotransposons and 355 Mb (7.41%) were DNA transposons. The distribution of LTR retrotransposons was uneven for each chromosome, with *Copia* elements enriched at both telomeric–subtelomeric regions, and *Gypsy* retrotransposons enriched in the pericentromeric–centromeric regions. Intact retroelements (tens of thousands) suggest a burst of *Gypsy* elements more than 1 Mya and of *Copia* elements less than 1 Mya, both after the divergence of A and B genomes.

High density SNP analyses were carried for 147 (Ling et al. 2018) and 298 *T. urartu* accessions (Brunazzi et al. 2018) from across the fertile crescent, spanning Jordan, Lebanon, Syria, Turkey, Armenia, Iraq and Iran. Genetic diversity was correlated with geographical distance, $R^2 = 019$ (Brunazzi et al. 2018). Interestingly, a principal component analysis performed on climatic variation across the collection of accessions showed that 52% of the genetic variation could be accounted for by temperature indices and to a lesser extent by altitude and rainfall parameters. Moreover, using a GWAS analysis, 57 markers showed association with environmental parameters, including loci associated with frost resistance and dormancy.

3.3.4 *Aegilops tauschii* Coss. (Donor of Bread Wheat D Subgenome)

The genome of *Aegilops tauschii*, the diploid donor of the D subgenome to hexaploid wheat, was first sequenced as a draft, in year 2013, by Jia et al. (2013), Luo et al. (2013). Then in 2017, three high-quality whole genome sequences were published. A high-quality genome, from *Ae. tauschii* ssp. *strangulata,* which is closely related to subgenome D of hexaploid wheat, was sequenced by Luo et al. (2017) using ordered BAC clones sequencing, shotgun sequencing and BioNano optical genome mapping and merging with a previous map (Zimin et al. 2017) that included long Pacific Biosciences (PacBio) reads. In addition, Zhao et al. (2017) used a combination of high coverage Illumina sequencing with the DeNovoMAGIC2 assembly software and PacBio long reads to sequence the same accession of ssp. *strangulata* (AL8/78) as Luo et al. 2017. The estimated genome size of *Ae. tauschii* is ~ 4.3–4.5 Gb, of which 4.0–4.3 Gb (depending on the studies) were assembled into super-scaffolds representing ~ 95% of the total genome. The total of high-confidence genes varied in the different studies, ~ 42,828 (Zhao et al. 2017) compared to 39,622 HC genes according to Luo et al. (2017) eventhough the same accession was used in both studies. This is probably due to the annotation method and the large amount of low confidence genes and pseudogenes that could be mis-classified. For some unclear reasons, *Ae. tauschii* had longer genes and transcripts than previously sequenced plant species, essentially due to longer exons (Luo et al. 2017; Zhao et al. 2017). The high-confidence genes can be clustered into gene families; a total of 12,607 clusters were shared between other grass genomes (Barley, *Brachypodium*, Rice and *Sorghum).* The same comparison with wheat subgenomes showed 15,180 shared gene family clusters. Interestingly, duplicated genes were the most abundant category of the *Ae. tauschii* genes (71.5%) with 4001 tandem (10.3%) and 23,722 dispersed (61.2%) duplicates, only 5050 (13%) genes were single-copy. Overall, the ratio of duplicated vs single copy genes is higher than in other species.

Repetitive DNA was highly abundant with transposable elements representing 84% (Luo et al. 2017) or ~ 86% (Zhao et al. 2017) of the genome. LTR retroelements were the most abundant (65.9% of the whole genome), with *Gypsy* elements constituting the highest component of the retroelements and among DNA transposons, *CACTA*

elements were the most abundant. The density of the *Gypsy* superfamily showed a gradient of density, lowest near telomeres and higher in centromeric regions. One particular family, *Gypsy12*, homologous to Barley *cerebra* elements, was clustered in the centromere, constituting part of the centromere core. While the LTR retroelements showed a peak of transposition burst ~ 1Mya, there was evidence for multiple bursts for specific sub-families for the past 3 M years. Beyond that TEs could not be identified due to fast turnover (Luo et al. 2017). 80% of the pseudogenes were somehow disrupted by TEs, suggesting TEs roles in pseudogenization (Zhao et al. 2017). Moreover, ~ half of the genes contained TEs and on average such genes had a lower expression than genes free of TEs, presumably due to increased cytosine methylation (Zhao et al. 2017).

3.3.5 Species of *Aegilops* Section Sitopsis (Jaub. and Spach) Zhuk.: *Ae. speltoides* Tausch; *Ae. bicornis* (Forssk.) Jaub. and Spach; *Ae. sharonensis* Eig; *Ae. longissima* Schweinf. and Muscl.; *Ae. searsii* Feldman and Kislev ex Hammer

While the origin of the A and D subgenomes of wheat is well established (see above Sects. 3.3.3 and 3.3.4), the origin of the B subgenome has remained elusive. It has been proposed, on the basis of morphological and karyotypic similarities between *Triticum* and the five diploid Sitopsis species (2n = 2x = 14), that the donor of the B genome could belong to a species from the Sitopsis section, most probably from *Ae. speltoides* (Sarkar and Stebbins 1956; Riley et al. 1958). However, the analysis of meiotic pairing between a low-pairing genotype of *Ae. speltoides* and bread wheat showed very little or no synapsis (Kimber and Athwal 1972), and C-banding pattern fwas different (Gill and Kimber 1974; Ruban and Badaeva 2018). Early molecular analyses showed that while *Ae. speltoides* was the known species with the highest similarity to the B subgenome, it remained quite divergent at both molecular and cytogenetical levels (Huang et al. 2002; Petersen et al. 2006; Gornicki et al. 2014; Miki et al. 2019). This led to various hypotheses regarding the B subgenome progenitor. First, is the monophyletic origin, with a direct progenitor, either *Ae. speltoides* itself, or an extinct or an undiscovered species. Second, is the polyphyletic scenario, whereby the B subgenome evolved rapidly in the polyploid background through hybridization and introgressions from other Sitopsis species (Glémin et al. 2019; Bernhardt et al. 2020) or other related allopolyploid species (Zohary and Feldman 1962; Natarajan and Sharma 1974; El Baidouri et al. 2017). Whole genome sequencing of the five Sitopsis species has helped better understand their genomic structure as well as the phylogenetic relationships within the section, and with the B subgenome (Li et al. 2022, Avni et al. 2022).

High-quality de novo reference genome sequencing of the five Sitopsis species was assembled by a combination of Oxford Nanopore Technologies, single-molecule real-time technology and Hi-C based scaffolding strategy, followed by Illumina short read-based sequencing (Li et al. 2022). Genome sizes of the five species ranged from 4.11 to 5.89 Gb, which is lower from the size range estimated by flow cytometry, namely, 5.81–7.52 pg (Table 3.1). *Ae. speltoides* (4.11 Gb) had the smallest genome among the five species. It is close in size to the bread wheat D subgenome and its donor *Ae. tauschii* (Table 3.1). In contrast, the remaining four Sitopsis species, *Ae. bicornis* (5.64 Gb), *Ae. longissima* (5.80 Gb), *Ae. searsii* (5.34 Gb) and *Ae. sharonensis* (5.89 Gb), all have much larger genomes, similar in size to the B subgenomes (5.11–5.18 Gb) (Table 3.1).

Avni et al. (2022) combined Illumina 250-bp paired-end reads, 150-bp mate-pair reads, 10X Genomics and Hi-C libraries for whole genome sequencing and chromosome-level assembly of *Ae. longissima* and *Ae. speltoides* genomes. They also compared these genomes with the recently assembled *Ae. sharonensis* genome (Yu et al. 2021). The genome of *Ae. longissima* has an assembly size of 6.70 Gb, highly similar to that of *Ae. sharonensis* (6.71 Gb) and substantially larger than the 5.13-Gb assembly of *Ae. speltoides*. These values are in agreement with nuclear DNA quantification that showed 1C-values of ~ 7.5, ~ 7.5, and ~ 5.8 pg for *Ae. longissima*, *Ae. sharonensis*, and *Ae. speltoides*, respectively (Eilam et al. 2007) (Table 3.1). Quality control tools showed a high level of genome completeness with 97.8% for *Ae. sharonensis*, 97.5% for *Ae. longissima*, and 96.4% for *Ae. speltoides*. The predicted chromosome sizes in *Ae. longissima* and *Ae. sharonensis* were similar for all chromosomes, except for chromosome 7, which is much smaller in *Ae. longissima* due to a translocation to chromosome 4 (Friebe et al. 1993; Zhang et al. 2001).

The differences in genome size were mostly due to the activity of transposable elements (Li et al. 2022). The difference in genome size among the five Sitopsis species can be mainly attributed to the total length of repetitive DNA (86.13–88.11% of the total), including 2.94–4.21 Gb (66.48–71.47%) retrotransposons and 0.52–1.03 Gb (12.54–19.21%) DNA transposons (Table 3.1). A general feature of the five Sitopsis species and of the bread wheat subgenome B, is that *copia-like* retrotransposons tend to cluster at telomeric regions of all chromosomes while *gypsy-like* retrotransposons cluster in pericentric regions (Li et al. 2022).

A total of 37,201–40,222 HC protein-coding genes were predicted in the five Sitopsis genome (Table 3.1). The average transcript size of the Sitopsis genes encode transcripts of 1193–1319 bp in length, which are comparable to the bread wheat subgenomes.

Consistently with other sequenced Triticeae species, gene density was higher in the distal than the proximal chromosomal regions in all Sitopsis species. These Sitopsis genes could be classified into ~ 24,000 gene families, out of which ~ 17,600 are shared with other diploid or allopolyploid *Triticum/Aegilops* species. A total of 419–1086 genes were Sitopsis-specific genes. A total of 38,994 structural variants were identified in the five Sitopsis species (ranging from 37,039 to 37,721 in each species), 18,153 of which are shared with the two diploid species, *Ae. tauschii* (DD) and *T. urartu* (AA).

The Sitopsis species represent a very broad germplasm from which valuable agronomic genes could be transferred to the domestic wheat backgrounds. Of particular interest are disease resistance genes which are widely available within the Sitopsis (Anikster et al. 2005; Olivera et al. 2007; Scott et al. 2014). The Mining of the Sitopsis genomes has enabled to identify the spectrum of multiple and new resistance candidate genes. Several hundreds of NBS-LRR genes were detected in each species (Li et al. 2022; Avni et al. 2022), some genes are homoeologous to known resistance genes which can provide resistance to different races of diseases such as Stripe rust or powdery mildew, and other are species-specific loci.

3.3.6 *Secale cereale* L. (Domesticated Rye)

Rye, *Secale cereale,* is an important crop in several countries, in particular in Northern Europe, being used for food and feed. It distinguishes itself from other Triticeae species through a strong allogamy. It is notorious for its climate resilience, being able to grow in poor soils and under biotic and abiotic stress. It is a diploid species (2n = 2x = 14; genome RR), belonging to the Triticineae subtribe, that can be crossed to both tetraploid and hexaploid wheat. The resulting F_1 hybrid is sterile but fertility can be restored by genome doubling. The resulting synthetic amphiploid shows heterotic features that have been exploited to generate Triticale, a new man-made crop. Moreover, hybridization with wheat can be used to introgress genes into the wheat germplasm, as was done to transfer the powdery mildew resistance gene from chromosome 1RS of rye to the wheat background. Rye therefore constitutes an important resource by itself, or as a parent of new crops or for wheat improvement. Sequencing of its genome was achieved in two inbred lines, a Chinese elite line Weining (Li et al. 2021) and Lo7, an inbred cultivar (Rabanus-Wallace et al. 2021). These genome sequence are a valuable resource for breeding in Rye and in the Triticineae. Rye turns out to have one of the largest genome, among the diploids of the Triticeae estimated at 7.9 Gb via flow cytometry, essentially due to transposable elements which constitute ~ 90% of the genome. To construct a high-quality genomic sequence, Li et al. (2021) integrated data generated by long-range PacBio and Illumina sequencing, and from chromatin conformation capture (Hi-C), genetic mapping and BioNano analysis. A genome assembly of 7.74 Gb was achieved, i.e. 98.47% of the total estimated genome size. The genome assembly by Rabanus-Wallace et al. (2021) using short reads together with genetic maps, Hi-C and BioNano analysis, assembled 6.74 Gb, i.e. ~ 78% of the estimated size. Li et al. (2021) annotated 45,596 high-confidence (HC) and 41,395 low-confidence genes. HC genes include 4217 single copy genes, 23,753 dispersed duplicated genes, 6659 proximal duplications, 7077 tandemly duplicated genes and 1866 segmentally duplicated genes. They also annotated 34,306 microRNA, 14,226 long non-coding RNA, 11,486 transfer RNA and 1956 small nucleolar RNA species. The average intron length of HC genes was the longest among sequenced grass genomes, but exons and coding sequences were similar in size compared to other genomes. TEs content (~ 90%) was higher than for other genomes, with retroelements being the dominant TEs (~ 84.5%) and among these, the *Gypsy*-like elements were responsible for most of the genome size expansion in the Weining genome. One burst of TE amplification (mostly that of *Gypsy*-like TEs) occurred ~ 0.5 MYA while an earlier peak took place ~ 1.7 MYA. Interestingly, TE transpositions was associated to a high proportion of dispersed gene duplications (10,357 out of 23,753) and to neofunctionalization of genes selected for breeding.

The contribution of large-scale structural variants to genome evolution was assessed by comparing two rye cultivars 'Lo7' and 'Lo225', representing two distinct heterotic gene pools (Rabanus-Wallace et al. 2021). Megabase-scale inversions were identified on four of the seven rye chromosomes, including a 50-Mb inversion on chromosome 5R, which coincides with a region lacking genetic recombination. Considering that 382 HC genes are present in this inversion, this can hinder breeding progress in that region.

3.3.7 *Hordeum vulgare* L. (Domesticated Barley)

Domestic barley, *Hordeum vulgare,* is an ancient and important crop used mostly as food, feed and for alcoholic beverages. Its genome is diploid (2n = 2x = 14; genome HH). It was one of the first crops of the Triticeae to have a high-quality physical map constituted of single BACs assembly

(International Barley Genome Sequencing Consortium 2012). This map, together with a genetic map, an optical map, a population sequencing (POPSEQ), Hi-C data and Illumina reads, enabled to assemble a high-quality full genome sequence (Mascher et al. 2017). A length of 4.79 Gb of non-redundant sequences was assembled, out of a total genome size estimated by flow cytometry at ~ 5.1 Gb (Dolezel et al. 1998). A total of 39,734 high-confidence genes and 41,949 low-confidence genes were identified. In addition, 19,908 long non-coding RNAs and 792 microRNA precursor loci were predicted. 29,944 genes were part of families with multiple copies.

Hi-C data was useful not only in joining contigs, but also into probing the three-dimensional structure of the chromatin in the nucleus. Data on adjacency of chromosomal regions suggested that the genome is organized into a Rabl-configuration, with centromeres clustered at one pole and telomeres at the opposite pole. This structure was supported by fluorescence in situ hybridization when using telomere or centromere probes (Mascher et al. 2017).

In a followup study from the Gatersleben team, Jayakodi et al. (2020) reported on the de-novo assembly of 20 varieties of barley, including modern cultivars, landraces and a wild barley. In addition, a shotgun analysis was done for 300 genebank accessions. This analysis enabled to identify the abundance and size of structural variants. Large inversions (> 5 Mb) were frequent in the genome of the 20 lines, with one inversion being 141 Mb-long. Structural variants were also found. It was estimated that on average, each of the 20 lines contained 2.9 Mb of single-copy sequence not present in the other lines.

Mascher et al. (2017) reported that a total of 3.7 Gb, or 80.8% of the total assembled sequence, was derived of transposable elements, with a majority of retroelements. A similar proportion of TEs was found in all 20 genomes sequenced by Jayakodi et al. (2020), however, their location was variable in the different genomes. TEs had typical insertion patterns along the chromosome (Mascher et al. 2017). For examples, MITEs and LINEs were found mostly in gene-rich distal regions, the pericentric regions were rich in *Gypsy* retroelements while *Copia* TEs were in distal or interstitial regions. Mariner elements were located ~ 1 kb upstream or downstream of genes while Helitrons and Harbinger elements had a preference for promoter regions. LTR elements were more distant from genes.

3.4 Gene Order, Comparative Genomics and Karyotypic Evolution

Grasses have a monophyletic origin dating back at an estimated ~ 70 MYA (Linder 1986; Clark et al. 1995), although the discovery of grass phytoliths into dinosaurs coprolites suggests it might have been even older (Prasad et al. 2005, 2011). One of the fascinating discoveries emerging from genome mapping with DNA markers in the late 80's to early 90's was that despite their long period of divergence, the differences in morphology, habitat, number of chromosomes and genome sizes, grass genomes show a very significant collinearity (synteny) in gene order. This was shown first when comparing Sorghum and maize (Hulbert et al. 1990). This was not totally unexpected as they diverged only 15–20 MYA. However, subsequent comparative mapping showed that long collinear segments are also found between rice and maize which diverged ∼ 70 MYA and also between wheat and rice (Ahn et al. 1993; Kurata et al. 1994). Collinearity is also conserved when comparing extremes in diploid genome size, such as *Brachypodium distachyon* (272 Mb) and the ∼ 30 fold larger rye genome (7.9 Gb), (Li et al. 2021). This led to the building of collinearity maps aimed at integrating data from the various grasses (Bennetzen and Freeling 1993) and to the very useful circular viewing configuration of syntenic genomes (Moore et al. 1995) which altogether provided a new understanding of the ancestral grass karyotype and of karyotypic evolution.

Recent sequencing of several Triticeae genomes and construction of high-resolution gene-based genetic maps, enabled comparison of modern Triticeae genomes with their reconstructed founder ancestral genomes (Salse et al. 2008; Murat et al. 2010, 2014a, b; Salse 2012). Such a comparison has revealed the evolutionary history of the present-day Triticeae genomes and has provided insight into how the wheat, barley, and rye genomes are organized today compared to their grass relatives (rice, sorghum, millet and maize) (Murat et al. 2014a, b).

Such comparisons show that grasses are derived from an ancestor having n = 7 protochromosomes containing 8581 ordered protogenes, dating back to ∼ 90 MYA. This ancestor went through a whole-genome duplication to reach the n = 14 ancestral grass karyotypes (Murat et al. 2014b). This chromosome number was then decreased to n = 12, as a result of two telomeric/centromeric fusions, three inversions and two translocations (Murat et al. 2014a). The genomes of maize, millet and sorghum were proposed to have derived from this duplicated intermediate, through distinct ancestral chromosome fusion patterns (Murat et al. 2014a). According to this scenario, the modern rice genome retained the n = 12 chromosome profile, making this the ancestral genome reference karyotype for comparative genomics investigation in grasses (Murat et al. 2014a, b). In fact, it is estimated that all the diploid grass species are paleopolyploids, since their ancestor underwent, at least one event of whole-genome duplication (Salse et al. 2008; Jiao et al. 2014; Wang et al. 2015; Murat et al. 2017). This polyploidization was followed by cytological and genetic diploidization, which explains the divergence of the diploid

grass genomes during their evolution from a common ancestor (Bolot et al. 2009).

The ancestral Triticeae karyotype has been proposed to have derived from the n = 12 ancestral grass karyotype (Salse et al. 2008; Luo et al. 2009). Comparative genetics showed that the reduction from n = 12 to n = 7, without the loss of genes, was accomplished through four centromeric ancestral chromosome fusions (leading to functional monocentric neochromosomes), one fission and two telomeric ancestral chromosome fusions (Salse et al. 2008; Luo et al. 2009). These changes involved translocation of the euchromatic portion of each arm of rice chromosomes to one of the seven Triticeae chromosomes; the heterochromatic centromeric region was subsequently lost (Luo et al. 2009). The Triticeae basic chromosome number of n = 7 evolved via the loss of five functional centromeres, four of which correspond to those of rice chromosomes Os4, Os5, Os6, and Os9 and the fifth to either that of Os3 or Os11 (Luo et al. 2009). Elucidation of these changes in the basic chromosome number in Triticeae was made possible by locating the sites of the seven active and five lost Triticeae centromeres, the present and past chromosome termini, and their relationships to those in rice (Luo et al. 2009).

The modern genome of barley, *Hordeum vulgare*, retained the original ancestral Triticeae n = 7 karyotype (The International Barley Genome Sequencing Consortium 2012), turning it into a reference karyotype for comparative genomics investigation within the Triticeae tribe (Mayer et al. 2011; International Barley Genome Sequencing Consortium 2012). The genome of modern rye, *Secale cereale*, derived from the n = 7 ancestral Triticeae karyotype. A comparison of the rye genome (cv Weining) to an ancestral grass karyotype (similar to the rice genome), showed 23 large syntenic blocs that encompass 10,949 orthologous gene pairs (Li et al. 2021). Chromosome 3R was mostly derived from rice chromosome Os1. Other blocs were more complex, with nested insertions and translocations. When comparing the Weining assembly to that of bread wheat, each subgenome had ∼ 50% conserved orthologous gene pairs. The order of these genes in chromosomes 1R, 2R, 3R was entirely colinear with that of wheat group 1, 2 and 3, respectively. Chromosome 4R, showed collinearity with parts of groups 4, 7 or 6. Chromosome 5R was entirely collinear with 5A and partly collinear with 5B and 5D due to translocation of 4B or 4D segments at the ends of 5BL or 5DL. Chromsome 6R showed collinearity with groups 6, 3 and 7 of wheat. Chromosome 7R was mostly colinear with group 7 with non co-linear regions due to translocations (Li et al. 2021). The genome of common wheat, *Triticum aestivum,* shares the ancestral reciprocal translocation (between A4 and A5 on the A subgenome) characterized in rye, and underwent two allopolyploidization events, first involving species with subgenomes A and B, and later between AB and D, as well as an additional lineage-specific translocation (between chromosomes 4A and 7B) to reach the modern 21 chromosomes (Salse et al. 2008; Luo et al. 2009; Murat et al. 2014a, b). Interestingly, ryegrass, *Lolium perenne*, which is not a Triticeae, presents a modern karyotype of n = 7 chromosomes that is close to the ancestral Triticeae karyotype (Pfeifer et al. 2013), suggesting that this ancestral Triticeae karyotype structure already existed before the formation of the Triticeae (Murat et al. 2014a, b). One of the most prominent large-scale karyoptypic variant that occurred in wheat is a translocation between chromosomes 5B and 7B. The translocation was found in the majority of the lines tested by Walkowiak et al. (2020). Sequence data showed that the translocation between chromosomes 5B and 7B, which are ∼ 737 and 762 Mb long, respectively, gave rise to translocated chromosomes of 488 Mb (5BS/7BS) and 993 Mb (7BL/5BL) in length. The translocation breakpoint was mapped to a ∼ 5 kb GAA microsatellite when comparing cv. ArinaLrFor and SY Mattis. Interestingly, the *Ph1* locus is near the breakpoint but remained syntenic among all the lines tested regardless of the presence or absence of the translocation.

The *T. urartu* genome was compared to the *T. aestivum* A and B subgenomes and to the rice genome (Ling et al. 2018). Overall, syntenic blocs A *versus* A or A *versus* B are well conserved with unaligned regions resulting mostly from retroelements insertions. Some rearrangements of syntenic blocks were observed, such as a reciprocal translocation at the distal end of the long arms of *T. urartu* chromosomes 4 and 5; a non-reciprocal translocation from *T. aestivum* 7B to *T. aestivum* 4A; and a pericentric inversion on *T. aestivum* 4A involving most of the long and short arms. The comparison with rice enabled fine mapping of the regions that underwent the chromosomal fusions that enabled the karyotypic transition from 12 (in rice) to 7 chromosome pairs in wheats.

Synteny was also assessed in more recent divergence events, such as between wild emmer and durum wheat. Comparison between the durum (Svevo) and *dicoccoides* (Zavitan) genomes, as described by Maccaferri et al. (2019) indicated a high synteny for high-confidence genes and a similarity in high-confidence genes copy numbers, namely, 66,559 versus 67,182 for durum and *dicoccoides* respectively. Out of 36,434 HC unigene groups, 12,842 had a syntenic copy between Svevo and Zavitan for both A and B genomes (with all homoeologs present and conserved); 6793 had a synthenic copy in A only or in B only; 4313 HC genes were only found in Svevo and a similar number (4227) was in Zavitan only. Intriguingly, the length of the coding sequence (CDS) of the genes that were not present in all sub-genomes or had no corresponding allele in durum vs *dicoccoides*, i.e. genes that tended to be eliminated during

evolution, had a significantly smaller CDS (∼ 750 bp), compared to genes retained in all subgenomes (∼ 1250 bp). Several syntenic LTR transposons were also identified, suggesting that there was no major burst of transposition during the time of divergence from a common ancestor.

In bread wheat, synteny is highly conserved in the three subgenomes. Synteny was more prominent, with larger syntenic blocks, in interstitial regions compared to subtelomeric regions or pericentric regions (IWGSC 2018). Higher gene duplication in the sub-telomeric regions was in part responsible for the reduction in synteny. Only 55% of the genes had a syntenic homoeolog in all three subgenomes; 15% had at least one missing homoeolog but had a paralog. The percent of missing homoeologs (gene loss) was similar, i.e. ∼ 10%, in each subgenome (IWGSC 2018).

When analyzing the genomic sequence of different bread lines (Walkowiak et al. 2020), it was found that overall, gene order on homologous chromosomes revealed a high collinearity and overall genome size was similar (Walkowiak et al. 2020). Nevertheless, ∼ 12% of the genes showed a "Presence/Absence Variation" (PAV) between the different cultivars and 26% of the predicted genes were found in tandem duplications. This suggests a dynamic gene copy number variation (CNV) that could have been stimulated by allopolyploidy following bread wheat formation and further shaped by selection. Several evidences suggest that CNV contributes to agronomic adaptation. For example, analysis of the repertoire of genes that control disease resistance, such as the nucleotide-binding leucine-rich repeat (NLR) coding genes, showed that only ∼ 1/3 of ∼ 2500 NLR loci are shared between all sequenced cultivars, the remaining 2/3 being partly shared between some of the cultivars and tens or hundreds of unique cultivar-specific NLR signatures were found (Walkowiak et al. 2020). Considering that hexaploid wheat was formed under domestication only ∼ 9000 years ago, this very rapid diversification of disease resistance genes likely points out to a response to selective pressure caused by diseases under wheat cultivation.

On a micro-scale, gene distribution was also found to be non-random. Pre-genome studies of Gill et al. (1996a, b), Sandhu and Gill (2002), Erayman et al. (2004) proposed that common wheat genes are clustered into a limited number of very large gene islands that are separated from each other by inter-genic regions regularly containing transposable elements. However, studies on gene distribution in the fully sequenced chromosome 3B of common wheat (Rustenholz et al. 2011) and in the seven chromosomes of *Aegilops tauschii* (Luo et al. 2013), failed to support this claim. Sequencing of single wheat BAC clones and BAC clone contigs suggested that gene clustering into many small islands is a more likely scenario (Feuillet and Keller 1999; Wicker et al. 2005; Choulet et al. 2010). Similarly, sequences obtained from a few BACs of *Ae. tauschii* genome, namely, loci from Glu-D1 (Anderson et al. 2003), Ha (grain hardness/softness) (Chantret et al. 2005), and equivalent region of *Lr21* (leaf rust resistance) (Brooks et al. 2002) have revealed a structure of gene islands within highly repetitive DNA (Ogbonnaya et al. 2006). Comparison between the *Ha* locus region in *Ae. tauschii* with that of the D subgenome of hexaploid wheat, showed a conserved gene order of the *Gsp-1* (grain softness protein) and the puroindoline genes *Pina* and *Pinb*, with the main differences confined to the insertion of various retroelements (Chantret et al. 2005). In order to gain a better understanding of the gene arrangement and its evolution in grass genomes, Gottlieb et al. (2013) studied gene distribution in orthologous regions of the genomes of four grass species differing in genome size, *Sorghum bicolor*, *Oryza sativa*, *Brachypodium distachyon*, and *Ae. tauschii*. *Aegilops tauschii*, like all members of the tribe Triticeae, has a large genome (Table 3.1), while the genomes of *O. sativa* and *B. distachyon*, are more than an order of magnitude smaller (Table 3.1). The sorghum genome, at 730 Mb (Paterson et al. 2009), is about one-sixth the length of the *Ae. tauschii* genome. By comparing these four genomes, Gottlieb et al. (2013) found that genes are distributed rather uniformly in the small genomes of *O. sativa* and *B. distachyon*, but in the larger *S. bicolor* and *Ae. tauschii* genomes, genes tended to cluster into islands.

Gottlieb et al. (2013) proposed to call the gene islands "gene insulae" to distinguish them from other types of gene clustering that have been proposed earlier, namely gene-rich regions and gene islands. An average insula is estimated to contain 3.7–3.9 genes, with an average inter-genic distance within an insula of 2.1 kb in *S. bicolor* and 16.5 kb in *Ae. tauschii*. The average inter-insular distances are 15.1 and 205 kb, in *S. bicolor* and *Ae. tauschii*, respectively. Gene density in many grass genomes generally increases from the proximal towards the distal regions of chromosome arms (Gottlieb et al. 2013). Using regression analysis, Gottlieb et al. (2013) showed that gene number per insula and gene density within an insula, were similar along *Ae. tauschii* chromosomes but that inter-insular distances were shorter in distal, high-recombination regions compared to proximal, low-recombination regions. Therefore, the increase in gene density toward the distal regions of *Ae. tauschii* chromosomes (Luo et al. 2013) can be ascribed to shortening of inter-insular distances. The variation in inter-insulae distances is mostly determined by LTR retroelements insertion or deletion. This is consistent with the hypothesis that the accumulation of LTR retroelements occurs principally in regions already containing LTR retroelements (Bennetzen and Kellogg 1997). Data from whole genome analysis confirmed these earlier studies showing that genes tend to be organized in small clusters localized in-between repetitive DNA (Luo et al. 2017).

In contrast, to the insulae relative fixed size along the chromosome reported in *Ae. tauschii*, insulae in the distal gene-rich regions of wheat chromosome 3B were reported to contain more genes than in insulae in the proximal, gene-poor chromosome regions (Rustenholz et al. 2011). The higher gene density in the distal regions of chromosome 3B of common wheat was related to a higher incidence of duplicated genes in these regions (Luo et al. 2013; Dvorak and Akhunov 2005) and these duplicated genes are not syntenic with *Oryza sativa* and *Brachypodium distachyon* (Rustenholz et al. 2011). For example, when comparing the *Ae. tauschii* and bread wheat D subgenome, only 87.4% of the genes were present at the expected orthologous position, highlighting a dynamic gene copy-number variation that is not related to whole genome duplication (Zhou et al. 2021). Some gene families, such as prolamine genes (including glutenins and gliadins) and disease-resistance genes were among those most prone to be non-orthologous with other grasses (Zhou et al. 2021). This is also consistent with the findings of Walkowiak et al. (2020) that CNV tend to be more frequent in subtelomeric regions.

In bread wheat, a genome-wide average of 27% of the high-confidence genes are organized in arrays of tandem duplicates (IWGSC 2018). These gene families show a trend of expansion as determined by comparing the number of family members in each subgenome to their progenitor or related species. Gene expansion had generally occurred in the wild progenitor or the common ancestor of the subgenomes: out of 8592 expanded families, 6216 expanded in all three A, B, D subgenomes while 1109 expanded in only one of the subgenomes (IWGSC 2018). Only 78 gene families contracted. Interestingly, when the families that expanded were assigned to functions, there were significant differences between subgenomes. There was an over-representation of seed-related genes (embryo and endosperm) in the A subgenome and of vegetative growth and development in the B subgenome. Families that expanded in all three genomes were enriched in genes that play an important role in wheat breeding such as yield or biotic and abiotic stress resistance (IWGSC 2018). Disease resistance–related NLR loci and WAK (wall-associated receptor)—like genes were clustered in high numbers at the distal regions of all chromosome arms. The restorer-of-fertility-like (RFL) subclade of P class penta-tricopeptide repeat (PPR) proteins, potentially of interest for hybrid wheat production, comprised 207 genes, nearly threefold more per haploid subgenome than have been identified in any other plant genome analyzed to date. They localized mainly as clusters of genes in regions on the group 1, 2, and 6 chromosomes, which carry fertility-restoration QTLs in wheat. Within the dehydrin gene family, implicated with drought tolerance in plants, 25 genes that formed well- defined clusters on chromosomes 6A, 6B, and 6D showed early increased expression under severe drought stress. From these few examples, it is evident that flexibility in gene copy numbers within the wheat genome has contributed to the adaptability of wheat to produce high-quality grain in diverse climates and environments (Feldman and Levy 2015). Knowledge of the complex picture of the genome-wide distribution of gene families, which needs to be considered for selection in breeding programs in the context of distribution of recombination and allelic diversity, can now be applied to wheat improvement strategies.

3.5 Meiotic Recombination in the Wheat Genome

Meiotic recombination between homologous chromosomes is a major engine of diversity in wheat evolution and breeding. In bread wheat, Gill et al. (1996a, b), Sandhu and Gill (2002) noted that recombination was suppressed in the centromeric regions and was mainly confined to the distal gene-rich regions. In accordance with these findings, Lukaszewski and Curtis (1993) and Akhunov et al. (2003) noticed that crossover (CO) events locate in the distal chromosomal regions. Dvorak (2009) reviewed data indicating that Triticeae genomes have a steep recombination gradient along the centromere-telomere axis and gene content is structured in a similar pattern along this axis, namely, the number of genes increases from the proximal towards the distal region. In their detailed study of the pattern of COs along chromosome 3B of bread wheat, Saintenac et al. (2009) found, like Dvorak (2009), that the crossover frequency increases gradually from the centromeres to the telomeres. These findings have been reinforced by physical mapping of wheat chromosomes, which revealed small chromosome segments of high gene density and frequent recombination interspersed with relatively large regions of low gene density and infrequent recombination (Saintenac et al. 2009). Multiple COs occurred within these gene-dense regions, and the degree of recombination in these regions is at least 11-fold greater than the genomic average (Faris et al. 2000). In chromosome 3B, most of the crossovers occurred in the distal regions, representing 40% of the chromosome, which contain most of the gene-rich regions (Saintenac et al. 2009). In contrast, the proximal regions, representing about 27% of chromosome 3B, showed a very weak crossover frequency, with only three crossovers found in the 752 gametes studied (Saintenac et al. 2009).

Fine mapping of crossover in chromosome 3B showed that 82% of the COs occurred in 19% of the chromosome length in the sub-telomeric regions which carry 60–70% of the genes (Valenzuela et al. 2013; Darrier et al. 2017). The remaining ~ 35% of the genes are located in recombinationally-poor chromosomal regions (Erayman et al. 2004; Darrier et al. 2017). This hinders the elimination of deleterious mutations or the introgression of beneficial ones during evolution and breeding. Recombination tends to occur in or near genes, often in promoter regions (Darrier et al. 2017). Most recombination events take place in hotspots that are characterized by specific sequence motifs, such as CCN or CTT repeats, or A-rich regions (Darrier et al. 2017) and are associated with typical chromatin modifications (Liu et al. 2021). A new member of the *RecQ* helicase gene family was recently shown to be associated with high CO frequency in hexaploid wheat (Gardiner et al. 2019). In tetraploid wheat, silencing of methylation-controlling genes *MET1* and *DDM1* affected the distribution of COs in sub-telomeric regions, while silencing of *XRCC2* (a *RAD51* paralog) lead also to an increase of CO rates in pericentric regions of several chromosomes (Raz et al. 2021). Altering the landscape of meiotic recombination might become a useful tool for wheat breeding. Recombination between homoeologous chromosomes is very rare in wheat due to the suppressing effect of the *Pairing homoeologous 1* (*Ph1*) locus (see Chap. 12). Nevertheless, it can happen, in particular in allopolyploids lacking the *Ph1* locus (Feldman 1965; Zhang et al. 2020). Another suppressor of homoeologous recombination, the *Ph2* locus, was isolated (Serra et al. 2021). It encodes for the wheat homolog of DNA mismatch repair gene *MSH7*, *TaMSH7*, located on chromosome 3DS. Homoeologous recombination events were shown to occur preferentially within exons, generating novel hybrid transcripts and proteins (Zhang et al. 2020). The ability to control the rate and location of homologous and homoeologous recombination would facilitate classical breeding processes as well as the gene transfer from wild relatives of wheat to the wheat background.

References

Aguilar M, Prieto P (2020) Sequence analysis of wheat sub-telomeres reveals a high polymorphism among homoeologous chromosomes. The Plant Genome 13:e20065

Ahn SN, Anderson JA, Sorrels ME, Tanksley SD (1993) Homoeologous relationships of rice, wheat and maize chromosomes. Mol Gen Genet 241:483–490

Akhunov ED, Goodyear AW, Geng S, Qi LL, Echalier B, Gill BS, Miftahudin J, Gustafson JP, Lazo G, Chao SM et al (2003) The organization and rate of evolution of wheat genomes are correlated with recombination rates along chromosome arms. Genome Res 13:753–763

Akhunov ED, Akhunova AR, Anderson OD, Anderson JA, Blake N, Clegg MT, Coleman-Derr D, Conley EJ, Crossman CC, Deal KR et al (2010) Nucleotide diversity maps reveal variation in diversity among wheat genomes and chromosomes. BMC Genom 11:702

Akpinar BA, Biyiklioglu S, Alptekin B et al (2018) Chromosome-based survey sequencing reveals the genome organization of wild wheat progenitor *Triticum dicoccoides*. Plant Biotechn J 16:2077–2087

Amado L, Abrabches R, Neves N, Viegas W (1997) Development-dependent inheritance of 5-azacytidine-induced epimutations in triticale: analysis of rDNA expression patterns. Chromosome Res 5:445–450

Anderson OD, Rausch C, Moullet 0, Lagudah ES (2003) The wheat D genome HMW glutenin locus: BAC sequencing, gene distribution, and retrotransposon clusters. Fuction Integr Genomics 3:56–68

Anikster Y, Manisterski J, Long DL (2005) Resistance to leaf rust, stripe rust, and stem rust in *Aegilops* spp. in Israel. Pl Dis 89:303–308

Appels R (1982) The molecular cytology of wheat-rye hybrids. Int Rev Cytol 80:93–132

Appels R, Dvorak J (1982) The wheat ribosomal DNA spacer region: its structure and variation in populations and among species. Theor Appl Genet 63:337–348

Appels R, Honeycutt RL (1986) rDNA: evolution over a billion years. In: Dutta SK (ed) DNA systematics. CRC Press, Florida, pp 81–135

Appels R, Gerlach WL, Dennis ES, Swift H, Peacock WJ (1980) Molecular and chromosomal organization of DNA sequences coding for the ribosomal RNAs in cereals. Chromosoma 78:293–311

Appels R, Moran LB, Gustafson JP (1986) The structure of DNA from the rye (*Secale cereale*) NOR R1 locus and its behaviour in wheat backgrounds. Can J Genet Cytol 28:673–685

Appels R, Reddy P, McIntyre CL, Moran LB, Frankel OH, Clarke BC (1989) The molecular-cytogenetic analysis of grasses and its application to studying relationships among species Triticeae. Genome 31:122–133

Avni R, Nave M, Barad O, Sven BK, Twardziok SO et al (2017) Wild emmer genome architecture and diversity elucidate wheat evolution and domestication. Science 357:93–97

Avni R, Lux T, Minz-Dub A, Millet E et al (2022) Genome sequences of *Aegilops* species of section Sitopsis reveal phylogenetic relationships and provide resources for wheat improvement. Plant J 110:179–192

Axtell MJ, Meyers BC (2018) Revisiting criteria for plant microRNA annotation in the era of big data. Plant Cell 30:272–284

Badaeva ED, Dedkova OS, Gay G, Pukhalkyi VA, Zelenin AV, Bernard S, Bernard M (2007) Chromosome rearrangements in wheat: their types and distribution. Genome 50:907–926

Badaeva ED, Amosova AV, Goncharov NP, Macas J, Ruban AS, Grechishnikova IV et al (2015) A set of cytogenetic markers allows the precise identification of all A-genome chromosomes in diploid and polyploid wheat. Cytogenet Genome Res 146:71–79

Badaeva E, Dedkova OS, Pukhalskyi VA, Zelenin V (2019) Chromosomal changes over the course of polyploid wheat evolution and domestication. In: Ogihara Y, Takumi S, Handa H (eds) Advances in wheat genetics: from genome to field, proceedings of 12th international wheat genetic symposium. Springer, Tokyo, Heidelberg, New York, Dordrecht, London, pp 83–92

Baum BR, Feldman M (2010) Elimination of 5S DNA unit classes in newly formed allopolyploids of the genera *Aegilops* and *Triticum*. Genome 53:430–438

Bebeli PJ, Kaltsikes PJ (1985) Karyotypic analysis of two durum wheat varieties. Can J Genet Cytol 27:617–621

Bedbrook JR, Jones J, O'Dell M, Thompson RD, Flavell RB (1980) A molecular description of telomeric heterochromatin in *Secale* species. Cell 19:545–560

Bennett MD, Leitch IJ (2005) Plant DNA C-values database (release 6.0). http://www.rbgkew.org.uk/cval/homepage.html

Bennetzen JL (2000) Transposable elements contributions to plant gene and genome evolution. Plant Mol Biol 42:251–269

Bennetzen JL (2002) Mechanisms and rates of genome expansion and contraction in flowering plants. Genetica 115:29–36

Bennetzen JL (2005) Transposable elements, gene creation and genome rearrangement in flowering plants. Curr Opin Genet Dev 15:621–627

Bennetzen JL, Freeling M (1993) Grasses as a single genetic system: genome composition, collinearity and compatibility. Trends Genet 9:259–261

Bennetzen JL, Kellogg EA (1997) Do plants have a one-way ticket to genomic obesity? Plant Cell 9:1509–1514

Bennetzen JL, Ma J, Devos KM (2005) Mechanisms of recent genome size variation in flowering plants. Ann Bot 95:127–132

Bernhardt N, Brassac J, Dong X, Willing E-M, Poskar CH, Kilian B, Blattner FR (2020) Genome-wide sequence information reveals recurrent hybridization among diploid wheat wild relatives. Plant J 102:493–506

Bhowal JG (1972) Nucleolar chromosomes in wheat. Zeitschrift Für Pflanzenzuchtung 68:253–257

Birchler JA, Han FP (2009) Maize centromeres: structure, function, epigenetics. Annu Rev Genet 43:287–303

Birchler JA, Han FP (2013) Centromere epigenetics in plants. J Genet Genome 40:201–204

Bird A (2002) DNA methylation patterns and epigenetic memory. Genes Dev 16:6–21

Birnstiel ML, Chipchase M, Speirs J (1971) The ribosomal RNA cistrons. Prog Nucleic Acid Res Mol Biol 11:351–389

Blackburn EH (1986) Structure and formation of telomeres in *Holotrichous ciliates*. Intern Rev Cytol 99:29–47

Bolot S, Abrouk M, Masood-Quraishi U, Stein N, Messing J, Feuillet C, Salse J (2009) The 'inner circle' of the cereal genomes. Curr Opin Plant Biol 12:119–125

Brettell RIS, Pallotta MA, Gustafson JP, Appels R (1986) Variation at the Nor loci in triticale derived from tissue culture. Theor Appl Genet 71:637–643

Brooks SA, Huang L, Gill BS, Fellers JP (2002) Analysis of 106 kb of contiguous DNA sequence from the D genome of wheat reveals high gene density and a complex arrangement of genes related to disease resistance. Genome 45:963–972

Brunazzi A, Scaglione D, Talini RF, Miculan M et al (2018) Molecular diversity and landscape genomics of the crop wild relative *Triticum urartu* across the fertile crescent. Plant J 94:670–684

Budak H, Kaya SB, Cagirici HB (2020) Long non-coding RNA in plants in the era of reference sequences. Front Plant Sci 11:276

Caetano-Anollés G (2005) Evolution of genome size in the grasses. Crop Sci 45:1809–1816

Cermeño MC, Lacadena JR (1985) Noclealar organizer competition in *Aegilops*–rye hybrids. Genome 27:479–483

Cermeño MC, Orellana J, Santos JL, Lacadena JR (1984a) Nucleolar organizer activity in wheat, rye and derivatives analyzed by a silver-staining procedure. Chromosoma 89:370–376

Cermeño MC, Orellana J, Santos JL, Lacadena JR (1984b) Nucleolar activity and competition (amphiplasty) in the genus *Aegilops*. Heredity 53:603–611

Cesare AJ, Quinney N, Willcox S, Subramanian D, Griffith JD (2003) Telomere looping in *P. sativum* (common garden pea). Plant J 36:271–279

Chague V, Just J, Mestiri I, Balzergue S, Tanguy AM, Huneau C, Huteau V, Belcram H, Coriton O, Jahier J, Chalhoub B (2010) Genome-wide gene expression changes in genetically stable synthetic and natural wheat allohexaploids. New Phytol 187:1181–1194

Chantret N, Salse J, Sabot F, Rahman S, Bellec A, Laubin B, Dubois I, Dossat C, Sourdille P, Joudrier P, Gautier MF, Cattolico L, Beckert M, Aubourg S, Weissenbach J, Caboche M, Bernard M, Leroy P, Chalhoub B (2005) Molecular basis of evolutionary events that shaped the hardness locus in diploid and polyploid wheat species (Triticum and Aegilops). Plant Cell 17:1033–1045

Chen ZJ, Pikaard CS (1997) Epigenetic silencing of RNA polymerase I transcription: a role for DNA methylation and histone modification in nucleolar dominance. Genes Dev 11:2124–2136

Cheng Z-J, Murata M (2003) A centromeric tandem repeat family originating from a part of Ty3/gypsy retroelement in wheat and its relatives. Genetics 164:665–672

Cheng Z, Dong F, Langdon T, Ouyang S, Buell CR, Gu M, Blattner FR, Jiang J (2002) Functional rice centromeres are marked by a satellite repeat and a centromere-specific retrotransposon. Plant Cell 14:1691–1704

Cheung WY, Money TA, Abbo S, Devos KM, Gale MD, Moore G (1994) A family of related sequences associated with (TTTAGGG)n repeats are located in the interstitial regions of wheat chromosomes. Mol Gen Genet 245:349–354

Choulet F, Wicker T, Rustenholz C, Paux E, Salse J, Leroy P, Schlub S, Le Paslier M-C, Magdelenat G, Gonthier C, Couloux A, Budak H, Breen J, Pumphrey M, Liu S, Kong X, Jia J, Gut M, Brunel D, Anderson JA, Gill BS, Appels R, Keller B, Feuillet C (2010) Megabase level sequencing reveals contrasted organization and evolution patterns of the wheat gene and transposable element spaces. Plant Cell 22:1686–1701

Choulet F, Alberti A, Theil S et al (2014) Structural and functional partitioning of bread wheat chromosome 3B. Science 345 (6194):1249721

Clark LG, Zhang W, Wendel JF (1995) A phylogeny of the grass family (Poaceae) based on ndhF sequence data. Syst Bot 20 (436):460

Crosby AR (1957) Nucleolar activity of lagging chromosomes in wheat. Amer J Bot 44:813–822

Crosby-Longwell AR, Svihla G (1960) Specific chromosomal control of the nucleolus and the cytoplasm in wheat. Exp Cell Res 20:294–312

Curtis CA, Lukaszewski AJ (1991) Genetic linkage between C-bands and storage protein genes in chromosome 1B of tetraploid wheat. Theor Appl Genet 81:245–252

Darrier B, Rimbert H, Balfourier F et al (2017) High-resolution mapping of crossover events in the hexaploid wheat genome suggests a universal recombination mechanism. Genetics 206:1373–1388

Darvey NL, Driscoll CJ (1972) Nucleolar behaviour in Triticum. Chromosoma 36:131–139

de Lange T (2004) T-loops and the origin of telomeres. Nat Rev Mol Cell Biol 5:323–329

de Rop V, Padeganeh A, Maddox PS (2012) CENP-A: the key player behind centromere identity, propagation, and kinetochore assembly. Chromosoma 121:527–538

Dennis ES, Gerlach WL, Peacock WJ (1980) Identical polypyrimidine-polypurine satellite DNAs in wheat and barley. Heredity 44:349–366D

Devos KM, Brown LKM, Bennetzen J (2002) Genome size reduction through illegitimate recombination counteracts genome expansion in Arabidopsis. Genome Res 12:1075–1079

Dubcovsky J, Dvorak J (1995) Ribosomal RNA loci: nomads in the Triticeae genomes. Genetics 140:1367–1377

Dubcovsky J, Luo MC, Zhong GY, Bransteitter R, Desai A, Kilian A, Kleinhofs A, Dvorak J (1996) Genetic map of diploid wheat, Triticum monococcum L., and its comparison with maps of Hordeum vulgare L. Genetics 143:983–999

Dvorackova M, Fojtova M, Fajkus J (2015) Chromatin dynamics of plant telomeres and ribosomal genes. Plant J 83:18–37

Dvorak J (2009) Triticeae genome structure and evolution. In: Feuiller C, Muehlbauer GJ (eds) Genetics and genomics of the Triticeae, plant genetics and genomics: crops and models, vol 7. Springer, Berlin, pp 685–711

Dvorak J, Akhunov ED (2005) Tempos of gene locus deletions and duplications and their relationship to recombination rate during diploid and polyploid evolution in the Aegilops-Triticum alliance. Genetics 171:323–332

Dvorak J, Appels R (1982) Chromosomal and nucleotide sequence differentiation in genomes of polyploid Triticum species. Theor Appl Genet 63:349–360

Dvorak J, Chen K-C, Giorgi B (1984) The C-band pattern of a Ph mutant of durum wheat. Can J Genet Cytol 26:360–363

Dvorak J, Zhang H-B, Kota RS, Lassner M (1989) Organization and evolution of the 5S ribosomal RNA gene family in wheat and related species. Genome 32:1003–1016

Eilam T, Anikster Y, Millet E, Manisterski J, Feldman M (2007) Genome size and genome evolution in diploid Triticeae species. Genome 50:1029–1037

Eilam T, Anikster Y, Millet E, Manisterski J, Feldman M (2008) Nuclear DNA amount and genome downsizing in natural and synthetic allopolyploids of the genera Aegilops and Triticum. Genome 51(8):616–627

Eilam T, Anikster Y, Millet E, Manisterski J, Feldman M (2009) Genome size in natural and synthetic autopolyploids and in a natural segmental allopolyploid of several Triticeae species. Genome 52:275–285

Ekwall K (2007) Epigenetic control of centromere behavior. Annu Rev Genet 41:63–81

El Baidouri M, Murat F, Veyssiere M, Molinier M, Flores R, Burlot L, Alaux M, Quesneville H, Pont C, Salse J (2017) Reconciling the evolutionary origin of bread wheat (Triticum aestivum). New Phytol 213:1477–1486

Endo TR, Gill BS (1984) Somatic karyotype, heterochromatin distribution, and nature of chromosome differentiation in common wheat, Triticum aestivum L. em Thell. Chromosoma 89:361–369

Erayman M, Sandhu D, Sidhu D, Dilbirligi M, Baenziger PS, Gill KS (2004) Demarcating the gene-rich regions of the wheat genome. Nucleic Acids Res 32:3546–3565

Fajkus J, Sykorova E, Leitch AR (2005) Telomeres in evolution and evolution of telomeres. Chromosome Res 13:469–479

Faris JD, Haen KM, Gill BS (2000) Saturation mapping of a gene-rich recombination hot spot region in wheat. Genetics 154:823–835

Fedoroff NV (2000) Transposons and genome evolution in plants. Proc Natl Acad Sci USA 97:7002–7007

Fedoroff NV (2012) Transposable elements, epigenetics, and genome evolution. Science 338:758–767

Feldman M (1965) Chromosome pairing between differential genomes in hybrids of tetraploid Aegilops species. Evolution 19:563–568

Feldman M (2001) Origin of cultivated wheat. In: Bonjean AP, Angus WJ (eds) The world wheat book. Lavoisier, Paris, pp 3–56

Feldman M, Levy AA (2015) Origin and evolution of wheat and related Triticeae species. In: Molnar-Lang M, Ceoloni C, Dolezel J (eds) Alien introgression in wheat, cytogenetics, molecular biology, and genomics. Springer, Cham, Switzerland, pp 21–76

Feldman M, Levy AA, Fahima T, Korol A (2012) Genome asymmetry in allopolyploid plants—wheat as a model. J Exp Bot 63:5045–5059

Feng S, Jacobsen SE, Reik W (2010) Epigenetic reprogramming in plant and animal development. Science 330:622–627

Feng C, Liu YL, Su HD, Wang HF, Birchler J, Han FP (2015) Recent advances in plant centromere biology. Sci China Life Sci Rev Special Top Plant Biol Chromatin Small RNA Signal 58(3):240–245

Feschotte C, Jiang N, Wessler S (2002) Plant transposable elements: where genetics meets genomics. Nat Rev Genet 3:329–341

Feuillet C, Keller B (1999) High gene density is conserved at syntenic loci of small and large grass genomes. Proc Natl Acad Sci USA 96:8265–8270

Fitzgerald MS, McKnight TD, Shippen DE (1996) Characterization and developmental patterns of telomerase expression in plants. Proc Natl Acad Sci USA 93:14422–14427

Flavell RB (1986) Repetitive DNA and chromosome evolution in plants. Phil Trans R Soc Lond B 312:227–242

Flavell RB, O'Dell M (1976) Ribosomal RNA genes on homoeologous chromosomes of groups 5 and 6 in hexaploid wheat. Heredity 37:377–385

Flavell RB, O'Dell M (1979) The genetic control of nucleolus formation in wheat. Chromosoma 71:135–152

Flavell RB, O'Dell M, Thompson WF, Vingentz M, Sardana R, Barker RF (1986) The differential expression of ribosomal RNA genes. Philos Trans R Soc Lond B 314:385–397

Flemming W (1882) Zellsubstanz. Kern und Zeltheilung. F.C.W Vogel, Leipzig

Frankel OH, Gerlach WL, Peacock WJ (1987) The ribosomal RNA genes in synthetic tetraploids of wheat. Theor Appl Genet 75:138–143

Friebe B, Tuleen N, Jiang J, Gill BS (1993) Standard karyotype of *Triticum longissimum* and its cytogenetic relationship with *T. aestivum*. Genome 36:731–742

Fukui KN, Suzuki G, Lagudah ES, Rahman S, Appels R, Yamamoto M, Mukai Y (2001) Physical arrangement of retrotransposon-related repeats in centromeric regions of wheat. Plant Cell Physiol 42:189–196

Gardiner LJ, Wingen LU, Bailey P et al (2019) Analysis of the recombination landscape of hexaploid bread wheat reveals genes controlling recombination and gene conversion frequency. Genome Biol 20:69

Gerlach WL (1977) N-banded karyotypes of wheat species. Chromosoma 62:49–56

Gerlach WL, Appels R, Dennis ES, Peacock WJ (1979) Evolution and analysis of wheat genomes using highly repeated DNA sequences. In: Ramanujan S (ed) Proceedings of 5th international wheat genetic symposium. The Indian Society of Genetics and Plant Breeding, Indian Agricultural Research Institute, New Delhi, India, pp 81–91

Gerlach WL, Miller TE, Flavell RB (1980) The nucleolus organizers of diploid wheats revealed by in situ hybridization. Theor Appl Genet 58:97–100

Gill BS (1987) Chromosome banding methods, standard chromosome band nomenclature, and applications in cytogenetic analysis. In: Heyne EG (ed) Wheat and wheat improvement, 2nd edn. American Society of Agronomy Inc., Madison, Wisconsin, USA, pp 243–254

Gill BS, Kimber G (1974) Giemsa C-banding and the evolution of wheat. Proc Natl Acad Sci USA 71:4086–4090

Gill KS, Gill BS, Endo TR, Boyko EV (1996a) Identification and high-density mapping of gene-rich regions in chromosome group 5 of wheat. Genetics 143:1001–1012

Gill KS, Gill BS, Endo TR, Taylor T (1996b) Identification and high-density mapping of gene-rich regions in chromosome group 1 of wheat. Genetics 144:1883–1891

Glémin S, Scornavacca C, Dainat J, Burgarella C, Viader V, Ardisson M, Sarah G, Santoni S, David J, Ranwez V (2019) Pervasive hybridizations in the history of wheat relatives. Sci Adv 5:eaav9188

González-García M-P, Pavelescu I, Canela A, Sevillano X, Leehy KA et al (2015) Single-cell telomere-length quantification couples telomere length to meristem activity and stem cell development in *Arabidopsis*. Cell Rep 11:977–989

Gorbunova V, Levy AA (1997) Non-homologous DNA end joining in plant cells is associated with deletions and filler DNA insertions. Nucleic Acids Res 25:4650–4657

Gornicki P, Zhu H, Wang J, Challa GS, Zhang Z, Gill BS, Li W (2014) The chloroplast view of the evolution of polyploid wheat. New Phytol 204:704–714

Gottlieb A, Müller H-G, Massa AN, Wanjugi H, Deal KR et al (2013) Insular organization of gene space in grass genomes. PLoS ONE 8: e54101

Grandbastien M-A, Lucas H, Morel J-B, Mhiri C, Vernhettes S, Casacuberta JM (1997) The expression of the tobacco *Tnt1* retrotransposon is linked to plant defense responses. Genetica 100:241–252

Greider CW, Blackburn EH (1985) Identification of a specific telomere terminal transferase activity in *Tetrahymena* extracts. Cell 43:405–413

Griffith JD, Comeau L, Rosenfield S, Stansel RM, Bianchi A, Moss H, de Lange T (1999) Mammalian telomeres end in a large duplex loop. Cell 97:503–514

Guo X, Han F (2014) Asymmetric epigenetic modification and elimination of rDNA sequences by polyploidization in wheat. Plant Cell 26:4311–4327

Guo W, Xin M, Wang Z, Yao Y et al (2020) Origin and adaptation to high altitude of Tibetan semi-wild wheat. Nat Commun 1:5085

Gupta PK, Varshney RK (2000) The development and use of microsatellite markers for genetic analysis and plant breeding with emphasis on bread wheat. Euphytica 113:163–185

Gupta P, Balyan H, Edwards K et al (2002) Genetic mapping of 66 new microsatellite (SSR) loci in bread wheat. Theor Appl Genet 105:413–422

Gustafson JP, Flavell RB (1996) Control of nucleolar expression in triticale. In: Guedes-Pinto H, Darvey N, Carnide VP (eds) Triticale: today and tomorrow. Kluwer Academic Publishers, The Netherlands, pp 119–125

Handa H, Kanamori H, Tanaka T et al (2018) Structural features of two major nucleolar organizer regions (NORs), *Nor-B1* and *Nor-B2*, and chromosome-specific rRNA gene expression in wheat. Plant J 96:1148–1159

Harper L, Golubovskaya I, Cande WZ (2004) A bouquet of chromosomes. J Cell Sci 117:4025–4032

Heller K, Kilian A, Piatyszek MA, Kleinhofs A (1996) Telomerase activity in plant extracts. Mol Gen Genet 252:342–345

Henikoff S, Furuyama T (2010) Epigenetic inheritance of centromeres. Cold Spring Harb Symp Quant Biol 75:51–60

Hirochika H, Sugimoto K, Otsuki Y, Tsugawa H, Kanda M (1996) Retrotransposons of rice involved in mutations induced by tissue culture. Proc Natl Acad Sci USA 93:7783–7788

Hou G, Du C, Gao H et al (2020) Identification of microRNAs in developing wheat grain that are potentially involved in regulating grain characteristics and the response to nitrogen levels. BMC Plant Biol 20:87

Houchins K, O'Dell M, Flavell R, Gustavson PJ (1997) Cytosine methylation and nucleolar dominance in cereal hybrids. Mol Gen Genet 255:294–301

Huang S, Sirikhachornkit A, Su X, Faris JD, Gill BS et al (2002) Genes encoding plastid acetyl-CoA carboxylase and 3-phosphoglycerate kinase of the *Triticum/Aegilops* complex and the evolutionary history of polyploid wheat. Proc Natl Acad Sci USA 99:8133–8138

Hudakova S, Michalek W, Presting GG, TenHoopen R, Dos Santos K, Jasencakova Z, Schubert I (2001) Sequence organization of barley centromeres. Nucl Acid Res 29:5029–5035

Hulbert SH, Richter TE, Axtell JD, Bennetzen JL (1990) Genetic mapping and characterization of sorghum and related crops by means of maize DNA probes. Proc Natl Acad Sci USA 87:4251–4255

International Brachypodium Initiative (2010) Genome sequencing and analysis of the model grass *Brachypodium distachyon*. Nature 463:763–768

International Barley Genome Sequencing Consortium (2012) A physical, genetic and functional sequence assembly of the barley genome. Nature 491:711–716

International Wheat Genome Sequencing Consortium (IWGSC) (2018) Shifting the limits in wheat research and breeding using a fully annotated reference genome. Science 361(6403):eaar7191

Ito H, Gaubert H, Bucher E, Mirouze M, Vaillant I, Paszkowski J (2011) An siRNA pathway prevents transgenerational retrotransposition in plants subjected to stress. Nature 472:115–119

Jaiswal S, Iquebal MA, Arora V et al (2019) Development of species specific putative miRNA and its target prediction tool in wheat (*Triticum aestivum* L.). Sci Rep 9:3790

Jayakodi M, Padmarasu S, Haberer G et al (2020) The barley pan-genome reveals the hidden legacy of mutation breeding. Nature 588:284–289

Jia J, Zhao S, Kong X, Li Y, Zhao G et al (2013) *Aegilops tauschii* draft genome sequence reveals a gene repertoire for wheat adaptation. Nature 496:91–95

Jiang J, Gill BS (1994) New 18S–26S ribosomal RNA gene loci: chromosomal landmarks for the evolution of polyploid wheats. Chromosoma 103:179–185

Jiao Y, Li J, Tang H, Paterson AH (2014) Integrated syntenic and phylogenomic analyses reveal an ancient genome duplication in monocots. Plant Cell 26:2792–2802

Jiao W, Yuan J, Shan Jiang S et al (2018) Asymmetrical changes of gene expression, small RNAs and chromatin in two resynthesized wheat allotetraploids. Plant J 93:828–842

Jin W, Melo JR, Nagaki K, Talbert PB, Henikoff S, Dawe RK, Jiang J (2004) Maize centromeres: organization and functional adaptation in the genetic background of oat. Plant Cell 16:571–581

Kanazin V, Ananiev E, Blake T (1993) The genetics of 5S rRNA encoding multigene families in barley. Genome 36:1023–1028

Kashkush K, Feldman M, Levy AA (2002) Gene loss, silencing and activation in a newly synthesized wheat allotetraploid. Genetics 160:1651–1659

Kashkush K, Feldman M, Levy AA (2003) Transcriptional activation of retrotransposons alters the expression of adjacent genes in wheat. Nat Genet 33:102–106

Kaszás E, Birchler JA (1996) Misdivision analysis of centromere structure in maize. EMBO J 15:5246–5255

Keidar D, Doron C, Kashkush K (2018) Genome-wide analysis of a recently active retrotransposon, *Au* SINE, in wheat: content, distribution within subgenomes and chromosomes, and gene associations. Plant Cell Rep 37:193–208

Keidar-Friedman D, Bariah I, Khalil Kashkush K (2018) Genome-wide analyses of miniature inverted-repeat transposable elements reveals new insights into the evolution of the *Triticum-Aegilops* group. PLoS ONE 13:e020497

Kenan-Eichler M, Leshkowitz D, Tal L, Noor E, Cathy Melamed-Bessudo C, Feldman M, Levy AA (2011) Wheat hybridization and polyploidization results in deregulation of small RNAs. Genetics 188:263–272

Kidwell MG, Lisch DR (2000) Transposable elements and host genome evolution. Trends Ecol Evol 15:95–99

Kim CM, Je B, Piao HL, Park SJ, Kim MJ et al (2002) Reprogramming of the activity of the activator/dissociation transposon family during plant regeneration in rice. Mol Cells 14:231–237

Kimber G, Athwal RS (1972) A reassessment of the course of evolution of wheat. Proc Nat Acad Sci USA 69:912–915

Kumar A, Bennetzen JL (1999) Plant retrotransposons. Annu Rev Genet 33:479–532

Kurata N, Moore G, Nagamura Y, Foote T, Yano M, Minobe Y, Gale MD (1994) Conservation of genome structure between rice and wheat. Biotechnology 12:276–278

Lassner M, Anderson O, Dvorak J (1987) Hypervariation associated with 12-nucleotide direct repeat and inferences on intragenomic homogenization of ribosomal RNA gene spacer based on the DNA sequence of a clone from the wheat Nor-D3 locus. Genome 29:770–781

Law JA, Jacobsen SE (2010) Establishing, maintaining and modifying DNA methylation patterns in plants and animals. Nat Rev Genet 11:204–220

Lawrence GJ, Appels R (1986) Mapping the nucleolus organizing region, seed protein loci and isozyme loci on chromosome 1R in rye. Theor Appl Genet 71:742–749

Leitch IJ, Heslop-Harrison JS (1992) Physical mapping of the 18S–5.8S–26S rRNA genes in barley by in situ hybridization. Genome 35:1013–1018

Levy AA (2013) Transposons in plant speciation. In: Fedoroff NV (ed) Plant transposons and genome dynamics in evolution. Wiley, pp 164–179

Li B, Choulet F, Heng Y, Hao W, Paux E, Liu Z, Yue W, Jin W, Feuillet C, Zhang X (2013) Wheat centromeric retrotransposons: the new ones take a major role in centromeric structure. Plant J 73:952–965

Li A, Liu D, Wu J, Zhao X et al (2014) mRNA and small RNA transcriptomes reveal insights into dynamic homoeolog regulation of allopolyploid heterosis in nascent hexaploid wheat. Plant Cell 26:1878–1900

Li A-L, Geng S-F, Zhang L-Q et al (2015) Making the bread: insights from newly synthesized allohexaploid wheat. Mol Plant 8:847–859

Li G, Wang L, Yang J et al (2021) A high-quality genome assembly highlights rye genomic characteristics and agronomically important genes. Nat Genet 53:574–584

Li L-F, Zhang Z-B, Wang Z-H et al (2022) Genome sequences of the five Sitopsis species of Aegilops and the origin of polyploid wheat B subgenome. Mol Plant 15:488–503

Linde-Laursen I, Baden C (1994) Comparison of the Giemsa C-banded karyotypes of the three subspecies of *Psathyrostachys fragilis*, subspp. *villosus* (2x), *secaliformis* (2x, 4x), and *fragilis* (2x) (Poaceae), with notes on chromosome pairing. Pl Syst Evol 191:183–198

Linder HP (1986) The evolutionary history of the poales/resti-onales—a hypothesis. Kew Bull 42:297–318

Ling H-Q, Zhao S, Liu D, Wang J, Sun H et al (2013) Draft genome of the wheat A-genome progenitor *Triticum urartu*. Nature 496:87–90

Ling H-Q, Ma B, Shi X et al (2018) Genome sequence of the progenitor of wheat A subgenome *Triticum urartu*. Nature 557:424–428

Liu H, Able A, Able JA (2020) Transgenerational effects of water-deficit and heat stress on germination and seedling vigour—new Insights from durum wheat microRNAs. Plants 9:189

Liu Y, Yuan J, Jia G, Ye W, Chen ZJ, Song Q (2021) Histone H3K27 dimethylation landscapes contribute to genome stability and genetic recombination during wheat polyploidization. Plant J 105:678–690

Lu C, Wei Y, Wang X et al (2020) DNA-methylation-mediated activating of lncRNA SNHG12 promotes temozolomide resistance in glioblastoma. Mol Cancer 19:28

Lukaszewski AJ, Curtis CA (1993) Physical distribution of recombination in B-genome chromosomes of tetraploid wheat. Theor Appl Genet 84:121–127

Lundblad V (2000) DNA ends: maintenance of chromosome termini versus repair of double strand breaks. Mutatation Res 451:227–240

Luo M-C, Deal KR, Akhunov ED, Akhunova AR, Anderson OD, Anderson JA, Blake N, Clegg MT, Coleman-Derr D, Conley EJ et al (2009) Genome comparisons reveal a dominant mechanism of chromosome number reduction in grasses and accelerated genome evolution in Triticeae. Proc Natl Acad Sci USA 106:15780–15785

Luo M-C, Gu YQ, You FM, Deal KR, Ma Y, Hu Y, Huo N, Wang Y, Wang J, Chen S, Jorgensen CM, Zhang Y, McGuire PE, Pasternak S, Stein JC, Ware D et al (2013). A 4-gigabase physical map unlocks the structure and evolution of the complex genome of *Aegilops tauschii*, the wheat D-genome progenitor. Proc Natl Acad Sci USA 110:7940–7945

Luo M-C, Gu YQ, Puiu D, Wang H, Twardziok SO, Deal KR et al (2017) Genome sequence of the progenitor of the wheat D genome *Aegilops tauschii*. Nature 551:498–502

Maccaferri M, Ricci A, Salvi S, Milner SG, Noli E et al (2015) A high-density, SNP-based consensus map of tetraploid wheat as a bridge to integrate durum and bread wheat genomics and breeding. Plant Biotechnol J 13:648–663

Maccaferri M, Harris NS, Twardziok SO, Pasam RK, Gundlach H et al (2019) Durum wheat genome highlights past domestication signatures and future improvement targets. Nat Genet 51:885–895

Madlung A, Tyagi AP, Watson B, Jiang H, Kagochi T, Doerge RW, Martienssen R, Comai L (2005) Genomic changes in synthetic *Arabidopsis* polyploids. Plant J 41:221–230

Makino T, Sasaki M, Morris R (1977) Misdivision of homoeologous group 5 univalent chromosomes in hexaploid wheat. I. Univalents derived from japanese cultivars. Cytologia 42:73–83

Mao L, Devos KM, Zhu L, Gale MD (1997) Cloning and genetic mapping of wheat telomere-associated sequences. Mol Gen Genet 254:584–591

Martínez P, Blasco MA (2010) Role of shelterin in cancer and aging. Aging Cell 9:653–666

Martini G, Flavell RB (1985) The control of nucleolus volume in wheat, a genetic study at three developmental stages. Heredity 54:111–120

Mascher M, Gundlach H, Himmelbach A et al (2017) A chromosome conformation capture ordered sequence of the barley genome. Nature 544:427–433

Mayer KFX, Martis M, Hedley PE, Šimková H, Liu H et al (2011) Unlocking the barley genome by chromosomal and comparative genomics. Plant Cell 23:1249–1263

McClintock B (1934) The relation of a particular chromosomal element tp the development of the nucleoli in Zea mays. Zeit Zellforsch Mik Anat 21:294–328

McClintock B (1942) The fusion of broken ends of chromosomes following nuclear fusion. Proc Natl Acad Sci USA 28:458–463

McClintock B (1984) The significance of responses of the genome to challenges. Science 226:792–801

Mehrotra S, Goyal V (2014) Repetitive sequences in plant nuclear DNA: types, distribution, evolution and function. Genom Proteom Bioinform 12:164–171

Middleton C, Stein N, Keller B, Kilian B, Wicker T (2013) Comparative analysis of genome composition in Triticeae reveals strong variation in transposable element dynamics and nucleotide diversity. Plant J 73:347–356

Miki Y, Yoshida K, Mizuno N et al (2019) Origin of wheat B-genome chromosomes inferred from RNA sequencing analysis of leaf transcripts from section Sitopsis species of *Aegilops*. DNA Res 26:171–182

Miller TE, Hutchinson J, Reader SM (1983) The identification of the nucleolus organizer chromosomes of diploid wheat. Theor Appl Genet 65:145–147

Mohan J, Flavell RB (1974) Ribosomal RNA cistron multiplicity and nucleolar organizers in hexaploid wheat. Genetics 76:33–44

Moore G, Devos KM, Wang Z, Gale MD (1995) Cereal genome evolution: grasses, line up and form a circle. Curr Biol 5:737–739

Morgante M, Brunner S, Pea G et al (2005) Gene duplication and exon shuffling by helitron-like transposons generate intraspecies diversity in maize. Nat Genet 37:997–1002

Morris R, Taira T, Schmidt J, Sasaki M (1977) Misdivision of homoeologous group 5 univalent chromosomes in hexaploid wheat. II. Univalents derived from American and European cultivars. Cytologia 42:85–99

Mukai Y, Endo TR, Gill BS (1991) Physical mapping of the 18S–26S rRNA multigene family in common wheat: identification of a new *locus*. Chromosoma 100:71–78

Murat F, Xu J-H, Tannier E, Abrouk M, Guilhot N, Pont C, Messing J, Salse J (2010) Ancestral grass karyotype reconstruction unravels new mechanisms of genome shuffling as a source of plant evolution. Genome Res 20:1545–1557

Murat F, Zhang R, Guizard S, Flores R, Armero A, Pont C, Steinbach D, Hadi Quesneville H, Cooke R, Salse J (2014a) Shared subgenome dominance following polyploidization explains grass genome evolutionary plasticity from a seven protochromosome ancestor with 16K protogenes. Genome Biol Evol 6:12–33

Murat F, Pont C, Salse J (2014b) Paleogenomics in Triticeae for translational research. Curr Plant Biol 1:34–39

Murat F, Armero A, Pont C, Klopp C, Salse J (2017) Reconstructing the genome of the most recent common ancestor of flowering plants. Nat Genet 49:490–496

Naranjo T (1990) Chromosome structure of durum wheat. Theor Appl Genet 79:397–400

Natarajan AT, Sharma NP (1974) Chromosome banding patterns and the origin of the B genome in wheat. Genet Res 21:103–108

Navashin MS (1928) Amphiplastie—eine neue karyologische Erscheinung. Proc Int Conf Genet 5:1148–1152

Navashin MS (1934) Chromosomal alterations caused by hybridization and their bearing upon certain general genetic problems. Cytologia 5:169–203

Nave M, Avni R, Çakır E, Portnoy V, Sela H, Pourkheirandish M, Ozkan H, Hale I, Takao Komatsuda T, Dvorak J, Distelfeld A (2019) Wheat domestication in light of haplotype analyses of the Brittle rachis 1 genes (BTR1-A and BTR1-B). Plant Sci 285:193–199

Neves N, Heslop-Harrison JS, Viegas W (1995) rRNA gene activity and control of expression mediated by methylation and imprinting during embryo development in wheat x rye hybrids. Theor Appl Genet 91:529–533

Ogbonnaya FC, Halloran GM, Lagudah ES (2006) D-genome of wheat —60 years on from Kihara, Sears and McFadden. Wheat Inf Serv 100:205–220

Olivera PD, Anikster Y, Kolmer JA, Steffenson BJ (2007) Resistance of sharon goat grass (*Aegilops sharonensis*) to fungal diseases of wheat. Plant Dis 91:942–950

Palmer DK, O'Day K, Trong HL, Charbonneau H, Margolis RL (1991) Purification of the centromere-specific protein CENP-A and demonstration that it is a distinctive histone. Proc Natl Acad Sci USA 88:3734–3738

Petersen G, Seberg O, Yde M, Berthelsen K (2006) Phylogenetic relationships of *Triticum* and *Aegilops* and evidence for the origin of the A, B, and D genomes of common wheat (*Triticum aestivum*). Mol Phylogenet Evol 39:70–82

Paterson AH, Bowers JE, Bruggmann R, Dubchak I, Grimwood J et al (2009) The *Sorghum bicolor* genome and the diversification of grasses. Nature 457:551–556

Pestsova E, Ganal MW, Röder MS (2000) Isolation and mapping of microsatellite markers specific for the D genome of bread wheat. Genome 43:689–697

Pfeifer M, Martis M, Asp T, Mayer KF, Lübberstedt T, Byrne S, Frei U, Studer B (2013) The perennial ryegrass genome zipper: targeted use of genome resources for comparative grass genomics. Plant Physiol 161:571–582

Pikaard CS (1999) Nucleolar dominance and silencing of transcription. Trend Pl Sci Rev 4:478–483

Pikaard CS (2000) The epigenetics of nucleolar dominance. Trends Genet 16:495–500

Pikaard CS, Mittlestein-Scheid O (2014) Epigenetic regulation in plants. Cold Spring Harb Perspect Biol 6:a019315

Prasad V, Strömberg CAE, Alimohammadian H, Sahni A (2005) Dinosaur coprolites and the early evolution of grasses and grazers. Science 310:1177–1180

Prasad V, Strömberg CAE, Leaché AD et al (2011) Late cretaceous origin of the rice tribe provides evidence for early diversification in Poaceae. Nat Commun 2:480

Presting GG, Malysheva L, Fuchs J, Schubert IZ (1998) A *TY3/GYPSY* retrotransposon-like sequence localizes to the centromeric regions of cereal chromosomes. Plant J 16:721–728

Pumphrey M, Bai J, Laudencia-Chingcuanco D, Anderson O, Gill BS (2009) Nonadditive expression of homoeologous genes is established upon polyploidization in hexaploid wheat. Genetics 181:1147–1157

Rabanus-Wallace MT, Hackauf B, Mascher M et al (2021) Chromosome-scale genome assembly provides insights into rye biology, evolution and agronomic potential. Nat Genet 53:564–573

Ramírez-González RH, Borrill P, Lang D et al (2018) The transcriptional landscape of polyploid wheat. Science 361:eaar6089

Raz A, Dahan-Meir T, Melamed-Bessudo C, Leshkowitz D, Levy AA (2021) Redistribution of meiotic crossovers along wheat chromosomes by virus-induced gene silencing. Front Plant Sci 11:635139

Riha K, Shippen DE (2003) Telomere structure, function and maintenance in *Arabidopsis*. Chromosome Res 11:263–275

Riha K, Fajkus J, Siroky J, Vyskot B (1998) Developmental control of telomere lengths and telomerase activity in plants. Plant Cell 10:1691–1698

Riley R, Unrau J, Chapman V (1958) Evidence on the origin of the B genome of wheat. J Hered 49:91–98

Röder MS, Lapitan NLV, Sorrells ME, Tanksley SD (1993) Genetic and physical mapping of barley telomeres. Mol Gen Genet 238:294–303

Röder MS, Korzun V, Wendehake K, Plaschke J, Tixier MH, Leroy P, Ganal MW (1998) A microsatellite map of wheat. Genetics 149:2007–2023

Rodríguez S, Perera E, Maestra B, Díez M, Naranjo T (2000) Chromosome structure of *Triticum timopheevii* relative to *T. turgidum*. Genome 43:923–930

Rustenholz C, Choulet F, Laugier C, Safar J, Simkova H et al (2011) A 3000-loci transcription map of chromosome 3B unravels the structural and functional features of gene islands in hexaploid wheat. Plant Physiol 157:1596–1608

Ruban AS, Badaeva ED (2018) Evolution of the S-genomes in *Triticum-Aegilops* alliance: evidences from chromosome analysis. Front Plant Sci 9:1756

Sabot F, Guyot R, Wicker T, Chantret N, Laubin B, Chalhoub B, Leroy P, Sourdille P, Bernard M (2005) Updating of transposable element annotations from large wheat genomic sequences reveals diverse activities and gene associations. Mol Genet Gen 274:119–130

Saghai-Maroof MA, Soliman KM, Jorgensen RA, Allard RW (1984) Ribosomal DNA spacer-length polymorphism in barley: mendelian inheritance, chromosomal location, and population dynamics. Proc Natl Acad Sci USA 81:8014–8018

Saintenac C, Falque M, Martin OC, Paux E, Feuillet C, Sourdille P (2009) Detailed recombination studies along chromosome 3B provide new insights on crossover distribution in wheat (*Triticum aestivum* L.). Genetics 181:393–403

Salse J (2012) In silico archeogenomics unveils modern plant genome organisation, regulation and evolution. Curr Opin Pl Biol 15:122–130

Salse J, Bolot S, Throude M, Jouffe V, Piegu B, Quraishi UM, Calcagno T, Cooke R, Delseny M, Feuillet C (2008) Identification and characterization of conserved duplications between rice and wheat provide new insight into grass genome evolution. Plant Cell 20:11–24

Sandhu D, Gill KS (2002) Gene-containing regions of wheat and the other grass genomes. Pl Physiol 128:803–811

Sarkar P, Stebbins GL (1956) Morphological evidence concerning the origin of the B genome in wheat. Amer J Bot 43:297–304

Scherthan H (2007) Telomere attachment and clustering during meiosis. Cellular Mol Life Sci 64:117–124

Schwarzacher T, Heslop-Harrison JS (1991) *In situ* hybridization to plant telomeres using synthetic oligomers. Genome 34:317–323

Scott JC, Manisterski J, Sela H, Ben-Yehuda P, Steffenson BJ (2014) Resistance of *Aegilops* species from Israel to widely virulent African and Israeli races of the wheat stem rust pathogen. Pl Dis 98:1309–1320

Seal AG (1982) C-banded wheat chromosome in wheat and triticale. Theor Appl Genet 63:38–47

Sears ER (1944) Inviability of intergeneric hybrids involving *Triticum monococcum* and *T. aegilopoides*. Genetics 29:113–127

Sears ER (1952) Misdivision of univalents in common wheat. Chromosoma 4:535–550

Sears ER (1954) The aneuploids of common wheat. Missouri Agric Exp Stn Res Bull 572:1–58

Sears ER, Miller TE (1985) The history of Chinese spring wheat. Cereal Res Commun 13:261–263

Senerchia N, Wicker T, Felber F, Parisod C (2013) Evolutionary dynamics of retrotransposons assessed by high throughput sequencing in wild relatives of wheat. Genome Biol Evol 5:1010–1020

Serra H, Svačina R, Baumann U et al (2021) *Ph2* encodes the mismatch repair protein MSH7-3D that inhibits wheat homoeologous recombination. Nat Commun 12:803

Shalev G, Levy AA (1997) The maize transposable element *Ac* induces recombination between the donor site and a homologous ectopic sequence. Genetics 146:1143–1151

Sharopova N (2008) Plant simple sequence repeats: distribution, variation, and effects on gene expression. Genome 51:79–90

Shcherban AB, Badaeva ED, Amosova AV, Adonina IG, Salina EA (2008) Genetic and epigenetic changes of rDNA in a synthetic allotetraploid, *Aegilops sharonensis* x *Ae. umbellulata*. Genome 51:261–271

Shirasu K, Schulman AH, Lahaye T, Schulze-Lefert P (2000) A contiguous 66 kb barley DNA sequence provides evidence for reversible genome expansion. Genome Res 10:908–915

Simon SA, Meyers BC (2011) Small RNA-mediated epigenetic modifications in plants. Curr Opin Pl Biol 14:148–155

Slotkin RK, Martienssen R (2007) Transposable elements and the epigenetic regulation of the genome. Nat Rev Genet 8:272–285

Somers DJ, Isaac P, Edwards K (2004) A high-density microsatellite consensus map for bread wheat (*Triticum aestivum* L.). Theor Appl Genet 109:1105–1114

Steinitz-Sears LM (1966) Somatic instability of telocentric chromosomes in wheat and the nature of the centromere. Genetics 54:241–248

Steinitz-Sears LM (1973) Misdivision of five different 3B monosomes in Chinese spring wheat. In: Sears ER, Steinitz-Sears LM (eds) Proceedings of 4th international wheat genetic symposium. University of Missouri, Agricultural Experiment Station, Columbia, MO, USA, pp 739–743

Su H, Liu Y, Liu C, Shi Q, Huang Y, Han F (2019) Centromere satellite repeats have undergone rapid changes in polyploid wheat subgenomes. Plant Cell 31:2035–2051

Tirosh I, Reikhav S, Levy AA, Barkai N (2009) A yeast hybrid provides insight into the evolution of gene expression regulation. Science 324:659–662

Valente LP, Silva MCC, Jansen LET (2012) Temporal control of epigenetic centromere specification. Chromosome Res 20:481–492

Valenzuela NT, Perera E, Naranjo T (2013) Identifying crossover-rich regions and their effect on meiotic homologous interactions by partitioning chromosome arms of wheat and rye. Chromosome Res 21:433–445

Vega JM, Feldman M (1998) Effect of the pairing gene *Ph1* on centromere misdivision in common wheat. Genetics 148:1285–1294

Vicient CM, Souniemi A, Anamthawat-Jonsson K et al (1999) Retrotransposon *BARE-1* and its role in genome evolution in the genus *Hordeum*. Plant Cell 11:1769–1784

Vicient CM, Jääskeläinen J, Kalendar R, Schulman AH (2001) Active retrotransposons are a common feature of grass genomes. Plant Physiol 125:1283–1292

Vieira R, Queiroz A, Morais L, Barão A, Melo-Sampayo T, Viegas WS (1990) 1R chromosome nucleolus organizer region activation by 5-azacytidine in wheat x rye hybrids. Genome 33:707–712

Wagenaar E, Bray D (1973) The ultrastructure of kinetochores of unpaired chromosomes in a wheat hybrid. Can J Genet Cytol 15:801–806

Walkowiak S, Gao L, Monat C et al (2020) Multiple wheat genomes reveal global variation in modern breeding. Nature 588:277–283

Wang J, Liu D, Guo X et al (2011) Variability of gene expression after polyhaploidization in wheat (*Triticum aestivum* L.). G3: genes. Genom Genet 1:27–33

Wang X, Wang J, Jin D et al (2015) Genome alignment spanning major Poaceae lineages reveals heterogeneous evolutionary rates and alters inferred dates for key evolutionary events. Mol Plant 8:885–898

Wang D, Ling L, Zhang W et al (2018) Uncovering key small RNAs associated with gametocidal action in wheat. J Exp Bot 69:4739–4756

Watson JM, Riha K (2010) Comparative biology of telomeres: where plants stand. FEBS Lett 584:3752–3759

Wendel JF, Cronn RC, Johnson JS, Price HJ (2002) Feast and famine in plant genomes. Genetica 115:37–47

Werner JE, Kota RS, Gill BS, Endo TR (1992) Distribution of telomeric repeats and their role in the healing of broken chromosome ends in wheat. Genome 35:844–848

Wessler SR (2006) Transposable elements and the evolution of eukaryotic genomes. Proc Natl Acad Sci USA 103:17600–17601

Wicker T, Zimmermann W, Perovic D, Paterson AH, Ganal M et al (2005) A detailed look at 7 million years of genome evolution in a 439 kb contiguous sequence at the barley *Hv-eIF4E* locus: recombination, rearrangements and repeats. Plant J 41:184–194

Wicker T, Gundlach H, Spannagl M et al (2018) Impact of transposable elements on genome structure and evolution in bread wheat. Genome Biol 19:103

Xiang D, Quilichini TD, Liu Z et al (2019) The transcriptional landscape of polyploid wheats and their diploid ancestors during embryogenesis and grain development. Plant Cell 31:2888–2911

Yaakov B, Kashkush K (2012) Mobilization of stowaway-like *MITEs* in newly formed allohexaploid wheat species. Plant Mol Biol 80:419–427

Yang CW, Zhao L, Zhang HK et al (2014) Evolution of physiological responses to salt stress in hexaploid wheat. Proc Natl Acad Sci USA 111:11882–11887

Yu G, Matny O, Champouret N et al (2021) Reference genome-assisted identification of the stem rust resistance gene *Sr62* encoding a tandem kinase. Nat Commun. https://doi.org/10.21203/rs.3.rs-1198968/v1

Yuan J, Guo X, Hu J, Lv Z, Han F (2015) Characterization of two *CENH3* genes and their roles in wheat evolution. New Phytol 206:839–851

Zaratiegui M, Irvine DV, Martienssen RA (2007) Noncoding RNAs and gene silencing. Cell 128:763–776

Zhang H, Zhu JK (2011) RNA-directed DNA methylation. Curr Opin Plant Biol 14:142–147

Zhang H, Reader SM, Liu X, Jia JZ, Gale MD, Devos KM (2001) Comparative genetic analysis of the *Aegilops longissima* and *Ae. sharonensis* genomes with common wheat. Theor Appl Genet 103:518–525

Zhang W, Qu L, Gu H, Gao W, Liu M, Chen J, Chen Z (2002) Studies on the origin and evolution of tetraploid wheats based on the internal transcribed spacer (ITS) sequences of nuclear ribosomal DNA. Theor Appl Genet 104:1099–1106

Zhang Z, Gou X, Kun H, Bian Y, Ma X, Lij LM, Gong L, Feldman M, Liu B, Levy AA (2020) Homoeologous exchanges occur through intragenic recombination generating novel transcripts and proteins in wheat and other polyploids. Proc Natl Acad Sci USA 117:14561–14571

Zhao G, Zou C, Li K et al (2017) The *Aegilops tauschii* genome reveals multiple impacts of transposons. Nat Plants 3:946–955

Zhou Y, Bai S, Li H et al (2021) Introgressing the *Aegilops tauschii* genome into wheat as a basis for cereal improvement. Nat Plants 7:774–786

Zimin AV, Puiu D, Luo MC et al (2017) Hybrid assembly of the large and highly repetitive genome of *Aegilops tauschii*, a progenitor of bread wheat, with the MaSuRCA mega-reads algorithm. Genome Res 27:787–792

Zinkowski RP, Meyne J, Brinkley BR (1991) The centromere-kinetochore complex: a repeat subunit model. J Cell Biol 113:1091–1110

Zohary D, Feldman M (1962) Hybridization between amphidiploids and the evolution of polyploids in the wheat (*Aegilops-Triticum*) group. Evolution 16:44–61

B Chromosomes

4

4.1 Introduction

B chromosomes (Bs, also known as supernumerary or accessory chromosomes) are dispensable chromosomes, presenting in addition to the standard complement (A chromosomes) and occur in a wide range of species from fungi to higher eukaryotes, including plants and animals (Jones and Rees 1982). B chromosomes were first discovered in *Secale cereale* by Gotoh (1924) and in Maize, by Kuwada (1925), Longley (1927). Longley (1927) called them supernumeraries, but Randolph (1928) used the term B chromosomes to distinguish them from the chromosomes of the basic complement. The term B chromosomes was later simplified to Bs (Jones et al. 2008a, b). Bs are present in about 15% of all eukaryotes (Beukeboom 1994), have been detected in more than 1500 plant species, and their properties have been well documented (Jones and Rees 1982; Jones 1995; Puertas 2002; Jones and Houben 2003; Jones et al. 2008a, b).

The distribution of Bs among different groups of angiosperms is not random (Jones et al. 2008a). The Poaceae is the plant family with the largest number of species containing Bs (Levin et al. 2005). There is virtually no difference in its frequency among diploids versus polyploids (Jones and Rees 1982; Palestis et al. 2004; Trivers et al. 2004). Some have suggested that Bs have a higher frequency in families with a large genome size (Trivers et al. 2004), ascribed to the larger amounts of noncoding DNA that may create a more conducive, or more tolerant, environment for the origin of Bs.

Bs vary in size, structure and chromatin properties, but also share certain characteristics that make them unique and distinguishable from other types of chromosome polymorphisms, e.g., aneuploidy (Banaei-Moghaddam et al. 2015). According to Jones and Houben (2003) and Jones et al. (2008a), Bs can be recognized by the following criteria: (i) they are dispensable and can be present or absent from individuals within a population; (ii) they pair only among themselves at meiosis and do not pair or recombine with any members of the standard set of A chromosomes. (iii) their inheritance is non-Mendelian and irregular, mainly due to nondisjunction at the first mitosis in the gametophytes; (iv) they lack any known major gene loci but have adverse and quantitative effects on phenotype, especially on fertility, when present in high numbers; (v) they significantly contribute to intraspecific variation in genome size; and (vi) they have no obvious adaptive properties.

Because of their non-Mendelian mode of inheritance, B chromosomes have a tendency to accumulate in gametes, resulting in an increase of B counts over generations. However, the large number of Bs is counterbalanced by selection against infertility. Hence, B-chromosome frequencies in populations result from a balance between their transmission rates and their effects on host fitness. In spite of this balance, it seems unlikely that Bs would persist in a species unless there was some positive adaptive advantage, which in a few cases, has been identified (see below).

One of the main features of Bs is that they are not essential for the life of a species and are not necessary for its normal growth and development (Jones et al. 2008a, b). Because of their dispensable nature, Bs can be present or absent among individuals of the same population in a species and can vary in number. In *Secale cereale*, for instance, Bs counts vary from zero to eight per plant, with a mean frequency of Bs in a number of populations of *Secale cereale* ranging from 6.6 to 54.0% plants having Bs (Jones and Rees 1982). In *Secale cereale*, Bs can be found in every region where the species grows in the wild, under semi-wild conditions or under cultivation (Jones and Puertas 1993). In most plant species, however, Bs are found in low numbers (0–5) in natural populations. In many species, several morphological types of Bs may exist within a single species (Houben et al. 2014). Several cases of B structural polymorphisms have been reported in plants, e.g., *Aegilops speltoides* (Belyayev and Raskina 2013) and *Secale cereale* (Müntzing (1946). However, in most populations of domesticated and weedy *Secale cereale*, Bs exhibit a similar molecular and cytological structure, suggesting that after a

4

© The Author(s) 2023
M. Feldman and A. A. Levy, *Wheat Evolution and Domestication*,
https://doi.org/10.1007/978-3-031-30175-9_4

period of rapid B development, the process of chromosome modification has slowed. Regarding the size of Bs, there is no known species in which the Bs exceed the size of the largest A chromosome (Jones et al. 2008a).

The existence and evolution of B chromosomes has been a topic of considerable discussion and speculation for decades, in particular their persistence in natural populations and conflicting opinions regarding their 'selfish' and/or 'adaptive' nature (Jones 1975, 2012; Jones and Rees 1982; Jones et al. 2008a). Bs equilibrium models depend on their effects on fitness as well as their transmission ratio (accumulation mechanism) in comparison to the regular chromosome complement, which together, dictate their evolutionary significance (Camacho et al. 2000). Although natural *Secale cereale* populations cover a broad range of geographical regions (Jones and Puertas 1993), there is, so far, no solid indication of positive fitness provided by Bs (Pereira et al. 2017). In most cases, Bs do not confer any advantage to the host and can even be detrimental if they exceed a certain number (Klemme et al. 2013). For instance, *S. cereale* plants are sterile when they harbor eight Bs (Rees and Ayonoadu 1973).

B-chromosomes are only known to occur spontaneously in cross-pollinating taxa (Müntzing et al. 1969) and are fully absent from self-pollinators (Jones et al. 2008b). In the sub-tribe Triticineae, B chromosomes exist in plants of the cross-pollinated species, namely, in *Agropyron cristaum* (Knowles 1955; Baenziger 1962; McCoy and Law 1965; Assadi 1995; Asghari et al. 2007), *Secale cereale* (Emme 1928; Darlington 1933; Hasegawa 1934; Müntzing 1944, 1950; Kranz 1963; Jones and Rees 1969; Niwa et al. 1990), *Amblyopyrum muticum* (Mochizuki 1957, 1960, 1964; Ohta 1995) and *Aegilops speltoides* (Simchen et al. 1971; Mendelson and Zohary 1972).

The sub-telocentric Bs of rye, *Secale cereale,* are roughly half the length of the As, consisting of approximately 580-Mbp sequences (Martis et al. 2012). In this species, Bs are found in both domesticated (*Secale cereale* ssp. *cereale*) and in wild and weedy (*S. cereale* ssp. *ancestrale*) subspecies. Their cytological morphology is virtually invariant throughout geographical regions (Jones and Puertas, 1993), and meiotic pairing of Bs in F_1 hybrids derived from weedy and cultivated rye lines of different origins, indicated that the Bs of wild and domesticated rye have a monophyletic origin (Niwa and Sakamoto 1995, 1996). No Bs are known from the older *Secale* cross-pollinating species, *Secale strictum* (formerly *S. montanum)* (Niwa et al. 1990). Thus, the origin of the Bs in *S. cereale* might be linked to the divergence of *S. cereale* from *S. strictum* (Martis et al. 2012).

The sub-metacentric Bs of *Ae. speltoides* are about 2/3 of the average length of the A chromosomes (Simchen et al. 1971). Chromosome counts and flow cytometric analysis of *Aegilops speltoides* revealed a tissue type-specific distribution of the roughly large 570 Mbp B chromosomes. The Bs of this species are absent in the roots but stably present in the aerial tissue of the same individual (Mendelson and Zohary 1972), with a maximum number of eight Bs per cell reported (Raskina et al. 2004). Comparable tissue type-specific B chromosome distribution is also known for *Agropyron cristatum* (Baenziger 1962), and *Amblyopyrum muticum* (Ohta 1995). Bs of related species are unique and nonhomologous. The B chromosomes of *Amblyopyrum muticum* did not pair with the B chromosomes of *Aegilops speltoides* in F_1 hybrids between these two species (Vardi and Dover 1972).

Bs often accumulate by a 'drive' mechanism (Jones 1991; Jones and Houben 2003), as best demonstrated in Rye (Gonzalez-Sanchez et al. 2004). Because they do not participate in meiotic recombination with As, Bs take a distinct evolutionary path and their sequence composition may differ from that of the As. (Houben et al. 2014). Since Bs are under little or no selection pressure, various transposable elements, retrotransposons and other DNA sequences may insert, spread, or amplify in Bs, altering their composition from that of the As (Klemme et al. 2013).

4.2   Origin and Molecular Characterization

Our knowledge of the origin and sequence composition of B chromosomes was limited until recently. Technological advances in sequencing and genome analysis have shed considerable light on these aspects (Houben et al. 2014). It is widely accepted that B chromosomes derive from A chromosomes, either of the same or of related species (Camacho et al. 2000). B chromosomes can derive from A chromosomes of the same species, often in progeny of trisomic (2n + 1) plants. After going through several rapid structural changes, it finally stabilized as a heterochromatic chromosome with features of a B chromosome (Jones et al. 2008b). Alternatively, Bs can arise from the same species as a result of chromosomal rearrangements. However, there is also evidence suggesting that Bs can spontaneously arise following interspecific hybridization or polyploidization (Jones and Rees 1982; Jones and Houben 2003; Houben et al. 2013a, b).

The B chromosomes of *Aegilops speltoides* likely originated from the standard set of A chromosomes as a consequence of interspecific hybridization or, more likely, from trisomic (2n + 1) plants. Several lines of *Ae. speltoides* occasionally form unreduced gametes (Feldman M, unpublished), leading to the production of triploid progeny upon fertilization of a reduced gamete. Self-pollination of such triploids yields, trisomic plants, aside from other aneuploids. Proposed potential donors of the Bs are the A chromosomes 1S, 4S and 5S of the *Ae. speltoides* genome (Friebe et al.

1995; Belyayev and Raskina 2013). The *Ae. speltoides* Bs are also characterized by a number of A chromosome-localized repeats like *Spelt1*, pSc119.2 tandem repeats, 5S rDNA and *Ty3-gypsy* retroelements (Friebe et al. 1995; Raskina et al. 2011; Hosid et al. 2012; Belyayev and Raskina 2013).

The age of *Secale cereale* Bs was estimated to be *c*. 1.1–1.3 MYA (Martis et al. 2012). Considering the assumed age of the species *Secale cereale* (1.7 MYA), it is tempting to speculate that B chromosomes originated as a by-product of a chromosome rearrangement event during the development of the annual *S. cereale* from the perennial *S. strictum* (Martis et al. 2012). In fact, the genome of *Secale cereale* differs from that of *S. strictum* by several rearrangements (Stutz 1972; Koller and Zeller 1976; Shewry et al. 1985; Naranjo et al. 1987; Naranjo and Fernández-Rueda 1991; Liu et al. 1992; Rognli et al. 1992; Devos et al 1993; Schlegel 2013).

Early attempts to elucidate the DNA composition of Bs were mainly based on comparative studies of − B versus + B genomic DNA (Rimpau and Flavell 1975; Timmis et al. 1975; Sandery et al. 1990; Wilkes et al. 1995). Later, microdissection (Houben et al. 2001) and flow-sorting (Martis et al. 2012) enabled more reliable isolation of B-derived DNA. The use of next-generation sequencing (Martis et al. 2012), sophisticated bioinformatics tools, e.g., genome zipper (Mayer et al. 2011) and repeat clustering analysis (Novak et al. 2010) shed light on the origin and composition of rye B chromosomes (Martis et al. 2012; Klemme et al. 2013). These studies showed that Bs of rye contain sequences that originated from one or more A chromosomes (Houben et al. 2001; Martis et al. 2012). Even sequences considered as B-specific are also present on As but in low copy numbers, indicating an intraspecific origin of the Bs. Comparison of sequences of flow-sorted B and A chromosomes showed that rye Bs originated from multiple As, most likely by a pathway involving partial genome duplication and chromosome rearrangements (Martis et al. 2012; Klemme et al. 2013). Their subsequent molecular evolution involved gene silencing, heterochromatinization and the accumulation of repetitive DNA and transposons (Camacho et al. 2000). Klemme et al. (2013) showed that rye Bs contain a similar proportion of high-copy sequences as A chromosomes, but differ substantially in repeat composition. They found a massive accumulation of B-enriched repeats, mostly in the nondisjunction control region at the terminal part of the long arm (see below), which is transcriptionally active and very late replicating, as well as in the extended pericentromeric region.

Sequencing of rye B chromosomes showed that the Bs descended from chromosome arm 3RS and chromosome 7R (Klemme et al. 2013), with subsequent accumulation of repeats and genic fragments from other A chromosome regions (Martis et al. 2012; Banaei-Moghaddam et al. 2015). The multi-chromosomal origin of B-chromosome sequences is further supported by the many short sequences that are similar to other regions of the rye A chromosomes (Martis et al. 2012). Thus, the rye Bs represent a conglomerate of mainly tandem repeat sequences derived from different A chromosome sites and could therefore not have originated by a single excision of an A fragment (Houben et al. 2001). Jones et al. (2008b) proposed that B-founder sequences were 'released' from a polymorphic A chromosome region and were then stabilized by the addition of other sequences such as organellar DNA and sequences necessary for their function as chromosomes (e.g., telomeric and centromeric sequences). Indeed, it should be noted that Bs contain coding and non-coding repeats similar to those found in organellar DNA of various organisms (Cohen et al. 2003). Indeed, studying the molecular structure of rye Bs revealed their origin as a mosaic of nuclear and organellar DNA sequences (Martis et al. 2012) that contain functional domains including centromeric and telomeric sequences homologous to those found on A chromosomes (Jones 2012). However, several B-specific regions exist as well (Timmis et al. 1975; Tsujimoto and Niwa 1992; Wilkes et al. 1995; Houben et al. 1996).

Different A chromosome sequences may enter Bs via trafficking, that occurs during double-strand break repair or via hitchhiking of genomic fragments with transposable elements (Scholz et al. 2011). Alternatively, Bs may represent an evolutionary mechanism aimed at sequestering additional copies of genes that are generated at the chromosome breakpoints associated with speciation. In addition to this basic A-derived architecture, rye Bs display large amounts of B-specific repeats and cytoplasmic organellar DNA. It seems that the B acts like a "genomic sponge" (or garbage can) that collects and maintains sequences of diverse origins (Martis et al. 2012).

Analysis of the composition and distribution of rye B-located, high-copy sequences revealed that Bs contain a similar proportion of repeats as A chromosomes but differ substantially in repeat composition (Klemme et al. 2013). The most abundant mobile elements (*Gypsy*, *Copia*) in the genome of rye are similarly distributed along As and Bs, while the ancient retroelement *Sabrina* (Shirasu et al. 2000), is less abundant on Bs than on As. In contrast, the active element *Revolver* (Tomita et al. 2008), as well as the predicted *Copia* retrotransposon *Sc36c82*, are disproportionately abundant on the Bs.

Houben et al. (2014) proposed a multi-step model for the origin of B chromosomes as autonomous elements. Initially, a proto-B chromosome was derived from a segmental (trisomy) or whole-genome duplication, followed by reductive chromosome rearrangements, unbalanced segregation of a translocation chromosome, and subsequent sequence insertions. Soon after, pairing and recombination with donor A chromosomes became restricted. This restriction is considered as the starting point for the independent evolution of the B chromosome. The development of functional centromere and de novo telomeres, combined with fast-evolving repetitive sequences, along with reduced selective pressure on gene integrity, then predisposed nascent Bs to rapidly accumulate further structural modifications. The development of a preferential transmission system and the relative absence of phenotypic effects, enabled the persistence of Bs in a species. However, this apparently neutral "hitchhiking" is probably limited, as suggested from the deleterious effects of large Bs number (Rees and Ayonoadu 1973; Klemme et al. 2013).

Hence, one of the first requirements for the independent existence of Bs is to achieve meiotic isolation, i.e., to develop a system that prevents pairing and recombination with the A chromosomes. Prevention of meiotic pairing between B and A chromosomes is due to structural and epigenetic changes in Bs as well as to development of B-specific sequences, resulting in Bs homoeologous rather than homologous to the As. In addition, it led to the development of a homoeologous-pairing suppressor system that prevents homoeologous pairing between Bs and As. Because of such a system, in almost all cases studied, there is an absolute barrier to recombination between A and B chromosomes (Jones et al. 2008b). Examples of pairing between B chromosomes and A chromosomes are very rare in plants; when such pairing was observed, it was not due to synapsis but, rather, to end-to-end associations (Battaglia 1964). Yet, Vardi and Dover (1972) reported that a plant of *Amblyopyrum muticum* with one B chromosome, showed one cell in which the B chromosome formed a trivalent with two A chromosomes. On the other hand, *Secale cereale* Bs frequently pair with each other and themselves in pachytene (Diez et al. 1993), but bivalents are less connected by chiasmata than A chromosomes (Jiménez et al. 2000).

Rye B chromosomes have accumulated significantly greater amounts of chloroplast- and mitochondrion-derived sequences than the A chromosomes (Martis et al. 2012). All parts of the chloroplast and mitochondrial genomes are found on the Bs, indicating that all sequences are transferable. The higher amount of organelle-derived DNA inserts in B, as compared to A chromosomes, and the increased mutation frequency of B-located organellar DNA, suggests a reduced selection against the insertion of organellar DNA in Bs. Insertion of organellar sequences into B DNA has fewer deleterious genetic consequences than their insertion into As. This may reflect the generally silent B chromosome, which may enable tolerance to essentially uncontrolled DNA insertions of all sorts. In contrast, insertions into A chromosomal DNA may disrupt gene expression with lethal consequences, particularly when they become homozygous.

Transfer of organellar DNA to the nucleus is very frequent (Timmis et al. 2004), but much of the organellar DNA is also rapidly lost again within one generation by a partially counterbalancing, but largely elusive, removal process (Sheppard and Timmis 2009).

The presence of a disproportionately large amount of organellar DNA on the rye B chromosomes suggests a long-term evolutionary role of the Bs. The considerable amount of B-specific accumulation of chloroplast- and mitochondria-derived sequences is due to a reduced selection against the insertion of organellar DNA in B chromosomes, whereas there is probably a considerable selection against the insertion of organellar DNA in genetically active A chromosomes (Jones et al. 2008b, a; Martis et al. 2012). Natural selection has certainly played an important role in determining the Bs DNA composition, however alternative mechanisms might be also at play. While there is no evidence for active capturing of mobile DNAs into B chromosomes, it cannot be ruled out, as B chromosomes being repeat-rich, might undergo more DNA breaks, due to stalled replication or transposons activity, with such breaks serving as entry points for extrachromosomal DNAs. In addition, recent works suggest that heterochromatin can serve as a preferred insertion site for certain types of mobile DNA (Shilo et al. 2017).

Klemme et al. (2013) provided detailed insight into the changes that high-copy rye sequences underwent in B chromosomes. Although most repeats are similarly distributed along As and Bs, several transposons are either amplified or depleted on the Bs. Accumulation of B-enriched high-copy sequences was found mostly in the nondisjunction control region of the Bs, which is transcriptionally active and late-replicating (see below).

Because any increased gene dosage may affect gene expression, the expression of paralogues on B chromosomes might be reprogrammed (potentially through epigenetic mechanisms) early during the evolution of the Bs (Klemme et al. 2013). Thus, proto-B genes, derived from A-chromosomal regions and cytoplasmic organellar genomes, might first be suppressed and then degenerate due to mutations. Exceptions could include those sequences that promote preferential transmission of Bs, an advantage for the maintenance of B chromosomes in populations. Thus, B chromosomes provide a kind of safe harbor for genes and sequences without immediate selective benefit (Klemme et al. 2013).

Detailed analysis of rye B-located high copy sequences revealed that Bs contain a similar proportion of repeats as A chromosomes, but differ substantially in their repeat composition (Martis et al. 2012; Klemme et al. 2013; Marques et al. 2013). More specifically, B-specific accumulation of *Gypsy* retrotransposons and other repeated sequences have been reported for rye (Sandery et al. 1990).

The accumulation of active retrotransposons on Bs might be rooted in a relaxed selection pressure. Reduced crossing-over in Bs might further facilitate retroelement accumulation.

As mentioned above, B-located genic sequences originated from A chromosomes or from organellar DNA. Because Bs are dispensable, it is expected that they are prone to mutation accumulation as they undergo pseudogenization (Banaei-Moghaddam et al. 2015). Pseudogenes can encode endo-siRNAs and regulate the expression of their parental A chromosome genes (Johnsson et al. 2014). Thus, as B-specific transcripts would might be aberrant due to less selective pressure, they can serve as a substrate for RNA-directed RNA polymerase, which would make double strand RNA (dsRNA) (Banaei-Moghaddam et al. 2015). The dsRNA is then processed to small regulatory RNA. This model is consistent with the cumulative effects of an increasing number of Bs (Banaei-Moghaddam et al. 2015). Alternatively, pseudogene transcripts can function as indirect post-transcriptional regulators. For example, they might act as miRNA sponges. Due to high similarity between parental and pseudogene transcripts, both could compete for miRNA, leading to degradation of parental gene transcripts (Muro et al. 2011). Further, it has been shown that some pseudogene transcripts translate and produce short peptides or truncated proteins of functional importance (Johnsson et al. 2014). Pseudogene transcripts could also act as a source of siRNAs via pathways involving RNA-directed RNA polymerases (Banaei-Moghaddam et al. 2015). It is not yet known whether B-derived siRNAs acting as regulators exist.

Little information is available on the chromatin composition of Bs. Preliminary classical cytological observations (e.g., Giemsa-banding) suggested that the Bs in about half of the plant species carrying them are heterochromatic (Jones 1975). In rye, the sub-terminal heterochromatic domain of the B is characterized by a unique combination of histone methylation marks (Carchilan et al. 2007; Marques et al. 2013). Contrary to the heterochromatic regions of the A chromosomes, this domain is simultaneously marked by tri-methylated histone H3K4 and tri-methylated H3K27. In addition, this domain shows a dark Giemsa band at mitosis, but undergoes decondensation during interphase and transcription of B-specific high copy repeat families (Banaei-Moghaddam et al. 2015).

4.3 Preferential Transmission (Accumulation Mechanism) of B Chromosomes

B chromosomes dispensability, i.e., non-essentiality for normal development of their host, remains undetermined. In addition, while they may have negative effects on nuclear physiology and phenotype, it remains unclear if Bs have any selective advantages and how they manage to persist in specific populations. Certainly, their maintenance requires means of survival against selection. To counteract elimination, Bs of many species have evolved a 'drive' mechanisms, which ensures their transmission to the next generation at frequencies that are higher than expected according to Mendelian rules (Jones 1991; Jones et al. 2008a). Such transmission enables the maintenance of Bs in natural populations. The variety of mechanisms, including segregation failure, by which B chromosomes gain heritable advantage in transmission, are known as accumulation or drive mechanisms. Depending on the species, B chromosome drive can be pre-meiotic, meiotic, or post-meiotic, but the underlying molecular process remains unclear (Jones 1991; Burt and Trivers 2006). Post-meiotic drive is frequent in flowering plants during gametophyte maturation.

Survival in populations is achieved by various mechanisms of mitotic and meiotic drive (Jones 1991, 1995). The more common drive process for Bs in plants, especially in the family Poaceae, is based on directed nondisjunction in the gametophyte phase of the life cycle. In rye, it takes place both at first pollen mitosis and at first egg cell mitosis, in maize, it occurs at the second pollen mitosis, and at other time points in other species (Jones and Rees 1982).

Matthews and Jones (1982, 1983) assumed that the differential transmission of Bs in rye is strong enough to overcome their negative effects on plant vigor and fertility. Moreover, they suggested that the main factor enabling variation in population equilibrium for B-frequency is the level of bivalent or multivalent pairing of Bs with themselves at meiosis. Confirmation of the predicted variation in pairing levels was achieved when selection for high and low transmission genotypes was found to correlate with the level of B pairing at first meiotic metaphase in lines of Korean rye (Jiménez et al. 1997). It was found that plants with 2B of the low line formed bivalents in only 20% of first meiotic metaphase cells, whereas, in the high line, there were more than 90% of bivalents. Puertas et al. (1998, 2000) later proposed that what Jiménez et al. (1997) assumed were 'genes' dictating transmission rate, were actually the sites of chiasma formation, or the binding sites, in the Bs themselves.

In most species carrying Bs, the mitotic transmission of Bs during growth and development is normal and hence, all cells within the individual carry the same number of Bs (Houben et al. 2014). However, there are some exceptions in which the Bs show mitotic instability and are therefore present in variable numbers, sometimes characterizing specific tissues and/or organs. For example, in the grasses *Aegilops speltoides* and *Amblyopyrum muticum*, Bs exist in aerial organs but not in roots (Mendelson and Zohary 1972; Ohta 1996).

The behavior of rye Bs during pollen mitosis was first studied by Hasegawa (1934), who described how the two chromatids of the B chromosome do not separate at anaphase of the first pollen grain mitosis and, in most cases, are included in the generative nucleus. In the second pollen grain mitosis, the generative nucleus divides to produce two sperm nuclei, each with an unreduced number of Bs. A similar nondisjunction process may occur in the female gametophytes of rye as well (Håkansson 1948).

Banaei-Moghaddam et al. (2012) proposed a model describing the B chromosome accumulation mechanism in rye. The model is based on analysis of the cellular mechanism of B chromosome drive in the male gametophyte of rye. At all mitotic stages of microgametogenesis, the As and Bs centromeres are active. However, at first pollen mitosis, sister chromatid cohesion differs between As and Bs. The B-specific pericentromeric repeats are involved in the formation of pericentric heterochromatin, which plays a critical role in the cohesion of sister chromatids and in their non-disjunction at the first pollen mitosis. Failure to resolve the pericentromeric cohesion is under the control of the B-specific nondisjunction control element. The asymmetry of this division plays a critical role in the determination and subsequent fate of the two unequal mitotic products: the vegetative and the generative cells. Due to unequal spindle formation, joined B chromatids preferentially migrate towards the generative pole. Thus, at first pollen mitosis of rye, the mitotic spindle asymmetry seems to play a central role in B accumulation. In the second pollen mitosis, the generative nucleus divides to produce two sperm nuclei, each with an unreduced number of Bs. Hence, a combination of nondisjunction and of unequal spindle formation at first pollen mitosis, results in the directed accumulation of Bs to the generative nucleus, which consequently ensures their transmission at a higher than Mendelian rate to the next generation.

The B centromeres demonstrate standard behavior at anaphase of the first pollen mitosis, and can be seen separated and pulling to opposite poles. In contrast, the B chromatids appear to be transiently held together at sensitive sticking sites on either side of the centromere (the receptors), and since the spindle is asymmetric, the equator is closer to the pole which will include the B chromatids in the generative nucleus. The question to be answered is how the B-specific region signals the receptors to remain conjoined just long enough to facilitate directed nondisjunction? This is a fundamental question in terms of genome evolution, since the mechanism had to arise de novo, and then become rapidly established in a highly conserved way to allow the rye Bs to survive following their origin. Furthermore, nondisjunction works equally well when the rye B is introduced as an additional chromosome into hexaploid wheat (Lindström 1965; Müntzing 1970; Niwa et al. 1997; Endo

et al. 2008), into hypo-pentaploid Triticale (Kishikawa and Suzuki 1982), or into *Secale vavilovii* (Puertas et al. 1985). Thus, the B autonomously controls the process of nondisjunction (Matthews and Jones 1983; Romera et al. 1991).

Müntzing (1946) demonstrated in *Secale cereale* that in the standard B chromosome, as well as in isochromosomes consisting of the large and small arms of the standard B chromosome, the centromeres divided normally at anaphase of the pollen mitosis. However, in the standard and large isochromosomes, there were sticking sites on either side of the centromere which prevented normal anaphase separation of the chromatids, thereby causing nondisjunction of these two types. The chromosomal region that carries the element controlling nondisjunction is comprised of a concentration of B-specific sequences from two families, E3900 and D1100, assembled from a variety of repetitive elements, some of which are also represented in the A genome (Sandery et al. 1990; Blunden et al. 1993; Houben et al. 1996; Langdon et al. 2000). No genes have been found in the region, which raises questions regarding the genetic process controlling nondisjunction (Matthews and Jones 1983; Ortiz et al. 1996).

Deficient Bs lacking the heterochromatic terminal region of the long arm, undergo normal disjunction at first pollen anaphase. Therefore, it seems that the accumulation mechanism of the B by nondisjunction requires factors located at the end of the long arm (Müntzing 1948; Håkansson 1959; Endo et al. 2008). This factor can act in trans because, if a standard B (Lima-de-Faria 1962) or the terminal region of the long arm of the B (Endo et al. 2008) is present in the same cell containing a deficient B, nondisjunction occurs for both the standard and the deficient B. The nondisjunction control region is enriched in B-specific repeats, which are highly transcriptionally active in anthers (Banaei-Moghaddam et al. (2012). In addition, the distal heterochromatin of the long arm is marked with the euchromatin-specific histone modification mark H3K4me3 (Carchilan et al. 2007).

Lima-de-Faria (1962) initially proposed that the rye B drive mechanism is controlled by the sub-telomeric domain of the B long arm, where two sequence families, D1100 (Sandery et al. 1990) and E3900 (Blunden et al. 1993), accumulate (Langdon et al. 2000). Later, analysis of chromosome behavior of wheat lines with introgressed fragments of rye Bs, established that non-disjunction is in fact dependent on the D1100 and E3900 sequence families (Endo et al. 2008). More recently, a B-specific chromatid adhesion site involving the pericentromeric repeat ScCl11, was implicated in the delay of sister chromatid separation (Banaei-Moghaddam et al. 2012). Langdon et al. (2000) suggested that E3900 and D1100 repeats evolved via amplification of ancestral A-located sequences within the dynamic nondisjunction control region on rye B. The B-enriched tandem repeats could have been amplified via unequal crossover (Smith 1976).

It has been previously suggested that Bs of the two subspecies of rye have a monophyletic origin (Niwa and Sakamoto 1995; Marques et al. 2013), and that the organization of D1100 and E3900 is highly conserved (Klemme et al. 2013). The complex organization of the E3900 and D1100 sequence families as high copy repetitive DNA is specific to Bs (Klemme et al. 2013). FISH showed that D1100 accumulates in two zones in the sub-telomeric region of the long B arm, physically separated by an interstitial less labeled space, while E3900 has a more homogeneous and distal signal that overlaps with the D1100 domain closer to the telomere (Wilkes et al. 1995). Besides the 3.9-kb form of E3900, a shorter 2.7-kb E3900-related sequence, which also accumulates on the B-specific domain, has been identified (Pereira et al. 2009). Estimates of copy number have shown that E3900 sequences are highly conserved and are present in 100–150 copies on Bs and in single or low copy numbers in A chromosome (Pereira et al. 2009). Importantly, E3900 sequences are differentially expressed in a tissue- and developmental stage-specific manner in plants with and without B chromosomes (Pereira et al. 2009). While the expression levels of E3900 do not vary in leaves from plants with and without Bs, they are significantly upregulated during meiosis exclusively in plants with Bs, maintaining a high level of transcription in the gametophyte (Pereira et al. 2009).

Transcripts of the recently identified D1100 and E3900 tandem repeats were only observed in + B plants, mostly restricted to anthers, where post-meiotic nondisjunction of rye Bs takes place (Carchilan et al. 2007). Although it remains to be directly demonstrated, anther-specific transcripts of sequences residing within the terminal nondisjunction control region might be related to the non-Mendelian accumulation of Bs, for example, by mediating stickiness of sister pericentromeric regions.

The discovery that some of the nondisjunction control region-specific repeats produce noncoding RNA, predominantly in anthers of rye (Prestel et al. 2010), suggests an intriguing possibility that the nondisjunction of Bs occurs because the control region somehow maintains cohesion in key regions of B-sister chromatids. Failure to resolve the pericentromeric heterochromatin during first pollen mitosis leads to the question: in which aspect does the first pollen mitosis differ from other mitotic events in other cell types? It is argued that either a haploid tissue type-specific expression of nondisjunction controlling transcripts (Carchilan et al. 2007) and/or the formation of a contrasting chromatin composition during first pollen mitosis (Houben et al. 2011) ensures this non-disjunction that results in specific accumulation of B chromosomes.

4.4 Effect on Morphology, Fitness and Meiotic Chromosomal Pairing in Species and Hybrids

4.4.1 Effect on Morphology and Fitness

The presence of Bs is associated with mild or no obvious change in phenotype in many species. This feature led to the conclusion that Bs are depleted of functional genes. However, an excessive number of Bs can cause phenotypic effects and may reduce host fertility (Jones and Rees 1982). The maximum number of Bs tolerated by individuals varies among different species. The harmful effects of Bs on fitness were charged to the energy cost of their maintenance and their potential interference with the proper assortment of A chrooosomes during meiosis (Jones and Rees 1982). Besides reduction of fitness, further phenotypic effects have been associated with the presence of Bs in several plant species (Jones and Rees 1982).

The phenotypic effects of the presence of B chromosomes are usually cumulative, depending upon the number of Bs, with a positive correlation between severity of effects and Bs counts (Jones and Rees 1982; Jones 1995; Bougourd and Jones 1997; Carlson 2009; Houben et al. 2013a).

A significant amount of information is available regarding the effects of rye Bs upon sporophyte and gametophyte fitness and viability, from seed germination to seed set (Jones and Rees 1982). As the number of Bs increases, negative effects on fertility (Müntzing 1943), seed germination timing (Moss 1966) and vigor (Müntzing, 1963) have been described. Bs also induce nuclear physiological effects in proportion to their number, such as increased cell cycle length (Evans et al. 1972) and decreased nuclear protein and RNA levels (Kirk and Jones 1970). These extra chromosomes have various effects on mitotic and meiotic A chromosome behavior. For example, Bs alter rDNA condensation patterns in mitotic cells (Morais-Cecílio et al. 1997; Delgado et al. 2004) and induce alterations in the frequency and distribution of A chromosome chiasmata at meiosis (Jones and Rees 1982). A dosage-dependent increase in the frequency of anomalous adherences between sister chromatids at anaphase and metaphase cells in the first mitosis of pollen grains provided more direct evidence that Bs affect A chromosome behavior (Pereira et al. 2009). Pereira et al. (2017) provided detailed cytogenetic and molecular insight into the effects of heat stress during reproductive development on meiosis in rye plants with 0 and 2B chromosomes. Their findings are the first indication that rye B chromosomes have implications in heat tolerance and protection against heat stress-induced damage at early stages of meiosis.

4.4.2 Effect on Meiotic Chromosome Pairing in Species and Hybrids

B chromosomes have been reported to have an effect on meiotic chromosome pairing in specie. Using the C-banding technique, Alvarez et al. (1991) studied the effect of Bs on homologous pairing, by examining first metaphase association of *S. cereale* chromosomes both in normal plants ($2n = 14$) and in plants with B chromosomes ($2n = 14 + Bs$). They noted a promoting effect on homologous meiotic chromosome pairing by B chromosome, particularly by its short arm.

Müntzing et al. (1969) studied meiosis in common wheat plants bearing one to several *S. cereale* B chromosomes that were transferred to common wheat from *Secale cereale* by Lindström in 1965. The effect of these B chromosomes on meiotic pairing of common wheat chromosomes was insignificant, whereas the meiotic pairing of the Bs was poor and the frequency of B chromosome univalents was much higher than in the corresponding strain of rye. The reason for this difference must involve an influence of the wheat chromosomes or the wheat cytoplasm on the rye chromosomes. Likewise, the influence of B-chromosomes of *Secale cereale* on homologous chromosome pairing at first meiotic metaphase of the inter-varietal common wheat hybrid, Chinese Spring x Lindström (carrying B chromosomes of *S. cereale*) was studied by Viegas (1979) in the presence and absence of chromosome 5D of common wheat, that carries a promoter of homologous pairing (Feldman 1966). The presence of these B chromosomes did not change the normal pattern of chromosome pairing in disomic plants, although it slightly increased chiasma frequency in monosomic 5D plants at 20 °C. When chromosome 5D was absent, e.g., nullisomic 5D, the increase in chiasma frequency was more pronounced, especially the number of ring bivalents. In nullisomic 5D plants, at 10 °C, where a high degree of asynapsis was observed, the addition of B chromosomes increased chromosome pairing, but no increase in the frequency of monosomics was observed, which at 10 °C showed a slight reduction in pairing.

The effect of B chromosomes on homoeologous pairing in hybrids with common wheat is ambiguous. B chromosomes of *Amblyopyrum muticum* and *Aegilops speltoides* suppress homoeologous meiotic pairing of A chromosomes in intergeneric hybrids with common wheat lacking the *ph1* gene, which suppresses homoeologous pairing (Mochizuki 1964; Vardi and Dover 1972; Dover and Riley 1972; Ohta and Tanaka 1982). Thus, the effect of Bs is similar to that of *Ph1* of wheat on homoeologous pairing in wheat hybrids with alien species. The B chromosomes of *A. muticum* do not affect homologous pairing but suppress homoeologous pairing in interspecific hybrids (Mochizuki 1964; Dover and Riley 1972; Vardi and Dover 1972). Similarly, studies on meiotic chromosomal pairing in F_1 hybrids between *A. muticum* and most of the diploid species of *Aegilops* and *Triticum* containing B-chromosomes of *A. muticum*, showed notably reduced chromosomal pairing (Vardi and Dover 1972; Ohta and Tanaka 1983; Ohta 1990, 1991).

Vardi and Dover (1972) assumed that B chromosomes of *Ae. speltoides* and *A. muticum* interact with specific gene loci of the A chromosome complement that inactivate the suppression of *Ph1*. Furthermore, they suggested that Bs affect chromosomal pairing by causing disturbances in the mitotic and meiotic spindle. The similarity between the effects of *Ph1* and B chromosomes on chromosomal pairing suggests that the *ph1* gene of polyploid wheat was transferred from B chromosomes of *A. muticum* or *Ae. speltoides* to chromosome 5B of common wheat, via an ancestral translocation between a B chromosome and an A chromosome, presumably 5B (Vardi and Dover 1972). However, the absence of meiotic pairing between the B chromosome and chromosome 5B of common wheat refutes this idea. Alternatively, since B chromosomes derived from one of the A chromosomes in these two diploid species and contain many DNA sequences of other A chromosomes, one of the most essential prerequisites for the establishment of the B chromosome as an independent entity was to prevent pairing and recombination between B and A chromosomes. Thus, the development of a genetic system suppressing homoeologous pairing was a necessary event for the independent existence and evolution of B chromosomes.

On the other hand, conflicting results were obtained concerning the effect of *Secale cereale* B chromosomes on homoeologous pairing in hybrids with common wheat. Viegas (1980) reported that B chromosomes of *S. cereale* suppressed homoeologous pairing in hybrids with common wheat, irrespective of the presence or absence of chromosome *Ph1*, while Cuadrado et al. (1988) found that *cereale* B chromosomes only suppressed homoeologous pairing in hybrids with common wheat in the absence of *Ph1*. Romero and Lacadena (1980) found that *cereale* B chromosomes also suppress homoeologous pairing in hybrids with common wheat lacking the pairing suppressors on wheat chromosomes 3A and 3D, although they increased the level of pairing when a chromosome with a promoter effect (3B, 5A or 5D) was absent. Roothaan and Sybenga (1976) found that B chromosomes of *S. cereale* do not compensate for the absence of *Ph1* in hybrids with common wheat lacking *Ph1*. Estepa et al. (1993) studied the effect of different numbers of B chromosomes on homoeologous pairing in common wheat x *S. cereale* hybrids and found no significant quantity-related effect, but the variance of distribution of means of bivalent and paired chromosome complements was significantly increased when odd numbers (3 or 5 B-chromosomes) were present. Jenkins and Jones (2004) concluded that B-chromosomes of *S. cereale* carry genes that act together with the pairing control genes of common wheat.

Kousaka and Endo (2012) studied the effect of a rye B chromosome and its segments (B-9 and B-10) on homoeologous pairing in hybrids between common wheat and *Aegilops peregrina*. The B-9 and B-10 chromosome segments are derived from reciprocal translocations between a wheat and B chromosome; B-9 had the B pericentromeric segment, whereas B-10 had the B distal segment. Kousaka and Endo (2012) found that both the complete B chromosome and the B-9 segment suppressed homoeologous pairing when chromosome 5B was absent. On the other hand, the B-9 and B-10 segments promoted homoeologous pairing when 5B was present. The mean chiasma frequency (10.23/cel) in the hybrid of common wheat x *Ae. Peregrina* possessing 5B and one B-9, was considerably higher than that of a hybrid possessing 5B alone (2.78/cell), and was comparable to that of a hybrid lacking 5B (14.09/cell). The results suggested that the effect of the B chromosomes on homoeologous pairing was not confined to a specific region, i.e., it resulted from a polygenic system, and that the intensity of the effect varied and depended on the presence or absence of *Ph1* and on the dose of the B chromosome and the B segments.

4.5 Transcriptional Activity of B Chromosomes

Considering the intra- or interspecific origin of Bs and the above-listed B-associated effects, many have sought to determine whether Bs carry genes. Studies have led to different conclusions regarding the transcriptional activity of Bs. In contrast to single- or low-copy genes that were rarely found on Bs in early studies, rRNA genes have been frequently identified on Bs of many species (Green 1990; Bougourd and Jones 1997). This is most likely due to the fact that their detection is rather easy by cytogenetic techniques in contrast to unique genes. In contrast to the prevalent view that Bs do not harbor genes, recent analysis revealed that Bs of sequenced species are rich in gene-derived sequences (Houben et al. 2014). Recent application of next generation sequencing-based approaches revealed that rye Bs contain more than 4000 putative genic sequences (Martis et al. 2012), many of which are partly transcriptionally active (Banaei-Moghaddam et al. 2013). Some of the rye B gene sequences had lower similarity to their A-located counterparts, reflecting their faster degeneration or earlier insertion in Bs (Banaei-Moghaddam et al. 2013). These studies suggest that B chromosomes carry transcriptionally active genic sequences that could affect the transcriptome profile of their host genome (Banaei-Moghaddam et al. 2015). Thus, the view that B chromosomes are genetically inert, selfish

elements without any functional genes, is gradually changing. This can partly explain the deleterious effects associated with their presence as well as the possible advantages that Bs confer on their host.

A comparative cDNA-AFLP analysis indicated that rye Bs can modulate the transcription of corresponding gene copies on A chromosomes (Carchilan et al. 2009) and, from these studies, regulatory interactions between A- and B-located coding sequences have been proposed (Banaei-Moghaddam et al. 2015). It is likely that Bs influence A-localized sequences through epigenetic mechanisms, such as homology-dependent RNA interference pathways (Slotkin and Martienssen 2007), as has been proposed for the modulation of gene-activity in newly formed hybrids and allopolyploids (Comai 2005; Kenan-Eichler et al. 2011). Bs may also exert control over A chromosomes via the spatial organization of As in interphase nuclei, and it has been suggested that spatial positioning of genes and chromosomes can influence gene expression (Misteli 2007).

As Bs are dispensable, it is expected that they are prone to accumulate mutations as they undergo pseudogenization (Banaei-Moghaddam et al. 2013). Indeed, next-generation sequencing technologies have shown that B chromosomes of *S. cereale* harbor many pseudogenes originating from the A chromosomes and organellar DNA sequences (Houben et al. 2013b). It has been further demonstrated that some of the A-derived sequences are transcribed in a genotype-specific manner (Banaei-Moghaddam et al. 2013). This could explain the apparently contradictory findings and complex interactions associated with the effect of B chromosomes on homoeologous pairing in hybrids with common wheat.

It is possible that only part of the B-located genes is inactive (Houben et al. 2014). If Bs share many, almost identical, genic sequences with As, why is the presence of Bs not associated with more severe phenotypes, particularly assuming that some sequence variants may still have a biological effect? Bearing in mind that the relative dosage of a chromosome is critical for normal development, it is striking that organisms with an additional B are little affected. It is probable that during early evolution of a proto B, A-derived genes are likely to be downregulated by dosage compensation.

Transcripts coming from a B chromosome in combination with their related A-located genes, provide additional complexity to the transcriptome of their host and this may partially explain the phenotypes and effects associated with the presence of Bs. The transcriptional activity of Bs could form regulatory transcripts such as siRNAs which have the potential to modulate the level of A-derived transcripts or to change the chromatin status of a target region by DNA or chromatin modification (Dalakouras and Wassenegger 2013; Filipowicz et al. 2005). In addition, transcripts from Bs similar to pseudogenes may lead to indirect effects by competing with A-derived transcripts for regulatory factors such as miRNAs (Pink et al. 2011). If truncated proteins or overexpression of functional proteins are produced in the presence of Bs, they can cause overloading of the proteasome machinery, required to process these unfolded, misfolded, aggregated, and/or uncomplexed proteins, thereby imparting an energetic burden in the host (Gordon et al. 2012). But Bs may also produce functional proteins that may have some role in maintenance of B chromosomes.

Banaei-Moghaddam et al. (2015) postulated the potential activity modulation of A chromosome-located genes by homologous sequences on B chromosomes. When the B chromosome is absent, each gene on A has a defined level and pattern of expression. In the presence of a B chromosome, when the B-located gene is transcribed, transcripts can serve as a substrate for RNA-directed RNA polymerase, which would generate dsRNA, which is then processed to small RNAs. These small RNAs could cause mRNA degradation or epigenetic changes of A-located sequences. Alternatively, the B-transcripts may compete for regulatory factors responsible for regulation of A-located genes, yielding up- or down-regulation of the counterpart A gene. In the presence of B, if the corresponding B-located gene is inactive, the expression level and pattern of an A-located gene will not change. Likewise, when both A- and B-located genes express from the same strand, the expression level of that gene may remain unchanged compared to that of plants lacking B, due to dosage compensation.

Transcription of B-enriched repetitive sequences have also been demonstrated in rye. A comparative analysis of RNA-seq reads obtained from rye plants with and without Bs, was recently performed to assess the transcriptional activity of repetitive sequences. In this species, several B-repeats are active in a tissue-dependent manner (Klemme et al. 2013; Carchilan et al. 2007). Transcripts derived from vegetative (root and leaf) and generative (anthers) tissues, represented by 26–151 million RNA-seq reads, were screened for their sequence similarity to a complete set of previously identified rye repetitive elements (Martis et al. 2012). Low levels of transcription were found for most repetitive elements, which represented a combined total of 3.1–4.9% of transcripts in analyzed tissues (excluding rRNAs). Repeat expression profiles were similar for plants with and without B chromosomes, with the exception of highly expressed B-specific satellite E3900, which was found in both vegetative tissues and anthers. This finding confirms previously reported data based on Northern blot experiments (Carchilan et al. 2007). On the other hand, none of the other satellites enriched on or specific to B chromosomes (Martis et al. 2012; Klemme et al. 2013) were expressed at comparable levels.

Dosage-sensitive genes are less likely a part of an evolving B, as duplication of a B chromosome-donor fragment containing such genes could result in a detrimental phenotype. In contrast, dosage-insensitive genes, such as structural genes bearing no regulatory roles in transcription or translation, could be B-hosted and eventually undergo loss or pseudogenization. Duplicated genes are often associated with detrimental effects and are removed by natural selection. As most of the mutations are degenerative, it is more likely that duplicated gene underwent inactivation rather than acquiring a new function. Nevertheless, evidence of the beneficial role of duplicated genes, especially in stressful conditions, exists. In some circumstances, selective retention of duplicated genes could occur, e.g., when their redundancy protects corresponding parental genes from immediate detrimental mutations, or when overdominance exists between their products. In this case, duplicated genes could convert to new genes by achieving a beneficial mutation. Therefore, B-located duplicated genes may accelerate the evolution of their carriers.

Could a B-located genic sequence with the ability to modulate the activity of a corresponding A-located gene cause long-term evolutionary effects? If the product of the B-located gene would become beneficial, positive selection would act to preserve the respective B present in the population. This could release the pressure on a B-specific differential transmission, namely, a former selfish B chromosome would transform into a beneficial genome component. It would be intriguing to test whether Bs affect the epigenetic status of As, by comparing epigenetic modifications and expression levels of parental A genes in the presence and absence of Bs, over several generations.

References

Alvarez MT, Fominaya A, Perez de la Vega M (1991) A possible effect of B-chromosomes on metaphase I homologous chromosome association in rye. Heredity 67:123–128

Asghari A, Agayev Y, Fathi SAA (2007) Karyological study of four species of wheat grass (*Agropyron* sp.). Pakistan J Biol Sci 10:1093–1097

Assadi M (1995) Meiotic configuration and chromosome number in some iranian species of *Elymus* L. and *Agropyron* Gaertner (Poaceae), Triticeae). Bot J Linn Soc 117:159–168

Baenziger H (1962) Supernumerary chromsomes in diploid and tetraploid forms of crested wheatgrass. Can J Bot 40:549–561

Banaei-Moghaddam AM, Schubert V, Kumke K, Weib O, Klemme S, Nagaki K, Macas J, González-Sánchez M, Heredia V, Gomez-Revilla D et al (2012) Nondisjunction in favor of a chromosome: the mechanism of rye B chromosome drive during pollen mitosis. Plant Cell 24:4124–4134

Banaei-Moghaddam AM, Meier K, Karimi-Ashtiyani R, Houben A (2013) Formation and expression of pseudogenes on the B chromosome of rye. Plant Cell 25:2536–2544

Banaei-Moghaddam AM, Martis MM, Macas J, Gundlach H, Himmelbach A, Altschmied L, Mayer KFX, Houben A (2015) Genes on B chromosomes: old questions revisited with new tools. BBA 1849:64–70

Battaglia E (1964) Cytogenetics of B-chromosomes. Caryologia 17:245–299

Belyayev A, Raskina O (2013) Chromosome evolution in marginal populations of *Aegilops speltoides*: causes and consequences. Ann Bot 111:531–538

Beukeboom LW (1994) Bewildering Bs: an impression of the 1st B-chromosome conference. Heredity 73:328–336

Blunden R, Wilkes TJ, Forster JW, Jimenez MM, Sandery MJ, Karp A, Jones RN (1993) Identification of the E3900 family, a 2nd family of rye chromosome-B specific repeated sequences. Genome 36:706–711

Bougourd SM, Jones RN (1997) B chromosomes: a physiological enigma. New Phytol 137:43–54

Burt A, Trivers R (2006) Genes in conflict: the biology of selfish genetic elements. Cambridge, Belknap

Camacho JPM, Sharbel TF, Beukeboom LW (2000) B-chromosome evolution. Philos Trans R Soc Lond B 355:163–178

Carchilan M, Delgado M, Ribeiro T, Costa-Nunes P, Caperta A, Morais-Cecilio L, Jones RN, Viegas W, Houben A (2007) Transcriptionally active heterochromatin in rye B chromosomes. Plant Cell 19:1738–1749

Carchilan M, Kumke K, Mikolajewsk S, Houben A (2009) Rye B chromosomes are weakly transcribed and might alter the transcriptional activity of A chromosome sequences. Chromosoma 118:607–616

Carlson W (2009) The B chromosome of maize. In: Bennetzen JL, Hake S (eds) Maize handbook. Genetics and genomics, vol II. Springer, Heidelberg, pp 459–480

Cohen S, Yacobi K, Segal D (2003) Extrachromosomal circular DNA of tandemly repeated genomic sequences in *Drosophila*. Genome Res 13:1133–1145

Comai L (2005) The advantages and disadvantages of being polyploid. Nat Rev Genet 6:836–846

Cuadrado MC, Romero C, Lacadena JR (1988) Interaction between wheat chromosome arms controlling homoeologous pairing and rye B-chromosomes. In: Miller TE, Koebner RMD (eds) Proceedings of 7th international wheat genetic symposium. England, Cambridge, pp 237–241

Dalakouras A, Wassenegger M (2013) Revisiting RNA-directed DNA methylation. RNA Biol 10:453–455

Darlington CD (1933) The origin and behavior of chiasmata. VIII. *Secale cereale*. Cytologia 4:444–452

Delgado M, Caperta A, Ribeiro T, Viegas W, Jones RN, Morais-Cecilio L (2004) Different numbers of rye B chromosomes induce identical compaction changes in distinct A chromosome domains. Cytogenet Genome Res 106:320–324

Devos KM, Atkinson MD, Chinoy CN, Francis HA, Harcourt RL, Koebner RMD, Liu CJ, Masoje P, Xie DX, Gale MD (1993) Chromosomal rearrangements in the rye genome relative to that of wheat. Theor Appl Genet 85:673–680

Diez M, Jiménez MM, Santos JL (1993) Synaptic patterns of rye B chromosomes. II. The effect of the standard B chromosomes on the pairing of the A set. Theor Appl Genet 87:17–21

Dover GA, Riley R (1972) Prevention of pairing of homoeologous meiotic chromosomes of wheat by an activity of supernumerary chromosomes of Aegilops. Nature 240:159–161

Emme HK (1928) Karyologie der Gattung *Secale* L. Zschr Ind Abst Vererb L 47:99–124

Endo TR, Nasuda S, Jones N, Dou Q, Akahori A, Wakimoto M, Tanaka H, Niwa K, Tsujimoto H (2008) Dissection of rye B chromosomes, and nondisjunction properties of the dissected segments in a common wheat background. Genes Genet Syst 83:23–30

Estepa MA, Cuadrado C, Romero C (1993) Odd-even effect of rye B-chromosomes on homoeologous pairing in wheat-rye hybrids. Caryologia 46:17–23

Evans GM, Rees H, Snell CL, Sun S (1972) The relationship between nuclear DNA amount and the duration of the mitotic cycle. Chromosomes Today 3:24–31

Feldman M (1966) The effect of chromosomes 5B, 5D and 5A on chromosomal pairing in *Triticum aestivum*. Proc Natl Acad Sci USA 55:1447–1453

Filipowicz W, Jaskiewicz L, Kolb FA, Pillai RS (2005) Posttranscriptional gene silencing by siRNAs and miRNAs. Curr Opin Struct Biol 15:331–341

Friebe B, Jiang J, Gill B (1995) Detection of 5S rDNA and other repeated DNA on supernumerary B-chromosomes of *Triticum* species (*Poaceae*). Plant Syst Evol 196:131–139

Gonzalez-Sanchez M, Chiavarino M, Jimenez G, Manzanero S, Rosato M, Puertas MJ (2004) The parasitic effects of rye B chromosomes might be beneficial in the long term. Cytogenet Genome Res 106:386–393

Gordon DJ, Resio B, Pellman D (2012) Causes and consequences of aneuploidy in cancer. Nat Rev 13:189–203

Gotoh K (1924) Über die Chromosomenzahl von *Secale cereale* L. Botan Mag Tokyo 38:135–152

Green DM (1990) Muller`s Rachet and the evolution of supernumerary chromosomes. Genome 33:818–824

Håkansson A (1948) Behaviour of accessory rye chromosomes in the embryo sac. Hereditas 34:35–59

Håkansson A (1959) Behaviour of different small accessry rye chromosomes at pollen mitosis. Hereditas 45:623–631

Hasegawa N (1934) A cytological study on 8-chromosome rye. Cytologia 6:68–77

Hosid E, Brodsky L, Kalendar R, Raskina O, Belyayev A (2012) Diversity of long terminal repeat retrotransposon genome distribution in natural populations of the wild diploid wheat *Aegilops speltoides*. Genetics 190:263–274

Houben A, Kynast RG, Heim U, Hermann H, Jones RN, Forster JW (1996) Molecular cytogenetic characterization of the terminal heterochromatic segment of the B-chromosome of rye (*Secale cereale*). Chromosoma 105:97–103

Houben A, Field BL, Saunders VA (2001) Microdissection and chromosome painting of plant

Houben A, Kumke K, Nagaki K, Hause G (2011) CENH3 distribution and differential chromatin modifications during pollen development in rye (*Secale cereale* L.). Chromosome Res 19:471–480

Houben A, Banaei-Moghaddam AM, Klemme S (2013a) Biology and evolution of B chromosomes. In: Greilhuber J, Dolezel J, Wendel JF (eds) Plant genome diversity, vol 2. Springer Vienna, pp 149–166

Houben A, Banaei-Moghaddam AM, Klemme S, Timmis JN (2013b) Evolution and biology of supernumerary B chromosomes. Cell Mol Life Sci 71:467–478

Houben A, Banaei-Moghaddam AM, Klemme S, Timmis JR (2014) Evolution and biology of supernumerary B chromosomes. Cell Mol Life Sci 71:467–478

Jenkins G, Jones N (2004) B chromosomes in hybrids of temperate cereals and grasses. Cytogenet Genome Res 106:314–319

Jiménez MM, Romera F, González-Sánchez M, Puertas MJ (1997) Genetic control of the rate of transmission of rye B chromosomes. III. Male meiosis and gametogenesis. Heredity 78:636–644

Jiménez G, Manzanero S, Puertas M (2000) Relationship between pachytene synapsis, metaphase I associations, and transmission of 2B and 4B chromosomes in rye. Genome 43:232–239

Johnsson P, Morris KV, Grander D (2014) Pseudogenes: a novel source of trans-acting antisense RNAs. Methods Mol Biol 1167:213–226

Jones RN (1975) B-chromosome systems in flowering plants and animal species. Inter Rev Cytol 40:1–100

Jones RN (1991) B-chromosome drive. Am Nat 137:430–442

Jones RN (1995) B chromosomes in plants. New Phytol 131:411–434

Jones RN (2012) B chromosomes in plants. Plant Biosyst 146:727–737

Jones RN, Houben A (2003) B chromosomes in plants: escapeees from the A chromosome genome? Trends Plant Sci 8:417–423

Jones RN, Puertas MJ (1993) The B chromosomes of rye (Secale cereale L.). In: Dhir KK, Sareen TS (eds) Frontiers in plant science research. Bhagwati Enterprises, Delhi, pp 81–112

Jones RN, Rees H (1969) An anomalous variation due to B chromosomes in rye. Heredity 24:265–271

Jones RN, Rees H (1982) B chromosomes. Academic Press, London

Jones RN, Viegas W, Houben A (2008a) A century of B chromosomes in plants: so what? Ann Bot 101:767–775

Jones RN, González-Sánchez M, González-García M, Vega JM, Puertas MJ (2008b) Chromosomes with a life of their own. Cytogenet Genome Res 120:265–280

Kenan-Eichler M, Leshkowitz D, Tal L, Noor E, Cathy Melamed-Bessudo C, Feldman M, Levy AA (2011) Wheat hybridization and polyploidization results in deregulation of small RNAs. Genet 188:263–272

Kirk D, Jones RN (1970) Nuclear genetic activity in B chromosome rye, in terms of quantitative interrelationships between nuclear protein, nuclear RNA and histone. Chromosoma 31:241–254

Kishikawa H, Suzuki A (1982) Cytological study on hypopentaploid Triticale with four B chromosomes of rye. Jpn J Genet 57:17–24

Klemme S, Banaei-Moghaddam AM, Macas J, Wicker T, Novák P et al (2013) High-copy sequences reveal distinct evolution of the rye B-chromosome. New Phytol 199:550–558

Knowles RP (1955) A study of variability in crested wheatgrass. Can J Bot 33:534–546

Koller OL, Zeller FJ (1976) The homoeologous relationships of rye chromosomes 4R and 7R with wheat chromosomes. Genet Res Camb 28:177–188

Kousaka R, Endo ER (2012) Effect of a rye B chromosome and its segments on homoeologous pairing in hybrids between common wheat and Aegilops variabilis. Genes Genet Syst 87:1–7

Kranz AR (1963) Beitrage zur cytologischen und genetischen Evolutionsforschung an dem Roggen. Zeitschrift für Pflanzenzüchtung 50:44–58

Kuwada Y (1925) On the number of chromosomes in maize. Botan Mag Tokyo 39:227–234

Langdon T, Seago C, Jones RN, Ougham H, Thomas H, Forster JW, Jenkins G (2000) De novo evolution of satellite DNA on the rye B chromosome. Genetics 154:869–884

Levin DA, Palestis BG, Jones RN, Trivers R (2005) Phyletic hot spots for B chromosomes in angiosperms. Evolution 59:962–969

Lima-de-Faria A (1962) Genetic interaction in rye expressed at chromosome phenotype. Genetics 47:1455–1462

Lindström J (1965) Transfer to wheat of accessory chromosomes from rye. Hereditas 54:149–155

Liu CJ, Atkinson MD, Chinoy CN, Devos KM, Gale MD (1992) Non-homoeologous translocations between group 4, 5 and 7 chromosomes within wheat and rye. Theor Appl Genet 83:305–312

Longley AE (1927) Supernumerary chromosomes in Zea mays. J Agric Res 35:769–784

Marques A, Banaei-Moghaddam AM, Klemme S, Blattner FR, Niwa K et al (2013) B chromosomes of rye are highly conserved and accompanied the development of early agriculture. Ann Bot 112:527–534

Martis MM, Klemme S, Banaei-Moghaddam AM, Blattner F, Macas J et al (2012) Selfish supernumerary chromosome reveals its origin as a mosaic of host genome and organellar sequences. Proc Natl Acad Sci USA 109:13343–13346

Matthews RB, Jones RN (1982) Dynamics of the B chromosomes polymorphism in rye. I. Simulated populations. Heredity 48:345–369

Matthews RB, Jones RN (1983) Dynamics of the B chromosome polymorphism in rye. II. Estimates of parameters. Heredity 50:119–137

Mayer KFX, Martis M, Hedley PE, Šimková H, Liu H et al (2011) Unlocking the barley genome by chromosomal and comparative genomics. Plant Cell 23:1249–1263

McCoy GA, Law AG (1965) Satellite chromosomes in crested wheatgrass (Agropyron desertorum (Fisch), Schult.). Crop Sci 5:283

Mendelson D, Zohary D (1972) Behaviour and transmission of supernumerary chromosomes in Aegilops speltoides. Heredity 29:329–339

Misteli T (2007) Beyond the sequence: cellular organization of genome function. Cell 128:787–800

Mochizuki A (1957) B chromosomes in Aegilops mutica Boiss. Wheat Inf Serv 5:9–11

Mochizuki A (1960) A note on the B-chromosomes in natural populations of Aegilops mutica Boies in central Turkey. Wheat Inf Serv 11:31

Mochizuki A (1964) Further studies on the effect of accessory chromosomes on chromosome pairing. Jpn J Genet 39:356–362 (in Japanese)

Morais-Cecílio L, Delgado M, Jones RN, Viegas W (1997) Interphase arrangement of rye B chromosomes in rye and wheat. Chromosome Res 5:177–181

Moss JP (1966) The adaptive significance of B chromosomes in rye. Chromosomes Today 1:15–23

Müntzing A (1943) Genetical effects of duplicated fragment chromosomes in rye. Hereditas 29:91–112

Müntzing A (1944) Cytological studies of extra fragment chromosomes in rye. I. Iso-fragments produced by misdivision. Hereditas 30:231–248

Müntzing A (1946) Different chromosome numbers in root tips and pollen mother cells in a sexual strain of Poa alpina. Hereditas 32:127–129

Müntzing A (1948) Cytological studies of extra fragment chromosomes in rye. V. A new fragment type arisen by deletion. Hereditas 34:435–442

Müntzing A (1950) Accessory chromosomes in rye populations from Turkey and Afghanistan. Hereditas 36:507–509

Müntzing A (1963) Effects of accessory chromosomes in diploid and tetraploid rye. Hereditas 66:279–286

Müntzing A (1970) Chromosomal variation in the Lindström strain of wheat carrying accessory chromosomes in rye. Hereditas 66:279–286

Müntzing A, Jaworska H, Carlbom C (1969) Studies of meiosis in the Lindström strain of wheat carrying accessory chromosomes of rye. Hereditas 61:179–207

Muro EM, Mah N, Andrade-Navarro MA (2011) Functional evidence of post-transcriptional regulation by pseudogenes. Biochimie 93:1916–1921

Naranjo T, Fernández-Rueda P (1991) Homoeology of rye chromosome arms to wheat. Theor Appl Genet 82:577–586

Naranjo T, Roca A, Goicoecha PG, Giraldez R (1987) Arm homoeology of wheat and rye chromosomes. Genome 29:873–882

Niwa K, Sakamoto S (1995) Origin of B-chromosomes in cultivated rye. Genome 38:307–312

Niwa K, Sakamoto S (1996) Detection of B chromosomes in rye collected from Pakistan and China. Hereditas 124:211–215

Niwa K, Ohta S, Sakamoto S (1990) B chromosomes of Secale cereale L. and S. montanum Guss. from Turkey. Jpn J Breed 40:147–152

Niwa K, Horiuchi G, Hirai Y (1997) Production and characterization of common wheat with B chromosomes of rye from Korea. Hereditas 126:139–146

Novak P, Neumann P, Macas J (2010) Graph-based clustering and characterization of repetitive sequences in next-generation sequencing data. BMC Bioinform 11:378

Ohta S (1990) Genome analysis of *Aegilops mutica* Boiss based on the chromosome pairing in interspecific and intergeneric hybrids. Ph.D. thesis, submitted to the University of Kyoto, pp 1–267

Ohta S (1991) Phylogenetic relationship of *Aegilops mutica* Boiss with the diploid species of congeneric *Aegilops-Triticum* complex, based on the new method of genome analysis using its B-chromosomes. Mem Coll Agric Kyoto Univ 137:1–116

Ohta S (1995) Distinct numerical variation of B-chromosomes among different tissues in *Aegilops mutica* Boiss. Jap J Genet 70:93–101

Ohta S (1996) Mechanisms of B-chromosome accumulation in *Aegilops mutica* Boiss. Genes Genet Syst 71:23–29

Ohta S, Tanaka M (1982) The effects of the B-chromosomes of *Aegilops mutica* Boias. on meiotic chromosome pairing. Rep Plant Germ-Plasm Inst Kyoto Univ 5:36–52

Ohta S, Tanaka M (1983) Genome relationships between *Ae. mutica* and the other diploid *Aegilops* and *Triticum* species, based on the chromosome pairing in the hybrids with or without B-chromosomes. In: Sakamoto S (ed) Proceedings of 6th international wheat genetic symposium, Kyoto, pp 983–991

Ortiz M, Puertas MJ, Jimeniz M, Romera F, Jones RN (1996) B chromosomes in inbred lines of rye (*Secale cereale* L.). 2. Effects on metaphase I and first pollen mitosis. Genetica 97:65–72

Palestis BG, Burt A, Jones RN, Trivers R (2004) B chromosomes are more frequent in mammals with acrocentric karyotypes: support for the theory of centromeric drive. Proc R Soc Lond B Biol Sci 271 (Suppl. 3):S22–S24

Pereira HS, Barão A, Caperta A, Rocha J, Viegas W, Delgado M (2009) Rye Bs disclose ancestral sequences in cereal genomes with a potential role in gametophyte chromatid segregation. Mol Biol Evol 26:1683–1697

Pereira HS, Delgado M, Viegas W, Rato JM, Barão A, Caperta AD (2017) Rye (*Secale cereale*) supernumerary (B) chromosomes associated with heat tolerance during early stages of male sporogenesis. Ann Bot 119:325–337

Pink RC, Wicks K, Caley DP, Punch EK, Jacobs L, Carter DR (2011) Pseudogenes: pseudo-functional or key regulators in health and disease? RNA 17:792–798

Prestel M, Feller C, Becker PB (2010) Dosage compensation and the global rebalancing of aneuploid genomes. Genome Biol 11:216

Puertas MJ (2002) Nature and evolution of B chromosomes in plants: a non-coding but information-rich part of plant genomes. Cytogenet Genome Res 96:198–205

Puertas MJ, Romera F, Delapena A (1985) Comparison of B-Chromosome effects on *Secale cereale* and *Secale vavilovii*. Heredity 55:229–234

Puertas MJ, González-Sánchez M, Silvia Manzanero S, Félix Romera F, Jiménez MM (1998) Genetic control of the rate of transmission of rye B chromosomes. IV. Localization of the genes controlling B transmission rate. Heredity 80:209–213

Puertas MJ, Jiménez G, Manzanero S, Chiavarino AM, Rosato M, Naranjo CA, Poggio L (2000) Genetic control of B chromosome transmission in maize and rye. Chromosomes Today 13:79–92

Randolph LF (1928) Types of supernumerary chromosomes in maize. Anat Rec 41:102

Raskina O, Belyayev A, Nevo E (2004) Activity of the *En/Spm*-like transposons in meiosis as a base for chromosome repatterning in a small, isolated, peripheral population of *Aegilops speltoides* Tausch. Chromosome Res 12:153–161

Raskina O, Brodsky L, Belyayev A (2011) Tandem repeats on an eco-geographical scale: outcomes from the genome of *Aegilops speltoides*. Chromosome Res 19:607–623

Rees H, Ayonoadu U (1973) B-chromosome selection in rye. Theor Appl Genet 43:162–166

Rimpau J, Flavell RB (1975) Characterization of rye B chromosome DNA by DNA-DNA hybridization. Chromosoma 52:207–217

Rognli OA, Devos KM, Chinoy CN, Harcourt RL, Atkinson MD, Gale MD (1992) RFLP mapping of rye chromosome 7R reveals a highly translocated chromosome relative to wheat. Genome 35:1026–1031

Romero C, Lacadena JR (1980) Interaction between rye B chromosomes and wheat genetic systems controlling homoeologous pairing. Chromosoma 80:33–48

Romera F, Jimenez MM, Puertas MJ (1991) Factors controlling the dynamics of the B-chromosome polymorphism in Korean rye. Heredity 67:189–195

Roothaan M, Sybenga J (1976) No 5-B compensation by rye B-chromosomes. Theor Appl Genet 48:63–66

Sandery MJ, Forster JW, Blunden R, Jones RN (1990) Identification of a family of repeated sequences on the rye B chromosome. Genome 33:908–913

Schlegel R (2013) Rye—genetics, breeding, and cultivation. CRC Press, Taylor and Francis Group, Roca Raton, FL, USA, pp 1–344

Scholz U, Simkova H, Kubalakova M, Choulet F, Taudien S, Platzer M, Feuillet C, Fahima T, Budak H, Dolezel J, Keller B, Stein N (2011) Frequent gene movement and pseudogene evolution is common to the large and complex genomes of wheat, barley, and their relatives. Plant Cell 23:1706–1718

Sheppard AE, Timmis JN (2009) Instability of plastid DNA in the nuclear genome. PLoS Genet 5:e1000323

Shewry PR, Parmer S, Miller TE (1985) Chromosomal location of the structural genes for the M_r 75,000 γ-secalins in *Secale montanum* Guss: evidence for a translocation involving chromosome 2R and 6R in cultivated rye (*Secale cereale* L.). Heredity 54:381–383

Shilo S, Tripathi P, Melamed-Bessudo C, Tzfadia O, Muth TR, Lev AA (2017) T-DNA-genome junctions form early after infection and are influenced by the chromatin state of the host genome. PLoS Genet 13:e1006875

Shirasu K, Schulman AH, Lahaye T, Schulze-Lefert P (2000) A contiguous 66 kb barley DNA sequence provides evidence for reversible genome expansion. Genome Res 10:908–915

Simchen G, Zarchi Y, Hillel J (1971) Supernumerary chromosomes in second outbreeding species of wheat group. Chromosoma 33:63–69

Slotkin RK, Martienssen R (2007) Transposable elements and the epigenetic regulation of the genome. Nat Rev Genet 8:272–285

Smith GP (1976) Evolution of repeated DNA sequences by unequal crossover. Science 191:528–535

Stutz HC (1972) The origin of cultivated rye. Am J Bot 59:59–70

Timmis JN, Ingle J, Sinclair J, Jones N (1975) The genomic quality of rye B chromosomes. J Exp Bot 26:367–378

Timmis JN, Ayliffe MA, Huang CY, Martin W (2004) Endosymbiotic gene transfer: organelle genomes forge eukaryotic chromosomes. Nat Rev Genet 5:123

Tomita M, Shinohara K, Morimoto M (2008) Revolver is a new class of transposon-like gene composing the Triticeae genome. DNA Res 15:49–62

Trivers R, Burt A, Palestis BG (2004) B chromosomes and genome size in flowering plants. Genome 47:1–8

Tsujimoto H, Niwa K (1992) DNA structure of the B-chromosome of rye revealed by in situ hybridization using repetitive sequences. Jap J Genet 67:233–241

Vardi A, Dover GA (1972) The effect of B chromosomes on meiotic and pre-meiotic spindles and chromosome pairing in *Triticum/Aegilops* hybrids. Chromosoma (berl.) 38:367–385

Viegas WS (1979) The effect of B-chromosomes of rye on chiasma frequency in *Triticum aestivum*. Genetica 51:69–75

Viegas WS (1980) The effect of B-chromosomes of rye on the chromosome association in F$_1$ hybrids *Triticum aestivum* x *Secale cereale* in the absence of chromosomes 5B or 5D. Theor Appl Genet 56:193–198

Wilkes TM, Francki MG, Langridge P, Karp A, Jones RN, Forster JW (1995) Analysis of rye B-chromosome structure using fluorescence in situ hybridization (FISH). Chromosome Res 3:466–472

5.1 General Description of the Subtribe

Whereas the economically important cereals, *Triticum* L., *Secale* L., *Hordeum* L., and *Aegilops* L., have been subjected to intensive taxonomic, cytogenetic, molecular, and evolutionary studies, several other Triticeae genera received less attention. These "orphan" genera are *Agropyron* Gaertner, *Eremopyrum* (Ledeb.) Jaub. & Spach, *Henrardia* C. E. Hubbard, *Dasypyrum* (Coss. & Dur.) Dur., *Heteranthelium* Jaub. & Spach., *Taeniatherum* Nevski, and *Crithopsis* Jaub. & Spach. Several diploid *Elymus* species, having the St and E genomes, are also included in this group. They are small genera containing few species, most of which diploids that are characterized by a distinct morphology, and grow in different regions of the tribe distribution area, some in more arid environments. Their genetic relationships to the other well-studied genera of the tribe are vaguely known. Studies of these small genera may provide additional knowhow on the range of genetic diversity in the tribe, on processes that have led to diverge evolutionary developments as well as on the phylogenetic relationships among members of the tribe. As relatives of the crops, species of these orphan genera may contain valuable genes that, may be transferred to the crops and enhance greater tolerance to biotic and abiotic stresses, improve quality and performance. As such, these orphan genera deserve greater attention.

A number of interspecific and intergeneric hybrids involving those genera were produced during the years (e.g., Cauderon 1966; Sakamoto 1967, 1968, 1969, 1972, 1973, 1974, 1979; Sakamoto and Muramatsu 1963; Dewey 1969, 1970, 1984; Frederiksen 1991a, b, 1993; Frederiksen and von Bothmer 1986, 1989, 1995). Successful production of the hybrids suggests fairly good genetic or cytoplasmic compatibility among those species. However, there is very little chromosome pairing in F_1 hybrids between them as well as between wheat and these species, indicating limited homology between their genomes. Hence, these genera seem

to be highly differentiated from the taxonomic and the genetic viewpoints.

5.2 *Elymus* Species with St or E Genome

5.2.1 Group Description

The delimitation of the genera *Elymus* L. and *Agropyron* Gaertner has been the subject of controversy over the years (Assadi and Runemark 1995), primarily due to the absence of clear-cut generic characters and from the presence of numerous intergeneric hybrids that gave rise to conflicting results as discussed below (Melderis 1978). Their delimitation using different taxonomic treatments was changed several times over the last eight decades. Nevski (1933), Bor (1968), Tzvelev (1976), and Sakamoto (1974) kept these two genera separated, while Gould (1947) and Runemark and Heneen (1968), assuming that the traditional subdivision into *Elymus* s. l. and *Agropyron* s. l. (including *Pseudoroegneria, Elytrigia,* and *Thinopyrum*) is artificial, united them into a single genus, *Elymus*.

The species within the *Elymus-Agropyron* group have traditionally been referred to as *Agropyron*, if the spikelets are solitary, and as *Elymus*, if they are arranged in pairs or larger numbers at each rachis node. Yet, this division is not very distinct and Runemark and Heneen (1968) and Melderis (1978) pointed out that the number of spikelets at each node has a limited taxonomic value since several *Agropyron* species contain pairs of spikelets at several rachis nodes, especially in the lower or in the middle part of the spike, while several species of *Elymus* contain only one spikelet on each rachis node. Also, with regard to leaf anatomy, no difference was found between *Elymus* and *Elytrigia* that was included in *Agropyron* (Runemark and Heneen 1968). Dewey (1969, 1970) found homology between the genomes of several *Elymus* and *Elytrigia* species and Runemark and Heneen (1968) noted similar chromosome morphology in *Agropyron elongatum* (now *Elymus elongatus*) and *Elymus*

© The Author(s) 2023
M. Feldman and A. A. Levy, *Wheat Evolution and Domestication*,
https://doi.org/10.1007/978-3-031-30175-9_5

caninus. In reality, the two genera only represent different levels in the reduction of a paniculate inflorescence (Runemark and Heneen 1968). Cytogenetic studies (e.g., Cauderon 1966; Sakamoto 1973; Dewey 1984, and reference therein) contributed to a better understanding of the genomic relationships among species of *Elymus* and *Agropyron* and, as a result, to modification of the delimitation of the species in these two genera. Based on the above information, as well as on the absence of morphological discontinuities between the taxa *Pseudoroegneria, Elytrigia, Thinopyrum*, and *Elymus*, Melderis (1978, 1980) included these taxa in the genus *Elymus* s. l., while retaining *Agropyron* s. str. as a separate genus for the crested wheatgrasses, that contains only species with a solitary spikelet at each rachis node. The restricted *Agropyron* genus contains diploid and polyploid species that are based on the P genome (Table 2.4) and are morphologically distinct from other genera in Triticeae. The genus *Elymus* is treated by Melderis (1978, 1980, 1985a, b) in a broad sense, as comprising the genera *Elytrigia* Desv., *Pseudoroegneria* (Nevski) A. Löve, *Thinopyrum* A. Löve, *Lophopyrum* A. Löve, and *Trichopyrum* A. Löve (Table 2.2). Wang (1989) supported Melderis' classification which viewed separation of *Pseudoroegneria* from *Elymus* as unjustified for both evolutionary and morphological reasons, since several *Elymus* species include St genomes from different *Pseudoroegneria* diploids, and *Pseudoroegneria* and *Elymus* can hardly be distinguished from each other. This taxonomic classification makes the discrimination between these two genera more straightforward and has been accepted by various taxonomists, e.g., Clayton and Renvoize (1986), Assadi and Runemark (1995), Watson and Dallwitz (1992), Watson et al. (1985). Therefore, this book follows the Melderis classification.

A large number of hybrids within and between the *Elymus* s. l. and *Agropyron* s. str. genera have spontaneously emerged in nature. Many hybrids are sterile, but a considerable number are more or less fertile, at least upon backcross to one of the parents. Apparently, introgressive hybridization has played an important role in the evolution of these two genera.

Melderis (1978, 1980) transferred the following two sections from *Agropyron* s. l. to *Elymus* s. l.: Caespitosae (Rouy) Melderis, comb. nov. (Syn.: *Agropyron* sect. Caespitosa Rouy; *Elytrigia* sect. Caespitosae (Rouy) Tzvelev) and Junceae (Prat) Melderis, comb. nov. [Syn.: *Agropyron* sect. Junceae (Prat) Tzvelev]. The subdivision of these two sections was mainly based on caespitose or rhizomatous habit. The constituent species contain the St, Ee (=E) and the Eb (=J) genomes (Table 5.1), three genomes that occur in species that were included by Nevski's classification (1933), in the genus *Elytrigia*, but, later on, Nevski himself (1934a, b) included *Elytrigia* as a section in *Agropyron*. Tzvelev (1973)

maintained the generic status of *Elytrigia*, but pointed out that species of this genus are close to species of *Elymus*.

The *Elymus* L. section Caespitosae is characterized by caespitose plant, lax and erect spikes, tough rachis, solitary or sometimes two spikelets on the node at the lower part of the spike, usually with 6–13 florets, and glumes 5–8 mm long, unkeeled with 5–9 veins. In some species, the rachilla is fragile and disarticulates above the glumes and beneath each floret (floret-type disarticulation). This section contains about 13 species (Table 5.1) comprising a polyploid series (2n = 14, 28, 56, and 70), and most of them are allogamous, and have long anthers.

The *Elymus* L. section Junceae is characterized by rhizomatous or caespitose plants, lax, erect, and sometimes curved spikes, a fragile rachis, disarticulating at maturity into spikelets with the rachis segment below them (Fig. 2.3; wedge type dispersal unit), solitary spikelets on the rachis, with 2–9 florets and keeled glumes, 5–18 mm long. This section comprises only three species, *E. farctus* (Viv.) Runemark ex Melderis, occurring in Europe and the Middle East, *E. curvifolius* (Lange) Melderis, occurring in south and central Spain, and *E. distichus* (Thumb.) Melderis, native to South Africa. *Elymus farctus* and *E. distichus* grow on maritime coasts.

The species of sections Caespitosa and Junceae presumably originate in Europe or west Asia. Genome St is found in several diploid and polyploid species, Ee exists in the diploid taxon *E. elongatus* (Host) Runemark subsp. *elongatus* [formerly *Agropyron elongatum* (Host) Beauv.] and in several auto- and allo- polyploids, whereas genome Eb occurs in the diploid taxon *E. farctus* (Viv.) Runemark ex Melderis subsp. *bessarabicus* (Savul. & Rayss) Melderis [formerly *Agropyron junceum* subsp. *bessarabicum* Savul. & Rayss; *Thinopyrum bessarabicum* (Savul. & Rayyss) A. Löve], in the allopolyploids of this species and of *E. distichus* (Thunb.) Melderis and in the autopolyploid *E. curvifolius* (Lange) Melderis (Table 5.1).

Löve (1984) used the genome symbols J for the genome of *Elymus farctus* and E for that of *Elymus elongatus*. Endo and Gill (1984) questioned the equivalence of J and E and based on differences in C-banding patterns, justified the separation of these two genomes. However, Dewey (1984) and Dvorak et al. (1984b), based on evidence from karyotype and genome analyses, considered the J and E genomes as the same basic genome. Previous studies of chromosome pairing in hybrids carrying these two genomes had already shown that they are closely related (Cauderon and Saigne 1961; Heneen and Runemark 1972; Dvorak 1981a; McGuire 1984) and more recent studies supported Dewey's consideration (Wang 1985b; Wang and Hsiao 1989). However, Jauhar (1988) reached a different conclusion by studying chromosome pairing in the hybrids analyzed by Wang (1985b) and Wang and Hsiao (1989). Since a recent

Table 5.1 Species of *Elymus* having the St or E genomes

Section	Species and subspecies[a]	Genome[b]	Geographical distribution
Elymus L. section Caespitosae (Rouy) Melderis	*reflexiaristatus* (Nevski) Melderis ssp. *strigosus* (M. Bieb.) Melderis	StSt	The Crimean Peninsula
	reflexiaristatus (Nevski) Melderis ssp. *reflexiaristatus*	StSt	The Crimean Peninsula, Russia
	libanoticus (Hack.) Melderis	StSt	South and southeastern Anatolia, Lebanon, Syria, Israel, north Iraq, northwestern Iran, Caucasus
	spicatus (Pursh) Gould	StSt, StStStSt	Western North America
	stipifollius (Czern. ex Nevski) Melderis	StSt, StStStSt	Southeastern Ukraine and Southwestern Russia
	tauri (Boiss. & Bal.) Melderis	StSt, StStPP	Turkey, Iran
	panormiitanus (Parl.) Tzvelev	StStPP	South Spain, South Italy, Jugoslavia, Romania, South-east Russia, Turkey, Lebanon, Syria, Israel, Iraq, Iran,
	elongatus (Host) Runemark ssp. *elongatus*	EeEe	The Mediterranean basin
	elongatus (Host) Runemark ssp. *flaccidiffolius* (Boiss. & Heldr.) Runemark	EeEeEeEe	The Mediterranean basin
	elongatus (Host) runemark ssp. *turcicus* (P.E. McGuire) Melderis	EeEeEeEeStStStSt	Greece, Turkey, Georgia and northern Iran
	elongatus (Host) Runemark ssp. *ponticus* (Podp) Melderis	EeEeEeEeEeEeStStStSt[c]	Southeastern Europe, Turkey, near the Black Sea and southern Russia
	nodosus (Nevski) Melderis ssp. *nodosus*	EeEeStSt	The Crimean Peninsula
	nodosus (Nevski) Melderis ssp. *caespitosus* (K. Koch) Melderis	EeEeStSt	Corsica
	bungeanus (Trin.) Melderis	EeEeStSt	The Crimean Peninsula, Russia
	hispidus (Opiz) Melderis	EeEeEeEeStSt	Europe
Elymus L. section Junceae (Prat) Melderis	*farctus* (Viv.) Runemark ex Melderis ssp. *bessaribicus* (Savul. & Rayss) Melderis	EbEb	Coasts o Coasts of Black Sea from Bulgaria to Crimea, Sea of Azov, Aegean and N.E. Mediterranean
	farctus ssp. *rechingeri* (Runemark) Melderis	EbEbEbEb	West Turkey, the Aegean islands. coasts of Greece, Crete, and the Mediterranean coast of Egypt
	farctus ssp. *boreali-atlanticus* (Simonet & Guinochet) Melderis	EbEbEeEe	North and western Europe
	farctus var. *sartorii* (Boiss. & Heldr.) Melderis	EbEbEeEe	Western and southern Europe
	farctus ssp. *farctus* (Viv.) Runemark ex Melderis	EbEbEbEbEeEe	Coasts of Mediterranean Sea
	distichus (Thunb.) Melderis	EbEbEbEb	South Africa
	curvifolius (Lange) Melderis	EbEbEbEb	South-central Spain

[a] These species were transferred from *Agropyron* to *Elymus* by Melderis (1978, 1980)
[b] Genome symbols according to Wang et al. (1995)
[c] Genome symbol according to Zhang et al. (1996) and Li and Zhang (2002)

literature review indicated that most studies regarded J and E genomes as members of the same cluster (see Table 1 in Wang and Lu 2014), it is now generally accepted to regard them as very closely related genomes, supporting the use of a common basic genome symbol, E (Seberg and Frederiksen 2001; Yen et al. 2005; Fan et al. 2007; Liu et al. 2008; Sha et al. 2010; Yan et al. 2011; Wang and Lu 2014). Thus, these genomes were designated Ee for *Elymus elongatus* and Eb for *Elymus farctus*, respectively, as proposed by Dvorak (1981a) and McGuire (1984).

While the use of one basic genome symbol for these two species was rejected by some researchers (Jauhar 1988, 1990a, b; Jarvie and Barkworth 1992; Jauhar et al. 2004), several studies using different methodologies, have further confirmed the close relationship between genomes Ee and Eb. The studies included chromosome pairing (de V Pienaar et al. 1988; Forster and Miller 1989; Wang and Hsiao 1989), random amplified polymorphic DNA (RAPD) and sequence-tagged site (STS) markers (Wei and Wang 1995; Li et al. 2007), genomic in situ hybridization (GISH) (Kosina and Heslop-Harrison 1996; Chen et al. 1998a, b, 2003), chloroplast DNA sequences (Mason-Gamer et al. 2002; Liu et al. 2008), sequences of a gene encoding plastid acetyl-CoA carboxylase (Fan et al. 2007, 2009), and nuclear rDNA internal transcribed spacer (ITS) sequences (Hsiao et al. 1995; Liu et al. 2008; Yu et al. 2008). At the second International Triticeae Symposium, the Genome Designation Committee (Wang et al. 1995) adopted a system for the application of nuclear genome symbols in the tribe Triticeae. This system is based mainly on prevailing symbols, but since the number of basic nuclear genomes in the Triticeae exceeds the number of single letters in the Roman alphabet, some basic genomes are designated with an uppercase letter followed by a lowercase letter, e.g., Ee or Eb, for the genome in *Elymus elongatus* and *E. farctus*, respectively. An uppercase letter followed by a superscript in small letters are used when modified versions of a basic genome is referred to, e.g., A^m for the genome found in *Triticum monococcum*.

Melderis transferred the Asiatic diploid species of *Elymus*, namely, *libanoticus*, *reflexiaristatus* (subsp. *reflexiaristatus* and subsp. *strigosus*), and the diploid cytotypes of *stipifolius* and *tauri* from *Agropyron* (=*Elytrigia*) to *Elymus* (1978, 1980). These species carry the St genome (Wang et al. 1995) that also exists in the *Elymus* polyploid species containing the Ee genome, i.e., *elongatus, nodosum, bungeanus,* and *hispidus* (Table 5.1). The St genome is related to the Ee and Eb genomes (Wang 1989). Bieniek et al. (2015) found that nucleotide sequences of the diploid Ee, E^b and St taxa are almost identical, with only one substitution within the *matK* gene, differentiating genome Eb from the Ee and St genomes. Petersen and Seberg (1997) and Wang and Lu (2014) confirmed this very close relationship between the three genomes. The St genome almost always has a dominant influence on the morphology of the taxa of which it is a component (Assadi and Runemark 1995) and since exists in more primitive *Elymus* species, it is reasonable to assume that Ee and Eb evolved from St.

5.2.2 *Elymus* Species with St Genome

5.2.2.1   Species Description

The St-genome species of *Elymus* were previously recognized as a biological unit and placed as a separate section, *Elytrigia*, in the traditional *Agropyron* s. l. (Nevski 1934a). Due to the fact that all these species contain one genome, Love (1980), treated them as a separate genus, *Pseudoroegneria*. However, due to the absence of morphological discontinuities between *Pseudoroegneria* and *Elymus,* Melderis (1978, 1980) included *Pseudoroegneria* in the genus *Elymus* s. l.

The *Elymus* species bearing the St genome include approximately 15 different taxa that consists of about equal numbers of diploids and tetraploids. The type species of this group is *E. reflexiaristatus* (Nevski) Melderis subsp. *strigosus* (M. Bieb.) Melderis [formerly *Pseudoroegneria strigosa* (M. Bieb.) Á. Löve] (Dewey 1984; Löve 1984; Watson and Dallwitz 1992; Yan and Sun 2011). Interspecific hybrids between the St diploid species exhibit almost complete chromosome pairing at fist meiotic metaphase, but with high or complete sterility, indicating divergence of the same basic genome in each diploid (Stebbins and Pun 1953; Dewey 1975). Some of the species, e.g., *stipifolius*, and *spicatus*, have diploid and tetraploid cytotypes and the tetraploids behave cytologically as autotetraploids or near autoploids (Dewey 1975). The tetraploid taxa of two other species, *tauri* and *panormitanus*, are allopolyploids containing the St and P subgenomes.

A large amount of the allopolyploid species of *Elymus* s.l. share a common St genome with diploid *Elymus* species in different combinations with H, Y, P, and W subgenomes (Table 2.4). The maximum likelihood tree constructed, using nuclear ribosomal internal transcribed spacer region (nrITS) data, showed that diploid *Elymus*, *Hordeum* and *Agropyron* species served as the St, H and P subgenomes donors, respectively, for the *Elymus* allopolyploids (Dong et al. 2015). The maximum likelihood tree for the chloroplast genes (*matK* and the intergenic region of *trnH-psbA*) suggests that the *Elymus* diploid donors of the St genome to *Elymus* allopolyploids served, in most cases, as the maternal donor. Moreover, the chloroplast genes data suggest that diploid St *Elymus* species from Central Asia and Europe are more ancient than those in North America (Dong et al. 2015). Thus, it was hypothesized that the *Elymus* s. l. species originated in Central Asia and Europe, and then spread to North America.

The St genome species are perennials, caespitose, and cross-pollinating, with culms between 30 and 90 cm tall, narrow, linear spikes with single, distantly spaced spikelets, 5–8 mm long glumes of equal length in *E. reflexiaristatus*, or unequal *in E. spicatus, tauri* and *libanoticus*, 8–30 mm long glume awns, absent in *E. tauri* and *libanoticus*, and long anthers. These species grow in the northern Hemisphere, from southwestern and southeastern Europe, the Middle East, Transcaucasia across Central Asia and Northern China to Western North America (Dewey 1984). They occur on open rocky hillsides, are exceptionally drought and salt tolerant and have excellent quality forage that is palatable to animals (Dewey 1984).

The *Elymus libanoticus*-related species, that exemplify all St genome diploid species, are described below.

5.2.2.2 *Elymus libanoticus* (Hack.) Melderis—A Representative Example

Morphological and Geographical Notes
Elymus libanoticus (Hackel) Melderis [Synonym: *Agropyron libanoticum* Hackel; *Pseudoroegneria libanotica* (Hackel) D.R. Dewey; *Elytrigia libanotica* (Hackel) Holub); *Pseudoroegneria tauri* ssp. *libanotica* (Hackel) Á. Löve; *Agropyron sosnovskyi* Hackel; *Elytrigia sosnovskyi* (Hackel) Nevski; *Elymus sosnovskyi* (Hack.) Melderis; *Pseudoroegneria sosnovskyi* (Hackel) A. Love; *Agropyron gracillimum* Nevski; *Elytrigia gracillima* (Nevski) Nevski; *Pseudoroegneria gracillima* (Nevski) Á. Löve], is perennial, caespitose, with short rhizomes, 45–85 cm high culms, 5–15 cm long linear spikes, with 4–7 spikelets, each 10–15 mm long, one

per node, tough rachis, 3–6 florets, unequal, lanceolate, 3–5-veined glumes, the lower ones 6–8 mm long, typically 3/4 or nearly as long as lower floret, and upper ones 7–9 mm long, 8–9 mm long, lanceolate, 3-veined, unawned, lemma, palea shorter than lemma, sparsely ciliate on keels, 4–5 mm long anthers, and caryopsis adherent to palea and lemma. Chromosome number 2n = 2x = 14 (Dewey 1972) (Fig. 5.1a).

Unlike *E. libanoticus*, *E. sosnovskyi* (Hack.) Melderis [=*Agropyron sosnovskyi* Hack.: *Elytrigia sosnovskyi* (Hack.) Nevski] bears acuminate glumes with 3 veins. *Agropyron gracillimum* Nevski differs from *E. libanoticus* by their smaller leaf thickness. These differences, however, fall within the variation of the Iranian *E. libanoticus* material (Assadi 1996). Moreover, hybrids between *E. sosnovskyi* or *A. gracillimum* with *E. libanoticus* were highly fertile, with regular meiotic metaphase. Therefore, the three names are considered synonymous (Assadi 1996). *E. libanoticus* is closely related to *E. tauri* subsp. *libanoticus*, differing only in several morphological traits (Assadi 1996).

Elymus libanoticus grows in Lebanon, Syria, northern Israel, south and southeastern Anatolia, northern Iraq, Iran, and Caucasus. It thrives on dry mountain slopes and limestone ravines, usually on more xeric habitats, 1000–3050 m a.s.l. It is an Irano-Turanian element.

Cytology, Cytogenetics and Evolution
Hsiao et al. (1986) analyzed the karyotype of diploid St genome species, including *Elymus spicatus*, *E. Reflexiaristatus subsp. Strigosus*, *E. libanoticus*, and *E. stipifolius*. All four species possess similar karyotypes and chromosomal

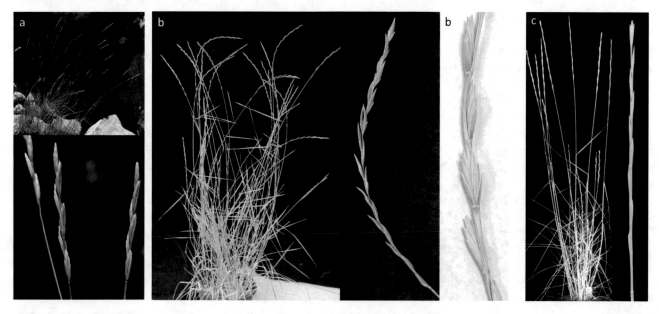

Fig. 5.1 Mediterranean *Elymus* species; **a** A plant and spikes of *E. libanoticus* (Hack.) Melderis (2n = 2x = 14), (Photographed by the late Prof. Avinoam Danin); **b** A plant and spike of *E. elongatus* (Host) Runemark ssp. *elongatus* (2n = 2x = 14); **c** A plant and spike of *E. farctus* (Viv.) Runemark ex Melderis ssp. *farctus* Runemark ex Melderis (2n = 6x = 42)

lengths. The karyotypes of all species have one pair of small and one pair of large satellites on the short arms of chromosomes 2 and 5, respectively (Hsiao et al. 1986). The karyotypes are symmetric; most chromosomes are metacentric and a few are sub-metacentric (Wang et al. 1985; Hsiao et al. 1986; Deng et al. 2004). The St genome consists of smaller chromosomes than those of the R, P, and Eb genomes. Despite their wide geographical distribution, the karyotype patterns of the St genome species have not been dramatically altered. The karyotype of *E. spicatus* has been reported previously (Schulz-Schaeffer and Jurasits 1962; Dvorak et al. 1984a, b).

Endo and Gill (1984), using the acetocarmine-Giemsa C-banding technique, studied heterochromatin distribution in somatic chromosomes of diploid *Elymus* and *Agropyron* species. With the exception of *E. elongatus,* which is moderately self-fertile, all other species are cross-pollinating and self-sterile. The cross-pollinating species showed large terminal C-bands and a high level of C-band polymorphism, whereas *E. elongatus* showed small terminal and interstitial bands and a minimal C-band polymorphism. C-banding patterns show that the Eb genome of diploid *E. farctus* appears to be distinct from the Ee genome of diploid *E. elongatus* and may constitute an intermediate link between the Ee and St genomes (Endo and Gill 1984).

E. spicatus, *E. libanoticus*, and *E. stipifolius* have similar C-band patterns, although C-bands were less prominent in *E. stipifolius* than in the others. Thus, the C-banding patterns and morphology of satellite chromosomes supported previous evidence that *E. spicatus, E. libanoticus,* and *E. stipifolius* share a common St genome. Variation in the intensity of terminal C-bands was observed in *E. stipifolius*, which is to be expected in a basic genome of species with worldwide distribution (Dewey 1981).

Wang (1989) produced the tetraploid hybrid (genome StStStH) from crossing the hexaploid *Elymus transhyrcanus* (genome StStStStHH) with *E. libanoticus* (genome StSt). This F$_1$ hybrid exhibited at first meiotic metaphase 13.94 univalents, 0.16 rod and 6.78 ring bivalents (6.94 total bivalents) and 0.06 trivalents. The reciprocal hybrid showed an average of 10.22 univalents, 2.34 rod and 5.24 ring bivalents (7.58 total bivalents), 0.74 trivalents and 0.10 quadrivalents. The amount of pairing in the hybrid and particularly that of trivalent cnfiguraton was much less than expected in the case of three fully homologous St genomes. Hence, either the two St sugenomes of the hexaploid had diverged from one another or both had diverged from the St genome of the diploid.

In the F$_1$ tetraploid hybrid (genome StStEeEe) between *Elymus libanoticus* (genome StSt) and *E. hispidus* (=*Thinopyrum intermedium*; genome StStEeEeEeEe) Wang

(1989) observed at first meiotic metaphase an average of 6.68 univalents, 4.96 rod and 3.66 ring bivalents (8.62 total bivalents), 1.06 trivalents, 0.20 quadrivalents and 0.03 pentavalents. These data show that, in addition to the autosyndetic pairing in the form of bivalents between the Ee subgenomes, the presence of multivalents indicates some allosyndetic pairing between St and Ee chromosomes, indicating that the two genomes are related.

Crosses with Other Triticineae Species

Studies of meiotic chromosome pairing in F$_1$ hybrids between diploid *Agropyron cristatum* (genome PP) and several different diploids species of *Elymus* with St genome (genome of all hybrids was PSt), showed that the two genomes, P and St, are related (Wang 1985a, 1986, 1987a, b, 1988, 1989, 1990, 1992; Wang et al. 1985). Size differences between *Agropyron* (large) and St genome *Elymus* (small) chromosomes facilitated interpretation of chromosome pairing in the F$_1$ hybrids. The average chromosome pairing at first meiotic metaphase of the diploid hybrid *A. cristatum* x *E. libanoticus* included 7.71 univalents, 2.77 bivalents, 0.22 trivalents, 0.01 quadrivalents and 0.01 pentavalents (Wang 1986), while that between *A. cristatum* and *E. stipifolius* displayed a similar amount and pattern of pairing (Wang 1985a). These pairing data indicate allosyndetic pairing between the homoeologous chromosomes of the two genomes, demonstrating a close relation between the St and the P genomes.

Meiotic chromosome pairing in the F$_1$ hybrid *Elymus spicatus* (genome StSt) x *Secale strictum* (genome RR) exhibited an average of 12.97 univalents, 0.49 bivalents and 0.01 trivalent (Wang 1987b). The F$_1$ hybrid A*gropyron. mongolicum* x *S. strictum*, which had the PR genome, showed an average of 12.86 univalents, 0.51 bivalents, 0.03 trivalents and 0.004 quadrivalents. The hybrid between *Elymus spicatus* and *A. mongolicum* (genome StP) had a mean configuration of 8.05 univalents, 2.86 bivalents, 0.07 trivalents and 0.01 quadrivalents. All hybrids were sterile. The meiotic pairings of these hybrids indicated that chromosome homology between the St and P genomes is higher than between St and R and between P and R. The degree of meiotic pairing in the *E. spicatus* x *A. mongoicum* hybrid was similar to those in other diploid hybrids bearing the same genome constitution, i.e., *A. cristatum* x *E. stipifolius* and *A. cristatum* x *E. libanoticus* (Wang et al. 1985; Wang 1986).

Following hybridization of the diploid *Elymus Stipifolius* (genome StSt) with tetraploid *Elymus elongatus* (genome EeEeEeEe), Dvorak (1981a) obtained a triploid hybrid (genome StEeEe), that exhibited 7.8 univalents, 5.9 bivalents and 0.41 trivalents at first meiotic metaphase. This pattern of pairing was attributed primarily to autosyndesis between homologous chromosomes of the Ee genomes. Stebbins and Pun (1953) had speculated that the Ee and St genomes might be variations of the same basic genome, yet the hypothesis was contradicted by Dvorak's (1981a) data, which showed that the Ee and St genomes are distinctly different.

Wang (1989) crossed *Elymus libanoticus* (genome StSt) with the tetraploid cytotype of *Agropyron cristatum* (genome PPPP) and observed an average meiotic pairing profile of 11.30 univalents, 3.40 rod and 1.50 ring bivalents (4.90 total bivalents) in the resulting triploid hybrid. Most pairing in this triploid hybrid was autosyndetic, indicating a difficulty in learning about the relationship between two genomes when one of the genomes exists in two doses.

Interpretation of chromosome pairing in St-*Elymus* and *Agropyron* hybrids is aided by size differences between the *Agropyron* (large) and St-*Elymus* (small) chromosomes. Chromosomes of autotetraploid *Elymus spicatus* (genome StStStSt) paired only rarely with chromosomes of diploid *Agropyron cristatum* (genome PP) in their triploid hybrids (StStP) (Dewey 1964). In the tetraploid hybrids (PPStSt) of *A. desertorum* (genome PPPP) and tetraploid *E. spicattus* (genome StStStSt), all chromosome pairing was attributed to autosyndesis between the PP and StSt genomes (Dewey 1967). Wang et al. (1985) crossed *Agropyron desertorum* (2n = 4x = 28; genome PPPP) with the autotetraploid cytotype of *Elymus stipifolius* (2n = 4x = 28; genome StStStSt). The tetraploid hybrid averaged 3.09 bivalents, most of which resulted from autosyndetic pairing between the P or the St genomes (Wang et al. 1985). In this tetraploid hybrid, because of the presence of homologous chromosomes, the P genome chromosomes rarely paired with the St genome chromosomes.

Wang (1989) used the mean C-values (the ratio between the number of chiasmata and the number of chromosome arms) to assess the relationships between genomes in diploid hybrids of the perennial Triticeae. He found that a C-value of 0.55 in diploid hybrids can serve as a critical value (in conjunction with other evidence, e.g., karyotype characteristics) to separate intergenomic from intragenomic divergence. Using this rule, he found that the *Secale* R genome, the *Hordeum* H genome and the *Psathyrostachys* N genome are distinct from each other and from other Triticeae genomes (C-values 0.03–0.17), while the St, Ee, Eb, and P genomes show considerable homoeology (C-values 0.24–0.36). Thus, the Eb and St genomes, despite considerable differences in total genome size (Hsiao et al. 1986), show considerable homoeology, with a mean C-value of 0.35 in the diploid hybrid between them (Wang 1989). Similar homoeology was recorded by Liu and Wang (1993) in the triploid hybrids (genomes StStEb and StEeEe).

5.2.3 *E. elongatus* (Host) Runemark (Based on Ee Genome)

5.2.3.1 Species Description

Elymus elongatus [syn. *Triticum elongatum* Host; *Agropyron elongatum* (Host) Beauv.; *Agropyron elongatum* subsp. *scirpeum*; *Elytrigia elongata* (Host) Nevski; *Lophopyrum elongatum* (Host) Á. Löve; *Thinopyrum elongatum* (Host) D. R. Dewey] is a perennial, caespitose, 30–100 cm high, with robust, glabrous culms, 10–25 cm long lax and erect spikes, tough rachis, 10–25 mm long solitary spikelets on each rachis node, sometimes two spikelets on one node, with 6–13 (9–25) awnless florets, glumes shorter than spikelet, 6–8 mm, 5–9 veined, without keels, 7–10 mm long, keeled lemma, keeled palea, 4–4.5 mm long anthers, caryopsis with adherent pericarp. The rachilla is fragile and disarticulates above the glumes and beneath each floret (floret-type disarticulation). This type of seed dispersal is characteristic of the Arctic-Temperate group and especially of species of *Elymus*.

The cytotaxonomy of *E. elongatus* was studied by several researchers, e.g., Peto (1930), Simonet (1935), Cauderon (1958, 1966), Schulz-Schaeffer and Jurasits (1962), Schulz-Schaeffer et al. (1971), Evans (1962), Runemark and Heneen (1968) Heneen (1972), Heneen and Runemark (1972, 1977), and Luria (1983), who showed that *E. elongatus* comprises a polyploid complex of diploid, tetraploid, octoploid, and decaploid taxa (Table 5.1). The diploid and decaploid taxa are well documented in the literature and evidence desmonstrates that the autotetraploid taxon also belongs to this group (Heneen and Runemark 1972). Hexaploid chromosome number was also found in material collected from Istria (Heneen and Runemark unpubl.). Schulz-Schaeffer and Jura (1967) reported the existence of hexaploid types in plants collected from Turkey. However, this hexaploid was not recognized as a valid subspecies. In addition to the diploid and tetraploid subspecies, an octoploid subsp. of *Elongatus*, subsp. *turcicus* (P. E. McGuire) Melderis, from Turkey, was described (McGuire 1984). Thus, it appears that the *Elymus elongatus* complex is represented in nature by types that form a complete polyploid series, ranging from diploids to decaploids.

Several authors described variants of this species as separate species or as subordinate taxa. Since the morphological differences between these taxa are not clear, Melderis (1980) recognized only two taxa that merit the subspecies status, namely, subsp. *elongatus*, a diploid, and subsp. *ponticus*, a decaploid. Later, Melderis (1985a) recognized an additional subspecies, subsp. *turcicus* (P. E. McGuire) Melderis, an octoploid taxon. Heneen and Runemark (1972) included an autotetraploid taxon from Cyprus and the

Aegean islands in *E. elongatus* as ssp. *flaccidifolius*. Breton-Sintes and Cauderon (1978) classified an accession from Sicily of Heneen and Runemark (1972) autotetraploid subspecies as *Agropyron elongatum* (Host) ssp. *scirpeum* (C. Presl.) Cifferi et Giacom. The taxon ssp. *flaccidifolius* was elevated by Melderis (1978) to the specific rank *Elymus flaccidifolius* (Boiss. & Heldr.) Melderis. However, since there are only minor morphological differences (mainly quantitative) between this species and other autotetraploids of *E. elongatus*, it is more appropriate to classify it, along with all the other autoetraploids of *E. elongatus*, as a subspecies of *elongatus*. Hence, in this book, all the autotetraploid taxa of *E. elongatus* (=*Agropyron elongatum* var. *flaccidifolium* Boiss. & Heldr.; *Agropyron flaccidifolium* (Boiss. & Heldr.) Candargy; *Elymus flaccidifolius* (Boiss. & Heldr.) Melderis; *Agropyron elongatum* Host subsp. *scirpeum* (C. presl.) Ciferri & Giacom.; *Lophopyrum scirpeum* (C. Presl) Á. Löve; *Thinopyrum scirpeum* (Presl) D. R. Dewey; *Agropyron scirpeum* C. Presl; *Elytrigia scirpea* (C. Presl) Holub) are referred to as *Elymus elongatus* (Host) Runemark ssp. *flaccidiffolius* (Boiss. & Heldr.) Runemark, and were grouped together as one subspecies.

The diploid and the tetraploid subspecies of *E. elongatus* exhibit wide morphological variation in the number of spikelets per spike, number of florets per spikelet, hairiness, and plant color. The decaploid subspecies, subsp. *ponticus*, exhibits wider variation than subsp. *elongatus* and *flaccidifolius*.

E. elongatus is found in all parts of the Mediterranean basin, in southwestern, southeastern, and eastern Europe, in North Africa, and the Middle East, Caucasus, western Asia, and Arabia. It was introduced or invaded Australasia, South America, and North America. It is a Mediterranean element (chorotype) and grows among Mediterranean plant communities. This species grows in salt marshes and near salty springs and is salt tolerant (Moxley et al. 1978; Dewey 1960; McGuire and Dvorak 1980).

5.2.3.2 Ssp. *elongatus* (2n = 2x = 14)

Morphological and Geographical Notes
E. elongatus (Host) Runemark subsp. *elongatus* [*Agropyron elongatum* (Host) Beauv.; *Lophopyrum elongatum* (Host) Á. Löve; *Thinopyrum elongatum* (Host) D.R. Dewey] is a diploid subspecies with tall stems (50–80 cm high); 10–25 cm long spikes, with 9–26 spikelets per spike, internodes at the base of the spike as long as the spikelets; 10–17 mm long spikelets with 7–8 florets, usually one spikelet at each rachis node, 7–10 mm long glumes with 5–9 veins, where the lower glume is shorter (about 2/3–3/4) than the lower floret, 9–10 mm long lemmas, 4–4.5 mm long anthers and 4 mm long caryopsis (Fig. 5.1b).

Subsp. *elongatus* grows in the Mediterranean basin. In Israel, the diploid taxon grows in the Coastal Plain from the Shfela (Einot Gibton) and northwards (Acre plain). It grows in salt marshes, near salty springs and on maritime sands, from sea level to 100 m a.s.l., throughout the range of the species. These saltmarsh habitats are characterized by high underground water that forms floods in the winter and salty soil with salty crust in the summer. The subspecies also grows on sandy soils near river mouths, on silt near river's banks or springs or on clay soil. When growing in wet soils, the amount of annual rainfall is not a limiting factor.

Cytology, Cytogenetics and Evolution
Matsumura and Sakamoto (1956), Cauderon (1958), Evans (1962), Schulz-Schaeffer and Jurasits (1962), Runemark and Heneen (1968), Heneen (1972) and Luria (1983) described the karyotype of the diploid subspecies. The karyotype is symmetric, consisting of four metacentric pairs and three sub-metacentric pairs. The differences in length and arm ratio among the chromosomes of this subspecies are relatively small (Dvorak and Knott 1974). However, the homologous chromosomes can be visually identified (Evans 1962). Two chromosome pairs have satellites, with one metacentric pair carrying a large satellite and one sub-metacentric pair bearing a small satellite. The constrictions between the satellites and the chromosome arms carrying them are the nucleolar organizing regions (NORs). The NORs contain a set of argyrophilic proteins which are selectively stained by silver. After silver staining, the NORs can be easily identified as black dots that are called Ag-NORs. Thus, in agreement with the number of satellite (SAT)-chromosomes mentioned above, four Ag-NORs are regularly observed in somatic cells of diploid *E. elongatus* (Lacadena et al. 1984). Giemsa C-banding analysis of the chromosomes of several accessions of this subspecies revealed small terminal and interstitial bands and a minimal C-band polymorphism (Endo and Gill 1984).

Heneen and Runemark (1972) observed karyotype differences between plants of ssp. *elongatus* collected from different locations, the major difference lying in the appearance of the SAT-chromosomes. Runemark and Heneen (1968), comparing the karyotype of subsp. *elongatus* with that of diploid *E. farctu* (genome EbEb), found that the two karyotypes resemble one another, but the chromosomes of *E. elongatus* are somewhat smaller than those of *E. farctus*. In addition, differences in morphology of the SAT-chromosomes exist between the two taxa (Heneen 1962); the pair with large satellites in *E. elongatus* has more median centromeres than the equivalent pair in *E. farctus*. The constriction in this SAT-chromosome divides the short arm in *E. elongatus* into two unequal parts, with the part proximal to the centromere being longer than the satellite, which is not the case in *E. farctus*.

Chromosomal pairing at meiosis in the diploid subspecies is regular (0.08–0.15 univalents and 6.92–6.95 bivalents per meiocyte; 12.04–12.86 chiasmata/cell (Cauderon 1958; Luria 1983). However, in accordance with the karyological observations, meiotic analysis of several inter-varietal hybrids showed the existence of structural heterozygosity among several accessions of this subspecies (Heneen and Runemark 1972). Similarly, in one inter-varietal cross, Luria (1983) observed a quadrivalent, suggesting the existence of a reciprocal translocation between the two accessions. These findings may indicate the occurrence of initial steps of karyotype divergence among and within accessions of the diploid subspecies of *E. elongata*.

Crosses with Other Triticineae Species

Crosses with Diploids Species
Cross of *Elymus farctus* ssp. *bessarabicus* (genome EbEb), as female, with *E. elongatus* ssp. *elongatus* (genome EeEe) was successful, while the reciprocal cross failed (Wang 1985a). Karyotypes of mitotic chromosomes in the parental species revealed that three of the seven chromosomes in the Eb and Ee genomes were similar in length and arm ratio. Meiosis in the F_1 hybrids substantiated this observation, but four chromosomes had undergone some structural rearrangements such as reciprocal translocations (Wang 1985b). Chromosomal pairing at meiotic first metaphase of the F_1 hybrid averaged 2.68 univalents, 4.68 bivalents, 0.27 trivalents, 0.27 quadrivalents, and 0.01 pentavalents (Wang 1985b). The F_1 hybrids were completely sterile upon self-pollination. From the relatively high pairing, Wang (1985a) concluded that the Eb and Ee genomes are closely related, supported the transfer of *Lophopyrum elongatum* to the genus *Thinopyrum*, as was suggested by Dewey (1984), as opposed to keeping them as two separate genera, as suggested by Löve (1984). GISH studies substantiated this conclusion by showing that genomes Ee and Eb are closely similar in their repetitive DNA (Kosina and Heslop-Harrison 1996).

Considering the suppression of pairing by the *Ph1* gene that inhibits homoeologous pairing between the chromosomes of ssp. *bessarabicus* and ssp. *elongatus* in the tri-generic hybrid with durum wheat (genome ABEbEe), Jauhar (1988, 1990a, b) argued that the genomes Eb and Ee are homoeologues rather than homologues and should be assigned distinct genome symbols (J and E, respectively).

Jauhar et al. (2004) later analyzed chromosomal pairing in meiosis of the tri-generic hybrids between durum wheat, with and without the *Ph1* gene, and the amphidiploid *E. farctus* ssp. *bessarabicus-E. elongatus* ssp. *elongatus*. Meiotic chromosome pairing was studied using both conventional staining and fluorescent genomic in situ hybridization (fl-GISH). As expected, the *Ph1*-intergeneric hybrids (genome ABEbEe) showed low chromosome pairing (23.86% of the total chromosome complement paired), whereas 49.49% of the trigeneric hybrids without *Ph1* showed pairing. Fl-GISH analysis provided insight to the study of the specificity of chromosome pairing: wheat with *Elymus* (AB with Ee and/or Eb), wheat with wheat (A with B), or *E. elongatus* with *E. farctus* (Ee with Eb). The analysis revealed that without the *Ph1* gene in the tri-generic hybrid, there were 3.97 chiasmata/cell between chromosomes of the Eb and Ee genomes, 2.29 chiasmata/cell between wheat chromosomes, and 2.6 chiasmata/cell between wheat–*Elymus* chromosomes. Thus, the two E genomes are more closely related to each other than A and B to one another.

Similarly, Forster and Miller (1989) reported that the chromosomes of ssp. *bessarabicus* and subsp. *elongatus* rarely paired in the presence of the *Ph1* gene, i.e., in the hybrid between the two amphiploids *Triticum aestivum*-diploid *E. farctus* x *T. aestivum*-diploid *E. elongatus*. However, they concluded that, because of the relative high frequency of pairing between chromosomes of these two species at the diploid level, their genomes warrant a common genome symbol. Yet, since the two genomes do not pair in a wheat genetic background, their differentiation should also be indicated. Therefore, Forster and Miller (1989) proposed that the genome symbol of *E. elongatus* be E and that of ssp. *bessarabicus* Eb, as suggested by Dvorak (1981a) and McGuire (1984).

Crossess with other Triticeae diploids revealed very little homology. Dvorak (1981b) succeeded in crossing *Aegilops tauschii* (=*Ae. squarrosa*) with ssp. *elongatus*, while crosses between ssp. *elongatus* and *Ae. speltoides* or *Triticum monococcum ssp. aegilopoides* (=*T. boeoticum*), were not successful. Mean chromosome pairing at first meiotic metaphase of the diploid hybrid yielded 10.7 univalents, 1.5 (0–5) bivalents, 0.027 trivalents per cell, indicating a certain degree of homoeology between the genomes of the two species (Dvorak 1981b). The F_1 hybrid plants were sterile, with very low pollen fertility.

Crosses with Tetraploid Species
Cauderon (1958) and Cauderon and Saigne (1961) crossed the allotetraploid *Elymus farctus* ssp. *boreo-atlanticus* (=*Agropyrum junceum boreo-atlanticum*) (genome EbEbEeEe) with *Elymus elongatus* ssp. *elongatus* (genome EeEe) and studied chromosome pairing at first meiotic metaphase of the triploid F_1 hybrid (genome EbEeEe). Meiotic pairing showed 3.40 univalents, 4.50 bivalents, 2.76

trivalents, and 0.08 quadrivalents. From the relatively high frequency of trivalents at the hybrid meiosis, they concluded that the Eb and Ee genomes are closely related. A similar conclusion was drawn following karyotype analysis (Caudereon 1958).

When diploid *E. elongatus* was crossed with tetraploid (durum) wheat (genome BBAA) (Jenkins and Mochizuki 1957; Mujeeb-Kazi and Rodriguez 1981), the F$_1$ hybrids showed very little pairing (0.3–2.6 bivalents per cell), suggesting that in the presence of one dose of *Ph1*, the Ee genome chromosomes of *elongatus* showed little, if any, pairing with those of the subgenomes A and B of durum wheat. However, the level of chromosomal pairing reported by Jenkins and Mochizuki (1957), i.e., 2.6 bivalents/cell in the hybrid durum wheat x diploid *E. elongatus* was significantly higher than expected on the basis of pairing in haploid durum wheat, i.e., 0.37 bivalents/cell, as reported by Kihara (1936) and Lacadena and Ramos (1968).

In a later study, Mochizuki (1960, 1962) studied chromosomal pairing between individual *E. elongatus* chromosomes and tetraploid wheat chromosomes in monosomic addition lines, where single *elongatus* chromosomes were added to the durum complement. No chromosome associations were observed between wheat and *elongatus* chromosomes in three lines, while a high frequency of trivalent associations was noted in the remaining four lines. From these results, he concluded that four *elongatus* chromosomes are partially homologous to durum chromosomes. However, Dvorak and Knott (1974) assumed that the trivalents resulted from translocations between the durum and *elongatus* chromosomes that occurred during the production of the monosomic addition lines and actually, in the presence of two doses of *Ph1* of durum wheat, there was no pairing between the *elongatus* and the durum chromosomes. Following this controversy, Ono et al. (1983) re-examined Mochizuki (1962) *durum-elongatus* addition lines and found that apart from 5Ee, no *elongatus* chromosomes paired with wheat chromosomes. Evans (1962) found that the nucleolar of *Elymus* were suppressed in the amphiploid *Triticum durum*-diploid *Elymus elongatus* by the durum NORs.

Crosses with Hexaploid Species

The F$_1$ hybrid between hexaploid wheat (BBAADD) and diploid *E. elongatus* (EeEe) exhibited very little chromosomal pairing at meiosis (Jenkins 1957). A low level of pairing between Ee genome chromosomes of diploid *E. elongatus* and those of common wheat was also observed in *elongatus* addition lines to common wheat (Dvorak and Knott 1974). Study of pairing of single diploid *E. elongatus* chromosomes with common wheat chromosomes in monosomic addition lines, in the presence of two doses of the homoeologous-pairing suppressor *Ph1*, showed that *elongatus* chromosomes do not pair with wheat chromosomes, with the exception of chromosome IV [assigned later to homoeologous group 3, and designated 3Ee by Dvorak (1980)], that very rarely paired with a wheat chromosome (Dvorak and Knott 1974). The researchers thus concluded that *elongatus* genomes did not play any role in the evolution of the polyploid series of *Aegilops* and *Triticum*.

When ten ditelosomic addition lines, comprising of diploid *E. elongatus* telosomes added to the common wheat complement, were crossed to *Aegilops speltoides*, that suppresses the activity of the *Ph1* gene, all ten *elongatus* telosomes paired with common wheat chromosomes (Dvorak 1979). But, because this pairing only occurred when *Ph1* was not active, Dvorak concluded that none of the ten *elongatus*-chromosome arms has a homologous partner among the three common wheat subgenomes A, B, and D, and the involved *speltoides* genome.

Likewise, in crosses between *elongatus* substitution lines, where the activity of *Ph1* was suppressed, *elongatus* chromosome 6Ee paired, to some extent (4.6%), with wheat chromosomes of homoeologous group 6 (Dvorak 1979). Similarly, pairing between *elongatus* and wheat chromosomes was also observed by Johnson and Kimber (1967), Dvorak (1979, 1981b), and Sears (1973) in hybrids between *elongatus* and common wheat, when the *Ph1* gene of wheat was suppressed or absent.

If the interpretation of these data, as well as that of Jenkins and Mochizuki (1957) on pairing in the hybrid *T. durum* x diploid *E. elongatus,* is correct, then there must be considerable homology between *Elymus elongatus* and common wheat chromosomes. However, Dvorak and Knott (1974) assumed that this degree of pairing does not result from chromosomal homology but, rather, from the presence of *E. elongatus* genes that promote homoeologous pairing. Indeed, Dvorak and Knott (1974) found that chromosome IV (designated later 3Ee by Dvorak 1980) and chromosome I (1Ee; Dvorak 1980) increased significantly the pairing of wheat chromosomes, i.e., they carry genes that promote pairing of homoeologous chromosomes. If this is the case, then, the pairing reported by Jenkins and Mochizuki (1957) between el*ongatus* and durum wheat chromosomes, resulted presumably from homoeologous pairing between chromosomes of the A and B subgenomes of durum in addition to that between *elongatus* and durum.

Dvorak (1987) assumed that genes promoting or suppressing pairing of homoeologous chromosomes are ubiquitous among Triticeae diploid species. To identify such genes in diploid *E. elongatus,* he crossed common wheat lines with added or substituted *E. elongatus* chromosomes with *Hordeum bulbosum* to obtain haploids, and with *Triticum urartu* to obtain interspecific hybrids. Studies of chromosome pairing at first meiotic metaphase in the resulting haploids and hybrids and in the parental addition and substitution lines revealed genes affecting homologous or homoeologous chromosome pairing. Genes promoting pairing were found on the short and long arms of chromosome 3Ee, on the short arms of 4Ee and 5Ee, and on chromosome 6Ee of *E. elongatus*. Genes suppressing pairing of homoeologous chromosomes were found on the long arms of chromosomes 4Ee and 7Ee (Dvorak 1987). That may explain why different results were found when using lines that may contain different alleles of these pairing genes.

While eight Ag-NORs were observed in many cells of the amphiploid common wheat–diploid *E. elongatus*, four on the wheat chromosomes 1B and 6B and on four on two *elongatus* chromosomes, in some cells the Ag-NORs of *elongatus* were suppressed by the wheat chromosomes (Lacadena et al. 1984).

5.2.3.3 Ssp. *flaccidifolius* (Boiss. & Heldr.) Runemark (2n = 4x = 28)

Morphological and Geographical Notes

Ssp. *flaccidifolius* [Syn.: Boiss. & Heldr.) Runemaks [Syn.: *Agropyron scirpeum* C. Presl; *Agropyron scirpeum* var. *flaccidifolium* Boiss. & Heldr.; *Agropyron elongatum* var. *flaccidifolium* (Boiss. & Heldr.) Boiss. & Heldr.; *Agropyron flaccidifolium* (Boiss. & Heldr.) Candargy; *Agropyron elongatum* Host ssp. *scirpeum* (C. Presl.) Ciferri & Giacom.; *Elymus elongatus* ssp. *flaccidifolius* (Boiss. a Heldr.) Runemark; *Elytrigia scirpea* (C. Presl) Holub; *Lophopyrum scirpeum* (C. Presl) Á. Löve; *Thinopyrum scirpeum* (C. Presl) D. R. Dewey} is a perennial caespitose, more or less glaucous grass with erect culms, 70–115 cm high, with 11–23 cm long spikes, with 5–17 spikelets per spike, 10–22 mm long spikelets, with 5–10 florets and glumes with 5–6 veins, 4–6 mm long anther and 5 mm long caryopsis. In several accessions, two spikelets are located at each rachis node in the lower part or the center of the spike.

The diploid and tetraploid subspecies of *E. elongatus* cannot be morphologically distinguished with certainty, primarily because they differ mainly in quantitative traits (Breton-Sintes and Cauderon 1978; Luria 1983). Luria (1983) found several tetraploid accessions of *E. elongatus* ssp. *flaccidifolius* in Israel, in addition to diploid accessions of ssp. *elongatus*. The tetraploid accessions morphologically resemble the tetraploid subsp *flaccidifolius* (Breton-Sintes and Cauderon 1978; Luria 1983). The Israeli tetraploid differs from the diploid cytotype of ssp. *elongatus* only in its somewhat taller plants, shorter flag leaf, larger stomata, larger pollen grains and longer caryopses.

The tetraploid subspecies grows in the Mediterranean basin (Heneen and Runemark 1972; Luria 1983; Gabi and Dogan 2010). In Israel, the tetraploid subspecies grows from Einot Gibton, Shfela, southwards (was found in Nahal-Zin springs in the Negev). The distribution of this taxon is fragmentary and the populations are isolated from one another. In the salty spring of Einot Gibton, the only site in Israel where the diploid and the tetraploid subspecies occur together, the diploid grows near the spring and the tetraploid in the outer ring (Luria 1983). Hence, the diploid subspecies can tolerate higher concentrations of salt than the tetraploid subspecies.

Cytology, Cytogenetics and Evolution

Based on karyomorphological data, Heneen and Runemark (1972) assumed that the tetraploid subspecies (2n = 4x = 28; genome $E^e E^e E^e E^e$) is an autotetraploid, derived from the diploid subspecies by chromosome doubling, or rather, via inter-varietal hybridizations followed by chromosome doubling. An inter-varietal origin of natural autopolyploids in different groups of plants is a widespread phenomenon, as discussed by Stebbins (1950, 1971).

Heneen and Runemark (1972), Breton-sintes and Cauderon (1978), and Luria (1983) arranged the chromosomes of the tetraploid subspecies in seven groups of four. These groups morphologically correspond to the seven pairs of the diploid subspecies, supporting the likelihood of an autopolyploid origin of the tetraploid (Heneen and Runemark 1972; Breton-Sintes and Cauderon 1978). Yet, detailed karyomorphological studies showed the existence of small differences between pairs within groups of four (Heneen and Runemark 1972; Breton-Sintes and Cauderon 1978; Luria 1983). Consequently, it was proposed that this taxon originated from hybridization between diploid varieties that underwent some chromosomal divergence and therefore, possess two partially diverged genomes, namely, Ee1Ee1Ee2Ee2 (Breton-Sintes and Cauderon 1978).

Chromosome pairing in the F1 triploid hybrid between the tetraploid and the diploid subspecies, is only slightly lower than that expected for a hybrid between autotetraploid and its diploid progenitor (Dvorak 1981b; Charpentier et al. 1986). The assessment of the homology between the two genomes of the tetraploid showed that differentiation had occurred in all chromosome arms that could be tested. From the pairing frequencies of individual telosomes of the diploid subspecies of *E. elongatus* with chromosomes of the tetraploid subspecies, Dvorak (1981a) concluded that slight differentiation occurred in every chromosome of the two subspecies. Thus, both genomes of the tetraploid subspecies appear to be a slightly modified version of the genome of the diploid subspecies (Dvorak 1981b).

To account for this genomic divergence, Heneen and Runemark (1972) suggested that the autotetraploid originated from hybridization(s) between different diploid lines whose karyotype underwent some structural chromosomal differentiation. On the other hand, Dvorak (1981b), Dvorak and Scheltgen (1973) and Dvorak and McGuire (1981)

Fig. 5.2 Nuclear DNA content in diploid species and their polyploid derivatives, natural (black) and synthetic (gray): **a** Diploid and autotetraploid cytotypes of *Elymus elongatus* (taken from Eilam et al. 2009), and **b** Diploids, allotetraploid and allo-auto-hexaploid of *Elymus farctus* (taken from the Angiosperm C-value database (http://www.rbgkew.org.uk.cval/) at Kew Botanic Gardens. *The expected value, for each polyploid derivative, is the sum of the observed values of its parents

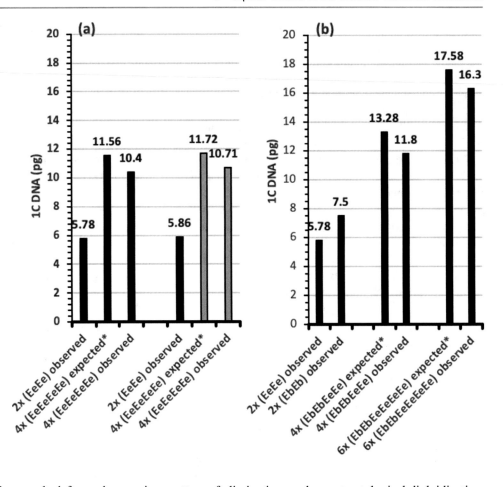

proposed that this differentiation resulted from changes in nucleotide sequences, rather than chromosomal aberrations such as inversions, translocations and other structural rearrangements. Alternatively, Eilam et al. (2009, 2010) suggested that the tetraploid subspecies underwent some cytological diploidization at the tetraploid level due to elimination of DNA sequences from two chromosomes in each group of four. Indeed, the tetraploid subspecies contained a significantly smaller amount of nuclear DNA (about 10% less) than the expected additive value of the diploid parent (Eilam et al. 2009, 2010). Also, a newly synthesized autotetraploid line of *E. elongatus*, produced by Charpentier et al. (1986), had significantly less DNA (8.57%) than the expected additive value (Eilam et al. 2009) (Fig. 5.2). The similarity in nuclear DNA content between the synthesized and the natural autotetraploids of *E. elongatus* indicates that the reduction in DNA content in the natural autotetraploid occurred immediately after its production, with only small changes in genome size over the history of the autotetraploid. Elimination of DNA sequences from two out of the four homologous chromosomes in each set of four, or elimination of sequences from one pair and other sequences from the second pair, augments the differentiation between the constituent subgenomes. Hence, the two subgenomes that became slightly divergent as a consequence of this

pattern of elimination, underwent cytological diploidization. This reduction in nuclear DNA may lead to exclusive bivalent pairing between fully homologous chromosomes and consequently, disomic inheritance. The eliminated sequences are likely to include those that participate in homologous recognition and initiation of meiotic pairing.

The chromosomes in other tetraploid *Elymus* species, such as *E. farctus* ssp. *boreo-atlanticus* (Heneen 1962) and *E. rechingeri* (Heneen and Runemark 1962), could not be grouped into groups of four. These species most likely have an allopolyploid origin. An autopolyploid origin of the tetraploid subspecies of *E. elongatus* is also indicated by the occasional formation of quadrivalents at meiosis (Heneen and Runemark 1972). This conclusion was also supported by genome analysis; Dvorak (1981b) and Charpentier et al. (1986) observed extensive pairing at the first meiotic metaphase in the triploid hybrid of these two subspecies and concluded that the three genomes of the triploid (Ee, Ee1, and Ee2) are closely related.

Meiosis is generally regular in subsp. *flaccidifolius* (Heneen and Runemak 1972; Dvorak 1981b; Luria 1983; Charpentier et al. 1986). For instance, Charpentier et al. (1986) observed 0.12 univalents, 13.9 bivalents 0.025 quadrivalents; 26.55 chaismata/cell at first meiotic metaphase. The majority of the cells had all the chromosomes paired as ring bivalents, indicating a high degree of homology within chromosome pairs. Multivalents, represented mainly by quadrivalents, occurred very rarely. The preferential bivalent pairing and the rarity of multivalent pairing in the autotetraploid subspecies, indicate either that this taxon is an autotetraploid that underwent cytological diploidization (Eilam et al. 2009, 2010) or that the tetraploid subspecies is a segmental allopolyploid (Breton-sintes and Cauderon 1978).

An induced autotetraploid of *E. elongatus*, produced by colchicine treatment of a diploid plant, was found morphologically indistinguishable from the natural tetraploid (Charpentier et al. 1986). The F_1 hybrid between the natural and the induced autotetraploid had almost complete chromosome pairing, with an average of 1.0 univalents, 7.9 bivalents, 2.8 quadrivalents and 23.8 chiasmata per cell, nearly similar to the chromosomal pairing observed in the induced autotetraploid parent (Charpentier et al. 1986). This pairing pattern further supports the autopolyploid nature of the natural tetraploid subspecies. Because of this slight genomic divergence, Dvorak (1981a, b) suggested classifying the tetraploid and the diploid taxa in two separate species. However, since autotetraploids and their diploid progenitors are usually included in the same species (Stebbins 1950), diploid and tetraploid *elongatus* are two cytotypes and were classified as members of a single biological species (Heneen and Runemark 1972; Breton-Sintes and Cauderon 1978).

Crosses with Other Triticineae Species

Homology between genomes Eel and Ee2 was also inferred from the meiotic behavior of the F_1 hybrid between tetraploid *E. elongatus* and common wheat. Despite the presence of one dose of *Phl*, the F_1 hybrid (2n = 5x = 35; genome BADEe1Ee2) exhibited at the first meiotic metaphase five to seven bivalents, interpreted as autosyndetic pairing of *elongatus* Eel and Ee2 chromosomes (El Gawas and Khalil 1973; Dvorak 1981b; Sharma and Gill 1983; Charpentier et al. 1988a). This number of bivalents shows that most, if not all, chromosomes of the two Ee subgenomes were involved in pairing. Pairing of the *elongatus* Ee1 and Ee2 chromosomes in the presence of one dose of the *Phl* gene indicated that these two genomes still retained their homology, further supporting the autoploid nature of tetraploid *E. elongatus*.

Indications that genomes Ee1 and Ee2 are closely related were also reported by Han and Li (1993). In two crossing combinations, *Triticum timopheevii ssp. timopheevii* (2n = 4x = 28; genome GGAA) x tetraploid *E. elongatus* and *T. turgidum* ssp. *durum* (2n = 4x = 28; genome BBAA) x tetraploid *E. elongatus*, chromosome pairing at first meiotic metaphase included a mean 9.10 univalents, 9.11

bivalents, 0.20 trivalents and 13.78 univalents, 6.87 bivalents, and 0.15 trivalents, respectively (Han and Li 1993). Since pairing between A and B wheat subgenomes is very low in haploids of tetraploid wheat containing the *Phl* gene (Kihara 1936; Lacadena and Ramos 1968), pairing in the hybrid containing the BAEe1Ee2 genomes was likely due to autosyndesis between Ee1 and Ee2 chromosomes of tetraploid *E. elongataus*. Similar homologous relationships were observd between the two Eb genomes of hexaploid *E. farctus* (Charpentier 1992) and between the two Eb genomes of tetraploid *E. farctus* (de V Pienaar et al. 1988).

The hybrid formed between tetraploid *E. elongatus* and common wheat, with zero dose of *Phl*, exhibited a relatively high degree of autosyndetic pairing between *elongatus* two-subgenome chromosomes and between wheat-subgenome chromosomes and allosyndetic pairing between wheat and *elongatus* chromosomes (range of chromosomal pairing was 9.6–11.2 bivalents and 1.2–1.9 trivalents per cell; quadrivalents, and some pentavalents were also observed) (Charpentier et al. 1988a). In contrast, a drastic reduction in pairing was observed in hybrids carrying one dose of *Phl*. Altogether they showed a means 4.6–7.7 bivalents per cell, multivalents were rare. The number of chiasmata/cell dropped from 19–20 in *Phl*-deficient hybrids to 6–10 in hybrids with one dose of this gene (Charpentier et al. 1988a).

Phl-deficient haploid bread wheat was found to form at first meiotic metaphase 3.2–4.2 bivalents, 0.9–2.0 trivalents and very few quadrivalents (0.02–0.12) and pentavalents per cell (Riley 1960). Assuming a similar level of pairing between the wheat chromosomes in hybrids generated from the tetraploid subspecies of *E. elongatus* and *Phl*-deficient bread wheat, then 6–7 bivalents of the observed 9.6–11.2 should be the results of pairing between Eel and Ee2 *elongatus* chromosomes. The number of quadrivalents and pentavalents that were observed in these hybrids indicate allosyndetic pairing. The relatively high level of allosyndetic pairing in F_1 hybrids between tetraploid *E. elongatus* and *Phl*-deficient bread wheat, and the low level of allosyndetic pairing in the presence of *Phl*, indicates that the subgenomes of tetraploid *elongatum* do not have genes that suppresses or promote homoeologous pairing.

Interestingly, in contrast to the effect of Ph-suppressors or homoeologous pairing promoters on diploid *E. elongatus* (Dvorak 1987), such an effect was not observed in hybrids between common wheat and tetraploid *E. elongatus* (Dvorak 1981b).

To determine the chromosomal location of these and other genes that control pairing in diploid *E. elongatus*, disomic addition lines of chromosomes derived from the diploid subspecies of *E. elongatus* in the background of Chinese Spring, were crossed with the tetraploid subspecies of *E. elongatus*, and pairing was then compared to those observed in hybrids between Chinese Spring and tetraploid *E. elongatus*, whose Ee1 and Ee2 chromosomes were previously defined as homologues (Charpentier et al. 1986, 1988a). The resultant F_1 hybrids ($2n = 5x = 36$), each carrying three doses of a given *elongatus* chromosome, enabled evaluation of the effect of each *elongatus* chromosome on pairing of homologues (Eel with Ee2) and homoeologues (A, B, D, and Ee). The study of chromosomal pairing in these hybrids enabled classification of the *elongatus* chromosomes into those that suppress (6Ee), promote (5Ee, 3Ee, and possibly also 1Ee), or have no effect on pairing (4Ee). The effect of chromosomes 2Ee and 7Ee was not studied. Chromosomes 5Ee and 3Ee differed in their effect on the degree and pattern of chromosome pairing, namely, the effect of 5Ee was stronger than that of 3Ee. Pairing analysis in such addition lines and in substitution lines, in their haploid derivatives and in hybrids between these lines and *Triticum urartu*, led Dvorak (1987) to allocate genes that promoted homologous or homoeologous pairing to chromosome arms 3EeS, 3EeL, 4EeS, 5Eep and to chromosome 6Ee of diploid *E. elongatus*. Genes suppressing homoeologous pairing were allocated to chromosome arms 4EeL and 7Eeq. In accord with Charpentier et al. (1988b), chromosomes 3Ee and 5Ee of diploid *elongatus* promoted homoeologous pairing.

In the presence of an extra dose of chromosome 6Ee of *elongatus*, the number of bivalents per cell was reduced, indicating suppression of pairing between the Eel and Ee2 chromosomes (Charpentier et al. 1988b), bringing Charpentier et al. to conclude that chromosome 6Ee of diploid *E. elongatus* carries gene(s) that inhibit(s) pairing or chiasma formation. This is in contrast to the finding of Dvorak (1987), who assigned a pairing-promoting effect to this chromosome. Chromosome 4E had no effect on pairing in hybrids with wheat (Charpentier et al. 1988b). This is in accord with Dvorak (1987), who reported pairing suppression by the long arm of chromosome 4Ee, but assumed the presence of a pairing promoter on the short arm of 4Ee, thus accounting for the lack of pairing effect by the entire 4E chromosome. He also found a suppressive effect of chromosome arm 7Eq, a chromosome arm that was not studied by Charpentier et al. (1988b).

Charpentier et al. (1988b) also studied chromosome pairing at first meiotic metaphase in hybrids between the bread wheat cultivar Chinese Spring and a synthetic autotetraploid line derived from diploid *E. elongatus*. The hybrids exhibited a high level of homoeologous pairing. Apparently, the genome of the diploid, from which the autotetraploid was synthesized, promoted pairing even in the presence of *Phl*. A similar effect was reported for gene(s) derived from another diploid accession of *E. elongatus* (Dvorak 1981b).

Promotion of homoeologous pairing by diploid *E. elongatus* was also observed in an amphiploid between allotetraploid *E. farctus* (subsp. *boreali-atlanticus*; genome EbEbEeEe) and diploid *E. elongatus* (genome EeEe) (Yvonne Cauderon, personal communication). The amphiploid had genome (EbEbEeEeEeEe). While tetraploid and hexaploid *E. farctus* exhibited mostly bivalents at meiosis, the amphiploid had several multivalents per cell, mostly quadrivalents but also some hexavalents. Evidently, in this amphiploid, genome Ee of dipoid *E. elongatus* promoted pairing between the homologues Ee genomes and between homoeologues Eb with Ee genoes. Three different accessions of diploid *E. elongatus* were found to promote homoeologous pairing: the accession used by Jenkins (1957) to produce the initial hybrid from which Dvorak and Knott (1974) derived their disomic addition lines, an accession from south France used by Cauderon in the cross with the tetraploid form of *E. farctus* and the Israeli accession, from which the induced autotetraploid was derived. Thus, the ability to promote homoeologous pairing may be a common feature of many accessions of diploid *E. elongatus*. Promotion of homoeologous pairing in the presence of *Phl* was described in several diploid Triticinae, viz. *Aegilops speltoides* (Riley 1960; Riley et al. 1961; Dvorak 1972), *Amblyopyrum muticum* (Riley 1966a, b; Dover and Riley 1972), *Ae. longissima* (Mello-Sampayo 1971b), *Secale cereale* (Riley et al. 1973; Lelley 1976; Dvorak 1977) and *Dasypyrum villosum* (Blanco et al. 1988b) (Table 5.2). However, chromosomal allocation of the promoters was only determined in rye (Lelley 1976) and in *E. elongatus* (Dvorak 1987; Charpentier et al. 1988b). In rye (Lelley 1976) chromosome 3R, and possibly also 5R, 4R, and 7R, were found to carry genes that promote homoeologous pairing in hybrids with wheat. This finding corresponds to the allocation of pairing promoters in diploid *E. elongatus* (Dvorak 1987; Charpentier et al. 1988b); the genes of 3Ee and 5Ee are presumably homoeoalleles to those of rye (Lelley 1976), as well as to those of homoeologous groups 3A, 3B and 3DL and 5A, 5D, and 5BS in bread wheat (Sears 1976).

5.2.3.4   Ssp. *turcicus* (P. E. McGuire) Melderis (2n = 8x = 56)

Morphological and Geographical Notes

Ssp. *turcicus* [=*Elytrigia turcica* P. E. Maguire; *Elytrigia elongata* ssp. *turcica* (P. E. McGuire) Valdés & H. Scholz; *Elytrigia pontica* ssp. *turcica* (P. E. McGuire) Jarvie & Barkworth; *Lophopyrum turcicum* (P. E. McGuire) McGuire ex Löve; *Thinopyrum turcicum* (P. E. Maguire) Cabi & Dogan], is a perennial caespitose, more or less glaucous grass with erect culms, 70–115 cm high, with 10–20 cm long spikes, spikelets with 7–9 florets and glumes with 7–9 veins. Anthers are 2.5–3.5 mm long.

Within the polyploid complex of *E. elongatus*, ssp. *turcicus* most resembles the decaploid ssp. *ponticus*. It differs from *ponticus* by the laxer leaves, less prominent ligules, more rounded apex of glumes, lack of hairs inside the glumes at the apex, smaller anthers, and in chromosome number (McGuire 1983). Although morphologically similar, the octoploid and the decaploid taxa were treated as separate species by McGurie (1983), but Melderis (1978), Dewey (1984), and Moustakas (1989) considered the morphological differences between the two taxa insufficient for separation on the specific level and consequently, classified them as two subspecies.

This subspecies distributes in Thassos Island, Greece (Moustakas 1993), Turkey, Georgia, and northern Iran (Jarvie 1992). It grows on dry calcareous, saline land from sea level to dry and saline mountain habitats, 1800 m above sea level, in low rainfall areas.

Cytology, Cytogenetics and Evolution

Chromosome counts in accessions of *Elymus elongatus* from eastern Turkey and northern Iran, showed the presence of octoploid plants with $2n = 8x = 56$ (Lorenz and Schulz-Schaeffer 1964; Sculz-Schaeffer et al. 1971; McGuire 1983), implying that the octoploids are not just sporadic individuals, arising in populations with other ploidy levels, but represent established populations (McGuire 1983).

Moustakas (1993) performed computer-aided karyotype analysis and found that the karyotype of the octoploid taxon is asymmetric, namely, it is composed of 10 metacentric chromosome pairs, 15 sub-metacentric chromosome pairs and 3 sub-telocentric chromosome pairs. Only two chromosome pairs have secondary constrictions, implying that a number of NORs are inactive in this octoploid.

All the chromosomes pairs of the octoploid can be matched to the chromosome pairs of the decaploid. Only the sub-telocentric satellited chromosome pairs differ slightly (Moustakas 1993). Yet, the karyotype analysis (Moustakas 1989, 1991, 1993) indicated that the chromosomes of the present-day diploid *elongatus* and those of the octoploid and decaploid diverged from each other.

Analysis of seed protein polymorphism patterns (Moustakas 1989) revealed that the octoploid originated from a speciation event more recent than that associated with the decaploid from which it presumably evolved. Moustakas (1989) found that the patterns of seed-protein electrophoresis of ssp. *turcicus* were qualitatively similar to those of ssp. *ponticus*.

Taking into consideration the results of the karyotype analysis, Moustakas (1993) concluded that ssp. *turcicus* is a segmental allopolyploid, with genome designation JjJjJjJjJeJeJeJe. (Genome designations Jj and Je represent the same genome but with some structural nuances.) On the other hand, Jarvie (1992) thought that the genome of subsp. *turcicus* is EEEEJJJJ is an auto-allo polyploid. Yet, if subsp. *turcicus* derived from the decaploid subsp. *ponticus*, its genome designation should be EeEeEbEbStStStSt.

5.2.3.5 Ssp. *ponticus* (Podp) Melderis (2n = 10x = 70)

Morphological and Geographical Notes

Ssp. *ponticus* (commonly known as tall wheatgrass and rush wheatgrass) [Syn.: *Triticum ponticum* Podp; *Agropyron elongatum* ssp. *ponticum* (Podp.) Senghas; *Agropyron incrustatum* Adamovic; *Elymus ponticus* (Podp.) N. Snow; *Elytrigia pontica* (Podp.) Holub; *Elytrigia elongata* ssp. *pontica* (Podp.) Gamisans; *Elytrigia ruthenica* (Griseb.) Prokudin; *Lophopyrum ponticum* (Podp.) Á. Löve; *Thinopyrum ponticum* (Podp.) Barkworth & D. R. Dewey] is a perennial, caespitose, tall plant with a 50–100 cm high stem. Its leaves are green or glaucous bluish with flat to curling blades that are often covered with short, stiff hairs, its lower sheaths usually ciliate, spikes are 10–35 cm long, lower internodes are usually much longer than the spikelets, rachis is not fragile, spikelets are 17–25 mm long, with 8–18 florets, glumes are thick and hardened, and 9–11 mm long, with 5–7 veins, lemmas are also thick and hardened, and 10–13 mm long, palea with cilia are seen along the entire length of keels and anthers are 4–7 mm long. Under certain conditions, this perennial grass can grow up to 2 m tall, and spikelets up to 3 cm long, each containing up to 12 flowers.

Ssp. *ponticus* is native to southeastern Europe, Turkey near the Black Sea and southern Russia. It grows well in dry and saline habitats, especially alkaline soils, as well as in disturbed habitats, such as waste ground and roadsides. This subspecies is found generally 360–1740 m above sea level.

Cytology, Cytogenetics and Evolution

Peto (1930), Simonet (1935), and Vakar (1935) reported that accessions of *E. elongatus* from Russia are decaploid, with $2n = 10x = 70$ chromosome number. Heneen and Runemark (1972) superficially described the karyotype of the decaploid subspecies. The large number of chromosomes rendered it difficult for them to construct the karyotype and to identify all the SAT-chromosomes. Generally, chromosome morphology and SAT-chromosome type seemed similar to those of the diploid and tetraploid subspecies of *E. elongatus*, indicating that ssp. *ponticus* is interrelated to these subspecies (Heneen and Runemark 1972), supporting the view that different subspecies of *E. elongatus* are involved in the origin of the decaploid subspecies.

On the diploid and tetraploid levels, there is a correlation between the number of SAT-chromosomes and degree of ploidy. This correlation is not obvious at the decaploid level,

Table 5.2 Promoters and suppressors of chromosomal pairing in Triticineae

	Species	Chromosome or chromosomal arm	References
Promoters	*Triticum aestivum*	5BS	Riley and Chapman (1967), Feldman and Mello-Sampayo (1967), Dvorak (1976), Cuadrado et al. (1991)
		5AS	Feldman (1966), Riley et al. (1966), Cuadrado et al. (1991)
		5DS	Feldman (1966), Riley et al. (1966)
		5AL	Feldman (1966)
		5DL	Feldman (1966)
		3BL	Sears (1954), Kempanna and Riley (1962)
		3DL	Driscoll (1972), Cuadrado et al. (1991)
		3AL	Mello-Sampayo (1971a), Mello-Sampayo and Canas (1973)
		2AS	Sears (1954, 1976), Riley et al. (1960), Ceoloni et al. (1986)
		2BS	Ceoloni et al. (1986)
		2DS	Ceoloni et al. (1986)
		6A (*ph KL*)	Liu et al. (1997, 2003)
	Amblyopyrum muticum		Riley and Law (1965), Dover and Riley (1972)
	Aegilops speltoides	3S (*Su1-Ph1*), 7S (*Su2-Ph1*), 5S (*QPh.ucd-5S*)	Riley et al. (1960), Riley et al. (1961), Dvorak (1972), Dvorak et al. (2006)
	Aegilops longissima		Mello-Sampayo (1971a), Feldman (1978)
	Aegilops caudata		Kihara (1959), Upadhya (1966)
	Aegilops umbellulata	5U	Riley et al. (1973)
	Aegilops peregrina		Driscoll and Quinn (1970), Farooq et al. (1990), Fernandez-Calvin and Orellana (1991)
	Aegilops geniculata		McGuire and Dvorak (1982)
	Elymus elongatus 2x,	3EeS, 3EeL, 4EeS, 5Eep,	Mochizuki (1962), Dvorak (1987), Charpentier et al. (1988b)
	Agropyron cristatum 2x and 4x	1P, 3P, 4PS, 5PL, 6PS	Chen et al. (1989, 1992a), Jubault et al. (2006), Jauhar (1992)
	Agropyron fragile		Ahmad and Comeau (1991)
	Dasypyrum villosum		Blanco et al. (1988b)
	Secale cereale	3R	Riley et al. (1973), Lelley (1976), Dvorak (1977)
	Taeniatherum caput-medusae		Frederiksen (1994), Jauhar et al. (1991)
Suppressors	*Triticum aestivum*	5BL (*Ph1*)	Riley and Chapman (1958), Sears and Okamoto (1958)
		5DL (*Ph3*)	Viegas et al. (1980)
		3DS (*Ph2*)	Mello-Sampayo (1971a), Mello-Sampayo and Canas (1973)
		3AS	Driscoll (1972a), Mello-Sampayo and Canas (1973), Cuadrado et al. (1991)
		3BS	Mello-Sampayo (1971a), Miller et al. (1983), Cuadrado et al. (1991)
		2DL	Ceoloni et al. (1986)
		2AL	Ceoloni et al. (1986)
		2BL	Ceoloni et al. (1986)
		4D	Driscoll (1973)
	Elymus elongatus 2x, 4x	4EeL, 6Ee, 7Eeq	Dvorak (1987), Charpentier et al. (1988b)

since the number of the barely detectable SAT-chromosomes in the decaploid subspecies was not proportional to the degree of ploidy (Heneen and Runemark 1972).

Up to ten nucleoli were recorded in pre-meiotic cells of the decaploid subspecies (Schulz-Schaeffer and Jura 1967). Brasileiro-Vidal et al. (2003), using silver nitrate staining to determine the number of nucleoli and NORs, revealed 17 AG-NOR sites on mitotic metaphase cells—a number similar to that of 45S rDNA detected via FISH. However, the mean number of nucleoli per interphase nucleus was much lower; in most cells, the number ranged from four to nine, indicating that at interphase, the active Ag-NOR sites tend to coalesce, as suggested by Lacadena et al. (1988).

Li and Zhang (2002) used FISH to study the distribution of the 18S-5.8S-26S rDNA in the decaploid subspecies and in its related diploid taxa, *E. elongatus* ssp. *elongatus* (genome EeEe), *E. farctus* subsp. *bessarabicus* (genome EbEb) and *E. stipifolius* (=Pseudoroegneria stipifolia) (genome StSt). The distribution of rDNA genes was similar in all three diploid taxa, i.e., two pairs of loci were observed in each somatic cell at metaphase and interphase. The first pair was located near the terminal end and the second in the interstitial regions of the short arms of a pair of chromosomes. The maximum number of major rRNA loci detected on metaphase spreads of the decaploid subspecies was 20, which corresponded to the additive sum of that of its progenitors. However, in the decaploid, all of the major loci were located on the terminal end of the short arms of the chromosomes. Apparently, the interstitial loci that exist in the possible diploid donors of genomes to the decaploid, changed their position during the formation and evolutionary history of the decaploid. These results suggest that there has been distinct differentiation between *ponticus* and its diploid relatives during the evolutionary process (Li and Zhang 2002). Positional changes of 18S-5.8S-26S rDNA loci between ssp. *ponticus* and its candidate genome donors, indicate that it is almost impossible to find a genome in the decaploid that is completely identical to that of its diploid donors (Li and Zhang 2002). The interstitial position is likely an ancestral trait, whereas the terminal position is probably a later-derived trait (Dubcovsky and Dvorak 1995). During polyploidization of ssp. *ponticus*, all of the interstitial loci have been either deleted and novel loci have been positioned on terminal regions of the chromosomes, or, alternatively, migrated to terminal positions. A similar phenomenon has been observed in other Triticeae species (Gill and Apples 1988; Dubcovsky and Dvorak 1995), although the exact underlying mechanism remains unknown.

Using FISH to determine the number and position of 45S and 5S rDNA sites in another accession of ssp. *ponticus*, Brasileiro-Vidal et al. (2003) detected both 45S and 5S rDNA sites on the short arms of 17 chromosomes, while on three other chromosomes, only the 5S rDNA site was observed. In ssp. *ponticus*, the 45S rDNA loci were always distally located in relation to the 5S rDNA loci. The occurrence of these sites in 17 instead of 20 chromosomes, as observed by Li and Zhang (2002), most likely indicates a reduction in the number of 45S rDNA sites in the accession used by Brasileiro-Vidal et al. (2003).

In ssp. *elongatus*, the 5S rDNA sites were associated with chromosomes 1Ee and, possibly, 5Ee (Scoles et al. 1988; Dvorák et al. 1989). Considering the distribution of these sites in diploid *elongatus*, the chromosomes carrying the 5S rDNA in the decaploid might also belong to homoeologous groups 1 and 5 (Brasileiro-Vidal et al. 2003).

Meiosis in the decaploid was less ordered. Cauderon (1958) observed 1.04 univalents, 19.9 bivalents, and a number of chain and ring multivalents (0.76 trivalents, 2.71 quadrivalents, 0.81 pentavalents, 0.52 hexavalents, 0.76 heptavalents, 0.19 octovalents, 0.05 ennevalents (=nine valents), and 0.04 decavalents). Similar patterns of chromosomal pairing were also observed by Zhang et al (1993, 1996) and Muramatsu (1990). The high frequency of multivalents in the decaploid subspecies may be the results of the activity of pairing promoters that exist in the ssp. *ponticus* genome (Zhang et al. 1993, 1995; Cai and Jones 1997).

Genomic relationships between the genomes of ssp. *ponticus* and its related taxa, have been the subject of several studies, and, due to chromosome pairing complexity, different genome formulae have been proposed for the decaploid. Peto (1936), assuming that the decaploid is an auto-allo-polyploid, tentatively assigned the genome formula AAXXXXYYYY, whereas Matsumura (1949) proposed the genome formula BBXXXXYYYY. Both researchers assumed that the A and B subgenomes of *Triticum* exist in ssp. *ponticus*. Muramatsu (1990) and Wang et al. (1991) concluded from the high frequency of multivalents in pollen mother cells (PMCs) of both the decaploid and the polyhaploid of ssp. *ponticus*, that this taxon contains several closely related or identical genomes, and therefore, is an autodecaploid with the genomic formula $J_1J_1J_2J_2J_3J_3J_4J_4J_5J_5$ and JJJJJJJJJJ, respectively. The J genome is from *E. farctus* ssp. *bessarabicus* (currently designated as Eb), is closely related to the Ee genome of ssp. *elongatus* and possesses modified versions of the same basic genome, namely, E (Forster and Miller 1989; Wang and Hsiao 1989; Wang 1990). Dvorak (1975) and Wang et al. (1991) regarded ssp. *ponticus* to be an autodecaploid that behaves as an allodecaploid, due to a bivalentization system.

Konarev (1979) assumed that the diploid subspecies of E. *elongatus* contributed to the karyotype of the decaploid taxon. Moustakas (1993), based on karyotype analysis, concluded that ssp. *ponticus* is a segmental allopolyploid, with genome formula JjJjJjJjJjJjJjJeJeJeJe, where genome designations Jj and Je represent the same genome but with structural differences. Dvorak (1975, 1981a) showed pairing of the chromosomes of diploid *elongatus* with some of the chromosomes of the decaploid, but the pairing was poor in every case, indicating that differentiation of the chromosomes had occurred. Consequently, Dvorak (1975) postulated that the decaploid evolved from an ancestral *elongatus*-like diploid taxa, by primary and secondary chromosome doubling of inter-ecotypic or inter-specific hybrids. In accordance, the karyotype analysis performed by Moustakas (1989, 1991), indicated that the chromosomes of the present-day diploid *elongatus* and those of the decaploid diverged from one another.

According to Dvorak (1981b), the decaploid appears to have one group of three closely related genomes and another group of two closely related genomes. Moreover, he suggested that the chromosomes of the diploid *elongatus* are more closely related to the doublet of the decaploid genome than to the triplet. Thus, Zhang and Dvorak (1990) suggested the genome designation ExExExExEyEyEyEyEyEy for subsp. *ponticus*. Jarvie (1992), assuming that ssp. *ponticus* is an auto-allo-polyploid, proposed the genome symbol EEEEEEJJJJ. Zhang et al. (1996), on the basis of GISH studies and genome specific markers, also suggested that this subspecies is an auto-allo-polyploid, but with the genome symbol StStStStEeEeEbEbExEx, where the St genome is homologous to the St genome of *Elymus stipifolius*. Moreover, GISH revealed that the centromeric region might be the critical area for discrimination between the St and E subgenomes (either Ee or Eb) in ssp. *ponticus*. Mitotic cells of several accessions of ssp. *ponticus*, when hybridized with the St probe and blocked by E genomic DNA, had 28 chromosomes strongly hybridized by the St probe at regions near the centromere (Zhang et al. 1996). When Ee or Eb was labeled as the probe and St was used as the blocker, all 70 chromosomes were labeled with FITC (fluorescein isothiocyanate). However, there were about 28 chromosomes lacking hybridization signals at the centromeric regions (Zhang et al. 1996). These consistent results were interpreted by Zhang et al. (1996) to mean that ssp. *ponticus* has 28 St genome chromosomes and 42 E genome chromosomes. The chromosome pairing data of Wang (1992) and the molecular studies of Hsiao et al. (1995) showed that the St, Ee, and Eb genomes are very closely related. The GISH results of Zhang et al. (1996) also revealed the close relationships between these three genomes and that GISH cannot distinguish between Eb and Ee. Taken together, the centromere and the region nearby may be the critical areas that discriminate the St from the E genomes in ssp. *ponticus*.

The 70 chromosomes of ssp. *ponticus* all fluoresced bright yellow when probed either with DNA from the Ee genome of *E. elongatus* ssp. *elongatus* or from the Eb genome of *E. farctus* ssp. *bessarabicus* (Chen et al. 1998a, b). This demonstrated that a substantial affinity exists between these probes and the subgenomes present in *ponticus*. Conversely, no obvious hybridization signal was detected in *ponticus* when probing either with DNA from the Ee genome and blocking with Eb genome DNA, or in the reverse analysis, using Eb genome DNA as probe and Ee genome DNA as blocker. Since this is expected when an effective DNA probe is used to block itself, the results suggest that the Ee and Eb genomes are closely related to one another and to the chromosomes of ssp. *ponticus*.

Chen et al. (1998a, b) also performed GISH using genomic DNA probes from *E. elongatus* ssp. *elongatus* (genome Ee), *E. farctus* ssp. *bessarabicus* (genome Eb), and *E. strigisus* (=*Pseudoroegneria strigosa*) (genome St), to investigate the genomic constitution of ssp. *ponticus*. Their findings indicated that the decaploid subspecies had only the two basic genomes Eb and Ebs (=Js). The Ebs genome of *ponticus* is homologous with E (Ee and Eb) genomes, but is quite distinct at the centromeric regions, which strongly hybridize with the St genomic DNA probe. This may indicate that the Ebs genome is a modified Eb (=J) genome whose chromosomes exchanged St segments via translocations between the two (Chen et al. 1998a, b). Support of this hypothesis also came from lack of centromeric hybridization signals upon hybridization of mitotic chromosomes of the diploid subspecies of *elongatus* and *farctus* with St genome DNA in the presence of Eb or Ee genome blocker (Chen et al. 1998a, b). Likewise, mitotic chromosomes of *E. strigosus*, probed with Eb genome DNA and blocked with St genome DNA, showed no hybridization signal. It appears that the chromosomes of *ponticus*, which show hybridization affinity with the centromeres of *E. strigosus* DNA, were not simply derived from any of these three diploid species, but rather, have a more complicated origin (Chen et al. 1998a, b).

Consequently, the group proposed that ssp. *ponticus* contains only segments of the St genome rather than any intact St genome or chromosomes. Based on the GISH results, namely, that all 70 chromosomes of ssp. *ponticus* hybridized extensively with Eb or Ee genome DNA probes, even in the presence of St genome blocker, Chen et al. (1998a, b) redesignated the genomic formula of ssp. *ponticus* as EEEEEEEbsEbsEbsEbs, where E refers to the Ee- or Eb-type chromosomes closely related to the genomes of ssp. *elongatus* and ssp. *bessarabicus*, respectively, while Ebs refers to a modified Ee- or Eb-type chromosomes distinguished by the presence of St genome-specific sequences close to the centromere.

The major disagreement between Zhang et al. (1996) and Chen et al. (1998a, b) centered around the explanation of the GISH results of ssp. *ponticus* probed by St genomic DNA and blocked by E genomic DNA. The St genomic probe hybridized all 70 chromosomes, but more strongly hybridized with 28 chromosomes at their centromeres and nearby regions. In the reverse GISH analysis, these 28 chromosomes were also hybridized by the E genomic probe, except for their centromeric and nearby regions, that were completely blocked by the St genomic DNA (Zhang et al. 1996). Zhang et al. opined that the unexpected signals appearing beyond the probe genome chromosomes were mainly caused by cross-hybridization between St and E genomes, arising from their close relationship in ssp. *ponticus*. Therefore, Zhang et al. (1996) proposed that the centromeres and nearby regions might be critical in the discrimination of St and E genomes.

Li and Zhang (2002) and Liu et al. (2007) accepted this interpretation, and used StStStStEeEeEbEbExEx as the genome formula for ssp. *ponticus*. Accordingly, the candidate donors of genomes to ssp. *ponticus* have been narrowed down to a few species, including the diploid species of the genus *Elymus*, namely, *E. elongatus* ssp. *elongatus*, *E. farctus* ssp. *bessarabicus* and *E. stipifolius*.

Since cytogenetic data indicate that the St, Ee, Eb (=J), and Ebs (=Js) genomes are very closely related (Wang 1992), the latest genomic designations are consistent with the earlier autopolyploid designation for these subspecies (Fedak et al. (2000). The study of Chen et al. (1998a, b) indicates that ssp. *ponticus* is not a characteristic autodecaploid and its five subgenomes are most likely modified versions of the Eb or Ee genomes. Since 28 chromosomes of ssp. *ponticus* containing centromeric region of St chromosomes are recombinant chromosomes, they concluded that *ponticus* can be regarded as a segmental autodecaploid with three sets of the E genome (Ee or Eb) genomes plus two sets of Ebs genome (Chen et al. 1998a, b). In contrast, Zhang et al. (1996) and Li and Zhang (2002) found that the decaploid is an auto-allo-polyploid containing six Ee genomes and four St genomes. The existence of multivalents in meiosis of the decaploid subspecies indicates that intergenomic recombination occurs quite frequently in this subspecies. Therefore, the genomes in the decaploid are recombinant genomes and differ from those of their donors.

Crosses with Other Triticineae Species

Peto (1936) suggested that the relatively high number of paired chromosomes in F_1 hybrids between *Triticum turgidum* ssp. *dicoccon* (2n = 4x = 28; genome BBAA) x *Elymus elongatus* ssp. *ponticus* (2n = 10x = 70; genome EeEeEbEbEbEbStStStSt) indicates homology between some wheat and *E. elongatus* chromosomes. Although B-subgenome chromosomes of wheat occasionally paired with other chromosomes at first meiotic metaphase of these F_1 hybrids, aceto-carmine Giemsa N-banding analysis of chromosomes in root tip cells and PMCs of the F_1 hybrid between *Triticum aestivum* cv. Fukuhoc and ssp. *ponticus*,

showed that the latter does not contain the B subgenome of wheat (Zhang et al. 1993).

The average pairing in first meiotic metaphase of the F_1 hybrid between tetraploid wheat, *Triticum turgidum* subsp. *durum* and *E. Elongatus* ssp. *ponticu*s included 14.93 (9–25) univalents, 11.92 (5–18) bivalents, 2.14 (0–5) trivalents, 0.54 (0–3) quadrivalents, and 0.42 (0–20) pentavalents (Zhang et al. 1993). The average chromosome pairing in the F_1 hybrid between hexaploid wheat, *Triticum aestivum* cv. Fukuhoc and ssp. *ponticus* displayed 10.87 (5–19) univalents, 16.40 (6–22) bivalents, 2.78 (0–6) trivalents, 0.55 (0–3) quadrivalents, and 0.23 (0–3) pentavalents (Zhang et al. 1993). Likewise, the F_1 hybrid *T. aestivum* cv. Chinese Spring x ssp. *ponticus* had 11.19 (4–17) univalents, 14.73 (7–20) bivalents, 3.12 (0–6) trivalents, 0.67 (0–2) quadrivalents, and 0.63 (0–2) pentavalents (Zhang et al. 1993). Cai and Jones (1997) used GISH to distinguish autosyndetic from allosyndetic pairing in the hybrid between ssp. *ponticus* and *Triticum aestivum* cv. Chinese Spring. Chromosome pairing in this hybrid occurred mainly among wheat chromosomes and among ssp. *ponticus* chromosomes, whereas allosyndetic pairing between wheat and *ponticus* chromosomes was very low. These results showed that the relationships among *T. aestivum* subgenomes and among ssp. *ponticus* subgenomes are much closer than the relationship between the subgenomes of the two species. The higher frequencies of autosyndetic pairing among the chromosomes of *ssp. ponticus* than among bread wheat chromosomes in the hybrid, indicated that the relationships between the five subgenomes of ssp. *ponticus* are closer than those between the three subgenomes of *T. aestivum*.

Comparing the observations in *aestivum* x *ponticus* to those in *durum* x *ponticus*, shows that adding the D subgenome leads to pairing of more than eleven chromosomes. This may indicate either that *ponticus* contains gene(s) that promote homoeologous pairing in the presence of *Ph1*, located in several different subgenomes of *ponticus* (Zhang et al. 1993, 1995), or that *ponticus* has a subgenome(s) related to the D subgenome of bread wheat (Zhang et al. 1993). Cai and Jones (1997) reported relatively high autosyndetic pairing frequencies among bread wheat chromosomes in the hybrid of subsp. *ponticus* x *Triticum aestivum* cv. Chinese Spring. The mean autosyndetic pairing frequency between wheat chromosomes in the hybrid was much higher than between those of euhaploids Chinese Spring with *Ph1* (Jauhar et al. 1991). Since this hybrid carries the *Ph1* gene, Cai and Jones (1997) suggested that ssp. *ponticus* carries gene(s) that can promote homoeologous chromosome pairing in the presence of *Ph1*.

Jauhar (1995) analyzed the F_1 hybrid between *ssp. ponticus* and bread wheat, which proved perennial and morphologically resembled the *Elymus* parent. The hybrid (2n = 8x = 56; genome EeEbEbStStBAD) showed high chromosome pairing (average pairing included 9.24 univalents, 6.23 rod- and 11.68 ring-bivalents, 2.07 trivalents, 1.09 quadrivalents, 0.02 pentavalents, and 0.05 hexavalents). Like Cai and Jones (1997), also Jauhar (1995) suggested that this high pairing was due to the inactivation of *Ph1* by ssp. *ponticus* promoters.

Addition lines of bread wheat bearing chromosomes of diploid *E. elongatus,* which are homoeologous with wheat chromosomes of groups 6 and 7, were crossed with addition lines of bread wheat carrying chromosomes of the decaploid subspecies of *E. elongatus,* which are also homoeologous with wheat chromosomes of groups 6 and 7 (Dvorak 1975). The chromosomes of the two *elongatus* subspecies paired with one another in the presence of the *Ph1* gene of common wheat. Since pairing between diploid *E. elongatus* and decaploid *E. elongatus* chromosomes in the presence of the *Ph1* gene, was generally low, it was suggested that the chromosomes assigned to the same group are not homologous, but rather, closely homoeologous (Dvorak 1975). In contrast to the low pairing between chromosomes of the two *elongatus* subspecies in the presence of *Ph1*, Dvorak (1975) assumed that these chromosomes could pair quite regularly in the absence of *Ph1*.

Johnson and Kimber (1967) produced complex hybrids bearing 29 chromosomes, including one telocentric chromosome (in different hybrids, different telocentric chromosomes represent a different chromosome arm) and twenty complete chromosomes of *T. aestivum* (2n = 6x = 42), seven complete chromosomes of *Aegilops. speltoides* (2n = 2x = 14) and one telocentric chromosome derived from *E. elongatus ssp. ponticus,* corresponding to homoeologous group 6. The presence of the *Ae. speltoides* genome induced pairing between homoeologous chromosomes at meiosis, even in the presence of *Ph1*. The *elongatus* telocentric chromosome paired with wheat chromosomes homoeologous to group 6. There was no evidence that it paired with chromosomes of any other group.

5.2.4 *E. farctus* (Viv.) Runemark Ex Melderis (Based on Eb Genome)

5.2.4.1 Species Description

E. farctus (Viv.) Runemark ex Melderis [Syn. *Triticum farctum* Viv.; *Triticum junceum* L.; *Agropyron junceum* (L.) Beauv.; *Agropyron junceum* ssp. *mediterraneum* Simonet; *Elytrigia juncea* (L.) Nevski; *Agropyron farctum* (Viv.) Rothm.; *Elymus multinodus* Gould; *Elymus farctus* (Viv.) Runemark; *Elytrigia juncea* ssp. *mediterranea* (simonet) Hyl.; *Thinopyrum* junceum (L.) Á. Löve; *Thinopyrum farctum* (Viv.) Cabi & Dogan comb. nov.] is rhizomatous, perennial, with rigid and glabrous 30–60 cm high stems, 2–5 mm broad, glaucous-green, leaves, densely pubescent on ribs of upper surface, with no auricles; 15–25 cm long spikes, fragile rachis (wedge type disarticulation), 5–12 veined and keeled 10–18 mm long glumes, 10–18 mm long lemmas and 6–12 mm long anthers, Caryopsis with adherent pericarp, 9.0 mm long.

This species (commonly known as sand couch-grass) comprises a polyploid complex represented by diploid, tetraploid, hexaploid, and possibly also octoploid subspecies, some of which are not well defined morphologically (Simonet and Guinochet 1938; Heneen 1972, 1977; Cauderon 1979). The diploid (2n = 2x = 14), tetraploid (2n = 4x = 28), and hexaploid (2n = 6x = 42) subspecies carry the genomes EbEb, Eb1Eb1Eb2Eb2, and Eb1Eb1Eb2Eb2EeEe (=JJ, $J_1J_1J_2J_2$, and $J_1J_1J_2J_2$EE), respectively (Cauderon 1958). While the tetraploid subspecies arose as an autopolyploid whose two genomes underwent some differentiation, it is assumed that the hexaploid was derived from a cross between the tetraploid subspecies and diploid *E. elongatus* (genome EeEe), through alloploidy (Cauderon 1958). The two Eb subgenomes (Eb1 and Eb2) of the tetraploid and the hexaploid subspecies are still very closely related, as is evident by the almost regular autosyndetic pairing of their chromosomes in hybrids with other polyploid *Elymus* species (Ostergen 1940a; Cauderon 1958). The Eb and Ee subgenomes are also related, as demonstrated by hybrids between tetraploid *E. farctus* and diploid *E. elongatus*, which show an average of 2.8 trivalents per cell (Cauderon and Saigne 1961). In fact, based on the high frequency of trivalents, Dvorak (1981a, b) and Dewey (1984) suggested that the Eb and Ee subgenomes be regarded as variations of the same genome. Yet, despite the close relationship between these subgenomes and the ability of their corresponding chromosomes to pair with each other, the hexaploid is characterized by almost complete bivalent pairing at the first meiotic metaphase; multivalents are rarely found (Charpentier et al. 1986).

All the subspecies are facultative cross-fertilizing and are capable of self-fertilization (Melderis 1978; Luria 1983). Nearly all subspecies possess long-creeping rhizomes, except for subsp. *rechingeri*, which grows in tufts on maritime rocks. All subspecies share a common smooth spike-rachis, readily disarticulating at maturity (wedge-type disarticulation), long anthers and flat, often convolute leaves, with densely and minutely hairy prominent ribs.

Simonet (1935) and Simonet and Guinochet (1938) were the first to notice the intraspecific differentiation of *E. farctus*. On the basis of karyological, morphological and chorological studies, they divided *E. farctus* (then *Agropyron junceum* (L.) P. Beauv.) into two subspecies: northwestern European (Atlantic) 2n = 28 (named *A. junceum* ssp. *boreali-atlanticum*) and southern European

(Mediterranean) 2n = 42 (named *A. junceum* subsp. *mediterraneum*). Prokudin (1954) divided *Elytrigia juncea* (L.) Nevski (=*Agropyron junceum*) into three separate species: *E. juncea* (L.) nevski s.str., and *E. mediterranea* (simonet et Guinochet) Prokudin, both having a south-Europe distribution range, and *E. junceiformis* A. & D. Löve, occurring in the northern coasts of Europe. However, Melderis (1978, 1980) transferred the *Elytrigia* genus to *Elymus* and named the species *Elymus farctus*, which he then further divided into the following four subspecies: ssp. *boreali-atlanticus*, occurring in the northern and western coasts of Europe, ssp. *farctus*, occurring in western Europe and the Mediterranean basin, ssp. *bessaribicus* and ssp. *rechingerii*, occurring in the southern part of Europe. Hence, *Elymus farctus* occurs in Europe and the Middle East and grows in maritime sands, near sea level.

5.2.4.2 Ssp. *bessaribicus* (Savul. & Rayss) Melderis (2n = 2x = 14)

Morphological and Geographical Notes
Ssp. *bessaribicus* (Savul. & Rayss) Melderis [=*Agropyron bessarabicum* Savul. & Rayss; *Elytrigia juncea* ssp. *bessarabicum* (Savul. & Rayss) Tzvelev; *Thinopyrum bessarabicum* (Savul. & Rayyss) Á. Löve]) is perennial, with shortly creeping or absent rhizomes, 50–80 cm high plants, with rigid, fairly thick, culms, usually not swollen at base, 15–35 cm long, erect or slightly curved spikes, fragile rachis, breaking at maturity above each spikelet (wedge type disarticulation), where rachis internodes are usually longer than lower spikelets; 10–25 mm long 5–9 flowered spikelets, appressed to rachis and laterally compressed, 10–18 mm long 6–12 veined, asymmetrically keeled, and unawned glumes, 10–20 mm long unawned lemma, keeled towards apex, and 10–12 mm long anthers.

Ssp. *bessaribicus* distributes in coasts of the Black Sea from Bulgaria to Crimea, Sea of Azov, Aegean and N.E. Mediterranean Sea. Corotype of subspecies is Mediterranean. Habitat: Seashores, sandy soil or sandy loam.

Cytology, Cytogenetics and Evolution
Cauderon (1958), Moustakas and Coucoli (1982), and Moustakas (1993) studied the chromosome number and karyotype of ssp. *bessarabicus*. All accessions studied had 2n = 2x = 14, with a symmetric karyotype, consisting of four metacentric pairs and three sub-metacentric pairs. Two chromosome pairs are SAT-chromosomes, one metacentric pair carries a large satellite and one sub-metacentric pair has a small satellite. The karyotype of ssp. *bessarabicus* is similar to the karyotype of diploid *E. elongatus* (Cauderon 1958; Runemark and Heneen 1968; Moustakas and Coucoli 1982; Wang 1985b; Hsiao et al. 1986). Yet, the chromosomes of *E.*

farctus are larger than those of *E. elongatus* and differences in morphology of the SAT-chromosomes exist between the two taxa (Heneen 1962); the pair with large satellites in *E. elongatus* has more median centromeres than the equivalent pair in *E. farctus*, whose short arm is divided into two equal parts by the secondary constriction. The second pair of ssp. *bessarabicus* has small satellites, which are somewhat larger than those of the equivalent pair in diploid *E. elongatus*. Endo and Gill (1984) also found a distinction between the two species in C-banding patterns. From their cytological analyses, Heneen and Runemark (1972), Moustakas and Coucoli (1982), and Moustakas (1993) concluded that the seven pairs of *E. bessarabicus* show striking similarity in chromosome size and centromere positions with the seven largest pairs of tetraploid *E. boreali-atlanticus* and the fourteen pairs of subsp. *rechingeri*.

Crosses with Oter Triticineae Species
Wang (1985a) crossed *E. bessarabicus* with *E. elongatus*, and, from the relatively high chromosomal pairing in their F$_1$ hybrid, concluded that the chromosomes of the two taxa show a high degree of homology, which counterweighed the C-banding differences reported by Endo and Gill (1984). Hsiao et al. (1986) assumed that the C-banding differences of the two taxa are due to structural rearrangements.

McGuire (1984) studied chromosomal pairing at meiosis of F$_1$ triploid hybrids between *Elymus curvifolius* (2n = 4x = 28; genome Eb1Eb1Eb2Eb2) x ssp. *bessarabicus* and observed an average of 3.71 univalents, 2.29 rod- and 1.82 ring-bivalents, 2.64 trivalents, and 0.29 quadrivalents. These pairing data indicate that the two subgenomes of *E. curvifolius* are closely related to the genome of ssp. *bessarabicus* and that the tetraploid had an autopolyploid origin.

Similarly, McGuire (1984) studied chromosomal pairing at meiosis of the F$_1$ triploid hybrid between *Elymus elongatus* ssp. *flaccidifolius* (=*Elymus scirpeus*; *Elytrigia scirpea*) (2n = 4x = 28; genome Ee1Ee1Ee2Ee2) and ssp. *bessarabicus* (genome EbEb). Mean chromosome pairing included 5.14 univalents, 1.28 rod- and 3.86 ring-bivalents, 1.47 trivalents, 0.11 quadrivalents, and 0.1 pentavalents. Also this hybrid provides evidence that the two Ee subgenomes of *E. flaccidifolius* are homologues and are related to genome Eb of ssp. *bessarabicus*. This is in accord with the finding of Wang (1985a, b), who demonstrated that genome Eb of ssp. *bessarabicus* is closely related to the genome Ee of diploid *E. elongatus*.

Wang (1988) produced diploid hybrids between ssp. *bessarabicuus* and *Elymus spicatus* (=*Pseudoroegneria spicata*) (genome StSt), as well as with *Secale strictum* (formerly *S. montanum*) (genome RR). Meiotic chromosome pairing of the F$_1$ hybrid *spicatus* x *bessarabicus* averaged

4.34 univalents, 2.77 rod- and 1.42 ring-bivalents, 0.24 trivalents, and 0.14 quadrivalents. On the other hand, chromosome pairing of the F_1 hybrid *bessarabicus* x *strictum* included 11.05 univalents, 1.22 rod- and 0.04 ring-bivalents, 0.13 trivalents, and 0.01 quadrivalents. These meiotic data suggest that the ST genome of *E. spicatus* and the Eb genome of ssp. *bessarabicus* are more closely related to each other than Eb is with the R genome of *Secale*.

Alonso and Kimber (1980) and Sharma and Gill (1983) produced an F_1 hybrid between *Triticum aestivum* cv. CS and diploid ssp. *bessarabicus*. The morphology of the hybrid was closer to wheat, but showed intermediate expression of some traits. The hybrid seemed not to be perennial, since it produced few tillers. Since hybrids between bread wheat and tetraploid or hexaploid *E. farctus* exhibit chromosomal pairing that result from autosyndesis of *Elymus* chromosomes, the study of the relationships between the Eb genome of ssp. *bessarabicus* and the subgenomes of bread wheat is incomprehensible. Hence, the relationships between the two species can be studied directly in hybrids with diploid *E. farctus*. A mean 0.2 bivalents per cell (always rod) was observed at meiosis of the hybrid; most of the cells had 28 univalents (Alonso and Kimber 1980). Somewhat higher chromosomal pairing was observed in such a F_1 hybrid by Sharma and Gill (1983), with a mean 0.83 rod bivalents, 0.04 ring bivalents, and 0.01 trivalents per cell. The very low pairing in these hybrids indicates that genome Eb of diploid *E. farctus* is only distantly related to the A, B, and D subgenomes of bread wheat. Moreover, the Eb genome of diploid *E. farctus* did not induce homoeologous pairing of wheat chromosomes in the hybrids containing the *Ph1* gene.

5.2.4.3 Ssp. *rechingeri* (Runemark) Melderis (2n = 4x = 28)

Ssp. *rechingeri* (Runemark) Melderis [Syn.: *Agropyron rechingeri* Runemark; *Elymus rechengeri* Runemark in Runemark & Heneen; *Elytrigia rechingeri* (Runemark) Holub] is perennial, without rhizomes, with culms usually swollen at base, short ligule (c. 0.5 mm), 2–10 cm long spikes, rachis internodes that are usually shorter than the lower spikelets, 5–12 mm, 4–5 veined glumes, and palea ciliate in upper half of keel only.

Ssp. *rechingeri* is a tetraploid (2n = 4x = 28) cross-fertilizing taxon. The karyotype of ssp. *rechingeri* shows great similarities to the karyotype of *E. furctus* ssp. *boreali-atlanticus* (2n = 4x = 28) (Heneen 1977). Meiosis is generally normal, but some asynapsis and multivalent formation at first meiotic metaphase and separation difficulties at first anaphase have been noted (Heneen 1977). Chromosomal polymorphism, manifested by changes in the three pairs of satellite chromosomes, was observed both between and within populations (Heneen 1977). Some offspring plants also show structural and numerical chromosome deviations. Fertility is lower in crosses between populations, indicating some genetic differences between populations (Heneen 1977).

This subspecies grows in small, isolated populations, mainly on the Aegean islands. It also occurs in west Turkey, coasts of Greece, Crete, and the Mediterranean coast of Egypt. Corotype: E. Mediterranean element.

5.2.4.4 Ssp. *boreali-Atlanticus* (Simonet & Guinochet) Melderis (2n = 4x = 28)

Morphological and Geographical Notes

Ssp. *boreali-atlanticus* (Simonet & Guinochet) Melderis [Syn.: *Agropyron junceum* subsp. *boreali-atlanticum* Simonet & Guinochet; *A. junceiforme* (Á. & D. Löve) Á. & D. Löve; *Elytrigia juncea* subsp. *boreo-atlantica* Hyl.; *Thinopyrum junceum* (L.) Á. Löve] is perennial with long rhizomes, up to 55 cm high, fragile, glabrous culms, erect or slightly curved, 5.0–14.0 cm long, spike with 4–10 nodes and 16–23 mm long internodes, spikelets longer than internodes, with 3–5 flowers, rachis breaking up between each spikelet at maturity (wedge-type disarticulation), 10–16 mm long glumes, 10–17 mm long awned lemma, sometimes with very short awns.

Ssp. *boreali-atlanticus* occurs in the northern part of the distribution of the species, from Portugal to Finland, however, its stands are concentrated mainly between Portugal and Germany, whereas in the east, their numbers decrease rapidly.

Cytology, Cytogenetics and Evolution

The chromosome number given for ssp. *boreali-atlanticus* from different areas of distribution is 2n = 4x = 28 (Peto 1930; Simonet and Guinochet 1938; Östergren 1940a; Moustakas et al. 1986; de V Pienaar et al. 1988). The karyotype is symmetric, exhibiting only two, out of the four possible, secondary constrictions in somatic metaphases (Simonet 1935). The genome of ssp. *boreali-atlanticus* was formulated as Eb1Eb1Eb2Eb2 (=$J_1J_1J_2J_2$), assuming an autopolyploidy origin that later sustained some differentiation in the two subgenomes (Cauderon 1958). Alternatively, Moustakas et al. (1986) and de V Pienaar et al. (1988), on the basis of genome analysis of the whole polyploid complex of *E. farctus*, concluded that ssp. *boreali-atlanticus* is a segmental allopolyploid containing a basic genome E (=J) (more or less modified at different polyploid levels of the entire complex). In line with this suggestion, Liu and Wang (1993) proposed EbEbEeEe as the genomic formula of this subspecies.

Östergren (1940a) reported that meiosis in ssp. *boreali-atlanticus* is regular; 14 bivalents were observed in almost all cells, whereas quadrivalents were very rare. Few individuals were found to be heterozygous for a paracentric inversion (Östergren 1940a).

Crosses with Other Triticineae

McGuire (1984) analyzed the F_1 hybrid between ssp. *boreali-atlanticus* (genome EbEbEeEe) x *E. curvifolius* (genome Eb1Eb1Eb2Eb2) and observed mean chromosomal pairing of 3.00 univalents, 0.93 rod- and 1.57 ring-bivalents, 1.36 trivalents, 1.79 quadrivalents (Open), 1.14 quadrivalents (close), and 0.79 pentavalents. The high frequency of quadrivalents and pentavalents resulted presumably, from the close relatedness of Eb and Ee. Östergren (1940a) studied chromosomal pairing at meiosis of the F_1 hybrid between ssp. *boreali-atlanticus* and *Elymus repens* (2n = 6x = 42; genome StStStStHH). Average chromosomal pairing included 11.8 univalents and 11.6 bivalents. From the pairing data, Östergren (1940a) concluded that ssp. *boreali-atlanticus* is not entirely autopolyploid.

Östergren (1940b) crossed *T. turgidum* subsp. *turgidum* (2n = 4x = 28; genome BBAA) with *ssp. boreali-atlanticus*, and obtained at the first meiotic metaphase of the F_1 hybrid an average chromosomal pairing of 18.4 univalents and 4.8 bivalents; most of which were rod shaped. Chromosome pairing in haploid tetraploid wheat was found by Kihara (1936), Lacdena and Ramos (1968) and Jauhar et al. (1999), to be very low and therefore, most of the pairing in the hybrid (genome BAEbEe) was autosyndetic of Eb and Ee chromosomes. This pairing, that took place in the presence of the *Ph1* gene of ssp. *turgidum,* again indicates that the two subgenomes of ssp. boreali-atlanticus, namely, Eb and Ee, are closely related.

5.2.4.5 Ssp. *farctus* (Viv.) Runemark Ex Melderis (2n = 6x = 42)

Morphological and Geographical Notes

Ssp. *farctus* (Viv.) Runemark ex Melderis [syn.: *Triticum farctum* Viv.; *Agropyron farctum* Viv.; *Agropyron junceum* (L.) Beauv.; *Agropyron junceum* ssp. *mediterraneum* Simonet & Guinochet; *Thinopyrum junceum* (L.) Á. Löve; *Elytrigia juncea* (L.) Nevski] is perennial, usually with long-creeping rhizomes, 50–80 cm high plants, with rigid, thick culms, glaucous, usually rolled leaves, 15–35 cm long, erect spike, more or less fragile rachis, breaking at maturity above each spikelet (wedge type disarticulation), 10–25 mm long, 5–9 flowered, glabrous, laterally compressed, awnless spikelets, appressed to rachis, 10–18 mm long, lanceolate, 6–12 veined, asymmetrically keeled glumes, 10–20 mm long lemma, keeled towards the apex, palea ciliate nearly along the entire length of keels, 10–12 mm long anthers (Fig. 5.1c).

Distribute along the coasts of the Mediterranean Sea. Grows near seashores, on sandy soil or sandy loam. Chorotype: Mediterranean element.

Cytology, Cytogenetics and Evolution

Simonet (1935) and Simonet and Guinochet (1938) reported that the number of chromosomes in this subspecies is 2n = 6x = 42, and its genome was formulated by Cauderon (1958) as Eb1Eb1Eb2Eb2EeEe (=$J_1J_1J_2J_2$EE). Charpentier et al. (1986) studied the degree and pattern of chromosomal pairing at the first meiotic metaphase of lines of ssp. *farctus* collected from several sites along the Israeli Mediterranean coast. Despite the relatedness between its subgenomes, ssp. *farctus* exhibited almost complete bivalent pairing, i.e., 21 bivalents; multivalents were rarely found.

A similar pattern of bivalent pairing with rare multivalents is characteristic of other autopolyploid or segmental allopolyploid species of *Elymus*, e.g., *E. hispidus* (=*Agropyron intermedium*) (genome EeEeEeEeStSt) (Cauderon 1958, 1966), and of tetraploid *E. elongatus* (genome Ee1Ee1Ee2Ee2) (Charpentier et al. 1986). The almost strict bivalent pairing in these auto-allopolyploids may be brought about by a gene system that induces bivalentization by restricting pairing to fully homologous chromosomes. A similar gene system, determining a bivalent rather than a multivalent pattern of pairing, was described for the autotetraploid *Avena barbata* (Ladizinsky 1973) and in an autotetraploid line of *Aegilops longissima* (Avivi 1976).

Crosses with Other Triticineae Species

Charpentier et al. (1986) analyzed chromosomal pairing at the first meiotic metaphase of F_1 hybrids generated from lines of ssp. *farctus* and the bread wheat cultivar Chinese Spring. In the presence of the wheat *Ph1* gene, chromosome pairing in the hybrid (genome BADEb1Eb2Ee) included 25.88 univalents, 5.43 rod- and 0.87 ring-bivalents, 1.08 trivalents, and 0.06 quadrivalents. Most of the pairing was autosyndetic of the Eb1 and Eb2 chromosomes of ssp. *farctus*. Based on this pattern of pairing, Charpentier et al. (1986) concluded that the corresponding chromosomes of the Eb1 and Eb2 subgenomes are distant homologues.

The seed set in the crosses CS x ssp. *farctus* was rather low (1.8–7.0%), and most seeds were very shriveled, with a poorly developed endosperm. Embryos were well differentiated, and when cultured on Orchid agar, 19–33% of the embryos germinated. The F_1 plants exhibited strong heterosis, i.e., they were tall (1.20–1.60 m) and had vigorous tillering. All plants were perennials with a non-brittle rachis.

The small seed set in crosses between bread wheat (as female) and hexaploid *E. farctus*, shrivelling of most of the seeds, and their poor germination, as well as the fact that the F_1 plants were completely male sterile with anthers that failed to dehisce, attest to the existence of significant strong chromosomal and/or genetic barriers between these two taxa.

Eleven disomic addition lines, and nine partial amphiploids have been obtained from hybridization between bread wheat cv. CS and the hexaploid subspecies of *E. farctus* (Charpentier 1992). The genomic structure of the disomic addition lines consists of 21 pairs of wheat chromosomes, plus an additional pair of ssp. *farctus*.

McArthur et al. (2012) used FISH to characterize thirteen disomic addition lines of chromosome pairs of ssp. *farctus* that were added to bread wheat cv. CS. Several lines were those produced by Charpentier (1992). Five disomic addition lines (AJDAj5, 7, 8, 9, and HD3508) were identified to contain a *farctus* chromosome that corresponded to homoeologous group 1. Addition lines AJDAj2, 3, and 4 contained a *farctus* chromosome that corresponded to homoeologous group 2, HD3505 to group 4, AJDAj6 and AJDAj11 to group 5, and AJDAj1 probably to group 6. Several ssp.. *farctus* chromosomes in the addition lines were found to contain genes for resistance to *Fusarium* head blight, tan spot, *Septoria nodorum* blotch, and stem rust (Ug99 races).

5.2.5   Phylogenetic Relationships of St, Ee and Eb Genome *Elymus* Species with Other Triticineae Species

Elymus s. l. is the largest, most morphologically diverse and most widely distributed genus in the Triticeae, consisting of approximately 150 diploid and polyploid species (Table 2.1). Cytogenetic analyses have shown that there are eight basic genomes within *Elymus,* namely, St, E (either Ee or Eb), L, Xp, H, P, W, and Y. The St, L, and Xp genomes exist in diploid species that grow in the Temperate-Arctic region whereas other diploids with St genome and those with E genome grow in the Mediterranean-Central Asiatic region.

The St genome of several diploid species of *Elymus* (Table 5.1) is closely related to Ee and Eb genomes of diploid *E. elongatus* and *E. farctus*, respectively (Wang 1989, 1992). Bieniek et al. (2015) found that the nucleotide sequences at three chloroplast loci (*matK*, *rbcL*, *trnH-psbA*) are almost identical in the diploid Ee, Eb and St taxa, with only one substitution within the *matK* gene, differentiating genome Eb from the Ee and St genomes. Petersen and Seberg (1997), based on *rpo*A sequence data, and Wang and Lu (2014), based on a literature review, corroborated the very close relationship between the Ee, Eb and St genomes. This close relationship was also substantiated by study of the sequences of a gene encoding plastid acetyl-CoA carboxylase (Fan et al. 2007). A study using 5S rDNA further strengthened the reported close relationship between St and Eb (Shang et al. 2007). The St genome is also closely related to the P genome of *Agropyron* species (Wang 1992). The St and Eb genomes were shown to be more closely related to the R genome of

Secale than to the Vv genome of *Dasypyrum villosum* (Shang et al. 2007). The Ns, H, and R genomes are remotely related to the E-St-P cluster (Wang 1992). Since the St genome exists in primitive diploid species of *Elymus* and in many allopolyploid species of this genus, it is assumed that genomes Ee and Eb evolved from St.

The St and E genomes are two important basic genomes in the perennial species of the Triticeae. In addition to their existence in diploids, the St and E genomes also exist in almost all polyploid (allopolyploid and autopolyploids) *Elymus* species, whereas the H, P, W, and Y genomes exist only in allopolyploid species (Dewey 1984; Wang et al. 1995). The H genome originated from one of the species of *Hordeum*, the P from *Agropyron*, the W from *Australopyrum*, and the origin of the Y genome is unknown (Dewey 1984; Yan et al. 2011; Petersen et al. 2011). Liu et al. (2006) and Okito et al. (2009) postulated a common origin of St and Y, whereas Dewey (1984) assumed that the donor of the Y genome is a distinct, yet undiscovered or extinct, diploid Asian species. Sun et al. (2008), Sun and Komatsuda (2010), and Yan et al. (2011)) supported Dewey's (1984) view, and separated the St and Y genomes into distinct groups. Studies on the chloroplast DNA of *Elymus* showed that the St genome of diploid *Elymus* species is the maternal donor of the St genome in all the allopolyploid *Elymus* species (Mason-Gamer et al. 2002; Liu et al. 2006; Hodge et al. 2010; Yang et al. 2017).

Recent molecular studies (Mason-Gamer et al. 2010; Sun et al. 2008; Yu et al. 2008; Sun and Komatsuda 2010; Yan and sun 2011) showed the presence of several distinct clades within the diploid St genome *Elymus* species, but analysis of various DNA sequences suggested different combinations of clades. For instance, *E. libanoticus* and *E. tauri* have similar sequences and form one clade, while the other diploid *St* genome species form another clade (Mason-Gamer et al. 2010; Sun et al. 2008; Sun and Komatsuda 2010; Yan and Sun 2011). Yu et al. (2008) found that *E. libanoticus*, diploid *E. tauri* and diploid *E. spicatus* are more closely to one another than they are to *E. stipifolius* and *E. reflexiaristatus*. Yan et al. (2011) grouped the Eurasian St genome species *E. libanoticus*, *E. reflexiaristatus* and diploid *E. tauri* into one clade and the North American *E. spicatus* into a separate clade. Evidently, there are great discrepancies between various phylogenetic studies performed on this group of diploid species.

By analyzing the sequence diversity in the internal transcribed spacer (ITS) region of nuclear ribosomal DNA, Dïzkirici et al. (2010) found that all the species of *Elymus* s. l. fall into one clade. Fan et al. (2013), using two single-copy nuclear gene (*Acc1* and *Pgk1*) sequences, found that that the St genome *Elymus* species are closely related to the Ee genome of *E. elongatus*. Yang et al. (2017) using nuclear internal-transcribed spacer and the chloroplast *trnL-F* sequences, analyzed phylogenetic relationships among

Elymus and related genera, and obtained four major clades: (1) the St/E clade, comprised all of the St and E genome species of *Elymus*; (2) the P/W clade, including *Agropyron* and *Australopyrum*; (3) the Ns clade which included *Psathyrostachys*; and (4) the H clade, which consisted of *Hordeum* species. The results suggested that: (a) diploid St genome species were the maternal donors of St in allopolyploid species in *Elymus* s.l. and that the *trnL-F* sequences are highly similar among these species; (b) the *trnL-F* sequences of *Agropyron* species and *Australopyrum* species are similar, and the P genomes are closely related to the W genome; and (c) the trnL-F sequences of species with the H or Ns genomes diverged considerably from that of species with the St, E, P, or W genomes.

Petersen and Seberg (1997) and Escobar et al. (2011) classified the diploid Triticeae into five major clades: (1) *Psathyrostachys*; (2) *Hordeum*; (3) *Elymus;* (4) *Agropyron* (includes *Australopyrum*)–*Eremopyrum*: and (5) *Aegilops*–*Triticum*–*Secale*–*Taeniatherum*. These major clades were defined recently also on the basis of the nuclear phosphoglycerate kinase (PGK) gene that codes for plastid PGK isozyme (Adderley and Sun 2014). Similarly, Bieniek et al. (2015), studying phylogenetic relationships among the Triticeae diploid species through analysis of three chloroplastic genes, also supported the finding of the above five clades. The *Elymus* clade is the largest clade, containing the diploid species of the genus *Elymus* s. l. (Melderis 1980), namely, the St-genome species, *E. elongatus,* and *E. farctus.*

Hsiao et al. (1995) and Kellogg et al. (1996) considered *T. monococcum* to be the sister group to *Elymus elongatus.* Based on internal transcribed spacers (ITS) of the nuclear rDNA sequences, Hsiao et al. (1995) reported that Ee and Eb jointly clustered with subgenomes A, B, and D. In accord with this finding, Liu et al. (2007), using genomic hybridization (both Southern and in situ hybridization), also showed that the St and Eb genomes are very closely related to the A, B and D subgenomes of common wheat, but are more closely related to the D subgenome than to the A and B subgenomes. These observations provide a possible explanation as to why most of spontaneous translocations and substitutions occurring in the common wheat—*E. elongatus* ssp. *ponticus*, usually take place in the D genome, some in the A subgenome and rarely in the B subgenome. In accord with the above findings, both genomes Ee and Eb were found to be closely related to subgenomes A, B, and D of the wheat group (reviewed in Wang and Lu 2014) and thus, the latter genomes may derive from the E genome(s). Interestingly, from their study of two single-copy nuclear gene (*Acc1* and *Pgk1*) sequences, Fan et al. (2013) concluded that the relationship between *Elymus farctus* subsp. *bessaribicus* (genome Eb) and *Triticum/Aegilops* is closer than between *Elymus elongatus* (genoe Ee) and *Triticum/Aegilops.*

5.3 *Agropyron* Gaertner *Senso Stricto*

5.3.1 Taxonomic Notes

The genus *Agropyron* Gaertner s.l. (the name *Agropyron* is derived from the Greek terms 'agros' meaning field and 'puros' meaning wheat) was one of the largest genera in the Triticeae, including more than 100 species (Gaertner 1770). However, Nevski (1933, 1934a) divided the perennial species of *Agropyron* to four genera, *Agropyron* s. str., *Elytrigia* Desv., *Roegneria* C. Koch and *Anthosachne* Steud. Gould (1947) included *Agropyron* s. str. in *Elymus*, while Runemark and Heneen (1968) further expanded the generic concept of *Elymus* to also include *Elytrigia, Roegneria, Aneurolepidium, Terella, Hystrix* (*Asperella*) and *Sitanion.* Melderis (1978, 1980, 1985a) followed this expansion of the generic concept of *Elymus* but retained *Agropyron* s. str. as a separate genus for the crested wheatgrasses.

Agropyron s. str. is currently considered to be a small genus, including only 10–15 species (Dewey 1984; Sakamoto 1991), all of which are the crested wheat grasses (Nevski 1933; Hitchcock 1951; Tzvelev 1983). The genus is morphologically well characterized by its distinctly keeled glumes, its short rachis internodes, spikelets divergent from the rachis at an angle of more than 45° and typically pectinate spikes. It should be emphasized that Nevski (1933, 1934a) treated the genus *Agropyron* on the basis of morphological features only; later cytogenetic analyses, mainly genome analyses, confirmed the validity of Nevski's treatment. Genome analyses revealed that *Agropyron* is genomically homogeneous with diploid, tetraploid, and hexaploid taxa, all displaying the P genome (Löve 1984), thus rendering it the only complex with just one genome and genomically distinct from other taxa of the *Agropyron-Elymus* complex (Melderis 1978; Löve 1984). The P genome also exists in several polyploid species of *Elymus*, but has a negligible influence on the morphology of these species (Assadi and Runemark 1995). Therefore, it seems reasonable to maintain *Agropyron* as a separate genus. The generic concept of Nevski was accepted by many taxonomists (e,g., Tzvelev 1973, 1976; Melderis 1978, 1980, 1985a; Clayton and Renvoize 1986; Watson and Dallwitz 1992; Assadi and Runemark 1995).

The species of *Agropyron* s. str. are perennials, rhizomatous or caespitose and cross-fertilizing (Melderis 1978). They have firm, straight, 20–70 cm long culms, spikes with hairy and tough rachis (rarely fragile), that do not disarticulate at maturity, with short internodes, solitary, sessile, pectinate and strongly laterally compressed, spikelets with 2–12 florets at each node of the rachis. The rachilla disarticulates at maturity above the glumes and beneath the florets (floret-type disarticulation). In this type of disarticulation, lemma and palea fall off, while glumes persist, firmly attached to the rachis. The glumes are boat-shaped, with a prominent keel, 1–2 inconspicuous lateral veins and a wide margin, lemma are membranous, 5-vined and keeled and both glumes and lemmas are awnless, or lemma have a short awn. Their anthers are 3.5–5.0 mm long.

The *Agropyron* species form an autopolyploid series consisting of diploids, tetraploids and hexaploids (Dewey and Asay 1975; Dewey 1967, 1982, 1983; Melderis 1978; Assadi 1995; Knowles 1955; Jensen et al. 2006). The autopolyploid series is based on the P genome of diploids *A. cristatum* and *A. mongolicum* (Knowles 1955; Dewey 1967, 1969, 1984; Löve 1982).

Much confusion prevailed in regard to the number of species in this genus (Cabi 2010). Tzvelev (1976) recognized 10 species in *Agropyron*, Yilmaz et al. (2014) reported that this genus contains ± 15 species, and Sakamoto (1991) assumed that it contains 19 species, with a distribution of 5 diploids, 13 tetraploids and 1 hexaploid. The taxonomy of the species has been complicated due to extensive interspecific introgressive hybridizations, as well as by the fact that many of the interspecific hybrids are fertile (Knowles 1955; Dewey 1983; Asay and Dewey 1979).

Thirteen *Agropyron* species, their synonyms, common name, ploidy level, genome formula, and geographic distribution, are presented in Table 5.3. The species grow in a wide range of habitats, including steppe-like habitats, mountains, saltmarshes, and seashores, on sands or stony mountain slopes, but primarily in the grasslands of Eurasia, at altitudes ranging from a few meters to more than 5000 m above sea level (Dewey and Asay 1975; Tzvelev 1983; Yang et al. 2014). Due to their environmental adaptability, tolerance to aridity and infertile soils, resistance to pest and disease damage, and palatability, several *Agropyron* species have been used extensively as an ecological resource and for feed research (Dewey 1984).

5.3.2 *Agropyron cristatum*—The Genus Type

The typification of the genus *Agropyron* was based on *A. cristatum* (L.) Gaertn. (Tzvelev 1976), which was first described as *Triticum cristatum* (Linnaeus 1753) and, later classified as *Agropyron* by Gaertner (1770). This species (Fig. 5.3a) comprises a polyploid complex of diploid, tetraploid and hexaploid cytotypes. It is a very polymorphic species, and was subdivided on the intraspecific level to the following eight subspecies: Ssp. *cristatum* (contains diploid, tetraploid and hexaploid cytotypes); Ssp. *pectinatum* (M. Bieb.) Tzvelev (contains diploid and tetraploid cytotypes); Ssp. *sabulosum* Lavrenko; Ssp. *brandzae* (Panfu & Solacolu) Melderis; Ssp. *ponticum* (Nevski) Tzvelev;

Ssp. *sclerophyllum* Novopokr. ex Tzvelev; Ssp. *bulbosum* (Boiss.) Á. Löve; Ssp. *incanum* (Nábělek) Melderis (hexaploid cytotype). The species *A. deweyi* Á. Löve, *A. incanum* (Nábělek) Tzvelev, *A. imbricatum* Roem. & Schult., and *A. bulbosum* Boiss. are considered synonyms of *A. cristatum*.

A. cristatum is facultative allogamous (Cabi 2010). It is native to Europe and Asia, growing from Portugal in the west to China in the east. It was introduced from Russia and Siberia to North America in the first half of the twentieth century, where it was often used as forage and in erosion control. However, currently it is considered a weed in the USA and Canada. *Agropyron cristatum* is the most widely distributed species of *Agropyron*, it exhibits significant inter- and intra-population variation in maturity time, height, texture, rhizome development, fertility, and seed size. It is a xerophytic species, which probably originated from central Asia, and is indigenous to this area, including parts of the former USSR, China, Afghanistan, Turkey, and Iran (Dewey and Asay 1975; Tzvelev 1976; Cabi 2010). It grows in a variety of steppes and steppe-like habitats. In its native range, it is frequently found on carbonate slopes in the forest steppe belt, on dry terraces, and in steppe woodlands (Cabi 2010). It grows from 1500 to about 2200 m above sea level and prefers well-drained, deep, loamy soils. It tolerates frost, drought, and salinity and prefers moderately alkaline conditions (Cabi 2010). It is best adapted to areas with poor precipitation (200–400 mm annual rainfall).

The broad pectinate spiked *A. cristatum* (L.) Gaertner contains three cytotypes (2n = 14, 28 and 42 chromosomes) (Araratian 1938); Dewey 1982, 1983; Dewey and Asay, 1982; Yang et al. 2014), all three of which occur in Iran (Dewey and Asay (1975). Tetraploids are the most common, exhibit high morphological variation, and are found throughout the entire distribution area. Hexaploid populations occur only in the Azerbaijan province in northwestern Iran. The diploid cytotype is rare in northwestern Iran but is known from Europe and other regions. The polyploid races behave cytologically as autoploids. Heterozygous chromosome interchanges are common in the tetraploids, and aneuploidy is uncommon.

Tetraploid *A. cristatum* has been used in wheat breeding for many years, has been hybridized and chromosomal addition, substitution and translocation lines were formed (Li and Dong 1991; Jensen and Bickford 1992; Yang et al. 2010).

5.3.3 Cytology, Cytogenetics and Evolution

Agropyron cristatum and *A. mongolicum* are the only diploid taxa in the genus *Agropyron* s. str. Both are cross-pollinating species, but differ morphologically: *A. cristatum* (an Eurasian species) has broad pectinate spikes, whereas *A.*

Table 5.3 Species of *Agropyron senso stricto*, their synonyms, common name, ploidy level, genome formula and geographic distribution

Species	Synonyms	Common name	Ploidy level	Genome	Geographical distribution
cristatum (L.) Gaertn	*Eremopyrum cristatum* (L.) Willk.; *Bromus cristatus* L.; *Bromus distichus* Georgi; *Triticum pumilum* L	Crested wheatgrass	2x, 4x, 6x	PP, PPPP, PPPPPP	Eurasia from Spain to Korea, North Africa; naturalized in western and central North America (USA, Canada, northern Mexico)
mongolicum Keng	–	Mongolian wheatgrass	2x	PP	Mongolia and China
fragile (Roth) P. Candargy	*Agropyron sibiricum* (Willd.) Beauv.; *Triticum fragile* Roth; *T. sibiricum* Willd	Siberian wheatgrass	2x, 4x	PP, PPPP	Caucasus to Mongolia; naturalized in western United States and Canada
desertorum (Fisch. ex Link) Schultes	*Agropyron cristatum* ssp. *desertorum* (Fisch. ex Link) Á. Löve; *Agropyron cristatum* var. *desertorum* (Fisch. ex Link) Dorn	Desert wheatgrass	4x	PPPP	Crimea, Southeast Russia, Caucasus, Central Asia
cimmericum Nevski	*Agropyron dasyanthum* ssp. *birjutczense* (Lavr.) Lavr	Kerch Wheatgrass	–	–	Southeast Ukraine, Crimea, South Russia
dasyanthum Ledeb	*Triticum dasyanthum* (Ledeb.) Sprengel	–	4x	PPPP	Ukraine, Russia
tanaiticum Nevski	–	–	4x	PPPP	East Ukraine and Southeast Russia
michnoi Roshevitz	–	–	4x	PPPP	Mongolia
thomsonii Hook	*Elymus nayarii* Karthik.: *Elymus thomsonii* (Hook.) Melderis	–	–	–	Western Himalayas
krylovianum Schischk. ex Krylov	*Elytrigia kryloviana* (Schischk.) Nevski	–	–	–	Central Asia, Siberia
pumilum (Steud.) P. Candargy	*Triticum pumilum* Steud. *Agropyron cristatum* subsp. *pumilum* (Steud.) Á. Löve	–	–	–	Siberia
pectinatum (Labill) Beauv	*Australopyrum pectinum* (Labill.) Á. Löve; *Agropyron bowne* (Kunth) Tzvelev	–	2x	WW	Australia, New Zealand

mongolicum (an East Asian species) has narrow linear ones (Hsiao et al 1986). Schulz-Schaeffer et al. (1963), McCoy and Law (1965), Taylor and McCoy (1973), Endo and Gill (1984) Hsiao et al. (1986) and Yang et al. (2014) described the karyotype of A. *cristatum*; that of A. *mongolicum* was reported by Hsiao et al. (1986). The chromosomes of these two diploids are all metacentric or sub-metacentric, relatively large and with a symmetric karyotype (Hsiao et al. 1986; Yang et al. 2014). The karyotypes are very similar and differ slightly in the centromere positions of chromosomes 5 and 7. Despite differences in plant morphology, the two species hybridize readily (Dewey and Hsiao 1984). The F_1 hybrids showed reasonably good chromosome pairing at meiosis, with an average of five to six bivalents per cell. They probably differ only in minor structural rearrangements of certain chromosomes (Hsiao et al. 1986).

Yang et al. (2014) studied the karyotype of six *Agropyron cristatum* populations distributed from Northern Europe (Sweden) to Southwest Asia (Iran). The European (Swedish and Bulgarian) populations were diploids, the two populations from the Russian Federation and one from Iran were tetraploids, while the second Iranian population was hexaploid. Differences in the centromere position in a number of chromosomes and in the relative length of the longest chromosome indicated the existence of karyological variation among these populations (Knowles 1955; Yang et al. 2014). Satellites were not observed in all populations (Yang et al. 2014). The karyotypes of the two Iranian populations were different from those reported by Hsiao et al. (1986) and Hsiao et al. (1989), who showed that the diploid cytotype had two small satellites on the fourth and the sixth chromosomes. The minute satellites appear as small dots visible only at early metaphase. The occurrence of minute satellites on two chromosome pairs of *A. cristatum*, was also reported by Knowles (1955), McCoy and Law (1968), Watson and Dallwitz (1992) and Endo and Gill (1984). The results reported by Yang et al. (2014) supported the relationship between distribution and ploidy levels (Dewey 1984; Dewey and Asay 1975; Yen and Yang 2006), with diploids distributed in small and scattered areas, tetraploids showing a universal distribution, and hexaploids distributed narrowly in Northeastern Turkey and Northwest Iran.

The karyotype of tetraploid *A. cristatum* was compared to that of a colchicine-induced autotetraploid of diploid *A. cristatum*. The idiograms of the two tetraploid taxa were strikingly similar, suggesting that the tetraploid cytotype evolved through autopolyploidy (Taylor and McCoy 1973).

Peto (1930) determined the chromosome number of *A. desertorum* as 2n = 4x = 28. Knowles (1955) determined chromosome numbers in *A. desertorum*, *A. sibiricum*, *A. fragile*, and *A. michnoi*, all of which are tetraploids (2n = 4x = 28). Although several works reported no satellite chromosomes in *A. desertorum* (Sarkar 1956; Schultz-Schaeffer and Jurasits 1962), McCoy and Law (1965) reported the existence of four to six such chromosomes in a number of clones of *A. desertotum*.

Endo and Gill (1984) used the acetocarmine-Giemsa C-banding technique to study heterochromatin distribution in somatic chromosomes of two diploid *Agropyron* taxa, *A. cristatum*, and *A. imbricatum* (a synonym of *A. cristatum*). While most cross-pollinating Triticeae species show large terminal C-bands and a high level of C-band polymorphism, *A. cristatum* exhibited only small to medium terminal bands in most of the chromosomes with low C-band polymorphism. Both *A. cristatum* and *A. imbricatum* showed gross similarity in C-banding patterns, although small differences were discernible. This confirms Dewey's (1983) assumption that *A. imbricatum* may carry a genome similar to the P genome of *A. cristatum* (Endo and Gill 1984).

Yousofi and Aryavand (2004) used flow cytometry to determine the ploidy levels of six different populations of *A. cristatum* in Iran. The mean nuclear 2C DNA content ranged from 26.41 to 27.56 pg for two varieties of *A. cristatum* ssp. *pectinatum* (five populations), and 43.47 pg for *Agropyron cristatum* ssp. *incanum* (one population). These results were supported by chromosome counting; chromosome number in the tetraploid populations varied from 28 to 31, and in the hexaploid population from 35 to 44. The frequency of aneuploidy was lower (3–4%) in tetraploids and much higher (about 18.9%) in the hexaploid population (Yousofi and Aryavand 2004).

Mean 2C DNA content was 26.26 pg for three tetraploid *A. cristatum* populations and 27.50 pg for the other two tetraploid populations of this species, a difference of 1.24 pg. Small differences in DNA content at the intraspecific level may be due to the presence or absence of accessory chromosomes (B-chromosome) (Vogel et al. 1999), or due to the aneuploidy observed within these populations. Yousofi and Aryavand (2004) argued that the differences in DNA content between the sub-specific taxa exceeds the probable DNA content of accessory chromosomes, since the average 2C DNA content of an *Agropyron* chromosome has previously been reported to be about 1 pg (Vogel et al. 1999).

Vogel et al (1999) determined the mean DNA content of three diploid accessions of *A. cristatum*, but did not analyze the tetraploid and hexaploid cytotypes. The data of Vogel et al. (1999) and those of Yousofi and Aryavand (2004) show that compared to other Triticeae species, the size of the haplome genome of *A. cristatum*, the P genome, and probably of the other *Agropyron* species, is intermediate to small.

Fig. 5.3 Species of the *Agropyron-Eremopyrum-Henrardia* clade: **a** A spike of *Agropyron cristatum* (L.) Gaertn.; **b** A plant and a spike of *Eremopyrum* distans (K. Koch) Nevski; **c** A plant and a spike of diploid *Eremopyrum buonapartis* (Spreng.) Nevski

Dewey and Asay (1982) hybridized three morphologically distinct taxa of diploid *A. cristatum*. Mean chromosome pairing at meiotic first metaphase of the three F_1 hybrid combinations included a range of 1.38–2.25 univalents, 5.05–5.83 bivalents, 0.03–0.52 trivalents, and 0.005–0.18 quadrivalents. The pairing data indicated that the three diploids contain the same basic genome, which differ by structural rearrangements of some chromosomes. The moderately high sterility in the F_1 hybrids serves as a genetic barrier but does not preclude gene flow among the diploids. Hence, the diploid taxa were identified as three different subspecies of *A. cristatum* (Dewey and Asay 1982).

Agropyron mongolicum was hybridized with the diploid cytotype of *A. cristatum* (Hsiao et al. 1989; Chen et al. 1992b). Chromosome pairing at first meiotic metaphase in the F_1 hybrid averaged 1.40 univalents, 5.59 bivalents, 0.35 trivalents, and 0.09 quadrivalents per cell (Hsiao et al. 1989) and 0.22 (0–2) univalents, and 0.79 (0–3) rod- and 6.10 (4–7) ring-bivalents (Chen et al. 1992b). The F_1 hybrids were partially fertile. The presence of seven bivalents in many pollen mother cells (PMCs) of the F_1 hybrid *A. cristatum* x *A. mongolicum* indicated that the two diploid species contain the same basic P genome. However, the occurrence of multivalents revealed that the genomes of these two diploids differ by a reciprocal translocation(s). These two diploids are the likely source of morphological and cytological variation in the tetraploid species of *Agropyron* (Hsiao et al. 1989). In accord with this view, Mellish et al. (2002), using AFLP markers, concluded that *A. desertorum* is an allopolyploid of *A. cristatum* and *A. mongolicum*.

Assadi (1995) analyzed chromosome numbers and meiotic behavior in *A. cristatum* ssp. *incanum* (2n = 4x = 42). The existence of multivalents at the first meiotic metaphase of this subspecies, which averaged 2.73 quadrivalents and 0.64 hexavalents per cell, indicated that this taxon is an autohexaploid (Assadi 1995).

Hybrids between the diploid cytotype of *A. cristatum* and tetraploid *A. desertorum* showed a high frequency of trivalents at the first meiotic metaphase (Knowles 1955), showing considerable homology between the *cristatum* and *desertorum* genomes, thus indicating an autoploid origin of *A. desertorum* (Knowles 1955). All crosses between the tetraploid species *A. sibiricum*, *A. fragile*, and *A. michnoi* with *A. desertorum* produced fertile hybrids (Knowles 1955), implying phylogenetic closeness and likely autoploid origin of these tetraploid species. Myers and Hill (1940) observed an average quadrivalent frequency of 3.8 (3.4–4.5) per PMC of tetraploid *A. cristatum*, suggesting the autotetraploid derivation of this cytotype.

Artificial crosses between diploid and hexaploid cytotypes of *A. cristatum* were not successful (Dewey 1969), but those between tetraploid *A. desertorum* and hexaploid *A. cristatum* produced viable and highly fertile seeds.

Dewey and Pendse (1968) crossed *Agropyron desertorum* and an induced-tetraploid derived from diploid *A. cristatum*. The *A. desertorum* used had 2n = 4x = 31 (the three extra *A. desertorum* chromosomes were believed to be B chromosomes), and the induced-tetraploid *A. cristatum* had 2n = 4x = 28. From a cytological aspect, the parents behaved as autoploids. Chromosome pairing at diakinesis of the 28-chromosome F_1 hybrids included an average of 0.02 univalents, 8.54 bivalents, 0.02 trivalents, 2.25 quadrivalents, 0.22 hexavalents, and 0.06 octavalents. Hexavalent and octavalent associations at diakinesis and bridge-fragment formations at first and second anaphase signified structural heterozygosity between the *A. cristatum* and *A. desertorum* genomes. However, the F_1 hybrids were fertile.

Dewey (1969) crossed doubled-diploid *Agropyron cristatum* and tetraploid *A. desertorum* with hexaploid *A. cristatum*. Meiosis in the parent plants was typical of that in autoploids, and the 35-chromosome F_1 hybrids exhibited pentavalent associations, with up to five per cell. Occasional higher multivalent associations and bridge-fragment formations at first anaphase indicated the existence of some structural heterozygosity. The pentaploid hybrids were surprisingly fertile. The high pentavalent pairing indicated close homology between the parental genomes. These results and others led Dewey (1969) to assume that all crested wheatgrasses, whether diploid, tetraploid, or hexaploid, contain one basic genome that has undergone some structural rearrangements including both translocations and inversions. Hence, autopolyploidy has played an important evolutionary role in the evolution of this genus.

Likewise, Knowles (1955) suggested that *Agropyron desertorum* is an autopolyploid of diploid *A. cristatum*. Sarkar (1956) concluded that evolution occurred primarily through autopolyploidy, followed by structural and genic changes in the chromosomes. However, since the degree of morphological variation among the tetraploid species implied the contribution of more than one genome, Sarkar (1956) suggested that segmental allopolyploidy must have been involved in the evolution of these species. Likewise, Schulz-Schaeffer et al. (1963) suggested segmental allopolyploidy in evolution of the tetraploid species, while the hexaploid cytotype of *A. cristatum* originated from autopolyploidy only. They suggested the possibility that *A. desertorum* is an allopolyploid involving diploid *A. cristatum* and an unknown diploid. The karyotype analysis performed by McCoy and Law (1965) supported the assumption of a segmental alloploidic nature of the tetraploids.

In line with these works, the chromatographic study of Lorenz and Schulz-Schaeffer (1964) showed that the tetraploid species *A. desertorum*, *A. pectinatum* (currently *A. cristatum*) and *A. sibiricum* contained more phenolic compounds than the diploid *Agropyron* species. They concluded that doubling of a single genome would not lead to proliferation of phenolic compounds, and, consequently, the tetraploids may have been derived through allopolyploidy, with the second diploid parent still unidentified.

The phenolic profile of the tetraploid species also brought Taylor and McCoy (1973) to conclude that while tetraploid *Agropyron cristatum* ssp. *pectinatum* (formerly *A. pectiniforme*) is a natural autopolyploid, another tetraploid subspecies of *A. cristatum*, namely, ssp. *imbricatum*, as well as *A. desertorum*, *A. fragile*, and *A. sibiricum* are segmental allopolyploids that derived from hybridization of different diploid subspecies of *A. cristatum*. To further support their conclusion, Taylor and McCoy (1973), produced a colchicine-induced autopolyploid from two different clones of *A. cristatum*, and confirmed that autopolyploidy does not, in itself, result in the production of phenolic compounds absent in the diploid progenitor.

While segmental allopolyploidy that resulted from intergeneric hybridizations and introgressions, e.g., from tetraploid *Elymus* species, can not be ruled out, the possibility exists, as suggested by Dewey (1969), that the tetraploids originated through autoploidy of different diploids of *A. cristatum* and *A. mongolicum*, that later hybridized with each other.

Yousofi and Aryavand (2004) found a genome size of 26.4 and 27.6 pg 2C DNA in different tetraploid lines of *A. cristatum*, respectively 7% and 3% less than the additive amount of the diploid ($14.2 \times 2 = 28.4$) (Vogel et al. 1999). These reductions in DNA content in the tetraploids, as well as structural chromosomal rearrangements, may have led to reduced multivalent formation and increased bivalent pairing and consequently, to a disomic mode of inheritance (Eilam et al. 2009). Autotetrapoids contain duplications of most of their gene loci. While the activity of most duplicated genes might be of adaptive value, the activity of some of the duplicated genes may lead to overproduction of proteins and other chemical compounds and consequently, to a disadvantageous or even deleterious effect (Birchler and Veitia 2007). Natural selection will favor changes leading to sub-functionalization or neo-functionalization in these loci. Sub-functionalization may occur when an ancestral gene with two functions becomes duplicated and each of the duplicated genes specializes in one of the ancestral gene functions, while neo-functionalization describes gain of a new, nonancestral function in a duplicated locus. Neo-functionalization of duplicated genes in autotetraploids may generate the formation of new gene products, such as new phenolic compounds that are not present in the diploid progenitor. Actually, new phenotypes often arise with polyploid formation and can contribute to the success of polyploids (Osborn et al. 2003, and reference therein).

Chromosome numbers in the tetraploid ssp. *pectinatum* of *A. cristatum* vary from 28 to 33 (Assadi 1995). Different PMCs within the same anther can display chromosome counts ranging from 28 to 32 in one plant and from 32 to (rarely) 33 in another plant. Aneuploid chromosome numbers have been reported in various tetraploid collections of *A. cristatum* (Myers and Hill 1940; Dewey and Asay 1975), a phenomenon that may have derived from cytologically unstable pentaploid hybrids between tetraploid and hexaploid cytotypes of *A. cristatum* (Dewey 1974). Hence, the plants that had the somatic chromosome numbers 2n = 32 and 33, were probably a derivative of such unstable pentaploid hybrids (Assadi 1995). The variable chromosome number in different PMCs may be caused by elimination of chromosomes in archesporial division or at an early stage of the meiotic cycle. Alternatively, the extra chromosomes may be B chromosomes, which were frequently observed in PMCs of *A. cristatum* (Knowles 1955; McCoy and Law 1965; Assadi 1995; Asghari et al. 2007). Baenziger (1962) found no B chromosomes in adventitious root-tips of diploid *A. cristatum* but reported the presence of these chromosomes in stem meristems, in primary roots, and in PMCs but they were absent in adventitious roots of the tetraploid *A. desertorum* and diploid *A. cristatum*. The B chromosomes are usually smaller than the basic (A) chromosomes, not heterochromatic, and show sub-terminal centromeres. At meiosis, there is good pairing between the B chromosomes but not between B and A chromosomes (Baenziger 1962). A B chromosome with a sub-terminal constriction, which is either a centromere or a secondary constriction, was also observed in *A. desertorum* (McCoy and Law 1965). The group reported a mitotic chromosome count of 2n = 28, whereas the meiotic chromosome count was 2n = 32. Evidently, as noted by Knowles (1955), the mitotic chromosome number may not agree with the meiotic number.

5.3.4 Crosses with Other Triticineae Species

Studies of meiotic chromosomal pairing in F_1 hybrids between diploid *A. cristatum* and several different diploid species *of Elymus* bearing genome StSt, (genome of all hybrids was PSt) showed that the P and St genomes are related (Wang 1985a, b, 1986, 1992). Size differences between *Agropyron* (large) and *Elymus* (small) chromosomes facilitated interpretation of chromosome pairing in these hybrids. Average chromosome pairing at first meiotic metaphase of the diploid hybrid between *Agropyron cristatum* and *Elymus stipifollus* included 7.65 univalents, 2.88 rod- and 0.21 ring-bivalents (total 3.09 bivalents), 0.04 trivalents and 0.01 quadrivalents (Wang 1985b). These pairing configurations indicate allosyndetic pairing between the homoeologous chromosomes of the two genomes, showing that the St and the P genomes are related. Later, Wang (1986) produced hybrids between diploi*d Agropyron* *cristatum* and *Elymus libanoticus*. Chromosomal pairing at first meiotic metaphase of the F_1 hybrid (genome PSt) averaged 7.71 univalents, 2.77 bivalents, 0.22 trivalents, 0.01 quadrivalents, and 0.01 pentavalents per cell. As with the hybrid of *A. cristatum* x *E. stipifolius* (Wang 1985b), chromosome pairing in the hybrid between diploid *A. cristatum* and *E. libanoticus* was mainly between the P genome (large) and the S-genome (small) chromosomes, i.e., allosyndetic. Meiotic pairing also in this hybrid suggests that the P and the St genomes are related.

Wang (1992) produced hybrids between diploid species of *Agropyron* (genome P), *Elymus* (genomes St, Ee and Eb), *Psathyrostachys* (genome Ns), *Hordeum* (genome H), and *Secale* (genome R). Chromosome pairing patterns in these diploid hybrids enabled the estimation of genomic similarity between the various genomes. The results showed that Ee of *Elymus elongatus* and Eb of *E. farctus* are the most closely related genomes, followed by the St and P genomes. The N, H, and R genomes are remotely related to the E-St-P cluster. These relationships are also reflected in hybrids of higher ploidy levels, when genes controlling chromosome pairing are kept in check (Wang 1992). Similarly, the average meiotic chromosome pairing in the intergeneric diploid hybrid *E. spicatus* (genome StSt) x *Secale strictum (formerly montanum)* (genome RR) included 12.97 univalents, 0.49 bivalents, and 0.01 trivalents (Wang 1987b). The hybrid *A. mongolicum* x *S. strictum*, which have the PR genomes, had an average of 12.86 univalents, 0.51 bivalents, 0.03 trivalents, and 0.004 quadrivalents. The hybrid between *E. spicatus* and *A. mongolicum* (genome StP) had a mean configuration of 8.05 univalents, 2.86 bivalents, 0.07 trivalents, and 0.01 quadrivalents. All hybrids were sterile. The meiotic pairings of these hybrids indicated that chromosome homology between the St and P genomes is higher than between both St and R or between P and R. The degree of meiotic pairing in the *E. spicatus* x *A. mongoicum* hybrid was similar to that observed in other diploid hybrids bearing the same genome constitution, i.e., *A. cristatum* x *E. stipifolius* and *A. cristatum* x *E. libanoticus* (Wang et al. 1985; Wang 1986). Interestingly, mitotic preparations of root-tip cells of these hybrids suggested that the chromosomes of different genomes were spatially separated (Wang 1987b). As was found in other plant species (Finch et al. 1981; Avivi et al. 1982), the separated genome distribution in the *A. cristatum* x *Elymus* species in the majority (50–67%) of root-tip cells, suggested that each genome in these hybrids occupy a different part of the nucleus.

Monoploids and hybrids were obtained from the cross of diploid *Elymus elongatus* (genome EeEe) and *A. mongolicum*. The monoploid was a result of gradual and eventually complete elimination of *A. mongolicum* chromosomes in the hybrid. About 95% of the root-tip cells, and nearly all of the pollen mother cells, had only seven chromosomes

(Wang 1987a). The genome in the monoploid cells was identified as Ee, by its characteristic satellited chromosomes (Wang 1987a). This was the first report of chromosome elimination following intergeneric hybridization in the Triticeae that did not involve species of *Hordeum* or *Critesion*. The monoploid plant had only a few root-tip cells that contained as many as seven additional chromosomes, whereas none of the PMCs had more than eight chromosomes. These observations indicated that chromosome elimination commenced some time after zygote formation and was nearly complete in the PMCs. Chromosome elimination in the hybrids between tetraploid *H. vulgare* and *H. bulbosum* begins at maximum, 3–5 days after pollination and was frequently complete 9 days after pollination (Fukuyama and Hosoya 1983).

Very little autosyndesis between chromosomes within the Ee genome occurred in the monoploid. On the other hand, extensive chromosome pairing was observed at first meiotic metaphase of the F_1 hybrid *E. elongatus* x *A. mongolicum* (genome EeP), averaging 6.42 univalents, 2.53 rod- and 0.85 ring-bivalents, 0.25 trivalents, and 0.02 quadrivalents. Bridges and fragments were present in many first anaphase cells. The hybrid was sterile and had non-dehiscent anthers. This pairing profile revealed a degree of chromosome homology between Ee and P, indicating a close phylogenetic relationship between these two species. The amount of pairing between the Ee and the P genomes, especially the occurrence of three ring bivalents in some cells, suggests a close relation between the two. Clark et al. (1986), who studied the spacer region of rDNA units, also found a close relationship between Ee and P genomes. However, the number of univalents in this intergeneric hybrid (6.42) exceeds that (2.89) observed in the interspecific hybrids of diploid *E. farctus* (genome Eb) and diploid *E. elongatus* (genome Ee) (Wang 1985b). Despite the capacity of the Ee and P genome chromosomes to pair, they have differentiated to a degree that the two genomes have different chromosome lengths and karyotypic patterns (Hsiao et al. 1986). These karyotypic differences, as well as morphological differences between *E. elongatus* and *A. mongolicum*, justify their classification into separate genera (Wang 1987a).

Wang (1985b) crossed tetraploid *A. desertorum* (genome PPPP) x tetraploid *E. stipifolius* (genome StStStSt) and found that most F_1 hybrid (genome PPStSt) pairing was autosyndetic, namely, pairing between chromosomes of the St genome and pairing between those of the P genome. Average pairing configurations included 4.48 univalents, 5.79 rod- and 5.07 ring-bivalents (total 10.86 bivalents), 0.53 trivalents and 0.05 quadrivalents (Wang 1985b). These data imply that the two parental tetraploid species are autoploids, that several chromosomes in each genome underwent some structural rearrangements, and that the P and St chromosomes tended to pair with their homologues (autosyndesis) rather than with their homoeologues (allosyndesis). Hence, chromosomal pairing in such hybrids cannot disclose the degree of relatedness between the P and St genomes.

Assadi and Runemark (1995) crossed *E. libanoticus*) (genome StSt) with a tetraploid cytotype of *A. cristatum, namely,* ssp. *pectinatum* (genome PPPP). Average meiotic configurations in the F_1 hybrid (genome StPP) displayed: 11.30 univalents, 3.40 rod- and 1.50 ring-bivalents (4.90 total bivalents). The hybrid was sterile. Most chromosomal pairing, if not all, was autosyndetic between P chromosomes. The preferential pairing between the homologous P chromosomes precluded the assessment of the relationships between genomes P and St.

Dewey (1963a) reported that meiosis in plants of *Elymus hispidus* (formerly *Agropyron trichophorum*) (2n = 6x = 42; EeEeEeEeStSt] was basically regular; average chromosomal associations at the first meiotic metaphase showed 0.09 univalents, 20.56 bivalents, 0.05 trivalents, and 0.16 quadrivalents, and was therefore described as an auto-allohexaploid. Dewey (1963a) analyzed chromosomal pairing at first meiotic metaphase of F_1 hybrids between *Elymus hispidus* and *A. desertorum* (genome PPPP) and between *E. hispidus* and hexaploid cytotype of *A. cristatum* (genome PPPPPP). The hexaploid *A. cristatum* parent averaged 0.18 univalents, 7.44 bivalents, 0.81 trivalents 2.86 quadrivalents, 0.08 pentavalents, and 2.11 hexavalents at diakinesis and consequently, was described as an autohexaploid (Dewey 1963b). Chromosome pairing at the fist meiotic metaphase of the F_1 hexaploid hybrid (genome EeEeStPPP) presented 5.08 univalents, 8.94 bivalents, 4.33 trivalents, 1.11 quadrivalents, 0.27 pentavalents, and 0.05 hexavalents per cell. On the basis of chromosome pairing in the parent species and their hybrids, it was concluded that one of the *E. hispidu* genomes was partially homologous with the P genomes of hexaploid *A. cristatum* and tetraploid *A. desertorum.*

Martín et al. (1999) reported the production of the amphiploid *Aegilops tauschii–Agropyron cristatum*, obtained by crossing an induced autotetraploid of *Ae. tauschii* (genome DDDD) with tetraploid *A. cristatum* (genome PPPP). They used multicolor fluorescence in situ hybridization (FISH), using total genomic DNA probes, to distinguish between the chromosomes of *Ae. tauschii* and those of *A. cristatum* at meiosis of the amphiploid. Analysis of chromosomal pairing at first meiotic metaphase of the amphiploid (genome DDPP) showed the presence of multivalents (trivalents, quadrivalents and pentavalents), which were of *A. cristatum* origin. Moreover, pairing between *Ae. tauschii* chromosomes was higher than between *A. cristatum* chromosomes. The high frequency of multivalents, presumably due to translocations, plus the reduced pairing of the *A. cristatum* chromosomes, indicated that the two P

genomes of the latter underwent some structural changes. FISH analysis also showed a rare event of pairing between *Ae. tauschii* and *A. cristatum* chromosomes. The presence of homologous chromosomes (DD and PP) competed with the homoeologous pairing between D and P and thus, the low level of homoeologous pairing between these two genomes is not indicative of the relationships between these two genomes. In the absence of homologous chromosomes, this pairing could be presumably higher.

Martin et al. (1998) studied chromosome pairing at meiosis between the diploid cytotype of *A. cristatum* (genome PP), *Aegilops tauschii* (genome DD) and *Hordeum chilense* (genome $H^{ch}H^{ch}$) in the trigeneric hybrid *Ae. tauschii-A. cristatum* x *H. chilense*. Since this trigeneric hybrid (genome DPH^{ch}) had a single dose of each genome and thus, lacks homologous chromosomes, analysis of its pairing pattern can reveal the level of affinity between the genomes of these species. Using FISH, the pairing of these hybrids at first meiotic metaphase showed higher pairing between the D and H^{ch} genomes than between each of them and the P genome.

Hybridization of *Agropyron* species with *Triticum* species has been difficult. White (1940) was unsuccessful in producing hybrids between *Triticum* and diploid *A. cristatum*. Smith (1942a, b) reported one hybrid plant produced from 882 *Triticum aestivum* florets pollinated with *A. cristatum*, and Mujeeb-Kazi et al. (1987) reported that no viable embryos were obtained from a cross of *T. aestivum* with tetraploid *Agropyron cristatum*. Chen et al. (1989) succeeded to produce two hybrid plants from the pollination of 952 *T. aestivum* florets with diploid *A. cristatum* pollen, but in both instances, the hybrid plants died before reaching maturity. Nevertheless, Chen et al. (1989, 1990), Li and Dong 1990, 1991), and Ahmad and Comeau (1991) recently succeeded to obtain hybrid plants between *T. aestivum* and various tetraploid *Agropyron* species, namely, *A. desertorum*, *A. michnoi*, and *A. fragile*. Studies of chromosomal pairing at first meiotic metaphase of these F_1 pentaploid hybrids showed higher levels of meiotic pairing than expected. But, the genomic relationships of the P genome of *Agropyron* with the A, B, and D subgenomes of *T. aestivum* could not be clearly assessed due to the presence of the wheat *Ph1* homoeologous-pairing suppressor and two homologous genomes in the tetraploid *Agropyron* species.

Ahmad and Comeau (1991) analyzed chromosome pairing in ten F_1 hybrids between *T. aestivum* cv. Fukuho and *A. fragile*. Mean chromosome configurations at first meiotic metaphase of the 10 pentaploid hybrids (genome BADPP) included 17.29 univalents, 6.57 rod- and 1.97 ring-bivalents, 0.18 trivalents, 0.03 quadrivalents, and 0.002 hexavalents per PMC. However, there was a considerable intra-hybrid variation in the mean number of bivalents per PMC, ranging from 5.88 to 11.03. Since one would expect to find up to seven bivalents in the polyhaploid of *A. fragile* and up to three bivalents (Ahmad and Comeau 1991) in the polyhaploid of *T. aestivum* cv. 'Fukuho', the expected maximum bivalents in these F_1 hybrids should be 10. The higher number of bivalents in some of the hybrids, which presumably occurred between wheat chromosomes, was attributed to the *a* pairing-promoter gene(s) present in *A. fragile* (Ahmad and Comeau 1991). Allosyndetic pairing between wheat and *A. fragile* chromosomes are not expected in these hybrids, since the *A. fragile* chromosomes tend to pair preferentially with their own homologues, rather than with wheat homoeologues. Such a pairing-promoter gene system has been previously reported in tetraploid *A. cristatum* (Chen et al. 1989), as well as in other Triticeae species, e.g., *Amblyopyrum muticum*, *Aegilops speltoides*, *Secale cereale*, and *Dasypyrum villosum* (Dover and Riley 1972; Chen and Dvorak 1984; Lelley 1976; Blanco et al. 1988b; Table 5.2). Such a gene system was not found in diploid *A. cristatum* (Limin and Fowler 1990). Since the pairing promotion system in tetraploid *Agropyron* species is determined by a polygenic system (Jubault et al. 2006), and different accessions may have different alleles of the system, the use of bulk pollen from many different plants to pollinate wheat (Ahmad and Comeau 1991), might produce hybrids that differ in the allelic composition of the *Ph* suppressors.

Jauhar (1992) produced and analyzed intergeneric hybrids between *T. aestivum* and tetraploid *Agropyron cristatum*. The F_1 pentaploid hybrids (genome BADPP) were perennial like the male wheatgrass parent and morphologically intermediate between the two parents. Two types of hybrids were obtained: a low-pairing (LP) hybrid and a high-pairing (HP) hybrid. The LP hybrid, with an apparently functional *Ph1* (the suppressor of homoeologous pairing), had a mean display of 25.91 univalents, 3.17 rod- and 1.16 ring-bivalents, and 0.14 trivalents. If *A. cristatum* were a true autotetraploid, its haploid complement (PP) in the hybrid should form approximately 7 bivalents. The mean 4.33 bivalents (of which about 1.0 bivalent probably involved the A, B and D subgenomes of wheat) suggests a certain degree of divergence between the two P genomes (Jauhar 1992). The degree of divergence between chromosomes that is required for the *Ph1* suppressor to operate on is not known. It would appear, however, that the degree of similarity between the two P genomes is inadequate to pass the discrimination limits of *Ph1* (Jauhar 1992).

The HP hybrid had 15.73 univalents, 5.89 rod- and 2.98 ring-bivalents, 0.47 trivalents, and 0.03 chain quadrivalents, a pairing profile that likely involved both autosyndesis (pairing within the BAD component and within the PP component of the BADPP hybrid) and allosyndesis (pairing between the parental complements), as indicated by the frequent formation of heteromorphic bivalents and asymmetrical trivalents (Jauhar 1992). This kind of pairing could

have occurred only if *Ph1* was partially suppressed by genes of tetraploid *A. cristatum*. The existence of LP and HP plants among the F_1 pentaploid hybrids presumably results from segregation of the *Ph1* suppressors of *A. cristatum*.

Limin and Fowler (1990) reported the first successful hybridization of *T. aestivum* cv. Chinese Spring with diploid *A. cristatum*. Average chromosomal pairing per PMC at the first meiotic metaphase of the F_1 tetraploid hybrid (genome BADP) included 27.69 univalents, 0.15 rod- and 0.00 ring-bivalents, and 0.003 trivalents (0.16 chiasmata per cell) (Limin and Fowler 1990). Haploid Chinese Spring wheat had 0.24–0.27 chiasmata per cell (Miller and Chapman 1976; McGuire and Dvorak 1982). Therefore, the chiasma frequency observed in the study of Limin and Fowler (1990) could be explained on the basis of autosyndesis pairing between the wheat subgenomes. The pairing data of the tetraploid hybrid indicates that pairing-promoting gene(s) does not exist in the diploid accession of *A. cristatum* used as parent in the described cross.

Triticum aestivum cv. Chinese Spring was crossed with diploid *Agropyron* species from Inner Mongolia, *A. cristatum* and *A. mongolicum*, with or without B chromosomes, generating intergeneric F_1 tetraploid hybrids with 2n = 27, 28, 32, and 33 chromosomes (Chen et al. 1992a, b). The extra chromosomes in the hybrids with 2n = 32 and 33 were assumed to be B chromosomes. Average meiotic pairing in the euploid hybrid (2n = 4x = 28; genome BADP), derived from the cross of Chinese Spring x (*A. cristatum* x *A. mongolicum*), included 14.38 univalents, 3.56 rod- and 1.36 ring-bivalents, and 1.26 trivalents, and the maximum number of bivalents was seven. This level of pairing is higher than expected and was likely due to homoeologous pairing between wheat chromosomes. Hence, the data indicated that the P genome of diploid *Agropyron* originated from Inner Mongolia, as those of the tetraploid *Agropyron* species, possess a genetic system that suppresses the *Ph1* genes of wheat. The chromosome pairing observed in hybrids of CS x *A. cristatum/A. mongolicum* included 4.92 bivalents and 1.26 trivalents on average, and was much higher than that previously reported for the hybrid CS x diploid *A. cristatum* produced by Limin and Fowler (1990), which had an average of 27.69 univalents, 0.15 bivalents and 0.003 trivalents per cell. This difference between the hybrids of common wheat x diploid *A. cristatum* could be attributed to the use of different accessions of *A. cristatum* used in the study of Chen et al. (1992a, b) versus that of Limin and Fowler (1990).

The observed pattern of pairing in the tetraploid hybrid (genome BADP) (Chen et al. 1992a, b) was very similar to that of the pentaploid hybrid CS x tetraploid *A. cristatum* (genome BADPP), previously produced by Chen et al. (1989). Both diploid and tetraploid *A. cristatum*, used in these crosses, have genes that suppress the *Ph*-system of wheat. In the pentaploid hybrids, average chromosome pairing included 8.18 univalents, 11.88 bivalents, 0.97 trivalents, and 0.03 quadrivalents (Chen et al. 1989). However, as the tetraploid *A. cristatum* used was a true autotetraploid (Dewey 1984; Chen et al. 1992a, b), in most PMCs, seven bivalents representing autosyndesis pairing of the 14 *Agropyron* chromosomes, was observed. Thus, the meiotic behavior of wheat chromosomes shows 8.18 univalents, 4.88 bivalents, 0.97 trivalents, and 0.03 quadrivalents, corresponding to the level of pairing in a wheat haploid deficient for the *Ph* gene (7.93 univalents, 5.20 bivalents, 0.53 trivalents, and 0.02 qiadrivalents; Kimber and Riley 1963). Hence, the high level of pairing in the pentaploid may be ascribed to the suppression of the *Ph* gene of wheat (Chen et al. 1989).

In the study of Chen et al. (1992a, b), the mean chromosome pairing observed in the tetraploid hybrids dislayed 14.38 univalents, 4.92 bivalents, and 1.26 trivalents. If we assume that the seven chromosomes of the P genome of *Agropyron* did not pair with those of wheat in the hybrid, and consequently, were univalents, the meiotic behavior of wheat chromosomes in the hybrid would therefore be 7.38 univalents, 4.92 bivalents, and 1.26 trivalents. This level of pairing is very similar to that of the wheat chromosome in the pentaploid hybrid. Hence, the P genome of both diploid and tetraploid *A. cristatumn* has a gene that suppress the *Ph* gene system of wheat. The fact that the P genome of another accession of diploid *A. cristatum* did not suppress the *Ph* effect (Limin and Fowler 1990), indicates the existence of variability within diploid *Agropyron* concerning the *Ph* suppressors.

Li and Dong (1990) produced intergeneric hybrids between *T. aestivum* cv. Chinese Spring and tetraploid *Agropyron desertorum*. Average meiotic chromosome pairing at the first metaphase of the F$_1$ hybrid (genome BADPP) showed 6.62 univalents, 4.16 rod- and 8.20 ring-bivalents, 0.57 trivalents, 0.35 quadrivalents, 0.06 pentavalents and 0.03 hexavalents. The number of bivalents and multivalents in the F$_1$ hybrid was higher than the expected seven bivalents between the PP genomes of *Agropyron* and one bivalent between the wheat genomes.

Li and Dong (1991) also produced intergeneric hybrids between *T. aestivum* cv. Chinese Spring and *Agropyron michnoi*. The average meiotic chromosome pairing at the first meiotic metaphase of F$_1$ pentaploid hybrid (genome BADPP) included 6.39 univalents, 3.75 rod- and 8.64 ring-bivalents, 0.81 trivalents, 0.30 quadrivalents and 0.04 pentavalents; the bivalent and multivalent formation was much higher than expected.

Chen et al. (1989) assumed that the higher pairing in hybrids between bread wheat and tetraploid *Agropyron* species resulted from *Agropyron* genes that suppress the wheat *Ph* effect and thus lead to wheat homoeologous

pairing. Li and Dong (1990) suggested that the duplicated dosage of the P genome induced pairing between the homoeologues. Their conclusion was inspired by the report of Riley et al. (1973) who showed that an extra dose of the rye R genome increased the level of pairing in hybrids between common wheat and tetraploid rye, i.e., the mean number of wheat-chromosome bivalents increased from 0.24 to 2.25. Li and Dong suggested a similar effect of the P genome, when present in an extra dose. However, the lack of dosage effect of the P genome of *Agropyron* on suppression of the *Ph* gene of wheat is indicated from the similar level of pairing between the wheat chromosomes in the tetraploid and the pentaploid hybrids (Chen et al. 1992a, b). It appears that wheat-*Agropyron* allosyndetic associations, if any, are rare, even if the level of chromosome pairing observed in the hybrid is high.

Hybrids between disomic addition lines of *A. cristatum* chromosomes or chromosome arms to the complement of bread wheat x *Aegilops peregrina* (=*Ae. variabilis*) (2n = 4x = 28; genome S^{v}S^{v}UU), can be used for studying the effect of individual *Agropyron* chromosomes and chromosome arm on homoeologous pairing between wheat and *Ae. peregrina* chromosomes and to assess the *Ph*-suppressing effect of different P genome chromosomes. Jubault et al. (2006) used five disomic addition lines (1P, 3P, 4P, 5P and 6P) and five ditelosomic addition lines (2PS, 2PL, 4PS, 5PL and 6PS) of wheat—*A. cristatum* addition lines, produced by Chen et al. (1994, 1992a, b), in crosses with *Ae. peregrina*. Chromosome configurations in each hybrid, which had either 2n = 36 or 35 + t, were recorded, and the pairing level for each of them was compared with that of the control hybrid *T. aestivum* CS–*Ae. peregrina*. All the genotypes, except those with 2PS and 2PL chromosomes, displayed a significantly higher level of homoeologous pairing than the control. Consequently, all the P chromosomes tested, with the exception of chromosomes arms 2PS and 2PL, seemed to promote homoeologous pairing. The *A. cristatum Ph*-suppressing system appeared polygenic. However, the pairing-promoting effect of every *Agropyron* chromosome was weaker than the effect of the absence of *Ph1*.

In addition, Jubault et al. (2006) assessed the level of pairing between individual *A. cristatum* chromosomes and those of common wheat, in hybrids lacking *Ph1*. Allosyndetic pairing between P and BAD chromosomes were very rare even in the absence of *Ph1*. Only telosome 5PL paired, at a very low frequency, with wheat chromosomes (Jubault et al. (2006). Since the addition lines did not provide any evidence for structural rearrangements between the P and the A, B, D subgenomes, it is assumed that the lack of ability of P chromosomes to pair with wheat chromosomes stems from divergence of the DNA sequences that are involved in homology recognition and initiation of meiotic pairing.

5.3.5 Phylogenetic Relationships of *Agropyron* with Other Triticineae

The study of Escobar et al. (2011) showed that the clade of *Agropyron, Astralopyrum, Eremopyrum* and *Henrardia* was not affected by introgression and/or incomplete lineage sorting. The analyses of 5S DNA sequences using Wagner parsimony and NJ distance methods (Baum and Appels 1992), placed consistently *Agropyron* (genome P), *Pseudoroegneria* (currently *Elymus*; genome St) and *Australopyrum* (genome W) in one clade. *Agropyron* evolved, most probably, from *Elymus* species having the St, Ee or Eb genomes that are moderately related to genome P of *Agropyron* (Wang 1989). The close phylogenetic relationship between *Agropyron* and *Ermopyrum* is supported by the data of Escobar et al. (2011); the latter might have evolved from the former.

The internal transcribed spacer (ITS) region of nuclear ribosomal DNA sequence phylogeny indicated that the endemic Australian grasses *Australopyrum pectinatum* (genome W) are closely related to species of *Agropyron* (genome P) (Hsiao et al 1995). Species of the W and P genomes share certain gross morphological similarities and *Australopyrum* was once treated as a member of *Agropyron* (Löve 1984). The karyotypes of P and W genome species are also similar, but the chromosomes of the W genome are smaller (Hsiao et al. 1986). The differences in chromosome size could simply be due to a low copy number of the repetitive DNA, because the chromosomes of *Australopyrum pectinatum* ssp. *velutinum* contain much less C-banded heterochromatin than do those of *Agropyron cristatum* (Endo and Gill 1984).

The P genome of the allohexaploid *Elymus* species most likely derived from a diploid species of *Agropyron* (Petersen et al. 2011; Fan et al 2013; Dong et al. 2015). Refoufi et al. (2001) found that *Elymus pycnanthus* (Godr.) Melderis (=*Elytrigia pycnantha* or *Thinopyrum pycnanthum*) is a hexaploid containing genomes St, Ee, and P. Using genomic in situ hybridization (GISH) techniques, they also proposed that the P genome of *E. pycnanthus* is closely related to that of *A. cristatum*. In accord with the above, Dizkirici et al. (2010), constructing a phylogenetic tree by the maximum parsimony method, based on sequence diversity in the internal transcribed spacer (ITS) region of nuclear ribosomal DNA, revealed that *Elymus pycnanthus* clustered with species of *Agropyron*. Molecular diversity statistics also indicated that *E. pycnanthus* is close to *Agropyron* species (Dizkirici et al. 2010).

The studies of Fan et al. (2009) and Sha et al. (2010) indicated that the Xm genome of the allopolyploid Leymus species might have originated from an ancestral lineage of *Agropyron* (genome P) and *Eremopyrum triticeum* (genome F).

5.4 *Eremopyrum* (Ledeb.) Jaub. & Spach.

5.4.1 Morphological and Geographical Notes

Eremopyrum (*eremia* 'desert', and *pyros* 'wheat' in Greek) was described by Ledebour (1853) and included in *Triticum* sect. *Eremopyrum*. However, since *Eremopyrum* was morphologically similar to the P genome-bearing *Agropyron* crested wheatgrasses, Bentham and Hooker (1883) included them in *Agropyron* Gaertner. But, due to its annual habit, Jaubert and Spach (1851) distinguished this taxon from *Agropyron* as a separate genus. Since then, a number of different *Eremopyrum* species have been described. Currently, there are five universally accepted species in the genus (Gabi and Dogan 2010), namely, *E. bonaepartis* (Spreng.) Nevski, *E. confusum* Melderis, *E. distans* (C. Koch) Nevski, *E. orientale* (L.) Jaub. et Spach, and *E. triticeum* (Gaertn.) Nevski (Figs. 5.3b and c).

In her review of the taxonomy of the genus *Eremopyrum*, Frederiksen (1991b) recognized only the following four species: *E. triticeum, E. orientale, E. distans*, and *E. bonaepartis*. Due to the absence of clear-cut delimitation between three previously considered species, she included them in *E. bonaepartis*: *E. confusum*, characterized by awned glumes and lemmas, *E. bonaepartis* s. str., with sharp pointed glumes and lemmas, and *Triticum sinaicum* Steud., characterized by gradually tapering glumes and lemmas on 1–3 lower spikelets, but distinctly awned lemmas on upper spikelets. Moreover, Frederiksen (1991b) determined the chromosome number of these taxa and found that *E. confusum* and *E. bonaepartis* s. str. are tetraploids, while *Triticum sinaicum* is a diploid. In contrast, Gabi and Dogan (2010) indicated clear differences between the three taxa. More specifically, they confirmed that *E. confusum* Melderis is a valid species and consequently, like Melderis (1985b), recognized five species in *Eremopyrum*. Since tetraplois *E. bonaepartis* is an allotetraploid (Sakamoto 1979), and differs morphologically from diploid *E. bonaepartis* (Gabi and Dogan 2010), they concluded that the two taxa should be treated as separate species.

All the five species are annual, short plants (30–40 cm high), with a short, compact, laterally compressed spike, rachis with very short internodes, solitary spikelets, seated distichously at a wide angle to the rachis, and with spikes that disarticulate at maturity at each rachis node beneath each spikelet (wedge-type disarticulation), but only in *E. triticeum* is the disarticulation at the base of each floret (floret-type disarticulation). The *Eremopyrum* species feature spikelets with 2–5 bisexual florets, distal or no sterile florets, and very short anthers (0.4–1.3 mm) indicating facultative self-pollination. Based on differences in disarticulation of spike and spikeletes, Nevski (1936) divided *Eremopyrum*

Table 5.4 Sections and species of *Eremopyrum*, their disarticulation type, ploidy level, genome formula and geographic distribution

Section	Species	Disarticulation type	Ploidy level	Genome[a]	Geographical distribution
Micropuryum Nevski	*triticeum* (Gaertn.) Nevski	Floret type	2x	FF	From southeastern Europe and Turkey in the west, to China in the east
Eremopyrum Nevski	*orientale* (L.) Jaub. et Spach	Wedge type	4x	XdXdFF	From Morocco and Algeria in the west, to China in the east
	distans (C. Koch) Nevski	Wedge type	2x	XdXd	From the east Mediterranean Turkey to Afghanistan
	bonaepartis (Spreng.) Nevski	Wedge type	2x	XbXb	From Morocco in the west through the Middle East, the Arabian Peninsula, Iran and Central Asia, to Afghanistan, Pakistan, and the Chinese province Xinjiang in the east
		Wedge type	4x	XbXbXdXd	
	confusum Melderis	Wedge type	4x	XbXbXdXd[b]	West Asia

[a] Genome symbol according to Wang et al. (1995)
[b] Assumed

into two sections, *Micropuryum* Nevski (includes *E. triticeum*) and *Eremopyrum* Nevski (includes the other four species) (Table 5.4).

Eremopyrum species grow in steppes and semi-desert regions, from the Balkan, through the East Mediterranean to Asia (Balkan, Turkey, Syria, Jordan, Israel, Sinai Peninsula, Caucasia, Turkmenistan, Iraq, Saudi Arabia, Iran, Afghanistan, Pakistan, and China) (Bor 1968, 1970; Davis et al. 1988). In their native ranges, they serve as valuable fodder on ephemeral spring pastures. *E. triticeum, E. bonaepartis,* and *E. orientale* have been found in North America.

The distribution of *E. triticeum* extends from southeastern Europe and Turkey in the west, to China in the east. The species appears to be widespread in the northern region of the genus distribution area. *E. distans* is an Asiatic species, widely distributed from the East Mediterranean and Eastern Turkey to Afghanistan in the east. *E. orientale* is a widespread species that distributes from Morocco and Algeria in the west, to China in the east. *E. bonaepartis* is the most common of the species, and widely distributed, growing from Morocco in the west through the Middle East, the Sinai Peninsula, the Arabian Peninsula, Iran and Central Asia, to Afghanistan, Pakistan, and the Chinese province Xinjiang in the east. It is a variable species that has been divided into several subspecies (Table 5.4).

5.4.2 Cytology, Cytogenetics and Evolution

Studies of somatic chromosome number showed that *Eremopyrum* includes both diploid and tetraploid taxa; *E. triticeum* and *E. distans,* are diploids (2n = 2x = 14), *E. ponaepartis* contains diploid and tetraploid cytotypes, and *E. confusum* and *E. orientale* are tetraploids (2n = 4x = 28)

(Avdulov 1931; Sakamoto and Muramatsu 1965; Frederiksen 1991b). Interspecific *Eremopyrum* hybrids were produced and studies of their chromosome pairing at meiosis showed: (1) very little pairing between the diploid species, indicating that their genomes are remarkably diverged from one another; (2) that the tetraploid species are allotetraploids, each containing two different genomes; tetraploid *E. bonaepartis* contains one genome of diploid *E. bonaepartis* and a second genome of *E. distans*, while tetraploid *E. orientale* contains one genome of *E. triticeum* and a second genome of *E. distans* (Sakamoto 1972). Since tetraploid *E. bonaepartis* is allotetraploid, the two ploidy types of *E. bonaepartis* should be separated into two different species. Based on these findings, Sakamoto (1979) classified the genomes of the *Eremopyrum* species as follows: diploid *E. bonaepartis* (genome AA), *E. distans* (genome BB), *E. triticeum* (genome CC), tetraploid *E. bonaepartis* (genome AABB) and tetraploid *E. orientale* (genome BBCC). Since genomic symbols A, B, and C, were previously given to *Triticum* and *Aegilops* species, Dewey (1984) changed the genome symbol of *Eremopyrum* species given by Sakamoto (1979). Later, Wang et al. (1995) suggested the presence of two different genomes in *Eremopyrum*, F and X. Following the genome analysis performed by Sakamoto (1979), the genome symbol of the *Eremopyrum* species are as follows: *E. triticeum* FF, diploid *E. bonaepartis* XbXb, tetraploid *E. bonaepartis* XbXbXdXd, *E. distans* XdXd, *E. orientale* XdXdFF (Table 5.4). The genome of *E. confusum* was not determined, but it is probably similar to that of tetraploid *E. bonaepartis*, i.e., XbXbXdXd.

The diploid species *E. triticeum* and *E. distans*, are more easily distinguishable than the other species. The allotetraploid species often grow in mixed populations with their diploid progenitors and exhibit wider morphological variation

(Sakamoto 1979). The three allotetraploid species are of recent origin, and most probably arose in the dry steppe zones of northwestern Iran, the assumed distribution center of this genus (Sakamoto 1979).

Intergeneric hybridizations showed that there are strong sterility barriers between *Eremopyrum* species and those of other Triticeae genera (Sakamoto 1967, 1968, 1972, 1974; Frederiksen 1991b, 1993, 1994; Frederiksen and von Bothmer 1995). Sakamoto (1974) succeeded in producing *Heteranthelium piliferum* x diploid *Eremopyrum bonaepartis* and *E. bonaepartis* x tetraploid *Hordeum depressum* hybrids. The diploid *H. piliferum* x *E. bonaepartis* hybrid exhibited abnormal growth and very little chromosomal pairing at first meiotic metaphase (average of 13.93 univalents, and 0.04 bivalents per cell). Growth of the triploid hybrid *E. bonaepartis* x *H. depressum* was highly vigorous and chromosomal pairing (averaged 9.97 univalents, 5.50 bivalents, 0.01 trivalents, and 0.00 quadrivalents) resulted mainly from autosyndesis of chromosomes derived from the autotetraploid *Hordeum* parent. From the very little pairing between the chromosomes of *E. bonaepartis* and the other two species, Sakamoto (1974) concluded that there is no homology among the genomes of the three species.

5.4.3 Phylogenetic Relationships with Other Triticineae Species

Phylogenetic studies, based on morphology (Seberg and Frederiksen 2001), chloroplast DNA (Mason-Gamer et al. 2002; Hodge et al. 2010), chloroplast, mitochondrial, and nuclear DNA sequences (Seberg and Petersen 2007; Escobar et al. 2011) and nuclear genes (Hsiao et al. 1995; Mason-Gamer et al. 2010), included *Eremopyrum* species in the same clade with species of *Agropyron* s. str.

5.5 *Henrardia* C. E. Hubbard

5.5.1 Morphological and Geographical Notes

Henrardia is a small genus containing two species, *H. persica* (Boiss.) C. E. Hubbard and *H. pubescens* (Bertol.) C. E. Hubbard (Hubbard 1946). The two species, having a characteristic morphology, differ from other genera of the tribe. Both species are annuals short plants with relatively long, cylindrical spike (5–15 cm long), anthers are small (1.5–2.2 mm in *H. persica* and 0.7 mm in *H. pubescens*) indicating a mating system of facultative self-pollination. Rachis harbors a solitary spikelet at each node, fragile, disarticulating at maturity just below the nodes (barrel type), so that each

spikelet falls with the rachis segment beside it. Caryopsis is free but tightly enclosed by the glumes (Fig. 5.3c).

Both species are distributed in Turkey (Anatolia) and from there have dispersed eastwards through Armenia and Transcaucasia to Central Asia and southwards to Iraq, Iran, Afghanistan and Baluchistan (Hubbard 1946). *H. persica* is not common than *H. pubescens* (Bor 1968). The latter species may be found also in Syria (Bowden 1966).

5.5.2 Cytology, Cytogenetics and Evolution

The two species are diploids ($2n = 2x = 14$); Sakamoto and Muramatsu (1965) and Sakamoto (1972) reported $2n = 14$ in *H. persica* and Bowden (1966) observed $2n = 14$ in *H. pubescens*. The karyotype of both species is extremely asymmetric, consisting of large chromosomes, of which four pairs have sub-telocentric and three pairs have telocentric chromosomes (Asghari-Zakaria et al. 2002). One of the chromosomes has a small satellite located at the end of its long arm (Asghari-Zakaria et al. 2002). *Henrardia* species have a most asymmetric karyotype; all other species have metacentric or sub-metacentric chromosomes, except *Aegilops caudata*, *Ae. umbellulata*, and *Ae. uniaristata* that have several sub-telocentric chromosomes (Chennaveeraiah 1960) and *Eremopyrum* triticeum (all seven pairs are sub-telocentric), *E. bonaepartis* (five pairs are sub-telocentric) and *E. distans* (two pairs are sub-telocentric) (Frederiksen 1991b). In all other Triticeae species the NOR region and the satellite are located on the short arm whereas in the *Henradia* species (and in *Eremopyrum distans*) they are located on the long arm (Asghari-Zakaria et al. 2002). Study of the C-banded karyotype of *H. persica* showed that each chromosome has a unique, easily recognizable C-banding pattern (Asghari-Zakaria et al. 2002). The karyotype of the *Henrardia* species is unique, differing from those of all other Triticeae.

In an attempt to study the genetic relationships between *Henrardia* and other genera of the tribe, Sakamoto (1972) crossed *H. persica*, as either female or male parent, with a number of species from different Triticeae genera. Hybrids were obtained only in the cross of tetraploid *Eremopyrum orientale* x *H. persica*. These hybrids were intermediate in spikes morphology but their spikelets were of *Eremopyrum* type. Disarticulation of ripe spikelets of *E. orientale* is of the wedge-type and that of *H. persica* is of the barrel-type. The F_1 showed the wedge-type disarticulation of the *Eremopyrum* parent. Chromosomal pairing at first meiotic metaphase was very low (13–21 univalents and 0–4 bivalents per cell), indicating reduced homology between the genomes of these two species. Another hybrid of *H. persica* x diploid *Eremopyrum distans* was obtained by Frederiksen (1993) but the plant was very weak, did not develop normal roots and died within a short time.

5.5.3 Phylogenetic Relationships with Other Triticineae Species

Because of its very peculiar morphology, *Henrardia* was earlier included in genera outside the Triticeae. Yet, in his taxonomical revision of these taxa, Hubbard (1946) noticed that *Henrardia* shares several diagnostic traits with the Triticeae, e.g., ovary and caryopsis hairy at the apex, lodicule hairy, lemma three or more nerved and seed longitudinally grooved with simple starch grains of the Triticeae-type in the endosperm (Tateoka 1962; Seberg et al. 1991). Consequently, since these characters have been regarded to be of diagnostic value in distinguishing the tribe Triticeae from other Poaceae tribes, Hubbard (1946) transferred this taxon to the Triticeae as a new genus, *Henrardia* C. E. Hubbard.

The deviating morphology of *Henrardia* led first to its classification in a separate sub-tribe, Henrardiinae Pilger, within Triticeae (Tzvelev 1976; Löve 1984). Later on, Clayton and Renvoize (1986) considered *Henrardia* as an offshoot of *Aegilops* and included it in the sub-tribe Triticineae. This is in accordance with Kellogg (1989) and Frederiksen and Seberg (1992) who, based on a cladistics analysis, concluded that *Henrardia* and the diploid species of *Aegilops* form a clade. As a consequence Kellogg (1989) suggested inclusion of *Henrardia* into *Aegilops* s. lat. As *Henrardia* possess a unique morphology and exceptional karyotype as well as low crossability with *Aegilops* species (Sakamoto 1972), it seems at present most convenient to consider it a part of the Triticineae, but to maintain it as a separate genus.

Seberg and Frederiksen (2001) performed a cladistic analysis, primarily based on morphology, of the monogenomic diploid genera of the Triticeae, and found that the large *Aegilops* clade consists of taxa traditionally included in *Aegilops* (van Slageren 1994), the closely related *Amblyopyrum muticum*, and *Henrardia persica*. But, using β-amylase gene sequences, Mason-Gamer (2005) carried out a phylogenetic study of the monogenomic Triticeae and found that *Henrardia persica* is close to *Eremopyrum bonaepartis* and both are in the same clade with *Psathyrostachys*. *Eeremopyrum species* grouped with *Elymus* (=*Agropyron*) and *Henrardia* on the chloroplast DNA tree (Mason-Gamer et al. 2002). A dual placement of *Eremopyrum* and *Henrardia* with Elymus (=*Agropyron*) has also been supported by other data sets (reviewed in Mason-Gamer 2005). Also Hodge et al. (2010), using chloroplast gene encoding ribosomal protein S16, found that *Eremopyrum bonaepartis* and *Henrardia persica* formed a

well supported clade. Phylogenies based on chloroplast DNA contradict the phylogeny based on morphological data of Petersen and Seberg (1997) and Seberg and Frederiksen (2001). The agreement of the rps16 data (Hodge et al. 2010) and the results of Mason-Gamer et al. (2002) support the placement of *Henrardia* in the *Henrardia–Eremopyrum* clade. A similar conclusion was reached by Hsiao et al. (1995) who analyzed nuclear DNA sequences. Escobar et al. (2010) determined the mating system of Triticeae species and combined the data with those obtained from molecular analysis of 27 protein-coding loci. They found that *Henrardia persica* is very close to *Eremopyrum bonaepartis* and form a clade (clade III) with *Eremopyrum triticeum* and *Agropyron mongolicum*. Using most comprehensive molecular data set of one chloroplastic and 26 nuclear genes, Escobar et al. (2011) found two well-supported clades, the first is formed by *Australopyrum* (clade IIIA), *Henrardia* and *Eremopyrum bonaepartis* (clade IIIB), and *Agropyron mongolicum* and *E. triticeum* (clade IIIC).

Seberg and Frederiksen (2001), based on morphology, placed *Elymus farctus* subsp. *bessaribicum* (=Thinopyrum bessarabicum) and *Henrardia persica* at the bases of their respective tree. However, the accumulated trees indicate that either *Psathyrostachys* or *Hordeum* is basal to the rest of the tribe and *Henrardia* is a more advanced type (Mason-Gamer 2005; Escobar et al. 2011).

5.6 *Dasypyrum* (Coss. & Durieu) T. Durand

5.6.1 Morphological and Geographical Notes

The taxonomy of and relationships of the *Dasypyrum* species have been the subject of controversy. Originally, they were placed in *Secale* L. and later in various other Triticeae genera until Schur (1866) recognized that this taxon was morphologically distinct from *Secale*, *Triticum* and other genera in the tribe. Schur (1866) placed it in a new genus, *Haynaldia* Schur, named in acknowledgement of Cardinal Haynald (1816–1891) and his interest in science and botany (Bor 1970). To avoid confusion with other *Haynaldia* genera, the genus name was later changed to *Dasypyrum* Cosson et Durieu and its generic rank was validated by Durand in 1888 (de Pace et al. 2011). According to Löve (1984), the genus name derives from the Greek words *dasy* (bushy, hairy) and *pyros* (wheat), and was selected to reflect the distinctive hairy keels of the glumes (de Pace et al. 2011). At first, three species were included in *Dasypyrum* (Candargy 1901): *D. villosum*, *D. hordeaceum*, and *D. sinaicum*, the latter being an annual species (Humphries 1978) occurring in eastern Mediterranean environments (Durand 1888) and was recognized as a species by Candargy as well (1901). However, in her taxonomical revision of the genus *Dasypyrum*,

Frederiksen (1991a) noted that *Dasypyrum sinaicum* (Steudel) Candargy is based on *Triticum sinaicum* Steudel, whose lectotype belonged to *Eremopyrum bonaepartis* (Sprengel) Nevski, thereby rendering it inappropriately assigned to *Dasypyrum*. Thus, she recognized only two species in the genus: the annual diploid *D. villosum* (L.) Candargy [=*Haynaldia villosa* (L.) Schur] and the perennial tetraploid *D. breviaristatum* (=*D. hordeaceum*), and demonstrated that the inclusion of *D. hordeaceum* was based on a later homonym, and for that reason, changed the name to *D. breviaristatum*. *D. breviaristatum* is commonly known as a perennial tetraploid, 2n = 4x = 28. However, Sarkar (1957) isolated both a tetraploid and a diploid cytotype from a 1954 collection of *D. breviaristatm* assembled by G. L. Stebbins in Morocco. Later on, Ohta et al. (2002) also reported on the existence of a diploid cytotype of *D. breviaristatum* among populations of the tetraploid cytotype in the Atlas Mountains. Morphologically, the diploids were similar to but smaller than the tetraploids in plant height, spike length, and spikelet number (Ohta et al. 2002). A distinct difference between the two cytotypes was only found in the number of trichomes on the leaf surfaces (Ohta et al. 2002). The two cytotypes are perennial, the diploid being slower in growth than the tetraploid and also with smoother leaves (Sarkar 1957).

The genus *Dasypyrum* bears distinctive two-keeled glumes with tufts of bristles along the keels, rendering them easily distinguishable from other genera in the Triticeae. Its plants are annuals or perennials, with 20–100 cm high culms, and terminal spikes that are 4–12 cm long, including the awns, with 1 spikelet per node. Their rachis disarticulates above each spikelet (wedge-type disarticulation). Spikelets are more than three times the length of the rachis internodes, and laterally compressed, with 2–4 florets. The lower two florets are usually fertile, the terminal florets are sterile, glumes are awned with two hairy keels, lemmas are awned and anthers are 4–7 mm long (Fig. 5.4a).

Dasypyrum (Cosson and Durieu) T. Durand (=*Haynaldia* Schur) comprises two allogamous (predominantly outcrossing) species: the annual *D. villosum* (L.) Candargy and the perennial *D. breviaristatum* (Lindb. f.) Frederiksen. In the recent literature, *Haynaldia villosa* (L.) Schur is most commonly known as *D. villosum*, although the former name is still occasionally used (Gradzielewska 2006a, b; de Pace and Qualset 1995).

Dasypyrum villosum (L.) P. Candargy–mosquito grass– [syn.: *Agropyron villosum* (L.) Link, *Haynaldia villosa* (L.) Schur, *Secale villosum* L., *Pseudosecale villosum* (L.) Degan; *Triticum villosum* (L.) Link] is an annual species with 20–100 cm long culms. Blades are light green, spikes

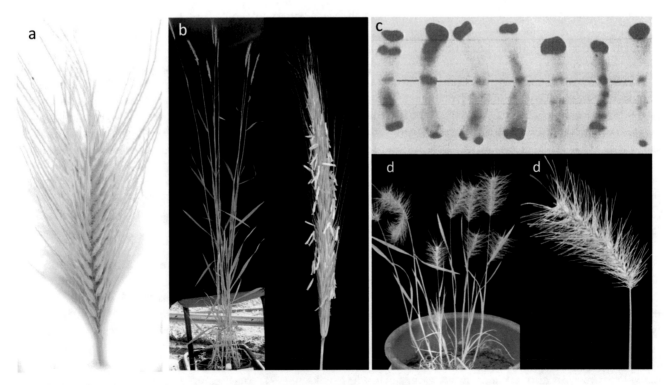

Fig. 5.4 Species of the *Dasypyrum-Secale-Heteranthelium* clade; **a** A spike of *D. villosum* (L.) Candargy; **b** A plant and spike of *S. cereale* L. ssp. cereale; **c** Giemsa C-banding of *S. Cereale* L. ssp. cereale chromosomes. The seven chromosomes are from left to right, 1R to 7R. (From Gill and Kimber 1974); **d** A plant and a spike of *Heteranthelium piliferum* Hochst

are 4–12 cm long, and glumes have tufts of hair on the two keels, with a tuft of stiff hairs below the awns, which are straight and 15–60 mm long; anthers are 4–7 mm long. In contrast, *Dasypyrum breviaristatum* (H. Lindb.) Frederiksen–[syn. *D. hordeaceum* (Cosson & Durieu) Candargy; *Haynaldia hordeacea* (Coss. & Durieu) Hack] is perennial, has short rhizomes, and features dark green leaves, glume keels with hairs that are not in tufts, and awns that are ~ 15 mm long.

Morphologically, the most conspicuous evolutionary divergence between diploid and tetraploid *D. breviaristatum* and *D. villosum* is apparent in the vegetative propagation device, with presence of rhizomes in diploid and tetraploid *D. breviaristatum* and absence of rhizomes in *D. villosum* (de Pace et al. 2011). These differences clearly reflect the major trends of adaptive radiation between *D. villosum* and the two cytotypes of *D. breviaristatum*, that occurred during colonization of high altitude habitats and further adaptation and differentiation of the tetraploid cytotype of *D. breviaristatum* to the environmentally disturbed habitats in forests and pastures at high altitude (de Pace et al. 2011). The F_1 hybrids between *D. villosum* and *D. breviaristatum* produce rhizomes, indicating that the perenniality trait is dominant (Ohta and Morishita 2001; Blanco and Simeone 1995).

All *D. villosum* and *D. breviaristatum* plants show dimorphism for kernel color, with a yellow and dark red kernel within every spikelet (Onnis 1967). Yellow kernels are more frequent on the second floret, are heavier than dark-red kernels, and germinate faster (de Pace et al. 1994). Similar dimorphism in kernel color exists in wild emmer, *Triticum turgidum* subsp. *dicoccoides*, where the light-colored kernels, develop in the second floret, germinate in the first year after seed dispersal, and the dark–colored kernels develop on the lower floret, in the second year. The inheritance of the kernel color does not show any Mendelian segregation differences are reported between the kernels color classes: the dark-red seeds have longer seed dormancy than the yellow ones and maintain longer germination ability (after eight years of storage) than that of the yellow seeds (Stefani et al. 1998).

The distribution area of the genus *Dasypyrum* is in the southwestern part of the distribution region of the sub-tribe Triticineae, and it is known more as the region of *D. villosum* than as the geographical range of *D. breviaristatum* (Sarkar 1957). The core distributional center of *D. villosum* is in the Mediterranean Basin of southern Europe, e.g., Italy (including Sicily and Sardinia), Slovenia, Croatia, Bosnia-Erzegovinia, Serbia, Albania, Macedonia, Greece, (including Crete). It also sporadically grows in Spain (the Baleares Islands), southern France, southern Switzerland, Austria, Hungary, Romania, Bulgaria, Moldova, Ukraine (Krym), Turkey, Caucasus, Armenia, Azerbaijan, Georgia,

Southeastern Russia, and western Turkmenistan (Maire 1952: Frederiksen 1991a; de Pace et al. 2011). In its core distributional centers in southern Europe, *D. villosum* is a vigorous plant that grows at low altitudes and is absent in habitats above 1350 m (de Pace et al. 2011). It is common in open herbaceous plant formations, often in dense stands, and also occupies disturbed habitats (de Pace et al. 2011). Genetic studies revealed lower inter-population and higher intra-population genetic diversity, as expected of an out-crosser species (de Pace et al. 2011). However, there is evidence of a positive relationship between spatial distance and genetic distance (de Pace et al. 2011). Therefore, to capture more genetic variation of *D. villosum*, samples should be collected from distant populations.

The distribution of *Dasypyrum breviaristatum* is rather restricted, and is mainly outside the range of the annual *D. villosum*. It grows in two isolated mountainous regions, each located over 1000 m above sea level, i.e., the Atlas Mountains of Morocco and Algeria, and Mt. Taygetos, in the Peloponnisos, Greece (Frederiksen 1991a; Ohta and Morishita 2001). Both the diploid and the autotetraploid grow in mixed populations, but the majority of the plants are tetraploids (Ohta et al. 2002). Recent investigations show that the diploid cytotype of *D. breviaristatum* is found only in Morocco and Algeria, while the tetraploid cytotype grows in Greece as well (Sarkar 1957; Ohta and Morishita 2001; Ohta et al. 2002).

Contrary to *D. villosum*, *D. breviaristatum* is common in the pastures and forests of the mountains of Algeria and Morroco, at an altitude ranging from 1000 to 2200 m above sea level. In Mt. Taygetos, Greece, it was found at an altitude of 1080 m (Frederiksen 1991a). The habitats of diploid *D. breviaristatum* in Morocco are disturbed oak forests and calcareous bedrock (Ohta et al. 2002). The distribution of the diploid is more restricted than that of the tetraploid. These ecological aspects of local, narrow distribution of the diploid cytotype and more expansive geographic distribution of the tetraploid cytotype, match the trends observed for other diploid–tetraploid taxa in the Triticeae (e.g., Zohary and Feldman 1962).

5.6.2 Cytology, Cytogenetics and Evolution

5.6.2.1 Karyotype and Genome Size

The karyotype of the *Dasypyrum* species is symmetric. *D. villosum* contains five pairs of metacentric (of which two are satellited (SAT)-chromosomes) and two pairs of sub-metacentric chromosomes (Linde-Laursen and Frederiksen 1991). de Pace et al. (2011) reported on a similar karyotype, but with only one SAT-chromosome, in *D. villosum*. The diploid cytotype of *D. breviaristatum* contains six pairs of metacentric chromosomes, one being satellited, and one pair of sub-metacentric chromosomes (Ohta et al. 2002), while the tetraploid cytotype has 13 pairs of meta-centric chromosomes (of which three pairs are SAT-chromosomes—two with large and one with small satellites), and one pair of sub-metacentric chromosomes (Linde-Laursen and Frederiksen 1991). Another line of this species contains only two pairs of SAT-chromosomes with large satellites (Linde-Laursen and Frederiksen 1991), indicating some degree of polymorphism regarding the number of SAT-chromosomes in tetraploid *D. breviaristatum*. In addition, nucleolar dominance of the *breviaristatum*-NOR region on the *villosum* NOR and wide C-band karyotype differences between the genomes of the two species were observed by Linde-Laursen and Frederiksen (1991).

The use of different banding techniques, including staining with fluorochromes, C-banding, and Ag-NOR (Gill 1981; Linde-Laursen and Frederiksen 1991; Blanco et al. 1996), and chromosomal localization of a species-specific 380 bp long satellite DNA sequence (de Pace et al. 1992), allowed for reliable identification of each chromosome pair of *D. villosum*, in different genomic backgrounds after interspecific or intergeneric hybridization. In contrast, chromosomal identification was not possible in *D. breviaristatum* because of the overall similarity of banding patterns and chromosome morphology (Linde-Laursen and Frederiksen 1991).

The 1C genome size of *D. villosum* is 5.065 pg (Obermayer and Greilhuber 2005; Eilam et al. 2007), although large intraspecific variation, either between or within populations, has been detected (Greilhuber 2005). Genome size of tetraploid *D. breviaristatum* is twice that of *D. villosum* (Blanco et al. 1996).

5.6.2.2 Cytogentic Relationship Within and Between the Two Cytotypes of *D. breviaristatum*

To elucidate the cytogenetic relationship between the diploid and the tetraploid cytotypes of *Dasypyrum breviaristatum*, the two cytotypes were reciprocally crossed with one another and chromosome pairing at first metaphase of meiosis and fertility were studied in the F_1 hybrids (Ohta and Morishita 2001). The researchers used two diploid and nine tetraploid plants of *D. breviaristatum* for within and between cytotype crossings. The diploids were collected from one population in Morocco, while two of the tetraploids were collected from the same population as the diploid plants, five from other regions of the Atlas Mountains and two from Greece (Ohta and Morishita 2001). F_1 hybrid between the two diploid plants exhibited almost complete chromosomal pairing at MI (0.14 univalents, 6.92 bivalents, and 0.003 quadrivalents per cell; 12.70 chiasmata/cell.) and pollen fertility of the F_1 hybrid was high (82.3%). The F_1 hybrids between the different ecotypes of the tetraploid cytotype from Morocco,

displayed chromosomal configurations typical of autotetraploid (0.14–0.75 univalents, 3.90–5.81 bivalents, 0.06–0.78 trivalents, 2.54–4.45 quadrivalents, and 0.02–0.26 hexa- or octo-valents) configurations. The range of chiasmata/cell was 23.75–25.44 and pollen fertility was high (69.2–77.9%). The F_1 hybrid between the tetraploid plants from Morocco and Greece showed somewhat more univalents, bivalents, and trivalents per cell (1.01–1.61 univalents, 6.21–6.30 bivalents, 0.46–0.87 trivalents), fewer quadrivalents and higher configurations per cell (2.77–3.28 quadrivalents, 0.01–0.02 higher configurations), and somewhat fewer chiasmata/cell (22.00–23.22). Nevertheless, pollen fertility was as high as that of the F_1 generated tetraploid cytotypes from different ecotypes in Morocco. The pattern of chromosomal pairing at meiosis in hybrids between tetraploid Moroccan ecotypes from different site of Morocco, indicated that this hybrid was an autotetraploid and, therefore, the tetraploid cytotype is an autotetraploid.

The presence of higher multivalents (hexa- and octo-valents) in several crosses between tetraploid *D. breviaristatum* ecotypes indicated the occurrence of reciprocal translocations between these ecotypes (Ohta and Morishita 2001). Ohta and Morishita further revealed the presence of aneuploid plants in natural populations of the tetraploid cytotype. Such chromosomal variation is also maintained in the natural populations of tetraploid *D. breviaristatum* by vegetative propagation. These researchers also observed the following mean pairing configurations in *D. breuiaristatum* (4x) x *D. breuiaristatum* (2x) F_1 hybrid: 3.38 univalents, 3.20 bivalents, 3.74 trivalents, and 0.005 quadrivalents per cell. The mean arm pairing frequency and relative affinity were 0.915 and 0.641, respectively, indicating homology of the diploid genome to the two genomes of the tetraploid cytotype. The seven trivalents, observed in many meiocytes of the F_1 hybrid between the diploid and the tetraploid cytotypes of *D. breviaristatum*, supported this conclusion.

5.6.2.3 Origin of the Tetraploid Cytotype of *D. breviaristatum*

The origin and genomic composition of the tetraploid cytotype of *D. breviaristatum* is under debate (de Pace et al. 2011). Several authors suggested an autoploid origin of the tetraploid cytotype (Sarkar 1957; Sakamoto 1986; von Bothmer and Claesson 1990; Galasso et al. 1997; Ohta and Morishita 2001; Ohta et al. 2002). Sarkar (1957) studied chromosomal pairing at meiosis of the tetraploid cytotype and found 11.7 (2–14) bivalents and 1.1 (0–6) quadrivalents per cell. The occurrence of up to six quadrivalents in a cell may indicate an autoploid derivation of the tetraploid *D. breviaristatum*, making the diploid cytotype is the most likely ancestral type (Sarkar 1957).

The pattern of chromosomal pairing at meiosis observed in F_1 hybrids between the tetraploid and the diploid cytotypes (Ohta and Morishita 2001), and particularly the existence of seven trivalents in several meiocytes of the F₁ triploid hybrid, supported the autotetraploidy hypothesis of Sarkar (1957) regarding the speciation event that led to formation of the tetraploid *D. breviaristatum*. A similar conclusion, based on the similarity in karyotype and in plant morphology of the two cytotypes of *D. breviaristatum*, was reached by Ohta et al. (2002), who also suggested that the diploid cytotype is the most probable ancestral form of the tetraploid cytotype. Indirect evidence for this hypothesis came from experiments of Nakajima (1960), where the diploid cytotype of *D. breviaristatum* produced several unreduced gametes. Also, the fact that diploid *D. breviaristatum* is perennial, as is the tetraploid cytotype, designates the diploid cytotype as the most likely progenitor in which the genome duplication event occurred. This hypothesis is also supported by the studies of Sakamoto (1986), and von Bothmer and Claesson (1990).

On the other hand, using DNA fragment analyses and isozymes, as well as FISH of structural genes sequences, Blanco et al. (1996) concluded that tetraploid *D. breviaristatum* originates from *D. villosum*. Other authors proposed an allopolyploid origin for tetraploid *D. breviaristatum*, most likely with *D. villosum* as one of the parents and diploid *D. breviaristatum* as the other parent (Frederiksen 1991a; Linde-Laursen and Frederiksen 1991). However, a direct derivation from *D. villosum* was ruled out following molecular cytogenetic analyses (Galasso et al. 1997), RAPD analyses of genomic DNAs (Yang et al. 2006), and studies of the meiosis in reciprocal crosses of the two species (Sakamoto 1986; Ohta and Morishita 2001). Following failure to detect *D. villosum*-specific DNA sequences in tetraploid *D. breviaristatum* fluorescent in situ hybridization (FISH), Uslu et al. (1999) concluded that *D. villosum* is not related to tetraploid *D. breviaristatum*. Therefore, the hypothesis that the tetraploid cytotype of *D. breviaristatum* was derived from the diploid cytotype of this species by autopolyploidy, is the only one, to date, that benefits from concurrent support (Ohta et al. 2002; de Pace et al. 2011).

5.6.2.4 Cytogenetic Relationships Between *D. breviaristatum* and *D. villosum*

To clarify the genomic relationships between the two species of *Dasypyrum*, Ohta and Morishita (2001) reciprocally crossed the two cytotypes of *D. breviaristatum* with *D. villosum* and examined chromosome pairing at the first metaphase of meiosis and fertility in the F₁ hybrids. Some seed setting (36.1%) was obtained in *D. villosum* x diploid *D. breviaristatum*; 30% of the shriveled caryopses germinated. The reciprocal cross did not produce any seeds. The mean pairing configurations and mean arm pairing frequency per ell in the diploid *D. villosum* x *D. breviaristatum* (2x) hybrids were 11.12 univalents and 1.44 bivalents, and 0.107

mean arm pairing, and the hybrids were almost completely sterile. Based on these results, Ohta and Morishita (2001) concluded that the genome of diploid *D. breviaristatum* is only distantly related to that of *D. villosum*. They, therefore, proposed, in accordance with Wang et al. (1995), to use of the symbol Vb for the haploid genome of the diploid cytotype of *D. breviaristatum* and Vv for the haploid genome of *D. villosum*. Furthermore, since they concluded that tetraploid *D. breviaristatum* is an autotetraploid, with diploid *D. breviaristatum* as the immediate ancestor, they proposed the genome symbol VbVb for the haploid genome of tetraploid *D. breviaristatum*.

A similar pattern of seed set and chromosome homoeology was observed after hybridization of *D. villosum* (as female) with tetraploid *D. breviaristatum*. About 50% and 12% F₁ hybrids seed setting was reported by Ohta and Morishita (2001) and Blanco et al. (1996), respectively, and over 80% of them germinated. The reciprocal combination did not produce any seed. Chromosome pairing at meiosis of the F₁ hybrid between *D. villosum* and tetraploid *D. breviaristatum* from Morocco contained 8.75–9.49 univalents, 5.51–5.68 bivalents, and 0.05–0.25 trivalents per cell, as well as 8.70–8.76 chiasmata/cell (Ohta and Morishita 2001). In contrast, the F₁ hybrid between *D. villosum* and tetraploid *D. breviaristatum* from Greece showed somewhat higher chromosomal pairing (7.28 univalents, 6.64 bivalents, 0.14 trivalents, and 0.004 quadrivalents per cell; 11.58 chiasmata/cell). The higher chromosomal pairing in the hybrid between *D. villosum* and the Greece ecotype of tetraploid *D. breviaristatum* may indicate exchange of some chromatin between these two taxa (Ohta and Morishita 2001). The very low pairing between the F₁ hybrid chromosomes of *D. villosum* and diploid *D. brevisaristatum* clearly indicated no homology between their genomes, which is supported by the almost complete sterility of the F₁ hybrid (Ohta and Morishita (2002). This conclusion coincides with previous results reached following karyotype analysis (Ohta et al. 2002).

Sakamoto (1986) also achieved interspecific hybridization between *D. villosum*, used as female, and tetraploid *D. breviaristatum*, as male. At meiosis, the triploid F₁ hybrid displayed chromosome configurations with an average of 6.5 bivalents and 7.9 univalents per cell. As the tetraploid parents contained many quadrivalents, the bivalents observed in the interspecific hybrid were ascribed to autosyndesis of the tetraploid *D. breviaristatum* chromosomes (Sarkar 1957; Ohta and Morishita 2001).

Because in the pollem mother cells (PMCs) of F₁ of *D. villosum* x tetraploid *D. breviaristatum*, the observed number of trivalents per cell was 0.14, while 3.74 were observed in the cross of the diploid cytotype and the tetraploid cytotype of *D. breviaristatum*, the contribution of *D. villosum* genome to the tetraploid was ruled out. In accord with this conclusion, Galasso et al. (1997) observed seven bivalents of *D. breviaristatum* and seven univalents of *D. villosum* after simultaneously labeling F_1 hybrid *D. breviaristatum* (4x) and *D. villosum* chromosomes in first meiotic metaphase with fluorescein (FITC)-labeled *D. breviaristatum* DNA. Yang et al. (2005, 2006) confirmed the divergence of the Vv and Vb genomes using GISH and RAPD markers. When the entire Vb genomic DNA was labeled to hybridize a somatic mitotic metaphase of a partial amphiploid with 42 chromosomes (genome BBAAVbVb), generated from the selfing population of the amphiploid bread wheat (cv. Chinese Spring)-*D. breviaristatum*, 14 chromosomes were strongly and uniformly labeled to Vb (Yang et al. 2005). In contrast, when *D. villosum* was used as a probe, many arms of the 14 Vb chromosomes displayed large regions in their distal half with less intense labeling. These differentiated GISH patterns not only reflected the large genomic divergence between the Vb and Vv genomes, as described by Galasso et al. (1997), but also helped identify the chromosome pairs of *D. breviaristatum* in wheat background labeled with Vv-DNA.

5.6.2.5 Cytogenetic Relationships Between *Dasypyrum* Species and Species of Other Triticineae Genera

Several researchers crossed *D. villosum* with various species from other Triticeae genera, in pursuit of identification of cytogenetic and evolutionary relationships between these species. A complete list of hybrids between *Dasypyrum* species and Triticeae species and their level of pairing at first meiotic metaphase, is provided in Tables 4.7 and 4.8 of the publication of de Pace et al. (2011). Little chromosome homology was observed between *Dasypyrum* species and any other species of the Triticeae. Similarly, hybridizations performed by Lucas and Jahier (1988) between *D. villosum* and other diploid Triticeae species, indicated a low average number of chromosome pairings between homoeologous arms.

Crosses between diploid wheat (wild and domesticated *T. monococcum* and *T. urartu*) with *D. villosum* produced F_1 hybrids that displayed very low chromosomal pairing at first meiotic metaphase (Sando 1935; Kihara 1937; Sears 1941; Lucas and Jahier 1988; von Bothmer and Claesson 1990). Similarly, studies of chromosome pairing in F_1 hybrids between tetraploid wheat, *T. turgidum* ssp. *durum,* and *D. villosum*, revealed very low pairing at meiosis, demonstrating that the Vv genome is not homologous to the A and B subgenomes of tetraploid wheat (Kihara and Nishiyama 1937; Nakajima 1966; von Blanco et al. 1983a, b, 1988b). Using the C-banding technique on PMCs of the F_1 *T. turgidum* ssp. *durum* cv. Capelli x *D. villosum* hybrid, Blanco et al. (1988b) estimated that of the 194 observed bivalents, 82% involved A-B chromosome associations, 11.3% A-Vv, 1%

B-Vv, 4.1% A-A, 1.6% B-B, and 0% VvVv, and deduced that A and B subgenomes are related, while the Vv genome is more distant, but closer to the A than to the B genome.

Yu et al. (1998, 2001) analyzed chromosome pairing in F_1 hybrids between the common wheat cultivar Chinese Spring and *D. villosum*. On average, they observed 1.61 bivalents per cell. Chen and Liu (1982) reported a cytogenetic study of *T. aestivum* x *D. villosum* F_1 plants, in which they were able to identify the *villosum* chromosomes in the *T. aestivum* background and found very little pairing at first meiotic metaphase, i.e., 22.5–27.8 univalents, 0.11–0.37 bivalents, and 0.10 trivalents per cell. The F_1 hybrid had reduced vigor and tillering ability, as compared to Chinese Spring (Chen and Liu 1986).

Nakajima (1953) studied meiosis in the F_1 hybrid of *T. timopheevii* x *D. villosum*, and noted, on average, one bivalent per cell. He later reported (Nakajima 1960) meiotic chromosome pairing in F_1 hybrids between *Triticum turgidum* and tetraploid *D. breviaristatum*, and concluded that there is no homology between the genomes of *D. breviaristatum* and the two subgenomes of *T. turgidum*. The mean bivalent frequency of 8.06 per cell in the tetraploid F_1 hybrid of this cross, and 8.9 in the pentaploid F_1 *T. aestivum* x tetraploid *D. breviaristatum* hybrid, resulted from autosyndesis of the *breviaristatum* chromosomes. The combination *T. turgidum* subsp. *durum* x tetraploid *D. breviaristatum* produced an average of 7.86 bivalents and 11.5 chiasmata/cell (Blanco and Simeone 1995). Seven out of the 7.86 bivalents resulted from autosyndesis of the *breviaristatum* chromosomes.

The *T. aestivum* x tetraploid *D. breviaristatum* F_1 hybrid seed set was 2.9% (von Bothmer and Claesson 1990). Two hybrid plants displayed a high level of pairing, with up to 12 bivalents and an average of 7.8 bivalents. The pairing of seven bivalents was attributed to autosyndesis of *breviaristatum* chromosomes.

Oehler (1933, 1935), Sando (1935), Kihara and Lilienfeld (1936), von Berg (1937), Sears (1941), Lucas and Jahier (1988), and Deng et al. (2004) crossed diploid and tetraploid species of *Aegilops* with *D. villosum*. The F_1 hybrids displayed very low chromosomal pairing and were completely sterile. An intergeneric triploid hybrid between *Aegilops tauschii* and the tetraploid cytotype of *D. breviaristatum* was produced and at meiosis, displayed 8.80 univalents and 6.15 bivalents per cell (Sakamoto 1986).

Sando (1935) produced hybrids between *Secale fragile* and *D. villosum*. In general, the F_1 plants resembled the *Secale* parent. Kostoff and Arutiunova (1937) analyzed pairing in a trigeneric hybrid (*T. turgidum* subsp. *dicoccon* x *D. villosum*) x *S. cereale* (genome BAVvR) and found that Vv chromosomes were not homologous with the *Secale* chromosomes (genome R) and both Vv and R chromosomes were not homologous to the wheat A or B chromosomes.

Similarly, Nakajima (1951) observed very low pairing in MI of the F$_1$ hybrid between *D. villosum* and *Secale cereale* and concluded that no homology existed between the Vv and R genomes. Extremely low pairing ability was also observed between Vv and R chromosomes in PMCs of hybrid plants generated by crossing two amphiploids: *Ae. Uniaristata–D. villosum* (2n = 28; genome NNSvSv) and *Ae. uniaristata–S. cereale* (2n = 28; genome NNRR) (Jahier et al. 1988). However, Jahier et al. (1988) did not reject the working hypothesis that Vv and R chromosomes share homologous sequences. Rather, they attributed the Vv and R asynapsis to factors such as asynchronous meiotic rhythm between R and Vv genomes. Similar causes can explain the lack of pairing between Vv and Vb reported above.

5.6.3 Phylogeny and Time of Origin

The shared peculiar spike morphology, unilocus molecular and biochemical markers (Blanco et al. 1996), and signals on the chromosomes of tetraploid *D. breviaristatum*, but not on most other Triticeae species, hybridization with the pHv62 *D. villosum*-species-specific repeated sequence (Uslu et al. 1999), seen between the two species of *Dasypyrum*, are indicative of common ancestry. The genomic distance between *D. villosum* and *D. breviaristatum*, as determined by 301 RAPD loci, was smaller than their distance from *Secale* species (Yang et al. 2006). Therefore, the formation of the *dasypyrum* species and their biological and taxonomical status may be explained by a cascade of events, which began in the earlier stages of Triticineae separation from the Hordeineae (13–15 MYA), and continued through the reproductive isolation of the lower-altitude *D. villosum* ecotypes from the high–altitude diploid *D. breviaristatum* prototype, followed by the autopolyploidization event of the tetraploid cytotype from the diploid one, and incipient reproductive isolation between the two cytotpes. Such divergence has not occurred for other syntenic and gene-rich DNA segments of genomes Vv and Vb, as suggested by the strong similarity between *D. villosum* and *D. breviaristatum* genomes in restriction fragment patterns of genomic DNA, the phenotypes for some isozyme systems, and the location of gliadin genes (Blanco et al. 1996).

The phyletic relationships within the *Dasypyrum* genus and among *Dasypyrum* species and other Triticineae species, has been assessed at the levels of morphology, protein, chromosome, chloroplast and nuclear fragments and nucleotide sequences. Morphology-based phylogenetic analyses showed that *Dasypyrum* branched in a sister group of *Secale* within the same clade (Baum 1978a, b, 1983; Kellogg 1989; Frederiksen and Seberg 1992; Seberg and Frederiksen 2001). Kellogg (1989) placed *Dasypyrum* near *Agropyron* and *Triticum monococcum*, and Baum (1978a, b,

1983) considered *Secale cereale* and *D. villosum* as evolutionarily more contiguous to *Triticum* and *Aegilops* than to the rest of the Triticineae.

Phylogenetic studies based on molecular data suggested that *Secale* is the closest relative of the *Triticum-Aegilops* genera (Kellogg et al. 1996; Huang et al. 2002a; Mason-Gamer et al. 2002). When *Hordeum* and *Dasypyrum* were analyzed with *Secale, Triticum* and *Aegilops*, they were positioned at the base of the tree topology, as out-groups (Yamane and Kawahara 2005; Kawahara et al. 2008), implying a much earlier divergence between *D. villosum* and the common ancestor of *Triticum-Aegilops*.

The analysis from two cpDNA data sets, one based on restriction site variation (Mason-Gamer and Kellogg 1996b) and the other on sequences encoding the *rpoA*-subunit of the RNA-polymerase (Petersen and Seberg 1997), placed *Elymus* species possessing the E genome, e.g., *E. elongatus* and *E. farctus,* and *Dasypyrum* together on cpDNA cladograms. Sequencing of the nuclear starch synthase gene also revealed a close affinity between *E. farctus* and *Dasypyrum* (Mason-Gamer and Kellogg 2000). The cpDNA tree contained a well-supported clade, including *Dasypyrum* and diploid *Elymus* with St genome. The *Dasypyrum-Elymus* with St genome monophyly was also observed in cladograms obtained from similar RFLP profiles of 14 cloned fragments covering the entire cpDNA of *T. aestivum* (Kellogg 1992), morphological data (Kellogg 1989), and 5S RNA (Appels and Baum 1991).

However, the nuclear DNA data (Hsiao et al. 1995; Kellogg and Appels 1995) are incongruent with the cpDNA data, in that they suggest different affinities of diploid *Elymus* with St genome and *Dasypyrum* within Triticineae. The *Heteranthelium* element of the transposon *Stowaway* is present in *Dasypyrum*, but absent in other Triticeae species (Petersen and Seberg 2000). *Heteranthelium piliferum* and *D. villosum* stand at one extreme of the phylogenetic relationships, determined by variation in the PCR sequences of *6-SFT* (sucrose:fructan 6-fructosyltransferase), whereas diploid *Secale, Triticum* and *Aegilops* species are at the other extreme (Wei et al. 2000). Molecular phylogeny of the *RPB2* (the second-largest subunit of RNA polymerase II) gene sequence revealed that the Vv genome of *D. villosum* is sister to the St genome of *Elymus* and that both diverged from the H-genome of barley (Sun et al. 2008). These findings fall in line with the phylogenetic relationships of mono-genomic species of Triticeae inferred from nuclear rDNA (internal transcribed spacer) sequences, where *Heteranthelium* and *Dasypyrum* demonstrated close relation to diploid *Elymus* with St genome.

DNA/DNA hybridization of the genomes of rye and *D. villosum* with labeled nuclear DNA from wheat and rye, revealed greater homology between the Vv- and R- than between Vv and A-, B-, and D-subgenomes of wheat (Lucas and Jahier 1988). Fish analysis involving hybridization of the genomes of different Triticeae species with species-specific molecular probes prepared from tandem repeated DNA sequences of *D. villosum* (pHv62) and *S. cereale* (pSc119.2), demonstrated a greater homology between the R- and Vv-genomes than between R- or Vv-genomes and between those of *Triticum* and *Aegilops* (Uslu et al. 1999).

Escobar et al. (2011; Fig. 2.2) obtained the most comprehensive molecular dataset to date in Triticeae, including one chloroplast and 26 nuclear genes. They found that *Dasypyrum, Heteranthelium* and genera of clade V, grouping *Secale, Taeniatherum, Triticum* and *Aegilops*, evolved in a reticulated manner. Their evidence supported the following clades (Escobar et al. 2011): The first includes *Australopyrum* (clade IIIA), *Henrardia* and *Eremopyrum bonaepartis* (clade IIIB), and *Agropyrum* and *E. triticeum* (clade IIIC), while the second consists of *Dasypyrum* and *Heteranthelium* (clade IV), on the one hand, and *Secale, Taeniatherum, Triticum* and *Aegilops* (clade V), on the other hand. St genome *Elymus* does not group with *Hordeum* but is sister to *Dasypyrum*. Consequently, *Heteranthelium* branches at the base of clade V and these two newly inferred clades (*Elymus-Dasypyrum* and *Heteranthelium*-clade V) are closely related to each other.

Lucas and Jahier (1988) concluded that the differentiation of *D. villosum* (and *Secale cereale*) from the genera *Aegilops* and *Triticum* occurred earlier than speciation in the latter two genera. They assumed that the closest *Aegilos-Triticum* species to *D. villosum* is wild diploid wheat, *T. monococcum* ssp. *aegilopides*, and not diploid species of *Aegilops*.

5.6.4 Use of *Dasypyrum* in Wheat Improvement

5.6.4.1 Production of Cytogenetic Lines Facilitating Identification of Useful Genes, Their Allocation to Chromosomes and Construction of Genetic Maps

D. Villosum has been crossed to *Secale, Aegilops, Agropyron*, and *Triticum* species, but permanent introgression of its chromosomal segments occurs only in wheat, following controlled backcrosses (de Pace et al. 2011). At the beginning of the twentieth century, Nazareno Strampelli was the first to show the feasibility of crossing wheat species with *D. villosum* and of transferring genes from the wild species to tetraploid and hexaploid wheat, by crossing, backcrossing and then implementing proper selection rules (Strampelli 1932). The *T. aestivum* cv. Rieti x *D. villosum* cross produced an F_1 hybrid, from which (after a putative process of backcrossing to cv. Rieti) the "gigas" winter bread wheat cultivar Cantore was derived (Strampelli 1932). Strampelli succeeded to release several interesting cultivars of bread

wheat, such as the so-called *Triticum giganteum* (which produced large spikes and kernels as large as a coffee seed) or lines with sweet kernels.

Following the pioneering work of Strampelli, several researchers used *D. villosum* in intergeneric hybridizations to study the morphology, chromosomal pairing at meiosis, and fertility of the hybrids. The studies were of interest to the evolutionist, biosystematist, and cytogeneticist, but they rarely went further in the backcrossing and selection programs for releasing cultivars. However, several cytogenetic lines have facilitated meaningful assessment and use of the potential of *D. villosum* for wheat improvement.

Production of wheat (tetraploid and hexaploid)—*D. villosum* amphiploids was the first methodological approach used to assess the potential of *Dasypyrum* species to contribute genes that may improve wheat. Tschermak-Seysenegg (1934) produced the first hexaploid amphiploid (2n = 6x = 42; genome BBAAVvVv) from the cross of *T. turgidum* with *D. villosum*. McFadden and Sears (1947) produced the hexaploid amphiploid *T. turgidum* ssp. *dicoccoides–D. villosum* and Jan et al. (1986) produced the octoploid amphiploid *T. aestivum* cv. CS–*D. villosum* with 2n = 8x = 56 (genome BBAADDVvVv). Several other amphiploid BBAADDVvVv were produced by Mini et al. (1988), but all octoploid amphiploids had agronomically poor plant type.

In contrast to the octoploid amphiploids, hexaploid amphiploids (*T. turgidum* ssp. *durum–D. villosum*) proved more promising. Such amphiploids were produced by several researchers (see review in de Pace et al. 2011), and exhibited a plant habitus and spike morphology resemblant of wheat, good seed quality, but brittle rachis. As a primary amphiploid, the overall performance, with the exception of brittle rachis, of the hexaploid BBAAVvVv was equal to or better than, most hexaploid primary triticale. Thus, the BBAAVvVv amphiploids deserve consideration as new crop plants, much as triticale did in its early stages of development (de Pace et al. 2011).

Dasypyrum addition and substitution lines in wheat are suitable material for studying the role of single *Dasypyrum* chromosomes in determining various traits in the wheat background. Sears (1953) and Hyde (1953) produced the first set of six out of seven possible monosomic addition lines, using *D. villosum* chromosomes to complement the standard laboratory bread wheat cultivar Chinese Spring host genome. Later, Lukaszewski (1988) completed the set of seven monosomic addition lines. Using C-banding, plant morphology, and molecular markers, Lukaszewski (1988) assessed the homoeology of the added Vv chromosomes to those of wheat, and assigned each chromosome to the wheat homoeologous groups. Similarly, Blanco et al. (1987) produced a set of six monosomic addition lines by donating *D. villosum* chromosomes to the durum wheat cultivar Creso. Each added Vv chromosome had a specific effect on plant morphology and fertility. Liu et al. (1995) described six *D. villosum* substitution lines in bread wheat and reported the homoeology assignment of the Vv chromosomes to the six different wheat homoeologous groups.

Various novel disease-resistance genes have been identified on specific Vb chromosomes of the perennial tetraploid *D. breviaristatum*. Addition and substitution lines were isolated in the progeny of wheat—*D. breviaristatum* amphiploids crossed with cultivated wheat, including different addition lines carrying genes for stripe rust (Yang et al. 2008), as well as stem rust and powdery mildew (Liu et al. 2011a, b) resistance. Marker data indicated that the Vb chromosomes in the latter two addition lines were rearranged with respect to wheat homoeologous groups. On the other hand, various molecular markers confirmed a group 2 homoeology for the Vb chromosome substituted into a Chinese bread wheat in place of chromosome 2D, able to confer stripe rust resistance at the adult plant stage (Li et al. 2014). Interestingly, FISH, C-banding, and PCR-based molecular marker analyses indicated that the 2Vb of *D. breviaristatum* was completely different from 2Vv of *D. villosum*, in line with the current view about the origin of $4 \times D.$ breviaristatum.

5.6.4.2 Production of Translocation Lines of *D. villosum* Chromosomal Segments in Wheat Chromosomes

Production of Translocations via Induction of Homoeologous Pairing

Halloran (1966) crossed *D. villosum* with the bread wheat cultivar Chinese Spring monosomic for chromosome 5B, and obtained two types of hybrids, with (2n = 28) and without (2n = 27) chromosome 5B. Very low pairing (0.25 bivalent/cell) was observed in meiosis of the 28-chromosome hybrid containing 5B, leading to the conclusion that *D. villosum* does not possesses gene(s) that remove the inhibition to homoeologous pairing due to the *Ph1* gene in chromosome 5B (Halloran 1966). The 27-chromosome hybrid (deficient for chromosome 5B) showed much higher chromosome pairing (9.6 univalents, 3.8 rod and 1.06 ring bivalents, 0.86 trivalents and 0.7 quadrivalents) than the 28-chromosome hybrid (Halloran 1966). Yu et al. (1998, 2001) also noted enhanced homoeologous pairing between wheat and *D. villosum* chromosomes in the absence of *Ph1*. These observations indicated that it is possible to induce transfer of chromosomal segments from *D. villosum* to bread wheat by promoting homoeologous pairing in the absence of *Ph1* or in the presence of its mutant *ph1ph1*.

In contrast to Halloran (1966), von Bothmer and Claesson (1990) suggested that *D. villosum* genotyopes might influence the pairing frequency in the F_1 hybrids between *D. villosum* and *Triticum-Aegilops* species. Likewise, Blanco et al. (1983a, b) concluded that *D. villosum* contains genes that promote homoeologous pairing in the presence of *Ph1*, explaining the similar proportions of homoeologous pairings observed in F_1 hybrids between *T. turgidum* and *D. villosum*, in the presence of *Ph1* or its mutant allele *ph1*. To explain the discrepancy between Halloran's results and theirs, Blanco et al. (1988b) presumed that ecotypes of *D. villosum* may vary in their ability to promote homoeologous pairing in the presence of *Ph1*. Genetic variation for the promotion of homoeologous pairing has been demonstrated in various accessions of several Triticeae species, e.g., *Amblyopyrum muticum* (Dover and Riley 1972), *Ae. speltoides* (Dvorak 1972), *Ae. longissima* (Mello-Sampayo 1971a), *Elymus elongatus* (Mochizuki 1962), and *Secale cereale* (Dvorak 1977) (Table 5.2).

Yu et al. (1998, 2001) also analyzed chromosome pairing in F_1 hybrids of the bread wheat cultivar Chinese Spring (with *Ph1*) and its *ph1b* mutant (a defciency for Ph1) with *D. villosum*. On average, 1.61 chromosomes per cell paired in the hybrid with *Ph1*, but 14.43 in the hybrid with *ph1b*. GISH revealed three types of homoeologous associations between wheat (W) and *D. villosum* (D) chromosomes (W-D, D-W-W and D-W-D) in PMCs of the CSph1 x *D. villosum* hybrid, and only one type (W-W) in the CS*Ph1* x *D. villosum* hybrid. Translocations of chromosome segments or entire arms, were detected by GISH in the BC_1 plants from the backcross of CS*ph1* x *D. villosum* to CS*ph1b*.

Production of Translocations via Irradiation

Irradiating mature female or male gametes of plans having addition or substitution of whole chromosomes or chromosome arm of *D. villosum* in either bread or durum wheat background, with ^{60}Co-gamma-ray, is a new and highly efficient means of eliciting small segment structural changes in chromosomes, especially interstitial translocations of *D. villosum* segments in wheat chromosomes (Chen et al. 2008; Cao et al. 2009; Bie et al. 2007).

5.6.4.3 Allocation of Useful Genes to Chromosomes

The various genetic lines mentioned above allowed allocation of useful genes of Vv to chromosomes of Vv, namely, genes for morphological (e.g., Sears 1982a, b; Mariani et al. 2003; Chen et al 2008*)*, biochemical (e.g., Resta et al. 1987; Shewry et al. 1987, 1991; Montebove et al. 1987; Blanco et al. 1991; de Pace et al. 1988, 1992, 2011), molecular (e.g., Gil and Appels 1988; de pace et al. 1992; Liu et al. 1995; Galasso et al. 1997), disease resistance (e.g., Pasquini et al. 1978; Panayotov and Todorov 1979; Chen et al. 1997;Yildirim et al 1998, 2000; Oliver et al. 2005; Huang et al. 2007;

Bizzarri et al. 2009), and abiotic stresses tolerance (e.g., Scarascia-Mugnozza et al. 1982; Schlegel et al. 1998) genes. An alternative approach in which genes are assigned to specific chromosomes of *D. villosum* using nullisomic amphiploids, has been proposed (Zhong and Qualset 1990).

5.6.4.4 Contribution of *Dasypyrum* Genes to Wheat Improvement

In an evaluation of forage crop potential, the amphiploid *T. turgidum* ssp. *Durum–D. villosum* showed high biomass quality (N yield) and quantity (biomass) (de Pace et al. 1990). The role of *D. villosum* chromosome segments introgressed in hexaploid wheat in pre-breeding and primary population mapping for complex genetic trait analysis, was evidenced by Mariani et al. (2003). Vaccino et al. (2007) found that wheat lines with chromosome 1BL containing chromatin introgressed from *D. villosum*, were early-heading, good grain yielders, improved bread making quality and were environmentally stable over the years. Analysis of several hexaploid lines derived from the backcross of the F_1 hybrid *T. turgidum* ssp. *durum* x *D. villosum* to bread wheat, exhibited good agronomic performance in field trials, in term of yield, kernel weight, and bread-making quality when compared to the best standards (Vaccino et al. 2009). Successful transfer of the *D. villosum* gene for powdery mildew resistance was described by several researchers, e.g., Blanco et al. 1988a; Shi et al. (1996) Chen et al. (2008), Liu et al. (1996), and Qi et al. (1995).

5.7 *Heteranthelium* Hochst

Heteranthelium is a monotypic genus containing the species *H. piliferum* (Banks et Sol.) Hochst. This species is annual, consisting of short plants with a very peculiar spike morphology that is different from that of other Triticeae genera (Fig. 5.4d). Its spike consists of two kinds of interspersing spikelets, fertile and sterile, both strikingly different in appearance. In the fertile spikelets the lower florets are bisexual and the upper ones are barren and scale-like, whereas the sterile spikelets are composed of barren scale-like florets only. The glumes and the lower parts of the awns are hairy. The spikelets are solitary at each rachis node. Rachis only partially fragile and consequently, ripe spikes of this species do not disarticulate between individual spikelets, but break up into a number of sections, each section contains at least three fused rachis-segments diminishing in length from below upwards, the lowest spikelet is fertile, the next above is smaller, fertile or sterile with a reduced lemma, and succeeding spikelet(s) reduced to a bunch of awned glumes. Thus, *H. piliferum* has an exceptional dispersal unit consisting of three spikelets of which 1–2 are sterile. The small size of the anthers (1 mm) indicates mating system of facultative self-pollination; bagged spikes were fertile almost as non-bagged spikes (Luria 1983). There is some variation in plants size, leaves size, spike color (from green to red), number of seeds in the fertile spikelets and in seed size.

Heteranthelium piliferum is native to the Eastern Mediterranean region and central Asia (Israel, Jordan, Syria, Lebanon, Turkey, Transcaucasia, Iraq, Iran, Pakistan, Afghanistan, Turkmenistan, Tajikistan and Kyrgyzstan). It is an Irano-Turanian element growing on rocky and dry slopes of foothills in semi-arid steppes as well as at the edges of the Mediterranean region. Its distribution is limited by drought (desert) and cold (higher mountainous regions). In the sub-Mediterranean regions it grows in 550–750 m above sea level in annual rainfall of 250–300 mm while in mountainous area (e.g., Hermon Mt.) it grows at elevation of 1400–1700 m above sea level, with annual rainfall of 1000–1300 mm, in plant community consisting of sub-Mediterranean and many Irano-Turanian plants.

H. piliferum is a diploid (2n = 2x = 14) (Sakamoto and Muramatsu 1965; Bowden 1966), with symmetric karyotype with median or sub-median centromere, one chromosome pair is satellite with a small satellite (Chennaveeraiah and Sarkar 1959; Ferederiksen 1993; Bowden 1966; Sakamoto 1974). Each chromosome had a distinct C-banding pattern and this technique provided adequate information to identify all of *H. piliferum* chromosomes (Asghari-Zakaria 2007). Homologous chromosomes were identified based on position of the centromere and similarities of C-banding patterns (Asghari-Zakaria 2007). Its genome symbol is Q (Wang et al. 1995).

In an attempt to elucidate the genetic relationships to other genera of the tribe, Sakamoto (1974) crossed *H. piliferum*, as the female parent, with a number of Triticeae species. Yet, only the inter-generic crosses of *H. piliferum* with 2 x *Eremopyrum bonaepartis* and autotetraploid *hordeum depressum* were successful. The F_1 hybrid H. piliferum x *E. bonaepartis* showed poor growth and the morphology of the spikes was similar to that of *Eremopyrum*, while the spikelets were intermediate. Very little chromosome pairing was observed at first meiotic metaphase of the hybrid (0.04 bivalents per cell), indicating that the genome of *H. piliferum* is only distantly related that of *E. bonaepartis* (Sakamoto 1974). Growth of the F_1 hybrid *H. piliferum* x tetraploid *Hordeum depressum* was vigorous and the spike morphology was intermediate between the parents, i.e., a solitary spikelet exists at each rachis node like the *Heteranthelium* parent but no sterile spikelets characteristic of *Heteranthelium* were found. The *Heteranthelium* characteristic of one spikelet per rachis node dominated the multiple spikelet character of *Hordeum*. Similar dominance of the solitary spikelet trait over multiple spikelets at each rachis node was also found in bread wheat x *Hordeum vulgare* hybrids (Muramatsu 2009). The triploid F_1 hybrid

H. piliferum x tetraploid *Hordeum depressum* exhibited 5.5 bivalents and 0.01 trivalents per cell, resulting from autosyndesis of the *Hordeum* parent chromosomes and indicating lack of homology between *H. piliferum* and *H. depressum*.

Considering the unique morphology of the spike and the dispersal unit, distribution in the east and southeast periphery of the distribution area of the sub-tribe Triticineae, inter-generic cross-ability and cytogenetic relationships of *H. piliferum*, Sakamoto (1974) concluded that the monotypic genus *Heteranthelium* is a distinctive entity, representing a specialized group that occupies an isolated position in in the tribe Triticeae. This taxon has evolved as an annual during the process of adaptation to rather dry habitats of the Mediterranean climatic regions.

In different classifications of Triticeae *Heteranthelium* is supposed to be related to either *Triticum/Aegilops* complex (Nevski 1934a; Tzvelev 1976) or *Hordeum* (Love 1984; Clayton and Renvoize 1986; Kellogg 1989). Clayton and Renvoize (1986) regarded it as an advanced offshoot of *Critopsis*. Yet, *Heteranthelium* has one spikelet per node like the sub-tribe Triticineae and awnlike glumes like *Hordeum* (Frederiksen 1993). Thus, the phylogenetic relationships of *Heretanthelium* are still ambiguous. Studies of Hodge et al. (2010) on the chloroplast gene encoding ribosomal protein S16, showed that *H. piliferum* is in the same clade with *Triticum monococcum*, *Secale cereale*, and all the *Aegilops* species. Their results are consistent with the finding of Mason-Gamer et al. (2002), based on combined cpDNA sequences of tRNA genes, spacer sequences, rpoA genes, and restriction sites. Phylogenetic relationships based on mating systems showed that *H. piliferum* is in a clade with *Dasypyrum Villosum* (Escobar et al. 2010). This is in spite of the fact that *H. piliferum* is facultative self-pollinated whereas *D. Villosum* is an out-crosser. Escobar et al. (2011) studied one chloroplastic and 26 nuclear genes and found that *H. piliferum* comprises a clade with *Dasypyrum villosum* (clade IV) that branches at the base of clade V (*Triticum*, *Aegilops*, *Secale* and *Taeniatherum*). Cllades IV and V are closely related to each other (Escobar et al. 2011).

5.8 *Taeniatherum* Nevski

5.8.1 Morphological and Geographical Notes

Because of its diverse morphological features, *Taeniatherum* Nevski has previously been included in several different genera. Linnaeus (1753) considered it as a single species belonging to the genus *Elymus* (*E. caput-medusae* L.), while Schreber (1772) concluded that this taxon consists of two *Elymus* species (*E. caput-medusae* and *E. crinitum* Schreb.), and Link (1827) defined three species (*E. cput-medusae*, *E.

crinitum* and *E. platatherus* Link). Other authors thought that *Taeniatherum* belongs either to the genus *Cuviera* Koeler (Simonka 1897) [now *Hordelymus* (Jessen) Harz] or *Hordeum* L. (Cosson and Durieu 1855; Ascherson and Graebner 1902). However, Nevski (1934a) pointed out that *Taeniatherum* differs from the above genera in several principal features, e.g., it differs from *Elymus* in its one-flowered spikelets with connate, subulate glumes, and annual life cycle. *Taeniatherum* differs from *Hordelymus* in its sessile spikelets with connate glumes, flattened lemma awns and annual life cycle, and differs from *Hordeum* in its tough rachis, rigid spike with a terminal spikelet and sessile spikelets in pairs (Frederiksen 1986). Consequently, Nevski (1934a) considered it a separate genus, *Taeniatherum* (Fig. 5.5a).

The genus *Taeniatherum* is characterized by a broad morphological variation, making the classification of its specific rank difficult. Consequently, the taxonomic literature contains different classifications of the specific rank in this genus. In his monograph on *Taeniatherum*, Nevski (1934a) defined three species, *T. caput-medusae*, *T. crinitum*, and *T. asperum*, while Bor (1968) recognized only two: *T. crinitum* and *T. asperum*, and noted that they only differ very little from one another in several morphological traits. The distinguishing characters between the taxa that Nevski (1934a) and Bor (1968) defined, were glume length and spreading of glumes after ripening, length of the lemma, and width of the lemma awn base. The culm length to spike length ratio was sometimes used to distinguish between the taxa Nevski 1934a, 1936; Maire and Weiller 1955; Humphris 1978). However, all these traits show gradual transition between the taxa and merely indicate quantitative, rather than qualitative differences, rendering them impractical to use for identification purposes. In his taxonomical treatment of the genus, Humphris (1978) combined the above-mentioned species of Nevski (1934a) and Bor (1968) to one, *T. caput-medusae*, which contains several morphological variants. He argued that the morphological characteristics that distinguished between those defined by Nevski and Bor were unstable and variable, and did not justify a split of the species.

While revising the taxonomy of *Taeniatherum*, Frederiksen (1986) thought that the morphological characters that led Nevski to split the species into three, were indistinct, with many intermediates occurring between them, rendering it rather difficult to define these taxa morphologically. Since, morphological features, geographical extension and crossing experiments (Sakamoto 1969; Frederiksen and von Bothmer 1986) indicate some taxonomic differentiation, Frederiksen (1986) treated the three Nevski species as subspecies of *T. caput-medusae* (L.) Nevski, namely, subsp. *caput-medusae*, subsp. *crinitum* (Schreb.) Melderis, and subsp. *asperum* (Simk.) Melderis. She stated that the absence of discontinuities in the morphology supports a sub-specific rank.

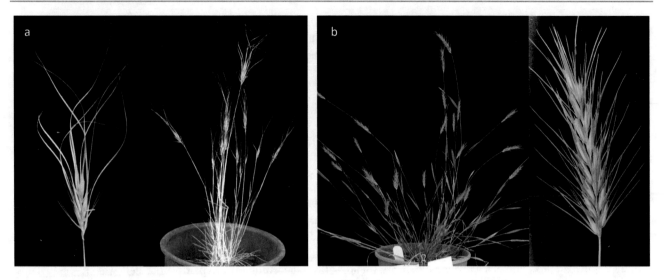

Fig. 5.5 Species of *Taeniatherum-Crithopsis* clade; **a** A spike and a plant of *T. caput-medusae* (L.); **b** A plant and spike of *C. delileana* (Schult.) Rozhev

Taeniatherum caput-medusae (L.) Nevski (common name medusa-head) is a small genus of annual, short plants (20–61 cm tall) with dense spikes, bearing one terminal spikelet and otherwise paired sessile spikelets, and a tough rachis, that does not disarticulate at maturity. It has two-flowered spikelets, with a lower hermaphrodite floret, and an upper floret reduced to a scale-like rudiment. At maturity, glumes are divergent from the rachis, forming an acute or obtuse angle. The glumes and lemma are awned. Glumes 3–5 cm long (including awns). Lemma's awns are flattened (0.4–1.5 mm wide) at the base, 6.0–13.5 cm long (Incl. lemma), and twisted when dry. The palea is a little longer than the body of the lemma. Anthers are small (about 1 mm in length), and the mature caryopsis is firmly adhered to the lemma and palea, which are released from the glumes, that remain on the rachis. The very small anthers that shed most of the pollen inside the floret indicate a facultative autogamous species (Frederiksen and von Bothmer 1986) (Fig. 5.5a).

The species distributes from West Mediterranean to Central Asia, i.e., Portugal, Spain, Southern France, Morocco, Algeria, Italy (incl. Sardinia and Sicily), Croatia, Serbia, Bosnia-Herzegovina, Macedonia, Albany, Turkey, Greece (incl. Crete), Romania, Bulgaria, Southern Hungary, Southwestern Russia, Crimea Caucasia, Cyprus, Tunisia, Libya, Egypt (incl. Sinai), Israel, Jordan, Syria, Lebanon, Iraq, Iran, West Pakistan, Afghanistan, Turkmenistan, Uzbekistan, Tajikistan, and Kirgizstan. Outside this area, it has been introduced as a weed in the northern and northwestern parts of Europe, in North and South America and Australia. *T. caput-medusae* inhabits low mountains and plateau areas, growing on altitudes from 600 to 800 m in Yatir, Israel, to 1850 m in south Sinai, Egypt, in a range of annual precipitation from 100 mm in Sinai to 1000–

1300 mm in Mt. Hermon. All sites it grows in are arid, somewhat extreme habitats, in sub-Mediterranean small-shrub formations and Irano-Turanian steppes, and is usually found on Nubian sandstone, basalt, and calcareous soil, as well as on stony or gravelly soils.

T. caput-medusae ssp. *caput-medusae* [(Syn.: *Elymus caput-medusae* L.; *Hordeum caput-medusae* (L.) Coss. & Dur.; *Cuviera caput-medusae* (L.) Simk.; *Hordeum caput-medusae* subsp. bobartii Asch. & Graebn.; *Elymus caput-medusae* var. *typicus* Halácsy; *Elymus caput-medusae* subsp. *bobartii* (Asch. & Graebn.) Maire; *Hordelymus caput-medusae* (L.) Pign.; *Taeniatherum caput-medusae* var. *caput-medusae* Humphries] bear glumes that are 3.5–8.0 mm long and horizontal. Paleae are 5.0–8.5 mm long. The awn base is 0.5–0.8 mm wide. This subspecies is restricted to the western part of the species distribution area, i.e., Portugal, Spain and southernmost France in Europe, and Morocco and Algeria in Africa.

T. caput-medusae ssp. *crinitum* (Schreb.) Melderis (1984) [syn.: *Elymus crinitus* Schreb.; *Hordeum crinitum* (Schreb.) Desf.; *Elymus caput-medusae* var. *crinitus* (schreb.) Ball.; *Elymus caput-medusae* subsp. *crinitus* (Schreb.) Nyman; *Hordeum caput-medusae* subsp. crinitum (Schreb.) Asch. & Graebn.; *Taeniatherum crinitum* (Schreb.) Nevski; *Hordelymus caput-medusae* subsp. *crinitus* (Schreb.) Pign; *Taeniatherum caput-medusae* var. *crinitum* (Schreb.) Humphries] bear glumes that are 1.5–4.0 mm long, erect or curved. Glume awns are 0.4–1.0 mm wide at base. Lemmas are 8.5–14.5 mm long. Lemma awns are 0.6–1.0 mm wide at base. This subspecies is found from Italy and eastwards into Asia. Within a large part of the distribution area, morphological transitions are found between this subspecies and subsp. *asperum*.

T. caput-medusae ssp. *asperum* (Simk.) Melderis [syn.: *Cuviera caput-medusae* var. *aspera* Simk; *Hordeum caput-medusae* subsp. *asperum* (Simk.) Degen in Asch. & Graebn.; *Cuviera aspera* (Simk.) Simk.; *Eltmus caput-medusae* var. *asper* (Simk.) Halácsy; *Elymus asper* (Simk.) Brand in Hallier & Brand; *Hordeum caput-medusae* var. *asperum* (Simk.) Fom. & Wor. Ex Fedtsch.; *Taeniatherum asperum* (Simk.) Nevski; *Elymus caput-medusae* subsp. *critinus* var. *asper* (Simk.) Maire; *Hordelymus asper* (Simk.) Beldie in Savulescu; *Hordelymus caput-medusae* subsp. *asper* (Simk.) Pign.] bears glumes that are 1.5–4.0 mm long, erect or curved. Lemmas are 5.0–14.5 mm long. The awn base is 0.4–1.0 mm wide. This subspecies has a very broad distribution, as it is found in most parts of the distribution area of the species. Morphological intermediary specimens between this subspecies and the others are found.

5.8.2 Cytology, Cytogenetics and Evolution

Chromosome number was determined by Sakamoto and Muramatsu (1965), Sakamoto 1969), Bowden (1966), Coucoli and Symeonides (1980), Luria (1983), Frederiksen (1986) and Linde-Laursen and Frederiksen (1989), who found that all taxa are diploids (2n = 2x = 14). The genome symbol of *Taeniatherum caput-medusae* is Ta (Wang et al. 1995). The three subspecies of *T. caput-medusae* have the same karyotype, with no interspecific variation (Frederiksen 1986). The karyotype is symmetric, with six metacentric pairs (one of which is a SAT-chromosome with a small satellite) and one sub-metacentric. All seven pairs of chromosomes were of nearly equal size, so that it was difficult to distinguish between them with certainty.

In an attempt to characterize the karyotype of each subspecies, Linde-Laursen and Frederiksen (1989) studied the C-banding patterns of somatic metaphases in plants from several populations of all three taxa. The C-banding patterns of all three subspecies were rather similar and characterized by a majority of small or very small bands. The number of bands per chromosome varied from 2 to 12 and had no preferential disposition. The banding pattern polymorphism was narrow but was sufficient for identifying the homologues of each of the seven chromosome pairs. No larger polymorphism was found within the sub-species, indicating that this trait is of no diagnostic value in distinguishing between the subspecies, and supported the view of a close relationship between them (Frederiksen 1986).

Frederiksen and von Bothmer (1986) hybridized the three subspecies of *Taeniatherum* and studied chromosomal pairing at meiotic first metaphase and fertility in the F$_1$ intra-specific hybrids. Crosses within a subspecies were as difficult to perform as crosses between subspecies. The intra-specific hybrid plants were vigorous but completely sterile (Frederiksen and von Bothmer 1986). Meiotic pairing in the parental plants was high, predominantly in the form of ring bivalents and 13.10–14.27 chismata/cell. Chromosomal pairing in the F$_1$ intra-specific hybrids also showed a very high degree of chromosomal pairing (average of 11.0 chiasmata /cell; no multivalents were observed), indicating that all three sub-species have the same basic genome. The almost complete lack of multivalents in the hybrids also shows that the basic genome has not been subjected to any larger structural rearrangements. However, this finding is in contradiction with the observations reported by Sakamoto (1969), who observed in several pollen mother cells of a sterile F$_1$ hybrid of ssp. *asperum* x ssp. *crinitum*, a trivalent and in several others, a quadrivalent, in addition to PMCs with seven bivalents or six bivalents and two univalents (average chromosome pairing per cell was 1.9 univalents, 5.8 bivalents, 0.1 trivalents, and 0.2 quadrivalent). From these observations, Sakamoto (1969) concluded that the two taxa had very similar but structurally differentiated genomes. However, it is most likely that one of the accessions used by Sakamoto (1969) contained a small reciprocal translocation that did not exist in the accessions used by Frederiksen and von Bothmer (1986). Consequently, Frederiksen and von Bothmer (1986) concluded from the pairing data, that the three subspecies share a high degree of genome homology, and thus, supported Frederiksen's (1986) treatment of these taxa as subspecies. This conclusion fails to account for the complete sterility of the intra-specific F$_1$ hybrids, indicating that certain genetic barriers exist between the subspecies. Frederiksen and von Bothmer (1986) assumed that these genetic barriers might have resulted from the processes of adaptation to the different conditions of the large variety of ecological habitats occupied by this species, that together with the effect on adaptive gene complexes, also selected genes involved in the intra-specific barriers.

The crossability of *T. caput-medusae* with other Triticeae species has been low, making it difficult to produce intergeneric hybrids, and consequently, the cytogenetic relationships between this species and other Triticeae species are poorly known. For instance, Sakamoto (1973) tried to cross *T. caput-medusae* with *Crithopsis delileana* and did not succeed to obtain F$_1$ seeds. However, several intergeneric crosses between *Taeniatherum caput-medusae* and other Triticeae species were successfully performed. Schooler (1966) crossed *T. caput-medusae* with *Aegilops cylindrica* and obtained a highly sterile triploid F$_1$ hybrid. At meiosis, univalents were the most common configuration, while rod bivalents and especially ring bivalents were only rarely observed (Schooler 1966). Thus, he concluded that the three genomes, Ta of *T. caput-medusae* and CD of *Ae. cylindrica*, are only distantly related.

Sakamoto (1991) analyzed intergeneric hybrids between *T. caput-medusae* ssp. *crinitum* and *Eremopyrum orientale*

(4x) and *Agropyron tsukushiense*. Chromosome pairing in several plants of the F₁ triploid hybrid *E. orientale* (4x) x *T. caput-meusae* ssp. *crinitum* (2x) was low (average of 19.3–20.6 univalents and 0.2–0.8 bivalents per cell) and the hybrid was completely sterile. Chromosome pairing in the F₁ tetraploid hybrid between *Agropyron tsukushiense* (6x) x *T. caput-medusae* ssp. *crinitum* (2x) was also low (average of 26.1 univalents and 0.9 bivalents per cell), and no seed set was obtained. No genomic homology was found between these three genera (Sakamoto 1991).

Frederiksen and von Bothmer (1989) crossed the 3 subspecies of *Taeniatherum caput-medusae* with 30 different species representing 11 genera of the Triticeae. A seed set were observed in 15 combinations only. However, most of the seeds lacked an embryo or the embryo was unable to form a vigorous seedling, and consequently, only six resulted in adult plants. In the combination of *Taeniatherum caput-medusae* with *Hordeum bulbosum* (2x), a haploid of *T. caput-medusae* was obtained as a result of selective elimination of the *H. bulbosum* chromosomes. All other hybrids included the following five combinations: *T. caput-medu*sae ssp. *caput-medusae* x *Psathyrost*achys *fragilis*, *T. caput-medu*sae ssp. *crinitum* x *Psathyrost*achys *juncea*, ssp. *crinitum* x *Dasypyrum villosum*, ssp. *asperum* x *Hordeum brevisubualatum*, and ssp. *crinitum* x *Eremopyron. orientale* (4x). All F₁ hybrids were morphologically intermediate between the parents. Perenniality dominated over annuality in the combinations with the Hordeinae species, and annuality in hybrids with *Dasypyrum* and *Eremopyrum*. All hybrids were highly sterile due to very low chromosome pairing at meiosis. The pairing data supported the doctrine that *Taeniatherum* is a distinct genus within the Triticeae, but slightly related to the genomes of *Psathyroatachys*, *Dasypyrum*, *Eremopyrum* or *Hordeum*.

Löve (1984) assumed that the tetraploid genus *Hordelymus* possesses one genome from *Hordeum* and another from *Taeniatherum*. As no hybrids between these two genera survived (Frederiksen and von Bothmer 1989), homology between the genomes of these two genera could not be proven nor rejected. Yet, von Bothmer et al. (1994) ruled-out the presence of the H genome in *Hordelymus* and demonstrated the presence of the Ns genome instead. Likewise, Bieniek et al. (2015) showed that *Psathyrostachys juncea* (genome NsNs) contributed its genome as a maternal parent to the allotetraploid *Hordelymus eyuropaeus* and not *Taeniatherum caput-medusae*, and suggested a genomic formula of NsNsXrXr for *Hordelymus*.

Frederiksen (1994) crossed *Taeniatherum caput-medusae* with *Triticunm aestivum* and obtained an F₁ hybrid. Morphologically, the hybrid looked like *T. aestivum*, with broad leaves and one spikelet per node, with three florets. The hybrid was completely sterile. Analysis of meiotic pairing at first meiotic metaphase showed a high number of univalents

(20.02), an unexpected high number of bivalents (3.52) (rod 3.44 and ring 0.08), and trivalents (0.08), resulting in 1–10 chiasmata/cell. The relatively high pairing was ascribed to pairing between the homoeologous chromosomes of the A, B, and D subgenomes of wheat. It is concluded that the genome of *Taeniatherum* eases the restriction on homoeologous pairing in *T. aestivum*, imposed by the *Ph1* gene.

As *Taeniatherum* seems to be a distant relative of *Triticum*, it is reasonable to suppose that the observed pairing in the hybrid only occurred between wheat chromosomes. The mean number of chiasmata in euhaploid *T. aestivum* containing *Ph1* was reported to be 1.06, while it was 8.48 in euhaploid mutants lacking *Ph1* (Jauhar et al. 1991). The number of chiasmata per cell in the hybrid studied by Frederiksen (1994) was 4.26 and thus, fell between the range referred to above. Thus, the genotype of *Taeniatherum* seems to have some repressive effect on the *Ph1* gene, resulting in increased pairing of homoeologous wheat chromosomes.

5.8.3 Phylogeny

In most morphological trees, *Taeniatherum* is linked to the *Hordeum* group, mainly because it shares the characteristic of multiple spikelets per node (Baum 1983; Baum et al. 1987; Kellogg 1989; Frederiksen and Seberg 1992). In contrast, phylogenetic studies based on molecular analyses have placed *Taeniatherum* close to *Secale*, *Triticum* and *Aegilops* (e.g., Mason-Gamer et al. 2002, based on chloroplast DNA, and Hsiao et al. 1995, Seberg and Petersen 2007, and Escobar et al. 2011, based on nuclear DNA sequences).

Hsiao et al. (1995) used the sequences of the internal transcribed spacer (ITS) region of nuclear ribosomal DNA and sequences of tRNA to estimate phylogenetic relationships among 30 diploid Triticeae species representing 19 genomes. They found that most of the annuals of Mediterranean origin, i.e., species of *Triticum*, *Aegilops*, *Crithopsis*, *Taeniatherum*, *Eremopyrum*, *Henrardia*, *Secale*, and two perennials, *Elymus farctus* and *Elymus elongatus*, constitute a monophyletic group. In a more restricted species sampling, the two perennial parsimony tree species *Elymus farctus* and *E. elongatus*, formed a sister group with *Triticum monococcum*, *Aegilops speltoides*, and *Ae. tauschii*. *Crithopsis*, *Taeniatherum*, *Eremopyrum*, and *Henrardia* were close to *Secale*. Based on this finding, Hsiao et al. (1995) suggested that the *Triticeae* apparently independently evolved similar sorts of characters in two parallel lineages, one in the Mediterranean and one in the Arctic-temperate region. Several conspicuous morphological characters appear in both groups, for example, a similar number of spikelets per node, a similar number of florets per spikelet, linear spikes, and keeled glumes. Species in the Mediterranean group that formed a monophyletic lineage in the ITS tree, were grouped with various perennials of Arctic-temperate origin on the morphology trees (Kellogg 1989; Frederiksen and Seberg 1992; and following Seberg Frederiksen, unpublished data). This apparent agreement between the molecular and bio-geographic data, and conflict of both of these lines of evidence with the morphological trees, suggest that large-scale morphological parallelism occurred in the evolutionary history of the Triticeae. The results of Hsiao et al. (1995) supported the suggestion of Sakamoto (1973) to classify Triticeae as two major groups, a Mediterranean group and an Arctic-temperate group (Table 2.3), with the Mediterranean lineage evolving from the Arctic-temperate species.

Mason-Gamer et al. (2002) analyzed new and previously published chloroplast (cp) DNA data from *Elymus* and from most of the mono-genomic genera of the Triticeae, and presented additional cp DNA data from Elymus and from mono-genomic genera and constructed the phylogeny for the mono-genomic genera. Their analysis was in agreement with previous cpDNA studies with regard to the close relationship between *Secale*, *Taeniatherum*, and *Triticum–Aegilops*. Further, their analysis provided moderate support for some additional still unresolved relationships or for those that were very weakly supported in the previous cpDNA studies. These included (i) the sister relationship between *Taeniatherum* and *Triticum–Aegilops*; (ii) the placement of *Heteranthelium* with a *Secale* + *Taeniatherum* + *Triticum–Aegilops* clade; and (iii) the placement of the *Elymus farctus* and *E. elongatus* + *Dasypyrum* clade with the *Secale* + *Taeniatherum* + *Aegilops–Triticum* + *Heteranthelium* clade. A close relationship between *Taeniatherum* and *Triticum–Aegilops* is completely at odds with the DMC1 tree (Petersen and Seberg 2000) and the morphology-based cladogram (Seberg and Frederiksen 2001). The ITS data left the relationship unresolved (Hsiao et al. 1995), and *Taeniatherum* was not included in the 5S long spacer data set (Kellogg and Appels 1995). But, Escobar et al. (2011) found that the clade containing *Taeniatherum* and *Triticum–Aegilops* is also seen on the 5S short spacer data tree, but only if *Elymus farctus* and *E. elongatus* are included in the clade. Some conflict appears in the position of *Secale* in the starch synthase tree, in which a close relationship among *Triticum*, *Aegilops*, and *Heteranthelium* was detected, while *Secale* was placed in a *Secale–Taeniatherum–Elymus–Dasypyrum* group (Mason-Gamer and Kellogg 2000).

The above phylogenetic studies suggest that *Taeniatherum* is closer to the species of the sub-tribe Triticineae than to those of the Hordeineae. Hence, use of the number of spikelets on each node, as a diagnostic marker for placing genera in one of the two sub-tribes (Tzvelev 1976; Clayton and Renvoize 1986), is not sufficient for such a grouping.

5.9 *Crithopsis* Jaub. & Spach.

5.9.1 Morphological and Geographical Notes

The genus *Crithopsis* was previously included in one of the *Elymus*, *Hordeum* and *Eremopyrum* genera, however, Jaubert and Spach (1851) recognized it as a separate genus. *Crithopsis* is monotypic and is represented by the species *C. delileana* (Schult.) Rozhev.

Crithopsis delileana [syn.: *Elymus delileana* Schult.; *Elymus geniculatus* Del.; *Elymus aegyptiacus* Spreng.; *Elymus rhachitrichus* Hochst. ex Jaub. & Spach; *Elymus subulatus* Forssk.; *Hordeum delileanum* (Schult.) Hack.; *Hordeum geniculatum* (Delile) Thell.; *Eremopyrum cretense* (Coustur. & Gand.) Nevski; *Crithopsis rhachitricha* Jaub. & Spach; *Crithopsis brachytricha* Walp] is annual, culms 10–30 cm high, spike dense and bristly, 2–5 cm long (excl. awns), with paired sessile spikelets at each node of rachis. The rachis is densely hairy and fragile and disarticulates above each node (wedge-type disarticulation). Spikelets contain two florets, where the lower is hermaphrodite and the upper rudimentary. Glumes are equal in size and linear, tapering to an awn. Lemma taper to form a long flat awn, equal in length to the glume awn. Anthers are about 0.8 mm long. The caryopsis is firmly adherent to lemma and palea. The dispersal unit is the rachis internode, which carries two spikelets with a single grain in each (Fig. 5.5b).

The species is characterized by a wide morphological variation, mainly in the color of the leaf sheath, hairiness of the leaves and glumes and lemmas, size of rachis internodes, glumes and lemmas. Luria (1983) found no difference in the seed sets of bagged and unbagged spikes, indicating that the pollination system is autogamous. This finding was corroborated by Frederiksen (1993) who, on account of the short anthers, assumed this species to be predominantly autogamous.

The distribution of the species is from North Africa, through the southeastern Mediterranean basin to Central Asia [Morocco, Algeria, Tunisia, Egypt, Libya, Greece (incl. Grete), Turkey, Cyprus, Syria, Lebanon, Jordan, Israel, Iraq, Iran, Afghanistan, and west Pakistan (Baluchistan)]. The species grows in a wide range of habitats, on a variety of climatic conditions and on different soils [terra rossa, rendzina (dark and light), basalt, grumusol, sandy soil, loess, and desert lithosol]. It is found at 100–950 m above sea level, from xeric climates with 100 mm annual rainfall to more mesic conditions, with 800 mm annual rainfall. *Crithopsis delileana* grows in dry steppe grassland and batha (Mediterranean small shrub formations), as well as in semi-disturbed habitats, in plant communities consisting of Irano-Turanian and sub-Mediterranean plants.

5.9.2 Cytology, Cytogenetics and Evolution

All analyzed accessions of *Crithopsis delileana* are diploids, 2n = 2x = 14 (Sakamoto and Muramatsu 1965; Bowden 1966; Sakamoto 1991, 1973; Luria 1983; Frederiksen 1993; Linde-Laursen et al. 1999). Löve (1984) designated its haploid genome as K. The karyotype of all analyzed accessions was similar, i.e., symmetric, consisting of five pairs of metacentric chromosomes, one pair of sub-metacentric chromosomes, and one pair of metacentric SAT-chromosomes, with rather small satellites (Frederiksen 1993; Linde-Laursen et al. 1999). Only the SAT-chromosome pair and the sub-metacentric pair could be identified reliably by morphology, whereas the morphological differences between the other five pairs of metacentric chromosomes were insufficient for safe identification. Linde-Laursen et al. (1999) made use of Giemsa C-banding, Giemsa N-banding and silver nitrate staining to discriminate between the chromosome pairs of *C. delileana*. The Giemsa C-banding patterns included a few small to very small, mainly centromeric or telomeric bands (Linde-Laursen et al. 1999). The banding patterns of the different populations were polymorphic, but within populations, the variation in banding patterns was sufficient for identifying the homologous chromosomes of each of the seven pairs. N banding produced no or few weakly developed bands in the chromosomes at the same positions as C bands. Silver nitrate staining identified two nucleolar organizer regions (NORs), at the nucleolar constriction of the pair SAT-chromosomes. The number of NORs was confirmed by observing a maximum of two nucleoli at interphase (Linde-Laursen et al. 1999).

Intergeneric hybridization between *Crithopsis* and other members of the Triticeae are very difficult. Attempts to produce intergeneric hybrids between *C. delieana* and *Aegilops peregrina* (=*Ae. variabilis*), *Eremopyrum bonaepartis*, *E. triticeum Taeniatherum caput-medusae* ssp. *caput-medusae* and *Agropyron tsukushiensis* have always failed (Sakamoto (1973). Likewise, Frederiksen and von Bothmer (1989) had no success in crossing *C. delileana* with *Taeniatherum caput-medusae*. However, a single weak hybrid was produced in the cross of *C. delileana* and *Eremopyrum distans* (Frederiksen 1993).

Crithopsis delileana and *Taeniatherum caput-medusae* are considered close taxonomically to each other (Clayton and Renvoize 1986) and morphologically similar. Both feature spikes with two sessile spikelets at each rachis node and a single hermaphroditic floret at each spikelet. In contrast to *Crithopsis*, which has a fragile rachis, *Taeniatherum* bears a tough rachis and a disarticulating rachilla. The karyotype of *C. delileana* is morphologically similar to that of *T. caput-medusae*, as shown by Linde-Laursen and Frederiksen (1989). However, the C-banding patterns of the two species exhibit differences in the distributions of the C-bands, with *C. delileana* having more telomeric and fewer intercalary bands than *T. caput-medusae*. Yet, altered distribution of C-bands is a weak diagnostic characteristic, as activity of transposable elements, that can affect the distribution and quantity of the C-banding, may be different in closely related species and even within a species.

5.9.3 Phylogeny

The genera *Crithopsis* and *Taeniatherum* are traditionally considered related to *Hordeum* and *Psathyrostchys* and therefore, are included in the Hordeineae (Tzvelev 1976; Clayton and Renvoize 1986). All four genera have more than one spikelet per rachis node and only one hermaphroditic floret per spikelet. Morphologically-based phylogenetic analyses supported inclusion of the four genera within the same, albeit not fully resolved, clade (Kellogg 1989; Frederiksen and Seberg 1992). Following her cladistics analysis, Kellogg (1989) even proposed to include *Crithopsis* and *Taeniatherum* in *Hordeum* in a single clade. Similar analysis performed by Frederiksen and Seberg (1992) placed *Crithopsis*, *Taeniatherum*, *Hordeum* and *Psathyrostachys* in a single clade.

Yet, the cytological studies of Linde-Laursen et al. (1999) indicate a closer relationship between the genera *Crithopsis* and *Taeniatherum* than between *Hordeum* or *Psathyrostachys*. In contrast to the comparatively minor differences in chromosome morphology distinguishing between the karyotypes of *C. delileana* and *T. caput-medusae*, the karyotypes of members of the genera *Pasthyrostachys* and *Hordeum* show several distinct characteristics (Linde-Laursen et al. 1999). Species of *Psathyrostachys* have significantly larger chromosomes, including SAT-chromosomes with generally minute satellites, as compared to the species of the other three genera, while species of *Hordeum* have chromosomes that produce bands after N-banding (Linde-Laursen 1981; Morris and Gill 1987; Xu and Kasha 1992). This banding pattern, unobserved in the other genera, indicates a significant qualitative difference in the composition of the constitutive heterochromatin (Gill 1987). The above differences support the taxonomic approach relating to them as individual genera.

There is also a discrepancy between the phylogenetic trees derived from morphological analysis versus those generated following molecular analysis of either chloroplast, mitochondrial or nuclear DNA sequences (Seberg and Frederiksen 2001; Seberg and Petersen 2007; Mason-Gamer 2005; Petersen et al. 2006; Escobar et al. 2011). In the morphological trees, Kellogg (1989), Frederiksen and Seberg (1992), Seberg and Frederiksen (2001), and Seberg and Petersen (2007) included *Crithopsis* and *Taeniatherum* in the *Hordeum* group, mainly because they share the characteristic of multiple spikekets per rachis node. Seberg and Petersen (2007) showed an incongruence between morphological evaluations and nucleotide sequence analysis of two plastid genes (*rbc*L, *rpo*A), one mitochondrial gene (*cox*II), and two single-copy nuclear genes (*DMC1*, *EF-G*). In addition, they found that data derived from chloroplast genes were somewhat incongruent with those from the mitochondrial and nuclear sequences. Similar limited agreement was found between gene trees derived from nuclear genes (the two 5S rDNA arrays and ITS) and those derived from the chloroplast (cpDNA RFLPs and *rpo*A; Kellogg et al. 1996; Petersen and Seberg 1997), both of which deviated considerably from the morphologically based phylogenies.

Based on the chloroplast data, Seberg and Petersen (2007) placed *Crithopsis* close to *Secale*, both of which were close to *Aegilops/Triticum* and *Taeniatherum*. Similar results showing that *Crithopsis* is close to *Secale* were obtained by Petersen and Seberg (1997), who analyzed the plastid genome spanning the entire *rpo*A gene, and by Petersen et al. (2006), who analyzed the plastid gene *ndh*F. They found that *Crithopsis* and Secale were sister species to a clade that includes *Aegilops*, *Triticum*, and *Taeniatherum*. Yet, based on the mitochondrial and the nuclear data, Seberg and Petersen (2007) found that *Crithopsis* was linked to the *Aegilops/Triticum* group. This finding is in agreement with that of Mason-Gamer and Kellogg (1996a), whose nuclear data supported classification of the clade of *Ae. tauschii*, *Triticum monococcum*, *Taeniatherum caput-medusae*, *Elymus farctus* ssp. *bessarabicus*, *Crithopsis delileana* and *Elymus elongatus*. Similar results were also obtained by Petersen et al. (2006), who studied the nuclear gene *DMC1*, and found that *Crithopsis* was close to *Aegilops/Triticum* and *Elymus elongatus*. Likewise, Hsiao et al. (1995) analysed the internal transcribed spacer (ITS) region of nuclear ribosomal DNA and found that *Crithopsis* is close to *Secale* and included both in the *Aegilops/Triticum* clade.

Mason-Gamer (2005) constructed a phylogenetic tree based on analysis of the β-amylase genes of the Triticeae. This tree consists on an evidence-supported clade included *Aegilops*, *Triticum*, *Crithopsis*, and *Taeniatherum*. Placement of *Crithopsis* (and *Taeniatherum*) within or very near *Aegilops* was supported by most of the molecular data sets that included *C. delileana* (see Table 5 of Mason-Gamer 2005), including the analyses of cpDNA restriction sites (Mason-Gamer et al., 2002), the *rpo*A gene (Petersen and Seberg 1997), the integrated highly repetitive genes (Kellogg et al. 1996), and the *DMC1* gene (Petersen and Seberg 2002). In full agreement with the above, Seberg and Petersen (2007) presented a highly resolved, strongly supported,

consensus phylogenetic tree, based on all their data, which included morphological assessments, and analyses of two chloroplast genes, one mitochondrial gene, and two nuclear genes. As a rule, the tree positions *Crithopsis* and *Taeniatherum* in the same clade, which includes several other taxa as well, mostly *Aegilops*, but invariably excludes *Psathyrostachys* and *Hordeum*.

Close homology between DNA sequences of *Crithopsis delileana* and *Taeniatherum caput-mesdusae* with *Elymus elongatus* was found by Arterburn et al. (2011). Diploid *E. elongatus* and *C. delileana* were most closely related. *E. elongatus* was equally close to *E. farctus* ssp. *bessarabicus* and *T. caput-medusae*, but ssp. *bessarabicus* was shared less similarity with *T. caput-medusae* and *C. delileana* than with *E. elongatus*.

The traditional taxonomic subdivision of the genera of the Triticeae into two sub-tribes, the Hordeineae and the Triticineae (e.g., Tzvelev 1976; Clayton and Renvoize 1986), is not supported by the phylogenetic schemes derived from molecular analyses. Sakamoto (1973, 1991) classified the Triticeae into two groups based on their geographical distribution: the Arctic-Temperate group and the Mediterranean-Central Asiatic group (Table 2.3). In their phylogenetic analysis of ITSs of nuclear rDNA, Hsiao et al. (1995) recovered the Mediterranean group. The Arctic-Temperate group includes most of the Hordeineae (characterized by three spikelets per rachis node), while the Mediterranean-Central Asiatic group includes all of the Triticineae (characterized by solitary spikelets at each rachis node), as well as species of *Hordeum*, *Crithopsis* and *Taeniatherum* (characterized by two to three spikelets per rachis node). The latter group is younger and assumed to have developed from the Arctic-Temperate group (Runemark and Heneen 1968; Sakamoto 1973). DNA sequence-based phylogenetic analyses, indicate that the two annual genera, *Crithopsis* and *Taeniatherum*, that are classified in the Hordeineae on the basis of morphological features, are linked to the genera of the sub-tribe Triticineae. Their inclusion in the Mediterranean-Central Asiatic group is not merely due to their geographical distribution, but also to close phylogenetic relationships between the two genera, *Crithopsis* and *Taeniatherum*. Their relationships with genera of the Triticinae may indicate their evolvement from the same ancestral group and suggests that evolutionary development of diagnostic morphological characteristics, e.g., number of spikelets on each rachis node and number of fertile florets in each spikelet, occurs at varying rates. The dispersal unit of many species belonging to the Mediterranean-Central Asiatic contains two seeds, either one in each of the two spikelets or two in a single spikelet, indicating different routes for achieving analogous adaptive traits to brace the long, dry summer of the Mediterranean and Central Asiatic regions.

5.10 Concluding Comments

5.10.1 Evolution of the Sub-tribe Triticineae

The cold climate that prevailed during the Oligocene geological era (33.7–23.8 MYA; Table 2.5) channeled the evolution of the early Triticeae toward the development of perennial and allogamous genera adapted to mesophyllic habitats that are characteristic to the temperate-arctic zones. During the Miocene era (23.8–5.3 MYA; Table 2.5), the climate warmed up and the Tethys Sea, that covered a large area of the Mediterranean basin and southwest Asia, disappeared and the lands in these regions rose. The climate in the east Mediterranean and central Asia became seasonal, namely, with cold and humid winters and hot and dry summers. These geological and climatic changes, characterized by a relatively short growth period in the winter and long inactivity in the summer, have led to the development of diversified ecosystems, expansion of grasslands and opening of new ecological niches (Table 2.5). Such ecological changes were suitable for the development of the annual, autogamous Triticineae genera. It is therefore assumed that the radiation of species in the Triticeae tribe might have been triggered by the middle to late Miocene climate (Fan et al. 2013).

Indeed, several molecular studies indicated that the radiation of genera and species in the sub-tribe Triticineae, the Mediterranean and central Asiatic lineage, occurred during the later part of the Miocene era. Sequencing of chloroplast genes indicated that the divergence of the sub-tribe Triticineae from the subtribe Hordeineae took place either 15 (Marcussen et al. 2014), 10.6 (Gornicki et al. 2014), or 8–9 (Middleton et al. 2014) MYA, whereas sequencing of nuclear genes showed that this divergence occurred 11.6 (Chalupska et al. 2008), 11.0 (Huang et al. 2002a, b) or 10.1 (Dvorak and Akhunov 2005) MYA (Table 2.6). The molecular data and biogeography of the sub-tribe suggest that the Mediterranean lineage derived from the Arctic-temperate lineage and that the two lineages have since evolved in parallel (Hsiao et al. 1995).

5.10.2 Appearance of Advanced Traits in Different Genera

Ancestral trait is a trait that exists in a group of taxa that are all descendent from a common ancestor in which the trait first developed. Primitive trait represents the original condition of the trait in the common ancestor, whereas an advanced trait or a derived trait signifies an important change from the original condition.

Major evolutionary changes that facilitated rapid adaptation of the sub-tribe Triticineae to the newly opened Mediterranean and central Asiatic habitats, included the transition from perennial to annual growth habit, from allogamy to autogamy, from tall to short plants, from a few to many tillers, from long to short spikes, from several to a single spikelet on each rachis nod, and from un-awned to awned spikelets. In the Mediterranean and Central Asiatic climate, annual growth habit is an adaptive trait, since annual plants can pass the long, dry and hot summer as seeds. The evolution of species with an annual growth habit occurred independently several times in the Triticineae, either between genera (*Agropyron* and *Eremopyrum*), or within genera (in *Dasypyrum* and *Secale*) (Table 5.5). Perennial growth habit is a dominant trait controlled by a small number of genes (Charpentier et al. 1986; Lammer et al. 2004), where mutations in these genes lead to the development of annual plants. Allogamy is considered a primitive character since most perennial species in the tribe Triticeae are allogamous. Stebbins (1957) pointed out that in many plant groups, autogamy derived from allogamy and therefore, autogamy might be considered a more advanced trait. Indeed, autogamy developed independently several times from the cross-pollinating species in the Triticineae (Escobar et al. 2010; Table 5.5). While annual growth habit requires reestablishment of the populations each year, autogamy is well fit for mass reproduction of plants with similar well-adapted genotypes, that can occupy successfully the same habitats, and facilitates the rapid colonization of newly opened habitats (Mac Key 2005). Since autogamy in the Triticineae species is not an obligatory trait, occasional out-crossing may result in the production of sufficient genetic flexibility that effectively used by the rapid generation shift (Allard and Kannenberg 1968; Allard et al. 1968; Allard 1975). The annual plants pass the dry and hot summers as dormant seeds that are buried and stored in the soil until autumn rains begin and growth starts again. The change from dormancy to germination is a fundamental process superimposed by a series of genetic trigger mechanisms, which selectively respond to environmental factors like water, temperature, light, and oxygen (Mac Key 1987, 1989). Selection for ability to compete against other species in dense stands may be combined with lower competition within the species itself (Mac Key 2005). Short plants with a large number of tillers can better tolerate drought and hot winds. Awned spikelets are better protected from drought and herbivores and more efficient in seed dispersal and self-sowing. Not all these changes occurred in all the genera and presumably did not happen simultaneously (Table 5.5). The Triticineae genera in Table 5.5 are classified in five clades, each containing genus or genera with ancestral and advanced traits. Genera (or species) having more advanced traits than ancestral ones, e.g., in clade 1, *Eremopyrum* and *Henrardia* have more advanced traits than *Agropyron*, in clade 2, *Heteranthelium* has more advanced traits than *Dasypyrum* and *Secale,* in clade 3, *Aegilops* is more advanced than *Amblyopyrum* and *Triticum,* and in clade 4, *Crithopsis* is more advanced than *Taeniatherum.* The genus *Aegilops* contains species with ancestral traits and species with advanced traits displaying evolutionary changes within the genus. Interestingly, genera or species exhibiting more advanced traits, namely, species of *Eremopyrum, Taeniatherum, Crithopsis, Heteranthelium* and *Henrardia,* occupy more xeric habitats in the peripheries of the Triticineae distribution area.

5.10.3 Chromosomal Pairing Level in Intergeneric F₁ Hybrids

Despite the fact that the various Triticineae genera are relatively young and still maintain a great deal of genetic relatedness, their chromosomes are homoeologous rather than homologous, due to genetic and structural changes that have occurred during their divergence. Consequently, F_1 intergeneric hybrids display minimal meiotic pairing between the homoeologous chromosomes of the parental genera. For instance, pairing between *Secale* species and other Triticineae is affected by chromosomal rearrangements, such as translocations and inversions, that drove the evolution of *Secale* genomes. Also, the accumulation of large amounts of subtelomeric heterochromatin can change the relative position of the telomeric regions at the beginning of meiosis, hindering pairing initiation of *Secale* and other Triticineae chromosomes (Devos et al. 1995; Lukaszewski et al. 2012; Megyeri et al. 2013). Furthermore, genetic and epigenetic changes, such as mutations, inactivation or elimination of DNA sequences that are involved in homology recognition and initiation of pairing at the beginning of meiosis, may cause pairing failure in hybrids between various Triticineae genera (Feldman et al. 1997; Ozkan et al. 2001).

Sybenga (1966) postulated that zygomeres control meiotic chromosomal pairing. The zygomeres may be made of homologous-specific DNA sequences that are clustered in "pairing initiation sites". Clustering of telomeres in the bouquet at the leptotene–early zygotene meiotic stages may facilitate the approaching of terminal regions of homologous chromosomes (Naranjo and Corredor 2008; Naranjo and Benaveste 2015). These terminal regions may contain homologous-specific DNA sequences that become active at these early meiotic stages and play a role in homology recognition and initiation of pairing. Pairing between chromosomes in interspecific and intergeneric hybrids only occurs when the chromosomes of the two parents share the same homologous-specific DNA sequences. It seems likely that the sequences initiating chromosome pairing between homologous chromosomes of one Triticineae genus may be different from those of another genus, thus hindering the alignment of homoeologous chromosomes at the beginning of meiosis and the initiation of chromosome pairing at early zygotene. Moreover, meiotic pairing in interspecific or intergeneric F_1 hybrids may either be promoted or suppressed by various pairing genes and B chromosomes that exist in these taxa (Table 5.2). Consequently, the reduced meiotic chromosomal pairing in intergeneric Triticineae hybrids, brought about by homoeologous pairing suppressors, cannot be used as an indication of the degree of evolutionary differentiation between the genomes of these genera.

5.10.4 Phylogenetic Scheme of the Triticineae Genera

5.10.4.1 *Elymus* Species with St, Ee and Eb Genomes

Escobar et al. (2011) classified the diploid Triticeae in five major clades: (1) *Psathyrostachys*; (2) *Hordeum*; (3) *Elymus;* (4) *Aegilops–Triticum–Secale–Taeniatherum*; and (5) *Eremopyrum–Agropyron.* The *Elymus* clade is the largest clade, consisting, among others, of representatives of the St, Ee and Eb genomes. These genomes are three important basic genomes from which, presumably, the subtribe Triticineae evolved. The St genome of several diploid species of *Elymus* (Table 5.1) is moderately related to the Ee genome of diploid *E. elongatus* and to the Eb genome of diploid *E. farctus* (Wang 1989). Petersen and Seberg (1997), based on *rpo*A sequence data, and Wang and Lu (2014), based on a literature review of chromosomal pairing data, confirmed the very close relationship among the St, Ee, and Eb genomes. The close relationship between St, Eb and Ee genomes was also substantiated by analysis of the sequences of a gene encoding plastid acetyl-CoA carboxylase (Fan et al. 2007), and between St and Eb by the study using 5S rDNA (Shang et al. 2007). Bieniek et al. (2015) found that the nucleotide sequences at three chloroplast loci (*matK, rbcL, trnH-psbA*) are almost identical in the diploid Ee, Eb and St taxa, with only one substitution within the *matK* gene, differentiating genome Eb from the Ee and St genomes. Since the St genome exists in more primitive diploid species of *Elymus,* it is assumed that genomes Ee and Eb evolved from St. Other Triticineae genera can be classified in the following four, semi-independent clades (Table 5.5).

Table 5.5 Possession of ancestral (bold) and advanced traits in Triticineae genera

An ancestral trait in all species ■ ; Several species exhibit an ancestral trait, others with an advanced trait ▨ ; All species exhibit an advanced trait □ .

Clade / Genus / Ancestral vs. advanced	Base[a]		1[b]			2[c]			3[d]			4[e]	
	Cae	Jun	Agr	Ere	Hen	Das	Sec	Het	Amb	Tri	Aeg	Tae	Cri
Diploidy vs. polyploidy	▨	▨	▨	▨	■	▨	■	■	■	▨	▨	■	■
Symmetric vs. asymmetric karyotype	■	■	■	□	□	▨	■	■	■	▨	▨	■	■
Caryopsis adhering to chaff vs. free	■	■	--	■	□	□	□	--	■	□	▨	■	■
Awns on lemma vs. awns on glumes	--	--	■	■	--	▨	■	--	--	■	▨	□	□
Awnless vs. awned spikelets	■	■	■	■	■	■	■	■	■	■	■	□	□
Spikelet vs. whole spike as dispersal unit	■	■	■	■	■	■	■	■	■	■	■	■	■
Floret vs. spikelet/spike disarticulation	■	□	■	▨	■	■	■	■	▨	■	■	□	□
Tough vs. fragile rachis	■	■	■	■	■	■	▨	■	■	■	■	□	□
Multi-floret vs. few-floret spikelets	■	■	▨	□	□	■	■	■	■	▨	▨	□	□
Multiple spikelets vs. solitary on each node	▨	□	□	□	□	□	□	□	□	■	■	■	■
Many vs. few spikelets per spike	■	■	□	□	■	■	□	□	■	□	□	□	□
Long vs. short spike	■	■	□	□	□	■	■	□	□	□	□	□	□
Few vs. many tillers	■	■	■	■	□	□	■	□	□	□	□	■	□
Tall vs. short plants	■	■	▨	□	□	■	■	□	■	▨	▨	□	□
Allogamy vs. autogamy	▨	■	■	□	□	▨	■	□	■	▨	▨	□	□
Perennial vs. annual growth habit	■	■	■	□	□	▨	▨	□	□	□	□	□	□
Rhizomatous vs. Caespitosa growth	□	■	▨	□	□	▨	□	□	□	□	□	□	□

[a] **Base**— *Elymus* sections: Caespitosae (Cae) & Junceae (Jun)

[b] **Clade-1:** *Agropyron* (Agr), *Eremopyrum* (Ere) & *Henrardi*a (Hen)

[c] **Clade-2:** *Dasypyrum* (Das), *Secale* (Sec) & *Heteranthelium* (Het)

[d] **Clade-3:** *Amblyopyrum* (Amb), *Triticum* (Tri) & *Aegilops* (Aeg)

[e] **Clade-4:** *Taeniatherum* (Tae) & *Crithopsis* (Cri)

5.10.4.2 The Clade *Agropyron–Eremopyrum–Henrardia*

The genus *Agropyron* most probably evolved from *Elymus* species bearing the St genome that is moderately related to genome P of *Agropyron* (Wang 1989). This assumption is supported by the analysis of 5S DNA sequences that consistently placed *Elymus* species with an St genome and *Agropyro*n (genome P) in one clade (Baum and Appels 1992). Additional phylogenetic studies, based on morphology (Seberg and Frederiksen 2001), chloroplast DNA (Mason-Gamer et al 2002; Hodge et al. 2010), chloroplast, mitochondrial, and nuclear DNA sequences (Seberg and Petersen 2007) and nuclear genes (Hsiao et al. 1995; Mason-Gamer et al. 2010), included *Eremopyrum* species in the same clade as species of *Agropyron*. Likewise, the close phylogenetic relationship between *Agropyron* and *Ermopyrum* is also evident from the data of Escobar et al. (2011), who placed these two genera in the same clade. It is assumed therefore, that *Eremopyrum*, whose species are self-pollinating annuals that exhibit many advanced traits (Table 5.5), may have evolved from diploids of the genus *Agropyron*.

Eeremopyrum species were grouped with *Agropyron* and *Henrardia* on the chloroplast DNA tree (Mason-Gamer et al. 2002). Similarly, using β-amylase gene sequences,

Mason-Gamer (2005) found that *Henrardia persica* is close to *Eremopyrum bonaepartis*. Dual placement of *Eremopyrum* and *Henrardia* with *Agropyron* has also been supported by other data sets (reviewed in Mason-Gamer 2005). Hsiao et al. (1995) reached a similar conclusion upon analysis of nuclear DNA sequences. Upon analysis of the chloroplast gene encoding ribosomal protein rps16, Hodge et al. (2010) also placed *Eremopyrum bonaepartis* and *Henrardia persica* in a single clade. Upon combination of Triticeae species mating system observations and data obtained from molecular analysis of 27 protein-coding loci, Escobar et al. (2010) found that *Henrardia persica* is very close to *Eremopyrum bonaepartis* and form a clade with *Eremopyrum triticeum* and *Agropyron mongolicum*. Likewise, Escobar et al. (2011) found two well-supported sub-clades, the first formed by *Henrardia* and *Eremopyrum bonaepartis*, and the second by *Agropyron mongolicum* and *E. triticeum*. The study of Escobar et al. (2011) showed that the genera *Agropyron*, *Eremopyrum* and *Henrardia* were not affected by introgression and/or incomplete lineage sorting.

5.10.4.3 The Clade *Dasypyrum–Secale–Heteranthelium*

Two cpDNA data sets, one based on restriction site variation (Mason-Gamer and Kellogg 1996b) and the other on sequences encoding the *rpoA* subunit of the RNA polymerase (Petersen and Seberg 1997), placed *Elymus* species possessing the E genome, e.g., *E. elongatus* and *E. Farctus*, together with *Dasypyrum* on cpDNA cladograms. Sequencing of the nuclear starch synthase gene also revealed a close relation between *Elymus* St species and *Dasypyrum* (Mason-Gamer and Kellogg 2000). The cpDNA tree contained a well supported clade that included *Dasypyrum*, with *Elymus* (St and Ee species). The *Dasypyrum-Elymus* (St species) monophyly was also observed in cladograms obtained from RFLP similarity patterns obtained from 14 cloned fragments covering the entire cpDNA of *T. aestivum* (Kellogg 1992), morphological data (Kellogg 1989) and 5S RNA (Appels and Baum 1991).

The nuclear DNA data (Hsiao et al. 1995; Kellogg and Appels 1995) are incongruent with the cpDNA data, as they suggest different affinities of St species of *Elymus* and *Dasypyrum*. However, molecular phylogeny of the *RPB2* (the second largest subunit of RNA polymerase II) gene sequence reveals that the *Dasypyrum villosum* genome is sister to the St genome of *Elymus* and that both diverged from the H genome of barley (Sun et al. 2008). This finding is in line with the phylogenetic relationships of monogenomic species of Triticeae inferred from nuclear rDNA (internal transcribed spacer) sequences, which established a close relation of *Dasypyrum* and *Heteranthelium* to the St genome species of *Elymus*.

The phyletic relationships among *Dasypyrum* and other Triticineae genera have been assessed at the level of morphology, protein and chloroplast and nuclear DNA sequences. Morphology-based phylogenetic analyses showed that *Dasypyrum* branched from a sister group of *Secale* within the same clade (Baum 1978a, b, 1983; Kellogg 1989; Frederiksen and Seberg 1992; Seberg and Frederiksen 2001). *Dasypyrum villosum* is morphologically similar to *Triticum*, in general (Baum 1978a, b), and to *Triticum monococcum* (Seberg and Frederiksen 2001), in particular. Kellogg (1989) placed *Dasypyrum* near *Agropyron* and *Triticum monococcum*, and Baum (1978a, b, 1983) considered *Secale cereale* and *D. villosum* as evolutionarily more contiguous to *Triticum* and *Aegilops* than to the rest of the Triticineae. Yang et al. (2006), studing genome relationship based on species-specific PCR markers, concluded that the formation of the *Dasypyrum* species started at the earlier stages of the separation of the sub-tribe Triticineae from the sub-tribe Hordeineae (13–15 MYA).

Phylogenetic studies based on molecular data suggested that *Secale* is the closest relative of the *Triticum-Aegilops* genera (Kellogg et al. 1996; Huang et al. 2002a; Mason-Gamer et al. 2002). When *Hordeum* and *Dasypyrum* were assessed with *Secale, Triticum* and *Aegilops*, they stood at the base of the tree topology as out-groups (Yamane and Kawahara 2005; Kawahara et al. 2008), implying a much earlier divergence between *D. villosum* and the common ancestor of *Triticum-Aegilops*.

DNA/DNA hybridization experiments, in which the genomes of *Secale cereale* and *D. villosum* were hybridized with labeled nuclear DNA from *Triticum aestivum* and *S. cereale*, revealed greater homology between *Dasypyrum* and *Secale* than between *Dasypyrum* and *Triticum* (Lucas and Jahier 1988). FISH analysis, in which the genomes of different species of the Triticeae were hybridized with species-specific molecular probes prepared from tandem repeated DNA sequences of *D. villosum* (pHv62) and *S. cereale* (pSc119.2), exhibited greater homology than the homology observed between *Secale* or *Dasypyrum villosum* genomes and those of *Triticum* and *Aegilops* (Uslu et al. 1999).

Escobar et al. (2011) included *Dasypyrum* and *Heteranthelium* in the same clade, and *Secale, Taeniatherum, Triticum* and *Aegilops* in another clade. St genome species of *Elymus* are sister to *Dasypyrum*. Lucas and Jahier (1988) concluded that the differentiation of *D. villosum* and *Secale cereale* from the other Triticineae genera occurred before speciation in *Aegilops* and *Triticum*. They assumed that *T. monococcum* subsp. *aegilopides* is the closest species to *D. villosum* from the *Aegilos-Triticum* group, and not diploid species of *Aegilops*. Shang et al. (2007) concluded that the St and Eb genomes of *Elymus* were also more closely related to the genome of *Secale cereale* than to the genome of *Dasypyrum villosum*.

The *Heteranthelium* transposon *Stowaway* is present in *Dasypyrum* but absent in other Triticeae species (Petersen and Seberg 2000). Variations in the PCR sequences of 6-SFT (sucrose-fructan 6-fructosyltransferase) placed *Heteranthelium piliferum* and *D. villosum* at one extreme of the phylogenetic relationships, whereas diploid species of *Secale, Triticum* and *Aegilops* were at the other extreme (Wei et al. 2000).

Taken together, the reported phylogenetic relationships of *Heretanthelium* remain ambiguous. Assessments of the chloroplast gene encoding ribosomal protein rps 16, performed by Hodge et al. (2010), placed *H. piliferum* in the same clade with *Triticum monococcum, Secale cereale*, and all the *Aegilops* species. Their results are consistent with the conclusions of Mason-Gamer et al. (2002), which were based on combined cpDNA sequences of tRNA genes, spacer sequences, rpoA genes and restriction sites. Phylogenetic relationships determined by mating systems showed that *Heteranthelium piliferum* is in a clade with *Dasypyrum villosum* (Escobar et al. 2010), despite the fact that *H. piliferum* is a facultative self-pollinated species, whereas *D. villosum* is an out-crosser. Similarly, Escobar et al. (2011) found that *Heteranthelium piliferum* comprises a clade with *Dasypyrum villosum* that branches at the base of the clade of *Triticum, Aegilops, Secale* and *Taeniatherum*. Thus, these two clades are closely related to each other.

Escobar et al. (2011) concluded that *Dasypyrum, Heteranthelium, Secale, Taeniatherum, Triticum* and *Aegilops*, evolved in a reticulated manner. The ambiguous reports concerning the phylogenetic relationships of *Dasypyrum, Secale, Heteranthelium, Triticum* and *Aegilops* may result from introgression between these genera and/or incomplete lineage sorting (Escobar et al. 2011).

5.10.4.4 The Clade *Amblyopyrum–Triticum–Aegilops*

Most morphological and molecular trees included *Amblyopyrum* in the *Aegilops* clade (Seberg and Petersen 2007); several molecular trees also included several species of *Elymus* in this clade (Mason-Gamer et al., 1998). In phylogenetic analyses based on nuclear DNA sequences, *Amblyopyrum* is a monophyletic taxon, although, the relationships of *Amblyopyrum* within the *Aegilops* clade remain unresolved (Frederiksen and Seberg 1992; Frederiksen 1993). Nevertheless, its position as an intermediate between *Elymus elongatus* and Sitopsis species of *Aegilops* was suggested by Eig (1929b). Numerical analysis (Baum 1977, 1978a, b; Schultze-Motel and Meyer 1981) indicated the close relationship of *Amblyopyrum* with *Aegilops* and *Triticum*, but also confirmed the morphological differences (Baum 1977).

Hsiao et al. (1995) and Kellogg et al. (1996) considered diploid wheat, *Triticum monococcum*, to be a sister group to *Elymus elongatus*. Upon studying internal transcribed spacers (ITS) of the nuclear rDNA sequences, Hsiao et al. (1995) reported that Ee and Eb jointly clustered with subgenomes A, B, and D of *Triticum aestivum*. In accord with this finding, Liu et al. (2007), using genomic hybridization (both Southern and in situ hybridization), demonstrated that the St and Eb genomes are very closely related to the A, B and D subgenomes of bread wheat, with the closest relation being to the D subgenome. Their findings provide a possible explanation as to why most of the spontaneous bread wheat—*E. elongatus* ssp. *ponticus* translocations and substitutions occur in the D subgenome, while only some occur in the A subgenome and rarely any in the B subgenome. Taken together, the diploid donors of subgenomes A, B, and D of the wheat group may have derived from the E genome of *Elymus* (reviewed in Wang and Lu 2014).

Studies of nuclear DNA sequences (genes or repetitious DNA) and chloroplast DNA sequences of the wheat group (the genera *Amblyopyrum, Aegilops* and *Triticum*), have shown significant inconsistencies, possibly due to both ancient and recent inter-specific and inter-generic hybridizations and introgressions (Kawahara 2009) that usually involve distal chromosomal region much more than proximal ones. Incongruence between chloroplast and nuclear genomic data has often been reported (Sasanuma et al. 2004; Kawahara 2009; Li et al. 2014). A monophyletic origin of *Aegilops* and *Triticum* was inferred from some of analyses (e.g., Hsiao et al. 1995; Kellogg and Appels 1995; Kellogg et al. 1996; Huang et al. 2002a, b), whereas a polyphyletic origin was deduced from others (Petersen and Seberg 1997, 2000; Seberg and Frederiksen 2001; Sallares and Brown 2004; Mason-Gamer 2005; Petersen et al. 2006). It is probable that intergeneric hybridizations and introgressions from other genera of the Triticineae e.g., *Elymus, Secale* and others, blurred the monophyletic origin of the wheat group.

Hammer (1980) assumed that allogamous, self-incompatible plants, resembling species of *Amblyopyrum muticum* and *Aegilops speltoides*, were ancestral types to the autogamous genera of the wheat group. He reached this assumption by combining anther length and amount of pollen produced, with reductions in anther size and/or amount of pollen, which suggested increasing autogamy, for which he found strong positive correlations. Long anthers that produce substantial amounts of pollen are an indication of allogamy or of a transitional state to autogamy. Based on these findings, Hammer (1980) produced an evolutionary model delineating the sequence of origin of the various species. Hammer's (1980) phylogenetic model assumes that *Amblyopyrum* separated from the ancestral *Aegilops* lineage at an early stage, of which *Ae. speltoides* is thought to be the most primitive representative of *Aegilops*. Phylogenetic speculation that assumes a change from allogamy (as in *A. muticum*) towards facultative autogamy (as in species of *Aegilops, Eremopyrum, Heteranthelium, Henrardia* and *Triticum*), coinciding with divergent morphological development, suggests separation from the ancestral *Aegilops* lineage at an early stage (Hammer 1980).

Dvorak and Zhang (1992a, b), analyzing repeated DNA sequences, concluded that *A. muticum* is close to *Ae. caudata, Ae. comosa, Ae. uniaristata* and *Ae. umbellulata*. Wang et al. (2000) compared the internal transcribed spacer (ITS) region of the ribosomal DNA in the diploids of the wheat group (including *A. muticum*) and observed wide divergences of this sequence between species. The highest divergence was between *Ae. speltoides* and *A. muticum*. In situ hybridization with repeated DNA markers and C-banding patterns, suggest that *A. muticum* occupies an isolated position, which is relatively closer to the Sitopsis species (*Ae. speltoides, Ae, bicornis, Ae. sharonensis, Ae. longissima* and *Ae. searsii*) than to other species of *Aegilops* (Badaeva et al. 1996a, b). Sallares and Brown (2004), who analyzed the transcribed spacers of the 18S ribosomal RNA genes, reached a similar conclusion, namely, that *A. muticum* has a basal position and that it is close to *Ae. speltoides*.

5.10.4.5 The Clade *Taeniatherum–Crithopsis*

Taxonomists assigned the two genera *Taeniatherum* and *Crithopsis* to the subtribe Hordeineae. In most morphological trees, *Taeniatherum* and *Crithopsis* were linked to the *Hordeum* group, mainly because they all present multiple spikelets per rachis node (Baum 1983; Baum et al. 1987; Kellogg 1989; Frederiksen and Seberg 1992). Yet, phylogenetic studies based on molecular analyses placed *Taeniatherum* and *Crithopsis* closer to *Secale*, *Triticum* and *Aegilops* than to Hordeum [e.g., Mason-Gamer et al. (2002), based on chloroplast DNA, and Hsiao et al. (1995) and Escobar et al. 2011), based on nuclear DNA sequences]. Hsiao et al. (1995) used the sequences of the ITS region of nuclear ribosomal DNA and sequences of tRNA to estimate phylogenetic relationships among 30 diploid Triticeae species representing 19 genomes. They found that most of the annuals of the Mediterranean origin, i.e., species of *Triticum*, *Aegilops*, *Crithopsis*, *Taeniatherum*, *Eremopyrum*, *Henrardia*, and the perennials, *Elymus farctus*, *E. elongatus*, and *Secale strictum*, comprise a monophyletic group. In the parsimony tree from a more restricted species sampling, the two perennial species *Elymus farctus* and *E. elongatus* formed a sister group to *Triticum monococcum*, *Aegilops speltoides*, and *Ae. tauschii*, whereas *Crithopsis*, *Taeniatherum*, *Eremopyrum*, and *Henrardia*, were close to *Secale*. Based on this finding, Hsiao et al. (1995) supported Sakamoto (1973), who suggested that the Triticeae should be classified as two major groups, a Mediterranean group and an Arctic-temperate group, where the Mediterranean lineage evolved from Arctic-temperate species.

Mason-Gamer et al. (2002) analyzed new and previously published chloroplast DNA data from *Elymus* and from most of the mono-genomic genera of the Triticeae, and presented additional cpDNA data to construct the phylogeny for the mono-genomic genera. They concluded that their analysis was in agreement with previous cpDNA studies with regard to the close relationship between *Secale*, *Taeniatherum*, and *Triticum–Aegilops*. Further, their analysis provided moderate support for some relationships that were unresolved or very weakly supported in earlier cpDNA studies. These included (i) the sister relationship between *Taeniatherum* and *Triticum–Aegilops*; (ii) the placement of *Heteranthelium* in a *Secale* + *Taeniatherum* + *Triticum–Aegilops* clade; and (iii) the placement of the St- and Ee-genome *Elymus* species + *Dasypyrum* clade in the *Secale* + *Taeniatherum* + *Aegilops–Triticum* + *Heteranthelium* clade. Yet, a close relationship between *Taeniatherum* and *Triticum–Aegilops* is completely at odds with the DMC1 tree (Petersen and Seberg 2000) and the morphology-based cladogram (Seberg and Frederiksen 2001). The relationship was unresolved by the ITS data (Hsiao et al. 1995), and *Taeniatherum* was not included in the 5S long spacer data set (Kellogg and Appels 1995). But, Escobar et al. (2011) reported that the clade

containing *Taeniatherum* and *Triticum–Aegilops* is also seen on the 5S short spacer data tree, but only if *Elymus farctus* and *E. elongatus* are included in the clade. Some conflict exists with regard to the position of *Secale* in the starch synthase tree, in which a close relationship among *Triticum*, *Aegilops*, and *Heteranthelium* was detected, while *Secale* was put in a *Secale–Taeniatherum–Elymus* (Ee genome species)–*Dasypyrum* group (Mason-Gamer and Kellogg 2000).

The above phylogenetic studies suggest that *Taeniatherum* is closer to the species of the subtribe Triticineae than to those of the Hordeinae. Hence, the number of spikelets on each node (Tzvelev 1976; Clayton and Renvoize 1986), is not an adequate feature for placing such genus in one of the two subtribes.

5.10.5 Evolutionary Changes in the Polyploid Triticineae

Several Triticineae genera contain polyploid species or subspecies that developed in the Mediterranean-Central Asiatic region (Table 2.8). These genera are *Elymus*, *Agropyron*, *Eremopyrum*, *Dasypyrum*, *Triticum* and *Aegilops*. The perennial polyploid taxa of *Agropyron* and *Dasypyrum* are typical autopolyploids exhibiting multivalents at meiosis due to pairing between the homologous chromosomes of the multiple genomes in the tetraploid and hexaploid cytotypes of these species (Dewey 1961; Ohta and Morishita 2001). In these taxa, although it causes some sterility, multivalent pairing increases plant heterozygosity, that, together with the perennial growth habit and allogamy, emphasizes the evolutionary principle of genetic flexibility. On the other hand, autopolyploids of the subspecies of *Elymus elongatus* and *E. farctus*, are diploidized autopolyploids exhibiting almost exclusive bivalent formation at meiosis due to strict intra-subgenomic pairing between fully homologous chromosomes (Cauderon 1958, 1966; Charpentier et al. 1986). This type of autopolyploidy, combined with facultative self-pollination, leads to full fertility, homozygosity and permanent heterozygosity between subgenomes, emphasizing the evolutionary principle of immediate fitness.

Eilam et al. (2009) determined genome size in several diploid and autotetraploid cytotypes of several Triticeae species. DNA content in the typical autotetraploids did not undergo downsizing, namely, it did not deviate significantly from the expected sum of the two diploid genomes. In contrast, the DNA content of the diploidized autotetraploids underwent downsizing, which was presumably required for the bivalent pattern of pairing due to restriction of pairing to intra-subgenomic homologs. Studies of Eilam et al. (2009) show that the DNA content of the diploidized autotetraploids, *E. elongatus* and *E. farctus* underwent downsizing, whereas that of the typical autopolyploids of *Agropyron* showed the expected DNA content (Table 2.4).

All the annual polyploid species of the genera *Eremopyrum*, *Triticum* and *Aegilops* are allopolyploids, which facilitates adaptation to the Mediterranean and Central Asiatic climate. Leitch and Bennett (2004) showed that there is genome downsizing in polyploids as compared to their diploid parents. Eilam et al. (2008) determined DNA content in allopolyploid species of *Aegilops* and *Triticum* and found that most contain less DNA than expected. Feldman et al. (1997), Ozkan et al. (2001) Shaked et al. (2001), Kashkush et al. (2002) and Salina et al. (2004) showed that DNA sequences are eliminated in newly formed allopolyploids at the same times as or very soon after chromosome doubling. It was assumed that elimination of these DNA sequences may ameliorate the harmonious co-existence of the two or more homoeologous genomes that reside in one allopolyploid nucleus. Furthermore, it is assumed that the homologous-specific DNA sequences play a role in homologous recognition at the beginning of meiosis and initiation of pairing, thus, their elimination from one of the subgenome, leads to diploid-like meiotic behavior in the allopolyploids, i.e., exclusive intra-subgenomic pairing of homologous chromosomes. This pattern of pairing ensures full fertility and prevents intergenomic gene exchanges, thereby sustaining plausible inter-subgenomic genetic interactions that may lead to permanent positive heterosis between subgenomes.

Allopolyploid patterns of gene expression might be intermediate (between that of the two parents), dominant (similar to one of the parents) or overdominant (greater than that of the parents). Overdominance can produce novel traits not found in the parents. It can be caused by novel cis–trans interactions between regulatory elements of the different subgenomes that harbor the same nucleus, as shown in yeast (Tirosh et al. 2009). The genetic system of the allopolyploids may contribute to the build-up of genetic variability and creation of populations with archipelagoes of genotypes via interspecific introgression (Zohary and Feldman 1962), thereby increasing their adaptability, fitness, competitiveness and capacity to colonize rapidly newly-opened ecological niches.

References

Adderley S, Sun G (2014) Molecular evolution and nucleotide diversity of nuclear plastid phosphoglycerate kinase (PGK) gene in Triticeae (Poaceae). Gene 533:142–148

Ahmad F, Comeau A (1991) A new intergeneric hybrid between *Triticum aestivum* L. and *Agropyron fragile* (Roth) Candargy: variation in *A. fragile* for suppression of the wheat *Ph*-locus activity. Plant Breeding 106:275–283

Allard RW (1975) The mating system and microevolution. Genetics 79 (Suppl):115–126

Allard RW, Kannenberg LW (1968) Population ptudies in predominantly self-pollinated species. XI. Genetic divergence among the members of the *Festuca microstachys* complex. Evolution 22:517–528

Allard RW, Jain SK, Workman PL (1968) The genetics of inbreeding population. Adv Genet 14:55–131

Alonso LC, Kimber G (1980) A hybrid between diploid *Agropyron Alonso junceum* and *Triticum aestivum*. Cereal Res Commun 80:355–358

Appels R, Baum B (1991) Evolution of the Nor and 5S DNA loci in the Triticeae. In: Soltis PS, Soltis DE, Doyle JJ (eds) Molecular plant systematics. Chapman and Hall, New York, USA, pp 92–116

Araratian NG (1938) The chromosome numbers of certain species and forms of *Agropyron*. Sovet Bot 6:109–111

Arterburn M, Kleinhofs A, Murray T, Jones S (2011) Polymorphic nuclear gene sequences indicate a novel genome donor in the polyploid genus *Thinopyrum*. Hereditas 148:8–27

Asay KH, Dewey DR (1979) Bridging ploidy differences in crested wheatgrass with hexaploid x diploid hybrids. Crop Sci 19:519–523

Ascherson P, Graebner P (1898–1902) Synopsis der mitteleuropäischen Flora, Bd. 2, Abt. 1. Wilhelm Engelmann, Leipzig

Asghari A, Agayev Y, Fathi SAA (2007) Karyological study of four species of wheat grass (*Agropyron* sp.). Pakistan J Biol Sci 10:1093–1097

Asghari-Zakaria R (2007) karyotype and C-banding patterns of mitotic chromosomes in *Hereranthelium piliferum*. Pak J Biol Sci 10:4160–4163

Asghari-Zakaria R, Kazemi H, Aghayev YM, Valizadeh M, Moghaddam M (2002) Karyotype and C-banding patterns of mitotic chromosomes in *Henrardia persica* (Boiss.) C.E. Hubb. Caryologia 55:289–293

Assadi M (1995) Meiotic configuration and chromosome number in some Iranian species of *Elymus* L. and *Agropyron* Gaertner (Poaceae), Triticeae. Bot J Linn Soc 117:159–168

Assadi M (1996) A taxonomic revision of *Elymus* sect. Caespitosae and sect. *Elytrigia* (Poaceae, Triticeae) in Iran. Willdenowia 26:251–271

Assadi M, Runemark H (1995) Hybridization, genomic constitution and generic delimitation in *Elymus s.l.* (Poaceae, Triticeae) Pl. Syst Evol 194:189–205

Avdulov N (1931) Karyo-systematisce Untersuchung der Familie Gramineen. Bull Appl Bot Genet PL Breed, Leningrad, Suppl 43:1–428

Avivi L (1976) The effect of gene controlling different degrees of homoeologous pairing on quadrivalent frequency in induced autotetraploid lines of *Triticum longissimum*. Can J Genet Cytol 18:357–364

Avivi L, Feldman M, Brown M (1982) An ordered arrangement of chromosomes in the somatic nucleus of common wheat, *Triticum aestivum* L. II. Spatial relationships between chromosomes of different genomes. Chromosoma 86:17–26

Badaeva ED, Friebe B, Gill BS (1996a) Genome differentiation in *Aegilops*. Distribution of highly repetitive DNA sequences on chromosomes of diploid species. Genome 39:293–306

Badaeva ED, Friebe B, Gill BS (1996b) Genome differentiation in *Aegilops*. 2. Physical mapping of 5S and 18S–26S ribosomal RNA gene families in diploid species. Genome 39:1150–1158

Baenziger H (1962) Supernumerary chromsomes in diploid and tetraploid forms of crested wheatgrass. Can J Bot 40:549–561

Barkworth ME (1992) Morphological variation and genome constitution in some perennial Triticeae. Bot J Linn Soc 108:167–180

Baum BR (1977) Taxonomy of the tribe Triticeae (Poaceae) using various numerical techniques. I. Historical perspectives, data, accumulation and character analysis. Can J Bot 55:1712–1740

Baum BR (1978a) Taxonomy of the tribe Triticeae (Poaceae) using various numerical techniques. II. Classification. Can J Bot 56:27–56

Baum BR (1978b) Taxonomy of the tribe Triticeae (Poaceae) using various numerical techniques. III. Synoptic key to the genera and synopses. Can J Bot 56:374–385

Baum BR (1983) A phylogenetic analysis of the tribe Triticeae (Poaceae) based on morphological characters of the genera. Can J Bot 61:518–535

Baum BR, Appels R (1992) Evolutionary change at the *5S Dna* loci of species in the Triticeae. Pl Syst Evol 183:195–208

Baum BR, Estes JR, Gupta PK (1987) Assessment of the genomic system of classification in the Triticeae. Amer J Bot 74:1388–1395

Bentham G, Hooker JD (1883) Genera plantarum, 2nd ed. Reeve, London, 1886, pp 1–468

Bie TD, Cao YP, Chen PD (2007) Mass production of intergeneric chromosomal traslocations through pollen irradiation of *Triticum durum-Haynaldia villosa* amphiploid. J Integr Plant Biol 49:1619–1626

Bieniek W, Mizianty M, Szklarczyk M (2015) Sequence variation at the three chloroplast loci (*matK*, *rbcL*, *trnH-psbA*) in the Triticeae tribe (Poaceae): comments on the relationships and utility in DNA barcoding of selected species. Pl Syst Evol 301:1275–1286

Birchler JA, Veitia RA (2007) The gene balance hypothesis: from classical genetics to modern genomics. Plant Cell 19:395–402

Bizzarri M, Pasquini M, Matere A, Sereni L, Vida G, Sepsi A, Molnar-Lang M, de Pace C (2009) *Dasypyrum villosum* 6V chromosome as source of adult plant resistance to *Puccinia triticina* in wheat. In: Proceedings of the 53rd Italian society of agricultural genetics annual congress, Torino, Italy, pp 16–19

Blanco A, Simeone R, Tanzarella OA (1983a) Morphology and chromosome pairing of a hybrid between *Triticum durum* Desf. and *Haynaldia villosa* (L.) Schur. Theor Appl Genet 64:333–337

Blanco A, Simeone R, Tanzarella OA, Greco B (1983b) Cytogenetic of the hybrid of *Triticum durum* Desf. x *Haynaldia villosa* (L.) Schur. Genetica Agraria 37:149

Blanco A, Simeone R, Resta P (1987) The addition of *Dasypvrum villosum* (L.) Candargy chromosomes to durum wheat (*Triticum durum* Desf.). Theor Appl Genet 74:328–333

Blanco A, Resta P, Simeone R, Parmar S, Shewry PR, Sabelli P, Lafiandra D (1991) Chromosomal location of seeds storage protein genes in the genome of *Dasypvrum villosum* (L.) Candargy. Theor Appl Genet 82:358–362

Blanco A, Simeone R, Resta P, de Pace C, Delre V, Caccia R, Scarascia Mugnozza GT, Frediani M, Cremonini R, Cionini PG (1996) Genomic relationships between *Dasypyrum villosum* (L.) Candargy and *D. hordeaceum* (Cosson et Durieu) Candargy. Genome 39:83–92

Blanco A, Simeone R (1995) Chromosome pairing in hybrids and amphiploids between durum wheat and the tetraploid *Dasypyrum hordeaceum*. In: Li ZS, Xin ZY (eds), Proceedings of the 8th international wheat genetics symposium, Beijing, China, pp 305–309

Blanco A, Perrone V, Resta P, Simeone R, Urbano M (1988a) The incorporation of powdery mildew resistance from *Dasypyrum villosum* (L.) Candargy to durum wheat. Genet Agrar 42:62

Blanco A, Perrone V, Simeone R (1988b) Chromosome pairing variation in *Triticum turgidum* L. x *Dasypyrum villosum* (L.) Candargy hybrids and genome affinities. In: Miller TE, Koebner RMD (eds) Proceedings of the 7th international wheat genetics symposium, vol 1. Institute of Plant Science Research, Cambridge, pp 63–67

Bor NL (1968) Gramineae. In: Towsend CC, Guest E, El-Rawi A (eds) Flora of Iraq, vol 9, pp 210–263

Bor NL (1970) Gramineae. In: rechiger KH (ed) Flora Iranica, vol 70, Graz, Austria: Akademische Druk-Und Verlagsanstalt, Wien

Bowden WM (1966) Chromosome numbers in seven genera of the tribe Triticeae. Can J Genet Cytol 8:130–136

Brasileiro-Vidal AC, Cuadrado A, Brammer SP, Zanatta ACA, Prestes AM, Moraes-Fernandes MIB, Guerra M (2003) Chromosome characterization in *Thinopyrum ponticum* (Triticeae, Poaceae) using in situ hybridization with different DNA sequences. Genet Mol Biol 26:505–510

Breton-Sintes S, Cauderon Y (1978) Étude cytotaxonomique de l'*Agropyron scirpeum* C. Presl et De l'*A. elongatum* (Host) PB. Bull Soc Bot Fr 125:443–455

Cabi E (2010) Taxonomic revision of the tribe Triticeae Dumortier (Poaceae) in Turkey. A Ph.D. thesis submitted to the Graduate School of Natural and Applied Sciences of Middle East Technical University, Ankara, Turkey, pp 1–364

Cai X, Jones S (1997) Direct evidence for high level of autosyndetic pairing in hybrids of *Thinopyrum intermedium* and *Th. ponticum* with *Triticum aestivum*. Theor Appl Genet 95:568–572

Candargy P (1901) Arch Biol Vég Athénes Fasc.1:35 in clavi 62

Cao Y, Bie T, Wang X, Chen P (2009) Induction and transmission of wheat-Haynaldia villosa chromosomal translocations. J Genet Genomics 36:313–320

Cauderon Y (1958) Etude cytogenetique des *Agropyrum* francais et de leurs hybrides avec les bles. Ann Amelior Plant 8:389–567

Cauderon Y (1966) Genome analysis in the genus *Agropyron*. Hereditas (suppl) 2:218–234

Cauderon Y (1979) Use of *Agropyron* species for wheat improvement. In: Zeven AC, van Harten AM (eds) Proceedings of the conference on broadening the genetic base of crops, Wageningen, Netherlands, 3–7 July 1978, Centre for Agricultural Publishing and Documentation, Wagening, pp 175–186

Cauderon Y, Saigne B (1961) New interspecific and intergeneric hybrids involving *Agropyron*. Hereditas 12:13–14

Ceoloni C, Strauss I, Feldman M (1986) Effect of different doses of group-2 chromosomes on homoeologous pairing in intergeneric wheat hybrids. Can J Genet Cytol 28:240–246

Chalupska D, Lee HY, Faris JD, Evrard A, Chalhoub B, Haselkorn R, Gornicki P (2008) ACC homoeoloci and the evolution of wheat genomes. Proc Natl Acad Sci USA 105:9691–9696

Charpentier A (1992) Production of disomic lines and partial amphiploids of Thinopyrum junceum on wheat. Comptes Rendus de l'Academie des Sciences. Series 3. Sciences 315:551–557

Charpentier A, Feldman M, Cauderon Y (1986) Genetic control of meiotic chromosome pairing in tetraploid *Agropyron elongatum*. I. Pattern of pairing in natural and induced tetraploids and in F_1 triploid hybrid. Can J Genet Cytol 28:783–788

Charpentier A, Cauderon Y, Feldman M (1988a) The effect of different doses of Ph1 on chromosome pairing in hybrids between tetraploid *Agropyron elongatum* and common wheat. Genome 30:974–977

Charpentier A, Cauderon Y, Feldman M (1988b) The effect of different *Agropyron elongatum* chromosomes on pairing in *Agropyron*-common wheat hybrids. Genome 30:978–983

Chen KC, Dvorak J (1984) The inheritance of genetic variation in *Triticum speltoides* affecting heterogenetic chromosome pairing in hybrids with *Triticum aestivum*. Can J Genet Cytol 26:279–287

Chen PD, Liu DJ (1982) Cytogenetic studies of hybrid progenies between *T. aestivum* and *H. villosa*. Nanjing Agric Coll Bull 4:1–16

Chen PD, Liu DJ (1986) Identification of *H. villosa* chromosomes in alien wheat addition of *T. aestivum-H. villosa*. In: Li ZS (ed) Proceedings of the 1st international symposium on chromosome engineering in plants, Xian, China, pp 31–33

Chen Q, Jahier J, Cauderon Y (1989) Production and cytogenetical studies of hybrids between *Triticum aestivum* L. Thell. and *Agropyron cristatum* (L.) Gaertn. C.R. Acad. Sci. Paris, 308, Ser. III, pp 425–430

Chen Q, Jahier J, Cauderon Y (1990) Intergeneric hybrids between *Triticum aestivum* and three crested wheatgrasses: *Agropyron mongolicum, A. michnoi*, and *A. desertorum*. Genome 33:663–667

Chen Q, Jahier J, Cauderon Y (1992a) Enhanced meiotic chromosome pairing in intergeneric hybrids between *Triticum aestivum* and diploid Inner Mongolian *Agropyron*. Genome 35:98–102

Chen Q, Jahier J, Cauderon Y (1992b) Production and cytogenetic analysis of BC$_1$, BC$_2$, and BC$_3$ progenies of an intergeneric hybrid between *Triticum aestivum* (L.) Thell. and tetraploid *Agropyron cristatum* (L.) Gaertn. Theor Appl Genet 84:698–703

Chen Q, Lu YL, Jahier J, Bernard M (1994) Identification of wheat–*Agropyron cristatum* monosomic addition lines by RFLP analysis using a set of assigned DNA probes. Theor Appl Genet 89:70–75

Chen X, Shi AN, Shang LM (1997) The resistance reaction of *H. villosa* to powdery mildew isolates and its expression in wheat background. Acta Phytopathol Sin 27:17–22

Chen Q, Conner RL, Laroche A, Thomas JB (1998a) Genome analysis of *Thinopyrum intermedium* and *Thinopyrum ponticum* using genomic in situ hybridization. Genome 41:580–586

Chen Q, Conner RL, Laroche FAA, Fedak G, Thomas JB (1998b) Molecular characterization of the genome composition of partial amphiploids derived from *Triticum aestivum* x *Thinopyrum ponticum* and *T. aestivum* x *Th. intermedium* as sources of resistance to wheat streak mosaic virus and its vector *Aceria tosichella*. Theor Appl Genet 97:1–8

Chen Q, Conner RL, Li HJ, Sun SC, Ahmad F, Laroche A, Graf RJ (2003) Molecular cytogenetic discrimination and reaction to wheat streak mosaic virus and the wheat curl mite in Zhong series of wheat – *Thinopyrum intermedium* partial amphiploids. Genome 46:135–145

Chen PD, Chen SW, Cao AZ, Xing LP, Yang XM, Zhang SZ, Wang XE, Qi LL, Liu DJ (2008) Transferring, mapping, cloning of powdery mildew resistance gene of *Haynaldia villosa* and its utilization in common wheat. In: Appels R, Eastwood R, Lagudah E, Langridge P, Mackay M, McIntyre L, Sharp P (eds) Proceedings of the 11th internataional wheat genetics symposium. University Press, Sydney, Australia, pp 727–729

Chennaveeraiah MS (1960) Karyomorphologic and cytotaxonomic studies in *Aegilops*. Acta Horti Gotoburgensis 23:85–178

Chennaveeraiah MS, Sarkar P (1959) Chromosomes of *Heteranthelium piliferum* Hochst. Wheat Inf Serv 9–10:42

Clark B, McIntyre CL, Appels R (1986) Evolutionary change at the rDNA locus in the Triticeae. In: Li ZJ (ed) First international symposium on chromosome engineering in plants, Xian, China, pp 34–35 (Abstr.)

Clayton SD, Renvoize SA (1986) Genera graminum, grasses of the world, distributed for royal botanic gardens, Kew bulletin. Additional series; 13, Kew, London, pp 1–389

Cosson E, Durieu de Maisonneuve MC (1855) Exploration Scientifique de l'Algerie. 2. –Paris

Coucoli HD, Symeonidis L (1980) Karyotype analysis on some Greek wild species of *Hordeum* (*marinum* group) and *Taeniatherum*, vol 20A. Sci Ann Fac Phys & Mathem, Univ Thessaloniki, pp 77–90

Cuadrado C, Romero C, Lacadena JR (1991) Meiotic pairing control in wheat-rye hybrids. I. Effect of different wheat chromosome arms of homoeologous groups 3 and 5. Genome 34:72–75

Davis PH, Mill RR, Tan K (1988) Flora of Turkey and The East Aegean islands, vol 10 (supplement). Esinburgh University Press, Edinburgh

Deng GB, Chen J, Ma XR, Pan ZF, Yu MQ, Li XF (2004) Morphology, cytogenetics of intergeneric hybrid between *Aegilops tauschii* and *Dasypyrum villosum*. Hereditas/yi Chuan 26:189–194

de Pace C, Qualset CO (1995) Mating system and genetic differentiation in *Dasypyrum villosum* (Poaceae) in Italy. Plant Syst Evol 197:123–147

de Pace C, Benedettelli S, Qualset CO, Kart GE, Scarascia-Mugnozza GT, Delre V, Vittori D (1988) Biochemical markers in *Triticum* x *Dasypyrum* amphiploids and derived disomic addition lines. In: Miller TE, Koebner RMD (eds) Proceedings of the 7th international wheat genetics symposium, vol 1. Institute of Plant Science Research, Cambridge, UK, pp 503–590

de Pace C, Paolini R, Scarascia Mugnozza GT, Qualset CO, Delre V (1990) Evaluation and utilization of *Dasypyrum villosum* as a genetic resource for wheat improvement. In: Srivastava JP, Damania AB (eds) Wheat genetic resources: meeting diverse needs. Wiley, West Sussex, England, pp 279–289

de Pace C, Delre V, Scarascia Mugnozza GT, Qualset CO, Cremonini R, Frediani M, Cionini PG (1992) Molecular and chromosomal characterization of repeated and single-copy DNA sequences in the genome of *Dasypyrum villosum*. Hereditas 116:55–65

de Pace C, Qualset CO, Scarascia Mugnozza GT (1994) Somatic dimorphism of caryopsis color in *Dasypyrum villosum* (L.) Candargy: some reproductive and ecological relationships. In: Wang RRC, Jensen KB, Jaussi C (eds) Proceedings of the 2nd international Triticeae symposium, Logan, UT, USA, pp 234–246

de Pace C, Vaccino P, Cionini PG, Pasquini M, Bizzarri M, Qualset CO (2011) *Dasypyrum*. In: Kole C (ed) Wild crop relatives: genomic and breeding resources, cereals. Springer, Berlin, pp 185–291

Devos KM, Dubcovsky J, Dvorak J, Chinoy CN, Gale MD (1995) structural evolution of wheat chromosomes 4A, 5A, and 7B and its impact on recombination. Theor Appl Genet 91:282–288

de V Pienaar R, Littlejohn GM, Sears ER (1988) Genomic relationships in *Thinopyrum*. S Afr J Bot 54:541–550

Dewey DR (1960) Salt tolerance of 25 strains of *Agropyron*. Agron J 52:631–635

Dewey DR (1961) Polyhaploids of crested wheatgrass. Crop Sci 1:249–254

Dewey DR (1963a) Morphology and cytology of synthetic hybrids of *Agropyron trichophorum* x *Agropyron cristatum*. Amer J Bot 50:1028–1034

Dewey DR (1963b) Cytology and morphology of a synthetic *Agropyron trichophorum* x *Agropyron desertorum* hybrid. Amer J Bot 50:552–562

Dewey DR (1964) Synthetic hybrids of New World and Old World Agropyrons. I. Tetraploid *Agropyron spicatum* X diploid *Agropyron cristatum*. Amer J Bot 51:763–769

Dewey DR (1967) Synthetic hybrids of new world and old world Agropyrons. IV. Tetraploid *Agropyron spicatum* f. Inerme x Tetraploid *Agropyron desertorum*. Amer J Bot 54:403–409

Dewey DR (1969) Hybrids between tetraploid and hexaploid crested wheatgrasses. Crop Sci 9:787–791

Dewey DR (1970) Hybrids and induced amphiploids of *Agropyron dasytachyum* x *Agropyron caninum*. Bot Gaz 131:342–348

Dewey DR (1972) Cytogenetics of tetraploid *Elymus cinereus, E. triticoides, E. multicaulis, E. karataviensis*, and heir F$_1$ hybrids. Bot Gaz 133:51–57

Dewey DR (1974) Cytogenetics of Elymus sibiricus and its hybrids with *Agropyron tauri, Elymus canadensis*, and *Agropyron caninum*. Bot Gaz 135:80–87

Dewey DR (1975) Genome relations of diploid *Agropyron libanoticum* with diploid and autotetraploid *Agropyron stipifolium*. Bot Gaz 136:116–121

Dewey DR (1981) Cytogenetics of *Agropyron ferganense* and its hybrids with six species of *Agropyron, Elymus*, and *Sitanion*. Amer J Bot 68:216–225

Dewey DR (1982) Genomic and phylogenetic relationships among North American perennial Triticeae. In: Estes JR, Tyrl RJ, Brunken JN (eds) Grasses and grasslands: systematics and ecology. University of Oklahoma Press, Norman, pp 51–88

Dewey DR (1983) Historical and current taxonomic perspectives of *Agropyron, Elymus*, and related genera. Crop Sci 23:637–642

Dewey DR (1984) The genomic system of classification as a guide to inter-generic hybridization with the perennial Triticeae. In: Gustafson JP (ed) Gene manipulation in plant improvement. Plenum Press, New York, pp 209–279

Dewey DR, Asay KH (1975) The crested wheatgrasses of Iran. Crop Sci 15:844–819

Dewey DR, Asay KH (1982) Cytogenetic and taxonomic relationships among three diploid crested wheatgrasses. Crop Sci 22:645–650

Dewey DR, Hsiao CI (1984) The source of variation in tetraploid crested wheatgrass. Agron Abstr 198464

Dewey DR, Pendse PC (1968) Hybrids between *Agropyron desertorum* and induced-tetraploid *Agropyron cristatum*. Crop Sci 8:607–611

Dizkirici A, Kaya Z, Cabi E, Doğan M (2010) Phylogenetic relationships of *Elymus* L. and related genera (Poaceae) based on the nuclear ribosomal internal transcribed spacer sequences. Turk J Bot 34:467–478

Dong ZZ, Fan X, Sha LN, Wang Y, Zeng J, Kang HY, Zhang HQ, Wang XL, Zhang L, Ding CB, Yang RW, Zhou YH (2015) Phylogeny and differentiation of the St genome in *Elymus* L. sensu lato (Triticeae; Poaceae) based on one nuclear DNA and two chloroplast genes. BMC Plant Biol 15:179

Dover GA, Riley R (1972) Variation at two loci affecting homoeologous meiotic chromosome pairing in *Triticum aestivum* × *Aegilops mutica* hybrids. Nature New Biol 235:61–62

Driscoll CJ (1972) Genetic suppression of homoeologous chromosome pairing in hexaploid wheat. Can J Genet Cytol 14:39–42

Driscoll CJ, Quinn CJ (1970) Genetic variation in *Triticum* affecting the level of chromosome pairing in intergeneric hybrids. Can J Genet Cytol 12:278–282

Dubcovsky J, Dvorak J (1995) Ribosomal RNA loci: nomads in the Triticeae genomes. Genetics 140:1367–1377

Durand T (1888) Index Generum Phanerogamorum. Brux-Elles

Dvorak J (1972) Genetic variability in *Aegilops speltoides* affecting homoeologous pairing in wheat. Can J Genet Cytol 14:371–380

Dvorak J (1975) Meiotic pairing between single chromosomes of diploid *Agropyron elongatum* and decaploid *A. elongatum* in *Triticum aestivum*. Can J Genet Cytol 17:329–336

Dvorak J (1976) The relationship between the genome of *Triticum urartu* and the A and B genomes of *Triticum aestivum*. Can J Genet Cytol 18:371–377

Dvorak J (1977) Effect of rye on homoeologous chromosome pairing in wheat x rye hybrids. Can J Genet Cytol 19:549–556

Dvorak J (1979) Metaphase I pairing frequencies of individual *Agropyron elongatum* chromosome arms with *Triticurn* chromosomes. Can J Genet Cytol 21:243–254

Dvorak J (1980) Homoeology between *Agropyron elongatum* chromosomes and *Triticum aestivum* chromosomes. Can J Genet Cytol 22:237–259

Dvorak J (1981a) Genome relationships among *Elytrigia (=Agropyron)* elongata, *E. stipifolia*, "*E. elongata* 4x," *E. caespitosa*, *E. intermedia* and "*E. elongata* 10x." Can J Genet Cytol 23:481–492

Dvorak J (1981b) Chromosome differentiation in polyploid species of *Elytrigia*, with special reference to the evolution of diploid-like chromosome pairing in polyploid species. Can J Genet Cytol 23:287–303

Dvorak J (1987) Chromosomal distribution of genes in diploid *Elytrigia elongata* that promote or suppress pairing of wheat homoeologous chromosomes. Genome 29:34–40

Dvorak J, Akhunov ED (2005) Tempos of gene locus deletions and duplications and their relationship to recombination rate during diploid and polyploid evolution in the *Aegilops-Triticum* alliance. Genetics 171:323–332

Dvorak J, Knott DR (1974) Disomic and ditelssomic additions of diploid *Agropyron elongatum* chromosomes to *Triticum aestivum*. Can J Genet Cytol 16(399–4):17

Dvorak J, McGuire PE (1981) Nonstructural chromosome differentiation among wheat cultivars, with special reference to differentiation of chromosomes in related species. Genetics 97:391–414

Dvorak J, Scheltgen E (1973) Phylogenetic relationships among DNAs of wheat, rye and *Agropyron*. In: Sears ER, Sears LM (eds) Proceedings of the 4th international wheat genetics symposium, Columbia, Missouri, pp 73–79

Dvorak J, Zhang H-B (1992a) Application of molecular tools for study of the phylogeny of diploid and polyploid taxa in Triticeae. Hereditas 116:37–42

Dvorak J, Zhang H-B (1992b) Reconstruction of the phylogeny of the genus *Triticum* from variation in repeated nucleotide sequences. Theor Appl Genet 84:419–429

Dvorak J, Chen K-C, Giorgi B (1984a) The C-band pattern of a *Ph* mutant of durum wheat. Can J Genet Cytol 26:360–363

Dvorak J, Lassner MW, Kota RS, Chen KC (1984b) The distribution of the ribosomal RNA genes in the *Triticum speltoides* and *Elytrigia elongata* genomes. Can J Genet Cytol 26:628–632

Dvorak J, Zhang H-B, Kota RS, Lassner M (1989) Organization and evolution of the 5S ribosomal RNA gene family in wheat and related species. Genome 32:1003–1016

Dvorak J, Deal KR, Luo M-C (2006) Discovery and mapping of wheat *Ph1* suppressors. Genetics 174:17–27

Eig A (1929) *Amblyopyrum* Eig. A new genus separated from the genus *Aegilops*. PZE Institute for Agricultural Natural History and Agricultural Research 2:199–204

Eilam T, Anikster Y, Millet E, Manisterski J, Feldman M (2007) Genome size and genome evolution in diploid Triticeae species. Genome 50:1029–1037

Eilam T, Anikster Y, Millet E, Manisterski J, Feldman M (2008) Nuclear DNA amount and genome downsizing in natural and synthetic allopolyploids of the genera *Aegilops* and *Triticum*. Genome 51(8):616–627

Eilam T, Anikster Y, Millet E, Manisterski J, Feldman M (2009) Genome size in natural and synthetic autopolyploids and in a natural segmental allopolyploid of several Triticeae species. Genome 52:275–285

Eilam T, Anikster Y, Millet E, Manisterski J, Feldman M (2010) Genome size in diploids, allopolyploids, and autopolyploids of Mediterranean Triticeae. J Bot 210: Article ID 341380. https://doi.org/10.1155/2010/341380

El Ghawas NI, Khalil HA (1973) The production and evaluation of a *Triticum-Agropyron* hybrid. Egypt J Bot 16:483–499

Endo TR, Gill BS (1984) The heterochromatin distribution and genome evolution in diploid species of *Elymus* and *Agropyron*. Can J Genet Cytol 26:669–678

Escobar JS, Cenci A, Bolognini J, Haudry A, Laurent A, David J, Glémin S (2010) An integrative test of the dead-end hypothesis of selfing evolution in Triticeae (Poaceae). Evolution 64:2855–2872

Escobar JS, Scornavacca C, Cenci A, Guilhaumon C, Santoni S, Douzery EJ, Ranwez V, Glémin S, David J (2011) Multigenic phylogeny and analysis of tree incongruences in Triticeae (Poaceae). BMC Evol Biol 11:181–198

Evans LE (1962) Karyotype analysis and chromosome designations for diploid *Agropyron elongatum* (HOST) P. B. Can J Genet Cytol 4:267–271

Fan X, Zhang HQ, Sha LN, Zhang L, Yang RW, Ding CB, Zhou YH (2007) Phylogenetic analysis among *Hystrix*, *Leymus* and its affinitive genera (Poaceae: Triticeae) based on the sequences of a gene encoding plastid acetyl-CoA carboxylase. Plant Sci 172:701–707

Fan X, Sha LN, Yang RW, Zhang HQ, Kang HY, Zhang L, Ding CB, Zheng YL, Zhou YH (2009) Phylogeny and evolutionary history of *Leymus* (Triticeae; Poaceae) based on a single-copy nuclear gene encoding plastid acetyl-CoA carboxylase. BMC Evol Biol 9:247

Fan X, Sha LN, Yu SB, Wu DD, Chen XH, Zhuo XF, Zhang HQ, Kang HY, Wang Y, Zheng YL, Zhou YH (2013) Phylogenetic reconstruction and diversification of the Triticeae (Poaceae) based on single-copy nuclear *Acc1* and *Pgk1* gene data. Biochem Syst and Ecol 50:346–360

Farooq S, Iqbal N, Shah TM (1990) Intergeneric hybridization for wheat improvement. III. Genetic variation in *Triticum* species affecting homoeologous chromosome pairing. Cereal Res Commun 18:233–237

Fedak G, Chen Q, Conner RL, Laroche A, René Petroski R, Armstrong KW (2000) Characterization of wheat–*Thinopyrum* partial amphiploids by meiotic analysis and genomic in situ hybridization. Genome 43:712–719

Feldman M (1966) The effect of chromosomes 5B, 5D and 5A on chromosomal pairing in *Triticum aestivum*. Proc Natl Acad Sci USA 55:1447–1453

Feldman M, Mello-Sampayo T (1967) Suppression of homeologous pairing in hybrids of polyploid wheats × *Triticum speltoides*. Can J Genet Cytol 9:307–313

Feldman M, Liu B, Segal G, Abbo S, Levy AA, Vega JM (1997) Rapid elimination of low-copy DNA sequences in polyploid wheat: a possible mechanism for differentiation of homoeologous chromosomes. Genetics 147:1381–1387

Feldman M (1978) New evidence on the origin of the B genome of wheat. In: Ramanujam RS (ed) Proceedings of the 5th international coference on wheat genetic symposium, New Delhi, pp 120–132

Fernández-Calvín B, Orellana J (1991) Metaphase I bound arms frequency and genome analysis in wheat-*Aegilops* hybrids. 1. *Ae. variabilis*-wheat and *Ae. kotschyi*-wheat hybrids with low and high homoeologous pairing. Theor Appl Genet 83:264–272

Finch RA, Smith JB, Bennett MD (1981) *Hordeum* and *Secale* genomes lie apart in a hybrid. J Cell Sci 52:391–403

Forster P, Miller TE (1989) Genome relationship between *Thinopyrum bessarabicum* and *Thinopyrum elongatum*. Genome 32:930–931

Frederiksen S (1986) Revision of *Taeniatherum* (Poaceae). Nord J Bot 6:389–397

Frederiksen S (1991a) Taxonomic studies in *Dasypyrum* (Poaceae). Nord J Bot 11:135–142

Frederiksen S (1991b) Taxonomic studies in *Eremopyrum* (Poaceae). Nord J Bot 11:271–285

Frederiksen S (1993) Taxonomic studies in some annual genera of the Triticeae. Nordic J Bot 13:490–492

Frederiksen S (1994) Hybridization between *Taeniatherum caput-medusa* and *Triticum aestivum*. Nord J Bot 14:3–6

Frederiksen S, Seberg O (1992) Phylogenetic analysis of the Triticeae (Poaceae). Hereditas 116:15–19

Frederiksen S, von Bothmer R (1995) Intergeneric hybridization with *Eremopyrum* (Poaceae). Nord J Bot 15:39–47

Frederiksen S, von Bothmer R (1986) Relationships and differentiation in *Taeniatherum* (Poaceae). Can J Bot 64:2343–2347

Frederiksen S, von Bothmer R (1989) Intergeneric hybridization between *Taeniatherum* and different genera of Triticeae, Poaceae. Nord J Bot 9:229–240

Fukuyama T, Hosoya H (1983) Genetic control and mechanism of chromosome elimination in the hybrids between *Hordeum bulbosum* (4x) and *H. vulgare* (4x). Jpn J Genet 58:241–250

Gabi E, Dogan M (2010) Taxonomic study on the genus *Eremopyrum* (Ledeb.) Jaub. et Spach (Poaceae) in Turkey. Pl Syst Evol 287:129–140

Gaertner J (1770) Observationes et descriptiones botanicae. Novi Comment Acad Sci Imper Petrop 14:531–547

Galasso I, Blanco A, Katsiotis A, Pignone D, Heslop-Harrison JS (1997) Genomic organization and phylogenetic relationships in the genus *Dasypyrum* analysed by Southern and in situ hybridization of total genomic and cloned DNA probes. Chromosoma 106:53–61

Gill BS (1981) Evolutionary relationships based on heterochromatin bands in six species of the Triticinae. J Hered 72:391–394

Gill BS (1987) Chromosome banding methods, standard chromosome band nomenclature, and applications in cytogenetic analysis. In: Heyne EG (ed) Wheat and wheat improvement, 2nd edn. American Society of Agronomy Inc., Madison, WI, USA, pp 243–254

Gill BS, Appels R (1988) Relationships between *Nor* loci from different Triticeae species. Pl Syst Evol 160:77–89

Gornicki P, Zhu H, Wang J, Challa GS, Zhang Z, Gill BS, Li W (2014) The chloroplast view of the evolution of polyploid wheat. New Phytol 204:704–714

Gould FW (1947) Nomenclaturial changes in *Elymus* with a key to the Californian species. Madroño 9:120–128

Gradzielewska A (2006a) The genus *Dasypyrum*-part 1. The taxonomy and relationships within *Dasypyrum* and with *Triticum* species. Euphytica 152:429–440

Gradzielewska A (2006b) The genus *Dasypyrum*-part 2. *Dasypyrum villosum*—a wild species used in wheat improvement. Euphytica 152:441–454

Greilhuber J (2005) Intraspecific variation in genome size in Angiosperms: identifying its existence. Ann Bot 95:91–98

Halloran GM (1966) Hybridization of *Haynaldia villosa* with *Triticum aestivum*. Aust J Bot 14:355–359

Hammer K (1980) Vorarbeiten zur monographischen Darstellung von Wildpflanzensortimenten: *Aegilops* L. Kulturpflanze 28:33–180

Han FP, Li J (1993) Morphology and cytogenetics of intergeneric hybrids of crossing *Triticum durum* and T. timopheevi with tetraploid *Elytrigia elongata*. Acta Genet Sin 20:44–49

Heneen WK (1962) Karyotype studies in *Agropyron junceum*, A. repens and their spontaneous hybrids. Hereditas 48:471–502

Heneen WK (1972) Separation difficulties during anaphase I in *Elymus* (Agropyron) species. Bot Notiser 125:430–438

Heneen WK (1977) Chromosomal polymorphism in isolated populations of *Elymus* (Agropyron) in the Aegean. II. *Elymus rechingeri*. Hereditas 86:211–224

Heneen WK, Runemark H (1962) Chromosomal polymorphism and morphological diversity in *Elymus rechingeri*. Hereditas 48:545–564

Heneen WK, Runemark H (1972) Cytology of the *Elymus* (Agropyron) *elongatus* complex. Hereditas 70:155–164

Hitchcock AS (1951) Manual of the grasses of the United States. 2nd ed. rev. by Agnes Chase. Dover Publications, US, pp 230–280

Hodge CD, Wang H, Sun G (2010) Phylogenetic analysis of the maternal genome of tetraploid StStYY *Elymus* (Triticeae: Poaceae) species and the monogenomic Triticeae based on rps16 sequence data. Plant Sci 178:463–468

Hsiao C, Wang RR-C, Dewey DR (1986) Karyotype analysis and genome relationships of 22 diploid species in the tribe Triticeae. Can J Genet Cytol 28:109–120

Hsiao C, Asay KH, Dewey DR (1989) Cytogenetic anlysis of interspecific hybrids and amphiploids between two diploid crested wheatgrasses *Agropyron mongolicum* and A. *cristatum*. Genome 32:1079–1084

Hsiao C, Chatterton NJ, Asay KH, Jensen KB (1995) Phylogenetic relationships of the mono genomic species of the wheat tribe, Triticeae (Poaceae), inferred from nuclear rDNA (internal transcribed spacer) sequences. Genome 38(2):11–223

Huang S, Sirikhachornkit A, Faris JD, Su X, Gill BS et al (2002a) Phylogenetic analysis of the acetyl-CoA carboxylase and 3-phosphoglycerate kinase loci in wheat and other grasses. Plant Mol Biol 48:805–820

Huang S, Sirikhachornkit A, Su X, Faris JD, Gill BS et al (2002b) Genes encoding plastid acetyl-CoA carboxylase and 3-phosphoglycerate kinase of the Triticum/Aegilops complex and the evolutionary history of polyploid wheat. Proc Natl Acad Sci, USA 99:8133–8138

Huang DH, Lin ZS, Chen X, Zhang ZY, Chen CC, Cheng SH, Xin ZY (2007) Molecular characterization of a Triticum durum-Haynaldia villosa amphiploid and its derivatives for resistance to Gaeumannomyces graminis var. tritici. Agric Sci Chin 6:513–521

Hubbard CE (1946) Henrardia, a new genus of the Gramineae. Blumea, Suppl 3:10–21

Humphries CJ (1978) Dasypyrum (Cosson & Durieu) T. Durand. In: Heywood VH (ed) Flora Europaea Notulae Systematicae ad Floram Europaeam spectantes No. 20. Bot J Linn Soc 76:361–362

Hyde BB (1953) Addition of individual Haynaldia villosa chromosomes to hexaploid wheat. Am J Bot 40:174–182

Jahier J, Tanguy AM, Lucas H (1988) Pairing between Dasypyrum villosum (L.) Candargy and Secale cereale L. chromosomes. In: Miller TE, Koebner RMD (eds) Proceedings of the 7th international wheat genetics symposium, vol I. Institute of Plant Sciences Research, Cambridge, UK, pp 315–321

Jan CC, de Pace C, McGuire PE, Qualset CO (1986) Hybrids and amphiploids of Triticum aestivum L. and T. turgidum L. with Dasypyrum villosum (L.) Candargy. Z Pflanzenzuchtg 96:97–106

Jarvie JK (1992) Taxonomy of Elytrigia sect Caespitosae and sect. junceae (gramineae: Triticeae). Nordic J Bot 12:155–169

Jaubert C, Spach E (1851) Ilustrationes plantarum orientalium, vol 4. Paris

Jauhar PP (1988) A reassessment of genome relationships between Thinopyrum bessarabicum and T. elongatum of the Triticeae. Genome 30:903–914

Jauhar PP (1990a) Multidisciplinary approach to genome analysis in the diploid species, Thinopyrum bessarabicum and Th. elongatum (Lophopyrum elongatum) of the Triticeae. Theor Appl Genet 80:523–536

Jauhar PP (1990b) Dilema of genome relationship in the diploid species Thinopyrum bessarahicum and Thinopyrum elongatum (Triticeae: Poaceae). Genome 33:944–946

Jauhar PP (1992) Chromosome pairing in hybrids between hexaploid bread wheat and tetraploid crested wheatgrass (Agropyron cristatum). Hereditas 116:107–109

Jauhar PP (1995) Meiosis and fertility of F₁ hybrids between hexaploid bread wheat and decaploid tall wheatgrass (Thinopyrum ponticum). Theor Appl Genet 90:865–871

Jauhar PP, Riera-Lizarazu O, Dewey WG, Gill BS, Crane CF, Bennet JH (1991) Chromosome pairing relationships among the A, B and D genomes of bread wheat. Theor Appl Genet 81:441–449

Jauhar PP, Almouslem AB, Peterson TS, Joppa LR (1999) Inter- and intragenomic chromosome pairing in haploids of durum wheat. J Hered 90:437–445

Jauhar PP, Dogramaci M, Peterson TS (2004) Synthesis and cytological characterization of trigeneric hybrids of durum wheat with and without Ph1. Genome 47:1173–1181

Jenkins BC (1957) The addition of an Agropyron genome to the common wheat variety Chinese Spring. Wheat Inf Serv 5:14

Jenkins BC, Mochizuki A (1957) A new amphiploid from a cross between Triticum durum and Agropyron elongatum (2n=14). Wheat Inf Serv 5:15

Jensen KB, Bickford IW (1992) Cytology of intergeneric hybrids between Psathyrostachys and Elymus with Agropyron (Poaceae: Triticeae). Genome 35:676–680

Jensen KB, Larson SR, Waldron BL, Asay KH (2006) Cytogenetic and molecular characterization of hybrids between 6x, 4x, and 2x ploidy levels in crested wheatgrass. Crop Sci 46:105–112

Johnson R, Kimber G (1967) Homoeologous pairing of a chromosome from Agropyron elongatum with those of Triticum aestivum and Aegilops speltoides. Genet Res 18:63–71

Jubault M, Tanguy A-M, Abélard P, Coriton O, Dusautoir J-C, Jahier J (2006) Attempts to induce homoeologous pairing between wheat and Agropyron cristatum genomes. Genome 49:190–193

Kashkush K, Feldman M, Levy AA (2002) Gene loss, silencing and activation in a newly synthesized wheat allotetraploid. Genetics 160:1651–1659

Kawahara T (2009) Molecular phylogeny among Triticum-Aegilops species and of the tribe Triticeae. Breed Sci 59:499–504

Kawahara T, Yamane K, Imai T (2008) Phylogenetic relationships among Aegilops-Triticum species based on sequence data of chloroplast DNA. In: Appels R, Eastwood R, Lagudah E, Langridge P, Mackay M, Mcintyre L, Sharp P (eds) Proceedings of the 11th international wheat genetics symposium, University Press, Sydney, Australia. http://hdl.handle.net/2123/3332

Kellogg EA (1989) Comments on genomic genera in the Triticeae (Poaceae). Amer J Bot 76:796–805

Kellogg EA (1992) Restriction site variation in the chloroplast genomes of the monogenomic Triticeae. Hereditas 116:43–47

Kellogg EA (1996) When the genes tell different stories: the diploid genera of Triticeae (Gramineae). Syst Bot 21:321–347

Kellogg EA, Appels R (1995) Intraspecific and interspecific variation in 5S RNA genes are decoupled in diploid wheat relatives. Genetics 140:325–343

Kellogg EA, Appels R, Mason-Gamer RJ (1996) When genes tell different stories: the diploid genera of Triticeae (Gramineae). Syst Bot 21:321–347

Kempanna C, Riley R (1962) Relationships between the genetic effects of deficiencies for chromosomes III and V on meiotic pairing in Triticum aestivum. Nature 195:1270–1273

Kihara H (1936) Ein diplo-haploides Zwillingspaar bei Triticum durum. Agric Hortic 11:1425–2143

Kihara H (1937) Genomanalyse bei Triticum und Aegilops. VII. Kurze ~ bersicht iiber die Ergebnisse der Jahre 1934–36. Mem Coll Agric Kyoto Imp Univ 41:1–61

Kihara H (1959) Fertility and morphological variation in the substitution and restoration backcrosses of the hybrids, Triticum vulgare x Aegilops caudata. In: Proceedings 10th international congress of genetics, Montreal, vol 2, pp 142–171

Kihara H, Lilienfeld FA (1936) Riesenpollenkorner bei den F1-bastarden Aegilops squarrosa x Haynaldia villosa und Aegilops caudata x Aegilops speltoides. Jpn J Genet 12:239–256

Kihara H, Nishiyama I (1937) Possibility of crossing-over between semihomologous chromosomes from two different genomes. Cytologia (Fujii Jub Vol):654–666

Kimber G, Riley R (1963) Haploid angiosperms. Bot Rev 29:480–531

Knowles RP (1955) A study of variability in crested wheatgrass. Can J Bot 33:534–546

Konarev AV (1979) Genome composition of Agropyron elongatum (Host) P. B. (The data of immunochemical analysis of grain proteins). Genetika (USSR) 15:510–517 (In Russian with English summary)

Kosina R, Heslop-Harrison JS (1996) Molecular cytogenetics of an amphiploid trigeneric hybrid between Triticum durum, Thinopyrum distichum and Lophopyrum elongatum. Ann Bot 78:583–589

Kostoff D, Arutiunova N (1937) Studies on the polyploid plants. Triticum-Haynaldia hybrids with special reference to the amphidiploid Triticum dicoccum x Haynaldia villosa. Curr Sci 5:414–415

Lacadena JR, Ramos LP (1968) Meiotic behaviour in a haploid plant of Triticum durum Desf. Genética Ibérica 20:55–71

Lacadena JR, Cermeño MC, Orellana J, Santos JL (1984) Analysis of nucleolar activity in Agropyron elongatum, its amphiploid with

Triticum aestivum and the chromosome addition lines. Theor Appl Genet 68:75–80

Lacadena JR, Cermeño MC, Orallana J, Santos JL (1988) Nucleolar competition Triticeae. In: Brandham PE (ed) Proceedings of third Kew Chrom Conf, 1987. HMSO, London, pp 151–165

Ladizinsky G (1973) Genetic control of bivalent pairing in the *Avena strigosa* polyploid complex. Chromosoma 42:105–110

Lammer D, Cai X, Arterburn M, Chatelain J, Murray T, Jones S (2004) A single chromosome addition from Thinopyrum elongatum confers a polycarpic, perennial habit to annual wheat. J Exp Bot 55:1715–1720

Ledebour CF (1853) Flora Rossica, vol 4, Stuttgart

Leitch IJ, Bennett MD (2004) Genome downsizing in polyploid plants. Biol J Linnean Soc 82:651–663

Lelley T (1976) Induction of homoeologous pairing in wheat by genes of rye suppressing chromosome 5B effect. Can J Genet Cytol 18:485–489

Li LH, Dong YS (1990) Production and study of intergeneric hybrids between *Triticum aestivum* and *Agropyron desertorum*. Sci China (Series B) (in Chinese) 5:492–496 (in English, 1991, 34:45–51)

Li LH, Dong YS (1991) Hybridization between *Triticum aestivum* L. and *Agropyron michnoi* Roshev. 1. Production and cytogenetic of F$_1$ hybrids. Theor Appl Genet 81:312–316

Li DY, Zhang XY (2002) Physical localization of the 18S–5.8S-26S rDNA and sequence analysis of ITS regions in *Thinopyrum ponticum* (Poaceae: Triticeae): implications for concerted evolution. Ann Bot 90:445–452

Li XM, Lee BS, Mammadov AC, Koo BC, Mott IW, Wang RRC (2007) CAPS markers specific to E[b], E[e], and R genomes in the tribe Triticeae. Genome 50:400–411

Li A, Liu D, Wu J, Zhao X et al (2014) mRNA and small RNA transcriptomes reveal insights into dynamic homoeolog regulation of allopolyploid heterosis in nascent hexaploid wheat. Plant Cell 26:1878–1900

Limin AE, Fowler DB (1990) An interspecific hybrid and amphiploid produced from *Triticum aestivum* crosses with *Agropyron cristatum* and *Agropyron desertorum*. Genome 33:581–584

Linde-Laursen I (1981) Giemsa banding patternsof the chromosomes of cultivated and wild barleys. In: Asher MJC, Ellis RP, Hayter AM, Whitehouse RNH (eds) Barley Genetics IV. Proceedings of the 4th international barley genetics symposium. Edinburgh University Press, Edinburgh, pp 786–795

Linde-Laursen I, Frederiksen S (1989) Giemsa C-banded karyotypes of three sub-species of *Taeniatherum caput-medusae* and two intergeneric hybrids with *Psathyrostachys* spp. (Poaceae). Hereditas 110:283–288

Linde-Laursen I, Frederiksen S (1991) Comparison of the Giemsa C-banded karyotypes of *Dasypyrum villosum* (2x) and *D. breviaristatum* (4x) from Greece. Hereditas 114:237–244

Linde-Laursen I, Frederiksen S, Seberg O (1999) The Giemsa C-banded karyotype of Crithopsis delileana (Poaceae; Triticeae). Hereditas 130:51–55

Link HF (1827) Hortus Regius Botanicus Berolinensis. I. Berolini

Linnaeus C (1753) Species Plantarum, Tomus I, May 1753: i-xii, 1–560; Tomus II, Aug 1753: 561–1200, plus indexes and addenda, 1201–1231

Liu ZW, Wang RRC (1993) Genome analysis of *Elytrigia caespitosa*, *Lophopyrum nodosum*, *Pseudoroegneria geniculate* ssp. scythica, and *Thinopyrum intermedium*. Genome 36:102–111

Liu DJ, Chen PD, Raupp WJ (1995) Determination of homoeologous groups of *Haynaldia villosa* chromosomes. In: Li ZS, Xin ZY (eds) Proceedings of the 8th international wheat genetics symposium, vol 1, China Agricultural Scientech, Beijing, China, pp 181–185

Liu DJ, Qi LL, Chen PD, Zhou B, Zhang SZ (1996) Precise identification of an alien chromosome segment introduced in wheat and stability of its resistance gene. Acta Genet Sin 23:18–23

Liu DC, Luo MC, Yang JL, Yan J, Lan XJ, Yang WY (1997) Chromosome location of a new paring promoter in natural populations of common wheat. SW China J Agric Sci 10:10–15

Liu DC, Zheng YL, Yan ZH, Zhou YH, Wei YM, Lan XJ (2003) Combination of homoeologous pairing gene *phKL* and *Ph2*-deficiency in common wheat and its mitotic behaviors in hybrids with alien species. Acta Bot Sin 45:1121–1128

Liu QL, Ge S, Tang HB, Zhang XL, Zhu GF, Lu BR (2006) Phylogenetic relationships in *Elymus* (Poaceae: Triticeae) based on the nuclear ribosomal internal transcribed spacer and chloroplast trnL-F sequences. New Phytol 170:411–420

Liu Z, Li D, Zhang X (2007) Genetic relationships among five basic genomes St, E, A, B and D in Triticeae revealed by genomic Southern and in situ hybridization. J Integr Plant Biol 49:1080–1086

Liu ZP, Chen ZY, Pan J, Li XF, Su M, Wang LJ, Li HJ, Liu GS (2008) Phylogenetic relationships in *Leymus* (Poaceae: Triticeae) revealed by the nuclear ribosomal internal transcribed space and chloroplast trnL-F sequences. Mol Phylogenet Evol 46:278–289

Liu D, Xiang Z, Zhang L, Zheng Y, Yang W, Chen G, Wan C, Zhang H (2011a) Transfer of stripe rust resistance from Aegilops variabilis to bread wheat. African J Biotechnol 10:136–139

Liu W, Rouse M, Friebe B, Jin Y, Gill BS, Pumphrey MO (2011b) Discovery and molecular mapping of a new gene conferring resistance to stem rust, Sr53, derived from *Aegilops geniculate* and characterization of spontaneous translocation stocks with reduced alien chromatin. Chromosome Res 19:669–682

Lorenz H, Schulz-Schaeffer J (1964) Biosystematic investigations in the genus *Agropyron* Gaertn. II. A chromatographic approach. Z Pflanzenz, Berlin 52:13–26

Löve Á (1980) IOPB chromosome number reports. LXVI. Poaceae-Triticeae- Amaricanae. Taxon 29:163–169

Löve Á (1982) Generic evolution of the wheatgrasses. Biol Zentralbl 101:199–212

Löve Á (1984) Conspectus of the Triticeae. Feddes Repert 95:425–521

Lucas H, Jahier J (1988) Phylogenetic relationships in some diploid species of Triticineae: cytogenetic analysis of interspecific hybrids. Theor Appl Genet 75:498–502

Lukaszewski AJ (1988) A comparison of several approaches in the development of disomic alien addition lines of wheat. In: Miller TE, Koebner RMD (eds) Proceedings of the 7th international wheat genetics symposium, vol 1. Institute of Plant Science Research, Cambridge, UK, pp 363–367

Lukaszewski AJ, Kopecky D, Linc G (2012) Inversions of chromosome arms 4AL and 2BS in wheat invert the patterns of chiasma distribution. Chromosoma 121:201–208

Luria Y (1983) Morphological, ecogeographical and cyto-taxonomic survey of five Israeli Triticeae genera. M.Sc. thesis submitted to the Hebrew University of Jerusalem (In Hebrew with English summary), pp 1–116

Mac Key J (1987) Dormancy in the wild and weedy relatives of modern cereals. In: Mares DJ (ed) Proceedings of the 4th international symposium on pre-harvest spouting in cereals 1986. Port Macquarie, NSW: Westview Press, Boulder, CO, pp 414–424

Mac Key J (1989) Seed dormancy in wild and weedy relatives of cereals. In: Derera NF (ed) Preharvest field sprouting in cereals. CRC Press, Boca Raton, pp 15–25

Mac Key J (2005) Wheat: its concept, evolution and taxonomy. In: Royo C et al (eds) Durum wheat, current approaches, future strategies, vol 1. CRC Press, Boca Raton, pp 3–61

Maire R (1952) Flore de l''afrique du Nord. Vol I. Pteridophyta, Gymnospermae, Monocotyledonae. P. Lechevalier Editeur Paris, France, pp 333–337

Maire R, Weiller M (1955) Flore de l'Afrique du nord. Vol III Monocotyledonae: Glumiflorae Gramineae: sf. Pooideae, pp 357, 371–372

Marcussen T, Sandve SR, Heier L, Spannagl M, Pfeifer M,The International Wheat Genome Sequencing Consortium, Jakobsen KS, Wulff BBH, Steuernagel B, Klaus FX, Mayer KFX, Olsen OA (2014) Ancient hybridizations among the ancestral genomes of bread wheat. Science 345(6194):288–291

Mariani M, Minelli S, Ceccarelli M, Cionini PG, Qualset CO, de Pace C (2003) Dasypyrum villosum chromosome segments introgressed in hexaploid wheat provide opportunities for prebreeding and preparing primary mapping populations for analyzing complex genetic traits. In: Pogna NE, McIntosh R (eds) Proceedings of the 10th international wheat genetics symposium, Inst Sperimentale per la Cerealicoltura, Rome, Italy, vol 2, pp 613–615

Martín A, Rubiales D, Cabrera A (1998) Meiotic pairing in a trigeneric hybrid Triticum tauschii-Agropyron cristatum-Hordeum chilense. Hereditas 129:113–118

Martín A, Cabrera A, Esteban E, Hernández P, Ramírez MC, Rubiales D (1999) A fertile amphiploid between diploid wheat (Triticum tauschii) and crested wheatgrass (Agropyron cristatum). Genome 42:519–524

Mason-Gamer RJ (2005) The B-amylase genes of grasses and a phylogenetic analysis of the Triticeae (Poaceae). Am J Bot 92:1045–1058

Mason-Gamer RJ, Kellogg EA (1996a) Chloroplast DNA analysis of the monogenomic Triticeae: phylogenetic implications and genome-specific markers. In: Jauhar PP (ed) Methods of Genome Analysis in Plants. CRC Press, Boca Raton, FL, pp 301–325

Mason-Gamer RJ, Kellogg EA (1996b) Testing for phylogenetic conflict among molecular data sets in the tribe Triticeae (Gramineae). Syst Bio l 45:524–545

Mason-Gamer RJ, Kellogg EA (2000) Phylogenetic analysis of the Triticeae using the starch synthase gene, and a preliminary analysis of some North American Elymus species. In: Jacobs SWL, Everett J (eds) Grasses, systematics and evolution. CSIRO Publishing, Collingwood, Victoria, Australia, pp 102–109

Mason-Gamer RJ, Weil CF, Kellogg EA (1998) Granule-bound starch synthase: structure, function, and phylogenetic utility. Mol Biol Evol 15:1658–1673

Mason-Gamer RJ, Orme NL, Anderson CM (2002) Phylogenetic analysis of North American Elymus and the monogenomic Triticeae (Poaceae) using three chloroplast DNA data sets. Genome 45:991–1002

Mason-Gamer RJ, Burns MM, Naum M (2010) Reticulate evolutionary history of a complex group of grasses: phylogeny of Elymus StStHH allotetraploids based on three nuclear genes. PLoS ONE 5: e10989

Matsumura S (1949) Genomanalyse bei Agropyrum, als verwandte Gattung von Triticum. II. Gattungsbastarde zwischen Triticum und Agropyrum. Jap J Genet, Suppl 2:35–44 (In Japanese with German summary)

Matsumura S, Sakamoto S (1956) Karyotypes of diploid Agropyron species. Ann Rep Nat Inst Genet (Japan) 6:49–50

McArthur RI, Zhu X, Oliver RE, Klindworth DL, Xu SS, Stack RW, Wang RR-C, Cai X (2012) Homoeology of Thinopyrum junceum and Elymus rectisetus chromosomes to wheat and disease resistance conferred by the Thinopyrum and Elymus chromosomes in wheat. Chromosome Res 20:699–715

McCoy GA, Law AG (1965) Satellite chromosomes in crested wheatgrass (Agropyron desertorum (Fisch.) Schult.). Crop Sci 5:283

McCoy GA, Law AG (1968) Satellite chromosomes in two diploid species of crested wheatgrass (Agropyron cristatum (L.) Gaertn. or Agropyron cristatum Sarkar and Agropyron imbricatum (M.B.) Roem. Et Schult.). Agronomy Abstr 13:221–223

McFadden ES, Sears ER (1947) The Genome approach in radical wheat breeding. Agron J 39:1011–1026

McGuire PE (1983) Elytrigia turcica sp. nova, an octoploid species of the E. elongate complex. Folia Geobot Phytotax, Praha 18:107–109

McGuire PE (1984) Chromosome pairing in triploid and tetraploid hybrids in Elytrigia (Triticeae; Poaceae). Can J Genet Cytol 26:519–522

McGuire PE, Dvorak J (1980) High salt-tolerance potential in wheatgrasses. Crop Sci 21:702–705

McGuire PE, Dvorak J (1982) Genetic regulation of heterogenetic chromosome pairing in polyploid species of the genus Triticum sensu lato. Can J Genet Cytol 24:57–82

Megyeri M, Molnár-Láng M, Molnár I (2013) Cytomolecular identification of individual wheat-wheat chromosome arm associations in wheat-rye hybrids. Cytogenet Genome Res 139:128–136

Melderis A (1978) Taxonomic notes on the tribe Triticeae (Gramineae), with special reference to the genera Elymus L. senso lato, and Agropyron Gaertbner senso lato. In: Tutin TG, Heywood VA, Burges NA, Moore DM, Valentine DH, Walters SM (eds) Flora Europea, vol 5, pp 369–384

Melderis A (1980) Taxonomic notes on the tribe Triticeae (Graminae), with special reference to the genera Elymus and Agropyron. Fl Europ Not sys, No 20

Melderis A (1984) New taxa and combinations in Gramineae tribe Triticeae from Turkey. Notes Roy Bot Gard Edinburgh 42:77–82

Melderis A (1985a) Elymus. In: Davis PH (ed) Flora of Turkey. Edinburgh University Press, Edinburgh, pp 206–227

Melderis A (1985b) Eremopyrum (Ledeb.) Jaub. & Spach. In: Davis PH (ed) Flora of Turkey and the East Aegean Islands, vol 9. University Press, Edinburgh, Scotland, pp 227–231

Mellish A, Coulman B, Ferdinandez Y (2002) Genetic relationships among selected crested wheatgrass cultivars and species determined on the basis of AFLP markers. Crop Sci 42:1662–1668

Mello-Sampayo T (1971a) Promotion of homoeologous pairing in hybrids of Triticum aestivum x Aegilops longissima. Genet Iber 23:1–9

Mello-Sampayo T (1971b) Genetic regulation of meiotic chromosome pairing by chromosome 3D of Triticum aestivum. Nat New Biol 230:22–23

Mello-Sampayo T, Canas AP (1973) Suppressors of meiotic chromosome pairing in common wheat. In: Sears ER, Sears LMS (eds) Proceedings of the 4th international wheat genetics symposium, Agricultural ExperimentStation, College of Agriculture and University of Missouri, Columbia, Missouri, pp 709–713

Middleton CP, Senerchia N, Stein N, Akhunov ED, Keller B, Wicker T, Kilian B (2014) Sequencing of chloroplast genomes from wheat, barley, rye and their relatives provides a detailed insight into the evolution of the Triticeae tribe. PLoS ONE 9: e85761

Miller TE, Chapman V (1976) Aneuhaploids in bread wheat. Genet Res 28:37–45

Miller TE, Reader SM, Gale DM (1983) The effect of homoeologous group 3 chromosomes on chromosome pairing and crossability in Triticum aestivum. Can J Genet Cytol 25:634–641

Mini P, de Pace C, Scarascia Mugnozza GT, Delre V, Vittori D (1988) Morphological and isozyme divergence in CS x v amphiploids (Triticum aestivum cv Chinese Spring x Dasypyrum villosum, AABBDDVV, 2n=8x=56) and derived lines from CS x v. In: 31st annual meeting A.I.G.A., Como 30/9-2/10, 1987. Genet Agrar 42:84–85

Mochizuki A (1960) Addition of individual chromosomes of *Agropyron* to *durum* wheat. Wheat Inform Serv 11:22–23

Mochizuki A (1962) *Agropyron* addition lines of *durum* wheat. Seiken Zihô 13:133–138

Montebove L, de Pace C, Jan CC, Qualset CO, Scarascia Mugnozza GT (1987) Chromosomal location of isozyme and seed storage protein genes in *Dasypyrum villosum* (L.) Candargy. Theor Appl Genet 73:836–845

Morris KLD, Gill BS (1987) Genomic affinities of individual chromosomes based on C-and N-banding analyses of tetraploid Elymus species and their diploid progenitor species. Genome 29(2):247–252

Moustakas M (1989) Biosystematic study of the polyploid complexes *Agropyron junceum* (L.) P. Beauv. and *A. elongatum* (Host) P. Beauv. of N. Greece. Ph.D. thesis, School of Biology, University of Thessaloniki, Greece, Scientific Annals, Faculty of Physics and Mathematics, University of Thessaloniki, Suppl 20, pp 1–148 (In Greek with English summary)

Moustakas M (1991) Further evidence of the genome relationships between *Thinopyrum hessarabicum* and *T. elongatum*. Cytobios 68:197–206

Moustakas M (1993) Genome relationships between octoploid and decaploid *Thinopyrum ponticum*. Bot J Linn Soc 112:149–157

Moustakas M, Coucoli H (1982) Karyotype and seed protein profile determination of *Agropyron striatulum* natural Greek populations. Wheat Inf Serv 55:27–31

Moustakas M, Symeonidis L, Coucoli H (1986) Seed protein electrophoresis in *Agropyron junceum* (L.) P. B. complex. Ann Bot 57:35–40

Moxley MG, Berg WA, Barrau EM (1978) Salt tolerance of five varieties of wheatgrass during seedling growth. J Range Manage 3 (I):54–55

Mujeeb-Kazi A, Rodriguez (1981) Cytogenetics of intergeneric hybrids involving genera within the Triticeae. Cereal Res Commun 9:39–45

Mujeeb-Kazi A, Roldan S, Suh DY, Sitch LA, Farooq S (1987) Production and cytogenetic analysis of hybrids between Triticum aestivum and some caespitose Agropyron species. Genome 29:537–553

Muramatsu M (1990) Cytogenetics of decaploid *Agropyron elongatum* (*Elytrigia elongate*) (2n=70) I. Frequency of decavalent formation. Genome 33:811–817

Muramatsu M (2009) A presumed genetic system determining the number of spikelets per rachis nodein the tribe Triticeae. Breed Sci 59:617–620

Myers WM, Hill HD (1940) Studies of chromosomal association and behavior and the occurrence of aneuploidy in autotetraploitl qrass species, orchard grass, tall oat grass, and crested wheatgrass. Bot Gaz 102:236–255

Nakajima G (1951) Cytogenetical studies on intergeneric hybrids between *Haynaldia* and *Secale*. I. Characteristics of an F₁ plant of *Haynaldia villosa* (n = 7) x *Secale cereale* (n=7) and maturation division in PMC's. La Kromosomo 9-10:364–369

Nakajima G (1953) F₁ plants of *Triticum timopheevi* x *Haynaldia villosa*. Cytologia 18:251–252

Nakajima G (1960) Karyogenetical studies on the intergeneric F₁ hybrids raised between *Triticum turgidum* and *Haynaldia hordeacea*. Cytologia 25:208–213

Nakajima G (1966) Caryogenetical studies on F₁ intergenic hybrids raised from crossing between *Triticum* and *Haynaldia*. La Kromosomo 41:2083–2100

Naranjo T, Benaveste E (2015) The mode and regulation of chromosome pairing in wheat-Alien hybrids (*ph* genes, an updated view). In: Molnár-Láng M, Ceoloni C, Doležel J (eds) Alien introgrssion in wheat, cytogentics, molecular biology and genomics. Springer, pp 133–162

Naranjo T, Corredor E (2008) Nuclear architecture and chromosome dynamics in the search of the pairing pattern in meiosis in plants. Cytogenet Genome Res 120:320–330

Nevski SA (1933) Agrostologische studien. IV. Uber das System der Tribe Hordeae Benth. Acta Inst Bot Sci, U. R. SS, Ser II

Nevski SA (1934a) Hordeae Bentb. In: Komarov VL (ed) Flora of the USSR II: 590

Nevski SA (1934b) Schedae ad Herbarium Flora Asiae Mediae. Acta Univ Asiae Med VIII b Bot 17:1–94

Nevski SA (1936) Conspectus Loliearum, Nardearum, Leptuearum hordecarumque florae Unionis Rerum publicarum Sovieticarum Socialisticarum. Trudy Bot Inst Akad Nauk SSSR 1–2:33–90

Obermayer R, Greilhuber J (2005) Does genome size in *Dasypyrum villosum* vary with fruit colour. Heredity 95:91–95

Oehler E (1933) Untersuchungen über Ansatzverhältnisse, Morphologie und Fertilität bei *Aegilops*-Weizenbastarden. I. Teil: Die F₁ Generation. Z Induk Abst Vererbgsl 64:95–153

Oehler E (1935) Untersuchungen an *Aegilops-Haynaldia* und *Triticum-Haynaldia*-Bastarden. Z Induk Abst Vererbgsl 68:187–208

Ohta S, Morishita M (2001) Relationships in the genus *Dasypyrum* (Gramineae). Hereditas 135:101–110

Ohta S, Koto M, Osada T, Matsuyama A, Furuta Y (2002) Rediscovery of a diploid cytotype of *Dasypyrum breviaristatum* in Morocco. Genet Resour Crop Evol 49:305–312

Okito P, Mott IW, Wu Y, Wang RRC (2009) A Y genome specific STS marker in *Pseudoroegneria* and *Elymus* species (Triticeae: Gramineae). Genome 52:391–400

Oliver RE, Cai X, Xu SS, Chen X, Stack RW (2005) Wheat-alien species derivatives: a novel source of resistance to Fusartrium Head Blight in wheat. Crop Sci 45:1353–1360

Onnis A (1967) La dormienza nelle cariossidi di Haynaldia villosa Schur in relazione allo stadio di maturazione. Giorn Bot Ital 101:135–137

Ono H, Nagayoshi T, Nakamura C, Shinotani K (1983) *Triticum durum–Elytrigia elongata* addition line—cytological, morphological and biochemical characteristics. In: Sakamoto S (ed) Proceedings of the 8th international wheat genetics symposium, Kyoto, Japan, pp 1049–1053

Osborn TC, Pires JC, Birchler JA, Auger AL, Chen ZJ, Lee HS, Comai L, Madlung A, Doerge RW, Colot V, Martinssen AR (2003) Understanding mechanisms of novel gene expression in polyploids. Trends Genet 19:141–147

Östergren G (1940a) Cytology of *Agropyron junceum*, *A. repens* and their spontaneous hybrids. Hereditas 26:305–316

Östergren G (1940b) A hybrid between *Triticum turgidum* and *Agropyron junceum*. Hereditas 26:395–398

Ozkan H, Levy AA, Feldman M (2001) Allopolyploidy—induced rapid genome evolution in the wheat (*Aegilops-Triticum*) group. Plant Cell 13:1735–1747

Panayotov I, Todorov I (1979) Study of sources of resistance to brown and black rusts and powdery mildew in the Triticinae. Genet Selekt 12:366–377

Pasquini M, de Los Angeles Gras M, Vallega J (1978) *Haynaldia villosa* (L.) Shur. come fonte di resistenza alle ruggini e all'oidio, da incorporare nelle specie di frumento coltivate. In: Atti Giornate Fitopatologiche, Ist Sper Cereal, Roma, Italy, pp 349–353

Petersen G, Seberg O (1997) Phylogenetic analysis of the Triticeae (Poaceae) based on *rpo*A sequence data. Mol Phylogenet Evol 7:217–230

Petersen G, Seberg O (2000) Phylogenetic evidence for the excision of Stowaway miniature inverted-repeat transposable elements in Triticeae (Poaceae). Mol Biol Evol 17:1589–1596

Petersen G, Seberg O (2002) Molecular evolution and phylogenetic application of DMC1. Mol Phylogent Evol 22:43–50

Petersen G, Seberg O, Yde M, Berthelsen K (2006) Phylogenetic relationships of *Triticum* and *Aegilops* and evidence for the origin of the A, B, and D genomes of common wheat (*Triticum aestivum*). Mol Phylogenet Evol 39:70–82

Petersen G, Seberg O, Salomon B (2011) The origin of the H, St, W, and Y genomes in allotetraploid species of *Elymus* L, and *Stenostachys* Turcz. (Poaceae: Triticeae). Plant Syst Evol 291:197–210

Peto FH (1930) Cytological studies in the genus *Agropyron*. Can J Res 3:428–448

Peto FH (1936) Hybridization of Triticum and Agropyron. II. Cytology of the male parents and F_1 generation. Can J Resh, Sect C 14:203–214

Prokudin YN (1954) K sistematike pyreev ryada *Junceae* Nevski [De serie *Junceae* Nevski generis *Elytrigia* Desv. notae systematicae]. Bot Mat Gerb Bot Inst VL Komarova ANSSSR 16:59–64 (in Russian)

Qi LL, Chen PD, Liu DJ, Zhou B, Zhang SZ (1995) Development ot translocation lines of *Triticum aestivum* with powdery mildew resistance introduced from *Haynaldia villosa*. In: Li ZS, Xin ZY (eds) Proceedings of the 8th international wheat genetics symposium, vol 1. China Agriculture Scientech Press. Beijing, China, pp 333–337

Refoufi A, Jahier J, Esnault MA (2001) Genome analysis of *Elytrigia pycnantha* and *Thinopyrum junceiforme* and of their putative natural hybrid using GISH technique. Genome 44:708–715

Resta P, Lafiandra D, Blanco A (1987) *Dasypyrum villosum* (L.) Cand. chromosomes affecting durum wheat proteins. In: Lasztity R, Bekes F (eds) Proceedings of the 3rd international workshop on gluten proteins. World Scientific, Budapest, Hungary, pp 299–313

Riley R (1960) The diploidization of polyploid wheat. Heredity 15:407–429

Riley R (1966a) The genetic regulation of meiotic behavior in wheat and its relatives. In: Mac Key J (ed) Proc Second Inter Wheat Genet Symp, Genetic Institute, University of Lund, Sweden, August 1963, Hereditas (suppl) 2:395–406

Riley R (1966b) Genetics and the regulation of meiotic chromosome behavior. Sci Progr (London) 54:193–207

Riley R, Chapman V (1958) Genetic control of the cytologically diploid behaviour of hexaploid wheat. Nature 182:713–715

Riley R, Chapman V (1967) Effect of 5BS in suppressing the expression of altered dosage of 5BL on meiotic chromosome pairing in *Triticum aestivum*. Nature 216:60–62

Riley R, Law CN (1965) Genetic variation in chromosome pairing. Adv Genet 13:57–114

Riley R, Chapman V, Kimber G (1960) Position of the gene determining the diploid-like meiotic behaviour of wheat. Nature 186:259–260

Riley R, Kimber G, Chapman V (1961) Origin of genetic control of diploid-like behavior of polyploid wheat. Hered 52:22–25

Riley R, Chapman V, Young RM, Belfield AM (1966) Control of meiotic chromosome pairing by the chromosomes of homoeologous group 5 of *Triticum aestivum*. Nature 212:1475–1477

Riley R, Chapman V, Miller TE (1973) The determination of meiotic chromosome pairing. In: Sears ER, Sears LMS (eds) Proceedings of the 4th international wheat genetics symposium, Columbia, Missouri, pp 731–738

Runemark H, Heneen WK (1968) *Elymus* and *Agropyron*, a problem of generic delimitation. Bot Notiser 121:51–79

Sakamoto S (1967) Cytogenetic studies in the tribe Triticeae. V. Intergeneric hybrids between two *Eremopyrum* species and *Agropyron tsukushiense*. Seiken Ziho 19:19

Sakamoto S (1968) Cytogenetic studies in the tribe Triticeae. VI. Intergeneric hybrid between *Eremopyrum orientale* and *Aegilops squarrosa*. Jpn J Genet 43:167–171

Sakamoto S (1969) Interspecific hybrid between the two species of the genus *Taeniatherum* of tbe tribe Triticeae. Wheat Inf Serv 28:27

Sakamoto S (1972) Intergeneric hybridization between *Eremopyrum orientale* and *Henrardia persica*, an example of polyploidy species formation. Heredity 28:109–115

Sakamoto S (1973) Patterns of phylogenetic differentiation in the tribe Triticeae. Seiken Zihô 24:11–31

Sakamoto S (1974) Intergeneric hybridization amomg three species of *Heteranthelium*, *Eremopyrum* and *Hordeum* and its significance for the genetic relationships within the tribe Triticeae. New Phytol 73:341–350

Sakamoto S (1979) Genetic relationships among four species of the genus *Eremopyrum* in the tribe Triticea, Gramineae. Mem Coll Agric, Kyoto Univ 114:1–27

Sakamoto S (1986) Genome analysis of a polyploid form of *Haynaldia hordeacea* in the tribe Triticeae, Gramineae. In: Li Z, Swaminathan MS (eds) Proceedings of 1st international symposium on chromosome engineering in plants, Xian, China, pp 52–53

Sakamoto S (1991) The cytogenetic evolution of Triticeae grasses. In: Tsuchiya T, Gupta PK (eds) Chromosome Engineering in plants: genetics, breeding, evolution. Elsvier Science Publications B.V, Amsterdam, The Netherlands, pp 469–482

Sakamoto S, Muramatsu M (1963) Preliminary studies on the relationship pf diploid *Eremopyrum distans* and *E. triticeum* (Gramineae). Can J Genet Cytol 5:433–436

Sakamoto S, Muramatsu M (1965) Morphological and cytological studies on various species of Gramineae collected in Pakistan, Afghanistan and Iran. In: Results of the Kyoto University Scientific Expedition to the Karakoram and Hindiikush, 1955, vol I, pp 119–140

Salina EA, Numerova OM, Ozkan H, Feldman M (2004) Alterations in sub-telomeric tandem repeats during early stages of allopolyploidy in wheat. Genome 47:860–867

Sallares R, Brown TA (2004) Phylogenetic analysis of complete 5′ external transcribed spacers of the 18S ribosomal RNA genes of diploid *Aegilops* and related species (Triticeae, Poaceae). Genet Resour Crop Evol 51:701–712

Sando WJ (1935) Hybrids of wheat, rye, *Aegilops* and *Haynaldia*. A series of 122 intra- and inter-generic hybrids shows wide variations in fertility. J Hered 26:229–232

Sarkar P (1956) Crested wheatgrass complex. Can J Bot 34:328–345

Sarkar P (1957) A new diploid form of *Haynaldia hordeacea* Hack. Wheat Inform Serv 6:22

Sasanuma T, Chabane K, Endo TR, Valkoun J (2004) Characterization of genetic variation in and phylogenetic relationships among diploid *Aegilops* species by AFLP: incongruity of chloroplast and nuclear data. Theor Appl Genet 108:612–618

Scarascia-Mugnozza GT, de Pace C, Tanzarella OA (1982) *Haynaldia villosa* (L.) Schur.: una specie di poteenziale valore per il miglioramento genetico del frumento. I. Analisi di alcumi caratteri morfologici. Genet Agrar 36:191

Schlegel R, Cakmak I, Torun B, Eker S, Tolay I, Ekiz H, Kalayci M, Braun HJ (1998) Screening for zinc efficiency among wheat relatives and their utilisation for alien gene transfer. Euphytica 100:281–286

Schooler AB (1966) *Elymus caput-medusae* L. crosses with *Aegilops cylindrica* Host. Crop Sci 6:79–82

Schreber JCD (1772) Beschreibung der Gräser. 2,1. Leipzig

Schultze-Motel J, Meyer D (1981) Numerical taxonomic studies in the genera *Triticum* L. and *Pisum* L. Kulturpflanze 29:241–250

Schulz-Schaeffer J, Jura P (1967) Biosystematic investigations in the genus *Agropyron*. IV. Species karyotype analysis, phytogeographic and other biosystematic studies. Z Pflanzenzücht 57:146–166

Schulz-Schaeffer J, Jurasits P (1962) Biosustematic investigatios in the genus *Agropyron*. I. Cytologic studies of species caryotypes. Amer J Bot 49:940–953

Schulz-Schaeffer J, Allderdice PW, Creel GC (1963) Segmental allopolyploidy in tetraploid and hexaploid *Agropyron* species of the crested wheatgrass complex (section *Agropyron*). Crop Sci 3:525–530

Schulz-Schaeffer J, Chapman SR, Yuan M (1971) Ploidy level distribution of tall wheatgrass in Turkey. Crop Sci 11:592–593

Schur PJF (1866) Enumeratio plantarum Transsilvaniae: exhibens: stripes phanerogamas sponte crescents atque frequentius cultas, crtptogamas vasculares, characeas, etiam muscos hepaticasque. Sumptibus C. Graeser, Vindobonae, p 984

Scoles GJ, Gill BS, Xin Z-Y, Clarke BC, McIntyre CL, Chapman C, Appels R (1988) Frequent duplication and deletion events in the 5S RNA genes and the associated spacer regions of the Triticeae. Pl Syst Evol 160:105–122

Sears ER (1941) Amphidiploids in the seven-chromosome Triticinae. Univ Missouri Agric Exper Stn Res Bull 336:1–46

Sears ER (1953) Addition of the genome of *Haynaldia villosa* to *Triticum aestivum*. Am J Bot 40:168–174

Sears ER (1954) The aneuploids of common wheat. Missouri Agric Exp Stn Res Bull 572:1–58

Sears ER (1973) Agropyron-wheat transfers induced by homoeologous pairing. In: Sears ER, Sears LMS (eds) Proceedings of the 4th international wheat genetics symposium, Columbia, MO, USA, pp 191–199

Sears ER (1976) Genetic control of chromosome pairing in wheat. Ann Rev Genet 10:31–51

Sears ER (1982a) A wheat mutation conditioning an intermediate level of homoeologous chromosome pairing. Can J Genet Cytol 24:715–719

Sears ER (1982b) Activity report. Ann Wheat Newsl 28:121

Sears ER, Okamoto M (1958) Intergenomic chromosome relationships in hexaploid wheat. In: Proceedings of the 10th international congress of genetics, Montreal, Quebec, vol 2, pp 258–259

Seberg O, Frederiksen S (2001) A phylogenetic analysis of the monogenomic Triticeae (Poaceae) based on morphology. Bot J Linn Soc 136:75–97

Seberg O, Petersen G (2007) Phylogeny of Triticeae (Poaceae) based on three organelle genes, two single-copy nuclear genes, and morphology. Aliso J Syst Evol Botany 23:362–371

Seberg O, Frederiksen S, Baden C, Linde-Lauresen I (1991) *Peridictyon*, a new genus from the Balkan peninsula, and its relationship with Festucopsis (Poaceae). Willdenowia 21:87–104

Sha LN, Fan X, Yang RW, Kang HY, Ding CB, Zhang L, Zheng YL, Zhou YH (2010) Phylogenetic relationships between Hystrix and its closely related genera (Triticeae; Poaceae) based on nuclear Acc1, DMC1 and chloroplast trnL-F sequences. Mol Phylogenets and Evol 54:327–335

Shaked H, Kashkush K, Ozkan H, Feldman M, Levy AA (2001) Sequence elimination and cytosine methylation are rapid and reproducible responses of the genome to wide hybridization and allopolyploidy in wheat. Plant Cell 13:1749–1759

Shang H-Y, Baum BR, Wei Y-M, Zheng Y-L (2007) The 5S rRNA gene diversity in the genus *Secale* and determination of its closest haplomes. Genet Resour Crop Evol 54:793–806

Sharma HC, Gill BS (1983) New hybrids between *Agropyron* and wheat. 2. Production, morphology and cytogenetic analysis of F_1 hybrids and backcross derivatives. Theor Appl Genet 66:111–121

Shewry PR, Parmar S, Pappin DJC (1987) Characterization and genetic control of the prolamins of *Haynaldia villosa*: relationship to cultivated species of the Triticeae (rye, wheat, and barley). Biochem Genet 25:309–325

Shewry PR, Sabelli PA, Parmar S, Lafiandra D (1991) α-type prolamins are encoded by genes on chromosome 4Ha and 6Ha of *Haynaldia villosa* Schur (syn. *Dasypyrum villosum* (L.). Biochem Genet 29:207–211

Shi AN, Leath L, Chen X, Murphy JP (1996) Transfer of resistance to wheat powdery mildew from *Dasypyrum villosum* to *Triticum aestivum*. Phytopathology 86:S46

Simonet M (1935) Observations sur quelques espèces et hybrides d'*Agropyrum*. 1. Revision de l'*Agropyrum junceum* (L) P. B. et de l'*A. elongatum* (HOST) P. B. d'après l'étude cytologique. Bull Soc Bot France 82:624–632

Simonet M, Guinochet M (1938) Observations sur quelques especes et hybrides d'*Agropyrum*. II. Sur la repartition geographique des races caryologiques de l'*Agropyrum junceum* (L.) P. B. Bull Soc Bot France 85:175–179

Simonka L (1897) *Cuviera caput-medusae*. Pótfüz Termésettud Közl 24:230–232

Smith DC (1942a) Intergeneric hybridization of cereals and other grasses. J Agric Res 64:33–47

Smith SG (1942b) Polarization and progression in pairing. II. Premeiotic orientation and the initiation of pairing. Can J Res 20:221–229

Stebbins GL (1950) Variation and evolution in plants. Columbia University Press, New York

Stebbins GL (1957) Self-fertilization and population variability in the higher plants. Amer Nat 91:337–354

Stebbins GL (1971) Chromosomal evolution in higher plants. Edward Arnold Ltd., London

Stebbins GL, Pun FT (1953) Artificial and natural hybrids in the Gramineae, tribe Hordeae. V. Diploid hybrids of *Agropyron*. Amer J Bot 40:444–449

Stefani A, Meletti P, Onnis A (1998) Dormancy and ageing in caryopses of the experimental amphidiploid *Triticum durum* x *Dasypyrum villossum* in comparison with parental forms. Agric Med 128:250–254

Strampelli N (1932) Origini, sviluppi, lavori e risultati Instituto Nazionale di Genetica per la Cerealicultura in Rome. Italy, Rome

Sun GL, Komatsuda T (2010) Origin of the Y genome in *Elymus* and its relationship to other genomes in Triticeae based on evidence from elongation factor G (EF-G) gene sequences. MolPhylogenet Evol 56:727–733

Sun GL, Ni Y, Daley T (2008) Molecular phylogeny of RPB2 gene reveals multiple origin, geographic differentiation of H genome, and the relationship of the Y genome to other genomes in *Elymus* species. Mol Phylogenet Evol 46:897–907

Sybenga J (1966) The zygomere as hypothetical unit of chromosome pairing initiation. Genetica 37:186–198

Tateoka T (1962) Starch grains of endosperm in grass systematics. Bot Mag Tokyo 75:377–383

Taylor RJ, McCoy GA (1973) Proposed origin of tetraploid species of crested wheatgrass based on chromatographic and karyotype analyses. Am J Bot 60:576–583

Tirosh I, Reikhav S, Levy AA, Barkai N (2009) A yeast hybrid provides insight into the evolution of gene expression regulation. Science 324:659–662

Tschermak-Seysenegg E (1934) Weitere studien am fertilen, kanstanten Artbastard *T. turgidovillosum* und seinen-Verwandten. I. Teil. Z Induk Abst Vererbgsl 66:180–218

Tzvelev NN (1973) Conspectus specierum tribus *Triticeae* Dum. Familiae *Poaceae* in Flora URSS. Nov Syst Pl Vasc 10:19–59

Tzvelev NN (1976) Poaceae URSS. Tribe III. Triticeae Dum. USSR. Academy of Sciences Press, Leningrad, pp 105–206

Tzvelev NN (1983) Taxonomic review of the USSR. Grasses of the Soviet Union. Oxonian Press, New Delhi, pp 190–224

Upadhya MD (1966) Altered potency of chromosome 5B in wheat-*caudata* hybrids. Wheat Inf Serv 22:7–9

Uslu E, Reader SM, Miller TE (1999) Characterization of *Dasypyrum villosum* (L.) Candargy chromosomes by fluorescent in situ hybridization. Hereditas 131:129–134

Vaccino P, Corbellini M, Cattaneo M, Negri S, Pasquini M, Cionini PG, Caceres E, Vittori D, Ciofo A, de Pace C (2007) Analysis of genotype-by-environment interaction in wheat using aneuploidy lines with chromatin introgressed from *Dasypyrum villosum*. In: Proceedings of the 56th Italian Society of Agricultural Genetics annual congress, Palazzo dei Congrssi, Riva del Garda (TN), Abstract a23

Vaccino P, Corbellini M, de Pace C (2009) Optimizing low input production systems using improved and stable wheat inbred lines arising from a new breeding scheme. In: Farming systems design 2009. In: Intern Symp Methodl Integr Anal Farm Production Systems, 23–26 Aug 2009, Monterey, CA, USA

Vakar BA (1935) Cytologische Untersuchungen der ersten Generation der Weizen-Quecken-grasbastarde. Züchter 7:199–206

van Slageren MW (1994) Wild wheats: a monograph of *Aegilops* L. and *Amblyopyrum* (Jaub. and Spach) Eig (Poaceae). Agricultural University, Wageningen, The Netherlands.

Viegas WS (1980) The effect of B-chromosomes of rye on the chromosome association in F₁ hybrids *Triticum aestivum* x *Secale cereale* in the absence of chromosomes 5B or 5D. Theor Appl Genet 56:193–198

Vogel KP, Arumuganathan K, Jensen KB (1999) Nuclear DNA content of perennial grasses of Triticeae. Crop Sci 39:661–667

von Berg KH (1937) Beitrag zur Genomanalyse in der Getreide-gruoppe. Der Züchter 9:157–163

von Bothmer R, Claesson L (1990) Production and meiotic pairing of intergeneric hybrids of *Triticum* x *Dasypyrum* species. Euphytica 51:109–117

von Bothmer R, Lu B-R, Linde-Laursen I (1994) Intergeneric hybridization and C-banding patterns in *Hordelymus* (Triticeae, Poaceae). Pl Syst Evol 189:259–266

Wang RR-C (1985a) Genome analysis of *Thinopyrum bessarabicum* and *T. elongatum*. Can J Genet Cytol 27:722–728

Wang RR-C (1985b) Identification of intergeneric hybrids in the tribe Triticeae by karyotype analysis. Agron Abstr Amer Soc Agron, Madison, WI, p 74

Wang RR-C (1986) Diploid perennial intergeneric hybrids in the tribe Triticeae. 1. *Agropyron cristatum* x *Pseudoroegneria libanotica* and *Critesion violaceum* x *Psathyrostachys juncea*. Crop Sci 26:75–78

Wang RR-C (1987a) Diploid perennial intergeneric hybrids in the tribe Triticeae. III. Hybrids among *Secale montanum*, *Pseudoroegneria spicata*, and *Agropyron mongolicum*. Genome 29:80–84

Wang RR-C (1987b) Progenies of *Thinopyrum elongatum* X *Agropyron mongolicum*. Genome 29:738–743

Wang RR-C (1988) Diploid perennial intergeneric hybrids in the tribe Triticeae. IV. Hybrids among *Thinpyrum bessarabicum*, *Pseudorogneria spicata*, and *Secale montanum*. Genome 30:356–360

Wang RR-C (1989) An assessment of genome analysis based on chromosome pairing in hybrids of perennial Triticeae. Genome 32:179–189

Wang RR-C (1990) Intergeneric hybrids between *Thinopyrum* and *Psathyrostachys* (Triticeae). Genome 33:845–849

Wang RR-C (1992) Genome relationships in the perennial Triticeae based on diploid hybrids and beyond. Hereditas 116:133–136

Wang RR-C, Hsiao CT (1989) Genome relationship between *Thinopyrum bessarabicum* and *T. elongatum*: revisited. Genome 32:802–809

Wang RR-C, Lu BR (2014) Biosystematics and evolutionary relationships of perennial Triticeae species revealed by genomic analyses. J Syst Evol 52:697–705

Wang RR-C, Dewey DR, Hsiao C (1985) Intergeneric hybrids of *Agropyron* and *Pseudoroegneria*. Bot Gaz (chicago) 146:268–274

Wang RR-C, Marburger JE, Hu C-J (1991) Tissue-culture-facilitated production of aneupolyploid *Thinopyrum ponticum* and amphidi-ploid *Hordeum violaceum* H. bogdanii and their uses in phyloge-netic studies. Theor Appl Genet 81:151–156

Wang RR-C, von Bothmer R, Dvorak J, Fedak G, Linde-Laursen I, Muramatsu M (1995) Genome symbols in the Triticeae (Poaceae). In Wang RR-C, Jensen KB, Jaussi C (eds) Proceedings of 2nd international Triticeae symposium, June 20–24. Logan, UT, pp 29–34

Wang J-B, Wang C, Shi S-H, Zhong Y (2000) ITS regions in diploids of *Aegilops* (Poaceae) and their phylogenetic implications. Hered-itas 132:209–213

Watson L, Clifford HT, Dallwitz MJ (1985) The classification of the Poaceae: subfamilies and supertribes. Austral J Bot 33:433–484

Watson L, Dallwitz MJ (1992) The grass genera of the world. C.A.B. International, Wallingford, p 1038

Wei JZ, Wang RR-C (1995) Genome- and species-specific markers and genome relationships of diploid perennial species in Triticeae based on RAPD analyses. Genome 38:1230–1236

Wei JZ, Chatterton NJ, Larson SR, Wang RR (2000) Linkage mapping and nucleotide polymorphisms of the 6-SFT gene of cool-season grasses. Genome 43:931–938

White WJ (1940) Intergeneric crosses between *Triticum* and *Agropy-ron*. Sci Agric 21:198–232

Xu J, Kasha KJ (1992) Identification of a barley chromosomal interchange using N-banding and in situ hybridization techniques. Genome 35:392–397

Yamane K, Kawahara T (2005) Intra- and interspecific phylogenetic relationships among diploid *Triticum-Aegilops* species (Poaceae) based on base-pair substitutions, indels, and microsatellites in chloroplast noncoding sequences. Am J Bot 92:1887–1898

Yan C, Sun GL (2011) Nucleotide divergence and genetic relationships of *Pseudoroegneria* species. Bioch Syst Ecol 39:309–319

Yang ZJ, Li GR, Feng J, Jiang HR, Ren ZL (2005) Molecular cytogenetic characterization and disease resistance observation or wheat-*Dasypyrum breviaristatum* partial amphiploid and its deriva-tives. Hereditas 142:80–85

Yang ZJ, Liu C, Feng J, Li GR, Zhou JP, Deng KJ, Ren ZL (2006) Studies on genome relationship and species-specific PCR marker for *Dasypyrum breviaristatum* in Triticeae. Hereditas 143:47–54

Yang ZJ, Zhang T, Liu C, Li GR, Zhou JP, Zhang Y, Ren ZL (2008) Identification of wheat Dasypyrum breviaristatum addition lines with stripe rust resistance using C-banding and genomic in situ hybridization. In: Appels R, Eastwood R, Lagudah E, Langridge P, Mackay M, McIntyre L, Sharp P (eds) Proceedings of the 11th international wheat genetics symposium. University Press, Sydney, Australia

Yang L, Wang XG, Liu WH, Li CY, Zhang JP, Gao AN, Wang YD, Yang XM, Li LH (2010) Production and identification of wheat-*Agropyron cristatum* 6P translocation lines. Planta 232:501–510

Yang CT, Fan X, Wang XL, Gu MX, Wang Y, Sha LN, Zhang HQ, Kang HY, Xiao X, Zhou YH (2014) Karyotype analysis of *Agropyron cristatum* (L.) Gaertner. Caryologia 67:234–237

Yang Y, Fan X, Wang L et al (2017) Phylogeny and maternal donors of *Elytrigia* Desv. sensu lato (Triticeae; Poaceae) inferred from nuclear internal-transcribed spacer and *trn*L-F sequences. BMC Plant Biol 17:207

Yen C, Yang JL (2006) Biosystematics on Triticeae (Poaceae). Agr Pr China Beijing 3:78–273

Yen C, Yang JL, Yen Y (2005) Hitoshi Kihara, Áskell Löve and the modern genetic concept of the genera in the tribe Triticeae (Poaceae). Acta Phytotaxonomica Sinica 43:82–93

Yildirim A, Jones SS, Murray TD (1998) Mapping a gene conferring resistance to *Pseudocercosporella herpotrichoides* on chromosome 4V of *Dasypyrum villosum* in a wheat background. Genome 41:1–6

Yildirim A, Jones SS, Murray TD, Line RF (2000) Evaluation of *Dasypyrum villosum* populations for resistance to cereal eyespot and stripe rust pathogens. Plant Dis 84:40–44

Yilmaz R, Cabi E, Dogan M (2014) Molecular analysis of the genera *Eremopyrum* (Ledeb) Jaub & Spach and *Agropyron* Gaertner (Poaceae) by PCR methods. Pak J Bot 46:769–774

Yousofi M, Aryavand A (2004) Determination of ploidy levels of some populations of *Agropyron cristatum* (Poaceae) in Iran by flow cytometry. Iranian J Sci Techn, Trans A 28:137–144

Yu Y, Yang WY, Hu XR (1998) The effectiveness of *ph1b* gene on chromosome association in the F$_1$ hybrid of *T. aestivum* x *H. villosa*. In: Slinkard AE (ed) Proceedings of the 9th international wheat genetics symposium, vol 2. University Extension Press, University of Saskatchewan, Saskatoon, Canada, pp 125–126

Yu MQ, Deng GB, Zhang XP, Ma XR, Chen J (2001) Effect of the ph1b mutant on chromosome pairing in hybrids between *Dasypyrum villosum* and *Triticum aestivum*. Plant Breed 120:285–289

Yu H-Q, Fan X, Zhang C, Ding CB, Wang XL, Zhou YH (2008) Phylogenetic relationships of species in *Pseudoroegneria* (Poaceae: Triticeae) and related genera inferred from nuclear rDNA ITS (internal transcribed spacer) sequences. Biologia 63:498–505

Zhang H-B, Dvorak J (1990) Isolation of repeated DNA sequences from *Lophopyrum elongatum* for detection of *Lophopyrum* chromatin in wheat genomes. Genome 33:283–293

Zhang X-Y, Dong Y-S, Li Z-W (1993) Cytogenetic research on a hybrid of *Triticum* with *Thinopyrum ponticum* as well as its derivatives: 1. Chromosome pairing in *Th. ponticum* and its hybrids with both *T. aestivum* and *T. durum*. Acta Genet Sin 20:439–449

Zhang X-Y, Dong Y-S, Yang X (1995) Cytogenetic research on hybrids of *Triticum* with both *Thinopyrum ponticum* and *Th. intermedium* as well as their derivatives. III. Primary detection of genetic base for introgression of useful genes from the two-alien species to wheat. Acta Genetica Sinica 22:217–222 (Chinese with English summary)

Zhang X-Y, Dong Y-S, Wang RR-C (1996) Characterization of genomes and chromosomes in partial amphiploids of the hybrid *Triticum aestivum* x *Thinopyrum ponticum* by in situ hybridization, isozyme analysis and RAPD. Genome 39:1062–1071

Zhong GY, Qualset CO (1990) An alternative method for assigning genes to specific chromosomes of *Dasypyrum villosum* (L.) Candargy. In: Kimber G (ed) Proceedings of 2nd international symposium on chromosome engineering in plants. University Extension Press, University of Missouri, Columbia, USA, pp 302–309

Zohary D, Feldman M (1962) Hybridization between amphidiploids and the evolution of polyploids in the wheat (*Aegilops-Triticum*) group. Evolution 16:44–61

6.1 The Genus *Secale*—Taxonomic Survey

Secale L. is a small genus including one perennial and two annual species (Frederiksen and Petersen 1998; Table 6.1). The species are characterized by two hermaphroditic florets in each spikelet. Species with more florets in each spikelet, that were previously included in *Secale*, were later separated as *Eremopyrum* (Ledeb.) Jaub. and Spach (Frederiksen 1991a) or *Dasypyrum* (Coss. and Durieu) T. Durand (Frederiksen 1991b). The taxonomy of *Secale* has been complicated by the presence of many weedy, annual intermediates between the wild and the domesticated species, as well as by many wild perennial races. Roshevitz (1947) described 14 species in *Secale*, in addition to over 20 intra-specific taxa. He categorized several of the annual, weedy intermediates as separate species, namely, *S. ancestrale* (Zhuk.) Zhuk., *S. dighoricum* (Vav.) Roshev., *S. segetale* (Zhuk.) Roshev., *S. afghanicum* (Vav.) Roshev. and *S. vavilovii* Grossh. The similar cytology between these weedy taxa and its resemblance to that of domesticated *Secale*, with whom the weeds are inter-fertile, led Sencer (1975), Kobyljanskij (1983), Hammer et al. (1987), Evans (1995) and Frederiksen and Petersen (1998) to consider them subspecies of *S. cereale,* rather than independent species. This group of weeds is virtually confined to agricultural lands, currently widespread as cereal crops in northeastern Iran, Afghanistan and Transcaspian (Zohary 1971). In addition, there is a group of wild perennial races that are widely distributed from Morocco eastwards through the Mediterranean countries and the plateau region of central and eastern Turkey to northern Iraq and Iran. Roshevitz (1947) separated these races into several distinct species, e.g., *S. ciliatoglume* (Boiss.) Grossh., *S. dalmaticum* Vis., *S. kuprijanovii* Grossh., *S. anatolicum* Boiss, *S. daralagesi* Tum., and *S. montanum* Guss. Members of this group are morphologically highly similar, show high cytogenetic affinity to each other and are highly inter-fertile. They differ from the *S. cereale* complex by two major reciprocal translocations involving three pairs of chromosomes (Riley

1955). These races are most probably best described as subspecies of *S. strictum* (Sencer 1975; Kobyljanskij 1983; Hammer et al. 1987; Evans 1995; Frederiksen and Petersen 1998). Consequently, the number of *Secale* species was reduced, e.g., Schiemann (1948) recognized five species, Kobyljanskij (1983), and Hammer et al. (1987) recognized four species and Sencer (1975), Sencer and Hawkes (1980) and Frederiksen and Petersen (1997, 1998) recognized only three species in this genus, namely, *S. strictum* (C. Persl.) C. Presl. (Formerly *S. montanum* Guss.), *S. sylvestre* Host, and *S. cereale* L. (Table 6.1). *Secale strictum* includes all the wild perennial taxa that show high morphological resemblance and cytogenetic affinity to each other. *S. sylvestre* is a wild, annual species which is isolated geographically, ecologically, and reproductively from *S. strictum*, although the two-show cytogenetic affinity under experimental conditions. Rye, *S. cereale,* contains the annual wild, weedy and domesticated types (Sencer 1975; Sencer and Hawkes 1980; Frederiksen and Petersen 1998).

The general taxonomic relationships between the *Secale* species, determined based on morphological and cytological studies, were supported by studies of thin-layer chromatographic patterns of 41 fluorescent compounds in young leaves of 11 *Secale* taxa (Dedio et al. 1969). They showed that domesticated rye grouped together with weedy ryes, the perennial wild *Secale* taxa grouped together with each other, *S. sylvestre* had a distinct chromatogram of its own and *S. vavilovii* (currently included in *S. cereale*) was distinguished by the presence of four compounds not present in any other taxa. Sencer (1975), studying thin-layer chromatographic patterns of phenolic compounds in mature leaves of nine *Secale* taxa, also supported the general taxonomic relationships of the species by demonstrating high resemblance between *S. cereale* and *S. afghanicum* (currently included in *S. cereale*) and between *S. anatolicum* and *S. vavilovii* (both are currently included in *S. cereale*).

Frederiksen and Petersen (1998) performed a critical taxonomic review of the genus *Secale*, studying specimens primarily obtained from natural habitats. They claimed

M. Feldman and A. A. Levy, *Wheat Evolution and Domestication*,
https://doi.org/10.1007/978-3-031-30175-9_6

Table 6.1 The species and subspecies of *Secale* L. according to Frederiksen and Petersen (1998)

Species	Synonyms	Subspecies	Synonyms	Spike rachis	Pollination mode	Growth habit	Karyotype
Strictum (C. Presl) C. Presl	*Triticum strictum* C. Presl; *Secale montanum* Guss.; *T. cereale* var. *montanum* (Guss.) Kuntze; *S. cereale* var. *montanum* (Guss.) Fiori; *Frumentum Secale* E. H. L. Krause	*Strictum*	*Secale anabolism* Boiss.; *S. dalmaticum* Vis.: *S. serbicum* PanÈic ex Griseb.; *S. kuprijanovii* Grossh.; *S. chaldicum* Fed.; *S. daralagesi* Tumanian; *S. rhodopaeum* Delip.; *S. perenne* Hortor in Fisch. & C. A. Mey.	Fragile	Allogamous	Perennial	Strictum type
		Africanum (Stapf) K. Hammer	*S. africanum* Stapf; *S. montanum* ssp. *africanum* (Stapf) Kobyl.	Fragile	Autogamous	Perennial	Strictum type
Sylvestre Host	*Triticum silvestre* (Host) Asch. & Graebn.; *T. campestre* (Kit.) Kit. ex Roem. & Schult.; *Secale fragile* M. Bieb.; *S. cereale* M. Bieb.; *S. glaucum* d'Urv.; *S. spontaneum* Fisch.	–	–	Fragile	Autogamous	Annual	Strictum type
Cereale L.		*Cereale*	*S. cereale* ssp. *indo-europaeum* Antropov & Antropova in Roshev; *S. trijlorum* P. Beauv.	Tough	Mostly allogamous	Annual	Cereale type
		Ancestrale Zhuk.	*S. ancestrale* (Zhuk.) Zhuk.	Fragile or tough	Mostly allogamous	Annual	Cereale type

priority of the taxonomical name *Secale strictum* (C. Presl) C. Presl over *S. montanum* Guss. They also divided this species into two subspecies, subsp. *strictum* and ssp. *africanum* (Stapf) Hammer, that are separated geographically. *Secale cereale* was also treated as having two subspecies; the domesticated taxa, marked by their tough rachises, were placed in subsp. *cereale* and the wild or weedy taxa, that have either tough or varying degrees of fragile rachis, were placed in subsp. *ancestrale* Zhuk. (Table 6.1). *S. sylvestre*, although morphologically distinct from the wild, perennial, cross-pollinating *S. strictum,* has a similar karyotype to that of *S. strictum* (Sencer and Hawkes 1980; Evans 1995). Although it is geographically, ecologically and reproductively isolated from *S. strictum,* the two-show cytogenetic affinity under experimental conditions (Sencer and Hawkes 1980). The classification of the genus *Secale* suggested by Frederiksen and Petersen (1998) is currently accepted by many taxonomists and is used in this book.

Roshevitz (1947) recognized *S. vavilovii* Grossh. as a valid species. *S. vavilovii* is an annual, self-pollinating taxon, with short culms, of limited geographical distribution; it grows in eastern Turkey and northern Iran. Its chromosome arrangement is identical to that of *S. cereale* and therefore, different from *S. strictum* and *S. sylvestre* by two reciprocal translocations involving three pairs of chromosomes (Stutz 1972). Although several studies recognized *S. vavilovii* as a valid species (e.g., Khush and Stebbins 1961; Kranz1961; Khush 1962, 1963a, b; Singh and Robbelen 1975, 1977; Vences et al. 1987; Hammer et al. 1987), Kobyljanskij (1983) included it as a subspecies in *S. cereale*. Similarly, following morphological analyses of many characteristics, Sencer (1975) and Frederiksen and Petersen (1998) also included it in *S. cereale*. It differs from *S. cereale* by its autogamous habit, but self-fertile strains are also known within both weedy and domesticated *S. cereale* (Jain 1960; Pérez de la Vega and Allard 1984; Voylokov et al. 1993; Meier et al. 1996).

Secale iranicum Kobyl. was described by Kobyljanskij (1975, 1983), based on material collected by H. Kuckuck in Hamadan, Iran. *S. iranicum* was initially identified by Kranz (1957) as *S. vavilovii* and, like *S. vavilovii*, it is a self-pollinating taxon. Morphological analyses based on several characters, as well as other evidence, led Frederiksen and Petersen (1998) to include it in *S. cereale*.

6.2 *Secale* Species

6.2.1 Introduction

The *Secale* species are perennials or annuals, with 25–120-cm-long culms and 6–15-cm-long spikes. The spikes are dense, laterally compressed, with solitary spikelets at each rachis node. In the wild species and in a number of the weedy types, spikes are brittle and disarticulate at maturity into spikelets with the rachis internode below each spikelet (wedge-type dispersal unit) or rarely, beneath the florets (floret-type dispersal unit); in the domesticated types and in most of the weedy types the spikes are not brittle. The number of spikelets per spike ranges from 15 to 25, with each spikelet containing two hermaphroditic florets and a third sterile, male or very rarely hermaphroditic floret. Rachises are fragile or tough, and rachis segments are densely covered with white hairs on the edges. The glumes are sub-equal, 8–18-mm-long, shorter than the adjacent lemmas, linear, 1-veined, keeled, and either awnless or taper to a straight, short awn (up to 35-mm-long). The lemmas are lanceolate, 10–19-mm-long, 5-veined, prominently keeled and terminate in a long, straight scabrous awn. The pales are nearly as long as the lemmas, are membranous, and 2-keeled. Anthers are 2.5–12-mm-long. The caryopsis is free.

The *Secale* species are cross-pollinating or self-pollinating. *S. strictum* subsp. *africanum* and *S. sylvestre* are self-compatible and have about 40% self-fertility, while most lines of *S. cereale* and *S. strictum* subsp. *strictum* are self-incompatible (Jain 1960; Kranz 1963; Kuckuck and Peters 1967; Stutz 1972). *S. vavilovii and S. iranicum,* which were included in *S. cereale* by Frederiksen and Petersen (1998), are autogamous (Kuckuck and Peters 1967) show about 50% self-fertility (Kranz 1963). In fact, Kranz (1963) observed many intermediate forms in *S. vavilovii and S. iranicum,* from allogamous through autogamous to cleistogamous. Assuming that the ancestor of the genus *Secale* was self-incompatible, then the transition from self-incompatible to self-compatible occurred independently several times in the genus, i.e., to *S. sylvestre,* to *S. strictum* subsp. *africanum,* and to *S. cereale* subsp. *vavilovii.*

A correlation between anther length and breeding habit exists in *Secale* species. Self-sterile, allogamous taxa feature long anthers (5–12 mm), whereas autogamous taxa have short anthers (2.5–5.0 mm) (Schiemann and Nürnberg-Krüger 1952; Nürnberg-Krüger 1960; Stutz 1957, 1972; Khush and Stebbins 1961; Khush 1963b; Kranz 1961,

1973; Kobyljanskij 1975, 1983; Hammer 1990). However, the actual degree of self-sterility has only been thoroughly studied in *S. cereale* (incl. *S. vavilovii*), which shows a continuous variation between self-sterility and self-fertility (Lundquist 1954, 1956, 1958a, b; Voylokov et al. 1993; Meier et al. 1996). The cross-pollinating taxa of *Secale,* i.e., the wild *S. strictum* subsp. *strictum,* the domesticated *S. cereale* subsp. *cereale,* and most of the weedy types of *S. cereale* subsp. *ancestrale,* exhibit broad morphological variation, both within and between populations.

The *Secale* species are native to the Mediterranean region and western Asia. The genus distributes from central Europe and the western Mediterranean through the Balkans, Anatolia, the Levant, and the Caucasus to Central Asia. An isolated population also exists in South Africa.

6.2.2 *Secale sylvestre* Host

Secale sylvestre Host [Syn.: *Triticum silvestre* (Host) Asch. and Graebn.; *Secale campestre* Kit. ex Schult.; *Triticum campestre* (Kit.) Kit. ex Roem. and Schult.; *Secale fragile* M. Bieb.; *S. cereale* M. Bieb.; *Triticum fragile* (M. Bieb.) Link; *Secale sylvestre* var. *fragile* (M. Bieb.) Fritsch ex Nevski; *S. glaucum* d'Urv.; *S. sylvestre* var. *glaucum* (d'Urv.) Fritsch ex Nevski; *S. sylvestre* f. *glaucum* (d'Urv.) Roshev.: *Secale spontaneum* Fisch.] is an annual and self-pollinating species, possible with cleistogamous flowers. Culms are 25–50-cm-long, the rachis is highly fragile, and disarticulates at maturity into dispersal units, each consisting of a single spikelet and the rachis segment below it (wedge-type disarticulation). Glume awns are 15–35-mm-long, lemma awns are up to 65-mm-long, bristles are 0.5–0.7-mm-long on lemma keel, anthers are 2.5–3.5-mm-long.

Secale sylvestre is characterized by its long awns. Morphologically, *S. sylvestre* is well separated from the other two species of *Secale.* It also differs from *S. cereale* by three chromosomal translocations and from *S. strictum* by a single one. However, it is largely inter-fertile with both these species (Sencer and Hawkes 1980).

Secale sylvestre distributes in the area ranging from Hungary in the west, to the Altai mountains of Central Asia in the east. It is a psammophyte (a plant that thrives in sandy conditions) and grows near riversides, in river deltas and seashores, sand dunes, sandy pastures, steppes and semi-deserts (Roshevitz 1947).

6.2.3 *Secale strictum* (C. Presl) C. Presl

Secale strictum (C. Presl) C. Presl [Syn.: *Triticum strictum* C. Presl; *S. montanum* Guss.; *T. cereale* var. *montanum*

(Guss.) Kuntze; *S. cereale* var. *montanum* (Guss.) Fiori in Fiori and Paol.; *Frumentum Secale* E. H. L. Krause; *S. anatolicum* Boiss.; *S. cereale* var. *anatolicum* (Boiss.) Regel; *S. montanum* var. *anatolicum* (Boiss.) Boiss.; *T. cereale* var. *anatolicum* (Regel) Kuntze; *T. cereale* ssp. *montanum* var. *anatolicum* (Boiss.) Asch. and Graebn.; *S. montanum* ssp. *anatolicum* (Boiss.) Tzvelev; *S. strictum* ssp. *anatolicum* (Boiss.) K. Hammer in Hammer, Skolimowska and Knüpffer; *S. dalmaticum* Vis.; *T. cereale* ssp. *montanum* var. *dalmaticum* (Vis.) Asch. and Graebn.; *S. montanum* var. *dalmaticum* (Vis.) E. Schiem.; *S. strictum* ssp. *strictum* K. Hammer in Hammer, Skolimowska and Knüpffer; *S. serbicum* PanÈic ex Griseb.; *S. kuprijanovii* Grossh.; *S. montanum* Guss. ssp. *kuprijanovii* (Grossh.) Tzvelev; *S. strictum* ssp. *kuprijanovii* (Grossh.) K. Hammer in Hammer, Skolimowska and Knüpffer; *S. kuprijanovii* ssp. *ciscaucasica* A. P. Ivanov and Yakovlev; *S. kuprijanovii* ssp. *transcaucasica* A. P. Ivanov and Yakovlev; *S. chaldicum* Fed.; *S. montanum* ssp. *chaldicum* (Fed.) Tzvelev; *S. kuprijanovii* var. *chaldicum* (Fed.) Sinskaya and Bork.; *S. cereale* var. *perennans* Dekapr. in Grossh.: *S. daralagesi* Tumanian; *S. rhodopaeum* Delip.; *S. anatolicum* Delip.; *S. perenne* Hortor in Fisch. and C. A. Mey.]. According to Frederiksen and Petersen (1998), *Secale strictum* contains two subspecies, ssp. *strictum* and ssp. *africanum* (Stapf) K. Hammer.

6.2.3.1 Ssp. *strictum*

The plants are perennial and allogamous, with 40–100-cm-high culms that are glaucous, tufted, and glabrous below the spikes. Spikes are 6–14-cm-long (excl. awns), the rachis is densely hairy along the margins, and fragile, and disarticulating with rachis internode below spikelets (wedge-type dispersal units). Spikelets are 2-flowered, glumes are keeled and 15–23-mm-long (incl. awns that are 0–6-mm-long), lemma are 15–24-mm-long (excl. awns), with a strongly developed keel, which features strongly developed bristles, and end in a scabrous awn that is 4–50-mm-long, and with 5–12-mm-long anthers.

Ssp. *strictum* contains two varieties, var. *strictum* and var. *c iliatoglume* (Boiss.) Frederiksen and Petersen comb. nov. (Frederiksen and Petersen 1998). Var. *strictum* is morphologically very variable, comprising a group of perennial cross-pollinated taxa, such as *S. dalmaticum* Vis., *S. anatolicum* Boiss., and *S. kupriyanovii* Grossh.

Var. *ciliatoglume* [Syn.: *Secale montanum* var. *ciliatoglume* Boiss.; *S. ciliatoglume* (Boiss.) Grossh.; *Triticum cereale* ssp. *montanum* var. *ciliatoglume* (Boiss.) Asch. and Graebn.; *S. anatolicum* var. *ciliatoglume* (Boiss.) A. P. Ivanov and Yakovlev; *S. strictum* ssp. *ciliatoglume* (Boiss.) K. Hammer in Hammer, Skolimowska and Knüpffer], differs from var. *strictum* by its dense layer of hairs over the internodes, leaf sheaths, and blades. Var. *ciliatoglume* also differs from var. *strictum* by three restriction site mutations in the plastid DNA (Petersen and Doebley 1993).

Var. *strictum* is an Irano-Turanian and Mediterranean element, extending from the western Mediterranean to the Caspian Sea, namely, Morocco southeast Spain, Sicily, south Balkans, Greece, Syria, Lebanon, Mt. Hermon and Golan Heights, central and eastern Turkey, southern Armenia, Caucasus, northern Iraq, and northwestern Iran. Var. *strictum* is native to elevated plateaus and mountain systems and mainly grows in primary habitats (meadows, rangelands, open oak-park forests, shrub formations) on calcareous slopes, on rocky mountain slopes, in dry mountain areas, in the sub-alpine and alpine regions of the mountains, and as a weed in segetal habitats, such as along roadsides or the edges of cultivated fields (Roshevitz 1947; Sencer and Hawkes 1980). Var. *ciliatoglume* is restricted to eastern Turkey, southern Armenia, north-western Iran and northern Iraq. It grows on dry, stony or sandy mountain slopes.

6.2.3.2 *Secale strictum* ssp. *africanum* (Stapf) K. Hammer

Secale strictum ssp. *africanum* (Stapf) K. Hammer [Syn.: *S. africanum* Stapf; *S. montanum* ssp. *africanum* (Stapf) Kobyl.] is perennial and self-pollinating, with culms up to 100-cm-high. Its spike is 80–120-mm-long, linear, very dense and laterally compressed, its rachis has short hairs and disarticulates at maturity, the spikelets are solitary, 10–15-mm-long, and laterally compressed. Its two unequal glumes are shorter than the lemmas, and awned; lemmas are 5-nerved, and keeled, with an awn up to 20-mm-long, with less developed bristles on the keel (only about 0.3-mm-long). Anthers are approximately 5-mm-long.

Ssp. *africanum* is a single taxon found only in South Africa. This subspecies inhabits natural vegetation on stony ground among shrubs, but only in a very small area of the country; even in this area it appears to be rare (Schiemann and Schweickerdt 1950). In fact, ssp. *africanum* is only known from a single locality in the Sutherland District of the Western Province of South Africa. This disjuncted location, at a long distance from the distribution of other *Secale* taxa, is peculiar and has been explained by either human activities or as the remain of an originally much larger continuous distribution area (Schiemann and Schweickerdt 1950; Khush 1962). Alternatively, grains (or spikelets) of this subspecies might have been brought from the Mediterranean-central Asiatic region to South Africa by migrating birds, either through droppings that contain undigested grains or through spikelets that were attached to their feathers. The enigma has not yet been satisfactorily solved.

6.2.4 *Secale cereale* L.

Secale cereale L. [Syn.: *Triticum cereale* (L.) Salisb., *T. Secale* Link, *Triticum cereale* ssp. *eu-cereale* Asch. & Graebn., *Frumentum secale* var. *cereale* E. H. L. Krause] contains the domesticated varieties and a group of weedy, semi-wild forms. All are annual, self-incompatible or sporadically self-compatible, chromosomally homologous, and fully inter-fertile with one another. The species was divided into two subspecies, ssp. *cereale*, containing all the domesticated forms, and ssp. *ancestrale*, containing all the weedy and semi-wild forms (Frederiksen and Petersen 1998).

6.2.4.1 Ssp. *cereale*

Secale *cereale* L. subsp. *cereale* [Syn.: *S. cereale* ssp. *indo-europaeum* Antropov and Antropova in Roshev., *S. cereale* var. *eligulatum* Vavilov in Majssurjan, *S. cereale* var. *compositum* Lilj., *S. cereale* var. *vulgare* Kom., *Triticum cereale* ssp. *eucereale* Asch. & Graebn., *S. trijlorum* P. Beauv., *S. cereale* var. *trijlorum* (P. Beauv.) Peterm.] is an annual, mostly allogamous (occasionally autogamous) plant, with a tough (occasionally brittle) rachis, and with characteristic large and plump grains (Fig. 5.4b). It is grown as a cereal in most temperate regions, tolerating cold climates and poor soils. Ssp. *cereale* is mainly cultivated in Europe (Germany, Poland, Russia, Belarus Ukraine, Denmark and Spain), China, Turkey, Caucasus, Central Asia, North and South America (mainly Canada), and Australia (Table 6.2). It is also cultivated in high elevations in the tropics and subtropics, i.e., in the highlands of East Africa. Rye is also grown in Morocco, Algeria, Egypt and South Africa.

It probably originated in northeastern Turkey as the weedy derivative of wild *S. strictum*, infesting wheat and barley fields, and eventually taken into cultivation when agriculture spread northwards and eastwards into colder climates (see Sect. 5.6 on rye domestication). Currently, subsp. *cereale* is grown as a grain crop and in many areas as forage. Rye is an important food plant, especially in north and east Europe. It includes a great number of varieties.

6.2.4.2 Ssp. *ancestrale* Zhuk.

This subspecies includes *S. ancestrale* (Zhuk.) Zhuk.; *S. cereale* ssp. afghanicum (Vavilov) K. Hammer in Hammer Skolimowska and Knüpffer.; *S. afghanicum* (Vavilov) Roshev.; *S. segetale* ssp. *segetale* var. *afghanicum* (Vavilov) Tzvelev; *S. segetale* ssp. *afghanicum* (Vavilov) Bondar. ex O. Korovina; *S. cereale* ssp. *dighoricum* Vavilov; *S. dighoricum* (Vavilov) Roshev.; *S. segetale* ssp. *dighoricum* (Vavilov) Tzvelev; *S. segetale* (Zhuk.) Roshev.; *S. cereale* ssp. *segetale* Zhuk.; *S. kasakorum* Roshev.; *S. cereale* ssp. *vavilovii* (Grossh.) Kobyl.

Ssp. *ancestrale* is an annual weedy or semi-wild taxon always found as weeds in cultivated fields and field borders in Anatolia, the Caucasus and Central Asia (Vavilov 1926; Zhukovsky 1933; Roshevitz, 1947; Sencer 1975). It contains forms with different degrees of spike brittleness, including non-brittle, partly brittle (usually the upper part of the spike) and fully brittle. The forms with non-brittle spikes are weeds that infest wheat and barley fields in southwestern Asia, Caucasia and Transcaucasia and in southeastern Europe. Weeds with partly brittle spikes are common in Armenia, northeastern Iran, Afghanistan and central Asia. Usually, the upper part of the mature spike disarticulates upon maturity, while the lower part remains intact and is harvested together with the wheat or the barley crop. Forms of subsp. *ancestrale* with fully brittle spikes are relatively rare and have been found in western Turkey and in western Iran. *S. cereale* ssp. *ancestrale* infests cultivated fields as well as secondary habitats, such as edges of cultivation and roadsides.

6.3 Cytology, Cytogenetics and Evolution

All three *Secale* species are diploids, with 2n = 2x = 14 (Sakamura 1918; Stolze 1925; Aase and Powers 1926; Thompson 1926; Emme 1927; Lewitsky 1929, 1931; Jain 1960; Bowden 1966; Love 1984; Petersen 1991a, b); their genome has been designated R (Love 1984; Wang et al. 1995). B-chromosomes occur in a low frequency in *S. cereale* and in a few populations of *S. strictum* subsp. *strictum* (Emme 1928; Darlington 1933; Hasegawa 1934; Popoff 1939; Müntzing 1944, 1950; Kranz 1963; Jones and Rees 1982; Niwa et al. 1990). Several synthetic autotetraploid lines of the domesticated subspecies *S. cereale* ssp. *cereale*, also exist. The karyotype of all *Secale* species is symmetric; three chromosome pairs are metacentric and four are sub-metacentric, of which one pair, chromosome 1R, carries a satellite (Bennett et al. 1977; Fig. 5.4c).

With the exception of the SAT-chromosome, the similar arm ratios and relative length of *Secale* chromosomes render it difficult to distinguish between chromosomes using conventional cytological methods. Yet, the presence of massive blocks of sub-telomeric heterochromatin characteristic to *Secale* chromosomes (Lima-de-Faria 1953; Vosa 1974; Gill and Kimber 1974; Singh and Röbbelen 1975), has enabled recognition of individual *Secale* chromosomes via C-banding techniques (Vosa 1974; Gill and Kimber 1974; Singh and Röbbelen 1975; Fig. 5.4c). Giemsa banding of domesticated *Secale cereale* chromosomes, not only enables identification of individual chromosome, but also shows large telomeric heterochromatic bands in most of the telomeres and a number of weaker bands in centromeric and interstitial positions (Vosa 1974; Gill and Kimber 1974;

Table 6.2 World's total and top ten rye producer countries in 2020 (from FAOSTAT 2020)

Country	Hectare	Production in metric tons	Productivity (tons/ha)
Germany	636,000	3,513,400	5.524
Poland	843,620	2,904,580	3.442
Russia	975,435	2,377,629	2.437
Belarus	360,409	1,050,702	2.913
Denmark	115,370	699,370	6.062
China	166,601	523,759	3.144
Canada	153,000	487,800	3.188
Ukraine	137,800	456,780	3.320
Spain	137,590	407,620	2.963
Turkey	104,211	295,681	2.837
World total	4,446,927	15,022,273	3.378

Singh and Röbbelen 1975). In fact, all *S. cereale* chromosomes possess thin centromeric bands and a number of bands adjacent to the secondary constriction on chromosome 1R (Vosa (1974). As an allogamous species, *Secale cereale* exhibits great variability in Giemsa banding patterns (Singh and Röbbelen 1975; Giraldez et al. 1979). The few specimens of *S. strictum* ssp. *strictum* that have been studied, also exhibit large telomeric heterochromatic bands (Singh and Röbbelen 1975; Cuadrado and Jouve 1995). In contrast, *S. sylvestre* bears small telomeric bands (Singh and Röbbelen 1975).

The constitutive heterochromatin in somatic chromosomes, detected by C-banding, has been shown by Gill and Kimber (1974) to be equivalent to the classical heterochromatic structures observed by Lima-de-Faria (1953) in pachytene chromosomes of *S. cereale*. The pattern and relative size of the terminal C-bands, pattern of minor interstitial bands and arm ratios enable individual chromosome identification. Since common wheat, *Triticum aestivum*, chromosomes do not have the dense band in the terminal position on the short arms, Gill and Kimber (1974) used the distinct traits of *Secale* chromosomes to identify the seven disomic additions of Imperial rye chromosomes to common wheat. On the basis of their homoeology with wheat chromosomes, Gill and Kimber (1974) arranged the seven rye chromosomes in the homoeologous groups of wheat chromosomes and designated them 1R to 7R. Imperial rye chromosomes are considered to be the standard for karyotype arrangement of *S. cereale* (Sybenga 1983).

The *Secale* genome differs from that of wheat in both size and structure (Gill and Friebe 2009). Its size of approximately 9 pg 1C DNA is 33% larger than the genome of diploid wheat ($\sim$6 pg 1C DNA), and is among the larger genomes of the Triticeae (Table 6.3). 1C DNA content ranges from 7.21 pg in *S. sylvestre* to 7.4 pg in *S. strictum* subsp. *africanum*, 8.28–8.65 pg in *S. cereale* ssp. *cereale* and 8.20–9.45 pg in *S. strictum* subsp. *strictum* (Table 6.3).

Intraspecific and interspecific variations in DNA content among *Secale* genomes are mainly due to different amounts of sub-telomeric heterochromatin (Bennett et al. 1977). In fact, Bennett et al. (1977) measured the length of sub-telomeric C-bands and found that the proportion of sub-telomeric heterochromatin in the genome ranged from about 6% in *S. Silvestre* and *S. strictum* ssp. *africanum* to about 9% in *S. strictum* ssp. *strictum* and to 12.24% in *S. cereale subsp. cereale* (Table 6.3). Thus, one of the major evolutionary changes in chromosome structure in *Secale* has involved the addition of heterochromatin close to the telomeres (Bennett et al. 1977). Bedbrook et al. (1980) succeeded in cloning six families of non-homologous repeat heterochromatin-specific DNA sequences of *S. cereale* ssp. *cereale*. The six sequences are predominantly located within the blocks of constitutive sub-telomeric heterochromatin that can be observed on all seven *ssp. cereale* chromosome pairs by Giemsa staining. Four of these sequence unrelated families, having repeating units of 120, 480, 610, and 630 bp, account for most, if not all, of the difference in sub-telomeric heterochromatin DNA content between the three *Secale* species (Bedbrook et al. 1980; Table 6.3). The four repeats are present in high copy numbers in *S. cereale* DNA and in somewhat smaller amounts in *S. strictum* ssp. *strictum* whereas only one repeat is detectable in *S. sylvestre* DNA and two repeats in *S. strictum* ssp. *africanum* (Table 6.3). The four families of ssp. *cereale*-specific sequences, accounting for most of the sub-telomeric heterochromatin, are arranged in tandem arrays, are complex and contain simple sub-repeats interspersed with an unrelated sequence without sub-repeats. Bedbrook et al. (1980) suggested that each of the ssp. *cereale*-specific repeats evolved by the insertion of DNA elements into an array of simple repeats, followed by amplification of the portion of the array containing the inserted sequence.

The structure, copy number and chromosomal location of arrays of the four families of highly repeated sequences have

Table 6.3 Nuclear DNA content, percentage of sub-telomeric heterochromatin, and percentage of repeated sub-telomeric DNA sequence families in different *Secale* species

Species	Subspecies	1C DNA (pg)	% Of sub-telomeric heterochromatin[a]	% Of repeated sub-telomeric sequence families[b]			
				120 bp	480 bp	610 bp	630 bp
S. sylvestre	–	7.21[c]	6.14	2.4	0.0	0.00	0.00
S. strictum	*strictum*	8.37[c] 9.45[d]	8.91	1.5–3.2	1.2–4.3	0.50	0.16
	africanum	7.42[c]	6.11	2.4	0.0	0.04	0.00
S. cereale	*cereale*	8.53[c] 8.65[d]	12.24	2.4	6.1	2.70	0.60

[a] Determined from the length of C-bands (Bennett et al. 1977)
[b] Bedbrook et al. (1980) and Jones and Flavell (1982b)
[c] Bennett et al. (1977)
[d] Eilam et al. (2007)

been investigated in *Secale* species (Jones and Flavell 1982a, b). Each species was found to be unique in its complement and/or chromosomal distribution of the sequence families. For example, *S. strictum* subsp. *strictum* and *S. cereale* ssp. *cereale* accessions show the same complement of repeated sequences, but differ substantially in the number of repeats they contain of the 480-bp, 610-bp, and 630-bp sequences (Table 6.3). The structure of the 480-bp repeating unit also varies across accessions of ssp. *strictum*. In this outbreeding subspecies, hetero-morphisms are frequent, and are particularly conspicuous in hybridization analyses detecting the 480-bp sequence.

The relationship between the chromosomal location of heterochromatin C-bands and of the four non-homologous repeat sequence families, constituting 12% or more of total ssp. *cereale* DNA, has been investigated by in situ hybridization in ssp. *cereale* chromosomes (Jones and Flavell 1982a, b). Only centromeric and nucleolar organizer region-associated C-bands failed to hybridize with at least one of the sequences, whereas many sub-telomeric blocks of heterochromatin contained all four repeat sequence families.

The repeat sequences are mostly ssp. *cereale*-specific and serve as convenient markers for ssp. *cereale* chromatin in wheat-rye hybrids (Appels 1982; Lapitan et al. 1986). Upon study of the distribution of families of repeats in triticale and in *Secale*-addition lines to bread wheat, Jones and Flavell (1982a) found that some of the families of repeats, present in rye but not in wheat sub-telomeric heterochromatin, were selected against in the wheat genetic background.

Using FISH, Hutchinson et al. (1981) studied the relationship between sub-telomeric C-bands and the four families of repeat sequences. They used the DNA probes, produced by Bedbrook et al. (1980) and Jones and Flavell (1982a) that represent the repeat sequences, namely, pSc119.2 representing the120 bp repeat, pSc74 the 350–480 bp repeat, and pSc34 the 630 bp repeat. Similar to Jones

and Flavell (1982b), also Hutchinson et al. (1981) found that these repeats account for most, if not all, of the differences in sub-telomeric heterochromatic DNA quantities between the *Secale* species. The proportion of DNA in the genome represented by these sequence families in different species of *Secale* is different (Hutchinson et al. 1981). The 120-bp family is equally abundant in all *Secale* species and is the only repeat family found in *S. sylvestre*, in which a correlation has also been observed between low C-heterochromatin content and the absence of the other three repeat sequences (Hutchinson et al. 1981; Jones and Flavell 1982b). The relative high DNA amount and the higher proportion of heterochromatin in ssp. *cereale* and ssp. *strictum*, is the result of accumulation of several repeat families that constitute the major components of sub-telomeric heterochromatin and are absent in *S. sylvestre* (Table 6.3).

In accord with the above, Cuadrado and Jouve (1997) used FISH with the probes pSc119.2, pSc74, and pSc34 to study the quantity and distribution of the highly repeated sub-telomeric DNA sequences that are located in the sub-telomeric heterochromatin. In accord with the above, they showed that *S. sylvestre* had considerably fewer repeat DNA sequences in the sub-telomeric heterochromatin than the other two species and proposed that it was due to its autogamous nature.

The fairly constant C-banding pattern of all seven chromosomes in various samples of *S. strictum* ssp. *strictum* were reported by Gustafson et al. (1976). The chromosomes of this subspecies are easily identifiable and can be distinguished from the seven *S. cereale* chromosome pairs by their distinct C-banding pattern (Gustafson et al. 1976). FISH analysis of the karyotype of ssp. *strictum* was performed using the probes pSC119.2, pSc74, and pSc34, enabling the identification of each ssp. *strictum* chromosome (Cuadrado and Jouve 1995). Moreover, this labeling allowed the comparison of the karyotype of *ssp. strictum* to that of ssp.

cereale since labeling of a combination of two different repetitive DNA sequences enabled identification of all chromosomes of both species (Cuadrado and Jouve 1995). Using this approach, it was found that the physical locations of the repetitive 120-, 480-, and 610-bp repeats are rather similar in the chromosomes of ssp. *strictum* and ssp. *cereale* and that the main qualitative difference lies in the rarity of a combination of the 610-bp and 480-bp repeats in the same telomeres in ssp. *strictum*, in contrast to ssp. *cereale* (Cuadrado and Jouve 1995). This finding is in accord with that of Jones and Flavell (1982b), who reported that both species have the same complement of repetitive DNA sequences, but differed significantly in the amount of telomeric 610-bp repetitive sequences and in the pattern of distribution of the 480-bp repetitive sequences along the chromosomes. Differences in the distribution of hybridization sites of the 120-bp family (pSc119.2) probe in both subspecies were limited to chromosome arms 2RS, 2RL and 7RL. The 610-bp repetitive sequence probe hybridized exclusively to the sub-telomeric regions predominantly of chromosome arms 1RS, 1RL, 4RS, 5RS, 6RS and 7RS, and was almost always absent in the most prominent heterochromatic blocks of sub-telomeric 2RS, 2RL and 7RL in ssp. *strictum*. Jones and Flavell (1982b) reported a major difference in the hybridization responses of ssp. *strictum* and ssp. *cereale* to the pSc34 (610 bp) probe, with hybridization being clearly more intense in ssp. *cereale*. Finally, use of the pSc74 probe, demonstrated the presence of the 480-bp sequence in almost all arms (Cuadrado and Jouve 1995). Interestingly, interstitial hybridization sites were found in the short arm of 6R in ssp. *strictum* and in the long arm of the same chromosome in *S. cereale* (Jones and Flavell 1982a; Lapitan et al. 1988; Mukai et al. 1992; Cuadrado and Jouve 1994).

González-García et al. (2006) used FISH to compare the morphology of the sub-telomeric heterochromatin during the transition from zygotene to second meiotic telophase in meiocytes of *Secale cereale* ssp. *cereale*. At zygotene, pachytene and diplotene, the sub-telomeric heterochromatin formed clumps, often featuring two or more bivalent ends, strongly suggesting ectopic recombination. The high variability between homologous chromosomes and the frequent nonhomologous bindings of sub-telomeric heterochromatin, strongly suggest that rye sub-telomeric heterochromatin is in a dynamic state and that its position along the chromosome changes frequently during meiosis (González-García et al. 2006).

Little is known about the mechanism driving the dynamics of telomeric chromatin. Insertions of DNA sequences into an array of single repeats, followed by saltatory amplification and the occurrence of other chromosomal rearrangements, may have played an important role in the process. These could result from saltatory amplification events at telomeres that were initially responsible for each

large increase in DNA quantity. Subsequently, unequal crossing-over between homologues may have played an important secondary role, by extending the range of variation in the amount of heterochromatin at a given telomere, while crossing-over between non-homologues may have driven an increase in the DNA quantity at one telomere to be distributed between chromosomes (Bennett et al. 1977). In accord with this, Evtushenko et al. (2016) found that the evolution of sub-telomeric heterochromatin appears to have involved a significant contribution of illegitimate recombination. They suggested that the large blocks of sub-telomeric heterochromatin arose from the combined activity of transposable elements (TEs) and the expansion of the tandem repeats, likely as a result of a highly complex network of recombination mechanisms. The abundance of TEs and repeats associated with heterochromatin in sub-telomeric regions raises questions as to their potential role in telomere activity. It is unlikely that this heterochromatin can substitute for telomeres as in *Drosophila* (Pardue and DeBaryshe 2003), because plants have both the telomerase and the array of simple telomeric repeats necessary for telomere maintenance. However, it is possible that the heterochromatin packaging and dynamics at the chromosome ends affects telomeric functions and chromosome behavior in cereals, as it does in other species (Schoeftner and Blasco 2009).

The occurrence and distribution of the sub-telomeric tandem arrays in the different *Secale* species suggest that *S. sylvestre* may be of ancient origin, while *S. strictum* and *S. cereale* may have a more recent origin (Jones and Flavell 1982b). The fact that the 120-bp repeat is the only repeat family that exists in *S. sylvestre*, may indicate that this repeat was the first to be amplified in the evolution of the genus *Secale* (Zeller and Cermeño 1991). On the other hand, if *S. strictum* ssp. *strictum* is more ancient than *S. sylvestre*, then repeat sequences may have also been lost during evolution of the species.

Murai et al. (1989) used several restriction endonucleases to study chloroplast DNA variation in all the species of *Secale*. The chloroplast genome size was estimated to be 136 kbp, which is very close to its size in *Triticum* and *Aegilops*, and produced identical patterns with all the restriction enzymes applied in *S. strictum* ssp. *strictum* and *S. cereale* ssp. *cereale*, but not in *S. sylvestre*. The restriction fragment patterns of *S. sylvestre* showed up to two differences from those of the other *Secale* taxa, suggesting early separation of *S. sylvestre* from the rest of the species.

There is a general agreement that the Triticeae tribe is monophyletic and therefore, assumed to be derived from a common ancestor (Watson et al. 1985; Kellogg 1989; Soreng et al. 1990; Hsiao et al. 1995a). Thus, genetic similarities, i.e., synteny (the conservation of order of loci on homoeologous chromosomes), might be expected between the homoeologous chromosomes of the various Triticeae species and those of the common wheat cultivar Chinese Spring, whose seven homoeologous chromosome groups serve as a standard (Sears 1954, 1966). The homoeology of the chromosomes of a given Triticeae species is determined by the degree to which a pair of chromosomes that is added to a nullisomic line of common wheat compensates for the missing wheat chromosome pair for vigor and fertility. Accordingly, the nomenclature of the chromosomes of the *Secale* species is based on the homoeologous system of the Triticeae. Good genetic compensation by *Secale* chromosomes to common wheat chromosomes has been indicated by the ability of the former to satisfactorily compensate for the missing bread wheat chromosomes when substituted into wheat. Over the years, individual chromosomes of both, domesticated rye, *Secale cereale* ssp. *cereale*, and wild type, *S. strictum* ssp. *strictum*, have been added to bread wheat, and substituted of wheat chromosomes (Miller 1984). Evidence from the compensating effect of *Secale* chromosomes in the absence of their common wheat homoeologues in wheat-*Secale* (*S. cereale* and *S. strictum*) addition and substitution lines implied homoeology relationships between *Secale* and bread wheat chromosomes (Zeller and Hsam 1983; Miller 1984). In fact, Gill and Kimber (1974), Zeller and Hsam (1983), and Miller (1984) designated the seven pairs of rye chromosomes 1R-7R relating to the seven homoeologous groups of common wheat.

Supporting evidence of the homoeology between *Secale* and wheat chromosomes can be obtained by comparing the chromosomal location of genes and of various biochemical and molecular markers. Studies of gene localization in wheat-*Secale* (*S. cereale* and *S. strictum*) addition and substitution lines, have contributed considerably to the know-whow on the position and order of genes in chromosomes of *Secale* relative to wheat. Devos et al. (1993) constructed an RFLP-based genetic map of *S. cereale* ssp. *cereale* that provided evidence that the genetic synteny between *Secale* and wheat chromosomes has been affected by chromosomal rearrangements that occurred during the evolution of the various *Secale* genomes. Their work revealed multiple evolutionary translocations in the *Secale* genome relative to that of bread wheat. DNA clones indicated that chromosome arms 2RS, 3RL, 4RL, 5RL, 6RS, 6RL, 7RS and 7RL have all been involved in at least one translocation. Moreover, Devos et al. (1993) identified the translocated chromosomal segments, and suggested a possible evolutionary pathway that could account for the present-day structure of *Secale* genomes relative to one another and to those of bread wheat.

Secale primary trisomics and telo-trisomics lines (Zeller et al. 1977) as well as rye addition and substitution lines (O'Mara 1940; Driscoll and Sears 1971; Miller 1984; Schlegel et al. 1986; Zeller and Cermeño 1991; Mukai et al. 1992) have been used for chromosome and arm mapping of genes. Schlegel (1982) even produced monosomic additions

of wheat chromosomes to rye. However, because all of these aneuploids do not breed true and are highly sterile, researchers preferred to use monosomic or telosomic additions of *Secale* chromosomes to wheat (Mukai et al. 1992) for rye (*S. cereale* ssp. *cereale*) genome mapping. Wheat-rye recombinant chromosome stocks provide further opportunities for chromosome mapping (Rogowsky et al. 1993; Lukaszewski 2000; Lukaszewski et al. 2004).

Translocations have played an important role in karyotype evolution and speciation of many plant groups (Stebbins 1958; Rieseberg 2001; Faria and Navarro 2010) and, as such, are thought to have played an important role in the evolution of the genus Secale (Stutz 1972; Koller and Zeller 1976; Shewry et al. 1985; Naranjo et al. 1987; Naranjo and Fernández-Rueda 1991; Liu et al. 1992; Rognli et al. 1992; Devos et al. 1993; Schlegel 2013). The chromosomes that were involved in the various translocations characterizing *Secale* species, were identified by studying the ability of trisomics of rye chromosomes (Heemert and Sybenga 1972) or chromosome arms to successfully substitute nullisomy for homoeologous wheat chromosomes (Koller and Zeller 1976; Miller 1984). This was achieved by studying pairing between rye and wheat chromosome arms in the presence and absence of the homoeologous-pairing suppressor gene, *Ph1* (Naranjo et al. 1987; Naranjo and Fernández-Rueda 1991), and by using various kinds of genetic markers (Shewry et al. 1985; Liu et al. 1992; Devos et al. 1993).

Chromosome pairing at first meiotic metaphase in PMCs of F$_1$ hybrids between *Secale* species, has shown that they differ from each other by a number of translocations (Riley 1955; Stutz 1957; Khush and Stebbins 1961; Khush 1962; Kranz 1973). The perennial wild *S. strictum* and the annual wild *S. sylvestre* differ from *S. cereale* by two reciprocal translocations that involve three pairs of chromosomes (Schiemann and Nürnberg-Krüger 1952; Riley 1955; Stutz 1957; Khush and Stebbins 1961; Khush 1962; Kranz 1963; Singh and Röbbelen 1977). These translocations appear as a hexavalent configuration in the F$_1$ hybrids of *S. strictum* and *S. sylvestre* with *S. cereale*. Heemert and Sybenga (1972) used a standard tester set of reciprocal translocations in *S. cereale* and several primary trisomics to identify the chromosomes involved in the two translocations between *S. cereale* and *S. strictum*. They reported that the two translocations involved chromosomes I (=2R), V (=6R) and III (=7R). Later, these translocations were found to involve the short arm of 2R (2RS), and the long arms of 6R and 7R (6RL and 7RL, respectively) (Naranjo and Fernández-Rueda 1991; Devos et al. 1993). Another translocation that differs between these two species of *Secale*, was discovered by Shewry et al. (1985), who located the *Sec2* gene on chromosome 6RS of *S. strictum* and on 2BS of *S. cereal*. They concluded that there was a translocation between 2 and 6RS in *S. cereale* relative to *S. strictum* (and probably to wheat, where homologous genes are located on 6AS, 6BS, and 6DS). This has been presented as evidence of a translocation between part of the 2R short arm and part of the 6R short arm (Devos et al. 1993). The short arm of chromosome 4R of *S. cereale* is homoeologous to wheat chromosome arms 4AS, 4BS and 4DL, whereas the proximal region of the long arm of 4R was translocated with a region of 7RS and is homoeologous to the short arm of wheat chromosomes 7A, 7B and 7D (Koller and Zeller 1976), and the distal region with 6RS (Devos et al. 1993). On the other hand, the proximal region of the short arm of *S. cereale* chromosome 7R is homoeologous to wheat chromosome arms 4AL, 4BL, and 4DS, while the distal region of 7RS is homoeologous to the distal region of chromosomes of wheat group 5 (Devos et al. 1993). The proximal region of the long arm of 7R is homoeologous to the long arm of wheat 7AL, 7BL, and 7DL, while the distal region was translocated with 2RS (Koller and Zeller 1976; Devos et al. 1993). Chromosomes 4R and 7R of *S. strictum* did not undergo rearrangements and are homoeologous to wheat homoeologous groups 4 and 7, respectively. The translocation between chromosomes 4R and 7R is in addition to the two translocations between 2R, 6R, and 7R in *S. cereale* (Koller and Zeller 1976; Devos et al. 1993). Only chromosome 1R of *S. cereale* was not rearranged and is homoeologous to wheat homoeologous group 1, whereas all other six chromosome pairs of this species were involved in translocations (Devos et al. 1993). Additional smaller translocations distinguish *S. sylvestre* from *S. cereale* (Khush and Stebbins 1961; Khush 1962; Kranz 1963), and a single minor translocation separates *S. sylvestre* from *S. strictum* (Khush 1962). *S. strictum*, subsp. *africanum* was found to differ from subsp. *strictum* by a small translocation (Khush 1962; Singh and Röbbelen 1977). It is probable that the chromosomal rearrangements separating *S. sylvestre* from the other two species and separating *S. cereale* from *S. sylvestre* and *S. strictum*, were instrumental in speciation within the genus.

The translocations characterizing the *Secale* genomes can be classified into five categories: (1) Translocations that are common to all *Secale* species, wheat and other Triticineae. King et al. (1994) reported the existence of a 4AL/5AL translocation in *T. urartu*. Devos et al. (1995), and Dubcovsky et al. (1996) identified the same translocation in diploid wheat, *Triticum monococcum*, and in hexaploid wheat, *T. aestivum*, which both showed identical breakpoints (Li et al. 2016), and assumed that it occurred at the diploid level. A similar 4RL/5RL translocation was found in *S. cereale* (Naranjo et al. 1987; Naranjo and Fernández-Rueda 1991; Devos et al. 1993). Homoeologous chromosome pairing studies between bread wheat and *S. cereale*, showed that the 4RL/5RL translocation underwent an additional rearrangement in which the 4RL arm was translocated to

7RS (Naranjo et al. 1987; Naranjo and Fernández-Rueda 1991; Liu et al. 1992; Rognli et al. 1992; Devos et al. 1993; King et al. 1994). Presence of the 4L/5L translocation has also been demonstrated in *Ae. umbellulata* and diploid *Elymus farctus* subsp. *bessarabicus* (King et al. 1994), but has not been demonstrated in diploid *Elymus elongatus* and in barley. Li et al. (2016) identified the breakpoints of the 4L/5L translocation in bread wheat and rye. The wheat translocation joined the ends of breakpoints downstream of a *WD40* gene on 4AL and a gene of the *PMEI* family on 5AL. While rye shares the same position for the 4L breakpoint, its 5L breakpoint position differs, although very close to that of wheat, indicating the recurrence of 4L/5L translocations in wheat and rye. These findings suggest that if this translocation occurred recurrently in various Triticeae species, then the translocation breakpoints are the two fragile sites on the chromosome arms 4L and 5L of diploid Triticeae species and that the 4L/5L translocation breakpoints represent two hotspots of chromosomal rearrangement recurrently used during Triticeae evolution. (2) Translocations that differ between the genera *Secale* and *Triticum*. Several small translocations interfering with the homoeology between rye and wheat chromosomes but that do not separate *S. cereale* from *S. strictum*, were found by the use of various genes and DNA markers. The presence of genes belonging to wheat homoeologous group 3 on chromosome 6R, indicates possible transfer of a small 3R segment to 6R (Zeller and Cermeño 1991, and reference therein). The use of cDNA clones showed that some markers of 4R of both *S. cereale* and *S. strictum*, show phylogenetic relationship to Triticeae group 7 chromosomes (Chao et al. 1989). Likewise, substitution experiments in wheat-rye substitution lines revealed that the short arm of 4R (4RS) compensates well for group 4 of wheat chromosomes and 4RL compensates well for group 7 chromosomes (Koller and Zeller 1976). The 4RL/7RS and 3RL/6RL translocations are present in both the *S. strictum* and the *S. cereale* genomes. The evidence concerning 4RL/7RS derives from the similar plant morphology of the CS/4R of *S. cereale* and CS/4R of *S. strictum* disomic addition lines, that include the presence of the purple culm gene in both additions, which is carried on wheat 7S (cited in Devos et al. 1993). 3RL/6RL translocation exists in both *S. strictum* and *S. cereale* (Devos et al. 1993). These translocations occurred after the separation of *Secale* from *Triticum*. (3) A single minor translocation that occurred between *S. sylvestre* and the other two *Secale* species (Khush 1962). (4) Two translocations that occurred between *S. cereale* and *S. strictum*. Since these translocations characterize all examined races of *S. cereale*, they define its genome rather than reflect inter-varietal differences of more recent origin, and are of evolutionary significance. These translocations, involving chromosomes 2R, 6R and 7R (Heemert and Sebenga 1972), characterize only the genome of *S. cereale*.

Another small translocation involving a fourth pair of chromosomes and characterizing the *S. cereale* genome, also occurred between *S. strictum* and *S. cereale* (Riley 1955). These translocations presumably happened during the evolvement of cereale from *strictum*. (5) A small intraspecific translocation occurring between the two sub-species of *S. strictum* (Khush 1962; Singh and Röbbelen 1977) as well as intraspecific translocations between races of *S. cereale* (Darlington 1933; Müntzing and Prakken 1941).

The F_1 hybrids of the *S. cereale* x *S. strictum* cross are heterozygotes for the two translocations involving chromosomes 2R, 6R and 7R exhibiting hexavalent at first meiotic metaphase (Riley 1955). These hybrids had low fertility and resembled the wild parent, *S. strictum,* in that they had ears with brittle rachis, closely invested grains and perennial habit (Riley 1955). Translocation heterozygotes produce functional gametes consisting of the parental chromosomal combination. Hence, the F_2 of the crosses between *S. cereale* and *S. strictum* segregated into three types, in terms of chromosome structure and plant morphology: those like *S. cereale*, those like the F_1, and those like *S. strictum*, and with frequency of 1:2:1, respectively (Riley 1955). This pattern of segregation results from a correlation between the phenotypic expression of important taxonomic characters and the constitution of the plants with regard to three chromosome pairs involved in the two large translocations that occurred during the evolution of *S. cereale* from *S. strictum*. It seems probable, therefore, that the proximal regions of these three chromosome pairs, which are protected from cross-overs by more distal chiasma formations, and which always segregate in parental combinations, contain genes that might be responsible for the development of the differentiating characters of the species. Genes present in the proximal regions of these three chromosomes, probably control characters responsible for the adaptation of the two species. The genes segregate undisturbed, since gametes are only viable when all three chromosomes occur in parental combinations. In this way, the proximal regions of these three chromosomes may be regarded as a 'super genes' (Darlington and Mather 1949). Hence, a new arisen translocation might, by chance, have isolated a series of selectively advantageous genes from recombinations, and the newly structural condition would be of immediate adaptive value and become fixed. These two translocations presumably occurred simultaneously, rather than successively, since no race of *S. cereale* was found to have only one of the two translocations.

Are the two gene systems that control the major traits separating *S. cereale* from *S. strictum*, namely, annual growth habit and tough rachis, located on one of the translocated chromosomes? The F_1 hybrid between common wheat (cv. Chinese Spring) and *Elymus Farctus* subsp. *farctus* (=6x *Agropyron junceum*) was perennial (Charpentier et al. 1986). Likewise, Dvorak and Knott (1974) noted that the amphiploid

Chinese Spring—diploid *Elymus elongatus* (=*Agropyron elongatum*), combining the A, B, and D subgenomes of hexaploid wheat with the Ee genome of the perennial *E. elongatus*, is perennial. In this regard, Lammer et al. (2004) pointed out that hybrids between bread wheat and its wild perennial relatives, as well as many genetically stable amphiploids and partial amphiploids derived from these hybrids, exhibit the perennial trait, indicating that annual growth habit is a recessive trait. In hybrids derived from crosses between perennial and annual parents, the perennial habit seems to dominate (Stebbins and Pun 1953; Fedak and Armstrong 1986; Petersen 1991a). This is also the case in other interspecific and intergeneric hybrids in the tribe (Sakamoto 1967; von Bothmer et al. 1985; Frederiksen and von Bothmer 1989, 1995). Consistent with the notion that annual growth habit is a derived trait in the Triticeae, the evolvement of annual *S. cereale* from the perennial *S. strictum*, required a recessive mutation in the locus or loci determining this trait. Using a complete series of diploid *Elymus elongatus* (genome EeEe) chromosome addition lines in a Chinese Spring background, Lammer et al. (2004) found that chromosome 4Ee [most probably the short arm 4Ee (4EeS)] conferred the perennial growth habit. Assuming that the recessive gene(s) conferring annual growth-habit is located in *S. cereale* on 4RS, a chromosome arm that is not involved in translocations, and therefore, can segregate in hybrids between the two species only if it is located on the distal region.

The genetic control of non-brittle rachis in domesticated taxa was studied in several Triticeae species. Takahashi (1955) and Takahashi and Hayashi (1964) reported that two recessive, complementary and tightly linked genes, that later were designated *btr1* and *btr2* and located on chromosome arm 3HS of barley (Komatsuda and Mano 2002; Komatsuda et al. 2004), determine tough rachis in domesticated barley. Likewise, Sharma and Waines (1980) found that tough rachis is determined in diploid wheat *T. monococcum* subsp. *monococcum* by the two recessive complementary genes. Also in domesticated tetraploid and hexaploid wheat the tough rachis trait is controlled by recessive genes, *br-A2* and *br-A3*, on chromosomes 3A and 3B (Levy and Feldman 1989a; Watanabe and Ikebata 2000; Watanabe et al. 2002, 2005a, b; Nalam et al. 2006; Millet et al. 2013). Comparative mapping analyses suggested that both *br-A2* and *br-A3* are present in homoeologous regions on their respective chromosomes. Furthermore, *br-A2* and *br-A3* from wheat and *btr1/btr2* on chromosome 3H of barley are also homoeologous, suggesting that the location of major determinants of the tough rachis trait in these species has been conserved (Nalam et al. 2006). Assuming that the homoeologous genes are located in *S. cereale* on the same arm, namely, 3RS, then the genes involved in the non-shattering trait of *S. cereale* can only segregate in hybrids between *S. cereale* and *S. strictum* if it is located on the distal region.

Levan (1942) observed bivalents, trivalents and quadrivalents in low frequency in the first meiotic metaphase of haploid *S. cereale*, averaging 0.08–0.83 chiasmata per cell. Likewise, Neijzing (1982) observed several associations of two or more chromosomes in meiosis of haploid rye. Heneen (1963) noted a bivalent of rye chromosomes in a single cell of the pentaploid hybrid octoploid *Elymus arenarius* x *S. cereale*. Intragenomic chromosome pairing of *S. cereale* chromosomes, indicated by the presence of homomorphic large bivalents, was also observed in the intergeneric hybrids between diploid and polyploid *Hordeum species* and *Secale cereale* (Gupta and Fedak 1987a, b; Wagenaar 1959; Thomas and Pickering 1985; Petersen 1991b) and in hybrids between *Aegilops* species and *S. cereale* (Melnyk and Unrau 1959; Majisu and Jones 1971; Su et al. 2016). In addition, although occurring at a lower frequency than in haploid plants of *S. cereal* (Müntzing 1937; Nordenskiöld 1939; Levan 1942; Heneen 1965; Puertas and Giraldez 1979), pairing between *S. cereale* chromosomes is a usual feature in wheat-rye hybrids (Dhaliwal et al. 1977; Schlegel and Weryszko 1979). In wheat-rye hybrids, the preferential homoeologous pairing between wheat and rye chromosomes competes with that of rye-rye chromosomes (Schlegel and Weryszko 1979). No rye-rye recombinant chromosomes were observed at first anaphase, therefore rye-rye associations at first metaphase could be considered non-chiasmatic (Orellana 1985). Yet, Neijzing (1982), using Giemsa C-banding, detected chromatid exchanges between differently marked chromosome arms at first anaphase, verifying that chromosome associations observed at first metaphase of haploid rye were due to chiasma formations, which were non-randomly located and probably occurred in homologous segments. Ten to twelve sets of homologous segments were found to be present in the haploid rye genome (Neijzing 1985), likely as a result of ancient duplications. The association between C-banded and un-banded arms indicated that the telomeric heterochromatic does not act as promoter or suppressor of homologous pairing in haploid rye (Neijzing 1982).

Martis et al. (2013), using high-throughput transcript mapping, chromosome survey sequencing and integration of conserved synteny information of three sequenced model grass genomes [*Brachypodium distachyon*, rice (*Oryza sativa*), and sorghum (*Sorghum bicolor*)], established a virtual linear gene order model (genome zipper) comprising 22,426 or 72% of the detected set of 31,008 *S. cereale* genes. This enabled a genome-wide, high-density, comparative analysis of rye/barley/model grass genome synteny. Seventeen syntenic linkage blocks conserved between the model grass genomes and the rye and barley genomes, were identified. Strikingly, differences in the degree of conserved syntenic gene content, gene sequence diversity signatures and phylogenetic networks were found between individual rye syntenic blocks. This indicates that introgressive hybridizations (diploid or polyploid hybrid speciation) and/or a series of whole-genome or chromosome duplications played a role in rye speciation and genome evolution (Martis et al. 2013).

6.4 Crosses with Other Triticineae Species

6.4.1 Opening Remarks

Spontaneous intergeneric hybrids between *Secale* species and other Triticeae seem to be rare (Frederiksen and Petersen 1998), with the octoploid *Elymus arenarius* L. being one of the few reported (Heneen 1963). On the other hand, many artificial intergeneric hybrids between *Secale* species and other Triticeae have been produced. These hybrids have primarily involved diploid and polyploid taxa of *Triticum, Aegilops,* and *Hordeum* (e.g., Majisu and Jones 1971; Hutchinson et al. 1980; Pagniez and Hours 1986; Gupta and Fedak 1986; Petersen 1991a). In addition, artificial hybrids with several other diploid and polyploid Triticeae, e.g., diploid *Elymus spicatus* (Pursh) Gould, diploid *Agropyron mongolicum* Keng, tetraploid *A. cristatum* (L.) Gaertn., tetraploids *Elymus pseudonutans*, *E. shandongensis*, and *E. semicostatus*, and hexaploid *Elymus hispidus* (Opiz) Melderis, have also been studied (Stebbins and Pun 1953; Fedak and Armstrong 1986). In all combinations, *Secale* has been used as the pollen donor, as crossing experiments using *Secale* as the female parent have always failed (Majisu and Jones 1971; Petersen 1991a). The hybrids are generally described as morphologically more or less intermediate between the parents (Stebbins and Pun 1953; Wang 1987, 1988; Petersen 1991a).

The chromosomes of *Secale* are larger than the chromosomes of most species of the Triticeae, enabling simple identification of the parental origin of the chromosomes in artificial hybrids (von Berg 1931; Stebbins and Pun 1953; Melnyk and Unrau 1959; Wagenaar 1959; Bhattacharyny et al. 1961; Heneen 1963; Majisu and Jones 1971; Wang 1987; Fedak and Armstrong 1986; Petersen 1991b; Gill and Friebe 2009). At mitosis and meiosis, rye chromosomes can be easily distinguished from chromosomes of other Triticeae species in intergeneric hybrids, also by the presence of massive blocks of terminal heterochromatin (Lima-de-Faria 1953; Gill and Kimber 1974). In rye, C-heterochromatin is predominantly located in terminal blocks, whereas in *Triticum, Aegilops* and other related species, it is mostly situated in centromeric and pericentromeric regions or dispersed throughout the chromosomes. Thus, techniques such as Giemsa C-banding (Fedak and Armstrong 1986; Gill and Friebe 2009) and in situ hybridization (Hutchinson et al. 1980) have also been used to identify parental chromosomes.

The ability to identify the parental chromosomes in these hybrids is of great help in determining whether the type of pairing is auto- or allo-syndetic.

Hybridizations of *Secale* species with species of over ten Triticeae genera have been reported (e.g., Crasniuk 1935; Stebbins and Pun 1953; Dvorak 1977; Hutchinson et al. 1980; Gupta and Fedak 1987a, b; Wang 1987, 1988; Lu et al. 1990; Lu and von Bothmer 1991; Petersen 1991b). Most authors have been able to clearly demonstrate very low levels of allosyndetic pairing between the chromosomes of such species, irrespective of the ploidy level (Stebbins and Pun 1953; Heneen 1963; Majisu and Jones 1971; Hutchinson et al. 1980; Gupta and Fedak 1987a, b; Fedak and Armstrong 1986; Wang 1987, 1988; Lu et al. 1990; Lu and von Bothmer 1991; Petersen 1991b).

6.4.2 Chromosome Pairing in Hybrids Between *Elymus* or *Agropyron* Species and *Secale* Species

Meiotic chromosomal pairing was analyzed in several intergeneric diploid hybrids between diploid *Elymus* or *Agropyron* species and *Secale strictum* (Wang 1987). Hybrids of *Elymus spicatus* x *Secale strictum*, having the genome formula StR, presented an average of 12.97 univalents, 0.49 bivalents, and 0.01 trivalents at first meiotic metaphase. The hybrid of *Agropyron mongolicum* x *S. strictum*, which have the PR genomes, had an average of 12.86 univalents, 0.51 bivalents, 0.03 trivalents and 0.004 quadrivalents. The hybrid between *E. spicatus* and *Agropyron mongolicum* (genome StP) had a mean configuration of 8.05 univalents, 2.86 bivalents, 0.07 trivalents and 0.01 quadrivalents. Chromosome pairing in PMCs of additional diploid hybrids, contained an average of 11.05 univalents, 1.22 rod bivalents, 0.04 ring bivalents, 0.13 trivalents and 0.01 quadrivalents in *E. farctus* subsp. *bessarabicus* x S. strictum (genome EbR), while in *E. spicatus* x *E. farctus* subsp. *bessarabicus* (genome StEb), average pairing included 4.34 univalents, 2.77 rod bivalents, 1.42 ring bivalents, 0.24 trivalents and 0.14 quadrivalents (Wang 1988). All hybrids had intermediate spike morphology, compared to their parents, and were sterile (Wang 1987, 1988). The meiotic pairings of these hybrids indicated that chromosome homology between the St and P genomes is higher than between the St and R and between the P and R. In addition, it demonstrated that the St genome of *Elymus spicatus* and the Eb genome of *E. farctus* subsp. *bessarabicus* are more closely related to each other than they are to the R genome of *Secale*. The R genome is slightly closer to the Eb genome than to the St genome (Wang 1987, 1988). Mitotic preparations of root-tip cells of these diploid hybrids

suggested that the chromosomes of the different genomes were spatially separated (Wang 1987).

Lu et al. (1990) analyzed chromosomal pairing at first meiotic metaphase in the following three intergeneric triploid hybrids, *Elymus pseudonutans* (2n = 4x = 28; genome StStYY) x *S. cereale*, *E. shandongensis* (2n = 4x = 28; genome StStYY) x *S. cereale* and *E. semicostatus* (2n = 4x = 28; genome StStYY) x *S. strictum*. Meiotic configurations of the F_1 triploid hybrids (2n = 3x = 21; genome StYR) included a mean 14.64 univalents, 2.82 bivalents, 0.20 trivalents, and 0.01 quadrivalents and 16.38 univalents, 2.02 bivalents and 0.16 trivalents, for two combinations of *E. pseudonutans* x *S. cereale*, 15.59 univalents, 2.62 bivalents and 0.08 trivalents for *E. shandongensis* x *S. cereale* and 19.63 univalents and 0.65 bivalents for *E. semicostatus* x *S. strictum*. A large number of chromosomes were involved in secondary associations. Most pairing between chromosomes of the *Elymus* St and Y genomes was autosyndetic. The pairing data showed higher pairing (2.02–2.82 bivalents per cell) between the St and Y genomes of *Elymus* in the presence of the *S. cereale* genome than in the presence of the *S. strictum* genome (0.65 bivalents per cell). Moreover, the bivalent frequency in haploid *E. pseudonutans* (averaging 0.55 bivalents/cell) and in haploid *E. shandongensis* (averaging 0.69 bivalents/cell) (Lu et al. 1990) was much lower than that in the hybrids of these species with *S. cereale*. These observations suggest that the genome of *S. cereale* promotes meiotic pairing of homoeologous chromosomes in *Elymus* species (Lu et al. 1990).

Lu and von Bothmer (1991) crossed *Secale cereale* with three polyploid *Elymus* species, namely, *E. caninus* (2n = 4x = 28; genome StStHH), *E. brevipes* (2n = 4x = 28; genome StStYY) and *E. tsukushiensis* (2n = 6x = 42; genome StStHHYY). Chromosome pairing at first meiotic metaphase in the F_1 hybrids included 20.74 univalents and 0.14 bivalents for the triploid *E. caninus* x *S. cereale* (genome StHR), 16.35 univalents, 2.17 bivalents and 0.09 trivalents for the triploid hybrid *E. brevipes* x *S. cereale* (genome StYR), and 25.84 univalents, 1.10 bivalents and 0.02 trivalents for the tetraploid hybrid *E. tsukushiensis* x *S. cereale* (StHYR). Several secondary associations were also observed in the hybrids. The researchers concluded that (1) the homoeologous relationship between "St", "H" and "Y" genomes in the investigated *Elymus* species, differ, namely, St is closer to Y than to H; (2) low homoeology exists between genomes of *Elymus* (either St, H or Y) and rye (R); (3) the *Secale* genome affects homoeologous chromosome pairing between different genomes in *E. brevipes* and *E. tsukushiensis*. This is in accord with earlier findings suggesting that the *Secale* genome can promote pairing of homoeologous chromosomes (Lelley 1976; Dvorak 1977; Gupta and Fedak 1985, 1987a, b; Lu et al. 1990).

Fedak and Armstrong (1986) produced hybrids between hexaploid *Elymus hispidus* (2n = 6x = 42; genome EeEeEeEeStSt) and *Secale cereale*. Mean chromosome pairing at first meiotic metaphase of the hybrid (2n = 4x = 28; genome EeEeStR) included 18.80 univalents, 3.71 bivalents, and 0.56 trivalents. Most of the pairing was autosyndetic. Indicating very low homology between the Elymus genomes (Ee and St) and the R of rye.

Heneen (1963) analyzed mitotic and meiotic behavior of the natural pentaploid hybrid between *Elymus arennrius* (2n = 8x = 56; genome EeEeEeEeStStStSt) and *Secale cereale*. The nucleolar organizer of rye was suppressed in this hybrid and no secondary constriction was observed in the short arm of chromosome 1R. Most of the hybrid PMCs (2n = 5x = 35; genome EeEeStStR) had 14 bivalents and 7 univalents. The univalents, which are large in size, represent the 7 chromosomes of rye. Pairing is thus autosyndetic. Various meiotic aberrations and irregularities, including spontaneous chromosome breakage, polyploid cells and plasmodia, occurred frequently in this hybrid (Heneen 1963).

6.4.3 Chromosome Pairing in Hybrids Between *Hordeum* and *Secale* Species

Meiotic chromosomal pairing in hybrids between various diploid and polyploid *Hordeum* species and *Secale* species was studied by a number of researchers (e.g., Fedak 1979, 1986; Finch and Bennett 1980; Fedak and Armstrong 1981; Thomas and Pickering 1985; Gupta and Fedak 1985, 1987a, b; Lu et al. 1990; Petersen 1991a, b). In all hybrids, pairing between *Hordeum* and *Secale* chromosomes was very low, indicating a very distant relationship between the two genera.

Several hybridizations between *Hordeum* and *Secale* were performed at the diploid level. Fedak (1979) obtained a diploid hybrid by crossing *H. vulgare* with *S. cereale*, and observed very low chromosome pairing, with an average chiasma frequency of 0.22 per cell. Phenotypically, the hybrid resembled rye, the pollen parent, but only the nucleolar organizers of barley were active. Petersen (1991a, b) produced 41 intergeneric hybrids between diploid, tetraploid and hexaploid cytotypes of several *Hordeum species* and *Secale* species and analyzed their chromosome pairing at first meiotic metaphase. *H. vulgare* and *H. bulbosum* have the same genome (genome H, while *H. marinum* and *H. murinum* each have one distinct genome, Xa and Xu, respectively; all other diploid species have the H-genome (Wang et al. 1995). The same four genomes are found in the polyploid species of *Hordeum* (Table 2.8). Differences between the chromosome sizes of *Hordeum* and *Secale* enabled distinction between auto- and allo-syndetic pairing.

Allosyndetic pairing between chromosomes of *Hordeum* and *Secale* was very rare, indicating very little homology between any of the four basic genomes of *Hordeum* (H, I, Xa, and Xu) and the R-genome of *Secale*. Similarly low allosyndetic pairing between *Hordeum* and *Secale* chromosomes (0.40 bivalents per cell) was reported in several other diploid *Hordeum* x *Secale* hybrids (Thomas and Pickering 1985; Gupta and Fedak 1987a). Very little chromosomal pairing was also observed at first meiotic metaphase of the diploid hybrid *H. chilense* x *Secale cereale* (Finch and Bennett 1980). This hybrid combination was also studied by Thomas and Pickering (1985), who observed 12.64 univalents, 0.62 rod bivalents, and 0.02 trivalents per cell. Among the bivalents, there were 0.13 large homomorphic rye-rye bivalents, 0.23 small homomorphic *chilense-chilense* bivalents, and 0.64 heteromorphic *chilense*-rye bivalents. Only the nucleolar organizers of *H. chilense* only were expressed. Gupta and Fedak (1987b) studied chromosome pairing at first meiotic metaphase in the diploid *Hordeum califormcum* x *S. vavilovii* (now included in *S. cereale*) hybrid and observed a mean 11.74 univalents and 1.13 bivalents per cell, most of which (86.7%) were autosyndetic.

Triploid hybrids between tetraploid species of *Hordeum* and diploid *Secale* also exhibited very little pairing between the chromosomes of the species of these two genera. Finch and Bennett studied chromosomal pairing at first meiotic metaphase of the triploid hybrid between tetraploid *H. jubatum* ssp. *breviaristatum* and *S. strictum* subsp. *africanum*. This hybrid had up to six bivalents between *H. jubatum* chromosomes and almost no pairing between *H. jubatum* and *Secale* chromosomes (Finch and Bennett 1980). These results are in accord with previous reports of hybridization between *S. cereale* and *H. jubatum* (Quincke 1940; Brink et al. 1944; Wagenaar 1959). Brink et al. (1944) and Wagenaar (1959) typically found five to six bivalents between *jubatum* chromosomes in the triploid hybrid, while the S. *cereale* chromosomes were mostly univalents. However, several autosyndetic bivalent and multivalent associations between the rye chromosomes, as well as some allosyndetic associations with *H. jubatum* chromosomes, were also seen in this triploid hybrid. The fairly strong autosyndesis displayed by the *H. jubatum* chromosomes is due to the two closely related genomes of the tetraploid *H. jubatum*. Similar results were obtained by Schlegel et al. (1980), who studied meiotic chromosomal pairing in the triploid hybrid *Hordeum jubatum* x *S. kuprijanovii* (now included in *S. Strictum* subsp. *Strictum*) and found that, apart from barley-barley and rye-rye, homoeologous barley-rye chiasmatic associations were also evident in about 1% of the PMCs.

Studies of chromosome pairing at first meiotic metaphase in the triploid hybrid *H. parodii* (4x) x *S. anatolicum* (included now in *S. strictum* subsp. *strictum*) showed a mean

14.74 univalents, 6.31 bivalents, 0.22 trivalents and 0.02 quadrivalents (Gupta and Fedak 1987b). Most pairing was attributed to autosyndetic among *Hordeum* chromosomes and among several *Secale* chromosomes. The scarcity of allosyndetic pairing in the form of heteromorphic bivalents between *Secale* large and barley small chromosomes, indicated a distant relationship between the parental genomes. In addition, in the triploid hybrid *H. depressum* (4x) x *S. cereale*, Morrison and Raihathy (1959) reported mean chromosomal pairing of 1.7 bivalents per cell, mostly between *H. depressum* chromosomes. They decided that there is no homology between the genomes of *H. depressum* and *S. cereale*. Studies in a triploid hybrid between tetraploid *H. marinum* ssp. *gussoneanum* and *S. cereale*, Staat et al. (1985) reported a frequency of 0.02 heteromorphic bivalents per cell. Hence, the triploid hybrids described above exhibited a strong preferential pairing between the two genomes of *Hordeum,* whereas heteromorphic bivalents involving *Hordeum* and *Secale* chromosomes were very rare.

Similar results were obtained in hybrids with higher ploidy level. The tetraploid hybrid *H. lechleri* (6x) x *S. cereale* had 0.30–0.56 heteromorphic rod bivalents per cell (Fedak and Armstrong 1981) and the trigeneric hybrid containing the genomes of *Triticum aestivum, Hordeum chilense* and *Secale cereale* showed similar low allosyndetic pairing (Fernández-Escobar and Martin 1989).

Preferential intragenomic chromosome pairing, involving *Hordeum-Hordeum* chromosomes, indicated by the presence of homomorphic small bivalents, and *Secale-Secale* chromosomes, indicated by the presence of homomorphic large bivalents, prevails in the diploid hybrids (Gupta and Fedak 1987a, b; Petersen 1991b). Autosyndetic intergenomic pairing among *Hordeum* chromosomes occurred in the hybrids involving polyploid species of *Hordeun* and *Secale.* In hybrids with *S. cereale*, the polyploids *H. brachyantherum, H. jubatum, H. lechleri, H. arionicum, H. prorerum, H. capense,* and *H. secalinum* exhibited a number of bivalents resulting from autosyndetic pairing. In general, the level of autosyndetic pairing among the *Hordeum* chromosomes was higher in the intergeneric combinations with *S. cereale* than in the corresponding *Hordeum* polyhaploids. It was therefore concluded that the *S. cereale* genome promotes homoeologous chromosome pairing between the genomes of *Hordeum* (Gupta and Fedak 1985). *S. cereale* promoted more homoeologous pairing than *S. strictum* subsp. *africanum* (Gupta and Fedak 1985). This is in accord with the finding of Lu et al. (1990), who noted a differential effect of the two *Secale* genomes on chromosome pairing; while *S. cereale* promoted homoeologous pairing in hybrids with *Elymus semicostatus, S. strictum* subsp. *strictum* did not promote such pairing.

Haploids of *Hordeum vulgare* were recovered at a frequency of less than 1% from pollinations of five strains of *H. vulgare* with cultivar Prolific of *S. cereale* (Gupta and Fedak 1987b). This is the first report of barley haploids obtained from barley x rye crosses. Preliminary studies suggest that haploids arose through a process of elimination of rye chromatin in the seedlings.

6.4.4 Chromosome Pairing in Hybrids Between *Aegilops* Species and *Secale* Species

The nucleolar organizer activity in several allopolyploid *Aegilops* species (*Ae. triuncialis, Ae. variabilis, Ae. biuncialis, Ae. juvenalis*) x *S. cereale* F_1 hybrids was analyzed using a highly reproducible silver-staining procedure (Cermeño and Lacadena 1985). All the *Aegilops* allopolyploids share the U subgenome that derived from the diploid species, *Ae. umbellulata.* The *Aegilops* 1U and 5U chromosomes showed strong nucleolar activity, which suppressed the NOR activity of the 1R rye chromosome in all the hybrid combinations. Low activity of the nucleolar-organizer chromosomes of the other subgenomes of the allopolyploid *Aegilops* parents, namely, C, S^v, M^o, and D^cX^c, was also observed. These findings confirm earlier reports of predominant nucleolar-organizer activity of the 1U and 5U chromosomes and the suppression of the NOR activity of 1R chromosome from rye.

Several early investigations on chromosomal pairing in F_1 hybrids between *Aegilops* and *Secale* were reported (von Berg 1931; Kagawa and Chizaki 1934; Karpechenko and Sorokina 1929; Kihara 1937). Because the chromosomes of *Secale* are larger than those of *Aegilops*, the parental members of the paired chromosomes can be identified at first metaphase of meiosis in the F_1 hybrids. This facilitated easy identification of the wheat and rye chromosomes. A low level of chromosome pairing between *Aegilops* and *Secale* chromosomes has been reported in the intergeneric F_1 hybrids between both diploid and tetraploid *Aegilops* species and *S. cereale* (Karpechenko and Sorokina 1929; von Berg 1931; Kagawa and Chizaki 1934; Melnyk and Unrau 1959; Majisu and Jones 1971).

Majisu and Jones (1971), using embryo culture, produced diploid hybrids between species of *Secale* and four diploid species of *Aegilops* section Sitopsis and *Amblyopyrum muticum.* Studies of chromosomal pairing at first meiotic metaphase of these hybrids showed that most *Aegilops* and *Secale* chromosomes were univalents, but some paired as normal chiasmatic bivalents. The pairing data suggested that the *Secale* species genome shows very little homology with the genomes of the *Aegilops* species (Majisu and Jones 1971). Likewise, there is no evidence that the chromosomes of *Secale* are homologous with those of another group of *Aegilops* diploid species, namely, *Ae. caudata*, *Ae. comosa* and *Ae. umbellulata* (Hutchinson et al. 1980), but does not preclude the presence of homologous segments between these species.

Crosses between *Ae. squarrosa* (currently *Ae. tauschii*), the donor of the D subgenome to hexaploid wheat, and *S. cereale*, were performed by several researchers (e.g., Melnyk and Unrau 1959; Kawakubo and Taira 1992; Su et al. 2015). Due to their large size, the chromosomes of *S. cereale* were easily distinguished from those of *Ae. tauschii*, and therefore, all observed heteromorphic bivalents were considered to be due to allosyndetic pairing between Rye and *Aegilops* chromosomes. Melnyk and Unrau (1959) observed chromosomal pairing with an average of 9.67 univalents, 1.61 rod-bivalents, 0.01 ring bivalents, 0.25 trivalents and 0.05 quadrivalent in the hybrid *Ae. squarrosa* var. *typica* x *S. cereale* (genome DR). Only 0.4% of the bivalents were formed by *Secale-Aegilops* pairing, and trivalents were composed of 1 or 2 *cereale* chromosomes, indicating that the chromosomes of these two species are distantly homoeologous but still capable of some intergeneric pairing.

Su et al. (2016) produced and analyzed hybrids from crosses of *Ae. tauschii* and *S. cereale*. The hybrids showed an average meiotic pairing configuration of 10.84 univalents, 1.57 bivalents and 0.01 trivalents. Genomic in situ staining revealed three types of bivalent associations, namely, D-D, R-R and D-R, at frequencies of 8.6, 8.2 and 83.3%, respectively. Trivalents consisting of D-R-D and of R-D-R associations were also found. These results suggested that both intra- and intergenomic chromosome homology contribute to chromosome pairing.

Kawakubo and Taira (1992) produced hybrid plants with different combinations of D and R genomes, by crossing *Ae. tauschii* (2x and 4x) and *Secale cereale* (2x and 4x). Amphidiploids were obtained directly from the cross between tetraploid parents. Diploid and triploid hybrids were completely seed-sterile, whereas the amphidiploid had an average self-fertility of 4.5%, ranging between 0 and 45%. At first meiotic metaphase, the mean chromosome associations included 13.4 univalents, 0.26 bivalents and 0.01 trivalents in the diploid hybrids, 7.1 univalents and 6.96 bivalents in the triploid hybrids, and 7.9 univalents and 10.5 bivalents in the amphidiploids. In diploids, end-to-end type pairing between D and R chromosomes was observed, presumably without chiasmata. Homologous pairing between D chromosomes was predominant in plants bearing two sets of D genomes, and homoeologous D-R pairing was scarcely observed in either triploids or amphidiploids.

The conclusion that there is very little homology between *Aegilops* and *Secale* chromosomes was also supported by the pattern of meiotic pairing in the triploid hybrid between synthetic autotetraploid *Ae. tauschii* and *S. cereale* (Majisu and Jones 1971). No trivalents were observed in the triploid hybrid, and bivalent formation was restricted to *Ae. tauschii* chromosomes, with the *Secale* chromosomes occurring as univalents.

Meiotic pairing behavior in hybrids generated between *Aegilops* and *S. cereale* was examined by in situ hybridization, using a 480-bp repeat probe, purified by Bedbrook et al. (1980), which is specific to rye and absent from *Aegilops* species, thus enabling the rye chromosomes to be clearly identified (Hutchinson et al. 1980). It was shown that in diploid hybrids between *Ae. caudata* or *Ae. comosa* and *S. cereale*, the small amount of pairing was mostly of the allosyndetic, *Aegilops*-rye type, although some autosyndetic associations also occurred. The two diploid hybrids showed a similar level of pairing, which, although fairly low, was slightly higher than previously reported (e.g., Majisu and Jones 1971). Hutchinson et al. (1980) concluded that the *Aegilops* chromosomes are more likely to pair with *Secale* chromosomes in diploid *Aegilops* x *Secale* hybrids, than with other *Aegilops* chromosomes in inter-genomic crosses. This suggests that there is partial homology between *Secale* and *Aegilops* chromosomes, and lies in agreement with the conclusions reached by Melnyk and Unrau (1959) based on studies of the meiotic pairing between *Ae. tauschii* and *S. cereal*, but contradicted the results of Majisu and Jones (1971). It should be pointed out, however, that the overall level of pairing obtained in the experiments of Majisu and Jones (1971) was lower than that reported by Hutchinson et al. (1980), and may reflect a lower level of allosyndetic pairing in the tested material. Yet, in the triploid hybrid between *Ae. columnaris* (4x; genome UUX^{n}X^n) and *S. cereale*, pairing and chiasma formation was largely limited to chromosomes of the two *Aegilops* subgenomes. This suggests that, as expected, the *Ae. columnaris* genomes U and X^n are more closely related to each other than they are to the R genome of *S. cereale* (Hutchinson et al. 1980).

Lucas and Jahier (1988) conducted a diallel crossing program, which included ten diploid species from the genera *Triticum*, *Aegilops*, *Dasypyrum* and *Secale*. Forty-one different interspecific hybrids were analyzed in this program. The number of associations between chromosome arms at first meiotic metaphase of the F$_1$ hybrids was taken as an indication of the degree of homology between the parental genomes. Lucas and Jahier (1988) found little affinity between the genomes of *Triticum* or *Aegilops* and *S. cereale*, and concluded that *S. cereale* and *Dasypyrum villosum* are only distantly related to the *Triticum and Aegilops* species.

Chromosomal pairing at first meiotic metaphase of *Ae. ovata* (currently *Ae. geniculata*; 2n = 4x = 28; genome M^{o}M^{o}UU) x *S. cereale* triploid hybrid, was studied by several researchers (e.g., Leighty et al. (1926), Kagawa and

Chizaki (1934), Khalilov and Kasumov (1989), Sechnyak and Simonenko (1991), Cuñado (1992) and Wojciechowska and Pudelska (2002). The studies showed a pairing frequency ranging from 0.40 to 0.86 for rod bivalents (Wojciechowska and Pudelska 2002) to 2–3 for rod bivalents (Kagawa and Chizaki 1934), 1.46 rod and 0.3 ring bivalents, 0.12 trivalents (Sechnyak and Simonenko (1991) and to 1–5 rod bivalents and 1 trivalent (Khalilov and Kasumov 1989). Cuñado (1992) crossed *Ae. uniaristata* (2x) with *S. cereale* and observed 0.38 rod bivalents per cell at meiosis. Using the C-banding technique, he was able to discern between autosyndetic *Aegilops–Aegilops* pairing (0.11 rod bivalents per cell), autosyndetic rye-rye pairing (0.04 rod bivalents per cell) and allosyndetic *Aegilops*-rye pairing (0.23 rod bivalents per cell). Cuñado (1992) also crossed several allotetraploid and allohexaploid *Aegilops* species with *S. cereale* and, using C-banding technique, found relatively little allosyndetic pairing between *Aegilops* and rye chromosomes (from 0.04 rod bivalents/cell in *Ae. ventricosa* x *S. cereale* to 0.30 rod bivalents/cell in *Ae. triuncialis* x *S. cereale*). On the other hand, the autosyndetic pairing between *Aegilops-Aegilops* chromosomes in hybrids with the allotetraploid *Aegilops,* ranged from 2.01 rods in *Ae. geniculata* x *S. cereale* to 6.59 in *Ae. cylindrica* x *S. cereale*. In hybrids with the allohexaploid *Aegilops*, it ranged from 4.41 rods in *Ae. juvenalis* x *S. cereale* to 8.77 rods in *Ae. crassa* x *S. cereale*. Autosyndetic pairing between rye-rye chromosomes also occurred in low frequency. Clearly, very little homoeologous pairing occurred between the R genome of *S. cereale* and the genomes of the polyploid *Aegilops* species. The pairing data indicate that subgenomes of the allopolyploid Aegilops species are much closer to each other than to the R genome of *S. cereale*.

Gupta and Fedak (1985) studied meiosis in hybrids between allohexaploid *Ae. crassa* (2n = 6x = 42; genome D^cX^cD) and species of *Secale*. Their results provide evidence that promotion of homoeologous pairing was observed in these hybrids. The increase in pairing mainly affected autosyndetic *Aegilops–Aegilops* pairing (Gupta and Fedak 1985). They concluded that a meiotic pairing control system that promotes homoeologous pairing operates in *Ae. crassa*. Different levels of homoeologous pairing were obtained in hybrids of 6x *Ae. crassa* x *Secale* species, depending on the rye DNA and C-heterochromatin content. For instance, the *S. strictum* genotype suppressed the function of this system in a manner that was inversely related to its heterochromatin content and total DNA content. Similar effects were observed in *S. cereale* x wheat and *S. cereale* x *Ae. ventricosa* hybrids, where C-heterochromatin appeared to affect homoeologous pairing involving both the chromosome arms carrying it as well as also involving other chromosomes (Gustafson 1983; Cuñado et al. 1986).

6.4.5 Chromosome Pairing in Hybrids Between *Triticum* and *Secale* Species

The first attempt to cross wheat with rye was made by the Scottish botanist A. Stephen (Wilson 1876), who pollinated common wheat with *S. cereale* pollen. The resulting hybrid plants were sterile. Wilson presented his results on April 8, 1875, in a communication to the Botanical Society of Edinburgh. His attempt inspired, the American plant breeder Elbert S. Carmann, who also obtained a sterile hybrid from a cross between common wheat and rye (Carman, E. S.: Rural New Yorker, August 30, 1884). Later on, in 1888, the German plant breeder Wilhelm Rimpau, crossed common wheat with *S. cereale* and obtained several sterile and one fertile hybrid. The fertile hybrid presumably originated from a spontaneous chromosome doubling, thus producing the first octoploid Triticale, which yielded fertile progeny (Rimpau 1891). In 1921, the Russian plant breeder G. K. Meister, observed spontaneous pollinations of wheat plants with rye pollens from neighboring breeding plots (Meister 1921). Since these pioneering attempts in the end of the nineteenth century and beginning of the twentieth one, that aimed mainly to transfer cold hardiness from rye to wheat, numerous hybridizations between *Triticum* species and *Secale* species have been produced and their chromosome behavior at meiosis has been analyzed. In most cases, the allohexaploid *T. aestivum* was used, whereas in a minority of cases, several subspecies of the allotetraploid wheat, *T. turgidum,* were crossed with *Secale* species.

6.4.5.1 Chromosome Pairing in Hybrids Between Diploid Wheat and *Secale* Species

Very few successful hybridizations involving diploid wheat and *Secale* species were reported. For example, Lucas and Jahier (1988) did not obtain viable hybrids when they crossed diploid wheats, *T. urartu* and wild *T. monococcum* with *S. cereale*. One of these crosses was performed by Sodkiewics (1982) who observed during meiosis of the diploid hybrid of domesticated *T. monococcum* (genome A^mA^m) x *S. cereale* little affinity between chromosomes of these two species.

6.4.5.2 Chromosome Pairing in Hybrids Between Tetraploid Wheat and *Secale* Species

Longley and Sando (1930) produced hybrids between wild emmer wheat, *Triticum turgidum* subsp. *dicoccoides,* the progenitor of domesticated *T. turgidum* (2n = 4x = 28; genome BBAA), and *S. montanum* (currently *S. strictum*), and found a mean 0–1 bivalent pair/cell at first meiotic metaphase in the triploid hybrid (genome BAR). In the same year, Plotnikowa (1930), Aase (1930), Oehler (1931), and Vasiljev (1932), produced hybrids from the crosses of domesticated *T. turgidum,* ssp. *durum* and ssp. *persicum,* with *S. cereale.* In all these triploid hybrids (genome (BAR) pairing was scarce (0–4 bivalents/cell). Kagawa and Chizaki (1934) observed 0–5 bivalents in the ssp. *durum* x *S. cereale* hybrid, and Liljefors (1936) observed somewhat higher pairing (0–6 bivalents and rare trivalents and quadrivalents) in the *T. turgidum* ssp. *turgidum* x *S. cereale* hybrid. Nakajima (1955) crossed *T. turgidum* subsp. *polonicum* with *S. strictum* subsp. *africanum* and observed 0–3 bivalents at meiosis, most of which resulted from autosyndesis between the chromosomes of BA genomes of ssp. *polonicum.* Nakajima (1956a) analyzed the F_1 hybrid progeny of the cross between domesticated emmer, *T. turgidum* subsp. *dicoccon* and *S. vavilovii* (currently included in *S. cereale* subsp. *ancestrale*). The hybrid resembled domesticated emmer in most traits, rather than being an intermediate between the two parents. Chromosomal pairing at first meiotic metaphase included 13–21 univalents/cell and 0–4 bivalents/cell. The bivalents were rod-, and rarely, ring-shaped and homomorphic, indicating autosyndesis of the *Triticum* B and A subgenome chromosomes.

Giorgi and Cuozzo (1980) crossed *S. cereale* with the *ph1c* mutant line of durum wheat cv. Cappelli, which has a deficiency of a segment on chromosome arm 5BL that includes the *Ph1* locus, and thus, enables high homoeologous pairing in wheat hybrids. In meiosis of the F_1 hybrid, an average of 11.4 univalents, 4.3 bivalents and 0.3 trivalents were observed per cell which is much higher count than that of the control hybrid Cappelli x rye possessing *Ph1,* in which only 0.36 bivalents per cell were observed (Giorgi and Cuozzo 1980). The increase in chromosome pairing, consequent to the absence of *Ph1,* was mainly due to autosyndetic pairing between the wheat B and A genome chromosomes. Relatively low pairing between wheat and rye chromosomes was also inferred from the frequency and size of the trivalents. Thus, the deficiency of *Ph1* only slightly increased the pairing between durum and *cereale* chromosomes.

All the hybrids between the subspecies of *T. turgidum* and *S. cereale* were completely sterile. In rye, there are only two flowers in each spikelet whereas in *T. turgidum* there are always more than two flowers. The average number of flowers in the hybrids was greater than 4, indicating dominance of a larger number of flowers per spikelet (Müntzing 1935). However, rye was dominant in determination of the number of spikelets per ear; the number of spikelets for rye, *T. turgidum* and F_1 was 31.6, 20.4 and 31.5, respectively (Müntzing 1935).

Naranjo (1982) analyzed meiotic pairing in the tetraploid hybrid (genome BARR) generated from hexaploid Triticale (2n = 6x = 42; genome BBAARR) and *S. cereale* (genome RR) and found that, in addition to the homologous pairing between chromosomes of the two R genomes,

homoeologous pairing took place preferentially between homoeologous chromosomes of group 1. 1A–1R associations were more frequent than 1B–1R associations, although, in both cases, pairing was mostly restricted to the long arms. In tetraploid wheat and rye hybrids, most of the wheat-rye homoeologous pairing was restricted to chromosomes 1RL/1BL or 1AL and 1DL, at low frequency (Naranjo 1982).

6.4.5.3 Chromosome Pairing in Hybrids Between Hexaploid Wheat and *Secale* Species

Most crosses between hexaploid *Triticum* and *Secale* have been made between common wheat, *T. aestivum* subsp. *aestivum* (genome BBAADD) and *S. cereale.* Longley and Sando (1930) reported the presence of 28 univalents and zero bivalents in such tetraploid hybrids (genome BADR). Kihara (1924), Thompson (1926), and Aase (1930), also observed very little pairing in hybrids between bread wheat and *S. cereale* (0–3 bivalents/cell), whereas Bleier (1930) and Kattermann (1934) observed somewhat more pairing (0–4 to 0–6 bivalents/cell). Hybrids with *S. strictum* subsp. *strictum* as parent exhibited similar results. In the hybrid of *T. aestivum* subsp. *aestivum,* x *S. strictum* subsp. *strictum,* Longley and Sando (1930) observed 0–1 bivalents/cell and in the hybrid *T. aestivum* subsp. *spelta* x subsp. *strictum,* they reported 0–3 bivalents/cell. In the latter hybrid, Aase (1930), Kagawa and Chizaki (1934), and Nakajima (1956b) observed 0–4 bivalents, whereas, in subsp. *spelta* x *S. strictum* subsp. *africanum,* Nakajima (1956b) observed 0–5 bivalents/cell. In the hybrid *T. aestivum* subsp. *compactum* x *S. cereale,* Kagawa and Chizaki (1934) observed 0–3 bivalents. The number of occasional bivalents observed in all these hybrids did not exceed that seen in haploids of common wheat. Therefore, they may be regarded as a result of autosyndesis between wheat-wheat chromosomes, and not as an indication of homology between wheat and rye chromosomes. The rye genome (R) is, therefore, not close to the A, B or D subgenomes of allohexaploid wheat (Thompson 1931).

Megyeri et al. (2013) studied chromosome pairing in first meiotic metaphase of bread wheat-rye F_1 hybrids, using sequential genomic and fluorescent in situ hybridization techniques. These methods enabled both discrimination of wheat and rye chromosomes, and identification of the individual wheat and rye chromosome arms involved in the chromosome associations. Mean chromosomal pairing at first meiotic metaphase of this hybrid included 25.74 univalents, 1.07 rod bivalents, 0.012 ring bivalents and 0.025 trivalents/cell. The majority of associations (93.8%) were observed between the wheat chromosomes, 5.2% between wheat and rye chromosomes and 1.0% between rye chromosomes (Megyeri et al. 2013). The largest number of wheat-wheat chromosome associations (53%) was detected between the A and D subgenomes, while the frequency of B-D and A-B associations was significantly lower (32% and 8%, respectively). Among the A-D chromosome associations, pairing between the 3AL and 3DL arms was observed at the highest frequency, while 3DS-3BS was the most frequent of all chromosome associations (0.113/cell). Pairing between the A and B chromosomes was found rarely, with 2AL and 2BL displaying the highest pairing affinity (pairing frequency was 0.011). Only four wheat chromosome arms (4AS, 5AL, 6BL, and 4DS) did not pair with other chromosome arms in the examined cells (Megyeri et al. 2013). Pairing between wheat and rye chromosomes was low. The frequency of A-R associations was 0.007, of B-R was 0.015 and of D-R was 0.015.

Megyeri et al. (2013) reported similar levels of wheat-wheat chromosome pairing in the wheat x rye hybrid and in the haploid of common wheat (Jauhar et al. 1991), suggesting that the *S. cereale* genome does not significantly influence the pairing of wheat-wheat chromosomes in the hybrid (as discussed below).

Hutchinson et al. (1983), using the C-banding technique, studied meiotic chromosomal pairing in hybrids between bread wheat cv. Chinese Spring aneuploids (nulli 3A-tetra 3B and nulli 5B tetra 5D), and *S. cereale*. The hybrids lacking chromosome 3A or 5B, displayed higher pairing than that observed in hybrids between euploid Chinese Spring and *S. cereale*. Pairing was more frequent between A and D subgenome chromosomes than between chromosomes of A or D subgenome and those of B subgenome and between B subgenome and R genome chromosomes.

Miller et al. (1994) used genomic in-situ hybridization (GISH) to compare the amount of wheat-rye chromosome pairing in *Triticum aestivum* x *Secale cereale* hybrids bearing the 5B chromosome (a), versus hybrids lacking the 5B chromosome (b) or those in which it was replaced by an extra dose of chromosome 5D (c). The mean number of chromosome arm associations per PMC in the three hybrid genotypes was as follows: (a) W-W 0.48, W-R 0.08, R-R 0.02 and total 0.58; (b) W-W 5.11, W-R 0.43, R-R 0.07 and total 5.61; and (c) W-W 4.19 (not including the 5D bivalent), W-R 0.18, R-R 0.04 and total 4.41. As expected, both of the chromosome 5B-deficient hybrid genotypes showed significantly higher pairing than the euploid wheat hybrid. The increase in pairing due to the absence of chromosome 5B was mainly between the wheat chromosomes, with much less between wheat and *Secale* and almost none between the *Secale* chromosomes.

In order to further establish the arm homoeology of common wheat and *S. cereale* chromosomes, Naranjo et al. (1987) studied chromosomal pairing at first meiotic metaphase in three different bread wheat (Chinese Spring) x *S. cereale* hybrid combinations (either bearing both 5B and 3D, 5B-deficient and 3D-deficient). The majority of individual

wheat chromosomes and their arms, as well as the arms of chromosomes 1R and 5R, were identified by means of C-banding. Chromosome arm 1RL paired with 1AL, 1BL and 1DL, while 5RL was homoeologous to 5AL and partially homoeologous to 4AL and 4DL. It was thus concluded that 5RL carries a translocated segment from 4RL (see Devos et al. 1993).

Chromosome pairing at first meiotic metaphase in hybrids between bread wheat and rye, bearing the *ph1b* mutation in bread wheat, were analyzed by Naranjo and Fernández-Rueda (1991, 1996), in efforts to establish the frequency of pairing between individual chromosomes of wheat and rye in the absence of the homoeologous pairing suppressor, *Ph1*. Diagnostic C-bands and other cytological markers, such as telocentrics or translocations, were used to identify each one of the rye chromosomes and wheat arms. Both the amount of telomeric C-heterochromatin and the structure of the rye chromosomes relative to wheat affected the level of wheat-rye pairing. The degree to which rye chromosomes paired with their wheat homoeologues varied with each of the three wheat subgenomes. In most homoeologous groups, the B-R association was more frequent than the A-R and D-R associations. Recombination between arms 1RL and 2RL and their wheat homoeologues possessing a different telomeric c-banding pattern, was detected and quantified at first anaphase. The frequency of recombinant chromosomes obtained supports the premise that recombination between wheat and rye chromosomes may be estimated from wheat-rye pairing patterns.

Cuadrado et al. (1997) performed fluorescence in situ hybridization (FISH), with multiple probes, to meiotic chromosome spreads derived from hybrid plants of the bread wheat *ph1b* mutant, i.e., has a deficiency of a region on chromosome arm 5BL that includes the *Ph1* gene, x *S. cereale*. Homoeologous pairing was expected to be increased due to the absence of *Ph1*. The probes used allowed for unequivocal identification of all of the rye and most of the wheat chromosomes, in both unpaired and paired configurations. Thus, it was possible to identify the pairing partners and to determine the frequency of wheat-wheat and wheat-rye associations. Most of the wheat-rye pairs, which averaged about 7–11% of the total pairs detected in the hybrids, involved B subgenome chromosomes (about 70%), and to a much lesser degree, D (almost 17%) and A (14%) subgenome chromosomes. In these pairs, rye arms 1RL and 5RL showed the highest pairing frequency (over 30%), followed by 2RL (11%) and 4RL (about 8%), and much lower values for all the other arms. 2RS and 5RS were never observed in pairings in the analysed sample. Chromosome arms 1RL, 1RS, 2RL, 3RS, 4RS and 6RS were exclusively bound to wheat chromosomes of the same homoeologous group. The opposite was true for 4RL, which paired with 6BS and 7BS and for 6RL, which paired with 7BL. 5RL, on the other hand, paired with 4WL arms or segments, in more than 80% of the cases, and with 5WL in the remaining 20% of pairings. Additional cases of pairings involving wheat chromosomes belonging to more than one homoeologous group, occurred with 3RL, 7RS and 7RL. These results support previous evidence (Naranjo and Fernández-Rueda 1991; Devos et al. 1993) of the existence of several translocations in the rye genome relative to that of wheat.

Meiotic pairing in euploid rye-triticale hybrids (2n = 4x = 28; genome BARR) was compared with that of three hypo-aneuploid BARR hybrids, two with 2n = 26 chromosomes and one with 2n = 27, also obtained from the triticale x *S. cereale* cross (Naranjo and Palla 1982). The aneuploid with 2n = 27 chromosomes was identified, by C-banding, as mono-5R; its meiotic pairing indicated that chromosome 5R of *S. cereale* is a strong promoter of homologous rye pairing and is also able to influence homoeologous wheat chromosome pairing. The results show that the application of hypo-aneuploidy compensated by homoeologous chromosomes, is useful in the study of genetic control of meiotic pairing in diploid species.

The ability to promote and suppress homologous and homoeologous chromosomal pairing in wheat x rye hybrids by *S. cereale* genes, is an important question and has been investigated by several researchers. Miller and Riley (1972), Naranjo et al. (1979), Jouve et al. (1980) Naranjo (1982) reported that hybrids between tetraploid or hexaploid wheat and rye bearing an extra dose of the rye R genome, resulting in hybrid genomes of BARR or BADRR, displayed higher homoeologous pairing between the chromosomes of the wheat subgenomes than in hybrids with a single R genome. A dosage effect of rye genomes on homoeologous pairing has also been shown by comparing BARR, BARRR, BADR combinations; however, BADRR hybrids frequently showed lower homoeologous wheat-rye pairing than BADR (Naranjo 1982). Similar results have been obtained by comparing *Hordeum-Secale* hybrids with bearing various genome ratios (Gupta and Fedak 1985). Thus, it appears that not only the number of rye genomes but also depending, at least partly, on rye DNA and C-heterochromatin, the genome ratio can affect the extent of homoeologous pairing. Naranjo et al. (1979) also noted an effect of rye promoters on homologous pairing between tetraploid and hexaploid wheat chromosomes of genomes BBAA in hybrids (genome BBAADRR) between hexaploid (genome BBAARR) and octoploid (genome BBAADDRR) triticale.

The increased homoeologous pairing in these hybrids has been explained by Cuñado et al. (1986), as an inhibition of the effect of a single dose of the *Ph1* locus on chromosome 5B by two or more doses of chromosome 5R (Jouve et al. 1980). However, Miller and Riley (1972) ascribed the promotion of homoeologous chromosome pairing by the rye genotypes to the activities of the complete genome and not

necessarily to the homoeologous group 5 chromosome. It seems clear that the rye genome may modulate the wheat homologous and homoeologous pairing in wheat x rye hybrids, but the mechanism of this control is unknown; further studies combining genomes in different doses are still required.

Studies performed by Lelley (1976), Dvorak (1977), Romero and Lacadena (1982) and Cuadrado and Romero (1984), in bread wheat x rye hybrids with one dose of the R genome, provided evidence of the possible existence of rye pairing promoting and suppressing gene(s) affecting both wheat-wheat and wheat-rye pairing (Dvorak 1977). These findings are in contrast to those of Megyeri et al. (2013), who reported that the *S. cereale* genome does not significantly influence wheat-wheat chromosomes pairing in the hybrid. The promotion of homoeologous pairing is greater in wheat-wheat than in wheat-rye homoeologous pairing. However, significant variation of wheat-rye pairing frequencies has been detected in different genotypes with the same genome constitution (Naranjo et al. 1979; Naranjo and Palla 1982).

Evidence for the existence of variation in the effect of such genes on homoeologous pairing among *Secale* taxa, was obtained by Cuadrado and Romero (1984). They analyzed meiotic pairing in wheat x rye hybrids obtained by crossing *T. aestivum* cv. Chinese Spring (CS) with two cultivars of *S. cereale* subsp. *cereale* (Elbon and Ailés) and *S. vavilovii* (currently included in *S. cereale* subsp. *ancestrale*). The results showed that the level of homoeologous pairing in hybrids was affected by the genotype of the rye cultivar used. The CS x *S. cereale* cv. Elbon hybrids exhibited fewer pairing (0.27 mean bivalents/cell) than *T. aestivum* haploids, CS x *S. vavilovii* hybrids had a similar number of bivalent pairings than did bread wheat haploids, and CS x *S. cereale* cv. Ailés hybrids had more pairings (0.44 mean bivalents/cell) than haploids of common wheat. Consequently, Cuadrado and Romero (1984) assumed that rye cultivar Ailés possesses genes that promote homoeologous pairing in hybrids with common wheat, while cv. Elbon possesses suppressor(s) of such pairing. The hybrids generated with *S. cereale* cultivars, that are allogamous, displayed much greater variation in the amount of pairing than hybrids with *S. vavilovii*, which is an autogamous taxon. This variation was interpreted by Cuadrado and Romero (1984) as being due to the different effect of the rye genotypes on the homoeologous pairing in relation with the reproductive systems (allogamy or autogamy) of the particular rye.

Similar evidence was obtained by Gupta and Fedak (1986) who studied the effect of two rye cultivars on chiasmata frequency in bread wheat-rye hybrids. Chiasmata frequencies ranging from 0.07 to 10.40 per cell were recorded in hybrid plants derived from the cross of bread wheat x F_1 between two cultivars of *S. cereale* (Gupta and Fedak 1986). These included one group of plants from the cross *Triticum aestivum* 'Chinese Spring' x *Secale cereale* F_1 ('Petkus' x 'Prolific') and another group from the cross 'Chinese Spring' x F_1 ('Prolific' X 'Puma'). These hybrids used to study the inheritance of genetic variation in rye affecting homoeologous chromosome pairing. In the progeny of the cross 'Chinese Spring' x *Secale cereale* F_1 ('Petkus' x 'Prolific') segregation for major genes affecting pairing was evident, since a bimodal distribution was observed and chiasmata frequencies ranging from 6.11–10.40 and 3.0–6.0 chiasmata/cell. In the second cross involving F_1 rye plants derived from 'Prolific' x 'Puma', a smaller-sample hybrids gave a continuous distribution of chiasmata with a single mode, and the chiasmata frequency never exceeded 2.70/cell. Gupta and Fedak (1986) concluded that the genetic system in 'Petkus' differs from that in 'Puma', and that both genes with major effects and minor effects on chromosome pairing, may be simultaneously present in different cultivars of rye. This could be due to a difference in genetic systems found in 'Puma' and 'Petkus', since 'Prolific' was a common parent in both crosses.

Direct involvement of rye chromosomes in homoeologous pairing is very low in wheat-rye hybrids (Hutchinson et al. 1983) and in *Aegilops*-rye hybrids (Gupta and Fedak 1985; Cuñado et al. 1986). Miller and Riley (1972) stated that rye chromosome arm 5RL, like 5BL of wheat, suppresses homoeologous pairing. But pairing analysis of hybrids involving disomic and monosomic substitution individuals of bread wheat, in which chromosome 5B of hexaploid wheat is replaced by the long arm of chromosome 5R of rye x *Aegilops peregrina*, quantitatively demonstrated that chromosome 5RL lacks the pairing regulator gene of chromosome 5BL. Lelley (1976), Cuadrado and Romero (1984), and Gupta and Fedak (1985) suggested that a polygenic system in rye that affects homoeologous pairing. The involvement of major genes was also suggested by Gupta and Fedak 1985).

Orellana et al. (1984) studied meiosis in wheat-rye addition and substitution lines and suggested that *S. cereale* and bread wheat chromosomes affect each other's homoeologous pairing. They also found that chromosomes 5R and 3R are pairing suppressors and that the effect of 5R in the decrease of wheat chromosome pairing was the strongest and that of 3R the weakest. Gupta and Fedak (1986) studied the effect of individual rye chromosomes on meiotic homoeologous pairing in hybrids between hexaploid and tetraploid wheat, as well as the effect of loss of specific heterochromatin blocks on meiotic pairing in these hybrids. In the pentaploid hybrids between tetraploid wheat, *Triticum turgidum* subsp. *turgidum* cv. 'Ma', and *T. aestivum* cv. 'Chinese Spring', an average 10.30% reduction in chiasmata frequency was observed, while addition lines with rye

chromosome 6R reduced chiasmata frequencies only by an average of 7.4%. Hybrids involving the 4R-bearing rye addition line showed a 25.04% reduction in chiasmata frequency. Hence, chromosome 6R promotes homologous pairing and 4R suppresses such pairing in these hybrids. Chromosome 2R was also found to impart a promotional effect on chromosome pairing (Gupta and Fedak 1986). Telomeric heterochromatin of 7RL did not influence homologous pairing in the pentaploid hybrids while a slight increase in chiasmata counts was observed upon loss of telomeric 6R heterochromatin.

Riley and Law (1965) reported more pairing in a nulli-5B haploid of *T. aestivum* as compared to the hybrid between nulli-5B of *T. aestivum* and *S. cereale*, and suggested that the rye genotype has a suppressing influence on chromosome pairing similar to that of chromosome 5B. However, Bielig and Driscoll (1970) showed that chromosome 5R of *S. cereale* did not compensate for the absence of chromosome 5B, indicating that it does not carry a homoeologous pairing suppressor.

The number of genes transferred from rye to wheat in breeding programs is usually small despite the amount of homoeologous pairing observed in bread wheat x *S. cereale* hybrids, indicating that there is a discrepancy between pairing and recombination in these hybrids. Since insufficient data are available on the relationship between wheat-rye homoeologous pairing and wheat-rye recombination in such hybrids, Orellana (1985) was able to distinguish between three types of wheat-rye associations (end-to-end extremely distal, end-to-end distal and interstitial) between homoeologous chromosomes at different first meiotic metaphase stages (early, middle and late) by analyzing telomeric C-bands. In addition, he estimated the actual recombination frequencies for such associations at first anaphase of wheat-rye hybrids. In all plants analyzed, only open bivalents were found. There was a decrease in the frequency of the end-to-end associations during first metaphase progressed, whereas the number of interstitial associations remained without significant change in all metaphase stages. Assuming a maximum of one chiasma per bond, a good correlation was found between the frequencies of interstitial associations at metaphase and the number of recombinant chromosomes at anaphase. In addition, rye-rye homologous pairing was observed at metaphase, but no evidence for rye-rye recombination was found at anaphase. Moreover, wheat-rye hybrids lacking chromosome 5B (either nulli-5B or nulli-5B tetra-5D) and the hybrids with suppressed *Ph1* activity, in which homoeologous pairing between wheat-wheat and wheat-rye chromosomes was increased, there was a clear reduction in the number of bound arms during the progression of first metaphase for all bivalents identified. Thus, even in the absence of *Ph1*, end-to-end homoeologous and nonhomologous associations are actually non-chiasmatic and are a remnant of prophase pairing (Orellana 1985). In contrast, Naranjo et al. (1989) found a good correspondence between the levels of pairing at first metaphase and those expected from the frequency of first anaphase recombination for the C-band-marked wheat chromosomal arm 1BL in wheat-rye hybrids with a *phl* mutant. Yet, upon analysis of several C-banded marked chromosomal arms, in addition to 1BL, Orellana (1985) reported a significant excess of wheat-rye metaphase associations as compared to the frequency of anaphase recombinant chromosomes.

Benavente et al. (1996) made similar observations, upon genomic in situ hybridization of first metaphase and first anaphase stages of meiosis of bread wheat x *S. cereale* hybrids carrying the *phlb* mutation. The frequency of associations between wheat and rye chromosomes greatly exceeded the level of wheat-rye recombination found in the examined hybrids. Extremely distal associations, accounting for about 50% of the total wheat–rye metaphase chromosomal pairing, were non-chiasmatic, which can explain the discrepancy between metaphase and anaphase recordings. If non-chiasmatic wheat-rye chromosomal pairing were the source of the excess metaphase associations, then it can be said that the very distal associations do not reflect chiasma formation, whereas interstitial and subterminal associations may be chiasmatic and lead to wheat-rye recombination.

In hybrids with bread wheat, the rye nucleolar organizer on 1R became inactive and its secondary constriction was no longer distinguishable. Silver staining showed that when the SAT chromosomes 1B or 6B of wheat or 1U or 5U of *Aegilops umbellulata* were present, the nucleolar activity of the rye SAT chromosome 1R appeared to be suppressed (Cermeño and Lacadena 1985). Suppression of the NORs of *Secale* genome R was reversed in wheat x rye hybrids and triticale (a synthetic allopolyploid between tetraploid wheat and rye, *Secale cereale*) upon treatment with the demethylating agent 5-aza-cytosine, suggesting that nucleolar suppression is triggered by cytosine methylation (Vieira et al. 1990; Neves et al. 1995; Amado et al. 1997).

Sallee and Kimber (1976) studied chromosome pairing in the hybrid between the auto-allo-hexaploid *Triticum timopheevii* var. *zhukovskyi* (currently *T. zhukovskyi*) (2n = 6x = 42; genome GGAAA^{m}A^m) and *S. cereale*. Mean chromosomal pairing per cell in the tetraploid hybrid (2n = 4x = 28; genome GAA^{m}R) included 12.65 univalents, 5.50 rod bivalents and 2.65 ring bivalents, 0.95 trivalents and 0.05 quadrivalents. Pairing between the A and A^m genomes and in multivalents between these two genomes and the G genome, was mainly autosyndetic.

The data above show that all diploid hybrids between *Secale* species and Triticeae species exhibit very little chromosomal pairing at first meiotic metaphase. In the triploid and tetraploid hybrids, there is somewhat higher pairing, most of which is autosyndetic between the chromosomes of the polyploid Triticeae species, a small amount is between the rye chromosomes, and only a small part of the pairing is allosyndetic between the *Secale* and the Triticeae species chromosomes. Evidently, *Secale* chromosomes underwent significant changes during the evolution of the genus, that affected their ability to homoeologously pair with other Triticeae species. Such alterations include chromosomal rearrangements, such as translocations and inversions, and accumulation of large amounts of telomeric heterochromatin. As a result, the relative position of the telomeric regions at the beginning of meiosis may shift, impairing pairing initiation of rye and other Triticeae chromosomes (Devos et al. 1995; Lukaszewski et al. 2012; Megyeri et al. 2013). Moreover, genetic and epigenetic changes, such as mutations or elimination of DNA sequences that are involved in homology recognition and pairing initiation, may underlie this restricted pairing.

Sybenga (1966) postulated that pairing is controlled by specific units or zygomeres, which, like centromeres and nucleolar organizers, become active at specific meiotic stages. Pairing between chromosomes in hybrids only occurs when certain zygomeres are common to both parental chromosomes. A genuine reduction of pairing between homoeologous chromosomes in hybrids containing the rye genotype, may thus indicate the presence of factors similar to those on chromosome 5B of wheat, that restrict pairing to chromosomes with very similar zygomeres. Evidence for such an influence of the rye genotype is not convincing, as autosyndetic pairing in hybrids between the polyploid species of *Aegilops* and *Secale* is unaffected by the rye genome (Majisu and Jones, unpublished). It seems more likely that the zygomeric system that initiates chromosome pairing in rye is different from those of *Aegilops* or *Triticum*. The disturbed alignment of chromosomes in pre-meiotic nuclei, possibly due to inactivity of zygomeres, could be responsible for the absence of homoeologous pairing in the hybrids. In conclusion, at present, differential pairing with *Secale* chromosomes cannot serve as evidence of evolutionary relationships with other genera in the Triticineae.

6.5 Phylogeny of *Secale*

6.5.1 Phylogenetic Relationships Within the Genus *Secale*

The genus *Secale* has evolved monophyletically (e.g., Hsiao et al. 1995b; Mason-Gamer et al. 2002; Petersen et al. 2004; Bernhardt 2016). Since six major translocations shaped the modern *Secale* genome, differing it from a putative Triticeae ancestral genome, the ancestor of the genus may have undergone chromosomal rearrangements (Martis et al.

2013). In addition, introgressive hybridizations from other Triticeae species and/or a series of whole-genome or chromosome duplications, may have played a role in *Secale* speciation and genome evolution (Martis et al. 2013).

Cytogenetic analyses showed that *S. sylvestre* is isolated from the other *Secale* species (Khush and Stebbins 1961; Khush 1962; Singh 1977). Crossability between *S. sylvestre* and other taxa of *Secale* is low and the hybrid plants have highly irregular meiosis, and thus, low fertility (Khush and Stebbins 1961; Khush 1962). Its isolation may have resulted from its autogamous habit (Schiemann and Nürnberg-Krüger 1952; Khush and Stebbins 1961) or from its characteristically low telomeric heterochromatin content, which results in unsynchronized mitotic cycles in embryos of hybrids with other *Secale* species that have larger amounts of telomeric heterochromatin (Singh 1977). In contrast, hybrids between *S. strictum* and *S. cereale* are easily made and the hybrid seeds are easily grown, even without embryo rescue techniques. The F$_1$ hybrids exhibit somewhat reduced fertility, possibly because they are heterozygous for the two chromosomal translocations that distinguish *S. cereale* from *S. strictum* (Khush and Stebbins 1961; Khush 1962; Singh 1977). However, in areas where the two species are sympatric, introgression is believed to occur quite frequently (Stutz 1957; Khush 1962; Perrino et al. 1984; Hammer et al. 1985; Zohary et al. 2012), and it is difficult to distinguish even the first-generation hybrids from their parental species (Frederiksen and Petersen 1997). Hybrids between *S. strictum* subsp. *strictum* and *S. strictum* subsp. *africanum* are highly fertile (Khush 1962). A larger number of taxa was recognized within *S. strictum* previously (e.g., Roshevitz 1947), yet, although minor chromosomal differences have been reported between some of these taxa cytogenetic analyses have shown high chromosome pairing in meiosis of the hybrids, as well as high pollen fertility and large seed set (Schiemann and Nürnberg-Krüger 1952; Riley 1955; Nürnberg-Krüger 1960; Khush 1962). Thus, no strong genetic isolation barrier exists between these intra-specific taxa. Hybridization between domesticated and weedy *S. cereale* resulted in vigorous plants, generally with normal meiosis and high fertility (Nürnberg-Krüger 1960; Khush 1963a). Although some meiotic irregularities have been observed in hybrid plants (Stutz 1976), only *S. vavilovii* (currently included in *S. cereale*; Sencer and Hawkes 1980; Frederiksen and Petersen 1998) seems to have a karyotype of its own (Khush and Stebbins 1961; Kranz 1961, 1963; Khush 1962, 1963b; Singh and Robbelen 1977). All other specimens of *S. cereale* seem to possess very similar karyotypes (Stutz 1972).

The phylogenetic relationships between species within the genus *Secale* have been studied via isozyme electrophoretic patterns (Jaaska 1975; Vences et al. 1987), thin-layer chromatography (Dedio et al. 1969; Sencer 1975), ribosomal DNA spacer lengths (Reddy et al. 1990), characterization of the internal transcribed spacer of the rDNA (de Bustos and Jouve 2002), distribution of other repeated DNA sequences (Jones and Flavell 1982a, b), and plastid RFLPs (Murai et al. 1989). All these studies clearly showed that *S. sylvestre* occupies an isolated position within the genus and differs substantially from both *S. strictum* and *S. cereale*. It was also evident that the difference between *S. cereale* and *S. strictum* is not extensive. Thus, morphological and cytogenetic evidence suggest that *S. sylvestre* is the most ancient *Secale* species (Reddy et al. 1990).

Likewise, a phylogenetic analysis, based on RFLP of the plastid genome, showed that *S. sylvestre* is the sister group of the rest of the genus and the only well separated taxon (Petersen and Doebley 1993). Similarly, using amplified fragment length polymorphism (AFLP), Chikmawati et al. (2005) showed that *S. sylvestre* is the most distantly related taxa in the *Secale* genus. They found the annual forms of *S. cereale* and the perennial forms of *S. strictum*, as more closely related to each other than to *S. sylvestre*. The data of Chikmawati et al. (2005) confirmed that *S. sylvestre* is the most ancient species, whereas *S. cereale* is the most recently evolved species. AFLP analysis clearly separated all *Secale* species into three major species groups: *S. sylvestre*, *S. strictum* for perennial forms, and *S. cereale* for annual forms. In contrast to the above studies, Del Pozo et al. (1995) failed to clearly distinguish *S. sylvestre* from *S. strictum* by PCR of amplified DNA fragments. However, in the phenogram constructed based on PCR amplified band polymorphisms at the species level, *S. sylvestre* separated early from the other two species, while *S. strictum* and *S. cereale* appeared relatively close. Del Pozo et al. (1995) also found that *S. vavilovii* (currently included in *S. cereale*) is closer to *S. strictum* than to *S. cereale*. This contradicts the accepted taxonomic classification and cytogenetic and molecular data that placed *S. vavilovii* within *S. cereale*. Similarly, Achrem et al. (2014), using inter-simple sequence repeat (ISSR) and inter-retrotransposon amplified polymorphism (IRAP) techniques, reported the highest value of similarity between *S. cereale* and *S. vavilovii*, thus, justifying the classification of the latter in *S. cereale*. Shang et al. (2006), using 24 *Secale cereale* microsatellite markers, found that the *S. sylvestre* accessions were clearly divergent from the accessions of other species and that the *S. vavilovii* accessions were closely related to the *S. cereale* accessions.

S. cereale and *S. strictum* specimens are intermingled on the phylogenetic tree and *S. strictum* ssp. *africanum* cannot be distinguished from the other *S. strictum* specimens. Yet, the tested specimens of *S. strictum* ssp. *strictum* var. *ciliatoglume* had unique restriction sites (Petersen and Doebley 1993). However, a multivariate analysis describing the variation in a number of morphological characters of *S. cereale* and *S. strictum*, showed no clear distinction between the two taxa (Frederiksen and Petersen 1997).

Ren et al. (2011), using inter-simple sequence repeat (ISSR) markers to analyze phylogenetic relationships among wild and domesticated *Secale* taxa, found that the annual weedy *S. cereale* subsp. *ancestrale* evolved from *S. strictum* subsp. *strictum*. These results support the division of the genus *Secale* into three species: the annual wild species *S. sylvestre*, the perennial wild species *S. strictum*, including several differential subspecies forms such as *strictum*, *africanum*, and *anatolicum*, and *S. cereale*, which includes domesticated and weedy rye as subspecies forms. No differences were found between the weedy and the domesticated forms of *S. cereale* (de Bustos and Jouve 2002).

Al-Beyroutiová et al. (2016) evaluated genetic diversity and phylogenetic relationships using 13,842 DArTseqTM polymorphic markers. Extracted genomic DNA samples were sent to Diversity Arrays Technology Pty Ltd. (http://www.diversityarrays.com) whole genome genotyping service for *Secale* analysis. The model-based clustering (STRUCTURE software) separated the 84 samples into three main clusters: perennial cluster, annual cluster, and *S. sylvestre* cluster. The same result was obtained using Neighbor-Joining tree and self-organizing maps. Their data confirm the taxonomic classification of Sencer and Hawkes (1980) and Frederiksen and Petersen (1998), which defined *Secale sylvestre*, *S. strictum*, and *S. cereale* as the three main species of the genus *Secale*. Several authors (Reddy et al. 1990; de Bustos and Jouve 2002; Chikmawati et al. 2005; Shang et al. 2006) considered *S. sylvestre* the oldest species, from which all other species evolved, while Hammer (1987) claimed that *S. sylvestre* evolved separately, and its evolution might have begun very early. All bioinformatical tools used by Al-Beyroutiová et al. (2016) confirmed the antiquity of *S. sylvestre* and showed that it is the most diverged species of all the *Secale* taxa. Three of the 84 *Secale* samples they studied (MON1, MON2 and MON3) were in basal positions in phylogenetic trees. MON3 is the oldest of all the *Secale* accessions and likely the ancestor of *S. sylvestre*. MON1 and MON2 show an ancestral position in the phylogeny of the *Secale* genus. The three accessions share ancient morphological characters and are probably the ancestors of different lineages within *Secale*. The three accessions do not belong to the same lineage (i.e., species), indicating that they are probably the ancestors of lineages leading to the formation of *S. sylvestre*, S. strictum and *S. cereale*. Furthermore, Al-Beyroutiová et al. (2016) found that var. *ciliatoglume* of *S. strictum* subsp. *strictum* and the semi-perennial taxon of *S. cereale* subsp. *ancestrale*, namely, var. *multicaule*, are genetically the most closely related to the annual forms of *S. cereale*. Chikmawati et al. (2005) affirmed that var. *ciliatoglume* is an ancestral type, being the second taxon diverging after *S. sylvestre*. The observations of Frederiksen and Petersen (1997), based on morphometric analyses, suggested that var. *ciliatoglume*

should be given an intraspecific rank. Concerning var. *multicaule* of *S. cereale* subsp. *ancestrale*, Hammer et al. (1987) suggested its hybrid origin from the cross *S. strictum* x *S. cereale*, but the results of Al-Beyroutiová et al. (2016) did not show a reticulated origin of this variety. The definitive status of these two varieties cannot be solved by DArTseq polymorphism only. In addition, the results of Al-Beyroutiová et al. (2016) confirmed that *S. vavilovii* could not be considered a separate species but rather, a subspecies of *S. cereale*. Overall, the phylogenetic relationships on the infrageneric level are still poorly understood. Bernhardt (2016) showed that *S. strictum* is somewhat different from *S. cereale*. Similarly, Petersen et al. (2004) stated that "an increased amount of evidence indicates that *S. cereale* and *S. strictum* are not exclusive lineages".

According to Kobyljanskij (1982), the oldest ancestor of the genus, *Protosecale*, appeared in the Oligocene (33.7–23.8 MYA) when early Triticeae appeared, and later evolved into a *Protosylvestre* and *Protostrictum*, from which *S. sylvestre* and *S. strictum* developed during the Pliocene epoch (5.3–1.8 MYA). In accord with this, Middleton et al. (2014), based on sequencing of chloroplast genomes, estimated that *Secale* diverged from *Triticum* approximately 3–4 million years ago.

6.5.2 Phylogenetic Relationships Between *Secale* and Other Triticineae Genera

The phylogenetic relationships of *Secale* with other Triticeae genera have been studied through morphological traits, genome analysis, isozymes, and cytoplasmic and nuclear DNA sequences, which have yielded contradictory results regarding the position of *Secale*. Aase (1935), assuming that the degree of chromosome pairing is a true test of genetic relatedness, concluded that *Secale* species are more closely related to *Aegilops* and *Agropyron* than to *Triticum*. In accord with this conclusion, Favorsky (1935) reported on the existence of a certain degree of relationship between *S. cereale* and *Agropyron cristatum*. Lucas and Jahier (1988) executed a diallel-crossing program, which included ten diploid species from the genera *Triticum*, *Aegilops*, *Dasypyrum* and *Secale*. The number of associations between chromosome arms in the hybrid PMCs at first meiotic metaphase was taken as an indication of the degree of homology between the parental genomes. They found little affinity between the genomes of *Aegilops* and *S. cereale*. In accord with these results, Lucas and Jahier (1988) noted that the affinity they found between the *Aegilops* genomes and the rye genome was as low as that reported by Sodkiewicz (1982) for the A^m genome of *Triticum monococcum* and the R genomes of *S. cereale*. Lucas and Jahier (1988) concluded that *S. cereale* and *Dasypyrum villosum* are more distantly related than to the *Aegilops* and *Triticum* species. However, as stressed by Frederiksen and Seberg (1992), genome analysis is incapable of revealing phylogenetic relationships of the diploid mono-genomic taxa of the Triticeae.

Taxonomical treatments by several well-known taxonomists (e.g., Nevski 1933; Melderis 1953; Hubbard 1959; Tzvelev 1973, 1976) placed the genus *Secale* close to the genus *Dasypyrum*. In accord with this taxonomical treatment, Baum (1983), on the basis of a phylogenetic analysis of Triticeae by means of numerical methods, grouped *Secale* and *Dasypyrum* close to one another. Further analysis of morphological characteristics suggested that *Secale* is the sister group of a clade consisting of *Dasypyrum villosum*, *Triticum monococcum* and *Aegilops* species (Frederiksen and Seberg 1992; Seberg and Frederiksen 2001; Seberg and Petersen 2007).

Incongruence exists between the phylogenetic trees constructed based on morphological versus molecular data, as well as between trees constructed using various kinds of molecular data (Seberg and Frederiksen 2001; Seberg and Frederiksen 2007; Escobar et al. 2011). According to several morphological trees, *T. monococcum* is included in the *Secale* clade (Seberg and Frederiksen 2001), but, in some molecular studies, is a sister clade to the *Aegilops* clade (Kellogg and Appels 1995; Mason-Gamer and Kellogg 1996). Molecular data from the plastid genome suggest a relationship with *Taeniatherum*, *Triticum* and *Aegilops* (Mason-Gamer and Kellogg 1996; Petersen and Seberg 1997), while DNA sequence data from various parts of the nuclear genome have given diverse results. Data from internal transcribed spacers (ITS) of the rDNA suggested that *Secale* is the sister group of *Eremopyrum* and *Henrardia* (Hsiao et al. 1995a, b), whereas data from the spacers between the 5S RNA genes suggested a rather basal position for *Secale* within Triticeae (Kellogg and Appels 1995; Kellogg et al. 1996). Thus, the position of *Secale* remains uncertain.

Monte et al. (1993) used RFLP in combination with other approaches, to reconstruct evolutionary events, which revealed a high degree of polymorphism both between and within the species examined. The RFLP data were used to generate a cladogram and a phenogram and the results of both methods were consistent with each other and with the general taxonomic information provided by earlier morphological studies, meiotic pairing analysis, isozyme tests, and sequence alignment in the rDNA and 5S DNA loci. Both the cladogram and the phenogram showed very close associations between the genera *Secale* and *Agropyron* and the Ee and Eb genomes of *Elymus*. The cladogram and the phenogram also clustered *Secale* and *Agropyron* together with the genomes of *Triticum*.

Mason-Gamer (2005) performed a phylogenetic analysis of sequences from a portion of the tissue-ubiquitous *β-amylase* gene in a broad range of the mono-genomic Triticeae. The results showed close relationships among *Secale*, *Australopyrum*, and *Dasypyrum*. Yet, no other molecular data sets support a *Secale* + *Australopyrum* + *Dasypyrum* clade; in general, there is little agreement with regard to the placement of any of these taxa. Furthermore, the morphological data (Seberg and Frederiksen 2001) showed that *Secale* is sister to a *Dasypyrum* + *Triticum* clade, while *Australopyrum* forms a paraphyletic grade at the base of a large clade containing over half of the remaining taxa. Sequencing of the ITS region of nuclear ribosomal DNA of diploid Triticeae species, brought Hsiao et al. (1995a, b) to conclude that *Secale* is close to *Taeniatherum* and sister clade to *Elymus bessarabicus, E. elongatus* and *Triticum monococcum*.

Seberg and Petersen (2007) studied the phylogeny of diploid Triticeae by combining morphological observations with nucleotide sequence data from two plastid genes (*rbcL*, *rpoA*), one mitochondrial gene (*coxII*) and two single-copy nuclear genes (*DMC1*, *EF-G*). Their data indicate incongruence between the four data sets, partitioning morphology and the three -genome-bearing compartments, was observed. Only the mitochondrial and nuclear sequences were mutually incongruent. They concluded that *S. strictum* is close to *Taeniatherum caput-medusae*, followed by *Dasypyrum villosum*, *Elymus elongatus Elymus bessarabicus*, *Crithopsis delileana* and the genera *Aegilops*, *Triticum* and *Amblyopyrum*.

Escobar et al. (2011), using one chloroplastic and 26 nuclear genes, obtained a comprehensive molecular dataset of phylogenetic value, on the diploid species of the Triticeae. They grouped Secale, including *Taeniatherum*, *Triticum* and *Aegilops* in Clade V, and the genera *Dasypyrum*, *Heteranthelium* as a sister clade. Clade V is retrieved, although branching within this clade changes relative to the super-matrix tree: *Secale* and *Taeniatherum* branched together, *T. monococcum* branched sister to *Ae. tauschii*, and *Ae. speltoides* and *Ae. longissima* grouped together. *Heteranthelium* branched at the base of clade V and these two newly inferred clades (*Pseudoroegneria-Dasypyrum* and *Heteranthelium*-clade V) were closely related to each other. According to Escobar et al. (2011), *Dasypyrum*, *Heteranthelium* and genera of clade V, grouping *Secale*, *Taeniatherum*, *Triticum* and *Aegilops*, evolved in a reticulated manner. In conclusion, with respect to the placement of the genus *Secale* in the tribe-wide phylogeny, virtually all genera of the Triticeae have been suggested—either alone or in combination with other genera—as a sister group to *Secale* (Petersen et al. 2004).

Whole genome sequencing data has enabled to better understand the phylogenetic relationship of rye to other species from the Triticeae as described in Chap. 3. The new genome sequences have shed light on rye evolution, which so far, had been not completely resolved regarding the proximity to wheat and barley as well as to its wild relatives. It was thought that multiple introgressions might generate a mosaic genome with reticulate evolution patterns and discordance in the phylogeny of various chromosomal segments. When comparing the genomes of wheat (cv. Chinese Spring) and barley (cv. Morex) to Lo7, Rabanus-Wallace et al. (2021) found no major discordance and showed that rye is more closely related to bread wheat than to barley across the whole genome. Furthermore, Li et al. (2021), analyzing single-copy conserved orthologous loci, estimated that rye separated from wheat 9.6MYA while barley separated from wheat 15MYA. By contrast when comparing sequence clusters of k-mers from 955 cultivated and wild rye lines, Rabanus-Wallace et al. (2021) found that reticulate evolution with multiple inter-species introgression of genomic clusters played a major role in rye evolution. For example, Lo7 was found to contain mostly clusters related to *S. cereale* and *S. vavilovii* and less related to *S. strictum* and *S. sylvestre*. In summary, while there was limited reproductive isolation between rye species, inter-genera exchanges did not play a major role in rye evolution.

6.6 Domestication

Secale cereale (rye) is the only domesticated species of the genus *Secale*. Vavilov (1917, 1926) considered rye a classical example of a secondary crop. Secondary crops evolved first as weeds that infested cultivated fields and only later, established as crops, whereas primary crops, developed from wild progenitors that were cultivated. When a crop is introduced into areas with harsh climatic conditions, to which the weed plant is more adapted, the latter may become more successful than the crop, forcing the farmer to domesticate it as a replacement for the crop. In this respect, rye is hardier than wheat and can withstand harsher climatic conditions and poorer soils. Vavilov (1917, 1926) observed that weedy rye is common in wheat and barley fields of southwest Asia and Central Asia, but in the mountainous areas of eastern Turkey, Caucasia, and Central Asia, at altitudes of 2000–2500 m asl, where rye always succeeded more than wheat, it gradually replaced wheat as a domesticated crop. In such areas, pure stands of domesticated rye were common. As wheat moved northward and eastward to areas with harsher climatic conditions, the weedy rye was consequently, domesticated. Vavilov (1926) and Khush (1963a, b) assumed that rye was domesticated in several places independently and at different times.

Having first arisen as a weed in fields of wheat and barley, rye may have been adopted as a crop at a somewhat later date than that of wheat and barley (Ladizinsky 1998). It is also reasonable to assume that these weedy races would

have only evolved with the development of agriculture in western Asia. There is also general agreement that the original ancestor of these weedy races, and hence of domesticated rye as well, was *S. strictum* (Vavilov 1926). Sencer and Hawkes (1980) assumed that wild populations of *S. cereale*, currently classified in *S. cereale* subsp. *ancestrale* (Frederiksen and Petersen 1998), which have presumably evolved from *S. strictum* (Khush and Stebbins 1961), invaded wheat and barley fields during the early days of cultivation and gave rise to weedy ryes with varying degrees of rachis brittleness. It is assumed that the impressive evolvement of weedy rye, as well as variation build-up in domesticated rye, may have been considerably enhanced by introgressive hybridization with the perennial *Secale strictum*, the wild rye species distributed over the elevated continental parts of Anatolia and adjacent areas in south-west Asia (Zohary et al. 2012).

It is not difficult to imagine the kind of selection necessary for the weedy rye to adapt to wheat fields. A more upright culm enhances competitive capacities. In addition, genes controlling non-shattering of the grains would be advantageous. Cultivation prior to sowing tended to destroy seedlings originating from seeds dropped by shattering spikes the previous year. Seeds of non-shattering genotypes would have been harvested with the main cereal crop and subjected to the same cultural procedures during the next season. At the same time, larger grains are positively selected, since even the most primitive winnowing procedure favors the retention of grains approaching the size of those of wheat and barley (Sencer and Hawkes 1980; Ladizinsky 1998; Zohary et al. 2012).

Stutz (1972), based on extensive cytological, ecological and morphological studies, hypothesized that *S. cereale* originated from hybridization between the perennial cross-pollinated *S. strictum* and the annual, self-pollinated *S. vavilovii* (currently included in *S. cereale* subsp. *ancestrale*; Sencer and Hawkes 1980; Frederiksen and Petersen 1998), the latter derived from *S. sylvestre* as a consequence of chromosomal translocations. *S. sylvestre* was, in turn, derived from *S. strictum* or a common ancestor. However, Nürnberg-Krüger (1960) and Khush and Stebbins (1961) considered this hypothesis improbable and assumed that *S. cereale* evolved from *S. strictum* as a result of progressive cytological and morphological differentiation, which was likely facilitated by adaptive superiority of translocation heterozygotes and rearrangement homozygotes. Zohary (1971) ascribed sympatric speciation of *S. cereale* from *S. strictum* to disruptive selection.

Almost all cytogenetic studies have implicated two instantaneous chromosomal rearrangements as the origin of *S. cereale* from *S. strictum*. All perennial rye taxa included in *S. strictum* and the annual *S. sylvestre* have the same chromosome arrangement and closer affinity to each other

(Stutz 1972). On the other hand, annual weedy and domesticated rye taxa of *S. cereale* have the same chromosome arrangement and display cytogenetic affinity to each other and differ from the other two species by two translocations involving three of the seven basic chromosome sets (Riley 1955; Khush and Stebbins 1961; Stutz 1972). Thus, *S. strictum* subsp. *strictum* is a perennial, self-incompatible, species with a *strictum* chromosome type. *S. strictum* subsp. *africanum* is a perennial, self-fertile species with a *strictum* chromosome type, *S. Sylvestre* is an annual, self-fertile, species with a *strictum* chromosome type and *S. cereale* is an annual, self-incompatible, (rarely self-compatible), species with a *cereale* chromosome type. From evolutionary and phylogenetic perspectives, the annual weedy taxa of *S. cereale* subsp. *ancestrale* are considered younger than the perennial wild ones and the domesticated rye, *S. cereale* subsp. *cereale*, is thought to be the youngest of all (Sencer and Hawkes 1980).

Wild forms of *S. cereale*, i.e., subsp. *ancestrale*, with complete brittle rachis, likely invaded wheat and barley fields and gave rise, through mutations, to weedy types with annual growth habits and varying degrees of rachis brittleness. The mature ears of weeds with non-brittle rachis, that do not shatter and also tend to mimic wheat in grain size and weight, have a great adaptive advantage over types with brittle rachis, since they are harvested and threshed together with wheat. Since traditional winnowing does not separate grains of rye from those of wheat, rye seed is included in the harvest and planted with the wheat in the subsequent year (Ladizinsky 1998). Farmers in the elevated plateau of eastern Turkey, Armenia and Central Asia, tolerate some rye-weed infestation in their crop, because in years with extreme cold and dry weather, the rye weed survives when wheat does not, ensuring a supply of cereal grains (Zohary et al. 2012).

The weedy races of *S. cereale*, occurring in west and central Asia, were presumably interfertile, differing from one another mainly in their seed dispersal pattern, ranging from races with a brittle rachis to non-brittle rachis (Ladizinsky 1998; Zohary et al. 2012). These races are common weeds in wheat fields and those with non-brittle rachis are not separated from wheat by harvesting, threshing and winnowing (Ladizinsky 1998). Farmers domesticated rye under cultivation, by selecting for the taxa with non-brittle rachis and bigger caryopsis (Sencer and Hawkes 1980). Thus, the immediate progenitors of the domesticated form, *S. cereale* subsp. *cereale*, arose from one or more of the weedy races belonging to *S. cereale* subsp. *ancestrale*.

Most of the weedy taxa were also cross-pollinated. *Secale strictum* subsp. *strictum*, the progenitor of the weedy types of *S. cereale*, is a cross-pollinated species that has a perennial growth habit and brittle rachis. Hence, domesticated *S. cereale* is unique among cereals, being the only cross-pollinated cereal taken into cultivation in southwest Asia.

While it seems highly likely that the agriculturally dependent *S. cereale* evolved directly from the wild *S. strictum*, the actual course of events is still obscure (Sencer and Hawkes 1980; Evans 1976, 1995). When considering its morphology and breeding system, *S. cereale* is close to *S. strictum*. Khush and Stebbins (1961) proposed gradual accumulation of the chromosomal difference between the two, but intermediate chromosome types have not been found, suggesting that the two translocations occurred simultaneously (Ladizinsky 1998). If no other distinct intermediate species were involved, then the question arises as to how major structural rearrangements (involving a double translocation) could have become established, particularly in light of the initial handicap of reduced fertility which would certainly have resulted. Moreover, the hypothesis claiming direct emergence of *S. cereale* from *S. strictum* fails to explain how a homozygous type for two chromosomal translocations could emerge in a population of a self-incompatible plant such as *S. strictum* (Ladizinsky 1998). The main obstacle is not so much the low fertility of a spontaneous heterozygote for two chromosomal rearrangements, but the constant pollination of the new *cereale*-chromosomal type by *S. strictum*. Such continuous pollination seems necessary for the accumulation of a sufficient number of alleles in the self-incompatible gene in the *cereale*-chromosome type. Yet, more rapid establishment could be achieved if the heterozygous genotypes and the homozygous *cereale*-type had some adaptive advantage over the *strictum*-type (Riley 1955; Khush and Stebbins 1961; Ladizinsky 1998). The partial fertility barrier between the new and old chromosome arrangements could have served as an isolating mechanism preventing swamping of the evolving weedy races of *S. cereale* by their progenitor, *S. strictum*. Eventually, the partial isolation enabled *S. cereale* to establish itself and to sympatrically develop as a separate species.

Evolvement of the annual growth habit of *S. cereale*, via mutations in the perennial *S. strictum*, is highly probable. Perennial growth habit is a dominant trait presumably controlled by a small number of genes. Charpentier et al. (1986) found that perennial growth habit was a dominant trait in F_1 hybrids between *Elymus farctus* subsp. *farctus* (=6x *Agropyron junceum*) and common wheat cv. Chinese Spring. In BC_1 plants, derived from backcrossing the F_1 hybrids to Chinese Spring, perennialism was less marked. They concluded that gene(s) located on two or more chromosomes of *E. farctus* determine perennial growth habit. Lammer et al. (2004) found that chromosome arm 4EeS of diploid *Elymus elongatus* confers perennial growth habit when added as a monosome or disome to Chinese Spring background. This may suggest that the gene(s) determining perennial growth habit is located on rye chromosome arm 4RS. Mutations in these genes presumably led to the formation of annual weedy races of rye. Annual growth habit is better adapted for successful infestation of cultivated wheat and barley fields and therefore, a strong selection pressure was exerted on the weedy rye, in favor of the annual forms. Therefore, it is highly probable that the weedy races of *S. cereale* evolved directly from *S. strictum* as was proposed by Riley (1955) and Khush and Stebbins (1961), via accommodation of the change from perenniality to annuality.

Takahashi (1972) described two complementary dominant genes, *Btr* and *Btr2*, located on chromosome 3H, that control spike fragility in wild barley, *Hordeum vulgare* L. subsp. *spontaneum* (C. Koch) Thell. Domesticated barley, which has a non-fragile spike due to a tough rachis, contains the recessive alleles *btr1btr2* of these loci. Similarly, Sharma and Waines (1980) crossed domesticated einkorn *Triticum monococcum* subsp. *monococcum*, which has a tough rachis, with wild einkorn *T. monococcum* subsp. *aegilopoides*, which has a brittle rachis, and found in F_2 and backcross generations, that tough rachis is controlled by two complementary recessive genes. Two recessive genes, located on the short arms of chromosomes 3A and 3B, control the tough rachis trait in domesticated allotetraploid and allohexaploid wheats (Levy and Feldman 1989a, b; Rong 1999; Watanabe and Ikebata 2000; Watanabe et al. 2002, 2005a, b; Nalam et al. 2006; Millet et al. 2013). Comparative mapping analyses suggest that the genes controlling tough rachis in wheat, are located on the short arm of chromosomes of homoeologous group 3 (Nalam et al. 2006). Thus, it is reasonable to assume that the genes determining tough rachis in *S. cereale* are located on the same arm, namely, 3RS, a chromosome that is not involved in the two translocations distinguishing *S. cereale* from *S. strictum*, in terms of tough rachis and annual growth habit.

The chromosomal location of the three characters distinguishing *S. cereale* from *S. strictum*, namely, annual growth habit, non-fragile rachis, and large caryopsis, presumably are not associated with the two chromosomal rearrangements existing between these two species. Such a conclusion has been drawn from genetic analysis of F_2 of *S. strictum* x *S. cereale* (Stutz 1957). Thus, it is assumed that while perennial *Secale* infested wheat and barley fields in northeastern Turkey, Armenia and northwestern Iran, mutation(s) supporting annual growth habit gave adaptive advantage to the weedy *Secale*. Shifts to tough rachis and large grains, mimic the wheat crops and increase the adaptation to harvest and threshing.

Alternatively, Sencer and Hawkes (1980) assumed that the switch to annual growth habit led to the development of early wild populations of *S. cereale*, which presumably evolved from *S. strictum*. This change improved their ability to invade wheat and barley fields during the early days of cultivation, and gave rise to the development of weedy ryes.

Weedy rye presumably originated in eastern Turkey and Armenia (Zohary et al. 2012) and evolved with the development of agriculture in this region. From this area, rye spread as a weed in wheat and barley fields, towards the north, east and west and imposed itself as a secondary crop under conditions unfavorable for wheat and barley. It is probable that rye became a crop in its own right in several places independently (Sencer and Hawkes 1980). Vavilov (1917, 1926), based on the fact that the greatest accumulation of genetic diversity of *S. cereale* was found in southwest Asia, considered this region to be the primary center of origin of domesticated rye. At the same time, he considered Afghanistan and Tadjikistan to be a secondary center of variation of this crop (Vavilov 1917, 1926). Sencer and Hawkes (1980), based on a synthetic analysis that considered evidence from the fields of morphology, taxonomy, ecology, phytogeography, reproductive biology, genetics, cytology, palaeoethnobotany, philology, phylogeny and evolution, suggested Mt. Ararat and Lake Van area in eastern Turkey as the geographic origin of domesticated rye. Linguistic evidence suggests early acquaintance with rye by the people living in the Caucasus and the northeastern Black Sea region (Sencer and Hawkes 1980). The name 'rye' in the Caucasian languages was retained by the Greeks and Celts, who spread the crop during their migrations (Sencer and Hawkes 1980). These are revealed by the facts that the Greek, Celtic and Latin names of rye are derivatives of its name in the Caucasian languages.

In summary, the available evidence points to the implicates the early weedy, annual, brittle rye types as the ancestors of the weedy non-brittle types, from which domesticated rye was picked up. Despite of the two translocations that exist between these two species, genes from *S. strictum* introgressed into *S. cereale*, enriching its gene pool. Indeed, partial fertile hybrids between annual *S. cereale* and perennial *S. strictum* can be produced in experimental fields and natural hybridizations of this sort occur quite frequently near cultivated fields in eastern Turkey (Zohary et al. 2012). It is assumed that variation build-up in domesticated rye and the impressive evolvement of weedy rye could have been considerably enhanced by introgressive hybridization with the perennial *S. strictum*.

Archaeological records support the proposal that rye was domesticated in eastern Turkey and Armenia, but much later than wheat. Due to the scarcity of rye remains in the Neolithic and the Bronze settlements of southwest Asia, it is difficult to define the exact time of rye domestication (Zohary et al. 2012). The earliest rye remains come from Epi-Palaeolithic sites in the Upper Euphrates valley in northern Syria. Numerous charred grains, later identified as a mixture of both wild rye and wild einkorn wheat, were unearthed in ca. 11,800–11,300 cal BP Tel Mureybit (van Zeist and Casparie 1968; van Zeist and Bakker-Heeres 1985;

Willcox and Fornite 1999). The narrow shape of the kernels indicates that they represent wild forms. Similar narrow rye grains, either of *S. cereale* subsp. *ancestrale* or of *S. strictum*, were found in the ca. 12,700–11,100 cal BP Epi-Palaeolithic Tel Abu Hureyra. Hillman (1975, 2000) and Hillman et al. (1989, 2001) suggested that these were domesticated forms, a view that was later criticized because of lack of chaff and proper dating (e.g., Nesbitt 2002; Colledge and Conolly 2010). Wild rye grains were also discovered in two Pre-Pottery Neolithic A (PPNA) northern Syria sites, ca. 11,500–11,000 cal BP J Ahmar (Willcox 2002; Willcox et al. 2008) and ca. 10,700–10,400 cal BP Djade el Mughara (Willcox et al. 2008). On the other hand, grains of domesticated rye were first found in ca. 9450–8450 cal BP PPNB Can Hasan III (Hillman 1972, 1978) and in very small quantities at the nearby ca. 9350–8950 cal BP site of Aceramic Neolithic Çatalhöyük in East Turkey (Helbaek 1964a, b; Fairbairn et al. 2002, 2005, 2007). At Can Hasan III, relatively plump grains were discovered, together with some non-brittle rachis segments. As argued by Hillman (1978), these finds suggest that rye had already entered cultivation in East Turkey in early Neolithic times, either as a non-brittle weed infesting wheat fields, or as a full-fledged domesticated cereal crop. Yet, no additional rye remains have been discovered in other Neolithic southwest Asian sites. The next record comes from ca. 4000 BP Bronze Age levels of Alaca Höyük in north-central Anatolia (Hillman 1978). There, a pure hoard of carbonized large grains of *S. cereale* was discovered, indicating that, at that time, rye was grown as a crop in its own right. Thus, a reasonable estimate for domestication is about 5000 BP.

Early archaeological evidence is also fragmentary in regions outside the Fertile Crescent and thus, the spread of rye cultivation in Europe is difficult to sketch (Behre 1992). Rye grains were found in Europe in Early Neolithic sites (ca. 7550–6459 cal BP) in northern Italy, from Middle Neolithic sites (ca. 6950–6650 cal BP) in Slovakia, and from several late Neolithic sites in Poland. Further evidence of rye remains in Europe came from several Bronze Age settlements in several countries of Central and East Europe (for a detailed description of the discovery of rye remains in Europe see Zohary et al. 2012). Only a limited number of rye grains were found in these sites, in contrast to the abundance of wheat or barley grains discovered, indicating that at that time, rye contaminated wheat and barley fields. In later Iron Age settlements in Germany (Hopf 1982), Denmark (Helbaek 1954), Poland (Willerding 1970), and Crimea (Januševič 1978), rye also typically appeared admixed with barley or wheat. Rye as a main crop, was part of the Roman grains-agriculture and was grown in the cooler northern provinces (Sencer and Hawkes 1980). Carbonized rye grains have been retrieved from several Roman frontier sites along the Rhine and the Danube (Hillman 1978), as well as from

the British Isles (Jessen and Helbaek 1944). Yet, wide cultivation of rye in Central and Eastern Europe only became marked from the Middle Ages. Renfrew (1973) refers to excavations dating rye in Austria, Poland, Czechoslovakia and eastern Crimea to the third-fourth centuries AD, where rye was found as a weed in crops of wheat and barley. Since then, it served as the main bread cereal in most areas east of the French-German border and north of Hungary, while in Southern Europe, it was cultivated on marginal lands (Sencer and HawSkes 1980).

Rye was probably originally grown in the same areas in Europe as the other temperate cereals, but its tolerance of low rainfall, cold winters and poor light soils made it particularly suitable for large areas of northern and eastern Europe, which are less suitable for wheat and barley. By the end of the eighteenth century, it had become the major cereal of the region. Since then, however, it has been gradually replaced by wheat.

Rye introduction into Europe, as a minor admixture with wheat and barley, probably occurred via two separate routes. One route was northwards, through the Caucasus from its primary center of origin. According to Engelbrecht (1916, 1917), rye grains were transported from north-eastern Anatolia to the north of the Black Sea by Greek traders, since they had close trade relations with the Scythians (on the river Dnjepr), who cultivated cereals during the fifth century BC. A second route was westwards, via the Aegean Basin and the south Balkan. Cultivation of rye spread eastwards to Iran and Central Asia, the secondary center of variation, and later on, further east. *S. cereale* appeared at the end of the fourth millennium BP, in Hasanlu, Iran, and seems to have been a staple crop in this region throughout the Iron Age (Tosi 1975).

Rye (*Secale cereale* subsp. *cereale*) is an important cereal crop in the cooler parts of northern and central Europe and Russia, cultivated up to the Arctic Circle and up to 4000 m above sea level (Zohary et al. 2012). The main rye belt stretches from northern Germany through Poland, Ukraine, Belarus, Lithuania and Latvia, into central and northern Russia. Rye is also grown in North America (Canada and the United States), in South America (Argentina, Brazil and Chile), in Oceania (Australia and New Zealand), in Turkey, in Kazakhstan and in northern China. It is particularly appreciated in these regions because of its winter hardiness, resistance to drought, and its ability to grow on acidic, sandy soils (Evans 1995). Its capacity to produce an economical crop in areas of cold winters and hot summers, renders it superior to the other temperate cereals. Although spring and winter biotypes of rye exist, most of the world supply is obtained from winter varieties.

Rye is grown extensively as a grain, a forage crop and as a cover crop (Evans 1976, 1995). Its grain is used for flour, rye bread, crisp bread, rye beer, some whiskeys, some vodkas, and for animal fodder. The area under cultivation is substantially smaller than that of wheat and barley and the world rye production is lower than any of the other major cereals. In 2012, the total production of rye grains was estimated at 14,615,719 metric tons (FAO 2015; Table 6.2). Although the production levels of rye have fallen in the last several years, in most of the producing nations, rye is still an important food plant in many areas of northern and eastern Europe and central Asia. Most rye is consumed locally or exported to neighboring countries only, rather than being shipped worldwide.

Rye bread, including pumpernickel, is a widely consumed food in Northern and Eastern Europe. Rye flour is high in gliadin, but low in glutenin, with a lower gluten content than wheat flour. It also contains a higher proportion of soluble fibers than wheat. Its grains contain appreciable amounts of proteins and therefore, its flour can be baked into dark-colored rye bread. In spite of the poor baking quality of the flour, much of the present world production of rye is consumed in the form of bread, appreciated for its flavor and distinctive dense texture. It has been estimated that in the early part of the twentieth century, rye bread was the main cereal food of a third of the European population. Rye grain is also used to make alcoholic drinks, like rye whiskey and rye beer, i.e., rye is used to make whiskey in the United States and Canada, gin in the Netherlands, and beer in Russia. In all rye-producing countries, more than 50% of the grain is used in animal feed and the young green plants are also commonly used for livestock fodder. The mature straw is too tough for animal fodder, but can be used for bedding, thatching, paper making, and straw hats.

Feral plants are plants that are derived, in part or fully, from crop plants that have become partly or fully undomesticated, and are no longer dependent on managed cultivation. Feral rye, a weed of wheat and barley fields, poses a serious threat to annual winter grain yields in western and central United States. Feral rye likely spread as a contaminant in the seed of domesticated cereals, as they were introduced into new areas.

By the time conscious plant breeding began, in the later part of the nineteenth century, the early plant breeders relied heavily on locally adapted land races as their immediate source of variability (Evans 1976). Breeding methods have been influenced by the outbreeding nature of the crop. Early breeding techniques are best described as forms of simple recurrent selection. With recently improved knowledge of the genetic structure of outbreeding populations, more sophisticated methods have been used (Evans 1976, 1995).

In contrast to most grain crops that are self-pollinating, rye is a cross-pollinated cereal. It has a gametophytic two-locus incompatibility system (Lundquist 1956). Consequently, rye yields depend, among other factors, on effective wind pollination. Rye shows inbreeding depression but

inbred lines of acceptable vigor can be isolated and used in the construction of synthetic varieties, following suitable progeny tests for combining ability (Evans 1976, 1995).

Unlike of the objectives in wheat and barley breeding, aspects of disease resistance have not dominated rye breeding. Improvement of grain yield, protein content and quality, together with cold tolerance and shorter straw, have been the main aims of recent breeding. Utilization of rye as forage has led to the breeding of varieties bred solely for this purpose and emphasis is then placed on characteristics other than grain production, namely, total dry matter, growth in winter and early spring and digestibility. Ergot (*Claviceps purpurea*) is a disease which has caused some trouble from time to time. Ergot is a fungus that parasitizes rye and is poisonous to humans and livestock. The poisonous sclerotia, which fully replace the grain and occasionally penetrate into flour, reportedly cause hallucinations in humans and abortion in farm animals. Thus, breeding for ergot resistance is also an important aim.

References

Aase HC (1930) The cytology of Triticum, Secale, and Aegilops hybrids with reference to phylogeny. Res Stud State Coll Wash 2:3–60

Aase HC (1935) Cytology of cereals. Bot Rev 1:467–496

Aase HC, Powers LR (1926) Chromosome numbers in crop plants. Amer J Bot 13:367–372

Achrem M, Kalinka A, Rogalska SM (2014) Assessment of genetic relationships among *Secale* taxa by using ISSR and IRAP markers and the chromosomal distribution of the AAC microsatellite sequence. Turk J Bot 38:213–225

Al-Beyroutiová M, Sabo M, Sleziak P, Dušinsky R, Birčák E, Hauptvogel P, Kilian A, Švec M (2016) Evolutionary relationships in the genus *Secale* revealed by DArTseq DNA polymorphism. Plant Syst Evol 302:1083–1091

Amado L, Abrabches R, Neves N, Viegas W (1997) Development-dependent inheritance of 5-azacytidine-induced epimutations in triticale: analysis of rDNA expression patterns. Chromosome Res 5:445–450

Appels R (1982) The molecular cytology of wheat-rye hybrids. Int Rev Cytol 80:93–132

Baum BR (1983) A phylogenetic analysis of the tribe Triticeae (Poaceae) based on morphological characters of the genera. Can J Bot 61:518–535

Bedbrook JR, Jones J, O'Dell M, Thompson RD, Flavell RB (1980) A molecular description of telomeric heterochromatin in Secale species. Cell 19:545–560

Behre K-E (1992) The history of rye cultivation in Europe. Veg Hist Archaeobot 1:141–156

Benavente E, Fernández-Calvín B, Orellana J (1996) Relationship between the levels of wheat-rye metaphase I chromosomal pairing and recombination revealed by GISH. Chromosoma 105:92–96

Bennett MD, Gustafson JP, Smith JB (1977) Variation in nuclear DNA in the genus Secale. Chromosoma 61:149–176

Bernhardt N (2016) Analysis of phylogenetic relationships among diploid Triticeae grasses. Ph.D. Thesis submitted to the Fakultät für Mathematik und Naturwissenschaften der Carl von Ossietzky Universität Oldenburg

Bhattacharyny AK, Evans LE, Jenkins BC (1961) Karyotype analysis of the individual "Dakold" fall rye chromosome additions to "Kharkov" winter wheat. Nucleus 4:25–38

Bielig LM, Driscoll CJ (1970) Substitution of rye chromosome 5RL for chromosome 5B of wheat and its effect on chromosome pairing. Genetics 65:241–247

Bleier H (1930) Cytologie von Art- und Gattungbastarden des Getreides. Züchter 2:12–22

Bowden WM (1966) Chromosome numbers in seven genera of the tribe Triticeae. Can J Genet Cytol 8:130–136

Brink RA, Cooper DC, Ausherman LE (1944) A hybrid between Hordeum jubatum and Secale cereale. J Hered 35:67–75

Cermeño MC, Lacadena JR (1985) Nocleolar organizer competition in Aegilops–rye hybrids. Genome 27:479–483

Chao S, Sharp PJ, Worland AJ, Warham EJ, Koebner RMD, Gale MD (1989) RFLP-based genetic maps of wheat homoeologous group 7. Theor Appl Genet 78:495–504

Charpentier A, Feldman M, Cauderon Y (1986) Chromosomal pairing at meiosis of F_1 hybrid and backcross derivatives of Triticum aestivum x hexaploid Agropyron junceum. Can J Genet Cytol 28:1–6

Chikmawati T, Skovmand B, Gustafson JP (2005) Phylogenetic relationships among Secale species revealed by amplified fragment length polymorphisms. Genome 48:792–801

Colledge S, Conolly J (2010) Reassessing the evidence for the cultivation of wild crops during the Younger Dryas at Tell Abu Hureyra, Syria. Environ Archaeol 15:124–138

Crasniuk AA (1935) The hybrid Secale cereale x Agropyron cristatum. Soc Grain Farm 5:106–114

Cuadrado A, Jouve N (1994) Mapping and organization of highly-repeated DNA sequences by means of simultaneous and sequential FISH and C-banding in 6x-Triticale. Chromosome Res 2:331–338

Cuadrado A, Jouve N (1995) Fluorescent in situ hybridization and C-banding analyses of highly repetitive DNA sequences in the heterochromatin of rye (Secale montanum Guss.) and wheat incorporating S. montanum chromosome segments. Genome 38:795–802

Cuadrado A, Jouve N (1997) Distribution of highly repeated DNA sequences in species of the genus Secale. Genome 40:309–317

Cuadrado MC, Romero C (1984) Interaction between different genotypes of allogamous and autogamous rye and the homoeolo-gous pairing control of wheat. Heredity 52:323–330

Cuadrado A, Vitellozzi F, Jouve N, C. Ceoloni C (1997) Fluorescence in situ hybridization with multiple repeated DNA probes applied to the analysis of wheat-rye chromosome pairing. Theor Appl Genet 94: 347–355

Cuñado N (1992) Genomic analysis in the genus Aegilops. 3. Intergeneric hybrids between different species of Aegilops and Secale cereale. Theor Appl Genet 85:309–316

Cuñado N, Cermeño MC, Orellana J (1986) Interactions between wheat, rye and Aegilops ventricosa chromosomes on homologous and homoeologous pairing. Heredity 56:219–226

Darlington CD (1933) The origin and behavior of chiasmata. VIII. Secale cereale. Cytologia 4:444–452

Darlington CD, Mather K (1949) The elements of genetics. London

de Bustos A, Jouve N (2002) Phylogenetic relationships of the genus Secale based on the characterization of rDNA ITS sequences. Pl Syst Evol 235:147–154

Dedio W, Kaltsikes PJ, Larter EN (1969) Numerical chemotaxonomy in the genus Secale. Can J Bot 47:1175–1180

Del Pozo JC, Figueiras AM, Benito C, de La Peña A (1995) PCR derived molecular markers and phylogenetic relationships in the Secale genus. Biologia Plantarum 37:481–489

Devos KM, Atkinson MD, Chinoy CN, Francis HA, Harcourt RL, Koebner RMD, Liu CJ, Masoje P, Xie DX, Gale MD (1993) Chromosomal rearrangements in the rye genome relative to that of wheat. Theor Appl Genet 85:673–680

Devos KM, Dubcovsky J, Dvorak J, Chinoy CN, Gale MD (1995) structural evolution of wheat chromosomes 4A, 5A, and 7B and its impact on recombination. Theor Appl Genet 91:282–288

Dhaliwal HS, Gill BS, Waines JG (1977) Analysis of induced homoeologous pairing in a ph mutant wheat x rye hybrid. J Hered 68:206–209

Driscoll CJ, Sears ER (1971) Individual additions of the chromosomes of 'Imperial' rye to wheat. Agron Abstr, p 6

Dubcovsky J, Luo MC, Zhong GY, Bransteitter R, Desai A, Kilian A, Kleinhofs A, Dvorak J (1996) Genetic map of diploid wheat, Triticum monococcum L., and its comparison with maps of Hordeum vulgare L. Genetics 143:983–999

Dvorak J (1977) Effect of rye on homoeologous chromosome pairing in wheat x rye hybrids. Can J Genet Cytol 19:549–556

Dvorak J, Knott DR (1974) Disomic and ditelssomic additions of diploid Agropyron elongatum chromosomes to Triticum aestivum. Can J Genet Cytol 16(399–4):17

Eilam T, Anikster Y, Millet E, Manisterski J, Feldman M (2007) Genome size and genome evolution in diploid Triticeae species. Genome 50:1029–1037

Emme HK (1927) Zur Cytologie der Gattung Secale L. Bull Appl Bot 17:73–100

Emme HK (1928) Karyologie der Gattung Secale L. Zschr Ind Abst Vererb L 47:99–124

Engelbrecht TH (1916) On the origin of some cultivated field crops. Geographische Zeilschrift 22:328–334

Engelbrecht TH (1917) Über die Entstehung des Kulturroggens. Edvart Hahn Festschrift, zum 60. Geburtstag, pp 17–21

Escobar JS, Scornavacca C, Cenci A, Guilhaumon C, Santoni S, Douzery EJ, Ranwez V, Glémin S, David J (2011) Multigenic phylogeny and analysis of tree incongruences in Triticeae (Poaceae). BMC Evol Biol 11:181–198

Evans GM (1976) Rye Secale cereale (Gramineae-Triticineae). In: Simmonds NW (ed) Evolution of crop plants. Longman Scientific and Technical, London and New York, pp 108–111

Evans GM (1995) Rye Secale cereale (Gramineae-Triticinae). In: Smartt J, Simmonds NW (eds) Evolution of crop plants, 2nd edn. Longman Scientific and Technical, London and New York, pp 166–170

Evtushenko EV, Levitsky VG, Elisafenko EA, Gunbin KV, Belousov AI, Šafář J, Doležel J, Vershinin AV (2016) The expansion of heterochromatin blocks in rye reflects the co-amplification of tandem repeats and adjacent transposable elements. BMC Genomics 17:337

Fairbairn A, Asouti E, Near J, Martinoli D (2002) Macro-botanical evidence for plant use at Neolithic Çatalhöyük south-central Anatolia, Turkey. Veg Hist Archaeobot 11:41–54

Fairbairn A, Martinoli D, Butler A, Hillman GC (2007) Wild plant seed storage at Neolithic Çatalhöyük East, Turkey. Veg Hist Archaeobot 16:467–479

Fairbairn A, Near J, Martinoli D (2005) Macrobotanical investigation-sof the North, South and KOPAL areas at Çatalhöyük. In: Hodder I (ed) Inhabiting Çatalhöyük reprts from the 1995–1999 seasons. Cambridge/Ankara, NcDonald Institute for Archaeological Research/British Institute of Archaeology at Ankara, pp 137–201

FAOSTAT (2015) FAO database. Food and Agricultural Organization of the United Nations, Statistics Division

FAOSTAT (2020) FAO Database. Food and Agriculture Organization of the United Nations. Available at: http://faostat.fao.org

Faria R, Navarro A (2010) Chromosomal speciation revisited: rearranging theory with pieces of evidence. Trends Ecol Evol 25:660–669

Favorsky NV (1935) On the reduction division in the hybrid of Secale cereale L. x Agropyron cristatum (L.) Gaertn. in connection with the question of its sterility. Social Grain Farming 1935:115–125

Fedak G (1979) Cytogenetics of a barley x rye hybrid. Can J Genet Cytol 21:543–548

Fedak G (1986) Wide crosses in *Hordeum*. In: Barley Agronomy Monograph, vol 26. ASA-CSSA-SSSA, Madison, pp 155–186

Fedak G, Armstrong KC (1981) Hybrids of *Hordeum parodii* and *Hordeum lechleri* with *Secale cereale*. In: Barley genetics IV: proceedings of the fourth international barley genetics symposium, Edinburgh, pp 740–745

Fedak G, Armstrong KC (1986) Intergeneric hybrids between *Secale cereale* (2x) and *Thinopyrum intermedium* (6x). Can J Genet Cytol 28:426–429

Fernández-Escobar J, Martin A (1989) A self-fertile trigeneric hybrid in the Triticeae involving *Triticum*, *Hordeum*, and *Secale*. Euphytica 42:291–296

Finch RA, Bennett MD (1980) Mitotic and meiotic chromosome behaviour in new hybrids of *Hordeum* with *Triticum* and *Secale*. Heredity 44:201–209

Frederiksen S (1991a) Taxonomic studies in *Eremopyrum* (Poaceae). Nord J Bot 11:271–285

Frederiksen S (1991b) Taxonomic studies in *Dasypyrum* (Poaceae). Nord J Bot 11:135–142

Frederiksen S, Petersen G (1997) Morphometrical analyses of *Secale* L. (Triticeae, Poaceae). Nord J Bot 17:1–14

Frederiksen S, Petersen G (1998) A taxonomic revision of *Secale* (Triticeae, Poaceae). Nord J Bot 18:399–420

Frederiksen S, Seberg O (1992) Phylogenetic analysis of the Triticeae (Poaceae). Hereditas 116:15–19

Frederiksen S, von Bothmer R (1989) Intergeneric hybridization between *Taeniatherum* and different genera of Triticeae, Poaceae. Nord J Bot 9:229–240

Frederiksen S, von Bothmer R (1995) Intergeneric hybridization with *Eremopyrum* (Poaceae). Nord J Bot 15:39–47

Gill BS, Friebe B (2009) Cytogenetic analysis of wheat and rye genomes. In: Feuillet C, Muehlbauer GJ (eds) Genetics and genomics of the Triticeae. Springer, pp 121–135

Gill BS, Kimber G (1974) The Giemsa C-banded karyotype of rye. Proc Natl Acad Sci, USA 71:1247–1249

Giorgi B, Cuozzo L (1980) Homoeologous pairing in a ph mutant of tetraploid wheat crossed with rye (x). Cereal Res Comm 8:485–490

Giraldez R, Cermeño MC, Orellana J (1979) Comparison of C-banding pattern in the chromosomes of inbred lines and open pollinated varieties of rye *Secale cereale* L. Z Pflanzenzüecht 83:40–48

González-García M, González-Sánchez M, Puertas MJ (2006) The high variability of subtelomeric heterochromatin and connections between nonhomologous chromosomes, suggest frequent ectopic recombination in rye meiocytes. Cytogenet Genome Res 115:179–185

Gupta PK, Fedak G (1985) Genetic control of meiotic chromosome pairing in the genus *Hordeum*. Can J Genet Cytol 27:515–530

Gupta PK, Fedak G (1986) The inheritance of genetic variation in rye (*Secale cereale*) affecting homoeologous chromosome pairing in hybrids with bread wheat (*Triticum aestivum*). Can J Genet Cytol 28:844–851

Gupta PK, Fedak G (1987a) Preferential intragenomic chromosome pairing in two new diploid intergeneric hybrids between Hordeum and Secale. Genome 29:594–597

Gupta PK, Fedak G (1987b) Meiosis in two new intergeneric hybrids between *Hordeum* and *Secale*. Plant Bree 99:155–158

Gustafson JP (1983) Cytogenetics of triticale. In: Swaminathan MS, Gupta PK, Sinha U (eds) Cytogenetics of crop plants. MacMillan India, Delhi, pp 225–250

Gustafson JP, Evans LE, Josifek K (1976) Identification of chromosomes in *Secale montanum* and individual Secale montanum chromosome addition to 'Kharkov' wheat by heterochromatin bands and chromosome morphology. Can J Genet Cytol 18:339–343

Hammer K (1987) Resistenzmerkmale und Reprodktionssystem als Indikatoren für evolutionäreTEndenzen der Gattung *Aegilops* L. Biol Zentralbl 106:264–273

Hammer K (1990) Breeding system and phylogenetic relationships in *Secale* L. Biol Zbl 109:45–50

Hammer K, Lehmann CO, Pemno P (1985) Die in den Jahren 1980, 1981 und 1982 in Süditalien gesammelten Getreide-Landsorten – botanische Ergebnisse. Kulturpflanze 33:237–267

Hammer K, Skolimowska E, Knüpffer H (1987) Vorarbeiten zur monographischen Darstellung von Wildpflanzensortimenten: *Secale* L. Kulturpflanze 35:135–177

Hasegawa N (1934) A cytological study on 8-chromosome rye. Cytologia 6:68–77

Helbaek H (1954) Prehistoric food plants and weeds in Denmark. A survey of archaeobotanical research 1923–1954. Danmark Geologjsk Undersøgelse 11:250–261

Helbaek H (1964a) Early Hassunan vegetable food at es-Sawwan near Samarra. Sumer 20:45–48

Helbaek H (1964b) First impressions of the Çatal Höyük plant husbandry. Anatol Stud 14:121–123

Heneen WK (1963) Cytology of the intergeneric hybrid *Elymus arenarius* x *Secale cereale*. Hereditas 49:61–77

Heneen WK (1965) On the meiosis of haploid rye. Hereditas 52:421–424

Hillman GC (1972) Plant remains in French DH. Excavations at Can Hassan III 1969–1970. In: Higgs ES (ed) Papers in economic prehistory. Cambridge University Press, Cambridge, pp 182–190

Hillman GC (1975) The plant remains from Tell Abu Huretra: a preliminary report. Proc Prehistory Society 41:70–73

Hillman GC (1978) On the origins of domestic rye-*Secale cereale*: the finds from Aceramic Can Hasan III in Turkey. Anatolian Stud 28:157–174

Hillman G (2000) Plant food economy of Abu Hureyra. In: Moore A, Hillman G, Legge T (eds) Village on the Euphrates, from foraging to farming at Abu Hureyra. Oxford Univ Press, New York, pp 372–392

Hillman GC, Colledge SM, Harris DR (1989) Plant–food economy during the Epipalaeolithic period at Tell Abu Hureyra, Syria: dietary diversity, seasonality, and modes of exploitation. In: Harris DR, Hillman GC (eds) Foraging and farming: the evolution of plant exploitation. Unwin Hyman, London, pp 240–268

Hillman GC, Hedges R, Moore A, Colledge S, Petitt P (2001) New evidence of Late Glacial cereal cultivation at Abu Hureyra on the Euphrates. The Holocene 11:383–393

Hopf M (1982) Vor- und frühgeschichtliche kulturpflanzen aus dem nördlichen Deutschland, Zentralmus, Nainz

Hsiao C, Chatterton NJ, Asay KH, Jensen KB (1995a) Phylogenetic relationships of the mono genomic species of the wheat tribe, Triticeae (Poaceae), inferred from nuclear rDNA (internal transcribed spacer) sequences. Genome 38(2):11–223

Hsiao C, Chatterton CJ, Asay KH, Jensen KB (1995b) Molecular phylogeny of the Pooideae (Poaceae) based on nuclear rDNA (ITS) sequences. Theor Appl Genets 90:389–398

Hubbard CE (1959) Gramineae. In: Hutchinson J (ed) The families of flowering plants, vol 2, Monocotyledons, 2nd edn. University Press, Oxford, pp 710–741

Hutchinson J, Chapman V, Miller TE (1980) Chromosome pairing at meiosis in hybrids between *Aegilops* and *Secale* species: a study by in situ hybridization using cloned DNA. Heredity 45:245–254

Hutchinson J, Flavell RB, Jones J (1981) Physical mapping of plant chromosomes by in situ hybridization. In: Setlow JK, Hollaender A (eds) Genetic engineering, vol III. Plenum Press, New York, pp 207–222

Hutchinson J, Miller TE, Reader SM (1983) C-banding at meiosis as a means of assessing chromosome affinities in the Triticeae. Can J Genet Cytol 25(3):19–33

Jaaska V (1975) Evolutionary variation of enzymes and phylogenetic relationships in the genus *Secale* L. Izv Akad Nauk Estonsk SSR, Ser Biol 24:179–198

Jain SK (1960) Cytogenetics of rye (*Secale* spp.). Bibliogr Genet 19:1–86

Januševič ZV (1978) Prehistoric food plants in the south-west of the Soviet Union. Berichte Der Deutschen Botanichen Gesellchaft 91:59–66

Jauhar PP, Riera-Lizarazu O, Dewey WG, Gill BS, Crane CF, Bennet JH (1991) Chromosome pairing relationships among the A, B and D genomes of bread wheat. Theor Appl Genet 81:441–449

Jessen K, Helbaek H (1944) Cereals in Great Britain and Ireland in prehistoric and early historic times. Det Kongelige Danske Videnskabernes Selskab Biologiske Skrifter 3:1–68

Jones JDG, Flavell RB (1982a) The mapping of highly-repeated DNA families and their relationship to C-bands in chromosomes of *Secale cereale*. Chromosoma 86:595–612

Jones JDG, Flavell RB (1982b) The structure, amount and chromosomal localization of defined repeated DNA sequences in species of the genus *Secale*. Chromosoma 86:613–641

Jones RN, Rees H (1982) B chromosomes. Academic Press, London

Jouve N, Diez N, Rodríguez M (1980) C-banding in 6x Triticale x *Secale cereale* L. hybrids cytogenetics. Theor Appl Genet 57:75–79

Kagawa F, Chizaki Y (1934) Cytological studies on the genus hybrids among *Triticum*, *Secale* and *Aegilops*, and the species hybrids in *Aegilops* Jap. J Bot 7:1–32

Karpechenko GD, Sorokina ON (1929) The hybrids of *Aegilops triuncialis* L. with rye. Bull Appl Bot Genet Pl Breed (USSR) 20:536–584

Kattermann G (1934) Die cytologischen Verhältnisse einiger Weizenroggenbastarde und ihrer Nachkommenschaft ("F2"). Der Züchter 6:97–107

Kawakubo J, Taira T (1992) Intergeneric hybrids between *Aegilops squarrosa* and *Secale cereale* and their meiotic chromosome behaviour. Pl Breed 109:108–115

Kellogg EA (1989) Comments on genomic genera in the Triticeae (Poaceae). Amer J Bot 76:796–805

Kellogg EA, Appels R (1995) Intraspecific and interspecific variation in 5S RNA genes are decoupled in diploid wheat relatives. Genetics 140:325–343

Kellogg EA, Appels R, Mason-Gamer RJ (1996) When genes tell different stories: the diploid genera of Triticeae (Gramineae). Syst Bot 21:321–347

Khalilov VG, Kasumov AG (1989) Some cytogenetic features of first-generation *Aegilops*-rye hybrids. Tsitologiya i Genet 23:66–68

Khush GS (1962) Cytogenetic and evolutionary studies in *Secale*. II. Interrelationships of the wild species. Evolution 16:484–496

Khush GS (1963a) Cytogenetic and evolutionary studies in *Secale*. III. Cytogenetics of weedy ryes and origin of cultivated rye. Econ Bot 17:60–71

Khush GS (1963b) Cytogenetic and evolutionary studies in *Secale*. IV. *Secale vavilovii* and its biosystematic status. Z Pflanzenzücht 50:34–43

Khush GS, Stebbins GL (1961) Cytogenetic and evolutionary studies in Secale. I. Some new data on the ancestry of *S. cereale*. Amer J Bot 48:721–730

Kihara H (1924) Cytologische und genetische studien bei wichtigcn getreidearten mit besonderer rücksicht ouf das verhalten der chromosomen und die sterilitat in den bastarden. Kyoto Imp Univ Bl, Mem Cell Sci, pp 1–200

Kihara H (1937) Genomanalyse bei *Triticum* und *Aegilops*. VII. Kurze ~ bersicht iiber die Ergebnisse der Jahre 1934–36. Mem Coll Agric Kyoto Imp Univ 41:1–61

King IP, Purdie KA, Liu CJ, Reader SM, Orford SE, Pittaway TS, Miller TE (1994) Detection of interchromosomal translocations within the Triticeae by RFLP analysis. Genome 37:882–887

Kobyljanskij VD (1975) K sistematike i filogenii roda *Secale* L. Byull Vses Nauchno-Issled Inst Rast 48:64–71

Kobyljanskij VD (1982) Rozh. Geneticheskije osnovy selekcii. Kolos, Moskva

Kobyljanskij VD (1983) The system of the genus *Secale* L. Trudy Prikl Bot 79:24–38

Koller OL, Zeller FJ (1976) The homoeologous relationships of rye chromosomes 4R and 7R with wheat chromosomes. Genet Res Camb 28:177–188

Komatsuda T, Mano Y (2002) Molecular mapping of the intermedium spike-c (*int-c*) and non-brittle rachis 1 (*btr1*) loci in barley (*Hordeum vulgare* L.). Theor Appl Genet 105:85–90

Komatsuda T, Maxim P, Senthil N, Mano Y (2004) High-density AFLP map of nonbrittle rachis 1 (*btr1*) and 2 (*btr2*) genes in barley (*Hordeum vulgare* L.). Theor Appl Genet 109:986–995

Kranz AR (1957) Populationsgenetische Untersuchungen am iranischen Primitivroggen. Z. Pflanzenzüchtg 38:101–146

Kranz AR (1961) Cytologische Untersuchungen und genetische Beobachtungen an den Bastarden zwischen *Secale cereale* L. und *Secale vavilovii* Grossh. Der Züchter 31:219–225

Kranz AR (1963) Beitrage zur cytologischen und genetischen Evolutionsforschung an dem Roggen. Zeitschrift Für Pflanzenzüchtung 50:44–58

Kranz AR (1973) Wildarten und primitivformen des Roggens (*Secale*). Fortschr Pflzücht 3:1–60

Kuckuck H, Peters R (1967) Genetische Untersuchungen über die Selbstfertilitat bei *S. vavilovii* Grossh. und *S. cereale* L. var. Dakold im Hinblick auf Probleme der Züchtung und Phylogenie. Zeitschrift zürPflanzenzüchtung 57:167–188

Ladizinsky G (1998) Plant evolution under domestication. Kluwer Academic Publishers, Dordrecht, The Netherlands

Lammer D, Cai X, Arterburn M, Chatelain J, Murray T, Jones S (2004) A single chromosome addition from *Thinopyrum elongatum* confers a polycarpic, perennial habit to annual wheat. J Exp Bot 55:1715–1720

Lapitan NLV, Sears RG, Rayburn AL, Gill BS (1986) Wheat-rye translocations. J Hered 77:415–419

Lapitan NLV, Sears RG, Gill BS (1988) Amplification of repeated DNA sequences in wheat x rye hybrids regenerated from tissue culture. Theor Appl Genet 75:381–388

Leighty CE, Sando WJ, Taylor JW (1926) Intergeneric hybrids in *Aegilops*, *Triticum* and *Secale*. J Agric Res 23:101–141

Lelley T (1976) Induction of homoeologous pairing in wheat by genes of rye suppressing chromosome 5B effect. Can J Genet Cytol 18:485–489

Levan A (1942) Studies on the meiotic mechanism of haploid rye. Hereditas 28:177–211

Levy AA, Feldman M (1989a) Genetics of morphological traits in wild wheat *Triticum turgidum* var. dicoccoides. Euphytica 40:275–281

Levy AA, Feldman M (1989b) Location of genes for high grain protein percentage and other quantitative traits in wild wheat *Triticum turgidum* var. dicoccoides. Euphytica 41:113–122

Lewitsky GA (1929) Investigations on the morphology of chromosomes. Proc USSR Cong Genet Pl and Anim Breed 2:87–105

Lewitsky GA (1931) The morphology of chromosomes. Bull Appl Bot Genet Pl Breed 27:19–174

Li W, Challa GS, Zhu H, Wei W (2016) Recurrence of chromosome rearrangements and reuse of DNA breakpoints in the evolution of the Triticeae genomes. G3 6:3837–3847

Li G, Wang L, Yang J et al (2021) A high-quality genome assembly highlights rye genomic characteristics and agronomically important genes. Nat Genet 53:574–584

Liljefors A (1936) Zytologische studien über den F_1 bastard *Triticum turgidum* x *Secale cereale*. Hereditas 21:240–262

Lima-de-Faria A (1953) Chromomere analysis of the chromosome complement of rye. Chromosoma 5:1–68

Liu CJ, Atkinson MD, Chinoy CN, Devos KM, Gale MD (1992) Non-homoeologous translocations between group 4, 5 and 7 chromosomes within wheat and rye. Theor Appl Genet 83:305–312

Longley AE, Sando WJ (1930) Nuclear divisions in the pollen mother cells of *Triticum*, *Aegilops*, and *Secale* and their hybrids. J Agr Res 40:683–719

Löve Á (1984) Conspectus of the Triticeae. Feddes Repert 95:425–521

Lu BR, von Bothmer R (1991) Cytogenetic studies of the intergeneric hybrids between *Secale cereale* and *Elymus caninus*, *E. brevipes*, and *E. tsukushiensis* (Triticeae: Poaceae). Theor Appl Genet 81:524–532

Lu B-R, Salomon B, von Bothmer R (1990) Cytogenetic studies of progeny from the intergeneric crosses *Elymus* x *Hordeum* and *Elymus* x *Secale*. Genome 33:425–432

Lucas H, Jahier J (1988) Phylogenetic relationships in some diploid species of Triticineae: cytogenetic analysis of interspecific hybrids. Theor Appl Genet 75:498–502

Lukaszewski AJ (2000) Manipulation of the 1RS.1BL translocation in wheat by induced homoeologous recombination. Crop Sci 40:216–225

Lukaszewski AJ, Rybka K, Korzun V, Malyshev SV, Lapinski B, Whitkus R (2004) Genetic and physical mapping of homoeologous recombination points involving wheat chromosome 2B and rye chromosome 2R. Genome 47:36–45

Lukaszewski AJ, Kopecky D, Linc G (2012) Inversions of chromosome arms 4AL and 2BS in wheat invert the patterns of chiasma distribution. Chromosoma 121:201–208

Lundquist A (1954) Studies on self-sterility in rye, *Secale cereale* L. Hereditas 40:278–294

Lundquist A (1956) Self-incompatibility in rye. I. Genetic control in the diploid. Hereditas 42:293–348

Lundqvist A (1958a) Self-incompatibility in rye. III. Homozygosity for incompatibility factors in relation to viability and vegetative development. Hereditas 44:174–188

Lundqvist A (1958b) Self-incompatibility in rye. IV. Factors related to self-seeding. Hereditas 44:193–256

Majisu BN, Jones JK (1971) *Aegilops* x *Secale* hybrids: the production and cytology of diploid hybrids. Genet Res, Camb 17:17–31

Martis MM, Zhou R, Haseneyer G, Schmutzer T, Vrána J, Kubaláková M, König S, Kugler KG, Scholz U, Hackauf B, Korzun V, Schön CC, Dolezel J, Bauer E, Mayer KF, Stein N (2013) Reticulate evolution of the rye genome. Plant Cell 25:3685–3698

Mason-Gamer RJ (2005) The B-amylase genes of grasses and a phylogenetic analysis of the Triticeae (Poaceae). Am J Bot 92:1045–1058

Mason-Gamer RJ, Kellogg EA (1996) Chloroplast DNA analysis of the monogenomic Triticeae: phylogenetic implications and genome-specific markers. In: Jauhar PP (ed) Methods of genome analysis in plants. CRC Press, Boca Raton, FL, pp 301–325

Mason-Gamer RJ, Orme NL, Anderson CM (2002) Phylogenetic analysis of North American *Elymus* and the monogenomic Triticeae (Poaceae) using three chloroplast DNA data sets. Genome 45:991–1002

Megyeri M, Molnár-Láng M, Molnár I (2013) Cytomolecular identification of individual wheat-wheat chromosome arm associations in wheat-rye hybrids. Cytogenet Genome Res 139:128–136

Meier S, Kunzmann R, Zeller FJ (1996) Genetic variation in germplasm accessions of *Secale vavilovii* Grossh. Genet Resourc Crop Evol 43:91–96

Meister GK (1921) Natural hybridization of wheat and rye in Russia. J Hered 12:467–470

Melderis A (1953) Generic problems within the tribe *Hordeae*. In: Osvald H, Aberg E (eds) Proceedings of the 7th international botanical congress, Stockholm 1950, Uppsala, Almquist & Wiksells, pp 1450–1485

Melnyk J, Unrau J (1959) Pairing between chromosomes of *Aegilops squarrosa* L. var. typica and *Secale cereale* L. var. prolific. Can J Genet Cytol 1:21–25

Middleton CP, Senerchia N, Stein N, Akhunov ED, Keller B, Wicker T, Kilian B (2014) Sequencing of chloroplast genomes from wheat, barley, rye and their relatives provides a detailed insight into the evolution of the Triticeae tribe. PLoS ONE 9: e85761

Miller TE (1984) The homoeologous relationship between the chromosomes of rye and wheat. Current Status. Can J Genet Cytol 26:578–589

Miller TE, Riley R (1972) Meiotic chromosome pairing in wheat-rye combinations. Genet Iber 24:241–250

Miller TE, Reader SM, Purdie KA, King IP (1994) Determination of the frequency of wheat-rye chromosome pairing in wheat x rye hybrids with and without chromosome 5B. Theor Appl Genet 89:255–258

Millet E, Rong JK, Qualset CO, McGuire PE, Bernard M, Sourdille P, Feldman M (2013) Production of chromosome-arm substitution lines of wild emmer in common wheat. Euphytica 190:1–17

Monte JV, McIntyre CL, Gustafson JP (1993) Analysis of phylogenetic relationships in the Triticeae tribe using RFLPs. Theor Appl Genet 86:649–655

Morrison JW, Rajhathy T (1959) Cytogenetic studies in the genus *Hordeum*. III. Pairing in some interspecific and intergeneric hybrids. Can J Genet Cytol 1:65–77

Mukai Y, Friebe B, Gill BS (1992) Comparison of C-banding patterns and in situ hybridization sites using highly repetitive and total genomic rye DNA probes of 'Imperial' rye chromosomes added to 'Chinese Spring' wheat. Jpn J Genet 67:71–83

Müntzing A (1935) Triple hybrids between rye and two wheat species. Hereditas 20:137–160

Müntzing A (1937) Note on a haploid rye plant. Hereditas 23:401–404

Müntzing A (1944) Cytological studies of extra fragment chromosomes in rye I. Iso-fragments produced by misdivision. Hereditas 30:231–248

Müntzing A (1950) Accessory chromosomes in rye populations from Turkey and Afghanistan. Hereditas 36:507–509

Müntzing A, Prakken R (1941) Chromosomal aberrations in rye populations. Hereditas 27:273–308

Murai K, Naiyu X, Tsunewaki K (1989) Studies on the origin of crop species by restriction endonuclease analysis of organellar DNA. III. Chloroplast DNA variations and interspecific relationships in the genus *Secale*. Jpn J Genet 64:36–47

Nakajima G (1955) A cytogenetical study on the intergeneric F_1 hybrid between *Triticum polonicum* and *Secale africanum*. Cytologia 20:273–279

Nakajima G (1956a) Cytogenetical studies on interspecific hybrids of rye. II. Meiosis in pollen mother cells of F_1 between *Secale cereale* on one hand and *S. vavilovii*, *S. africanum* and *S. montanum* on the other. Jap J Breed 6:171–174

Nakajima G (1956b) Cytological studies on the intergeneric F₁ hybrids between *Triticum spelta* and two species of *Secale*. Jap J Breed 6:101–106

Nalam VJ, Vales MI, Watson CJW, Kianian SF, Riera-Lizarazu O (2006) Map-based analysis of genes affecting the brittle rachis character in tetraploid wheat (*Triticum turgidum* L.). Theor Appl Gen 112:373–381

Naranjo T (1982) Preferential occurrence of wheat-rye meiotic pairing between chromosomes of homoeologous group 1. Theor Appl Genet 63:219–225

Naranjo T, Fernández-Rueda P (1991) Homoeology of rye chromosome arms to wheat. Theor Appl Genet 82:577–586

Naranjo T, Fernández-Rueda P (1996) Pairing and recombination between individual chromosomes of wheat and rye in hybrids carrying the ph1b mutation. Theor Appl Genet 93:242–248

Naranjo T, Palla O (1982) Genetic control of meiotic pairing in rye. Heredity 48:57–62

Naranjo T, Lacadena JR, Giraldez R (1979) Interaction between wheat and rye genomes on homologous and homoeologous pairing. Z Pflan 82:289–305

Naranjo T, Roca A, Goicoecha PG, Giraldez R (1987) Arm homoeology of wheat and rye chromosomes. Genome 29:873–882

Naranjo T, Fernández-Rueda P, Goicoechea PG, Roca A, Giráldez R (1989) Homoeologous pairing and recombination between the long arms of group 1 chromosomes in wheat x rye hybrids. Genome 32:293–301

Neijzing MG (1982) Chiasma formation in duplicated segments of the haploid rye genome. Chromosoma 85:287–298

Neijzing MG (1985) Number and localization of sets of homologous segments in the genome of rye on the basis of Giemsa-banded metaphase-I chromosome associations in a haploid. Genetica 66:41–52

Nesbitt M (2002) When and where did domesticated cereals first occur in southwest Asia? In: Cappers RTJ, Bottema S (eds) The dawn of farming in the Near East. Studies in Near Eastern Production, Subsistence and Environment, 6, 1999. Ex Oriente, Berlin, pp 113–132

Neves N, Heslop-Harrison JS, Viegas W (1995) rRNA gene activity and control of expression mediated by methylation and imprinting during embryo development in wheat x rye hybrids. Theor Appl Genet 91:529–533

Nevski SA (1933) Agrostologische studien. IV. Uber das System der Tribe Hordeae Benth. Acta Inst Bot Sci, U. R. SS, Ser II

Niwa K, Ohta S, Sakamoto S (1990) B chromosomes of *Secale cereale* L. and *S. montanum* Guss. from Turkey. Jpn J Breed 40:147–152

Nordenskiöld H (1939) Studies of a haploid rye plant. Hereditas 25:204–210

Nürnberg-Krüger U (1960) Cytogenetische Untersuchungen der Gattung *Secale* L. Z. Pflzücht. 44:63–72

Oehler E (1931) Untersuchungen über Ansatzverhältnisse, Morphologie und Fertilität bei Weizen-Roggenbestanden. Zeitschr Pflanzenzg 36:357–393

O'Mara JG (1940) Cytogenetic studies on triticale. I. A method of determining the effects of individual Secale chromosomes on Triticum. Genetics 25:401–408

Orellana J (1985) Most of the homoeologous pairing at metaphase I in wheat-rye hybrids is not chiasmatic. Genetics 111:917–931

Orellana J, Santos JL, Lacadena JR, Cermeño MC (1984) Nucleolar competition analysis in Aegilops ventricosa and its amphiploids with tetraploid wheats and diploid rye by the silver-staining procedure. Can J Genet Cytol 26:34–39

Pagniez M, Hours C (1986) Compatibilitees de *Triticum tauschii* (Coss.) Schmal. et *Triticum ventricosum* Ces. Avec *Secale cereale* L. et *Hordeum bulbosum* L. 2x et 4x. Z Pflanzenzücht 96:15–24

Pardue M-L, DeBaryshe PG (2003) Retrotransposons provide an evolutionarily robust non-telomerase mechanism to maintain telomeres. Annu Rev Genet 37:485–511

Perez de la Vega M, Allard RW (1984) Mating system and genetic polymorphism in populations of *Secale cereale* and *S. vavilovii*. Can J Genet Cytol 26:308–317

Perrino P, Hammer K, Hanelt P (1984) Collection of landraces of cultivated plants in South Italy 1983. Kulturpflanze 32:207–216

Petersen G (1991a) Intergeneric hybridization between *Hordeum* and *Secale* (Poaceae). I. Crosses and development of hybrids. Nordic J Bot 11:253–270

Petersen G (1991b) Intergeneric hybridization between *Hordeum* and *Secale*. II. Analysis of meiosis in hybrids. Hereditas 114:141–159

Petersen G, Doebley JF (1993) Chloroplast DNA variation in the genus *Secale* (Poaceae). Pl Syst Evol 187:115–125

Petersen G, Seberg O (1997) Phylogenetic analysis of the Triticeae (Poaceae) based on *rpo*A sequence data. Mol Phylogenet Evol 7:217–230

Petersen G, Seberg O, Aagesen L, Frederiksen S (2004) An empirical test of the treatment of indels during optimization alignment based on the phylogeny of the genus *Secale* (Poaceae). Mol Phylogenet Evol 30:733–742

Plotnikowa TW (1930) Cytological investigations on hybrids between 28-chromosome wheat and rye. Planta 12:167–183

Popoff A (1939) Untersuchungen über den Formenreichtum und die Schartigkeit des Roggens. Ang Bot 21:325–356

Puertas MJ, Giraldez R (1979) Meiotic pairing in haploid rye. Genét Ibér 39:30–34

Quincke FL (1940) Interspecific and intergeneric crosses with *Hordeum*. Can J Res 18:372–373

Rabanus-Wallace MT, Hackauf B, Mascher M et al (2021) Chromosome-scale genome assembly provides insights into rye biology, evolution and agronomic potential. Nat Genet 53:564–573

Reddy P, Appels R, Baum BR (1990) Ribosomal DNA spacer-length variation in *Secale* spp. (Poaceae). Pl Syst Evol 171:205–220

Ren TH, Chen F, Zou YT, Jia YH, Zhang HQ, Yan BJ, Ren ZL (2011) Evolutionary trends of microsatellites during the speciation process and phylogenetic relationships within the genus *Secale*. Genome 54:316–326

Renfrew JM (1973) Palaeoethnobotany: the prehistoric food plants of the near east and Europe. Columbia University Press, New York

Rieseberg LH (2001) Chromosomal rearrangements and speciation. Trends Ecol Evol 16:351–358

Riley R (1955) The cytogenetics of the differences between some *Secale* species. J Agric Sci 46:277–283

Riley R, Law CN (1965) Genetic variation in chromosome pairing. Adv Genet 13:57–114

Rimpau W (1891) Kreuzungsprodukte Landwirtschaftlicher Kulturpflanzen. Land- Wirtschaftl Jahrb 20:335–371

Rognli OA, Devos KM, Chinoy CN, Harcourt RL, Atkinson MD, Gale MD (1992) RFLP mapping of rye chromosome 7R reveals a highly translocated chromosome relative to wheat. Genome 35:1026–1031

Rogowsky PM, Sorrells ME, Shepherd KW, Langridge P (1993) Characterization of wheat-rye recombinants with RFLP and PCR probes. Theor Appl Genet 83:489–494

Romero C, Lacadena JR (1982) Effect of rye B-chromosomes on pairing in *Triticum aestivum* × *Secale cereale* hybrids. Z Pflan 89:39–46

Rong JK (1999) Mapping and tagging by DNA markers of wild emmer alleles that affect useful traits in bread wheat. Ph.D. Thesis, submitted to the Scientific Council of The Weizmann Institute of Science, Rehovot, Israel, April 1999

Roshevitz RJ (1947) A monograph of the wild, weedy and cultivated species of rye. Acta Inst Bot Acad Sci USSR., ser 1, 6:105–163

Sakamoto S (1967) Cytogenetic studies in the tribe Triticeae. V. Intergeneric hybrids between two *Eremopyrum* species and *Agropyron tsukushiense*. Seiken Ziho 19:19

Sakamura T (1918) Kurze mitteilung über die chromosomenzahalen und die verwandtschaftsverhältnisse der Triticum Arten. Bot Mag 32:151–154

Sallee PJ, Kimber G (1976) The meiotic analysis of the hybrid *T. timopheevii* var. zhukovskyi x *Secale cereale*. Wheat Inf Serv 41–42:9–11

Schiemann E (1948) Weizen, Roggen, Gerste. Systematik, Geschichte und Verwendung. Verlag Gustav Fischer, Jena, pp 1–102

Schiemann E, Nürnberg-Krüger U (1952) Neue Untersuchungen an Secale africanum Stapf. Die Naturwiss 6:136–137

Schiemann E, Schweickerdt HG (1950) Neue Untersuchungen an *Secale africanum* Stapf. Bot Jahrb Syst 75:196–200

Schlegel R (1982) First evidence for rye-wheat additions. Biol Zbl 101:641–646

Schlegel R (2013) Rye—genetics, breeding, and cultivation. CRC Presss, Taylor and Francis Group, Roca Raton, FL, USA, pp 1–344

Schlegel R, Weryszko E (1979) Intergeneric chromosome pairing in different wheat-rye hybrids revealed by Giemsa banding technique and some implications on karyotype evolution in the genus *Secale*. Biol Zentralbl 98:399–407

Schlegel R, Numerova OM, Pershina LA (1980) Confirmation of chiasmatic barley-rye chromosome pairing by means of the Giemsa banding technique. Cereal Res Comm 8:509–514

Schlegel R, Melz G, Mettin D (1986) Rye cytology, cytogenetics and genetics—current status. Theor Appl Genet 72:721–734

Schoeftner S, Blasco MA (2009) A 'higher order' of telomere regulation: telomere heterochromatin and telomeric RNAs. The EMBO J 28:2323–2336

Sears ER (1954) The aneuploids of common wheat. Missouri Agric Exp Stn Res Bull 572:1–58

Sears ER (1966) Nullisomic-tetrasomic combinations in hexaploid wheat. In: Riley R, Lewis KR (eds) Chromosome manipulations and plant genetics. Springer, Boston, MA, USA, pp 29–45

Seberg O, Frederiksen S (2001) A phylogenetic analysis of the monogenomic Triticeae (Poaceae) based on morphology. Bot J Linn Soc 136:75–97

Seberg O, Petersen G (2007) Phylogeny of Triticeae (Poaceae) based on three organelle genes, two single-copy nuclear genes, and morphology. Aliso J Syst Evol Botany 23:362–371

Sechnyak AL, Simonenko VK (1991) Effect of the rye genome R^{ce} on homoelogous chromosome pairing in hybrids of *Secale cereale* L. with polyploid specie of *Triticum* L. and *Aegilops* L. Tsitologiya i Genet 25:20–23

Sencer HA (1975) Study of variation in the genus *Secale* L. and on the origin of the cultivated rye. Ph.D. Thesis, The University of Birmingham

Sencer HA, Hawkes JG (1980) On the origin of cultivated rye. Biol J Linn Soc 13:299–313

Shang H-Y, Wei Y-M, Wang X-R, Zheng Y-L (2006) Genetic diversity and phylogenetic relationships in the rye genus *Secale* L. (rye) based on *Secale cereale* microsatellite markers. Genet Mol Biol 29:685–691

Sharma HC, Waines JG (1980) Inheritance of tough rachis in crosses of *Triticum monococcum* and *T. boeoticum*. J Hered 71:214–216

Shewry PR, Parmer S, Miller TE (1985) Chromosomal location of the structural genes for the M_r 75,000 γ-secalins in *Secale montanum* Guss: evidence for a translocation involving chromosome 2R and 6R in cultivated rye (*Secale cereale* L.). Heredity 54:381–383

Singh RJ (1977) Cross compability, meiotic pairing and fertility in 5 Secale species and their interspecific hybrids. Cereal Res Commun 5:67–75

Singh RJ, Robbelen G (1975) Comparison of somatic Giemsa banding pattern in several species of rye. Z. Pflanzenzücht 75:270–285

Singh RJ, Robbelen G (1977) Identification by Giemsa technique of the translocations separating cultivated rye from three wild species of *Secale*. Chromosoma 59:217–225

Sodkiewicz W (1982) Hybrids between diploid wheat and rye. I. F_1 hybrids of *Triticum monococcum* L. x *S. cereale* L. Genet Pol 23:123–131

Soreng RJ, Davis JI, Doyle JJ (1990) A phylogenetic analysis of chloroplast DNA restriction site variation in Poaceae subfam Pooideae. Pl Syst Evol 172:83–97

Staat B, Pohler W, Clauss E (1985) Unerwartet starke Homöologenpaarung bei Bastarden zwischen *Hordeum geniculatum* All, und *Secale cereale* L. im vergleich zum Wildger-stenelter. Arch. Züchtungsforsch. (Berlin), pp 19–25

Stebbins GL (1958) The inviability, weakness and sterility of interspecific hybrids. Adv Genet 9:147–215

Stebbins GL, Pun FT (1953) Artificial and natural hybrids in the Gramineae tribe Hordeae. VI. Chromosome pairing in *Secale cercale–Agropyron intermedium* and the problem of genome homologies in the Triticineae. Genetics 38:600–608

Stolze K (1925) Die Chromosomenzahlen der hauptsächlichsten Getreide- arten nebst allgemeinen Betrachtungen iiber Chromosomenzahl und Chromosomengrosse im Pflanzenreich. Bibliotheca Genet 8:8–71

Stutz HC (1957) A cytogenetic analysis of the hybrid *Secale cereale* L. x *Secale montanum* Guss. and its progeny. Genetics 42:199–222

Stutz HC (1972) The origin of cultivated rye. Amer J Bot 59:59–70

Stutz HC (1976) Genetically controlled chromosome breakage as an isolation barrier in the origin and maintenance of *Secale ancestrale*. Can J Genet Cytol 18:105–109

Su Y, Zhang D, Li Y, Li S (2016) Nonhomologous chromosome pairing in *Aegilops–Secale* hybrids. Cytogenet Genome Res 147:268–273

Sybenga J (1966) The zygomere as hypothetical unit of chromosome pairing initiation. Genetica 37:186–198

Sybenga J (1983) Rye chromosome nomenclature and homoeology relationships. Workshop report. Z Pflanzenzücht 90:297–304

Takahashi R (1955) The origin and evolution of cultivated barley. Adv Genet 7:227–266

Takahashi R (1972) Non-brittle rachis 1 and non-brittle rachis 2. Barley Genet Newsletter 2:181–182

Takahashi R, Hayashi J (1964) Linkage study of two complementary genes for brittle rachis in barley. Ber Ohara Inst Landw Biol, Okayama Univ 12:99–105

Thomas HM, Pickering RA (1985) Comparisons of the hybrid *Hordeum chilense* x *H. vulgare*, *H. chilense* x *H. bulbosum*, *H. chilense* x *Secale cereale* and the amphidipioid of *H. chilense* x *H. vulgare*. Theor Appl Genet 69:519–522

Thompson WP (1926) Chromosome hehavior in a cross between wheat and rye. Genetics 11:317–332

Thompson WP (1931) Chromosome homologies in wheat, rye and *Aegilops*. Can J Res 4:624–634

Tosi M (1975) Hasanlu project 1974: paleobotanical survey. Iran 8:185–186

Tzvelev NN (1973) Conspectus specierum tribus *Triticeae* Dum. Familiae *Poaceae* in Flora URSS. Nov Syst Pl Vasc 10:19–59

Tzvelev NN (1976) Poaceae URSS Tribe III. Triticeae Dum. USSR. Academy of Sciences Press, Leningrad, pp 105–206

van Heemert C, Sybenga J (1972) Identification of the three chromosomes involved in the translocations which structurally differentiate the genome of *Secale cereale* L. from those of *Secale montanum* Guss. and *Secale vavilovii* Grossh. Genetica 43:387–393

van Zeist W, Bakker-Heeres JAH (1985) Archaeobotanical studies in the Levant 1. Neolithic sites in the Damascus basin: Aswad, Choraife. Ramad. Palaeohistoria 24:165–256

van Zeist W, Casparie WA (1968) Wild einkornwheat and barley from Tell Mureybit in northern Syria. Acta Botanica Beerlandica17:44

Vasiljev B (1932) Wheat-rye hybrids. 1. An analysis of the first generation of various wheat-rye hybrid combinations. Bull Lab Genet 9:69–87

Vavilov NI (1917) On the origin of cultivated rye. Bull Appl Bot, Genet and Pl Breed 10:561–590

Vavilov NI (1926) Studies on the origin of cultivated plants. Bull Appl Bot, Genet, Pl Breed 16:1–248 (In Russian and English summary)

Vences FJ, Vaquero F, Garcia P, Perez d e La Vega M (1987) Further studies on the phylogenetic relationships in *Secale*: on the origin of its species. Pl Breed 98:281–291

Vieira R, Queiroz A, Morais L, Barão A, Melo-Sampayo T, Viegas WS (1990) 1R chromosome nucleolus organizer region activation by 5-azacytidine in wheat x rye hybrids. Genome 33:707–712

von Berg KH (1931) Autosyndese in *Aegilops triuncialis* L. x *Secale cereale* L. Zeitschr f Züchtung: A Pflanzenzüchtung17:55–69

von Bothmer R, Jacobsen N (1985) Origin, taxonomy and related species. In: Rasmusson D (ed) Barley, Agronomy Monograph 26, pp 19–56

Vosa CG (1974) The basic karyotype of Rye (*Secale cereale*) analysed with Giemsa and fluorescence methods. Heredity 33:403–408

Voylokov AV, Fuong FT, Smirnov VG (1993) Genetic studies of self-fertility in rye (*Secale cereale* L.). 1. The identification of genotypes of self-fertile lines for the Sf alleles of self-incompatibility genes. Theor Appl Genet 87:616–618

Wagenaar EB (1959) Intergeneric hybrids between *Hordeum jubatum* L. and *Secale cereale* L. J Hered 50:195–202

Wang RR-C (1987) Diploid perennial intergeneric hybrids in the tribe Triticeae. III. Hybrids among *Secale montanum*, *Pseudoroegneria spicata*, and *Agropyron mongolicum*. Genome 29:80–84

Wang RR-C (1988) Diploid perennial intergeneric hybrids in the tribe Triticeae. IV. Hybrids among *Thinpyrum bessarabicum*, *Pseudorogneria spicata* and *Secale montanum*. Genome 30:356–360

Wang RR-C, von Bothmer R, Dvorak J, Fedak G, Linde-Laursen I, Muramatsu M (1995) Genome symbols in the *Triticeae* (Poaceae). In Wang RR-C, Jensen KB, Jaussi C (eds) Proceedings of the 2nd international Triticeae symposium, June 20–24. Logan, UT, pp 29–34

Watanabe N, Ikebata N (2000) The effects of homoeologous group 3 chromosomes on grain colour dependent seed dormancy and brittle rachis in tetraploid wheat. Euphytica 115:215–220

Watanabe N, Sugiyama K, Yamagishi Y, Sakata Y (2002) Comparative telosomic mapping of homoeologous genes for brittle rachis in tetraploid and hexaploid wheats. Hereditas 137:180–185

Watanabe N, Takesada N, Fujii Y, Martinek P (2005a) Comparative Mapping of Genes for Brittle Rachis in *Triticum* and *Aegilops*. Czech J. Genet. Plant Breed 41:39–44

Watanabe N, Takesada N, Shibata Y, Ban T (2005b) Genetic mapping of the genes for glaucous leaf and tough rachis in *Aegilops tauschii*, the D-genome progenitor of wheat. Euphytica 144:119–123

Watson L, Clifford HT, Dallwitz MJ (1985) The classification of the Poaceae: subfamilies and supertribes. Austral J Bot 33:433–484

Willcox G (1999) Agrarian change and the beginnings of cultivation in the Near East: evidence from wild progenitors, experimental cultivation and archaeobotanical data. In: Hather J, Gosden C (eds) The prehistory of food. Routledge, London, pp 479–500

Willcox G (2002) Geographical variation in major cereal components and evidence for independent domestication events in the Western Asia. In: Cappers RT, Bottema S (eds) The dawn of farming in the Near East. Studies in Near Eastern Production, Subsistence and Environment, 6, 1999. Ex Oriente, Berlin, pp 133–140

Willcox G, Fornite S, Herveux L (2008) Early Holocene cultivation before domestication in Northern Syria. Veg Hist Archaeol 17:313–325

Willerding U (1970) Vor- und frühgeschichtliche kulturpflanzen in Mitteleuropa. Neue Ausgrabungen Und Forschungen in Niedersachsen 5:287–375

Wilson AS (1876) On wheat and rye hybrids. Trans Proc Bot Soc, Edinburgh 12:286–288

Wojciechowska B, Pudelska H (2002) Hybrids and amphiploids of *Aegilops ovata* L. with *Secale cereale* L.: production, morphology and fertility. J Appl Genet 43:415–422

Zeller FJ, Cermeño M-C (1991) Chromosome manipulations in *Secale* (rye). In: Gupta PK, Tsuchiya T (eds) Chromosome engineering in plants: genetics, breeding, evolution, Part A. Elsevier, Amsterdam, pp 313–333

Zeller FJ, Hsam SLK (1983) Broadening the genetic variability of cultivated wheat by utilizing rye chromatin. In: Sakamoto S (ed) Proceedings of 6th international wheat genetic symposium, Kyoto, Japan, pp 161–173

Zeller FJ, Kimber G, Gill BS (1977) The identification of rye trisomics by translocations and Giemsa staining. Chromosoma 62:279–289

Zhukovsky PM (1933) Roggen (*La Turquie Agricole*). Vsesoiusn Akad Selsk Khoz Nauk, Lenina Inst Rastenievod, Moscow- Leningrad, pp 274–275

Zohary D (1971) Origin of south-west Asiatic cereals: wheats, barley, oats and rye. In: Davis PH et al (eds) Plant life of south-west Asia. Bot Soc Edinburgh, pp 53–258

Zohary D, Hopf M, Weiss E (2012) Domestication of Plants in the old world—the origin and spread of domesticated plants in south-west Asia, Europe and the Mediterranean Basin, 4th edn. Oxford University Press, pp 1–243

Classification of the Wheat Group (The Genera *Amblyopyrum*, *Aegilops* and *Triticum*)

7.1 Dilemmas Concerning the Classification of the Wheat Group

The wheat group, comprising the genera *Amblyopyrum*, *Aegilops* and *Triticum,* represents a congeneric complex, since the allotetraploid wheats, *T. turgidum* and *T. timopheevii*, contain one subgenome of diploid *Triticum* and one subgenome of diploid *Aegilops*, while allohexaploid wheat, *T. aestivum*, contains two subgenomes of *Aegilops* and one of *Triticum* diploid species. For this reason, Bowden (1959) suggested to unite the two genera into one genus, *Triticum*. Yet, despite their close genetic relationship, most taxonomists (e.g., Zhukovsky 1928; Eig 1929; Mac Key 1966, 1981; Hammer 1980; van Slageren 1994) prefer, based on morphological differences, to regard them as two separate genera.

Classification of the genera *Aegilops* and *Triticum* is complicated due to: (1) allopolyploid species, which have bi-phyletic or tri-phyletic origin, resulting from intergeneric hybridization between species from two different genera; (2) occurrence of both domesticated and wild forms within a single biological species; (3) the relatively young age of the group, with possibilities for formation of hybrid species, and allopolyploids, as well as inter-specific and inter-generic hybridization, leading to introgression, that blurs boundaries between species. These complications raise the following dilemmas and complexities: (1) definition of the genera, particularly regarding preservation of *Aegilops* and *Triticum* as separate genera or classifying them in a single genus; (2) consider of the wild and domesticated forms of inter-breeding taxa (taxonomic vs. biological species); (3) handling of new genera or species of inter-generic or inter-specific synthetic allopolyploids; (4) nomenclature of synthetic allopolyploids and mixoploids. Neither the International Code of Botanical Nomenclature (ICBN) (McNeill et al. 2006), nor the International Code of Botanical Nomenclature for Cultivated Plants (Brickell et al. 2009) address the specific problems relating to the classification of such a group of cultivated plants.

7.2 Taxonomic Species Concept Versus Biological Species Concept

The taxonomic species concept defines a species as a taxon comprised of individuals placed together because of overall likeness, where a degree of morphological homogeneity is the first requirement (Heslop-Harrison 1963). All individuals in a species must have more morphological characteristics in common with other members of the same species than with an individual of a different species (Heslop-Harrison 1963). Hence, the basic taxonomical categories are based on morphological similarity, as was defined by Linnaeus and his successors, who still adhered to the basic ideas of pre-Darwinian thinking (Löve 1964).

In contrast, the biological species concept considers a species as groups of actually or potentially inter-breeding populations that are reproductively isolated from other groups (Mayr 1940, 1942). Hence, the evolutionary principles of classification below the generic level, demand genetic isolation of biological species from other species. Morphological differences are less important in this concept, as "to be a different species is not a matter of difference but of distinctness" (Mayr 1963). The biological concept is based on biological connections and grouping of taxa and thus, expresses evolutionary and phylogenetic relationships.

Van Slageren and Payne (2013) discussed the difference between these two taxonomic concepts, the narrow, morphology-based concept (the taxonomical species concept) and the much broader, genome-based one (the biological species concept) in regard to the classification of the wheat group. The taxonomical concept endorses a detailed specific and intraspecific classification that aims to describe and appreciate the richness of morphological variation found in the species, whether wild or domesticated (Hammer et al. 2011). According to Van Slageren and Payne (2013), this classification concept originated from the so-called 'German-Russian School' that proposed wheat classifications of increasing complexity. Van Slageren and Payne (2013) presented several examples demonstrating the

M. Feldman and A. A. Levy, *Wheat Evolution and Domestication*, https://doi.org/10.1007/978-3-031-30175-9_7

confusion and contradiction that a hierarchical, detailed system can easily generate while attempting to classify each and every slightly different cultivar.

In contrast to this detailed system, Thellung (1918a, b) classified the wheats based on phylogeny and practical use. Thellung recognized only the following three species: *T. monococcum*, *T. turgidum,* and *T. aestivum*, and classified all other species that were formerly suggested on the basis of the taxonomical species concept at the sub-specific rank. In his classification, Thellung laid the ground for a simple, flexible and phylogenetically informative system. More recently, Mac Key (1945a, b, 1966, 1975, 1981, 1988, 2005), taking in account that the wheats consist of three ploidy levels (Sakamura 1918), used very similar principles to those used by Thellung, and suggested a classification that accepted only a small number of taxa at the species level and classified all others at the subspecies rank. Van Slageren (1994), van Slageren and Payne (2013) supported this system.

The biological species approach is particularly relevant to the classification of domesticated plants that are subjected to intensive and dynamic evolution. The gene pool of domesticated wheat has been repeatedly and periodically stirred through recurrent conscious or unconscious selection in the polymorphic fields of traditional farmers and through genetic introductions or migrations, followed by natural or artificial hybridization, in breeder fields (Harlan 1970). Farmers and breeders, sometimes in diverse directions, continuously exerted very strong selection pressures on wheat. The end result is an enormous degree of conspicuous gene pool variation among very closely related forms of domesticated wheat. Moreover, spontaneous hybridizations between wild or between domesticated taxa, as well as between wild and domesticated taxa, followed by recombination and sorting of parental genes, as well as introgression, may lead to continuous variation among taxa, and blur taxonomic markers and specific boundaries. This dynamic evolution is obviously more pronounced in domesticated plants when the farmer imposes rigorous evolutionary processes. Thus, a classification system should account for the dynamic evolutionary processes that bring about genetic changes during the life of a species and affect inter- and intra-specific variation. In this regard, crop taxonomy must be simple and flexible and complex hierarchical classification methods, often used in connection with cultivated plants, must be avoided owing to their inherent rigidity (Mac Key 1981). Consequently, the biological species classification approach (Mayr 1940, 1942; Löve 1964) is most suitable for classification of domesticated plants (Mac Key 1981, 2005).

The biological species classification approach has cytogenetic and evolutionary significance, that is of particular relevance when considering crop breeding and use of various genetic resources for crop improvement. This method corresponds with the primary gene pool (GP-1) approach

developed by Harlan and de Wet (1971)."This gene pool corresponds with the traditional concept of the biological species. Among forms of this gene pool, crossing is easy; hybrids are generally fertile with good chromosomal pairing; gene segregation is approximately normal and gene transfer is generally easy. The biological species almost always includes spontaneous races (wild and/or weedy) as well as cultivated races. The species be divided into two subspecies: subspecies A to include the cultivated races, and subspecies B to include the spontaneous races" (Harlan and de Wet 1971). The genus *Triticum* contains four different GP-1 groups based around einkorn wheat (*T. monococcum* and *T. urartu*)), turgidum wheat (*T. turgidum*), timopheevii wheat (*T. timopheevii*), and bread wheat (*T. aestivum* L.) (Harlan and de Wet 1971).

Since this book focuses on genetic and evolutionary aspects of wheat and its relatives, the biological species concept is used throughout the book. Consequently, and following the classifications of Mac Key (1954a,b, 1966, 1975, 1981, 2005), Van Slageren (1994), the inter-fertile wild and domesticated forms of *Triticum* are included as sub-species in a single species, despite their disparate selection pressures and subsequently different evolutionary directions. Such differing trends between wild and domesticated plants, do not necessitate different basic principles in taxonomy.

7.3   Designation and Classification of Synthetic Allopolyploids

The absence of uniform principles for designating natural and synthetic allopolyploids introduces confusion. Gupta and Baum (1986) proposed to restrict the classification of synthetic taxa to primary amphiploids and F_1 hybrids. They claimed that both the artificially synthesized taxa such as triticale, designated as x *Triticosecale*, and the naturally evolved bread wheat, are established biological units and should be treated as botanical taxa. Van Slageren (1994) disagreed and suggested to designate hybrids or allopolyploids by 'naturally evolved' versus 'artificially created' taxa. When applying this principle, bread and tetraploid wheats remain under the botanical taxon name *Triticum* L, while the triticales are under x *Triticosecale* Wittm. ex A. Camus. Consequently, van Slageren (1994) suggested that synthetic allopolyploids should be excluded from the genus *Triticum* and comprise a separate genus.

In contrast, Mac Key (1968, 1981, 1989, 2005) reckoned that natural taxonomic treatment of the triticales illustrates the need to give *Triticum* a type of open status. Consequently, he suggested that well-established, man-made inter-generic and inter-specific hybrids, e.g., triticale and *T. kiharae*, (2n = 6x = 42; genome GGAADD) must be regarded in the same manner as their natural counterparts

(Mac Key 1981). Thus, he claimed no fundamental evolutionary differences between *Triticum* and *Triticosecale* Wittm. ex A. Camus (Mac Key 2005). *Triticum* and triticale are both annual and autogamous and have the basic morphological design that comes from the same progenitor, diploid wheat. Intergeneric crosses occur spontaneously and artificially between these two taxa in cultivation, resulting in an ongoing convergence of the two crops. These activities render it unrealistic to keep them apart by a generic border (Mac Key 2005). Thus, if *Triticum* is not treated as a nothogenus (a genus denoting a hybrid origin), neither is *Titicosecale* (see ICBN H.3, Note 1) (Mac Key 2005).

Mac Key's (1981, 2005) arguments for inclusion of the synthetic species of triticale in the genus *Triticum*, are based on the fact that (i) some strains of *Triticum* cross more easily with *Secale* than with *Aegilops*, (ii) the basic morphology of triticale is similar to that of *Triticum monococcum*, and (iii) both wheat and triticale crops undergo convergent evolution under domestication. Therefore, it seems most reasonable to classify them in the same generus. Consequently, Mac Key (1981, 2005) suggested to include the three allopolyploid species of triticale in section Triticosecale (Wittm. ex Camus) Mac Key, under the *Triticum* as *T. semisecale* Mac Key, (Subtriticale; $2n = 4x = 28$; genome AARR), *T. neoblaringhemii* (Wittm. ex Camus) Mac Key (Triticale; $2n = 6x = 42$; genome BBAARR), and *T. rimpaui* (Wittm.) Mac Key, (Eutriticale; $2n = 8x = 56$; genome BBAADDRR).

The taxonomic classification of the wheats used in this book concerns only with naturally-formed allopolyploids and does not refer to that of synthetic allopolyploids. The objective of the book is mainly to describe evolutionary events occurring in nature and under cultivation. Description of man-driven evolution will be discussed in the chapter on evolution under cultivation (Chap. 13).

7.4 The Taxonomic Relationships Between *Aegilops* and *Triticum*

The question whether the two genera, *Aegilops* and *Triticum*, should be united or kept separate has been addressed by several taxonomists and cytogeneticists (reviewed in van Slageren 1994). The two genera were treated as separate genera by Linnaeus (1753). However, Godron (1845), Grenier and Godron (1856) were the first to classify *Aegilops* as a section of *Triticum*. Several eminent botanists at the end of the 19th and the beginning of the twentieth centuries, e.g., Hackel (1887), Ascherson and Graebner (1898–1902), accepted this unification of the two genera. The tendency to unite the two genera became more widely accepted [e.g., Bowden (1959), Chennaveeraiah (1960), Yamane and Kawahara (2005)] when genome analyses showed that all the allopolyploid species of *Triticum* contain *Aegilops*

genome(s). Moreover, Krause (1898), recognizing the presence of hybrids between *Aegilops* and *Triticum* and their subsequent hybridization with *Elymus*, *Secale,* and *Hordeum*, suggested classification of the five genera as subspecies of a single species, *Frumentum*. In this approach, he regarded *Aegilops* and *Triticum* as two separate subspecies. However, in most floras, the two genera remained distinct, but in several cases, the species of section Sitopsis of *Aegilops,* were included in *Triticum* because of this section resemblance to wild wheats.

Cytogenetic studies, mainly genome analysis, performed by Kihara and his associates (e.g., see review of Lilienfeld 1951; Kihara 1954; Kihara and Tanaka 1970), Sax (1921a, b, 1927), and others, led researchers to consider *Aegilops* and *Triticum* as two different genera, despite their very close relation and discovery of genomes of *Aegilops* in allopolyploid wheat. These studies led Stebbins (1956) to propose merging *Aegilops* and *Triticum* into one genus, due to the many genetic links between the two. Stebbins (1956) noted that only diploid wheat does not contain chromosomes derived from *Aegilops*. Bowden (1959) made the same proposal based on cytogenetic evidence (Table 7.1). Bowden (1959) accepted Thellung's classification but ignored new discoveries of the cytogenetic relationship between the newly discovered species, namely, the wild and domesticated *T. timopheevii* and the hexaploid *T. zhukovskyi*. He included the wild and domesticated forms of *T. timopheevii* and hexaploid *T. zhukovskyi* in *T. turgidum*, despite their genetic isolation from the latter. Bowden's scheme adhered to the nomenclatural rule classifying inter-generic hybrids together, e.g., allopolyploid *Triticum* species that are of hybrid origin and their progenitors are found in two different genera. However, it did not consider the numerous allopolyploids of the tribe Triticeae that form a network of genetic inter-connections. If the nomenclatural rule for inter-generic hybrids would be strictly followed, the entire tribe would have to be lumped into a single genus (Stebbins 1956; Mac Key 1966, 1975, 1981, 2005; Runemark and Heneen 1968). No realistic taxonomist is willing to be that compliant.

The merge of *Aegilops* and *Triticum* was later adopted by Morris and Sears (1967), with some modifications [e.g., separation of the *timopheevii* wheats (genome GA) from *T. turgidum* (genome BA)], and was used by several cytogeneticists, e.g., Kimber and Feldman (1987). However, taxonomists as Mac Key (1966, 1968, 1981, 2005), Hammer (1980), Gupta and Baum (1989), van Slageren (1994), justified the separation of the two genera on the basis of morphological and genetic differences. The unification of *Aegilops* sensu lato with *Triticum* was also criticized by Miller (1987).

Another classification approach, based on karyomorphological studies, includes only the Sitopsis section of *Aegilops*, namely, the species *Ae. speltoides*, *Ae. bicornis*,

Table 7.1 Subdivision in wheat species: *Triticum. L.* em. Bowden (According to Bowden 1959)[a]

Wheat species	Goat grass species
I. Diploid species *T. monococcum* (A^b)	*T. bicorne* (S^b) *T. speltoides* (S) *T. comosum* (M) *T. uniaristatum* (M^u) *T. longissimum* (S^l) *T. umbellulatum* (C^u) *T. tripsacoides* (M^t) *T. dichasians* (C) *T. aegilops* (D)
IIa. Allotetraploid wheats *T. turgidum* ($A^uB/A^uG/A^uA^bG$)	
IIb. Allohexaploid wheats T. x *aestivum* (A^uBD)	
IIc. Allopolyploid species of interspecific origin	*T. ovatum* (C^uM^o) *T. triaristatum* (C^uM^t) *T. kotschyi* (C^uS^l) *T. triunciale* (C^uC) *T. cylindricum* (CD) *T. macrochaetum* (C^uM^b) *T. crassum* (DM^{cr}/DD^2M^{cr}) *T. turcomanicum* *T. juvenale* (DC^uM^j) *T. ventricosum* (DM^v)
III. Other artificial and natural interspecific hybrids	

[a] Bowden's classification was rejected by most taxonomists and as such not followed in the present text

Ae. longissima, *Ae. sharonensis*, and *Ae. searsii*, in *Triticum* (Chennaveeraiah 1960). This researcher, inspired by the close resemblance of the Sitopsis species to diploid wheats, assumed that, aside from this section, *Aegilops* is very distinct from *Triticum*. He transferred the Sitopsis section of *Aegilops* to *Triticum* on the basis of the presence of both a rudimentary keel on the glume, and a sub-median centromere on the chromosomes of diploid wheats and the Sitopsis species. Chennaveeraiah's classification has not been accepted for the obvious reason that it is illogical to transfer only one of the *Aegilops* genome donors (the B genome) to *Triticum*, without the second donor (the D genome of *Ae. tauschii*) (Mac Key 1968). In addition, there is no rudimentary keel on the glumes of the Sitopsis species, and Chennaveeraiah must have mistaken the ridge, which sometimes develops on the outer surface of the glumes of *Ae. speltoides*, for keels, as they can be found in the same location as the genuine keel in all *Triticum* species (van Slageren 1994). Thus, Chennaveeraiah's classification was rejected by Hammer (1980), Clayton and Renvoize (1986), van Slageren (1994), who transferred Sitopsis back from *Triticum* to *Aegilops*.

Maintenance of a generic separation between *Aegilops* and *Triticum*, as suggested by Zhukovsky (1928), Eig

(1929), Kihara (1954), Mac Key (1966, 1968, 1975, 1981, 2005), Kihara and Tanaka (1970), Hammer (1980), Miller (1987), Clayton and Renvoize (1986), Gupta and Baum (1989), van Slageren (1994), was rooted in morphological and genetic differences. *Aegilops* is morphologically highly distinct from *Triticum*, with rounded rather than keeled glumes.

Baum (1977, 1978a, b) analyzed the variation in the Triticeae by various numerical techniques, using as many morphological characters as possible. The results led him to maintain *Aegilops* in its traditional sense. Similarly, Schultze-Motel and Meyer (1981) investigated the *Aegilops–Triticum* group using a number of morphological characters, and in almost all cases, *Aegilops* was clearly distinct from *Triticum*.

Löve (1982, 1984) suggested classifying the wheat group on the basis of genomic similarities, and conditioned generic status to a genomically homogeneous taxon. His classification recognized genomically defined genera in *Aegilops* 13 and in *Triticum* 3. However, the high incidence of inter-generic and inter-specific hybridizations in this group indicate that Löve's scheme is not adequately sound; it is a single-character classification method, grouping genome types that are generally not connected via morphologically recognizable units and ignores taxonomy and nomenclature (Gupta and Baum 1986). Subsequently, the approach leads to too many monotypic genera, and is unstable as genome designation and combinations continue to change (Baum et al. 1987; Kellogg 1989).

In view of the various options, van Slageren concluded that the best and most practical approach is to maintain *Aegilops* and *Triticum* as separate genera. Van Slageren approach is followed in this book.

References

Ascherson P, Graebner P (1898–1902) Synopsis der mitteleuropäischen Flora, Bd. 2, Abt. 1. Wilhelm Engelmann, Leipzig

Baum BR (1977) Taxonomy of the tribe Triticeae (Poaceae) using various numerical techniques. I. Historical perspectives, data, accumulation and character analysis. Can J Bot 55:1712–1740

Baum BR (1978a) Taxonomy of the tribe Triticeae (Poaceae) using various numerical techniques II. Classification. Can J Bot 56:27–56

Baum BR (1978b) Taxonomy of the tribe Triticeae (Poaceae) using various numerical techniques. III. Synoptic key to the genera and synopses. Can J Bot 56:374–385

Baum BR, Estes JR, Gupta PK (1987) Assessment of the genomic system of classification in the Triticeae. Amer J Bot 74:1388–1395

Bowden MW (1959) The taxonomy and nomenclature of the wheats, barleys and ryes and their wild relatives. Can J Bot 37:657–684

Brickell CD, Brickell CD, Alexander C, David JC, Hetterscheid WLA, Leslie AC, Malecot V, Jin X (2009) International code of nomenclature for cultivated plants, 8th edn. Scripta Hort 10:1–204

Chennaveeraiah MS (1960) Karyomorphologic and cytotaxonomic studies in *Aegilops*. Acta Horti Gotoburgensis 23:85–178

Clayton SD, Renvoize SA (1986) Genera Graminum, grasses of the world, distributed for royal botanic gardens, Kew bulletin. Additional series, 13, Kew, London, pp 1–389

Eig A (1929) Monographisch-Kritische Ubersicht der Gattung Aegilops. Reprium nov. Spec Regni Veg 55:1–288

Godron DA (1845) De la fecondation naturelle et artificelle des Aegilops par le Triticum. Ann Sci Nat Bot 4(2):215–222

Grenier J, Godron DA (1855–56) Flore de France, vol 3 (Triticum on pp 598–603 from 1856)

Gupta PK, Baum BR (1986) Nomenclature and related taxonomic issues in wheats, triticales and some of their wild relatives. Taxon 35:144–149

Gupta PK, Baum BR (1989) Stable classification and nomenclature in the Triticeae: desirability, limitations and prospects. Euphytica 41:191–197

Hackel E (1887) Triticum L. In: Engler A, Prantl K (eds) Die natuerlichen Pflanzenfamilien nebst ihren Gattungen und wichtigeren Arten insbesondere den Nutzpflanzen. Wilhelm Engelmann, Leipzig

Hammer K (1980) Vorarbeiten zur monographischen Darstellung von Wildpflanzensortimenten: Aegilops L. Kulturpflanze 28:33–180

Hammer K, Filatenko AA, Pistrick K (2011) Taxonomic remarks on Triticum L. and ×Triticosecale Wittm. Genet Resour Crop Evol 58:3–10

Harlan JR (1970) Evolution of cultivated plants. In: Frankel OH, Bennet E (eds) Genetic resources in plants—their exploration and conservation. Blackwell Scientific Publications, Oxford, pp 19–32

Harlan JR, de Wet JMJ (1971) Towards a rational classification of cultivated Plants. Taxon 20:509–517

Heslop-Harrison J (1963) Species concept: theoretical and practical aspects. In: Swain T (ed) Chemical plant taxonomy. Academic Press, London, pp 17–40

Kellogg EA (1989) Comments on genomic genera in the Triticeae (Poaceae). Amer J Bot 76:796–805

Kihara H (1954) Considerations on the evolution and distribution of Aegilops species based on the analyzer-method. Cytologia 19:336–357

Kihara H, Tanaka M (1970) Addendum to the classification of the genus Aegilopsby means of genome-analysis. Wheat Inf Serv 30:1–2

Kimber G, Feldman M (1987) Wild wheats: an introduction. Special Report 353. College of Agriculture, Columbia, Missouri, USA, pp 1–142

Krause EHL (1898) Floristische Notizen II. Gräser. 1. Zur Systematik Und Synonymik Botanisches Centralblatt (jena) 73:337–343

Lilienfeld F (1951) H. Kihara: Genome analysis in Triticum and Aegilops X. concluding review. Cytologia 16:101–123

Linnaeus C (1753) Species Plantarum, Tomus I, May 1753: i–xii, 1–560; Tomus II, Aug 1753: 561–1200, plus indexes and addenda, 1201–1231

Löve Á (1964) The biological species concept and its evolutionary structure. Taxon 13:33–45

Löve Á (1982) Generic evolution of the wheatgrasses. BioI Zentralbl 101:199–212

Löve Á (1984) Conspectus of the Triticeae. Feddes Repert 95:425–521

Mac Key J (1954a) The taxonomy of hexaploid wheat. Svensk Bot Tidskr 48:579–590

Mac Key J (1954b) Neutron and X-ray experiments in wheat and a revision of the speltoid problem. Hereditas 40:65–180

Mac Key J (1966) Species relationship in Triticum. In: Proc 2nd international wheat genetic symposium, vol 2. Lund, Sweden, Hereditas, Suppl, pp 237–276

Mac Key J (1968) Relationships in the Triticinae. In: Finlay KW, Shepherd KW (eds) Proceedings of 3rd international wheat genetic symposium. Australian Academy Science, Canberra, pp 39–50

Mac Key J (1975) The boundaries and subdivision of the genus Triticum. Proc 12th Inter Botl Congress, Leningrad: Abstract 2:509 (complete manuscript available at Vavilov Institute of Plant Industry, Leningrad, pp. 1–23

Mac Key J (1981) Comments on the basic principles of crop taxonomy. Kulturpflanze 29(S):199–207

Mac Key J (1988) A plant breeder's perspective on taxonomy of cultivated plants. Biol Zentralblatt 107:369–379

Mac Key J (1989) Seed dormancy in wild and weedy relatives of cereals. In: Derera NF (ed) Preharvest field sprouting in cereals. CRC Press, Boca Raton, pp 15–25

Mac Key J (2005) Wheat: its concept, evolution and taxonomy. In: Royo C et al. (eds) Durum wheat, current approaches, future strategies, vol 1. CRC Press, Boca Raton, pp 3–61

Mayr E (1940) Speciation phenomena in birds. Amer Nat 74:249–278

Mayr E (1942) Systematics and the origin of species. Columbia Univ Press, New York

Mayr E (1963) Animal species and evolution. Mass, Cambridge

McNeill J, Barrie FR, Burdett HM, Demoulin V, Hawksworth DL Marhold K, Nicolson DH, Prado J, Silva PC, Skog JE, Wiersema JH, Turland NJ (eds) (2006) International code of botanical nomenclature (Vienna code), adopted by the 7th international botanical congress. http://ibot.sav.sk/icbn/main.htm

Miller TE (1987) Systematics and evolution. In: Lupton FGH (ed) Wheat breeding. Its scientific basis. Chapman & Hall, London, pp 1–30

Morris R, Sears ER (1967) The cytogenetics of wheat and its relatives. In: Quisenberry KS, Reitz LP (eds) Wheat and wheat improvement. American Society of Agronomy, Madison, Wisconsin, USA, pp 19–87

Runemark H, Heneen WK (1968) Elymus and Agropyron, a problem of generic delimitation. Bot Notiser 121:51–79

Sax K (1921a) Chromosome relationships in wheat. Science 54:413–415

Sax K (1921b) Sterility in wheat hybrids. I. Sterility relationships and endosperm development. Genetics 6:399–416

Sax K (1927) Chromosome behavior in Triticum hybrids. Verhandlungen des V Int Kongresses für Vererbungswissenchaft 2:1267–1284

Sakamura T (1918) Kurze mitteilung über die chromosomenzahalen und die verwandtschaftsverhältnisse der Triticum Arten. Bot Mag 32:151–154

Stebbins GL (1956) Taxonomy and evolution of genera with special reference to the family Gramineae. Evolution 10:235–245

Thellung A (1918a) Neuere Wege und Ziele der botanischen Systematik, erläutert am Beispiele unserer Getreidearten. Naturw Wochenschr Neue Folge 17(449–458):465–474

Thellung A (1918b) Neuere Wege und Ziele der botanischen Systematik, erläutert am Beispiele unserer Getreidearten. Mitt Naturwiss Ges Winterthur 12:109–152

van Slageren MW (1994) Wild wheats: a monograph of Aegilops L. and Amblyopyrum (Jaub. and Spach) Eig (Poaceae). Agricultural University, Wageningen, The Netherlands

van Slageren MW, Payne T (2013) Concepts and nomenclature of the Farro wheats, with special reference to Emmer, Triticum turgidum subsp. dicoccum (Poaceae). Kew Bull 68:477–494

Yamane K, Kawahara T (2005) Intra- and interspecific phylogenetic relationships among diploid Triticum-Aegilops species (Poaceae) based on base-pair substitutions, indels, and microsatellites in chloroplast noncoding sequences. Am J Bot 92:1887–1898

Zhukovsky PM (1928) A critical systematical survey of the species of the genus Aegilops L. Bull Appl Bot Genet and Pl Breed 18: 417–609

8.1 Separation of *Amblyopyrum* from *Aegilops*

Amblyopyrum is a monotypic genus of the sub-tribe Triticineae represented by the species *A. muticum* (Jaub. and Spach) Eig that was separated from *Aegilops* by Eig (1929b). The name *Amblyopyrum* is derived from the Greek 'amblyos' (=blunt) and 'pyros' (=wheat) (Van Slageren 1994). The taxonomic rank of this taxon is controversial. While describing and designating this species as *Aegilops mutica*, Boissier (1844a, b) assumed it resembles *Ae. Aucheri* Boiss. (a synonym of *Ae. speltoides* ssp. *speltoides* Tausch) in morphology, and therefore, regarded it as a primitive species of *Aegilops*. Yet, Boissier (1844a, b) considered it to be an intermediate between *Aegilops* and several species of *Agropyron* (currently *Elymus* species). Eig (1929a) in his detailed and comprehensive monograph on the genus *Aegilops*, placed *Ae. mutica* too within the genus *Aegilops* but very distant from the rest of the species; he classified it as subgenus *Amblyopyrum*, whereas all the other species were placed under subgenus *Eu-Aegilops*. The inclusion of *Ae. mutica* within the genus *Aegilops* was accepted by taxonomists such as Zhukovsky (1928), Hammer (1980), and Clayton and Renvoize (1986). However, later, after studying species of *Elymus* (formerly *Agropyron*), particularly *E. elongatus*, and realizing that *Aegilops mutica* is an intermediate in several basic morphological features between *Aegilops* and several species of *Elymus*, Eig (1929b) decided to separate it from *Aegilops* as a monotypic genus, *Amblyopyrum*, that includes the species *A. muticum*. This decision was based on the following morphological traits of *A. muticum*: long (up to 30 cm) linear awnless spikes with many multi-floret spikelets, without any rudimentary spikelets. Glumes widest at apex with divergent venation. The rachis is fragile and after ripening disarticulates into single spikelets that fall with the rachis internode below the spikelets (wedge type disarticulation), and fragile rachillae (the axis of the spikelet) that disarticulate into florets that fall separately especially in the upper part of each spikelet (floret

type disarticulation). These morphological features of *A. muticum*, particularly the floret type of disarticulation, are characteristics of the older Arctic-temperate group and especially to species of *Elymus*, thus corroborating its intermediate position between *Elymus* and *Aegilops*. It is assumed that *A. muticum* is morphologically similar to the putative ancestor of the wheat group while maintaining several features of *Elymus*. By separating this taxon, Eig (1929b) emphasized its intermediate position between *Aegilops* and *Elymus*. Numerical analysis (Baum 1977, 1978a, b; Schultze-Motel and Meyer 1981) indicated the close relationship of *Amblyopyrum* with *Aegilops* and *Triticum*, while, especially in Baum's work, confirming the morphological differences at the same time.

The morphological contrast between *Amblyopyrum muticum* and all other *Aegilops* species brought taxonomists, e.g., Bor (1968), Baum (1978a, b), Schultze-Motel and Meyer (1981), Löve (1982) Watson et al. 1985; Davis et al. 1988; Tzvelev (1989), and van Slageren (1994), to agree with Eig's (1929b) separation. Van Slageren (1994) pointed out that not only the morphological differences but also the early separation of *Amblyopyrum* from the *Aegilops* lineage requires an independent generic status.

8.2 Morphological and Geographical Notes

Amblyopyrum muticum (Jaub. and Spach) Eig (Syn.: *Aegilops mutica* Boiss.; *Triticum muticum* (Boiss.) Hack. in Fraser.; *Aegilops tripsacoides* Jaub. and Spach.; *Triticum tripsacoides* Bowden) is annual, 50–80 cm high (excluding spikes), culms usually few, mostly upright, and are sparsely foliated. Leaves glaucous-green, 8–25 cm long; Ligule short, membranous, up to 1 mm long. Spikes 15–35 (sometimes up to 45) cm long, thin, cylindrical, one-rowed, with many spikelets (15–20, sometimes more), awnless. Rachis fragile and disarticulates at maturity into individual spikelets each with a rachis segment below it (wedge type dispersal unit), and fragile rachillae (the axis of the spikelet)

M. Feldman and A. A. Levy, *Wheat Evolution and Domestication*, https://doi.org/10.1007/978-3-031-30175-9_8

that disarticulate into florets that fall separately especially in the upper part of each spikelet (floret type disarticulation). Spikelets 8–15 mm long, linear to linear-elliptical, equally large or slowly decreasing in size to the tip of the spike, usually shorter than the adjacent rachis segment and diverging from it. The apical spikelet is at right angle to the lateral spikelets. Florets 5–9, the lowest 2–3 fertile. Glumes 7–10 mm long, trapezoid with the upper edge being larger than the base. Upper margin with 2–4 short blunt teeth separated by notches. Lemmas 7–10 mm long, leathery, about the same length as the glumes. Anthers 4–5 mm long. Caryopsis 4 mm long, adherent to lemma and palea (Fig. 8.1).

A variable species, however, only awnless forms are known. Variation exists in the length and width of the spike and in number of spikelets. Glumes are either hairy or glabrous, and floral parts and rachis segregates for red, black and colorless as well as for width. Different morphological forms grow in mixed stands and interbreed freely.

Amblyopyrum muticum contains two varieties: var. *muticum* and var. *loliaceum*. In var. *muticum.* the glumes and apical parts of the lemmas are covered with short, stiff hairs while in var. *loliaceum* glumes and lemmas are glabrous.

Amblyopyrum muticum is native to west Asia (Anatolian Plateau, southeastern Turkey, Turkish Armenia, Caucasus, western Iran and north-east Syria). It may also occur in northern Iraq. Grows on sandy, stony or steppical-grey soils in abandoned fields, edges of wheat fields or roadsides. In the center of its distribution (Anatolian Plateau) it forms dense stands, but in other sites it occurs more sporadically in

Fig. 8.1 A spike and part of a spike of *Amblyopyrum muticum* (Jaub. and Spach) Eig

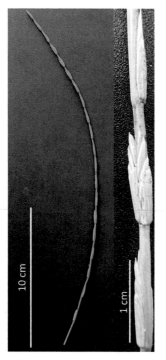

wadis and lower slopes. Alt: 700–1200 m asl. Its distribution in this area is considered as the probable center of origin of the group, and may indicate its primitive status (Hammer 1980).

Amblyopyrum muticum has a relatively limited distribution in the central region of the distribution of the wheat group (the genera *Amblyopyrum, Aegilops* and *Triticum*). It is a sub-steppical (Irano-Turanian) element; restricted mainly to steppical areas with 300 mm annual rainfall. It grows in many secondary, disturbed habitats. Sympatric with the following species: *Ae. speltoides, Ae. caudata, Ae. umbellulata, T. monococcum* ssp. *aegilopoides, T. urartu, Ae. geniculata Ae. biuncialis, Ae. neglecta, Ae. columnaris, Ae. triuncialis,* and *Ae. cylindrica*. Allopatric with *Ae. tauschii, T. timopheevii* ssp. *armeniacum* and. *T. turgidum* ssp. *dicoccoides.*

Othmemi et al. (2019) reviewed information showing that *A. muticum* carries several useful genes that, upon transfer to bread wheat, may improve its performance. It has been reported that this species tolerates environmental stresses (Iefimenko et al. 2015) and is resistant to powdery mildew (Eser 1998), and leaf rust (Dundas et al. 2015). In recent years, attempts are made to transfer some of these useful genes to bread wheat (King et al. 2017).

8.3 Cytology, Cytogenetics and Evolution

Amblyopyrum muticum is a diploid species (2n = 2x = 14) whose genome is designated T (Kimber and Tsunewaki 1988; Dvorak 1998) (formerly Mt; Kihara and Lilienfeld 1935). Its nuclear DNA amount is relatively small, i.e., 1C DNA is 5.82 pg (Eilam et al. 2007), similar to that of diploid *Elymus elongatus, Aegilops speltoides* and *Ae. uniaristata,* but significantly smaller than that of the other Sitopsis species and larger than that of *Ae. tauschii, Ae. caudata, Ae. umbellulata,* and *Ae. comosa* (Eilam et al. 2007). *A. muticum* has a symmetric karyotype with large metacentric or submetacentric chromosomes with two large satellites (Chennaveeraiah 1960). By means of in situ hybridization with pTa71 (18S-26S rDNA) and pTa794 (5S rDNA) DNA probes from *T. aestivum,* Badaeva et al. (1996a, b) determined the distribution of the 18S-5.8S-26S (18S-26S) and 5S ribosomal DNA gene families on chromosomes of *A. muticum.* It was found that *A. muticum* has major NOR loci, having the 18S-26S rDNA gene family on the short arm of chromosomes of groups 1 and 6 T, and a moderate-sized NOR locus was observed on the long arm of chromosome 7 T. Two 5S rDNA loci were observed, one on the short arm distal to the NOR locus of chromosome 1 T and a small locus on the short arm of chromosome 5 T. The distribution of the major NORs in *A. muticum* is as in *Ae. spelto*ides and that of the 5S rDNA is similar to that of *Ae. bicornis.*

Both, *A. muticum* and *Aegilops speltoides are* allogamous species; *A. muticum* is obligatory allogamous, namely, self-incompatible (Ohta 1990), whereas *Ae. speltoides* is facultative allogamous and, at least several accessions, produce seeds upon bagging (Feldman M, unpublished). Both species have genes that promote pairing between homoeologous chromosomes in hybrids with allopolyploid wheat by counteracting the effect of the homoeologous-pairing suppressor of wheat, *Ph1* (Riley 1960, 1966; Dover and Riley 1972a). There is high pairing at meiosis in hybrids between *A. muticum* and tetraploid wheat (Kihara and Lilienfeld 1935), and hexaploid wheat (Riley 1966). The high pairing is presumably due to the occurrence of homoeologous pairing through the suppression of the activity of *Ph1* of allopolyploid wheats.

Dover and Riley (1972a) found that there is genetic variation in *Ae. mutica* affecting homoeologous chromosomal pairing at meiosis. From their study of hybrids between *T. aestivum* and different accessions of *Ae. mutica*, they concluded that *A. muticum* has two loci with alternative alleles that affect homoeologous pairing in hybrids with *T. aestivum*. Both the low pairing alleles, when together, condition little or no homoeologous pairing. The high pairing alleles are *epistatic* to the low-pairing ones at the other locus.

Two such genes inducing homoeologous pairing in hybrids with allopolyploid wheat containing the *Ph1* gene, were also found in *Ae. speltoides* (Dvorak 1972). In *Ae. speltoides* they were allocated to chromosomes 3S and 7S (Dvorak et al. 2006). In addition, a QTL with a minor effect was allocated to the short arm of chromosome 5S (Dvorak et al. 2006). The *speltoides* genes did not affect the level of pairing in the inter-specific diploid hybrids *Ae. speltoides* x *Ae. tauschii* and *Ae. speltoides* x *Ae. caudata* (Chen and Dvorak 1984). In contrast, studies of meiotic chromosomal paring in hybrids between *A. muticum* and diploid species of *Aegilops* and *Triticum* show relatively high paring with almost each of them, presumably due to the promotion of pairing by the *muticum* genes (Ohta 1990, 1991).

Kihara and Lilienfeld (1935) reported a mode of 7 bivalents (a range of 3 to 7 bivalents of which 3 to 4 were ring bivalents; mean bivalents 5.86) in F_1 hybrids of *Ae. comosa* x *Ae. mutica*. These data indicated to Kihara and Lilienfeld that *Ae. mutica* had close genomic homology with *Ae. comosa* and other M-genome diploids. Consequently, they assigned the symbol Mt to the genome of *Ae. mutica* to distinguish this species from others having M or M-derivatives ($M°$, M^t, M^u, M^{cr}, M^v, M^b, and M^c), because *Ae. mutica* has a distinctive spike morphology (Kihara 1947, 1954; Lilienfeld 1951). However, Chennaveeraiah (1960) found that the karyotype of *A. muticum* is very similar to that of *Ae. speltoides* and differ from the karyotypes of the M-genome species. In this regard, Jones and Majisu (1968) reported that chromosome pairing in *Ae. tauschii* x *A.*

muticum hybrids was almost regular and exceeded the pairing in hybrids between *A. muticum* and Sitopsis species of *Aegilops* as well as between *A. muticum* and diploid *Triticum* species. They assumed that the chromosomes of *A. muticum* appear to have considerable homoeology with the D genome of *Ae. tauschii*.

Riley (1966) crossed *Ae. speltoides*, *Ae. longissima*, *Ae. caudata* and the wild and domesticated forms of *T. monococcum* with *A. muticum* and found high chromosome pairing in meiosis of all the F_1 hybrids. He suggested that *A. muticum* is cytogenetically close to *Ae. speltoides,* not only because of the high pairing, but also for the reasons that no chromosomal rearrangements were found between the two species and the similarity existing in the two species in pairing control as well as in their karyotypes (Chennaveeraiah 1960). Riley (1966) pointed out that the absence of any translocation difference between *Ae. speltoides* and *A. muticum* confirms the phylogenetic proximity that is indicated by their similarities in karyotype and pairing control of the two species. However, pairing in the *Ae. speltoides* x *A. muticum* hybrids is not so high as that in hybrids between *Ae. speltoides* and other members of the Sitopsis section of *Aegilops* (Kimber 1961).

Ohta (1990, 1991) performed a very detailed study on the cytogenetic relationships of *A. muticum* and diploids of the wheat group. He crossed *A. muticum* with 11 diploid species of the genera *Aegilops* and *Triticum*. The crossability was good and F_1 seeds were successfully obtained in all reciprocal cross combinations. However, in combinations where *A. muticum* was the female parent the seeds did not germinate. The cross *Ae. tauschii* x *A. muticum* yielded shriveled seeds while those of the reciprocal combination germinated regularly. In the cross *Ae. searsii* x *A. muticum*, the F_1 seeds did not germinate in the two reciprocal combinations. Previously, Jones and Majisu (1968) obtained similar results, i.e., shriveled seeds which did not germinate in the cross *Ae. tauschii* x *A. muticum* when *Ae. tauschii* was the female parent and normal size seeds that germinated in the reciprocal cross. They concluded that the difference in germination between the reciprocal crosses involving *Ae. tauschii* and *A. muticum* was not attributed to cytoplasmic difference between the parental species. Dhaliwal (1977) suggested that the difference in germination between reciprocal crosses might be attributed to different ratios of the parental genes in the triploid endosperm.

Most of the F_1 plants obtained by Ohta (1990) from the inter-generic crosses involving *A. muticum* and diploids of the wheat group grew normally and vigorously. From morphological features of the hybrid spikes, Ohta (1990) concluded that *A. muticum* is most similar to *Ae. speltoides*. In contrast to the F_1 hybrids between *A. muticum* and the diploid species of the wheat group that were completely sterile (except plants that formed unreduced gametes), partially fertile F_1 hybrids were obtained from the crosses between *Ae. speltoides* and *A. muticum* (Ohta 1990). The fact that several functional male and female gametes with seven chromosomes were produced, in spite of a high frequency of inter-genomic recombination, clearly indicates that the two parental species, *Ae. speltoides* and *A. muticum*, are very closely related to each other. In the Anatolian Plateau these two species sometimes grow sympatrically. However, no natural hybrids between these two species were reported though they are out-crossing species (Ohta 1990).

Although they exhibit a high level of chromosome pairing at meiosis, the complete or partial sterility of the F_1 hybrids can result from cryptic structural hybridity (chromosomal sterility) or from genes that cause hybrid incompatibility (genic sterility). The fact that pollen grains containing unreduced chromosome complements are viable and functional, suggests that the sterility of the F_1 hybrids is not genic but chromosomal.

In most F_1 hybrids Ohta (1990, 1991) reported very high meiotic pairing (Table 8.1). Only in hybrids from the crosses of *Ae. caudata, Ae. uniaristata* and *Ae. umbellulata* x *A. muticum* there was a lower level of chromosome pairing than in the other hybrids. The relatively high pairing in most hybrids does not indicate very close relationship between the genome of *A. muticum* and the genomes of the different diploid species of *Aegilops* and *Triticum* that have diverged considerably from one another (Kihara 1954). In accord with the above, Kimber (1982), based on his numerical analysis of chromosome pairing in meiosis of F_1 hybrids between five allopolyploid *Aegilops* species and *A. muticum*, decided that the genome of *A. muticum* is non-homologous to the A, B, and D subgenomes of hexaploid wheat nor with U genomes of *Ae. umbellulata* and with the M-genome species, i.e., *Ae. comosa* and *Ae. uniaristata*. Consequently, Kimber and Tsunewaki (1988) while changing the two letter designations of the genome symbols in some *Aegilops* species to single capital letters, proposed the genome symbol T for *A. muticum* to distinguish it from the M-genome group and from the other diploids of the wheat group.

Amblyopyrum muticum and *Ae. speltoides* are the only species that contain B-chromosomes [Mochizuki (1957, 1960) in *muticum* and Simchen et al. (1971) in *speltoides*]. The Bs in *A. muticum* are euchromatic, metacentric and smaller than the A chromosomes, and their number in different individuals ranged from one to five (Mochizuki 1960). The Bs, in both *A. muticum* and *Ae. speltoides*, were stably found in the shoot apices and PMCs, while they were almost entirely absent from the seminal and adventitious roots (Ohta 1995).

The B-chromosomes of *A. muticum* do not affect homologous pairing but suppress homoeologous one in interspecific hybrids (Mochizuki 1964; Dover and Riley 1972b; Vardi and Dover 1972; Ohta and Tanaka 1982; Table 8.2). Studies on meiotic chromosomal pairing in F_1 hybrids

between *A. muticum* and most of the diploid species of *Aegilops* and *Triticum*, containing B- chromosomes of *A. muticum*, showed much reduced chromosomal pairing, i.e., 14 univalents or 12 univalents with a rod-shaped bivalent of A-chromosomes (Vardi and Dover 1972; Ohta and Tanaka 1983; Ohta 1990, 1991; Table 8.1). Only the F_1 hybrids *Ae. speltoides* x *A. muticum* and *Ae. tauschii* x *A. muticum* exhibited somewhat higher pairing (Ohta 1990). However, in hybrids between autotetraploid *Ae. tauschii* and *A. muticum* only seven bivalents were observed, presumably between the *tauschii* chromosomes, indicating that the genomes of *muticum* and *tauschii* are not closely related (Ohta 1990).

Genome analysis in F_1 hybrids between *A. muticum* and ten different species of the wheat group, having zero or two B chromosomes of *A. muticum*, enabled Ohta (1990) to classify the various species of *Aegilops* and *Triticum* into the following three groups, based on their cytogenetic relationships to the genome of *A. muticum*: (i) the genomes of *Ae. caudata, Ae. uniaristata* and *Ae. umbellulata* are distantly related to that of *A. muticum;* (ii) the genomes of *Ae. bicornis, Ae. longissima, Ae. sharonensis, Ae. comosa,* and *T. monococcum* are homoeologous with that of *Ae. mutica;* and (iii) the genomes of *Ae. speltoides* and *Ae. tauschii* are closely related to that of *A. muticum.* Yet, based on plant morphology, chromosome pairing and fertility of the F_1 hybrids, Ohta (1990, 1991) concluded that the genome of *Ae. speltoides* is closer to that of *A. muticum* than the genome of *Ae. tauschii*, and that *A. muticum* and *Ae. speltoides* are the basal species in the group from which all other species have diverged.

Maan (1977) produced alloplasmic lines of bread wheat, *T. aestivum* ssp. *aestivum*, and *T. turgidum* ssp. *durum* in which their nuclear genomes were substituted into the cytoplasm of the diploid species, *Amblyopyrum muticum*, *Ae. comosa* ssp. *heldreichii*. (genome MM), *Ae. uniaristata* (genome NN), and allotetraploid *Ae. geniculata* (genome M^oM^oUU), to identify the M-genome diploid cytoplasm donor of *Ae. geniculata*. Substitution of the ssp. *durum* genome into *Ae. uniaristata* cytoplasm resulted in a large proportion of shriveled inviable seeds. A few plump viable seeds were obtained all of which produced male-sterile plants. The ssp. *aestivum* plants having *Ae. uniaristata* or *A. muticum* cytoplasms were fertile. *A. muticum* was similar to *Ae. geniculata* in the induction of delayed maturity and tall robust growth habit to the ssp. *durum* and ssp. *aestivum* plants. Cytoplasms of the other U- and M-genome diploids, *Ae. umbellulata* and ssp. *heldreichii* had been shown to differ from that of *Ae. geniculata*. Therefore, Maan (1977) concluded that *A. muticum* is the most likely cytoplasm donor to *Ae. geniculata* and its plasma type was designated "T". Panayotov and Gotsov (1973) and Panayotov (1983) found some variation in the effect of the cytoplasm of *A. muticum* on the phenotype of bread wheat and therefore, designated the *muticum* cytoplasm T^2. Thus, *Ae mutica*,

Table 8.1 Mean chromosomal pairing at first meiotic metaphase of F_1 hybrids between *Amblyopyrum muticum* (without B chromosomes) and species of *Triticum* and *Aegilops*

Hybrid combination	Genome	Univalents	Bivalents			Multivalents			Reference
			Rod	Ring	Total	III	IV	V	
Ssp. *monococcum* x *A. muticum*	A^mT	2.52	–	–	4.96	0.28	0.18	–	Riley (1966)
Ssp. *aegilopoides* x *A. muticum*	A^mT	3.70	–	–	4.88	0.10	0.06	–	
Ssp. *monococcum* x *A. muticum*	A^mT	0.63	2.70	2.83	5.53	0.10	0.50	–	Ohta (1990)
Ssp. *monococcum* x *A. muticum*	A^mT.	1.20	2.58	2.48	5.06	0.28	0.46	–	
Ssp. *monococcum* x *A. muticum*	A^mT	2.03	2.80	2.23	5.03	0.23	0.30	–	
Ssp. *dicoccoides* x *A. muticum*	ABT	6.52	–	–	4.30	1.88	0.06	–	Riley (1966)
Ssp. *durum* x *A. muticum*	ABT	6.38	–	–	4.68	1.70	0.04	–	
Ssp. *aestivum* x *A. muticum*	ABDT	5.32	–	–	5.14	2.06	1.52	–	
Ae. speltoides x *A. muticum*	ST	2.84	–	–	5.58	–	–	–	
Ae. speltoides x *A. muticum*	ST	0.35	1.94	4.88	6.82	0.005	–	–	Ohta (1990)
Ae. speltoides x *A. muticum*	ST	1.12	2.99	3.43	6.43	0.003	0.003	–	
Ae. speltoides x *A. muticum*	ST	1.99	3.27	2.64	5.91	0.04	0.02	–	
Ae. bicornis x *A. muticum*	S^bT	0.20	1.50	5.40	6.90	–	–	–	
Ae. bicornis x *A. muticum*	S^bT	0.80	2.73	3.87	6.60	–	–	–	
Ae. sharonensis x *A. muticum*	$S^{sh}T$	0.96	2.04	4.48	6.52	–	–	–	
Ae. sharonensis x *A. muticum*	$S^{sh}T$	1.42	3.06	3.16	6.22	0.02	0.02	–	
Ae. longissima x *A. muticum*	S^lT	3.52	–	–	5.06	0.12	–	–	Riley (1966)
Ae. longissima x *A. muticum*	S^lT	1.83	3.96	1.30	5.26	0.40	0.07	–	Ohta (1990)
Ae. longissima x *A. muticum*	S^lT	0.67	1.77	3.67	5.44	0.33	0.37	–	
Ae. tauschii x *A. muticum*	DT	0.16	1.78	4.11	5.89	0.02	0.59	–	
Ae. tauschii x *A. muticum*	DT	1.09	2.62	3.78	6.40	0.01	0.02	–	
Ae. caudata x *A. muticum*	CT	3.53	–	–	3.33	1.27	–	–	
Ae. caudata x *A. muticum*	CT	3.03	2.83	0.70	3.53	1.30	–	–	
Ae. comosa x *A. muticum*	MT	0.33	2.87	3.97	6.84	–	–	–	
Ae. comosa x *A. muticum*	MT	1.93	3.37	2.67	6.04	–	–	–	
Ae. uniaristata x *A. muticum*	NT	4.44	3.54	0.34	3.88	0.60	–	–	
Ae. uniaristata x *A. muticum*	NT	4.48	4.10	0.13	4.23	0.35	–	–	
Ae. umbellulata x *A. muticum*	UT	2.60	2.15	0.06	2.21	2.02	0.04	0.15	
Ae. umbellulata x *A. muticum*	UT	4.00	2.52	0.08	2.60	1.54	0.02	0.02	
Ae. crassa 6x x *A. muticum*	D^cX^cDT	11.14	4.22	2.24	6.46	1.30	0.60	0.04	Melnyk and McGinnis (1962)
Ae. juvenalis x *A. muticum*	D^cX^cUT	14.21	4.66	0.08	4.74	1.08	0.25	0.02	

Table 8.2 Mean chromosomal pairing in F_1 hybrids between species of *Aegilops* or *Triticum* and *Amblyopyrum muticum* with and without B chromosomes (calculated from the data of Ohta and Tanaka 1983)

A. Pairing in hybrids lacking B- chromosomes

Hybrid combination	Hybrid genome	Bivalents		Multivalents		Mean arm-pairing/cell
		Rod	Ring	III	IV	
Ae. speltoides x A. muticum	ST	2.75	3.06	0.03	0.04	0.65
Ae. bicornis x A. muticum	S^bT	2.80	4.00	–	–	0.77
Ae. longissima x A. muticum	S^lT	3.33	1.87	0.37	0.30	0.59
Ae. comosa x A. muticum	MT	2.79	3.93	–	–	0.77
Ae. caudata x A. muticum	CT	2.83	0.70	1.30	–	0.44
T. monococcum ssp. aegilopoides x A. muticum	A^mT	2.69	2.58	0.23	0.36	0.68

B. Pairing in hybrids containing B chromosomes

Ae. speltoides x A. muticum	ST	3.05	0.85	0.06	0.01	0.35
Ae. bicornis x A. muticum	S^bT	0.04	–	–	–	0.003
Ae. longissima x A. muticum	S^lT	0.40	–	–	–	0.03
Ae. comosa x A. muticum	MT	0.50	–	–	–	0.04
Ae. caudata x A. muticum	CT	0.70	–	–	–	0.05
T. monococcum ssp. aegilopoides x A. muticum	A^mT	0.58	0.06	0.01	–	0.05

being an obligatory allogamous species, comprises intraspecific plasmon differentiation (Tsunewaki 2009). The plasmon of the allotetraploid *Ae. geniculata*, designated M^0, did not have any close relatives in the diploids, although their phenotypic effects to wheat characters were close to those of the T^2 plasmon of *A. muticum* (Tsunewaki 2009). Therefore, Ogihara and Tsunewaki (1988) and Tsunewaki (2009) concluded that the plasmon type of *Ae. geniculata* closely resembles plasmon types of *Ae. umbellulata* and *A. muticum*, and, therefore, either of those diploids or an unknown species related to them seems to be the cytoplasm donor of *Ae. geniculata*.

8.4 Phylogenetic Relationships of *A. muticum* to *Aegilops* and *Triticum*

Amblyopyrum muticum and *Aegilops speltoides* are the only allogamous species in the wheat group. Allogamy is considered a primitive character since most perennial species in the tribe Triticeae are allogamous. Stebbins (1957) pointed out that in many plant-groups autogamy derived from allogamy and therefore, autogamy may be considered as a more advanced trait. Hammer (1980) assumed that allogamous self-incompatible plants, resembling *Amblyopyrum muticum* and facultative allogamous like *Aegilops speltoides*, were ancestral types to the other genera of the wheat group, i.e., *Aegilops* and *Triticum*. He based this assumption on the combination of anther length and amount of pollen produced, with reductions in anther size and/or amount of pollen pointing at increasing autogamy, for which he found

strong positive correlations. Long anthers that produce substantial amount of pollen are an indication of allogamy or transitional state to autogamy. On this, Hammer (1980) produced an evolutionary model explaining the sequence of origin of the various species.

Hammer (1980) assumed that the origin of the ancestral genera of the wheat group was in Transcaucasia. Awing to increase drought in the Pleistocene (1.8–0.01 MYA), many grass species, including the ancestral stocks of *Amblyopyrum–Aegilops*, started spreading in western and southwestern directions. The areas of the distribution of the diploids can thus be explained: the more primitive the closer to the center of origin. Thus, *A. muticum* and *Ae. speltoides*, the former restricted to Turkey, the latter in Turkey, reaching Bulgaria and along the Fertile Crescent might be considered as the most primitive species of the group. Speciation of the Sitopsis group apparently happened mainly on the western arc of the Fertile Crescent, with the species with the smallest anthers (*Ae. bicornis*) reaching the furthest (the Coast of Cyrenaica, Libya). Other diploids reached the eastern Mediterranean (*Ae. uniaristata* and *Ae. comosa*), or only partly so (*Ae. caudata and Ae. umbellulata*); one diploid spread mainly to the east (*Ae. tauschii*). In this old group speciation is relatively strong as is shown by the reported sterility of artificial hybrids of the Sitopsis species.

Coinciding with the speciation has been a gradual change in fertilization mechanism within the diploids from allogamy (*A. muticum* and *Ae. speltoides*), to facultative autogamy (*Ae. longissima, Ae. caudata, Ae. comosa*). Higher levels of autogamy are associated with lower values for anther length x width (as scored by Hammer 1980). This development happened at the lowest, most primitive ploidy level. This model of distribution, speciation, and change to autogamy can also be applied to the wild taxa of *Triticum*. Hammer (1980) notes that the anther length of diploid *Triticum* is shorter than those of Sitopsis species, while other characters could be interpreted as reductions or subsequent changes in any of the two groups, e.g., two versus only one keel, hairy versus glabrous rachillae, and 1–2 versus 2–3 kernels per spikelet. This, as well as the distribution patterns, underline the development of both groups out of a common ancestor, with according to Hammer, the flower biology of the diploid *Triticum* considered more derived and of *Aegilops* more-closer to the common ancestor.

At the early stage during the late Pleistocene, it is supposed that *A. muticum*, an obligate allogamous species, separated from the common stock during the westward migration through Asia minor (Hammer 1980). Eig (1929b) considered this an old species, being most closely related to the oldest section of Aegilops (Sitopsis) and showing relatively little plasticity in its morphology. Many morphological characters place this species apart from all *Aegilops* species while karyotype analysis showed similarity with what Hammer (1980) considered the most primitive *Aegilops*, i.e., *Ae. speltoides* (Chennaveeraiah 1960).

Out of the initial distribution of the diploids, the allotetraploids spread further westwards along the Mediterranean basin as well as in more northern, southern and eastern directions. This process continued until halted by natural boundaries and lack of suitable environments, such as the Saharan and Arabian deserts, the central Asian steppes, the Tian Shan and Himalayan Mountains or the coldness of the continental climate affecting the spread to the north and east.

The complicated interactions that happened in the process of allopolyploidization resulted in an intergrading network of forms and reproduction strategies that make it impossible to point at any direction of the evolution at this stage. The tetraploid stage is apparently dominant and a further development into hexaploids has been limited.

Amblyopyrum separated from the ancestral *Aegilops* lineage at an early stage, of which *Ae. speltoides* is thought to be the most primitive representative of *Aegilops*. Phylogenetic speculation that assumes a change from obligatory allogamy (as in *A. muticum*) towards almost complete autogamy (as in species of *Aegilops*, *Eremopyrum*, *Heteranthelium*, *Henrardia* and *Triticum*), coinciding with a divergent development in morphology, makes a separation at an early stage from the ancestral *Aegilops* lineage plausible (Hammer 1980).

Based on karyomorphological studies, Senjaninova-Korczagina (1932) concluded that the karyotype of *A. muticum* is similar to those of *Ae. tauschii* and *Ae. comosa* but is very different from those of *Ae. caudata* and *Ae. umbellulata*. Chennaveeraiah (1960), found that *A. muticum* has large chromosomes with median or sub-median centromeres, of which two pairs have fairly large satellites on their short arms. He argued that the karyotype of *A. muticum* is different from those of *Ae. comosa* and *Ae. tauschii*, and is more similar to those of the species of section Sitopsis, especially to that of *Ae. speltoides*. The karyotype analysis of these species by Giorgi and Bozzini (1969) confirmed the finding of Chennaveeraiah (1960), displaying a similar though not identical karyotype of *Ae. speltoides* and *A. muticum*. Thus, also from a karyomorphological view, a close phylogenetic relationship exists between *A. muticum* and *Ae. speltoides*. However, according to Jones and Majisu (1968), this karyotype similarity does not seem to be indicative of any greater homology between these genomes. In spite of that, since most species of the Triticeae, and particularly those that belong to the Arctic-temperate group, have large chromosomes with median or sub-median centromeres, it is assumed that this type of karyotype is characteristic of the prototype of the tribe. Hence, *A. muticum* having large chromosomes and symmetric karyotype is one of the primitive species in the sub-tribe Triticineae.

Analysis of the cytoplasm of *A. muticum* pointed at a close genetic relationship with the cytoplasm of the allotetraploid species *Ae. geniculata* (Maan 1977). Ohsako et al. (1996), studying variation in chloroplast and mitochondrial DNA by PCR-SSCP analysis, found that the level of intraspecific variation in *A. muticum* was lower than that in *Ae. speltoides* and, consequently, suggested that *A. muticum* is not older than *Ae. speltoides*. This suggestion is in contrast to the hypothesis of Hammer (1980) and Ohta (1990, 1991) that *A. muticum* is an ancestral species in the group. In the phylogenetic trees, *A. muticum* was included in different cluster than the other *Aegilops* and *Triticum* species (Ohsako et al. 1996). Yamane and Kawahara (2005) conducted phylogenetic studies by analyzing chloroplast DNA sequences from four regions of the diploid species of the wheat group and found that *A. muticum* was included in the most terminal clade close to *Ae. umbellulata*. Likewise, Terachi et al. (1984) and Murai et al. (1989) studied chloroplast DNA and suggested that *A. muticum* is close to *Ae. umbellulata* while Terachi and Tsunewaki (1992), based on mitochondrial RFLP analysis, suggested that it is close to *Ae. tauschii*. This poses a discrepancy between molecular phylogeny and classification based on morphology.

Sasanuma et al. (2004) pointed out that discrepancy also exists between results of chromosomal pairing in interspecific hybrids and molecular data from plasmon analysis. They studied intra- and inter-specific variation in seven diploid *Aegilops* species (including *A. muticum*) using AFLP technique. Of the seven species, the cross-pollinating *Ae. speltoides* and *A. muticum* showed the highest levels of intraspecific variation. In their study, *A. muticum* did not form a clear cluster with any other *Aegilops* species.

Dvorak and Zhang (1992), analyzing repeated DNA sequences, concluded that *A. muticum* is close to *Ae. caudata*, *Ae. comosa*, *Ae. uniaristata* and *Ae. umbellulata*. Wang et al. (2000) compared the internal transcribed spacer (ITS) region of the ribosomal DNA in the diploids of the wheat group (including *A. muticum*) and observed wide divergences of this sequence between species. The highest divergence was between *Ae. speltoides* and *A. muticum*. In-situ hybridization with repeated DNA markers and C-banding patterns suggest that *A. muticum* occupies an isolated position but relatively closer to the Sitopsis species than to other species of *Aegilops* (Badaeva et al. 1996a). Sallares and Brown (2004), who analyzed the transcribed spacers of the 18S ribosomal RNA genes, reached a similar conclusion, namely, that *A. muticum* has a basal position and that it is close to *Ae. speltoides*. Yet, in several phylogenetic analyses based on nuclear DNA sequences *Amblyopyrum* is a good, monophyletic genus (Frederiksen and Seberg 1992; Frederiksen 1993). Most morphological and molecular trees included *Amblyopyrum* in the *Aegilops* clade (e.g., Seberg and Petersen 2007) and in other molecular trees several species of *Elymus* (=Agropyron) are also included in this clade (Mason-Gamer et al. 1998), even though, the relationships of *Amblyopyrum* within the *Aegilops* clade remains at present somewhat vague.

Based on their molecular phylogenetic studies, Marcussen et al. (2014) concluded that the A and B (=S) lineages, the early *Triticum* and *Aegilops speltoides* forms, diverged from a common ancestor about 7 MYA, and that these ancestral forms gave rise to the D-genome lineage through homoploid hybrid speciation 1–2 million years later. Since then, more complex evolutionary scenarios with several rounds of hybridization have been proposed (Li et al. 2015; El Baidouri et al. 2017). However, these studies involved only a number of the diploids of the wheat group. In contrast, Glémin et al. (2019) obtained and analyzed a comprehensive genomic dataset including all extant diploid species of *Amblyopyrum*, *Aegilops* and *Triticum,* and developed a new framework to test intricate hybridization scenarios. Owing to these new developments, Glémin et al. (2019) were able to propose a core reference scenario for the history of diploid *Aegilops/Triticum* species. They confirmed the occurrence of an ancient hybridization event that gave rise to the D lineage, but showed (i) that this lineage includes 9, not only 5, of the 13 diploid species of the wheat group, and (ii) that the hybridization scenario involved a different parental species, *A. muticum* instead of *Ae. speltoides*. Glémin et al. (2019) pointed out that *A. muticum* has been an overlooked species with a debated phylogenetic position, and that their results plead for reconsideration and extensive study of this key species in the history of wheat relatives. To reconcile with the hypothesis of Marcussen et al. (2014) . with that of Glémin et al. (2019), it is suggested that the homoploid hybridization scenario involved the ancestral genome S/T before the divergence of *Ae. speltoides* from *A. muticum.*

Similar to Glémin et al. (2019), also Bernhardt et al. (2020) hypothesize that most of the diploid species of the wheat group were shaped by a primordial homoploid hybrid speciation event, that is the *Triticum* lineage merged with the ancestor of *A. muticum* to form all other species of *Aegilops* except *Ae. speltoides*. These results highlight the pivotal role of *A. muticum*, instead of *Ae. speltoides*, in the formation of the wild diploids of the wheat group. This hybridization event was followed by multiple introgressions affecting all taxa except *Triticum*. Mostly progenitors of the extant species were involved in these processes, while, according to Bernhardt et al. (2020), recent interspecific gene flow seems insignificant.

The results of Glémin et al. (2019) and Bernhardt et al. (2020) highlight the key role of *A. muticum*, instead of *Ae. speltoides*, in the formation of the diploid species of *Aegilops*. This hybridization event, was estimated to have occurred about 5.5 MYA based on whole genome sequences (Li et al. 2022). It was followed by multiple ancient introgressions affecting all taxa except *Triticum*. In contrast with Glémin et al. (2019), Bernhardt et al. (2020) do not find introgression of *Triticum* into *A. muticum*, instead their results indicated that *A. muticum* may have been introgressed by the *unbellulata/caudata* (U/C) group. Introgression to *A. muticum* from *Ae. umbellulata* (genome U) was also suggested from chloroplast phylogenetic research (Yamane and Kawahara 2005; Bordbar et al. 2011; Bernhardt et al. 2017). Hence, the maternal lineage of *A. muticum* does not group with *Ae. speltoides*, although both are sister taxa in nuclear phylogenies, but it shares a common ancestor with *Ae. umbellulata*.

References

Badaeva ED, Friebe B, Gill BS (1996a) Genome differentiation in *Aegilops*. 2. Physical mapping of 5S and 18S–26S ribosomal RNA gene families in diploid species. Genome 39:1150–1158

Badaeva ED, Friebe B, Gill BS (1996b) Genome differentiation in *Aegilops*. Distribution of highly repetitive DNA sequences on chromosomes of diploid species. Genome 39:293–306

Baum BR (1977) Taxonomy of the tribe Triticeae (Poaceae) using various numerical techniques. I. Historical perspectives, data, accumulation and character analysis. Can J Bot 55:1712–1740

Baum BR (1978a) Taxonomy of the tribe Triticeae (Poaceae) using various numerical techniques. II. Classification. Can J Bot 56:27–56

Baum BR (1978b) Taxonomy of the tribe Triticeae (Poaceae) using various numerical techniques. III. Synoptic key to the genera and synopses. Can J Bot 56:374–385

Bernhardt N, Brassac J, Kilian B, Blattner FR (2017) Dated tribe-wide whole chloroplast genome phylogeny indicates recurrent hybridizations within Triticeae. BMC Evol Biol 17:141

Bernhardt N, Brassac J, Dong X, Willing E-M, Poskar CH, Kilian B, Blattner FR (2020) Genome-wide sequence information reveals recurrent hybridization among diploid wheat wild relatives. Plant J 102:493–506

Boissier PE (1844a) Diagnoses plantarum orientalium novarum. Sér 1:73–74

Boissier PE (1884b) Flora orientalis, vol 5. Geneva, Basel & Lyon, pp 673–679

Bor NL (1968) Gramineae. In: Towsend CC, Guest E, El-Rawi A (eds) Flora of Iraq, vol 9, pp 210–263

Bordbar F, Rahiminejad MR, Saeidi H, Blattner FR (2011) Phylogeny and genetic diversity of D-genome species of *Aegilops* and *Triticum* (Triticeae, Poaceae) from Iran based on microsatellites, ITS, and *trn*L-F. Pl Syst Evol 291:117–131

Chen KC, Dvorak J (1984) The inheritance of genetic variation in *Triticum speltoides* affecting heterogenetic chromosome pairing in hybrids with *Triticum aestivum*. Can J Genet Cytol 26:279–287

Chennaveeraiah MS (1960) Karyomorphologic and cytotaxonomic studies in *Aegilops*. Acta Horti Gotoburgensis 23:85–178

Clayton SD, Renvoize SA (1986) Genera graminum, grasses of the world. Distributed for Royal Botanic Gardens, Kew Bull Addit Ser 13. Kew, London, pp 1–389

Davis PH, Mill RR, Tan K (1988) Flora of Turkey and The East Aegean islands, vol 10 (supplement). Esinburgh University Press, Edinburgh

Dhaliwal HS (1977) Basis of difference between reciprocal crosses involving *Triticum boeoticum* and *T. urartu*. Theor Appl Genet 49:283–286

Dover GA, Riley R (1972a) Variation at two loci affecting homoeologous meiotic chromosome pairing in *Triticum aestivum×Aegilops mutica* hybrids. Nature New Biol 235:61–62

Dover GA, Riley R (1972b) Prevention of pairing of homoeologous meiotic chromosomes of wheat by an activity of supernumerary chromosomes of Aegilops. Nature 240:159–161

Dundas I, Verlin D, Islam R (2015) Chromosomal locations of stem and leaf rust resistance genes from *Ae. caudata*, *Ae. searsii* and *Ae. mutica*. In: BGRI Workshop

Dvorak J (1972) Genetic variability in *Aegilops speltoides* affecting homoeologous pairing in wheat. Can J Genet Cytol 14:371–380

Dvorak J, Zhang H-B (1992) Reconstruction of the phylogeny of the genus *Triticum* from variation in repeated nucleotide sequences. Theor Appl Genet 84:419–429

Dvorak J (1998) Genome analysis in the *Triticum-Aegilops* alliance. In: Slinkard AE (ed) Proc 9th Inter Wheat Genet Symp, University Extension Press, University of Saskatoon, Saskatoon, Saskatchewan, Canada, pp 8–11

Dvorak J, Deal KR, Luo M-C (2006) Discovery and mapping of wheat *Ph1* suppressors. Genetics 174:17–27

Eig A (1929a) Monographisch-Kritische Ubersicht der Gattung *Aegilops*. Reprium Nov. Spec Regni Veg 55:12–88

Eig A (1929b) *Amblyopyrum* Eig. A new genus separated from the genus *Aegilops*. PZE Inst Agric Nat Hist Agric Res 2:199–204

Eilam T, Anikster Y, Millet E, Manisterski J, Feldman M (2007) Genome size and genome evolution in diploid Triticeae species. Genome 50:1029–1037

El Baidouri M, Murat F, Veyssiere M, Molinier M, Flores R, Burlot L, Alaux M, Quesneville H, Pont C, Salse J (2017) Reconciling the evolutionary origin of bread wheat (*Triticum aestivum*). New Phytol 213:1477–1486

Eser V (1998) Characterisation of powdery mildew resistant lines derived from crosses between *Triticum aestivum* and *Aegilops speltoides* and *Ae. mutica*. Euphytica 100:269–272

Frederiksen S, Seberg O (1992) Phylogenetic analysis of the Triticeae (Poaceae). Hereditas 116:15–19

Frederiksen S (1993) Taxonomic studies in some annual genera of the Triticeae. Nordic J Bot 13:490–492

Giorgi B, Bozzini A (1969) Karyotype analysis in *Triticum*. III. Analysis of the presumed diploid progenitors of polyploid wheats. Caryologia 22:279–287

Glémin S, Scornavacca C, Dainat J, Burgarella C, Viader V, Ardisson M, Sarah G, Santoni S, David J, Ranwez V (2019) Pervasive hybridizations in the history of wheat relatives. Sci Adv 5 (eaav9188):1–10

Hammer K (1980) Vorarbeiten zur monographischen Darstellung von Wildpflanzensortimenten: *Aegilops* L. Kulturpflanze 28:33–180

Iefimenko TS, Fedak YuG, Antonyuk MZ, Ternovska TK (2015) Microsatellite analysis of chromosomes from the fifth homoeologous group in the introgressive *Triticum aestivum/Amblyopyrum muticum* wheat lines. Cytol Genet 49:183–191

Jones JK, Majisu BN (1968) The homoeology of *Aegilops mutica* chromosomes. Can J Genet Cytol 10:620–626

Kihara H, Lilienfeld FA (1935) Genomanalyse bei *Triticum* und *Aegilops*. VI. Weitere Untersuchungen an *Aegilops* x *Triticum* und *Aegilops* x *Aegilops*-Bastarden. Cytologia 6:195–216

Kihara H (1947) The genus Aegilops classified on the basis of genome analysis. Seiken Zjho 3:7–25

Kihara H (1954) Considerations on the evolution and distribution of *Aegilops* species based on the analyzer-method. Cytologia 19:336–357

Kimber G (1961) Cytogenetics of haploidy in *Gossypium* and *Triticum*. PhD thesis, Univ. Manchester, UK, pp 1–297

Kimber G (1982) The genome relationships of *Triticum tripsacoides*. Z Pflanzenzuchtg 89:289–294

Kimber G, Tsunewaki K (1988) Genome symbols and plasma types in the wheat group. In: Miller TE, Koebner RMD (eds) Proc 7th Int Wheat Genet Symp, Cambridge, pp 1209–1210

King J, Grewal S, Yang C, Hubbart S, Scholefield D, Ashling S et al (2017) A step change in the transfer of interspecific variation into wheat from *Amblyopyrum muticum*. Plant Biotech J 15:217–226

Li L-F, Liu B, Olsen KM, Wendel JF (2015) A re-evaluation of the homoploid hybrid origin of *Aegilops tauschii*, the donor of the wheat D-subgenome. New Phytol 208:4–8

Li L-F, Zhang Z-B, Wang Z-H, et al (2022) Genome sequences of the five *Sitopsis* species of *Aegilops* and the origin of polyploid wheat B subgenome. Mol Plant 15:488–503

Lilienfeld F (1951) H. Kihara: genome analysis in *Triticum* and *Aegilops*. X. Concluding review. Cytologia 16:101–123

Löve Á (1982) Generic evolution of the wheatgrasses. BioI Zentralbl 101:199–212

Maan SS (1977) Cytoplasmic homology between *Aegilops mutica* Boiss. and *Ae. ovata* L. Euphytica 26:601–613

Marcussen T, Sandve SR, Heier L, Spannagl M, Pfeifer M, The International Wheat Genome Sequencing Consortium, Jakobsen KS, Wulff BBH, Steuernagel B, Klaus FX, Mayer KFX, Olsen OA (2014) Ancient hybridizations among the ancestral genomes of bread wheat. Science 345(6194):288–291

Mason-Gamer RJ, Weil CF, Kellogg EA (1998) Granule-bound starch synthase: structure, function, and phylogenetic utility. Mol Biol Evol 15:1658–1673

Melnyk JH, McGinnis RC (1962) Analysis of chromosome pairing in interspecific and intergeneric F1 hybrids involving hexaploid *Aegilops crassa*. Wheat Inf Serv 14:24–25

Mochizuki A (1957) B chromosomes in *Aegilops mutica* Boiss. Wheat Inf Serv 5:9–11

Mochizuki A (1960) A note on the B-chromosomes in natural populations of *Aegilops mutica* Boies. in Central Turkey. Wheat Inf Serv 11:31

Mochizuki A (1964) Further studies on the effect of accessory chromosomes on chromosome pairing (in Japanese). Jap J Genet 39:356–362

Murai K, Xu NY, Tsunewaki K (1989) Studies on the origin of crop species by restriction endonuclease analysis of organellar DNA. III. Chloroplast DNA variation and interspecific relationships in the genus *Secale*. Jpn J Genet 64:35–47

Ogihara Y, Tsunewaki K (1983) The diversity of chloroplast DNA among *Triticum* and *Aegilops* species. In: Sakamoto S (ed) Proc 6th Intern Wheat Genet Symp, Kyoto, pp 407–413

Ogihara Y, Tsunewaki K (1988) Diversity and evolution of chloroplast DNA in Triticum and Aegilops as revealed by restriction fragment analysis. Theor Appl Genet 76:321–332

Ohsako T, Wang G-Z, Miyashita NT (1996) Polymerase chain reaction—single strand conformational polymorphism analysis of intra- and interspecific variations in organellar DNA regions of *Aegilops mutica* and related species. Genes Genet Syst 71:281–292

Ohta S, Tanaka M (1982) The effects of the B-chromosomes of *Aegilops mutica* Boias. on meiotic chromosome pairing. Rep Plant Germ-Plasm Inst, Kyoto Univ 5:36–52

Ohta S, Tanaka M (1983) Genome relationships between Ae. *mutica* and the other diploid *Aegilops* and *Triticum* species, based on the chromosome pairing in the hybrids with or without B-chromosomes. In: Sakamoto S (ed) Proc 6th Int Wheat Genet Symp, Kyoto, pp 983–991

Ohta S (1990) Genome analysis of *Aegilops mutica* Boiss. Based on the chromosome pairing in interspecific and intergeneric hybrids. PhD thesis, University of Kyoto, pp 1–267

Ohta S (1991) Phylogenetic relationship of *Aegilops mutica* Boiss with the diploid species of congeneric *Aegilops-Triticum* complex, based on the new method of genome analysis using its B-chromosomes. Mem Coll Agric Kyoto Univ 137:1–116

Ohta S (1995) Distinct numerical variation of B-chromosomes among different tissues in *Aegilops mutica* Boiss. Jap J Genet 70:93–101

Othmeni M, Grewal S, Hubbart-Edwards S, Yang C, Scholefield D, Ashling S, Yahyaoui A, Gustafson P, Singh PK, King IP, King J (2019) The use of pentaploid crosses for the introgression of Amblyopyrum muticum and D-genome chromosome segments into durum wheat. Front Plant Sci 10(1110):1–11. https://doi.org/10.3389/fpls.2019.01110

Panayotov I, Gotsov K (1973) Interactions between nucleus of *Triticum aestivum* L. and cytoplasm of certain species of *Triticum* and *Aegilops*. In: Sears ER, Sears LMS (eds) Proc Fourth Int Wheat Genet Symp. Columbia, MO, USA, pp 381–383

Panayotov I (1983) The cytoplasms in Triticineae. In: Sakamoto S (ed) Pro 6th Int Wheat Gener Symp, Kyoto, Japan, pp 481–497

Riley R (1960) The diploidization of polyploid wheat. Heredity 15:407–429

Riley R (1966) The genetic regulation of meiotic behavior in wheat and its relatives. In: Mac Key J (ed) Proc Second Inter Wheat Genet Symp, suppl 2, Genetic Institute, University of Lund, Sweden, Hereditas, pp 395–406

Sallares R, Brown TA (2004) Phylogenetic analysis of complete 5' external transcribed spacers of the 18S ribosomal RNA genes of diploid *Aegilops* and related species (Triticeae, Poaceae). Genet Resour Crop Evol 51:701–712

Sasanuma T, Chabane K, Endo TR, Valkoun J (2004) Characterization of genetic variation in and phylogenetic relationships among diploid *Aegilops* species by AFLP: incongruity of chloroplast and nuclear data. Theor Appl Genet 108:612–618

Schultze-Motel J, Meyer D (1981) Numerical taxonomic studies in the genera *Triticum* L. and *Pisum* L. Kulturpflanze 29:241–250

Seberg O, Petersen G (2007) Phylogeny of Triticeae (Poaceae) based on three organelle genes, two single-copy nuclear genes, and Morphology. Aliso J Syst Evol Bot 23:362–371

Senjaninova-Korczagina MV (1932) Karyo-systematical investigations of the genus *Aegilops* L. (Russian with English summary) Bull Appl But Genet and Pl Breed Ser 2:1–90

Simchen G, Zarchi Y, Hillel J (1971) Supernumerary chromosomes in second outbreeding species of wheat group. Chromosoma 33:63–69

Stebbins GL (1957) Self-fertilization and population variability in the higher plants. Amer Nat 91:337–354

Terachi T, Ogihara Y, Tsunewaki K (1984) The molecular basis of genetic diversity among cytoplasms of *Triticum* and *Aegilops*. III. Chloroplast genomes of the M and modified M genome-carrying species. Genetics 108:681–695

Terachi T, Tsunewaki K (1992) The molecular basis of genetic diversity among cytoplasms of *Triticum* and *Aegilops*. VIII. Mitochondrial RFLP analyses using cloned genes as probes. Mol Biol Evol 9:911–931

Tsunewaki K (2009) Plasmon analysis in the *Triticum-Aegilops* complex. Breed Sci 59:455–470

Tzvelev NN (1989) The system of the grasses (Poaceae) and their evolution. Bot Rev 55:14–203

van Slageren MW (1994) Wild wheats: a monograph of *Aegilops* L. and *Amblyopyrum* (Jaub. and Spach) Eig (Poaceae). Agricultural University, Wageningen, The Netherlands

Vardi A, Dover GA (1972) The effect of B chromosomes on meiotic and pre-meiotic spindles and chromosome pairing in *Triticum/Aegilops* hybrids. Chromosoma (berl.) 38:367–385

Wang J-B, Wang C, Shi S-H, Zhong Y (2000) ITS regions in diploids of *Aegilops* (Poaceae) and their phylogenetic implications. Hereditas 132:209–213

Watson L, Clifford HT, Dallwitz MJ (1985) The classification of the Poaceae: subfamilies and supertribes. Austral J Bot 33:433–484

Yamane K, Kawahara T (2005) Intra- and interspecific phylogenetic relationships among diploid *Triticum-Aegilops* species (Poaceae) based on base-pair substitutions, indels, and microsatellites in chloroplast noncoding sequences. Am J Bot 92:1887–1898

Zhukovsky PM (1928) A critical systematical survey of the species of the genus *Aegilops* L. Bull. Appl Bot Genet Pl Breed 18:417–609

9.1 Classification of the Genus *Aegilops* L.

Aegilops is the name of a grass mentioned in Theophrastus' botanical treatise "Enquiry into Plants", that was a major source for botanical knowledge during antiquity and the Middle Ages. The name *Aegilops* comes from the Greek *aegilos*, which could mean "a herb liked by goats", or "a goat-like herb", and refers to the whiskery-awned spikelets of some of its species (Bor 1968; Watson and Dallwitz 1992). Since the taxonomic treatment of the *Aegilops* genus by Linnaeus (1753), various taxonomists provided different definitions of the species and sub-genus ranks of the genus. Zhukovsky (1928) described 20 species in the genus, which he classified into nine sections, while Eig (1929a) grouped 22 species into two sub-genera and six sections. In his review summarizing results from a genome analysis of the genus, Kihara (1954) recognized 21 species of which the tetraploid and hexaploid taxa of *Ae. triaristata* and *Ae. crassa*, were considered one species. He grouped the 21 species into six sections. Based on his karyomorphological study, Chennaveeraiah (1960) separated *Ae. vavilovii* as a new hexaploid species from *Ae. crassa*. Following Chennaveeraiah (1960), Kihara and Tanaka (1970) separated also the hexaploid taxon *Ae. recta* from the tetraploid *Ae. neglecta*, and accepted the separation of the hexaploid *Ae. vavilovii* from *Ae. crassa*. Consequently, Kihara and Tanaka (1970) grouped a total of 22 species into six sections. Later, Feldman and Kislev (1977) described *Ae. searsii* as a new diploid species belonging to section Sitopsis. While accepting *Ae. searsii* as a valid species, Hammer (1980) continue to consider *Ae. vavilovii* as a subspecies of *Ae. crassa* and grouped 22 species into three subgenera and four sections. All of the above classifications included *Ae. mutica* either as a separate section (Zhukovsky 1928; Kihara 1954; Kihara and Tanaka 1970) or as a sub-genus (Eig 1929a; Hammer 1980). Yet, Eig (1929b) removed *Ae. mutica* from the genus *Aegilops* and included it as a monotypic species in a new genus *Amblyopyrum*. Lastly, van Slageren (1994), in his recent comprehensive taxonomic classification of the genus *Aegilops*, recognized five sections containing 22 species. It included *Ae. vavilovii* and *Ae. searsii* as species, but kept *Ae. recta* in *Ae. neglecta*. This book recognizes *Ae. recta* as a species and, consequently, 23 *Aegilops* species are considered here: 10 diploids, 9 tetraploids, one species *Ae. crassa* containing two cytotypes, tetraploid and hexaploid, and 3 hexaploids (Table 9.1). A 23-species genus, instead of the usual 1–5, is exceptional in the sub-tribe Triticineae.

An entirely different tendency exists regarding the sub-species level classification taxa (van Slageren 1994). Hammer (1980) suggested to regard taxa with different chromosome numbers, or differing in their morphology and/or geographical distribution, as subspecies. Van Slageren (1994), like Mac Key (1981), disagreed and suggested a drastic consolidation at the intra-specific level, i.e., to maintain only groups exhibiting obvious discontinuities in several characters. Kihara (1954) included *Ae. kotschyi*, *Ae. heldreichii*, *Ae. aucheri* and *Ae. sharonensis* in *Ae. variabilis* (now *Ae. peregrina*), *Ae. comosa*, *Ae. speltoides* and *Ae. longissima*, respectively, because they shared very similar genomes. However, like Eig (1929a), Hammer (1980) considered *Ae. kotschyi* a separate species, because it differs from *Ae. peregrina* in its exclusive geographical distribution, i.e., the two species are good vicariad species. Similarly, van Slageren (1994) accepted Eig's (1929a) definition of *Ae. sharonensis* as a valid species due to its morphological and ecological distinction from *Ae. longissima*. On the other hand, van Slageren (1994) maintained the sub-specific rank of *Ae. heldreichii* and *Ae. aucheri* (Table 9.1).

The genus *Aegilops* L. [Syn.: *Agicon* Adans; *Triticum* L. Sect. *Aegilops* (L.) Godr. & Gren. in Grenier & Godron; *Triticum* L. Subg. *Aegilops* (L.) Schmahlh.; *Frumentum* E.H. L. Krause subg. *Aegilops* (L.) E.H.L. Krauase; *Aegilops* L. subg. *Eu-Aegilops* Eig; *aegilopoides* Á. Löve] consists of wild annual, mostly autogamous 15–100-cm-high grasses, with few to many tillers that are usually geniculate at the base then turning upright, and sparsely foliated in the lower parts. The leaf blades are flat, with short ligules. The leaves are linear and up to 15-cm long and one cm wide. The spikes

© The Author(s) 2023
M. Feldman and A. A. Levy, *Wheat Evolution and Domestication*,
https://doi.org/10.1007/978-3-031-30175-9_9

Table 9.1 Sections and species of *Aegilops* and their synonyms

Section[a]	Species[a]	Subspecies and varieties	Synonyms
Sitopsis (Jaub. & Spach) Zhuk.	*speltoides* Taush	var. *speltoides* var. *ligustica* (Savign) Fiori	*Triticum speltoides* (Tausch) Gren. ex Richt.; *Ae. aucheri* Boiss.; *Ae. ligustica* (Savign.) Coss.; *Ae. agropyroides* Godr.; *T. ligusticum* Bert.; *Agropyron ligusticum* Savign
	bicornis (Forssk.) Jaub. & Spach	Var. *bicornis* Var. *anathera* Eig	*Triticum bicorne* Forssk
	longissima Schweinf. & Muscl.		*T. longissimum* (Schweinf. & Muschl.) Bowden
	sharonensis Eig	Var. *sharonensis* Var. *mutica* (Post) Eig	*T. sharonense* (Eig) Bowden; *T. longissimum* (Eig) Bowden ssp. sharonensis (Eig) Chennav.; *Ae. longissima* Schweinf. & Muscl. ssp. sharonensis (Eig) Chennav
	searsii Feldman & kislev ex Hammer		*T. searsii* Feldman & kislev
Vertebrata Zhuk. emend. Kihara	*tauschii* Coss	ssp. *tauschii* ssp. *strangulata* (Eig) Tzvel	*Ae. squarrosa* L.; *T. tauschii* (Coss.) Schmal; *T. aegilops* P. Beauv. ex R. & S.; *T. squarrosum* Raspail
	ventricosa Tausch		*T. ventricosum* Ces., Pass. & Gib.,
	crassa Boiss.		*T. crissum* (Boiss.) Aitch. & Hemsl
	vavilovii (Zhuk.) Chennav.		*T. syriacum* Bowden; *Ae crassa* ssp. *palestina* Eig; *Ae. crassa* ssp. *vavilovii* Zhuk
	juvenalis (Thell.) Eig		*T. juvenile* Theil.; *T. turcomanicum* (Rosh.) Bowden, *Ae. turcomanica* Rosh
Cylindropyrum (Jaub. & Spach) Zhuk.	*caudata* L.	Var. *typica* Var. *polyathera* Boiss	*Triticum caudatum* (L.) Godr. & Gren.; *T. dichasians* (Zhuk.) Bowden; *Ae. dichasians* (Bowden) Humphries; *T. markgrafii* Greuter & Rechinger; *Ae. markgrafii* (Greuter) Hammer
	cylindrica Host	Var. *cylindrica* Var. *pauciaristata* Eig Var. *aristulata* (Zhuk.) Tzvel Var. *prokhanovii* Tzvel	*T. cylindricum* Ces., Pass. & Gib., *Ae. caudata* L. var cylindricum Fiori., *Ae. nova* Win
Comopyrum (Jaub. & Spach) Zhuk.	comosa Sm. In Sibth. & Sm.	ssp. *comosa* ssp. *heldreichii* (Boiss.) Eig	*Triticum comosum* (Sm. In Sibth. &Sm.) K. Richt.; *Comopyrum comosum* (Sm. In Sibth. &Sm.) Á. Löve; *Ae. heldreichii* Holzm.; *Ae. turcica* Azn.; *Ae. connate* Steud.; *Ae. ambigua* Haussk;
	uniaristata Vis.		*T. uniaistatum* (Vis.) Richt., *Ae. notarisii* Clem
Aegilops	*umbellulata* Zhuk.		*T. umbellulatum* (Zhuk.) Bowden
	geniculata Roth	ssp. *geniculata* (formerly *euovata*) ssp. *gibberosa* (Zhuk.) Hammer	*Ae. ovata* L.; *T. ovatum* (L.) Raspail.; *Ae. echinum* Godr.; *Ae. fausii* Sennen.; *Ae. neglecta* Req.; *Ae. divaricata* Jodr. & Fourr.; *Ae. eratica* Jodr. and Fourr.; *Ae. erigens* Jodr. and Fourr.; *Ae. microstachys* Jodr. and Fourr.; *Ae. parvula* Jodr. and Fourr.; *Ae. procera* Jodr. and Fourr.; *Ae. publiglumis* Jodr. and Fourr.; *Ae. sicula* Jodr. and Fourr.; *Ae. nigricans* Jodr. and Fourr
	biuncialis Vis.		*T. machrochaetum* (Shuttl. & Huet.) Richt.; *Ae. lorentii* Hochst.; *T. biunciale* Richt
	columnaris Zhuk.		*T. columnare* (Zhuk.) Morris et Sears; *Ae. mixta* Sennen., *Ae. neglecta* Req.; *Ae. fausii* Sennen
	neglecta Req. ex Bertol.	*	*Ae. triaristata* (4x) Willd.; *T. aristatum* (4x) (wild.) Godr. & Gren.; *Ae. geniculate* Roth
	Recta (Zhuk.) Chennv.	'	*T. triaristatum* (6x) (Willd.) Godr. & Gren.; *Ae. neglecta* Req. ex Bertol
	kotschyi Boiss.		*T. kotschyi* (Boiss.) Bowden; *Ae. divaricata* Jodr. & Fourr

(continued)

Table 9.1 (continued)

Section[a]	Species[a]	Subspecies and varieties	Synonyms
	peregrina (Hack. In J. Fraser) Maire &Weiller	ssp. *peregrina* (ssp. *euvariabilis*) Eig ssp. *cylindrostachys* Eig	*Ae. variabiis* Eig; *T. peregrinum* Hack.; *T. kotschyi* (Boiss.) Bowden
	triuncialis L.	ssp. *triuncialis* ssp. *persica* (Boiss.) Zhuk	*T. triunciale* (L.) Raspail.: *Ae. persica* Boiss

[a] Sections and species according to van Slageren (1994); hexaploids *Ae. vavilovii* and *Ae. recta* were given here the specific rank

are linear, or ovate to lanceolate, wholly or partly awned or awnless, with 2–20 solitary spikelets at each rachis node, with each spikelet up to 1.2-cm-long, sessile, or sub-sessile. Spikes contain 1–4 rudimentary spikelets at the base or the top. The spikes break off at maturity above the rudimentary spikelets and fall entire or disarticulate into single spikelets. Spikelets contain 2–8 florets, the upper often being staminate or sterile. The two glumes are more or less equal, shorter than the adjacent lemmas, or almost as long as the adjacent lemmas, with one or more teeth or awns, round on back, rarely keeled. The lemmas are papery or membranous with 1–3 teeth, or awns. The two palea are keeled. The caryopsis either adheres to the lemma and/or to the palea or is free from both lemma and palea. In some species, all the spikelets are fully awned, whereas in others, only the terminal spikelets are awned. Certain lines of several species (e.g., *Ae. bicornis*, *Ae. sharonensis*, *Ae. peregrina*) are awnless. Three types of dispersal units exist in the different species: wedge (the spikes disarticulate at maturity into spikelets with the rachis internode that belong to them), barrel (the spikes disarticulate at maturity into spikelets with the rachis internode that belong to the spikelet above them), and umbrella (the spikes fall entire at maturity) types.

The genus *Aegilops* is distinguished from the genus *Triticum* by the absence of well-developed keel on the glumes, causing the sharp angle in the glume outline of both the wild and domesticated *Triticum* taxa (van Slageren 1994). It also differs from *Triticum* by its glabrous rachis and a larger number of grains per spikelet.

9.2   Geographical Distribution and Ecological Affinities

The genus is a Mediterranean–western Asiatic element (Eig 1929a, 1936; Sakamoto 1973; Feinbrun-Dothan 1986a; van Slageren 1994; Hegde et al. 2002), containing species that are distributed from the Iberian Peninsula in the west, through the Mediterranean basin, southern Ukraine, Crimea, the Caucasus, the Middle East, and to central Asia and

western China (Zhukovsky 1928; Eig 1929a, 1936; Kihara 1954; Miller 1987; Kimber and Feldman 1987; van Slageren 1994; Table 9.2), i.e., from about 10°W to 82°E and about 24°S to 47°N (van Slageren 1994). The area of distribution is rectangular in shape, with its width being about four times its length. In central, northern, and eastern Europe, the distribution of the genus is bordered by the cold climate, in North Africa by the Saharan desert, in the southwest by the deserts of Sinai and the Arabian peninsulas, in the central-northern Asia by the steppes of Turkmenistan and Uzbekistan, and in the east by the Tian Shan and Himalaya mountain ranges and in the south east by the banks of the Indus river (Zhukovsky 1928; Eig 1929a, 1936; Kihara 1954; Kimber and Feldman 1987; van Slageren 1994). Several allotetraploid species introduced in the USA, of which *Ae. cylindrica* (jointed goat grass) is widespread and reduces wheat yield due to its severe infestation of wheat fields. Two other allopolyploid species (*Ae. geniculata*, and *Ae. triuncialis*) are locally spread, and the others are adventive with a few locations only. Several species are adventive in Canada, in the Canary Islands, in northern and northwestern Europe and in China (van Slageren 1994).

Most of the species grow in the central part of the genus distribution, i.e., in the Fertile Crescent arc (Israel, Jordan, Lebanon, Syria, southeastern Turkey, northern Iraq, and northwestern Iran) (Table 9.1). Countries like Iran, Iraq, Syria and Turkey contain many (13–17) of the 23 species in the genus, while in peripheral countries, like Afghanistan and Pakistan in the east and those of south-western Europe and North Africa in the west, contain few (2–6) species. The Fertile Crescent arc contains 11–17 species and can be considered as the center of origin and development of most of the species. From this primary center that is characterized by sub-Mediterranean ecological conditions, the various species spread westward to more typical Mediterranean conditions or southward and eastward to steppical areas comprising more extreme environments.

In many parts of the distribution area, the genus has a massive, broad and almost continuous distribution. Species of *Aegilops* are found in almost every place except for high

Table 9.2 The occurrence of species of *Aegilops* in different countries

A. Diploid species[a]

Confirmed presence is designated by the sign+; and possible occurrence by the sign?

Country	Spe	Bic	Sha	Lon	Sea	Tau	Cau	Com	Uni	Umb	Species/Country
China (Himal.)						+					1
India (Kashmir)						+					1
Pakistan						+					1
Afghanistan						+	?				1(1)
Kyrgyzstan						+					1
Tajikistan						+					1
Kazakhstan						+					1
Uzbekistan						+					1
Turkmenistan						+					1
Iran	+					+	+			+	4
Iraq	+					+	+			+	4
Armenia						+				+	2
Azerbaijan						+				+	2
Georgia						+				+	2
Cis-Caucasia						+				+	2
Turkey	+					+	+	+	+	+	6
Syria	+		+	+		?	+			+	5(1)
Lebanon	+		+	+	+		+			+	6
Jordan	+	+		+	+						4
Israel	+	+	+	+	+						5
Egypt		+		+							2
Libya		+									1
Cyprus		+					+	+			3
Crimea	?					+					1(1)
Ukraine						?					0(1)
Greece + Crete	+						+	+	+	+	5
Albania							+		+		2
Serbia + Montenegro							+	+	+		3
Croatia							+		+		2
Slovenia							+	?	+		2(1)
North Macedonia							+	+	+		3
Bulgaria	+					+	+	+			4
Italy + Sicily									+		1
No. of countries	9(1)	5	2	5	4	18(2)	12(1)	7(1)	8	10	80(5)

B. Tetraploid species[b]

Confirmed presence is designated by+; and possible occurrence by?

Country	Ven	Cra	Cyl	Biu	Gen	Col	Neg	Kot	Per	Tri	Species/Country
Pakistan			+					+		+	3
Afghanistan		+	+					+		+	4
Kyrgyzstan		+	+							+	3
Tajikistan		+	+							+	3

(continued)

Table 9.2 (continued)

B. Tetraploid species[b]

Confirmed presence is designated by+; and possible occurrence by**?**

Country	Ven	Cra	Cyl	Biu	Gen	Col	Neg	Kot	Per	Tri	Species/Country
Kazakhstan		+								+	2
Uzbekistan		+	+					+		+	4
Turkmenistan		+	+				+	+		+	5
Iran		+	+	+	+	+	+	+	+	+	9
Iraq		+	+	+	+	+	+	+	+	+	9
Kuwait								+		+	2
Saudi Arabia								+	?	?	1(2)
Armenia		+	+	+	+	+	+			+	7
Azerbaijan		+	+	+	+	+	+	+		+	8
Georgia		+	+	+	+	+	+			+	7
Cis-Caucasia		+	+	+	?	?	+			+	5(2)
Syria		+	+	+	+	+	+	+	+	+	9
Lebanon		?	+	+	+	+	+	+	+	+	8(1)
Jordan		+	?	+	+			+	+	+	6(1)
Israel				+	+			+	+	+	5
Egypt					?			+	+		2(1)
Libya	+			+	+			+	+		5
Cyprus				+	+			?	+	+	4(1)
Crimea			+	+	+					+	4
Ukraine			+	+	?		+			+	4(1)
Greece + Crete			+	+	+	+	+		+	+	7
Albania			+	+	+		+			+	5
North Macedonia				+	+		+			+	4
Serbia + Montenegro				+	+		+			+	4
Croatia				+	+		+			+	4
Slovenia				+	+		+			+	4
Bulgaria			+	+	+		+			+	5
Romania			+	+	+					+	4
Hungary			+		+		?				2(1)
Italy + Sicily	+		+	+	+		+		+	+	7
France	+		+	+	+		?			+	5(1)
Spain	+			+	+					+	4(1)
Portugal	+		?		+					+	3(1)
Tunisia	+			+	+			?	+	+	5(1)
Algeria	+			+	+				+	+	5
Morocco	+			+	+				+	+5	5
No. of countries	8	14(1)	23(1)	28(1)	28(3)	8(1)	18(2)	15(2)	14(2)	36(1)	204(14)

(continued)

Table 9.2 (continued)

C. Hexaploid species

Confirmed presence is designated by+; and possible occurrence by?

Country	recta	Crassa 6X	vavilovii	juvenalis	Country/species
Uzbekistan				?	0(1)
Turkmenistan				+	1
Afghanistan		+			1
Iran		+		+	2
Iraq				+	1
Saudi Arabia			?		0(1)
Turkey	+			?	1(1)
Syria			+	?	1(1)
Lebanon			?		0(1)
Jordan			+		1
Israel			+		1
Egypt			+		1
Greece + Crete	+				1
Italy	+				1
France	+				1
Spain	+				1
Portugal	+				1
No. of countries	6	2	4(2)	3(3)	15(5)

After Kimber and Feldman (1987) and van Slageren (1994)

[a] *Speltoides* (Spe); *bicornis* (Bic); *sharonensis* (Sha); *longissima* (Lon); *searsii* (Sea); *tauschii* (Tau); *caudata* (Cau); *comosa* (Com); *uniaristata* (Uni); *umbellulate* (Umb)

[b] *Ventricosa* (Ven); *crassa* 4x (Cra); *cylindrica* (Cyl); *biuncialis* (Biu); *geniculata* (Gen); *columnaris* (Col); *neglecta* (Neg); *kotschyi* (Kot); *peregrina* (Per); *triuncialis* (Tri)

mountains and deserts (*Ae. bicornis, Ae. kotschyi, Ae. longissima,* and *Ae. crassa* even penetrate into semi-deserts areas). All the islands of the Mediterranean Sea are also inhabited by some species. Several species grow only in the Mediterranean region (e.g., *Ae. ventricosa, Ae. comosa, Ae. uniaristata, Ae. geniculata, Ae. biuncialis, Ae. sharonensis* and *Ae. peregrina*), others are restricted to the Irano-Turanian region (west Asiatic-central Asiatic regions; *Ae. crassa, Ae. kotschyi, Ae. vavilovii,* and *Ae. juvenalis)* while still others (*Ae. speltoides, Ae. caudata, Ae. cylindrica, Ae. umbellulata, Ae. triuncialis,* and *Ae. columnaris)* grow in both regions.

The genus has very flexible adaptation capabilities and it occupies a large number of the habitats existing in its distribution area. The altitudinal distribution of the genus is from 400 m below sea level (Dead Sea area) to 2700 m above sea level (asl) with a great variation among species (van Slageren 1994). The climate of many parts of the genus distribution area, especially in the Fertile Crescent arc, where presumably the genus originated and developed, has a short, mild and rainy winter and a long, hot and dry summer. The genus has adapted itself to the conditions characterizing this climate in that all the species are annuals (grow in the winter and pass the dry, hot summer as dormant seeds) and the species are predominantly self-pollinated and have large, well-protected grains for the safe and rapid reestablishment of the stand (Sakamoto 1973; Feldman 1976, 2001). The self-pollination trait enables rapid colonization of newly disturbed habitats as well as maintenance of colonized sites by adaptive genotypes.

All of the species have a, more or less, continuous distribution and usually occupy open habitats in the edges and openings of Mediterranean plant formations, in herbaceous park-forest formations (in which some of the species are natural components), in pastures, abandoned fields, edges of cultivation and roadsides (Zohary and Feldman 1962; Feldman 1963). Some of the habitats are primary in well-defined and balanced ecological conditions, while many are secondary, i.e., in disturbed and degraded areas. Many of the species also grow as weeds in cultivated fields.

In disturbed and newly opened habitats, some of the species (particularly the allotetraploids) can form massive and very dense stands, usually consisting of several species. Their genetic system provides them with the ability to colonize such newly opened areas quickly and efficiently.

The diversity in plant habitus and spike morphology enables adaptation of the various *Aegilops* species to a broad range of habitats. In addition, as has already been pointed out by Stebbins (1956), the adaptive specialization of the various *Aegilops* species is also reflected in their mode of seed dispersal: they have evolved complex and distinct fruiting spikes, which constitute highly efficient methods of fruit dissemination. Consequently, *Aegilops* species occupy a variety of primary and secondary habitats.

The distribution of the diploid species is as follows: three species (*Ae. speltoides*, *Ae. caudata* and *Ae. umbellulata*) are distributed in the central part of the genus distribution area. The remaining species of the Sitopsis section (*Ae. longissima*, *Ae. searsii*, *Ae. sharonensis* and *Ae. bicornis*) grow south of the center, and the species of section Comopyrum (*Ae. comosa* and *Ae. uniaristata*) are found west of the center, while *Ae. tauschii* is in the eastern part of the genus distribution (Table 9.1). Several diploid species have a relatively large distribution area (*Ae. umbellulata*, *Ae. caudata* and *Ae. tauschii*). The distribution of *Ae. tauschii* is very wide, due to its weediness and segetal growth habit (van Slageren 1994). Another diploid, *Ae. speltoides*, has a medium-sized distribution area, while others have smaller ones (*Ae. bicornis*, *Ae. searsii*, *Ae. sharonensis*, *Ae. longissima*, *Ae. uniaristata* and *Ae. comosa*). *Ae. sharonensis* is endemic to the coastal plain of Israel and south Lebanon. The pattern of the geographical distribution of the various diploid species indicates that the genus already underwent extensive differentiation in its early stages of development.

The allotetraploid species have, in general, a broader distribution than the diploids (Zohary and Feldman 1962; Feldman 1963; Kimber and Feldman 1987). Several allotetraploids, *Ae. neglecta*, *Ae. geniculata*, *Ae. biuncialis*, *Ae. triuncialis*, *Ae. cylindrica* and *Ae. crassa*, have a very broad distribution, other allotetraploids, *Ae. columnaris*, *Ae. ventricosa*, *Ae. peregrina* and *Ae. kotschyi*, have an intermediate distribution, while the hexaploid species, *Ae. recta*, *Ae. vavilovii*, *Ae. juvenalis*, and hexaploid *Ae. crassa*, have a somewhat more restricted distribution. The polyploids of the U-genome group (section Aegilops) are distributed in the central and western parts of the genus distribution (except for *Ae. triuncialis* and *Ae. columnaris*, which extend to the east, and *Ae. kotschyi*, which is found only in the south-east), those of the D-genome group (section Vertebrata) are in the eastern part (except for *Ae. ventricosa*, which is found in the western part), and the allotetraploid species of section Cylindropyrum, *Ae. cylindrica*, is distributed all over the central and northern part (Table 9.1). In most cases, the distribution of the allotetraploid species overlaps, completely or partly, with that of their putative diploid parents. Exception is *Ae. ventricosa*, which does not overlap with the distribution of either of its parents, the N genome and the D genome donors, namely, *Ae. uniaristata* and *Ae. tauschii*, respectively. In some cases, the donor of one of the allotetraploid subgenomes is unknown, e.g., the diploid donor of the X^c subgenome to *Ae. crassa*, *Ae. vavilovii* and *Ae. juvenalis* and the diploid donor of the X^n subgenome to *Ae. neglecta*, *Ae. recta* and *Ae. columnaris* (Dvorak 1998).

The distribution areas of the allotetraploid species of the U-genome group and of *Ae. cylindrica* from the D-genome group, are larger than those of each of their diploid parents (only that of *Ae. columnaris* and *Ae. peregrina* is equal or somewhat smaller than that of *Ae. umbellulata*). The distribution area of the allotetraploids and allohexaploids of the D-genome group, is smaller than that of *Ae. tauschii*, the D-genome donor. The distribution of *Ae. ventricosa* is larger than that of the diploid donor of its second genome, N.

There is a large difference in morphological variation in diploid versus tetraploids. While the diploids have clear-cut boundaries or morphological discontinuities, the tetraploids are characterized by blurred morphological boundaries. This feature of the tetraploid species was already described by Zhukovsky (1928) and Eig (1929a). Eig, in particular, reported on overlapping variation ranges and presence of intermediate linking forms between the various tetraploid species.

There are also striking differences between the patterns of geographical distribution and ecological affinities of the diploids and allotetraploid species (Zohary and Feldman 1962; Feldman 1963). All the diploids are distributed either in or around the center of the genus distribution area (except for *Ae. tauschii*, which grows in the eastern part). They are relatively restricted in their distribution (Table 9.1) and are much more specialized than the tetraploids in their ecological requirements, usually occupying well-defined habitats with specific edaphic or climatic conditions. Some of the diploids (*Ae. tauschii*, *Ae. umbellulata*, *Ae. caudata* and, to some extent, *Ae. speltoides*) show wider ecological amplitudes, which correlate with their weedy and segetal tendency.

In contrast, the allotetraploids have a larger distribution area and wider ecological amplitude than the diploids. *Ae. triuncialis*, *Ae. geniculata*, *Ae. neglecta*, *Ae. biuncialis* and *Ae. cylindrica* occupy large parts of the distribution area of the genus. The tetraploids do not show the marked ecological specificity of the diploids, as evidenced by their growth in a very wide array of edaphic and climatic conditions. Their weedy nature is reflected in the ability to rapidly and efficiently colonize a variety of newly disturbed and secondary habitats. Undoubtedly, the expansion of agriculture and the opening up of many segetal habitats (in cultivated areas), played a key role in the massive distribution of these tetraploid species throughout the range of the genus.

Many tetraploid species are sympatric and tend to grow in mixed stands, usually with several species in each population (Feldman 1965a). However, in various parts of the genus distribution, one tetraploid is the dominant species. *Ae. cylindrica* is such a species in the northern part of the genus distribution, *Ae. triuncialis* in the central and western parts, *Ae. peregrina* in the southern part, *Ae. kotschyi* in the southeastern part and *Ae. crassa* in the eastern part. This interregional kind of vicarism reflects the ability of the various tetraploids to adapt themselves to different climatic conditions.

The diploid species tend to grow in separate habitats, sometimes mixed with tetraploid species. The allotetraploids, in sharp contrast, usually grow intermingled with other tetraploid *Aegilops* species. In most of the localities studied in Israel, Turkey and Greece, the tetraploid species tend to form mixed populations (Zohary and Feldman 1962; Feldman 1965a). This phenomenon is especially apparent in Turkey, where many tetraploid species occur sympatrically. There, it is possible to find mixed populations which consists of five or even six tetraploid species (*Ae. triuncialis, Ae. biuncialis, Ae. neglecta, Ae. columnaris, Ae. geniculate* and *Ae. cylindrica*). In such mixed populations, each species typically exhibits variation in morphological traits and represents by several lines which differ morphologically from one another. The number of distinct morphological lines of each species in a given mixed population is generally related to the number of its individuals, with the most prominent species also tending to be the most variable.

One may assume that the tendency of the allotetraploid species to form polymorphic mixed populations consisting of several species, increases the frequency of genetic contact between them and facilitates interspecific hybridizations and gene flow. Indeed, detailed analyses of several mixed populations demonstrated that the intraspecific variation characterizing each species was partly the result of introgression: some of the lines of each species represented established hybrid derivatives that differed from one another (Zohary and Feldman 1962; Feldman 1965a).

In contrast to the broad distribution of the tetraploid species, the distribution area of the hexaploid species of *Aegilops* (*Ae. vavilovii, Ae. juvenalis, Ae. recta* and hexaploid *Ae. crassa*) is smaller than that of their tetraploid and diploid parents. Only *Ae. recta* have a larger distribution area than that of its diploid parent, *Ae. uniaristata*. In addition, the ecological amplitudes of the hexaploids are much more restricted than those of their ancestral tetraploids and even than those of their diploid parents. They grow in a smaller number of habitats and often only sporadically. The morphological variation of the hexaploids is also relatively limited.

According to van Slageren (1994), the genus *Aegilops* L. is subdivided into the following five sections: Sitopsis (Jaub. & Spach) Zhuk., Vertebrata Zhuk. emend. Kihara, *Cylindropyrum (Jaub. & Spach) Zhuk., Comopyrum (Jaub. & Spach) Zhuk.*, and *Aegilops*.

9.3 Cytology and Cytogenetics

9.3.1 General Description

The species of *Aegilops* comprise an allopolyploid series with diploids (2n = 2x = 14), allotetraploids (2n = 4x = 28) and allohexaploids (2n = 6x = 42) (Kihara 1954). Genome size (1C DNA amount) in the diploid species ranges from 4.84 pg in *Ae. caudata* to 7.52 pg in *Ae. sharonensis*, in the tetraploid species, it ranges from 9.59 pg in *Ae. cylindrica* to 12.64 pg in *Ae. kotschyi*, and in the hexaploid species, from 16.22 pg in *Ae. recta* to 17.13 pg in *Ae. vavilovii* (Eilam et al. 2007, 2008; Table 9.3). The karyotype of most species is symmetric, with median or submedian centromeres, with the exception of *Ae. caudata, Ae. umbellulata, Ae. comosa, Ae. uniaristata* and allopolyploids containing subgenomes that derived from these diploids and have an asymmetric karyotype (Senyaninova-Korchagina 1932; Chennaveeraiah 1960). All species contain two chromosome pairs with a satellite (SAT chromosomes), except for *Ae. tauschii* and *Ae. uniaristata* that have only one satellite pair (Chennaveeraiah 1960). Several allopolyploid species contain the sum of the SAT chromosomes of their diploid progenitors, but many more exhibit a smaller number due to amphiplasty, i.e., nucleolar dominance (Chennaveeraiah 1960).

The satellite is a chromosome segment that is separated from the rest of the chromosome by a constriction, called the secondary constriction, whose region is active in nucleolus formation and referred to as nucleolar organizer region (NOR) (McClintock 1934). This NOR contains ribosomal (rDNA) genes that code for the 18S-5.8S-26S (18S-26S). Dubcovsky and Dvorak (1995b) and Badaeva et al. (1996b) presented evidence showing that the major NOR loci are located in homoeologous groups 1, 5, and 6 of the diploid *Aegilops* species. Using in situ hybridization, additional minor 18S-26S rDNA loci were detected in the genomes of several diploid *Aegilops* species (Badaeva et al. 1996b).

The 5S rDNA loci have been mapped on chromosomes of the diploid *Aegilops* species using in situ hybridization (Appels et al. 1980; Castilho and Heslop-Harrison 1995; Friebe et al. 1995c; Badaeva et al. 1996b). Each of the diploid *Aegilops* species had either one or two 5S rDNA loci on chromosomes of groups 1 and (or) 5 (Badaeva et al. 1996b), either located on the same chromosome arm as the 18S-26S rDNA loci, but unlinked to them, or on different chromosomes (Dvorak et al. 1989; Badaeva et al. 1966b).

Table 9.3 Nuclear and organellar genome, and genome size of the species of *Aegilops*

Species[a]	Genome[b]		Genome size[c] (Mean ± SD of 1C DNA in pg)
	Nuclear	organellar	
speltoides Taush	SS	S, G, G^2	5.81 ± 0.123
bicornis (Forssk.) Jaub. & Spach	$S^b S^b$	S^b	6.84 ± 0.097
longissima Schweinf. & Muscl	$S^l S^l$	S^{l2}	7.48 ± 0.082
sharonensis Eig	$S^{sh} S^{sh}$	S^l	7.52 ± 1.000
searsii Feldman & kislev ex Hammer	$S^s S^s$	S^v	6.65 ± 0.091
tauschii Coss	DD	D	5.17 ± 0.087
ventricosa Tausch	DDNN	D	10.64
crassa Boiss	$D^c D^c X^c X^c$; $D^c D^c X^c X^c$DD	D^2	10.86; 15.90^d
Vavilovii (Zhuk.) Chennav	$D^c D^c X^c X^c S^s S^s$	D^2	17.13 ± 0.139
juvenalis (Thell.) Eig	$D^c D^c X^c X^c$UU	D^2	–
caudata L	CC	C	4.84 ± 0.089
cylindrica Host	DDCC	D	9.59
comosa Sm. In Sibth. & Sm	MM	M, M^h	5.53 ± 0.052
uniaristata Vis	NN	N	5.82 ± 0.105
umbellulata Zhuk	UU	U	5.38 ± 0.073
geniculata Roth	$M^o M^o$UU	M^o	10.29 ± 0.008
biuncialis Vis	$UUM^b M^b$	U	10.37 ± 0.037
columnaris Zhuk	$UUX^n X^n$	U'	10.86
neglecta Req. ex Bertol	$UUX^n X^n$	U	10.64 ± 0.404
recta	$UUX^n X^n$NN	U	16.22
kotschyi Boiss	$S^v S^v$UU	S^v	12.64 ± 0.183
peregrina (Hack. In J. Fraser) Maire & Weiller	$S^v S^v$UU	S^v	12.52 ± 0.181
triuncialis L	UUCC; CCUU	U, C^2	9.93 ± 0.041

[a] Species according to van Slageren (1994)

[b] Nuclear genome designations according to Dvorak (1998), and Badaeva et al. (2004); the genome of the maternal parent in the allopolyploids is given first; organellar genome designation according to Wang et al. (1997), Ogihara and Tsunewaki (1988), and Tsunewaki (2009); X^n genome is not homologous to any diploid genome; its origin is currently unknown; X^c genome corresponds to that of an extinct ancestor of section Sitopsis (designated S^{sit})

[c] Genome size from Eilam et al. (2007, 2008)

[d] Genome size of *Ae. crassa* 6x from Naghavi et al. (2013)

The variation in chromosomal location and position of major NORs and 5S rDNA loci and the number and distribution of minor NOR loci are characteristic for each diploid species (Badaeva et al. 1996b). The rDNA and the 5S rDNA loci are mobile (Dubcovsky and Dvorak 1995b).

Among the species of section Sitopsis, *Aegilops longissima*, *Ae. sharonensis*, *Ae. searsii* and *Ae. bicornis* have major NOR loci on chromosomes of groups 5 and 6 and a variable number of minor loci on chromosomes of groups 1, 3, 5, and 6 (Badaeva et al. 1996b). The 5S rDNA loci were observed on chromosomes of groups 1 and 5. On the other hand, *Ae. speltoides* have a different distribution pattern of NOR and 5S rDNA loci, similar to that of *Amblyopyrum muticum* (Badaeva et al. 1996b). The major NOR loci are

located on chromosomes of groups 1 and 6 and only one 5S rDNA locus was found on the short arm of chromosomes of group 5. Likewise, the distribution of major NOR and 5S rDNA loci in *Ae. comosa* was similar to that in *Ae. speltoides* (and in *A. muticum*), except that minor NOR loci were observed in all seven chromosome pairs. The distribution patterns of NOR and 5S rDNA loci in *Ae. umbellulata* and *Ae. caudata* were identical; both loci were located on the short arm of chromosomes of groups 1 and 5 (Gerlach et al. 1980; Miller et al. 1983). However, an additional pair of NOR sites was observed in *Ae. umbellulata* (Badaeva et al. 1996b).

The distribution of NOR and 5S rDNA loci in *Ae. uniaristata* and *Ae. tauschii* is distinct, with only one major

NOR locus in the short arm of chromosome 5N and 5D, respectively, and several minor NORs on the short and long arms of chromosomes 1, 6 and 7 in *Ae. uniaristata*, and on the long arm of chromosome 7 in *Ae. tauschii* (Badaeva et al. 1996b). Two 5S rDNA loci exist in both species—one in the long arm of chromosomes of group 1 and the other in the short arm of chromosomes of group 5 of *Ae. uniaristata*. In *Ae. tauschii*, these two loci are located on the short arm of chromosomes of groups 1 and 5. This mode of distribution of the two 5S rDNA loci is similar in most other diploid *Aegilop*s species (Badaeva et al. 1996b).

In many *Aegilops* allopolyploid species, the rDNA genes of one parental set are transcribed, while most or all of the rDNA genes inherited from the other parent are silent or absent, a phenomenon known as a nucleolar dominance (Navashin 1928, 1934; Pikaard 1999, 2000). Nucleolar dominance occurs in almost all allopolyploid species of *Aegilops* (Cermeño et al. 1984b; Feldman et al. 2012). In the allopolyploid species containing the U subgenome of *Ae. umbellulata*, the U genome completely suppresses the NOR activity of the M, S and D subgenomes of the allopolyploids (Cermeño et al. 1984b).

Since most chromosomes of *Aegilops* species are morphologically indistinguishable from one another at mitotic metaphase, several methods were developed to identify individual chromosomes, for studying chromosome structure and organization and for genome analysis. Two widely used methods are C-banding and fluorescence in situ hybridization (FISH). The C-banding technique detects heterochromatic regions of the chromosomes, whose distribution can be chromosome- and species-specific. The C-banding patterns of chromosomes at mitotic metaphase were studied in all diploids (Teho and Hutchinson 1983; Teho et al. 1983; Friebe et al. 1992a, 1993, 1995a; Badaeva et al. 1996a, b) and polyploid species (Badaeva et al. 2002, 2004, 2011) of *Aegilops*. Chromosomes of all species show a distinctive and characteristic C-banding pattern, enabling the identification of their individual chromosomes. The results of the above-mentioned studies indicated that the total amount and the type of distribution of the heterochromatic regions in chromosomes were species-specific. In addition to differences in chromosome morphology (Chennaveeraiah 1960), karyotypes of all species could be distinguished by the distribution of heterochromatic regions in all their chromosomes.

The FISH method exploits repetitive DNA sequences, e.g., probe pSc119 isolated by Bedbrook et al. (1980) from the genome of *Secale cereale*, or probe pAs1 isolated by Rayburn and Gill (1986) from *Aegilops tauschii*. This method facilitates the precise location of such sequences on chromosomes may be chromosome-, species-, and probe-specific (Jiang and Gill 2006). A comparative analysis of in situ hybridization with the highly repetitive DNA

sequences pSc119 and pAs1 in all the diploid *Aegilops* species confirmed significant differentiation of their genomes (Badaeva et al. 1996a). In addition to interspecific differences, significant intraspecific polymorphism was also detected in the distribution of repetitive DNA sequences (Badaeva et al. 1996a).

9.3.2 Structure and Distribution of Repetitious DNA

The genomes of diploid and allopolyploid *Aegilops* species are very large (Eilam et al. 2007, 2008; Table 9.3), and are comprised of about 85–90% repeated nucleotide sequences (Flavell et al. 1979), most of them being transposable elements (TEs), primarily families of retrotransposons (Li et al. 2004; Wicker and Buell 2009; Yaakov et al. 2013; Senerchia et al. 2013; Jia et al. 2013). TEs have the potential to affect genome structure, function and size through transposition (Bennetzen 2005; Slotkin and Martienssen 2007; Fedoroff 2012) and, so, differential proliferation of TEs is considered to be one of the main driving forces of genome size variation in the Triticeae (Charles et al. 2008). It is therefore likely that the large differences in genome size between the various *Aegilops* species (Eilam et al. 2007, 2008; Table 9.3), derived from differential proliferation of TEs that were active during the speciation processes of these species and that they played an important role in their genomic evolution (Yaakov et al. 2013).

Middleton et al. (2013) found that the abundance of several TE families significantly differs between the Triticeae species, indicating that TE families can thrive extremely successfully in one species while going virtually extinct in another. In this regard, Senerchia et al. (2013) found that ancestral TE families followed independent evolutionary trajectories in several *Aegilops* species, highlighting the evolution of TE populations as a key factor of genome differentiation. Already in 1979, Flavell et al. (1979) showed that DNA of different *Aegilops* species hybridized to differing extents with a repetitive probe that derived from *Ae. speltoides*. These results are consistent with the hypothesis that speciation has been accompanied by quantitative changes in the repeated sequence complements of genomes (Flavell et al. 1979).

Badaeva et al. (1996a), using in situ hybridization with two highly repetitive DNA sequences, pSc119 from *Secale cereale* (Bedbrook et al. 1980) and pAs1 from *Ae. tauschii* (Rayburn and Gill 1986), studied genome differentiation in all diploid *Aegilops* species. While chromosomes of all the diploid species hybridized with the pSc119 probe, the level of hybridization and labeling patterns differed among genomes. Only three species, *Ae. tauschii*, *Ae. comosa*, and *Ae.*

uniaristata, showed distinct hybridization with pAs1. The labeling patterns were species-and chromosome-specific, confirming significant differentiation of their genomes.

Similar conclusion was reached by Yaakov et al. (2013), who assessed the relative copy number of 16 TE families in *Aegilops* species of section Sitopsis and in *Ae. tauschii*. They reported on a wide variation and genome-specificity of TEs in these species. Likewise, Senerchia et al. (2013) investigated genome restructuring and assessed the evolutionary trajectories of 17 long-terminal repeat (LTR) retrotransposon families after allopolyploidization events. Comparisons between these retrotransposons of the diploid progenitors and the allopolyploids highlighted the proliferation of several TE families and the predominant sequence deletion in others, indicating species-specific and TE-specific evolutionary trajectories following allopolyploidy.

9.3.3 Gametocidal Chromosomes in the Genus *Aegilops*

9.3.3.1 Opening Remarks

During recurrent backcrossing to produce addition lines of chromosomes from several *Aegilops* species to common wheat and during recurrent backcrossing to produce alloplasmic lines of wheat containing cytoplasm of *Aegilops* species substituting the wheat cytoplasm, it was found that certain alien chromosomes from a number of *Aegilops* species were preferentially transmitted to the offspring (Endo 2007, 2015). When introduced in a single dose to durum or common wheat, as in F_1 hybrids, in backcrossed progeny to wheat, in monosomic addition or monosomic substitution, these chromosomes, ensured their endurance in wheat by inducing severe chromosomal breakage in gametes lacking them, thus, causing their abortion and consequently, leading to their preferential transmission to the offspring of gametes possessing the gametocidal chromosome (Endo 1982, 1985, 1990; Maan 1975; Finch et al. 1984). In consequence, a severe reduction in the fertility of both sexes in wheat plants having a monosomic addition or substitution of one of the gametocidal chromosomes was observed (Endo 1985). Self-pollination of lines carrying such a chromosome yields offspring predominantly bearing a disomic addition or disomic substitution, and restored fertility. These *Aegilops* chromosomes are termed gametocidal (*Gc*) chromosomes (Endo 1979, 1982, 2007) or "cuckoo" chromosomes (Miller et al. 1982), and the genes that are responsible for the gametocidal action are called *Gc* genes (Endo 1982). Using FISH with a probe of a repetitive DNA sequence that marks the Gametocidal (*Gc*) gene, Friebe et al. (2003) directly demonstrated that chromosome breakage in pollen mitosis occurred only in gametes lacking this gene. The Gc chromosomes derived

from different *Aegilops* genomes (C, S, S^{sh}, S^l and M^o) and belong to four different homoeologous groups: 2, 3, 4, and 6 (Endo 2007, 2015). Currently known species possessing gametocidal chromosomes are: *Ae. sharonensis* (genome $S^{sh}S^{sh}$) (Maan 1975; Miller et al. 1982; Tsunewaki and Tsujimoto 1983; Tsujimoto and Tsunewaki 1985b; Endo 1990, 2007, 2015), *Ae. longissima* (genome S^lS^l) (Maan 1975; Endo 1990), *Ae. speltoides* (genome SS) (Tsujimoto and Tsunewaki 1984, 1988), *Ae. caudata* (genome CC) (Endo and Katayama 1978; Endo 1985), *Ae. triuncialis* (genome CCUU) (Endo and Tsunewaki 1975), *Ae. cylindrica* (genome CCDD) (Endo 1979) and *Ae. geniculata* (genome UUM^{o}M^o) (Friebe et al. 1999).

The *Ae. sharonensis* and *Ae. longissima* Gc chromosomes were first discovered by Maan (1975). Three such chromosomes were identified in different *Ae. sharonensis* accessions, (1) by Endo 1982, (2) by Miller et al. 1982, and (3) by Maan 1975, and two such chromosomes in different *Ae. longissima* accessions, (1) by Maan 1975, and (2) by Panayotov, cited in Endo (1985). The cytological features, homoeology and interrelation of the different Gc chromosomes were studied (Endo 1985). One of the *sharonensis* Gc chromosomes (no. 1) and one of the *longissima* Gc chromosomes (no. 2) have the same gametocidal action, and both have an N-banding pattern that resembles that of wheat chromosome 2B, and thus, are homoeologous to wheat group 2, and successfully substitute for any wheat chromosome of wheat homoeologous group 2. The other two *sharonensis* Gc chromosomes (nos. 2 and 3) and the *longissima* Gc chromosome (no. 1) are homoeologous to wheat group 4 (Miller et al. 1982), and exhibit relatively similar N-banding patterns. The group 4 chromosomes showing the same gametocidal action, all showed an N-banding pattern rather similar to that of wheat chromosome 4A. Dvorak (1983) presented the similarity between the C-banding pattern of the *sharonensis* Gc chromosome (no. 2) and wheat chromosome 4A, as evidence supporting his view that chromosome 4A of *T. aestivum* was contributed by a species of the section Sitopsis and, consequently, belongs to the B genome. The gametocidal chromosomes of group 2 were designated Gc1 and those of group 4 designated Gc2 (Endo 1985, 2007).

Two *Gc* genes, derived from two different strains, were found in *Ae. speltoides* (Tsujimoto and Tsunewaki 1984, 1988). These two genes are allelic, located on the Gc chromosome 2S, which is homoeologous to wheat group 2, and consequently, designated *Gc1a* and *Gc1b* (Tsujimoto and Tsunewaki 1984, 1988). The two alleles when present in monosomic addition of the Gc chromosome, differ in their ability to induce damage to the offspring of plants lacking the Gc chromosome; *Gc1a* causes endosperm degeneration and chromosome aberrations, whereas *Gc1b* results in abnormal seed lacking the shoot primordium. No correlation between embryo or endosperm degeneration and chromosome breakage was observed (Tsujimoto and Tsunewaki 1984, 1988).

The Gc chromosomes of *Ae. caudata* and *Ae. triuncialis* are homoeologous to wheat chromosomes of group 3 (Endo and Tsunewaki 1975). The morphology, pairing homology, selective gametocidality, and effects on plant growth of gametocidal chromosomes of natural and of synthetic *Ae. triuncialis* are almost the same as those of *Ae. caudata* (Endo 1979). On the other hand, the *Ae. cylindrica* Gc chromosome is homoeologous to group 2 (Endo 1979). It differs from the Gc chromosomes of *Ae. triuncialis* and *Ae. caudata* in its characteristic appearance and many aborted seeds. In addition, the centromere of the *cylindrica* Gc chromosome is not so extremely subterminal as that of the *caudata* and *triuncialis* Gc chromosome, and the selective gametocidal action of the *cylindrica* Gc chromosome is not effective in the types of common wheat where the *caudata* and *triuncialis* chromosomes exert their gametocidal effect. In respect of the selective gametocidal chromosome, therefore, the C genome of *Ae. cylindrica* is farther differentiated than that of *Ae. caudata* and of *Ae. triuncialis*.

Chromosome $4M^o$, which is homoeologous to wheat group 4 chromosomes, is the Gc chromosome of *Ae. geniculata* (Kynast et al. 2000). When transferred to cv. Chinese Spring of bread wheat as a monosomic addition, it induces chromosome breakage and anaphase bridges at anaphase and telophase of the first and second pollen mitosis. Gc-induced multicentric and ring chromosomes, among other chromosomal aberrations, can be transmitted to the offspring and initiate breakage-fusion-bridge cycles in dividing root tip meristem cells of the derived sporophytes.

9.3.3.2 Interaction Between Gametocidal (*Gc*) Genes

In double monosomic additions of common wheat with the gametocidal chromosomes of *Ae. sharonensis*, $2S^{sh}$ and $4S^{sh}$, only gametes carrying the alien chromosome $4S^{sh}$ were functional. Hence, there are two types of gametocidal chromosomes in *Ae. sharonensis*, with the $2S^{sh}$ chromosome being weaker than the $4S^{sh}$ (Endo 1985).

When observing double monosomic addition lines derived from three different Gc chromosomes, Endo (1982) found that the Gc gene of *Ae. triuncialis* does not interact with the Gc gene of *Ae. longissima* or *Ae. sharonensis*. In addition, he found that the activity of the Gc gene of *Ae. longissima* dominated over that of *Ae. sharonensis*. Endo (1985) further reported that Gc genes located on chromosome $4S^l$ of *Ae. longissima* or 4^{sh} of *Ae. sharonensis* are epistatic to those on chromosome $2S^l$, irrespective of species. Hence, the Gc chromosomes derived from *Ae. triuncialis*, *Ae. sharonensis* and *Ae. longissima* were found to differ in gametocidal action, as well as in morphology and homoeology (Endo 1982). The mode of action of the Gc genes of *Ae. longissima*, *Ae. sharonensis* and *Ae. speltoides* differs from that of Gc genes in *Ae. triuncialis* and *Ae. cylindrica* (Tsujimoto and Tsunewaki 1985b; Tsujimoto and Noda 1989; Endo 1988b).

Using plants with two different Gc genes, Tsujimoto (1995) investigated the functional relationship between six Gc genes and concluded that there are three functional groups. The first group included Gc genes located on the chromosomes of homoeologous group 2; the *Gc1* genes of *Ae. speltoides* showed similar function to those on chromosome $2S^{sh}$ of *Ae. sharonensis*. The second group included the Gc genes on chromosomes $4S^{sh}$ of *Ae. sharonensis* and $4S^l$ of *Ae. longissima*, i.e., the *Gc2* genes. These genes were epistatic to the *Gc1* genes in the first group in terms of gamete abortion and preferential transmission, as indicated by Endo (1985). Although, by themselves, the Gc genes in the first group only cause chromosome breakage at a low frequency (Tsujimoto and Tsunewaki 1985b; Tsujimoto and Noda 1989), they highly enhance breakage by Gc genes of the second group. Conversely, the Gc genes in the second group may enhance breakage induced by those in the first group.

The third group included the Gc gene on chromosome 3C of *Ae. triuncialis*, proved to have activity independent of that of the Gc genes of the first or second group. The function of the *triuncialis* Gc gene is suppressed by an inhibitor, *Igc1*, located on chromosome 3B of some common wheat lines. Based on the interactions between the different Gc genes, Tsujimoto (1995) proposed re-designation of the gene symbols following the rules for gene symbolization in wheat (McIntosh 1988). Tsujimoto (1995) proposed the name *Gc1* for the Gc genes in the first group, *Gc2* for the Gc genes in the second group, and *Gc3* for the Gc genes in the third group, with each designation followed by the name of the genome carrying the gene. The relationship between these Gc genes and those on chromosome 2C of *Ae cylindrica*, chromosome $4M^o$ of *Ae. geniculata* and chromosome 6S of *Ae. speltoides* has not yet been examined.

9.3.3.3 Mode of Action of Gc Genes

The explanation to the phenomenon of differential transmission of Gc chromosomes in monosomic additions or disomic substitutions of *Aegilops* Gc chromosomes to common wheat, is that meiospores (microspores and megaspores) with the alien chromosome develop into normal gametophytes, while meiospores lacking the alien chromosome exhibit a wide range of chromosome and chromatid aberrations at first gametophytic mitosis (Endo 2015, and reference therein). This would explain the partial male and female sterility in the monosomic addition or substitution lines. Hence, Gc genes have a dual function, i.e., to induce chromosomal mutations in gametes that lack them and to suppress such mutations in gametes that carry them. Indeed, Endo (1990) and Tsujimoto (2006) hypothesized that two genetic factors are associated with the preferential transmission of the Gc chromosome, the breaker (*GcB*) that induces chromosome breakage, and the inhibitor that prevents chromosome breakage. Chromosome aberrations do not occur in gametes carrying both elements, as the inhibitor neutralizes the gametocidal action. This hypothesis is supported by the discovery of, *Igc1*, an inhibitor of the *Gc* of *Ae. cylindrica* on 2C (Tsujimoto and Tsunewaki 1985a), and by the isolation of a knockout mutation of the breaker gene of the *Gc* of *Ae. sharonensis* on $4S^{sh}$, which renders the breakage function ineffective, while having no influence on the latter inhibition function (Friebe et al. 2003). The *Ae. sharonensis* $4S^{sh}$ breaker element (*Gc1B*) has been mapped, by C-banding, to the distal end of the long arm of chromosome $4S^{sh}$ (Endo 2007). Knight et al. (2015) confirmed this reported location and more specifically defined its location in a region proximal to the sub-telomeric heterochromatin of this chromosome arm. However, the molecular mechanism of the effect of the GcB has not yet been elucidated (Tsujimoto 2006).

It is not known if *Gc* genes are functional in their species of origin, i.e., in intraspecific hybridization between lines bearing and those lacking Gc chromosomes, nor in interspecific hybrids between various diploid and allopolyploid *Aegilops* species bearing with those lacking Gc chromosomes. Also, it is not known if some lines of diploid *Aegilops* and *Triticum* species possess a gene(s) that suppresses the function of the *GcB* gene.

9.3.3.4 Modification of *Gc* Action

The *Gc* genes action varies, depending on the common wheat cultivar into which a GC chromosome is introduced (Endo 1988b). Chromosome 2C of *Ae. cylindrica*, for instance, has complete Gc action and is therefore exclusively transmitted to progeny in the common wheat cultivar Jones Fife (JF), whereas its Gc action becomes incomplete in the common wheat cultivar Chinese Spring (CS), where chromosome 2C is lost in part of the progeny (Endo 1988a). Chromosome 3C of *Ae. triuncialis* has severe Gc action in CS and some other common wheat cultivars, but it displays almost no Gc action in Norin 26 (N26), which possesses the *Igc1* Gc-inhibitor gene on chromosome 3B (Tsujimoto and Tsunewaki 1985a). In both cases of incomplete Gc action, semi-lethal chromosomal mutations occur in gametes lacking the Gc chromosome, and structurally rearranged chromosomes are transmitted to the progeny.

Endo (1978) described evidence for the existence of suppressor(s) of Gc genes in certain cultivars of common wheat. He reported that monosomic addition of Gc chromosome 3C of *Ae. triuncialis* correlated with male and female semi-sterility in the genetic backgrounds of the common wheat cultivars JF and CS, whereas semi-sterility did not appear in the background of cultivar N26. Chromosome 3C was preferentially transmitted to the next generation of both sides in JF but only of the female side in CS. Although Endo (1978) did not mention preferential transmission of Gc chromosome 3C in N26, recovery of fertility in the background of this cultivar indicated that gametes lacking chromosome 3C were also normally transmitted.

Tsujimoto and Tsunewaki (1985a) analyzed the genetic factor in N26 that suppresses the Gc action of *Ae. triuncialis* chromosome 3C. They crossed the disomic addition line of Chinese Spring carrying chromosome 3C (21″w + 1″ae) with the F_1 progeny of the hybrid Chinese Spring × N26. In the resultant monosomic addition lines, fertile and semi-sterile plants segregated 1:1, indicating that a dominant suppressor gene, termed *Igc1*, inhibits the action of the *Gc* gene on chromosome 3C. By monosomic analysis Tsujimoto and Tsunewaki (1985a) localized *Igc1* to chromosome 3B of N26. The facts that both the *Gc1* gene and its suppressor are located on chromosomes of the same homoeologous group, may indicate relationships between these two genetic factors, i.e., that *Igc1* is an antimorph allele of a Gc gene, acting antagonistically to the *triuncialis Gc1* gene. Since *Igc1* exists only in a number of common wheat cultivars (see below), it is reasonable to assume that it did not derive from the diploid donor of the B genome, but rather, evolved at the polyploid level to counteract the action of *Gc* genes.

In the JF genetic background, both male and female gametes without chromosome 3C were abortive, whereas in the CS background, pollen without the Gc chromosome functioned and transmitted to the progeny (Tsujimoto and Tsunewaki 1985a). This result suggests the existence of an incomplete suppression in the CS background. In addition, no suppressors for the *Ae. longissima*, *Ae. sharonensis*, or *Ae. speltoides* Gc chromosome actions were discovered among the hundreds of common wheat cultivars tested so far (Tsujimoto, unpublished, cited in Tsujimoto and Tsunewaki 1985a).

Tsujimoto and Tsunewaki (1985a) and Tsujimoto and Tsunewaki (1988) studied the distribution of the inhibitor gene, *Igc1*, by crossing many cultivars of common wheat with CS following a disomic addition of *Ae. triuncialis* chromosome 3C, and classifying them into *Igc1* carriers and non-carriers, based on the seed fertility of the F_1 progeny. They found that the *Igc1* gene exists in Japanese and East and Southwest China cultivars but not in American, African, and Asiatic cultivars. Interestingly, cultivar CS, originally from the Sichuan Basin in China (Yen et al. 1988), is an *Igc1* non-carrier in spite of the fact that most Sichuan landraces are carriers. The Japanese cultivar Norin 10, which was used as the source of the semi-dwarf genes *Rht1* and *Rht2* in modern wheat breeding, carries *Igc1*, but the suppressor gene did not pass into the high-yielding cultivars containing these dwarfing genes.

Tsujimoto and Tsunewaki (1985b) made note of the fact that the phenomena associated with Gc genes in wheat are similar to those observed in association with hybrid dysgenesis in the fruit fly, *Drosophila*. These include sterility, lethality, mutation, chromosome breakage, male recombination and segregation distortion, all of which appeared only in the F$_1$ progeny of a cross between P or I lines of *Drosophila* males and M or R strain females (Crow 1983; Bregliano and Kidwell 1983). Later, Tsujimoto and Noda (1989) noted the similarity between the nature of Gc genes and the restriction-modification systems found in many bacteria. In bacteria, a restriction endonuclease in the host cuts alien DNA at or around a particular base sequence. The host DNA, by contrast, is protected from digestion by methylation. This restriction-modification system provides a mechanism that could explain chromosome breakage in gametogenesis and in zygotic cells in wheat (Tsujimoto and Tsunewaki 1985b). Tsujimoto and Tsunewaki (1985b) proposed a model for Gc action in which a *Gc* gene produces both a restriction enzyme (RE) and a modification enzyme (ME), e.g., a methylase. The RE cleaves the specific restriction sites that it recognizes. But, when these sites are protected by DNA methylation, caused by the ME, the RE cannot cleave them. This would be the case in homozygotes for the Gc gene, where no chromosome breakage occurs in any gametes. If ME function is incomplete and cannot protect all of the restriction sites, chromosome breakage may appear at some frequency in all gametes, regardless of presence of the *Gc* genes.

In plants hemizygotic for a *Gc* gene, haploid cells without the *Gc* gene are generated after meiosis. Since DNA replication occurs prior to the first mitotic division in the gametogenesis, cells lacking the *Gc* gene and therefore, the ME, will not contain modified restriction sites on one of the strands of the replicated DNA. If the RE remains in the cell longer than the ME, or if RE can be supplied by other cells (for example, egg or pollen mother cells), the unmodified restriction sites are broken by the RE. In the following mitosis, unmodified DNA is broken in the same manner. Thus, the gametes without Gc become abortive.

Interestingly, de las Heras et al. (2001) observed that treatment of plants carrying the *Ae. sharonensis* Gc gene, with the hypomethylation agent 5-azacytidine, induced chromosome breakage in root tip cells. This result supports that the process of chromosome breakage in early seed development was repressed by DNA methylation.

9.3.3.5 Evolutionary Significance of *Gc* Genes

At the diploid level, *Gc* genes can restrict intraspecific gene exchange between two different populations that possess non-compensating *Gc* genes (Endo 2015). If these *Gc* genes are alleles, located on homologous chromosomes, the intraspecific hybrid between these two populations will be completely sterile, as all gametes will include one of these incompatible *Gc* alleles. If the two different *Gc* genes are on non-homologous chromosomes, one-fourth of the gametes produced by the hybrid will be fertile. Thus, sexual isolation will be established within a species between two populations that easily cross-fertilize. As a result, the two populations can develop independent of one another and gradually diverge to two different species. Similarly, two closely related species can be sexually isolated and undergo independent evolutionary development.

An example of sexual isolation within a species is seen in hybrids between two allopatric accessions of *Ae. caudata*, which have normal meiotic chromosomal pairing, but produce completely sterile pollen (Ohta 1992). This sterility might be explained as the result of the occurrence of two different alleles at the Gc loci on homologous chromosomes of the allopatric accessions.

While all studied accessions of *Ae. sharonensis* possess gametocidal chromosomes, several lines of *Ae. longissima* do not and, therefore, addition and substitution lines of chromosomes of these lines were produced in common wheat. Feldman (1975) used *Ae. longissima* line TL01, from Revivim, central Negev, Israel, as a source for successful production of six different addition lines to common wheat cv. CS. A second complete set of disomic chromosome addition lines was successfully obtained by NA Tuleen (described in Friebe et al. 1993), by crossing *Ae. longissima* line 4 (an accession collected by G. Hart in Israel, and which is different from TL01), with common wheat cv. CS. Thus, two different lines of *Ae. longissima* do not possess gametocidal chromosomes. Several lines of *Ae. speltoides* also lack a gametocidal chromosome (Tsujimoto and Tsunewaki 1984), as also evident by the successful production of a complete series of seven alien addition lines with a low-pairing accession (Friebe et al. 2000).

To date, gametocidal chromosomes have not been found, neither in common wheat nor in the allotetraploids *Ae. peregrina* and *Ae. kotschyi*, which contain the S^l subgenome that derived from *Ae. longissima*. This can be the result of elimination or suppression of the gene(s) causing the gametocidal action during the evolution of the allopolyploid species. Alternatively, these genomes derived from accessions lacking gametocidal chromosomes. The *Gc* genes of the C and M^o genomes in the allopolyploids *Ae. cylindrica*, *Ae. triuncialis* and *Ae. geniculata*, respectively, induce mild, or semi-lethal, chromosome mutations in alien addition lines of common wheat (Endo 2007). The weak or complete absence of *Gc* gene activity in allopolyploid species, in contrast to their strong effect in diploid species, suggests that allowance of interspecific hybridization and gene exchange in allopolyploids has a great evolutionary advantage (Zohary and Feldman 1962). This is in contrast to the situation at the diploid level, where restriction of interspecific gene exchange provides an evolutionary advantage.

The presence of incomplete Gc action of some gametocidal genes suggests that the Gc system might also be involved in the evolution of the karyotype in the genus *Aegilops* (Endo 2015). Incomplete Gc action induces chromosomal rearrangements in hybrids heterozygous for a *Gc* gene, and gametes with rearranged chromosomes will survive and self-fertilize. The karyotype of the selfed progeny will stabilize when the *Gc* gene becomes homozygous, and some well-balanced karyotypes might be established in separate populations.

9.3.3.6 Use of *Gc* Genes in the Production of Deletion and Dissection Lines

The Gc genes of the C and M^o subgenomes in from allotetraploid *Aegilops* induce only mild, or semi-lethal, chromosome mutations in alien addition lines of common wheat. Consequently, induced chromosomal rearrangements have been identified and established in wheat stocks carrying deletions of wheat and alien chromosomes or wheat-alien translocations. Thus, gametocidal chromosomes may serve as a tool to produce cytogenetic stocks for cytogenetic manipulations (Endo 2007).

Monosomic addition of *Ae. cylindrica* chromosome 2C showed preferential transmission of 2C in the background of the common wheat cultivar JF but not in the background of CS (Endo 1979, 1988a). The disappearance of Gc action in CS is similar to the case of the *Ae. triuncialis* chromosome 3C in the background of cultivar N26 (Tsujimoto and Tsunewaki 1985a). Chromosome aberrations caused by chromosome 2C appeared most often in offspring without the alien chromosome. Endo (1988a) suggested that when the gametocidal action is mild, gametophytes without the alien chromosome are fertilized, suffer slight chromosome damage, and develop into plants with chromosome aberrations.

Tsujimoto et al. (1990) observed chromosome fragments, bridges and micronuclei in the first and the second pollen mitoses of monosomic 3C addition to CS and Nasuda et al. (1998) observed similar chromosome aberrations in monosomic 2C addition to CS. The breakpoints of the chromosome aberrations do not appear to be distributed randomly in wheat chromosomes and may be restricted to specific chromosome structures or DNA sequences (Endo and Gill 1996). In their effort to produce a large-scale collection of 1B deletion chromosomes, Tsujimoto et al. (2001) recognized breakage 'hot spots'. However, the distribution pattern of the breakage hot spots in the studies of Endo and Gill (1996) and Tsujimoto et al. (2001) did not coincide with each other, despite the fact that both studies used the same *Gc* gene (Tsujimoto et al. 2001; Friebe et al. 2001). The

broken end gradually acquired repetitive telomere sequences, indicating that incomplete (or perhaps undetected) telomere sequences were sufficient to heal the broken ends (Tsujimoto 1993). Using the telomere sequence as a primer for PCR, the DNA sequences at the broken ends were amplified and then analyzed. However, no specific sequences were identified (Tsujimoto et al. 1997, 1999).

The abnormal chromosomes induced by the *Gc* gene of *Ae. triuncialis* or *Ae. cylindrica* can be transmitted to the next generation. Because the breakage only occurs in the gametes without the Gc chromosome in monosomic addition lines, the offspring with a chromosome deletion in the next generation were stable and did not induce additional chromosome aberrations. Thus, these deletion lines were useful for mapping and were maintained as the standard for mapping genes to specific chromosome regions in common wheat (Endo and Gill 1996; Tsujimoto et al. 2001).

Using mostly chromosome 2C of *Ae. cylindrica*, Endo and Gill (1996) produced approximately 350 homozygous deletion lines of CS wheat that contain deletions of various size in specific chromosomes. These lines are useful in cytologic mapping (deletion mapping) of genes and especially of DNA markers to the missing chromosomal regions (Werner et al. 1992; Qi et al. 2004). Most of the CS deletion lines, together with the Gc chromosomes, are available at NBRP-wheat website (http://www.shigen.nig.sc.jp/wheat/komugi/strains/aboutNbrpL.gku.jsp).

In addition, the Gc genes can be usefully applied to induce translocations between alien chromosomes introduced into common wheat and wheat chromosomes. This is of particular relevance in the case of alien chromosomes from species distantly related to wheat, and which show little tendency to undergo homoeologous recombination with wheat chromosomes, even under genetically permissive conditions (Endo 2015). As an example, the *Ae. cylindrica* Gc chromosome 2C was introduced into CS wheat having all barley chromosomes as disomic addition lines, except for 1H (Shi and Endo 1997). Chromosomal rearrangements were induced by the 2C gametocidal system for each barley chromosome, including 2H (Joshi et al. 2011), 3H (Sakai et al. 2009), 4H (Sakata et al. 2010), 5H (Ashida et al. 2007), 6H (Ishihara et al. 2014) and 7H (Schubert et al. 1998; Serizawa et al. 2001; Masoudi-Nejad et al. 2005; Nasuda et al. 2005). The Gc system was similarly proven to be effective in inducing structural rearrangements in rye chromosome 1R introduced into common wheat (Endo et al. 1994; Masoudi-Nejad et al. 2002; Gyawali et al. 2009, 2010; Li et al. 2013). Since both terminal deletions and wheat-alien translocations enable cytological mapping of alien chromosomes, Endo (2015) has been developing many common wheat lines carrying deletions and translocations of alien chromosomes, collectively named "dissection lines". Comparative studies of cytological and genetic maps obtained in the above studies revealed that crossing-over is generally more frequent in the distal region than in the proximal region for all the wheat, barley and rye chromosomes that were analyzed.

9.4 Section Sitopsis (Jaub. & Spach) Zhuk.

9.4.1 General Description

Section Sitopsis (Jaub. & Spach) Zhuk. [Syn.: subgen. Sitopsis Jaub. & Spach; sect. Platystchys Eig; *Triticum* L. sect. Sitopsis (Jaub. & Spach) Chennav.; Sitopsis (Jaub. & Spach) Á. Löve] is characterized by long spikes, of lengths that are at least 20 times their width, with either two-rowed, lemmas of lateral spikelets are awned and disarticulating wedge-type (*Ae. speltoides* var. *ligustica, Ae. bicornis, Ae. sharonensis*), or with one-rowed, only lemmas of apical spikelets with awns, disarticulating as one unit, umbrella type (*Ae. speltoides* var. *speltoides* and *Ae. searsii* disarticulate above a basal rudimentary spikelet and *Ae. longissima* above several fertile spikelets that remain on the culm). Caryopsis is adherent to lemma and palea or free (in *Ae. searsii*) (Fig. 9.1).

The Sitopsis species can be distinguished by their morphology, and more easily by their specific habitat, climatic adaptation or area of distribution. Zhukovsky (1928) included four species in section Sitopsis, while Eig (1929a) recognized five, grouping them in two subsections, Truncata and Emarginata. Subsection Truncata, containing one species, *Ae. speltoides* Taush (Table 9.1), is characterized by many-flowered spikelets and relatively short glumes (about half the length of the florets), terminating in a thick margin,

with or without a small tooth on one side. Species of this subsection are found in the central, eastern and northern parts of the Sitopsis distribution area and grow on heavy and moist soils. According to Eig (1929a), subsection Truncata includes two species, namely, *Ae. speltoides* Tausch and *Ae. ligustica* (Savigny) Cosson. Yet, Sears (1941b) and Kihara (1954), based on cytogenetic studies, concluded that *Ae. ligustica* should be included in *Ae. speltoides*. Zohary and Imber (1963) supported this conclusion by presenting genetic evidence indicating that *Ae. speltoides* and *Ae. ligustica* are two genetic forms of the same species. Molecular analysis by Goryunova et al. (2008) also supported this conclusion.

Subsection Emarginata is characterized by few-flowered spikelets and relatively longer glumes than those of subsection Truncata, about 2/3 or more the length of the florets, terminating generally in two teeth (occasionally 0–3), separated by an angle. The species of this subsection are found in the central and southern parts of the distribution area of the section, on light, sandy soils. Subsection Emarginata includes four species: *Ae. sharonensis* Eig, *Ae. longissima* Schweinf. et Muschl., *Ae. bicornis* (forssk.) Jaub. et Sp., and *Ae. searsii* Feldman and Kislev ex Hammer (Table 9.1). Based on chromosome pairing at meiosis of F_1 hybrids between *Ae. sharonensis* and *Ae. longissima*, Kihara (1954) suggested including *Ae. sharonensis* in *Ae. longissima*. This suggestion was accepted by several taxonomists and cytogeneticists (e.g., Bowden 1959; Morris and Sears 1967; Mackey 1968). However, Ankori and Zohary (1962) and Waines and Johnson (1972) were inclined to accept Eig's recognition of *Ae. sharonensis* as a separate species. Later on, taxonomists (e. g., Hammer 1980; van Slageren 1994)

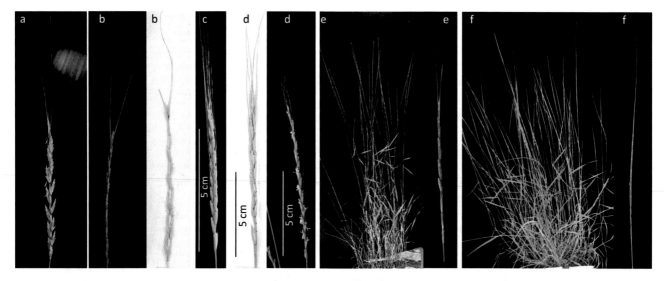

Fig. 9.1 *Aegilops* species of section Sitopsis; **a** A spike of *Ae. speltoides* Tausch var. *ligustica* (Savign.) Fiori in Fiori & Paoletii; **b** A spike of *Ae. speltoides* var. *speltoide*s Tausch; **c** A spike of *Ae. bicornis* (Forssk.) Jaub. & Spach; **d** A spike of *Ae. sharonensis Eig;* **e** A plant and a spike of *Ae. longissima* Schweinf. & Muschl.; **f** A plant and a spike of *Ae. searsii* Feldman and Kislev ex Hammer

and cytogeneticists (e. g., Teoh and Hutchinson 1983; Yen and Kimber 1990b) presented evidence justifying the specific rank of *Ae. sharonensis*.

Aegilops searsii was described by Feldman and Kislev (1977) as a taxon with unique habit and habitats, possessing sufficient new characteristics to justify treating it as an independent species. This new species, that was included in subsection Emarginata, differs from the other members of this subsection in morphological, eco-geographical and karyotypic characteristics.

Mendlinger and Zohary (1995) assessed the extent and structure of genetic variation in 21 populations covering the five species of the Sitopsis section, by electrophoretically analyzing water-soluble leaf proteins. All loci were polymorphic across the five species. Over 40% of the alleles were found in all five species and only three rare alleles were species-specific. Genetic diversity was high (D = 0.267), with 51% of the total diversity contributed by within-population diversity, 16% by diversity between populations within a species and 33% by diversity between species. *Ae. speltoides* was genetically distant from the other four species. *Ae. sharonensis* was found to be equally close to *Ae. longissima* and *Ae. bicornis*, whereas *Ae. searsii* was equally distant from *Ae. longissima*, *Ae. sharonensis* and *Ae. bicornis*.

Similar results were obtained by Giorgi et al. (2002) who used restriction-fragment-length polymorphism (RFLP) analysis to investigate phylogenetic relationships among the Sitopsis species. A dendrogram derived from a cluster analysis of the complete RFLP dataset showed subdivision of the species into two groups, one comprising the species of the Truncata subsection and the other comprised of the four species of the Emarginata subsection. The findings indicated that *Ae. speltoides* is the most divergent species within the *Sitopsis* section, and that *Ae. sharonensis* and *Ae. longissima* are closely related species and form a separate subgroup within subsection Emarginata. *Ae. bicornis* and *Ae. searsii* also form separate subgroups, where that of *Ae. bicornis* is closer to the *sharonensis-longissima* subgroup, and that of *Ae. searsii* is more distant from the other two subgroups. Similar results were obtained by Goryunova et al. (2008), who used random amplification of polymorphic DNA (RAPD) analysis to study the intraspecific variation and phylogenetic relationships of the Sitopsis species. They found that *Ae. speltoides* formed the most isolated species in the section, justifying its classification into a separate subsection. In the other subsection, *Ae. longissima* and *Ae. sharonensis* were the closest species, *Ae. bicornis* and *Ae. searsii* formed separate subgroups, that of *Ae. searsii* being the most distant. Their findings showed that the extent of intraspecific polymorphism considerably varies among the Sitopsis species: *Ae. speltoides* is the most polymorphic species of the group, *Ae. bicornis* and *Ae. searsii* display the

lowest diversity, while *Ae. longissima* and *Ae. sharonensis* are intermediate.

That *Ae. speltoides* differs significantly from the *Emarginata* species is also evident from studies that used different methodological approaches, e.g., Giemsa C-banding (Teoh and Hutchinson 1983; Friebe and Gill 1996; Friebe et al. 2000), in situ hybridization patterns with DNA probes (e.g., Talbert et al. 1991; (Yamamoto 1992a; Jiang and Gill 1994; Badaeva et al. 1996a, b; Salina et al. 2006; Raskina et al. 2011; Belyayev and Raskina 2013), analysis of the restriction patterns of chloroplast and mitochondrial DNA (Ogihara and Tsunewaki (1988), of biochemical markers (Bahrman et al. 1988), or of repeated sequences (Dvorak and Zhang 1990, 1992a, b; Giorgi 1996) and RFLP analysis (Sasanuma et al. 1996). The separate position of Ae. speltoides within the Sitopsis section and, in contrast, the similarity of the Emarginata species was also shown by Dvorák and Zhang (1992b) who studied variation of repeated nucleotide sequences (RNS). RAPD- and AFLP analyses revealed that *Ae. speltoides* forms a cluster with polyploid wheats, which is separated from other *Sitopsis* species (Kilian et al. 2007, 2011; Goryunova et al. 2008). Likewise, study of organellar DNAs by PCR-single-strand conformational polymorphism (PCR-SSCP) revealed high similarity of *Ae. bicornis—Ae. sharonensis—Ae. longissima* plasmons and their distinctness from plasmon of *Ae. speltoides* (Wang et al. 1997).

Similar results were obtained by Ruban and Badaeva (2018). These authors studied the relationships between the S-genome species using Giemsa C-banding and fluorescence in situ hybridization (FISH) with several DNA probes. To correlate the C-banding and FISH patterns, they used the microsatellites $(CTT)_{10}$ and $(GTT)_9$, which are major components of the C-banding positive heterochromatin in wheat. Their results justify the classification of the Sitopsis species into the two subsections, *Truncata* and *Emarginata*, which differ in the C-banding patterns, and distribution of DNA repeats. Evolution of *Emarginata* species was associated with an increase of C-banding and $(CTT)_{10}$-positive heterochromatin, as well as amplification of the DNA probe Spelt-52.

In accord with the evidence that *Ae. speltoides* has less constitutive heterochromatin, as seen from smaller amount of C-banding, than the Emarginata species, the amount of nuclear DNA in subsection Truncata is significantly smaller (5.81 pg 1C DNA) than that in species of subsection Emarginata (6.65 pg 1C DNA in *Ae. searsii* and 7.52 pg in *Ae. sharonensis*) (Furuta et al. 1977; Eilam et al. 2007; Table 9.3). No significant intraspecific variation in nuclear DNA size was detected (Furuta et al. 1977; Eilam et al. 2007). Likewise, Li et al. (2022) showed that the five *Sitopsis* species have variable genome sizes (4.611–6.22 Gb) with high proportions of repetitive sequences

(85.99–89.81%); nonetheless, they retain high collinearity with other wheat genomes. Li et al. (2022) concluded that differences in genome size are primarily due to independent post-speciation amplification of transposons rather than to inter-specific genetic introgression.

Hence, although *Ae. speltoides* has been considered by most taxonomists and cytogeneticists a member of the Sitopsis section, sequencing of chloroplast DNA by Middleton et al. (2014), and Gornicki et al. (2014), and nuclear DNA by Marcussen et al. (2014), showed that *Ae. speltoides* forms a phylogenetic clade with the B and G subgenomes of allopolyploid wheats and not with the other Sitopsis species. Similar results were presented by Ruban and Badaeva (2018) and by Li et al. (2022). Assembly of chromosome-level genome sequences of all the five *Sitopsis* species enable Li et al. (2022) to propose that the diploid species and B and G subgenomes of allopolyploid wheats fall into two independent clades, with *Ae. speltoides* being clustered with the B-subgenome of allopolyploid wheat (B-lineage) while the rest four *Sitopsis* species being grouped with the D-subgenome (D-lineage) and its diploid donor *Ae. tauschii* (D-lineage).

Subsection Truncata and subsection Emarginata are genetically isolated from one another, as shown by sterility of hybrids between them. On the other hand, hybrids between Emarginata species are fertile or partially fertile. The F_1 hybrid between *Ae. longissima* and *Ae. sharonensis* and the reciprocal hybrid had an 81–85% seed set (Tanaka 1955a; Ankori and Zohary 1962), that between *Ae. sharonensis* and *Ae. bicornis* had a 62% seed set (Tanaka 1955a), that between *Ae. bicornis* and *Ae. longissima* had a 40% seed set (Waines G, personal communication), and that between *Ae. longissima* and *Ae. searsii* had a 35% seed set in one hybrid combination and 7% in another (Feldman et al. 1979).

Ae. speltoides differs from the four species of subsection Emarginata in several additional important features: (i) *Ae. speltoides* is the only predominantly cross-pollinated species in the section. (ii) Its distribution area is in the central part of the genus area, i.e., in and around the Fertile Crescent, on terra rossa or alluvial soils. In contrast, the species of subsection Emarginata reside in the southern part of the western wing of the Fertile Crescent, with *Ae. sharonensis* and *Ae. bicornis* growing on sandy soils, and *Ae. longissima* on red sandy loam in the coastal plain, on sand derived from Nubian sandstone and on grey calcareous steppe soil or loess in the inland steppe and desert. *Ae. searsii* grows on terra rossa or basalt soil (Feldman and Kislev 1977; Kimber and Feldman 1987).

The five Sitopsis species differ in their genome size (Eilam et al. 2007), that is primarily due to independent post-speciation amplification of transposons (Yaakov et al. 2013). The sum of repetitive sequences in the different species ranges from 85.99 to 89.81% (Li et al. 2022). *Ae. speltoides* has the smallest genome, *Ae. bicornis* and *Ae. searsii* have a larger genome, and *Ae. sharonensis* and *Ae. longissima* have the largest genome.

Ae. speltoides is the only Sitopsis species found in the entire Fertile Crescent and in most parts of Turkey, with both varieties (*speltoides* and *ligustica*) often occurring sympatrically and with similar ecological tendencies (Hammer 1980; Kimber and Feldman 1987). Notably, the area of the other four species of the section (*Ae. bicornis*, *sharonensis*, *longissima*, and *searsii*) occurs within a very limited region in the southeastern corner of the Mediterranean, which marginally overlaps with the southwestern part of the area of *Ae. speltoides*. It seems as if these species replace *Ae. speltoides* in the southern climatically and edaphically special environments. Of these, the closely related *Ae. sharonensis* and *Ae. bicornis* are an example of bioregional vicariance (Zohary 1962), with the former confined to the coastal plain of the Mediterranean region in Israel and southern Lebanon, and the latter replacing it southwardly up to Lybia and also spreading towards the Saharo-Arabian region in desert sands of the Israeli Negev and Sinai Peninsula. *Ae. longissima* occurs in two different phyto-geographical regions, the Israeli and Egyptian coastal plain (Mediterranean region) and the Israeli Negev, southern Jordan and Sinai Peninsula (Irano-Turanian and Saharo-Arabian regions). *Ae. searsii* grows in sub-Mediterranean areas in Israel, Jordan, southwestern Syria and southeastern Lebanon and also occurs in the high elevations of the Israeli Negev and southern Jordan. It should also be noted that the distribution of *Ae. searsii*, similar to *Ae. vavilovii*, has only recently been defined, and may be larger than hitherto supposed.

9.4.2 *Aegilops speltoides* Tausch

9.4.2.1 Morphological and Geographical Notes

Ae. speltoides Tausch [Syn.: *Triticum speltoides* (Tausch) Gren. ex Richter; *Sitopsis speltoides* (Tausch) Á. Löve] is a predominantly allogamous, annual plant, with 40–70-cm high culms (excluding spikes), and leaves that are usually hairy, and sometimes pendant. Its spike is linear, narrow, tapers to the tip, either two- or one-rowed, and is (6-) 7-11(-15)-cm long (excluding awns). At maturity, its spike disarticulates above the basal rudimentary spikelet, either into single spikelets, with the rachis internode immediately below them (wedge-type dispersal unit), or falling entire as a unit (umbrella-type dispersal unit). Spikelets 7–11, 11–15 mm long, lanceolate or linear, sessile in hollows of rachis, and longer or shorter than the adjacent rachis internode. Each contains 4–8 (usually 4–6) florets, with the upper 1–3 being sterile. There is usually 1 rudimentary spikelet at the base of the spike. Glumes are 5–7 mm long, truncate at the apex, asymmetrical, somewhat keeled, and about 2/3 the length of the lemma. Lemmas are 7–10 mm long and boat-shaped, those of lateral spikelets have either short triangular awns or are not awned at all, while those of terminal spikelets are always have longer awns than those on the lateral spikelets. Anthers are 5–6 mm long. The caryopsis is adherent to lemma and palea (Fig. 9.1a, b).

Ae. speltoides has a median-sized distribution in the central region of the distribution of the genus. It is an eastern Mediterranean element, extending into the steppical (Irano-Turanian) region. It occupies primary and secondary habitats. *Ae. speltoides* grows in Israel (in the Coastal Plain and Esdraelon Plain on humid, alluvial soil; in Mt. Carmel and western slopes of the Samaria Mts., on terra rossa and Rendzina soils), Jordan, Lebanon, Syria, Turkey, northern Iraq and northwestern Iran. It is a weed in central and western Turkey, Greece, Bulgaria and possibly also Crimea. The center of variation of this species is in north Syria-southeastern Turkey, which is likely the center of its origin.

Ae. speltoides grows sympatrically with *Ae. caudata*, *Ae. umbellulata*, *Ae. triuncialis*, *Ae. geniculata*, *Ae. neglecta*, *Ae. biuncialis*, *Ae. peregrina*, *Ae. cylindrica*, wild *T. monococcum* (ssp. *aegilopoides*), *T. urartu*, wild *T. timopheevii* and wild *T. turgidum*, and allopatrically with *Ae. sharonensis*, *Ae. longissima* and *Ae. searsii* in its southern distribution area and with *Ae. tauschii* in its northeastern distribution area.

*Ae. speltoide*s has limited morphological variation, mainly in spike characteristics. It contains two main morphological types that differ markedly in the structure of their fruiting spike and consequently in their mode of seed dispersal. Taxonomists (see van Slageren 1994) regard these two types as varieties, var. *speltoides* Tausch and var. *ligustica* (Savign.) Fiori in Fiori & Paoletii.

Var. *speltoides* [Syn.: *Ae. speltoides* ssp. *speltoides* Hammer; *Ae. speltoides* var. *aucheri* (boiss.) Fiori in Fiori & Paoletii; *Ae. aucheri* Boiss.] is characterized by a one-rowed, long and cylindrical spike, with relatively widely interspersed spikelets. Only the apical spikelet is awned and rarely has awns on some lateral spikelets and then shorter and thinner than the awns on the lateral spikelet. The rachis of the spike is tough, except for a brittle node at its base, above a rudimentary spikelet. The spike falls at maturity as a unit (umbrella-type dispersal unit) (Fig. 9.1b).

Var. *ligustica* [Syn.: *Ae. ligustica* (Savign.) Coss.; *Ae. speltoides* ssp *ligustica* (savign,) Zhuk.; *Agropyron ligusticum* Savign.] is characterized by a denser, two-rowed spike, in which the lateral spikelets are also awned; awns of terminal spikelet are somewhat sturdier than those of lateral spikelets. The rachis is brittle at every joint, so that the mature spike disarticulates into single spikelets, each with the rachis internode immediately below it (wedge-type dispersal unit). Each individual spikelet thus serves as an independent dissemination unit; it is mechanically adapted to insert itself into the ground and usually buries itself soon after detachment (Fig. 9.1a). The two varieties grow, in most sites, in mixed stands and are cross-fertile; few intermediates are usually found in mixed populations. Presumably, the *ligustica* form, having the wedge-type dispersal unit typical of many Triticineae species, is the ancestral form. Alternatively, the *speltoides* form, having one-rowed spike similar to that of several *Elymus* species, may be the ancestral type.

The morphological differences between var. *speltoides* and var. *ligustica* are restricted to the fruiting spike alone. In all other traits, such as vegetative characters and phenological behavior, the types are strikingly similar to each other (Miczynski 1926; Schiemann 1928; Eig 1929a). In addition, F_1 hybrids between the two types are fully fertile and show complete pairing of chromosomes at meiosis (Kihara and Lilienfeld 1932; Sears 1941b; Zohary and Imber 1963). Further indications of the close relationship between *speltoides* and *ligustica* were obtained from the usual occurrence of mixed stands of the two types (Eig 1929a). There are wide fluctuations in the proportions between *ligustica* and *speltoides* in different sites and in different plant formations, but both types were almost universally present in the sites examined. In many stands, *ligustica* and *speltoides* plants were, more or less, evenly mixed. Yet, *ligustica* plants predominated the stands in higher altitudes and *speltoides* in warmer sites (Zohary and Imber 1963). These observations led to the conclusion that mixed populations are the rule; *speltoides* and *ligustica* are apparently spatially interconnected almost throughout their distribution area (Zohary and Imber 1963).

Rare intermediate types, e.g., plants with a laterally awned but tough rachis (var. *polyathera* according to Eig 1929a) or also, very occasionally, plants with brittle rachis but laterally awnless spikes, occur in very low frequencies (less than 1%) in mixed *speltoides-ligustica* populations (Zohary and Imber 1963). Progeny of *speltoides* x *ligustica* crosses revealed that the main differences in spike morphology are inherited in an almost monohybrid Mendelian fashion (dominance of *ligustica* over *speltoides* in F_1, and 3:1 segregation in F_2) (Miczynsky 1926; Schieman 1928; Kihara and Lilienfeld 1932). These data were first interpreted as indicating the presence of only a single pleiotropic gene, but Sears (1941b), based on occasional occurrence of recombinants or intermediate plants, proposed the operation of a block of closely linked genes. This block should possess at least three genes: one determining the length of rachis internodes (short in *ligustica* bringing about two-rowed spikes and long in *speltoides* leading to one-rowed spikes); a second gene that determines awns on the lateral spikelets; and a third gene that determines the type of dispersal unit. Luo et al. (2005) mapped the closely linked genes controlling the *ligustica/speltoides* spike dimorphism to the centromeric region of chromosome 3S. (the genome symbol of *Ae. speltoides* is S). The location near the centromere, being a chromosome region with rare recombination events, may explain the rare recombination between these three genes.

Progeny tests performed in plants sampled from natural mixed populations, indicated that *speltoides* and *ligustica* plants are genetically interconnected by virtue of their mating system of predominantly cross-pollination. Several *ligustica* plants, that apparently were F_1, segregated into 3 *ligustica*: 1 *speltoides* and several *speltoides* plants, that were presumably pollinated by *ligustica*, yielded *ligustica* progeny (Zohary and Imber 1963). Accordingly, Zohary and Imber (1963) suggested not to regard *speltoides* and *ligustica* types as two independent varieties, but as two constituents of dimorphic populations. The mixed stands of the two types are looked upon as a case of genetically determined fruit dimorphism, where two types of seed dispersal apparatus function in the same population in a complementary way (Zohary and Imber 1963). In the *speltoides-ligustica* pair, the two kinds of dispersal units are borne by different individuals: two types of plants, which differ with regard to their fruit structure, form a common population. This is an instance of genetically determined fruit dimorphism, with a population as its unit of operation.

A model of fruit dimorphism can account for the peculiar mode of inheritance of the many differences between the two types: they are inherited in an almost monohybrid fashion, with only rare cases of recombinants. As already proposed by Sears (1941b), this mode of inheritance suggests the presence of a single block of closely linked genes. A system based on such a single block is essential for the establishment of population dimorphism. The parallel evolvement of fruit dimorphism in two Sitopsis groups, *ligustica* vs. *speltoides* on the one hand, and *bicornis-sharonensis* versus *longissima-searsii* on the other hand, is an interesting evolutionary phenomenon. The genetic change(s) determining the *speltoides* and the *longissima*-spike morphology occurred independently in the two groups, or alternatively, happened once and genes of this system were transferred from one group to the other via interspecific introgressive hybridization. However, it seems there is a difference between the gene system in the two groups; while the genes in *ligustica-speltoides* are closely linked, those in *sharonensis-longissima* are not linked, and segregate independently from one another, as is indicated in the scatter diagram of F_2 artificial generation in Ankori and Zohary (1962).

9.4.2.2 Cytology, Cytogenetics, and Evolution

Aegilops speltoides is a diploid species containing seven pairs of homologous chromosomes (2n = 2x = 14), all of which have a median or sub-median centromere and lack distinctive morphological features, except for the SAT chromosomes (Riley et al. 1958; Chennaveeraiah 1960). In contrast, Dong et al. (2017), studying karyotypic polymorphism of *Ae. speltoides*, found intraspecific variation in the centromere position. In *Ae. speltoides* var. *ligustica*, there is one pair with a median centromere, while the rest of the pairs in both taxa have submedian centromeres (Chennaveeraiah 1960). The difference between the karyotypes of the two taxa seems to be very small. However, in var. *ligustica*, the short arms in the satellite pairs are more or less of the same length. In var. *speltoides*, there seems to be a difference in the lengths of the short arms of the satellite pairs (Chennaveeraiah 1960). The karyotype of both varieties of *Ae. speltoides* contains two pairs of SAT chromosomes with fairly large satellites (Pathak 1940; Riley et al. 1958; Chennaveeraiah 1960).

In *Ae. speltoides*, the genes coding for the 18S and 26S ribosomal RNA (rDNA genes) are located in the NOR regions on the short arms of chromosomes of groups 1 and 6 (Dvorak et al. 1984). There is only one 5S rDNA locus on the short arm of group 5 chromosomes that codes for 5S rRNA (Badaeva et al. 1996b). In contrast, *Ae. longissima*, *Ae. sharonensis*, *Ae. bicornis* and *Ae. searsii*, exhibit a different distribution pattern of NORs and 5S DNA loci. They contain major NOR loci on the short arms of chromosomes of groups 5 and 6 and two 5S rDNA loci on the short arms of chromosomes of groups 1 and 5 (Badaeva et al. 1996b). Thus, the distribution pattern of the NORs and 5S loci in the five Sitopsis species support the subdivision of this section into the two subsections.

The genome of *Ae. speltoides* is the smallest among the five Sitopsis species; 1C DNA is 5.81 pg (Furuta et al. 1977; Eilam et al. 2007) and 5.1 pg (The Angiosperm C-value database at Kew Botanic gardens) (Tables 2.4 and 9.3). Li et al. (2022) assessed the size of *Ae. speltoides* genome after genome sequencing and assembly to be somewhat smaller amount, i.e., 4.6 Gb (=4.5 pg), and Avni et al. (2022), also after sequencing and assembly, reported 5.13 Gb (=5.02 pg). No significant difference was noted in the genome size of the two taxa (Furuta et al. 1977; Eilam et al. 2007). Teoh and Hutchinson (1983), used an improved C-banding technique, found that all five Sitopsis species have telomeric, interstitial, and centromeric bands, varying in size and staining intensity. However, *Ae. speltoides* has exceptionally distinctive and extensive centromeric bands. The extent of polymorphic variation was found to vary between chromosomes of *Ae. speltoides* and plants were found to be heterozygous for the banding pattern (Teoh and Hutchinson 1983). Yet, the intraspecific variation in C-banding patterns neither differentiated between var. *ligustica* and var. *speltoides*, nor corresponded with impaired meiotic chromosomal pairing, since hybrids between the two varieties exhibited complete pairing and full fertility

(Teoh and Hutchinson 1983). Badaeva et al. (1996a, b) and Ruban and Badaeva (2018) also found that *Ae. speltoides* have a different C-banding pattern than those of the other species of section Sitopsis.

Friebe and Gill (1996) established the homoeology between chromosomes of *Ae. speltoides* and those of *T. aestivum*, on the basis of similarities in chromosome morphology and C-banding patterns. Maestra and Naranjo (1998) determined the homoeologous chromosome relationship by analyzing homoeologous pairing at meiotic first metaphase of F_1 hybrids between *Triticum aestivum* and *Ae. speltoides* carrying the homoeologous–pairing suppressor gene of common wheat, *Ph1*, or deficiencies *ph1b* for *Ph1*, and *ph2b* for *Ph2*. The chromosomes and their arms were identified, in both species, by C-banding. Data from relative pairing affinities were used to determine homoeologous relationships between *Ae. speltoides* chromosomes and bread wheat. All arms of the seven chromosomes of the *speltoides* genome showed normal homoeologous pairing with the wheat chromosomes, implying that no apparent chromosome rearrangements occurred in the evolution of *Ae. speltoides* relative to chromosomal structure in wheat. There was agreement between Friebe and Gill (1996) and Maestra and Naranjo (1998) in the assignment of five *speltoides* chromosomes to homoeologous groups 1, 4, 5, 6 and 7, but discrepancy with regards to chromosomes 2S and 3S.

Using in situ hybridization (ISH), Badaeva et al. (1996a) found that all Sitopsis species were similar to each other and to *Amblyopyrum muticum* in the distribution of the hybridization sites of two highly repetitive DNA sequences, pSc119 and pAsl. Yet, chromosomes of *Ae. speltoides* contain rich repetitive DNA sequences highly homologous to the *pSc119* probe and, on the other hand, lack such homology to probe *pTa53* (Dong et al. 2017). The distribution of *pSc119* on chromosomes of *Ae. speltoides* show differences between accessions, between plants of one accession and even between homologous chromosomes in one plant.

Ae. speltoides comprises many of TEs, particularly long terminal repeat (LTR) retrotransposons, of which Ty1-*copia* superfamilies (*Angela*, *Barbara*, and *Wis-A*) and Ty3-*gypsy* superfamilies (*Fatima* and *Erika*) make up the main fraction (Middleton et al. 2013; Yaakov et al. 2013). Non-LTR retrotransposons (*Ramona* and *Paula*) and DNA transposons (*Baldwin*, *Rong*, and *Charon*) also exist in *Ae. speltoides*. All these TEs families vary greatly in copy number among different accessions of *Ae. speltoides* (Yaakov et al. 2013). Likewise, Hosid et al. (2012) analyzed intraspecific variation of four LTR retrotransposons (*WIS2*, *Wilma*, *Daniela*, and *Fatima*) in 13 different populations of *Ae. speltoides* from all over the distribution area of this species and found significant diversity in retrotransposon distribution. The various genotypes significantly differ with respect to the patterns of the four explored LTR retrotransposons, indicating a constant ongoing process of LTR retrotransposon fraction restructuring among and within populations of *Ae. speltoides*. Maximum changes were recorded in genotypes from small, stressed populations. The data of Hosid et al. (2012) revealed dynamic changes in LTR retrotransposon fractions in the *Ae. speltoides* genome, that are continually reshaping the genome of *Ae. speltoides*, particularly in stressful environments (Raskina et al. 2004a, b, 2008).

In *Ae. speltoides* euchromatin, widely interspersed TEs are clustered non-randomly, and may affect chromosomal structure. For instance, Raskina et al. (2004a) reported on the involvement of the transposable element system *En/Spm* [Enhancer (*En*) and Suppressor-Mutator (*Spm*)] in ongoing chromosomal repatterning in a small, isolated, peripheral population of *Ae. speltoides*. Cytogenetic analysis of the dynamics of *En/Spm* transposons in meiosis indicated that this transposon is active during male gametogenesis, changing the position of the rDNA sites (Raskina et al. 2004a). Such findings may indicate the importance of TEs in genome architecture (Belyayev et al. 2001, 2005; Altinkut et al. 2006), as well as their most important role in intraspecific divergence (Belyayev et al. 2010). The variability of TE content in *Ae. speltoides*, as well as in other *Aegilops* and *Triticum* species, might has a great impact on the dynamic, ongoing evolution of their genomes (Charles et al. 2008; Yaakov and Kashkush 2012).

Repetitive DNA sequences form chromosome-specific heterochromatin patterns in *Ae. speltoides* (Salina et al. 2006; Raskina et al. 2008). The species-specific tandem-repeat *Spelt1* forms distinct clusters almost exclusively in subtelomeric chromosome positions, while *Spelt52* clustering was most prominent at interstitial sites (Salina et al. 2006; Raskina et al. 2011). These two repetitive DNA sequences exhibit high intraspecific variation in both number and size (Salina et al. 2006; Raskina et al. 2011). Using fluorescence in situ hybridization, Salina et al. (2006) observed considerable polymorphisms in the hybridization patterns of *Spelt1* between and within the studied lines of *Ae. speltoides*. There was a distinct ecogeographical gradient in the abundance of *Spelt1* and *Spelt52* blocks in *Ae. speltoides*; in marginal populations, the number of Spelt 1 chromosomal blocks was, at times, 12–14 times lower than in the center of the species distribution (Raskina et al. 2011). Likewise, *Ae. speltoides* had distinct Spelt52 hybridization patterns in the studied lines, but several distal Spelt52 sites, namely, on chromosome arms 3SL, 6SS, and 7SS, were common to all.

Aegilops speltoides contains a genetic mechanism that, in hybrids with allopolyploid wheats, suppresses the activity of the homoeologous-pairing suppressor (*Ph1*) gene of allopolyploid wheats, *Triticum aestivum* and *T. turgidum*. Riley and Law (1965) and Dover and Riley (1977) pointed out that increased homoeologous chromosome pairing at meiosis of common wheat x *Ae. speltoides* occurs only if *Ph1* of common wheat is present and therefore proposed that the *Ae. speltoides* genome can promote homoeologous chromosome pairing by overpowering the activity of *Ph1*. Chen and Dvorak (1984) and Dvorak et al. (2006) found that the *speltoides* genes were ineffective in diploid hybrids lacking a gene like *Ph1*. Consequently, in accord with the above conclusion of Riley and Law (1965) and Dover and Riley (1977), they concluded that *Ae. speltoides* genes act on the *Ph1* locus by suppressing its expression in common wheat x *Ae. speltoides* hybrids. Dvorak et al. (2006) mapped these, *Ae. speltoides* suppressor genes and demonstrated their significant effects as Mendelian loci on the long arms of chromosomes 3S and 7S. The chromosome *3S* locus was designated *Su1-Ph1* and the chromosome-*7S* locus was designated *Su2-Ph1* (Dvorak et al. 2006). A QTL with a minor effect was mapped on the short arm of chromosome *5S* and was designated *QPh.ucd-5S* (Dvorak et al. 2006). The expression of *Su1-Ph1* and *Su2-Ph1* increased homoeologous chromosome pairing in common wheat × *Ae. speltoides* hybrids by 8.4 and 5.8 chiasmata/cell, respectively. *Su1-Ph1* was fully epistatic to *Su2-Ph1*, and, when acting together, the two genes increased homoeologous chromosome pairing in common wheat × *Ae. speltoides* hybrids to the same level when *Su1-Ph1* acting alone. *QPh. ucd-5S* expression increased homoeologous chromosome pairing by 1.6 chiasmata/cell in common wheat × *Ae. speltoides* hybrids and was additive to the effect of *Su2-Ph1*. It is hypothesized that the products of *Su1-Ph1* and *Su2-Ph1* affect pairing between homoeologous chromosomes by regulating the expression of *Ph1*, while the product of *QPh. ucd-5S* primarily regulates recombination between homologous chromosomes.

The *Ae. speltoides* genes that suppress *Ph1* activity were transferred from a high-pairing accession of *Ae. speltoides* to common wheat (Chen et al. 1994). It appeared that two genes, presumably *Su1-Ph1* and *Su2-Ph1*, were transferred. These suppressor genes were transferred to several lines of common wheat and used to induce homoeologous pairing in hybrids between these lines and wild relatives of wheat and, consequently, enhanced the gene transfer from wild relatives to common wheat (Li et al. 2011).

Yet, it is puzzling to assume that the homoeologous-pairing promoters of *Ae. speltoides* (and also of *Amblyopyrum muticum*) evolved at the diploid level as a preadaptation, to repress the expression of the pairing suppressors at the polyploid level. These promoters may also promote homologous and homoeologous pairing per se or counteract the activity of suppressors that exist at the diploid level. Meiotic chromosome pairing data, collected by Sears (1941b), from hybrids of various diploid Triticineae species were interpreted by Waines (1976) to indicate that

intraspecific genotypic differences control the amount of homoeologous chromosome pairing in diploid inter-specific hybrids. He suggested that homoeologous-pairing suppressors analogous to the *Ph1* allele in bread wheat, are already present in genotypes of diploid species and do not necessarily have to arise by mutation or translocation after allopolyploid formation. He suggested that these homoeologous-pairing suppressors reinforce the isolating mechanism among the diploid species. The existence of *Ph1*-like genes in diploid species of the wheat group was also suggested by Okamoto and Inomata (1974) and by Maan (1977a). Feldman (1978) presented evidence for the existence of homoeologous-pairing suppressors in a low-pairing line of *Ae. longissima*.

There is also evidence that the promoters of the diploid species promote homoeologous pairing per se, not necessarily via the repression of *Ph1* expression. Feldman and Mello-Sampayo (1967) concluded that, *Ae. speltoides* carries a promoter(s) that directly promotes homoeologous pairing in bread wheat x *Ae. speltoides* hybrids lacking the *Ph1* gene. Similarly, all the promoters that exist in bread wheat promote pairing rather than suppress the expression of *Ph1* (Feldman 1966). Indication that *Ae. speltoides* (and also *A. muticum*) contain genes that promote homoeologous pairing in the absence of *Ph1* may be deduced from the following pairing data. The F_1 hybrids between *Ae. speltoides* and *T. monococcum* and between *Ae. speltoides* and *Ae. comosa* analyzed by Sears (1941b), showed a similar amount of chromosome pairing to that observed by Ohta (1990, 1991) in the F_1 hybrids between *T. monococcum* and *A. muticum*, between *Ae. comosa* and *A. muticum*, and between *Ae. speltoides* and *A. muticum*.

Massive chromosome restructuring was observed during the substitution of chromosome 6B of cultivar Chinese Spring of bread wheat by chromosome 6S from *Ae. speltoides* (Kota and Dvorak 1988). The chromosome rearrangements including deletions, translocations, ring chromosomes, dicentric chromosomes and a paracentric inversion in both euchromatic and heterochromatic regions of both wheat and *Ae. speltoides* chromosomes. The frequency of chromosome rearrangements was high among the B-subgenome chromosomes, moderate among the A-subgenome chromosomes, and low among the D-subgenome chromosomes (Kota and Dvorak 1988). In the B subgenome, the rearrangements were nonrandom, and occurred most frequently in chromosomes 1B and 5B. These observations indicate that wheat genomes can be subject to uneven rates of structural chromosome differentiation, even when within the same nucleus. Other examples of massive chromosome aberrations in wheat and *Aegilops* species are rare but were observed in root-tip cell of the Brazilian semi-dwarf cultivar IAS-54 of bread wheat (Dos Santos Guerra et al. 1977), in mitotic and meiotic cells of an individual plant of *Elymus farctus* (=*Agropyron junceum*) (Heneen 1963), and in first meiotic metaphase of an *Ae. longissima* line (TL02) (Feldman and Strauss 1983). Chromosome instability was also associated with gametocidal effects caused by several alien chromosomes of homoeologous group 4 in the monosomic state (Maan 1975; Finch et al. 1984; Tsujimoto and Tsunewaki 1984; Endo 1985).

Aegilops speltoides and *Amblyopyrum muticum*, the two-primitive species of the wheat group, are predominantly cross-pollinated, and contain genes promoting homoeologous pairing. On the other hand, these two species contain B chromosomes that suppress such pairing (Simchen et al. 1971; Mendelson and Zohary 1972; Zarchi et al. 1974; Mochizuki 1957, 1960, 1964; Ohta 1995a). B chromosomes are only known to occur spontaneously in cross-pollinating taxa (Müntzing et al. 1969) and are fully absent in self-pollinators (Jones et al. 2008). In both species, *Ae. speltoides* and *A. muticum*, the B chromosomes are absent in the roots but stably present in the aerial tissue (Mendelson and Zohary 1972; Ohta 1995a); a maximum of eight Bs per cell has been reported in *Ae. speltoides* (Raskina et al. 2004b). A comparable situation of tissue type-specific B chromosome distribution is also known for *Agropyron cristatum* (Baenziger 1962). The B chromosomes of *Ae. speltoides* and *A. muticum* are not homologous since they do not pair in F₁ hybrids between these two species (Vardi and Dover 1972). The presence of one to three Bs has a positive effect on the plant, whereas a higher number of Bs reduces fertility and vigor (Mendelson and Zohary 1972; Belyayev et al. 2010).

The B chromosomes of *Ae. speltoides* are also characterized by a few A-chromosome-localized repeats, like *Spelt1*, pSc119.2 tandem repeats, 5S rDNA and *Ty3-gypsy* retroelements (Friebe et al. 1995c; Raskina et al. 2011; Hosid et al. 2012; Belyayev and Raskina 2013). It is assumed that these B chromosomes originated from the standard set of A chromosomes as a consequence of interspecific hybridization or, more likely, from trisomic (2n + 1) plants. Several lines of *Ae. speltoides* form a low frequency of unreduced gametes as a result of meiotic disorders or spontaneous non-disjunction of the entire chromosome complement, leading to the formation of triploid plants upon the fusion of the 2n gamete with 1n gamete (Belyayev and Raskina 2013; Feldman M, unpublished). Self-pollination of such triploids yields aneuploids and trisomic plants. Several trisomic *Ae. speltoides* plants were occasionally found in plants that originated from different populations (Feldman M, unpublished). Potential donors of the B chromosome of *Ae. speltoides* are the A chromosomes 1S, 4S and 5S of the *Ae. speltoides* genome (Friebe et al. 1995c; Belyayev and Raskina 2013). All Bs in *Ae. speltoides* have a single intercalary *Spelt1* tandem repeat cluster and a 5S rDNA cluster in both arms (Raskina et al. 2011). In addition, a large intercalary cluster of Ty3-gypsy elements was found in close proximity to the 5S rDNA and *Spelt1* blocks (Belyayev and Raskina 2013). Since chromosome 4S is the only *Ae. speltoides* chromosome that carries the intercalary Spelt1 cluster, and chromosome 5S, which is an exclusive source of 5S rDNA in the *Ae. speltoides* genome, it may be involved in the heterologous synapses and recombination resulting in the formation of Bs in this species (Raskina et al. 2011). Support of the idea that B derived from A chromosome(s) also comes from the observation that Bs pair at meiosis, although very rarely, with A chromosome(s) (Belyayev and Raskina 2013). The similarity in the B chromosome structures throughout the species range indicated that they were generated from a similar heterologous recombination of certain A chromosomes.

Ae. speltoides B chromosomes suppress homoeologous meiotic pairing of A chromosomes in intergeneric hybrids with bread wheat lacking the *ph1* gene (Vardi and Dover 1972; Dover and Riley 1972). The effect of B chromosomes is like that of wheat *Ph1* on homoeologous pairing in wheat hybrids with alien species. If *Ae. speltoides* B chromosomes derived from one of the A chromosomes, then one of the first essential prerequisites for the establishment of B chromosome as an independent entity should be to prevent pairing and recombination between the newly formed B and the ancestral A chromosome(s). Thus, the development of a genetic system suppressing homoeologous pairing was necessary for the independent existence and evolution of B chromosomes.

B chromosomes suppress homoeologous pairing in interspecific hybrids and thus, counteract the activity of the pairing promoters of *Ae. speltoides*. In a predominantly cross-pollinated species, the presence of genes that promote pairing is important, either to assure complete pairing in intraspecific hybrids or to increase pairing and recombination in interspecific ones. Yet, the presence of B chromosomes in some individuals of this species helps to conserve the integrity of their genomes.

9.4.2.3 Crosses with Other Species of the Wheat Group

Riley (1966a), Vardi and Dover (1972), and Ohta and Tanaka (1983) determined mean chromosome pairing at first meiotic metaphase in F₁ hybrids of *Ae. speltoides* x *Amblyopyrum muticum* with and without B chromosomes (Tables 8.1 and 8.2). The hybrid *speltoides* x *muticum* without B chromosomes had mean pairing much higher than hybrid with four B chromosomes. Seed fertility of the F₁ hybrids was very low (about 2.0%). The hybrids with B chromosomes, that suppress homoeologous pairing, indicating that the high pairing in hybrids without B chromosomes results mainly from homoeologous pairing promoted by the homoeologous-pairing promoters that exist in these diploids. Hence, the above pairing data show that the genomes of these two species have substantially diverged from one another.

In general, chromosome pairing at meiotic first metaphase in F_1 hybrids between species of the Sitopsis section has been found to be high (Table 9.4) and consequently, Kihara (1949, 1954) grouped the Sitopsis species all together with the same primary genome symbol, S. The species of subsection Emarginata, namely, *Ae. sharonensis*, *Ae. longissima*, *Ae. bicornis* and *Ae. searsii*, form fertile F_1 hybrids with one another, whereas F_1 hybrids of any of these species with high- or low-pairing types of *Ae. speltoides*, the only species of subsection Truncata, show slightly reduced pairing and are sterile (Sears 1941b; Kihara 1949; Tanaka 1955a; Roy 1959; Riley et al. 1961; Kimber 1961; Ankori and Zohary 1962; Feldman et al. 1979). Thus, it is clear that the *Ae. sharonensis-longissima-bicornis-searsii* are of close affinity, while *Ae. speltoides* is somewhat more distant.

Hybrids between high-pairing lines of *Ae. speltoides* and other diploid species of *Aegilops*, i.e., *Ae. caudata*, *Ae. comosa*, *Ae. uniaristata*, and *Ae. umbellulata* exhibit about 10.70–11.26 (0.76–0.80%) paired chromosomes, while hybrids between low-pairing type of *Ae. speltoides* and *Ae. caudata*, *Ae uniaristata, and Ae. umbellulata* show only

4.92–7.30 (35–60%) paired chromosomes (Table 9.4). Evidently, the pairing data indicate that *Ae. speltoides* is more distant to these diploid species than to species of subsection Emarginata.

Chromosome pairing between the allotetraploid species *Ae. kotschyi* and high-pairing and low-pairing types of *Ae. speltoides* exhibit somewhat reduced pairing (Table 9.5). The F_1 hybrid between *Ae. kotschyi* and high pairing line of *Ae. speltoides* show 14.65 paired chromosomes (69.8%), whereas hybrids between *Ae. kotschyi* and low-pairing type of *Ae. speltoides* show only 13.8 (65.7%) paired chromosomes (Rubenstein and Sallee 1973). Hybrid between *Ae. peregrina* with low pairing type of Ae. speltoides showed 8.02 (38.2%) paired chromosomes (Yu and Jahier 1992). Also, F_1 hybrid between other allotetraploid species of section Aegilops, namely, *Ae. biuncialis*, *Ae. columnaris*, *Ae. neglecta* and *Ae. triuncialis*, with *Ae. speltoides*, presumably a high-pairing type, showed similar levels of reduced pairing (Kihara 1949). In all the above diploid and triploid hybrids, the *speltoides* high-pairing genotypes promoted higher pairing between the homoeologous chromosomes of the studied species than the low-pairing genotypes. Data of chromosome pairing between allohexaploids of

Table 9.4 Chromosome pairing at first meiotic metaphase of F_1 hybrids between diploid *Aegilops* species

Combination	Genome	Univalents	Bivalents			Multivalents		References
			Rod	Ring	Total	III	IV	
Ae. speltoides I[a] x *Ae. sharonensis*	SS^{sh}	0.32	1.50	5.35	6.85	–	–	Sears (1941b)
Ae. speltoides II[b] x *Ae. sharonensis*	SS^{sh}	0.68	3.00	3.66	6.66	–	–	Sears (1941b)
Ae. longissima x *Ae. speltoides*	S^{l}S	0.54	2.32	2.86	5.18	0.34	0.52	Riley et al. (1961)
Ae. bicornis x *Ae. longissima*	S^{b}S^l	0.28	–	–	5.78	0.16	0.42	Kimber (1961)
Ae. sharonensis x Ae. longissima	S^{sh}S^l	0.26	–	–	5.36	0.18	0.62	Kimber (1961)
Ae. longissima x *Ae. searsii*	S^{l}S^s	0.44	1.41	3.56	4.97	0.30	0.68	Feldman et al. (1979)
		1.15	1.84	3.13	4.97	0.81	0.12	Feldman et al. (1979)
Ae. caudata x *Ae. speltoides* I[a]	CS	2.74	3.26	0.48	3.74	1.26	–	Sears (1941b)
Ae. speltoides I[a] x *Ae. caudata*	SC	3.30	3.10	0.42	3.52	1.22	–	Sears (1941b)
Ae. caudata x *Ae. speltoides* II[b]	CS	6.70	2.78	0.12	2.90	0.50	–	Sears (1941b
Ae. sharonensis x *Ae. caudata*	S^{sh}C	6.34	3.26	0.12	3.36	0.30	–	Sears (1941b)
Ae. speltoides I[a] x *Ae. comosa*	SM	2.68	3.80	1.12	4.92	0.44	0.04	Sears (1941b)
Ae. speltoides II[b] x *Ae. uniaristata*	SN	5.54	3.70	0.26	3.96	0.18	–	Sears (1941b)
Ae. sharonensis x *Ae. uniaristata*	S^{sh}N	4.88	3.70	0.38	4.08	0.34	–	Sears (1941b)
Ae. speltoides I[a] x *Ae. umbellulata*	SU	3.22	2.98	–	2.98	1.58	0.02	Sears (1941b)
Ae. speltoides II[b] x *Ae. umbellulata*	SU	9.08	1.98	–	1.98	0.32	–	Sears (1941b)
Ae. sharonensis x *Ae. umbellulata*	S^{sh}U	6.74	2.78	–	2.78	0.20	0.02	Sears (1941b)
Ae. comosa x *Ae. uniaristata*	MN	2.88	3.98	0.56	4.54	0.68	–	Sears (1941b)
		2.94	5.08	0.33	5.41	0.08	–	Cuñado and Santos (1999)

[a] High pairing type of *Ae. speltoides*
[b] Intermediate pairing type of *Ae. speltoides*

Table 9.5 Chromosome pairing at first meiotic metaphase of F_1 hybrids between allotetraploids and diploids of the genus *Aegilops*

Hybrid combination	Genome	Univalents	Bivalents			Multivalents				References
			Rod	Ring	Total	III	IV	V	Above V	
Ae. peregrina x *Ae. longissima*	US^{v}S^l	6.66	1.00	5.66	6.66	0.34	–	–	–	Feldman (1963)
Ae. peregrina x *Ae. longissima*	US^{v}S^l	6.77	2.11	2.98	5.09	0.30	0.73	0.04		Yu and Jahier (1992)
Ae. kotschyi x *Ae. longissima*	US^{v}S^l	5.48	2.48	3.34	5.82	0.92	0.28	–	–	Feldman (1963)
Ae. kotschyi x *Ae. sharonensis*	US^{v}S^{sh}	5.41	2.26	2.02	4.28	1.50	0.46	0.09	–	Rubenstein and Sallee (1973)
Ae. peregrina x *Ae. bicornis*	US^{v}S^b	9.13	3.39	0.89	4.28	0.53	0.25	0.14	–	Yu and Jahier (1992)
Ae. peregrina x *Ae. searsii*	US^{v}S^s	10.97	3.48	0.33	3.81	0.54	0.16	0.03	–	Yu and Jahier (1992)
Ae. peregrina x *Ae. speltoides* (LP)^a	US^{v}S	12.98	3.26	0.06	3.32	0.45	0.01	–	–	Yu and Jahier (1992)
Ae. kotschyi x *Ae. speltoides* (HP)^b	US^{v}S	6.35	4.60	–	4.60	1.55	0.20	–	–	Rubenstein and Sallee (1973)
Ae. kotschyi x *Ae. speltoides* (IP)^c	US^{v}S	9.82	4.38	0.02	4.40	0.76	–	–	–	Rubenstein and Sallee (1973)
Ae. kotschyi x *Ae. speltoides* (LP)^a	US^{v}S	7.20	4.70	0.20	4.90	1.00	0.25			Rubenstein and Sallee (1973)
Ae. crassa (4x) x *Ae. tauschii*	D^{c}X^{c}D	7.67	3.07	1.96	5.03	0.91	0.07	0.03	0.01 VI	Kimber and Zhao (1983)
Ae. ventricosa x *Ae. tauschii*	DND	6.16	1.29	4.73	6.02	0.85	0.04	–	–	Kimber and Zhao (1983)
Ae. cylindrica x *Ae. caudata*	DCC	6.70	0.20	6.76	6.96	0.13	–	–	–	Kimber and Abu-Baker (1981)
Ae. geniculata x *Ae. umbellulata*	M^{o}UU	6.42	3.32	2.37	5.69	0.92	0.11	–	–	Kimber and Abu-Baker (1981)
Ae. peregrina x *Ae. umbellulata*	US^{v}U	6.95	1.59	5.41	6.99	0.02	–	–	–	Yu and Jahier (1992)
Ae. kotschyi x *Ae. umbellulata*	US^{v}U	7.04	2.90	3.66	6.56	0.29	–	–	–	Kimber and Abu-Baker (1981)
Ae. columnaris x *Ae. umbellulata*	UX^{n}U	6.17	1.87	4.30	6.17	0.83	–	–	–	Kimber and Abu-Baker (1981)
Ae. crass 4 x x *Ae. uniaristata*	D^{c}X^{c}N	11.11	3.46	0.43	3.89	0.61	0.07	–	–	Kimber et al. (1983)
Ae. ventricosa x *Ae. uniaristata*	DNN	7.55	3.70	2.05	5.75	0.65	–	–	–	Kimber et al. (1983)
Ae. comosa x *Ae geniculata*	MM^{o}U	6.15	2.65	3.05	5.70	1.15	–	–	–	Kimber et al. (1983)
Ae. geniculata x *Ae. uniaristata*	M^{o}UN	10.40	4.00	0.10	4.10	0.80	–	–	–	Kimber et al. (1983)
Ae. crassa 6 x x *Ae. speltoides*	D^{c}X^{c}DS	8.00	3.05	3.42	6.47	1.42	0.38	0.05	–	Melnyk and McGinnis (1962)
Ae. crassa 6 x x *Ae. speltoides*	D^{c}X^{c}DS	10.57	2.54	3.32	5.86	1.11	0.36	–	0.18 VII	Melnyk and McGinnis (1962)
Ae. crassa 6 x x *Ae. sharonensis*	D^{c}X^{c}DS^{sh}	10.88	4.44	2.34	6.78	0.94	0.15	0.04	–	Melnyk and McGinnis (1962)
Ae. juvenalis x *Ae. speltoides*	D^{c}X^{c}US	8.32	3.38	0.20	3.58	1.89	0.84	–	0.62 XI	McGinnis and Melnyk (1962)
Ae. juvenalis x *Ae. speltoides*	D^{c}X^{c}US	10.24	4.60	0.92	5.52	1.57	0.40	0.08	–	McGinnis and Melnyk (1962)
Ae. juvenalis x *Ae. sharonensis*	D^{c}X^{c}US^{sh}	13.14	4.17	0.42	4.59	1.40	0.35	–	0.03 VI	McGinnis and Melnyk (1962)
Ae. juvenalis x *Ae. longissima*	D^{c}X^{c}US^l	15.86	3.23	0.08	3.31	1.28	0.36	0.01	–	McGinnis and Melnyk (1962)
Ae. crassa 6 x x *Ae. tauschii*	D^{c}X^{c}DD	10.21	2.61	4.42	7.03	1.03	0.24	–	0.03 VI	Melnyk and McGinnis (1962)
Ae. vavilovii x *Ae. tauschii*	D^{c}X^{c}S^{s}D	13.89	3.89	1.14	5.03	0.75	0.37	0.05	–	Kimber and Zhao (1983)
Ae. juvenalis x *Ae. tauschii*	D^{c}X^{c}UD	11.73	4.42	2.13	6.55	0.92	0.10	–	–	McGinnis and Melnyk (1962)

(continued)

Table 9.5 (continued)

Hybrid combination	Genome	Univalents	Bivalents			Multivalents				References
			Rod	Ring	Total	III	IV	V	Above V	
Ae. juvenalis x *Ae. tauschii*	D^{c}X^{c}UD	13.55	3.73	1.38	5.11	1.07	0.26	–	–	Kimber and Zhao (1983)
Ae. crassa 6 x x *Ae. caudata*	D^{c}X^{c}DC	13.28	4.05	2.02	6.07	0.76	0.05	0.02	–	Melnyk and McGinnis (1962)
Ae. juvenalis x *Ae. caudata*	D^{c}X^{c}UC	14.70	4.88	0.08	4.96	0.11	–	–	–	McGinnis and Melnyk (1962)
Ae. juvenalis x *Ae. caudata*	D^{c}X^{c}UC	15.10	4.42	0.07	4.49	1.13	0.13	0.02	–	Kimber and Abu-Baker (1981)

[a] LP—Low paring type of *Ae. speltoides*
[b] HP—High paring type of *Ae. speltoides*
[c] IP—Intermediate paring type of *Ae. speltoides*

Aegilops section Vertebrata, *Ae. crassa* 6 x and *Ae juvenalis*, and *Ae. speltoides* are presented in Table 9.6. These hybrids had higher pairing because of the effect of the homoeologous-pairing promoters in *Ae. speltoides*.

Data of chromosome pairing between allohexaploids of *Aegilops* section Vertebrata, *Ae. crassa* 6x, *Ae juvenalis*, and *Ae. vavilovii*, and diploid species of *Aegilops* are presented in Table 9.6. The hybrid *Ae. crassa* 6x x *Ae. tauschii* had

Table 9.6 Chromosome pairing at first meiotic metaphase of F$_1$ hybrids between allohexaploids and diploids of the genus *Aegilops*

Combination	Genome	Univalents	Bivalents			Multivalents				References
			Rod	Ring	Total	III	IV	V	Above V	
Ae. crassa 6x x *Ae. speltoides*	D^{c}X^{c}DS	8.00	3.05	3.42	6.47	1.42	0.38	0.05	–	Melnyk and McGinnis (1962)
Ae. crassa 6x x *Ae. speltoides*	D^{c}X^{c}DS	10.57	2.54	3.32	5.86	1.11	0.36	–	0.18 VII	Melnyk and McGinnis (1962)
Ae. crassa 6x x *Ae. sharonensis*	D^{c}X^{c}DS^{sh}	10.88	4.44	2.34	6.78	0.94	0.15	0.04	–	Melnyk and McGinnis (1962)
Ae. juvenalis x *Ae. speltoides*	D^{c}X^{c}US	8.32	3.38	0.20	3.58	1.89	0.84	–	0.62 XI	McGinnis and Melnyk (1962)
Ae. juvenalis x *Ae. speltoides*	D^{c}X^{c}US	10.24	4.60	0.92	5.52	1.57	0.40	0.08	–	McGinnis and Melnyk (1962)
Ae. juvenalis x *Ae. sharonensis*	D^{c}X^{c}US^{sh}	13.14	4.17	0.42	4.59	1.40	0.35	–	0.03 VI	McGinnis and Melnyk (1962)
Ae. juvenalis x *Ae. longissima*	D^{c}X^{c}US^l	15.86	3.23	0.08	3.31	1.28	0.36	0.01	–	McGinnis and Melnyk (1962)
Ae. crassa 6x x *Ae. tauschii*	D^{c}X^{c}DD	10.21	2.61	4.42	7.03	1.03	0.24	–	0.03 VI	Melnyk and McGinnis (1962)
Ae. vavilovii x *Ae. tauschii*	D^{c}X^{c}S^{s}D	13.89	3.89	1.14	5.03	0.75	0.37	0.05	–	Kimber and Zhao (1983)
Ae. juvenalis x *Ae. tauschii*	D^{c}X^{c}UD	11.73	4.42	2.13	6.55	0.92	0.10	–	–	McGinnis and Melnyk (1962)
Ae. juvenalis x *Ae. tauschii*	D^{c}X^{c}UD	13.55	3.73	1.38	5.11	1.07	0.26	–	–	Kimber and Zhao (1983)
Ae. crassa 6x x *Ae. caudata*	D^{c}X^{c}DC	13.28	4.05	2.02	6.07	0.76	0.05	0.02	–	Melnyk and McGinnis (1962)
Ae. juvenalis x *Ae. caudata*	D^{c}X^{c}UC	14.70	4.88	0.08	4.96	0.11	–	–	–	McGinnis and Melnyk (1962)
Ae. juvenalis x *Ae. caudata*	D^{c}X^{c}UC	15.10	4.42	0.07	4.49	1.13	0.13	0.02	–	Kimber and Abu-Baker (1981)
Ae. juvenalis x *Ae. uniaristata*	D^{c}X^{c}UN	18.50	2.80	0.02	2.82	1.15	0.08	0.02	–	McGinnis and Melnyk (1962)

somewhat higher pairing than most other hybrids indicating high homology between the second D subgenome of the hexaploid and the D genome of the diploid. The remaining hybrid exhibited much less pairing. Chromosome pairing in F_1 hybrids between allopolyploid species of Aegilops indicates that also several chromosomes of the differential subgenomes are involved in pairing in addition to that of he shared subgenome (Table 9.7).

Mean chromosomal pairing at meiosis in F_1 hybrids between high-pairing type of *Ae. speltoides* and domesticated *T. monococcum*, ssp. *monococcum*, was very high, almost complete (Table 9.8). Similar high pairing was observed in the F_1 hybrid between *Ae. speltoides* (presumably a high-pairing line) and wild *T. monococcum*, ssp. *aegilopoides*, was reported by Shang et al. (1989). However, mean chromosomal pairing at meiosis of F_1 hybrids between the low-pairing type of *Ae. speltoides* and *T. monococcum* ssp. *monococcum* was low (Table 9.8) These results indicate that the *speltoides* homoeologous-pairing promoter genes bring about a high degree of homoeologous pairing in *speltoides* x *T. monococcum* hybrids, and, when these genes are absent, as in the low-pairing genotype, the low level of pairing reveals excessive divergence of the genomes of these two species.

Table 9.7 Chromosome pairing at first meiotic metaphase of F_1 hybrids between allopolyploids of the genus *Aegilops*

Hybrid combination	Genome	Univalents	Bivalents			Multivalents				References
			Rod	Ring	Total	III	IV	V	Above V	
Ae. ventricosa x *Ae. crassa* 4x	DND^cX^c	14.08	3.97	0.76	4.73	1.16	0.22	0.01	–	Kimber and Zhao (1983)
Ae. geniculata x *Ae. biuncialis*	M^oUUM^b	10.25	4.20	2.30	6.50	1.25	0.25	–	–	Kimber et al. (1988)
Ae. geniculata x *Ae. columnaris*	M^oUUX^n	9.20	4.45	2.70	7.15	1.10	0.30	–	–	Kimber et al. (1988)
Ae. geniculata x *Ae. kotschyi*	M^oUUS^v	10.65	3.20	3.25	6.45	1.35	0.10	–	–	Kimber et al. (1988)
Ae. geniculata x *Ae. kotschyi*	M^oUUS^v	11.70	1.60	4.42	6.02	1.42	–	–	–	Feldman (1963)
Ae. kotschyi x *Ae. peregrina*	US^vUS^v	0.94	1.94	9.04	10.98	0.20	0.54	0.18	0.24 VI	Feldman (1963)
Ae. geniculata x *Ae. peregrina*	M^oUUS^v	11.44	2.02	5.00	7.02	0.84	–	–	–	Feldman (1963)
Ae. biuncialis x *Ae. peregrina*	UM^bUS^v	12.74	1.22	5.14	6.36	0.74	0.08	–	–	Feldman (1963)
Ae. columnaris x *Ae. kotschyi*	UX^nUS^v	12.10	4.05	3.00	7.05	0.60	–	–	–	Kimber et al. (1983)
Ae. cylindrica x *Ae. crassa* 4x	DCD^cX^c	8.51	3.32	2.68	6.00	1.84	0.40	0.04	–	Kimber and Zhao (1983)
Ae. cylindrica x *Ae. ventricosa*	$DCDN$	8.24	3.04	4.26	7.30	1.44	0.16	0.03	–	Kimber and Zhao (1983)
Ae. ventricosa x *Ae. cylindrica*	$DNDC$	8.08	3.10	3.98	7.08	1.33	0.35	0.05	0.03 VI	Kimber and Zhao (1983)
Ae. juvenalis x *Ae. ventricosa*	D^cX^cUDN	20.39	4.41	0.45	4.86	1.25	0.25	0.04	–	Kimber and Zhao (1983)
Ae. vavilovii x *Ae. ventricosa*	$D^cX^cS^sDN$	19.45	4.08	0.73	4.81	1.69	0.20	0.02	–	Kimber and Zhao (1983)
Ae. ventricosa x *Ae. crassa* 6x	DND^cX^cD	15.61	4.39	2.27	6.66	1.77	0.20	–	–	Kimber and Zhao (1983)
Ae. juvenalis x *Ae. cylindrica*	D^cX^cUDC	10.97	5.14	1.43	6.57	2.03	0.77	0.17	0.14 VI	Kimber and Zhao (1983)
Ae. juvenalis x *Ae. cylindrica*	D^cX^cUDC	14.35	5.97	1.09	7.06	1.53	0.39	–	0.11 VII	McGinnis and Melnyk (1962)
Ae. cylindrica x *Ae. juvenalis*	DCD^cX^cU	14.36	5.18	1.09	6.27	2.45	0.18	–	–	Kimber and Zhao (1983)
Ae. vavilovii x *Ae. cylindrica*	$D^cX^cS^sDC$	12.15	6.07	1.27	7.34	2.08	0.39	0.08	–	Kimber and Zhao (1983)
Ae. crassa 6x x *Ae. cylindrica*	D^cX^cDDC	13.12	4.39	3.73	8.12	1.42	0.32	0.02	–	Melnyk and McGinnis (1962)
Ae. crassa 6x x *Ae. geniculata*	$D^cX^cDM^oU$	23.50	3.73	0.80	4.53	0.77	0.03	–	–	Kimber et al. (1988)
Ae. juvenalis x *Ae. geniculata*	$D^cX^cUM^oU$	13.05	5.30	1.53	6.83	1.95	0.60	–	–	Kimber et al. (1988)
Ae. geniculata x *Ae. juvenalis*	$M^oUD^cX^cU$	13.70	4.50	1.90	6.40	2.10	0.55	–	–	Kimber et al. (1988)
Ae. juvenalis x *Ae. columnaris*	$D^cX^cUUX^n$	12.95	4.24	3.57	7.81	1.79	0.22	–	0.04 VI	McGinnis and Melnyk (1962)
Ae. juvenalis x *Ae. triuncialis*	D^cX^cUUC	12.94	5.37	2.27	7.64	1.70	0.34	–	0.02 VII	McGinnis and Melnyk (1962)
Ae. crassa 6x x *Ae. peregrina*	$D^cX^cDS^vU$	19.67	4.29	1.50	5.79	0.98	0.17	–	0.03 VI	McGinnis and Melnyk (1962)

Table 9.8 Chromosome pairing at first meiotic metaphase of F_1 hybrids between diploid species of *Triticum* and *Aegilops*

Combination	Genome	Univalents	Bivalents			Multivalents		References
			Rod	Ring	Total	III	IV	
Ae. speltoides I[a] x ssp. *monococcum*	SA^m	0.82	3.84	2.44	6.28	0.18	0.02	Sears (1941b)
Ae. speltoides II[b] x ssp. *monococcum*	SA^m	8.42	2.34	0.44	2.78	0.15	–	Sears (1941b)
Ae. speltoides II[b] x ssp. *monococcum*	SA^m	6.04	3.28	0.38	3.66	0.16	0.04	Sears (1941b)
Ae. speltoides x ssp. *monococcum*	SA^m	7.25	–	–	3.37	–	–	Riley et al. (1958)
Ae. speltoides x ssp. *aegilopoides*	SA^m	5.52	3.26	0.88	4.14	0.07	–	Shang et al. (1989)
Ae. speltoides x ssp. *aegilopoides*	SA^m	2.92	3.49	1.76	5.25	0.15	0.03	Shang et al. (1989)
Ae. longissima (IP)[c] x ssp. *monococcum*	S^{l}A^m	10.77	1.47	0.02	1.49	0.07	–	Feldman (1978)
Ae. sharonensis x ssp. *monococcum*	S^{sh}A^m	13.35	–	–	0.30	0.01	–	Kushnir and Halloran (1981)
Ae. bicornis x ssp. *monococcum*	S^{b}A^m	8.62	2.28	0.23	2.51	0.11	0.01	Sears (1941b)
Ssp. *monococcum* x *Ae. tauschii*	A^{m}D	3.36	3.28	1.58	4.86	0.20	0.08	Sears (1941b)
Ssp. *aegilopoides* x *Ae. tauschii*	A^{m}D	6.56	2.88	0.28	3.16	0.34	0.02	Sears (1941b)
Ssp. *monococcum* x *Ae. caudata*	A^{m}C	4.80	3.69	0.03	3.72	0.59	–	Sears (1941b)
Ssp. *monococcum* x *Ae. comosa*	A^{m}M	7.44	3.00	0.22	3.22	0.04	–	Sears (1941b)
Ssp. *aegilopoides* x *Ae. comosa*	A^{m}M	7.06	2.48	0.08	2.56	0.42	0.14	Sears (1941b)
Ssp. *monococcum* x *Ae. uniaristata*	A^{m}N	10.15	1.86	0.02	1.88	0.03	–	Sears (1941b)
Ssp. *aegilopoides* x *Ae. uniaristata*	A^{m}N	7.66	2.98	–	2.98	0.02	–	Sears (1941b)
Ssp. *aegilopoides* x *Ae. umbellulata*	A^{m}U	5.78	2.78	0.10	2.88	0.82	–	Sears (1941b)

[a] High pairing type of *Ae. speltoides*
[b] Intermediate pairing type of *Ae. speltoides*
[c] Intermediate pairing type of *Ae. longissima*

McFadden and Sears (1947) reported that mean chromosomal pairing in the triploid hybrid between wild tetraploid wheat *T. turgidum* ssp. *dicoccoides* x *Ae. speltoides* was high, about two-third of the chromosomes paired (Table 9.9). Similar results were obtained by Riley et al. (1958) in hybrids between several subspecies of tetraploid wheat *T. turgidum* and *Ae. speltoides*. Yet, F_1 hybrids between *T. turgidum* ssp. *durum* x low-pairing type of *Ae. speltoides* showed much lower pairing, involving very few paired chromosomes (Shands and Kimber 1973).

The triploid hybrid between the second tetraploid species of wheat, *T. timopheevii* and *Ae. speltoides* showed somewhat higher pairing than that between *T. turgidum* and *Ae. speltoides* (Table 9.9). Hybrids between *T. timopheevii* with the high-pairing *Ae. speltoides* genotype showed pairing of more than ten chromosomes, while hybrids with the low-pairing *Ae. speltoides* genotype showed pairing of seven bivalents only (Shands and Kimber 1973). These data indicate greater homology between *Ae. speltoides* and one of the subgenomes of *T. timopheevii* than in T. turgidum x *Ae. speltoides* hybrids.

Mean chromosome pairing at first meiotic metaphase in hybrids between hexaploid wheat, *T. aestivum* ssp. *aestivum* (bread wheat) and the high-pairing line of *Ae. speltoides* is very high (Table 9.10). On the other hand, hybrids with the low-pairing *Ae. speltoides* genotype, showed much reduced chromosomal pairing (Kimber and Athwal (1972).

Maestra and Naranjo (1998) analyzed homoeologous chromosome pairing at first meiotic metaphase in F_1 hybrids derived from crosses of bread wheat, either carrying the homoeologous-pairing suppressor, *Ph1*, or the deficiency *ph1b*, lacking this gene, with a high-pairing genotype of *Aegilops speltoides*. Data from relative pairing affinities were used to predict homoeologous relationships of *Ae. speltoides* chromosomes to wheat. Chromosomes of both species, and their arms, were identified by C-banding. The *Ae. speltoides* genotype carried genes that induced a high level of homoeologous pairing in the two types of hybrids analyzed. All arms of the seven chromosomes of the *speltoides* S genome showed normal homoeologous pairing that implies that no apparent chromosome rearrangements occurred in the evolution of the genome of *Ae. speltoides* relative to wheat evolution. A pattern of preferential pairing of two types, A-D and B-S, confirmed that the S genome is very closely related to the B subgenome of wheat. Although this pairing pattern was also reported in hybrids of wheat with *Ae. longissima* and *Ae. sharonensis*, a different behavior was found in group 5 chromosomes. In the hybrids

Table 9.9 Chromosome pairing at first meiotic metaphase of F_1 hybrids between allotetraploid species of *Triticum* and diploid species of *Aegilops*

Combination	Genome	Univalents	Bivalents			Multivalents		References
			Rod	Ring	Total	III	IV	
Ssp. *dicoccoides* x *Ae. bicornis*	BAS[b]	17.37	1.78	–	1.78	0.02	–	Riley et al. (1958)
Ae. sharonensis x ssp. *dicoccoides*	S[sh]BA	16.62	2.20	0/03	2.23	–	–	Kushnir and Halloran (1981)
Ssp. *turgidum* x *Ae. sharonensis*	BAS[sh]	16.48	2.22	–	2.22	0.02	–	Riley et al. (1958)
Ssp. *turgidum* x *Ae. sharonensis*	BAS[sh]	16.06	2.50	–	2.50	–	–	Roy (1959)
Ae. longissima x ssp. *dicoccon*	S[l]BA	17.43	–	–	1.60	0.01	–	Riley et al. (1958)
Ssp. *durum* x *Ae. longissima*	BAS[l]	18.6	–	–	1.20	–	–	Vardi (1973)
Ssp. *dicoccoides* x *Ae. speltoides*	BAS	7.44	–	–	4.66	1.41	–	McFadden and Sears (1947)
Ae. speltoides x ssp. *dicoccoides*	SBA	5.94	–	–	5.21	1.59	0.06	Riley et al. (1958)
Ae. speltoides x ssp. *dicoccon*	SBA	6.70	–	–	6.22	0.62	–	Riley et al. (1958)
Ae. speltoides x ssp. *georgicum*[a]	SBA	8.00	–	–	4.96	1.08	0.08	Riley et al. (1961)
Ae. speltoides x ssp. *durum*	SBA	5.90	–	–	5.96	1.06	–	Riley et al. (1961)
Ssp. *durum* x *Ae. speltoides* (HP)[b]	BAS	6.60	3.00	1.00	4.00	2.10	0.5	Feldman and Mello-Sampayo (1967)
Ssp. *durum* x *Ae. speltoides* (HP)[b]	BAS	8.80	3.50	0.40	3.90	1.40	0.10	Shands and Kimber (1973)
Ssp. *durum* x *Ae. speltoides* (IP)[c]	BAS	12.10	2.70	1.45	4.15	0.20	–	Shands and Kimber (1973)
Ssp. *durum* x *Ae. speltoides* (LP)[d]	BAS	20.20	0.40	–	0.40	–	–	Shands and Kimber (1973)
Ssp. *durum* x *Ae. speltoides*	BAS	5.00	–	–	4.50	2.50	–	Vardi (1973)
Ae. speltoides x ssp. *turgidum*	SBA	7.92	–	–	5.94	0.40	–	Riley et al. (1961)
Ssp. dicoccoides x Ae. tauschii	BAD	20.89	–	–	0.05	–	–	McFadden and Sears (1947)
Ae. speltoides x ssp. *timopheevii*	SGA	7.60	–	–	6.28	0.28	–	Riley et al. (1961)
Ssp. *timopheevii* x *Ae. speltoides*	GAS	5.34	3.00	1.64	4.64	1.94	0.14	Sears (1941b)
Ssp. *timopheevii* x *Ae. speltoides* (HP)[b]	GAS	3.77	2.87	1.30	4.17	2.70	0.20	Shands and Kimber (1973)
Ssp. *timopheevii* x *Ae. speltoides* (IP)[c]	GAS	6.05	3.45	1.18	4.63	1.70	0.15	Shands and Kimber (1973)
Ssp. *timopheevii* x *Ae. speltoides* (LP)[d]	GAS	7.22	3.34	1.57	4.91	1.30	0.20 + 0.02 V	Shands and Kimber (1973)
Ssp. *timopheevii* x *Ae. bicornis*	GAS[b]	17.53	1.61	0.05	1.66	0.04	0.01	Sears (1941b)
Ssp. *timopheevii* x *Ae. tauschii*	GAD	17.37	–	–	1.71	0–1	–	McFadden and Sears (1947)

[a] *Triticum turgidum* ssp. *georgicum* is currently *T. turgidum* ssp. *paleocolchicum* (Menabde) Á. löve & D. Löve;
[b] HP-High paring type of *Ae. speltoides;* [c] IP-Intermediate paring type of *Ae. speltoides*;
[d] LP-Low paring type of *Ae. speltoides*

of bread wheat with *Ae. speltoides*, chromosome 5B-5S pairing was much more frequent than 5D-5S, while these chromosome associations reached similar frequencies in the hybrids of *Ae. longissima* and *Ae. sharonensis*. These results are in agreement with the hypothesis that the B subgenome of wheat is closely related to the B subgenome of *Ae. speltoides*.

9.4.3 *Aegilops bicornis* (Forssk.) Jaub. & Spach

9.4.3.1 Morphological and Geographical Notes
Aegilops bicornis (Forssk.) Jaub. & Spach [Syn.: *Triticum bicorne* Forssk.; *Sitopsis bicornis* (forssk.) Á. Löve] is predominantly an autogamous annual plant. Culms are slender, erect, 15–45-cm-high. Spikes are 4–5(-8)-cm-long (excluding awns), two-rowed, mostly awned, with 8–15 (-19) spikelets, disarticulating at maturity into spikelets, each spikelet falls with the rachis internode immediately below it (Fig. 2.3; wedge-type dispersal unit). Several lower spikelets often remain attached at the tip of culm. Spikelets are 5.5–8.5-mm-long (excluding awns), mostly 3-flowered (2 lower florets are fertile). The attached rachis internode is half the length of the spikelet. Rudimentary spikelets at the base of the spike are absent. Glumes are 4.5–5.5-mm-long, emarginated or with an angle between teeth, asymmetrical and keeled. Lemmas are 4–6-mm-long, boat-shaped, mostly ending in a slender, 4.5–6.0-cm-long awn (in lower

Table 9.10 Chromosome pairing at first meiotic metaphase of F_1 hybrids between allohexaploid species of *Triticum* and diploid species of *Aegilops*

Combination	Genome	Univalents	Bivalents			Multivalents		References
			Rod	Ring	Total	III	Above III	
Ssp. *aestivum* x Ae. *longissima*	BADS[l]	23.90	1.96	–	1.96	–	–	Riley et al. (1961)
Ssp. *aestivum* x Ae. *longissima*	BADS[l]	24.20	–	–	1.78	0.08	–	Riley and Chapman (1963)
Ssp. *aestivum* x Ae. *longissima*	BADS[l]	24.04	–	–	1.96	0.01	–	Riley (1966a)
Ssp. *aestivum* x Ae. *longissima*	BADS[l]	24.52	–	–	1.68	0.04	–	Riley (1966a)
Ssp. *aestivum* x Ae. *longissima* (LP)[a]	BADS[l]	25.40	1.43	0.02	1.45	0.02	–	Ceoloni et al. (1986)
Ssp. *aestivum* x Ae. *longissima* (LP)[a]	BADS[l]	26.33	0.80	0.02	0.82	0.01	–	Yu and Jahier (1992)
Ssp. *aestivum* x Ae. *longissima* (LP)[a]	BADS[l]	24.55	1.59	0.06	1.65	0.05	0.003 IV	Naranjo and Maestra (1995)
Ssp. *aestivum* x Ae. *longissima* (IP)[b]	BADS[l]	17.20	–	–	4.86	0.31	0.04 IV	Mello-Sampayo (1971b)
Ssp. *aestivum* x Ae. *longissima* (IP)[b]	BADS[l]	15.35	–	–	5.49	0.39	0.06 IV	Feldman (1978)
Ssp. *aestivum* x Ae. *longissima* (IP)[b]	BADS[l]	17.77	4.40	0.39	4.79	0.19	0.02 IV	Ceoloni et al. (1986)
Ssp. *aestivum* x Ae. *sharonensis* (LP)[a]	BADS[sh]	24.99	–	–	1.50	0.02	–	Mello-Sampayo (1971b)
Ssp. *aestivum* x Ae. *sharonensis* (LP)[a]	BADS[sh]	25.21	1.18	0.03	1.21	0.03	–	Maestra and Naranjo (1997)
Ssp. *aestivum* x Ae. *bicornis*	BADS[b]	26.70	0.61	–	0.61	0.03	–	Yu and Jahier (1992)
Ssp. *aestivum* x Ae. *searsii*	BADS[s]	24.62	–	–	1.57	0.07	–	Feldman (1978)
Ssp. *aestivum* x Ae. *searsii*	BADS[s]	26.61	0.68	–	0.68	0.01	–	Yu and Jahier (1992)
Ssp. *aestivum* x Ae. *speltoides*	BADS	6.04	4.60	2.04	6.64	1.88	0.76 IV	Riley et al. (1961)
Ssp. *aestivum* x Ae. *speltoides*	BADS	3.40	4.14	2.00	6.14	2.20	1.38 IV + 0.04 V	Riley et al. (1961)
Ssp. *aestivum* x Ae. *speltoides* (line G)	BADS	8.44	–	–	6.22	1.33	0.74 IV	Riley (1966a, b)
Ssp. *aestivum* x Ae. *speltoides* (line E)	BADS	6.60	–	–	6.60	1.44	0.92 IV + 0.04 V	Riley (1966a)
Ssp. *aestivum* x Ae. *speltoides* (line M)	BADS	9.82	–	–	6.02	1.38	0.50	Riley (1966a)
Ssp. *aestivum* x Ae. *speltoides* (HP)[c]	BADS	5.40	5.10	2.50	7.60	1.00	1.10	Feldman and Mello-Sampayo (1967)
Ssp. *aestivum* x Ae. *speltoides* (IP)[b]	BADS	5.83	–	–	5.27	1.47	1.47	Mello-Sampayo (1971b)
Ssp. *aestivum* x Ae. *speltoides* (HP)[c]	BADS	6.80	5.00	1.80	6.80	1.80	0.50 IV + 0.05 VI	Kimber and Athwal (1972)
Ssp. *aestivum* x Ae. *speltoides* (IP)[b]	BADS	16.70	5.20	–	5.20	0.03	–	Kimber and Athwal (1972)
Ssp. *aestivum* x Ae. *speltoides* (LP)[a]	BADS	26.60	0.70	–	0.70	–	–	Kimber and Athwal (1972)
Ssp. *aestivum* x Ae. *speltoides* (LP)[a]	BADS	23.37	2.17	0.08	2.25	0.04	–	YU and Jahier (1992)
Ssp. *aestivum* x Ae. *tauschii*	BADD	15.38	2.42	3.75	6.17	0.07	0.02 IV	Kimber and Riley (1963)
Ssp. *aestivum* x Ae. *tauschii*	BADD	14.66	1.94	4.44	6.38	0.12	0.02 IV	Kimber and Riley (1963)
Ssp. *aestivum* x Ae. *caudata*	BADC	24.12	–	–	1.82	0.08	–	Riley (1966a)
Ssp. *aestivum* x Ae. *comosa*	BADM	22.44	–	–	2.46	0.06	–	Riley (1966a)
Ssp. *aestivum* x Ae. *umbellulata*	BADU	24.50	–	–	1.60	0.10	–	Riley (1966a)
Ssp. *aestivum* x Ae. *umbellulata*	BADU	23.83	1.93	0.02	1.95	0.09	0.01 IV	Yu and Jahier (1992)

[a]LP-Low paring type of Ae. speltoides; [b]IP-Intermediate paring type of Ae. speltoides; [c]HP-High paring type of Ae. speltoides.

spikelets, awns are short or absent); the awn is usually not flanked by lateral teeth. The caryopsis adheres to lemmas and palea (Fig. 9.1c).

The species includes two varieties: var. *bicornis* (var. *typica* Eig) and var. *anathera* [var. *mutica* (Aschers.) Eig; *Triticum bicorne muticum* (Aschers.) Eig]. In var. *bicornis*, all spikelets are awned, sometimes, the lowest is awnless, whereas, in var. *anathera*, all spikelets, except for the uppermost 1–3, are awnless. Var. *anathera* is rare and sporadic, grows together with var. *bicornis*.

Ae. bicornis exhibits limited morphological variation, involving mainly spike and spikelet size, plant size, and degree of awn development. It is close in spike morphology to *Ae. sharonensis*, *Ae. speltoides* var. *ligustica*, *Triticum monococcum* ssp. *aegilopoides* and *T. urartu*.

Ae. bicornis is a semi-desert element, distributing in southeastern Mediterranean and in Saharo-Arabian regions. It occurs in the coastal regions of Libya (Cyrenaica)), Egypt (lower Egypt and Sinai) and southern Israel (western Negev), as well as in inland desert area in Israeli Negev and in Southern Jordan (Edom). Few populations were found in northeastern Cyprus. It grows from sea level to 200 m above sea level (in southern Jordan, it is found at 200–900 m), in areas with an annual rainfall of 75–275 mm (the driest part of the Mediterranean), usually on stable sandy soils, in open dwarf shrubs or herbaceous steppe-like or desert-like formations, xeric coastal and desert plains, in plantations, edges of cultivation and roadsides. It is common in coastal regions, sometimes in dense populations in the coastal plain of southern Israel, Sinai, and lower Egypt, and sporadic in inner sandy deserts. It is a very early maturing type. *Ae. bicornis* may have originated in the southern part of the Fertile Crescent. Currently, its center of variation is lower Egypt and the Sinai Peninsula. In the Israeli Negev and south Jordan (and possibly also in Egypt), *Ae. bicornis* grows sympatrically with *Ae. longissima*, and *Ae. kotschyi*. In south Israel, it grows allopatrically with *Ae. sharonensis* and *Ae. peregrina*.

9.4.3.2 Cytology, Cytogenetics, and Evolution

Ae. bicornis is a diploid (2n = 2x = 14) whose nuclear genome is a modified S genome (designated S^b by Kihara 1954, Kimber and Tsunewaki 1988; Dvorak 1998) and its organellar genome is unique (designated S^b by Ogihara and Tsunewaki 1988). Its nuclear 1C DNA content is relatively high (6.84 ± 0.097 pg; Eilam et al. 2007; 7.1 pg; The Angiosperm C-value database at Kew Botanic gardens) (Tables 2.4 and 9.3). Li et al. (2022) assessed the size of *Ae. bicornis* genome after genome sequencing and assembly to be 5.73 Gb (=5.6 pg), a value that is smaller than those above that were obtained by flow cytometry measurement.

Ae. bicornis has a symmetric karyotype (one chromosome pair has an almost median centromere and the rest of the pairs have submedian centromeres), with one pair with large and another with small satellites on the short arms (Senyaninova-Korchagina 1932; Riley et al. 1958; Chennaveeraiah 1960). The two SAT-chromosomes belong to homoeologous groups 5 and 6 (Friebe and Gill 1996). According to Riley et al. (1958), the karyotype of *Ae. bicornis* is similar to that of *Ae. sharonensis* and *Ae. longissima*.

Ae. bicornis, like the other species of subsection Emarginata, namely, *Ae. longissima*, *Ae. sharonensis*, and *Ae. searsii*, has major NOR loci on chromosomes of groups 5 and 6 and a variable number of minor loci on chromosomes of homoeologous groups 1, 3, 5 and 6 (Badaeva et al. 1996b). The 5S rDNA loci were observed in these species on chromosomes of groups 1 and 5, distal to minor NOR loci in the short arm of chromosome $1S^b$ and proximal to a major NOR locus in the short arm of chromosome $5S^b$. *Ae. bicornis* differs from the other species of subsection Emarginats, with the exception of *Ae. searsii*, by having a lower heterochromatin content. In addition, *Ae. bicornis* differs from the other three Emarginata species in the size of the two 5S rDNA sites, and by the presence of a minor polymorphic NOR locus in a distal part of the long arm of chromosome $5S^b$ (Badaeva et al. 1996b).

The karyotype of *Ae. bicornis* differs from that of *Ae. longissima*, *Ae. sharonensis*, and *Ae. searsii*, in both its C-banding pattern and by the size of C-bands detected in the karyotype, all of which are small (Teho and Hutchinson 1983; Teho et al. 1983a; Badaeva et al. 1996a), and most resembled that of *Amblyopyrum muticum* (Friebe et al. 1996). *Ae. longissima* and *Ae. sharonensis* are highly heterochromatic species; C-bands are present in intercalary, telomeric, and proximal regions of the chromosomes, while *Ae. searsii* and *Ae. bicornis* have much less C-heterochromatin and have relatively smaller centromeric bands (Teoh and Hutchinson 1983; Badaeva et al. 1996a).

Strong hybridization with the noncoding, highly repetitive DNA sequence, pSc119, derived from *Secale cereale*, was observed at the telomeres and at some subtelomeric regions of several chromosomes (Badaeva et al. 1996a). A strong signal with the pScll9 probe was detected at both telomeres of chromosomes $1S^b$ and one polymorphic site was found in the subterminal region of the long arm of $1S^b$. Chromosomes $2S^b$, $3S^b$, and $6S^b$ (a SAT chromosome) had similar labeling patterns with pSc119 but differed in chromosome morphology. The hybridization pattern of chromosome $4S^b$ included one telomeric site in the short arm and a double site in the long arm. An additional site was detected in the subterminal region of the long arm. Chromosome $5S^b$, the second SAT chromosome, had one pSc119 site at the telomere of the short arm. Two telomeric pScll9 sites and one subterminal site were detected in the short arm of chromosome $7S^b$. The distribution of pScll9 hybridization sites in the *Ae. bicornis* genome was like that of *Ae. longissima, Ae. sharonensis, Ae. searsii* and *A. muticum.*

Using FISH in *Ae. bicornis*, Salina et al. (2006) did not find any detectable FISH signal with the two tandem repeated sequences Spelt1 and Spelt52 probes. This confirms earlier studies that concluded that Spelt52 is absent in *Ae. bicornis* (and in *Ae. searsii*), while it occurs in *Ae. sharonensis* and *Ae. longissima* (Anamthawat-Jonsson and Heslop-Harrison 1993; Zhang et al. 2002; Salina et al. 2004a; Belyayev and Raskina 1998; Raskina et al. 2011). *Ae. sharonensis* and *Ae. longissima* have similar pSc119 labeling patterns as those seen in *Ae. bicornis* and *Ae. searsii* but differ from them in having more heterochromatic chromosomes (Badaeva et al. 1996a; Friebe and Gill 1996).

9.4.3.3 Crosses with Other Species of the Wheat Group

The chromosomes of *Ae. bicornis* pair relatively well with those of the other Sitopsis species, but only the F₁ hybrid with *Ae. sharonensis* was fertile (Tanaka 1955a). Tanaka (1955a) reported that in the F₁ hybrid of a biotype of *Ae. sharonensis* with one of *Ae. bicornis*, meiotic chromosome pairing was fairly regular, with six to seven bivalents. Pollen and seed fertility in this hybrid was high, even though the plants were dwarf-like and weak. Similarly, high chromosomal pairing was observed in meiosis of a hybrid between a biotype of *Ae. longissima* and *Ae. bicornis* (Table 9.4) indicated that these species differed by a reciprocal translocation and the hybrid was more or less self-sterile (Kimber 1961). However, a backcross to either parent set seed when the hybrid was used as the female parent (Waines and Johnson 1972). Kihara (1949) reported 5–7 bivalents (seven bivalents were the mode) in first meiotic metaphase of F₁ hybrids between *Ae. speltoides* and *Ae. bicornis*, however, the hybrids were sterile.

Few studies have been performed on crosses with other *Aegilops* species. Meiocytes of the F₁ hybrid between *Ae. bicornis* and *Ae. tauschii* exhibited from zero to six bivalents and a rare one trivalent (Kihara 1949). Those between *Ae. bicornis* and tetraploid *Ae. crassa* as well as with *Ae. uniaristata*, had similar levels of chromosomal pairing, i.e., one to six bivalents and few multivalents (Kihara 1949). The F₁ hybrid between *Ae. bicornis* and *Ae. umbellulata, Ae. biuncialis* and *Ae. columnaris* had an even lower level of pairing (0–5 with 3 bivalents as mode and few trivalents; Kihara 1949), while that with *Ae. peregrina* had higher pairing (Table 9.5), indicating that the genome of *Ae. bicornis* is related to one of the subgenomes of *Ae. peregrina*, presumably subgenome Sᵛ.

Ohta (1990) produced F₁ hybrids between *Ae. bicornis* and *Amblyopyrum muticum* with and without B chromosomes (Tables 8.1 and 8.2). In studies of chromosomal pairing at first meiotic metaphase of the hybrids that did not

have B chromosomes, he observed very little pairing (0.003 mean arm pairing per cell) whereas in hybrid with B chromosomes the pairing was significantly higher (0.77 mean arm pairing per cell; 82% of the cells had seven bivalents). No cells had multivalents and the frequency of univalents was very low. This high pairing in this hybrid presumably resulted from the activity of the homoeologous pairing promoters that exist in *A. muticum* rather than from the high degree of homology between the chromosomes of the two species.

Crosses between *Ae. bicornis* and *Triticum monococcum* ssp. *monococcum* was studied by Sears (1941b) (Table 9.8). At meiosis of this F₁ hybrid, 5.38 chromosomes paired on average, most of which in the form of rod bivalent. In other crosses between *Ae. bicornis* and *T. monococcum* ssp. *aegilopoides*, the offspring were inviable or poorly viable, whereas hybrid with another variety of ssp. *aegilopoides* was viable (Sears 1944a). This lack of viability is apparently mono-factorially determined (Sears 1944a). Poor pairing was observed in the triploid hybrids between wild tetraploid wheat, *T. turgidum* subsp. *dicoccoides* x *Ae. bicornis*, and *T. timopheevii* ssp. *timopheevii* and *Ae. bicornis* (Table 9.9), and between hexaploid wheat, *T. aestivum* subsp. *aestivum* and *Ae. bicornis* (Table 9.10). Evidently, very little homology exists between the chromosomes of diploid wheat and *Ae. bicornis*; the triploid and tetraploid hybrids had even less pairing, due to the presence of the *Ph1* gene of the allopolyploid wheats.

9.4.4 *Aegilops sharonensis* Eig

9.4.4.1 Morphological and Geographical Notes

Aegilops sharonensis Eig, common name: Sharon goat grass, [Syn.: *Aegilops bicornis* (Forssk.) Jaub. & Spach var. *major* Eig; *Triticum sharonnse* (Eig) Feldman & Sears; *T. longissimum* (Schweinf. & Muschl.) Bowden ssp. *sharonensis* (Eig) Chennav.; *Aegilops longissima* Schweinf. & Muschl. ssp. *sharonensis* (Eig) Chennav.; *Ae. longissima* Schweinf. & Muschl. var. *major* (Eig) Hammer; *Sitopsis sharonensis* (Eig) Á. Löve] is a predominantly autogamous annual plant. Its culm is (40-) 50–70 (-100)-cm-high (excluding spikes). Spikes are more or less broad, linear, two-rowed, usually awned (except in var. *mutica*), and 7–13-cm-long (excluding awns). Rachis is zig-zagged, with each segment being bow-shaped, disarticulating to individual spikelets at maturity, each spikelet with the rachis internode immediately below it (wedge-type dispersal unit; Fig. 2.3). The lower-most spikelet or the few lowest spikelets remain attached at the tip of the culm. Spikelets are 8–13-mm-long (excluding awns), linear, elliptical, grow smaller toward the tip, and are more or less flattened. There are 3–5 florets per spikelet, with the upper 1–3 being sterile. The attached rachis internode is shorter than the spikelets and curved. Glumes are 6–7-mm-long, with two small points, one of which is sometimes elongated into a small awn. Lemmas are 8–11-mm-long, with a 40–60-mm-long awn, there are two short broad teeth at the base. Awns increase in length toward the tip of the ear. The 1–3 lowest spikelets are usually almost awnless. Anthers are 5–6-mm-long. The caryopsis adheres to the lemma and palea (Fig. 9.1d).

Since its discovery, the status of *Ae. sharonensis* as an independent species has been in dispute and has been changed on several occasions. Zhukovsky (1928) considered *Ae. sharonensis* as a subspecies of *Ae. longissima*, while Eig (1928a) described it as a variety of *Ae. bicornis* and named it *Ae. bicornis* (forssk.) Jaub. & Spach. var. *major* Eig. Later, Eig (1928b), based on a re-examination of the plant morphology, his advanced knowledge of the various habitats, and the absence of hybrid swarms or intermediate forms in the contact zones of *Ae. sharonensis* and *Ae. bicornis*, elevated the former to the species rank and named it *Ae. sharonensis* Eig. However, its validity as a biological species had since been doubted. Based on complete chromosome pairing of the F$_1$ hybrids at meiosis and their high fertility, Kihara (1937, 1940a, b, 1949, 1954), and Lilienfeld 1951) treated it as a subspecies or variety of *Ae. longissima* and assigned the two taxa the same genome formula (S^l). This treatment was supported by studies of the fertility and chromosome pairing in pollen mother cells (PMCs) of hybrids between these two taxa (Tanaka 1955a; Roy 1959; Kimber 1961). Also, Bowden (1959), while integrating the genus *Aegilops* into *Triticum*, considered *Ae. sharonensis* a subspecies of *T. longissimum*. Chennaveeraiah (1960), based on karyomorphology and plant morphology, also considered *Ae. sharonensis* to be a subspecies of *Ae. longissima*. The merging of *Aegilops* into *Triticum* was adopted by several cytogeneticists (e.g., Morris and Sears 1967; Mac Key 1968; Kimber and Sears 1983; Kimber and Feldman 1987), but these scientists considered *T. sharonense* as a separate species, named *T. sharonense* (Eig) Kimber and Feldman. While reorganizing the tribe Triticeae in genera based on genome homology, Löve (1984), recognized section Sitopsis as a genus including five diploid species bearing the S genome. Under this taxonomic treatment, *Aegilops sharonensis* was named *Sitopsis sharonensis* (Eig) A. löve. However, taxonomists (e.g., Baum 1977, 1978a, b; Hammer 1980; Feinbrun-Dothan 1986a; van Slageren 1994) kept *Aegilops* as a separate genus. Ankory and Zohary (1962) reported that the two species, *Ae. sharonensis* and *Ae. longissima*, are well-differentiated in nature and were apparently separated by seasonal and ecological factors for natural hybridization and, therefore, treated them as different species. Using the same arguments, Kimber and Feldman (1987) also treated them as separate species. Using C-banding pattern of mitotic metaphase chromosomes, Teoh

and Hutchinson (1983) found that the differences between *Ae. sharonensis* and *Ae. longissima* are large enough to justify their treatment as two separate species. Likewise, studies of electrophoretic mobility of isoenzymes (Brody and Mendlinger 1980; Nakai and Tsuji 1984) and analysis of water-soluble leaf proteins (Mendlinger and Zohary 1995) showed *Ae. sharonensis* to be a valid species. Waines and Johnson (1972), studying the electrophoretic pattern of ethanol-extracted seed proteins, found that *Ae. sharonensis* was intermediate between *Ae. longissima* and *Ae. bicornis* and concluded that, *Ae. sharonensis* is genetically different from these two species and should be considered a separate species. Finally, Yen and Kimber (1990b) hybridized *Ae. sharonensis* with autotetraploid lines of *Ae. speltoides*, *Ae. longissima* and *Ae. bicornis* and analyzed chromosomal pairing at meiotic first metaphase of these hybrids. They found that *Ae. sharonensis* is almost equally related to *Ae. speltoides* and *Ae. longissima*, while it is distant from *Ae. bicornis*, and concluded, therefore, that *Ae. sharonensis* should not be treated as a subspecies or variety of *Ae. longissima*.

Ae. sharonensis includes two varieties: var. *sharonensis* (var. *typica* Eig), in which lemmas are awned in all spikelets except in the lowest, and var. *mutica* (Post) Eig, where lemmas are awnless in all spikelets or awned in the upper-most spikelets only. The latter is rare and occurs in mixed populations in Acre Plain and Sharon Plain of Israel.

Morphological variation involves differences in spike color and size, spikelet number and size, degree of expression of two-rowed nature and awn development. Studies on genetic and phenotypic diversity indicated that *Ae. sharonensis* is a diverse species, in spite of its limited geographic distribution and not highly variable environments (Olivera and Steffenson 2009). Morphologically, *Ae. sharonensis* is like to *Ae. bicornis* by having two-rowed spikes, awns in all the spikelets, spike disarticulation into single spikelets, and the shape of the lemmas (Eig 1929a), but have larger plant and spikelet sizes, larger grains and a somewhat laxer rachis. Edaphically, *Ae. sharonensis* is closer to *Ae. bicornis* than to *Ae. longissima*, though sympatric with the latter in several localities and allopatric with the former.

Olivera et al. (2010) used microsatellites to study genetic diversity and population structure of *Ae. sharonensis* from different sites in Israel and identified the Sharon Plain as the region exhibiting the highest level of allele richness and average gene diversity. Because it is located in the center of the geographic distribution of the species and includes sites with the highest level of diversity, the authors suggested that the Sharon Plain might be the center of origin and center of variation of *Ae. sharonensis*. There are two possibilities to explain the origin of *Ae. sharonensis*. First, it derived from *Ae. bicornis* that penetrated the coastal plain of southern Israel and absorbed genes from *Ae. longissima* via introgressive hybridization. Indeed, Waines and Johnson (1972), based on studies of seed protein patterns, proposed that, *Ae. sharonensis* derived from a hybridization between *Ae. bicornis* and *Ae. longissima*. Alternatively, Raskina et al. (2004b), based on studies of intrapopulation variability of rDNA in marginal populations of *Ae. speltoides* and *Ae. sharonensis*, suggested that *Ae. sharonensis* derived from *Ae. speltoides*.

Phenologically, *Ae. sharonensis* and *Ae. bicornis* are the earliest heading species of the genus *Aegilops*. Both species flower from March to May, albeit *Ae. sharonensis* starts flowering two weeks later than *Ae. bicornis* (Eig 1929a). *Ae. speltoides*, *Ae. longissima* and *Ae. searsii* head later; these species flower from April to June and in mountainous areas, even up to July (Feinbrun-Dothan 1986a; van Slageren 1994).

Ae. sharonensis is an east Mediterranean element. It has a very limited distribution in the south-central distribution region of the genus. It is endemic to the coastal plain of Israel (the Acre Plain, Sharon Plain, and Philistean Plain) and south Lebanon (Eig 1928b, 1929a, 1936; Millet 2006; Olivera and Steffenson 2009). The name *Ae. sharonensis* refers to the Sharon Pain in Israel, where this taxon was first described (Eig 1928a). It grows at 0–100 m above sea level, locally common, often in dense stands, on well-drained sandy soils, consolidated sand dunes, and on marine dilluvial rocks (kurkar rocks), in open park-, shrub-, and herbaceous-plant formations, abandoned fields, disturbed habitats and roadsides (Eig 1928b, 1929a; Post 1933; Ankori and Zohary 1962; Witcombe 1983; Kimber and Feldman 1987; van Slageren 1994). It grows sympatrically with *Ae. longissima* (in few sites in the Israeli Coastal Plain), where intermediate types between these two species in mixed populations, mainly in disturbed habitats in the Sharon Plain, were reported by Eig (1929a) and Ankori and Zohary (1962), and with *Ae. peregrina And Ae. geniculata*, and allopatrically with *Ae. longissima* (in most sites in the Sharon and Philistean Plains), *Ae. bicornis*, and *Ae. speltoides*. *Aegilops sharonensis* has a rich source of genes providing resistance to important wheat diseases and abiotic stresses (Olivera and Steffenson 2009 and references therein). Some forms grow in salt-marshes (e.g., Na'aman salt-marsh, north of Haifa) and may contain genes for salt tolerance. With the most limited distribution of any species in the genus *Aegilops*, *Ae. sharonensis* is rapidly losing its habitats, owing to the combined effects of modern agricultural intensification and expansion of urban and industrial areas (Millet 2006; Olivera and Steffenson 2009).

9.4.4.2 Cytology, Cytogenetics, and Evolution

Ae. sharonensis is diploid (2n = 2x = 14), having a modified S genome, designated S^l by Kihara (1954) and Dvorak (1998), or S^{sh} by Theo and Hutchinson (1983) (from here

on, the genome symbol of *Ae. sharonensis* will be S^{sh}). Its organellar genome is quite similar to that of *Ae. longissima*, and consequently, is designated S^l (Ogihara and Tsunewaki 1988). According to Eilam et al. (2007), this species has the largest genome in comparison to all other diploid *Aegilops* species (1C DNA = 7.52 ± 0.100 pg; 7.1 pg; The Angiosperm C-value database at Kew Botanic gardens (Tables 2.4 and 9.3). Li et al. (2022) assessed the size of *Ae. sharonensis* genome after genome sequencing and assembly to be 6.07 Gb (=5.93 pg), and Avni et al. (2022), also after sequencing and assembly, reported 6.71 Gb (=6.56 pg). Furuta et al. (1986) described the genome size of *Ae. sharonensis* as equal to that of *Ae. longissima* and larger than those of *Ae. searsii* and *Ae. speltoides*. The karyotype of *Ae. sharonensis* is symmetric; all chromosome pairs have submedian centromere (Chennaveeraiah 1960). Riley et al. (1958) and Chennaveeraiah (1960) found that the karyotype of *Ae. sharonensis* is similar to that of *Ae. longissima* and, in both species, there is one pair with large and another pair with distinctly small satellites. The smaller satellites are slightly less than half the size of the larger ones. The karyotypes of *Ae. sharonensis* and *Ae. longissima* are quite similar to each other except for minor differences in the arm ratios of the non-satellited pairs (Chennaveeraiah 1960).

Ae. sharonensis, like the other three species of subsection Emarginata, has major NOR loci on chromosomes of groups 5 and 6 and a variable number of minor loci on chromosomes of groups 1, 3, 5, and 6 (Badaeva et al. 1996b). The 5S rDNA loci were observed on chromosomes of groups 1 and 5, distal to minor NOR loci in the short arm of chromosome $1S^{sh}$ and proximal to a major NOR locus in the short arm of chromosome $5S^{sh}$ (Badaeva et al. 1996b).

Teoh and Hutchinson (1983) and Teoh et al. (1983) studied the C-banding patterns of chromosomes at mitotic metaphase in all 10 diploid *Aegilops* species and found that all diploid species exhibit characteristically different patterns that enable the chromosomes of any complement to be individually identified. *Ae. sharonensis* has a unique C-banding pattern with telomeric, interstitial and exceptionally distinctive centromeric bands, varying in size and staining intensity. Hence, from the specific C-banding pattern, Teoh and Hutchinson (1983) concluded that the genome of *Ae. sharonensis* is different from that of *Ae. longissima*, and, consequently, gave it the symbol S^{sh}. The C-banding pattern of *Ae. sharonensis* indicates that this species is much more closely related to *Ae. speltoides* and *Ae. longissima* than to *Ae. bicornis* and *Ae. searsii*.

A repetitive DNA sequence thought to be a noncoding, highly repeated, 260-bp DNA fragment derived from the B subgenome of bread wheat (Hutchinson and Lonsdale 1982), was used in FISH experiments on genomes of all the diploid species of *Aegilops* (Teoh et al. 1983). This repetitive sequence was found in variable amounts in all diploid species but was restricted to specific regions of the chromosomes. *Ae. sharonensis*, *Ae. speltoides* and *Ae. longissima* possess many copies of the sequence and its distribution is correlated with their respective C-banding patterns. The strongest labelling was observed in *Ae. longissima* and *Ae. sharonensis*, both showing localization near the centromeres and also interstitially along the chromosomes.

Salina et al. (2006), using two telomere-associated tandem repeat sequences of *Ae. speltoides*, *Spelt1* and *Spelt52*, found that the in-situ hybridization patterns of these probes in *Ae. speltoides*, *Ae. longissima* and *Ae. sharonensis* are different. The FISH signal with *Spelt1* was only observed in *Ae. speltoides*, while the FISH signal with *Spelt52* was found, in addition to *Ae. speltoides*, in *Ae. sharonensis* and *Ae. longissima* as well. Similar results were obtained by Anamthawat-Jonsson and Heslop-Harrison (1993), Zhang et al. (2002), Salina et al. (2004b) and Raskina et al. (2011). Thus, *Spelt1* was completely absent from the genomes of the four Emarginata species, *Ae. sharonensis*, *Ae. longissima*, *Ae. searsii* and *Ae. bicornis*.

Badaeva et al. (1996a), using C-banding and in situ hybridization with two highly repetitive DNA sequences, pSc119, and pAs1, found that *Ae. sharonensis* was, together with *Ae. speltoides* and *Ae. longissima*, the very heterochromatic species of the diploid *Aegilops* species. While they did not observed hybridization with the pAs1 probe on *Ae. sharonensis*, *Ae. speltoides* or *Ae. longissima* chromosomes, *Ae. sharonensis*, as well as the other Emarginata species (*Ae. longissima*, *Ae. searsii* and *Ae. bicornis*), showed strong labeling with the pSc119 probe in the telomeric chromosomal regions. Yet, the karyotypes of these species exhibited different distributions of C-heterochromatin (Badaeva et al. 1996b). *Ae. sharonensis*, like *Ae. longissima*, was highly heterochromatic; C-bands were present in intercalary, telomeric, and proximal regions of the chromosomes. *Ae. searsii* and *Ae. bicornis* had much fewer C-bands in these regions.

Yaakov et al. (2013) utilized quantitative real-time PCR to assess the relative copy numbers of 16 TE element families in various *Triticum* and *Aegilops* species. They found that the *Latidu* family of TEs showed specific proliferation in *Ae. sharonensis*, with more *Latidu* than in *Ae. speltoides* and *Ae. longissima*, and much more than in *Ae. searsii*. The *Rong* family was also observed in relatively high quantities in *Ae. sharonensis*.

9.4.4.3   Crosses with Other Species of the Wheat Group

Crosses between *Ae. sharonensis* and three other Sitopsis species, i.e., *Ae. longissima*, *Ae. bicornis* and *Ae. speltoides*, produced F_1 hybrids that exhibited complete or almost complete chromosomal pairing at first meiotic metaphase.

Thus, F_1 hybrids between *Ae. sharonensis* and *Ae. longissima* and their reciprocal hybrid, had 5 bivalents (mostly ring bivalents) and a quadrivalent (or trivalent and univalent), indicating complete homology between the genomes of these two species and the existence of a reciprocal translocation between them (Tanaka 1955a; Kimber 1961; Ankori and Zohary 1962). Only Roy (1959) observed 6.90 bivalent and no multivalents in these hybrids. However, when assessing the photographs presented by Roy, Ankori and Zohary (1962) noticed that the *longissima* line used was already highly introgressed with *Ae. sharonensis*, which explains why Roy did not encounter a translocation configuration in his hybrids. Hybrid between *Ae. sharonensis* and *Ae. bicornis* exhibited 6–7 bivalents (Tanaka 1955a) and the hybrid between *Ae. sharonensis* and *Ae. speltoides* had similar levels of pairing (6.85 bivalents per cell of which 5.35 were ring bivalents; Kihara 1949; Sears 1941b; Kimber 1961), whereas Tanaka (1955a) reported on somewhat less pairing in this hybrid (4–7 bivalents). The F_1 hybrids between *Ae. sharonensis* and *Ae. longissima* was almost fully fertile (more than 80% seed set; Tanaka 1955a; Ankori and Zohary 1962) and the hybrid *Ae. sharonensis* x *Ae. bicornis* had an approximate 62% seed set (Tanaka 1955a), while the hybrid between *Ae. sharonensis* and *Ae. speltoides* was sterile (Sears 1941b).

Data of chromosome pairing in F_1 hybrids between *Ae. sharonensis* and other diploid *Aegilops* species are presented in Table 9.4. These hybrids exhibited low pairing, indicating that the genome of *Ae. sharonensis* is more distant from the genomes of these species than from those of Sitopsis species. The F_1 hybrid between the tetraploid species *Ae. kotschyi* (genome S^vS^vUU) and *Ae. sharonensis* exhibited 4.28 bivalents (of which 2 were ring), 1.50 trivalents, 0.46 quadrivalents and 0.09 pentavalents (Rubenstein and Sallee 1973), indicating high homology between the genome of *Ae. sharonensis* and one of the subgenomes of *Ae. kotschyi*, most probably genome S^v.

In analyzing F_1 hybrids between *Ae. sharonensis* and *Ambliopyrum muticum* without B chromosomes, Ohta (1990) found chromosome pairing at first meiotic metaphase ranging from 6.22 to 6.50 bivalents (of which 3.10 to 4.48 were ring bivalents) (Table 8.1). Several hybrids had few multivalent configurations. These hybrids showed very good pairing, presumably due to the promotion of homoeologous pairing by the genome of *A. muticum* but were completely sterile (Ohta 1990).

Ae. sharonensis was crossed with diploid, tetraploid, and hexaploid *Triticum* species (Tables 9.8, 9.9 and 9.10). The F_1 hybrid between *Ae. sharonensis* and *T. monococcum* ssp. *monococcum* had very little pairing and were completely sterile (Kushnir amd Halloran 1981), indicating little homology between the S^{sh} and A^m genomes. The F_1 hybrid between *Ae. sharonensis* and tetraploid wild wheat *T.*

turgidum ssp. *dicoccoides* had 2.23 bivalents (of which 0.03 were ring bivalents) (Kushnir and Halloran 1981), whereas McFadden and Sears (1947) reported 5.18 bivalents and 1.05 trivalents in the reciprocal hybrid. Hybrids between domesticated tetraploid wheat *T. turgidum* subsp. *turgidum* and *Ae. sharonensis* exhibited 2.22 bivalents and 0.02 trivalents (Riley et al. 1958). A similar level of chromosomal pairing in this hybrid combination, i.e., 2.50 bivalents per cell, was reported by Roy (1959). These F_1 triploid hybrids had very little chromosome pairing at meiosis, indicating that, in the presence of the homoeologous-pairing suppressor, *Ph1*, of allopolyploid wheats, *Ae. sharonensis* chromosomes rarely pair with the wheat chromosomes and, likewise, the wheat chromosomes rarely pair with each other. Similar low chromosomal pairing exists in F_1 hybrids between hexaploid wheat *T. aestivum ssp. aestivum* and *Ae. sharonensis* (Table 9.10). Interestingly, the amount of chromosome pairing in hybrids with hexaploid wheat is lower than that measured in hybrids with tetraploid wheat.

The homoeologous relationship between *Ae. sharonensis* and chromosomes of *T. aestivum* was determined by Friebe and Gill (1996), by comparing the C-banding pattern of *Ae. sharonensis* chromosomes, described by Teoh and Hutchinson (1983), with those of *Ae. longissima* chromosomes, whose homoeologous relationships to *T. aestivum* chromosomes were previously established by Friebe et al. (1993). Then, Friebe and Gill (1996) assigned the seven chromosomes of *Ae. sharonensis* to each homoeologous group of *T. aestivum*. Later, Maestra and Naranjo (1997) confirmed the homoeologous relationship between *Ae. sharonensis* and *T. aestivum* chromosomes, by C-banding analysis of specific pairing at first meiotic metaphase in F_1 hybrids between *Ae. sharonensis* and *T. aestivum* ssp. *aestivum*. They analyzed chromosomal pairing in hybrids between three different genotypes of *Triticum aestivum*, each carrying (*Ph1*), or lacking it (*ph2b* mutant), and *Ae. sharonensis*, in order to establish the homoeologous relationships of *Ae. sharonensis* chromosomes to those of hexaploid wheat. Since *Ae. sharonensis* chromosomes show a distinctive C-banding pattern (Teoh and Hutchinson 1983), and thus, could be distinguished from those of wheat, C-banding was used by Maestra and Naranjo (1997) to identify the chromosomes of both species and their arms. Normal homoeologous relationships for the seven chromosomes of the S^{sh} genome of *Ae sharonensis*, and their arms, were revealed in this study. The pattern of pairing between chromosomes of *Ae. sharonensis* and ssp. *aestivum* indicated that no apparent chromosome rearrangement occurred during the evolution of the *Ae. sharonensis* genome relative to the subgenomes of hexaploid wheat. Thus, the chromosome structure of the ancestral genome from which subgenomes A, B, and D of allohexaploid wheat derived, was also preserved in genome S^{sh}. All three types of hybrids with *Ph1*,

demonstrating low pairing level, however, with *ph2b*, demonstrating intermediate pairing level was observed, showing preferential pairing between A-D and B-S^{sh}. A close relationship between the S^{sh} genome and the B subgenome of bread wheat was confirmed, but the results provided no evidence that the B subgenome was derived from *Ae. sharonensis*. Similar results were obtained by Fernández-Calvín and Orellana (1993).

On the basis of the results of homoeologous pairing between chromosomes of *Ae. sharonensis* and *T. aestivum*, Maestra and Naranjo (1997) were able to identify the homoeologous relationships between all of the chromosomes of the S^{sh} genome and bread wheat. Their homoeologous pairing findings confirmed the homoeology of chromosome 4S^{sh} to wheat chromosomes of group 4, reported by Miller et al. (1982), as well as the assignment of the S^{sh} genome chromosomes to the seven homoeologous groups of wheat, and the arm designation, suggested by Friebe and Gill (1996), based on chromosomal morphology and C-banding.

9.4.5 *Aegilops longissima* Schweinf. & Muschl.

9.4.5.1 Morphological and Geographical Notes
Aegilops longissima Schweinf. & Muschl. [Syn.: *Aegilops longissima* (Schweinf. & Muschl.) ssp. *longissima* Hammer; *Triticum longissimum* (Schweinf. & Muschl.) Bowden; *Sitopsis longissima* (Schweinf. & Muschl.) Á. Löve] is a predominantly autogamous annual plant. Its culm is 40–110-cm-high (excluding spikes). Spike is narrow, linear, one-rowed, tapering slightly to the tip, and 10–20-cm-long (excluding awns). At maturity, they are fragile near, but usually not at, the base, with the greater part of the spike falling entire, while lower spikelets remain on the culm or fall later singly or in 2–3 pieces. Spikelets 8–15 are 12–14-mm-long (excluding awns), become thinner and shorter toward the tip of the spike, and are appressed to the rachis segment. There are 3–5 florets in each spikelet, the upper 1–3 being sterile. The rachis internodes are nearly as long as spikelets in the middle of the spike. Glumes are 6–8-mm-long, tough, and usually with two teeth separated by a membranous edge. Glumes of the terminal spikelet sometimes have three teeth, with the center one sometimes elongated into a very short awn. Lemmas of the lateral spikelets are canoe-shaped and awnless. Lemmas of the terminal spikelet have a 7–12-cm-long, broad, convex awn, often with a small unequal tooth on each side. Anthers are 5–6-mm-long. The caryopsis adheres to both lemma and palea (Fig. 9.1e).

Limited morphological variation is mainly seen in spike length, spikelet number, rachis form (zig-zagged or straight) and awn length. Like *Ae. speltoides* and *Ae. searsii*, *Ae. longissima* ripens later than *Ae. sharonensis* and *Ae.*

bicornis. The center of variation is in steppic regions of Israel and Jordan.

Morphologically, *Ae. longissima* is taller than *Ae. sharonensis* and differs from it by its long, narrow, linear, and one-rowed spike, by the absence of awns on the lemmas of the lateral spikelets, and by the presence of long and broad awns on the terminal spikelet. At maturity, the spike breaks near the base and falls almost entire and, at the end of the summer, the spike may disarticulate into single spikelets with the rachis internode beside them (barrel-type disarticulation). Ankori and Zohary (1962) reported that the two species are well-differentiated in nature and apparently separated by seasonal and ecological factors for natural hybridization. Cytologically, the two taxa have minor differences in their karyotype, and even though the hybrid between them exhibits complete chromosomal pairing at meiosis, the two taxa differ by a reciprocal translocation. The occurrence of a reciprocal translocation between the distal regions of chromosome arms $4S^lL$ and $7S^lL$ in *Ae. longissima* was later confirmed by Friebe et al. (1993). This translocation is not present in other Sitopsis species (Kihara 1949; Tanaka 1955a; Riley et al. 1961; Kimber 1961; Feldman et al. 1979).

Ae. longissima has a relatively limited distribution in the south-central region of the distribution area of the genus. A steppical (Irano-Turanian) element extends into sub-Mediterranean and desert (Saharo-Arabian) regions. It grows in Egypt (lower Egypt and Sinai), Israel (coastal plain, northern, and central Negev and eastern Judea and Samarian Mountains up to the east to Ein Gev, the eastern shore of the Lake of Galilee), Jordan, Lebanon (southern coastal plain), and Syria. Alt: 0–900 m above sea level (Table 9.1). It grows on sandy loams derived from the sandstone of the coastal plain, rarely on somewhat heavier soil, in the coastal plain of Egypt, Israel and Lebanon, and on sand derived from Nubian sandstone, in grey calcareous steppe soil or loess in the inland steppe or desert regions, in open dwarf shrub or herbaceous steppe-like or desert-like formations, plains, abandoned fields, edges of cultivation and roadsides. It is common, and often abundant in the coastal plain and in several steppic habitats bordering on the Mediterranean region in Israel, Jordan, and Syria.

Like *Ae. bicornis*, it grows in the hot and dry parts of the south-east Mediterranean, but its distribution area is closer to the center of the distribution area of the genus. The center of variation is in the steppic regions of Israel and Jordan, where it presumably originated and from which it invaded the coastal plain. It is sympatric with *Ae. sharonensis* [hybrid swarms with *Ae. sharonensis* can be found in the coastal plain (Ankori and Zohary 1962)] and *Ae. peregrina* in the coastal plain and with *Ae.*, *bicornis*, *Ae. kotschyi* and *Ae. vavilovii* in the steppical region. It is allopatric with *Ae. speltoides* in the coastal plain, with *Ae. searsii* in southern Judea Mts., with *Ae. umbellulata*, *Ae. geniculata* and *Ae. biuncialis* in the sub-Mediterranean area in Lebanon and Syria and with *Ae. crassa* in the steppical regions of Syria.

9.4.5.2   Cytology, Cytogenetics, and Evolution

Ae. longissima is diploid (2n = 2x = 14) with a modified S genome, designated S^l by Kihara (1949, 1954, 1963, 1970). Its organellar genome was designated S^{l2}, as a subtype of the S^l plasmon (Wang et al. 1997; Table 9.3. This species has the second largest genome among diploid *Aegilops* species (1C DNA = 7.48 ± 0.082 pg) (Eilam et al. 2007), 6.0 pg (The Angiosperm C-value database at Kew Botanic gardens) (Tables 2.4 and 9.3). Li et al. (2022) assessed the size of *Ae. longissima* genome after genome sequencing and assembly to be somewhat smaller amount, i.e., 6.22 Gb (=6.07 pg), and Avni et al. (2022), also after sequencing and assembly, reported 6.70 Gb (=6.55 pg). Furuta et al. (1986) reported that the genome size of *Ae. longissima* is equal to that of *Ae. sharonensis* but larger than those of *Ae. searsii* and *Ae. speltoides*.

The karyotype of *Ae. longissima* is symmetric; the satellite chromosomes have median centromeres, while all the non-satellited chromosomes have sub-median centromeres (Senyaninova-Korchagina 1932; Chennaveeraiah 1960). Riley et al. (1958) reported that the karyotypes in *Ae. longissima* and *Ae. sharonensis* are similar, and that in both species, there is one pair with large and another with distinctly small satellites. These findings were confirmed by Chennaveeraiah (1960), who stated that the smaller satellites are slightly less than half the size of the larger ones, and that there are minor differences in in the karyotypes of *Ae. longissima* and *Ae. sharonensis* in the arm ratios of the non-satellited pairs.

Based on the presence of secondary constrictions (Chennaveeraiah 1960; Chen and Gill 1983; Teoh and Hutchinson 1983), in situ hybridization analysis using a radioactive rDNA probe (Miller et al. 1983; Teoh et al. 1983), and analysis of nucleolar activity by Ag-NOR banding (Cermeño et al. 1984a, b), two pairs of NORs were identified in the secondary constrictions of *Ae. longissima*. Yet, Friebe et al. (1993) and Badaeva et al. (1996b), using in situ hybridization with a 18S–26S rDNA probe, which enabled high sensitivity, detected, in addition to the two major-active NORs on the short arms of chromosomes $5S^l$ and $6S^l$, an additional minor NOR on chromosomes $1S^l$, and a polymorphic minor NOR on chromosome $3S^l$. The minor NORs usually do not form secondary constriction, indicating that these NORs are not transcribed. The 5S rDNA loci were detected in *Ae. longissima* on chromosomes $1S^l$ and $5S^l$, distal to minor NOR loci in the short arm of chromosome $1S^l$ and proximal to a major NOR locus in the short arm of chromosomes $5S^l$ (Badaeva et al. 1996b).

In situ hybridization (ISH) patterns with both pTa71 (rDNA) and pTa794 (5S rDNA) probes revealed similar distribution of the two probes in *Ae. longissima* and *Ae. sharonensis* (Badaeva et al. 1996a). However, while ISH patterns were not polymorphic in *Ae. sharonensis*, intraspecific polymorphism was found in *Ae. longissima*. On the other hand, *Ae. longissima* chromosomes were similar to those of *Ae. searsii* in hybridization patterns with both probes (Badaeva et al. 1996a). This confirms the high degree of similarity of their genomes, as first observed by the distribution of highly repetitive DNA sequences (Teoh et al. 1983; Badaeva et al. 1996b). *Aegilops bicornis* differed from *Ae. longissima* by its lower heterochromatin content, differences in size between the two 5S rDNA sites, and the presence of a minor polymorphic NOR locus in a distal part of the long arm of chromosome $5S^b$ (Badaeva et al. 1996b).

The C-banding patterns of *Ae. longissima* chromosomes were determined at mitotic metaphase by Teoh and Hutchinson (1983), who showed that all seven pairs of *Ae. longissima* chromosomes can be identified by their characteristic C-banding patterns. Like all other Sitopsis species, *Ae. longissima* also had telomeric, interstitial and centromeric bands, that varied in size and staining intensity. The proximal and telomeric bands were smaller than the centromeric ones (Teoh and Hutchinson 1983).

The karyotypes of *Ae. longissima*, *Ae. sharonensis*, *Ae. searsii* and *Ae. bicornis* have different C-banding patterns (Badaeva et al. 1996a). *Ae. longissima* and *Ae. sharonensis* are highly heterochromatic species; C-bands are present in intercalary, telomeric, and proximal regions of the chromosomes. *Ae. searsii* and *Ae. bicornis* have much less C-heterochromatin compared with the two other Emarginata species.

Teoh et al. (1983) used a noncoding repetitive sequence, derived from *T. aestivum*, as a probe in in situ hybridization experiments on the genome of diploid *Aegilops* species. *Ae. longissima*, like *Ae. speltoides* and *Ae. sharonensis*, possesses many more copies of the noncoding sequence than *Ae. bicornis* and *Ae. searsii*. The distribution of the copies in each species correlated with their respective C-banding patterns.

Friebe et al. (1993) analyzed C-banding polymorphism in 17 accessions of *Ae. longissima* from Israel and Jordan and established a generalized idiogram of this species. Polymorphism for C-band size and C-band location was observed between different accessions but did not prevent chromosome identification. The C-banding patterns of *Ae. longissima* chromosomes they reported were similar to the N- and C-banding patterns reported earlier for this species (Chen and Gill 1983; Jewell and Driscoll 1983; Teoh and Hutchinson 1983; Kota and Dvorak 1985; Hueros et al. 1991). However, in most earlier reports, only one accession was analyzed and, therefore, no data were available on

C-band polymorphisms in this species. Moreover, not all *Ae. longissima* chromosomes assigned correctly according to their homoeologous groups (Friebe et al. 1993).

A complete series of *Ae. longissima* addition and substitution lines were produced in the *T. aestivum* background (Hart and Tullen 1983). Sporophytic and gametophytic compensation tests were used to determine the homoeologous relationships of *Ae. longissima* chromosomes (Friebe et al. 1993). All *Ae. longissima* chromosomes compensated rather well and fertility was restored even in substitution lines involving wheat chromosomes 2A, 4B and 6B that contain major fertility genes.

The homoeologous relationships between *Ae. longissima* and wheat were determined for several chromosomes by their ability to compensate for the absence of wheat chromosomes in substitution lines (Jewell and Driscoll 1983; Kota and Dvorak 1985) and by all the chromosomes by C-banding analysis of *Ae. longissima* chromosomes in disomic and ditelosomic addition lines of *Ae. longissima* to *T. aestivum* and in substitution lines of *Ae. longissima* chromosomes for their *T. aestivum* homoeologues (Friebe et al. 1993).

The homoeologous relationships between *Ae. longissima* and wheat chromosomes have been established earlier by isozyme, storage protein, and morphological markers, as well as by analyzing their compensating ability in substitution lines (Hart and Tuleen 1983; Jewell and Driscoll 1983; Netzle and Zeller 1984; Kota and Dvorak 1985; Levy et al 1985; Millet et al. 1988; Hueros et al. 1991). The results obtained by Friebe et al. (1993) largely agreed with those reported previously. However, there were some discrepancies caused by chromosome misidentification in earlier studies.

C-banding patterns and morphology of *Ae. longissima* chromosomes $1S^l$, $3S^l$ and $5S^l$ are very similar to those of chromosomes 1B, 3B and 5B of *T. aestivum* (Gill et al. 1991a). The C-banding pattern of the short arm of $4S^l$ is almost identical to that of the short arm of wheat chromosome 4B. The remaining *Ae. longissima* chromosomes showed differences in C-banding patterns and arm ratio with their wheat homoeologues.

Salina et al. (2006) and Raskina et al. (2001), using fluorescence in situ hybridization (FISH), observed considerable polymorphisms in the hybridization patterns of two tandemly repeated sequences, *Spelt1* and *Spelt52*, among *Aegilops* species. While there was no detectable *Spelt1* FISH signal in any species of section Sitopsis, except for *Ae. speltoides*, hybridization patterns of *Spelt52* were species-specific in *Ae. speltoides*, *Ae. longissima* and *Ae. sharonensis* (Salina et al. 2006). Two very small and dim *Spelt52* blocks were detected in *Ae. searsii*, whereas *Ae. bicornis* did not contain any *Spelt52* repeats in its genome (Raskina et al. 2011). On the other hand, *Ae. longissima* and *Ae. sharonensis* showed that probe pSc119*ffff* (from *Secale cereale*) labeling patterns similar to *Ae. bicornis* and *Ae. searsii* but differed from them by having more heterochromatic chromosomes (Badaeva et al. 1996a; Friebe and Gill 1996).

Badaeva et al. (1996a), using both in situ hybridization (ISH) with the highly repetitive DNA sequences pSc119, from *Secale cereale*, and pAsl, from *Ae. tauschii*, as probes, and C-banding, analyzed genome differentiation in all diploid *Aegilops* species. The level of hybridization and labeling patterns differed among genomes. All five Sitopsis species had different C-banding pattern but they were similar to each other and to *Amblyopyrum muticum* in the distribution of pSc119 hybridization sites. On the other hand, no hybridization was observed with the pAsl probe on *Ae. speltoides*, *Ae. sharonensis* and *Ae. longissima* chromosomes, whereas a few minor pAsl sites were observed in *Ae. searsii* and *Amblyopyrum muticum* (Badaeva et al. 1996a). The ISH Patterns showed only minor intraspecific variations. *Ae. bicornis* and *A. muticum* had a low amount of C-heterochromatin, whereas *Ae. longissima* and *Ae. sharonensis* were the most heterochromatic species, and *Ae. searsii* possessed intermediate amounts of C-heterochromatin (Badaeva et al. 1996a). *Ae. longissima*, like the other three species of subsection Emarginata, *Ae. sharonensis*, *Ae. searsii*, and *Ae. bicornis*, had strong labeling in the telomeric chromosomal regions with the pSc119 probe (Badaeva et al. 1996a).

The C-banding pattern of the *A. muticum* genome chromosomes was different from that of the chromosomes of *Ae. speltoides*, and most resembled those of *Ae. bicornis* (Friebe et al. 1996). Distribution of pScll9 hybridization sites in the *A. muticum* genome also differed from that of *Ae. speltoides* but was similar to those of *Ae. longissima*, *Ae. sharonensis*, *Ae. searsii* and *Ae. bicornis*. Yet, the *A. muticum* genome was easily morphologically differentiated from the genome of the Sitopsis species.

During its evolution, *Ae. longissima* suffered a translocation involving the long arm of chromosome $4S^l$ and the long arm of chromosome $7S^l$ (Hart and Tuleen 1983; Friebe et al. 1993; Naranjo 1995). Analysis of chromosomal pairing at first meiotic metaphase showed a quadrivalent or a trivalent plus univalent in F_1 hybrids between *Ae. longissima* and all other members of Sitopsis, indicating that *Ae. longissima* differs from the other four species by a reciprocal translocation (Feldman et al. 1979 and references therein). Since F_1 hybrids between these four species form only bivalents at meiosis, it is reasonable to conclude that all four Sitopsis species, except *Ae. longissima*, have the ancestral chromosome structure.

Ae. longissima is probably involved in the parentage of the allotetraploid species *Ae. peregrina* (genome S^lS^lUU) and *Ae. kotschyi* (genome S^lS^lUU) (Kihara 1954; Feldman 1963; Kimber and Feldman 1987; Kimber and Sears 1987;

Yen and Kimber 1989; Zhang et al. 1992; Dvorak 1998). *Ae. longissima* is a good source for genes conferring resistance to mildew, leaf and stem rust, heat and drought tolerance, and determination of high grain protein.

Some forms of *Ae. longissima* have genetic mechanisms capable of partially suppressing the *Ph1* gene of polyploid wheat in hybrids with *T. aestivum* and *T. turgidum*. Mello-Sampayo (1971b) reported that a line of *Ae. longissima* promoted homoeologous pairing in meiosis of the F_1 hybrid with *T. aestivum*, thereby, partially counteracting the suppressive effect of wheat-*Ph1*. This line was designated an intermediate-pairing (IP) line to distinguish it from the more common low-pairing (LP) lines. Upadhya and Swaminathan (1967) reported on another IP line of *Ae. longissima* that promoted pairing in hybrids with common wheat. In studies of chromosome pairing in F_1 hybrids between ditelosomic lines of *T. aestivum* cv. Chinese Spring and the IP line of *Ae. longissima*, discovered by Mello-Sampayo (1971b), an intermediate amount of pairing was observed (5.49 bivalents, 0.39 trivalents and 0.06 quadrivalents per pollen mother cells (PMCs), most of which was between chromosomes of the wheat B subgenome and those of *Ae. longissima* (Feldman 1978).

To assess the effect of LP and IP genotypes on homologous pairing, Avivi (1976) produced autotetraploids from both lines of *Ae. longissima* and followed chromosomal pairing at first metaphase of meiosis. While the two induced autotetraploids did not differ in chiasma frequency or in the number of paired chromosomal arms, they differed significantly in multivalent frequency; the IP autotetraploid exhibited the same multivalent frequency as that expected on the basis of random pairing between the four homologues, namely, 4.7 multivalents per/cell. In contrast, the LP autotetraploid exhibited a significantly lower than expected frequency of multivalents. Avivi (1976) assumed that the LP genotype in the autotetraploid does not affect meiotic pairing per se but modifies the pattern of homologous association by separating the four sets of homologous chromosomes in somatic and premeiotic cells into two groups of two. In contrast, the IP genotype does not affect the spatial arrangement of the chromosomes. Accordingly, the gene *Ph1* of common wheat suppresses homoeologous pairing in the *aestivum* x *longissima* hybrid by separating the wheat genomes from that of the *longissima* genome, thereby preventing pairing of distantly located chromosomes, the IP genotype does not do it and, consequently, enables some pairing between common wheat and *longissima* chromosomes (Avivi 1976).

9.4.5.3 Crosses with Other Species of the Wheat Group

Several intraspecific crosses were performed between lines of *Ae. longissima* from different habitats and geographical regions (Feldman et al. 1979 and Feldman M, unpublished). All the hybrids showed seven bivalents, most of which were ring bivalents. Seed set was normal except in one combination between a line from the Israeli Negev (steppe area) and a line from the Israeli coastal plain (Mediterranean region), in which seed fertility was low (53%) (Feldman et al. 1979).

Crosses between *Ae. longissima* and all of the other four Sitopsis species, i.e., *Ae. sharonensis*, *Ae. bicornis*, *Ae. searsii* and *Ae. speltoides*, produced F_1 hybrids that exhibited complete or almost complete chromosomal pairing at first meiotic metaphase (Table 9.4). Thus, F_1 hybrids between *Ae. longissima* and *Ae. sharonensis* and the reciprocal hybrids exhibited almost complete chromosome pairing at first meiotic metaphase, with 5 bivalents, one quadrivalent or trivalent and univalent, indicating the existence of high homology and the presence of a reciprocal translocation between these two species (Tanaka 1955a; Kimber 1961; Ankori and Zohary 1962). The F_1 hybrids are almost fully fertile, with an 82–86% seed set (Tanaka 1955a). Similarly, high pairing indicating great chromosomal homology and the presence of a reciprocal translocation was observed in all the hybrids of *Ae. longissima* and the other four Sitopsis species (Table 9.4). Nevertheless, the F_1 hybrids involving *Ae. longissima* and the Emarginata species, namely, *Ae. sharonensis*, *Ae. bicornis* and *Ae. searsii*, were highly or only partially fertile. In contrast, the F_1 hybrid *Ae. longissima* x *Ae. speltoides* was completely sterile in contrast to the high pairing (Riley et al. (1961), implying that the chromosomes of the two species differ by cryptic structural hybridity or by genetic barriers. Thus, *Ae. longissima* diverges from the other Sitopsis species by a translocation and is totally isolated from *Ae. speltoides* by partial internal chromosomal or genetic barriers which are superimposed upon predominant self-fertilization and ecological specialization of these two species. Since all intraspecific hybrids of *Ae. longissima* studied had seven bivalents, i.e., they were homozygous for the translocation, it is reasonable to assume that the translocation originated during the formation of this species.

Kihara (1949) crossed *Ae. longissima* with *Ae. caudata*, *Ae. comosa*, *Ae. uniaristata* and *Ae. umbellulata* analyzed chromosome pairing in the F_1 hybrids and in the reciprocal combinations. In *longissima* x *caudata* Kihara observed 3–6 bivalents and 0–2 trivalents, in *longissima* x *comosa* up to 6 bivalents and 0–1 trivalents, in *longissima* x *uniaristata* 2–5 bivalents and in *Ae. longissima* x *Ae. umbellulata* 0–4 bivalents (with mode of 2) and 0–1 trivalent These hybrids were completely sterile. These findings indicate that these four-specie are not close to *Ae. longissima* and have diverged quite considerably from it.

Chromosome pairing in F_1 hybrids between *Ae. longissima* and *Amblyopyrum muticum* (without B chromosomes) was relatively high (Tables 8.1 and 8.2). These hybrids showed very good pairing, presumably due to the promotion of homoeologous pairing by the genome of *A. muticum*, but were completely sterile (Ohta 1990). The presence of multivalents indicated that the two species differ by a reciprocal translocation. There was a slight difference in the amount of pairing between these two hybrids. Since the same accession of *A. muticum* was used in the two hybrids, Ohta (1990) suggested that the frequency and configuration of chromosome pairing observed in these two hybrid plants might be caused by small differences in the chromosomal structure or in the genotypes affecting homoeologous chromosome pairing between the two accessions of *Ae. longissima*. In hybrids *Ae. longissima* x *A. muticum* with B chromosome pairing was considerably reduced (Table 8.2).

Data from analyses of chromosomal pairing at first meiotic metaphase in F_1 triploid hybrids between *Ae. longissima* and a number of allotetraploid species of section Aegilops are presented below. Kihara (1949) reported the presence of 6–8 bivalents (with mode of 7) and 0–1 trivalents in the F_1 *Ae. peregrina* x *Ae. longissima*. Similar data were also reported by other researchers (Table 9.5) indicating that one of the two subgenomes of *Ae. peregrina* is homologous to the genome of *Ae. longissima*. Similar results were observed in the F_1 triploid *Ae. kotschyi* x *Ae. longissima* (Table 9.5). Chromosomal pairing in the F_1 *Ae. biuncialis* x *Ae. longissima* exhibited 0–5 bivalents and 0–2 trivalents, and the F_1 *Ae. geniculata* x *Ae. longissima* hybrid had 2–6 bivalents (with mode of 4), and 0–1 trivalents (Kihara 1949). Evidently, there is reduced homology between the genomes of these species.

Several crosses were performed between *Ae. longissima* and diploid, tetraploid and hexaploid *Triticum* species (9.8, 9.9 and 9.10). Kihara (1949) studied chromosome pairing in the F_1 *Ae. longissima* x *T. monococcum* ssp. *monococcum* and found 0–3 bivalents at meiosis. Feldman (1978) reported 1.49 bivalents, of which 0.02 were ring bivalents, and 0.07 trivalents in F_1 *longissima* x ssp. *monococcum* hybrid. These data indicate that the two species have diverged quite considerably from one another.

Chromosome pairing was analyzed in several *Ae. longissima* x *T. turgidum* hybrids (Table 9.9). In all these hybrids pairing was very low, indicating that very little homology exists between the chromosomes of *Ae. longissima* and those of *T. turgidum*. Kihara (1949) also crossed *Ae. longissima* with the second tetraploid species of wheat, *T. timopheevii*, and also found very little pairing in this hybrid (0–7 bivalents and 0.01 trivalents), indicating that the chromosomes of this tetraploid species are not homologous to those of *Ae. longissima*.

Fig. 9.2 *Aegilops* species carrying the D genome; **a** A. spike of *Ae. tauschii* Coss. ssp. *tauschii* (formerly ssp. *eusquarrosa* Eig); **b** A spike of *Ae. tauschii* Coss. ssp. *strangulata* (Eig) Tzvelev; **c** A spike of *Ae.* *ventricosa* Tausch; d A plant and a spike of 4x *Ae. crassa* Boiss.; **e** A spike of *Ae. vavilovii* (Zhuk.) Chennav.; **f** A spike of *Ae. juvenalis* (Thell.) Eig; **g** A spike of *Ae. cylindrica* Host

Table 9.1 in McFadden and Sears (1947) includes data on chromosomal pairing in the F₁ hybrid between *Triticum turgidum* subsp. *dicoccoides* and *Aegilops sharonensis*, which showed the formation of 5.18 bivalents 0–3 multivalents. But, on page 1016 of their article Fig. 9.2i is showing *Ae. longissima* and not *Ae. sharonensis*. Therefore, the pairing data in Table 9.1 is between subsp. *dicoccoides* and *Ae. longissima*, and, judging from the level of pairing in this hybrid, as well as in hybrids between *T. aestivum* and intermediate-pairing (IP) type *Ae. longissima* (Mello-Sampayo 1971b; Feldman 1978), it is most likely that the *longissima* parent in the McFadden and Sears (1947) paper is an IP type of *Ae. longissima*, and not of *Ae. sharonensis*.

Several crosses were performed between low-pairing (LP) type and IP type *Ae. longissima* and hexaploid wheat, *T. aestivum*. Hybrids involving LP types of *Ae. longissima* yielded very low pairing, with 1.45–1.96 bivalents and 0.01–0.08 trivalents (Riley et al. 1961; Riley and Chapman 1963; Riley 1966a; Ceoloni et al. 1986; Yu and Jahier 1992; Naranjo and Maestra 1995). Chromosomal pairing in these hybrids also showed little homology between the genomes of *Ae. longissima* and those of *T. aestivum*. On the other hand, hybrids involving IP types of *Ae. longissima*, that contains homoeologous pairing promoter(s), exhibited higher chromosomal pairing, namely, 4.79–5.49, 0.19–0.39

trivalents and 0.04–0.07 quadrivalents (Mello-Sampayo 1971; Feldman 1978; Ceoloni et al. 1986).

9.4.6 *Aegilops searsii* Feldman & Kislev Ex Hammer

9.4.6.1 Morphological and Geographical Notes

Aegilops searsii Feldman & Kislev ex Hammer [Syn.: *Triticum searsii* Feldman & Kislev; *Sitopsis searsii* (Feldman & Kislev ex Hammer) Á. Löve] is a predominantly autogamous annual plant. Its culm is 20–40 (-50)-cm-high (excluding spikes). The spike is narrow, linear, 5–11-cm-long (excluding awns), one-rowed, and tapers slightly towards the tip. At maturity, it disarticulates above the lowest spikelet and falls as a unit. There are 8–12 spikelets, which are linear, and generally adorned with 3 florets per spikelet in middle of the spike, of which 2 are fertile, 1 in the lower and one in the upper part of spike. The uppermost floret in the middle of spike is sterile. The rachis internode is more or less as long as the spikelets. Glumes feature two teeth, separated by a membranous edge. Glumes are ¾ the length of the spikelet. Lemmas of the terminal spikelet are awned. One floret has a short awn, triangular in cross section, while the other floret has a very long (equal to or longer than the spike), flat awn.

Awns are flanked at the base by 1–2 unequal, short–articulate teeth; the middle tooth at glumes in terminal spikelet sometimes lengthened into an up to 1-cm-long awn. Anthers are short (2–3 mm). The caryopsis is more or less free at maturity (Fig. 9.1f).

A unique character of *Ae. searsii* is the unequal (one very long and the second short or sometimes absent) awns in the terminal spikelet, which undoubtedly results from the presence of only one fertile floret in the terminal spikelet. Generally, in the related species of the section, two developed awns form at the top of the ear, due to the occurrence of two fertile florets.

It is assumed (Feldman and Kislev 1977) that *Ae. searsii* is a young species which returned from the relatively dry habitats of the other Emarginata species to more mesophilic habitats. The very low stature of the plant, as well as its short ear, the almost equal length between the glumes and the florets, as well as the free caryopsis, are considered characteristics of advanced members of the genus (Eig 1929a).

Ae. searsii has limited morphological variation, involving mainly spike size, spikelet number and awn length. Israel and Jordan are the center of variation, where it probably originated. Morphologically, *Ae. searsii* is close to *Ae. longissima*, but differs from it in habitat and in the following characteristics: shorter plant, shorter spike, fewer spikelets, fewer fertile florets, larger glumes to spikelet ratio, awns of lemma in the terminal spikelet are very unequal, shorter anthers and free caryopsis.

Analysis of electrophoretically discernible water-soluble leaf proteins (Mendlinger and Zohary 1995) and random amplified polymorphic DNA (RAPD) (Goryunova et al. 2008), showed that *Ae. searsii* (and *Ae. bicornis*) displays the lowest diversity among the Sitopsis species. This lies in agreement with earlier studies (Dvorak and Zhang 1992b; Sasanuma et al. 1996; Giorgi et al. 2002; Goryunova et al. 2008), which have shown that *Ae. searsii* appears to be remote from the other three Emarginata species, equally distant from *Ae. longissima*, *Ae. bicornis* and *Ae. sharonensis* (Mendlinger and Zohary 1995). Likewise, meiotic pairing analyses in triploid hybrids of *Ae. searsii* with autotetraploid *Ae. longissima*, autotetraploid *Ae. speltoides* and autotetraploid *Ae. bicornis*, indicated the S^s genome of *Ae. searsii* is equally distant to the S genome of *Ae. speltoides*, the S^l genome of *Ae. longissima* and the S^b genome of *Ae. bicornis* (Yen and Kimber 1990a).

Ae. searsii has a relatively limited distribution region in the south-central area of the distribution of the genus. It occupies east Mediterranean primary and secondary habitats. It is a sub-Mediterranean element, extending into the Irano-Turanian region. Alt: 200–1000 m above sea level. *Ae. searsii* is limited to Israel (Judea, Samaria, the Golan Heights, and the higher Negev), Jordan (Gilead, Ammon, Moav, and Edom), southeast Lebanon and southwest Syria.

It grows on terra rossa, sometimes mixed with loess, or on basalt soil in the destroyed sub-Mediterranean habitat of *Sarcopoterium spinosum*, in open-park herbaceous formations, in the degraded deciduous steppe-maquis, in small shrub (Batha) formations, abandoned fields and edges of cultivation. Locally common, sometimes dense stands in Judea.

Ae. searsii grows sympatrically with *Ae. peregrina*, *Ae. geniculata*, *Ae. biuncialis* and *T. turgidum* subsp. *dicoccoides* (in Judea and Samaria), with *Ae. triuncialis*, *T. monococcum* subsp. *aegilopoides* and *T. urartu* (in Lebanon and Syria), and with *Ae. kotschyi* and *Ae. vavilovii* (in southern Jordan and Israeli Negev). It grows allopatrically with *Ae. speltoides*, *Ae. longissima*, *Ae. caudata*, *Ae. umbellulata* and *Ae. neglecta*. *Ae. searsii* is involved in the parentage of *Ae. vavilovii* (mixed populations of these two species occur in the higher central Negev in Israel and in southern Jordan), and it presumably introgressed with wild tetraploid wheat, *T. turgidum* subsp. *dicoccoides*, *Ae. peregrina* and *Ae. kotschyi*. This is the only diploid *Aegilops* species that grows in mixed populations with wild *T. turgidum* in the southern Fertile Crescent. It also has massive contact with *T. urartu* in southeastern Lebanon and southwestern Syria. Variation in the ability to suppress the *Ph1* gene of polyploid wheat has not yet been demonstrated in this species.

9.4.6.2 Cytology, Cytogenetics, and Evolution

Ae. searsii is a diploid (2n = 2x = 14), with a modified S genome, designated S^s by Feldman et al. (1979). Its organellar genome was designated S^v (Ogihara and Tsunewaki 1988; Wang et al. 1997) since it differs from those of the other Sitopsis species (Wang et al. 1997). The genome of *Ae. searsii* is the smallest genome in the Emarginata subsection (1C DNA = 6.65 ± 0.091 pg; Eilam et al. 2007), 5.9 pg (The Angiosperm C-value database at Kew Botanic gardens) (Tables 2.4 and 9.3). Li et al. (2022) assessed the size of *Ae. searsii* genome after genome sequencing and assembly to be 5.55 Gb (=5.42 pg). This genome size is somewhat smaller than that of *Ae. bicornis* and considerably smaller than those of *Ae. sharonensis* and *Ae. longissima*, but larger than that of *Ae. speltoides*. The karyotype of *Ae. searsii* is symmetric, with two metacentric chromosome pairs and five sub-metacentric pairs (Feldman and Kislev 1977; Feldman et al. 1979). One of the metacentric pairs and one of the submedian pairs are satellited chromosomes; the metacentric pair has large satellites and the submetacentric one has a pair of medium-size satellites. *Ae. searsii* differs from the other three Emarginata species in that it has a pair of the medium-size satellites, whereas the other species contain a pair of smaller satellites.

Karyotype analysis showed the presence of two secondary constrictions in *Ae. searsii* (Feldman and Kislev

1977; Feldman et al. 1979; Teoh and Hutchinson 1983). In situ hybridization analysis, using a radioactive rDNA probe (Teoh et al. 1983), revealed that two pairs of NORs are located in the secondary constriction of *Ae. searsii*. Yet, hybridization with a 18S–26S rDNA probe, which enabled high sensitivity, revealed that the two-major active NORs are located on the short arms of chromosomes 5S^s and 6S^s (Badaeva et al. 1996b). The numbers and chromosomal locations of the NORs corresponded to the number of satellited (SAT) chromosomes (Feldman and Kislev 1977; Feldman et al. 1979; Teoh and Hutchinson 1983), indicating that these regions are actively transcribed. An additional pair of minor NORs was detected on the short arm of chromosome 1S^s (Badaeva et al. 1996b). The minor NORs do not usually form secondary constriction, indicating that these NORs are not transcribed. In situ hybridization with the pTa794 (5S rDNA) probe showed that the 5S rDNA loci are located in *Ae. searsii* on the short arms of chromosomes 1S^s and 5S^s (Badaeva et al. 1996b).

Ae. longissima chromosomes were similar to *Ae. searsii* chromosomes in hybridization patterns with both the pTa71 and pTa794 DNA probes, confirming the high degree of similarity between their genomes (Teoh et al. 1983; Badaeva et al. 1996b).

Teoh and Hutchinson (1983), using an improved C-banding technique, found that the C-banding patterns of mitotic metaphase chromosomes of *Ae. searsii* exhibited a characteristically different pattern from those of the Sitopsis species. Moreover, the C-banding pattern enables the chromosomes of this species to be identified individually. *Ae. searsii*, like other Sitopsis species, has telomeric, interstitial, and centromeric bands. However, *Ae. searsii* and *Ae. bicornis* have relatively smaller centromeric bands (Teoh and Hutchinson 1983).

Teoh and Hutchinson (1983) analyzed only one accession per species. C-banding polymorphism in a larger number of accessions of the diploid *Aegilops* species was studied by Friebe et al. (1992b, 1993, 1995b) and Friebe and Gill (1996). C-banding polymorphism was analyzed in 14 accessions of *Ae. searsii* from Israel, enabling establishment of a generalized idiogram of the species (Friebe et al. 1995b). All seven *Ae. searsii* chromosome pairs were individually identified on the basis of their C-banding patterns. No variation was observed within the accessions, but C-band polymorphism was detected between the different accessions. *Ae. searsii* was easily distinguished from *Ae. longissima* by their distinct C-banding patterns; *Ae. searsii* had fewer and smaller C-bands, and, in this respect, was similar to *Ae. bicornis* (Friebe et al. 1995b). Isozyme studies (Pietro et al. 1988) showed that there is no translocation 4/7 in *Ae. searsii*, also differing in this respect, from *Ae. longissima*.

C-banding analysis was also used to identify the seven disomic addition lines of *Ae. searsii* chromosomes to T. *aestivum* cv. Chinese Spring, 14 ditelosomic chromosome addition lines, 21 disomic substitution of whole chromosome and 31 ditelosomic chromosome substitution lines, produced by NA Tuleen. The identity of these lines was further confirmed by meiotic pairing analysis. Sporophytic and gametophytic compensation tests were used to determine the homoeologous relationships of the *Ae. searsii* chromosomes. The results show that the *Ae. searsii* chromosomes do not compensate well for their wheat homoeologues. The short arm of *Ae. searsii*, 3S^{s}S, causes spike fragility; the spike breaks between the first and the sixth spikelets. Chromosome arm 2S^{s}S causes tenacious glumes.

In situ hybridization with the highly repetitive DNA sequences pSc119 and pAsl, and C-banding analysis, enabled the genome differentiation in all the diploid species of *Aegilops*. The ISH patterns of chromosomes with pSc119 showed only minor intraspecific variations in the subsection Emarginata species. Chromosomes of all these diploid species hybridized with the pSc119 probe; however, the level of hybridization and labeling patterns differed among genomes (Badaeva et al. 1996a). All these species had strong pSc119 hybridization sites located mainly in the telomeres. *Ae. longissima* and *Ae. sharonensis* are highly heterochromatic species; C-bands were present in intercalary, telomeric and proximal regions of the chromosomes. *Ae. searsii* and *Ae. bicornis* have much less C-heterochromatin compared with the other two Emarginata species and hence, these species form a third subgroup (Badaeva et al 1996a).

On the other hand, the Sitopsis species differ in their ability to hybridize with the pAs1 probe (Badaeva et al. 1996a). No hybridization was observed with the pAs1 probe on *Ae. speltoides*, *Ae. sharonensis* and *Ae. longissima* chromosomes. A few minor pAs1 sites were observed in *Ae. searsii* and *Ae. bicornis*.

The evolution of two tandemly repeated sequences, Spelt1 and Spelt52, was studied in species of *Aegilops* sect. Sitopsis (Salina et al. 2006). Fluorescence in situ hybridization showed considerable polymorphisms in the hybridization patterns of Spelt1 and Spelt52 repeats between and within *Aegilops* species. Hybridization patterns of Spelt52 in *Ae. speltoides*, *Ae. longissima* and *Ae. sharonensis* were species-specific. There was no detectable Spelt1-associated FISH signal in Section Sitopsis, with the exception of *Ae. speltoides*. Spelt52 and its analogues pGc1R-1 and pAesKB52, were found in *Ae. speltoides*, *Ae. longissima* and *Ae. sharonensis*, but not in *Ae. bicornis* or *Ae. searsii* (Anamthawat-Jonsson and Heslop-Harrison 1993; Zhang et al. 2002; Salina et al. 2004a).

Yaakov et al. (2013) assessed the relative copy numbers of a number of TE families in five accessions of *Ae. searsii* and in several other Sitopsis species. The analysis of six *Gypsy* families revealed that *Fatima* is very abundant in *Ae. searsii* and in *Ae. speltoides*. On the other hand, several elements, i.e., *Latidu*, *Sabrina*, *BAGY2* (all *Gypsy* retrotransposon) and *Charon* (*Mutator* DNA transposon), had specific proliferation in the *Ae. searsii* genome, in comparison to the other Sitopsis species.

9.4.6.3 Crosses with Other Species of the Wheat Group

Very few hybrids between *Ae. searsii* and other *Aegilops* and *Triticum* species were produced and analyzed. Feldman et al. (1979) noticed that the crossability between *Ae. longissima* with *Ae. searsii* is apparently very low. Over 380 pollinated florets (about 15 spikes) in both directions yielded about 10 seeds, and then, only when *Ae. longissima* was the female parent. Most of the seeds were shriveled and their embryos poorly developed; they either did not germinate, even after placing the embryo on culture medium, or the seedlings died very young. Only three seeds, representing two hybrid combinations, germinated and survived the seedling stage. When studying chromosome pairing at first meiotic metaphase of these F_1 hybrids, Feldman et al. (1979) found that meiosis was irregular and chromosomal pairing was somewhat reduced (Table 9.4). One of the rod bivalents was heteromorphic, indicating that the total length and the arm ratio of the two pairing partners were not the same. Moreover, most meiocytes exhibited one asymmetrical multivalent, indicating that the two species differ in one asymmetric reciprocal translocation. The asymmetrical configuration implies that the translocated segments were unequal. Further indication of chromosomal differentiation between *Ae. searsii* and *Ae. longissima* was the occurrence of a chromatid bridge and an acentric fragment at first anaphase of the two hybrids, indicating heterozygosity for a paracentric inversion. A comparison of the karyotypes of the two species confirms the existence of several structural differences between the chromosomes of the two species (Feldman et al. 1979). Pollen and seed fertility of the F_1 hybrids was low; pollen fertility was 30–37% and seed fertility 6.7–35%. The significant reduction in meiotic pairing in these hybrids, as manifested by reduced numbers of chiasmata, shows the incomplete homology between the *Ae. searsii* and *Ae. longissima* genomes. Clearly, in one or both species, several chromosomes have undergone independent re-patterning, as indicated by the asymmetrical translocation, heteromorphic bivalent and the paracentric inversion.

Ae. searsii and *Ae. longissima* are presumably isolated via several different mechanisms that include low crossability, partial hybrid sterility and differences in ecological requirements and geographical distribution. In addition to these, hybridization between the two species in nature is also greatly reduced because of the predominance of self-pollination. The multiplicity of isolating mechanisms is probably responsible for the rarity of hybrids between these species in the contact zone between them.

Yu and Jahier (1992) produced F_1 hybrid between the allotetraploid *Ae. peregrina* and *Ae. searsii*. Analysis of chromosome pairing at first meiotic metaphase (Table 9.5) revealed that the S^v subgenome of *Ae. peregrina* is not completely homologous to that of *Ae. searsii*. From the level of pairing in the triploid hybrid, Yu and Jahier (1992) concluded that subgenome S of *Ae. peregrina* is very close to the genome of *Ae. longissima*, and relatively more distant from that of *Ae. searsii*.

Hybrids between hexaploid wheat, *Triticum aestivum* ssp. *aestivum* and *Ae. searsii* were produced and analyzed by Feldman (1978) and Yu and Jahier (1992) (Table 9.10). The low pairing observed in these hybrids indicated that very little homology exists between the S^s genome of *Ae. searsii* and all of the three subgenomes of *T. aestivum*. Hence, *Ae. searsii* is not the source of the B subgenome of allopolyploid wheat.

9.5 Section Vertebrata Zhuk. Emend. Kihara

9.5.1 General Description

Section Vertebrata Zhuk. emened Kihara [Syn.: Pachystachys Eig, Gastropyrum (Jaub. & Spach) Zhuk.; Polyploids Zhuk.; Aegilonearum Á. Löve] consists of annual, predominantly autogamous species. The plants are robust, with a more or less long spike, (3-) 6–10 (-15)-cm-long, mostly thick, 3–7 mm wide, cylindrical or moniliform, mostly awned and disarticulating at maturity into spikelets, each with the laterally adjacent rachis-internode (barrel-type dispersal unit). Spikelets are more or less ventricose or linear and glumes are mostly awnless. Lemmas of apical spikelets, sometimes of several upper spikelets, with one well-developed broad and mostly flat awn. The caryopsis adheres to lemmas and palea (Fig. 9.2).

Section Vertebrata contains five species, one diploid (*Ae. tauschii* Coss., genome DD) and two tetraploids (*Ae. ventricosa* Tausch (genome DDNN) and *Ae. crassa* Boiss. (genome $D^cD^cX^cX^c$). The latter also has a hexaploid cytotype (genome $D^cD^cX^cX^cDD$)), and two hexaploids (*Ae. juvenalis* (Thell.) Eig (genome $D^cD^cX^cX^cUU$) and *Ae. vavilovii* (Zhuk.) Chennav. (Genome $D^cD^cX^cX^cS^sS^s$). Kimber and Zhao (1983) and Zhao and Kimber (1984), based on analysis of meiotic chromosome pairing, decided that the allopolyploid species containing the D subgenome can be divided into three clusters: (1) *T. aestivum*, *Ae. cylindrica* and *Ae. ventricosa*, in which the D subgenome has undergone little modification from the genome of the diploid progenitor, *Ae. tauschii*; (2) tetraploid and hexaploid *Ae. crassa*, in which the D subgenome is somewhat modified; (3) *Ae. juvenalis* and *Ae. vavilovii*, in which the D subgenome is substantially modified. Dubcovsky and Dvorak

(1995a), Dvorak (1998) and Dvorak et al. (2012) arrived at a similar conclusion but pointed out that the D subgenome of *Ae. crassa* differentiated considerably from that of the diploid parent and consequently, it was designated D^c (Dvorak 1998). Analyses of the plasmon of these species clearly showed that *Aegilops tauschii* is the maternal parent of the two allotetraploid species, *Ae. crassa* and *Ae. ventricosa*, and that the tetraploid cytotype of *Ae. crassa* is the maternal parent of three hexaploids, *Ae. juvenalis, Ae. vavilovii* and 6x *Ae. crassa* (Tsunewaki 1993, 2009; Ogihara and Tsunewaki 1988; Wang et al. 1997). These studies suggested that the D-genome containing allotetraploids originated at three different times from *Ae. tauschii* in the following order: *Ae. crassa, Ae. cylindrica,* and *Ae. ventricosa. Ae. crassa* is the oldest allotetraploid containing the D genome, while *Ae. cylindrica* and *Ae. ventricosa* originated later. (Tsunewaki 1993, 2009; Ogihara and Tsunewaki 1988; Wang et al. 1997). Thus, 4x *Ae. crassa* appears to be an ancient allotetraploid that originated from hybridization of primitive *Ae. tauschii* with an ancient species in the evolutionary lineage leading to the section Sitopsis (Dubcovsky and Dvorak 1995a; Badaeva et al. 2002).

The Vertebrata species are characterized by, more or less, high morphological resemblance. All the allopolyploid species contain the D subgenome or a modified D subgenome that derived from the genome of the diploid species, and the impact of this genome is clearly apparent in the morphology and dispersal unit type of all the allopolyploid species. In addition, the organellar genome of all the allopolyploid species derived from that of the diploid species.

The central and northeastern part of the Fertile Crescent (i.e., eastern Turkey, Syria, Iraq and Iran) is the region with the greatest diversity for the Vertebrata species (Table 9.1). Only one species, *Ae. ventricosa*, occurs exclusively west of the Fertile Crescent, while other three, *Ae. tauschii, Ae. crassa* and *Ae. juvenalis*, have spread eastward, and *Ae. vavilovii* grows in the southeastern part of the genus distribution area. With the exception of *Ae. juvenalis* and, to a lesser extent, *Ae. vavilovii* (both hexaploids), all Vertebrata species are widespread.

The distribution of *Ae. ventricosa* is the most difficult to explain in an evolutionary sense, as it distinctly does not overlap with the areas of any of its putative parents. *Ae. tauschii* is the donor of one subgenome to *Ae. ventricosa* (Kihara 1949) and the second genome is derived from *Ae. uniaristata* (Yen and Kimber 1992b). There seems only a theoretical overlap on the Istrian peninsula of Croatia of *uniaristata* and *ventricosa*, but *Ae. tauschii* is still far away.

Some Vertebrata species are also found at higher altitudes and may show better adaptation to cold than most species of *Aegilops*. Several species grow in areas with relatively little annual rainfall and may be drought-tolerant. On the other hand, the D-genome of these species may be responsible for the overall poor performance of species with this genome in resisting rust infection (Hammer 1987).

9.5.2 *Aegilops tauschii* Coss.

9.5.2.1 Morphological and Geographical Notes

Ae. tauschii Coss. [Syn: *Ae. squarrosa* L.; *Triticum squarrosum* (L.) Rasp.; *Triticum aegilops* P. Beauv. ex Roem. & Schult.; *Triticum tauschii* (Coss.) Schmahlh.; (Coss.) Á. Löve], also known as Tausch's goat grass and rough-spike hard grass, is a predominantly autogamous, annual plant. It is tuft, few- to many-tillered plant, its culms are 20–45-cm-high (excluding spikes) and it defoliates in lower parts at maturity. The spike is cylindrical, thick or thin, one-rowed, tapers slightly to the tip, and is relatively long (4–10-cm-long, excluding awns). At maturity, the spike disarticulates into individual spikelets, each with its adjacent rachis segment (barrel-type dispersal unit). The number of spikelets per spike is 5–13, cylindrical, barrel-shaped, usually equal in length to the adjacent rachis segment. Basal rudimentary spikelets are absent, or rarely, there are 1 or 2. There are 3–5 florets, usually 4, with the upper 1–3 being sterile. Glumes are almost rectangular, narrow, equally spaced and with small nerves. The tip of the glume is truncated, with a clearly thickened edge, with one or no teeth. Lemmas are membranous, with a keel that terminates with a small tooth or awn, up to 4-cm long, and sometimes accompanied by 1–2 short, wide lateral teeth. The lemmas of the apical spikelet have longer awns, that up to 5.5-cm-long. The awns are triangular in cross section and are shorter on the lower spikelets. The caryopsis adheres to lemmas and palea (Fig. 2.1).

Ae. tauschii possesses a very wide morphological variation, mainly in spike shape, length and width, awned to awnless, straight or winding rachis, cylindrical to square spikelets, number of spikelets per spike and length of awns. Several accessions have a tough rachis (Waines et al. 1982; Knaggs et al. 2000). The morphological variation and ecological amplitude of *Ae. tauschii* exceed those of any other diploids of the *Triticum-Aegilops* group (Zohary et al. 1969). This variation led Eig (1929a) to classify the various morphological forms in two subspecies, ssp. *eusquarrosa* Eig (now ssp. *tauschii*), which has elongated, cylindrical spikelets (Fig. 9.2a), and ssp. *strangulata* (Eig) Tzvelev, which have a curved rachis segment, noticeably longer and narrower than the adjacent spikelets, giving the spike a markedly moniliform appearance, as well as square spikelets, equally long as wide (Fig. 9.2b). The latter subspecies is morphologically distinct, being taller with greater seed weight and rounded seed shape than in ssp. *tauschii*. Despite its variability, ssp. *strangulata* has been considered a discrete taxon (Jaaska 1981). Ssp. *tauschii* has been found throughout the geographic range of the species while ssp. *strangulata* mainly occurs in narrow belts along the southeastern Caspian Sea in Iran (Zohary et al. 1969; Tanaka 1983; van Slageren 1994).

Ssp. *tauschii* was further divided into three varieties: *typica* Eig (now var. *tauschii)*, *anathera* (Eig) Hammer (awnless), and *meyeri* (Griesb.) Tzvelev (has a slender spike). var. *anathera* is rather easily distinguishable by awnless type and is shorter than the other varieties, whereas var. *tauschii* and var. *meyeri* proved difficult to identify and were not easily distinguished due to many intermediates between them (Knaggs et al. 2000). In spite of this difficulty, subdivision of *Ae. tauschii* on the basis of morphology appears to be reasonably valid (Knaggs et al. 2000).

Eig's classification was based primarily on variation in spike morphology. Hammer (1980) accepted Eig's intraspecific classification of *Ae. tauschii* and added var. *paleidenticulata* (Gandiljan) Hammer to ssp. *tauschii*. Hammer (1980) recommended using *Ae. tauschii* as the correct name for this species, renamed ssp. *eusquarrosa* as ssp. *tauschii* and var. *typica* as var. *tauschii*. Knaggs et al. (2000), based on morphological study of a large number of *Ae. tauschii* accessions, also identified the two subspecies, *tauschii* and *strangulata*, and the three varieties of ssp. *tauschii*. Kihara and Tanaka (1958) adopted Eig's (1929a) intraspecific classification of this species and presented a detailed account of the morphological and genetic variation of *Ae. tauschii*. Yet, Kihara and Tanaka (1958) and Kihara et al. (1965) showed all three varieties of ssp. *tauschii* are interfertile and described many intermediates between them. Only ssp. *strangulata* is more distinct, and its occurrence is limited to a narrow belt on the southern shores of the Caspian Sea (Kihara and Tanaka 1958; Zohary et al. 1969) and in Armenia, Azerbaijan, and Turkmenistan (Kim et al. 1992). In the southern shores of the Caspian Sea, in northern Iran, ssp. *strangulata* grows next to many robust types of *Ae. tauschii* with large, thick spikes that presumably have segetal growth (Zohary et al.1969). Kim et al. (1992) pointed out that intermediate forms exist among all morphologically distinct forms of *Ae. tauschii* and that gene flow takes place quite frequently between the two subspecies. Moreover, they found that a ribosomal DNA genotype of an accession of ssp. *strangulata* from Armenia was different from other *strangulata* genotypes and similar to genotypes of the *tauschii* subspecies. Consequently, Kim et al. (1992) considered intraspecific classification of *Ae. tauschii* on morphological grounds inadequate. Similarly, van Slageren (1994) thought it justified not to classify the observed intraspecific variation of *Ae. tauschii*. To this day, the status of subspecies and varieties is still under discussion (Hammer 1980; van Slageren 1994; Dudnikov 2000; Knaggs et al. 2000).

Nevertheless, the morphologically-based intraspecific classification of *Ae. tauschii* has been used by many cytogeneticists.

Since the discovery that *Ae. tauschii* donated its genome (genome D) to hexaploid wheat (McFadden and Sears 1944, 1946; Kihara 1944), *Ae. tauschii* has been a subject of intensive genetic, cytogenetic, and molecular studies which seek answers to many questions, including the center of genetic diversity of *Ae. tauschii* and the place of origin of hexaploid wheat (Gill 2013). A large portion of these studies have concentrated on assessment of the intraspecific genetic diversity of this species in order to understand the factors leading to the evolutionary success of *Ae. tauschii* and to exploit its rich gene pool for wheat improvement.

Very wide intraspecific variation in *Ae. tauschii* has been demonstrated through morphological (Kihara and Tanaka 1958; Hammer 1980; Knaggs et al. 2000), isozyme (Nishikawa et al. 1980; Jaaska 1981; Dudnikov 2014 and reference therein), restriction fragment length polymorphism (RFLPs) (Lubbers et al. 1991; Dvorak et al. 1998c), amplified fragment length polymorphism (AFLPs) (Saeidi et al. 2008) and microsatellite (Pestsova et al. 2000; Saeidi et al. 2006) analyses. Diversity was studied in accessions of *Ae. tauschii* that were collected from the western (Syria, Turkey, and Georgia) to the eastern (Central Asia and western China) range of its distribution. The above studies showed that accessions from the region along the Southern Caspian Sea exhibited the highest genetic variation, suggesting that this region is the center of variation of *Ae. tauschii*. Genetic variability within the D subgenome of wheat is much lower than it is within *Ae. tauschii* (Appels and Lagudah 1990; Lagudah et al. 1991a; Lubbers et al. 1991), so the wild progenitor offers great potential for wheat improvement. Utilization of *Ae. tauschii* for wheat improvement is further aided by the ability of the chromosomes of *Ae. tauschii* and the D subgenome chromosomes of bread wheat to naturally recombine.

Isozyme analysis conducted by Jaaska (1981) revealed intraspecific differentiation of *Aegilops tauschii* into two groups of biotypes, which essentially correspond to its two-morphological subspecies, *tauschii* and *strangulata*. Dudnikov (2014) pointed out that the isozymic variation in *Ae. tauschii* reflects adaptive intraspecies divergence: ssp. *strangulata* favors the habitats of the Caspian seaside climate, with warm and moist winters, while ssp. *tauschii* mostly occupies the habitats, which have a rather continental climate, with relatively cold and dry winters.

RFLP marker analysis of genetic diversity, performed by Tsunewaki et al. (1991) and Lubbers et al. (1991), found close similarities between ssp. *tauschii* var. *meyeri* and ssp. *strangulata*. A molecular study by Dvorak et al. (1998c) found evidence of gene migration between the different divisions in accessions from the southwest Caspian area of Iran. The greatest amount of variation was found in *Ae. tauschii* accessions collected in Iran and western Transcaucasia (Lubbers et al. 1991; Tsunewaki et al. 1991). Var. *meyeri*, although formally placed in ssp. *tauschii*, was found to be genetically closer to *ssp. strangulata* than to *ssp. tauschii, and consequently*, Lubbers et al. (1991) concluded that var. *meyeri* actually belongs to ssp. *strangulata*.

Actually, Lubbers et al. (1991) identified two genetically diverse groups, one consisting of ssp. *tauschii* vars. *tauschii* and *anathera*, and the other of ssp. *strangulata* and ssp. *tauschii* var. *meyeri*. Their analysis strongly supported the suggestion the that Caspian Sea region is the center of genetic diversity and origin of *Ae. tauschii*. Likewise, Dvorak et al. (1998c), in a large study of RFLP markers, confirmed the existence of two genetically diverse groups in *Ae. tauschii*, which crosscut taxonomic groupings, but, in contrast to Lubbers et al. (1991), proposed Armenia as the center of genetic diversity and origin of hexaploid wheat. However, Wang et al. (2013b), using 7815 single nucleotide polymorphisms (SNPs) previously mapped by Luo et al. (2009), that provide complete coverage of the genome, to analyze 402 accessions of *Ae. tauschii*, 75 hexaploid wheats, and seven tetraploid wheats, concluded that southwestern Caspian Iran is the center of genetic diversity of *Ae. tauschii* and the center of origin of hexaploid wheat.

As seen from the above, the intraspecific botanical classification of *Aegilops tauschii* agrees poorly with the genetic relationships (Dvorak et al. 1998c). The most apparent contradiction is encountered with var. *meyeri*, which is assigned to ssp. *tauschii* on the basis of morphology but is genetically closely related to ssp. *strangulata* (Lubbers et al. 1991; Dvorak et al. 1998c). In genetic studies, therefore, the use of categories based on genetic subdivision of *A. tauschii* is preferable to those based on formal taxonomy (Wang et al. 2013b). In fact, recent genetic and molecular studies have shown that two major lineages exist in *Ae. tauschii* (Takumi et al. 2008; Mizonu et al. 2010; Sohail et al. 2012; Wang et al. 2013b). These two lineages were named by Takumi et al. (2008) and Mizuno et al. (2010) lineage 1 (L1) and lineage 2 (L2). L1 consisted of accessions of ssp. *tauschii* from the eastern habitats, mainly Afghanistan and Pakistan, whereas L2 included ssp. *tauschii* accessions of the western habitats and all accessions of ssp. *strangulata*. Accordingly, both lineages included accessions of ssp. *tauschii*, whereas ssp. *strangulata* belonged only to lineage 2. The two varieties of ssp. *tauschii, meyeri* and *anathera* were classified into L2 and L1, respectively.

Nucleotide sequence variations of 10 nuclear genes were used to construct a phylogenetic tree for each gene, which were compared to the SSR phylogenetic tree (Takumi et al. 2008). Although some discrepancies were found between the trees, the results supported the subdivision of *Ae. tauschii*

into two major lineages and suggested that ssp. *strangulata* derived from only one of the two lineages, i.e., lineage 2. This phylogenetic tree was consistent with a previous report of a similar SSR analysis (Pestsova et al. 2000).

Mizuno et al. (2010) conducted an AFLP analysis to study population structure of 122 accessions of *Ae. tauschii*. Their phylogenetic and principal component analyses revealed, similar to the finding of Takumi et al. (2008), two major lineages in *Ae. tauschii*, i.e., lineages one (L1) and two (L2). The results of the study of Wang et al. (2013b) also supported the subdivision of *Ae. tauschii* into two lineages (L1 and L2) that have been reproductively isolated in nature. Particularly informative was a comparison of the SNPs tree, obtained by Wang et al. (2013b), with the AFLP tree reported by Mizuno et al. (2010). In both trees, branches in L2 were longer than those in L1, indicating that L2 was more diverse than L1. Wang et al. (2013b) found that each lineage consists of two closely related sub-lineages that appear to be geographically isolated. Within L1, sub-lineage 1W is located in Turkey, Transcaucasia, and western Iran, whereas sub-lineage 1E is located from central Iran to China. Sub-lineages 1W and 1E are predominantly found at high elevations (400–3000 m above sea level). Within L2, sub-lineage 2W occupies elevations between 400 and 1500 m in Transcaucasia (Armenia and Azerbaijan), whereas sub-lineage 2E occupies elevations ≤ 25 m and is distributed across Azerbaijan and Caspian Iran. The 2E sub-lineage is morphologically heterogeneous, including both the typical moniliform *Ae. tauschii* ssp. *strangulata* in southern Caspian Iran and morphologically intermediate forms classified as ssp. *tauschii* vars. *meyeri* and *tauschii* in southern and southwestern Caspian Sea.

The poor agreement between morphological and genetic relationships among *A. tauschii* accessions (Lubbers et al. 1991; Dvorak et al. 1998c; Mizuno et al. 2010; Sohail et al. 2012; Wang et al. 2013b) was reconciled by two mutually exclusive hypotheses (Wang et al. 2013b): (1) the morphological traits of *Ae. tauschii* and the subsequent classification are trustworthy, but the taxa are genetically heterogeneous as a result of gene flow between them; or (2) *Ae. tauschii* is genetically clearly subdivided but the subdivision is not faithfully reflected by morphology and taxonomic classification. Clear genetic separation of lineages 1 and 2 and the paucity of intermediate genotypes, led Wang et al. (2013b) to favor the second alternative.

Ae. tauschii is the donor of the D genome to hexaploid wheat, *T. aestivum* (McFadden and Sears 1944, 1946; Kihara 1944), to which it confers many important traits, including bread making quality (Kerber and Tipples 1969; Orth and Bushuk 1973), cold hardiness (Limin and Fowler 1981; Le et al. 1986) and salt tolerance (Schachtman et al. 1992). In addition, many accessions of *Ae. tauschii*, particularly those of subsp. *strangulata*, showed resistance to various diseases (Yildirim et al. 1995; Cox et al. 1995; Appels and Lagudah 1990; Knaggs et al. 2000). Limin and Fowler (1981) rated cold hardiness in a large number of accessions from species which share a common genome with hexaploid wheat and found that more than half of the *Ae. tauschii* accessions survived the coldest temperatures of the Canadian winter and had a hardiness level which approached that of the hardiest winter cultivars of hexaploid wheat. Le et al. (1986) obtained similar results. These results support Tsunewaki's (1968) suggestion that the addition of the D genome to tetraploid wheat made spread of the cultivation of the resulting hexaploid to colder northern countries possible. Indeed, study of intervarietal substitution lines, in which a chromosome of a cold hardiness cultivar of hexaploid wheat substituted its homologous chromosome in a spring cultivar, showed that chromosomes 4D and 5D accounted for much of the difference in cold hardiness between these two cultivars (Law and Jenkins 1970; Cahalan and Law 1979).

The differentiation of *A. tauschii* into two lineages brings forth several questions relevant to the origin of wheat and shaping of its diversity (Wang et al. 2013b). Recurrent hybridization and introgression between wheat and *Ae. tauschii* were known to have played a role in the origin of wheat D-subgenome diversity, although the magnitude is unknown (Dvorak et al. 1998a, b, c; Talbert et al. 1998; Caldwell et al. 2004; Akhunov et al. 2010). Did only lineage 2 contribute germplasm to the wheat D subgenome? If so, why lineage 2 and not lineage 1? Diversity is uneven among and along the wheat D-subgenome chromosomes (Akhunov et al. 2010). So, does the distribution of diversity along wheat chromosomes have anything to do with its distribution along the *A. tauschii* chromosomes, and what is the cause of this pattern?

The wheat D subgenome, unlike the other two subgenomes of hexaploid wheat, shows great fluctuation in diversity among chromosomes (Akhunov et al. 2010). Similar uneven distribution of diversity among all D-genome chromosomes was noted by Wang et al. (2013b) in *Ae. tauschii*. They also showed that diversity correlated with recombination rates along the chromosomes. Similar correlations were also observed along each chromosome in both lineages of *Ae. tauschii*. Likewise, RFLP was shown to correlate with recombination rates in *Ae. tauschii* (Dvorak et al. 1998a).

Previous genetic studies placed the origin of wheat in Transcaucasia and southwestern Caspian Iran (Tsunewaki 1966; Nakai 1979; Jaaska 1980; Dvorak et al. 1998c) or southeastern Caspian Iran (Nishikawa et al. 1980). The consensus has been that ssp. *strangulata* was the wheat progenitor (Nishikawa 1973; Nakai 1979; Jaaska 1980; Hammer 1980; Nishikawa et al. 1980; Lagudah et al. 1991a; Lubbers et al. 1991; Dvorak et al. 1998c, 2012).

Jaaska (1981), by way of isozyme analysis, identified ssp. *tauschii* as the contributor of the D genome to the allotetraploid *Ae. cylindrica* (genome DDCC) and the third subgenome to allohexaploid *Ae. crassa* Boiss. (genome $D^cD^cX^cX^cDD$), and subsp. *strangulata* as the contributor of the D genome to the allohexaploid *Triticum aestivum*, to the allotetraploids *Ae. crassa* (genome $D^cD^cX^cX^c$) to *Ae. ventricosa* (genome DDNN), and to the allohexaploid *Ae. juvenalis* (genome $D^cD^cX^cX^cUU$).

Wang et al. (2013b) identified 12 *Ae. tauschii* accessions that are closely related to the D-subgenome of wheat. All 12 accessions belonged to sub-lineage 2E and were members of populations located in southwestern and southern Caspian Iran. In a surprising departure from the belief that ssp. *strangulata* was the source of the wheat D subgenome, only one of these 12 accessions had been classified as ssp. *strangulata* on the basis of its morphology. Eleven of the 12 accessions were classified as ssp. *tauschii* var. *tauschii* or var. *meyeri*. However, if it is accepted that morphology does not reflect genetic relationships, as in the second hypothesis of Wang et al. (2013b), this conflict with the previous conclusions regarding the progenitor of the wheat D subgenome becomes irrelevant, as, genetically, these accessions are members of the 2E sub-lineage. Of the 7185 SNP sites in the wheat D subgenome studied by Wang et al. (2013b), 0.8% appeared to originate by introgression from the L1 lineage, while 99% of the D subgenome was contributed by *A. tauschii* lineage 2. A population within lineage 2E in the southwestern and southern Caspian appears to be the main source of the wheat D subgenome (Wang et al. 2013b).

Given the extensive opportunity for natural hybridization between wheat and *Ae. tauschii* (Kihara et al. 1965), why does hexaploid wheat appear monophyletic and why the preference for sub-lineage 2E? An answer to this question may reside in the geography of cultivation of tetraploid wheat by early farmers. *Aegilops tauschii* readily hybridizes with tetraploid wheat, and triploid hybrids often produce so many unreduced gametes, that they are fertile (Zhang et al. 2010). Spontaneous chromosome doubling via union of unreduced (2n) gametes has been thought to be the way that hexaploid wheat originated from the hybridization of *T. turgidum* with *Ae. tauschii*. Previous works have observed unreduced gametes in F_1 hybrids of *Ae. tauschii* with six of the eight *T. turgidum* subspecies tested (Zhang et al. 2008, 2010). By contrast, hybridization of *Ae. tauschii* with hexaploid wheat is arduous, and hybrids can only be obtained with the aid of embryo rescue. Introgression from *A. tauschii* into hexaploid wheat should therefore only be expected in the areas where tetraploid wheat was farmed in mixed populations with hexaploid wheat (Dvorak et al. 1998c).

Today, farming of tetraploid durum wheat is limited to a few mountainous regions in northern Iran (Matsuoka et al.

2008), but the situation could have been different in the past. If wheat farming was predominantly in low elevations in Caspian Iran, and if the distribution of *Ae. tauschii* was similar to its present-day distribution, the only possible source of the D genome was sub-lineage 2E, as only sub-lineage 2E is found at low elevations.

Among the diploid *Aegilops* species, *Ae. tauschii* has the widest geographic distribution. It grows in a very large area in the eastern part of the distribution of the genus. This species is the only diploid that spread eastward from the center of origin of the genus. It grows in the following countries: Crimea (possibly), Ciscaucasia (northern Caucasia) (possibly), Transcaucasia (Georgia, Armenia, and Azerbaijan), Uzbekistan, Turkmenistan, Kazakhstan, Tagikistan, Kyrgyzstan, Afghanistan, China (western slopes of Himalaya), India (Kashmir), Pakistan, Iran, Iraq, Turkey, and northeastern Syria (Table 9.2). Its distribution is limited in the east by the Himalaya mountains, in the south by the deserts of southern Pakistan and southern Iran, in the north by the cold steppes of Central Asia, and in the west by the Mediterranean climate. Zohary et al. (1969) assumed that the southern shores of the Caspian Sea, Turkmenistan, and northern Afghanistan, include the 'primary habitat' of the species, while the peripheral localities in e.g., Central Asia, Pakistan, Armenia, Iraq, and Syria, are 'secondary' in being always associated with weedy growth, often in (irrigated) wheat fields. They pointed out that the center of diversity and abundance comprises a belt around the southern shores of the Caspian Sea, across northern Iran, Turkmenistan, and northern Afghanistan. In this area, *Ae. tauschii* attains its widest ecological amplitude and morphological diversity. From this center, it spread westward to the Turkish border and the Syrian steppes, and eastwards to Pakistan, approaching the Chinese border. It is mentioned as a rare plant in the Caucasus and the Crimea, but it is doubtful that it is a native plant there (Zohary et al. 1969). In its broadly scattered distribution area, *Ae. tauschii* is also a weed and often a weed in cereal fields.

Ssp. *strangulata* is distributed from Transcaucasia to eastern Caspian Iran (Kihara and Tanaka 1958; Kihara et al. 1965; Zohary et al. 1969; Jaaska 1980). Ssp. *tauschii* var. *meyeri* grows in the southwestern and southern Caspian Iran, where there is some overlap with ssp. *strangulata*. Ssp. *tauschii* var *tauschii*, var. *anathera*, and their intermediate types, have a broad distribution and were found in almost all parts of the distribution of the species (Kihara and Tanaka 1958). *Ae. tauschii* is also found in several Chinese locations near the Yellow River in Shaanxi and Henan provinces (Yen et al. 1983). These authors assume that *Ae. tauschii* was introduced to China with emmer wheat (Yen et al. 1983). Adventive presence of *Ae. tauschii* is reported from USA (New York area) and various countries in south, west and central Europe (van Slageren 1994).

Ae. tauschii occupies a variety of different habitats, from steppes and margins of deserts, to the very wet and more temperate forests of the southern Caspian seashores in Iran and to the cool and dry central Asian steppes. This diploid occurs over a strikingly wide range of climatic conditions, from the dry *Artemisia* steppes and margins of deserts to the rain-soaked temperate hyrcanic forest belt of the southern coast of the Caspian Sea, and from the hot plains of southern Iran to the extreme continental climate of the Central Asiatic steppes (Zohary et al. 1969). It grows on a variety of soils, i.e., grey-calcareous steppe, marl, alluvial and sandy soils and at altitudes of 200–1800 m above sea level (Hodgkin et al. 1992), infrequently at higher elevation (up to 3000 m above sea level; Wang et al. 2013b). It grows in primary habitats, being a common component of several types of plant formations, such as open areas of deciduous steppe maquis, dwarf-shrub steppe-like formations and steppical plains, as well as in secondary disturbed habitats such as abandoned fields, edges of cultivation, and roadsides, from which it has developed as a successful weed of cultivated cereals (Zohary et al. 1969).

Ae. tauschii grows sympatrically with *Ae. umbellulata*, *Ae. columnaris*, *Ae. triuncialis*, *Ae. cylindrica*, *Ae. crassa* and *Ae. juvenalis*, and allopatrically with *Amblyopyrum muticum*, *Ae. speltoides*, *Ae. caudata*, *T. monococcum* ssp. *aegilopoides*, *T. urartu*, *T. timopheevii* ssp. *armeniacum*, *T. turgidum* ssp. *dicoccoides*, *Ae. geniculata*, *Ae. neglecta*, *Ae. biuncialis* and *Ae. kotschyi*.

9.5.2.2 Cytology, Cytogenetics, and Evolution

Ae. tauschii is a diploid species (2n = 2x = 14) bearing the D genome (Kihara 1949, 1954). Its organellar genome was designated D by Tsunewaki (1993, 2009) and Wang et al. (1997). It has a small genome, the second smallest in the genus (1C DNA = 5.17 ± 0.087 pg) (Eilam et al. 2007; Table 9.2). Similar results on genome size were previously obtained by Furuta et al. (1986) and Rees and Walters (1965). Smaller amount of 1C DNA (4.17 pg) was reported by Arumuganathan and Earle (1991). The DNA amount reported by Eilam et al. (2007) is equivalent to 5.056 Mbp, since 978 Mbp exist in 1 pg DNA (Doležel et al. 2003). The main part (about 80% or more) was estimated to be repetitive DNA (You et al. 2011).

The karyotype of *Ae. tauschii* is symmetric, consisting of four types of chromosomes (Senyaninova-Korchagina 1932). There is only one pair with satellites on the short arm, and the centromeres in the rest of the pairs are submedian. Later studies by Riley et al. (1958) and Chennaveeraiah (1960), confirmed the findings of Senyaninova-Korchagina (1932), but Chennaveeraiah (1960) recognized five types of chromosomes in the *Ae. tauschii* karyotype. In all analyzed accessions of *Ae. tauschii*, the size of the satellites was very much the same, varying from 0.7 to 0.8 μ (Chennaveeraiah

1960). The short arm that bears the satellite being about 3 times larger than the satellite (Chennaveeraiah (1960). There is, however, a negligible difference in the lengths of the long arms of this satellite pair. This long arm is roughly twice the size of the short arm that bears the satellite. The karyotypic differences between the varieties seem to be very small; there is more uniformity than there are differences.

Teoh et al. (1983) used a DNA sequence that codes for ribosomal RNA in in situ hybridization experiments on the genome of *Ae. tauschii*. This sequence consists of 18S and 25S rRNA genes, with associated spacer DNA (Gerlach and Bedbrook 1979). They found a complete fit between the number of satellited chromosomes and the number of rRNA sites; one pair of chromosomes exhibited one such site. Likewise, Badaeva et al. (1996b) studied the distribution of the 5S and 18S-26S ribosomal RNA gene families on *Ae. tauschii* chromosomes by in situ hybridization with pTa71 (18S-26S rDNA) and pTa794 (5S rDNA) DNA clones. The distribution of hybridization sites with pTa71 in *Ae. tauschii* was unique. Only one major 18S-26S rDNA locus, that was found in the NOR of *Ae. tauschii*, was located on the short arm of chromosome 5D. Similar results were obtained by Appels et al. (1980), Lawrence and Appels (1986) and Lassner et al. (1987). In addition, a minor NOR locus was located on the short arm of chromosome 7D (Badaeva et al. 1996b). Hybridization with pTa794 showed the presence of two 5S rDNA loci in the short arms of chromosomes of group 1 and 5 (Badaeva et al. 1996b). This is similar to the location of these loci in most other diploid *Aegilops* species. The 5S rDNA loci were not associated with NORs.

The chromosomal locations of major and minor NORs and 5S rDNA loci in the D genome of *Ae. tauschii* were identical to those of the D subgenome of hexaploid wheat (Friebe et al. 1992a; Yamamoto 1992b), with the minor NOR on chromosome 7D being polymorphic among different lines.

The first complete investigation of C-banded chromosomes of *Ae. tauschii* was carried out by Teoh and Hutchinson (1983), using an improved C-banding technique. They found that *Ae. tauschii* exhibited a unique pattern, different from that of other *Aegilops* species. In this pattern, each of the seven chromosomes comprising the haploid set of *Ae. tauschii* displayed its own characteristic banding pattern and could be identified individually and clearly (Iordansky et al. 1978; Teoh and Hutchinson 1983). The banding pattern described by Teoh and Hutchinson (1983) is not comparable to that reported by Gill and Kimber (1974) but bears some resemblance to that of Iordansky et al. (1978).

Teoh and Hutchinson (1983) studied C-banding in only one accession of *Ae. tauschii* and therefore, Friebe et al. (1992a) and Badaeva et al. (1996a) extended the study to include more accessions. A generalized C-banded karyotype of *Ae. tauschii* was established based on chromosome

analysis of 15 accessions of *Ae. tauschii* of diverse origins, including the two subspecies, *tauschii* (with the varieties *tauschii*, *anathera* and *meyeri*) and *strangulata* (Friebe et al. 1992a). Whereas only minor variation in C-banding patterns was observed within accessions, a larger amount of polymorphic variation was found between accessions. Yet, this polymorphic variation did not prevent chromosome identification in these accessions. All chromosomes of *Ae. tauschii* contain centromeric bands of almost equal staining intensity and size, while they differ in the amount and distribution of interstitial, subtelomeric and telomeric bands (Teoh and Hutchinson 1983; Friebe et al. 1992a). One accession (TA 2462) was found to be homozygous for a reciprocal translocation involving the complete arms of chromosomes 1D and 7D (Friebe et al. 1992a). In situ hybridization using the D-genome-specific probe, pAs1, confirmed the presence of this translocation in the accession TA 2462. The C-banding pattern of *Ae. tauschii* chromosomes was similar to that of the D-subgenome chromosomes of hexaploid wheat, thus permitting their unequivocal identification and homoeologous groups designations (Friebe et al. 1992a).

Badaeva et al. (1996a) studied genome structure through in situ hybridization with *Ae. tauschii*-derived pAsl probe. As expected, *Ae. tauschii* showed heavy labeling with this probe, as was found previously (Rayburn and Gill 1986; Cabrera et al. 1995). The labeling pattern in this species was chromosome-specific. Yet, intraspecific variation in the distribution of the pAs1 probe was observed (Badaeva et al. 1996a). The labeling pattern in one accession was similar to that of the D subgenome of hexaploid wheat (Rayburn and Gill 1986, 1987; Mukai et al. 1991), whereas two accessions were similar to the second unmodified D subgenome of hexaploid *Ae. crassa*. Comparison of the labeling pattern of pAs1 in *Ae. tauschii* with the distribution of heterochromatin detected by C-banding in this species, showed that *Ae. tauschii* possessed an intermediate amount of C-heterochromatin. On the other hand, using a highly repetitive, 260-bp, non-coding DNA sequence derived from the B subgenome of hexaploid wheat (Hutchinson and Lonsdale 1982), *Ae. tauschii* was shown to have very little labeling with this probe, indicating a small amount of heterochromatin (Teoh et al. 1983).

Genetic diversity of *Ae. tauschii* was further assessed using fluorescence in situ hybridization (FISH) with eleven DNA probes representing satellite and microsatellite DNA sequences as well as the 45S and 5S rRNA gene families and by electrophoretic (EF) analysis of seed storage proteins (gliadins) (Badaeva et al. 2019). A clear genetic differentiation of accessions into groups *strangulata* and *tauschii* was observed. These two groups differ in the presence of microsatellite repeats GAAn and ACTn and in the distribution of satellite DNA families, especially pAs1. Based on similarities of labeling patterns of DNA probes used in the study, they concluded that the *strangulata* group was phylogenetically closest to the D subgenome of common wheat. A comparison of spectra of gliadins revealed the highest similarity of Armenian and Azerbaijani accessions of *Ae. tauschii* to common wheat, which may indicate a contribution of Transcaucasian members of the *strangulata* group to the genetic pool of common wheat.

Since the D subgenome of hexaploid wheat and the genome of *Ae. tauschii* are still homologous, the genome of *Ae. tauschii* serves as an invaluable reference for wheat genetics and genomics as well as an important resource for wheat improvement. Undoubtedly, the usefulness of *Ae. tauschii* as a reference for the structure of wheat D subgenome and for the ability to identify and use important *Ae. tauschii* genes would be further enhanced by a high-quality sequencing of its entire D genome. Still, development of high-quality physical and genetic maps was a necessary step in progressing towards complete sequencing of the *Ae. tauschii* genome. These maps of *Ae. tauschii* are continuously updated, with more sequences and particularly, with detailed information about functional elements.

Genetic maps are based on meiotic recombination frequency between different pairs of genetic or molecular markers, with distances between markers expressed in crossover units. Physical maps use molecular biology techniques to align DNA sequences to construct maps showing the positions of sequences, including genes, relative to one another along the DNA helix axis, with actual physical distance between markers expressed in base pairs (bp). Physical distances between markers have been determined by techniques such as radiation hybrid mapping, fluorescence in situ hybridization (FISH) or, ideally, by automated DNA sequencing. Genome assembly involves a multi-step procedure, in which DNA fragments were cloned, sequenced and, on the basis of the markers they were found to contain, ordered relative to each other and to the genetic map. Obtaining sufficient coverage of the genome involved generating much physical and genetic data so that the two maps could be reconciled. Following assembly, the physical and genetic maps were continuously updated with detailed information about functional elements. For the physical sequence map, the primary annotation task was identification of genes for the genetic map. High-resolution genetic and physical maps serve as the framework for genome sequence assembly. Accession AL8/78 of *Ae. tauschii* subsp. *strangulata* was chosen as the standard accession for the construction of physical and genetic maps.

DNA markers and coding sequences such as cDNAs and ESTs (expressed sequence tag) have been used for construction of genetic maps of *Ae. tauschii*. Kam-Morgan et al. (1989) used restriction fragment length polymorphisms (RFLPs) as genetic markers and determined linkage relationships between RFLP loci. In addition, they demonstrated the use of segregating populations of *Ae. tauschii* for linkage measurements and the use of wheat aneuploid lines to allocate markers to chromosome arms in the D subgenome of *T. aestivum*. Gill et al. (1991b), Lagudah et al. (1991a) and Boyko et al. (1999, 2002) used this strategy to enrich the genetic map of *Ae. tauschii* and that of the D subgenome of *T. aestivum* by a large number of loci. Boyko et al. (2002) carried on building a high-density map of the *Ae. tauschii* by inserting additional loci, including retrotransposon loci and microsatellite and ISSR loci. Comparison of the genetic maps of *Ae. tauschii* with those of the D subgenome of hexaploid wheat confirmed the conserved collinearity between the two D genomes. Yet, accessions of *Ae. tauschii* revealed greater polymorphism than that observed in the D subgenome of common wheat.

Physical map construction necessitates the production of a large number of bacterial artificial chromosome (BAC) clones, sequencing with a next-generation DNA sequencing platform, assembly into long contiguous sequences and anchoring the contigs on a genetic map (Luo et al. 2013). Genes and transposable elements in the assembled sequences are annotated. However, due to the lack of recombination in certain chromosomal regions, genetic mapping alone is not sufficient to develop high-quality marker scaffolds for a sequence ready physical map. Radiation hybrid mapping has proven to be a successful approach for developing marker scaffolds for sequence assembly in the genome of *Aegilops tauschii*. This method offers much higher and more uniform marker resolution across the length of the chromosome, compared to genetic mapping, and does not require marker polymorphism per se, as it is based on a presence vs. absence marker assay (Kumar et al. 2012, 2015). Kumar et al. (2012) reported the development of high-resolution radiation hybrid maps for the genome of *Ae. tauschii* accession AL8/78, the standard accession for the construction of physical and genetic maps, which were then used for anchoring unassigned sequence scaffolds. Their study demonstrated how radiation hybrid mapping, which offers high and uniform resolution across the length of the chromosome, can facilitate complete sequence assembly of large and complex plant genomes.

Several physical maps of *Ae. tauschii* were constructed in the last decades (e.g., Fleury et al. 2010; Massa et al. 2011; Kumar et al. 2012, 2015; Luo et al. 2013; Hastie et al. 2013; Zhu et al. 2016). The assembled scaffolds of high-quality sequences in these maps represent more than 80% of the D genome, of which about 66% of the sequences are comprised of repetitive elements. More than 43,000 protein-coding genes were identified, 71.1% of which were uniquely anchored to chromosomes. Genes, pseudogenes and transposable elements were annotated.

The Department of Plant Sciences, University of California, Davis and Genomics and Gene Discovery Unit, USDA/ARS Western Regional Research Center, Albany,

California, runs a database which accumulates and releases up-to-date information on genome mapping and sequencing, genetic and physical maps, genetic markers and genomic sequences of *Ae. tauschii*. Gaining sufficient coverage of the *Ae. tauschii* genome may pave the way to the assembly of a high-quality sequence of the entire genome.

Sequencing of more than 90% of the *Ae. tauschii* genome was recently achieved by Jia et al. (2013). They identified 43,150 protein-coding genes and found that more than 66% of the *Ae. tauschii* genome was composed of 410 different transposable element families, of which the 20 most abundant comprised more than 50% of the genome.

However, even though sequencing whole plant genomes has advanced rapidly during the last decades, with the development of next-generation sequencing (NGS) technologies and bioinformatics (Bierman and Botha 2017), complete sequence assembly of the *Ae. tauschii* genome was problematic due to its large size and about 80% repetitive DNA (Jia et al. 2013). Nevertheless, using an array of advanced technologies, Luo et al. (2017) succeeded to obtain a reference-quality genome sequence for *Ae. tauschii* ssp. *strangulata*, accession AL8/78 which is closely related to the wheat D subgenome (see Chap. 3, Sect. 3.7). They show that the *Ae. tauschii* genome contains unprecedented amounts of very similar repeated sequences, a greater number of dispersed duplicated genes than other sequenced genomes, and its chromosomes have been structurally evolving an order of magnitude faster than those of other grass genomes. The decay of colinearity with other grass genomes correlates with recombination rates along chromosomes. These authors propose that the vast amounts of very similar repeated sequences cause frequent errors in recombination and lead to gene duplications and structural chromosome changes that drive fast genome evolution.

Zhao et al. (2017) generated a chromosome-scale, high-quality reference genome of *Ae. tauschii*, in which 92.5% sequences have been anchored to chromosomes. Using this assembly, they accurately characterized genic loci, gene expression, pseudogenes, methylation, recombination ratios, microRNAs and especially TEs on chromosomes. In addition to the discovery of a wave of very recent gene duplications, the authors revealed that TEs occurred in about half of the genes, and found that such genes are expressed at lower levels than those without TEs, presumably because of their elevated methylation levels. All wheat molecular markers and mapped allowing the authors to construct a high-resolution integrated genetic map corresponding to genome sequences, thereby placing previously detected agronomically important genes/ quantitative trait loci (QTLs) on the *Ae. tauschii* genome for the first time.

9.5.2.3 Crosses with Other Species of the Wheat Group

Intraspecific F_1 hybrids between accessions of *Ae. tauschii* that were collected from different regions, had regular chromosome pairing at meiosis (seven bivalents) and high fertility in most cases, but some hybrids were partially sterile (Kihara et al. 1965). Yet, F_1 hybrids involving one accession from Iran, had a ring of four chromosomes and five bivalents, indicating the presence of a reciprocal translocation in this accession (Kihara et al. 1965). Lagudah et al. (1991b) and Hohman and Lagudah (1993) found that this reciprocal translocation involves chromosomes 1DS-7DL and 7DS-1DL.

Chromosomal pairing at first meiotic metaphase of F_1 hybrids between *Ae. tauschii* and other species of the wheat group are characterized by relatively low pairing. The F_1 hybrid *Ae. tauschii* x *Ae. bicornis* had a low to intermediate level of chromosomal pairing at meiosis (0–6 bivalents and 0–1 trivalents; Kihara 1949). Evidently the D genome of *Ae. tauschii* has diverged quite considerably from the S^b genome of *Ae. bicornis*. Likewise, chromosomal pairing at first meiotic metaphase of F_1 hybrids between *Ae. tauschii* and *Ae. caudata* had 3–5 bivalents (Kihara 1949) or 3.60 bivalents, of which 0.60 were ring, 1.06 trivalents and 0.04 quadrivalents (Sears 1941b), indicating homoeology rather than homology of the genomes of these two diploid species. Divergence of the genome of *Ae. tauschii* from that of *Ae. uniaristata* (genome NN) is obvious from the low pairing in the F_1 hybrids between these two species. Kihara (1949) observed 1–5 bivalents and 0–1 trivalents and Sears (1941b) reported 3.82 bivalents, of which 0.20 were ring bivalents, and 0.14 trivalents. The F_1 hybrid between *Ae. umbellulata* (Genome UU) and *Ae. tauschii* had 3–5 bivalents with a mode of 3 bivalents and 1–3 trivalents (Kihara 1949), indicating low homology between the genomes of these two species. Pairing in the F_1 hybrid between *Ae. geniculata* (genome M^oM^oUU) and *Ae. tauschii* was also low, with 5–8 bivalents and 0–1 trivalents, (Kihara 1949), indicating that the two subgenomes of *Ae. geniculata* are not homologous to that of *Ae. tauschii*.

Chromosome pairing in *Ae. tauschii* x *A. muticum* hybrids was almost regular and exceeded the pairing in hybrids between *Ae. mutica* and Sitopsis species and between *A. muticum* and diploid *Triticum* species (Jones and Majisu 1968). Similarly, Ohta (1990) produced F_1 hybrids between *Ae. tauschii* and *A. muticum* (presumably without B chromosomes) and observed a relatively high amount of chromosomal pairing at first meiotic metaphase (Table 8.1). This high pairing presumably resulted from the homoeologous-pairing promoters of *A. muticum*. And indicates *that* the chromosomes of *A. muticum* appear to have considerable homoeology with the D genome of *Ae. tauschii*.

Data of chromosomal pairing in F$_1$ hybrids between *Ae. tauschii* and the allotetraploid species of section Vertebrata, namely, *Ae. crassa* 4x and *Ae. ventricosa*, are presented in Table 9.5. These pairing data and that of Kihara (1949) clearly show that the two allotetraploids contain one subgenome that is very closely related to the genome of *Ae. tauschii*. Moreover, the similarity of the spike morphology of these allotetraploids to that of *Ae. tauschii* further supports the conclusion that *Ae. tauschii* is the donor of one of the two subgenomes of these allotetraploids. Chromosomal pairing in the F$_1$ hybrids between *Ae. tauschii* and the allohexaploid species of section Vertebrata, namely, *Ae. crassa* 6x, *Ae. juvenalis*, and *Ae. vavilovii*, indicates that hexaploid *Ae. crassa* (genome D^{c}D^{c}X^{c}X^{c}DD) contains one unaltered and one modified subgenomes of *Ae. tauschii*, as suggested by Kihara (1949), whereas the other two allohexaploids, *Ae. vavilovii* (genome D^{c}D^{c}X^{c}X^{c}S^{s}S^s) and *Ae. juvenalis* (genome D^{c}D^{c}X^{c}X^{c}UU), contain a modified subgenome of *Ae. tauschii* (Kimber and Zhao. 1983) (Table 9.6). On the other hand, chromosomal pairing in the F$_1$ hybrid of *Ae. cylindrica* (genome DDCC) x *Ae. tauschii* showed 8–9 bivalents and up to 5 trivalents (Kihara 1949), indicating that one of the two genomes of *Ae. cylindrica* was donated by *Ae. tauschii*.

Chromosomal pairing in F$_1$ hybrids with the two subspecies of diploid wheat, *T. monococcum* (genome A^{m}A^m) domesticated ssp. *monococcum* and wild ssp. *aegilopoides*, showed that the genome of *Ae. tauschii* diverged from those of diploid wheat although somewhat less than its divergence from the genomes of other diploid *Aegilops* species (Table 9.8). Chromosomal pairing in F$_1$ hybrids between *Ae. tauschii* and the two allotetraploid species of *Triticum*, namely, the wild subspecies of *T. turgidum*, ssp. *dicoccoides* (genome BBAA), and the domesticated subspecies of *T. timopheevii* ssp. *timopheevii* (genome GGAA) showed that the subgenomes of the *Triticum* allotetraploids are not homologous to that of *Ae. tauschii*, although the pairing in the *timopheevii* x *tauschii* hybrid was somewhat higher than in the *dicoccoides* x *tauschii* hybrid (Table 9.9). On the other hand, chromosomal pairing in the F$_1$ hybrid with allohexaploid wheat, *T. aestivum* ssp. *aestivum*, showed that the allohexaploid contains one subgenome homologous to that of *Ae. tauschii* (Table 9.10).

9.5.3 *Aegilops ventricosa* Tausch

9.5.3.1 Morphological and Geographical Notes

Ae. ventricosa Tausch [Syn: *Triticum ventricosum* (Tausch) Ces., Pass. & Gibelli; *Gastropyrum ventricosum* (Tausch) Á. Löve] is an annual plant with few to many rather thick culms, that are 20–40-cm-high (excluding spikes). Its entire length is foliated with broad linear glabrous, seldom hairy, leaves. Spikes are long, moniliform, 4–6 (rarely up to 12)-cm-long, excluding awns, with 5–10 spikelets, more or less rough, and usually awned. Rudimentary spikelets are absent, rarely 1–2 are found. The spike disarticulates into individual spikelets, each falling with its adjacent rachis segment (barrel-type dispersal unit). Sometimes, the entire spike disarticulates at the base. Spikelets are oval, equal in length to the rachis segment, and suddenly become inflated in the lower parts (urn–shaped). There are 4–5 florets; the upper 1–3 are sterile. The glumes are strongly overlapping, with curved nerves, tips somewhat thickened with two teeth separated with a broad sinus; one tooth may be lengthened into an awn (up to 3-cm-long), that is usually shorter than the lemma awn (up to 4-cm-long). Tips of the glumes of the terminal spikelet are 3-toothed, the center one usually elongated into an awn. The lemmas are membranous, thickened in the upper parts, with weak keel at the tip, which is elongated into an awn, at the base of which are 1–2 small teeth. The awns are triangular, with lemma awns stronger than glume awns. All awns become longer toward the tip of the spike. The caryopsis adheres to the lemma and palea (Fig. 9.2c).

Ae. ventricosa exhibits limited morphological variation, mainly in narrowness of the upper part of the spikelets, spike length and width, spikelet number and awning. Some of the spike and spikelet characters are similar to those of *Ae. uniaristata*, one of the two diploid progenitors of this allotetraploid, but basic characters of the spikes and spikelets indicate that *Ae. ventricosa* morphology derived from *Ae. tauschii*, the second diploid parent of this allotetraploid species.

Ae. ventricosa is a Mediterranean element. It grows in the western and northern parts of the Mediterranean Sea, namely, Portugal, Spain, South France, Corsica, and Sardinia, rarely Italy and Sicily, Egypt (near Alexandria, rare), Libya, Tunisia, Algeria, and Morocco (Table 9.2). *Ae. ventricosa* grows on terra rossa, rendzina and light, sandy soils in the edges and openings of deciduous and sclerophyllous Mediterranean forests and maquis, in degraded dwarf-shrub and semi-steppical formations, pastures, abandoned fields, edges of cultivation, disturbed habitats and roadsides. It is common in a wide array of habitats and invades wheat fields, vineyards, and olive groves as weed. *Ae. ventricosa* develops well in areas with 200–350 mm annual rainfall but was also found in areas with less than 100 mm and up to 600 mm annual rainfall. It is found in a range of altitudes, from sea level to 1850 m above sea level.

Ae. ventricosa has a medium-sized distribution in the western part of the distribution of the genus. It is relatively isolated from the rest of the group. Interestingly, its distribution does not overlap with that of either of its diploid progenitors (*Ae. tauschii* and *Ae. uniaristata*), and it is larger than that of *Ae. uniaristata*, but smaller than that of *Ae.*

tauschii. Its distribution area is closer to the distribution area of *Ae. uniaristata* than to that of any other diploid species (excluding *Ae. bicornis*, with which it may have contact in Libya). *Ae. ventricosa* distributes sympatrically with other *Aegilops* tetraploid species, i.e., *Ae. geniculata*, *Ae. neglecta*, *Ae. biuncialis*, *Ae. triuncialis*, *Ae/ perigrina* and *Ae. cylindrica*, and allopatrically with *Ae. bicornis*, *Ae. recta* and *Ae. kotschyi* (in Libya and Tunisia). Natural hybridizations are known with many species, including domesticated tetraploid and hexaploid wheat. Several economically important traits, such as resistance to eyespot, stripe-, leaf- and stem-rust, and tolerance to aluminum, were discovered in *Ae. ventricosa* (e.g., Maia 1967; Dosba et al. 1980), some of which were already transferred to common wheat (e.g., Jahier et al 1978, 1996; Bariana and Mcintosh 1993; Tanguy et al. 2005).

9.5.3.2 Cytology, Cytogenetics, and Evolution

Ae. ventricosa is an allotetraploid species (2n = 4x = 28; genome DDNN) that evolved as a result of hybridization between two different diploid *Aegilops* species, followed by chromosome doubling. The existence of a D subgenome in *Ae. ventricosa* was first directly proved by the observation by Kihara, in 1938, of seven ring bivalents in the *Ae. tauschii* x *Ae. ventricosa* F$_1$ hybrid (see Kihara 1949). The second subgenome of this allotetraploid was first believed to be a modified M genome, designated M^v, derived from *Ae. comosa*, which underwent changes at the polyploid level (Kihara 1949, 1954, 1963; Kihara and Lilienfeld 1932). However, Chennaveeraiah (1960), Kimber et al. (1983), Kimber and Zho (1983), Zhao and Kimber (1984) and Yen and Kimber (1992b) demonstrated that the second subgenome in *Ae. ventricosa* is actually a N genome from *Ae. uniaristata*. Thus, it is currently accepted that the genomic constitution of *Ae. ventricosa* is DDNN (Kimber and Tsunewaki 1988; Dvorak 1998). Its organellar genome is similar to that of *Ae. tauschii*, which was the female parent in the formation of *Ae. ventricosa*, and consequently, was designated D by Ogihara and Tsunewaki (1988) and Wang et al. (1997).

Matsumoto and Kondo (1942) synthesized an amphiploid of *Ae. tauschii* – *Ae. uniaristata*, and Matsumoto et al. (1957) synthesized the amphiploid of the reciprocal combination. Morphologically, both synthetic amphiploids were very similar to *Ae. ventricosa*. The synthetic amphiploids exhibited almost regular pairing at meiosis and had about 60% seed fertility. Hybrids between the synthetic and natural *Ae. ventricosa* exhibited chromosomal pairing at meiosis that was quite good (average of 12 bivalents and 0–3 trivalents or quadrivalents per cell). In spite of the high pairing, seed fertility was very low (3–5%) (Matsumoto et al. 1957).

Ae. ventricosa has 10.64 pg 1C DNA (Eilam et al. 2008; Table 9.3). This allotetraploid species has 3.18% less DNA than expected from the DNA sum of its two diploids parents, i.e., 10.99 pg (*Ae. tauschii* contains 5.17 pg and *Ae.*

uniaristata contains 5.82 pg; Eilam et al. 2007). The loss of DNA in the allotetraploid was confirmed by Badaeva et al. (2012) who, based on differential C-banding and in situ hybridization, found that *Ae. ventricosa* exhibits substantial structural chromosome rearrangements, including deletion of chromosomal segments and reduction of heterochromatin content.

Earlier karyomorphological studies (Emme 1924; Sorokina 1928; Senyaninova-Korchagina 1932) did not find satellites in *Ae. ventricosa*. Since there is no species of *Aegilops* that is without at least one pair of satellites, it is quite unlikely that this species is without any satellites. Indeed, the karyotype of this allotetraploid species contains one chromosome pair with a satellite on the short arm (Chennaveeraiah 1960). In agreement with this finding, Orellana et al. (1984), using a silver-staining procedure, analyzed the activity of the nucleolar organizer of *Ae. ventricosa* and detected only one pair of Ag-NORs, indicating that natural amphiplasty occurs in this allotetraploid species.

The karyotype of *Ae. ventricosa* is asymmetric, consisting of five pairs with submedian-subterminal centromeres and the rest with submedian centromeres (Chennaveeraiah 1960). One half of the set consist of seven pairs with submedian centromeres, resembling the chromosome set of *Ae. tauschii*, whereas the other seven pairs closely resemble the set in *Ae. uniaristata* (Chennaveeraiah 1960). Hence, the karyotype of the D subgenome chromosomes of *Ae. ventricosa* is symmetric, consisting of metacentric and submetacentrics chromosomes, whereas the karyotype of the N subgenome is asymmetric; five chromosomes are subtelocentrics (Chennaveeraiah 1960). Though this chromosome set agrees mostly with the chromosome set of *Ae. uniaristata*, it differs from it in the absence of a satellite pair (Chennaveeraiah 1960; Badaeva et al. 2012). Therefore, the N subgenome of *Ae. ventricosa* is not completely homologous with the genome of its diploid progenitor *Ae. uniaristata*, but it is slightly modified.

Meiotic pairing analysis revealed only minor modification of the D and N subgenomes in *Ae. ventricosa* (Kimber et al. 1983; Kimber and Zhao 1983; Zhao and Kimber 1984; Yen and Kimber 1992b). Yet, Badaeva et al. (2002, 2011, 2012), using the C-banding technique and fluorescence in situ hybridization with clones pTa71 (18S-5.8S-26S rDNA), pTa794 (5S rDNA), both clones from *T. aestivum*, and pAs1 (non-coding repetitive DNA sequence from *Ae. tauschii*) as probes, found that the N subgenome of *Ae. ventricosa* substantially differed from that of *Ae. uniaristata*. The D subgenome also differed, albeit to a lesser extent, from the genome of *Ae. tauschii*. They observed polymorphism for the presence and size of C-bands and for chromosome aberrations in these subgenomes, including minor changes in the C-banding and pAs1-FISH patterns, complete deletion of the NOR on chromosome 5D and the loss of several minor

18S-5.8S-26S rDNA loci on N subgenome chromosomes. In addition, several chromosomal translocations involving different chromosomes and a pericentric inversion of chromosome 5N, were identified in several accessions of *Ae. ventricosa* (Bardsley et al. 1999; Badaeva et al. 2012).

Cuñado et al. (1986) pointed out that two chromosome groups in *Ae. ventricosa* have a distinct C-banding pattern, i.e., 16 chromosomes with prominent C-heterochromatin located in centromeric and pericentromeric regions and 12 almost entirely euchromatic chromosomes. These differences may distinguish between the N and D subgenomes, respectively, although at least one of the heterochromatic chromosomes should belong to the D genome. These different C-banding patterns allowed for estimation of the frequencies of homologous and homoeologous pairing in the hybrids. The N subgenome chromosomes of *Ae. ventricosa* are also characterized, in addition to their distinct asymmetry, by large pericentric heterochromatin complexes, like in *Ae. uniaristata* (Badaeva et al. 2012). In contrast, the D subgenome chromosomes of *Ae. ventricosa* have only a few small, predominantly intercalary C bands, like the chromosomes of *Ae. tauschii*. Yet distinct reproducible changes associated with chromosome modification that accompanied the formation of the allopolyploid species, were observed in the two *Ae. ventricosa* subgenomes. As apparent from the appearance or disappearance of C-bands and from changes in band size, the heterochromatin content decreased in some N subgenome chromosomes (1N, 3N, and 7N) and increased in others (2N, 4N, 5N, and 6N). The greatest changes in C banding pattern were observed for chromosome 3N, which lost the large pericentric heterochromatin block in the short arm and the distal C-band marker in the long arm, while several new bands appeared instead (Badaeva et al. 2012). In situ hybridization with the *pAs1* clone revealed changes in the N subgenome of *Ae. ventricosa*, manifested by fewer hybridization sites or less intense signals (Badaeva et al. 2012). The greatest changes were observed for chromosomes 1N, 4N, 6N, and 7N. In addition, the minor NORs of chromosomes 1N and 7N were partly lost, but new, earlier unidentified minor 45S rDNA sites, appeared in the subterminal regions of the short and long arms of *Ae. ventricosa* chromosome 3D (Badaeva et al. 2011).

Bardsley et al. (1999) analyzed lines of *Ae. ventricosa* using fluorescent in situ hybridization with probes including rDNA, repeated sequences from wheat and rye, simple-sequence repeats (SSRs) and total genomic DNA. They found that the banding patterns could be used to distinguish most chromosome arms of this allotetraploid. All lines had a single major 18S-25S rDNA site, the nucleolar organizing region (NOR) in chromosome 5N and several minor sites of 18S-25S rDNA and 5S rDNA. A 1NL.3DL-1NS.3DS translocation was identified, and other minor differences were found between the lines.

9.5.3.3 Crosses with Other Species of the Wheat Group

Crosses between *Ae. ventricosa* with its parental diploids, namely, *Ae. ventricosa* x *Ae tauschii* and *Ae. ventricosa* x *Ae uniaristata*, showed that the allotetraploid contains one subgenome similar to that of *Ae. tauschii* and another alike the genome of *Ae. uniaristata* (Table 9.5). The F$_1$ hybrid *Ae. ventricosa* x *Ae. tauschii* had 6 bivalents and 1 trivalent at meiosis (Kihara 1949; Kimber and Zhao 1983) and the F$_1$ hybrid between *Ae. ventricosa* and *Ae. uniaristata* also had a similar level of pairing, i.e., 6 bivalents and a trivalent (Kimber et al. 1983). Since the F$_1$ hybrid between the two diploid progenitors of *Ae. ventricosa*, *Ae. tauschii* and *Ae. uniaristata*, exhibited reduced pairing, i.e., 3.82 bivalents (Sears 1941b), indicating homoeologous relationships between the D and the N genomes, the pairing in the triploid hybrids between *Ae. ventricosa* and its parental diploids was rather homologous. The chromosome pairing in the tetraploid hybrid between *Ae. ventricosa* and *Ae. crassa* (4x) (genome D^{c}D^{c}X^{c}X^c) (about 5 bivalents and up to 2 trivalents; Kimber and Zhao 1983) also indicates that the two allotetraploid species share one subgenome, most probably D (Table 9.7).

Chromosome pairing in the F$_1$ hybrid between tetraploid wheat (*T. turgidum* subsp. di*coccoides*) (genome BBAA) and *Ae. ventricosa* was very low [0–2 (mode of 0) bivalents per cell] (Kihara 1949). On the other hand, the F$_1$ hybrid between hexaploid wheat, *T. aestivum* ssp. *aestivum* (genome BBAADD), and *Ae. ventricosa* showed 5 bivalents and up to 2 trivalents (Kimber and Zhao 1983), indicating that allohexaploid wheat has one subgenome homologous to one of the subgenomes of *Ae. ventricosa*. Since the hybrid with tetraploid wheat had very little pairing, the good pairing in the hybrid with hexaploid wheat is between the D subgenomes of ssp. *aestivum* and that of *Ae. ventricosa*.

9.5.4 *Aegilops crassa* Boiss.

9.5.4.1 Morphological and Geographical Notes

Ae. crassa Boiss. (Persian goat grass) [Syn: *Triticum crassum* (Boiss.) Aitch. & Hemsl.; *Triticum syriacum* Bowden; *Ae. platyathera* Jaub. & Spach; *Castropyrum crassum* (Boiss.) Á. Löve] is an annual robust plant with many-jointed culms, sometimes thick culms, that are 20–40-cm high (excluding spikes) and often foliated along the entire length. The spikes are usually long (4–10-cm, excluding awns), with 6–10 spikelets, thick, cylindrical, somewhat like a string of beads (moniliform), tapering toward the tip, usually hairy, with barrel-shaped disarticulation, and each spikelet falling with its adjacent rachis segment. Rudimentary spikelets are absent, or rarely, there are 1–2. There are 3–5 florets, with the upper 1–3 being sterile. Glumes are somewhat or not fully overlapped, covered with fine silvery hairs, with a tip blunt, thickened, with 1–4, usually 2, teeth, that are separated by a broad and shallow sinus, and usually not awned or seldom with a short awn. Glume awns are slender, triangular and weaker than the lemma awns of the same spikelet. Lemmas are membranous or cartilaginous, somewhat keeled, usually tipped with a tooth or an awn, flanked by 1–2 lateral teeth. The lemma awns are mostly broad and more strongly developed in the upper spikelets. The caryopsis adheres to lemmas and palea (Fig. 9.2d).

Ae. crassa is morphologically closer to *Ae. tauschii* than *Ae. ventricosa*. It is a most variable species, varying in all the diagnostic elements, such as spike and spikelet size, form, structure, color, glume and lemma awn development, form, and place of attachment. The wide morphological variation of *Ae. crassa* led Eig (1929a) to recognize two varieties within this species: var. *typica* and var. *palaestina* (= ssp. *vavilovii* Zhuk.). [The later was elevated to the specific rank by Chennaveeraiah (1960), who designated it *Ae. vavilovii* (zhuk.) Chennav.]. Hammer (1980) split *Ae. crassa* into two subspecies: ssp. *crassa*, with three varieties (var. *crassa* (=var. *typica*), var. *glumiaristata* Eig. and var. *macrathera* Boiss.) and subsp. *vavilovii* Zhuk. Var. *typica* was subdivided by Hammer (1980) into f. *crassa*, f. *rubiginosa* (Popova) Hammer and f. *fuliginosa* (Popova) Hammer. Indeed, a very wide variation in *Ae. crassa* from Iran was found in morphological elements (Ranjhar et al. 2007; Bordbar and Rahiminejad 2010), in simple sequence repeat (SSR) DNA markers (Naghavi et al. 2009a, b), in SSR and ISSR markers (Moradkhani et al. 2015), in nuclear microsatellite loci, nuclear rDNA ITS, and in chloroplast *trn*L-F sequences (Bordhar et al. 2011). The wide variation of *Ae. crassa* may has accumulated throughout the lifetime of this species, which is considered very old (Tsunewaki 1993, 2009).

Ae. crassa is a steppical (Irano-Turanian) element, penetrating into semi-desert regions. It has a relatively large distribution area in the eastern part of the distribution of the genus, i.e., in western Asia. It grows in Transcaucasia (Georgia, Armenia, and Azerbaijan), southern Turkmenistan, southern Uzbekistan, southernmost part of Kazakhstan, northern Tajikistan, western Kyrgyzstan, Afghanistan, Iran, central and northern Iraq, northern and northeastern Syria and southeastern Turkey. It is rarely present in Jordan and Lebanon. *Ae. crassa* inhabits a wide range of primary and secondary habitats. It grows on grey-calcareous, loess, and alluvial soils and on stony slopes and gravel, in degraded deciduous steppe maquis, Juniperous forests, dwarf-shrub steppe-like formations, steppical plains, wadis, edges of cultivation, disturbed habitats and roadsides. It is a common weed of cultivation.

Ae. crassa is considered a drought-tolerant species because it grows in areas with 150–350 mm rainfall (Kilian

et al. 2011). In this regard, Harb and Lahham (2013) reported that several accessions of *Ae. crassa* from semiarid and arid areas in Jordan, are drought-tolerant and can be used to improve this trait in domesticated wheat. (A likely possibility is that these authors refer to *Ae. vavilovii*).

The distribution area of *Ae. crassa* is somewhat smaller than that of *Ae. tauschii*, but much larger than that of the M-genome species (Kihara 1963). It grows sympatrically with the following species: *Ae. columnaris*, *Ae. triuncialis*, *Ae. tauschii*, *Ae. cylindrica* and *Ae. juvenalis*, and allopatrically with *Ae. speltoides*, *Ae. caudata*, *Ae. umbellulata*, *Ae. geniculata*, *Ae. biuncialis*, *Ae. neglecta*, *Ae. vavilovii*, wild *T. monococcum* (subsp. *aegilopoides*), and wild *T. timopheevii* (subsp. *armeniacum*).

Ae. crassa contains two cytotypes, an allotetraploid (2n = 4x = 28; genome $D^cD^cX^cX^c$) and an auto-allohexaploid (2n = 6x = 42; genome $D^cD^cX^cX^cDD$) (Table 9.3), that are very similar morphologically. Yet, there is a morphological difference between the two cytotypes; the allotetraploid has more robust spikes and has a greater tendency to display moniliform spikes, while the auto-allohexaploid has more or less cylindrical spikes (Kihara 1963).

The allotetraploid cytotype was found throughout the species distribution area, whereas the distribution of the auto-allohexaploid cytotype is restricted to northern Afghanistan and northeastern Iran (Kihara 1963). This difference in the size of the distribution areas presumably results from the age of the two cytotypes; the tetraploid is an old polyploid, while the hexaploid is much younger (Tsunewaki 1993, 2009). In northern Afghanistan and northeastern Iran, there are mixed populations of the hexaploid and the tetraploid cytotypes. It is assumed therefore, that the hexaploid cytotype was formed there. Variation analysis of the restriction profiles of nuclear repeated nucleotide sequences support this assumption (Dubcovsky and Dvorak 1995a). The allotetraploid *Ae. crassa* appears to be an ancient allotetraploid (Tsunewaki 1993, 2009) that originated from hybridization of primitive *Ae. tauschii* with an ancient species in the evolutionary lineage, leading to the section Sitopsis (Dubcovsky and Dvorak 1995a). The auto-allohexaploid cytotype is a young taxon. According to Kihara (1954), the tetraploid originated in Asia Minor and the hexaploid in northern Afghanistan.

9.5.4.2 Cytology, Cytogenetics, and Evolution

The allotetraploid cytotype of *Ae. crassa* (2n = 4x = 28; genome $D^cD^cX^cX^c$) evolved as a result of hybridization between two different diploid *Aegilops* species, and the auto-allohexaploid cytotype (2n = 6x = 42; genome $D^cD^cX^cX^cDD$) was formed as a result of hybridization between the allotetraploid cytotype and *Ae. tauschii*. The F_1 hybrids of both hybridizations, i.e., between the two diploid parents of the allotetraploid and between the allotetraploid and *Ae. tauschii*, are sterile, but chromosome doubling led to the production of fertile polyploids. The existence of a D subgenome in the allotetraploid *Ae. crassa* was demonstrated by the observation of seven ring bivalents in the F_1 hybrid between *Ae. crassa* 4x and *Ae. tauschii* (Kihara 1949, 1957). The C-banding and ISH results reported by Badaeva et al. (1998) support the genome analysis data of Kihara (1949), demonstrating that one of the two subgenomes of 4x *Ae crassa* was derived from the D genome of *Ae. tauschii*. Yet, the D subgenome of the allotetraploid differentiated from that of the diploid progenitor (Dubcovsky and Dvorak 1995a) and consequently, was designated D^c by Dvorak (1998). The second subgenome of this allotetraploid was first believed to be a modified M genome of *Ae. comosa*, designated M^{cr} (Kihara 1940a, 1957, 1963; Lilienfeld 1951; Kihara et al. 1959; Kihara and Tanaka 1970; Kimber and Feldman 1987), but analysis of variation in nuclear repeated nucleotide sequences (Dvorak 1998 and reference therein) showed this subgenome to differ from the M genome and instead, to be related to genome S of the Sitopsis species. It is assumed therefore, that the X^c genome derived from an extinct ancestor of section Sitopsis and was tentatively designated X^c until its equivalence to a Sitopsis genome will be verified by further studies (Dubcovsky and Dvorak 1995a; Dvorak 1998). Yet, the X^c subgenome exhibits a large number of intra-subgenomic rearrangements, making it difficult to identify the origin of this subgenome (Badaeva et al. 1998). Hexaploid *Ae. crassa* derived from hybridization between the tetraploid cytotype of *Ae. crassa* and *Ae. tauschii*. The speciation of 6x *Ae. crassa* was accompanied by a reciprocal translocation which is specific for the 6x cytotype. The distinct genetic differences between the tetraploid and hexaploid cytotypes of *Ae. crassa* would justify classifying them as different subspecies (Badaeva et al. 1998) or even different species.

The D^c subgenomes of 4x and 6x *Ae. crassa* are highly modified compared with the D genome of the progenitor species *Ae. tauschii* (Kihara 1940a, 1957; Lilienfeld 1951; Siddique and Jones 1967; Kihara and Tanaka 1970; Chapman and Miller 1978; Nakai 1982; Kimber and Zhao 1983; Zhao and Kimber 1984; Kimber and Feldman 1987; Zhang and Dvorak 1992; Tsunewaki 1993). In situ hybridization with the D-genome-specific DNA clone pAs1 confirmed that the D^c subgenome of tetraploid *Ae. crassa* is significantly different from the ancestral D genome of *Ae. tauschii* (Badaeva et al. 1998). The D subgenome of 6x *Ae. crassa* is different from the D^c subgenome and is similar to the D-genome of *Ae. tauschii*.

The organellar genome of the allotetraploid cytotypes of *Ae. crassa* is a subtype of the organellar genome D of *Ae. tauschii*, designated D^2 (Ogihara and Tsunewaki 1988; Wang et al. 1997). It is similar to that of *Ae. tauschii*, which was the female parent in the formation of 4x *Ae. crassa*, but

diverged from it somewhat. The organellar genome of the hexaploid cytotype is similar to that of the allotetraploid and, consequently, designated D^2 (Ogihara and Tsunewaki 1988; Wang et al. 1997). This indicates that the allotetraploid was the female parent in the hybridization that led to the formation of the auto-allo-hexaploid.

Analysis of meiotic chromosome pairing in F_1 hybrids between 4x and 6x *Ae. crassa* confirmed that the hexaploid forms were derived from a hybridization between 4x *Ae. crassa* and *Ae. tauschii* (Kihara et al. 1959; Kihara 1963; Kimber and Zhao 1983; Zhao and Kimber 1984). The F_1 hybrid between the two cytotypes showed up to 14 bivalents and several multivalents (Kihara 1963), indicating that genome $D^cD^cX^cX^c$ of the tetraploid is homologous to the $D^cD^cX^cX^c$ subgenomes of the hexaploid cytotype, and that the D^c subgenome of the hexaploid is related to the D^c subgenome of the tetraploid.

A synthetic auto-allohexaploid was produced by crossing the allotetraploid cytotype of *Ae. crassa* with *Ae. tauschii* (Shigenobu and Sakamoto 1977). This synthetic hexaploid (genome $D^cD^cX^cX^cDD$) closely resembles the natural auto-allohexaploid cytotype of *Ae. crassa* (Kihara 1963). The F_1 hybrid between the synthetic and the natural hexaploid had 19 bivalents at meiosis and had high fertility (89%) (Kihara 1963). Since there are mixed populations of tetraploid *Ae. crassa* and *Ae tauschii* in northern Afghanistan and northeastern Iran, they had many opportunities to hybridize and to form the hexaploid cytotype (Kihara (1963).

The allotetraploid cytotype of *Ae. crassa* has 10.86 pg 1C DNA (Eilam et al. 2008; Naghavi et al. 2013) and the auto-allohexaploid has 15.90 pg 1C DNA (Naghavi et al. 2013) (Table 9.3). The expected amount of 1C DNA in the tetraploid cannot be calculated since the donor of the X^c subgenome has yet to be identified. The expected amount of 1C DNA in the hexaploid cytotype is 16.03 pg (10.86 pg of 4x + 5.17 pg of *Ae. tauschii;* Eilam et al. 2007), 0.82% higher than the content in the natural hexaploid Ae. crassa. The loss of DNA in the two cytotypes was confirmed by Badaeva et al. (1998) who, based on differential C-banding and in situ hybridization, found that *Ae. crassa* exhibits substantial structural chromosome rearrangements, including deletion of chromosomal segments and reduction of heterochromatin content.

The correct number of chromosomes in the allotetraploid *Ae. crassa* was first reported by Emme (1924), but this report did not mention the presence of satellites. Chennaveeraiah (1960) described the correct karyotype, which is symmetric, consisting of two pairs with satellites of different sizes on short arms, one pair with a median centromere, and one pair with an almost median centromere. Three chromosome pairs with satellites exist in the hexaploid *Ae. crassa*. The remaining chromosomes in both cytotypes have submedian centromeres. There are no subterminal centromeres in the set.

With the aim of analyzing the activity of the nucleolar organizer regions (NORs), Cermeño et al. (1984b performed a comparative analysis of somatic metaphase chromosomes in 4x and 6x *Ae. crassa* by phase contrast, C-banding and Ag-staining. They found that the nucleolar activity of chromosome "C" of the X^c subgenome was much higher than that of chromosome "B" of the D^c subgenome, indicating the incidence of partial amphiplasty. The NORs activity of the D subgenome chromosomes are also inhibited in the 6x *Ae. crassa* (Cermeño et al. 1984b). ISH analysis using the pTa71 clone, containing 18S, 5.8S, and 26S rDNA loci, detected a third NOR locus in 4x *Ae. crassa*, but this locus is not actively transcribed (Yamamoto and Mukai 1995).

Using C-banding and in situ hybridization (ISH) analyses, Badaeva et al. (1998) studied the distribution of highly repetitive DNA sequences on chromosomes of the tetraploid and the hexaploid cytotypes of *Ae. crassa*. The ISH studies were carried out with the pSc119 [120-bp sequence from *Secale cereale* (Bedbrook et al. 1980)], pAs1 [1 kb from *Ae. tauschii* (Rayburn and Gill 1986)] and pTa794 [410-bp from *T. aestivum* containing the 5S rRNA gene unit (120-bp) separated by a 290-bp spacer (Gerlach and Dyer 1980)] DNA clones. All chromosomes were identified by their C-banding and ISH pattern with the pAs1 clone, and the position of C-bands generally coincided with the location of the pAs1 sequence (Badaeva et al. 1998). Only a few pSc119 hybridization sites were observed in the telomeric regions of several chromosomes. Since pSc119 hybridizes with all S-genome chromosomes of Sitopsis species (Badaeva et al. 1996a), the few minor pSc119 ISH sites that were detected in 4x *Ae. crassa*, weaken the assumption that X^c subgenome originates from a Sitopsis species. Three pTa794 ISH sites (5S rDNA) were detected in tetraploid *Ae. crassa* and five such sites were identified in the hexaploid cytotype, three having derived from 4x *Ae. crassa*, and two contributed by *Ae. tauschii* (Badaeva et al. 1998). All the hexaploid accessions differed from the tetraploids by a reciprocal non-centromeric translocation. Several additional intraspecific translocations were detected between 4 and 6X accessions (Badaeva et al. 1998).

Using C-banding and fluorescence in situ hybridization (FISH) with ten DNA probes, Badaeva et al. 2021b) studied genome structure of 4x *Ae. crassa*. They confirmed that the D^c subgenome of allotetraploid *Ae. crassa* was contributed by *Ae. tauschii*, although the retention of minor NORs on chromosomes $1D^c$ and $6D^c$ indicated that *Ae. crassa* probably emerged prior to the loss of the respective loci in the diploid progenitor. Subgenome X^c might have originated from an ancestral S-genome species of subsection Emarginata.

9.5.4.3 Crosses with Other Species of the Wheat Group

Upon crossing tetraploid *Ae. crassa* with *Ae. tauschii* the F_1 triploid hybrids exhibited 7 bivalents (Kihara 1949) or 5.03 bivalents, 0.91 trivalents and 0.07 quadrivalents at first meiotic metaphase (Kimber and Zhao 1983). This high chromosomal pairing indicates that tetraploid *Ae. crassa* contains a subgenome that is related to the *Ae. tauschii* genome. Another triploid hybrid involving tetraploid *Ae. crassa* and *Ae. bicornis* showed 2–7 (mode of 5) bivalents, 0–2 trivalents, and 0–2 quadrivalents (Kihara 1949). A similar low level of pairing was also observed in the hybrid between *Ae. crassa* 6x and *Ae. bicornis*, namely, 3–6 bivalents (Kihara 1949), indicating that the two subgenomes of tetraploid and the three subgenomes of hexaploid *Ae. crassa* are not homologous to the S^b genome of *Ae. bicornis*. Chromosomal pairing between tetraploid *Ae. crassa* and *Ae. uniaristata* [3.89 bivalents, 0.61 trivalents, and 0.07 quadrivalent (Kimber et al. 1983)] and between hexaploid *Ae. crassa* and *Ae. uniaristata* [6–9 bivalents and 3 trivalents (Kihara 1949)], show that the genomes of the two species are not homologous. On the other hand, Kimber and Zhao (1983) studied chromosomal pairing in the hybrid between *Ae. cylindrica* and tetraploid *Ae. crassa* (Table 9.7), and hound that the two-allotetraploid species share one homologous subgenome, subgenome D.

Gupta and Fedak (1985) studied meiotic chromosomal pairing in F_1 hybrids between hexaploid *Ae. crassa* and species of *Secale*. The chiasmata frequency per cell ranged from 6.86 in hybrids with *S. cereale* to 9.93 in hybrids with *S. strictum*. These results provide evidence that a homoeologous-pairing control system operates in *Ae. crassa*. However, Cuñado (1992) and Cuñado et al. (2005) reported that chromosomal pairing at first meiotic metaphase of tetraploid *Ae. crassa* was as follows: 0.01 univalents, 3.02 rod bivalents, 10.88 ring bivalents and 24.78 chiasmata/cell. That of 6x *Ae. crassa*: 0.12 univalents, 4.27 rod bivalents, 15.78 ring bivalents, 0.39 multivalents and 36.98 chiasmata/cell. The multivalent pairing in 6x *Ae. crassa* presumably resulted from the partial homology between some chromosomes of D^c and D subgenomes, indicates that the cytologically-diploidizing genetic system is not fully effective in hexaploid *crassa*.

9.5.5 *Aegilops vavilovii* (Zhuk.) Chennav.

9.5.5.1 Morphological and Geographical Notes

Aegilops vavilovii (Zhuk.) Chennav. [Syn.: *Ae. crassa* Boiss. subsp. *vavilovii* Zhuk.; *Ae. crassa* var. *palaestina* Eig; *Triticum syriacum* Bowden; *Gastropyrum vavilovii* (Zhuk.) Á. Löve] is an annual robust plant, with thick 20–40-cm-high culms (excluding spikes). Spikes are 10–15-cm-long (excluding awns), with a cylindrical or slightly zigzag-shaped, tapering towards the tip, disarticulating into individual spikelets at maturity, each with its adjacent rachis segment (barrel-type dispersal unit). There are 5–10 spikelets, with 1–2, and usually 1, basal rudimentary spikelets. Spikelets are 12–14-mm-long, with 4–5 florets, linear, cylindrical to slightly inflated at the base, and slightly overlapping glume edges. Glumes are nearly truncate, covered with fine silvery hairs, and membranous at the tip, and are approximately 2/3 to ¾ as long as the lemmas, with 2–3 teeth (or weak awn) and a shallow sinus between the teeth and with veins equal in width. Lemmas are thickened in the upper part and usually keeled, ending with a tooth, which may be extended into a small awn. Lemmas of the terminal spikelet end with a strong, broad 5–6-cm-long awn, flanked with two small teeth, and has a prominent central nerve. The caryopsis adheres to lemma and palea (Fig. 9.2e).

Ae. vavilovii has limited morphological variation. It differs from *Ae. crassa* by characteristics derived from its S^s subgenome parent, i.e., *Ae. searsii*, including its long cylindrical and somewhat zig-zagged spike, that tapers in the upper half, and by the absence of awns on the lateral spikelets, while those of the uppermost spikelet are very long.

Ae. vavilovii is an Irano-Turanian (steppical) element. It has relatively limited distribution in the south-central part of the distribution area of the genus, namely, in the southeastern Mediterranean. Tetraploid *Ae. crassa* and *Ae. searsii*, the two parental species, may have contact in eastern Syria, where *Ae. vavilovii* presumably originated and spread southwards to high elevations of the southeastern Mediterranean steppes and dry and semi–desert habitats, where it is relatively isolated geographically from most of the other species of the group. Its distribution partially overlaps with that of its diploid parent *Ae. searsii*, and is south to the distribution of its tetraploid parent, *Ae. crassa*.

Ae. vavilovii grows in Egypt (Santa Katerina, Sinai Peninsula), south and southeastern Israel, Jordan, Syria, possibly also in eastern Lebanon, rarely in southeastern Turkey (close to the Syrian border) and in western Iraq (Table 9.1). It grows on grey calcareous, rendzina, alluvial or sandy soils, rarely on basalt, in the edges of dwarf-shrub steppe-like formations, wadis, stony slopes, fallows, grasslands, roadsides, edges of cultivation, and disturbed habitats. Populations of *Ae. vavilovii* may vary from small and scattered to large and dense stands, sometimes intermingled with *Ae. searsii* (in the higher elevations of the Israeli Negev and in southern Jordan) or with *Ae. crassa* in northeastern Syria. It grows at altitudes of 275–1550 m, in areas with 100–275 mm annual rainfall (and up to 550 mm in some higher locations). As such, it might be a drought-tolerant species. *Ae. vavilovii* grows sympatrically with *Ae. searsii*, *Ae. longissima* and *Ae. kotschyi*, and allopatrically with *Ae. peregrina*, *Ae. crassa*, and possibly *Ae. triuncialis*.

9.5.5.2   Cytology, Cytogenetics, and Evolution

Ae. vavilovii is an allohexaploid species (2n = 6x = 42: genome $D^cD^cX^cX^cS^sS^s$) (Dvorak 1998). The $D^cD^cX^cX^c$ subgenomes derived from 4x *Ae. crassa* (Kihara 1957; Chennaveeraiah 1960; Chapman and Miller 1978; Kimber and Zhao 1983; Zhang and Dvorak 1992). Analysis of variation in 27 repeated nucleotide sequences and the 5S rRNA demonstrated that the D^c subgenome was contributed by ancient *Ae. tauschii*, and the X^c subgenome by an extinct species, possibly a species that was ancestral to the entire genus *Aegilops*. The origin of the third subgenome was a source of controversy; Kihara (1957), Kihara et al. (1959) and Nakai (1982) proposed that it derived from the D genome of *Ae. tauschii*, whereas Kihara (1963) Kihara and Tanaka (1970), Talbert et al. (1991) and Yen and Kimber (1992a) thought that it derived from *Ae. longissima*. However, recent molecular studies by Zhang and Dvorak (1992) and Dubcovsky and Dvorak (1995a) showed that the third subgenome of *Ae. vavilovii* derived from *Ae. searsii*. Similar conclusions were reached following C-banding analysis (Badaeva et al. 2002). In addition, FISH analysis confirmed the relationship between *Ae. vavilovii*, 4x *Ae. crassa* and *Ae. searsii* (Badaeva et al. 2002). The organellar genome is D^2, similar to that of *Ae. crassa*. (Ogihara and Tsunewaki 1988; Wang et al. 1997). Hence, *Ae. vavilovii* was formed through hybridization of 4x *Ae. crassa* (as female) and *Ae. searsii* (Zhang and Dvorak 1992; Dubcovsky and Dvorak 1995a; Badaeva et al. 2002). Slight variations in the C-banding pattern relative to that of the parental species *Ae. crassa*, were detected in *Ae. vavilovii*, as was the existence of intraspecific variation in *Ae. vavilovii* due to a translocation between chromosomes $3X^c$ and $3D^c$ (Badaeva et al. 2002). While Gong et al. (2006), using ISSR markers, reported that the D^c subgenome of *Ae. vavilovii* changed only slightly from the D genome of *Ae. tauschii*, studies of chromosomal pairing showed that the D^c subgenome of *Ae. vavilovii* may have been substantially modified from that of the D genome of *Ae. tauschii* (Zhao and Kimber 1984).

Karyotypically, Chennaveeraiah (1960) noticed that the morphology of the chromosomes of two subgenomes of *Ae. vavilovii* (formerly *Ae. crassa* subsp. *vavilovii* Zhuk. Or *Ae. crassa* var. *palaestina* Eig) resembled that of 4x *Ae. crassa*, and that that of the third subgenome was different from that of the D genome that exists in 6x *Ae. crassa*. Therefore, he separated this taxon from *Ae. crassa* and elevated it to the species rank, designated *Ae. vavilovii*. Kihara (1963) reported that *Ae. vavilovii* has features quite different from those of 6x *Ae. crassa*, and consequently, supported Chennaveeraiah's decision to consider it as a separate species. A similar conclusion, based on cytological studies, was reached by Kihara and Tanaka (1970).

The *Ae. vavilovii* karyotype is symmetric, consisting of three pairs with median centromeres, and the rest with submedian centromeres (Chennaveeraiah 1960). While Sorokina (1928) noticed the presence of a single pair with a satellite, Senyaninova-Korchagina (1932) and Pathak (1940) observed three pairs with satellites. The presence of three pairs with satellites was also confirmed by Chennaveeraiah (1960). Two of the three satellite pairs are large and somewhat similar to each other, whereas those on the third pair are smaller and different. It is the smaller type that corresponds to the satellite of *Ae. tauschii* and is located on chromosome pair 5D^c.

Multicolor FISH analysis with probes pTA794 [410-bp-long sequence containing 5S rDNA from *T. aestivum* (Gerlach and Dyer 1980)] and pTa71 [9 kb of 18S-5.8S-26S rDNA from *T. aestivum* (Gerlach and Bedbrook 1979)] revealed six 5S rDNA loci and four 18S-5.8S-26S rDNA sites in *Ae. vavilovii*. Since three pairs of satellited chromosomes were observed in *Ae. vavilovii* (Chennaveeraiah 1960), the existence of only four 18S-5.8S-26S rDNA loci show that not all three pairs of NORs are active in organizing nucleoli (Badaeva et al. 2002). The reduction in the number of active NORs in *Ae. vavilovii* suggests that this locus was inactivated as a result of amphiplasty.

Usually, 21 bivalents per cell are formed at first meiotic metaphase, of which 4–5 are rod bivalents and the rest are ring bivalents. In some cells, two univalents were formed, one forming a ring univalent, indicative of an isochromosome (Chennaveeraiah 1960). Chapman and Miller (1978) observed 20.50 bivalents, of which 0–6 were rod bivalents, and a very low frequency of trivalents and quadrivalents. A similar frequency of bivalents and no multivalents were found in this species by Cuñado (1992).

9.5.5.3 Crosses with Other Species of the Wheat Group

The F$_1$ hybrid between *Ae. vavilovii* and *Ae. tauschii* (2n = 28; hybrid-genome D^{c}X^{c}S^{s}D) had at first meiotic metaphase 13.89 univalents, 5.03 bivalents, of which only 1.14 were ring bivalents, and several multivalents (0.75 trivalents, 0.37 quadrivalents and 0.05 pentavalents) (Table 9.6). This pattern of meiotic chromosome pairing confirms that the D^c subgenome of *Ae. vavilovii* is not completely homologous with the D genome of *Ae. tauschii*. Evidently, the D^c subgenome is a modified D genome. Likewise, the chromosomal pairing in the F$_1$ hybrid *Ae. vavilovii* x *Ae. ventricosa* (2n = 5x = 35; hybrid-genome D^{c}X^{c}S^{s}DN) and in the reciprocal hybrid, had 4.81 bivalents in one combination and 5.67 bivalents in another, of which only very few were ring bivalents, and several multivalents indicating again that D^{c}D subgenomes of the two species are not completely homologous (Table 9.7). A somewhat higher number of bivalents was observed in the F$_1$ hybrid *Ae. vavilovii* x *Ae. cylindrica* (2n = 5x = 35; hybrid-genome D^{c}X^{c}S^{s}CD), showing that the D^c subgenome of *Ae. vavilovii*

has better pairing with the D subgenome of *Ae. cylindrica* than with D subgenome of *Ae. ventricosa* and the D genome of *Ae. tauschii*. On the other hand, the F$_1$ hybrid hexaploid *Ae. crassa* x *Ae. vavilovii* (2n = 6x = 42; hybrid-genome D^{c}D^{c}X^{c}X^{c}DS^s) had 11 bivalents and several multivalents (2.63 trivalents and 0.43 quadrivalents, indicating good homology between the D^{c}X^c subgenomes of the two species (Table 9.7). This type of chromosomal pairing justifies the conclusion that tetraploid *Ae. crassa* was one of the parents of *Ae. vavilovii*. The good homology between the D^{c}X^c subgenomes of the two species suggests that the formation of *Ae. vavilovii* from 4x *Ae. crassa* was relatively recent.

The subgenomes of *Ae. vavilovii* and those of tetraploid wheat, *T. turgidum* ssp. turgidum and ssp. *durum* (2n = 5x = 35; hybrid-genome D^{c}X^{c}S^{s}BA), have almost no chromosomal pairing in the F$_1$ hybrid between these two species (Chapman and Miller 1978; Melnyk and McGinnis 1962). Similar low chromosomal pairing was observed in the F$_1$ hybrid between *Ae. vavilovii* and *T. aestivum* (2n = 6x = 42; hybrid-genome D^{c}X^{c}S^{s}BAD) (Chapman and Miller 1978). In spite of the presence of the D^c subgenome of *Ae. vavilovii* and the D of *T. aestivum*, this low pairing is due to the action of the *Ph1* gene of *T. aestivum* that suppresses pairing of homoeologous chromosomes.

9.5.6 Aegilops juvenalis (Thell.) Eig

9.5.6.1 Morphological and Geographical Notes

Aegilops juvenalis (Thell.) Eig [Syn.: *Triticum juvenalis* Thell.; *Ae. turcomanica* (Roshev.) Roshe.: *Aegilonearum juvenale* (Thell.) Á. Löve] is an annual robust plant, 15–40-cm tall (excluding spikes), jointed in the lower parts and then upright, and nearly or fully glabrous. Its leaves are broad. The spike is medium-sized, 3–7-cm-long (excluding awns), one rowed, cylindrical to slightly moniliform, becomes narrower toward the tip, is hairy, and has 3–7 (usually 5) spikelets. There are 1–3 and seldom no rudimentary spikelets. Spikelets disarticulate into individual spikelets, each with its adjacent rachis segment (barrel-type dispersal unit). Spikelets are elliptical, weakly inflated in the lower parts, and somewhat incised above. Glumes are hairy and overlapping in the upper parts. The tips of the glumes of lateral spikelets have 1–4 small, narrow and flat awns, separated from each other by an interval. Glumes of the terminal spikelet have a flat awn and two flanking teeth, or with 2 or 3 awns. The lemma is leathery, hairy at the top and about one-third longer than the glume, with 1 flat awn, 2–5-mm-long in the lower spikelets but up to 2–4 mm and two lateral teeth in the upper spikelet, or seldom with 3 awns with 2 lateral teeth, one of which may develop into a short awn. Lemma awns are more developed than glume awns. The caryopsis adheres to the lemma and palea (Fig. 9.2f).

The plant shows limited morphological variation mainly in length of spike, number of spikelets, glume and lemma. Thellung, who first described this species, thought it was a hybrid between *Ae. crassa* and *Ae. triuncialis*. It resembles its one parent *Ae. crassa* in many features, but differs from it by its many flat awns on the glumes and lemmas and by shorter and wider glumes, all of which are characteristics of its second parent, *Ae. umbellulata*.

It is a western and central Asiatic species occurring at rather dispersed locations in Turkmenistan, Uzbekistan, Iran, Iraq, north-east Syria and eastern Turkey. It grows on grey calcareous and alluvial soils, stony ground, gravel and open steppical habitats, edges of cultivation and roadsides. *Ae. juvenalis* is a weed of cultivation. It shows sporadic distribution, mainly in secondary disturbed habitats, throughout the warm steppes of central Asia. It is adventive in south France (Port Juvenal-from where it receives its name) and grows at altitudes of 150–1000 m. The plant is drought-tolerant in areas with 250–350 mm annual rainfall.

Ae. juvenalis has a medium-sized distribution in the eastern part of the distribution of the genus and is a steppical (Irano-Turanian) element. The two progenitors have contact in Iran, Northern Iraq, and eastern Turkey, where this species presumably originated. *Ae. juvenalis* is sympatric with the following species: *Ae. columnaris*, *Ae. triuncialis*, *Ae. tauschii*, *Ae. cylindrica* and *Ae. crassa* and allopatric with *Ae. speltoides*, *Ae. caudata*, *T. monococcum* subsp. a*egilopoides*, *T. timopheevii* subsp. *armeniacum*, *Ae. umbellulata*, *Ae. geniculata*, *Ae. neglecta* and *Ae. biuncialis*.

9.5.6.2 Cytology, Cytogenetics, and Evolution

Ae. juvenalis is an allohexaploid species (2n = 6x = 42; genome $D^cD^cX^cX^cUU$), (Dvorak 1998), originating from hybridization of tetraploid *Ae. crassa* (2n = 4x = 28; genome $D^cD^cX^cX^c$), as female, and *Ae. umbellulata* (2n = 2x = 14; genome UU) (Dubcovsky and Dvorak 1995a; Wang et al. 1997). Morphological analysis of *Ae. juvenalis* led to its classification into section Vertebrata. Some of the morphological characteristics that differentiate between it and tetraploid *Ae. crassa* are features derived from the U genome of *Ae. umbellulata* (Kihara et al. 1959), indicating that the U genome is the third subgenome of *Ae. juvenalis*, in addition to the two subgenomes of tetraploid *Ae. crassa* (Kihara et al. 1959).

Dubcovsky and Dvorak (1995a), using restriction fragments of nuclear repeated nucleotide sequences, found that *Ae. juvenalis* contains all restriction fragments of tetraploid *Ae. crassa* (except one) and six of the seven marker bands of *Ae. umbellulata*. All *Ae. juvenalis* bands of nuclear repeated nucleotide sequences were shared with either tetraploid *Ae. crassa* or *Ae. umbellulata*. However, the U subgenome of *Ae. juvenalis* has somewhat diverged from that of *Ae. umbellulata*. Divergence of a similar magnitude may have

also occurred between *Ae. juvenalis* and tetraploid *Ae. crassa*. Differentiation of the U subgenome of *Ae. juvenalis* from the genome of *Ae. umbellulata* was also suggested by Kimber and Yen (1989) from their investigation of chromosome pairing in interspecific hybrids.

Ae. juvenalis was studied by C-banding and FISH using clones pTa71 (18S-5.8S-26S rDNA), pTa794 (5S rDNA), and pAs1 (non-coding repetitive DNA sequence) as probes (Badaeva et al. 2002). Their data confirm previous conclusions from genome analysis (Kihara 1957; Kihara et al. 1959) that tetraploid *Ae. crassa* and *Ae. umbellulata* are the parental species of *Ae. juvenalis*. Previously, the genomic constitution of *Ae. juvenalis* was $DDM^jM^jC^uC^u$ (Kihara 1957; Kihara et al. 1959). The D subgenome is a modified D genome, designated D^c (Dvorak 1998), the C^u subgenome is currently designated U (Kimber and Tsunewaki 1988), and the M^j subgenome is not a modified M genome but rather close to the S genome of section Sitopsis and tentatively designated X^c (McGinnis and Melnyk 1956; Dubcovsky and Dvorak 1995a; Dvorak 1998). The presence of the D^c and U subgenomes was confirmed by genome analysis (McGinnis 1956; McGinnis and Melnyk 1962). The D^c subgenome of *Ae. crassa*, *Ae. vavilovii* and *Ae. juvenalis* are very similar since those of the latter two species derived from *Ae. crassa*. Indeed, studies of meiotic chromosome pairing showed that the D subgenome in *Ae. juvenalis* is substantially modified from the D genome of *Ae. tauschii* (Kimber and Zhao 1983; Zhao and Kimber 1984). Likewise, Bordbar et al. (2011), using allelic diversity at 25 nuclear microsatellite loci, nuclear rDNA ITS, and chloroplast *trn*L-F sequences, found a close phylogenetic relationship between the D^c subgenomes of *Ae. crassa*, *Ae. vavilovii* and *Ae. juvenalis*. In these three species, cloned sequences revealed high diversity at the nuclear rDNA ITS region (Bordbar et al. 2011). Using inter-simple sequence repeat (ISSR), Gong et al. (2006) also found that the D^c subgenome of *Ae. juvenalis* is a modified D genome of *Ae. tauschii*. McGinnis (1956) analyzed meiotic chromosome pairing in a number of interspecific and intergeneric F_1 hybrids involving species of *Triticum* and *Aegilops*, in order to determine the subgenomes present in *Ae. juvenalis*. It was concluded that *Ae. juvenalis* has the D genome of *Ae. squarrosa*, which is somewhat altered from the D of *T. aestivum*, and the U genome of *Ae. umbellulata*. The third genome was not determined. McGinnis and Melnyk (1962) presented evidence that the third subgenome of *Ae. juvenalis* is close to the S genome of section Sitopsis and not close to the M genome, as suggested by Kihara et al. (1959). Jaaska (1981), studying the electrophoretic genotype of the enzyme aspartate aminotransferase alcohol dehydrogenase in several D-genome species, concluded that *Ae. tauschii* subsp. *strangulata* contributed the D genome to hexaploid wheat, to tetraploids *Ae. crassa*, and *Ae. ventricosa*, and to hexaploid *Ae. juvenalis*. Its organellar genome is D^2, similar to that of its female parent (Ogihara and Tsunewaki 1988; Wang et al. 1997).

Kihara (1963) wrote that Tanaka (unpublished data) synthesized an allohexaploid from hybridization of tetraploid *Ae. crassa* x *Ae. umbellulata* and found that the synthetic hexaploid morphologically resembles *Ae. juvenalis*. Tanaka crossed the synthetic allohexaploid with *Ae. juvenalis* and chromosome pairing and fertility in the F_1 hybrid were as follows: the synthetic allohexaploid (2n = 6x = 42; genome $D^cD^cX^cX^cUU$) had 2.40 univalents, 17.00 bivalents, 0.70 trivalents, 0.90 quadrivalents and 0.10 pentavalents, and a 28.9% seed set (by selfing). The F_1 synthetic x natural *Ae. juvenalis* (2n = 6x = 42; genome $D^cD^cX^cX^cUU$) had 4.50 univalents, 14.30 bivalents, 0.50 trivalents, 1.00 quadrivalents and 0.40 pentavalents, and a 15.9% seed set. These data indicate that the synthesized allohexaploid and natural *Ae. juvenalis* are closely related, although not identical (Kihara 1963).

Parts of the distribution area of *Ae. juvenalis* in Iran and Iraq overlap with those of its two-parental species. This prompted Kihara et al. (1959) to suggest that these areas may be the place of origin of *Ae. juvenalis*. Its distribution overlaps that of its tetraploid female parent *Ae. crassa*, but is much more restricted.

The karyotypic study of Chennaveeraiah (1960)) supported the reported existence of the D^c and U subgenomes in *Ae. juvenalis*. The karyotype of *Ae. juvenalis* consists of two types, the subgenomes D^c and X^c that derived from diploid species with symmetric karyotype, whereas the subgenome U derived from *Ae. umbellulata* is asymmetric. Chennaveeraiah (1960) observed three chromosome pairs with satellites, the pairs are of dissimilar sizes. In one satellited pair, there was an interstitial minute chromosome segment between the satellite and the short arm, probably caused by a supernumerary constriction. Others included one pair with extreme subterminal centromeres, four with submedian-subterminal centromeres, one with a median centromere, and the remaining with submedian centromeres. The chromosomes corresponding to the third subgenome did not have satellites and all the pairs had submedian centromeres (Chennaveeraiah 1960).

A comparative analysis of somatic metaphase chromosomes by phase contrast, C-banding and Ag-staining was performed to analyze the activity of the NORs (Cermeño et al. (1984b). It was found that the U subgenome completely suppresses the NOR activity of the D subgenome of *Ae. juvenalis* and that of one pair of the nucleolar organizer chromosomes of the X^c subgenome of *Ae. juvenalis*. This corresponds to the observation of only three pairs of satellited chromosomes in *Ae. juvenalis* Chennaveeraiah 1960) and to the fact that two active NORs belong to subgenome U and the third to subgenome X^c (Badaeva et al. 2002).

Ae. juvenalis had 21 bivalents at meiosis, of which 3–4 are rod bivalents (Chennaveeraiah 1960). In few cells, however, a quadrivalent was formed, indicating the presence of a reciprocal translocation (Chennaveeraiah 1960). C-banding analysis of chromosome pairing at first meiotic metaphase was studied in diploid and polyploid *Aegilops* species (Cuñado 1992). Most of the polyploid *Aegilops* species showed a diploid-like meiotic behavior, although multivalents involving homoeologous associations were occasionally observed in *Ae. biuncialis*, *Ae. juvenalis* and *Ae. crassa*(6x); therefore, the *Aegilops* diploidizing genetic system is not equally effective in all polyploid species (Cuñado 1992).

Badaeva et al. (2021b) recognized two karyotypic groups in *Ae. juvenalis*: juv-I and juv-II. All genomes of juv-I were significantly modified, whereas juv-II was karyotypically similar to 4x *Ae. crassa* and *Ae. umbellulata*. Probably, juv-II originated independently of juv-I, from more recent hybridization of the same parental species.

9.5.6.3 Crosses with Other Species of the Wheat Group

Chromosomal pairing in F_1 hybrids between *Ae. juvenalis* and its two diploid parents, *Ae. tauschii* and *Ae. umbellulata* confirm the suggestion made based on morphological characteristics, that *Ae. juvenalis* contains subgenomes that derived from these two diploids (Table 9.6). The F_1 hybrid *Ae. juvenalis* x *Ae. tauschii* had 6.55 bivalents, 0.92 trivalents and 0.10 quadrivalents (McGinnis and Melnyk 1962) or 5.11 bivalents, 1.07 trivalents and 0.26 quadrivalents (Kimber and Zhao 1983). The data from both hybrids show that *Ae. juvenalis* contains one subgenome that is closely related, although not completely homologous, to that of *Ae. tauschii*. This subgenome is a modified D genome and therefore, was designated D^c. On the other hand, the F_1 hybrid *Ae. juvenalis* x *Ae. umbellulata* had 8.05 bivalents, 1.42 trivalents, 0.28 quadrivalents and 0.03 hexavalents (McGinnis and Melnyk 1962), indicating that subgenome U of *Ae. juvenalis* underwent relatively few changes at the hexaploid level and still is homologous to that of its diploid parent.

Chromosomal pairing in F_1 hybrids between *Ae. juvenalis* and tetraploid species having the D genome, namely, *Ae. ventricosa* (genome DDNN) and *Ae. cylindrica* (genome DDCC), indicated that the D^c subgenome of *Ae. juvenalis* is not so close to the D subgenome of the two tetraploid species. The hybrid *Ae. juvenalis* x *Ae. ventricosa*) had 4.86 bivalents, 1.25 trivalents, 0.25 quadrivalents and 0.04 pentavalents (Kimber and Zhao 1983; Table 9.7) and the hybrid *Ae. juvenalis* x *Ae. cylindrica* had 6.57 bivalents, 2.03 trivalents, 0.77 quadrivalents, 0.17 pentavalents and 0.14 hexavalents (Kimber and Zhao 1983). The higher pairing in the latter hybrid results from pairing between the U subgenome of the hexaploid and the C subgenome of the tetraploid.

Chromosome pairing in F_1 hybrids between *Ae. juvenalis* and hexaploid species having the D subgenome, namely, *Ae. crassa* 6x (genome $D^cD^cX^cX^cDD$) and *Triticum aestivum* (genome BBAADD) show that the D^c subgenome of *Ae. juvenalis* is somewhat diverged from the D subgenome of either hexaploid *Ae. crassa* or *T. aestivum* (Table 9.7). The hybrid 6x *Ae. crassa* x *Ae. juvenalis* had 8.90 bivalents, 2.01 trivalents, 0.59 quadrivalents and 0.28 higher multivalents (McGinnis and Melnyk 1962) and the hybrid *T. aestivum* x *Ae. juvenalis* had 4.33 bivalents, 0.48 trivalents and 0.32 quadrivalents (Riley 1966a). The higher pairing in the former hybrid results from the presence of two homologous subgenomes ($D^cD^cX^cX^c$) in the hybrid.

Hybrids between *Ae. juvenalis* and several other *Aegilops* diploid species, namely, *Ae. speltoides*, *Ae. sharonensis*, *Ae. longissima*, *Ae. caudata* and *Ae. uniaristata*, showed reduced pairing (McGinnis and Melnyk 1962), indicating that all three subgenomes of *Ae. juvenalis* are homoeologous to the genomes of the diploid species. Similarly, hybrids between *Ae. juvenalis* and tetraploid species of section Aegilops having the U subgenome, namely, *Ae. geniculata*, *Ae. columnaris* and *Ae. triuncialis*, showed that the U subgenome of *Ae. juvenalis* is closely related to that of the tetraploid species, while the other subgenomes are homoeologous (McGinnis and Melnyk 1962; Kimber et al.1988).

Hybrids between *Ae. juvenalis* and *Amblyopyrum muticum* (2n = 2x = 14; genome TT) had 6.46 bivalents, 1.30 trivalents, 0.60 quadrivalents and 0.02 pentavalents (McGinnis and Melnyk 1962). The relatively higher pairing in this hybrid results from the activity of homoeologous pairing promoters that exists in *A. muticum* (Dover and Riley 1972a).

9.6   Section Cylindropyrum (Jaub. & Spach) Zhuk

9.6.1   General Description

Section Cylindropyrum (Jaub. & Spach) Zhuk. (Syn.: Monoleptathera Eig) consists of annual and predominantly autogamous two species. The plants are slender with upright, 20–40-cm-high culms (excluding spikes), with narrow and cylindrical spikes, more than 10–20 times as long as wide but sometimes somewhat shorter. At maturity, the whole spike falls entire or disarticulates into individual spikelets, each falling with its segment beside it (a barrel-type dispersal unit). All spikelets have a similar shape but become smaller towards the tip of the spike. There are 4–10 spikelets per

spike, all equal in length to the adjacent rachis segment. There are 1–2, and seldom no rudimentary spikelets. Each spikelet contains 3–5 florets, the upper 1–2 being sterile. Glumes of apical spikelets are with one or three awns, that are at least 3-cm-long, broad, with or without small lateral teeth at the base. Glumes of lateral spikelets may be awned with short awns or awnless. The caryopsis adheres to lemmas and palea.

Two monographs on the genus *Aegilops* that were published by Zhukovsky (1928) and Eig (1929a), morphologically classified the *Aegilops* species into several sections. Zhukovsky (1928) included only one species in section Cylindropyrum, *Ae. cylindrica* Host and put *Ae. caudata* in section Comopyrum together with *Ae. comosa*, *Ae. heldreichii* and *Ae. uniaristata*. Likewise, Eig (1929a) included only *Ae. cylindrica* in section Monoleptathera (=Cylindropyrum) and put *Ae. caudata* in section Macrathera together with *Ae. comosa*, *Ae. heldreichii* and *Ae uniaristata*. Yet, Kihara (1954), classifying the *Aegilops* species on the basis of genome analysis, noting that the allotetraploid *Ae. cylindrica* contains a subgenome that derived from *Ae. caudata*, included *Ae. caudata* in section Cylindropyrum, together with *Ae. cylindrica*. This classification was also adopted, on a morphological basis, by van Slageren (1994).

In light of the above, section Cylindropyrum contains two species—one diploid, *Ae. caudata* L. (2n = 2x = 14; genome CC) and one allotetraploid, *Ae. cylindrica* Host (2n = 4x = 28; genome DDCC). The *Ae. tauschii* contributed the D genome to the tetraploid, which was responsible for the barrel-shape disarticulation of many types in this species.

Species of this section distribute in the center of the distribution area of the genus, namely, Turkey, northern Iraq, western Iran, Syria and Lebanon, (Israel (rare) and the diploid species spread from this center to Cyprus and westwards into the Aegean Islands, Greece (incl. Crete), Albania and Bulgaria, and eastwards to Afghanistan. The tetraploid species spread westwards [Greece, Albania Bulgaria, Italy (incl. Sicily)], northwards (Macedonia, Serbia, Croatia, Hungary, Ukraine, Crimea, Georgia, Armenia, and Azerbaijan) and eastwards (Pakistan, Afghanistan, Kyrgyzstan, Tajikistan, Uzbekistan and Turkmenistan). The spread of the tetraploid species northwards and eastwards is presumably the influence of its D genome.

The diploid species grows mainly on the red Mediterranean terra rossa soil, while the tetraploid species grows on a variety of soils from terra rossa to grey-calcareous steppe soil as well as on stony slopes. Both species grow at the edge of and openings in deciduous and sclerophyllous oak forests and maquis, and open herbaceous park formations, and in secondary disturbed habitats such as abandoned fields, edges of cultivation and roadsides. The tetraploid species also grows in open dwarf shrub steppe-like formations, at altitudes of almost sea level to 1750 m above sea level. The two species are common throughout most of the distribution area of the section.

9.6.2   *Aegilops caudata* L.

9.6.2.1   Morphological and Geographical Notes

Aegilops caudata L. [Syn.: *T. markgrafii* Greuter in Greuter & Rechinger; *Ae. markgrafii* (Greuter) K. Hammer; *T. caudatum* (L.) Godr. & Gren. in Grenier & Godron; *T. dichasians* Bowden; *Ae. dichasians* (Bowden) Humphries; *Orrhopygium caudatum* (L.) Á. Löve] is annual, predominantly autogamous, and tufted, with many tillers, upright, 20–45-cm-tall culms (excluding spikes) and hairy leaves. The spike is linear, narrow, tapering a little toward the tip, 3–10-cm-long (excluding awns), and disarticulates as entire spike at maturity. There are 3–7 cylindrical spikelets, equal in length to the adjacent rachis segment, each with 3–4 florets, the upper 1–2 being sterile. There are 1–3 (usually 2) rudimentary spikelets at the base of the spike. The glumes are rough, with the upper parts overlapping. Glumes of lower spikelets have two teeth, or a sharp tooth and a short, thin awn, separated from the tooth by an acute angle. Glumes of the terminal spikelet gradually taper into a long (4–12-cm), broad awn, longer than the entire spike, with a small adjacent tooth separated by a gap (no angle). The awns diverge sharply from each other. The lemma is membranous, upper parts thickened, with 2–3 teeth, from which weak, short awns can develop in the terminal spikelet. The caryopsis adheres to the lemma and palea (Fig. 9.3a).

Several taxonomists (e.g., Greuter, Bowden) included *Ae. caudata* in the genus *Triticum* while others maintained it in *Aegilops* (e.g., Zhukovsky, Eig, van Slageren). Consequently, the name of *Ae. caudata* has been a matter of controversy (Scholz and van Slageren 1994). Greuter (1976) included it in *Triticum* under the name *T. markgrafii* Greuter. But Hammer (1980) brought it back to *Aegilops* while endorsing the name *Aegilops markgrafii* (Greuter) Hammer. Bowden (1959) included the genus *Aegilops* in *Triticum* and named it *T. dichasians*, which later, when *Aegilops* was separated from *Triticum*, was changed to *Aegilops dichasians* (Bowden) Humphries. In spite of these suggestions, the name *Ae. caudata* has been widely used by taxonomists and geneticists, e.g., Zhukovsky (1928), Eig (1929a), Kihara (1954), Badaeva et al. (1996a, b), Ohta (2000), van Slageren (1994), Tsunewaki (1993), and many others. To maintain the common and widely used name *Aegilops caudata* L. in its traditional sense, Scholz and van Slageren (1994) proposed neotypification of this species endorsing to maintain the usage of the well-known name *Ae. caudata*.

Fig. 9.3 Spikes of diploid *Aegilops* species of sections Cylindropyrum and Comopyrum; **a** *Ae. caudata* L.; **b** *Ae. comosa* Sm. in Sibth. & Sm. ssp. *eucomosa*; **c** *Ae. uniaristata* Vis.

Ae. caudata resembles *Ae. cylindrica* (2n = 4x = 28; genome DDCC) in its spike morphology. The main morphological differences between these two species of section Cylindropyrum are caused by the D genome in *Ae. cylindrica*. According to Scholz and van Slageren (1994), these differences are: (1) in *Ae. caudata*, the awns of the apical glumes are equal or usually longer than the entire spike and with a base which is a continuation of the apex of the glume, whereas in *Ae. cylindrica*, these awns are only one-third to one-half of the spike length; (2) absence of small lateral teeth at the base of the apical glume's awn in *Ae. caudata;* (3) the apical lemma awns in *Ae. caudata* are short and narrowly linear, with one or two lateral teeth at the base, whereas they are well developed, equal in length but usually longer than the apical glume awns in *Ae. cylindrica*. The presence of well-developed awns on the lateral glumes in some forms of *Ae. caudata*, namely in var. *polyathera*, superficially makes this species look similar to *Ae. cylindrica*.

In some spike characteristics, *Ae. caudata* resembles *Ae. comosa*. However, the spike of *Ae. caudata* is more primitive, less specialized and shows a greater resemblance to the basic spike structure of the group. Symeonidis et al. (1979) studied the esterase and peroxidase patterns in five varieties of *Aegilops caudata* and *Ae. comosa* in order to elucidate the phylogenetic relationships within and between the two groups. In spite of considerable isozyme polymorphism, closer relationships in the banding patterns were found between different varieties of a single species than between varieties of the two different species. Esterase and peroxidase patterns of the two *Ae. caudata* varieties, *typica* and *polyathera*, were very similar and prove their close phylogenetic relationship. Overall, the electrophoretic data agree well with morphological and cytological findings (Zhukovsky 1928; Eig 1929a; Chennaveeraiah 1960; Kihara 1954).

Ae. caudata has limited morphological variation, mainly involving spike length, number and size of spikelets, awn length and development on the lateral spikelets. On a morphological basis, *Ae. caudata* was subdivided by Eig (1929a) into two varieties, var. *typica* (=var. *dichasians*; var. *margrafii*), which has awnless lateral spikelets, and var. *polyathera* Boiss., which has awned lateral spikelets. Var. *typica* is found in Greece, western and central Turkey, and neighboring Aegean Islands, while var. *polyathera* extends over the whole area reaching Syria and North Mesopotamia (Kihara 1954).

Ae. caudata is an East Mediterranean and Western Asiatic element. It has a relatively large distribution in the central and eastern regions of the distribution of the genus. It is a bi-regional species of the Mediterranean and steppical (Irano-Turanian) regions. *Ae. caudata* grows in Croatia, Serbia, Bosnia-Hercegovina, Macedonia, Albania, South Bulgaria, Greece (incl. Crete, Rhodes, and the Aegean region), Turkey, Cyprus, Lebanon, Syria, N. Iraq, Iran and Afghanistan (rare). *Ae. caudata* occupies a variety of primary and secondary habitats. It grows on a variety of soils from terra rossa to grey-calcareous steppe soil, as well as on stony slopes at the edge of and openings in sclerophyllous and deciduous oak forests and maquis, open herbaceous park formations, open dwarf shrub steppe-like formations, abandoned fields, edges of cultivation and roadsides. It is common throughout most of its distribution area. It grows from almost sea level to 1850 m, in areas with annual rainfall that range from 399–799 mm. *Ae. caudata* is adventive in the Genoa region and in Sardinia in Italy. It is also found near Marseille, France, and in Scotland.

Ae. caudata can form dense stands, often together with other *Aegilops* species. It grows sympatrically with *A. muticum, Ae. speltoides, Ae. comosa, Ae. uniaristata, Ae. umbellulata, Ae. tauschii, T. monococcum* subsp. *aegilopoides, T. urartu, T. timopheevii* subsp. *armeniacum, T, turgidum* subsp. *dicoccoides, Ae. geniculata, Ae. neglecta, Ae. recta, Ae. biuncialis, Ae. columnaris, Ae. triuncialis* and *Ae. cylindrica,* and allopatrically with *Ae. searsii, Ae. peregrina, Ae. crassa,* and *Ae. juvenalis. Ae. caudata* is involved in the parentage of *Ae. cylindrica* and *Ae. triuncialis* (section Aegilops).

Ae. caudata possesses genes that confer resistance to leaf rust, stem rust, and powdery mildew, as well as genes for higher grain protein and lysine content (Baldauf et al. 1992; Gong et al. 2017).

9.6.2.2   Cytology, Cytogenetics and Evolution

Ae. caudata is a diploid (2n = 2x = 14), bearing the C genome (Kihara and Lilienfeld 1932; Kihara 1949; Kimber and Tsunewaki 1988; Dvorak 1998). Its organellar genome was designated C by Tsunewaki (1993, 2009) and Wang et al. (1997). This species has the smallest genome in the genus (1C DNA = 4.84 ± 0.089 pg) (Eilam et al. 2007; Table 9.3). Eilam et al. (2007) determined DNA content in seven different accessions of *Ae. caudata* collected from different regions of the distribution area of the species and very little, if any, intraspecific variation in DNA content was noticed. The karyotype of *Ae. caudata* is highly asymmetric. Sorokina (1928), Senyaninova-Korchagina (1932), and Chennaveeraiah (1960) described the karyotype of *Ae. caudata.* They recognized two pairs with satellites on short arms, one pair with sub-median centromere, one pair with a sub-median-sub-terminal centromere and three pairs with subterminal centromeres. Chennaveeraiah (1960) found that the three pairs with sub-terminal centromeres differ from one another with respect to their long arms, while their short arms were not different. *Ae. caudata* was considered by Eig (1929a) to be a more advanced species than the Sitopsis species and the question is if the formation of this advance species was accompanied by a decrease in DNA content and in the change to an asymmetric karyotype.

In in situ hybridization experiments using a noncoding repetitive DNA sequence derived from *T. aestivum* (designated TC22b) on genomes of diploid *Aegilops* species, Teoh et al. (1983) found that *Ae. caudata* had much less heterochromatin than other species, as shown by a moderate amount of labeling, as well as the intensity of the hybridization sites amount on some chromosomes. The relatively small amount of heterochromatin is in agreement with the low DNA content in this species. In addition, Teoh et al. (1983) and Badaeva et al. (1996b), using a cloned repetitive DNA that consists of 18S and 25S rRNA genes, found that has two chromosome pairs had nucleolar organizer regions. This finding is in complete correspondence with the number of satellites in this species. Badaeva et al. (1996b) found that the two pairs of satellited chromosomes in *Ae. caudata* belong to homoeologous groups 1 and 5.

Teoh and Hutchinson (1983), using an improved C-banding technique, described a characteristic C-banding pattern in *Ae. caudata*, which enables the identification of individual chromosomes. All chromosomes have at least one intensely stained centromeric band and most have variable numbers of interstitial bands, while prominent telomeric bands are absent. Likewise, Badaeva et al. (1996a), studying the C-banding pattern of *Ae. caudata*, found that C-bands in this species are mainly medium or small in size with predominantly interstitial locations. They, like Teoh and Hutchinson (1983), concluded that *Ae. caudata* possesses an intermediate amount of C-heterochromatin. In situ hybridization with the highly repetitive DNA sequences pSc119 (from *S. cereale*) and pAsl (from *Ae. tauschii*), showed that only a few minor pAs1 sites were observed in *Ae. caudata*, whereas the pSc119 probe hybridized with telomeres of one or both arms of all chromosomes and several interstitial sites (Badaeva et al. (1996a).

Friebe et al. (1992b) analyzed the C-banding patterns of 19 different accessions of *Ae. caudata* from Turkey, Greece and west Asia, and established a generalized C-banded karyotype. Chromosome-specific C-bands are present in all C-genome chromosomes, allowing the identification of each of the seven chromosome pairs of *Ae. caudata*. While only minor variations in the C-banding pattern were observed within the accessions, a large amount of polymorphic variation was found between different accessions (Friebe et al. 1992b). C-banding analysis was also carried out to identify *Ae. caudata* chromosomes in six chromosome addition lines to common wheat. C-banding patterns of the added *Ae. caudata* chromosomes were identical to those of the ancestor species, indicating that these chromosomes were not structurally rearranged.

In an attempt to produce addition lines of *Ae. caudata* in common wheat, one chromosome was found to have been selectively retained in common wheat (Endo and Katayama 1978). The gametocidal chromosome was sub-telocentric and related to homoeologous group 3 (Endo 1990; See Sect. 9.3.3.).

Kihara and Lilienfeld (1935), Kihara (1959) and Upadhya (1966) studied chromosomal pairing in F_1 hybrids between common wheat and *Ae. caudata* and found somewhat higher pairing than in other hybrids between common wheat and diploid species, e.g., with *Ae. longissima* or *Secale cereale*. The number of bivalents in common wheat x *Ae. caudata* hybrids ranged from 3 to 5, with a mean of 4 bivalents/cell and trivalents varying ranged from 0 to 2 (Kihara and Lilienfeld 1935), 0–5 bivalents (mode 3) and trivalents (Kihara 1959), and the number of paired chromosomes ranged from 2 to 13, with a mean of 5.74 chromosomes per cell; the mean chiasma frequency per cell was found to be 2.98 (Upadhya 1966). Evidently, the *Ae. caudata* genome partially counteracts the suppression of pairing between homoeologous chromosomes in meiosis of common wheat x diploid hybrids, which is driven by the *Ph1* gene of common wheat. The *Ae. caudata* genome seems to modify the potency of the *Ph1* gene, resulting in a lower threshold of homoeologous chromosome pairing inhibition (Upadhya 1966). This altered threshold is not of the same magnitude as that brought about by *Ae. speltoides* or *A. muticum* genotypes. Kihara's (1959) report of the occurrence of higher frequencies of multivalents in the *Ae. caudata* x common

wheat hybrids as compared to in the reciprocal common wheat x *Ae. caudata* hybrids, indicates that the effect of the *caudata* genome on the *Ph1* gene in the F_1 hybrids is enhanced by the *caudata* cytoplasm.

Kihara and Lilienfeld (1932) thought that the genomes of *Ae. caudata* and *Ae. umbellulata* are related and therefore grouped them in the same genomic group, i.e., the C group, and designated them C and C^u, respectively. The opinion that *Ae. triuncialis* (2n = 4x = 28; genome CCC^uC^u) is nearly autotetraploid, has been expressed by several authors (e.g., von Berg 1931; Karpechenko and Sorokina 1929; Kihara 1929, 1937). It was based on the fact that the hybrid between *Secale cereale* and *Ae. triuncialis* has 3–7 bivalents (or sometimes even 6–7), which should be ascribed to autosyndesis of the *triuncialis* chromosomes. This high pairing led (Kihara 1937) to assume that *Ae. triuncialis* was formed when the two genomes of the diploid parents, C and C^u, were not significantly differentiated (Kihara and Lilienfeld 1932).

However, Sears (1948), as well as Kihara (1954) himself, felt the inadequacy of placing the genomes of these two species in one genomic group. The two species are morphologically different and are placed in different sections by taxonomists, e.g., Zhukovsky (1928) and Eig (1929a), and by the cytologists Senyaninova-Korchagina (1932) and Chennaveeraiah (1960), who found that the karyotypes of *Ae. caudata* and *Ae. umbellulata* are vastly different. Likewise, Teoh and Hutchinson (1983) and Badaeva et al. (1996a) described differences in the C-banding patterns of *Ae. caudata* and *Ae. umbellulata*, thus providing additional evidence supporting separation of these two species to different genomic group. The C and C^u genomes also differ in heterochromatin content (Friebe et al. 1992b), which lies in agreement with the low chromosomal pairing observed in meiosis of F_1 hybrids between these two species (Kimber and Abu-Baker 1981). Consequently, Chennaveeraiah (1960) and Kimber and Abu-Baker (1981) suggested to separate *Ae. umbellulata*'s genome from the C group and to designate a new genomic symbol, U.

Molnár et al. (2016) described the distribution of GAA and ACG microsatellite repeats on chromosomes of the U, M, S and C genomes of *Aegilops*, and the use of microsatellite probes to label the chromosomes in suspension by fluorescence *in situ* hybridization (FISH). Purified chromosome fractions enabled them to investigate the structure and evolution of the *Aegilops* genomes, by comparing the positions of conserved orthologous set (COS) markers on the purified *Aegilops* chromosomes with known positions on common wheat A, B and D subgenomes. Such comparisons revealed that the distribution of GAA and ACG hybridization signals differs within the U, M, S and C genomes, namely, significant rearrangements had occurred in the U and C genomes, while the M and S genomes exhibited structure similar to wheat.

At the whole-genome level, the structures of the S genome chromosomes of *Ae. speltoides* and the M genome chromosomes of *Ae. comosa* were the most similar to wheat, followed by the U genome of *Ae. umbellulata*, while the structure of the C genome in *Ae. caudata* differed considerably (Molnár et al. 2016). These results are in line with the findings of previous phylogenetic studies in which *Ae. umbellulata* and *Ae. caudata* formed a closer sub-cluster on the *Aegilops-Triticum* clade, indicating greater genetic similarity, relative to *Ae. comosa* and *Ae. speltoides* (Petersen et al. 2006; Mahelka et al. 2011).

Danilova et al. (2017) used molecular cytogenetic techniques and next-generation sequencing to explore the genome organization of *Ae. caudata*. Fluorescence *in situ* hybridization with a set of common wheat cDNAs showed that only two chromosomes of *Ae. caudata* maintained collinearity with wheat, whereas the remaining were highly rearranged as a result of inversions and inter- and intra-chromosomal translocations. Danilova et al. (2017) used sets of barley and wheat orthologous gene sequences to compare discrete parts of the *Ae. caudata* genome involved in the rearrangements. Analysis of sequence identity profiles and phylogenic relationships grouped chromosome blocks into two distinct clusters. Chromosome painting revealed the distribution of transposable elements and differentiated chromosome blocks into two groups consistent with the sequence analyses. Danilova et al. (2017) suggested that introgressive hybridization accompanied by gross chromosome rearrangements might have had an impact on karyotype evolution and speciation in *Ae. caudata*.

Tanaka et al. (1967) discovered the occurrence of intraspecific sterility in *Ae. caudata*, and Ohta (1992, 1995b, 2000; Ohta and Yasukawa 2015) further studied this phenomenon in crossing experiments. In two series of crossing experiments, Ohta (1992) first made reciprocal crosses between the two varieties, var. *typica* and var. *polyathera*, collected from sympatric populations and found that all F₁ hybrids showed normal fertility. Second, he crossed accessions collected from different geographical regions, namely, from the western distribution of the species (Aegean area and Greece) and the eastern distribution areas (eastern Turkey, Syria and northern Iraq) and found that the F₁ hybrids from these crosses were completely sterile. In contrast, F₁ hybrids from crosses between the Aegean accessions and the accession from Greece showed normal fertility, and F₁ hybrids from crosses between the accession from northern Iraq and those from Syria and northern Turkey showed high fertility. Accordingly, Ohta (1992) concluded that the intra-specific hybrid sterility in *Ae. caudata* did not correlate with morphological differences, i.e., it is not related to the varietal subdivision of the species, but rather, to geographical differentiation.

A similar conclusion was reached by Ohta (2000), who further studied the intra-specific hybrid sterility in *Ae. caudata*. Based on his results, the distribution area of *Ae. caudata* was divided into two geographical regions effectively isolated by the mountainous region lying between West Anatolia and Central Anatolia. The western region includes Greece, the Aegean Islands, and West Anatolia, while the eastern region consists of Central Anatolia, South Anatolia, East Anatolia, Syria, and Northern Iraq.

The results of Ohta (2000) and the findings from recent palaeopalynological works suggest that during the maximum glacial period from 18,000 BP to 16,000 BP, *Ae. caudata* occurred in the two isolated regions, i.e., the region surrounding the Aegean Sea and the western Levant or some sheltered habitats in the East Taurus/Zagros mountains arc. It then migrated into Central and East Anatolia from the latter regions, as the climate became warmer (Ohta 2000). Furthermore, it is also suggested that the Levant populations now occur in the eastern region of the distribution, while those occurring in the Aegean Sea region during the last glacial period now occupy the western region of the distribution (Ohta 2000).

Further elucidation of the geographical differentiation pattern of *Ae. caudata*, was reported by Ohta and Yasukawa (2015). They crossed 35 accessions derived from the entire distribution area, with four tester lines. It became clear that the present distribution area of *Ae. caudata* can be divided into the western and eastern regions, with the border in the mountains lying between West Anatolia and Central Anatolia: the western and eastern accessions are isolated not only geographically but also reproductively by genetic hybrid sterility.

Ohta (1995b), studied chromosome pairing and segregation at meiosis as well as fertility in sterile F₁ hybrids, a tetraploid derivative induced from one of the sterile hybrids, and their parental lines. The F₁ hybrids showed normal configurations and frequency of chromosome pairing at first meiotic metaphase, but was completely sterile. At first anaphase, chromosomes consisting of two sister chromatids of different lengths were observed. The induced tetraploid was shown to be an autotetraploid based on the configuration and frequency of chromosome pairing at first metaphase, and it showed partial restoration of fertility. Ohta (1995b) suggested that the intraspecific hybrid sterility observed in *Ae. caudata* is caused by chromosomal cryptic structural hybridity. The differences in chromosomal structure between the parental lines are presumably not large enough to cause preferential pairing in the induced autotetraploid. However, *Ae. caudata* represents intraspecific divergence that might be an initial step towards speciation due cryptic chromosomal rearrangements (Ohta 1995b).

9.6.2.3 Crosses with Other Species of the Wheat Group

Chromosomal pairing at first meiotic metaphase of *Ae. caudata* exhibited 7 bivalents, of which 2.87 were rod and 4.13 ring (Cuñado 1992). The relatively high number of rod bivalents in this species results from lack of chiasmata in the short arms of the three pairs of sub-telocentric chromosomes (Chennaveeraiah 1960).

Chapman and Riley (1964) reported chromosomal pairing at meiosis of a haploid individual of *Ae. caudata.* They observed that all the chromosomes were univalents in most cells, but one bivalent was occasionally formed. The low frequency of this bivalent was interpreted to mean that there is a small duplication of genetic material within the genome of *Ae. caudata.* This information may be of value to those concerned with the interpretation of meiotic chromosome pairing in hybrids involving *Ae. caudata,* in genome analysis and evolutionary studies.

Analysis of chromosome pairing at first meiotic metaphase of the F_1 triploid hybrid *Ae. caudata* x *Ae. cylindrica* showed the presence of 6–7 bivalents and 0–1 trivalent (Kihara 1954 and Table 9.5), indicating that *Ae. cylindrica* contains a subgenome that derived from *Ae. caudata.* [The second subgenome of this allotetraploid derived from *Ae. tauschii* (Kihara 1949)].

Data of meiotic chromosomal pairing in F_1 hybrids between *Ae. caudata* and other diploid *Aegilops* species are presented in Table 9.4. Sears (1941b) crossed two different lines of *Ae. speltoides* var. *ligustica* with *Ae. caudata,* one of which (designated *speltoides* I) is a high-pairing line that induces high homoeologous pairing in interspecific hybrids and the second (designated *speltoides* II) is an intermediate-pairing line. In the hybrid *Ae. caudata* x *speltoides* I, Sears (1941b) found 3.74 bivalents, of which only 0.48 were ring bivalents, and 1.26 trivalents. The reciprocal hybrids had a similar level of pairing. The *Ae. caudata* x *speltoides* II hybrid had 2.90 bivalents, of which only 0.12 were ring bivalents, and 0.50 trivalents. A similar level of chromosomal pairing was observed in F_1 hybrids *Ae. caudata* x *Ae. sharonensis,* and somewhat lower level of pairing was observed in the *Ae. caudata* x *Ae. longissima* hybrid. These data indicate that the genome of *Ae. caudata* and those of the studied Sitopsis species have diverged considerably from one another and are homoeologous. Hybrids between *Ae. caudata* x *Ae. tauschii* had 3.60 bivalents, of which only 0.60 were ring bivalents, 1.06 trivalents and 0.04 quadrivalents (Sears 1941b; Kihara 1949). This low level of chromosomal pairing shows that the genomes of *Ae. caudata* and *Ae. tauschii* also diverged considerably from one another. Somewhat higher chromosomal pairing was observed in F_1 hybrids between *Ae. caudata* and the species of section Comopyrum, *Ae. comosa* and *Ae. uniaristata.* The *Ae. caudata* x *Ae. comosa* hybrid had 4–6 bivalents with a mode of 5 and 0–1 trivalents (Kihara 1949), and the *Ae. caudata* x *Ae. uniaristata* hybrid had 3.94 bivalents, of which 0.44 were ring bivalents, 0.46 trivalents and 0.06 quadrivalents (Sears 1941b) or 2–5 bivalents (Kihara 1949). A similar level of chromosomal pairing was observed in the F_1 hybrid *Ae. caudata* x *Ae. umbellulata,* namely, 3.58–3.68 bivalents, of which 0.12–0.28 were ring bivalents, and 0.40–1.12 trivalents (Sears 1941b) or 3–6 bivalents with a mode of 5 bivalents and 0–1 trivalents (Kihara 1949). Evidently, the genomes of *Ae. caudata* and those of the Comopyrum species and *Ae. umbellulata* have diverged from each other, but less than that of *Ae. caudata* and the Sitopsis species.

Chromosome pairing at first meiotic metaphase of F_1 hybrids *Ae. caudata* x *Amblyopyrum muticum,* lacking B chromosomes, showed 3.03 univalents, 2.83 rod bivalents, 0.70 ring bivalents, 1.30 trivalents and 7.70 chiasmata/cell (Ohta 1990; Table 8.2). This low pairing, in spite of the presence of homoeologous-pairing promoters in *A. muticum* (Dover and Riley 1972a), indicates little homology between the chromosomes of the two species. Moreover, hybrids with two B chromosomes of *A. muticum* showed a drastically low frequency of A chromosome pairing (Table 8.2). The F_1 hybrids with or without B chromosomes, were completely sterile (Ohta 1990).

Hybrids between *Ae. caudata* and *Ae. triuncialis* had 7 bivalents, mostly ring bivalents (Kihara 1949), indicating that this allotetraploid contains a subgenome that derived from *Ae. caudata* [the second subgenome of *Ae. triuncialis* derived from *Ae. umbellulata* (Kihara 1954)]. The hybrids between other tetraploids of section Aegilops, i.e., *Ae. biuncialis* and *Ae. columnaris,* with *Ae. caudata,* had approximately 7 bivalents (Kihara 1949) as a result of pairing between the chromosomes of *Ae. caudata* with either those of the U subgenome of the allotetraploids or the second subgenome, M^b or X^n, as well as autosyndetic pairing between the two subgenomes of the allotetraploids.

The hybrid *Ae. caudata* x diploid wheat, *T. monococcum,* had 3.72 bivalents, of which only 0.03 were ring bivalents, and 0.59 trivalents (Table 9.8). The hybrid *Ae. caudata* x tetraploid wheat, *T. timopheevii,* had 0–5 bivalents (Kihara 1949), and the hybrid *Ae. caudata* x hexaploid wheat, *T. aestivum* ssp. *aestivum,* had 1.82 bivalents and no trivalents (Table 9.10). Evidently, there is very little pairing between *Ae. caudata* and the wheat species. Interestingly, there is more pairing with diploid wheat than with polyploid wheats, presumably due to the suppressive effect of the *Ph1* gene of polyploid wheat.

9.6.3 *Aegilops cylindrica* Host

9.6.3.1 Morphological and Geographical Notes

Ae. cylindrica Host, jointed goat grass, [Syn.: *Triticum cylindricum* (Host) Ces., Pass & Gibelli; *Triticum caudatum* (L.) Godr. & Gren.; *Cylindropyrum cylindricum* (Host) Á. Löve] is a predominantly autogamous, annual species, tufted with many tillers, with a 20–40(-80)-cm tall (excluding spikes) culm is prostrate near the ground and then upright. Leaves are narrow, linear, smooth or hairy. The spike is 5–8 (-12)-cm-long (excluding awns), cylindrical, one-rowed, and more or less awned. There are 6–10(-12) spikelets, that become smaller toward the tip of the spike. The spike disarticulates into individual spikelets, each with the adjacent rachis segment (barrel-type dispersal unit) or falls entire from the culm. The number of rudimentary spikelets is 1–2, seldom zero. Spikelets are cylindrical, three or more times longer than broad, mostly equal in length and appressed to the rachis segment. Glumes have a short triangular awn (up to 3.5-cm-long) and an associated short, broad, blunt tooth forming an obtuse or acute angle with the awn. Awns on glumes of lower spikelets are shorter than awns on glumes of upper spikelets, which are 3–6-cm-long. Lemmas are solid in their upper parts, nerved, 2–3-toothed, always with 3 teeth in the terminal spikelet, with the middle tooth elongated into an awn thicker and longer (4–8-cm-long) than the awns on the glumes. Awns of lemmas of lower spikelets are shorter (or absent) than associated glume awns. Apical glume and lemma awns are always shorter than the length of the spike. The caryopsis adheres to the lemmas and palea (Fig. 9.2g).

Ae. cylindrica exhibits very wide variation in the length of culms, length of spike, number, size, hairiness and color of spikelets, and in development of the lateral glume awns. There are more awns on the glumes but the lemma awns are always more developed. This wide morphological variation led Zhukovsky (1928), Eig (1929a) and Hammer 1980) to subdivide the species into several intraspecific categories. Hammer (1980) subdivided *Ae. cylindrica* into four varieties: var. *cylindrica*, var. *aristulata* (Zhuk.) Tzvel., var. *pauciaristata* Eig, and var. *prokhanovii* Tzvel. Var. *cylindrica* was further divided by Hammer (1980) into three forma: f. *cylindrica*, f. *ferruginea* (Popova) Hammer, and f. *brunnea* (Popova) Hammer. Var. *prokhanovii* was also further divided to three formas: f. *prokhanovii* (Tsvel.) Hammer, f. *rubiginosa* (Popova) Hammer, and f. *fuliginosa* (Popova) Hammer. Chennaveeraiah (1960), based on karyotypic studies, elevated the varieties to subspecies and recognized only two subspecies in *Ae. cylindrica*, subsp. *cylindrica* and subsp. *pauciaristata* (Eig) Chennav.

Genetic variation of *Ae. cylindrica* was assessed on various levels and using different methodological approaches, morphological studies, isozyme analyses, storage protein comparisons and molecular markers. Using quantitative and qualitative morphological traits, Arabbeigi et al. (2015) evaluated genetic variation of 66 *Aegilops cylindrica* genotypes, collected from western and northwestern Iran. They noted high genetic variation in the quantitative and qualitative traits of Iranian *Ae. cylindrica*, with lemma length and glume color showing the highest variation. Cluster analysis divided the studied genotypes into three groups, each occupying a different region of Iran.

Isozymes analyses showed little variation between accessions of *Ae. cylindrica* collected from various locations in the distribution area of the species. Thus, Nakai (1981), studying isoelectric focusing patterns of esterase isozymes, Watanabe et al. (1994), studying alpha-amylase isozymes, and Hedge et al. (2002), studying 10 isozymes, revealed high uniformity in the isozyme patterns of the studied accessions. Likewise, analysis of glutenin subunits in different accessions of *Ae. cylindrica* did not disclose variation in the high-molecular weight glutenin subunits (Farkhari et al. 2007; Khabiri et al. 2012), whereas some variation was noticed in the low-molecular weight (LMW) glutenin subunits (Khabiri et al. 2012). While glutenin subunits exhibit low diversity, gliadin subunits are much more polymorphic in *Ae. cylindrica*. Khabiri et al. (2013), using polyacrylamide gel electrophoresis, assessed genetic diversity in banding patterns of gliadin protein in seventeen populations of *Aegilops cylindrica* from northwestern Iran. Their results showed that most bands were related to the ω type of gliadins, whereas the smallest number of bands pertained to the β type gliadins. Genetic diversity between populations was greater than within populations. Assessment of total variation for the three gliadin types indicated that the highest total variation was related to the β type, while the lowest belonged to the ω type. Cluster analysis using the complete linkage method divided populations into two separate groups, in which genetic diversity does not follow geographical distribution.

Molecular markers such as random amplified polymorphic DNA (RAPD) assays (Okuno et al. 1998; Pester et al. 2003; Goryunova et al. 2004; Farkhari et al. 2007), and amplified fragment length polymorphism (AFLP) (Pester et al. 2003) revealed a certain degree of variation. RAPD analysis of *Ae. cylindrica* accessions collected in Central Asia and North Caucasia showed that while accessions of this species from north Caucasia were genetically uniform, those from Central Asia were slightly more diverse (Okuno et al. 1998). Yet the authors did not find associations between altitudinal variation and variability of RAPD markers. Pester et al. (2003) used RAPD and AFLP to study genetic variation in *Ae. cylindrica* and found that, although AFLP produced more scorable bands as compared to RAPD, both methods revealed limited genetic diversity in *Ae.*

cylindrica. Similarly, Goryunova et al. (2004), using RAPD, found that the intraspecific level of genetic variation in *Ae. cylindrica* was considerably lower than in either parent, *Ae. tauschii* or *Ae. caudata.* Yet, in contrast to the above findings, Farkhari et al. (2007), using RAPD, evaluated genetic variation of 28 populations of *Ae. cylindrica*, collected from different parts of Iran, and found that primers of RAPD generated 133 reproducible fragments, 69% of which were polymorphic. There was little relationship between genetic divergence and geographical origin (Farkhari et al. 2007).

The results of Farkhari et al. (2007) showed that RAPD is suitable for genetic diversity assessment in *Ae. cylindrica* populations. This is expected, as protein markers reflect only variation in the coding parts of the genome, which is, by nature, more conservative and thus less polymorphic. RAPDs, on the other hand, can detect variation in both coding and non-coding sequences, and the length of the primers allows the amplification of a large number of fragments with a single primer (Guadagnuolo et al., 2001). This technique has the advantage of requiring very small quantities of template DNA and no prior sequencing knowledge of the target genome (Pester et al. 2003).

Simple sequence repeats (SSRs) are another type of molecular marker used to study intraspecific genetic diversity. Naghavi et al. (2008. 2009b), using SSRs to assess genetic diversity in *Ae. cylindrica* accession, collected from 13 different sites in Iran, found that intraspecific genetic diversity in this species is not as high as it was in *Ae. tauschii* and *Ae. crassa.* Moreover, SSR markers showed that the genetic distance between *Ae. tauschii* and *Ae. cylindrica* is very small. Mohammadi et al. (2014), using 17 inter-simple sequence repeat (ISSR) primers, investigated genetic diversity among 35 accessions of *Ae. cylindrica*, collected from different Iranian locations. Out of 190 alleles that were amplified, 188 (98.95%) were polymorphic. Cluster and principal coordinate analysis (PCA) showed no association between molecular diversity and geographic diversity of genotypes, indicating that there is high genetic diversity within populations (Mohammadi et al. 2014). Based on their findings, Mohammadi et al. (2014) concluded that the center of diversity and origin of *Ae. cylindrica* might be the western region of Iran and this species migrated from this region to other parts of its distribution area.

Ae. cylindrica is a Mediterranean and steppical (Irano-Turanian) element, which distributes in the north Mediterranean and central Asiatic parts of the distribution of the genus. It grows in the west in Greece (rare), including Crete (rare), Albania, Macedonia, Serbia, Croatia, Hungary, Romania, Bulgaria, south Ukraine, Crimea, Ciscaucasia, Georgia, Armenia, Azerbaijan, southern Turkmenistan, southern Uzbekistan, northern Tajikistan, western Kyrgyzstan, Afghanistan (rare), Iran, northern Iraq (rare), Turkey, Syria, Lebanon (rare), Jordan (rare) and Israel (rare).

Ae. cylindrica inhabits a wide variety of primary and secondary habitats. It grows at the edges and in openings of deciduous Mediterranean and steppical oak forest and maquis, dwarf shrub formations, dwarf shrub steppe-like shrub formations, dry hill and mountain slopes, plains, pastures, abandoned fields, ruderal and disturbed sites, wastelands, edges of cultivation, disturbed habitats and roadsides, railway sides, grasslands, and close by or within cultivation, such as orchards, vineyards, and wheat fields. Near common wheat fields, *Ae. cylindrica* forms natural hybrids with wheat. It grows mainly on calcareous and basaltic soils, less frequently on sands, in areas with annual rainfall from 450 to 800 mm, indicating a preference for more humid environments than most *Aegilops* species. *Ae. cylindrica* is usually found at altitudes from 300 to 2000 m, rarely at lower attitudes (around the Caspian Sea).

The species is common throughout most of its range and locally abundant. *Ae. cylindrica*, more than most species of the genus *Aegilops*, shows weedy behavior, occupying large stands after recent disturbances, i.e., it is a successful colonizing species (van Slageren 1994). It grows in mixed populations with many *Aegilops* species, with which it may introgress. Its two diploid parents overlap in east Turkey, and Iran, where *Ae. cylindrica* may have originated and then spread both westwards and northwards. Its distribution area is larger than those of its diploid parents. It grows sympatrically with *Ae. caudata, Ae. tauschii, Ae. umbellulata, Ae. geniculata, Ae. neglecta, Ae. biuncialis, Ae. columnaris, Ae. triuncialis, Ae. ventricosa, Ae. crassa, Ae. juvenalis, Amblyopyrum muticum*, wild *Triticum monococcum* ssp. *aegilopoides, T. urartu*, and wild *T. timopheevii*, and allopatrically with *Ae. comosa, Ae. uniaristata, Ae. speltoides, Ae. peregrina, Ae. vavilovii* and wild *T. turgidum.*

Ae. cylindrica is an adventive in north Italy, France, Germany, Switzerland, and in many other countries of central, northwestern, northern and eastern Europe. It was first introduced to Kansas, USA, in the late nineteenth century, as a contaminant in Turkey winter wheat seed, and since then, it was introduced again and again and spread to many different additional states, mainly to western and northwestern states where it has become very common (Hitchcock 1951; Barkley 1986; Donald and Ogg 1991). Moreover, changing wheat production practices during the second half of the twentieth century have encouraged the spread and increase of this weed (Donald and Ogg 1991).

No variation was revealed when isozyme studies were carried out on *Ae. cylindrica* introduced to North America and Eurasian accessions (Watanabe et al. 1994; Hedge et al. 2002). Likewise, Pester et al. (2003), using two DNA molecular marker techniques, RAPD and AFLP, found only low variation between North American and Eurasian accessions. Cluster analysis showed small genetic distances between all *Ae. cylindrica* accessions from North America, with the greatest distance observed between an accession from Washington and a group of three other U.S. accessions from Nebraska, Oklahoma and Utah. These results suggest either-multiple introductions of *Ae. cylindrica* into the United States or some genetic divergence within U.S. populations since their introduction. These introductions may have originated from different Eurasian geographic locations and included different *Ae. cylindrica* genotypes. The Eurasian accessions, which, although somewhat more genetically varied than those from the United States, are still much less diverse than might be expected in populations from the center of origin of the species (Okuno et al. 1998; Watanabe et al. 1994).

Ae. cylindrica has similar growth habits as hexaploid winter wheat and consequently, has developed into a serious pest in winter wheat fields, especially in northwestern states, where it can lower winter wheat yield by competing for growth requirements, thereby reducing harvesting efficiency, and lowering crop quality by contaminating harvested grain (Donald and Ogg 1991). In 1999, it was reported that *Ae. cylindrica* infested an estimated 2 million hectares in the US alone, and caused annual losses of $145 million in crop yield and quality, and that the infested area increases annually at a rate of about 20,000 hectares (Kennedy and Stubbs 2007). In addition, it can easily hybridize with winter wheat and form viable hybrids that can produce seeds via backcrossing to either parent (Gandilyan and Jaaska 1980; Guadagnuolo et al. 2001; Hegde and Waines 2004; Galaev and Sivolap 2005; Gandhi et al. 2006). Because *Ae. cylindrica* contains the D subgenome, which is homologous to the D subgenome of common wheat, it is very difficult to control it in commercial fields. Aside from its spikelets (the dispersal units) that are similar in shape and size to the grains of winter wheat, making it difficult to separate them from wheat using conventional methods of grain cleaning, the shared genetics makes it practically impossible to selectively kill off *Ae. cylindrica* without harming the winter wheat. This poses problems for farmers who have to suffer through reduced yields and poorer quality winter wheat (Donald and Ogg 1991). However, it is hoped that breeding of winter wheat lines resistant to imidazolinone will allow the use of imazamox to selectively kills *Ae. cylindrica*. Yet, Seefeldt et al. (1998) mentioned the concerns about the possibility of transferring the resistant genes from common wheat to *Ae. cylindrica* through hybridization and backcrossing to the wild parent. Indeed, Guadagnuolo et al. (2001) found that introgression of wheat DNA into *Ae. cylindrica* is possible. Gandhi et al. (2006) obtained similar results when studying the occurrence of wheat nuclear microsatellite markers in BC_1 plants obtained from *Ae. cylindrica* x common wheat backcrossed to the wild parent.

Ae. cylindrica possesses a large number of desirable genes that can be used for wheat improvement. It contains genes for resistance to various fungal diseases (e.g., Bai et al. 1995; Singh et al. 2004) for cold hardiness and salt tolerance (e.g., Farooq et al. 1992; Arabbeigi et al. 2014; Kiani et al. 2015), and for storage protein genes (Wan et al. 2000).

9.6.3.2   Cytology, Cytogenetics, and Evolution

Ae. cylindrica is an allotetraploid species (2n = 4x = 28; genome DDCC). Its genome constitution was determined by the analyses of chromosome pairing (Kihara 1930; Kihara and Matsumura 1941; Sears 1944b; McFadden and Sears 1946; Kimber and Zhao 1983; Zhao and Kimber 1984), karyotype structure (Chennaveeraiah 1960), storage proteins (Johnson 1967; Masci et al. 1992), isozymes (Jaaska 1981; Nakai 1981), and differences in restriction length patterns of repeated nucleotide sequences (Dubcovsky and Dvorak 1994). All these studies identified the diploid species *Ae. caudata* as the donor of the C subgenome and *Ae. tauschii* as the donor of the D subgenome. Hence, *Ae. cylindrica* originated from spontaneous chromosome doubling of an F_1 hybrid between *Ae. tauschii* and *Ae. caudata* (Kihara 1944), the former being the plasmon donor to *Ae. cylindrica* (Tsunewaki 1996).

Amphiploid *Ae. caudata-Ae. tauschii* (2n = 4x = 28; genome CCDD) was synthesized from its putative parents (Sears 1941b). The synthetic amphiploid closely morphologically resembled the natural *Ae. cylindrica* and chromosomal pairing at meiosis of the F_1 hybrid between the synthetic and the natural *Ae. cylindrica* was almost regular, i.e., 0.54 univalents, 13.14 bivalents, 0.02 trivalents, and 0.28 quadrivalents per cell (Sears 1941b). It thus appears that the synthetic amphiploid has no significant chromosome differences from natural *Ae. cylindrica*, except possibly a reciprocal translocation. Concerning seed set, the hybrid was less fertile (44%) than the natural *Ae. cylindrica* (76%), but more fertile than the synthetic amphiploid (27%).

Johnson (1967) showed that *Ae. tauschii* subsp. *tauschii* (former subsp. *eusquarrosa* Eig) var. *tauschii* (former var. *typica* Eig) + *Ae. caudata* var. *polyathera* had similar seed protein pattern as that of Ae. cylindrica. Based on isozyme studies, Jaaska (1981) also proposed subsp. *tauschii* of *Ae. tauschii* as the donor of the D genome to *Ae. cylindrical*, and subsp. *strangulata* as the D subgenome donor to the hexaploid wheats. Nakai (1981) found no zymogram variation in studies of the isoelectric focusing patterns of esterase isozymes in 30 strains of *Ae. caudata*. Consequently, it was impossible to identify the line that contributed the C subgenome to *Ae. cylindrica*. Since *Ae. cylindrica* had an isozyme pattern which corresponded to a mixture of esterases from *Ae. caudata* and one genotype of *Ae. tauschii*, Nakai (1981) concluded that *Ae. cylindrica* originated with a single amphiploidy event, and that the C and D subgenomes have remained remarkably constant regarding esterase isozyme composition. Similar to Jaaska (1981), Nakai (1981) also found that the D subgenome of common wheat derived from a genotype of *Ae. tauschii* different than the one donated to *Ae. cylindrica*. Masci et al. (1992) confirmed the presence of the alpha-gliadin genes of *Ae. caudata* in *Ae. cylindrica*, in addition to those of *Ae. tauschii*.

Maan (1976) obtained fully fertile plants of normal vigor, with a normal chromosome number, and normal meiotic pairing in alloplasmic lines of common wheat bearing the cytoplasm of *Ae. tauschii* or *Ae. cylindrica*. The similar nucleo-cytoplasmic interactions indicate cytoplasmic homology between *Ae. tauschii* and *Ae. cylindrica*. Thus, Maan (1976) concluded that the cytoplasm of *Ae. cylindrica* was contributed by *Ae. tauschii*, a conclusion that was later verified by Tsunewaki (1989, 1996). *Ae. tauschii* is the maternal parent of *Ae. cylindrica* and of two other allotetraploids, *Ae. ventricosa* and *Ae. crassa*. The average distances between pairs of the D-genome tetraploids and *Ae. tauschii* are 0.015 with *Ae. cylindrica*, 0.050 with *Ae. crassa* and 0.006 with *Ae. ventricosa* (Wang et al. 1997). This suggests that the allotetraploids with the D genome originated at three different times from *Ae. tauschii* as female, in the following order: *Ae. crassa*, *Ae. cylindrica*, and *Ae. ventricosa* (Wang et al. 1997).

Tsunewaki et al. (2014) produced an alloplasmic *Ae. cylindrica* with the plasmon of *Ae. caudata* and found that the genetic effect of the *Ae. caudata* plasmon on the manifestation of *Ae. cylindrica* genomes was limited to male fertility: the alien plasmon caused pollen and self-pollinated seed sterility, without exerting any detectable effects on the other morphological and physiological characters investigated. Further studies (Tsunewaki et al. 2014) indicated that the male sterility expressed by the alloplasmic line was due to genetic incompatibility between the *Ae. cylindrica* genome and *Ae. caudata* plasmon. Previously, Tsunewaki et al. (2002) investigated the genetic effects of the *Ae. caudata* plasmon on the genome manifestation of 12 genotypes of common wheat and found that the *caudata* plasmon caused pollen and seed sterility in nine of the 12 genotypes.

Yet, a recent investigation comparing chloroplast and nuclear microsatellite loci of 36 *Ae. cylindrica* accessions with those of seven accessions of *Ae. caudata* and 17 accessions of *Ae. tauschii*, revealed that one of the examined *Ae. cylindrica* accessions possessed a plastom of the C type, the *caudata* plastom, whereas all others had the D type, the *tauschii* plastom (Gandhi et al. 2009). The accession with the C cytoplasm was originally collected in Turkey. Their result suggests that *Ae. cylindrica* arose from multiple hybridizations between *Ae. caudata* and *Ae. tauschii*, presumably along the Fertile Crescent, where the geographic distributions of its diploid progenitors overlap. In most cases, *Ae. cylindrica* originated from a cross between *Ae. tauschii* as female and *Ae. caudata* as male, as proposed by Kihara (1944) and Tsunewaki (1996). However, their result

also suggests a diphyletic origin of *Ae. cylindrica* from the reciprocal crosses between *Ae. tauschii* and *Ae. caudata*, although its diphyletic origin must be confirmed by other means (Tsunewaki 2009). In this case, *caudata* cytoplasm did not cause male sterility in the allotetraploid.

There is 9.59 pg nuclear 1C DNA in *Ae. cylindrica* (Eilam et al. 2008). This value deviates by −4.20% from the expected cumulative amount of DNA of the two diploid parents, 4.84 pg 1C DNA of *Ae. caudata* and 5.17 pg of *Ae. tauschii* (Eilam et al. 2007). Bakhshi et al. (2010) determined the DNA peak mode of 100 accessions of *Ae. cylindrica* using flow cytometry and found variations among accessions from different locations. There was no significant difference in morphological traits of accessions with high and low DNA peak mode. Accessions with lower and higher DNA content had shorter and longer chromosome lengths, respectively (Bakhshi et al. 2010). The observed intraspecific differences in DNA size in *Ae. cylindrica* are predominantly associated with differences in the amounts of repetitive sequences. Changes in copy number of certain DNA sequences, as a response to different environments, may be responsible for the observed changes in DNA size.

The karyotype of *Ae. cylindrica* is asymmetric. Senyaninova-Korchagina (1932) reported that the karyotype of *Ae. cylindrica* contains four pairs with satellites on short arms, three pairs with extreme subterminal centromeres, two pairs with submedian-subterminal centromeres, one pair with an almost median centromere, and the rest with submedian centromeres. Pathak (1940) observed one pair with satellites and another pair with secondary constrictions and four nucleoli. In other words, he observed two pairs with satellites. Chennaveeraiah (1960) studied the karyotype in two subspecies of *Ae. cylindrica*, ssp. *cylindrica* and ssp. *pauciaristata*. In neither of them did he observe four satellited pairs. *Ae. cylindrica* ssp. *cylindrica* showed a karyotype including three satellite pairs, three pairs with extreme subterminal centromeres, one pair with a submedian-subterminal centromere, and the rest with submedian centromeres. The karyotype of *Ae. cylindrica* ssp. *pauciaristata* differs from that of subsp. *cylindrica* in that it has only one pair with satellites.

Senyaninova-Korchagina (1932), based on her karyotypic studies, thought that one subgenome of *Ae. cylindrica* corresponds to the karyotype of *Ae. caudata*, whereas the other subgenome is closely related to the karyotype of *Ae. bicornis*. She made another wrong decision by depicting four satellited pairs in *Ae. cylindrica*. In contrast, Chennaveeraiah (1960) observed only three satellited pairs in ssp. *cylindrica*, two on the C subgenome and one on the D subgenome. According to him, one subgenome of *Ae. cylindrica* resembles the karyotype of *Ae. caudata* whereas the other subgenome resembles the karyotype of *Ae. tauschii*. The single pair of satellited chromosomes in *Ae.*

cylindrica ssp. *pauciaristata* corresponds to the satellited pair in *Ae. tauschii*. The second subgenome of this subspecies, the C subgenome, does not have the two satellited pairs. It is possible that the differences between the two subspecies reflect some variation in the karyotype of the parental species, *Ae. caudata* (Chennaveeraiah 1960).

Teoh et al. (1983) used a repetitive DNA sequence, derived from *T. aestivum*, coding for ribosomal RNA, as a probe in an in-situ hybridization experiment on genome of *Ae. cylindrica*. They found only two pairs of rRNA sites, one less than the expected three from the presence of three pairs of satellited chromosomes (Chennaveeraiah 1960).

The genomic constitution of *Ae. cylindrica* was analyzed by C-banding, genomic in situ hybridization (GISH), and fluorescence in situ hybridization (FISH), using the DNA clones pSc119, pAs1, pTa71 and pTA794 (Linc et al. 1999). The C-banding patterns of the D and C subgenome chromosomes were similar to those of the D and C genome chromosomes of the diploid progenitor species *Ae. tauschii* and *Ae. caudata*, respectively. These similarities permitted the identification at first meiotic metaphase of the chromosomes of the D and the C subgenomes of *Ae. cylindrica* (Cuñado 1992). The chromosomes of the D subgenome were almost without C-banding, just like the chromosomes of *Ae. tauschii*, whereas the chromosomes of the C subgenome were smaller than those of the D subgenome and showed thin interstitial bands (Cuñado 1992).

FISH analysis detected one major 18S-5.8S-25S rDNA locus in the short arm of chromosome 1C (Linc et al. 1999). Minor 18S-5.8S-25S rDNA loci were mapped in the short arms of 5D and 5C. 5S rDNA loci were identified in the short arm of chromosomes 1C, 5D, 5C and 1D. GISH analysis detected inter-genomic translocation in three of the five *Ae. cylindrica* accessions studied. The breakpoints in all translocations were non-centromeric, with similar-sized segment exchanges (Linc et al. 1999).

Cermeño et al. (1984b) analyzed the activity of the nucleolar organizer regions (NORs) in somatic metaphase chromosomes of *Ae. cylindrica* by phase contrast, C-banding and Ag-staining. They reported that the nucleolar activity of the D subgenome is completely suppressed by the C subgenome.

A recent FISH analysis, using repetitive DNA probes (Afa family, pSc119.2, and 18S rDNA) on root tip mitotic metaphase spreads, enabled identification of the entire set of chromosomes in *Ae. cylindrica* (Molnar et al. 2015). In addition, use of two-color GISH that differentially labeled total genomic DNA from *Ae. caudata* and *Ae. tauschii*, discriminated the constituent C and D subgenomes of *Ae. cylindrica*. Yet, certain differences were found between the C genome of *Ae. caudata* and that of *Ae. cylindrica* (Molnar et al. 2015).

Chromosome 2C has a gametocidal effect when added as a monosome to common wheat cv. Chinese Spring (Endo 1996,). Yet the effect of this gametocidal chromosome is relatively mild and there are viable progeny that lack the *Ae. cylindrica* chromosome. Approximately half of these progeny exhibit structural changes, such as deletions and translocations in various regions of all the wheat chromosomes. Thus, the ability of the gametocidal chromosome 2C to induce deletions of a variety of chromosomal segments observed in the viable progeny of the monosomic addition line that lacks the gametocidal chromosome, has been exploited to induce deletions in various wheat stocks (Endo 2007). The deletions that were stably transmitted to the offspring that became homozygous lines, were identified by C-banding (Endo and Gill 1996). The deletion stocks showed variations in morphological, physiological and biochemical traits, depending on the size of their chromosomal deficiency, and constitute powerful tools for physical mapping of wheat chromosomes (Endo and Gill 1996).

9.6.3.3   Crosses with Other Species of the Wheat Group

Chromosome pairing at first meiotic metaphase of *Ae. cylindrica* was regular, i.e., 0.09 univalents, 3.61 rod bivalents, 10.30 ring bivalents and 24.21 chiasmata per cell (Cuñado 1992). With the ability to identify the chromosomes of the D and C subgenomes at first meiotic metaphase, Cuñado (1992) determined chromosomal pairing of each subgenome. Average pairing of D subgenome chromosomes was 0.03 univalents, 1.58 rod bivalents, 5.39 ring bivalents and 12.31 chiasmata. That of the C subgenome chromosomes was 0.65 rod bivalents, 6.34 ring bivalents and 13.34 chiasmata. Interestingly, in spite of the occurrence of three pairs of chromosomes with extremely subterminal centromeres in the C subgenome (Chennaveeraiah 1960), there are more ring bivalents in this subgenome than in the D subgenome.

Chromosomal pairing at first meiotic metaphase was analyzed in F_1 hybrids between *Ae. cylindrica* and each of its two parental diploids. The triploid hybrid *Ae. cylindrica* x *Ae. caudata* (genome DCC) had 6–7 bivalents and 0–1 trivalents (Kihara 1954) and the triploid hybrid *Ae. cylindrica* x *Ae. tauschii* (genome DCD) had 8–9 bivalents and 0–5 trivalents (Kihara 1949). Evidently, one of the two subgenomes of *Ae. cylindrica*, subgenome C is homologous to the genome of *Ae. caudata* while the second, subgenome D, is homologous to that of *Ae. tauschii*. The genomes of the two diploids are not homologous, as evident from the average chromosomal pairing in the hybrid between them, that had only 3.60 bivalents, of which only 0.60 were ring bivalents, and 1.06 trivalents (Sears 1941b; Table 9.4).

Data of chromosomal pairing in F_1 hybrids between *Ae cylindrica* and allopolyploids of Section Vertebrata are presented in Table 9.7. The hybrid *Ae. cylindrica* x tetraploid *Ae. crassa* exhibited 6.00 bivalents and several multivalents and the reciprocal hybrid had 7.30 bivalents (Kimber and Zhao 1983). The hybrid *Ae. ventricosa* x *Ae. cylindrica* also had seven bivalents (Kimber and Zhao 1983). Likewise, the hybrids between *Ae. cylindrica* and the hexaploid species *Ae. juvenalis* x *Ae. cylindrica*, and the hybrid *Ae. vavilovii* x *Ae. cylindrica*, and hexaploid *Ae. crassa* x *Ae. cylindrica*, had about 7 bivalents at meiosis (McGinnis and Melnyk 1962; Melnyk and McGinnis 1962; Kimber and Zhao 1983). These data indicate that the D subgenomes in all these species are still closely related.

Average chromosome pairing in the hybrid *T. turgidum* ssp. *durum* x *Ae. cylindrica* (genome BADC) was 27.50 univalents and 0.50 bivalents (Sears 1944b), indicating that these two species do not share a common subgenome. In contrast the *T. aestivum* x *Ae. cylindrica* hybrid (genome BADDC) had about 7 bivalents (Riley 1966a; Kimber and Zhao 1983), indicating that the third subgenome of hexaploid wheat is homologous to one of the subgenomes of *Ae. cylindrica*.

9.7 Section Comopyrum (Jaub. & Spach) Zhuk.

9.7.1 General Description

Section Comopyrum (Jaub. & Spach) Zhuk. (Syn.: Macrathera Eig) consists of annual, predominantly autogamous two species. Plants are many-tillered with thin, short (10–30-cm-tall; excluding spikes) culms, with narrow, usually hairy leaves, and narrow, cylindrical or moniliform linear-lanceolate or narrow-elliptical, short (1.5–7.0-cm-long) spikes that gradually taper toward the tip. At maturity, the entire spike falls as one unit. There are 2–3 long narrow to oval spikelets that are shorter than the adjacent rachis segment, with the terminal spikelet sometimes being sterile. There are 1–3 rudimentary spikelets at the base. Spikelets contain 3–4 florets, the upper 1–2 being sterile. Glumes of apical spikelets have either 3 awns (a well-developed central one and 2 more slender, lateral ones which are sometimes reduced to teeth), or only 1 awn. Glumes of lateral spikelets are rough or hairy, with the upper parts overlapping, with two teeth with an angle between; sometimes one of the teeth develops into a small awn. Lemmas are seldom awned in the upper spikelet. The caryopsis adheres to lemmas and palea (Fig. 9.3).

Section Comopyrum contains two diploid species, *Ae. comosa* Sm. in Sibth. & Sm. (genome M), and *Ae. uniaristata* Vis. (genome N). The former has a cylindrical and the latter a moniliform spike. Section Comopyrum was first described by Jaubert and Spach in 1850–1851, who

included in it only one species, *Ae. comosa*, that contained one variety, var. *subventricosa*. Zhukovsky (1928), based on taxonomical criteria, classified *Ae. caudata* L., *Ae. comosa* Sm. in Sibth. & Sm., *Ae. heldreichii* Holzm. and *Ae. uniaristata* Vis. in section Comopyrum. Eig (1929a) also included *Ae. caudata*, together with *Ae. comosa* and *Ae. uniaristata*, in section Macrathera (currently Comopyrum), but included *Ae. heldreichii* in *Ae. comosa*. Based on studies of chromosomal pairing in the interspecific hybrids, Kihara (1937, 1940b, 1954) transferred *Ae. caudata* from section Comopyrum to section Cylindropyrum. Moreover, Kihara (1937, 1940b) thought that *Ae. comosa* and *Ae. uniaristata* are very closely related species and designated their genomes M and M^u, respectively. However, Chennaveeraiah (1960), based on a karyotypic analysis, suggested to designate the genome of *Ae. uniaristata* N. This suggestion was accepted by Kimber and Sears (1987), Kimber and Feldman (1987), Kimber and Tsunewaki (1988), and Dvorak (1998). Wang et al. (2000) claimed that the classification of section Comopyrum should be reconsidered based on their study of internal transcribed spacers (ITS) of nuclear ribosomal DNA sequences. Badaeva et al. (1996b) also postulated that *Ae. uniaristata* should be removed from section Comopyrum based on in situ hybridization and chromosome morphology data. However, Yamane and Kawahara (2005) reported that the unrooted strict consensus tree based on BPS + indels + microsatellites showed that *Ae. uniaristata* and *Ae. comosa* are clustered together. This finding is in agreement with previous molecular studies, based on chloroplast DNA RFLP analysis, that showed monophyletic origin of the two species of section Comopyrum (Terachi et al. 1984; Ogihara and Tsunewaki 1988; Tsunewaki 2009).

The species of this section grow in the western part of the distribution area of the diploid *Aegilops* species, i.e., western Turkey, Greece (incl. Crete) Albania, Macedonia, Serbia, Bosnia-Herzegovina, and Croatia. They grow on terra rossa soil in the edges of sclerophyllous and deciduous oak forests and maquis, in open and degraded dwarf shrub formations, pastures, abandoned fields, edges of cultivation and roadsides. It is common and forms dense stands in many localities and grows as altitudes of almost sea level to 1000 m.

9.7.2 *Aegilops comosa* Sm. in Sibth. & Sm.

9.7.2.1 Morphological and Geographical Notes

Ae. comosa Sm. in Sibth & Sm. [Syn.: *Ae. subventricosa* Jaub & Spach ex Bornm.; *Triticum heldreichii* (Holzm. ex Boiss.) K. Richt.; *Ae. turcica* Azn.; *Triticum comosum* (Sm.) K. Richt.; *Ae. ambigua* Hausskn.; *Comopyrum comosum* (Sm.) Á. Löve] is a predominantly autogamous, annual, tufted, multi-tillered plant. It has thin 10–35-cm-tall (excluding spikes) culms, that geniculate at the base then rising upright, with narrow, and usually hairy leaves. Its spike is linear-lanceolate to narrow-elliptical to narrow-oval, short (2–3.5-cm-long), usually rough or hairy, and gradually tapering towards the tip. The entire spike disarticulates at maturity (umbrella-type of dispersal unit). There are 3–4(-5), long, narrow to oval spikelets, that are shorter than the adjacent rachis segment. There are 1 and rarely 2 rudimentary spikelets at base. Each spikelet contains 3–4 florets, with the upper 1–2 being sterile. Glumes are more or less rough or hairy, with the upper parts overlapping. Glumes of lateral spikelets have two teeth, with an angle between them; sometimes one of the teeth develops into a short awn. The glumes of the terminal spikelet have 3 (seldom 1) large, 4–11-cm-long awns. The awns on each glume spread laterally from each other and the central awn of each glume diverges laterally from the central awn of the other glume of the spikelet. Lemmas are membranous, cartilaginous in the upper parts, and seldom awned in the upper spikelets. The caryopsis adheres to lemma and palea (Fig. 9.3b).

Ae. comosa is a highly polymorphic species. Variation involves spike and spikelet structure and number and length of awns. Eig (1929a) divided the species into two subspecies, ssp. *eucomosa* (currently *comosa*) and ssp. *heldreichii* (Boiss.) Eig. Ssp. *eucomosa* contains three varieties, var. *comosa* (=var. *typica* Eig), var. *thessalica* Eig, and var. *ambigua* Eig [=*Ae. ambigua* (Hausskn.) Eig]. Ssp. *heldreichii* contains four varieties: var. *subventricosa* Boiss., var. *achaica* Eig, var. *biaristata* Eig, and var. *polyathera* Hausskn. Hammer (1980), accepted Eig's division, but changed the name of subsp. *eucomosa* to subsp. *comosa*.

The question if the taxa *comosa* and *heldreichii* is composed of two subspecies of *Ae. comosa* or of two different species, *Ae. comosa* and *Ae. heldreichii*, was a matter of dispute. Zhukovsky (1928), on a taxonomical ground, and Kihara (1940b), on a cytogenetic basis, regarded ssp. *heldreichii* as a separate species, *Ae. heldreichii* (Holzm.) Halac. Badaeva et al. (1999) supported this view of separation of these two taxa into two separate species, arguing that *Ae. comosa* and *Ae. heldreichii* were well-differentiated (though similar genomes) and consequently, must be considered separate species. This argument was based on the finding that the two taxa differ in heterochromatin content, morphology and C-banding pattern of several chromosomes, as well as in the number of the 5S rRNA gene loci (Badaeva et al. 1996a, 1999).

Teoh et al. (1983) also observed differences in the C-banding patterns of the two taxa: *comosa* had predominantly interstitial C-bands, whereas *heldreichii* had centromeric and telomeric bands. Yet, since this sort of change has no effect on chromosome pairing and fertility of the F_1 hybrids between the two taxa, they saw no justification to regard them as two separate species. The two taxa had similar karyotype structure, similar pattern of hybridization with pSc119 and pAs1, and similar distribution of major and minor NORs (Badaeva et al. 1999). Moreover, based on the observation of Kihara (1940b), that five bivalents and one quadrivalent are regularly formed at meiosis of the F_1 hybrid between the two subspecies and that the hybrid is fertile, Lilienfeld (1951) concluded that the two taxa should be at the intraspecific rank rather than at the specific one. In agreement to this, Chennaveeraiah (1960) reported only minor karyotypic differences between the two subspecies. Wang et al. (2000) analyzed nucleotide sequences of the internal transcribed spacer (ITS) of nuclear ribosomal DNA in *Ae. comosa*, and found that ssp. *comosa* and ssp. *heldreichii* were closely related and formed a sister-group relationship. The two subspecies grow sympatrically in the same geographic area and in the same ecological niches (van Slageren 1994). Since there is no noteworthy ecogeographical difference between them, the existence of a reciprocal translocation between the two taxa is not sufficient to separate them into two species.

Van Slageren (1994) considered the two intraspecific taxa as varieties, var. *comosa* and var. *subventricosa* Boiss. [syn: subsp. *heldreichii* Eig; *Ae. subventricosa* Jaub. & Spach ex Bornm.; *Triticum heldreichii* (Holzm. ex Boiss.) K. Richt.; *Ae. turcica* Azn.], which followed Art 35.3 of the International Code of Botanical Nomenclature (ICBN), adopted by the 14th International—Botanical Congress of Berlin 1987 (Greuter et al. 1988). According to this code, when the two taxa are classified at the variety level, the final epithet must be *subventricosa*, and when classified at the subspecific level, the epithet must be *heldreichii*. Since the epithet *heldreichii* is usually used by researchers of this group, it is adopted in this book and presented as ssp. *heldreichii*.

According to van Slageren (1994), the two taxa differ from one another by the following morphological characteristics: *comosa* has a cylindrical spike, is not inflated, and bears 3–4(-5) spikelets. The apical glume has 1 strongly developed, diverging, 4–11-cm-long central awn, and 2 lateral awns, more-slender, 3–7.5(-10) long awns. *Heldreichii* is shorter than *comosa*, with a monoliform spike that contains 1–2(-3) fertile spikelets; glumes of the apical spikelet have 1 strongly developed, diverging, 3–5.5-cm-long central awn, and 2 lateral, more-slender, 2–2.5-cm-long awns that are frequently reduced to teeth or absent. The spikelets of *heldreichii* are more inflated than those of *comosa*, and *heldreichii* has 1 + 3 awns on the terminal spikelet, while *comosa* has 3 + 3. The awns of *heldreichii* are usually shorter than those of *comosa* and the uppermost spikelet is smaller. Intermediate forms between the two taxa are frequent. The two taxa grow sympatrically in Greece, Albania and in western Turkey. *Heldreichii* does not grow in Cyprus and Bulgaria and is rare throughout its range, but possibly more common in Greece.

Ae. comosa distributes in the northeastern Mediterranean: in western Turkey, northern Cyprus, Greece (including

Aegean Islands and Crete), Albania and southern Bulgaria. Greece is its main area of distribution. It grows on terra rossa soil and on dry, sandy, grassy and rocky places, in the edges of sclerophyllous and deciduous oak forests and maquis, in edges of and opening in garrigue, open and degraded dwarf shrub formations, cleared areas, pastures, abandoned fields, edges of cultivation and roadsides. It is common in Greece, frequently abundant and forms dense stands with restricted ranges. It is uncommon to rare in other regions of its distribution area. It grows at altitudes of almost sea level to 500 m, rarely 800 m.

Ae. comosa has a relatively limited distribution in the central-western region of the distribution of the genus. It is a Mediterranean element. Together with *Ae. uniaristata*, the other species of section Comopyrum, it occupies the western part of the distribution region of the diploid species of this group. It grows sympatrically with *Ae. caudata*, *Ae. uniaristata*, *Ae. umbellulata*, *Ae. geniculata*, *Ae. biuncialis*, *Ae. neglecta*, *Ae. recta* and *Ae. triuncialis*, and allopatrically with *Ae. speltoides*, *Ae. peregrina*, *Ae. cylindrica* and wild *T. monococcum*.

Ae. comosa is resistant to all the physiologic races of yellow rust (*Puccinia striiformis*) for which it has been tested (Riley et al. 1968).

9.7.2.2 Cytology, Cytogenetics, and Evolution

Ae. comosa is a diploid species (2n = 2x = 14), bearing the M genome (Kihara 1937, 1940a, b, 1954; Kimber and Tsunewaki 1988; Dvorak 1998). The organellar genome of its two subspecies, subsp. *comosa* and subsp. *heldreichii*, are designated M and M^h, respectively (Tsunewaki 1993; Ogihara and Tsunewaki 1988). The nuclear genome of *Ae. comosa* is small (1C DNA = 5.53 ± 0.052 pg) (Eilam et al. 2007 and Table 9.3). Eilam et al. (2007) determined DNA content in 3 different accessions of *Ae. comosa*, two belonging to subsp. *comosa* and one to subsp. *heldreichii*, that were collected from different regions of the distribution area of the species. Very little intraspecific variation in DNA content was found between the accessions.

According to Senyaninova-Korchagina (1932), the karyotype of *Ae. comosa* consists of 5 types of chromosomes including one pair with a large satellite on the short arm, one pair with an almost median centromere, and the rest with submedian centromeres. The karyotype described by Chennaveeraiah (1960) is asymmetric and differs in some respects from that described by Senyaninova-Korchagina (1932). It has two pairs with secondary constrictions, one pair with almost median centromere, three pairs with submedian centromeres and one pair with a somewhat submedian-subterminal centromere. One of the two pairs with the secondary constrictions has a large satellite, while the second

pair has a small satellite; the satellites in both pairs are on the short arm.

Chennaveeraiah (1960) reported that the karyotype of ssp. *heldreichii* is almost similar to that of ssp. *comosa*. The differences in the karyotypes of these two subspecies are that ssp. *heldreichii* has two pairs with a submedian-subterminal centromere in contrast to one in ssp. *comosa*. In addition, two pairs in *heldreichii* versus one in *comosa* have almost median centromeres, and only one pair in *heldreichii* has a submedian centromere. The difference between the two subspecies lies mainly in the arm ratio of these pairs.

Similar results were obtained by Karataglis (1975), who studied the karyotypes of three varieties of *Aegilops comosa* from Greece, two (var. *biaristata* and var. *subventricosa*) of ssp. *heldreichii*, and one (var. *thessalica*) of ssp *comosa*. He found that the two varieties of ssp. *heldreichii* had the same chromosome morphology, but differed in their total chromosome length, in the length of the long arms and in the length of the satellites. The karyotype of var. *thessalica* was slightly different from those of *heldreichii*; the longest satellite of *heldreichii* occurred in the longest SAT chromosome, whereas the contrary was observed in the shortest SAT in the case of *thessalica*. From the two varieties of *heldreichii*, var. *biaristata* stands closer to var. *thessalica* than var. *thessalica* does to var. *subventricosa*. The great similarity between the karyotypes of the two varieties of *heldreichii* indicates their close phylogenetic relationship whereas the slight differences between the karyotypes of the varieties of *heldreichii* and that of *comosa* show some remoteness between the two subspecies, albeit insufficient to elevate them to the specific rank.

Teoh et al. (1983) used a repetitive DNA sequence, pTa71, that codes for ribosomal RNA, and which was derived from *T. aestivum* by Gerlach and Bedbrook (1979), as a probe in an *in-situ* hybridization experiment on the genome of *Ae. comosa*. They observed two pairs of rRNA sites in *Ae. comosa*, as expected from the presence of two pairs of SAT chromosomes (Chennaveeraiah 1960). An extra pair of rRNA gene cluster located terminally in some cells, but not on others, was also observed (Teoh et al. 1983). This third pair is not in correspondence with the number of satellites in this species.

C-banding studies also showed that the karyotypes of ssp. *comosa* and ssp. *heldreichii* are asymmetric, possessing metacentric, submetacentric and sub-telocentric chromosomes (Badaeva et al. 1996a). The distribution of the 18S-26S and 5S ribosomal RNA gene families on chromosomes of *Ae. comosa* was studied by in situ hybridization with pTa71 (18S-26S rDNA) and pTa794 (5S rDNA) DNA clones (Badaeva et al. 1996b). Two major 18S-26S rDNA loci were found in the nucleolar organizer region (NOR) of chromosomes 1 M and 6 M. In addition to the major NORs, minor loci were observed in all the chromosomes, except for 6 M of ssp. *comosa* and 7 M of ssp. *heldreichii* (Badaeva et al. 1996b). Some minor loci were polymorphic, whereas others were conserved. All major and minor 18S-26S rDNA loci were on the short arms, except for chromosome 4 M (in both subspecies) and 3 M (in ssp. *heldreichii*), where the minor loci were on the long arm. One major 5S rDNA was located on the short arm of chromosome 1 M and a minor locus on the short arm of 5 M. The minor 5S rDNA on 5 M was absent in ssp. *heldreichii*.

The genomes of ssp. *comosa* and ssp. *heldreichii* had similar highly repetitive DNA sequence (Badaeva et al. 1996a) and NOR probe labelling patterns (1996b). However, they differed from each other in the amount and distribution of heterochromatin, the number of 5S rDNA loci, and the chromosomal locations of some minor 18S-26S rDNA hybridization sites (Friebe et al. 1996; Badaeva et al. 1996b).

The chromosomal locations of the major NORs were similar in *Amblyopyrum muticum*, in *Ae. speltoides* and in both subspecies of *Ae. comosa* (Badaeva et al. 1996b). However, *Ae. comosa* differs from *A. muticum* by the presence of several minor NORs, C-banding patterns, and the distribution of pSc119 and pAs1 DNA clones (Badaeva et al. 1996b). The phylogenetic proximity between *A. muticum* and *Ae. comosa* is evident from the data obtained from a comparative analysis of repeated nucleotide sequences in the *Triticum–Aegilops* group (Dvorak and Zhang 1992b). Thus, it seems that *A. muticum* contributed to the development of *Ae. comosa* via hybridization.

Cermeño et al. (1984b) analyzed the nucleolar activity in *Ae. comosa* via comparative analysis of somatic metaphase chromosomes by phase contrast, C-banding and Ag-staining. There were four Ag-NORs in *Ae. comosa*, and the number of nucleoli at interphase registered were one in 38, two in 121, three in 42, and four in 7 cells. Accordingly, the maximum number of nucleoli fits the expectation from the number of pairs of Ag-NORS and SAT chromosomes.

Teoh and Hutchinson (1983), using an improved C-banding technique, found that *Ae. comosa* has very faint and small centromeric bands in all chromosomes. The NORs can be clearly seen but are not banded. Moreover, there is an intraspecific variation in C-banding patterns differentiating the two subspecies of *Ae. comosa*, i.e., *comosa* and *heldreichii* (Teoh et al. 1983). Ssp. *comosa* has predominantly interstitial C-bands, in contrast to centromeric and telomeric bands in ssp. *heldreichii*. Yet, these polymorphic differences between the two subspecies have no effect on chromosome pairing or fertility of the F$_1$ hybrids between these two taxa (Kihara 1940a). Polymorphic differences in C-bands patterns appear to be widespread within and between the two subspecies of *Ae. comosa* and their non-random distribution seems to suggest that these differences could be intimately

associated with the processes of subspeciation as suggested by Teoh et al. (1983).

Friebe et al. (1996) analyzed the C-banding pattern and polymorphisms in several accessions of the two subspecies of *Ae. comosa* and established the standard karyotypes of these subspecies. Although variation in C-band size and location was observed between different accessions, the different chromosomes were still distinct. The homoeologous relationships of these chromosomes were established by comparison of chromosome morphologies and C-banding patterns of *comosa* chromosomes to those of other diploid *Aegilops* species with known chromosome homoeology (Friebe et al. 1996). In addition, in situ hybridization analysis with a 5S rDNA probe was used to identify homoeologous groups 1 and 5 chromosomes. The present analysis enabled assignment of all the chromosomes of ssp. *comosa* and ssp. *heldreichii* to their homoeologous groups.

Teoh et al. (1983) used a non-coding repetitive DNA sequence, TC22b, which was derived from the B subgenome of hexaploid wheat by HaeIII restriction endonuclease digest (Hutchinson and Lonsdale 1982). *Ae. comosa* exhibited an intermediate amount of labeling with this probe. Comparisons between the in-situ hybridization and C-banding results indicated that, *Ae. comosa*, which has little heterochromatin, also shows little labeling with the TC22b probe.

Badaeva et al. (1996a), using in-situ hybridization with the highly repetitive DNA sequences pSc119 and pAs1, combined with C-banding, found that *Ae. comosa* possessed an intermediate amount of C-heterochromatin. The genome of ssp. *heldreichii* was characterized by a very low C-heterochromatin content, distinguishing it from ssp. *comosa*. Only few small C-bands were detected in this subspecies and they were located mainly at the telomeric regions.

Five chromosomes of *Ae. comosa* had hybridization sites with pSc119, whereas all chromosomes were labeled with pAs1. One to four pAs1 hybridization sites per chromosome were observed in *Ae. comosa* and each chromosome had a distinctive labeling pattern (Badaeva et al. 1996a). Yet, moderate levels of labeling were detected in ssp. *comosa* and ssp. *heldreichii* and the distribution of hybridization sites of both the pSc119 and pAS1 probes on chromosomes of these two subspecies was similar but not identical; the distribution of pAS1 sites in several chromosomes of ssp. *heldreichii* was different from the distribution in their homologous chromosomes of ssp. *comosa*.

The two species of section Comopyrum, *Ae. comosa* and *Ae. uniaristata*, display very different C-banding patterns (Teoh and Hutchinson 1983; Badaeva et al. 1996a). Kihara (1937, 1940a, b) designated the genome of *Ae. uniaristata* M^u because, based on chromosome pairing in the F$_1$ hybrid between *Ae. comosa* and *Ae. uniaristata*, he thought that the genome of *Ae. uniaristata* is related to genome M of *Ae.*

comosa. However, Chennaveeraiah (1960) found major karyotypic differences between the two species and suggested to designate the genome of *Ae. uniaristata* N. Support for this suggestion came from the observation, made by Maan and Sasakuma (1978), of high univalent frequencies in amphiploid combinations containing these two species, indicating the homoeology level of their genomes. The different C-banded patterns of *Ae. comosa* and *Ae. uniaristata* reinforces this conclusion and the adoption of genome symbol N for *Ae. uniaristata* ((Teoh and Hutchinson 1983). Indeed, Kimber et al. (1983), based on meiotic chromosome pairing configurations in triploid, tetraploid and pentaploid hybrids involving *Ae. comosa* and *Ae. uniaristata*, found different genomes between the two, and consequently, proposed the genome symbol U^n for *Ae. uniaristata*. Later, Kimber and Sears (1987) and Kimber and Feldman (1987) adopted the symbol N for the genome of *Ae. uniaristata*. In accord with the above, Yamane and Kawahara (2005) clustered *Ae. comosa* and *Ae. uniaristata* as sister species in one clade in the trees based on SSRs and cpDNA, and Haider et al. (2010) constructed two phylogenetic trees based on DNA analysis using RAPDs and ISSRs and found that *Ae. comosa* and *Ae. uniaristata* are sister species. Earlier studies had also revealed such a close genetic relationship between the two species based on cpDNA analysis (Terachi et al. 1984; Ogihara and Tsunewaki 1988; Haider 2003) and RFLPs (Dvorak and Zhang 1992b).

Ogihara and Tsunewaki (1988) and Haider and Nabulsi (2008) mentioned that the cpDNA in the two species showed identical restriction profiles using restriction enzymes. Similarly, the two species were clustered together based on the analysis of the 5S rDNA sequence (Appels et al. 1992) and restriction of repeated nucleotide sequences (Ogihara and Tsunewaki 1988). In another study carried out by Cuñado and Santos (1999), authors confirmed that there is a close relationship between the genomes of both species based on chromosome pairing in a hybrid of the two species. All these data agree with the classification of Eig (1929a), who treated these two species as sister species in the section Comopyrum.

Genome analysis studies (Kihara 1949) have shown that *Ae. comosa* was involved in the parentage of the allotetraploid species, *Ae. geniculata* (genome UUM^oM^o) and *Ae. biuncialis* (genome UUM^bM^b) of section Aegilops.

9.7.2.3 Crosses with Other Species of the Wheat Group

Meiosis in *Ae. comosa* was found to be very regular (Chennaveeraiah 1960); seven bivalents, mostly ring bivalents, were found in all 67 PMCs studied. Similar data were obtained by Cuñado (1992) who, using C-banding to study chromosome pairing at first meiotic metaphase of *Ae. comosa* ssp. comosa, observed an average of 0.89 rod

bivalents and 6.11 ring bivalents (13.11 chiasmata/cell) and in ssp. *heldreichii*, an average of 0.02 univalents, 1.19 rod bivalents and 5.79 ring bivalents (12.77 chiasmata/cell).

Yamada and Suzuki (1941) studied chromosome pairing in F$_1$ hybrids between ssp. *heldreichii* and ssp. *comosa* and observed in most (about 90%) PMCs, an average of 5.00 bivalents and 1.00 quadrivalent. Metaphase cells with 7.00 bivalents or a univalent, 5.00 bivalents and 1 trivalent were also observed in about 10% of the PMCs. F$_2$ segregated to plants with two chromosome types in a ratio of 1:1, one with 7.00 bivalents and the other with 5.00 bivalents and 1 quadrivalent. The formation of a quadrivalent in these generations indicates the occurrence of reciprocal translocation between the two subspecies.

Kihara and Yamada (1942) and Kihara (1949) observed 1.00 univalent, 5.39 bivalents and 1.00 trivalent in the F$_1$ hybrid of *Ae. comosa* ssp. *comosa* x *Ae. uniaristata*, and a similar level of pairing in the ssp. *heldreichii* x *Ae. uniaristata* hybrid. Similar data in *Ae. comosa* x *Ae. uniaristata* hybrids were obtained by Sears (1941b) and Cuñado and Santos (1999) (Table 9.4), validating that the two species of section Comopyrum are closely related.

In the F$_1$ hybrid between *Ae. speltoides* x *Ae. comosa* Kihara (1949) reported on the presence of 1–7 bivalents (a mode of 6 bivalents), and 0–1 trivalent, whereas in the *Ae. comosa* x *Ae. longissima* hybrid, he found lower pairing, i.e., 0–6 bivalents (a mode of 3 bivalents), and 0–1 trivalent. The higher pairing in the hybrids that involve *Ae. speltoides* is presumably due to the action of the pairing promoters of *Ae. speltoides*. Nevertheless, the above pairing data show that the genome of *Ae. comosa* are homoeologous rather than homologous to those of the Sitopsis species. Similar conclusion was obtained from chromosome pairing in the F$_1$ hybrid between *Ae. comosa* or *Ae. uniaristata* and Sitopsis species (Table 9.4). Sears (1941b) observed higher pairing in the hybrid *Ae speltoides* (HP type) x *Ae. comosa* than in the hybrid *Ae. speltoides* (IP type) x *Ae. uniaristata*. Similar reduced level of pairing was observed in the hybrid *Ae. sharonensis* x *Ae. uniaristata*. The hybrid *Ae. caudata* x *Ae. comosa* had 4–6 bivalents (with a mode of 5 bivalents), and 0–1 trivalents, while the hybrid *Ae. umbellulata* x *Ae. comosa* had a lower level of pairing, i.e., 2–5 bivalents and 0–2 trivalents (Kihara 1949).

The hybrids between the allotetraploid species of section Aegilops bearing a modified M genome, namely, *Ae. geniculata* (genome M^oM^oUU) and *Ae. biuncialis* (genome UUM^bM^b) with *Ae. comosa* (genome MM), had 7 bivalents (Kihara 1949; Kimber et al. 1988) or 6 bivalents and a trivalent (Kimber et al. 1988), indicating that one of the subgenomes of these allotetraploids, subgenome M^o, is very closely related to the genome of *Ae. comosa*. The hybrid between *Ae. comosa* and another allotetraploid species of section Aegilops, i.e., *Ae. columnaris* (genome UUX^nX^n), had much less pairing [3–7 bivalents and 0–3 trivalents (Kihara 1949)], indicating that none of the subgenomes of *Ae. columnaris* are closely related to the genome of *Ae. comosa*.

The hybrids with diploid wheat *T. monococcum*, either domesticated ssp. *monococcum* or wild ssp. *aegilopoides*, exhibited relatively low chromosomal pairing (Table 29). The hybrid with hexaploid wheat *T. aestivum* ssp. *aestivum* x *Ae. comosa* also showed very little chromosome pairing (Table 9.10). Evidently, the genome of *Ae. comosa* is only homoeologous to the subgenomes of allopolyploid wheats.

The cytogenetic relationships between *Ae. comosa* and *Amblyopyrum muticum* were studied by Ohta (1990). The F_1 hybrid *Ae. comosa* x *A. muticum* without B chromosomes of *A. muticum* showed a high frequency of chromosome pairing, whereas pairing in the hybrid with B chromosomes was much reduced (Table 8.2). All the hybrids with 0, 1, and 2 B chromosomes were male sterile (Ohta 1990).

B chromosomes of *A. muticum* effectively suppress pairing between homoeologous chromosomes but do not affect pairing of fully homologous ones (Mochizuki 1964; Dover and Riley 1972; Vardi and Dover 1972). Thus, the pairing in hybrids with to the genome of *Ae. comosa*. The hybrids *Ae. caudata*, *Ae. uniaristata* and *Ae. umbellulata* x *A. muticum* showed a lower frequency of A chromosome pairing than those from the hybrids *Ae. comosa* x *A. muticum*. This may indicate that the genome of *A. muticum* is closer to that of *Ae. comosa* than to those of these three species.

The hybrids between each of the two subspecies of *Ae. comosa* and *A. muticum* (without B chromosomes) showed high frequency of pairing (Ohta 1990). Yet, they differed in the frequency of multivalents. The hybrids involving ssp. *comosa* showed few or no quadrivalents, while those involving ssp. *heldreichii* frequently formed a quadrivalent or a trivalent. The occurrence of a reciprocal translocation between the two subspecies of *Ae. comosa* [Kihara (1937) and Yamada and Suzuki (1941)], causes the difference in the frequency of multivalents in hybrids involving these two subspecies. The high frequency of multivalents in the hybrid with ssp. *heldreichii* may indicate that this translocation was formed in ssp. *heldreichii*.

9.7.3 *Aegilops uniaristata* Vis.

9.7.3.1 Morphological and Geographical Notes

Aegilops uniaristata Vis. [Syn.: *Triticum uniaristatum* (Vis.) K. Richt.: *Chennpyrum uniaristatum* (Vis.) Á. Löve] is predominantly an allogamous, annual plant, tufted with many culms, which are 15–35-cm-tall (excluding spikes), and usually prostrate before turning upwards. Leaves are narrow, linear and usually hairy in the upper pars. The spike is moniliform, short, 1.5–4.0-cm-long (excluding awns), lanceolate to oval-lanceolate, glabrous, and tapering rapidly to the tip. The spike disarticulates entirely at maturity (umbrella-type dispersal unit). There are 3–5 number of spikelets, the terminal one usually being sterile, and 2–3 basal rudimentary spikelets. There are 4 florets, the upper 1–2 sterile. Glumes of the lateral spikelets have a conspicuous triangular tooth separated from an awn (up to 4-cm-long) by an acute angle. Glumes of the terminal spikelet end stepwise or abruptly in a broad, flat awn (3–5-cm-long), sometimes with an accompanying tooth. The middle nerve of the awn of the terminal glume is strongly projecting and is a continuation of the projecting nerve of the glume. Lemmas of the lateral spikelets have small teeth which sometimes elongate into small awns. The lemmas of the terminal spikelet have a weakly developed awn (up to 2-cm-long) and 1–2 adjacent teeth. The caryopsis adheres to lemma and palea (Fig. 9.3c).

Ae. uniaristata has limited morphological variation, primarily manifesting in spike and spikelet size. Kawahara (2000) analyzed genetic variation at 21 enzyme loci of *Ae. comosa* and *Ae. uniaristata*, the two species exhibited different levels of genetic variation; in *Ae. comosa*, the mean number of alleles per locus was 2.00 and the proportion of polymorphic loci was 0.667, while in *Ae. uniaristata*, they were 1.19 and 0.143, respectively. No heterozygotes were found, confirming that *Ae. uniaristata* and *Ae. comosa* are self-pollinating species, as was also reported by Hammer (1980). Since *Ae. uniaristata* occupies a relatively limited geographical area and is uncommon or rare throughout its distribution area (van Slageren 1994), it is probable that this species consists of small populations that are susceptible to the loss of genetic variation due to genetic drift during its evolution (Kawahara 2000).

Morphologically, *Ae. uniaristata* is close to *Ae. comosa*, particularly to several forms of *Ae. comosa* ssp. *heldreichii*, from which it differs in the following characters: the glumes of the terminal spikelet carry only one awn and those of the lateral spikelets have an awn and a large triangular tooth.

The taxonomists Zhukovsky (1928) and Eig (1929a), classified *Ae. uniaristata* and *Ae. comosa* as sister species: in section Comopyrum by Zhukovsky and in section Macrathera by Eig. Hammer (1980), and van Slageren (1994) gave priority, on the basis of earliest designation, to Comopyrum. The genome analysis studies of Kihara (1937, 1949, 1954) also arranged *Ae. uniaristata* and *Ae. comosa* in section Comopyrum. This classification was confirmed by the study of Yamane and Kawahara (2005) on the phylogenetic relationships among the diploid *Aegilops-Triticum* species. They found that SSRs and cpDNA clustered *Ae. comosa* and *Ae. uniaristata* as sister species in one clade in the phylogenetic trees. Likewise, the two phylogenetic trees that were constructed by Haider et al. (2010), based on DNA analysis using RAPDs and ISSRs, categorized *Ae. uniaristata* and

Ae. comosa as sister species. Previous cpDNA analyses (Terachi et al. 1984; Ogihara and Tsunewaki 1988; Haider and Nabulsi 2008), and RFLPs (Dvorak and Zhang 1992b) also reported such close genetic relationship between these species. Ogihara and Tsunewaki (1988) and Haider and Nabulsi (2008) mentioned that the cpDNA of these species showed identical restriction profiles using restriction enzymes. Similarly, the two species were clustered together based on an analysis of the 5S rDNA sequence (Appels et al. 1992). In another study, Cuñado and Santos (1999) confirmed that there is a close relationship between the genomes of both species based on chromosome pairing in hybrids between the two species.

Ae. uniaristata has limited distribution in the central-western part of the distribution of the genus. It is a Mediterranean element that grows in European Turkey, Greece (including Crete and the Aegean Islands), Albania, Bosnia-Herzegovina, Serbia, and Croatia, and rarely in southeastern Italy. It is rare in European Turkey, sporadic in Greece, and more common in the Adriatic region of Bosnia-Herzegovina and Croatia.

Ae. uniaristata grows in grasslands and bushy slopes, mainly on rocky, calcareous soils (terra rossa), more rarely on sandstone, in edges of sclerophyllous Mediterranean oak forest and maquis, open or degraded dwarf shrub formations, pasture, disturbed habitats, edges of cultivation and roadsides, at altitudes of 0–750 m.

Ae. uniaristata occupies the northwestern corner of the distribution area of the diploid species of *Aegilops*. It also occupies the most mesophytic of all the habitats of the diploid species. It grows sympatrically with *Ae. geniculata*, *Ae. biuncialis*, *Ae. neglecta*, *Ae. recta*, *Ae. triuncialis*, and *Ae. comosa* (in Greece and European Turkey). It grows allopatrically with *Ae. caudata*, *Ae. umbellulata* and *Ae. cylindrica*.

Ae. uniaristata is tolerant to high levels of soil aluminum (Berzonsky and Kimber 1986, 1989), which remains effective when transferred to wheat (Miller et al. 1995, 1997). It is highly resistant to wheat stem and leaf rusts (Valkoun et al. 1985), and to some other fungal diseases (Gong et al. 2014).

9.7.3.2 Cytology, Cytogenetics, and Evolution

Ae. uniaristata is a diploid species (2n = 2x = 14), with the nuclear genome N (Kimber and Tsunewaki 1988; Dvorak 1998) and organellar genome N (Ogihara and Tsunewaki 1988; Wang et al. 1997). Haider and Nabulsi (2008) showed that *Ae. uniaristata* and *Ae. comosa* have very similar chloroplasts. This finding is in accord with Ogihara and Tsunewaki (1988), who reported that the cytoplasm of both species displayed identical restriction patterns when using 13 endonucleases. In contrast, there has been dispute concerning the nuclear genome of the two species. Originally, Kihara (1937, 1940a, b, 1949), based on genome analysis,

gave to *Ae. uniaristata* the genomic symbol M^u, implying a genome modified from the M genome of *Ae. comosa*. Yet, Chennaveeraiah (1960) observed that the karyotype of *Ae. uniaristata* is unique and considerably different from that of *Ae. comosa* and, consequently, suggested to change its genome symbol to N. This conclusion was further supported by the data of Maan and Sasakuma (1978), who observed high univalent frequencies in amphiploid combinations containing *Ae. uniaristata* and *Ae. comosa*, indicating non-homology of their genomes. Kimber et al. (1983) examined meiotic chromosome pairing in triploid, tetraploid, and pentaploid hybrids involving the genomes of these two species, looking for values of relative affinity, and concluded that there is no preferential pairing between the chromosomes of these genomes. Consequently, they suggested that the M^u genome of *Ae. uniaristata* be changed to Un, and later, it was changed to N (Kimber and Sears 1987; Yen and Kimber 1992b). C-banding and FISH patterns of chromosomes (Teoh and Hutchinson 1983; Teoh et al. 1983; Badaeva et al. 1996a, b; Friebe et al. 1996; and Iqbal et al. 2000a, b) and isozyme studies (Kawahara 2000) confirmed that genome N is different from M. *Ae. uniaristata* is involved in the parentage of the allotetraploid *Ae. ventricosa* (genome DDNN) and the allohexaploid *Ae. recta* (genome UUX^nX^nNN).

Ae. uniaristata has the largest genome in section Comopyrum; its 1C nuclear DNA content = 5.82 ± 0.105 pg is significantly larger than that of *Ae. comosa* (1C DNA = 5.53 ± 0.052 pg) (Eilam et al. 2007). The karyotype of *Ae. uniaristata* was first described by Chennaveeraiah (1960) and was compared with that of *Ae. comosa*. It is an asymmetric karyotype containing only one pair with satellites, whereas in *Ae. comosa* there are two pairs with satellites. Of its six other pairs, two have sub-median centromeres, two have sub-median-sub-terminal centromeres, and the other two have sub-terminal centromeres. None of the distinctive karyotypic characteristics of *Ae. comosa* are seen in *Ae. uniaristata* (Chennaveeraiah 1960). A pair with a secondary constriction and an arm ratio of 1:1:1.5 characteristic to *Ae. comosa* was never found in *Ae. uniaristata*. In *Ae. uniaristata*, there are two pairs with sub-terminal centromeres which are not present in *Ae. comosa*. On the whole, the karyotype of *Ae. uniaristata* differs from, more than it resembles, the karyotypes of the two subspecies of *Ae. comosa*. The karyotypic changes in *Ae. uniaristata* may have occurred by inversions (Chennaveeraiah 1960).

The C-banding pattern at mitotic metaphase of *Ae. uniaristata* chromosomes was first described by Teoh and Hutchinson (1983), and was found to be unique and characteristically different from that of *Ae. comosa*. However, one common feature shared by both species is the absence of prominent telomeric bands. In *Ae. uniaristata*, all the centromeric regions have large and intensely stained bands. There are very few interstitial bands in both arms, which makes identification of chromosomes easier. Badaeva et al. (1996a) also found that large proximal C bands and some medium-sized interstitial bands were typical of the *Ae. uniaristata* N genome.

C-band findings in eight different accessions of *Ae. uniaristata* reported by Friebe et al. (1996) supported the observation of Teoh and Hutchinson (1983). Friebe et al. (1996) observed intraspecific variation in C-band size and location between different accessions, but this did not prevent chromosome identification and establishment of standard karyotypes. Similar to the report by Chennaveeraiah (1960), Friebe et al. (1996) found that the karyotype of *Ae. uniaristata* consisted of two sub-metacentric and five more or less acrocentric chromosome pairs, all with diagnostic markers.

Teoh et al. (1983) used two different cloned repetitive DNA sequences as probes in *in-situ* hybridization experiments on the genome of *Ae. uniaristata*. The TC22b probe is a highly repeated, noncoding 260-bp DNA segment, derived from the B subgenome of hexaploid wheat (Hutchinson and Lonsdale 1982). The rRNA probe they used, described by Gerlach and Bedbrook (1979), consists of 18S and 25S rDNA with an associated spacer DNA. *Ae. uniaristata* chromosomes exhibited little labeling with TC22b and the large centromeric heterochromatic bands showed hardly any labeling with this highly repetitive probe. *In-situ* hybridization with the rDNA probe revealed only one pair of rRNA sites. The location of these sites was on the one pair of satellited chromosomes that was described by Chennaveeraiah (1960).

Badaeva et al. (1996a) performed *in-situ* hybridization with the highly repetitive noncoding DNA sequences, pSc119, derived from *Secale cereale*, and pAsl, derived from *Ae. tauschii*, and observed strong labeling with both pSc119 and pAs1, indicating that the karyotype of this species was highly heterochromatic. The pSc119 probe hybridized to the telomeres of all chromosomes and the distribution of its hybridization sites was similar to that seen in *Ae. umbellulata* and *Ae. caudata*, whereas the strong and unique hybridization sites with pAs1 differed from labeling in *Ae. umbellulata*, *Ae. caudata* and *Ae. comosa* (Badaeva et al. 1996a). This difference, combined with differences in chromosome morphology, karyotypic features, and C-banding, indicates again that *Ae. uniaristata* and *Ae. comosa* have different genomes.

The distribution of the 5S and 18S-26S ribosomal RNA gene families on *Ae. uniaristata* chromosomes was studied by *in-situ* hybridization with pTa71 (18S-26S rDNA) and pTa794 (5S rDNA) DNA clones (Badaeva et al. 1996b). Similar to the finding of Teoh et al. (1983), only one major 18S-26S rDNA locus was found in the single pair of the

nucleolar organizer region (NOR). This chromosome pair was identified as chromosomes 5N by hybridization with rDNA. In addition to the major NORs, several minor loci were also observed in *Ae. uniaristata*, some of which were polymorphic, while others were conserved. The presence of only one pair of chromosomes with major NORs in *Ae. uniaristata* (and in *Ae. tauschii*), and their location on the short arms of homoeologous group 5, is in sharp contrast to the presence of two such pairs, on homoeologous groups 1 and 6, in all other diploid *Aegilops* species.

Two 5S rRNA loci were observed in *Ae. uniaristata*, one major locus located in the distal part of the long arm of chromosome 1N and one minor locus located in the short arm of chromosome 5N. The 5S rDNA loci were not associated with NORs. The different distribution of 5S rRNA loci in *Ae. uniaristata* and *Ae. comosa* also confirm that these two species have distinct genomes.

The nucleolar activity of the NORs in somatic chromosomes of *Ae. uniaristata* was analyzed using a highly reproducible silver-staining procedure (Cermeño et al. 1984b. Two Ag-NORs were identified, one nucleolus was seen in 262 cells at interphase and two nucleoli were seen in 269 cells. All diploid *Aegilops* species analyzed, except for *Ae. uniaristata* and *Ae. tauschii*, showed 4 nucleolus organizer chromosomes, further supporting the difference of *Ae. uniaristata* and *Ae. tauschii* in this respect from the rest of the diploid *Aegilops* species.

The homoeologous relationships of the chromosomes of *Ae. uniaristata* were established by comparison of their morphologies and C-banding patterns to chromosomes of other diploid *Aegilops* species with known chromosome homoeology (Friebe et al. 1996). In addition, *in-situ* hybridization analysis with a 5S rDNA probe was used to identify chromosomes of homoeologous groups 1 and 5. This analysis permitted the assignment of three *Ae. uniaristata* chromosomes to their homoeologous groups. The homoeology of the remaining four acrocentric chromosomes could not be determined. Chromosome 5N is sub-metacentric with a secondary constriction and a small satellite in the short arm. The middle of the short arm of this chromosome has a 5S rDNA in-situ hybridization site. The other sub-metacentric chromosome pair was identified as 4N. The least sub-telocentric chromosome pair had a 5S rDNA in situ hybridization site in a distal region of the long arm, suggesting that this chromosome is at least partially homoeologous to group 1 chromosomes, and was designated accordingly. One accession of *Ae. uniaristata* was homozygous for a whole-arm translocation involving chromosomes 1N and 5N.

Iqbal et al. (2000b) performed RFLP analyses on wheat-*Aegilops uniaristata* addition lines and translocation lines to confirm the identity of the added N-genome chromosomes. Complete 1N, 3N, 4N, 5N and 7N chromosome additions were identified, while the complete long arm and only part of the short arm was identified for chromosome 2N. There were no wheat-like 4/5 and 4/7 translocations in the *Ae. uniaristata* chromosomes (Iqbal et al. 2000a, b). Chromosome 3N carried an asymmetric pericentric inversion. Chromosome-specific RAPD and microsatellite markers were also identified all the added *Ae. uniaristata* chromosomes available in this set of addition lines.

Hybridization sites of the repetitive DNA sequences pAs1, pSc119.2 and pTa71 were identified on the N-genome chromosomes of *Ae. uniaristata* using the FISH technique (Iqbal et al. 2000a, b). Like Badaeva et al. (1996a), Iqbal et al. (2000a, b) also found six pairs of *Ae. uniaristata* chromosomes showing strong hybridization signals with pAs1, while an additional pair showed weak signals with this probe. pAs1 is a D genome specific probe isolated from *Ae. tauschii* (Rayburn and Gill 1986) and shows strong hybridization sites with the D subgenome of common wheat, indicating the relatedness of the N genome to the D subgenome of wheat. For pSC119.2, three pairs of chromosomes showed one signal at the telomeres of their short arms, while two pairs had signals at both the short- and the long-arm telomeres. Two pairs of chromosomes showed three sites on each chromosome, one at the telomere of the short arm and two on the long arm.

Construction of the genetic map of *A. uniaristata* chromosome 3N revealed the important role that asymmetric pericentric inversions played in the evolution of the N-genome (Iqbal et al. 2000a). This mechanism led to the formation of all four sub-telocentric chromosomes of *A. uniaristata* and caused transposition of the 5S rDNA locus from the short arm of chromosome 1N to its long arm.

Gong et al. (2014) performed a C-banding analysis on *Ae. uniaristata* chromosomes present in the amphiploid *Triticum turgidum–Ae. uniaristata* and on the available set of *T. aestivum* cv. Chinese Spring–*Ae. uniaristata* addition lines (6N addition was lacking in this set). Their study showed easily recognizable C-banding patterns for chromosomes 1N–5N and 7N that were distinguishable from wheat chromosomes based on chromosome size and the position of the C-bands. Moreover, chromosome 6N was easily recognized by comparing the C-bands of the six N-genome chromosomes (1N–5N and 7N) to the seven N-genome chromosomes in the amphiploid *Triticum turgidum–Ae. uniaristata*. Similar to the results of Teoh and Hutchinson (1983) and those of Badaeva et al. 1996a), Gong et al. (2014) found that the C-bands mainly exist in the centromeric regions of *Ae. uniaristata* and rarely at the distal ends.

Moreover, FISH on mitotic metaphase chromosomes of the amphiploid *T. turgidum–Ae. uniaristata*, using SSR $(GAA)_8$ as a probe, showed that the hybridization signals of all *Ae. uniaristata* chromosomes differ from those of *T.*

turgidum. Thus, $(GAA)_8$ can be used to identify all *Ae. uniaristata* chromosomes in wheat background simultaneously (Gong et al. 2014). In addition, a total of 42 molecular markers specific for *Ae. uniaristata* chromosomes were developed by screening expressed sequence tag—sequence tagged site (EST-STS), expressed sequence tag—simple sequence repeat (EST-SSR), and PCR-based landmark unique gene (PLUG) primers (Gong et al. 2014) These markers were subsequently localized using the *T. aestivum–Ae. uniaristata* addition lines and different wheat cultivars as controls (Gong et al. 2014).

Evolution of the N-genome has also been associated with amplification, elimination and re-distribution of the different families of repetitive DNA sequences (Badaeva et al. 2011). The most interesting example is the emergence of pericentromeric heterochromatin bands on *Ae. uniaristata* chromosomes, which was probably caused by massive amplification of certain classes of highly repetitive DNA sequences. Large pericentromeric C-bands are also found on the chromosomes of *A. speltoides*; however, the results of a comparative study of several *Aegilops* species using a *TC22b* probe, (Teoh et al. 1983), suggest that the molecular composition of pericentromeric heterochromatin differ between *Ae. uniaristata* and *Ae. speltoides* species. Bardsley et al. 1999) also found that probes for the SSR repeats $(AAC)_5$, $(ACG)_{10}$ and $(CGT)_{10}$, which are the major components of C-bands in wheat and many *Aegilops* species, hybridized poorly to the N-genome chromosomes. Pericentromeric C-bands on *A. uniaristata* chromosomes also did not contain the pAs1 sequence. Moreover, the long arm of chromosome 7N, which displayed the highest amount of hybridization with the pAs1 sequence, contained only few very faint interstitial C-bands, and conversely, the most heterochromatic chromosome, 4N, possessed only two very small pAs1 sites. Thus, the molecular composition and organization of heterochromatin in the *A. uniaristata* genome is distinct from other species and remains to be characterized.

9.7.3.3 Crosses with Other Species of the Wheat Group

Chromosome pairing at first meiotic metaphase of *Ae. uniaristata* was studied by Cuñado (1992), using the C-banding technique. Chromosome pairing was regular with average configurations per PMC of 0.03 univalents, 4.07 rod bivalents, 2.90 ring bivalents and 9.88 chiasmata. Seven bivalents existed in most cells; the relatively large frequency of rod bivalents resulted from the presence of four chromosome pairs with sub-median–subterminal and subterminal centromeres (Chennaveeraiah 1960).

Chromosome pairing in F_1 hybrids between the two species of section Comopyrum, *Ae. uniaristata* and *Ae. comosa*, showed a high number of bivalents, with most being rod bivalents (Table 25). This can result partly from the presence of several chromosome pairs with sub-median–subterminal and sub-terminal centromeres in *Ae. uniaristata* and partly from reduced homology. Cuñado and Santos (1999) studied chromosome pairing in this hybrid by electron microscopy in surface-spread-first meiotic prophase nuclei and compared the results with light-microscopic observations of first metaphase cells after C-banding and FISH. At first prophase, the hybrid showed extensive synapsis and complex multivalents, involving up to 14 chromosomes, but at first metaphase, most associations were in the form of bivalents between homoeologous chromosomes as follows: the average pairing between the M and the N genomes was 5.08 rod bivalents, 0.33 ring bivalents and 0.08 trivalents. Yet, there was also autosyndesis pairing between chromosomes of genome M (0.01 rod bivalents and 0.03 trivalents). In the hybrid *comosa* x *uniaristata*, the mean bivalent frequency at first prophase (3.07) was lower than that observed at first metaphase (5.41). Therefore, a considerable number of bivalents formed by homoeologous chromosomes must be involved in the complex multivalent associations observed at first prophase (Cuñado and Santos 1999).

Chromosomal pairing in hybrids was more or less the same in hybrids between *Ae. uniaristata* and each of the two subspecies of *Ae. comosa*. Average chromosome pairing in the hybrid involving ssp. *comosa* contained 1.00 univalent, 5.00 bivalent and 1.0 trivalents and with ssp. *heldreichii*, 5.48 bivalents (Kihara and Yamada 1942). Percival (1932) observed 0–4 univalents, 5–7 bivalents (mostly rod), and an occasional trivalent in the hybrid *Ae. uniaristata* x ssp. *heldreichii*. The above data show that ssp. *comosa* differs by a reciprocal translocation from *Ae. uniaristata*, whereas such translocation does not exist between ssp. *heldreichii* and *Ae. uniaristata*, indicating that the latter may evolved from ssp. *heldreichii*.

Chromosome pairing in hybrids involving two Sitopsis species, *Ae. speltoides* (IP type) and *Ae. sharonensis*, with *Ae. uniaristata* (Table 9.4) showed fa reduced number of bivalents, most of which were rod, indicating that the genomes of these species have diverged considerably. Likewise, *Ae. uniaristata* x *Ae. bicornis* had 1–6 bivalents and 0–2 trivalents (Kihara 1949), *Ae. longissima* x *Ae. uniaristata* had 0–6 bivalents with a mode of 2, and 0–1 trivalents (Kihara 1949). The diploid hybrid *Ae. tauschii* x *Ae. uniaristata* had 5.94 univalents, 3.63 rod bivalents, 0.08 ring bivalents and 0.14 trivalents (Sears 1941b). In the reciprocal hybrid, Cuñado and Santos (1999) observed a similar level of pairing, i.e., 3.08 rod bivalents, 0.08 ring bivalents and 0.14 trivalents. Evidently, the genomes of these two species have diverged considerably from one another. Chromosomal pairing in hybrids of *Ae. uniaristata* with the diploid species *Ae. caudata* and *Ae. umbellulata* was also relatively low, indicating divergence of the genome of *Ae. uniaristata* from those of *Ae. caudata* and *Ae. umbellulata*. More specifically,

the pairing in the hybrid *Ae. caudata* x *Ae. uniaristata* had 4.50 univalents, 3.50 rod bivalents, 0.44 ring bivalents, 0.46 trivalents and 0.06 quadrivalents (Sears 1941b), or 2–5 bivalents (Kihara 1949). In the hybrid with *Ae. umbellulata*, Percival (1932) observed 6–14 univalents and 0–4 bivalents (all rod) and Sears (1941b) found 8.30 univalents, 2.70 bivalents (all rod) and 0.10 trivalents.

Chromosome pairing with the allotetraploid *Ae. crassa* (genome $D^cD^cX^cX^c$) was similar to the amount of pairing with *Ae. tauschii*, but had a somewhat higher frequency of trivalents (Table 9.5). On the other hand, chromosomal pairing with *Ae. ventricosa* (genome DDNN), that possesses subgenome N from *Ae. uniaristata*, was significantly higher (Table 9.5). Chromosome pairing in hybrids with the allohexaploid cytotype of *Ae. crassa* (genome $D^cD^cX^cX^cDD$) showed higher pairing, most of which was presumably autosyndesis between chromosomes of the two D subgenomes of hexaploid *Ae. crassa*. In contrast, the hybrid with the allohexaploid species *Ae. juvenalis* (genome $D^cD^cX^cX^cUU$), that contains only one D subgenome, showed a significantly low level of pairing (McGinnis and Melnyk 1962).

Similarly, chromosome pairing in the hybrid *Ae. geniculata* (genome M^oM^oUU) x *Ae. uniaristata* (genome NN) indicated low homology and great divergence between the genomes of these species (Table 9.5). The low level of pairing in this hybrid is in spite of the presence of the modified M subgenome of *Ae. geniculata* and the N genome of *Ae. uniaristata*. A similarly low level of chromosome pairing was observed in the hybrids of *Ae. columnaris* (genome UUX^nX^n) x *Ae. uniaristata* (4–6 bivalents and 0–1 trivalents) and of *Ae. neglecta* (genome UUX^tX^t) x *Ae. uniaristata* (2–6 bivalents with mode of 4) (Kihara 1949).

Sears (1941b) observed a low pairing in the hybrids between *T. monococcum*, either domesticated or wild, and *Ae. uniaristata* (Table 9.8), indicating great divergence between the genomes of these two diploid species. Likewise, chromosome pairing in F1 hybrids *Ae. uniaristata* x *Amblyopyrum muticum* without B chromosomes was relatively low, indicating a considerable divergence of the genomes of the two species (Table 8.1). The hybrids were sterile (Ohta 1990).

9.8 Section Aegilops

9.8.1 General Description

Section Aegilops (Syn.: sect. Surcullosa Zhuk.; sect. Polyides Zhuk.; sect. Pleionathera Eig) contains annual, predominantly autogamous species. The plants are slender, with short spikes, which are more or less ovate, elliptic or lanceolate in outline, rarely elongate-linear, mostly awned,

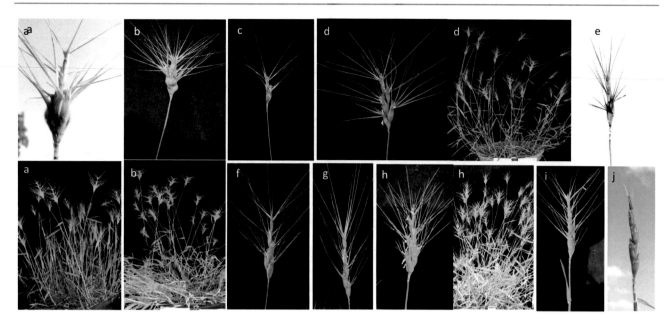

Fig. 9.4 Plants and spikes of *Aegilops* species carrying the U genome; **a** *Ae. umbellulata* Zhuk.; **b** *Ae. geniculata* Roth; **c** *Ae. biuncialis* Vis.; **d** *Ae. neglecta* Req. ex Bertol.; **e** *Ae. recta* (Zhuk.) Chennv.(from Kimber and Feldman 1987); **f** *Ae. columnaris* Zhuk.; **g** *Ae. triuncialis* L.; **h** *Ae. kotschyi* Boiss. **i** *Ae. peregrina* (Hack. in J. Fraser) Maire & Weiler ssp. *euvariabilis* Eig; and **j** *Ae. peregrina* ssp. *cylindrostachys* Eig

and fall as a unit at maturity (umbrella-type dispersal unit). There are 2–5 spikelets, which are ventricose or elliptic, and rarely linear. mostly awned; Spikelets are reduced and not inflated at the upper part of the spike, and usually less than 10 times long as wide, rarely more. The glumes of apical spikelets have 3–4 awns, those of lateral spikelets with 2–4 (-5) awns and or teeth. Glume awns are always stronger than lemma awns. There are 1–4 rudimentary spikelets at the base of the spike. The caryopsis is free or adhered to lemmas and palea (Fig. 9.4).

Section Aegilops contains nine species: one diploid [*Ae. umbellulata* Zhuk. (genome UU), seven allotetraploids [*Ae. geniculata* Roth (genome M^oM^oUU), *Ae. biuncialis* Vis. (genome UUM^bM^b), *Ae. neglecta* Req. ex Bertol. (Genome UUX^nX^n), *Ae. columnaris* Zhuk. (Genome UUX^nX^n), *Ae. triuncialis* L. (genome UUCC), *Ae. kotschyi* Boiss. (genome S^lS^lUU) and *Ae. peregrina* (Hack. in J. Fraser) Maire & Weiler (genome S^lS^lUU)] and one allohexaploid, [*Ae. recta* (Zhuk.) Chennv. (Genome UUX^nX^nNN)]. The spike of all the allopolyploid species resembles that of *Ae. umbellulata*, the U genome donor (Fig. 9.4).

Section Aegilops was divided by Eig into two subsections: subsection Libera Eig, which contains all species with free caryopsis, and subsection Adhaerens Eig, which contains the two species, *Ae. kotschyi* and *Ae. peregrina*, which have an adherent caryopsis.

The distribution of the species of section Aegilops is very wide, one of them, *Ae. triuncialis*, distributes in almost all the distribution area of the genus whereas several others distribute from western Mediterranean to West Asia. The distribution of the single diploid species in the section, *Ae. umbellulata*, is relatively wide, falling in the central part of the genus distribution area, namely, from Greece to west and north Iran. The distribution of the allopolyploid species, on the other hand, is, in most cases, very wide, extending beyond that of their diploid progenitors, and covering a large part of the genus distribution area. While the distribution of *Ae. triuncialis* is extremely large almost overlaps the genus distribution area, that of the three species, *Ae. geniculata*, *Ae. biuncialis* and *Ae. neglecta*, is more in the Mediterranean part of the genus distribution area. The distribution of *Ae. columnaris* extends from the center eastwards up to Iran, *Ae. peregrina* grows in the south and east Mediterranean parts, whereas *Ae. kotschyi*, the steppical species, grows from the south (Libya) toward the east, up to Pakistan. The distribution of the single allohexaploid species, *Ae. recta*, is more limited than that of its allotetraploid progenitor, *Ae. neglecta*; it was found in Portugal, Spain, France, Italy, Croatia, Greece and Turkey. All species of this section (except *Ae. kotschyi*) grow on a great variety of soils (e.g., terra rossa, basalt, alluvial, stand sands and sandy loams), *Ae. kotschyi* grows on loess and grey calcareous steppe soil. All species occupy a large number of habitats and often infest cultivated fields as weeds. The allotetraploids are great colonizers, strive well in disturbed habitats, and often form very dense, mixed populations.

9.8.2 *Aegilops umbellulata* Zhuk.

9.8.2.1 Morphological and Geographical Notes

Aegilops umbellulata Zhuk. [Syn.: *Triticum umbellulatum* (Zhuk.) Bowden; *Ae. ovata* var. *anatolica* Eig; *Kiharopyrum umbellulatum* (Zhuk.) Á. Löve] is a predominantly autogamous, annual plant, tufted with many, upward-bent, 10–30-cm-tall (excluding spikes) culms. Leaves are linear, 2–5-cm-long, and more or less hairy. The spike is lanceolate-ovoid, 1.5–4-cm-long, and typically rough. The entire spike disarticulates at maturity (umbrella-type dispersal unit). Spikelets 5–6, rarely 3, the upper 1–3 being sterile, so the ear suddenly becomes narrow, rudimentary basal spikelets 3, rarely 2. The rachis of the lower spikelets is much shorter than the adjacent spikelet. There are four florets, the upper two being sterile. The rachis segment of the upper spikelets is much longer than the adjacent spikelet causing the narrow upper part of the spike to protrude from the lower wider part. Glumes are similar, shorter than the spikelets, and suddenly inflated above the middle, above which they narrow to a deeply incised margin. Glumes of the lower spikelets have 4–5 (3–6) awns, while glumes of the upper spikelets have 3–5 awns, all of which are similar in shape. Lemmas have 1–3 awns, all resembling the glume awns. All awns diverge at maturity, producing a characteristic umbel shape. The lemma of the lower spikelets has 8–12 veins near the upper margin. The caryopsis is free (Fig. 9.4a).

Morphological variation involves mainly spike size, color and hairiness. There are two forms which differ in spike size, and sometimes grow in mixed populations and in such stands, intermediates may be found. According to Hammer (1980), *Ae. umbellulata* contains two subspecies, subsp. *umbellulata* (glumes with 5–7 awns) and subsp. *transcaucasua* Dorof. et Migusch. (glumes with 3–4 awns), the former contains two varieties, var. *umbellulata* (glabrous glumes) and var. *pilosa* Eig (hairy glumes).

Ae. umbellulata is an East Mediterranean/Western Asiatic element extending into the steppical (Irano-Turanian) region. It is predominantly occurring in Turkey, but also present in Greece (including Rhodes and the Aegean Islands), Syria, Lebanon, North Iraq, West and Northwest Iran, Armenia, and Azerbaijan. It is common and locally abundant, growing at altitudes from sea level to 1800 m, on terra rossa, basalt, alluvial and grey calcareous steppe soils in the edges and openings of sclerophyllous or deciduous oak forest or maquis, degraded dwarf shrub formations, open dwarf shrub semi-steppe and steppe-like formations, abandoned fields, edges of cultivation and roadsides. It often grows as a weed in cultivated areas. *Ae umbellulata* prefers more humid conditions than many other *Aegilops* species, with annual rainfall from 350 to 700 mm.

Ae. umbellulata has a wide distribution in the central region of the distribution of the genus. It occupies a large variety of primary and secondary habitats. In the central part of the genus distribution area, *Ae. umbellulata* occurs in mixed populations with many other species of *Aegilops* and has sporadic contact with others. It grows sympatrically with *Amblyopyrum muticum*, *Ae. speltoides*, *Ae. caudata*, *Ae. comosa*, *Ae. tauschii*, wild *T. monococcum*, wild *T. timopheevii*, *Ae. geniculata*, *Ae. biuncialis*, *Ae. neglecta*, *Ae. recta*, *Ae. columnaris*, *Ae. triuncialis*, *Ae. peregrina*, and *Ae. cylindrica*, and allopatrically with *Ae. longissima*, *Ae. searsii*, *Ae. uniaristata*, wild *T. turgidum*, *Ae. kotschyi*, *Ae. crassa*, and *Ae. juvenalis*.

Ae. umbellulata is resistant to leaf rust (Sears 1956), stem rust (Ozgen et al. 2004), stripe rust (Bansal et al. 2017), powdery mildew, hessian fly and green bug (Gill et al.1985). Sears (1956), using a radiation treatment, pioneered the transfer of a gene resistant to leaf rust, *Lr9*, from chromosome 6U of *Ae. umbellulata* to common wheat chromosome 6B (Sears 1941a).

9.8.2.2 Cytology, Cytogenetics and Evolution

Ae. umbellulata is the only diploid species in the section Aegilops having the basic chromosome number of the section ($2n = 2x = 14$). Both its nuclear genome and organellar genome were designated U (Kimber and Tsunewaki 1988; Dvorak 1998). It is involved in the parentage of the seven allotetraploids of section Aegilops, namely, *Ae. geniculata* (genome M^oM^oUU), *Ae. biuncialis* (genome UUM^bM^b), *Ae. neglecta* (genome UUX^nX^n), *Ae. columnaris* (genome UUX^nX^n), *Ae. triuncialis* (genome UUCC), *Ae. peregrina* (genome S^lS^lUU), and *Ae. kotschyi* (genome S^lS^lUU), and of two allohexaploids (*Ae. recta* (genome UUX^nX^nNN) from section Aegilops and *Ae. juvenalis* (genome $D^cD^cX^cX^cUU$) from section Vertebrata (Kihara 1954, 1957; Kihara and Tanaka 1970; Kimber and Feldman 1987; Kimber and Tsunewaki 1988; Dvorak 1998). As the donor of the U genome to all the allopolyploids of the section, *Ae. umbellulata* contributed many of its morphological features (glume awns, sudden narrowing of the spike, free caryopsis, inflation of the glumes and whole-spike disarticulation) to its related allopolyploids.

Kihara and Lilienfeld (1932) designated the genome of *Ae. umbellulata* C^u and considered it to be a modified genome derived from the C genome of *Ae. caudata* Consequently, Kihara (1954) placed it, together with *Ae. caudata*, in the C-genomic group, based on the assumption that *Ae. triuncialis* ($2n = 4x = 28$; genome UUCC), containing the genomes of *Ae. umbellulata* and *Ae. caudata*, was considered to be an autotetraploid by von Berg (1931), who observed the autosyndetic behavior of the C and C^u genomes of *Ae. triuncialis* in a hybrid with *Secale cereale*. Kihara

(1954) also based his grouping of *Ae. umbellulata* in the same genomic group with *Ae. caudata*, on the formation of some multivalents in a synthetic amphiploid of *Ae. triuncialis* (Karpechenko and Sorokina (1929). Yet, Kihara (1954) himself expressed the inaccuracy of placing these two species in one group, since the view of autopolyploidy of *Ae. triuncialis* was rejected by several cytogenetic studies. Also, there was little or no support of the assumption of Kihara and Lilienfeld (1932) that the two genomes were not widely differentiated at the time of their incorporation into the tetraploid *Ae. triuncialis*. Sears (1948) questioned the genomic similarity of the C and C[u] genomes on both cytological and morphological grounds, pointing out that both the C- and C[u]-genome chromosomes paired with M-genome chromosomes of *Ae. comosa* at about the same frequency as they paired with each other, and further, that the spike morphology of the two species is very different. Such weak chromosomal affinity in meiosis of hybrids between these two species was also found by Sears (1941b) and Kihara (1949). Kimber and Abu-Baker (1981) reached the same conclusion by investigating chromosome pairing in a series of hybrids involving *Ae. umbellulata* and *Ae. caudata*. These findings agree with the taxonomic treatment of *Ae. umbellulata* and *Ae. caudata* that, based on their morphology, placed the two species in two different sections (Zhukovsky 1928), Eig (1929a), Hammer (1980), and van Slageren (1994). In addition, the karyotypes of *Ae. umbellulata* and *Ae. caudata* are widely different (Chennaveeraiah 1960). Consequently, it was concluded that the C[u] genome of *Ae. umbellulata* is not a modified C genome of *Ae. caudata* and should, therefore, be given a separate genomic status (Chennaveeraiah 1960; Kimber and Abu-Baker 1981). Thus, the symbol U was assigned to the genome of *Ae. umbellulata* and is also used in the genomic descriptions of allopolyploid species with *Ae. umbellulata* in their parentage (Chennaveeraiah 1960; Kimber and Abu-Baker 1981). The major differences in the C-banding patterns of *Ae. caudata* and *Ae. umbellulata* provide additional support for the differential designation of the *Ae. umbellulata* and *Ae. caudata* genomes (Teoh and Hutchinson 1983).

The nuclear genome of *Ae. umbellulata* is significantly larger than that of *Ae. caudata*, equal to that of *Ae. comosa* and smaller than the genome of *Ae. uniaristata* (Eilam et al. 2007; Table 9.3). Its 1C DNA size, determined in 9 accessions that were collected from different regions of the distribution area of the species, is 5.53 ± 0.052 pg (Eilam et al. 2007). Very little intraspecific variation in DNA content was found between the accessions.

The karyotype of *Ae. umbellulata* is asymmetric. Senyaninova-Korchagina (1932) recognized several types of chromosomes that included one pair with large satellites on the short arm, one pair with a median centromere, four pairs with sub-median-sub-terminal centromeres, and one pair

with a characteristic knob due to an extremely sub-terminal centromere. The karyotype described by Chennaveeraiah (1960), although essentially similar to that observed by Senyaninova-Korchagina (1932), differed from it in one respect. One of the pairs with a sub-median-sub-terminal centromere also featured a satellite on the shorter arm, but the satellite was considerably smaller, approximately 1/3 of the large satellite. Thus, the karyotype consists of 6 types of chromosomes, one pair with a large satellite on the short arm, one pair with a small satellite on the short arm, one pair with a sub-median centromere, two pairs with sub-median-sub-terminal centromeres, another smaller pair with a sub-median centromere and one pair with the characteristic knob due to an extreme sub-terminal centromere. The shortest chromosome is about ¾ the size of the largest one. A similar description of the *Ae. umbellulata* karyotype was presented by Al-Mashhadani et al. (1978), who analyzed the karyotype of several *Aegilops* species native to Iraq.

Ae. umbellulata has a complex C-banding pattern with intensely stained pericentromeric, telomeric and interstitial bands (Teoh and Hutchinson 1983). Telomeric bands occur on most of the long arms. Chromosomes 2U, 4U, and 5U according to the designation of Friebe et al. (1995b), [chromosomes D, F, and C according to Kimber's (1967) designation] are easily distinguishable by their C-banding patterns, but the other chromosomes also exhibit specific C-banding patterns. Chromosomes 1U and 5U are the SAT chromosomes that are clearly differentiated by their banding patterns (Teoh and Hutchinson 1983). Badaeva et al. (1996a) also found that *Ae. umbellulata* is among the most heterochromatic species with large centromeric, intercalary and terminal C-bands present on all chromosomes.

While Teoh and Hutchinson (1983) analyzed only one accession of *Ae. umbellulata*, Friebe et al. (1995b) studied ten different accessions, collected from different geographic regions, *and established* a standard karyotype and a generalized ideogram of *Ae. umbellulata*, based on C-banding of these accessions. In addition, these authors identified, by C-banding and GISH, the individual *Ae. umbellulata* chromosomes in *T. aestivum* cv. CS– *Ae. umbellulata* chromosome addition lines, produced by Kimber (1967). These six *umbellulata* chromosomes were designated by Kimber (1967) as follows: A (subterminal chromosome), B (SAT chromosome), C (SAT chromosome), D (subterminal chromosome), E (subterminal chromosome), G (subterminal chromosome). Friebe et al. (1995a, b, c) determined the homoeology of these six *umbellulata* chromosomes in the addition lines as follows: 1U = B, 2U = D, 4U = F, 5U = C, 6U = A, and 7U = E). [The missing chromosome, F (=3U), is sub-median; Kimber (1967)].

Ae. umbellulata has two pairs of SAT-chromosomes, 1U and 5U, that were described earlier by the presence of secondary constrictions (Chennaveeraiah 1960). These pairs were shown by *in-situ* hybridization (ISH) using a ribosomal DNA probe (Teoh et al. 1983), and by Ag-Nor banding (Cermeño et al. 1984b), to possess NORs. The NOR in 1U is usually more prominent than that in 5U. Chromosome 6U is the shortest chromosome with the extreme sub-terminal centromere. Some polymorphism for C-band size and position was observed between the different accessions. However, this did not prevent chromosome identification (Friebe et al. 1995b). No large structural rearrangements detectable by C-banding analysis were found in any of the accessions. Yet, Badaeva et al. (1996b) reported that line TA1965 of *Ae. umbellulata* from Iran has a reciprocal translocation of chromosomes 1U and 5U.

The homoeology of chromosome 6U of *Ae. umbellulata* was also determined by studying its pairing relationships with wheat telocentric lines in hybrids having the chromosome constitution of 20 wheat chromosomes + one wheat telocentric chromosome + seven *Ae. speltoides* chromosomes that promote homoeologous pairing + chromosome 6U (Athwal and Kimber 1972). It has been found that chromosome 6U is homoeologous to the group-6 chromosomes of wheat. Castilho and Heslop-Harrison (1995) performed ISH with the 5S and the 18S–5.8S–25S rRNA genes and the repetitive DNA sequence pSc119.2, on *Ae. umbellulata* chromosomes. The pSc119.2 probe hybridized with all *Ae. umbellulata* chromosomes at the telomeres, except for the short arm of chromosome 6U, and showed intercalary sites on the long arms of chromosomes 6U and 7U. The 5S and 18S–25S rDNA only mapped physically on the short arms of chromosomes 1U and 5U. On chromosome 1U, 5S rDNA was shown to be sub-terminal and 18S–25S rDNA more proximal, while on chromosome 5U, the position of the genes was reversed. Similar result was obtained by Badaeva et al. (1996b), using in situ hybridization with pTa71 (18S-26S rDNA) and pTa794 (5S rDNA) DNA probes. Also, Martini et al. (1982), studying common wheat plants into which 1U and 5U chromosomes of *Ae. umbellulata* had been separately introduced, found that these two chromosomes possess ribosomal RNA genes. ISH analysis showed that the two addition lines with 1U and 5U have more ribosomal RNA genes and rDNA clusters than the recipient wheat plants. The repeating rDNA unit in *Ae. umbellulata* is longer than most of the units in the standard laboratory wheat cultivar Chinese Spring (Martini et al. 1982). The additional DNA is probably in the non-transcribed spacer, as probably suggested by restriction endonuclease maps of rDNA. In Chinese Spring plants possessing 1U or 5U chromosomes, the largest nucleoli formed on 1U or 5U chromosomes and the wheat NORs formed only micronucleoli. This is not because the NORs on chromosomes 1U and 5U have many more rRNA genes than the wheat NORs, rather, Martini et al. (1982) suggested that they compete more effectively for some limiting factor. The partial inactivation of the wheat NORs by chromosomes 1U or 5U does not result in reduced total nucleolus volume in root tip or PMCs, because of the compensation by the NORs of chromosomes 1U or 5U.

Lacadena and Cermeño (1985) analyzed the influence of U genome chromosomes on NOR activity of *common wheat*, in the complete set of the chromosome addition lines of *Ae. umbellulata* to common wheat. Chromosomes 1U and 5U induced partial inactivation of wheat NORs of chromosome 6B, 1B and 5D. Chromosomes 2U and 3U, which are not SAT-chromosomes, also influenced the activity of wheat NORs. The predominant status of the U genome with respect to nucleolar competition in *common wheat was in accord with its predominance in the allopolyploid species of Aegilops that contains the U genome* (Cermeño et al. 1984b).

Teoh et al. (1983) probed the *Ae. umbellulata* genome with two different cloned repetitive DNA sequences derived from common wheat, one, TC22b, being a noncoding sequence and the other coding for ribosomal RNA. Chromosomes exhibited very little labelling with TC22b. Like in *Ae. uniaristata*, *Ae. umbellulata* also had large centromeric heterochromatic bands which did not correlate with heavy labelling by the TC22b probe. On the other hand, ISH with the repetitive DNA sequence that codes for ribosomal RNA, confirmed the presence of two rDNA sites in *Ae. umbellulata*.

Using the probe pAs1 from *Ae. tauschii* in ISH experiments with *Ae. umbellulata*, revealed only a few minor pAs1 sites on chromosomes 1U and 6U (Badeva et al. 1996a). Faint inconsistent signals were detected in several other U-genome chromosomes. In contrast, the probe pSc119 from *Secale cereale* hybridized to telomeric regions and some interstitial sites of most *Ae. umbellulata* chromosomes (Badaeva et al. 1996a).

The C genome of *Ae. caudata* and the U genome of *Ae. umbellulata* differ significantly in heterochromatin content (Friebe et al. 1995b), but minimally in C-band patterns (Badaeva et al. 1996a). The number and chromosomal location of hybridization sites with the pAs1 and pSc119 probes were similar in the two species (Badaeva et al. 1996a). This kind of similarity in ISH patterns suggests that the C and U genomes are related. Yet, both genomes were significantly rearranged during speciation and their chromosomes show little affinity in meiosis of hybrids between the two species (Kimber and Abu-Baker 1981).

Zhang et al. (1998), using RFLP probes that detect homoeoloci previously mapped in hexaploid wheat, constructed a comparative genetic map of *Ae. umbellulata* with wheat. It was found that all seven *Ae. umbellulata* chromosomes displayed one or more rearrangements relative to wheat, changes that are consistent with the sub-terminal morphology of chromosomes 2U, 3U, 6U and 7U. Comparison of the chromosomal locations assigned by mapping and those obtained by hybridization to wheat –*Ae.*

umbellulata single chromosome addition lines verified the composition of the added *Ae. umbellulata* chromosomes and indicated that no further cytological rearrangements had taken place during the production of the alien-wheat aneuploid lines. Relationships between *Ae. umbellulata* and wheat chromosomes were also confirmed, based on homoeology of the centromeric regions.

Edae et al. (2017) developed a framework consensus genetic map for *Ae. umbellulata* comprising 3009 genotype-by-sequence SNPs with a total map size of 948.72 cM. On average, there were three SNPs per centimorgan for each chromosome. Chromosome 1U was the shortest (66.5 cM), with only 81 SNPs, whereas the remaining chromosomes had between 391 and 591 SNP markers. A total of 2395 unmapped SNPs were added to the linkage maps through a recombination frequency approach, and brought the number of SNPs placed on the consensus map to a total of 5404 markers. Segregation distortion was disproportionally high for chromosome 1U for both populations used to construct component linkage maps, and thus, segregation distortion may be one of the reasons for the exceptionally reduced linkage size of chromosome 1U. From comparative analysis, all *Ae. umbellulata* chromosomes, with the exception of 4U, showed moderate to strong collinearity with corresponding homoeologous chromosomes of hexaploid wheat and barley. The present consensus map may serve as a reference map in QTL mapping and validation projects, and in genome assembly of a reference genome sequence for *Ae. umbellulata*.

Zymogram analysis was used to identify the *Ae. umbellulata* chromosomes that carry the structural genes for particular isozymes (Benito et al. 1987). It was found that *umbellulata* chromosome 6U carries a structural gene for 6-phosphogluconate dehydrogenase, chromosome 1U carries structural genes for glucose phosphate isomerase and phosphoglucose mutase, chromosome 2U carries genes for leaf peroxidases, chromosome 7U carries structural genes for endosperm peroxidases, acid phosphatases and leaf esterases, chromosome 4U carries a gene for embryo plus scutellum peroxidases, and chromosome 3U carries structural genes for endosperm alkaline phosphatases, leaf alkaline phosphatases and leaf esterases. The results obtained indicate that chromosome 1U is partially homoeologous to the common wheat chromosomes of group 1 and 4, and chromosome 7U is partially homoeologous to common wheat chromosomes of groups 7 and 4.

9.8.2.3 Crosses with Other Species of the Wheat Group

Meiosis in *Ae. umbellulata* is regular; seven bivalents were formed in every pollen mother cell (PMC) of which 3–4 were rod and 3–4 ring (Chennaveeraiah 1960). Likewise, Cuñado (1992) observed an average of 3.07 rod bivalents, 3.93 ring bivalents, (total of seven bivalents), and 10.94 chiasmata per cell. Obviously, *Ae. umbellulata*, having several sub-telocentric chromosomes, displays less ring bivalents than species with symmetric karyotype.

Data on average chromosome pairing between allotetraploid species of section Aegilops containing a U subgenome with *Ae. umbellulata*, signifies the homology between this subgenome and that of the diploid (Table 9.5). Average chromosome pairing in meiotic metaphase of F_1 hybrids between Sitopsis Species and *Ae. umbellulata* was low. Thus, hybrids involving two Sitopsis species, *Ae. speltoides* and *Ae. sharonensis*, with *Ae. umbellulata* shows that their genomes are very distantly related (Table 9.4). More specifically, the hybrid *Ae. speltoides* var. *ligustica* I (HP type) x *Ae. umbellulata* had 3.22 univalents, 2.98 rod bivalents, 0.0 ring bivalents, 1.58 trivalents and 0.03 quadrivalents, whereas the hybrid *Ae. speltoides* var. *ligustica* (IP type) x *Ae. umbellulata* had 9.08 univalents, 1.98 rod bivalents, 0.0 ring bivalents, and 0.32 trivalents. Similar to the hybrid with the HP type of *Ae. speltoides*, also the hybrid *Ae. sharonensis* x *Ae. umbellulata* had low pairing. Similarly, Tanaka (1955a) observed 5–11 univalents, 1–5 bivalents with a mode of 3, 0–1 trivalents and 0–1 quadrivalents in the hybrid *Ae. sharonensis* x *Ae. umbellulata*. Chromosome pairing in the hybrid *Ae. bicornis* x *Ae. umbellulata* had 0–5 bivalents (with a mode of 3) and 0–3 trivalents (Kihara 1949), and in the hybrid *Ae. longissima* x *Ae. umbellulata* 0–4 bivalents (with mode of 2) and 0–1 trivalents (Kihara 1949).

Average chromosomal pairing in the F_1 hybrid *Ae. umbellulata* x *Ae. tauschii* displayed 3.36 rod bivalents, 0.0 ring bivalents and 0.66 trivalents (Cuñado and Santos 1999). Similar data were obtained by Kihara (1949). Average chromosome pairing in hybrids between *Ae. caudata* and *Ae. umbellulata* included 5.64 univalents, 3.46 rod bivalents, 0.12 ring bivalents and 0.40 trivalents, and in another hybrid of *Ae. caudata* x *Ae. umbellulata* 5.64 univalents, 3.46 rod bivalents, 0.12 ring bivalents and 0.40 trivalents (Sears 1941b). Somewhat higher pairing in the hybrid *Ae. caudata* x *Ae. umbellulata* was reported by Kihara (1949), with 3–6 bivalents (with mode of 5) and 0–1 trivalents. Nevertheless, the above data indicate low affinity between the genomes of these two species. In the F_1 hybrid *Ae. umbellulata* x *Ae. comosa*, Kihara (1949) observed 2–5 bivalents and 0–2 trivalents. In the hybrid *Ae. uniaristata* x *Ae. umbellulata*, Percival (1932) observed 0–4 bivalents, while Sears (1941b) observed 8.30 univalents, 2.70 rod bivalents, 0.0 ring bivalents and 0.10 trivalents). Evidently, the genome of *Ae. umbellulata* is closer, although not very close, to the genome of *Ae. caudata* than to those of *Ae. comosa* and *Ae. uniaristata*.

Chromosomal pairing between hexaploid *Ae. crassa* (genome $D^cD^cX^cX^cDD$) x *Ae. umbellulata* included 13.94 univalents, 4.06 rod bivalents, 1.46 ring bivalents (5.52 total), 0.81 trivalents, 0.09 quadrivalents and 0.03 hexavalents (Melnyk and McGinnis 1962). Most of the pairing is presumably autosyndetic between chromosomes of genomes D^c and D of *Ae. crassa*. The rest of the pairing indicate very little affinity between the U genome and the subgenomes of hexaploid *Ae. crassa*. In contrast, two reports of chromosome pairing in the hybrid *Ae. juvenalis* (genome $D^cD^cX^cX^cUU$) and *Ae. umbellulata* had 6.39 univalents, 3.59 rod bivalents, 4.46 ring bivalents (8.05 total), 1.42 trivalents, 0.28 quadrivalents and 0.03 hexavalents (McGinnis and Melnyk 1962) and 6.93 univalents, 4.80 rod bivalents, 4.42 ring bivalents (9.22 total), 0.27 trivalents and 0.08 quadrivalents (Kimber and Abu-Baker 1981), indicating that a subgenome in *Ae. juvenalis* is homologous to the genome of *Ae. umbellulata*.

Average chromosomal pairing in the F_1 hybrid between *T. monococcum* subsp. *aegilopoides* and *Ae. umbellulata* had low pairing (Table 9.8). Evidently, the genomes of these two species are very distantly related. The hybrid between hexaploid wheat, *T. aestivum* ssp. *aestivum* and *Ae. umbellulata* was very low (Table 9.10) Obviously, the subgenomes of hexaploid wheat are not closely related to the genome of *Ae. umbellulata*.

In certain lines of domesticated diploid wheat, *T. monococcum* subsp. *monococcum*, Sears (1944a) identified two alleles which act as dominant-lethal that are responsible for unviability of hybrids with *Ae. umbellulata*, but without an effect in *T. monococcum* itself. The alleles differ in the time at which they cause death. A third normal allele is present in the wild subspecies of *T. monococcum*, i.e., subsp. *aegilopoides*.

Chromosome pairing in the F_1 hybrid *Ae. umbellulata* x *Amblyopyrum muticum* without *muticum* B chromosomes, showed intermediate level of pairing and many multivalents (Table 8.1). Hence, pairing was low in spite of the presence of homoeologous pairing promoter genes in *A. muticum*. The range of chromosome pairing in hybrids with two B chromosomes, which suppress homoeologous pairing, was much lower. Evidently, the genome of *Ae. umbellulata* is only distantly related to that of *A. muticum*.

9.8.3 *Aegilops geniculata* Roth

9.8.3.1 Morphological and Geographical Notes

Ae. geniculata Roth, commonly known as ovate goat grass, [Syn.: *Ae. ovata* L.; *T. ovatum* (L.) Raspail:] is a predominantly autogamous, annual, tufted, multi-tillered plant. It has thin, 10–40-cm-tall (excluding spikes) culms, that geniculate at the base and then rise upright, with narrow, hairy or glabrous leaves. The upper 1/3 or 1/4 of the culms is defoliated. Leaf blades are usually short, 2–5-cm-long. Its spike is broad-oval to narrow-elliptical, short (1.5–3.0-cm-long), does not become suddenly narrow, and is usually awned. The entire spike disarticulates at maturity (umbrella-type dispersal unit). There are 3(2–4) oval or urceolate spikelets that are appressed to the rachis, all potentially fertile, aside from the terminal, which is usually sterile. The lower rachis segment is usually shorter than the adjacent spikelet. There is one basal rudimentary spikelet, and rarely 2 or none. There are 5 florets in each spikelet, with the upper 3 being sterile. Glumes are 7–8-mm-long, with veins (5–7) that are unequal in width and unequally spaced, and are inflated at the middle, usually with 3–4(-5) awns (2.0–3.5-cm-long), equal in width. Upper spikelets usually have more awns (4–5), which are as long as or shorter than lateral spikelet awns. There are usually two lemma awns on lower spikelets, and three on upper spikelets. All awns are widely spread at maturity. There are 5–7 veins near the upper margin of the lemma of the lower spikelets. The caryopsis is free (Fig. 9.4b).

Ae. geniculata exhibit very large variation in spike shape, size, hairiness, and compactness, spikelet shape, site of glume inflation, and awn count, length and structure. In the southwestern part of its distribution, there is a compact form with irregular awns. The wide morphological variation led taxonomists to subdivide the species into two subspecies, each containing several varieties, albeit, with some disagreement on the sub-specific classification. Thus, Zhukovsky (1928) subdivided *Ae. geniculata* into four subspecies, namely, *gibberosa*, *umbonata*, *globulosa* and *planiuscula*. Eig (1929a) divided *Ae. geniculata* into two subspecies, *euovata* (currently subsp. *geniculata*), that contains four varieties, and subsp. *atlantica*, that contains three varieties. Hammer (1980) also subdivided the species into two subspecies: *geniculata* and *gibberosa*. Ssp. *geniculata* includes six varieties and is characterized by loose arrangement of spikelets in the ear, awns that have the same width, and often by more than 3 spikelets per spike, whereas subsp. *gibberosa* includes three varieties and is characterized by more compact arrangement of spikelets in the ear, and rarely with more than 3 spikelets per spike.

Medouri et al. (2015) evaluated morphological polymorphism among *Ae. geniculata* accessions from Algeria and found a great variation in most of the studied traits. A weak relationship between morphological traits and ecological factors was found. They also revealed high polymorphism among accessions in the high molecular weight (HMW) glutenin subunits and discovered several new subunits. Mahjoub et al. (2009, 2010), used morphological traits and RAPD markers to assess genetic diversity in *Ae. geniculata* from North and Central Tunisia. Both morphological traits and RAPD markers showed a high degree of variation within and between populations. Yet, gene diversity was attributable mostly to diversity within populations. The morphological variation was associated with environmental (climatic) change. A non-significant correlation was found between morphological and RAPD variations. Similar results of high variation in European and Tunisian *Ae. geniculata* accessions, determined using RAPD, was also reported by Zhang et al. (1996) and Mahjoub et al. (2016), respectively. On the other hand, Monte et al. (2001) reported only slight variation in *Ae. geniculata* accessions from Spain, whereas, when using RAPD and ISSR, Thomas and Bebell (2010) found great genetic diversity in accessions of this species from Greece, the center of its distribution. Similarly, AFLP analysis of *Ae. geniculata* accessions from Turkey revealed high polymorphism in this species (Kaya et al. 2011).

Ae. geniculata is a widespread Mediterranean and Western Asiatic element. It grows in Macaronesia (Madeira, Tenerife), Portugal, Spain, South France (including Corsica), Italy (including Sardinia and Sicily), Malta, Slovenia, Croatia, Serbia, Bosnia-Herzegovina, North Macedonia, Albania, Bulgaria, Greece, (including Crete and the Aegean Islands), South Ukraine, South Russia (including Crimea and Ciscaucasia), rare in Transcaucasia (Georgia, Armenia, and Azerbaijan), Turkey (Northwest and Mediterranean), Northern Iraq, Western Iran, Syria, Lebanon, Cyprus, Israel, Jordan, Egypt, Libya, Tunisia, Algeria and Morocco. In several regions of North-Africa, it penetrates into the Sahara. *Ae. geniculata* grows on terra-rossa, basalt, rendzina, calcareous sandstone and alluvial soils. It is found in edges and openings of sclerophyllous oak forests, garrigue and maquis, shrub and herbaceous formations, fallow fields, roadsides, disturbed habitats, and often on the edges of and within cultivated plantations and wheat fields (with which it may form natural hybrids). It grows at altitudes from almost sea level to 1750 m. *Ae. geniculata* is adventive in parts of central and northwestern Europe (van Slageren 1994) and was introduced into parts of United States, e.g., California, where it is a noxious weed invading dry land pastures (Hitchcock 1935).

Ae. geniculata grows with its two diploid parents in Turkey and Greece. However, its distribution is larger than that of its diploid parents and it occupies more types of habitats that its diploid parents. It is common, locally abundant, forming dense stands throughout its range in both primary and secondary habitats. Yet, it is sporadic in marginal and semi-steppical regions. Like other *Aegilops* allotetraploids, *Ae. geniculata* is a typical colonizer species. The two diploid parents grow together in western Turkey and in Greece and this is presumably the center of origin of the species.

Ae. geniculata is very widely distributed in the central and western part of the distribution of the genus. With the exclusion of *Ae. bicornis* and *Ae. vavilovii*, both of which grow in xeric habitats, and *Ae. juvenalis*, *Ae. geniculata* has contact with all the species of the wheat group. *Ae. geniculata* grows sympatrically with most of the specpes of *Aegilops* and wild, and allopatrically with *A. muticum*, *Ae. bicornis*, *Ae. tauschii*, *Ae. kotschyi*, and *Ae. juvenalis*.

Ae. geniculata usually grows in mixed stands with other *Aegilops* species, particularly with those belonging to the U-genome cluster, i.e., species of section *Aegilops*, with which it may form natural hybrids and introgressed derivatives. In many mixed populations of *Ae. geniculata*, *Ae. peregrina* and *Ae. biuncialis* in Israel, natural hybrids, hybrid derivatives and many highly introgressed types involving *Ae. geniculata* have been found, indicating a wide occurrence of gene flow between these three species, resulting in increased variation (Zohary and Feldman 1962; Feldman 1965a, b, c). It is assumed, therefore, that in many parts of the *Ae. geniculata* distribution area, mixed populations of *Ae. geniculata* and other allopolyploid species of section *Aegilops*, such interspecific genetic connections leading to introgression that blurs, to some extent, the specific boundaries. Evidence indicating gene flow that occurred between *Ae. geniculata* and common wheat, when the former grew near or in wheat fields, was presented by Zaharieva and Monneveux (2006), Loureiro et al. (2006), Loureiro and Escorial (2007), and by Arrigo et al. (2011). Spontaneous hybridization between tetraploid wheat (*T. turgidum ssp. durum*) and *Ae. geniculata* was also regularly observed in sympatric populations, resulting in spontaneous formation of *durum-geniculata* fertile amphiploids that arose through unreduced gametes of the F_1 hybrids (David et al. 2004).

Ae. geniculata contains genes that confer resistance to several fungal diseases, such as powdery mildew (Gill et al. 1985; Stoilova and Spetsov 2006), leaf rust (Gill et al. 1985; Zaharieva et al. 2001b; Aghaee-Sarbarzeh et al. 2002; Anikster et al. 2005), stem rust (Zaharieva et al. 2001b; Liu et al. 2011a, b), and yellow (stripe) rust (Zaharieva et al. 2001b; Aghaee-Sarbarzeh et al. 2002; Anikster et al. 2005), resistance to barley yellow dwarf virus (BYDV) (Zaharieva et al. 2001b), Hessian fly (Gill et al. 1985), common root rot (Bailey et al. 1993), and cereal cyst nematodes (CCN) (Zaharieva et al. 2001b). It also contains a gene that may improve tolerance to abiotic stresses related to water status, chlorophyll content and plant thermal regulation under Mediterranean field conditions (Zaharieva et al. 2001a). The genetic potential of *Ae. geniculata* to improve resistance to biotic stresses and tolerance to abiotic stresses in wheat harbors invaluable potential for wheat improvement.

9.8.3.2 Cytology, Cytogenetics and Evolution

Ae. geniculata is an allotetraploid (2n = 4x = 28) species. Its nuclear genome designation is M^{o}M^{o}UU and organellar genome is M^o (Kimber and Tsunewaki 1988; Dvorak 1998). Kihara (1937, 1954) designated the genome of *Ae. geniculata* $C^uC^uM^oM^o$ (C^u is currently U) primarily due to the chromosomal pairing observed in hybrids with other *Aegilops* allotetraploids. Kihara (1929) and Kihara and Lilienfeld (1932) observed 5–11 bivalents in the F_1 hybrid *Ae. triuncialis* (genome C^uC^uCC; currently UUCC) and *Ae. geniculata* and in the reciprocal combination and concluded that the C^u (U) subgenome of *Ae. triuncialis* also exists in *Ae. geniculata*. Likewise, Kihara (1949) observed 6–9 bivalents and 0–3 trivalents in the F_1 *Ae. geniculata* x *Ae. columnaris* hybrid (genome $C^uC^uM^cM^c$; currently UUX^nX^n), confirming the inclusion of the U genome in *Ae. geniculata*. On the other hand, Kihara (1929) observed 5–10 bivalents in the F_1 hybrid of *Ae. geniculata* x *Ae. ventricosa* (genome DDM^uM^u; currently DDNN) and configurations of 10 bivalents + 8 univalents were frequently observed in the F_1 of the reciprocal cross (Kihara and Lilienfeld 1932). Therefore, it was concluded that one of the two subgenomes in *Ae. geniculata* is close to the M genome *of Ae. uniaristata* or *Ae. comosa*. This conclusion was accepted because pairing in the F_1 hybrid *Ae. geniculata* x *T. aestivum* (genome BBAADD) had only 2–3 bivalents (Kihara and Lilienfeld 1932; Riley 1966a), indicating that *Ae. geniculata* does not carry the D subgenome. Genome analysis following direct crosses between *Ae. geniculata* and its putative diploid progenitors was performed by Kihara (1937), who observed 7 bivalents in the *Ae. geniculata* x *Ae. umbellulata* (genome UU) hybrid. Kimber and Abu-Baker (1981) reported the presence of 6 bivalents and one trivalent in the F_1 of this hybrid combination, thus, presenting direct evidence that one of the subgenomes of *Ae. geniculata* is U of *Ae. umbellulata*. Likewise, Kimber et al. (1988) reported the presence of 7 bivalents or 6 bivalents and one trivalent in the F_1 *Ae. geniculata* x *Ae. comosa* (genome MM) hybrid, verifying the presence of the M subgenome in *Ae. geniculata*. Yet, based on meiotic analysis of a number of hybrids, Kimber et al. (1988) concluded that, while the U subgenome of *Ae. geniculata* is much closer to the U genome of *Ae. umbellulata* than the M subgenome is to the M genome of *Ae. comosa*, and suggested that the M subgenome underwent substantial modifications at the tetraploid level. By this conclusion, they reinforced Kihara's (1937, 1954) view that the M subgenome of *Ae. geniculata* is a modified genome. The presence of a modified M-subgenome in *Ae. geniculata* was further confirmed by cytogenetic and molecular data (Talbert et al. 1993; Friebe et al. 1999; Resta et al. 1996; Badaeva et al. 2002, 2004).

Resta et al. (1996) investigated the origin of the *Ae. geniculata* subgenomes by examining specific restriction fragments of repeated nucleotide sequences in the DNA of this allotetraploid species. The analysis showed that *Ae. geniculata is* closely related to *Ae. biuncialis*; in both

species, one subgenome was closely related to the genome of *Ae. umbellulata* and the other was a modified genome of *Ae. comosa*. Consequently, they proposed the same genome formula, UUM^oM^o, for *Ae. geniculata and Ae. biuncialis. Ae. neglecta* and *Ae. columnaris*, which are also closely related to each other and have the same genomes, share the U genome with *Ae. geniculata and Ae. biuncialis*, but their second pair of subgenomes is unrelated to the M^o subgenome.

A similar conclusion was reached by Gong et al. (2006) who, using 31 ISSR primers, found that the genome constituents of *Ae. geniculata* had considerably changed as compared to the genomes of its ancestral diploid species. The genetic similarity index between *Ae. geniculata* and *Ae. umbellulata* was much higher than that with *Ae. comosa*, indicating that the U subgenome of *Ae. geniculata* had changed relatively minimally while its M^o subgenome was significantly modified.

Karyological studies also showed that one chromosomal set of *Ae. geniculata* corresponds to that of *Ae. umbellulata* (Senjaninova-Korczagina 1932; Pathak 1940; Chennaveeraiah 1960). The second chromosomal set is different, bearing median or sub-median centromeres and with no satellites or secondary constrictions (Senvaninova-Korchaginova 1932; Chennaveeraiah 1960). Since the karyotype of the second set differed from that of all the diploid species of *Aegilops*, Chennaveeraiah (1960) supported Kihara's (1937, 1954) claim that the second chromosomal set of *Ae. geniculata*, the M^o genome, became modified at the tetraploid level.

Maan (1977b) reported that the cytoplasm of *Ae. geniculata* is similar to that of *Amblyopyrum muticum*. Similarly, Tsunewaki (1996) and Wang et al. (1997) suggested that a form of *A. muticum* might be the maternal and *Ae. umbellulata* the paternal parent of *Ae. geniculata*. Yet, Terachi et al. (1984), based on restriction fragment patterns of chloroplast DNA, concluded that the chloroplast genome of *Ae. geniculata* is unique, being distinctly different from that of all species of sections Aegilops and Comopyrum as well as from that of *Amblyopyrum muticum*. No diploid species with a chloroplast genome closely related to that of *Ae. geniculata* has been found in the *Triticum, Aegilops* or *Amblyopyrum* genera (Ogihara and Tsunewaki 1982; Terachi et al. 1984). Even *Ae. umbellulata* and *A. muticum*, whose chloroplast genomes are closest to that of *Ae. geniculata*, showed seven differences in ctDNA restriction fragments when compared to that of *Ae. geniculata*. It was therefore suggested that the cytoplasm of *Ae. geniculata* was derived from the M^o genome donor (Mukai and Tsunewaki 1975; Tsunewaki 1980). Consequently, the plasma type of *Ae. geniculata* was designated M^o (Tsunewaki 1980; Tsunewaki and Tsujimoto 1983; Kimber and Tsunewaki 1988; Dvorak 1998), and is different from that of the related allotetraploid species, *Ae. biuncialis*, *Ae. neglecta* and *Ae. columnaris*, that have the cytoplasm of *Ae. umbellulata*.

Tsunewaki (2009), reviewing the results of his and his coworkers on plasmon (plastom and chondrion) analysis of the *Aegilops-Triticum* group, maintained the conclusion that the origin of the M° plasmon of *Ae. geniculata* is unclear. This allotetraploid species shares a genome constitution, UM, with *Ae. biuncialis*, and has a genome constitution related to that of *Ae. columnaris* and *Ae. neglecta*. Its plasmon, however, greatly differed from those of these three species in most respects, although similar to the *Ae. columnaris* plasmon in its phenotypic effects, and to the T, T2 and U plasmons of *A. muticum* and *Ae. umbellulata*, respectively, in the plastom.

Assuming the degree of plasmon dissimilarity between two species parallels differences in time of origin, Wang et al. (1997) and Tsunewaki (2009) speculated relative time of the origin of related polyploids by comparing their plasmon similarity with their closest diploid relatives. Thus, plasmon differences were compared between *Ae. geniculata* and its phylogenetically related species of section *Aegilops* (*Ae. biuncialis*, *Ae. neglecta* and *Ae. columnaris*) and the plasmons of the closest diploid relatives. The results suggest that the origin of *Ae. geniculata* is more ancient than that of the other related allotetraploid species. A similar conclusion on the early origin of *Ae. geniculata* was reached by Terachi et al. (1984) on the basis of plastom analysis.

Ae. geniculata contains 10.29 ± 0.008 pg 1C DNA (Eilam et al. 2008), which is 5.68% less DNA than that expected from the sum of the DNA of its two diploid parents, i.e., 10.91 pg (*Ae. comosa* contains 5.53 pg and *Ae. umbellulata* contains 5.38 pg; Eilam et al. 2007). The loss of DNA in *Ae. geniculata* was confirmed by Badaeva et al. (2002, 2004) who, based on differential C-banding and in situ hybridization, found that this allotetraploid exhibits substantial structural chromosome rearrangements, including deletion of chromosomal segments and reduction of heterochromatin content.

Early studies of the karyotype of *Ae. geniculata* revealed only one pair of SAT chromosomes on the somatic complement (Senyaninova-Korchagina (1932) or two such pairs on short arms (Matsumura 1940), and Pathak (1940) noted the presence of one pair with a satellite and another pair with a secondary constriction. Chennaveeraiah (1960) confirmed the presence of two SAT pairs on short arms, and also reported one pair with a knob due to extreme sub-terminal constriction, one pair with a median centromere, eight pairs with sub-median centromeres and four pairs with sub-median-sub-terminal centromeres.

Chennaveeraiah (1960) analyzed the karyotype of two taxa of *Ae. geniculata*, var. *brachyathera* (Pomel) Eig [=*Ae. geniculata* ssp. *gibberosa* (Zhuk.) Hammer] and subsp. *globulosa* Zhuk., and found no significant differences

between them. In both, two satellite pairs were observed, confirming the observations of Matsumura (1940) and Pathak (1940). Both had a pair with a median centromere, another pair with an almost median centromere, and a pair with the extreme sub-terminal centromere. The rest of the pairs in both taxa had sub-median and sub-median-sub-terminal centromeres. The only difference between the two taxa was that ssp. *globulosa* had more pairs with sub-median-sub-terminal centromeres than var. *brachyathera*.

Chromosome morphology and variation in *N*-band distribution along the *Ae. geniculata* chromosomes were studied in 13 accessions by Landjeva and Ganeva (2000). The *N*-banding technique differentially stained all 14 chromosome pairs, enabling their identification. The most prominent bands were observed near the centromeres and in the intercalary regions of both arms, whereas telomeric bands were found in only seven chromosome pairs. Polymorphism for presence or absence of particular bands was observed among different accessions. The variable bands were predominantly located at the terminal and subterminal chromosome regions, and were also unevenly distributed over chromosomes. In their work, the researchers present a generalized idiogram of *Aegilops geniculata* (Landjeva and Ganeva 2000).

Badaeva et al. (2002, 2004), using C-banding technique, analyzed heterochromatin banding patterns in the somatic metaphase chromosomes of several accessions of *Ae. geniculata*. Chromosome morphology and C-banding patterns in most studied accessions were similar to those reported previously (Pathak 1940; Chennaveeraiah 1960; Friebe et al. 1999), allowing chromosome designations according to the standard nomenclature (Friebe et al. 1999). Chromosome modifications were mostly found in Turkish accessions involving a high frequency of chromosomal rearrangements, represented by paracentric inversions and intragenomic and intergenomic translocations. Some modifications were the result of either Robertsonian translocations or translocations with interstitial breakpoints (Badaeva et al. (2002, 2004). This is in accord with the finding of Furuta (1981a), who reported a high frequency of chromosomal rearrangements in Turkish *Ae. geniculata* accessions. On the other hand, Yen and Kimber (1990c) did not observe modifications of the U genome chromosomes in five Turkish *Ae. geniculata* accessions.

The findings reported by Badaeva et al. (2002, 2004) confirmed that *Ae. geniculata* evolved as a result of hybridization between *Ae. umbellulata* and *Ae. comosa*. Comparing *Ae. geniculata* with its diploid ancestors revealed differences in morphology and C-banding patterns of many chromosomes belonging both to the M° and U genomes, indicating that both genomes had been modified, which agrees with previous data (Kimber and Abu-Baker 1981; Kimber et al. 1988; Kimber and Yen 1989; Yen and Kimber 1990d).

Talbert et al. (1993) cloned three repetitive DNA sequences found primarily in the U genome and two repetitive DNA sequences found primarily in the M genome and used these to monitor variation in species containing both genomes. The U genome of Ae. umbellulata and the U subgenome of the allotetraploid U-genome species were similar with regards to hybridization patterns observed with the U genome probes. Much more variation was found both among diploid Ae. comosa accessions and allopolyploids containing M subgenomes. The observed variation supports the cytogenetic evidence that the M subgenome in the allotetraploid species is more variable than the U subgenome. It also raises the possibility that part of the differential nature of the M subgenome of the allotetraploids may be due to variation within the diploid Ae. comosa, while other variations may have been generated at the tetraploid level.

Badaeva et al. (2004) also studied the distribution of hybridization sites in Ae. geniculata, by performing FISH with the repetitive DNA probes pSc119 and pAs1, as well as the distribution of NOR and 5S DNA loci, using the pTa71 (18S-26S rDNA), and pTa794 (5S rDNA) probes. FISH with pSc119 revealed signals on 10 chromosome pairs, while five chromosome pairs had pAs1 FISH sites. Large- or medium-sized pSc119 FISH sites were observed in telomeric regions of either one or both chromosome arms. Interstitial pSc119 FISH sites were located in the long arms of 6U and 7U, which were identified on the basis of their similarity with the corresponding chromosomes of Ae. umbellulata (Badaeva et al. 1996a; Castilho and Heslop-Harrison 1995). On the other hand, the pAs1 FISH sites were faint and located in interstitial chromosome regions and, in general, resembled the pAs1–labeling pattern of Ae. comosa chromosomes (Badaeva et al. 1996a, 1999).

Badaeva et al. (2004) observed two pairs of SAT chromosomes in Ae. geniculata, confirming previous reports (Pathak 1940; Chennaveeraiah 1960; Cermeño et al. 1984b; Yamamoto 1992b; Yamamoto and Mukai 1995). Multicolor FISH with the probes pTa71 and pTa794 exposed the inactivation of major NORs on the M-subgenome chromosomes, and redistribution of 5S rDNA sites and loss of some minor 18S-26S rDNA loci in both subgenomes (Badaeva et al. (2004). Their study detected two major and four minor 18S-26S rDNA and three 5S rDNA sites, whose locations and presence were different from those in the parental species; chromosome 1U had a major NOR but lacked the 5S rDNA site and 1M^o only had the 5S rDNA locus. Small 5S rDNA loci were detected in the short arms of all group-5 chromosomes, which also had either major (5U) or minor (5M^o) 18S-26S rDNA loci. Minor pTa71 FISH sites of various intensities were detected in telomeric regions of two large sub-metacentric chromosome pairs, presumably 2 and 3 Mg, and in a small metacentric chromosome pair,

presumably a derivative of chromosome 6 M of Ae. comosa that lost the NOR during the formation of Ae. geniculata. A weak hybridization signal was occasionally observed in an interstitial region of the long arm of chromosome 6U, which is also present in chromosome 6U of Ae. umbellulata (Badaeva et al. 1996b). The data presented by Badaeva et al. (2004) substantiated that Ae. geniculata speciation was accompanied by modification of both parental genomes as a result of amplification, deletion, and re-distribution of various classes of repetitive DNA sequences and chromosomal rearrangements.

Chromosome 1U is one of the two SAT chromosomes of diploid Ae. umbellulata and is one of the most conserved chromosomes among the different allopolyploid species containing the U subgenome. Loss of the 5S rDNA locus on 1U of Ae. geniculata was probably caused by a species-specific translocation. Badaeva et al. (2004) confirmed that the second subgenome of Ae. geniculata, the M subgenome, is a modified subgenome that underwent many modifications because of amplification, elimination and redistribution of highly repetitive DNA sequences, as well as chromosomal rearrangements. Part of the significant intraspecific heterogeneity observed in Ae. geniculata was presumably derived from independent hybridization and introgression with other species.

Cermeño et al. (1984b) analyzed the activity of the NORs in somatic metaphase chromosomes of Ae. geniculata using a highly reproducible silver-staining procedure. Similar to the finding of Badaeva et al. (2004), Cermeño et al. (1984b) found that the U subgenome completely suppressed the NOR activity of the M^o subgenome. They identified two pairs of SAT chromosomes and two pairs of Ag-NORs, although four Ag-NORs pairs were expected, based on their count in their diploid progenitors. The number of nucleoli at interphase was as follows: 22 cells had 1 nucleolus, 175 cells had 2 nucleoli, 99 had 3, and 15 had 4.

The production and identification of a complete set of an intact Ae. geniculata chromosome additions to common wheat has been described (Friebe et al. 1999). C-banding and meiotic pairing analyses revealed that all added Ae. geniculata chromosomes were structurally identical to the Ae. geniculata parent accession. Consequently, all U and M^o subgenome chromosomes were tentatively assigned to their homoeologous groups based on C-banding, first meiotic metaphase pairing analyses and plant morphologies.

FISH signals were generally associated with constitutive heterochromatin regions corresponding to C-band-positive chromatin, including telomeric, pericentromeric, centromeric and interstitial regions of all the 14 Ae. geniculata chromosome pairs. The newly identified satellite DNA CL36, used by Koo et al. (2016), produced localized M^o-subgenome chromosome-specific FISH signals in Ae. geniculata and in the M genome of Ae. comosa ssp. heldreichii var. *subventricosa* but not in *Ae. comosa* ssp. *comosa*, suggesting that the M° subgenome of *Ae. geniculata* derived from var. *subventricosa*. Friebe et al. (1999) suggested a different source of the *Ae. geniculata* M° subgenome, based on comparison with the C-banding patterns that were established for *Ae. umbellulata* and *Ae. comosa* by Badaeva et al. (1996a, b). They confirmed that the U genome of *Ae. geniculata* was contributed by *Ae. umbellulata*, but claimed that the M° genome derived from *Ae. comosa* ssp. *comosa*.

9.8.3.3 Crosses with Other Species of the Wheat Group

Chromosomal pairing at first meiotic metaphase of an Israeli accession of *Ae. geniculata* was regular: 0.06 univalents, 1.20 rod bivalents, 12.77 ring bivalents and 30.66 chiasmata per cell (Feldman 1963). The single pair with a sub-terminal centromere, belonging to the U subgenome, appeared as rod bivalents but, in many cells, it had two chiasmata. A higher number of rod bivalents (3.01 rod bivalents, 10.99 ring bivalents) and lower number of chiasmata per cell at first meiotic metaphase (24.99) was reported by Cuñado (1992) for *Ae. geniculata* from Turkey. Similar pattern of chromosomal pairing at first metaphase of meiosis of *Ae. geniculata* was reported by Cuñado et al. (1996a, b), who also analyzed the pattern of pachytene pairing by whole-mount surface-spreading of synaptonemal complexes under the electron microscope. Their data indicated that in more than 90% of the cells at the pachytene stage, all the chromosomes were associated as bivalents, while in the rest of the cells, they observed 12 bivalents and one quadrivalent, presumably the result of inter-subgenomic homoeologous pairing.

Furuta (1981a) studied meiosis in intraspecific hybrids of *Ae. geniculata* and found that many Turkish accessions differed by one or two reciprocal translocations. On the other hand, Yen and Kimber (1990c) did not observe any rearrangement in the U genome chromosomes in five Turkish accessions of *Ae. geniculata*.

Data on chromosomal pairing in F_1 hybrids between *Ae. geniculata* and its diploid parents, *Ae. umbellulata and Ae. comosa*, are presented in Table 9.5. Kihara (1937) reported that the *Ae. geniculata* x *Ae. umbellulata* hybrid had 7 bivalents, and similar observation was made by Kimber and Abu-Baker (1981) observed 6.42 univalents, 5.69 bivalents, 0.92 trivalents and 0.11 quadrivalents. The *Ae. geniculata* x *Ae. comosa* hybrid had 6.15 univalents, 2.65 rod bivalents, 3.05 ring bivalents and 1.15 trivalents (Kimber et al. 1988). The triploid hybrids pairing data substantiated the Kihara's (1937, 1949, 1954) claim that *Ae. geniculata* originated from a hybridization between *Ae. comosa* and *Ae. umbellulata*, followed by chromosome doubling. Yet, the meiotic analysis of Kimber and Yen (1988) showed that the M genome of *Ae. geniculata* had undergone substantial modification; the U

subgenome is much closer to the U genome of *Ae. umbellulata* than the M° subgenome is to the M genome of *Ae. comosa*. However, the possibility exists that the U subgenome has also been somewhat modified (Kimber and Yen 1988).

Meiotic data from other F_1 triploid hybrid involving *Ae. geniculata* and *Ae. uniaristata* (Table 9.5) displayed low chromosomal pairing, indicating that there is low homology between the subgenomes of *Ae. geniculata* and the genomes of *Ae. uniaristata* (Table 9.5). The triploid hybrid *Ae. geniculata* x *Ae. longissima* had 2–6 bivalents, with mode of 4, and 0–1 trivalents (Kihara 1949). The *Ae. geniculata* x *T. monococcum* hybrid had 13.50 univalents, 3.42 rod bivalents, 0.06 ring bivalents and 0.18 trivalents (Bell and Sachs 1953). A similarly low level of pairing [0–5 (usual 1–3) bivalents] were reported earlier in this hybrid combination by Aase (1930), Bleier (1930 and Percival (1930).

Chromosome pairing at meiosis of F_1 hybrids between *Ae. geniculata* and allotetraploids sharing the U subgenome but differing in the second subgenome, showed that, in addition to pairing between the chromosomes of the shared subgenome, that pairing also occurred between several chromosomes of the non-shared subgenomes (Table 9.7). For instance, the F_1 tetraploid hybrid *Ae. geniculata* x *Ae. biuncialis* (hybrid genome M°M^{b}UU) had 10.25 univalents, 4.20 rod bivalents, 2.30 ring bivalents, 1.25 trivalents and 0.25 quadrivalents (Kimber et al. 1988). The *Ae. geniculata* x *Ae. columnaris* (hybrid genome M°X^{n}UU) hybrid had 9.20 univalents, 4.45 rod bivalents, 2.70 ring bivalents, 1.10 trivalents and 0.30 quadrivalents (Kimber et al. 1988). The *Ae. geniculata* x *Ae. kotschyi* (hybrid genome M°S^{v}UU) hybrid had 11.70 univalents, 1.60 rod bivalents, 4.42 ring bivalents and 1.42 trivalents (Feldman 1963), and Kimber et al. (1988) observed 10.65 univalents, 3.20 rod bivalents, 3.25 ring bivalents, 1.35 trivalents and 0.10 quadrivalents in this hybrid combination, and 11.95 univalents, 3.50 rod bivalents, 2.15 ring bivalents, 1.25 trivalents and 0.25 quadrivalents in the reciprocal hybrid. The *Ae. geniculata* x *Ae. peregrina* hybrid had 11.44 univalents, 2.02 rod bivalents, 5.00 ring bivalents and 0.84 trivalents (Feldman 1963). The *Ae. geniculata* x *Ae. triuncialis* hybrid had 5–11 (usually 7–8) bivalents (Kihara 1929).

Hybrids between tetraploid species with different genomic constitution had significantly less pairing than hybrids between tetraploid species that share one subgenome. For example, the *Ae. cylindrica* x *Ae. geniculata* (hybrid genome DCM°U) hybrid had 3–8 bivalents (Aase 1930) and the *Ae. geniculata* x *Ae. ventricosa* (hybrid genome M°UDN) hybrid had 3–10 bivalents (usually 7) (Kihara 1929) or 3–7 bivalents (all rod bivalents) (Percival 1930).

Chromosomal pairing was also studied in hybrids between *Ae. geniculata* and tetraploid and hexaploid wheat (Tables 9.9 and 9.10). The tetraploid hybrid *Ae. geniculata* x

T. turgidum sp. *turgidum* (hybrid genome M°UBA) had 0–3 rod bivalents (Aase 1930; Bleier 1928; Gaines and Aase 1926 Yen and Kimber (1990d) Genomic relationships of *Triticum searsii* to other S-genome diploid *Triticum* species. Genome 33: 369–374.

Kagawa 1929; Kihara 1929; Percival 1930; Sax and Sax 1924), and the *Ae. geniculata* x *T. turgidum* subsp. *durum* (genome M°UBA) hybrid had 0–4 rod bivalents (Aase 1930). The *T. aestivum* x *Ae. geniculata* (genome BAD-M°U) hybrid had 29.67 univalents, 2.62 bivalents and 0.03 trivalents (Riley 1996a), or 0–3 rod bivalents (Bleier 1928), indicating very low homology between the subgenomes of these two species. Similarly, low levels of pairing in the *T. aestivum* x *Ae. geniculata* hybrid was observed by Cifuentes and Benavente (2009), who examined chromosome pairing at meiotic metaphase in this F$_1$ pentaploid hybrid by FISH, which enabled simultaneous discrimination of the chromosomes of the A, B, D, U and M° subgenomes. They observed 32.86 univalents, 1.04 bivalents, and 0.02 trivalents, of which more than 60% represented allosyndetic pairing between wheat and *geniculata* chromosomes. The average ratio of *aestivum-geniculata* associations was 5:1:12 for those involving the A, B and D subgenomes, respectively, indicating that somewhat higher homology exists between D subgenome chromosomes and those of *Ae. geniculata* than between A or B subgenomes chromosomes and *Ae. geniculata* chromosomes.

Fernandez-Calvin and Orellana (1992) analyzed meiotic associations at first metaphase in the pentaploid hybrid *Ae. geniculata* x *T. aestivum*, bearing the *Ph1* gene (suppressor of homoeologous pairing) and the mutant *phlb* (a deficiency of *Ph1* that enables high homoeologous pairing), using the C-banding technique. The observed associations revealed the same relative order: AD-M°U > A-D > U–M° > AD-B > UM°-B in both low- and high-homoeologous-pairing hybrids.

Benavente et al. (2001) analyzed karyotypes of offspring of two different *T. turgidum–Ae. geniculata* amphiploids (2n = 8x = 56; genome BBAAM°M°UU) carrying *Ph1* or lacking it (*ph1c deletion*), by GISH. The offspring, obtained after two generations of amphiploid selfing, had, on average, fewer chromosomes than expected. Most of the lost chromosomes belonged to *Ae. geniculata*. The two families differed greatly in the number of intergenomic translocations. The *ph1c* family showed nine translocations over 12 plants, while only one translocation was observed in the *Ph1* family. All exchanges involved either the M° and U chromosomes or the M° and wheat chromosomes. The results suggest an epistatic effect of the *ph1c* deletion on the genetic diploidizing system that operates in *Ae. geniculata*, since translocated chromosomes are most likely derived from homoeologous recombination.

In summary, *Ae. geniculata* has very little homology with the subgenomes of either tetraploid or hexaploid wheat, as expressed by the minimal pairing in its hybrids with wheat species of these ploidy levels. Aase (1930) suggested that *Ae. geniculata* may be an autotetraploid. But, if so, more bivalents would be expected in its hybrids both with tetraploid and hexaploid wheat, and Kagawa's studies (1929) on chromosome morphology in *Ae. geniculata* unambiguously rule out an autotetraploid interpretation of *Ae. geniculata*.

9.8.4 *Aegilops biuncialis* Vis.

9.8.4.1 Morphological and Geographical Notes

Ae. biuncialis Vis., common name Mediterranean *Aegilops*, [Syn.: *Ae. lorentii* Hochst.; Ae. *machrochaeta* Shuttl. & Huet. ex Duval-Jouve; *T. macrochaetum* (Shuttl. & Huet. ex Duval-Jouve) Richter; *T. biuncialis* Vill.; *Ae. connata* Ateud.] is a predominantly autogamous, annual, multi-tillered plant, 15–40(-50) cm tall (excluding spikes). The uppermost 1/3 or 1/4 of the culms is defoliated. Leaf blades are glabrous or ciliate, seldom hairy, short (2–5-cm-long) and narrow-linear. Its spike is narrow-lanceolate to narrow-elliptical, lax, 2.0–3.5-cm-long (excluding awns), awned, usually with two spikelets, rarely 3–4, the uppermost of which is not significantly smaller than the lower ones. The entire spike disarticulates at maturity and falls as a unit (umbrella-type dispersal unit). The lower rachis internode is shorter than the adjacent spikelet. There is one basal rudimentary spikelet, seldom two. The spikelets are narrow to broad elliptical, and their lower parts are sometimes slightly inflated. There are 4–5 florets, the two lower one being fertile. The glumes are 8–10-mm-long, awned, with unequally broad nerves that are unequally spaced. Glume awns are usually smooth underneath, all with the same breadth, but with unequal lengths in different spikelets. Terminal spikelet have three glume awns, which are 4–7-cm-long, and much longer and broader than those of the lateral spikelets. The central awn of the terminal spikelet is sometimes longer than its lateral awns. There are 2–3 awns on the lateral spikelets, and when there are three, the central awn is shorter than the lateral awns. The lemma is membranous, usually awned, but the awns are more poorly developed than the glume awns, plainly shorter and fewer, but always more than one per spikelet, usually 2 in lateral spikelets, and 3–4 in the terminal ones. All awns diverge at maturity. The caryopsis is free (Fig. 9.4c).

There is a difference of opinion among taxonomists concerning the valid name of this species. The name *Ae. biuncialis* was given by Visiani (1842), accompanied by an illustration but without a description. His description of the species appeared only ten years later (Visiani 1852). In the meantime, Hochstetter (1845) described the same species, collected by Lorent (1845), and named it *Ae. lorentii*. Zhukovsky (1928), Eig (1929a), Nevski (1934a, b), Tzvelev (1976) and Gandilian (1980) used the name *Ae. biuncialis*, whereas other taxonomists, e.g., Bor (1968, 1970), Tutin and Humphries (1980) and Hammer (1980), used the name *Ae. lorentii*. Yet, according to Article 44 of the International Code of the Botanical Nomenclature, the former name, which was published earlier and accompanied by an illustration with analysis, is the valid name; and so, it is, in practice (Mattatia and Feinbrun-Dothan 1986).

Ae. biuncialis exhibits a very wide morphological variation mainly involving the spike shape (elliptical to lanceolate), size (2 or 3, seldom 4 fertile spikelets), color, hairiness and awn width and length. This variation is reflected in the adaptation of *Ae. biuncialis* to different environmental conditions throughout its distribution area (van Slageren 1994). Because of this morphological variation, Zhukovsky (1928), Eig (1929a), and Hammer (1980) subdivided *Ae. biuncialis* into three varieties that mainly differed in the shape of the spikelets, hairiness and length of glumes and awns.

A high degree of intraspecific molecular variation was also revealed in *Ae. biuncialis*; analysis of several populations of this species from the Iberian-peninsula and the Balearic-islands, using AFLP DNA markers, showed high genetic variation within and between the studied populations (Monte et al. 1999, 2001). Similarly, Thomas and Bebell (2010), using RAPD and ISSR markers, revealed a significant amount of genetic variability in *Ae. biuncialis* from Greece. Rabokon et al. (2019), using intron-specific DNA polymorphism of β-tubulin gene family members as a molecular marker, uncovered DNA polymorphism sufficiently high to distinguish between different accessions of *Ae. biuncialis*.

Ae. biuncialis distributes in the Mediterranean and western Asiatic regions. It grows in Portugal (possibly), Spain, South France (including Corsica), Italy (including Sicily and Sardinia), Malta, Greece (including Crete and the Aegean Islands), Albania, Macedonia, Serbia, Bosnia-Herzegovina, Croatia, Romania, Bulgaria, South Ukraine, South Russia (Crimea and Cis-Caucasia), Trans-Caucasia (Armenia, Azerbaijan, Georgia), Turkey, northern Iraq, western Iran, Syria, Lebanon, Cyprus, Israel, Jordan, Libya, Tunisia, Algeria, and Morocco. Ae. biuncialis is common in southern Europe and the Aegean, Greece, Turkey, Bulgaria and Cyprus. It is well represented in the western arc of the Fertile Crescent but virtually rare in the central part and the eastern arc, in Cis- and Trans- Caucasia, southern Crimea and southern Ukraine. It is less common in North Africa. *Ae. biuncialis* is adventive in parts of central and northwestern Europe such as Germany, Switzerland and the Netherlands. *Ae. biuncialis* grows on a variety of soils, e.g., terra rossa, basalt, and rendzina soils. It is prevalent in the edges and

openings of the sclerophyllous and deciduous oak forests and maquis, in degraded shrub formations, in semi-steppe herbaceous formations, in stony hillsides, abandoned fields, edges and within cultivation such as olive groves, vineyards, fruit tree plantations, and within or near barley and wheat fields, in disturbed and eroded areas and roadsides. It is common and locally abundant in generally dry, somewhat disturbed habitats, such as fallow wastelands, roadsides, and dry rocky slopes of hills and mountains. It grows at altitudes of 200–1750 m, rarely up to 2100 m in Lebanon. As a typical colonizer, the species can form massive stands, especially in regularly disturbed places. It grows in areas with annual rainfall varying from less than 100 mm up to 1100 mm, with most growing in areas with a range of 200–700 mm rain.

The distribution area *of Ae. biuncialis* is in the western and central parts of the distribution of the genus. It is a Mediterranean element extending into semi-steppical (West Irano-Turanian) region, occupying a wide variety of primary and secondary habitats. *Ae. biuncialis* usually grows in mixed stands with other *Aegilops* species, particularly those belonging to the U-genome cluster, i.e., the allopolyploids of section Aegilops, with which it may introgress. Interspecific hybrids and hybrid derivatives were frequently found in mixed populations of *Ae. biuncialis*, *Ae. geniculata*, and *Ae. peregrina* in Israel, indicating continuous genetic connections between these three species (Zohary and Feldman 1962; Feldman 1965a, b, c). Hybrids between *Ae. biuncialis* and common wheat, when the wild species grew in or close to the wheat fields, were reported by Loureiro et al. (2006), Loureiro and Escorial (2007), who estimated the field hybridization rate under central Spain conditions. Their study showed that the hybrid *Ae. biuncialis* x common wheat can be partially fertile by pollinating it by the wheat parent and thus, can introgress with the domesticated parent.

The distribution of the two diploid parents of *Ae. biuncialis* overlaps in western Turkey and Greece. It is assumed therefore, that this area is the center of origin of *Ae. biuncialis*. The distribution of *Ae. biuncialis* overlaps with and is much larger than that of its two parents. It has a sympatric distribution with the following species: *A. muticum*, *Ae. speltoides*, *Ae. searsii*, *Ae, caudata*, *Ae. comosa*, *Ae. uniaristata*, *Ae. umbellulata*, wild *T. monococcum*, wild *T. timopheevii*, wild *T. turgidum*, *Ae. geniculata*, *Ae. neglecta*, *Ae. recta*, *Ae. columnaris* and *Ae. ventricosa*, and an allopatric distribution with *Ae. longissima*, *Ae. kotschyi*, *Ae. tauschii*, *Ae. crassa*, *Ae. vavilovii* and *Ae. juvenalis*.

Ae. biuncialis contains genes that confer resistance to powdery mildew (Gill et al. 1985; Zhou et al. 2014), to leaf rust (Gill et al. 1985; Marais et al. 2003), yellow (stripe) rust (Damania and Pecetti 1990; Marais et al. 2003; Zhou et al. 2014), barley yellow dwarf virus (BYDV) (Makkouk et al. 1994), and Hessian Fly (Gill et al. 1985). It also contains genes that may improve tolerance to abiotic stresses such as drought and salt (Molnár et al. 2004; Colmer et al. 2006; Dulai et al. 2014). In this regard, Molnáret al. (2004) found that *Ae. biuncialis* genotypes originating from a dry habitat have better drought tolerance than wheat, rendering them good candidates for improving the drought tolerance of wheat through intergeneric crossing. In addition, it was found that one accession of *Ae. biuncialis* had a high gluten and grind quality because of a high percentage of γ-45.31 and γ-43.5 gliadins, while another accession had low gluten content, because of a low percentage of these gliadins (Ahmadpoor et al. 2014). A disomic addition line of a pair of chromosomes 1U of *Ae. biuncialis* in the background of common wheat, improved the end-product quality of wheat (Zhou et al. 2014). Also, chromosome 3 MB of *Ae. biuncialis* improved the grain micronutrient content, namely, higher K, Zn, Fe, and Mn contents in wheat (Farkas et al. 2014).

9.8.4.2 Cytology, Cytogenetics and Evolution

Ae. biuncialis is an allotetraploid (2n = 4x = 28) species. Its nuclear genome designation is UUM^bM^b and that of its organellar genome is U. Kihara (1937, 1954) designated the genome of *Ae. biuncialis* $C^uC^uM^oM^o$ (current designation C^u = U and M^o = M^b), and Lilienfeld (1951) concluded that the M^o subgenome of *Ae. biuncialis* is a modified M genome, that was derived from either the M genomes of *Ae. comosa* or from the M^u (currently N) genome of *Ae. uniaristata*. The designation $C^uC^uM^oM^o$ was based, in addition to genome analyses pairing data, on morphological characteristics and on karyotype (Senyaninova-Korchagina 1932). However, a later karyotypic study showed that one chromosome set corresponded very well with the chromosomes of *Ae. umbellulata*, whereas the second chromosome set did not correspond to any chromosomes of the diploid species (Chennaveeraiah 1960).

Sasanuma et al. (2006) presented evidence that the U subgenome of *Ae. biuncialis* had multiple origins deriving from different accessions of *Ae. umbellulata*. This conclusion was based on PCR–RFLP of the U genome-specific U31 fragment, developed by Kadosumi et al. (2005). Sasanuma et al. (2006) investigated the PCR–RFLP of this fragment in 48 accessions of *Ae. biuncialis* and found that most accessions possessed one allele of this fragment, whereas other accessions had a second allele. Since these two alleles exist in different accessions of *Ae. umbellulata*, Sasanuma et al. (2006) concluded that the U subgenome in *Ae. biuncialis* had multiple origins. The multiple origin of the M subgenome of *Ae. biuncialis* has been suggested by Chee et al. (1995),

The origins of the subgenomes of *Ae. biuncialis and Ae. geniculata were* investigated by examining the presence of specific restriction fragments of repeated nucleotide sequences in the DNA of the allopolyploid species (Resta et al. 1996). The analysis showed that *Ae. biuncialis* and *Ae. geniculata* are closely related, that the U subgenome of both species is closely related to the genome of *Ae. umbellulata*, whereas the second subgenome, M^b in *Ae. biuncialis* and M^o in *Ae. geniculata*, is a modified genome of *Ae. comosa*. Modification of the M^b and M^o subgenomes could be attributed to hybridization of allotetraploids sharing the U subgenome but differing in their second subgenome (Zohary and Feldman 1962; Feldman 1965a, b, c). C-banding and FISH studies also confirmed that the U subgenome of *Ae. biuncialis* derived from *Ae. umbellulata* and that the second subgenome, M^b, is a modified M genome of *Ae. comosa* (Badaeva et al. 2004). Like Resta et al (1996), Badaeva et al. (2004) also assumed that intraspecific divergence of *Ae. biuncialis* involved introgression of genetic material from other species. In addition to sharing the U and M subgenomes with the allotetraploid species *Ae. geniculata, Ae. biuncialis* shares the U subgenome with the allopolyploids *Ae. columnaris* (UUX^nX^n), and *Ae. neglecta* (UUX^nX^n), but differ from them in the second subgenome (Resta et al. 1996). In accord with this, Badaeva et al. (2004) found significant differences in the karyotype structure, in the total amount and distribution of C-heterochromatin and in the number and location of 5S and 18S-26S rDNA loci between *Ae. geniculata* and *Ae. biuncialis* on the one hand, and between *Ae. columnaris* and *Ae. neglecta*, on the one hand, evidence that these two groups of species contain the U genome of *Ae. umbellulata* but differ in the source of the second subgenome.

Several investigations were performed to identify the cytoplasm donor of *Ae. biuncialis*. Using restriction endonucleases, Ogihara and Tsunewaki (1982) found no differences in the chloroplast genome of two *Ae. biuncialis* accessions. Terachi et al. (1984) and Ogihara and Tsunewaki (1988) discovered that the chloroplast genome of *Ae. biuncialis* is almost identical with that of *Ae. umbellulata* and therefore concluded that *Ae. umbellulata* was the cytoplasm donor to *Ae. biuncialis*. Consequently, the designation U was given to the plasma type of *Ae. biuncialis*. These results were also confirmed by further studies on the chloroplast and mitochondrial DNA (Wang et al. 1997; Tsunewaki 2009). In contrast, based on studies of cytoplasm-substituted wheats, Maan (1975, 1978) and Panayotov and Gotsov (1975, 1976) proposed that *Ae. uniaristata* or *Ae. comosa* were the cytoplasm donors of *Ae. biuncialis*. However, the results of Terachi et al. (1984), Ogihara and Tsunewaki (1988), Wang et al. (1997), and Tsunewaki (2009) clearly confirmed the conclusion that *Ae. umbellulata* was the cytoplasm donor to *Ae. biuncialis*. Based on plasmon differences between the cytoplasms of the allotetraploids and those of *Ae. umbellulata*, Terachi et al. (1984), Wang et al. (1997), and Tsunewaki (2009) proposed that *Ae. biuncialis* is a young species whose origin was later than that of the origin of *Ae. geniculata*.

Ae. biuncialis contains 10.37 ± 0.039 pg 1C DNA (Eilam et al. 2008), which is 4.95% less DNA than that which is expected from the sum of the DNA of its two diploid parents, i.e., 10.91 pg (*Ae. umbellulata* contains 5.38 pg and *Ae. comosa* contains 5.53 pg; Eilam et al. 2007). The loss of DNA in *Ae. biuncialis* was also found by Badaeva et al. (2002, 2004) who, based on differential C-banding and FISH, found that *Ae. biuncialis* exhibits substantial structural chromosome rearrangements, including deletion of chromosomal segments and reduction of heterochromatin content.

Senyaninova-Korchagina (1932) was the first to describe the karyotype of *Ae. biuncialis*. She recognized seven types of chromosomes that have only primary constrictions and observed no SAT chromosomes. In contrast, Chennaveeraiah (1960) reported on three chromosome pairs with satellites on short arms, two pairs with markedly larger satellites, whereas those on the third pair were very small. All in all, he recognized 12 types of chromosomes, where one pair had an extreme subterminal centromere, three pairs had submedian-subterminal centromeres, and the rest of the pairs had submedian centromeres. No chromosome had a median centromere.

Teoh et al. (1983), using FISH with a repetitive clone that codes for rRNA, confirmed that there are three chromosome pairs in *Ae. biuncialis* exhibiting rRNA sites, whereas the expected number of pairs of rRNA is 4. Cermeño et al. (1984b) analyzed nucleolar activity in *Ae. biuncialis* using a highly reproducible silver-staining procedure and found that the U subgenome suppressed one pair of the NORs of the M^b subgenome.

Badaeva et al. (2002, 2004) studied the karyotype structure of *Ae. biuncialis by* analyzing heterochromatin banding patterns of their somatic metaphase chromosomes, as revealed by C-banding and FISH with the heterochromatin-limited repetitive DNA probes pSc119, pAs1, as well as the distribution of NOR and 5S DNA loci, revealed by pTa71 (18S-26S rDNA) and pTa794 (5S rDNA) probes. The C-banding studies displayed a wide polymorphism resulting from chromosomal rearrangements represented by paracentric inversions and intra- and inter-subgenomic translocations. The results obtained confirmed that the allotetraploid species *Ae. biuncialis* was formed as a result of hybridization of the diploids *Ae. umbellulata* and *Ae. comosa*. The dissimilarity of the C-banding patterns and FISH sites in several chromosomes of this allotetraploid species and those of its ancestral diploid species indicated that chromosomal changes occurred at the tetraploid level. Badaeva et al. (2004) found that *Ae. geniculata* and *Ae. biuncialis* differed from each other and from the putative diploid progenitors in

the inactivation of major NORs on the M-subgenome chromosomes, in the redistribution of 5S rDNA sites, and loss of some minor 18S-26S rDNA loci in *Ae. geniculata* and *Ae. biuncialis*. These differences indicate that various types of chromosomal alterations occurred during the formation and evolution of these allopolyploid species.

When analyzing the karyotype of three different accessions of *Ae. biuncialis*, Wang et al. (2013b) found that they all exhibited a similar karyotype. All *Ae. biuncialis* chromosomes had identifiable C-bands and FISH sites, which allowed for simultaneous discrimination of all U and M^b chromosomes. The U subgenome of *Ae. biuncialis* resembled the U genome of the diploid species *Ae. umbellulata*, whereas the M^b subgenome had some differences compared to the M genome of *Ae. comosa*. The C-banding pattern of the three studied accessions was similar to that reported by Badaeva et al. (2004).

In contrast to Wang et al. (2013a), Schneider et al. (2005), using FISH with the two repetitive DNA sequences pSc119.2 and pAs1 on root-tip metaphases, observed differences in the FISH patterns of all chromosomes among four *Ae. umbellulata* accessions, four *Ae. comosa* accessions, and three *Ae. biuncialis* accessions. The hybridization patterns of the analyzed *Ae. biuncialis* accessions were more variable, and differences were observed not only among the *Ae. biuncialis* lines but also between *Ae. biuncialis* and its diploid progenitors. The genetic variability of *Ae. biuncialis* was manifested by the different locations of the repetitive sequences, which led to differences in the FISH pattern. FISH polymorphism was detected in both the U and M subgenome chromosomes, but, as also reported by Chee et al (1995), the level of repetitive DNA variation in the M subgenome was much higher than in the U subgenome. Similar results were obtained by Gong et al. (2006) who, using ISSR markers, found that the U-subgenome of the allopolyploids of the U subgenome group, was very similar to that of *Ae. umbellulata* and was practically unchanged, while the other subgenomes were greatly altered in the allopolyploids, as was suggested by Zohary and Feldman (1962). Likewise, Kimber and Yen (1988), analyzing chromosome pairing in the hybrid between *Ae. biuncialis* and autotetraploid *Ae. umbellulata*, found that the U subgenome was relatively unmodified.

In addition, Schneider et al. (2005), produced and identified five different *T. aestivum–Ae. biuncialis* disomic addition lines. To differentiate between the added *Ae. biuncialis* chromosomes and those of common wheat, they used genomic in situ hybridization and detected no chromosome interchanges involving wheat and *Ae. biuncialis* chromosomes. Schneider et al. (2005) used three repetitive DNA clones (pSc119.2, pAs1, and pTa71) and identified the *Ae. biuncialis* disomic additions as 2M, 3M, 7M, and 3U.

9.8.4.3 Crosses with Other Species of the Wheat Group

Feldman (1963) studied chromosomal pairing at first meiotic metaphase of two *Ae. biuncialis* accessions and found regular behavior, i.e., an average of 0.97 rod bivalents and 13.03 ring bivalents per cell. Cuñado (1992) observed more rod bivalents These authors also analyzed the pattern of early meiotic zygotene pairing in *Ae. biuncialis* by whole-mount surface-spreading of synaptonemal complexes under the electron microscope. Their data indicate that at the zygotene stage, almost all cells (93%) had 14 bivalents, whereas only a few cells (7%) had 12 bivalents and 1 quadrivalent.

Chromosomal pairing at first meiotic metaphase of F_1 hybrids between *Ae. biuncialis* and its putative diploid parents, *Ae. umbellulata* and *Ae. comosa*, was studied by Kihara (1937, 1949). The *Ae. biuncialis* x *Ae. umbellulata* hybrid had an average of 7 bivalents (Kihara 1937) and the *Ae. biuncialis* x *Ae. comosa* hybrid had 5–8 bivalents with a mode of 7 and 0–2 trivalents (Kihara 1949). Kihara (1949) also studied hybrids between *Ae. biuncialis* and diploid species of section Sitopsis and noted much less pairing than in the hybrids between *Ae. biuncialis* and its putative diploid parents. *Ae. biuncialis* x *Ae. speltoides* had 3–7 bivalents and 0–3 trivalents, *Ae. biuncialis* x *Ae. longissima* had 0–5 bivalents and 0–2 trivalents, and the hybrid *Ae. biuncialis* x *Ae. bicornis* had 0–6 bivalents and 0–1 trivalents. On the other hand, the F_1 hybrid *Ae. biuncialis* x *Ae. caudata* had 5–7 bivalents with a mode of 7, and 0–3 trivalents (Kihara 1949), indicating a greater homology with *Ae. caudata* than with the Sitopsis species.

Hybrids between *Ae. biuncialis* and allopolyploid species sharing the U subgenome were studied by Kihara (1937), Feldman (1965c), and Kimber et al. (1988). Kihara observed 7 bivalents in the F_1 hybrid *Ae. biuncialis* x *Ae. peregrina* (hybrid genome UM^bS^vU) while Feldman reported 12.16 univalents, 1.16 rod bivalents, 5.38 ring bivalents, 0.76 trivalents, and 0.12 quadrivalents in this hybrid. The range of the bivalents in this hybrid was 5–9, indicating that in addition to the 7 bivalents between the chromosomes of the U subgenomes, some bivalents were also formed between the chromosomes of the differential genomes M^b and S^v. The F_1 *Ae. geniculata* x *Ae. biuncialis* hybrid (hybrid genome M^oUUM^b) had 10.24 univalents, 4.20 rod bivalents, 3.39 ring bivalents, 1.25 trivalents and 0.25 quadrivalents (Kimber et al. 1988). This hybrid also had pairing between chromosomes of the differential subgenomes M^o and M^b. Lindschau and Oehler (1936) observed very low chromosome pairing in the F_1 hybrid *Ae. biuncialis* x *Ae. cylindrica* (hybrid genome UM^bDC), indicating that Ae. *biuncialis* has no subgenome in common with *Ae. cylindrica*.

9.8.5 *Aegilops neglecta* Req. ex Bertol.

9.8.5.1 Morphological and Geographical Notes

Ae. neglecta Req. ex Bertol., common name tri-awn goatgrass, [Syn.: *Ae. ovata* L. emend. Roth; *T. ovatum* (L.) Raspail; *Ae. triaristata* Willd.; *T. triaristatum* Willd.) Godr. & Gren. in Grenier & Godron; *T. neclectum* (req. ex Bertol.) Greuter in Greuter & Rechinger] is a predominantly autogamous, annual, tufted, multi-tillered, 20–40(-50)-cm-tall (excluding spikes) plant. Its culms have few joints, are upright, or somewhat jointed near the ground, with the uppermost 1/4 or 1/3 being defoliated. Leaves are more or less hairy, often with ciliate margins. Its spike is 2.0–4.5-cm-long (excluding awns), lanceolate or narrowly ovoid and compact, with the lower parts inflated and ellipsoid, becoming suddenly narrow-cylindrical in the upper part, and awned. The entire spike disarticulates at maturity and falls as a single unit (umbrella-type dispersal unit). There are 3 basal rudimentary spikelets, seldom 2. There are 3–6 spikelets, usually 4, with the two lowest being narrow to broad-elliptical, proportionally large and lying against each other, and the upper two being narrow, sterile, and projecting conspicuously from the lower spikelets. The lower rachis internodes are shorter than the adjacent spikelet, whereas the upper rachis internodes are longer than the spikelets. There are 4 florets, the upper 2 being sterile. The glumes have curved, unequally broad, flattened nerves, with 2–3 awns that are equal in length to those of the lemma. There are usually three, 3.5–4.5-cm-long glume awns 2–3 on the lower spikelets and almost always 3 on the upper spikelets. There are 2–4 (usually 2) lemma awns on lower spikelets, usually none on the upper ones, and if they exist, they are weakly developed. Often the awns in the upper spikelets decrease in length so that all awns of the spike end at the same height. The awns of the mature spike are generally weak, seldom strong. The caryopsis is free (Fig. 9.4d).

Ae. ovata (senso lato) was presented in the species Plantarum (Linnaeus 1753) as containing two taxa, *ovata* (senso stricto) and *triaristata*. The taxon *triaristata* was separated from *ovata* and elevated to the specific rank by von Willdenow in (1806), but the name *Ae. triaristata* was illegitimate since it was already given to a part of the original *Ae. ovata*. *Ae. neglecta* Req. ex Bertol. was described by Bertoloni in 1834. Since it is the oldest available name to replace the *triaristata*-part of the old *ovata* of Linneaus, it is the legitimate name for this species (van Slageren 1994). Yet, a number of taxonomists and cytogeneticists [e.g., Zhukovsky (1928), Eig (1929a), Chennaveeraiah (1960), Kihara (1940a, b, 1954, 1963), Kihara and Tanaka (1970)] used the name *triaristata*, while others, e.g., Hammer (1980), used the name *neglecta*.

Ae. triaristata included tetraploid and hexaploid forms. The two forms were considered by Zhukovsky (1928), and Kihara 1954, 1963), as subspecies of *Ae. triaristata*, whereas Hammer (1980) considered them subspecies of *Ae. neglecta*. Hammer (1980) and van Slageren (1994) designated the tetraploid forms as ssp. *neglecta* and the hexaploid ones as ssp *recta*. Chennaveeraiah (1960), based on karyotype differences, elevated the hexaploid forms to the species rank and named it *Ae. recta* (Zhuk.) Chennav., but kept the tetraploid as *Ae. triaristata*.

Chennaveeraiah (1960) described few morphological differences between *Ae. neglecta* and *Ae. recta*, such as in the number of spikelets, with 2 to 3 in *neglecta* and 5 to 6 in *recta*. The awns in *neglecta* are spread and in *recta*, are slanted. *Neglecta* glumes are glabrous and in *recta* are hairy. In addition, Kihara (1963) noted that *neglecta* spikes suddenly taper, whereas those of *recta*, gradually taper. Yet, there are many morphological intermediates between the two species and sometimes it is difficult to differentiate between them (Kihara 1963).

Ae. neglecta has a wide morphological variation mainly in spikelet number, shape and size, hairiness, awn number and development. It is sometimes confused with its closest relatives *Ae. recta* and *Ae. columnaris*. Thomas and Bebell (2010), using RAPDs and ISSRs, reported on the occurrence of wide molecular diversity in Greek *Ae. neglecta*.

Ae. neglecta is an East Mediterranean and West-Asiatic element. It distributes in the northeastern Mediterranean region, that is in western Turkey, Greece (including the Aegean Islands and Crete), the Balkan (Croatia, Bosnia-Herzegovina, Serbia-Montenegro, North Macedonia, and Albania), Southern Crimea, and Syria, and in West Asia, i.e., northern Iraq, Cis-Caucasus, Trans-Caucasus, Northern Iran (rare) and Turkmenistan (Kihara 1963; van Slageren 1994; Ohta et al. 2016). Although *Ae. neglecta* was described from France and other West Mediterranean countries, there are no chromosome counts to assure that all these collections are tetraploids.

Ae. neglecta grows on a variety of soils, e.g., terra rossa, basalt, rendzina, and alluvial soils. It is common in the edges and openings of sclerophyllous and deciduous oak forests and maquis, in openings of shrub formations, degraded dwarf-shrub formations, semi-steppe herbaceous formations, pastures, stony hillsides, abandoned fields, edges of cultivation, disturbed and eroded areas and roadsides, within cultivations, such as olive groves, vineyards, fruit tree plantations, and within or near barley and wheat fields. It is also common, locally abundant in the Eastern Mediterranean but somewhat less common in its eastern distribution area. It grows at altitudes from almost sea level to 1600 m, rarely, up to 2000 m. As a typical colonizer, the species can form massive stands, especially in regularly disturbed places. It

grows in areas with annual rainfall varying from less than 450 mm to 750 mm.

Ae. neglecta is an adventive in central and northwestern Europe, e.g., in Scotland, Belgium, Germany, Switzerland and in the Netherlands. It was introduced to the USA and reported as a "weed in fields of California and Virginia" by Hitchcock and Chase in their 1951 Manual of the Grasses of the United States. *Ae. neglecta* is still very restricted in North America, known in California, Oregon, New York and Virginia.

Ae. neglecta has a wide distribution in the central region of the genus distribution. It occupies a large variety of primary and secondary habitats. *Ae. neglecta* grows sympatrically with *Ae. caudata*, *Ae. comosa*, *Ae. uniaristata*, *Ae. umbellulata*, *Ae. biuncialis*, *Ae. triuncialis*, *Ae. cylindrica* and *Ae. recta* and allopatrically with *A. muticum*, *Ae. speltoides*, *Ae. peregrina* and *Ae. columnaris*. Hybrids and intermediate forms between *Ae. neglecta* and other species of the U-subgenome group, which usually grow in mixed stands, are occasionally found (Feldman 1965a).

Ae. neglecta contains genes that confer resistance to powdery mildew (Gill et al. 1985; Worthington et al. 2015), leaf rust (Gill et al. 1985; Marais et al. 2009), yellow (stripe) rust (Marais et al. 2009), barley yellow dwarf virus (BYDV) (Makkouk et al. 1994), and to Hessian Fly (Gill et al. 1985; El Bouhssini et al. 1998). Damania and Pecetti (1990) reported that *Ae. neglecta* is tolerant to drought and frost.

9.8.5.2 Cytology, Cytogenetics, and Evolution

Ae. neglecta is an allotetraploid ($2n = 4x = 28$), with a UUX^nX^n genome designation (modified by Badaeva et al. 2008 from Dvorak 1998; Tsunewaki 2009). The designation of its plasmon is U, similar to that of *Ae. umbellulata* (Kimber and Tsunewaki 1988; Dvorak 1998; Tsunewaki 2009). Originally, *Ae. neglecta* was given the genomic formula $C^uC^uM^tM^t$ (Kihara 1937, 1949; 1954; Morris and Sears 1967). The C^u subgenome (currently U) was derived from *Ae. umbellulata* and the M^t subgenome (Currenly X^n) was thought to be a modified form of the *Ae. comosa* M genome (Kihara 1937, 1949). While Kihara did not analyze hybrids between *Ae. neglecta* and its assumed diploid parents, he determined the genomic formula of *Ae. neglecta* from chromosomal pairing in hybrids between *Ae. neglecta* and other allotetraploid species bearing U and M subgenomes. The *Ae. neglecta* (genome UUX^nX^n) x *Ae. biuncialis* (genome UUM^bM^b) hybrid had 5–10 bivalents, with a mode of 8, the hybrid between *neglecta* and *columnaris* (genome UUX^nX^n) had more than 12 bivalents (Kihara 1936, 1940a, b), and another work on *neglecta* x *columnaris* hybrid reported 1.98 univalents, 12.54 bivalents, 0.26 trivalents and 0.04 quadrivalents (Furuta and Tanaka 1970). Consequently, Kihara (1936, 1940a, b, 1954) assigned the genomic formulas $C^uC^uM^tM^t$ to *Ae. neglecta*. Kimber and Yen (1989) and Yen and Kimber (1992b) reported that all the accessions of *Ae. neglecta* they used had an essentially unchanged U subgenome, which reinforced Kihara's conclusion concerning the origin of the U subgenome. Tsuchiya (1956) produced hybrids between *Ae. neglecta* and its assumed diploid parent *Ae. comosa*, but most of the seeds did not germinate. In one hybrid that did germinate, the meiotic behavior of the chromosomes was not studied in detail, as the anthers degenerated. However, a few configurations observed at first metaphase of several meiotic cells, included 2 or 3 bivalents and many univalents. This may indicate that the second subgenome of *Ae. neglecta* is either different or greatly modified from the M genome of *Ae. comosa*. Additional hybrids between *neglecta* and *comosa* were not studied. Hybrids between *Ae. neglecta* and autotetraploid *Ae. uniaristata* showed that there is no N genome in *neglecta* (Kimber and Yen 1989).

Upon analysis of the karyotype of *Ae. Neglecta*, Chennaveeraiah (1960), like Senyaninova-Korchagina (1932), reported that there is only one pair of chromosomes with satellites on short arms, one pair with an extreme subterminal centromere, four pairs with submedian-subterminal centromeres, one pair with median centromeres and the rest with submedian centromeres. Senyaninova-Korchagina (1932) has called attention to the resemblance of one half of the chromosomes in *Ae. neglecta* to the chromosomes of *Ae. umbellulata*. The most distinguished character of the U genome is the single pair with the extremely subterminal centromere. *Ae. umbellulata* has two chromosome pairs with satellites, whereas *Ae. neglecta* has only one pair with satellites. It seems certain, however, that the U subgenome is present in *Ae. neglecta* either in a pure or in a slightly modified form. The second set of chromosomes includes five types, one pair with a median centromere and the rest with submedian centromeres. No secondary constrictions or satellites which are characteristic of the M genome are present here. The M^t genome of *Ae. neglecta* is either considerably modified from the basic type or it must be a genome foreign to *Aegilops* or at least to the known diploid analyzers of Kihara (1954).

Ae. neglecta and *Ae. columnaris* are closely related to each other and have the same genomes (Resta et al. 1996). Variations in restriction fragments of repeated nucleotide sequences showed that their second subgenome did not derive from the M genome of *Ae. comosa*. While they share the U subgenome with *Ae. biuncialis* and *Ae. geniculata*, their second subgenome is unrelated to the M^o or M^b subgenomes of these species. No relationship was found between this subgenome and the genome of any extant diploid species of *Aegilops* or any phylogenetic lineage leading to the extant diploid species (Resta et al. 1996; Dvorak 1998). This unknown genome was designated X^t by

Resta et al. (1996) and Dvorak (1998), but was recently changed to X^n, (X of *neglecta*) and consequently, the proposed genomic formula for *Ae. neglecta* and *Ae. columnaris* is UUX^nX^n.

Analysis of the restriction fragment pattern of chloroplast DNA of *Ae. neglecta* and of *Ae. umbellulata* using restriction endonucleases (Ogihara and Tsunewaki 1982; Terachi et al. 1984), showed that the chloroplast genome (the plastom) of *Ae. neglecta* arose from *Ae. umbellulata*. Studies on the chondrion genome also showed that the cytoplasm (the plasmon) of *Ae. neglecta* derived from *Ae. umbellulata* (Terachi and Tsunewaki 1986). Comparison the plasmon of several U-subgenome allotetraploid species with that of *Ae. umbellulata* indicated that *Ae. neglecta* is of more recent origin than *Ae. geniculata* (Tsunewaki 2009). On the other hand, the four-tetraploid species, *Ae. biuncialis*, *Ae. columnaris Ae. neglecta* and *Ae. triuncialis* arose from *Ae. umbellulata* as female (Tsunewaki 1996, 2009). The distance between *Ae. neglecta* and *Ae. umbellulata* is 0.01 but is zero between the other three tetraploid and *Ae. umbellulata*. This result suggests that *Ae. neglecta* is the oldest of the four tetraploids.

Kadosumi et al. (2005), studying variation of the genome-specific PCR primer set U31 in *Ae. umbellulata* and *Ae. neglecta*, found three alleles of this DNA sequence in *Ae. umbellulata*, and two in *Ae. neglecta*. This result indicated that the U genome had at least two, probably more, independent origins in *Ae. neglecta*. Similarly, Meimberg et al. (2009) showed that *Ae. neglecta* originated from two independent hybridizations as indicated by presence of at least two different *Ae. umbellulata* chloroplast haplotypes. This multiple origin could have introduced genetic variability that increased the ecological amplitude and evolutionary success of *Ae. neglecta*.

Ae. neglecta contains 10.64 ± 0.404 pg 1C DNA (Eilam et al. 2008). The genome of *Ae. neglecta*, and that of its closest relative *Ae. columnaris*, are larger than the genomes of *Ae. biuncialis* and *Ae. geniculata* (Eilam et al. 2008). This may indicate that the genome of the, yet unknown, diploid donor of the subgenome X^n is larger than the M genome of *Ae. comosa*, the putative donor of the M subgenome to *Ae. biuncialis* and *Ae. geniculata*.

Badaeva (2002) identified all the chromosomes of *Ae. neglecta* on the basis of morphology and C-banding patterns, while Badaeva et al. (2004) also used FISH. All accessions they analyzed possessed three pairs of NORs, two of which were attributed to the U and one to the X^n subgenome. Yet, Senyaninova-Korchagina (1932) and Chennaveeraiah (1960) found only one pair of SAT chromosomes in *Ae. neglecta*. The values obtained by Badaeva (2002) and Badaeva et al. (2004) corresponded to the number of active NORs detected by Teoh et al. (1983) and Ag-NOR staining (Cermeño et al. 1984b). The absence of secondary constriction on two chromosome pairs carrying NORs indicated that the respective NORs were inactivated. Nevertheless, C-banding analysis showed the presence of active NORs on chromosomes 1U, 5U and a X^n chromosome pair.

Comparison of 10 *Ae. neglecta* accessions from diverse geographical regions, revealed low variability in the C-banding patterns and translocation polymorphism (Badaeva 2002, and Badaeva et al. 2004). Badaeva et al. (2004) performed FISH with the heterochromatin-limited repetitive DNA probes pSc119, pAs1, and studied the distribution of NORs (18S-26S rDNA) and 5S rDNA loci using the pTa71 (18S-26S rDNA), and pTa794 (5S rDNA) probes. The data obtained confirmed significant differences in karyotype structure, in the total amount and distribution of heterochromatin, and in the number and location of 5S and 18S-26S rDNA loci between *Ae. neglecta* -*Ae. columnaris*, and *Ae. geniculata*—*Ae. biuncialis*. *Ae. geniculata* and *Ae. biuncialis* showed moderate heterochromatin content, with small- or medium-sized bands located in telomeric and interstitial chromosome regions. The C-banding patterns of these two-species corresponded with that of their diploid ancestors *Ae. umbellulata* and *Ae. comosa*. In contrast to the C-banding patterns of *Ae. geniculata* and *Ae. biuncialis*, *Ae. neglecta* and *Ae. columnaris* chromosomes were characterized by a high heterochromatin content, and large C-bands or C-band complexes located in pericentromeric, interstitial, and telomeric chromosome regions. Similarities in C-banding and FISH patterns of most *Ae. columnaris* and *Ae. neglecta* chromosomes suggest that they likely had a common ancestral diploid species. The differences in three chromosome pairs may indicate that the divergence of these two species was probably associated with chromosomal rearrangements and/or introgressive hybridization. Three intergenomic translocations between U and X^n chromosomes exist in *Ae. neglecta*.

These data reinforced the conclusion of Resta et al. (1996) and Dvorak (1998) that *Ae. neglecta* and *Ae. columnaris* (genomes (UUX^{n}X^n) had a different origin than *Ae. geniculata* and *Ae. biuncialis* (genomes M^{o}M^{o}UU and UUM^{b}M^b, respectively). The similarity of the C-banding patterns of *Ae. neglecta* and *Ae. columnaris* chromosomes suggest that they originated from a common ancestor, with *Ae. umbellulata* being the U-subgenome donor, while the donor of the second subgenome is yet to be determined. Moreover, Badaeva (2002) assumed that *Ae. neglecta* and *Ae. columnaris* exchanged genetic material through processes of introgressive hybridization with other tetraploid species bearing the U subgenome.

9.8.5.3 Crosses with Other Species of the Wheat Group

Using C-banding to study chromosome pairing at meiotic metaphase of the two subgenomes of *Ae. neglecta*, (Cuñado

1992) showed diploid-like meiotic behavior, namely, only bivalents were formed between fully homologous chromosomes. They analyzed the pattern of chromosome pairing at the meiotic zygotene and pachytene stages in *Ae. neglecta* by whole-mount surface-spreading of synaptonemal complexes under the electron microscope. The observations indicated that already at these early meiotic stages, the chromosomes were almost exclusively associated as bivalents.

There are no data on chromosomal pairing in hybrids between *Ae. neglecta* and its diploid parent *Ae. umbellulata*, the donor of the U subgenome. However, chromosomal pairing in the hybrid *Ae. neglecta* x autotetraploid *Ae. umbellulata* supported the assumption that *Ae. neglecta* contains the U subgenome and that this subgenome is fully homologous to the U genome of *Ae. umbellulata* (Kimber and Yen 1989; Yen and Kimber 1992b). On the other hand, there are no data on chromosomal pairing in the hybrid between *Ae. neglecta* and *Ae. comosa*, that was assumed to be the donor of the second subgenome of *neglecta*. Nevertheless, there are data on meiotic chromosomal pairing in hybrids between *Ae. neglecta* and other U-subgenome-bearing allotetraploids. The F$_1$ *Ae. neglecta* x *Ae. columnaris* hybrid had 11–14 bivalents with a mode of 12 bivalents (Kihara 1949), indicating that these two allotetraploids are very close to each other and share close genomes. The hybrid *neglecta* x *geniculata* had 5–10 bivalents, with a mode of 8 bivalents, 0–2 trivalents and 0–1 quadrivalents (Kihara and Nishiyama 1937), and the hybrid *neglecta* x *biuncialis* had similar pairing, i.e., 5–10 bivalents, with a mode of 8 bivalents, 0–1 trivalents and 0–1 quadrivalent (Kihara 1949). This degree of chromosomal pairing indicates that *geniculata* and *biuncialis* are close to *neglecta* but, since all three contain the U subgenome, the reduced pairing in the hybrids between them indicates that they differ in their second subgenome. The *Ae. neglecta* x *Ae. triuncialis* hybrid (genome UX^{n}UC) had 0–9 bivalents, with a mode of 5 bivalents, and 0–1 trivalents (Lindschau and Oehler 1936; Kihara 1949), the hybrid *Ae. neglecta* x *Ae. peregrina* or *Ae. neglecta* x *Ae. kotschyi* (hybrids genome UX^{n}S^{v}U) had 7–10 bivalents, with a mode of 9 bivalents, and 0–2 trivalents, indicating that *Ae. neglecta* differs in its second subgenome from *Ae. triuncialis*, *Ae. peregrina* and *Ae. kotschyi*. On the other hand, chromosome pairing in the pentaploid hybrid *Ae. neglecta* x *Ae. recta* (genome UX^{n}UX^{n}N) showed 14 bivalents and 7 univalents (Kihara 1937) indicating that the two species share two subgenomes and that the third subgenome of *Ae. recta* differ from the two subgenomes of *Ae. neglecta*.

Chromosomal pairing in hybrids between *Ae. neglecta* and diploid species having the S, C, and N genomes showed that the two subgenomes of *Ae. neglecta* are not homologous to those of the diploids. More specifically, the *Ae. neglecta* x *Ae. speltoides* hybrid (genome UX^{n}S) had 2–8 bivalents, with a mode of 4–5 bivalents, the *Ae. neglecta* x *Ae. caudata* hybrid (genome UX^nC) had 6.30 univalents, 4.48 rod bivalents, 0.17 ring bivalents, 1.73 trivalents and 0.05 quadrivalents (Kimber and Abu-Baker 1981), and the *Ae. neglecta* x *Ae. uniaristata* hybrid (genome UX^nN) had 2–6 bivalents, with a mode of 4 bivalents (Kihara 1949).

The frequency of homoeologous pairing at first meiotic metaphase in F_1 hybrids between *T. turgidum* subsp. *durum* cv. Langdon and *Ae. neglecta* was determined by a genomic in situ hybridization (GISH) procedure that allowed simultaneous discrimination of *durum* and. *neglecta* chromosomes (Cifuentes et al. 2010). Chromosomal pairing was low in this hybrid; average pairing contained 22.71 univalents, 2.37 rod bivalents, 0.01 ring bivalents, 0.16 trivalents and 0.02 quadrivalents, with a total of 2.75 associations per cell. Some of the associations were auto syndetic, namely, 0.19 associations/cell were between A and B chromosomes of *durum* and 0.67 associations/cell were between U and X^n chromosomes of *neglecta*. However, most of the associations were allosyndetic, with 1.39 associations/cell between A and UX^n chromosomes and 0.5 between B and UX^n chromosomes. Hence, interspecific *durum-neglecta* associations account for 69% of total first meiotic metaphase pairing. Chromosomes of the A subgenome were most frequently involved in pairing with their *neglecta* homoeologues than the B subgenome chromosomes (Cifuentes et al. 2010).

9.8.6   *Aegilops recta* (Zhuk.) Chennav.

9.8.6.1   Morphological and Geographical Notes

Ae. recta (Zhuk.) Chennav. [Syn.: *Ae. triaristata* Willd.; *Ae. triaristata* subsp. *recta* Zhuk.; *Ae. neglecta* Req. ex Bertol.; *Ae. neglecta* subsp. *tecta* (Zhuk.) Hammer; *T. neglectum* (Req. ex Bertol.) Greuter; *T. rectum* Bowden] is an annual, predominantly autogamous, 20–35-cm-tall (excluding spikes) plant. The uppermost 1/4 or 1/3 of the culms is defoliated. Leaf blades are more or less hairy, often with ciliate margins. Its spike is lanceolate, compact, 2.0–3.5-cm-long (excluding awns), awned, and generally becoming narrow in the upper parts. The entire spike disarticulates at maturity and falls as a unit (umbrella-type dispersal unit). The lower rachis internode is shorter than the adjacent spikelet. There are three basal rudimentary spikelets, seldom two. There are 3–6 spikelets, usually 4, with the two lowest being narrow to broad-elliptical, proportionally larger than the upper ones and lying against each other, while the upper two are usually fertile, seldom sterile, and project from the lower part of the spike. There are 4–5 florets, the two lower ones being fertile. The glumes are awned and broad, with curved, unequally broad, flattened nerves. There are usually 3 glume awns on the upper spikelets, and 2–3 on the lower spikelets. There are 2–4 lemma awns, usually 2, on lower

spikelets and usually none on upper ones, and if they exist, they are poorly developed. Often the awns in the upper spikelets decrease in length so that all awns of the spike end at the same height. The awns of the mature spike are generally weak, seldom strong. The caryopsis is free (Fig. 9.4e).

Originally, *Ae. recta* has been included in *Ae. neglecta* (formerly *Ae. triaristata*) which contained tetraploid and hexaploid cytotypes. Zhukovsky (1928), and Kihara (1954, 1963) considered the two forms subspecies of *Ae. triaristata*, whereas Hammer (1980) and van Slageren (1994) considered them subspecies of *Ae. neglecta*, i.e., the tetraploid forms as ssp. *neglecta* and the hexaploid ones as ssp. *recta* (Zhuk,) Hammer.

The tetraploid and hexaploid subspecies of *Ae. neglecta* posed a taxonomic problem in the sense that they are almost undistinguishable, yet, the different ploidy level causes some reproductive isolation and subsequent speciation (van Slageren 1994). The association 'recta' with the hexaploid level was made for the first time by Senyaninova-Korchagina (1930) in her karyosystematical overview of *Aegilops*. Since then, the epithet 'recta' has been associated with the hexaploid level (e.g., Chennaveeraiah 1960; Bowden 1966; Löve 1984; Kimber and Feldman 1987). Yet, the F_1 hybrid between the tetraploid and the hexaploid subspecies had 14 bivalents and 7 univalents (Kihara 1937), indicating that the third subgenome in Ae. *recta* differ from the two subgenomes of Ae. *neglecta*. The F_1 hybrid is sterile (Chennaveeraiah 1960) or partially fertile (Kihara 1963). Therefore, according to Chennaveeraiah (1960), the two taxa should be treated as separate species. Consequently, Chennaveeraiah (1960) elevated the hexaploid forms to the species rank and named the new species *Ae. recta* (Zhuk.) Chennav.

Separation of *Ae. neglecta* and *Ae. recta* to two different species would be easy if morphological characters linked to the ploidy level can be identified. In this regard, Chennaveeraiah (1960) described few morphological differences between *Ae. neglecta* and *Ae. recta*, namely, there are 2 to 3 spikelets in *Ae. neglecta*, whereas *Ae. recta* has 5 to 6. *Ae. neglecta* awns are spread at maturity, while they are slanted in *Ae. recta*. Glumes are glabrous in *Ae. neglecta* and usually hairy in *Ae. recta*. In addition, Kihara (1963) noted that the spikes of *Ae. neglecta* suddenly taper, whereas those of *Ae. recta* taper gradually. Kimber and Feldman (1987) suggested that the hexaploid forms possess fertile terminal spikelet(s) while the tetraploid forms have sterile ones. Yet, there are many morphological intermediates between the two species, at times rendering it difficult to differentiate between them (Kihara 1963).

Aryavand et al. (2003) reported on a significant difference in stomatal counts and size in the leaves between the two species. Higher ploidy in *Ae. recta* is associated with fewer but larger stomata per unit leaf area, showing a highly negative correlation between stomatal count and size. Hence, this trait can assist in the identification of the two species. Additionally, Giraldo et al. (2016) developed two chloroplast DNA-based molecular markers that accurately discriminate *Ae. recta* from *Ae. neglecta*. The use of these markers, in addition to chromosome counting, facilitates further the ability to differentiate *Ae. recta* from *neglecta*.

Ae. recta exhibit relatively limited morphological variation involving spike size, color and hairiness. Its variation is relatively low compared with that of other polyploid species of *Aegilops*. Monte et al. (2001) used AFLP DNA markers to characterize the genetic diversity in *Ae. recta* populations distributed in the Iberian Peninsula and Balearic Islands. *Ae. recta* exhibited low variation in contrast with *Ae. biuncialis*, that presented a high degree of polymorphism.

Originally, there was confusion concerning the distribution of *Ae. neglecta* and *Ae. recta* (Ohta et al. 2016). The distribution of *Ae. neglecta* senso lato was usually described as Mediterranean and West Asiatic, namely, from Morocco and Portugal in the west to Transcaucasia, western Iran and Turkmenistan in the east. There was almost no attempt to determine the distribution of the two cytotypes of *Ae. neglecta*. Also, after the separation of the tetraploid forms to a different species, the geographical distributions of the two-species remained unclear. To determine more accurately the distribution of these two species, Ohta et al. (2016) analyzed the chromosome numbers of accessions of these two species from 137 populations, located in the western area of the species distribution from the Aegean Islands to Morocco. Taken together with data from previous studies, Ohta et al. (2016) revealed a difference in the geographical distribution of *Ae. neglecta* and *Ae. recta*: *Ae. neglecta* is distributed in the eastern part of the species area, whereas *Ae. recta* predominantly occurs in the western part, with the border between these two species on the western margin of the Aegean Sea. Near the border, *Ae. neglecta* and mixed populations of both species were sporadically found among populations of *Ae. recta* in the Balkan and Peloponnesus Peninsulas, while a few *Ae. recta* and mixed populations were found among populations of *Ae. neglecta* in the East Aegean Islands and West Anatolia (Ohta et al. 2016).

Support of the conclusions reached by Ohta et al. (2016) was provided by Baik et al. (2017), who carried out a karyological study of several populations of *Ae. neglecta* from different eco-geographical sites in North Algeria. Chromosome counting showed that all accessions were hexaploids, that is to say, *Ae. recta*. Similar results were obtained by Belkadi et al. (2003), who determined chromosome counts in two accessions of *Ae. neglecta* from Morocco and found them to be hexaploids.

Hence, *Ae. recta* is a West Mediterranean element. It grows in Algeria, Morocco, Portugal, Spain, southern France, Italy, Croatia, Bosnia-Herzegovina, Serbia, Bulgaria, Montenegro, North Macedonia, Albania, Greece, and western Turkey. It grows on terra rossa soil in edges and openings of sclerophyllus oak forests and maquis and dwarf-shrub formations, in abandoned fields, edges of cultivation, and in disturbed and eroded areas and roadsides. It is relatively common.

Ae. recta has a medium-sized distribution in the western part of the distribution of the genus. It grows sympatrically with its two parents, *Ae. neglecta* and *Ae. uniaristata*, in the Balkan, the Aegean islands and western Turkey. Since its two parents grow sympatrically in West Turkey and the Balkan, this is presumably the region where *Ae. recta* originated and from which it spread westward (Kihara 1963).

Ae. recta usually grows in mixed stands, with other species with which it introgresses. It grows sympatrically with *Ae. umbellulata*, *Ae. comosa*, *Ae. uniaristata*, *Ae. caudata*, *Ae. neglecta*, *Ae. geniculata*, *Ae. biuncialis*, *Ae. triuncialis*, and *Ae. ventricosa*, and allopatrically with *Ae. columnaris*.

In Europe, *Ae. recta* grows sympatrically with common wheat and spontaneous hybridization between these two species is known to occur (Zaharieva and Monneveux 2006). Arrigo et al. (2011) investigated introgression between common wheat and *Ae. recta* and compared wheat field borders to areas isolated from agriculture. All *Ae. recta* had $2n = 42$. Individuals were characterized with AFLP fingerprinting, analyzed through two computational approaches (i.e., Bayesian estimations of admixture and fuzzy clustering), and sequences marking wheat-specific insertions of transposable elements. With this combined approach, Arrigo et al. (2011) detected substantial gene flow between wheat and *Ae. recta*, and noted significantly more admixed individuals close to wheat fields than in locations isolated from agriculture. Arrigo et al. (2011) concluded that reproductive barriers have been regularly bypassed during the long history of sympatry between wheat and *Ae. recta*.

Gill et al. (1985) reported that *Ae. recta* confers resistance to powdery mildew, leaf rust and Hessian Fly. El Bouhssini et al. (1998) also found that *Ae. recta* from Morocco showed resistance to Hessian Fly.

9.8.6.2 Cytology, Cytogenetics and Evolution

Ae. recta are an allohexaploid species ($2n = 6x = 42$). Its nuclear genome is designated as UUX^nX^nNN (modified from Dvorak 1998) and the plasmon genome is designated U, similar to that of *Ae. umbellulata* (Tsunewaki 2009). Kihara (1937), based on the observation of 14 ring bivalents in the pentaploid hybrid between tetraploid *Ae. neglecta* and hexaploid *Ae. recta*, concluded that *Ae. recta* contain the two subgenomes of *Ae. neglecta*, U and M^t (currently X^n), and consequently, regarded *Ae. recta* as a cytotype of *Ae.*

neglecta (Kihara 1963). Karyotype analyses confirmed that *Ae. recta* contain the two subgenomes of *Ae. neglecta* (Senyaninova-Korchagina 1930, 1932; Chennaveeraiah 1960). Kihara (1957), using indirect pairing data obtained from hybrids between *Ae. recta* and other allotetraploid species bearing the M subgenome, as well as from the morphological similarity between *Ae. recta* and *Ae. neglecta*, believed that the third subgenome of *Ae. recta* is a modified form of genome M of *Ae. comosa*. So, Kihara (1963) designated the genome of *Ae. recta* $C^u C^u M^t M^t M^{t2} M^{t2}$. This genomic formula indicated that *Ae. recta* are, in fact, an allo-auto-hexaploid containing two modified M subgenomes. But, Kihara's assumption has never been confirmed by direct evidence from hybrids between *Ae. recta* and *Ae. comosa*.

As was pointed out by Senyaninova-Korchagina (1930, 1932) and Chennaveeraiah (1960), fourteen chromosome pairs of *Ae. recta* are identical to those of *Ae. neglecta*; thus, karyotypic results also revealed that two of the subgenomes of *Ae. recta* derived from *Ae. neglecta*. The third chromosomal set resembles the chromosomes of *Ae. uniaristata*, although *Ae. uniaristata* contains no chromosome with a median centromere. Bowden (1959) suggested that one subgenome of *Ae. neglecta* and *Ae. recta* derived from *Amblyopyrum muticum*, yet Chennaveeraiah (1960) rejected this suggestion because the karyotypes of *Ae. neglecta* and *Ae. recta* contained no *A. muticum* genome.

Evidence of the presence of the U subgenome in *Ae. recta* has been indirect, based largely on studies of hybrids between allopolyploid species, of which one species, *Ae. geniculata*, had previously been shown to have the *Ae. umbellulata* genome (von Berg 1937; Kihara 1937, 1940a, b, 1949; Kimber et al. 1988). Indirect evidence also came from the analysis of chromosome pairing in the F_1 hybrid between *Ae. recta* and a synthetic tetraploid *Ae. umbellulata* – *Ae. uniaristata* (hybrid genome $UX^n NUN$), which showed good chromosome pairing; the average number of bivalents was 13.4 (Kihara 1963). On the other hand, the hybrid between *Ae. neglecta* (genome $UUX^n X^n$) and this synthetic tetraploid (hybrid genome $UX^n UN$) showed an average of only 8.3 bivalents. Hence, it was concluded that *Ae. neglecta* and *Ae. recta* share the U subgenome, while the third subgenome of *Ae. recta* (subgenome N) is entirely different from the second subgenome of *Ae. neglecta* (X^n). Kihara (1963) assumed that the third subgenome of *Ae. recta* is closely related the genome of *Ae. uniaristata*, which was designated M^u, since Kihara (1954) believed it to be a modified M genome, and consequently, designated it M^{12} (Kihara (1963). Accordingly, Kihara (1963) concluded that *Ae. recta* was formed via hybridization of *Ae. neglecta* x *Ae. uniaristata*. Since these two species grow in mixed stand in western Turkey and the southern Balkan, *Ae. recta* was formed there and then spread westward. But, data on chromosome pairing in interspecific hybrids do not agree with the assumption that the second and third subgenomes of *Ae recta* are modified M (Kimber et al. 1983; Resta et al. 1996).

The presence of the U subgenome in *Ae. recta* was also confirmed by crossing this species with an induced autotetraploid of *Ae. umbellulata* ($2n = 4x = 28$; genome UUUU) (Kimber and Yen 1989; Yen and Kimber 1992b). Chromosome pairing was analyzed at first meiotic metaphase in this pentaploid hybrid that had an average of 9.30 univalents, 4.76 rod bivalents, 3.16 ring bivalents, 2.83 trivalents, 0.33 quadrivalents and 0.10 pentavalents. The optimization analysis of the meiotic data fit the 3:2 model, indicating that there are three homologous genomes in the hybrid (Yen and Kimber 1992b). Since two U genomes were introduced from the autotetraploid *Ae. umbellulata*, it is quite clear that there is a U subgenome in the hexaploid Ae. *recta* which is homologous to genome U of *Ae. umbellulata* (Yen and Kimber 1992b).

Likewise, the origin of the third subgenome of *Ae. recta* was determined by studying chromosome pairing in the hybrid *Ae. recta* x an induced autotetraploid of *Ae. uniaristata* ($2n = 4x = 28$; genome NNNN), which showed an average of 9.85 univalents, 5.05 rod bivalents, 2.30 ring bivalents, 2.95 trivalents and 0.40 quadrivalents (Yen and Kimber 1992b). Also in this hybrid, the optimization analysis of the meiotic data fit the 3:2 model, indicating that there are three homologous genomes in the hybrid. Since two were introduced from the autotetraploid *Ae. uniaristata*, it was concluded that the third subgenome of the hexaploid *Ae. recta* was homologous to genome N of *Ae. uniaristata* (Yen and Kimber 1992b). However, since the pairing data of another hybrid involving a different line of *Ae. recta* and the autotetraploid of *Ae. uniaristata* indicated that some differences may exist between the N genomes, and since no data were available on the pairing of the *Ae. recta* chromosomes with *Ae. comosa* chromosomes, the existence of an N genome in *Ae. recta* was not unequivocally established. If *Ae. uniaristata* had contributed chromosomes to *Ae. recta*, they must have undergone substantial modification (Yen and Kimber 1992b).

The origin of the *Ae. recta* genome was also investigated by examining the presence of specific restriction fragments of repeated nucleotide sequences in its DNA (Resta et al. 1996). The data showed that all bands of *Ae. neglecta* and *Ae. columnaris*, both bearing genome $UUX^n X^n$, were present in *Ae. recta* (Resta et al. 1996) indicating that one of these species is the tetraploid parent of *Ae. recta*. *Ae. neglecta* is morphologically very similar to *Ae. recta* and is therefore more likely a tetraploid ancestor of *Ae. recta* than *Ae. columnaris* (Resta et al. 1996).

Yen and Kimber (1992b) considered their study concerning the origin of the third subgenome of *Ae. recta* as inconclusive since they did not investigate the genetic connection between *Ae. recta* and *Ae. comosa*. This relationship was investigated by Resta et al. (1996), who showed that *Ae. comosa* did not contribute any of the three *Ae. recta* subgenomes. They determined the value of the repeated nucleotide sequence correspondence (RSC), that is, the fraction of marked bands shared between *Ae. uniaristata* and *Ae. recta* and found it to be 1.00, meaning that all marker bands of *Ae. uniaristata* were encountered in *Ae. recta*. This finding provided strong evidence that *Ae. uniaristata* donated the third subgenome to *Ae. recta*. Thus, the most probable origin of *Ae. recta* was the hybridization of *Ae. neglecta* with *Ae. uniaristata*. Resta et al. therefore proposed, to revise Kihara's (1963) formula $C^{u}C^{u}M^{t}M^{t}M^{t2}M^{t2}$ for *Ae. recta* to $UUX^{t}X^{t}NN$ (currently $UUX^{n}X^{n}NN$).

Tsunewaki (2009) reviewed his team investigations on the origins of the plasmons of the various diploid and polyploid species in the *Aegilops-Triticum* group. RFLP analyses of chloroplast (cp) DNA (Terachi et al. 1984; Ogihara and Tsunewaki 1988; Tsunewaki 1996) and mitochondrial (mt) DNA (Terachi and Tsunewaki 1986), as well as PCR–single-strand conformational polymorphism analyses of chloroplast and mitochondrial DNAs (Wang et al. 1997) showed that the plasmon of *Ae. recta* was identical to that of *Ae. neglecta* and that both were similar to that of *Ae. umbellulata* and therefore, were designated U. A similar conclusion from recent analyses of chloroplast DNA was reached by Bernhardt et al. (2017). The relative times of origin of the allopolyploid species of section Aegilops were inferred from the genetic distances of their plasmon from that of their putative maternal parent, *Ae. umbellulata* (Terachi et al. 1984; Wang et al. 1997; Tsunewaki 2009). Thus, it was suggested that the origin of *Ae. geniculata* was older than that of *Ae. neglecta*, which, in turn, was older than that of *Ae. biuncialis* and *Ae. columnaris* (Terachi et al. 1984; Wang et al. 1997; Tsunewaki 2009). Obviously, the origin of *Ae. recta*, which derived from hybridization of *Ae. neglecta*, as female parent, and *Ae. uniaristata*, was more recent than that of *Ae. neglecta*.

Ae. recta contain 16.22 pg 1C DNA (Eilam et al. 2008), which is 1.46% less DNA than expected from the sum of the DNA content of its two parents, i.e., 16.46 pg (*Ae. neglecta* contains 10.64 pg and *Ae. uniaristata* contains 5.82 pg; Eilam et al. 2007, 2008). The loss of some DNA in *Ae. recta* was also found by Badaeva et al. (2002, 2004) who, based on differential C-banding and FISH, found that *Ae. recta* exhibited several minor structural chromosome modifications, including deletion of chromosomal segments.

According to Senyaninova-Korchagina (1930, 1932), the hexaploid species, *Ae. recta*, had all the chromosomes present in the tetraploid species, *Ae. neglecta*, plus 7 chromosome pairs which had only submedian centromeres. Yet, the karyotype described by Chennaveeraiah (1960) showed that the description by Senyaninova-Korchagina (1930, 1932) of

seven additional chromosome pairs in *Ae. recta* was incorrect. According to Chennaveeraiah (1960), the karyotype of *Ae. recta* consisted of two pairs with satellites of different sizes on short arms, one pair with an extreme subterminal centromere, seven pairs with submedian-subterminal centromeres, two pairs with median centromeres and the rest with submedian centromeres. The extra seven pairs in the hexaploid, therefore, consisted of one pair with satellites, one pair with median centromere, four pairs with submedian-subterminal centromeres and one pair with a submedian centromere. Therefore, this third set in *Ae. recta* is not in any way a duplicated complete set of either of the other two sets in the tetraploid parent, *Ae. neglecta*. In all, there are 16 types of chromosomes in *Ae. recta* (Chennaveeraiah 1960).

In accord with the findings of Senyaninova-Korchagina (1932) and Chennaveeraiah (1960), Teoh et al. (1983), using FISH with a repetitive DNA that codes for rRNA, found two pairs of SAT chromosomes and four pairs of rRNA sites in *Ae. recta*. Badaeva (2002) and Badaeva et al. (2004) observed that all analyzed accessions of *Ae. recta* possessed three SAT chromosomes and four major NORs on chromosomes 1U, 5U, $1X^{n}$, and 5N, only three of which were active. These values correspond to the number of SAT chromosomes and active NORs detected by Ag-NOR staining (Cermeño et al. 1984b). All in all, the results of C-banding analysis and FISH reported by Badaeva et al, (2004) in *Ae. recta* showed the presence of active NORs on chromosomes 1U, 5U and $1X^{n}$ and inactivation of NOR on chromosome 5N.

Badaeva et al. (2004) analyzed the heterochromatin banding patterns of somatic metaphase chromosomes revealed by C-banding in several accessions of *Ae. recta*, and revealed low variation in C-banding patterns and translocation polymorphism. Four accessions from Spain, Portugal and Morocco had a Robertsonian translocation between chromosomes $6X^{n}$ and 3N. Another Robertsonian translocation between $2X^{n}$ and 3N was identified in an accession from Bulgaria. Badaeva et al. (2004, 2011) claimed that this relatively low variation of C-banding demonstrated that the formation of this allohexaploid involved only minor functional modifications of the parental genomes, while intraspecific divergence was accompanied by genome rearrangements, namely, translocations involving the total chromosome arms of all three subgenomes (Badaeva et al. 2011).

The distribution of C-bands on chromosomes of the U and X^{n} subgenomes of *Ae. recta* was mostly identical and similar to that of chromosomes of *Ae. neglecta*. Likewise, the C-banding patterns of the N subgenome chromosomes of *Ae. recta* were similar to those of *Ae. uniaristata* chromosomes. It was possible to identify all *Ae. recta* chromosomes on the basis of morphology and C-banding patterns. Similar to its tetraploid parent, *Ae. neglecta*, the karyotype of *Ae.*

recta was characterized by a high heterochromatin content (Badaeva 2002).

In addition, Badaeva et al. (2004) used FISH with the heterochromatin-limited repetitive DNA probes pSc119 and pAs1, and the pTa71 probe, to reveal the 18S-26S rDNA sites, and pTa794, to reveal the 5S rDNA sites. The pSC119 and pAs1 FISH patterns on U and X^n subgenome chromosomes of *Ae. recta* were nearly identical to those of *Ae. neglecta*. Also, no differences in the labeling pattern of the N subgenome chromosomes were observed between *Ae. recta* and *Ae. uniaristata* (Badaeva et al. 2004). Therefore, in contrast to the conclusion of Yen and Kimber (1992b), the N subgenome of *Ae. recta* appeared to be only slightly structurally modified from that of *Ae. uniaristata*. Hence, the UX^n and the N subgenome chromosomes of *Ae. recta* are similar to those of *Ae. neglecta* and *Ae. uniaristata*, respectively, with regard to the distribution of C-bands, 45S and 5S rDNA loci and FISH sites of pSc119 and pAs1 (Badaeva 2002; Badaeva et al. 2004, 2011).

Four major (1U, 5U, $1X^n$, 5N) and two minor ($7X^n$ and 1N) NORs and seven 5S rDNA loci were detected on all group-1 and group-5 chromosomes in *Ae. Recta*, using pTa71 and pTa794. Thus, the number of NORs and 5S rDNA in *Ae. recta* was similar to the total number of these loci in the parental species. However, the number of minor 18S-26S rDNA sites was smaller than expected.

9.8.6.3 Crosses with Other Species of the Wheat Group

Chromosome associations at first meiotic metaphase were studied in the allohexaploid species, *Ae. recta*, using the C-banding technique (Cuñado 1992). Mean chromosomal pairing included 0.12 univalents, 5.31 rod bivalents, 15.57 ring bivalents, and 36.45 mean chromosome associations per metaphase cell. Use of C-banding enabled the analysis of associations between homologous chromosomes of each of the three subgenomes. Average homologous associations between chromosomes of the UX^n subgenomes showed 0.06 univalents, 3.16 rod bivalents, 10.78 ring bivalents and 24.09 mean chromosome associations per metaphase cell, and between chromosomes of the N subgenomes showed 0.05 univalents, 2.13 rod bivalents, 4.81 ring bivalents and 11.75 mean chromosome associations for a metaphase cell (Cuñado 1992). No difference between the UX^n and N subgenomes was noted in regards to the mean associations per bivalent.

The allohexaploid *Ae. recta* regularly form bivalents at first meiotic metaphase (Cuñado 1992). Cuñado et al. (2005) analyzed the pattern of synapsis at late zygotene and pachytene in *Ae. recta* using whole-mount surface-spreading of synaptonemal complexes under an electron microscope. It was revealed that the chromosomes were mostly associated as bivalents in these early meiotic stages, with a mean 0.17 multivalents per nucleus. It can be concluded that the mechanism controlling bivalent formation in this species acts mainly at zygotene, by restricting synapsis to homologous chromosomes, but also acts at pachytene, by preventing chiasma formation in the homoeologous associations (Cuñado et al. 2005).

Chromosomal pairing was studied in F_1 hybrids of *Ae. recta* x several allotetraploid species bearing the U subgenome. Only the *Ae. neglecta* x *Ae. recta* hybrid (genome UX^nUX^nN) had 14 ring bivalents and 7 univalents (Kihara 1937), indicating that only U and X^n subgenomes are homologous to those of *Ae. neglecta*. On the other hand, the other hybrids exhibited a reduced level of pairing. More specifically, the hybrid *Ae. geniculata* x *Ae. recta* (genome M^oUUX^nN) had 6–7 bivalents (Kihara 1937), *Ae. recta* x *Ae. biuncialis* (genome UX^nNUM^b) had 6–11 bivalents and 0–1 trivalent (Kihara 1949), the hybrid *Ae. recta* x *Ae. triuncialis* (genome UX^nNUC) had + 7(-9) bivalents (Kihara 1937) and the hybrid *Ae. recta* x *Ae. peregrina* (genome (UX^nNS^vU) had 7–11 bivalents and 0–1 trivalent Kihara 1949). The amount of chromosomal pairing in these four hybrids showed that *Ae. recta* shared only one subgenome, i.e., U, with *Ae. geniculata*, *Ae. biuncialis*, *Ae. triuncialis*, and *Ae. peregrina*.

The hexaploid hybrid between *Ae. recta* and hexaploid *Ae. crassa* (genome $UX^nND^cX^cD$) had 7–8 bivalents, with a mode of 8 bivalents, and 0–2 trivalents (Kihara 1949). Since most of the pairing in this hybrid was presumably autosyndetic between subgenomes D^c and D of *Ae. crassa*, the little remaining pairing was allosyndetic pairing between chromosomes of the homoeologous subgenomes.

9.8.7 *Aegilops columnaris* Zhuk.

9.8.7.1 Morphological and Geographical Notes

Ae. columnaris Zhuk. [Syn.: *Triticum columnare* (Zhuk.) Morris and Sears] is an annual, predominantly autogamous, multi-tillered plant. Its culms are 20–50-cm-tall (excluding spikes) and slightly geniculate at base. The leaves are narrow, linear-lanceolate, 2–7-cm-long and generally hairy. The spike is lanceolate or narrow, ovoid to oblong, becoming suddenly narrow in the upper half, generally awned, and 2.5–7.0 (usually 3.5–5.5)-cm-long (excluding awns). The entire spike disarticulates at maturity and falls as a unit (umbrella-type dispersal unit). There are 4–6 spikelets (usually 5) in each spike. The lowest two (seldom 3) are elliptical and longer than the adjacent rachis segments, the upper spikelets are equally long or shorter than the adjacent rachis segment. The upper spikelets bear small grains. There are 2–4 (usually 3) basal rudimentary spikelets. The glumes of the lower spikelets are elliptical, 7–11-mm-long, with two 3–5-cm-long awns, one being much wider than the other, with a deep cleft between. Glumes on upper spikelets often have 3 awns, which are shorter than the awns of the lower spikelets. Lemmas have 2–3 awns, are more slender and shorter than glume awns, and are generally present in all spikelets. The caryopsis is free Fig. 9.4f.

Ae. columnaris has relatively little morphological variation involving spike size, number of fertile and rudimentary spikelets and awn development. Sometimes it is difficult to distinguish it from *Ae. neglecta* and *Ae. recta* bearing only two awns per glume. According to van Slageren (1994), the main differences between *Ae. columnaris* and *Ae. neglecta* are: *Ae. columnaris* glumes of the lower 2–3 spikelets are elliptic-oblong, the apex usually has 2 awns–one large, 1.5–2.5-mm-wide at the base and often bifurcating above, and one very small and linear, 1 mm or less at the base–its spike is ovoid in the lower part, and more linear in the upper part, with 3–4(-6) spikelets, all fertile. *Ae. neglecta* glumes of the lower, fertile spikelets are obovate-elliptical, the apex usually has 3 awns of equal length and width at the base, its spike is ovoid-ellipsoid and inflated in the lower part, then abruptly constricted and almost linear, with 3–6 spikelets, of which the upper 1–3 are sterile.

Ae. columnaris is a Mediterranean-Western Asiatic element, occurring mainly in Turkey and the western arc of the Fertile Crescent, but scattered in the eastern part of the arc as well. The area of distribution extends westwards to Crete and eastwards to Transcaucasia and northwestern Iran (rare). It is uncommon throughout its range. *Ae. columnaris* grows in Greece (including Crete), Turkey, Syria, East Lebanon, North Iraq (rare), Iran (rare), Armenia, and Azerbaijan. It is found as an adventive plant in southern France. *Ae. columnaris* is mainly found on limestone, less frequently on basalt or grey-calcareous steppe soils. The soil textures are predominantly stony, with additional clay, (clay)-loam and occasionally sand. It grows in areas with a range of annual rainfall 450–1250 mm, indicating that it prefers a generally wetter environment than most *Aegilops* species. It occurs in degraded deciduous oak forests, deciduous steppe maquis, degraded dwarf-shrub formations, open dwarf-shrub steppe-like formations, pastures, abandoned fields, edges of cultivation, disturbed areas and roadsides. It grows at altitudes of 450–1990 m.

Ae. columnaris has a medium-sized distribution in the central part of the distribution of the genus. It occupies mainly open, secondary habitats and usually grows in mixed stands with *Ae. biuncialis, Ae. neglecta, Ae. triuncialis, Ae. cylindrica* and other species, with which it introgresses, particularly with other U-subgenome allopolyploids. These mixed populations contained different types of intermediate forms; some of which are rich in hybrids and hybrid derivatives, while others contain only a small number of introgressed lines. Series of intermediates between *Ae.*

neglecta and *Ae. columnaris* were sampled at the edges of cultivation, near Malatya, Turkey (Zohary and Feldman 1962), and intermediates between *Ae. triuncialis* and *Ae. columnaris* were collected west of Malatya, Turkey, at roadsides, and north of Ankara, central Turkey, at roadsides and edges of wheat fields (Feldman 1965a). Karagoz (2006) presented evidence of spontaneous hybridization between *Ae. columnaris* and domesticated wheat.

Ae. columnaris grows sympatrically with *A. muticum, Ae. speltoides, Ae. caudata, Ae. umbellulata, Ae. tauschii,* wild *T. monococcum, T. urartu,* wild *T. timopheevii, Ae. neglecta, Ae. triuncialis, Ae. cylindrica, Ae. crassa,* and *Ae. juvenalis,* and allopatrically with *Ae. searsii, Ae. geniculata, Ae. kotschyi* and *Ae. peregrina.*

Accessions of *Ae. columnaris* confer resistance to powdery mildew, leaf rust and Hessian Fly (Gill et al. 1985). Chromosome $5X^n$ of *Ae. columnaris* is resistant to leaf rust (Badaeva et al. 2018). Several accessions of *Ae. columnaris* are tolerant to drought and frost (Damania and Pecetti 1990).

9.8.7.2 Cytology, Cytogenetics and Evolution

Ae. columnaris is an allotetraploid species (2n = 4x = 28). Its nuclear genome is designated UUX^nX^n (similar to the genome of *Ae. neglecta*; modified from Dvorak 1998) and the plasmon genome is designated U', a sub-type of the U plasmon of *Ae. umbellulata* (Tsunewaki 2009). The origin of *Ae. columnaris* is not yet well understood (Resta et al. 1996; Yen et al. 1996; Dvorak 1998). Originally, the nuclear genome of *Ae. columnaris* was designated by Kihara (1937, 1949, 1954) as $C^uC^uM^cM^c$ (C^u is currently U and M^c is modified M), closely related to the genome of *Ae. Umbellulata,* based on indirect evidence, namely, on analysis of chromosome pairing in hybrids between *Ae. columnaris* and other allotetraploid species that had previously been carry the *Ae. umbellulata* genome (von Berg 1937; Kihara 1937, 1940a, b, 1949). A direct cross between *Ae. columnaris* and *Ae. umbellulata* was analyzed by Kimber and Abu-Bakar (1981), who observed 6.17 univalents, 1.87 rod bivalents, 4.30 ring bivalents, 6.17 total number of bivalents and 0.83 trivalents at first meiotic metaphase of this triploid hybrid (hybrid genome UX^nU). This pairing data indicated that *Ae. columnaris* indeed contains one subgenome that is homologous to genome U of *Ae. umbellulata.* Analysis of chromosome pairing in hybrids between *Ae. columnaris* and an induced autotetraploid line of *Ae. umbellulata* again showed that one subgenome of *Ae. columnaris* is homologous to that of *Ae. umbellulata* (Kimber and Yen 1988, 1989). The conclusion that one subgenome of *Ae. columnaris* originated from *Ae. umbellulata* was later confirmed by several authors (e.g., Resta et al. 1996; Badaeva 2002; Badaeva et al. 2004, 2018). Both genomes have similar chromosome morphologies, distribution of 5S and 45S rDNA loci, labeling patterns of pSc119.2 and pAs1 probes and weak hybridization with the (GTT)$_{10}$ probe (Badaeva et al. 2018).

Evidence for multiple origins of the U subgenome of *Ae. columnaris* was obtained by Kadosumi et al. (2005). Using the genome-specific PCR primer set U31, they analyzed variation of the U genome in 48 accessions of *Ae. columnaris* and 72 accessions of its diploid progenitor *Ae. umbellulata*. Three alleles were distinguishable by the length of the amplified sequence, namely, allele I (normal size with an *Msp*I site), allele II (normal size without an *Msp*I site), and allele III (shorter size caused by a 123 bp deletion). Sequence comparison indicated the inheritance of alleles I and III from the diploid to the tetraploid, suggesting multiple origins of the U subgenome of the tetraploid. Regarding allele II, however, the sequence comparison indicated that parallel mutations at the *Msp*I site produced allele II several times. The phylogenetic tree based on the sequences of the U31 region, demonstrated the presence of a third lineage of the U genome from *Ae. umbellulata* to *Ae. columnaris*. Consequently, Kadosumi et al. (2005) concluded that the *Ae. columnaris* U subgenome had at least three independent origins.

The origin of the second subgenome of *Ae. columnaris* has not been satisfactorily determined. Initially, the second genome was designated by Kihara (1937, 1949, 1954) as M^c, i.e., a modified M genome, since, based on morphological comparisons and meiotic analysis, Kihara assumed that this subgenome was closely related to the M genome of *Ae. comosa* or to the M^u genome of *Ae. uniaristata*. Yet, the hybrids between the diploid species *Ae. comosa* or *Ae. uniaristata* and *Ae. columnaris* displayed relatively low pairing, indicating that there was no homology between any of the subgenomes of *Ae. columnaris* and the genomes of these diploid species. The *Ae. comosa* x *Ae. columnaris* hybrid showed three to seven bivalents per first meiotic metaphase cell, of which only one was a ring bivalent, and the hybrid *Ae. uniaristata* x *Ae. columnaris* showed four to six rod bivalents per cell (Kihara 1949). To resolve this incongruity, Kihara ((1963) suggested that the incomplete pairing between the M subgenomes of *Ae. columnaris* with the M or the M^u genomes of the diploid species, resulted from either divergence of the M at the diploid level, followed by extinction of some of the diploid taxa, or genomic modification at the tetraploid level. But, while resemblance between one chromosome set of *Ae. columnaris* to that of *Ae. umbellulata* was observed, karyotype analysis failed to find a genome resembling to that of *Ae. comosa* or *Ae. uniaristata* in *Ae. columnaris* (Chennaveeraiah 1960). Moreover, Chennaveeraiah noted that no genome in the diploid species of the *Aegilops-Triticum* group matched the second subgenome of *Ae. columnaris*.

Resta et al. (1996) employed variation in randomly selected families of repeated nucleotide sequences to study the origin of the subgenomes of *Ae. columnaris*. They determined the fraction of marker bands of these repeated sequences of a diploid shared with a polyploid which they called repeated nucleotide sequence correspondence (RSC). RSC varies from 0.00, if no marker band of a diploid is encountered in a polyploid, to 1.00, if all are encountered. They found that fifteen marker bands of *Ae. umbellulata* were shared with *Ae. columnaris* (RSC = 0.88), and that two marker bands of *Ae. speltoides* (RSC = 0.05) and one marker band of *Ae. caudata* (RSC = 0.03) were found in *Ae. columnaris*. Thus, only the RSC of *Ae. umbellulata* with *Ae. columnaris* differed significantly from 0.00, indicating that *Ae. columnaris* has a subgenome similar to that of *Ae. umbellulata*. The analysis also showed that *Ae. columnaris* is closely related to *Ae. neglecta*. Both species share the U subgenome with *Ae. geniculata* and *Ae. biuncialis*, but their second subgenome differs from those of *Ae. geniculata* and *Ae. biuncialis*, in other words, is not a modified M subgenome. As concluded by Chennaveeraiah (1960), they found no relationship between the second subgenome of *Ae. columnaris* and *Ae. neglecta* and a genome of any extant diploid species of *the Aegilops-Triticum group*. Consequently, the second subgenome of *Ae. columnaris* and *Ae. neglecta* was designated X^t by Resta et al. (1996) and the proposed genome formula for *Ae. columnaris and Ae. neglecta was* UUX^{t}X^t (Resta et al. 1996; Dvorak 1998). This genomic formula was later modified to UUX^{n}X^n for *Ae. columnaris* and for *Ae. neglecta* (Badaeva 2002; Badaeva et al. 2004).

Differences in C-banding patterns, and the distribution of 5S and 45S rDNA loci also contradicted the conclusion that the second subgenome of *Ae. columnaris* is a modified M genome (Badaeva 2002; Badaeva et al. 2004). Considering data from C-banding and FISH analyses, Badaeva (2002) and Badaeva et al. (2004) suggested that the putative progenitor of the X^n subgenome should have a relatively symmetric karyotype, with highly heterochromatic chromosomes visualized by C-banding. In hybridization patterns of pSc119.2 and pAs1 probes, the progenitor species should be similar to S, T, U, or C genomes of diploid *Aegilops*. An abundance of (GTT)$_n$ microsatellite repeats is an important diagnostic feature of the X^n subgenome. The X^n subgenome is also unique with respect to the number and distribution of 5S rDNA loci, because other diploid or polyploid species of *Aegilops* and *Triticum*. except *Ae. neglecta*, carry two or, rarely, one, 5S rDNA locus per haploid genome (Dvorak et al. 1989; Badaeva et al. 1996b). In all cases, these loci are located on different chromosomes. *Ae. columnaris* has four 5S rDNA loci, two of which are located in one chromosome.

None of the extant diploid species of *Aegilops* fit the criteria listed above. Consequently, none of them could be the direct progenitor of the X^n genome, which agrees with previous results (Resta et al. 1996; Dvořák 1998). The X^n subgenome was probably derived from an extinct yet an undiscovered extant diploid species of *Aegilops*.

While Resta et al. (1996) detected considerable differences between *Ae. geniculata* and *Ae. biuncialis*, on one side, and *Ae. columnaris and Ae. neglecta*, on the other, almost no divergence was found between *Ae. columnaris* and *Ae. neglecta*. This close relationship between the genomes of the latter two species is reflected in the high chromosomal pairing in the hybrid *Ae. neglecta* x *Ae. columnaris*, i.e.,12–14 bivalents per cell with a mode of 12 bivalents, and in the 20% pollen viability (Kihara 1949). In contrast, the hybrid *Ae. biuncialis* x *Ae columnaris* showed a mode of seven to eight bivalents per cell (Kihara 1937), *Ae. biuncialis* x *Ae neglecta* showed a mode of eight bivalents (Kihara 1937), and *Ae. geniculata* x *Ae. neglecta* showed a mode of eight bivalents, ranging from 6–12 bivalents per cell (Percival 1932; Kihara 1937). Hence, the data of chromosome pairing and the studies of Resta et al. (1996) showed that the species within each group (*Ae. geniculata* and *Ae. biuncialis* vs. *Ae. columnaris* and *Ae. neglecta*) have two subgenomes in common, whereas the species between the groups have only one subgenome in common, i.e., subgenome U. This contradicts the conclusion of Kihara (1937, 1949, 1954, 1963) who assigned the same basic genome formula to all these four species.

According to Mukai and Tsunewaki (1975), Tsunewaki (1993, 1996, 1980), Tsunewaki and Tsujimoto (1983), and Ikeda and Tsunewaki (1996), the cytoplasm of *Ae. columnaris* causes variegation (variegated yellowing of leaf color) in midwinter and growth depression in most common wheat lines bearing *Ae. columnaris* cytoplasm. In addition, it causes haploidy and haplo-diplo twinning in two particular common wheats and complete male sterility in about half of the common wheats tested. Since these genetic effects are similar to those caused by the cytoplasm of *Ae. umbellulata*, Mukai and Tsunewaki (1975), Tsunewaki (1980) and Tsunewaki and Tsujimoto (1983), classified *Ae. columnaris* cytoplasm as C^u (currently U) plasma type and concluded that *Ae. umbellulata* as the cytoplasm donor to *Ae. columnaris*. Ogihara and Tsunewaki (1988) used restriction enzymes to investigate variation due to fragment size mutations in the chloroplast DNAs of these species and classified the chloroplast DNAs of the *Triticum* and *Aegilops* species. This classification was in agreement with that of the plasma types assigned according to phenotypes arising from nucleus-cytoplasm interactions. On the basis of these studies, Ogihara and Tsunewaki (1988) confirmed the *Ae. umbellulata* origin of the *Ae. columnaris* cytoplasm. However, based on studies of the phenotypes in cytoplasm-substituted wheats, Maan (1975, 1978) and Panayotov and Gotsov (1975, 1976) proposed that *Ae. uniaristata* or *Ae. comosa* was the cytoplasm donor to *Ae. columnaris*. Yet, the results of Terachi et al. (1984), Ogihara and Tsunewaki (1988) and Wang et al. (1997) unequivocally showed that the chloroplast genome of *Ae. columnaris* is nearly identical to that of *Ae. umbellulata*.

Terachi and Tsunewaki (1986) reported that the mitochondrial genome diversity in the allopolyploid species of the wheat group is far more extensive than the chloroplast genome diversity. In this respect, they found that of the mitochondrial DNAs isolated from the allotetraploids with U-type cytoplasms, that of *Ae. columnaris* was the most differentiated. Whereas the mitochondrial DNA of the other allotetraploids was almost identical to that of *Ae. umbellulata*, that of *Ae. columnaris* differed to some extent from that of *Ae. umbellulata* (Terachi and Tsunewaki 1986; Wang et al. 1997). In agreement with the above findings, Ikeda et al. (1994) and Ikeda and Tsunewaki (1996) reported that alloplasmic wheats having *Ae. columnaris* cytoplasm, that are characterized by growth inhibition, show impaired mitochondrial cytochrome c oxidase. Hence, *Ae. columnaris* differs only by a small number of mutations from that of *Ae. umbellulata*, whereas almost no mutations exist in the chloroplast genome of *Ae. columnaris*, the allotetraploid undoubtedly originated from *Ae. umbellulata* as a female (Tsunewaki 2009). Because of the mitochondrial changes the cytoplasm of *Ae. columnaris* was designated U' (Tsunewaki 2009).

The relative times of origin of the four allopolyploid species bearing U cytoplasm, i.e., *Ae. neglecta*, *Ae. columnaris*, *Ae. biuncialis*, and *Ae. triuncialis*, were inferred from genetic distances between their cytoplasms and that of *Ae. umbellulata*, their putative maternal parent. The number of differences between the plasmon of these allotetraploids and that of *Ae. umbellulata* indicates that *Ae. neglecta* is older than *Ae. columnaris*, *Ae. biuncialis* and *Ae. triuncialis* (Tsunewaki 2009).

Ae. columnaris contains 10.86 pg 1C nuclear DNA (Eilam et al. 2008). Its genome, and that of is closest relative *Ae. neglecta*, are larger than the genomes of *Ae. biuncialis*, *Ae. geniculata, and Ae. triuncialis* and smaller than those of *Ae. peregrina* and *Ae. kotschyi* (Eilam et al. 2008). This may indicate that the genome of the, yet unknown, diploid donor of the subgenome X^n to *Ae. columnaris* is larger than both the M genome of *Ae. comosa*, the putative M subgenome donor to *Ae. biuncialis* and *Ae. geniculata* and the C genome of *Ae. caudata*, the putative C subgenome donor to *Ae. triuncialis*, but smaller than the S^l genome of *Ae. longissima*, the putative S^l subgenome donor to *Ae. peregrina* and *Ae. kotschyi*.

The karyotype of *Ae. columnaris* has three pairs of chromosomes with satellites on short arms (Chennaveeraiah

1960). The satellites on two chromosome pairs are fairly large and those on the third pair are comparatively smaller. One chromosome pair with large and another pair with smaller satellites resemble those of *Ae. umbellulata* (Chennaveeraiah 1960). There is a single pair with the characteristic extreme subterminal centromere similar to that of *Ae. umbellulata*; all the other pairs have submedian centromeres. (Chennaveeraiah (1960). Thus, the karyotype of *Ae. columnaris* differs considerably from those of *Ae. geniculata* and *Ae. biuncialis* (Chennaveeraiah 1960).

Three pairs of SAT chromosomes were also found in *Ae. columnaris* by Teoh et al. (1983). Using FISH with a repetitive DNA sequence coding for rRNA, they noted the presence of three pairs of rRNA sites, agreeing with the number of SAT chromosomes pairs. NORs activity studied in somatic metaphase chromosomes of *Ae. columnaris* by phase contrast, C-banding and Ag-staining (Cermeño et al. 1984b), showed that the U subgenome suppressed the activity of one pair of NORs of subgenomes X^n. In accord with the finding of Chennaveeraiah (1960) and Teoh et al. (1983), Cermeño et al. (1984b) found three pairs of SAT chromosomes as well as three pairs of active NORs. Of the active three NOR pairs, two pairs were of the U subgenome and only one pair of the X^n subgenome. The X^n NOR showed reduced activity as was evident by the small size of the Ag-NORs of this chromosome pair (Cermeño et al. 1984b).

The heterochromatin banding patterns of somatic metaphase chromosomes of *Ae. columnaris*, as revealed by C-banding, contained broad C-banding polymorphism and chromosomal rearrangements, represented by paracentric inversion and intra- and inter-subgenomic translocations (Badaeva (2002) and Badaeva et al. (2004). Significant differences in karyotype structure and total amount and distribution of C-heterochromatin were observed between *Ae. columnaris* and *Ae. neglecta*, on the one hand, and *Ae. geniculata* and *Ae. biuncialis*, on the other, evidence supporting claims of different origins of these two groups of species. In turn, similarity of the C-banding patterns of *Ae. columnaris* and *Ae. neglecta* chromosomes suggested that they were derived from a common ancestor (Badaeva (2002). *Ae. umbellulata* was shown to be the *U*-subgenome donor of *Ae. columnaris* and *Ae. neglecta*, but the donor of the second subgenome of these two species was not determined by the C-banding pattern. Badaeva (2002) assumed that the second subgenome of these two tetraploid species resulted from introgressive hybridization.

In addition, Badaeva et al. (2004) used FISH with the heterochromatin-limited repetitive DNA probes pSc119 and pAs1 to study the distribution of these hybridization sites, as well as the distribution of NOR and 5S DNA loci revealed by pTa71 (18S-26S rDNA) and pTa794, (5S rDNA), respectively. Similar to the finding of Resta et al. (1996) that differentiated between *Ae. columnaris* and *Ae. neglecta*, on

one hand, and *Ae. geniculata* and *Ae. biuncialis*, on the other hand, the studies of Badaeva et al. (2004) revealed significant differences between these two groups of species in the total amount and distribution of heterochromatin, and in the number and location of 5S and 18S-26S rDNA loci, indicating that these two groups of species have different origins. Similarities in C-banding and FISH patterns of most *Ae. columnaris* and *Ae. neglecta* chromosomes suggest that they were probably derived from a common ancestor, whereas distinct differences of three chromosome pairs may indicate that the divergence of these species was probably associated with chromosomal rearrangements and/or introgressive hybridization (Badaeva et al. 2004). *The contribution of the U subgenome to Ae. columnaris and Ae. neglecta by Ae. umbellulata was again confirmed, however, the source of their second genomes remains unknown.* FISH with the probes pSc119 and pAs1 detected polymorphic hybridization patterns in *Ae. columnaris*. Large to medium PSc119 sites were observed in telomeric regions of one or both arms of 11 of 14 chromosome pairs, and interstitial sites were observed in the long arms of chromosomes 6U and 7U. FISH with pAs1 revealed signals of different intensities on several chromosomes.

Ae. columnaris is characterized by a unique distribution of the hybridization sites of the two ribosomal probes, pTa71 and pTa794. (Badaeva et al. 2004). Two major NORs were detected on chromosomes 1U and 5U. These chromosomes also had 5S rDNA sites distal and proximal to the major NORs. The third major NOR was detected in the short arm of chromosome $1X^n$, which also had two 5S rDNA loci; one distal and the other proximal to the NOR. Thus, *Ae. columnaris* has three major MORs and four 5S rDNA loci, two of which are located in one chromosome arm. Such distribution of 18S-26S and 5S rDNA loci has no analogs in diploid or polyploid *Aegilops/Triticum* species, except for *Ae. neglecta* (Badaeva et al. 1996b). No minor NORs were observed in *Ae. columnaris*. The genetic nomenclature of *Ae. columnaris* chromosomes have not been established, and consequently, Badaeva et al. (2004) tentatively assigned chromosomes to homoeologous groups and genomes on the basis of their similarity with chromosomes of other *Aegilops* species.

According to Badaeva (2002), of all allopolyploid *Aegilops* species, *Ae. columnaris* showed the highest level of intraspecific chromosome variation. Two major chromosomal types, A and B, differing in karyotype structure, in the number of SAT chromosomes, and in the total amount and distribution of heterochromatin, could be distinguished in this species (Badaeva 2002). Broad C-banding polymorphism was revealed within each chromosomal type. Chromosome type A was found in 13 of 17 *Ae. columnaris* accessions (Badaeva 2002). These accessions had three pairs of SAT chromosomes, two of which were assigned to the U and one to the X^n

subgenomes. The length of secondary constrictions on the satellite chromosomes of the U subgenome (1U and 5U) significantly exceeded the length of the secondary constriction on the X^n subgenome. The type A chromosomes of *Ae. columnaris* showed significant polymorphism due to both variability of the C-banding patterns and chromosomal rearrangements. A paracentric inversion of chromosome 7U was the most common type of chromosomal rearrangements found in six out of 13 lines with type A chromosomes. Two unrelated accessions possessed centromeric translocation between chromosomes 1U and 5U, accompanied by inactivation of one NOR locus (Badaeva 2002).

Type B chromosomes were found in four accessions (Badaeva 2002). These accessions contained only two pairs of SAT chromosomes, i.e., they lost the satellite on chromosome 1U. At the same time, the secondary constriction on chromosome $1X^n$, which is poorly expressed in accessions of group A, was clearly extended. In addition, the accessions with type B chromosomes possessed less heterochromatin compared to accessions of group A. They also showed different morphology and C-banding patterns than those of seven chromosome pairs belonging to both U and X^n subgenomes. The observed differences were so significant that Badaeva (2002) could not explain them by simple chromosome translocation and/or inversions. She thus suggested that intraspecific chromosome diversity of *Ae. columnaris* is accompanied not only by chromosomal aberrations, but, to a great extent, by introgression of genetic material from other species. It may be assumed that type B chromosomes derived from type A, since the latter are similar to the karyotype of *Ae. neglecta*.

Data from C-banding, FISH, nuclear and chloroplast (cp) DNA analyses, and gliadin electrophoresis, enabled Badaeva et al. (2021a) to confirm the division of *Ae. columnaris* into two distinctive groups, C-I (equivalent to group A) and C-II (equivalent to group B). C-I group was more similar to *Ae. neglecta* than C-II, and less polymorphic than C-II. Most C-II accessions were collected from a very narrow geographic region, and they might have originated from a common ancestor. The authors suggest that C-II emerged from C-I relatively recently, probably due to introgression from another *Aegilops* species, and might be at the initial stage of a speciation process.

Badaeva et al. (2018) reported the development of 57 common wheat-*Ae. columnaris* introgression lines covering 8 of the 14 *Aegilops* chromosomes. Based on the compensating capability in substitution lines and the results of FISH analysis of the parental *Ae. columnaris* line with seven DNA probes, Badaeva et al. (2018) determined the genetic nomenclature of 11 of 14 chromosomes. Each of these 11 *Ae. columnaris* chromosome was characterized on the basis of C-banding pattern and the distribution of the following seven DNA sequences: two wheat rDNAs pTa71, and pTa794, two

microsatellite sequences $(GAA)_{10}$ and $(CTT)_{9,}$ $(GAA/CTT)_n)$, $(GTT)_{10}$, and three tandemly repeated DNA families, pSc119.2 pAs1, and pAesp_SAT86 (Badaeva et al. 2015).

GISH on *Ae. columnaris* chromosomes failed to clearly discriminate between genomes. Therefore, Badaeva et al. (2018) tested different probes to find appropriate genome-specific markers. Distribution of the $(GAA/CTT)_n$ sequence was similar to the C-banding pattern but, by contrast, the $(GTT)_{10}$ probe showed different labelling patterns on chromosomes of the two *Ae. columnaris* subgenomes. Poorly labelled chromosomes were assigned to the U subgenome, based on their similarity with *Ae. umbellulata* chromosomes, which display weak hybridization with this sequence. Heavily labelled chromosomes were assigned to the X^n subgenome.

9.8.7.3 Crosses with Other Species of the Wheat Group

Chromosome pairing at first meiotic metaphase of *Ae. columnaris* was studied by Cuñado (1992) using C-banding. The lower C-heterochromatin content of the U subgenome differentiated them from the chromosomes of the X^n subgenome. Mean chromosomal pairing per cell included 0.03 univalents, 2.23 rod bivalents, 11.73 ring bivalents and 25.70 chiasmata. Thus, *Ae. columnaris* exhibits diploid-like meiotic behavior. Pairing between homologous chromosomes of the U subgenome included 0.03 univalents, 1.58 rod bivalents, 5.39 ring bivalents and 12.36 chiasmata, while the X^n genome included 0.00 univalents, 0.65 rod bivalents, 6.34 ring bivalents and 13.34 chiasmata. The somewhat higher pairing between chromosomes of the X^n subgenome than between chromosomes of the U subgenome may be due to higher C-heterochromatin content in the latter subgenome (Cuñado 1992).

The pattern of zygotene and pachytene pairing in *Ae. columnaris* was analyzed by whole-mount surface-spreading of synaptonemal complexes, as viewed under an electron microscope (Cuñado et al. 1996b). The data indicated that at the early meiotic zygotene and pachytene stages the chromosomes were almost exclusively associated as 14 bivalents; only rarely as 12 bivalents and one trivalent.

Data of chromosomal pairing in the F_1 hybrid *Ae. columnaris* x *Ae. umbellulata* (genome UX^nU) showed that *Ae. columnaris* contains one subgenome that is homologous to the genome of *Ae. umbellulata*. In contrast, the triploid hybrids between *Ae. columnaris* and the M-and N-genome diploid species, *Ae. comosa* and *Ae. uniaristata*, showed a much lower level of pairing, indicating that none of the subgenomes of *Ae. columnaris* are homologous to the M or N genomes. More specifically, *Ae. columnaris* x *Ae. comosa* (genome UX^nM) had 3–7 bivalents and 0–3 trivalents and *Ae. columnaris* x *Ae. uniaristata* (genome UX^nN) had 4–6 bivalents and 0–1 trivalents (Kihara 1949). The triploid hybrid *Ae. columnaris* x *Ae caudata* (genome UX^{n}C) showed somewhat higher pairing than the hybrids of *Ae. columnaris* with the M and N genome species. Kihara (1949) observed 7–8 bivalents and 0–3 trivalents in *Ae. columnaris* x *Ae. caudata*, while Kimber and Abu-Bakar (1981) observed 7.47 univalents, 5.06 rod bivalents, 0.14 ring bivalents, 5.20 bivalents, 0.99 trivalents and 0.04 quadrivalents. This level of pairing indicates that one subgenome of *Ae. columnaris*, most probably subgenome U, is closer to the C genome than to the M and N genomes. The triploid hybrid *Ae. columnaris* x *Ae. speltoides* (genome UX^{n}S) showed 5–7 bivalents with a mode of 7 bivalents and 0–4 trivalents (Kihara 1949). In contrast, the triploid hybrid *Ae. columnaris* x *Ae. bicornis* (genome UX^{n}S^b) had 1–7 bivalents and 0–2 trivalents (Kihara 1949). No homology between the subgenomes of *Ae. columnaris* and *Ae. bicornis* was observed, and the high pairing of the hybrid with *Ae. speltoides* resulted presumably from the ability of the *speltoides* genome to promote homoeologous pairing.

The *Ae. neglecta* x *Ae. columnaris* hybrid (genome UX^{n}UX^n) had 11–14 bivalents, with a mode of 12 bivalents (Kihara 1949). This level of pairing shows that the genomes of the two species are closely related; both contain the U genome and a slightly modified X^n genome. Chromosomal pairing of other hybrids between *Ae. columnaris* and allotetraploid species having the U-subgenome and differing in the second subgenomes, usually had more than seven bivalents, indicating that, in addition to pairing between the chromosomes of the shared U subgenome, some pairing also occurs between chromosomes of the differential, unshared subgenomes. Thus, chromosome pairing data between *Ae. triuncialis* and *Ae. columnaris* hybrid (genome UCUX^n) (6–12 bivalents, with a mode of 10 bivalents, and 0–2 trivalents) shown by Lindschau and Oehler (1936), indicate that there is one subgenome common to both species, whereas the second subgenome is different. The *Ae. geniculata* x *Ae. columnaris* hybrid (genome M^{o}UUX^n) had 7.15 bivalents, 1.10 trivalents and 0.30 quadrivalents (Kimber et al. 1988) or 6–9 bivalents and 0–3 trivalents (Kihara 1949). The *Ae. columnaris* x *Ae. biuncialis* hybrid (hybrid genome UX^{n}UM^b) had 7–9 bivalents, with a mode of 8 bivalents, and 0–2 trivalents (Kihara and Nishiyama 1937). The *Ae. columnaris* x *Ae. kotschyi* hybrid (genome UX^{n}S^{v}U) had 7.05 bivalents and 0.90 trivalents (Kimber et al. 1983). The pentaploid hybrid between the hexaploid species *Ae. juvenalis* and *Ae. columnaris* (genome D^{c}X^{c}UUX^n) had 7.81 bivalents, 1.79 trivalents and 0.22 quadrivalents (McGinnis and Melnyk 1962), indicating that the two species share only one subgenome, i.e., U, and differ in the other subgenomes. The subgenome X^c of *Ae. juvenalis*, like X^c of *Ae. crassa*, is different from the X^n of *Ae. columnaris*; the symbol 'X^n' means that the origin of this subgenome is unknown (Dvorak 1998).

9.8.8 *Aegilops triuncialis* L.

9.8.8.1 Morphological and Geographical Notes

Ae. triuncialis L., common name Barbed Goatgrass [Syn.: *Triticum triunciale* (L.) Raspail; *Aegilopoides triuncialis* (L.) Á. Löve; *Ae. aristata* Req. ex Bertol.; *Ae. echinata* C. Presl: *Ae. persica* Boiss.] is an annual, predominantly autogamous, tufted, many–tillered plant. Its culms are without many joints, usually prostrate and then turning upright, (15-)20–45 (-60)-cm-tall (excluding spikes), and defoliated in the upper quarter. The leaves are usually hairy, narrow-linear and 2.0–6.0-cm-long. The spike is narrow-lanceolate, 2.5–6.0-cm-long (excluding awns), usually awned, tapering to the tip, and well above the flag leaf on a long peduncle. The entire spike disarticulates at maturity and falls as a unit (umbrella-type dispersal unit), seldom as individual spikelets, each with its adjacent rachis segment (barrel-type dispersal unit). The rachis internode is generally approximately as long as the adjacent spikelet. There are three, and seldom 2, basal rudimentary spikelets. There are 3–8 (usually 4–5) spikelets, narrowly elliptical, usually all of which are potentially fertile. Each spikelet contains 4 florets, with the upper 2 being sterile. Glumes feature curved, unequally wide nerves, often covered with short, silvery hairs. There are 3 glume awns, unequal in length, which are usually smooth underneath, gradually tapering to the tip. The central glume awn of lower spikelets is shorter than its laterals. The central awn of glumes of the terminal spikelet is longer (4.5–7.0-cm-long) and wider than its laterals and is often the longest awn on the spike, and diverges almost at a right angle to the spike axis. Lemma awns are poor developed or absent. The caryopsis is free (Fig. 9.4g).

Ae. triuncialis is a polymorphic species exhibiting very wide morphological variation that mainly involves awn development (presence or absence), length and count, spike color and hairiness. This large morphological variation led taxonomists to subdivide the species into several intraspecific taxa. Zhukovsky (1928) classified *Ae. triuncialis* in a separate section, Surculosa Zhuk., and divided it into two subspecies, *brachyathera* Boiss. and *kotschyi* Boiss. Eig (1929a) also divided *Ae. triuncialis* into two subspecies: *eu-triuncialis* and *orientalis* Eig, with subsp. *eu-triuncialis* containing two varieties: typical and *constantinopolitana* Eig, and subsp. *orientalis* containing three varieties: *assyriaca* Eig, persica (Boiss.) Eig, and var. *anathera* Hausskn. et Bornm. Hammer (1980) also divided the species into two subspecies: *triuncialis* and *persica* (Boiss.) Zhuk, with subsp. *triuncialis* containing three varieties: *triuncialis*, *flavescens* Popova, and *constantinopolitana* Eig, and subsp. *persica* containing three varieties: *persica* (Boiss.) Eig, *assyriaca* Eig, and *anathera* Hausskn. et Bornm. In contrast to the above taxonomists who divided *Ae.*

triuncialis into subspecies, van Slageren (1994) divided *Ae. triuncialis* into two varieties: *triuncialis* and *persica* (Boiss) Eig. Hammer's subdivision (1980) into two subspecies, *triuncialis* and *persica*, will be used in this book.

Ssp. *triuncialis* has lateral spikelets glumes with 2–3 well-developed, 1.5-cm-long awns, while apical glumes have a well-developed, 5–8-cm central awn which is the longest awn of the spike, as well as 1–3-cm-long lateral awns. The spike always falls entire at maturity. Ssp. *persica* lateral spikelets have glumes with 1 or 2 teeth and a short, up to 1.5-cm-long awn. Its apical glumes have a well-developed, 2–5-cm-long central awn, and 2 short, 1–2-cm-long lateral awns, or lateral awns reduced to teeth. Lemmas of the terminal spikelet are awn-free, while the glumes bear 1 (1–3) awn. The spike sometimes disarticulates into individual spikelets, each with its adjacent rachis segment (barrel-type dispersal unit).

A new subspecies of *Aegilops triuncialis* L., ssp. *bozdagensis* Cabi & Doğan, confined to Denizli, Acıpayam, Bozdağ in southwestern Anatolia, was recently described (Cabi et al. 2018). It differs from the other two subspecies of *Ae. triuncialis*, *triuncialis* and *persica*, by the unawned glumes of the lateral spikelets, the three, up to 0.5-cm-long glume apex, with the middle one shorter than the others. The distribution map and notes on the biogeography and ecology of this subspecies, as well as an identification key of the three subspecies of *Ae. triuncialis*, are provided in Cabi et al. (2018).

Intraspecific variation was also detected on the molecular level. Nakai (1981), studying banding patterns of esterase isozymes in *Ae. triuncialis* and its putative parental diploid species, *Ae. umbellulata* and *Ae. caudata*, reported that zymogram phenotypes of both parents were quite uniform, whereas seven zymogram phenotypes were found among the 260 lines of *Ae. triuncialis* examined. Of these seven phenotypes, one (phenotype 3) had all bands of both parents, while the other six phenotypes differed greatly from phenotype 3. Thus, the zymogram phenotype in which the isozymes of both parental species are present, was considered by Nakai (1981) to be the most primitive of the seven types. Whether the phenotypes other than type 3 were due to introgressive hybridization or chromosomal rearrangement, is yet not determined.

Likewise, a RAPD analysis with 21 primers revealed more than 80% polymorphism between any two accessions of *Ae. triuncialis* collected in central Asia and North Caucasia (Okuno et al. 1998). Similarly, Monte et al. (1999, 2001), using RAPD and AFLP DNA markers, detected high polymorphism among Spanish populations of *Ae. triuncialis* and Thomas and Bebell (2010), using RAPD and ISSR analyses, observed significant genetic variability among 13 accessions of *Ae. triuncialis* from Greece. In contrast, Goryunova et al. (2010), using RAPD analysis, observed relatively low intraspecific polymorphism among 23 accessions of *Ae. triuncialis* that were collected from several areas of the species range.

Ae. triuncialis is a widespread Mediterranean/Western Asiatic/circumboreal element, occurring all over southern Europe and the Near East, extending eastwards into central Asia, and well-represented along most of the entire Fertile Crescent arc. It grows in Morocco, Algeria, Portugal, Spain, France (including Corsica), Italy (including Sardinia and Sicily), Slovenia, Croatia, Serbia, Bosnia-Herzegovina, Bulgaria, North Macedonia, Albania, Greece (including the Aegean Islands), Turkey, Cyprus, Syria, Lebanon, Israel (north), Iraq, Iran, Ukraine, Southern Crimea, Ciscaucasia, Transcaucasia (Georgia, Armenia and Azerbaijan), Kuwait, Saudi Arabia, Turkmenistan, Kazakhstan, Uzbekistan, Tajikistan, Kyrgyzstan, Afghanistan, and Pakistan. *Ae. triuncialis* grows in open areas and degraded forest and maquis, dwarf-shrub formations, steppe-like formations, pastures, roadsides and other disturbed habitats. It is also present in edges and within cultivation such as olive groves, vineyards, fruit tree plantations, and cereal crops, such as barley and wheat (with which it may form natural hybrids). *Ae. triuncialis* is also found in steppes but not in deserts, and, more rarely, in humid pastures, river terraces, and even at the seaside, apparently tolerating saline conditions. It grows on a variety of soils such as Mediterranean terra rossa, basalt and sandstone. It is also found on clay-and sandy loam, (sandy) clay, and gravel, and more rarely on loess, pure sands, and marly soils. It grows at altitudes from sea level up to 2700 m, usually in the 150–1800 m range. *Ae. triuncialis* grows under broad amplitudes of annual rainfall, varying from 125 mm up to 1400 mm per annum. Most data are, however, from the range 350–700 mm.

Ae. triuncialis has very large distribution (the largest in the genus), which overlaps with almost all of the distribution of the genus, except for the southeastern corner of the Mediterranean basin. It is common throughout its range and the most massive species in its distribution area. It has a very wide ecological amplitude, occupying a very large number of primary and secondary habitats. As a typical colonizer species, *Ae. triuncialis* can be found in massive stands and dominate a vegetation. It usually grows in mixed stands with other allotetraploid *Aegilops* species, with which it may introgress. According to Kihara (1954), *Ae. triuncialis* ssp. *triuncialis* with its var. *typica* covers the whole area of the species, while var. *constantinopolitana* is found in a very limited locality. The three varieties of ssp. *persica* are mainly in the Asiatic part of the *Aegilops* area. Among them one variety, *persica*, has the largest area extending to Iran, Transcaspian region and Afghanistan, while var. *anathera* has the same distribution except for not occurring in Afghanistan; var. *assyrica* has a very small habitat in Assyria.

The two diploid parents of *Ae. triuncialis*, *Ae. umbellulata* and *Ae. caudata*, have massive contact throughout Greece, Turkey, Syria, and Iran, where hybridization between them could have occurred, yielding *Ae. triuncialis* (Kihara 1954). However, the present distribution of *Ae. triuncialis* is very wide, much wider than that of its two diploid parents.

Ae. triuncialis grows sympatrically with almost all the species of the wheat group, except for *Ae. bicornis*, *Ae. sharonensis*, *Ae. longissima*, *Ae. kotschyi* and *Ae. vavilovii*. Intermediates between *Ae. triuncialis* and other allotetraploid species sharing the U genome are quite common in such mixed populations. Zohary and Feldman (1962) and Feldman (1965a) described intermediates between *Ae. triuncialis* and *Ae. biuncialis* and intermediates between *Ae. triuncialis* and *Ae. neglecta* in several locations in Turkey, and intermediates between *Ae. peregrina* and *Ae. triuncialis* in northern Israel. There is also evidence of hybridizations and introgression between *Ae. triuncialis* and common wheat (Hegde and Waines 2004). Arrigo et al. (2011) detected substantial gene flow between wheat and *Ae. triuncialis* growing close to wheat fields. Zaharieva and Monneveux (2006) assessed the probability of introgression between wheat and *Ae. triuncialis* in Europe, in areas where *Ae. triuncialis* occurred near or within wheat fields, and found that hybridization of *Ae. triuncialis* with wheat was quite common. Parisod et al. (2013), using 12 EST-SSR markers mapped on wheat chromosomes, obtained evidence of gene flow between common wheat and *Ae. triuncialis* growing in proximity to cultivated fields. Loci from the A subgenome of wheat were significantly less introgressed than sequences from the two other subgenomes, indicating differential introgression into *Ae. triuncialis*.

Ae. triuncialis is found as an adventive in central and northern France, Germany, Switzerland, Belgium, the Netherland, and the UK (van Slageren 1994). It was introduced with domesticated wheat into the USA (California, Oregon, Nevada, the New England area, and Pennsylvania) in the twentieth century and became a troublesome weed mainly on rangeland in California and Pennsylvania (Hitchcock 1951). Over the last several decades, *Ae. triuncialis* has rapidly spread into many annual grasslands and serpentine soil sites within California (Rice et al. 2013). It is unclear whether genetic differentiation, phenotypic plasticity, or both have allowed this species to invade competitive (i.e., high productivity, non-serpentine, annual grassland) and edaphically stressful (i.e., low productivity serpentine) environments (Rice et al. 2013). Despite its significant presence in northern California and in Pennsylvania, there have been no records of hybrids forming between common wheat and *Ae. triuncialis* in the USA (Watanabe and Kawahara 1999; Hedge and Waines 2004).

Accessions of *Ae. triuncialis* confer resistance to powdery mildew (Gill et al. 1985). leaf rust (Gill et al. 1985; Aghaee-Sarbarzeh et al. 2002), and to Hessian fly (Gill et al. 1985; Martín-Sánchez et al. 2003). *Ae. triuncialis* was highly resistant to Spikstersh, French and Swedish populations of cereal cyst nematode (Romero et al. 1998), and accessions of *Ae. triuncialis* were highly resistant to barley yellow dwarf luteovirus (BYDV) (Makkouk et al. 1994). Damania and Pecetti (1990) reported that several accessions of *Ae. triuncialis* were tolerant of drought and frost.

9.8.8.2 Cytology, Cytogenetics and Evolution

Ae. triuncialis is an allotetraploid species ($2n = 4x = 28$). Its nuclear genome is designated UUCC and CCUU (Kimber and Tsunewaki 1988; Dvorak 1998) and the plasmon genome of some forms is designated U, similar to that of *Ae. umbellulata*, while others contain plasmon C', a subtype of the cytoplasm of *Ae. caudata* (Tsunewaki 2009). The origin of the subgenomes of *Ae. triuncialis* remains an enigma (Kihara 1954). Karyotype studies reported by Senyaninova-Korchagina (1932) and by Chennaveeraiah (1960) reached considerably different conclusions. While they agreed on the occurrence in this species of a subgenome similar to that of *Ae. umbellulata*, they differed in the observed nature of the second subgenome. Senyaninova-Korchagina (1932) described the karyotype of *Ae triuncialis* ssp. *persica* as composed of two sets of chromosomes, namely, one similar to that of *Ae. umbellulata* and the other to that of *Ae. caudata*, and the karyotype of ssp. *triuncialis* as composed of an *Ae. umbellulata*–chromosome set and another one different from that of *Ae. caudata*. Based on this, Senyaninova-Korchagina (1932) separated ssp. *persica* from *Ae. triuncialis* and elevated it to the specific rank, *Ae. persica* Boiss. In contrast, Chennaveeraiah (1960) reported that *Ae. triuncialis* ssp. *persica* had one set of chromosomes which corresponded to the genome of *Ae. umbellulata*, and a second set, though resembled, in most respects, the genome of *Ae. caudata*, differed from it in in having only one pair of SAT chromosomes. In *Ae. triuncialis* ssp. *triuncialis* he found, again, one chromosome set corresponding to that of *Ae. umbellulata*, whereas the second set differed not only from the typical genome of *Ae. caudata* but also from that of ssp. *persica*. Thus, Chennaveeraiah (1960) concluded that the *Ae. triuncialis* complex possesses chromosomal sets similar to those of *Ae. umbellulata* and *Ae. caudata*, but that either the chromosome set derived from *Ae. caudata* had undergone significant modifications or some other genome is involved.

Kihara (1940b) pointed out that karyotype analysis and genome analysis may lead to different results and, based on genome analysis, he included *Ae. persica* in *Ae. triuncialis* (Kihara 1954). Genome analysis (reviewed in Kihara 1954) showed that *Ae. triuncialis* originated from hybridization between the two diploids *Ae. umbellulata* and *Ae. caudata* and consequently, assigned to it the genome formula C^uC^uCC (currently C^u is U). Only one cross combination, *Ae. caudata* x *Ae. triuncialis*, was successful, displaying 7 bivalents and 7 univalents at first meiotic metaphase, thus, indicating unequivocally that *Ae. triuncialis* possesses the C subgenome that derived from *Ae. caudata* (Kihara 1949).

Since attempts to produce the triploid hybrid *Ae. umbellulata* x *Ae. triuncialis* were unsuccessful, the artificially produced amphidiploid of *Ae. caudata* x *Ae. umbellulata* (genome CCUU) was employed to determine whether the second subgenome of *Ae. triuncialis* is homologous to that of *Ae. umbellulata* (Kihara and Kondo 1943). The resulting amphidiploid was morphologically similar to ssp. *triuncialis*, whereas its karyotype was identical with that of ssp. *persica* (Kihara 1954), as described by Senyaninova-Korchagina (1932). Both the synthetic amphidiploid and ssp. *persica* had the umbrella-type disarticulation, but at the same time, the ear was easily broken in a barrel-type fashion in the upper part of the spike, a feature that is seldom found in ssp. *triuncialis* (Kihara 1954).

Kihara and Kondo (1943) used their own synthetic amphidiploid, and that of Sears (1939), to study chromosome pairing in hybrids between the synthetic amphidiploid and *Ae. triuncialis*. Chromosome pairing in the F_1 hybrid between the amphidiploid and the two subspecies of *Ae. triuncialis* was almost regular; the hybrid ssp. *triuncialis* x the amphidiploid had 11–12 bivalents (mode 12) and its reciprocal had 10–13 bivalents (mode 12); the hybrid ssp. *persica* x the amphidiploid had 11–13 bivalents (mode 12) and its reciprocal had 11–14 bivalents (mode 12) (Kihara and Kondo 1943). Pollen and seed fertility of these hybrids were low when the amphidiploid (C cytoplasm) was the female parent and higher in the reciprocal crosses when the natural subspecies bearing the U cytoplasm was the female parent (Kihara and Kondo 1943). The almost complete pairing in the hybrids between the synthetic amphidiploid and the two subspecies of *Ae. triuncialis* confirmed the karyotype studies which showed that the second subgenome of *Ae. triuncialis* is very close or even almost identical to the U genome of *Ae. umbellulata*. Thus, from the observation on chromosome pairing and fertility of the F_1 hybrid, Kihara and Kondo (1943) concluded that the genome-types of the two subspecies of *Ae. triuncialis*, *triuncialis* and *persica*, are identical or almost identical.

Additional evidence that *Ae. triuncialis* contains the U subgenome was presented by Kimber and Yen (1988, 1989), who observed 8.00 univalents, 2.08 rod bivalents, 2.37 ring bivalents, 2.87 trivalents, 0.52 quadrivalents, 0.07 pentavalents, and 0.02 hexavalents in the hybrid *Ae. triuncialis* x an induced autotetraploid *Ae. umbellulata* (hybrid genome UUUC). Using the numerical methods of chromosome

pairing, Kimber and Yen (1988, 1989) determined that this type of chromosomal pairing best fits the 3:1 model, namely, three U and one C, indicating that *Ae. triuncialis* indeed contains a subgenome homologous to genome U of *Ae. umbellulata*. Studies on variation of repeated nucleotide sequences also confirmed the presence of U and C subgenomes in *Ae. triuncialis* (Dubcovsky and Dvorak 1994). They found that the number of bands of the repeated nucleotide sequences shared between each diploid species and *Ae. triuncialis*, was significantly higher than those shared with any other diploid species.

Early cytogeneticists were of the opinion that *Ae. triuncialis* is nearly an autotetraploid species (von Berg 1931; Karpechenko and Sorokina 1929; Kihara 1929, 1937). This opinion was based on the fact that the hybrid between *Secale cereale* and *Ae. triuncialis* had 3–7 bivalents (or sometimes even 6–7), which were attributed to autosyndesis of the *triuncialis* chromosomes. This level of pairing led Kihara and Lilienfeld (1932) to assume that *Ae. triuncialis* is not an autotetraploid, but rather, a segmental allopolyploid that was formed at a time when genomes U and C of the parental diploid species had not been widely differentiated.

Currently, the accepted view is that *Ae. triuncialis* is a typical allopolyploid (Kihara 1954, 1963). This view also gained support from the analysis of chromosomal pairing at first meiotic metaphase of a haploid plant of *Ae. triuncialis* (Chapman and Miller 1977), which showed 9.17 univalents, 2.10 bivalents, 0.17 trivalents, and 0.03 quadrivalents in the haploid plant. This low level of pairing indicated that the two subgenomes of *Ae. triuncialis* are less closely related than previously suggested (Chapman and Miller 1977).

Several accessions of *Ae. triuncialis* have the U plasmon type that was contributed by *Ae. umbellulata* (Tsunewaki 1980; Tsunewaki and Tsujimoto 1983). The chloroplast genome of these accessions showed identical restriction fragment patterns to those of *Ae. umbellulata* (Ogihara and Tsunewaki 1982, 1983; Terachi et al. 1984). Restriction fragment patterns of mitochondrial DNA isolated from one accession of *Ae. triuncialis*, have been analyzed using five restriction endonucleases and confirmed the presence of U cytoplasm (Terachi and Tsunewaki 1986).

Yet, by comparing morphological and physiological characters of alloplasmic wheat lines bearing *Ae. triuncialis* cytoplasm, Mukai et al. (1978) found that two accessions of *Ae. triuncialis* ssp. *triuncialis* have genetically different cytoplasms. One had a cytoplasm almost identical to that of *Ae. umbellulata* and the other had one similar to that of *Ae. caudata*. This finding was supported by restriction endonuclease analysis of the chloroplast DNA of these accessions (Ogihara and Tsunewaki 1982, 1983). Together, these observations indicated a possible diphyletic origin of *Ae. triuncialis* from the reciprocal crosses between *Ae. caudata* and *Ae. umbellulata* (Ogihara and Tsunewaki 1982). Murai and Tsunewaki (1986), aiming to clarify (a) the extent of chloroplast DNA variation within *Ae. triuncialis*, (b) the geographical distribution of accessions bearing different chloroplast genomes, and (c) the distribution of each chloroplast genome type among different taxa, further studied the diphyletic origin of *Ae. triuncialis*. These authors, using restriction endonuclease analysis of chloroplast DNA of 21 accessions of *Ae. triuncialis*, found that 13 accessions had the type 2a chloroplast genome derived from *Ae. caudata*, eight possessed the type 3 chloroplast genome of *Ae. umbellulata*, and the remaining five contained a new chloroplast genome (named type 2b), which differed from the 2a type, by a 0·3 kbp insertion. The accessions with type 2a (*caudata*) and type 3 (*umbellulata*) chloroplast genomes distribute in wide geographical areas, and occur in both of its subspecies, *triuncialis* and *persica*, whereas those with the type 2b chloroplast genome occurred only in Azerbaijan. Waines and Barnhart (1992) mistakenly suggested that *Ae. umbellulata* is the female parent of ssp. *triuncialis* (the typical subspecies), while *Ae. caudata* is the female parent ssp. *persica*. From these results, Murai and Tsunewaki (1986) drew the following two conclusions: (a) *Ae. triuncialis* has a diphyletic origin from the reciprocal crosses between *Ae. caudata* and *Ae. umbellulata*, and (b) the type 2b chloroplast genome arose from type 2a chloroplast genome by a 0·3 kbp insertion.

Likewise, Ogihara and Tsunewaki (1988), studying restriction fragment patterns of chloroplast DNAs from two alloplasmic lines of common wheat carrying cytoplasm of two different accessions of *Ae triuncialis*, discovered that one accession (code no. 26) showed identical restriction fragment patterns to those of *Ae. umbellulata*, whereas the second accession (code no. 38) had patterns identical to that of *Ae. caudata*. This finding corroborated the proposal that *Ae. triuncialis* was produced diphyletically by the reciprocal crosses between *Ae. caudata* and *Ae. umbellulata*.

Wang et al. (1997) performed PCR–single-strand conformational polymorphism (PCR-SSCP) analyses of 14.0-kb chloroplast and 13.7-kb mitochondrial DNA regions that were isolated from alloplasmic wheat lines carrying *Ae. triuncialis* cytoplasm. Their study, showing that *Ae. triuncialis* contains two plasmon types, U and C^2, supported the claimed dimaternal origin of *Ae. triuncialis* from reciprocal crosses between *Ae. umbellulata* and *Ae. caudata*. The plasmon of these species pointed to zero genetic distance between *Ae. triuncialis* (line 26) and *Ae. umbellulata*, and a large distance (average 0.074) between *Ae. triuncialis* and *Ae. caudata*. On the other hand, *Ae. triuncialis* (line 38) was a short distance (average 0.019) from *Ae. caudata* as compared to *Ae. umbellulata* (0.081). If intraspecific variation of *Ae. caudata* is subtracted from the distance between *Ae. triuncialis* (38) and *Ae. caudata*, the distance between them is small, suggesting that both types of *Ae. triuncialis* arose

recently and at approximately the same time. Since the C cytoplasm of *Ae. triuncialis* differs slightly from that of *Ae. caudata*, Tsunewaki (2009) designated it as C' (a subset of C).

In accord with the above, Chee et al. (1995), using genome-specific primer sets, reported that variation in the C subgenome of *Ae. triuncialis* resulted from variability in *Ae. caudata* rather than from frequent introgression events after formation of the allopolyploid, thus, suggesting a polyphyletic origin for *Ae. triuncialis*. Meimberg et al. (2009) concluded that *Ae. triuncialis* was formed in four independent origins, affecting differences in ecological tolerance among the independent origins of this species. Similarly, Vanichanon et al. (2003) claimed that the number of independent origins of an allopolyploid species was traditionally underestimated. They screened 84 primer sets to identify genome-specific primer sets for *Aegilops triuncialis* and its diploid progenitors; some of the primers were U genome-specific and others C genome-specific. A DNA sequence comparison revealed at least two or three independent formations of *Ae. triuncialis*.

Ae. triuncialis contains 9.93 ± 0.041 pg 1C DNA (Eilam et al. 2008), the smallest genome among the allotetraploid species of section Aegilops and, together with *Ae. cylindrica*, the smallest genome among all the allotetraploid species of the wheat group (Eilam et al. 2008). The 1C DNA content of *Ae. triuncialis* is 2.84% smaller than that which is expected from the sum of the DNA of its two diploid parents, namely, 10.22 pg (*Ae. umbellulata*, the donor of the U subgenome, contains 5.38 pg and *Ae. caudata*, the donor of the C subgenome, contains 4.84 pg; Eilam et al. 2007). The loss of DNA in *Ae. triuncialis* was also noted by Badaeva (2002) and Badaeva et al. (2004) who, based on differential C-banding and FISH, found that *Ae. triuncialis* exhibits substantial structural chromosome rearrangements, including deletion of chromosomal segments and reduction of heterochromatin content.

Early karyomorphological studies in *Ae. triuncialis* (Emme 1924; Sorokina 1928) did not show any presence of SAT chromosomes. According to Senyaninova-Korchagina (1930, 1932) the karyotype consists of 9 types of chromosomes, including only one pair with a satellite on a short arm, one pair with an extreme subterminal centromere, two pairs with submedian-subterminal centromeres and the rest with submedian centromeres. Chennaveeraiah (1960) analyzed the karyotype of the two subspecies of *Ae. triuncialis* and found marked differences between that described by Senyaninova-Korchagina (1930, 1932). The karyotype of ssp. *persica* studied by Chennaveeraiah (1960), was in agreement with that described by Senyaninova-Korchagina (1930, 1932) for this subspecies. However, Chennaveeraiah (1960) reported the existence of three chromosome pairs with satellites on short arms, with the satellites differ from each other in size. In addition, there are four pairs with extreme subterminal centromeres, two pairs have submedian-subterminal centromeres, and the rest with submedian centromeres. In all, there are 12 types of chromosomes (Chennaveeraiah 1960). Ssp. *triuncialis* (*brachyathera* Boiss. in Chennaveeraiah) has a different karyotype in that it has only two pairs with satellites on short arms, only three pairs with extreme subterminal centromeres (whereas there are four in ssp. *persica*), three pairs with submedian-subterminal centromeres, and the rest with submedian centromeres. In all, there are 12 types of chromosomes in this subspecies as well (Chennaveeraiah 1960). Because of the diversity of morphological traits within the species, it was suggested that different accessions of *Ae. triuncialis* show different types of karyotypes (Chennaveeraiah 1960).

According to Chennaveeraiah (1960), *Ae. triuncialis* ssp. *persica* has one set of chromosomes which corresponds to the U genome of *Ae. umbellulata*. The second set, which in most respects resembles the C genome of *Ae. caudata*, differs in that it has a only one pair with satellites. In *Ae. triuncialis* ssp. *triuncialis*, one set corresponds to the U genome, whereas the second set differs not only from the typical C genome set but also from that of ssp. *persica*. In this case, the second set does not even contain a single satellited pair. From this it appears that the C genome underwent tremendous modification.

In the light of the present knowledge of the karyotypes of these subspecies, the synthesized amphidiploid of *caudata-umbellulata* (Kihara and Kondo 1943) is expected to have 4 pairs with satellites, as both C and U genome sets have two SAT-pairs each. Yet, ssp. *persica* has three SAT chromosomes and ssp. *triuncialis* has two. In spite of this and other differences, Chennaveeraiah (1960) believed in the presence of both the C and the U subgenomes in the *triuncialis* complex, but in some, either the C genome had undergone significant modification or there was some involvement of a foreign genome instead of the C genome. The U genome has remained unchanged.

Teoh et al. (1983), using FISH with a cloned repetitive DNA sequence derived from common wheat coding for ribosomal RNA, observed three SAT pairs (like the report above), and three pairs of rRNA sites on the genome of *Ae. triuncialis*. On the other hand, two SAT chromosomes were observed in *Ae. triuncialis* by phase contrast of metaphase cells (Cermeño et al. 1984b), corresponding to the report by Chennaveeraiah (1960) for ssp. *triuncialis*. This number is in quite good agreement with the two Ag-NORs (2) and maximum number of four nucleoli observed by Cermeño et al. (1984b). Apparently, the U subgenome fully suppresses NOR activity of the C subgenome in *Ae. triuncialis* (Cermeño et al. 1984b).

In accord with the above, Al-Mashhadani et al. (1980) found two pairs of SAT chromosomes in *Ae. triuncialis*.

They also reported the presence of four metacentric pairs, three submetacentric pairs and five subtelocentric pairs, whereas Tanaka and Matsumoto (1965) found four metacentric chromosome pairs, three submetacentric pairs and seven subtelocentric pairs. Ahmadabadi et al. (2002), analyzing 13 populations of *Ae. triuncialis*, collected from northwest regions of Iran, found that karyological characters such as total length of chromosomes, chromosome arm ratio, number and length of satellites, showed high variation. Sadeghian et al. (2015) also recorded intraspecific karyotype divergence in *Ae. triuncialis* from northwestern Iran. Studying samples of *Ae. triuncialis* from different eco-geographical sites in northern Algeria, Baik et al. (2017) revealed two cytotypes which differed in chromosome lengths, karyotype symmetry and in the presence or absence of the satellites. One cytotype had 7 metacentric, 4 submetacentric, and 3 subtelocentric pairs, whereas the second cytotype had 9 metacentric, 11 submetacentric, and 2 subtelocentric pairs. One cytotype, consisting larger chromosomes, was sampled from populations at low altitudes under a humid bioclimate, whereas the second cytotype, having smaller chromosomes, was found in populations of high steppe plains under a semi-arid bioclimate.

Badaeva et al. (2004) examined karyotypes of 21 accessions of *Ae. triuncialis* (13 of ssp. *triuncialis* and 8 of ssp. *persica*) by C-banding and of four accessions (three of ssp. *triuncialis* and one of ssp. *persica*) by FISH with the pSc119, pAs1, pTa71 (18S-26S rDNA), and pTa794 (5S rDNA) probes. The karyotype structure of the accessions studied was similar to that described by Chennaveeraiah (1960) for *Ae. triuncialis* var. *persica*. All *Ae. triuncialis* chromosomes had distinct C-banding patterns, similar to those of the parental species *Ae. umbellulata* and *Ae. caudata* that were described by Friebe et al. (1992a, b, 1995a, c) and Badaeva et al. (1996a). Only limited intraspecific polymorphism was observed, manifested by the presence or absence of certain bands and variation in C-band size. Chromosomal arrangements found in three of the 21 accessions were represented by single or multiple translocation breakpoints and by a paracentric inversion. The origin of additional modified chromosomes (chromosome 1U in line k-940; chromosome 5U in lines k-146 and k-1965) remains unknown. FISH with pSc119 revealed distinct hybridization sites in the telomeric regions of one or both arms of 12 chromosome pairs of *Ae. triuncialis*. Interstitial FISH sites were observed in the long arms of chromosomes 7U and one C genome chromosome. Comparison of the pSc119 labeling patterns with the parental species showed that both the number and intensity of hybridization signals on *Ae. triuncialis* chromosomes decreased.

An extended secondary constriction was always observed on chromosome 1C, whereas the satellites on the 1U-subgenome chromosomes were usually not visible. Two active NORs were observed using FISH, confirming previous observations. The signal size decreased in the order $1C > 5U > 1U >> 5C$. A faint signal was occasionally observed in the middle of the long arm of chromosome 6U. Four approximately equal 5S rDNA sites were detected in all chromosomes of homoeologous groups 1 and 5. Thus, the distribution of ribosomal RNA gene families on *Ae. triuncialis* chromosomes is similar to those of the parental species (see Badaeva et al.1996b), suggesting that speciation in *Ae. triuncialis* was not associated with large genomic modifications. The low C-banding polymorphism and the low frequency of chromosomal rearrangements suggest a relatively recent origin of *Ae. triuncialis*. The study of Badaeva et al. (2004), while confirming the origin of the two subgenomes in *Ae. triuncialis*, also detected a certain level of modification of the C subgenome in this species.

Different repetitive sequences, namely, pSc119.2–1, pTa535-1, pAs1-1, $(CTT)_{10}$ and the 45S rDNA clone from wheat (pTa71), were hybridized to chromosomes of *Ae. triuncialis* and compared to the patterns of its diploid progenitors *Ae. umbellulata* and *Ae. caudata* (Mirzaghaderi et al. (2014). Like Badaeva et al. (2004), also Mirzaghaderi et al. (2014) found that the FISH patterns of the U and C subgenomes of *Ae. triuncialis* were, in general, similar to those of U and C genomes of *Ae. umbellulata* and *Ae. caudata*, respectively, although some differences were observed.

In contrast, Gong et al. (2006), using 31 ISSR primers, found that subgenome U showed few alterations, while subgenome C had undergone changes after allopolyploidization. Compared with their ancestral diploid genomes, the genetic similarity index between *Ae. triuncialis* and *Ae. umbellulata* was 0.6842, which was higher than that between *Ae. triuncialis* and *Ae. caudata*.

9.8.8.3 Crosses with Other Species of the Wheat Group

Kihara and Kondo (1943) reported that *Ae. triuncialis* ssp. *triuncialis* showed regular pairing (14 bivalents) at first meiotic metaphase. Fertility of this subspecies was high (94.8% pollen fertility and 88.4% seed fertility). Similar data were obtained in ssp. *persica*, namely, 14 bivalents, 96.7% pollen fertility and 92.1% seed fertility. Cuñado (1992), using C-banding, studied metaphase-I chromosome associations in *Ae. triuncialis* and observed 0.09 univalents, 5.45 rod bivalents and 8.46 ring bivalents, culminating to a mean 22.36 chromosome associations per cell. The pattern of zygotene and pachytene pairing in *Ae. triuncialis*, analyzed by whole-mount surface-spreading of synaptonemal complexes under the electron microscope (Cuñado et al. 1996b), indicated that the chromosomes were exclusively associated as bivalents at the zygotene stage, while at pachytene, one multivalent was observed in 7% of the cells.

Chromosomal pairing in the hybrid between *Ae. triuncialis* and one of its two diploid parents, *Ae. caudata* (genome UCC), included 6–7 bivalents, with a mode of 7, and 0–2 trivalents (Kihara 1949), indicating that *Ae. triuncialis* possesses one subgenome, C, that derived from *Ae. caudata*. On the other hand, the hybrid between *Ae. triuncialis* and another diploid species, *Ae. speltoides* (genome UCS) had only 4–7 bivalents (Kihara 1949). The lower pairing in the latter hybrid, in spite of the presence of *Ae. speltoides* genes promoting homoeologous pairing, showed that none of the subgenomes of *Ae. triuncialis* are close to the genome of *Ae. speltoides*.

Chromosome pairing in hybrids between *Ae. triuncialis* and other allotetraploid species sharing the U subgenome but differing in the second subgenome, showed that, in addition to pairing between chromosomes of the U subgenomes, some pairing also occurred between chromosomes of the differential subgenome. More specifically, the hybrid *Ae. neglecta* x *Ae. triuncialis* (genome UX^nUC) had 0–9 bivalents, with a mode of 5, and 0–1 trivalents (Lindschau and Oehler 1936), or up to 9 bivalents (Kihara 1949). The hybrid *Ae. geniculata* x *Ae. triuncialis* (genome M^oUUC) had 5–11 bivalents (Kihara 1929), and the hybrid A*e. triuncialis* x *Ae columnaris* (h genome $(UCUX^n)$ had 6–12 bivalents, with a mode of 10, and 0–2 trivalents (Lindschau and Oehler 1936). Chromosome pairing between the hexaploid species bearing the U subgenome, *Ae. juvenalis*, and *Ae. triuncialis* (genome D^cX^cUUC) was 12.94 univalents, 5.37 rod bivalents, 2.27 ring bivalents (total 7.64 bivalents), 1.70 trivalents, 0.34 quadrivalents and 0.02 heptavalents (McGinnis and Melnyk 1962). This level of chromosome pairing showed the homology between the U subgenomes as well as some pairing between chromosomes of the differential subgenomes. Chromosome pairing with *Ae. cylindrica*, which shares the C subgenome with *Ae. triuncialis* (genome DCUC), included 3–12 bivalents (Percival 1930), indicating that up to three bivalents were formed between the chromosomes of the D and the U subgenomes.

Chromosome pairing in hybrids between *Ae. triuncialis* and several subspecies of tetraploid wheat, *Triticum turgidum* (genome UCBA), was very low, indicating no homology between the subgenomes of these species. More specifically, the hybrid *Ae. triuncialis* x ssp. *dicoccoides* had 0–7 bivalents (usually 2–4) (Kihara 1929) and 1–3 rod bivalents (Percival 1930). The hybrid *Ae. triuncialis* x ssp. *dicoccon* had 1–7 rod bivalents, usually 4–5 (Kihara 1929). The hybrid *Ae. triuncialis* x ssp. *durum* had 0–8 rod bivalents, usually 6 (Kihara 1929), and 1–6 bivalents (Percival 1930). The hybrid *Ae. triuncialis* x ssp. *polonicum* had 3–8 bivalents, usually 5–6 (Kihara 1929), and the hybrid *Ae. triuncialis* x subsp. *turgidum* had 1–3 rod bivalents (Percival 1930).

Cifuentes et al. (2010) studied chromosome pairing at meiotic first metaphase in hybrids between *T. turgidum* ssp. *durum* and *Ae. triuncialis*, using a genomic in situ hybridization (GISH) procedure that allows simultaneous discrimination between A, B and U, C subgenomes. Chromosome pairing in the hybrid presented 25.6 univalents, 1.16 rod bivalents, 0.002 ring bivalents and 0.024 trivalents, culminating to 1.21 associations/cell. The general picture that was drawn showed that A and B wheat subgenomes paired with each other less than U and C did in the examined hybrid. Interspecific wheat-*triuncialis* (AB-UC) pairing accounted for 57% of the total pairing in the hybrid; the A subgenome was always the wheat partner most frequently involved in pairing with the *Ae. triuncialis* homoeologues. Pairing between the U and C chromosomes in *durum* x *Ae. triuncialis* showed 0.55 bivalents/cell (45.8%), whereas pairing between chromosomes of subgenome A of durum and UC of *triuncialis* included 0.49 bivalents/cell (40.8) and that of subgenome B chromosomes and *triuncialis* 0.08 bivalents/cell (6.67%) (Cifuentes et al. 2010). Pairing between A and B chromosomes of *durum* was at an incidence of 0.08 bivalents/cell (6.67%). Evidently, there was more autosyndetic than allosyndetic pairing between chromosomes of *Ae. triuncialis*. These results support the suggestion that U and C subgenomes show a higher pairing affinity for each other, and for the wheat A genome, than any of the other pairs of the constituent subgenomes present in the hybrid. Thus, the U and C subgenomes of *Ae. triuncialis* are more closely related than any combination of the other subgenomes present in this hybrid.

Chromosome pairing in hybrids between hexaploid wheat, *T. aestivum*, and *Ae. triuncialis* (genome BADUC) presented 26.97 univalents, 3.77 bivalents, and 0.17 trivalents (Riley 1966a), indicating that no homology exists between the subgenomes of these species. A similarly low level of pairing was observed in this hybrid combination by Aase (1930) [0–3 bivalents] and by Kihara (1929) [0–5 bivalents, usually 1–3].

9.8.9 *Aegilops kotschyi* Boiss.

9.8.9.1 Morphological and Geographical Notes

Ae. kotschyi Boiss. [Syn.: *Ae. triuncialis* L. var. *kotschyi* (Boiss.) Boiss.; *Ae. triuncialis* ssp. *kotschyi* (Boiss.) Zhuk.; *Triticum kotschyi* (Boiss.) Bowden; *Aegilemma kotschyi* Á. Löve] is an annual, predominantly autogamous, tufted, 15–30 (-40)-cm-tall (excluding spikes) plant. It is usually bushy, with many tillers, of which the upper parts are upright. The leaves are usually glabrous. The spike is narrow-lanceolate to ovate-lanceolate, 2.0–3.0(-4)-cm-long (excluding awns), which narrows toward the tip, and is awned. The entire spike

disarticulates at maturity (umbrella-type dispersal unit). There are (3-)4–5(-6) spikelets, usually linear and narrow, appressed to the rachis or to each other, with the two lower spikelets usually longer than the adjacent rachis segment. There are 2–4 (usually 3) basal rudimentary spikelets. There are 3–4 florets in the two lower spikelets, 2–3 of which are fertile; the upper spikelets are only partly fertile with small seed. The glume is usually shorter than the lemma, with narrow, equally wide, parallel nerves. There are usually 3, glume awns, and when there are 2, the central awn is replaced with a tooth or a gap. Glume awns are flat, gradually tapering to the tip. There are 1–3 lemma awns, which are equal to or slightly shorter than glume awns. In total, there are 8–14 awns on the spike, which tend to spread at maturity. The caryopsis adheres to the lemma and palea (Fig. 9.4h).

The original specimen of this species was collected by the botanist Kotschy in Iran. Boissier first described it as a species in 1846, but later, in 1884, he changed his mind about this species and listed it as a variety of *Ae. triuncialis* L. (Eig 1929a). Indeed, *Ae. kotschyi* shows some morphological similarity with *Ae. triuncialis*, and consequently was also treated by Zhukovsky (1928) as a subspecies of the latter, namely, *Ae. triuncialis* L. ssp. *kotschyi* (Boiss.) Zhuk. However, according to Eig (1929a), Boissier erred by classifying *Ae. kotschyi* as a variety of *Ae. triuncialis*.

Genome analysis studies led Kihara (1937, 1949, 1954) designated the genome of *Ae. triuncialis* as C^uC^uCC and that of *Ae. kotschyi* as $C^uC^uS^vS^v$, indicating that these two allotetraploids are separate species. Moreover, *Ae. triuncialis* and *Ae. kotschyi* are distinguished by their acid phosphatase isoenzyme patterns (Jaaska 1978a). *Ae. triuncialis* has the acid phosphatase isoenzyme doublet characteristic of *Ae. caudata* (genome C), whereas this isoenzyme was not observed in *Ae. kotschyi*. These results, like those of Kihara (1949, 1954, 1937), show the absence of the C genome in the latter, implying that *Ae. kotschyi* differs from *Ae. triuncialis* in its genome composition and should be treated taxonomically as an independent species.

Based on almost complete chromosome pairing in interspecific F_1 hybrids between *Ae. kotschyi* and *Ae. peregrina* (Lindschau and Oehler 1936; Kihara 1937), Kihara (1954, 1957) considered these two species very closely related and therefore, included *Ae. kotschyi* in *Ae. peregrina*. While the inclusion of *Ae. kotschyi* in *Ae. peregrina* was accepted by some authors, e.g., Bowden (1959) and Morris and Sears (1967), that followed Bowden' classification, included *Ae. peregrina* in *Ae. kotschyi*. However, other wheat scientists treated them as separate species (e.g., Chennaveeraiah 1960; Furuta 1981b; Kimber and Feldman 1987; Waines and Burnhart 1992). On taxonomical grounds, the taxonomists Eig (1929a), Hammer (1980) and van Slageren (1994) considered them separate species.

The two species are vicarious taxa that grow in two different phytogeographical regions; *Ae. kotschyi* is an Irano-Turanian element and *Ae. peregrina* is a Mediterranean element. The two species also differ by the following morphological characters: *Ae. kotschyi* has linear, appressed spikelets and both glumes of the lowest fertile spikelet always have 3 fine awns that are equally wide at the base. The awns of all glumes and lemmas being more or less of the same length in *Ae. kotschyi*. In *Ae. peregrina*, the spikelets are urn-shaped to elliptical and not appressed. The glumes are coarse, one glume of the lowest fertile spikelet has 3 awns that are equally wide at the base, the other glume has only 2 awns that are unequally wide at the base. The lemma awns are poorly developed and often absent. The spike is stout, with an irregular appearance, caused by variation in glume and lemma awn development. Glume awns are 2–4-cm-long, and lemmas have one or two 0.3–3-cm-long awns, which are always shorter than the glume awns and with 1 or 2 teeth.

Hammer (1980) followed Eig (1929a) in recognizing the following five varieties in *Ae. kotschyi*, var. *kotschyi* (=var. *tipica* Eig), var. *leptostachys* (Bornm.) Eig, var. *palaestina* Eig, var. *caucasica* Eig, and var. *hirta* Eig. Yet, in part four of Flora Palaestina, Feinbrun-Dothan (1986b) recognized only two varieties in *Ae. kotschyi*, var. *kotschyi* and var. *brachyathera*. The spike of var. *kotschyi* is oval-lanceolate to lanceolate in outline, and awned. It commonly grows in steppes. The spike of var. *brachyathera* is narrow-lanceolate; all spikelets are awnless, sometimes with the exception of the upper one, which is shortly and irregularly awned. The variety is rare and grows in Nubian sandstone hills together with var. *kotschyi*.

Ae. kotschyi has some morphological variation involving spike shape (ratio of length to width), compactness due to rachis segment length, spikelet size, awn length and development. Eig (1929a) did not encounter a form that can be described as a subspecies. Of note, *Ae. kotschyi* is fairly uniform in Israel and surrounding countries and seems to be so also in other parts of its area (Feinbrun –Dothan 1986a; b). The predominant taxon in all the distribution area is var. *kotschyi*.

Goryunova et al. (2010) used RAPD analysis to study genetic variation in 15 accessions of *Ae. kotschyi* and its phylogenetic relationships to other allopolyploid *Aegilops* species having the U subgenome. The majority of *Ae. kotschyi* accessions displayed a moderate variation; two accessions however, were distinct from the other accessions. The maximum genetic distance between *Ae. kotschyi* accessions was 0.20. The greatest separation within the U-subgenome cluster was observed for the US genome species *Ae. kotschyi* and *Ae. peregrina*. This result is in agreement with Eig (1929a), who noticed that these two species differ morphologically from the other U-subgenome species. Based on their morphological characters, Eig (1929a) isolated *Ae. kotschyi* and *Ae. peregrina* into a separate subsection, Adhaerens, in which the caryopsis is fused to floral bracts, while the other U-subgenome species were assigned to subsection Libera, in which the caryopsis is free.

The study reported by Zhang et al. (1992) indicated that repeated nucleotide sequence families were more variable in *Ae. peregrina* than in *Ae. kotschyi* . This agrees with the morphological studies of Eig (1929a), which also indicated greater variation in *Ae. peregrina* than in *Ae. kotschyi* .

Ae. kotschyi occurs mainly along the coast of eastern North Africa and the western arc of the Fertile Crescent but is rarer and displays a scattered presence in the central and eastern parts of the Fertile Crescent and central Asia. It grows in Uzbekistan, Turkmenistan, Afghanistan, Pakistan, southern Iran, Iraq (Mesopotamia), Kuwait, eastern Saudi Arabia, Azerbaijan, southeastern Turkey, Syria, Lebanon, Cyprus (rarely), Israel, Jordan, Egypt (lower Egypt and Sinai), Libya, and Tunisia. *Ae. kotschyi* grows on grey-calcareous steppe soil, white rendzina, loess, sandy-clay and sandy soils, in deciduous steppe maquis, in dwarf-shrub steppe-like formations, marginal dwarf-shrub formations, steppical plains, wadis, edges of cultivation, disturbed habitats and roadsides. It is very common and locally abundant, forming very dense populations in open disturbed habitats of the warm steppes. It is sporadic in primary, stable habitats. It grows at altitudes of 100–1100 m. *Ae. kotschyi* is one of the few *Aegilops* species that clearly extends into the Saharo-Arabian region.

Ae. kotschyi has a relatively large distribution in the south, central and eastern part of the distribution of the genus. It is a steppical (Irano-Turanian) element, penetrating into the desert (Saharo-Arabian) region. It is the southern-most species at many sites. In various parts of its distribution, it grows sympatrically with *Ae. longissima*, one of the putative donors of the S subgenome. In other parts of its distribution, *Ae. kotschyi* has mainly allopatric contact with *Ae. umbellulata*, the donor of its second subgenome. As its putative parents, *Ae. longissima* and *Ae. umbellulata*, have sporadic contact in semi-steppical, steppical or sub-Mediterranean regions of Syria, it is possible that this region was the center of origin of *Ae. kotschyi*, from which it spread both south- and eastward. Currently, *Ae. kotschyi* grows sympatrically in some sites with *Ae. bicornis*, *Ae. longissima*, tetraploid *Ae. crassa*, and *Ae. vavilovii* and allopatrically with *Ae. searsii*, *Ae. caudata*, *Ae. umbellulata*, *Ae. geniculata*, *Ae. biuncialis*, *Ae. triuncialis*, *Ae. peregrina* and *Ae. tauschii*.

The Mediterranean *Ae. peregrina* and the more arid species, *Ae. kotschyi*, come in contact in the southern coastal plain of Israel, in the border area between the steppe of the Negev and the Mediterranean territory. In this border belt, stands of both species are commonly in contact, and hybrids and hybrid derivatives were readily detected in several such mixed stands (Zohary and Feldman 1962; Feldman 1963). Triploid hybrids and hybrid derivatives between *Ae. kotschyi* and *Ae. longissima* were found in several site in the northern Israeli Negev (Feldman M, unpublished), indicating the possibility of gene flow from a diploid species to the allotetraploid. Similarly, spontaneous, introgression and stabilization of a DNA sequence from *Ae. searsii* into *Ae. kotschyi was described by* Weissman et al. (2005).

Ae. kotschyi contains gene(s) conferring resistance to leaf and stripe rust (Marais et al. 2005) and several accessions of this allotetraploid were resistant to widely virulent races of stem rust (Scott et al. 2014). *Ae. kotschyi* accumulates high concentrations of iron and zinc in its grains and it is a promising source for increasing the amount of these chemical elements in the grains of common wheat (Chhuneja et al. 2006). It is assumed that *Ae. kotschyi* contains genes for drought, heat, and salt tolerance (Kimber and Feldman 1987).

9.8.9.2 Cytology, Cytogenetics and Evolution

Ae. kotschyi is an allotetraploid species (2n = 4x = 28). Its nuclear genome was designated $C^u C^u S^v S^v$ (Kihara 1963; Kihara and Tanaka 1970) and its plasmon genome as S^v (identical to the plasmon of *Ae. searsii*) (Tsunewaki 2009). Subgenome C^u is currently designated U and subgenome S^v is closely related to genome S^l of *Ae. longissima* and genome S^{sh} of *Ae. sharonensis* (Zhang et al. 1992; Dvorak 1998). Since the plasmon of *Ae. kotschyi* derived from the S^v donor, the genomic formula of this species is $S^v S^v UU$. Genome analyses reported the formation of about seven bivalents in the F_1 hybrid *Ae . kotschyi* x *Ae. umbellulata* (von Berg 1937; Kimber and Abu-Bakar 1981) and approximately seven bivalents in the hybrid between *Ae. longissima* and *Ae. kotschyi* (Kihara 1949). On the basis of these data, Kihara (1949) concluded that *Ae. kotschyi* contains the C^u (=U) subgenome from *Ae. umbellulata* and the S subgenome from a *Sitopsis* species and consequently, designated the genome of *Ae. kotschyi* as $C^u C^u S^v S^v$. While several authors (Talbert et al. 1991; Zhang et al. 1992; Friebe et al.1996) considered *Ae. longissima*, or the immediate precursor of *Ae. longissima*, and *Ae. sharonensis* (Zhang et al. 1992) as the source of the S subgenome of *Ae. kotschyi*, Kihara (1949) noted that the genome of *Ae. longissima* was not truly homologous to the S subgenome of this allotetraploid. Similarly, Feldman (1963) observed 5.48 univalents, 5.82 bivalents, 0.92 trivalents and 0.28 quadrivalents in *Ae. kotschyi* x *Ae. longissima*, showing that the S^v subgenome of *Ae. kotschyi* differs from the S^l genome of *Ae. longissima* by two reciprocal translocations. Therefore, the donor species of the S genome in *Ae. kotschyi* still remains uncertain (Kimber and Feldman 1987).

Sears (1941b) produced an amphidiploid (2n = 4x = 28; genome $S^{sh} S^{sh} UU$) from the cross of *Ae. sharonensis* and *Ae. umbellulata*. This amphidiploid was not very similar in its morphology neither to *Ae. kotschyi* nor to *Ae. peregrina* (Kihara 1954). Tanaka (1955b) produced another amphidiploid from the reciprocal cross, *Ae. umbellulata* x *Ae. sharonensis*. He then produced hybrids from reciprocal crosses between the resulting amphidiploid and *Ae. peregrina*. Chromosome pairing at first meiotic metaphase of the F_1 hybrids revealed almost complete pairing, i.e., 9–14 (mode of 13) bivalents and two multivalents that were formed in some of the PMCs. Pollen and seed fertility was relatively high. Tanaka (1955b) concluded that *Ae. peregrina* and *Ae. kotschyi* must have arisen as allotetraploids from a cross between *Ae. umbellulata*, with the U-genome, and *Ae. sharonensis* with S^{sh}-genome. The amphidiploid resembles *Ae. kotschyi* in several morphological characters, but differed from it by other characters, such as in the number of spikelets per spike, awn shape, and others.

Rubenstein and Sallee (1973) analyzed chromosome pairing in hybrids between *Ae. kotschyi* and different pairing genotypes of *Ae. speltoides* as well as with *Ae. sharonensis* and their result supported the view that the second subgenome of *Ae. kotschyi* showed a high degree of chromosome pairing with the S^{sh} genome of *Ae. sharonensis*, and less so with that of *Ae. speltoides*.

Jaaska (1978a, b), studying the acid phosphatase isoenzyme (isophosphatase) variation in *Ae. kotschyi*, found that the isophosphatases characteristic of *Ae. umbellulata* are expressed in *Ae. kotschyi*, thus confirming the presence of the U subgenome in this allotetraploid species. In addition, the observations did not contradict the proposition that *Ae. kotschyi* contains the S^v subgenome that either derived from *Ae. bicornis*, *Ae. longissima*, or *Ae. sharonensis*. However, the isophosphatase data did not provide any further information on the similarity of the S^v subgenome of *Ae. kotschyi* to one of the genomes of these diploids. On the other hand, the data clearly argue against the presumed involvement of the genome of *Ae. speltoides* in *Ae. kotschyi*, since the former species is distinct from *Ae. bicornis*, *Ae. longissima, and Ae. sharonensis*, by having isophosphatase of higher electrophoretic mobility not encountered in *Ae. kotschyi*.

Kimber and Yen (1988) and Yen and Kimber (1990d) used a new cytogenetic approach to examine the origin of the U and S^v subgenomes of *Ae. kotschyi*. Kimber and Yen (1988) studied chromosome pairing in F_1 hybrids between *Ae. kotschyi* and an induced autotetraploid line of *Ae. umbellulata* (hybrid genome $S^v UUU$), The pattern of pairing in this tetraploid hybrid best fit the 3:1 model, indicating that the U subgenome of *Ae. kotschyi* paired fairly well with the two U genomes of the *Ae. umbellulata* autotetraploid. Thus, they concluded that the U subgenome of *Ae. kotschyi* is homologous to the U genome of *Ae. umbellulata*, confirming that *Ae. kotschyi* contains a subgenome that derived from *Ae. umbellulata*. To determine the origin of the S^v subgenomes of *Ae. kotschyi*, Yen and Kimber (1990d) analyzed chromosome pairing in F_1 of the hybrid between *Ae. kotschyi* and an induced autotetraploid lines of three *Aegilops* species of section Sitopsis, namely, *Ae. longissima*, *Ae. speltoides*, and *Ae. bicornis*, as well as with the diploid *Ae. speltoides*. Their pairing data showed that the hybrid involving the autotetraploid *Ae. longissima* (hybrid genome $S^vUS^lS^l$) best fit the 3:1 model, i.e., subgenome S^v is homologous to S^l, while those involving the autotetraploid *Ae. speltoides* (hybrid genome S^vUSS) and autotetraploid *Ae. bicornis* (hybrid genome $S^vUS^bS^b$) best fit the 2:2 and 2:1:1 model, respectively. The triploid hybrid of *Ae. kotschyi* x diploid *Ae. speltoides* (hybrid genome S^vUS) best fit the 3:0 model. Thus, they concluded that although modifications occurred in the S^v subgenome of *Ae. kotschyi*, they were not extensive enough to clearly distinguish the S^v subgenome of *Ae. kotschyi* from the S^l genome of *Ae. longissima*. Consequently, Yen and Kimber (1990d) decided that the genome of *Ae. longissima* is more similar to the S^v subgenome of *Ae. kotschyi* than those of *Ae. speltoides* and *Ae. bicornis*. However, the absence of the remaining two Sitopsis species, *Ae. sharonensis* and *Ae. searsii*, from their study precluded determination of the exact source of the S^v subgenomes. Additional hybrids between *Ae. kotschyi* and an autotetraploid line of *Ae. searsii* need to be made to study their genomic relationship (Yen and Kimber 1990d).

The genomes of *Ae. longissima*, *Ae. sharonensis*, *Ae. bicornis*, and *Ae. searsii*, are very close to one another, rendering it difficult to identify the diploid donor of the S^v subgenome. To resolve this problem, Talbert et al. (1991) used repeated nucleotide sequences to investigate the four species, but failed to distinguish the genomes of these four diploids from one another because only two repeated nucleotide sequences were used in their study.

To reinvestigate the origin of the two subgenomes of the closely-related allotetraploid species, *Ae. kotschyi* and *Ae. peregrina*, and the extent of their genome modification, Zhang et al. (1992) employed a developed phylogenetic technique (Dvorak and Zhang 1990; Zhang and Dvorak 1991) based on variation in repeated nucleotide sequences (Dvorak et al. 1988). They made use of twenty-seven randomly selected clones of repeated nucleotide sequences and one 5S rRNA gene clone to identify diagnostic bands and diagnostic hybridization intensities in the restriction profiles of the diploid *Aegilops* and *Triticum* species. The presence of each diagnostic band was then determined in *Ae. kotschyi* and in its closely related species *Ae. peregrina*. One subgenome in both allotetraploid species was found to be almost identical to the U genome of *Ae. umbellulata* and the other to the genome of *Ae. longissima*, *Ae. sharonensis*, or the internode in the phylogenetic tree of *Aegilops/Triticum*, immediately preceding the divergence of *Ae. longissima* and *Ae. sharonensis*. Twenty diagnostic bands were

found to characterize the internode, 19 of which were encountered in *Ae. kotschyi* and *Ae. peregrina*, indicating that the S^v subgenome of the two allotetraploids was contributed by the *Ae. longissima—Ae. sharonensis* evolutionary lineage. Since the shared number of the diagnostic bands between the remaining Sitopsis species and the allotetraploids was close to zero, it is improbable that any of them contributed to the second subgenome of *Ae. kotschyi* and *Ae. peregrina*. Taken together, since *Ae. kotschyi* and *Ae. peregrina* originated from hybridization between *Ae. umbellulata* and a species in the lineage of *Ae. longissima* and *Ae. sharonensis*, the genome formulae of the two allotetraploids should be S^lS^lUU or $S^{sh}S^{sh}UU$ (Zhang et al. 1992). However, these conclusions disagree with those reached following studies of the cpDNA and mtDNA of these species (Tsunewaki and Ogihara 1983; Ogihara and Tsunewaki 1988; Siregar et al. 1988; Wang et al. 1997; Tsunewaki 2009), which indicated that the cytoplasms of both allotetraploids were contributed by *Ae. searsii*. Whether this conflict indicates introgression of the cytoplasm from *Ae. searsii* into *Ae. kotschyi* and *Ae. peregrina* or some other causes is not known and requires further investigation. Until the unequivocal identification of the donor of the second subgenome of *Ae. kotschyi* and *Ae. peregrina* this subgenome will be designated S^v.

Both *Ae. kotschyi* and *Ae. peregrina* have the same S^v type cytoplasm and are thought to have received their cytoplasm from a genome donor with this type cytoplasm (Mukai and Tsunewaki 1975; Tsunewaki et al. 1978). The genetic characteristics of *Ae. kotschyi* and *Ae. variabilis* cytoplasms are identical, and differ greatly from those of *Ae. umbellulata*, and more or less resemble those of *Ae. speltoides*, *Ae. longissima* and *Ae. bicornis* (the cytoplasm of *Ae. searsii* was not studied (Mukai and Tsunewaki 1975; Tsunewaki et al. 1978; Tsunewaki 1980)). In an attempt to identify the donor of the cytoplasm to the *Ae. kotschyi-Ae. peregrina* complex, the restriction fragment patterns of chloroplast DNAs of the two allotetraploids were compared with those of several diploids of section Sitopsis (Tsunewaki and Ogihara 1983). The large subunit peptide composition of the chloroplast fraction I protein of *Ae. kotschyi* and *Ae. peregrina* is identical with that of *Ae. bicornis*, *Ae. sharonensis* and *Ae. searsii* (Hirai and Tsunewaki 1981), and the ctDNA restriction patterns of the former are identical to those of *Ae. bicornis* and *Ae. searsii* (Ogihara and Tsunewaki 1982; Tsunewaki and Ogihara 1983). Thus, Tsunewaki and Ogihara (1983) suggested that the cytoplasm donor of *Ae. kotschyi* and *Ae. peregrina* is *Ae. bicornis* or *Ae. searsii*. But, Tsunewaki and Ogihara (1983) also pointed out that the cytoplasm of *Ae. bicornis* differs from those of *Ae. kotschyi* and *Ae. peregrina* in the following two respects: first, *Ae. kotschyi-Ae. peregrina* cytoplasm induced complete male sterility in three of the 12 alloplasmic common wheats tested, whereas *Ae. bicornis* cytoplasm did not, and, secondly, the cytoplasm of *Ae. kotschyi* and *Ae. peregrina* induced haploids in the common wheat cultivar Salmon, but the *Ae. bicornis* cytoplasm did not. Obviously, *Ae. bicornis* has different cytoplasm DNA than *Ae. peregrina* and *Ae. kotschyi*, although Tsunewaki and Ogihara (1983) did not reveal any differences between the ctDNAs of these allotetraploids and that of *Ae. bicornis*. This might mean that their mitochondrial DNAs differ (Tsunewaki and Ogihara 1983).

To more accurately detect the diploid donor of the cytoplasm to *Ae. kotschyi* and *Ae. peregrina*, Terachi and Tsunewaki (1986), analyzing restriction fragment patterns of mitochondrial (mt) DNA isolated from *Ae. kotschyi*, *Ae. peregrina*, *Ae. bicornis*, and *Ae. searsii*, revealed that whereas the mitochondrial genomes of *Ae. kotschyi* and *Ae. peregrina* was identical to that of *Ae. searsii*, that of *Ae. bicornis* was somewhat different. Siregar et al. (1988) used alloplasmic lines of common wheat containing cytoplasm of several Sitopsis species, to compare the effect of the cytoplasm of *Ae. bicornis*, *Ae. searsii* and *Ae. sharonensis* to those of *Ae. kotschyi* and *Ae. peregrina*, on the fertility spectrum, haploid and twin induction and restriction fragment patterns of chloroplast and mitochondrial DNAs. In all these respects, the cytoplasm of *Ae. searsii* most closely resembled the cytoplasm of *Ae. kotschyi* and *Ae. peregrina*. Similar results were obtained by Terachi and Tsunewaki (1992), who used RFLPs among mtDNA digests of *Triticum* and *Aegilops* species that were analyzed by Southern blot hybridization with four cloned mitochondrial genes, as probes. With all of the probes used, mtDNA from *Ae. searsii* gave profiles identical to those of *Ae. kotschyi*. Thus, *Ae. searsii* was proposed as the cytoplasm donor to these two-allotetraploid species (Terachi and Tsunewaki 1986, 1992; Siregar et al. 1988; Tsunewaki 2009). Similarly, Wang et al. (1997), analyzing PCR–single-strand conformational polymorphism (PCR-SSCP) of 14.0-kb chloroplast (ct) and 13.7-kb mitochondrial mtDNA regions that were isolated from 46 alloplasmic wheat lines, concluded that *Ae. searsii* was the maternal ancestor of *Ae. kotschyi* and *Ae. peregrina*.

Due to the presence of three chloroplast haplotypes, two of which are shared with S-genome diploids and one with *Ae. umbellulata*, Meimberg et al. (2009) suggested that *Ae. kotschyi* evolved through three independent hybridizations between the donor of the S^v subgenome and that of the U subgenome.

In order to more deeply understand the cytogenetic relationships between the two-closely related allotetraploid species, *Ae. peregrina* and *Ae. kotschyi*, Feldman (1963), Furuta (1981b) and Cuñado (1993b) studied chromosome pairing in F$_1$ hybrids between these two allotetraploid species. These studies reported the presence of almost complete pairing in these hybrids, i.e., 0.8–1.0 univalents, 11–12

bivalents and 1–2 multivalents. Evidently, the genome of *Ae. peregrina* differs, by at least two reciprocal translocations from that of *Ae. kotschyi*. It was pointed out by Furuta (1981a) that the genome of *Ae. kotschyi* significantly differs from that of *Ae. peregrina*. Pairing among chromosomes of the U subgenomes in *Ae. peregrina* x *Ae. kotschyi* hybrids was significantly higher than between chromosomes of the S^v subgenomes (Cuñado 1993b). Thus, the differences between *peregrina* and *kotschyi* could be exclusively attributed to the S^v subgenomes (Cuñado 1993b).

Based on genome analysis, Kihara (1937, 1949, 1954) concluded that the S^v subgenome of *Ae. kotschyi* is modified relative to those of the diploid species of A*egilops* section Sitopsis and, accordingly, designated it S^V. To explain the occurrence of modified subgenomes in most allopolyploid *Aegilops* species, Kihara (1954) suggested that either presently extinct species were the donors of the modified subgenomes or that the subgenomes were significantly rearranged during formation and evolution of the allopolyploids. The degree of S^v-subgenome modification in *Ae. kotschyi* was estimated differently in different studies; some authors reported that it is modified (Kihara 1940a, 1946, 1949, 1954; Chennaveeraiah 1960; Zohary and Feldman 1962; Kimber and Yen 1989), while others observed only minor differences compared to that of *Ae. longissima* (Zhang et al. 1992; Friebe et al. 1996).

In an attempt to explain the evolution of allopolyploid species of the wheat (*Aegilops-Triticum*) group, Zohary and. Feldman (1962) hypothesized that interspecific hybridization between allotetraploid species sharing one subgenome and differing in the second subgenome, also led to some pairing between the differential subgenomes, resulting in introgression and subsequently to modification of these differential genomes. The results of Zhang et al. (1992) showed no difference in divergence between the U subgenome of *Ae. kotschyi* and the genome of *Ae. umbellulata*, and only a minor difference was observed between the S^v subgenome of the allotetraploid and the genome of the *Ae. longissima–Ae. sharonensis* lineage. One of nine bands diagnostic for *Ae. bicornis* and 1 of 15 bands diagnostic for *Ae. searsii*, were detected in *Ae. kotschyi*. Whether these bands introgressed into the allotetraploid or their absence in *Ae. longissima* and *Ae. sharonensis* reflects evolution of the *Ae. longissima— Ae. sharonensis* lineage after the origin of the allotetraploids, remains to be determined (Zhang et al. 1992). The question of modification of the S^v subgenome by chromosome introgression (Zohary and Feldman 1962) may require further investigation, since the introgression of small chromosome segments might have escaped detection in the work of Zhang et al. (1992).

Following C-banding analysis, Badaeva et al. (2004) reported that the U subgenome in *Ae. kotschyi* is similar to the U subgenomes of all other polyploids bearing the U subgenome and to that of the parental diploid species *Ae. umbellulata*. On the other hand, the second subgenome of these allopolyploid species were modified compared to the donor chromosomes, as was suggested by Kihara (1954, 1963), Chennaveeraiah (1960), Kimber and Abu-Bakar (1981), Kimber and Zhao (1983), Kimber and Feldman (1987), and Kimber and Yen (1989), with the extent of subgenome modification varying between species. Similarly, Gong et al. (2006), using 31 inter-simple sequence repeat (ISSR) primers, compared the genome of *Ae. kotschyi* with those of its ancestral diploids. Based on the genetic similarity index between *Ae. kotschyi* and *Ae. umbellulata*, they reported that the allotetraploid contains one subgenome similar to the U genome of *Ae. umbellulata*, while the second subgenome was altered greatly in the allotetraploid.

Feldman (1963), Kawahara (1986, 1988), and Fernández-Calvín and Orellana (1991) found that most interchanges between *Ae. kotschyi* and *Ae. peregrina* involved chromosomes of the S^v subgenome. This suggests that genome rearrangement occurs more frequently in the S^v subgenome than in the U subgenome. Kawahara (1986, 1988) proposed that the modified genome of *Ae. kotschyi* probably evolved through its high structural variability rather than through introgression with other species. Yet, Furuta and Tanaka (1970) concluded from their experimental hybridization between *Ae. peregrina* and *Ae. columnaris*, that it may have been exchange of genetic material between the differential subgenomes of these allotetraploids, namely, S^v and X^n, that modified them. Thus, introgression occurring between species sharing one subgenome and differing in the second subgenome can lead to formation of recombinant (modified) subgenomes. Since the S^v subgenomes of *Ae. kotschyi* and *Ae. peregrina* are close to each other, it is more difficult to obtain cytological and molecular evidence for modification of their S^v subgenomes through introgression, although morphological evidence indicated that such introgression indeed takes place (Feldman 1963). Assuming that the S^v subgenome *of Ae. kotschyi* derived from *Ae. sharonensis* and that of *Ae. peregrina* from *Ae. longissima*, as was suggested by Badaeva et al. (2004), the latter differs from the former diploid parent by a translocation (Ankori and Zohary 1962), or alternatively, one of the allopolyploids originated as a result of introgressive hybridization with *Ae. searsii* (Tsunewaki 2009 and references therein), which differs from *Ae. longissima* by a translocation (Feldman et al. 1979), such courses of events may explain the changes in the S^v subgenome of these allotetraploids.

Ae. kotschyi contains 12.64 ± 0.183 pg 1C DNA (Eilam et al. 2008), the largest genome among the allotetraploid species of *Aegilops* (Eilam et al. 2008). The 1C DNA content of *Ae. kotschyi* is 1.71% smaller than the DNA expected from the sum of the DNA of its two putative diploid parents, i.e., 12.86 pg (*Ae. umbellulata*, the donor of the U subgenome, contains 5.38 pg and *Ae. longissima*, the donor of the S^v subgenome, contains 7.48 pg; Eilam et al. 2007). If assuming that the S^v subgenome donor is *Ae. sharonensis*, then the 1C DNA content of *Ae. kotschyi* is 2.02% smaller than the DNA expected from the sum of the DNA of its two diploid parents, i.e., 12.90 pg (*Ae. umbellulata*, the donor of the U subgenome, contains 5.38 pg and *Ae. sharonensis*, the donor of the S subgenome, contains 7.52 pg; Eilam et al. 2007).

The karyotype figured by Senyaninova-Korchagina (1930, 1932) for *Ae. triuncialis* subsp. *kotschyi* Boiss. (Currently *Ae. kotschyi*) was similar to the one she figured for *Ae. peregrina*. In genome analytical studies, Kihara (1954, 1957) included *Ae. kotschyi* under *Ae. peregrina*, perhaps, partly due to the karyotypic results of Senyaninova-Korchagina (1930, 1932). However, the karyotypes studied by Chennaveeraiah (1960) revealed considerable differences between the karyotypes of *Ae. kotschyi* and *Ae. peregrina*. *Ae. kotschyi* has only two pairs with satellites on short arms whereas *Ae. peregrine* has three. The rest of the pairs in *Ae. kotschyi* consist of one with an extreme subterminal centromere, one with a median centromere, one with an almost median centromere, and the others with submedian centromeres. Thus, in all, there are 12 types of chromosomes in *Ae. kotschyi* (Chennaveeraiah 1960).

One half of the chromosomes of *Ae. kotschyi* corresponds to one half of the chromosomes of *Ae. peregrina*, which all correspond to the U genome of *Ae. umbellulata* (Chennaveeraiah 1960). Distinct differences in the second half of chromosomes in the two species were noted (Chennaveeraiah 1960). The second set in *Ae. kotschyi* consisted of chromosomes with only primary constrictions which were median or submedian. There were no pairs with satellite or secondary constriction in th S^v subgenome of *Ae. kotschyi*, whereas *Ae. peregrina* did have a pair with satellite in the second set of chromosomes. There was no chromosome with a median centromere in the entire set of *Ae. peregrina*, whereas *Ae. kotschyi* had one such pair. The chiasma frequency in meiotic first metaphase was also different in the two species; more rod bivalents formed in *Ae. peregrina* than in *Ae. kotschyi* (Chennaveeraiah 1960). Judging from karyotypic observations, Chennaveeraiah (1960) proposed that the second subgenome of *Ae. kotschyi* is similar to the M-genomes of species of section Comopyrum. It could even be foreign to *Aegilops* but never the same as the second subgenome in *Ae. peregrina*. Thus, also the study of Chennaveeraiah (1960) revealed that *Ae. kotschyi* and *Ae. peregrina* are separate species.

Like Chennaveeraiah (1960), Al-Mashhadani et al. (1980) also observed two satellite chromosome pairs in *Ae. kotschyi* from Iraq. The satellites in one pair were larger than in the other. In addition to the SAT chromosomes, five metacentric chromosome pairs, four submetacentric pairs and three subtelocentric pairs were observed. These observations are quite similar to those of Tanaka and Matsumoto (1965), and clearly differ from those of Chennaveeraiah (1960).

Badaeva et al. (2004) investigated the heterochromatin structure in the karyotype of *Ae. kotschyi* using C-banding and FISH with the heterochromatin-specific DNA probes pSc119 and pAs1 (non-coding, highly repeated DNA sequences), as well as of rDNA loci, using pTa71 (18S-26S rDNA) and pTa794 (5S rDNA) probes. Ten accessions of *Ae. kotschyi* were examined by C-banding and one accession was analyzed by FISH. Although *Ae. kotschyi* has the same genome formula as *Ae. peregrina*, the C-banding patterns were different. Comparison of *Ae. kotschyi* with the diploid ancestors *Ae. umbellulata* and *Ae. sharonensis* revealed a higher degree of S^v subgenome modification in *Ae. kotschyi* than in *Ae. peregrina*. Chromosomes 4 Sv and 7 Sv of *Ae. peregrina* are nearly identical to $4S^l$ and $7S^l$ of *Ae. longissima*, whereas these chromosomes of *Ae. kotschyi* are similar to $4S^{sh}$ and $7S^{sh}$ of *Ae. sharonensis*. Previous studies performing isozyme (Hart and Tuleen1983) and C-banding analyses (Friebe et al. 1993, 1996; Friebe and Gill 1996) revealed that chromosomes $4S^l$ and $7S^l$ of *Ae. longissima* are involved in a species-specific reciprocal translocation that is absent in *Ae. sharonensis*. The presence of this translocation in *Ae. peregrina* and its absence in *Ae. kotschyi* suggests that *Ae. peregrina* originated from the hybridization of *Ae. umbellulata* and *Ae. longissima* and that *Ae. kotschyi* originated from the hybridization of *Ae. umbellulata* and *Ae. sharonensis* (Badaeva et al. 2004). Thus, the C-banding analysis showed that the S^v subgenome of *Ae. kotschyi* was derived from either the S^{sh} genome of *Ae. sharonensis* or its immediate precursor. The high frequency of chromosomal aberrations and reduction in the number and size of 18S-26S rDNA loci observed in the S^v subgenome compared to the S^{sh} and S^l of *Ae. peregrina*, suggested that *Ae. kotschyi* is an older species than *Ae. peregrina*.

FISH with clone pSC119 revealed signals of various sizes in telosomic regions of either the short or both arms of 12 chromosome pairs of *Ae. kotschyi* (Badaeva et al. 2004). Interstitial pSc119 FISH sites were detected in the long arm of chromosome 7U, while chromosome 6U had no such sites; distinct telomeric and interstitial FISH sites were present in *Ae. umbellulata* (Badaeva et al. 1996a). However, in the allotetraploid species, chromosome 6U has telomeric and interstitial pAs1 FISH sites that are absent in *Ae. umbellulata*.

Similar to Chennaveeraiah (1960) and Al-Mashhadani et al. (1980), Badaeva et al. (2004) found two pairs of satellite chromosomes in *Ae. kotschyi*, which coincides with the number of active NORs detected by Ag-NOR staining (Cermeno et al. 1984b) and in situ hybridization (Yamamoto

1992a). Like in *Ae. peregrina*, the NORs on *Ae. kotschyi* S^v-subgenome chromosomes were inactivated and accompanied with a decrease or loss of rDNA sequences. Consequently, the major NORs were observed on group 1U and 5U chromosomes. *Ae. kotschyi* also have several minor 18S-26S rDNA sites. Two consistent minor loci detected on group 5 Sv and 6 Sv chromosomes were associated with a significant reduction in copy number of 18S-26S rRNA genes. The hybridization patterns with pTa794 revealed four similar-sized 5S rDNA loci that were located on chromosomes of homoeologous groups 1 and 5.

9.8.9.3 Crosses with Other Species of the Wheat Group

Chennaveeraiah (1960) reported that meiosis is very regular in *Ae. kotschyi* exhibiting 14 bivalents at first meiotic metaphase of every PMC. Likewise, Feldman (1963) reported that all PMCs at first meiotic metaphase of *Ae. kotschyi* had 14 bivalents and 29.40 chiasmata/cell. Chromosome pairing in triploid hybrids between *Ae. kotschyi* and *Ae. umbellulata* indicated that the former contains one subgenome that is homologous to the genome of the latter (von Berg 1937; Kimber and Abu-Bakar 1981). Kimber and Abu-Bakar reported the presence of 7.04 univalents, 2.90 rod bivalents, 3.66 ring bivalents, 6.56 total number of bivalents and 0.29 trivalents (Table 9.5). Kihara (1949) reported 7 bivalents, five of which were rings, and up to one multivalent per cell in the triploid hybrid *Ae. longissima* x *Ae. kotschyi*. However, in the footnote to his Table 9.5, Kihara indicated that the genome of *Ae. longissima* was not truly homologous to the S^v subgenome of *Ae. kotschyi*. Likewise, Feldman (1963) studied chromosome pairing in the F_1 triploid hybrid *Ae. kotschyi* x *Ae. longissima* and observed 5.48 univalents, 5.82 bivalents, (of which 1.52 were heteromorphic), 0.92 (0–2) trivalents, 0.28 (0–1) quadrivalents and 11.92 chiasmata/cell. This pattern of chromosome pairing indicates the existence of homology between one subgenome of *Ae. kotschyi* and the S^l genome of *Ae. longissima*, although some structural differences (two reciprocal translocations) exist between the two species. On the other hand, chromosome pairing in the triploid hybrid *Ae. kotschyi* x *Ae. caudata* (genome S^vUC) had 10.70 univalents, 3.45 rod bivalents, 0.35 ring bivalents (3.80 total number of bivalents) and 0.90 trivalents (Kimber and Abu-Bakar 1981), indicating that the genome of *Ae. caudata* is homoeologous to the subgenomes of the allotetraploid.

Hybrids between *Ae. kotschyi* and both *Ae. speltoides* and *Ae. sharonensis* were produced and examined cytologically (Table 9.5). Chromosome pairing in the F_1 triploid hybrids between *Ae. kotschyi* and different pairing types of *Ae. speltoides* (genome S^vUS) reveals that the allotetraploid contains one genome that is related to the genome of the diploid. Interestingly, the high pairing and the intermediate pairing genotypes of *Ae. speltoides* did not promote pairing in the above hybrids (Rubenstein and Sallee 1973). Chromosome pairing in the F_1 hybrid *Ae. kotschyi* x *Ae. sharonensis* was somewhat higher than in the hybrid *kotschyi* x *speltoides*, confirming the presence of a subgenome in *Ae. kotschyi* that is homologous to the genome of *Ae. sharonensis*, but which differs from it by at least two reciprocal translocations.

Chromosome pairing in the F_1 tetraploid hybrid between several lines of the two subspecies of *Ae. peregrina* and *Ae. kotschyi*, was studied by Feldman (1963). The hybrids involving lines of *Ae. peregrina* ssp. *peregrina* and *Ae. kotschyi* had 0.84–0.94 univalents, 10.98–11.28 bivalents, 0.14–0.20 trivalents, 0.26–0.54 quadrivalents, 0.14–0.18 pentavalents, and 0.24–0.36 hexavalents. Hybrids involving lines of *Ae. peregrina* ssp. *cylindrostachys* and *Ae. kotschyi* had 0.84–1.02 univalents, 11.42–11.72 bivalents, 0.18–0.24 trivalents, 0.18–0.68 quadrivalents, 0.02–0.30 pentavalents, and 0.06–0.20 hexavalents. Evidently, the genomes of both subspecies of *Ae. peregrina* are homologous to that of *Ae. kotschyi*, but differ from it by, at least, two reciprocal translocations.

Similarly, Furuta (1981b) crossed 21 lines of *Ae. kotschyi*, collected in Egypt, Jordan, and Syria, with a line of *Ae. peregrina* that served as the common pollen parent. Chromosome pairing in the F_1 hybrids also displayed the presence of one or two reciprocal translocations between these two species. Furuta (1981b) studies revealed that variable and continuous chromosomal differentiation occurred between the chromosomes of *Ae. peregrina* and *Ae. kotschyi*. Based on these results, Furuta (1981b) concluded that the genome of *Ae. kotschyi* significantly differs from that of *Ae. peregrina*.

Comparison of chromosomal pairing of intraspecific hybrids in *Ae. peregrina* with hybrids between *Ae. peregrina* and *Ae. kotschyi*, enabled assessment of whether the U and S^v subgenomes had been altered during their evolution in one or both species (Cuñado 1993b). Chromosome pairing in the intraspecific hybrid *peregrina* x *peregrina* showed 0.18 univalents, 4.32 rod bivalents, 9.56 ring bivalents, and 23.44 associations/cell. Pairing in the U subgenome included 0.02 univalents, 2.14 rod bivalents, 4.84 ring bivalents, and 11.83 associations/cell, and pairing in the S^v subgenome showed 0.16 univalents, 2.18 rod bivalents and 4.72 ring bivalents (11.61 associations/cell (Cuñado 1993b). The number of associations/cell in the intraspecific hybrid of *peregrina* was lower than that of the parental lines *Ae. peregrina* (23.44 vs. 25.30; Cuñado 1993b). This decrease in association frequency cannot be attributed to structural changes since multivalents were not observed (Cuñado 1993b). Chromosome pairing in the hybrid *peregrina* x *kotschyi* included 0.99 univalents, 3.67 rod bivalents, 8.43

ring bivalents 0.47 trivalents, 0.35 quadrivalents, and 21.71 associations/cell. Pairing in the U subgenome showed 0.04 univalents, 2.05 rod bivalents, 4.93 ring bivalents, and 11.93 associations/cell, and pairing in the S^v subgenome had 0.95 univalents, 1.62 rod bivalents, 3.50 ring bivalents, 0.47 trivalents, 0.35 quadrivalents, and 9.78 associations/cell. Evidently, the S^v subgenome of the two species differs in a reciprocal translocation. These results are in agreement with those reported by Furuta (1981b) and Feldman (1963), although Furuta and Feldman found a higher number of interchanges in some hybrids involving other lines. The mean number of associations/cell between chromosomes of the U subgenomes were similar in the two hybrids. However, the frequencies of chromosome associations of the S^v subgenomes differed significantly in the *peregrina* x *kotschyi* as compared to the intraspecific hybrid. Thus, the differences between *peregrina* and *kotschyi* could be exclusively attributed to the S^v subgenomes (Cuñado 1993b).

Chromosome paring in tetraploid hybrids between *Ae. kotschyi* and several allotetraploid species bearing the U subgenome were studied by a number of groups. The hybrid *Ae. geniculata* x *Ae. kotschyi* (genome M^oUS^vU) had 11.70 univalents, 6.02 bivalents (of which 1.76 were heteromorphic) and 1.42 trivalents (Feldman 1963), while others reported 10.65 univalents, 3.20 rod bivalents, 3.25 ring bivalents, 6.45 total number of bivalents, 1.35 trivalents and 0.10 quadrivalents (Kimber et al. 1988). The hybrid *Ae. columnaris* x *Ae. kotschyi* (genome UX^nS^vU) had 7–10 bivalents with mode of 9, 0–3 trivalents and 0–2 quadrivalents (Kihara 1949). The hybrid *Ae. neglecta* x *Ae. kotschyi* (genome UX^nS^vU) had 7–10 bivalents with mode of 10, and 0–2 trivalents (Kihara 1949). The hybrid *Ae. kotschyi* x *Ae. triuncialis* (genome S^vUUC) had 6–9 bivalents (Kihara 1937). In all these hybrids, there were more than seven bivalents, indicating that, in addition to homologous pairing between the shared U subgenomes, homoeologous pairing also occurred between some chromosomes of the differential subgenomes.

Chromosome pairing in the hybrid common wheat x *Ae. kotschyi* (genome $ABDS^vU$) included 30.92 univalents, 1.98 rod bivalents, 0.02 ring bivalents, and 0.03 trivalents (Fernández-Calvín and Orellana 1991). Of the rod bivalents, 0.54 were between A-D chromosomes, 0.74 between AD-US^v chromosomes, 0.05 between AD-B chromosomes, 0.38 between US^v-B chromosomes and 0.25 between U-S^v chromosomes. The data indicated that more than 40% of the pairing in these three hybrids was autosyndetic (A-D + AD-B = 25%, and U-S^v = 16%) and that more than 50% was allosyndetic (AD-US^v + US^v-B = 51%) (Fernández-Calvín and Orellana 1991). The allosyndetic pairing between AD chromosomes and US^v chromosomes was twice as high as that between US^v and B chromosomes.

9.8.10 *Aegilops peregrina* (Hackel) Maire & Weiller

9.8.10.1 Morphological and Geographical Notes

Ae. peregrina (Hackel) Maire & Weiller [Syn.: *Ae. variabilis* Eig; *Ae. peregrina* (Hack. In Fraser) Eig; *T. peregrinum* Hack. In J. Fraser; *Aegilemma peregrina* (Hack.) Á. Löve] is a predominantly autogamous, annual, multi–tillered, 15–40-cm-tall (excluding spikes) plant. It branched and prostrate near the ground but upright at its upper parts. Its leaves are hairy or glabrous and spike is broad oval, linear to cylindrical, 1.2–7.5-cm-long (excluding awns), disarticulating entirely at maturity (umbrella-type dispersal unit), and usually awned. There are 2–4 (usually 3, exceptionally 1) basal rudimentary spikelets and 2–7, (usually 3–5), urn-shaped to elliptical spikelets, not appressed to the rachis or to each other. The spikelets become smaller to the tip of the spike, with the uppermost spikelet seldom becoming suddenly smaller. There are 3–6, usually 4–5 florets, with the upper 1–3 being sterile. The glume is usually tough and rough, 6–8-mm–long and 4–6-mm-wide, with weak, narrow, parallel, and equally long and wide nerves. There are 3 glume awns of the terminal spikelet, and 2–3 awns on lateral spikelets, with the central one replaced by a tooth or gap when there are only two. The number of awns never exceeds 3. Glume awns are narrow and flat at the base, strongly polymorphic and variable in number, width and length, and spread out at maturity. They are either equally broad, broader at the lower spikelets, or one is considerably broader (particularly in the lower spikelets). Their lengths are either equal, or differ (by 4–8 mm), or shorter in the middle as compared to its laterals. Lemma awns are weakly developed, and often missing, and when present, there is 1, very seldom 3, and always shorter than glume awns, often with more or less long teeth. The caryopsis adheres to the lemma and palea (Fig. 9.4i, j).

The species shows the most extraordinary variability (Fig. 9.5) and was therefore called *Ae. variabilis* by Eig (1929a), although he was perfectly aware that Hackel had already given the name *Triticum peregrinum* to a specimen of this species which had been casually introduced into Scotland. Eig preferred to give a new name to the species because Hackel's specimen was atypical, calling for abandonment of the specific epithet. Maire and Weiller (1955) seem to be the first authors to have validly published the combination of *Ae. peregrina*, and transferred most of the intraspecific taxa of Eig from the species epithet *variabilis* Eig to the correct one, *peregrina* Hack.

Eig (1929a) described two subspecies in this species, *eu-variabilis* (containing 7 varieties) and *cylindrostachys* (containing 3 varieties). Hammer (1980), accepting the transfer of Eig's intraspecific taxa from *variabilis* to *peregrina*, designated Eig's two subspecies as *peregrina* and *cylindrostachys* (Eig et Feinbrun) Hammer. In contrast to Eig and Hammer, van Slageren (1994) ranked the intraspecific taxa of *Ae. peregrina* as varieties: var. *peregrina* and var. *brachyathera*. The sub-classification of Hammer (1980) into subspecies will be used in this book.

Fig. 9.5 Spikes of *Ae. peregrina* (=*Ae. variabilis* Eig) representing portion of the morphological variation of this species (From Feldman 1963)

The description of the two subspecies is as follows: Ssp. *peregrina* (*Ae. variabilis* ssp. *eu-variabilis* Eig & Feinbr.; Fig. 9.4i) has 1.5–4-cm–long (excl. awns), broadly ovate to lanceolate spikes, with 3–4 spikelets. The lower spikelets are longer than the adjacent rachis internodes. The glumes have 2–3 awns, with terminal spikelet glumes bearing three awns. The lemmas of all spikelets have 1–2 awns of notably uneven length (0.3–3-cm), flanked by 1–2 teeth. This subspecies is most variable in awn length. Eig (1929a) described seven varieties within this subspecies. Ssp. *cylindrostachys* (Eig & Feinbr.) Hammer (*Ae. variabilis* Eig ssp. *cylindrostachys* Eig & Feinbr.; Fig. 9.4j) has 3.5–7.5-cm–long (excl. awns), narrowly lanceolate to linear spikes, with (3-)5(-7) spikelets. The glumes of the lateral spikelets feature 2–3 sharp teeth, 1 or 2 of which may develop into a short awn (up to ± 7 mm-long, increasing to 1.5 mm subapically). The glumes of the apical spikelet bear 1–3, 1–3-cm-long awns. In the case of only one, the awn is flanked by acute teeth of up to 6-mm in length. All spikelet lemmas have 2–3 teeth. The rachis internodes are generally about as long as those of the adjacent spikelets. Awned variants are rare, and, when present, the awns on glumes are short whereas awns on lemmas are present only in the terminal spikelet. Eig (1929a) described three varieties within this subspecies. Subsp. *cylindrostachys* is less common than subsp. *peregrina*.

Ae. peregrina also exhibits wide variation at the biochemical and molecular levels. Nakai and Tsunewaki (1971) found variations in the zymograms of esterase isozymes in four *Ae. peregrina* accessions, analyzed using the gel electrofocusing method. Similarly, Nakai and Tsuji (1984), after examining four accessions of *Ae. peregrina* for acid phosphatase isozymes, using gel electrofocusing technique, reported on intraspecific variation of two variant phenotypes. Such intraspecific variation may be the result of adaptation to different environment conditions, as shown by Nevo et al. (1984), who electrophoretically analyzed allozymic diversity in two polymorphic esterase loci in 70 *Ae. peregrina* plants, collected from a microsite at Tabigha, north of the Sea of Galilee, Israel. The test involved a 100 m-transect, equally subdivided into basalt and terra-rossa soil types. Significant genetic differentiation across soil type was found over very short distances. The results suggested that allozyme polymorphisms in *Ae. peregrina* are adaptive and differentiate primarily by soil selection, probably through aridity stress.

RAPD analysis was used to study genetic variation and phylogenetic relationships among allopolyploid *Aegilops* species sharing the U-subgenome (Goryunova et al. 2010). In total, the group examined 115 DNA samples of eight allopolyploid species containing the U subgenome and of the diploid species *Ae. umbellulata* (genome U). Substantial interspecific polymorphism was observed in the majority of these allpolyploids. As with most other species, the 12 accessions of *Ae. peregrina* exhibited wide intraspecific variation. When establishing the phylogenetic relationships for the U-subgenome species, the authors noted the greatest separation within this group between the S^VU-subgenome species *Ae. peregrina and Ae. kotschyi*.

Ae. peregrina is a Mediterranean element, growing in Southern Italy (including Sicily), South Greece (including Crete and Rhodes), South Turkey, Iraq (lower Mesopotamia), Iran (northwest and south), Azerbaijan, Syria, Lebanon, Cyprus, Israel, Jordan, Egypt (lower), Libya, Tunisia, Algeria, and Morocco. In this region, *Ae. peregrina* thrives on a large variety of soils, in edges and openings of sclerophyllous oak forest, maquis, dwarf shrub formations, herbaceous formations, pastures abandoned fields, edges of cultivation, disturbed areas and roadsides. It grows at altitudes of 0–1600 m. The species is very common and locally abundant. It rapidly colonizes deserted fields as well as open, unstable, secondary habitats.

Ae. peregrina has a medium-sized distribution in the southwestern part of the distribution of the genus. Like *Ae. triuncialis* in the central and northern parts of the genus area, *Ae. peregrina* is the massive species in the southwestern part of the genus area. Its ecological amplitude is exceptionally large. It differs from its closely related species *Ae. kotschyi*, in that, in some areas, it grows sympatrically with both of its putative diploid parents, *Ae. umbellulata* and either *Ae. longissima and, Ae. sharonensis or Ae. searsii*. *Ae. peregrina* usually grows in mixed populations with other species, with which it introgresses. It may have contact with its two putative diploid parents in semi-steppical, steppical and sub-Mediterranean regions of Syria or Israel where *Ae. peregrina* might have originated and then spread southwards and westwards. Its distribution area is larger than those of its two putative parents.

Ae. peregrina grows sympatrically with *Ae. speltoides*, *Ae. sharonensis*, *Ae. longissima*, *Ae. searsii*, *Ae. caudata*, *Ae. umbellulata*, wild tetraploid wheat, *T. turgidum* ssp. *dicoccoides*, *Ae. geniculata*, *Ae. biuncialis* and *Ae. triuncialis*, and allopatrically with *Ae. bicornis*, *Ae. comosa*, the wild subspecies of *T. monococcum*, *Ae. urartu*, *Ae. neglecta*, *Ae. columnaris*, and *Ae. kotschyi*. *Ae. peregrina* forms mixed populations with *Ae. geniculata* and *Ae. biuncialis* in many parts of the Mediterranean phytogeographic region of Israel. In such populations, interspecific hybrids and hybrid derivatives between *Ae. peregrina* and the other two allotetraploids are quite common (Zohary and Feldman 1962; Feldman 1965a; Pazy and Zohary 1965). *Ae. peregrina* and its close relative *Ae. kotschyi* are vicarious species, the former growing in the Mediterranean phytogeographic region and the latter in semi-steppical and steppical (Irano-Turanian) regions. In the transition zone between southern Israel's Mediterranean and steppical regions where *Ae. peregrina* and *Ae. kotschyi* have massive contact, there are many indications for gene flow between these two species (Feldman 1963). In northern Israel, there are mixed populations of four allotetraploids, namely, *Ae. peregrina, Ae. geniculata, Ae. biuncialis* and *Ae. triuncialis*, with intermediates between *Ae. peregrina* and *Ae. triuncialis* found in several such populations (Zohary and Feldman 1962; Feldman 1965a). In addition, hybrids between *Ae. peregrina* and wild and domesticated tetraploid wheat, as well as domesticated hexaploid wheat, were repeatedly found (Percival 1921; Feldman M, unpublished). In this regard, Weissman et al. (2005) described a spontaneous DNA introgression from domesticated hexaploid wheat into *Ae. peregrina* and the stabilization of this introgression in wild populations. Vardi and Zohary (1967) described triploid hybrids and hybrid derivatives in mixed populations of tetraploid Ae. peregrina and diploid *Ae. sharonensis* or *Ae. longissima*, indicating the possibility of gene flow between these species.

Accessions of *Ae. peregrina* carry genes that confer resistance to powdery mildew (Gill et al. 1985; Spetsov et al. 1997), leaf rust (Gill et al. 1985; Marais et al. 2008), stem rust (Anikster et al. 2005; Scott et al. 2014) and strip rust (Anikster et al. 2005; Liu et al. 2011a, b; Zhao et al. 2016). Accessions of this species were also found resistant to Hessian fly (Gill et al. 1985), and green bug (Gill et al. 1985), as well as to cereal cyst nematode (Coriton et al. 2009) and to root-Knot nematodes (Yu et al. 1990; Coriton et al. 2009). Several accessions of *Ae. peregrina* were salt tolerant (Farooq et al.1989).

9.8.10.2 Cytology, Cytogenetics, and Evolution

Ae. peregrina is an allotetraploid species (2n = 4x = 28). Its nuclear genome is designated S^vS^vUU (modified from Dvorak 1998) and its plasmon genome as S^v (identical to S^s of *Ae. searsii*) (Tsunewaki 2009). Early genome analysis studies showed that *Ae. peregrina* has the U subgenome from *Ae. umbellulata* (von Berg 1937; Kihara 1940a; Kimber and Yen 1989), findings that were later substantiated by a karyological study (Chennaveeraiah 1960), cytogenetic studies (Yu and Jahier 1992), biochemical studies (Jaaska (1978a, b) and subsequent molecular studies (Zhang et al. 1992; Badaeva et al. 2004). Yet, in contrast to the consensus concerning the nature and origin of the U subgenome of *Ae. peregrina*, that of the second subgenome is still enigmatic. Chennaveeraiah (1960), based on karyotypic considerations, suggested that the donor of the second subgenome is a species of the M-genome group. From his cytogenetic studies, Kihara (1946, 1949) proposed that the second subgenome of *Ae. peregrina* was contributed by a species of the Sitopsis group, possibly *Ae. longissima*. This proposal was supported by Talbert et al. (1991) and Friebe et al. (1996), whereas Zhang et al. (1992) assumed that either *Ae.*

longissima, Ae. sharonensis or, more likely, the immediate precursor of these two-diploid species, was the donor of the S^v subgenome to *Ae. peregrina*. Badaeva et al. (2004), using C-banding method, revealed that *Ae. peregrina* contains a reciprocal translocation involving chromosomes 4 and 7 of the S subgenome, similar to the translocation that exists in *Ae. longissima*. Since this translocation does not present in *Ae. sharonensis*, they concluded that *Ae. longissima* was the source of the second subgenome of *Ae. peregrina*, and that *Ae. peregrina* arose after the separation of *Ae. longissima* from *Ae. sharonensis*. In contrast to the above, Ogihara and Tsunewaki (1988) and Siregar et al. (1988), studying restriction patterns of chloroplast DNA, proposed *Ae. searsii* as the cytoplasm donor to *Ae. peregrina*.

Sears (1941b) produced an *Ae umbellulata—Ae sharonensis* amphidiploid (2n = 4x = 28; genome $UUS^{sh}S^{sh}$), which was not similar to *Ae. peregrina*, while the F_1 of the hybrid *Ae. bicornis* x *Ae. umbellulata* (Genome S^bU) had a similar ear-form to *Ae. peregrina* (Kihara 1954). Tanaka (1955b) also produced an amphidiploid from the *Ae. sharonensis* x *Ae. umbellulata* cross, that reportedly resembled *Ae. kotschyi* in many morphological characters, but differed from *Ae. kotschyi* and *Ae. peregrina* with respect to spikelet count and awn shape. Tanaka (1955b) reported that chromosome pairing in the F_1 hybrid of the amphidiploid x *Ae. peregrina* was almost regular, i.e., 0–4 univalents, with a mode of 2, 9–14 bivalents, with a mode of 13, and 0–2 trivalents or quadrivalents; pollen fertility was 71.5% and seed set 50.0%. These data implied that *Ae. peregrina* originated from the hybridization between *Ae. sharonensis*, or another closely related species, and *Ae. umbellulata*. Kihara (1954) regarded the S subgenome of *Ae. peregrina* as a modified subgenome and therefore, formulated it as S^v.

These findings were reinforced by Zhang et al. (1992), who identified diagnostic bands in Southern blots hybridized with repeated nucleotide sequences and one 5S rRNA gene. Their study confirmed that one subgenome in *Ae. peregrina* was identical to the U genome of *Ae. umbellulata* and that the other was identical to the S^1 genome of *Ae. longissima* or to the S^{sh} genome of *Ae. sharonensis* or, more likely, to the internode in the phylogenetic tree of *Triticum* immediately preceding the divergence of *Ae. longissima* and *Ae. sharonensis*. Their data indicated that the second subgenome of *Ae. peregrina* was contributed by the *Ae. longissima-Ae. sharonensis* evolutionary lineage, and not by any other Sitopsis species.

When the data of Zhang et al. (1992) are interpreted in the context of the phylogenetic tree of the wheat group, it appears that *Ae. peregrina* and *Ae. kotschyi* are of recent origin and evolved after the differentiation of *Ae. umbellulata* from *Ae. caudata*, and *Ae. longissima* and *Ae. sharonensis* from *Ae. bicornis*.

The results of the cytogenetic (Tanaka 1955b) and molecular (Zhang et al. 1992) studies on the source of the S^v subgenome in *Ae. peregrina* and *Ae. kotschyi* disagree with the inference based on chloroplast (cp) DNA and mitochondrial (mt) DNA (Ogihara and Tsunewaki 1988; Siregar et al. 1988; Wang et al. 1997; Tsunewaki 2009), which indicated that the cytoplasm of both allotetraploids was contributed by *Ae. searsii* and not by *Ae. longissima*.

Both *Ae. peregrina* and *Ae. kotschyi* are allotetraploid with the same S^v-type cytoplasm and are thought to have received their cytoplasm from a genome donor with the S^v-type cytoplasm (Mukai and Tsunewaki 1975; Tsunewaki et al. 1978). Studies on the large subunits of the chloroplast Fraction I protein (Hirai and Tsunewaki 1981) and chloroplast DNA restriction pattern (Ogihara and Tsunewaki 1982; Tsunewaki and Ogihara 1983), suggested that the S^v cytoplasm derived either from *Ae. searsii* or *Ae. bicornis*. While it is obvious that *Ae. bicornis* has different phenotypic effects in alloplasmic lines of bread wheat than the cytoplasm of *Ae. searsii* (Siregar et al. 1988), studies of the cytoplasm of *Ae. peregrina* and *Ae. kotschyi*, did not reveal any differences between their ctDNAs and that of *Ae. bicornis*, suggesting that these species differ in their mitochondrial genomes (Tsunewaki and Ogihara 1983).

Indeed, while aiming to more accurately determine the diploid donor of the cytoplasm to *Ae. peregrina* and *Ae. kotschyi*, Terachi and Tsunewaki (1986), analyzing restriction fragment patterns of mtDNA isolated from various *Aegilops* species, revealed that whereas the mitochondrial genomes of *Ae. peregrina* and *Ae. kotschyi* are identical to that of *Ae. searsii*, that of *Ae. bicornis* is somewhat different. Thus, *Ae. searsii* was proposed as the cytoplasm donor to these two allotetraploid species (Tsunewaki 2009 and references therein). Whether this conclusion indicates introgression of the cytoplasm from *Ae. searsii* into *Ae. peregrina* (and *Ae. kotschyi*) or some other cause, is currently not known and requires further investigation.

Wang et al. (1997) analyzed PCR–single-strand conformational polymorphism (PCR-SSCP) of 14.0-kb ct and 13.7-kb mt DNA regions that were isolated from 46 alloplasmic wheat lines. In accord with the above, they found that the genetic distances between *Ae. searsii* and both *Ae. peregrina* and *Ae. kotschyi* were moderate (0.008 and 0.011, respectively) and between the two allotetraploids was small (0.005). Therefore, Wang et al. (1997) designated the cytoplasm of *Ae. searsii* S^v, like the designation of the cytoplasm of *Ae. peregrina* and *Ae. kotschyi*. Apparently, *Ae. peregrina* and *Ae. kotschyi* had a monophyletic origin (Wang et al. 1997). In this respect, it is interesting to note that Meimberg et al. (2009) assumed a single origin for *Ae. peregrina*.

In his attempt to explain the origin of the modified genomes in the *Aegilops* allotetraploid species, Kihara (1963) assumed subgenome donors to be either extinct or yet unknown diploid species, or that independent chromosome differentiation occurred within the subgenome. Kawahara (1986, 1988) proposed an alternative and more simple explanation for the existence of modified subgenomes in the *Aegilops-Triticum* group. He assumed that this group contains several diploid species with stable genomes, while other diploids have less stable and more variable genomes. In an allotetraploid species with the genomic combination of one stable and one variable subgenomes, structural differentiation or segmental rearrangements would accumulate in the variable genome. Since one genome is stable, serving as a genetic buffer, the chromosome structure of the second genome would change far more rapidly than that of the corresponding genome of the diploid species. To verify this hypothesis, Kawahara (1986, 1988) identified the breakpoints of spontaneous reciprocal translocations in each subgenome. In *Ae. peregrina* and *Ae. kotschyi* he identified seven translocations, three being between the S^v subgenome chromosomes, two between the U and the S^v subgenomes and one between the U subgenome chromosomes. The breakpoint of the remaining translocation was assumed to be on a S^v subgenome chromosome but was not identified (Kawahara T, unpublished). Evidently, the number of breakpoints on the modified S^v is clearly about twice that on the pivotal U subgenome. Therefore, Kawahara (1986, 1988) concluded that genome rearrangement occurs more frequently in the modified genomes than in the pivotal ones, and that the modified genomes likely evolved through their high structural variability.

The coexistence of a modified and unchanged subgenome within the same nucleus of an allotetraploid *Aegilops* species, led to the hypothesis that interspecific hybridization between allotetraploid species sharing one subgenome, but differing in the other subgenome, results in the modification of the differential subgenomes via hybridization and gene exchange (Zohary and Feldman 1962; Feldman 1965a). These authors studied this evolutionary process using *Ae. peregrina* as a model, and argued that the extensive morphological variation in this species was indicative of genome modification. They presented evidence for the existence of hybrids and hybrid derivatives between *Ae. peregrina* and several other allotetraploid species that share the U subgenome and grow with *Ae. peregrina* in mixed populations. Moreover, F_1 hybrids between *Ae. peregrina* and other U-subgenome allopolyploids exhibit, in addition to pairing of the homologous U-chromosomes, also pairing between several chromosomes of the differential genomes (Feldman 1965c), even though such pairing is usually precluded (Riley 1966b). Upon backcrossing to either parent, this pairing may lead to the formation of introgressed (modified) subgenomes.

Furuta and Tanaka (1970) examined whether introgression occurs between tetraploid species belonging to the U-subgenome group of *Aegilops*, by carrying out cytological and morphological analyses on hybrid progenies of three cross-combinations: *Ae. neglecta* (UUX^{n}X^n) x Ae. columnaris (UUX^{n}X^n), *Ae. peregrina* (UUS^{v}S^v) x *Ae. columnaris*, and *Ae. biuncialis* (UUM^{b}M^b) x *Ae. columnaris*. In all cases, the U subgenome was the common buffer subgenome and subgenomes X^n, S^{v}and M^b form 5, 3.5 and 2 bivalents, respectively, with the X^n subgenome. In the most closely related combination, *Ae. neglecta* x *Ae. columnaris*, introgression was cytologically and morphologically confirmed. In *Ae. peregrina* x *Ae. columnaris*, only introgression between two homoeologous chromosomes was observed, whereas in the third distinctly related combination, *Ae. biuncialis* x *Ae. columnaris*, no introgression took place. These observations suggest that introgression is a function of the relationship between species and impacts the resulting modified genomes in the *Aegilops* allotetraploid species (Furuta and Tanaka 1970).

Support for the above hypothesis came from Nakai and Tsuji (1984) who, using the gel electrofocusing technique, examined acid phosphatase isozymes in four accessions of *Ae. peregrina* and reported that the S^v subgenome of *Ae. peregrina* and *Ae. kotschyi*, had been modified by introgressive hybridization from *Ae. geniculata* and *Ae. columnaris*. In contrast, Jaaska (1978a, b) noted from his acid phosphatase data, no indication of the occurrence of subgenome recombination in Aegilops species sharing a common subgenome. However, his results did not necessarily overthrow Zohary and Feldman's (1962) hypothesis, as the banding profiles of the allotetraploids presented may reflect complex evolutionary processes. Likewise, C-banding and FISH analyses performed by Badaeva et al. (2004), strongly suggested that the S^v subgenome of *Ae. peregrina* was derived from *Ae. longissima*, and that it is not structurally altered relative to that of the parental species. This agrees with molecular data of Zhang et al. (1992), who obtained no evidence for an extensive modification of the S^v subgenome relative to that of its diploid donor, *Ae. longissima*.

In contrast to the above, Gong et al. (2006) used 31 ISSR (inter-simple sequence repeat) primers to study genomic evolutions among 23 species of *Aegilops*. The results indicated that the genome constituents of the allopolyploid species had considerably changed through evolution compared with their ancestral diploid species. Genome U showed little alterations in U-containing allopolyploids, while others had undergone changes after allopolyploidization. Gong et al. (2006), found that *Ae. peregrina* and *Ae. kotschyi* were more similar to *Ae. umbellulata* than to *Ae. searsii*, despite the fact that *Ae. searsii* was their cytoplasmic donor (Terachi et al. 1990). They concluded that their findings supported the pivotal-differential hypothesis proposed by Zohary and Feldman (1962).

Ae. peregrina contains 12.52 ± 0.181 pg 1C DNA (Eilam et al. 2008), one of the largest genomes among the allotetraploid species of *Aegilops* (Eilam et al. 2008). The 1C DNA content is 2.64% less DNA than that expected from the sum of the DNA of its two diploid parents, namely, 12.86 pg (*Ae. umbellulata*, the donor of the U subgenome, contains 5.38 pg and *Ae. longissima*, the assumed donor of the S^v subgenome, contains 7.48 pg; Eilam et al. 2007).

The karyotype of *Ae. peregrina* described by Sorokina (1928) did not show the presence of satellites. However, according to Senyaninova-Korchagina (1930, 1932), there are 8 types of chromosomes, including one pair with satellites. Chennaveeraiah (1960) revealed three pairs with satellites on short arms, two pairs with large satellites and the third pair with smaller satellites. One pair has an extreme subterminal centromere, two pairs have submedian-subterminal centromeres, and the rest have submedian centromeres. No pairs have median centromere. In all, there are 12 types of chromosomes. One set of chromosomes, consisting of a chromosome pair with small satellites, a pair with large satellites, a pair with extreme subterminal centromere, two pairs with submedian centromeres and two pairs with submedian-subterminal centromeres, corresponds to the karyotype of the U genome of *Ae. umbellulata*. The second set of chromosomes contains one pair with large satellites, and the rest with submedian centromeres but with different arm ratio. Based on these karyotypic features, Chennaveeraiah (1960) concluded that this set falls into the M-group rather than the S-group.

A repetitive DNA sequence, derived from *T. aestivum*, coding for ribosomal RNA, was used as a probe in FISH analysis of *Ae. peregrina* and *Ae. kotschyi* genomes (Teoh et al. 1983). Similar to the finding of Chennaveeraiah (1960), Teoh et al. (1983) also found three pairs of SAT chromosomes and three pairs of rRNA sites. In contrast, Cermeño et al. (1984b) found only two SAT chromosomes and two pairs of Ag-NORs in *Ae. peregrina*. In parallel, they found up to four nucleoli in interphase cells and reported that the U subgenome completely suppressed the NOR activity of the *Ae. peregrina* S^v subgenome.

Badaeva et al. (2004) studied structure of the *Ae. peregrina* genome by analyzing heterochromatin banding patterns of its somatic metaphase chromosomes, as revealed by study of 20 accessions by C-banding and three accessions by FISH with the heterochromatin-limited repetitive DNA probes pSc119 and pAs1, as well as the distribution of NOR and 5S DNA loci revealed by pTa71 (18S-26S rDNA) and pTa794 (5S rDNA) probes. All *Ae. peregrina* chromosomes were highly heterochromatic and with distinct C-banding patterns, allowing their easy identification. Similar to Badaeva et al. (2004), Zhao et al. (2016) identified each of the 14 pairs of *Ae. peregrina* chromosomes using a FISH probe combination of pSc119.2, pTa71 and pTa-713. Using N Banding, Jewell and Driscoll (1983) succeeded to identify nine of the 14 possible chromosomes of *Ae. peregrina* that were added to common wheat as a monosomic addition.

Significant C-banding polymorphism was detected in *Ae. peregrina* (Badaeva et al. 2004). However, the frequency of chromosomal aberrations was comparatively low, which agrees with meiotic pairing data in intraspecific hybrids of this species (Feldman 1963; Furuta 1981b; Kawahara 1986, 1988; Yu and Jahier 1992). Rearranged chromosomes were only found in four accessions. These modified chromosomes likely arose as a result of introgression of genetic material from a related species, followed by meiotic recombination. (Badaeva et al. 2004).

Studies based on isozyme (Hart and Tuleen1983) and C-banding analyses (Friebe et al. 1993, 1996; Friebe and Gill 1996) revealed that chromosomes 4S^l and 7S^l of *Ae. longissima* are involved in a species-specific reciprocal translocation that is absent in *Ae. sharonensis*. The presence of this translocation in *Ae. peregrina* and its absence in *Ae. kotschyi* suggests that *Ae. peregrina* originated from the hybridization of *Ae. umbellulata* with *Ae. longissima* and that *Ae. kotschyi* originated from the hybridization of *Ae. umbellulata* with *Ae. sharonensis*; both hybridizations occurred after the differentiation of *Ae. longissima* from *Ae. sharonensis*.

While the U subgenome of *Ae. peregrina* is similar to that of *Ae. umbellulata* (Friebe et al. 1995b, 1996; Badaeva et al. 1996a), Badaeva et al. (2004) found some differences in the size and position of C-bands of the corresponding chromosomes. On the other hand, the S^v subgenome was nearly identical to the S^l genome of *Ae. longissima*, as also reported by Friebe et al. (1993), Friebe and Gill (1996), and Badaeva et al. (1996a), indicating that *Ae. peregrina* derived from hybridization of *Ae. umbellulata* with *Ae. longissima*. Little modification occurred in the S^v subgenome at the tetraploid level, contradicting earlier findings (Chennaveeraiah 1960; Zohary and Feldman 1962; Feldman 1965a).

Although *Ae. peregrina* has a genome constitution similar to that of *Ae. kotschyi*, the C-banding patterns of these two species were different (Badaeva et al. 2004). Comparison of *Ae. kotschyi* with the diploid ancestors *Ae. umbellulata* and *Ae. sharonensis*, revealed a higher degree of genome modification in *Ae. kotschyi* as compared to *Ae. peregrina*. Chromosomes 4 Sv and 7S^{v}of *Ae. peregrina* are nearly identical to 4S^l and 7S^l of *Ae. longissima*, whereas chromosomes 4S^{v}and 7S^{v}of *Ae. kotschyi* are similar to 4S^{sh} and 7S^{sh} of *Ae. sharonensis*. The high frequency of chromosomal aberrations and reduction in the number and size of 18S-26S rDNA loci observed in the S^v subgenome of *Ae. kotschyi*, compared to the S^{sh} genome of *Ae. sharonensis* and

S^v of *Ae. peregrina*, suggest that *Ae. kotschyi* is an older species than *Ae. peregrina*.

FISH with clone pSC119 revealed signals of various sizes in telosomic regions of either the short or both arms of 13 *Ae. peregrina* chromosome pairs. An interstitial pSc119 FISH site was detected in the long arm of *Ae. peregrina* chromosome 7U, whereas chromosome 6U had no pSc119 FISH site, despite the fact that distinct telomeric and interstitial FISH sites are present in *Ae. umbellulata* (Badaeva et al. 1996a). Chromosome 6U of *Ae. peregrina* has a telomeric and an interstitial pAs1 FISH site, which are absent in *Ae. umbellulata*. Another a pAs1 FISH site was detected in the middle of the satellite of chromosome 1U of *Ae. peregrina*.

In contrast to Chennaveeraiah (1960), Badaeva et al. (2004) found only two pairs of satellite chromosomes in *Ae. peregrina* and *Ae. kotschyi*, which coincides with the number of active NORs detected by Ag-NOR staining (Cermeno et al. 1984b) and in situ hybridization (Yamamoto 1992a). In both species, major NORs were observed on group 1 and 5 from the U-subgenome chromosomes. *Ae. peregrina* and *Ae. kotschyi* also have several minor 18S-26S rDNA sites. Two consistent minor loci detected on group 5 and 6 of the S^v-subgenome chromosomes were associated with a significant reduction in copy number of 18S-26S rRNA genes. The hybridization patterns with pTa794 were similar in both species; four similar-sized rDNA loci were located on chromosomes of homoeologous groups 1 and 5. As was found by Cermeño et al. (1984b), the NORs on S^v subgenome chromosomes were inactivated and were accompanied with a decrease or loss of rDNA sequences.

9.8.10.3 Crosses with Other Species of the Wheat Group

Chennaveeraiah (1960) reported that all PMCs of *Ae. peregrina* had 14 bivalents at first meiotic metaphase, two of which were usually rod bivalents. No multivalents were observed. Likewise, Feldman (1963), who studied chromosome pairing in different lines of both *Ae. peregrina* subspecies, noted that all lines had 14 bivalents and 27.86 to 32.34 chiasmata/cell. Feldman (1963) also studied chromosome pairing at first meiotic metaphase in F$_1$ hybrids between different lines of *Ae. peregrina*. Average chromosome pairing in hybrids between lines of ssp. *peregrina* included 0.0–0.14 univalents, 10.0–14.0 bivalents, 0.0–0.04 trivalents, 0.0–1.96 quadrivalents and 26.10–26.74 chiasmata. Average chromosome pairing in the hybrid between two lines of ssp. *cylindrostachys* contained 0.12 univalents, 11.92 bivalents, of which one was heteromorphic, 0.04 trivalents, 0.98 quadrivalents and 26.80 chiasmata. Average chromosome pairing in F$_1$ hybrids between lines of ssp. *peregrina* x lines of ssp. *cylindrostachys*, presented 0.06–0.56 univalents, 12.02–12.304 bivalents, of which

0.02–0.46 were heteromorphic, 0.02–0.08 trivalents, 0.78–0.96 quadrivalents, and 26.04–26.46 chiasmata. Average chromosome pairing in the intraspecific hybrids was somewhat less regular than pairing in the parental lines, namely, though they had a low frequency of univalents, they had fewer bivalents, some multivalents and a lower number of chiasmata per cell.

Cuñado (1992) used C-banding to analyze chromosome pairing at first meiotic metaphase of *Ae. peregrina*. This technique facilitated the pairing of the whole complement as well as of specific-subgenome chromosomes. *Ae. peregrina* had an average of 0.04 univalents, 2.61 rod bivalents, 11.34 ring bivalents and 25.30 chromosome associations per cell (Cuñado 1992). U subgenome chromosome pairings showed 0.04 univalents, 1.41 rod bivalents, 5.54 ring bivalents, and 12.50 chromosome associations per cell, while S^v subgenome chromosome pairings included 1.20 rod bivalents, 5.80 ring bivalents and 12.80 chromosome associations per cell. Slight differences were observed between the pairing data of the two subgenomes, presumably because of the presence of one chromosome pair with a subterminal centromere in the U subgenome and higher C-heterochromatin content in the S^v subgenome. Cuñado et al. (1996b) also studied chromosome pairing in early stages of meiotic prophase of *Ae. peregrina* and found, in zygotene, 14 bivalents in 9 cells and 12 bivalents and one multivalent in one cell, whereas, in pachytene, all cells had 14 bivalents. These data indicate that bivalent pairing in *Ae. peregrina* occurs already at early stages of meiosis.

Chromosome pairing in F_1 triploid hybrids of *Ae. peregrina* with its putative diploid progenitors, *Ae. umbellulata* and *Ae. longissima*, are presented in Table 9.5. Pairing in *peregrina* x *umbellulata* included 6.95 univalents, 1.59 rod bivalents, 5.41 ring bivalents, 0.02 trivalents and 12.44 associations/cell. Pairing in *peregrina* x *longissima* showed 6.77 univalents, 2.11 rod bivalents, 2.98 ring bivalents, 0.30 trivalents, 0.73 quadrivalents, 0.04 other multivalents and 11.42 associations/cell. Similar results were obtained by Feldman (1963), who studied chromosome pairing in F_1 hybrids between several lines of *Ae. peregrina* and *Ae. longissima*. Average chromosome pairing included 6.58–6.72 univalents, 6.52–6.72 bivalents, 0.28–0.34 trivalents, and 14.36–15.12 chiasmata/cell. One hybrid combination had 6.82 univalents, 4.90 bivalents, 0.16 trivalents, 0.90 quadrivalents, 0.06 pentavalents and 13.08 chiasmata/cell. The above data substantiated that *Ae. umbellulata* and *Ae. longissima* are the parental species of *Ae. peregrina*. The S^v subgenome in two lines of *Ae. peregrina* (one studied by Yu and Jahier 1992 and the second by Feldman 1963) differed from the S^l genome of *Ae. longissima* by a reciprocal translocation. Thus, the U subgenome of *Ae. peregrina* remained nearly unchanged from that of *Ae. umbellulata*,

whereas the S^v subgenome of *Ae. peregrina* underwent some changes compared to the S^l genome of *Ae. longissima* and is therefore structurally differentiated by at least one interchange (Yu and Jahier 1992).

Yu and Jahier (1992) also studied chromosome pairing in F_1 triploid hybrids between *Ae. peregrina* and three other diploid species of section Sitopsis, *Ae. bicornis*, *Ae. searsii* and a low-pairing line of *Ae. speltoides* (Table 9.5). Pairing in *peregrina* x *bicornis* exhibited 7.57 associations/cell, pairing in *peregrina* x *searsii* had 5.82 associations/cell, and pairing in the *peregrina* x *speltoides* hybrid had 4.31 associations/cell. The relatively low level of pairing in these three hybrids showed that the genomes of these three Sitopsis species are only homoeologous to the S^v subgenome of *Ae. peregrina*.

Feldman (1965c) studied chromosome pairing in hybrids between *Ae. peregrina* and two other allopolyploid species of *Aegilops* sharing the U-subgenome, *Ae. geniculata* and *Ae. biuncialis* (Table 9.7). The F_1 of the *Ae. biuncialis* x *Ae. peregrina* ssp. *peregrina* hybrid (genome UM^bUS^v) showed 12.16 univalents, 6.54 bivalents, 0.76 trivalents and 0.12 quadrivalents, and the F_1 *Ae. biuncialis* x *Ae. peregrina* ssp. *cylindrostachys* hybrid showed 11.82 univalents, 6.60 bivalents, 0.70 trivalents, and 0.22 quadrivalents. The F_1 *Ae. geniculata* x *Ae. peregrina* ssp. *peregrina* hybrid (genome M^oUUS^v) had 11.44 univalents, 7.02 bivalents and 0.84 trivalents, and the F_1 *Ae. geniculata* x *Ae. peregrina* ssp. *cylindrostachys* hybrid had 11.42 univalents, 6.63 bivalents, 1.00 trivalent and 0.08 quadrivalent. Kihara (1937) observed 6–8 bivalents in the F_1 hybrid *Ae. geniculata* x *Ae. peregrina* and 7 bivalents in the F_1 hybrid *Ae. biuncialis* x *Ae. peregrina*, and Kihara (1949) observed 7–11 bivalents in the *Ae. neglecta* x *Ae. peregrina* hybrids, Lindschau and Oehler (1936) observed 2–12 bivalents in *Ae. peregrina* x *Ae. triuncialis* (genome S^vUUC) and Kihara (1937) observed 7–9 bivalents in the reciprocal combination. All these hybrids had more than 7 bivalents, indicating that pairing also took place between chromosomes of the differential genomes of these species.

Cuñado (1993a) also analyzed chromosome pairing in the F_1 *Ae. triuncialis* x *Ae. peregrina* hybrid and observed 12.74 univalents, 3.82 rod bivalents, 2.66 ring bivalents, 0.66 trivalents, 0.08 quadrivalents and 10.74 association/cell. The S^v—subgenome chromosomes were distinguishable from the U and C chromosomes by their larger size and higher C-heterochromatin content, which facilitated determination of the homomorphic and heteromorphic pairing in the hybrid. Average homomorphic pairing between the homologous chromosomes of the U subgenome, showed 2.04 rod bivalents, 2.50 ring bivalents and 6.66 associations/cell, and heteromorphic pairing between the homoeologous chromosomes of the C and S^v differential subgenomes included 1.78

rod bivalents, 0.16 ring bivalents, 0.66 trivalents, 0.08 quadrivalents and 3.58 associations/cell.

Meiotic pairing was analyzed by Fernández-Calvín and Orellana (1991) at first meiotic metaphase of the F_1 pentaploid hybrids common wheat x *Ae. peregrina* ssp. *peregrina*, common wheat x ssp. *cylindrostachys*, and common wheat x *Ae. kotschyi*. The hybrids had either the *Ph1* gene of common wheat that induces low pairing in hybrids or the *ph1b* mutant that allows homoeologous pairing in hybrids. The use of C-banding technique enabled identification of the various chromosomes in the different pairing configurations. Chromosome pairing in the hybrid common wheat x *Ae. peregrina* ssp. *peregrina* var. *typica* (hybrid genome BADUS^v) included 30.99 univalents, 1.69 rod bivalents, 0.10 ring bivalents, 0.03 trivalents and 0.01 quadrivalents. Among the rod bivalents, 0.33 were between A and D chromosomes, 0.67 were between AD-US^v chromosomes, 0.01 were between AD-B chromosomes, 0.37 were between US^v-B chromosomes and 0.31 were between U-S^v chromosomes. Chromosome pairing in the hybrid common wheat x *Ae. peregrina* ssp. *cylindrostachys* presented 28.99 univalents, 2.81 rod bivalents, 0.06 ring bivalents and 0.10 trivalents. Among the rod bivalents, 0.60 were between A-D chromosomes, 1.20 between AD-US^v chromosomes, 0.14 between AD-B chromosomes, 0.42 between US^{v}B chromosomes, and 0.45 between U-S^v chromosomes. Chromosome pairing in the hybrid common wheat x *Ae. kotschyi* (hybrid genome BADUS^v) had 30.92 univalents, 1.98 rod bivalents, 0.02 ring bivalents and 0.03 trivalents. Among the rod bivalents, 0.54 were between A-D chromosomes, 0.74 between AD-US^v chromosomes, 0.05 between AD-B chromosomes, 0.38 between US^v-B chromosomes and 0.25 between U-S^v chromosomes. These data indicated that more than 40% of the pairing in these three hybrids was autosyndetic (A-D + AD-B = 25%, and U-S^v = 16%) and more than 50% was allosyndetic (AD-US^v + US^v-B = 51%) (Fernández-Calvín and Orellana 1991). The allosyndetic pairing between AD chromosomes and US^v chromosomes was twice as frequent as that between US^{Iv}and B chromosomes. The genotype of *Ae. peregrina* ssp. *cylindrostachys* seemed to promote homoeologous pairing in the hybrid with common wheat (an increase of approximately a bivalent/cell). Its effect was detectable in the low pairing, but not in the high pairing hybrid (Fernández-Calvín and Orellana 1991).

Evidence for the existence of a gene(s) promoting pairing in a line of *Ae. peregrina* was obtained by Farooq et al. (1990) who analyzed chromosome pairing in hybrids of common wheat x three different accessions (A, B, and E) of *Ae. peregrina*. These authors found significant differences in the frequencies of homoeologous chromosome pairing at first meiotic metaphase. Hybrids between common wheat and *Ae. peregrina* accessions A and B showed very little pairing, as indicated by a chiasma frequency of 1.0 and 1.5 per cell, respectively. On the other hand, the hybrid between common wheat and *Ae. peregrina* accession E, showed significantly more homoeologous pairing (mean chiasma frequency was 12.6/cell). The level of such pairing was essentially the same as that between the hybrids of common wheat *ph1b* x *Ae. peregrina* accessions A and B. However, when the *ph1b* mutant was hybridized with accession E, the level of chromosome pairing significantly increased further (mean chiasma frequency was 17.52/cell), indicating the presence of pairing promoter gene(s) in *Ae. peregrina* accession E, which are epistatic to the wheat *Ph1* allele, and which positively interact with its mutant form to further increase the *ph1b* ceiling to homoeologous pairing in wheat.

References

Aase HC (1930) The cytology of Triticum, Secale, and Aegilops hybrids with reference to phylogeny. Res Stud State Coll Wash 2:3–60

Aghaee-Sarbarzeh M, Ferrahi M, Singh S, Singh H, Friebe B, Gill BS, Dhaliwal HS (2002) *Ph1*-induced transfer of leaf and stripe rust-resistance genes from *Aegilops triuncialis* and *Ae. geniculata* to bread wheat. Euphytica 127:377–382

Ahmadabadi M, Tehrani PA, Omidi M, Davoodi D (2002) Intraspecific karyotype divergence of *Aegilops triuncialis* in the northwest regions of Iran. Cytologia 67: 443-447

Ahmadpoor F, Asghari-Zakaria R, Firoozi B, Shahbazi H (2014) Investigation of diversity in *Aegilops biuncialis* and *Aegilops umbellulata* by A-PAGE. Natural Product Res 28:1626–1636

Akhunov ED, Akhunova AR, Anderson OD, Anderson JA, Blake N, Clegg MT, Coleman-Derr D, Conley EJ, Crossman CC, Deal KR et al (2010) Nucleotide diversity maps reveal variation in diversity among wheat genomes and chromosomes. BMC Genomics 11:702

Al-Mashhadani AN, Soliman AS, Al-Shehbaz IA (1978) Karyotype analysis of some diploid Aegilops species native to Iraq. Caryologia 31:299–303

Al-Mashhadani AN, Al-ShehbazI A, Soliman AS (1980) Karyotype analysis of five tetraploid Aegilops species native to Iraq. Caryologia 33:495–502

Altinkut A, Kotseruba V, Kirzhner VM, Nevo E, Raskina O, Belyayev A (2006) Ac-like transposons in populations of wild diploid Triticeae species: comparative analysis of chromosomal distribution. Chromosome Res 14:307–317

Anamthawat-Jonsson K, Heslop-Harrison JS (1993) Isolation and characterization of genome-specific DNA sequences in Triticeae species. Mol Gen Genet 240:151–158

Anikster Y, Manisterski J, Long DL (2005) Resistance to leaf rust, stripe rust, and stem rust in Aegilops spp. in Israel. Pl Disease 89:303–308

Ankori H, Zohary D (1962) Natural hybridization between Aegilops sharonensis and Ae. longissima: a morphological and cytological study. Cytologia, Tokyo 27:314–324

Appels R, Lagudah ES (1990) Manipulation of chromosomal segments from wild wheat for the improvement of bread wheat. Austral J Pl Physiol 17:253–266

Appels R, Gerlach WL, Dennis ES, Swift H, Peacock WJ (1980) Molecular and chromosomal organization of DNA sequences coding for the ribosomal RNAs in cereals. Chromosoma 78:293–311

Appels R, Baum BR, Clarke BC (1992) The 5S DNA units of bread wheat (Triticum aestivum L.). Plant Syst Evol 183:183–194

Arabbeigi M, Arzani A, Majidi MM, Kiani R, Tabatabaei BES, Habibi F (2014) Salinity tolerance of Aegilops cylindrica genotypes collected from hyper-saline shores of Uremia Salt Lake using physiological traits and SSR markers. Acta Physiol Plant 36:2243–2251

Arabbeigi M, Arzani A, Majidi MM, Habibi F, Rakhzadi A, Sayed-Tabatabaei BE (2015) Genetic diversity of *Aegilops cylindrica* species from west of Iran using morphological and phenological traits. Tulīd va Farāvarī-i Maḥṣūlāt-i Zirāī va Bāghī 5:123–133

Arrigo N, Guadagnuolo R, Lappe S, Pasche S, Parisod P, Felber F (2011) Gene flow between wheat and wild relatives: empirical evidence from Aegilops geniculata, Ae. neglecta and Ae. triuncialis. Evol Appl 4:685–695

Arumuganathan K, Earle ED (1991) Nuclear DNA content of some important plant species. Pl Mol Biol Rep 9:208–218

Aryavand A, Ehdaie B. Tran B, Waines JG (2003) Stomatal frequency and size differentiate ploidy levels in *Aegilops neglecta*. Genet Resourc Crop Evol 50:175–182

Ashida T, Nasuda S, Sato K, Endo TR (2007) Dissection of barley chromosome 5H in common wheat. Genes Genet Syst 82:123–133

Athwal RS, Kimber G (1972) The pairing of an alien chromosome with homoeologous chromosomes of wheat. Can J Genet Cytol 14:325–333

Avivi L (1976) The effect of gene controlling different degrees of homoeologous pairing on quadrivalent frequency in induced autotetraploid lines of Triticum longissimum. Can J Genet Cytol 18:357–364

Avni R, Lux T, Minz-Dub A, Millet E et al (2022) Genome sequences of Aegilops species of section Sitopsis reveal phylogenetic relationships and provide resources for wheat improvement. The Plant J 110:179–192

Badaeva ED (2002) Evaluation of phylogenetic relationships between five polyploid Aegilops L. Species of the U-genome cluster by means of chromosome analysis. Russian J Genet 28:664–675

Badaeva ED, Friebe B, Gill BS (1996a) Genome differentiation in Aegilops. Distribution of highly repetitive DNA sequences on chromosomes of diploid species. Genome 39:293–306

Badaeva ED, Friebe B, Gill BS (1996b) Genome differentiation in Aegilops. 2. Physical mapping of 5S and 18S–26S ribosomal RNA gene families in diploid species. Genome 39:1150–1158

Badaeva ED, Friebe B, Zoshchuk SA, Zelenin AV, Gill BS (1998) Molecular cytogenetic analysis of tetraploid and hexaploid Aegilops crassa. Chromosome Res 6:629–637

Badaeva ED, Chikida NN, Filatenko AA, Zelenin AV (1999) Comparative analysis of the M genome in Aegilops comosa and Ae. heldreichii by means of C-banding technique and in situ hybridization. Russian J Genet 35:670–677

Badaeva ED, Amosova AV, Muravenko OV, Samatadze TE, Chikida NN, Zelenin AV, Friebe B, Gill BS (2002) Genome differentiation in Aegilops. 3. Evolution of the D-genome cluster. Plant Syst Evol 231:163–190

Badaeva ED, Amosova AV, Samatadze TE, Zoshchuk SA, Shostak NG, Chikida NN, Zelenin AV, Raupp WJ, Friebe B, Gill BS (2004) Genome differentiation in Aegilops. 4. Evolution of the U-genome cluster. Plant Syst Evol 246:45–76

Badaeva ED, Dedkova O, Koenig J et al (2008) Analysis of introgression of Aegilops ventricosa Tausch. Genetic material in a common wheat background using C-banding. Theor Appl Genet 117:803–811

Badaeva ED, Dedkova OS, Zoshchuk SA, Amosova AV, Reader SM, Bernard M, Zelenin AV (2011) Comparative analysis of the N-genome in diploid and polyploid Aegilops species. Chromosome Res 19:541–548

Badaeva ED, Dedkova OS, Pukhalskiy VA, Zelenin AV (2012) Cytogenetic comparison of N-genome Aegilops L. Species. Russ J Genet 48:522–531

Badaeva ED, Amosova AV, Goncharov NP, Macas J, Ruban AS, Grechishnikova IV et al (2015) A set of cytogenetic markers allows the precise identification of all A-genome chromosomes in diploid and polyploid wheat. Cytogenet Genome Res 146:71–79

Badaeva ED, Ruban AS, Shishkin AA, Sibikeev SN, Druzhin AE, Surzhikov SA, Dragovich AYu (2018) Genetic classification of *Aegilops columnaris* Zhuk. ($2n=4x=28$, $U^cU^cX^cX^c$) chromosomes based on FISH analysis and substitution patterns in common wheat × Ae. columnaris introgressive lines. Genome 61:131–143

Badaeva ED, Fisenko AV, Surzhikov SA et al (2019) Genetic heterogeneity of a diploid grass Aegilops tauschii revealed by chromosome banding methods and electrophoretic analysis of the seed storage proteins (gliadins). Russ J Genet 55:1315–1329

Badaeva ED, Chikida NN, Fisenko AN, Surzhikov SA, Belousova MK, Özkan H, Dragovich AY, Kochieva EZ (2021a) Chromosome and molecular analyses reveal significant karyotype diversity and provide new evidence on the origin of Aegilops columnaris. Plants 10:956. https://doi.org/10.3390/plants10050956

Badaeva ED, Chikida NN, Belousova MK et al (2021b) A new insight on the evolution of polyploid Aegilops species from the complex Crassa: molecular-cytogenetic analysis. Plant Syst Evol 307:3. https://doi.org/10.1007/s00606-020-01731-2

Baenziger H (1962) Supernumerary chromsomes in diploid and tetraploid forms of crested wheatgrass. Can J Bot 40:549–561

Bahrman N, Zivy M, Thiellement H (1988) Genetic relationships in the Sitopsis section of Triticum and the origin of the B genome of polyploid wheats. Heredity 61:473–480

Bai D, Scoles GJ, Knott DR (1995) Rust resistance in Triticum cylindricum Ces. (4x, CCDD) and its transfer into durum and hexaploid wheats. Genome 38:8–16

Baik N, Maamri F, Bandou H (2017) Karyological study and meiotic analysis of four species of Aegilops (Poaceae) in Algeria. Caryologia 70:324–337

Bailey KL, Harding H, Knott DR (1993) Transfer to bread wheat of resistance to common root rot (Cochliobolus sativus) identified in Triticum timopheevii and Aegilops ovata. Can J Pl Pathol 15:211–219

Bakhshi B, Aghaei MJ, Bihamta MR, Darvish F, Zarifi E (2010) Ploidy determination of Aegilops cylindrica Host accessions of Iran by using flow cytometry and chromosome counting. Iran J Bot 16:258–266

Baldauf F, Schubert V, Metzlaff M (1992) Repeated DNA sequences of Aegilops markgrafii (Greuter) Hammer var. markgrafii: cloning, sequencing and analysis of distribution in Poaceae species. Hereditas 116:71–78

Bansal M, Kaur S, Dhaliwal HS, Bains NS, Bariana HS, Chhuneja PUK, Bansal K (2017) Mapping of Aegilops umbellulata-derived leaf rust and stripe rust resistance loci in wheat. Pl Pathology 66:38–44

Bardsley D, Cuadrado A, Jack P et al (1999) Chromosome markers in the tetraploid wheat Aegilops ventricosa analysed by in situ hybridization. Theor Appl Genet 99:300–304

Bariana HS, McIntosh RA (1993) Cytogenetic studies in wheat. XV. Location of rust resistance genes in VPM1 and their genetic linkage with other disease resistance genes in chromosome 2A. Genome 36:476–482

Barkley TM (1986) Flora of the great plains. Kansas University Press, Lawrence, Great Plains Flora Association

Baum BR (1977) Taxonomy of the tribe Triticeae (Poaceae) using various numerical techniques. I. Historical perspectives, data, accumulation and character analysis. Can J Bot 55:1712–1740

Baum BR (1978a) Taxonomy of the tribe Triticeae (Poaceae) using various numerical techniques. II. Classification. Can J Bot 56:27–56

Baum BR (1978b) Taxonomy of the tribe Triticeae (Poaceae) using various numerical techniques. III. Synoptic key to the genera and synopses. Can J Bot 56:374–385

Bedbrook JR, Jones J, O'Dell M, Thompson RD, Flavell RB (1980) A molecular description of telomeric heterochromatin in Secale species. Cell 19:545–560

Belkadi B, Assali N, Benlhabib (2003) Variation of specific morphological traits and ploidy level of five Aegilops L. species in Morocco. Acta Botanica Malacitana 28:47–58

Bell GDH, Sachs L (1953) Investigations in the Triticinae. II. The cytology and fertility of intergeneric and interspecific F_1 hybrids and their derived amphidiploids. J Agric Sci 43:105–115

Belyayev A, Raskina O (1998) Heterochromatin discrimination in Aegilops speltoides by simultaneous genomic in situ hybridization. Chromosome Res 6:559–566

Belyayev A, Raskina O (2013) Chromosome evolution in marginal populations of Aegilops speltoides: causes and consequences. Ann Bot 111:531–538

Belyayev A, Raskina O, Nevo E (2001) Chromosomal distribution of reverse transcriptase-containing retroelements in two Triticeae species. Chromosome Res 9:129–136

Belyatev A, Raskina O, Nevo E (2005) Variability of Ty3-gypsy retrotransposons chromosomal distribution in populations of two wild Triticeae species. Cytogen Genome Res 109:43–49

Belyayev A, Kalendar R, Brodsky L, Nevo E, Schulman AH, Raskina O (2010) Transposable elements in a marginal plant population: temporal fluctuations provide new insights into genome evolution of wild diploid wheat. Mob DNA 1:6. https://doi.org/10.1186/1759-8753-1-6

Benavente E, Alix K, Dusautoir J-C, Orellana J, David JL (2001) Early evolution of the chromosomal structure of Triticum turgidum–Aegilops ovata amphiploids carrying and lacking the Ph1 gene. Theor Appl Genet 103:1123–1128

Benito C, Figueiras AM, González-Jaén MT (1987) Location of genes coding isozyme markers on Aegilops umbellulata chromosomes adds data on homoeology among Triticeae chromosomes. Theor Appl Genet 73:581–588

Bennetzen JL (2005) Transposable elements, gene creation and genome rearrangement in flowering plants. Curr Opin Genet Dev 15:621–627

Bernhardt N, Brassac J, Kilian B, Blattner FR (2017) Dated tribe-wide whole chloroplast genome phylogeny indicates recurrent hybridizations within Triticeae. BMC Evol Biol 17:141

Bertoloni A (1834) Flora Italica 1:786–792

Berzonsky WA, Kimber G (1986) Tolerance of Triticum species to aluminium. Pl Breed 97:275–278

Berzonsky WA, Kinmber G (1989) The tolerance of aluminium of Triticum N-genome amphiploids. Pl Breed 103:37–42

Bierman A, Botha A (2017) A review of genome sequencing in the largest cereal genome, Triticum aestivum L. Agric Sci 8:194–207

Bleier H (1928) Zytologische Untersuchungen an Seltenen Getreide– und Rübenbastarden. Ztschr f Induktive Abstam u Vererbungslehre, Supp 1:447–452

Bordbar F, Rahiminejad MR (2010) A taxonomic revision of Aegilops crassa Boiss. (Poaceae) in Iran. Iran. J. Bot. 16:30–35

Bleier H (1930) Cytologie von Art- und Gattungbastarden des Getreides. Züchter 2:12–22

Bor NL (1968) Gramineae. In: Towsend CC, Guest E, El-Rawi A (eds) Flora of Iraq, vol 9, pp 210–263

Bor NL (1970) Gramineae. In: Rechiger KH (ed) Flora Iranica, vol 70. Akademische Druk-Und Verlagsanstalt, Wien, Graz, Austria

Bordbar F, Rahiminejad MR, Saeidi H, Blattner FR (2011) Phylogeny and genetic diversity of D-genome species of Aegilops and Triticum (Triticeae, Poaceae) from Iran based on microsatellites, ITS, and trnL-F. Pl Syst Evol 291:117–131

Bowden MW (1959) The taxonomy and nomenclature of the wheats, barleys and ryes and their wild relatives. Can J Bot 37:657–684

Bowden WM (1966) Chromosome numbers in seven genera of the tribe Triticeae. Can J Genet Cytol 8:130–136

Boyko EV, Gill KS, Mikelson-Young L, Nasuda S, Raupp WJ, Ziegle JN, Singh S, Hassawi DS, Fritz AK, Namuth D, Lapitan NL, Gill BS (1999) A high-density genetic linkage map of Aegilops tauschii the D-genome donor of bread wheat. Theor Appl Genet 99:16–26

Boyko E, Kalendar R, Korzun V, Fellers J, Korol A, Schulman AH, Gill BS (2002) A high-density cytogenetic map of the Aegilops tauschii genome incorporating retrotransposons and defense-related genes: insights into cereal chromosome structure and function. Plant Mol Biol 48:767–789

Bregliano JC, Kidwell MG (1983) Hybrid dysgenesis determinants. In: Shapiro JA (ed) Mobile genetic elements. Academic Press, New York, pp 363–410

Brody T, Mendlinger S (1980) Species relationships and genetic variation in the diploid wheats (Triticum, Aegilops) as revealed by starch gel electrophoresis. Pl Syst Evol 136:247–258

Cabi E, Ekici B, Doğan M (2018) Aegilops triuncialis subsp. bozdagensis (Poaceae), a new subspecies from South-Western Turkey. Acta Biologica Turcica 31:55–61

Cabrera A, Friebe B, Jiang J, Gill BS (1995) Characterization of Hordeum chilense chromosomes by C-banding and in situ hybridization using highly repeated DNA probes. Genome 38:435–442

Cahalan C, Law CN (1979) The genetical control of cold resistance and vernalization requirement in wheat. Heredity 42:125–132

Caldwell KS, Dvorak J, Lagudah ES, Akhunov E, Luo M-C, Wolters P, Powell W (2004) Sequence polymorphism in polyploid wheat and their D-genome diploid ancestor. Genetics 167:941–947

Castilho A, Heslop-Harrison JS (1995) Physical mapping of 5S and 18S–25S rDNA and repetitive DNA sequences in Aegilops umbellulata. Genome 38:91–96

Ceoloni C, Strauss I, Feldman M (1986) Effect of different doses of group-2 chromosomes on homoeologous pairing in intergeneric wheat hybrids. Can J Genet Cytol 28:240–246

Cermeño MC, Orellana J, Santos JL, Lacadena JR (1984a) Nucleolar organizer activity in wheat, rye and derivatives analyzed by a silver-staining procedure. Chromosoma 89:370–376

Cermeño MC, Orellana J, Santos JL, Lacadena JR (1984b) Nucleolar activity and competition (amphiplasty) in the genus Aegilops. Heredity 53:603–611

Chapman V, Miller TE (1977) Haploidy in the genus Aegilops. Wheat Inf Serv 44:21–22

Chapman V, Miller TE (1978) The relationship of the D genomes of hexaploid Ae. crassa, Ae. vavilovii and hexaploid wheat. Wheat Inf Serv 47:17–20

Chapman V, Riley R (1964) Haploid Aegilops Caudata. Wheat Inf Serv 17–18:16

Charles M, Belcram H, Just J, Huneau C, Viollet A, Couloux A, Segurens B, Carter M, Huteau V, Coriton O, Appels R, Samain S, Chalhoub B (2008) Dynamics and differential proliferation of transposable elements during the evolution of the B and A genomes of wheat. Genetics 180:1071–1086

Chee PW, Talbert LE, Lavin M (1995) Molecular analysis of evolutionary patterns in U genome wild wheats. Genome 38:290–297

Chen KC, Dvorak J (1984) The inheritance of genetic variation in Triticum speltoides affecting heterogenetic chromosome pairing in hybrids with Triticum aestivum. Can J Genet Cytol 26:279–287

Chen PD, Gill BS (1983) The origin of chromosome 4A, and genomes B and G of tetraploid wheat. In: Sakamoto S (ed) Proceedings of 6th international wheat genetics symposium, pp 39–48

Chen PD, Tsujimoto H, Gill BS (1994) Transfer of Ph1 gene promoting homoeologous airing from Triticum speltoides into common wheat and their utilization in alien genetic introgression. Theor Appl Genet 88:97–101

Chennaveeraiah MS (1960) Karyomorphologic and cytotaxonomic studies in Aegilops. Acta Horti Gotoburgensis 23:85–178

Chhuneja P, Dhaliwal HS, Bains NS, Singh K (2006) Aegilops kotschyi and Aegilops tauschii as sources for higher level of grain Iron and Zinc. Pl Breed 125:529–531

Cifuentes M, Benavente E (2009) Wheat-alien metaphase I pairing of individual wheat genomes and D genome chromosomes in interspecific hybrids between Triticum aestivum L. and Aegilops geniculata Roth. Theor Appl Genet 119:805–813

Cifuentes M, Garcia-agüero V, Benavente E (2010) A Comparative analysis of chromosome pairing at metaphase I in interspecific hybrids between durum wheat (Triticum turgidum L.) and the most widespread Aegilops Species. Cytogenet Genome Res 129:124–132

Colmer TD, Flowers TJ, Munns R (2006) Use of wild relatives to improve salt tolerance in wheat. J Exp Bot 57:1059–1078

Coriton O, Barloy D, Virginie H, Jahier J (2009) Assignment of Aegilops variabilis Eig chromosomes and translocations carrying resistance to nematodes in wheat. Genome 52:338–346

Cox TS, Sears RG, Bequette RK (1995) Use of winter wheat x Triticum tauschii backcross populations for germplasm evaluation. Theor Appl Genet 90:571–577

Crow JF (1983) Hybrid dysgenesis and P factor in Drosophila. Jpn J Genet 58:621–625

Cuñado N (1992) Analysis of metaphase I chromosome association in species of the genus Aegilops. Theor Appl Genet 85:283–292

Cuñado N (1993a) Genomic analysis in the genus Aegilops. I. Interspecific hybrids between tetraploid species sharing a common genome. Heredity 70:9–15

Cuñado N (1993b) Genomic analysis in the genus Aegilops. II. Interspecific hybrids between polyploid species sharing two common genomes. Heredity 70:16–21

Cuñado N, Santos JL (1999) On the diploidization mechanism of the genus Aegilops: meiotic behaviour of interspecific hybrids. Theor Appl Genet 99:1080–1086

Cuñado N, Callejas S, García MJ, Fernández A, Santos JL (1996a) The pattern of zygotene and pachytene pairing in allotetraploid Aegilops species sharing the U genome. Theor Appl Genet 93:1152–1155

Cuñado N, Callejas S, García MJ, Santos JL, Fernández A (1996b) Chromosome pairing in the allotetraploid Aegilops biuncialis and a triploid intergeneric hybrid. Genome 39:664–670

Cuñado N, Cermeño MC, Orellana J (1986) Interactions between wheat, rye and Aegilops ventricosa chromosomes on homologous and homoeologous pairing. Heredity 56:219–226

Cuñado N, Blazquez S, Melchor L, Pradillo M, Santos JL (2005) Understanding the cytological diploidization mechanism of polyploid wild wheats. Cytogenet Genome Res 109:205–209

Damania AB, Pecetti L (1990) Variability in a collection of Aegilops species and evaluation for yellow rust resistance at two locations in Northern Syria. J Genet Breed 44:97–102

Danilova TV, Akhunova AR Akhunov ED, Friebe B, Gill BS (2017) Major structural genomic alterations can be associated with hybrid speciation in Aegilops markgrafii (Triticeae). The Plant J 92:317–330

David JL, Benavente E, Brès-Patry (2004) Are neopolyploids a likely route for a transgene walk to the wild? The Aegilops ovata × Triticum turgidum durum case. Biol J Linnean Soc 82:503–510

de Las Heras JI, King IP, Parker JS (2001) 5-azacytidine induces chromosomal breakage in root tips of wheat carrying the cuckoo chromosome 4S^L from Aegilops sharonensis. Heredity 87:474–479

Doležel J, Bartoš J, Greilhuber J (2003) Nuclear DNA content and genome size of trout and human. Cytometry 51A:127–128

Donald WW, Ogg AG (1991) Biology and control of jointed goatgrass (Aegilops cylindrica), a review. Weed Technol 5:3–17

Dong L, Dong Q, Zhang W, Hu X, Wang H, Wang Y (2017) Karyotypic analysis of Aegilops speltoides revealed by FISH. Sci Agric Sin 50:1378–1387

Dos Santos Guerra M, Irene M, de Moraes Fernandes B (1977) Somatic instability in the Brazilian semi-dwarf wheat IAS 54. Can J Genet Cytol 19:225–230

Dosba F, Tanguy A-M, Douaire G (1980) Study of the characteristics linked to an M^v chromosome of Aegilops ventricosa in an addition line wheat × Aegilops. Cereal Res Commun 8:501–507

Dover GA, Riley R (1972) Prevention of pairing of homoeologous meiotic chromosomes of wheat by an activity of supernumerary chromosomes of Aegilops. Nature 240:159–161

Dover GA, Riley R (1977) Inferences from genetical evidence on the course of meiotic chromosome pairing in plants. Philos Trans R Soc Lond Ser. B 277:313–326

Dubcovsky J, Dvorak J (1994) Genome origins of Triticum cylindricum, Triticum triunciale, and Triticum ventricosum (Poaceae) inferred from variation in restriction patterns of repeated nucleotide sequences: a methodological study. Amer J Bot 81:1327–1335

Dubcovsky J, Dvorak J (1995a) Genome identification of the Triticum crassum complex (Poaceae) with the restriction patterns of repeated nucleotide sequences. Am J Bot 82:131–140

Dubcovsky J, Dvorak J (1995b) Ribosomal RNA loci: nomads in the Triticeae genomes. Genetics 140:1367–1377

Dudnikov AJ (2000) Multivariate analysis of genetic variation in Aegilops tauschii from the world germplasm collection. Genet Resour Crop Evol 47:185–190

Dudnikov AJ (2014) Aegilops tauschii Coss.: allelic variation of enzyme-encoding genes and ecological differentiation of the species. Genet Resour and Crop Evol 61:1329–1344

Dulai S, Molnár I, Szopko D, Darkó E, Vojtkó A, Sass-Gyarmati A, Molnár-Láng M (2014) Wheat-Aegilops biuncialis amphiploids have efficient photosynthesis and biomass production during osmotic stress. J Plant Physiol 171:509–517

Dvorak J (1983) The origin of wheat chromosomes 4A and 4B and their genome reallocation. Can J Genet Cytol 25:210–214

Dvorak J (1998) Genome analysis in the Triticum-Aegilops alliance. In: Slinkard AE (ed) Proc 9th Inter Wheat Genet Symp, University Extension Press, University of Saskatoon, Saskatoon, Saskatchewan, Canada, pp. 8–11

Dvorak J, Zhang H-B (1990) Variation in repeated nucleotide sequences sheds light on the phylogeny of the wheat B and G genomes. Proc Natl Acad Sci, USA 87:9640–9644

Dvorak J, Zhang H-B (1992a) Application of molecular tools for study of the phylogeny of diploid and polyploid taxa in Triticeae. Hereditas 116:37–42

Dvorak J, Zhang H-B (1992b) Reconstruction of the phylogeny of the genus Triticum from variation in repeated nucleotide sequences. Theor Appl Genet 84:419–429

Dvorak J, Lassner MW, Kota RS, Chen KC (1984) The distribution of the ribosomal RNA genes in the Triticum speltoides and Elytrigia elongata genomes. Can J Genet Cytol 26:628–632

Dvorak J, McGuire PE, Cassidy B (1988) Apparent sources of the A genomes of wheats inferred from polymorphism in abundance and restriction fragment length of repeated nucleotide sequences. Genome 30:680–689

Dvorak J, Zhang H-B, Kota RS, Lassner M (1989) Organization and evolution of the 5S ribosomal RNA gene family in wheat and related species. Genome 32:1003–1016

Dvorak J, Luo M-C, Yang Z-L (1998a) Restriction fragment length polymorphism and divergence in the genomic regions of high and low recombination in self-fertilizing and cross- ferti lizing *Aegilops* species. Genetics 148:423–434

Dvorak J, Luo M-C, Yang ZL, Zhang HB (1998b) Genetic evidence on the origin of *T. aestivum* L. In: Damania AB, Valkoun J, Willcox G, Qualset CO (eds) The origins of agriculture and crop domestication. The Harlan Symposium. ICARDA, IPGRI, FAO, UC/GRCP, Aleppo, Syria, pp 235–251

Dvorak J, Luo M-C, Yang ZL, Zhang H-B (1998c) The structure of the *Aegilops tauschii* gene pool and the evolution of hexaploid wheat. Theor Appl Genet 67:657–670

Dvorak J, Akhunov ED, Akhunov AR, Deal KR, Luo M-C (2006) Molecular characterization of a diagnostic DNA marker for domesticated tetraploid wheat provides evidence for gene flow wild tetraploid wheat to hexaploid wheat. Mol Biol Evol 23:1386–1396

Dvorak J, Deal KR, Luo M-C, You FM, von Borstel K, Dehghani H (2012) The origin of spelt and free-threshing hexaploid wheat. J Hered 103:426–441

Edae EA, Olivera PD, Jin Y, Rouse MS (2017) Genotyping-by-sequencing facilitates a high-density consensus linkage map for Aegilops umbellulata, a wild relative of cultivated wheat. G3: Genes Genomes, Genet 7:1551–1561

Eig A (1928a) Aegilops sharonensis n. sp. Notitzbl Bot Gart Mus Berlin 10:490–491

Eig A (1928b) Notes sur la gener *Aegilops* Bull Soc Botanique de Geneve, Ser 3, vol XIX fasc 2

Eig A (1929a) Monographisch-Kritische Ubersicht der Gattung *Aegilops*. Reprium nov. Spec Regni veg 55:1–288

Eig A (1929b) Amblyopyrum Eig. A new genus separated from the genus Aegilops. PZE Inst Agric Nat Hist Agric Res 2:199–204

Eig A (1936) *Aegilops* L. In: Hannig E, Winkler H (eds) Die Pflanzenareale 4. Reihe e 4, Jena, Fischer, Germany, pp 43–50

Eilam T, Anikster Y, Millet E, Manisterski J, Feldman M (2007) Genome size and genome evolution in diploid Triticeae species. Genome 50:1029–1037

Eilam T, Anikster Y, Millet E, Manisterski J, Feldman M (2008) Nuclear DNA amount and genome downsizing in natural and synthetic allopolyploids of the genera Aegilops and Triticum. Genome 51(8):616–627

El Bouhssini M, Benlhabib O, Nachit MM, Houari A, Bentika A, Nsarellah N, Lhaloui S (1998) Identification in Aegilops species of resistant sources to Hessian fly (Diptera: Cecidomyiidae) in Morocco. Genet Resourc Crop Evol 45:343–353

Emme HK (1924) Resultate der zytologischen Untersuchungen einiger Aegilops-Arten. Zeitschr Russ Bot Gesell 8:193–197

Endo TR (1978) On the *Aegilops* chromosomes having gametocidal action on common wheat. In: Ramanujan RS (ed) Proceeding of 5th international wheat genetics symposium, vol 1, New Delhi, pp 306–314

Endo TR (1979) Selective gametocidal action of a chromosome of Aegilops cylindrica in a cultivar of common wheat. Wheat Lnf Serv 50:24–28

Endo TR (1982) Gametocidal chromosome of three Aegilops species in common wheat. Can J Genet Cytol 24:201–206

Endo TR (1985) An Aegilops longissima chromosome causing chromosome aberrations in common wheat. Wheat Inf Serv 60:29

Endo TR (1988a) Induction of chromosomal structural changes by a chromosome of Aegilops cylindtica L. in common wheat. J Hered 79:366–370

Endo TR (1988b) Chromosome mutations induced by gametocidal chromosomes in common wheat. In: Miller TR, Koebner RMD (eds) Proceedings of 7th international wheat genetics sympsoium, vol 1, Cambridge, pp 259–263

Endo TR (1990) Gametocidal chromosomes and their induction of chromosome mutations in wheat. Jpn J Genet 65:135–152

Endo TR (1996) Allocation of a gametocidal chromosome of Aegilops cylindrica to wheat homoeologous group 2. Genes Genet Syst 71:243–246

Endo TR (2007) The gametocidal chromosome as a tool for chromosome manipulation in wheat. Chromosome Res 15:67–75

Endo TR (2015) Gametocidal genes. In: Molnár Láng M, Ceoloni C, Doležel J (eds) Alien introgression in wheat, cytogenetics, molecular biology, and genomics. Springer Internatinal Publishing, Switzerland, pp 121–131

Endo TR, Gill BS (1996) The deletion stocks in common wheat. J Hered 87:295–307

Endo TR, Katayama Y (1978) Finding of a selectively retained chromosome of Aegilops caudata L. in common wheat. Wheat Inf Serv 47–48:32–33

Endo TR, Tsunewaki K (1975) Sterility of common wheat with Aegilops triuncialis cytoplasm. J Hered 66:13–18

Endo TR, Yamamoto M, Mukai Y (1994) Structural changes of rye chromosome 1R induced by a Gc chromosome. Jpn J Genet 69:13–19

Farkas A, Molnár I, Dulai S, Rapi S, Oldal V, Cseh A, Kruppa K, Molnár-Láng M (2014) Increased micronutrient content (Zn, Mn) in the 3M^b(4B) wheat—Aegilops biuncialis substitution and 3M^b.4BS translocation identified by GISH and FISH. Genome 57:61–67

Farkhari M, Naghavi MR, Pyghambari SA, Sabokdast (2007) Genetic variation of Jointed Goatgrass (Aegilops cylindrica Host) from Iran using RAPD-PCR and SDS-Page of seed proteins. Pakistan J Biol Sci 10:2868–2873

Farooq S, Niazi MLK, Iqbal N, Shah TM (1989) Salt tolerance potential of wild resources of the tribe Triticeae. II. Screening of species of the genus Aegilops. Plant Soil 119:255–260

Farooq S, Iqbal N, Shah TM (1990) Promotion of homoeologous chromosome pairing in hybrids of Triticum aestivum x Aegilops variabilis. Genome 33:825–828

Farooq S, Iqbal N, Asghar M, Shah TM (1992) Intergeneric hybridization for wheat improvement. IV. Expression of salt tolerance gene(s) of Aegilops cylindrica in hybrids with hexaploid wheat. Cereal Res Commun 20:111–118

Fedoroff NV (2012) Transposable elements, epigenetics, and genome evolution. Science 338:758–767

Feinbrun-Dothan N (1986a) *Aegilops*, Flora Palaestina, part four, The Israel Academy of Sciences and Humanities, Jerusalem, pp. 168–177

Feinbrun-Dothan N (1986b) *Aegilops kotschyi* Boiss, Flora Palaestina, Part Four. The Israel Academy of Sciences and Humanities, Jerusalem, p 175

Feldman M (1963) Evolutionary studies in the *Aegilops-Triticum* group with special emphasis on causes of variability in the polyploid species of section *Pleionathera*. Thesis submitted for the degree of "Doctor of Philosophy" to the Senate of the Hebrew University of Jerusalem, Oct 1963, pp 1–216 (in Hebrew with English summary)

Feldman M (1965a) Further evidence for natural hybridization between tetraploid species of Aegilops section Pleionathera. Evolution 19:162–174

Feldman M (1965b) Fertility of interspecific F$_1$ hybrids and hybrid derivatives involving tetraploid species of Aegilops section Pleionathera. Evolution 19:556–562

Feldman M (1965c) Chromosome pairing between differential genomes in hybrids of tetraploid Aegilops species. Evolution 19:563–568

Feldman M (1966) The effect of chromosomes 5B, 5D and 5A on chromosomal pairing in Triticum aestivum. Proc Natl Acad Sci USA 55:1447–1453

Feldman M (1975) Alien addition lines of common wheat containing *Triticum longissimum* chromosomes. In: Proceedings of 12th international botanical congress, vol 2, Leningrad, USSR, p 506

Feldman M (1976) Wheats Triticum spp. (Gramineae-Triticineae). In: Simmonds NW (ed) Evolution of crop plants. Longman Group Ltd., London, pp 120–128

Feldman M (1978) New evidence on the origin of the B genome of wheat. In: Ramanujam RS (ed) In: Proceedings of 5th international wheat genetics symposium, New Delhi, pp 120–132

Feldman M (2001) Origin of cultivated wheat. In: Bonjean AP, Angus WJ (eds) The world wheat book. Lavoisier, Paris, pp 3–56

Feldman M, Kislev M (1977) Aegilops searsii, a new species of section Sitopsis (Platystachys). Isr J Bot 26:190–201

Feldman M, Mello-Sampayo T (1967) Suppression of homeologous pairing in hybrids of polyploid wheats × Triticum speltoides. Can J Genet Cytol 9:307–313

Feldman M, Strauss I (1983) A genome-restructuring gene in *Aegilops longissima*. In: Sakamoto S (ed) Proceedings of 6th international wheat genetics symposium, Kyoto, Japan, pp 309–314

Feldman M, Strauss I, Vardi A (1979) Chromosome pairing and fertility of F_1 hybrids of Aegilops longissima and Ae. searsii. Can J Genet Cytol 21:261–272

Feldman M, Levy AA, Fahima T, Korol A (2012) Genome asymmetry in allopolyploid plants—wheat as a model. J Exp Bot 63:5045–5059

Fernández-Calvín B, Orellana J (1991) Metaphase I bound arms frequency and genome analysis in wheat-Aegilops hybrids. 1. Ae. variabilis-wheat and Ae. kotschyi-wheat hybrids with low and high homoeologous pairing. Theor Appl Genet 83:264–272

Fernandez-Calvin B, Orellana J (1992) Relationship between pairing frequencies and genome affinity estimations in Aegilops ovata × Triticum aestivum hybrid plants. Heredity 68:165–172

Fernández-Calvín B, Orellana J (1993) Metaphase-I bound-arm frequency and genome analysis in wheat-Aegilops hybrids. 2. Cytogenetical evidence for excluding Ae. sharonensis as the donor of the B genome of polyploid wheats. Theor Appl Genet 85:587–592

Finch RA, Miller TE, Bennett MD (1984) "Cuckoo" Aegilops addition chromosome in wheat ensures its transmission by causing chromosome breaks in meiospores lacking it. Chromosoma 90:84–88

Flavell R, O'Dell O, Smith D (1979) Repeated sequence DNA comparisons between Triticum and Aegilops species. Heredity 42:309–332

Fleury D, Luo M-C, Dvorak J, Ramsay L, Gill BS, Anderson OD, You FM, Shoaei Z, Deal KR, Langridge P (2010) Physical mapping of a large plant genome using global high-information-content-fingerprinting: the distal region of the wheat ancestor Aegilops tauschii chromosome 3DS. BMC Genomics 11:382

Friebe B, Gill BS (1996) Chromosome banding and genome analysis in diploid and cultivated polyploid wheats. In: Jauhar PP (ed) Methods in Genome analysis in plants: their merits and pitfalls, Boca Ration, FL; New York, NY; London; CRC Press, Tokyo, pp 39–60

Friebe B, Mukai Y, Gill BS (1992a) C-banding polymorphism in several accessions of Triticum tauschii (Aegilops squarrosa). Genome 35:192–199

Friebe B, Schubert V, Blüthner WD, Hammer K (1992b) C-banding pattern and polymorphism of Aegilops caudata and chromosomal constitutions of the amphiploid T. aestivum—Ae. caudata and six derived chromosome addition lines. Theor Appl Genet 83:589–596

Friebe B, Tuleen N, Jiang J, Gill BS (1993) Standard karyotype of Triticum longissimum and its cytogenetic relationship with T. aestivum. Genome 36:731–742

Friebe B, Jiang J, Gill B (1995a) Detection of 5S rDNA and other repeated DNA on supernumerary B-chromosomes of Triticum species (Poaceae). Plant Syst Evol 196:131–139

Friebe B, Jiang J, Tuleen N, Gill BS (1995b) Standard karyotype of Triticum umbellulatum and the characterization of derived chromosome addition and translocation lines in common wheat. Theor Appl Genet 90:150–156

Friebe B, Tuleen NA, Gill BS (1995c) Standard karyotype of Triticum searsii and its relationship with other S-genome species and common wheat. Theor Appl Genet 91:248–254

Friebe B, Badaeva ED, Hammer K, Gill BS (1996) Standard karyotypes of Aegilops uniaristata, Ae. mutica, Ae. comosa subspecies comosa and heldreichii (Poaceae). Plant Syst Evol 202:199–210

Friebe B, Tuleen NA, Gill BS (1999) Development and identification of a complete set of Triticum aestivum—Aegilops geniculata chromosome addition lines. Genome 42:374–380

Friebe B, Qi L, Nasuda S et al (2000) Development of a complete set of Triticum aestivum-Aegilops speltoides chromosome addition lines. Theor Appl Genet 101:51–58

Friebe B, Kynast RG, Zhang P, Qi L, Dhar M, Gill BS (2001) Chromosome healing by addition of telomeric repeats in wheat occurs during the first mitotic divisions of the sporophyte and is a gradual process. Chromosome Res 9:137–146

Friebe B, Zhang P, Nasuda S, Gill BS (2003) Characterization of a knockout mutation at the Gc2 locus in wheat. Chromosoma 111:509–517

Furuta Y (1981a) Chromosome structural variation in Aegilops ovata L. Jpn J Genet 56:287–294

Furuta Y (1981b) Intraspecific variation in Aegilops variabilis and Ae. kotschyi revealed by chromosome pairing in F_1 hybrids. Jpn J Genet 56:495–504

Furuta Y, Tanaka M (1970) Experimental introgression in natural tetraploid Aegilops species. Jpn J Genet 45:129–145

Furuta Y, Nishikawa K, Kimizuka T (1977) Quntitative comparison of the nuclear DNA in section Sitopsis of the genus Aegilops. Jpn J Genet 52:107–115

Furuta Y, Nishikawa K, Yamaguchi S (1986) Nuclear DNA content in diploid wheat and its relatives in relation to the phylogeny of tetraploid wheat. Jpn J Genet 61:97–105

Gaines EF, Aase HC (1926) A haploid wheat plant. Amer J Bot 13:373–385

Galaev AV, Sivolap YM (2005) Molecular-genetic analysis of wheat (T. aestivum L.) genome with introgression of Ae. cylindrica Host genetic elements. Cytol Genet 39:57–66

Gandhi H, Mallory-Smith CA, Watson CJW, Vales MI, Zemetra RS, Riera-Lizarazu O (2006) Hybridization between wheat and jointed goatgrass (Aegilops cylindrica) under field conditions. Weed Sci 54:1073–1079

Gandhi HT, Vales MI, Mallory-Smith C, Riera-Lizarazu O (2009) Genetic structure of Aegilops cylindrica Host in its native range and in the United States of America. Theor Appl Genet 119:1013–1025

Gandilian PA (1980) Key to wheats, *Aegilops*, rye and barley. Academy of Sciences of the Armenian SSR, Erevan (in Russian)

Gandilian PA, Jaaska VE (1980) A stable introgressive hybrid from hybridization between Aegilops cylindrica Host and Triticum aestivum L. Genetika 16:1052–1058

Gerlach WL, Bedbrook JR (1979) Cloning and characterization of ribosomal RNA genes from wheat and barley. Nucleic Acids Res 7:1869–1885

Gerlach WL, Dyer TA (1980) Sequence organization of the repeating units in the nucleus of wheat which contain SS rRNA genes. Nucleic Acids Res 8:485–4865

Gerlach WL, Miller TE, Flavell RB (1980) The nucleolus organizers of diploid wheats revealed by in situ hybridization. Theor Appl Genet 58:97–100

Gill BS (2013) SNPing Aegilops tauschii genetic diversity and the birthplace of bread wheat. New Phytol 198:641–642

Gill BS, Kimber G (1974) Giemsa C-banding and the evolution of wheat. Proc Natl Acad Sci, USA 71:4086–4090

Gill BS, Sharma HC, Raupp WJ, Browder LE, Hatchet J et al (1985) Evaluation of Aegilops species for resistance to wheat powdery mildew, wheat leaf rust and Hessian fly and green bug. Pl Dis 69:314–316

Gill BS, Friebe B, Endo TR (1991a) Standard karyotype and nomenclature system for description of chromosome bands and structural aberrations in wheat (Triticum aestivum). Genome 34:830–839

Gill KS, Lubbers EL, Gill BS, Raupp WJ, Cox TS (1991b) A genetic linkage map of Triticum tauschii (DD) and its relationship to the D genome of bread wheat (AABBDD). Genome 34:362–374

Giorgi D (1996) Caratterizzazione molecolare delle specie di Aegilops appartenenti alla sezione Sitopsis e loro affinita con il genoma B dei frumenti coltivati, ph.d. Ph.D. dissertation. University of Tuscia, Viterbo, Italy

Giorgi D, D'Ovidio R, Tanzarella OA, Porceddu E (2002) RFLP analysis of Aegilops species belonging to the Sitopsis section. Genet Res Crop Evo 49:145–151

Giraldo P, Ruiz M, Rodriguez-Quijano M, Benavente E (2016) Development and validation of chloroplast DNA markers to assist Aegilops geniculata and Aegilops neglecta germplasm management. Genet Resourc Crop Evol 63:401–407

Gong H-Y, Liu A-H, Wang J-B (2006) Genomic evolutionary changes in Aegilops allopolyploids revealed by ISSR markers. Acta Phytotaxonomica Sinica 44:286–295

Gong WP, Li GR, Zhou JP, Li GP, Liu C, Huang CY, Zhao ZD, Yang ZJ (2014) Cytogenetic and molecular markers for detecting Aegilops uniaristata chromosomes in a wheat background. Genome 57:489–497

Gong W, Han R, Li H, Song J, Yan H, Li G, Liu A, Vao X, Guo J, Zhai S, Cheng D, Zhao Z, Liu C (2017) Agronomic traits and molecular marker identification of wheat–Aegilops caudata addition lines. Front Plant Sci 8:1743

Gornicki P, Zhu H, Wang J, Challa GS, Zhang Z, Gill BS, Li W (2014) The chloroplast view of the evolution of polyploid wheat. New Phytol 204:704–714

Goryunova SV, Kochieva EZ, Chikida NN, Pukhalskyi VA (2004) Phylogenetic relationships and intraspecific variation of D-genome Aegilops L. as revealed by RAPD analysis. Russian J Genet 40:515–523

Goryunova SV, Chikida NN, Kochieva EZ (2008) Molecular analysis of the phylogenetic relationships among the diploid Aegilops species of the section Sitopsis. Russ J Genet 44:115–118

Goryunova SV, Chikida NN, Kochieva EZ (2010) RAPD analysis of the intraspecific and interspecific variation and phylogenetic relationships of Aegilops L. species with the U genome. Russian J Genet 46:841–854

Greuter W (1976) The flora of Psara (E. Aegean islamds, Greece)—an annotated catalogue. Candollea 31:191–242

Greuter W, Burdet HM, Chaloner WG, Demoulin V, Grolle R et al (1988) International code of botanical nomenclature, adopted by the fourteenth international botanical congress, Berlin, July–Aug 1987. Regnum Veg, p 118

Guadagnuolo R, Savova-Bianchi D, Felber F (2001) Gene flow from wheat (Triticum aestivum L.) to jointed goatgrass (Aegilops cylindrica Host.), as revealed by RAPD and microsatellite markers. Ther Appl Genet 103:1–8

Gupta PK, Fedak G (1985) Variation in induction of homoeologous chromosome pairing in 6x Aegilops crassa by genomes of six different species of Secale. Can J Genet Cytol 27:531–537

Gyawali YP, Nasuda S, Endo TR (2009) Cytological dissection and molecular characterization of chromosome 1R derived from 'Burgas 2' common wheat. Genes Genet Syst 84:407–416

Gyawali YP, Nasuda S, Endo TR (2010) A cytological map of the short arm of rye chromosome 1R constructed with 1R dissection stocks of common wheat and PCR-based markers. Cytogenet Genome Res 129:224–233

Haider N (2003) Development and use of universal primers in plants. A Ph.D. thesis, Reading University, Reading, UK

Haider N, Nabulsi I (2008) Identification of Aegilops L. species and Triticum aestivum L. based on chloroplast DNA. Genet Resour Crop Evol 55:537–549

Haider N, Nabulsi I, MirAli N (2010) Comparison of the efficiency of A-PAGE and SDS–PAGE, ISSRs and RAPDs in resolving genetic relationships among Triticum and Aegilops species. Genet Resour Crop Evol 57:1023–1039

Hammer K (1980) Vorarbeiten zur monographischen Darstellung von Wildpflanzensortimenten: Aegilops L. Kulturpflanze 28:33–180

Hammer K (1987) Resistenzmerkmale und Reprodktionssystem als Indikatoren für evolutionäreTEndenzen der Gattung Aegilops L. Biol Zentralbl 106:264–273

Harb AM, Lahham JN (2013) Response of three accessions of Jordanian Aegilops crassa Boiss. And durum wheat to controlled drought. Jordanian J Biol Sci 6:151–158

Hart GE, Tuleen NA (1983) Characterizing and selecting alien genetic material in derivtives of wheat-alien species hybrids by analysis of isozyme variation. In: Sakamoto S (ed) In: Proceedings of 6th international wheat genetics symposium, Maruzen, Kyoto, pp 377–385

Hastie AR, Dong L, Smith A, Finkelstein J, Lam ET, Huo N et al (2013) Rapid genome mapping in nanochannel arrays for highly complete and accurate de novo sequence assembly of the complex Aegilops tauschii genome. PLoS ONE 8:e55864

Hegde SG, Waines JG (2004) Hybridization and introgression between bread wheat and wild and weedy relatives in North America. Crop Sci 44:1145–1155

Hegde SG, Valkoun J, Waines JG (2002) Genetic diversity in wild and weedy Aegilops, Amblyopyrum, and Secale species—a preliminary survey. Crop Sci 42:608–614

Heneen WK (1963) Meiosis in the interspecific hybrid Elymus farctus × A. repens (=Agropyron junceum x A. repens). Hereditas 49:107–118

Hirai A, Tsunewaki K (1981) Genetic diversity of the cytoplasm in Triticum and Aegilops. VIII. Fraction I protein of 39 cytoplasms. Genetics 99:487–493

Hitchcock AS (1935) Manual of the grasses of the United States, 1st edn. USDA Misc Publ 200, pp 1–1040

Hitchcock AS (1951) Manual of the grasses of the United States, 2nd edn, rev. by Agnes Chase. Dover Publications, US, pp 230–280

Hochstetter CF (1845) In Lorent JA, Waderungen im Moregenlande waehrend der Jahren: 1842–1843 (Extract). Flora (regensb.) 28:24–32

Hodgkin T, Adham YJ, Powell KS (1992) A preliminary survey of wild Triticum and Aegilops species in the world' genebanks. Hereditas 116:155–162

Hohman U, Lagudah ES (1993) C-banding polymorphism and linkage of non-homologous RFLP loci in the D genome progenitor of wheat. Genome 36:235–243

Hosid E, Brodsky L, Kalendar R, Raskina O, Belyayev A (2012) Diversity of long terminal repeat retrotransposon genome distribution in natural populations of the wild diploid wheat Aegilops speltoides. Genetics 190:263–274

Hueros G, Gonzales JM, Sanz JC, Ferrer E (1991) Gliadin gene location and C-banding identification of Aegilops longissima chromosomes added to wheat. Genome 34:236–240

Hutchinson J, Lonsdale DM (1982) The chromosomal distribution of cloned highly repetitive sequences from hexaploid wheat. Heredity 48:371–376

Ikeda TM, Tsunewaki K (1996) Deficiency of cox1 gene expression in wheat plants with Aegilops columnaris cytoplasm. Curr Genet 30:509–514

Ikeda TM, Shibasaka M, Tsunewaki K (1994) Respiratory activities, cytochrome composition from alloplasmic wheats having Aegilops cytoplasms. Plant Cell Physiol 35:779–784

Iordansky AB, Zurabishvili TB, Badaev NS (1978) linear differentiation of cereal chromosomes. I. Common wheat and its supposed ancestors. Theor Appl Genet 51:145–152

Iqbal N, Reader SM, Caligari PDS et al (2000a) Characterization of Aegilops uniaristata chromosomes by Comparative DNA marker analysis and repetitive DNA sequence in situ hybridization. Theor Appl Genet 101:1173–1179

Iqbal N, Reader SM, Caligari PDS, Miller TE (2000b) The production and characterization of recombination between chromosome 3N of Aegilops uniaristata and chromosome 3A of wheat. Heredity 84:487–492

Ishihara A, Mizuno N, Islam AKMR, Doležel J, Endo TR, Nasuda S (2014) Dissection of barley chromosomes 1H and 6H by the gametocidal system. Genes Genet Syst 89:203–214

Jaaska V (1978a) Electrophoretic study of acid phosphatase isoenzymes in the grass genus Aegilops L. Biochemie Physiologie Pflanzen 172:133–153

Jaaska V (1978b) NADP-dependent aromatic alcohol dehydrogenase in polyploid wheats and their diploid relatives. On the origin and phylogeny of polyploid wheats. Theor App Genet 53:209–217

Jaaska V (1980) Electrophoretic survey of seedling esterases in wheats in relation to their phylogeny. Theor App Genet 56:273–284

Jaaska V (1981) Aspartate aminotransferase and alcohol dehydrogenase isoenzyme: intraspecific differentiation in Aegilops tauschii the origin of the D genome polyploids in the wheat group. Plant Syst Evol 137:259–273

Jahier J, Doussinault G, Dosba F, Bourgeois F (1978) Monosomic analysis of resistance to eyespot in the variety Roazon. In: Ramanujan S (ed) Proceedings of 5th international wheat genetics symposium, New Delhi, India, 23–28 Feb 1978. Indian Society of Genetics Plant Breeding, New Delhi, India, pp 437–440

Jahier J, Tanguy AM, Abélard P, Rivoal R (1996) Utilization of deletions to localize a gene for resistance to the cereal cyst nematode, Heterodera avenae, on an Aegilops ventricosa chromosome. Plant Breed 11:282–284

Jewell DC, Driscoll CJ (1983) The addition of Aegilops variabilis chromosomes to Triticum aestivum and their identification. Can J Genet Cytol 25:76–84

Jia J, Zhao S, Kong X, Li Y, Zhao G et al (2013) Aegilops tauschii draft genome sequence reveals a gene repertoire for wheat adaptation. Nature 496:91–95

Jiang J, Gill BS (1994) Different species-specific chromosome translocations in Triticum timopheevii and T. turgidum support the diphyletic origin of polyploid wheats. Chromosome Res 2:59–64

Jiang J, Gill BS (2006) Current status and the future of fluorescence in situ hybridization (FISH) in plant genome research. Genome 49:1057–1068

Johnson BL (1967) Confirmation of the genome donors of Aegilops cylindrica. Nature 216:859–862

Jones JK, Majisu BN (1968) The homoeology of Aegilops mutica chromosomes. Can J Genet Cytol 10:620–626

Jones RN, González-Sánchez M, González-García M, Vega JM, Puertas MJ (2008) Chromosomes with a life of their own. Cytogenet Genome Res 120:265–280

Joshi GP, Nasuda S, Endo TR (2011) Dissection and cytological mapping of barley chromosome 2H in the genetic background of common wheat. Genes Genet Syst 86:231–248

Kadosumi S, Kawahara T, Sasanuma T (2005) Multiple origins of U genome in two UM genome tetraploid Aegilops species, Ae. columnaris and Ae. triaristata, revealed based on the polymorphism of a genome-specific PCR fragment. Genes Genet Syst 80:105–111

Kagawa F (1929) A study on the phylogeny of some species in Triticum and Aegilops, based upon comparison of the chromosomes. J Coll Agric Tokyo 10:173–228

Kam-Morgan LNW, Gill BS, Muthukrishnan S (1989) DNA restriction fragment length polymorphisms: a strategy for genetic mapping of D genome of wheat. Genome 32:724–732

Karagoz I (2006) Hybridization in Turkish Aegilops L. species. Pakistan J Biol Sci 9:2243–2248

Karataglis SS (1975) Karyotype analysis on some diploid native Greek Aegilops species. Caryologia 28:99–110

Karpechenko GD, Sorokina ON (1929) The hybrids of Aegilops triuncialis L. with rye. Bull Appl Bot Genet Pl Breed (USSR) 20:536–584

Kawahara T (1986) Difference in structural variability of genomes in Triticum and Aegilops. Wheat Inf Serv 63:42–43

Kawahara T (1988) Variation in chromosome structure in Aegilops Kotschyi Boiss. and Ae. variabilis Eig. In: Miller TE, Koebner RMD (eds) Proceedings of 7th international wheat genetics symposium, Cambridge, UK, Bath Press, Bath, pp 99–104

Kawahara T (2000) Isozyme variation in species of the section Comopyrum of Aegilops. Genet Resourc Crop Evol 47:641–645

Kaya I, Kirişözü AC, Ersoy FY, Dere S, Akkaya MS (2011) Genetic diversity and relationship analysis among accessions of Aegilops ssp. in Turkey using amplified fragment length polymorphism (AFLP) markers. Afr J Biotechnol 10:16167–16174

Kennedy AC, Stubbs TL (2007) Management effects on the incidence of jointed goatgrass inhibitory rhizobacteria. Biol Control 40:213–221

Kerber ER, Tipples KH (1969) Effects of the D-genome on milling and baking properties of wheat. Can J Pl Sci 49:255–263

Khabiri T, Zakaria RA, Zareh N, Sofalian O (2012) Analysis of genetic diversity based on HMW and LMW glutenin subunits in Aegilops cylindrica from Nourthwest of Iran. Int J Agric Crop Sci 4:529–533

Khabiri T, Asghari-Zakaria R, Zare N, Sofalian O (2013) Assessing genetic diversity based on gliadin proteins in Aegilops cylindrica populations from northwest of Iran. Notulae Sci Biol 5:109–113

Kiani R, Ahmad Arzani A, Habibi F (2015) Physiology of salinity tolerance in Aegilops cylindrica. Acta Physiol Plant 37:article number 135

Kihara H (1929) Conjugation of homologous chromosomes in genus hybrids Triticum x Aegilops and species hybrids of Aegilops. Cytologia 1:1–15

Kihara H (1930) Genomanalyse bei Triticum und Aegilops II. Aegilotricum Und Aegilops CylindRica. Cytologia 2:106–156

Kihara H (1936) Ein diplo-haploides Zwillingspaar bei Triticum durum. Agric Hortic 11:1425–2143

Kihara H (1937) Genomanalyse bei Triticum und Aegilops. VII. Kurze ~ bersicht iiber die Ergebnisse der Jahre 1934–36. Mem Coll Agric Kyoto Imp Univ 41:1–61

Kihara H (1940a) Verwandtschaft der Aegilops-Arten im Lichte der Genomanalyse. Ein Überblick. Der Züchter 12:49–62

Kihara H (1940b) Anwendung der Genomanalyse fur die Systematik von Triticurn und Aegilops. Jpn J Genet 16:309–320

Kihara H (1944) Discovery of the DD-analyser, one of the ancestors of Triticum vulgare. Agric Hortic 19:13–14

Kihara H (1946) Genomanalyze bei Triticum und Aegilops. IX. Systematischer Aufbau der Gattung Aegilops auf genomanalytischer Grundlage. Cytologia 14:135–144

Kihara H (1949) Genomanalyse bei Triticum und Aegilops. IX. Systematischer Aufbau der Gattung Aegilops auf genomanalytischer Grundlage. Cytologia 14:135–144

Kihara H (1954) Considerations on the evolution and distribution of Aegilops species based on the analyzer-method. Cytologia 19:336–357

Kihara H (1957) Completion of genome-analysis of three 6x species of Aegilops. Wheat Inf Serv 6:11

Kihara H (1959) Fertility and morphological variation in the substitution and restoration backcrosses of the hybrids, *Triticum vulgare* x *Aegilops caudata*. In: Proceedings of 10th international congress genetics, Montreal, vol 2, pp 142–171

Kihara H (1963) Interspecific relationship in Triticum and Aegilops. Seiken Zihô 15:1–12

Kihara H, Kondo N (1943) Studies on amphidiploids of Aegilops caudata x Ae. umbellulata induced by colchicine. Seiken Ziho 2:24–42

Kihara H, Lilienfeld FA (1932) Genomanalyse bei Triticum und Aegilops. IV. Untersuchungen an Aegilops x Triticum—und Aegilops x Aegilops Bastarden. Cytologia 3:384–456

Kihara H, Lilienfeld FA (1935) Genomanalyse bei Triticum und Aegilops. VI. Weitere Untersuchungen an Aegilops x Triticum und Aegilops x Aegilops-Bastarden. Cytologia 6:195–216

Kihara H, Matsumura S (1941) Genomanalyse bei Triticum und Aegilops. VIII. Rückkreuzung des Bastards Ae. caudata x Ae. cylindrica zu den Eltern und seine Nachkommen. Cytologia 11:493–506

Kihara H, Nishiyama I (1937) Possibility of crossing-over between semihomologous chromosomes from two different genomes. Cytologia (Fujii Jub Vol):654–666

Kihara H, Tanaka M (1958) Morphological and physiological variation among Aegilops squarrosa strains collected in Pakistan, Afghanistan and Iran. Preslia 30:241–251

Kihara H, Tanaka M (1970) Addendum to the classification of the genus *Aegilops* by means of genome-analysis. Wheat Inf Serv 30:1–2

Kihara H, Yamada Y (1942) Ein Fall von Plasmonwirkung auf das artfremde Genom. Seiken Ziho 1:1–38

Kihara H, Yamashita H, Tanaka M (1959) Genomes of 6x species of Aegilops. Wheat Inf Ser 8:3–5

Kihara H, Yamashita K, Tanaka M (1965) Morphological, physiological, genetical and cytological studies in *Aegilops* and *Triticum* collected from Pakistan, Afghanistan and Iran. In: Yamashita K (ed) Cultivated plants and their relatives, Koei Printing Comp, Japan, pp 1–118

Kilian B, Özkan H, Deusch O, Effgen S, Brandolini A, Kohl J, Martin W, Salamini F (2007) Independent wheat B and G genome origins in outcrossing Aegilops progenitor haplotypes. Mol Biol Evol 24:217–227

Kilian B, Mammen K, Millet E, Sharma R, Graner A, Salamini F, Hammer K, Özkan H (2011) Aegilops. In: Kole C (ed) Wild crop relatives: genomic and breeding resources, cereals, vol 24. Springer. Berlin, Germany, pp 1–76

Kim WK, Innes RL, Kerber ER (1992) Ribosomal DNA repeat unit polymorphism in six Aegilops species. Genome 35:510–514

Kimber G (1961) Cytogenetics of haploidy in *Gossypium* and *Triticum*. Ph.D. thesis, University Manchester, UK, pp 1–297

Kimber G (1967) The addition of the chromosomes of Aegilops umbellulata to Triticum aestivum (var. Chinese Spring). Genet Res Camb 9:111–114

Kimber G, Abu-Bakar M (1981) Genomic relationships of Triticum dichasians and Triticum umbellulatum. Z. Pflanzenzuchtg 87:255–273

Kimber G, Athwal RS (1972) A reassessment of the course of evolution of wheat. Proc Nat Acad Sci USA 69:912–915

Kimber G, Feldman M (1987) Wild wheats: an introduction. Special Report 353. College of Agriculture, Columbia. Missouri, USA, pp 1–142

Kimber G, Riley R (1963) The relationships of the diploid progenitors of hexaploid wheat. Can J Genet Cytol 5:83–88

Kimber G, Sears ER (1983) Assignment of genome symbols in the Triticeae. In: Sakamoto S (ed) Proceedings of 6th international wheat genetics symposium, Kyoto, Japan, pp 1195–1196

Kimber G, Sears ER (1987) Evolution in the genus Triticum and the origin of cultivated wheat. In: Heyne EG (ed) Wheat and wheat improvement, 2nd edn. American Society of Agronomy, Madison, Wisconsin, USA, pp 154–164

Kimber G, Tsunewaki K (1988) Genome symbols and plasma types in the wheat group. In: Miller TE, Koebner RMD (eds) Proceedings of 7th international wheat genetics symposium, Cambridge, pp 1209–1210

Kimber G, Yen Y (1988) Analysis of pivotal-differential evolutionary patterns. Proc Natl Acad Sci, USA 85:9106–9108

Kimber G, Yen Y (1989) Hybrids involving wheat relatives and autotetraploid Triticum umbellulatum. Genome 32:1–5

Kimber G, Zhao YH (1983) The D genome of the Triticeae. Can J Genet Cytol 25:581–589

Kimber G, Pignone D, Sallee PJ (1983) The relationships of the M and Mu genomes of Triticum. Can J Genet Cytol 25:509–512

Kimber G, Sallee PJ, Feiner MM (1988) The interspecific and evolutionary relationships of Triticum ovatum. Genome 30:218–221

Knaggs P, Ambrose MJ, Reader S, Miller TE (2000) Morphological characterization and evaluation of the subdivision of Aegilops tauschii Coss. Wheat Inf Serv 91:15–19

Knight E, Binnie A, Draeger T, Moscou M, María-Dolores M, Rey MD, Sucher J, Mehra S, King L, Moore G (2015) Mapping the 'breaker' element of the gametocidal locus proximal to a block of sub-telomeric heterochromatin on the long arm of chromosome 4S^{sh} of *Aegilops sharonensis*. Theor Appl Genet 128:1049–1059

Koo D-H, Tiwari VK, Hřibová E, Doležel J, Friebe Gill BS (2016) Molecular cytogenetic mapping of satellite DNA sequences in Aegilops geniculata and wheat. Cytogenet Genome Res 148:314–321

Kota RS, Dvorak J (1985) A rapid technique for substituting alien chromosomes into Triticum Aestivum and determining their homoeology. Can J Genet Cytol 27:549–558

Kota RS, Dvorak J (1988) Genomic instability in wheat induced by chromosome 6B^s of Triticum speltoides. Genetics 120:1085–1094

Kumar A, Bassi FM, Paux E, Al-Azzam O, Michalak de Jimenez M, Denton AM et al (2012) DNA repair and crossing over favor similar chromosome regions as discovered in radiation hybrid of Triticum. BMC Genomics 13:339

Kumar A, Seetan R, Mergoum M, Tiwari VK, Iqbal MJ, Wang Y, Al-Azzam O, Simkova H, Luo MC, Dvorak J, Gu YQ, Denton A, Kilian A, Lazo GR, Kianian SF (2015) Radiation hybrid maps of the D-genome of Aegilops tauschii and their application in sequence assembly of large and complex plant genomes. BMC Genomics 16:800

Kushnir U, Halloran GM (1981) Evidence for Aegilops sharonensis Eig as the donor of the B genome of wheat. Genetics 99:495–512

Kynast RG, Friebe B, Gill BS (2000) Fate of multicentric and ring chromosome induced by a new gametocidal factor located on chromosome 4M^g of Aegilops geniculata. Chromosome Res 8:133–139

Lacadena JR, Cermeño MC (1985) Nucleolus organizer competition in Triticum aestivum—Aegilops umbellulata chromosome addition lines. Theor Appl Genet 71:278–283

Lagudah ES, Appels R, McNeil R (1991a) The Nor-D3 locus of Triticum tauschii: natural variation and genetic linkage to markers in chromosome 5. Genome 36:387–395

Lagudah ES, Apples R, Brown AHD, McNeil D (1991b) The molecular-genetic analysis of Triticum tauschii the D genome donor to hexaploid wheat. Genome 34:375–386

Landjeva SP, Ganeva GD (2000) Chromosome N-banding polymorphism in Aegilops geniculata Roth. Genet Resourc Crop Evol 47:53–41

Lassner M, Anderson O, Dvorak J (1987) Hypervariation associated with 12-nucleotide direct repeat and inferences on intragenomic homogenization of ribosomal RNA gene spacer based on the DNA sequence of a clone from the wheat Nor-D3 locus. Genome 29:770–781

Law CN, Jenkins G (1970) A genetic study of cold resistance in wheat. Genet Res Camb 15:197–208

Lawrence GJ, Appels R (1986) Mapping the nucleolus organizing region, seed protein loci and isozyme loci on chromosome 1R in rye. Theor Appl Genet 71:742–749

Le HT, Reicosky DA, Olien CR, Cress CE (1986) Freezing hardiness of some accessions of Triticum tauschii, Triticum turgidum var. durum. Can J Plant Sci 66:893–899

Levy AA, Galili G, Feldman M (1985) The effect of additions of Aegilops longissima chromosomes on grain protein in common wheat. Theor Appl Genet 69:429–435

Li W, Zhang P, Fellers JP, Friebe B, Gill BS (2004) Sequence composition, organization, and evolution of the core Triticeae genome. Plant J 40:500–511

Li H, Gill BS, Wang X, Chen PD (2011) A Tal-Ph1 wheat genetic stock facilitates efficient alien introgression. Genet Res Crop Evol 58:667–678

Li C, Chen M, Chao S, Yu J, G Bai G (2013) Identification of a novel gene, H34, in wheat using recombinant inbred lines and single nucleotide polymorphism markers. Theor Appl Genet 126:2065–2071

Li L-F, Zhang Z-B, Wang Z-H et al (2022) Genome sequences of the five Sitopsis species of Aegilops and the origin of polyploid wheat B subgenome. Mol Plant 15:488–503

Lilienfeld F (1951) H. Kihara: genome analysis in Triticum and Aegilops. X. Concluding review. Cytologia 16:101–123

Limin AE, Fowler DB (1981) Cold hardiness of some relatives of hexaploid wheat. Can J Genet Cytol 59:572–573

Linc G, Friebe BR, Kynast RG, Molnar-Lang M, Köszegi B, Sutka J, Gill BS (1999) Molecular cytogenetic analysis of Aegilops cylindrica Host. Genome 42:497–503

Lindschau M, Oehler E (1936) Cytologische Untersuchungen an tetraploiden Aegilops-Artbastarden. Der Züchter 8:113–117

Linnaeus C (1753) Species Plantarum, Tomus I, May 1753: i–xii, 1–560; Tomus II, Aug 1753: 561–1200, plus indexes and addenda, 1201–1231

Liu D, Xiang Z, Zhang L, Zheng Y, Yang W, Chen G, Wan C, Zhang H (2011a) Transfer of stripe rust resistance from Aegilops variabilis to bread wheat. African J Biotechnol 10:136–139

Liu W, Rouse M, Friebe B, Jin Y, Gill BS, Pumphrey MO (2011b) Discovery and molecular mapping of a new gene conferring resistance to stem rust, Sr53, derived from Aegilops geniculata and characterization of spontaneous translocation stocks with reduced alien chromatin. Chromosome Res 19:669–682

Lorent JA (1845) Wanderungen im Morgenlander der waehrend Jahren 1842–1843. Verlag von T Loffler, Mannheim

Loureiro I, Escorial MC (2007) Hybridization between wheat (Triticum aestivum) and the wild species Aegilops geniculata and A. biuncialis under experimental field conditions. Agriculture, Ecosystems & Environment 120:384–390

Loureiro I, Escorial MC, García-Baudín JM, Chueca MC (2006) Evidence of natural hybridization between Aegilops geniculata and wheat under field conditions in Central Spain. Environ Biosafety Res 5:105–109

Löve Á (1984) Conspectus of the Triticeae. Feddes Repert 95:425–521

Lubbers EL, Gill KS, Cox TS, Gill BS (1991) Variation of molecular markers among geographically diverse accessions of Triticum tauschii. Genome 34:354–361

Luo M-C, Deal KR, Yang Z-L, Dvorak J (2005) Comparative genetic maps reveal extreme crossover localization in the Aegilops speltoides chromosomes. Theor Appl Genet 111:1098–1106

Luo M-C, Xu K, Ma Y, Deal KR, Nicolet CM, Dvorak J (2009) A high-throughput strategy for screening of bacterial artificial chromosome libraries and anchoring of clones on a genetic map constructed with single nucleotide polymorphisms. BMC Genomics 10:28

Luo M-C, Gu YQ, You FM, Deal KR, Ma Y, Hu Y, Huo N, Wang Y, Wang J, Chen S, Jorgensen CM, Zhang Y, McGuire PE, Pasternak S, Stein JC, Ware D et al (2013) A 4-gigabase physical map unlocks the structure and evolution of the complex genome of Aegilops tauschii, the wheat D-genome progenitor. Proc Natl Acad Sci, USA 110:7940–7945

Luo M-C, Gu YQ, Puiu D, Wang H, Twardziok SO, Deal KR et al (2017) Genome sequence of the progenitor of the wheat D genome Aegilops tauschii. Nature 551:498–502

Maan SS (1975) Cytoplasmic variability and speciation in Triticineae, In; Wali MK (ed) Prairie. Univ North Dakota Press, Grand Forks, pp 255–281

Maan SS (1976) Alien chromosome controlling sporophytic sterility in common wheat. Crop Sci 16:581–583

Maan SS (1977a) Cytoplasmic control of cross incompatibility and zygotic and sporophytic sterility in interspecific hybrids. In: Proceedings 8th congress Eucarpia, Madrid, Spain, pp 201–214

Maan SS (1977b) Cytoplasmic homology between Aegilops mutica Boiss. and Ae. ovata L. Euphytica 26:601–613

Maan SS (1978) Cytoplasmic relationships among the D and M genome of Aegilops species. In: Ramanujan S (ed) Proceedings of 5th internnational wheat genetics symposium, vol 1, pp 231–260

Maan SS, Sasakuma T (1978) Chromosome pairing relationships among the D and M genomes of Triticum and Aegilops species. In: Ramanujam S (ed) Proceedimgs of 5th international wheat genetics symposium, Indian Society Genetics Plant Breeding, Indian Agric Res Institute, New Delhi, India, pp 322–331

Mac Key J (1968) Relationships in the Triticinae. In: Finlay KW, Shepherd KW (eds) Proceedings of 3rd international wheat genetics symposium, Canberra, Aust Acad Sci, Canberra, pp 39–50

Mac Key J (1981) Comments on the basic principles of crop taxonomy. Kulturpflanze 29(S):199–207

Maestra B, Naranjo T (1997) Homoeologous relationships of Triticum sharonense chromosomes to T. aestivum. Theor Appl Genet 94:657–663

Maestra B, Naranjo T (1998) Homoeologous relationships of Aegilops speltoides chromosomes to bread wheat. Theor Appl Genet 97:181–186

Mahelka V, Kopecky D, Paštová L (2011) On the genome constitution and evolution of intermediate wheatgrass (Thinopyrum intermedium: Poaceae, Triticeae). BMC Evol Biol 11:127

Mahjoub A, El Gharbi MS, Mguis K, El Gazzah M, Ben Brahim N (2009) Evaluation of genetic diversity in Aegilops geniculata Roth accessions using morphological and RAPD markers. Pakistan J Biol Sci 12:994–1003

Mahjoub A, Abdellaoui R, Ben Naceur M, Brahim NN (2010) Genetic diversity of Tunisian accessions of Aegilops geniculata Roth and durum wheats (Triticum durum Desf.) using RAPD markers. Acta Bot Gallica 157:3–12

Mahjoub A, Khaled M, Ben Brahim N (2016) Genetic variability and phylogenetic relationships studies of Aegilops L. using some molecular markers. Int J Agron Agruc Res 8:14–25

Maia N (1967) Obtention de blés tendres résistants au piétinverse (Cercosporella herpotrichoides) par croisements interspécifiques. CR Acad Agric Fr 53:149–154

Maire R, Weiller M (1955) Flore de l'Afrique du nord. Vol III Monocotyledonae: Glumiflorae, sf. Pooideae, Gramineae, pp 357, 371–372

Makkouk KM, Ghulam W, Comeau A (1994) Resistance to barley yellow dwarf luteovirus in Aegilops species. Can J Pl Sci 74:631–634

Marais GF, Pretorius ZA, Marais AS, Wellings CR (2003) Transfer of rust resistance genes from Triticum species to common wheat. S Afr J Pl Soil 20:193–198

Marais GF, McCallum B, Snyman JR, Pretorius ZA, Marais AS (2005) Leaf rust and stripe rust resistance genes Lr54 and Yr37transferred to wheat from Aegilops. Plant Breed 124:525–614

Marais GF, McCallum B, Marais AS (2008) Wheat leaf rust resistance gene Lr59 derived from Aegilops peregrina. Pl Breed 127:340–345

Marais F, Marais A, McCallum BD, Pretorius ZA (2009) Transfer of leaf rust and stripe rust resistance genes Lr62 and Yr42 from Aegilops neglecta Req. ex Bertol. to common wheat. Crop Sci 49:871–879

Marcussen T, Sandve SR, Heier L, Spannagl M, Pfeifer M. The International Wheat Genome Sequencing Consortium, Jakobsen KS, Wulff BBH, Steuernagel B, Klaus FX, Mayer KFX, Olsen OA (2014) Ancient hybridizations among the ancestral genomes of bread wheat. Science 345(6194):288–291

Martini G, O'Dell M, Flavell RB (1982) Partial inactivation of wheat nucleolus organisers by the nucleolus organiser chromosomes from Aegilops umbellulata. Chromosoma 84:687–700

Martín-Sánchez JA, Gómez-Colmenarejo M, Del Moral J, Sin E, Montes MJ, González-Belinchón C, López-Braña I, Delibes A (2003) A new Hessian fly resistance gene (H30) transferred from the wild grass Aegilops triuncialis to hexaploid wheat. Theor Appl Genet 106:1248–1255

Masci S, D'Ovidio R, Lafiandra D, Tanzarella OA, Porceddu E (1992) Electrophoretic and molecular analysis of alpha-gliadins in Aegilops species (Poaceae) belonging to the D genome cluster and in their putative progenitors. Pl Syst Evol 179:115–128

Masoudi-Nejad A, Nasuda S, McIntosh RA, Endo TR (2002) Transfer of rye chromosome segments to wheat by a gametocidal system. Chromosome Res 10:349–357

Masoudi-Nejad A, Nasuda S, Bihoreau M-T, Endo TR (2005) An alternative to radiation hybrid mapping for large-scale genome analysis in barley. Mol Genet Genomics 274:589–594

Massa AN, Wanjugi H, Deal KR, O'Brien K, You FM, Maiti R, Chan AP, Gu YQ, Luo M-C, Anderson OD, Rabinowicz PD, Dvorak J, Devos KM (2011) Gene space dynamics during the evolution of Aegilops tauschii, Brachypodium distachyon, Oryza sativa, and Sorghum bicolor genomes. Mol Biol Evol 28:2537–2547

Matsumoto K, Kondo N (1942) Two new amphiploids in Aegilops. Jpn J Genet 18:130–133

Matsumoto K, Shimotsuma M, Nezu M (1957) The amphiploid $M^{u}M^{u}DD$ and its hybrid with Aegilops ventricosa. Wheat Infor Serv 5:12–13

Matsumura S (1940) Induced haploidy and autotetraploidy in Aegilops Ovata L. Bot Mag Tokyo 54:404–413

Matsuoka Y, Aghaei MJ, Abbasi MR, Totiaei A, Mozafari J, Ohta S (2008) Durum wheat cultivation associated with Aegilops tauschii in northern Iran. Genet Resour Crop Evol 55:861–868

Mattatia J, Feinbrun-Dothan N (1986) The nomenclature of Aegilops biuncialis Vis. Israel J Bot 35:53–54

McClintock B (1934) The relation of a particular chromosomal element tp the development of the nucleoli in Zea mays. Zeit Zellforsch Mik Anat 21:294–328

McFadden ES, Sears ER (1944) The artificial synthesis of Triticum spelta. Records Genet Soc Amer 13:26–27

McFadden ES, Sears ER (1946) The origin of Triticum spelta and its free-threshing hexaploid relatives. J Hered 37(81–89):107–116

McFadden ES, Sears ER (1947) The Genome approach in radical wheat breeding. Agron J 39:1011–1026

McGinnis RC (1956) Genome analysis of Aegilops juvenalis. Can J Agruc Sci 36:284–291

McGinnis RC, Melnyk JH (1956) Cytological evidence for the S genome in Aegilops juvenalis. Wheat Inf Serv 4:8

McGinnis RC, Melnyk JH (1962) Analysis of chromosome pairing in interspecific F_1 hybrids involving Aegilops Juvenalis. Wheat Inf Serv 14:22–23

McIntosh RA (1988) Catalogue of gene symbols for wheat. In: Miller TE, Koebner RMD (eds) Proceedings of 7th international wheat genetics symposium, vol 2, Cambridge, pp 1225–1323

Medouri A, Bellil I, Douadi K (2015) Polymorphism at high molecular weight glutenin subunits and morphological diversity of Aegilops geniculata Roth collected in Algeria. Cereal Res Comm 43:272–283

Meimberg H, Rice KJ, Milan NF, Njoku CC, McKay JK (2009) Multiple origins promote the ecological amplitude of allopolyploid Aegilops (Poaceae). Am J Bot 96:1262–1273

Mello-Sampayo T (1971) Promotion of homoeologous pairing in hybrids of Triticum aestivum x Aegilops longissima. Genet Iber 23:1–9

Melnyk JH, McGinnis RC (1962) Analysis of chromosome pairing in interspecific and intergeneric F_1 hybrids involving hexaploid Aegilops crassa. Wheat Inf Serv 14:24–25

Mendelson D, Zohary D (1972) Behaviour and transmission of supernumerary chromosomes in Aegilops speltoides. Heredity 29:329–339

Mendlinger S, Zohary D (1995) The extent and structure of genetic variation in species of the Sitopsis group of Aegilops. Heredity 74:616–627

Miczynski K Jr (1926) Études génétiques sur le genre Aegilops. I. Expériences avec l' Aegilops speltoides Jaub.et Spach. Acta Soc Bot Pol 4:20–40

Middleton C, Stein N, Keller B, Kilian B, Wicker T (2013) Comparative analysis of genome composition in Triticeae reveals strong variation in transposable element dynamics and nucleotide diversity. Plant J 73:347–356

Middleton CP, Senerchia N, Stein N, Akhunov ED, Keller B, Wicker T, Kilian B (2014) Sequencing of chloroplast genomes from wheat, barley, rye and their relatives provides a detailed insight into the evolution of the Triticeae tribe. PLoS ONE 9:e85761

Miller TE (1987) Systematics and evolution. In: Lupton FGH (ed) Wheat breeding. Its scientific basis. Chapman & Hall, London, pp 1–30

Miller TE, Hutchinson J, Chapman V (1982) Investigation of a preferentially transmitted Aegilops sharonensis chromosome in wheat. Theor Appl Genet 61:27–33

Miller TE, Hutchinson J, Reader SM (1983) The identification of the nucleolus organizer chromosomes of diploid wheat. Theor Appl Genet 65:145–147

Miller TE, Reader SM, Mahmood A, Purdie KA, King IP (1995) Chromosome 3N of Aegilops uniaristata—a source of tolerance to high levels of aluminium for wheat. In: Li ZS, Xin ZY (eds) Proceedings of 8th international wheat genetics symposium, Beijing. China Agricultural Scientech Press, Beijing, China, pp 1037–1042

Miller TE, Iqbal N, Reader SM, Mahmood A, Cant KA, King IP (1997) A cytogenetic approach to the improvement of aluminum tolerance in wheat. New Phytol 137:93–98

Millet E (2006) Rescue collections of Aegilops sharonensis in Israel. Theor Appl Genet 61:27–33

Millet E, Avivi Y, Zaccai M, Feldman M (1988) The effect of substitution of chromosome 5S[l] of Aegilops longissima for its wheat homoeologues on spike morphology and on several quantitative traits. Genome 30:473–478

Mirzaghaderi G, Houben A, Badaeva ED (2014) Molecular-cytogenetic analysis of Aegilops triuncialis and identification of its chromosomes in the background of wheat. Mol Cytogenet 7:91

Mizuno N, Yamasaki M, Matsuoka Y, Kawahara T, Takumi S (2010) Population structure of wild wheat D-genome progenitor Aegilops tauschii Coss.: implications for intraspecific lineage diversification and evolution of common wheat. Mol Ecol 19:999–1013

Mochizuki A (1957) B chromosomes in Aegilops mutica Boiss. Wheat Inf Serv 5:9–11

Mochizuki A (1960) A note on the B-chromosomes in natural populations of *Aegilops mutica* Boies. in central Turkey. Wheat Inf Serv 11:31

Mochizuki A (1964) Further studies on the effect of accessory chromosomes on chromosome pairing (in Japanese). Jap J Genet 39:356–362

Mohammadi S, Mehrabi AA, Arminian A, Fazeli A (2014) Genetic diversity structure of Aegilops cylindrica accessions revealed by genomic ISSR markers. J Pl Genet Res 1:13–26

Molnár I, Gáspár L, Sárvári E et al (2004) Physiological and morphological responses to water stress in Aegilops biuncialis and Triticum aestivum genotypes with differing tolerance to drought. Functional Pl Biol 31:1149–1159

Molnár I, Vrána J, Farkas A, Kubaláková M, Cseh A, Molnár-Láng M, Doležel J (2015) Flow sorting of C-genome chromosomes from wild relatives of wheat Aegilops markgrafii, Ae. triuncialis and Ae. cylindrica, and their molecular organization. Ann Bot 116:189–200

Molnár I, Jan Vrána J, Burešová V, Cápal P, Farkas A, Darkó E, Cseh A, Kubaláková A, Molnár-Láng M, Doležel J (2016) Dissecting the U, M, S and C genomes of wild relatives of bread wheat (Aegilops spp.) into chromosomes and exploring their synteny with wheat. The Plant J 88:452–467

Monte J-V, Casanova C, Soler C (1999) Genetic variation in Spanish populations of the genus Aegilops revealed by RAPDs. Agronomie 19:419–427

Monte JV, de Nova PJG, Soler C (2001) AFLP-based analysis to study genetic variability and relationships in the Spanish species of the genus Aegilops. Hereditas 135:233–238

Moradkhani H, Mehrabi AA, Etminan A, Pour-Aboughadareh A (2015) Molecular diversity and phylogeny of Triticum-Aegilops species possessing D genome revealed by SSR and ISSR markers. Pl Breed Seed Sci 71:81–95

Morris R, Sears ER (1967) The cytogenetics of wheat and its relatives. In: Quisenberry KS, Reitz LP (eds) Wheat and wheat improvement. Amer Soc Agron, Madison, Wisconsin, USA, pp 19–87

Mukai Y, Tsunewaki K (1975) Genetic diversity of the cytoplasm in Triticum and Aegilops. II. Comparison of the cytoplasms between four 4x Aegilops Polyeides species and their 2x relatives. Seiken Ziho 25–26:67–78

Mukai Y, Maan SS, Panayotov I, Tsunewaki K (1978) Comparative studies of the nucleus-cytoplasm hybrids of wheat produced by three research groups. In: Ramanujan S (ed) Proceedings of 5th International wheat genetics symposium, vol 1, pp 282–292

Mukai Y, Endo TR, Gill BS (1991) Physical mapping of the 18S–26S rRNA multigene family in common wheat: identification of a new locus. Chromosoma 100:71–78

Müntzing A, Jaworska H, Carlbom C (1969) Studies of meiosis in the Lindström strain of wheat carrying accessory chromosomes of rye. Hereditas 61:179–207

Murai K, Tsunewaki K (1986) Molecular basis of genetic diversity among cytoplasms of Triticum and Aegilops species. IV. CtDNA Variation in Ae. Triuncialis. Heredity 57:335–339

Naghavi MR, Aghaei MJ, Taleei AR, Omidi M, Hassani ME (2008) Genetic diversity of hexaploid wheat and three *Aegilops* species using microsatellite markers. In: Rudi A, Eastwood R, Lagudah E, Langridge P, Lynne MM (eds) Proceedings of 11th international wheat genetics symposium, Sydney University Press, Sydney

Naghavi MR, Mardi M, Pirseyedi S, Tabatabaei S (2009a) Evaluation of genetic diversity in the subspecies of Aegilops tauschii using microsatellite markers. Cereal Res Comm 37:159–167

Naghavi MR, Aghaei MJ, Taleei AR, Omidi M, Mozafari J, Hassani ME (2009b) Genetic diversity of the D-genome in T. aestivum and Aegilops species using SSR markers. Genetic Resour Crop Evol 56:499–506

Naghavi MR, Ranjbar M, Hassani MH, Aghaee MJ, Bamneshin M (2013) Characterization of Iranian accessions of Aegilops crassa Boiss. using flow cytometry and protein analysis. J Agr Sci Tech 15:811–818

Nakai Y (1979) Isosyme variation in Aegilops and Triticum. IV. The origin of the common wheats revealed from the study on esterase isozymes in synthesized wheats. Jpn J Genet 54:175–189

Nakai Y (1981) D genome doners for Aegilops cylindrica (CCDD) and Triticum aestivum (AABBDD) deduced from esterase isozyme analysis. Theor Appl Genet 60:11–16

Nakai Y (1982) D genome donors for Aegilops crassa (DDM^{cr}M^{cr}, DDD^{2}D^{2}M^{cr}M^{cr}) and Ae. Vavilovii (DDM^{cr}M^{cr}S^{p}S^p) deduced from esterase analysis by isoelectric focusing 1. Jpn 57:349–360

Nakai Y, Tsuji S (1984) A polyacrylamide gel isoelectrofocusing study of acid phosphatase isozymes of Aegilops species with reference to their numerical taxonomy. Seiken Zihô 32:1–8

Nakai Y, Tsunewaki K (1971) Isozymes variations in Aegilops and Triticum. I. Esterase isozymes in Aegilops studied using the gel isoelectrofocusing method. Jpn J Genet 46:321–336

Naranjo T (1995) Chromosome structure of Triticum longissimum relative to wheat. Theor Appl Genet 91:105–109

Naranjo T, Maestra B (1995) The effect of ph mutations on homoeologous pairing in hybrids of wheat with Triticum longissimum. Theor Appl Genet 91:1265–1270

Nasuda S, Friebe B, Gill BS (1998) Gametocidal genes induce chromosome breakage in the interphase prior to the first mitotic cell division of the male gametophyte in wheat. Genetics 49:1115–1124

Nasuda S, Kikkawa Y, Ashida T, Rafiqul Islam AKMR, Sato K, Endo TR (2005) Chromosomal assignment and deletion mapping of barley EST markers. Genes Genet Systems 80:357–366

Navashin MS (1928) Amphiplastie—eine neue karyologische Erscheinung. In: Proceedings of international conference on genetics, vol 5, pp 1148–1152

Navashin MS (1934) Chromosomal alterations caused by hybridization and their bearing upon certain general genetic problems. Cytologia 5:169–203

Netzle S, Zeller FJ (1984) Cytogenetic relationship of Aegilops longissima chromosomes with common wheat chromosomes. Plant Syst Evol 145:1–13

Nevo E, Krugman T, Beiles A (1984) Edaphic natural selection of allozyme polymorphisms in Aegilops peregrina at a Galilee microsite in Israel. Heredity 72:109–112

Nevski SA (1934a) Hordeae Bentb. In: Komarov VL (ed) Flora of the USSR II, p 590

Nevski SA (1934b) Schedae ad Herbarium Flora Asiae Mediae. Acta Univ Asiae Med VIII b Bot 17:1–94

Nishikawa K (1973) Alpha-amylase isozymes and phylogeny of hexaploid wheal. In: Sears ER, Sears LMS (eds) Proceedings of 4th international wheat genetics symposium, Columbia MO, pp 851–855

Nishikawa K, Furuta Y, Wada T (1980) Genetic studies on α-amylase isozymes in wheat. III. Intraspecific variation in Aegilops squarrosa and birthplace of hexaploid wheat. Jpn J Genet 55:325–336

Ogihara Y, Tsunewaki K (1982) Molecular basis of the genetic diversity of the cytoplasm in Triticum and Aegilops. I. Diversity of the chloroplast genome and its lineage revealed by the estriction pattern of chloroplast DNAs. Jpn J Genet 57:371–396

Ogihara Y, Tsunewaki K (1983) The diversity of chloroplast DNA among *Triticum* and *Aegilops* species. In: Sakamoto S (ed) Proceedings of 6th international wheat genetics symposium, Kyoto, pp 407–413

Ogihara Y, Tsunewaki K (1988) Diversity and evolution of chloroplast DNA in Triticum and Aegilops as revealed by restriction fragment analysis. Theor Appl Genet 76:321–332

Ohta S (1990) Genome analysis of *Aegilops mutica* Boiss. based on the chromosome pairing in interspecific and intergeneric hybrids. Ph.D. Thesis, submitted to the University of Kyoto, pp 1–267

Ohta S (1991) Phylogenetic relationship of Aegilops mutica Boiss with the diploid species of congeneric Aegilops-Triticum complex, based on the new method of genome analysis using its B-chromosomes. Mem Coll Agric Kyoto Univ 137:1–116

Ohta S (1992) Intraspecific hybrid sterility in Aegilops caudata L. Hereditas 116:247–251

Ohta S (1995a) Distinct numerical variation of B-chromosomes among different tissues in Aegilops mutica Boiss. Jap J Genet 70:93–101

Ohta S (1995b) Cytogenetic evidence for cryptic structural hybridity causing intraspecific hybrid sterility in Aegilops caudata L. Jpn J Genet 70:355–364

Ohta S (2000) Genetic differentiation and post-glacial establishment of the geographical distribution in Aegilops caudata L. Genes Genet Syst 75:189–196

Ohta S, Tanaka M (1983) Genome relationships between *Ae. mutica* and the other diploid *Aegilops* and *Triticum* species, based on the chromosome pairing in the hybrids with or without B-chromosomes. In: Sakamoto S (ed) Proceedings of 6th international wheat genetics symposium, Kyoto, pp 983–991

Ohta S, Yasukawa N (2015) Genetic variation and its geographical distribution in Aegilops caudata L.: morphology, hybrid sterility and gametocidal genes. In: Ogihara Y, Takumi S, Handa H (eds) Advances in wheat genetics: from genome to field. Springer, Tokyo, pp 53–61

Ohta S, Nosaka Y, Yamagata H (2016) Allopatric distributions of tetraploid and hecapoid forms of Aegilops neglecta Req. ex Bertol. Genet Resour Crop Evol 63:193–197

Okamoto M, Inomata N (1974) Possibility of 5B-like effect in diploid species. Wheat Inf Serv 38:15–16

Okuno K, Ebana K, Voov B, Yoshida H (1998) Genetic divers of central Asian and north Caucasian Aegilops species as revealed by RAPD markers. Gen Resour Crop Evol 45:389–394

Olivera PD, Steffenson BJ (2009) Aegilops sharonensis: origin, genetics, diversity, and potential for wheat omprovement. Botany 87:740–756

Olivera PD, Anikster Y, Steffenson BJ (2010) Genetic diversity and population structure in Aegilops sharonensis. Crop Sci 50:636–648

Orellana J, Santos JL, Lacadena JR, Cermeño MC (1984) Nucleolar competition analysis in Aegilops ventricosa and its amphiploids with tetraploid wheats and diploid rye by the silver-staining procedure. Can J Genet Cytol 26:34–39

Orth RA, Bushuk W (1973) Studies of glutenin: III. Identification of subunits coded by the D-genome and their relation to breadmaking quality. Cereal Chem 50:680–687

Ozgen M, Yildiz M, Ulukan KN (2004) Association of gliadin protein pattern and rust resistance derived from Aegilops ubmbellulata Zhuk. in winter Triticum durum Desf. Breed Sci 54:287–290

Panayotov I, Gotsov K (1975) Results of nucleus substitution in Aegilops and Triticum species by means of successive backcrossing with common wheat. Wheat Inf Serv 40:20–22

Panayotov I, Gotsov K (1976) Interactions between Aegilops cytoplasms and Triticum species. Cereal Res Commun 4:297–306

Parisod C, Definod C, Sarr A, Arrigo N, Felber F (2013) Genome-specific introgression between wheat and its wild relative Aegilops triuncialis. J Evol Biol 26:223–228

Pathak GN (1940) Studies in the cytology of cereals. J Genet 39:437–467

Pazy B, Zohary D (1965) The process of introgression between Aegilops polyploids: natural hybridization between Ae. variabilis, Ae. ovata, and Ae. biuncialis. Evolution 19:385–394

Percival J (1921) The wheat plant: a monograph. Duckworth, London, pp 1–46

Percival J (1930) Cytological studies of some hybrids of Aegilops sp. x wheats, and of some hybrids between different species of Aegilops. J Genet 22:201–278

Percival J (1932) Cytological studies of some wheat and Aegilops hybrids. Ann Bot 46:479–501

Pester TA, Ward SM, Fenwick AL, Westra P, Nissen SJ (2003) Genetic diversity of jointed goatgrass (Aegilops cylindrica) determined with RAPD and AFLP markers. Weed Sci 51:287–293

Pestsova E, Korzun V, Goncharov NP, Hammer K, Ganal MW, Röder MS (2000) Microsatellite analysis of Aegilops tauschii germplasm. Theor Appl Genet 101:100–106

Petersen G, Seberg O, Yde M, Berthelsen K (2006) Phylogenetic relationships of Triticum and Aegilops and evidence for the origin of the A, B, and D genomes of common wheat (Triticum aestivum). Mol Phylogenet Evol 39:70–82

Pietro ME, Tuleen NA, Hart GE (1988) Development of wheat-*Triticum searsii* disomic chromosome addition lines. In: Miller TE, Koebner RMD (eds) Proceedings of 7th international wheat genetics symposium, Bath Press, Bath, UK, pp 409–413

Pikaard CS (1999) Nucleolar dominance and silencing of transcription. Trend Pl Sci Rev 4:478–483

Pikaard CS (2000) The epigenetics of nucleolar dominance. Trends Genet 16:495–500

Post GE (1933) Flora of Syria, Palestine and Sinai, vol II, 2nd ed, revised and enlarged by Dinsmore JE. American Press, Beirut, p 788

Qi LL, Echalier B, Chao S, Lazo GR, Butler GE, Anderson OD, Akhunov ED, Dvorak J, Linkiewicz AM, Ratnasiri A, Dubcovsky J, Bermudez-Kandianis CE, Greene RA, Kantety R, La Rota CM, Munkvold JD, Sorrells SF, Sorrells ME, Dilbirligi M, Sidhu D, Erayman M, Randhawa HS, Sandhu D, Bondareva SN, Gill KS, Mabmoud AA, Ma XF, Miftahudin GJP, Conley EJ, Nduati V, Gonzalez-Hernandez JL, Anderson JA, Peng JH, Lapitan NL, Hossain KG, Kalavacharla V, Kianian SF, Pathan MS, Zhang DS, Nguyen HT, Choi DW, Fenton RD, Close TJ, McGuire PE, Qualset CO, Gill BS (2004) A chromosome bin map of 16,000 expressed sequence tag loci and distribution of genes among the three genomes of polyploid wheat. Genetics 168:701–712

Rabokon A, Demkovych A, Sozinov A, Kozub N, Sozinov I, Pirko Y, Blume Y (2019) Intron length polymorphism of β-tubulin genes of Aegilops biuncialis Vis. Cell Biol Int 43:1031–1039

Ranjbar M, Naghavi MR, Zali A, Aghaei MJ (2007) Multivariate analysis of morphological variation in accessions of Aegilops crassa from Iran. Pakistan J Biol Sci 10:1126–1129

Raskina O, Betyayev A, Nevo E (2001) Repetitive DNA of the wild emmer wheat *Triticum dicoccoides* Korn: hybridization analysis of the relatedness of the S-genome B-genome. No. Rep-11261, CIMMYT

Raskina O, Belyayev A, Nevo E (2004a) Activity of the En/Spm-like transposons in meiosis as a base for chromosome repatterning in a small, isolated, peripheral population of Aegilops speltoides Tausch. Chromosome Res 12:153–161

Raskina O, Belyayev A, Nevo E (2004b) Quantum speciation in Aegilops: molecular cytogenetic evidence from rDNA cluster variability in natural populations. Proc Natl Acad Sci USA 101:14818–14823

Raskina O, Barber J, Nevo E, Belyayev A (2008) Repetitive DNA and chromosomal rearrangements: Speciation-related events in plant genome. Cytogenet Genome Res 120:351–357

Raskina O, Brodsky L, Belyayev A (2011) Tandem repeats on an eco-geographical scale: outcomes from the genome of Aegilops speltoides. Chromosome Res 19:607–623

Rayburn AL, Gill BS (1986) Isolation of a D-genome specific repeated DNA sequence from Aegilops squarrosa. Plant Mol Biol Rep 4:104–109

Rayburn AL, Gill BS (1987) Molecular analysis of the D-genome of the Triticeae. Theor Appl Genet 73:385–388

Rees H, Walters MR (1965) Nuclear DNA and the evolution of wheat. Heredity 20:73–82

Resta P, Zhang HB, Dubkovsky J et al (1996) The origin of the genomes of Triticum biunciale, T. ovatum, T. neglectum, T. columnare, and T. rectum based on variation in repeated nucleotide sequences. Am J Bot 83:1556–1565

Rice KJ, Gerlach J, Dyer AR, Mckay JK (2013) Evolutionary ecology along invasion fronts of the annual grass Aegilops triuncialis. Biol Invasions 15:2531–2545

Riley R (1966a) The genetic regulation of meiotic behavior in wheat and its relatives. In: Mac Key J (ed) Proceedings of second international wheat genetics symposium. Genetic Institute, University of Lund, Sweden, Aug 1963, Hereditas (suppl) 2:395–406

Riley R (1966b) Genetics and the regulation of meiotic chromosome behavior. Sci Progr (London) 54:193–207

Riley R, Chapman V (1963) The effects of the deficiency of chromosome V (5B) of Triticum aestivum on the meiosis of synthetic amphiploids. Heredity 18:473–484

Riley R, Law CN (1965) Genetic variation in chromosome pairing. Adv Genet 13:57–114

Riley R, Unrau J, Chapman V (1958) Evidence on the origin of the B genome of wheat. J Hered 49:91–98

Riley R, Kimber G, Chapman V (1961) Origin of genetic control of diploid-like behavior of polyploid wheat. Hered 52:22–25

Riley R, Chapman V, Johnson R (1968) The incorporation of alien disease resistance in wheat by genetic interference with the regulation of meiotic chromosome synapsis. Genet Res 12:199–219

Romero MD, Montes ML, Sin E, Lopez-Braña I, Duce A, Martín-Sanchez JA, Andrés MF, Delibes A (1998) A cereal cyst nematode (Heterodera avenae Woll.) resistance gene transferred from Aegilops triuncialis to hexaploid wheat. Theor Appl Genet 96:1135–1140

Roy RP (1959) Genome analysis of Aegilops sharonensis. Genetica 29:331–357

Ruban AS, Badaeva ED (2018) Evolution of the S-genomes in Triticum-Aegilops alliance: evidences from chromosome analysis. Front Plant Sci 9:1756

Rubenstein JM, Sallee PJ (1973) The genome relationships of Triticum kotschyi to the S-genome diploids. In: Sears ER, Sears LMS (eds) Proceedings of 4th international wheat genetics symposium, Agricultural Experiment Station, College of Agriculture, University of Missouri, pp 95–99

Sadeghian A, Mohsen SHH, Hatami A (2015) Investigation of chromosome variation in four Aegilops L. (Poaceae) species and populations in Iran. IUFS J Biol 74:9–16

Saeidi H, Rahiminejad MR, Vallian S, Heslop-Harrison JS (2006) Biodiversity of D-genome Aegilops tauschii Coss. in Iran measured using microsatellites. Genet Resour Crop Evol 53:1477–1484

Saeidi H, Tabatabaei BES, Rahimmalek M, Talebi-Badaf M, Rahiminejad MR (2008) Genetic diversity and gene-pool subdivisions of diploid D-genome Aegilops tauschii Coss. (Poaceae) in Iran as revealed by AFLP. Genet Resour Crop Evol 55:1231–1238

Sakai K, Nasuda S, Sato K, Endo TR (2009) Dissection of barley chromosome 3H in common wheat and comparison of 3H physical and genetic maps. Genes Genet Syst 84:25–34

Sakamoto S (1973) Patterns of phylogenetic differentiation in the tribe Triticeae. Seiken Zihô 24:11–31

Sakata M, Nasuda S, Endo TR (2010) Dissection of barley chromosome 4H in common wheat by the gametocidal system and cytological mapping of chromosome 4H with EST markers. Genes Genet Syst 85:19–29

Salina EA, Adonina IG, Vatolina T, Kurata N (2004a) A comparative analysis of the composition and organization of two sub-telomeric repeat families in Aegilops speltoides Tausch. and related species. Genetica 122:227–237

Salina EA, Numerova OM, Ozkan H, Feldman M (2004b) Alterations in sub-telomeric tandem repeats during early stages of allopolyploidy in wheat. Genome 47:860–867

Salina EA, Lim KY, Badaeva ED, Shcherban AB, Adonina IG, Amosova AV, Samatadze TE, Vatolina TY, Zoshchuk SA, Leitch AR (2006) Phylogenetic reconstruction of Aegilops section Sitopsis and the evolution of tandem repeats in the diploids and derived wheat polyploids. Genome 49:1023–1035

Sasanuma T, Miyashita NT, Tsunewaki K (1996) Wheal phylogeny determined by RFLP analysis of nuclear DNA. 3. Intra- and interspecific variations of five Aegilops Sitopsis species. Theor Appl Genet 92:928–934

Sasanuma T, Kadosumi S, Kawahara T (2006) Multiple origins of U genome in two UM genome tetraploid Aegilops species, Ae. biuncialis and Ae. ovata. Wheat Inf Serv 101

Sax K, Sax HJ (1924) Chromosome behavior in a genus cross. Genetics 9:454–464

Schachtman DP, Lagudah ES, Munns R (1992) The expression of salt tolerance from Triticum tauschii in hexaploid wheat. Theor Appl Genet 84:714–719

Schiemann E (1928) Zytologische und pflanzen-geographische Beiträge zur Gatung Aegilops (II. Miteilung). Ber Deutsch Bot Ges 46:107–123

Schneider A, Linc G, Molnár I, Molnár-Láng M (2005) Molecular cytogenetic characterization of Aegilops biuncialis and its use for the identification of 5 derived wheat—Aegilops biuncialis disomic addition lines. Genome 48:1070–1082

Schubert I, Shi F, Fuchs J, Endo TR (1998) An efficient screening for terminal deletions and translocations of barley chromosomes added to common wheat. The Plant J 14:489–495

Scott JC, Manisterski J, Sela H, Ben-Yehuda P, Steffenson BJ (2014) Resistance of Aegilops species from Israel to widely virulent African and Israeli races of the wheat stem rust pathogen. Pl Dis 98:1309–1320

Sears ER (1939) Amphidiploids in the Triticinae induced by colchicine. J Hered 30:38–43

Sears ER (1941a) Identification of the wheat chromosome carrying leaf-rust resistance from Aegilops umbellulata. Wheat Inf Serv 12:12

Sears ER (1941b) Chromosome pairing and fertility in hybrids and amphidiploids in the Triticinae. Univ Missouri Agric Exp Stn Res Bull 337:1–20

Sears ER (1944a) Inviability of intergeneric hybrids involving Triticum monococcum and T. aegilopoides. Genetics 29:113–127

Sears ER (1944b) The amphidiploids *Aegilops cylindrica* x *Triticum durum* and *A. ventricosa* x *T. durum* and their hybrids with *T. aestivum*. J Agric Res 68:135–144

Sears ER (1948) The cytology and genetics of the wheats and their relatives. Adv Genet 2:239–270

Sears ER (1956) The transfer of leaf-rust resistance from *Aegilops umbellulata* to wheat. In: Genetics in plant breeding Brook-haven symposia in biology, pp 1–22

Seefeldt SS, Zemetra R, Young FL, Jones SS (1998) Production of herbicide-resistant jointed goatgrass (Aegilops cylindrica) × wheat (Triticum aestivum) hybrids in the field by natural hybridization. Weed Sci 46:632–534

Senerchia N, Wicker T, Felber F, Parisod C (2013) Evolutionary dynamics of retrotransposons assessed by high throughput sequencing in wild relatives of wheat. Genome Biol Evol 5:1010–1020

Senyaninova-Korchagina MV (1930) Karyo-systematical investigation of the genus *Aegilops*. In: Proceedings of USSR congress genetics plant animal breed Leningrad 1929. Genetics 2:453–466

Senjaninova-Korczagina MV (1932) Karyo-systematical investigations of the genus *Aegilops* L. (Russian with English summary) Bull Appl But Genet Pl Breed Ser 2:1–90

Serizawa N, Nasuda S, Shi F, Endo TR, Prodanovic S, Schubert I, Künzel G (2001) Deletion-based physical mapping of barley chromosome 7H. Theor Appl Genet 103:827–834

Shands H, Kimber G (1973) Reallocation of the genomes of *Triticum timopheevii* Zhuk. In: Sears ER, Sears LMS (eds) Proceedings of 4th international wheat genetics symposium, Agricultural Experiment Station, College of Agriculture, University of Missouri, pp 101–108

Shang XM, Jackson RC, Nguyen HT, Huang JY (1989) Chromosome pairing in the Triticum monococcum complex: evidence for pairing control genes. Genome 32:216–226

Shi F, Endo TR (1997) Production of wheat-barley disomic addition lines possessing an Aegilops cylindrica gametocidal chromosome. Genes Genet Syst 74:49–54

Shigenobu T, Sakamoto S (1977) Production of a polyhaploid plant of Aegilops crassa 6x pollinated by Hordeum bulbosum. Jpn J Genet 52:397–402

Siddique AK, Jones JK (1967) The D genomes of Aegilops crassa Boiss. Bol Genet Inst Fitotec Castelar 4:29–31

Simchen G, Zarchi Y, Hillel J (1971) Supernumerary chromosomes in second outbreeding species of wheat group. Chromosoma 33:63–69

Singh S, Franks CD, Huang L, Brown-Guedira GL, Marshall DS, Gill BS, Fritz A (2004) Lr41, Lr39, and a leaf rust resistance gene from Aegilops cylindrica may be allelic and are located on wheat chromosome 2DS. Theor Appl Genet 108:586–591

Siregar UJ, Ishii T, Tsunewaki K (1988) *Aegilops searsii* is a possible cytoplasm donor to *Ae. kotschyi* and *Ae. variabilis*. In: Miller TE, Koebner RMD (eds) Proceedings of 7th international wheat genetics symposium, Cambridge, England, Bath Press, Bath, Avon, UK, pp 145–152

Slotkin RK, Martienssen R (2007) Transposable elements and the epigenetic regulation of the genome. Nat Rev Genet 8:272–285

Sohail Q, Shehzad T, Kilian A, Eltayeb AE, Tanaka H, Tsujimoto H (2012) Development of diversity array technology (DArT) markers for assessment of population structure and diversity in Aegilops tauschii. Breed Sci 62:38–45

Sorokina ON (1928) On the chromosomes of Aegilops species. Bull Appl Bot Gen Pl Br 19:523–532

Spetsov P, Mingeot D, Jacquemin JM, Samardjieva K, Marinova E (1997) Transfer of powdery mildew resistance from Aegilops variabilis into bread wheat. Euphytica 93:49–54

Stebbins GL (1956) Taxonomy and evolution of genera. with special reference to the family Gramineae. Evolution 10:235–245

Stoilova T, Spetsov P (2006) Chromosome 6U from Aegilops geniculata Roth carrying powdery mildew resistance in bread wheat. Breed Sci 56:351–357

Symeonidis L, Karataglis S, I. Tsekos I (1979) Electrophoretic variation in esterases and peroxidases of native Greek diploid *Aegilops* species (*Ae. caudata* and *Ae. comosa*, Poaceae). Plant Syst Evol 13:1–15

Takumi S, Mizuno N, Okumura Y, Kawahara T, Matsuoka Y (2008) Two major lineages of *Aegilops tauschii* Coss. revealed by nuclear DNA variation analysis. In: Appels R, Eastwood R, Lagudah E, Langridge P, Mackay M, McIntyre L, Sharp P (eds) Proceedings of 11th international wheat genetics symposium, Sydney, Australia, Sydney University Press, pp 312–314

Talbert LE, Magyar GM, Lavin M, Blake TK, Moylan SL (1991) Molecular evidence for the origin of the S-derived genomes of polyploid Triticum species. Amer J Botany 78:340–349

Talbert LE, Kimber G, Magyar GM, Buchanan CB (1993) Repetitive DNA variation and pivotal–differential evolution of wild wheats. Genome 36:14–20

Talbert LE, Smith LY, Blake NK (1998) More than one origin of hexaploid wheat is indicated by sequence comparison of low-copy DNA. Genome 41:402–407

Tanaka M (1955a) Chromosome pairing in hybrids between Aegilops sharonensis and some species of Aegilops and Triticum. Wheat Infor Serv 2:7–8

Tanaka M (1955b) A new amphidiploid from the hybrid Ae. sharonensis x Ae. umbellulata. Wheat Inf Serv 2:8–9

Tanaka M (1983) Geographical distribution of *Aegilops* species based on the collections at the Plant Germ-plasm Institute Kyoto University. In: Sakamoto S (ed) Proceedings of 6th international wheat genetics symposium, Kyoto, Japan, pp 1009–1024

Tanaka S, Matsumoto K (1965) Karyotype analysis in the genus Aegilops. I. Karyotypes of C, Cu and D genomes. Mem Osaka Uni Lib Arts Edu, B 14:212–219

Tanaka M, Suemoto H, Ichikawa S (1967) The awn characters and sterility in Aegilops caudata L. I. Jpn J Breed 17(Suppl 2):155–156 (in Japanese)

Tanguy A-M, Coriton O, Abélard P, Dedryver F, Jahier J (2005) Structure of Aegilops ventricosa chromosome 6N^v, the donor of wheat genes Yr17, Lr37, Sr38 and Cre5. Genome 48:541–546

Teho SB, Hutchinson J (1983) Interspecific variation in C-banded chromosomes of diploid Aegilops species. Theor Appl Genet 65:31–40

Teho SB, Hutchinson J, Miller TE (1983) A comparison of the chromosomal distribution of cloned repetitive DNA sequences in different Aegilops species. Heredity 51:635–641

Terachi T, Tsunewaki K (1986) The molecular basis of genetic diversity among cytoplasms of Triticum and Aegilops. 5. Mitochondrial genome diversity among Aegilops species having identical chloroplast genome. Theor Appl Genet 73:175–181

Terachi T, Tsunewaki K (1992) The molecular basis of genetic diversity among cytoplasms of Triticum and Aegilops. VIII. Mitochondrial RFLP analyses using cloned genes as probes. Mol Biol Evol 9:911–931

Terachi T, Ogihara Y, Tsunewaki K (1984) The molecular basis of genetic diversity among cytoplasms of Triticum and Aegilops. III. Chloroplast genomes of the M and modified M genome-carrying species. Genetics 108:681–695

Terachi T, Ogihara Y, Tsunewaki K (1990) The molecular basis of genetic diversity among cytoplasms of Triticum and Aegilops. 7. Restriction endonuclease analysis of mitochondrial DNAs from polyploid wheats and their ancestral species. Theor Appl Genet 80:366–373

Thomas KG, Bebell PJ (2010) Genetic diversity of Greek Aegilops species using different types of nuclear genome markers. Mol Phylogenet Evol 56:951–961

Tsuchiya T (1956) Hybrids berween Aegilops triaristata (4x) and Ae. comosa, Heldreichii and Uniaristata. Wheat Inf Serv 3:22–23

Tsujimoto H (1993) Molecular cytological evidence for gradual telomere synthesis at the broken chromosome ends in wheat. J Plant Res 106:239–244

Tsujimoto H (1995) Gametocidal genes in wheat and its relatives. IV. Functional relationship between six gametocidal genes. Genome 38:283–289

Tsujimoto H (2006) Gametocidal genes in wheat as the inducer of chromosome breakage. Wheat Inf Serv 100:33–48

Tsujimoto H, Noda K (1989) Structure of chromosome 5A of wheat speltoid mutants induced by the gametocidal genes of Aegilops speltoides. Genome 32:1085–1090

Tsujimoto H, Tsunewaki K (1984) Gametocidal genes in wheat and its relatives. I. Genetic analyses in common wheat of a gametocidal gene derived from Aegilops speltoides. Can J Genet Cytol 26:78–84

Tsujimoto H, Tsunewaki K (1985a) Gametocidal genes in wheat and its relatives. II. Suppressor of the chromosome 3C gametocidal gene of Aegilops triuncialis. Can J Genet Cytol 27:178–185

Tsujimoto H, Tsunewaki K (1985b) Hybrid dysgenesis in common wheat caused by gametocidal genes. Jpn J Genet 6:565–578

Tsujimoto H, Tsunewaki K (1988) Gametocidal genes in wheat and its relatives. III. Chromosome location and effects of two Aegilops speltoides-derived gametocidal genes in common wheat. Genome 30:239–244

Tsujimoto H, Tsujimoto A, Tanaka M (1990) Abnormal pollen development in wheat lines carrying the gametocidal genes. Jpn J Breed 40(suppl. 2):396–397 (in Japanese)

Tsujimoto H, Yamada T, Sasakuma T (1997) Molecular structure of a wheat chromosome end healed after gametocidal-gene-induced breakage. Proc Natl Acad Sci USA 94:3140–3144

Tsujimoto H, Usami N, Hasegawa K, Yamada T, Nagaki K, Sasakuma T (1999) De novo synthesis of telomere sequence at the healed breakpoints of wheat deletion chromosomes. Mol Gen Genet 262:851–856

Tsujimoto H, Yamada T, Hasegawa K, Usami N, Kojima T, Endo TR, Ogihara Y, Sasakuma T (2001) Large scale selection of lines with deletions in chromosome 1B in wheat and applications for fine deletion mapping. Genome 44:501–508

Tsunewaki K (1966) Comparative gene analysis of common wheat and Its ancestral species. III. Glume hairiness. Genetics 53:303–311

Tsunewaki K (1968) Origin and phylogenetic differentiation of bread wheat revealed by comparative gene analysis. In: Finley KW, Shepherd KW (eds.) Proceedings of 3rd International wheat genetics symposium, Canberra, Australia, pp 71–85

Tsunewaki K (ed) (1980) Genetic diversity of the cytoplasm in Triticum and Aegilops. Jpn Soc Prom Sci, Tokyo, pp 1–290

Tsunewaki K (1989) Plasmon diversity in Triticum and Aegilops and its implication in wheat evolution. Genome 31:143–154

Tsunewaki K (1993) Genome-plasmon interactions in wheat. Jpn J Genet 68:1–34

Tsunewaki K (1996) Cytoplasmic genome analysis: phylogenetic inferences: plasmon analysis as the counterpart of genome analysis. In: Jauhar PP (ed) Methods of genome analysis in Plants. CRC, Boca Raton, FL, pp 271–298

Tsunewaki K (2009) Plasmon analysis in the Triticum-Aegilops complex. Breed Sci 59:455–470

Tsunewaki K, Ogihara Y (1983) The molecular basis of genetic diversity among cytoplasms of Triticum and Aegilops species. II. On the origin of polyploid wheat cytoplasms as suggested by chloroplast DNA restriction fragment patterns. Genetics 104:155–171

Tsunewaki K, Tsujimoto H (1983) Genetic diversity of the cytoplasm in Triticum and Aegilops. In: Sakamoto S (ed) Proceedings of 6th international wheat genetics symposium Kyoto, pp 1139–1144

Tsunewaki K, Mukai Y, Endo TR (1978) On the descent of the cytoplasms of polyploid species of Triticum and Aegilops. In: Ramanujan S (ed) Proceedings of 5th international wheat genetics symposium, Indian Society Genetics Plant Breeding, vol 1, Indian Agricultural Research Institute, New Delhi, India, pp. 261–272

Tsunewaki K, Takumi S, Mori N, Achiwa T, Liu YG (1991) Origin of polyploid wheats revealed by RFLP analysis. In: Sasakuma T, Kinoshita T (eds) Nuclear and organellar genomes of wheat species. Kihara Memo Found, Yokohama, pp 31–39

Tsunewaki K, Wang GZ, Matsuoka Y (2002) Plasmon analysis of Triticum (wheat) and Aegilops. 2. Characterization and classification of 47 plasmons based on their effects on common wheat phenotype. Genes Genet Syst 77:409–427

Tsunewaki K, Mori N, Takumi S (2014) Genetic effect of the Aegilops caudata plasmon on the manifestation of the Ae. cylindrica genome. Genes Genet Syst 89:195–202

Tutin TG, Humphries CJ (1980) Aegilops. In: Tutin TG et al.(eds) Flora Europaea, vol 5, Cambridge, p 201

Tzvelev NN (1976) Poaceae URSS. Tribe III. Triticeae Dum. USSR. Academy of Sciences Press, Leningrad, pp 105–206

Upadhya MD (1966) Altered potency of chromosome 5B in wheat-caudata hybrids. Wheat Inf Serv 22:7–9

Upadhya MD, Swaminathan MS (1967) Mechanism regulating chromosome pairing in Triticum. Biol Zentral (suppl) 86:239–255

Valkoun J, Hammer K, Kučerová D, Bartoš P (1985) Disease resistance in the genus Aegilops L.—stem rust, leaf rust, stripe rust, and powdery mildew. Genet Resour Crop Evol 33:133–153

van Slageren MW (1994) Wild wheats: a monograph of Aegilops L. and Amblyopyrum (Jaub. and Spach) Eig (Poaceae). Agricultural University, Wageningen, The Netherlands

Vanichanon A, Blake N, Sherman J, Talbert L (2003) Multiple origins of allopolyploid Aegilops triuncialis. Theor Appl Genet 106:804–810

Vardi A (1973) Introgression between different ploidy levels in the wheat group. In: Sears ER, Sears LMS (eds) Proceedings of 4th international wheat genetics symposium, Agricultural Experiment Station, College of Agriculture, University of Missouri, pp 131–141

Vardi A, Dover GA (1972) The effect of B chromosomes on meiotic and pre-meiotic spindles and chromosome pairing in Triticum/Aegilops hybrids. Chromosoma (berl.) 38:367–385

Vardi A, Zohary D (1967) Introgression in wheat via triploid hybrids. Heredity 22:541–560

Visiani R (1842) Flora dalmatica, vol 1, Lipisiae

Visiani R (1852) Flora dalmatica, vol 3, Lipsiae

von Berg KH (1931) Autosyndese in Aegilops triuncialis L. x Secale cereale L. Zeitschr f Züchtung: A Pflanzenzüchtung 17:55–69

von Berg KH (1937) Beitrag zur Genomanalyse in der Getreide-gruoppe. Der Züchter 9:157–163

von Willdenow CL (1806) Species plantarum (ed) 4:943–944

Waines JG (1976) A model for the origin of diploidizing mechanisms in polyploid species. Amer Naturalist 110:415–430

Waines JG, Johnson BL (1972) Genetic differences between Aegilops longissima, A. sharonensis, and A. bicornis. Can J Genet Cytol 14:411–415

Waines JG, Barnhart D (1992) Biosystematic research in Aegilops and Triticum. Hereditas 116:207–212

Waines JG, Hilu K, Sharma H (1982) Species formation in Aegilops and Triticum. In: Estes JR, Tyrl RJ, Brunken JN (eds) Grasses and Grasslands. Systematic and ecology, pp 94–95, 97

Wan Y, Liu K, Wang D, Shewry PR (2000) High-molecular-weight glutenin subunits in the Cylindropyrum and Vertebrata section of the Aegilops genus and identification of subunits related to those encoded by the Dx alleles of common wheat. Theor Appl Genet 101:879–884

Wang G-Z, Miyashita N, Tsunewaki K (1997) Plasmon analyses of Triticum (wheat) and Aegilops: PCR–single-strand conformational polymorphism (PCR-SSCP) analyses of organellar DNAs. Proc Natl Acad Sci USA 94:14570–14577

Wang J-B, Wang C, Shi S-H, Zhong Y (2000) ITS regions in diploids of Aegilops (Poaceae) and their phylogenetic implications. Hereditas 132:209–213

Wang J, Zhang W, Zhao H (2013a) Molecular cytogenetic characterization of the Aegilops biuncialis karyotype. Genet Mol Res 12:683–692

Wang J, Luo M-C Chen Z,You FM, Wei Y, Zheng Y, Dvorak J (2013b) *Aegilops tauschii* single nucleotide polymorphisms shed light on the origins of wheat D-genome genetic diversity and pinpoint the geographic origin of hexaploid wheat. New Phytol 198:925–937

Watanabe N, Kawahara T (1999) Aegilps species collected in California and Oregon, USA. Wheat Inf Serv 89:33–36

Watanabe N, Mastui K, Furuta Y (1994) Uniformity of the alpha-amylase isozymes of *Aegilops cylindrica* introduced into North America: comparisons with ancestral Eurasian accessions. In: Wang K, Jensen B, Jaussi C (eds) Proceedings of 2nd international wheat symposium, Utah State University, Logan, USA, pp 215–218

Watson L, Dallwitz MJ (1992) The grass genera of the world. C.A.B. International, Wallingford, p 1038

Weissman S, Feldman M, Gressel J (2005) Sequence evidence for sporadic inter-generic DNA introgression from wheat into wild Aegilops species. Mol Biol Evol 22:2055–2062

Werner JE, Kota RS, Gill BS, Endo TR (1992) Distribution of telomeric repeats and their role in the healing of broken chromosome ends in wheat. Genome 35:844–848

Wicker T, Buell CR (2009) Gene and repetitive sequence annotation in the Triticeae. In: Muehlbauer GJ, Feuillet C (eds) Genetics and genomics of the Triticeae. Springer, New York, pp 407–425

Witcombe JR (1983) A guide to the species of *Aegilops* L. Their taxonomy, morpholog and distribution. International Board for Plant Genetic Resources (IBPGR), Rome, pp 1–74

Worthington M, Lyerly J, Petersen S, Brown-Guedira G, Marshall D, Cowger C, Parks R, Murphy JP (2015) MlUM15: an Aegilops neglecta-derived powdery mildew resistance gene in common wheat. Crop Sci 54:1397–1406

Yaakov B, Kashkush K (2012) Mobilization of Stowaway-like MITEs in newly formed allohexaploid wheat species. Plant Mol Biol 80:419–427

Yaakov B, Meyer K, Ben-David S, Kashkush K (2013) Copy number variation of transposable elements in Triticum-Aegilops genus suggests evolutionary and revolutionary dynamics following allopolypoliduzation. Plant Cell Rep 32:1615–1624

Yamada I, Suzuki E (1941) The behavior of tetravalent chromosomes and its bearing on sterility in Aegilops Heldreichii × Ae. comosa and its offsprings. Jpn J Genet 17:83–96

Yamamoto M (1992a) Detection of ribosomal RNA genes in Aegilops by in situ hybridization. Bull Osaka Priv Coll Assoc 29:77–82

Yamamoto M (1992b) Distribution of ribosomal RNA genes in Aegilops and Triticum chromosomes. Bull Kansai Women Coll 2:25–37

Yamamoto M, Mukai Y (1995) Physical mapping of ribosomal RNA genes in *Aegilops* and *Triticum*. In: Li S, Xin ZY (eds) Proceedings of 8th international wheat genetics symposium, Bejing, China, 20–25 July 1993, pp 807–811

Yamane K, Kawahara T (2005) Intra- and interspecific phylogenetic relationships among diploid Triticum-Aegilops species (Poaceae) based on base-pair substitutions, indels, and microsatellites in chloroplast noncoding sequences. Am J Bot 92:1887–1898

Yen Y, Kimber G (1989) Triploid hybrid between autotetraploid Triticum longissimum and Triticum speltoides. Cereal Res Commun 17:259–264

Yen Y, Kimber G (1990a) Genomic relationships of Triticum searsii to other S-genome diploid Triticum species. Genome 33:369–374

Yen Y, Kimber G (1990b) Genomic relationships of Triticum sharonense with other S-genome diploid Triticum species. Plant Breed 104:53–57

Yen Y, Kimber G (1990c) Production and meiotic analysis of autotriploids Triticum speltoides and T. bicorne. Theor Appl Genet 79:525–528

Yen Y, Kimber G (1990d) Reinvestigation of the S genome in Triticum kotschyi. Genome 33:521–524

Yen Y, Kimber G (1992a) The S genome in *Triticum syriacum*. Genome 35:709–713

Yen Y, Kimber G (1992b) Genomic relationships of N-genome *Triticum* species. Genome 35:962–966

Yen C, Yang JL, Liu XD, Li LR (1983) The distribution of *Aegilops tauschii* Cosson in China and with reference to the origin of the chinese common wheat. In: Sakamoto S (ed) Proceedings of 6th international wheat genetics symposium, Kyoto, Japan, pp 55–58

Yen C, Luo MC, Yang JL (1988) The origin of the Tibetan weedrace of hexaploid wheat, Chinese Spring, Chengdu-guang-tou and other landraces of the white wheat complex from China. In: Miller TE, Koebner RMD (eds) Proceedings of 7th international wheat genetics symposium, Cambridge, England, pp 175–179

Yen J, Baenziger PS, Morris R (1996) Genomic constitution of bread wheat: current status. In: Jauhar PP (ed) Methods in genome analysis in plants: their merits and pitfalls, CRC, pp 359–373

Yildirim A, Jones SS, Murray TD, Cox TS, Line RP (1995) Resistance to stripe rust and eyespot diseases of wheat in Triticum tauschii. Plant Dis 79:1230–1236

You FM, Huo NX, Deal KR, Gu YQ, Luo MC, McGuire PE, Dvorak J, Anderson OD (2011) Annotation-based genome-wide SNP discovery in the large and complex Aegilops tauschii genome using next-generation sequencing without a reference genome sequence. BMC Genomics 12:59

Yu MQ, Jahier J (1992) Origin of S^v genome of Aegilops variabilis and utilization of the S^v as analyser of the S Genome of the Aegilops Species in the Sitopsis Section. Plant Breed 108:290–295

Yu MQ, Person-Dedryver P, Jahier J, Pannetier D, Tanguy AM, Abelard P (1990) Resistance to root knot nematode, Meloidogyne naasi (Franklin) transferred from Aegilops variabilis Eig to bread wheat. Agronomie 10:451–458

Zaharieva M, Monneveux P (2006) Spontaneous hybridization between bread wheat (Triticum aestivum L.) and its wild relatives in Europe. Crop Sci 46:512–527

Zaharieva M, Gaulin E, Havaux M, Acevedo E, Monneveux P (2001a) Drought and heat responses in the wild wheat relative Aegilops geniculata Roth: Potential interest for wheat improvement. Crop Sci 41:1321–1329

Zaharieva M, Monneveux P, Henry M, Rivoal R, Valkoun J, Nachit MM (2001b) Evaluation of a collection of wild wheat relative Aegilops geniculata Roth and identification of potential sources for useful traits. Euphytica 119:33–38

Zarchi Y, Hillel J, simchen G (1974) Supernumerary chromosomes and chiasma distribution in *Triticum speltoides*. Heredity 33:173–180

Zhang H-B, Dvorak J (1991) The genome origin of tetraploid species of Lemus (Poaceae: Triticeae) inferred from variation in repeated nucleotide sequences. Amer J Bot 78:871–884

Zhang H-B, Dvorak J (1992) The genome origin and evolution of hexaploid Triticum crassum and Triticum syriacum determined from variation in repeated nucleotide sequences. Genome 35:806–814

Zhang H-B, Dvorak J, Waines JG (1992) Diploid ancestry and evolution of Triticum kotschyi and T. peregrinum examined using variation in repeated nucleotide sequences. Genome 35:182–191

Zhang X-Y, Wang RR-C, Dong Y-S (1996) RAPD polymorphisms in *Aegilops geniculata* Roth (*Ae. ovata* auct. Non L.) Genet Resourc Crop Evol 43:429–433

Zhang H, Jia J, Gale M, Devos K (1998) Relationships between the chromosomes of Aegilops umbellulata and wheat. Theor Appl Genet 96:69–75

Zhang W, Qu L, Gu H, Gao W, Liu M, Chen J, Chen Z (2002) Studies on the origin and evolution of tetraploid wheats based on the internal transcribed spacer (ITS) sequences of nuclear ribosomal DNA. Theor Appl Genet 104:1099–1106

Zhang LQ, Yan ZH, Dai SF, Chen QJ, Yuan ZW, Zheng YL, Liu DC (2008) The crossability of Triticum turgidum with Aegilops tauschii. Cereal Res Commun 37:417–427

Zhang LQ, Liu DC, Zheng YL, Yan ZH, Dai SF, Li YF, Jiang Q, Ye YQ, Yen Y (2010) Frequent occurrence of unreduced gametes in Triticum turgidum-Aegilops tauschii hybrids. Euphytica 174:285–294

Zhao YH, Kimber G (1984) New hybrids with D genome wheat relatives. Genetics 106:509–515

Zhou JP, Yao CH, Yang EN, Yin MQ, Liu C, Ren ZL (2014) Characterization of a new wheat-Aegilops biuncialis addition line conferring quality-associated HMW glutenin subunits. Genet Mol Res 13:660–669

Zhao L, Ning S, Yu J, Hao M, Zhang L, Yuan Z, Zheng Y, Liu D (2016) Cytological identification of an Aegilops variabilis chromosome carrying stripe rust resistance in wheat. Breed Sci 66:522–529

Zhao G, Zou C, Li K et al (2017) The Aegilops tauschii genome reveals multiple impacts of transposons. Nat Plants 3:946–955

Zhu T, Rodriguez JC, Deal KR, Van S, Dvorak J, Luo M-C (2016) Evolution of genome structure in polyploid wheat revealed by comparison of wheat and *Aegilops tauschii* whole-genome BioNano maps. In: Proceedings of 24th plant and animal genome conference, San Diego, 9–13 Jan 2016, pp 1–2

Zhukovsky PM (1928) A critical systematical survey of the species of the genus Aegilops L. Bull Appl Bot Genet and Pl Breed 18:417–609

Zohary M (1962) Plant life of palestine: Israel and Jordan. Ronald Press Co., pp 1–262

Zohary D, Feldman M (1962) Hybridization between amphidiploids and the evolution of polyploids in the wheat (Aegilops-Triticum) group. Evolution 16:44–61

Zohary D, Imber D (1963) Genetic dimorphism in fruit types in Aegilops speltoides. Heredity 18:223–231

Zohary D, Harlan JR, Vardi A (1969) The wild diploid progenitors of wheat and their breeding value. Euphytica 18:58–65

10.1 Description of the Genus

10.1.1 Taxonomic Complexities

At all times human strived to recognize, define and control the plants and animals around them and, as the first step in this endeavor, they named and classified them. This is well reflected in the biblical story in which God asked Adam to name all animals and plants, … "And the Lord God … brought them unto Adam to see what he would call them: and whatsoever Adam called every living creature, that was the name thereof" (Genesis 2, 19). Already in ancient times, at the end of the second millennium BC, the domesticated wheats were divided into two major groups: free-threshing wheats and hulled wheats, referred to in the bible (e.g., Exodus 9: 32) as wheat (probably a free-threshing form of tetraploid wheat) and as emmer (hulled form of tetraploid wheat; spelt was not grown in ancient Egypt), respectively. This classification was also accepted by the early Greek taxonomists of the fourth century BC, Aristoteles and Theophrastus, and by the first century Latin agronomist Columella, who classified the domesticated wheats in two sections, namely: *Triticum*—wheats whose spikes have a tough rachis and grain so loosely invested by the chaff that they fall out when the spikes are threshed (free-threshing types), and *Zea*—wheats whose spikes have a semi-fragile rachis, which, when pressed, breaks into spikelets, and whose grains are so firmly enclosed by the glumes that they are separated with difficulty (hulled wheats). The *Zea* section includes the present-day "spelt" wheats, namely, small spelt (*Triticum monococcum* ssp. *monococcum*), Emmer (*T. turgidum* ssp. *dicoccon*) and common spelt (*T. aestivum* ssp. *spelta*).

This classification was more or less in use until the eighteenth century, when Linnaeus (1753) suggested the classification of the wheats based on the binaric system. He was the first to place all the domesticated wheats under a single genus, *Triticum,* and included the following five different species in this genus: *T. aestivum* (bearded spring wheat), *T. hybernum* (bearded winter wheat), *T. turgidum, T. spelta* and *T. monococcum.* In the third edition of his Species Plantarum (Linnaeus 1764), he added *T. polonicum.* All the six species were domesticated forms.

Since then, the scientific name of bread (common) wheat has been changed by several taxonomists, e.g., *T. sativum* Lamarck (1786), *T. vulgare* Villars (1787), *T. cereale* Schrank (1789), *T. vulgare* subsp. *vulgare* Körnicke (1885), *Frumentum triticum* Krause (1898), *T. vulgare* Percival (1921), *T. aestivum* Schiemann (1948), *T. aestivum* cultivar group *aestivum* Bowden (1959), and *T. aestivum* ssp. *aestivum* Mac key (1966, 1988), van Slageren (1994) (for details see Bálint et al. 2000).

The discovery of the wild one-grained wheat, *T. monococcum* ssp. *aegilopoides,* and the two-grained *T. turgidum* ssp. *dicoccoides*, in the nineteenth century and in the beginning of the twentieth century, respectively, enabled Schultz (1913b) to assemble the first natural classification of the wheats (Table 10.1). He divided the genus *Triticum* into three major taxonomic groups, namely, einkorn, emmer and dinkel. Each group was subdivided further into wild and domesticated species and the domesticated species further separated into hulled and naked (free-threshing) types. Schultz (1913b) assumed that the domesticated naked types derived from the domesticated hulled species, which, in turn, derived from the wild prototypes. He postulated the existence of three different types of wild wheats, i.e., prototypes of the einkorn, the emmer and the dinkel. The prototypes of einkorn and emmer were known, and Shultz believed that the wild progenitor of the dinkel wheats would also be found.

The data collected by von Tschermak (1914) on the fertility of the various interspecific F_1 hybrids of wheats supported Schultz's classification of three major groups of wheat. Schultz's grouping is also in accord with the serological relationships as determined by Zade (1914). Thellung (1918a, b) disagreed with the morphological species classification approach and used a broader definition of species in the genus *Triticum*, a definition that was based on genetic

© The Author(s) 2023
M. Feldman and A. A. Levy, *Wheat Evolution and Domestication*,
https://doi.org/10.1007/978-3-031-30175-9_10

Table 10.1 Schulz's (1913b) phylogenetic classification of species and domesticated groups of *Triticum*

	Wild progenitors		Domesticated wheats		
			Spelt (hulled) wheat		Naked (free-threshing) wheat
Einkorn series (one-grained wheat)	*T. aegilopoides*	→	*T. monococcum*		None
Emmer series (two-grained wheat)	*T. dicoccoides*	→	*T. dicoccon*	→	*T. durum*
					T. turgidum
					T. polonicum
Dinkel series	Not known		*T. spelta*	→	*T. compactum*
					T. vulgare

relatedness. Thus, Thellung recognized only three wheat species, namely, *T. monococcum*, *T. turgidum* and *T. aestivum*. The natural classifications of Schultz and Thellung were also supported by the pioneering cytological studies of Sakamura (1918) who showed that the three groups of wheat differ in their chromosome number: the einkorns (*T. monococcum*) are diploids (2n = 14), the emmers (*T. turgidum*) are tetraploids (2n = 28), and the dinkels (*T. aestivum*) are hexaploids (2n = 42). Thus, the wheat species form a polyploid series based on X = 7 chromosome number. Of note, Sakamura (1918) was the first to determine the correct chromosome number of the wheats. Additional significant support for Schultz's and Thellung's classification came from the cytogenetic analysis of meiotic chromosome pairing in interspecific F_1 hybrids between the various wheats, e.g., the studies of Sax (1918, 1921, 1922), Kihara (1919, 1924), Sax and Sax (1924). These studies showed that the polyploid species of the wheats are allopolyploids, i.e., containing subgenomes that derived from different species, and that the tetraploid wheat contains a subgenome of diploid wheat plus an additional subgenome and that allohexaploid wheat comprises the genome of tetraploid wheat plus an additional subgenome.

Since then, several new domesticated and wild taxa were discovered and a better understanding of the taxonomical, as well as of the cytogenetic and phylogenetic relationships among the various wheat taxa, has been gained; the classification of wheat was then modified accordingly. Not only the species names, but also their classification have been a matter of disagreement. The lack of agreement among wheat taxonomists have led to various classification proposals of the genus *Triticum*, each comprising a different number of sections and species. Percival (1921) described only two species in this genus, while Bowden (1959) listed three, Mac Key (1966) six, Miller (1987) 22, Jackubziner and Dorofeev (1968) 24, Dorofeev et al. (1980) 27, and Goncharov (2011) and Goncharov et al. 2009) 29. Obviously, one proper and

acceptable classification of the *Triticum* species is essential for the study of their genetic structure, origin and evolutionary relationships as well as for efficient use of their gene pools to improve the domesticated forms (van Slageren 1994).

Classification of the *Triticum* species is complicated for several reasons: (i) The allopolyploid species of wheat contain subgenomes originating from two different genera, *Triticum* and *Aegilops* (e.g., hexaploid wheat, *Triticum aestivum*, contains one subgenome from *Triticum* and two from *Aegilops*). (ii) Both cultivated and wild forms occur in a single biological taxon. (iii) The genus *Triticum* and its related genera are relatively young, with possibilities of inter-specific and inter-generic hybridization leading to introgression. (iv) Over the 10,000 years of cultivation, numerous forms of domesticated wheats have evolved under human selection. This great diversity has led to much confusion in the taxonomical treatment of wheats. The above taxonomical complications have raised the following problems: (a) Is separation of the *Triticum* and *Aegilops* genera taxonomically justified? (b) Is it appropriate to group wild and domesticated forms under one biological species or to separate them into different taxonomic species? (c) How should new intergeneric hybrids and synthetic amphiploids be regarded?

In spite of the fact that *Aegilops* species contributed genome(s) to allopolyploid species of *Triticum*, their separation into two different genera is justified on the basis of morphological differences and the different evolutionary trajectories of these two genera. Mac Key (1966, 1968, 1981) claimed that *Aegilops*, especially the allopolyploid species, are developing towards increased weediness, whereas the domesticated taxa of *Triticum*, are taking a different evolutionary route, due to selection by man towards tough rachis, naked grains, rapid and uniform germination, reduced competition with neighboring wheat plants, long grain filling period, higher yield, and other traits. This evolutionary route has caused the domesticated wheat to be completely dependent on man for its survival.

The traditional taxonomists (e.g., Dorofeev et al. 1980; Goncharov et al. 2009; Goncharov 2011) do not consider the presence of reproductive barriers as an essential criterion for definition of a species, whether cultivated or wild. They claim that even though some wheat species readily cross with each other and form fertile progeny under experimental conditions, they fail to do so in nature because they do not grow in the same locations. Moreover, wild forms are under different selection pressure than domesticated ones, and consequently, have a different evolutionary trajectory. This means, according to the traditional taxonomists, that wild and domesticated forms cannot be included in the same species. The traditional taxonomist tends to overclassify; they find conspicuous characters, often without intermediates, and frequently use them to define new species (Mac Key 1981, 1989). The characters may be controlled by one or a few genes of little biological significance. As a result, too many species are defined, and then, to accommodate the enormous variability, unreasonable numbers of intra-specific categories are often established (Mac Key 1981, 1989; van Slageren 1994). Several classifications of the genus *Triticum*, such as those of Dorofeev et al. (1980), Goncharov (2011), subdivide it into a large number of species defined on the basis of a single major gene that affects the morphology of the spike. These classifications were built upon Körnicke's (1885) morphological research, and were strongly influenced by Flaksberger's classification (1935), both of which emphasized morphological differences. The classification of Dorofeev et al. (1980) (Table 10.2) divides the genus into two subgenera, *Triticum* and *Boeoticum* Migusch. et Dorof., each containing three sections; one subgenus comprises 19 and the other, eight different species. Within each section, crossability between species is perfect and the hybrids are fully fertile, borders are often indistinct and clearly only depend on individual major genes that affect spike morphology (Mac Key 1966, 1968, 1981). Among the species recognized by Dorofeev et al. (1980), there were several that are best regarded as mutant forms of other species. For instance, *T. militinae* Zhuk. & Migush. is a mutant form of *T. timopheevii*, *T. jakubzineri* (Udachin & Shakhm.) of *T. turgidum*, and *T. sinskajae* Filat. & Kurk. of *T. monococcum* (Goncharov 2011; van Slageren 1994). Moreoer, the division into two subgenera is based on the wrong concept that *T. monococcum* ssp. *aegilopoides* (= *boeoticum*; genome A[b)] is the donor of the A subgenome to the *T. timopheevii* and *T. kiharae* instead of *T. urartu*. The classification proposed by Goncharov (Goncharov 2011; Goncharov et al. 2009) (Table 10.3) includes 29 species in five sections (the fifth section is of synthetic amphiploids).

The concept "biological species" has a cytogenetic and evolutionary (biosystematic) meaning and bears great relevance when genetic resources are considered (van Slageren 1994). On the other hand, "taxonomic species" is of great relevance when classification and nomenclature are considered. On this basis, and since this book adopts the cytogenetic and evolutionary approaches, the wild and domesticated forms will be included as sub-species under one species, even though they are under different selection pressure and consequently, different evolutionary direction.

According to Mac Key (1989), wheat taxonomy should take the dynamic process of speciation into account. In particular, when working with domesticated forms where breeders accelerate the creating forces, taxonomic decisions must not only be evolutionarily retrospective but also foresighted. Van Slageren and Payne (2013) maintained that a narrow, morphology-based concept is contrasted with a much wider, genome-based one, leading to profound differences in the recognition of taxa at species level and below. The latter concept accepts far fewer taxa. Considerations regarding the nomenclature of taxa are presented by van Slageren and Payne (2013), applying both the International Code for the Nomenclature of Cultivated Plants (ICNCP) and the International Code of Nomenclature for algae, fungi and plants (ICN or the 'Melbourne Code').

With the genome types being of obvious importance in any classification of *Triticum*, and when considering the biphyletic origin of the allopolyploids in this genus, Mac Key (1981, 1989, 2005) proposed a classification of the *Triticum* species that recognized three sections of wheat and only six biological species two in each section, and includes, in addition to the three wheat sections, also a section of the manmade crop, triticale, named *Triticoseale* (Wittm. ex Camus) MK, section nov. comprising two species (Table 10.4). The recognition of only six biological species of wheat by Mac Key (1981, 1989) was acceptable by van Slageren (1994) (Table 10.5) who omitted the section of Triticosecale since the triticale species were artificially created. This classification is used in this book (Table 10.5). The six species evolved from the five previously analyzed and enumerated by Mac Key (1966), who originally included *T. urartu* in *T. monococcum* ssp. *aegilopoides* (ssp. *aegilopoides* was under the name ssp. *boeoticum* by Mac Key 1966), which was later shown to be a biological species separable from *aegilopoides* by various morphological characters and reproductive isolation (Johnson and Dhaliwal 1976). Mac Key (1981) rearranged both the wild and cultivated species of the genus into species defined by their genetic relationships. Interspecific genome constitutions were identical in all six species of the genus *Triticum*. Consequently, species sharing the same genome constitution were ranked subspecies (van Slageren 1994) [convarieties by (Mac Key 1981)] within the same species.

The classification of the genus *Triticum* suggested by Mac Key (1966, 1975, 1981, 2005, 1954b) and modified by

Table 10.2 Classification of *Triticum* (after Dorofeev et al. 1980)

Subgenus	Section	Species group	Species	Chromosome number	Genome
Triticum	Urartu Dorof. et A. Filat.	Einkorn	*T. urartu* Thum. Ex Gandil	2n = 14	A^uA^u
	Dicoccoides Flaksb.	Emmer	*T. dicoccoides* (Koern. Ex Aschers. Et Graeb.) Schweif.	2n = 28	A^uA^uBB
			T. dicoccum (Schrank) Schuebl.	2n = 28	A^uA^uBB
			T. karamyschevii Nevski.	2n = 28	A^uA^uBB
			T. ispahanicum Heslot.	2n = 28	A^uA^uBB
	Triticum.	Naked tetraploids	*T. turgidum* L.	2n = 28	A^uA^uBB
			T. jakubzineri Udacz. et Schachm.	2n = 28	A^uA^uBB
			T. durum Desf.	2n = 28	A^uA^uBB
			T. turanicum Jakubz.	2n-28	A^uA^uBB
			T. polonicum L.	2n = 28	A^uA^uBB
			T. aethiopicum Jakubz.	2n = 28	A^uA^uBB
			T. carthlicum Nevski (syn. T. persicum Vav.)	2n = 28	A^uA^uBB
		Spelt	*T. macha* Dekapr. et Menabde.	2n = 42	A^uA^uBBDD
			T. spelta L.	2n = 42	A^uA^uBBDD
			T. vavilovii (Tum.) Jakubz.	2n = 42	A^uA^uBBDD
		Naked hexaploids	*T. compactum* Host.	2n = 42	A^uA^uBBDD
			T. aestivum L.	2n = 42	A^uA^uBBDD
			T. sphaerococcum Percv.	2n = 42	A^uA^uBBDD
			T. petropavlovsky Udacz. et Migufch.	2n = 42	A^uA^uBBDD
Boeoticum Migusch. et Dorof.	Monococcum Dum.	Einkorn	*T. boeoticum* Boiss.	2n = 14	A^bA^b
			T. monococcum L.	2n = 14	A^bA^b
		Naked einkorn	*T. sinskajae* A. Filat. et Kurk.	2n = 14	A^bA^b
	Timopheevii A. Filat. et Dorof.	Emmer	*T. araraticum* Jakubz.	2n = 28	A^bA^bGG
			T. timopheevii (Zhuk.) Zhuk.	2n = 28	A^bA^bGG
			T. zhukovskyi Menabde et Erizjan.	2n = 42	$A^bA^bA^bA^b$GG
		Naked tetraploid	*T. militinae* Zhuk. et Migusch.	2n = 28	A^bA^bGG
	Kiharae Dorof. et Migusch.	Spelt	*T. kiharae* Dorof. Et Migusch.	2n = 42	A^bA^bGGDD

van Slageren (1994), is based on the biological species approach. It includes three sections, each with a different ploidy level, six species, two in each section, and 19 sub-species (Table 10.5). Taxa that differ by a major gene(s) are classified as sub-species. Elevation of Mac Key's subspecies to a species rank, as suggested by Dotofeev et al. (1980), Miller (1987), Goncharov (2011), is not preferred, as it obscures the genetic relationships. With the exception of *aestivum* and *durum*, the other wheats are, in fact, relics with only local importance, e.g., *dicoccon*, *macha*, *spelta*, *compactum* and *sphaerococcum*. This taxonomy technique accounts for genetic barriers and phylogenetic relationships.

The Mac Key's classification is elegant, simple, flexible, and based on a genetic concept (van Slageren 1994). It stands in remarkable contrast to the highly hierarchical, strictly morphological system of Dorofeev et al. (1980), Goncharov (2011). It is therefore, not surprising that Mac Key's approach was adopted by Mansfeld's Encyclopedia of Agricultural and Horticultural Crops (Hanelt 2001) and was used by the (former) International Board for Plant Genetic Resources (IBPGR; now IPGRI).

Van Slageren (1994) preferred the adoption of the sub-species status for the intraspecific taxa instead of the 'cv. Group' designation used by Bowden (1959), Morris and

Table 10.3 *Triticum* classification (after Goncharov et al. 2009)

Section	Group of species	Species	2n	Genomes
Monococcum Dum.	Hulled	*T. urartu* Thum. ex Gandil.	14	A^u
		T. boeoticum Boiss.	14	A^b
		T. monococcum L.	14	A^b
	Naked	*T. sinskajae* A. Filat. et Kurk.	14	A^b
Dicoccoides Flaksb.	Hulled	*T. dicoccoides* (Korn. ex Aschers. et Graebn.) Schweinf.	28	BA^u
		T. dicoccum (Schrank) Schuebl.	28	BA^u
		T. karamyschevii Nevski.	28	BA^u
		T. ispahanicum Heslot.	28	BA^u
	Naked tetraploids	*T. turgidum* L.	28	BA^u
		T. durum Desf.	28	BA^u
		T. turanicum Jakubz.	28	BA^u
		T. polonicum L.	28	BA^u
		T. aethiopicum Jakubz.	28	BA^u
		T. carthlicum Nevski.	28	BA^u
Triticum	Hulled	*T. macha* Dekapr. et Menabde.	42	BA^{u}D
		T. spelta L.	42	BA^{u}D
		T. vavilovii (Thum.) Jakubz.	42	BA^{u}D
	Naked hexaploids	*T. compactum* Host.	42	BA^{u}D
		T. aestivum L.	42	BA^{u}D
		T. sphaerococcum Perciv.	42	BA^{u}D
Timopheevii A. Filat. et Dorof.	Hulled	*T. araraticum* Jakubz.	28	GA^u
		T. timopheevii (Zhuk.) Zyuk.	28	GA^u
		T. zhukovskyi Menabde et Erizjan.	42	GA^{u}A^b
Compositum N. P. Gontsch	Hulled	*T. palmovae* G. Ivanov (syn. T. erebuni Gandil.)	28	DA^b (DA^u)
		T. dimococcum Schieman et Staudt.	42	BA^{u}A^b
		T. kiharae Dorof. et Migusch.	42	GA^{u}D
		T. soveticum Zhebrak.	56	BA^{u}GA^u
		T. borisii Zhebrak.	70	BA^{u}DGA^u
	Naked octoploid	*T. flaksbergerii* Navr.	56	GA^{u}BA^u

Sears (1967). Subspecies designation expresses their classification as a botanical taxon, similar to species and genus (van Slageren 1994). Cultivars should be designated as, for example, *T. turgid*um L. ssp. *durum* (Desf.) Husn. 'Cappelli', an Italian cultivar.

According to van Slageren (1994), species described only on the basis of a discovered or induced mutation, subsequently selected and multiplied and found to be stable, but never released as a commercial cultivar, should be made synonyms under the cultivated (sub)species from which they were isolated. This applies to the following species:

(1) *T. sinskajae* A. Filat. & Kurk.—this diploid species (genome A^{m}A^m) is a free-threshing mutant of domesticated *T. monococcum* developed in the Vavilov Institute in Daghestan, Russia; (2) *T. militinae* Zhuk. & Migush.—this tetraploid species (genome GGAA) is a free-threshing mutant selected from a single specimen (Miller 1987) of cultivated *T. timopheevii*; (3) *T. jakubzineri* Udachin & Schachm.—this tetraploid species (genome BBAA) appears to be a form of *T. turgidum* (Dorofeev and Korovina 1979); (4) *T. petropavlovskyi* Udachin & Migush.—this hexaploid species (genome BBAADD) is a Chinese landrace, originated from hybridization of hexaploid wheat, ssp. *aestivum* and tetraploid wheat, ssp. *polonicum* (Chen et al. 1985; Watanabe and Imamura 2002; (5) *T. aethiopicum* Jakubz. (Syn: T. abyssinicum Steud.)—this tetraploid

Table 10.4 Species of *Triticum* (after Mac Key 2005)

Sect. Monococca Dumort			
	T. monococcum (Genome A^b)		
		ssp. *Aegilopoides* (Link) Thell.	
			Var. thaoudar (Reut.) Perc.
		ssp. *Monococcum*	
			Var. sinskajae (Filat. Et Kurk.) MK, comb. Nov.
	T. urartu Thum. ex. Gandil. (Genome A^u)		
Sect. Dicoccoidea Flaksb.			
	T. timopheevi (Zhuk.) Zhuk. (Genome GA^u)		
		ssp. *armeniacum* (Jakubz.) MK	
		ssp. *timopheevii*	
			Var. militinae (Zhuk. et Migusch.) Zhuk. et Migusch.
	T. turgidum (L.) Thell. (Genome BA^U)		
		ssp. *dicoccoides* (Körn. ex Aschers. Et Graebn.) Thell.	
		ssp. *dicoccon* (Schrank) Thell.	
		ssp. *georgicum* (Dek. et Men.) MK	
		ssp. *turgidum*	
		ssp. *durum* (Desf.) Husn.	
		ssp. *turanicum* (Jakubz.) Löve et Löve	
		ssp. *polonicum* (L.) Thell.	
		ssp. *carthlicum* (Nevski) Löve et Löve	
Sect. Triticum L. (Speltoidea Flaksb.)			
	T. zhukovskyi Men. et Erizan. (Genome GA^{u}A^b)		
	T. aestivum (L.) Thell. (Genome BA^{u}D)		
		ssp. *spelta* (L.) Thell.	
		ssp. *macha* (Dekapr. & Menabde) MK	
		ssp. *compactum* (Host) MK	
		ssp. *sphaerococcum* (Perc.) MK	
		ssp. *aestivum* ['vulgare' (Vill.)] MK	
Sect. Triticosecale (Wittm. ex Camus) MK, section nov.			
	T. semisecale MK (B/A^{u}R)		
	T. neoblaringhemii (Wittm. ex Camus) Mk, comb. nov. (Genome BA^{U}R)		
	T. rimpaui (Wittm.) MK, comb. Nov. (Genome BA^{U}DR)		

species (genome BBAA) presumably arose as a free-threshing mutant from Ethiopian emmer, and was included by Miller (1987) in *T. durum*. It was classified by Mac Key (1966) in *T. turgidum*; (6) *T. isphahanicum* Heslot.—this tetraploid species (genome BBAA) is, according to Mac Key (1966), a form of *T. polonicum,* and shows a tendency towards a more standard *Triticum* glume morphology. This taxon may have been cultivated in the Isfahan region of Iran; (7) *T. pyramidale* Perciv.—this tetraploid species (genome BBAA) was considered by Mac Key (1966) as a special form of *T. turgidum* ssp *durum,* but by Miller (1987) as a form of *T. turgidum* ssp. *turgidum.* This taxon may have been or is may still be cultivated in Egypt; (8) *T. vavilovii* (Tumanian) Jakubz—

this hexaploid species (genome BBAADD) is a branching mutant found by Tumanian during the 1929–30 autumn–winter season as an admixture in a stand of bread wheat landrace called 'Dir', northeast of lake Van, at an altitude of 1780 m. In 1970, it was also found as an admixture in Azerbaijan, by Mustfaev, and in Armenia, by Gandilyan. Plants with branched, vavilovii-type spikes were produced by Mac Key (1966), among others, in progenies of wide interspecific crosses of cultivated wheat groups.

van Slageren (1994) objected to the inclusion of inter-generic synthetic allopolyploids under *Triticum*. Thus, the synthetic allopolyploids *T. kiharae* Dorof. & Migush. (Genome GGAADD), resulting from the cross of *T. timopheevii* x *Ae. tauschii, T. x boeoticourarticum*

Table 10.5 Classification of the wheats (after van Slageren 1994)

Species and subspecies		Common name
Section Monococcon Dumort. (2n = 14)		
Triticum monococcum L. (Genome A^m)		
	ssp. *aegilopoides* (Link) Thell.	Wild einkorn
	ssp. *monococcum*	Domesticated einkorn or small spelt
Triticum urartu Tum. ex Gand. (Genome A)		None (wild form)
Section Dicoccoidea Flaksb. (2n = 28)		
Triticum timopheevii (Zhuk.) Zhuk. (Genome GA)		
	ssp. *armeniacum* (Jakubz.) van Slageren	Wild timopheevii
	ssp. *timopheevii*	Domesticated timopheevii
Triticum turgidum L. (Genome BA)		
	ssp. *dicoccoides* (Körn. ex Asch. & Graebn.) Thell.	Wild emmer
	ssp. *dicoccon* (Schrank ex Schübl.) Thell.	Domesticated emmer
	ssp. *paleocolchicum* (Men.) A. Löve & D. Löve	Georgian wheat
	ssp. *parvicoccum* Kislev.	Domesticated (currently extinct)
	ssp. *durum* (Desf.) Husn.	Macaroni or hard wheat
	ssp. *turgidum*	Rivet, cone or pollard wheat
	ssp. *polonicum* (L.) Thell.	Polish wheat
	ssp. *turanicum* (Jakubz.) Löve & Löve	Khorassan wheat
	ssp. *carthlicum* (Nevski) Löve & Löve	Persian wheat
Section Triticum (2n = 42)		
Triticum zhukovskyi Men. & Er. (Genome GAA^m)		None
Triticum aestivum L. (Genome BAD)		
	ssp. *spelta* (L.) Thell.	Dinkel or large spelt
	ssp. *macha* (Dek. & Men.) MK	None
	ssp. *aestivum*	Common or bread wheat
	ssp. *compactum* (Host.) MK	Club wheat
	ssp. *sphaerococcum* (Percival) Mk	Indian dwarf or shot wheat
	ssp. *tibetanum*	Semi-wild wheat

Gandilyan et al. (genome A^{m}A^{m}AA), resulting from the cross *T. monococcum* ssp. *aegilopoides* x *T. urartu*, *T.* x *boeoticotayschicum* (genome A^{m}A^{m}DD), resulting from the cross *T. monococcum* ssp. *aegilopoides* x *Ae. tauschii*, should not be enumerated under *Triticum*. Likewise, Mac Key (1968) included two 'subgroups' in *Triticum*: Triticale (Tscherm. seys. ex Müntzing) Mac Key, representing the triticales and Trititrigia Mac Key), representing the Triticum x Elytrigia hybrids. However, according to van Slageren (1994), both subgroups cannot be included in *Triticum*.

10.1.2 Morphology

The *Triticum* L species plants are annual, predominantly autogamous and 30–100-cm-tall. Their leaves are flat with a membranous ligule. The spike is determinate or indeterminate, linear, two-rowed, laterally compressed, 8–20-cm-long, and either awned or awnless. The spikelets are solitary at nodes, laterally compressed. There are 2–6(–9) florets in every spikelet. The florets are hermaphrodites, with the 1–2 upper usually sterile. The glumes are sub-equal, ovate or oblong, mostly shorter than the spikelet, veined, more or less keeled, and either with 1–2 teeth on the tip or awned. The lemma is sub-ventricose, boat-shaped, coriaceous, keeled towards the apex, and either with 1–2 teeth or one awn. The palea are membranous, 2-veined, 2-keeled, and ciliated along the keels. The caryopsis is either tightly enclosed by the tough glumes, or free, oblong-elliptic, hairy at the apex, deeply grooved along its adaxial side, and with an embryo about 1/5 its length; the hilum is linear, and as long as the caryopsis.

In contrast to the species of *Aegilops* that exhibit wide morphological differences, those of *Triticum* are more or less uniform in their gross morphology. The two diploid wheat species are similar in their spike morphology and all the allopolyploid wheats carry the pivotal A subgenome that determines plant morphology and thus, they resemble diploid wheats in their basic morphology (stature, leaf shape, spike and spikelet morphology, grain shape, free caryopsis, keeled glumes, plant habitus, and growth habit) and in the structure of the seed dispersal unit (Mac Key 2005; Feldman et al. 2012). The differential B, G, and D subgenomes are primarily responsible for the eco-geographical adaptation of the allopolyploids and their resistance to biotic and abiotic stresses (Feldman et al. 2012). The A subgenome also controls the autogamous naure of the allopolyploid wheats (assuming that the donor of the B subgenome is an allogamous species, similar to *Ae. speltoides*) and harbors many domestication genes, such as the genes for non-brittle spike on 3AS (Rong et al. 1999; Nalam et al. 2006), free-threshing on 5AL (Sears 1954), QTIs for kernel size predominantly on A subgenome, i.e., on chromosomes 1A, 2A, 3A, 4A, 7A, 5B, and 7B) (Elias et al. 1996), and for grain size and grain form on chromosomes 1A, 3A, 4B, 5A, and 6A (Gegas et al. 2010), and a number of domestication-related QTLs (Peng et al. 2003a, b).

There are four wild wheat taxa, whereas all the other taxa are domesticated (Table 10.5). The wild and domesticated taxa are inter-fertile and only a small number of genes control the traits that distinguish them. The wild wheats, namely, diploid *T. monococcum* ssp. *aegilopoides* and *T. urartu*, and tetraploids *T. timopheevii* ssp. *armeniacum* and *T. turgidum* ssp. *dicoccoides*, have a brittle rachis that disarticulates into arrowhead-shaped spikelets upon maturity, each with the rachis segment below it (wedge-type dispersal unit). This type of seed dispersal unit facilitates self-burial in the soil, hence, protection during the hot summer and successful germination after the beginning of the rainy season.

10.1.3 Geographic Distribution and Ecological Affinities

Following the discovery of wild diploid wheat during the second half of the nineteenth century and wild tetraploid wheats in the beginning of the twentieth century, the putative progenitors of the domesticated diploid and tetraploid wheats were identified and their distribution area and ecological affinities have become well known. It was then relatively simple to establish the geographical regions from which the various wheat taxa were taken into cultivation.

The wild progenitors of domesticated wheats are natural constituents of some of the open oak-park belts and the herbaceous plant formations in southwest Asia. Their center of origin, proposed based on ample archaeological evidence (Feldman and Kislev 2007), and current center of distribution and diversity is in the "Fertile Crescent" arc—a hilly and mountainous region extending from the foothills of the Zagros Mountains in south-western Iran, through the Tigris and Euphrates basins in northern Iraq and southeastern Turkey, continuing southwestward over Syria to the Mediterranean, and extending to Israel, Palestinian Territory and Jordan. The four wild wheat taxa and 17 species of the closely-related genus *Aegilops,* as well as several genera and species of other Triticeae, are endogenous to this region.

The Fertile Crescent is bound by the Mediterranean in the west, by chains of large and high mountain ranges in the north and east (the Amanos in north-western Syria, the Taurus in southern Turkey, Ararat in north-eastern Turkey and the Zagros in western Iran), and in the East and south by the Syrio-Arabian desert, with its western extension (e.g., Paran desert) in the Sinai Peninsula. Situated between the sea, mountains, and the desert, the Fertile Crescent is under the influence of several different climates: on the one hand, it enjoys the temperate Mediterranean climate with a short, mild and rainy winter and long, hot and dry summer. On the other hand, it is influenced by the more extreme steppical climate of the Iranian and Anatolian plateaus in the northeast and north, and by the desert climate in the east and south. Consequently, the Fertile Crescent encompasses two different phytogeographical regions, the Mediterranean in the western part and the steppical (Irano-Turanian) in the northeastern part, and is affected by two other regions, the Saharo-Arabian in the south and the Euro-Siberian in the north. The Mediterranean part of the Fertile Crescent includes Israel, the Palestinian Territory, Jordan, Lebanon, Syria and the western part of southeastern Turkey and it centers around the Syrio-African rift (the Jordan rift valley in the south, the Beqa Valley of Lebanon and the Orentos valley in Syria). The Irano-Turanian part includes the eastern part of southeastern Turkey, northern Iraq, and southwestern

Iran, and is influenced by the continental climate of the Iranian and Central Asiatic steppes.

It is no wonder, therefore, that this region is very ecologically diversified, comprising a wide array of different habitats. These versatile ecological conditions are manifested by the wide display of different plant formations, ranging from well-developed Mediterranean forests and maquis, through open parks, shrubs (*garrigue*) and herbaceous (*batha*) formations, to small shrub and steppical plant formations. Many annual grasses and legumes occupy the open habitats in these formations, which presumably served in the pre-agricultural era as the main pasture area for wild sheep, goats and gazelles—the game of the pre-Neolithic hunter and plant collector.

Like de Candolle (1886), Braidwood (1960), Harlan and Zohary (1966), Zohary and Hopf (2000), Zohary et al. (2012), and others thought that Near Eastern agriculture originated within or near the distribution area of the wild progenitors in the Fertile Crescent region. New geological, climatic and archaeological data from the east Mediterranean region indicate that the Younger Dryas climatic event, which was characterized by a cold and dry climate, occurred from 12,900 to 11,700 years Cal-BP (for details see Hillman 1996; Bar-Yosef 1998). Hillman (1996), using palaeobotanical data, reconstructed the phytogeographical belts of this region during the Younger Dryas and concluded that the habitats of the annual cereals lie mainly in the open areas of the oak-park maquis, in a relatively narrow strip of the east Mediterranean. This narrow strip, called the "Levantine Corridor", begins in the Taurus foothills (Diyarbakir area) in southeastern Turkey and extends along the Mediterranean southward, incorporating the middle Euphrates through the Damascus basin, the Lebanese mountains, the two sides of the Jordan Rift Valley into the Sinai Peninsula (Bar-Yosef 1998). In the central-southern part of the corridor (from north of Damascus to southern Judea), tetraploid wild wheat, *T. turgidum* ssp. *dicoccoides*, has massive stands, while diploid wild wheats, *T. monococcum* ssp. *aegilopoides*, and *T. urartu,* and also another tetraploid wild wheat, *T. timopheevi* ssp. *armeniacum,* occupy the northern part. The diploid species also grow in the central part of the corridor. These wild taxa mainly grow on terra-rossa and basalt soils and are adapted to a variety of habitats and a wide range of altitudes (from 100 to 1600 m asl) (Kimber and Feldman 1987; Zohary and Hopf 2000; Mac Key 2005).

The accumulating archaeological data indicate that agriculture originated in the Levantine Corridor of the Fertile Crescent. This is clearly apparent from the distribution of the earliest Neolithic sites 10,300–9500 BP: Tell Mureybit and Tell Abu Hureyra in the Middle Euphrates in the northern part of the corridor, Tell Aswad and Tell Ghoraife in the Damascus basin, and Netiv Hagdud, Dhra, Gilgal, and Jericho in the Jordan Valley, between the Lake of Galilee and the Dead Sea (Bar-Yosef and Kislev, 1989; Hillman, 1996; Bar-Yosef, 1998). All these settlements were established along the ecotone between the relatively temperate Mediterranean phytogeographical region and the steppical Irano-Turanian region. The predominant plant formations in this transitional zone are open parks and herbaceous covers containing a large number of annual grasses and legumes.

The present distribution of the wild forms of wheat could provide important information on the place(s) of origin of Neolithic agriculture. For example, emmer was probably taken into agriculture in the upper Jordan watershed and einkorn was domesticated in southeast Turkey (Harlan and Zohary 1966; Feldman and Kislev 2007). It seems likely therefore, that, while these wild wheats were domesticated within the Fertile Crescent, each was domesticated in a different sub-region of the zone. Yet, any interpretation of modern distribution must take into account (i) the possibility that the climate likely changed in the last 12,000 years, and (ii) the possibility that the wild progenitors themselves changed location, producing weed races whose ranges expanded after the spread of agriculture (Harlan and Zohary 1966).

10.1.4 Preadaptation for Domestication

In many respects, wheat was well preadapted for domestication. Its massive stands in some regions and large seeds that are nutritious and storable, rendered it attractive to the ancient collector. Its annual growth habit, by which it escaped the dry season, made it suitable for dryland farming. In addition, wheat's predominant self-pollination could have helped in the fixation of desirable mutants and of recombinants resulting from rare outcrossing events. Moreover, while the wild wheats occupy poor, thin, rocky soils in their natural sites, they respond well when transferred to richer habitats.

Of the various species of cereals that grew in the Fertile Crescent and which were harvested by the pre-neolithic (Natufian) people, only wild emmer, einkorn, rye and barley (and possibly also wild *timopheevii*) were cultivated by the early farmers (van Zeist and Bakker-Heeres 1985; Kislev et al. 1986). Assuming that the amount of grain collected per unit time was the most important criterion (Evans 1981), wild stands of wheat and barley were preferred over other cereals, because their large, heavy grains borne in spikes, facilitated their harvesting. Indeed, Harlan (1967) in wild einkorn and Ladizinsky (1975) in wild emmer, succeeded in gathering considerable amounts of grains per hour of harvesting. Wild emmer, having larger grains and two grains in each dispersal unit, was a better candidate for domestication than barley, which only has one grain. Indeed, over a 3-h period, Ladizinsky (1975) harvested twice as many wild emmer grains as wild barley from a site in the upper Jordan

Valley, Israel, (1950 g vs. 1040 g). More specifically, 247 wild emmer grains per hour were collected as compared to 107 barley grains per hour. Being a tetraploid, wild emmer exhibits greater and more rapid adaptability to cultivation conditions than barley. The taste of wheat, its high nutritional quality, and its large grain might have also contributed to its preference over barley. In addition, judging from today's pattern of distribution, barley was probably very common and grew in abundance within a short distance from the early Neolithic settlements, while the dense stands of wild emmer were somewhat more distant. This might have created additional pressure to cultivate wheat, as sufficient quantities of barley could have been harvested from nearby wild stands.

10.2 Section Monococcon Dumort. (2n = 2x = 14)

10.2.1 Description of the Section

This section [Syn.: *Crithodium* Link; *Triticum* L. sect. *Crithodium* (Link) Nevski; *Triticum* L. 'congregatio' Diploidea Flaksb.] contains two species, *T. monococcum* L. and *T. urartu* Tum. ex Gand. Both species are morphologically similar, diploids (2n = 2x = 14), with closely related genomes. The *T. monococcum* genome is designated A^m (modified A) and that of *T. urartu* is designated A. *T. monococcum* contains two subspecies, ssp. *aegilopoides* and ssp. *monococcum*, with the former being a wild form with a wide distribution in the northeastern Mediterranean and western Asia, while the latter is a domesticated form which is still cultivated in certain regions in southern Europe, North Africa and Asia. ssp. *aegilopoides* is the wild ancestor of the

domesticated ssp. *monococcum*. The second diploid species, *T. urartu*, grows sympatrically with ssp. *aegilopoides* in southeastern Turkey, northern Iraq, northwestern Iran and Armenia.

10.2.2 *T. monococcum* L. (Genome A^{m}A^m)

10.2.2.1 Description of the Species

T. monococcum L., commonly known as einkorn, one-grained wheat, and small spelt, is an annual, predominantly autogamous, 30–70(–80)-cm-tall (excluding spikes) plant. It has a small number of tillers of which the upper parts are upright or erect. The spike is indeterminate, bilaterally compressed, tow-rowed, 8–12-cm-long (excluding awns) and awned. In the wild form the entire spike disarticulates at maturity into individual spikelets, each with its associated rachis segment (wedge–type dispersal unit) while in the domesticated form the spike remains intact on the culm. The rachis internodes are covered with hairs. The spikelets are compressed, with the top spikelet being fertile and generally in the same plane as those below. There are 8–16 spikelets, and 2–3 basal rudimentary spikelets. Each fertile spikelet contains 3 florets, the upper 1–2 are sterile. Usually, there is one grain per spikelet, sometimes two in the center of the spike or, more rarely, in the whole spike. The glumes with well-developed keels, two unequal teeth which usually do not develop into awns. The lemma tapers into a long awn, with a lateral tooth. The palea is membranous split along the keel at maturity. The caryopsis is free and laterally compressed (Fig. 10.1).

T. monococcum contains two subspecies, wild ssp. *aegilopoides* (Fig. 10.1a) and domesticated ssp. *monococcum* (Fig. 10.1b). The two subspecies are morphologically similar

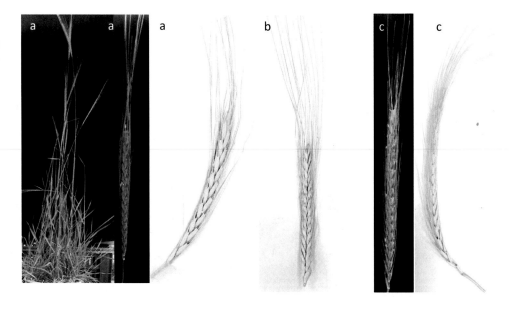

Fig. 10.1 Plants and spikes of diploid species of *Triticum;* **a** *T. monococcum L.* ssp. *aegilopoides* (Link) Thell. (Wild einkorn); **b** *T. monococcum* L. ssp. *monococcum* (domesticated einkorn); **c** *T. urartu* Tum. Ex Gand

and genetically closely related (Smith 1936). Genetic divergence between these species has been accompanied by few or minor rearrangements in chromosome structure, and most characters differentiating between the two subspecies are dominant in the wild versus the domesticated form (Smith 1936).

The most distinguishing traits between the wild and domesticated subspecies are spike fragility, which does not disarticulate at maturity in the domesticated form, and the seeds, which are larger and wider in the domesticated as compared to the wild form. The wild forms have a brittle rachis, and the individual spikelets disarticulate at maturity to disperse the seed. The spikelets have one or two awns and a sharp rachis segment below each spikelet. These arrow-head like structure are very effective devices for seed dispersal and self-planting under wild conditions (Zohary et al. 1969). In domesticated *monococcum*, the mature spike remains intact and breaks into individual spikelets only on pressure. A mutation(s) for a tough rachis, that may occasionally occur in the wild, prevents the dispersal of seeds and thus, has a negative adaptive value and consequently, is soon selected against. In contrast, under cultivation, it has a positive value, facilitating easy harvesting, and has thus been preferred by the early farmers.

Non-brittleness in the domesticated form, ssp. *monococcum*, is determined by two complementary recessive genes (Sharma and Waines 1980). Pourkheirandish et al. (2018) identified *non-brittle rachis 1* (*btr1*) and *non-brittle rachis 2* (*btr2*) in einkorn as homologous to those of barley. Re-sequencing of the dominant and the recessive alleles of the *Btr1* and *Btr2* genes in a collection of 53 lines of wild and domesticated *T. monococcum*, showed that a single non-synonymous amino acid substitution (alanine to threonine) at position 119 of *btr1*, is responsible for the non-brittle rachis trait in domesticated einkorn. However, the rachis is not very tough and when pressure is applied, e. g., during threshing, it breaks into segments similar to the breakage obtained with the wild form. Evans and Dunstone (1970) found that increases in grain and leaf size accompanied the development of domesticated *T. monococcum* as a result of selection for increased yield, and that larger grains led to faster seedling development.

T. monococcum is a polymorphic taxon with variation involving spike and spikelet size, color, hairiness, number of grains per spikelet, and awn count, color and length. It features delicate spikes and spikelets, and has either spring or winter lines, with heading in the latter delayed until the plant experiences vernalization, usually a period of 30–60 days of cold winter temperatures (0°–5 °C). Both the wild and the domesticated forms are hulled types, i.e., their grains are tightly invested by the tough glumes, and therefore, the product of threshing is spikelets rather than grains. While several authors reported wide variation among

domesticated varieties of *T. monococcum* (e.g., Smith 1936; Sharma and Waines 1980; Empilli et al. 2000), Dhaliwal (1977b) found that accessions from different sources were remarkably similar in their growth habit and some morphological characteristics. Likewise, Kuspira et al. (1989) investigated 460 true-breeding lines and reported on observations that supported Dhaliwal's conclusion that phenotypic and genetic variability in *T. monococcum* is limited.

Smith-Huerta et al. (1989), using starch gel electrophoresis of extracts of young leaves, found low genetic diversity in the wild subspecies of *T. monococcum* and in the second diploid wheat, *T. urartu*. Both taxa had a small number of alleles per locus, as well as a low percentage of polymorphic loci, and mean gene diversity. The intra- and inter-populations affinity of both taxa, computed using Nei's identity index (NI), were highly uniform on a genetic level. However, using starch gel electrophoresis of extracts of young leaves, but with several different enzymes Kuspira et al. (1989) detected a higher level of genetic diversity in these two wild taxa.

In contrast to the above, study of variations at the gliadin loci, *Gli-1* and *Gli-2,* of the diploid wheats, revealed a remarkable allelic variation in these two loci (Ciaffi et al. 1997). The gliadin patterns of the wild and domesticated subspecies of *T. monococcum* were very similar but differed substantially from those of *T. urartu*. The gliadin composition of *T. urartu* resembled that of the A subgenome of polyploid wheats, supporting the hypothesis that *T. urartu* is the donor of this subgenome to the polyploid wheats. Similarly, Waines and Payne (1987), Ciaffi et al. (1998) studied variation in the high-molecular weight (HMW) glutenin subunit composition in wild and domesticated diploid wheats and found that all the taxa analyzed were characterized by high intraspecific variation. Yet, the biochemical characteristics of the HMW-glutenin subunits of wild and domesticated *T. monococcum* were very similar but distinctly different from those of *T. urartu*, which is consistent with their classification as two different species. High intraspecific allelic variation at the *Glu-A1* locus of *T. monococcum* was also reported by Li et al. (2015a).

Using PCR with random primers, Vierling and Nguyen (1992) detected polymorphisms in ssp. *monococcum*, namely, out of 103 amplified products 41 showed polymorphism. Nasernakhaei et al. (2015), using single-strand conformation polymorphism to evaluate the nucleotide diversity in the *Acc-1* and *Pgk-1* loci of Iranian wild diploid wheats, detected a meaningful inter-population variation in wild *T. monococcum*. Mizumoto et al. (2002) studied genetic diversities of nuclear and chloroplast genomes using AFLP and simple sequence length polymorphism (SSLP) analyses, in the wild and domesticated subspecies of *T. monococcum* as well as in the second wild diploid species of wheat, *T. urartu*. The diploid wheats showed high nuclear and chloroplast DNA variation. In addition, nuclear AFLP and chloroplast SSLP analyses demonstrated a clear distinction between *T. urartu* and the wild and domesticated forms of *T. monococcum*.

10.2.2.2   Ssp. *aegilopoides* (Link) Thell. (Wild Einkorn)

T. monococcum ssp. *aegilopoides* (Link) Thell. [Syn.: *T. boeoticum* Boiss. ssp. *aegilopoides* (Link) E. Sciem.; *Crithodium aegilopoides* Link; *Crithodium monococcum* (L.) Á. Löve; *T. monococcum* ssp. *boeoticum* (Boiss.) Hayek; *T. aegilopoides* (Link) Balansa ex Körn.; *T. spontaneum* ssp. *aegilopoides* (Link) Flaksb.; *T. thaoudar* Reut.] was first described under the name *Crithodium aegilopoides* by Link, who found it in Greece between Nauplia and Corinth in 1833 (see Boissier 1884). In 1854, a similar specimen was discovered by Balanza on Mount Siphylus in Anatolia, Turkey, and was identified by Gay in 1860 as a wild *T. monococcum aegilopoides* (see Boissier 1884). Boissier collected another specimen of this taxon in Boeotica, Greece, and named it *T. baeoticum* Boiss. [Boissier's original spelling of the new species was '*bæoticum*' which is transcribed as '*baeoticum*' (see van Slageren 1994)]. Thellung (1918a, b), however, decided that there is no justification to rank it at the species level and named it *T. monococcum* ssp. *aegilopoides*.

This wild subspecies is found in the northeastern Mediterranean and west Asia. It grows in primary habitats in the Fertile Crescent arc, namely, in Syria, Lebanon, southeastern Turkey, northern Iraq and northwestern Iran. Aaronsohn (1910) found ssp. *aegilopoides* on Mount Hermon, in southwestern Syria, and in southeastern Lebanon and later, it was also found in Georgia by Zhukovsky in 1923, in Armenia by Tumanian in 1930, and in Nakhichevan by Jakubziner in 1932 (Jakubziner 1932a).

From the Fertile Crescent region, it spread as weed with the expansion of the Neolithic agriculture into central, western, and northern Turkey, Greece (Thessaly and Achaia), Albania, southern Bulgaria, and southern Serbia, Crimea, and Ciscaucasia. In this region, it mainly grows as a segetal plant in somewhat disturbed, secondary habitats. ssp. *aegilopoides* thrives on terra rossa, basalt and several types of alluvial soils in degraded deciduous oak forests and maquis, deciduous steppe maquis, open herbaceous and dwarf shrub formations, pastures, abandoned fields, edges of cultivation and roadsides. Over much of its present range, it is a weedy plant growing along roadsides, field margins, and paths, and often invading wheat fields in quantity (Harlan and Zohary 1966). It is very common and locally abundant in the northern part of the Fertile Crescent. ssp. *aegilopoides* grows at altitudes of 600–2000 m. It is more mesophytic and tolerant of cold than wild emmer wheat and occurs in massive stands at altitudes of as high as 2000 m in

southeastern Turkey and Iran. Its center of distribution is in the Taurus-Zagros arc. In southeast Turkey, there are almost pure stands of ssp. *aegilopoides* in the Karacadag Mountains and several other areas in southeastern Turkey (Harlan and Zohary 1966). These massive stands range from elevations of 2000 m down to the edge of the plains in Urfa and Gaziantep, in southeastern Turkey (about 600 m). Massive stands of ssp. aegilopoides also occur, but less extensively, in northern Iraq and here and there in the Zagros of Iran.

Wild *monococcum* has a relatively large distribution in the central part of the distribution of the genus. It is an east Mediterranean element extending to the steppical (Irano-Turanian) region. It grows sympatrically with *Amblyopyrum muticum, Ae. speltoides, Ae. searsii, Ae, caudata, Ae. umbellulata, T. urartu,* the wild form of *T. timopheevii,* the wild form of *T. turgidum, Ae. geniculata, Ae. biuncialis, Ae. neglecta, Ae. columnaris, Ae. triuncialis* and *Ae. cylindrica.* It grows allopatrically with *Ae. comosa, Ae. uniaristata, Ae, perigrina, Ae. kotschyi, Ae. tauschyii, Ae. crassa* and *Ae. juvenalis.*

This subspecies contains two varieties, var. *boeoticum* that is characterized by a relatively small spike, spikelets mostly with one grain and with one awn, its second flower, ontogenetically the first, may be developed but is usually sterile and awnless lemma. This variety prefers less-extreme climates and grows more in the northern and northwestern part of the endemic region. The second variety, var. *thaoudar,* is more robust, has larger spikes, spikelets usually with two grains and two awns. The seed developing in the second floret is usually larger than the seed of the first floret, with the latter generally being half as big, richer in protein, and darker in color. While the larger seeds germinate in the first winter, the germination of the smaller darker seeds is delayed to the second winter (Harlan and Zohary 1966; Mac Key 1975). Var. *boeoticum* is common to the Balkan, the Anatolian Plateau and the parts of the Fertile Crescent arc, while var. *thaoudar* is found mainly in southeastern Turkey, Iraq and Iran. In central Anatolia and in Transcaucasia, many intermediates between the two races occur.

10.2.2.3   Ssp. *monococcum* (Domesticated Einkorn or Small Spelt)

Soon after its discovery, it became apparent that ssp. *aegilopoides* is the progenitor of domesticated einkorn (Schulz 1913b). This idea, that was mainly based on morphological evidence, was later supported by Blaringhem (1927), Kihara et al. 1929), who reported on complete fertility of hybrids between the wild and domesticated forms of *T. monococcum,* as well as by cytological data, namely, formation of seven ring bivalents at first meiotic metaphase of the F_1 hybrids between these forms (Smith 1936). In addition, a survey of variation in alpha- and beta–amylase isosymes of dry and germinating seeds showed that ssp. *monococcum,* derived from its wild form ssp. *aegilo-poides,* either from var. *boeoticum* or var. *thaoudar* (Nishikawa 1983). Further evidence from morphology and cross-compatibility of diploid wheats and fertility of the F_1 hybrids among them, suggested that domesticated *T. monococcum* derived only once from a population of ssp. *aegilopoides* and underwent limited introgression from *T. urartu* (Dhaliwal 1977a). Experimental evidence showed that introgression could have only been possible from *T. urartu* to *T. monococcum* ssp. *aegilopoides* but not in the opposite direction (Dhaliwal 1977a).

Kilian et al. (2007) investigated haplotype variation among > 12 million nucleotide sequences at 18 loci across 321 wild and 92 domesticated lines of *T. monococcum.* Studies of the wild lines revealed that this taxon underwent a process of genetic differentiation prior to domestication, which led to formation of three genetically distinct races, designated α, β, and γ (Kilian et al. 2007). Only β was domesticated by the Neolithic farmers. Nucleotide and haplotype diversity in domesticated *T. monococcum* was higher than in its wild ancestral race, the β race, indicating that the domesticated form did not undergo reduction of diversity during domestication.

The domesticated subspecies was obviously preferred by the ancient farmers. Heun et al. (1997, 2008) analyzed domesticated and wild *T. monococcum* from the Fertile Crescent and beyond, and found that *T. monococcum* was domesticated in the Karacadağ Mountains in southeast Turkey. The findings also supported the assumption of its monophyletic origin. These findings were recently supported by Kilian et al. (2007).

It was only natural that the domestication of diploid wheat took place in the northern part of the Fertile Crescent arc, the distribution center of the wild progenitor during the Late Epipalaeolithic–early Neolithic period, as has been confirmed by archaeological and genetic data (review in Feldman 2001). Wild *monococcum* was collected by plant gatherers throughout its distribution area long before it was cultivated. The earliest known carbonized grains of brittle einkorn wheat were found in the prehistoric settlement of Tell Mureybit, the northern Levantine Corridor, about 10,000 years BP, where it was apparently collected or harvested from wild stands (Renfrew 1973). Indeed, the phylogenetic analysis by Heun et al. (1997) pinpoints the Karacadağ region of southeastern Turkey, in the northern part of the Levantine Corridor, as the region where einkorn was domesticated. At somewhat later archaeological sites in this area, non-brittle *monococcum* occurred side by side with a brittle type and gradually replaced it: in Ali Kosh in Iranian Khuzistan, 9500–8750 BP, Jarmo in Iraqi Kurdistan, ca. 8750 BP, Cayonu Tepesi in southeastern Turkey, ca. 9000 BP, Tell Abu Hureyra, northern Syria, ca. 9000 BP, and Hacilar in west-central Anatolia, ca. 9000 BP (for

review see Hillman 1975, 1996; Kislev 1984, 1992; Bar-Yosef 1998; Bar-Yosef and Kislev 1989).

Harlan (1967) reported that, today, much of the Karacadağ Mountains is covered with vast stands of wild diploid wheats, together with a few wild barleys and other grasses. Most abundant are var. *thaoudar* of the wild einkorn and *Ae. speltoides.* Wild emmer was found scattered as a minor component in patches among the vast seas of the diploid wild wheats.

It is assumed that the natural massive stands of wild einkorn were attractive to the pre-Neolithic plant-gatherer. To obtain an idea of the quantity of grains of wild einkorn that the ancient gatherer could harvest in a day or a season, Harlan (1967), using a reconstructed sickle with flint sickle blades, was able to harvest 2.45 kg grains per hour. The actual grain content, free of glumes, palea, lemma, rachis segment, and awns, of the harvested material was 46% by weight, i.e., approximately 1 kg. Chemical analysis of the harvested grains showed that they contained 7.91% water, 2.77% ash, 2.64% other extract, 2.33% crude fiber, 22.83% crude protein and 60.04% nitrogen-free extract (Harlan 1967). The wild wheat is far higher in protein content than modern bread wheat cultivars that contain 12–14% protein. Hence, according to Harlan (1967), the ancient man would have had no difficulty in collecting about 227 g (1/2 lb) of protein per hour during the wild wheat harvesting season. One of the conspicuous advantages of obtaining food from wild cereal harvest is that grain stored in a dry place can be kept several years and still preserve its nutritive value.

ssp. *monococcum* is one of the earliest cultivated forms of wheat, alongside domesticated emmer wheat (*T. turgidum* ssp. *dicoccon*). According to archaeological evidence from the Fertile Crescent region (e.g., Arranz-Otaegui et al. 2018), hunter gatherers in the Fertile Crescent may have started harvesting einkorn as long as 30,000 years ago. Although gathered from the wild for thousands of years, einkorn wheat was first domesticated approximately 10,000 years BP in the Pre-Pottery Neolithic A (PPNA) or B (PPNB) periods (Zohary et al. 2012).

The predominantly autogamous nature of *T. monococcum* made domestication of this species easier. Desirable gene combinations of domesticated forms could be selected and maintained due to self-fertilization. Since autogamy is not absolute, inter-genotypic hybridization, resulting in the formation of numerous new recombinant genetic combinations, ensures sufficient genetic flexibility. The annual growth habit also aided in the domestication of *T. monococcum* since it enables passing the hot, dry season as seeds. The combination of annual growth habit and autogamy facilitated fixation and effective exploitation of combination with desirable genes by the rapid generation shift (Mac Key 2005).

Zohary (1999) assumed that comparison of domesticated *T. monococcum* with its wild relative may provide clues for discriminating between monophyletic and polyphyletic origins of the domesticated form. After such a comparison, he arrived at the conclusion that it is very likely that domesticated *T. monococcum* was taken into cultivation only once or, at most, very few times. Dhaliwal (1977a) concluded from morphological and cytological evidence that *T. monococcum* was domesticated only once from a population of ssp. *aegilopoides*, with a limited introgression from the second diploid wild wheat, *T. urartu*, that is found in southeastern Turkey, the site of *T. monococcum* domestication. In contrast, Kilian et al. (2007), based on molecular studies and archaeological findings from the Fertile Crescent, concluded that *T. monococcum* was domesticated in several independent events.

The cultivation of einkorn wheat spread from the northern part of the Fertile Crescent into several other regions in Eurasia. It spread westwards to central, northern and western Turkey, the Balkans, Italy, Spain, central, western and northern Europe, Morocco, north of Caucasia, and Southern Russia, and eastwards to Iran, Turkmenistan, and India. The cultivation of einkorn was never extensive in most of these regions. No remains of either wild or domesticated *T. monococcum* have been found in irrigated settlements in the lowlands of the Mesopotamian plains or in the Nile valley, to which emmer wheat farming spread in the 8th and 6th millennium BP, respectively (Feldman 2001). This might be due to the cooler and moister climate required by einkorn compared to emmer. On the other hand, domesticated einkorn spread into central and Western Europe through the Danube and Rhine valleys. In subsequent eras (Bronze Age), einkorn wheat attained a wide distribution in Europe and in the Near East. However, later on it was replaced by free-threshing polyploid wheats (Zaharieva and Monneveux 2014). Nowadays, traditional einkorn crops can still be found in very small areas on poor soil in marginal mountain areas of Turkey, Balkan countries, southern Italy, southern France, Spain and Morocco (Brandolini and Hidalgo 2011; Zaharieva and Monneveux 2014) and in few locations in central Asia.

An additional domesticated type of *T. monococcum* was recently found in a restricted area in Daghestan, central Asia. This type, var. *sinskajae* A. Filat & Kurk. differs from ssp. *monococcum* and *aegilopoides* by its softer glumes and free-threshing habit, a character controlled by a single recessive allele (Waines 1983). The glumes of this variety, however, are very long, resulting in somewhat more difficult threshing.

Domesticated *T. monococcum* contains lines with spring and winter growth habit. Kuspira et al. (1986) studied the mode of inheritance of this trait in *T. monococcum* and found that only one major gene determined growth habit in this species; the allele determining spring growth habit is dominant over that determining winter growth habit. They

suggested that this locus is homoeoallelic to the *VrnI* locus of the A subgenome of *T. aestivum*.

The domesticated forms of *T. monococcum* are low-yielding but produce a reasonable quantity of grains when grown on poor, marginal soils where other wheats usually fail (Castagna et al. 1995; Borghi et al. 1996). Most domesticated einkorn varieties produce one grain per spikelet, hence its name, but cultivars with two grains exist as well (Schiemann 1948; Harlan 1981). Domesticated *T. monococcum* has a higher percentage of protein than most modern domesticated wheats and is considered more nutritious because it has also higher levels of fat, phosphorus, potassium, pyridoxine, vitamin A, beta-carotene, lutein and more riboflavin than other wheats. Because its flour lacks the rising characteristics desirable for bread, einkorn is primarily consumed as boiled whole grains, porridge, bulgur or as animal feed.

Yet, interest in *T. monococcum* is growing due to the demand for high quality wheat and the awareness of its agronomic potential and nutritional qualities. Attempts have been made to improve it to a more prolific and nutritious crop (Waines 1983; Vallega 1992; Castagna et al. 1995; Borghi et al. 1996; Brandolini and Hidalgo 2011;). Photosynthesis rates in diploid wheats exceed those of polyploid wheat (Austin et al. 1982) and there is considerable variation in several traits such as flag-leaf size, plant-height, and seed protein content and composition, that may be exploited for the improvement of diploid wheats. Many new sources of disease and pest resistance have been identified in diploid wheats which would be immediately usable at the diploid level. *T. monococcum* could become an important crop for the production of baked foods rich in carotenoids and proteins (Waines 1983; Borghi et al. 1996). Its adaptation to low-input agriculture and high level of resistance to pests and diseases represent advantages for organic farming and its gene pools may serve as a valuable reservoir of desirable genes for improvement of durum and bread wheat (Sharma et al. 1981; Megyeri et al. 2012; Zaharieva and Monneveux 2014). Since diploid wheats may be easier to manipulate than polyploid wheats in conventional breeding programs, it has gradually been recognized as an attractive diploid model for exploitation of useful traits, discovery of novel genes and variant alleles, and functional genomics (Jing et al. 2009). Currently, natural and artificially mutagenized *T. monococcum* are being used to identify and map genes of agronomic importance (Bullrich et al. 2002; Kuraparthy et al. 2007). Furthermore, new technologies for studying functional genomics, e.g., TILLING (Targeting Induced Local Lesions in Genomes), VIGS (Virus-Induced Gene Silencing) and DArT (Diversity Arrays Technology), are currently in use in *T. monococcum* (Jing et al. 2009).

Lines of *T. monococcum* are resistant to powdery mildew (Saponaro et al. 1995; Lebedeva and Peusha 2006), to leaf rust (Gill et al. 1983; Dyck and Bartos 1994; Saponaro et al. 1995; Anker et al. 2001), to stem rust (Gerechter-Amitai et al. 1971; Kerber and Dyck 1973; Chen et al. 2018), to eyespot (Cadle and Murray 1997), green bug (Gill et al. 1983), Russian aphid (Potgieter et al. 1991), Hessian fly (Gill et al. 1983), and wheat streak mosaic virus (Gill et al. 1983). *T. monococcum* has genes conferring resistance to pre-harvest sprouting (Sodkiewicz 2002).

10.2.2.4 Cytology, Cytogenetics and Evolution

T. monococcum is a diploid species (2n = 2x = 14). Its nuclear genome is a modified A genome and designated A^m (Kimber and Tsunewaki 1988; Dvorak 1998), and its organellar genome is designated A in certain lines and A^2, a subset of the A organellar genome, in others (Ogihara and Tsunewaki 1982, 1988; Wang et al. 1997; Tsunewaki 2009). The organellar genome of *T. monococcum* differs considerably from the organellar genomes of all the diploid and polyploid *Aegilops* species, as well as from those of the polyploid wheats (Tsunewaki 2009).

Hybrids between the two subspecies of *T. monococcum,* the wild ssp. *aegilopoides* and the domesticated ssp. *monococcum,* showed seven ring bivalents at meiotic first metaphase and were fully fertile (Kihara et al. 1929; Percival 1932; Smith 1936), and no differences in the restriction profiles of repeated nucleotide sequences (Dvorak et al. 1988). On the other hand, genetic, cytological and molecular evidence indicated that the two diploid wheat species, *T. monococcum* and *T. urartu,* diverged from one another. Studies of male and female fertility in hybrids and backcrosses showed that the interspecific F_1 hybrid between either wild or domesticated *T. monococcum* (as male), and *T. urartu* (as female) were completely sterile and backcrosses were completely to partially sterile, indicating that *T. urartu* is a separate species (Sharma and Waines 1981; Waines and Payne 1987; Ciaffi et al. 1998; Castagna et al. 1994).

Takumi et al. (1993) performed RFLP analyses of the nuclear DNAs of diploid wheats and calculated the genetic distances between all the pairs of accessions from the RFLP data. Using the UPGMA method, all the accessions of *T. urartu* were found to cluster in one group, whereas those of wild and domesticated *T. monococcum* were in a second group. Similar results were obtained by Le Corre and Bernard (1995). Hammer et al. (2000) noted that microsatellite markers differentiated the wild form of *T. monococcum* from *T. urartu.* The existence of genome divergence between *T. monococcum* and *T. urartu* is also evident from extensive differences in the restriction profiles of repeated nucleotide sequences and the promoter region of the 18S-5.8S-26S rRNA genes, which show very little intraspecific variation in the *Triticum* species (Dvorak et al. 1993). This indication of divergence between the genome of *T. monococcum* and the

A genome of *T. urartu* led Dvorak et al. (1993) to propose re-designating the genome of *T. monococcum* as A^m.

The divergence between the two diploid wheat species, *T. monococcum* and *T. urartu,* at the molecular level is also apparent from the pattern of chromosome pairing in hybrids between them. Johnson and Dhaliwal (1978) observed a mean 6.97 bivalents at first meiotic metaphase of the F_1 hybrid between *T. monococcum* and *T. urartu* and Shang et al. (1989) described 5.03 bivalents and 0.94 quadrivalents per cell in hybrids between different lines of these species, indicating that some chromosome differentiation occurred between the two species.

Meiotic pairing between *T. monococcum* chromosomes, when individually substituted in common wheat, and the wheat chromosomes of the A subgenome, is low in the presence of the homoeologous pairing suppressor *Ph1* (Paull et al. 1994; Dubcovsky et al. 1995). Paull et al. (1994) reported on a very low level of recombination between chromosome $7A^m$ of *T. monococcum* ssp. *aegilopoides* and 7A of *T. aestivum.* Likewise, Dubcovsky et al. (1995) noted that recombination between *T. aestivum* chromosome 1A and its closely related homoeologous chromosome $1A^m$ of ssp. *monococcum* was low in the presence of the *Ph1* gene. Chromosomes 1A and $1A^m$ were shown to be colinear, and consequently, Dubcovsky et al. (1995) concluded that the A^m genome of *T. monococcum* and subgenome A of hexaploid wheat diverged from one another by small chromosomal segments, and this amount of chromosomal differentiation is already recognized by the *Ph1* gene. In the absence of *Ph1*, the distribution and frequencies of crossing over between the 1A and $1A^m$ homoeologues were similar to the distribution and frequencies of crossover between 1A homologues. This indicates that some distinction exists between the chromosomes of the A^m genome of *T. monococcum* and those of the A subgenome of common wheat.

Additional evidence on divergence between chromosome $1A^m$ and chromosome 1A of *T. aestivum* was reached by Wicker et al. (2003), who sequenced and compared two-large physical contigs of 285 and 142 kb, covering orthologous low molecular weight (LMW) glutenin loci on the short arm of chromosome $1A^m$ of domesticated *T. monococcum* and on the short arm of chromosome 1A of tetraploid wheat, *T. turgidum* ssp. *durum.* Sequence conservation between the two species was restricted to small regions containing the orthologous LMW glutenin genes, whereas > 90% of the compared sequences were not conserved. Dramatic sequence rearrangements occurred in the regions rich in repetitive elements. Dating of long terminal repeat retrotransposon insertions revealed different insertion events occurring in the past 5.5 million years in both genomes. These insertions are partially responsible for the lack of homology between the intergenic regions. In addition, the gene space was conserved only partially, as demonstrated by several predicted genes identified on both contigs. Duplications and deletions of large fragments that might be attributable to illegitimate recombination, also contributed to the differentiation of this region in both genomes. The striking differences in the intergenic landscape of the A subgenome versus the A^m genomes, that diverged 1.28 million years ago (Li et al. 2011), provide evidence of dynamic and rapid genome evolution in wheat species.

Despite domestication of diploid wheat, both the wild and domesticated forms of *T. monococcum* contain a similar amount of 1C DNA (6.45 ± 0.103 pg and 6.48 ± 0.043 pg, respectively; Eilam et al. 2007). *T. monococcum* features one of the large genomes among the diploid species of the wheat group (the genera *Ambliopyrum*, *Aegilops* and *Triticum*). Only Emarginata species of *Aegilops*, namely, *Ae. bicornis*, *Ae searsii*, *Ae. sharonensis* and *Ae. longissima,* have larger genomes than that of *T. monococcum* (Eilam et al. 2007). *T. urartu*, the second diploid species of the genus *Triticum*, has a significantly smaller genome (1C DNA = 6.02 ± 0.062 pg) than that of *T. monococcum* (Eilam et al. 2007). The estimated amount of 1C DNA of the A subgenome of wild tetraploid wheat, *T. turgidum* ssp. *dicoccoides,* is 4.9 Gbp (= 4.79 pg) (Avni et al. 2017) and that of *T. aestivum* is 5.95 Gbp (=5.81 pg) (IWGSC 2018) (see Table 3.1). The A subgenome of polyploid wheat is thus smaller than that of *T. urartu.*

The karyotype of wild and domesticated forms of *T. monococcum* was studied by many cytologists. Kagawa (1929) classified the chromosomes in five groups, two-chromosome pairs had a secondary constriction. Subsequent works described only one chromosome pair with satellites, e.g., Levitsky et al. (1939), Riley et al. (1958), Upadhya and Swaminathan (1963a), Coucoli and Skorda (1966), or two pairs of satellited chromosomes, e.g., Smith (1936), Pathak (1940), Camara (1943), Oinuma (1953), Giorgi and Bozzini (1969b). Yet, it is possible that some lines of *T. monococcum* have only one satellited pair, as was found by Waines and Kimber (1973). These authors surveyed six biotypes of wild and domesticated *T. monococcum* from Europe and Iran and found variations in the number and size of the satellites. Biotypes with both one and two pairs of satellites were found, and the satellite size varied both within and between biotypes.

Giorgi and Bozzini (1969b) reported that the karyotypes of *T. monococcum* ssp. *aegilopoides* and *T. urartu* are very similar. Both have two chromosome pairs with small satellites. These chromosome pairs are the most hetero-brachial in the complement, with SAT 1 bearing a slightly more subterminal centromere than SAT 2. Three submedian and two median chromosome pairs were found in both species (Giorgi and Bozzini 1969b). There is good agreement between the ideogram drawn by Giorgi and Bozzini and that

reported by Pathak (1940). The presence of the satellites, together with the differences in total length, relative arm length, and position of secondary constrictions, make the identification of individual chromosomes possible (Smith 1936; Giorgi and Bozzini 1969b).

Gerlach et al. (1980) hybridized labelled RNA, transcribed in vitro from wheat ribosomal DNA cloned in a bacterial plasmid, to metaphase chromosomes of accessions of diploid wheats, including the two varieties of wild *T. monococcum* (var. *boeoticum* and var. *thaoudar*), domesticated *T. monococcum* and *T. urartu*. Autoradiography of the chromosomes provided unequivocal evidence that these taxa possess two pairs of nucleolus organizer (NOR) chromosomes. Yet, the diploid wheat accessions used possessed widely differing numbers of ribosomal RNA genes. The two chromosomes carrying the (NORs) of *T. monococcum* were identified as $1A^m$ and $5A^m$ by the combination of in situ hybridization and cytological markers (Miller et al. 1983). Later, Dvorak et al. (1989) found that the short arms of chromosomes $1A^m$ and $5A^m$ of both subspecies of *T. monococcum* also carry the 5S rRNA genes. The locus on chromosome $1A^m$ contains the 5S DNA subfamily with short spacers, while the locus on chromosome $5A^m$ contains 5S DNA subfamily with long spacers. Thus, in *T. monococcum*, $1A^m$ contains 360-bp units which belong to the short-unit subfamily, whereas $5A^m$ contains 500-bp units, belonging to the long-unit subfamily. The location of the 5S DNA in the short arms of chromosomes $1A^m$ and $5A^m$ suggests an ancestral linkage between the NORs and the 5S DNA loci.

Evidence that the loci carrying the 5S rRNA and 18S + 26S rRNA genes are located close to one another was also obtained by Kim et al. (1993) who performed FISH with the pScT7 (5S rDNA probe from *Secale cereale*) and pTa80 (18S + 26S rDNA probe from *T. aestivum*) and observed that both probes hybridized to the sub-terminal regions of the short arms of chromosomes $1A^m$ and $5A^m$ in *T. monococcum*. Both probes labelled the pair of $1A^m$ more heavily than those on $5A^m$, indicating that in *T. monococcum*, chromosomes $1A^m$ carry more copies of rRNA genes than chromosomes 5Am (Kim et al. 1993). The close linkage between 18S + 26S and 5S rDNA genes in diploid wheats, (as well as in several other Triticeae species), may be due to the fact that RNAs specified by these two types of genes, which are transcribed by different RNA polymerases, are required in approximately the same amount for ribosomal formation. Hence, one advantage of them being in close proximity may be that a common regulatory mechanism ensures such equality (Kim et al. 1993).

No differences were found in the 5S rRNA restriction sites among the 12 accessions of *T. monococcum* and *T. urartu* studied with several restriction enzymes (Kim et al. 1993). Similar to the finding of Dvorak et al. (1989),

Southern analysis of 5S rRNA with *Bam*HI-digested DNA from the 12 accessions, yielded two superimposed ladders of approximate sizes of 500 and 330 bp (Kim et al. 1993). The 500-bp ladder derived from chromosome $5A^m$ and the 330-bp ladder from chromosome $1A^m$.

Megyeri et al. (2012), performing FISH with the repetitive DNA probes pSc119.2, Afa family and pTa71 (rDNA probe) on mitotic chromosomes of *T. monococcum*, showed that the pSc119.2 probe was not suitable for the identification of *T. monococcum* chromosomes. On the other hand, all chromosomes were distinguishable by their Afa family signals that were observed on all chromosomes in the intercalary and distal regions, albeit, with different intensities. Strong fluorescent pTa71 signals were observed on the sub-telomeric region of the short arms of chromosomes $1A^m$ and $5A^m$. A similar FISH hybridization pattern was observed in *T. urartu* (Molnar et al. 2014). Comparison of the hybridization pattern of the chromosomes of *T. monococcum* and *T. urartu* with those of the A subgenome chromosomes of tetra- and hexa-ploid wheat, using the same hybridization probes, showed that the chromosomes of the diploid species have more complex Afa family and pTa71 hybridization patterns than the A-subgenome chromosomes of the polyploid wheats (Molnar et al. 2014). On the other hand, the pSc119.2 signals located on chromosomes 4A and 5A of tetra-and hexa-ploid wheat were not observed on these chromosomes of the diploid wheats.

The results of Megyeri et al. (2012), Molnar et al. (2014) showed that the entire set of *T. monococcum* chromosomes (especially chromosomes 1, 4, 5 and 7) could be discriminated by their hybridization patterns of pTa71 and Afa family. In situ hybridization with the microsatellite motifs GAA, CAG, AAC and AGG, demonstrated that these SSRs represented additional landmarks for the identification of *T. monococcum* chromosomes. The most promising SSR probes were the GAA and CAG motifs, which when used in combination with the Afa family and pTa71 probes, allowed for reliable identification of the entire set of *T. monococcum* chromosomes and for their discrimination from the A subgenome chromosomes of polyploid wheat background (Megyeri et al. 2012).

In order to improve the identification of and localization to chromosomes useful genes, a series of primary trisomics of *T. monococcum* was generated from autotriploids derived from crosses between induced autotetraploids and diploids (Friebe et al. 1990). All trisomics differed phenotypically from their diploid progenitor, but only two of the seven possible trisomic types exhibited a distinct morphology enabling their identification. C-banding, Ag–NOR staining and FISH, using rDNA probes, were employed to identify the chromosomes and the trisomics of *T. monococcum*. A comparison of the C-banding patterns of the chromosomes of *T. monococcum* with those of the A subgenome

of *T. aestivum*, enabled identification of five *monococcum* chromosomes, viz., $1A^m$, $2A^m$, $3A^m$, $5A^m$, and $7A^m$. The two remaining chromosomes, $4A^m$ and $6A^m$, showed C-banding patterns that were not equivalent to those of any of the chromosomes in the A subgenome of bread wheat. When one of these undesignated chromosomes from ssp. *aegilopoides* var. *boeoticum* was substituted for chromosome 4A of *T. turgidum*, it compensated well phenotypically for the loss of chromosome 4A in the recipient species. Because this *T. monococcum* chromosome appeared to be homoeologous to the group 4 chromosomes of polyploid wheats, it was designated $4A^m$, and the second undesignated chromosome in *T. monococcum* was designated $6A^m$ (Friebe et al. 1990).

Haploid plants (1n = 1x = 7) may be present among *T. monococcum* progeny (Smith 1936). Kihara and Katayama (1932, 1933), Chizaki (1934) studied meiosis in haploids of this species and found that the chromosomes often become attached end-to-end, especially during the meiotic phase of diakinesis. Kostoff and Arutiunian (1938) assumed that this phenomenon is due to associations between heterochromatic regions of non-homologous chromosomes rather than to the presence of duplications or interchanges. Indeed, Kostoff (1938) showed that the distal ends of almost all chromosomes of *T. monococcum* stained very dark with Newton's gentian violet, while all the other chromosomal parts were stained much lighter. This differential staining indicates the presence of heterochromatin at the chromosome ends.

The presence of genes that either prevent crossability of lines of wild and domesticated *T. monococcum* with other species or that cause death of interspecific or intergeneric hybrids, was demonstrated with several hybridization events. The F_1 hybrids from the *T. monococcum* ssp. *aegilopoides (as female)* x *T. urartu* (as male) cross yielded small, plump and viable seeds, while the reciprocal cross had long, shriveled and non-viable seeds (Johnson and Dhaliwal 1976; Dhaliwal 1977a; Fricano et al. 2014). Alloplasmic lines, where a nucleus of one species was introduced into the cytoplasm of another species, were developed through repeated backcrossing, which were then crossed as female parents with respective non-recurrent parents that were the cytoplasm donors (Dhaliwal 1977a). It was concluded that the difference between the reciprocal crosses was presumably attributable to different ssp. *aegilopoides-urartu* genomic ratios in the triploid endosperm rather than to cytoplasmic differences between the diploid wheats. The endosperm with two doses of the *aegilopoides* and one of the *urartu* genome resulted in small, plump and viable seeds while the endosperm of the reciprocal cross with two doses of the *urartu* and one of the *aegilopoides* genome, developed into large but shriveled and non-viable seeds irrespective of the cytoplasmic type (Dhaliwal 1977a).

Gill and Waines (1978) performed diallel crosses among four lines of *T. monococcum* ssp. *aegilopoides* from different geographical areas, with *T. urartu, Ae. tauschii* and *Ae. speltoides* and observed reciprocal differences in hybrid seed morphology, endosperm development, and embryo viability. *T. urartu and Ae. tauschii* as females, crossed with ssp. *aegilopoides* and *Ae. speltoides*, led to the production of shriveled inviable seed. ssp. *aegilopoides* accessions as female, crossed with *Ae. speltoides* also led to shriveled seeds. The reciprocal crosses produced plump seeds which either resembled the maternal parent or showed size differences. Altering the endospermic genome ratios was achieved by crossing an autotetraploid line of ssp. *aegilopoides* (2n = 4x = 28), either as female or as male, with diploid *T. urartu, Ae. tauschii* or *Ae. speltoides*. When the autotetraploid ssp. *aegilopoides* was the male parent and any of the diploid species was the female parent, the hybrid endosperms contained two doses from the female parent genome and two doses from the male parent genome and consequently, the seeds showed extreme shriveling. On the other hand, when the autopolyploid was the female parent and each of the diploid species was the male parent, the hybrid endosperm contained four doses from the female parent genome and one dose from the male parent genome, and consequently, the seeds were moderately shriveled to plump. Genetic experiments involving hybrids of *T. monococcum* (wild and domesticated forms)*, and T. urartu* showed that a factor showing a dosage effect is present in male gametes, which, by interacting with the maternal genome, leads to endosperm abortion (Gill and Waines 1978).

Endosperm development and embryo lethality were studied in hybrids of different accessions of *Ae. tauschii* (as female) and the diploid wheats, i.e., wild and domesticated *T. monococcum* and *T. urartu* (Gill et al. 1981), that produced shriveled and non-viable seed. *Ae. tauschii x T. urartu* hybrids showed good seed development but embryos were semi-lethal and seedling death resulted. *Ae. tauschii x T. monococcum* ssp. *monococcum* hybrids showed endosperm abortion and embryos were lethal or semi-lethal. *Ae. tauschii x T. monococcum* ssp. *aegilopoides* hybrid seeds were dead by day 14. However, using embryo culture from day 10 of seed development, Gill et al. (1981) obtained viable F_1 *tauschii x* ssp. *aegilopoides* plants. These authors found that the block that leads to hybrid endosperm abortion is effective between days 5 and 10 of seed development. Seed abortion in *tauschii* x ssp. *aegilopoides* crosses were seemingly primarily the result of faster, rather than abnormal, nuclear division in the hybrid endosperm. The lack of storage protein synthesis indicates that abortion of the hybrid endosperm is complete by day 10. This rapid endosperm degeneration almost certainly adversely affects the viability of hybrid embryos (Gill et al. 1981).

Sears (1944a) identified two alleles in a domesticated line of *T. monococcum* which acted as dominant lethals in hybrids with *Ae. umbellulata*, but which induced no effect in *T. monococcum* itself. The two alleles differed in the earliness with which they caused death. A third, normal allele that did not cause death in hybrids with *Ae. umbellulata*, was present in the wild form, ssp. *aegilopoides* . Semi-lethality occurred in hybrids of *Ae. bicornis* with certain varieties of domesticated *T. monococcum,* while hybrids with another variety of *T. monococcum* were viable (Sears 1944a). This effect on viability was apparently mono-factorially determined. Non-crossability of domesticated *T. monococcum* with *Dasypyrum villosum* also appears to be simply inherited (Sears 1944a).

The modes of inheritance of 12 morphological characters were investigated in domesticated *T. monococcum* by crosses, complementation studies, and observations of phenotypes of F_1s and F_2s from crosses between lines expressing the different traits (Kuspira et al. 1989). All studied traits were found to be controlled by single genes. The genes for six of these 12 characters fall into two closely linked groups; *Bg* (glume color) and *Hg* (Hairy glume) are the same distance apart in *T. monococcum* as in in *T. aestivum*, indicating that this segment has been highly conserved. The genes *Sg* (glume hardness), *La* (lemma awn length), *Fg* (false glume), and *Lh* (head type) were also very closely linked, with the outside markers being only 4 map units apart. Tentative assignments of genes and linkage groups identified in this investigation, to specific chromosomes of *T. monococcum* have been made on the basis of known chromosomal locations in A subgenome genes of *T. aestivum* (Kuspira et al. 1989).

Since cultivated and wild genotypes of *T. monococcum* show high levels of restriction fragment length polymorphism (RFLP) (Castagna et al. 1994; Le Corre and Bernard 1995), *T. monococcum* can be used to produce high-density RFLP maps that would complement the genetic maps of the A subgenome of *T. aestivum*. RFLP and AFLP markers have been developed and used to generate genetic linkage maps, for map-based cloning, and genome synteny comparisons in *T. monococcum* (Dubcovsky et al. 1995, 1996; Faris et al. 2008).

Singh et al. (2007) produced an integrated molecular linkage map of the A^m genome based on 93 recombinant inbred lines (RILs) derived from a cross of wild x domesticated *T. monococcum*. The parental lines were analyzed with SSRs and RFLP markers, and bin-mapped ESTs. The polymorphic markers, that were assayed on the RILs, mapped on the seven linkage groups with a total map length of 1262 cM. About 58 loci, mostly mapping on chromosome $2A^m$, showed distorted segregation. With a few exceptions, the position and order of the markers was similar to the ones in maps of the A subgenome of polyploid wheat.

Chromosome 1A^m of ssp. *monococcum* and ssp. *aegilopoides* showed a small paracentric inversion relative to the A subgenome of hexaploid wheat.

Partial genetic maps of chromosomes 1A^m and 5A^m, constructed based on crosses between winter and spring lines of ssp. *aegilopoides* and between ssp. *aegilopoides* and ssp. *monococcum*, showed the same order of markers and similar interval lengths between markers (Dubcovsky et al. 1995). A map of the ssp. *aegilopoides* chromosome 1A^m, constructed based on a winter x spring F2 population, has a similar genetic length as a map of chromosome 1A of *T. aestivum* (Dubcovsky et al. 1995). Additional genetic map of F$_2$ progeny of a cross between domesticated and wild subspecies of *T. monococcum* involving 335 markers, including RFLP DNA markers, isozymes, seed storage proteins, rRNA, and morphological loci, was constructed by Dubcovsky et al. (1996). These authors reported that *T. monococcum* and barley linkage groups are remarkably conserved. They differ by a reciprocal translocation involving the long arms of chromosomes *4* and *5*, and paracentric inversions in the long arm of chromosomes *1* and *4*; the latter is in a segment of chromosome arm *4L* translocated to *5L* in *T. monococcum*.

Jing et al. (2007, 2009) developed Diversity Arrays Technology (DArT), consisting of 2304 hexaploid wheat, 1536 tetraploid wheat, 1536 domesticated and 1536 representative wild *T. monococcum* genomic clones, to assess genetic diversity in *T. monococcum*, and to construct a genetic linkage map integrating DArT and microsatellite markers. In total, 846 polymorphic DArT markers were identified and used to fingerprint 16 *T. monococcum* accessions of diverse geographical origins. The fingerprinting data showed a partial correlation between the geographic origin of *T. monococcum* accessions and their genetic variation. Using DArT and SSR markers, Jing et al. (2007, 2009) constructed a linkage map in F2 progeny from a cross between two different accessions of *T. monococcum*. In total, 356 (274 DArTs and 82 SSRs) molecular markers were mapped and formed nine linkage groups. Two morphological traits, namely awn color and leaf hairiness, were found to segregate in a 1:3 ratio in the *T. monococcum* mapping population. These two traits, *Ba* (black awn) and *Hl* (hairy leaf), each controlled by a single gene, were also included in the linkage analysis. The linkage map derived from the combined data set spanned 1062.72 cM, with an average length of 151.82 cM per chromosome and an average density of one marker per 2.97 cM. Each of the seven chromosomes contained both DArT and SSR markers. Six of the linkage groups corresponded to six *T. monococcum* chromosomes, but chromosome 4A^m was formed by three linkage groups.

10.2.2.5 Crosses with Other Species of the Wheat Group

Chromosomal pairing at meiosis of both wild and domesticated forms of *T. monococcum*, is complete, namely, seven ring bivalents. A similar level of pairing was observed in F$_1$ hybrids between the two subspecies of *T. monococcum* (Kihara et al. 1929; Perecival 1932; Smith 1936), indicating that their genomes are fully homologous. On the other hand, F$_1$ hybrids between wild *T. monococcum*, ssp. *aegilopoides* x *T. urartu* (hybrid genome A^{m}A) (Table 10.6) had somewhat reduced pairing indicates a small scale of chromosomal differentiation between the chromosomes of these two diploid wheats. Chromosome pairing in hybrids between wild or domesticated forms of *T. monococcum* with diploid *Aegilops* species of section Sitopsis (Table 9.8), showed greater genomic differentiation. The F$_1$ hybrid between a high-pairing type of *Ae. speltoides* and domesticated *T. monococcum* (hybrid genome SA^m) has more pairing than the hybrid with intermediate pairing type (Sears 1941b). This difference resulted from the promotion of homoeologous pairing by the high-pairing gene(s) of *Ae. speltoides*. Chromosome pairing in the F$_1$ hybrids *Ae. longissima* x ssp. *monococcum* (hybrid genome S^{l}A^m), *Ae. sharonensis* x ssp. *monococcum* (hybrid genome S^{sh}A^m), and *Ae. bicornis* x ssp. *monococcum* (hybrid genome S^{b}A^m) exhibited much reduced pairing than the hybrid with the intermediate - pairing type of *Ae. speltoides* (Table 9.8). While genome A^m of *T. monococcum* diverged relatively little from the S genome of *Ae. speltoides*, it differed quite considerably from those of the other Sitopsis species.

Sears (1941b) studied chromosome pairing in F$_1$ hybrids between domesticated and wild *T. monococcum* x several

Table 10.6 The wild and domesticated subspecies of *Triticum turgidum*. (Classification of subspecies after van Slageren 1994; common names in parentheses)

Wild	Domesticated	
	Hulled wheat	Free-threshing wheat
dicoccoides (wild emmer)	*dicoccon* (domesticated emmer) *paleocolchicum* (Georgian wheat)	*parvicoccum*[a] *durum* (macaroni or hard wheat) *turgidum* (rivet or pollard wheat) *polonicum* (polish wheat) *turanicum* (khorasan wheat) *carthlicum* (Persian wheat)

[a] Extinct, described by Kislev (1979/1980)

other diploid *Aegilops* species (Table 9.8). The domesticated *T. monococcum* x *Ae. tauschii* hybrid (hybrid genome A^mD), exhibited more pairing than the hybrid with wild *monococcum*. The lower level of chromosomal pairing in the hybrid with wild *T. monococcum* may result from the presence of gene(s) affecting pairing in the wild line used for this cross. The F_1 hybrids between ssp. *monococcum* x *Ae. caudata*, ssp. *aegilopoides* x *Ae. caudata* (hybrid genome A^mC), ssp. *monococcum* x *Ae. comosa*, ssp. *aegilopoides* x *Ae. comosa* (hybrid genome A^mM), a ssp. *monococcum* x *Ae. uniaristata*, ssp. *aegilopoides* x *Ae. uniaristata* (hybrid genome A^mN), and ssp. *aegilopoides* x *Ae. umbellulata* (hybrid genome A^mU) had reduced pairing (Table 9.8). The above data showed that genome A^m of *T. monococcum*, either domesticated or wild, had extensively diverged from these genomes of *Aegilops*.

Chromosome pairing between alltetraploid wheat, i.e., various subspecies of *T. turgidum* and *T. monococcum*, either ssp. *monococcum* or ssp. *aegilopoides*, (hybrid genome BAA^m), domesticated *T. timopheevii* x ssp. *monococcum* (hybrid genome GAA^m), and ssp. *aestivum* x ssp. *monococcum* (hybrid genome $BADA^m$) was studied by several cytogeneticists (Table 10.6). Pairing in all these F_1 hybrids show that the allotetraploids and the allohexaploid wheats contain a subgenome related to the A^m genome of *T. monococcum*.

10.2.3 *T. urartu* Tum. ex Gand. (Donor of the A Subgenome to Allopolyploid Wheats)

10.2.3.1 Description of the Species

T. urartu Tum. ex Gand., common names red wild einkorn and urartu wheat [Syn.: *T. boeoticum* ssp. *urartu* Dorof.; *T. monococcum* ssp. *urartu* (Tum. ex Gand.) Á. Löve and D. Löve; *Crithodium urartu* (Gandilyan) Á. Löve], is an annual, predominantly autogamous, 60–90-cm-tall (excluding spikes) plant. Its culms are first geniculated and then ascending upright. Culm internodes are distally glabrous. The leaf sheaths are pubescent, their auricles are curved, and the ligule is membranous. Leaf blades are 30–45-cm-long. The spike is indeterminate, bilaterally compressed, two-rowed, 7–11(–13) cm-long (excluding awns) and awned. The entire spike disarticulates at maturity into individual spikelets, each with its associated rachis segment (wedge–type dispersal unit). The rachis is densely hairy on margins and sparsely hairy at nodes. The spikelets are laterally compressed, 17–20-mm-long, the top one being fertile and generally in the same plane as those below it. There are 8–16 fertile spikelets per spike, and 1–2 basal rudimentary spikelets. There are 3 florets, the upper one being sterile. The spikelets usually have two grains and two 2.5–5.5 cm-long

awns. The glumes are shorter than the spikelet, 11.2–11.6-mm-long, glabrous, with two well-developed keels, and 3–5 -veins. There are two unequal teeth, which usually do not develop into awns, with one tooth being well developed and the second less. The lemma are glabrous, 13–15-mm-long, keeled, taper into an awn, and bear a lateral small tooth. The palea is 11–13-mm-long, membranous and splits along the keel at maturity. The anthers are small, 2.0–2.5-mm-long. The caryopsis has a reddish color, is 7.3–9.2-mm-long, and is free but laterally compressed by the tough glumes (Fig. 10.1c).

T. urartu was discovered by Tumanian (1937) in Armenia, who distinguished it from the other wild diploid wheat *T. monococcum* ssp. *aegilopoides*, by the following morphological features: its smaller second tooth of the glume, smaller anthers, two-awned two-grained spikelets and red grains (Gandilian 1972; Dorofeev et al. 1980). Johnson (1975), who discovered *T. urartu* in northern Lebanon, southeastern Turkey, and southwestern Iran, found that this species also differs from ssp. *aegilopoides* by the electrophoretic pattern of seed Beta albumin. He described additional morphological differences between *T. urartu* and ssp. *aegilopoides*, namely, the awns of *T. urartu* are shorter and slenderer than those of ssp. *aegilopoides*, they are nearly equal and spread at maturity. *T. urartu* is most easily identified by the presence of an awn up to 1-cm-long on the lemma of the third floret (Johnson 1975). In several locations, this awn was often reduced to an inconspicuous bristle, but in Johnson's fields and a greenhouse at Riverside, California, it was discernible in all accessions of this species. With respect to plant and spikelet size, the Turkish populations provide a transition between the Lebanese biotypes and the more robust one of Transcaucasia, Iraq and Iran. Both in the wild and under cultivation, *T. urartu* from all parts of its range, except Transcaucasia, was found to generally mature earlier than ssp. *aegilopoides* (Johnson 1975).

The two diploid wheats, *T. urartu* and the wild and domesticated forms of *T. monococcum*, also differ by biochemical and molecular markers such as isoenzymes of ACP (Jaaska 1974), EST (Jaaska 1980), SOD (Jaaska 1982), by the composition of the seed storage proteins, namely, the high-molecular weight (HMW) glutenin subunits (Waines and Payne 1987; Ciaffi et al. 1998; Castagna et al. 1994), and Gliadin genes (Ciaffi et al. 1997). In addition, AFLP (Sasanuma et al. 2002; Heun et al. 2008) and RFLP (Takumi et al. 1993; Le Corre and Bernard 1995) analyses of nuclear DNAs of *T. urartu* and *T. monococcum* revealed differences between these two species. AFLP and simple sequence length polymorphism (SSLP) analyses clearly showed that *T. urartu* was greatly differentiated from both the wild and domesticated forms of *T. monococcum* (Sasanuma et al. 2002). A similar conclusion was reached following chloroplast SSLP analysis (Mizumoto et al. 2002). Comparable results were also obtained by Hammer et al. (2000), who found that microsatellite markers (SSRs) differentiated the wild form of *T. monococcum* from *T. urartu*. Furthermore, Baum and Baily (2004) found that *T. urartu* differs from *T. monococcum* in the short and long units of the 5S DNA; *T. monococcum* contained the long A1 and the short A1 unit classes, whereas *T. urartu* had the long A1 and the short G1 unit classes. Likewise, Dvorak et al. (1988, 1993) found extensive differences between these two species in the restriction profiles of repeated nucleotide sequences and the promoter region of the 18S-5.8S-26S rRNA genes.

Phase-contrast microscopy and scanning electron microscopy measuring the pollen grain diameter and examining the exine (the outer layer of the pollen grain) sculpturing of *T. urartu* and *T. monococcum* ssp. *aegilopoides* var. *thaoudar* (López-Merino et al. 2015), found that *T. urartu* pollen is smaller on average than that of var. *thaoudar*, and its exine sculpturing differs from that of var. *thaoudar*.

The divergence between *T. urartu* and *T. monococcum* is also apparent from the cytogenetic data; chromosome pairing in the F_1 hybrids between the two species was somewhat reduced (Johnson and Dhaliwal 1978; Shang et al. 1989) and the hybrid was completely sterile when *T. urartu* served as the female parent (Johnson and Dhaliwal 1976). Hence, gene flow via hybridization between these two species can occur only from *T. urartu* into *T. monococcum* (Johnson and Dhaliwal 1976). Tsunewaki et al. (1999) found that *T. urartu*, when used as female, showed strong cross incompatibility to emmer wheat, differing, in this respect, from *T. monococcum* ssp. *aegilopoides,* which crosses relatively easily with emmer wheat. All the above biochemical, molecular and cytogenetic data revealed the existence of a profound genome divergence between *T. urartu* and *T. monococcum* and thus, substantiating their classification into two separate species.

T. urartu exhibits wide morphological variation in spike and spikelet size, color and hairiness (Johnson 1975). This variation is also expressed at the biochemical and molecular levels. While Smith-Huerta et al. (1989), using isoenzymes, observed low genetic diversity, Yaghoobi-Saray (1979), Moghaddam et al. (2000), Hedge et al. (2000) reported a high level of isoenzyme diversity, either within or among populations. Studies at the molecular level also revealed high intra- ad inter-population variation in *T. urartu*. Sasanuma et al. (2002), Singh et al. (2006), using AFLP, revealed variation in all analyzed accessions of *T. urartu*. Similar results were obtained by Brunazzi et al. (2018), who genotyped a collection of 352 accession of *T. urartu*, sampled from Armenia, Iran, Iraq, Jordan, Lebanon, Syria and Turkey, with a large number of high-quality genome-wide single nucleotide polymorphisms (SNPs), and revealed broad molecular variation across the sampled populations. In addition, Brunazzi et al. (2018) obtained phenotypic data that highlighted a wide variation for flowering time and plant height. Storage protein analysis indicated the presence of highly polymorphic protein bands, while SDS sedimentation tests showed broad variability in the dough volume, even including some accessions approaching good bread-making quality (Brunazzi et al. 2018). These researchers used 1.3 million genome-wide SNPs to assess variation in a large collection of *T. urartu* accessions. They found a correlation between the amount of genetic diversity and the geographical distance existing between samples from different regions. Using a genome-wide association approach, they identified several marker-environment associations, such as association of molecular markers with altitude, temperature, and/or association with rainfall measures. The most significant marker–environment associations were observed with genomic loci with adaptive potential, including dormancy and frost resistance loci.

A high level of genetic diversity was found in low-molecular weight (LMW) glutenin subunits from *T. urartu*, that are coded by the *Glu-A3* locus on chromosome 1A, including detection of 11 novel alleles (Cuesta et al. 2015). Wang et al. (2017) investigated the morphological and genetic diversity and population structure of 238 *T. urartu* accessions collected from different geographic regions. They found wide variation in SSR markers and in HMW-glutenin subunits. Their analysis indicated that the 238 *T. urartu* accessions could be classified into two subpopulations, of which Cluster I contained accessions from the Eastern Mediterranean coast, while those from Mesopotamia and Transcaucasia belonged to Cluster II. Significant associations were observed between SSRs or HMW-GSs and six morphological traits: heading date, plant height, spike length, spikelet number per spike, tiller angle and grain length.

T. urartu distributes abundantly in southeastern Turkey, Syria and Lebanon and sporadically in northern Iraq, western Iran and Armenia (Johnson 1975). Recently, it was also discovered in the high mountains of southern Jordan. *T. urartu* is not found west of the Fertile Crescent arc. In this respect, it also differs from ssp. *aegilopoide*s, which spread westwards as a weed. The chorotype (the general range of distribution of a species characterizing its phytogeographical nature) of *T. urartu* is sub-Mediterranean-west Irano-Turanian. It grows in a relatively large area, in a wide range of altitudes (500–1600 m), occurs naturally in openings of oak park-forests, in open herbaceous formations, and in steppe-like formations. It thrives well on terra rossa, basalt, and on several types of alluvial soils. *T. urartu* is more adapted to dry climates than ssp. *aegilopoides* (Mac Key 2005). Studies on heat and cold tolerance in wild and domesticated wheat forms by Damania and Tahir (1993), revealed that *T. urartu* was significantly more cold-tolerant than other wild wheat taxa.

T. urartu often grows mixed with *T. monococcum* ssp. *aegilopoides* and wild emmer, *T. turgidum* ssp. *dicoccoides*, in southwestern Syria, southeastern Lebanon and in some sites in southeastern Turkey, and with wild *T. timopheevii* in southeastern Turkey, northern Iraq, southwestern Iran and Transcaucasia. It also grows sympatrically with *Ae. speltoides, Ae. searsii, Ae. caudata, Ae. umbellulata, Ae. geniculata, Ae. biuncialis, Ae. triuncialis* and *Ae. columnaris* and allopatrically with *Ae. longissima, Ae. peregrina, Ae. tauschii, Ae cylindrica* and *Ae. crassa.*

Being the donor of the A subgenome to allopolyploid wheats (Chapman et al. 1976; Dvorak 1976), many of the chromosomes of *T. urartu* pair regularly with their homologous chromosomes of allopolyploid wheats in the F_1 hybrids between them, and therefore, desirable genes can be transferred from the wild diploid species to the domesticated allopolyploid species via conventional plant breeding procedures. Thus, searching the germplasm of *T. urartu* for economically useful traits is very important.

T. urartu contains genes conferring resistance to powdery mildew (Hovhannisyan et al. 2011; Qiu et al. 2005; Ling et al. 2018), leaf rust (Hovhannisyan et al. 2011), stem rust (Rouse and Jin 2011), yellow rust (Xiao et al. 2018), and cereal aphids (Radchenko 2011). *T. urartu* might provide some glutenin subunit genes to improve the quality of common wheat dough (Luo et al. 2015; Cuesta et al. 2015).

10.2.3.2   Cytology, Cytogenetics and Evolution

T. urartu is a diploid species (2n = 2x = 14). Its nuclear genome is designated A, which differs from the A^m genome of *T. monococcum* (Dvorak 1998). Based on the restriction fragment pattern of its chloroplast DNA, its organellar genome was assigned A (Tsunewaki and Ogihara 1983). The cytoplasm of *T. urartu* and *T. monococcum* have identical chloroplast DNA (Tsunewaki and Ogihara 1983).

T. urartu contains 6.02 ± 0.062 pg 1C DNA (Eilam et al. 2007). While there is little variation in the amount of 1C DNA at the intraspecific level, the genome of *T. urartu* is significantly smaller than that of *T. monococcum, Ae. bicornis, Ae. searsii, Ae. sharonensis* and *Ae. longissima* and significantly larger than that of *Amblyopyrum muticum, Ae. tauschii, Ae. caudata, Ae. comosa, Ae. uniaristata* and *Ae. umbellulata* (Eilam et al. 2007).

The karyotype of *T. urartu* is very similar to that of *T. monococcum*, with small differences in the length of the short and long arms of several chromosome pairs (Giorgi and Bozzini 1969b). It consists of two chromosome pairs with small satellites, and three pairs with submedian and two with median centromeres. A similar karyotype of *T. urartu* was reported by Kerby and Kuspira (1988). *T. urartu* possess two pairs of nucleolus organizer (NOR) chromosomes (Gerlach et al. 1980), which were identified as 1A and 5A by the combination of in situ hybridization and cytological

markers (Miller et al. 1983). The number of rRNA genes in *T. urartu* was estimated to be somewhat smaller than 4500 (Gerlach et al. 1980).

The karyotypes of *T. urartu* and *T. monococcum* were examined using C-banding and FISH, with DNA probes representing 5S and 45S rDNA families, the microsatellite sequences GAA_n and GTT_n, the pSc119.2, Spelt52, *Fat*, pAs1 and pTa535 probes, and a newly identified repeat called Aesp_SAT86 (Badaeva et al. 2015). The C-banding pattern of *T. urartu* was similar to that of *T. monococcum*, except for differences in chromosomes 4A and 6A. Besides two major 45S rDNA loci on the short arms of chromosomes 1A and 5A, two-minor polymorphic NORs were observed in the terminal part of the long arm of chromosomes 5A and in the distal part of the short arm of chromosomes 6A in *T. urartu* and in *T. monococcum*. An additional minor locus was found in the distal part of the long arm of chromosomes 7A of *T. monococcum*, but not in *T. urartu*. Two 5S rDNA loci were observed in the short arms of chromosomes 1A and 5A. The pTa535 probe displayed species- and chromosome-specific hybridization patterns, enabling full identification of all *T. urartu* chromosomes. The distribution of pTa535 on the chromosomes of *T. urartu* was more similar to its distribution on the A-subgenome chromosomes of wild *T. turgidum* and wild *T. timopheevii*, confirming the origin of these subgenomes from *T. urartu*. The probe pAs1 allowed for the identification of four chromosomes of *T. urartu* and two of *T. monococcum*. The Aesp_SAT86-derived patterns were polymorphic; main clusters were observed on chromosomes 1A and 3A of *T. urartu*. The study of Badaeva et al. (2015) showed that a set of the above probes proved to be most informative for the analysis of A genomes in diploid and allopolyploid wheat species.

Johnson and Dhaliwal (1976) demonstrated that *T. urartu* is isolated *from T. monococcum* through a genetic barrier. ssp. *aegilopoides* (as female) x *T. urartu* (as male) produced seeds with germination capacity, whereas the F_1 hybrid from the reciprocal cross, produced inviable seeds (Johnson and Dhaliwal 1976; Dhaliwal 1977a). Indeed, the *T. urartu* x *T. monococcum* hybrids, involving either wild or domesticated forms, were completely self-sterile (Johnson and Dhaliwal 1976). These authors proposed that the reproductive isolation between *T. urartu* and *T. monococcum* is cytoplasmic, but Dhaliwal (1977a), suggested that the difference between the reciprocal crosses can be attributed to different *monococcum-urartu* genomic ratios in the triploid endosperm of the F_1 hybrid rather than to cytoplasmic difference between these diploid wheats.

Johnson (1975) discovered that the electrophoretic pattern of seed proteins of allotetraploid wheat contains B albumin bands similar to those of *T. urartu*. He mistakenly proposed that *T. urartu* is the donor of the B subgenome of the allopolyploid wheats. Chapman et al. (1976), Dvorak

(1976), based on chromosome pairing in hybrids between ditelosomic lines of *T. aestivum* and *T. urartu*, concluded that the genome of T. urartu is homologous to the A subgenome of allopolyploid wheats, and that it did not correspond, as had been proposed by Johnson (1975), to the B subgenome. In contrast, chromosome pairing in the hybrid ditelosomic lines of *T. aestivum* x *T. monococcum* ssp. *aegilopoides* showed that the genome of *T. monococcum* is homoeologous, rather than homologous, to the A subgenome of allohexaploid wheat (Chapman et al. 1976). Moreover, from the higher trivalent frequencies observed in the latter hybrids, Chapman et al. (1976) concluded that the genotype of ssp. aegilopoides has the capacity to partly suppress the activity of the Ph1 locus of common wheat, allowing for some homoeologous pairing to occur. Dvorak (1976) noted that pairing of *T. urartu* chromosomes was significantly reduced in hybrids lacking chromosome arms 5AS or 5BS of *T. aestivum*. He suggested that this reduction in chromosome pairing resulted from the absence of genes which promote pairing and which are normally present on chromosome arms 5AS and 5BS in cultivar Chinese Spring of *T. aestivum.*

Further evidence supporting the idea that *T. urartu* is the donor of the A subgenome of the allopolyploid wheats, came from biochemical and cytogenetic studies. Konarev et al. (1979), Konarev (1983) concluded from immunochemical and electrophoretic studies of seed proteins, that the A subgenome of *T. turgidum* was contributed by *T. urartu,* whereas the A subgenome of *T. timopheevii* was contributed by *T. monococcum.* Yet, Nishikawa (1983), based on variation in esterases, showed that the A subgenome in both *T. turgidum* and *T. timopheevii* was contributed by *T. urartu.* Kerby et al. (1988) compared the amino acid sequence of *T. urartu* purothionin to the amino acid sequences of the purothionins in *T. monococcum, T. turgidum* and *T. aestivum,* and found that the sequence of the purothionin from *T. urartu* is identical to the β form specified by a gene in the A subgenome of the allpolyploid wheats and differs by five and six amino acid substitutions, from the α_1 and α_2 forms coded for by genes in the B and D subgenomes, respectively. Their results showed that *T. urartu,* rather than *T. monococcum,* is the source of the A subgenome in *T. turgidum* and *T. aestivum.* Phylogenetic analyses based on whole genome sequences suggest that the wheat A sub-genome diverged from T. urartu 1.28 MYA (Li et al. 2022). It might be therefore that an *urartu*-closely related species, unknown or now extinct, is in fact the progenitor of the A subgenome that hybridized with the B-donor ∼ 0.8 MYA.

The data of Dvorak et al (1988, 1993), from variation in repeated nucleotide sequences, substantiated Nishikawa's (1983) hypothesis, that the A subgenome in both tetraploid and hexaploid wheat is more related to the A genome of *T. urartu* than to the A^m genome of *T. monococcum.* In

T. zhukovskyi, one A subgenome was contributed by *T. urartu* and the other by *T. monococcum* (Dvorak et al. 1993). Takumi et al. (1993), using RFLP analysis of diploid and polyploid wheats, also concluded that *T. urartu* is the donor of the A subgenome to allopolyploid wheats, and Baum and Bailey (2004), based on 5S DNA unit classes, supported the view that the A subgenome of *T. turgidum* and *T. aestivum* was donated by *T. urartu.* Similarly, Badaeva et al. (2015) found that the pattern of FISH sites with the probe pTa535 confirmed the *T. urartu* origin of the A subgenome in *T. turgidum* and *T. timopheevii.* As the donor of one of the subgenomes of polyploid wheat, recognition of the genome structure, function, and diversity of T. urartu may provide important information for understanding the genomes of tetraploid and hexaploid wheats.

Ling et al. (2018), using technologies like bacterial artificial chromosome (BAC)-by-BAC sequencing, single-molecule, real-time, whole-genome shotgun sequencing, linked read sequencing and optical mapping, assembled a high-quality sequence of the *T. urartu* genome. This genome sequencing very much improved the draft genome of this species published earlier by Ling et al. (2013). See Sect. 3.3 in Chap. 3 and Table 3.1 for details on *T. urartu* genome structure. The genome of *T. urartu* includes *37,516* HC genes and 3991 LC genes. On average, the genes have a transcript length of 1453 bp, protein length of 332 amino acids and 4.5 exons per transcript, which is comparable to genes in other Triticeae species. Throughout the genome, Ling et al. (2018) identified 31,269 miRNAs, 5810 lncRNAs), 3620 tRNAs, 80 rRNAs and 2,519 snRNAs. A total of 3.90 Gb (81.42%) of genome sequences was identified as repetitive elements, including 3.44 Gb (71.83%) retrotransposons and 355 Mb (7.41%) DNA transposons. Among long-terminal repeat (LTR) retrotransposons, the *Gypsy* and *Copia* super families comprised 42.71% and 24.30% of the genome, respectively. The distribution of *Copia* elements was enriched at both telomeric and sub-telomeric regions, whereas *Gypsy* retrotransposons were enriched in the pericentromeric–centromeric regions. Comparative analyses with genomes of other grasses showed gene loss and amplification in the number of transposable elements in the *T. urartu* genome. *T. urartu*-specific or wheat-specific amplification of gene families was associated with stress response or vernalization. Large-scale retrotransposon-mediated structural rearrangements occurred during A-genome evolution, as revealed by comparing the A genomes among *T. urartu* and tetraploid and hexaploid wheat.

Ling et al. (2018) found substantially higher gene density and recombination rates, as well as lower densities of transposable elements and tandem repeats, in the sub-telomeric regions of each chromosome. The accumulated gene expression level was higher in the sub-telomere than in the centromere regions. Analyses of genes in the

T. urartu genome, together with those from rice, maize, sorghum and *Brachypodium,* clustered the genes into 24,860 gene families. Of these, 10,681 families were shared among the five examined cereal genomes, representing a core set of genes across these grass genomes. There were 4610 genes from 1567 gene families that were specific to *T. urartu,* many of which have functional gene ontology annotations relating to responses to stimulus and stress.

Upon comparison of the *T. urartu* genome to the draft sequences of three subgenomes of hexaploid wheat, Ling et al. (2018) identified three large structural variations, with clearly defined boundaries, that occurred in either *T. urartu* or *T. aestivum.* They aligned the *T. urartu* genome with sequences from six BACs of the A subgenome of *T. turgidum* and eleven BACs from the A subgenome of *T. aestivum,* and found that the unaligned regions between the BAC and the *T. urartu* genomic sequences resulted from the insertion of LTR retrotransposons in either *T. urartu* or *T. turgidum* and/or *T. aestivum.* Furthermore, they compared the chromosome 7 assembly of the A subgenome of *T. aestivum* to *T. urartu* chromosome 7 and found that 655 Mb (91.03%) and 536 Mb (90.06%) of *T. urartu* 7 and *T. aestivum* 7A sequences, respectively, were aligned to each other at a minimum identity of 90% or lower, with many unaligned retrotransposon regions. These results show that the different wheat A genomes underwent large-scale rearrangements with other genomes, and experienced independent gain or loss of LTR retrotransposons after the allopolyploidization event.

10.2.3.3 Crosses with Other Species of the Wheat Group

Most F_1 hybrids between different accessions of *T. urartu* exhibited seven bivalents at first meiotic metaphase, of which the greater part was ring bivalents (Shang et al. 1989). As was already stated above, the F_1 hybrid between ssp. *aegilopoides,* the wild form of *T. monococcum,* and *T. urartu* (hybrid genome A^mA) displayed almost complete chromosomal pairing at meiosis (Table 10.6). Chromosome pairing in the F_1 hybrids *T. urartu* x *Ae. tauschii* (hybrid genome AD), *Ae. comosa* x *T. urartu* (hybrid genome MA), and *Ae. umbellulata* x *T. urartu* (hybrid genome UA) was much lower (Table 9.8). The hybrid with *Ae. tauschii* had somewhat higher pairing that the other two hybrids.

Data on chromosomal pairing in the F_1 hybrid between allotetraploid wheat *T. turgidum* ssp. *dicoccon* and *T. urartu,* and between hexaploid wheat *T. aestivum* ssp. *aestivum* and *T. urartu* are presents in Table 10.6. The formation of about six bivalents in these hybrids, that contain the *Ph1* gene of the allopolyploid wheats, shows, as was found by Chapman et al. (1976), Dvorak (1976), that most of the chromosomes of *T. urartu* are homologous to those of the A subgenome of the allopolyploid wheats.

10.3 Section Dicoccoidea Flaksb. (2n = 4x = 28)

10.3.1 Description of the Section

Section Dicoccoidea Flaksb. (Syn.: *Triticum* L. sect. Spelta Dumort; *Triticum* L sect. Orthatherum Nevski; *Triticum* L. 'congregatio' Tetraploidea Flaksb.) contains two species, *T. turgidum* L. and *T. timopheevii* (Zhuk.) Zhuk., which are morphologically similar (Tanaka and Ishii 1973) but genetically isolated (Lilienfeld and Kihara 1934; Kostoff 1937a; Wagenaar 1961a). The species are allotetraploids, with a shared subgenome A, and differ by the other subgenome, B in *T. turgidum* and G in *T. timopheevii* (Lilienfeld and Kihara 1934). Hence, the genome of *T. turgidum* is designated BBAA and that of *T. timopheevii* is GGAA. The cytoplasms of the allotetraploids derived from the diploid donors of the B or G subgenome, namely, that of *T. turgidum* derived from one genotype of *Ae. speltoides,* or rather, from a species that is closely related to *Ae. speltoides,* and that of *T. timopheevii* derived from another genotype of *Ae. speltoides* (Gornicki et al. 2014). Both species contain wild and domesticated subspecies. *T. turgidum* contains one wild subspecies, ssp. *dicoccoides* (Körn. ex Asch. & Graebn.) Thell., one domesticated fossil taxon, ssp. *parvicoccum* Kislev, and seven domesticated subspecies, including the important crop ssp. *durum.* *T. timopheevii* contains one wild subspecies, ssp. *armeniacum* (Jakubz.) van Slageren, and one domesticated subspecies, ssp. *timopheevii.* The wild subspecies of *T. turgidum* grows in the southwestern, central and northeastern part of the Fertile Crescent arc, whereas the wild subspecies of *T. timopheevii* grows in the northeastern part of the Fertile Crescent arc and in Transcaucasia (Kimber and Feldman 1987; van Slageren 1994). The two-wild subspecies grow sympatrically in some sites in southeastern Turkey, northern Iraq and western Iran (Harlan and Zohary 1966). The domesticated form of *T. timopheevii* is grown as a small crop in western Georgia. In contrast, ssp. *durum* of *T. turgidum,* is the second most cultivated crop of wheat, constituting 5–8% of global wheat production (Boyacioglu 2017). It is cultivated in many regions of the world and mainly used for pasta. The other extant six subspecies of *T. turgidum* are locally grown.

10.3.2 *T. turgidum* (L.) Thell. (Genome BBAA)

10.3.2.1 Description of the Species

Triticum turgidum (L.) Thell. is an annual, predominantly autogamous, 40–160-cm-tall (excluding spikes) plant. Culms are erect, stiff, and glabrous, with nodes that are sometimes hairy, hollow lower internodes, and generally solid uppermost nodes. Leaf blades are flat, linear, pointed,

Fig. 10.2 A natural stand, plant and spike of wild emmer *T. turgidum* L. ssp. *dicoccoides* (Körn. Ex Asch. & Graebn.) Thell. and spikelets disseminated on the soil

and up to 60-cm-long, with short auricles and membranous ligule. Spikes are bilaterally compressed, dense, determinate, two-rowed, parallel, 3–14-cm-long (excluding awns) and awned. In the wild form, the rachis is fragile and the entire spike disarticulates at maturity into individual spikelets, each with its associated rachis segment (wedge-type dispersal unit). In the domesticated forms, the rachis is tough and, consequently, the spike remains intact on the culm. In the wild form, the rachis has hairs on its margins, with a tuft of hair up to 5 mm long at each node, whereas in the domesticated forms, the hairs are shorter or entirely absent. The spikelets are 14–15-mm-long, solitary at nodes, lanceolate, appressed to the rachis, and glabrous or hairy, with the top spikelet being fertile, and at right angles to the plane of the lateral spikelets. There are 5–18 fertile spikelets, with 3 florets in the wild form, and up to 6 florets in several domesticated forms; the upper floret is usually sterile. There are 2–3 basal rudimentary spikelets. The glumes are oblong, similar in size, shorter than the spikelet, usually glabrous, 8–13-mm-long, with 2 strong keels and 5–7 veins, and two teeth on the upper margin, one larger and pointed and separated from the other by an acute angle. Lemma is 10–12-mm-long, without keels, with 9–11 veins, and a central vein prolonged to an awn, 15–20-cm–long, flattened, straight, and with a small basal tooth. The palea is membranous, and splits along the keel at maturity. Usually there are two grains per spikelet, but

in several domesticated forms, there are up to 5 grains. Caryopsis is 5–11-mm-long, and free but adherent to the lemma and palea (hulled). In several domesticated forms, the caryopsis is not adherent to the lemma and palea (free-threshing, naked), and is hairy at the apex. The embryo approximately 1/5 the length of the caryopsis (Fig. 10.2).

The *T. turgidum* species is subdivided into nine subspecies: one wild, ssp. *dicoccoides* (Körn. ex Asch. & Graebn.) Thell. (Known as wild emmer), and eight domesticated subspecies, namely, two are primitive hulled forms, ssp. *dicoccon* (Schrank) Thell. (Known as domesticated emmer) and ssp. *paleocolchicum* (Menabde) Á. Löve and D. Löve (known as Georgian wheat), one currently extinct free-threshing form, ssp. *parvicoccum* Kislev, one major commercial free-threshing form, ssp. *durum* (Defs.) Husn. (Known as macaroni, hard wheat, or durum wheat), and four free-threshing forms that are locally cultivated, ssp. *turgidum* (rivet, cone, or pollard wheat), ssp. *polonicum* (L.) Thell. (Polish wheat), ssp. *turanicum* (Jakubz.) Á. Löve and D. Löve (Khorassan wheat), and ssp. *carthlicum* (Nevski) Á. Löve and D. Löve (Persian wheat). The wild subspecies of *T. turgidum* is morphologically similar and genetically closely related to the domesticated subspecies, and the F_1 hybrids between them are fully or almost fully fertile. Therefore, all these wheats are included in a single biological species, *T. turgidum*.

Both the wild and the domesticated forms exhibit wide morphological, cytological and molecular variation. The morphological variation involves mainly spike, spikelet size and shape, glume and awn color and hairiness, grain color and size, plant height, and leaf width. In many habitats in north Israel, Jordan, and in south Syria, wild emmer wheat displays a large number of forms assembling conspicuously polymorphic populations that are easily noted by their variation in glume hairiness, spike color, spikelet size, and leaf shape (Poyarkova 1988; Poyarkova and Gerechter-Amitai 1991; Zohary et al. 2012). Hybrids between wild emmer and most of the domesticated subspecies of *T. turgidum*, are fertile; the chromosomes pair regularly or almost regularly and give every indication of a close relationship (e.g., von Tschermak 1914; Percival 1921; Rao and Smith 1968; Tanaka and Ichikawa 1972; Rawal and Harlan 1975; Dagan and Zohary 1970).

10.3.2.2 Ssp. *dicoccoides* (Körn. ex Asch. & Graebn.) Thell. (Wild Emmer)

Description of the Subspecies

T. turgidum ssp. *dicoccoides*, known as wild emmer wheat, and two-grained wild wheat (called in Hebrew the mother of wheat), [Syn.: *T. vulgare* Vill. var. *dicoccoides* Körn.; *T. dicoccoides* (Körn. ex Asch. & Graebn.) Schweinf.; *T. sativum* Lam.; *T. dicoccon* (Schrnk) Schübler var. *dicoccoides* Körn. ex Ascher. & Graebn.; *T. hermonis* Cook; *T. turgidum* L. var. *dicoccoides* (Körn in Schweinf.) Bowden] is an annual, predominantly autogamous, 65–100-cm-tall (excluding spikes) plant. Culms are prostrated or, rarely, erect, stiff, with a glabrous or pubescent sheath, and with nodes that are sometime hairy, hollow lower internodes and generally solid uppermost internodes. Leaf blades are flat, linear, pointed, and up to 60-cm-long, with short auricles and a membranous ligule. The spike is rigid, bilaterally compressed, dense, determinate, two-rowed, parallel, 3–10-cm-long (excluding awns) and awned. The rachis is brittle and the entire spike disarticulates at maturity into individual spikelets, each with its associated rachis segment (wedge-type dispersal unit). The rachis of the ripe spike disarticulates on the slightest shake, with the spikelets near the apex detaching first, and the others breaking off in orderly succession towards the base. The rachis has hairs on its margins, with a tuft of white, yellow or brown hair, up to 5-mm-long at each node. Spikelets are 14–15-mm-long, solitary at the nodes, lanceolate, appressed to the rachis, and glabrous or hairy, with the top spikelet being fertile and at right angle to the plan of the lateral spikelets. There are 5–15 fertile spikelets, with 3 florets, the upper floret usually being sterile, or rarely fertile. There are 2–3 basal rudimentary spikelets. The glumes are rigid, similar in size, 10–

13-mm-long, shorter than the spikelet, with a strong keel and 5–7 veins; the strongest vein converges towards the base of the apical tooth, where it ends in a secondary tooth which, in some specimens is 5 mm long, very short in others. The glume color is either yellow, white, red, uniformly black, or striped along the margins or spotted irregularly with dark brown, and either glabrous or pubescent. The lemma is boat-shaped,10–13-mm-long, without keels, with 9–11 veins, and membranous and slightly divided near the apex, with a central vein prolonged as a strong and long awn, 10–20-cm–long, flattened, straight, and a small basal tooth. Awns are always present on the two lemmas of the fertile spikelets and are often of nearly equal dimensions. The epidermis of the awns is covered with hard and sharp silicified hairs that form a rugose structure, protecting the grains from grazers. Moreover, these hairs serve as a sort of rachet that propels the dispersal unit, the spikelet, into the ground upon cyclic bending of the awns caused by changes in humidity during the day (Elbaum et al. 2007). The palea is as long as the lemmas, and membranous and splits at the tip at maturity, and with two veins and two keels. The flowers have purple or yellow anthers. Usually there are two, rarely three, grains per spikelet. The caryopsis is 7–11-mm-long, free but adherent to the lemma and palea (hulled), hairy at the apex, with white 1–1.5-mm-long hairs. The embryo approximately 1/5 the length of the caryopsis. In each spikelet, the grain of the lower flower is smaller than that of the second flower, and with a somewhat darker color. The upper grain in each spikelet germinates in the first fall, but the lower one remains dormant for one year. This dispersal in time of germination prevents competition between seedlings that derived from seeds of the same spikelet (Fig. 10.2).

Glume color and pubescence led researchers to subdivide wild emmer into several intra-specific forms. Aaronsohn (1909) noted the following forms: white-eared, black-eared with black awns, red-eared, and black-eared with white awns. Flaksberger (1915) subdivided wild emmer into three varieties, but Percival (1921) defined the following five varieties: var. *kotschyanum* Percival (glumes white and glabrous), var. *fulvovillosum* Percival (glumes white and pubescent), var. *aaronsohni* (Flaksb.) Percival (glumes pinkish-red and glabrous), var. *spontaneonigrum* (Flaksb.) Percival (glumes uniformly black or striped and glabrous), and var. *spontaneovillosum* (Flaksb.) Percival (glumes black or striped and pubescent). The latter variety is rare. All the varieties grow sympatrically but some populations containing only one or two varieties.

Geographical Distribution and Ecological Requirements

Geological, climatic, and archaeological data from the east Mediterranean region indicate the presence of wild emmer wheat during periods of changing climates. Indeed, it was found during the Last Glacial Maximum period ~ 23,000 years ago on the shores of the lake of Galilee (Snir et al. 2015). Following this cold period, temperatures raised and dropped again ~ 12,000–10,300 years ago, during the Younger Dryas or "Big Freeze", an extreme event of rapid change of climate characterized by cold and dry weather, returning to present-day temperatures ~ 10,300 years ago (for details see Hillman 1996; Bar-Matthews et al. 1997; Bar-Yosef 1998). Hillman (1996), using paleo-botanical data, reconstructed the phytogeographic belts of this region during the Younger Dryas, and concluded that the habitats of the annual cereals lay mainly in the open areas of the oak-park forest, in a relatively narrow strip of the east Mediterranean. This narrow strip, known as the "Levantine Corridor", begins in the southern Taurus foothills (Diyarbakir area) in southeastern Turkey and extends along the Mediterranean southward, incorporating the middle Euphrates through the Damascus basin, the Lebanese mountains, and the two sides of the Jordan Rift Valley into the Sinai Peninsula (Bar-Yosef 1998). At that time, wild emmer was a natural constituent of this corridor, but was more widespread in its central-southern part than in its northern part (if it was there at all).

Currently, wild emmer is a natural constituent of several open oak-forest belts and herbaceous plant formations in southwest Asia. Its distribution area is in the Fertile Crescent —a hilly and mountainous region extending from the foothills of the Zagros Mountains in southwestern Iran, through the Tigris and Euphrates basins in northern Iraq and southeastern Turkey, continuing southwestward over Syria and Lebanon to the Mediterranean, and extending to Israel and Jordan (Aaronsohn 1910; Harlan and Zohary 1966; Dagan and Zohary 1970; Tanaka and Ishii 1973; Johnson 1975). The current distribution area of wild emmer wheat is discontinuous (Zohary and Hopf 2000; Zohary et al. 2012). The "southwestern" (Israel, Jordan, southwestern Syria, and southeastern Lebanon) populations of wild emmer are geographically semi-isolated from the "northeastern" (northern Syria, southeastern Turkey, northern Iraq, and western Iran) populations. There are presumably only sporadic connections (if any) in central-western Syria between populations of the southern and northern wild emmer (Kimber and Feldman 1987; Valkoun et al. 1998; Nevo 2001).

Wild emmer is an east-Mediterranean element extending into marginal sub-Mediterranean regions. It grows in a wide range of ecological conditions, from 200 m below sea level (the Jordan Valley) to 1600 m above sea level (Mt. Hermon). It occurs as a common annual component in the herbaceous cover of the deciduous open oak park-forest belt, as well as in evergreen dwarf shrub formations, in steppe-like herbaceous plant formations, in pastures, abandoned fields, and on the edges of cultivation (Kimber and

Feldman 1987; Feldman 2001; Nevo 2001; Zohary et al. 2012). It is a calciferous plant and does not grow on calcareous soils, but thrives well on soils that are formed on hard limestone bedrock (terra rossa soil), on basalt bedrock (basaltic soils), and on soil formed on Nubian sandstone. In rocky places that have not been severely overgrazed, *dicoccoides* wheat often grows in large stands; with wild barley *Hordeum spontaneum* and wild oat *Avena sterilis*, they form 'fields of wild cereals' (Zohary et al. 2012).

In Israel, wild emmer grows on Mt. Hermon, in the Golan Heights, the Jordan Valley, eastern Upper and Lower Galilee, Gilboa Mts., Mt. Carmel, the eastern and western slopes of Samaria and Judean Mountains, and southwards up to the Yattir region, in southern Judea. In Jordan, it grows in the Gilead highlands (Irbid Plateau), and southwards, to Moab and Edom Mountains. In Southern Syria, wild emmer grows on the northeastern slopes of Mt. Hermon, extending eastwards the region of Hauran, Daraa, and the Druze Mt. In southeastern Lebanon, it grows on the northwestern slopes of Mt. Hermon and in the southern part of the Beqaa Valley. It is quite common and locally abundant in the catchment area of the upper Jordan Valley, in some sites in northern Jordan and southwestern Syria, where it occupies a variety of primary and secondary habitats. Wild emmer is less frequent in Turkey, Iraq and Iran (Harlan and Zohary 1966; Rawal and Harlan 1975; Johnson 1975; Valkoun et al. 1998). Its distribution in the Euphrates basin is limited because most of the soils there are calcareous (Willcox 2005). In most of its distribution area, it grows in patches, in mixed stands with wild barley, oat, and several legumes; in the northeastern region of the Fertile Crescent, it also grows in mixed stands with a second wild, tetraploid wheat, *T. timopheevii* ssp. *armeniacum*, with wild diploid wheat, *T. monococcum* ssp. *aegilopoides* and with its putative diploid parent, *T. urartu*. In this region, it also grows sympatrically with the diploids *Ae. speltoides*, *Ae. caudata*, and *Ae. umbellulata*, and the allotetraploids *Ae. geniculata*, *Ae. biuncialis*, *Ae. triuncialis*, and *Ae. peregrina*. It has an allopatric distribution with *Ae. neglecta*, *Ae. columnaris* and *Ae. cylindrica*. In the southwestern range of its distribution, wild emmer grows sympatrically with the diploid *Ae. searsii*, and allotetraploids *Ae. geniculata*, *Ae. biuncialis*, *Ae. triuncialis* and *Ae. peregrina*.

Diversity

Upon discovery of wild wheat in nature, Aaronsohn (1909, 1910) pointed out that this taxon exhibits wide variation in spike size, glume size, apical tooth, hairiness, and had different spike colors (i.e., white, red and black). Indeed, in its native habitats, this wild wheat exhibits great morphological diversity in spike and spikelet size, shape, color, hairiness, glume shape and size, color and hairiness, awn shape and size, keel and secondary teeth size and prominence, grain color and size, plant height, leaf sheath pigmentation and hairiness, leaf shape and width, and growth habit (Kimber and Feldman 1987). Phenological differences occur in heading time (Percival 1921; Anikster et al. 1991). The larger size of the ear and the spikelet, the form and size of the grain, and the character of the pubescence of the leaves distinguish it from wild diploid wheat, *T. monococcum* ssp. *aegilopoides*, while the exceptionally easy disarticulation of the spike, length and form of the spikelets, the striking abundance of hair on the rachis, and the shape and size of the grain, distinguish it from domesticated emmer (ssp. *dicoccon*), to which it has the closest affinity (Percival 1921).

Detailed genetic studies of populations of wild emmer indicated that this subspecies of *T. turgidum* is highly polymorphic (Rawal 1971; Nevo et al. 1982, 1984, 1986; Levy et al. 1988; Nevo and Beiles 1989; Anikster et al. 1991; Felsenburg et al. 1991; Huang et al, 1999; Ozbek et al. 2007a). Electrophoretic studies of nonspecific esterases in germinating seeds of accessions of wild emmer revealed genetic polymorphism, with accessions from Turkey showing heterogeneous isoenzyme patterns (Rawal 1971). Similarly, Nevo et al. (1984, 1986, 1982) studying 457 wild emmer samples, taken from 12 populations across its eco-geographical range in Israel, observed a large allozyme variation both among and within populations. Later, Nevo and Beiles (1989) extended their investigation and studied isosymes encoded by 42 gene loci in 1815 plants representing 37 populations, 33 of which were from Israel and 4 from Turkey. Their results showed that wild emmer is highly polymorphic [15 loci (36%) were locally polymorphic, and 21 loci (50%) were regionally polymorphic], with a mean 1.252 alleles per locus (range: 1.050–1.634); the proportion of polymorphic loci per population averaged 0.220 (range: 0.050–0.415), and genic diversity averaged 0.059 (range: 0.002–0.119). Altogether, there were 119 alleles at the 42 putative loci tested—114 in Israel, and only five in Turkey. Genetic differentiation was primarily regional and local, and not clinal; 70% of the variant alleles were common and rather localized or sporadic, displaying an "archipelago" population genetics and ecology structure (Nevo and Beiles 1989). The coefficients of genetic distance between populations were high and averaged D = 0.134 (0.018–0.297), an indication of sharp genetic differentiation over short distances. Discriminant analyses differentiated Israeli from Turkish populations, and within Israel, between central and three marginal regions, as well as between different soil-type populations. Allozyme diversity, overall and at single loci, was significantly correlated with, and partly predictable by, climatic and edaphic factors. The results of Nevo and Beiles (1989) suggest that during the evolutionary history of wild emmer, diversification of natural selection through climatic and edaphic factors was a major agent of genetic structure

and differentiation at both the single and multi-locus levels. In addition, they indicate that wild emmer harbors large amounts of genetic diversity exploitable as genetic markers in sampling as well as abundant genetic resources utilizable for wheat improvement.

Polymorphism of the high molecular weight (HMW) glutenin subunits was studied in 456 accessions of wild emmer wheat, originating from 21 different Israeli populations (Levy and Feldman 1988; Levy et al. 1988). A total of 50 different SDS-PAGE migration patterns were observed, resulting from the combinations of 15 subunit patterns of the A subgenome and 24 subunit patterns of the B subgenome. Migration patterns consisted of between 3 and 6 subunits, with most containing five. The migration patterns of the A subgenome had 0–3 subunits, with most containing two. The migration patterns of the B subgenome had 1–3 subunits, with three being most common. The polymorphism of the HMW glutenin genes found in wild emmer wheat is much higher than that of domesticated wheats. Marginal populations tended to be more uniform than those at the center of distribution. The various HMW glutenin alleles tended to be clustered, both at a regional level and within a single population. Significant correlations were found between the molecular weight of subunits encoded by *Glu-A1-1* and population altitude, average temperature and rainfall.

Felsenburg et al. (1991) studied variation in the electrophoretic mobility pattern of the HMW glutenin subunits by different genotypes of the Ammiad wild emmer population studied during a five year period (1984–1988). One-dimensional SDS-PAGE of seed extracts showed that the population was highly polymorphic. The spatial distribution of these genotypes was nonrandom, with each of the 11 habitats characterized by different genotype frequencies. Yearly changes in genotypes had little effect on the total frequencies of the various genotypes. A high affinity was found between specific HMW glutenin genotypes and certain habitats.

Ozbek et al. (2007a) estimated the spatio-temporal genetic variation in populations of wild emmer wheat and assessed the contribution of spatial versus temporal factors to the maintenance of genetic variation in a population. Single spikes were collected in the years 1988 and 2002, from plants that grew in the same sampling points, from six different habitats in the Ammiad conservation site, Eastern Galilee, Israel. DNA was extracted from each plant and analyzed by the AFLP method. Fourteen primer combinations yielded 1545 bands, of which 50.0% and 48.8% were polymorphic in the years 1988 and 2002, respectively. Genetic diversity was much larger within populations than between populations and the temporal genetic diversity was considerably smaller than the spatial genetic diversity. Nevertheless, population genetic structure may vary to some degree in different years, mainly due to fluctuations in population size because of yearly rainfall variations. This may lead to predominance of different genotypes in different years. Clustering the plants by their genetic distances grouped them according to their habitats and demonstrated the existence of genotype-environment affinities.

The characterization of the Ammiad population has continued, starting in 1984 until 2020, sampling plants at the same location during all the years of collection (Dahan-Meir et al. 2022). During these 36 years, temperatures have raised by almost 2 °C and CO_2 concentration increased from 340 to 410 ppm. Dense genotyping of 832 individuals along the transect of collections, provided thousands of genetic markers mapped along the reference genome sequenced by Avni et al. (2017). The study by Dahan-Meir et al. (2022) showed that the population was highly variable and that genotypes tended to be clustered to the same ecological microhabitats over the 36 years of collection. Simulations, using realistic demographic parameters of gene flow through outcrossing and seed mobility, and population density, indicate that it is unlikely that neutral processes alone can explain the observed spatial and temporal stability of the population. These results suggested that natural selection together with limited gene flow, explain the remarkable stability of the population. The resilience of this wild emmer wheat population emphasizes the potential importance of such gene pool for the breeding of domesticated wheat in the face of a changing climate. It also shows the importance of in situ conservation over long periods. In fact, the Ammiad population has been declared as a natural reserve for wild-wheat conservation by the Israel nature and parks authority in 2006, which protects it from extinction due to overgrazing or other anthropogenic activities.

RFLP diversity in the nuclear genome was estimated within and among wild emmer wheat populations from several Israeli locations (Huang et al. 1999). Use of 55 enzyme-probe combinations showed high levels of genetic diversity. Population genetic structure in this wild taxon appears to have been influenced by historical founder events as well as selective factors. Multivariate analyses indicated that individuals tend to cluster together according to their population of origin, and that there is little geographical differentiation among populations.

Avivi (1979), Avivi et al. (1983) analyzed 47 different accessions of wild emmer, representing almost the entire eco-geographical range in Israel, for their grain protein percentage (GPP). The GPP range from 17.0 to 27.3%, indicating wide variation between genotypes. The observed GPP was much higher than that of all the domesticated subspecies of *T. turgidum*. Within a given genotype, the small grains of the first florets exhibited almost the same amount of protein as the large grains of the second florets.

Comparison between different genotypes revealed a positive correlation between protein content and grain size.

To study the effect of various habitats on GPP in wild emmer, the trait was determined in 910 accessions collected from 22 different populations representing different eco-geographical conditions in Israel (Levy and Feldman 1988). High values of GPP were found, with population means ranging from 19.7 to 28.0%, and single accession means ranging from 14.1 to 35.1%. Marginal geographical populations usually had a lower GPP and smaller intra-population variations than central ones. Repeated sampling of several central populations for four consecutive years revealed relatively large intra-population GPP fluctuations. No correlation was found between GPP and ecological factors, except for soil type; accessions growing on terra-rossa soil had higher GPP than those growing on basaltic soil. Accessions with black glumes, glabrous auricles, or large grains exhibited high GPP values. Their association with morphological and biochemical markers enabled mapping of genes for high GPP to six chromosomes, namely, 1AS, 1BS, 5A, 5B, 7A and 7B (Levy et al. 1988).

In sharp contrast to the above data on diversity in wild emmer, Haudry et al. (2007) concluded from the mean nucleotide diversity in wild emmer that this subspecies of *T. turgidum* is not a highly polymorphic taxon. They suggested several possible reasons for this low diversity. First, wild emmer arose through a relatively recent allopolyploidy event (0.7–0.9 MYA; Gornicki et al. 2014; Marcussen et al. 2014; Middleton et al. 2014), that may have resulted in a large decrease in diversity in the new allopolyploid species with respect to its diploid ancestors, and since nucleotide mutation rate is low (Lande and Barrowclough 1987), the time from its formation has not been sufficiently long to restore diversity. Second, the small effective population size of the current population of wild emmer may account for the low level of nucleotide diversity.

Cytological Variation

Several studies demonstrated the wide occurrence of reciprocal translocations among accessions of wild emmer (Rao and Smith 1968; Dagan and Zohary 1970; Tanaka and Ishii 1973; Rawal and Harlan 1975; Kawahara and Tanaka 1978, 1981, 1983; Kawahara 1984, 1986, 1987; Joppa et al. 1995). Rao and Smith (1968), Rawal and Harlan (1975) reported the presence of one or two reciprocal translocations in hybrids between Turkish accessions as well as between Turkish and Israeli accessions of wild emmer. Cytological analysis of meiosis in the F_1 hybrids of two accessions from Iran crossed with Israeli wild emmer showed that the wild Iranian wheats were fully inter-fertile with the Israeli *dicoccoides* line (Dagan and Zohary 1970). Chromosome pairing was normal and the presence of two reciprocal translocations was observed, between the two Iranian accessions and between one of the Iranian accession7 and the Israeli accession.

Mixed stands of wild emmer and wild *timopheevii* wheats were found in several sites in southeastern Turkey, northern Iraq and western Iran (Tanaka and Ishii 1973). Morphological differences between plants of wild emmer and those of wild *timopheevii* were not clear, except that the leaf surface of the former was exclusively glabrous, while that of wild *timopheevii* was pubescent (Tanaka and Ishii 1973). The F₁ hybrids between several accessions of wild emmer and a tester line of wild emmer, exhibited one or two reciprocal translocations, indicating that chromosomal structural differences between accessions is quite common in this wild subspecies of *T. turgidum*.

A relatively large number of genomes which differ from each other by one or two reciprocal translocations, were identified in wild emmer through hybridization experiments (Kawahara and Tanaka 1978, 1981, 1983; Kawahara 1984, 1986, 1987). Variation in chromosome structure was the highest in Turkey, followed by Israel, whereas wild emmer from Iraq and Iran showed little variation. Consequently, Kawahara (1987) concluded that the center of diversity in chromosome structure is in southeastern Turkey. Nishikawa et al. (1994), using telocentric lines of domesticated emmer wheat, identified the chromosomes involved in seven translocation kinds in wild emmer. Likewise, Joppa et al. (1995), analyzing Israeli accessions of wild emmer, determined 119 genotypes of wild emmer with translocations (as compared to the ordinary chromosome arrangement typified by that in the standard laboratory common wheat cultivar Chinese Spring) in an investigated sample of 171 genotypes (70%). The frequency of translocations in different Israeli populations observed by Joppa et al. (1995) varied from 0.27 to 1.00, and all populations had 1 or more genotypes with one or more translocations. A sample of 17 genotypes from 12 populations were crossed with the Langdon D-genome disomic substitutions to determine the identity of the chromosomes involved in the translocations. There were nine genotypes with translocations and with the exception of a 2A/2B translocation, none of them involved the same homoeologous chromosomes (Joppa et al. 1995). The B-subgenome chromosomes were involved in translocations more frequently than the A-subgenome chromosomes. Translocation frequencies of the various populations were correlated with environmental variables, primarily with water availability and humidity, and possibly also with soil type. In general, translocation frequency was higher in peripheral populations in the ecologically heterogeneous frontiers of wild emmer distribution than in the central populations located in the catchment area of the upper Jordan valley.

To identify accessions of tetraploid wheat in mixed populations of ssp. *dicoccoides* and wild *T. timopheevii*, ssp. *armeniacum,* collected by Jack R. Harlan in Turkey, Rao and Smith (1968) crossed six of the accessions with four Israeli accessions of ssp. *dicoccoides* and then morphologically and cytogenetically analyzed the F₁ hybrids. Both, the accessions of ssp. *dicoccoides* and ssp. *armeniacum* were also crossed with a number of domesticated tetraploid wheats, including *T. timopheevii* ssp. *timopheevii*, *T. turgidum* ssp. *turgidum*, and ssp. *dicoccon*. Cytogenetically, the four Israeli accessions of wild emmer were similar and exhibited very close relationships with ssp. *dicoccon* and ssp. *turgidum,* but their hybrids with ssp. *timopheevii* showed poor chromosome pairing and were fully sterile. Four of the six Turkish accessions were similar to the Israeli group in pairing relationships and seed set percentages (Rao and Smith 1968). The remaining two Turkish accessions showed considerable cytogenetic differentiation. Turkish accession 11,189 showed a close pairing relationship and some fertility with ssp. *timopheevii* and exhibited poor pairing and complete sterility in crosses with ssp. *dicoccon* and the Israeli ssp. *dicoccoides* group. Surprisingly, Turkish accession 11,191 exhibited almost complete chromosome pairing and some fertility in crosses with both ssp. *timopheevii* and ssp. *dicoccon* (Rao and Smith 1968).

Rawal and Harlan (1975) studied chromosome pairing in meiosis of F₁ hybrids involving three Israeli accessions of ssp. *dicoccoides*, six Turkish accessions of tetraploid wheat, and one line of domesticated *T. timopheevii*. Like Rao and Smith (1968), they found that four of the six Turkish accessions were cytologically similar to the Israeli *dicoccoides* accessions, and one Turkish accession (# 189) was cytologically similar to *T. timopheevii*, whereas the other Turkish accession (# 191) showed good chromosome pairing with both all three accessions of ssp. *dicoccoides* and ssp. *timopheevii*. Interestingly, the four Turkish accessions that were identified as ssp. *dicoccoides* had somewhat better chromosomal pairing with *T. timopheevii* (average number of chromosome association/cell was 22.0–22.8) than did the Israeli accessions (average number of chromosome association/cell was 18.8–21.5).

Studies of chromosomal pairing in F₁ hybrids of accessions of either the northern race (the Turkish race) or the southern race (the Israeli race) of wild emmer and wild *T. timopheevii* showed somewhat better pairing between the northern race and *T. timopheevii* than between the southern one and *T. timopheevii* (Rao and Smith 1968; Rawal and Harlan 1975; Tanaka and Kawahara 1976; Tanaka et al. 1978). Actually, the four Turkish accessions with ssp. *dicoccoides*-like behavior had chromosomes that paired slightly better with ssp. *timopheevii* as compared to the accessions from Israel (Rawal and Harlan 1975); the differences were small but seemed to be consistent. It is possible that with more collections, forms with a truly intermediate behavior might be found. This indicates that the northern race of wild emmer introgressed with *T. timopheevii*. According to the hypothesis set forth by Zohary and Feldman (1962), wild emmer and wild *T. timopheevii*, which share the A subgenome, could have acquired adaptive traits through introgressive hybridization, leading to recombination of their B and G subgenomes. Cytologically intermediate types predicted by this hypothesis were indeed discovered by Sachs (1953), Wagenaar (1961a, 1966), Rao and Smith (1968), Rawal and Harlan (1975), Tanaka and Kawahara (1976), Tanaka et al. (1978). In accord with this, Gornicki et al. (2014) provided molecular evidence that evolution of these two allotetraploid wheats was also accompanied by chloroplast introgression. One accession of *T. turgidum* ssp. *dicoccoides* (G4991), for example, which showed high chromosome pairing with both *T. turgidum* and *T. timopheevii* (Rawal and Harlan 1975), carries the *T. timopheevii* chloroplast haplotype (*H09*) as a result of a cross between wild emmer and wild *timopheevii*. Conversely, wild *T. timopheevii* accession TA976 carries the emmer-lineage chloroplast haplotype (*H04*) (Gornicki et al. 2014).

The pairing homoeologous (*Ph1*) gene of common wheat has long been considered the main factor responsible for the diploid-like meiotic behavior of polyploid wheat (Riley 1960; Sears 1977). This dominant gene, located on the long arm of chromosome 5B (5BL), suppresses pairing of homoeologous chromosomes in allopolyploid wheat and in their hybrids with related species. Ozkan and Feldman (2001) reported on the existence of genotypic variation among wild emmer wheat in the control of homoeologous pairing, most probably in the *Ph1* locus. Compared with the level of homoeologous pairing in hybrids between *Aegilops peregrina* and the bread wheat cultivar Chinese Spring (CS), significantly higher levels of homoeologous pairing were obtained in hybrids between *Ae. peregrina* and CS substitution lines in which chromosome 5B of CS was replaced by 5B of several lines of wild emmer. Searching for variation in the control of homoeologous pairing among lines of wild emmer showed that hybrids between *Aegilops peregrina* and different lines of this wild wheat exhibited three different levels of homoeologous pairing: low, low-intermediate, and intermediate-high. The genotypes with low-intermediate and intermediate-high pairing may possess weak alleles of *Ph1*. The three different ssp. *dicoccoides* pairing genotypes were collected from different geographical regions in Israel, indicating that this trait may have an adaptive role (Ozkan and Feldman 2001).

Intra-Subspecific Differentiation
Wild emmer contains two main races that are morphologically, ecologically, and genetically distinct (Harlan and

Zohary 1966; Nishikawa et al. 1994; Joppa et al. 1995; Kawahara and Nevo 1996; Ozkan et al. 2002, 2005; Mori et al. 2003; Luo et al. 2007). The southern race grows in Israel, Jordan, southwest Syria and southeast Lebanon, whereas the norther race grows in southeastern Turkey, northern Iraq and western Iran. The two races are geographically separated by a conspicuous discontinuity in central Syria (Kimber and Feldman 1987; Zohary et al. 2012). In general, the plants of the northern race are characterized by relatively compact heads, fine-textured awns, often hairy spikelets, with sparse pubescence on the rachis and spikelet base. In contrast, the southern race is large and robust, with spikelets featuring coarse awns, dense pubescence on the rachis internode edge and spikelet base, large seeds, wide leaves and thick stems (Harlan and Zohary 1966; Rawal and Harlan 1975). Robust early-maturing types occupy the winter-warm basin around the Sea of Galilee, to altitudes as low as 100 m below sea level, whereas more-slender, late-blooming forms occur higher up in the Galilee mountains, reaching elevations of 1600 m on the east- and south-facing slopes of Mt. Hermon (Zohary et al. 2012). The southern race may occur in massive stands over considerable areas on basaltic and hard limestone slopes of the oak woodland belt of the region. But it was not until the current state of Israel was established and grazing became regulated, that the abundance of these stands was recognized. Where grazing is controlled, non-arable sites support stands as dense as cultivated wheat fields (Harlan and Zohary 1966). In contrast, the northern race is never really abundant and occurs in sporadic, isolated patches and thin, scattered stands in the lower oak-woodland belt, often in association with wild diploid wheat and wild barley (Harlan and Zohary 1966). It is never the dominant species of the grassland flora and is usually found only as a minor component among other cereals. Since it is not a weedy plant, the range and abundance of the northern race may well have become restricted, since the land was disturbed by agriculture (Harlan and Zohary 1966).

Large-grain forms of wild emmer were first noted by Cook (1913) in northeastern Israel (then Palestine). In 1926, Vavilov found in Israel (then Palestine) wild emmer with large spikes and large grains that resemble domesticated *T. turgidum*, mainly ssp. *durum*, that was grown in Israel for, at least, 2300 years. Consequently, Vavilov subdivided wild emmer into two major groups: a narrow-spike form and a wide-spike one (Vavilov et al. 1931). Jakubziner (1932) maintained Vavilov's division of wild emmer, but named the narrow-spike form grex *horanum* Vav., since it was collected in Hauran, Syria, and the wide-spike form grex *judaicum* Vav., since it was collected in Israel (then Palestine). He reported that *judaicum* in several regions in Israel, namely, the Upper Jordan Valley, Mt. Gilboa, and Mt. Hermon, in Syria, Jordan and even in the Cilician Taurus,

Turkey, always occurs close to cultivated wheat fields. He also described some wild emmer and ssp. *durum* hybrids that had been collected in the Upper Jordan Valley. These observations suggest introgression.

Jakubziner's subdivision of wild emmer was supported by morphological studies of Israeli accessions (Poyarkova 1988), which, following Jakubziner, classified the slender type as *horanum* and the robust type as *judaicum*. Poyarkova et al. (1991) drew attention to the morphological, phenological, and geographical differences between the narrow-spiked and the wide-spiked types of Israeli wild emmer accessions. A tendency for increased grain number to three in a spikelet was observed in the both variants, but was more strongly expressed in the wide-spiked accession (Poyarkova et al. 1991). The narrow-spiked form, *horanum*, is widely distributed in Israel, whereas the wide-spiked form, *judaicum*, is restricted to the vicinity of the Sea of Galilee and to Mt. Gilboa (Poyarkova et al. 1991). Intermediate morphological forms are abundant in natural habitats where the two variants are sympatric (Poyarkova et al. 1991). Therefore, it was concluded by Anikster et al. (1988) that the two forms of wild emmer are extremes of a continuum.

Genetic variation in several Israeli (the southern race) and Turkish (the northern race) populations of wild emmer was assessed by RFLP analysis (Ozbek et al. 2007b). Frequencies of polymorphic loci and gene diversity were significantly higher in the southern than in the northern populations. The southern populations contained more unique alleles than northern populations. Genetic distance was larger between Israeli and Turkish populations than between populations within each country, indicating that the Israeli and Turkish populations are considerably diverged. Similarly, AFLP analysis showed that the southern race is clearly separated from the northern one (Ozkan et al. 2002, 2005). However, genetic studies (Tanaka and Sakamoto 1979; Saito and Ishida 1979; Nakai 1978a, b; Nishikawa et al. 1979) revealed that the northern race of wild emmer also showed wide variation and that it is differentiated into several populations (Luo et al. 2007). Luo et al. (2007), who performed RFLP analysis at 131 loci of accessions of wild emmer, showed that this taxon consists of a southern population (in Israel, Jordan, southwestern Syria, and southeastern Lebanon) and northern population (southeastern Turkey, northern Iraq, and western Iran), each which can be further subdivided. The southern race consists of two distinct groups, the robust *judaicum* group located north and northwest of the Sea of Galilee, and the slender group, grown in other Israeli regions (Luo et al. 2007).

Luo et al. (2007) found that gene flow between wild and domesticated tetraploid wheat occurred across the entire area of wild emmer distribution, but failed to show that the *judaicum* group originated from hybridization between wild emmer and ssp. *durum*. Feldman and Millet (Feldman M, Millet E, unpublished data) found robust plants of wild emmer on the edges of wheat fields in several northwestern sites of the Sea of Galilee and assumed that they resulted from introgression with domesticated wheats, either tetraploid ssp. *durum* or hexaploid *T. aestivum* ssp. *aestivum*. Similarly, Blumler (1998) studying the *judaicum* type in the upper Jordan Valley, concluded that it originated relatively recently through hybridization of wild emmer with durum wheat. The occurrence of spontaneous hybrids between wild emmer and ssp. *dicoccon, durum* or common wheat, was already reported by Cook (1913), Percival (1921), Jakubziner (1932a), Zohary and Brick (1962). Indeed, in a number of wild emmer accessions sown in experimental fields in southern Russia, natural crossing with ssp. *durum* were frequently observed (Jakubziner 1932a). These spontaneous hybridizations indicate that hybridization between wild emmer and domesticated wheats are not an isolated phenomenon. Such hybridization is doubtlessly due to the numerous intermediate and extraordinarily diverse morphological forms found in its native habitats (Percival 1921). Percival (1921) held the opinion that most of the "large-seeded" forms are of hybrid origin, and contain a trace of ssp. *durum*. Some of these forms are likely hybrids of ssp. *dicoccoides* with the domesticated wheats *T. turgidum* ssp. *durum* and *T. aestivum* ssp. *aestivum*, and with the wild *T. monococcum* ssp. *aegilopoides*, which is often found growing sympatrically with wild emmer (Percival 1921).

While typical wild emmer (*horanum*) is not at all similar to durum wheat, plants from the Israeli upper Jordan Valley occasionally contain several traits of durum wheat (Blumler 1994, 1998) such as plant robustness, grain shape and size, glume shape, first glume tooth, glume pubescence, spikelet width and glutenin A1-1 allele, and early maturing. In general, wild emmer from the upper Jordan valley is highly variable, containing intermediates between durum wheat and *horanum* wild emmer, as one would expect of products of hybridization (Blumler 1994). Indeed, electrophoretic studies of Nevo et al. (1982), Nishikawa et al. (1994) showed that the wild emmer populations in the upper Jordan Valley are genetically differentiated from all other investigated Israeli populations. As one travels west or east from this site, this race of emmer is replaced by populations of typical wild emmer (Golenberg 1988; Blumler 1994). Domesticated individuals should have had opportunities to come in contact with wild plants especially along rocky, untillable field margins and hybridize with them.

Likewise, the glutenin data of Levy and Feldman (1988) are particularly informative. Glutenin A1-1 is present in domesticated emmer, ssp. *dicoccon*, but absent in ssp. *durum*. It is almost always present in wild emmer, but is generally absent in accessions from the upper Jordan Valley, which suggests potentially massive introgression from ssp. *durum*. Of the 19 wild populations that Levy and

Feldman (1988) examined, glutenin A1-1 was present in all but two: a population just south of the upper Jordan Valley, and in Majdal-es-Shams on Mt. Hermon. Jakubziner (1932a) reported on *judaicum* from Majdal es–Shams. Both populations are in agricultural areas. Interestingly, the glutenin A1-2 locus presented a different pattern, as it is absent in most wild emmer populations, in domesticated emmer and ssp. *durum*, but present in bread wheat and in some wild emmer plants from the upper Jordan Valley.

Time and Place of Origin of Wild Emmer

The time of origin of wild emmer was estimated in several studies (Table 10.7; see also Chap. 12). Dvorak and Akhunov (2005), using locus duplications as a clock for estimation of the age of this allotetraploid, suggested that it was formed about 0.360 million years ago (MYA), while Huang et al. (2002), analyzing Acc-1 (plastid acetyl-CoA carboxylase) and Pgk-1 (plastid 3-phosphoglycerate kinase) genes, found that the A subgenome of wild emmer diverged from the genome of *T. urartu* less than half a MYA, indicating an origin 0.500 MYA. On the other hand, studies of nucleotide sequences of hundreds of nuclear genes showed that wild emmer formed about 0.800 MYA (Marcussen et al. 2014). Gornicki et al. (2014), Middleton et al. (2014), based on sequencing of the entire genomic chloroplast DNA, found that the cytoplasm of the emmer lineage diverged from that of *Ae. speltoides* 0.700–0.900 MYA, respectively, meaning that the allotetraploidization event that formed wild emmer occurred within the last 0.700–0.900 MYA. This date is slightly above earlier estimates of Huang et al. (2002), Dvorak and Akhunov (2005).

Wild emmer presumably originated in the southwestern part of the Fertile Crescent, i.e., in the vicinity of Mt. Hermon and the catchment area of the Jordan River. This is inferred from the wider morphological, phenological, biochemical, and molecular variation of wild emmer in Israel, Jordan, southern Syria, and southern Lebanon, as opposed to its more limited variation in southeastern Turkey, northern Iraq, and southwestern Iran (Nevo and Beiles 1989; Ozbek et al. 2007b). AFLP analyses showed that the pattern of *T. urartu* from Mt. Hermon is the closest to that of the A subgenome of wild emmer, suggesting that this allotetraploid formed in the vicinity of Mt. Hermon (Dvorak and Luo 2007). In accordance, the geographic distribution of chloroplast haplotypes of wild emmer and of *Ae. speltoides* (the assumed cytoplasm donor to allotetraploid wheat) illustrates the possible geographic origin of the emmer lineage in the southern Levant (Gornicki et al. 2014).

From this region, wild emmer spread southward and northward. Wild emmer could presumably have spread into the northern and northeastern parts of the Fertile Crescent during the last part of the Pleistocene (10,000–400,000 years ago), which was characterized by climatic fluctuations that might have facilitated its northward spread. Alternatively, wild emmer may have moved northward much later, in the beginning of the Holocene (somewhat around 10,000 years ago), with the expansion of the cultivation of wild emmer from the Jordan Rift Valley to southeastern Turkey, in the Pre-Pottery Neolithic A (PPNA) 10,300–9500 years ago. Its escape from cultivation in the northern region of the Fertile Crescent could have laid the foundation for "feral" populations there. In addition, the morphological resemblance of wild emmer from the north-eastern region of the Fertile Crescent to wild *T. timopheevii* and the somewhat better chromosomal pairing in hybrids between wild emmer and wild *T. timopheevii* (see above) also suggests that introgressive hybridization from the local wild tetraploid wheat, *T. timopheevii* ssp. *armeniacum*, might have helped in the establishment of wild emmer in the northeastern region of the Fertile Crescent. The lack of archaeological remains of wild emmer in southeastern Turkey prior to 9500 BP (all dates are uncalibrated) (Nesbitt 2002) supports this second possibility.

Table 10.7 Time of formation of the allopolyploid species of *Triticum* in million years ago

Species	Time of formation	Method of study	References
Triticum turgidum subsp. *dicoccoides* (wild emmer)	< 0.500	Nucleotide sequences of two nuclear genes	Huang et al. (2002)
	0.360	Nucleotide sequences of four nuclear genes	Dvorak and Akhunov (2005)
	< 0.800	Nucleotide sequences of hundreds nuclear genes	Marcussen et al. (2014)
	0.700	Sequencing chloroplast DNA and nuclear genome sequencing	Gornicki et al. (2014) Li et al. (2022)
Triticum timopheevii subsp. *armeniacum* (wild timopheevii)	0.300–0.050	RFLP analysis of nuclear DNA	Mori et al. (1995)
	0.400	Sequencing chloroplast DNA	Gornicki et al. (2014)
	< 0.400	Nuclear genome sequencing pay attention to the addition in this cell	Li et al. (2022)
Triticum aestivum subsp. *aestivum* (common wheat)	0.008	Nucleotide sequences of two genes	Huang et al. (2002)
	0.008	Nucleotide sequences of four nuclear genes	Dvorak and Akhunov (2005)

This idea is shared by Civáň et al. (2013), who used super-networks with datasets of nuclear gene sequences and novel markers detecting retrotransposon insertions in ribosomal DNA loci, to reassess the evolutionary relationships among tetraploid wheats. The observed diversity and reticulation patterns indicate that wild emmer evolved in the southern Levant, and that the wild emmer populations in south-eastern Turkey and the Zagros Mountains are relatively recent reticulate descendants of a subset of the wild southern Levantine populations.

Economically Important Genes in Wild Emmer

Already Aaronsohn (1910), who discovered in 1906 wild emmer in nature, was impressed by the adaptation of this taxon to a wide range of climatic and edaphic conditions, by its large grain size, its high resistance to rust, and its ability to grow in relatively dry habitats. Consequently, he recommended to transfer its desirable traits to domesticated wheats, particularly, to improve their resistance to biotic stresses, and tolerance to extreme climatic and soil conditions. Aaronsohn believed that "the cultivation of wheat might be revolutionized by the utilization of wild wheat. Such utilization might facilitate the formation of many new varieties, some of which will be hardy and able to grow in dry and warm habitats or in areas with poor soil and can thus expand the wheat growing area" (Aaronsohn, 1910, p. 52).

Aaronsohn's belief that wild emmer can be utilized in the improvement of domesticated wheats was shared by Schweinfurth (1908), von Tschermak (1914), Percival (1921), Vavilov (1932), and others. The fertile F_1 hybrids between ssp. *durum* and wild emmer, produced by von Tschermak (1914), showed that gene transfer from wild into domesticated forms is possible. Consequently, selected specimens of subsp. *dicoccoides* were introduced to various research stations in Europe and the United States for observation and crosses with domesticated wheats. Yet, these early attempts to utilize wild emmer in breeding programs were met with very little success, and Aaronsohn's vision of using this wild gene resource for the improvement of domesticated wheats was soon neglected. This mainly resulted from lack of genetic and cytogenetic knowledge of the wheats, coupled with a poor understanding of relationships between wild and domesticated wheats. Moreover, wheat breeders were discouraged from utilizing wild wheat in breeding programs because of difficulties to select against undesirable characters derived from wild emmer.

This situation has changed radically in the second half of the twentieth century, leading to a renewed interest in the germplasm of wild emmer as a source of agronomically important traits. The renewed interest in the germ plasm of wild wheat has been stimulated by the large genetic erosion that occurred in domesticated wheat, mainly to common wheat, because of the replacement in parts of the world of a huge number of land races by high-yielding varieties without preserving the land races, resulting in a drastic narrowing of the genetic basis of common wheat. During this period, much information has been accumulated on the genetic relationships between wild emmer and domesticated wheats. The availability of aneuploidy lines of *T. aestivum* (Sears 1954) and substitution lines in which wild emmer chromosomes or chromosome arms are substituted for their durum wheat homologues (Joppa 1993) or bread wheat homologous arms (Millet et al. 2013, 2014), enabled the genetic analysis of individual wild emmer chromosomes or chromosome arms on the genetic background of domesticated wheat, and thus, facilitated the transfer of selected wild chromosomal segments to domesticated wheat. In particular, the recent sequencing of the wild emmer genome (Avni et al. 2017) opens the possibility of identifying and utilizing beneficial wild genes.

Wild emmer contains an invaluable rich gene resource for wheat improvement that, at present, has been hardly exploited (Feldman and Millet 1995; Feldman et al. 1994, 1996). The gene pool of wild emmer, which is larger and richer than that of the domesticated wheats, contains many agronomically-important genes (for review see Feldman and Millet 1995; Huang et al. 2016). This taxon is genetically very close to durum and common wheat and the F_1 hybrids between them are fertile. Since the chromosomes of wild emmer are homologous to those of durum wheat and to those of the A and B subgenomes of common wheat, it is possible to transfer desirable traits from wild emmer into domesticated wheat by simple, conventional plant breeding procedures.

The screening of wild emmer for economically valuable characteristics is only in its initial stages. Recent surveys of samples of wild emmer collected throughout its distribution area have shown that this wild wheat contains many agronomically important genes, such as those conferring resistance to pests and diseases (e.g., Anikster et al. 2005), greater tolerance to drought and heat (Peleg et al. 2005, 2009), higher grain protein content and quality (Avivi 1979; Avivi et al. 1983), higher zinc and iron in the grains (Cakmac et al. 2004), and larger grains (Cook 1913). Moreover, this wild gene pool contains many alleles that do not exist in the domesticated gene pool such as those coding for different subunits of storage proteins (glutenin and gliadins) (Levy and Feldman 1987). Despite its breeding potential, this gene pool has been utilized relatively minimally in wheat improvement (Gerechter-Amitai and Grama 1974; Grama and Gerechter-Amitai 1974; Feldman 1977; Feldman and Sears 1981; Levy and Feldman 1987; Feldman et al. 1994; Feldman and Millet 1995). Simply inherited traits have been transferred from wild emmer into domesticated wheats and the value of these traits has been widely documented. Yet, gene transfer of complexly inherited traits, for which the controlling genes may be located on several different chromosomes of wild emmer, is more difficult, and the presence of undesirable wild alleles may musk the effect of the desirable ones.

Wild emmer contains many genes conferring adaptation populations to biotic and abiotic stresses. (Huang et al. 2016). It comprises resistance to fungal and pest diseases, such as resistance to powdery mildew (Moseman et al. 1984; Gerechter-Amitai and van Silfhout 1984; Dinoor et al. 1991; Rong et al. 2000; Yahiaoui et al. 2009; Li et al. 2020), leaf rust (Moseman et al. 1985; Dinoor et al. 1991; The et al. 1993; Anikster et al. 2005), stem rust (Nevo et al. 1991; Dinoor et al. 1991; The et al. 1993; Moseman et al. 1985; Anikster et al. 2005), stripe rust (yellow rust) (Gerechter-Amitai and Stubbs 1970; Gerechter-Amitai and Grama 1974; Gerechter-Amitai, 1980; Reinhold et al. 1983; Valkoun 2001; Klymiuk et al. 2018), *Septoria nodorym* (Dinoor et al. 1991), tan spot (Faris et al. 2020), and take-all (Dinoor et al. 1991). It has a variety of glutenins and gliadins subunits (Galili and Feldman 1983; Levy and Feldman 1988; Felsenburg et al. 1991), a high percentage of grain protein Avivi et al. 1983; Mansur-Vergara et al. 1984; Nevo et al. 1986; Levy and Feldman 1987, 1989; Millet et al.1992; Peleg et al. 2008; Chatzav et al. 2010), a high concentration of micronutrients such as zinc (Zn), iron (Fe) and manganese (Mn) (Cakmac et al. 2004; Uauy et al. 2006; Distelfeld et al. 2006; Peleg et al. 2008; Chatzav et al. 2010), and tolerance to abiotic stresses, such as drought (Peleg et al. 2005, 2009), heat (Ullah 2016), and salinity (Nevo and Chen 2010). Several accessions of wild emmer have large grains (Avivi, 1979), and the potential to increase grain yield and grain protein yield (Feldman and Millet, 1995; Millet et al. 2013, 2014), and improve photosynthetic efficiency (Nevo et al. 1991). In a *T. durum* x *T. dicoccoides* mapping population, Peng et al. (2003a, b) identified quantitative trait loci (QTLs) contributing to early flowering time, higher spike number and weight, higher kernel number and higher yield.

Chromosome 6B of wild emmer wheat was previously reported to be associated with high grain protein content (Joppa and Cantrell 1990), zinc, iron, and manganese (Distelfeld et al. (2006). This chromosome also carries the wild type allele of *Gpc-B1*, causing earlier senescence of flag leaves (Uauy et al. 2006). To explain the pleiotropic effect of the *Gpc-B1* gene on the high concentration of protein, zinc, iron, and manganese in the grain, Distelfeld et al. (2006) suggested that this locus is involved in more efficient remobilization of nutrients from the leaves to the grains, in addition to its effect on earlier senescence of the green tissues. Yet, Uauy et al. (2006) claimed that the *Gpc-B1* gene confers

a short duration of grain fill time due to an earlier flag leaf senescence, thus ceasing synthesis of carbohydrates and their translocation to the grains. This results in a relatively smaller amount of grain carbohydrates and consequently, in a higher concentration of grain protein and several micronutrients.

10.3.2.3 Ssp. *dicoccon* (Schrank) Thell. (Domesticated Emmer)

Description of the Subspecies

T. turgidum ssp. *dicoccon* (Schrank) Thell., known as domesticated emmer or two-grained domesticated wheat [Syn.: *T. spelta* Host; *T. spelta* (L.) var. *dicoccon* Schrank; *T. dicoccum* (Schrank) Schübl.; *T. vulgare dicoccum* Alef.; *T. vulgare* ssp. *dicoccum* Körn.; *T. sativum dicoccum* Hack.; *T. sativum* Lam. ssp. d*icoccum* Ascher.et Graebn.; *T. aestivum* L. var. *dicoccon* (Schrank) Fiori; *T. farrum* Bayle-Barelle; *T. amyleum* Seringe; *T. zea* Wagini; *T. dicoccum* Körn.; *Gigachilon polonicum* (L.) Seidl ssp. *dicoccon* (Schrank) Á. Löve] is an annual, predominantly autogamous, 55–125-cm-tall (excluding spikes) plant. Culms are usually erect, and hollow in some varieties, while solid at the upper internode in others. The spike, whose color is white, red, or black, is bilaterally compressed, dense, determinate, two-rowed, 5–10-cm-long (excluding awns), with 15–30 spikelets, and awned. The rachis is short, flat and smooth, more or less hairy along the margins, with a tuft of hair in its base, and usually non-brittle, but breaks into spikelets upon threshing. Spikelets are oval or lanceolate, 10–16-mm-long, solitary at nodes, with the top spikelet being fertile, and at right angles to the plane of the lateral spikelets. There are 3–4 florets, the lower two being fertile, while the upper ones are sterile. The glumes are rigid, oval and boat-shaped, similar in size, shorter than the spikelet, 10–13-mm-long, with a strong keel ending in a straight or curved teeth that varies in length and form in different varieties, and with 7 veins; the strongest vein converges towards the base of the apical tooth, where it ends in a secondary tooth, which, in some specimens, is 5-mm-long, while in others, is very short. Some varieties have pubescent glumes. The lemma is boat shaped, with 9–11 veins, but without a keel, its central vein is prolonged as a strong and long, 5–14-cm–long awn. The awn is flattened, straight, and with a small basal tooth. Awns are always present on the two lemmas of the fertile spikelets and are often of nearly equal in dimensions. The palea is equal to, or slightly longer than the lemma, ovate-lanceolate, with a narrow apex and two ciliate keels. The palea is membranous and splits at the tip at maturity, and has two veins and two keels. There are 2, rarely 3, caryopses in each spikelet, firmly enclosed by the palea and lemma. The color of the grain is white, yellowish,

Fig. 10.3 A plant and spikes of domesticated subspecies of *T. turgidum* L.; **a** Domesticated emmer, ssp. *dicoccon* (Schrank ex Schübl.) Thell. **b** ssp. *durum* (Desf.) Husn. cv. Inbar; **c** ssp. *polonicum* (L.) Thell.; **d** ssp. *carthlicum* (Nevski) Löve and Löve

or red, rarely purple. The grains are comparatively narrow and pointed at both ends and more or less laterally compressed. Each grain has a brush of hairs on the apex. Well-developed grains measure 7–9 mm in length, and 2.85–3.4 mm in width (Fig. 10.3a).

Domesticated emmer closely matches wild emmer in almost all its morphological characteristics (Percival 1921). In both taxa, the hairs on the surface of the leaf blades are similar in form, length and arrangement; the spikes are flat, with narrow spikelets containing two grains in each, and the glumes alike in shape and texture. The conspicuous fringe of silky hairs on the margins of the rachis is missing or much reduced in the domesticated emmer and the grains of wild emmer are longer than those of domesticated emmer. The main difference between wild and domesticated emmer is rachis brittleness; in wild emmer, the rachis is brittle, and the spike disarticulates at maturity into individual spikelets, whereas, in domesticated emmer, the rachis is tough (or semi tough) and non-brittle, and the spike remains intact at maturity. Like wild emmer, domesticated emmer is also hulled wheat, i.e., it has rigid glumes that firmly enclose the grains. This requires pounding to release the grains from the glumes. On threshing, a hulled wheat spike breaks up into spikelets.

Being an ancient crop adapted to diverse climatic and edaphic conditions that exist in many parts of the world as well as to different agricultural practices, domesticated emmer is characterized by numerous landraces and varieties. Percival recognized two groups in domesticated emmer: (1) Indo-Abyssinian emmers—all are early forms, with four-to six–nerved coleoptiles, yellowish-green leaves, short straw (55–65-cm-tall), paler yellowish-green culm leaves, which have fewer and shorter hairs on the ridges, frequently rather glabrous auricles, and ears with a brittle or tough rachis;

(2) European emmers—later in growth, with two nerved coleoptiles, glaucous leaves, taller straw (100–125-cm-tall), leaf-blades covered with soft hairs and bluish-green in color, large auricles and a somewhat fragile rachis.

Dorofeev et al. (1980), following Vavilov, distinguished four taxa in domesticated emmer on morphological and geographical bases: *abyssinicum* Vav. (Abyssinian emmer), *asiaticum* Vav. (Eastern emmer), *dicoccum* (European emmer) and *maroccanum* Flaksb. (Moroccan emmer). These categories were further subdivided by Dorofeev et al. (1980), who distinguished between 8, 13, 40 and 3 botanical varieties within the *abyssinicum*, *asiaticum*, *dicoccum*, and *maroccanum* taxa, respectively. A detailed description of these botanical varieties is provided by Dorofeev et al. (1980).

Domesticated emmer is the primitive domesticated form of *T. turgidum* and one of the most ancient cultivated cereals. It was one of the principle wheats in summer-dry, relatively warm south-west Asia, central Asia, the Mediterranean basin, temperate Europe, and Egypt, from the Neolithic time to the second half of the third millennium BP, when it was replaced by the more advanced, free-threshing tetraploid form, *T. turgidum* ssp. *durum* (Schrank) Thell (macaroni or hard wheat) (Nesbitt and Samuel 1996; Zohary et al. 2012). In the eighth millennium BP, it was taken from the hilly and mountainous areas of the Fertile Crescent to the lowlands of Mesopotamia, and during the seventh and sixth millennia, to Egypt, the Mediterranean basin, Europe, and central Asia. It was taken to Ethiopia some 5000 years ago. Since its replacement by ssp. *durum* and by bread wheat, domesticated emmer has been grown as a relic crop in Ethiopia, India, Iran, Transcaucasia, the Volga Basin, eastern Turkey, the Balkans, Italy, Spain and central Europe (Nesbitt and Samuel 1996; Perrino et al. 1996).

Today, domesticated emmer covers only 1% of the total world wheat area (Zaharieva et al. 2010). It has also been introduced to the USA, and presently, limited amounts of spring varieties of domesticated emmer are grown in scattered areas throughout Montana and North Dakota (Stallknecht et al. 1996).

Genetic Diversity

Domesticated emmer contains very early and late accessions (Zaharieva et al. 2010). Percival (1921), Dorofeev et al. (1980) noted that most emmer wheat landraces are spring types, with the exception of some from western Europe, which are winter type. Damania et al. (1990) found a wide variation in tillering traits, grain protein content, and resistance to common bunt and yellow rust among accessions from Ethiopia, Jordan and Turkey. Wang et al. (2007) reported a wide range of diversity in eight agronomic characteristics among 91 accessions of this subspecies.

Domesticated emmer wheat is characterized by high protein and mineral concentrations; its grain protein concentration can reach 18–23% (Blanco et al. 1990; Dhaliwal 1977a; Perrino et al. 1993; Damania et al.1992). A high variation was found for gluten strength (Blanco et al. 1990; Perrino et al. 1993), and Grausgruber et al. (2004) reported a large variation in rheological properties of this wheat.

Current Uses of Domesticated Emmer

Domesticated emmer is currently primarily used as a human food, and also for animal and chicken feed (Zaharieva et al. 2010). Ancient Egyptians mainly used domesticated emmer for making a variety of breads (Samuel 1994) and beer (Kemp 1989). In Roman Italy, emmer wheat was used for making bread, as well as porridge and groats (Braun 1995), and is still used in Tuscany as whole grains (farricello) in traditional soups. Its use in the pasta industry is a recent response to the health food market.

Domesticated Emmer as Health Food

The advantages of domesticated emmer as a health food was recently reviewed by Zaharieva et al. (2010). This subspecies contains an amino acid composition similar to that of bread wheat (Cubadda and Marconi 1996), with some varieties having a higher lysine content (up to 3.65%) (Stehno 2007). Crude fiber content is higher in domesticated emmer than in durum wheat. Mineral and ash content were found in different varieties to be lower, similar, or higher than in durum (Hanchinal et al. 2005; Cubadda and Marconi 1996; Piergiovanni et al. 2009). Domesticated emmer also has a higher concentration of selenium, an important antioxidant factor. Genc and MacDonald (2008) identified domesticated emmer wheat accessions with greater grain

zinc concentration than modern durum and bread wheat genotypes.

The nutritional value of domesticated emmer is mainly due to its high fiber and antioxidant concentrations (Piergiovanni et al. 1996), high protein digestibility (Hanchinal et al. 2005) and high resistant-starch content, as well as its slower in vitro carbohydrate digestibility (Mohan and Malleshi 2006). The low glycemic index value and high satiating value (the degree at which food gives a human the sense of food gratification) of this wheat make it particularly suitable for diabetes patients (Buvaneshwari et al. 2003). Most of these qualities are related to a higher total dietary fiber (Yenagi et al. 1999; Annapurna 2000; Hanchinal et al. 2005), which is associated with a reduced rate of starch digestion (Jenkins et al. 1984). Substitution of bread wheat by emmer wheat in the diet for 6 weeks led to a significant reduction in total lipids, triglycerides and LDL (low-density lipoprotein) cholesterol (Zaharieva et al. 2010).

Because of its high value as a health food, nutritional value, and the unique taste of its products, current interest in domesticated emmer is increasing (Zaharieva et al. 2010). Such a renewed interest has its origin mainly in countries with well-developed intensive agriculture (Hammer and Perrino 1995; Nielsen and Mortensen 1998; Olsen 1998).

Economically Important Genes in Domesticated Emmer

Accessions of domesticated emmer are suited to warm, dry climates, can grow on poor soils, tolerate drought and heat stress, and are resistant to several fungal diseases. As such, they constitute a valuable genetic resource for improving durum and bread wheats. Several accessions of domesticated emmer from Italy were found resistant to stem rust, leaf rust and yellow rust (Corazza et al. 1986). Stem rust resistance was found in the Indian variety Khapli (Jakubziner 1969), in an Israeli accession (Rondon et al. 1966) and in a landrace from Ethiopia (Lebsock et al. 1967). Resistance to yellow rust has also been reported in 18 accessions of emmer wheat belonging to the ICARDA gene bank collection (Damania and Srivastrava 1990). Jakubziner (1969) identified accessions resistance to powdery mildew. A dominant gene for resistance to powdery mildew, designated as Pm4, was transferred by Briggle (1966) from Khapli into the Chancellor variety of bread wheat. Domesticated emmer resistance to loose smut was reported by Michalikova (1970). Similarly, there have been reports of resistance to fusarium head blight (Oliver et al. 2008), tan spot and Septoria blotch (Chu et al. 2008), Russian Wheat Aphid (Robinson and Skovmand 1992) and Hessian Fly (Zhukovsky 1964). Some accessions showed tolerance to drought (Zhukovsky 1964; Damania et al. 1992; Al Hakimi and Monneveux 1997) and heat (Hanchinal et al. 2005).

Domesticated emmer crosses easily with durum wheat and desirable genes can be transferred from the former to the latter. But hybrid necrosis and hybrid chlorosis are frequently met within crosses between domesticated emmer and durum (and also bread wheat), creating significant gene transfer barriers. Hybrid necrosis is governed by two complementary genes, *Ne1* and *Ne2*, located on chromosomes 5B and 2B, respectively (Tsunewaki 1960), and hybrid chlorosis is controlled by two complementary genes, *Ch1* located on 2A (Hermsen and Waninge 1972) and *Ch2* on 3D (Tsunewaki and Kihara 1961).

Several synthetic hexaploid wheats that were produced at CIMMYT using domesticated emmer as a female parent and *Ae. tauschii* as a male, were backcrossed to elite cultivars of bread wheat. Several backcross-derived lines were found to have a high level of resistance to green bug (Lage et al. 2003) and Russian wheat aphid (Lage et al. 2004), as well as good grain quality (Lage et al. 2006). Emmer-based synthetic backcross-derived lines also showed higher yield under drought-prone conditions in Mexico, Pakistan and eastern India compared to those using durum wheat (Trethowan and Mujeeb-Kazi 2008).

10.3.2.4 Ssp. *paleocolchicum* (Men.) A. Löve and D. Löve (Georgian Wheat)

T. turgidum ssp. *paleocolchicum* (Menabde) A. Löve and D. Löve, known as Kolkhuri Asli, Colchian emmer, or Colchis emmer, kolchis wheat, [Syn.: ssp. *georgicum* (Dek. et Men.) MK; *T. paleocolchicum* Men.; *T. georgicum* (Dek. et Men.) MK: *T. dicoccum* Schrank grex *georgicum* Dek. et Men.; *T. dicoccum* ssp. *georgicum* (Dek. et Men.) Flaksb.; *Triticum dicoccon* (Schrank) Schübl.; *T. karamyschevii* Nevski; *T. paleocolchicum* Men.; *T. turgidum* grex *paleocolchicum* (Men.) Bowden; *Gigachilon polonicum* (L.) Seidi ssp. *paleocolchicum* (Men.) Löve; *T. macha* Dek et Men. ssp. *paleocolchicum* (Men.) Cai; *T. turgidum* L. em. Thell. ssp. *georgicum* (Dek. et Men.) Hanelt]. Van Slageren (1994) maintained that at the species level, the older name *T. karamyschevii* Nevski should be adopted over the apparently *T. paleocolchicum,* but Löve and Löve's combination at the subspecies level is valid.

ssp. *paleocolchicum* is the second domesticated hulled form of *T. turgidum* (Table 10.5). This subspecies differs from ssp. *dicoccon* by its broad, compact, flat spike. The spike has a relatively large number of spikelets (34–36), and 4–5 florets per spikelet. It flowers relatively late, from June to July, and the seeds ripen from August to September. The taxonomic classification and rank have been a matter of much dispute. In light of its distinct morphological traits, it was first included in *T. dicoccum* as a variety, var. *chvamlicum* Supat (Supatashvili 1929). Later, Dekaprelevich and Menabde (1932) classified it as a subspecies, namely, *T. dicoccum*

ssp. *georgicum,* and later, Menabde (1948) (in Jorjadze et al. 2014) ranked it as the species *T. paleocolchicum* Men. Mac Key (1988) classified it as *T. Turgidum* ssp. *georgicum* (Dekapr. & Menabde) Mac Key, whereas van Slageren (1994) considered it as *T. turgidum* subsp. *paleocolchicum* (Menabde) A. Love and D. Love. This subspecies contains four varieties (Mosulishvili et al. 2017).

Wang et al. (2019) sequenced and assembled the complete chloroplast (cp) genome sequence of ssp. *paleocolchicum*, using Illumina sequencing. The assembled cp genome is 136,445 bp in size, and consists of four parts, namely LSC (79,993 bp), SSC (12,832 bp) and two IRs (21,815 bp). Gene annotation found that it encodes 109 non-redundant genes, including 76 protein-coding genes, 29 tRNA genes and 4 rRNA genes, 19 of which are located in the IR region with two copies. An evolutionary tree constructed based on the whole chloroplast genome sequence showed that ssp. *paleocolchicum* and other species with the BBAA genome were clustered together, while separated from *T. timopheevii* wheat (GGAA). This study enriched the sequence resources of ssp. *paleocolchicum* and also provided important data for its molecular identification, marker development and phylogenetic studies.

ssp. *paleocolchicum* is a small endemic, relic taxon of observed in Georgia (Jorjadze et al. 2014). Currently, it is cultivated on a limited scale in western Georgia (Colchis), often in a mixture with hexaploid wheat *T. aestivum* ssp. *macha*. But in the past, it was grown more widely, in eastern Georgia as well. The reason for its maintenance in west Georgia stems from its well performance under the humid conditions prevailing in this part of Georgia. Under these conditions, it has successfully replaced ssp. *dicoccon* that did not thrive in humid environments. It grows well on sandy, loamy, and clay soils, but prefers well-drained beds. While this subspecies adapts well to the humid climate of west Georgia, it also performs well under the dry, hot climate conditions of east Georgia. It is usually ground into a flour and used for making bread, biscuits, etc. Its grain protein content reaches 18.8%, lysine content is 2.9%, and its high-quality gluten enables the production of a good bread and offers desirable baking properties (Jorjadze et al. 2014). ssp. *paleocolchicum* is resistant to several fungal disease and can serve as a good source for desirable traits. This subspecies is thought to have been generated by inter-ploidy hybridization between wild emmer wheat, ssp. *dicoccoides*, and a form of bread wheat (Dvorak and Luo 2001).

10.3.2.5 Ssp. *parvicoccum* Kislev (Currently Extinct)

T. turgidum L. ssp. *parvicoccum* (Kislev) Kislev, known as a small-grain wheat (Syn.: *T. parvicoccum* Kislev sp. nov.), is an archaeobotanical, free-threshing, tetraploid wheat taxon

(Kislev 1979, 1980, 2009). It seemed justifiable (Kislev 1979/1980) to combine the grains and rachis fragments found in several Near Eastern sites into one taxon since (a) they were located (e.g., in Tel Aphek) within close proximity, at distances of a few millimeters or centimeters; (b) they both belonged to a new taxon; and (c) their morphology was complementary, viz. small grains and compact rachis.

In the description of the species, efforts were made by Kislev (1979/1980, 2009) to reconstruct ssp. *parvicoccum* as fully as possible. Some features were highly correlated with the habitat or with other plant qualities which survived in the same archaeological sites. The validity of such a description is, of course, not completely satisfactory and the reader may judge each characteristic for himself.

ssp. *parvicoccum* has a tough rachis with very short and narrow internodes. The grain is very short (about 5-mm-long), oval to elliptic, widest in the middle, thickest in the lower third, with a wide, rounded or truncate apex, and wide, short-haired, and sometimes collared brush, especially in small grains. The embryo is small and oval, the radicle is prominent or slightly so, the plumule is slightly prominent or not at all, the ventral side is flat, ascending at the base, sometimes also at the apex, the crease is narrow or moderately wide and cheeks are rounded.

The following additional characteristics were deduced from circumstantial evidence or correlation with other characteristics: the plant has a domesticated annual, spring habit, and short, dense, compressed and oblong or oval ears. Spikelets are two-grained, and the third floret is reduced. Glumes are very short, diagonally oriented, grains are free, and loosely invested by chaff (Fig. 10.4a–c).

Kislev (2009) investigated well preserved archaeological remains of ssp. *parvicoccum* wheat from Late Bronze Timnah (Tel Batash), Israel, and added the following to its morphological description: (1) the culm upper internode is solid, with striate and a slightly rough surface; (2) the ear is

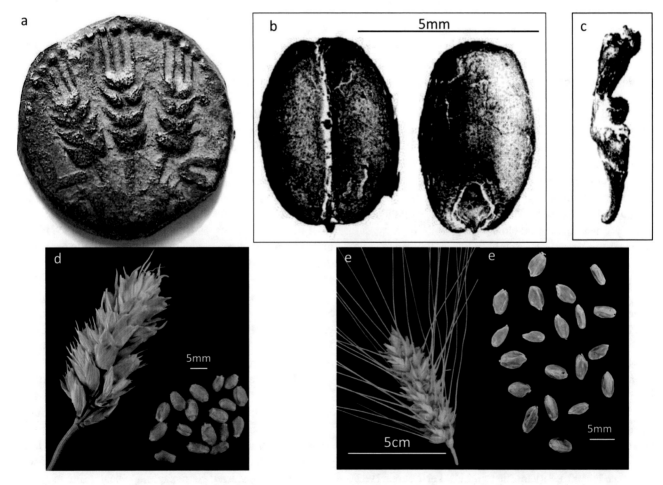

Fig. 10.4 a A coin coined by Agripas 1, king of Judea in the first century AD, showing three wheat spikes, presumably of *T. turgidum* ssp. *parvicoccum*, that was cultivated in Israel at least until 130 AD (Kislev 1979/1980), and **b** grains; and **c** rachis of *T. turgidum* L. ssp. *parvicoccum* Kislev (From Kislev 1979/1980); **d** A spike and grains of an extracted tetraploid [extracted by Kerber (1964) from the hexaploid cultivar Canthath], that resemble ssp. *parvicoccum*; **e** A spike and grains of *T. turgidum* L ssp. *durum* (Desf.) Husn. cv. Hourani that according to Kislev (1979/1980) may have derived from ssp. *parvicoccum*

bearded, laterally compressed, and of medium to high density; (3) the rachis is fringed, with long hairs along the margins, and bears a frontal tuft of long hairs at the base of each spikelet; (4) the internodes are relatively thick; (5) the spikelets are two-flowered; (6) the outer face of the hairy glumes is somewhat flat, with a prominent keel that runs from the base to the tip; (7) a pair of prominent lumps is present on the rachis node, beneath its glume bases; and (8) after threshing, the basal part of the glume is usually retained on the rachis node.

The archaeological samples of free-threshing (naked), small, and somewhat spherical grains (< 5 mm) and of short internodes found in Near Eastern sites from 8900 to 7000 BP, were considered by Helbaek (1959) to be club wheat, the hexaploid *T. aestivum* ssp. *compactum* (= *T. compactum antiquorum*). However, it is difficult to accept that these Near Eastern wheat remnants from such an early period (9000–7000 years BP) are hexaploids, as explained by Kislev (1979): (1) Modern native strains of hexaploid wheats are more adapted to the northern latitudes (Vavilov 1926) than to the Near East, where Helbaek found his remnants, the latter area being dominated by tetraploid taxa. In the western flank of the Fertile Crescent, however, hexaploid wheat was grown, but only as a crop of little importance (Jakubziner 1932a). (2) Bread wheat (*T. aestivum* L.) which is sometimes short-grained (c. 5 mm) is lax-eared, while the dense-eared club wheat, *T. aestivum* ssp. *compactum*, has grains longer than 5.7 mm (Percival 1921). (3) If Helbaek's remnants are indeed hexaploid, this implies that hexaploid naked wheat, found in the earliest periods in the Near East, preceded cultivation of tetraploid naked wheats. However, on an evolutionary basis, one would expect that the tetraploids developed first. (4) The earliest samples are probably from too early a period for hexaploid wheats to have come into existence; the spread of agriculture had not yet made geographic contact between the domesticated tetraploid and *Ae. tauschii*, the wild diploid progenitors of hexaploid wheat.

Based on the morphological characteristics of this archaeological material and the difficulty in defining ancient Near Eastern naked wheat as hexaploid, Kislev (1979/1980) concluded that these wheat remnants belong to an extinct tetraploid species, *T. parvicoccum* Kislev. According to current classification, this taxon should be referred to as *T. turgidum* ssp. *parvicoccum* (Table 10.5).

The first records of the naked tetraploid wheat ssp. *parvicoccum* were found in the early PPNB. Naked grains were unearthed in phase II of Tell Aswad, near Damascus, dated 8900–8600 BP (Hillman 1996), in Ramad, southwestern Syria, dated 8300 BP (van Zeist and Bakker-Heeres 1982); and in Can Hassan III, south Anatolia, dated 8400 BP (Hillman 1972, 1978). Remnants of naked, small-grain wheat were also reported in Atlit Yam, Israel, the late PPNB

(8000–7500 BP) (Kislev et al. 2004) and in various sites in the Pottery Neolithic period (7500–6200 BP), including western Iran, Anatolia, Iraq, Syria, Israel, Georgia, and the Balkans (Kislev 1979, 1980, 1981; Schultze-Motel 2019). Archaeological excavations in Israel dated it to as late as 1870 years BP (Kislev 1979/1980, 1981). Schulz–Motel (2019) found ssp. *parvicoccum* that was presumably grown in Georgia about 800 years BP. It was abundantly grown from the 8th millennium BP onwards in south-east Asia, Transcaucasia and the Balkans (see list of sites in Kislev 1979/1980, 2009). Schultze-Motel (2019) detected it recently in two places in Georgia, broadening the area of this species to the north.

The great advantage of ssp. *parvicoccum* was its comparatively delicate glumes and tough rachis nodes, which facilitated easy threshing of its grains. The fact that both hulled and naked wheat were found contemporaneously in the Near East for such a long period may indicate that the overall advantage of the latter (free-threshing, but with small grains) must not have been sufficiently great to replace the former (bearing large but hulled grains) as a major crop. In biblical times, they were grown in separate fields and treated as two different, albeit closely related, crops (Kislev 1979/1980, 1981). ssp. *parvicoccum* was replaced by the related taxon, ssp. *durum*, that became more abundant, since it combines the free-threshing feature of *parvicoccum* and the large grains of *dicoccon*.

The origin of ssp. *parvicoccum* is obscure. Three major morphological characters distinguish it from domesticated emmer, namely, nakedness (free-threshability), ear compactness, and small grain size. The free-threshing trait could have been derived from a mutation that reduces the toughness of the glumes and increase the rigidity of the rachis (McFadden and Sears 1946; Morris and Sears 1967). The hulled, non-free-threshing emmer contains the *q* gene, which determines both a semi-fragile rachis and tough, thick glumes. All the extant free-threshing tetraploids contain the *Q* factor, dictating the free-threshing trait (Muramatsu 1986; Simons et al. 2006). This gene was recently isolated and characterized (Faris et al. 2003; Jantasuriyarat et al. 2004; Simons et al. 2006). Consequently, it is assumed that the ancient free-threshing ssp. *parvicoccum* also contained this gene. The *Q* factor, located on chromosome arm 5AL of common wheat (Sears 1954), has a pleiotropic expression pattern, affecting free-threshing as well as several other characteristics that are related to spike and spikelet structure, such as tough rachis and loose glumes. The mutation from *q* to *Q* occurred at the tetraploid level, from where it was presumably transferred to hexaploid wheat (Simons et al. 2006). The two alleles differ in a single amino acid, and *Q* is more abundantly transcribed than *q* (Simons et al. 2006). The higher expression of *Q* is in accord with the finding of

Muramatsu (1963), who noted that extra doses (five or six) of *q* mimic the effect of *Q* in bread wheat. There is a wide variation in the phenotype of *QQ* lines, which is presumably due to different genetic backgrounds.

In addition to *q*, there is also an extra dose of group-2 chromosomes in bread wheat, which increases glume toughness (Sears 1954). In accord with this, Kerber and Rowland (1974) found the *Tg* (tenacious glumes) gene, conferring tough glumes, on chromosome arm 2DS of *Ae. tauschii*. Likewise, addition of the short arm of chromosome $2S^l$ ($2S^lS$) of *Ae. longissima* to common wheat resulted in a similar effect (Feldman unpublished). Chromosome arm 2BS of emmer also contains the *Tg* gene which determines tough glumes, whereas chromosome arm 2AS does not contain such a gene (Simonetti et al. 1999). Thus, this character is determined in tetraploid wheat by, at least, two complementary genes, *Q* and *tg*. Thus, at least two mutations were required to produce the free-threshing character in ssp. *parvicoccum* (Jantasuriyarat et al. 2004).

Wild and domesticated emmer have long grains, and almost no transitional forms between types with long and short grains are known in the archaeological literature. Yet, in PPNB of Çayönü, southeastern Turkey, a few mid-long grains of domesticated emmer were unearthed (van Zeist 1972), suggesting that the missing link between long and short grains may have been a compact, short-grained domesticated emmer. Outside the Near East, two early Neolithic findings of mid-long grains of naked wheat from Azmak, south Bulgaria, dated to the beginning of the 7th millennium BP, were found (Hopf 1973). Their mean dimensions are 5.6 × 3.3 × 2.9 mm and 5.8 × 3.2 × 2.7 mm. The formation of small-grained wheat presumably occurred as a result of a series of mutations, such as an *s*-like mutation, which shortened the grain and made the ear dense, as found in other tetraploids (Schmidt and Johnson 1963).

ssp. *parvicoccum* has been found hitherto in the Near East and the Balkans (Kislev 1979/1980). Recently, Schultze-Motel (2019) found grains of ssp. *parvicoccum* in several sites in Georgia, dated to about 2300 and 800 BP. According to Kislev (1979/1980), ssp. *parvicoccum* was grown side by side with domesticated emmer, ssp. *dicoccon*, in the East Mediterranean and Near Eastern regions over several millennia. The last evidence for cultivation of this subspecies in the Levant is from up 1900 years ago (Kislev 1979/1980, 1981). ssp. *parvicoccum* was grown in Georgia up to 800 years ago (Schultze-Motel 2019).

Changes in spike compactness and grain size probably occurred as a result of a series of mutations, including an *s*-like (*sphaerococcum*-like) mutation that shortened the grain and increased ear density. Such a mutation creating a *sphaerococcum*-like tetraploid wheat, was indeed described by Schmidt and Johnson (1963). The finding of a few medium-sized grains of wild and/or domesticated emmer

In PPNB Çayönü, led Kislev (1979/1980) to suggests that the missing link may have been a compact, short-grained emmer. Alternatively, these traits developed in the free-threshing wheat. In addition, one cannot rule out the introgression of genes conferring compact spikes and short grains from tetraploid *Aegilops* species, such as *Ae. peregrina* (= *Ae. variabilis*) or *Ae. geniculata* (= *Ae. ovata*).

10.3.2.6 Ssp. *durum* (Desf.) Husn. (Macaroni or Hard Wheat)

Description of Subspecies

T. turgidum ssp. *durum* (Desf.) Husn., known as macaroni wheat, hard wheat or simply durum wheat, [Syn.: *T. durum* Desf.; *T. alatum* Peterm.; *T. vulgare* var. grex *durum* Alef.; *T. sativum durum* Hack.; *T. sativum* ssp. *durum* (Desf.) K. Richt.; *T. aestivum* ssp. *durum* (Desf.) Thell.; *T. turgidum* ssp. *sementivum* Rasse *durum* (Desf.) Thell.; *T. turgidum* grex *durum* (Desf.) Bowden; *T. turgidum* ssp. *durum* Löve and Löve; *T. turgidum* ssp. *turgidum* convr. *durum* (Desf.) MK; *Gigachilon polonicum* (L.) Seidl ssp. *durum* (Desf.) Á. Löve] is a predominantly autogamous, plant. Its culms, usually 3–4 per plant, are erect, 60–160-cm-high, with 5 or 6 internodes above ground, and generally solid throughout or in the upper internodes, while hollow with thick walls in some varieties. The leaves are linear, flat, 16–25-cm-long, yellowish-green or bluish-green, glabrous in most varieties, and with a transparent ligule and ciliate auricle. The spikes are 4–11-cm-long (excluding awns), two-rowed, determinate, almost always awned, laterally compressed, oblong, or square in cross-section, depending upon the laxness of the spike, and the number and size of the grains in each spikelet, and bear an average of 20 spikelets per spike, at a density varying from about 20 in lax spikes to 47 in compact ones per 10 cm length of rachis. The rachis is usually tough and non-brittle, although in some varieties, it disarticulates more or less easily, especially near the base of the spike; it is fringed with hairs along the margins and bears a frontal tuft at the base of each spikelet, and its internodes are 2.5–5-mm-long, flattened, narrow and wedge-shaped. The spikelets are 10–15-mm-long, with 5–7 florets, the upper 2 or 3 being sterile, and those of square spikes containing three or four grains, while those of compressed spikes usually contain only two. The glumes are yellow, red, or blue-black in color, loosely appressed to the lower florets, glabrous in some varieties and pubescent in others, with considerable variation in hair length and amount, glumes of the lateral spikelets being 8–12-mm-long, asymmetrical, with one prominent keel running from the base to the tip, and 5–7 nerves running from the base to a point close to the apical tooth, with the latter being acute or blunt, and of variable length, and the lateral secondary tooth usually being short or missing. The glumes of the terminal spikelet are ovate and more or less symmetrical. The lemmas are thin and pale, rounded on the back, 10–12-mm-long, with 9–15 nerves which converge at the tip into a terminal, firm awn. As a rule, only the two lower lemmas of each spikelet bear long awns, measuring, in some cases in the lower spikelets, as long as 20–23 cm, longer than those of any other wheats. The awns of the flowers of the upper spikelet are usually 5–4 cm long, i.e., all awns of a spike usually terminate at the same height. Some varieties are awnless or have short (6.0 cm) or medium (6.1–9.0 cm) length awns. The awns are white, red, or black, almost smooth near the base, thus differing from those of ssp. *turgidum* which are usually rough throughout their entire length, generally straight and more or less parallel to the sides of the spike. The palea is membranous and shorter than the corresponding lemma, and does not split at maturity. The caryopsis is larger (45–60 mg) than those of other wheats, glabrous, free of lemma and palea, white, amber, yellow, or red in color, generally somewhat narrow, tapering towards both ends, more or less laterally compressed, with a narrow dorsal ridge, and wanting in plumpness, and with a shallow furrow usually exhibiting flattish sloping sides; the cross-section is more or less triangular. The embryo is large, with an elongated oval scutellum. Measurements of grains taken from the middle of the spike of flinty forms were of an average length of 8.30 (7.0–9.7) mm, average breadth of 3.48 (2.8–4.1) mm, and an average thickness of 3.61 (3.2–4.25) mm. However, some cultivars, e.g., Hurani, have smaller (about 5 mm long) grains. Most varieties grow in the spring, while several grow in the winter (Fig. 10.3b).

Durum in Latin means "hard" and reflects the hard-flinty nature of the plant grain, the hardest of all other wheat grains. Yet, the name Hard Wheat is confusing, as it is applied to forms of North American bread wheat. It is advisable therefore, to use the common name durum wheat or macaroni wheat for this subspecies.

Because of the similarity between plant remains of tetraploid and hexaploid free-threshing wheats, it was very difficult, if not impossible, to identify the species of the Pre-Pottery Neolithic B (PPNB; 9500–7500 BP) wheat remnants, and consequently, they were referred to as *T. turgidum-T. aestivum* or as 'aestivo- compactum' wheats (Zohary et al. 2012). In fact, almost all naked (free-threshing) wheat remnants described so far have been compared to a free-threshing hexaploid wheat that was grown by the Neolithic Lake dwellers of Switzerland during the Bronze Age, described by Heer (1886) as *T. vulgare antiquorum*. This wheat, now believed to be extinct, was very similar to *T. aestivum* ssp. *compactum* (club wheat); it was a dwarf plant, with extremely small, stubby grains and compact, awnless spikes. Because of its resemblance to *T. aestivum* ssp. *compactum*, it has generally been assumed to have been a hexaploid. In fact, palaeoethnobotanists have

already described very early naked wheat remnants as belonging to hexaploid species. For example, Helbaek (1959) identified archaeological samples of small grains (less than 5 mm in length) and short internodes, found in very early Near Eastern sites, as hexaploid *T. compactum*.

Yet, Kislev (1979/1980) found it difficult to explain these remnants as hexaploids for the following reasons: (i) Modern genotypes of hexaploid wheats are more adapted to the northern latitudes than to the Near East, the latter area being dominated by tetraploid forms. In the western flank of the Fertile Crescent, however, hexaploid wheats are grown, but only as a crop of little importance (Jakubziner 1932a) (ii) The most ancient samples are probably from too early a period for hexaploid wheats to have come into existence (van Zeist 1976), i.e., the beginning of the 8th millennium BP; the spread of agriculture had not yet made geographic contact between domesticated tetraploid wheats and *Aegilops tauschii*, the wild diploid donor of the third subgenome (subgenome D) to hexaploid wheat. Since hexaploid wheat originated toward the middle or even the end of the 8th millennium BP (Kislev 1984), in northwestern Iran, south of the Caspian Sea (Dvorak et al. 1998a, b), outside the Fertile Crescent region, it has to be assumed that most of the PPNB remnants of naked wheat in the Fertile Crescent are of tetraploid forms rather than of hexaploid wheats.

The idea that wheat grown in the ancient Near East was tetraploid rather than hexaploid, has been recently suggested by several researchers (Kislev 1973; van Zeist and Bakker-Heeres 1973; Zohary 1973). Yet, because of the small grains (less than 5 mm in length) and short internodes characterizing these wheat remnants, Kislev (1979/1980) suggested that these wheats belong to an extinct tetraploid taxon, named *T. parvicoccum* Kislev (currently *T. turgidum* L. ssp. *parvicoccum* Kislev). Practically, all prehistoric naked wheat grains in the Fertile Crescent and East Mediterranean regions are less than 5 mm on average (Kislev 1979/1980).

Ssp. *durum* evolved only later, presumably from hybridization of domesticated emmer and ssp. *parvicoccum*. Renfrew (1973) indicated that archeological studies suggest that durum wheat may have originated later than hexaploid wheat. Material ascribed to ssp. *durum* has been found, at first sporadically, among prehistoric plant remnants, already from the 8th millennia BP in Syria. It was unearthed by Hillman (1978) in layers of Pottery Neolithic Can Hassan III (7500–6200 BP). During later Neolithic periods, this wheat gradually gained prominence, until it became the main wheat in the Mediterranean countries, first in its western part, and only in the Helenistic time, in the east Mediterranean and Egypt. It was suggested that ssp. *durum* was derived from domesticated emmer by a series of mutations that reduced the toughness of the glumes until a free-threshing form was attained (McFadden and Sears 1946; Morris and Sears 1967; Sears 1969). Alternatively, it could have derived from the free-threshing ssp. *parvicoccum*, or more likely, from crosses between *parvicoccum* and *dicoccon*. Hybridization between *parvicoccum* and *dicoccon* may lead to the formation of progeny with the free-threshing trait from *parvicoccum* and the large grains from *dicoccon*.

Studies of chloroplast markers (cpSSRs) indicated that wild and domesticated emmer and ssp. *durum* share a common maternal ancestral gene pool (Oliveira et al. 2012). Zohary et al. (2012) proposed that the early naked forms of tetraploid wheats evolved from domesticated emmer in the Fertile Crescent. The data of Oliveira et al. (2012) support this hypothesis. If this event occurred only once or several limited number of times, it is reasonable to assume that only a small number of cp-haplotypes, from a broader gene pool present in domesticated emmer, would appear in the early naked tetraploid forms, and later in ssp. *durum*, as a consequence of a bottleneck effect. This is apparent from the strong difference between the hulled wild and domesticated emmer and the naked tetraploid wheat, ssp. *durum*, suggests that the latter subspecies evolved from only a small number of hulled tetraploid genotypes, and have been in relative reproductive isolation since their spread into Europe and North Africa during the Neolithic Age. Oliviera et al. (2012) also detected cp-haplotypes in accessions of both wild emmer and ssp. *durum*. This suggested an alternate scenario in which free-threshing tetraploid wheats were domesticated de novo from lines of wild emmer, independently acquiring, by mutation, the tough-rachis trait of domesticated emmer, plus the free-threshing trait.

The fact that ssp. *durum*, for which evidence dates back to 6500–7500 BP, became established as a prominent crop in the east Mediterranean only 2000–2300 years ago, requires an explanation. More specifically, it is perplexing that in spite of its relatively early origin, ssp. *durum*, featuring both large grains like those of ssp. *dicoccon*, and a free-threshing trait like that of *ssp. parvicoccum*, was only established as a major crop in the Mediterranean basin and the Near East during the Late Bronze Age and Hellenistic period (Nesbitt, 2002; Feldman and Kislev 2007). Perhaps the early *durum* types were not actively cultivated because small-grained wheat and domesticated emmer were better adapted to the semi-arid conditions of the region. Adapted genotypes evolved only later, through a series of mutations, or through introgression of genes from the other two tetraploid subspecies. Kislev (2009) suggested that small grains may have been more competitive than large grains because they could not be attacked by *Sitophilus granarius*, the most destructive pest beetle of stored cereals. Because the size of its larva would take up half of the *parvicoccum* grain, it would fail to provide sufficient food for the pest's normal development. Only upon construction of air-sealed granaries in the classical period, did

larger grains become more resistant to pest beetle. Taken together, development of sealable granaries may explain the establishment of ssp. *durum* in the Near East and the impoverishment of ssp. *parvicoccum* (Kislev 2009).

The large grain of ssp. *durum* was probably preferred over the small grains of *parvicoccum*, and likely lead to the prominence of *durum* as a tetraploid wheat and to the extinction of *parvicoccum*. Most other subspecies of *T. turgidum* with a naked grain are probably of a relatively more recent origin; they deviate from ssp. *durum* in only a few characteristics (Mac Key 1966; Morris and Sears 1967) and share the genetic system for the free-threshing habit. The cultivation of ssp. *parvicoccum* alongside ssp. *durum* in the Near East may explain the morphology of some peculiar taxa of naked tetraploid wheats, such as the compact cv. *pyramidale* Perc. in Egypt and Horan wheat, cv. *horanicum* (Jakubz.) Flaksb. in the Levant, with compact ears and small-plump grains (Fig. 10.4e). Also, the extracted tetraploids, derived from *T. aestivum* ssp. *aestivum* by Kerber (1964), resemble very much ssp. parvicoccum (Fig. 10.4d). Crosses between *durum* and *parvicoccum* might have resulted in such intermediate forms. These new taxa, some of which are adapted to semi-arid conditions, took the place of the old small-grained wheat.

In southwest Asia and in the Mediterranean basin, hulled and *parvicoccum* wheats were replaced by more modern free-threshing types. Up to the present time, the hulled wheats, emmer and einkorn, continued to occur as minor components. In south-eastern Turkey emmer was replaced by more modern free-threshing wheats at the beginning of the Early Bronze Age (ca. 5000 BP) and have been nearly absent from the archaeobotanical record thereafter (van Zeist and Bakker-Heeres 1975; Nesbitt 1995). In central Turkey, einkorn and emmer appear to be minor crops in Middle Bronze Age samples (3900–3700 BP) (Nesbitt 1993). In the Levant, emmer became a minor component in the 5th millennium BP and almost disappeared in the 4th millennium BP (Miller 1991), whereas in Egypt, emmer was grown until the Hellenistic time (ca. 2300 BP), when it was replaced by the free-threshing ssp. *durum* (Crawford 1979; Bowman 1990). For further details, see Nesbitt and Samuel (1996). Towards the Late Bronze Age, southwest Asia and the Mediterranean basin both showed a high prevalence of naked wheats. The large quantities of naked wheat unearthed in the Levantine Bronze Age villages in the Levant are impressive (Zohary et al. 2012).

Current Cultivation of Durum Wheat

The earliest historical reference to durum wheat was made by Dodoens in his *Historia frumentorum*, published in 1566. Only near the end of the eighteenth century was ssp. *durum* distinguished from the Mediterranean forms of ssp. *turgidum* and ssp. *aestivum* by Desfontaines, who described it in his *Flora Atlantica*, mentioning solid straw, pubescent glumes, and long flinty grain as its specific characteristics. Today, ssp. *durum* is the principal tetraploid wheat and next to bread wheat, *T. aestivum* ssp. *aestivum,* the various varieties of ssp. *durum* are the most widely cultivated tetraploid wheats, constituting about 5–9% of global wheat production. The area planted annually is approximately 20 million hectares, averaging about 38 million metric tons annually (FAOSTAT 2018), and current trends are upwards. ssp. *durum* developed in the hot and dry climate of the Near East and Mediterranean basin. Currently, it grows as a major crop in the Mediterranean basin (Portugal, Spain, Italy, Morocco, Algeria, Tunisia, Egypt, Greece, Turkey, Syria, Lebanon, Israel, and Jordan), the Near East (Iraq and Iran), trans- and cis-Caucasia, Afghanistan, Pakistan, Turkestan, Kazakhstan, the southern portion of East Siberia, Bulgaria, Ukraine, southern Russia, Ethiopia, India, China, and in low rainfall areas of the great plains of the United States and Canada, Mexico, Chili, Argentina, South Africa and Australia (Feldman 2001). The European Union (mainly Italy, Spain, and Greece) is the largest durum wheat producer, Canada is the second largest producer, followed by Turkey and the USA.

Genetic Analysis of Durum Wheat

A recent major achievement in the study of genome structure of durum wheat is the assembly of 10.45 gigabase (Gb) of the genome of cultivar Svevo by Maccaferri et al. (2019) (see Sect. 3.3.2.2. in Chap. 3). This assembly facilitated the comparison between the genome of ssp. *durum* and that of its wild ancestor, ssp. *dicoccoides,* that was sequenced earlier (Avni et al. 2017) (see Sect. 3.3.2.1. in Chap. 3). The comparison revealed, and will continue to do so, changes imposed by the domestication process, by thousands of years of unconscious and conscious selection under cultivation, and by modern, scientific based, breeding. Regions exhibiting strong signatures of genetic divergence associated with domestication and breeding were widespread in the *durum* genome with several major diversity losses in the pericentromeric regions that occurred during domestication of wild emmer. The reduction of diversity continued more moderately, but spread over the genome, during the evolution of domesticated emmer wheat and that of *durum* landraces, and, more recently, in modern cultivars as a consequence of the breeding activity.

The study of the assembled genome of the modern Italian durum cultivar Svevo by Maccaferri et al. (2019) showed little evidence of a polyphyletic origin of this subspecies. Principal component analysis showed a close relationship between Svevo and a specific group of durum landrace populations. The observations indicated that two domesticated emmer populations from the southern Levant showed the closest relationship to all durum wheat landrace populations, while the modern *durum* cultivars were mostly related to two durum wheat landrace populations from North Africa and Transcaucasia. Landraces of ssp. *durum* from Ethiopia, were genetically isolated.

Assembly of the genomes of *durum* and *dicoccoides* may provide a better understanding of how the two durum subgenomes, A and B, interact and coordinate their activities. It may clarify the nature and mode of action of the genetic mechanism(s) governing cytological and genetic diploidization in allopolyploid wheats. These are fundamental biological questions that may elucidate the reasons for the evolutionary success of this allotetraploid as well as of other *Triticum* and *Aegilops* allopolyploids.

Assembly of the *durum* genome will facilitate the investigation of the genetic control of many agricultural and nutritional properties in this important crop, via analysis of genes, their structure, order, control of expression, and communication with one another. It may lead to the identification of genes that are responsible for useful traits such as yield, disease resistance, and nutritional properties, allowing for their selected for breeding programs. For instance, Maccaferri et al. (2019) has already identified the gene most likely responsible for high cadmium (Cd) accumulation in the grains of modern cultivars, an undesirable gene located on chromosome 5B, and the recovery of an allele for a low Cd build up.

During the last several millennia, durum wheat has spread all over the world, which has been accompanied by increased diversity driven by unintentional and conscious selection of adaptive genotypes for different climatic and edaphic conditions as well as different human tastes. The main centers of variation of ssp. *durum* are in southwest Asia, the Mediterranean basin and Ethiopia. Genetic improvement of durum wheat began very late and has proceeded more slowly than that of common wheat. Breeding of durum wheat is currently concentrating on the simultaneous improvement of grain yield, disease and insect resistance, tolerance to abiotic stresses (cold, heat, drought and salt), grain quality traits, and processing quality traits, including protein concentration, yellow pigment concentration (high carotenoids and flavonoids content), gluten strength, semolina milling properties, and pasta cooking quality.

The increase in genetic diversity of ssp. *durum* is manifested by the large number of varieties and land races. Percival (1921) described more than 60 varieties, based on spike characteristics, such as awn presence or absence, glume color (white, red or black), pubescence or glabrousness, awn color, and grain color and texture. In the monograph by Dorofeev et al. (1980), durum wheat (as *T. durum*) is subdivided into the ssp. *durum* and ssp. *horanicum* Vav., and about 140 botanical varieties were described, mainly on the basis of spike and grain characteristics.

A consensus genetic map presented by Maccaferri et al. (2014), providing nearly complete genome coverage, as well as marker density, was used as a reference for genetic diversity and mapping analyses of ssp. *durum*. The consensus map provides the basis for high-density single nucleotide polymorphic (SNP) marker implementation in durum wheat. Markers previously mapped in hexaploid wheat constitute a strong link between the two species. But, differences in marker order and local recombination rate were observed between the durum and hexaploid wheat consensus maps.

Seventy SSRs and 234 AFLPs were used to profile a collection of 58 durum wheat accessions representing the most important extant breeding programs (Maccaferri et al. 2007). In addition, 42 phenotypic traits, including the morphological characteristics recommended for the distinctness, uniformity, and stability tests, were recorded. The correlation between the genetic similarities obtained with the two marker classes was high ($r = 0.81$), whereas lower correlations were observed between molecular and phenotypic data ($r = 0.46$ and 0.56 for AFLPs and SSRs, respectively). Morphological data, even if sampled in high numbers, largely failed to describe the pattern of genetic similarity, according to known pedigree data and the indications provided by molecular markers.

Levels of genetic diversity and population genetic structure of a collection of 230 accessions of seven subspecies of *T. turgidum* were investigated by Laidò et al. (2013), using six morphological, nine seed storage protein loci, 26 SSRs and 970 DArTs. As expected, genetic diversity of the morphological traits and seed storage proteins was always lower in durum wheat compared to those of wild and domesticated emmers. The two sets of molecular markers distinguished durum cultivars from the other free-threshing subspecies of *T. turgidum*. The genetic diversity of morphological traits and seed storage proteins was always lower in the improved versus older durum cultivars. This marked effect on diversity was not observed for molecular markers, where there was only a weak reduction. The SSR markers identified a greater number of groups within each subspecies as compared to DArT.

Kabbaj et al. (2017) investigated population structure and genetic diversity among elites and landraces of durum wheat collected from 32 countries. A total of 10 sub-populations were identified, with six bearing modern germplasm and four constituting landraces of different geographical origins. Interestingly, genomic comparison between groups indicated that the Middle East and Ethiopia had the lowest level of allelic diversity, while breeding programs and landraces collected outside these regions were the richest in rare alleles. Further, phylogenetic analysis of landraces indicated that Ethiopia represents a second center of durum wheat diversity. Overall, the analyses performed by Kabbaj et al.

(2017) provided a global picture of the current genetic diversity for this crop and shall guide its targeted use by breeders.

Durum Products

The large, hard-textured grains of ssp. *durum* yield low-gluten flour that creates plastic doughs, contrasting with the strong elastic doughs obtained from flour of bread wheat. As a result, durum wheat is less commonly used in bread making and is especially suitable for making pasta and other semolina products (Hanelt 2001). Durum breads do exist, but, in most instances, the doughs contain only a portion of durum flour and are substantially supplemented with white bread wheat flours; often, bread wheat flour is high in gluten, and is necessary to offset the poor contribution of durum flour to the gluten network. Pure durum breads are often dense, containing few air bubbles, with relatively little elastic structure. Most of the durum grown today is amber durum, the grains of which are amber-colored and larger than those of other types of wheat. Durum has a yellow endosperm that gives pasta its color. When milled, the endosperm is ground into a granular product called semolina, which is used for premium pastas. In southern Europe, it is mainly used for pasta, whereas in North Africa, is used for couscous and in the Levant, for dishes such as tabbouleh, kubbeh, frikeh and bulgur for pilafs. The Israeli variant of couscous involves larger pearls of durum called ptitim in Hebrew. In many Mediterranean countries, it forms the basis of many soups, gruels, stuffings, puddings and pastries. When ground as fine as flour, it is used for making bread; in the Middle East, it is used for flat round breads (pita), and in Europe and elsewhere, it is used for pizza.

Sayaslan et al. (2012) found that several durum landraces have the potential to improve high-quality pasta processing of modern durum cultivars. The seed protein composition may affect the pasta and semolina quality. SDS-PAGE analyses were performed to characterize the four seed protein fractions (albumins, globulins, gliadins, and glutenin subunits), as well as several proteins from each of the four subunits of gliadin (α, β, γ, and ω) in the grains of five bread wheat and five durum wheat genotypes (Žilić et al. 2011). In addition, content of tryptophan and wet gluten were analyzed. It was found that gliadin and glutenin subunits comprised from 58.17% to 65.27% and 56.25% to 64.48%, respectively, of total proteins and, as such, account for both quantity and quality of the bread and durum wheat grain proteins. The analysis demonstrated that bread wheat genotypes had a higher concentration of $\alpha + \beta + \gamma$-subunits of gliadin (on average, 61.54% of extractable proteins) as compared to durum wheat (above 55% of extractable proteins). Low concentrations of the ω-subunit were found in both bread (0.50–2.53% of extractable proteins) and durum

(3.65–6.99% of extractable proteins) wheat genotypes. On average, durum wheat contained significantly higher amounts of tryptophan and wet gluten (0.163% dry weight (dw) and 26.96% dw, respectively) than bread wheat (0.147% dw and 24.18% dw, respectively).

High polymorphism of gliadin alleles was described by Dukic et al. (2008) in 21 durum wheat cultivars, each exhibiting different gliadin allelic composition. Gliadin alleles at the *Gli-B1* locus showed the highest positive connection with gluten contents associating with good gluten quality and water absorption of flour. A great variation in high molecular weight (HMW) and low molecular weight (LMW) glutenin subunit was found in landraces compared to very low one in modern cultivars (Nazco et al. 2014a, b). The large variation found in landraces proved their potential value in breeding for gluten quality improvement.

The carotenoid concentration in grains of durum wheat is a criterion for the assessment of semolina quality, and it is of particular importance in determining the color of pasta (Beleggia et al. 2011). Among carotenoids controlling yellow color, the presence of β-carotene is also important as precursors of vitamin A. Blanco et al. (2011) detected the amount of individual carotenoid compounds (lutein, zeaxanthin, β-kryptoxanthin, α-carotene and β-carotene) in different durum lines and in segregating populations from inter-lines crosses. Total yellow pigment concentration among the durum genotypes was variable in all environments, but the genotype x environment interaction was not significant. Carotenoid concentration amounted to 37% of the yellow pigments, indicating the existence of unknown color-producing compounds in the durum extracts. Lutein was the most abundant carotenoid, followed by zeaxanthin, α-carotene and β-carotene, while β-Kryptoxanthin was a minor component. The phytoene synthase marker *Psy-A1*, 150 SSR and EST-SSR markers, and 345 DArT® markers, were used in search for QTLs affecting the various carotenoid compounds. Clusters of QTL for total and/or one or more carotenoid compounds were detected on chromosomes 2A, 3B, 5A and 7A, where major QTLs for yellow pigment concentration and yellow index were identified. The molecular markers associated with major QTLs would be useful in marker-assisted selection programs, to enhance the concentrations of carotenoid compounds with high nutritional value in wheat grain.

Grain characteristics that affect the technological quality of durum wheat, namely, semolina yield and its ability to be processed into pasta, were defined by Porceddu (1995). Semolina yield is influenced by a fusion of grade, intrinsic properties and ash content. Pasta quality can be considered either from the visual or cooking point of view. The visual aspect considers pasta color, which is a combination of yellowness and brownness, the former determined by the carotenoid content and lipoxygenase activity, while the latter

is attributed to peroxidase and polyphenoxydase. Cooking quality is associated with gluten properties. Protein content may account for 30–40% of the variability in cooking quality, but the protein ratio, i.e., the of gliadin and glutenin components, has a strong effect on quality.

The very hard texture of durum grains is due to the loss of the *puroindoline* genes that were eliminated during the allopolyploid formation of ssp. *dicoccoides,* 0.7–0.9 million years ago (Marcussen et al. 2014). Morris et al. (2011) described transfer of the *puroindoline* genes from chromosome 5D of *T. aestivum* cv. Chinese Spring to ssp. *durum* cv. Langdon, using a Langdon 5D(5B) disomic substitution line, which allowed homoeologous pairing due to the absence of chromosome 5B that carries the *Ph1* gene. *Puroindoline a* and *puroindoline b* were successfully transferred to durum cv. Svevo, which segregated to soft: heterozygous: very hard in a 1:2:1 ratio. The final backcross (BC₃) Svevo line produced uniformly soft grains. The transfer of this fundamental grain property to durum wheat will undoubtedly have an expansive and profound effect on the way that durum grain is milled as well as on the products derived from it (Royo and Abio 2003; Morris et al. 2011).

Importance of Durum Germplasm

Ssp. *durum* usually has higher yields than other wheats in areas of low precipitation (250–350 mm per annum). In districts too dry to support the cultivation of bread wheat, ssp. *durum* will yield 2000–4000 kg per hectare. Its growth is most prosperous, however, in areas with higher annual rainfall. ssp. *durum* does not tiller much but grows rapidly and succeeds best as a spring crop. In regions with mild winters, it may be sown in autumn, whereas in regions with freezing winters, it is sown in the spring.

Lines of ssp. *durum* contain genes conferring resistance to biotic stresses and tolerance to abiotic stresses. Several lines show good resistance to powdery mildew (Marone et al. 2013), stem rust (Klindworth et al. 2007; Aghaee-Sarbarzeh et al. 2013; Lemma et al. 2014; Nirmala et al. 2017; Miedaner et al. 2019), leaf rust (Maccaferri et al. 2008; Marone et al. 2009), stripe (yellow) rust (Yahyaoui et al. 2000; Lin et al. 2018), *Septoria tritici* (van Ginkel and Scharen 1988; Ferjaoui et al. 2015; Berraies et al. 2014; Kidane et al. 2017), *Septoria nodorum* (Nelson and Gates 1982; Cao et al. 2001), Hessia fly (Amri et al. 1990; El Bouhssini et al. 1999; Ratcliffe et al. 2002; Bassi et al. 2019), and soil-borne cereal mosaic virus (Vallega and Rubies-Autonell 1985; Maccaferri et al. 2012). Lines of durum wheat show tolerance to abiotic stresses such as drought (Saleeem 2003; Golabadi et al. 2006; Diab et al. 2008; Nouri et al. 2011; Kacem et al. 2017), salt (Royo and Abió 2003; Borrelli et al. 2011; Kim et al. 2016; Blumenthal et al. 1995; Sall et al. 2018), and cold (Szűcs et al. 2003; Longin et al. 2013).

Yield loss from sawfly (*Cephus cinctus* Norton) can be prevented by growing solid-stemmed durum wheat. When crossing solid-stemmed with hollow-stemmed durum lines, Clarke et al. (2002a) found that the F_1 were solid-stemmed, and the F_2 had three solid-stem to 1 hollow stem plants, showing that the expression of stem solidness fit the expected segregation ratios for a single dominant gene.

Clarke et al. (2002b) assessed the effect of low grain cadmium concentrations on uptake of other elements and on economic traits, such as yield. Grain yield, test weight, kernel weight and protein concentration were determined in five pairs of near-isogenic high/low cadmium durum wheat lines and their parents that were grown in a randomized complete block trial, with three replications. Average grain cadmium concentration differed across years within a given locations, across locations within a given year, and among genotypes. The average grain cadmium concentration of the high-cadmium isolines was approximately double that of the low-cadmium isolines. Grain concentrations of the other tested elements were associated with significant genotypic differences, but the differences were not associated with the high or low cadmium traits. The low-cadmium allele seemed to be specific for cadmium, lowering cadmium without altering concentrations of other elements or affecting economic traits. The low-cadmium trait had no significant effect on average yield, grain protein concentration, test weight, or kernel weight, as indicated by comparison of the high- and low-cadmium isolines.

Triticum pyramidale Percival

T. pyramidale Percival (Egyptian cone wheat) was considered by Miller (1987) as a special form of *T. turgidum senso stricto*, but by Mac Key (1966, 1975), as a dense-eared form of ssp. *durum*. It is a small race, found only in Egypt and Ethiopia, and among the earliest wheats (Percival 1921), with features including short straw, characteristic yellow-green culm leaves, and pointed grains. Egyptian Cone Wheat is a small and distinct race confined to Egypt and Ethiopia. Percival (1921) regarded it as an endemic dense-spiked mutation derived from the Ethiopian form of ssp. *dicoccon*. It is similar to the latter in the pubescence of its young leaves, yellow-green culm-leaves, short culms, very early habit, and the shape of its grain, and only differs from it in having short, dense spikes, with a tough rachis and free-threshing grain.

10.3.2.7 Ssp. *turgidum* (Rivet, Cone, or Pollard Wheat)

T. turgidum ssp. *turgidum*, known as rivet, cone, pollard, or branched wheat (Syn.: *T. turgidum* L.; *T. sativum* Hackel.), is a predominantly autogamous plant. It is the tallest of all wheats, with a culm reaching an average height of about 150 cm (120–180 cm) or more. It is slender but strong, having six or seven internodes above ground, the upper internode being curved and, in many forms, solid or filled with pith, while, when hollow, the culm wall is thick. The young shoot is prostrate or semi-erect and the culms are semi-erect or erect. The culm leaves are long and broad, with both sides covered with short, soft, white hairs. The auricles are usually fringed, with a few long hairs. The spikes are simple or branched at base, compact or relatively lax, 7.0–11.5-cm-long (excluding awns), bearing 19–33 spikelets, glabrous or pubescent, large, heavy, bending over at maturity, two-rowed, almost always awned, and square or oblong in section depending upon the laxness of the spike and the number and size of the grains in each spikelet. The rachis is tough, non-brittle, smooth, but fringed along its edges, with white hairs and with white a tuft with 1–2-mm-long hairs at the base of the spikelets. The individual rachis internodes are 2.5–4.7-mm-long, flattened, narrow and wedge-shaped. The spikelets are 10–13-mm-long, with 4–7 florets, with the upper 2 or 3 being sterile. The glumes are broadly lanceolate or oblong-lanceolate, white, yellow, red, or dark-bluish, glabrous in some forms and pubescent in others. The glumes of the lateral spikelets are 8–11-mm-long, shorter or longer than the lemma, inflated unsymmetrical, and with 5–7 nerves running from the base to a point close to the apical tooth. The glumes of the terminal spikelet are ovate and more or less symmetrical. All glumes are keeled from apex to base, with an apical tooth 1–1.5-mm-long. In some forms, the apical tooth is short and blunt, and the lateral nerve in some forms terminates in a short secondary tooth. The lemmas are thin, fragile, pale, oval, inflated, without a keel and with 9–15 fine nerves, which converge at the tip into a terminal, firm, yellowish-white, red, or black, awn, 10–19-cm-long, and triangular in section. Several forms shed their awns when the grain is ripe. As a rule, only the two lower lemmas of the spikelets bear long awns, while those of the third and higher flowers are much shorter or altogether missing. The palea is membranous and shorter than the corresponding lemma. The caryopsis is free from lemma and palea, white, yellow, or red in color, large, broad, plump (6.7–8.37-mm-long, 3.26–4.43-mm-broad, and 3.23–4.1-mm thick), blunt or truncate at the apex, and with a high dorsal arch or hump behind the embryo. The embryo is small and the endosperm is generally opaque and starchy; in few forms it is flinty.

Ssp. *turgidum* contains many varieties, which mainly differ in glume and awn color, glume hairiness, awn presence or absence, and in spike branching. Genetic diversity of 313 accessions of ssp. *turgidum*, obtained from different countries in Europe, Asia, USA, and Australia, was assessed by analyzing morphological traits and high-molecular weight glutenin subunits (Carmona et al. 2010a, b). A high level of variability was observed; 20 allelic variants were observed, five in the *Glu-A1* loci, two of which were new, and 15 alleles in the *Glu-B1* loci, six of these being novel. Genetic diversity among accessions of different countries was considerable, whereas diversity observed within countries was relatively low. The data indicated a clear decrease of morphological variability, along with an asymmetric distribution of the alleles and seed storage protein patterns.

Up until 1950, ssp. *turgidum* was widely grown in China. However, later its cultivation areas have been gradually reduced and, currently, most *turgidum* landraces are only preserved in germplasm banks. C-banding analysis revealed more genetic variations in Chinese landraces of ssp. *turgidum* than in those of common wheat and in other tetraploid wheats (Dou and Chen 2003). Likewise, Li et al. (2006), using SSR markers, investigated the genetic diversity and genetic relationships among 48 accessions of *T. turgidum*, including 30 ssp. *turgidum*, 7 ssp. *durum*, 4 ssp. *carthlicum*, 3 ssp. *paleocolchicum*, 2 ssp. *turanicum*, and 2. ssp. *polonicum*. A total of 97 alleles were detected at 16 SSR loci. The genetic diversity among the ssp. *turgidum* accessions from Gansu, China, was higher than among those from Sichuan and Shanxi, China. Similarly, Wei et al. (2008), using EST-SSR markers and analysis of molecular variance (AMOVA), investigated 68 accessions of ssp. *turgidum* landraces, originating from four geographic areas in China. They noted that 92.5% of the total variations was attributed to genetic variations between accessions from the same area, whereas only 7.5% of the variations were among accessions from different areas.

Oliveira et al. (2012) used nuclear SSRs (nuSSRs), chloroplast SSRs (cpSSRs), insertion site-based polymorphisms (ISBPs) and functional markers in expressed genes to investigate genetic diversity and population structure in landraces of several subspecies of *T. turgidum* in the Mediterranean basin, including wild and domesticated emmer, ssp. *turgidum* and *durum*. Wild emmer was the most diverse of all subspecies (gene diversity of 0.833), followed by domesticated emmer (gene diversity of 0.708) and ssp. *turgidum* (gene diversity of 0.682), while ssp. *durum* from northwest African was the least diverse (gene diversity of 0.546).

Most naked tetraploid wheats are likely of relatively recent origin and deviate in only a few characteristics from ssp. *durum,* its closely-related subspecies (Mac Key 1966; Morris and Sears 1967). It differs from ssp. *durum* by its softer grain, its somewhat taller stature, and branched spikes, in several forms. Transitional forms between the two subspecies also exist. The time of origin of ssp. *turgidum* is unclear. Due to the difficulty of distinguishing grains of ssp. *turgidum* from those of ssp. *durum* and bread wheat, grains described as those of ssp. *turgidum* found in deposits

of Neolithic and Bronze sites, should be considered with caution.

One of the first accurate descriptions of ssp. *turgidum* was given by botanists in the first half of the sixteenth century, when this taxon was grown in small areas in southern European countries. Because of its relatively high yield when grown under suitable conditions (Percival 1921), its popularity increased and during the 16th, 17th, and eighteenth centuries, it was also cultivated in England and Central Europe (Germany and northern France). During this period, ssp. *turgidum* was designated Rivet, Cone, or Pollard wheats. However, in the second half of the nineteenth century, its cultivation area was greatly reduced. Currently, ssp. *turgidum* is mainly grown in the countries bordering the Mediterranean, namely, Spain, Portugal, and Italy. Several, forms are grown, to a lesser degree, in Bulgaria, Greece, Turkey, northern Iraq, southern Iran, western Georgia, Azerbaijan, and in smaller amounts in Central Asia, and possibly also in Ethiopia and India. It is grown on small scales in Algeria, South Africa, Australia, China, Canada, USA, Chili, and Argentina.

The productive power of most varieties of ssp. *turgidum* is greater than that of any other subspecies of tetraploid wheat when growing conditions are suitable and climate allows a long growing period for the crop. Their high grain-yielding capacity is correlated with a long vegetative period, leading to an abundance of green-assimilating tissue. Moreover, the number of spikelets on each spike is usually greater among varieties of ssp. *turgidum* than among the varieties of bread wheat. Their tall straw is strong, the crop rarely lodges, and the stiff awned ears are not readily damaged by birds, deer, and other herbivores. ssp. *turgidum* contains some varieties that can be cultivated as winter wheat, but the majority of its varieties are cultivated as spring wheats. It grows well on light (sandy), medium (loamy) and heavy (clay) soils and prefers well-drained soil. ssp. *turgidum* also has the capacity to grow in poor soils, shows a strong weed competitiveness and a good resistance to diseases. Yet, compared to modern cultivars of *durum* and bread wheat, the varieties of ssp. *turgidum* occupy the fields during a relatively long period, have a late heading and a modest yield (from 2 to 2.5 tons per hectare).

Ssp. *turgidum* is less suitable for bread making because of its poor physical quality (elasticity and extensibility) and reduced gluten content. Loaves of bread made from ssp. *turgidum* flour are more or less dense and non-porous, and of small volume when compared with loaves made from the same weight of dough prepared from the "strong" flour of certain varieties of *T. aestivum*. Most ssp. *turgidum* varieties have a soft, opaque, starchy endosperm, which yields "weak" flour, suited to the requirements of the biscuit, porridge, and pasta industries.

Percival (1921) assumed that ssp. *turgidum* has the characteristics of a hybrid race. According to him, the long growing period, number of culm leaves, velvety leaf surface, plant height, surface, and pithy interior of its straw, spike density, as well as the frequent occurrence of branched spikes, indicate a close relationship between ssp. *turgidum* and the tall European forms of ssp. *dicoccon*. Several forms suggest affinity with ssp. *durum,* while others appear to be allied with *T. aestivum*. Its affinity with the European emmer can be traced by the morphological characteristics and habit. The two taxa agree in the characteristic pubescence of their young leaves. Both have tall, solid, or nearly solid, culms, and spikes with spikelets very regularly arranged along the rachis. They tiller very little and have a similar late-ripening period. Moreover, the tendency to produce branched spikes is strongly evident in these two taxa and rare in others. The square spike of ssp. *turgidum*, its many-flowered spikelets, and plump, blunt-ended grains are characteristics derived from the dense-spiked ssp. *compactum* or ssp. *aestivum* parent, while the dorsal hump of the grain is derived from the emmer parent.

Despite this, Oliveira et al. (2012), using several types of molecular markers, found that while wild and domesticated emmer is genetically distinct from ssp. *durum* and *turgidum*, the latter two share a common gene pool and are almost genetically indistinguishable. Differences in key genes between *turgidum* and *durum* have not been identified or quantified and it is debatable if the differences in phenotype are sufficient to classify them as different taxa. Consequently, Oliveira et al. (2012) classified *turgidum* and *durum* as varieties of the same subspecies, each with distinct morphological characteristics. Likewise, Morris and Sears (1967) assumed that the turgidum group differs little from the durum group, and that there exist transitional forms between the two types, as reported by Watkins (1940). However, following van Slageren (1994), who thought that the morphological differences between the two justify their separation, we prefer to maintain ssp. *turgidum* and ssp. *durum* as separate subspecies of *T. turgidum*.

Considering the strong genetic similarity between *turgidum* and *durum*, demonstrated in all the marker systems used, Oliveira et al. (2012) suggested that the two subspecies originated from a common domesticated ancestor. Their distinct adaptation to specific conditions as they were introduced into Europe could have yielded landrace varieties with distinct morphological characteristics, such as the distinct head form in *turgidum* or its higher tolerance to cold and humidity in comparison with ssp. *durum*. Yet, Oliveira et al. (2012) thought that these selective pressures were apparently not sufficiently strong to create a distinct genetic pool between the two. They proposed, based on the considerable difference between the hulled emmer and the naked *turgidum* and *durum,* that naked wheats evolved from a small number of hulled tetraploid genotypes, and have been in relative reproductive isolation since their spread into Europe and North Africa during the Neolithic Age. Nevertheless, the data of Oliveira et al. (2012) do not negate the possibility that ssp. *turgidum* developed from ssp. *durum* under different cultivation conditions.

10.3.2.8   Ssp. *polonicum* (L) Thell. (Polish Wheat)

T. turgidum ssp. *polonicum* (L) Thell., known as Polish wheat, (Syn.: *T. polonicum* L.; *T. levissimum* Haller; *T. maximum* Vill.; *T. glaucum* Moench: *T, turgidum* grex *polonicum* Bowden; *T. turgidum* ssp. *turgidum* convar. *polonicum* Mk; *Gigachilon polonicum* Seidl.; *Deina polonica* Alef.) is a subspecies that was discovered by the scientific world last of all the subspecies of *T. turgidum*. Evidence of its existence was not obtained before the first half of the seventeenth century (Percival 1921). Its name, *T. polonicum*, was given to this taxon since it was first described after a specimen obtained from the Botanic Garden at Leyden, Poland, but the origin of the name and the early connection of the wheat with Poland is obscure. There is no evidence that it was grown in Poland before 1870 (Percival 1921). In fact, it would be more appropriate to call it Galician wheat. When giving the name *T. polonicum*, Linnaeus (1753) confused Galicia (a region of Spain) with Galicia (a region of Poland and Ukraine).

Ssp. *polonicum* is a predominantly autogamous plant. It is one of the tallest wheats, with culms usually reaching a height of 100–160 cm. The upper internode is generally solid, while the lower ones are hollow. The plants tiller very little (three or four tillers per plant), and the shoots are erect. The culm leaves are glabrous or pubescent, bluish-green in color, and with smooth surfaces, like those of ssp. *durum*. The upper culm leaves are 2-cm-broad, with small auricles. The spikes are very long, 10–16-cm-long (excluding awns), narrow, very lax, almost square in cross-section, with a tough, non-brittle rachis, two-rowed, and awned. It bears 19–23 spikelets, with three or four of the lower ones being rudimentary. In some forms, the spike is compact, short, 7–9-cm-long, and oblong in cross-section. The rachis is flat, narrow and wedge-shaped, each about 1.5-mm-wide at the base and 3-mm-wide at the top, fringed with hairs on the sides, and with a frontal tuft of hairs 2–2.5-mm-long at the base of each spikelet. The spikelets are large, flat, 3–4-cm-long, and consisting of four or five flowers, of which two (or three) produce grain, while the others are sterile. The glumes are delicate, with a glaucous color, glabrous or pubescent, long, lanceolate and narrow, 2–4-cm-long, extending beyond the rest of the spikelet, keeled from the apex to the base, and with a short apical tooth; the second tooth is very short or absent. The lemmas of the two lowest flowers are 2–3-cm-long, the upper ones being much shorter, boat-shaped, rounded on the back and with 15–17 nerves.

Their membranous edges are fringed with short hairs, bear a coarse awn, which is 7–12-cm-long, and white or black. The awns of the upper lemmas are either very short, not more than about 5-mm-long, or absent. The palea are only 1.2–1.5-cm-long, lanceolate, with a slightly divided tip, and possess 4 nerves. The grains, are free-threshing, long and narrow, 11–12-mm-long and 4-mm-broad, i.e., the largest of all wheats, narrow, and yellowish-white or pale red, with a flinty endosperm (Fig. 10.3c).

The typical form of ssp. *polonicum* is strikingly different from all other subspecies of *T. turgidum*, in that it possesses large ears with long, narrow, papery, somewhat loose glumes of a glaucous color, which extend beyond the rest of the spikelet. A large number of ssp. *polonicum* varieties have been described. The majority are not currently cultivated and do not appear to have been grown, except for in Botanic Gardens. Percival (1921) recognized only three varieties that are currently found in cultivated fields, namely, *levissimum* Körn. (white glumes, glabrous, long awn, quadrate ear, white grain), *villosum* Körn. (white glumes, pubescent, long awns, quadrate ear), and *Martinari* Körn. (white glumes, pubescent, long awns, flattened ears, white grain).

According to Percival (1921) ssp. *polonicum* is a mutation of ssp. *durum*. He considered the only point of difference, namely the excessively long, thin glumes, as a hereditary teratological variation. Percival reported on several specimens of ssp. *durum* from India with elongated glumes, suggestive of incipient *polonicum*. Körnicke (in Percival 1921) claimed that he obtained from Upper Egypt a specimen that looked like a transition type between *durum* and *polonicum*.

Ssp. *polonicum* is a rather minor, primitive, *durum*-like, spring wheat subspecies. It is sporadically Kwiatek grown, not prolific, and of marginal importance in the contemporary grain market (Kwiatek et al. 2016). ssp. *polonicum* requires hot climate and well-drained soils for satisfactory growth. It is cultivated on small scales in several countries bordering the Mediterranean (Algeria, Spain, Italy, and Turkey), in eastern and southern Europe (European Russia, and Ukraine), in Transcaucasia (Armenia, Azerbaijan, Georgia), in central Asia (Kazakhstan, Kyrgyzstan, and Turkmenistan) and in Ethiopia. It is also grown to a limited degree in the USA and Argentina. In several countries, it is grown principally for plant breeding, while in others, it is grown for its edible seed, and in the Mediterranean, also for cereals, and macaroni but not for bread. The straw is used for fuel, thatching, and as a mulch in the garden. Wiwart et al. (2013) compared morphometric parameters such as plant height, spike length, spike density, grain weight per spike and single kernel weight, and some chemical properties of the grain, i.e., protein, ash, fat, crude fiber, minerals and mycotoxins content, in nine lines of ssp. *polonicum* with Kamut wheat (a variety of *T. turgidum* ssp. *turnicum*) and two common wheat cultivars. The average height of the *polonicum* lines was 109.8 cm, considerably greater than that of common wheat (88.8 cm), and Kamut wheat (89.3 cm). The average length and density of *polonicum* spikes did not differ significantly from those of bread wheat, yet the differences between lines were considerable. The shortest spikes were observed in Kamut (5.98 cm), and the most-dense spikes were reported in the *polonicum* P-4 line (22.7 spikelets per 10 cm of rachis). The highest grain weight per spike was reported in one cultivar of bread wheat (2.17 g), in Kamut (1.90 g), and in the *polonicum* lines (1.35–1.78 g) and exceeded those of the high-yielding second cultivar of bread wheat (1.31 g). Like Oliver et al. (2008), Wiwart et al. (2013) also found that lines of ssp. *polonicum* possess moderate to high levels of resistance against *Fusarium* head blight.

Concerning the chemical properties of the grain, Wiwart et al. (2013) reported that the grains of ssp. *polonicum* were characterized by a significantly higher average protein (16.61%) and ash (2.14%) content than the grains of bread wheat (13.87% and 1.73%, respectively). The common wheat cultivars had considerably higher concentrations of fat and dietary fiber than *polonicum* and Kamut. In comparison with bread wheat, the grains of the examined *polonicum* lines had significantly more phosphorous, sulfur, magnesium, copper, calcium, zinc, iron and molybdenum and contained significantly less aluminum and strontium. Similarly, Bieńkowska et al. (2019) found that ssp. *polonicum* grain is characterized by a significantly high content of phosphorus (4.55 g/kg), sulfur (1.82 g/kg), magnesium (1.42 g/kg), zinc (49.5 mg/kg), iron (39.1 mg/kg) and boron (0.56 mg/kg). Hence, the nutrient profile of most ssp. polonicum lines differs completely from that of bread and durum wheat. More specifically, seven lines of ssp. *polonicum* had the highest content of copper, iron and zinc, and the lowest concentrations of strontium, aluminum and barium, which are undesirable in food products, providing a particularly beneficial micronutrient profile (Bieńkowska et al. 2019). These data show that ssp. *polonicum* may constitute an important genetic resource for improving the nutritive value and resistance to *Fusarium* head blight of durum and bread wheat (Wiwart et al. 2013; Bieńkowska et al. 2019).

Biffen (1905) was the first to study the genetic control of glume length by making reciprocal crosses between ssp. *polonicum* (long glumes) and ssp. *turgidum* (short glumes). In F_2, he observed a long:intermediate:short glumes ratio of 1:2:1, showing that only one gene determines this trait and that the long-glume allele is not dominant over the short glume allele. These observations were corroborated by data collected by Engledow (1920), crossing ssp. *polonicum* x ssp. *durum* var. *Kubanka,* and also showed that glume length bears some definite relationship with grain length, and thus, is of economic significance.

In accord with the above, Matsumura (1950) observed in F_2 of crosses involving ssp. *polonicum, dicoccoides, and carthlicum* (formerly *T. persicum*), that the long glume trait is controlled by one allele (*P*), which is dominant on the short glume allele (*p*). He reported that the *P locus* is linked to the red coleoptile locus (*Rc*), with a cross-over value of 20.3%. *Rc* genes are known to be located on the homoeologous group 7 chromosomes (McIntosh et al. 1998). In agreement with these results, Watanabe et al. (1996), Watanabe (1999), used telosomic mapping to locate *P*, designated by them as P_1, on the long arm of chromosome 7A, 9.8 cM from the centromere. Moreover, Watanabe et al. (1996) noted a link between the P_1 gene and the chlorina gene *CDd6* (*cn-A1d*; Klindworth et al. 1995, 1997), further suggesting that P_1 is located on chromosome 7A. Furthermore, crosses between a near isogenic line of ssp. *durum* cv. LD222 carrying P_1, with the durum cultivar Langdon (LDN) and the LDN D genome substitution lines, LDN 7D (7A) and LDN 7D(7B), corroborated the location of P_1 on 7A. Segregation for the long glume trait in the F_2 of LDN/P-LD222 and LDN 7D(7B)/P-LD222 was normal (3:1) and indicated that the P_1 gene was not on chromosome 7B. The location of P_1 on 7AL was confirmed by Koval (1999), Efremova et al. (2001).

The Chinese wheat landrace, Xinjiang rice wheat (*T. aestivum* ssp. *aestivum* var. *petropavlovskyi* (=*T. petropavlovskyi* Udacz. et Migusch.), was found in 1948, in the agricultural areas of the west part of Talimu basin, Xinjiang, China. This taxon is characterized by long glumes. Watanabe and Imamura (2002) introduced the gene for long glume from var. *petropavlovskyi* into the LD222 cultivar of ssp. *durum* and found that it is controlled by a gene located on chromosome 7A. The gene was located approximately 12.45 cM from the centromere on the long arm of 7A. Consequently, Watanabe and Imamura (2002) considered the gene for long glume from var. *petropavlovskyi* an allele of the P_1 locus and designated it P_1a. It was suggested that var. *petropavlovskyi* originated from either a natural hybrid between *T. aestivum* and ssp. *polonicum* or a natural point mutation of *T. aestivum* (Chen et al. 1985; Watanabe and Imamura 2002).

A second gene for elongated glumes (P_2) has been identified in var. *ispahanicum* (Watanabe 1999). The Ispahan emmer wheat, ssp. *dicoccon* var. *ispahanicum*, (= *T. ispahanicum* Heslot), is a hulled taxon that was discovered in Ispahan, Iran, in 1957, by the French expedition of Vinnot-Bourgen, and was considered by Miller (1987) *to be a variety of* ssp. *dicoccon. This wheat* has a long glume and a more-slender spike than ssp. *polonicum*. Watanabe (1999), using a near-isogenic line that carries the P_2 gene of var. *ispahanicum,* located this gene on chromosome arm 7BL, approximately 36.5 cM from the *cn-B1* locus, which controls the chlorina trait (Klindworth et al. 1995, 1997) and

approximately 40 cM from the centromere. The location of P_2 approximately 29.6 cM from the purple color gene (*Pc*) (McIntosh et al. 1998), provided additional evidence that the order of loci was *cn-B1*, and that var. *ispahanicum* originated following a mutation of a gene affecting glume length on chromosome 7B of ssp. *dicoccon*.

The most pronounced elongated glumes are present in ssp. *polonicum*, reaching a length of up to 40 mm, compared to 5–8 mm in normal wheats (Dorofeev et al. 1980). Less pronounced glume elongation is present in the tetraploid subspecies of *T. turgidum, turanicum* and *durum* convar. *falcatum* and in the hexaploid taxon *T. aestivum* ssp. *aestivum var, petropavlovskyi*. Wang et al. (2002) mapped the respective genes that determine glume length in these taxa, using wheat microsatellite markers. In ssp. *polonicum* and hexaploid *var. petropavlovskyi,* loci conferring long glume were mapped near the centromere on the long arm of chromosome 7A. These two loci were designated by Wang et al. (2002), $P\text{-}A^{pol}1$ (currently P_1) and $P\text{-}A^{pet}1$ (currently P_1a), respectively. It was shown that both are probably homoeoallelic to each other and to the P_2 gene of var. *ispahanicum* on the long arm of chromosome 7B. The loci determining elongated glumes in ssp. *turanicum* and ssp. *durum* conv. *falcatum* are not homoeologous to the *P* loci in the centromeric region of group 7 chromosomes (Wang et al. 2002).

Pan et al. (2007), using SDS-PAGE, assessed genetic diversity in ssp. *polonicum* by analyzing gliadins and high molecular weight glutenin subunits in 72 accessions from 23 countries. High genetic variability was observed in both types of storage proteins. Diversity indices (H) at *Glu-B1* loci (0.659) were much higher than in *Glu-A1* loci (0.271). Variation in these proteins was associated with their geographic origins.

Watanabe (1994) studying near-isogenic lines of ssp. *durum* cv. LD222 carrying the P_1 gene, derived from crosses involving ssp. *polonicum*, reported that the long glume trait, controlled by the P_1 gene, which resulted in a large photosynthetic area, tended to increase the main culm dominance, and plant height, resulting in declined grain yield and harvest index.

Biffen (1905), Engledow (1920), crossing ssp. *polonicum* x ssp. *turgidum* and *durum*, showed that glume length has some definite relation to grain length, and thus, appeared to be of economic significance. In this respect, Millet and Pinthus (1984) assumed that long glumes may allow grain growth by preventing the effect of penetrating light to the flower cavity. They found that removal of organs that exposed the developing wheat grain to increased light intensity, resulted in a reduction in grain size. The restriction in grain growth was already apparent two weeks after anthesis. Covering the treated spikes with opaque bags restored grain growth.

Like all other subspecies of *T. turgidum*, ssp. p*olonicum* is also a tetraploid taxon, with 2n = 4x = 28 chromosomes and genome BBAA. Nakajima (1955) observed 0–3 bivalents and 15–21 univalents in meiotic first metaphase of F$_1$ hybrids from the cross ssp. p*olonicum* x *Secale africanum*. He concluded that the bivalents resulted from autosyndesis between the chromosomes of the AB subgenomes of ssp. *polonicum* and not from pairing between the chromosomes of the two taxa.

10.3.2.9 Ssp. *turanicum* (Jakubz.) Löve and Löve (Khorassan Wheat)

T. turgidum ssp. *turanicum* (Jakubz.) Löve & Löve, known as Khorassan wheat or Oriental wheat [Syn.: *T. turanicum* Jakubz.; *T. orientale* Percival; *T. percivalii* Hubb. ex Schiem.; *T. percivalanum* Parodi; *T. turgidum* grex *turanicum* (Jakubz.) Bowden; *T. turgidum* ssp. *turgidum* conva. *turanicum* (Jakubz.) MK; *T. georgicum* convar. *turanicum* (Jakubz.) Mandy; *T. durum* ssp. *turanicum* L. B. Cai; *Gigachilon polonicum* ssp. *turanicum* (Jakubz,) Á. Löve] is a predominantly autogamous, 66–110-cm-tall plant, with an upper internode that is either solid or hollow with thick walls. The young shoots are erect, with very narrow pubescent leaves. The plants tiller very little and the straw is thin. The spikes are very long, narrow, very lax, almost square in cross-section, with a tough non-brittle rachis, 10–11.5-cm-long (excluding awns), two-rowed, and awned. The rachis sides are fringed with white hairs, and there is a conspicuous frontal tuft below each spikelet. The rachis internodes are narrow and wedge-shaped, each about 1.5-mm-wide at the base and 3-mm-wide at the top. The spikelets (15–20 per spike) are 15–17-mm-long, and with 2–3 grains. The glumes are delicate, white, pubescent, large, 12–15 mm-long, keeled from the apex to the base, with a short apical tooth, and prominent lateral nerve. The lemma bears a coarse awn, which scabrid to the base, 14–16-cm-long, and white or black. The grains are free-threshing, very long (10.5–12 mm), narrow, twice the size of modern wheat kernels, with a thousand-kernel weight up to 60 g, white to amber in color and flinty, with a short brush, resembling those of ssp. *polonicum.* This wheat grown in spring, is early in maturity and resistant to several fungal diseases.

ssp. *turanicum,* Khorasan wheat (Khorasan refers to a region in northeast of Iran, Afghanistan and central Asia, where this subspecies was cultivated) contains two varieties, differing only in the color of their awns (white or black). It is a small, ancient race of the free-threshing subspecies of *T. turgidum.* Khorasan wheat was probably continuously cultivated on small scales and for personal use, in Near East, Central Asia, and northern Africa (Vavilov 1951). In these areas, it was grown mostly as an admixture in durum wheat fields, seldom in pure sowing. Currently, Khorasan wheat is grown on a small scale in Europe and the Middle East, mainly for special bread, and in central Asia, mainly as food for livestock (camels). Approximately 6500 ha of Khorasan wheat, mostly cultivar Kamut, were cultivated worldwide in 2006, mainly in north-central Montana, USA, and southern Saskatchewan and southeast Alberta, Canada (Brester et al. 2009).

The origin of the cultivar Kamut is accompanied by a mystical story, namely, a USA airman claimed to have taken a handful of grains from a stone box in a tomb near Dashare, Egypt (Quinn 1999). Wheat grains usually lose their germination ability few years after harvest, and therefore, it is unfeasible that seeds several thousands of years old will germinate. It is more likely that someone put fresh seeds in the tomb or that making of such a story furnishes some mysterious sense to the origin of Kamut. In any case, Kamut (an ancient Egyptian word for wheat) was developed in Montana, USA, and appeared on the market about 35 years ago. In 1990, the US Department of Agriculture recognized the grain as a protected cultivar, which was given the official name 'QK-77'. Kamut yield is less than that of standard cultivars of spring wheats (Stallknecht et al. 1996).

Cultivar Kamut contains some nutritional, health, and taste advantages over modern wheat varieties, but lacks some of their agronomic advantages (Brester et al. 2009). The average yield of cv. Kamut is very low (1.1–1.3 t/ha), approximately 1/3 that of durum wheat. Its grain is twice the size of that of modern-day wheat and is known for its rich nutty flavor. Grains of Khorasan wheat contain a high protein content (15%), several B vitamins and minerals, such as iron, zinc, magnesium, manganese, phosphorus and potassium, and soluble and insoluble fibers (Brester et al. 2009; Abdel-Haleen et al. 2012). Kamut grains can be either directly consumed or milled into flour, which can be found in breads, bread mixes, breakfast cereals, cookies, waffles, pancakes, bulgur, baked goods, pastas, drinks, beer and snacks (Brester et al. 2009). Sofi et al. (2013) examined the effect of a Kamut diet on cardiovascular risk parameters by conducting a randomized, single-blinded cross-over trial with two intervention phases on 22 healthy subjects (14 females; 8 males). Their results suggested that a diet with Kamut products could be effective in reducing metabolic risk factors, markers of both oxidative stress and inflammatory status.

ssp. *turanicum* is a tetraploid taxon, a free-threshing subspecies of *T. turgidum.* It is considered to be an ancient taxon of this species. With respect to ear form and grain length, it closely resembles some of the varieties of ssp. *polonicum* (Percival 1921). But, although it is long-grained, ssp. *turanicum* lacks the long glumes of ssp. *polonicum.* Khlestkina et al. (2006), studying the taxonomic classification of Kamut and ssp. *turanicum,* presented genetic evidence from DNA fingerprinting that ssp. *turanicum* is perhaps a natural hybrid between ssp. *durum* and ssp. *polonicum*. They investigated the taxonomic placement of Kamut using micro-satellite genotyping. In total, 89 accessions of 13 tetraploid wheat species, including ssp. *turanicum* and Kamut, were genotyped. Kamut clustered together with three accessions of, ssp. *polonicum* and three of ssp. *durum*, indicating the genetic proximity of ssp. *turanicum* (including Kamut) to ssp. *durum* and ssp. *polonicum*.

10.3.2.10 Ssp. *carthlicum* (Nevski in Kom.) Á. Löve and D. Löve (Persian Wheat)

T. turgidum ssp. *carthlicum* (Nevski) Á. Löve and D. Löve, known as dika wheat, Persian wheat, Persian black wheat, [Syn.: *T. carthlicum* Nevski; *T. persicum* Vavilov ex Zhuk.; *T. dicoccum* var. *persicum* Perciv.; *T. persicum* (Boiss.) Aitch. & Hemsl.; *T. ibericum* Menabde; *T. paradoxum* Parodi; *T. carthlicum* grex *carthlicum* (Nevski) Bowden; *T. turgidum* ssp. *carthlicum* (Nevski) MK; *T. turgidum* convar. *carthlicum* (Nevski) Morris & Sears: *T. georgicum* convar. *carthlicum* (Nevski) Mandy: *Gigachilon polonicum* ssp. *carthlicum* (Nevski) Á. Löve] is an autogamous plant. Its young shoots are erect and young leaves pubescent, its straw is thin, of medium height (104 cm), solid or hollow with thick walls and with hairy nodes. Its rachis is narrow, spike lax, narrow, 9-cm-long, square in cross-section, and with 19–20 spikelets, each with 2–3 red grains. Glumes bear a single awn, 3–4 cm in length. The lemma is white at the bottom and dark brown at the top, with an 8–11-cm-long awn. Grains are flinty and reddish, the apex is blunt, 6.5-mm-long, 3-mm-broad, and 2.9-mm-thick. The plant flowers very late and grows in the spring (Fig. 10.3d).

Ssp. *carthlicum* was found in Transcaucasia and was first referred to as a separate species, *T. persicum* Vav, by Vavilov (1918), who believed it originated in Iran. This peculiar species morphologically resembles hexaploid wheat, *T. aestivum* ssp. *aestivum,* but is biologically and cytologically related to tetraploid wheats (Jakubziner 1959). Later, it was found that this tetraploid taxon was not exclusive to Iran, as it was discovered by Zhukovsky (1923) in Georgia, and by other in Armenia, Azerbaijan, Dagestan, and Turkey (Jakubziner 1959).

The epithet was renamed by Nevski as *Triticum carthlicum* Nevski (Nevski 1934), with original publication in Komarov VL (ed.), Fl. URSS 2: 688 (1934), replacing *T. persicum* Vavilov ex Zhuk., which is illegitimate because the name *T. persicum* (Boiss.) Aitch. & Hemsl. was awarded to another taxon now included in the genus *Aegilops*, i.e., *Ae. persicum* Boiss. (Now recognized as a variety of *Ae. triuncialis* namely, var. *persica* (Boiss.) Eig). More recent names given to this taxon such as *T. ibericum* Men. (1940) and *T. paradoxum* Parodi (1940), ought to be rejected (Jakubziner 1959).

ssp. *carthlicum* is an endemic domesticated wheat from the Caucasus-Trans-Caucasus region, grown in pure stands or, more commonly, in mixture with domesticated emmer, durum, or bread wheat, confined to mountainous regions in Georgia, Armenia, Azerbaijan, Dagestan, northern Ossetia, northwestern Iran, northern Iraq and northeastern Turkey (Dorofeev 1968; Hanelt 2001).

The study of Dekaprelevich (1925) showed that this subspecies, which was previously considered more or less ecologically homogenous, is actually divided into two different ecotypes: the mountainous-forest ecotype (forma *caucasionis* dika) and the mountainous-steppe ecotype (forma *dzhavachetica* dika). The former is distributed in elevations of 900–1400 m above sea level and is represented by a late-ripening, moisture-loving, black-spiked variety. The latter is adapted to a higher elevation of 1400–2100 m above sea level and is chiefly represented by early-ripening, and white- or red-spiked varieties. Dorofeev (1968) described two varieties of ssp. *carthlicum* in Armenia, vars. *Stramineum* and *rubiginosum.*

Variation in spike color of ssp. *carthlicum*, namely, black, red and white, was found in Georgia and Daghestan, but, in Armenia, only types with white or red ears occur (Dorofeev 1968). The various forms have a spring habit, and are tolerant to low temperatures (Dorofeev 1968).

A sample of 74 accessions of ssp. *carthlicum* were scored for 46 characteristics using numerical analysis (Vieira 1985). The study enabled the recognition of three morphological distinguishable groups in this subspecies. The data provide supportive evidence for the recognition of var. *rubiginosum* Zhuk. and var. *fuliginosum* Zhuk. as distinct taxa. However, recognition of var. *stramineum* Zhuk. based on spike color could not be achieved. Variation patterns in ssp. *carthlicum* suggested that geographical distance and taxonomic distance between populations are associated. SDS-PAGE analysis of the HMW glutenin subunits in the 74 *carthlicum* accessions provided identical HMW glutenin subunits profile, indicating a monophyletic origin of ssp. *carthlicum*. The lack of variability also suggests its recent origin. This interpretation seems to favor the hypothesis that ssp. *carthlicum* is a young form derived from a cross between a free-threshing and a tetraploid wheat.

Interspecific relationship studies involving the subspecies of *T. turgidum* were conducted using iso-enzymatic characters located in dry mature seeds (Asins and Carbonell 1986b). Studies of peroxidases in the embryo and alkaline phosphatase in the endosperm showed no variation in these two enzymes in ssp. *carthlicum* which, in turn, showed no new enzymatic pattern for that had not been observed in wild emmer, ssp. *dicoccoides*. Hence, from all the subspecies of *T. turgidum*, the closest relationship was found between ssp. *carthlicum* and ssp. *dicoccoides*, suggesting the origin of the former from the latter (Asins and Carbonell 1986b).

ssp. *carthlicum* is very morphologically distinct from the other free-threshing subspecies of *T. turgidum,* namely, *durum, turgidum, turanicum,* and *polonicum.* Watkins (1928, 1940) reported that while ssp. *carthlicum* is characterized by non-keeled round glumes, and possesses the recessive *k* allele, all the other tetraploid wheats had keeled glumes, and the dominant *K* allele. Following Watkins (1928, 1940), many wheat geneticists, e. g., Mac Key (1954a, 1966, 1975; Morris and Sears 1967) assumed that only in this tetraploid subspecies is the free-threshing trait determined by the *Q* factor, as in hexaploid wheats, while in other tetraploids, this trait is determined by another genetic system. Mac Key (1954a), using irradiation experiments, found that the free-threshing, non-keeled glumes and rachis toughness traits of ssp. *carthlicum* are, in fact, controlled by one factor, i.e., the same as the square-head gene *Q*, while all other free-threshing tetraploids possess the q allele (Mac Key 1954a, 1966, 1975). These traits in the other tetraploid wheats were assumed to be controlled by a different gene system than *Q*, i.e., a polygenic system (Mac Key 1966, 1975). Another striking feature of ssp. *carthlicum* is its awned glumes, with all the spikelets displaying four awns, one on each of the two glumes and one on each of the lemmas of the two lower flowers. As it looks very much like hexaploid *T. aestivum* ssp. *aestivum,* ssp. *carthlicum* was classified at first as a hexaploid species, *T. persicum* Vav. (Vavilov 1918). Only later, on account of its chromosome number (2n = 28), and the high sterility of its F_1 hybrid with *T. aestivum,* ssp. *carthlicum* was recognized to be a tetraploid taxon (Schiemann 1948). The very restricted distribution area of ssp. *carthlicum* to the mountainous regions of Caucasia and Trans-Caucasia and surroundings countries (Zhukovsky 1923; Vavilov and Jakushkina 1925) align with its recent origin, presumably by hybridization between an unknown tetraploid and a hexaploid of the *aestivum* group (Mac Key 1954a, 1966, 1975; Morris and Sears 1967). Alternatively, it may be an offspring of a mutation in an emmer–like ancestor (Hanelt 2001). Mac Key (1966) assumed that the *Q* gene was transferred to hexaploid wheat from *carthlicum*, with the latter being involved in the formation of hexaploids via hybridization with *Ae. tauschii*, the donor of the D subgenome. Alternatively, spelt-type hexaploid wheat was first formed and then crossed with *carthlicum*, producing the free-threshing hexaploids. On the other hand, if *Q* evolved on the hexaploid level, *carthlicum* must be considered a much younger type, originating from a cross of *Q*-hexaploid x q-emmer (Vavilov 1926).

Free-threshing tetraploid types evolved from the *dicoccoides-dicoccum* group but through two different genic systems (Mac Key 1966 1975). *Carthlicum* (cf. Jakubziner 1959) is the only free-threshing tetraploid endemic to Transcaucasia. It resembles the 6x ssp. *aestivum* wheat, with which it consistently grows with and with which it shares *Q*.

This gene, located on chromosome 5A (Sears 1954, 1959), appears to be lethal at the diploid level (Mac Key 1966). If *Q* arose already at the tetraploid level, *carthlicum* may have been the key for the development of the naked hexaploid wheats. However, it seems more likely that *carthlicum* arose through 6x *aestivum* x 4x *dicoccon* than formation of 6x *aestivum* through 4x c*arthlicum* x 2x *Ae. tauschii* or (4x d*icoccum* x *Ae. tauschii*) x *carthlicum*. All three possibilities have been proven to yield 6x *T. aestivum* (Mac Key 1966).

Kuckuck (1979) found in the border region of Iran, Turkey and Transcaucasia, accessions of free-threshing hexaploid wheat exhibiting the subsp. *carthlicum*-like morphology, i.e., four awns on each spikelet; however, the glume awns of ssp. *carthlicum* are longer than those of the hexaploid taxon. Since chromosome number was 2n = 42, he named these accessions *T. aestivum* subsp. *carthlicoides.* This taxon is not mentioned among the subspecies of *T. aestivum* in recent taxonomical treatments of the genus *Triticum*, e, g, by Bowden (1959), Löve (1984), Dorofeev et al. (1980), Mac key (1988, 2005), van Slageren 1994. Hence, it will be referred to from here on as var. *carthlicoides.* Kuckuck (1979) proposed that ssp. *carthlicum* originated from a spontaneous hybridization between var. *carthlicoides* and an accession of domesticated emmer. The 6x var. *carthlicoides* should be considered as the original and older genotype from which genes for the particular morphology of the spike were transferred together with the Q-factor, to ssp. *carthlicum.* The prolongation of the glume awns in ssp. *carthlicum*, relative to that of var. *carthlicoides*, might be due to the lack of the D subgenome (Kuckuck 1979).

Kerber and Bendelow (1977), based on similarity in several milling and baking properties, and Bushuk and Kerber (1978), based on similarity in the gliadin electrophoretic mobility, concluded that ssp. *carthlicum* cannot be rejected as a possible source of the BBAA component of bread wheat, nor did the evidence exclude the hypothesis that this tetraploid is merely a segregate from a hexaploid wheat x tetraploid wheat hybrid.

Haque et al. (2011) found that the tetra-aristatus recessive allele, *t,* controlling the glume awns in ssp. *carthlicum*, is located on chromosome arm 5AL, and is among the *b1* genes that determine awn development on the lemma. This conclusion was reached because the semi-dwarf *Rht12* gene is linked to the *b1* gene and the *t* locus is approximately 11 cM units from *Rht12*. Gandilyan (1972) pointed out the significance of the *t* gene for wheat domestication along with the *Q* gene, because the four-awned-hexaploid accessions were more easily threshed than other genotypes. Thus, a supposed spontaneous mutation at the *T* locus in *T. aestivum* (*TTQQ*) led to the formation of var. *carthlicoides* (*ttQQ*), which then hybridized with the hulled tetraploid ssp. *dicoccon* (*TTqq*); subsequent recombination and fixation produced ssp. *carthlicum* (*ttQQ*). The phenotype "tetra-aristatus" has not been found in tetraploid wheat species other than ssp. *carthlicum*.

Takumi and Morimoto (2015) supported Kuckuck's (1979) idea that ssp. *carthlicum* evolved through inter-ploidy hybridization between a tetraploid form and hexaploid var. *carthlicoides*. They provided evidence for the origin of ssp. *carthlicum* based on the discovery of a new allele for the 5th-to-6th exon region of the *Wknox1b KNOTTED1*-type homeobox gene in the common wheat var. *carthlicoides*. In this *Wknox1b* region, var. *carthlicoides* contains an inverted duplication mutation in the 3' flanking region of a 157-bp MITE insertion site. This structural mutation resulted in the suppression of *Wknox1b* expression in var. *carthlicoides*, but no structural mutation was observed in the same region of ssp. *carthlicum*. In addition, the ssp. *carthlicum Wknox1b* 5th-to-6th exon region exhibited the same sequence as that of the wild emmer wheat subsp. *dicoccoides*. These observations support the suggestion that ssp. *carthlicum* originated from inter-ploidy hybridization between wild emmer and var. *carthlicoides*.

However, Muramatsu (1978, 1979, 1985, 1986) found out that all the free-threshing subspecies of *T. turgidum* possess the *Q* factor. By substituting chromosome 5A of the tetraploids for 5A of hexaploid ssp. *aestivum* cv. Chinese Spring (that carries the *Q* factor), he found out that not only does ssp. *carthlicum* have the *Q* factor, but also all other free-threshing tetraploids. Even ssp. *dicoccon* var. *liguliforme*, a hulled subspecies with a compact spike, with a brittle rachis and keeled glumes, has the Q factor, while other hulled cultivars of ssp. *dicoccon, farrum*, Large White, and Vernal, possess the *q* allele. Muramatsu (1986) concluded that there is wide phenotypic variation of characteristics in different *QQ* lines, and suggested that the range of variation of these characteristics is very narrow in the absence of *Q*, but when *Q* is present, they express an obvious phenotype. The discovery that all free-threshing tetraploids carry the Q factor may indicate that ssp. *carthlicum* could originate from an hybridization between free-threshing tetraploid with wild or domesticated emmer.

Nevertheless, crucial questions concerning this conclusion have been raised, and stand in the way of a full understanding of the tetraploid wheat phenotype. That the expression of the *Q* factor may be modified by the genetic background, seemed to decisively clear up the difficulty in explaining the relation between spike morphology and wheat evolution. So, if the effect is only due to the genetic background, and if the genes making up this background are minor in effect, then why are there not many transitional types between brittle and tough rachis in *qq KK* tetraploids? Or, is there a major gene with such a strong effect that it is equivalent to *Q*? Such a gene would have been highly important, but it has not been discovered. Besides, there are

some varieties of tetraploid wheat that have square-headed spikes. Because squareheadedness is one of the pleiotropic effects of *Q*, it is unlikely in a plant with genotype *qq*. Analysis of one such typical variety of ssp. *polonicum*, namely, var. *vestitum*, showed that, despite having keeled glumes, it has the *aestivum* gene *Q* (Muramatsu 1978). A preliminary result similar to this was obtained even with ssp. *dicoccon* var. *liguliforme* (Muramatsu 1979).

Because the rest of ssp. *dicoccon* differs from var. *liguliforme* and ssp. *carthlicum* in being speltoid, its chromosome 5A is presumed to carry the spelta gene *q*, and, therefore, this 5A should affect speltoidy in ssp. *aestivum* cv. Chinese Spring background (Mac Key 1954a; Muramatsu 1963). This has been confirmed with two series of aneuploid progeny involving ssp. *dicoccon* cv. *Large White* emmer, and cv. *Vernal-squarrosa* amphiploid (Muramatsu 1985).

The round glume of ssp. *carthlicum* is ascribed to lack of the proper genotype to develop strong keels. However, it is also assumed that *carthlicum* does not completely lack such genes because its phenotype resembles that of the hexaploid cultivars with round glumes, in which removal of *Q* led to keeled glumes (Muramatsu 1979, 1985, 1986).

McFadden and Sears (1946) mentioned that addition of the D subgenome to a tetraploid wheat from *Ae. tauschii* tends to increase the size of the grains. This is especially noticeable when the small-grained ssp. *carthlicum* is the tetraploid involved. It has also been observed that free-threshing hexaploid segregates from crosses between ssp. *carthlicum* and 6x ssp. *spelta* invariably have larger seeds than the free-threshing tetraploid parent.

The grains of ssp. *carthlicum* have a high protein content, but a low baking quality (Hanelt 2001). ssp. *carthlicum* is mainly used as a cereal. The seed is low in gluten, therefore bread made from it will not rise very well. The straw has many uses, as a biomass for fuel, for thatching, and as a mulch in the garden. The fibers obtained from the stems are used for paper-making. The stems are harvested in the late summer, after the seed has been harvested, they are cut into usable pieces and soaked in clear water for 24 h. They are then cooked for 2 h in lye or soda ash and then beaten in a ball mill for 1.5 h. The fibers make a green-tan paper. The starch from the seed is used for laundering.

Ssp. *carthlicum* is marked by a high resistance to fungal diseases. As such, they are considered to be of high value for wheat improvement (Dorofeev 1968). Already Vavilov (1914, 1926) found it resistant to downy mildew. Oliver et al. (2008) evaluated reactions to Fusarium head blight (FHB) in 376 accessions of five cultivated subspecies of *T. turgidum*, including *carthlicum, dicoccon, polonicum, turanicum*, and *turgidum*. Preliminary data showed that 16 *carthlicum* and 4 *dicoccon* accessions consistently exhibited full resistance or moderate resistance to FHB. These accessions likely carry genetic resistance to FHB and can be used directly in breeding programs to enhance FHB resistance in durum wheat.

10.3.2.11 Diversity of Domesticated *T. turgidum*

Given that only a relatively small number of wild emmer genotypes were taken into cultivation, the genetic basis of the cultivated wild emmer was relatively narrow, representing only a fraction of the large variation that exists in the wild form. This narrow genetic basis was further reduced during the formation of domesticated emmer, as mutations from fragile to non-fragile rachis presumably occurred only in a small number of wild emmer genotypes. These two phenomena, referred to as a "genetic bottleneck", is characteristic of many domesticated crops (Stebbins 1950). Moreover, a large fraction of the ancient gene pool of domesticated tetraploid wheat was lost in the Near East about 2500 years ago, or even earlier, when ssp. *durum* replaced domesticated emmer and the small-grained, free-threshing tetraploid wheat, ssp. *parvicoccum* (Nesbitt 2002). In addition, modern plant breeding practices have further eroded the genetic basis of domesticated tetraploid wheat due to the replacement, in many countries, of lots of traditional varieties (landraces) by a small number of high-yielding cultivars. The current, relatively narrow genetic basis of the domesticated tetraploid wheats decreases their adaptability to abiotic stresses, increases their susceptibility to biotic pressures, and considerably limits the ability to further improve their performance. This has further boosted the interest of wheat geneticists and breeders in the wild relatives of wheat, mainly in wild emmer, in an attempt to exploit their broad gene pool for the improvement of domesticated tetraploid wheat.

Several studies have estimated changes in diversity between wild emmer and its domesticated descendants. Most studies used isozymes, SSRs, or RFLPs, but recently, studies based on nucleotide diversity, that are being more comparable between laboratories and experimental systems, have been performed (Buckler et al. 2001).

Thus, Haudry et al. (2007) analyzed nucleotide diversity at 21 loci in a sample of 28 wild emmer accessions, collected from all the distribution areas of this taxon, in 12 ssp. *dicoccon* lines, 20 ssp. *durum* lines, and 41 *T. aestivum* ssp. *aestivum* lines. As expected, their results showed that the diversity of the domesticated subspecies of *T. turgidum, dicoccon* and *durum*, and of *T. aestivum* was a subset of that observed in wild emmer, namely, nucleotide diversity levels were found to be much lower in the domesticated forms than in the wild progenitor. Assuming that the sample of wild emmer studied by Haudry et al. (2007) accurately reflected the diversity of wild emmer 10,000 years ago, initial diversity was reduced by 84% in ssp. *durum* and 69% in ssp. *aestivum*. The loss of nucleotide diversity during domestication of the wheats is one of the largest reported

thus far for a crop species (Haudry et al. 2007). Most crops have nucleotide diversities about 30% lower than those of their wild progenitor, but wheat and barley lost significant and similar amounts of diversity (Kilian et al. 2006).

Akhunov et al. (2010) assessed the distribution of diversity in 2114 genes among and within the bread wheat subgenomes of *T. aestivum* ssp. *aestivum* and wild emmer. Of the analyzed loci, 305 (52%) and 296 (51%) were polymorphic in the A and B subgenomes of ssp. *aestivum*, respectively, and 316 (54%) and 338 (59%) were polymorphic in the A and B subgenomes of wild emmer, respectively. The estimates of nucleotide diversity were similar between the A and B subgenomes of bread wheat and between the A and B subgenomes of wild emmer, which showed higher diversity than the corresponding genomes in ssp. *aestivum* (Akhunov et al. 2010). Chromosome 5A of ssp. *aestivum* and chromosomes 2A and 7A of wild emmer had higher diversity than the genome-wide average. With the sole exception of ssp. *aestivum* chromosome 2A, diversity was low in genes in proximal chromosomal regions and high in genes in distal chromosomal regions. In the B subgenome of ssp. *aestivum*, chromosome 2B had higher diversity and chromosome 4B had lower diversity than the rest of the chromosomes (Akhunov et al. 2010).

Thuillet et al. (2005) reported a series of bottleneck effects in the population history of ssp. *durum* landraces, detected as decreases in the effective population size, one of these being in the transition from emmer to free-threshing wheat. In accordance with these findings, maternal lineages in emmer and wild emmer also seem to be more diversified (Oliveira et al. 2012). The number of unique chloroplast (cp) haplotypes detected in wild emmer (7) was higher than in domesticated emmer (3), ssp. *durum* (3) or ssp. *turgidum* (2). Out of the 14 cp haplotypes found in *durum* and *turgidum*, 4 were also present in domesticated emmer accessions and 3 were present in wild emmer. Of the 7 cp haplotypes found in domesticated emmer, 2 were also found in wild emmer. This suggests a scenario in which all three subspecies share a common maternal ancestral gene pool that later became distinct between them, due to different population histories.

The recent assembly of the genome of cultivar Svevo of ssp. *durum* by Maccaferri et al. (2019), facilitated the comparison between the genome of ssp. *durum* and that of its wild ancestor, ssp. *dicoccoides* (Avni et al. 2017), revealing changes imposed by the domestication process during thousands of years of unconscious and conscious selection under cultivation, and by modern, scientific-based, breeding programs. Regions exhibiting strong signatures of genetic divergence associated with domestication and breeding were widespread in the *durum* genome, with several major diversity losses in the pericentromeric regions, which occurred during the domestication of wild emmer.

The reduction of diversity continued more moderately, but spread over the genome, during the evolution of domesticated emmer wheat and that of *durum* landraces, and, more recently, in modern cultivars as a consequence of breeding activity.

In accordance, Avni et al. (2017) examined DNA variation in regions of the emmer genome that were under domestication selection. To identify these regions, they characterized the genomic diversity of 31 domesticated accessions and 34 wild emmer accessions, using a whole exome capture assay. Depending on the method used, between 32 and 154 genomic regions, spanning 0.6% (68 Mb)–3.1% (373 Mb) of the domesticated emmer genome, emerged as regions potentially affected by selection. These regions in the domesticated emmer genome were significantly enriched (> 95th percentile) with nonsynonymous SNPs. In these regions, Avni et al. (2017) found only a minor loss of genetic diversity among domesticated emmer genotypes (mean nucleotide diversity in domesticated emmer was $\pi_D = 1.1 \times 10^{-3}$) as compared to wild emmer (mean nucleotide diversity $\pi_D = 1.3 \times 10^{-3}$). This minor loss of genetic diversity indicating that selection under domestication preferentially enriched variants with possible functional effects in coding regions. The enrichment under domestication included genes involved in response to auxin stimulus.

It is likely that, the long-mixed cultivation of wild and domestic forms of emmer in many sites in the Levantine Corridor (Kislev 1984) has provided ample opportunities for some gene flow from wild to domesticated emmer. Consequently, domesticated emmer evolved in many sites as a polymorphic population, rather than as single genotypes (Feldman and Levy 2015). The increase in the genetic basis of the young crop, reduced its vulnerability to biotic and abiotic stresses. Moreover, the spread of wheat culture to different countries with different climatic and edaphic conditions, created different kinds of abiotic stresses. These stresses might have triggered activities of retrotransposons and MITEs as well as of transcription factors that, in turn, turned on many silent genes or suppressed the activity of others, and also caused new genetic variation via mutations. Selection under domestication by different farmers for different useful traits in different climatic and edaphic regions, further increased the diversity of the domesticated forms. Moreover, the tendency of traditional farmers in many parts of the world to grow in a single field (polymorphic fields), a mixture of genotypes that hybridized and recombined, enabled the selection of genotypes that were more desirable to the farmers. Selection pressure was thus, exerted consistently, but in different directions, by different farmers. These efforts resulted in many landraces that had a better adaptation to a wider range of climatic and edaphic conditions and to diverse farming regimes.

Furthermore, even after the complete replacement of wild emmer by domesticated emmer, the latter could continue to absorb genes from wild genotypes that grew on the edges of many cultivated fields or among the stones and rocks within fields, which further broadened its genetic basis (Percival 1921; Huang et al. 1999; Dvorak et al. 2006; Luo et al. 2007). This is in agreement with the comparatively high level of RFLP recently found in a sample of lines of domesticated tetraploid wheat, i.e., in *T. turgidum* ssp. *dicoccon* and ssp. *durum* (Huang et al. 1999; Luo et al. 2007). Genetic diversity of domesticated tetraploid wheat was only somewhat smaller than that of wild emmer (Huang et al. 1999), indicating that the domesticated types absorbed a significant portion of the genetic variation that exists in the wild forms.

Thus, the study of He et al. (2019) revealed that the genome-wide single-nucleotide polymorphism (SNP) diversity in hexaploid wheat was strongly influenced by gene flow from its tetraploid wild ancestor. Regions of introgression in the wheat A and B subgenomes showed increased levels of genetic diversity and reduced genetic differentiation from wild emmer. Both patterns were consistent with wild-relative introgression into domesticated wheat, which offset the effects of allopolyploidization and domestication bottlenecks on diversity in these genomes (Akhunov et al. 2010).

10.3.2.12   Cytology, Cytogenetics and Evolution

The discovery of wild diploid wheat *T. monococcum* ssp. *aegilopoides* and wild tetraploid wheat *T. turgidum* ssp. *dicoccoides*, made it possible for Schulz (1913b) to assemble, on the basis of plant morphology, the first natural classification of the wheats (Table 10.1). He divided the genus *Triticum* into three major groups: einkorn, emmer, and dinkel. Each group was subdivided into wild and domesticated species; the domesticated species were separated further into hulled and naked (free-threshing) types. Schulz (1913b) assumed correctly that the domesticated naked types were derived from the domesticated hulled forms, which, in turn, were derived from the wild progenitors. Schulz's classification was supported by further studies of taxonomic classifications (von Tschermak 1914; Percival 1921), by serological relationships, as determined by Zade (1914), by Vavilov's (1914) classification in respect to reaction to the pathogens rust and mildew, and by Sax's (1921) studies of sterility in interspecific wheat hybrids. Table 10.6 presents the modern classification of the wild and domesticated subspecies of the emmer series, i.e., *Triticum turgidum*.

At the same time, a comparatively large number of investigators performed cytological studies of the chromosome number in the wheats. Several cytologists reported 8 haploid chromosomes in *T. monococcum*, while others reported 40 in *T. vulgare* (currently *T. aestivum* ssp. *aestivum*) (reviewed in Sax 1922). However, the pioneering cytological studies of Sakamura (1918), Sax (1918) revealed the correct chromosome number of the wheats. Sakamura (1918), analyzing root-tip cells, obtained the following results: *T. monococcum* had 14 chromosomes, *T. dicoccon*, *T. durum*, *T.* turgidum and *T. polonicum* (currently all are subspecies of *T. turgidum*) had 28, and *T. vulgare, T. compactum* and *T. spelta* (currently all are subspecies of *T. aestivum*) had 42. Sax (1918) determined in meiotic cells the correct chromosome number of tetraploid wheat, *T. turgidum* ssp. *durum*. It then became obvious that Schultz's three groups of wheats also differ in their chromosome number and represent a polyploid series in which the einkorn are diploids (2n = 14), the emmer are tetraploids (2n = 28), and the dinkel are hexaploids (2n = 42).

But, since Sakamura (1918) did not present any illustrations to support his counts, his results were questioned by Percival (1921). Yet, the studies of Kihara (1919, 1924) and Sax (1921, 1922) regarding the cytogenetic relationships between the three wheat groups, diploid, tetraploid and hexaploid, showed that Sakamura's results were definitely correct and the wheats comprise a polyploid series, based on sets of seven chromosome pairs.

In 1930, Kihara upgraded the definition of "genome", that was termed earlier by Winkler (1920), as a haploid chromosome set, to a chromosome set that acts as a fundamental genetic and physiological system whose complete gene content is indispensable for the normal development and activity of an organism. One pair of homologous genomes must be present in a fully viable fertile diploid organism, and at least one in polyploid. Two genomes are strictly homologous, if the chromosomes forming bivalents at first meiotic metaphase are identical, similar in length, gene position, and centromere location, whereas two genomes are semi-homologous (= homoeologous; Huskins 1931) if they are phylogenetically similar but not strictly homologous and all or a part of their pairing chromosomes have only similar segments in common (Kihara 1930).

Since the discovery that the polyploid species of *Triticum* comprise an allopolyploid (having two or three different subgenomes) series, attempts have been made to identify the diploid donors of the two subgenomes to allotetraploid wheat. In this endeavor, studies were extended to the wild relatives of wheat, particularly to the closely related genus *Aegilops* that also comprises an allopolyploid series with diploid, tetraploid and hexaploid species (Lilienfeld 1951, and reference therein). The species of these two genera have been subjected to extensive taxonomic, cytogenetic, genetic, biochemical, molecular, and evolutionary studies by numerous scientists (see review s of Kihara 1954; Mac Key 1966; Morris and Sears 1967; Kimber and Sears 1987; Feldman et al. 1995; Feldman 2001; Gupta et al. 2005; Dvorak 2009).

Based on the concept of genome stability, and on the assumption that the genomes of the allopolyploid species remain similar to those of their parental forerunners, Kihara (1930) developed the "genome analysis" method for the identification of the genomic constitution of the wheat allopolyploid species. This method has been one of the most extensively used in attempts to identify the diploid ancestors of the allotetraploid wheats. This method is based on the use of diploid species as analyzers in crosses with allopolyploid species whose genome constitution had to be ascertained. The first step in this method involves the analysis of the chromosomal-pairing relations between the genomes of all available diploid species of the group in question. Kihara broadened his investigation to include species of the closely related genera *Aegilops* and *Ambliopyrum* (Kihara 1929, 1930, 1937, 1940, 1947, 1949, 1954). In these studies, Kihara and colleagues studied chromosomal pairing at first meiotic metaphase and fertility of all the possible combinations of interspecific diploid hybrids. These studies has provided the most consistent recognition of genomic similarities in the wheat group (Lilienfeld 1951). Genome analysis of wheat and its relatives also provided insight into the evolutionary past of these species. With this method. Kihara used the diploid species of *Aegilops* and *Amblyopyrum* as analyzers of the genomes in the allopolyploids of the genera *Aegilops* and *Triticum* (Lilienfeld 1951). Nine diploid analyzers were established: one from *Amblyopyrum* (*A. muticum*), and eight from *Aegilops* (*caudata, umbellulata, comosa, uniaristata, squarrosa* (currently *tauschii*), *bicornis, longissima* (including *sharonensis*) and *speltoides*). All allopolyploid *Aegilops* and *Triticum* species are comprised of contributions of the genomes of these diploid analyzers, except for *A. muticum, Ae. bicornis, Ae. sharonensis*, and *T. monococcum* (Kihara 1954). [*Ae. searsii* was discovered later on, in 1976 (Feldman and Kislev 1977), and was not included in Kihara's genome analysis]. The lack of participation of these diploids in the formation of allopolyploid species of *Triticum* and *Aegilops* requires an explanation.

The working hypothesis of Kihara (1930), Sax (1935) was that cytogenetics of interspecific hybrids, especially between species of different ploidy levels, might have great theoretical and applied aspects; they may shed light on the origin and mode of evolution of the relevant allopolyploid species and offer the possibility to synthesize them from different lines (genotypes) of the parental species and thus, to augment the genetic basis of the pertinent allopolyploid species (Lilienfeld 1951). The cytological analysis of interspecific hybrids has been of value in determining the relationships and origin of many species of plants (Sax 1935). Species which produce hybrids with regular meiotic pairing and normal fertility appear to be distinguished primarily by differentiation of genetic factors. Such species retain their identity only solely by geographic or physiological isolation.

Hybrids showing irregular meiotic pairing and reduced fertility indicate that the genomes of their parental species have diverged.

Kihara (1919, 1924), Sax (1922) provided the first evidence indicating that the polyploid wheats were allopolyploids. This conclusion was based on the mode of meiotic chromosome behavior observed in F₁ hybrids between diploid and tetraploid wheat species and between tetraploid and hexaploid wheat species. Seven bivalents and seven univalents were observed in the majority of the meiocytes in triploid hybrids derived from crosses between *T. turgidum* (2n = 28) and *T. monococcum* (2n − 14). Thus, Kihara (1919, 1924), Sax (1922) concluded that the chromosomes in one of the two subgenomes in *T. turgidum* were homologous with the chromosomes of the genome in *T. monococcum*. The pollen mother cells in the pentaploid hybrids derived from crosses between *T. aestivum* (2n = 42) and *T. turgidum* contained 14 bivalents and 7 univalents. This indicated that the chromosomes in two of the three subgenomes of *T. aestivum* were homologous with those of the two subgenomes of *T. turgidum*. These studies indicated that *T. turgidum* evolved as a result of hybridization between two different diploid species, a

diploid wheat and another yet unknown diploid species, followed by chromosome doubling, and that *T. aestivum* is an allohexaploid that resulted from hybridization of *T. turgidum* with a yet unknown diploid species (Kihara 1924, 1925, 1932, 1938; Percival 1921; Sax 1921, 1927). Consequently, Kihara (1924) designated the genome of diploid wheat AA, that of allotetraploid wheat AABB, and the genome of hexaploid wheat AABBDD. However, since the International Code for Botanical Nomenclature dictates indication of the genome of the female parent in hybrids and allopolyploids first, then, as the donor of the B subgenome was the female parent, the correct designation of the genome of *T. turgidum* should be BBAA and that of *T. aestivum* BBAADD.

Following van Slageren (1994), modern classification for the genus *Triticum* recognizes two diploid species, *T. monococcum* L. and *T. urartu* Tum. ex Gand., two tetraploid species, *T. turgidum* L. and *T. timopheevii* (Zhuk.) Zhuk., and two hexaploid species, *T. aestivum* L. and *T. zhukovskyi* Men. & Er. (Table 10.5). The evolution of the wheats is illustrated in Fig. 10.5 (for details see Levy and Feldman 2022). The economically important wheats are *T. aestivum* ssp. *aestivum* (bread or common wheat,

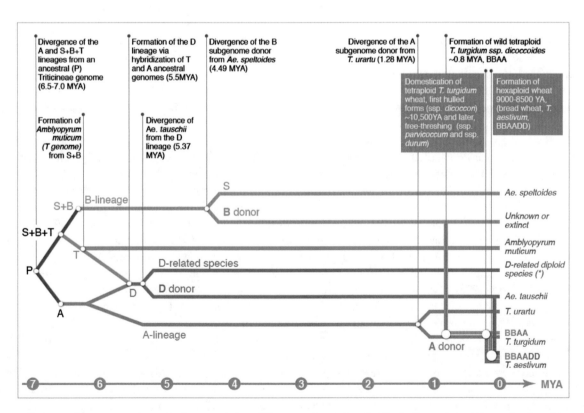

Fig. 10.5 Phylogenetic representation of wheat evolution. Wheat evolution is shown starting ∼ 7 MYA from a progenitor that gave rise to the A, B and D lineages that merged to form bread wheat. The relative timing of the major speciation events is shown in the horizontal axis and described in the boxes above. The tree is adapted from Glémin et al. 2019; Avni et al. 2022; Li et al. 2022. [Fig. 1 from Levy and Feldman (2022)]

comprising 95% of the global wheat production) and *T. turgidum* ssp. *durum* (macaroni wheat).

The polyploid species of wheat are a classic example of evolution through allopolyploidy (Kihara 1924; Sax 1921, 1927; Sears 1948, 1969; Kihara et al. 1959; Morris and Sears 1967). They behave like typical genomic allopolyploids; that is, their chromosomes pair in a diploid-like fashion and the mode of inheritance is disomic. Many attempts were made to identify the diploid donors of the B subgenomes to *T. turgidum*, and the third subgenome of *T. aestivum*. Most of these attempts used the cytogenetic approach of genome analysis, developed by Kihara (1919, 1924, 1930). However, the accumulating cytogenetic and molecular evidence has indicated that, while one subgenome remained relatively unchanged, the second subgenomes of allopolyploid wheat and *Aegilops*, changed considerably from those of their parental diploids. These genomes were termed modified genomes by Kihara (1954) and other wheat cytogeneticists. Thus, almost every allotetraploid species of *Aegilops* and *Triticum* contains an unchanged genome alongside a modified one whose diploid origin has been difficult to trace (Zohary and Feldman 1962). It is therefore more difficult to identify the diploid donor of the B subgenome of the allopolyploid wheats than the donor of the A subgenome.

10.3.2.13 Origin of the A Subgenome

From the time when only one diploid wheat species, *T. monococcum*, was known, and although the correspondence of the A subgenome of the tetraploid subspecies of *T. turgidum* with the A genome of the wild and domesticated *T. monococcum* is not perfect (Kihara and Lilienfeld 1932), it was generally accepted by wheat cytogeneticists that this diploid wheat specie is the donor of the A subgenome to allotetraploid wheat (Sax 1922; Kihara 1924; Lilienfeld and Kihara 1934; Sears 1948). The discovery in 1937 in Armenia (Tumanian 1937) of a second wild diploid wheat species, *T. urartu*, suggested the presence of another potential donor of the A subgenome. *T. urartu* differs from wild *T. monococcum,* ssp. *aegilopoides*, by several morphological features (Gandilian 1972; Dorofeev et al. 1980; Johnson 1975). Johnson (1975) also found *T. urartu* growing abundantly in Lebanon, southeastern Turkey, southwestern Iran and Transcaucasia. Although Giorgi and Bozzini (1969b) showed that the karyotypes of *T. urartu* and *T. monococcum* are identical, the nuclear DNA content is significantly different in the two species (Furuta et al. 1986; Eilam et al. 2007; Table 2.4). Moreover, *T. urartu* and *T. monococcum* differ in biochemical and molecular features, namely, isozymes (Jaaska 1974, 1980, 1982), HMW glutenin subunits (Waines and Payne 1987; Ciaffi et al. 1998; Castagna et al. 1994*)*, Gliadin genes (Ciaffi et al. 1997), AFLP (Sasanuma et al. 2002; Heun et al. 2008), RFLP

(Takumi et al. 1993; Le Corre and Bernard 1995), SSRs (Hammer et al. 2000), and AFLP and simple sequence length polymorphism (SSLP) (Sasanuma et al. 2002). A similar conclusion that the two diploid species differ from each other was also reached following chloroplast SSLP analysis (Mizumoto et al. 2002). Furthermore, Baum and Baily (2004) found that *T. urartu* differs from *T. monococcum* in the short and long units of the 5S DNA, and Dvorak et al. (1988, 1993) found extensive differences between these two species in the restriction profiles of repeated nucleotide sequences and the promoter region of the 18S-5.8S-26S rRNA genes. The divergence between *T. urartu* and *T. monococcum* is also apparent from the cytogenetic data; chromosome pairing in the F_1 hybrids between the two species was somewhat reduced (Johnson and Dhaliwal 1978; Shang et al. 1989) and the F_1 hybrid was completely sterile when *T. urartu* served as the female parent (Johnson and Dhaliwal 1976). Although they grow sympatrically in many parts of their distribution area, the two species are partly genetically isolated and have diverged to some extent from each other on the morphological, cytogenetic and molecular levels. Yet, in spite of these differences, it is generally accepted that the two-diploid *Triticum* species are closely related and presumably, of monophyletic origin (Dvorak and Zhang 1992). Hence, Dvorak (1998) designated the nuclear genome of *T. urartu* A and that of *T. monococcum* A^m.

As it became apparent that diploid wheat comprises two different species, it was important to reexamine the sources of the A subgenome in the allopolyploid wheats. Studies of chromosome pairing in F_1 hybrids between ditelosomic lines of *T. aestivum* and *T. urartu* or *T. monococcum* showed somewhat better pairing with *T. urartu* chromosomes than with *T. monococcum,* indicating that the A genome of *T. urartu* is closer to the A subgenome of polyploid wheat than to that of *T. monococcum* (Chapman et al. 1976; Dvorak 1976). To determine whether the *T. urartu* genome is more closely related to the A or B subgenome of the polyploid wheats, Chapman et al. (1976), Dvorak (1976) crossed *T. urartu* with lines of *T. aestivum* that were ditelosomic for the A and B subgenome chromosomes. In both studies, only the telocentrics of the A subgenome paired with *T. urartu* chromosomes, unequivocally showing that the chromosomes of *T. urartu* are homologous to the chromosomes of the A subgenome.

Similarly, by means of serological storage protein markers, Konarev and his associates (Konarev 1983; Konarev et al. 1976, 1979) were able to distinguish between the genomes of *T. urartu* and *T. monococcum*, a discovery that enabled them to show that the genome of *T. urartu* is more similar to the A subgenome of the wild and domesticated subspecies of *T. turgidum* than to the genome of *T. monococcum*. Additional evidence that subgenome A of allopolyploid wheats derived from *T. urartu* were obtained by polyacrylamide gel electrophoresis (PAGE) and by differential staining of seed albumins and globulins (Caldwell and Kasarda 1978). Likewise, Nishikawa (1983), using isozyme studies, showed that emmer wheat received its A subgenome from *T. urartu*. Similar results were obtained by Takumi et al. (1993), who used DNA clones, known to hybridize with the DNA of the A subgenome chromosomes of common wheat, in RFLP analyses of nuclear DNAs of diploid and polyploid wheats. They calculated genetic distances between all the pairs of accessions determined using RFLP data and clustered the species using the UPGMA method. All the accessions of *T. urartu*, *T. turgidum* ssp. *durum* and *T. aestivum* ssp. *aestivum* were clustered in one group. Those of wild and domesticated *T. monococcum* were in a different group. So, they concluded that the A subgenomes of tetraploid and hexaploid wheats originated from *T. urartu*. Dvorak et al. (1988), analyzing polymorphism in repeated nucleotide sequences, confirmed that the A subgenome derived from *T. urartu* and not from *T. monococcum*. Similarly, variation in 16 repeated nucleotide sequences showed that the A subgenomes of *T. turgidum*, *T. timopheevii*, and *T. aestivum* were contributed by *T. urartu* (Dvorak et al. 1993). Still little divergence in the repeated nucleotide sequences of the A subgenomes of these allopolyploid species from the genome of *T. urartu* was detected (Dvorak et al. 1993). In accordance with the data of Dvorak et al. (1993), recent whole genome sequencing showed that the donor of the A subgenome to allotetraploid wheat diverged from the genome of *T. urartu* ∼ 1.28 MYA (Li et al. 2022).

The donor of the A subgenome to allotetraploid wheat diverged from the genome of *T. urartu* ∼ 1.26 MYA (Li et al. 2022).

10.3.2.14 Origin of the B Subgenome

In contrast to the A subgenome donor, the identity of the B subgenome donor has so far not been conclusively defined. On the basis of chromosome pairing at meiosis in F_1 hybrids involving tetraploid and hexaploid wheats with polyploid species of *Agropyron*, researchers like Wakar (1935), Peto (1936), Matsumura (1951) believed that the B subgenome must have been contributed by an *Agropyron* species. In accord with this, McFadden and Sears (1946) suggested that the species involved might be the diploid *A. triticeum*. Yet, it soon became apparent that the diploid donor of the B subgenome to allopolyploid wheats is a more closely related species, possibly from the section Sitopsis of the genus *Aegilops*. Studies were then extended to the five species of *Aegilops* section Sitopsis, namely *Ae. speltoides*, *Ae. bicornis*, *Ae. sharonensis*, *Ae. longissima* and *Ae. searsii*, since they appear to possess the complementary requisite morphological characteristics of the B subgenome donor (Kerby and Kuspira 1987). Consequently, the five Sitopsis species

have been subjected to intense morphological, geographical, cytogenetic, genetic, biochemical, molecular, and evolutionary studies by numerous researchers, who employed various experimental approaches (see reviews of Kihara 1954; Riley et al. 1958; Mac Key 1966; Morris and Sears 1967; Kerby and Kuspira 1987; Kimber and Sears 1987; Feldman et al. 1995; Feldman 2001; Gupta et al. 2005; Dvorak 2009). These studies implicated six different diploid species as putative B subgenome donors: the five species of section Sitopsis and *T. urartu*. Yet, despite these countless efforts, the presented evidence obtained by the above-mentioned studies still failed to unequivocally identify the B subgenome donor and its identity remains ambiguous and controversial.

Morphological Evidence

Based on morphological comparison, Sarkar and Stebbins (1956) concluded that wild *T. turgidum* arose as an allotetraploid between *T. monococcum* and another species that is morphologically close to *Ae. speltoides* var. *ligustica*. Consequently, they suggested that *Ae. speltoides*, or a closely related species, is the donor of the B subgenome. On the other hand, Sears (1956a, 1956b) observed that the amphiploid *Ae. bicornis*—*T. monococcum* more closely morphologically resembles *T. turgidum* ssp. *dicoccon* than the amphiploid *Ae. speltoides*—*T. monococcum*. Moreover, at meiosis of the F_1 hybrid between the amphiploid *Ae. bicornis*—*T. monococcum* and ssp. *dicoccon*, an average of almost 9 bivalents or its equivalent was observed, indicating some pairing affinity between subgenome B and the *Ae. bicornis* genome. From these morphological and cytological data, Sears (1956a) concluded that *Ae. bicornis* or a closely related species, is the donor of the B subgenome.

Tanaka (1956) pointed out that both *Ae. longissima* and *Ae. sharonensis* possess morphological traits expected of the B subgenome donor. The amphiploid derived from crossing *Ae. longissima* with *T. monococcum* resembled *T. turgidum* with respect to several morphological properties. However, from the low pairing (average of 7 bivalents) in the hybrid between them, he concluded that there is very little chromosomal homology between the chromosomes of *Ae. longissima* and those of subgenome B and thus, it can not be considered the B subgenome donor. On the other hand, Kushnir and Halloran (1981) observed that the amphiploid derived from the hybrid *Ae. sharonensis* x *T. monococcum* had a spike, spikelet, and grain morphology similar to that of wild emmer, *T. turgidum* ssp. *dicoccoides*, and can be considered the B subgenome donor. On the basis of its morphology, *Ae. searsii* was also suggested as a possible B subgenome donor (Feldman 1978). Hence, based on morphological traits, each of the five *Aegilops* species of section Sitopsis were proposed as the donor of the B subgenome.

Geographical Evidence

Currently, *Ae. speltoides* is in contact with *T. urartu* in Lebanon, Syria, Turkey, Iraq, and Iran. In this region, *Ae. speltoides* is also in contact with wild *T. turgidum* and wild *T. timopheevii*. *Ae. searsii*, on the other hand, is native to the southern Levant (Feldman and Kislev 1977), where it has a sympatric distribution with wild *T. turgidum*. In southern Syria and southeastern Lebanon, *Ae. searsii* has massive contact with *T. urartu* (Feldman and Kislev 1977; Feldman 1978). Dvorak and Luo (2007), based on RFLP studies, found that *T. urartu* from the vicinity of Mount Hermon is the closest to the A subgenome of wild *T. turgidum* and assumed that this allotetraploid formed in this region. Assuming that the present-day distribution of the Sitopsis species reflects their distribution 700,000–900,000 years ago, when wild *T. turgidum* formed (Gornicki et al. 2014; Marcussen et al. 2014; Middleton et al. 2014), then, *Ae. searsii* can be considered the donor of the B subgenome to allotetraploid *T. turgidum*.

Evidence from Karyotypic Studies

T. turgidum and *T. aestivum* contain two pairs of satellited (SAT) chromosomes (Pathak 1940; Riley et al. 1958). Using monosomic lines, Okamoto (1957b) determined that these satellited chromosomes belong to the B subgenome; one satellite is located on the short arm of chromosome 1B and the second on the short arm of chromosome 6B. Riley et al. (1958) found that two similar pairs of SAT chromosomes are found in *Ae. speltoides* and not in *Ae. bicornis*, *Ae. longissima*, and *Ae. sharonensis*, thus showing that *Ae. speltoides* is the only Sitopsis species that can be considered the B subgenome donor. Kushnir and Halloran (1981) reported that the SAT chromosomes in *Ae. sharonensis* are similar to those observed in *T. turgidum and* concluded that this *Aegilops* species could be the B subgenome donor. Feldman (1978), upon observing that the chromosomes of *Ae. searsii* with large and medium sized satellites resembled chromosomes 1B and 6B of *T. aestivum*, respectively, suggested that *Ae. searsii* could be the B subgenome donor to the allopolyploid wheats.

Giorgi and Bozzini (1969a, b, c) described the karyotypes of *T. turgidum*, *T. monococcum*, *T. urartu*, *Ae. bicornis*, *Ae. speltoides*, and an *Ae. speltoides* x *T. monococcum* amphiploid. They found that the karyotypes of *T. urartu*, *Ae. bicornis*, and *Ae. speltoides* show that these three species could not be donors of the B subgenome to *T. turgidum*. Kerby and Kuspira (1988) compared the karyotypes of T. *turgidum, T. monococcum,* and all the putative B subgenome donors. The chromosomes of the A subgenome of *T. turgidum* were identified via comparison to the karyotype of *T. monococcum*. Comparisons of the chromosomes of the B subgenome of *T. turgidum* with the karyotypes of the

putative B subgenome donors showed that only the karyotype of *Ae. searsii* was similar to the one deduced for the donor of the B subgenome, suggesting that *Ae. searsii* is, therefore, the most likely donor of the B subgenome to the allopolyploid wheats.

Natarajan and Sarma (1974) studied the distribution of large blocks of constitutive heterochromatin and found that *Ae. speltoides,* of the Sitopsis species, had a pattern of distribution of heterochromatin that most resembled that of the B subgenome chromosomes. Contrarily, Gill and Kimber (1974b) observed that the C-banding pattern of *Ae. speltoides* chromosomes differs substantially from that of the B subgenome chromosomes of *T. aestivum* and therefore, casts doubt on the validity of the satellite and heterochromatin distribution evidence.

Evidence from Meiotic Chromosome Pairing

One of the most extensive searches for the B subgenome donor involved analysis of chromosome pairing at first meiotic metaphase of F_1 hybrids between *T. turgidum* and different diploid species considered to be the B subgenome donor. Many such studies indicated *Ae. speltoides* as a possible donor of the B subgenome, since hybrids between *Ae. speltoides* and *T. turgidum* exhibited a relatively high level of pairing, namely, from a range of 4.66 to 6.22 bivalents per meiocyte (McFadden and Sears 1946; Riley et al. 1958), to 7 bivalents per meiocyte in most pollen mother cells (Jenkins 1929). Since F_1 hybrids between *Ae. speltoides* and *T. monococcum* showed little or no pairing (Kihara 1940), it was concluded that *Ae. speltoides* and *T. monococcum* possessed different genomes, and that most of the pairing in *T. turgidum* x *Ae. speltoides* hybrids was between the B subgenome and *speltoides* chromosomes (Riley et al. 1958).

But allopolyploid wheats contain mechanisms restricting homoeologous pairing, including the *Ph1* gene on chromosome arm 5BL of *T. aestivum* (Okamoto 1957a; Riley and Chapman 1958; Sears and Okamoto 1958), and, therefore, pairing in hybrids between allopolyploid wheats and the four Sitopsis species (excluding *Ae. speltoides*), namely, *Ae. bicornis*, *Ae. sharonensis*, *Ae. longissima*, and *Ae. searsii*, is very low. On the other hand, *Ae. speltoides* contains genes that suppress the *Ph1* effect and thus, brings about homoeologous pairing (Riley et al. 1961; Dvorak 1972). Indeed, in 1961, Riley et al. crossed *T. aestivum* monosomic for chromosome 5B, the chromosome that carries the *Ph1* gene, with *Ae. speltoides*. Regardless of whether chromosome 5B (*Ph1*) was present, the F_1 hybrids showed a high incidence of chromosome pairing, which included the formation of bivalents, trivalents, and quadrivalents. These observations indicated that all or most of the pairing was between and among homoeologous chromosomes. From this result, it was concluded that the genotypes of *Ae. speltoides* used in this and previous experiments suppressed the action of *Ph1*, allowing homoeologous as well as homologous chromosomes to pair. Thus, it was impossible from the original cytological data to determine the extent of homology between the chromosomes of *Ae. speltoides* and those in the B subgenome of the allopolyploid wheats.

Dvorak (1972), in studying chromosome pairing of different accessions of *Ae. speltoides* in hybrids with bread wheat, recognized three *speltoides* types of genotypes: those showing high pairing, intermediate pairing and low pairing. On the basis of heteromorphic bivalents observed in hybrids derived from crosses between ditelosomic lines of *T. aestivum* and low-, intermediate-, and high-pairing genotypes of *Ae. speltoides*, Kimber and Athwal (1972) concluded that the pairing in these hybrids was between homoeologous chromosomes. Chromosomal pairing in the F_1 hybrid with the low-pairing genotype of *Ae. speltoides* was very low, with average 0.7 rod bivalents per cell. The conclusion that *Ae. speltoides* contains only homoeologous chromosomes rather than homologous ones to the chromosomes of *T. aestivum* was further supported by the observation that only homologous chromosomes paired in the amphiploids *T. aestivum*—low-pairing type of *Ae. speltoides* (Kimber and Athwal 1972). These cytological data preclude *Ae. speltoides* as the source of the B subgenome (Kimber and Athwal 1972).

Ae. longissima, *Ae. sharonensis*, and *Ae. bicornis* have been ruled out as the potential donors of the B subgenome, mainly on the grounds of the low meiotic pairing in F_1 hybrids between them and allopolyploid wheats (Riley et al. 1958; Riley 1965). Recent studies, however, attributed the low pairing to a *Ph*-like gene, existing in these three *Aegilops* species, that, together with the *Ph1* of allopolyploid wheats, suppresses homoeologous pairing in the hybrids between allopolyploid wheat and these species (Feldman 1978). Mello-Sampayo (1971) discovered one genotype of *Ae. longissima* that induced an intermediate level of pairing (IP) (five to six bivalents) in hybrids with *T. aestivum*. It was assumed that this IP genotype lacks the *Ph*-like gene (Avivi 1976; Feldman 1978). The existence of *ph*-like genes in diploid species of *Aegilops* has already been suggested by several researchers (Okamoto and Inomata 1974; Waines 1976; Maan 1977). Feldman (1978) crossed this IP genotype of *Ae. longissima* with ditelosomic lines of the A and B subgenomes of *T. aestivum* and found that most of the pairing involved chromosomes of the B subgenome and those of *Ae. longissima*, while the chromosomes of the A and D subgenomes paired relatively little. The chromosomes of the B subgenome paired at a much higher frequency in hybrids with *Ae. longissima* as compared to hybrids with *Ae. speltoides*. Based on these observations, Feldman (1978) concluded that the genome of *Ae. longissima* and probably

of other Sitopsis species (e.g., *Ae. searsii*) is closer to the B subgenome than that of *Ae. speltoides*.

The meiotic behavior of *Triticum aestivum* x *Aegilops speltoides*, *T. aestivum* x *Ae. sharonensis* and *T. aestivum* x *Ae. longissima* (hybrid genome constitution ABDS, ABDS^{sh}, and ABDS^l, respectively) has been analyzed by Fernandez–Calvin and Orellana (1994), using the C-banding technique. In all of the hybrids analyzed, the mean number of bound arms per cell for associations between the A and D subgenomes chromosomes was significantly higher than the mean number of associations between the B and S, the B and S^{sh} and the B and S^l genomes. These results indicated that the genomes of *Ae. speltoides*, *Ae. sharonensis* and *Ae. longissima* show a similar affinity with the genomes of hexaploid wheat; therefore, none of these species can be considered to be a distinct donor of the B subgenome of wheats (Fernandez–Calvin and Orellana 1994).

A different conclusion was reached by Maestra and Naranjo (1998), who analyzed homoeologous pairing at first meiotic metaphase of F_1 hybrids between *T. aestivum*, bearing the *Ph1* and the *Ph2* genes, or its mutants *ph1b* or *ph2b*, with a high-pairing genotype of *Ae. speltoides*. They applied the C-banding technique to identify the chromosome arms of both species. All chromosome arms of *Ae. speltoides* showed normal homoeologous pairing, implying that no apparent chromosome rearrangements occurred in the evolution of *Ae. speltoides* relative to wheat. A pattern of preferential pairing of A-D and B-S, confirmed that the S genome is closely related to the B subgenome of allopolyploid wheat. Consequently, they concluded that their results sustain the hypothesis that the B subgenome of the allopolyploid wheats was derived from *Ae. speltoides*.

Rodriguez et al. (2000a) also set out to assess the relationships between the genome of *Ae. speltoides* and those of the allopolyploid wheats. To this end, they used C-banding to analyze chromosome pairing at first meiotic metaphase of F_1 hybrids involving *Ae. speltoides* (genome SS), *T. timopheevii* (genome GGAA), *T. turgidum* (genome BBAA), and *T. aestivum* (genome BBAADD). Pairing between chromosomes of the G and S genomes in *T. timopheevii* x *Ae. speltoides* (GAS) hybrids reached a frequency much higher than pairing between chromosomes of the B and S in *T. turgidum* x *Ae. speltoides* (BAS) hybrids and *T. aestivum* x *Ae. speltoides* (BADS) hybrids and pairing between B and G genome chromosomes in *T. turgidum* x *T. timopheevii* (BGAA) hybrids or *T. aestivum* x *T. timopheevii* (BGAAD) hybrids. These results support a higher degree of closeness of the G and S genomes to each other than to the B subgenome. Such relationships are consistent with independent origins of tetraploid wheats *T. turgidum* and *T. timopheevii* and with a more recent formation of the *timopheevii* lineage.

Evidence from Seed-Storage Proteins

The electrophoretic pattern of the water-soluble endosperm proteins was determined for *T. monococcum, T. aestivum* and *T. turgidum,* (Johnson and Hall 1965; Johnson 1972), *T. timopheevii* (Johnson et al. 1967), *Ae. tauschii* (Johnson et al. 1967), and *Ae. speltoides, Ae. longissima,* and *Ae. sharonensis* (Johnson 1972). The three *Aegilops* species of section Sitopsis were found to have electrophoretic band patterns that were inconsistent with those expected of a B subgenome donor, thereby ruling them out as potential donors (Johnson 1975). Johnson then mixed equimolar amounts of proteins from selected lines of *T. monococcum* and *T. urartu* and found that the electrophoretic pattern of the mixture included all the bands found in *T. turgidum.* This led him (Johnson 1975) to suggest that *T. urartu* could be the B subgenome donor to polyploid wheats. However, the pairing data of Chapman et al. (1976), Dvorak (1976) showed unequivocally that *T. urartu* is very close to the A subgenome and cannot be the B subgenome donor.

One component of water-soluble endosperm proteins is amylase inhibitors. Vittozzi and Silano (1976) compared the molecular weight and activity of the α-amylase inhibitors from the polyploid wheats with those from all the putative B subgenome donors, excluding *Ae. searsii* and *Ae. sharonensis. Ae. bicornis, Ae. speltoides* and *T. urartu* were rejected as possible B subgenome donors on the basis of differences in α-amylase inhibitor content. *Ae. longissima* was the only species found to possess all the α-amylase inhibitors contained in *T. turgidum,* but the specific activities of these proteins were not identical, implying that *Ae. longissima* might play a role in the speciation of *T. turgidum.*

Konarev et al. (1976, 1979), Peneva and Konarev (1982), Konarev (1983) compared albumins and gliadins, that are species- and genus-specific, from all the putative B subgenome donors with those from allopolyploid wheats. They found that both *Ae. longissima* and *Ae. searsii* possess proteins that are identical to one another as well as to those specified by the B subgenome in the allopolyploid wheats. Many of the proteins of the other potential donors were different from those coded for by the B subgenome. On the basis of this information, Konarev and colleagues concluded that both *Ae. longissima* and *Ae. searsii* are equally probable B subgenome donors.

Purothionins are small, basic, highly conserved endosperm proteins (Jones et al. 1982). Jones and Mak (1977) established that there were two forms of the α-purothionin fraction in *T. turgidum* and *T. aestivum* isolates, which they designated α and α1, the latter form being specific to the B subgenome. Kerby (1986) determined the amino acid sequence of purothionins from all the putative B subgenome donors and compared these sequences with those of α1-purothionin of *T. turgidum* and *T. aestivum.* None of the purothionins isolated from these species had an amino acid sequence identical to that of α1-purothionin. Each of the proteins from *Ae. searsii* and *Ae. bicornis* differed from α1 purothionins by a single amino acid substitution, whereas the purothionins from the remaining putative B subgenome donors differed from the α1 sequence by two to five amino acid substitutions. On the basis of these protein comparisons, Kerby (1986) concluded that either *Ae. bicornis* or *Ae. searsii* is the most likely donor of the B subgenome. Similarly, Dass (1972) compared the chromatographic profiles of the phenolic compounds from *T. turgidum, T. monococcum, Ae. bicornis,* and *Ae. speltoides.* His comparisons revealed *Ae. bicornis* to possess a profile that most closely resembled the B subgenome profile of *T. turgidum,* and therefore, he concluded that *Ae. bicornis* is the more likely B subgenome donor.

Evidence from Isozymes

Jaaska (1974, 1976, 1980), surveying variation in several enzymes in wheat and diploid *Aegilops* species, found that, among the contemporary species of section Sitopsis, only *Ae. speltoides* proved suitable to be the B-subgenome donor to allopolyploid wheats. Diploids of the Emarginata subsection of section Sitopsis, namely, *Ae. longissima, Ae. sharonensis, Ae. searsii* and *Ae. bicornis,* are unsuitable for the role of the wheat B genome donors. Nakai (1979), analyzing esterase isozymes, concluded that wild *T. turgidum* originated from the hybrid of wild einkorn and a member of the Sitopsis section of *Aegilops.* Likewise, Nishikawa (1983), studying variation in α- and- ß- amylase isozymes in allopolyploid wheats and in their putative diploid ancestors, revealed that the B subgenome of *T. turgidum* is a recombinant subgenome comprising, at least, chromosome 6S[l] from *Ae. longissima* and 7S from *Ae. speltoides.* Nishikawa et al. (1992) further sustained the conclusion that *T. urartu,* the A subgenome donor, first gave rise to a hybrid with *Ae. speltoides* as the donor of the second subgenome, B, and cytoplasm, and later, a species of subsection Emarginata, most likely *Ae. longissima,* was involved in repatterning the second subgenome, which resulted in subgenome B of allotetraploid wheat.

Evidence from Studies on Nuclear DNA

Nuclear DNA Content

Based on nuclear DNA content of *T. monococcum, T. turgidum, T. aestivum, Ae. tauschii, Ae. speltoides, Ae. longissima,* and *Ae. bicornis,* Rees (1963), Rees and Walters (1965), Pegington and Rees (1970) supported the notion that *Ae. speltoides* is the donor of B subgenome to allopolyploid wheats. In contrast, Nishikawa and Furuta (1978) found that the DNA content of *T. turgidum* is very close to the sum of the content of the diploid species *T. monococcum* and *Ae. longissima*, or *Ae. bicornis*, while the sum DNA of *T. monococcum* and *Ae. speltoides* is considerably less than that of *T. turgidum*. They suggested that *Ae. longissima* or *Ae. bicornis*, rather than *Ae. speltoides*, contributed the B subgenome. Furuta et al. (1984) determined the DNA content of the B subgenome of hexaploid wheat by summing up the DNA contents of the individual B subgenome chromosomes. They found that the nuclear DNA content of *Ae. speltoides*, is only about 87% that of the B subgenome of hexaploid wheat, whereas the DNA contents of *Ae. bicornis*, *Ae. searsii*, *Ae. longissima*, and *Ae. sharonensis* are comparable to that of the B subgenome (Furuta et al. 1984). This was consistent with the results they obtained from the synthetic tetraploids *T. monococcum–Ae. longissima*, *T. monococcum–Ae. sharonensis*, and *T. monococcum–Ae. bicornis*, which had DNA content very similar to that of natural *T. turgidum*. From this, Furuta et al. (1984) concluded that *Ae. speltoides* was not the donor of the B subgenome.

The results of Eilam et al. (2008) also showed that tetraploid wheat, *T. turgidum*, contains much more DNA than the sum of DNA content of *T. urartu* and *Ae. speltoides*. Consequently, the B subgenome either derived from a species containing more DNA or, if it derived from *Ae. speltoides*, its DNA content was increased at the polyploid level. If so, the B subgenome is unique in that it underwent DNA content upsizing, as all other genomes of *Aegilops* and *Triticum* allopolyploids underwent downsizing, including the synthetic allopolyploids containing the S genome of *Ae. speltoides* (Eilam et al. 2008), as also shown by Salina et al. (2004), Han et al. (2005), who described elimination of repetitive sequences from genome S of *Ae. speltoides* in newly formed allopolyploids.

In Vitro DNA:DNA Hybridizations

To identify the donor of the B subgenome of the allopolyploid wheats, Nath et al. (1983, 1984) hybridized ^{3}H-*Triticum aestivum* DNA to the unlabeled DNAs of the six putative donors of the B subgenome, namely, *T. urartu*, *Ae. speltoides*, *Ae. sharonensis*, *Ae. bicornis*, *Ae. longissima*, and *Ae. searsii*, and found that while the genome of *T. urartu* was more closely related to the A subgenome than to the B subgenome, that of *Ae. searsii* was most closely related to the B subgenome of *T. aestivum*. Consequently, they concluded that *Ae. searsii* was the B-subgenome donor to the allopolyploid wheats or a major chromosome donor, if the B subgenome is, in fact, polyphyletic in origin. Similarly, Thompson and Nath (1986) performed DNA:DNA hybridization between unique and repeated-sequence fractions of labeled *T. turgidum* ssp. *durum* DNA and the corresponding fractions of unlabeled DNAs of *Ae. searsii*, *Ae. speltoides*, *Ae. longissima*, *Ae. sharonensis*, and *Ae. bicornis*,

and found *Ae. searsii* fractions to be the most closely related to the B subgenome of *T. turgidum* ssp. *durum*.

Evidence from RFLP Patterns of Genes Coding for rRNAs

Hybridization of total DNA from the tetraploid and hexaploid wheats, treated with the BamHl restriction enzyme, to a labelled 5S rRNA probe, showed that the repeating unit of the 5S rRNA genes comprise two fragments, 420- and 500-base pairs-long (Peacock et al. 1981). They found that the 420-base pair repeat unit was located on chromosome 1B. While neither *T. monococcum* nor *Ae. speltoides* contained the 420-base pair repeat unit, whereas *Ae. longissima*, *Ae. sharonensis* and *Ae. searsii* did. On this basis, they concluded that the 5S gene pattern excludes *Ae. speltoides* but is consistent with the other species of Sitopsis section, being closely related to the B subgenome.

A different conclusion was reached by Ruban and Badaeva (2018) who, using Giemsa C-banding and fluorescence in situ hybridization (FISH) with DNA probes representing 5S (pTa794) and 18S-5.8S-26S (pTa71) rDNAs, as well as the following nine tandem repeats: pSc119.2, pAesp_SAT86, Spelt-1, Spelt-52, pAs1, pTa-535, and pTa-s53, found that the B and G subgenomes of polyploid wheat are most closely related to the S genome of *Ae. speltoides*. Likewise, Zhang et al. (2002) cloned, sequenced and compared the internal transcribed spacer (ITS) sequences of nuclear rDNA of *T. turgidum*, *T. timopheevii*, and of the following diploid species, *T. monococcum*, *T. urartu*, the five Sitopsis species, and *Ae. tauschii*. Phylogenetic analysis demonstrated that the ITS sequences of *Ae. speltoides*, that were distinct from those of other Sitopsis species, were similar to the B and G subgenomes of the allotetraploid wheats.

Sallares and Brown (2004) amplified the entire 5′ external transcribed spacer (ETS) region of the 18S rRNA gene of all the diploid species of *Aegilops* and polyploid wheat. The phylogenetic analysis performed on the complete set of ETS sequences showed that the B and G subgenomes of tetraploid wheats form a clade with *Ae. speltoides*, in which the B subgenome diverged first and the G subgenome more recently.

Evidence from Hybridization with Single Copy Nuclear Genes

Petersen et al. (2006), using a sophisticated extension of the PCR technique, successfully isolated two single-copy nuclear genes, *DMC1* and *EF–G*, from each of the three subgenomes of *T. aestivum* and from the two subgenomes of the tetraploid progenitor *T. turgidum*. Phylogenetic analysis of these sequences showed that the B subgenome derived from *Ae. speltoides*. The B subgenome donor was associated with a higher diversification rate of the B subgenome as compared to the diversification rate seen in A subgenome in the polyploid wheats. Petersen et al. (2006) extended the phylogenetic hypothesis suggesting that neither *Triticum* and *Aegilops* are monophyletic.

Daud and Gustafson (1996) cloned a genome-specific DNA sequence pSP89.XI from *Ae. speltoides* which was barely detected in any of the other Sitopsis genomes, while Southern blot analyses established that this sequence was present in the B subgenome of allotetraploid and allo-hexaploid wheat, though its relative abundance seemed to decrease at the allopolyploid levels. Hence, Daud and Gustafson (1996) concluded that the B subgenome derived from *Ae. speltoides* but that it was somewhat modified compared with that of modern *Ae. speltoides*.

Salse et al. (2008) compared sequences of the storage protein activator (*SPA*) locus region of the S genome of *Ae. speltoides* to those of the A, B and D subgenomes of hexaploid wheat. They concluded that the S genome of *Ae. speltoides* is more evolutionary related to the B subgenome of *T. aestivum and* had diverged very early from the progenitor of the B subgenome which remains to be identified.

Evidence from Hybridization with Repeated DNA Sequences

Flavell et al. (1979) found that the B subgenome of hexaploid wheat is highly enriched with repeated sequences homologous to those of *Ae. speltoides*, thus supporting the notion that the B subgenome of hexaploid wheat is closely related to an *Ae. speltoides*-like genome.

Dennis et al. (1980) isolated and hybridized a single highly repeated satellite DNA with chromosomes of various *Triticum* and *Aegilops* species and found that this satellite DNA hybridized to specific major sites in all B subgenome chromosomes, as well as to chromosomes 4A and 7A of *T. aestivum*. Observations of such in situ hybridization patterns in chromosomes of all putative B subgenome donors (except *Ae. searsii*) showed that none had a satellite distribution identical to that of the chromosomes of the B subgenome of hexaploid wheats. However, three chromosomes of *Ae. longissima* had a satellite distribution pattern similar to chromosomes those of 2B, 3B, and 5B in *T. aestivum*. Thus, *Ae. longissima* was selected as the most likely source of the B subgenome (Dennis et al. 1980). Peacock et al. (1981) extended these studies and found that *Ae. tauschii*, *T. monococcum, and T. urartu* had no major sites of hybridization, eliminating them as possible contributors to the B subgenome. They also showed that the diploid species with greatest similarity to the B subgenome of the allopolyploid wheats was *Ae. longissima*. On the other hand, whereas every chromosome in the complement of *Ae. searsii* possessed a specific pattern of major hybridization sites, the patterns did not resemble those of the B set of chromosomes of *T. aestivum* (Peacock et al. (1981). Dvorak and Zhang (1990) developed a general method, based on variation in repeated nucleotide sequences, for the

identification of the diploid species most closely related to the B and G subgenomes of the allopolyploid wheats. Using this method, they demonstrated that *Ae. speltoides* is the most closely related species to both the B and G subgenomes of allotetraploid wheats.

Evidence from Cytoplasmic Analyses

Maan and Lucken (1967, 1968b, 1970) observed that hybrids derived from crosses between *T. monococcum* as female parent and *T. turgidum* and *T. aestivum* as male parents, were sterile males and lacked vigor. This led them to conclude that only the B subgenome donor could have contributed the cytoplasm to the allopolyploid wheats. Suemoto (1968), studying F_1 hybrids and backcross derivatives of *T. monococcum* x *T. turgidum* and *Ae. speltoides* x *T. turgidum*, found that both pollen and seed fertility were significantly greater among the progeny of the latter cross, and consequently suggested that *Ae. speltoides* or a close relative in the Sitopsis section, was the female parent of *T. turgidum*. Fertile offspring were obtained after replacing the cytoplasms of *T. turgidum* and *T. aestivum* with cytoplasms of any of the *Aegilops* species from the Sitopsis section (Hirai and Tsunewaki 1981; Tsunewaki and Ogihara 1983). These studies confirmed Suemoto's findings that the cytoplasm of the polyploid wheats was derived from a diploid *Aegilops* species of section Sitopsis.

Based on this conclusion, comparisons have been made between the cytoplasmic components of the allopolyploid wheats with those of *T. monococcum*, *Ae. tauschii*, and all the putative B subgenome donors from the Sitopsis section. Chen et al. (1975) determined the isoelectric points of the polypeptides comprising the large and small subunits of ribulose-1–5, biphosphate carboxylase-oxygenase (RuBisCO or fraction I protein). The small subunit was composed of one polypeptide chain and was coded for in the nuclear genome, whereas the large subunit was composed of three polypeptide chains encoded in the chloroplast genome. Isoelectric focusing of the polypeptides that comprise the large subunit revealed two patterns: *T. aestivum*, *T. turgidum*, and *Ae. speltoides* had an identical and higher isoelectric pattern, while *T. monococcum*, *T. urartu*, and *Ae. tauschii* had identical and lower isoelectric points. Thus, they concluded that *T. turgidum* is the female parent of *T. aestivum* and not *Ae. tauschii*. Moreover, the data indicated that neither *T. monococcum* nor *T. urartu* provided the genetic information for the large subunits in *T. turgidum*, thereby precluding both as the B subgenome donor and female parent of *T. turgidum*. On the other hand, the data unequivocally show that *Ae. speltoides*, or another species containing an identical large subunit pattern, was the B subgenome donor to *T. turgidum*. Hirai and Tsunewaki (1981) extended this study by using different alloplasmic lines, each of which contained the nuclear chromosome complement of *T. aestivum* and the cytoplasm from a specific *Aegilops* species in the Sitopsis section. Their results, supporting those of Chen et al. (1975), showed that the large subunit polypeptides of fraction 1 protein in alloplasmic lines containing *Ae. speltoides* or *Ae. longissima* cytoplasms had the same migration patterns (H type) as those of allopolyploid wheats. On the other hand, the large subunit polypeptides in the alloplasmic lines containing *Ae. bicornis* or *Ae. sharonensis* cytoplasms were of the L type found in *T. monococcum*, *T. urartu*, and *Ae. tauschii*. The authors concluded that either *Ae. speltoides or Ae. longissima* provided the cytoplasm to the allopolyploid wheats and, therefore, could have been the B subgenome donor to these species. Of note, *Ae. searsii* was not included in this study.

Comparison of the effect of various cytoplasms on several traits in euplasmic and alloplasmic lines (bearing the wheat nucleus in alien cytoplasm) made by Hori and Tsunewaki (1967), Suemoto (1968), Maan and Lucken (1971), and Tsunewaki et al. (1976, 1978), showed that the cytoplasm of several Sitopsis species can be considered the cytoplasm donor to the allopolyploid wheats. These results were supported by the results of Hirai and Tsunewaki (1981) regarding the Fraction I protein, as *Ae. speltoides* and *Ae. longissima* had a plasma gene encoding the H-type large subunit in all the polyploid wheats. *Ae. bicornis* and *Ae. sharonensis* could not have been the cytoplasm donor because they bear a plasma gene for the L-type large subunit.

Assuming that little divergence of potential B subgenome donors and the polyploid wheats has occurred since the origin of *T. turgidum*, several studies have compared the restriction fragment pattern of chloroplast DNA from the polyploid wheats with those of a number of *Triticum* and *Aegilops* species (Vedel et al. 1976; Ogihara and Tsunewaki 1982; Tsunewaki and Ogihara 1983). The most complete study reported to date has been that of Tsunewaki and Ogihara (1983), who used seven restriction enzymes. When compared to the chloroplast restriction fragment patterns of *T. aestivum* and *T. turgidum*, *T. monococcum* differed by 11 fragments, *Ae. bicornis*, *Ae. searsii*, and *Ae. speltoides* by 10 fragments and *T. urartu* by 8 fragments. Yet, the fragment pattern of chloroplast DNA from *Ae. longissima* was identical to those of *T. turgidum* and *T. aestivum*, prompting the authors to conclude that it is the most likely B subgenome donor.

Ogihara and Tsunewaki (1988), assessing variation in the chloroplast (cp) DNAs of 35 *Triticum* and *Aegilops* species, by analyzing restriction fragments that were obtained by applying 13 different restriction enzymes, classified the chloroplast genomes of these species into 16 types. This classification of cpDNAs was principally in agreement with that of the plasma types assigned according to phenotypes

arising from nucleus-cytoplasm interactions. The chloroplast genome of the *Aegilops* diploid species of section Sitopsis separated into two distinct groups, one consisting of *Ae. speltoides* and the other comprising *Ae. bicornis*, *Ae. sharonensis* and *Ae. searsii* (*Ae. longissima* was not included in this study). The *speltoides* cpDNA was closely related to the that of *T. timopheevii* and *T. aestivum*, but somewhat more distant from the latter. Similar results were obtained by Miyashita et al. (1994), who, using a battery of four-cutter restriction enzymes, investigated restriction map variation in two 5–6-kb chloroplast DNA regions of the five Sitopsis species and of two wild allotetraploid wheats, *T. turgidum* ssp. *dicoccoides* and *T. timopheevii* ssp. *armeniacum*, as well as a single accession each of *T. turgidum* ssp. *durum*, *T. timopheevii* ssp. *timopheevii* and *T. aestivum* ssp. *aestivum*. Whereas low polymorphism was found in *Ae. speltoides*, no restriction site polymorphisms were detected in any of the other diploid and allopolyploid species. One accession of *Ae. speltoides* had a plastotype identical to those of wild and domesticated *T. timopheevii*. On the other hand, no diploid species had the plastotype of *T. turgidum*. Three of the plastotypes found in the Sitopsis species were very similar, but not identical, to those of *T. turgidum* and *T. aestivum*. It was concluded that *T. timopheevii* and *T. turgidum* have a diphyletic origin, evolving at different times and originating from two different, but closely related, maternal parents.

Provan et al. (2004) utilized polymorphic chloroplast microsatellites to analyze cytoplasmic relationships between *Triticum* and *Aegilops* species. Phylogenetic analyses revealed three distinct groups of accessions; one group contained all the non-*Ae. speltoides* S-type cytoplasm species, another comprised almost exclusively A, C, D, M, N, T and U cytoplasm-type accessions, and the third contained the allopolyploid *Triticum* species and all the *Ae. speltoides* accessions. These results further confirm that *Ae. speltoides*, or a closely related species, was the original B subgenome donor of allopolyploid wheat. Similarly, Haider (2012) compared the polymorphism of chloroplast DNA between *T. aestivum* and 8 different *Aegilops* species, using cleaved amplified polymorphic sequence (CAPS) and sequencing of 28 chloroplast loci and concluded that *Ae. speltoides* is B subgenome and the cytoplasm donor to *T. aestivum*.

Studies of mitochondrial (mt) DNA variation in *Triticum* and *Aegilops* may provide a sensitive assay for assessing cytoplasmic relationships between the diploid *Triticum* and *Aegilops* species and the allopolyploid wheats. Breiman (1987) found that the *Sitopsis* species exhibited wide intra- and inter-specific variation, with *Ae. speltoides* showing the most extensive intraspecific diversity, whereas no variation was detected among the cytoplasms of the allopolyploid *Triticum* species sharing the BBAA genome. In an attempt to identify the donor of the B subgenome to *T. turgidum* and *T. aestivum*, the restriction endonuclease profiles of two regions around the mitochondrial cytochrome oxidase subunit I gene were compared with those of *Ae. speltoides, Ae. bicornis, Ae. sharonensis, Ae. longissima, Ae. searsii, Ae. tauschii* and *T. monococcum* (Graur et al. 1989). The results indicated that none of these diploid species were likely to have either donated the B subgenome or to be closely related to the donor.

Terachi et al. (1990) examined the mitochondrial genomes of three Sitopsis species (*Ae. bicornis, Ae. sharonensis,* and *Ae. speltoides*), three subspecies of *T. turgidum* (*dicoccoides, dicoccon,* and *durum*), three subspecies of *T. aestivum* (*spelta, aestivum,* and *compactum*), two subspecies of the *timopheevii* group (*armeniacum and timopheevii*), and the species *T. zhukovskyi*. mtDNAs from the subspecies *dicoccon, durum, aestivum,* and *compactum* yielded identical restriction fragment patterns which differed from those of the subspecies *dicoccoides* and *spelta* in only 2.3% of their fragments. The fragment patterns of ssp. *timopheevii* and *T. zhukovskyi* were identical, and both differed from the ssp. *armeniacum* mtDNA pattern by only one fragment. The differences in the mitochondrial genome of *T. turgidum* and *T. aestivum* from those of *T. timopheevii* and *T. zhukovskyi* suggest a diphyletic origin of the two groups. Whereas the mtDNAs of *Ae. bicornis, Ae. sharonensis,* and *Ae. searsii* were relatively similar, that of *Ae. speltoides* differed greatly from the other three, and was identical, or nearly so, to the mtDNAs of *T. timopheevii* and *T. zhukovskyi*. Terachi et al. (1990) could not determine with precision the cytoplasm donor to *T. turgidum* and *T. aestivum*, as their results revealed that the ctDNA underwent smaller evolutionary divergence than the mtDNAs from these same accessions.

To investigate phylogenetic relationships among plasmons (the whole cytoplasmic genome) in *Triticum* and *Aegilops*, Wang et al. (1997), performed PCR–single-strand conformational polymorphism (PCR-SSCP) analyses on 14.0-kb chloroplast and 13.7-kb mitochondrial DNA regions isolated from 46 alloplasmic wheat lines and one euplasmic line. The phylogenetic trees of plasmons indicated *Ae. speltoides* as the cytoplasm donor to the allopolyploid wheat, suggesting that this species is the B and G subgenomes donor of all allopolyploid wheats. Mori et al. (1997) studied variation in mitochondrial DNA of *Triticum* and *Aegilops* species by PCR-aided RFLP analysis of a 1.3 kb region containing the intron of *coxII*. All but one accession of *Ae. speltoides* possessed a haplotype common to *T. timopheevii* wheat, thus supporting the hypothesis that *Ae. speltoides* donated the G subgenome to *T. timopheevii*. However, these findings did not agree with the hypothesis that *Ae. speltoides* was the B subgenome donor to the allotetraploid and allohexaploid wheats.

Gornicki et al. (2014) sequenced 25 chloroplast genomes and genotyped 1127 accessions of 13 *Triticum* and *Aegilops*

species. They found that *Ae. speltoides* diverged before the divergence of *T. urartu, Ae. tauschii* and the *Aegilops* species of section Sitopsis. *Ae. speltoides* had formed a monophyletic clade with the allopolyploids *T. turgidum* and *T. timopheevii*, which originated within the last 0.7 and 0.4 million years ago, respectively. The geographic distribution of chloroplast haplotypes of the wild tetraploid wheats and *Ae. speltoides* illustrates the possible geographic origin of wild *T. turgidum* in the southern Levant and of wild *T. timopheevii* in northern Iraq. Chloroplast haplotypes were often shared by species or subspecies within major allopolyploid lineages and between the lineages, indicating the contribution of introgression to the evolution of the allopolyploid wheats.

Concluding Remarks on the Origin of the B Subgenome

As seen from the above data, the identification of the donor of the B subgenome to allopolyploid wheats has so far lacking conclusive results, despite the many attempts that were made during the last century to identify the diploid donor(s) of the B subgenome of the allopolyploid wheats. The morphological, geographical, cytological, genetic, and molecular data reviewed above implied that one of the species of *Aegilops* section Sitopsis, can be the donor of the B subgenome albeit unequivocally. Indeed, over the years, all five species of section Sitopsis have been proposed, by various authors, as the putative donors of the B subgenome to *T. turgidum* and *T. aestivum*.

As was pointed out by Kerby and Kuspira (1987), already Sarkar and Stebbins (1956), Sears (1948) ascribed these failures to the great antiquity of wild *T. turgidum* and hence, to the many changes that must have occurred in the B subgenome since its incorporation into the allotetraploid. Chromosomal and genetic changes could have also occurred in the diploid donor of the B subgenome since its hybridization with *T. urartu*.

On the other hand, evidence for a polyphyletic origin of the B subgenome was presented by Giorgi and Bozzini (1969a) on the basis of karyotype analysis, by Vittozzi and Silano (1976) from studies of enzyme systems, and by Dennis et al. (1980) on the basis of satellite distribution patterns. Support for such an origin of the B subgenome was also obtained from isozyme studies by Nishikawa (1983). His results implied that chromosome 6B of allohexaploid wheat derived from *Ae. longissima* and 7B from *Ae. speltoides*. Further evidence for a recombinant B subgenome was presented by Nishikawa et al. (1992), suggesting that the B subgenome of *T. turgidum* and *T. aestivum* is a recombinant subgenome, comprising segments that derived from various Sitopsis species. Chromosomes 2B and 3B contain segments of *Ae. longissima* (Gerlach et al. 1979), and chromosome 3B also has segments of *Ae. speltoides* (Vittozzi and Silano 1976; Jaaska 1980). Chromosome 4B (formerly 4A) derived from *Ae. speltoides* (Dvorak 1983; Chen and Gill 1983), and contains segments from *Ae. sharonensis* (Rayburn and Gill 1985). Chromosome 5B contains segments from *Ae. longissima* (Gerlach et al. 1979), and *Ae. speltoides* (Jaaska 1978). The long arm of chromosome 6B derived from *Ae. longissima* (Nishikawa 1983), and contains segments from *Ae. sharonensis* or *Ae. bicornis* and *Ae. speltoides* (Nishikawa et al. 1992). The long arm of chromosome 7B derived from *Ae. speltoides* (Nishikawa 1983; Nishikawa et al. 1992) and contains segments from *Ae. sharonensis* or *Ae. searsii* (Nishikawa et al. 1992). Similarly, Zhang et al. (2017), using an integrative cytogenetic and genomic approach, assessed the homology of the wheat B subgenome with the S genome of *Ae. speltoides* and revealed noticeable homology between wheat chromosome 1B and *Ae. speltoides* chromosome 1S, but not between other chromosomes in the B subgenome and S genome. Evidently, *Ae. speltoides* had been involved in the origin of the wheat B subgenome but should not be considered an exclusive donor of the subgenome. To elucidate the origin of wheat B subgenome, Kong et al. (Kong XY, Dong YS, Jia JZ, personal communication) isolated four B subgenome-specific repetitive sequences from *T. turgidum* ssp. *dicoccon*. A Southern hybridization analysis with these clones showed that four species, *Ae. speltoides, Ae. longissima, Ae. sharonensis and Ae. searsii*, contained different B subgenome-specific repetitive sequences. Hence, their results implied that the B-subgenome of allopolyploid wheat is a recombined subgenome comprising chromosomal segments from several Sitopsis species. Thus, one possibility to explain the above data is to assume that the wheat B subgenome might have a polyphyletic origin with multiple ancestors involved.

Blake et al. (1999) used the B subgenome of allopolyploid wheat as a model system to test hypotheses that bear on the monophyly or polyphyly of the individual constituent subgenomes. By using aneuploid wheat stocks, combined with PCR-based cloning strategies, they cloned and sequenced two single-copy DNA sequences from each of the seven chromosomes of the wheat B subgenome and the homologous sequences from representatives of the five diploid species in section *Sitopsis*. Phylogenetic comparisons of sequence data suggested that the B subgenome of wheat diverged from a diploid B subgenome donor. The extent of genetic diversity among the *Sitopsis* diploids and the failure of any of the *Sitopsis* species to group with the wheat B subgenome, indicated that these species also diverged from the ancestral B subgenome donor. Their results support monophyletic origin of the wheat B subgenome.

Thus, a plausible possibility therefore, is that the B-subgenome donor is a distinct diploid species different from the current Sitopsis species, which is either extinct or

extant that still remain to be discovered (Feldman et al. 1995). The genome of this species is closely related to that of *Ae. speltoides* and other Sitopsis species, and it was estimated, based on whole genome sequencing of Sitopsis species, to have diverged from the S genome of *Ae speltoides* ~ 4.49 (4.31–4.67) MYA (Li et al. 2022) and to have introgressed with the Sitopsis species of the D-lineage before its hybridization with *T. urartu* leading to formation of *T. turgidum*. Hence, hybridization between the B subgenome donor as female and *T. urartu* as male, 700,000–900,000 years ago (Gornicki et al. 2014; Marcussen et al. 2014; Middleton et al. 2014), led to the formation of wild emmer *T. turgidum* ssp. *dicoccoides*. That such allopolyploidization occurred over and over again is conceivable, possibly involving somewhat different genotypes of the concerned diploid parents. The accumulated cytogenetic and molecular evidence has indicated that assumption of genome stability, that is, that the genomes of the allopolyploid species remain similar to those of their diploid parents, is not always correct, and that subgenome(s) may undergo considerable changes at the polyploid level. Indeed, the formation of allotetraploid wheat was followed by revolutionary (occurring during allopolyploid formation) and evolutionary (occurring during the life of the allopolyploid) genetic, and epigenetic changes that brought about cytological and genetic diploidization (Feldman et al. 1997; Liu et al. 1998a, b; Ozkan et al. 2001, 2002; Shaked et al. 2001; Kashkush et al. 2002, 2003; Salina et al. 2004; Han et al. 2003, 2005; Ma and Gustafson 2005, 2006; Baum and Feldman 2010; Guo and Han 2014; Cheng et al. 2019) as well as introgression with other diploid and allopolyploid species (Zohary and Feldman, 1962; Vardi 1973; Rao and Smith 1968; Rawal and Harlan 1975; Gornicki et al. 2014; El-Baidouri et al. 2017: Glemin et al. 2019; Bernhardt et al. 2020).

10.3.2.15   DNA Content of *T. turgidum*

The subspecies of *T. turgidum* contained from 12.52 to 12.91 pg 1C DNA (Eilam et al. 2008). Sixteen ssp. *dicoccoides* accessions (35 plants) had a mean 12.91 ± 0.194 pg 1C DNA; the three ssp. *dicoccon* cultivars (8 plants) analyzed had mean 1C DNA content of 12.87 ± 0.093 pg, the eleven ssp. *durum* cultivars (22 plants) analyzed had a mean 1C DNA content of 12.84 ± 0.175 pg, the one ssp. *polonicum* cultivar (2 plants) analyzed had a mean 1C DNA content of 12.52 pg, the two ssp. *turgidum* cultivars (4 plants) analyzed had a mean 1C DNA content of 12.75 ± 0.085 pg and the one in ssp. *carthlicum* cultivar (3 plants) analyzed had a mean 1C DNA content of 12.87 pg. Domestication seemingly had no effect on genome size in this species. The genome of *T. turgidum* is larger than the genomes of *T. timopheevii* and of most *Aegilops* allopolyploids (Eilam et al. 2008). The DNA content reported for

T. turgidum by Eilam et al. (2008) is in accord with the findings of Rees (1963) and Rees and Walters (1965), who determined DNA content in tetraploid wheats and found significantly less nuclear DNA in *T. timopheevii* as compared to *T. turgidum* ssp. *durum*. The 1C DNA content of *T. turgidum* is significantly larger than the additive amount of 1C DNA of *Ae. speltoides* (5.81 pg), and *T. urartu* (6.02 pg) (Eilam et al. 2007). On the other hand, the amount of DNA in *T. turgidum* is about equal to the sum of the amphiploids of any of the other four Sitopsis species together with *T. urartu*, i.e., *Ae. bicornis* 6.84 pg, *Ae. searsii* 6.65 pg, *Ae. longissima* 7.48 pg, *Ae. sharonensis* 7.52 pg and *T. urartu* 6.02 pg (Eilam et al. 2007). It is deduced therefore, that 1C DNA amount of the B-subgenome donor was around 7.50 pg. Estimates of genome size in the Sitopsis from whole genome sequences (Avni et al. 2022) are in same range as those of Eilam et al. (2007; Table 9.3) while those of Li et al. (2022) are a bit lower.

Similar results were obtained by Furuta et al. (1984), who determined the DNA content of the three subgenomes of hexaploid wheat by summing the DNA contents of the seven individual chromosomes of each subgenome. They found that the DNA content of the B subgenome was significantly higher than those of the A and D subgenomes, and that the nuclear DNA content of *Ae. speltoides* is only about 87% of that of the B subgenome of hexaploid wheat. On the other hand, the DNA contents of *Ae. bicornis*, *Ae. searsii*, *Ae. longissima, and Ae. sharonensis* were comparable to that of the B subgenome (Furuta et al. 1984). This was consistent with the results they obtained for the synthetic allotetraploids *T. monococcum—Ae. longissima*, *T. monococcum—Ae. sharonensis*, and *T. monococcum—Ae. bicornis*. All three synthetic allotetraploids had DNA content very similar to that of natural *T. turgidum*.

10.3.2.16 Chromosome Morphology

The different subgenomes of *T. turgidum* are composed of chromosomes with median or submedian centromeres, which lack distinctive morphological features (Riley et al. 1958). For this reason, there are no distinct diagnostic markers in the chromosome complement of this species, except for the satellited (SAT) chromosomes and some differences in total length and, to some extent, in arm length ratios.

Kagawa (1929), one of the first cytologists to study chromosome morphology of domesticated emmer, *T. turgidum* ssp. *dicoccon*, observed that its genome in mitotic metaphase consisted of two different chromosomal sets which were designated A and B by Kihara (1924). Kagawa recognized ten chromosome types in this subspecies, seven of which were with secondary constrictions. Bhatia (1938) found that the morphology of all the chromosomes in ssp. *dicoccon* was quite clear, and that the homologous pairs

were easily recognizable. In contrast to Kagawa, Bhatia observed only two chromosome pairs that were satellited, and which differed in their morphology from one another. Of the remaining 24 chromosomes, 22 chromosomes had only one constriction each and the remaining pair had two constrictions each. All the 28 chromosomes fell into fourteen pairs and no two pairs were alike in the position of the centromere. Pathak (1940) observed two pairs of satellited chromosomes in *T. turgidum* ssp. *durum*. Similarly, Waines and Kimber (1973) observed in one line of wild emmer, *T. turgidum* ssp. *dicoccoides*, two pairs of SAT chromosomes. Morrison (1953) identified the two satellited pairs most commonly observed as chromosomes I and X of Chinese Spring (the old designation system of bread wheat chromosomes; Sears 1954). Okamoto (1957b, 1962) confirmed that both of these chromosomes belong to the B subgenome; chromosome I was identified as 1B and chromosome X as 6B (Okamoto 1957b).

Riley et al. (1958) noted that the short arm of chromosome 1B carried the smallest satellite and was larger than the second SAT chromosome and more hetero-brachial, whereas the short arm of chromosome 6B carried the longer satellite and was shorter than 1B. The satellites of chromosome 1B were less than 1/3 the length of the adjacent arm, whereas, in 6B, the satellites were 1/3 to ½ the length of the adjacent arm. Similarly, Coucoli and Skorda (1966) analyzing the karyotype of *T. turgidum* ssp. *durum* var. *leucurum*, found two pairs of satellited chromosomes. But, in contrast to what was found by Riley et al. (1958), Coucoli and Skorda (1966) reported that the longer SAT pair, on chromosome 6B, also had the longest satellite (2.29 μ), corresponding to approximately 1/5 of the total chromosome length and less than the half of the adjacent short arm. The SAT pair on chromosome 1B possessed the shorter satellite (approximately 1/7 the total length), which is almost equal to one third of the adjacent short arm and was more heterobrachial. Chromosomes length was found to range from 12.97 μ (the longest pair) to 8.80 μ (the shortest). All chromosomes of ssp. *durum* are metacentric or sub-metacentric, but more hetero-brachial chromosomes are present in the genome of ssp. *durum* than in *T. monococcum*.

Giorgi and Bozzini (1969a) performed a detailed analysis of the karyotypes of several subspecies, cultivars and varieties of *T. turgidum,* including ssp. *durum* (cvs. Cappelli and Aziziah), ssp. *dicoccoides*, ssp. *dicoccon*, ssp. *aethiopicum* Jakubz. (currently a variety of ssp. *durum*), ssp. *carthlicum*, ssp. *turgidum* (var. *plinianum*) and ssp. *ispahanicum* (currently a variety of ssp. *dicoccon*). Analyzing mitotic metaphase in root-tip cells, they showed that all the studied taxa had a basically similar karyotype. They then classified the chromosomes of *T. turgidum* as follows: (i) Two satellited (SAT) chromosomes, SAT1, presumably on chromosome 1B, can be distinguished from SAT2, presumably on chromosome 6B, mainly due to its shorter satellite and greater arm ratio. (ii) Two subterminal (ST) chromosomes, one is a little longer and with a slightly more subterminal centromere than the second. (iii) seven submedian (SM) chromosomes (of which some are nearly median, while others are almost subterminal. SM 7 is the shortest submedian chromosome in the complement and has the highest arm ratio in the group, which makes it easily distinguishable. (iv) Three median (M) chromosomes of which one, M1, is the longest, usually slightly submedian and, therefore, sometimes difficult to separate from submedian chromosomes. Chromosome M2 is invariably a true median chromosome, shorter than M1. Chromosome M3 is the shortest in the M group. Chromosomes of this group are fairly easily distinguishable. In summary, at least 6 out of the 14 chromosome pairs present in *T. turgidum* are distinguishable with sufficient accuracy (SAT1 and SAT2, SM7, and M1, M2 and M3).

The measurement of B and A subgenome chromosome length in the common wheat cultivar Chinese Spring (Sears 1954; see Sect. 10.4.2.8 in Chap. 10) indicated that B subgenome chromosomes are longer than A chromosomes (total subgenome length is 43.65 μ vs. 41.01 μ, respectively, with an average chromosome length of 6.24 μ vs. 5.86 μ, respectively). Chromosome 3B is the longest and 2B is the shortest in the B subgenome, whereas chromosome 2A is the longest and 1A is the shortest in the A subgenome. Chromosome 5B is the most brachial (arm ratio 2.65:1) and 6B is the least brachial, almost metacentric, in the B subgenome (arm ratio 1.05:1), while chromosome 1A is the most brachial (arm ratio 1.91:1) and 6A the least brachial (arm ratio 1.12:1) in the A subgenome. Sasaki et al. (1963) measured chromosome length in the bread wheat cv. Cheyenne, and Gill et al. (1963 in cv. Wichita and their results sustain, for the most part, those of Sears (1954). In all three studies, chromosomes 1B and 6B were found to carry satellites on their short arm.

As most chromosomes of all subspecies of *T. turgidum*, as well as all other wheat species, lack morphological features, it was necessary to look for other means of identifying individual chromosomes. This necessity has led to the development of chromosome banding techniques, especially Giemsa C-banding and N-banding, that facilitated the identification of individual chromosomes in diploid, allotetraploid and allohexaploid wheats, and thus, has advanced cytogenetic studies in wheat and related species (Gill 1987). In addition to chromosome size, centromere position and arm ratio, the position, size, and intensity of individual C- or N-bands are also important criteria for chromosome identification.

The Giemsa stain interacts specifically with constitutive heterochromatin and thus, the C-bands expose the position of this type of chromatin in the chromosomes. The dark (stained) bands and light (unstained) bands represent heterochromatic and euchromatic regions, respectively (Gill 1987). The C-banding technique stains all classes of constitutive heterochromatin and identifies each of the 21 chromosomes in *T. aestivum* (Endo and Gill 1984), and each of the 14 chromosomes of *T. turgidum* (Seal 1982; Bebeli and Kaltsikes 1985; Badaeva et al. 2015).

As in *T. aestivum* (Endo and Gill 1984), subgenome B in *T. turgidum* is also more heavily banded than subgenome A; chromosomes 1B, 3B, 5B, 6B and 7B of the B subgenome and chromosome 4A of the A subgenome are most heavily banded in durum wheat (Seal 1982; Bebeli and Kaltsikes 1985; Badaeva et al. 2015). The bands in both A and B subgenomes were concentrated in the centromeric, distal and terminal regions. Yet, there is widespread banding polymorphism among different lines of wild and domesticated *T. turgidum* (Badaeva et al. 2015). For example, these authors found that karyotypes of wild and domesticated emmer showed an extremely high diversity of C-banding patterns. B subgenome chromosomes were more polymorphic than A subgenome chromosomes. The lowest diversity of C-banding patterns was found for chromosome 3A, while chromosomes 2A and 4A proved to be most variable among the A subgenome chromosomes. On the B subgenome, the lowest polymorphism was observed for chromosome 4B and the highest, for chromosomes 3B and 7B, respectively.

The C-banding method has been used in studies of several aspects of cytogenetics and evolution of *T. turgidum*. For example, using C-bands, Seal (1982) identified all 14 *T. turgidum* chromosome pairs in hexaploid triticales (genome BBAARR). Similarly, Bebeli and Kaltsikes (1985) constructed the karyotypes of Capeiti and Mexicali, two durum wheat cultivars, on the basis of C-banding of their chromosomes.

In contrast to C-banding, the N-banding technique specifically reveals heterochromatin containing polypyrimidine DNA sequences (Dennis et al. 1980). Nine of the 21 chromosome pairs of the hexaploid wheat *Triticum aestivum* cv. Chinese Spring showed distinctive N-banding patterns (Gerlach 1977). These nine chromosomes were 4A, 7A and all of the B subgenome chromosomes. The remaining chromosomes showed either faint bands or no bands at all. Wild tetraploid wheat, *T. turgidum* ssp. *dicoccoides,* showed banded chromosomes similar to those observed in the A and B subgenomes of hexaploid wheat. Of the diploid species, only *Ae. speltoides* had several N-banded chromosomes similar to, but somewhat different than those of the B subgenome. Endo and Gill (1984) identified 16 of the 21 chromosomes of common wheat using an improved N-banding technique.

C-banding has been used quite extensively in the identification of intraspecific chromosomal rearrangements in the different subspecies of *T. turgidum* (Badaeva et al. 2007, 2015, 2019). During a C-banding survey of a large collection of cultivars of domesticated emmer, ssp. *dicoccon*, Rodríguez et al. (2000b), Badaeva et al. (2015) identified different types of chromosomal rearrangements, some of which were novel to *T. turgidum*. Chromosomal rearrangements were represented by single translocations and or multiple translocations, as well as by paracentric and pericentric inversions. The use of C-banding enabled the detection of the position of translocation breakpoints, which was either at or near the centromere, or interstitial. Centromeric translocations significantly prevailed in tetraploid wheat (Badaeva et al. 2015).

Use of the C-banding technique in the identification of individual chromosomes and chromosome arms at first meiotic metaphase of F_1 hybrids involving *T. turgidum* and related species, enabled determination of the type of pairing of each T. *turgidum* arm, i.e., homologous versus homoeologous pairing. Thus, for example, Naranjo (1990) analyzed meiotic pairing in the hybrid tetraploid triticale (genome BARR x rye) and identified the arm homoeology of A-B chromosomes of *T. turgidum* using the C-banding technique. Results confirmed that the homoeologous relationships between chromosome arms of the A and B subgenomes in *T. turgidum* are the same as in *T. aestivum*, and that a double translocation involving *4AL*, *5AL*, and *7BS*, and a pericentric inversion involving a substantial portion of chromosome 4A, are present in *T. turgidum* as in *T. aestivum*.

C-banding was also applied in *T. turgidum* to determine the approximate location of the *Ph* gene in the 5BL arm. Dvorak et al. (1984) C-banded the long arm of chromosome 5B of a mutant line of the Italian cultivar Cappelli of *T. turgidum* ssp. *durum* deficient for the *Ph* gene and of another Cappelli line bearing a duplication of part of the long arm of 5B that carries *Ph*. Compared with arm 5BL of the parental cultivar, the 5B long arm of the *Ph* mutant was shorter, owing to a deletion of one of two inter-band regions in the middle of the arm. In the line suspected to have a duplication, the 5BL arm was longer than in Cappelli and the interband region that was absent in the *Ph* mutant was twice as long.

Genetic mapping of polymorphic C-bands also enables direct comparisons between genetic and physical maps (Curtis and Lukaszewski 1991). More specifically, Curtis and Lukaszewski used eleven C-bands and two seed storage protein genes on chromosome *1B*, polymorphic between Langdon durum and four accessions of ssp. *dicoccoides*, to study the distribution of recombination along the entire length of the chromosome. The genetic maps obtained from the four individual *T. dicoccoides* chromosomes were combined to yield a consensus map of 14 markers (including the centromere) for chromosome 1B.

10.3.2.17 Development and Use of Aneuploid Lines in Durum Wheat

Several series of aneuploid lines were produced in *T. turgidum* ssp. *durum* cultivar Langdon by Joppa and colleagues (Joppa 1987, 1993). These aneuploids have been used to identify and locate genes on chromosomes, to map gene-to-centromere distances, to transfer chromosomes from one cultivar or species to another, and to identify chromosome homoeology (Joppa 1987). In contrast to hexaploid aneuploids that were produced in the 50 s of the previous century (Sears 1954), aneuploids in the tetraploid *T. turgidum* only became available during the late 1970s. Several kinds of aneuploids are not vigorous and fertile in tetraploid wheat, particularly those with reduced chromosomes number, i.e., monosomes, monotelosomes and nullisomes. On the other hand, increases in chromosome numbers in the form of trisomics, is tolerated in tetraploid wheat. Indeed, a complete set of primary trisomics that are vigorous and fertile was developed in the Italian durum cultivar Cappelli by Simeone et al. (1983).

A complete set of double-ditelosomics, dimonotelosomics for the A and B subgenome chromosomes has been developed in cultivar Langdon of durum wheat (Joppa 1987, 1993; Joppa and Williams 1988). These lines are fully fertile and vigorous and resemble Langdon in morphology. The lines have been used to identify chromosome substitutions.

Sears (1966) classified the chromosomes of hexaploid wheat into seven homoeologous groups and showed that the chromosomes of different subgenomes within the same group can compensate for each other in nullisomic-tetrasomic combinations. Thus, for example, four doses of chromosome 1D compensated for the absence of 1A or 1B. Joppa (1987, 1993), Joppa and Williams (1988) developed a more useful material for cytogenetic and genetic analyses in tetraploid wheat, namely, disomic substitutions, in which a pair of chromosomes of another subspecies of tetraploid wheat or of the D subgenome of hexaploid wheat, substituted for a homoeologous durum pair. Such inter-cultivar chromosome substitution lines, D-subgenome disomic substitutions and also homozygous recombinant lines were produced in the durum cultivar Langdon.

Joppa and colleagues produced complete sets of 14 different disomic-substitution lines bearing D subgenome chromosomes of hexaploid wheat or A and B subgenome chromosomes of several subspecies of *T. turgidum*, which substituted their durum homoeologues or homologues, respectively. These lines enabled the evaluation of the genetic contribution of each of the substituted chromosomes on the genetic background of durum, as well as *identification of useful genes and their location on specific chromosomes* (Joppa and Williams 1983; Joppa and Cantrell 1990; Cantrell and Joppa 1991; Joppa et al. 1991).

10.3.2.18 Crosses with Other Species of the Wheat Group

Chromosome pairing at first meiotic metaphase of F_1 hybrids between wild and several domesticated subspecies of *T. turgidum* is complete, or practically so, exhibiting 14 bivalents, or close to this value, per pollen mother cell (PMC) (Table 10.8). This pattern of pairing implies that the chromosomes of the wild and domesticated subspecies of *T. turgidum* are homologous and that domestication did not cause any conspicuous genomic changes. The existence of reciprocal translocations between accessions of wild emmer (Rao and Smith 1968; Rawal and Harlan 1975; Kawahara and Tanaka 1978, 1981, 1983; Kawahara 1984, 1986, 1987; Joppa et al. 1995; Badaeva et al. 2007, 2015) is not sufficient for setting up a genetic barrier(s) and preventing gene flow between them.

Hybrids between *T. turgidum* and the wild and domesticated subspecies of diploid wheat, *T. monococcum*, (hybrids genome BAA^m), had approximately 4.61–5.70 bivalents per PMC (Table 10.9), indicating that one of the two subgenomes of *T. turgidum*, i.e., subgenome A, is close but not fully homologous with the A^m genome of *T. monococcum*. The F_1 hybrid between *T. turgidum* ssp. *dicoccon* and the second diploid wheat, *T. urartu* (hybrid genome BAA), had somewhat higher pairing (6.10 bivalent per PMC; Table 10.8), but, still, subgenome A of the tetraploid albeit close to but not completely homologous with the A genome of *T. urartu*.

Hybrids between wild and domesticated *T. turgidum* and wild and domesticated *T. timopheevii* (hybrids genome BAGA) exhibited reduced chromosome pairing at first meiotic metaphase (Table 10.8). This pattern of pairing showed that the two allotetraploid wheats share one subgenome and differ in the second subgenome. However, the more than 7 bivalents that existed in these hybrids indicate that sporadic hybridization between these two allotetraploid species and backcross to one of the parents can produce progeny with recombinant B/G subgenomes. Indeed, such recombinant subgenomes were found in natural population where the two species grows sympatrically (Sachs 1953; Rao and Smith 1968; Wagenaar 1966).

Hybrids between domesticated *T. turgidum* ssp. *durum* and *T. aestivum* ssp. *aestivum* (hybrid genome BABAD) exhibited almost 14 bivalents or 12 bivalents and one translocation (Table 10.8), indicating that the two allopolyploid species share two subgenomes and that the allohexaploid contains an additional non-homologous subgenome.

Hybrids between wild and domesticated subspecies of *T. turgidum* with three species of *Aegilops* section Sitopsis, namely, *Ae. bicornis*, *Ae. sharonensis*, and *Ae. longissima*, showed very little pairing at meiotic first metaphase, with between 0.00 and 2.50 bivalents per PMC (Table 9.9). This very low pairing indicates either that there is no homology between the subgenomes of the allotetraploid wheat and the genomes of these three Sitopsis species or, alternatively, that these species contain a gene(s) that, together with *ph1* of *T. turgidum*, suppresses homoeologous pairing (Feldman 1978). On the other hand, there are two types of hybrids between *T. turgidum* and *Ae. speltoides,* the first, involving the high-pairing type of *Ae. speltoides,* had higher pairing than the second hybrid with the low-pairing type of *Ae. speltoides* (Table 9.9). Evidently, the high-pairing genotype of *Ae. speltoides* carries a gene(s) that promotes homoeologous pairing, even in the presence of the wheat *Ph1*. Hybrids between wild *T. turgidum* and *Ae. tauschii* (hybrid genome BAD) had very little pairing, 0.0–3.0 bivalents per PMC (Table 9.9) indicating no homology between the subgenomes of *T. turgidum* and the genome of *Ae. tauschii*.

10.3.3 *T. timopheevii* (Zhuk.) Zhuk. (Genome GGAA)

10.3.3.1 Description of the Species

Triticum timopheevii (Zhuk.) Zhuk., commonly known as timopheev's wheat [Syn.: *T. dicoccoides* Körn. var. *timopheevii* Zhuk.; *T. dicoccoides* Körn. subsp. *armeniacum* Jakubz.; *T. turgidum* L. var. *timopheevii* Zhuk.; *T. turgidum* L. var. *timopheevii* (Zhuk.) Bowden; *T. turgidum* var. *tumanianii* (Jakubz.) Bowden; *T. dicoccoides* var. *nudiglumis* Nabalek*; T. araraticum* Jakubz.; *T. armeniacum* (Jakubz.) Magush. & Dorof.; *T. chaldicum* Menabde; *T. miguschovae* Zhirov; *T. timonovum* Heslot & Ferrary; *Gigachilon timopheevii* (Zhuk.) Á. Löve] is an annual, predominantly autogamous, 70–90(–100)-cm-tall (excluding spikes) plant. Culms are erect. The entire plant is clothed with stiff hairs which can be up to 3-mm-long on the leaf sheaths. Leaf blades are 20–45-cm-long, and hairy on both sides. The spike is indeterminate, strongly bilaterally compressed, two-rowed, 8–12-cm-long (excluding awns), ovoid, tapering to both the base and tip, hairy and awned. In the wild form, the entire spike disarticulates into individual spikelets at maturity, each with its associated rachis segment (wedge-type dispersal unit), while in the domesticated form, the rachis is not fragile and consequently, the spike remains intact on the culm. In both subspecies, wild *armeniacum* and domesticated timopheevii, the rachis internodes are covered with dense, white hairs and the spikelets are compressed and ovoid, with the top spikelet being fertile and generally in the same plane as those below. There are 8–15 spikelets per spike, and the 2–3 basal spikelets are rudimentary. The fertile spikelet is 10–12-mm-long and has 3 florets, the upper one usually being sterile. The glumes are similar to one another, 7–10-mm-long, very hairy, with 2 keels and 5–9

Table 10.8 Chromosome pairing at first meiotic metaphase of F$_1$ hybrids between allopolyploid species of *Triticum*

Hybrid combination	Genome	Univalents	Bivalents			Multivalents			References
			Rod	Ring	Total	III	IV	Above IV	
ssp. *dicoccoides* (T[a]) x ssp. *dicoccoides* (I)	BABA	–	–	–	13.34	–	0.34	–	Rao and Smith (1968)
ssp. *dicoccoides* (I[a]) x ssp. *dicoccon*	BABA	–	–	–	14.00	–	–		Rao and Smith (1968)
ssp. *dicoccoides* (T) x ssp. *dicoccon*	BABA	–	–	–	13.80	–	0.10	–	Rao and Smith (1968)
ssp. *dicoccoides* (I) x ssp. *turgidum*	BABA	–	–	–	14.00	–	–	–	Rao and Smith (1968)
ssp. *dicoccoides* (T) x ssp. *turgidum*	BABA	–	–	–	13.87	–	0.07	–	Rao and Smith (1968)
ssp. *dicoccoides* x ssp. *durum*	BABA	0.35	–	–	13.20	0.15	0.20	–	Tanaka and Ichikawa (1972)
ssp. *armeniacum* x ssp. *timopheevii*	GAGA	0.44	–	–	12.44	0.22	0.44	–	Tanaka and Ichikawa (1972)
ssp. *dicoccoides* (a[b]) x ssp. *timopheevii*	GAGA	0.20	–	–	13.90	–	–	–	Sachs (1953)
ssp. *dicoccoides* (b[b]) x ssp. *timopheevii*	BAGA	5.62	–	–	10.00	0.74	0.04	–	Sachs (1953)
ssp. *dicoccoides* x ssp. *timopheevii*	BAGA	5.62	–	–	10.00	0.74	0.04	–	Sachs (1953)
ssp. *dicoccoides* (T) x ssp. *timopheevii*	BAGA	7.87	–	–	8.13	0.18	0.80	0.02 V	Rao and Smith (1968)
ssp. *dicoccoides* (I) x ssp. *timopheevii*	BAGA	6.62	–	–	7.62	–	1.53	–	Rao and Smith (1968)
ssp. *armeniacum* x ssp. *dicoccoides*	GABA	10.15	–	–	7.05	1.05	0.15	–	Tanaka and Ichikawa (1972)
ssp. *armeniacum* x ssp. *dicoccoides*	GABA	7.40	–	–	8.90	0.80	0.15	–	Tanaka and Ichikawa (1972)
ssp. *armeniacum* x ssp. *dicoccoides*	GABA	5.20	–	–	9.55	0.90	0.25	–	Tanaka and Ichikawa (1972)
ssp. *armeniacum* x ssp. *durum*	GABA	5.85	–	–	8.40	1.45	0.25	–	Tanaka and Ichikawa (1972)
ssp. *armeniacum* x ssp. *carthlicum*	GABA	6.47	–	–	7.40	1.60	0.40	–	Tanaka and Ichikawa (1972)
ssp. *timopheevii* x ssp. *durum*	GABA	7.97	5.65	3.52	9.17	0.51	0.04	–	Shands and Kimber (1973)
ssp. *timopheevii* x ssp. *durum*	GABA	3.90	–	–	10.55	1.00	–	–	Sachs (1953)
ssp. *dicoccon* x ssp. *timopheevii*	BAGA	4.98	–	–	10.30	0.78	0.02	–	Sachs (1953)
ssp. *dicoccon* x ssp. *timopheevii*	BAGA	6.25	–	–	9.00	1.18	0.05	–	Wagenaar (1966)
ssp. *timopheevii* x ssp. *turanicum*	GABA	5.78	–	–	9.44	1.06	0.04	–	Sachs (1953)
ssp. *timopheevii* x ssp. *turgidum*	GABA	6.32	–	–	9.04	1.12	0.06	–	Sachs (1953)
ssp. *carthlicum* x ssp. *timopheevii*	BAGA	6.16	–	–	9.44	0.96	0.02	–	Sachs (1953)
ssp. *timopheevii* x ssp. *polonicum*	GABA	7.76	–	–	8.66	0.92	0.04	–	Sachs (1953)
ssp. *timopheevii* x ssp. *dicoccon*	GABA	7.42	–	–	8.18	0.18	0.87	0.04 V	Rao and Smith (1968)
ssp. *aestivum* X ssp. *armeniacum*	BADGA	12.25	5.35	3.80	9.15	1.35	0.10	–	Dhaliwal (1977a)
ssp. *aestivum* x ssp. *durum*	BADBA	7.37	2.80	10.91	13.71	0.08	-	–	Mello-Sampayo (1968)

(continued)

Table 10.8 (continued)

Hybrid combination	Genome	Univalents	Bivalents			Multivalents			References
			Rod	Ring	Total	III	IV	Above IV	
ssp. aestivum x T. Kiharae[c]	BADGAD	8.92	5.48	9.50	14.98	0.99	0.12	–	Feldman (1966b)
T. zhukovskyi x ssp. aestivum	GAA^mBAD	11.27	5.53	3.47	9.00	3.13	0.40	0.07 V + 0.07 VI	Shands and Kimber (1973)
T. zhukovskyi x ssp. spelta	GAA^mBAD	15.12	–	–	10.70	1.39	0.33	–	Upadhya and Swaminathan (1963b)

[a] Dicoccoides (I) is from Israel and dicoccoides (T) is from Turkey
[b] T. timopheevii subsp. armeniacum is T. dicoccoides (a) var. nudiglumis which is a wild form of T. timopheevii
[c] T. kiharae Dorof. & Migush. is a synthetic amphiploid having genome AAGGDD

Table 10.9 Chromosome pairing at first meiotic metaphase of F_1 hybrids between diploid species of *Triticum* and between allopolyploids and diploids of the genus *Triticum*

Combination	Genome	Univalents	Bivalents			Multivalents			References
			Rod	Ring	Total	III	IV	V	
ssp. monococcum x ssp. aegilopoides	A^mA^m	–	–	–	7.00	–	–	–	Smith (1936)
ssp. aegilopoides x T. urartu	A^mA	0.06	0.90	6.07	6.97	–	–	–	Johnson and Dhaliwal (1978)
ssp. aegilopoides x T. urartu	A^mA	0.14	0.89	4.14	5.03	0.02	0.94	–	Shang et al. (1989)
ssp. dicoccon x T. urartu	BAA	8.90	2.00	4.10	6.10	0.04	–	–	Johnson and Dhaliwal (1978)
ssp. dicoccon x ssp. aegilopoides var. thaoudar	BAA^m	9.77	–	–	5.47	0.10	–	–	Riley et al. (1958)
ssp. dicoccon x ssp. aegilopoides var. boeoticum	BAA^m	10.63	–	–	4.93	0.17	–	–	Riley et al. (1958)
ssp. dicoccon x ssp. aegilopoides var. boeoticum	BAA^m	9.00	2.30	3.40	5.70	0.19	–	–	Johnson and Dhaliwal (1978)
ssp. dicoccon x ssp. monococcum	BAA^m	11.37	–	–	4.67	0.01	–	–	Riley et al. (1958)
ssp. dicoccon x ssp. monococcum	BAA^m	11.72	–	–	4.61	0.02	–	–	Wagenaar (1961c)
ssp. durum x ssp. monococcum	BAA^m	9.56	–	–	5.65	0.05	–	–	Wagenaar (1961c)
ssp. durum x ssp. monococcum	BAA^m	9.40	1.80	3.90	5.70	0.07	–	–	Feldman and Mello-Sampayo (1967)
ssp. durum x ssp. aeglopoides	BAA^m	9.80	–	–	5.40	0.10	–	–	Vardi (1973)
ssp. timopheevii x ssp. monococcum	GAA^m	7.80	–	–	6.09	0.34	–	–	Sachs (1953)
ssp. timopheevii x ssp. monococcum	GAA^m	7.14	–	–	5.61	0.44	0.24	0.08	Wagenaar (1961b)
ssp. aestivum x T. urartu	BADA	16.16	1.58	4.35	5.93	–	–	–	Dhaliwal (1977c)
ssp. aestivum x ssp. monococcum	BADA^m	19.25	3.11	1.05	3.99	0.14	–	–	González et al. (1993)

veins, and with two teeth on the upper margin, one larger and pointed and separated from the other by an acute angle. Lemma is elliptic, hairy, 10–12-mm-long, without keels, and with 9–11 veins, and features a central vein prolonged as a narrow 50–60(–90)-mm-long awn. In addition to awn, the lemma has a lateral tooth. At maturity, the palea is membranous and split along the keel. Usually, there are two grains per spikelet. The plant is caryopsis-free but adhered to

Fig. 10.6 Plants and spikes of allopolyploid species of *Triticum* with the G subgenome; **a** *T. timopheevii* (Zhuk.) Zhuk. ssp. *armeniacum* (Jakubz.) van Slageren (wild timopheevii); **b** *T. timopheevii* ssp. *timopheevii* (Zhuk.) Löve and Löve (domesticated timopheevii)

the lemma and palea (hulled type). It is laterally compressed, and hairy at the apex (Fig. 10.6).

Wild *T. timopheevii* was discovered several years after the discovery of wild *T. turgidum*. In 1910, Theodor Strauss, a botanist who served as the British vice-consul in Sultanabad, Iran, found several specimens of two-grained wild wheat in Noa-kuh, near the city of Kerind, in the mountainous region of western Iran. Schulz (1913a) identified the material collected by Strauss as *T. dicoccoides* (Kcke) form. *Straussiana,* distinguishing it from the Syrio-Palestinian two-grained wild wheat, which he named *T. dicoccoides* (Kcke) form. *kotschyana.* Schulz assumed that *dicoccoides* had a continuous range from the Hermon region in northern Israel to Iran, via eastern Turkey.

In 1923, Zhukovsky found a unique form of hulled, two-grained, domesticated wheat in western Georgia. He first classified it as *T. dicoccon* Schr. ssp. *timopheevii* Zhuk. (Zhukovsky 1923), but in 1928, he raised this taxon to the species rank and termed it *T. timopheevii* (Zhuk.) Zhuk. (Zhukovsky 1928b). Wild timopheevii was collected by M. G. Tumanyan and A. G. Araratyan in 1928 southeast of Erevan, Armenia, and in 1932 in several other locations in Armenia, Azerbaijan, Iran, Iraq, and Turkey (Jakubziner 1932b, 1959). The understanding of the natural distribution of wild *timopheevii* got bigger by the botanical expeditions to southwest Asia, namely, by Johnson to Turkey in 1965 and to the Fertile Crescent in 1972–1973, and by The Botanical Expedition of Kyoto University to the Northern Highlands of Mesopotamia in 1970 (Badaeva et al. 2021).

Wild *timopheevii was* found also in north-western Syria (Valkoun et al. 1998).

A specimen of two-grained wild wheat was also collected by J. B. Gillett in Rowanduz, northern Iraq, which Nabalek (1929) classified as *T. dicoccoides* ssp. *nudiglumis.* Sachs' classification (1953) coincided with Nabalek's, despite the fact that his studies of chromosome pairing in F_1 hybrids between this taxon and *T. turgidum* ssp. *dicoccoides* clearly showed that this wild wheat actually belongs to *T. timopheevii* ssp. *armeniacum.* Indeed, based on chromosome pairing and fertility, Tanaka and Ichikawa (1972) included this taxon in the *timopheevii* complex. Also, the specimen discovered by Strauss in 1910 in western Iran should be included in *T. timopheevii* ssp. *armeniacum a*nd Theodor Strauss should be credited as the first to have found ssp. *armeniacum* in nature.

Further studies on the distribution of ssp. *armeniacum* by Jakubziner and others (see Jakubziner 1959), showed that this wild wheat also grows in Azerbaijan and on the eastern slopes of the Akhsam Mountain in Caucasia. In addition, more recent studies (see Kimber and Feldman 1987) established that ssp. *armeniacum* also grows in the northeastern region of the Fertile Crescent arc, namely, southeastern Turkey, northern Iraq and western Iran. In the Fertile Crescent part of its distribution area, it is found in mixed populations with types that morphologically and cytologically resemble wild emmer, *T. turgidum* ssp. *dicoccoides* (Harlan and Zohary 1966; Rao and Smith 1968; Dagan and Zohary 1970; Rawal and Harlan 1975). Yet, it is quite difficult to distinguish *T. timopheevii* ssp. *armeniacum* from ssp. *dicoccoides*, although both subspecies are genetically isolated even when they grow sympatrically in the same population (Tanaka and Ichikawa 1972).

Wild *timopheevii* differs from wild *T. turgidum,* ssp. *dicoccoides* by its shorter stature, hairy leaves (blades and sheaths), culms and spikes, shorter and ovoid spike, less-developed keel, and delicate awns. Accordingly, Jakubziner (1932b) placed it in a separate sub-species, namely, *T. dicoccoides* ssp. *armeniacum.* Further studies by Makushina in 1938 (see Jakubziner, 1959) showed that this taxon is sufficiently distinct from the Syrio-Palestinian ssp. *dicoccoides* to warrant the rank of a separate species, and consequently, she called it *T. armeniacum* Mak. However, since this name had been briefly used for another taxon, Jakubziner changed it in 1947 to *T. araraticum* Jakub. (see Jakubziner 1959). Still, *T. araraticum* was included as a subspecies in *T. timopheevii* and therefore, cannot be used as a species name (Mac Key 1975). Tzvelev (1976, 1984), van Slageren (1994) correctly considered Mac Key's classification of *araraticum* as a subspecies of *T. timopheevii* illegitimate, since the older epithet *armeniacum* should have been used both at the subspecies and varietal levels. Accordingly, the wild form of this species should be referred to as *T. timopheevii* (Zhuk.) Zhuk. ssp. *armeniacum* (Jakubz.) van Slageren (van Slageren 1994).

A mutant of domesticated *T. timopheevii* with naked (free-threshing) seeds, was discovered by Zhukovsky and, in 1969, was given the species status, *Triticum militinae* Zhuk. & Migush. However, this naked wheat was later referred to by van Slageren (1994), Mac Key (2005) as a variety of ssp. *timopheevii,* namely, *var. militinae (Zhuk. et Migusch.) Zhuk.* et Migusch., because it is a free-threshing mutant that was selected from a single specimen of domesticated *T. timopheevii* (Miller 1987) and not commercially cultivated (Mac Key 2005).

Triticum timopheevii contains two subspecies, the wild ssp. *armeniacum (*former *araraticum)* and the domesticated ssp. *timopheevi.* These subspecies, are morphologically similar, and genetically are closely related. The F_1 hybrids between them are either sterile, semi-fertile or fully fertile, while they differ genomically and are reproductively isolated from the second tetraploid wheat, *T. turgidum* (Tanaka and Ichikawa 1972; Tanaka and Ishii 1975; Tanaka et al. 1978). The most differentiating trait between the wild and the domesticated subspecies is rachis brittleness at maturity; in the wild form, the rachis disarticulates, whereas, in the domesticated one it remains intact. In the wild form, the brittle rachis is an essential trait, leading at maturity, to disarticulation of the spike into individual spikelets. Each spikelet, equipped with two awns above and a sharp rachis segment below, is an arrow-like seed-dispersal unit that is very effective for seed dispersal and self-planting under wild

conditions (Zohary et al. 2012). In contrast, in the domesticated subspecies, the mature spike remains intact on the culm and breaks into individual spikelets only when slight mechanical pressure is applied at threshing (Dorofeev and Korovina 1979). This difference was caused by two recessive mutations yielding a tough rachis, that occasionally occur in the wild, preventing the dispersal of seeds. While such a mutation is nuisance in nature, it has a positive adaptive value under cultivation, facilitating harvesting, and has thus been preferred by farmers.

Domesticated *T. timopheevii* contains the recessive *br-A1* allele on the short arm of chromosomes 3A (Li and Gill 2006). It is likely that the domesticated form contains an additional recessive allele in the *Br-B1* locus on the short arm of chromosome 3G, since *Ae. speltoides*, (genome SS), the putative donor of the G subgenome, carries the dominant *Br-S1* allele for rachis brittleness on chromosome 3S (Li and Gill 2006).

It is difficult to distinguish wild *timopheevii*, ssp. *armeniacum*, from wild emmer, *T. turgidum* ssp. *dicoccoides,* by morphology (Tanaka and Sakamoto 1979). But, both wild taxa can easily be differentiated based on biochemical, immunological, cytological and molecular markers (Lilienfeld and Kihara 1934; Kawahara and Tanaka 1977; Konarev et al. 1976; Gill and Chen 1987; Jiang and Gill 1994b; Badaeva et al. 1994). From an archaeobotanical perspective, both species can be reliably identified based on several characteristics of charred spikelets (Jones et al. 2000). In blind tests, it was possible to distinguish modern representatives of the two wild taxa on the basis of the primary keel of the glume, which arises just below the rachis disarticulation scar, and the prominent vein on the secondary keel, observable at the base of the glume, which is the part of spikelet most commonly preserved by charring in archaeological material (Jones et al. 2000).

In order to differentiate between seeds of domesticated *T. timopheevii* from those of domesticated emmer, that were uncovered in archaeological excavations, Boscato et al. (2008) used ribosomal primers *ITS2* and *ITS2*, both from *T. timopheevii*, and the nuclear primer acetyl-coenzyme A from *timopheevii*, to discriminated the related DNA sequences of these two taxa. Moreover, Tanno et al. (2018) developed a multiplex PCR DNA marker for quick and easy identification of the genome of the *T. timopheevii* lineage, including the wild and the domesticated subspecies, and of the *T. turgidum* lineage, including the wild and domesticated subspecies. This multiplex PCR system is based on the simultaneous PCR amplification of two chloroplast regions, matK and rbcL. The matK region molecularly distinguishes between the two lineages with complete accuracy, whereas the rbcL region serves as a positive control amplicon.

Morphological variation in *T. timopheevii* involves mainly spike and spikelet sizes and shapes, glume and awn colors and hairiness, plant heights, and leaf widths. *T. timopheevii* is normally characterized by a dense and pyramidal spike. Yet, wider intraspecific variation was found on the biochemical and molecular levels. Asins and Carbonell (1986a), using the isoenzymes peroxidase and alkaline phosphatase, observed a wide intraspecific variability of these enzymes in wild and domesticated forms of *T. timopheevii*, with inter-population variability proving greater than the intra-population variability. Domesticated *T. timopheevii* showed higher variability than the wild subspecies. The interspecific variability data indicated that *T. timopheevii* is closer to *T. turgidum* than to any of the diploid species (Asins and Carbonell 1986b). These close relationships stem from the fact that both species share the A subgenome derived from *T. urartu*. Thus, the differences between these two species must be due to the second subgenome, G in *timopheevii* and B in *turgidum*, which derived from different, but closely related, diploid species.

Protein spectra from several subspecies of *T. turgidum* and from the two subspecies of *T. timopheevii,* were obtained by electrophoresis of seed extracts on polyacrylamide gels (Johnson et al. 1967). The subspecies of *T. turgidum* showed nine fast-moving albumin homologues, while those of *T. timopheevii* showed seven. The two species had only five albumin bands in common.

Badaeva et al. (2021) presented a comprehensive survey of genomic and cytogenetic diversity of the gene pool of *T. timopheevii*. Their study provided detailed insights into the cytogenetic composition of this wheat, revealed group-, and population-specific markers and show that chromosomal rearrangements play an important role in intraspecific diversity of *T. araraticum*. This study enabled Badaeva et al. (2021) to make several major remarks: (i) the extant *timopheevii* gene pool consists of three distinct lineages, two *wild,* designated as ARA-0 and ARA-1, and one domesticated; (ii) while ARA-0 was found to be geographically widespread, ARA-1 was restricted to southeastern Turkey and north-western Syria, where it grows sympatrically with wild emmer, *Triticum turgidum* ssp. *dicoccoides*; (iii) wild emmer is genetically more diverse, supporting a more recent origin of *wild timopheevii*; (iv) among the *timopheevii* lineages, ARA-0 harbors more genetic diversity than ARA-1; (v) Nei's genetic distance between lineages based on all polymorphic Sequence-Specific Amplification Polymorphism (SSAP) markers or considering only the *BARE-1* markers (distributed between A- and G-subgenome chromosomes) revealed that ARA-0 is phylogenetically more closely related to domesticated *timopheevii* than ARA-1. However, considering only the *Jeli* markers (target more the A subgenome chromosomes), ARA-1 was more closely related to domesticated *timopheevii;* and (vi) wild emmer is most closely related to the ARA-1 lineage.

An important outcome of the study of Badaeva et al. (2021) is the identification of three distinct lineages in *T. timopheevii:* one comprising all *domesticated* genotypes (including var. *militinae* and *T. zhukovskyi*), *and two,* ARA-0 and ARA-1, belong to the *wild form.* ARA-0 was found across the whole area of species distribution whereas ARA-1 was only detected in southeastern Turkey and in neighboring northwestern Syria.

Badaeva et al. (2021), concerning the characteristics, composition and geographical distribution of ARA-0 and ARA-1 lineages, are not in agreement with the taxonomic treatment of wild *timopheevii* by Dorofeev et al. (1980), who divided this taxon into two subspecies: subsp. *kurdistanicum* Dorof. et Migusch. and subsp. *araraticum*.

The origin of ssp. *timopheevii* remains unclear, but Badaeva et al. (2021) speculate that it was probably introduced from easternTurkey, on the grounds that ssp. *timopheevii.* is more closely related to ssp. *armeniacum* from Turkey or northern Iraq than to the Transcaucasian types.

ssp. *armeniacum* is known to have various degrees of structural chromosomal differentiations, involving several interchanges (Svetozarova 1939; Wagenaar 1966; Tanaka and Ichikawa 1968, 1972; Tanaka and Ishii 1975; Kawahara and Tanaka 1977; Jiang and Gill 1994b; Badaeva et al. 1994). In contrast, no distinct structural variations of chromosomes were found in ssp. *timopheevii*. Most *armeniacum* and all *timopheevii* lines in Transcaucasia have the same chromosome structure (Tanaka and Ishii 1975). Consequently, it was concluded that cultivated *T. timopheevii* in Transcaucasia derived from wild ssp. *armeniacum* bearing a similar chromosome architecture (Tanaka and Ishii 1975). Kawahara and Tanaka (1977), analyzing chromosome pairing in F_1 hybrids between different accessions of ssp. *armeniacum* collected in southeastern Turkey and northern Iraq, observed multivalents, indicating the wide occurrence of intra-sub-specific translocations. Seed fertility of hybrids between the various accessions of ssp. *armeniacum.* and domesticated *T. timopheevii* varied from almost sterile to fully fertile. These findings supported previous studies (Tanaka and Ichikawa 1972) by showing that chromosome differentiation in ssp. *armeniacum* is more abundant in northern Iraq than in Transcaucasia, indicating that the wild *T. timopheevii* originated in the north-eastern part of the Fertile Crescent and later spread northward to Transcaucasia.

Both subspecies of *T. timopheevii* contain genes conferring resistance to wheat fungal and insect pathogens, such as powdery mildew (Tomerlin et al. 1984; Brown-Guedira et al. 1996, 1997, 1999b; Jarve et al. 2000; Leonova et al. 2011; Timonova et al. 2013), leaf rust (Tomerlin et al. 1984; Brown-Guedira et al. 1996, 1997, 1999a, 2003; Leonova et al. 2010; Timonova et al. 2013; Singh et al. 2017), stem rust (Sawhney and Goel 1979; Dyck 1992; Brown-Guedira et al. 1996; Leonova et al. 2011; Timonova et al. 2013), stripe rust (Brown-Guedira et al. 1996) *Septoria nodorum* blotch (Tomerlin et al. 1984), *Septoria* blotch (Brown-Guedira et al. 1996), spot blotch (Leonova et al. 2011; Timonova et al. 2013), tan spot (Brown-Guedira et al. 1996), loose smut (Leonova et al. 2011; Timonova et al. 2013) and fusarium head blight (Malihipour et al. 2017), as well as resistance to Hessian fly and curl mite (Brown-Guedira et al. 1996; El Haddoury et al. 2005).

In the production of hybrid wheat lines of *T. aestivum*, that yield higher than pure, true-breeding lines, and exhibit improved quality and greater tolerance to environmental and biotic stresses (Briggle 1963; Wilson and Driscoll 1983; Bruns and Peterson 1998; Jordaan et al. 1999), alloplasmic lines having *T. timopheevii* cytoplasm that is different from common wheat cytoplasm (Maan 1975; Tsunewaki 1989) is commonly used to create male-sterile female lines (Wilson and Ross 1962). Yet, such male sterility requires restoration of fertility in the F_1 hybrids by the male parent having a restorer gene(s). Several such restorers were identified (Maan and Lücken 1968a; Chen 2003), but it is difficult to find such genes that are effective in a wide range of *T. aestivum* genotypes. Moreover, the system also requires breeding of the male parent to produce male lines that contain suitable restorer gene(s), thus rendering hybrid seed production more expensive and limiting of the number of male parents that can be tested for combining ability.

10.3.3.2 Ssp. *armeniacum (Jakubz.) Van Slageren* (Wild *timopheevii*)

The essential nomenclature of this subspecies is currently as follows: *T. timopheevii* (Zhuk.) Zhuk. ssp. *armeniacum* (Jakubz.) van Slageren, comb. nov. [Syn.: *T. dicoccoides* var. *nudiglumis* Nabalek; *T. dicoccoides* (Körn. ex Asch. & Graebn.) Schweinf. ssp. *armeniacum* Jakubz.; *T. araraticum* Jakubz.; *T. armeniacum* (Jakubz.) Makush.; *T. timopheevii* Zhuk. subsp. *araraticum* (Jakubz.) Mac Key; *T. turgidum* L. ssp. *armeniacum* (Jakubz.) Á. Löve and D. Löve.; *T. timopheevii* (Zhuk.) Zhuk. var. *araraticum* (Jakubz.) C. Yen; *T. timopheevii* (Zhuk.) Zhuk. ssp. *araraticum* (Jakubz.) Mac Key; *Gigachilon timopheevii* (Zhuk.) Á. Löve and D. Löve ssp. *armeniacum* (Jakubz.) Á. Löve] (Fig. 10.6a).

The geographical distribution of the wild subspecies of *T. timopheevii*, ssp. *armeniacum*, is west-Irano-Turanian phytogeographic region, including southeastern Turkey, northern Iraq, western Iran and Trans- and Cis-Caucasia. Compared to the second wild allotetraploid wheat, wild emmer (*T. turgidum* ssp. *dicoccoides*), the distribution area of subsp. *armeniacum* is more affected by the steppical conditions of the Irano-Turanian region. In this area, it thrives on terra rossa, basalt, and other soils that were produced from hard limestone bedrock, in the herbaceous cover of the deciduous oak-park forests, in evergreen dwarf shrub formations, and in steppe-like formations where it occupies primary habitats, as well as in secondary habitats such as abandoned fields, roadsides and edges of cultivation. In most of its habitats, ssp. *armeniacum* grows in patches and in mixed stands with other annual cereals and several legumes. It grows at altitudes of 300–1600 m.

Subsp. *armeniacum* is distributed in the north-central region of the distribution of the wild species of *Triticum*. In some locations in the northeastern part of the Fertile Crescent, it is abundant while in other is more sporadic. The distribution of ssp. *armeniacum* overlaps that of both of its putative diploid parents, *Ae. speltoides* and *T. urartu*, but it is more restricted. The putative diploid parents have massive contact in the northeastern part of the Fertile Crescent, which is the presumed center of origin of subsp. *armeniacum*. It grows sympatrically with *Ae. speltoides, Ae. caudata, Ae. umbellulata, T. urartu, T. monococcum* ssp. *aegilopoides* (mainly var. *thaoudar*), *T. turgidum* ssp. *dicoccoides* (in the northeastern part of the Fertile Crescent arc), *Ae. cylindrica, Ae. columnaris* and *Ae. triuncialis*, and allopatrically with *Amblyopyrum muticum, Ae. geniculata, Ae. biuncialis, Ae. neglecta, Ae. peregrina. Ae. tauschii, Ae. crassa* and *Ae. juvenalis*.

Ssp. *armeniacum* grows sympatrically with wild emmer in many sites in southeastern Turkey, northern Iraq and western Iran (Harlan and Zohary 1966), and morphological differences between the two wild wheats are not clear with respect to ear color, glume pubescence and shooting time. Distinguishing their ripe ears from one another by morphological examination is difficult (Tanaka and Ishii 1973) and requires crossing and/or by molecular tests. For these reasons, taxonomists dealing with the flora of south-west Asia (e.g., Bor 1968) frequently group all Kurdish wild tetraploid wheat material together into what they call *T. dicoccoides*, disregarding the fact that they are actually two reproductively well-isolated entities.

In contrast to the accessions of ssp. *dicoccoides* from the southwestern region of the Fertile Crescent arc, i.e., northeastern Israel, northwestern Jordan, southwestern Syria and southeastern Lebanon, that features larger and most robust plants with lax heads and heavy large awns, northeastern Fertile Crescent arc tetraploid wheat plants (ssp. *dicoccoides* mixed with ssp. *armeniacum*) are smaller and slender and characterized by somewhat small, compact heads, fine-textured awns, small spikelets, and are often hairy, with sparse pubescence on the rachis and spikelet base (Rao and Smith 1968; Rawal and Harlan 1975). These tetraploid wheats are found as a minor component of the annual herbaceous flora of the oak woodlands in the Taurus and Zagros Mountains (Rawal and Harlan 1975), where they are never dominant and occur in widely scattered patches on rocky slopes. In some areas in Turkey, e.g., near Gaziantep, the mixture of tetraploid wheats tends to converge toward the wild diploid wheats of the area (Rawal and Harlan 1975). However, the diploids always have more spikelets per head, but the spikelets are strikingly similar in size and appearance.

Differential production of semi-fertile and sterile hybrids have been reported when lines of ssp. *armeniacum* were crossed with lines of the domesticated form ssp. *timopheevii* (Tanaka and Ichikawa 1968, 1972; Tanaka and Ishii 1975). Two groups of ssp. *armeniacum* lines were found in crosses with a tester strain of ssp. *timopheevii*. The first group gave rise to semi-fertile F_1 hybrids that showed normal chromosome association at first meiotic metaphase, whereas the second group yielded sterile hybrids that exhibited several chromosome interchanges. The former had dark green leaves and a procumbent tillering habit, while the latter had light green leaves and a semi-erect tillering habit (Tanaka and Ishii 1975). It seems therefore that ssp. *armeniacum* has undergone extensive structural changes at the chromosomal, and possibly also at the genetic, level and that genetic isolation by hybrid sterility between lines of ssp. *armeniacum* as well as between some of them and ssp. *timopheevii* is, currently, partially established (Tanaka and Ishii 1975).

High intraspecific diversity of the karyotype was detected in ssp. *armeniacum* by several researchers by crossing experiments (Tanaka and Ichikawa 1968, 1972; Tanaka and Ishii 1973, 1975; Tanaka et al. 1968; Kawahara and Tanaka 1977; Kawahara et al. 1996), by C-banding (Badaeva et al. 1990), by nuclear (Nave et al. 2021; Shcherban et al. 2016) and chloroplast DNA markers (Mori et al. 2009; Gornicki et al. 2014).

Using C-banding techniques, Badaeva et al. (2021) analyzed the karyotypes of a large number of accessions of ssp. *armeniacum* from all over across its distribution area of this subspecies. They found that this wild taxon comprises two distinct lineages, ARA-0 encompassed 342 genotypes and is geographically widespread and characterized by a wide genetic diversity, and ARA-1, encompassed 49 genotypes that has limited in its distribution to southeastern Turkey and north-western Syria, as well as in its genetic variation.

A recent detailed analysis of karyotype structure and C-banding patterns highlighted significant differences between wild emmer and wild and domesticated *T. timopheevii* (Badaeva et al. 2021). Wild *timopheevii* is characterized by a wide diversity of C-banding patterns and broad translocation polymorphisms. The karyotype lacking chromosomal rearrangements (designated normal by the authors), found in 44.6% genotypes of wild form, was the most frequent karyotype variant shared by wild and domesticated *T. timopheevii*.

Cytogenetic data showed that the speciation of ssp. *armeniacum* has been accompanied by chromosomal rearrangements involving either intra- or inter-subgenome translocations (Kawahara and Tanaka 1977, 1981; Chen and Gill 1983; Jiang and Gill 1994b; Badaeva et al. 1990, 1994; Kawahara et al. 1996; Rodriguez et al. 2006b). Badaeva et al. (2021) reported that 216 out of 391 (55.4%) wild *timopheevii* accessions carried a translocated karyotype. One-hundred-forty-seven genotypes differed from the 'normal' karyotype by one, 45 genotypes by two (double translocations), 21 genotypes by three (triple translocations) and three genotypes by four chromosomal rearrangements. While most variants of chromosomal rearrangements were unique and identified in one or few genotypes, only four following variants, representing 16% of the whole analyzed material, were relatively frequent.

Translocations occurred more frequently in the chromosomes of the G subgenome than of the A subgenome. Individual chromosomes differed in the frequencies of their involvement in translocations and each geographical region contained a unique spectrum of translocations. Karyotypic diversity was the highest in Iraq, followed by Transcaucasia and Turkey; Iran showed little karyotypic variation.

Badaeva et al. (2021), using the six DNA probes pTa-535, pSc119.2, pAesp_SAT86, GAA_n, Spelt-1, and Spelt-52, estimated the intraspecific diversity of wild *timopheevii* and assessed its phylogenetic relationships with domesticated *timopheevii*. The distribution of pTa-535 was monomorphic among both the wild and domesticated forms, while the pSc119.2 site on 1AL discriminated ARA-1 and the domesticated form from the ARA-0 lineage. Whereas the distribution of the pAesp_SAT86 probe was similar in all domesticated *timopheevii* genotypes, the labeling patterns were highly polymorphic in wild *timopheevii*. Large pAesp_SAT86 sites were found only on some G-subgenome chromosomes. The A-subgenome chromosomes possessed several small, but genetically informative polymorphic sites, some of which were lineage-specific. Most obvious differences were observed for 3A, 4G and 7G chromosomes. Thus, all domesticated *timopheevii* and ARA-1 genotypes carried the pAesp_SAT86 signal in the middle of 3AS, while for ARA-0 it was located sub-terminally on the long arm. One large pAesp_SAT86 cluster was present on the long arm of 4G in ARA-0, but on the short arm in ARA-1 and domesticated *timopheevii*. Two large and adjacent pAesp_SAT86 clusters were detected on 7GS in ARA-1 and domesticated *timopheevii*, but they were split between opposite chromosome arms in all ARA-0 genotypes. Differences between ARA-0 and ARA-1 in pAesp_SAT86 cluster position on 4G and 7G could be caused by pericentric inversions.

FISH with Spelt-1 and Spelt-52 probes revealed high intraspecific diversity of wild. and low polymorphism in domesticated *timopheevii*. The broadest spectra of labeling patterns were found in genotypes from Dahuk and

Sulaymaniyah (Iraq) and in the ARA-1 group from Turkey, while material from Transcaucasia exhibited the lowest variation. The Spelt-1 signals in various combinations appeared in sub-telomeric regions of either one or both arms of 2A, 6A, and all G-subgenome chromosomes. The Spelt-52 signals were observed in various combinations on 2AS, 1GS, and 6GL chromosomes.

The existence of two lineages in ssp. *armeniacum* is supported by the study of Mori et al. (2009) that was based on 13 polymorphic chloroplast microsatellite markers (cpSSR). Their 'plastogroup G-2' was distributed in south-eastern Turkey and northern Syria and was closely related to domesticated *timopheevii*.

The high level of karyotype diversity in wild ssp. *armeniacum* collected in northern Iraq (Badaeva et al. 1994, 2021), suggests that this is the region where *T. timopheevii* originated. Indeed, studies cof whole chloroplast genomes of wild *T. timopheevii*, ssp. *armeniacum,* and *Ae. speltoides* indicated that the wild subspecies of the former originated within the last 0.4 MYA, likely in the northern Iraq region around Arbil (Gornicki et al. 2014). Two accessions of wild *T. timopheevii* collected near Arbil had the dominant haplotype of *Ae. speltoides* chloroplast, hinting that they are the ancestral forms of ssp. *armeniacum*.

The region around Dahuk in northern Iraq can be considered the center of origin, but also the center of diversity of *T. araraticum*.

Gornicki et al. (2014), based on whole chloroplast genome sequence information and large taxon sampling, showed that *T. timopheevii* most probably originated in northern Iraq. According to Badaeva et al. (2021) data, the wild lineage that exists in northern Iraq belong to the ARA-0 lineage as no ARA-1 occurs in Iraq. This was supported by Bernhardt et al. (2017) showing that some ARA-0 and domesticated *timopheevii* genotypes are most closely related. Haplotype analysis of the Brittle rachis 1 (BTR1-A) gene in a set of wild *timopheevii* in comparison with two domesticated timopheevii accessions (Nave et al. 2021) also showed closer relationships of domesticated to wild *timopheevii* from Iraq.

Lineage ARA-1 of ssp. *armeniacum* grows in mixed populations with wild emmer, ssp. *dicoccoides,* in the Northern Levant, and is phylogenetically most closely related to *it*. ARA-1 is morphologically more similar to ssp. *dicoccoides*.

Based on karyotype analyses, translocation spectra and distribution of DNA probes wild populations from Dahuk and Sulaymaniyah, Iraq, harbored the highest karyotypic diversity among all ssp. *armeniacum* populations studied (Badaeva et al. 2021). They therefore, consider the region around Dahuk in Northern Iraq as the center of diversity of wild *timopheevii*, and this is probably the region

where ssp. *armeniacum* originated. This is supported by Nave et al. (2021), who found the highest haplotype diversity among ssp. *armeniacum* from Iraq, and by Bernhardt et al. (2017), Gornicki et al. (2014) who traced chloroplast haplotypes from *Aegilops speltoides* growing in Iraq via ssp. *armeniacum* (ARA-0) to domesticated *timopheevii* and *T. zhukovskyi*.

However, some FISH patterns suggested that domesticated *timopheevii* probably originated in Turkey and probably from ARA-1 This is supported by the following observations: (i) TIM and ARA-1 carry the pSc119.2 signal in the middle of 1A long arm, while this site was absent from ARA-0; (ii) all ARA-0 and most ARA-1 possessed the Spelt-52 signal on 6GL, but it is absent in all TIM and five ARA-1 genotypes from Gaziantep-Kilis, Turkey. The distribution of Spelt-1 and Spelt-52 probes on chromosomes of these five genotypes was similar to, and in accession IG 116165 (ARA-1 from Gaziantep) almost identical with TIM; (iii) the pAesp_SAT86 patterns on chromosomes $3A^t$, 4G, and 7G are similar in TIM and ARA-1 but differed from ARA-0. Differences between ARA-1 and TIM based on FISH patterns of some other chromosomes as well as the results of C-banding and molecular analyses suggest that extant ARA-1 genotypes are not the direct progenitors of TIM but that the ARA-1 lineage is most closely related to it. Based on AFLP, C-banding, FISH and *Jeli* retrotransposon markers, TIM was genetically most closely related to ARA-1. Additional evidence for the close relationship between TIM and ARA-1 lineages comes from allelic variation at the *VRN-1* locus of genome A (Shcherban et al. 2016). This analysis revealed a 2.7 kb deletion in intron 1 of *VRN-A1* in three *T. timopheevii* and four *T. araraticum* accessions, which, according to our data, belong to the ARA-1 lineage. However, at *Vrn-G1*, TIM from Kastamonu in Turkey (PI 119442) shared the same haplotype (*Vrn1Ga*) with ARA-1 samples, while TIM from Georgia harbored haplotype *VRN-G1* as found in ARA-0. These results suggest multiple introgression events and incomplete lineage sorting as suggested by Bernhardt et al. (2017, 2020).

Iran occupies a marginal part of the distribution range of wild *timopheevii*. An abundance of the pericentric inversion of the $7A^t$ chromosome in the Iranian group indicates that it is derived from Iraq. The karyotypically 'normal' genotype was probably introduced to Transcaucasia via Western Azerbaijan (Iran). The low diversity of FISH patterns and the low C-banding polymorphism of wild *timopheevii* from Transcaucasia indicate that this wild taxon, ssp. *armeniacum*, was introduced as a single event. Interestingly, the AFLP data suggested some similarity between ARA-0 from Armenia and Azerbaijan and domesticated *timopheevii*.

10.3.3.3 Ssp. *timopheevii* (Zhuk.) Löve and Löve (Domesticated *timopheevii*)

Triticum timopheevii ssp. *timopheevii* (Zhuk.) Löve and Löve, commonly known as sanduri wheat or timopheev's wheat, is the domesticated form of its wild progenitor. It is an endemic crop restricted to Transcaucasia, particularly to western Georgia (Zohary et al. 2012) and mainly used as a cereal for making bread, biscuits and cookies. The grains are also used to feed livestock and poultry, whereas the straw is used as a biomass for fuel, for thatching, and for making mats, carpets, and baskets.

The plant has delicate spikes and spikelets, and like the wild subspecies also ssp. *timopheevii,* shows close morphological similarities to ssp. *dicoccon* and *paleocolchicum,* the hulled types of *T. turgidum.* The relatively limited genetic diversity manifested by morphological variation of the domesticated form of *T. timopheevii,* was also detected at the molecular level (Fig. 10.6b).

Wild T. timopheevii contains both winter and spring types, whereas domesticated *timopheevii* is considered a spring type (Shcherban et al. 2016). In order to clarify the origin of the spring growth habit in ssp. *timopheevii,* allelic variability of the vernalization *VRN-1* gene was investigated in a set of accessions of both subspecies of *T. timopheevii,* together with *Ae. speltoides,* presumed to be the donor of the G subgenome to this tetraploid species. Among accessions of ssp. *armeniacum,* two large mutations were found in both *VRN-A1* and *VRN-G1* loci, which were found to have no effect on vernalization requirements (Shcherban et al. 2016). Spring ssp. *timopheevii* had *VRN-G1a* allele in common for the two subspecies, and two alleles that were specific (*VRN-A1f-ins, VRN-A1f-del/ins*). These two alleles include mutations in the first intron of *VRN-A1* and also share a 0.4 kb MITE insertion near the start of intron 1. Hence, Shcherban et al. (2016) suggested that this insertion resulted in a spring growth habit in a progenitor of ssp. *timopheevii,* which was probably selected for during subsequent domestication.

Domesticated *T. timopheevii* is presumably evolved in isolation from the more common *T. turgidum,* and hybrids between it and *T. turgidum* are reportedly sterile (Lilienfeld and Kihara 1934; Sachs 1953; Tanaka and Ichikawa 1972). The wild wheat from which it could have been derived is obviously ssp. *armeniacum* (Jakubziner 1932b; Dorofeev et al. 1980), which is also scattered across Transcaucasia, and with which the domesticated types are inter-fertile and share identical genomic constitution. It was probably domesticated in Georgia, where it was mostly grown mixed with domesticated diploid wheat, *Triticum monococcum* ssp. *monococcum,* or as a weed in fields of *Triticum aestivum.*

Domesticated *timopheevii* was part of the spring landrace Zanduri (a mixture of ssp. *timopheevii* and *T. monococcum*) (Zeven 1980). This landrace was cultivated in the humid and moderately cool climate zone of western Georgia, 400–800 m above sea level (Dorofeev et al. 1980). Martynov et al. (2018) reported that ssp. *timopheevii* is not drought tolerant.

Currently, the cultivation of ssp. *timopheevii* is limited to western Georgia, but the area of cultivation might have been larger in the past. Badaeva et al. (2021), while screened all available passport data of *T. timopheevii,* found two accessions of ssp. *timopheevii* that were collected from Turkey in the first half of the twentieth century. The two accessions harbor the normal karyotype of *T. timopheevii* and were characterized as a typical domesticated *timopheevii.* Assuming that the passport data of these two accessions is correct, Badaeva et al. (2021) supposed that *T. timopheevii* may have been cultivated in Turkey during the first half of the twentieth century, indicating that the cultivation range of ssp. *timopheevi* at this time was wider than the current one. Badaeva et al. (2021) speculate that the two ssp. *timopheevii* accessions were probably introduced from Transcaucasia to Turkey and may have been left over from unsuccessful cultivation or breeding experiments.

From the point of view of domestication and spread of domesticated wheats, the role of the wild and domesticated forms of *T. timopheevii* is apparently negligible. It is likely that in the Neolithic days, both wild *T. turgidum* and wild *T. timopheevii* were taken into cultivation in southeastern Turkey, northern Iraq, and western Iran, and that the early non-brittle, hulled wheat remains found in these places represent both stocks (Zohary et al. 2012). Judging from the present geographical distribution of wild *T. timopheevii,* as well as from its assumed distribution at the end of the Younger Dryas cold period (ca 12,900–11,700 years before present), the carbonized grains, spikelets, and clay impressions found at Cayonu Tepesi in southeastern Turkey (ca. 9000 BP) and Jarmo northern Iraq on the foothills of the Zagros Mountains (ca. 8750 BP) could belong to this taxon rather than to wild *T. turgidum.* If this is the case, then *T. timopheevii* was first domesticated in the northeastern part of the Fertile Crescent. Eventually, the brittle form of *T. timopheevii* gave rise to only a restricted number of non-brittle cultivars, all of which belong to ssp. *timopheevii,* which are currently cultivated in a few localities in Transcaucasia. Yet, the question remains: if such alleged domesticated *T. timopheevii* were indeed produced in south-west Asia, why were they fully replaced by *T. turgidum,* even among the local landraces (Zohary et al. 2012)? It is speculated that when emmer cultivation spread to Transcaucasia, local populations of wild *T. timopheevii* colonized emmer crops fields as a weed, and, became domesticated by being incorporated into the agricultural cycle of harvest and sowing (Nesbitt and Samuel 1996).

The oldest known records of prehistoric ssp. *timopheevii* are from Turkey, ALıklı Höyük in Cappadocia and Cafer

Höyük in southeastern Turkey (Cauvin et al. 2011). At three Neolithic sites and one Bronze Age site in northern Greece, spikelet bases of a "new glume wheat" type of hulled wheat have been recovered (Jones et al. 2000). These spikelet bases are morphologically distinct from the typical hulled wheats, *T. monococcum* ssp. *monococcum* (einkorn), *T. turgidum* ssp. *dicoccon* (emmer) and *T. aestivum* ssp. *spelta* (spelt), previously observed in Greece as well as in Neolithic and Bronze Age sites in Turkey, and southern and central Europe. The taxonomic identification of this new type remains uncertain, but it seems likely that they are tetraploid, and have morphological features in common with *T. timopheevii* (Jones et al. 2000). At the northern Greek sites, at least, the new type may have been cultivated as a mixed crop with einkorn and emmer. At some other sites, this wheat was grown as a minor component in wheat fields (Ulaş and Fiorentino 2021). Yet, in some places, new glume wheat was a major crop in itself (Bogaard et al. 2017; Ergun 2018).

Czajkowska et al. (2020) detected DNA sequences from the G subgenome in two samples of archaeological new glume wheat, one from Turkey and the second from Poland, thus providing evidence that this wheat is indeed a domesticated type of *T. timopheevii*. This confirms that ssp. *timopheevii* was domesticated from ssp. *armeniacum* during early neolithic age and was widely cultivated in the prehistoric past.

Vavilov (1935) suggested that ssp. *timopheevii* of western Georgia was probably introduced from northeastern Turkey. The possibility of introduction of ssp. *timopheevii* into Georgia from northeastern Turkey is supported by Dorofeev et al. (1980).

10.3.3.4 Cytology, Cytogenetics and Evolution

T. timopheevii is an allotetraploid species (2n = 4x = 28) that exhibits a diploid-like behavior at meiosis, namely, formation of 14 bivalents of homologous chromosomes. The genomic designation of its nuclear genome is GGAA (Lilienfeld and Kihara 1934; Kihara 1963; Kihara and Tanaka 1970; Kimber and Tsunewaki 1988; Dvorak 1998) and its organellar genome is designated G (Tsunewaki et al. 1976; Ogihara and Tsunewaki 1988; Tsunewaki 1989; Wang et al. 1997) or S (Dvorak 1998). Lilienfeld and Kihara (1934) analyzed chromosome pairing in F_1 hybrids between different subspecies of *T. turgidum* and wild and domesticated subspecies of *T. timopheevii* and found that the hybrids involving *T. timopheevii* differed from those involving subspecies of *T. turgidum*. The *T. turgidum x T. timopheevii* hybrids had poor chromosome pairing at first meiotic metaphase, and were sterile, whereas hybrids between different subspecies of *T. turgidum* exhibited almost complete chromosome pairing and were fertile. On the basis of chromosome pairing in these hybrids, they agreed with Zhukovsky (1928) that *T. timopheevii* should be considered

a different species from *T. turgidum*, and designated the genome of *T. timopheevii* by the formula AAGG in contrast to the BB formula given to the genome of *T. turgidum*. [Note: The subgenome symbols G and B, donated by the female parents of the allotetraploid wheats (see below), should be on the left side of the genomic formula]. Moreover, Lilienfeld and Kihara (1934) placed *T. timopheevii* in a section of its own, thus suggesting a different origin for this species as compared to that of *T. turgidum*, a suggestion that was followed by Svetozarova (1939), Sears (1948, 1969). Kostoff (1937a, b, c) obtained similar results to those of Lilienfeld and Kihara, but interpreted them as showing partial homology between the genomes of *T. timopheevii* and *T. turgidum*, and therefore suggested to designate the genome of *T. timopheevii* by the formula ßßAA.

The formula GGAA for the genome of *T. timopheevii* shows that this species contains one subgenome similar to the A subgenome of *T. turgidum* (genome BBAA) and to that of *T. aestivum* (BBAADD) but differs in the second subgenome. *T. timopheevii* also contains a distinct cytoplasm which induces male sterility when combined with the nuclear genome of *T. turgidum* and *T. aestivum* (Maan and Lücken 1971; Maan 1973; Tsunewaki et al. 1976; Tsunewaki 1989).

The results reported by Sachs (1953) broadly agree with those obtained by Lilienfeld and Kihara (1934) and by Kostoff (1937b, c), i.e., the F_1 hybrids *T. timopheevii* x *T. turgidum* showed considerably less chromosome pairing. Yet, Sachs followed Nabalek (1929) in identifying an accession, that was collected in northern Iraq, as *T. dicoccoides* Körn. var. *nudiglumis* Nabalek, in spite of the fact that the F_1 hybrids between ssp. *dicoccoides* and var. *nudiglumis*, exhibited reduced chromosome pairing and were completely sterile. Notwithstanding, because of the inclusion of var. *nudiglumis* in *dicoccoides*, Sachs (1953) concluded that the latter is characterized by a wide chromosomal variation and postulated that *T. timopheevii* arose from *dicoccoides* and that cytologically intermediate genotypes between the two varieties should exist.

Wagenaar (1961a, 1966) studied chromosome pairing in F_1 hybrids between *T. timopheevii* and *T. turgidum*. While his results were exactly in line with those obtained by previous researchers, he interpreted them differently. Wagenaar suggested that the pairing failure in these hybrids could be attributed to genes affecting chromosome pairing and chiasma formation and less to structural changes of the chromosomes. Consequently, he credited gene action and not chromosome structural differentiation as the origin of *T. timopheevii* speciation processes. Wagenaar (1966) considered the G subgenome as a modified B and proposed that ssp. *armeniacum* evolved after becoming isolated from ssp. *dicoccoides* through mutations of the gene complex controlling chromosomal pairing.

To identify accessions of tetraploid wheat that were collected by Jack R. Harlan in mixed populations of ssp. *dicoccoides* and ssp. *armeniacum* in Turkey, Rao and Smith (1968) crossed six of the accessions with four Israeli accessions of ssp. *dicoccoides*; the F₁ hybrids were then morphologically and cytogenetically analyzed. Both groups were also crossed with a number of other tetraploid wheats including *T. timopheevii*, *T. turgidum* ssp. *turgidum*, and ssp. *dicoccon*. Cytogenetically, the four Israeli accessions were similar and exhibited very close relationships with *dicoccon* and *turgidum*, but their hybrids with *timopheevii* showed poor chromosome pairing and were completely sterile. Four of the six Turkish accessions were similar to the Israeli group in pairing relationships and seed set percentages (Rao and Smith 1968). The remaining two Turkish accessions showed considerable cytogenetic differentiation. The Turkish accession 11,189 showed a close pairing relationship and some fertility with *timopheevii* and exhibited poor pairing and complete sterility in crosses with *dicoccon* and the Israeli ssp. *dicoccoides* group. Surprisingly, the Turkish accession 11,191 exhibited almost complete chromosome pairing and some fertility in crosses with both *timopheevii* and *dicoccon* (Rao and Smith 1968).

Rawal and Harlan (1975) studied chromosome pairing in meiosis of F₁ hybrids involving three Israeli accessions of ssp. *dicoccoides*, the six Turkish accessions of tetraploid wheat that were used by Rao and Smith (1968), and one line of *T. timopheevii*. Like Rao and Smith (1968), Rawal and Harlan (1975) found that four of the six Turkish accessions were cytologically similar to the Israeli *dicoccoides* accessions, and one Turkish accession (# 189) was cytologically similar to *T. timopheevii*, whereas the other Turkish accession (# 191) showed good chromosome pairing with both ssp. *dicoccoides* and ssp. *timopheevii*. Interestingly, the four Turkish accessions that were identified as ssp. *dicoccoides* had somewhat better chromosomal pairing with *T. timopheevii* (average number of chromosome association/cell was 22.0–22.8) than did the Israeli accessions (average number of chromosome association/cell was 18.8–21.5 (Rawal and Harlan 1975).

The results of Rao and Smith (1968) and those of Rawal and Harlan (1975) tend to support Wagenaar's contention that the pairing behavior in these taxa is primarily under genetic control. It is difficult to visualize genomic differentiation based on alteration of chromosome structure that would account for this behavior. If the differences were due to structure only, then a chromosome of the B subgenome that pairs with a *timopheevii* chromosome, should not pair with a *dicoccoides* chromosome (Rawal and Harlan 1975). Yet, in accession # 191, chromosomes paired with both subspecies to approximately the same degree. The fact that there is little or no morphological differentiation between the plants of the Turkish population with different chromosome

behaviors, suggests that there is no intermediate race, but rather, different genotypes within the population (Rawal and Harlan 1975).

The pattern of chromosome pairing in F₁ hybrids between a synthetic amphiploid *T. timopheevii-Ae. tauschii* ($2n = 6x = 42$; genome GGAADD), used as male parent, and 19 monotelocentric lines of *T. aestivum* cv. Chinese Spring, each containing one dose of a known chromosome arm, instead of a pair of the relevant chromosomes, was studied by Feldman (1966b). Formation of a heteromorphic bivalent (pairing between a complete *timopheevii* chromosome with a telocentric Chinese Spring chromosome) at meiosis of the F₁ hybrids, indicated that the telocentric chromosome was homologous or partly homologous to the corresponding arm of the *timopheevii* chromosome. A multivalent involving the telocentric chromosome indicated the occurrence of chromosome structural modifications. The pairing of the telocentric chromosomes clearly showed a large difference between the A and B subgenome chromosomes. Feldman (1966b) reported that six out of the eight arms of the studied A subgenome chromosomes showed high pairing and only two showed poor pairing, whereas only one arm out of the 11 arms of the B subgenome showed good pairing, whereas the other 10 arms paired very poorly. Telocentrics of the B subgenome differed markedly in pairing frequency (Feldman 1966b), indicating that some *T. timopheevii* chromosomes are more significantly differentiated than others. The poor pairing was mainly between chromosomes of the B and G subgenomes (Feldman 1966b). That is, while the A subgenome of *T. aestivum* is still relatively closely related to the A subgenome of *T. timopheevii*, the B subgenome is greatly diverged from the G subgenome. Indeed, recent whole genome sequencing showed that the G subgenome donor to wild *T. timopheevii* diverged from *Ae. speltoides* 2.86 (3.24–2.51) MYA (Li et al. 2022).

Multivalent chromosomes associations were observed in ten out of the 19 hybrid lines; four from A and six from B subgenomes chromosomes were involved in multivalent associations, indicating the existence of at least five translocations between the parental species. This poor chromosome pairing in hybrids between *T. aestivum* and *T. timopheevii* might be due to chromosome structural differences rather than to desynaptic genetic factors causing asynapsis when heterozygous (Wagenaar 1961a, 1966). Indeed, no such genes were found on any of the 26 arms of *timopheevii* tested by being made hemizygous in various hybrids. According to Wagenaar, these genes are assumed to prevent pairing of chromosomes only when a single dose of each gene from each parent is present, thereby explaining why there is asynapsis in the hybrids but not in the amphiploids derived from this hybrid, as was found by Sachs (1953). If these genes are actually present, it is difficult to understand how they can specifically affect the pairing of the chromosomes of the B-G subgenomes and why they have differential effects on different chromosome arms.

Maestra and Naranjo (1999) used the C-banding technique to study chromosome pairing at first meiotic metaphase in the F_1 of hybrids *T. turgidum* ssp. *durum* x *T. timopheevii* (hybrid genome BGAA), *T. aestivum* x *T. timopheevii* (hybrid genome BGAAD), and *T. turgidum* ssp. *durum* x *T. aestivum* (hybrid genome BBAAD). C-banding enabled them to identify and differentiate between individual chromosome arms of *T. timopheevii* and those of *T. turgidum* ssp. *durum* and *T. aestivum*. Maestra and Naranjo (1999) found that homologous pairing between the A subgenome chromosomes was similar in the three hybrid types BGAA, BGAAD and BBAAD, but B-G chromosome associations were less frequent than B-B associations. Homoeologous associations were also observed, especially in the BGAAD hybrids.

Originally, it was thought that the A subgenome of *T. timopheevii* derived from the A^m genome of *T. monococcum* (Lilienfeld and Kihara 1934). In accord with this, Kostoff (1937a) reported that in the F_1 *T. timopheevii* x *T. monococcum* triploid hybrid, the chromosomes of the A^m genome of *monococcum* paired most frequently with 7 chromosomes of *timopheevii*, while the other 7 chromosomes of *timopheevii* remained as univalents. Consequently, Kostoff (1937a) concluded that *T. timopheevii* has one subgenome homologous with genome A^m. Similarly, following observations of meiotic pairing in hybrids between the two tetraploid species, Wagenaar (1961a) and Tanaka et al. (1978) suggested that *T. monococcum* contributed the A subgenome to both tetraploid wheats. However, Dvorak et al. (1993), studying variations in 16 repeated nucleotide sequences, detected relatively little divergence between the repeated nucleotide sequences of the A subgenome of *T. timopheevii* and the genome of *T. urartu* and noted more variation between the A subgenome of *T. timopheevii* and the genome of *T. monococcum*. They therefore, concluded that the A subgenome of *T. timopheevii* was contributed by *T. urartu*.

Following analysis of chromosome pairing in hybrids, Riley et al. (1958) proposed that *Ae. speltoides* was the most likely diploid species that donated the G subgenomes to *T. timopheevii*. This proposal was reinforced by the cytogenetic study of Shands and Kimber (1973). In agreement with the above, Dvorak and Zhang (1990), studying variations in repeated nucleotide sequences, concluded that *Ae. speltoides* is the closest extant species to the G subgenome of *T. timopheevii*. The incomplete homology between S and G is accounted for by assuming that the G subgenome has undergone modifications at the tetraploid level. Owing to uniparental transmission of the cytoplasm through the female parent, the maternal parent of the wheat species was

traced back to *Ae. speltoides*, which chloroplast and mitochondrial DNA studies have shown is the cytoplasm donor to both the *timopheevii* and the *turgidum* lineages (Ogihara and Tsunewaki 1988; Terachi et al. 1990; Golovnina et al. 2007; Wang et al. 1997). Likewise, Kilian et al. (2007), using nuclear and cytoplasm DNA in studies of a large collection of polyploid wheats and *Ae. speltoides*, obtained evidence supporting the suggestion that *Ae. speltoides* is the donor of the B and G subgenomes to the allotetraploid wheats. In accord with the above, Gornicki et al. (2014), studying whole chloroplast genomes of a large number of *Triticum* and *Aegilops* accessions, found that the cytoplasm of *Ae. speltoides* is the closest to the cytoplasms of *T. timopheevii* and *T. turgidum*. However, the chloroplast genome of *Ae. speltoides* is more closely related to that of *T. timopheevii* than to the chloroplast genome of *T. turgidum* (Gornicki et al. 2014). Consequently, Gornicki et al. (2014), like Dvorak and Zhang (1990), assumed that either the female donor of the cytoplasm and B genome to the *T. turgidum* is either a relative of *Ae. speltoides* (perhaps even an extinct species), or the allotetraploidization event of *T. turgidum* occurred much earlier (about 300,000 years earlier) than that of *T. timopheevii* (Gornicki et al. 2014). As an older species, the *T. turgidum* B subgenome and cytoplasm exhibit more divergence from the S genome and cytoplasm of *Ae. speltoides* than the G subgenome and the cytoplasm of *T. timopheevii*. In addition, *T. turgidum* is considerably more polymorphic than *T. timopheevii* (Gornicki et al. 2014).

In the *T. timopheevii* x *Ae. speltoides* hybrids, chromosome pairing at first meiotic metaphase occurred preferentially between chromosomes of the G subgenome and the S genome (Rodriguz et al. 2000a). Pairing between G and S chromosomes reached frequencies higher than 50% in most of the arms. The reduced arm length (Gill and Chen 1987; Maestra and Naranjo 1998) may account for the pairing decrease between the 3GS-3SS and 5GS-5SS arms. Structural differences existing between the 1SS and 1GS and between 4SS and 4GS arms (Gill and Chen 1987; Jiang and Gill 1994b; Maestra and Naranjo 1999) affecting pairing between them. The 1GS arm with two translocated segments, an intercalary segment from 6AS, and a terminal segment from 4GS, did not pair with 1SS.

In vitro DNA: DNA hybridizations and hydroxyapatite thermal-elution chromatography were employed by Nath et al. (1985) to identify the diploid donor of the G subgenome of *T. timopheevii*. Total genomic, unique-sequence, and repeated-sequence fractions of ^{3}H-*T. timopheevii* DNA were hybridized to the corresponding fractions of unlabeled DNAs of the five Sitopsis species. The heteroduplex thermal stabilities indicated that *Ae. speltoides* was the most closely related species to the G genome of *T. timopheevii*. Dobrovolskaya et al. (2011) transferred microsatellite (SSR) markers from *Ae. speltoides* to *T. timopheevii* and found that most of the SSRs were integrated into chromosomes of the G subgenome rather than to chromosomes of the A subgenome. This also indicates a close relationship between the G subgenome and the S genome.

Strangely enough, the results of Takahashi et al. (2010) indicated that the wild and domesticated subspecies of *T. timopheevii* might have arisen independently via allotetraploidization, with both subspecies obtaining the A subgenome from *T. urartu*, but ssp. *armeniacum* obtaining the G subgenome from *Ae. speltoides* and ssp. *timopheevii* obtaining it from one of the other *Sitopsis* species, most probably from *Ae. searsii*. The group compared the 19th intron (PI19) sequence of the *PolA1* gene, encoding the largest subunit of RNA polymerase I. Two different sized DNA fragments containing PI19 sequences (PI19A and PI19G) were amplified both in ssp. *timopheevii* and ssp. *armeniacum*. The shorter PI19A (112 bp) sequences of both subspecies were identical to PI19 sequences of *T. urartu* and *T. monococcum*. Interestingly, the longer PI19G (241–243 bp) sequences of ssp. *armeniacum* showed more similarity to PI19 sequences of *Ae. speltoides*, whereas those of ssp. *timopheevii* showed more similarity to PI19 sequences of *Ae. searsii*.

Aegilops speltoides is in contact with *T. urartu*, the donor of the A subgenome, in southeastern Turkey, northwestern Iraq and western Iran. In these regions, there are also numerous mixed populations of these two species with wild *T. timopheevii*.

Yet, an interesting outcome of whole genome sequence of the Sitopsis species is the new insight on their relationship to the B and G subgenomes of allopolyploid wheats (Li et al. 2022). As previously shown by Marcussen et al. (2014), Glémin et al. (2019), the genomes of the Sitopsis species, the B and G wheat subgenomes, and the *A. muticum* genome fall into three clades corresponding to the A, B and D lineages. The A lineage consists of the genomes of diploid wheat and the A subgenomes of allopolyploid wheat. The B lineage includes the B subgenomes of the allopolyploid wheats, *Ae. speltoides*, and *A. muticum* (Glémin et al. 2019) and the G subgenome of *T. timopheevii*. The D-lineage includes the ancestral genomes of the subsection *Emarginata* species of *Aegilops*, the D genome and those of the remaining *Aegilops* species. The G subgenome sequence is not yet available, however, a substitution of most of chromosome 2B by an introgression from *T. timopheevii* (450 Mb in length) in the bread wheat cultivar (LongReach Lancer) whose genome has been sequenced (Walkowiak et al. 2020), enabled to include it into the phylogenetic analysis of the B lineage (Li et al. 2022). The timing of divergence between these genomes was determined based on full or partial genomic sequences. Interestingly, *Ae. speltoides* diverged from the B

subgenome donor ~ 4.49 MYA. Considering that wild emmer wheat (genome BBAA) was formed ~ 0.8 MYA (Marcussen et al. 2014), this rules out that *Ae. speltoides* is the direct progenitor of the B subgenome *A. muticum* was confirmed as belonging to the B-lineage and diverged from *Ae. speltoides* and B-subgenome donor at a more ancient time (~6.37 MYA), supporting the proposal of Glémin et al. (2019) that *A. muticum* is the extant representative most directly related to the B-lineage ancestor. *Ae. speltoides* diverged from the *T. timopheevii* G-subgenome donor ~ 2.85 MYA, i.e., after its divergence from the B-subgenome progenitor. This makes the donor of the G-subgenome substantially older than the estimated allotetraploidization time (< 0.4 MYA) leading to speciation of ssp. *armeniacum*, the wild progenitor of *T. timopheevii* (Gornicki et al. 2014). From this analysis, Li et al. (2022) concluded that *Ae. speltoides* is also not the direct donor to the G-subgenome of *T. timopheevii*, although the latter is more closely related to the G- than to the B-subgenome. The donor of the G-subgenome is thus thought to be a distinct species related to *Ae. speltoides*, that is either extinct or yet undiscovered (Li et al. 2022). This is consistent with earlier reports (Ogihara and Tsunewaki 1988) showing that *Ae. speltoides* shares near identical cytoplasm with *T. timopheevii* but not with *T. turgidum* and *T. aestivum*.

Ssp. *timopheevii* contains 11.87 ± 0.630 pg 1C DNA (Eilam et al. 2008), similar to the amount of DNA in wild ssp. *armeniacum* (11.82 ± 0.071 pg 1C DNA); domestication seemingly had no effect on genome size in this species. The genome of *T. timopheevii* is larger than the genomes of most *Aegilops* allopolyploids but significantly smaller than that of *T. turgidum* (Eilam et al. 2008). This in accord with Rees and Walters (1965), who determined DNA amounts in tetraploid wheats and found significantly less nuclear DNA in *T. timopheevii* as compared to *T. turgidum* (ssp. *durum*). The 1C DNA content of the G-subgenome donor is expected to be close to that of *Ae. speltoides* since DNA amount in *T. timopheevii* is equal to the additive amount of 1C DNA of *Ae. speltoides* (5.81 pg) and that of *T. urartu* (6.02 pg) (Eilam et al. 2007).

The karyotypes of wild and domesticated forms of *T. timopheevii* have identical chromosome morphology (Bozzini and Giorgi 1969). Both contain two pairs of SAT chromosomes and two pairs of submetacentric chromosomes, while the rest of the pairs are metacentric chromosomes. The karyotype of *T. timopheevii* is quite different from the karyotype of *T. turgidum*; Bozzini and Giorgi (1969) observed differences in total chromosome length and arm ratio between the two species, particularly in the SAT chromosomes. Moreover, chromosomes with a median centromere are more frequent in *T. timopheevii* than in *T. turgidum*. The differences in chromosome morphology between the two tetraploid wheats, reinforce cytogenetic data showing that *T. timopheevii* and *T. turgidum* have different chromosome architectures.

Hutchinson et al. (1982), using C-banding technique, showed that the G subgenome chromosomes of the wild and domesticated forms of *T. timopheevii* were banded, whereas those of the A subgenome showed no bands. Only one of the two pairs of satellited chromosomes had strong heterochromatic bands. The G chromosomes paired less in the triploid hybrids between *timopheevii* and diploid wheats as compared to the A chromosomes. Upon assessment of the relationship between the genomes of *T. timopheevii* and *T. turgidum* ssp. *dicoccon* at meiosis in F_1 hybrids between these two taxa, Hutchinson et al. (1982) concluded that the two species differ in the amount and distribution of heterochromatin and by several intra- and inter-subgenome translocations.

All the somatic chromosomes of domesticated *T. timopheevii* and those of two varieties of *T. aestivum*, were identified by Giemsa staining (Badaeva et al. (1986). Similar to the finding of Hutchinson et al. (1982), Badaeva et al. (1986) also found that the chromosomes of the A subgenome of *T. timopheevii* contained a small amount of heterochromatin located mainly in the centromeric region, whereas the chromosomes of the G subgenome contained large amounts of heterochromatin located in the telomeric, centromeric and intercalary positions. While chromosome 6G possesses a nucleolar organizer, that of chromosome 1G was translocated to chromosome 6A (Badaeva et al. 1986). Comparison of the C-banding patterns of *T. timopheevii* chromosomes with those of the homoeologous chromosomes of the A and B subgenomes of *T. aestivum* showed similarity between these chromosomes, although the G subgenome chromosome contained a larger amount of heterochromatin than the B chromosomes and exhibited differences in the location and size of these heterochromatic blocks. Studies of N-banded mitotic and meiotic chromosomes of *T. timopheevii* and *T. turgidum* revealed that the satellites are on the short arms of chromosomes 6A and 6G of *T. timopheevii* and of 1B and 6B of *T. turgidum* (Gill and Chen 1987; Badaeva et al. 1986),

Studies of C-banded somatic chromosomes of *T. timopheevii* (Hutchinson et al. 1982) and N-banded chromosomes of *T. timopheevii* and *T. turgidum* and studies of chromosome pairing in *T. timopheevii* x *T. turgidum* hybrids (Gill and Chen 1987) uncovered differences between these two species in several translocations. Such differences do not include translocation 5AL/4AL (Jiang and Gill 1994b; Maestra and Naranjo 1999) that both allotetraploid species inherited from *T. urartu*, the donor of the A subgenome to both species. In the emmer lineage, i.e., *T. turgidum* and *T. aestivum*, chromosome 4A showed three more species-specific rearrangements, i.e., translocation 4AL/7BS, a large

pericentric inversion, and a paracentric inversion in the long arm. Neither translocation 4AL/7BS nor the pericentric inversion exists in *T. timopheevii*, and whether the paracentric inversion of 4AL is present in *T. timopheevii* could not be determined (Jiang and Gill 1994a, b; Gill and Chen 1987; Maestra and Naranjo 1999, 2000). C-banding analysis of meiotic pairing in interspecific hybrids, revealed four species-specific translocations in *T. timopheevii* (Maestra and Naranjo 1999). A double translocation involving the arms 1GS, 6AS, and 4GS is present in both domesticated and wild forms of *T. timopheevii*. Two more translocations between the 4GS/4AL and 3AL/4AL arms, which are present in domesticated lines, probably also exist in wild forms. In addition to the 1GS/6AS and 1GS/4GS double translocations, 4GS/4AL/3AL translocations were also identified (Jiang and Gill 1994b). Analysis of the wild and domesticated forms of *T. timopheevii* by sequential N-banding and genomic in situ hybridization, showed that chromosomes 6A, 1G and 4G were involved in A-G inter-genomic translocations in all six lines analyzed (Jiang and Gill 1994b). These chromosomes may have derived from a cyclic translocation that is species-specific to *T. timopheevii*. All the species-specific translocations of *T. timopheevii* and *T. turgidum* occurred at the tetraploid level (Maestra and Naranjo 1999, 2000).

Rodríguez et al. (2000b), using genomic in situ hybridization (GISH) and C-banding analysis of meiotic configurations in the F_1 *T. timopheevii* x *T. turgidum* hybrids, found that four species-specific translocations (6AS/1GS, 1GS/4GS, 4GS/4AL, and 4AL/3AL) exist in *T. timopheevii*, and that *T. timopheevii* and *T. turgidum* differ in the pericentric inversion of chromosome 4A and the paracentric inversion of 4AL. Rodríguez et al. (2000b) concluded that both tetraploid species had undergone independent and distinct evolutionary chromosomal rearrangements.

Using C-banding analysis, Kawahara et al. (1996) studied the karyotypes at meiosis of F_1 hybrids between different accessions of ssp. *armeniacum*, and found that out of 18 translocations, 12 were between the G-subgenome chromosomes, five were between G and A subgenomes and one was between A subgenome chromosomes. Within the G subgenome, chromosomes 4G and 6G had higher frequency of translocations than the other chromosomes. Evidently, the G chromosomes have cytologically diverged much more than the A chromosomes and that chromosome restructuring has played an important role in the formation of intraspecific diversity of ssp. *armeniacum* (Kawahara and Tanaka 1977, 1983; Badaeva et al. 1990, 1994; Kawahara et al. 1996). Seventy-nine accessions (58.5% of the studied accessions) had the same type of translocation, and the F_1 hybrids between them exhibited complete chromosome pairing and regular meiosis. This translocation, considered as the standard translocation type, was found in all geographical regions, while other translocation types were mostly restricted to a single locality (Kawahara et al. 1996).

Zoshchuk et al. (2007) studied the distribution of the Spelt1 and Spelt52 repetitive DNA sequences on chromosomes of seven accessions of *T. timopheevii* ssp. *armeniacum*, two accessions of *T. timopheevii* ssp. *timopheevii* and one accession of *T. kiharae* (a synthetic amphiploid *T. timopheevii-Ae. tauschii*). Sequences of both repetitive DNA families were found mostly in the sub-telomeric chromosome regions of the G subgenome. The number, location and size of Spelt1 differed among the seven accessions of ssp. *armeniacum,* but were identical in the two ssp. *timopheevii* accessions and in *T. kiharae*. Spelt52 was detected in the sub-telomeric regions of chromosomes 1–4 of the G subgenome and its sites did not coincide with the Spelt1 sites. As with Spelt1, the distribution and signal intensity of Spelt52 varied in ssp. *armeniacum,* but were identical in ssp. *timopheevii* and *T. kiharae*. Comparison of the distributions of the Spelt1 and Spelt52 repeats in the G subgenome chromosomes of *T. timopheevii* and *T. kiharae* with those of *Ae. speltoides*, the putative donor of the G subgenome, revealed a decrease in both the number and size of the sites in the G subgenome. The decrease was assumed to result from repeat elimination during allopolyploidization and subsequent evolution of the allopolyploid wheats (Zoshchuk et al. 2007).

Adonina et al. (2015), using pSc119.2 (120 bp repeat from rye), pTa71 (45S RNA genes from *T. aestivum*), and pTm30 containing (GAA)$_{56}$ microsatellite sequence from *T. monococcum*) as probes in FISH experiments, observed great differences in the presence and polymorphism in the sequences that bind these probes between domesticated *T. timopheevii* and wild and domesticated *T. turgidum.* ssp. *timopheevii* had pSc119.2 signals predominantly on the G subgenome chromosomes and on 1AL and 5AS of the A subgenome, whereas in *T. turgidum*, pSc119.2 hybridized to 1AS and 4AL in all lines and to 5AS, 5AL, and 2AL in some lines. The pTa71 probe detected the NOR regions on the short arm of chromosomes 6A and 6G in *T. timopheevii*, whereas in *T. turgidum*, they were located on 1BS and 6BS and smaller one on 1AS. The probe pTm30 hybridized to all *T. timopheevii* G subgenome chromosomes and only to one site, on 6AS, in the A subgenome, whereas among three wild accessions of *T. turgidum*, 4 conserved and 9 polymorphic (GAA)$_n$ sites were observed in the A subgenome. The (GAA)$_n$ loci found on chromosomes 2AS, 4AL, and 5AL in the wild forms, were retained in the domesticated form of *T. turgidum* (ssp. *durum*) and in *T. aestivum*.

Badaeva et al. (2016) developed a set of molecular cytogenetic landmarks, based on eleven DNA probes, to characterize the different chromosomes of *T. timopheevii*. They found that the pTa535 sequence, derived from *T.*

aestivum, enabled identification of all the A subgenome chromosomes, whereas the G subgenome and some of the A subgenome chromosomes could be identified using the (GAA/CTT)$_n$ and pSc119.2 probes. The pAsSAT86, pAs1, Spelt-1 and Spelt-52 probes, as well as 5S and 45S rDNA, discriminated particular chromosomes or chromosomal regions. The distribution of (GAA/CTT)$_n$, pTa-535 and pSc119.2 on *T. timopheevii* chromosomes was distinct from that of *T. turgidum*.

Several available DNA probes were tested in in situ hybridization experiments seeking to identify the chromosomes of *T. timopheevii*. Hybridization patterns generated by the (GAA)$_7$ probe was consistent with the C-banding patterns, and enabled identification of 11 chromosome pairs. As reported by Badaeva et al. (2016), probes pAs1 and pSc119.2 produced hybridization bands on all chromosomes; nine chromosome pairs were clearly distinguishable by pSc119.2, while pSc119.2 in combination with pAs1 identified all *T. timopheevii* chromosomes. Probing of the putative diploid progenitors with total genomic DNA clearly distinguished between both subgenomes.

In wheat, grain texture is mainly determined by the *Hardness* (*Ha*) locus consisting of the *Puroindoline a* (*Pina*) and *b* (*Pinb*) genes. These genes were conserved in all diploid species of *Aegilops* and *Triticum*, but were deleted from the A and B subgenomes of all *T. turgidum* lines studied (Li et al. 2008). In contrast to *T. turgidum*, *Pina* and *Pinb* were eliminated from the G subgenome, but were maintained in the A subgenome of *T. timopheevii*.

Upadhya and Swaminathan (1965) concluded that *T. timopheevii* does not contain the *Ph1* gene that suppresses homoeologous pairing in interspecific wheat hybrids. Yet, studies of meiotic chromosomal pairing in 41- and 42-chromosome F$_1$ hybrids between *T. aestivum* monosomic for chromosome 5B and the amphiplods *T. timopheevii–Ae. tauschii*, revealed that chromosome 5G of *T. timopheevii* compensated for the absence of 5B of *T. aestivum*, namely, it suppresses homoeologous pairing (Feldman 1966a). Thus, Feldman concluded that *T. timopheevii* contains a gene system, presumably *Ph1*, occurring on chromosome 5G, similar to that found in chromosome 5B of *T. aestivum*. Evidence that chromosome 5G of ssp. *timopheevii* carries a gene like *Ph1* was also presented by Ozkan and Feldman (2001).

10.3.3.5 Theories Concerning the Phylogenetic Relationships Between *T. timopheevii* and *T. turgidum*

Since the designation of the genome of *T. timopheevii* GGAA (Lilienfeld and Kihara 1934), implying a genome different from the BBAA genome of *T. turgidum*, several theories were proposed concerning the evolutionary relationships between these two-allotetraploid wheats. Despite the intensive cytogenetic, biochemical and molecular studies, the phylogenetic relationships of these two-allotetraploid wheats still remain controversial and the question is to what extent the B and the G subgenomes are related is currently unresolved. The main theories concerning the relationships between these two-wheat species are presented below.

Theories Concerning Close Relationships Between the B and the G Subgenomes

Recent molecular studies show that the two allotetraploid species were formed in different times: wild emmer, *T. turgidum* ssp. *dicoccoides*, was formed 700,000–900,000 years ago (Marcussen et al. 2014; Gornicki et al. 2014; Middleton et al. 2014), and wild *T. timopheevii*, ssp. *armeniacum*, 400,000 years ago (Marcussen et al. 2014; Li et al. 2022). Wild emmer was formed via allopolyploidization process involving the B-subgenome donor, that diverged from *Ae speltoides* 4.49 MYA (Li et al. 2022), and the A subgenome donor that diverged from the *T. urartu* 1.28 MYA (Li et al. 2022). Wild ssp. *armeniacum* was formed from the G subgenome donor, that diverged from *Ae. speltoides* much later, 2.85 MYA (Li et al. 2022) and the A-subgenome donor. Since the formation of wild *T. timopheevii* was 300,000–500,000 years after that of wild emmer, the question is how much the B and the G subgenomes donor, both diverged from *Ae. speltoides*, and to a lesser extent the A subgenome donor, have diverged from one another.

Close genetic relationships between the B and the G subgenomes was suggested following observations of meiotic pairing in hybrids between the two-allotetraploid species or between *T. timopheevii* and *T. aestivum* (Sachs 1953; Wagenaar 1961a; Feldman 1966b; Tanaka et al. 1978). Sachs (1953) concluded that all the allotetraploid wheats originated from a common allotetraploid progenitor and that the hybrid sterility between *T. timopheevii* and *T. turgidum* was due to cryptic structural hybridity. This conclusion was supported by Sears (1956b), Riley et al. (1958). Wagenaar (1966) assumed that wild *T. timopheevii* arose from wild *T. turgidum* in northern Iraq, through a series of mutations which occurred simultaneously or in quick succession in both species and induced a strong sterility barrier between them. Several F_1 hybrids between ssp. *armeniacum* and ssp. *dicoccoides* from a mixed stand in eastern Turkey and northern Iraq, showed considerably high chromosome pairing (Rao and Smith 1968; Rawal and Harlan 1975; Tanaka and Kawahara 1976; Tanaka et al. 1978), suggesting that the two wild allotetraploid wheats are phylogenetically closely related.

The possibility of close phylogenetic relationship between the two-allotetraploid wheat species was also supported by chromosome pairing in hybrids, reported by Upadhya and Swaminathan (1965), Feldman (1966b), Tanaka et al. (1978), and by isozyme studies performed by Jaaska (1974). Feldman (1966b) found that the G subgenome of *T. timopheevii* is sufficiently closely related to the B subgenome of T. *turgidum* and may have differentiated from the latter as a result of extensive chromosomal rearrangements. All the studied telocentric chromosomes of *T. aestivum* subgenome B paired to some extent with the corresponding chromosomes of *T. timopheevii* (Feldman 1966b), indicating that the second subgenome of *T. timopheevii* is close to subgenome B of *T. turgidum*. The relatively close relationship between these two subgenomes was especially conspicuous in the relatively high pairing in hybrids between *T. timopheevii* and some varieties of T. *turgidum*, as observed by Kostoff (1937a, b), Wagenaar (1961a), Rao and Smith (1968), Rawal and Harlan (1975). Likewise, Tanaka et al. (1978) suggested that these two allotetraploid wheat species had differentiated from a common ancestor as the result of extensive chromosomal differentiation. A similar conclusion was reached by Badaeva et al. (1986), following comparison of C-banding in *T. timopheevii* versus *T. aestivum*. They concluded that the G and B subgenomes have a common ancestor, from which they differentiated at the tetraploid level, due to chromosomal rearrangements and increased heterochromatinization in the chromosomes of the G subgenome of *T. timopheevii*.

None of the current diploid species of *Triticum* and *Aegilops*, from which wild *T. timopheevii* and *T. turgidum* presumably derived, possess a *Ph1* gene identical to that of *T. aestivum* (Dhaliwal 1977c). This suggests that the *Ph1* gene originated at the tetraploid level. The fact that this gene occurs in both allotetraploid species (Feldman 1966a; Dhaliwal 1977c; Ozkan and Feldman 2001)), may suggest that either the donors of subgenomes B and G possess the *Ph1* gene, or a more likely alternative, *Ph1* evolved in wild *T. turgidum* and was transferred to wild *T. timopheevii* via introgressive hybridization.

Theories Concerning More Distant Relationships Between the B and the G Subgenomes

Other researchers bearing in mind that *T. timopheevii* and *T. turgidum* originated in two independent events. The GGAA genomic formula assigned to *T. timopheevii* by Kihara and Lilienfeld (1934) and supported by Svetozarova (1939), Sears (1948), suggested that the G subgenome derived from a different species than the B subgenome of *T. turgidum*. The different species-specific chromosome rearrangements that occurred at the tetraploid level in both lineages, are consistent with a diphyletic origin (Tanaka and Ichikawa 1972; Tanaka and Ishii 1975; Kawahara and Tanaka 1977; Hutchinson et al. 1982; Badaeva et al. 1990, 1995; Naranjo et al. 1987; Jiang and Gill 1994a, b; Maestra and Naranjo 1999, 2000; Rodriguez et al. 2000b). The analyses of chloroplast DNA (Ogihara and Tsunewaki 1988; Miyashita et al. 1994), mitochondrial DNA (Terachi et al. 1990), and chloroplast and mitochondrial DNA (Wang et al. 1997), also reinforce the diphyletic origin. Phylogenetic relationships among the plastotypes (plastid genotypes) of the two-tetraploid wheat species and those of the five species of the *Aegilops* of section Sitopsis, suggested a diphyletic origin of ssp. *dicoccoides* and ssp. *armeniacum* (Miyashita et al. 1994). The plastotype of one *Ae. speltoides* accession was identical to that of ssp. *armeniacum*, while three of the plastotypes found in the other Sitopsis species were very similar, but not identical, to that of ssp. *dicoccoides*. These studies imply that two different, perhaps closely related, species were the donors of the cytoplasm to the two-tetraploid wheats.

Mori et al. (1995) studied intra- and inter-specific variations in nuclear DNA of 32 lines of ssp. *dicoccoides* and 24 lines of ssp. *armeniacum* by RFLP. The average genetic distance between the ssp. *dicoccoides* accessions was 0.0135 ± 0.0031 and between the ssp. *armeniacum* accessions was 0.0036 ± 0.0015, indicative of about a four-fold intraspecific variation in ssp. *dicoccoides* as compared to ssp. *armeniacum*. The genetic distance between the two species was 0.0482 ± 0.0022, and when corrected for intraspecific divergence was 0.0395, about three times that for ssp. *dicoccoides* and 11 times that for ssp. *armeniacum*. These results showed that in the wild state, ssp. *dicoccoides* and ssp. *timopheevii* are clearly differentiated and that ssp. *dicoccoides* has much greater variation than ssp. *armeniacum*, suggesting a relatively more recent origin for the latter and therefore, a diphyletic origin for these species.

Pairing between G subgenome and S genome chromosomes in *T. timopheevii* x *Ae. speltoides* hybrids reached a frequency much higher than pairing between B subgenome and S genome chromosomes in *T. turgidum* x *Ae. speltoides* hybrids and in *T. aestivum* x *Ae. speltoides* hybrids, and between B and G subgenome chromosomes in *T. turgidum* x *T. timopheevii* or *T. aestivum* x *T. timopheevii* hybrids (Maestra and Naranjo 1999; Rodriguez et al. 2000a). These results, demonstrating a higher degree of closeness between the G subgenome and S genome than with the B subgenome, are consistent with the claimed diphyletic origins of the two-allotetraploid species.

The diphyletic mode of origin theory is supported by the fact that B and G subgenomes derived from different, albeit related, diploid donors as female and from *T. urartu* as male. Subsequent introgressive hybridizations between the two allotetraploids, sharing the A subgenome and differing in the B and G subgenomes, may have led to further differentiation of the differential subgenomes (Zohary and Feldman 1962).

In such hybrids, the two differentiated genomes, which have partial homologous chromosomes, can exchange genetic material and recombine. As a result of such hybridizations, the related differential subgenomes could have become more modified, acquired adaptive genetic combinations, leading to recombination of the B and G subgenomes. An intermediate types predicted by this hypothesis, was previously found by Sachs (1953) Wagenaar (1961a, 1966), Rao and Smith (1968), Rawal and Harlan (1975), and Tanaka and Kawahara (1976). Gornicki et al. (2014) provided molecular evidence that evolution of these allotetraploid wheats was also accompanied by chloroplast introgression. One accession of *T. turgidum* ssp. *dicoccoides* (G4991), for example, which showed high chromosome pairing with both *T. turgidum* and *T. timopheevii* (Rawal and Harlan 1975), carries the *T. timopheevii* chloroplast haplotype (*H09*) as a result of a cross between wild emmer and wild *timopheevii*. Conversely, wild *T. timopheevii* accession TA976 carries the emmer-lineage chloroplast haplotype (*H04*) (Gornicki et al. 2014).

An Independent Origin from Closely Related Donors

Up till now, it was accepted (e.g., Dvorak and Zhang 1990; Gornicki et al. 2014) that both, the B and the G subgenomes of the allotetraploid wheats derived from *Ae. speltoides* and therefore, their origin cannot be considered biphyletic. Likewise, Gill and Friebe (2002), proposed that two different genotypes of *Ae. speltoides* contributed to the formation of the allopolyploid wheats, one genotype participated in the formation of the *T. turgidum*–*T. aestivum* lineage and another in the *T. timopheevii*–*T. zhukovskyi* lineage. Yet, the two subgenome donors diverged from *Ae. speltoides* in different times and may have diverged considerably from each other in chromosome structure and cytoplasm. Moreover, these two species arose independently from two different allotetraploidization events, and as such, their origin can barely be considered monophyletic.

Hybridization between the two allotetraploids might have led to exchange of chromosomal material such as chromosome 4B of *turgidum* that is present in *timopheevii*; Gill and Chen 1987) and chromosome segments, including the possible transfer of the *Ph1* gene from chromosome arm 5BL of *turgidum* to 5G of *timopheevii* (Feldman 1966a). Additional evidence for possible introgression between these two-species is evident from the high pairing (in 80% of the cells) between chromosome arm 2BL of Chinese Spring and the corresponding arm of *T. timopheevii,* and between chromosome arms 1BL and 5BL (in about 50% of the cells) with *timopheevii* chromosomes (Feldman 1966b). Hybridization with diploids, mainly *Ae. speltoides* or other Sitopsis species, might have led to further differentiation of the G and B subgenomes. This might align with the finding of Dvorak and Zhang (1990), who observed two bands of diagnostic restriction fragments of species of sub-section Emarginata of section Sitopsis in the allopolyploid wheats, which could be considered as evidence of introgression.

A similar conclusion was reached by Brown-Guedira et al. (1996), who studied the phylogenetic relationships between *T. turgidum* and *T. timopheevii* by assessing the genetic compensation capacity of individual *T. timopheevii* chromosomes that substituted missing chromosomes of *T. aestivum*. Six *T. timopheevii* chromosome substitutions were isolated: 6A (6A), 2G (2B), 3G (3B), 4G (4B), 5G (5B) and 6G (6B). The substitution lines had normal morphology and fertility. Their findings indicated a common origin for the two tetraploid wheat species, but because of the presence of a different spectrum of inter-genomic translocations, the authors concluded rightly that the two-species originated from separate hybridization events. Likewise, Kilian et al. (2007) using AFLP markers and haplotypes of several nuclear and chloroplast loci, revealed that both the B and the G subgenomes derived from *Ae. speltoides* at different times from different genotypes of *Ae. speltoides*.

Allopolyploidization in the wheat group was followed by rapid genetic and epigenetic changes (Feldman et al. 1997, 2013; Levy and Feldman 2002, 2004; Feldman and Levy 2005, 2009; 2012). The genetic changes, mainly elimination and duplication of various low-and high-copy DNA sequences, may have led to genomic rearrangements, which could have accelerated the cytological and genetic diploidization of the nascent allopolyploids. On the other hand, if recurrent allopolyploidization events occurred from different genotypes of the diploid parents, the magnitude of genomic reorganization could have also impeded the genetic flow between the newly formed allopolyploids (Maestra and Naranjo 2000).

T. turgidum is an older species than *T. timopheevii* and was suggested to have originated 700,000–900,000 years ago (Marcussen et al. 2014; Gornicki et al. 2014; Middleton et al. 2014), while wild *T. timopheevii* was formed 400,000 years ago (Marcussen et al. 2014; Li et al. 2022). The B subgenome of *T. turgidum* derived from a species that diverged from *Ae. speltoides* 4.49 MYA (Li et al. 2022) and the G subgenome of *T. timopheevii* from a species that diverged from *Ae. speltoides* 2.85 MYA (Li et al. 2022). Wild *T. turgidum* was formed in the southern Levant, whereas wild *T. timopheevii* originated in northern Iraq (Gornicki et al. 2014).

10.3.3.6   Intraspecific Hybrids Involving ssp. *armeniacum* and ssp. *timopheevii*

The F_1 hybrid between ssp. *armeniacum* (mistakenly referred to as *T. dicoccoides* var. *nudiglumus*) and *T. timopheevii* had almost regular pairing at first meiotic metaphase (Table 10.8). A similar level of chromosomal pairing was

observed in the F_1 hybrid between ssp. *armeniacum* and ssp. *timopheevii* (Table 10.8). The multivalents observed in this hybrid by Tanaka and Ichikawa (1972) most probably resulted from reciprocal translocations occurring between the two studied lines of the two subspecies, rather than from homoeologous pairing. The high level of chromosomal pairing indicates the great homology between the genomes of the two subspecies of *T. timopheevii*.

10.3.3.7   Crosses with Other Wheat Group Species

Meiotic chromosomal pairing in the F_1 hybrid ssp. *timopheevii x ssp. monococcum* (2n = 3x = 21; genome GAA^m) (Table 10.9) indicated that the A subgenome of ssp. *timopheevii* is closely related to that of ssp. *monococcum*.

Data of meiotic chromosomal pairing in F_1 hybrids between ssp. *timopheevii* and diploid species of *Aegilops,* namely, *Ae. speltoides* (hybrid genome GAS), *Ae. bicornis* (hybrid genome GAS^b), and *Ae. tauschii* (hybrid genome GAD) are presented in Table 9.9. Shands and Kimber (1973) studied chromosomal pairing in F_1 hybrids of ssp. *timopheevii* with high-, intermediate-, and low-pairing types of *Ae. speltoides*. In the hybrid with the high-pairing type they observed somewhat higher pairing than in the hybrids with the intermediate- and low-pairing types, but in the latter two hybrids pairing was also high. From these data, Shands and Kimber (1973) concluded that the G subgenome of *T. timopheevii* is very loosely related to the S genome of *Ae. speltoides* and most probably derived from it. Hybrids between ssp. *timopheevii* and the other two *Aegilops* species exhibited significantly lower pairing. A similar low level of pairing was reported in the hybrid ssp. *timopheevii* x *Ae. caudata* (0–5 bivalents; Kihara 1949). Low level of chromosomal pairing was also observed in the F_1 tetraploid hybrids between ssp. *armeniacum* and *Ae. peregrina* (hybrid genome GAS^{v}U). The F_1 hybrids between two lines of the wild subspecies of *T. timopheevii* and *Ae. peregrina* exhibited very low chromosomal pairing (Table 10.10). A hybrid with one line of ssp. *armeniacum* (line TIA28) had significantly lower pairing than the hybrid with another *armeniacum* line (TIA02). This pattern of pairing indicates that both subgenomes *of Ae. peregrina* (S^v and U) are not close to any of the two subgenomes of *T. timopheevii*.

Meiotic pairing in hybrids between *T. timopheevii* and different subspecies of *T. turgidum* indicated that only one subgenome in each species is closely related (Table 10.8). The level of chromosomal pairing in the F_1 hybrid between hexaploid wheat *T. aestivum* ssp. *aestivum* and *T. timopheevii* ssp. *armeniacum* (hybrid genome BADGA and in the hybrid *T. aestivum* ssp. *macha* x *T. timopheevii* ssp. *timopheevii* (hybrid genome BADGA) included 7–14 bivalents (Kihara 1949), shows that *T. timopheevii* shares

Table 10.10 Chromosome pairing at first meiotic metaphase of F_1 hybrids between Allopolyploid species of *Triticum* and *Aegilops*

Combination	Genome[a]	Univalents	Bivalents			Multivalents		References
			Rod	Ring	Total	III	IV	
Ae. peregrina x ssp. *dicoccoides* (TTD20)	S^{v}UBA	26.64	0.68	–	0.68	–	–	Ozkan and Feldman (2001)
Ae. peregrina x ssp. *dicoccoides* (TTD24)	S^{v}UBA	25.20	1.40	–	1.40	–	–	Ozkan and Feldman (2001)
Ae. peregrina x ssp. *dicoccoides* (TTD22)	S^{v}UBA	23.92	1.88	0.08	1.96	0.10	–	Ozkan and Feldman (2001)
Ae. peregrina x ssp. *dicoccoides* (TTD06)	S^{v}UBA	16.48	4.75	0.16	4.91	0.53	–	Ozkan and Feldman (2001)
Ae. peregrina x ssp. *timopheevii* (TIA28)	S^{v}UGA	19.04	3.90	0.10	4.00	0.32	–	Ozkan and Feldman (2001)
Ae. peregrina x ssp. *timopheevii* (TIA02)	S^{v}UGA	16.76	4.68	0.14	4.82	0.48	0.04	Ozkan and Feldman (2001)
ssp. *durum* x *Ae. neglecta*	BAUX^n	22.71	2.37	0.01	2.38	0.16	0.02	Cifuentes et al. (2010)
ssp. *aestivum* x *Ae. peregrina*	BADS^{v}U	30.86	2.02	0.02	2.04	0.02	–	Ceoloni et al. (1986)
ssp. *aestivum* x *Ae. peregrina*	BADS^{v}U	34.00	0.50	–	0.50	–	–	Ozkan and Feldman (2001)
ssp. *aestivum* x *Ae. peregrina*	BADS^{v}U	30.51	2.16	0.02	2.18	0.05	–	Yu and Jahier (1992)
ssp. *aestivum* x *Ae. geniculata*	BADM^{o}U	29.67	–	–	2.62	0.03	–	Riley (1966)
ssp. *aestivum* x *Ae. triuncialis*	BADUC	26.97	–	–	3.77	0.17	–	Riley (1966)
ssp. *aestivum* x *Ae. juvenalis*	BADD^{c}X^{c}U	–	–	–	4.33	0.98	0.32	Riley (1966)
Ae. vavilovii x ssp. *aestivum*	D^{c}X^{c}S^{s}BAD	–	–	–	0.60	0.37	–	Chapman and Miller (1978)
6x *Ae. crassa* x ssp. *aestivum*	D^{c}X^{c}DBAD	24.53	4.22	2.34	6.45	1.31	0.12	Melnyk and McGinnis (1962)

[a] Genome symbols after Dvorak (1998)

only one subgenome, subgenome A, with *T. aestivum*; the other three subgenomes, B, D, and G, are homoeologous.

10.4 Section Triticum (Speltoidea Flaksb.) (2n = 6x = 42)

10.4.1 Description of the Section

Section *Triticum* [Syn: *Triticum* L sect. *Spelta* Dumort; *Triticum* L. sect. *Speltoidea* Flaksb.; *Triticum* L. 'congregatio' Hexaploidea Flaksb.] contains two species: *T. aestivum* L. and *T. zhukovskyi* Menabde & Ericz. *T. aestivum* is an allohexaploid (2n = 6x = 42; genome BBAADD), whereas *T. zhukovskyi* is an allo-auto-hexaploid (2n = 6x; genome GGAAA^{m}A^m). The two species share subgenome A and differ by the other subgenomes. The cytoplasm of *T. aestivum* derived from the B-subgenome donor, a close relative of *Ae. speltoides*, and that of *T. zhukovskyi* derived from the G-subgenome donor which is closely related to *Ae. speltoides*. The donors of the B and G subgenomes are currently extinct or yet not discovered.

The two species contain only domesticated forms, except for var. *tibetanum* of *T. aestivum* ssp. *aestivum* which is a feral form that escaped from a domesticated cultivar of *T. aestivum*

due to a mutation in one of the *Br* (brittle rachis) genes, leading to plants with fragile rachis. This subspecies is grown near edges of wheat and barley fields in Tibet (Shao et al. 1983; Guo et al. 2020). *T. aestivum* contains five domesticated subspecies, ssp. *spelta*, ssp. *macha*, ssp. *aestivum*, ssp. *compactum*, and ssp. *sphaerococcum*. ssp. *aestivum* (common or bread wheat) is the most important crop of wheat, constituting about 95% of global wheat production. It is cultivated in all wheat-growing regions of the world and mainly used for bread. The other extant domesticated subspecies of *T. aestivum* are locally grown. *T. zhukovskyi* is a small species and only grown as a minor crop in western Georgia. Thus, among the hexaploid wheats, two natural groups can be observed, reflecting a biphyletic process of evolution (Mac Key 1975).

10.4.2 *T. aestivum* L. (Genome BBAADD)

10.4.2.1 Description of the Species

Triticum aestivum L., known as Dinkel wheat, [Syn.: *T. hybernum* L.; *T. sativum* Lamarck.; *T. vulgare* Vill.; *T. vulgare* Host] is an annual, predominantly autogamous plant. Culms are erect and glabrous, hollow, with thin walls, nodes are sometimes hairy, 65–150 cm tall (excluding spikes), and with 4 to 6 internodes. The young shoots are erect, semi-erect, or prostrate. Leaf blades are flat and linear, glaucous, or more rarely, yellowish-green in color, and pubescent in most forms, but glabrous in some. The spikes are awned with short awns of uniform length all over, or only on the upper spikelets, or, in few forms, are awnless. Spikes are determinate, squarehead in cross section or lax, dense or lax, 4 to 18 cm long (excluding awns), with 14–30 spikelets. The rachis is tough and smooth, and fringed, with short hairs along its margins, and a few hairs immediately below the base of each spikelet. In some forms, the rachis is semi-tough and breaks into individual spikelets upon threshing. The spikelets are 10–15 mm long, solitary at the nodes, and glabrous or hairy. The top spikelet is fertile and at right angle to the plane of lateral spikelets, each spikelet contains 3–9 florets, and those near the center of the spike bear 2–5 grains, whereas those near the base or the tip of the spike contain 1–2 grains only. The glumes are oblong, asymmetrical in size, shorter than or almost equal in length to the rest of the spikelet, glabrous or pubescent, with white, yellow, red, brown, or blue-black, 6–11 mm long, with 5–7 nerves, and terminate with a short tooth or, in some of the bearded forms, extend into a slender, short awn (1–3 cm long). The lemmas are thin, rounded, and with 7–11 nerves. In bearded forms, they carry awns 5–10 cm long, usually straight, which are yellowish-white, reddish, black, or brown. The caryopsis is 5–10 mm long, 3.0–4.2 mm wide and 2.5–4.0 mm thick, free and non-adherent to the lemma and palea (free-threshing), but in some forms, is adherent to lemma and palea (hulled). The grains are white, yellow, orange, or red, generally plump, usually with a shallow furrow, and with a brush of hairs at the apex. The endosperm is flinty, semi-flinty, or mealy. The embryo is about 1/5 the length of the caryopsis (Fig. 10.7).

Currently, *T. aestivum* is the main cultivated wheat species, including bread (common) wheat and a number of domesticated forms, all of which are hexaploids with the BBAADD genome, and inter-fertile when crossed with one another. Mac Key (1954b) revised the classification of the hexaploid wheats based on genetic and agronomic features. He drew attention to the fact that the main differences between the various forms of *T. aestivum* are due to a small number of genes affecting the morphology of the plant, mainly the spike. Consequently, Mac Key (1954b) suggested to classify these forms at the intraspecific level as subspecies of *T. aestivum* rather than to consider them as separate species, as was done by several others (e.g., Dorofeev et al. 1980; Goncharov 2011). More specifically, Mac Key (1954b) classified *T. aestivum* as having five subspecies, two of which are hulled (glumed), ssp. *spelta* (L.) Thell., and ssp. *macha* (Dek. & Men.) Mac Key, and three of which are free-threshing, ssp. *aestivum*, ssp. *compactum* (Host) Mac

Key, and ssp. *sphaerococcum* (Percival) Mac Key. The one feral form, semi-wild, ssp. *tibetanum*, presumably escaped from wheat cultivation of bread wheat in Tibet (Shao et al. 1983). Mac Key's classification was adapted by van Slageren (1994), who found that it expresses the various groups as a botanical taxon, similar to the higher ranks of species and genus (van Slageren 1994).

Except for ssp. *aestivum*, which is of global importance, the other subspecies are only locally cultivated. The products of threshing in the hulled subspecies are individual spikelets, whereas free grains in the free-threshing ones. Currently, the hulled subspecies are cultivated as relic crops, while the free-threshing ones are more in extensive use.

The principal differences between the major forms of *T. aestivum* are due to one or two genes that affect gross morphology (Mac Key 1954b). Studies of the speltoid mutations occurring in ssp. *aestivum* and changing the morphology of the spike from a squarehead to lax and glumes from loose (free-threshing) to adherent (hulled type), showed already that these two traits were found to be genetically associated with a simple pattern of inheritance (Nilsson-Ehle 1917). Philiptschenko (1934) proposed that these two traits are controlled in *T. aestivum* by the genes q, which regulates the squarehead-spike and k, which suppresses speltoidy. Following the genetic association reported by Nilsson-Ehle (1917), Watkins (1940) assumed that q and k are closely linked, and Watkins and Ellerton (1940) showed that this is indeed the case. Through studies of induced deficiency-duplication mutants, Mac Key (1954a) found that the recessive $q + k$ genes are in fact one codominant factor, which he named it Q. This factor was llocated to the long arm of chromosome 5A (Sears 1954). The phenotypic influence of Q proved to be much broader than merely suppression of its homoeologous counterpart q. It affects a large number of traits in roots, stems, leaves, and especially spikes. Investigations by Mac Key (1954a, b) showed that Q determines the main free-threshing and domestication-related traits distinguishing between free-threshing ssp. *aestivum* and the hulled ssp. *spelta*, and, as such, it is one of the most significant domestication loci. While Q on 5AL has the most significant contribution, other homoeoalleles (on 5BL and 5DL) were also shown to be involved in the domestication traits (Zhang et al. 2011).

Q is thought to be a central gene regulating floral development, encoding an AP2-like transcription factor that plays an important role in the activation of a number of genes in hexaploid wheat (Simons et al. 2006; Zhang et al. 2011). Actually, Q, being a transcription factor (Simons et al. 2006; Zhang et al. 2011), has a pleiotropically influence on several domestication-related traits in addition to free-threshing, such as spike density and length (square-head spike), fertility of the basal spikelets, glume shape and

Fig. 10.7 Plants and spikes of
Triticum aestivum L.;
a ssp. *spelta* (L.) Thell.;
b ssp. *aestivum* cv. Chinese
Spring on the left side and of an
additional cultivar with awns on
the right; **c** spikes of two cultivars
of ssp. *compactum* (Host) MK
(From Percival 1921); **d** ssp.
sphaerococcum (Percival) MK;
e ssp. *tibetanum* J. Z. Shao (feral
wheat)

tenacity, glume keel formation, rachis fragility, plant height,
spike emergence time, and chlorophyll pattern along nerves,
as well as grain size and shape.

The mutation from *q* to *Q* presumably occurred at the
tetraploid level, from where it was transferred to hexaploid
wheat (Muramatsu 1986; Simons et al. 2006). It has been
suggested (Dvorak et al. 2012) that the tetraploid parent of
hexaploid wheat was a free-threshing form. This was also
apparent from the fact that all the tetraploids with the genome BBAA, extracted from hexaploid cultivars, were
free-threshing (Kerber and Rowland 1974). Muramatsu
(1963) proposed that *Q* is a triplication of *q*, since five doses
of *q* conferred the same phenotype as two doses of Q.
However, *Q* was recently isolated and characterized (Faris

et al. 2003; Jantasuriyarat et al. 2004; Simons et al. 2006),
and was shown by Simons et al. (2006) to have likely arisen
through a gain-of-function mutation, and not from a duplication of q. *Q* and *q*, differ by a single nucleotide (GAG in
q and GCG in *Q*), leading to a change in one amino acid,
namely, all *q*-containing forms had valine in position 329,
whereas all *Q*-containing forms possessed an isoleucine at
this position (Simons et al. 2006). *Q* is more abundantly
transcribed than *q* (Simons et al. 2006). The higher expression of *Q* is in accord with the finding of Muramatsu (1963),
who showed that extra doses (five or six) of *q* mimic the
effect of *Q* in bread wheat. Increased transcription of *Q* was
most obviously associated with spike compactness and
reduced plant height, as in plants tetrasomic for chromosome

5A (Sears 1954). In fact, the increased transcription of *Q* is apparently not due to the valine by an isoleucine substitution at position 329, as originally thought (Simons et al. 2006), but to a SNP in miRNA 172 binding site (Debernardi et al. 2017). Subsequent work by Debernardi et al. (2017) showed that in fact the SNP within the miRNA binding site is the causal polymorphism for the functional difference between the *Q* and *q* alleles. This is consistent with Q dominance, and with the increased transcription due to the reduction of miRNA 172 suppressive effect (Debernardi et al. 2017).

The non-free-threshing trait in synthetic hexaploids, irrespective of whether their tetraploid parent carried *Q* or *q,* was linked to the *Tg* gene, derived from *Ae. tauschii* (Kerber and Rowland 1974). Monosomic and telosomic analysis of synthetic hexaploid lines revealed the presence of *Tg*, a partially dominant gene for tenacious glumes, on the short arm of chromosome 2D (Kerber and Rowland 1974). Some variation in the degree of glume tenacity was noted among the synthetic hexaploids; those having ssp. *dicoccon* as a parent and containing *q* and *Tg* were the most difficult to thresh. The interaction between *Tg* and *Q* was clearly demonstrated by extraction of the tetraploid component of a hexaploid wheat and then resynthesizing the hexaploid. The extracted tetraploids containing only the BBAA component of the original free-threshing hexaploids were also free-threshing. In further studies, crosses between the extracted free-threshing tetraploids and *Ae. tauschii* produced spelta-like (hulled) hexaploids. It was therefore concluded that the *Tg* gene of *Ae. tauschii* inhibits the expression of *Q*. The interaction between *tg* and *Q*, conferring the free-threshing character, is complementary (Kerber and Rowland 1974), with a requirement for presence of both *tg* and *Q* to obtain a free-threshing character in hexaploid wheat. Some variation in the degree of threshability occurs among the many lines of free-threshing hexaploid wheat (Kerber and Rowland 1974), which is assumed to reflect the different genetic backgrounds. The probability that the genotypes of *Ae. tauschii* which served as the progenitors of hexaploid wheat, possessed *Tg*—as apparently do all extant forms of this species—supports the above hypothesis that hulled, hexaploid wheats are more primitive than

free-threshing hexaploids; they carried the *Tg* gene and, therefore, were non-free threshing (Kerber and Rowland 1974). The mutation from *Tg* to *tg* is presumed to have occurred at the hexaploid level.

Simonetti et al. (1999) found that while chromosome arm 2BS of wild emmer contains the *Tg* gene that determines tough glumes, that of domesticated emmer contains the *tg* allele, determining soft glumes. Thus, the free-threshing trait in tetraploid wheat is determined by two complementary genes, *Q* on 5AL and *tg* on 2BS (Simonetti et al. 1999). Faris et al. (2014) found that, in wild emmer, both chromosome arms 2AS and 2BS, carry *Tg* alleles, thus requiring a change of *Tg* to *tg* on 2AS as well, in order to obtain a free-threshing form. Further fine genetic mapping by Sharma et al. (2019) showed that emmer wheat contains two dominant homoeoalleles on 2A and 2B (*Tg-A1* and *Tg-B1*), which together with the *q* wildtype allele are responsible for the non-free-threshing phenotype.

The key gene separating ssp. *aestivum* and ssp. *compactum* is *C*, dominant in *compactum* and recessive in *aestivum* (Nilsson-Ehle 1911), which is located on chromosome XX (currently 2D) (Unrau 1950). The major gene separating these two subspecies from ssp. *sphaerococcum* is *S*, which is dominant in *aestivum* and *compactum* and recessive in *sphaerococcum* (Ellerton 1939) and located on chromosome XVI (currently 3D) (Sears 1946). Hence, the major genes determining the differences between the subspecies of *T. aestivum* are described in Table 10.11.

T. vavilovii (Tumanian) Jakubz. (syn: *T. vulgare* Vill. var. *vavilovii* Tumanian; *T. aestivum* L. ssp. *vavilovii* (Tumanian) Sears), is a branching mutant discovered by Tumanian in 1930, as an admixture in a stand of a bread wheat landrace in Armenia. In the 1970s, it was also found as an admixture in Azerbaijan, by Mustafaev, and in Armenia, by Gandilyan. Plants with branched spikes of the *vavilovii* type were produced by Mac Key (1966), including in progenies of some intraspecific crosses. Stable mutants like *vavilovii,* that never were released as commercial cultivars, should be made synonyms under the cultivated subspecies from which they were isolated and not to regard them as a special species or subspecies (van Slageren 1994).

Table 10.11 Genetic characterization of the various subspecies of *T. aestivum*

Sub-species	Genotype[a]	Phenotype
spelta *macha*	*TgTgqqccSS* *TgTgQQccSS* *tgtgqqccSS*	Hulled; normal spike and grains
aestivum	*tgtgQQccSS*	Free-threshing; normal spike and grains
compactum	*tgtgQQCCSS*	Free-threshing; compact spike and normal grains
sphaerococcum	*tgtgQQccss*	Free-threshing; normal spike and spherical grains

[a] The four genes concerned are located as follows: *Tg* (tenacious glumes) on chromosome arm 2DL (Kerber and Rowland 1974); *Q* (free-threshing) on 5AL (Sears 1954); *C* (compact spike) on 2DL (Rao 1972); and *S* (spherical grain) on 3DS (Rao 1977)

There are no wild hexaploid wheats, although feral forms of ssp. *aestivum* and ssp. *macha* are found. A feral form of ssp. *aestivum*, namely, var. *tibetanum*, presumably escaped from wheat cultivation in Tibet (Shao et al. 1983). This form has brittle rachis and grows as weed in edges of wheat and barley fields, but not in well-defined primary habitats. A feral form of ssp. *macha*, that grows as weed in edges of wheat fields, was identified in Georgia (Dekaprelevich 1961). Both of these forms are presumably feral derivatives of domesticated wheats, rather than a truly-wild taxa.

The free-threshing subspecies of *T. aestivum*, i.e., *aestivum*, *compactum* and *sphaerococcum*, are considered to be more advanced than the hulled subspecies *spelta* and *macha*. The primitive status of hulled *T. aestivum* is supported by the observation that artificial hybridizations between almost all subspecies of *T. turgidum*, either free-threshing or hulled, with all known races of *Ae. tauschii*, gave rise to hulled types (McFadden and Sears 1946; Kerber and Rowland 1974). The hulled subspecies, being the primitive forms, are the predecessors of the more advanced, free-threshing types. Among the free-threshing forms, ssp. *aestivum* gave rise to ssp. *compactum* and ssp. *sphaerococcum* through mutation. While ssp. *aestivum* is grown world-wide, ssp. *compactum* is grown today in restricted areas of Europe, the Near East and the north-western USA, and ssp. *sphaerococcum* is grown in parts of India and central Asia.

The genetic data which show that the first hexaploid wheats were hulled, spelt-like, and more primitive than the free-threshing subspecies, do not agree with the archaeological chronology. While free-threshing forms of *T. aestivum* were found at the middle of the 9th millennium BP and were abundant in the pre-historic Near East from the 8th millennium onwards, thus far, archaeological evidence of ssp. *spelta* dates to only one thousand years later. Neolithic, Near Eastern ssp. *spelta* is very rare, but there is evidence of *spelta* grains from Yarim Tepe II, northern Iraq, dating back to the 7th millennium BP and probably also from Yarim Tepe I, about one thousand years earlier (Kislev 1984). Earlier evidence for the existence of ssp. *spelta* is still missing. This discrepancy between the genetic and archaeological data posed some difficulties in tracing the early history of the hexaploid wheats. However, assuming that the first hexaploid wheats were hulled, their very low occurrence in the prehistoric remains of the Near East suggest their lack of advantage over domesticated emmer in that area. ssp. *spelta* is grown today in the Near East in extreme environments, such as the high plateau of west-central Iran, eastern Turkey, and Transcaucasia. This cultivation is possibly of an ancient origin. In central Europe, *spelta* appeared at ca. 4000 BP, about 2000 years later than forms of free-threshing hexaploid wheat. It could have been brought to Europe, where it replaced the free-threshing type in many sites of the upper Rhine region, particularly at high altitudes

where extreme temperatures prevail. Alternatively, *spelta* could have arisen in the Rhine valley through mutation of *Q* to *q*, or *tg* to *Tg*, or as a result of a cross between a free-threshing hexaploid form and a hulled tetraploid form, ssp. *dicoccon*, both of which were grown in that area. The relatively wide distribution of ssp. *spelta* in central Europe in the past was presumably due to its winter hardiness and ability to out-yield the other crops on poor soils. It was also preferred for its good quality. Spelt wheat is still cultivated today in several areas of central Europe and in the plateau of western Iran.

Ssp. *macha* was derived from mutations in either ssp. *spelta* or from ssp. *aestivum*. Some forms of this subspecies may contain the *C* (compact spike) allele, but others contain the *c* allele for normal spike. It is cultivated in a restricted area in Transcaucasia.

Ssp. *aestivum* gave rise to ssp. *compactum* (club wheat) and *sphaerococcum*, through mutations. This assumption is based on the fact that no line of *Ae. tauschii* has been found to carry the *compactum* gene *C* or the *sphaerococcum* gene *s*, both of which are located on D-genome chromosomes (Rao 1972, 1977), indicating that these mutations appeared at the hexaploid level. The fact that hulled hexaploid wheat does not carry these genes, shows clearly that neither ssp. *compactum* nor ssp. *sphaerococcum* could have been the first free-threshing hexaploids. The lineage of ssp. *compactum* from ssp. *aestivum*, entails only a single mutation from *c* to *C*, believed to have occurred in the Near East. Subsequently, *compactum* was transported to Europe as an admixture with emmer-einkorn, and was established as the dominant form in several locations, such as in the Lake Dweller area in Switzerland. ssp. *compactum* is grown today in a few restricted areas of Europe, the Near East, and the northwestern United States. Similarly, ssp. *sphaerococcum* originated from a single mutation in *aestivum* (*S* to *s*). This mutation presumably occurred in an *aestivum* that had been carried eastward, since *sphaerococcum* has not been found in the prehistoric Near East and its culture nowadays is largely confined to India. ssp. *sphaerococcum* is known to have been in India as early as the 5th millennium BP, and currently grows, to some extent, in India and Pakistan.

T. aestivum originated southwest of the Caspian Sea (Wang et al. 2013). As man migrated to new areas, cultivated wheats encountered new environments, to which they responded with bursts of variation resulting in many endemic forms. Secondary centers of variation for tetraploids in the Ethiopian plateau and the Mediterranean basin and for hexaploids in the Hindu Kush area of Afghanistan, were described by Vavilov (1987). Transcaucasia is one such secondary center for both tetraploid and hexaploid types. Secondary centers of diversity are valuable to wheat breeders, as they present gene pools additional to those existing at the primary centers of variation. More recent studies are also pointing to the Caspian-sea origin of the D subgenome: whole genome analysis of a core collection of 278 accessions covering the eco-geographic distribution of *Ae. tauschii,* in arid and semi-arid habitats from central Asia, Transcaucasia to China, confirmed that the wheat D subgenome is mostly derived from the *strangulata* subgroup originating from the south Caspian-sea and further narrowing down the origin to accessions from the Mazandaran province (Zhou et al. 2021). A recent analysis of 242 *Ae. tauschii* accessions showed that a rare and distinct lineage (different from *strangulata*) from Transcaucasia also contributed $\sim 1\%$ on average to the current wheat D subgenome (Gaurav et al. 2021).

An enormous amount of variation has developed in *T. aestivum,* as reflected in its very wide morphological and ecological variation. This hexaploid wheat species exhibits a wide range of genetic flexibility, as shown in its adaptation to a great variety of environments. While tetraploid wheats, in keeping with their Near Eastern origin, are adapted to mild winters and rainless summers, the addition of the central Asiatic *tauschii* (genome D) must have contributed to the adaptation of hexaploid wheats to a more continental climate and northern latitudes. This could have greatly facilitated the spread of bread wheat into parts of Europe and through the highlands of Iran to central and eastern Asia.

10.4.2.2   Ssp. *spelta* (L.) Thell. (Dinkel or Large Spelt)

Description of Subspecies

Ssp. *spelta,* known as large spelt, Dinkel, or hulled 6 x wheat [Syn.: *T. spelta* L.; *T. zea* Host; *Spelta vulgare* Seringe; *T. vulgare spelta* Alef.; *T. aestivum* var. *spelta* (L.) L. H. Bailet; *T. aestivum* ssp. *transcaucasicum* Dorof. & Laptev], is an annual, predominantly autogamous, plant. Culms are erect, stiff, 60–120 cm long (excluding spikes), glabrous or pubescent, hollow and with thin walls. Leaf sheaths have few hairs or glabrous, auricles are very large, curved, and fringed, with long hairs, and ligules are membranous. Leaf blades are 30–60 cm long, glabrous or sparsely hairy, those of young plants are dark green, and relatively narrow, whereas the leaves of the older culms are pale greenish-yellow, and broader. The spikes are relatively long, 10–15 cm long, lax, straight or slightly curved, white, red, grey-blue or blue-black, determinate, and awned or awnless. Each spike possesses from 16 to 22 spikelets, which are usually well separated from each other on the rachis. The rachis is flattened and smooth, with hairy margins, with a very small or absent frontal tuft. It is semi-brittle at the nodes and breaks easily with pressure (as during threshing), with each spikelet carrying the rachis segment below it (wedge

type) and/or beside it (barrel type). The spikelets are oval, 12–16 mm long, solitary at nodes, and glabrous or hairy. The top spikelet is fertile and at right angle to the plan of lateral spikelets. Each spikelet contains 3–4(–5) florets, those near the center of the spike bear 2 (rarely 3) grains, whereas those near the base or the tip of the spike contain 1 grain only. The Two glumes of each spikelet have a similar length, 8–12 mm long, white, yellow, red, brown, or blue-black, shorter than the spikelet, broad, with a truncate apex, 2 keels that are keeled all along, 11 nerves, with one ending in a short tooth or, in some of the awned forms, extending into a slender, short awn (1–3 cm long). The lateral nerve ends in a blunt bulge, which is always far from the apical tooth. The lemma is boat-shaped relatively thin, 9–13-mm-long, keeled above, and with 9–11 nerves. In the awned varieties, it terminates in a stiff, 6–8 cm long awn. Sometimes the lemma of the third flower bears a short, 2-cm-long awn. The palea is about as long as the lemma, oval, with two keels, 2 veins, and short hairs on their fringe. The caryopses is reddish, 7–10 mm long, with a flinty endosperm, and long and pointed at both ends. The apex is covered with a brush of white hairs, adhered to the palea and lemma and enclosed by the firm glumes (hulled type). The furrow of the grain is shallow. The weight of one thousand grains is 50–58 g (Fig. 10.7a).

The three domesticated hulled wheats, einkorn (*T. monococcum* ssp. *monococcum*), emmer (*T. turgidum* ssp. *dicoccon*), and spelt (*T. aestivum* ssp. *spelta*), are known as *farro* (Szabó and Hammer 1996; van Slageren and Payne 2013). In early times, the European ssp. *spelta* was not distinguished from domesticated emmer, and only during the thirteenth century, a distinction was made between the two crops (Percival 1921).

The hulled character of the three crops results from the semi-brittle nature of the rachis, and the toughness of the glumes. Because of these features, the products of threshing are not grain but rather individual spikelets. While in einkorn and emmer the rachis segment below the spikelet remains attached to the threshed spikelet, in spelt, the segment attached is either below or beside the spikelets or both. After threshing, an additional grinding process is required to separate the grain from the closely investing glumes (Percival 1921). This enclosure in the glumes of hulled wheats gives excellent protection to the grains against birds and animals and ensures extended viability of the grains (Nesbitt and Samuel 1996).

Ssp. *spelta* was formed, presumably repeatedly, in cultivated fields, as a result of hybridization between domesticated tetraploid wheat *T. turgidum,* and a diploid *Aegilops* species, *Ae. tauschii,* that infested cultivated fields of tetraploid wheat as weed (McFadden and Sears 1944, 1946; Kihara 1944). Such infestation of wheat fields, either tetraploid or hexaploid, by the weedy *Ae. tauschii*, also exists today in northern Iran (Matsuoka et al. 2008). All synthetic amphiploids *turgidum–tauschii* are hulled, similarly to spelt and irrespective of the threshability of the tetraploid parent (McFadden and Sears 1946; Kerber and Rowland 1974). Thus, the first hexaploid wheat would, therefore, have been a hulled wheat.

ssp. *spelta* exhibits wide morphological variation, mainly in spike awnedness, glume pubescent, glume and awn color and grain color (Percival 1921; Szabó and Hammer 1996). It contains many botanical varieties, which are classified according to the presence or absence of awns, presence or absence of hairs, and spike color (Percival 1921; Szabó and Hammer 1996).

Spelt distributes over two large geographical areas, Europe and Asia. While the European spelts are known for several millennia (ca. 4000 years), the Asiatic group was discovered in the 1950s (Kuckuck and Schiemann 1957). In fact, ssp. *spelta* had already been discovered earlier in Iran. In 1877, Andre Michoux saw spelt wheat growing wild north of Hamadan, western Iran, and, in 1807, Olivier (1807) found wheat, barley, and spelt in uncultivated areas northwest of Anah, on the right bank of the Euphrates, and mentioned that he had already seen such wheat several times in northern Mesopotamia. Both these finds are quoted by de Candolle (1886).

The Asiatic type presumably originated from the hybridization of tetraploid x *Ae. tauschii* and yielded the free-threshing wheat, ssp. *aestivum,* through mutation(s), whereas the European *spelta* apparently descended from a cross between ssp. *aestivum* and ssp. *dicoccon* (Tsunewaki 1968, 1971). To distinguish between these two spelt types, Kislev (1984) designated the Asiatic *spelta* "*TgQ spelta*" and the European spelt, carrying the *tg* gene, "*tgq spelta*". Thus, a *TgQ spelta* arose as a result of crosses between *Ae. tauschii* and the free-threshing (naked) tetraploid wheat ssp. *parvicoccum* or ssp. *durum*, which presumably contained the Q factor (Muramatsu 1986). Only one mutation, from *Tg* to *tg*, was required to derive free-threshing hexaploid wheats from such *spelta* (Kerber and Rowland 1974). On the other hand, ssp. *spelta* formed from domesticated emmer and *Ae. tauschii* would have carried both the *Tg* and *q* alleles, supplying a double dosage for hulledness. Since the chance for the occurrence of mutations in both of these genes within several centuries is small, it is more likely that the first mutation, *q* to *Q*, occurred in domesticated emmer fields, forming naked tetraploid wheat, while the second mutation, *Tg* to *tg*, occurred in *TgQ spelta* fields, forming bread wheat. The synthetic *spelta* obtained by McFadden and Sears (1946), likely contained both factors responsible for hulledness, namely, *q* and *Tg*, as it was obtained by hybridization of the hulled domesticated emmer and *Ae. tauschii*.

After bread wheat and emmer were established in Europe, the European type of ssp. *spelta* appeared, mostly north of the Alps. This *spelta*, which carries *q*, could have been derived from bread wheat through a back-mutation of *Q* to *q* or, more likely, via hybridization with emmer wheat (genotype *Tgqq*), as suggested by Tsunewaki (1968, 1971). While the European ssp. *spelta* is of the *tgq spelta* type, some of the Iranian spelta contain Tg (Dvorak et al. 2012). Thus, on the basis of genetic data, one can explain how Kuckuck obtained free-threshing types, ssp. *aestivum,* among the progenies of crosses between Iranian and European spelt wheats (see Kuckuck 1964). A cross between a *TgQ spelta* and a *tgq spelta* yields, besides the two parental types, the two recombinants, *tgQ* and *Tgq*, with the former being a free-threshing type.

Flaksberger (1925) already proposed that the European spelt derived in Europe from bread wheat. Such a hypothesis suggesting an independent origin of European and Asian spelt, was confirmed by several biochemical studies. von Büren (2001), based on polymorphisms of two γ-gliadin genes in various diploid, tetraploid and hexaploid species of wheat, and Blattner et al. (2002), based on analysis of genes of high molecular weight glutenin subunits of European and Asian spelt, concluded that the European spelt originated from crosses of free-threshing hexaploid wheat and domesticated emmer. Support for this hypothesis was also presented by Poltoretsky et al. (2018), who reviewed literature dealing with the genetic evolution of the European and Asiatic groups of ssp. *spelta*.

Archaeological Evidence

Nesbitt and Samuel (1996) reviewed the early archaeological evidence of ssp. *spelta* in the Near East and in southeast Europe. As mentioned above, evidence of Neolithic Near Eastern spelt is very rare. One of the earliest pieces of evidence is from Yarim Tepe II, northern Iraq, dating back to the 7th millennium BP (Bakhteyev and Yanushevich 1980). Other forms of evidence are from the 7th millennium BP in Transcaucasia (Lisitsina 1984), from north of the Black Sea (Janushevich 1984). A large number of glume imprints of spelt from Moldavia date to between 6800 and 6500 BP (Körber-Grohne 1987), while Popova (1991) reports three minor Neolithic and Chalcolithic occurrences in Bulgaria. These findings led Zohary and Hopf (1993) to regard the archaeobotanical evidence from Transcaucasia and the Balkans as consistent with the hybridization of emmer and *Ae. tauschii* near the Caspian belt, and its travel to Europe by way of the north shore of the Black Sea. However, Nesbitt and Samuel (1996) questioned this conclusion. They drew attention to the facts that, in all the archaeological records from outside of Europe, spelt was usually present in small proportions as compared to the other wheats, and its identification criteria were poorly documented. The most common identification criterion of spelt in these wheat mixtures is the barrel-type of disarticulation. However, fragmentary remains of spelt spikelets can be difficult to distinguish from those of the *Aegilops* species that also have barrel-type disarticulation and are common weeds in wheat fields, namely. *Ae. tauschii, Ae. cylindrica, Ae. crassa* and *Ae. juvenalis.* While most of these *Aegilops* species grow in the Near East, *Ae. cylindrica* does grow north of the Black Sea and in the Balkan, which could account for some spelt identifications there. Hence, these records of spelt from outside of Europe are still doubtful (Nesbitt and Samuel 1996).

Spelt was discovered in the 1950s in central Iran, growing as a crop at elevations of 2000–2300 m (Kuckuck and Schiemann 1957). Both domesticated emmer and ssp. *aestivum* were also important crops in the same area (Kuckuck and Schiemann 1957). The presence of spelt in this Iranian region is either a remnant of the original spelt form, resulting from the ancient hybridization of *T. turgidum* and *Ae. tauschii,* or, alternatively, a relatively younger crop derived from a recent hybridization(s) of tetraploid wheat and *Ae. tauschii* or of ssp. *aestivum* and domesticated emmer.

The absence of spelt from archaeological remains of southwest and central Asia remains a mystery. While spelt is expected to be well adapted to the highlands of Turkey or Afghanistan, there are no ancient or modern records of its existence in these areas (Nesbitt and Samuel 1996). Perhaps spelt was replaced, shortly after its formation, by the free-threshing ssp. *aestivum* that derived from it by a mutation of *Tg* to *tg*, and, since it was free-threshing, it was preferred by the Near Eastern farmers.

In contrast to the Asiatic spelt, ssp. *spelta* appeared in central Europe about 1000 years later than forms of free-threshing wheat (Nesbitt and Samuel 1996). Spelt remains occur at later Neolithic sites (4500–3700 BP) in eastern Germany and Poland, Jutland and possibly two sites in southwest Germany (Körber-Grohne 1989). During the Bronze Age, it spread widely in northern Europe (Nesbitt and Samuel 1996).

Spelt wheat replaced the dense-ear, free-threshing type (thought to be ssp. *compactum* (Heer 1866) but suggested by Kislev (1979/1980) to a free-threshing tetraploid form) grown by the lake-dwellers in the upper Rhine region, particularly at high altitudes where temperatures were extreme. Alternatively, and more likely, it could have arisen in the Rhine valley as a result of a cross between a hexaploid dense-eared form and tetraploid ssp. *dicoccon,* both of which were grown in the area.

Spelt was not grown in Egypt, and all the currently available evidence suggest that it was not known to the ancient Greeks and Romans (Percival 1921; Nesbitt and Samuel 1996). No remains of spelt have been found in the Neolithic Age in Europe. Yet, spelt was an important staple in parts of Europe from the Bronze to medieval times. Currently, spelt wheat is a relic crop, with cultivation confined to small areas in Europe and Asia. In Europe, the largest volume is grown in Bavaria, where it has been cultivated from the earliest times instead of bread wheat. It is grown on a smaller scale in parts of Prussia, Hesse and Alsace. Some is also grown in Switzerland, Austria, northern parts of Spain, France, and Italy. Small amounts are also grown in north Africa. In Asia, it is grown in several regions in Iran, central Asia, and Transcaucasia (Kuckuck and Schiemann 1957; Dorofeev 1969; Zohary et al. 2012). Spelt was introduced to the United States in the 1890s, but during the twentieth century, it was replaced by bread wheat in almost all areas where it was grown. However, with the growing trend for organic agriculture and "health food", there is increasing interest in this wheat. In addition to its higher nutritional value as compared to bread wheat, spelt is more tolerant to poor soil conditions, and resistant to a range of fungal diseases, such as smut, bunt, and rust. It is one of the hardiest of cereals, rarely affected by cold and frosts, and also grows well at all elevations. Because of its stiff culms, it is also more resistant to lodging than the other subspecies of hexaploid wheat and tolerates sprouting.

Although spelt is slightly less productive than bread wheat and possesses the disadvantage of a semi-brittle rachis and tough glumes, from which the grain cannot be easily threshed, it has advantages which enable it to compete successfully with bread wheat in areas with more marginal conditions. Its greater winter hardiness is of greatest importance. Another point in its favor is its smaller loss from attack of birds. Most spelt cultivars in Europe are winter forms, although a few less hardy spring varieties are also cultivated.

Nutritional Quality of ssp. *spelta*

The nutritional value of spelt is somewhat higher than that of bread wheat, with larger amounts of grain protein, dietary fibers, B vitamins and minerals (e.g., Suchowilska et al. 2019). In addition, Ruibal-Mendieta et al. (2005) found that ash, copper, iron, zinc, magnesium, and phosphorus contents were higher in spelt samples than in bread wheat. On the other hand, phytic acid content was 40% lower in spelt than in bread wheat. Analysis of two spelt cultivars showed differences in protein, iron and Zinc content (Rodríguez-Quijano et al. 2019).

Spelt has a relatively high-grain protein content, although a great variation of protein content, presumably due to different growing conditions, has been documented. The average grain protein percentage ranges from 12.7 to 19.0% (Belitz et al. 1989; Graber and Kuhn 1992; Perrino et al. 1993; Ranhotra et al. 1995; Cubadda and Marconi 1996).

The gluten of spelt flour is suitable for pastry, pasta, muesli, flakes, puddings, and soups. It is also used to make special kinds of bread. In comparison to hard red winter wheat of ssp. *aestivum*, spelt has a more soluble protein matrix, characterized by a higher gliadin:glutenin ratio. The subunits composition of high molecular weight glutenin shows extensive polymorphism, covering a large number of different allelic combinations in all loci. The wet gluten content in spelt varieties is about 40% higher than in a compared cultivar of bread wheat, but the quality of gluten in spelt varieties is lower. The total yield of flours in spelt varieties was only 70% of that of a check bread wheat cultivar (Capouchova 2001). In mean quality parameters, a marked difference was observed between the bread wheat and spelt cultivars, indicative of weaker gluten and dough in the spelt varieties studied.

In Germany and Austria, spelt loaves and rolls (Dinkelbrot) are widely available in bakeries, as is spelt flour in supermarkets. In addition, the unripe spelt grains are dried and eaten as green grain. In Poland, spelt breads and flour are commonly available as health foods and easy to find in bakeries (Defrise and Jacqmain 1984; Graber and Kuhn 1992; Boller 1995). Beer brewed from spelt is sometimes seen in Bavaria and spelt is distilled to make vodka in Poland.

Organic agriculture and health food products have been gaining increasing popularity (Cubadda and Marconi 1996; Lacko-Bartošová et al. 2010) and has revived spelt popularity, as consumption of spelt-wheat-based products provides for increased intake of minerals, vitamins and dietary fibers. With an important role in decreasing the glycemic index of final products (Lacko-Bartošová et al. 2010), spelt has become a common wheat substitute for making breads, pastas, and flakes.

10.4.2.3   Ssp. *macha* (Dek. & Men.) MK

Triticum aestivum ssp. *macha*, known as macha wheat [Syn.: *T. Macha* Dekapr. & Menabde; *T. sativum spelta* Hackl.; *T. spelta* ssp. *macha* (Dek. & Men.) Dorof; *T. aestivum* L. group *macha* Bowden] is an annual, predominantly autogamous, plant. Culms are erect, 60–100 cm long (excluding spikes), glabrous or pubescent, hollow, and thin-walled. The leaf sheath hairy, auricles are curved, and ligules are membranous. Leaf blades are 20–60 cm long, and 10–15 mm wide, with a scabrous, glabrous, or pubescent surface. The spikes are relatively compact, 10–12.5 cm long, 8–10 mm wide, bilateral, vary in density from open to dense, determinate, solitary at nodes, and glabrous or hairy, and bear short awns. The top spikelet is fertile and lies at right angle to the plane of lateral spikelets. Each spike possesses 15–23 spikelets. The rachis is flattened, hairy on the margins, and with a very small or absent frontal tuft. Spikelets are densely packed broadside to the rachis. Rachis

internodes are oblong, 3–5 mm long, fragile (semi-brittle) at the nodes and break easily with pressure (as during threshing), with each spikelet carrying the rachis segment below it (wedge type). Spikelets are solitary, sessile, ellipsoid, laterally compressed, and 15–18 mm long, and are comprised of 2–3 fertile florets, of which the upper one is sterile. Glumes are similar in length and shorter than the spikelets. The lower glume is lanceolate, or elliptic; 10–11 mm long, with 2-keels, keeled all along. The surface of the lower glume is hairy, its apex bearing a unilateral tooth, truncate. The upper glume is lanceolate or elliptic, pubescent, and 10–11 mm long, with 2 keels, and lateral veins divergent at the apex, which has a truncate tooth. Lemma is ellipsoid, pubescent, 11–13 mm long, and keeled, with an acute, awned, 4.0–6.5 cm long. Palea are two –veined and keeled. The caryopsis is elliptical, red, intermediate in hardness, hairy at its apex, with a linear hilum, and adheres to the lemma and palea.

Ssp. *macha* was discovered in 1928 and described by Dekaprelevich and Menabde (1932). It is endemic to western Georgia, growing as a minor component in admixture with the hulled west Georgian tetraploid *T. turgidum* ssp. *paleocolchicum* (Dorofeev et al. 1980). The two taxa are morphologically similar and it is difficult to distinguish one from the other (Jakubziner 1959). The similarity is so great that Flaksberger (1938) considered them as one species. However, the leaves of ssp. *macha* are coarser than those of ssp. *paleocolchicum*. Being at two different ploidy levels, hybridization between ssp. *macha* and ssp. *paleocolchicum* is difficult and produces highly sterile hybrids, whereas it crosses relatively easily with different subspecies of *T. aestivum*, yielding a fertile progeny (Jakubziner 1959).

Bhaduri and Ghosh (1955) stated that although very restricted in distribution, ssp. *macha* shows remarkable diversity of forms, with as many as eight varieties already described. Mosulishvili et al. (2017) also noted the great morphological variation of ssp. *macha*, particularly in the brittleness of the rachis. They mention that 16 varieties were recognized in this subspecies.

Ssp. *macha* has been assumed to descend from ssp. *palaecolchicum* (Jakubziner 1959). Its discoverers, Dekaprelevich and Menabde (1932) regarded the two hulled subspecies of *T. aestivum*, ssp. *macha* and ssp. *spelta*, as closely related, resembling each other in the semi-brittleness of the rachis and the toughness of the glumes. Because of this variation, they consider ssp. *macha* as a basic polymorphic group from which ssp. *spelta* probably originated. However, it is distinct from ssp. *spelta* in the method of disarticulation of the rachis (Cao et al. 1997). In ssp. *macha*, disarticulation of the rachis, due to pressure during threshing, results in individual spikelets, with the rachis internode attached below each spikelet (like in emmer wheats; wedge type) (Kabarity 1966; Singh et al. 1957). In contrast, ssp. *spelta* frequently disarticulates into spikelets, each with the rachis segment beside each spikelet (like in *Ae. tauschii;* barrel type) (McFadden and Sears 1946). Another difference between ssp. *macha* and ssp. *spelta* is in the shape of the spike; in ssp. *macha*, the spike is dense, while ssp. *spelta* has a lax spike. By means of monosomic analysis, Joshi et al. (1980) allocated the gene which controls spike compactness in ssp. *macha* to chromosome 6B. A gene symbol C^m was proposed for this character, to distinguish it from the C gene of ssp. *compactum*, which is located on 2D. In addition, the F_1 hybrids between *spelta* and *macha* exhibit some pairing failure at meiosis (Chin and Chwang 1944). Hybrids between *macha* and *spelta* or free-threshing *T. aestivum* are weak and are sterile, indicating that ssp. *macha* also differs cytogenetically from the other hexaploid wheats.

Poltoretsky et al. (2018) mentioned that according to several Russian scientists, ssp. *macha* was regarded as the likely parent of the free-threshing hexaploid subspecies of *T. aestivum.* This hypothesis is based on the fact that free-threshing forms are obtained among progenies of crosses between ssp. *macha* and Asiatic ssp. *spelta.* Conversely, in view of its limited distribution and its status as a minor crop component, Nesbitt and Samuel (1996) regarded ssp. *macha* as a local form that evolved in isolation, rendering it unlikely to have played any role in the evolution of the other forms of *T. aestivum.*

Tsunewaki (1971) determined the necrosis and chlorosis genotypes (*Ne* and *Chl* genes) of 13 lines of ssp. *macha* and 105 cultivars of ssp. *spelta*, by crossing them with appropriate testers. Results of this and previous investigations (Tsunewaki 1968) indicated that common wheat, ssp. *aestivum*, differentiated into two geographical populations, i.e., Asian and western. The Asian population is characterized by a high frequency of *Ne1*-carriers, while the Western population has more *Ne2*-carriers. A strong isolation barrier was found between ssp. *macha* and the other subspecies of *T. aestivum*, due to the complementary chlorosis genes, *Chl* and *Ch2.* Tsunewaki (1968, 1971) assumed that European spelt originated from *Ne2*-carrying free-threshing ssp. *aestivum* and *q*-carrying hulled ssp. *dicoccon*, while the origin of ssp. *macha* was assumed to have derived from crosses between ssp. *aestivum* and *Chl*-carrying ssp. *dicoccon.* Accordingly, at least, two independent introgressions of genes from domesticated emmer wheat seem to have played important roles in subspecies differentiation in hexaploid wheat.

Comparative and molecular genetic analyses suggest that macha wheat is a segregant from a cross between wild emmer wheat, ssp. *dicoccoides*, and bread wheat, ssp. *aestivum* (Tsunewaki 1968, 1971). These researchers suggested that it is likely that ssp. *macha*, as well as other west Georgian wheats, are sibling cultivars that arose in a hybrid swarm involving ssp. *aestivum* and wild emmer wheat.

Corroborating this suggestion, Gornicki et al. (2014) found that the ssp. *macha* cytoplasm derived from the haplotype H04 of wild emmer cytoplasm.

The ssp. *macha* cytoplasm is of the B type, as that of the other subspecies of *T. aestivum* and *T. turgidum* (Wang et al. 1997). To assemble a phylogeny tree of Georgian allopolyploid wheats containing the B cytoplasm, Gogniashvili et al. (2018) determined complete nucleotide sequences of chloroplast DNA of 11 representatives of these wheats. According to the simplified scheme based on SNP and indel data, the predecessor of plasmon B (chloroplast DNA) is an unknown X taxon. Four lines were formed from this X taxon: one SNP and two inversions (38 and 56 bp) caused the formation of the *paleocolchicum* line, two SNPs formed the *macha* line, three SNPs formed the *durum* line, and four SNPs formed the *carthlicum* lines. The *carthlicum* line includes tetraploid subsp. *carthlicum* and hexaploid ssp. *aestivum.*

The hypothesis that the cytoplasm of ssp. *macha* derived from wild emmer, demands an explanation of how wild emmer and ssp. *aestivum* contacted each other. Wild emmer is native to the Fertile Crescent and is not found in Georgia and in other regions of the South Caucasus, and ssp. *aestivum* was formed northeast to the fertile Crescent, namely, in the southwestern belt of the Caspian Sea. One possible explanation is transfer of some wheat species and subspecies by current inhabitant of south Caucasus who lived several millennia ago in the Fertile Crescent area (Gogniashvili et al. 2018). It is also tempting to assume that early farmers (circa 1000–8000 years ago) brought various wheats into present-day Georgia and that wild emmer was brought as an admixture with domesticated emmer but did not established there.

Ssp. *macha*, one of the modern subspecies of *T. aestivum*, probably developed around 7000 years ago as a result of a cross between wild or domesticated emmer wheat, and ssp. *aestivum.* It is a late-maturing winter wheat and grows well in most well-drained soils. ssp. *macha* shows a good resistance to Fusarium head blight (FHB (Burt et al. 2015), and to common bunt (*Tilletia tritici*) and rust (*Puccinia* sp.) (Borgen 2010).

Ssp. *macha* is used as cooked seed, or ground into a flour and used as a cereal for making bread, and biscuits. Its straw is used as a biomass for fuel, for thatching, and as a mulch in the gardens. Fibers obtained from the stem are used for making paper.

10.4.2.4   Ssp. *aestivum* (Bread or Common Wheat)

Description of Subspecies

T. aestivum L. ssp. *aestivum*, also known as bread wheat or common wheat [Syn.: *Triticum aestivum* L. (bearded);

T. hybernum L. (beardless); *T. sativum* Lam.; *T. vulgare* Vill.; *T. sativum* Pers.; *T. cereale* Schrank.; *Frumentum triticum* E. H. L. Krause] is an annual, predominantly autogamous, plant. Culms are erect, vary in length from 60 to 150 cm (excluding spikes) (50 cm dwarf wheat; 80–100 cm semi-dwarf wheat, 140–150 cm standard-height wheat), usually hollow and with thin walls, but in several forms, they are solid, possessing 5–6, glabrous internodes, nodes sometime are hairy. The young shoots are erect, semi-erect, or prostrate. Leaf blades are 20–60 cm long, 10–20 mm wide, flat and linear, usually glaucous, or more rarely, yellowish-green, and pubescent in most forms, but glabrous in a few. The spikes are either awned all over with short awns of uniform length or, only awned on the upper spikelets in the forms called beardless; very few forms are truly awnless. Spikes are determinate, squarehead in cross section, dense or lax, 6–18 cm long (excluding awns), with 14–30 spikelets. The rachis is tough, not brittle, and smooth, and fringed with short hairs along its margins, with a few hairs immediately below the base of each spikelet. The spikelets are 10–15-mm-long, solitary at the nodes, and glabrous or hairy. The top spikelet is fertile and at right angle to the plane of the lateral spikelets. Each spikelet contains 4–9 florets, and those near the center of the spike bear 3–5 grains, whereas those near the base or the tip of the spike contain 1 or 2 grains only. The glumes are oblong, asymmetrical in size, shorter than or almost equal in length to the rest of the spikelet, glabrous or pubescent, with white, yellow, red, brown, or blue-black color, 6–11 mm long, with 5–7 nerves, terminating in a short tooth or, in some of the bearded forms, extend into a slender, short awn (1–3-cm-long). The lemmas are 10–15 mm long, thin and pale, rounded on the back, with 7–11 nerves, In bearded forms, they carry awns 5–10 cm long, are usually straight, and yellowish-white, reddish, black, or brown. The caryopsis is 5–9 mm long, 3.0–4.5 mm wide, 2.5–4.0 mm thick

and free and not adherent to lemma and palea (free-threshing). The grains are white, yellow, orange, or red, generally plump, usually with a shallow furrow, and with a brush of hairs at the apex. The endosperm is flinty, semi-flinty, or mealy. The embryo about 1/5 the length of the caryopsis (Fig. 10.7b).

Ssp. *aestivum,* bread wheat, is by far the most economically important wheat growing today on a world-wide scale. It is the most widely adapted crop, growing in diverse environments and climates, from 67° N in Norway, Finland and Russia, to 45° S in Argentina, but in the tropics and subtropics, its cultivation is restricted to higher elevations (Feldman et al. 1995). The world's main wheat-producing regions are north-central China, southern Russia and the Ukraine, the central plains of the USA and adjacent areas in Canada, northwest Europe, the Mediterranean basin, India, Argentina and south-western Australia. Based on growth habit, wheat is classified into spring wheat and winter wheat, covering about 65% and 35% of the total global wheat production area, respectively (Braun et al. 2010).

Bread wheat is high-yielding in a wide range of ecosystems, and, as such, accounts for about 95% of world wheat production (durum wheat comprises the other 5%). It is the most important staple food of about two billion people (35% of the world population), providing nearly 55% of the carbohydrates and 20% of the food calories consumed globally. It is grown on more land area than any other food crop (219 million hectares in 2021), and in 2021, worldwide annual wheat production reached a record of ∼ 770 mt with China and India as the top 2 producers (Table 10.12). Global demand for wheat is increasing due to the unique viscoelastic and adhesive properties of gluten proteins, which facilitate the production of processed foods, whose consumption is increasing as a result of the worldwide industrialization and adoption of western diet in the east (Shewry and Hey 2015; Day et al. 2006).

Table 10.12 World total and top ten ssp. *aestivum* (bread or common wheat) producer countries in 2020 (From FAOSTAT 2021)

Country	Hectare	Production in metric tones	Productivity (tons/ha)
China	23,382,215	134,254,710	5.742
India	31,357,000	107,590,000	3.431
Russia	28,864,312	85,896,326	2.976
United States	14,870,740	49,690,680	3.342
Canada	10,017,800	35,183,000	3.512
France	4,512,420	30,144,110	6.680
Pakistan	8,804,677	25,247,511	2.868
Ukraine	6,564,500	24,912,350	3.795
Germany	2,835,500	22,172,100	7.821
Turkey	6,914,632	20,500,000	2.964
World total	219,006,893	765,789,635	3.497

Genetic Diversity and Genetic Erosion

Having originated from a small number of genotypes of the parental species, the nascent bread wheat harbored only a small fraction of the genetic diversity of its parental species. Yet, recurrent formation over time from different genotypes of the parental species (Caldwell et al. 2004; Pont et al. 2019) and gene flow via introgressive hybridizations from different domesticated and wild wheat taxa, different genotypes of *Ae. tauschii,* and from various other wild relatives, have increased considerably the genetic base of ssp. *aestivum.* Moreover, from the time of its origin in west Asia ssp. *aestivum* was taken to many parts of the globe where it had to adapt to a wide range of new climatic, edaphic and biotic conditions on the one hand, and to different human demands on the other. This exerts strong stresses to which ssp. *aestivum* could adapt thanks to its exceptional capacity to sustain high dosage of mutations (Dvorak and Akhunov 2005; Akhunov et al. 2007), enabling its spread to new environments. The cultivation of this subspecies in many parts of the world in admixtures of genotypes, facilitated inter-genotypic hybridizations and establishment of new recombinant types. Selection under domestication by different farmers, seeking different nutritional profiles, for different end-uses in different regions of the globe, brought about additional variation.

The fast increase in genetic diversity of bread wheat has also aided by its genetic structure, reinforced by a diploid-like meiotic cytological behavior and predominantly self-pollination. These features have proven a very successful genetic system, facilitating a rapid build-up of genetic diversity. Being a hexaploid, most of the gene loci are present in six doses, and the accumulation of genetic variation through mutations or hybridizations is tolerated more readily than in tetraploid and diploid species. Moreover, allopolyploidy facilitates genetic diploidization—the process whereby genes existing in multiple doses can be diverted to new functions. Furthermore, its system of predominant self-pollination could have helped in the fixation of desirable mutants and recombinants resulting from rare outcrossing events. Thus, hexaploid wheat can accumulate a significant amount of genetic variability through mutations, hybridizations and introgressions. Mutability may have been triggered and accelerated by the activity of various factors, such as transposable elements, gametocidal genes, and genome-restructuring genes that induce various kinds of chromosomal rearrangements and mutations (cf. Kashkush et al. 2002, 2003; Endo 2007; Feldman and Strauss 1983). Activation of these elements by various stresses, such as new, unexpected and extreme environmental conditions might have been of critical significance in the build-up of the genetic diversity of this crop.

Hybridization has also played an important role in building up the genetic variation in ssp. *aestivum.* For thousands of years, farmers have been growing admixtures of different genotypes of land races in polymorphic fields. Such admixtures allow gene flow between different genotypes, creating new genetic combinations, some of which were better adapted to the farmer requirements in specific regions. Moreover, some admixtures included, in addition to bread wheat, other subspecies of *T. aestivum,* e.g., *compactum* and *spelta,* and even different representatives of *T. monococcum* and *T. turgidum* (Zeven 1980; Feldman 2001), thus facilitating massive intra- and inter-specific gene flow.

The B and A subgenomes of wild emmer are homologous to the B and A subgenomes of ssp. *aestivum* and the genome of *Ae. tauschii* is homologous to the D subgenome. Hybrids between wild emmer or *Ae. tauschii* and bread wheat are semi-fertile, or sterile, respectively, but a few seeds are produced upon backcrossing to the hexaploid parent. Further backcrossing may fully restore fertility. So, genes can be transferred from these wild forms into bread wheat chromosomes at hybrid meiosis simply through crossing over. The cultivation of hexaploid wheat in the Near East next to stands of wild emmer for more than 8000 years facilitated the production of many hybrids and hybrid swarms, resulting in almost constant gene flow from the wild into the domesticated background. Even today, spontaneous hybridizations between bread wheat and wild emmer have been described (Percival 1921; Zohary and Brick 1962).

Highly sterile hybrids have also been frequently obtained as a result of hybridizations between bread wheat and more distant diploid and tetraploid species of *Aegilops* (see Kimber and Feldman 1987), as well as species of *Secale, Elymus, Haynaldia* and other related Triticeae genera which grew within or near wheat fields. The few seeds that these hybrids produce upon backcrossing yield more-fertile plants. Such occasional intergeneric hybridizations may often result in successful introgression, thereby maintaining a weak but constant flow of genes into the hexaploid domesticated background.

Pont et al. (2019) used exome sequencing of a worldwide panel of almost 500 genotypes selected from across the geographical range of the wheat species complex, to explore how 9000 years of hybridization, selection, adaptation and plant breeding shaped the genetic makeup of modern bread wheats. They observed considerable genetic variation at the genic, chromosomal and subgenomic levels, and used this information to decipher the likely origins of modern-day wheats, the consequences of range expansion and the allelic variants selected since its domestication. Zhou et al. (2020a, b), evaluating nucleotide diversity of bread wheat and its progenitors, found that the genetic bottleneck of bread wheat, resulting from allopolyploidization, has been largely compensated by a massive gene flow from multiple groups of tetraploid wheat and wild relatives. According to them, the extent of introgression in bread wheat genome is much higher than it in other species; about 13–36% of bread wheat genome was directly contributed by introgressions from other wheats, including wild emmer, and various other wild relatives. The increase of genetic diversity from alien introgression has been essential to the global expansion of bread wheat. Their study confirmed earlier studies showed that bread wheat has asymmetric distribution of nucleotide diversity on the three subgenomes, with diversity in B being greater than in A, and in A being greater than in D. Reinforcement of these findings was recently provided by He et al. (2019), Cheng et al. (2019), who showed that introgression from wild emmer is one of the primary reasons for the relatively high diversity of the B and A subgenomes in bread wheat. Indeed, bread wheat adaptability to diverse environments and end-uses is surprising, given the diversity bottleneck expected from the allopolyploid speciation events (Dubcovsky and Dvorak 2007). Buckler et al. (2001) reviewed the present knowledge of molecular diversity among the grass crops and relate diversity mainly to genes involved in domestication and to yield gains.

As an outcome to increased mutability and introgression, bread wheat exhibits a large amount of morphological, physiological, biochemical, and molecular variations. This subspecies surpasses all other wheats in the number of forms and cultivars that have been classified on the basis of morphological, physiological, and quality traits; more than 25,000 different cultivars have been produced up to the 1990s of the 20th Century (Feldman et al. 1995). Already in 1921, Percival (1921) wrote, in his comprehensive book on wheat, that this wide variation resulted from a collection of mutants and hybrids, which originated from intraspecific (between races and cultivars), interspecific (with other wheat species) and intergeneric hybridization (with wild relatives, e.g., several closely related *Aegilops* species).

Under modern conditions, the wheat field has become genetically uniform so that spontaneous gene exchange between different genotypes is less likely. On the other hand, gene migration has been greatly increased by world–wide introduction and exchange of cultivars. At the same time, new techniques have become available for the identification and introduction of desirable genes from one cultivar to another, as well as from wild to domesticated type. But, until now, hybridizations have mainly been confined to intraspecific crosses and relatively little use has been made of diploid and tetraploid gene pools of wild and domesticated taxa in improvement of the bread wheat (Feldman and Sears 1981).

Today's selection techniques can achieve the objectives of the primitive farmer with much greater certainty. High-yielding cultivars owe their improved performance to genetic increases in the number of fertile florets in the spikelet, to the size of the ear and to the number of ears per plant. This is, to a large extent, determined by the harvest index, and the ratio of grain to straw weight, but is also much influenced by resistance to diseases and pests and to loss by lodging and shattering or by ability of the crop to utilize heavy doses of nitrogenous fertilizers, with the effects of these components largely interrelated.

Still, the diversity in bread wheat is only a fraction of that in its parental species. Haudry et al. (2007) reported that the diversity of bread wheat is only 69% of that of its tetraploid parent. Also, it has been estimated that the genetic diversity of the D subgenome of modern bread wheat cultivars only accounts for 15% of that of *Ae. tauschii* growing in the Transcaucasia area (Dvorak et al. 1998a, b). Talbert et al. (1995) observed that the amount of DNA sequence variability in wheat is low, although somewhat more variability existed in the B subgenome than in the D subgenome. Estimates of RFLP diversity at the RbcS loci have indicated that bread wheat has perhaps 30% of the diversity levels found in its diploid relatives, but there are substantial differences between the A, B, and D subgenomes; the A subgenome was somewhat more polymorphic than the B subgenome, while the D subgenome was the most conserved (Galili et al. 2000). Akhunov et al. (2010) estimated nucleotide diversity in 2114 wheat genes and found it to be similar in the A and B subgenomes and reduced in the D subgenome. Nucleotide diversity varies within a subgenome, and along chromosomes. Low diversity was always accompanied by an excess of rare alleles (Akhunov et al. 2010). Whole-genome shotgun sequencing of the bread wheat genome indicated that upon allopolyploidization and domestication, between 10,000 and 16,000 genes and several gene families were lost in hexaploid wheat, compared with its three diploid progenitors, (Brenchley et al. 2012). At the same time, several classes of gene families with predicted roles in defense, nutritional content, energy metabolism, and growth increased in size in the domesticated wheats, possibly as a result of selection during domestication (Brenchley et al. 2012).

A major advance in wheat productivity was achieved in the1960s, following a cross by an American breeder, Orville Vogel, of an economically unimportant dwarf Japanese cultivar Norin 10, with the North American winter wheat cultivar Brevor. This led to the release in 1961 of the high-yielding cultivar Gaines, which was widely used as a parent by breeders in US and western Europe. Under the leadership of Norman E. Borlaug at the International Maize and Wheat Improvement Centre in Mexico (CIMMYT),

Norin 10 was crossed with spring varieties, resulting in the production of high-yielding, semi-dwarf wheats which became the basis of the Green Revolution in India, Pakistan, Iran, and the Mediterranean basin. By the end of the twentieth century, 81% of the developing world's wheat area was grew semi-dwarf and dwarf wheats, giving both increased yields and better response to nitrogenous fertilizer.

Modern varieties of bread wheat have short stems (semi-dwarf wheat is about 80–100 cm tall and dwarf wheat is 50–60 cm tall), due to *Rht* dwarfing genes that reduce the plant's sensitivity to Gibberellic acid (Gale and Youssefian 1985; Lenton et al. 1987; Youssefian et al. 1992). Short stems are important because harvest index (the proportion of grains to the straw) is improved and they allow for the application of high levels of chemical fertilizers that in traditional tall wheat (140–150 cm tall), would cause lodging of the stems, and consequently, loss of yield (Brooking and Kirby 1981).

During the green revolution, in the 1960s, numerous polymorphic traditional admixtures were replaced in many parts of the globe by a small number of elite high-yielding cultivars, each grown mono-genotypically on large areas. A very large number of the land races was not preserved and their germ plasm was lost, causing a massive erosion in the bread wheat gene pool. Awareness among wheat geneticists and breeders of the negative effects of this erosion has been increasing (e.g., Feldman and Sears 1981; Feldman et al. 1995; Feldman 2001). Yet, based on the high mutation rates in hexaploid wheat and the buffering effects caused by allopolyploidy, Dubcovsky and Dvorak (2007) pointed out that the loss in bread wheat diversity can be reverted to some extent. In fact, Warburton et al. (2006) showed a recovery in genetic diversity among the most modern CIMMYT lines. The extent of this recovery is however not an absolute trend in all breeding programs and in spite of recent advances, some traits of interest may have been neglected (Nazco et al. 2014b).

Ssp. *aestivum* originated and entered cultivation only after the more or less simultaneous domestication of diploid and allotetraploid forms. In the past, archaeological remains of ssp. *aestivum* were difficult to separate from those of free-threshing forms of *T. turgidum*. Grain shape, that was considered a valuable diagnostic trait, proved to be problematic because tetraploid and hexaploid wheats overlap considerably in this trait (Zohary et al. 2012). This overlap is more prominent in archaeological remains, because of swelling and other deformations caused by charring (Hopf 1955; van Zeist 1976; Harlan 1981). Similar difficulties have been encountered with the shape of the scutellum, which, although it is somewhat morphologically different in tetraploid versus hexaploid free-threshing wheats, it exhibits some overlapping in grains of these to taxa. For these reasons, most archaeobotanists in the past did not attempt to distinguish between tetraploid and hexaploid free-threshing wheats and tended to group them together as '*aestivo-compactum*' or *T. turgidum-T. aestivum* finds (Zohary et al. 2012). Yet, Hillman (2001) showed that archaeological remains of free-threshing *T. turgidum* can be separated from free-threshing *T. aestivum* by the morphology of their rachis segments, enabling a more precise identification of the two forms in the archaeological remains.

Ssp. *aestivum* appears in archaeological data from the middle of the 9th millennium BP. The earliest finds were at Can Hasan, south Anatolia (see Hillman 1996). Free-threshing forms of ssp. *aestivum* were abundant in the prehistoric Near East, from the 8th millennium BP onwards. Indeed, finds from the 8th millennium BP, identified as ancestral forms of free-threshing ssp. *aestivum*, have been unearthed at Tepe Sabz in Iranian Khudistan, at Tell Sawwan in Iran, at Çatal Huyük in central Anatolia, in Haciar in west-central Anatolia and at Knossos on Grete (Kislev 1984). Between 8000 and 7000 BP, ssp. *aestivum* penetrated, together with domesticated emmer, into the irrigated agriculture of the plains of Mesopotamia and western Iran. In the 7th millennium, it appeared also in finds from the central and western Mediterranean basin. Dense forms of ssp. aestivum, sometimes mixed with ssp. *compactum*, were cultivated in central and western Europe at the end of the 6th millennium BP, where they were found associated, together with einkorn and emmer, with the first traces of agricultural activities. Bread wheat first reached North America with Spanish missions in the sixteenth century, and Australia in 1788, with the arrival of the colonists.

While tetraploid wheats, in keeping with their Near Eastern origin, are adapted to mild winters and rainless summers, the addition of subgenome D from the central Asiatic species *Ae. tauschii*, must have contributed to the adaptation of hexaploid wheats to a more continental climate and northern latitudes. This could have greatly facilitated the spread of bread wheat into parts of central, eastern and northern Europe and through the highlands of Iran, to central and eastern Asia.

The bread wheat grain contains most of the nutrients essential to man. It contains 13% water, 60–80% carbohydrates (mainly as starch), 1.5–2.0% fat, 8–15% protein, including adequate amounts of all amino acids except lysine, tryptophan and methionine which exist in relatively small amount, 1.5–2.0% minerals, and vitamins such as the B complex and vitamin E (Table 10.13). Of the protein, 75–80% is gluten. Each 100 g of bread wheat provides 327 cal and is a rich source of multiple essential nutrients, such as protein, dietary fiber, Manganese, Phosphorus, and Niacin. Significant quantities of several B vitamins and other dietary minerals are also present (Table 10.13).

Wheat is an important source of carbohydrates and the leading source of vegetable protein in human food (Shewry and Hey 2015) (Table 10.13). While wheat protein content is

Table 10.13 Nutritional value per 100 g of hard red winter wheat grains (USDA Nutrient Database)

Energy 327 calories

Nutrients	Quantity	% DV[a]
Total carbohydrates	**71.18 g**	–
Sugars	0.41 g	–
Dietary fiber	12.2 g	–
Total fat	**1.54 g**	–
Protein	**12.62 g**	–
Vitamins	–	–
Thiamine (B1)	0.383 mg	33%
Riboflavin (B2)	0.115 mg	10%
Niacin (B3)	5.464 mg	36%
Pantothenic acid (B5)	0.954 mg	19%
Vitamin B6	0.3 mg	23%
Folate (B9)	38 μg	10%
Choline	31.2 mg	6%
Vitamin E	1.01 mg	7%
Vitamin K	1.9 μg	2%
Minerals	–	–
Calcium	29 mg	3%
Iron	3.19 mg	25%
Magnesium	126 mg	35%
Manganese	3.985 mg	190%
Phosphorus	288 mg	41%
Potassium	363 mg	8%
Sodium	2 mg	0%
Zinc	2.65 mg	28%
Other constituents	–	–
Water	13.1 g	–
Selenium	70.7 μg	–

[a] Percent of Daily Value **for the main nutrients categories**. The % DV tells you how much a nutrient in a serving of food contributes to a daily diet (the percentages are roughly approximated using US recomandation for adults); 2000 calories a day is used for general nutrition advice

relatively high compared to that of other major cereals, it is of relatively low protein quality with low levels of essential amino acids. In addition, in a small part of the general population, gluten—the major part of wheat protein—can trigger Coeliac disease, non-celiac gluten sensitivity, gluten ataxia and Dermatitis herpetiformis (Ludvigsson et al. 2013).

Because of the high gluten content of its endosperm, bread wheat, especially its harder grained cultivars, are highly valued for bread making. The sticky gluten protein entraps the carbon dioxide formed during yeast fermentation and enables the leavened dough to rise. Nowadays, bread wheat is a central ingredient in foods such as bread, porridge, crackers, biscuit, muesli, pancakes, pie, pastries, Pizza, cake, cookies, muffins, rolls, doughnuts, gravy, beer, vodka, boza (a fermented beverage) and breakfast cereals. Worldwide, bread

wheat has proven highly adaptable to modern industrial baking, and has displaced many of the other wheat, barley, and rye species that were once commonly used for bread making, particularly in Europe. The most common forms of ssp. *aestivum* are white and red wheat. In North America, bread wheat is classified dark-colored or light-colored. The dark-colored bread wheats include Hard Red Winter, characterized by grain, brownish color, and high protein content, and used for bread, and other baked goods and the Hard Red Spring, characterized by hard grains, brownish color, and high protein content, used for bread and other baked goods. The light-colored bread wheats include the Soft Red Winter, which is soft, low-protein wheat, used for cakes, pie crusts, biscuits, and the Hard White, which is hard, light-colored, opaque, chalky, medium-protein wheat, used for bread and brewing, and the Soft White, which is soft, light-colored, very low-protein wheat grown in temperate moist areas, used for pie crusts and pastry.

In addition to its high nutritive value, the low water content, ease of processing and transport and good storage qualities of the bread wheat crop have made it the most important staple food of about 35% of the world's population. There is an increasing demand for wheat in countries undergoing urbanization and industrialization. As a result, over the last 70 years, the global wheat area has increased by 50%, reaching 220 Mha in 2014. During the same period, average yields increased, mainly due to broader use of fertilizers and improved cultivars. In 2017, global wheat production was 772 million tones (MT), with China (134.3 MT), India (98.5 MT), Russia (85.9 MT), United States (47.4 MT), and France (36.9 MT) being the main producers (FAOSTAT 218). This production accounted for more than 25 per cent of the total cereal crops consumed throughout the world.

10.4.2.5 Ssp. *compactum* (Host) MK (Club Wheat)

T. aestivum ssp. *compactum* (Host) Mac Key, known as club, dwarf, cluster or hedgehog wheat, [Syn.: *T. compactum* Host; *T. vulgare compactum* Alef.; *T. sativum compactum* Hackel.] is an annual, predominantly autogamous, plant. Culms are erect, vary in length from 70 to 140 cm long (excluding spikes), possess 5–6 internodes above ground, and are generally hollow, with thin walls, but in several varieties, are solid. In most varieties, the young shoots are erect, but in some, they are semi-erect or prostrate. Leaf blades are yellowish-green or blue-green, sparsely hairy, and 10 mm wide. The spikes are determinate, bilateral, square-head in cross section, awned (bearded) or awnless (beardless), and usually appear from the side of the leaf sheath and not from the apex, as in most wheats. They are short, stiff and compact, 3.5–6.0 cm long (excluding awns), 10–15 mm wide, and generally with 17–25 closely packed spikelets that are arranged almost at right angles to the rachis. Typical spikes are with a uniform density, while in some forms they are more crowded towards the apex and more loosely arranged at the base. The rachis is tough, flattened, and fringed, with short hairs along the side and across the upper part of each notch immediately beneath the points of insertion of the spikelets. Spikelets pack broadside to the rachis, are crowded, oblong to ovate, 10–13 mm long, 13–15 mm wide, solitary at the nodes, and usually hairy. The top spikelet is fertile and at right angle to the plane of lateral spikelets. Each spikelet contains 6–7 florets, with fewer florets at the apex, 2–4 of which frequently produce grain. Those near the center of the spike bear 2–4 grains, whereas those near the base or the tip of the spike contain 1 or 2 grains only. The glumes are stiff, oblong, symmetrical in size, shorter than or almost equally long as the rest of the spikelet,

white, yellow, red, brown, or blue-black, either glabrous or covered with soft hairs, 8–9-mm-long, and with 5 to 6 nerves. In beardless varieties, the apical tooth of the glume is blunted, 0.5–1.5 mm long, whereas in bearded varieties, it generally extends into a fine awn, usually not more than 0.5-3.0-cm-long. In most forms, there is no single prominent keel, except for in the upper half of the glumes; some Asiatic forms, however, are keeled from the tip to the base. The lemmas are elliptic, thin and pale, inflated, 10 mm long, with 9 to11 nerves, and without a keel. In bearded varieties, the lemmas terminate in a stiff awn, 50–90-cm-long, which in some forms the awns diverge widely. The palea is of the ordinary bicarinate form, two-veined, and keeled. The caryopsis is ellipsoid, small, oval, narrow towards the apex and generally plump and hairy at the apex, 5–7 mm long, 3.0–3.7 mm wide, and 2.8–3.6 mm thick, free (free-threshing), white, yellow, or red, usually shallow and generally with a filled furrow. The endosperm in the majority of forms is opaque and starchy, but in some, it is flinty (Fig. 10.7c).

Ssp. *compactum* contains many varieties that were classified on the basis of awn presence (bearded or beardless), glume color and hairiness, and grain color. Most cultivars are spring types, although there are several winter-type forms.

Ssp. *compactum* is similar to bread wheat (ssp. *aestivum*) but can be straightforwardly distinguished by its more compact spike, due to its shorter rachis segments, and smaller grains than those of most bread-wheat cultivars. Therefore, in 1807, Host first gave this taxon a specific rank, but Mac Key (1954b) later classified it as a subspecies of *T. aestivum*. Percival (1921) considered ssp. *compactum* as closely related to ssp. *aestivum*. In contrast, Mac Key (2005) stated that there are no genetic reasons to consider ssp. *compactum* more closely related to ssp. *aestivum* than to the other subspecies of *T. aestivum*. He was of the opinion that ssp. *compactum* is even more phylogenetically closely related to ssp. *spelta* than is ssp. *aestivum*. Mac Key based this opinion on the cytological evidence presented by Thompson and Robertson (1930), who found somewhat better pairing in the F_1 *compactum* × *spelta* hybrids than in *compactum* x *aestivum* or *aestivum* x *spelta* hybrids. The serological analysis of Zade (1914) provides additional supporting evidence.

Helbaek (1959) identified the archaeological samples of naked, small, somewhat spherical grains (< 5 mm) and short internodes found in Near Eastern sites from 8900 to 7000 BP, as club wheat, ssp. *compactum*. However, it is difficult to accept that these Near Eastern wheat remnants from such an early period are hexaploids, as explained by Kislev (1979/1980): (1) Modern forms of the hexaploid wheat *T. aestivum* are more adapted to the northern (Vavilov 1926) than to the

Near East latitudes, where Helbaek found his remnants, the latter area being dominated by tetraploid wheats. (2) In the south-western flank of the Fertile Crescent, hexaploid wheat was only grown as a crop of little importance (Jakubziner 1932a). (3) If Helbaek's remnants are indeed hexaploids, this implies that hexaploid naked wheat, found in the earliest periods in the Near East, preceded cultivation of tetraploid naked wheats. However, on an evolutionary basis, one would expect that the tetraploids developed first. (4) The earliest samples are probably from too early a period for hexaploid wheats to have come into existence; the spread of agriculture had not yet made geographic contact between the domesticated, free-threshing tetraploids and *Ae. tauschii*, the wild diploid progenitor of the D subgenomic of *T. aestivum*. Based on the morphological characteristics of this archaeological material and the difficulty in defining ancient Near Eastern naked wheat as hexaploid, Kislev (1979/1980), van Zeist and Bakker-Heeres (1973), Zohary (1973) concluded that these wheat remnants belong to an extinct free-threshing tetraploid form, *T. turgidum* ssp. *parvicoccum* Kislev. It is assumed that ssp. *compactum* originated later in the Near East and subsequently, was transported to Europe as an admixture with other wheats, and later established in several places as the dominant form (Feldman 2001).

According to Schiemann (1932, 1948), ssp. *compactum* is to be considered as an older wheat than ssp. *aestivum*. However, it is generally assumed that ssp. *compactum* derived from ssp. *aestivum* by a mutation of *c* to *C*, as was suggested by Nilsson-Ehle (1911), Mac Key (2005), Johnson et al. (2008). This gene is located on chromosome 2D (formerly XX) (Unrau 1950), and was either mutated on the hexaploid level or contributed by a line of *Ae. tauschii* containing the *C* allele. Since they grew in many places in mixture, hybridization and gene exchange between ssp. *compactum* and ssp. *aestivum* has repeatedly occurred.

Percival (1921) considered ssp. *compactum* one of the most ancient of free-threshing wheats widely grown by Neolithic man in many parts of Europe. According to him, ssp. *compactum* was found in many Neolithic and Bronze Ages deposits in Hungary, Germany, Switzerland, Italy, Spain, and Sweden. The grains of this wheat are small, and more or less hemi-spherical, with a blunt apex, and well-developed furrow. On average, they are 4.6 mm long, 3.4 mm broad, and 3–3.3 mm thick. Examples of ssp. *compactum*, with somewhat larger grains (5.5–7 mm long), similar to those of the common forms of the present day, have only been found in the Neolithic and Bronze Age deposits of Switzerland and northern Italy.

The small, naked wheat grains found in many sites of Neolithic Europe, were considered to belong to hexaploid wheat and were named by Heer (1866) *Triticum compactum antiquorum*. Because of the difficulty to distinguish between the fossil grains of ssp. *compactum* and those of many forms of ssp. *aestivum*, Schiemann (1932) and Bertsch and Bertsch (1949) referred the European remains of hexaploid wheat of this epoch as *aestivo-compactum*.

Yet, since it is difficult to distinguish between 4x and 6x seeds in archaeological finds (Nesbitt 2001), and since the spike morphology and grain size of *T. compactum antiquorum* resemble those of the extinct *T. turgidum* ssp. *parvicoccum* and the extant *T. turgidum* ssp. *durum* var. Hourani (Fig. 10.4e), Kislev (1979/1980) suggested that it may well be a tetraploid rather than a hexaploid taxon. That these primitive types may be of the same category as Kislev's (1979/1980) tetraploid, small-seeded ssp. *parvicoccum*, was also suggested by Mac Key (2005), Zohary et al. (2012).

Rao (1972) assigned the dominant allele, *C*, of ssp. *compactum* to the left (beta) arm (currently long arm; Sears and Sears 1979) of chromosome 2D, i.e., 2DL, close to the centromere. Johnson et al. (2008), studying the location of C on chromosome 2D, localized *C* to an interval flanked by the markers *Xwmc245* and *Xbarc145*, but could not unambiguously localize the *C* locus to a chromosome bin because markers that were completely linked to *C* or flanked this locus were localized to chromosome bins on either side of the centromere. Consequently, following Rao's work (1972) localizing *C* near the centromere on chromosome arm 2DL, and based on the locations of flanking and linked markers, Johnson et al. (2008) placed it in deletion bin C-2DL3, which is located on both sides of 2D centromere. On the other hand, *Tg1* was localized to a more distal position on the short arm of chromosome 2D (Nalam et al. 2007).

Johnson et al. (2008) also studied the relationship between *C* and another spike-compacting gene in wheat, namely, *soft glume* (*Sog*) in *T. monococcum* ssp. *monococcum* var. *sinskajae* (Lebedeva and Rigin 1994). This recessive factor on chromosome 2A^m yields a compact spike, as well as soft glumes that are longer and broader than those of *T. monococcum* ssp. *monococcum*. They suggested that *C* and *Sog* are present in homoeologous regions on chromosomes 2D and 2A^m, respectively. However, *Sog* is located on the short arm of 2A^m (Taenzler et al. 2002; Sood et al. 2009) and *C* on the long arm. Thus, it is possible that these loci are orthologous, but their positions with respect to the centromere, are under debate. Alternatively, *Sog* and *C* might not be orthologous, despite their similar locations. The *C* allele of ssp. *compactum* seems to be homologous with the gene causing spike-compactness in ssp. *macha* (Swaminathan and Rao 1961). Similarly, *C* is not allelic to the gene controlling the compact spike of spp. *sphaerococcum* (Goncharov and Gaidalenok 2005).

There have been a number of QTL analyses dealing with spike compactness in bread wheat (Sourdille et al. 2000;

Jantasuriyarat et al. 2004; Nalam et al. 2007; Ma et al. 2007). No QTLs affecting spike compaction have been found to coincide with the location of *C*. This suggests that there is no allelic variation at the *c* locus of ssp. *aestivum* and that variation in spike dimensions is due to other factors (Johnson et al. 2008). Ausemus et al. (1967) showed that genes affecting spike compaction are present in every chromosome except for chromosomes 2B, 4D, 5A and 5D, suggesting that spike compactness may be affected by many genes other than *C*. Thus, using the compact spike characteristic as a taxon-defining trait may not always be appropriate.

Club wheat, ssp. *compactum*, is grown today in a few restricted areas of Europe (south-east Russia, Germany, France, Italy, Switzerland, Spain, and Portugal), in Asia (Transcaucasia, Turkestan, Siberia, Mongolia, China and India), in the Near East (Turkey), North and South Africa, in the northwestern United States (California, Oregon, Washington, and Idaho), in South America (Chili), and in Australia. In many regions, ssp. *compactum* is currently predominantly distributed as a constituent in admixtures with ssp. *aestivum* or ssp. *durum*, with pure crops rarely seen. In North America, however, they are cultivated as a pure crop on a somewhat extensive scale. An extinct variety of ssp. *compactum*, California Club Wheat (*Triticum compactum erinaceum*), named for its appearance which resembles a hedgehog, had a compact (2–5-cm-long) and bearded spike, hairy rachis, red chaff, and small, soft, and red kernels. This variety, introduced to California by the Spaniards via Mexico in 1787, was farmed extensively during the beginning of California's agricultural history. However, most of it disappeared during the first half of the nineteenth century, whereas a small amount was grown until the middle of the twentieth century.

Most ssp. *compactum* varieties tolerate frost, drought, and several fungal diseases, and grow well on poor soils. Club wheat better adapts to certain agro-climatic regions (Johnson et al. (2008), and it was suggested that this subspecies may be more competitive than ssp. *aestivum* in dryland areas, where stand establishment is difficult (Gul and Allan 1972; Zwer et al. 1995). They are mainly spring forms, many of them exceptionally early; only a few late-ripening winter forms exist. The straw of most varieties is not liable to lodge. While the spikes have a large number of grains, the yield in volume or weight per unit area is comparatively lower than that of ssp. *aestivum*, as the individual grains are smaller. The grains are firmly held by the glumes, a character which renders these wheats particularly suited for cultivation in districts where it is the practice to leave the crop on the field for considerable periods of time before harvesting can be completed. The grain is of soft or medium hardness, and of quality resembling that of several varieties of bread wheat.

Ssp. *compactum* are used for cake, crackers, cookies, pastries, and flours. Some varieties are also used for the production of starch, paste, malt, dextrose, gluten, alcohol, and other products.

10.4.2.6 Ssp. *sphaerococcum* (Percival) MK (Indian, Dwarf or Shot Wheat)

Triticum aestivum ssp. *sphaerococcum* (Perc.) Mac Key, known as Indian dwarf wheat, and shot wheat [Syn.: *T. sphaerococcum* Percival; *T. compactum* A. et G. Howard; *T. aestivum* gr. *sphaerococcum* (Percival) Bowden; *T. aestivum* convar. *sphaerococcum* (Percival) Morris and Sears], is an annual, predominantly autogamous, plant. Culms are erect, short, 50–100 cm long (excluding spikes), very stiff, hollow, with 4–5 internodes, and frequently bent below the spike in a winding manner. In some varieties, the culm has a reddish-pink color. The young shoots are erect, with leaves as in ssp. *aestivum*. The culm leaf-blade are somewhat rigid, comparatively short, 10–16 cm long, 1.2–1.5 cm wide, taper towards the tip, and with scabrid upper surfaces, and a few coarse hairs on the ribs. The auricles are long, narrow, and fringed with a few hairs. The spikes are determinate, dense, short, 4–6 cm long (excluding awns), with 14–20 spikelets, and squarehead in cross-section. The rachis is tough, non-brittle, and fringed with very short white hairs along its margins, which also extend across the front of the rachis at the base of each spikelet. The spikelets are solitary at the nodes, about 10 mm long, 10 mm across, and 4 mm thick. The top spikelet is fertile and at right angle to the plane of lateral spikelets. Each spikelet contains 6 or 7 flowers, 4 or 5 of which may produce grain. The glumes have an inflated appearance, are shorter than the rest of the spikelet, 8–9 mm long, possess 6 or 7 nerves, glabrous or pubescent, white or red in color, are generally keeled in the upper part only, and terminate in a broad curved scabrid tooth. The lemmas are inflated, 8–9 mm long, and rounded on the back with 9 nerves. In the beardless varieties, the tip terminates in a short awn 3–4-mm-long, whereas the bearded forms possess very stiff short awns, 1.5–2 cm long, which are frequently bent near the base and which in ripe ears, spread outwards irregularly. The palea is 7–8 mm long and fringed with hairs along the two keels. The grains are of very characteristic form, distinctly shorter and rounder than those of other wheats. They are 4–5.5 mm long, 3–3.7 mm wide, and 3–3.7 mm thick, free and not adherent to lemma and palea (free-threshing), and often somewhat unsymmetrical due to pressure of the lemmas. The apex is truncate with a short "brush" at the apex, and the furrow is shallow. Grains color is white or red (Fig. 10.7d).

Ssp. *sphaerococcum* is a very early wheat; all varieties have a spring habit. There are several varieties that are classified on the basis of bearded or beardless spikes, white or red glume color, glabrous or pubescent glumes, and white or red grain color.

This subspecies was first described by Percival (1921). Based on its difference from other hexaploid wheats in its short stature (averaging 54–70 cm), small spikes, characteristically small, round grains and practically hemispherical, inflated glumes, Percival (1921) considered it as a separate species, but Mac Key (1954b) classified it as a subspecies.

Ssp. *sphaerococcum* is one of the more modern subspecies of *T. aestivum*, probably originating as a mutation in cultivated fields of bread wheat. Archaeological evidence dates the contribution of this wheat to ancient civilization in the Indus valley back to the 5th millennium BP (Feldman et al. 1995). According to Percival (1921), ssp. *sphaerococcum* only resembles ssp. *compactum* in their short dense spike, while it also has a tufted appearance and erect, stiff straw. The two glumes are different from each other in form and texture. The spikes are bearded or beardless, the former never having long awns, but irregularly spreading ones are 1.5–2.0 cm long at the apex of the spike and much shorter at the base. It has a red or white, glabrous or pubescent chaff, and the grain is very small and characteristically hemispherical in shape.

ssp. *aestivum* gave rise to *sphaerococcum* through a single mutation (Feldman et al. 1995; Feldman 2001). This mutation presumably occurred in an *aestivum* that had been carried eastward, since *sphaerococcum* has not been found in the prehistoric Near East and its culture nowadays is largely confined to India. This assumption is based on the fact that no line of *Ae. tauschii* has been found to carry the *sphaerococcum* allele *s*, indicating that these mutations appeared at the hexaploid level.

Miczynski (1930) crossed ssp. *sphaerococcum* with ssp. *aestivum* and concluded that the entire characters of ssp. *sphaerococcum* that distinguish it from ssp. *aestivum* is inherited as if it were determined by a single gene. He designated ssp. *aestivum*, bearing the dominant allele, as *SS*, and ssp. *sphaerococcum* as *ss*. While the *s* allele was recessive to the *S* allele, several intermediates were found, and the F_2 population classified as *aestivum* varied somewhat in its characters. Miczynski (1930) concluded that the *s* allele also determines short awns, which were 2 cm long in fully bearded forms.

Ellerton (1939) crossed a bearded form of ssp. *sphaerococcum* with a beardless variety of ssp. *aestivum*. The F_1 closely resembled *aestivum*, although dominance was not complete. The spikelets were rather shorter and more inflated than in the *aestivum* parent, tough the grains closely resembled those of *aestivum* in shape. F_2 segregated in a ratio of 1:2:1 indicating a difference of a single gene. The absence of crossover types in the above cross showed that the entire complex of characters differentiating between *sphaerococcum* and *aestivum* behaves as if determined by a single gene and not by a group of closely but incompletely linked genes. Ellerton (1939) assumed that the magnitude and diversity of the effects of this single Mendelian factor are much greater than one would normally associate with a recessive mutation at a single locus, particularly in a hexaploid species. Consequently, he suggested that the *s* mutation involved a structural change in a short chromosomal section. Since the general characteristic of the mutation suggests that it is the effect of gene deficiency, one likely hypothesis is that the two subspecies differ by a single deletion covering several loci (Ellerton 1939). Such a deletion would be conserved and inherited as a single Mendelian factor.

In contrast, Swaminathan and Prabhakara Rao (1961), Swaminathan et al. (1963) claimed that ssp. *sphaerococcum* could not have arisen through a deletion in ssp. *aestivum*, since back-mutations to ssp. *aestivum* readily occur in *sphaerococcum*. In addition, induced mutations via gamma rays from *sphaerococcum*-type to *aestivum*-type were also obtained by Josekutty (2008). Moreover, the *sphaerococcum* locus, which tends to behave as one Mendelian unit in recombination, can be broken up by irradiation, resulting in phenotypes lacking the compact growth habit and rigidity of leaves, but possessing hemispherical glumes and spherical grains (Swaminathan and Prabhakara Rao 1961; Swaminathan et al. 1963).

Sears (1947), using monosomic analysis, showed that the *sphaerococcum* phenotype is due to a hemizygous-ineffective recessive gene, which in two doses produces the *sphaerococcum* phenotype but in a single dose, is relatively ineffective. Sears located the gene on chromosome 3D (formerly XVI). Using the telomeric method, Prabhakara Rao (1977) mapped the *sphaerococcum* gene *s* 5.7 crossover units away from the centromere on the beta (short) arm of chromosome 3D of wheat. In contrast, Koba and Tsunewaki (1978) located the *s* gene on the very proximal region of the long arm of 3D, approximately 5.0 cM from the centromere. Singh (1987) observed that the 1030 F_2 plants of the cross between monosomic 3D of var. Pb. C 591 of ssp. *aestivum* and ssp. *sphaerococcum*, formed two distinct classes, parental-type (812 (78.8%) plants, morphologically *sphaerococcum* and *aestivum* types) and recombinant-type (218 (21.2%) plants having hybrid characters of *sphaerococcum* and *aestivum*). The occurrence of the recombinant sub-classes showed that the *sphaerococcum* phenotype is not governed by a single gene, but by at least three closely linked genes that are located very close to the centromere on the long arm of chromosome 3D.

Cheng et al. (2020) studied the genetic basis of the semispherical grain trait in ssp. *sphaerococcum*, by generating an F2 segregating population from a cross of the wheat line HeSheng 2 of ssp. *aestivum* with Nongda 4332 (ND4332, derived from a cross between ssp. *aestivum* and ssp. *sphaerococcum*. The corresponding segregation ratio fit a Mendelian model of 3:1, indicating that the semispherical grain trait is controlled by a single nuclear gene, which is consistent with previous findings (Miczynski 1930; Sears 1947). In accordance with the official nomenclature rules of gene designation in wheat, Cheng et al. (2020) named the gene determining semispherical grain *Tasg-D1*. This gene was mapped between markers Xgwm341 and Xgdm72 on the short arm of chromosome 3D, near the centromeric region. Fine mapping of *Tasg-D1* confirmed that the locus is located between markers 3DS-68 and 3DS-44. Cheng et al. (2020) narrowed the candidate region to a 1.01-Mb region between markers 3DS-68 and 3DS-94, a region containing 13 predicted high-confidence genes. Resequencing of these genes revealed only one single-nucleotide polymorphism (SNP) in the coding sequences. This SNP (A/G) is located in exon 9 of TraesCS3D01G137200, between ND4332 (*Tasg-D1*) and HS2 (*TaSG-D1*). This SNP in the coding region led to an amino acid substitution from lysine (286 K) to glutamic acid (286E). Studies of the expression profiles of *TaSG-D1* in different tissues showed that it was highly expressed in shoot meristem, root, spike, grain, shoot axis and ovary tissue. *TaSG1* has three homoeologs that share expression patterns, each located in a different subgenome.

Previous studies have revealed a discrepancy in the inheritance pattern of the *sphaerococcum* gene, which either has a hemizygous-ineffective recessive effect or an incompletely dominant effect (Sears 1947; Schmidt et al., 1963; Salina et al. 2000). Since it was determined that *Tasg-D1* is a gain-of-function allele in wheat line ND4332, Cheng et al. (2020), evaluated its genetic effects on grain shape and other traits in a segregating population. The grain length showed a 1:2:1 segregation ratio in the examined populations, supporting the notion that *Tasg-D1* shows incomplete dominance. Moreover, the phenotypes of heterozygous individuals were significantly different from those of homozygous ones, in traits such as grain length, plant height, spike length, spikelet density, and thousand-grain weight. This suggests that the effects of *Tasg-D1* on plant architecture associated with the *s* locus are indeed pleiotropic (is it a transcription factor?).

According to Cheng et al. (2020), the mutant allele *Tasg-D1* encodes a serine/threonine protein kinase glycogen synthase kinase 3 (STKc_GSK3) that negatively regulates brassinosteroid signaling. Expression of *TaSG-D1* and the mutant form *Tasg-D1* in *Arabidopsis thaliana,* suggested that a single amino acid substitution in the TREE domain of *TaSG-D1* enhances protein stability in response to brassinosteroids, likely leading to formation of round grains in wheat. This gain-of-function mutation has pleiotropic effects on plant architecture and exhibits incomplete dominance. Cheng et al. (2020) proposed that the *Tasg-D1* gene of

ssp. *sphaerococcum* might have originated at the hexaploid level from ssp. aestivum by a spontaneous mutation in the TREE domain of *TaSG-D1*.

Mutations determining the *sphaerococcum* phenotype were described in hexaploid, tetraploid and diploid wheats. Schmidt et al. (1963) reported the appearance of a drastic mutation in bread wheat, ssp. *aestivum*, simulating ssp. *sphaerococcum.* This effect was controlled by an incompletely dominant gene not allelic to the *sphaerococcum* gene. The chromosome carrying the gene could not be identified in monosomic crosses because of sterility interactions in 10 of the 21 chromosomes. Therefore, the chromosomal location of this mutant could not be assigned to a specific subgenome of bread wheat.

Further evidence that the *sphaerococcum* phenotype is not restricted to the D subgenome of hexaploid wheat, came from the report by Schmidt and Johnson (1963, 1966) of the same character in a tetraploid wheat. *Sphaerococcum*-like plants were seen in a plot of a *durum* introduction from China. Cytological studies showed that the *sphaerococcum*-like plants had 14 bivalents at first meiotic metaphase and regular pairing in crosses with other *durum* lines. One possibility is that a translocation of the gene from the D subgenome to one of the tetraploid subgenomes had occurred. However, an alternative hypothesis is that this variant may represent a mutation of a normal allele to a *sphaerococcum* allele in either the A or the B subgenome. In this respect, it is interesting that Georgiev (1979) reported that EMS treatment produced mutants with phenotype similar to ssp. *sphaerococcum* in diploid wheat, *T. monococcum.*

Likewise, treatment of ssp. *aestivum* with chemical mutagens, produced three independent mutants with morphological features resembling those of ssp. *sphaerococcum* (Maystrenko et al. 1998). A monosomic analysis situated the three mutant genes, designated *S1, S2,* and *S3,* on chromosome 3D, 3B, and 3A, respectively (Maystrenko et al. 1998). Salina et al. (2000) reached similar results using microsatellite markers from a homoeologous group 3 of ssp. *aestivum* and mapped the *S1, S2,* and *S3* genes of the induced sphaerococcoid mutation to chromosome 3D, 3B, and 3A, respectively. The *S1* locus was found to be closely linked to the centromeric marker *Xgwm456* of the long arm (2.9 cM) and mapped not far (8.0 cM) from the *Xgdm72* marker of the short arm of chromosome 3D. The *S2* gene was tightly linked to two centromeric markers (*Xgwm566* and *Xgwm845*) of chromosome 3B and *S3* was located between *Xgwm2* (5.1 cM), the marker of the short arm, and *Xgwm720* (6.6 cM), the marker of the long arm, of chromosome 3A. Thus, the *sphaerococcum* trait is not only restricted to the responsible gene in the D subgenome but can also be attributed to its homoeologs on the A and B subgenomes (Cheng et al. 2020).

Ssp. *sphaerococcum* is endemic to southern Pakistan and northwestern India (Elleton 1939; Josekutty 2008). It was one of the main crops grown by ancient Indian cultures. In modern times, it was grown as a major crop on a larger area in northwestern India and in southern Pakistan, but during the Green Revolution, in the 1960s, it was mostly replaced by high-yielding varieties of bread wheat (Mori et al. 2013). Currently, it is grown as a relic crop in northwestern India and southern Pakistan.

Indian dwarf wheat, ssp. *sphaerococcum*, presented the lowest nucleotide diversity among all *T. aestivum* subspecies (Zhou et al. 2020a). The extremely low diversity of Indian dwarf wheat is likely the result of its early migration to remote areas at the southwest of the Himalayas, which lacks wild relatives, and a consequential escape from alien introgressions.

Ssp. *sphaerococcum* resists drought well (Ellerton 1939) and is generally grown in areas of relatively little rain. Many varieties are resistant to yellow rust caused by *Puccinia striiformis* (Josekutty 2008). The grain has a high protein content compared to other hexaploid subspecies of *T. aestivum*, (Singh 1946; Josekutty 2008), but its yield is lower than that of ssp. *aestivum*. Its grains are usually ground into a flour and used as a cereal for making bread, biscuits, etc. The straw is used as a biomass for fuel, for thatching, or as a mulch in the garden. A fiber obtained from its stems is used the paper industry.

Kihara (1937) reported that Lilienfeld and Kihara in 1934 found a modal arrangement of 21 bivalents, most of which were ring bivalents, in both *aestivum* x *sphaerococcum* and the reciprocal cross. Occasional multivalents were also observed. The fertility of the F₁ hybrid was good. Similarly, in crosses made by Ellerton (1939), 21 bivalents (mostly ring) were observed in most cells. Percival (1930) described an *Ae. ovata* (currently *Ae. geniculata*) x ssp. *sphaerococcum* cross in which he found a maximum of four bivalents, and similar pairing in hybrids between *ovata* and other hexaploid wheats. These results show that *sphaerococcum* is very closely related to *aestivum*.

Vakar (1932) studied chromosome pairing at first meiotic metaphase in F₁ hybrids involving ssp. *sphaerococcum* and two forms of tetraploid wheat, i.e., two ssp. *sphaerococcum* x ssp. *turgidum* crosses and one *sphaerococcum* x var. *pyramidale* cross. All three hybrids exhibited fourteen bivalents and seven univalent. One of the *sphaerococcum* x *turgidum* hybrids showed a chromosome bridge and acentric fragment in several cells, indicating that the hybrid was heterozygous for a paracentric inversion. Baghyalakshmi et al. (2015) studied F₁ hybrids between *T. timopheevii* ssp. *timopheevii* and ssp. *sphaerococcum* and observed 11.36 (5–12) univalents, 4.48 (4–11) bivalents, 4.49 (1–5) trivalents, and 0.60 (0–1) quadrivalents at meiosis.

10.4.2.7 Ssp. *tibetanum* J. Z. Shao (Semi-Wild, Feral Wheat)

No wild hexaploid progenitors of *T. aestivum* are known, but the two distinguishing characteristics of wild *Triticum* species, i.e., fragile rachises breaking into wedge-shaped units and closely appressed glumes, are found in plants in Tibet and named *T. aestivum* ssp. *tibetanum* J. Z. Shao. It is an autogamous plant, whose height ranges from 90 to 130 cm, leaves are light green, and culm and nodes are glabrous. Spikes are square, somewhat denser at the upper portion, 9–15 cm long, and awnless or awned, with short straight or curved white awns. The rachis is brittle, and at maturity, the spike disarticulates into individual spikelets, each falling with the rachis segment below it (wedge-type disarticulation). Spikes contain 15–27 spikelets, with each spikelet containing 5–6 florets, rachilla is hairy. Glumes are ovate, stiff and rigid, glabrous or pubescent and red or white in color. Lemmas are ovate, glabrous and awned. The grains are adherent to the palea and lemma and threshing results in spikelets rather than naked grains, their number per spikelet in the middle portion of spike is 3–5. All forms have spring habit (Fig. 10.7e).

Morphological, physiological and genetic studies have shown that ssp. *tibetanum* is closely related to bread wheat landraces of the white wheat complex native to the Sichuan province of south-west China (Yen et al. 1988). It has a brittle rachis, different from that of ssp. *spelta* of west Asia. It closely resembles the Chinese white wheat complex by its thin leaves, light green color, square spike, with multi-floret spikelets, rounded glume, with hooded lemma and tipped or curly awn. The crossability genes and non-interchanged chromosomes phylogenetically connect ssp. *tibetanum* with the white wheat complex (Yen et al. 1988). Its chloroplast genome is similar to that of bread wheat, its mitochondrial genome is only slightly different from that of the latter, and its cytoplasm does not induce male sterility in all studied bread wheat genotypes (Tsunewaki et al. 1990). These facts suggest that it is an off-type of Tibetan bread wheat.

Ssp. *tibetanum* has a very primitive compliment of the D subgenome chromosomes since their D-subgenome chromosomes are structurally similar to the chromosomes of *Ae. tauschii*; at meiosis of F₁ hybrids between ssp. *tibetanum* and *Ae. tauschii*, no multivalents form and 7 ring bivalents and 14 univalents are always observed (Yen et al. 1988).

In 1974, this subspecies was collected in five counties in the Tibet plateau, by a Chinese scientific expedition (Shao et al. 1980, 1983). Based on its morphology and chromosome number, Shao et al. (1980, 1983) suggested to regard this semi-wild wheat as a new *T. aestivum* subspecies, namely, ssp. *tibetanum*. The new subspecies, considered to be an off-type of Tibetan bread wheat (Tsunewaki et al. 1990), usually grows as weed within or on edges of barley and wheat fields (Shao et al. 1983). It is a polymorphic taxon, exhibiting variation in almost all morphological characters, and based on this variation, Shao et al. (1983) classified the collected samples into three varieties.

Ssp. *tibetanum* has a brittle rachis, different from that of ssp. *spelta* of west Asia. Although it is a hulled form having brittle rachis and stiff glumes, significant differences exist between it and ssp. *spelta* and ssp. *macha*, the two other hulled subspecies of *T. aestivum*, (Shao et al. 1983; Tsunewaki et al. 1990). Seedlings of all three ssp. *tibetanum* lines studied have broad, light green leaves. In contrast, the two spelt-type wheats ssp. *spelta* and ssp. *macha*, have rather narrow leaves of dark green color (Tsunewaki et al. 1990). The spikes of the three ssp. *tibetanum* lines are like those of the two other hulled subspecies of *T. aestivum*, however, the spikes of ssp. *tibetanum* disarticulate spontaneously from the tip, as the spikelets mature, and the spikelets drop off the culm at the time of ear ripening. Such strong disarticulation is not observed in any other forms of *T. aestivum*, including *spelta* and *macha*. The mode of disarticulation of ssp. *tibetanum* is of the wedge-type, i.e., with the rachis segment below each spikelet, and is clearly different from the barrel-type, in which each spikelet contains the rachis segment beside it, a type exhibited by ssp. *spelta* (Shao et al. 1983; Tsunewaki et al. 1990).

Analysis of root tip cells revealed that ssp. *tibetanum* is a hexaploid with 2n = 42 chromosomes (Shao et al. 1983; Tsunewaki et al. 1990). It can easily be crossed with bread wheat, yielding fully fertile F_1 hybrids, with complete chromosomal pairing at meiosis, although three interchanges exist between the studied lines of ssp. *tibetanum* and cultivar Chines Spring of ssp. *aestivum* (Tsunewaki et al. 1990).

Shao et al. (1983) found that F_1 hybrids of a bread wheat cultivar (white glume, white grain, glabrous glumes, non-brittle rachis) with a line of ssp. *tibetanum* (red glume, red grains, pubescent glumes, fragile rachis) exhibited dominance of the four studied traits, i.e., the F_1 hybrids had a fragile rachis, red glumes and red grains. Data obtained by Tsunewaki et al. (1990) support the conclusion of Shao et al. (1980, 1983), i.e., that ssp. *tibetanum* is genetically very closely related to bread wheat. The observed mode of spike disarticulation, chromosomal instability and the genotype of hybrid chlorosis clearly favor the hypothesis that ssp. *tibetanum* originated rather recently as an off-type of domesticated forms of Tibetan bread wheat (Tsunewaki et al. 1990).

Sun et al. (1998) used RAPD analysis of seven accessions of ssp. *tibetanum*, 22 cultivars of ssp. *aestivum* from China, and 17 lines of European ssp. spelta, to study the genetic relationships between these three subspecies of *T. aestivum*, and to assess genetic diversity among and within these taxa. RAPD polymorphism was found to be much higher within ssp. *spelta* and ssp. *tibetanum* than within ssp. *aestivum*. The

genetic distance between ssp. *tibetanum* and ssp. *aestivum* was smaller than that between ssp. *tibetanum* and ssp. *spelta*. Cluster analysis clearly classified all the studied genotypes into two groups: one included all the European *spelta* lines, and the second included all cultivars of ssp. *aestivum* and the lines of ssp. *tibetanum*, thus supporting classification of the Tibetan wheat as a subspecies in *T. aestivum*. Similar results were obtained by Cao et al. (2000b), who used RAPD to assess the phylogenetic relationships among the subspecies of *T. aestivum*. Their RAPD data, are in-agreement with those based on morphological classification, suggesting that, of the subspecies of *T. aestivum*, ssp. *tibetanum* is most closely related to bread wheat.

Study of the genetic control of rachis fragility and glume tenacity in ssp. *tibetanum* was carried out by Cao et al. (1997), in an attempt to help establish the taxonomic status and genetic origin of semi-wild wheat. Progenies of crosses and backcrosses of semi-wild wheat with cultivar Columbus of bread wheat indicated that the fragile rachis and non-free-threshing character of ssp. *tibetanum* were dominant over the tough rachis and free-threshing character of bread wheat. F_2 and backcross data indicated that the rachis fragility and glume tenacity of the semi-wild wheat were each controlled by a single gene. On the other hand, in the cross between ssp. *tibetanum* and spp. *spelta*, the F_2 and F_3 populations did not segregate by glume tenacity, but did segregate by rachis fragility. These data suggest that three genes interact to control three types of rachis fragility in hexaploid wheat: in semi-wild wheat-type, in *spelta*-type and in the tough rachis of common wheat. Semi-wild wheat differs from bread wheat in rachis fragility and glume tenacity and from the hulled subspecies of *T. aestivum* (ssp. *spelta,* and ssp. *macha*) in the pattern and degree of rachis disarticulation. Cao et al. (1997) concluded that semi-wild wheat is likely a subspecies within *T. aestivum*, at the same taxonomic level as spp. *spelta* and *macha*.

Chen et al. (1998), using monosomic and ditelosomic lines of bread wheat, found that the gene controlling the brittle rachis of ssp. *tibetanum*, designated *Br1*, is dominant and located on the short arm of chromosome 3D i.e., on 3DS. Consequently, it was designated *Br-D1*. So, it seems that this gene in ssp. *tibetanum* derived from *Ae. tauschii*. It is interesting to note that the brittle rachis gene in ssp. *tibetanum*, like those in some related genera such as *Aegilops*, is also located on homoeologous group 3 chromosomes. Interestingly, the *Br* gene in *Ae. tauschii* determines barrel-type disarticulation in the diploid species and not wedge-type disarticulation but exhibits the wedge-type of disarticulation in hexaploid background. As was found earlier (Shao et al. 1983; Tsunewaki et al. 1990), *Br-D1* is different from the gene determining brittle rachis in ssp. *spelta*. In the progenies of the cross *spelta* x *tibetanum*, F_1 plants exhibited the wedge and barrel types of disarticulation indicating that wedge-type disarticulation in ssp. *tibetanum* is codominant with the barrel type in spelt wheat.

The genes for brittle rachis were mapped using aneuploid stocks in hexaploid and tetraploid wheat (Watanabe et al. 2002). Similar to Chen et al. (1998), also Watanabe et al. (2002) located the *Br-D1* in the Tibetan weed races on the short arm of chromosome 3D. The average distance from centromere was 20.6 cM. In accordance with the rule for the symbolization of genes in homoeologous sets, they propose to designate the group 3 brittle rachis genes, *Br-A1, Br-B1* and *Br-D1*.

Guo et al. (2020) present a draft genome sequence of a Tibetan semi-wild wheat accession Zang1817 and re-sequence 245 wheat accessions, including world-wide wheat landraces, cultivars as well as Tibetan landraces. They demonstrate that high-altitude environments can trigger extensive reshaping of wheat genomes, and also uncover that Tibetan wheat accessions accumulate high-altitude adapted haplotypes of related genes in response to harsh environmental constraints. Moreover, Guo et al. (2020) find that Tibetan semi-wild wheat is a feral form of Tibetan landrace, and identify two associated loci, including a 0.8-Mb deletion region containing *Brt1/2* homologs and a genomic region with *TaQ-5A* gene, responsible for rachis brittleness during the de-domestication episode. Gheir study provides confident evidence to support the hypothesis that Tibetan semi-wild wheat is de-domesticated from local landraces, in response to high-altitude extremes.

This feral wheat has an established growth habitat and a distinct morphology and should therefore be considered a sub-species. It grows as weed within and on edges of wheat and barley fields. A feral form of ssp. *macha*, that grows as weed in edges of wheat fields in Georgia, was also described (Dekaprelevich 1961). This form is also considered a feral derivative of domesticated wheats rather than a truly wild species.

Kuckuck (1964) suggested that the hexaploid wheat with a fragile rachis found by Dekaprelevich, may have originated as an amphiploid between wild emmer, ssp. *dicoccoides*, and *Ae. tauschii*, independently of the origin of domesticated hexaploids, which are believed to have involved free-threshing tetraploid wheat as their tetraploid parent. Sears (1976a) crossed wild emmer, ssp. *dicoccoides*, with *Ae. tauschii* and obtained a F₁ hybrid with a brittle rachis. Thus, it remains uncertain whether Dekaprelevich's brittled-rachis hexaploid wheat actually originated in this way or as a segregate from a cross of ssp. *macha* with wild tetraploid wheat, or as result of back-mutation of Q to q in the domesticated hulled-wheat ssp. *macha*.

Kuckuck (1970) suggested that a brittle-rachis form of ssp. *macha*, var. *megrelicum* Dek. et Men., could be a genuinely wild hexaploid wheat, and therefore a candidate ancestor species. However, its fully brittle-eared form, var. *megrelicum*, is not described as growing outside cultivated fields, and is therefore not a truly wild wheat.

10.4.2.8 Cytology, Cytogenetics, Genomics, and Evolution

Origin of *T. aestivum*

Hexaploid wheat, *T. aestivum* (2n = 6x = 42; genome BBAADD), evolved through two allopolyploidization events. The first event, involving two diploid species, *T. urartu*, the male donor of the A subgenome, and a species related to *Ae. speltoides*, the B-genome donor, the female parent, led to the formation of wild emmer, *T. turgidum* ssp. *dicoccoides*, about 700,000–900,000 years ago (Marcussen et al. 2014; Gornicki et al. 2014; Middleton et al. 2014). Subsequently, following mutations, several domesticated subspecies of *T. turgidum* evolved. The second allopolyploidization event that produced hexaploid wheat, occurring during the 9th millennium BP (Feldman 2001), presumably involved the free-threshing tetraploid wheat, ssp. *parvicoccum* as the female donor of the B and A subgenomes, and a diploid species, *Ae. tauschii*, the male donor of the D subgenome. This event first formed the hulled allohexaploid wheat, ssp. *spelta*, from which the more advanced free-threshing forms developed (Feldman et al., 1995; Feldman 2001). Almost all F₁ hybrids between the different subspecies of *T. aestivum* show complete chromosome pairing at meiosis and high fertility, justifying their categorization at the sub-specific rank (Mac Key 1954b).

Earlier chromosome counts of the various wheat species were wrong. Several cytologists reported the presence of 16 chromosomes in the diploid wheat, *T. monococcum*, while others reported 40 in the hexaploid, *T. vulgare* (= ssp. *aestivum*) (reviewed in Sax 1922). Conversely, Sakamura (1918), analyzing root tip cells of different wheats, obtained the following results: *T. monococcum* had 14 chromosomes, the subspecies of *T. turgidum*, namely, *dicoccon, durum, turgidum* and *polonicum* had 28, and the subspecies of *T. aestivum*, *compactum* and *spelta*, had 42. At the same time, Sax (1918) found 28 chromosomes in the first division of the fertilized egg of ssp. *durum*. These chromosome counts reinforced Schultz's (1913b) classification showing that his three wheat groups comprise a polyploid series; einkorn is a diploid (2n = 14), emmer is a tetraploid (2n = 28), and dinkel is a hexaploid (2n = 42).

Soon after the discovery of the correct chromosome count in wheats, Kihara (1919, 1924), Sax (1918, 1921, 1922, 1923, 1927), Sax and Sax (1924) started to cross representatives of the different ploidy levels to study the cytogenetic relationship between these groups. In F₁ hybrids involving representatives of the hexaploid species *T. aestivum*, and those of the tetraploid species *T. turgidum* (called pentaploid hybrids; 2n = 5x = 35), Kihara (1919) observed at first meiotic metaphase 14 bivalents (mostly ring) and 7 univalents, indicating that the hexaploid parent shares 14 chromosome pairs with tetraploids wheats and differs by an extra 7 pairs. Consequently, Kihara concluded that hexaploid wheat originated from a cross between a form(s) of *T. turgidum* that contributed 14 chromosomes, while an alien diploid species contributed the additional seven chromosomes. Such a cross yielded F_1 triploid hybrid (2n = 3x = 21) that underwent spontaneous chromosome doubling.

Since the indication that the third chromosome set (subgenome) of *T. aestivum* was donated by an alien diploid species, extensive attempts have been made to identify the diploid donor of this subgenome. In these endeavors, studies extended to the wild relatives of wheat, particularly to species of the closely related genera *Aegilops* that also comprises a polyploid series with diploid, tetraploid and hexaploid species (Percival 1923; Schiemann 1929; Sorokina 1937; Lilienfeld 1951, and reference therein). These works resulted in much speculation as to which species of *Aegilops* may have contributed the third subgenome of hexaploid wheat. In this regard, Percival (1921, 1923) expressed the belief that the free-threshing hexaploid wheats were segregants from crosses between wild emmer and *Ae. geniculata* (formerly *Ae. ovata*). However, cytological studies by Sax and Sax (1924) and others, including Percival (1930), have proven quite conclusively that *Ae. geniculata* could not have played a major role in the origin of the hexaploid wheats.

Morphological indication as to the donor of the third subgenome of *T. aestivum* was obtained from the comparison of the spikelet structure in the hulled hexaploid wheat ssp. *spelta*, with that of wild *Triticum* and *Aegilops* species. ssp. *spelta* spikelets show two types of rachis breaks upon maturity, one like that of wild emmer, ssp. *dicoccoides*, namely, rachis breaks yield spikelets having arrow-head shape, each with the rachis segment below it (wedge-type), and the second like that of several species of *Aegilops* from sections Vertebrata (including the diploid *Ae. tauschii* and the allotetraploids *Ae. crassa* and *Ae, ventricosa*) and from section Cylindropyrum (including the allotetraploid *Ae. cylindrica*). In these *Aegilops* species the brittle rachis disarticulates into spikelet containing the rachis segment beside them (barrel type). Hence, this type of rachis disarticulation implies that the donor of the third subgenome to hexaploid wheat is from one of these *Aegilops* species.

In fact, several years earlier, Stapf (1909) suggested, based on morphological characteristics, that ssp. *spelta* derived from *Ae. cylindrica*. Percival (1921, 1923) assumed that this species had contributed certain characters to ssp. *spelta*, but was not involved in its origin. Conversely,

Sax and Sax (1924) reported that the third chromosome set in *T. aestivum* is found in *Ae. cylindrica* (2n = 4x = 28), since F_1 hybrids between this species and *T. aestivum*, a pentaploid hybrid with 2n = 5x = 35, had 7 bivalents and 21 univalents, indicating that they share one subgenome. Also, chromosomal pairing at meiosis of the F_1 hybrid between the amphiploid *cylindrica–durum* (2n = 8x = 56) and *T. aestivum* (2n = 6x = 42), a heptaploid hybrid with 2n = 7x = 49, which showed 21 bivalents (14 bivalents between the A and B subgenomes of *aestivum* with those of *durum* and 7 bivalents between one of the subgenomes of *Ae. cylindrica* and that of *T. aestivum*) and seven univalents, indicated again that *Ae. cylindrica* and *T. aestivum* share one subgenome (Sears 1944b). The intensity of pairing and the absence of heteromorphic bivalents indicated fairly complete homology between one of the *aestivum* subgenomes and one of the *Ae. cylindrica* sets. Since average chromosome pairing in the F_1 hybrid *cylindrica* x *durum* (2n = 4x = 28) was only 0.50 bivalents per cell, the shared subgenome of *Ae. cylindrica* and *T. aestivum* should involve the third subgenome of *T. aestivum*. Taken together, *Ae. cylindrica* contains one subgenome which is homologous to one subgenome of ssp. *spelta* and ssp. *aestivum*.

An allopolyploid resulting from a cross between *Ae. cylindrica* and a tetraploid wheat, would have 56 chromosomes instead of the required 42. This, therefore, appears to eliminate *Ae. cylindrica*, and other allotetraploid species of *Aegilops* with barrel-type disarticulation, as possible parents of the hexaploid wheats. *Ae. cylindrica* has been shown by Sax and Sax (1924), Bleier (1928), Kihara (1937) to be an allopolyploid carrying the C subgenome from *Ae. caudata*, and the D subgenome from *Ae. tauschii*, and Sears (1941a) produced an amphiploid from the cross of *Ae. caudata* x *Ae. tauschii* that morphologically resembles *Ae. cylindrica*. In addition, Sears (1941a) produced a hexaploid from the cross ssp. *dicoccoides* x *Ae. caudata*, but the resulting amphiploid did not morphologically resemble ssp. *aestivum* or ssp. *spelta*, thus eliminating *Ae. caudata* (genome C) as the third subgenome donor and leaving *Ae. tauschii* (genome D) as the probable one.

In accord with the above, Pathak (1940), following an analysis of figures of *Aegilops* chromosomes prepared by Senjaninova-Korczagina (1932), suggested that, *Ae. tauschii* may have been the donor of the third subgenome of *T. aestivum*. This aligns with the morphological evidence of rachis disarticulation, suggesting that the third subgenome of ssp. *spelta* most likely derived from hybridization between tetraploid wheat and *Ae. tauschii* (Kihara 1944; McFadden and Sears 1944b, 1946). *Ae. tauschii* has both the barrel-type dispersal unit, and the square-shouldered glumes which characterize *T. aestivum* (Nesbitt and Samuel 1996).

Indeed, the F_1 of ssp. *dicoccon* x *Ae. tauschii* resembled the taxonomic characters of ssp. *spelta* but was completely sterile (McFadden and Sears 1946). In contrast, the amphiploid *dicoccon–tauschii*, derived from colchicine treatment of the F_1 hybrid, had the hexaploid number of chromosomes (2n = 6x = 42; genome BBAADD), exhibited regular pairing at meiosis and high fertility, and closely resembled ssp. *spelta* in most traits (McFadden and Sears 1946). Moreover, crosses involving the synthetic amphiploid and ssp. *spelta* or ssp. *aestivum* produced fertile F_1 hybrids with 21 bivalents at meiosis, and no multivalents (McFadden and Sears 1946), thus, demonstrating unequivocally that *Ae. tauschii* is the donor of the third subgenome of *T. aestivum*. For the origin of the A and B subgenomes of *T. aestivum*, see Sect. 10.3.2.12.

The cytoplasm of *T. aestivum*, designated B, is identical to that of its maternal parent, *T. turgidum*, and closely related to the S cytoplasm of *Ae. speltoides* (Terachi and Tsunewaki 1992; Wang et al. 1997; Tsunewaki 2009, and reference therein). Similarly, Provan et al. (2004), utilizing polymorphic chloroplast microsatellites to analyze cytoplasmic relationships in the genera *Triticum* and *Aegilops*, reported that the allopolyploid *Triticum* species have cytoplasm similar to that of *Ae. speltoides*. Similar results were obtained by Gornicki et al. (2014), who found that *Ae. speltoides* is the closest relative to the diploid donor of the chloroplast of the emmer lineage, that is, the allopolyploids containing the B subgenome. This further demonstrates that *Ae. speltoides* or a closely related specie was the B-subgenome donor of allopolyploid wheats.

Following identification of the donors of the subgenomes of *T. aestivum* and consequently, the three diploid species involved in the ancestry of this hexaploid species, i.e., *T. urartu*, an extinct or yet not discovered B-subgenome donor, and *Ae. tauschii*, it became possible to investigate the cytogenetic relationships and to estimate the cytological similarities between these three subgenomes of the hexaploid. Mochizuki and Okamoto (1961) studied chromosomal pairing at first meiotic metaphase of the 21-chromosome hybrid, *T. monococcum* ssp. *aegilopoides* (very close to the donor of the A subgenome), *Ae. speltoides* (close to the donor of the B subgenome), and *Ae. tauschii* (the donor of the D subgenome) and found more than 5 bivalents and several multivalents (an average of 12 chromosomes out of the 21 were involved in pairing). These results were directly comparable to those of Kimber and Riley (1963), who observed a similar level of pairing in haploids of bread wheat lacking chromosome 5B that carries the homoeologous-pairing suppressor gene, *Ph1*. Likewise, euhaploids of bread wheat (2n = 3x = 21; genome BAD), with deletion of *Ph1*, as in the mutant *ph1b*, exhibited extensive homoeologous pairing, with 1.53–1.74 ring bivalents, 2.90–3.57 rod bivalents, and 0.53–1.16 trivalents (Jauhar et al. 1991). The most reasonable conclusions to draw from these data is that the genomes of the three diploid

species still have not diverged considerably from each other, that chromosomes of the three subgenomes of *T. aestivum* have undergone little change during the evolution of the allopolyploid, and that the regular behavior in the allopolyploid is due to the presence of two doses of homologous chromosomes and to the action of the *Ph1* gene. Accordingly, despite it being a segmental allopolyploid (i.e., an allopolyploid that exhibits partial homology between chromosomes of its subgenomes that that derived from relatively closely related species), *T. aestivum*, underwent complete cytological diploidization), and behaves as a genomic allopolyploid (i.e., an allopolyploid having little homology between chromosomes of its subgenomes that derived from relatively distant species), with exclusive pairing of homologous chromosomes, i.e., 21 bivalents always form at first meiotic metaphase of this species.

Hexaploid wheat originated from spontaneous chromosome doubling of the triploid F_1 hybrid between the domesticated form of *T. turgidum* and *Ae. tauschii*. Indeed, Fukuda and Sakamoto (1992), Matsuoka and Nasuda (2004), Zhang et al. (2010) reported occasional production of unreduced gametes and consequently, fertility, in such hybrids. At first meiotic division of such F_1 hybrids, Fukuda and Sakamoto (1992) observed that unreduced gametes were formed as a result of restitution of the first meiotic division, and normal second division, followed by formation of dyads which developed into two fertile 2n pollen grains. Further studies (Matsuohka and Nasuda 2004) involved crossing of a durum wheat cultivar that carried a gene for meiotic restitution with a line of *Ae. tauschii*. Some of the F_1 hybrids were highly fertile and spontaneously set hexaploid F_2 seeds. Cytological analyses of F_1 male gametogenesis showed that meiotic restitution was responsible for the high fertility of the triploid F_1 hybrids.

Matsuoka et al. (2013) examined the genetic basis of the spontaneous genome doubling of triploid F_1 hybrids between *T. turgidum* ssp. *durum* and *Ae. tauschii*. They found six QTLs in *Ae. tauschii* that are involved in hybrid genome doubling, presumably through the production of unreduced gametes. In addition, Hao et al. (2014) detected a major QTL controlling the production of unreduced gametes in two F_2 populations that derived from F_1 *T. turgidum* x *Ae. tauschii* hybrids. The QTL, named *QTug.sau-3B*, is located in chromosome 3B of the *T. turgidum* parent and situated between the markers *Xgwm285* and *Xcfp1012*. *QTug.sau-3B* is a haploid-dependent QTL, as it was not detected in doubled haploid populations.

Farming of this durum wheat is limited today to several mountainous regions in northern Iran (Matsuoka et al. 2008), but the situation may have been different in the past. If tetraploid wheat farming was predominantly adopted in low elevations of Caspian Iran, and if the distribution of *Ae. tauschii* was similar to its present-day distribution, most likely sources of the D genome are genotypes of *Ae. tauschii* ssp. *strangulata* that grow in the area. Indeed, biochemical and molecular studies indicated that the birthplace of hexaploid wheat was in Transcaucasia and in Iran, southwest to the Caspian Sea (Tsunewaki 1966; Nakai 1979; Jaaska 1980; Dvorak et al. 1998a, b) or southeastern Caspian Iran (Nishikawa et al. 1980).

The accumulated data indicate that plants of *Ae. tauschii* native to the south-west part of the Caspian Sea (mainly forms taxonomically placed in ssp. *strangulata*) had genome similar to that of subgenome D found in the hexaploid wheats. This led to the consensus concept that *Ae. tauschii* ssp. *strangulata* was the wheat progenitor (Nishikawa 1973; Nakai 1979; Jaaska 1980; Hammer 1980; Nishikawa et al. 1980; Lagudah et al. 1991; Lubbers et al. 1991; Dvorak et al. 1998a, b, 2012; Wang et al. 2013). ssp. *strangulata* is distributed from Transcaucasia to eastern Caspian Iran (Kihara et al. 1965; Jaaska 1980). In the southwestern and southern Caspian Iran, subsp. *strangulata* overlaps with subsp. *tauschii* var. *meyeri* and var. *typica*.

Wang et al. (2013), using the 10 K Infinium single-nucleotide polymorphism (SNP) array, studied genetic relationships between 477 *Ae. tauschii* lines and the D subgenome of bread wheat. They found that *Ae. tauschii* consists of two lineages (designated 1 and 2), each consisting of two closely related sub-lineages. The distinct separation of lineages 1 and 2 from each other, the scarcity of intermediate genotypes between the two lineages, and the relative lengths of branches in the phylogenetic tree obtained, agreed with trees constructed with AFLP markers (Mizuno et al. 2010), RFLP markers (Dvorak et al. 2012), diversity arrays technology (DarT) markers (Sohail et al. 2012), and haplotype sequencing (Dvorak et al. 2012). A population within lineage 2 in the southwestern and southern Caspian appears to be the main source of the wheat D subgenome.

Ecogeographic and genetic evidence strongly favors the origin of hexaploid wheat from domesticated tetraploid wheat rather than from the wild emmer (*Triticum timopheevii*, either domesticated or wild, cannot be considered a putative parent because of its different genomic constitution). In the middle of the 9th millennium BP, when hexaploid types first appeared, there was no geographical contact between wild emmer and *Ae. tauschii*. Moreover, any *dicoccoides-tauschii* amphiploid would have had a brittle rachis (Sears 1976a) and, hence, little chance to be selected by ancient farmers. By the time hexaploid wheat evolved, domesticated tetraploid wheat was already grown in eastern Turkey and western and north-western Iran, and came into contact with *Ae. tauschii*, which presumably was growing as a weed within and at the edges of wheat fields. Therefore, the most likely area of origin of the hexaploid bread wheat is the south-western corner of the Caspian belt. Such association between domesticated *T. turgidum* and

weedy *Ae. tauschii* in cultivation can still be found in this area (Matsuoka et al. 2008). A recent analysis of 242 *Ae. tauschii* accessions showed that a rare and distinct lineage (different from *strangulata*) from Transcaucasia also contributed ∼ 1% on average of the current wheat D sub-genome (Gaurav et al. 2021), in accordance with earlier studies that analyzed allelic variation of high molecular weight (HMW) glutenins (Giles and Brown 2006).

Although several sources of evidence point to domesticated emmer, ssp. *dicoccon*, rather than ssp. *dicoccoides*, as the tetraploid parent (Tsunewaki 1966; Porceddu and Lafiandra 1986; Kimber and sears 1987), it is more likely that the donor of the BA subgenomes to hexaploid wheat was a free-threshing tetraploid wheat (Dvorak et al. 2012), or more specifically, ssp. *durum* (Matsuoka and Nasuda 2004; Pont et al. 2019). Indeed, based on the ability to induce the production of unreduced gametes in the F_1 hybrid, Matsuoka and Nasuda (2004) suggested that *T. turgidum* ssp. *durum* was the tetraploid parent of hexaploid wheat. While surveying current cultivation areas of ssp. *durum* in northern Iran, they observed that, *Ae. tauschii* occurred widely as a weed in the durum fields. This finding showed that the *T. turgidum*–*Ae. tauschii* association hypothesized in the theory regarding *T. aestivum* evolution, still exists in the area where bread wheat likely evolved.

Extraction of the BA subgenomes from hexaploid wheat has given an indication of the type of tetraploid that was involved in the synthesis of the hexaploid. This was done by Kerber (1964), who crossed *aestivum* cultivars with a tetraploid, and backcrossed the pentaploid hybrids to the hexaploid parents for several generations, each time using only those plants that were themselves pentaploids. Finally, he selfed the pentaploids and selected tetraploid progeny that contained the BBAA subgenomes of hexaploid wheat. These extracted tetraploids were similar in spike morphology to the primitive free-threshing tetraploid, ssp. *parvicoccum* (Kislev 1979/1980). In accord with this, at the time when *T. aestivum* was formed, during the ninth millennium BP, ssp. *durum* was scarce or possibly nonexistent, whereas ssp. *parvicoccum* was widely cultivated and presumably was in massive contact with *Ae. tauschii*. Later on, when ssp. *durum* replaced ssp. *parvicoccum*, additional hybridizations between ssp. *durum* and *Ae. tauschii* presumably occurred.

Since no wild prototype of the hexaploid group is known to exist, many theories have been proposed as to the time, place, and way of origin of the various subspecies of *T. aestivum*. The fact that ssp. *spelta* has a brittle rachis and hulled seeds, led de Candolle (1886), Hackel (1890), Schulz (1913b), Carleton (1916) to consider it more primitive than the free-threshing hexaploid forms, and thus, as the oldest form of *T. aestivum*. This conclusion is further supported by the fact that all crosses of either hulled or free-threshing tetraploid wheats with all used lines of *Ae. tauschii* yielded only hulled forms resembling ssp. *spelta*, indicating that this subspecies is the prototype of hexaploid wheat (McFadden and Sears 1946; Kerber and Rowland 1974), and therefore, the predecessor of the more advanced, free-threshing forms.

Already Schroder (1931), based on anatomical evidence, proposed that ssp. *aestivum* arose from ssp. *spelta*. With the understanding that ssp. *spelta* is the most primitive subspecies of *T. aestivum*, it was assumed that the free-threshing forms of *T. aestivum* derived from it as a result of mutations (McFadden and Sears 1946). Indeed, the principal differences between the major hexaploid taxa are due to one or two genes that affect gross morphology (Mac Key 1954b) (Table 10.11).

Yet, the genetic data suggesting that the first hexaploid wheats were hulled, spelt-type, and more primitive than the free-threshing forms, do not agree with the archaeological chronology. While free-threshing forms of *T. aestivum*, i.e., ssp. *aestivum*, were found at the middle of the 9th millennium BP and were abundant in the pre-historic Near East from the 8th millennium onwards, thus far there is archaeological evidence for ssp. *spelta* only a thousand years later (Kislev 1984). Neolithic, Near Eastern ssp. *spelta* is very rare and earlier evidence for the existence of ssp. *spelta* is still missing. There is evidence of *spelta* grains from Yarim Tepe II, northern Iraq, dating back to the 7th millennium BP and probably also from Yarim Tepe I, about one thousand years earlier (Kislev 1984). These discrepancies between the genetic and archaeological data pose some difficulties in tracing the early history of the hexaploids. Indeed, several researchers (see Tsunewaki 1968) postulated that spelt wheat could not be the progenitor of bread wheat, but rather, its derivative. On the other hand, assuming that the first hexaploids were hulled, their absence from the prehistoric remains of the Near East may indicate their lack of advantage over domesticated emmer and free-threshing forms of *T. turgidum* in that area. ssp. *spelta* is grown today in extreme environments of the Near East, such as the high plateau of west-central Iran, eastern Turkey, and Transcaucasia. This cultivation is possibly of an ancient origin.

The earliest hexaploid wheats, seemingly originating south-west of the Caspian Sea, were hulled, *spelta*-type, presumably carrying the *Q* factor contributed by the free-threshing tetraploid wheat parent, and *Tg* (tenacious glume) from *Ae. tauschii* (Kerber and Rowland 1974). So, only a single mutation from *Tg* to *tg* was necessary to produce a free-threshing form (Kerber and Rowland 1974). If free-threshing hexaploid wheats indeed derived from *spelta*, formed from domesticated emmer and *Ae. tauschii*, and thus, carrying both the *Tg* and *q* genes, they carried a double dosage for hulledness. Since the chance for the occurrence of mutations in both of these two genes within several centuries is small, it is more likely that the first mutation, *q* to *Q*,

occurred in domesticated emmer fields, forming naked tetraploid wheat (Muramatsu 1986), and the second mutation, *Tg* to *tg*, occurred in *TgQ spelta* fields, forming free-threshing wheat. The artificial *spelta* obtained by McFadden and Sears (1946) should contain both factors responsible for hulledness, namely, *q* and *Tg*, as they hybridized the hulled domesticated emmer, ssp. *dicoccon* and *Ae. tauschii*.

The non-free-threshing trait of the synthetic hexaploids, irrespective of carriage of *Q* or *q* by their tetraploid parent, was found to be due to the *Tg* gene derived from *Ae. tauschii* (Kerber and Rowland 1974). Some variation in the degree of glume tenacity was noted among the synthetic hexaploids; those having ssp. *dicoccon* as a parent and containing *q* and *Tg* were the most difficult to thresh (Kerber and Rowland 1974). The interaction between *Tg* and *Q* was clearly demonstrated by extraction of the tetraploid component of a hexaploid wheat and then resynthesizing the hexaploid (Kerber and Rowland 1974). The extracted tetraploids containing only the BBAA component of the original free-threshing hexaploids were also free-threshing. In later studies, crosses between the extracted free-threshing tetraploids and *Ae. tauschii* produced *spelta*-like hexaploids. It was concluded that the *Tg* gene of *Ae. tauschii* inhibits the expression of *Q* (Kerber and Rowland 1974). The interaction between *tg* and *Q* conferring the free-threshing character, is complementary (Kerber and Rowland 1974), namely, both *tg* and *Q* must be present for the expression of the free-threshing trait in hexaploid wheat. The probability that the genotypes of *Ae. tauschii* which served as the progenitors of hexaploid wheat possessed *Tg*—as apparently do all extant forms of this species—supports the above hypothesis that hulled, hexaploid wheats are more primitive than free-threshing hexaploids; they carried the *Tg* gene and, therefore, were non-free threshing (Kerber and Rowland 1974). The mutation from *Tg* to *tg* is presumed to have occurred at the hexaploid level.

The mutation from *q* to *Q* most probably occurred not so long after the creation of ssp. *spelta*. The free-threshing ssp. *aestivum* thus formed, was preferred by the early farmers of the region and quickly replaced the hulled forms. As man migrated to new areas, cultivated wheats encountered new environments, to which they responded with bursts of variation, resulting in many endemic forms. Secondary centers of variation for hexaploids in the Hindu Kush area of Afghanistan were described by Vavilov (1951). Transcaucasia is a secondary center for both tetraploid and hexaploid types. Such secondary centers of diversity provide valuable gene pools to wheat breeders, beyond those existing at the primary centers of variation.

ssp. *spelta* appears to be comprised of two genetic types: the Asiatic type which gave rise to the free-threshing ssp. *aestivum*, and the European *spelta*, which apparently descended from ssp. *aestivum* (Tsunewaki 1968). To distinguish between these two types, Kislev (1984) designated the Asiatic one "*TgQ spelta*" and the European type, carrying the *tg* gene, "*tgq spelta*". ssp. *spelta* appeared in central Europe at ca. 4000 BP, about 2000 years later than forms of free-threshing hexaploid wheat. It could have been brought to Europe, where it replaced the free-threshing type in many sites of the upper Rhine region, particularly at high altitudes, where extreme temperatures prevail. Alternatively, ssp. *spelta* could have arisen in the Rhine valley through back-mutation of Q to q, or, more likely, as a result of a spontaneous hybridization between a hexaploid, free-threshing form (genotype QQtgtg) and tetraploid *dicoccon* (genotype *qq*), both of which were grown in that area, as suggested by Schiemann (1929), Tsunewaki (1968). This hybridization yielded, among others, hulled, spelt-type hexaploid progenies (genotype qqtgtg). The relatively wide distribution of ssp. *spelta* in central Europe in the past was presumably due to its winter hardiness and ability to out-yield the other crops on poor soils. ssp. *spelta* was also preferred for its good quality and it is still cultivated today in several areas of central Europe.

The possibility that European spelt is a form of comparatively recent origin that originated independently of the Asiatic spelt, was already proposed by Flaksberger (1939), Bertsch (1943, 1950), Kuckuck and Schiemann (1957). This hypothesis claimed that after establishment of bread wheat and domesticated emmer in Europe, the second type of ssp. *spelta* originated, mostly north of the Alps. While the European ssp. *spelta* is of the *tgq spelta* type, the genetic structure of the Iranian spelta is not known. On the basis of these genetic data, one can explain how Kuckuck in 1959, obtained free-threshing types, ssp. *aestivum*, among the progenies of crosses between Iranian and European spelt wheats (see Kuckuck 1964). A cross between a *TgQ* Asiatic *spelta* and a *tgq* European *spelta* yields, besides the two parental types, the two recombinants, *tgQ* and *Tgq*, of which the former is a free-threshing type. An important contribution to that hypothesis was presented by Blatter et al. (2002), who supported the claim of a European origin of ssp. *spelta* by analyzing the glutenin subunit genes *B1-1* and *A1-2* in 58 accessions of hexa- and tetraploid wheats from Europe and Asia. Their findings suggested that European spelt originated by introgression of a tetraploid wheat into free-threshing hexaploid wheat, as a secondary evolution after the development of bread wheat.

Dekaprelevich and Menabde (1932) assumed that ssp. *macha* is the primary form of hexaploid hulled wheats, from which ssp. *spelta* branched off. However, based on the morphological similarity between synthetic hexaploid wheat, formed from a cross of tetraploid wheat with *Ae. tauschii*, and ssp. *spelta*, the accepted view is that ssp. *macha* derived from *spelta* through mutations. Dekaprelevich and Menabde

(1932) assumed that forms of ssp. *macha* contain the *C* allele, conferring a compact spike, while others contain the recessive *c* allele for normal spike. Cultivation of ssp. *macha* subspecies is currently limited to a restricted area in Transcaucasia. Another hulled form, ssp. *vavilovii,* characterized by branched spikes, has a restricted cultivation in Armenia and is considered to be a form of ssp. *spelta* (van Slageren 1994).

The advanced, free-threshing subspecies of *T. aestivum, aestivum, compactum* and *sphaerococcum,* differ from each other in only single genes (Mac Key 1954a, b). ssp. *aestivum,* bread wheat, is, by far, the most economically important wheat growing today on a world-wide scale. The earliest remnants of ssp. *aestivum* are from Can Hassan III, south Anatolia and Cafer Hoyuk from about the middle of the 9th millennium BP (see Hillman 1996; Bilgic et al. 2016). Finds of free-threshing hexaploids from the 8th millennium, have also been unearthed in western Iran, northern Iraq, eastern, central and western Anatolia, and other sites. Between 8000 and 7000 BP, ssp. *aestivum,* together with domesticated emmer, penetrated into the irrigated agricultural plains of Mesopotamia and, in the 6th millennium BP, into the Nile basin (Fig. 13.1). ssp. *aestivum* also appeared in archaeological finds of the 7th millennium BP, in the central and western Mediterranean basin. Forms of free-threshing hexaploid wheat appeared in central and western Europe at the end of the 6th millennium BP, associated (together with einkorn and emmer) with the first traces of agricultural activities. *T. aestivum* spread into central Asia and, by way of the highlands of Iran (8th millennium BP), to the Indus valley, where it appeared at the beginning of the 5th millennium BP (Fig. 13.1).

ssp. *aestivum* is assumed to have given rise to ssp. *compactum* (club wheat) and *sphaerococcum* (Indian dwarf wheat) through mutations. This assumption is based on the fact that no line of *Ae. tauschii* has been found to carry the *compactum* allele *C* or the *sphaerococcum* allele *s*, both of which are located on D-subgenome chromosomes (Rao 1972, 1977), indicating that these mutations appeared at the hexaploid level. The fact that ssp. *spelta* does not carry these genes clearly shows that neither ssp. *compactum* nor ssp. *sphaerococcum* could have been the first free-threshing hexaploids. The lineage of ssp. *compactum* from ssp. *aestivum,* entailed only a single mutation from *c* to *C*, believed to have occurred in the Near East. Subsequently, *compactum* was transported to Europe as an admixture with other wheats and was established in several places as the dominant form. ssp. *compactum* is grown today in a few restricted areas of Europe, the Near East, and the northwestern United States. ssp. *sphaerococcum* originated from *aestivum* by a single mutation as well (*S* to *s*), which presumably occurred in an ssp. *aestivum* that had been carried eastward, since *sphaerococcum* has not been found in the prehistoric Near East and its culture nowadays is largely confined to northwestern India and southern Pakistan. ssp. *sphaerococcum* has been documented in India as early as the 5th millennium BP, and currently grows, to some extent, in India and Pakistan.

A free-threshing wheat very similar to ssp. *compactum* was grown by the Neolithic Lake Dwellers of Switzerland at least 1000 years before ssp. *spelta* reached that part of Europe. This "Lake Dweller wheat", now believed to be extinct, was described by Heer (1866) as *T. vulgare antiquorum*. It was a dwarf wheat with extremely small, stubby grains and compact, awnless spikes. Because of its resemblance to ssp. *compactum*, it has generally been assumed to have been a hexaploid wheat. The similarity between the tetraploid ssp. *carthlicum* (formerly *T. persicum*) and the Lake Dweller wheat eliminates the only reason for assuming that the latter was a hexaploid (Kislev 1979/1980).

Hexaploid *T. aestivum* originated only after the domestication of diploid and tetraploid wheats. While there is no wild progenitor to domesticated hexaploid wheat, a feral, semi-wild weedy form of hulled and brittle hexaploid wheat, ssp. *tibetanum,* which grows near the edges of wheat and barley fields, was discovered in Tibet (Shao et al. 1983). Since wild tetraploid wheats are not grown in China, this emmer-type brittle wheat is considered a derivative of a domesticated plant that underwent back-mutations at brittle rachis loci (Yen et al. 1988). Another emmer-type brittle hexaploid wheat, growing wild in Georgia, was described by Dekaprelevich in 1961. Kuckuck (1964) suggested that this wheat was derived from hybridization between a wild tetraploid wheat, most probably *T. timopheevii* ssp. *armeniacum,* and *Ae. tauschii*. Alternatively, this wheat was derived by back-mutations either from the free-threshing form, ssp. *aestivum*, or the hulled forms, ssp. *spelta* or *macha*.

To retrace the origin of the genome of modern bread wheat, ssp. *aestivum*, El Baidouri et al. (2017) investigated the evolutionary dynamics of gene-based transposable elements (TEs) and of single-nucleotide mutations across homoeologs of the A, B and D subgenomes of ssp. *aestivum*, as well as across hexaploid, tetraploid and diploid wheats. Based on these studies, they proposed a novel concept clarifying the structural asymmetry observed between the A, B and D subgenomes in bread wheat. Their concept derives from the cumulative effect of diploid progenitor divergence, the hybrid origin of the D subgenome, as was suggested by Marcussen et al. (2014), and subgenome partitioning following allopolyploidization events. In this model, the evolution of the A subgenome appears quite simple, whereas that of the other two subgenomes is more complex than initially reported. According to El Baidouri et al. (2017), the B subgenome in tetraploid/hexaploid wheat derived from an ancient S-genome progenitor, from which the modern S

genome of *Ae. speltoides* had considerably diverged. The D subgenome of the progenitor *Ae. tauschii* hexaploid wheat derived from an ancient hybridization between A and S (Marcussen et al. 2014), as well as between other species (Li et al. 2015b), which that accounts for at least 19% of the origin of the modern D genome.

The Contribution of the D Subgenome to the Wide Adaptability of ssp. *aestivum*

The isozyme study conducted by Jaaska (1981) revealed intraspecific differentiation of *Aegilops tauschii* into two groups of biotypes, which essentially correspond to its two morphological subspecies, subsp. *tauschii*, with cylindrical spikes, and subsp. *strangulata*, with a bead-like arrangement of spikelets. Jaaska identified subsp. *tauschii* as the contributor of a D genome to the allotetraploid *Ae. cylindrica* (genome DDCC) and of the third subgenome, D, to the allohexaploid *Ae. crassa* ssp. *crassa* (genome $D^cD^cX^cX^cDD$), and ssp. *strangulata* as the contributor of a D subgenome to allohexaploid wheats, to the allotetraploids *Ae. crassa* subsp. *macrathera* (genome $D^cD^cX^cX^c$), to *Ae. ventricosa* (genome DDNN), and to the allohexaploid *Ae. juvenalis* (genome $D^cD^cX^cX^cUU$). Reinforcement of Jaaska's (1981) conclusion, came from additional isoenzyme studies (Jaaska 1993), as well as biochemical and molecular studies (Wang et al. 2013, and reference therein), which confirmed that ssp. *strangulata* is the donor of the D subgenome to hexaploid wheat. ssp. *strangulata* has a narrower distribution than ssp. *tauschii*, mainly growing in the southwest fringes of the Caspian Sea (Jaaska 1993). This reinforces the suggestion that hybridization of tetraploid wheat with *Ae. tauschii* occurred in the Caspian region.

Ae. tauschii grows in a wide range of ecological conditions. It occupies both primary and segetal habitats (Eig 1929; Kimber and Feldman 1987; van Slageren 1994; Zohary et al. 2012) and thrives in areas characterized by continental climatic conditions, from the dry sagebrush steppes of the elevated Iranian and Afghan plateaus, to desert margins, as well as in more temperate climates, such as the rain-soaked southern coastal plain of the Caspian Sea. At the same time, throughout this, area *Ae. tauschii* is a successful colonizer of secondary, manmade habitats, and a common weed in cereal fields. Towards the periphery of its distribution, it is almost exclusively a weed in cultivation (Zohary et al. 2012).

Consideration of the ecology and distribution of *Ae. tauschii* reveals that this wild grass contributed substantially to the adaptation and worldwide success of bread wheats (Zohary et al. 2012). This is the easternmost diploid species in the wheat group, with a center of distribution lying in continental or temperate central Asia. It is widespread and very common in northern Iran and adjacent Transcaucasia, Transcaspia, and Afghanistan (Eig 1929; van Slageren

1994). From this geographic center, *Ae. tauschii* spreads west to east Turkey and Syria, and east to Pakistan. *Ae. tauschii* is a variable species represented by a multitude of forms, from slender types with cylindrical spikes (ssp. *tauschii*), to more robust plants with thick, beaded spikes (ssp. *strangulata*).

The wheat D subgenome appeared anomalous among the three wheat subgenomes, in its great fluctuation chromosomal diversity (Akhunov et al. 2010). Gene flow from *A. tauschii* has been an important source of wheat genetic diversity and influenced its distribution along the D-subgenome chromosomes. Yet, despite its growth as a weed in bread wheat fields, and ample opportunities for hybridization between hexaploid wheat and its D-subgenome donor (Kihara et al. 1965), direct hybridization of *Ae. tauschii* with hexaploid wheat is arduous. On the other hand, *Aegilops tauschii* readily hybridizes with tetraploid wheat, and triploid hybrids often produce many unreduced gametes and are fertile (Matsuoka and Nasuda 2004; Zhang et al. 2010). This wide variation of forms may indicate recurrent formation of hexaploid wheat from many independent crosses, involving different genotypes of tetraploid wheat and *Ae. tauschii* (Kuckuck 1964; Mac Key 1966; Jakubziner 1959; Morris and Sears 1967; Feldman et al. 1995). This recurrent origin presumably occurred throughout the area where tetraploid wheat was farmed in the distribution area of *Ae. tauschii*, i.e., from eastern Turkey in the west up to western China in the east.

While tetraploid wheats, either hulled or free-threshing, in keeping with their Near Eastern origin, are adapted to the Mediterranean-type environments (with mild winters and warm, rainless summers), addition of the D subgenome of the central Asiatic *tauschii* greatly extended the range of adaptation of hexaploid wheats to a more continental climate and northern latitudes (Zohary 1969; Zohary et al. 2012; Feldman 2001). Incorporation of the *Ae. tauschii* subgenome rendered the hexaploid plants more capable of withstanding continental winters and humid summers, facilitating the spread of hexaploid bread wheat over the continental plateaus of Asia and the colder temperate areas in eastern, central, and northern Europe, explaining their prevalence in these regions.

The D subgenome confers many desirable bread wheat qualities, including bread making quality (Orth and Bushuk 1973), cold hardiness (Limin and Fowler 1981; Le et al. 1986), and salt tolerance (Schachtman et al. 1992). Bread wheat is the dominant crop in temperate countries and is used for human food and livestock feed. Its success, resulting from the addition of the D subgenome to the BA of tetraploid wheat, depends partly on its adaptability and high yield potential, but also on its gluten protein fraction, which confers the viscoelastic properties that allow dough to be processed into bread, pasta, noodles, and other food products (Shewry 2009). Bread wheat also contributes essential amino acids, minerals, and vitamins, and beneficial phytochemicals and dietary fiber components to the human diet and are particularly enriched in whole-grain products. However, wheat products are also known or suggested to be responsible for a number of adverse reactions in humans, including intolerances (notably celiac disease) and allergies (respiratory and food). Current and future concerns include sustaining wheat production and quality with reduced use of agrochemicals and developing lines with enhanced quality for specific end-uses, notably for biofuels and human nutrition (Shewry 2009).

In addition, accessions of *Ae. tauschii*, particularly those of subsp. *strangulata*, show resistance to many diseases (Yildirim et al. 1995; Cox et al. 1995; Appels and Lagudah 1990; Knaggs et al. 2000). The increase in cold hardiness ascribed to the D subgenome, supports Tsunewaki's (1968) suggestion that the addition of this subgenome to tetraploid wheat enabled the spread of the cultivation of the resulting hexaploid to colder northern countries. Analysis of inter-varietal substitution lines, in which a chromosome of a cold hardiness cultivar of hexaploid winter wheat substituted its homologous chromosome in a spring cultivar, showed that chromosomes 4D and 5D accounted for much of the difference in cold hardiness between these two cultivars (Law and Jenkins 1970; Cahalan and Law 1979).

Hexaploid wheat has greater tolerance to frost and other environmental extremes than tetraploid wheat, and cultivation of hexaploid wheat consequently became far more widespread than that of tetraploid wheat (Dubcovsky and Dvorak 2007). Because farming of tetraploid wheat has been very limited in the Far East, such as China, introgression from *Ae. tauschii* did not take place in the Far East, while it continued in west Asia. The absence of introgression in the Far East subdivided Asian hexaploid wheat into two populations, western and Far Eastern (Dvorak et al. 2006; Balfourier et al. 2007). Because of the importance of tetraploid wheat as a bridge in gene flow from *Ae. tauschii* to hexaploid wheat, and because of the paucity of tetraploid wheat in the eastern area of wheat distribution, Far Eastern hexaploid wheat more faithfully documents the original hexaploid wheat than the west Asian hexaploid wheat (Dvorak et al. 2006). Since the identification of the parental species of *T. aestivum* (McFadden and Sears 1946), many synthetic hexaploids were produced by various researchers and breeders using a variety of different lines of tetraploid wheat and *Ae. tauschii*. These synthetic hexaploids were crossed with cultivars of ssp. *aestivum*, enriching the genetic basis of this important crop (Mujeeb Kazi et al. 1996; Dreisigacker et al. 2008).

Karyotype and Chromosome Morphology

The chromosomes of bread wheat ssp. *aestivum* cv. Chinese Spring were numbered from I to XXI by Sears (1954). But, on the basis of resemblance between different nullisomics (plants deficient for one pair of homologous chromosomes, and from study of nullisomic-tetrasomic combinations, the 21 chromosomes have been placed in seven homoeologous groups of 3 (Sears 1952). Within these groups each tetrasome (plant with four homologous chromosomes) shows the ability to compensate to some degree for either of the two other two nullisomes. The placement of the various chromosomes to one of the subgenomes was followed. For subgenome D this determination involved crossing each momsome $(2n = 41;\ 20_{II} + 1_D)$ with tetraploid wheat $(2n = 14_{II})$ and observing whether the F_1 has 14 bivalents and 6 univalents or 13 bivalents and 8 univalents. Okamoto (1957b) identified the A- and B-subgenome chromosomes by the occurrence of a heteromorphic bivalent in F1 hybrids involving telocentrics of particular A- and B-subgenome chromosomes and the amphiploid AADD. The allocation of the various chromosomes to subgenomes and homoeologous groups made possible the assignment of each chromosome to its respective genome and homoeologous group and to suggest a more logical system of renumbering the chromosomes of bread wheat (Sears 1959). This renumbering assigned the 21 chromosome pairs toto their respective subgenomes and homoeologous groups (Table 10.14).

Previous efforts to identify the chromosomes of hexaploid wheat in somatic cells by their morphological characteristics (Levitsky et al. 1939; Camara 1943, 1944; Schulz-Schaeffer and Haun 1961; Khan 1963), failed to yield satisfactory results because of the similarity between some of the chromosomes. With monosomics (chromosomes that exist in a single dose rather than in two), it is possible to study chromosome morphology, because a monosome appears as a univalent at first meiotic metaphase and lags behind the other chromosomes during first and second meiotic anaphases (Sears 1954). After measuring the size of monosomic ssp. *aestivum* cv. Chinese Spring chromosomes at second meiotic telophase, Morrison (1953) noted two chromosomes with a secondary constriction that was assumed to contain the nucleolar-organizing regions (NORs), one in chromosome 1B and a second in 6B.

Giorgi and Bozzini (1970), using aneuploid lines of hexaploid wheat cv. Chinese Spring, succeeded to study, with sufficient accuracy, the morphology of somatic metaphase chromosomes in root tip cells and compared them with those of tetraploid wheat, *T. turgidum* ssp. *durum* cv. Cappelli, and of the diploid *Ae. tauschii*. Using relative chromosome length and arm ratio as criteria, they observed that, apart from minor differences, the chromosomes of the A and B subgenomes of hexaploid wheat are very similar to those of tetraploid wheat, and the chromosomes of the D subgenome are similar to those of *Ae. tauschii*, showing that there were no major structural changes at the hexaploid level. Whereas homoeologous chromosomes generally have similar arm ratios, no simple relationship exists between chromosome homoeology and chromosome lengths. Giorgi and Bozzini (1970), like Morrison (1953), identified two pairs of satellited chromosomes (1B and 6B), 5 medians (7B, 7A, 6A, 7D, and 6D), 10 submedian (3B, 2A, 3A, 2B, 4B, 4A, 1A, 3D, 2D, and 4D), and 4 subterminal (5B, 5A, 1D, and 5D) chromosomes in the complement of Chinese Spring.

Measurements presented by Sears (1954), of the monosomes of the bread wheat cultivar Chinese Spring at second meiotic telophase, were in reasonable agreement with Morrison's (1953) results. The measurements obtained at first meiotic metaphase and second meiotic telophase by Sears (1954) (Table 10.15) showed that average chromosome length at MI in the B subgenome is 6.43 μ, at TII is 10.14 μ, and total B subgenome length at MI is 44.98 μ and at TII 70.97 μ. Chromosome 5B is the most brachial and 6B is the least brachial, almost metacentric. Average chromosome length in the A subgenome at MI is 5.6 μ and at TII 8.15 μ, and total A subgenome length at MI is 39.56 μ and at TII 57.03 μ. Chromosome 1A is the most brachial and 6A is the least brachial, submetacentric and almost metacentric.

Table 10.14 New and old designation of of the chromosomes of *T. aestivum* ssp. *aestivum*, cv. Chinese Spring, and assignment to their respective subgenomes and homoeologous groups (From Sears 1959)

Homoeologous groups	Subgenome					
	A		B		D	
	New	Old	New	Old	New	Old
1	1A	XIV	1B	I	1D	XVII
2	2A	II[a]	2B	XIII[a]	2D	XX
3	3A	XII	3B	III	3D	XVI
4	4A	IV	4B	VIII	4D	XV
5	5A	IX	5B	V	5D	XVIII
6	6A	VI	6B	X	6D	XIX
7	7A	XI	7B	VII	7D	XXI

[a] After Chapman and Riley (1966)

Table 10.15 Total length and arm ratio of the chromosomes of *Triticum aestivum* ssp. *aestivum* cv. Chinese Spring (Studied in monosomic combinations, see Table 3 in Sears 1954)

Homoeologous group	Subgenome								
	A			B			D		
	Length (μ)		Arm ratio	Length (μ)		Arm ratio	Length (μ)		Arm ratio
	MI^a	TII^b	TII^b	MI^a	TII^b	TII^b	MI^a	TII^b	TII^b
1	4.67	7.34	1.91:1	6.47	10.42	1.38:1	5.02	5.55	1.82:1
2[c]	5.52	8.11	1.26:1	6.87	10.92	1.25:1	5.58	8.18	1.23:1
3	6.21	8.50	1.15:1	6.87	12.32	1.29:1	5.86	7.45	1.37:1
4[d]	5.85	7.91	1.55:1	5.52	8.11	1.13:1	4.90	6.85	1.80:1
5	6.43	9.81	1.79:1	6.71	11.34	2.65:1	4.83	5.77	1.82:1
6	4.71	6.26	1.12:1	6.61	9.10	1.05:1	4.22	5.90	1.11:1
7	6.27	9.10	1.21:1	5.93	8.76	1.24:1	6.16	9.06	1.17:1
Total genome length	39.56	57.03	–	44.98	70.97	–	36.57	48.76	–
Average chromosome length	5.65	8.15	–	6.43	10.14	–	5.22	6.97	–

[a] First meiotic metaphase
[b] Second meiotic telophase
[c] 2A was formerly 2B and 2B was formerly 2A (Chapmam and Riley 1966)
[d] 4A was formerly 4B and 4B was formerly 4A (Dvorak 1983)

Average chromosome length in the D subgenome at MI is 5.22 μ and at TII 6.97 μ, and total D subgenome length at MI is 36.57 μ and at TII 48.76 μ. Chromosomes 5D and 1D are the most brachial and chromosome 6D is the least brachial. Chromosomes 3B and 2B are the longest and 6D is the shortest in the complement.

Sasaki et al. (1963), measuring monosome length at first meiotic metaphase of the bread wheat winter cultivar Cheyenne, and Gill et al. (1963), measuring monosome length at first meiotic metaphase and at second meiotic anaphase of the bread wheat winter cultivar Wichita, found, similar to Sears's (1954) measurements in the spring cultivar Chinese Spring, that the B subgenome included the longest chromosomes, while the D subgenome had the shortest ones. Like Sears (1954), Sasaki et al. (1963), Gill et al. (1963) also Giorgi and Bozzini (1970) found that the chromosomes of the D subgenome are usually the smallest in the complement, i.e., the total chromosome length of the D subgenome is 28.2% of the total chromosome complement length of hexaploid wheat.

In the 1970s, modern staining techniques were used to analyze the structures of cereal chromosomes, and a cytogenetic karyotype of wheat was developed (Gill and Kimber 1974a; Gill et al. 1991; Jiang and Gill 1994a). Among the several staining methods, two techniques, namely, C-banding and N-banding, have been most useful in cytogenetic studies of wheat (Gill 1987). While C-bands mark constitutive heterochromatin and stain all somatic metaphase chromosomes in hexaploid wheat, N-bands reveal specialized heterochromatin on only nine of the 21 chromosome pairs, i.e., all of the B genome chromosomes and chromosomes 4A, and 7A (Gerlach 1977). The remaining chromosomes show either faint bands or no bands at all. Later, Endo and Gil (1984) identified 16 of the 21 chromosomes of hexaploid wheat using an improved N-banding technique.

Endo and Gill (1984), using N-banding, and Gill (1987), using C-banding, measured chromosome length at metaphase in somatic cells of the bread wheat cultivar Chinese Spring and found, like Sears (1954), Giorgi and Bozzini (1970), that the B subgenome included the longest chromosomes and the D subgenome the shortest chromosomes. Like Morrison (1953), Gill et al. (1963), Giorgi and Bozzini (1970), Endo and Gill (1984), Gill (1987) also observed secondary constrictions and satellites on the short arms of chromosomes 1B and 6B.

There are some discrepancies between the measurements of Sears (1954) and those of Endo and Gill (1984), Gill (1987) which presumably resulted from the sample preparation and measurements techniques. Yet, in all these studies, the B subgenome was the largest (38%), the A subgenome was intermediate (33%), and the D subgenome was the smallest (28%) of the chromosome complement. The chromosomes of the B and the D subgenomes were more heterobrachial than those of the A subgenome, with chromosome 5B being the most and 7A the least heterobrachial. implying perhaps that the chromosomes of subgenomes B and D underwent more chromosomal rearrangements than those of subgenome A. Since cv. Chinese Spring has been shown to possess a 'primitive' chromosome structure and has been extensively used for the development of cytogenetic stocks, and in genetic studies, its idiogram was adopted as the wheat standard.

Pedersen and Langridge (1997), using the *Ae. tauschii* clone pAs1, together with the barley clone pHvG38 (a GAA-satellite sequence) for two-color FISH, were able to identify the entire chromosome complement of hexaploid wheat, facilitating easy discrimination of the three genomes of wheat. A detailed idiogram was constructed, including 73 GAA bands and 48 pAs1 bands. Identification of the wheat chromosomes by FISH will be particularly useful in connection with the physical mapping of other DNA sequences to chromosomes, as well as for chromosome identification in general, as an alternative to C-banding.

Chromosomal Rearrangements in ssp. *aestivum*

Chromosomal rearrangements are abundant among accessions of allopolyploid wheats. Using the C-banding technique, Badaeva et al. (2007) detected chromosomal rearrangements in 70 of 208 accessions of tetraploid wheat, and 69 of 252 accessions of hexaploid wheat. Among all chromosomal aberrations identified in tetraploid and hexaploid wheats, single translocations were the most frequent type (39 types), followed by multiple rearrangements (9 types), pericentric inversions (9 types), and paracentric inversions (3 types). The breakpoints were located at or near the centromere in 60 rearranged chromosomes, while, in 52 cases, they were in interstitial chromosome regions. In the latter group, translocation breakpoints were often located at the border between constitutive heterochromatin (C-bands) and euchromatin or between two adjacent C-bands. According to Badaeva et al. (2007), some of these regions seem to be translocation "hotspots". Their results, as well as data published by other authors, indicate that translocations most frequently involve B-subgenome chromosomes, followed by A- and D-subgenome chromosomes. Individual chromosomes also differ in the frequencies of translocations. Other translocations seem to occur independently and their broad distribution can result from selective advantages of rearranged genotypes in diverse environmental conditions. Badaeva et al. (2007) found significant geographic variation in the spectra and frequencies of translocation in wheat: the highest proportions of rearranged genotypes were found in Central Asia, the Middle East, Northern Africa, and France. A low proportion of aberrant genotypes was characteristic of tetraploid wheat from Transcaucasia and hexaploid wheat from Middle Asia and Eastern Europe.

In hexaploid wheat, evolutionary translocation involving chromosome arms 4AL (formerly 4BL), 5AL, and 7BS were proposed by Naranjo et al. (1987), following study of induced homologous chromosome pairing with the aid of differential chromosome staining. A similar translocation exists in rye, also involving chromosomes 4R, 5R and 7R

(Naranjo et al. 1987). Since diploid wheat, *T. monococcum*, has a similar 4A^{m}L/5A^{m}L translocation (Miller et al. 1981), it was concluded that the translocation 4L/5L is an ancestral translocation involving many Triticineae species.

Genetic maps of wheat chromosome 4A, and the chromosomal locations of 70 sets of isozymes and molecular homoeologous loci, have been used to further define the structure of wheat chromosomes 4A, 5A and 7B (Liu et al. 1992). Evidence was provided showing that an interstitial segment on 4AL originated from 5AL. The construction of comparative genetic maps of chromosomes 4A^m and 5A^m of *T. monococcum*, and of chromosomes of homoeologous groups 4, 5 and 7 of ssp. *aestivum* has provided insight into the evolution of these chromosomes (Devos et al. 1995). As was shown by Naranjo et al. (1987), and Liu et al. (1992), wheat chromosome 4A is a translocated chromosome carrying segments derived from 4A, 5A, and 7B. This translocation existing in bread wheat, which can be explained by a 4AL/5AL translocation that already occurred at the diploid level, before the differentiation of *T. urartu* and *T. monococcum*, is an important chromosomal rearrangement incident (Luo et al. 2018). Three further rearrangements, a 4AL/7BS translocation, a pericentric inversion and a paracentric inversion, took place in the tetraploid progenitor of hexaploid wheat (Devos et al. 1995). The structurally rearranged chromosomes 4A, 5A, and 7B are an exception to other translocations. The first step in these rearrangements was fixation of a 4A/5A translocation in diploid ancestors of several Triticineae taxa. This translocation existed already in diploid wheat prior to the divergence of *T. monococcum* and *T. urartu*. The remaining rearrangements were fixed during the evolution of *T. turgidum* ssp. *dicoccoides*. These involved a pericentric inversion in 4A which converted the long arm into the short arm, a paracentric inversion in the 4AL arm, and a reciprocal translocation between 7BS and the rearranged 4AL arm. Translocations fixed during the evolution of *T. turgidum* ssp. *dicoccoides* differ from those fixed during the evolution of *T. timopheevii* ssp. *armeniacum*.

Although additional terminal translocations and inversions in the A, B, and D subgenomes may have been fixed during the evolution of the three subgenomes of *T. aestivum*, and escaped molecular detection, the order of loci in wheat homoeologous chromosomes is largely colinear.

Nucleolar-Organizing Regions (NORs) and rRNA Genes

Crosby (1957) studied the nucleolar activity of each of the 21 chromosomes of the complement of ssp. *aestivum* cv. Chinese Spring, by analyzing micronucleus formation at second meiotic anaphase and telophase of the lagged monosomic chromosomes. She found that four chromosomes can produce nucleoli in the pollen mother cells

(PMCs): chromosome 1B produced nucleoli in 15% of its micronuclei, chromosome 6B in 20%, chromosome 1A in 13% and chromosome 5D in 7%. The normal nuclei in PMCs of disomic plants had an average 1.5 nucleoli per cell. Lines with four doses of chromosome 1B ((tetra 1B) and four doses of chromosome 6B each had an average of 3.0 nucleoli per PMC; tetra 1A and tetra 5D had 2.1 and 2.0 nucleoli per PMC, respectively. Thus, Crosby (1957) concluded that chromosomes 1B and 6B have strong nucleolar-organizing capacity, while chromosomes 1A and 5D have a weaker capacity. Crosby-Longwell and Svihla (1960) further studied the effect of increased and decreased doses of the chromosomes bearing NORs on nucleoli number in PMCs of bread wheat, and found that removal of pairs of strong nucleolar chromosomes induced latent nucleolar chromosomes to compensate for almost 100% of the nucleolar activity of the missing pair of strong nucleolar chromosomes.

Nucleolar organizer regions (NORs), which are located in the secondary constrictions of satellited chromosomes, have been shown to be the sites of multiple rRNA genes in eukaryotes (Birnstiel et al. 1971). In accordance, studies of Flavell and Smith (1974), Flavell and O'Dell (1976), Miller et al. (1980) showed that the major rRNA gene clusters in ssp. *aestivum* and ssp. *spelta* are located in chromosome 1B and 6B. Additional smaller sites were found on chromosomes 1A and 5D. This was confirmed by Hutchinson and Miller (1982), who showed four pairs of sites in some cells of ssp. *spelta* and ssp. *aestivum*. Using in situ hybridization of cloned ribosomal DNA, Hutchinson and Miller (1982), established the numbers of NORs in a range of subspecies of *T. aestivum*. In all subspecies, two pairs of sites occurred on chromosome arms with marked satellites. However, ssp. *compactum* and ssp. *sphaerococcum* showed an additional pair of minor sites located on a chromosome without an obvious satellite.

Following Crosby (1957), also Flavell and Smith (1974) and Flavell and O'Dell (1976) considered chromosomes 1B and 6B as possessing "strong" and the 1A and 5D as "weak" nucleolar organizers. NORs on 1B and 6B of the cultivar Chinese Spring possess approximately 90% of the total rRNA gene complement. Chromosome 6B possesses approximately 5500 (60%) rRNA genes, chromosome 1B possesses 2700 (30%) rRNA genes, whereas chromosomes 1A and 5D possess 950 (10%) rRNA genes of the total 9150 rRNA genes of the cv. Chinese Spring (Flavell and Smith 1974). Yet, different cultivars of bread wheat have different numbers of ribosomal RNA genes, as indicated by rRNA/DNA hybridizations (Flavell and Smith 1974; Mohan and Flavell 1974). For Example, in the cultivar Holdfast, chromosome 6B possess approximately only 2000 ribosomal RNA genes (Flavell and Smith 1974).

In plants, the rRNA cistrons (18S-5.8S-26S) are organized as families of repeated units in tandem arrays, at the NORs of chromosomes. Each repeating unit consists of one rRNA sequence including transcribed (18S-5.8S-26S) and non-transcribed spacer regions. The number of repeat units and the size of the spacer vary across plants. Also in hexaploid wheat the ribosomal RNA genes are organized in tandem arrays at the nucleolus organizers (Flavell 1989). By the use of various aneuploid lines and other genetic stocks, and by in situ hybridization using high specific activity ^{125}I-rRNA, Appels et al. (1980) determined the location of the rRNA genes in wheat, rye and barley and studied the organization of the ribosomal RNA cistrons in these grasses. Their study confirmed the location of the 18S-5.8S-26S rRNA gene repeat units in ssp. *aestivum* (cv. Chinese Spring) on chromosomes 1B, 6B and 5D, and this repeating unit is approximately 9.5 kb long (Appels et al. 1980).

In situ hybridization in conjunction with deletion mapping (using chromosomal deficiencies described by Endo and Gill 1996) was employed to physically map the 18S.26S multigene rRNA gene family in ssp. *aestivum* cv. Chinese Spring (Mukai et al. 1991). These authors found a new locus in the long arm of chromosome 7D of Chinese Spring and *Aegilops squarrosa*, and also confirmed the location of the NOR locus at the telomeric end of the short arm of chromosome 1A. Moreover, they showed that the rDNA exists as condensed DNA (heterochromatic) at each end, while diffused rDNA was identified within the secondary constriction region of the NORs in 1B (*Nor*-B1), 6B (Nor-B2), and 5D (Nor- D3). On the basis of these observations, Mukai et al. (1991) supported the model of Hilliker and Appels (1989) which claimed that the usual state of rDNA is inactive (facultatively heterochromatic). A small fraction of rDNA at a specific location (usually in the middle of the NORs in wheat) exists as a diffuse region (active) in condensed metaphase chromosomes.

In ribosomal DNA, there are transcribed and non-transcribed spacers within and between gene clusters. The transcribed spacer contains both internal transcribed spacer (ITS) and external transcribed spacer (ETS) (Flavell 1980). The ITS is the spacer DNA situated between the small-subunit rRNA and the large-subunit rRNA genes in the polycistronic rRNA precursor transcript; ITS1 is located between 18S and 5.8S rRNA genes, while ITS2 is between 5.8S and 25S. Thus, each eukaryotic ribosomal cluster contains the 5′ ETS, the 18S rRNA gene, ITS1, the 5.8S rRNA gene, ITS2, the 25S rRNA gene, and finally the 3′ ETS. During rRNA maturation, ETSs and ITSs are excised (Flavell 1980) and rapidly degraded.

The non-transcribed spacer, termed the intergenic spacer (IGS) or non-transcribed spacer (NTS), separates the rRNA genes (and internal transcribed spacers) that are arranged in tandem repeats. Due to the non-coding nature of spacer DNA, its nucleotide sequence changes much more rapidly over time than nucleotide sequences coding for proteins that are subject to selective forces. Although spacer DNA might not have a function that depends on its nucleotide sequence, it may still have sequence-independent functions.

Although rRNA gene clusters were homogeneous within cultivars of ssp. *aestivum* and could be assigned to particular chromosomes, extensive polymorphism was observed for the spacer region in various cultivars (Appels and Dvorak 1982). May and Appels (1987) examined the length of the non-transcribed spacer separating rRNA genes in 25 ssp. *aestivum* cultivars, carrying up to 3000 such genes on chromosomes 1B and 6B. The data showed that there were three distinct alleles of the 1B locus, and at least five allelic variants of the 6B locus. Chromosome 5D had only one allelic variant. Whereas the major spacer variants of the 1B alleles apparently differed by the loss or gain of one or two of the 133 bp sub-repeat units within the spacer DNA, the 6B allelic variants showed major differences in their compositions and lengths. This may be related to the greater number of rRNA genes repeat units at this locus.

Genetic variation exists within a species for the number of the rRNA genes at a locus and also for the structure of the intergenic, regulatory DNA. This variation can affect the activity of a locus, relative to that of another in the same cell (Flavell 1989). The active loci are enriched with genes that are not methylated at specific CpG residues in the intergenic regulatory DNA. Flavell (1989) suggested a model that attempts to relate the structural variation between genes to the differential expression and nucleolar organization of the rRNA genes at different loci. The model is based on the affinity of proteins, at limited concentrations, to specific regulatory DNA sequences. These sequences are subject to change as a result of mechanisms that can spread mutations through a locus. The resulting variation, which may not be eliminated by selection, unless it is very deleterious and accounts for a large part of the total rDNA, may be one reason why plants maintain a large excess of ribosomal RNA genes.

Flavell and O'Dell (1979) investigated the genetic control of the various NORs in root tip cells of bread wheat by cytologically scoring the number of nucleoli per cell in (a) aneuploid derivatives of cv. Chinese Spring, each having a different dosage of a particular chromosome or chromosome arm, and (b) in substitution lines where nucleolus organizer chromosomes had been replaced by homologues possessing different NORs. They assumed that NOR activity correlates with nucleolus size and thus, with the intensity of a cytologically visible marker. The authors reported that the weak NORs on chromosomes 1A and 5D infrequently formed a visible nucleolus in the presence of the strong NORs on chromosomes 1B and 6B. When a major pair of NORs on chromosomes 1B or 6B was deleted, the smaller NORs formed a visible nucleolus more frequently. Similarly, when the major NORs were replaced by weak NORs, the smaller nucleolus organizers formed visible nucleoli more frequently. When a small nucleolus organizer was replaced by a strong NOR a larger nucleolus is formed. These and other findings led to the general conclusions that there is a correlation between the number of rRNA gene and nucleolus size, i.e., the relative size of the nucleolus formed depends principally upon the number of the total active rRNA genes in the cell. Varying the dosage of at least 13 non-nucleolus-organizer chromosomes also resulted in changes in the number of visible nucleoli per cell, implying that, in bread wheat, genetic control of the activity of individual nucleolus organizers is complex (Flavell and O'Dell 1979).

The number of nucleoli visualized at interphase by specific silver (Ag) staining, was also used to infer the activity of NORs. The silver staining procedure can be used to visualize gene activity at the ribosomal DNA sites with conventional light microscopy, since only NORs that are functionally active during the preceding interphase are stained by silver. Using this procedure, Lacadena et al. (1984) studied amphiplasty (suppression or dominance of NOR activity) in somatic metaphase cells of the synthetic allohexaploid triticale (tetraploid wheat-rye amphiploid) and found that while the NORs on 1B and 6B were active, rye NORs were suppressed. Cermeño et al. (1984) provided evidence for only four active NORs in hexaploid wheat, likely corresponding to the chromosome pairs 1B and 6B. However, some somatic metaphase cells also showed NOR activity on 1A and 5D. Their results clearly demonstrated the relative nucleolar activity of the four organizer chromosomes to be 6B > 1B > 5D > 1A in euploid, ditelosomic and nulli-tetrasomic plants of common wheat. Inclusion of a nucleolus organizer chromosome from *Ae. umbellulata* in the bread wheat genome, caused suppression of the wheat nucleolus organizers, while the *Ae. umbellulata* organizer remained active (Flavell and O'Dell 1979). Such suppression occurs in many interspecific plant hybrids and was described by Navashin (1928, 1931) as differential amphiplasty, i.e., dominance of one nucleolar organizer over another. Differential amphiplasty has been reported to exist in tetraploid and hexaploid wheats (Lacadena et al. 1984, and references therein).

Nucleolar differential amphiplasty is an epigenetic phenomenon that describes the formation of nucleoli, by chromosomes inherited from only one parent/progenitor of a genetic hybrid (Pikaard 2001). As only transcriptionally active rRNA genes give rise to nucleoli, the molecular basis for nucleolar dominance or amphiplasty is differential rRNA gene expression that might result from selective silencing of specific subsets of rRNA genes via changes in DNA methylation and histone posttranslational modifications (Pikaard 2001). Because rRNA genes are nearly identical in sequence, it has long been a mystery how rRNA genes destined for silencing can be discriminated from genes that remain active. However, recent genetic evidence indicates that selective rRNA gene silencing results from inactivation of entire NORs, and not through mechanisms that discriminate between individual rRNA genes (Pikaard 2001).

rRNA genes at different NOR loci in hexaploid wheat are expressed at different levels. Even in cv. Chinese Spring, the NOR on chromosome 1B is partially dominant to that on chromosome 6B, since the 1B locus is more active despite the smaller number of genes (Thompson and Flavell 1988). These authors have previously shown that these and other examples of nucleolar dominance in wheat are associated with undermethylation of cytosine residues in certain regions of the dominant rDNA. Thompson and Flavell (1988) showed that rRNA genes at dominant loci are organized in a chromatin conformation that renders them more sensitive to DNase I digestion than other rRNA genes.

Ribosomal RNA genes of *T. aestivum*, are known to be expressed at different levels and some loci exhibit full or partial dominance over others (Gustafson and Flavell 1996). Nucleolar dominance is best seen in hexaploid triticale where the wheat loci dominate over the rye NOR locus. A correlation appears to exist between the methylation of cytosine residues and the expression of a rRNA locus suggesting that a dominant locus would have a much-reduced methylation pattern over a recessive locus. The effect on rye NOR expression in tetraploid triticales containing various wheat and rye NOR loci was studied by Gustafson and Flavell (1996). The results showed that when both wheat chromosomes 1B and 6B were present, the rye NOR locus was methylated and suppressed. When either 1B or 6B were present, some minor rye activity was seen, and that a slightly higher degree of activity was observed in the absence of 1B versus 6B. When both 1B and 6B were absent, the rye locus was expressed as in rye. These data indicate that either one of the wheats rRNA gene loci suppress rye rRNA gene loci.

Flavell et al. (1988) studied cytosine methylation in wheat rRNA genes at nucleolar organizers displaying different activities. They observed that the methylation pattern within a specific multigene locus is influenced by the number and type of rRNA genes in other rRNA gene loci in the cell. Dominant, very active loci had a higher proportion of rRNA genes with nonmethylated cytosine residues in comparison with recessive and inactive loci. The authors concluded that cytosine methylation in rRNA genes is regulated and that the methylation pattern correlates with the transcription potential of an rRNA gene. Suppression of rRNA genes, originating from one parent, is often due to cytosine methylation (Thompson and Flavell 1988; Gustafson and Flavell 1996; Houchins et al. 1997). Neves et al. (1995)

showed that treatments during germination with the cytosine analogue 5-azacytidine, that causes cytosine demethylation, stably reactivated the expression of the suppressed rRNA genes of rye in triticale.

Wheat rRNA genes are methylated at CCGG sites that are present in the intergenic regions (Sardana et al. 1993). In all the genotypes of *T. aestivum* studied, the rRNA gene loci with larger intergenic regions between their genes, possessed a larger number of rRNA genes unmethylated at one or more CCGG sites in the intergenic regions as compared to loci with shorter intergenic regions. In four genotypes, rDNA loci with longer intergenic regions had larger secondary constrictions on metaphase chromosomes, a measure of relative locus activity, as compared to loci with shorter intergenic regions. The findings have been integrated into a model for the control of rDNA expression based on correlations between cytosine methylation patterns and the number of upstream 135 by repeats in intergenic regions (Sardana et al. 1993).

The reduction in the number of rye rRNA genes containing an unmethylated CCGG site in the promoter was associated in wheat x rye hybrids with the suppression of the rye nucleolus (Houchins et al. 1997). These results are consistent with a model in which promoter and upstream regulatory repeats of rRNA genes compete for limited concentrations of regulatory proteins, and genes that are methylated at key binding sites fail to engage these regulatory proteins and thus remain inactive.

Carvalho et al. (2010) studied the methylation patterns of the NOR regions in 18 Portuguese bread wheat lines, of which 10 presented six Ag-NORs per somatic metaphase and six nucleoli per interphase, and eight presented four Ag-NORs per metaphase and four nucleoli per interphase. Using Southern blot, with pTa71 as probe, which identified a complete rDNA unit of bread wheat, they noted that DNA digestions, performed by the restriction enzymes *Msp*I and *Hpa*II, resulted in different patterns, revealing the high level of cytosine methylation at their recognition sequences. The total percentage of NOR methylation indicated that wheat lines with a maximum of four Ag-NORs were more heavily methylated at the NOR region than lines with a maximum of six Ag-NORs.

Jointly with 25S and 5.8S rRNA, also *5S rRNA* is an integral component of the large ribosomal subunits in plants. The genes coding for the 5S rRNA (5S DNA) of *T. aestivum* have been extensively studied (e.g., Gerlach and Dyer 1980; Appels et al. 1980; Scoles et al. 1988; Dvorak et al. 1989). This species contains several thousands of repeated 5S DNA units that are arranged in tandem arrays and are located at one or two loci in the A, B, and D subgenomes of *T, aestivum* (Appels et al. 1992). The 5S DNA multigene family in bread wheat cv. Chinese Spring consist of 10,000 copies of the 5S rRNA coding sequence per nucleus (Appels et al.

1980). Sequencing of entire repeating units in ssp. *aestivum* showed that there are two size classes of repeating units, 410 and 500 bp, in the 5S rRNA genes (Gerlach and Dyer 1980). Each of the size classes carry 5S rRNA genes (120 bp) coding for a small 120-bp molecule that forms part of the ribosome, and a variable spacer region (290 bp and 380 bp, respectively) (Baum and Bailey 2001). The spacer region changes much faster in evolution than the gene region (Fedoroff 1979; Scoles et al. 1988; Sastri et al. 1992) and consequently, each 5S DNA locus consists of tandem arrays of sequences containing blocks that are highly conserved (gene regions), alternating with blocks that are highly divergent (spacer regions) (Scoles et al. 1988; Baum and Appels 1992; Sastri et al. 1992). The units at a given locus are very similar in DNA sequence but marked differences have been observed between loci. Nucleotide sequences of the two clones showed that their 120-bp coding regions were very similar, while the spacers showed very low homology, indicating that their divergence must be relatively ancient.

In *T. aestivum*, the short (410 bp) units were found in the short arm of chromosomes 1B and 1D, while the repeated 500-bp units were in the short arm of chromosomes 5A and 5B (Dvorak et al. 1989). This suggests that, although the two subfamilies evolved in concert in the coding regions, homogenization between the spacers of different subfamilies, is limited or does not occur. This lack of homogenization of spacers between 5S DNA loci of different genomes in polyploid wheat, parallels a similar lack of spacer homogenization of the Nor loci between the B and D genomes of bread wheat (Lassner et al. 1987).

The number of 5S DNA repeats in *T. aestivum* is estimated to be 4700–5200 copies for the short sequence and almost 3100 copies for the long one (Appels et al.1992; Sastri et al. 1992). The copy number, however, may differ among cultivars (Röder et al. 1992). The 5S DNA repeats are located in bread wheat on the short arms of chromosomes of homoeologous groups 1 and 5 (Dvorak et al. 1989); those of the short-size class are located on group 1 chromosomes, while those of the long-size class are located on group 5 chromosomes (Appels et al. 1992). The physical location of these loci was assigned more precisely by deletion mapping and in situ hybridization (Mukai et al. 1990). The six loci of 5S DNA units located on chromosome arms 1AS, 5AS, 1BS, 5BS, 1DS, and 5DS were designated by Dvorak et al. (1989), according to the rules for wheat genetic nomenclature, *5SDna-A1, 5SDna-B1,* and *5SDna-D1,* on chromosome arms 1AS, 1BS, and 1DS, respectively, and *5SDna-A2, 5SDna-B2*, and *5SDNA-D2* on chromosome arms 5AS, 5BS, and 5DS, respectively.

In wheat, the major 5S RNA gene sites are close to the secondary constrictions where the 18S-5.8S-26S repeating units are found. The repeat unit of the 5S RNA genes was approximately 0.5 kb in wheat (Appels et al. 1980). Dvorak et al. (1989) determined the location of 5S rRNA genes (5S DNA) in bread wheat on chromosome arms 1BS, 1DS, 5AS, 5BS, and tentatively 5DS.

In chromosome arm 1BS, the 5S DNA probe hybridized in situ to the middle of the satellite (Appels et al. 1980). Chromosome 1B shows a C-band at that position. This suggests equivalence between the subterminal C-band in the satellite of chromosome 1 and the physical location of the 5S DNA. Deletion mapping of chromosome 5B placed all 5S DNA copies into the distal half of the short arm (Kota and Dvorak 1986). As in the short arm of chromosome 1B, there is a subterminal C-band in this region. Hence, the subterminal C-bands of 1BS, 1DS, 5AS, and 5BS may be equivalent to the 5S DNA arrays (Dvorak et al. 1989).

These observations suggest that 5S DNA is located in a single locus in each chromosome arm. According to the rules for wheat genetic nomenclature and in adherence with the designations used for 5S DNA by Xin and Appels (1988), the loci on chromosome arms 1AS, 1BS, and1DS were designated 5SDna-A1, 5SDna-B1, and 5S Dna-D1, respectively. The loci on chromosome arms 5AS, 5BS, and 5DS were designated 5SDna-A2, 5SDna-B2, and 5SDNA-D2, respectively. The existence of several species in which there is no linkage between the NORs and 5S DNA loci.

Genome Size

The 1C DNA content in the subspecies of *T. aestivum* ranges from 17.67 pg in ssp. *aestivum* to 18.92 pg in ssp. *tibetanum* (Eilam et al. 2008). A similar estimate was provided by Bennett (1972), who reported on 18 pg 1C DNA in ssp. *aestivum*, and a somewhat lower estimate by Arumuganathan and Earle (1991), who reported on 16.5 pg in this subspecies. Two cultivars (4 plants) of ssp. *spelta* had a mean of 17.72 ± 0.039 pg, 12 cultivars (19 plants) of ssp. *aestivum* had 17.67 ± 0.311 pg, one cultivar (2 plants) of ssp. *compactum* had 17.78 pg, two cultivars (4 plants) of ssp. *sphaerococcum* had 17.99 ± 0.244 pg, and one line (2 plants) of ssp. *tibetanum* had 18.92 pg 1C DNA. The amount of DNA in the four domesticated subspecies of *T. aestivum* did not differ significantly, whereas ssp. *tibetanum* had a significantly higher DNA content than ssp. *aestivum* and ssp. *spelta* (but only one line was analyzed in ssp. *tibetanum*) (Eilam et al. 2008). The DNA content of *T. aestivum* did not differ from that of the other hexaploid wheat, *T. zhukovskyi* (Eilam et al. 2008). Nishikawa and Furuta (1968) found no significant differences in DNA content among American and Japanese cultivars of ssp. *aestivum*. Likewise, Nishikawa and Furuta (1969), Furuta et al. (1974), Eilam et al. (2008) reported no differences in DNA content neither between synthesized hexaploid wheat and cultivars of this subspecies, nor between the various subspecies of

T. aestivum. The synthesized hexaploids have nuclear DNA content equal to the sum of the DNA contents of their respective parents (Nishikawa and Furuta 1969; Furuta et al. 1974). Similarly, Rees (1963), Rees and Walters (1965), Pegington and Rees (1970) found no evidence of appreciable change in nuclear DNA subsequent hybridization and allopolyploidy that gave rise to the hexaploid wheats. Yet, Pegington and Rees (1970) found that chromosomes in *T. aestivum* are shorter than in its tetraploid and diploid ancestors, which they claim is not reflection of a diminution in the chromosome material but, rather, of its reorganization. In contrast, Pai et al. (1961) reported that a considerable degree of elimination of chromosomal material took place in hexaploid wheats subsequent their origin, and Upadhya and Swaminathan (1963a) reported that in two studied cultivars of ssp. *aestivum*, there is less DNA than the expected amount from the sum of its two putative parents. Pai et al. (1961), Upadhya and Swaminathan (1963a) suggested that this reduction in DNA content might have been an important factor in the conversion of allopolyploid wheats into functional diploids. The combination of the multivalent gene suppressor system, the *Ph1* gene, and DNA elimination, appears to have led to a synthesis of the advantageous features of allopolyploidy in wheats. Eilam et al. (2008) also found that the 1C DNA content of *T. aestivum* is significantly smaller than the expected additive amount of 1C DNA of its two putative parents, namely, 12.84 ± 0.175 pg in *T. turgidum* ssp. *durum,* one of the putative donors of the B and A subgenomes, and 5.17 ± 0.087 pg of *Ae. tauschii,* the donor of the D subgenome (Eilam et al. 2007, 2008).

Furuta et al. (1984) determined the DNA content of the three subgenomes of hexaploid wheat by summing the DNA contents of the seven individual chromosomes of each subgenome. They found that the DNA content of the B subgenome was significantly higher than those of the A and D subgenomes. In addition, DNA content of individual chromosomes varied from chromosome to chromosome in the studied lines. Lee et al. (1997) estimated the DNA content of individual chromosomes at the G1 stage using flow cytometry, by subtracting the readings of a monosomic line from those of euploid *T. aestivum*. The haploid 2C DNA content of individual wheat chromosomes at the G1 stage range from about 0.58 pg in single-chromosome 1D, to approximately 1.12 pg in single-chromosome 3A. The A subgenome (haploid 2C content = 6.15 pg) seems to contain more DNA than the B (haploid 2C = 6.09 pg) and D (haploid 2C = 5.05 pg) subgenomes. Analysis of variance showed significant differences ($\alpha = 0.01$) in DNA content both among homoeologous groups and among genomes. Yet, their estimates of interphase wheat chromosomes DNA content in monosomic lines correlated poorly with the chromosome sizes at metaphase ($r = 0.622, p \leq 0.01$).

Chromosome Arrangement in the Nucleus of *T. aestivum*

Arrangement of chromosomes in the eukaryotic nucleus and its impact on various aspects of chromosomal behavior and function, have been a focal point of interest in cytological research for over a century (for review see Avivi and Feldman 1980; Comings 1980; Hilliker and Appels 1989). Arrangement of chromosomes has two main elements: (i) arrangement of chromosomes with respect to nuclear polarity and to various nuclear components (e.g., Rabl orientation, attachment of centromeres to the spindle fibers, involvement of centromeres in the formation of the nuclear membrane, and attachment of centromeres and telomeres to the nuclear membrane), and (ii) arrangement of chromosomes with respect to one another (Comings 1968; Feldman and Avivi 1973a, b; Vogel and Schroeder 1974; Mosolov 1974; Horn and Waldman 1978; Avivi et al. 1982a, b). While there is consensus regarding the concept of polarized chromosomal arrangements in the interphase nucleus, there is still some controversy concerning the spatial relationships of chromosomes with respect to one another. This controversy stems from difficulties in performing direct observations of interphase chromosomal arrangements and from the fact that mitotic metaphase chromosomal distribution, i.e., arrangement of chromosomes on the metaphase plate by means of the spindle fibers, may be distinct from that of interphase. In most cases, therefore, interphase chromosomal arrangement is inferred from chromosomal positions at metaphase of physically or chemically treated cells. In such cells, the spindle system did not develop and, presumably, the chromosomes did not move much after the disintegration of the nuclear membrane; therefore, they are assumed to retain, more or less, their interphase location.

In diploid organisms the arrangement of chromosomes with respect to one another in the interphase nucleus comprises two kinds of spatial relationships: (a) those between homologous chromosomes, and (b) those between non-homologous chromosomes. In interspecific hybrids and in allopolyploids, spatial relationships also exist between homoeologous (partially homologous) chromosomes and non-homoeologous chromosomes of different subgenomes (Avivi and Feldman 1980).

Bread wheat, *T. aestivum* ssp. *aestivum*, is quite suitable for studies of such spatial relationships. First, as bread wheat is a hexaploid with three homoeologous subgenomes, A, B, and D, and six genetically related chromosomes in each of the seven homoeologous groups, it can be used to examine the relationships between chromosomes of the same or of different subgenomes. Second, the availability in bread wheat cv. Chinese Spring complete series of telocentric chromosomes (Sears 1954; Sears and Sears 1979), facilitates production of plants with different types of telocentric pairs, easily recognizable in mitosis and meiosis. Third, the deduction that chromosome arm 5BL affects spatial relationships between chromosomes (Feldman 1966c; Feldman and Avivi 1973b, 1984), allows studies of the genetic control and subcellular and molecular mechanisms involved in chromosomal arrangement.

When studying spatial relationships in hexaploid wheat, chromosomes can be grouped into telocentric pairs, whose members are either genetically related or unrelated. In allohexaploid wheat, related chromosomes can be either of the same subgenome (homologues) or of different subgenomes (homoeologues). Unrelated combinations of two chromosomes (non-homologues and non-homoeologues) can be divided into combinations whose members belong to the same subgenome and those whose members belong to different subgenomes.

The approximate two decades of study of chromosomal arrangement in the somatic and meiotic nucleus of bread wheat (Feldman et al. 1966; Feldman and Avivi 1973a, 1973b, 1984; Avivi et al. 1982a, b; Yacobi et al. 1985) have provided much information on spatial relationships between different types of chromosomes and on the genetic control of these relationships. These studies indicate that in root-tip and premeiotic cells and in first meiotic prophase and metaphase, chromosomes are distributed nonrandomly with respect to each other and occupy definite positions in the nucleus.

Spatial Relationships Between Chromosomes of the Different Subgenomes

Feldman and Avivi (1973b), Avivi et al. (1982b), measuring distances between different telocentric pairs in root-tip cells of bread wheat cv. Chinese Spring, found that the chromosomes of each of the three subgenomes are arranged at larger distances from each other than chromosomes of the same subgenome do. This might imply that each subgenome occupies a distinct region in the allohexaploid nucleus.

Spatial separation of parental genomes in somatic and meiotic metaphases of various plant hybrids has been reported by several researchers (Finch et al. 1981; Schwarzacher et al. 1989, 1992; Linde-Laursen and Jensen 1991). Others have made such observations in allopolyploid *Milium montianum* (Bennett and Bennett 1992) and allotetraploid cotton (Han et al. 2015). In allopolyploid plants, there appears to be subgenome separation of interphase chromosomes, with chromosomes of each subgenome tending to cluster together (Hilliker and Appels 1989). A similar genome separation phenomenon was found in synthesized tetraploid cotton (genome AAGG) (Han et al. 2015), indicating that genome separation established immediately after tetraploid cotton formation. Given the evidence of parental genome separation in other plants, Han et al. (2015) speculated that genome separation might be a normal phenomenon in diploid and allopolyploid species.

Concia et al. (2020) analyzed the whole genome interaction matrix in allohexaploid wheat and revealed three hierarchical layers of chromosome interactions, from strongest to lowest: (i) within chromosomes, (ii) between chromosomes of the same subgenome, and (iii) between chromosomes of different subgenomes. This organization indicates a non-random spatial distribution of the three subgenomes that could mirror the presence of functional "subgenome territories."

Concia et al. (2020) confirmed the presence of subgenome-specific territories using genomic in situ hybridization (GISH) in root meristematic cells. Similar to Avivi et al. (1982b), Concia et al. (2020) observed that the A and B subgenomes interact more frequently than do A and D or B and D. To determine whether other polyploid plants share the same large-scale nuclear organization, Concia et al. (2020) analyzed 14-day-old rapeseed seedlings (*Brassica napus*), and again revealed a three-layer hierarchy of chromosomal interactions identical to those in wheat, suggesting that this organization might be a general feature of allopolyploid plants.

Concia et al. (2020) showed, using two genome-wide complementary techniques, GISH and Hi-C, that the chromatin of hexaploid wheat is not uniformly distributed across the nucleus but, rather, occupies subgenome-specific nuclear compartments. This finding is consistent with previous cytological observations (Feldman and Avivi 1973b; Avivi et al. 1982a, 1982b), indicating that chromosomes of the same subgenome tend to be physically closer to each other than chromosomes of different subgenomes. Consequently, Concia et al. (2020) proposed that subgenome territories are the primary level of chromatin spatial organization in bread wheat. The establishment of subgenome territories may be a mechanism that favors the pairing of homologues versus homoeologues by creating territorial "boundaries" between the different subgenomes, i.e., either homoeologues or non-homologues (Concia et al. (2020).

Similar finding showing that the three subgenomes of bread wheat are located in three different territories in the interphase nucleus were presented by Li et al. (2000) and Jia et al. (2021). Using genomic in situ hybridization (GISH), Li et al. (2000) found that the three subgenomes of hexaploid wheat tend to localize to specific nuclear territories. Jia et al. (2021) investigated the mechanism affecting higher-order structure on both chromosome and subgenome levels in common wheat. Their data rsupports the existence of subgenome-specific territories and reveals the impact of the genetic sequence context on the higher-order chromatin structure and subgenome stability in hexaploid wheat.

Spatial Relationships Between Homologous and Homoeologous Chromosomes

In normal dosage, the long arm of chromosome 5B (5BL) of ssp. *aestivum* cv. Chinese Spring prevents pairing of homoeologues while allowing homologues to pair regularly at meiosis. In the absence of 5BL as in nullisomic 5B or in the *ph1b* mutant line, which is deficient for the *Ph1* gene, the suppressor of homoeologous pairing on 5BL, some homoeologous pairing superimposed on the homologous pairs can be observed (Sears 1976b). Six doses of 5BL caused partial asynapsis of homologous chromosomes at meiosis, and at the same time, some pairing of homoeologous chromosomes and interlocking of bivalents (Feldman 1966c). To explain this paradoxical effect of six doses of 5BL, it was assumed (Feldman 1966c, 1968) that this chromosome arm carries a gene, most probably *Ph1*, that neither interferes with the pairing process itself nor with those of recombination, but, rather, disrupts a premeiotic event which is a prerequisite for regular meiotic pairing. It was hypothesized that six doses of 5BL alter the arrangement of homologues and homoeologues in the premeiotic nucleus of bread wheat. According to this assumption, in plants lacking 5BL, the homoeologues as well as homologues lie close to each other, resulting in some homoeologous pairing, which is superimposed on the homologous pairs, and also in few interlocking of two or, more rarely, three bivalents. Two doses of 5BL, as in disomic 5B or mono-isosomic 5BL, while scarcely affecting the homologous chromosomes, maintain the homoeologues apart, thus, leading to exclusive homologous pairing without interlocking of bivalents. Six doses of 5BL (as in tri-isosomic 5BL) fully suppress premeiotic arrangement, leading to a random distribution of chromosomes, in which the homologues may presumably be separated by a distance of as much as several microns. The meiotic attraction forces which usually cause pairing of very closely oriented homologues are not sufficient to bring about pairing of such distantly separated homologues, and, as a result, many homologous chromosomes fail to pair. Since homoeologous chromosomes may, by chance, lie close to each other, homoeologous pairing can take place. Although multivalent associations were rare at first metaphase of tri-isosomic 5BL, due to the general reduction of pairing, heteromorphic rod bivalents occur, presumably also the result of homoeologous pairing. Finally, the wide occurrence of interlocking bivalents, in spite of the reduction in the total number of bivalents, shows clearly that some of the pairing is between widely separated partners.

The gene on 5BL is not a specific suppressor of homoeologous pairing, since in extra doses, it also affects the pairing of homologous chromosomes (Feldman 1966c; Wang 1990). In disomic 5B plants, the normal two doses of the 5BL gene are presumably counteracted by the presence of the homoeoallelic pairing promoters on chromosomal arms 5AL and 5DL and, therefore, only homoeologous pairing is suppressed, whereas in tri-isosomic 5BL plants, the extra dose of this gene also suppresses homologous

pairing. Indeed, it was found (Feldman, 1966c, 1968) that the long arms of 5A and 5D carry gene(s) that promote pairing. There are several other pairing genes that may also be involved in the control of premeiotic association between bread wheat chromosomes. These include weak suppressors on 3DS (Mello-Sampayo 1972), on 3AS (Driscoll 1973), on 2DL and possibly also on 2AL and 2BL (Ceoloni et al. 1986), and promoters on the short arms of chromosomes of group 5 (Feldman 1966c; Feldman and Mello-Sampayo 1967; Riley et al. 1966; Dvorak 1976), on the long arms of chromosomes of group 3 (Driscoll 1973), and on the short arms of chromosomes of group 2 (Ceoloni et al. 1986). Hence, the control of pairing in bread wheat stems from a balance between several suppressors and promoters.

Further evidence indicating that extra doses of 5BL interfere with premeiotic chromosomal arrangement came from their effect on pairing of isochromosomes in plants bearing tri-isosomic 5BL. Such bread wheat plants carry three isochromosomes 5BL, each with two homologous 5BL arms, which may undergo inter-chromosomal pairing between 5BL arms of different isochromosomes or intra-chromosomal pairing involving arms of the same isochromosome. While six doses of 5BL reduced the inter-chromosomal pairing between the three 5BL isochromosomes down to a frequency similar to that of conventional homologous chromosomes (to about 50% of the normal pairing), the degree of intra-chromosomal pairing of these isochromosomes was not affected at all. Hence, when the homologous arms are connected by a common centromere and cannot be spatially separated, their pairing is not suppressed by six doses of 5BL.

The effect of *Ph1* on intra- and inter-chromosomal pairing was compared to the effect of *the Syn-B1* gene (Vega and Feldman 1998) located on the long arm of chromosome 3B (3BL), whose activity is responsible for normal synapsis and chiasma formation (Sears 1954; Kempanna and Riley 1962; Kato and Yamagata 1982, 1983). In contrast to the effect of *Ph1*, intra-chromosome pairing was strongly reduced in the absence of the synaptic gene Syn-B1 (Vega and Feldman 1998).

Absence of 5BL also induces bivalent interlocking, but to a lesser extent than six doses. However, in zero dose of 5BL, interlocking configurations are composed of only two or three bivalents, as predicted for interlocking involving only homoeologous bivalents. In the presence of two doses of 5BL, no interlocking bivalents are formed. On the other hand, in plants with four and six doses of 5BL, more than three bivalents (up to seven) form interlocking configurations, indicating that in extra dose of 5BL, non-homologous bivalents, presumably from the same subgenome, may interlock with each other (Yacobi et al. 1982; Yacobi and Feldman 1983). Chromosomal arm 5BS affects interlocking in an opposite manner than 5BL; namely, two and four doses of 5BS markedly reduce interlocking frequency (Yacobi et al. 1982).

Although this arrangement of close association of homologues and distant positioning of homoeologues could conceivably be limited to the stage immediately preceding meiosis, a number of investigators (e.g., Kitani 1963) have maintained that homologues are associated throughout the life of the plant. The study of Feldman et al. (1966), Feldman and Avivi (1973a, b, 1984), Avivi et al. (1982a, b) was carried out to see whether chromosomes in bread wheat are arranged non-randomly also in root-tip cells.

For the study chromosomal arrangement in root tip cells of bread wheat, telocentric chromosome pairs of homologues, homoeologues, or of non-homologues of the same subgenome, and non-homoeologues of different subgenomes, were used as marked pairs. The telocentric chromosomes are easily distinguished from the normal two-arm chromosomes of bread wheat. Measures of distances between different types of telocentric pairs, showed that in root tip cells of bread wheat, chromosomes are distributed non-randomly with respect to one another (Feldman et al. 1966; Feldman and Avivi 1973a, b, 1984; Avivi et al. 1982a, b). Within each of the three subgenomes composing the bread wheat nucleus, homologous chromosomes exhibited a so-called somatic association, being located closer to each other than any other pair of non-homologues (Feldman et al. 1966). Approximately equal degrees of association were observed between the two telocentrics for opposite arms of the same chromosome and between homologous telocentrics, suggesting that the centromere is at least partly responsible for somatic association (Feldman et al. 1966; Mello-Sampayo 1968, 1973). On the other hand, chromosomes of the different subgenomes, either homoeologues or non-homoeologues, were spatially separated, indicating that each subgenome occupies a different territory in the interphase nucleus (Feldman and Avivi 1973b; Avivi et al 1982b). Similarly, Yacobi et al. (1985) determined distances between marked pairs of bivalents at first meiotic metaphase of bread wheat by tallying the number of bivalents intervening between two marked bivalents. Within each subgenome, the association of telocentric bivalents representing the two different arms of one chromosome was much more intimate than that of genetically unrelated bivalents. The data from meiotic cells indicate that a similar pattern of chromosomal arrangement exists in both meiotic and somatic cells. Moreover, the arrangement of chromosomes in cold-treated cells, in which mitotic spindle formation is inhibited, is maintained on the functioning spindle at meiotic metaphase.

The close association between homologous chromosomes in root-tip cells of bread wheat aligns well with data of Schulz-Schaeffer and Haun (1961), Singh and Joshi (1972), Singh et al. (1976) obtained in hexaploid wheat. Specific chromosome order within a set was found in a large number of plant species (reviewed in Avivi and Feldman 1980). On the other hand, Darvey and Driscoll (1972), Dvorak and Knott (1973) found no evidence that homologous chromosomes are closer to one another than are non-homologues in root tip cells of bread wheat. As Avivi et al. (1982a) demonstrated, experimental errors could account for this failure to detect somatic association.

Studies of C-banded first meiotic metaphase in plants with four doses of 5BL have shown that most of the unpaired chromosomes, as well as the interlocked bivalents, belong to the B subgenome (Lukaszewski and Feldman, unpublished). Likewise, evidence exists for a greater effect of the *Phl* gene, which is located on 5BL (Okamoto 1957a), on the pairing of B-subgenome chromosomes than on the A- and D-subgenome chromosomes in wheat-rye hybrids (Naranjo et al. 1987). This finding supports the notion that the hexaploid wheat nucleus still maintains some individual ancestral genomic organization, which, in turn, is recognized by *Phl*.

10.4.2.9   Crosses with Other Species of the Wheat Group

A large number of *T. aestivum* accessions exhibit one, or, more rarely, two, reciprocal translocations, some of which are of different types (Badaeva et al. 2007, and reference therein). Some of these translocations may have occurred at the tetraploid level, while most translocations and obviously, those involving chromosomes of the D subgenome, occurred after the formation of the allohexaploid. The B-subgenome chromosomes are most frequently involved in translocations, followed by the A- and D-subgenome chromosomes (Badaeva et al. 2007). Individual chromosomes also differ in the frequencies of translocations. Several types of pericentric inversions, and paracentric inversions have also been detected in different accessions (Badaeva et al. 2007). The occurrence of such rearrangements and presumably also some cryptic structural hybridity, undetectable by cytological studies, may have led to some pairing failure at meiosis. In fact, the frequency of meiotic cells with univalents was 4.0–4.4% in pure strains of several subspecies of *T. aestivum*, namely, ssp. *aestivum*, ssp. *compactum*, and ssp. *spelta* (Thompson and Robertson 1930), and some inter-varietal hybrids of ssp. *aestivum* had two or more univalents in few cells and rare trivalents and quadrivalents (Hollingshead 1932).

Most F_1 hybrids between the various subspecies of *T. aestivum* displayed complete or almost complete chromosomal pairing at first meiotic metaphase, indicating full homology between their genomes. Yet, the F_1 hybrids between ssp. *spelta* and ssp. *macha* showed some pairing failure at first meiotic metaphase (bivalents exhibited reduced chiasmata) and a bridge + an acentric fragment, frequently formed at first anaphase, indicating the occurrence of a paracentric inversion (Chin and Chwang 1944). At meiosis of ssp. *aestivum* x ssp. *macha*, a multivalent (either a quadrivalent or a trivalent plus univalent) was observed in many meiocytes (Chin and Chwang 1944).

Chapman et al. (1976), Dvorak (1976) presented evidence implying that the A subgenome of ssp. *aestivum* derived from the genome of *T. urartu*. Chapman et al. (1976) compared the pairing behavior of the ssp. *aestivum* x *T. urartu* hybrids with earlier results obtained from hybrids between ssp. *aestivum* and T. monococcum ssp. *aegilopoides* var. *boeoticum* (Chapman and Riley 1966) and concluded that pairing affinities of the chromosomes of *T. urartu* or *T. monococcum* ssp. *boeoticum* with the chromosomes of ssp. *aestivum* are essentially similar but the lhybrid ssp. *aestivum* x *T. monococcum* exhibited a higher frequency of trivalents (Chapman et al. 1976). Dvorak (1976) while investigating the relationships between the A and B subgenomes of ssp. *aestivum* and the genome of *T. urartu*, reported that the F$_1$ hybrids showed relatively high chromosome pairing with an average of 5.4 bivalents, and 0.05 trivalents per cell. The low frequency of trivalents suggests that the majority of ssp. *aestivum* A subgenome chromosomes do not differ structurally from the chromosomess of *T. urartu*. Dhaliwal (1977c) observed in the F$_1$ hybrid between ssp. *aestivum*, and diploid wheat, *T. urartu* 5.94 bivalents (of which 1.58 were rod and 4.36 ring). This high level of pairing, in the presence of the *Ph1* gene of hexaploid wheat, indicates great, but not complete homology, between the A subgenome of ssp. *aestivum* and the genome of *T. urartu*.

Chapman et al. 1976) analyzed fourteen distinct F$_1$ hybrids between di-telocentric lines of A and B subgenomes of ssp. *aestivum* cv. Chinese Spring and the wild diploid wheat *T. monococcum* ssp. *aegilopoides* var. *thaoudar,* and observed 3.06–5.30 (range from 1 to 7) bivalents and 0.02–0.34 (range from 0 to 2) trivalents. The trivalents occurring in these hybrids represent interchanges mainly between A subgenome chromosomes of ssp. *aestivum* and the A chromosomes of var. *thaoudar*. The meiotic pairing behavior at first meiotic metaphase of the hybrid ssp. *aestivum* cv. Courtot × *T. monococcum* ssp. *monococcum* has been studied by means of the C-banding technique, to ascertain the homology between the chromosomes in the A subgenome of the hexaploid and those in the genome of the diploid (González et al. 1993). The technique allowed the A and B subgenome chromosomes and the 2D, 3D and 5D chromosomes to be identified. Average chromosomal pairing was 3.99 (1–7) bivalents (of which 3.11 (1–6) were rod and 1.05 (0–3) ring), and 0.14 (0–2) trivalents. A similarly low level of pairing in this hybrid combination was reported by Miller and Reader (1980), while Melburn and Thompson (1927) found higher pairing (average of 5.0 bivalents).

The *T. monococcum* 4A chromosome did not pair with any of the ssp. *aestivum* chromosomes. Two reciprocal translocations 2B/2D and 2A/3D have been identified in ssp. *aestivum* cv. Courtot. Evidently, these translocations occurred at the hexaploid level.

Data of chromosome pairing in F$_1$ hybrids between ssp. *aestivum* diploid specie of *Aegilops* are presented in Table 9.10. The hybrids ssp. *aestivum* and *Ae. speltoides* show different results depending on the genotype of the latter. Kimber and Athwal (1972) produced F$_1$ hybrids between ssp. *aestivum* cv. Chinese Spring and a high-intermediate- and low-pairing types of *Ae. speltoides*. At first meiotic metaphase of these hybrids, they observed much higher pairing when the high-pairing type was used than when the intermediate pairing type *was* used, and much reduced pairing when the low-pairing type was used (Table 9.10). These results show that none of the subgenomes of ssp. *aestivum* are homologous to the genome of *Ae. speltoides* (Kimber and Athwal 1972).

Hybrids between ssp. *aestivum* and the other species of *Aegilops* section Sitopsis, namely, *Ae. sharonensis, Ae, bicornis, Ae. longissima,* and *Ae. searsii*, had very little pairing (observed average pairing was 1.50–1.78 bivalents, and 0.02- 0.08 trivalents; Table 9.10), showing that either very little homology exists between the chromosomes of these species or that these diploids contain a *Ph*-like gene that suppresses pairing of homoeologous chromosomes (Feldman 1978). On the other hand, a hybrid between ssp. *aestivum* and an intermediate pairing type of *Ae. longissima* showed more pairing than the hybrids with standard lines of Ae. longissima (Feldman 1978).

The meiotic behavior of ssp. *aestivum* x *Ae. speltoides*, ssp. *aestivum* x *Ae. sharonensis* and ssp. *aestivum* x *Ae. longissima* has been analyzed by the C-banding technique (Fernández-Calvín and Orellana 1994). These authors reported that in all the hybrids analyzed the mean number of bound arms per cell for the A-D type was significantly higher than the mean number of associations between the B subgenome and the S/S^{sh}/S^l genomes of the aegilops species. Usually, the relative contribution of each type of pairing is maintained among hybrids with different *Aegilops* species. These results indicate that the genomes of *Ae. speltoides, Ae. sharonensis* and *Ae. longissima* show a similar affinity with the genomes of hexaploid wheat; therefore, none of these species can be considered to be a distinct donor of the B subgenome of hexaploid wheat.

Using the C-banding technique, Maestra and Naranjo (1998) studied homoeologous pairing at first meiotic metaphase in the *Ph1, ph1b,* and *ph2b* hybrids of ssp. *aestivum* and a high-pairing type of *Ae. speltoides*. All arms of the seven chromosomes of the *speltoides* genome showed normal homoeologous pairing, which implies that no apparent chromosome rearrangements occurred in the evolution of

Ae. speltoides relative to that of bread wheat. A pattern of preferential A-D and B-S pairing confirmed that the S genome is closely related to the B subgenome of wheat. Although this pairing pattern was also reported in hybrids of wheat with *Ae. longissima* and *Ae. sharonensis* (Fernández-Calvín and Orellana 1994), somewhat more intimate pairing was observed between some chromosomes of the B subgenome and those of *Ae. speltoides*. These results are in agreement with the hypothesis that the B subgenome of wheat is derived from a species closely related to *Ae. speltoides*. Chromosomal pairing in the F₁ hybrid ssp. *aestivum* x *Ae. tauschii* (genome BADD) and in the reciprocal hybrid, was studied by (Kimber and Riley 1963; Table 9.10). This high level of pairing observed in these hybrids shows clearly that ssp. *aestivum* possess one subgenome (D) homologous to the genome of *Ae. tauschii*. Hybrids between ssp. *aestivum* and other diploid species of *Aegilops*, namely, *caudata*, *comosa*, and *umbellulata*, had little pairing (Table 9.10) Hence, the genomes of these diploid species have no homology to any of the three subgenomes of hexaploid wheat.

Crosses between ssp. *aestivum* x and tetraploid wheat, *T. turgidum* ssp. *durum* (Table 10.8), produced a pentaploid F₁ hybrid (genome BADBA) that had at meiosis, close to 14 bivalents that resulted from pairing between the A and B subgenomes of ssp. *aestivum* each with its homologous subgenome of ssp. *durum* (Table 10.8). Likewise, the F₁ hybrid ssp. *macha* x ssp. *durum* had 12.0 bivalents, and a quadrivalent (Kihara 1949). This high level of pairing corroborates the conclusion of Kihara (1924), who showed that two subgenomes of hexaploid wheat (B and A) are fully homologous with the two subgenomes of *T. turgidum*. On the other hand, the pentaploid F₁ hybrid between ssp. *aestivum* and *T. timopheevii* ssp. *armeniacum* (hybrid genome BADGA) had had reduced pairing (Table 10.8), showing that the G and B subgenomes of these two taxa are homoeologous rather than homologous.

Chromosomal pairing in F₁ hybrids between ssp. *aestivum* and the second hexaploid wheat species, *T. zhukovskyi* (hybrid genome BADGAAᵐ) (Table 10.8) implied that while the A subgenomes of these species are homologous, the Aᵐ subgenome is very closely related A, and B, G, and D subgenomes are homoeologous.

Data on meiotic chromosome pairing between ssp. *aestivum* and allotetraploid *Aegilops* species are presented in Table 10.10. The pentaploid hybrid between ssp. *aestivum* and *Ae. ventricosa* (genome BADDN) had 5.02 bivalents (of which 2.94 were rod, and 2.08 ring), and 0.13 trivalents and the reciprocal hybrid had 3.35 rod, 0.92 ring and 4.27 total bivalents, and 0.19 trivalents (Kimber and Zhao 1983). This level of chromosomal pairing shows that the D subgenome of hexaploid wheat is homoeologous rather than homologous to the D subgenome of *Ae. ventricosa*. The pentaploid

hybrid ssp. *aestivum* x *Ae. cylindrica* (hybrid genome BADDC) had 6.74 bivalents, 0.18 trivalents, and 0.02 quadrivalents (Riley 1966), 2.16 rod, 4.66 ring and 6.82 total bivalents and 0.08 trivalents (Kimber and Zhao 1983). Evidently, and as was shown previously by Sears (1944b), the D subgenome of *Ae. cylindrica* is homologous to the D subgenome of hexaploid wheat. In contrast, the pentaploid F₁ hybrids between ssp. *aestivum* and *Ae. peregrina* (hybrid genome BADSⱽU), *Ae geniculata* (hybrid genome BADMᵒU) and *Ae. triuncialis* (hybrid genome BADUC) had much reduced pairing (Table 10.6). The reduced pairing in these hybrids implies that the subgenomes of these allotetraploid species are only distantly homoeologous to those of hexaploid wheat.

Chromosome pairing was studied in in F₁ hexaploid hybrids between ssp. *aestivum* and allohexaploid species of *Aegilops*, *Ae. juvenalis* (genome hybrid BADDᶜXᶜU), *Ae. vavilovii* (hybrid genome BADDᶜXᶜSˢ), and hexaploid Ae. crassa (hybrid genome BADDᶜXᶜD) (Table 10.6). Ssp. *aestivum* x *Ae. juvenalis* and ssp. *aestivum* x *Ae vavilovii* had a number of paired chromosomes between the D subgenome of hexaploid wheat and the D subgenomes of these *Aegilops* allohexaploids showing that they are homoeologous rather than homologous. Presumably, the D subgenomes of *Aegilops* underwent some changes on the polyploid level and became modified subgenomes. On the other hand, the hybrid 6x *Ae. crassa* x ssp. *aestivum* had a higher pairing indicating homology between the D subgenome of ssp. *aestivum* and the unmodified subgenome D of *Ae. crassa*.

10.4.3   *T. zhukovskyi* Menabde & Ericz. (Genome GGAAᵐAᵐ)

10.4.3.1   Description of Species

Triticum zhukovskyi, known as zhukovsky's wheat, [Syn.: *T. timococcum* Kostov.; *T. timopheevii* ssp. *zhukovskyi* (Men. et Ericz.) L. B. Cai: *T. timopheevii* var. *zhukovskyi* (Men. et Ericz.) Morris et Sears; *T. turgidum* var. *zhukovskyi* (Men. & Er.) Bowden; *Gigachilon zhukovskyi* (Menabde & Erizin) Á. Löve] is an annual, predominantly autogamous, 90–130-cm-tall (excluding spikes), pubescent plant. Culms are erect and hollow throughout. The spike is indeterminate, strongly bilaterally compressed, two-rowed, 9–14-cm-long (excluding awns), ovoid or lanceolate, tapered towards the base and tip, hairy and awned. The rachis is semi-fragile and subsequently, the spike either remains intact on the culm or, during threshing, easily disarticulates at maturity into single spikelets with the rachis segment below it. The rachis internodes are covered with dense, white hairs and the spikelets are compressed and ovoid or lanceolate, with the top spikelet being fertile and generally in the same plane as those below. There are 10–15 spikelets per spike, and the 2–3 basal spikelets are rudimentary. The fertile spikelet is 10–12-mm-long and has 3 florets, the upper one usually being sterile. The glumes are similar to one another, 9–12-mm-long, very hairy, with 2 keels and 5–9 veins, and with two teeth on the upper margin, one larger and pointed and separated from the other by an acute angle. Lemma are elliptic, hairy, 9–12-mm-long, without keels, with 9–11 veins, and with a central vein, prolonged as a narrow 60–100-mm-long white or black awn. In addition to the awn, the lemma has a lateral tooth. At maturity, the palea is membranous and split along the keel. Usually, there are two grains per spikelet. The caryopsis is free but laterally compressed (hulled), hairy at the apex, long and red in color.

T. zhukovskyi was discovered recently, in 1957, in Western Georgia, by Menabde and Ericzjan (Jakubziner 1959). It only exists as a domesticated form and has never been cultivated alone. Zhukovsky's wheat is continually found in cultivated fields in West Georgia, containing, in addition to *T. zhukovskyi*, a mixture of diploid *T. monococcum* var. *hornemanni* and tetraploid *T. timopheevii*, a mixture called the *Zanduri* wheat (Jakubziner 1959; Dorofeev 1966). *T. zhukovskyi* has been a minor component of the endemic Zanduri wheat complex and is currently grown in extremely limited amounts in Western Georgia. The three components of the *Zanduri* complex are genetically effectively isolated by different ploidy levels (Mac Key 1966). *T. zhukovskyi* constitutes a minor part of the *Zanduri* complex which currently, is mainly consumed as porridge (Mac Key 1975).

T. zhukovskyi resembles *T. timopheevii* in its ear morphology and growth characteristics, but its spike is somewhat larger than that of *T. timopheevii* (Jakubziner 1959). Likewise, Upadhya and Swaminathan (1963b) pointed out that *T. zhukovskyi* resembles *T. timopheevii* in most plant characters and particularly in their hairs, which, in both species, are long, thin and unicellular. However, it has enlarged cells as compared to T. *timopheevii*. *T. zhukovskyi* and *T. timopheevii* have similar disease immunity (Jakubziner 1959) and seedlings of *T. zhukovskyi* are highly resistant to manty races of leaf rust and stem rust (Upadhya and Swaminathan 1963b). When sown in the fall, its development lag*s* several days behind *T. timopheevii*. *T. zhukovskyi* has a spring habit and exhibits very limited variation.

10.4.3.2 Cytology, Cytogenetics and Evolution

Its morphology and its constant association with *T. timopheevii* and *T. monococcum* suggest that *T. zhukovskyi* is a hexaploid species (2n = 6x = 42) that derived from a spontaneous hybridization between a domesticated subspecies of tetraploid *T. timopheevii* and a domesticated subspecies of diploid *T. monococcum* (presumably var. *hornemanni* which is a component of the Zanduri complex). Menabde and Ericzjan noted that the crossing between *T.*

timopheevii and *T. zhukovskyi* occurs readily, but the F_1 hybrid displays low fertility. If *T. zhukovskyi* originated from hybridization of the other two components of the *Zanduri* mixture, then, it is obviously of a more recent origin than its two domesticated parents.

Bowden (1959) was the first to suggest that *T. zhukovskyi* originated as a Allohexaploid from the cross of *T. timopheevii* and consequently, designated its genome AAAAGG *i*, implying the sum of the genomes of *T. timopheevii* (genome GGAA) and of *T. monococcum* ($A^m A^m$), the other two components of the Zanduri mixture. The idea that *T. zhukovskyi* is a natural polyploid derived from hybridization of *T. timopheevii* and *T. monococcum* was confirmed cytogenetically by Upadhya and Swaminathan (1963a, b). Upadhya and Swaminathan (1963) pointed out that *T. zhukovskyi* is the firbst hexaploid wheat to have a genomic constitution other than AABBDD. Thus, the allopolyploid species of the genus *Triticum* constitute two evolutionary lineages; the BBAA and BBAADD comprise one lineage and GGAA and $GGAAA^m A^m$ the second lineage.

Kimber and Sears (1983, 1987), following Bowden (1959), designated the genome of *T. zhukovskyi* AAAAGG, assuming that the genomes of *T. urartu* and *T. monococcum* are very closely related. However, Dvorak et al. (1993), studying variation in 16 repeated nucleotide sequences, showed that *T. zhukovskyi* contains one A subgenome, that was contributed by *T. urartu*, and another related subgenome, contributed by *T. monococcum*, thereby verifying that *T. zhukovskyi* originated from hybridization of *T. timopheevii* with *T. monococcum*. Dvorak et al. (1993), in view of the differences in the profiles of repeated nucleotide sequences and the reproductive isolation by hybrid sterility between *T. urartu* and *T. monococcum* (Johnson and Dhaliwal 1976), considered their genomes modified relative to each other. Consequently, they proposed to designate the genome of *T. urartu* AA and that of *T. monococcum* $A^m A^m$. Thus, the genome formula of *T. zhukovskyi* is $AAA^m A^m GG$ (Dvorak et al. 1993). Since the female parent in the formation of *T. zhukovskyi* was *T. timopheevii*, its genome should be designated $GGAAA^m A^m$. In addition, the repeated nucleotide sequence profiles in the A and A^m subgenomes of *T. zhukovskyi* showed reduced correspondence with those in the genomes of both ancestral species, *T. urartu* and *T. monococcum*. According to Dvorak et al. (1993), this differentiation is attributed to heterogenetic chromosome pairing and segregation among chromosomes of the two A genomes in *T. zhukovskyi*. This agrees with the formation of some multivalents at meiosis of *T. zhukovskyi*, indicating that *T. zhukovskyi* has four A related subgenomes (Upadhya and Swaminathan 1963b).

The karyotypes of *T. zhukovskyi, T. timopheevii* ssp. *timopheevii* and *T. monococcum* ssp. *monococcum* var.

hornemanni, were studied by Upadhya and Swaminathan (1963b). They found that all the chromosomes of *T. zhukovskyi* possess median or sub-median centromeres. The species contains three pairs of satellited chromosomes, two of which bear large satellites and one with small satellites. The two chromosome pairs with the large satellites resemble the two pairs present in *T. timopheevii*, whereas the other pair is similar in arm ratio to the pair found in *T. monococcum* var. *hornemanni*.

Electrophoretic seed protein patterns confirmed the origin of *T. zhukovskyi* as a product of hybridization between *T. timopheevii* and *T. monococcum* (Johnson 1968). The electrophoretic pattern of *T. zhukovskyi* proteins is similar to the pattern of *T. timopheevii*, specifically to that of ssp. *timopheevii*, but also contains bands of domesticated *T. monococcum*. Immunoelectrophoretic findings (Aniol 1973) also supported the conclusion that *T. zhukovskyi* originated from chromosome doubling of a hybrid between *T. timopheevii* and *T. monococcum*. Using a set of molecular cytogenetic landmarks, based on eleven DNA probes, that identified the chromosomes of *T. timopheevii*, Badaeva et al. (2016) demonstrated the existence of the *timopheevii* genome in *T. zhukovskyi*. This further corroborated the conclusion of Upadhya and Swaminathan (1963b) and others, that the latter contains the genome of the former, and presumably resulted from hybridization of *T. timopheevii* with *T. monococcum*. However, Badaeva et al. (2016) observed that the formation of *T. zhukovskyi* was accompanied by structural changes involving mostly A subgenome chromosomes, presumably, due to inter-subgenomic chromosomal pairing between the two closely related A- and A^m subgenomes in *T. zhukovskyi*.

Sallares and Brown (1999), using the sequence of the intergenic spacer regions of the NOR loci in diploid and polyploid species of wheat, showed that while *T. timopheevii* contains the spacer originating from *T. urartu*, *T. zhukovskyi* contains the spacer of both *T. urartu* and *T. monococcum*. This observation confirmed that one of the A subgenomes of *T. zhukovskyi* originated from *T. urartu* and the other from *T. monococcum*. On the other hand, Adonina et al. (2015), using the pTm30 probe cloned from *Triticum monococcum* genome and containing $(GAA)_{56}$ microsatellite sequence, found that (GAA)n sites observed in *T. monococcum* are undetectable in the A^m genome of *T. zhukovskyi*. They concluded that this site may have been eliminated during the process of allopolyploidization.

Using *T. timopheevi* as female and *T. monococcum* as male, synthetic amphiploids (genome $GGAAA^mA^m$) were produced, by several teams (Kostoff (1937b), Bell et al. (1955), Watanabe et al. 1956; Dhaliwal and Johnson 1976; Upadhya and Swaminathan 1963b; Cao et al. 2000a) and, in all cases, the morphological characteristics of the amphiploid closely resembled that of *T. zhukovskyi*. This further supports the conclusion that this species originated from a natural cross between *T. timopheevii* and *T. monococcum*, followed by chromosome doubling. However, the amphiploid is reported to be partially sterile.

Thus, the 14 *timopheevii* pairs (genome GGAA) plus the 7 pairs of *T. monococcum* (genome A^mA^m), constitute the 21 pairs of hexaploid *T. zhukovskyi*. The cytoplasm of *T. timopheevii* and of *T. zhukovskyi* derived from the donor of the G genome (Ogihara and Tsunewaki 1988; Wang et al. 1997; Gornicki et al. 2014), implying that *T. timopheevii* was the female parent. *T. zhukovskyi* is an auto-allo-hexaploid (genome $GGAAA^mA^m$), carrying two closely related subgenomes, A and A^m. Consequently, a maximum of 4 multivalents per cell was found at meiosis in *T. zhukovskyi*, the average being 1.35 per cell (Upadhya and Swaminathan 1963b). Wang et al. (1997) reported that while the average genetic distance between the A genome of emmer wheat and that of hexaploid wheat was 0.016, the average distance between *T. timopheevii* and *T. zhukovskyi* was only 0.004, suggesting that *T. zhukovskyi* arose quite recently.

Eilam et al. (2008) found the 1C DNA content of one line of *T. zhukovskyi* to be 17.74 pg. The 1C DNA content of *T. zhukovskyi* is less than 18.35 pg, the expected additive amount of 1C DNA of its two-parental species, namely 11.87 pg of domesticated *T. timopheevii*, and 6.48 pg of domesticated *T. monococcum*, (Eilam et al. 2007). Interestingly, Upadhya and Swaminathan (1963b) found that total chromatin length in μ in *T. zhukovskyi* is 461.52 ± 12.68 (smaller than the expected additive amount of its two parents: *T. timopheevii* 329.47 and *T. monococcum* var. *hornemanni* 146.23, together 475.70*)*.

On the basis of chromosome pairing in 41- and 42-chromosome hybrids between ssp. *aestivum* cv. Chinese Spring monosomic for chromosome 5B and *T. zhukovskyi*, Upadhya and Swaminathan (1965) suggested that *T. zhukovskyi* does not contain the *Ph1* gene, that suppresses pairing of homoeologous chromosomes. Since *T. zhukovskyi* arose from the cross of *T. timopheevii* and *T. monococcum*, they assumed that *T. timopheevii* also lacks *Ph1*. Joshi et al. (1970) presented pairing data indicating that the potency of *Ph1* in *T. timopheevii* is somewhat weaker than in ssp. *aestivum*. On the other hand, Feldman (1966a), Sallee and Kimber (1976) presented evidence implying that *T. timopheevii* contains a *ph1* as effective as that of ssp. *aestivum*. Thus, different cultivars of *T. timopheevii* may contain *Ph1* with different potencies, as was found by Ozkan and Feldman (2001).

10.4.3.3 Crosses with Other Species of the Wheat Group

Few hybrids were produced involving T. zhukovskyi and other species of the wheat group. Actually, only two such hybrids were reported. The F_1 hybrid *T. zhukovskyi* and *T. aestivum* ssp. *spelta,* had an average of 15.12 univalents, 10.70 bivalents, 1.39 trivalents and 0.33 quadrivalents per cell (Upadhya and Swaminathan 1963b). The F_1 hybrid *T. zhukovskyi* and *T. aestivum* ssp. *aestivum* had an average frequency of 12.27 univalents, 9.00 bivalents of which 5.53 were rod and 3.47 ring, 3.13 trivalents, 0.49 quadrivalents, 0.07 pentavalents, and 0.07 hexavalents (Shands and Kimber 1973). The partial failure of chromosome pairing in these two hybrids is similar to that usually found in hybrids between the G subgenome of *T. timopheevi* and the B subgenome of *T. turgidum* and *T. aestivum* o (Upadhya and Swaminathan 1963b). The multivalents, mainly trivalents and quadrivalents, are presumably the result of pairing between chromosomes of the A subgenome of the two specie and and A^m subgenome of *T. zhukovskyi.*

References

Aaronsohn A (1909) Contribution à l'histoire des céréales, le blé, l'orge et le seigle à l'état sauvage. Bull Soc Bot France 56:196–203, 237–245, 251–258

Aaronsohn A (1910) Agricultural and botanical explorations in Palestine. Bull Plant Industr 180:1–63

Abdel-Haleem AMH, Seleem HA, Galal WK (2012) Assessment of Kamut® wheat quality. World J Sci Technol Sustain Dev 9:194–203

Adonina IG, Goncharov NP, Badaeva ED, Sergeeva EM, Petrash NV, Salina EA (2015) (GAA)n microsatellite as an indicator of the A genome reorganization during wheat evolution and domestication. Comp Cytogenet 9:533–547

Aghaee-Sarbarzeh M, Kaviani R, Mohammad-Reza Bihamta M-R, Mohammadi M (2013) Identification of durum wheat (*Triticum turgidum ssp. durum* L.) germplasm from Iran, Italy, Argentina, and Bulgaria with resistance at the seedling stage to race Ug99 of *Puccinia graminis f. sp. tritici.* Can J Pl Pathol 35:251–255

Akhunov ED, Akhunova AR, Dvorak J (2007) Mechanisms and rates of birth and death of dispersed duplicated genes during the evolution of a multigene family in diploid and tetraploid wheats. Mol Biol Evol 24 (2):539–550. https://doi.org/10.1093/molbev/msl183

Akhunov ED, Akhunova AR, Anderson OD, Anderson JA, Blake N, Clegg MT, Coleman-Derr D, Conley EJ, Crossman CC, Deal KR et al (2010) Nucleotide diversity maps reveal variation in diversity among wheat genomes and chromosomes. BMC Genom 11:702

Al Hakimi A, Monneveux P (1997) Utilization of ancient tetraploid wheat species for drought tolerance in durum wheat (*Triticum durum* Desf.). In: Damania AB, Valkoun J, Willcox G, Qualset CO (eds) The origins of agriculture and crop domestication. ICARDA, Aleppo, pp 273–279

Amri A, Hatchett JH, Cox TS, Bouhssini M, Sears RG (1990) Resistance to hessian fly from North African durum wheat germplasm. Crop Sci 30:378–381

Anikster Y, Eshel A, Ezrati S, Horovitz A (1991) Patterns of phenotypic variation in wild tetraploid wheat at Ammiad. Isr J Bot 40:397–418

Anikster Y, Eshel A, Horovitz A (1988) Phenotypic variation. In Final report of the U.S.-Israel co-operative program: the biological structure of native populations of wild emmer wheat (*Triticum turgidum* var. *dicoccoides*) in Israel. I. Tel Aviv University, Ramat Aviv, Israel, pp 39–51

Anikster Y, Manisterski J, Long DL, Leonard KJ (2005) Leaf rust and stem rust resistance in *Triticum dicoccoides* populations in Israel. Plant Dis 89:55–62

Aniol A (1973) A serological investigation of wheat evolution. In: Sears ER, Sears LMS (eds) Proceedings of 4th international wheat genetics symposium, Columbia, Mo, USA, p 59

Anker C, Buntjer J, Niks R. (2001) Morphological and molecular characterisation confirm that *Triticum monococcum s.s.* is resistant to wheat leaf rust. Theor Appl Genet 103:1093–1098

Annapurna K (2000) Comparative study on protein and storage quality of supplemented uppuma of dicoccum and durum wheat. M.Sc. Thesis, University of Agricultural Sciences, Dharwad, Karnataka, India

Appels R, Dvorak J (1982) The wheat ribosomal DNA spacer region: its structure and variation in populations and among species. Theor Appl Genet 63:337–348

Appels R, Lagudah ES (1990) Manipulation of chromosomal segments from wild wheat for the improvement of bread wheat. Austral J Pl Physiol 17:253–266

Appels R, Gerlach WL, Dennis ES, Swift H, Peacock WJ (1980) Molecular and chromosomal organization of DNA sequences coding for the ribosomal RNAs in cereals. Chromosoma 78:293–311

Appels R, Baum BR, Clarke BC (1992) The 5S DNA units of bread wheat (*Triticum aestivum* L.). Plant Syst Evol 183:183–194

Arranz-Otaegui A, González Carretero L, Roe J, Richter T (2018) "Founder crops" v. wild plants: assessing the plant-based diet of the last hunter-gatherers in Southwest Asia. Quat Sci Rev 186:263–283

Arumuganathan K, Earle ED (1991) Nuclear DNA content of some important plant species. Pl Mol Biol Rep 9:208–218

Asins MJ, Carbonell EA (1986a) A comparative study on variability and phylogeny of *Triticum* species. 1. Intraspecific variability. Theo Appl Genet 72:551–558

Asins MJ, Carbonell EA (1986b) A comparative study on variability and phylogeny of *Triticum* species. 2. Interspecific variability. Theor Appl Genet 72:559–568

Ausemus ER, McNeal FH, Schmidt JW (1967) Genetics and inheritance. In: Quisenberry KS, Reitz LP (eds) Wheat and wheat improvement. ASA CSSA SSSA, Madison

Austin RB, Morgan CL, Ford MA, Bhagwat SG (1982) Flag leaf photosynthesis of *Triticum aestivum* and related diploid and tetraploid species. Ann Bot 49:177–189

Avivi L (1976) The effect of gene controlling different degrees of homoeologous pairing on quadrivalent frequency in induced autotetraploid lines of *Triticum longissimum* . Can J Genet Cytol 18:357–364

Avivi L (1979) High grain protein content in wild tetraploid wheat *Triticum dicoccoides* Korn. In: Procedings of 5th international wheat genetics symposium, New Delhi, India, vol 1, pp 372–380

Avivi L, Feldman M (1980) Arrangement of chromosomes in the interphase nucleus of plants. Human Genet 55:281–295

Avivi L, Feldman M, Brown M (1982a) An ordered arrangement of chromosomes in the somatic nucleus of common wheat, *Triticum aestivum* L. I. Spatial relationships between chromosomes of the same genome. Chromosoma 86:1–16

Avivi L, Feldman M, Brown M (1982b) An ordered arrangement of chromosomes in the somatic nucleus of common wheat, *Triticum aestivum* L. II. Spatial relationships between chromosomes of different genomes. Chromosoma 86:17–26

Avivi L, Levy AA, Feldman M (1983) Studies on high protein durum wheat derived from crosses with the wild tetraploid wheat *Triticum turgidum* var. *dicoccoides.* In: Sakamoto S (ed) Proceedings of 6th international wheat genetics symposium, Kyoto, Plant Germ-Plasm Inst, Kyoto, Japan, pp 199–204

Avni R, Nave M, Barad O, Baruch K Sven O, Twardziok SO, et al (2017) Wild emmer genome architecture and diversity elucidate wheat evolution and domestication. Science 357:93–97

Avni R, Lux T, Minz-Dub A, Millet E et al (2022) Genome sequences of *Aegilops* species of section Sitopsis reveal phylogenetic relationships and provide resources for wheat improvement. Plant J 110:179–192

Badaeva ED, Shkutina FM, Bogdevich IN, Badaev NS (1986) Comparison study of *Triticum aestivum* and *T. timopheevii* genomes using C-banding technique. Pl Syst Evol 154:183–194

Badaeva ED, Boguslavsky RL, Badaev NS, Zelenin AV (1990) Intraspecific chromosomal polymorphism of *Triticum araraticum* (Poaceae*)* detected by C-banding technique. Pl Syst Evo 169:13–24

Badaeva ED, Badaev NS, Gill BS, Filatenko AA (1994) Intraspecific karyotype divergence in *Triticum Araraticum* (Poaceae). Pl Sci Evol 192:117–145

Badaeva ED, Jiang J, Gill BS (1995) Detection of intergenomic translocations with centromeric and non-centromeric breakpoints in *Triticum araraticum*: mechanism of origin and adaptive significance. Genome 38:976–981

Badaeva ED, Dedkova OS, Gay G, Pukhalkyi VA, Zelenin AV, Bernard S, Bernard M (2007) Chromosome rearrangements in wheat: their types and distribution. Genome 50:907–926

Badaeva ED, Amosova AV, Goncharov NP, Macas J, Ruban AS, Grechishnikova IV et al (2015) A set of cytogenetic markers allows the precise identification of all A-genome chromosomes in diploid and polyploid wheat. Cytogenet Genome Res 146:71–79

Badaeva ED, Ruban A, Zoshchuk S et al (2016) Molecular cytogenetic characterization of *Triticum timopheevii* chromosomes provides new insight on genome evolution of *T. zhukovskyi*. Pl Syst Evol 302:943–956

Badaeva ED, Fisenko AV, Surzhikov SA et al (2019) Genetic heterogeneity of a diploid grass *Aegilops tauschii* revealed by chromosome banding methods and electrophoretic analysis of the seed storage proteins (gliadins). Russ J Genet 55:1315–1329

Badaeva ED, Konovalov FA, Knüpffer H et al (2021) Genetic diversity, distribution and domestication history of the neglected GGA^tA^t genepool of wheat. Theor Appl Genet 135:755–776

Baghyalakshmi K, Vinoth R, Ulaganathan V, Ramchander S (2015) Chromosomal behavior studies during meiosis: a cross between *Triticum timopheevii* and *Triticum sphaerococcum*. Intern J Agric Sci 7:853–857

Bakhteyev FK, Yanushevich ZV (1980) Discoveries of cultivated plants in the early farming settlementsof Yarym-Tepe II in northern Iraq. J Archaeol Sci 7:167–178

Balfourier F, Roussel V, Strelchenko P, Exbrayat-Vinson F, Sourdille P, Boutet G et al (2007) A worldwide bread wheat core collection arrayed in a 384-well plate. Theor Appl Genet 114:1265–1275

Bálint AF, Kovács G, Sutka J (2000) Origin and taxonomy of wheat in the light of recent research. Acta Agronomica Hungarica 48:301–313

Bar-Matthews M, Ayalon A, Kaufman A (1997) Late Quaternary paleoclimate in the eastern Mediterranean region from stable isotope analysis of speleothems at Soreq cave, Israel. Quat Res 47:155–168

Bar-Yosef O (1998) On the nature of transitions: the middle to upper Palaeolithic and the Neolithic revolution. Cambridge Archaeol J 8:141–163

Bar-Yosef O, Kislev ME (1989) Early farming communities in the Jordan Valley. In: Harris DR, Hillman GC (eds) Foraging and farming: the evolution of plant exploitation. Unwin-Hyman Ltd., London, pp 632–642

Bassi FM, Brahmi H, Sabraoui A, Amri A, Nsarellah N et al (2019) Genetic identification of loci for hessian fly resistance in durum wheat. Mol Breed 39:24–40

Baum BR, Appels R (1992) Evolutionary change at the *5S Dna* loci of species in the Triticeae. Pl Syst Evol 183:195–208

Baum BR, Bailey LG (2001) The 5S rRNA gene sequence variation in wheats and some polyploidy wheat progenitors (Poaceae: Triticeae). Genet Resour Crop Evol 48:35–51

Baum BR, Bailey LG (2004) Genetic diversity in the Red wild einkorn: *T. urartu* Gandilyan (Poaceae: Triticeae). Genet Resourc Crop Evol 60:77–87

Baum BR, Feldman M (2010) Elimination of 5S DNA unit classes in newly formed allopolyploids of the genera *Aegilops* and *Triticum*. Genome 53: 430-438

Bebeli PJ, Kaltsikes PJ (1985) Karyotypic analysis of two durum wheat varieties. Can J Genet Cytol 27:617–621

Beleggia R, Platani C, Nigro F, Papa R (2011) Yellow pigment determination for single kernels of durum wheat (*Triticum durum* Desf.). Cereal Chem 88:504–508

Belitz HD, Seilmeier W, Wieser H (1989) Problems of spelt wheat (*Triticum spelta*). Z fwer Lebensmittel Untersuchung und Forschung 189:1–5

Bell GDH, Lupton M, Riley R (1955) Investigations in the Triticinae. III. The morphological and field behaviour of the A_2 generation of interspecific and intergeneirc amphidiploids. J Agri Sci 46:199–231

Bennett MD (1972) Nuclear DNA content and minimumgeneration time in herbaceous plants. Proc Royal Soc Lond Ser B 181:109–135

Bennett ST, Bennett MD (1992) Spatial separation of ancestral genomes in the wild grass *Milium montianum* Parl. Ann Botany 70:111–118

Bernhardt N, Brassac J, Kilian B, Blattner FR (2017) Dated tribe-wide whole chloroplast genome phylogeny indicates recurrent hybridizations within Triticeae. BMC Evol Biol 17:141

Bernhardt N, Brassac J, Dong X, Willing E-M, Poskar CH, Kilian B, Blattner FR (2020) Genome-wide sequence information reveals recurrent hybridization among diploid wheat wild relatives. Plant J 102:493–506

Berraies S, Ammar K, Gharbi MS, Yahyaoui A, Rezgui S (2014) Quantitative inheritance of resistance to Septoria tritici blotch in durum wheat in Tunisia. Chilean J Agric Res 74:35–40

Bertsch F (1943) Der Dinkel. Landw Jahrbuch 92:241–252

Bertsch K (1950) Vom Ursprung der hexaploiden Weizen. Züchter 20:24–27

Bertsch K, Bertsch E (1949) Geschichte unserer Kulturpflanzen. Wissenschaftliche Verlagsgsellschaft M.B.H., Stuttgart, 2 Auflge

Bhaduri PN, Ghosh PN (1955) SAT-Chromosome of *Triticum macha*, a unique feature among *Triticum* species. Cytologia 20:148–149

Bhatia GS (1938) The cytology and genetics of some Indian wheats. II. The cytology of some Indian wheats. Ann Bot 2:335-371B

Bieńkowska T, Suchowilska E, Kandler W, Krska R, Wiwart M (2019) *Triticum polonicum* L. as potential source material for the biofortification of wheat with essential micronutrients. Pl Genet Resour Charact Utiliz 17:213–220

Biffen RH (1905) Mendel's laws of inheritance and wheat breeding. J Agric Sci 1:4–48

Bilgic H, Hakki EE, Pandey A et al (2016) Ancient DNA from 8400-year-old Çatalhöyük wheat: implications for the origin of neolithic agriculture. PLoS ONE 11:e015197

Birnstiel ML, Chipchase M, Speirs J (1971) The ribosomal RNA cistrons. Prog Nucleic Acid Res Mol Biol 11:351–389

Blake NK, Lehfeldt BR, Lavin M, Talbert LE (1999) Phylogenetic reconstruction based on low copy DNA sequence data in an allopolyploid: the B genome of wheat. Genome 42:351–360

Blanco A, Giorgi B, Perrino P, Simeone R (1990) Risorse genetiche e miglioramento della qualita del frumento duro. Agricoltura Ricerca 114:41–58

Blanco A, Colasuonno P, Gadaleta A, Mangini G, Schiavulli A, Simeone R, Digesù AM, de Vita P, Mastrangelo AM, Cattivelli L (2011) Quantitative trait loci for yellow pigment concentration and individual carotenoid compounds in durum wheat. J Cereal Sci 54:255–264

Blaringhem L (1927) Aftinites des blés sauvages *Triricum aegilopoides* Balansa et *Tr. monococcum* L. demontrées par leurs hybrids reciproques. C R Acad Sci 184:225–227

Blatter R, Jacomet S, Schlumbaum A (2002) Spelt-specific alleles in HMW glutenin genes from modern and historical European spelt (*Triticum spelta* L.). Theor Appl Genet 104:329–337

Bleier H (1928) Zytologische Untersuchungen an Seltenen Getreide– und Rübenbastarden. Ztschr f Induktive Abstam u Vererbungslehre 1:447–452

Blumenthal C, Bekes F, Gras PW, Barlow EWR, Wrigley CW (1995) Identification of wheat genotypes tolerant to the effects of heat stress on grain quality. Cereal Chem 72:539–544

Blumler MA (1994) Evolutionary trends in the wheat group in relation to environment, quaternary climatic change and human impacts. In: Millington AC, Pye K (eds) Environmental change in dryland: biogeographical and geomorphological perspectives. Wiley, Chichester, pp 253–269

Blumler MA (1998) Introgression of durum into wild emmer and the agricultural origin question. In: Damania AB, Valkoun J, Willcox G, Qualset CO (eds) The origin of agriculture and crop domestication, proceedings of the harlan symposium, 10–14 May, 1997, ICARDA, IPGRI, FAO and UC/GRCP, pp 252–268

Bogaard A, Filipović D, Fairbairn A et al (2017) Agricultural innovation and resilience in a long-lived early farming community: the 1500-year sequence at Neolithic to early Chalcolithic Çatalhöyük, central Anatolia. Anatol Stud 67:1–28

Boissier PE (1844) Diagnoses plantarum orientalium novarum. Sér 1:73–74

Boller S (1995) Conservation and promotion of spelt by breeding and research in Switzerland. Oral presentation given at the workshop "Farro: promoting the conservation and use of a valuable underutilized crop", IPGRI, 21 July 1995, Castelvecchio Pascoli, Lucca, Italy

Bor NL (1968) Gramineae, vol 9. In: Towsend CC, Guest E, El-Rawi A (eds) Flora of Iraq, pp 210–263

Borgen A (2010) Resistance to common bunt (*Tilletia tritici*) and rust (*Puccinia* sp.) in hulled wheat. Abstract for the XVIth Biennial Workshop on the Smuts and Bunts in Lethbridge, Alberta, Canada

Borghi B, Castagna R, Corbellini M, Heun M, Salamini F (1996) Breadmaking quality of einkorn wheat (*Triticum monococcum* ssp. *monococcum*). Cereal Chem 73:208–214

Borrelli GM, Ficco DBM, Giuzio L et al (2011) Durum wheat salt tolerance in relation to physiological, yield and quality characters. Cereal Res Commun 39:525–534

Boscato P, Carioni C, Brandolini A, Sadori L, Rottoli M (2008) Molecular markers for the discrimination of *Triticum turgidum* L. subsp. *dicoccum* (Schrank ex Schübl.) Thell. and *Triticum timopheevii* (Zhuk.) Zhuk. subsp. *timopheevii*. J Archaeol Sci 35:239–246

Bowden MW (1959) The taxonomy and nomenclature of the wheats, barleys and ryes and their wild relatives. Can J Bot 37:657–684

Bowman AK (1990) Egypt after the pharaohs. Oxford University Press, Oxford

Boyacioglu H (2017) Global durum wheat use trending upward. WORLD-GRAIN.com 23.10. 2017

Bozzini A, Giorgi B (1969) Karyotype analysis in *Triticum*: II—Analysis of *T. araraticum* Jakubz. and *T. timopheevii* Zhuk. and their relationships with other tetraploid wheats. Caryologia 22:261–268

Braidwood RJ (1960) The agricultural revolution. Sci Am 203:130–152

Brandolini A, Hidalgo A (2011) Einkorn (*Triticum monococcum*) flour and bread. In: Preedy VR, Watson RR, Patel V (eds) Flour and breads and their fortification in health and disease prevention. Academic Press, London, pp 79–88

Braun T (1995) Barley cakes and emmer bread. In: Wilkins J, Harvey D, Dobson M (eds) Food in antiquity. University of Exeter Press, Exeter, pp 25–37

Braun HJ, Atlin G, Payne T (2010) Multi-location testing as a tool to identify plant response to global climate change. In: Reynolds MP (ed) Climate change and crop production. CABI Publishers, Wallingford, UK, pp 115–138

Breiman A (1987) Mitochondrial DNA diversity in the genera of *Triticum* and *Aegilops* revealed by southern blot hybridization. Theor Appl Genet 73:563–570

Brenchley R, Spannagl M, Pfeifer M et al. (2012) Analysis of the bread wheat genome using whole-genome shotgun sequencing. Nature 491:705–710

Brester GW, Grant B, Michael A. Boland MA (2009) Marketing organic pasta from big sandy to Rome: it's a long Kamut®. Appl Econ Perspect Policy 31:359–369

Briggle LW (1963) heterosis in wheat—a review. Crop Sci 3:407–412

Briggle LW (1966) Transfer of resistance to erysiphe graminis f. sp. tritici from khapli emmer and yuma durum to hexaploid wheat. Crop Sci 6:459–461

Brooking I, Kirby ELM (1981) Interrelationships between stem and ear development in winter wheat: the effects ofa Norin 10 dwarfing gene, Gai/Rht2. J Agric Sci 97:373–381

Brown-Guedira GL, Gill BS, Bockus WW, Cox TS, Hatchett JH, Leath S, Peterson CJ, Thomas JB, Zwer PK (1996) Evaluation of a collection of wild timopheevii wheat for resistance to disease and arthropod pests. Pl Disease 80:928–933

Brown-Guedira GL, Gill BS, Cox TS, Leath S (1997) Transfer of disease resistance genes from *Triticum araraticum* to common wheat. PL Breed 116:105–112

Brown-Guedira GL, Cox TS, Gill BS, Sears RG (1999a) Registration of KS96WGRC35 and KS96WGRC36 leaf rust-resistant hard red winter wheat germplasm. Crop Sci 39:595

Brown-Guedira GL, Cox TS, Gill BS, Sears RG, Leath S (1999b) Registration of KS96WGRC37 powdery mildew-resistant hard white winter wheat germplasm. Crop Sci 39:596

Brown-Guedira GL, Singh S, Fritz AK (2003) Performance and mapping of leaf rust resistance transferred to wheat from *Triticum timopheevii* subsp. *armeniacum*. Phytopathology 93:784–789

Brunazzi A, Scaglione D, Talini RF, Miculan M et al (2018) Molecular diversity and landscape genomics of the crop wild relative *Triticum urartu* across the Fertile Crescent. Plant J 94:670–684

Bruns R, Peterson CJ (1998) Yield and stability factors associated with hybrid wheat. Euphytica 100:1–5

Buckler E, Thornsberry JM, Kresovich S (2001) Molecular diversity, structure and domestication of grasses. Genet Res 77:213–218

Bullrich L, Appendino .M, Tranquilli G et al (2002) Mapping of a thermo-sensitive earliness *per se* gene on *Triticum monococcum* chromosome 1A[m]. Theor Appl Genet 105:585–593

Bushuk and Kerber (1978) The role of *Triticum carthlicum* in the origin of bread wheat based on gliadin electrophoregrams. Can J Pl Sci 58:1019–1024

Burt C, Steed A, Gosman N, Lemmens M, Bird N, Ramirez-Gonzalez R, Holdgate S, Nicholson P (2015) Mapping a type 1 FHB resistance on chromosome 4AS of *Triticum macha* and deployment in combination with two type 2 resistances. Theor Appl Genet 128:1725–1738

Buvaneshwari G, Yenagi NB, Hanchinal RR, Naik RK (2003) Glycaemic responses to dicoccum products in the dietary management of diabetes. Ind J Nutr Dietet 40:363–368

Cadle MM, Murray TD (1997) Identification of Resistance to *Pseudocercosporella herpotrichoides* in *Triticum monococcum*. Pl Dis 81:1181–1186

Cahalan C, Law CN (1979) The genetical control of cold resistance and vernalization requirement in wheat. Heredity 42:125–132

Cakmak I, Torun A, Millet E, Feldman M, Fahima T, Korol A, Nevo E, Braun HI, Ozkan H (2004) *Triticum dicoccoides*: an important genetic resource for increasing zinc and iron concentration in modern cultivated wheat. Soil Sci Plant Nutr 50:1047–1054

Caldwell KA, Kasarda DD (1978) Assessment of genomic and species relationships in *Triticum* and *Aegilops* by PAGE and by differential staining of seed albumins and globulins. Theor Appl Genet 52:273–280

Caldwell KS, Dvorak J, Lagudah ES, Akhunov E, Luo M-C, Wolters P, Powell W (2004) Sequence polymorphism in polyploid wheat and their D-genome diploid ancestor. Genetics 167:941–947

Camara A (1943) Estudo comparative ce caritipos no género *Triticum*. Agronomia Lusitans 5:95–117

Camara A (1944) Cromosomas dos trigos hexaploides. Agron Lusiclna 6:221–251

Cantrell RG, Joppa LR (1991) Genetic analysis of quantitative traits in wild emmer (*Triticum turgidum* L. var. *dicoccoides*). Crop Sci 31:645–649

Cao W, Scoles GJ, Hucl P (1997) The genetics of rachis fragility and glume tenacity in semi-wild wheat. Euphytica 94:119–124

Cao W, Armstrong K, Fedak G (2000a) A synthetic zhukovskyi wheat. Wheat Inf Serv 91:30–32

Cao W, Scoles G, Hucl P, Chibbar RN (2000b) Phylogenetic relationships of five morphological groups of hexaploid wheat (*Triticum aestivum* L. em Thell.) based on RAPD analysis. Genome 43:724–727

Cao W, Hughes G, Ma H et al (2001) Identification of molecular markers for resistance to *Septoria nodorum* blotch in durum wheat. Theor Appl Genet 102:551–554

Capouchova I (2001) Technological quality of spelt (*Triticum spelta*) from ecological growing system. Agris Since 2002 32:307–322

Carleton MA (1916) The small grains. The Macmillian Co, New York

Carmona S, Caballero L, Martín LM, Alvarez JB (2010a) Genetic diversity in khorasan and rivet wheat by assessment of morphological traits and seed storage protein. Crop Pasture Sci 61:938–944

Carmona SMM, Alvarez JB, Fernández Caballero LF (2010b) Genetic diversity for morphological traits and seed storage proteins in Spanish rivet wheat. Biologia Plantarum 54:69–75

Carvalho A, Polanco C, Lima-Brito J et al (2010) Differential rRNA genes expression in hexaploid wheat related to NOR methylation. Plant Mol Biol Rep 28:403–412

Castagna R, Maga G, Perenzin M, Heun M, Salamini F (1994) RFLP-based genetic relationships of einkorn wheats. Theor Appl Genet 88:818–823

Castagna R, Borghi B, Di Fonzo N, Heun M, Salamini F (1995) Yield and related traits of einkorn (*T. monococcum* ssp. *monococcum*) in different environments. Euro J Agron 4:371–378

Cauvin J, Aurenche O, Cauvin M-C, Balkan-Atlı N (2011) The prepottery site of cafer Höyük. In: Özdogan M, Başgelen N, Kuniholm P (eds) The neolithic in Turkey. Archaeology and Art Publications, Istanbul, pp 1–40

Ceoloni C, Strauss I, Feldman M (1986) Effect of different doses of group-2 chromosomes on homoeologous pairing in intergeneric wheat hybrids. Can J Genet Cytol 28:240–246

Cermeño MC, Orellana J, Santos JL, Lacadena JR (1984) Nucleolar organizer activity in wheat, rye and derivatives analyzed by a silver-staining procedure. Chromosoma 89:370–376

Chapman V, Miller TE (1978) The relationship of the D genomes of hexaploid *Ae. crassa, Ae. vavilovii* and hexaploid wheat. Wheat Inf Serv 47:17–20

Chapman V, Riley R (1966) The allocation of the chromosomes of *Triticum aestivum* to the A and B genomes and evidence of genome structure. Can J Genet Cytol 8:57–63

Chapman V, Miller TE, Riley R (1976) Equivalence of the A genome of bread wheat and that of *Triticum uratru*. Genet Res 27:69–76

Chatzav M, Peleg Z, Ozturk L, Yazici A, Fahima T, Cakmak I, Saranga Y (2010) Genetic diversity for grain nutrients in wild emmer wheat: potential for wheat improvement. Ann Bot 105:1211–1220

Chen Q-F (2003) Improving male fertility restoration of common wheat for *Triticum timopheevii* cytoplasm. Plant Breeding 122:401–404

Chen PD, Gill BS (1983) The origin of chromosome 4A, and genomes B and G of tetraploid wheat. In: Sakamoto S (ed) Proceedings of 6th international wheat genetics symposium, pp 39–48

Chen KC, Gray JG, Wildman SG (1975) Fraction 1 protein and the origin of polyploid wheats. Science 190:1304–1306

Chen Q, Sun Y, Dong Y (1985) Cytogenetic studies on interspecific hybrids of Xinjiang wheat. Acta Agron Sin 11:23–28

Chen Q, Conner RL, Laroche A, Thomas JB (1998) Genome analysis of *Thinopyrum intermedium* and *Thinopyrum ponticum* using genomic in situ hybridization. Genome 41:580–586

Chen S, Guo Y, Briggs J et al. (2018) Mapping and characterization of wheat stem rust resistance genes SrTm5 and Sr60 from *Triticum monococcum*. Theor Appl Genet 131:625–635

Cheng H, Liu J, Wen J, Nie X, Xu L, Chen N, Li Z, Qilin Wang Q et al (2019) Frequent intra- and inter-species introgression shapes the landscape of genetic variation in bread wheat. Genome Biol 20:136

Cheng X, Xin M, Xu R, Chen Z, Cai W, Chai L et al (2020) A single amino acid substitution in STKc_GSK3 kinase conferring semi-spherical grains and its implications for the origin of *Triticum sphaerococcum* Perc. Plant Cell Advance Publication. Published on February 14, 2020. https://doi.org/10.1105/tpc.19.00580

Chin TG, Chwang CS (1944) Cytogenetic studies of hybrids with "mskha" wheat. Bull Torrey Bot Club 71:356–366

Chizaki Y (1934) Another new haploid plant in *Triticum monococcum* L. Shokubutsugaku Zasshi (published by the Bot Soc Japan) 48:621–628

Chu CG, Xu SS, Faris JD, Nevo E, Friesen TL (2008) Seedling resistance to tan spot and *Stagonospora nodorum* leaf blotch in wild emmer wheat (*Triticum turgidum* ssp. *dicoccoides*). Plant Dis 92:1229–1236

Ciaffi M, Dominici L, Lafiandra D (1997) Gliadin polymorphism in wild and cultivated einkorn wheats. Theor Appl Genet 94:68–74

Ciaffi M, Dominici L, Lafiandra D (1998) High molecular weight glutenin subunit variation in wild and cultivated einkorn wheats (*Triticum* spp., Poaceae). Pl Syst Evol 209:123–137

Cifuentes M, Garcia-agüero V, Benavente E (2010) A Comparative analysis of chromosome pairing at metaphase I in interspecific hybrids between durum wheat (*Triticum turgidum* L.) and the most widespread *Aegilops* Species. Cytogenet Genome Res 129:124–132

Civáň P, Ivaničová Z, Brown TA (2013) Reticulated origin of domesticated emmer wheat supports a dynamic model for the emergence of agriculture in the Fertile Crescent. PLoS ONE 8: e81955

Clarke FR, Clarke JM, Knox RE (2002a) Inheritance of stem solidness in eight durum wheat crosses. Can J Plant Sci 82:661–664

Clarke JM, Norvell WA, Clarke FR, Buckley WT (2002b) Concentration of cadmium and other elements in the grain of near-isogenic durum lines. Can J Anim Sci 82:27–33

Comings DE (1968) The rationale for an ordered arrangement of chromatin in the interphase nucleus. Amer J Hum Genet 20:440–460

Comings DE (1980) Arrangement of chromatin in the nucleus. Hum Genet 53:131–143

Concia L, Veluchamy A, Ramirez-Prado JS, Martin-Ramirez A, Huang Y et al (2020) Wheat chromatin architecture is organized in genome territories and transcription factories. Genome Biol 21:104

Cook OF (1913) Wild wheat in palestine. Bull Plant Industr 274:1–58

Corazza L. Pasquini N, Perrino P (1986) Resisttance to rusts and powdery mildew in some strains of *Triticum monococcum* L. and *Triticum dicoccum* Schubler cultivated in Italy. Genet Agr 40:243–254

Coucoli HD. and Skorda EA (1966) Further evidence on the karyotype of *Triticum monococcum* L. and *Triticum durum* Desf. Can J Genet Cytol 8:102–110

Cox TS, Sears RG and Bequette RK (1995) Use of winter wheat x *Triticum tauschii* backcross populations for germplasm evaluation. Theor Appl Genet 90:571–577

Crawford DJ (1979) Food: tradition and change in Hellenistic Egypt. World Archaeol 11:1136–1146

Crosby AR (1957) Nucleolar activity of lagging chromosomes in wheat. Amer J Bot 44:813–822

Crosby-Longwell AR, Svihla G (1960) Specific chromosomal control of the nucleolus and the cytoplasm in wheat. Exp Cell Res 20:294–312

Cubadda R, Marconi E (1996) Technological and nutritional aspects in emmer and spelt. In: Padulosi S, Hammer K, Heller J (eds) Proceedings of first international workshop on hulled wheats, 21–22 July,1995, Castelvecchio Pascoli, Tuscany, Italy, International Plant Genetic Resources Institute, Rome, Italy, pp 203–211

Cuesta S, Guzmán C, Alvarez JB (2015) Molecular characterization of novel LMW-i glutenin subunit genes from *Triticum urartu* Thum. ex Gandil. Theor Appl Genet 128:2155–2165

Curtis CA, Lukaszewski AJ (1991) Genetic linkage between C-bands and storage protein genes in chromosome 1B of tetraploid wheat. Theor Appl Genet 81:245–252

Czajkowska BI, Bogaard A, Charles M et al (2020) Ancient DNA typing indicates that the "new" glume wheat of early Eurasian agriculture is a cultivated member of the *Triticum timopheevii* group. J Arch Sci 123:105258

Dagan J, Zohary D (1970) Wild tetraploid wheats from west Iran cytogenetically identical with Israeli *T. dicoccoides*. Wheat Inf Serv 31:15–17

Dahan-Meir T, Ellis TJ, Sela H, Mafessoni F et al (2022) The genetic structure of a wild wheat population has remained associated with microhabitats over 36 years. https://doi.org/10.1101/2022.01.10.475641

Damania AB, Srivastrava JP (1990) Genetic resources for optimal input technology. ICARDA's perspectives. In: El-Bassam N, Dambrot M, Loughman BC (eds) Genetic aspects of plant mineral nutrition. Kluwer, Dordrecht, pp 425–430

Damania AB, Tahir M (1993) Heat and cold tolerance in wild relatives and primitive forms of wheat. In: Damania AB (ed) Biodiversity and wheat improvement. Wiley, Chichester, West Sssex, UK, pp 217–224

Damania AB, Pecetti L, Jana S (1990) Evaluation for useful genetic traits in primitive and wild wheats. In: Srivastava JP, Damania AB (eds) Wheat genetic resources: meeting diverse needs. Wiley, Chichester, pp 57–64

Damania AB, Hakim S, Moualla MY (1992) Evaluation of variation in *T. dicoccum* for wheat improvement in stress environment. Hereditas 116:163–166

Darvey NL, Driscoll CJ (1972) Evidence against somatic association in hexaploid wheat. Chromosoma 36:140–149

Dass HC (1972) Phylogenetic affinities in *Triticineae* studied by thin-layer chromatography. Can J Genet Cytol 14:703–712

Daud HM, Gustafson JP (1996) Molecular evidence for *Triticum speltoides* as B-genome progenitor of wheat (*Triticum aestivum*). Genome 39:543–548

Day L, Augustin MA, Batey IL, Wrigley CW (2006) Wheat-gluten uses and industry needs. Trends Food Sci Technol (rev) 17:82–90

de Candolle A (1886) Origin of cultivated plants. London: Kegan, Paul, Trench and CO Second Edition du ble et de l'orge. Ann Sci Nat Ser I (Paris) IX: 61–82

Debernardi JM, Lin H, Chuck G, Faris JD, Dubcovsky J (2017) MicroRNA172 plays a crucial role in wheat spike morphogenesis and grain threshability. Development 144:1966–1975

Defrise D, Jacqmain D (1984) Determination of spelt content of bread by electrophoretic analysis of gliadins. Revue Des Fermentations Et Des Industries Alimentaires 39:123–136

Dekaprelecich L (1961) Die Art *Triticum macha* Dek. et Men.im Lichte neuester Untersuchungen über die Herkuft der hexaploidenWeizen. Z fur Pflanzenzüchtung 45:17–30

Dekaprelevich LL (1925) The geographical centre of the culture of Persian wheat—*Triticum persicum* Vay. Bull Appl Bot Pl Breed 15:199–202

Dekaprelevich LL, Menabde VL (1932) Spelt wheats of Western Georgia (Western Transcaucasia). Bull Appl Bot Genet PI Breed 5:1–46

Dennis ES, Gerlach WL, Peacock WJ (1980) Identical polypyrimidine-polypurine satellite DNAs in wheat and barley. Heredity 44:349–366D

Devos KM, Dubcovsky J, Dvorak J, Chinoy CN, Gale MD (1995) structural evolution of wheat chromosomes 4A, 5A, and 7B and its impact on recombination. Theor Appl Genet 91:282–288

Dhaliwal HS (1977a) Basis of difference between reciprocal crosses involving *Triticum boeoticum* and *T. urartu*. Theor Appl Genet 49:283–286

Dhaliwal HS (1977b) Genetic control of seed proteins in wheat. Theor Appl Genet 50:235239

Dhaliwal HS (1977c) The *Ph* gene and the origin of tetraploid wheats. Genetica 47:177–182

Dhaliwal HS, Johnson BL (1976) Origin of *Triticum zhukovskyi*. Wheat Inf Serv 41–42:33–35

Diab AA, Ozturk NZ, Benscher D, Nachit MM, Sorrells ME (2008) Drought-inducible genes and differentially expressed sequence tags associated with components of drought tolerance in durum wheat. Sci Res Essays 3:9–26

Dinoor A, Eshed N, Ecker R, Gerechter-Amitai Z, Solel Z, Manisterski J, Anikster Y (1991) Fungal diseases of wild tetraploid wheat in a natural stand in northern Israel. Isr J Bot 40:481–500

Distelfeld A, Uauy C, Fahima T, Dubcovsky J (2006) Physical map of the wheat high-grain protein content gene Gpc-B1 and development of a high-throughput molecular marker. New Phytol 169:753–763

Dobrovolskaya O, Boeuf C, Salse J, Pont C, Sourdille P, Bernard M, Salina E (2011) Microsatellite mapping of *Ae. speltoides* and map-based comparative analysis of the S, G, and B genomes of Triticeae species. Theor Appl Genet 123:1145–1157

Dorofeev VF (1966) Geographical localization and gene center of hexaploid wheat in Transcaucasia. Genetika 3:16–33 (in Russian)

Dorofeev VF (1968) The variability and breeding value of Armenian wheats. Euphytica 17:451–461

Dorofeev VF (1969) Die Weizen Transkaukasiens und ihre Bedeutung in der Evolution der Gattung Triticum L. Z Pflanzenzücht 61:1–28

Dorofeev VF, Korovina ON (1979) Wheat. In: Dorofeev VF, Korovina ON (eds), Flora of Cultivated Plants (1979), Kolos, Leningrad (Russian)

Dorofeev VF, Filatenko AA, Miguschova EF (1980) Wheat. In: Dorofeev VF, Korovina ON (eds) Flora of Cultivated Plants in the USSR. "Kolos", Leningrad, Russia, Vol 1 (In Russian)

Dou QW, Chen PD (2003) C-band polymorphism in tetraploid wheat (*Triticum turgidum*) landraces from Qinghai and Gansu. Acta Bot Boteal-Occident Sin 23:335–338

Dreisigacker S, Kishii M, Lage J, Warburton M (2008) Use of synthetic hexaploid wheat to increase diversity for CIMMYT bread wheat improvement. Austral J Agric Res 59:413–420

Driscoll CJ (1973) Minor genes affecting homoeologous pairing in hybrids between wheat and related genera. Genetics 74:s66

Dubcovsky J, Dvorak J (2007) Genome plasticity a key factor in the success of polyploid wheat under domestication. Science 316:1862–1866

Dubcovsky J, Luo MC, Dvorak J (1995) Differentiation between homoeologous chromosomes 1A of wheat and 1A^m of *Triticum monococcum* and its recognition by the wheat *Ph1* locus. Proc Natl Acad Sci USA 92:6645–6649

Dubcovsky J, Luo MC, Zhong GY, Bransteitter R, Desai A, Kilian A, Kleinhofs A, Dvorak J (1996) Genetic map of diploid wheat, *Triticum monococcum* L., and its comparison with maps of *Hordeum vulgare* L. Genetics 143:983–999

Dukic N, Knežević D, Zečević V (2008) Genetic determination of technological quality in Triticum *durum*. Periodicum Biologorum 110:285–289

Dvorak J (1972) Genetic variability in *Aegilops speltoides* affecting homoeologous pairing in wheat. Can J Genet Cytol 14:371–380

Dvorak J (1976) The relationship between the genome of *Triticum urartu* and the A and B genomes of *Triticum aestivum*. Can J Genet Cytol 18:371–377

Dvorak J (1983) The origin of wheat chromosomes 4A and 4B and their genome reallocation. Can J Genet Cytol 25:210–214

Dvorak J (1998) Genome analysis in the *Triticum-Aegilops* alliance. In: Slinkard AE (ed) Proceedings of 9th international wheat genetics symposium, University Extension Press, University of Saskatoon, Saskatoon, Saskatchewan, Canada, pp 8–11

Dvorak J (2009) Triticeae genome structure and evolution. In: Feuiller C, Muehlbauer GJ (eds) Genetics and genomics of the Triticeae, plant genetics and genomics: crops and models, vol 7. Springer, Berlin, pp 685–711

Dvorak J, Akhunov ED (2005) Tempos of gene locus deletions and duplications and their relationship to recombination rate during diploid and polyploid evolution in the *Aegilops-Triticum* alliance. Genetics 171:323–332

Dvorak J, Knott DR (1973) A study of somatic association of wheat chromosomes. Can J Genet Cytol 15:411–416

Dvorak J, Luo M-C (2001) Evolution of free-threshing and hulled forms of *Triticum aestivum*: old problems and new tools. In: Caligari PDS, Brandham PE (eds) Wheat taxonomy: the legacy of John Percival. Academic Press, London, pp 127–136

Dvorak J, Luo M-C (2007) Domestication and evolution of tetraploid and hexaploid wheats. Abstracts of The Aaronsohn-ITMI conference, Lecture E1, 16–20 Apr 2007, Tiberias, Israel

Dvorak J, Zhang H-B (1990) Variation in repeated nucleotide sequences sheds light on the phylogeny of the wheat B and G genomes. Proc Natl Acad Sci 87:9640–9644

Dvorak J, Zhang H-B (1992) Reconstruction of the phylogeny of the genus *Triticum* from variation in repeated nucleotide sequences. Theor Appl Genet 84:419–429

Dvorak J, Chen K-C, Giorgi B (1984) The C-band pattern of a *Ph* mutant of durum wheat. Can J Genet Cytol 26:360–363

Dvorak J, McGuire PE, Cassidy B (1988) Apparent sources of the A genomes of wheats inferred from polymorphism in abundance and restriction fragment length of repeated nucleotide sequences. Genome 30:680–689

Dvorak J, Zhang H-B, Kota RS, Lassner M (1989) Organization and evolution of the 5S ribosomal RNA gene family in wheat and related species. Genome 32:1003–1016

Dvorak J, di Terlizzi P, Zhang H-B, Resta P (1993) The evolution of polyploid wheats: identification of the A genome donor species. Genome 36:21–31

Dvorak J, Luo M-C, Yang ZL, Zhang HB (1998a) Genetic evidence on the origin of *T. aestivum* L. In: Damania AB, Valkoun J, Willcox G, Qualset CO (eds) The origins of agriculture and crop domestication. The Harlan symposium. ICARDA, IPGRI, FAO, UC/GRCP, Aleppo, Syria, pp 235–251

Dvorak J, Luo M-C, Yang ZL, Zhang H-B (1998b) The structure of the *Aegilops tauschii* gene pool and the evolution of hexaploid wheat. Theor Appl Genet 67:657–670

Dvorak J, Akhunov ED, Akhunov AR, Deal KR, Luo M-C (2006) Molecular characterization of a diagnostic DNA marker for domesticated tetraploid wheat provides evidence for gene flow wild tetraploid wheat to hexaploid wheat. Mol Biol Evol 23:1386–1396

Dvorak J, Deal KR, Luo M-C, You FM, von Borstel K, Dehghani H (2012) The origin of spelt and free-threshing hexaploid wheat. J Hered 103:426–441

Dyck PL (1992) Transfer of a gene for stem rust resistance from *Triticum araraticum* to hexaploid wheat. Genome 35:788–792

Dyck PL, Bartos P (1994) Attempted transfer of leaf rust relstance lrom *Trilicum monococcurn* and durum wheat to hexaploid wheat. Can J Plant Sci 74:733–736

Efremova TT, Laikova LI, Arbuzova VS, Popova OM, Maystrenko OI (2001) Use of aneuploid analysis in chromosomal localisation and wheat genes mapping. In: Proceedings of 11th European wheat aneuploid conference, Novosibirsk, 2000, Russia, 115–118

Eig A (1929) Monographisch-Kritische Ubersicht der Gattung *Aegilops*. Reprium nov. Spec Regni veg 55:1–288

Eilam T, Anikster Y, Millet E, Manisterski J, Feldman M (2007) Genome size and genome evolution in diploid Triticeae species. Genome 50:1029–1037

Eilam T, Anikster Y, Millet E, Manisterski J, Feldman M (2008) Nuclear DNA amount and genome downsizing in natural and synthetic allopolyploids of the genera *Aegilops* and *Triticum*. Genome 51(8):616–627

El Baidouri M, Murat F, Veyssiere M, Molinier M, Flores R, Burlot L, Alaux M, Quesneville H, Pont C, Salse J (2017) Reconciling the evolutionary origin of bread wheat (*Triticum aestivum*). New Phytologist 213:1477–1486

El Bouhssini M, Nsarellah NN, Nachit MM, Bentika A, Benlahbib O, Lhaloui S (1999) First source of resistance in durum wheat to Hessian fly (Diptera: Cecidomyiidae) in Morocco. Genet Resourc Crop Evol 46:107–109

El Haddoury J, Amri A, Lhaloui S, Nsarellah N (2005) Use of interspecific hybridization for the transfer of hessian fly resistance from *Triticum araraticum* to durum wheat. Al Awamia 113:5–13

Elbaum R, Zaltzman L, Burgert I, Fratzl P (2007) The role of wheat awns in the seed dispersal unit. Science 316:884–886

Elias EM, Steiger DK, Cantrell RG (1996) Evaluation of lines derived from wild emmer chromosome substitutions: II. Agronomic traits. Crop Sci 36:228–233

Ellerton S (1939) The origin and geographical distribution of *Triticum sphaerococcum* Perc. and its cytogenetical behaviour in crosses with *T. vulgare* Vill. J Genet 38:307–324

Empilli S, Castagna R, Brandolini A (2000) Morpho-agronomic variability0f the diploid wheat *Triticum monococcum* L. Plant Genet Resourc Newsletter 124:36–40

Endo TR (2007) The gametocidal chromosome as a tool for chromosome manipulation in wheat. Chromosome Res 15:67–75

Endo TR, Gill BS (1984) Somatic karyotype, heterochromatin distribution, and nature of chromosome differentiation in common wheat, *Triticum aestivum* L. em Thell. Chromosoma 89:361–369

Endo TR, Gill BS (1996) The deletion stocks in common wheat. J Hered 87:295–307

Engledow FL (1920) The inheritance of glume-length and grain-length in a wheat cross. J Genet 10:109–134

Ergun M (2018) Where the wild things are. Contextual insights into wild plant exploitation at Aceramic Neolithic Aşıklı Höyük. Turkey Paléorient 44:9–28

Evans LT (1981) Yield improvement in wheat: empirical or analytical? In: Evans LT, Peacock WJ (eds) Wheat science—today and tomorrow. Cambridge University Press, Cambridge, pp 203–222

Evans LT, Dunstone RL (1970) Some physiological aspects of evolution in wheat. Austr J Biol Sci 23:725–742

FAOSTAT (2018) Food and Agriculture Organization of the United Nation (FAO), Statistical Database, Statistical Division. Rome. http://faostat3.fao.org/browse/Q/QC/

Faris JD, Fellers JP, Brooks SA, Gill BS (2003) A bacterial artificial chromosome contig spanning the major domestication locus *Q* in wheat and identification of a candidate gene. Genetics 164:311–321

Faris JD, Zhang Z, Fellers JP, Gill BS (2008) Micro-colinearity between rice, *Brachypodium*, and *Triticum monococcum* at the wheat domestication locus *Q*. Funct Integr Genomics 8:149–164

Faris JD, Zhang Z, Chao S (2014) Map-based analysis of the tenacious glume gene *Tg-B1* of wild emmer and its role in wheat domestication. Gene 542:198–208

Faris JD, Overlander ME, Kariyawasam GK, Carter A et al (2020) Identification of a major dominant gene for race-nonspecific tan spot resistance in wild emmer wheat. Theor Appl Genet 133:829–841

Fedoroff NV (1979) On spacers. Cell 16:697–710

Feldman M (1966a) The mechanism regulating pairing in *Triticum timopheevii*, Wheat Inf Serv 21:1–2

Feldman M (1966b) Identification of unpaired chromosomes in F_1 hybrids involving *Triticum aestivum* and *T. timopheevii*. Can J Genet Cytol 8:144–151

Feldman M (1966c) The effect of chromosomes 5B, 5D and 5A on chromosomal pairing in *Triticum aestivum*. Proc Natl Acad Sci USA 55:1447–1453

Feldman M (1968) Regulation of somatic association and meiotic pairing in common wheat. In: Finlay KW, Shepherd KW (eds) Proceedings of 3rd international wheat genet symposium, Canberra, Australia, pp 169–178

Feldman M (1977) Historical aspects and significance of the discovery of wild wheats. Stadler Symp 9:121–146

Feldman M (1978) New evidence on the origin of the B genome of wheat. In: Ramanujam RS (ed) Proceedings of 5th international wheat genetics symposium, New Delhi, pp 120–132

Feldman M (2001) Origin of cultivated wheat. In: Bonjean AP, Angus WJ (eds) The world wheat book. Lavoisier, Paris, pp 3–56

Feldman M, Avivi L (1973a) Non-random arrangement of chromosomes in common wheat, vol 4. In: Wahrman J, Lewis KR (eds) Chromosomes today. Wiley, New York, Toronto and Israel Universities Press, Jerusalem, pp 187–196

Feldman M, Avivi L (1973b) The pattern of chromosomal arrangement in nuclei of common wheat and its genetic control. In: Sears ER, Sears LMS (eds) Proceedings of 4th international wheat genetics symposium, Missouri Agr Exp Sta, Columbia, Missouri, pp 675–684

Feldman M, Avivi L (1984) Ordered arrangement of chromosomes in wheat. Chromosomes Today 8:181–189

Feldman M, Kislev M (1977) *Aegilops searsii*, a new species of section *Sitopsis (Platystachys)*. Isr J Bot 26:190–201

Feldman M, Kislev M (2007) Domestication of emmer wheat and evolution of free-threshing tetraploid wheat. Isr J Pl Sci 55:207–221

Feldman M, Levy AA (2005) Allopolyploidy—a shaping force in the evolution of wheat genomes. Cytogenet Genome Res 109:250–258

Feldman M, Levy AA (2009) Genome evolution in allopolyploid wheat—a revolutionary reprogramming followed by gradual changes. J Genet Genom 36:511–518

Feldman M, Levy AA (2012) Genome evolution due to allopolyploidization in wheat. Genetics 192:763–774

Feldman M, Levy AA (2015) Origin and evolution of wheat and related Triticeae species. In: Molnar-Lang M, Ceoloni C, Dolezel J (eds) Alien Introgression in wheat, cytogenetics, molecular biology, and genomics. Springer, Cham, Switzerland, pp 21–76

Feldman M, Mello-Sampayo T (1967) Suppression of homeologous pairing in hybrids of polyploid wheats x *Triticum speltoides*. Can J Genet Cytol 9:307–313

Feldman M, Millet E (1995) Methodologies of identification, allocation and transfer of quantitative genes from wild emmer into cultivated wheat. In: Li ZS, Xin ZY (eds) Proceedings of 8th international wheat genetics symposium, Beijing, 1993. China Agric. Scientech Press, Beijing, China, pp 19–27

Feldman M, Sears ER (1981) The wild gene resources of wheat. Sci Am 244:102–112

Feldman M, Strauss I (1983) A genome-restructuring gene in *Aegilops longissima*. In: Sakamoto S (ed) Proceedings of 6th international wheat genetics symposium, Kyoto, Japan, pp 309–314

Feldman M, Mello-Sampayo T, Sears ER (1966) Somatic association in *Triticum aestivum*. Proc Natl Acad Sci USA 56:1192–1199

Feldman M, Millet E, Abbo S (1994) Exploitation of wild emmer wheat to increase yield and protein content in durum and common wheat. In: Proceedings of EUCARPIA meeting, genetic resources section, March 15–18, 1994 Clermont-Ferrand, France

Feldman M, Lupton FGH, Miller TE (1995) Wheats, 2nd edn. In: Smart J, Simmonds NW (eds) Evolution of crop plants. Longman Scientific & Technical, pp 184–192

Feldman M, Millet E, Avivi Y (1996) Wild emmer wheat, *Triticum turgidum* var. *dicoccoides*—a rich genetic resource for the improvement of durum and common wheat. In: Ranieri R, Romano R (eds) Premio Barilla dal grano alla pasta, Barilla Alimentare, Parma, Italy, pp 39–45

Feldman M, Liu B, Segal G, Abbo S, Levy AA, Vega JM (1997) Rapid elimination of low-copy DNA sequences in polyploid wheat: a possible mechanism for differentiation of homoeologous chromosomes. Genetics 147:1381–1387

Feldman M, Levy AA, Fahima T, Korol A (2012) Genome asymmetry in allopolyploid plants—wheat as a model. J Exp Bot 63:5045–5059

Feldman M, Levy AA, Chalhoub B, Kashkush K (2013) Genome plasticity in polyploid wheat. In: Soltis PS, Soltis D (eds) Polyploidy and genome evolution. Springer-Verlag, Berlin/Heidelberg, pp 109–135

Felsenburg T, Levy AA, Galili G, Feldman M (1991) Polymorphism of high-molecular weight glutenins in wild tetraploid wheat: spatial and temporal variation in a native site. Isr J Bot 40:451–479

Ferjaoui S, M'Barek SB, Bahri B, Slimane RB, Hamza S (2015) Identification of resistance sources to *Septoria tritici* blotch in old Tunisian durum wheat germplasm applied for the analysis of the *Zymoseptoria tritici*-durum wheat interaction. J Pl Pathol 97:1–11

Fernández-Calvín B, Orellana J (1994) Metaphase I-bound arms frequency and genome analysis in wheat-*Aegilops* hybrids. 3. Similar relationships between the B genome of wheat and S or S^l genomes of *Ae. speltoides*, *Ae. longissima* and *Ae. sharonensis*. Theor Appl Genet 88:1043–1049

Finch RA, Smith JB, Bennett MD (1981) *Hordeum* and *Secale* genomes lie apart in a hybrid. J Cell Sci 52:391–403

Flaksberger CA (1915) Determination of wheats. Trudy prikl Bot Genet Selek VIII (1–2). *Triticum dicoccoides*, pp 14–15, 26–27, 48–52 (Russian, with English summary)

Flaksberger CA (1925) Pshenitsa odnozemyahki (wheat- einkorn). Bull Appl Bot Pl Breed (trudy Po Prikladnoi Botanike I Selektsii) 15:207–227

Flaksberger CA (1935) Psenica (wheat). Kulturnaja flora SSSR. Acad. USSR, vol 1. Moskva, Leningrad

Flaksberger CA (1938) Wheats. State Agricultural Publishing Company, Moscow, Leningrad

Flaksberger CA (1939) Key to true cereals. Academy of Science International Plant Cultural, Leningrado, USSR

Flavell RB (1980) The molecular characterization and organization of plant chromosomal DNA sequences. Ann Rev PL Physiol 31:569–596

Flavell RB (1989) Variation in structure and expression of ribosomal DNA loci in wheat. Genome 31:963–968

Flavell R, O'Dell O, Smith D (1979) Repeated sequence DNA comparisons between *Triticum* and *Aegilops* species. Heredity 42:309–332

Flavell RB, O'Dell M (1976) Ribosomal RNA genes on homoeologous chromosomes of groups 5 and 6 in hexaploid wheat. Heredity 37:377–385

Flavell RB, O'Dell M (1979) The genetic control of nucleolus formation in wheat. Chromosoma 71:135–152

Flavell RB, Smith DB (1974) Variation in nucleolar organiser rRNA gene multiplicity in wheat and rye. Chromosoma 47:327–334

Flavell RB, O'Dell M, Thompson WF (1988) Regulation of cytosine methylation in ribosomal DNA and nucleolus organizer expression in wheat. J Mol Biol 204:523–534

Fricano A, Brandolini A, Rossini L, Sourdille P et al (2014) Crossability of *Triticum urartu* and *Triticum monococcum* wheats, homoeologous recombination, and description of a panel of interspecific introgression lines. G3 Genes Genom Genet 4:1931–1941

Friebe B, Kim N-S, Kuspira J, Gill BS (1990) Genetic and cytogenetic analyses of the A genome of *Triticum monococcum*. VI. roduction and identification of primary trisomics using the C-banding technique. Genome 33:542–555

Fukuda K, Sakamoto S (1992) Studies on unreduced gamete formation in hybrids between tetraploid wheats and *Aegilops squarrosa* L. Hereditas 116:253–255

Furuta Y, Nishikawa K, Tanino T (1974) Stability in DNA content of AB genome component of common wheat during the past seven thousand years. Jpn J Genet 49:179–187

Furuta Y, Nishikawa K, Makino T, Sawai Y (1984) Variation in DNA content of 21 individual chromosomes among six subspecies in common wheat. Jpn J Genet 59:83–90

Furuta Y, Nishikawa K, Yamaguchi S (1986) Nuclear DNA content in diploid wheat and its relatives in relation to the phylogeny of tetraploid wheat. Jpn J Genet 61:97–105

Gale MD, Youssefian S (1985) Dwarfing genes in wheat. In: Russell GE (ed) Progress in plant breeding. Butterworths, London, pp 1–35

Galili G, Feldman M (1983) Genetic control of endosperm proteins in wheat. 2. Variation in high molecular weight glutenin and gliadin subunits of *Triticum aestivum*. Theor Appl Genet 66:77–86

Galili S, Avivi Y, Millet E, Feldman M (2000) RFLP-based analysis of three *RbcS* subfamilies in diploid and polyploid species of wheat. Mol Gen Gen 263:674–680

Gandilian PA (1972) On the wild-growing *Triticum* species of the Armenian SSR. Botanical Zhurnal 57:173–181

Gaurav K, Arora S, Silva P, Sanchez-Martin J et al (2021) Population genomic analysis of *Aegilops tauschii* identifies targets for bread wheat improvement. Nat Biotechnol. https://doi.org/10.1038/s41587-021-01058-4

Gegas VC, Nazari A, Griffiths S, Simmonds J, Fish L, Orford S, Sayers L, Doonan JH, Snape JH (2010) A genetic framework for grain size and shape variation in wheat. The Plant Cell 22:1046–1056

Genc Y, MacDonald GK (2008) Domesticated emmer wheat [*T. turgidum* L. subsp. *dicoccon* (Schrank) Thell.] as a source for improvement of zinc efficiency in durum wheat. Plant Soil 310:67–75

Georgiev S (1979) EMS-induced sphaerococcum mutation in *Triticum monococcum* L. Cereal Res Commun 7:183–189

Gerechter-Amitai ZK (1980) Specific and non-specific resistance to yellow rust in *Triticum dicoccoides*. Triennial Report of Research, Inst of Plant Protection, Agric Res Orgn, Bet Dagan, Israel

Gerechter-Amitai ZK, Grama A (1974) Inheritance of resistance to stripe rust (*Puccinia striiformis*) in crosses between wild emmer (*Triticum dicoccoides*) and cultivated tetraploid and hexaploid wheats. I. *Triticum durum*. Euphytica 23:387–392

Gerechter-Amitai ZK, Stubbs RW (1970) A valuable source of yellow rust resistance in Israeli populations of wild emmer, Triticum dicoccoides. Euphytica 19:12–21

Gerechter-Amitai ZK, Van Silfhout CH (1984) Resistance to powdery mildew in wild emmer (*Triticum dicoccoides* Körn). Euphytica 33:273–280

Gerechter-Amitai ZK, Wahl I, Vardi A et al (1971) Transfer of stem rust seedling resistance from wild diploid einkorn to tetraploid *durum* wheat by means of a triploid hybrid bridge. Euphytica 20:281–285

Gerlach WL (1977) N-banded karyotypes of wheat species. Chromosoma 62:49–56

Gerlach WL, Dyer TA (1980) Sequence organization of the repeating units in the nucleus of wheat which contain SS rRNA genes. Nucleic Acids Res 8:485–4865

Gerlach WL, Appels R, Dennis ES, Peacock WJ (1979) Evolution and analysis of wheat genomes using highly repeated DNA sequences. In: Ramanujan S (ed) Proceedings of 5th international wheat genetics symposium, Indian Soc Genet Pl Breed, Indian Agricultural Research Institute, New Delhi, India, pp 81–91

Gerlach WL, Miller TE, Flavell RB (1980) The nucleolus organizers of diploid wheats revealed by in situ hybridization. Theor Appl Genet 58:97–100

Giles RJ, Brown TA (2006) Glu Dy allele variations in *Aegilops tauschii* and *Triticum aestivum*: implications for the origin of hexaploid wheats. Theor Appl Genet 112:1563–1572

Gill BS (1987) Chromosome banding methods, standard chromosome band nomenclature, and applications in cytogenetic analysis. In: Heyne EG (ed) Wheat and wheat improvement, 2nd edn. American Society of Agronomy Inc., Madison, Wisconsin, USA, pp 243–254

Gill BS, Chen PD (1987) Role of cytoplasm-specific introgression in the evolution of the polyploid wheats. Proc Natl Acad Sci USA 84:6800–6804

Gill BS, Friebe B (2002) Cytogenetics, phylogeny and evolution of cultivated wheats. In: Curtis BC, Rajaram S, Gomez Macpherson H (eds) Bread wheat improvement and production. FAO, Rome, pp 71–88

Gill BS, Kimber G (1974a) The Giemsa C-banded karyotype of rye. Proc Natl Acad Sci 71:1247–1249

Gill BS, Kimber G (1974b) Giemsa C-banding and the evolution of wheat. Proc Natl Acad Sci 71:4086–4090

Gill BS, Waines JG (1978) Paternal regulation of seed development in wheat hybrids. Theor Appl Genet 51:265–270

Gill BS, Morris R, Schmidt JW, Maan SS (1963) Meiotic studies on chromosome morphology in the Wichita winter wheat variety by means of monosomics. Can J Genet Cytol 5:326–337

Gill BS, Waines JG, Sharma HC (1981) Endosperm abortion and the production of viable *Aegilops squarrosa* x *triticum boeoticum* hybrids by embryo culture.Plant Science Letters, 23:181–187

Gill BS, Browder LE, Hatchett JH, Harvey TL et al (1983) Disease and insect resistance in wild wheats. In: Sakamoto S (ed) Proceedings of 6th internaional wheat genetics symposium, Kyoto Japan, pp 785–792

Gill BS, Friebe B, Endo TR (1991) Standard karyotype and nomenclature system for description of chromosome bands and structural aberrations in wheat (*Triticum aestivum*). Genome 34:830–839

Giorgi B, Bozzini A (1969a) Karyotype analysis in *Triticum*. I. Analysis of *Triticum turgidum* (L.) Thell. and some related tetraploid wheats. Caryologia 22:249–258

Giorgi B, Bozzini A (1969b) Karyotype analysis in *Triticum*. III. Analysis of the presumed diploid progenitors of polyploid wheats. Caryologia 22:279–287

Giorgi B, Bozzini A (1969c) Karyotype analysis in Triticum. IV. Analysis of (*Aegilops speltoides* x *Triticum boeoticum*) amphiploid and a hypothesis on the evolution of tetraploid wheats. Caryologia 22:289–306

Giorgi B, Bozzini A (1970) Karyotype analysis in *Triticum*: V. Identification of chromosomes of bread and *durum* wheats using aneuploids of chinese spring. Caryologia 23:565–574

Glémin S, Scornavacca C, Dainat J, Burgarella C, Viader V, Ardisson M, Sarah G, Santoni S, David J, Ranwez V (2019) Pervasive hybridizations in the history of wheat relatives. Sci Adv 5:eaav9188

Gogniashvili M, Maisaia L, Kotorashvili A, Kotaria N, Beridze T (2018) Complete chloroplast DNA sequences of Georgian indigenous polyploid wheats (*Triticum* spp.) and B plasmon evolution. Genet Resour Crop Evol 65:1995–2002

Golabadi M, Arzani A, Mirmohammadi Maibody SAM (2006) Assessment of drought tolerance in segregating populations in durum wheat. African J Agric Res 1:162–217

Golenberg EM (1988) Outcrossing rates and their relationship to phenology in *Triticum dicoccoides*. Theor Appl Genet 75:937–944

Golovnina K, Glushkov S, Blinov A, Mayorov V, Adkison L, Goncharov N (2007) Molecular phylogeny of the genus *Triticum* L. Plant Syst Evol 264:195–216

Goncharov NP (2011) Genus *Triticum* L. taxonomy: the present and the future. Plant Syst Evol 295:1–11

Goncharov NP, Gaidalenok RF (2005) Localization of genes controlling spherical grain and compact ear in *Triticum antiquorum* Heer ex Udacz. Russ J Genet 41:1262–1267

Goncharov NP, Golovnina KA, Kondratenko EYA (2009) Taxonomy and molecular phylogeny of natural and artificial wheat species. Breed Sci 59:492–498

González JM, Bernard S, Bernard M (1993) Metaphase-I analysis of a *Triticum aestivum* x *T. monococcum* hybrid by the C-banding technique. Euphytica 68:187–192

Gornicki P, Zhu H, Wang J, Challa GS, Zhang Z, Gill BS, Li W (2014) The chloroplast view of the evolution of polyploid wheat. New Phytol 204:704–714

Graber S, Kuhn M (1992) Evaluation of the baking quality of different varieties of spelt. Getreide Mehl Und Brot 46:102–108

Grama A, Gerechter-Amitai ZK (1974) Inheritance of resistance to stripe rust (*Puccinia striiformis*) in crosses between wild emmer and cultivated tetraploid and hexaploid wheats. II. *Triticum aestivum*. Euphytica 23:393–398

Graur D, Bogher M, Breiman A (1989) Restriction endonuclease profiles of mitochondrial DNA and the origin of the B genome of bread wheat, *Triticum aestivum* . Heredity 62:335–342

Grausgruber H, Sailer C, Ghambashidze G, Bolyos L, Ruckenbauer P (2004) Genetic variation in agronomic and quantitative traits of ancient wheats. In: Grausgruber H, Ruckenbauer P (eds) Genetic variation for plant breeding: proceedings of the 17th EUCARPIA general congress, Vienna, Vollman J BOKU–University of Natural Resources and Applied Life Sciences, pp 19–22

Gul A, Allan RE (1972) Relation of the club gene with yield and yield components of near-isogenic wheat lines. Crop Sci 12:297–301

Guo X, Han F (2014) Asymmetric epigenetic modification and elimination of rDNA sequences by polyploidization in wheat. Plant Cell 26:4311–4327

Guo W, Xin M, Wang Z, Yao Y et al (2020) Origin and adaptation to high altitude of Tibetan semi-wild wheat. Nat Commun 1:5085

Gupta PK, Kulwal PL, Rustgi S (2005) Wheat cytogenetics in the genomics era and its relevance to breeding. Cytogenet Genome Res 109:315–327

Gustafson JP, Flavell RB (1996) Control of nucleolar expression in triticale. In: Guedes-Pinto H, Darvey N, Carnide VP (eds) Triticale: today and tomorrow. Kluwer Academic Publishers, The Netherlands, pp 119–125

Hackel E (1890) The true grasses. 228 p., illus. New York

Haider N (2012) Evidence for the origin of the B genome of bread wheat based on chloroplast DNA. Turkish J Agric for 36:13–25

Hammer K (1980) Vorarbeiten zur monographischen Darstellung von Wildpflanzensortimenten: *Aegilops* L. Kulturpflanze 28:33–180

Hammer K, Perrino P (1995) Plant genetic resources in South Italy and Sicily—studies toward *in situ* and on-farm conservation. IPGRI Plant Genet Resourc Newsletter 103:19–23

Hammer K, Filatenko AA, Korzun V (2000) Microsatellite markers: a new tool for distinguishing diploid wheat species. Genet Res Crop Evol 47:497–505

Han FP, Fedak G, Ouellet T, Liu B (2003) Rapid genomic changes in interspecific and intergeneric hybrids and allopolyploids of Triticeae. Genome 46:716–723

Han FP, Fedak G, Guo WL, Liu B (2005) Rapid and repeatable elimination of a parental genome specific DNA repeat (pGcIR-1a) in newly synthesized wheat allopolyploids. Genetics 170:1239–1245

Han J, Zhou B, Shan W, Yu L, Wu W, Wang K (2015) A and D genomes spatial separation at somatic metaphase in tetraploid cotton: evidence for genomic disposition in a polyploid plant. Plant J 84:1167–1177

Hanchinal R, Yenagi N, Bhuvaneswari G. Math K (2005). Grain quality and value addition of emmer wheat. Dharward: University of Agricultural Sciences Dharwad

Hanelt P (2001) *Triticum* L. In: Hanelt P, Institute of Plant Genetics and Crop Plant Research (eds) Mansfeld's encyclopedia of agricultural and horticultural crops, vol 5. Springer, Berlin, pp 2565–2591

Hao M, Luo J, Zeng D, Zhang L, Ning S, Yuan Z et al. (2014) *QTug.sau-3B* Is a Major Quantitative Trait Locus for Wheat Hexaploidization. G3 Genes Genom Genet 4:1943–1953

Haque MA, Takayama A, Watanabe N, Kuboyama T (2011) Cytological and genetic mapping of the gene for tour-awned phenotype in Triticum carthlicum Nevski. Genet Resourc Crop Evol 58:1087–1093

Harlan JR (1967) A wild wheat harvest in Turkey. Archaeology 20:197–201

Harlan JR (1981) The early history of wheat: earliest trace to the sack of Rome. In: Evans LT, Peacock WJ (eds) Wheat science—today and tomorrow. Cambridge University Press, pp 1–19

Harlan JR, Zohary D (1966) Distribution of wild wheats and barley. Science 153:1074–1080

Haudry A, Cenci A, Ravel C, Bataillon T et al (2007) Grinding up wheat: a massive loss of nucleotide diversity since domestication. Mol Biol Evol 24:1506–1517

He F, Pasam R, Shi F, Kant S, Keeble-Gagnere G et al (2019) Exome sequencing highlights the role of wild-relative introgression in shaping the adaptive landscape of the wheat genome. Nat Genet 51:896–904

Heer O (1866) Die Pflanzen der Pfahlbauten, Zurich

Hedge SG, Valkoun J, Waines JG (2000) Genetic diversity in wild wheats and goat grass. Theor Appl Genet 101:309–316

Helbaek H (1959) Domestication of food plants in the old world. Science 130:365–372

Hermsen JGT, Waninge J (1972) Attempts to localize the gene *Ch1* for hybrid chlorosis in wheat. Euphytica 21:204–208

Heun M, Schäfer-Pregl R, Klawan D, Castagna R et al (1997) Site of einkorn wheat domestication identified by DNA fingerprinting. Science 278:1312–1214

Heun MS, Haldorsen K. Vollan K (2008) Reassessing domestication events in the Near East: einkorn and *Triticum urartu*. Genome 51:444–451

Hilliker AJ, Appels R (1989) The arrangement of interphase chromosomes: structural and functional aspects. Exp Cell Res 185:297–318

Hillman GC (1972) Plant remains in French DH. Excavations at Can Hassan III 1969–1970. In: Higgs ES (ed) Papers in economic prehistory. Cambridge University Press, Cambridge, pp 182–190

Hillman GC (1975) The plant remains from Tell Abu Huretra: a preliminary report. Proc Prehistory Soc 41:70–73

Hillman GC (1978) On the origins of domestic rye-*Secale cereale*: the finds from Aceramic Can Hasan III in Turkey. Anatolian Stud 28:157–174

Hillman CG (1996) Late Pleistocene changes in wild plant foods available to hunter-gatherers of the northern Fertile Crescent: possible preludes to cereal cultivation. In: Harris D (ed) The origin and spread of agriculture and pastoralism in Eurasia. UCL Press, London, pp 159–203

Hillman G (2001) Archaeology, percival, and the problems of identifying wheat remains. In: Caligari PDS, Brandham PE (eds) Wheat taxonomy: the legacy of John Percival. The Linn Soc London, Burlington House, London, pp 27–36

Hirai A, Tsunewaki K (1981) Genetic diversity of the cytoplasm in *Triticum* and *Aegilops*. VIII. Fraction I protein of 39 cytoplasms. Genetics 99:487–493

Hollingshead L (1932) The occurrence of unpaired chromosomes in hybrids between varieties of *Triticum vulgare*. Cytologia 3:119–141

Hopf M (1955) Formveränderungen von Getreidekörnern beim Verkohlen. Berichte Der Deutschen Botanishen Gesellschaft 68:191–193

Hopf M (1973) Frühe Kulturflanzen aus Bulgarien. Jahrb Röm-German. Zentralmus Mainz- 20:1–47

Hori T, Tsunewaki K (1967) Study on substitution lines of several Emmer wheats having the cytoplasm of Triticum boeoticum. Seiken Ziho 19:55–59

Horn JD, Walden DB (1978) Affinity distance values among somatic metaphase chromosomes in maize. Genetics 88:181–200

Houchins K, O'Dell M, Flavell R, Gustavson PJ (1997) Cytosine methylation and nucleolar dominance in cereal hybrids. Mol Gen Genet 255:294–301

Hovhannisyan NA, Dulloo ME, Yesayan AH, Knüpffer H, Amri A (2011) Tracking of powdery mildew and leaf rust resistance genes in *Triticum boeoticum* and *T. urartu*, wild relatives of common wheat. Czech J Genet Plant Breed 47:45–57

Huang L, Millet E, Rong JK, Wendel JF, Anikster Y, Feldman M (1999) Restriction fragment length polymorphism in wild and cultivated tetraploid wheat. Isr J Pl Sci 47:213–224

Huang S, Sirikhachornkit A, Su X, Faris JD, Gill BS et al (2002) Genes encoding plastid acetyl-CoA carboxylase and 3-phosphoglycerate kinase of the *Triticum/Aegilops* complex and the evolutionary history of polyploid wheat. Proc Natl Acad Sci 99:8133–8138

Huang L, Raats D, Sela H, Klymiuk V et al (2016) Evolution and adaptation of wild emmer wheat populations to biotic and abiotic stresses. Ann Rev Phytopathol 54:279–301

Huskins CL (1931) A cytological study of Vilmorin's unfixable dwarf wheat. J Genet 25:113–124

Hutchinson J, Miller TE (1982) The nucleolar organisers of tetraploid and hexaploid wheats revealed by in situ hybridisation. Theor Appl Genet 61:285–288

Hutchinson J, Miller TE, Jahier J, Shepherd KW (1982) Comparison of the chromosomes of *Triticum timopheevi*with related wheats using the techniques of C-banding and in situ hybridization. Theor Appl Genet 64:31–40

International Wheat Genome Sequencing Consortium (IWGSC) (2018) Shifting the limits in wheat research and breeding using a fully annotated reference genome. Science 361(6403):eaar7191

Jaaska V (1974) The investigation of tetraploid wheat origin by method of enzyme electrophoretic study. Izv Acad Nauk EstSSR 23:201–220

Jaaska V (1976) Aspartate aminotransferase isoenzymes in the polyploid wheats and their diploid relatives. On the originof tetraploid wheats. Biochem Physiol Pflanzen 170:159–171

Jaaska V (1978) NADP-dependent aromatic alcohol dehydrogenase in polyploid wheats and their diploid relatives. On the origin and phylogeny of polyploid wheats. Theor App Genet 53:209–217

Jaaska V (1980) Electrophoretic survey of seedling esterases in wheats in relation to their phylogeny. Theor App Genet 56:273–284

Jaaska V (1981) Aspartate aminotransferase and alcohol dehydrogenase isoenzyme: intraspecific differentiation in *Aegilops tauschii* the origin of the D genome polyploids in the wheat group. Plant Syst Evol 137:259–273

Jaaska V (1982) Isoenzymes of superoxide dismutase in wheats and their relatives: allozyme variation. Biochem Physiol Pflanzen 177:747–755

Jaaska V (1993) Isoenzymes in the evaluation of germplasm diversity in wild diploid relatives of cultivated wheat. In: Damania AB (ed) Biodiversity and wheat improvement. John Wiley, Chichester, pp 247–257

Jakubziner MM (1932a) The wheats of Syria, Palestine and Transjordania, cultivated and wild. Trudy prikl Bot Genet Selek, Suppl 53, *Triticum dicoccoides*, pp 150–178 (Russian, with English summary)

Jakubziner MM (1932b) A contribution to the knowledge of the wild wheat of Transcaucasia Trudy prikl Bot Genet Selek. Ser 5, 1:147–198 (Russian, with English summary)

Jakubziner MM (1959) New wheat species. In: Jenkins BC (ed) Proceedings of 1st international wheat genetics symposium. Public Press Ltd., Winnipeg, Manitoba, Canada, pp 207–220

Jakubziner MM (1969) Immunity of different wheat species. Agric Biol 4:837–847 (in Russian)

Jakubziner MM, Dorofeev SF (1968) Word wheat resources in service of soviet breeding. Bull Appl Bot Genet Pl Breed 39:65–79

Jantasuriyarat C, Vales MI, Watson CJW, Riera-Lizarazu O (2004) Identification and mapping of genetic loci affecting the free-threshing habit and spike compactness in wheat (*Triticum aestivum* L.). Theor Appl Genet 108:261–73

Janushevich ZV (1984) The specific composition of wheat finds from ancient agricultural centers in the USSR. In: van Zeist W, Casparie WA (eds) Plants and ancient man: studies in palaeoethnobotany. Balkema, Rotterdam, pp 267–276

Järve K, Peusha HO, Tsymbalova J, Tamm S, Devos KM, Enno TM (2000) Chromosomal location of a *Triticum timopheevii*—derived powdery mildew resistance gene transferred to common wheat. Genome 43:377–381

Jauhar PP, Riera-Lizarazu O, Dewey WG, Gill BS, Crane CF, Bennet JH (1991) Chromosome pairing relationships among the A, B and D genomes of bread wheat. Theor Appl Genet 81:441–449

Jenkins JA (1929) Chromosome homologies in wheat and *Aegilops*. Amer J Bot 16:238–245

Jenkins DJA, Wolever TMS, Thorme MJ (1984) The relationship between glycaemic responses, digestibility and factors influencing habits of diabetes. Am J Clin Nutri 40:1175–1191

Jia J, Xie Y, Cheng J, Kong C, Wang M et al (2021) Homology-mediated inter-chromosomal interactions in hexaploid wheat leads to specific subgenome territories following polyploidization and introgression. Genome Biol 22:26

Jiang J, Gill BS (1994a) New 18S–26S ribosomal RNA gene loci: chromosomal landmarks for the evolution of polyploid wheats. Chromosoma 103:179–185

Jiang J, Gill BS (1994b) Different species-specific chromosome translocations in *Triticum timopheevii* and *T. turgidum* support the diphyletic origin of polyploid wheats. Chromosome Res 2:59–64

Jing HC, Kornyukhin D, Kanyuka K, Orford S et al (2007) Identification of variation in adaptively important traits and genome-wide analysis of trait-marker associations in *Triticum monococcum*. J Exp Bot 58:3749–3764

Jing HC, Bayon C, Kanyuka K et al (2009) DArT markers: diversity analyses, genomes comparison, mapping and integration with SSR markers in *Triticum monococcum*. BMC Genom 10:458

Johnson BL (1968) Electrophoretic evidence on the origin of Triticum zhukovskyi. In: Finlay KW, Shepherd KW (eds) Proceedings of third international wheat genetics symposium. Australian Academy of Science, Canberra, pp 105–110

Johnson BL (1972) Protein electrophoretic profiles and the origin of the B genome of Wheat. Proc Natl Acad Sci 69:1398–1402

Johnson BL (1975) Identification of the apparent B-genome donor of wheat. Can J Genet Cytol 17:21–39

Johnson BL, Dhaliwal HS (1976) Reproductive isolation of *Triticum boeoticum* and *Triticum urartu* and the origin of tetraploid wheats. Amer J Bot 63:1088–1094

Johnson BL, Dhaliwal HS (1978) *Triticum urartu* and genome evolution in the tetraploid wheats. Am J Bot 65:907–918

Johnson BL, Hall O (1965) Analysis of phylogenetic affinities in the Triticinae by protein electrophoresis. Am J Bot 52:506–513

Jones BL, Mak AS (1977) Amino acid sequences of the a-purothionins of hexaploid wheat. Cereal Chem 54:511–523

Johnson BL, Barnhart D, Hall O (1967) Analysis of genome and species relationships in the polyploid wheats by protein electrophoresis. Amer J Bot 54:1089–1098

Johnson EB, Nalam VJ, Zemetra RS, Riera-Lizarazu O (2008) Mapping the *compactum* locus in wheat (*Triticum aestivum* L.) and its relationship to other spike morphology genes of the Triticeae. Euphytica 163:193–201

Jones BL, Lookhart GL, Mak A, Cooper DB (1982) Sequences of purothionins and their inheritance in diploid, tetraploid, and hexaploid wheats. J Hered 73:143–144

Jones G, Valamoti S, Charles M (2000) Early crop diversity: a "new" glume wheat from northern Greece. Veg Hist Archaeobotany 9:133–146

Joppa LR (1987) Aneuploid analysis in tetraploid wheat, 2nd edn. In: Heyne EG (ed) Wheat and wheat improvement. Amercan Society of agronomy, Madison, WI, USA, pp 255–267

Joppa LR (1993) Chromosome engineering in tetraploid wheat. Crop Sci 33:908–913

Joppa LR, Cantrell RG (1990) Chromosomal location of genes for grain protein content of wild tetraploid wheat. Crop Sci 30:1059–1064

Joppa LR, Williams ND (1983) The Langdon durum D-genome disomic substitutions: development, characteristic, and uses. Agronomy Abstract American Society of Agronomy, Madison, WI, USA, p 68

Joppa LR, Williams D (1988) Langdon durum disomic substitution lines and aneuploid analysis in tetraploid wheat. Genome 30:222–228

Joppa LR, Hareland GA, Cantrell RG (1991) Quality characteristics of the Langdon *durum-dicoccoides* chromosome substitution lines. Crop Sci 31:1513–1517

Joppa LR, Nevo E, Beiles A (1995) Chromosome translocations in wild populations of tetraploid emmer wheat in Israel and Turkey. Theor Appl Genet 91:713–719

Jordaan JP, Engelbrecht SA, Malan J, Knobel HA (1999) The genetics and exploitation of heterosis in crops, Madison, WI, USA, pp 411–421

Jorjadze M, Berishvili T, Ahatberashvili E (2014) The ancient wheats of Georgia and their traditional use in the southern part of the country. Emir J Food Agric 26:192–202

Josekutty PC (2008) Defining the genetic and physiological basis of *Triticum sphaerococcum* Perc. A thesis submitted for the degree of master of science in cellular and molecular biology at the University of Canterbury, New Zealand, pp 1–133

Joshi BC, Upadhya MD, Swaminathan MS (1970) Aneuploid analysis of chromosome pairing in *Triticum timopheevii*. Wheat Inf Serv 31:2–4

Joshi BC, Sawhney RN, Singh D (1980) F_1 monosomic of *Triticum macha*. Euphytica 29:609–614

Kabarity A (1966) On the origin of the new cultivated wheat. II. Cytogenetical studies on the karyotypes of some *Triticum macha* varieties. Beitr Biol Pflanzen 42:339–346

Kabbaj H, Sall AT, Al-Abdallat A, Geleta M, Amri A, Filali-Maltouf A, Belkadi B, Ortiz R, Bassi FM (2017) Genetic diversity within a global panel of durum wheat (*Triticum durum*) landraces and modern germplasm reveals the history of alleles exchange. Front Plant Sci 8:1277

Kacem NS, Delporte F, Mukovski Y, Djekoun A, Watillon B (2017) In vitro screening of durum wheat against water-stress mediated through polyethylene glycol. J Genet Engineer Biotechnol 15:239–247

Kagawa F (1929) A study on the phylogeny of some species in *Triticum* and *Aegilops,* based upon comparison of the chromosomes. J Coll Agric Tokyo 10:173–228

Kashkush K, Feldman M, Levy AA (2002) Gene loss, silencing and activation in a newly synthesized wheat allotetraploid. Genetics 160:1651–1659

Kashkush K, Feldman M, Levy AA (2003) Transcriptional activation of retrotransposons alters the expression of adjacent genes in wheat. Nat Genet 33:102–106

Kato T, Yamagata H (1982) Effect of 3B chromosome deficiency on the meiotic pairing between the arms of an isochromosome in common wheat. Jpn J Genet 57:403–406

Kato T, Yamagata H (1983) Analysis of the action of 3B chromosome on meiotic homologous chromosome pairing in common wheat. In: Sakamoto S (ed) Proceedings of 6th international wheat genetics symposium. Faculty of Agriculture, Kyoto, Japan, pp 321–325

Kawahara T (1984) Studies on intraspecific structural differentiation of chromosomes in the wild tetraploid wheats. Doctorate Thesis, Kyoto University, Kyoto

Kawahara T (1986) Identification of reciprocal translocation chromosome types in the emmer wheats. II. Thirty-eight strains of *Triticum dicoccoides* Korn. with the fundamental chromosome structure. Wheat Inf Serv 63:2–7

Kawahara T (1987) Identification of reciprocal translocation chromosome types in the emmer wheats. III. Six chromosome types in *Triticum dicoccoides*. Jpn J Genet 62:197–204

Kawahara T, Nevo E (1996) Screening of spontaneous major translocations in Israeli populations of *Triticum dicoccoides* Koern. Wheat Inf Serv 83:28–30

Kawahara T, Tanaka M (1977) Six chromosome types in *Triticum araraticum* Jakubz. differing with reciprocal translocations. Jpn J Genet 52:261–267

Kawahara T, Tanaka M (1978) Identification of reciprocal translocation chromosome types in the emmer wheats. I. *Triticum dicoccoides* Korn. Wheat Inf Serv 45–46:29–31

Kawahara T, Tanaka M (1981) Intraspecific differentiation in chromosome structures of the wild tetraploid wheats. Wheat Inf Serv 52:33

Kawahara T, Tanaka M (1983) Chromosomal interchanges and the evolution of the B and G genomes. In: Sakamoto S (ed) Proceedings of 6th international wheat genetics symposium, Kyoto, Japan, pp 977–981

Kawahara T, Badaeva ED, Badaev NS, Gill BS (1996) Spontaneous translocations in *Triticum araraticum* Jakubz. Wheat Inf Serv 83:7–14

Kemp B J (1989) *Ancient Egypt: Anatomy of a civilization*. Routledge, London, 1989

Kempanna C, Riley R (1962) Relationships between the genetic effects of deficiencies for chromosomes III and V on meiotic pairing in *Triticum aestivum*. Nature 195:1270–1273

Kerber ER (1964) Wheat: reconstitution of the tetraploid component (AABB) of hexaploids. Science 143:253–255

Kerber ER, Bendelow VM (1977) The role of *Triticum carthlicum* in the origin of bread wheat based on comparative milling and baking properties. Can J Pl Sci 57:367–373

Kerber ER, Dyck PL (1973) Inheritance of stem rust resistance transferred from diploid wheat (*Triticum monococcum*) to tetraploid and hexaploid wheat and chromosome location of the gene involved. Can J Genet Cytol 15:397–409

Kerber ER, Rowland GG (1974) Origin of the free threshing character in hexaploid wheat. Can J Genet Cytol 16:145–154

Kerby K (1986) A cytological and biochemical characterization of the potential B genome donors to common wheat, *Triticum aestivum*. Ph.D. thesis, University of Alberta, Edmonton, Canada

Kerby K, Kuspira J (1987) The phylogeny of the polyploid wheats *Triticum aestivum* (bread wheat) and *Triticum turgidum* (macaroni wheat). Genome 29:722–737

Kerby K, Kuspira J (1988) Cytological evidence bearing on the origin of the B genome in polyploid wheats. Genome 30:36–43

Kerby K, Kuspira J, Jones BL (1988) Biochemical data bearing on the relationship between the genome of *Triticum urartu* and the A and B genomes of the polyploid wheats. *Genome* 30:576–581

Khan SI (1963) Karyotype analysis of «Holdfast» a cultivar of *Triticum aestivum*. Cellule 63:293–305

Khlestkina EK, Röder NS, Grausgrruber H, Börner A (2006) A DNA fingerorinting-based taxonomic allocation of Kamut wheat. Pl Genet Resourc 4:172–180

Kidane YG, Mancini C, Mengistu DK, Frascaroli E, Fadda C, Pè ME, Dell'Acqua M (2017) Genome wide association study to identify the genetic base of smallholder farmer preferences of durum wheat traits. Front Pl Sci 8:1–11

Kihara H (1919) Uber cytologische Studien bei einigen Getreidiarten. I. Species-bastarde des weizens und weizenroggen-bastarde. Bot Mag Tokyo 32:17–38

Kihara H (1924) Cytologische und genetische studien bei wichtigcn getreidearten mit besonderer rücksicht ouf das verhalten der chromosomen und die sterilitat in den bastarden. Mem Cell Sci, Kyoto Imp Univ Bl, pp 1–200

Kihara H (1925) Weitere Untersuchungen über die pentaploiden *Triticum*-Bastarde. I. Jpn J Bot II:299–304299

Kihara H (1929) Conjugation of homologous chromosomes in genus hybrids *Triticum* x *Aegilops* and species hybrids of *Aegilops*. Cytologia 1:1–15

Kihara H (1930) Genomanalyse bei *Triticum* und *Aegilops*. Cytologia 1:263–284

Kihara H (1932) Genomanalyse bei *Triticum* und *Aegilops*. IV. Untersuchungen an *Aegilops* x *Triticum* und *Aegilops*x *Aegilops*-Bastarden. Cytologia 3:384–456

Kihara H (1937) Genomanalyse bei *Triticum* und *Aegilops*. VII. Kurze ~ bersicht iiber die Ergebnisse der Jahre 1934–36. Mem Coll Agric Kyoto Imp Univ 41:1–61

Kihara H (1938) Cytogenetics of species hybrids. Curr Sci (Special Number Genet): 20–23

Kihara H (1940) Verwandtschaft der *Aegilops*-Arten im Lichte der Genomanalyse. Ein Überblick. Der Züchter12:49–62

Kihara H (1944) Discovery of the DD-analyser, one of the ancestors of *Triticum vulgare*. Agric Hortic 19:13–14

Kihara H (1947) The genus Aegilops classified on the basis of genome analysis. Seiken Zjho 3:7–25

Kihara H (1949) Genomanalyse bei *Triticum* und *Aegilops*. IX. Systematischer Aufbau der Gattung Aegilops auf genomanalytischer Grundlage. Cytologia 14:135–144

Kihara H (1954) Considerations on the evolution and distribution of *Aegilops* species based on the analyzer-method. Cytologia 19:336–357

Kihara H (1963) Interspecific relationship in *Triticum* and *Aegilops*. Seiken Zihô 15:1–12

Kihara H, Katayama Y (1932) On the progeny of haploid plants of *Triticum monococcum*. Kwagau 2:408–410

Kihara H, Katayama Y (1933) On the maturation division of haploid *Triticum monococcum*. Agr Hort (Tokyo) 8:2725–2738

Kihara H, Lilienfeld FA (1932) Genomanalyse bei *Triticum* und *Aegilops*. IV. Untersuchungen an *Aegilops* x *Triticum* – und *Aegilops* x *Aegilops* Bastarden. Cytologia 3:384–456

Kihara H, Lilienfeld F (1934) Genomanalyse bei *Triticum* und *Aegilops*. V. *Triticum Timopheevi* Zhuk. Cytologia 6:87–122

Kihara H, Tanaka M (1970) Addendum to the classification of the genus Aegilopsby means of genome-analysis. Wheat Inf Serv 30:1–2

Kihara H, Wakakuwa S, Nishiyama I (1929) Notes on species hybrids of *Tririrum*. Jpn J Genet 5:8

Kihara H, Yamashita H, Tanaka M (1959) Genomes of 6x species of *Aegilops*. Wheat Inf Ser 8:3–5

Kihara H, Yamashita K, Tanaka M (1965) Morphological, physiological, genetical and cytological studies in *Aegilops* and *Triticum* collected from Pakistan, Afghanistan and Iran. In: Yamashita K (ed) Cultivated plants and their relatives, Koei Printing Company, Japan, pp 1–118

Kilian B, Özkan H, Deusch O, Effgen S, Brandolini A, Kohl J, Martin W, Salamini F (2007) Independent wheat B and G genome origins in outcrossing *Aegilops* progenitor haplotypes. Mol Biol Evol 24:217–227

Kilian B, Ozkan H, Kohl J, von Haeseler A, Barale F, Deusch O, Brandolini A, Yucel C, Martin W, Salamini F (2006) Haplotype structure at seven barley genes: relevance to gene pool bottlenecks, phylogeny of ear type and site of barley domestication. Mol Genet Genom 276:230–241

Kim N-S, Kuspira J, Armstrong KC, Bhambhani R (1993) Genetic and cytogenetic analyses of the A genome of Triticum monococcum. VIII. Localization of rDNAs and characterization of 5S rRNA genes. Genome 36:77–86

Kim SH, Kim JY, Kim DY, Yoon JS et al (2016) Development of a SCAR marker associated with salt tolerance in durum wheat (*Triticum turgidum* ssp. *durum*) from a semi-arid region. Genes Genom 38:939–948|

Kimber G, Athwal RS (1972) A reassessment of the course of evolution of wheat. Proc Nat Acad Sci USA 69:912–915

Kimber G, Feldman M (1987) Wild wheats: an introduction. Special report 353. College of Agriculture, Columbia. Missouri, USA, pp 1–142

Kimber G, Riley R (1963) The relationships of the diploid progenitors of hexaploid wheat. Can J Genet Cytol 5:83–88

Kimber G, Sears ER (1983) Assignment of genome symbols in the Triticeae. In: Sakamoto S (ed) Proceedings of 6th international wheat genetics symposium, Kyoto, Japan, pp 1195–1196

Kimber G, Sears ER (1987) Evolution in the genus *Triticum* and the origin of cultivated wheat, 2nd edn. In: Heyne EG (ed) Wheat and wheat improvement. American Society of Agronomy, Madison, Wisconsin, USA, pp 154–164

Kimber G, Tsunewaki K (1988) Genome symbols and plasma types in the wheat group. In: Miller TE, Koebner RMD (eds) Proceedings of 7th international wheat genetics symposium, Cambridge, pp 1209–1210

Kimber G, Zhao YH (1983) The D genome of the Triticeae. Can J Genet Cytol 25:581–589

Kislev ME (1973) Hitta and Kussemet, notes on their interpretation. Leshonenu. 37:83–95

Kislev ME (1979/1980) *Triticum parvicoccum* sp. nov, the oldest naked wheat. Isr J Bot 28:95–107

Kislev ME (1981) The history of evolution of naked wheats. Z Archaeol 15:57–64

Kislev ME (1984) Emergence of wheat agriculture. Palaeorient 10:61–70

Kislev ME (1992) Agricultural situation in the Near East in the VIIth millennium BP. In: Anderson PC (ed) Prehistoire de l'Agriculture: Nouvelles Approches Experimentales et Ethnographiques. Édition du CNRS, Monographie du Centre Recherches Archéologiques No. 6, Paris, pp 87–93

Kislev ME (2009) Reconstructing the ear morphology of ancient small-grain wheat (*Triticum turgidum* ssp. *parvicoccum*). In: Fairbairn A, Weiss E (eds) From foragers to farmers: papers in honour of Gordon C Hillman. Oxbow Books, pp 235–238

Kislev ME, Bar-Yosef O, Gopher A (1986) Early neolithic domesticated and wild barley from the Netiv Hagdud region in the Jordan Valley. Isr J Bot 35:197–201

Kislev ME, Hartmann A, Galili E (2004) Archaeobotanical and archaeoentomological evidence from a well at Atlit-Yam indicates colder, more humid climate on the Israeli coast during the PPNC period. J Archaeol Sci: 1301–1310

Kitani Y (1963) Orientation, arrangement and association of somatic chromosomes. Jap J Genet 38:244–256

Klindworth DL, Klindworth MM, Williams ND (1997) Telosomic mapping of four morphological markers of durum wheat. J Hered 88:229–232

Klindworth DL, Williams ND, Duysen ME (1995) Genetic analysis of Chlorina mutants of durum wheat. Crop Sci 35:431–436

Klindworth DL, Miller JD, Jin Y, Xu SS (2007) Chromosomal locations of genes for stem rust resistance in monogenic lines derived from tetraploid wheat accession ST464. Crop Sci 47:1441–1450

Klymiuk V, Yaniv E, Huang L, Raats et al (2018) Cloning of the wheat Yr15 resistance gene sheds light on the plant tandem kinase-pseudokinase family. Nature Commun 9:3735

Knaggs P, Ambrose MJ, Reader S, Miller TE (2000) Morphological characterization and evaluation of the subdivision of *Aegilops tauschii* Coss. Wheat Inf Serv 91:15–19

Koba T, Tsunewaki K (1978) Mapping of the s and *Ch2* genes on chromosome 3D of common wheat. Wheat Inf Serv 45:18–20

Konarev VG (1983) The nature and origin of wheat genomes on the data of grain protein immunochemistry and electrophoresis. In: Sakamoto S (ed) Proceedings of 6th international wheat genetics symposium, Kyoto, Japan, pp 65–75

Konarev VG, Gavrilyuk IP, Peneva TI (1976) About nature and origin of wheat genomes on the data of biochemistry and immunochemistry of grain proteins. Skh Biol 11:656–665

Konarev VG, Gavrilyuk IP, Gubareva NK, Peneva TI (1979) Seed proteins in genome analysis, cultivar identification, and documentation of cereal genetic resources: a review. Cereal Chem 56:272–278

Körber-Grohne U (1987) Nutzpflanzen in Deutschland. Konrad Thesis, Stuttgart, Germany

Körber-Grohne U (1989) The history of spelt (Triticum spelta) on the basis of archaeological findings from Neolithic to Medieval times, and the data by written sources until today. In: Devroey JP, van Mol JJ (eds) L'épeautre (*Triticum spelta*): histoire et ethnologie, Dire, Treignes, pp 51–59

Körnicke F (1885) Die Arten und Varieta ten des Getreides, vol 1. In: Körnicke F, Werner H (eds) Handbuch des Getreidebaus. Berlin

Kostoff D (1937a) The genomes of *Triticum Timopheevi* Zhuk, *Secale cereale* L., and *Haynaldia villosa* Schr. Z Ind Abst Vererb 72:115–118

Kostoff D (1937b) Chromosome behavior in *Triticum* hybrids and allied genera. I. Interspecific hybrids with *Triticum timopheevii*. Proc Indian Acad Sci 5:231–236

Kostoff D (1937c) Studies on the polyploid plants. XI. Amphidiploid *Triticum timopheevi* ZHUK. x *Triticum monococcum* L. Z. Pflanzenzücht 21:41–45

Kostoff D (1938) Heterochromatin at the distal ends of the chromosomes in *Triticum monococcum*. Nature 141:690–691

Kostoff D, Arutiunian N (1938) Heterochromatic (inert) regions in the chromosomes of *Crepis capillaris*. Nature 141:514–515.

Kota RS, Dvořák J (1986) Mapping of a chromosome pairing gene and 5S rRNA genes in *Triticum aestivum* L. by a spontaneous deletion in chromosome arm 5Bp. Can J Genet Cytol 28:266–271

Koval SF (1999) Near-isogenic lines of spring common wheat Novosibirskaya 67 marked with short and long glume. Wheat Inf Ser 88:37–42

Krause EHL (1898) Floristische Notizen II. Gräser. 1. Zur Systematik Und Synonymik Botanisches Centralblatt (jena) 73:337–343

Kuckuck H (1964) Experimentelle Unersuchumgen zur Entstehungder Kulturweizen. I. Die Variation des Iranischen Speltzweizns und seine genetischen Beziehungen zu *Triticum aestivum* ssp. v*ulgare* (Vill-, HOSt) Mac Key, ssp. *spelta* (L.) Thell. und ssp. *macha* (Dek et Men.) Mac Key mit einem Beitrag zur Genetik den *Spelta*-Komplexes. Z Pfanzenzücht 51:97–140

Kuckuck H (1970) Primitive wheats. In: Frankel OH, Bennett E (eds) Genetic resources in plants—their exploration nd conservation, London, pp 249–266

Kuckuck H (1979) On the origin of *Triticum carthlicum* Neyski (= *Triticum persicum* Vav.). Wheat Inf Sev 50:1–3

Kuckuck H, Schiemann E (1957) Über das Vorkommen von Spelz und Emmer (*Triticum spelta* L. und *T. dicoccum* Schebl.) im Iran. Z Pflanzenzüchtg 38:383–396

Kuraparthy V, Sood S, Dhaliwal HS, *et al.* (2007) Identification and mapping of a tiller inhibition gene *(tin3)*in wheat. Theor Appl Genet 114:285–294

Kushnir U, Halloran GM (1981) Evidence for *Aegilops sharonensis* Eig as the donor of the B genome of wheat. Genetics 99:495–512

Kuspira J, Maclagakn J, Kerby K, Bhambhani N (1986) Genetic and cytogenetic analyses of the A genome of *Triticum monococcum*. II. The mode of inheritance of spring versus winter growth habit. Can J Genet Cytol 28:88–89

Kuspira J, Maclagan J, Bhambhani RN, Sadsivaiah RS, Kim N-S (1989) Genetic and cytogenetic analyses of the A genome of *Triticum monococcum* L. V. Inheritance and linkage relationships of genes determining the expression of 12 qualitative characters. Genome 32:869–881

Kwiatek M, Majka M, Majka J et al (2016) Intraspecific polymorphisms of cytogenetic markers mapped on chromosomes of *Triticum polonicum* L. PLoS One 11:e0158883

Lacadena JR, Cermeño MC, Orellana J, Santos JL (1984) Evidence for wheat-rye nucleolar competition (amphiplasty) in triticale by silver-staining procedure. Theor Appl Genet 67:207–213

Lacko-Bartošová M, Korczyk-Szabó J, Ražny R (2010) Triticum spelta —a specialty grains for ecological farming systems. Res J Agric Sci 42:143–147

Ladizinsky G (1975) Collection of wild cereals in the upper Jordan Valley. Econ Bot 29:264–267

Lage J, Skovmand B, Andersen SB (2003) Characterization of greenbug (*Homoptera: Aphidiae*) resistance in synthetic hexaploid wheats. J Econ Entomol 90:1922–1928

Lage J, Skovmand B, Andersen SB (2004) Field evaluation of emmer wheat-derived synthetic hexaploid wheat for resistance to Russian wheat aphid (*Homoptera: Aphidiae*). J Econ Entomol 97:1065–1070

Lage J, Skovmand B, Peña RJ, Andersen SB (2006) Grain quality of emmer derived synthetic hexaploid wheats. Genet Resour Crop Evol 53:955–962

Lagudah ES, Apples R, Brown AHD, McNeil D (1991) The molecular-genetic analysis of *Triticum tauschii* the D genome donor to hexaploid wheat. Genome 34:375–386

Laido G, Mangini G, Taranto F, Gadaleta A, Blanco A, Cattivelli L, Marone D, Mastrangelo AM, Papa R, de Vita P (2013) Genetic diversity and population structure of tetraploid wheats (*Triticum turgidum* L.) estimated by SSR, DArT and pedigree data. Plos One 8

Lamarck JBM de (1786) Encyclopédie Methodique: Botanique. Chez Panckoucke, Paris

Lande R, Barrowclough GF (1987) Effective population size, genetic variation, and their use in population management. In: Soule ME (ed) Viable populations for conservation. Cambridge University Press, New York, pp 87–123

Lassner M, Anderson O, Dvorak J (1987) Hypervariation associated with 12-nucleotide direct repeat and inferences on intragenomic homogenization of ribosomal RNA gene spacer based on the DNA sequence of a clone from the wheat Nor-D3 locus. Genome 29:770–781

Law CN, Jenkins G (1970) A genetic study of cold resistance in wheat. Genet Res 15:197–208

Le HT, Reicosky DA, Olien CR, Cress CE (1986) Freezing hardiness of some accessions of *Triticum tauschii, Triticum turgidum* var. *durum*. Can J Plant Sci 66:893–899

Le Corre V, Bernard M (1995) Assessment of the type and degree of restriction fragment length polymorphism (RFLP) in diploid species of the genus *Triticum*. Theor Appl Genet 90:1063–1067

Lebedeva TV, Peusha HO (2006) Genetic control of the wheat *Triticum monococcum* L. resistance to powdery mildew. Russ Genet 42:60–66

Lebedeva TV, Rigin BV (1994) Inheritance of some morphological characteristics, growth habit, and powdery mildew resistance in einkorn *Triticum monococcum* L. Russ J Genet 30:1383–1387

Lebsock KL, Gough FJ, Sibbitt LD (1967) Registration of 'Leeds' durum wheat. Crop Sci 7:169–170

Lee J, Yen Y, Arumuganathan K, Baenziger PS (1997) DNA content of wheat monosomics at interphase estimated by flow cytometry. Theor Appl Genet 95:1300–1304

Lemma A, Woldeab G, Semahegn Y, Dilnesaw Z (2014) Survey and virulence distribution of wheat stem rust (*Puccinia graminis f. sp. tritici*) in the major wheat growing areas of central ethiopia. Sci-Afric J Sci Issues Res Essays 2:474–478

Lenton JR, Hedden P, Gale MD (1987) Gibberellin insensitivity and depletion in wheat—consequences for development. In: Hoad GV, Lenton JR, Jackson MB, Atkin RK (eds) Hormone action in plant development—a critical appraisal. Butterworths, London, pp 145–160

Leonova IN, Budashkina EB, Flath K, Weidner A, Börner A Röder MS (2010) Microsatellite mapping of a leaf rust resistance gene transferred to common wheat from *Triticum timopheevii*. Cereal Res Commun 38:211–219

Leonova IN, Budashkina EB, Kalinina NP, Röder MS, Börner A, Salina EA (2011) *Triticum aestivum* x *Triticum timopheevii* introgression lines as a source of pathogen resistance genes. Czech Genet Pl Breed 47:S49–S55

Levitsky GA, Sizova MA, Poddubnaja-Arnoldi VA (1939) Comparative morphology of the chromosomes of wheat. C R Acad Sci USSR 25:142–145

Levy AA, Feldman M (1987) Increase in grain protein percentage in high yielding common wheat breeding lines by genes from wild tetraploid wheat. Euphytica 36:353–359

Levy AA, Feldman M (1988) Ecogeographical distribution of HMW glutenin alleles in populations of the wild tetraploid wheat *Triticum turgidum* var. *dicoccoides*. Theor Appl Genet 75:651–658

Levy AA, Feldman M (1989) Location of genes for high grain protein percentage and other quantitative traits in wild wheat *Triticum turgidum* var. *dicoccoides*. Euphytica 41:113–122

Levy AA, Feldman M (2002a) The impact of polyploidy on grass genome evolution. Plant Physiol 130:1587–1593

Levy AA, Feldman M (2022b) Evolution and origin of bread wheat. Plant Cell 34:2549–2567

Levy AA, Feldman M (2004) Genetic and epigenetic reprogramming of the wheat genome upon allopolyploidization. Biol J Linn Soc 82:607–613

Levy AA, Galili G, Feldman M (1988) Polymorphism and genetic control of high molecular weight glutenin subunit in wild tetraploid wheat *Triticum turgidum* var. *dicoccoides*. Heredity 61:63–72

Li W, Gill B (2006) Multiple genetic pathways for seed shattering in the grasses. Funct Integr Genomics 6:300–309

Li DY, Zhang XY, Yang J, Rao GY (2000) Genetic relationship and genomic in situ hybridization analysis of the three genomes in *Triticum aestivum*. Acta Bot Sin. 2000;42:957–64.

Li W, Zhang D-F, Wei Y-M, Yan Z-H, Zheng Y-L (2006) Genetic diversity of *Triticum turgidum* L. based on microsatellite markers. Russ J Genet 42:311–316

Li W, Huang L, Gill BS (2008) Recurrent deletions of puroindoline genes at the grain hardness locus in four independent lineages of polyploid wheat. Plant Physiol 146:200–212

Li H, Gill BS, Wang X, Chen PD (2011) A Tal-*Ph1* wheat genetic stock facilitates efficient alien introgression. Genet Res Crop Evol 58:667–678

Li A-L, Geng S-F, Zhang L-Q et al (2015a) Making the bread: insights from newly synthesized allohexaploid wheat. Mol Plant 8:847–859

Li L-F, Liu B, Olsen KM, Wendel JF (2015b) A re-evaluation of the homoploid hybrid origin of *Aegilops tauschii*, the donor of the wheat D-subgenome. New Phytol 208:4–8

Li M, Dong L, Li B, Wang Z et al (2020) A CNL protein in wild emmer wheat confers powdery mildew resistance. New Phytol 228:1027–1037

Li G, Wang L, Yang J et al (2021) A high-quality genome assembly highlights rye genomic characteristics and agronomically important genes. Nat Genet 53:574–584

Li L-F, Zhang Z-B, Wang Z-H et al (2022) Genome sequences of the five *Sitopsis* species of *Aegilops* and the origin of polyploid wheat B subgenome. Mol Plant 15:488–503

Lilienfeld F, Kihara H (1934) Genomanalyse bei *Triticum* und *Aegilops*. V. *Triticum timopheevii* Zhuk. Cytologia 6:87–122

Lilienfeld F, Kihara H (1951) Genome analysis in *Triticum* and *Aegilops*. X. Concluding review. Cytologia 16:101–123

Limin AE, Fowler DB (1981) Cold hardiness of some relatives of hexaploid wheat. Can J Genet Cytol 59:572–573

Lin X, N'Diaye A, Sean Walkowiak S, Nilsen KT, Cory AT et al (2018) Genetic analysis of resistance to stripe rust in durum wheat (*Triticum turgidum* L. var. durum). PLoS One 13:e0203283

Linde-Laursen I, Jensen J (1991) Genome and chromosome disposition at somatic metaphase in a *Hordeum* x *Psathyrostachys* hybrid. Heredity 66:203–210

Ling H-Q, Zhao S, Liu D, Wang J, Sun H et al (2013) Draft genome of the wheat A-genome progenitor *Triticum urartu*. Nature 496:87–90

Ling H-Q, Ma B, Shi X et al (2018) Genome sequence of the progenitor of wheat A subgenome *Triticum urartu*. Nature 557:424–428

Linnaeus C (1753) Species plantarum, Tomus I, May 1753: i–xii, 1–560; Tomus II, Aug 1753: 561–1200, plus indexes and addenda, 1201–1231

Linnaeus C (1764) Species plantarum, 3rd ed, Vindobonae [Vienna]: typis Joannis Thomae de Trattner. https://bibdigital.rjb.csic.es/idurl/1/12666

Lisitsina GN (1984) The Caucasus—a centre of ancient farming Eurasia. In: van Zeist W, Casparie WA (eds) Plants and ancient man: studies in palaeoethnobotany. Balkema, Rotterdam, pp 285–292

Liu B, Vega JM, Segal G, Abbo S, Rodova H, Feldman M (1998a) Rapid genomic changes in newly synthesized amphiploids of *Triticum* and *Aegilops*. I. Changes in low-copy noncoding DNA sequences. Genome 41:272–277

Liu B, Vega JM, Feldman M (1998b) Rapid genomic changes in newly synthesized amphiploids of *Triticum* and *Aegilops*. II. Changes in low-copy coding DNA sequences. Genome 41:535–542

Liu CJ, Atkinson MD, Chinoy CN, Devos KM, Gale MD (1992) Non-homoeologous translocations between group 4, 5 and 7 chromosomes within wheat and rye. Theor Appl Genet 83:305–312

Longin CFH, Sieber A-N, Reif JC (2013) Combining frost tolerance, high grain yield and good pasta quality in durum wheat. Plant Breed 132:353–358

López-Merino L, Leroy SAG, Haldorsen S, Heun M, Reynolds A (2015) Can *Triticum urartu* (Poaceae) be identified by pollen analysis? Implications for detecting the ancestor of the extinct two-grained einkorn-like wheat. Bot J Linnean Soc 177:278–289

Löve Á (1984) Conspectus of the Triticeae. Feddes Repert 95:425–521

Lubbers EL, Gill KS, Cox TS, Gill BS (1991) Variation of molecular markers among geographically diverse accessions of *Triticum tauschii*. Genome 34:354–361

Ludvigsson JF, Leffler DA, Bai JC, Biagi F, Fasano A, Green PH, Hadjivassiliou M, Kaukinen K, Kelly CP, Leonard JN, Lundin KE, Murray JA, Sanders DS, Walker MM, Zingone F, Ciacci C (2013) The oslo definitions for coeliac disease and related terms. Gut 62:43–52

Luo M-C, Yang Z-L, You FM, Kawahara T, Waines JG, Dvorak J (2007) The structure of wild and domesticated emmer wheat populations, gene flow between them, and the site of emmer domestication. Theor Appl Genet 114:947–959

Luo G, Zhang X, Zhang Y et al (2015) Composition, variation, expression and evolution of low-molecular-weight glutenin subunit genes in *Triticum urartu*. BMC Plant Biol 15:68

Luo W, Qin N, Mu Y, Tang H et al (2018) Variation and diversity of the breakpoint sequences on 4AL for the 4AL/5AL translocation in *Triticum*. Genome 61:635–641

Ma X-F, Gustafson JP (2005) Genome evolution of allopolyploids: a process of cytological and genetic diploidization. Cytogenet Genome Res 109:236–249

Ma X-F, Gustafson JP (2006) Timing and rate of genome variation in triticale following allopolyploidization. Genome 49:950–958

Ma Z, Zhao D, Zhang C et al (2007) Molecular genetic analysis of five spike-related traits in wheat using RIL and immortalized F2 populations. Mol Genet Genom 277:31–42

Maan SS (1973) Cytoplasmic and cytogenetic relationships among tetraploid *Triticum* species. Euphytica 22:287–300

Maan SS (1975) Cytoplasmic variability and speciation in Triticineae. In: Wali MK (ed) Prairie, Univ North Dakota Press, Grand Forks, pp 255–281

Maan SS (1977) Cytoplasmic control of cross incompatibility and zygotic and sporophytic sterility in interspecific hybrids. In: Proceedings of 8th congress Eucarpia, Madrid, Spain, pp 201–214

Maan SS, Lucken KA (1967) Additional cytoplasmic male sterility-fertility restoration systems in *Triticum*. Wheat Inf Serv 23–24:6–9

Maan SS, Lücken KA (1968a) Cytoplasmic male sterility and fertility restoration in *Triticum*L. I. Effects of aneuploidy. II. Male sterility-fertility restoration systems. Proceedings of 3rd international wheat genetics symposium, Canberra, pp 135–140

Maan SS, Lucken KA (1968b) Male sterile durum: Interaction of Triticum boeoticum and *T. monococcum* cytoplasms with *T. durum* nucleus. Wheat Inform Serv 26:5

Maan SS, Lucken KA (1970) Interaction of *Triticum boeoticum* cytoplasm and the genomes of *T. aestivum* and *T. durum*: restoration of male fertility and plant vigor. Euphytica 19:498–508

Maan SS, Lucken KA (1971) Nucleo-cytoplasmic interactions involving *Aegilops* cytoplasms and *Triticum* genomes. J Hered 62:149–152

Mac Key J (1954a) Neutron and X-ray experiments in wheat and a revision of the speltoid problem. Hereditas 40:65–180

Mac Key J (1954b) The taxonomy of hexaploid wheat. Svensk Bot Tidskr 48:579–590

Mac Key J (1966) Species relationship in *Triticum*. In: Proceedings of 2nd international wheat genetics symposium, Lund, Sweden, Hereditas, Suppl 2, pp 237–276

Mac Key J (1968) Relationships in the Triticinae. In: Finlay KW, Shepherd KW (eds) Proceedings of 3rd international wheat genetics symposium, Canberra, Aust Acad Sci, Canberra, pp 39–50

Mac Key J (1975) The boundaries and subdivision of the genus *Triticum*. In: Proceedings of 12th international Botl congress, Leningrad: Abstract 2:509 (complete manuscript available at Vavilov Institute of Plant Industry, Leningrad, pp 1–23)

Mac Key J (1981) Comments on the basic principles of crop taxonomy. Kulturpflanze 29(S):199–207

Mac Key J (1988) A plant breeder's perspective on taxonomy of cultivated plants. Biol Zentralblatt 107:369–379

Mac Key J (1989) Seed dormancy in wild and weedy relatives of cereals. In: Derera NF (ed) Preharvest field sprouting in cereals. CRC Press, Boca Raton, pp 15–25

Mac Key J (2005) Wheat: its concept, evolution and taxonomy, vol 1. In: Royo C et al (eds) Durum wheat, current approaches, future strategies. CRC Press, Boca Raton, pp 3–61

Maccaferri M, Stefanelli S, Rotondo F, Tuberosa R, Anguineti MC (2007) Relationships among durum wheat accessions. I. Comparative analysis of SSR, AFLP, and phenotypic data. Genome 50:373–384

Maccaferri M, Mantovani P, Tuberosa R, DeAmbroguo E, Giuliani S, Demontis A, Massi A, Sanguineti MC (2008) A major QTL for durable leaf rust resistance widely exploited in durum wheat breeding programs maps on the distal region of chromosome arm 7BL. Theor Appl Genet 117:1225–1240

Maccaferri M, Francia R, Ratti C, Rubies-Autonell C, Colalongo, C., Ferrazzano G, Tuberosa R, Sanguineti MC (2012) Genetic analysis of *Soil-Borne Cereal Mosaic Virus* response in durum wheat: evidence for the role of the major quantitative trait locus *QSbm.ubo-2BS* and of minor quantitative trait loci. Mol Breed 29:973–988

Maccaferri M, Cane MA, Sanguineti MC et al (2014) A consensus framework map of durum wheat (*Triticum durum* Desf.) suitable for linkage disequilibrium analysis and genome-wide association mapping. BMC Genom 15:873

Maccaferri M, Harris NS, Twardziok SO, Pasam RK, Gundlach H et al (2019) Durum wheat genome highlights past domestication signatures and future improvement targets. Nature Genet 51:885–895

Maestra B, Naranjo T (1998) Homoeologous relationships of *Aegilops speltoides* chromosomes to bread wheat. Theor Appl Genet 97:181–186

Maestra B, Naranjo T (1999) Structural chromosome differentiation between *Triticum timopheevii* and *T. turgidum* and *T. aestivum*. Theor Appl Genet 98:744–750

Maestra B, Naranjo T (2000) Genome evolution uin Triticeae. Chromosomes Today 13:155–167

Malihipour A, Gilbert J, Fedak G, Brûlé-Babel A, Cao W (2017) Mapping the A genome for QTL conditioning resistance to Fusarium head blight in a wheat population with *Triticum timopheevii* background. Pl Disease 101:11–19

Mansur-Vergara L, Konzak CV, Gerechter-Amitai ZK, Grama A, Blum A (1984) A computer assisted examination of the storage protein genetic variation in 841 accessions of *Triticum dicoccoides*. Theor Appl Genet 69:79–86

Marcussen T, Sandve SR, Heier L, Spannagl M, Pfeifer M (2014) The international wheat genome sequencing consortium; Jakobsen KS, Wulff BBH, Steuernagel B, Klaus FX, Mayer KFX, Olsen OA (eds) Ancient hybridizations among the ancestral genomes of bread wheat. Science 345(6194):288–291

Marone D, Del Olmo AI, Laidò G, Sillero JC, Emeran AA, Russo MA, Ferragonio P, Giovanniello V, Mazzucotelli E, de Leonardis AM, de Vita P, Blanco A, Cattivelli L, Rubiales D, Mastrangelo AM (2009) Genetic analysis of durable resistance against leaf rust in durum wheat. Mol Breed 24:25–39

Marone D, Russo MA, Laidò G et al (2013) Genetic basis of qualitative and quantitative resistance to powdery mildew in wheat: from consensus regions to candidate genes. BMC Genomics 14:562

Martynov SP, Dobrotvorskaya TV, Krupnov VA (2018) Analysis of the distribution of *Triticum timopheevii* Zhuk. genetic material in common wheat varieties (*Triticum aestivum* L.). Russ J Genet 54:166–175

Matsumura S (1950) The effect of the long glumegene in Triticum polonicum. Jap J Genet 25:36

Matsumura S (1951) Genetics of some cereals. Ann Rep Nat Hist Genet Japan 1:22–27

Matsuoka Y, Nasuda S (2004) Durum wheat as a candidate for the unknown female progenitor of bread wheat: an empirical study with a highly fertile F_1 hybrid with *Aegilops tauschii* Coss. Theor Appl Genet 109:1710–1717

Matsuoka Y, Aghaei MJ, Abbasi MR, Totiaei A, Mozafari J, Ohta S (2008) Durum wheat cultivation associated with *Aegilops tauschii* in northern Iran. Genet Resour Crop Evol 55:861–868

Matsuoka Y, Nasuda S, Ashida Y, Nitta M, Tsujimoto H, Takumi S, Kawahara T (2013) Genetic basis for spontaneous hybrid genome doubling during allopolyploid speciation of common wheat shown by natural variation analyses of the paternal species. PLoS ONE 8: e68310

May CE, Appels R (1987) Variability and genetics of spacer DNA sequences between the ribosomal-RNA genes of hexaploid wheat (*Triticum aestivum*). Theor Appl Genet 74:617–624

Maystrenko OI, Laikova LI, Arbuzova VS, Melnik VM (1998) The chromosomal location of the *S1*, *S2* and *S3* genes of induced sphaerococcoid mutations in common wheat. In: EWAC newsletter, proceedings of 10th EWAC meeting, University of Tuscia, Italy, pp 127–130

McFadden ES, Sears ER (1944) The artificial synthesis of *Triticum spelta*. Records Genet Soc Amer 13:26–27

McFadden ES, Sears ER (1946) The origin of *Triticum spelta* and its free-threshing hexaploid relatives. J Hered 37:81–89, 107–116

McIntosh RA, Hart GE, Devos KM, Gale MD, Rogers WJ (1998) Catalogue of gene symbols for wheat, vol 5. In: Slinkard SE (ed) Proceedings of 9th Intern Wheat Genetics Symp, Saskatoon, Canada. University Extension Press, University of Saskatchewan, pp 1–236

Megyeri M, Farkas A, Varga M et al (2012) Karyotypic analysis of *Triticum monococcum* using standard repetitive DNA probes and simple sequence repeats. Acta Agronomica Hungarica 60:87–95

Melburn MC, Thompson WP (1927) The cytology of tetraploid wheat hybrid (*Triticum spelta x T. monococcum*). Am J Bot 14:327–333

Mello-Sampayo T (1968) Somatic association between telocentrics for both arms of the same chromosome in *Triticum aestivum*. Proc XIIth Intern Congr Genet 1:163

Mello-Sampayo T (1971) Promotion of homoeologous pairing in hybrids of *Triticum aestivum* x *Aegilops longissima*. Genet Iber 23:1–9

Mello-Sampayo T (1972) Compensated monosomic 5B-trisomic 5A plants in tetraploid wheat. Can J Genet Cytol 14:463–475

Mello-Sampayo T (1973) Somatic association of telocentric chromosomes carrying homologous centromeres in common wheat. Theor Appl Genet 43:174–181

Melnyk JH, McGinnis RC (1962) Analysis of chromosome pairing in interspecific and intergeneric F_1 hybrids involving hexaploid *Aegilops crassa*. Wheat Inf Serv 14:24–25

Menabde VL, (1948) Wheats of Georgia (in Russian), Institute of Botany, Academy of Sciences of Georgian SSR. Publishing House of Academy of Sciences of Georgian SSR

Michalikova A (1970) The influence of stigma extracts of individual varieties of wheat on the germination of chlamydospores of *Ustilago tritici* (Pers.) Jens. Polnohospodarstvo 16:19–23

Miczynski K Jr (1930) 0 dziedziczeniu sie niektorych cech u pszenioy w krzyzowkach *Triticum pyramidale* x *T. durum* i *T. vulgare* x *T. sphaerococcum*. Roczn Nauk rol 23:27–62

Middleton CP, Senerchia N, Stein N, Akhunov ED, Keller B, Wicker T, Kilian B (2014) Sequencing of chloroplast genomes from wheat, barley, rye and their relatives provides a detailed insight into the evolution of the Triticeae tribe. PLoS ONE 9:e85761

Miedaner T, Rapp M, Flath K et al (2019) Genetic architecture of yellow and stem rust resistance in a durum wheat diversity panel. Euphytica 215:71

Miller TE (1987) Systematics and evolution. In: Lupton FGH (ed) Wheat breeding. Its scientific basis. Chapman & Hall, London, pp 1–30

Miller NF (1991) The Near East. In: van Zeist W, Wasylikova K, Behre K-E (eds) Progress in old world palaeoethnobotany. Balkema, Rotterdam, pp 133–160

Miller TE, Reader SM (1980) Variation in the meiotic chromosome pairing of hybrids between hexaploid and diploid wheats. Cereal Res Com 8:477–483

Miller TE, Gerlach WL, Flavell RB (1980) Nucleolus organizer variation in wheat and rye revealed by *in situ* hybridisation. Chromosoma 45:377–382

Miller TE, Shepherd KW, Riley R (1981) The relationship of chromosome 4A of diploid wheat. Clarification of an earlier study. Cereal Res Commun 9:327–329

Miller TE, Hutchinson J, Reader SM (1983) The identification of the nucleolus organizer chromosomes of diploid wheat. Theor Appl Genet 65:145–147

Millet E, Pinthus MJ (1984) Effects of removing floral organs, light penetration and physical constraints on the development of wheat grains. Ann Bot 53:261–269

Millet E, Zaccai M, Feldman M (1992) Paternal and maternal effects on grain weight and protein percentage in crosses between hexaploid and tetraploid high-and low-protein wheat genotypes. Genome 35:257–260

Millet E, Rong JK, Qualset CO, McGuire PE, Bernard M, Sourdille P, Feldman M (2013) Production of chromosome-arm substitution lines of wild emmer in common wheat. Euphytica 190:1–17

Millet E, Rong JK, Qualset CO, McGuire PE, Bernard M, Sourdille P, Feldman M (2014) Grain yield and grain protein percentage of common wheat lines with wild emmer chromosome-arm substitutions. Euphytica 195:69–81

Miyashita NT, Mori N, Tsunewaki K (1994) Molecular variation in chloroplast DNA regions in ancestral species of wheat. Genetics 137:883–889

Mizumoto K, Hirosawa S, Shigeo C, Takumi S (2002) Nuclear and chloroplast genome genetic diversity in the wild einkorn wheat, *Triticum urartu*, revealed by AFLP and SSLP analyses. Hereditas 137:208–214

Mizuno N, Yamasaki M, Matsuoka Y, Kawahara T, Takumi S (2010) Population structure of wild wheat D-genome progenitor *Aegilops tauschii* Coss.: implications for intraspecific lineage diversification and evolution of common wheat. Mol Ecol 19:999–1013

Mochizuki A, Okamoto M (1961) The role of *Aegilops speltoiifes* genome in chromosome pairing. Chromosome Inf Serv 2:12–11

Moghaddam M, Ehdaie B, Waines JG (2000) Genetic diversity in populations of wild diploid wheat *Triticum urartu* Tum. ex. Gandil. revealed by isozyme markers. Genet Resour Crop Evol 47:323

Mohan J, Flavell RB (1974) Ribosomal RNA cistron multiplicity and nucleolar organizers in hexaploid wheat. Genetics 76:33–44

Mohan BH, Malleshi NG (2006) Characteristics of native and enzymatically hydrolyzed common wheat (*Triticum aestivum*) and dicoccum wheat (*Triticum dicoccum*) starches. Eur Food Res Technol 223:355–361

Molnár I, Kubaláková M, Šimková H et al (2014) Flow cytometric chromosome sorting from diploid progenitors of bread wheat, *T. urartu*, *Ae. speltoides* and *Ae. tauschii*. Theor Appl Genet 127: 1091–1104

Mori M, Liu Y-G, Tsunewaki K (1995) Wheat phylogeny determined by RFLP analysis of nuclear DNA. 2. Wild tetraploid wheats. Theor Appl Genet 90:129–134

Mori N, Miyashita NT, Terachi T, Nakamura C (1997) Variation in *coxII* intron in the wild

Mori N, Ishi T, Ishido T, Hirosawa S, Watatani H et al (2003) Origins of domesticated emmer and common wheat inferred from chloroplast DNA fingerprinting. In: Pogna NE, M Romano M, Pogna EA, Galterio G (eds) Proceedings of 10th international eheat genetics symposium, Paestum, Italy. Rome: Instituto Sperimentale per la Cerealicoltura, pp 25–28

Mori N, Kondo Y, Ishii T, Kawahara T, Valkoun J, Nakamura C (2009) Genetic diversity and origin of timopheevi wheat inferred by chloroplast DNA fingerprinting. Breed Sci 59:571–578

Mori N, Ohta S, Chiba H, Takagi T, Niimi Y, Shinde V, Kajale MD, Osafjda T (2013) Rediscovery of Indian dwarf wheat (*Triticum aestivum* L. ssp. *sphaerococcum* (Perc.) MK an ancient crop of the Indian subcontinent. Genet Resourc Crop Evol 60:1771–1775

Morris R, Sears ER (1967) The cytogenetics of wheat and its relatives. In: Quisenberry KS, Reitz LP (eds) Wheat and wheat improvement. American Society of Agronomy, Madison, Wisconsin, USA, pp 19–87

Morris CF, Simeone MC, King GE, Lafiandra D (2011) Transfer of soft kernel texture from Triticum aestivum to durum wheat, *Triticum turgidum* ssp. Durum. Crop Sci 51:114–122

Morrison JW (1953) Chromosome behaviour in wheat monosomics. Heredity 7:203–217

Moseman JG, Nevo E, El-Morshidy MA, Zohary D (1984) Resistance of *Triticum dicoccoides* to infection with *Erysiphe graminis tritici*. Euphytica 33:41–47

Moseman JG, Nevo E, Gerechter-Amitai ZK, El-Morshidy MA, Zohary D (1985) Resistance of *Triticum dicoccoides* collected in Israel to infection with *Puccinia recondita* tritici. Crop Sci 25:262–265

Mosolov AN (1974) Evolutional conception of nucleus and chromosome structure. Nucleus 27:51–64

Mosulishvili M, Bedoshvili D, Maisaia I (2017) A consolidated list of *Triticum* species and varieties of Georgia to promote repatriation of local diversity from foreign gene banks. Ann Agrar Sci 15:61–70

Mujeeb-Kazi A, Rosas V, Roldan S (1996) Conservation of the genetic variation of *Triticum tauschii* (Coss.) Schmalh. (*Aegilops squarrosa* auct. non L.) in synthetic hexaploid wheats (*T. turgidum* L. x *T. tauschii*; $2n = 6x =42$, AABBDD) and its potential utilization for wheat improvement. Genet Resourc Crop Evol 43:129–134

Mukai Y, Endo TR, Gill BS (1990) Physical mapping of the 5S rRNA multigene family in common wheat. J Hered 81:290–295

Mukai Y, Endo TR, Gill BS (1991) Physical mapping of the 18S-26S rRNA multigene family in common wheat: identification of a new *locus*. Chromosoma 100:71–78

Muramatsu M (1963) Dosage effect of the spelta gene *q* of hexaploid wheat. Genetics 48:469–482

Muramatsu M (1978) Phenotypic expression of the *vulgare* gene *Q* in tetraploid wheat, vol 1. In: Ramanujan S (ed) Proceedings of 5th international wheat genetics symposium, New Delhi, pp 92–102

Muramatsu M (1979) Presence of the *vulgare* gene, *Q*, in a dense-spike variety of *Triticum dicoccum* Schubl. Report of the Plant Germ-Plasm Institute, Kyoto University, No. 4, pp 39–41

Muramatsu M (1985) Spike type in two cultivars of *Triticum dicoccum* with the *spelta* gene *q* compared with the *Q*-bearing variety *liguliforme*. Jpn J Breed 35:255–267

Muramatsu M (1986) The *vulgare* super gene, *Q*: its universality in *durum* wheat and its phenotypic effects in tetraploid and hexaploid wheats. Can J Genet Cytol 28:30–41

Nabalek F (1929) Inter turco-persicum pars. V. Publ Fac Sci Masaryk Univ Brno III:27–28

Nakai Y (1978a) Genetical studies on the variations of esterase isozymes in *Aegilops* and *Triticum*. Dr. thesis, Kyoto University, Kyoto

Nakai Y (1978b) The origin of the tetraploid wheats revealed from the study on esterase isozymes. In: Ramanujan S (ed) Proceedings of 5th international wheat genetics symposium, New Delhi, 108–119

Nakai Y (1979) Isosyme variation in *Aegilops* and *Triticum*. IV. The origin of the common wheats revealed from the study on esterase isozymes in synthesized wheats. Jpn J Genet 54:175–189

Nakajima G (1955) A cytogenetical study on the intergeneric F₁ hybrid between *Triticum polonicum and Secale africanum*. Cytologia 20:273–279

Nalam VJ, Vales MI, Watson CJW, Kianian SF, Riera-Lizarazu O (2006) Map-based analysis of genes affecting the brittle rachis character in tetraploid wheat (*Triticum turgidum* L.). Theor Appl Gen 112:373–381

Nalam VJ, Vales MI, Watson CJW, Johnson EB, Riera-Lizarazu O (2007) Map-based analysis of genetic loci on chromosome 2D that affect glume tenacity and threshability, components of the free-threshing habit in common wheat (*Triticum aestivum* L.). Theor Appl Genet 116:135–145

Naranjo T (1990) Chromosome structure of durum wheat. Theor Appl Genet 79:397–400

Naranjo T, Roca A, Goicoecha PG, Giraldez R (1987) Arm homoeology of wheat and rye chromosomes. Genome 29:873–882

Nasernakhaei F, Rahiminejad MR, Saeidi H, Tavassoli M (2015) Genetic structure and diversity of *Triticum monococcum* ssp. *aegilopoides* and *T. urartu* in Iran. Plant Genet Resour Charact Util 13:1–8

Natarajan AT, Sharma NP (1974) Chromosome banding patterns and the origin of the B genome in wheat. Genet Res 21:103–108

Nath J, McNAy JW, Paroda CM, Gulati SC (1983) Implication of *Triticum searsii* as the B-genome donor to wheat using DNA hybridizations. Biochem Genet 21:745–760

Nath J, Hanzel JJ, Thompson JP, McNay JW (1984) Additional evidence implicating *Triticum searsii* as the B-genome donor to wheat. Biochem Genet 22:38–49

Nath J, Thompson JP, Gulati SC (1985) Identification of the G-genome donor to *Triticum timopheevii* by DNA:DNA hybridizations. Bioch Genet 23:125–137

Navashin MS (1928) Amphiplastie — eine neue karyologische Erscheinung. Proc Int Conf Genet 5:1148–1152

Navashin MS (1931) Chromatin mass and cell volume in related species. Univ Calif Publ Agri Sci 6:207–230

Nave M, Tas M, Raupp J, Tiwari VK et al (2021) The independent domestication of timopheev's wheat: insights from haplotype analysis of the *Brittle rachis 1* (*BTR1-A*) Gene Genes 12:338

Nazco R, Peña RJ, Ammar K, Villegas D (2014a) Variability in glutenin subunit composition of Mediterranean durum wheat germplasm and its relationship with gluten strength. J Agric Sci 152:379–393

Nazco R, Peña RJ, Ammar K et al (2014b) Durum wheat (*Triticum durum* Desf.). Mediterranean landraces as sources of variability for allelic combinations at *Glu-1/Glu-3* loci affecting gluten strength and pasta cooking quality. Genet Resour Crop Evol 61:1219–1236

Nelson LR, Gates CE (1982) Genetics of host plant resistance of wheat to *Septoria nodorum*. Crop Sci 22:771–773

Nesbitt M (1993) Ancient crop husbandry at Kaman-Kalehöyük: 1991 archaeological report. In: Mikasa T (ed) Essays on anatolian archaeology, bulletin of the middle eastern culture center in Japan 7. Harrassowitz, Wiesbaden, pp 75–97

Nesbitt M (1995) Plants and peoplein ancient Anatolia. Biblic Archaeol 58:68–81

Nesbitt M (2001) Wheat evolution: integrating archaeological and biological evidence. In: Caligari PDS, Brandham PE (eds) Wheat taxonomy: the legacy of john Percival. The Linn Soc London, Burlington House, London, pp 37–60

Nesbitt M (2002) When and where did domesticated cereals first occur in southwest Asia? In: Cappers RTJ, Bottema S (eds) The dawn of farming in the Near East. Studies in Near Eastern production, subsistence and environment, 6, 1999. Ex Oriente, Berlin, pp 113–132

Nesbitt M, Samuel D (1996) From staple crop to extinction? The archaeology and history of hulled wheat. In: Padulosi S, Hammer K, Heller J (eds) First international workshop on hulled wheats. Castelvecchio Pascoli, Tuscany, Italy: International Plant Genetic Resources Institute, Rome, Italy, 41–100

Neves N, Heslop-Harrison JS, Viegas W (1995) rRNA gene activity and control of expression mediated by methylation and imprinting during embryo development in wheat x rye hybrids. Theor Appl Genet 91:529–533

Nevo E (2001) Genetic resources of wild emmer, *Triticum dicoccoides*, for wheat improvement in the third millennium. Isr J Plant Sci 49: S77–S91

Nevo E, Beiles A (1989) Genetic diversity of wild emmer wheat in Israel and Turkey: structure, evolution, and application in breeding. Theor Appl Genet 77:421–455

Nevo E, Chen G (2010) Drought and salt tolerances in wild relatives for wheat and barley improvement. Plant Cell Environ 33:670–685

Nevo E, Golenberg E, Beiles A, Brown AHD, Zohary D (1982) Genetic diversity and environmental associations of wild wheat, *Triticum dicoccoides*, in Israel. Theor Appl Genet 62:241–254

Nevo E, Beiles A, Gutterman Y, Starch N, Kaplan D (1984) Genetic resources of wild cereals in Israel and vicinity. I. Phenotypic variation within and between populations of wild wheat, *Triticum dicoccoides*. Euphytica 33:717–735

Nevo E, Grama A, Beiles A, Golenberg EM (1986) Resources of high protein genotypes in wild wheat *Triticum dicoccoides*, in Israel: predictive method by ecology and allozyme markers. Genetica 68:215–227

Nevo E, Gerechter-Amitai Z, Beiles A (1991) Resistance of wild emmer wheat to stem rust: ecological, pathological and allozyme associations. Euphytica 53:121–130

Nevski SA (1934) Hordeae Bentb. In: Komarov VL (ed) Flora of the USSR II, 590

Nielsen F, Mortensen JV (1998) Ecological cultivation of emmer and giant durum. 4:194–196

Nilsson-Ehle H (1911) Kreuzungsuntersuchungen an Hafer und Weizen. II. Lunds UNiv Arsskr NF Afd 2, Bd 7, N r 6, pp 84

Nilsson-Ehle H (1917) Untersuchungen über Speltoidmutationen bein Weizen. Botaniska Notise 1917:305–329

Nirmala J, Saini J, Newcomb M, Olivera P, Gale S, Klindworth D, Elias E, Talbert L, Chao S, Faris J, Xu S, Jin Y, Rouse MN (2017) Discovery of a novel stem rust resistance allele in durum wheat that exhibits differential reactions to Ug99 isolates. G3 Genes Genom Genet 7:3481–3490

Nishikawa K (1973) Alpha-amylase isozymes and phylogeny of hexaploid wheal. In: Sears ER, Sears LMS (eds) Proceedings of 4th international wheat genetics symposium Columbia MO, pp 851–855

Nishikawa K (1983) Species relationship of wheat and its putative ancestors as viewed from isozyme variation. In: Sakamoto S (ed) Proceedings of 6th int wheat genetics symposium. Maruzen, Kyoto, pp 59–63

Nishikawa K, Furuta Y (1968) Comparison of DNA content per nucleus between Japanese and U.S. varieties of common wheat. Jpn J Genet 18:94–95

Nishikawa K, Furuta Y (1969) DNA content per nucleus in relation to phylogeny of wheat and its relatives. Jpn J Genet 44:23–29

Nishikawa. K, Furuta Y (1978) DNA content and individual chromosomes and its evolutionary significance, vol 1. In: Ramanujam S (ed) Proceedings of 5th international wheat genetics symposium, New Delhi, Kapoor Art Press, New Delhi, pp 133–138

Nishikawa K, Furuta Y, Kudo S, Ujihara K (1979) Differentiation of tetraploid wheat in relation to DNA content of nucleus and alpha-amylase isozymes. Rep Plant Germ Plasm Inst Kyoto Univ 4:30–38

Nishikawa K, Furuta Y, Wada T (1980) Genetic studies on α-amylase isozymes in wheat. III. Intraspecific variation in *Aegilops squarrosa* and birthplace of hexaploid wheat. Jpn J Genet 55:325–336

Nishikawa K, Furuta Y, Yamada T, Kudo S (1992) Genetic studies of α-amylase isozymes in wheat VII. Variation in diploid ancestral species and phylogeny of tetraploid wheat. Genes Genet Syst 67:1–15

Nishikawa K, Mizuno S, Furuta Y (1994) Identification of chromosomes involved in translocations in wild Emmer. Jpn J Genet 69:371–376

Nouri A, Alireza E, da Silva T, Jaime A, Reza M (2011) Assessment of yield, yield-related traits and drought tolerance of durum wheat genotypes (*Triticum turjidum* var. *durum* Desf.). Australian J Crop Sci 5:8–16

Ogihara Y, Tsunewaki K (1982) Molecular basis of the genetic diversity of the cytoplasm in *Triticum* and *Aegilops*. I. Diversity of the chloroplast genome and its lineage revealed by the estriction pattern of chloroplast DNAs. Jpn J Genet 57:371–396

Ogihara Y, Tsunewaki K (1988) Diversity and evolution of chloroplast DNA in *Triticum* and *Aegilops* as revealed by restriction fragment analysis. Theor Appl Genet 76:321–332

Oinuma T (1953) Karyomorphology of cereals. IX. Karyotype alteration in *Aegilops* and *Triticum* and relationship between karyotype and genome. Jpn J Genet 28:219–226

Okamoto M (1957a) Asynaptic effect of chromosome V. Wheat Inf Serv 5:6

Okamoto M (1957b) Identification of the chromosomes of the A and B genome. Wheat Inf Serv 5:7

Okamoto M (1962) Identification of the chromosomes of common wheat belonging to the A and B genomes. Gan J Genet Cytol 4:31–37

Okamoto M, Inomata N (1974) Possibility of 5B-like effect in diploid species. Wheat Inf Serv 38:15–16

Oliveira HR, Campana MG, Jones H, Hunt HV, Leigh F, Redhouse DI, Lister DL, Jones MK (2012) Tetraploid wheat landraces in the Mediterranean basin: taxonomy, evolution and genetic diversity. PLoS ONE 7:e37063

Olivier (1807) Voyage dans l'Empire Ottoman, vol 3, p 46

Oliver RE, Cai X, Friesen TL, Halley S, Stack RW, Xu SS (2008) Evaluation of Fusarium head blight resistance in tetraploid wheat (*Triticum turgidum* L.). Crop Sci 48:213–222

Olsen CC (1998) Old cereal species, growing emmer and durum wheat without pesticides. Gron Viden. Markburg 4:196–199

Orth RA, Bushuk W (1973) Studies of glutenin: III. Identification of subunits coded by the D-genome and their relation to breadmaking quality. Cereal Chem 50:680–687

Ozbek O, Millet E, Anikster Y, Arslan O, Feldman M (2007a) Spatio-temporal genetic variation in populations of wild emmer wheat, *Triticum turgidum* ssp. *dicoccoides*, as revealed by AFLP analysis. Theor Appl Genet 115:19–26

Ozbek O, Millet E, Anikster Y, Arslan O, Feldman M (2007b) Comparison of the genetic structure of populations of wild emmer wheat, *Triticum turgidum* ssp. *dicoccoides*, from Israel and Turkey revealed by AFLP analysis. Genet Resour Crop Evol 54:1587–1598

Ozkan H, Feldman M (2001) Genotypic variation in tetraploid wheat affecting homoeologous pairing in hybrids with *Aegilops peregrine*. Genome 44:1000–1006

Ozkan H, Levy AA, Feldman M (2001) Allopolyploidy—induced rapid genome evolution in the wheat (*Aegilops-Triticum*) group. Plant Cell 13:1735–1747

Ozkan H, Brandolini A, Schäfer-Pregl R, Salamini F (2002) AFLP analysis of a collection of tetraploid wheats indicates the origin of emmer and hard wheat domestication in southeast Turkey. Mol Biol Evol 19:1797–1801

Ozkan H, Brandolini A, Pozzi C, Effgen S, Wunder J, Salamini F (2005) A reconsideration of the domestication geography of tetraploid wheats. Theor Appl Genet 110:1052–1060

Pai RA, Upadhya MD, Bhaskaran S, Swaminathan MS (1961) Chromosome diminution and evolution of polyploid species in *Triticum*. Chromosoma 12:398–409

Pan J, Zhu Y, Cao W (2007) Modeling plant carbon flow and grain starch accumulation in wheat. Field Crops Res 101:276–284

Pathak GN (1940) Studies in the cytology of cereals. J Genet 39:437–467

Paull JG, Pallotta MA, Langridge P et al (1994) RFLP markers associated with Sr22 and recombination between chromosome 7A of bread wheat and the diploid species *Triticum boeoticum*. Theoret Appl Genet 89:1039–1045

Peacock WJ, Gerlach WL, Dennis ES (1981) Molecular aspects of wheat evolution: repeated DNA sequences. In: Evans LT, Peacock WJ (eds) Wheat science-today and tomorrow. Cambridge University Press, Cambridge, pp 41–60

Pedersen C, Langridge P (1997) Identification of the entire chromosome complement of bread wheat by two-colour FISH. Genome 40:589–593

Pegington C, Rees H (1970) Chromosome weights and measures in the triticinae. Heredity 25:195–205

Peleg Z, Fahima T, Abbo S, Krugman T, Nevo E, Yakir D, Saranga Y (2005) Genetic diversity for drought resistance in wild emmer wheat and its ecogeographical associations. Plant Cell Environ 28:176–191

Peleg Z, Saranga Y, Yazici A, Fahima T, Ozturk L, Ismail Cakmak I (2008) Grain zinc, iron and protein concentrations and zinc-efficiency in wild emmer wheat under contrasting irrigation regimes. Plant Soil 306:57–67

Peleg Z, Fahima T, Krugman T, Abbo S, Yakir D, Korol AB, Saranga Y (2009) Genomic dissection of drought resistance in durum wheat x wild emmer wheat recombinant inbreed line population. Plant Cell Environ 32:758–779

Peneva TJ, Konarev VG (1982) The identification of *Ae. searsii* Feld. et Kislev genome on protein markers. Bull Appl Bot Gen Pl Breed 73:132–133

Peng J, Ronin Y, Fahima T, Röder MS, Li Y, Nevo E, Korol A (2003a) Genomic distribution of domestication QTLs in wild emmer wheat, *Triticum dicoccoides*. In: Pogna NE, Mcintosh RA (eds) Proceedings of 10th international wheat genetics symposium. Paestum, Italy, pp 34–37

Peng J, Ronin Y, Fahima T, Röder MS, Nevo E, Korol A (2003b) Domestication quantitative trait loci in *Triticum dicoccoides*, the progenitor of wheat. Proc Natl Acad Sci USA 100:2489–2494

Percival J (1921) The wheat plant: a monograph. Duckworth, London, pp 1–46

Percival J (1923) Chromosome numbers in *Aegilops*. Nature 111:810

Percival J (1930) Cytological studies of some hybrids of *Aegilops* sp. x wheats, and of some hybrids between different species of *Aegilops*. J Genet 22:201–278

Percival J (1932) Cytological studies of some wheat and *Aegilops* hybrids. Ann Bot 46:479–501

Perrino P, Infantino S, Basso P, Di Marzio A, Volpe N, Laghetti G (1993) Valutazione e selezione di farro in ambienti marginali dell'Appennino molisano (II nota). Informatore Agrario 43:41–44 (in Italian)

Perrino P, Laghetti G, D'Antuono LF, Al Ajlouni M, Kanbertay M, Szabò AT, Hammer K (1996) Ecogeographical distribution of hulled wheat species. In: Padulosi S, Hammer K and Heller J (eds) Hulled wheats. Promoting the conservation and use of underutilized and neglected crops. 4. Proceedings of the "first international workshop on hulled wheats, 21–22 July 1995,

Castelvecchio Pascoli, Tuscany, Italy, IPGRI, Rome, Italy, pp 101–119

Petersen G, Seberg O, Yde M, Berthelsen K (2006) Phylogenetic relationships of *Triticum* and *Aegilops* and evidence for the origin of the A, B, and D genomes of common wheat (*Triticum aestivum*). Mol Phylogenet Evol 39:70–82

Peto FH (1936) Hybridization of *Triticum* and *Agropyron*. II. Cytology of the male parents and F_1 generation. Can J Resh Sect C 14:203–214

Philiptschenko J (1934) Genetics of soft wheats. State Press Literature, State and College Farms, Moscow, Leningrad (in Russian)

Piergiovanni AR, Laghetti G, Perrino P (1996) Characteristics of meal from hulled wheats (*Triticum dicoccum* Schrank and *T. spelta* L.): an evaluation of selected accessions. Cereal Chem 73:732–735

Piergiovanni AR, Simeone R, Pasqualone A (2009) Composition of whole and refined meals of Kamut under southern Italian conditions. Chem Engin Trans 17:891–896

Pikaard CS (2001) Genomic change and gene silencing in polyploids. Trends Genet 17:675–677

Poltoretskyi S, Hospodarenko H, Liubych V, Poltoretska N, Demydas H (2018) Toward the theory of origin and distribution history of *Triticum spelta* L. Ukrainian J Ecol 8:263–268

Pont C, Leroy T, Seidel M, Tondelli A, Duchemin W et al (2019) Tracing the ancestry of modern bread wheats. Nat Genet 51:905–911

Popova C (1991) Palaeoethnobotanic studies of the Neolithic and Eneolithic epoch in the territory of Bulgaria. In: Hajnalová E (ed) Acta Interdisciplinaria Archaeologica 7. Archaeological Institute of the Slovak Acad 7: 167–178 emy of Sciences, Nitra, pp 269–272

Porceddu E (1995) Durum wheat quality in the Mediterranean countries. In: Di Fonzo N, Kaan F, Nachit M (eds) Durum wheat quality in the Mediterranean region. Zaragoza, CIHEAM, pp 11–21

Porceddu E, Lafiandra D (1986) Origin and evolution of wheats. In: Barigozzi C (ed) The origin and domestication of cultivated plants. Elsevier, Amsterdam, pp 143–178

Potgieter GF, Marais GF, Du Toit F (1991) The transfer of resistance to the Russian wheat aphid from *Triticum monococcum* L. to common wheat. Pl Breed 106:284–292

Pourkheirandish M, Dai F, Sakuma S, Kanamori H, Distelfeld A et al (2018) On the origin of the non-brittle rachis trait of domesticated einkorn wheat. Front Plant Sci 8:2031

Poyarkova H (1988) Morphology, geography and infraspecific taxonomics of *Triticum dicoccoides* Körn. A detailed retrospective of 80 years of research. Euphytica 38:11–23

Poyarkova H, Gerechter-Amitai ZK, Genizi A (1991) Two variants of wild emmer (*Triticum dicoccoides*) native to Israel: morphology and distribution. Can J Bot 69:2772–2789

Prabhakara Rao MVP (1977) Mapping of the sphaerococcum gene '*S*' on chromosome 3D of wheat. Cereal Res Commun: 15–17

Provan J, Wolters P, Caldwell KH, Powell W (2004) High-resolution organellar genome analysis of *Triticum* and *Aegilops* sheds new light on cytoplasm evolution in wheat. Theor Appl Genet 108:1182–1190

Qiu YC, Zhou RH, Kong XY et al (2005) Microsatellite mapping of a *Triticum urartu* Tum. derived powdery mildew resistance gene transferred to common wheat (*Triticum aestivum* L.). Theor Appl Genet 111:1524–1531

Quinn RM (1999) Kamut®: ancient grain, new cereal. In: Janick J (ed) Perspectives on new crops and new uses. ASHS Press, Alexandria, VA, pp 182–183

Radchenko EE (2011) Resistance of *Triticum* Species to cereal aphids. Czech J Genet Pl Breed 47:s67–s70

Ranhotra GS, Gelroth JA, Glaser BK, Lorenz KJ (1995) Baking and nutritional qualities of a spelt wheat sample. Lebensm Wiss Technol 28:118–122

Rao MVP (1972) Mapping of the compactum gene C on chromosome 2D of wheat. Wheat Inf Serv 35:9

Rao MVP (1977) Mapping of the sphaerococcum gene 's' on chromosome 3D of wheat. Cereal Res Commun 5:15–17

Rao PS, Smith EL (1968) Studies with Israeli and Turkish accessions of *Triticum turgidium* L. emend. var. *dicoccoides* (Korn.) Bowden. Wheat Inf Serv 26:6–7

Ratcliffe RH, Patterson FL, Cambron SE, Ohm HW (2002) Resistance in durum wheat: Sources to hessian fly (*Diptera: Cecidomyiidae*) populations in eastern USA. Crop Sci 42:1350–1356

Rawal KM (1971) Esterase polymorphism in wild *Triticum turgidurn* var. *dicoccoides* (Kern.) Bowden. W Afr J Biol Appl Chem 14:12–15

Rawal K, Harlan JR (1975) Cytogenetic analysis of wild emmer populations from Turkey and Israel. Euphytica 24:407–411

Rayburn AL, Gill BS (1985) Molecular evidence for the origin and evolution of chromosome 4A in polyploidy wheats. Can J Genet Cytol 27:246–250

Rees H (1963) Deoxyribonucleic acid and the ancestry of wheat. Nature 198:108–109

Rees H, Walters MR (1965) Nuclear DNA and the evolution of wheat. Heredity 20:73–82

Reinhold M, Sharp EL, Gerechter-Amitai Z (1983) Transfer of additive "minor effect" genes for resistance to *Puccinia striiformis* from *Triticum dicoccoides* into *Triticum durum* and *Triticum aestivum*. Can J Bot 61:2702–2708

Renfrew JM (1973) Palaeoethnobotany: the prehistoric food plants of the near East and Europe. Columbia University Press, New York

Riley R (1960) The diploidization of polyploid wheat. Heredity 15:407–429

Riley R (1965) Cytogenetics and evolution of wheat. In: Hutchinson J (ed) Essays in crop plant evolution. Cambridge University Press, London, pp 103–122

Riley R (1966) The genetic regulation of meiotic behavior in wheat and its relatives. In: Mac Key J (ed) Proceedings of second international wheat genetics symposium, Genetic Institute, University of Lund, Sweden, August 1963, Hereditas (suppl) 2, pp 395–406

Riley R, Chapman V (1958) Genetic control of the cytologically diploid behaviour of hexaploid wheat. Nature 182:713–715

Riley R, Unrau J, Chapman V (1958) Evidence on the origin of the B genome of wheat. J Hered 49:91–98

Riley R, Kimber G, Chapman V (1961) Origin of genetic control of diploid-like behavior of polyploid wheat. Hered 52:22–25

Riley R, Chapman V, Young RM, Belfield AM (1966) Control of meiotic chromosome pairing by the chromosomes of homoeologous group 5 of *Triticum aestivum*. Nature 212:1475–1477

Robinson J, Skovmand B (1992) Evaluation of emmer wheat and other Triticeae for resistance to Russian wheat aphid. Gen Res Crop Evol 39:159–163

Röder MS, Sorrells ME, Tanksley SD (1992) 5S ribosomal gene clusters in wheat: pulsed field gel electrophoresis reveals a high degree of polymorphism. Mol Gen Genet 232:215–220

Rodríguez S, Maestra B, Perera E, Díez M, Naranjo T (2000a) Pairing affinities of the B- and G-genome chromosomes of polyploid wheats with those of *Aegilops speltoides*. Genome 43:814–819

Rodríguez S, Perera E, Maestra B, Díez M, Naranjo T (2000b) Chromosome structure of *Triticum timopheevii* relative to *T. turgidum*. Genome 43:923–930

Rodríguez-Quijano M, Vargas-Kostiuk ME, Ribeiro M et al (2019) Triticum aestivum ssp. vulgare and ssp. spelta cultivars. 1. Functional evaluation. Eur Food Res Technol 245:1561–1570

Rondon MR, Gough FJ, Williams ND (1966) Inheritance of stem rust resistance in *Triticum aestivum* ssp. *vulgare* 'Reliance' and PI 94701 of *Triticum durum*. Crop Sci 6:177–179

Rong JK, Millet E, Feldman M (1999) Unequal RFLP among genomes, homoeologous groups and chromosome regions in wheat. In: Proceedings of 9th ITMI workshop, August 25–27, Viterbo, Italy

Rong JK, Millet E, Manisterski J, Feldman M (2000) A new powdery mildew resistance gene: introgression from wild emmer into common wheat and RFLP-based mapping. Euphytica 115:121–126

Rouse MN, Jin Y (2011) Stem rust resistance in A-genome diploid relatives of wheat. Plant Dis 95:941–944

Royo A, Abió D (2003) Salt tolerance in durum wheat cultivars. Spanish J Agric Res 1:27–35

Ruban AS, Badaeva ED (2018) Evolution of the S-genomes in *Triticum-Aegilops* alliance: evidences from chromosome analysis. Front Plant Sci 9:1756

Ruibal-Mendieta NL, Delacroix DL, Eric Mignolet E, Pycke JM, Marques C, Rozenberg R, Petitjean G, Habib-Jiwan J-L, Meurens M, Quetin-Leclercq J, Delzenne NM, Larondelle Y (2005) Spelt (*Triticum aestivum* ssp. *spelta*) as a source of bread-making flours and bran naturally enriched in oleic acid and minerals but not phytic acid. J Agric Food Chem 53:2751–2759

Sachs L (1953) Chromosome behavior in species hybrids with *Triticum timopheevii*. Heredity 7:49–58

Saito H, Ishida N (1979) Speciation of wild tetraploid wheats concerning susceptibility to leaf rust. Rep. Plant Germ-plasm Inst., Kyoto Univ. No. 4, pp 18–22

Sakamura T (1918) Kurze mitteilung über die chromosomenzahlen und die verwandtschaftsverhältnisse der Triticum Arten. Bot Mag 32:151–154

Saleem M (2003) Response of durum and bread wheat genotypes to drought stress: biomass and tield components. Asian J Pl Sci 2:290–293

Salina E, Börner A, Leonova I, Korzun V, Laikova L, Maystrenko O, Röder MS (2000) Microsatellite mapping of the induced sphaerococcoid mutation genes in *Triticum aestivum*. Theor Appl Genet 100:686–689

Salina EA, Numerova OM, Ozkan H, Feldman M (2004) Alterations in sub-telomeric tandem repeats during early stages of allopolyploidy in wheat. Genome 47:860–867

Sall AT, Cisse M, Gueye H, Kabbaj H, Ndoye I, Filati-Maltouf A, Belkadi B, El-Mourid M, Ortiz R, Bassi F (2018) Heat tolerance of durum wheat (*Triticum durum* Desf.) elite germplasm tested along the Senegal River. J Agric Sci 10:217–233

Sallares R, Brown TA (1999) PCR-based analysis of the intergenic spacers of the *Nor* loci on the A genomes of *Triticum* diploids and polyploids. Genome 42:116–128

Sallares R, Brown TA (2004) Phylogenetic analysis of complete 5' external transcribed spacers of the 18S ribosomal RNA genes of diploid *Aegilops* and related species (Triticeae, Poaceae). Genet Resour Crop Evol 51:701–712

Sallee PJ, Kimber G (1976) The meiotic analysis of the hybrid *T. timopheevii* var. *zhukovskyi* x *Secale cereale*. Wheat Inf Serv 41–42:9–11

Salse J, Bolot S, Throude M, Jouffe V, Piegu B, Quraishi UM, Calcagno T, Cooke R, Delseny M, Feuillet C (2008) Identification and characterization of conserved duplications between rice and wheat provide new insight into grass genome evolution. Plant Cell 20:11–24

Samuel D (1994) An archaeological study of baking and bread in New Kingdom Egypt (Doctoral thesis), Submitted to the Univ of Cambridge

Saponaro C, Pogna NE, Castagna R, Pasquini M, *et al.* (1995) Allelic variation at the *Gli-1ᵐ*, *Gli-A2ᵐ* and *Glu-A1ᵐ* loci and breadmaking quality in diploid wheat *Triticum monococcum*. Genet Res 66:127–137

Sardana RK, O'Dell M, Flavell RB (1993) Correlation between the size of the intergenic regulatory region, the status of cytosine methylation of rRNA genes and nucleolar expression in wheat. Mol Gen Genet 236:155–162

Sarkar P, Stebbins GL (1956) Morphological evidence concerning the origin of the B genome in wheat. Amer J Bot 43:297–304

Sasaki M, Morris R, Schmidt JW, Gill BS (1963) Metaphase I studies on F$_1$ monosomics from crosses between the Chinese Spring and Cheyenne common wheat varieties. Can J Genet Cytol 5:318–325

Sasanuma T, Chabane K, Endo TR, Valkoun J (2002) Genetic diversity of wheat wild relatives in the Near East detected by AFLP. Euphytica 127:81–93

Sastri DC, Hilu K, Appels R, Lagudah ES, Playford J, Baum BR (1992) An overview of evolution in plant 5S DNA. Pl Syst Evol 183:169–181

Sawhney RN, Goel LB (1979) Stem rust resistance in accessions of Triticum timopheevii and three Triticum aestivum lines with resistance from timopheevi. Wheat Inf Serv 50:56–58

Sax K (1918) The behavior of chromosomes in fertilization. Genetics 3:309–327

Sax K (1921) Chromosome relationships in wheat. Science 54:413–415

Sax K (1922) Sterility in wheat hybrids. II. Chromosome behavior in partially sterile hybrids. Genetics 7:513–552

Sax K (1923) The relation between chromosome number, morphological characters and rust resistance in segregates of partially sterile hybrids. Genetics 8:301–321

Sax K (1927) Chromosome behavior in Triticum hybrids, Verhandlungen des V Int. Kongresses für Vererbungswissenchaft, Berlin, 2, 1267–1284

Sax K (1935) The cytology analysis of species-hybrids. Bot Rev 1:100–117

Sax K, Sax HJ (1924) Chromosome behavior in a genus cross. Genetics 9:454–464

Sayaslan A, Koyuncu M, Yildirim A, Güleç TE, SönmeZoglu Ö, Kandemir N (2012) Some quality characteristics of selected durum wheat (Triticum durum) landraces. Turk J Agric 36: 749–756

Schachtman DP, Lagudah ES, Munns R (1992) The expression of salt tolerance from Triticum tauschii in hexaploid wheat. Theor Appl Genet 84:714–719

Schiemann E (1929) Zytologische Beitrage zur Gattung Aegilops. Chromosomenzahlen und Morphologie. III. Mitteilung. Ber Deutsch Bot Ges 47:164–181

Schiemann E (1932) Entstehung der Kulturpflanzen. In: Bauer E, Hartmann M (eds) Handbuch der Vererbungs-Wissenschaften Bornträger, Vol. III. Verlag Gebr Bornträger, Berlin

Schiemann E (1948) Weizen, Roggen, Gerste. Systematik, Geschichte und Verwendung. Verlag Gustav Fischer, Jena, pp 1–102

Schmidt JW, Johnson VA (1963) A sphaerococcum—like tetraploid wheat. Crop Sci 3:98–99

Schmidt JW, Johnson VA (1966) Inheritance of the sphaerococcum effect in tetraploid wheat. Wheat Inf Serv 22:5–6

Schmidt JW, Weibel DE, Johnson VA (1963) Inheritance of an incompletely dominant character in common wheat simulating Triticum sphaerococcum. Crop Sci 3:261–264

Schröder E (1931) Anatomische Untersuchungen an den Spindeln der Triticum- und Aegilops-Arten zur Gewinnung neuer Gesichtspunkte für die Abstammung und Systematik der Triticum-Arten. Beih Bot Zentralbl 48:333–403

Schultze-Motel J (2019) Triticum parvicoccum Kislev in Transcaucasia. Genet Resour Crop Evol 66:1363–1366

Schulz A (1913a) Über eine neue spontane Eutriticumform: Triticum dicoccoides Kcke. forma Straussiana, Berichte der Deutschen Bot. Gesselschaft 31:226–230

Schulz A (1913b) Die Geschichte der kultivierten Getreide, Nebert, Halle

Schulz-Schaeffer J, Haun CR (1961) The chromosomes of hexaploid common wheat, Triticum aestivum L. Z Pflanzenzücht 46:112–124

Schwarzacher T, Leitch AR, Bennett MD, Heslop-Harrison JS (1989) In situ localization of parental genomes in a wide hybrid. Ann Bot 64:315–324

Schwarzacher T, Heslop-Harrison JS, K. Anamthawat-Jónsson K, Finch RA, Bennett MD (1992) Parental genome separation in reconstructions of somatic and premeiotic metaphases of Hordeum vulgare x H. bulbosum. J Cell Sci 101:13–24

Schweinfurth G (1908) Über die von A. Aaronsohn ausgeführten Nachforschungen nach dem wilden Emmer (Triticum dicoccoides Kcke.). Ber Deutsch Bot Ges 26:309–324

Scoles GJ, Gill BS, Xin Z-Y, Clarke BC, McIntyre CL, Chapman C, Appels R (1988) Frequent duplication and deletion events in the 5S RNA genes and the associated spacer regions of the Triticeae. Pl Syst Evol 160:105–122

Seal AG (1982) C-banded wheat chromosome in wheat and triticale. Theor Appl Genet 63:38–47

Sears ER (1941a) Amphidiploids in the seven-chromosome Triticinae. Univ Missouri Agric Exper Stn Res Bull 336:1–46

Sears ER (1941b) Chromosome pairing and fertility in hybrids and amphidiploids in the Triticinae. Univ Missouri Agric Exp Stn Res Bull 337:1–20

Sears ER (1944a) Inviability of intergeneric hybrids involving Triticum monococcum and T. aegilopoides. Genetics 29:113–127

Sears ER (1944b) The amphidiploids Aegilops cylindrica x Triticum durum and A. ventricosa x T. durum and their hybrids with T. aestivum. J Agric Res 68:135–144

Sears ER (1946) The sphaerococcum gene in wheat. Rec Genet Soc Amer 15:65–66

Sears ER (1947) The sphaerococcum gene in wheat. Genetics 32:102–103

Sears ER (1948) The cytology and genetics of the wheats and their relatives. Adv Genet 2:239–270

Sears ER (1952) Homoeologous chromosomes in Triticum aestivum. Genetics 37:624

Sears ER (1954) The aneuploids of common wheat. Missouri Agric Exp Stn Res Bull 572:1–58

Sears ER (1956a) The B genome of Triticum. Wheat Inform Serv 4:8–10

Sears ER (1956b) Weizen (Triticum L.) I. The systematics cytology and genetics of wheat, 2nd edn. In: Kappert H, Rudorf W (eds) Handbuch der Pflanzenziichtung, Paul Parey, Berlin und Hamburg, 164–187

Sears ER (1959) Wheat cytogenetics. Ann Rev Genet 3:451–468

Sears ER (1966) Nullisomic-tetrasomic combinations in hexaploid wheat. In: Riley R, Lewis KR (eds) Chromosome manipulations and plant genetics. Springer, Boston, MA, USA, pp 29–45

Sears ER (1969) Wheat cytogenetics. Annu Rev Genet 3:451–468

Sears ER (1976a) A synthetic hexaploid wheat with fragile rachis. Wheat Inf Serv 41–42:31–33

Sears ER (1976b) Genetic control of chromosome pairing in wheat. Ann Rev Genet 10:31–51

Sears ER (1977) An induced mutant with homoeologous chromosome pairing. Can J Genet Cytol 19:585–593

Sears ER, Okamoto M (1958) Intergenomic chromosome relationships in hexaploid wheat. In: Proceedings of 10th international congress genetics, Montreal, Quebec 2, 258–259

Sears ER, Sears LMS (1979) The telocentric chromosomes of common wheat. In: Ramanuam S (ed) Proceedings of 5th international wheat genetics symposium, New Delhi, 23–28 Feb 1978, pp 389–407

Senjaninova-Korczagina MV (1932) Karyo-systematical investigations of the genus Aegilops L. Bull Appl But Genet and Pl Breed Ser 2:1–90 (Russian with English summary)

Shaked H, Kashkush K, Ozkan H, Feldman M, Levy AA (2001) Sequence elimination and cytosine methylation are rapid and reproducible responses of the genome to wide hybridization and allopolyploidy in wheat. Plant Cell 13:1749–1759

Shands H, Kimber G (1973) Reallocation of the genomes of *Triticm timopheevii* Zhuk. In: Sears ER, Sears LMS (eds) Proceedings of 4th international wheat genet symposium. Agricultural Experiment Station, College of Agriculture, University of Missouri, pp 101–108

Shang XM, Jackson RC, Nguyen HT, Huang JY (1989) Chromosome pairing in the *Triticum monococcum* complex: evidence for pairing control genes. Genome 32:216–226

Shao Q, Li C, Basang C (1983) Semi-wild wheat from Xizang (Tibet). In: Sakamoto S (ed) Proceedings of 6th international wheat genetics symposium. Kyoto, Japan, pp 11–114

Shao Q, Shao C, Li C, Basang C (1980) Semi-wild wheat from Xizang (tibet). Genet Sin 7:149–156 (in Chinese with English summary)

Sharma HC, Waines JG (1980) Inheritance of tough rachis in crosses of *Triticum monococcum* and *T. boeoticum*. J Hered 71:214–216

Sharma HC, Waines JG (1981) The relationships between male and female fertility and among taxa in diploid wheats. Botany 68:449–451

Sharma HC, Waines JG, Foster KW (1981) Variability in primitive and wild wheats for useful genetic characters. Crop Sci 21:555–559

Sharma JS, Running KLD, Xu SS, Zhang Q et al (2019) Genetic analysis of threshability and other spike traits in the evolution of cultivated emmer to fully domesticated durum wheat. Mol Genet Genom 294:757–772

Shcherban AB, Schichkina AA, Salina EA (2016) The occurrence of spring forms in tetraploid *Timopheevi* wheat is associated with variation in the first intron of the *VRN-A1* gene. BMC Pl Biol 16 (suppl 3):236

Shewry PR (2009) Wheat. J Exp Bot 60:1537–1553

Shewry PR, Hey SJ (2015) The contribution of wheat to human diet and health. Food Energy Secur 4:178–202

Simeone R, Blanco A, Giorgi B (1983) The primary trisomics of durum wheat (*Triticum durum* Desf.) In: Sakamoto S (ed) Proceedings of 6th international wheat genetics symposium, Kyoto, Japan, pp 1103–1107

Simonetti MC, Bellomo MP, Laghetti G, Perrino P, Simeone R, Blanco A (1999) Quantitative trait loci influencing free-threshing habit in tetraploid wheats. Genet Res Crop Evol 46:267–271

Simons KJ, Fellers JP, Trick HN, Zhang Z, Tai YS, Gill BS, Faris JD (2006) Molecular characterization of the major wheat domestication gene *Q*. Genetics 172:547–555

Singh R (1946). *Triticum sphaerococcum* Perc. (Indian dwarf wheat). Indian J Genet 6:34–37

Singh D (1987) Mapping of the complex T. *sphaerococcum* locus. Wheat Inf Serv 64:17–20

Singh D, Joshi BC (1972) Orientation of homoeologous chromosomes in the somatic cells of wheat. Wheat Inf Serv 33–34:9

Singh HB, Anderson E, and Pal BP (1957) Studies in the genetics of *Triticum vavilovii* Jakubz. Agron J 49:4–11

Singh RJ, Röbbelen G, Okamoto M (1976) Somatic association at interphase studied by giemsa banding technique. Chromosoma 56:265–273

Singh M, Chabane K, Valkoun J et al (2006) Optimum sample size for estimating gene diversity in wild wheat using AFLP markers. Genet Resour Crop Evol 53:23–33

Singh K, Ghai M, Garg M et al (2007) An integrated molecular linkage map of diploid wheat based on a *Triticum boeoticum* x *T. monococcum* RIL population. Theor Appl Genet 115:301–312

Singh AK, Sharma JB, Singh PK, Singh A, Mallick N (2017) Genetics and mapping of a new leaf rust resistance gene in *Triticum aestivum* L. x *Triticum timopheevii* Zhuk. derivative 'selection G12'. J Genet 96:291–297

Smith L (1936) Cytogenetic studies in *Triticum monococcum* L. and *T. aegilopoides* Bal. Res Bull Mo Agric Exp Sta 248:38

Smith-Huerta NL, Huerta AJ, Barnhart, D et al (1989) Genetic diversity in wild diploid wheats *Triticum monococcum* var. *boeoticum* and *T. urartu* (Poaceae). Theor Appl Genet 78:260–264

Snir A, Nadel D, Groman-Yaroslavski I, Melamed Y, Sternberg M, Bar-Yosef O, Weiss E (2015) The origin of cultivation and proto-weeds, long before Neolithic farming. PLoS ONE 10: e0131422

Sodkiewicz W (2002) Diploid wheat—*Triticum monococcum* as a source of resistance genes to preharvest sprouting of triticale. Cereal Res Commun 30:323–328

Sofi F, Whittaker A, Cesari F, Gori AM, Fiorillo C, Becatti M, Marotti I, DinelliG CA, Abbate R, Gensini GF, Benedettelli S (2013) Characterization of Khorasan wheat (Kamut) and impact of a replacement diet on cardiovascular risk factors: cross-over. Europ J Clinical Nutrition 67:190–195

Sohail Q, Shehzad T, Kilian A, Eltayeb AE, Tanaka H, Tsujimoto H (2012) Development of diversity array technology (DArT) markers for assessment of population structure and diversity in *Aegilops tauschii*. Breed Sci 62:38–45

Sood S, Kuraparthy V, Bai G, Gill BS (2009) The major threshability genes soft glume (*sog*) and tenacious glume (*Tg*), of diploid and polyploid wheat, trace their origin to independent mutations at non-orthologous loci. Theor Appl Genet 119:341–351

Sorokina ON (1937) Contribution to the synthesis of *Aegilops* species. Bull Appl Bot Genet Pl Br 7:151–160

Sourdille P, Tixier MH, Charmet G, Gay G, Cadalen T, Bernard S, Bernard M (2000) Location of genes involved in ear compactness in wheat (*Triticum aestivum*) by means of molecular markers. Mol Breed 6:247–255

Stallknecht GF, Gilbertson KM, Ranney JE (1996) Alternative wheat cereals as food grains: einkorn, emmer, spelt, kamut, and triticale. In: Janick J (ed) Progress in new crops. ASHS Press, Alexandria, pp 156–170

Stapf O (1909) History of the wheats. Rept Brit Assoc Adv Sci, Winnipeg, 799–807

Stebbins GL (1950) Variation and evolution in plants. Columbia University Press, New York

Stehno Z (2007) Emmer wheat Rudico can extend the spectraof cultivated plants. Czech J Genet Plant Breed 43:113–115

Suchowilska E, Szafrańska A, Słowik E, Wiwart M (2019) Flour from *Triticum polonicum* L. as a potential ingredient in bread production. Cereal Chm 96:554–563

Suemoto H (1968) The origin of the cytoplasm of tetraploid wheat. In: Finlay KW, Shepherd KW (eds) Proceedings of 3rd international wheat genetics symposium, Canberra, Australia, pp 141–152

Sun QX, Ni ZF, Liu ZY, Gao JW, Huang TC (1998) Genetic relationships and diversity among Tibetan wheat, common wheat and European spelt wheat revealed by RAPD markers. Euphytica 99:205–211

Supatashvili V (1929) Emmers of lechkhumi District. Bull Exp Agron Inst Ga 1:83–98

Svetozarova VV (1939) Second genome of *Triticum timopheevi* Zhuk. C R Acad Sci USSR 23:473–477

Swaminathan MS, Prabhakara Rao MV (1961) Macro-mutations and subspecific differentiation in *Triticum*. Wheat Inf Serv 13:9–11

Swamlnathan MS, Jagathesan D, Chopra VL (1963) Induced sphaerococcoid mutations in *Triticum aestivum* and their phylogenetic and breeding significance. Curr Sci 32:530–540

Szabó AT, Hammer K (1996) Notes on the taxonomy of farro: *Triticum monococcum, T. dicoccon* and *T. spelta*. In: Padulosi S, Hammer K, Heller J (eds) Proceedings of 1st international workshop on hulled wheats, 21–22 July, 1995, Castelvecchio Pascoli, Tuscany, Italy, Intern Pl GenetResourc Institute, Rome, Italy, pp 2–40

Szűcs P, Veisz O, Vida G, Bedő Z (2003) Winter hardiness of durum wheat in Hungary. Acta Agron Hung 51:389–396

Taenzler B, Esposti RF, Vaccino P, Brandolini A, EVgen S, Heun M, Schafer-Pregl R, Borghi B, Salamini F (2002) Molecular linkage map of einkorn wheat: mapping of storage-protein and soft-glume genes and bread-making quality QTLs. Genet Res 80:131–143

Takahashi H, Rai B, Kato K, Nakamura I (2010) Divergent evolution of wild and cultivated subspecies of *Triticum timopheevii* as revealed by the study of *PolA1* gene. Genet Resourc Crop Evol 57:101–109

Takumi S, Morimoto R (2015) Implications of an inverted duplication in the wheat *KN1*-type homeobox gene *Wknox1* for the origin of Persian wheat. Genes Genet Syst 90:115–120

Takumi S, Nasuda S, Liu Y-G, Tsunewaki K (1993) Wheat phylogeny determined by RFLP analysis of nuclear DNA. 1. Einkorn wheat. Jpn J Genet 68:73–79

Talbert LE, Blake NK, Storlie EW, Lavin M (1995) Variability in wheat based on low-copy DNA sequence comparisons. Genome 38:951–957

Tanaka M (1956) Chromosome pairing and fertility in the hybrid between the new amphidiploid S^bS^bAA and emmer wheat. Wheat Inf Serv 3:21–22

Tanaka M, Ichikawa S (1968) Cytogenetical examinations of *Triticum araraticum* Jakubz., a wild-type tetraploid wheat species. Genetics 60:229

Tanaka M, Ichikawa S (1972) Cytogenetical relationships of two types of *Triticum araraticum* Jakubz. to other tetraploid wheat species. Jpn J Genet 47:103–114

Tanaka M, Ishii H (1973) Cytogenetical evidence on the speciation of wild tetraploid wheats collected in Iraq, Turkey, and Iran. In: Sears ER, Sears LMS (eds) Proceedings of 4th international wheat genetics symposium, Columbia, Missouri, pp 115–121

Tanaka M, Ishii H (1975) Hybrid sterility and chromosomal interchanges found in the *Timopheevi* group of tetraploid wheat. Jpn J Genet 50:141–149

Tanaka M, Kawahara T (1976) Wild tetraploid wheat from northern Iraq cytogenetically closely related to each other. Wheat Inf Serv 43:3–4

Tanaka M, Sakamoto S (1979) Morphological and physiological variations in wild tetraploid wheats collected from the Zagros Mountains. Rep Plant Germ-Plasm Inst Kyoto Univ 4:12–17

Tanaka M, Ishii H, Ichikawa S (1968) Origin and differentiation of the tetraploid wheats, II. Two types of *Triticum araraticum* Jakubz. Jpn J Genet 43:442 (in Japan)

Tanaka M, Kawahara T, Sano J (1978) The origin and the evolution of wild tetraploid wheats. Wheat Inf Serv 47–48:7–9

Tanno K-I, Takeuchi A, Akahori E, Kobayashi K (2018) Multiplex PCR effectively identifies tetraploid *Triticum* AABB—or AAGG-genome species. Pl Genet Resour 16:279–283

Terachi T, Tsunewaki K (1992) The molecular basis of genetic diversity among cytoplasms of *Triticum* and *Aegilops*. VIII. Mitochondrial RFLP analyses using cloned genes as probes. Mol Biol Evol 9:911–931

Terachi T, Ogihara Y, Tsunewaki K (1990) The molecular basis of genetic diversity among cytoplasms of *Triticum* and *Aegilops*. 7. Restriction endonuclease analysis of mitochondrial DNAs from polyploid wheats and their ancestral species. Theor Appl Genet 80:366–373

The TT, Nevo E, McIntosh RA (1993) Responses of Israeli wild emmer to selected Australian pathotypes of *Puccinia* species. Euphytica 71:75–81

Thellung A (1918a) Neuere Wege und Ziele der botanischen Systematik, erläutert am Beispiele unserer Getreidearten, Naturw. Wochenschr. Neue Folge 17(449–458):465–474

Thellung A (1918b) Neuere Wege und Ziele der botanischen Systematik, erläutert am Beispiele unserer Getreidearten. Mitt Naturwiss Ges Winterthur 12:109–152

Thompson WF, Flavell RB (1988) DNase I sensitivity of ribosomal RNA Genes in chromatin and nucleolar dominance in wheat. J Mol Biol 204(535):548

Thompson JP, Nath J (1986) Elucidation of the B-genome donor to *Triticum turgidum* by unique- and repeated-sequence DNA hybridizations. Biochem Genet 24:39–50

Thompson WP, Robertson HT (1930) Cytological irregularities in in hybrids between species of wheat with the same chromosome number. Cytologia 1:252–262

Thuillet A-C, Bataillon T, Poirier S, Santoni S, David JL (2005) Estimation of long-term effective population sizes through the history of durum wheat using microsatellite data. Genetics 169:1589–1599

Timonova EM, Leonova IN, Röder MS et al (2013) Marker-assisted development and chracterization of a set of *Triticum aestivum* lines carrying different introgressions from the *T. timopheevii* genome. Mol Breed 31:123–136

Tomerlin JR, El-Morshidy MA, Moseman JG, Baenziger PS, Kimber G (1984) Resistance to *Erysiphe graminis* f. sp. *tritici*, *Puccinia recondita* f. sp. *tritici* and *Septoria nodorum* in wild Triticeae species. Pl Disease 68:10–13

Trethowan RM, Mujeeb-Kazi A (2008) Novel germplasm resources for improving environmental stress tolerances of hexaploid wheat. Crop Sci 48:1255–1265

Tsunewaki K (1960) Monosomic and conventional analysis in common wheat. III. Lethality. Jpn J Genet 35:71–75

Tsunewaki K (1966) Comparative gene analysis of common wheat and its ancestral species. III. Glume Hairiness. Genetics 53:303–311

Tsunewaki K (1968) Origin and phylogenetic differentiation of bread wheat revealed by comparative gene analysis. In: Finley KW, Shepherd KW (eds) Proceedings of 3rd international wheat genetics symposium, Canberra, Australia, pp 71–85

Tsunewaki K (1971) Distribution of necrosis genes in wheat v. *Triticum macha*, *T. spelta* and *T. vaviloviii*. Jpn J Genet 46:93–101

Tsunewaki K (1989) Plasmon diversity in *Triticum* and *Aegilops* and its implication in wheat evolution. Genome 31:143–154

Tsunewaki K (2009) Plasmon analysis in the *Triticum-Aegilops* complex. Breed Sci 59:455–470

Tsunewaki K, Kihara H (1961) F_1 monosomic analysis of *Triticum macha*. Wheat Inf Serv 12:1–3

Tsunewaki K, Ogihara Y (1983) The molecular basis of genetic diversity among cytoplasms of *Triticum* and *Aegilops* species. II. On the origin of polyploid wheat cytoplasms as suggested by chloroplast DNA restriction fragment patterns. Genetics 104:155–171

Tsunewaki K, Mukai Y, Endo TR, Tsjii S, Murata M (1976) Genetic diversity of the cytoplasm in *Triticum* and *Aegilops*. v. classification of 23 cytoplasms into eight plasma types. Jpn J Genet 51:175–191

Tsunewaki K, Mukai Y, Endo TR (1978) On the descent of the cytoplasms of polyploid species of Triticum and Aegilops, vol 1. In: Ramanujan S (ed) Proceedings of 5th international wheat genetics symposium Indian soceity genetics Pl breed. Indian Agricultural Research Institute, New Delhi, India, pp 261–272

Tsunewaki K, Yamada S, Mori N (1990) Genetical studies on a Tibetan semi-wild wheat *Triticum aestivum* ssp. *tibetanum*. Jpn J Genet 65:353–365

Tsunewaki K, Shimada T, Matsuoka Y (1999) Transfer of *Triticum urartu* cytoplasm to emmer wheat is difficult, if not impossible. Wheat Inf Serv 88:27–31

Tumanian MG (1937) The occurrence in nature of polyploid mutations in wild monococcal wheat. Compt Rend (Dok) Wcad Sci UrSS 16:325–327

Tzvelev NN (1976) Poaceae URSS. Tribe III. Triticeae Dum. USSR. Academy of Sciences Press, Leningrad, pp 105–206

Tzvelev NN (1984) Grasses of the Soviet Union, 2 Vols, Translated from Russian by Sharma BR, AA Balkema, Rotterdam

Uauy C, Distelfeld A, Fahima T, Blechl A, Dubcovsky J (2006) A *NAC* gene regulating senescence improves grain protein, zinc, and iron content in wheat. Science 314:1298–1301

Ulaş B, Fiorentino G (2021) Recent attestations of "new" glume wheat in Turkey: a reassessment of its role in the reconstruction of Neolithic agriculture. Veget Hist Archaeobot 30:685–701

Ullah AA (2016) Rationalizing migration decisions, labour migrants in East and South-East Asia. Routledge Taylor & Francis Group, London and New York

Unrau J (1950) The use of monosomes and nullisomes in cytogenetic studies of common wheat. Sci Agr 30:66–89

Upadhya MD, Swaminathan MS (1963a) Deoxyribonucleic acid and the ancestry of wheat. Nature 200:713–714

Upadhya MD, Swaminathan MS (1963b) Genome analysis in *Triticum zhukovskyi*, a new hexaploid wheat. Chromosoma 14:589–600

Upadhya MD, Swaminathan MS (1965) Studies on the origin of Triticum zhukovskyi and on the mechanism regulating chromosome pairing in *Triticum*. Indian J Genet Pl Breed 25:1–13

Vakar BA (1932) Cytological study of the interspecific hybrids of the genus *Triticum*. Bull Appl Bot Leningrad, Ser 2:180–241

Valkoun JJ (2001) Wheat pre-breeding using wild progenitors. Euphytica 119:17–23

Valkoun JJ, Waines JG, Konopka J (1998) Current geographical distribution and habitat of wild wheats and barley. In: Damania AB, Valkoun J, Willcox G, Qualset CO (eds) The origins of agriculture and crop domestication. Proceedings of Harlan symposium international center for Agric Res in the Dry Areas (ICARDA), Syria, pp 293–299

Vallega V (1992) Agronomical performance and breeding value of selected strains of diploid wheat, *Triticum monococcum.* Euphytica 61:13–23

Vallega V, Rubies-Autonell C (1985) Reactions of Italian *Triticum durum* cultivars to Soilborne wheat mosaic. Plant Dis 69:64–66

van Ginkel M, Scharen AL (1988) Host-pathogen relationships of wheat and *Septoria tritici*. Phytopathology 78:762–766

van Slageren MW (1994) Wild wheats: a monograph of *Aegilops* L. and *Amblyopyrum* (Jaub. and Spach) Eig (Poaceae). Agricultural University, Wageningen, The Netherlands

van Slageren MW, Payne T (2013) Concepts and nomenclature of the Farro wheats, with special reference to Emmer, *Triticum turgidum* subsp. *dicoccum* (Poaceae). Kew Bull 68:477–494

van Zeist W (1972) Palaeobotanical results of the 1970 season at Qayonii, Turkey. Helinium 12:3–19

van Zeist W (1976) On macroscopic traces of food plants in southwestern Asia (with some reference to pollen data). Phil Trans R Soc Lond B 275:27–41

van Zeist W, Bakker-Heeres JAH (1973) Paleobotanical Studies of Deir 'Alla, Jordan. Paléorient 1:21–37

van Zeist W, Bakker-Heeres JAH (1975) Prehistoric andearly historic plant husbandry in the Altinova Plain, southeastern Turkey. In: van loon MN (ed) Korucutepe 1, North–Holland, Amsterdam, pp 221–257

van Zeist W, Bakker-Heeres JAH (1982) Archaeobotanical studies in the Levant. 1. Neolithic sites in the Damascus Basin: Aswad, Ghoraifé, Ramad. Palaeohistoria 24:165–256

van Zeist W, Bakker-Heeres JAH (1985) Archaeobotanical studies in the Levant 1. Neolithic sites in the Damascus basin: Aswad, Choraife, Ramad. Palaeohistoria 24:165–256

Vardi A (1973) Introgression between different ploidy levels in the wheat group. In: Sears ER, Sears LMS (eds) Proceedings of 4th international wheat genetics symposium, agricultural experiment station. College of Agriculture, University of Missouri, pp 131–141

Vavilov NI (1914) Immunity to fungous diseases as a physiological test in genetics and systematics, exemplified in cereals. J Genet 4:49–65

Vavilov NI (1918) Immunity of plants to infectious diseases. News Petr Agrar Acad 1918; Rabushinski Press, Moscow (1919) (in Russian)

Vavilov NI (1926) Studies on the origin of cultivated plants. Bull Appl Bot Genet Pl Breed 16:1–248 (in Russian and English summary)

Vavilov NI (1932) The process of evolution in cultivated plants. Proc 6th Intern Congr Genet 1:331–342

Vavilov NI (1935) Botaniko-geograficheskie osnovy selektsii [Botanical-geographical foundations of breeding]. Teoreticheskie osnovy selektsii rastenii [Theoretical foundations of plant breeding], Moscow, Leningrad, pp 17–75 (in Russian)

Vavilov NI (1951) The origin, variation, immunity and breeding of cultivated species. Selectedwritings, translated from the Russian by Chester KS. Chronica Botanica 13: Chronica Botanica Co. Waltham, Mass. and Wm. Dawson and Sons, Ltd., London, pp 1–366

Vavilov NI (1987) Origin and geography of cultivated plants. Cambridge University Press, Cambridge, UK

Vavilov NI, Jakushkina OV (1925) A contribution to the phylogenesis of wheat and yjr interspecific hybridization in wheat. Bull Appl Bot PL Breed 15:159

Vavilov NI, Fortunatova OK, Jakubziner MM, Palmova ET, Nikolaenko EI, Stoletova EA, Verkhovshkiaya KA, Shreiber LL, Syrovatsky SG (1931) The Wheats of Abyssinia. Trudy prikl Bot Genet Selek (Suppl. 51) (in Russian)

Vega JM, Feldman M (1998) Effect of the pairing gene *Ph1* and premeiotic colchicine treatment on intra- and inter-chromosome pairing of isochromosomes in common wheat. Genetics 150:1199–1208

Vedel F, Quetier F, Bayen M (1976) Specific cleavage of chloroplast DNA from higher plants by *EcoR1* restriction nuclease. Nature 263:440–442

Vieira LGE (1985) Characterization of *Triticum turgidum* ssp. *carthlicum* (Nevski) Löve et Löve by discriminant analysis. Dissertation for the degree of Ph.D., Crop Sci Dept, Oregon State Univ, Corvalis, Oregon, USA

Villars D (1787) Nistoire des Plantes de Dauphiné. Grenoble, Lyon, and Paris, p 690

Vittozzi L, Silano V (1976) The phylogenesis of protein α-amylase inhibitors from wheat seed and the speciation of polyploid wheats. Theor Appl Genet 48:279–284

Vogel F, Schroeder TM (1974) The internal order of the interphase nucleus. Humangenetik 25:265–297

von Buren M (2001) Polymorphisms in two homeologous gammagliadin genes and the evolution of cultivated wheat. Genet Res Crop Evol 48:205–220

von Schrank FP (1789) Baiersche Flora. *Bd. 2.* J.B. Strohl, München, pp 703

von Tschermak E (1914) Die Verwertung der Bastardierung für phylogenetische Fragen in der Getreidegruppe. Zeitschr. Pflanzenzucht 2:291–312

Wagenaar EB (1961a) Studies on the genome constitution of *Triticum timopheevii* Zhuk. I. Evidence for genetic control of meiotic irregularities in tetraploid hybrids. Can J Genet Cytol 3:47–60

Wagenaar EB (1961b) Cytological studies of the development of metaphase I in *Triticum* hybrids. II. The behavior of univalents in meiotic cell division. Can J Genet Cytol 3:204–225

Wagenaar EB (1961c) Cytological studies of the development of metaphase I in *Triticum* hybrids. III. The lagging patterns in two triploids. Can J Genet Cytol 3:361–371

Wagenaar EB (1966) Studies on the genome constitution of *Triticum timopheevi* Zhuk. II. The *T. timopheevi* complex and its origin. Evolution 20:150–164

Waines JG (1976) A model for the origin of diploidizing mechanisms in polyploid species. Am Naturalist 110:415–430

Waines JG (1983) Genetic resources in diploid wheat: the case for diploid commercial wheats. In: Sakamoto S (ed) Proceedings of 6th international wheat genetics symposium, Kyoto, Japan, pp 115–122

Waines JG, Kimber G (1973) Satellite number and size in *Triticum monococcum* L. and the evolution of the polyploid wheats. Can J Genet Cytol 15:117–122

Waines JG, Payne PI (1987) Electrophoretic analysis of the high-molecular-weight glutenin subunits of *Triticum monococcum*, *T. urartu*, and the A genome of bread wheat (*Triticum aestivum*). Theor Appl Genet 74:71–76

Wakar BA (1935) Cytologische Untersuchungen der ersten Generation der Weizen queckenbastarde. Der Ziichter 7:199–206

Walkowiak S, Gao L, Monat C et al (2020) Multiple wheat genomes reveal global variation in modern breeding. Nature 588:277–283

Wang RR-C (1990) Intergeneric hybrids between *Thinopyrum* and *Psathyrostachys* (Triticeae). Genome 33:845–849

Wang G-Z, Miyashita N, Tsunewaki K (1997) Plasmon analyses of *Triticum* (wheat) and *Aegilops*: PCR–single-strand conformational polymorphism (PCR-SSCP) analyses of organellar DNAs. Proc Natl Acad Sci USA 94:14570–14577

Wang H-J, Huang XQ, Röder MS, Börner A (2002) Genetic mapping of loci determining long glumes in the genus *Triticum*. Euphytica 123:287–293

Wang X, Li W, Zheng Y (2007) Principal component and cluster analysis of agronomic characters in *Triticum dicoccum* Schrank. J Sichuan Agric Univ 25:239–248

Wang H, Yang H, Shivalila CS et al (2013) One-step generation of mice carrying mutations in multiple genes by CRISPR/Cas-mediated genome engineering. Cell 153:910–918

Wang X, Luo G, Yang W et al (2017) Genetic diversity, population structure and marker-trait associations for agronomic and grain traits in wild diploid wheat *Triticum urartu*. BMC Plant Biol 17:112

Wang T, Sun X, Zhang L, Zheng W, Chen C, Nie X (2019) Sequencing and characterization of the whole chloroplast genome of *Triticum paleocolchicum* Men. Hans J Comput Biol 9:22–31

Warburton ML, Crossa J, Franco J et al (2006) Bringing wild relatives back into the family: recovering genetic diversity in CIMMYT improved wheat germplasm. Euphytica 149:289–301

Watanabe N (1994) Near-isogenic lines of durum wheat: their development and plant characteristics. Euphytica 72:143–147

Watanabe N (1999) Genetic control of the long glume phenotype in tetraploid wheat by homoeologous chromosomes. Euphytica 106:39–43

Watanabe N, Imamura I (2002) Genetic control of long glume phenotype in tetraploid wheat derived from *Triticum petropavlovskyi* Udacz. et Migusch. Euphytica 128:211–217

Watanabe Y, Mukade K, Kokubun K (1956) Studies on the production of amphidiploids as the sources of resistance to leaf-rust in wheats. II. Cytogenetical studies on the F₁ hybrids and the amphidiploids. *T. timopheevi* ZHUK. x *T. monococcum* L. [Japanese.] Jpn J Breed 6:23–31

Watanabe N, Yotani Y, Furuta Y (1996) The inheritance and chromosomal location of a gene for long glume in durum wheat. Euphytica 91:235–239

Watanabe N, Sugiyama K, Yamagishi Y, Sakata Y (2002) Comparative telosomic mapping of homoeologous genes for brittle rachis in tetraploid and hexaploid wheats. Hereditas 137:180–185

Watkins AE (1928) The genetics of wheat species crosses. I J Genet 20:1–27

Watkins AE (1940) The inheritance of glume shape in *Triticum*. J Genet 39:240–264

Watkins AE, Ellerton S (1940) Variation and genetics of the awn in *Triticum*. J Genet 40:243–270

Wei L, Dong P, Wei Y-M, Cheng G-Y, Zheng Y-L (2008) Genetic variation in *Triticum turgidum* L. ssp. *turgidum* landraces from China assessed by EST-SSR markers. Agric Sci China 7:1029–1036

Wicker T, Yahiaoui N, Guyot R, Schlagenhauf E et al (2003) Rapid genome divergence at orthologous low molecular weight glutenin loci of the A and Aᵐ genomes of wheat. Plant Cell 15:1186–1197

Willcox G (2005) The distribution, natural habitats and availability of wild cereals in relation to their domestication in the Near East: multiple events, multiple centres. Veget Hist Archaeobot 14:534–541

Wilson P, Driscoll CJ (1983) Hybrid wheat. In: Frenkel R (ed) Heterosis, monographs on theoretical and applied genetics, vol 6. Springer-Verlag, pp 94–123

Wilson JA, Ross WM (1962) Male sterility interaction of the *Triticum aestivum* nucleus and *Triticum timopheevi* cytoplasm. Wheat Inf Serv 14:29

Winkler H (1920) Verbreilung und Ursache der Parthenogenesis im Pflanzen-un Tierreiche. Fischer, Jena

Wiwart M, Suchowilska E, Kandler W, Sulyok M, Groenwald P, Krska R (2013) Can Polish wheat (*Triticum polonicum* L.) be an interesting gene source for breeding wheat cultivars with increased resistance to *Fusarium* head blight? Genet Resour Crop Evol 60:2359–2373

Xiao J, Dong L, Jin H, Zhang K et al (2018) Reactions of Triticum urartu accessions to two races of the wheat yellow rust pathogen. Crop J 6:509–515

Xin Z, Appels R (1988) Occurrence of rye (*Secale cereale*) 350-family DNA sequences in *Agropyron* and other Triticeae. Pl Syst Evol 160:65–76

Yacobi YZ, Feldman M (1983) The control of the regularity and pattern of chromosome pairing in common wheat by the *Ph1* gene. In: Sakamoto S (ed) Proceedings of 6th international wheat genetics symposium, Kyoto, Japan, pp 1113–1118

Yacobi YZ, Mello-Sampayo T, Feldman M (1982) Genetic induction of bivalent interlocking in common wheat. Chromosoma 87:165–175

Yacobi YZ, Levanony H, Feldman M (1985) An ordered arrangement of bivalents at first meiotic metaphase of wheat. I. Hexaploid wheat. Chromosoma 91:347–354

Yaghoobi-Saray J (1979) An electrophoretic analysis of genetic variation within and between populations of five species in *Triticum–Aegilops* complex. Ph.D. Thesis, University of California, Davis, CA, USA

Yahiaoui N, Kaur N, Keller B (2009) Independent evolution of functional *Pm3* resistance genes in wild tetraploid wheat and domesticated bread wheat. Plant J 57:846–856

Yahyaoui A, Al Naimi M, Nachit M, Hakim S (2000) Multiple disease resistance in durum wheat (*Triticum turgidum* L. var. *durum*). In: Royo C, Nachit M, Di Fonzo N, Araus JL (eds) Durum wheat improvement in the mediterranean region: new challenges. CIHEAM, Zaragoza, pp 387–392

Yen C, Luo MC, Yang JL (1988) The origin of the Tibetan weedrace of hexaploid wheat, Chinese Spring, Chengdu-guang-tou and other landraces of the white wheat complex from China. In: Miller TE, Koebner RMD (eds) Proceedings of 7th international wheat genetics symposium, Cambridge, England, pp 175–179

Yenagi NB, Hanchinal R R, Suma C (1999) Nutritional quality of dicoccum wheat semolina and its use in planning therapeutic diets. Paper presented in 32nd annual meeting of Nutrition Society of India, Coimbatore, Tamil Nadu, India, pp 25–26

Yildirim A, Jones SS, Murray TD, Cox TS, Line RP (1995) Resistance to stripe rust and eyespot diseases of wheat in *Triticum tauschii.* Plant Disease 79:1230–1236

Youssefian S, Kirby EJM, Gale MD ((1992) Pleiotropic effects of the GA-insensitive *Rht* dwarfing genes in wheat. 2. Effects on leaf, stem, ear and floret growth. Field Crops Res 28:191–210

Yu MQ, Jahier J (1992) Origin of S^v genome of *Aegilops variabilis* and utilization of the S^v as analyser of the S genome of the *Aegilops* species in the sitopsis section. Plant Breed 108:290–295

Zade (1914) Serologische Studien an Leguminosen und Gramineen. Zeitschr Pflanzenzuchtung 2:101–151

Zaharieva M, Monneveux P (2014) Cultivated einkorn wheat (*Triticum monococcum* L. subsp. *monococcum*): the long life of a founder crop of agriculture. Genet Resour Crop Evol 61:677–706

Zaharieva M, Ayana NG, Al Hakimi A, Misra SC, Monneveux P (2010) Cultivated emmer wheat (*Triticum dicoccon* Schrank), an old crop with promising future: a review. Genet Resour Crop Evol 57:937–962

Zeven AC (1980) Polyploidy and domestication: the origin and survival of polyploids in cytotype mixtures. In: Lewis WH (ed) Polyploidy—biological relevance. Plenum Press, New York, pp 385–407

Zhang W, Qu L, Gu H, Gao W, Liu M, Chen J, Chen Z (2002) Studies on the origin and evolution of tetraploid wheats based on the internal transcribed spacer (ITS) sequences of nuclear ribosomal DNA. Theor Appl Genet 104:1099–1106

Zhang LQ, Liu DC, Zheng YL, Yan ZH, Dai SF, Li YF, Jiang Q, Ye YQ, Yen Y (2010) Frequent occurrence of unreduced gametes in *Triticum turgidum-Aegilops tauschii* hybrids. Euphytica 174:285–294

Zhang ZC, Belcram H, Gornicki P, Charles M, Just J, Huneau C, Magdelenat G, Couloux A, Samain S, Gill BS, Rasmussen JB, Barbe V, Faris JD, Chalhoub B (2011) Duplication and partitioning in evolution and function of homoeologous *Q* loci governing domestication characters in polyploid wheat. Proc Natl Acad Sci USA 108:18737–18742

Zhang W, Zhang M, ZhuX, CaoY, Sun Q, Ma G, Chao S, Yan C, Xu SS, Cai X (2017) Genome-wide homology analysis reveals new insights into the origin of the wheat B genome. bioRxiv 197640. https://doi.org/10.1101/197640

Zhou Y, Bai S, Li H et al (2021) Introgressing the Aegilops *tauschii* genome into wheat as a basis for cereal improvement. Nat Plants 7:774–786

Zhou Y, Zhao X, Li Y, Xu J, Aoyue Bi A et al (2020a) Convergence within divergence: insights of wheat adaptation from *Triticum* population sequencing. https://doi.org/10.1101/2020.03.21.001362

Zhou Y, Zhao X, Li Y et al (2020b) Triticum population sequencing provides insights into wheat adaptation. Nat Genet 52:1412–1422

Zhukovsky P (1923) "Persian wheat"—*Triticum persicum* Vav. in Transcaucasia. Bull Appl Bot 13:45–55 (in Russian with English summary)

Zhukovsky PM (1928) A new species of wheat. Bull Appl Bot Genet PL Breed 19:59–66

Zhukovsky P M (1964) Cultivated plants and their wild relatives, 2nd edn. Leningrad, pp 1–791

Žilić S, Barać M, Pešić M, Dodig D, Ignjatović-Micić D (2011) Characterization of proteins from grain of different bread and durum wheat genotypes. Int J Mol Sci 12:5878–5895

Zohary D (1969) The progenitors of wheat and barley in relation to domestication and agriculture dispersal in the old world. In: Ucko PJ, Dimbleby GW (eds) The domestication and exploitation of plants and animals. Duckworth, London, pp 47–66

Zohary D (1973) The origin of cultivated cereals and pulses in the Near East. Chromosomes Today 4:307–320

Zohary D (1999) Monophyletic vs. polyphyletic origin of the crops on which agriculture was founded in the Near East. Genet Resour Crop Evol 46:133–142

Zohary D, Brick Z (1962) *Triticum dicoccoides* in Israel: notes on its distribution, ecology and natural hybridization. Wheat Inf Serv 13:6–8

Zohary D, Feldman M (1962) Hybridization between amphidiploids and the evolution of polyploids in the wheat (*Aegilops-Triticum*) group. Evolution 16:44–61

Zohary D, Hopf M (1993) Domestication of plants in the old world. The origin ans Spread of Cultivated Plants in West Asia, Europe, and the Nile Valley, 2sd Ed, Clarendon Press, Oxford, England

Zohary D, Hopf M (2000) Domestication of plants in the old world, 3rd edn. Oxford Science Publications, Oxford

Zohary D, Harlan JR, Vardi A (1969) The wild diploid progenitors of wheat and their breeding value. Euphytica 18:58–65

Zohary D, Hopf M, Weiss E (2012) Domestication of plants in the old world-the origin and spread of domesticated plants in south-west Asia, Europe and the Mediterranean Basin, 4th edn. Oxford University Press, pp 1–243

Zoshchuk SA, Badaeva ED, Zoshchuk NV, Adonina IG, Shcherban AB, Salina EA (2007) Intraspecific divergence in wheats of the *Timopheevi* group as revealed by in situ hybridization with tandem repeats of the Spelt1 and Spelt52 families. Russian J Genet 43:636–645

Zwer PK, Sombrero A, Rickman RW, Klepper B (1995) Club and common wheat yield component and spike development in the Pacific Northwest. Crop Sci 35:1590–1597

Evolution of the Diploid Species of the Sub-tribe Triticineae

11.1 Introduction

The Triticineae, the wheat lineage of the Triticeae (Clayton and Renvoize 1986; Table 2.1; Fig. 2.1), diverged from the Hordeineae, the barley lineage, during the Miocene epoch, about 8–15 million years ago (MYA) (Wolfe et al. 1989; Ramakrishna et al. 2002; Chalupska et al. 2008; Huang et al. 2002a, b; Dvorak and Akhunov 2005; Gornicki et al. 2014; Middleton et al. 2014) (Table 2.6). During this time, the global climates became warm, the Tethys Sea, that covered a large part of the East Mediterranean and Southwest Asia, disappeared and the east Mediterranean region rose (Table 2.5). This brought about diversification of temperate ecosystems and opening of new ecological niches, allowing the expansion of grasslands. These geological and climatic changes triggered the appearance first of diploid *Elymus* species containing the St, Ee, or Eb genomes, and somewhat later, the diploid *Agropyron* species, in the East Mediterranean area and Southwest and central Asia (Tables 5.1, 5.3, and 11.1). In the late Miocene, about seven MYA, the ancestral forms of *Amblyopyrum*, *Triticum* and *Aegilops* appeared (Marcussen et al. 2014; Huynh et al. 2019) and diverged from the lineage of *Secale* (Marcussen et al. 2014). Further divergence of the diploid species of *Aegilops* and *Triticum* occurred during the Pliocene (5.3–1.8 MYA). During this period the climate of the East Mediterranean and West Asia regions became seasonal. i.e., cold and humid in the winters and hot and dry in the summers, leading to a relatively short growth period in the winter and long drought in the summer. Also, today landscapes developed during the Pliocene facilitating further spread of grasslands. The climate changes and the opening of new ecological niches, applied a selection pressure that accelerated speciation processes.

In reaction to the environmentally unstable conditions of the East Mediterranean and West Asia, annualism and autogamy developed, enabling rapid colonization of new habitats by new genotypes that were ecologically isolated from their ancestral forms. Annualism enabled plants to pass the dry summer as seeds. The evolution of annual species from perennials occurred independently several times in the Triticineae, e.g., in the lineage Agropyron–*Eremopyrum*, in the lineage *Dasypyrum–Secale,* and in the divergent of the *Amblyopyrum-Triticum-Aegilops linage* from diploid *Elymus* species, presumably *E. elongatus* ssp. *elongatus.* Perennial growth habit is a dominant trait controlled by a small number of genes (Charpentier et al. 1986; Lammer et al. 2004), and mutations in these genes led to the development of annual plants. Autogamy in the Triticineae species is not an obligatory trait; occasional out-crossing may result in the production of sufficient genetic flexibility that effectively used by the rapid generation shift (Allard and Kannenberg 1968; Allard et al. 1968; Allard 1975). Changes from multi-floret spikelets to spikelets with only a few florets improved seed dispersal and reduced competition between siblings, while increased grain size ensured rapid and successful germination. More efficient seed dispersal systems were brought about by the development of awns on glume and lemma ends, that either assisted in burying the spikelet in the soil, through movements of the awns due to changes in humidity (Elbaum et al. 2007), or helped disperse the spikelets by clinging to various parts of animals. The primitive genera, *Elymus, Agropyron*, and *Amblyopyrum*, lack these specializations. Interestingly, genera or species exhibiting more advanced traits, namely, species of *Eremopyrum, Taeniatherum, Crithopsis, Heteranthelium* and *Henrardia,* occupy more xeric habitats in the peripheries of the genus distribution area.

The very wide variation in the inflorescence traits and in the seed dispersal techniques among the genera and species of the Triticineae subtribe reflects rapid adaptation to the broad radiation that occurred in the habitats of the East Mediterranean and Central Asia regions (Sakamoto 1973). In accord with the above, several molecular studies proposed that a major radiation of the Triticineae occurred during a relatively narrow period of time in the late Miocene and early Pliocene (5.3 MYA).

The initial steps of such differentiation occurred at the diploid level, i.e., the newly-formed diploid species

M. Feldman and A. A. Levy, *Wheat Evolution and Domestication*,
https://doi.org/10.1007/978-3-031-30175-9_11

Table 11.1 Time of beginning divergence in million years ago of the Triticineae lineages

Lineages	Beginning divergence time	Sequencing of	Reference
Mediterranean diploid *Elymus* species (primary Triticineae) from the other Hordeineae *Elymus* species	10.0–14.0	Chloroplast genome	Wolfe et al. (1989)
	10.6	Chloroplast genome	Gornicki et al. (2014)
	11.6	Nuclear genes	Chalupska et al. (2008)
Divergence of *Agropyron-Eremopyrum–Henrardia*	> 10.0*	–	–
Dasypyrum-Secale–Heteranthelium	14.0–7.0	Nuclear and chloroplast genes	Yang et al. (2006), Marcussen et al. (2014)
Taeniatherum-Crithopsis	7.0–6.0*	–	–
Amblyopyrum-Triticum-Aegilops	6.5–7.0	275 nuclear genes and three chloroplast genes	Marcussen et al. (2014); Huynh et al. (2019)

* Assumed divergent time

underwent divergent evolution. This divergent evolution was accompanied by a convergent evolution, resulting from allopolyploidization of inter-generic and inter-specific hybrids, that was followed by a further divergence at the polyploid level, to new allopolyploid forms. Allopolyploidy considerably facilitates gene transfer between species and genera via hybridization and introgression (Zohary and Feldman 1962), further enhancing convergent evolution. Thus, the sub-tribe Triticineae developed in cycles of divergence at the diploid level, convergence followed by some divergence at the polyploid level and further convergence due to interspecific hybridizations. These cycles became an important factor in the evolution of the Triticineae.

Based on their geographical distributions, Sakamoto (1973, 1991) classified the Triticeae genera into two major groups: arctic-temperate group and east Mediterranean-central Asiatic group (Table 2.3), a classification that was supported by the studies of Hsiao et al. (1995a) and Fan et al. (2013). Most of the genera that developed in the east Mediterranean and central Asia regions belong to the subtribe Triticineae. This subtribe mainly contains annual species that have a solitary spikelet at each rachis node (except for *Crithopsis* and *Taeniatherum*, that have two spikelets at each rachis node). The genera *Elymus* and Agropyron have only perennial species, *Dasypyrum* and *Secale* have both perennial and annual species, and the remaining eight genera have only annual species (Table 2.1). *Aegilops* is the largest genus (24 species) and *Taeniatherum, Crithopsis, Heteranthelium* and *Amblyopyrum* are monotypic. Each genus of this group is morphologically distinct.

Speciation at the diploid level might have resulted from accumulation of mutations in coding and non-coding sequences and structural changes that led to the buildup of genetic barriers between the diverging taxa. Moreover, amplification or reduction in specific repetitive DNA

sequences, mainly transposons, and their mobilization and activity, activity of genome-restructuring genes (Heneen 1963a; Feldman and Strauss 1983), as well as introgressive hybridization between the diverging taxa, may have boosted the speciation processes. Many Triticineae species have genes that either promote or suppress homoeologous pairing in interspecific and intergeneric hybrids (Table 5.2). Thus, despite the fact that the various Triticineae species are relatively young and still maintain a great deal of genetic relatedness, their chromosomes are homoeologous, rather than homologous, due to genetic and structural changes that occurred during their evolution. Consequently, the chromosomes of one species show reduced pairing with the homoeologous chromosomes of another species curtailing intergeneric and interspecific gene flow.

The most effective genetic barrier is complete sterility or semi-sterility of the interspecific F_1 hybrids, resulting, in many cases, in cryptic structural hybridity. Stebbins (1945) defined cryptic structural hybridity as chromosomal sterility due to heterozygosity for structural differences too small to materially influence chromosome pairing at meiosis. Indeed, several F_1 interspecific and intergeneric hybrids that exhibit high chromosomal pairing were completely sterile and their anthers did not dehisce (Ohta 1990). It cannot be strictly determined whether the sterility observed in the Triticineae F_1 hybrids is chromosomal or genic. This decision can be made only after chromosome doubling of the sterile hybrid. Fertile allopolyploid indicates that the sterility of the F_1 hybrid is chromosomal and not genic.

A considerable number of morphological and molecular studies failed to reach a consensus concerning the phylogenetic relationships between the various diploid taxa of the Triticineae. This ambiguity is due either to a limited number of samples (Kellogg and Appels 1995; Kellogg et al. 1996; Mason-Gamer and Kellogg 1996a; Escobar et al. 2011) or to the small number of genes that were analyzed (Hsiao et al.

1995a, b; Kellogg and Appels 1995; Petersen and Seberg 1997; Helfgott and Mason-Gamer 2004; Mason-Gamer 2005). Relationships between the diploid species of the Triticineae have also been blurred by intergeneric and interspecific hybridizations and introgression events, as well as to incomplete lineage sorting of ancestral polymorphisms, indicating an intricate, reticulate pattern of evolution in this sub-tribe (Kellogg 1996; Komatsuda et al. 1999; Nishikawa et al. 2002; Mason-Gamer 2005; Kawahara 2009; Escobar et al. 2011). Such a reticulate pattern of evolution presents a considerable challenge in phylogenetic analyses, since different genes may exhibit conflicting genealogical histories (Escobar et al. 2011).

Yet, Escobar et al. (2011), using a comprehensive molecular dataset, succeeded to construct a comprehensive, multigenic phylogeny of the diploid taxa of the Triticeae tribe. The multigenic network structure (Escobar et al. 2011) highlights parts of the Triticineae history that did not evolve in a tree-like manner but rather in a reticulate pattern. Moreover, the results of Escobar et al. (2011) provided strong evidence of incongruence among single-gene trees, with different portions of the genome exhibiting different histories. They determined the role of recombination and gene location in the incongruence, and demonstrated that loci in close physical proximity are more likely to share a common history than distant ones, due to a low incidence of recombination in proximal chromosomal regions (Akhunov et al. 2003a, b; Luo et al. 2000, 2005).

Escobar et al. (2011) showed that despite strong tree conflicts, not all Triticineae clades are affected by introgression and/or incomplete lineage sorting. Notably, *Agropyron*, *Eremopyrum* and *Henrardia* diverge in a tree-like manner, whereas the evolution of *Elymus*, *Dasypyrum*, *Secale*, *Heteranthelium*, *Taeniatherum*, *Amblyopyrum*, *Triticum* and *Aegilops* is reticulated. There is no straightforward way to determine whether incongruence in Triticineae results from introgression or incomplete lineage sorting. Recombination could be an important evolutionary force in exacerbating the level of incongruence among gene trees.

11.2 Phylogenetic Relationships of the Diploid *Elymus* Species Having the St or E Genomes

Melderis (1978, 1980) transferred these species, having multiple spikelets per rachis node, from *Agropyron* to *Elymus*, leaving in *Agropyron* only species containing one spikelet per rachis node. Except *E. spicatus* that grows in western North America, all the other diploid species, that were transferred to *Elymus*, are native to southern Ukraine, the Mediterranean basin and Southwest Asia (Table 5.1).

This group contains diploid species, some of which from autopolyploids developed and allopolyploids containing subgenomes from diploid species of this group were formed. The diploid and polyploid *Elymus* species of this group are perennial, and with the exception of *E. elongatus*, which is moderately self-fertile (Melderis 1978; Luria 1983), all the other species are cross-pollinating and self-sterile.

Genome St occurs in several diploid *Elymus* species, Ee in the diploid taxon *E. elongatus* subsp. *elongatus*, whereas genome Eb occurs in the diploid taxon *E. farctus* subsp. *bessarabicus* (Table 11.1). Several cytogenetic and molecular studies showed that these three genomes are closely related (de V Pienaar et al. 1988; Forster and Miller 1989; Wang 1989; Wang and Hsiao 1989; Hsiao et al. 1995a; Wei and Wang 1995; Kosina and Heslop-Harrison 1996; Petersen and Seberg 1997; Chen et al. 1998, 2003; Mason-Gamer et al. 2002; Li et al. 2007, 2008; Fan et al. 2007, 2009; Shang et al. 2007; Liu et al. 2008; Yu et al. 2008; Wang and Lu 2014). Bieniek et al. (2015) found that the nucleotide sequences at three chloroplast loci (*matK*, *rbcL*, *trnH-psbA*) of genomes Ee, Eb and St are almost identical, with only one substitution within the *matK* gene differentiating genome Eb from Ee and St. Similarly, Fan et al. (2013), using two single-copy nuclear gene (*Acc1* and *Pgk1*) sequences, found that the St genome is closely related to the Ee genome of *E. elongatus*. Genome in situ hybridization (GISH) studies substantiated this conclusion by showing that genomes Ee and Eb are very similar in their repetitive DNA (Kosina and Heslop-Harrison 1996). Also, the almost complete chromosome pairing at meiosis of the F_1 hybrids between the diploid species of *Elymus* indicated close relatedness of these genomes (Wang 1985). The high or complete sterility of these F_1 hybrids results presumably from initial steps of divergence leading to cryptic structural hybridity.

The karyotypes of the diploid *Elymus* species are symmetric, the St genome consists of smaller chromosomes than genomes Ee and Eb, and Eb has larger chromosomes than Ee. These differences in chromosome size are also evident in DNA amount (Table 11.2). Since the St genome exists in the more primitive diploid species of *Elymus*, it is assumed that genomes Ee and Eb evolved from St. Their lager size resulted most probably from increase in repetitive DNA.

E. libanoticus grows in all parts of the Fertile Crescent and *E. tauri* grows in the northern part of this region only. Molecular studies have shown that these two species form one clade, while the other diploid St genome species form another clade (Mason-Gamer et al. 2010; Sun et al. 2008; Sun and Komatsuda 2010; Yan and Sun 2011). Likewise, Yu et al. (2008) found that *E. libanoticus* and diploid *E. tauri* are more closely related to one another than they are to *E. stipifollius* and *E. reflexiaristatus*. On the other hand, Yan et al. (2011) grouped the Eurasian St genome species *E.*

Table 11.2 Nuclear DNA amount in diploid species of *Elymus*

Species	1C DNA (pg)[*]
E. reflexiaristatus (Nevski) Melderis ssp. *strigosus* (M. Bieb.) Melderis	4.86
E. libanoticus (Hack.) Melderis	3.95
E. spicatus (Pursh) Gould	4.65
E. stipifollius (Czern. ex Nevski) Melderis	4.00
E. elongatus (Host) Runemark ssp. *elongatus*	5.98
E. farctus (Viv.) Runemark ex Melderis ssp. *bessaribicus* (Savul. and Rayss) Melderis	7.45

[*] From Vogel et al. (1999)

libanoticus, E. reflexiaristatus and diploid *E. tauri* into one clade and the North American *E. spicatus* into a separate clade. Evidently, there are some discrepancies between various phylogenetic studies performed on this group of diploid species. The diploid subspecies of *E. elongatus*, subsp. *elongatus*, grows in the Mediterranean basin and diploid *E. farctus*, subsp. *bessarabicus*, grows in the coasts of the Black Sea, Aegean and N.E. Mediterranean Sea. These two taxa, frequently have one spikelet at each rachis node, are presumably evolved from the St-genome species. The ancestral Triticineae lineages presumably evolved from St or E genomes *Elymus* species in the following four, semi-independent clades: *Agropyron, Eremopyrum* and *Henrardia* clade, *Dasypyrum, Secale* and *Heteranthelium* clade, *Amblyopyrum, Aegilops*, and *Triticum* clade, and *Crithopsis* and *Taeniatherum* clade (Table 11.1). A scheme of the evolution of the Triticineae genera is presented in Fig. 2.1.

11.3 The *Agropyron-Eremopyrum-Henrardia* Clade

11.3.1 Clade Description

This clade contains one genus (*Agropyron*) with many ancestral traits and two genera (*Eremopyrum* and *Henrardia*), with many advanced traits and, therefore, may be considered younger than the former (Table 5.5). The *Agropyron* species are perennials, cross-fertilizing (Melderis 1978), and have tough rachises that do not disarticulate at maturity. In contrast, the species of *Eremopyrum* and *Henrardia* are, annuals, facultative autogamous, with a disarticulating rachis (wedge-type in *Eremopyrum* and barrel-type in *Henrardia*). The diploid species of these two genera have presumably evolved from diploid *Agropyron*.

Phylogenetic studies, based on morphology (Seberg and Frederiksen 2001), chloroplast DNA (Mason-Gamer et al. 2002; Hodge et al. 2010), chloroplast, mitochondrial, and nuclear DNA sequences (Seberg and Petersen 2007) and nuclear genes (Hsiao et al. 1995a, b; Mason-Gamer et al. 2010), separated *Agropyron* from *Elymus*, and included the genus *Eremopyrum* in a clade with *Agropyron*. The close phylogenetic relationship between *Agropyron* and *Eremopyrum* is also evident from the data of Escobar et al. (2011), who placed these two genera in the same clade. *Eremopyrum* species were grouped with *Agropyron* and *Henrardia* on the chloroplast DNA tree (Mason-Gamer et al. 2002). Similarly, using β-amylase gene sequences, Mason-Gamer (2005) found that *Henrardia persica* is close to *Eremopyrum bonaepartis*. Placement of both *Eremopyrum* and *Henrardia* with *Agropyron* has also been supported by other data sets (reviewed in Mason-Gamer 2005). reached a similar conclusion in analysis of nuclear DNA sequences. Analysis of the chloroplast gene encoding ribosomal protein rps16, led Hodge et al. (2010) also to place *Eremopyrum bonaepartis* and *Henrardia persica* in a single clade. Upon combination of Triticineae species mating system observations and data obtained from molecular analysis of 27 protein-coding loci, Escobar et al. (2010) found that *Henrardia persica* is very close to *Eremopyrum bonaepartis* and forms a clade with *Eremopyrum triticeum* and *Agropyron mongolicum*. Likewise, Escobar et al. (2011) found two well-supported sub-clades, the first formed by *Henrardia* and *Eremopyrum bonaepartis*, and the second by *Agropyron mongolicum* and *E. triticeum*. The study of Escobar et al. (2011) showed that the genera *Agropyron* (*Astralopyrum* is included), *Eremopyrum* and *Henrardia* were not affected much by introgression and/or incomplete lineage sorting. Yet, a number of hybrids between *Agropyron* and *Elymus* species have spontaneously emerged in nature. While many hybrids are sterile, a considerable number are more or less fertile, at least upon spontaneous backcrossing to one of the parents. Apparently, introgressive hybridization may have played a role in the evolution of these two genera.

11.3.2 *Agropyron* Gaertn.

The genus *Agropyron* sensu stricto contains diploid and polyploid species that are based on the P genome (Table 5.3) and are morphologically distinct from other genera in Triticineae. The diploid cytotype of *Agropyron cristatum* and *A. mongolicum* are the only diploids in the genus *Agropyron s.*

str. Both are perennial and cross-pollinating species, but differ morphologically and in their geographical distribution (Table 5.3) The two species hybridize readily (Dewey and Hsiao 1984) and the F_1 hybrids showed reasonably good chromosome pairing at first meiotic metaphase, with an average of five to six bivalents per cell. Dewey (1969) concluded that all *Agropyron* species, whether diploids, autotetraploids, or autohexaploids, contain one basic genome, P, implying that autopolyploidy played an important evolutionary role in this genus.

Studies of meiotic chromosome pairing in hybrids between diploid *A. cristatum* and several different diploid species of *Elymus* showed that the P genome is closely related to the St genome and moderately related to the Ee and Eb genomes, respectively (Wang 1985, 1986, 1989, 1992). This may imply that diploid *Agropyron* derived from *Elymus* species possessing the St-genome. This assumption is supported by the analysis of 5S DNA sequences that consistently placed *Elymus* species with an St genome and *Agropyron* in one clade (Baum and Appels 1992).

The internal transcribed spacer (ITS) region of nuclear ribosomal DNA sequence phylogeny indicated that the endemic Australian grasses *Australopyrum pectinatum* (genome W) are closely related to species of *Agropyron* (genome P) (Hsiao et al. 1995a, b). Species of the W and P genomes share certain gross morphological similarities and *Australopyrum* was once treated as a member of *Agropyron* (Löve 1984). The karyotypes of P and W genome species are also similar, but the chromosomes of the W genome are smaller (Hsiao et al. 1986). The differences in chromosome size could simply be due to a low copy number of the repetitive DNA, because the chromosomes of *Australopyrum pectinatum* ssp. *velutinum* contain much less C-banded heterochromatin than do those of *Agropyron cristatum* (Endo and Gill 1984).

11.3.3   Eremopyrum (Ledeb.) Jaub. and Spach

The *Eremopyrum* genus includes both diploid and allotetraploid taxa, namely, diploids *E. bonaepartis*, *E. distans*, and *E. triticeum*, and allotetraploids *E. bonaepartis* (=*E. confusum*) and *E. orientale* (Sakamoto 1972; Table 5.4). Following the genome analysis performed by Sakamoto (1979), the genome symbol of the *Eremopyrum* species are as follows: *E. triticeum* FF, diploid *E. bonaepartis* XbXb, tetraploid *E. bonaepartis* XbXbXdXd, *E. distans* XdXd, *E. orientale* XdXdFF (Table 5.4).

All *Eremopyrum* species are annual, short plants, with solitary spikelets at each rachis node, and with spikes that disarticulate at maturity (wedge-type disarticulation). Only in *E. triticeum* the disarticulation is at the base of each floret (floret-type disarticulation). The two tetraploid species are of recent origin, and most probably arose in the dry steppe zones of northwestern Iran, the assumed distribution center of this genus (Sakamoto 1979).

Studies of chromosome pairing at meiosis of interspecific *Eremopyrum* hybrids showed very little pairing between the diploid species, indicating that their genomes diverged considerably from one another, and that the tetraploid species are allotetraploids (Sakamoto 1972). Intergeneric hybridizations showed that there are strong sterility barriers between *Eremopyrum* species and those of other Triticeae genera (Sakamoto 1967, 1968, 1972, 1974; Frederiksen 1991b, 1993, 1994; Frederiksen and von Bothmer 1995). Sakamoto (1974) succeeded in producing the hybrid *Heteranthelium piliferum* x diploid *Eremopyrum bonaepartis*, which exhibited abnormal growth and very little chromosomal pairing at first meiotic metaphase (an average of 0.04 bivalents per cell).

Phylogenetic studies, based on morphology (Seberg and Frederiksen 2001), chloroplast DNA (Mason-Gamer et al. 2002; Hodge et al. 2010), chloroplast, mitochondrial, and nuclear DNA sequences (Seberg and Petersen 2007; Escobar et al. 2011) and nuclear genes (Hsiao et al. 1995a, b; Mason-Gamer et al. 2010), included *Eremopyrum* species in the same clade with species of *Agropyron s. str.*

11.3.4   Henrardia C.E. Hubbard

Henrardia is a small genus containing two species, *H. persica* and *H. pubescens* (Hubbard 1946). The two species differ morphologically from other genera in the Triticineae subtribe. Both species are annuals, short plants, with facultative self-pollination. The rachis harbors a solitary spikelet at each node, which are fragile, and disarticulate at maturity with the rachis segment alongside it (barrel type).

The two species are diploids (Sakamoto and Muramatsu 1965; Sakamoto 1972; Bowden 1966) and have a unique, extremely asymmetric karyotype, comprised of large chromosomes, and different from those of all other Triticineae, (Asghari-Zakaria et al. 2002), indicating an advanced genus.

Sakamoto (1972) crossed *H. persica,* as either female or male parent, with a number of species from different Triticeae genera, but only obtained hybrids in the cross of tetraploid *Eremopyrum orientale* x *H. persica*. The F_1 hybrid showed the wedge-type disarticulation of the *Eremopyrum* parent. Chromosomal pairing at first meiotic metaphase was very low (0–4 bivalents per cell), indicating lack of homology between the genomes of these two species.

Because of its very peculiar morphology, *Henrardia* was earlier included in genera outside the Triticineae, but Hubbard (1946) noticed that *Henrardia* shares several diagnostic traits with the Triticineae that have been regarded to be of diagnostic value in distinguishing the Triticeae tribe from other Poaceae tribes. Consequently, Hubbard (1946) transferred this taxon as a new genus to Triticineae.

Clayton and Renvoize (1986) considered *Henrardia* an offshoot of *Aegilops*. This is in accordance with Kellogg (1989) and Frederiksen and Seberg (1992), who, based on a cladistics analysis, concluded that *Henrardia* and the diploid species of *Aegilops* form a clade. But, in a number of molecular studies, *Henrardia* was grouped in a clade with the *Eremopyrum* species (see above).

11.4 The *Dasypyrum-Secale-Heterantheliun* Clade

11.4.1 Clade Description

This group contains three genera, two of which, *Dasypyrum and Secale*, include perennial and annual species, and one, *Heteranthelium*, having only one annual species. It is assumed that *Dasypyrum* is the most ancient genus in this clade (Blanco et al. 1996) while *Secale* is a younger genus and *Heteranthelium* is the youngest (Table 11.1). *Dasypyrum* diverged very early from the ancestral Triticineae, about 14 MYA (Yang et al. 2006: Table 11.1), during the early stages of separation between Triticineae and Hordeineae. Marcussen et al. (2014) concluded that *Secale* diverged from the ancestors of the wheat group during the Miocene, about 7 MYA, and probably at that time or somewhat earlier from *Dasypyrum*. *Heteranthelium* is the youngest in the group since it has more advanced traits than *Dasypyrum* and *Secale* (Table 5.5).

11.4.2 Dasypyrum (Coss. and Durieu) T. Durand

The genus *Dasypyrum* differs from the other Triticineae genera by its distinctive two-keeled glumes, with tufts of bristles along the keels. This genus comprises two allogamous species with a fragile rachis (wedge-type disarticulation), *D. villosum* (=*Haynaldia villosa*) and *D. breviaristatum* (=*D. hordeaceum*). *D. villosum* is an annual diploid (Frederiksen 1991a), whereas D. *breviaristatum* contains two cytotypes, an annual diploid and a perennial autotetraploid (Sarkar 1957; Ohta et al. 2002).

Wang et al. (1995) used the symbol Vb for the haploid genome of the diploid cytotype of *D. breviaristatum* and Vv for the haploid genome of *D. villosum*. Indeed, meiotic chromosome pairing in the interspecific F$_1$ hybrid *D. villosum x diploid D. breviaristatum* was very poor (1.44 bivalents per cell) and the hybrid was almost completely sterile, supporting the notion that the genomes of these two species are only distantly related to each other (Ohta and Morishita 2001). Yet, the shared peculiar spike morphology,

indications of molecular and biochemical markers (Blanco et al. 1996), and hybridization with the species-specific repeated sequence pHv62 of *D. villosum* (Uslu et al. 1999), suggest a common ancestry for these two species. The genomic distance between *D. villosum* and *D. breviaristatum*, as determined by 301 RAPD loci, is smaller than their distance from *Secale* species (Yang et al. 2006).

Very low pairing was reported in F$_1$ hybrids involving *Dasypyrum* species and many other Triticeae species, showing little chromosome homology (Oehler 1933, 1935; Sando 1935; Kihara and Lilienfeld 1936; von Berg 1937; Kihara 1937; Kihara and Nishiyama 1937; Sears 1941b; Nakajima 1966; Chen and Liu 1982; Blanco et al. 1983a, 1983b, 1988; Lucas and Jahier 1988; von Bothmer and Claesson 1990; Yu et al. 1998, 2001; Deng et al. 2004; de Pace et al. 2011). Interestingly, low chromosome pairing was also observed in hybrid of *D. villosum* and *Secale cereale* that are included in the same phylogenetic clade. Jahier et al. (1988) crossed two amphiploids: *Ae. uniaristata–D. villosum* (2n = 28; genome NNVvVv) and *Ae. uniaristata–S. cereale* (2n = 28; genome NNRR). Despite of the low pairing in this hybrid, Jahier et al. (1988) did not reject the hypothesis that Vv and R chromosomes share homologous sequences. Rather, they attributed the Vv and R lack of pairing to factors such as asynchronous meiotic rhythm between R and Vv genomes. Similar causes can explain the lack of pairing between Vv and Vb reported above.

Morphology-based phylogenetic analyses showed that *Dasypyrum* and *Secale* form one clade (Baum 1978a, b, 1983; Kellogg 1989; Frederiksen and Seberg 1992; Seberg and Frederiksen 2001). Kellogg (1989) also placed *Dasypyrum* near *Agropyron* and *Triticum monococcum*, and Baum (1978a, b, 1983) considered *D. villosum* and *Secale cereale* as evolutionarily more contiguous to *Triticum* and *Aegilops* than to the rest of the Triticeae tribe. This is in accord with molecular phylogenetic studies which suggested that *Secale* is the closest relative of the *Triticum-Aegilops* genera (Kellogg et al. 1996; Huang et al. 2002a; Mason-Gamer et al. 2002). However, when *Hordeum* and *Dasypyrum* were analyzed with *Secale, Triticum* and *Aegilops*, they were positioned at the base of the tree topology, as out-groups (Yamane and Kawahara 2005; Kawahara et al. 2008), implying a much earlier divergence between *D. villosum* and the common ancestor of *Triticum-Aegilops*.

Analysis of different chloroplast and nuclear DNA sequences placed *Elymus* species possessing the E and St genomes and *Dasypyrum* species closely together (Kellogg 1992; Mason-Gamer and Kellogg 1996b, 2000; Petersen and Seberg 1997). Similarly, molecular phylogeny of the gene sequence encoding the second-largest subunit of RNA polymerase II revealed that the Vv genome of *D. villosum* is sister to the St genome of *Elymus* and that both diverged from the H-genome of barley (Sun et al. 2008). These findings fall in line with phylogenetic relationships inferred from nuclear rDNA sequences, showing a close relation of *Heteranthelium* and *Dasypyrum* to *Elymus* (Hsiao et al. 1995b). The *Heteranthelium* element of the transposon *Stowaway* is present in *Dasypyrum* but absent in other Triticeae species (Petersen and Seberg 2000). DNA/DNA hybridization of the genomes of *Secale* and *D. villosum* with labeled nuclear DNA from wheat and rye, revealed greater homology between the Vv genome of *D. villosum* and R genome of *Secale* than between Vv and the A-, B-, and D-subgenomes of wheat (Lucas and Jahier 1988). Similarly, FISH analysis involving hybridization of the genomes of different Triticeae species with species-specific molecular probes prepared from tandem repeated DNA sequences of *D. villosum* (pHv62) and *S. cereale* (pSc119.2), demonstrated greater homology between the R- and Vv-genomes than between R- or Vv-genomes and those of *Triticum* and *Aegilops* (Uslu et al. 1999).

Escobar et al. (2011) found that *Dasypyrum, Heteranthelium, Secale, Taeniatherum, Triticum* and *Aegilops*, evolved in a reticulated manner. They found that *Dasypyrum* and *Heteranthelium* form one clade, while *Secale, Taeniatherum, Triticum* and *Aegilops* formed another clade. *Elymus* forms a sister clade to *Dasypyrum* and *Heteranthelium*.

The formation of the *Dasypyrum* species was explained by a cascade of events, which began in the earlier stages of the separation of the Triticineae from the Hordeineae (13–15 MYA), and continued through the reproductive isolation of *D. villosum* from diploid *D. breviaristatum* prototype, and incipient reproductive isolation between the two cytotypes (Blanco et al. 1996). Such divergence did not occur for syntenic and gene-rich DNA segments of genomes Vv and Vb, as suggested by the strong similarity between the genomes of these two species in restriction fragment patterns of genomic DNA, the phenotypes of some isozyme systems, and the location of gliadin genes (Blanco et al. 1996). Thus, *Dasypyrum* was one of the earliest Triticineae genera that diverged from the basal *Elymus* clade. Likewise, Yang et al. (2006), studying genome relationships based on species-specific PCR markers, concluded that the formation of the *Dasypyrum* species began at the earlier stages of the separation of the sub-tribe Triticineae from the sub-tribe Hordeineae (13–15 MYA). Lucas and Jahier (1988) concluded that the differentiation of *D. villosum* (and *Secale cereale*) from the genera *Aegilops* and *Triticum* occurred earlier than speciation in the latter two genera. Indeed, Marcussen et al. (2014) proposed that the *Secale* lineage diverged from the *Aegilops-Triticum* lineage about 7 MYA, while Middleton et al. (2014) proposed it emerged 3–4 MYA. Lucas and Jahier (1988) assumed that the closest *Aegilops-Triticum* species to *D. villosum* is diploid wheat *T. monococcum* subsp. *aegilopoides* and not diploid species of *Aegilops*.

11.4.3 *Secale* L.

Secale is a small genus including one perennial, *S. strictum* and two annual species, *S. sylvestre* and *S. cereale* (Table 6.1). Most lines of the perennial *S. strictum* ssp. *strictum* and the annual *S. cereale* are self-incompatible, whereas the perennial *S. strictum* ssp. *africanum*, the annual *S. sylvestre* and some lines of *S. cereale* are self-compatible (Jain 1960; Kranz 1963; Kuckuck and Peters 1967; Stutz 1972). All three *Secale* species are diploids, (Sakamura 1918; Stolze 1925; Aase and Powers 1926; Thompson 1926; Emme 1927; Lewitsky 1929, 1931; Jain 1960; Bowden 1966; Love 1984; Petersen 1991a, b) and their genome has been designated R (Love 1984; Wang et al. 1995). B chromosomes occur in a low frequency in several lines of *S. cereale* and in a few populations of *S. strictum* subsp. *strictum* (Emme 1928; Darlington 1933; Hasegawa 1934; Popoff 1939; Müntzing 1944, 1950; Kranz 1963; Jones and Rees 1982; Niwa et al. 1990).

The *Secale* genus evolved monophyletically (Hsiao et al. 1995a; Mason-Gamer et al. 2002; Petersen et al. 2004; Bernhardt 2016). Its genome differs from that of wheat in both size and structure (Gill and Friebe 2009). Its size is the largest in the Triticineae (1C DNA of *S. cereale* = 8.65 pg, and that of *S. strictum* ssp. *strictum* = 9.45 pg; Table 6.3) and is 33 to 45% larger than the genome of diploid wheat (Table 3.1). This size difference is mainly due to the large amount of heterochromatin in *Secale*. Thus, one of the major evolutionary changes in chromosome structure in *Secale* has involved the addition of heterochromatin close to the telomeres (Bennett et al. 1977). The occurrence and distribution of the sub-telomeric heterochromatin in the different *Secale* species suggest that *S. sylvestre*, having a relatively small amount of heterochromatin, may be of ancient origin, while *S. strictum* and *S. cereale* may have a more recent origin (Jones and Flavell 1982).

In addition to increase in sub-telomeric heterochromatin, translocations have also played an important role in the evolution of the genus *Secale* (Stutz 1972; Koller and Zeller 1976; Shewry et al. 1985; Naranjo et al. 1987; Naranjo and Fernández-Rueda 1991; Liu et al. 1992; Rognli et al. 1992; Devos et al. 1993; Schlegel 2013).

Martis et al. (2013), using molecular data, suggested that also introgression from other Triticeae species may have played a role in *Secale* speciation and R genome evolution. However, the rarity of spontaneous formation of intergeneric hybrids involving *Secale* species and other Triticeae (Frederiksen and Petersen 1998), may indicates that introgression played a minor evolutionary role. Moreover, artificial intergeneric hybrids between *Secale* species and other Triticeae exhibited very low levels of pairing (Stebbins and Pun 1953; Heneen 1963b; Majisu and Jones 1971; Hutchinson et al.

1980; Gupta and Fedak 1985, 1987a, b; Fedak and Armstrong 1986; Wang 1987, 1988; Lu et al. 1990; Lu and von Bothmer 1991; Petersen 1991b). This low level of pairing indicates a very distant relationship between the R and the H genomes (Fedak 1979, 1986; Finch and Bennett 1980; Fedak and Armstrong 1981; Thomas and Pickering 1985; Gupta and Fedak 1985, 1987a, b; Lu et al. 1990; Petersen 1991a, 1991b), and that the R genome is slightly closer to the Eb genome than to the St genome (Wang 1987, 1988). Similarly, very little homology was observed between genomes of *Aegilops* and Secale species (Karpechenko and Sorokina, 1929; von Berg, 1931; Kagawa and Chizaki, 1934; Melnyk and Unrau, 1959; Majisu and Jones, 1971; Hutchinson et al. 1980; Lucas and Jahier 1988; Kawakubo and Taira 1992; Su et al. 2015), and between the genomes of diploid wheat and *S. cereale* (Sodkiewicz 1982), as well as between subgenomes of allopolyploid wheats and *S. cereale* (Longley and Sando 1930; Plotnikowa 1930; Oehler 1931; Vasiljev 1932). The absence of *Ph1* has a much smaller effect on pairing between wheat and Secale chromosomes than between chromosomes of wheat subgenomes (Miller et al. 1994). Naranjo and Fernández-Rueda (1991, 1996) and Cuadrado et al. (1997) found that most of the wheat-rye pairs in the absence of *Ph1* involved B subgenome chromosomes and, to a much lesser degree, D and A subgenome chromosomes.

The pairing data above show that Secale chromosomes underwent significant changes during the evolution of the genus, which affected their ability to homoeologously pair with other Triticineae species. Such alterations include, in addition to chromosomal rearrangements, also accumulation of large amounts of telomeric heterochromatin. As a result, the relative position of the telomeric regions that are involved in the commencement of meiotic pairing may shift, impairing pairing initiation of rye and other Triticineae chromosomes (Devos et al. 1995; Lukaszewski et al. 2012; Megyeri et al. 2013). Genetic and epigenetic changes, such as mutations or elimination of DNA sequences that are involved in homology recognition and pairing initiation, may also underlie this restricted pairing.

Biochemical and molecular studies have shown that *S. sylvestre* occupies an isolated position within the genus and differs substantially from both *S. strictum* and *S. cereale*. Its isolation may have resulted from its autogamous habit (Schiemann and Nürnberg-Krüger 1952; Khush and Stebbins 1961) or from its characteristically low sub-telomeric heterochromatin content, which results in unsynchronized mitotic cycles in embryos of hybrids with other Secale species that have larger amounts of sub-telomeric heterochromatin (Singh 1977). In contrast, hybrids between *S. strictum* and *S. cereale* are easily formed and exhibit somewhat reduced fertility, possibly because they are heterozygous for the two chromosomal translocations distinguishing *S. cereale* from *S. strictum* (Khush and Stebbins 1961; Khush 1962; Singh 1977). In areas where *S. strictum* grows near or even within *S. cereale* cultivated fields, plants resulting from hybridization and introgression between these two species were frequently observed (Stutz 1957; Khush 1962; Perrino et al. 1984; Hammer et al. 1985; Zohary et al. 2012). Biochemical and molecular studies showed that differences between *S. cereale* and *S. strictum* are not extensive (Jaaska 1975; Vences et al. 1987; Dedio et al. 1969; Sencer 1975; Reddy et al. 1990; de Bustos and Jouve 2002; Jones and Flavell 1982b; Murai et al. 1989). Yet, although *S. cereale* and *S. strictum* specimens are intermingled on the phylogenetic tree (Frederiksen and Petersen 1997), Bernhardt (2016) showed that *S. strictum* is somewhat different from *S. cereale*.

According to Kobyljanskij (1982), *Protosecale,* the oldest ancestor of the genus, appeared in the Oligocene Epoch (33.7–23.8 MYA) and later evolved in the Miocene (23.8–5.3 MYA) into a *Protosylvestre* and *Protostrictum* forms, from which *S. sylvestre* and *S. strictum* developed during the Pliocene epoch (5.3–1.8 MYA) (Table 2.6). The divergence of the *Hordeum* lineage and the *Secale/Triticum/Aegilops* lineage(s) occurred 8–15 MYA; Wolfe et al. (1989), Ramakrishna et al. (2002), and Marcussen et al. (2014) suggested that this event occurred 10–15 MYA, On the other hand, Huang et al. (2002a, b), Dvorak and Akhunov (2005), Chalupska et al. (2008), Gornicki et al. (2014), and Middleton et al. (2014) suggested that this divergence occurred 8–11 MYA (Table 2.6). The separation of the Secale lineage from the *Triticum/Aegilops* lineage occurred 7 MYA (Marcussen et al. 2014, based on nuclear genes), when the ancestral genomes of the wheat group (A. muticum, Ae. speltoides and diploid wheat) evolved.

The phylogenetic relationships between *Secale* and other Triticeae genera have been studied through morphological traits, genome analysis, isozymes, and cytoplasmic and nuclear DNA sequences, which have yielded contradictory results regarding the position of Secale. Taxonomical treatments by several well-known taxonomists (e.g., Nevski 1933; Melderis 1953; Tzvelev 1973, 1976) placed the genus Secale close to the genus *Dasypyrum*. In accord with this taxonomical treatment, Baum (1983), on the basis of a phylogenetic analysis of Triticeae by means of numerical methods, also grouped Secale and Dasypyrum close to one another. Further analysis of morphological characteristics suggested that Secale is the sister group of a clade consisting of *Dasypyrum villosum, Triticum monococcum* and *Aegilops* species (Frederiksen and Seberg 1992; Seberg and Frederiksen 2001; Seberg and Petersen 2007). In some molecular studies, Secale was classified as a sister clade to the *Aegilops* clade (Kellogg and Appels 1995; Mason-Gamer and Kellogg 1996a). Monte et al. (1993) place Secale in close associations with *Agropyron* and *Elymus* species bearing the E^e and E^b genomes.

Data from internal transcribed spacers (ITS) of the rDNA suggested that *Secale* is the sister group of *Eremopyrum* and *Henrardia* (Hsiao et al. 1995a), whereas data from the spacers between the 5S RNA genes suggested a rather basal position for *Secale* within Triticineae (Kellogg and Appels 1995; Kellogg et al. 1996). Mason-Gamer (2005), analyzing sequences from a portion of the tissue-ubiquitous β-amylase gene in a broad range of the mono-genomic Triticeae, found close relationships between *Secale*, *Australopyrum* (=*Agropyron*) and *Dasypyrum*. Sequencing of the ITS region of nuclear rDNA of diploid Triticeae species, brought Hsiao et al. (1995a) to conclude that *Secale* is close to *Taeniatherum* and sister clade to *Elymus farctus E. elongatus* and *Triticum monococcum*. The study of Seberg and Petersen (2007) indicated incongruence between morphological and molecular data sets. They concluded that *S. strictum* is close to *Taeniatherum caput-medusae*, followed by *Dasypyrum villosum*, *Elymus elongatus Elymus bessarabicus*, *Crithopsis delileana* and the genera *Aegilops*, *Triticum* and *Amblyopyrum*.

Thus, the phylogenetic position of *Secale* remains ambiguous. In conclusion, with respect to the placement of the genus *Secale* in the subtribe-wide phylogeny, virtually all genera of the Triticineae have been suggested–either alone or in combination with other genera–to be a sister group to *Secale* (Petersen et al. 2004).

11.4.4 *Heteranthelium* Hochst

Heteranthelium is a monotypic genus containing the species *H. piliferum* (Banks et Sol.) Hochst. It is an annual, facultative self-pollinating species (Luria 1983), with a very peculiar spike morphology that is different from that of other Triticeae genera. It is a diploid species (Sakamoto and Muramatsu 1965; Bowden 1966), with a symmetric karyotype (Chennaveeraiah and Sarkar 1959; Bowden 1966; Sakamoto 1974; Frederiksen 1993).

Chromosome pairing in F_1 hybrid showed that the genome of *H. piliferum* is distantly related to that of *E. bonaepartis* (Sakamoto 1974). Other artificial hybridizations involving *H. piliferum* and other Triticeae species were not successful. Considering the unique morphology of the spike and the dispersal unit, inter-generic crossability and the cytogenetic relationships of *H. piliferum*, Sakamoto (1974) concluded that the genus *Heteranthelium* is a distinctive entity, representing a specialized group that occupies an isolated position in the subtribe Triticineae.

In different taxonomic classifications of the Triticeae, *Heteranthelium* is supposed to be related to either *Triticum/Aegilops* complex (Nevski 1934b; Tzvelev 1976) or *Hordeum* (Love 1984; Clayton and Renvoize 1986; Kellogg 1989). Clayton and Renvoize (1986) regarded it as an advanced offshoot of *Crithopsis*. Yet, *Heteranthelium* has one spikelet per node like most species of the sub-tribe Triticineae, and awn-like glumes as in *Hordeum* (Frederiksen 1993). Thus, the phylogenetic relationships of *Heteranthelium* are still ambiguous. Studies of Hodge et al. (2010) of the chloroplast gene encoding ribosomal protein S16, showed that *H. piliferum* is in the same clade as *Triticum monococcum*, *Secale cereale*, and all the *Aegilops* species. Their results are consistent with the finding of Mason-Gamer et al. (2002), based on combined cpDNA sequences, of tRNA genes, spacer sequences, *rpoA* genes, and restriction sites. Phylogenetic relationships based on mating systems showed that *H. piliferum* is in a clade with *Dasypyrum villosum* (Escobar et al. 2010, 2011).

11.5 The *Taeniatherum–Crithopsis* Clade

The *Taeniatherum* and *Crithopsis* genera consist of annual, facultative autogamous, monotypic species, that have two sessile spikelets at each rachis node and contain only one hermaphrodite floret in each spikelet. *Taeniatherum caput-medusae* and *Crithopsis delileana* are considered taxonomically close to each other (Clayton and Renvoize 1986). The species of both genera are diploids (Sakamoto and Muramatsu 1965; Luria 1983) with a symmetric karyotype; the genome symbol of *Taeniatherum* is Ta (Wang et al. 1995) and that of *Crithopsis* is K (Löve 1984). The karyotype of *C. delileana* is similar to that of *T. caput-medusae*, as shown by Linde-Laursen and Frederiksen (1989). However, the C-banding patterns of the two species exhibit differences, with *C. delileana* having more telomeric and fewer intercalary bands than *T. caput-medusae*. Yet, altered distribution of C-bands is a weak diagnostic characteristic, as activity of repetitious sequences, that can affect the distribution and quantity of the C-banding, may be different in closely related species and even within a species. *Taeniatherum* and *Crithopsis* differ in their seed-dispersal system; *Taeniatherum* has a tough rachis and a disarticulating rachilla (like *E. elongatus*), so that the dispersal unit is a floret, whereas *Crithopsis* has a brittle rachis and wedge-type dispersal unit of spikelets.

The crossability of either *T. caput-medusae* or *C. delileana* with other Triticeae species has been difficult, and consequently, the cytogenetic relationships between these two species and other Triticeae species are poorly known. However, in the few successful intergeneric crosses, the F_1 hybrids exhibited scarce amounts of chromosomal pairing at meiosis (Sakamoto 1991; Frederiksen and von Bothmer 1989), indicating great divergence of the genomes of *Taeniatherum* and *Crithopsis* from those of the other Triticeae.

Arterburn et al. (2011) found close homology between DNA sequences of *Crithopsis delileana* and *Taeniatherum caput-medusae.* Moreover, these authors also found sizeable homology between these two species and diploid *Elymus elongatus* and *E. farctus*. The relationships between diploid *E. elongatus* and *C. delileana* were closer than those between *E. elongatus* and *T. caput-medusae.*

Based on the possession of two spikelets at each rachis node and one grain in each spikelet, taxonomists (e.g., Tzvelev 1976; Clayton and Renvoize 1986) assigned the two genera to the subtribe Hordeineae. Indeed, in most morphological trees, *Taeniatherum* and *Crithopsis* were linked to the *Hordeum* group (Baum 1983; Baum et al. 1987; Kellogg 1989; Frederiksen and Seberg 1992). Yet, the traditional taxonomic subdivision of the genera of the Triticeae into two sub-tribes, Hordeineae and Triticineae (Tzvelev 1976; Clayton and Renvoize 1986), is not supported by phylogenetic studies based on molecular analyses that placed *Taeniatherum* and *Crithopsis* closer to species of the Triticineae than to *Hordeum* (Hsiao et al. 1995a; Mason-Gamer and Kellogg 1996b; Petersen and Seberg 1997; Mason-Gamer et al. 2002; Seberg and Petersen 2007; Escobar et al. 2011). Mason-Gamer et al. (2002), analyzing new and previously published chloroplast DNA data from *Elymus* and from most of the mono-genomic genera of the Triticeae, concluded that their analysis agrees with previous cpDNA studies with regard to the close relationship between *Secale, Taeniatherum, Triticum,* and *Aegilops.* Escobar et al. (2011) reported that the clade containing *Taeniatherum* and *Triticum–Aegilops* is also seen on the 5S short-spacer data tree, but only if *Elymus farctus* and *E. elongatus* are included in the clade.

Sakamoto (1973, 1991) classified the Triticeae into two groups based on their geographical distribution: the Arctic-Temperate group and the Mediterranean-Central Asiatic group (Table 2.3). Hsiao et al. (1995a), analyzing ITSs of nuclear rDNA and sequences of tRNA in 30 diploid Triticeae species representing 19 genomes, found that most of the annuals of Mediterranean origin, i.e., species of *Triticum, Aegilops, Crithopsis, Taeniatherum, Eremopyrum, Henrardia, Secale,* and two perennials, *Elymus farctus* and *Elymus elongatus,* comprise a monophyletic group. In the parsimony tree from a more restricted species sampling, the two-perennial species *Elymus farctus* and *E. elongatus,* formed a sister group with *Triticum monococcum, Aegilops speltoides,* and *Ae. tauschii,* whereas *Crithopsis, Taeniatherum, Eremopyrum,* and *Henrardia,* were close to *Secale.* Based on this finding, Hsiao et al. (1995a) supported Sakamoto (1973), who suggested that the Triticeae should be classified as two major groups, a Mediterranean group and an Arctic-temperate group, where the Mediterranean lineage evolved from the Arctic-temperate species (Runemark and Heneen 1968; Sakamoto 1973). The inclusion of *Crithopsis* and *Taeniatherum* in the Mediterranean-Central Asiatic group is not merely due to their geographical distribution,

but also to close phylogenetic relationships between the two genera, *Crithopsis* and *Taeniatherum,* and other Triticineae species. Their relationships with genera of the Triticineae may indicate their evolvement from the same ancestral group and suggest that evolutionary development of diagnostic morphological characteristics, e.g., number of spikelets on each rachis node and number of fertile florets in each spikelet, occurs at varying rates in different taxa. The dispersal unit of many species belonging to the Mediterranean-Central Asiatic group contains two seeds, either one in each of the two spikelets or two in a single spikelet, indicating different routes for achieving analogous adaptive traits to brace the long, dry summer of the Mediterranean and Central Asiatic regions.

The above molecular phylogenetic studies showed that *Taeniatherum* and *Crithopsisis* are closer to the species of the subtribe Triticineae than to those of the Hordeinae. Consequently, these two genera are included in this book as members of the Triticineae. *Crithopsis* contains more advanced traits than *Taeniatherum* (Table 5.5) and can be considered a younger genus, more advanced than *Taeniatherum.*

11.6 The *Amblyopyrum-Aegilops-Triticum* Clade

11.6.1 Clade Description

This clade comprises three genera, *Amblyopyrum* (contains one diploid species), *Aegilops* (contains 10 diploid, 10 allotetraploid, and 4 allohexaploid species; Table 9.3), and *Triticum* (contains 2 diploid, 2 allotetraploid, and 2 allohexaploid species (Table 10.5). All species are annuals, and two of the three basal species, *A. muticum* and *Ae. speltoides* are allogamous, with the former being self-incompatible (Kimber and Feldman 1987), and the latter being a predominantly cross-pollinator (Zohary and Imber 1963). *Ae. longissima* has been described recently as a facultative outcrossing species (Escobar et al. 2010), whereas the remainder of the species are facultative autogamous. Many of the 13-diploid species of the wheat group are differentiated from each other by their unique spike and spikelet features and specialized dispersal units, namely, wedge, barrel, and umbrella types. The diploid species also differ in their eco-geographical requirements, and distinguished well-defined habitats. These species have distinct genomes, with different genome sizes (Table 9.3) and pairing patterns in inter-specific and inter-generic hybrids (Table 9.4).

The particular eco-geographical affinities of many of the diploid species may indicate that they had undergone extensive differentiation in their early stages of development. Southeast Turkey is the geographical center of the group distribution and thus, is presumably the center of origin of the genus (Kimber and Feldman 1987). Hammer (1980) suggested that the Fertile Crescent region, which currently maintains the greatest diversity of the wheat group genera, is not necessarily the region in which the group originated and developed during its evolution. He assumed that the primary center of origin of the group was in Transcaucasia, a region where several diploid species formed and developed and later, due to climatic change, migrated in western, southern and eastern directions.

On the basis of karyomorphological studies, Senyaninova-Korchagina (1932) and Chennaveeraiah (1960) separated the diploids into two categories: those having a symmetric karyotype (the T–, A–, S–, and D– genome species) and those with an asymmetric karyotype (the C–, M–, N–, and U– genome species). Avdulov (1931) and Stebbins (1950, 1971) considered an asymmetrical karyotype more advanced than a symmetric karyotype, since the former is found in diploids with increased specialization with respect to two morphological characteristics: the type of rachis fragility and the number of awns on the glumes. Indeed, in the more primitive species, i.e., *A. muticum,* wild *T. monococcum, Ae. speltoides* ssp. *ligustica, Ae. bicornis* and *Ae. sharonensis,* the dispersal unit is wedge-type and the glumes are awnless. In *Ae. tauschii,* the dispersal unit is barrel-type and the glumes are awnless. In *Ae. speltoides* ssp. *speltoides, Ae. longissima,* and *Ae. searsii,* the spike is long, with awns only on the lemmas of the uppermost spikelet and the dispersal unit is umbrella-type, but their glumes are awnless, whereas in the more advanced species, namely, *Ae. caudata, Ae. comosa, Ae. uniaristata,* and *Ae. umbellulata,* the spike is shorter, the dispersal unit is the umbrella type and the glumes are awned. Thus, since most species of the Triticeae have median or sum-median centromeres, the ancestral taxon(s) of the genera *Amblyopyrum, Aegilops* and *Triticum* presumably had a symmetric karyotype comprised only of chromosome pairs with median or submedian centromeres and awnless glumes.

Parisod and Badaeva (2020) studied the interplay between hybridization, chromosomal evolution and biological diversification of the diploid species of the wheat group. Comparative profiling of low-copy genes, repeated sequences and transposable elements among the divergent species, characterized by different karyotypes, highlighted high genome dynamics and shed light on the processes underlying chromosomal evolution in these wild diploid species. One of the hybrid clades (e.g., species of subsection Emarginata of *Aegilops* section Sitopsis and *Triticum*) presents upsizing of metacentric chromosomes, which paralleled the proliferation of specific repeats, thus leading to a large genome size (Eilam et al. 2007), whereas other species (e.g., *Ae. caudata* and *Ae. umbellulata*) showed stable, or

even reduced genome size (Eilam et al. 2007), which was associated with increasing chromosomal asymmetry.

Most morphological and molecular trees share the *Aegilops–Triticum* clade, while the morphological tree produced by Seberg and Petersen (2007) also included *Amblyopyrum* and *Henrardia* in this clade. Morphological trees produced by Kellogg and Appels (1995) and Mason-Gamer and Kellogg (1996a) included *T. monococcum* in the *Secale* clade, which is a sister clade to the *Aegilops* clade. Escobar et al. (2011), based on studies of a large number of nuclear genes, reported that *T. monococcum* is a branched sister of *Ae. tauschii, Ae. speltoides* and *Ae. longissima.* Hsiao et al. (1995a, b) and Kellogg et al. (1996) considered *T. monococcum* as a sister group of *Elymus elongatus.* In line with this, Hsiao et al. (1995b), based on the ITSs of the nuclear rDNA sequences, reported that the Ee and Eb genomes of *Elymus* clustered with the A, B, and D subgenomes of *T. aestivum.* Equally, Wang and Lu (2014) found that both genomes Ee and Eb are closely related to subgenomes A, B, and D of hexaploid wheat, and thus, the wheat diploid genomes may have derived from the E genome(s). In accord with these findings, also Liu et al. (2007), using genomic hybridization (both Southern and in situ hybridization), showed that the St and Eb genomes of *Elymus* are very closely related to the A, B and D subgenomes of *T. aestivum.* Interestingly, from their study of two single-copy nuclear gene (*Acc1* and *Pgk1*) sequences, Fan et al. (2013) concluded that the relationship between *Elymus farctus* subsp. *bessaribicus* (genome Eb) and *Triticum/Aegilops* is closer than between *Elymus elongatus* (genome Ee) and *Triticum/Aegilops.*

A considerable number of the above morphological and molecular studies failed to reach a consensus concerning the phylogenetic relationships between the various diploid species of the wheat group (Bernhardt et al. 2020). This ambiguity is due to either a limited number of samples or to the small number of genes that were analyzed (Hsiao et al. 1995a, b; Kellogg and Appels 1995; Mason-Gamer and Kellogg 1996; Petersen and Seberg 1997; Helfgott and Mason-Gamer 2004; Mason-Gamer 2005; Escobar et al. 2011; Bernhardt 2015; Glémin et al. 2019). Phylogenetic relationships between the diploid species of the wheat group have also been blurred by inter-generic hybridizations and introgression events, indicating an intricate, reticulate pattern of evolution in this group of species (Komatsuda et al. 1999; Nishikawa et al. 2002; Mason-Gamer 2005; Kawahara 2009; Escobar et al. 2011). In addition, an incomplete lineage sorting of ancestral polymorphisms also may lead to incongruent results. Such an intricate pattern of evolution presents a considerable challenge in phylogenetic analyses, since different genes may exhibit conflicting genealogical histories (Escobar et al. 2011).

Table 11.3 and Figs. 10.5 and 11.1 presents data on the time of beginning divergence of lineages and species of the wheat group in million years ago. The ancestral genomes of the group, T, S, and A, started diverging about 7.0–6.5 MYA from an ancestral Triticineae genome. From several morphological traits, shared by the T genome (*Ambliopyrum muticum*) and the Ee genome of *Elymus elongatus*, it can be assumed that the former derived from the latter. It is probable that also the S and the A ancestral genomes derived sequentially from diploid *E. elongatus* or from a closely-related species. Several hundred-thousand years later, the T and the S ancestral genomes started to diverge from one another. Homoploid hybridization involving ancestral S or T, or both, with an A ancestral genome, 6.0–5.0 MYA, formed the D-lineage that included the progenitors of *Ae. tauschii*, Emarginata species (*Ae. bicornis, Ae. searsii, Ae. longissima,* and *Ae sharonensis*), and the more advanced *Aegilops* species (*Ae. caudata, Ae. comosa, Ae. uniaristata,* and *Ae. umbellulata*). Soon after, *Ae. tauschii* (genome D) diverged from the D lineage. The donor of the B subgenome to allotetraploid and allohexaploid what, *T. turgidum* and *T. aestivum,* diverged from *Ae. speltoides* 4.5 MYA, the donor of the G subgenome to *T. timopheevii* diverged from *Ae. speltoides* about 2.85 MYA, the donor of the A subgenome to all allopolyploid wheats diverged from *T. urartu* 1.28 MYA, and the donor of the D subgenome to *T. aestivum* diverged from *Ae. tauschii* about 0.68 MYA. Interestingly, the four diploid progenitors that donated their genomes to allopolyploid wheat, the B, G, A, and D, are currently extinct or yet undiscovered. Similarly, the diploid donor of the X^n subgenome to the allotetraploids *Ae. neglecta* and *Ae. columnaris,* and to the allohexaploid *Ae recta,* and the donor of the X^c subgenome to allotetraploid *Ae. crassa,* and to the allohexaploid *Ae. crassa, Ae. vavilovii,* and *Ae. juvenalis,* are currently extinct or yet undiscovered. Li et al. (2022) explained the extinction of all these diploid donors by assuming that competition with their more fit allopolyploid derivatives, occupying the same habitats, caused their elimination.

Divergence of the diploid species of *Aegilops* started 3.73 MYA when the Emarginata species (*Ae. bicornis, Ae. sharonensis, Ae. longissima,* and *Ae. searsii*) diverged from the D-lineage. Divergence of the diploid species of *Aegilops* from one another started between 2.9 to 2.1 MYA. Within the Emarginata species, *Ae. bicornis* diverged from the lineage *longissima-sharonensis* 2.0 −1.0 MYA, *Ae. searsii* diverged from the other species, most probably *Ae. bicornis,* 1.4 MYA, and Ae. sharonensis diverged from *Ae. longissima* 0.4 MYA. The divergence of the more advanced diploid species of *Aegilops,* i.e., *Ae. caudata, Ae. comosa. Ae. uniaristata,* and *Ae. umbellulata,* from the basal *Aegilops* species, occurred presumably about 2.0 MYA (Marcussen et al. 2014).

Bernhardt et al. (2020) highlighted the contribution of multiple rounds of hybridization and introgression to the evolution of the diploid species of the wheat group. They analyzed DNA sequences of 244 nuclear low-copy genes, evenly distributed across all the chromosomes, as well as genome-wide single nucleotide polymorphisms (SNPs) for all the wild diploid species of the group. The use of a combination of different phylogenetic and network approaches together with advanced statistics revealed ancient complex reticulated processes partly involving many rounds of introgression as well as at least one homoploid hybrid speciation that occurred during the formation of the extant taxa. Based on a comprehensive taxon sampling, Bernhardt et al. (2020) were able to propose a detailed scheme of events that shaped the wild species of the wheat group and which seemed to best reflect the evolution of these species. This scheme of events is much more complex than previously suggested (Marcussen et al. 2014; Sandve et al. 2015; Li et al. 2015; El Baidouri et al. 2017; Huynh et al. 2019).

Marcussen et al. (2014) determined the divergence time of the A and B diploid ancestral genomes from a common ancestor $\sim$ 7 MYA and that these genomes gave rise to the D-lineage through homoploid hybrid speciation 1 to 2 million years later (Table 11.3). The A and B parental lineages contributed equally to the D lineage. This model of homoploid hybrid origin of the D lineage agrees with the fact that lineages A and B are more closely related to D individually than to each other and thus, contradicts a tree-like phylogeny. The majority of the analyses of Marcussen et al. (2014) show a slightly younger divergence of A and D lineages compared with B and D lineages, indicating that gene flow from A to D may have persisted after gene flow from B to D had ceased. Support for a homoploid hybrid origin of the D lineage is found in independent analyses using the genome sequence of bread wheat. Both at the base-pair level and in gene content [International Wheat Genome Sequencing Consortium (IWGSC) 2014], the A and B lineages are more similar to the D genome lineage than they are to each other.

However, Glémin et al. (2019) substantiated the origin of the D lineage through homoploid hybridization but suggested the involvement of the T ancestral genome instead the S ancestral genome. Bernhardt et al. (2020) were able to confirm the evolutionary scenario developed by Glémin et al. (2019), and also to uncover more complex patterns of interspecific gene flow. Their phylogenetic scheme is congruent with the proposed formation of the D lineage [refers to the progenitor of the entire S (Emarginata species) + D (*Ae. tauschii*) + M (*Ae. comosa*) clade]. through homoploid hybrid speciation as suggested by Marcussen et al. (2014) and Huynh et al. (2019), but also proposes, in agreement with Glémin et al. (2019), that ancestral *A. muticum,* rather

Table 11.3 Time of beginning divergence of lineages and species of the wheat group (*Amblyopyrum*, *Aegilops*, and *Triticum*)* in million years ago

Linages and species	Beginning divergence time	Method of Study	Reference
Divergence of ancestral T, S, and A genomes from an ancestral Triticineae genome	6.5–7.0	Nuclear DNA sequences	Marcussen et al. (2014) Huynh et al. (2019)
Divergence of *A. muticum* (genome T) from *Ae. speltoides* (genome S) and the B-subgenome donor	6.37 (6.79–5.97)	Whole genome sequencing	Li et al. (2022)
Formation of the D-lineage (the progenitors of the Emarginata species + *Ae. tauschii* + *Ae. comosa*) via homoploid hybridization of ancestral S and/or T with A genomes	6.0–5.0	Nuclear DNA sequences; Transcriptome data	Marcussen et al. (2014) Glémin et al. (2019) Bernhardt et al. (2020)
Divergence of *Ae. tauschii* from the D-lineage	5.37 (5.58–5.16)	Whole genome sequencing	Li et al. (2022)
Divergence of B-subgenome donor from *Ae. speltoides*	4.49 (4.67–4.31)	Whole genome sequencing	Li et al. (2022)
Divergence of Emarginata species (*Ae. bicornis, Ae. sharonensis, Ae. longissima,* and *Ae. searsii*) from the D-lineage	3.73 (3.88–3.58)	Whole genome sequencing	Li et al. (2022)
Divergence of *Ae. speltoides* from the G subgenome donor to wild *T. timopheevii*	2.85 (3.24–2.51)	Whole genome sequencing	Li et al. (2022)
Divergence of *T. urartu* from the A-subgenome donor	1.28 (1.33–1.22)	Whole genome sequencing	Li et al. (2022)
Divergence of *Ae. tauschii* from the D-subgenome donor	< 0.88	Whole genome sequencing	Li et al. (2022)
Divergence of the diploid *Aegilops* species from one another	4.5–2.5	Two nuclear genes	Huang et al. (2002b) Marcussen et al. (2014)
	2.7 (4.1–1.4)	Four nuclear genes	Dvorak and Akhunov (2005)
	2.9–2.1	Whole-chloroplast genome sequencing	Middleton et al. (2014) Gornicki et al. (2014)
Divergence of *Ae. searsii* from other Emaeginata species	1.4	Nuclear DNA sequences	Marcussen et al. (2014)
Divergence of *Ae. bicornis* from the ancestral lineage of *Ae. longissima* *Ae. sharonensis*	2.0–1.0	Nuclear DNA sequences	Marcussen et al. (2014)
Divergence of *Ae. sharonensis* from *Ae. longissima*	0.4	Nuclear DNA sequences	Marcussen et al. (2014)
Formation of wild emmer, *T. turgidum* ssp. *dicoccoides*	0.8	Nuclear DNA sequences	Marcussen et al. (2014)
	< 0.8	Whole genome sequencing	Li et al. (2022)
Formation of wild *timopheevii*	0.4	Nuclear DNA sequences	Marcussen et al. (2014)
	< 0.40	Whole genome sequencing	Li et al. (2022)
Formation of allotetraploid *Aegilops* species	1.6–0.18	Sequencing of nuclear genes	Huang et al. (2002b) Dvorak and Akhunov (2005) Marcussen et al. (2014)
		Sequencing of chloroplast DNA	Gornicki et al. (2014) Middleton et al. (2014)

* The divergence time of *T. monococcum* and *T. urartu* is currently unknown

Fig. 11.1 Phylogeny tree of seven diploid *Triticum/Aegilops* species, B-subgenome of bread wheat (cultivar Chinese Spring) and G-subgenome of *T. timopheevii* (an introgressed region on chromosome 2B of bread wheat cultivar "LongReach Lancer") based on the RRGRs on chromosome 2B of Chinese Spring. [From the supplementary Fig. 5A in Li et al. (2022)]

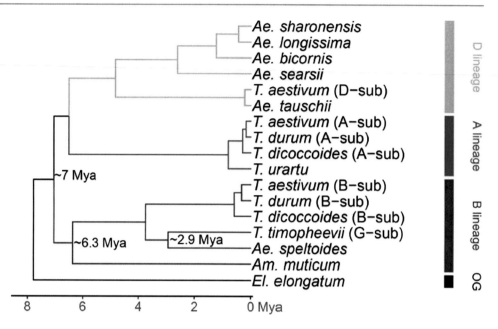

than ancestral *Ae. speltoides,* were together with the *Triticum* lineage, the progenitors of the group. The ancestors of *A. muticum* and the *Triticum* clade contributed approximately equal proportions (0.54 and 0.46, respectively) to the common ancestor of all other *Aegilops* species, except for *Ae. speltoides,* whose ancestral genome evolved, more or less, at the same time as the ancestral T and A genomes. Mostly progenitors of the extant diploid species of the wheat group were involved in further hybridizations and introgressions, but recent interspecific gene flow seems less significant, perhaps due to further divergence and build-up of strong genetic inter-specific isolating systems. Remarkably, despite the fact that several diploids have massive spatial eco-geographical contact (Kimber and Feldman 1987), present-day natural hybridization between diploid species of *Aegilops* is a rare phenomenon. Examples of current inter-specific hybridizations and introgressions were only reported between two of these species, *Ae. longissima* and *Ae. sharonensis* (Ankori and Zohary 1962).

In contrast to the current situation, Bernhardt et al. (2020) suggest that ancient inter-specific or even inter-generic hybridizations significantly contributed to the evolution of the various *Aegilops* species. Glémin et al. (2019) and Bernhardt et al. (2020), concluded that all the diploid *Aegilops* species, except *Ae. speltoides,* derived from an initial homoploid hybridization event involving the ancient A (*Triticum*) and T (*A. muticum*) lineages, highlighted the pivotal role of *A. muticum*, instead of *Ae. speltoides*, in the formation of the diploid *Aegilops* species. This hybridization event was followed by multiple introgressions affecting all taxa, except *Triticum*. Following the development of the diploid species of the group, Bernhardt et al. (2020) found strong signals of introgression from the *caudata-umbellulata*

group to *A. muticum*. This introgression seems to have occurred in both directions. Weaker signals of introgression of Emarginata species into *Ae. caudata* and *Ae. comosa*, as well as into *Ae. tauschii* were also found (Bernhardt et al. 2020). These authors proposed that an ancient, now extinct, lineage was introgressed by *Ae. longissima*, or another species of subsection Emarginata, and possibly also by an ancestor of the *caudata–umbellulata–comosa-uniaristata* clade, forming *Ae. tauschii*. Indeed, chloroplast phylogenies (Yamane and Kawahara 2005; Bernhardt et al. 2017) trace the maternal lineage of *Ae. tauschii* sister to the *caudata–umbellulata–comosa-uniaristata* (CUMNS) clade, suggesting that one of its ancestors is an ancient, perhaps extinct lineage (El Baidouri et al. 2017). This idea is in accord with its placement in nuclear phylogenies in which *Ae. tauschii* shows a moderately supported sister relationship to subsection Emarginata or members of CU(MN) clade. Moreover, the data of Bernhardt et al. (2020) provide evidence of gene flow between species of section Emarginata and the B genome lineage, a hypothesis also raised by El Baidouri et al. (2017) and Glémin et al. (2019). Their study also confirms the close relationships between the members of subsection Emarginata and *Ae. speltoides*. Among the members of subsection Emarginata, *Ae. longissima* appeared as a major introgressor of B genome donor. The close relationship between *Triticum* species and the *caudata-umbellulata-comosa-uniaristata-tauschii* (genomes C, U, M, N, D) clade was confirmed, although no direction could be inferred (Bernhardt et al. 2020).

Ae. searsii, which diverged from the other Emarginata species about 2.0 -1.0 MYA (Marcussen et al. 2014) and *Ae. caudata*, which diverged from the advanced *Aegilops* species, that diverged from the basal *Aegilops* species about 2.5

MYA (Marcussen et al. 2014), have identical chloroplast type (Alnaddaf et al. (2012) and both were found to be close to one another (Sliari and Amer 2011). This may imply that *Ae. caudata* introgressed with the prototype of *Ae. searsii*, as suggested by Bernhardt et al. (2020). Such introgression may explain the advanced trait that exists in *Ae. searsii*, namely, glume length close to the length of florets, that does not exist in other Emarginata species. Also, the adaptation of *Ae. searsii* to Mediterranean habitats may derive from such an introgression. Hence, the evolutionary scenarios of the evolution of the diploid species of the wheat group, proposed by El Baidouri et al. (2017) and Bernhardt et al. (2020), are highly reticulated.

It is generally accepted that the divergence of the T, A and S genomes from an ancestral Triticineae genome established the basal lineages of the wheat group. Comparison of chloroplast (Yamane and Kawahara 2005; Gornicki et al. 2014; Middleton et al. 2014; Bernhardt et al. 2017) and nuclear DNA sequences (Petersen et al. 2006; Salse et al. 2008; Kawahara 2009; Marcussen et al. 2014) confirmed the basal position of *Ae. speltoides* on the phylogenetic *Aegilops/Triticum* tree. *Ae. speltoides* likely diverged from the progenitor of the Triticineae earlier than the ancestral A genome and much earlier than the other *Aegilops* species (Yamane and Kawahara 2005; Salse et al. 2008; Gornicki et al. 2014; Middleton et al. 2014; Bernhardt et al. 2017). Estimates obtained from the analyses of nuclear DNA sequences placed the possible divergence time of the three basal genomes within the period between ~ 7 MYA (Marcussen et al. 2014). Estimates obtained from chloroplast DNA favored a more recent origin of *Ae. speltoides*, i.e., between 4.1–3.6 MYA (Bernhardt et al. 2017) and 2.67 ± 1.1. MYA (Middleton et al. 2014). On the other hand, the divergence of the other diploid species of the wheat group from one another occurred much later. Huang et al. (2002b) estimated that these diploid species began to diverge from one another at 4.5–2.5 MYA and Dvorak and Akhunov (2005) suggested that the divergence time of these species was about 2.7 (4.1–1.4) MYA.

It seems reasonable that the diploid species of the wheat group, other than the ancestral *A. muticum*, *Ae. speltoides* and diploid *Triticum*, evolved at different times–the primitive species about 4.5–2.5 MYA and the advanced ones later on, at about 2.5–1.5 MYA. This period corresponds to the geological epoch Pliocene (5.3–1.8 MYA; Table 2.5), that was characterized by development of seasonal climate (cold and humid winters and hot and dry summers) in the east Mediterranean and south west Asia, the presumed center of origin of the diploid species of this group. The adaptation to dry habitats with seasonal growth periods presumably led to the development of their annual growth habit, associated with increased self-fertilization and large grains.

Genomic divergence may result from the activity of transposable elements (TE) (McClintock 1984; Fedoroff 2012). Senerchia et al. (2013) suggested that ancestral TE families, mainly retrotransposons, followed independent evolutionary trajectories in related species, highlighting the evolution of TE populations as a key factor of genome differentiation in the diploid species of the wheat group. In accordance, Middleton et al. (2013) also found that several TE families differ strongly in their abundance across the diploid species of the wheat group, indicating that these families can thrive extremely successfully in one species, while going virtually extinct in another. Yaakov et al. (2013) also reported that several TE families have undergone either proliferation or reduction in abundance during species diversification at the diploid level. The balance between genome expansion through TE proliferation and contraction through deletion of TE sequences drives variation in genome size and organization (Bennetzen and Kellogg 1997). Hence, the large differences in genome size between the various diploid species of the wheat group (Eilam et al. 2007; Table 2.4) suggest that TE activity has played an important role in the genomic evolution of these species. Indeed, Yaakov et al. (2013) determined the relative copy numbers of TE families in diploid species of section Sitopsis of *Aegilops* and found high variation and genome-specificity of TEs, implying that the main genomic differences between these species are the results of differential activity of TEs. TEs, accounting for a very large fraction of the genomes of the diploid species of the wheat group [80% of well-annotated TEs, with a majority of LTR retrotransposons (Senerchia et al. 2013)], were found to be one of the main drivers of genome divergence and evolution in this group (Yaakov et al. 2013). Whole genome sequencing in the Sitopsis group confirmed that genome size variation could be largely associated with TEs proliferation (Li et al. 2022). Charles et al. (2008) estimated from the insertion dates of TEs that the majority of differential proliferation of TEs in the B and A subgenomes of bread wheat, occurred in these genomes already at the diploid level, prior to the allotetraploidization event that brought them together in *Triticum turgidum*, about 0.8 MYA (Marcussen et al. 2014; Gornicki et al. 2014; Middleton et al. 2014). Finally, rewiring of gene expression in hybrids might dysregulate the silencing of transposons, resulting in activation of transposons, and in reduction of the hybrid fitness or viability, thereby contributing to speciation (Levy 2013).

Another genetic system that can restructure the genome in the diploid species of the wheat group and thus lead to genomic divergence and speciation, is the activity of genome restructuring genes (McClintock 1978). These genes are normally in an inactive state and can be activated by severe stress, either physical, physiological or genetic. Upon activation, they induce a wide range of chromosomal rearrangements that lead to genome restructuring. Heneen (1963a) described an extensive chromosomal breakage occurring spontaneously in an *Elymus farctus* individual. Likewise, Feldman and Straus (1983) reported on a mutant line in *Ae. longissima* that carried a recessive gene causing a wide range of chromosomal rearrangements in meiotic and mitotic cells. None of the chromosome breaks were random, indicating that specific DNA sequences were affected. Other examples of massive chromosomal aberrations in higher plants are rare but were observed in root-tip cells of the Brazilian semi-dwarf wheat cultivar IAS-54 (Guerra et al. 1977), in the hybrid *Elymus arenarius* x *Secale cereale* (Heneen 1963b), and in the hybrid *Elymus farctus* x *Agropyron repens* (Heneen 1963c). Several other cases of spontaneous chromosome breakage in meiotic and mitotic cells of several plants and several intergeneric hybrids were reviewed by Heneen (1963a). Genome restructuring is an ongoing process in natural *Ae. speltoides* populations (Belyayev 2013). Indeed, numerical chromosomal aberrations, spontaneous aneuploidy and re-patterning and reduction in the species-specific tandem repeats have been detected in marginal populations of *Ae. speltoides* (Raskina et al. 2004; Belyayev et al. 2010).

The activation of genome restructuring genes by various stresses has important evolutionary significance. In addition to the generation of genetic variability, due to changes in small DNA sequences as well as formation of cryptic-structural hybridity that may bring about hybrid sterility even though chromosomal pairing looks complete. Genome restructuring also may lead to the formation of new linkage groups. Rapid chromosomal rearrangement can also contribute to the evolvement of isolating mechanisms between differentiating sympatric taxa. Indeed, Lewis (1966) pointed out that rapid chromosomal reorganization played a major role in the formation of many plant species. Activity of genome restructuring genes during wheat evolution may explain the wide occurrence of chromosomal rearrangements among wild as well as domesticated wheats.

11.6.2 *Amblyopyrum* (Jaub. and Spach) Eig

This genus contains only one species, *A. muticum*. Eig (1929b) regarded it as a primitive form, since he noted that this species is an intermediate in several basic morphological features between *Aegilops* and several species of *Elymus*. These traits are: a long, linear awnless spike, many cylindrical, multi-floret spikelets, absence of rudimentary spikelets, and a fragile rachilla that disarticulates into florets that fall separately, especially in the upper part of each spikelet.

Two of the three basal species of the wheat group, *A. muticum* and *Ae. speltoides,* are annual, allogamous and are

the only species that contain B-chromosomes. The B chromosomes of *A. muticum* do not affect homologous pairing, but suppress homoeologous pairing in intergeneric hybrids (Mochizuki 1964; Dover and Riley 1972; Vardi and Dover 1972; Ohta and Tanaka 1982, 1983). These two species possess genes that promote pairing between homoeologous chromosomes in hybrids involving allopolyploid wheat, by counteracting the effect of the homoeologous-pairing suppressor, *ph1,* of allopolyploid wheat (Riley 1960: Feldman and Mello-Sampayo 1967; Dover and Riley 1972a; Dvorak 1972). Ohta (1990, 1991) crossed *A. muticum* with all the diploid species of the wheat group, and reported that most F$_1$ hybrids were completely sterile. Partial fertility was observed only in the hybrid with *Ae. speltoides*, leading Ohta (1990) to conclude that *A. muticum* is most closely related to *Ae. speltoides.*

Numerical analysis (Baum 1977, 1978a, b; Schultze-Motel and Meyer 1981) indicated the close relationship between *Amblyopyrum* and *Aegilops* and *Triticum*. Likewise, several morphological and molecular trees included *Amblyopyrum* in the *Aegilops* clade (Seberg and Petersen 2007), and Mason-Gamer et al. (1998) also included species of *Elymus* in this clade. Hammer (1980) and Ohta (1990, 1991) proposed that *A. muticum* is the ancestral species in the group, whereas Ohsako et al. (1996) studying variation in chloroplast and mitochondrial DNA by single-strand conformational polymorphism (PCR-SSCP) analysis, suggested that *A. muticum* is not older than *Ae. speltoides.*

In some phylogenetic trees, e.g., Ohsako et al. (1996) and Sasanuma et al. (2004), *A. muticum* was included in a different cluster than the other species of the wheat group. *In-situ* hybridization with several repeated DNA markers, and C-banding patterns, suggest that *A. muticum* occupies an isolated position, closer to the Sitopsis species than to other species of *Aegilops* (Badaeva et al. 1996). Sallares and Brown (2004), who analyzed the ITSs of the rRNA genes, reached a similar conclusion, namely, that *A. muticum* has a basal position and that it is close to *Ae. speltoides*. Recent studies of Glémin et al. (2019) and Bernhardt et al. (2020) presented evidence indicating the basal position of *A. muticum* and its contribution, together with the ancestral A genome of *Triticum*, and later on, through introgression with the S genome of *Ae. speltoides*, to the evolvement of most other species of *Aegilops.*

11.6.3 *Aegilops* L.

The *Aegilops* genus contains 10 diploid species (Table 9.3), all of which are annual, and facultatively autogamous, except for *Ae. speltoides*, which is a predominantly allogamous plant (Kimber and Feldman 1987) and *Ae. longissima,* which has a high percentage of cross pollination (Escobar et al. 2010). The northern region of the Fertile Crescent is the geographical center of the group distribution and thus, is presumably the center of origin of the genus (Kimber and Feldman 1987).

Eig (1929a) described principles of evolutionary succession of the *Aegilops* species that were based on morphological characters of the plant and particularly of the spike in comparison with other Triticeae species. His view concerning the evolutionary trends of characters in the diploid species of *Aegilops* was presented in the following generalization (Table 11.4): (1) tall plants represent primitive species (*Ae. speltoides, Ae. sharonensis,* and *Ae. longissima*) whereas short plants represent more advanced species (all other *Aegilops* species); (2) plants with awnless glumes on the apical spikelets are primitive species (species of section Sitopsis and *Ae. tauschii*), while single-awned glumes reflect advanced species (*Ae. caudata* and *Ae. uniaristata*), and many-awned glumes are characteristic of the most advanced species (*Ae. comosa* and *Ae. umbellulata*); (3) awns only on lemmas are seen in primitive species (Sitopsis species and *Ae. tauschii*), whereas main awns on glumes are seen in advanced species (*Ae. caudata, Ae. comosa, Ae. uniaristata,* and *Ae. umbellulata*); (4) wedge-type disarticulation of the spike is a feature of primitive species (*Ae. speltoides var. ligustica, Ae. sharonensis,* and *Ae. bicornis*), barrel-type of advanced type to some extent (*Ae. tauschii*), whereas umbrella-type is seen in the advanced species (*Ae. longissima, Ae. speltoides* var. *speltoides, Ae. searsii, Ae. caudata, Ae. comosa, Ae. uniaristata,* and *umbellulata*); (5) short glumes in relation to the length of the lemmas (about ½ or 2/3 of the length of the lemmas) are characteristic of primitive species (Sitopsis species, except *Ae. searsii,* and *Ae. tauschii*), whereas long glumes of a length that is almost equal to that of the lemmas is a common trait of advanced species (*Ae. searsii, Ae. caudata, Ae. comosa, Ae. uniaristata* and *Ae. umbellulata*); (6) caryopsis adhering to lemma and palea is seen in primitive species (most species of *Aegilops*), whereas a free caryopsis is seen in advanced species (*Ae. searsii* and *Ae. umbellulata*).

Moreover, according to Eig (1929a), the advanced species of *Aegilops* are mainly characterized by the following four morphological characteristics: (a) spikes with many awns, (b) ovoid spikes, (c) ovate spikelets, and (d) spikes falling entire when ripped. He found these four characteristics in *Ae. umbellulata* and concluded that this species is most differentiated from the other types. In contrast, these four characteristics were least distinct in section Sitopsis), which possesses many morphological characteristics common to the other genera in the Triticeae tribe. Thus, Eig concluded that the species of section Sitopsis are the most similar to the ancestral form of *Aegilops*.

The diploid species can be classified into three groups: those having the S genome or modified S genome (*Ae. speltoides, Ae. bicornis, Ae. sharonensis, Ae.* longissima, and *Ae. searsii*), the species having the D genome (*Ae. tauschii*), and those having the C, M, N, and U genomes (*Ae. caudata, Ae. comosa, Ae. uniaristata, and Ae. umbellulata*). Section Sitopsis was sub-divided by Eig (1929a) into two sub-sections, Truncata, containing one species (*Ae. speltoides*), and *Emarginata*, containing the other four species. Hybrids between species of the two subsections, show high pairing but are sterile, while hybrids between Emarginata species also exhibit high pairing but are fertile or partially fertile (Sears 1941a; Kihara 1949; Tanaka 1955a; Kimber 1961; Riley et al. 1961; Roy 1959; Ankori and Zohary 1962; Feldman et al. 1979; Yen and Kimber 1990). The species of the two subsections also differ in karyotype structure (Riley et al. 1958), in C-banding patterns (Friebe and Gill 1996; Ruban and Badaeva 2018), in the number and distribution of certain molecular probes (Ruban and Badaeva 2018), and in the amount of nuclear DNA content (Eilam et al. 2007). On the other hand, the above cytological and molecular data imply a close relationship between the Emarginata species. Based on their studies and other reports, Ruban and Badaeva (2018) suggest the following scenario of the evolution of the five Sitopsis species: *Ae. speltoides* is the most distinct diploid *Aegilops*, that diverged from the common ancestor very early, prior to the split of the other *Aegilops* species (Salse et al. 2008; Gornicki et al. 2014; Marcussen et al. 2014). Divergence of *Ae. speltoides* from an ancestral form was not associated with major chromosomal rearrangements (Rodríguez et al. 2000; Dobrovolskaya et al. 2011). Subsequent evolution of *Ae. speltoides* was accompanied by several transposon insertions (Salse et al. 2008) and by the loss of the 5S rDNA locus on chromosome 1S of modern *Ae. speltoides* (Badaeva et al. 2016) which is present in the B subgenome of emmer and bread wheat (Mukai et al. 1990).

In this respect, Gornicki et al. (2014) and El Baidouri et al. (2017) showed that the female donor of the cytoplasm and B subgenome to *T. turgidum* and *T. aestivum* is not *Ae. speltoides,* but a relative of *Ae. speltoides,* that diverged from the latter 4.48 MYA (Li et al. 2022). The B-subgenome donor is currently either extinct or extant that yet has not been discovered. Homoploid hybridization involving the ancestral S and A genomes (Marcussen et al. 2014) or the ancestral T and A genomes (Glémin et al. 2019; Bernhardt et al. 2020) formed the D lineage, 6.0–5.0 MYA. The D lineage diverged later to the ancestral D genome (5.37 MYA (Li et al. 2022), to the Emarginata species 3.73 MYA (Li et al. 2022), and to the more advanced *Aegilops* species diverged from the basal *Aegilops* species about 2.5 MYA (Marcussen et al. 2014). The divergence of the Emarginata

Table 11.4 Eig's (1929a) definition of ancestral and advanced traits in the genus *Aegilops*

Ancestral traits	Advance traits
Tall plant	Short plants
Small number of tillers	Large number of tillers
Long spike	Short and compact spike
Large number of fertile spikelets per spike	Small number of fertile spikelets per spike
Large number of florets per spikelet	Small number of florets per spikelet
Glumes are shorter than the florets	Glumes are about the same size as the florets
Awns on lemmas	Awns on glumes
Glumes of the apical spikelets are awnless	Glumes on the apical spikelets are with many awns
The spike disarticulates at maturity into spikelets with the rachis internode that belong to them (wedge type)	The spike falls entire at maturity (umbrella type)
The grain is joined together with the chaff	The grain is free

species was associated with an increase of high-copy DNA sequences due to the activity of transposable elements (Yaakov et al. 2013), amplification of the CTT-repeat, re-distribution of C-bands, massive amplification of Spelt-52 and gradual elimination of the D-genome-specific sequences pAs1, pTa-535 and pTa-s53 (Ruban and Badaeva 2018). The data of Ruban and Badaeva (2018) show that most drastic changes probably occurred at the stage of radiation of *Ae. longissima/Ae. sharonensis*, and included massive amplification of Spelt-52 and CTT-repeats, resulting in the gain of heterochromatin in these two species, and in an approximately 12% increase of nuclear DNA content in *Ae. longissima/Ae. sharonensis* as compared to that of *Ae. searsii/Ae. bicornis* (Eilam et al. 2007). The similar distribution of all analyzed DNA sequences on chromosomes of *Ae. longissima* and *Ae. sharonensis* point to a rather recent divergence of these species 0.4 MYA (Marcussen et al. 2014), which was accompanied by the species-specific 4S^l/7S^l translocation in *Ae. longissima*.

Morphologically, *Ae. bicornis* is the most primitive species in the Emarginata group (Eig 1929a). Indeed, *Ae. bicornis* diverged from the ancestral lineage of *Ae. longissima/Ae. sharonensis* 1.4 MYA (Marcussen et al. 2014). It is more difficult to produce hybrids with *Ae. bicornis* than with other S-genome *Aegilops* species (Kimber and Feldman 1987).

Morphologically, *Ae. searsii* resembles *Ae. longissima*, but molecular studies of chloroplast DNA showed that it is closer to *Ae. bicornis* than to *Ae. longissima* (Tsunewaki and Ogihara 1983). *Ae. searsii* differs from *Ae. longissima* and from the other Emarginata species by several morphological traits which are considered as evolutionarily advanced, namely, short stature, length of glumes, and free kernels (Feldman and Kislev 1977). *Ae. longissima* x *Ae. searsii*

hybrids exhibit meiotic irregularities, including a reciprocal translocation, and are partial sterile (Feldman et al. 1979), and, similarly, *Ae. longissima* x *Ae. bicornis* had a reciprocal translocation, some pairing failure and the hybrid was highly sterile (Kimber 1961). By contrast, the F$_1$ *Ae. longissima* x *Ae. sharonensis* are fertile and show complete chromosome pairing in meiosis (five bivalents and one quadrivalent, due to a reciprocal translocation) (Tanaka 1955; Ankori and Zohary 1962). Isolation of these species is caused by different ecological requirements (Ankori and Zohary 1962; Kimber and Feldman 1987; Feldman and Levy 2015). The close relationships between the *Emarginata* species and the separate position of *Ae. speltoides* within the *Sitopsis* section were confirmed by molecular analyses of nuclear and cytoplasmic DNA. Based on variation of repeated nucleotide sequences (RNS), Dvorak and Zhang (1992b) showed that the *Sitopsis* species are phylogenetically similar, but *Ae. speltoides* is clearly separated from species of the *Emarginata* group. RAPD and AFLP analyses revealed that *Ae. speltoides* forms a cluster with polyploid wheats, which is separated from other *Sitopsis* species (Kilian et al. 2007, 2011). Likewise, the phylogenetic reconstructions of Middleton et al. (2014), showed that *Ae. speltoides* is not a member of Sitopsis, but together with *T. turgidum, T. aestivum,* and *T. timopheevii* lineages it forms a clade (B lineage).

Dvorak and Zhang (1992a) and Sasanuma et al. (1996, 2004) found a close relationship between *Ae. caudata* and *Ae. umbellulata.* Cytogenetic and phylogenetic studies of the four advanced species of *Aegilops*, i.e., *caudata, comosa, umbellulata* and *uniaristata*, showed that the N genome of *Ae. uniaristata* is one of the most advanced genomes in the group and is closer to the U genome of *Ae. umbellulata* than to the genomes of *Ae. caudata* and *Ae. comosa* (Sallares and

Brown 2004; Badaeva et al. 1996). PCR fragment polymorphism analyses of chloroplast genomes placed *Ae. umbellulata* and *Ae. comosa* closer to *Ae. tauschii* than to the *T. monococcum* and *Ae. speltoides* (Tsunewaki et al. 1996; Gandhi et al. 2005).

Molecular analyses have shown that the diploid genomes S, D and A are much more closely related to each other than to other genomes in the wheat group (Monte et al. 1993; Dvorak and Zhang 1990; Dvorak et al. 1998). Indeed, comparisons of a large number of nuclear genes indicated that an ancestral D lineage derived from hybridization between ancient A and S lineages, about 6–7 MYA (Marcussen et al. (2014). This finding spurred a discussion regarding the hybrid origin of the extant *Ae. tauschii* (Sandve et al. 2015; Li et al. 2015, reevaluating the origin of *Ae. tauschii* by using recently published data from nuclear DNA (Marcussen et al. 2014) and chloroplast DNA sequencing (Gornicki et al. 2014), as well as additional data of chloroplast DNA of their own, confirmed the hybrid origin of the extant D genome but concluded that this genome has a more complex origin, one that may have involved multiple rounds of hybridizations. El Baidouri et al. (2017), following analysis of sequences of homoeologous genes and transposable elements derived from *T. aestivum*, *T. turgidum* ssp. *durum*, *T. urartu*, *Ae. speltoides*, and *Ae. tauschii*, deduced that, about 6 MYA, an ancestral D genome introgressed into a homoploid hybrid of the ancestral A and B genomes. The ancestral D genome became extinct sometime later. Today's D genome, occurring in diploid *Ae. tauschii* and as one subgenome in *T. aestivum* and other allopolyploid species of *Aegilops*, is, therefore, a hybrid genome combining three genomes (El Baidouri et al. 2017). As the B subgenome of allopolyploid wheat is different from its closest extant relative *Ae. speltoides*, El Baidouri et al. (2017) assumed that the B genome itself might also have been introgressed by species of subsection Emarginata of section Sitopsis. Such introgression was also suggested by Bernhardt et al. (2020). Glémin et al. (2019), based on transcriptome data for all species of the group, proposed a complex scenario of hybridizations, and identified *A. muticum* (genome T), instead of *Ae. speltoides* (genome S), as an ancestor of the D genome lineage and of at least two more hybridization events. Bernhardt et al. (2020) also found that the ancestral T genome, and not B, was involved in the ancient hybridization with genome A.

Molecular findings relating to the chloroplast and mitochondrial genome (Tsunewaki 2009; Kawahara 2009) reinforced the studies on phylogenetic relationships of the diploid species. Tsunewaki (2009) reviewed such studies and concluded that the diploid species of the wheat group exhibit great diversification. *A. muticum* and *Ae. speltoides*, the two-outbreeding species, showed especially clear intra-specific chloroplast and mitochondrial differentiation. On the other hand, Gornicki et al. (2014) reveal low sequence variation of the chloroplast genome within *Ae. speltoides* as well as little haplotype variation in the Emarginata species. Yet, earlier studies (Chen et al. 1975; Hirai and Tsunewaki 1981) revealed two types of electromorphs of the Rubisco large subunit (the chloroplast subunit), H- and L-types, in these species. The H-type large subunit was found in the chloroplast of *Ae. speltoides* (and also in that of allopolyploid *Triticum* species) while the L-type large subunit exists in the chloroplast of all diploid *Aegilops* and *Triticum* species.

In many phylogenetic trees, *Ae. speltoides* forms a moderately supported clade with *A. muticum*, and, as in previous studies (Petersen et al. 2006), it was always clearly separate from the other species of *Aegilops*. All the analyses reported by Bernhardt et al. (2020) always classified *Ae. tauschii* as sister of subsection Emarginata. Marcussen et al. (2014) assumed that *Ae. sharonensis* is close to *Ae. tauschii* and is a hybrid involving the B genome lineage. The data of Bernhardt et al. (2020) showed that not only *Ae. sharonensis* is closely related to *Ae. tauschii* but that genome parts of the latter suggest the involvement of the entire subsection Emarginata, i.e., *Ae. bicornis*, *Ae. sharonensis*, *Ae. longissima*, and *Ae. searsii*. Yet, the absence of any relationship to the B genome clearly indicates a more complex evolutionary history than previously hypothesized, of the *Ae. tauschii* genome and perhaps also of the genome of subsection Emarginata.

11.6.4 *Triticum* L.

This genus contains two diploid, two tetraploid and two hexaploid species. Among the diploids, *T. monococcum* comprises two subspecies, wild ssp. *aegilopoides* and domesticated ssp. *monococcum*, also known as einkorn wheat, whereas *T. urartu* consists only of a single wild taxon. Both diploid species are annuals and facultative autogamous. The wild forms have a fragile rachis and disarticulate at maturity into wedge-type dispersal units, whereas the domesticated subspecies, that derived from ssp. *aegilopoides*, have a tough rachis so that at maturity the spike remains intact on the culm. The two wild taxa distribute in the northern part of the Fertile Crescent and in Transcaucasia, sympatrically in many sites. Following the spread of wheat cultivation, ssp. *aegilopoides* expanded its distribution as a weed, westward up to the Balkan.

Key (1966, 1968, 1981) and Hammer (1980) argued that different evolutionary tendencies exist in domesticated and wild wheat. The domesticated forms, under the selection pressure exerted by man, will further develop its greater ability to utilize fertile land and will follow a completely different trend in ear construction, namely, more specialization for tough rachis and free threshing. The wild forms will continue to occupy shrinking primary habitats or adapted as weed to various disturbed sites or cereal fields.

Hybrids between *T. monococcum* and the basal species of the clade, *Ae. speltoides* and *A. muticum,* had high chromosomal pairing at meiosis (Sears 1941b; Ohta 1990), which may either be due to effect of pairing promoters or indicate close phylogenetic relationships. On the other hand, hybrids between diploid wheat and the other Sitopsis species exhibit low pairing (Sears 1941b; Kushnir and Halloran 1981; Feldman 1978), indicating great divergence between the genomes of these species. Only the hybrid *T. monococcum* x *Ae. tauschii* had somewhat more pairing (4.86 bivalents and 0.21 trivalents per cell; Sears 1941b), indicating much closer cytogenetic affinities between the A and the D genomes than between the A genome and the genomes of the remaining diploid *Aegilops* species.

Following a morphological analysis, Baum (1983) included *Amblyopyrum, Aegilops, Triticum* and *Henrardia* in one cluster, close to species of *Elymus*. The findings of Escobar et al. (2011) were in agreement with previous works (Petersen and Seberg 1997), which included *Triticum, Aegilops, Secale,* and *Taeniatherum* in one clade. In some molecular studies, *Triticum* is a sister clade to the *Aegilops* clade (Kellogg and Appels, 1995; Mason-Gamer and Kellogg, 1996a), while others (Hsiao et al. 1995a, b; Kellogg et al. 1996; Liu et al. 2007) considered it to be the sister group to *Elymus elongatus*, assuming that *Triticum* may have derived from the *Elymus* E genome.

Yet, based on variation in repeated nucleotide sequences, Dvorak and Zhang (1992b) constructed a phylogenetic tree of the species of the *Aegilops* and *Triticum*. The tree obtained was consistent with many cyto-taxonomical data on species relationships in the two genera. Their studies clustered the two *Triticum* diploids, *monococcum* and *urartu,* that have been shown cytogenetically to have a common genome (Dvorak 1976; Chapman et al. 1976).

Glémin et al. (2019), Huynh et al. (2019), and Bernhardt et al. (2020) have proposed that the ancestral *A. muticum* and the ancestral *Triticum* genomes each contributed approximately equal proportions to the common ancestor of all other *Aegilops* species, with the exception of *Ae. speltoides*. Bernhardt et al. (2020), while highlighting the contribution of hybridization to the evolution of the species of *Aegilops*, stated that the evolution of the diploid *Triticum* species was not affected by inter-generic hybridization. It seems therefore, that after the ancient hybridization with the T genome, later hybridization between species of the wheat group and the A genome was restricted, presumably due to development of strong genetic barriers. As previously shown by Marcussen et al. (2014), Glémin et al. (2019), and

Li et al. 2022 the Sitopsis species, the wheat subgenomes A, B, D, and G, and the T genome of *A. muticum* fall into three clades corresponding to the A, B and D lineages (Figs. 10.5 and 11.1).

References

Aase HC, Powers LR (1926) Chromosome numbers in crop plants. Amer J Bot 13:367–372

Akhunov ED, Akhunova AR, Linkiewicz AM, Dubcovsky J, Hummel D, Lazo G, Chao SM, Anderson OD, David J, Qi LL, Gerard G et al (2003a) Synteny perturbations between wheat homoeologous chromosomes caused by locus duplications and deletions correlate with recombination rates. Proc Natl Acad Sci (USA) 100:10836–10841

Akhunov ED, Goodyear AW, Geng S, Qi LL, Echalier B, Gill BS, Miftahudin J, Gustafson JP, Lazo G, Chao SM et al (2003b) The organization and rate of evolution of wheat genomes are correlated with recombination rates along chromosome arms. Genome Res 13:753–763

Allard RW, Kannenberg LW (1968) Population ptudies in predominantly self-pollinated species. XI. Genetic divergence among the members of the Festuca microstachys complex. Evolution 22:517–528

Allard RW, Jain SK, Workman PL (1968) The genetics of inbreeding population. Adv Genet 14:55–131

Allard RW (1975) The mating system and microevolution. Genetics 79 (Suppl):115–126

Alnaddaf LM, Moualla MY, Haider N (2012) the genetic relationships among *Aegilops* L. and *Triticum* L. species. Asian J Agric Sci 4:352–367

Ankori H, Zohary D (1962) Natural hybridization between *Aegilops sharonensis* and *Ae. longissima*: a morphological and cytological study. Cytologia, Tokyo 27:314–324

Arterburn M, Kleinhofs A, Murray T, Jones S (2011) Polymorphic nuclear gene sequences indicate a novel genome donor in the polyploid genus *Thinopyrum*. Hereditas 148:8–27

Asghari-Zakaria R, Kazemi H, Aghayev YM, Valizadeh M, Moghaddam M (2002) Karyotype and C-banding patterns of mitotic chromosomes in *Henrardia persica* (Boiss.) C.E. Hubb. Caryologia 55:289–293

Avdulov N (1931) Karyo-systematisce Untersuchung der Familie Gramineen. Bull Appl Bot Genet PL Breed, Leningrad, Suppl 43:1–428

Badaeva ED, Friebe B, Gill BS (1996) Genome differentiation in *Aegilops*. Distribution of highly repetitive DNA sequences on chromosomes of diploid species. Genome 39:293–306

Badaeva ED, Ruban A, Zoshchuk S et al (2016) Molecular cytogenetic characterization of *Triticum timopheevii* chromosomes provides new insight on genome evolution of *T. zhukovskyi*. Pl Syst Evol 302:943–956

Baum BR (1977) Taxonomy of the tribe Triticeae (Poaceae) using various numerical techniques. I. Historical perspectives, data, accumulation and character analysis. Can J Bot 55:1712–1740

Baum BR (1978a) Taxonomy of the tribe Triticeae (Poaceae) using various numerical techniques. II. Classification. Can J Bot 56:27–56

Baum BR (1978b) Taxonomy of the tribe Triticeae (Poaceae) using various numerical techniques. III. Synoptic key to the genera and synopses. Can J Bot 56:374–385

Baum BR (1983) A phylogenetic analysis of the tribe Triticeae (Poaceae) based on morphological characters of the genera. Can J Bot 61:518–535

Baum BR, Estes JR, Gupta PK (1987) Assessment of the genomic system of classification in the Triticeae. Amer J Bot 74:1388–1395

Baum BR, Appels R (1992) Evolutionary change at the *5S Dna* loci of species in the Triticeae. Pl Syst Evol 183:195–208

Belyayev A, Kalendar R, Brodsky L, Nevo E, Schulman AH, Raskina O (2010) Transposable elements in a marginal plant population: temporal fluctuations provide new insights into genome evolution of wild diploid wheat. Mob DNA 1:6. https://doi.org/10.1186/1759-8753-1-6

Belyayev A (2013) Chromosome evolution in marginal populations of *Aegilops speltoides*: causes and consequences. Ann Bot 111:531–538

Bennett MD, Gustafson JP, Smith JB (1977) Variation in nuclear DNA in the genus *Secale*. Chromosoma 61:149–176

Bennetzen JL, Kellogg EA (1997) Do plants have a one-way ticket to genomic obesity? Plant Cell 9:1509–1514

Bernhardt N (2015) Taxonomic treatments of Triticeae and the wheat genus *Triticum*. In: Molnar-Lang M, Ceoloni C, Dolezel J (eds) Alien introgression in wheat, cytogenetics, molecular biology, and genomics. Springer, Cham, Switzerland, pp 1–19

Bernhardt N (2016) Analysis of phylogenetic relationships among diploid Triticeae grasses. PhD thesis, Fakultät für Mathematik und Naturwissenschaften der Carl von Ossietzky Universität Oldenburg

Bernhardt N, Brassac J, Kilian B, Blattner FR (2017) Dated tribe-wide whole chloroplast genome phylogeny indicates recurrent hybridizations within Triticeae. BMC Evol Biol 17:141

Bernhardt N, Brassac J, Dong X, Willing E-M, Poskar CH, Kilian B, Blattner FR (2020) Genome-wide sequence information reveals recurrent hybridization among diploid wheat wild relatives. Plant J 102:493–506

Bieniek W, Mizianty M, Szklarczyk M (2015) Sequence variation at the three chloroplast loci (*matK, rbcL, trnH-psbA*) in the Triticeae tribe (Poaceae): comments on the relationships and utility in DNA barcoding of selected species. Pl Syst Evol 301:1275–1286

Blanco A, Simeone R, Tanzarella OA (1983a) Morphology and chromosome pairing of a hybrid between *Triticum durum* Desf. and *Haynaldia villosa* (L.) Schur. Theor Appl Genet 64:333–337

Blanco A, Simeone R, Tanzarella OA, Greco B (1983b) Cytogenetic of the hybrid of *Triticum durum* Desf. x *Haynaldia villosa* (L.) Schur. Genetica Agraria 37:149

Blanco A, Perrone V, Simeone R (1988) Chromosome pairing variation in Triticum turgidum L. x Dasypyrum villosum (L.) Candargy hybrids and genome affinities. In: Miller TE, Koebner RMD (eds) Proceedings of the 7th international wheat genetics symposium, vol 1, Institute of Plant Science Research, Cambridge, pp 63–67

Blanco A, Simeone R, Resta P, de Pace C, Delre V, Caccia R, Scarascia Mugnoza GT, Frediani M, Cremonini R, Cionini PG (1996) Genomic relationships between *Dasypyrum villosum* (L.) Candargy and *D. hordeaceum* (Cosson et Durieu) Candargy. Genome 39:83–92

Bowden WM (1966) Chromosome numbers in seven genera of the tribe Triticeae. Can J Genet Cytol 8:130–136

Chalupska D, Lee HY, Faris JD, Evrard A, Chalhoub B, Haselkorn R, Gornicki P (2008) ACC homoeoloci and the evolution of wheat genomes. Proc Natl Acad Sci (USA) 105:9691–9696

Chapman V, Miller TE, Riley R (1976) Equivalence of the A genome of bread wheat and that of *Triticum urartu*. Genet Res 27:69–76

Charles M, Belcram H, Just J, Huneau C, Viollet A, Couloux A, Segurens B, Carter M, Huteau V, Coriton O, Appels R, Samain S, Chalhoub B (2008) Dynamics and differential proliferation of transposable elements during the evolution of the B and A genomes of wheat. Genetics 180:1071–1086

Charpentier A, Feldman M, Cauderon Y (1986) Chromosomal pairing at meiosis of F$_1$ hybrid and backcross derivatives of *Triticum aestivum* x hexaploid *Agropyron junceum*. Can J Genet Cytol 28:1–6

Chen KC, Gray JG, Wildman SG (1975) Fraction 1 protein and the origin of polyploid wheats. Science 190:1304–1306

Chen PD, Liu DJ (1982) Cytogenetic studies of hybrid progenies between *T. aestivum* and *H. villosa*. Nanjing Agric Coll Bull 4:1–16

Chen Q, Conner RL, Laroche A, Thomas JB (1998) Genome analysis of *Thinopyrum intermedium* and *Thinopyrum ponticum* using genomic in situ hybridization. Genome 41:580–586

Chen Q, Conner RL, Li HJ, Sun SC, Ahmad F, Laroche A, Graf RJ (2003) Molecular cytogenetic discrimination and reaction to wheat streak mosaic virus and the wheat curl mite in Zhong series of wheat–*Thinopyrum intermedium* partial amphiploids. Genome 46:135–145

Chennaveeraiah MS, Sarkar P (1959) Chromosomes of *Heteranthelium piliferum* Hochst. Wheat Inf Serv 9–10:42

Chennaveeraiah MS (1960) Karyomorphologic and cytotaxonomic studies in Aegilops. Acta Horti Gotoburgensis 23:85–178

Clayton WD, Renvoize SA (1986) Genera graminum, grasses of the world. Distributed for Royal Botanic Gardens, Kew bulletin. Additional series, 13. Kew, London, pp 1–389

Cuadrado A, Vitellozzi F, Jouve N, C. Ceoloni C (1997) Fluorescence in situ hybridization with multiple repeated DNA probes applied to the analysis of wheat-rye chromosome pairing. Theor Appl Genet 94:347–355

Darlington CD (1933) The origin and behavior of chiasmata. VIII. *Secale cereale*. Cytologia 4:444–452

de Bustos A, Jouve N (2002) Phylogenetic relationships of the genus *Secale* based on the characterization of rDNA ITS sequences. Pl Syst Evol 235:147–154

de Pace C, Vaccino P, Cionini PG, Pasquini M, Bizzarri M, Qualset CO (2011) Dasypyrum. In: Kole C (ed) Wild crop relatives: genomic and breeding resources, cereals. Springer-Verlag, Berlin, pp 185–291

de V Pienaar R, Littlejohn GM, Sears ER (1988) Genomic relationships in *Thinopyrum*. S Afr J Bot 54:541–550

Dedio W, Kaltsikes PJ, Larter EN (1969) Numerical chemotaxonomy in the genus *Secale*. Can J Bot 47:1175–1180

Deng GB, Chen J, Ma XR, Pan ZF, Yu MQ, Li XF (2004) Morphology, cytogenetics of intergeneric hybrid between *Aegilops tauschii* and *Dasypyrum villosum*. Hereditas/yi Chuan 26:189–194

Devos KM, Atkinson MD, Chinoy CN, Francis HA, Harcourt RL, Koebner RMD, Liu CJ, Masoje P, Xie DX, Gale MD (1993) Chromosomal rearrangements in the rye genome relative to that of wheat. Theor Appl Genet 85:673–680

Devos KM, Dubcovsky J, Dvorak J, Chinoy CN, Gale MD (1995) Structural evolution of wheat chromosomes 4A, 5A, and 7B and its impact on recombination. Theor Appl Genet 91:282–288

Dewey DR (1969) Hybrids between tetraploid and hexaploid crested wheatgrasses. Crop Sci 9:787–791

Dewey DR, Hsiao CI (1984) The source of variation in tetraploid crested wheatgrass. Agron Abstr, p 64

Dobrovolskaya O, Boeuf C, Salse J, Pont C, Sourdille P, Bernard M, Salina E (2011) Microsatellite mapping of *Ae. speltoides* and map-based comparative analysis of the S, G, and B genomes of Triticeae species. Theor Appl Genet 123:1145–1157

Dover GA, Riley R (1972) Prevention of pairing of homoeologous meiotic chromosomes of wheat by an activity of supernumerary chromosomes of Aegilops. Nature 240:159–161

Dvorak J (1972) Genetic variability in *Aegilops speltoides* affecting homoeologous pairing in wheat. Can J Genet Cytol 14:371–380

Dvorak J (1976) The relationship between the genome of *Triticum urartu* and the A and B genomes of *Triticum aestivum*. Can J Genet Cytol 18:371–377

Dvorak J, Zhang H-B (1990) Variation in repeated nucleotide sequences sheds light on the phylogeny of the wheat B and G genomes. Proc Natl Acad Sci, USA 87:9640–9644

Dvorak J, Zhang H-B (1992a) Reconstruction of the phylogeny of the genus *Triticum* from variation in repeated nucleotide sequences. Theor Appl Genet 84:419–429

Dvorak J, Zhang H-B (1992b) Application of molecular tools for study of the phylogeny of diploid and polyploid taxa in Triticeae. Hereditas 116:37–42

Dvorak J, Luo M-C, Yang Z-L (1998) Restriction fragment length polymorphism and divergence in the genomic regions of high and low recombination in self-fertilizing and cross–ferti lizing *Aegilops* species. Genetics 148:423–434

Dvorak J, Akhunov ED (2005) Tempos of gene locus deletions and duplications and their relationship to recombination rate during diploid and polyploid evolution in the *Aegilops-Triticum* alliance. Genetics 171:323–332

Eig A (1929a) *Amblyopyrum* Eig. A new genus separated from the genus *Aegilops*. PZE Inst Agric Nat Hist Agric Res 2:199–204

Eig A (1929b) Monographisch-Kritische Ubersicht der Gattung *Aegilops*. Reprium Spec Novarum Regni Veg 55:1–288

Eilam T, Anikster Y, Millet E, Manisterski J, Feldman M (2007) Genome size and genome evolution in diploid Triticeae species. Genome 50:1029–1037

El Baidouri M, Murat F, Veyssiere M, Molinier M, Flores R, Burlot L, Alaux M, Quesneville H, Pont C, Salse J (2017) Reconciling the evolutionary origin of bread wheat (*Triticum aestivum*). New Phytol 213:1477–1486

Elbaum R, Zaltzman L, Burgert I, Fratzl P (2007) The role of wheat awns in the seed dispersal unit. Science 316:884–886

Emme HK (1927) Zur Cytologie der Gattung *Secale* L. Bull Appl Bot 17:73–100

Emme HK (1928) Karyologie der Gattung *Secale* L. Zschr Ind Abst Vererb L 47:99–124

Endo TR, Gill BS (1984) The heterochromatin distribution and genome evolution in diploid species of *Elymus* and *Agropyron*. Can J Genet Cytol 26:669–678

Escobar JS, Cenci A, Bolognini J, Haudry A, Laurent A, David J, Glémin S (2010) An integrative test of the dead-end hypothesis of selfing evolution in Triticeae (Poaceae). Evolution 64:2855–2872

Escobar JS, Scornavacca C, Cenci A, Guilhaumon C, Santoni S, Douzery EJ, Ranwez V, Glémin S, David J (2011) Multigenic phylogeny and analysis of tree incongruences in Triticeae (Poaceae). BMC Evol Biol 11:181–198

Fan X, Zhang HQ, Sha LN, Zhang L, Yang RW, Ding CB, Zhou YH (2007) Phylogenetic analysis among *Hystrix*, *Leymus* and its affinitive genera (Poaceae: Triticeae) based on the sequences of a gene encoding plastid acetyl-CoA carboxylase. Plant Sci 172:701–707

Fan X, Sha LN, Yang RW, Zhang HQ, Kang HY, Zhang L, Ding CB, Zheng YL, Zhou YH (2009) Phylogeny and evolutionary history of *Leymus* (Triticeae; Poaceae) based on a single-copy nuclear gene encoding plastid acetyl-CoA carboxylase. BMC Evol Biol 9:247

Fan X, Sha LN, Yu SB, Wu DD, Chen XH, Zhuo XF, Zhang HQ, Kang HY, Wang Y, Zheng YL, Zhou YH (2013) Phylogenetic reconstruction and diversification of the Triticeae (Poaceae) based on single-copy nuclear *Acc1* and *Pgk1* gene data. Biochem Syst Ecol 50:346–360

Fedak G (1979) Cytogenetics of a barley x rye hybrid. Can J Genet Cytol 21:543–548

Fedak G, Armstrong KC (1981) Hybrids of Hordeum parodii and Hordeum lechleri with Secale cereale. In: Barley genet IV. Proc 4th Int Barley Genet Symp, Edinburgh, pp 740–745

Fedak G (1986) Wide crosses in Hordeum. In: Barley agronomy monograph, vol 26. ASA-CSSA-SSSA, Madiso, pp 155–186

Fedak G, Armstrong KC (1986) Intergeneric hybrids between *Secale cereale* (2x) and *Thinopyrum intermedium* (6x). Can J Genet Cytol 28:426–429

Fedoroff NV (2012) Transposable elements, epigenetics, and genome evolution. Science 338:758–767

Feldman M, Mello-Sampayo T (1967) Suppression of homeologous pairing in hybrids of polyploid wheats×*Triticum speltoides*. Can J Genet Cytol 9:307–313

Feldman M, Kislev M (1977) *Aegilops searsii*, a new species of section *Sitopsis (Platystachys)*. Isr J Bot 26:190–201

Feldman M (1978) New evidence on the origin of the B genome of wheat. In: Ramanujam RS (ed) Proc 5th Intern Wheat Genet Symp, New Delhi, pp 120–132

Feldman M, Strauss I, Vardi A (1979) Chromosome pairing and fertility of F₁ hybrids of *Aegilops longissima* and *Ae. searsii*. Can J Genet Cytol 21:261–272

Feldman M, Strauss I (1983) A genome-restructuring gene in *Aegilops longissima*. In: Sakamoto S (ed) Proc. 6th Intern Wheat Genet Symp, Kyoto, Japan, pp 309–314

Feldman M, Levy AA (2015) Origin and evolution of wheat and related Triticeae species. In: Molnar-Lang M, Ceoloni C, Dolezel J (eds) Alien introgression in wheat, cytogenetics, molecular biology, and genomics. Springer, Cham, Switzerland, pp 21–76

Finch RA, Bennett MD (1980) Mitotic and meiotic chromosome behaviour in new hybrids of *Hordeum* with *Triticum* and *Secale*. Heredity 44:201–209

Forster P, Miller TE (1989) Genome relationship between *Thinopyrum bessarabicum* and *Thinopyrum elongatum*. Genome 32:930–931

Frederiksen S, von Bothmer R (1989) Intergeneric hybridization between *Taeniatherum* and different genera of Triticeae, Poaceae. Nord J Bot 9:229–240

Frederiksen S (1991a) Taxonomic studies in *Eremopyrum* (Poaceae). Nord J Bot 11:271–285

Frederiksen S (1991b) Taxonomic studies in *Dasypyrum* (Poaceae). Nord J Bot 11:135–142

Frederiksen S (1993) Taxonomic studies in some annual genera of the Triticeae. Nordic J Bot 13:490–492

Frederiksen S (1994) Hybridization between *Taeniatherum caput-medusa* and *Triticum aestivum*. Nord J Bot 14:3–6

Frederiksen S, Seberg O (1992) Phylogenetic analysis of the Triticeae (Poaceae). Hereditas 116:15–19

Frederiksen S, von Bothmer R (1995) Intergeneric hybridization with *Eremopyrum* (Poaceae). Nord J Bot 15:39–47

Frederiksen S, Petersen G (1997) Morphometrical analyses of *Secale* L. (Triticeae, Poaceae). Nord J Bot 17:1–14

Frederiksen S, Petersen G (1998) A taxonomic revision of *Secale* (Triticeae, Poaceae). Nord J Bot 18:399–420

Friebe B, Gill BS (1996) Chromosome banding and genome analysis in diploid and cultivated polyploid wheats. In: Jauhar PP (ed) Methods in Genome Analysis in Plants: Their Merits and Pitfals. CRC Press, Boca Ration, FL; New York, NY; London; Tokyo, pp 39–60

Gandhi HT, Vales MI, Watson CJW, Mallory–Smith CA, Mori N et al (2005) Chloroplast and nuclear micro-satellite analysis of *Aegilops cylindrica*. Theor Appl Genet 111:561–572

Gill BS, Friebe B (2009) Cytogenetic analysis of wheat and rye genomes. In: Feuillet C, Muehlbauer GJ (eds) Genetics and genomics of the triticeae. Springer, pp 121–135

Glémin S, Scornavacca C, Dainat J, Burgarella C, Viader V, Ardisson M, Sarah G, Santoni S, David J, Ranwez V (2019) Pervasive hybridizations in the history of wheat relatives. Sci Adv 5(5): eaav9188

Gornicki P, Zhu H, Wang J, Challa GS, Zhang Z, Gill BS, Li W (2014) The chloroplast view of the evolution of polyploid wheat. New Phytol 204:704–714

Guerra MS, Moraes-Fernandes MIB (1977) Somatic instability in the Brazilian semi-dwarf wheat IAS 54. Can J Genet Cytol 19:225–230

Gupta PK, Fedak G (1985) Genetic control of meiotic chromosome pairing in the genus *Hordeum*. Can J Genet Cytol 27:515–530

Gupta PK, Fedak G (1987a) Preferential intragenomic chromosome pairing in two new diploid intergeneric hybrids between Hordeum and Secale. Genome 29:594–597

Gupta PK, Fedak G (1987b) Meiosis in two new intergeneric hybrids between *Hordeum* and *Secale*. Plant Bree 99:155–158

Hammer K (1980) Vorarbeiten zur monographischen Darstellung von Wildpflanzensortimenten: *Aegilops* L. Kulturpflanze 28:33–180

Hammer K, Lehmann CO, Pemno P (1985) Die in den Jahren 1980, 1981 und 1982 in Süditalien gesammelten Getreide-Landsorten–botanische Ergebnisse. Kulturpflanze 33:237–267

Hasegawa N (1934) A cytological study on 8-chromosome rye. Cytologia 6:68–77

Helfgott M, Mason-Gamer RJ (2004) The evolution of North American *Elymus* (Triticeae, Poaceae) allotetraploids: evidence from phosphoenolpyruvate carboxylase gene sequences. Syst Bot 29:850–861

Heneen WK (1963a) Extensive chromosome breakage occurring spontaneously in certain individuals of *Elymus farctus* (=*Agropyron junceum*). Hereditas 49:1–32

Heneen WK (1963b) Cytology of the intergeneric hybrid *Elymus arenarius* x *Secale cereale*. Hereditas 49:61–77

Heneen WK (1963c) Meiosis in the interspecific hybrid Elymus farctus×A. repens (=Agropyron junceum×A. repens). Hereditas 49:107–118

Hirai A, Tsunewaki K (1981) Genetic diversity of the cytoplasm in *Triticum* and *Aegilops*. VIII. Fraction I protein of 39 cytoplasms. Genetics 99:487–493

Hodge CD, Wang H, Sun G (2010) Phylogenetic analysis of the maternal genome of tetraploid StStYY Elymus (Triticeae: Poaceae) species and the monogenomic Triticeae based on rps16 sequence data. Plant Sci 178:463–468

Hsiao C, Wang RR-C, Dewey DR (1986) Karyotype analysis and genome relationships of 22 diploid species in the tribe Triticeae. Can J Genet Cytol 28:109–120

Hsiao C, Chatterton NJ, Asay KH, Jensen KB (1995a) Phylogenetic relationships of the mono genomic species of the wheat tribe, Triticeae (Poaceae), inferred from nuclear rDNA (internal transcribed spacer) sequences. Genome 38(2):11–223

Hsiao C, Chatterton CJ, Asay KH, Jensen KB (1995b) Molecular phylogeny of the Pooideae (Poaceae) based on nuclear rDNA (ITS) sequences. Theor Appl Genets 90:389–398

Huang S, Sirikhachornkit A, Faris JD, Su X, Gill BS et al (2002a) Phylogenetic analysis of the acetyl-CoA carboxylase and 3-phosphoglycerate kinase loci in wheat and other grasses. Plant Mol Biol 48:805–820

Huang S, Sirikhachornkit A, Su X, Faris JD, Gill BS et al (2002b) Genes encoding plastid acetyl-CoA carboxylase and 3-phosphoglycerate kinase of the *Triticum/Aegilops* complex and the evolutionary history of polyploid wheat. Proc Natl Acad Sci (USA) 99:8133–8138

Hubbard CE (1946) *Henrardia*, a new genus of the Gramineae. Blumea, Suppl 3:10–21

Hutchinson J, Chapman V, Miller TE (1980) Chromosome pairing at meiosis in hybrids between *Aegilops* and *Secale* species: a study by *in situ* hybridization using cloned DNA. Heredity 45:245–254

Huynh S, Marcussen T, Felber F, Parisod C (2019) Hybridization preceded radiation in diploid wheats. Mol Phylogenet Evol 139:106554

International Wheat Genome Sequencing Consortium (IWGSC) (2014) A chromosome-based draft sequence of the hexaploid bread wheat genome. Science 345(6194):1251788. https://doi.org/10.1126/science.1251788

Jaaska V (1975) Evolutionary variation of enzymes and phylogenetic relationships in the genus *Secale* L. Izv Akad Nauk Estonsk SSR, Ser Biol 24:179–198

Jahier J, Tanguy AM, Lucas H (1988) Pairing between *Dasypyrum villosum* (L.) Candargy and *Secale cereale* L. chromosomes. In: Miller TE, Koebner RMD (eds) Proceedings of the 7th international wheat genetics symposium, vol I. Institute of Plant Sciences Research, Cambridge, UK, pp 315–321

Jain SK (1960) Cytogenetics of rye (*Secale* spp.). Bibliogr Genet 19:1–86

Jones JDG, Flavell RB (1982) The structure, amount and chromosomal localization of defined repeated DNA sequences in species of the genus *Secale*. Chromosoma 86:613–641

Jones RN, Rees H (1982) B chromosomes. Academic Press, London

Kagawa F, Chizaki Y (1934) Cytological studies on the genus hybrids among *Triticum*, *Secale* and *Aegilops*, and the species hybrids in *Aegilops*. Jap J Bot 7:1–32

Karpechenko GD, Sorokina ON (1929) The hybrids of *Aegilops triuncialis* L. with rye. Bull Appl Bot Genet Pl Breed (USSR) 20:536–584

Kawahara T, Yamane K, Imai T (2008) Phylogenetic relationships among Aegilops-Triticum species based on sequence data of chloroplast DNA. In: Appels R, Eastwood R, Lagudah E, Langridge P, Mackay M, Mcintyre L, Sharp P (eds) Proc 11th intern wheat genet symp. Univ Press, Sydney, Australia, http://hdl.handle.net/2123/3332

Kawahara T (2009) Molecular phylogeny among *Triticum-Aegilops* species and of the tribe Triticeae. Breed Sci 59:499–504

Kawakubo J, Taira T (1992) Intergeneric hybrids between *Aegilops squarrosa* and *Secale cereale* and their meiotic chromosome behaviour. Pl Breed 109:108–115

Kellogg EA (1989) Comments on genomic genera in the Triticeae (Poaceae). Amer J Bot 76:796–805

Kellogg EA (1992) Restriction site variation in the chloroplast genomes of the monogenomic Triticeae. Hereditas 116:43–47

Kellogg EA, Appels R (1995) Intraspecific and interspecific variation in 5S RNA genes are decoupled in diploid wheat relatives. Genetics 140:325–343

Kellogg EA (1996) When the genes tell different stories: the diploid genera of Triticeae (Gramineae). Syst Bot 21:321–347

Kellogg EA, Appels R, Mason-Gamer RJ (1996) When genes tell different stories: the diploid genera of Triticeae (Gramineae). Syst Bot 21:321–347

Key JM (1966) Species relationship in Triticum. In Proc 2nd Int Wheat Genet Symp, Lund, Sweden, Hereditas, Suppl 2:237–276

Key JM (1968) Relationships in the Triticinae. In: Finlay KW, Shepherd KW (eds) Proc 3rd Int Wheat Genet Symp, Canberra, Aust Acad Sci, Canberra, pp 39–50

Key JM (1981) Comments on the basic principles of crop taxonomy. Die Kulturpflanze 29:199–207

Khush GS, Stebbins GL (1961) Cytogenetic and evolutionary studies in *Secale*. I. Some new data on the ancestry of *S. cereale*. Amer J Bot 48:721–730

Khush GS (1962) Cytogenetic and evolutionary studies in *Secale*. II. Interrelationships of the wild species. Evolution 16:484–496

Kihara H, Lilienfeld FA (1936) Riesenpollenkorner bei den F1-bastarden *Aegilops squarrosa* x *Haynaldia villosa* und *Aegilops caudata* x *Aegilops speltoides*. Jpn J Genet 12:239–256

Kihara H (1937) Genomanalyse bei *Triticum* und *Aegilops*. VII. Kurze ~ bersicht iiber die Ergebnisse der Jahre 1934–36. Mem Coll Agric Kyoto Imp Univ 41:1–61

Kihara H, Nishiyama I (1937) Possibility of crossing-over between semihomologous chromosomes from two different genomes. Cytologia Fujii Jub Vol, pp 654–666

Kihara H (1949) Genomanalyse bei *Triticum* und *Aegilops*. IX. Systematischer Aufbau der Gattung Aegilops auf genomanalytischer Grundlage. Cytologia 14:135–144

Kilian B, Özkan H, Deusch O, Effgen S, Brandolini A, Kohl J, Martin W, Salamini F (2007) Independent wheat B and G genome origins in outcrossing *Aegilops* progenitor haplotypes. Mol Biol Evol 24:217–227

Kilian B, Mammen K, Millet E, Sharma R, Graner A, Salamini F, Hammer K, Özkan H (2011) Aegilops. In: Kole C (ed) Wild crop relatives: genomic and breeding resources, cereals, vol 24. Springer. Berlin, Germany, pp 1–76

Kimber G (1961) Cytogenetics of haploidy in *Gossypium* and *Triticum*. PhD thesis, Univ. Manchester, UK, pp 1–297

Kimber G, Feldman M (1987) Wild wheats: an introduction. Special Report 353. College of Agriculture, Columbia, Missouri, USA, pp 1–142

Kobyljanskij VD (1982) Rozh. Geneticheskije osnovy selekcii. Kolos, Moskva

Koller OL, Zeller FJ (1976) The homoeologous relationships of rye chromosomes 4R and 7R with wheat chromosomes. Genet Res Camb 28:177–188

Komatsuda T, Tanno K, Salomon B, Bryngelsson T, von Bothmer R (1999) Phylogeny in the genus *Hordeum* based on nucleotide sequences closely linked to the *vrs1* locus (row number of spikelets). Genome 42:973–981

Kosina R, Heslop-Harrison JS (1996) Molecular cytogenetics of an amphiploid trigeneric hybrid between *Triticum durum, Thinopyrum distichum* and *Lophopyrum elongatum*. Ann Bot 78:583–589

Kranz AR (1963) Beitrage zur cytologischen und genetischen Evolutionsforschung an dem Roggen. Zeitschrift Für Pflanzenzüchtung 50:44–58

Kuckuck H, Peters R (1967) Genetische Untersuchungen über die Selbstfertilitat bei S. vavilovii Grossh. und S. cereale L. var. Dakold im Hinblick auf Probleme der Züchtung und Phylogenie. Z Pflanzenzüchtung 57:167–188

Kushnir U, Halloran GM (1981) Evidence for *Aegilops sharonensis* Eig as the donor of the B genome of wheat. Genetics 99:495–512

Lammer D, Cai X, Arterburn M, Chatelain J, Murray T, Jones S (2004) A single chromosome addition from *Thinopyrum elongatum* confers a polycarpic, perennial habit to annual wheat. J Exp Bot 55:1715–1720

Levy AA (2013) Transposons in plant speciation. In: Fedoroff NV (ed) Plant transposons and genome dynamics in evolution. Wiley, pp 164–179

Lewis H (1966) Speciation in flowering plants. Science 152:167–172

Lewitsky GA (1929) Investigations on the morphology of chromosomes. Proc USSR Cong Genet Pl and Anim Breed 2:87–105

Lewitsky GA (1931) The morphology of chromosomes. Bull Appl Bot Genet Pl Breed 27:19–174

Li W, Huang L, Gill BS (2008) Recurrent deletions of puroindoline genes at the grain hardness locus in four independent lineages of polyploid wheat. Plant Physiol 146:200–212

Li XM, Lee BS, Mammadov AC, Koo BC, Mott IW, Wang RRC (2007) CAPS markers specific to E^b, E^e, and R genomes in the tribe Triticeae. Genome 50:400–411

Li L-F, Liu B, Olsen KM, Wendel JF (2015) A re-evaluation of the homoploid hybrid origin of *Aegilops tauschii*, the donor of the wheat D-subgenome. New Phytol 208:4–8

Li L-F, Zhang Z-B, Wang Z-H, et al (2022) Genome sequences of the five Sitopsis species of Aegilops and the origin of polyploid wheat B subgenome. Mol Plant 15:488–503

Linde-Laursen I, Frederiksen S (1989) Giemsa C-banded karyotypes of three sub-species of *Taeniatherum caput-medusae* and two intergeneric hybrids with *Psathyrostachys* spp. (Poaceae). Hereditas 110:283–288

Liu CJ, Atkinson MD, Chinoy CN, Devos KM, Gale MD (1992) Non-homoeologous translocations between group 4, 5 and 7 chromosomes within wheat and rye. Theor Appl Genet 83:305–312

Liu Z, Li D, Zhang X (2007) Genetic relationships among five basic genomes St, E, A, B and D in Triticeae revealed by genomic Southern and *in situ* hybridization. J Integr Plant Biol 49:1080–1086

Liu ZP, Chen ZY, Pan J, Li XF, Su M, Wang LJ, Li HJ, Liu GS (2008) Phylogenetic relationships in *Leymus* (Poaceae: Triticeae) revealed by the nuclear ribosomal internal transcribed space and chloroplast trnL-F sequences. Mol Phylogenet Evol 46:278–289

Longley AE, Sando WJ (1930) Nuclear divisions in the pollen mother cells of *Triticum, Aegilops,* and *Secale* and their hybrids. J Agr Res 40:683–719

Löve Á (1984) Conspectus of the Triticeae. Feddes Repert 95:425–521

Lu B-R, Salomon B, von Bothmer R (1990) Cytogenetic studies of progeny from the intergeneric crosses *Elymus* x *Hordeum* and *Elymus* x *Secale*. Genome 33:425–432

Lu BR, von Bothmer R (1991) Cytogenetic studies of the intergeneric hybrids between *Secale cereale* and *Elymus caninus, E. brevipes,* and *E. tsukushiensis* (Triticeae: Poaceae). Theor Appl Genet 81:524–532

Lucas H, Jahier J (1988) Phylogenetic relationships in some diploid species of Triticineae. Cytogenetic analysis of interspecific hybrids. Theor Appl Genet 75:498–502

Lukaszewski AJ, Kopecky D, Linc G (2012) Inversions of chromosome arms 4AL and 2BS in wheat invert the patterns of chiasma distribution. Chromosoma 121:201–208

Luo M-C, Yang Z-L, Kota RS, Dvorak J (2000) Recombination of chromosomes 3A (m) and 5A (m) of *Triticum monococcum* with homoeologous chromosomes 3A and 5A of wheat: the distribution of recombination across chromosomes. Genetics 154:1301–1308

Luo M-C, Deal KR, Yang Z-L, Dvorak J (2005) Comparative genetic maps reveal extreme crossover localization in the *Aegilops speltoides* chromosomes. Theor Appl Genet 111:1098–1106

Luria Y (1983) Morphological, ecogeographical and cyto-taxonomic survey of five Israeli Triticeae genera. MSc thesis, Hebrew University of Jerusalem (In Hebrew with English summary), pp 1–116

Majisu BN, Jones JK (1971) *Aegilops × Secale* hybrids: the production and cytology of diploid hybrids. Genet Res Camb 17:17–31

Marcussen T, Sandve SR, Heier L, Spannagl M, Pfeifer M, The International Wheat Genome Sequencing Consortium, Jakobsen KS, Wulff BBH, Steuernagel B, Klaus FX, Mayer KFX, Olsen OA (2014) Ancient hybridizations among the ancestral genomes of bread wheat. Science 345(6194):288–291

Martis MM, Zhou R, Haseneyer G, Schmutzer T, Vrána J, Kubaláková M, König S, Kugler KG, Scholz U, Hackauf B, Korzun V, Schön CC, Dolezel J, Bauer E, Mayer KF, Stein N (2013) Reticulate evolution of the rye genome. Plant Cell 25:3685–3698

Mason-Gamer RJ, Kellogg EA (1996a) Testing for phylogenetic conflict among molecular data sets in the tribe Triticeae (Gramineae). Syst Bio l 45:524–545

Mason-Gamer RJ, Kellogg EA (1996b) Chloroplast DNA analysis of the monogenomic Triticeae: phylogenetic implications and genome-specific markers. In: Jauhar PP (ed) Methods of genome analysis in plants. CRC Press, Boca Raton, Florida, pp 301–325

Mason-Gamer RJ, Weil CF, Kellogg EA (1998) Granule-bound starch synthase: structure, function, and phylogenetic utility. Mol Biol Evol 15:1658–1673

Mason-Gamer RJ, Kellogg EA (2000) Phylogenetic analysis of the Triticeae using the starch synthase gene, and a preliminary analysis of some North American *Elymus* species. In: Jacobs SWL, Everett J (eds) Grasses, systematics and evolution. CSIRO Publishing, Collingwood, Victoria, Australia, pp 102–109

Mason-Gamer RJ, Orme NL, Anderson CM (2002) Phylogenetic analysis of North American *Elymus* and the monogenomic Triticeae (Poaceae) using three chloroplast DNA data sets. Genome 45:991–1002

Mason-Gamer RJ (2005) The B-amylase genes of grasses and a phylogenetic analysis of the Triticeae (Poaceae). Am J Bot 92:1045–1058

Mason-Gamer RJ, Burns MM, Naum M (2010) Reticulate evolutionary history of a complex group of grasses: phylogeny of *Elymus* StStHH allotetraploids based on three nuclear genes. PLoS ONE 5: e10989

McClintock B (1978) Mechanisms that rapidly reorganize the genome. Stadler Symp 10:25–47

McClintok B (1984) The significance of responses of the genome to challenges. Science 226:792–801

Megyeri M, Molnár-Láng M, Molnár I (2013) Cytomolecular identification of individual wheat-wheat chromosome arm associations in wheat-rye hybrids. Cytogenet Genome Res 139:128–136

Melderis A (1953) Generic problems within the tribe *Hordeae*. In Osvald H, Aberg E (eds) Proc 7th Intern Bot Congr, Stockholm 1950, Uppsala, Almquist & Wiksells, pp 1450–1485

Melderis A (1978) Taxonomic notes on the tribe Triticeae (Gramineae), with special reference to the genera *Elymus* L. *senso lato*, and *Agropyron* Gaertbner *senso lato*. In: Tutin TG, Heywood VA, Burges NA, Moore DM, Valentine DH, Walters SM (eds) Flora Europea, vol 5, pp 369–384

Melderis A (1980) Taxonomic notes on the tribe Triticeae (Graminae), with special reference to the genera *Elymus* and *Agropyron*. Fl Europ Not sys, No 20

Melnyk J, Unrau J (1959) Pairing between chromosomes of *Aegilops squarrosa* L. var. *typica* and *secale cereale* L. var. *prolific*. Can J Genet Cytol 1:21–25

Middleton C, Stein N, Keller B, Kilian B, Wicker T (2013) Comparative analysis of genome composition in Triticeae reveals strong variation in transposable element dynamics and nucleotide diversity. Plant J 73:347–356

Middleton CP, Senerchia N, Stein N, Akhunov ED, Keller B, Wicker T, Kilian B (2014) Sequencing of chloroplast genomes from wheat, barley, rye and their relatives provides a detailed insight into the evolution of the Triticeae tribe. PLoS ONE 9: e85761

Miller TE, Reader SM, Purdie KA, King IP (1994) Determination of the frequency of wheat-rye chromosome pairing in wheat x rye hybrids with and without chromosome 5B. Theor Appl Genet 89:255–258

Mochizuki A (1964) Further studies on the effect of accessory chromosomes on chromosome pairing (in Japanese). Jap J Genet 39:356–362

Monte JV, McIntyre CL, Gustafson JP (1993) Analysis of phylogenetic relationships in the Triticeae tribe using RFLPs. Theor Appl Genet 86:649–655

Mukai Y, Endo TR, Gill BS (1990) Physical mapping of the 5S rRNA multigene family in common wheat. J Hered 81:290–295

Müntzing A (1944) Cytological studies of extra fragment chromosomes in rye. I. Iso-fragments produced by misdivision. Hereditas 30:231–248

Müntzing A (1950) Accessory chromosomes in rye populations from Turkey and Afghanistan. Hereditas 36:507–509

Murai K, Naiyu X, Tsunewaki K (1989) Studies on the origin of crop species by restriction endonuclease analysis of organellar DNA. Ill. Chloroplast DNA variations and interspecific relationships in the genus *Secale*. Jpn J Genet 64:36–47

Nakajima G (1966) Caryogenetical studies on F_1 intergenic hybrids raised from crossing between *Triticum* and *Haynaldia*. La Kromosomo 41:2083–2100

Naranjo T, Roca A, Goicoecha PG, Giraldez R (1987) Arm homoeology of wheat and rye chromosomes. Genome 29:873–882

Naranjo T, Fernández-Rueda P (1991) Homoeology of rye chromosome arms to wheat. Theor Appl Genet 82:577–586

Naranjo T, Fernández-Rueda P (1996) Pairing and recombination between individual chromosomes of wheat and rye in hybrids carrying the *ph1b* mutation. Theor Appl Genet 93:242–248

Nevski SA (1933) Agrostologische studien. IV. Uber das System der Tribe Hordeae Benth. Acta Inst Bot Sci, U. R. S. S, Ser II

Nevski SA (1934) Schedae ad Herbarium Flora Asiae Mediae. Acta Univ Asiae Med VIII b Bot 17:1–94

Nishikawa T, Salomon B, Komatsuda T, von Bothmer R, Kadowaki K (2002) Molecular phylogeny of the genus Hordeum using three chloroplast DNA sequences. Genome 45:1157–1166

Niwa K, Ohta S, Sakamoto S (1990) B chromosomes of *Secale cereale* L. and *S. montanum* Guss. from Turkey. Jpn J Breed 40:147–152

Oehler E (1931) Untersuchungen über Ansatzverhältnisse, Morphologie und Fertilität bei Weizen-Roggenbestanden. Zeitschr Pflanzenzg 36:357–393

Oehler E (1933) Untersuchungen über Ansatzverhältnisse, Morphologie und Fertilität bei *Aegilops*-Weizenbastarden. I. Teil: Die F_1 Generation. Z Induk Abst Vererbgsl 64:95–153

Oehler E (1935) Untersuchungen an *Aegilops*– *Haynaldia* und *Triticum*-*Haynaldia*-Bastarden. Z Induk Abst Vererbgsl 68:187–208

Ohsako T, Wang G-Z, Miyashita NT (1996) Polymerase chain reaction —single strand conformational polymorphism analysis of intra- and interspecific variations in organellar DNA regions of *Aegilops mutica* and related species. Genes Genet Syst 71:281–292

Ohta S, Tanaka M (1982) The effects of the B-chromosomes of *Aegilops mutica* Boias. on meiotic chromosome pairing. Rep Plant Germ-Plasm Inst, Kyoto Univ 5:36–52

Ohta S, Tanaka M (1983) Genome relationships between *Ae. mutica* and the other diploid *Aegilops* and *Triticum* species, based on the chromosome pairing in the hybrids with or without B-chromosomes. In: Sakamoto S (ed) Proc 6th Int Wheat Genet Symp, Kyoto, pp 983–991

Ohta S (1990) Genome analysis of *Aegilops mutica* Boiss. Based on the chromosome pairing in interspecific and intergeneric hybrids. PhD Thesis, University of Kyoto, pp 1–267

Ohta S (1991) Phylogenetic relationship of *Aegilops mutica* Boiss with the diploid species of congeneric *Aegilops-Triticum* complex, based on the new method of genome analysis using its B-chromosomes. Mem Coll Agric Kyoto Univ 137:1–116

Ohta S, Morishita M (2001) Relationships in the Genus *Dasypyrum* (Gramineae) Hereditas 135:101–110

Ohta S, Koto M, Osada T, Matsuyama A, Furuta Y (2002) Rediscovery of a diploid cytotype of *Dasypyrum breviaristatum* in Morocco. Genet Resour Crop Evol 49:305–312

Parisod C, Badaeva ED (2020) Chromosome restructuring among hybridizing wild wheats. New Phytol 226:1263–1273

Perrino P, Hammer K, Hanelt P (1984) Collection of landraces of cultivated plants in South Italy 1983. Kulturpflanze 32:207–216

Petersen G (1991a) Intergeneric hybridization between *Hordeum* and *Secale* (Poaceae). I. Crosses and development of hybrids. Nordic J Bot 11:253–270

Petersen G (1991b) Intergeneric hybridization between *Hordeum* and *Secale*. II. Analysis of meiosis in hybrids. Hereditas 114:141–159

Petersen G, Seberg O (1997) Phylogenetic analysis of the Triticeae (Poaceae) based on *rpo*A sequence data. Mol Phylogenet Evol 7:217–230

Petersen G, Seberg O (2000) Phylogenetic evidence for the excision of *Stowaway* miniature inverted-repeat transposable elements in Triticeae (Poaceae). Mol Biol Evol 17:1589–1596

Petersen G, Seberg O, Aagesen L, Frederiksen S (2004) An empirical test of the treatment of indels during optimization alignment based on the phylogeny of the genus *Secale* (Poaceae). Mol Phylogenet Evol 30:733–742

Petersen G, Seberg O, Yde M, Berthelsen K (2006) Phylogenetic relationships of *Triticum* and *Aegilops* and evidence for the origin of the A, B, and D genomes of common wheat (*Triticum aestivum*). Mol Phylogenet Evol 39:70–82

Plotnikowa TW (1930) Cytological investigations on hybrids between 28-chromosome wheat and rye. Planta 12:167–183

Popoff A (1939) Untersuchungen über den Formenreichtum und die Schartigkeit des Roggens. Ang Bot 21:325–356

Ramakrishna W, Dubcovsky J, Park Y-J, Busso C, Emberton J, SanMiguel P, Bennetzen J (2002) Different types and rates of genome evolution detected by comparative sequence analysis of orthologous segments from four cereal genomes. Genetics 162:1389–1400

Raskina O, Belyayev A, Nevo E (2004) Activity of the *En/Spm*-like transposons in meiosis as a base for chromosome repatterning in a small, isolated, peripheral population of *Aegilops speltoides* Tausch. Chromosome Res 12:153–161

Reddy P, Appels R, Baum BR (1990) Ribosomal DNA spacer-length variation in *Secale* spp. (*Poaceae*). Pl Syst Evol 171:205–220

Riley R, Unrau J, Chapman V (1958) Evidence on the origin of the B genome of wheat. J Hered 49:91–98

Riley R (1960) The diploidization of polyploid wheat. Heredity 15:407–429

Riley R, Kimber G, Chapman V (1961) Origin of genetic control of diploid-like behavior of polyploid wheat. Hered 52:22–25

Rodríguez S, Maestra B, Perera E, Díez M, Naranjo T (2000) Pairing affinities of the B- and G-genome chromosomes of polyploid wheats with those of *Aegilops speltoides*. Genome 43:814–819

Rognli OA, Devos KM, Chinoy CN, Harcourt RL, Atkinson MD, Gale MD (1992) RFLP mapping of rye chromosome 7R reveals a highly translocated chromosome relative to wheat. Genome 35:1026–1031

Roy RP (1959) Genome analysis of *Aegilops sharonensis*. Genetica 29:331–357

Ruban AS, Badaeva ED (2018) Evolution of the S-genomes in *Triticum-Aegilops* alliance: evidences from chromosome analysis. Front Plant Sci 9:1756

Runemark H, Heneen WK (1968) *Elymus* and *Agropyron*, a problem of generic delimitation. Bot Notiser 121:51–79

Sakamoto S, Muramatsu M (1965) Morphological and cytological studies on various species of Gramineae collected in Pakistan, Afghanistan and Iran. In: Results of the Kyoto university scientific expedition to the karakoram and hindiikush, 1955, vol 1, pp 119–140

Sakamoto S (1967) Cytogenetic studies in the tribe Triticeae. V. Intergeneric hybrids between two *Eremopyrum* species and *Agropyron tsukushiense*. Seiken Ziho 19:19–27

Sakamoto S (1968) Cytogenetic studies in the tribe Triticeae. VI. Intergeneric hybrid between *Eremopyrum orientale* and *Aegilops squarrosa*. Jpn J Genet 43:167–171

Sakamoto S (1972) Intergeneric hybridization between *Eremopyrum Orientale* and *Henrardia persica*, an example of polyploidy species formation. Heredity 28:109–115

Sakamoto S (1973) Patterns of phylogenetic differentiation in the tribe Triticeae. Seiken Zihô 24:11–31

Sakamoto S (1974) Intergeneric hybridization amomg three species of *Heteranthelium*, *Eremopyrum* and *Hordeum* and its significance for the genetic relationships within the tribe Triticeae. New Phytol 73:341–350

Sakamoto S (1979) Genetic relationships among four species of the genus *Eremopyrum* in the tribe Triticea, Gramineae. Mem Coll Agric Kyoto Univ 114:1–27

Sakamoto S (1991) The cytogenetic evolution of Triticeae grasses. In: Tsuchiya T, Gupta PK (eds) Chromosome engineering in plants: genetics, breeding, evolution. Elsvier Science Publications B.V, Amsterdam, The Netherland, pp 469–482

Sakamura T (1918) Kurze mitteilung über die chromosomenzahalen und die verwandtschaftsverhältnisse der Triticum Arten. Bot Mag 32:151–154

Sallares R, Brown TA (2004) Phylogenetic analysis of complete 5' external transcribed spacers of the 18S ribosomal RNA genes of diploid *Aegilops* and related species (Triticeae, Poaceae). Genet Resour Crop Evol 51:701–712

Salse J, Bolot S, Throude M, Jouffe V, Piegu B, Quraishi UM, Calcagno T, Cooke R, Delseny M, Feuillet C (2008) Identification and characterization of conserved duplications between rice and wheat provide new insight into grass genome evolution. Plant Cell 20:11–24

Sando WJ (1935) Hybrids of wheat, rye, *Aegilops* and *Haynaldia*. A series of 122 intra- and inter-generic hybrids shows wide variations in fertility. J Hered 26:229–232

Sandve SR, Marcussen T, Mayer K, Jakobsen KS, Heier L, Steuernagel B, Wulff BB, Olsen OA (2015) Chloroplast phylogeny of *Triticum/Aegilops* species is not incongruent with an ancient homoploid hybrid origin of the ancestor of the bread wheat D-genome. New Phytol 208:9–10

Sarkar P (1957) A new diploid form of *Haynaldia hordeacea* Hack. Wheat Inform Serv 6:22

Sasanuma T, Miyashita NT, Tsunewaki K (1996) Wheal phylogeny determined by RFLP analysis of nuclear DNA. 3. Intra–and interspecific variations of five *Aegilops* Sitopsis species. Theor Appl Genet 92:928–934

Sasanuma T, Chabane K, Endo TR, Valkoun J (2004) Characterization of genetic variation in and phylogenetic relationships among diploid *Aegilops* species by AFLP: incongruity of chloroplast and nuclear data. Theor Appl Genet 108:612–618

Schiemann E, Nürnberg-Krüger U (1952) Neue Untersuchungen an *Secale africanum* Stapf. Die Naturwiss 6:136–137

Schlegel R (2013) Rye–genetics, breeding, and cultivation. CRC Presss, Taylor and Francis Group, Roca Raton, FL, USA, pp 1–344

Schultze-Motel J, Meyer D (1981) Numerical taxonomic studies in the genera *Triticum* L. and *Pisum* L. Kulturpflanze 29:241–250

Sears ER (1941a) Chromosome pairing and fertility in hybrids and amphidiploids in the Triticinae. Univ Missouri Agric Exp Stn Res Bull 337:1–20

Sears ER (1941b) Amphidiploids in the seven-chromosome Triticinae. Univ Missouri Agric Exper Stn Res Bull 336:1–46

Seberg O, Frederiksen S (2001) A phylogenetic analysis of the monogenomic Triticeae (Poaceae) based on morphology. Bot J Linn Soc 136:75–97

Seberg O, Petersen G (2007) Phylogeny of Triticeae (Poaceae) based on three organelle genes, two single-copy nuclear genes, and Morphology. Aliso J Syst Evol Bot 23:362–371

Sencer HA (1975) Study of variation in the genus *Secale* L. and on the origin of the cultivated rye. PhD thesis, The University of Birmingham

Senerchia N, Wicker T, Felber F, Parisod C (2013) Evolutionary dynamics of retrotransposons assessed by high throughput sequencing in wild relatives of wheat. Genome Biol Evol 5:1010–1020

Senjaninova-Korchagina MV (1932) Karyo-systematical investigations of the genus *Aegilops* L. (Russian with English summary) Bull Appl But Genet and Pl Breed Ser 2:1–90

Shang H-Y, Baum BR, Wei Y-M, Zheng Y-L (2007) The 5S rRNA gene diversity in the genus *Secale* and determination of its closest haplomes. Genet Resour Crop Evol 54:793–806

Shewry PR, Parmer S, Miller TE (1985) Chromosomal location of the structural genes for the M_r 75,000 γ-secalins in *Secale montanum* Guss: evidence for a translocation involving chromosome 2R and 6R in cultivated rye (*Secale cereale* L.). Heredity 54:381–383

Singh RJ (1977) Cross compability, meiotic pairing and fertility in 5 *Secale* species and their interspecific hybrids. Cereal Res Commun 5:67–75

Sliai AM, Amer SAM (2011) Contribution of chloroplast DNA in the biodiversity of some *Aegilops* species. Afr J Biotechnol 10:2212–2215

Sodkiewicz W (1982) Hybrids between diploid wheat and rye. I. F_1 hybrids of *Triticum monococcum* L. x *S. cereale* L. Genet Pol 23:123–131

Stebbins GL (1945) The cytological analysis of species hybrids. Bot Rev 11:463–486

Stebbins GL (1950) Variation and evolution in plants. Columbia University Press, New York

Stebbins GL, Pun FT (1953) Artificial and natural hybrids in the Gramineae tribe Hordeae. VI. Chromosome pairing in *Secale cercale–Agropyron intermedium* and the problem of genome homologies in the Triticineae. Genetics 38:600–608

Stebbins GL (1971) Chromosomal evolution in higher plants. Edward Arnold Ltd., London

Stolze K (1925) Die Chromosomenzahlen der hauptsächlichsten Getreide- arten nebst allgemeinen Betrachtungen iiber Chromosomenzahl und Chromosomengrosse im Pflanzenreich. Bibliotheca Genet 8:8–71

Stutz HC (1957) A cytogenetic analysis of the hybrid *Secale cereale* L. x *Secale montanum* Guss. and its progeny. Genetics 42:199–222

Stutz HC (1972) The origin of cultivated rye. Amer J Bot 59:59–70

Su Y, Zhang D, Li Y, Li S (2015) Nonhomologous chromosome pairing in *Aegilops–Secale* Hybrids. Cytogenet Genome Res 147:268–273

Sun GL, Ni Y, Daley T (2008) Molecular phylogeny of *RPB2* gene reveals multiple origin, geographic differentiation of H genome, and the relationship of the Y genome to other genomes in *Elymus* species. Mol Phylogenet Evol 46:897–907

Sun GL, Komatsuda T (2010) Origin of the Y genome in *Elymus* and its relationship to other genomes in Triticeae based on evidence from elongation factor G (*EF-G*) gene sequences. Mol Phylogenet Evol 56:727–733

Tanaka M (1955) Chromosome pairing in hybrids between *Aegilops sharonensis* and some species of *Aegilops* and *Triticum*. Wheat Infor Serv 2:7–8

Thompson WP (1926) Chromosome hehavior in a cross between wheat and rye. Genetics 11:317–332

Thomas HM, Pickering RA (1985) Comparisons of the hybrid *Hordeum chilense* x *H. vulgare*, *H. chilense* x *H. bulbosum*, *H. chilense* x *Secale cereale* and the amphidipioid of *H. chilense* x *H. vulgare*. Theor Appl Genet 69:519–522

Tsunewaki K, Ogihara Y (1983) The molecular basis of genetic diversity among cytoplasms of *Triticum* and *Aegilops* species. II. On the origin of polyploid wheat cytoplasms as suggested by chloroplast DNA restriction fragment patterns. Genetics 104:155–171

Tsunewaki K, Wang GZ, Matsuoka Y (1996) Plasmon analysis of *Triticum* (wheat) and *Aegilops*. 1. Production of alloplasmic common wheats and their fertilities. Genes Genet Syst 71:293–311

Tsunewaki K (2009) Plasmon analysis in the *Triticum-Aegilops* complex. Breed Sci 59:455–470

Tzvelev NN (1973) Conspectus specierum tribus *Triticeae* Dum. Familiae *Poaceae* in Flora URSS. Nov Syst Pl Vasc 10:19–59

Tzvelev NN (1976) Poaceae URSS. Tribe III. Triticea Dum. USSR. Academy of Sciences Press, Leningrad, pp 105–206

Uslu E, Reader SM, Miller TE (1999) Characterization of *Dasypyrum villosum* (L.) Candargy chromosomes by fluorescent in situ hybridization. Hereditas 131:129–134

Vardi A, Dover GA (1972) The effect of B chromosomes on meiotic and pre-meiotic spindles and chromosome pairing in *Triticum/Aegilops* hybrids. Chromosoma (berl.) 38:367–385

Vasiljev B (1932) Wheat-rye hybrids. 1. An analysis of the first generation of various wheat-rye hybrid combinations. Bull Lab Genet 9:69–87

Vences FJ, Vaquero F, Garcia P, de La Vega MP (1987) Further studies on the phylogenetic relationships in *Secale*: On the origin of its species. Pl Breed 98:281–291

Vogel KP, Arumuganathan K, Jensen KB (1999) Nuclear DNA content of perennial grasses of Triticeae. Crop Sci 39:661–667

von Berg KH (1931) Autosyndese in *Aegilops triuncialis* L. x *Secale cereale* L. Pflanzenzüchtung, vol 17. Zeitschrift fur Zuchtung A, pp 55–69

von Berg KH (1937) Beitrag zur Genomanalyse in der Getreidegruoppe. Der Züchter 9:157–163

von Bothmer R, Claesson L (1990) Production and meiotic pairing of intergeneric hybrids of *Triticum* x *Dasypyrum* species. Euphytica 51:109–117

Wang RR-C (1985) Genome analysis of *Thinopyrum bessarabicum* and *T. elongatum*. Can J Genet Cytol 27:722–728

Wang RR-C (1986) Diploid perennial intergeneric hybrids in the tribe Triticeae. 1. *Agropyron cristatum* x *Pseudoroegneria libanotica* and *Critesion violaceum* x *Psathyrostachys juncea*. Crop Sci 26:75–78

Wang RR-C (1987) Progenies of *Thinopyrum elongatum* X *Agropyron mongolicum*. Genome 29:738–743

Wang RR-C (1988) Diploid perennial intergeneric hybrids in the tribe Triticeae. IV. Hybrids among Thinpyrum bessarabicum, *Pseudorogneria spicata* and *Secale montanum*. Genome 30:356–360

Wang RR-C (1989) An assessment of genome analysis based on chromosome pairing in hybrids of perennial Triticeae. Genome 32:179–189

Wang RR-C, Hsiao CT (1989) Genome relationship between *Thinopyrum bessarabicum* and *T. elongatum*: Revisited. Genome 32:802–809

Wang RR-C (1992) Genome relationships in the perennial Triticeae based on diploid hybrids and beyond. Hereditas 116:133–136

Wang RR-C, von Bothmer R, Dvorak J, Fedak G, Linde-Laursen I, Muramatsu M (1995) Genome symbols in the Triticeae (Poaceae). In: Wang RR-C, Jensen KB, Jaussi C (eds). Proc 2nd Inter Triticeae Symp. Logan, UT, pp 29–34

Wang RR-C, Lu BR (2014) Biosystematics and evolutionary relationships of perennial Triticeae species revealed by genomic analyses. J Syst Evol 52:697–705

Wei JZ, Wang RR-C (1995) Genome- and species-specific markers and genome relationships of diploid perennial species in Triticeae based on RAPD analyses. Genome 38:1230–1236

Wolfe KH, Gouy M, Yang Y-W, Sharp PM, Li W-H (1989) Date of the monocot-dicot divergence estimated from chloroplast DNA sequence data. Proc Natl Acad Sci (USA) 86:6201–6205

Yaakov B, Meyer K, Ben-David S, Kashkush K (2013) Copy number variation of transposable elements in *Triticum-Aegilops* genus suggests evolutionary and revolutionary dynamics following allopolyploiduzation. Plant Cell Rep 32:1615–1624

Yamane K, Kawahara T (2005) Intra- and interspecific phylogenetic relationships among diploid *Triticum-Aegilops* species (Poaceae) based on base-pair substitutions, indels, and microsatellites in chloroplast noncoding sequences. Am J Bot 92:1887–1898

Yan C, Sun GL (2011) Nucleotide divergence and genetic relationships of *Pseudoroegneria* species. Bioch Syst Ecol 39:309–319

Yan C, Sun GL, Sun DF (2011) Distinct origin of the Y and St genome in *Elymus* species: Evidence from the analysis of a large sample of St genome species using two nuclear genes. PLoS ONE 6:e26853

Yang ZJ, Liu C, Feng J, Li GR, Zhou JP, Deng KJ, Ren ZL (2006) Studies on genome relationship and species-specific PCR marker for *Dasypyrum breviaristatum* in Triticeae. Hereditas 143:47–54

Yen Y, Kimber G (1990) Genomic relationships of *Triticum searsii* to other S-genome diploid *Triticum* species. Genome 33:369–374

Yu Y, Yang WY, Hu XR (1998) The effectiveness of *ph1b* gene on chromosome association in the F_1 hybrid of *T. aestivum* x *H. villosa*. In: Slinkard AE (ed) Proc 9th intern wheat genet symp, vol 2. University Extension Press, University of Saskatchewan, Saskatoon, Canada, pp 125–126

Yu MQ, Deng GB, Zhang XP, Ma XR, Chen J (2001) Effect of the *ph1b* mutant on chromosome pairing in hybrids between *Dasypyrum villosum* and *Triticum aestivum*. Plant Breed 120: 285–289

Yu H-Q, Fan X, Zhang C, Ding CB, Wang XL, Zhou YH (2008) Phylogenetic relationships of species in *Pseudoroegneria* (Poaceae:

Triticeae) and related genera inferred from nuclear rDNA ITS (internal transcribed spacer) sequences. Biologia 63:498–505

Zohary D, Feldman M (1962) Hybridization between amphidiploids and the evolution of polyploids in the wheat (*Aegilops-Triticum*) group. Evolution 16:44–61

Zohary D, Imber D (1963) Genetic dimorphism in fruit types in *Aegilops speltoides*. Heredity 18:223–231

Zohary D, Hopf M, Weiss E (2012) Domestication of plants in the old world-the origin and spread of domesticated plants in south-west asia, europe and the mediterranean basin. 4 edn. Oxford University Press, pp 1–243

Evolution of the Allopolyploid Species of the Sub-tribe Triticineae

12.1 Mode, Time, and Place of Origin of the Allopolyploids

Allopolyploidization is a biological process that has played a central role in plant speciation and evolution (Manton 1950; Stebbins 1950, 1971; Grant 1971; Soltis and Soltis 1993, 1995; Soltis et al. 2009; Masterson 1994, de Bodt et al. 2005; Tang et al. 2008), and has driven and shaped the evolution of vascular plants, perhaps more than any other evolutionary process (Feldman and Levy 2005, 2009). It constitutes a radical and rapid mode of speciation that produces a new species by means of inter-specific or inter-generic hybridization of two diverging diploid species, followed by chromosome doubling of the F_1 hybrids. Allopolyploidization produces a new species in a single step, a novel taxon that is immediately isolated genetically from its two parental species.

The pioneering discoveries of the accurate chromosome numbers of the different *Triticum* and *Aegilops* species (Sakamura 1918; Kihara 1919, 1924, 1937; Sax 1921a, b, 1922, 1927; Sax and Sax 1924; Percival 1921, 1923; Schiemann 1929; Sorokina 1937; Lilienfeld 1951, and reference therein) showed that these two genera comprise a polyploid series, containing diploids, tetraploids and hexaploid species. Subsequent studies of chromosomal pairing in hybrids between the allopolyploid and the diploid species of the group revealed the allopolyploid nature of the polyploids, namely, each polyploid species contains two or three different subgenomes that derived from diverging diploid-level genomes (Lilienfeld 1951, and reference therein). The conclusion that the polyploids of the wheat group are allopolyploids was also supported by the fact that only bivalents were formed at first meiotic metaphase of the polyploids (except that of the auto-allohexaploid *T. zhukovskyi*, which has the genomic constitution GGAAA^{m}A^m and therefore, produces few quadrivalents at meiosis). The diploid-like meiotic pairing pattern, i.e., regular bivalent formation due to exclusive homologous pairing, is characteristic of allopolyploids (Stebbins 1950, 1971). Hence, the two genera *Aegilops* and *Triticum* contains 18 allopolyploid species, 12 allotetraploids (10 of *Aegilops* and 2 of *Triticum*), and 6 allohexaploids, (4 of *Aegilops* and 2 of *Triticum*) (Tables 2.8, 9.3, and 10.5). These species have been subjected to extensive taxonomic, cytogenetic, genetic, biochemical, molecular, phylogenetic, and evolutionary studies by numerous scientists (see reviews of Kihara 1954; Mac Key 1966; Morris and Sears 1967; Kimber and Sears 1987; Feldman et al. 1995; Feldman 2001; Gupta et al. 2005; Dvorak 2009), which resulted in the identification of the diploids that were involved in the formation of most allopolyploids.

Whereas the diploid species of the wheat group evolved in the Pliocene [5.3–1.8 million years ago (MYA)] and early Pleistocene (1.8–0.01 MYA), the allotetraploids were produced afterwards, in the mid- or late Pleistocene (Tables 10.7 and 11.4). *Ae. speltoides* forms a monophyletic clade with the allopolyploid *Triticum* species, the emmer lineage (allotetraploid *T. turgidum* and allohexaploid *T. aestivum*) and the *timopheevii* lineage (allotetraploid *T. timopheevii* and auto-allohexaploid *T. zhukovskyi*) (Gornicki et al. 2014). The geographical distribution of chloroplast haplotypes of the wild allotetraploid wheats and *Ae. speltoides* illustrates the possible geographic origin of the emmer lineage in the southern Levant, the present-day chloroplast diversity center of wild emmer (Gornicki et al. 2014). This is in accord with the finding of Jan Dvorak (personal communication) suggesting that wild emmer was formed in the vicinity of Mt. Hermon. The origin of the *timopheevii* lineage was in northern Iraq, around Arbil, where many accessions of major haplotypes of wild *T. timopheevii* and *Ae. speltoides* were found (Gornicki et al. 2014). Two accessions of wild *T. timopheevii*, collected near Arbil, carry the dominant haplotype of *Ae. speltoides*, hinting that they are the ancestral state of the species. The high level of karyotype diversity in wild *T. timopheevii* collected in northern Iraq (Badaeva et al. 1994), is consistent with the

M. Feldman and A. A. Levy, *Wheat Evolution and Domestication*, https://doi.org/10.1007/978-3-031-30175-9_12

claims that the species originated there. Two chloroplast haplotypes, H08 and H10, were present in both domesticated *T. timopheevii* ssp. *timopheevii* and in *T. zhukovskyi*, either due to recurrent allohexaploidization events or outcrossing between these two species (Gornicki et al. 2014).

Most of the allopolyploid species of Aegilops were presumably produced in the east Mediterranean region and west Asia, the distribution center of their diploid parents. The *Aegilops* allotetraploids then spread further westwards along the Mediterranean basin, as well as in more northern and eastern directions (Kihara 1954). The distribution of the *Aegilops* allotetraploids has been halted by natural boundaries and lack of suitable environments, such as the Saharan and Arabian deserts, the central Asian steppes, the Tian Shan and Himalayan Mountains or the coldness of the continental climate affecting their spread to the north and east. In contrast to the *Aegilops* allotetraploids, wild *Triticum* allotetraploids remained in or near the site of their origin (Harlan and Zohary 1966). Wild emmer, originating in the southern Levant, only spread to the northern Levant, where several new chloroplast haplotypes were established (Gornicki et al. 2014).

Middleton et al. (2014), based on sequencing of the chloroplast genome of 12 Triticeae species, estimated that the B subgenome donor to allopolyploid wheat diverged from *Ae. speltoides* approximately 980,000 years ago, but Li et al. (2022), based on whole genome sequencing of the Sitopsis species, suggest that this divergence occurred much earlier, 4.49 MYA (Table 11.3). The divergence of the S genome of *Ae. speltoides* and the B genome of the assumed parent of allopolyploid wheat, might have been followed by a speciation. In accord with this possibility, none of the chloroplast haplotypes carried by the 391 accessions of the emmer lineage was found among the 450 *Ae. speltoides* accessions (Gornicki et al. 2014). Hence, the female donor of the cytoplasm and the B subgenome to the emmer lineage is either a yet undiscovered relative of *Ae. speltoides*, or perhaps even extinct. Alternatively, the time span from the formation of wild emmer allowed the *speltoides* accession(s) donor to undergo evolutionary changes, and consequently, the current *speltoides* accessions differ from it.

The beginning divergence between *T. urartu* and the A-subgenome donor to allopolyploid wheat is estimated to have occurred 1.28 MYA and between *Ae. tauschii* and the D-subgenome donor to *T. aestivum* to have occurred somewhat more than 880,000 years ago (Li et al. 2022; Table 11.3). Gornicki et al. (2014) based on chloroplast DNA sequencing, suggested, that wild emmer, *T. turgidum* ssp. *dicoccoides,* was formed 700,000 years ago, whereas Marcussen et al. (2014), based on sequencing of a several hundred nuclear genes, suggested that this event occurred

800,000 years ago, and Li et al. (2022), based on whole genome sequencing, considered it to be somewhat earlier.

Tsunewaki et al. (1991) and Wang et al. (1997), based on intraspecific levels of restriction fragment length polymorphism and single-strand configuration polymorphism in chloroplast DNA, concluded that formation of wild emmer is more ancient than that of *T. timopheevii*. Indeed, Mori et al. (1995), based on restriction fragment length polymorphism (RFLP) analysis of nuclear DNA, proposed that the wild form of *T. timopheevii*, ssp. *armeniacum*, formed 50,000–300,000 years ago. Yet, the data of Gornicki et al. (2014) of sequence-based divergence time of chloroplast DNA, showed that the *timopheevii* lineage diverged from *Ae. speltoides* 400,000 years ago. Thus, the allotetraploidization events of wild emmer and of wild *T. timopheevii* occurred within the last 800,000 and 400,000 years ago, respectively. The *Timopheevii* lineage and *Ae. speltoides* chloroplast genomes are very closely related, consistent with *Ae. speltoides* being the maternal donor of the cytoplasm and G subgenome to *T. timopheevii* (Shands and Kimber 1973; Kimber 1974; Dvorak and Zhang 1992a, 1992b; Wang et al. 1997). Yet, Li et al. (2022) reported that the G-subgenome donor to *T. timopheevii* is not *Ae. speltoides* but rather, a species that diverged from *Ae. speltoides* 2.85 MYA (Table 11.2).

The allopolyploid species of *the* wheat group were formed 1.0 MYA or later (Table 11.3). Tsunewaki (2009), based on plasmon comparison between related allotetraploid species and their closest diploid relatives, concluded that wild emmer wheat, *T. turgidum* ssp. *dicoccoides*, *Ae. crassa* and *Ae. geniculata* (formerly *Ae. ovata*), are the oldest allotetraploids in the wheat group, followed by other allotetraploids, some of which originated relatively recently. Middleton et al. (2014) reached similar conclusions, in their report that *Ae. geniculata* is one of the oldest allotetraploid *Aegilops* species and formed about 1.62 MYA. On the other hand, *Ae. cylindrica* is a younger allotetraploid whose D subgenome diverged from D genome of *Ae. tauschii*, approximately 180,000 years ago (Middleton et al. 2014).

Many allopolyploids with combinations similar to natural ones, were synthetically produced, indicating that these genomic combinations can be created in the lab (e.g., Ozkan et al. 2001). Yet, some genomic combinations that were formed in the lab, are absent in nature. The reasons for this can be partly explained by the current eco-geographical isolation of the corresponding parental species and possibly also by the low viability of some combinations that might have hybrid weakness and consequently, could not compete with their parental species and establish themselves in nature.

12.2 Cytological Diploidization

12.2.1 Genetic and Epigenetic Changes Due to Allopolyploidization

The newly formed allopolyploid species, are hybrid species containing two or three different subgenomes enveloped within one nucleus. This situation exerts significant genetic stress on the nascent allopolyploids that must overcome several immediate challenges in order to be able to successfully establish themselves and survive in nature (Levy and Feldman 2002, 2004; Feldman and Levy 2005, 2009, 2011, 2012). Overcoming these challenges is achieved through immediate triggering of a variety of cardinal genetic and epigenetic changes that affect genome structure and gene expression.

Genomic changes of newly formed allopolyploids of *Triticum* and *Aegilops* comprise chromosome rearrangements, elimination of coding and noncoding DNA sequences, transposable element (TEs) and tandem repeat elimination or amplification, and gene expression modifications (Feldman et al. 1997; Ozkan et al. 2001; Shaked et al. 2001; Ma and Gustafson 2005, 2006; Cheng et al. 2019). In addition, allopolyploidization triggers the activity of a variety of TEs, which may affect gene expression and induce many structural rearrangements, including deletions or duplications of chromosomal segments (Kashkush et al. 2002, 2003). Yaakov and Kashkush (2011a) and Bariah et al. (2020) reviewed the accumulated data on genetic and epigenetic dynamics of TEs, particularly in newly formed allopolyploid wheats, and discussed the underlying mechanisms and the potential biological significance of TE dynamics following allopolyploidization. Similarly, Parisod and Senerchia (2012) presented evidence that TEs play a central role in driving genome reorganization subsequent to allopolyploidization, predominantly involving deletion of DNA sequences, as opposed to transposition. Genome reorganization generally occurs in the first generations following allopolyploidization and involves extensive epigenetic changes in the vicinity of TEs. Since massive transpositional activation of TEs could be highly deleterious to the nascent allopolyploid, Parisod and Senerchia (2012) speculated that only allopolyploids with transposition controlled through substantial repatterning of epigenetic marks and/or having lost TE fragments, could be viable.

Genetic and epigenetic alterations brought about by allopolyploidization may be induced by small RNAs, which affect key cellular processes, including TE activity, chromatin acetylation, cytosine methylation, and gene expression. Kenan-Eichler et al. (2011), performing high-throughput sequencing of small RNAs of parental, intergeneric hybrids, and of synthetic allopolyploid plants of the wheat group, found that the percentage of small RNAs corresponding to miRNAs increased whereas the percentage of siRNAs corresponding to TEs decreased soon after allopolyploidization. The reduction in siRNAs, together with decreased CpG methylation of the Veju TE element, shown by the same group, represent hallmarks of TE activation. TE-siRNA downregulation in the newly-formed allopolyploids may contribute to their genome destabilization at the initial stages of speciation.

One of the major challenges of the nascent allopolyploids of the wheat group stems from the fact that the homoeologous chromosomes of the different subgenomes are still genetically very closely related, as shown by the ability of four doses of one chromosome to compensate for the deficiency of either of its two homoeologues in allohexaploid wheat (Sears 1952a, 1966). Moreover, molecular studies revealed a high level of gene synteny and collinearity in the homoeologous chromosomes of allopolyploid wheat (Gale et al. 1995; Gale and Devos 1998). This genetic relatedness should enable the homoeologues to pair and recombine during meiosis. However, since homoeologous pairing leads to reduced adaptiveness, namely, partial sterility and multisomic inheritance, mechanisms restricting meiotic chromosome pairing to fully homologous chromosomes, i.e., a rapid process leading to cytological diploidization, have been acquired in allopolyploids of this group.

Since the different subgenomes of the wheat allopolyploids are genetically and structurally closely related, allopolyploids of this group are, in fact, segmental allopolyploids, rather than genomic allopolyploids. Yet, cytologically, they behave as genomic allopolyploids, namely, there is only intra-genomic pairing in the form of bivalents between fully homologous chromosomes (diploid-like meiotic behavior). By restricting pairing to homologous chromosomes, and preventing inter-subgenomic pairing, i.e., between homoeologous chromosomes (chromosomes of different subgenomes that are partly homologous), the cytologically diploidizing systems ensure bivalent pairing at meiosis and, consequently, regular segregation of genetic material, complete fertility, genetic stability, and disomic, rather than polysomic, inheritance, sustaining one of the advantages of allopolyploidy, i.e., heterosis between subgenomes. The establishment of allopolyploids of the wheat group as successful competitive taxa in nature, required the development of cytological and genetic systems that prevent inter-subgenomic pairing, while allowing intra-genomic pairing. Meeting this challenge is presumably a critical prerequisite for the success of the newly formed allopoly-ploids, that ensures their increased fitness and successful establishment in nature as competitive entities.

One means of bringing about cytological diploidization in newly formed allopolyploids of the wheat group is via rapid differential elimination of a number of DNA sequences, either low-copy or high-copy, from one subgenome in allotetraploids and from the additional subgenome in allo-hexaploids, leaving the concerned sequences in only one homologous pair, and thus, rendering them homologous-specific sequences (Feldman et al. 1997; Liu et al. 1998a, b; Ozkan et al. 2001; Han et al. 2003, 2005; Salina et al. 2004; Baum and Feldman 2010; Guo and Han 2014). Such dif-ferential elimination, i.e., some sequences are eliminated from one subgenome, whereas others are eliminated from the second subgenome, brings to a rapid, further divergence of the homoeologous chromosomes. The subsequent homologous-specific sequences may determine chromosome homology and strengthen homology search and attraction at the commencement of meiosis. Such DNA sequences exist in all the diploid progenitors of the allopolyploid species of the wheat group, whereas in the natural allopolyploids, they exist in only one homologous pair of one subgenome, implying that they were eliminated during or soon after the allopolyploidization event. (Feldman et al. 1997; Ozkan et al. 2001). This rapid elimination, occurring during or soon after the formation of the allopolyploids, was designated as a **revolutionary change** (Feldman et al. 1997; Ozkan et al. 2001; Table 12.1). No noteworthy further elimination of sequences occurs in the subsequent generations of the allopolyploids. The elimination of sequences is reproducible, as shown by elimination of the same sequences in synthetic and natural allopolyploids bearing the same genomic com-binations (Ozkan et al. 2001; Han et al. 2005). It has been concluded that instantaneous elimination of DNA sequences in the first generation(s) of the newly formed allopolyploids of *Triticum* and *Aegilops* was one of the major and imme-diate responses to allopolyploidization.

Rapid elimination of DNA sequences in newly formed allopolyploids was also observed in *Triticale*, a synthetic allopolyploid derived from wheat and rye (Boyko et al. 1984, 1988; Ma and Gustafson 2005, 2006; Bento et al. 2011). Elimination of DNA sequences were also reported in newly formed allopolyploids of *Brassica* (Song et al. 1995), *Nicotiana* (Skalická et al. 2005), and in *Arabidopsis* (Mad-lung et al. 2005). It seems that sequence elimination in newly formed allopolyploids is a widespread phenomenon.

DNA elimination seems not to be random at the intra-chromosomal level. For example, Liu et al. (1997a)

found that the chromosome-specific sequences on chromo-some arm 5BL in allohexaploid wheat are not distributed along the chromosome arm but, rather, cluster in terminal (sub-telomeric), subterminal and interstitial regions of this arm, rendering these regions extremely homologous-specific. Hence, it is tempting to suggest that these chromosome-specific regions, the only regions that deter-mine homology, are equivalents to the classical "pairing-initiation sites" that play a critical role in homology search and initiation of chromosomal pairing at the begin-ning of meiosis (Feldman et al. 1997).

The extent of DNA elimination was estimated by deter-mining the amount of nuclear DNA in natural allopolyploids and in their diploid progenitors, as well as in newly syn-thesized allopolyploids and in their parental plants (Ozkan et al. 2003; Eilam et al. 2008, 2010). Allopolyploid species of the wheat group contain 2–10% less DNA than the additive sum of their diploid parents, and synthetic allopolyploids exhibit a similar loss, indicating that DNA elimination occurs soon after allopolyploidization (Nishi-kawa and Furuta 1969; Furuta et al. 1974; Eilam et al. 2008, 2010; Table 12.1). In addition, the narrow intra-specific variation in DNA content of the natural allopolyploids indicates that the loss of DNA occurs immediately after allopolyploid formation, and that there is almost no subse-quent change in DNA content during the evolution of the allopolyploid species (Eilam et al. 2008). Boyko et al. (1984, 1988) and Ma and Gustafson (2005) found that there was a major reduction in DNA content in the course of *Triticale* formation, amounting to about 9% for the octoploid and 28–30% for the hexaploid *Triticale*. In this synthetic allopoly-ploid, the various subgenomes were not equally affected; the wheat genomic sequences were relatively conserved, whereas the rye genomic sequences underwent a high level of variation and elimination (Ma et al. 2004; Ma and Gus-tafson 2005; Bento et al. 2011). Bento et al. (2011) reana-lyzed data concerning genomic analysis of octoploid and hexaploid *Triticale* and found that restructuring depended on parental genomes, ploidy level, and sequence type (repeti-tive, low copy non-coding and/or coding). Similarly, in hexaploid wheat, subgenome D underwent a considerable reduction in DNA, while the A and B subgenomes were not reduced in size (Eilam et al. 2010). The fact that DNA elimination in *Triticale* and in allohexaploid wheat hardly affects the allopolyploid 4x and 6x wheat parents may be explained by the assumption that the subgenomes of these parents already underwent elimination during their formation and the additional subgenome of the diploid parent had to be adjusted.

Table 12.1 Allopolyploid species of the wheat group

Ploidy level	Genus	Species[1] and subspecies	Genome[2]	Genome size[3] (mean ± SD of 1C DNA in pg)	Expected genomic size[4]	Natural groups[5]
Tetraploids	Aegilops	biuncialis	UM^b	10.37 ± 0.037	10.91	The U group
		geniculata (=ovata)	M^{o}U	10.29 ± 0.008	10.91	
		neglecta (=triaristata)	UX^n	10.64 ± 0.404	–	
		columnaris	UX^n	10.86	–	
		triuncialis	UC; CU	9.93 ± 0.041	10.22	
		kotschyi	S^{v}U	12.64 ± 0.183	12.86	
		peregrina(=variabilis)	S^{v}U	12.52 ± 0.181	12.86	
		cylindrica	CD	9.59	10.01	The D group
		crassa	D^{c}X^c	10.86	–	
		ventricosa	DN	10.64	10.99	
	Triticum	turgidum ssp. dicoccoides	BA	12.91 ± 0.199	–	The A group
		turgidum ssp. durum	BA	12.84 ± 0.175	–	
		timopheevii ssp. armeniacum	SA	11.82 ± 0.071	–	
		timopheevii ssp. timopheevii	SA	11.87 ± 0.630	–	
Hexaploids	Aegilops	recta (=triaristata)	UX^{n}N	16.22	16.46	The U group
		vavilovii	D^{c}X^{c}S^s	17.13 ± 0.139	18.34	The D group
		crassa	D^{c}X^{c}D	–	–	
		juvenalis	D^{c}X^{c}U	–	–	
	Triticum	aestivum ssp. spelta	BAD	17.72 ± 0.039	18.04	The A group
		aestivum ssp. aestivum	BAD	17.67 ± 0.311	18.04	
		zhukovsky	GAA^m	17.74	18.35	

[1]Species designation after van Slageren (1994)
[2]Genome designations according to Dvorak (1998); the first genome is the donor of the cytoplasm; subgenome B is related to S, the donors of subgenomes X^c and X^{tn} are yet unknown
[3]Genome size from Eilam et al. (2008)
[4]Expected genomic size from the sum of the DNA amounts of the two parental species (see Eilam et al. 2007 for 1C DNA amount in the known diploid progenitors)
[5]Groups according to Zohary and Feldman (1962)

12.2.2 Suppression of Homoeologous Pairing by the *Ph1* Gene

12.2.2.1 Discovery and Induction of Mutations in *Ph1*

Superimposed on the divergence of the homoeologous chromosomes due to differential sequence elimination, is a genetic system that contributes to the maintenance and reinforcement of the exclusive bivalent pairing in the allopolyploid *Triticum* species, ensuring that chromosome pairing is wholly restricted to homologous chromosomes (reviewed by Sears 1976b). The most potent suppressor of homoeologous pairing in allopolyploid wheats themselves and in their interspecific or intergeneric hybrids, is a gene located on the long arm of chromosome 5B (5BL). The first evidence that chromosome arm 5BL of cv. Chinese Spring (CS) of bread wheat, *T. aestivum* ssp. *aestivum*, carries a gene that suppresses chromosome pairing in hybrids came from the observation of Okamoto (1957), who showed that F$_1$ hybrids between CS, deficient for 5BL, and the amphiploid *T. monococcum* ssp. *aegilopoides-Ae. tauschii* (genome AADD), exhibited much higher pairing than the hybrids with 5BL. Soon after, it was independently deduced, by Sears and Okamoto (1958) in the USA and Riley and Chapman (1958) in the UK, that a gene on chromosome arm 5BL (the former designation of chromosome 5B is V), designated *Ph* (pairing homoeologues; Wall et al. 1971a), and *Ph1* by Sears (1982) and Jampates and Dvorak (1986), suppresses homoeologous chromosomes pairing, without imparting any effect on homologues chromosome pairing. Consequently, only bivalents are formed at meiosis and allohexaploid wheat, in spite of being segmental allopolyploid, behaves like a typical genomic allopolyploid. In allohexaploid wheat plants lacking chromosome 5B or chromosome arm 5BL, a low level of pairing occurs between the homoeologues of the A, B, and D subgenomes (Riley

1960; Sears 1976a). On the other hand, in haploids of allopolyploid wheats (Riley and Chapman 1958; Jauhar et al. 1999) or in hybrids between wheat lacking 5BL and related species, when homologous chromosomes do not exist, the otherwise normally suppressed homoeologous pairing occurs at a relatively high frequency (Riley 1960, 1966; Sears 1976b). Supporting evidence that in the absence of 5BL, meiotic pairing indeed involves homoeologous chromosomes, has come from the finding of Okamoto and Sears (1962), who showed that the pairing in haploids of hexaploid wheat is largely between chromosomes of different subgenomes belonging to the same homoeologous group. Additionally, the work of Riley and Kempanna (1963) showed that the absence of chromosome arm 5BL in bread wheat induced pairing of homeologues, in addition to that of homologues.

Chromosome 5B of the allotetraploid *T. turgidum* compensates for the absence of chromosome 5B in *T. aestivum*, indicating the presence of a *Ph* gene, similar to that on chromosome 5B of hexaploid wheat, at the tetraploid level (e.g., Riley 1960; Dhaliwal 1977; Dvorak et al. 1984). Also, it was shown (Feldman 1966a) that chromosome 5G of *T. timopheevii* compensates for the absence of 5B of *T. aestivum*, implying that this allotetraploid species contains a gene system similar to that in chromosome 5B of *T. aestivum*. Since *T. timopheevii* contributed the G and A subgenomes to hexaploid *T. zhukovskyi*, it is assumed that the latter also contains a *Ph*-like gene. Thus, all the four allopolyploid species of *Triticum* contain a *Ph*-like gene system suppressing homoeologous pairing, which drives cytological diploidization.

The discovery that homoeologous pairing is prevented by chromosome arm 5BL explained why the chromosomes of the three subgenomes of allohexaploid wheat are unable to pair with each other, neither in the allopolyploid *Triticum* species nor in their inter-generic hybrids (Riley 1960, 1966; Sears 1976b), even though they are genetically so closely related. The discovery that a deletion for 5BL provides the means of inducing the chromosomes of wheat to pair and recombine with those of related species and genera, has been used widely in the transfer of valuable genes from wild relatives to wheat (e.g., Sears 1976b, Feldman 1988). No wonder, therefore, that from the time of its discovery, *Ph1* has had a great impact on wheat cytogenetics, breeding and beyond, and its isolation, structure, and mode of action have been the subjects of intensive cytogenetic and molecular research.

The suppressive effect of *Ph1* on homoeologous pairing in inter-generic *Triticum* hybrids is absolute. In contrast, its effect on homoeologous pairing in hexaploid wheat itself might be dispensable, as plants deficient for this gene exhibit relatively little homoeologous pairing [less than one multivalent per cell, which results from inter-subgenomic pairing

(Sears 1976b)]. Interestingly, a *Ph*-like effect has not been found in any of the allopolyploid species of the closely related *Aegilops* genus (Riley and Law 1965; Sears 1976b). None of the allopolyploid *Aegilops* genotypes studied by McGuire and Dvorak (1982) compensated for the absence of *Ph1* in hybrids with *T. aestivum* lacking this gene. Likewise, Cünado (1992) reported that all the polyploid species of *Aegilops* display a strict bivalent pairing behavior, with the exception of allotetraploid *Ae. biuncialis* and allohexaploid *Ae. juvenalis*, which occasionally form a few multivalents at first meiotic metaphase. These findings substantiated the belief that the allopolyploid *Aegilops* species do not possess a gene that has a *Ph*-like activity. Nevertheless, these species also exhibit exclusive bivalent pairing of fully homologous chromosomes, presumably due to the structural changes that were created by the sequence-elimination system.

The use of bread wheat plants lacking chromosome 5B or chromosome arm 5BL to induce homoeologous pairing in hybrids between allopolyploid wheats and their wild relatives, encounters some difficulties. The deletion of chromosome 5B or chromosome arm 5BL in ssp. *aestivum* resulted in reduced vigor and fertility and was associated with difficulty maintaining them as laboratory lines. To offset these deleterious effects, an extra dosage of chromosome 5D must be provided to compensate for the absence of 5B. This complicated the use of such a line (nullisomic 5B-tetrasomic 5D) in hybrids with wild relatives that would contain a pair of 5D. What was really needed is a simple mutation or a deletion of *Ph1*.

The first attempt to produce a mutation in *Ph1* was made by Wall et al. (1971a), who induced a mutation, designated *10/13*, by ethylmethanesulfonate (EMS) treatment of ditelosomic 5BL plants of cv. CS of ssp. *aestivum*. The homozygous mutant had no effect on homoeologous pairing, but in F$_1$ hybrids with rye, one dose this mutation increased the level of homoeologous meiotic pairing. Wall et al. (1971b) considered *10/13* to be a recessive allele to *Ph1* and designated it *ph1a*. Yet, the *10/13* mutant did not entirely meet the requirements of being an allele of *Ph1*, as the level of pairing it induced in hybrids was not as high as that in nullisomic 5B (Wall et al. 1971a). *Ph1* deficiency is actually expected to cause higher pairing than 5B nullisomy, because the short arm of 5B, 5BS, carrying a promoter of pairing (Feldman 1966b), would still be present in a mutant line. Sears (1982, 1984), based on genetic studies (test for allelism), found that the *10/13* mutation is actually a mutant of another homoeologous-pairing suppressor, *Ph2*, located on chromosome arm 3DS, rather than of *Ph1*. Accordingly, the correct designation of the *10/13* mutation is *ph2a*, and not *ph1a* (Sears 1984).

A deletion in chromosome arm 5BL that includes the *Ph1* gene, was induced by Sears (1977) via X-irradiation of normal pollen and using it in pollination of on plants monosomic for a 5B chromosome carrying a morphological marker. The 438 monosomic progeny of the cross with an irradiated chromosome 5B, were crossed with *Ae. peregrina* and the obtained F₁ hybrids were analyzed for increased pairing. Only one mutation was obtained, which appeared to be a mutation of *Ph1*. The homozygote mutant differed appreciably from the wild type in morphology, and also exhibited somewhat reduced vigor and fertility. In addition, male transmission from the heterozygote was less than 40%, leading Sears (1977) to conclude that the mutation is a deficiency. The author supposed that on the long arm of 5B, distal to *Ph1*, there is at least one gene for male fertility, and therefore, the one fertile mutant recovered would then be an interstitial deficiency that includes *Ph1* but not the fertility gene. This lies in accord with the fact that *Ph1* was mapped near the middle of the 5BL arm (Jampates and Dvorak 1986), about 1.0 centimorgan (cM) from the centromere (Sears 1984). This mutation, designated *ph1b* (Sears 1982), exhibits some homoeologous pairing under homozygous conditions (about 0.66 multivalents per pollen mother cell (PMC), (Sears 1977) and therefore, its progeny should be selected against inter-subgenomic translocations that result from homoeologous pairing. On the other hand, in euhaploids of allopolyploid wheats (Jauhar et al. 1999), as well as in interspecific and intergeneric hybrids (Sears 1976b), *ph1b* allows for somewhat higher homoeologous pairing than obtained in hybrids deficient for 5BL, because of the presence of 5BS.

One year after the induction of *ph1b* in allohexaploid wheat, a mutation in the *ph1* gene of allotetraploid wheat was obtained via X irradiation of cv. Cappelli of *Triticum turgidum* ssp. *durum* (Giorgi 1978, 1983; Giorgi and Cuozzo 1980; Giorgi and Barbera 1981a, b). The mutant was originally identified by the altered morphology of chromosome 5B, which, in some plants, appeared to be shorter than in Cappelli, whereas, in other plants in the same progeny, the arm was longer than in Cappelli (Giorgi and Barbera 1981a). It was found that the plants bearing the short 5B had a deletion in the region containing *Ph1* while those possessing the long 5B, had a duplication of the chromosome segment that includes *Ph1* (Giorgi and Cuozzo 1980; Giorgi and Barbera 1981a, b).

This mutation, designated *ph1c* (Jampates and Dvorak 1986), is also an interstitial deletion for a segment of the long arm of chromosome 5B containing *ph1*. In addition to the deletion, plants with a tandem duplication of part of 5BL that includes *Ph1* were produced, thus, plants homozygous for this duplication have four doses of *Ph1*. The deletion and the duplication in 5BL of Cappelli have a common origin, probably occurred in the same cell and comprised the same chromosome segment (Dvorak et al. 1984). Hybrids between *ph1c* and wild wheat relatives exhibited high homoeologous pairing as almost hybrids with *ph1b*. On the other hand,

hybrids bearing the duplicated segment of 5BL and consequently, with two doses of *Ph1*, had less pairing than hybrids with wild type Cappelli that carry only one dose of *Ph1*.

To infer the approximate location of the *Ph1* gene in the 5BL arm, Dvorak et al. (1984) compared the C-banding of the 5BL arm of wild type Cappelli, the 5BL arm of the homozygous mutant *ph1c*, and the 5BL arm of the line with the duplication. Like Giorgi and Cuozzo (1980) and Giorgi and Barbera (1981a, b), they found that 5BL of *ph1c* was shorter than 5BL of the wild type, owing to a deletion of one of two inter-band regions in the middle of the arm. In the line with the duplication, the 5BL arm was longer than its Cappelli counterpart and the interband region that was absent in *ph1c* was twice as long in the line with the duplication. Hence, C-band patterns confirmed that the difference between *ph1c* and wild-type Cappelli was due to a deletion of a chromosome segment from 5BL of the mutant, whereas the same segment was duplicated in the line with the duplication. Jampates and Dvorak (1986) crossed the two mutant lines, *ph1c* and the line with the duplication, and cv. Cappelli with several allotetraploid species of *Aegilops*. Hybrids involving *ph1c* had higher levels of chromosome pairing than those involving cv. Cappelli, whereas those involving the duplication had lower levels of pairing than those involving cv. Cappelli. Jampates and Dvorak (1986) found that *ph1c* is a deletion of sub-region 5BL12.3 between C-bands 5BL12.2 and 5BL21 (The C-bands and euchromatin on the 5BL arm were designated according to the system proposed by Gill (1987) for cv. Chinese Spring).

Several translocations involving chromosome arm 5BL were studied with the objective of finding the approximate location of the *Ph1* locus (Driscoll and Quinn 1968; Makino 1970; Mello-Sampayo 1972). In each case, it was inferred that *Ph1* is in the proximal part of the arm. These data may not conflict with the reported location of the *Ph1* locus in the area of the first C-inter band region in the middle of the 5BL arm.

Gill and Gill (1991) presented direct evidence corroborating that the *ph1b* mutation is a submicroscopic deletion. The probe XksuS1-5, from a genomic library of *Ae. tauschii*, detects a single fragment of each of the long arms of chromosomes 5A, 5B and 5D. The specific chromosome 5B fragment it recognizes, which was present in Chinese Spring, was missing in *ph1b* and *ph1c* mutants. Therefore, Gill and Gill (1991) suggested that XksuS1-5 lies adjacent to *Ph1* on the same chromosome fragment that is deleted in *ph1b* and *ph1c*. Thus, XksuS1-5 can be used to tag the *Ph1* gene and might also be a useful marker in cloning *Ph1* by chromosome walking.

Gill and Gill (1996) also developed an PCR-based screening for the detection of *Ph1*. This assay is based on the 0.6 kb probe *pHvksu8*, from barley, which maps in the interstitial region where *Ph1* is located. The probe was sequenced and 20 bp forward and reverse primers were generated. Using PCR, these primers amplified a fragment of chromosomes 5A, 5B, and 5D. The specific 5B fragment was not amplified in the *phlb* mutant, and thus, this probe can be used as a diagnostic fragment to screen plants for the presence or absence of the *Phl* gene.

Likewise, in an effort to tag the specific chromosomal region where *Ph1* is located, Segal et al. (1997) micro-dissected bread wheat chromosome arm 5BL and produced a plasmid library by random PCR amplification and cloning. From this library, a 5BL-specific probe, *WPG90*, was isolated and mapped within the region corresponding to the interstitial deleted chromosome fragments carrying *Ph1* in bread and durum wheat. This *WPG90*-based PCR assay allows for easy identification of homozygous genotypes deficient for *Ph1*.

Deletions in bread wheat chromosomes, induced by the gametocidal chromosome of *Ae. cylindrica*, arose from a single break, followed by the loss of the distal chromosome region (Endo 1988; Gill et al. 1993a). The breakpoints of the deletions in chromosome arm 5BL, namely, 5BL-9, 11, 1 and 5, are at fraction lengths (FLs) 0.76, 0.59, 0.55 and 0.54, respectively. (FL values were calculated by dividing the arm ratio of the deletion line with that of the normal). To determine the presence and absence of *Ph1* in the various deletion lines of 5BL, Gill et al. (1993b) crossed the deletion lines with *Ae. peregrina* and studied chromosome pairing in the F₁ hybrids. Low pairing indicates the presence of *Ph1*, whereas high pairing indicates its absence. The hybrids derived from the deletion lines 5BL-9 and 5BL-11, had low pairing, i.e., these deletion lines possess *Ph1*. On the other hand, hybrids derived from deletion lines 5BL-1 and 5BL-5, had high pairing, indicating the absence of the *Ph1* in these deletion lines. Since deletion 5BL-1 was the smallest deletion lacking the *Ph1* gene and 5BL-11 was the largest deletion in which *Ph1* is present, it was concluded that *Ph1* is located in the chromosome region between FL 0.55 and 0.59 (the breakpoints of deletion 5BL-1 and 5BL-11), respectively.

The deletions in the mutations *phlb* and *phlc* are 1.05 μm and 0.89 μm, and located proximal to C-band 5BL2.1 (Gill and Gill 1991; chromosome-banding nomenclature is according to Gill et al. 1991). Thus, the deletion of *phlb* encompasses 73.5 million bp and that of *phlc* 62.3 million bp (estimated from genomic DNA content of 16 billion base pairs divided by 250 μm, the total length of the chromosome complements of bread wheat; Gill and Gill 1991). Similarly, Foote et al. (1997) identified the *Ph1* locus of *T. aestivum* in the 70 Mb region between the *Xrgc846* and *Xpsr150A* markers. Gyawali et al. (2019) delimited the *phlb* deletion to a genomic region of 60 Mb by chromosome walking.

Gill and Gill (1991) and Gill et al. (1993b), using physical mapping of DNA markers on the 5BL region corresponding to the *phlb* deletion, demarcated the *Ph1* gene to a submicroscopic chromosome region, i.e., the *Ph1* gene region, whose size is ∼ 2.4 Mb. Both Griffiths et al. (2006) and Sidhu et al. (2008) delimited the *Ph1* region to a 2.5-Mb region within the *phlb* deletion. Gill et al. (1993b) identified three probes, *XksuS1*, *Xpsr128* and *Xksu75*, that mapped in the *Ph1* gene region of Chinese Spring and were missing in the deletion of *phlb*. These authors reported that the *Ph1* gene region is bordered by the breakpoints of two deletions (5BL-1 and *phlc*) and is marked by the DNA probe *XksuS1*. Two other DNA probes, *Xpsr128* and *Xksu75*, flank the *Ph1* region, *Xpsrl28* being proximal and *Xksu75* being distal. These two probes map in the interstitial deletion of the *phlb* mutant, whereas only *Xksu128* maps in the *phlc* deletion, while *Xksu75* detects a DNA fragment distal to the deletion Gill et al. 1993b).

In summary, *Ph1*, along with the DNA marker *XksuS1*, is located between the breakpoints of deletion 5BL-1 and *phlc*. The breakpoint of *phlb* is distal to the *Ph1* region but proximal to the C-band 5BL2.1, the region that possesses *Xksu75*. *Xpsr128* is present proximal to the *Ph1* gene region. The chromosome region around *Ph1* is high in recombination, as the genetic distance of the region spanned by *XksuS1* and *Xksu75* is at least 9.3 cM. This chromosome region is also prone to breaks, as the breakpoints of the mutants *phlb* and *phlc* also map in the region.

12.2.2.2 Theories Concerning the Mode of Action of *Ph1*

At present, it is not known how *Ph1* prevents homoeologues from pairing at meiosis and what are the product(s) and the subcellular target(s) of its activity. Does *Ph1* recognize specific regions that differ between homoeologous chromosomes and thus enable to distinguish between homologous and homoeologous chromosomes through homology search or pairing initiation? In this respect, are the homologous-specific sequences, produced by the differential elimination during the allopolyploidization event, recognized by *Ph1*? Or does *Ph1* operate on the divergence of a large number of differing sequences spreading out along the homoeologous chromosomes? Does the fact that *Ph1* impacts a number of traits, as described below, imply that it has a pleiotropic effect, or does the mutation that includes *Ph1* comprise several genes?

Over the years, many studies attempted to elucidate the mode of action of *Ph1*. The accumulated evidence falls into two main categories: (i) those showing that this gene operates during meiotic prophase, affecting processes involved in synapsis and crossing over, and (ii) those suggesting that this gene exerts its effect during premeiotic stages, affecting the premeiotic alignment of homologous and homoeologous chromosomes, thereby controlling the regularity and pattern of meiotic pairing.

Among the first theories concerning the mechanism of action of *Ph1*, was that of Riley (1960), who suggested that *Ph1* reduces the long-range pairing forces that bring chromosomes together in meiotic prophase. Since the attraction between homoeologues can be assumed to be less than that between homologues, the reduced pairing forces were assumed to no longer bring homoeologues together, although they are still sufficient to unite homologues. This suggestion left too many questions unanswered, particularly, if chromosomes are distributed non-randomly in the premeiotic nucleus, how could pairing of homoeologues lying by chance close to one another be prevented if their homologues happened to be on the opposite side of the nucleus? Also, since pairing is believed to be initiated at chromosome ends, how can homologues coming together from different parts of the nucleus avoid interlocking with other pairs that were also finding each other from a distance?

A later suggestion by Riley (1968) was based on the possibility that *Ph1* shortens the period available for the chromosomes to pair, thus, preventing the lower-affinity homoeologues from synapsing but allowing the high-affinity homologues to fully pair. But, when this idea was tested (Bennett et al. 1974), no effect of *Ph1* on the duration of meiosis was found.

Upadhya and Swaminathan (1967) suggested that the absence of *Ph1* causes a decrease in the speed and degree of chromosome condensation, and that this, in turn, allows homoeologous pairing to occur. However, the amount of reduction in condensation observed in the absence of *Ph1* was evidently marginal. Furthermore, the suggestion that differences in condensation have an appreciable effect on homoeologous pairing is difficult to accept knowing that the substantial increase in condensation observed in the absence of chromosome 6A did not result in a decrease in homoeologous pairing.

Colas et al. (2008) showed that chromatin remodeling of homologues at the onset of meiosis in allohexaploid wheat, enabling intimate association and recombination, can only occur if the homologues are identical or nearly identical. Failure to undergo such remodeling results in reduced pairing between the homologues. In this respect, Knight et al. (2010) showed that *Ph1* delays chromosome condensation at premeiotic and early meiotic stages, while treatment of premeiotic interphase of interspecific hybrids involving allohexaploid wheat and wild relatives with okadaic acid, a drug known to induce chromosome condensation, induced homoeologous pairing even in the presence of *Ph1*. Thus, the timing of chromosome condensation during the onset of meiosis is an important factor in controlling chromosome pairing.

When studying three-dimensional reconstructions at the ultrastructural level, of late zygotene and early pachytene of *T. aestivum* cv. Chinese Spring, Hobolth (1981) observed multivalent configurations, indicating pairing of homologues and homoeologues, and bivalents of strict homologous pairing at pachytene. He proposed that the regular bivalent formation in allohexaploid wheat is due to a temporal delay of crossover by the *Ph1* gene until pairing correction is completed at early pachytene.

Likewise, Gillies (1987) studied synaptonemal complexes at zygotene-pachytene in spread nuclei of *T. aestivum* x *Ae. peregrina* hybrids with and without *Ph1* and concluded that this gene does not impact the ability of homoeologous chromosomes to form synaptonemal complexes, but rather, influences the rate of pairing or the *time of crossover*.

Holm et al. (1988), Holm and Wang (1988) and Wang and Holm (1988) studied the effect of *Ph1* on chromosome pairing of homologous and homoeologous chromosomes and on synaptonemal complex formation at zygotene-pachytene stages in spread nuclei of *T. aestivum* cv. Chinese Spring, of aneuploid lines of this cultivar, and of hybrids with wild relatives, all with different doses of chromosome arm 5BL. They found that plants lacking 5BL had a large increase in the number of pairing partners at these early stages of meiosis, but only plants with zero or six doses of 5BL contained crossovers between homoeologues. This lies in line with the finding that plants either without 5BL or with six doses exhibit homoeologous pairing (Feldman 1966b). Holm et al. (1988), Holm and Wang (1988) and Wang and Holm (1988) concluded that *Ph1* affects both synapsis and crossover.

Similarly, Martinez et al. (1996) analyzed the synaptic process at mid-zygotene, late-zygotene and pachytene in spread nuclei of *T. timopheevii*. Nuclei at pachytene showed a lower frequency of multivalents than did zygotene nuclei. The authors concluded that a pairing-correction mechanism at pachytene transforms quadrivalents into pairs of bivalents, possibly by the suppression of crossover between homoeologues in the synaptonemal complexes.

Martinez et al. (2001b) reported that the mean number of lateral elements involved in synaptonemal complex multivalent associations at mid-zygotene was relatively high in plants with zero, two, and four doses of *Ph1*. At pachytene, multivalents were transformed to bivalents. Multivalent correction was more efficient in the presence than in the absence of *Ph1*. These findings suggest that the main action of the *Ph1* locus on the diploidization mechanism is related to a process which checks for homology during first meiotic prophase.

In *ph1b* mutant plants, that are deficient for *Ph1*, the number of ring bivalents and chiasmata decreased, while the number of univalents, rod bivalents, trivalents and quadrivalents increased (Martín et al. 2014). Consequently, these authors proposed that *Ph1* has a dual effect at meiosis, namely, it promotes early synapsis between homologues and prevents sites on homoeologues from becoming crossovers.

Dubcovsky et al. (1996) and Luo et al. (1996) studied chromosome pairing between chromosome 1A of *T. aestivum* carrying interstitial segments of 1A^m of *T. monococcum*, and ordinary 1A of *T. aestivum*. While the 1A^m and 1A segments recombined very little in the presence of *Ph1*, in the absence of *Ph1*, they recombined practically as if they were homologues. Consequently, these researchers concluded that *Ph1* recognizes the structural differences between homoeologous chromosomes and ensures homologous pairing in allopolyploid wheat by processing homology along the entire length of the chromosomes.

All the above hypotheses, inferring that *Ph1* operates at first meiotic prophase by affecting synapsis and crossover, do not explain the mechanism of action. However, other lines of evidence support the hypothesis that this gene exerts its effect at presynaptic stages, controlling the premeiotic alignment of homologous and homoeologous chromosomes, and thereby controls the regularity and pattern of pairing at meiosis.

In an attempt to study if *Ph1* also affects homologous pairing, Feldman (1966b) studied chromosome behavior at meiosis of *T. aestivum* cv. Chinese Spring plants carrying six doses of *Ph1*, as in tri-isosomic 5BL plants. The rationale behind studying plants with higher dose of *Ph1* stemmed from the assumption that the normal two-dose *Ph1* effect is partly counteracted by its homoeoalleles on 5AL and 5DL that promote pairing. Indeed, six doses of chromosome arm 5BL caused partial asynapsis of homologues (Feldman 1966b), i.e., it reduced homologous pairing to about one half of the normal level. Hence, higher doses of *Ph1* can also act on homologous pairing. Wang (1990) also showed that *Ph1* is not an exclusive suppressor of homoeologous pairingand can also affect the pairing of homologous chromosomes.

Apart from the partial suppression of homologous pairing, six doses of 5BL allowed some pairing of homoeologous chromosomes and induced a high frequency of interlocking bivalents (Feldman 1966b). The seemingly contradictory effect of an extra dose of *Ph1*, namely, partial suppression of homologous pairing on the one hand and induction of homoeologous pairing and interlocking of bivalents on the other hand, suggests that *Ph1* does not simply suppress synapsis at meiotic prophase. In light of these phenomena, it has been proposed (Feldman 1966b) that *Ph1* affects the presynaptic alignment of both homologous and homoeologous chromosomes.

A phenocopy of the effects of six doses of *Ph1* on chromosomal pairing was observed following pre-meiotic treatment of *T. aestivum* with colchicine (Driscoll et al. 1967; Yacobi et al.1982; Feldman and Avivi 1988). Such treatments induced partial asynapsis of homologues and pairing of homoeologues in allohexaploid wheat (Driscoll et al. 1967). Induction of interlocking of bivalents by pre-meiotic treatment with colchicine, in addition to partial

asynapsis of homologues and homoeologous pairing, was observed in allotetraploid *Ae. kotschyi* (M. Feldman, unpublished).

A role for premeiotic interphase chromosome associations in homologous recognition was also reported in bread wheat by other researchers (Aragón-Alcaide et al. 1997a, b; Schwarzacher 1997; Mikhailova et al. 1998; Martínez-Pérez et al. 1999) and in *Saccharomyces cerevisiae* (Loidl 1990; Weiner and Kleckner 1994). It has been argued that such associations lead directly to meiotic homologue pairing during first prophase (Kleckner 1996). However, the different studies disagree with regards to the extent and role of premeiotic chromosome association, where they start and how long they last (e.g., see Schwarzacher 1997; Mikhailova et al. 1998; Martínez-Pérez et al. 1999). Chromosome arrangement in interphase, somatic as well as premeiotic, nuclei, is presumably accomplished through distribution of centromeres and telomeres during each telophase, at opposite sides of the nuclei into a Rabl configuration (Fussell 1987). This organization eases the homolog search and the subsequent alignment (Pernickova et al. 2019).

Assuming premeiotic chromosome alignment in *T. aestivum*, the effect of different doses of *Ph1* can be clearly explained. A model was thus proposed whereby *Ph1* exerts its effect at the end of each cell division, including in the last pre-meiotic mitosis, where it affects the alignment of homologous and homoeologous chromosomes in telophase, and consequently in interphase, and so shapes the pattern of synapsis commencement and, as a result, controls the regularity and pattern of chromosomal pairing (Feldman 1993). Accordingly, in the absence of *Ph1*, i.e., in nullisomic 5B plants of cv. Chinese Spring of *T. aestivum* or in the *ph1b* mutant, the three subgenomes coexist in the nucleus and consequently, homologues as well as homoeologues would be closely associated at premeiotic stages, albeit the latter to a lesser extent. This results in reduced pairing of homologues whose pairing initiation was interrupted by the presence of homoeologues. The reduced pairing of homologues is expressed by few univalents and increased numbers of rod bivalents, alongside low frequency of multivalent homoeologous pairing. Mingling of homologues and homoeologues may lead to interlocking of bivalents, mainly of homoeologous bivalents (Feldman 1966b; Yacobi et al. 1982). In contrast, in euploid *T. aestivum*, which carries the normal two doses of *Ph1*, the association of homologous chromosomes is barely affected, while the somatic and premeiotic association of homoeologues is suppressed to the extent that they no longer lie together, and therefore, are not able to pair at meiosis or to pair somewhat less intimately. With six doses of *Ph1*, or pre-meiotic treatment with colchicine, even homologues no longer associate somatically, resulting in more or less random distribution of all the chromosomes. Then, at meiosis, homologues still pair, provided they do not lie too far apart, but in coming together from a distance and beginning pairing at their ends, they frequently intermingle with other bivalents to form interlocking bivalents. Homoeologues that lie close enough to each other also pair, if neither has a homologue close enough to generate greater attraction. In tri-isosomic 5BL plants, multivalent associations at first meiotic metaphase were rare because of the general reduction of pairing, while heteromorphic rod bivalents resulting from homoeologous pairing were more frequent (Feldman 1966b). Finally, about 20% of the ring bivalents present were interlocked with one, two or more (up to seven) other bivalents (Feldman 1966b; Yacobi et al. 1982). The broad occurrence of interlocking bivalents, in spite of the reduction in the total number of bivalents, showed clearly that some of the pairing occurred between somewhat separated partners.

The pairing behavior of two or three homologous isochromosomes can distinguish between factors affecting premeiotic alignment and those affecting synapsis and crossover. An isochromosome, consisting of two homologous arms can undergo either intra-chromosomal pairing between the two homologous arms of the same isochromosome, to form a ring univalent at first meiotic metaphase, or inter-chromosome pairing in cells having two or three homologous isochromosomes, to form a bivalent or a trivalent. Factors that disrupt homologous alignment would reduce the frequency of inter-chromosome pairing without affecting intra-chromosome pairing because the two homologous arms of an isochromosome are connected by a common centromere and their relative position remains undisrupted. On the other hand, factors that prevent synapsis or crossover would affect both types of pairing.

In *T. aestivum*, Sears (1952a) and Driscoll and Darvey (1970) observed almost complete intra-chromosome pairing in a univalent isochromosome at a frequency similar to that of pairing between homologous arms of conventional chromosomes. Application of colchicine during the last premeiotic interphase resulted in pairing failure of conventional homologues at the first meiotic metaphase (Driscoll et al. 1967; Dover and Riley 1973), but did not affect pairing between the two arms of an isochromosome (Driscoll and Darvey 1970). Hence, these authors concluded that colchicine inhibits premeiotic association of homologues rather than their synapsis and crossover. Similarly, high-temperature treatment during the last premeiotic interphase, which considerably reduced pairing of conventional homologous chromosomes, did not interfere with synapsis and chiasma formation between the two arms of an isochromosome (Kato and Yamagata 1980). In contrast to the effect of colchicine and high temperature, absence of wheat chromosome 3B, known to possess a recessive gene whose activity is responsible for normal synapsis and chiasma-formation (Li et al. 1945; Sears 1944, 1954;

Kempanna and Riley 1962; Kato and Yamagata 1982, 1983), reduced homologous pairing not only between the arms of conventional chromosomes (Sears 1954) but also the intra-chromosomal pairing between the two arms of an isochromosome (Kato and Yamagata 1982). This suggests that the pairing gene located on 3B controls either synaptic or postsynaptic events.

To determine which of the processes involved in chromosome pairing is affected by *Ph1*, Feldman and Avivi (1988) and Vega and Feldman (1998a) studied the effect of different doses of *Ph1* on the pairing of two or three isochromosomes and of an isochromosome with a telocentric-chromosome homologous to the isochromosome. These researchers showed that in tri-isosomic 5BL plants, the six doses of *Ph1* suppressed the inter-chromosomal pairing of the three isochromosomes to the same extent as it affected conventional pairing, without reducing the intra-chromosomal pairing. Under these conditions, the extra doses of *Ph1* could not modify the premeiotic alignment of the arms that were connected to one another by a common centromere. Hence, the failure of inter-chromosomal pairing as well as that of conventional chromosome pairing reflected disturbances in the premeiotic alignment. Thus, *Ph1* does suppress presynaptic homologous alignment. The similar outcome of premeiotic colchicine treatment and *Ph1* on the pattern of isochromosome pairing supports the hypothesis claiming the premeiotic association of homologues (Feldman 1966b; Feldman and Avivi 1988; Vega and Feldman 1998a).

Similar to the effect of *Ph1*, premeiotic colchicine treatment, which drastically decreased pairing of conventional chromosomes, reduced inter-chromosome but not intra-chromosome pairing of isochromosomes (Driscoll and Darvey 1970; Dover and Riley 1973; Feldman and Avivi 1988; Vega and Feldman 1998a). Based on these genetic and chemical effects, Feldman and Avivi (1988) reasoned that the effect of *Ph1* is exerted before the onset of meiosis. They reviewed data showing that in somatic cells of many plant species, the chromosomes are not randomly arranged with respect to each other, but rather, found that homologues are already lying side by side and therefore do not have to find each other from a distance at the beginning of meiosis (Avivi and Feldman 1980). In other words, the hexaploid nucleus still maintains some organizational aspects of the individual subgenomes, i.e., each subgenome occupies a separate region in the nucleus (Avivi et al. 1982b; Feldman 1993) (Also see Chap. 10, Sect. 4.2.8 on relationships between chromosomes of the different subgenomes and between homologues and non-homologues of the same subgenome).

Moreover, application of colchicine to spikes of normal hexaploid wheat before the last premeiotic mitosis (8–9 days before meiosis) resulted in a doubling of the chromosome number, and in bivalent pairing of nearly all the 84 chromosomes during the subsequent meiosis, even though each homologue exists in four doses (Driscoll et al. 1967; Dover and Riley 1973). It is assumed that the paired chromosomes were sister chromatids in the last mitosis and because of the colchicine treatment, their centromeres remained attached to each other until sometime in interphase, and thus, they remained very close to each other, while the homologous partners were randomly distributed. Likewise, in F_1 hybrids between allohexaploid wheat and related species, where without treatment, only homoeologous pairing occurs, colchicine treatment before the last premeiotic mitosis induced chromosome doubling, leading to regular bivalent formation, with no detectable homoeologous pairing (Dover and Riley 1973). These results indicate that premeiotic colchicine treatments, like extra dosage of *Ph1*, do not alter the processes of synapsis or crossover, but, rather, the premeiotic alignment of homologous chromosomes, which appears to be a prerequisite for meiotic pairing regularity.

Aragón-Alcaide et al. (1997a), using FISH with centromeric and telomeric sequences as probes, investigated centromeric behavior in PMCs of allohexaploid wheat and wheat x alien hybrids carrying different combinations of pairing genes. Their study revealed that centromeres are associated in pairs in pre-meiotic interphase, irrespective of the presence or absence of the homoeologous-pairing suppressors, *Ph1* and *Ph2*. Moreover, they found a difference in centromeric structure in pre-meiotic interphase, pachytene, first meiotic metaphase and anaphase plants carrying versus missing *Ph1* and *Ph2*. In plants lacking *Ph1* or *Ph2*, the centromeres exhibited diffuse hybridization sites during pre-meiotic interphase, and from pachytene through anaphase, whereas in plants carrying *Ph1* and *Ph2*, the discrete hybridization sites at premeiotic interphase remained as dense sites at these meiotic stages. The authors suggested that after replication of centromeres in pre-meiotic interphase, centromeres in the presence of *Ph1* and *Ph2* can form a more condensed structure throughout first meiotic division as compared to plants lacking the homoeologous-pairing suppressor genes. A diffuse structure may result in increased exposure of the centromere during the pairing process, which could increase interactions at these sites via proteins or via DNA sequences. The increased accessibility could expose regions of the centromere that are more conserved in structure, thus reducing the stringency of pairing at the centromere. Similar mechanisms may also impact other chromosomal sites involved in pairing.

To examine the occurrence of homologous association in somatic cells, Feldman et al. (1966) took advantage of the availability of telocentric chromosomes in hexaploid wheat, which are identifiable in somatic metaphases. Chinese Spring plants having two doses of *Ph1* with two telocentric chromosomes, either homologous or non-homologous, were produced. In root tip cells, the two non-homologous telocentrics were found to be located at random with respect to each other, while the two homologous telocentrics were significantly closer together than expected on a random basis. Telocentrics for the opposite arms of the same chromosome were also found to be associated (Feldman et al. 1966; Mello-Sampayo 1973). Thus, it was concluded that the centromere, the DNA sequences shared by the two different arms of the same chromosome, determines the position of each chromosome in the nucleus.

In nulli-5B plants (lacking *Ph1*), telocentric chromosomes of the different subgenomes lay as close together as telocentrics of the same subgenome, and thus, homoeologues were close to each other as homologues were (Feldman and Avivi 1984).

In disomic 5B plants, carrying two doses of *Ph1*, chromosomes of the same subgenome lay closer to each other than chromosomes of the different subgenomes (Feldman and Avivi 1973, 1984; Avivi et al. 1982a, b). Thus, the location of homologues and homoeologues with respect to each other, at different levels of *Ph1*, conformed with expectations based on Feldman's (1966b) hypothesis.

The closer somatic association of unrelated chromosomes of the same subgenome as compared to chromosomes of different subgenomes, is interpreted as being the result of a tendency for the chromosomes to have fixed positions within the nucleus and for the chromosomes of each subgenome to be grouped together and occupy different region in the nucleus. Spatial segregation of parental genomes in somatic and meiotic metaphases of various plant hybrids has been reported by several researchers (Finch et al. 1981; Schwarzacher et al. 1989, 1992; Linde-Laursen and Jensen 1991). Others have made such observations in the wild grass allotetraploid *Milium montianum* (Bennett and Bennett 1992) and in allotetraploid cotton (Han et al. 2015). In allopolyploid plants, there appears to be subgenome separation of interphase chromosomes, with chromosomes of each subgenome tending to cluster together (Avivi et al. 1982b; Feldman and Avivi 1984; Hilliker and Appels 1989). A similar genome separation phenomenon was found in synthesized tetraploid cotton (genome AAGG) (Han et al. 2015). Given the evidence of parental genome separation in other plants, Han et al. (2015) speculated that genome separation might be a normal phenomenon in diploid hybrids and in allopolyploid species. Concia et al. (2020) analyzed the entire genome interaction matrix in allohexaploid wheat and revealed three hierarchical layers of chromosome interactions, presented here from strongest to lowest: (i) within chromosomes, (ii) between chromosomes of the same subgenome, and (iii) between chromosomes of different subgenomes. This organization may indicate a non-random spatial distribution of the three subgenomes that could mirror the presence of functional "genome territories."

Concia et al. (2020) confirmed the presence of subgenome-specific nuclear territories using genomic in situ hybridization (GISH) in root meristematic cells of hexaploid wheat. To determine whether other polyploid plants share the same large-scale nuclear organization, Concia et al. (2020) analyzed 14-day-old rapeseed seedlings (*Brassica napus*), and again revealed a three-layer hierarchy of chromosomal interactions identical to those in allopolyploid wheat, suggesting that this organization is a general feature of allopolyploid plants. They also showed, using two genome-wide complementary techniques, GISH and in situ Hi-C, that the chromatin of hexaploid wheat is not uniformly distributed across the nucleus but, rather, occupies subgenome-specific nuclear compartments. This finding is consistent with previous cytological observations (Feldman and Avivi 1973; Avivi et al. 1982a, b), indicating that chromosomes of the same subgenome tend to be physically closer than chromosomes of different subgenomes. Consequently, they proposed that genome territories are the primary level of chromatin spatial organization in allohexaploid wheat. Little is known about the mechanisms facilitating homologous pairing versus homoeologous pairing during the telomere bouquet stage at the start of meiosis. The establishment of genome territories may be a mechanism that favors the pairing of homologues versus homoeologues by creating territorial "boundaries" between the different subgenomes, i.e., either homoeologues or non-homologues (Concia et al. (2020).

Similarly, Jia et al. (2021) probed the three-dimensional chromatin architecture of a Chinese cultivar of *T. aestivum* and found that the three subgenomes occupy specific territories in the nucleus. This is in accord with previous studies revealing that the three subgenomes of hexaploid wheat tend to localize to specific nuclear territories (Avivi et al. 1982b; Li et al. 2000; Concia et al. 2020). Moreover, the data of Jia et al. (2021) suggested that transposable elements help promote the higher order subgenome affinity in allohexaploid wheat. Bhat et al. (2021) in a recent review, propose that DNA, RNA and proteins are organized within precise 3D compartments in the nucleus, affecting many aspects of gene regulation. Non-coding RNA contribute to such intra-nuclear organization.

Driscoll and Darvey (1970), Darvey and Driscoll (1971, 1972), and Darvey et al. (1973) were unable to confirm that homologues lie closer than homoeologues in root-tip nuclei of common wheat. Both homologous and non-homologous telocentrics showed the same distribution, and homologous nucleoli showed a similar tendency to fuse than did non-homologous nucleoli. The tendency of proximal nucleoli to fuse, was established by demonstrating that the number of nucleoli per nucleus decreases substantially during interphase (Crosby 1957).

Moreover, Driscoll et al. (1979) estimated the probability of premeiotic association of homologous chromosomes and of chiasma formation, from frequencies of different chromosome configurations at first meiotic metaphase of allohexaploid wheat, several aneuploid lines, and wheat hybrids. Based on their estimations, they suggested that 5BL did not appear to affect the premeiotic association of chromosomes, but rather, the number of chiasmata.

Despite the evidence that presynaptic alignment of homologues ensures the regularity of pairing, there is little agreement regarding the timing of the first alignment of homologous chromosomes (reviewed by Loidl 1990). While some assume that homologues are already associated at the last premeiotic interphase (e.g., Smith 1942; Feldman 1966b; Maguire 1967), others hold that homologues do not associate before the beginning of zygotene (e.g., John 1976; Rasmussen and Holm 1978). Although there have been indications in a number of organisms that premeiotic alignment is a characteristic feature of meiosis (Avivi and Feldman 1980), it was difficult to conclusively demonstrate this phenomenon because individual chromosomes could not be clearly distinguished. In several species, this problem has recently been circumvented by fluorescence in situ hybridization with DNA probes that detect a specific pair of homologous chromosomes or chromosome segments. In the budding yeast *Saccharomyces cerevisiae*, the homologues were found to be associated via multiple interstitial interactions during the last premeiotic interphase (Weiner and Kleckner 1994). Genomic in situ hybridization in a wheat line carrying a pair of homologues originating from barley, showed that the hybridization signals of the two barley homologues fused into a single fluorescent signal during the last premeiotic interphase, indicating their complete association (Aragón-Alcaide et al. 1997b). Moreover, Aragón-Alcaide et al. (1997a) observed that in the absence of *Ph1*, the barley homologues were not in contact along their length and proposed that the absence of *Ph1* disrupts premeiotic homologue association. Premeiotic association was also observed in a pair of homologous rye telocentrics added to common wheat (E. I. Mikhailova, T. Naranjo, K. Shepherd, J. Wennekes, C. Heyting and J. H. de Jong, unpublished results).

Martinez-Perez et al. (2001) observed that *Ph1* also acts somatically by reducing non-homologous centromere associations. This effect during premeiotic interphase leads to exclusive homologue association during the telomere-bouquet stage in meiosis. The authors proposed that non-homologously associated centromeres separate at the beginning of meiosis in the presence, but not in the absence, of *Ph1* and concluded that *Ph1* is not responsible for the induction of centromere association, but, rather, regulates its specificity. Likewise, Moore (2002) proposed that *Ph1* determines correct pairing of homologous chromosomes at premeiotic and early meiosis prophase by giving nonhomologous chromosomes an almost 'Teflon'-like status, that increases chromosome specificity in the pairing process. In the absence of *Ph1*, the homoeologous chromosomes might be able to pair as a result of the loss of such a coating. All these findings in yeast and wheat demonstrate that homologous chromosomes recognize each other and associate before meiosis, a process that leads to exclusive synapsis of homologues at first meiotic prophase.

Of importance to the somatic-association hypothesis (Feldman 1966b) is how much, if any, association there is during the last premeiotic interphase. Following the report by Stack and Brown, somatic pairing is expected in premeiotic cells, even though it did not occur in root tips. Walters (1970, 1972) found no evidence of somatic pairing in *Lilium longiflorum* at any premeiotic stage, but it seems doubtful that the loose and intermittent association expected during pre-meiosis and early meiotic prophase would be detectable cytologically in a plant with such large chromosomes. Indeed, Walters' excellent photographs of leptotene nuclei, in which the chromosomes are assembled into what Brown and Stack (1968) would call a ball of yarn, raise the question of how homologous strands located in different parts of the nucleus could possibly thread their way through the mass of other strands and align themselves precisely alongside each other in zygotene, with no entanglement, and all in a relatively short time. Scherthan et al. (1996) found no associated homologues until early meiotic prophase in mouse and humans, but failed to compare the distance between the hybridization signals of homologues with that between non-homologues, and, therefore, the results from premeiotic stages are inconclusive.

It can be concluded that the controversy concerning the occurrence of homologous association in root-tip cells should be resolved by further work, but what is really needed is a careful analysis of the last premeiotic mitosis and premeiotic interphase. Unfortunately, this division is not as easy to study as are root-tip mitoses. Unless lack of association in the premeiotic mitosis can be clearly shown, there is little choice but to accept the premeiotic association hypothesis, because it so simply explains several otherwise puzzling phenomena.

Avivi et al. (1972) argue that somatic association has a physiological advantage and therefore occurs throughout the life of the plant. They found a different pattern of activity of the triplicate series of alcohol dehydrogenase genes in plants lacking 5BL than in plants carrying this chromosome arm. This observation needs to be extended to other loci, especially since analysis of isozymes of glutamate oxaloacetate transaminase indicated that even in the presence of *Ph1*, monomers from genes on homoeologous chromosomes combine with each other as readily as do the monomers from

homologues (Hart et al. 1976). Yet, DNA organization in precise 3D compartments in the nucleus is one of prerequisite conditions for regular gene action (Bhat et al. 2021).

Considering the phenocopy of the effect of six doses of *Ph1* on meiotic pairing by premeiotic treatments with colchicine (Driscoll et al. 1967; Dover and Riley 1973; Yacobi et al. 1982; Feldman and Avivi 1988), and since colchicine binds specifically to tubulin subunits, thereby preventing them from polymerizing into microtubules (Borisy and Taylor 1967), it was concluded that the microtubule system is one of the subcellular targets of *Ph1* (Avivi and Feldman 1973a; Feldman 1993; Feldman and Avivi 1988) and that microtubules are involved in the process of intimate homologous association. This may involve attachment of centromeres, telomeres and other chromosome segments to the nuclear envelope at the end of each telophase, thereby stabilizing chromosome position in the nucleus throughout premeiotic interphase. The disruption of microtubules by colchicine would detach the chromosomes from the nuclear membrane, leading to their movement at interphase and disruption of the association between homologues. In agreement with this view, Vega and Feldman (1998b) showed that *Ph1* affects centromere-microtubule interactions at meiotic anaphases.

Spindle inhibitors, particularly in combination with different *Ph1* genotypes, have been used to help elucidate the mechanisms of *Ph1* actions. Colchicine treatment of root-tip cells disrupted association of homologues (Avivi et al. 1969). The sensitivity of root-tip mitosis to colchicine and other microtubules drugs, such as vinblastine and griseofulvin, decreased with increased doses (up to 4) of 5BL (Avivi et al. 1973b; Ceoloni et al. 1984; Gualandi et al. 1984). This led to the conclusion that *Ph1* somehow affects the binding of microtubule subunits to colchicine, thereby protecting the spindle against colchicine action. The differential sensitivity of different genotypes of *Ph1* to colchicine (Avivi et al. 1970, 1973b; Ceoloni et al. 1984) is not confined to hexaploid wheat. Carla Ceoloni (unpublished data) found that the mitotic spindle of a *phlc* mutant induced in the Italian durum cultivar Cappelli (Giorgi 1983), is much more sensitive to colchicine than that of plants with *Ph1*.

It has been reported that detyrosination or acetylation of a-tubulin reduces microtubule dynamics (Webster and Borisy 1989). Such microtubules are more stable and less sensitive to depolymerization by antimicrotubule drugs (Piperno et al. 1987; Kreis 1987; Khawaja et al. 1988). Most microtubule-associated proteins (MAPs), e.g., tau, MAP2 and MAP1, stimulate the assembly of tubulin into microtubules (see Olmsted 1986 for a review) and, consequently, increase the stability of the polymers and suppress microtubule dynamics (Murphy et al. 1977). Post-translational modifications of MAPs, mainly phosphorylation, may be involved in reduction of microtubule dynamics. Interaction of microtubules with each other or other cytoskeletal elements (mainly intermediate filaments) may also increase microtubule stability (Gelfand and Bershadsky 1991). In addition, microtubule stability is affected by kinetochores, which are highly differentiated structures at the centromeres that serve as microtubule attachment sites, and which cap the plus-ends of microtubules (McIntosh and Hering 1991). All these interactions are presumably mediated by MAPs. Hence, phosphorylation-dephosphorylation of these MAPs may be the main mechanism underlying the regulation of such interactions (Gelfand and Bershadsky 1991).

How is chromosomal arrangement in the telophase and interphase nuclei affected by the reduced microtubule dynamics induced by *Ph1*? According to the presynaptic hypothesis, *Ph1* controls chromosome arrangement in the somatic as well as premeiotic nucleus, by operating on the subcellular elements that are involved in chromosome positioning: microtubules and centromeres. Vega and Feldman (1998b) assume that *Ph1* action may target the interaction of centromeres with spindle microtubules—an interaction that is critical for the movement of chromosomes to their specific interphase positions. Consequently, Vega and Feldman (1998b) studied centromere behavior of univalents at meiosis of monosomic lines in the presence and absence of *Ph1* and found that the frequency of centromere misdivision (transverse division) of univalent chromosomes is affected by *Ph1*. In common wheat, the centromere of unpaired chromosomes may undergo precocious division at first anaphase or telophase (Sears 1952b). This division is either longitudinal, leading to the formation of two sister chromosomes, each consisting of one chromatid, or transverse (misdivision), leading to the formation of telocentric chromosomes and isochromosomes. Transverse division of the centromere of one-chromatid chromosomes may also occur at second anaphase. In the presence of *Ph1*, the frequency of centromere misdivision in both first and second meiotic divisions was much higher than in the absence of the gene (Vega and Feldman 1998b), suggesting a role for *Ph1* in the interaction between kinetochores and microtubules at anaphase.

The *ph1b* deletion may contain a number of genes that might affect centromere-microtubule interaction. However, the fact that premeiotic treatments with colchicine and other antimicrotubule drugs phenocopy the effect of extra doses of *Ph1* on chromosome pairing (Feldman and Avivi 1988) indicates that the effect on pairing and on centromere microtubule interaction is caused by *Ph1*.

Several lines of evidence support the involvement of the centromere in chromosomal arrangement in wheat, in somatic (Feldman et al. 1966; Mello-Sampayo 1973) and meiotic (Yacobi et al. 1985a, b) cells. There is also direct evidence of nonrandom distribution of centromeres in the interphase nucleus. Using anti-kinetochore antibodies in

interphase nuclei of rat-kangaroo and Indian muntjac, Hadlaczky et al. (1986) observed that centromeres were arranged in pairs. In a similar experiment, half the expected number of pre-kinetochores were detected at interphase in *Vicia faba* (Houben et al. 1995). These observations indicate at least transient association of homologous centromeres at interphase. Su et al. (2019) suggested that variation in centromere satellite sequences and copy number, and their structural rearrangements, result in asymmetries in allohexaploid wheat homoeologues, highlighting the role of centromeres in homolog pairing during meiosis. This asymmetry in centromere organization among the three subgenomes of allohexaploid wheat, was suggested to play a role in proper homologous pairing during meiosis.

Taken together, *Ph1* seems to affect several different traits in somatic, premeiotic and meiotic cells of allohexaploid wheat. These traits are: spindle sensitivity to antimicrotubules drugs, somatic association of homologues, separation of the subgenomes into different nuclear compartments, concise centromeres at premeiotic and meiotic stages, chromatin condensation, synapsis of homologues and homoeologues at zygotene, correction of pairing to homologues at pachytene, prevention of crossover between homoeologues in pachytene, and strength of centromere attachment to microtubules. *Ph1* also affects the number and shape of pollen grain pores in common wheat (Avivi and Feldman 1973b). This gene has a pronounced effect on the condensation of heterochromatin in hybrids of wheat with related species and on the condensation of euchromatin in wheat itself (Martinez-Perez et al. 2001; Prieto et al. 2005). Hence, *Ph1* influences a number of phenotypic traits and, as such, may be either a gene with a pleiotropic effect (a transcription factor?) or as a complicated locus comprising a cluster of genes.

12.2.2.3 Isolation of *Ph1*

It is generally accepted that *Ph1* represents a single locus, but the possibility that it consists of a cluster of tandemly arranged genes, has not been completely ruled out (Sears 1976b). The existence of two pairing loci on 5BL was already suggested by Mello-Sampayo (1972), who noted that the F$_1$ hybrids *T. turgidum* x *Ae. sharonensis* and *T. turgidum* x *S. cereale*, in which the distal part of 5BL had been replaced by an homoeologous portion of 5DL, had an intermediate, rather than high, level of pairing. Consequently, he concluded that *Ph1* was still present but that its effectiveness was reduced by the presence of the 5DL segment, that might carry a pairing promoter. The 5BL homoeoallele of this 5DL promoter could be a low-grade promoter, or, alternatively, a low-grade suppressor.

Some evidence against the assumption that *Ph1* contains more than one locus is the failure to obtain intermediate mutants on chromosome arm 5BL by irradiation (Sears

1977). On the other hand, the inability to induce mutations in *Ph1* exhibiting the *ph1b*-like phenotype by EMS treatment, may indicate that its phenotype is determined by more than one gene (Griffiths et al. 2006).

The fact that *Ph1* activity is unique to chromosome 5B, and not to its homoeologues, led Griffiths et al. (2006) to suggest that this gene arose through a post-allopolyploidization structural change on chromosome 5B. Support for this conclusion is provided by studies showing that *Ae. speltoides* and *Ae. longissima*, two diploids that are closely related to the B and G subgenomes of allotetraploid wheats, do not compensate for the absence of *Ph1*, while the B and G subgenomes of these allotetraploids do (Griffiths et al. 2006). Since *Ph1* has a pronounced effect on several traits, Griffiths et al. (2006) concluded that *Ph1* is likely to comprise a multigene family, heterochromatin or both. These authors used two approaches to analyze the genetic structure of the *Ph1* region. First, they used genes of the orthologous regions in rice and *Brachypodium sylvaticum* to obtain markers for the saturation of the *Ph1* region, and second, they used five deletions produced by Roberts et al. (1999), using fast-neutron irradiation, that overlapped the *ph1b* deletion, to physically dissect the *Ph1* region. These approaches enabled the narrowing down of the *Ph1* locus to a 2.5 Mb region on 5BL, which contains a chromosomal segment derived after allotetraploidization from the sub-telomeric region of chromosome arm 3BL of bread wheat, comprising a block of heterochromatin and a single *Zip4* gene, inserted within a cluster of defective cyclin-dependent kinase-(*cdk*) like genes, between *Cdk6* and 7 (Griffith et al. 2006; Al-kaff et al. 2008; Martín et al. 2014, 2017). Griffiths et al. (2006) assumed that this region fulfills all criteria of a *Ph1* candidate structure. On the basis of their sequence homology to genes of known function, Griffiths et al. (2006) singled out the *cdk*-like genes of the *Ph1* region as the best candidates for *Ph1* function. This is the only multigene cluster in the region, and at least one of its members (*cdk2-4*) is 5BL-specific. Similarly, Al-Kaff et al. (2008), who further characterized the *Ph1* region by exploiting new deletions in this region and conducting expression analysis studies, assigned the *Ph1* locus to the region containing the *cdk*-like cluster and the neighboring heterochromatin segment. In fact, there are seven cdk-like genes on 5BL compared with at least five on 5AL and two on 5DL, however, the sub-telomeric heterochromatin segment inserted between the two *cdk*-like genes, cdk-like B6 and B7, is unique to chromosome arm 5B (Al-Kaff et al. 2008). The *cdk*-like gene cluster in the *Ph1* locus, designated *Ta5B2*, shows some similarity to mammalian *Cdk2* (Yousafzai et al. 2010). Greer et al. (2012) showed that *Cdk2*-type phosphorylation plays a major role in determining chromosome specificity during meiosis. Deletion of *Ph1* leads to increased phosphorylation at *cdk2*-type sites during meiosis, implying that the presence of *Ph1*

decreases *Cdk2*-type phosphorylation. Consistent with this, treatment with okadaic acid, an inhibitor of phosphatase activity, increases *Cdk2*-type phosphorylation, and phenocopies the deletion of *Ph1* by inducing crossovers (COs) (Knight et al. 2010). Deleting *Ph1* or treating with okadaic acid, both of which increase *Cdk2*-type activity, increases the efficiency of *MLH1*, a gene involved in DNA mismatch repair, whose active sites on paired homoeologues may lead to crossing overs (Martín et al. 2014). These authors proposed that *Ph1* has a dual effect in wheat, namely, it brings about cytological diploidization by both promoting homologous synapsis during early meiosis and by preventing *MLH1* sites on synapsed homoeologues from becoming COs later in meiosis. The effect on synapsis occurs during the telomere bouquet stage, when *Ph1* promotes homologous synapsis, thereby reducing the chance of homoeologous synapsis (Martín et al. 2014, 2017). The effect on crossing over formation occurs later in meiosis, when *Ph1* prevents *MLH1* activity.

Later, Martín et al. (2017) explored these two effects and demonstrated that regardless of the presence or absence of *Ph1*, synapsis between homoeologues does not take place during the telomere bouquet stage; only homologous synapsis takes place during this stage. Furthermore, in wheat lacking *Ph1*, overall synapsis was delayed with respect to the telomere bouquet, with more synapsis occurring after the bouquet stage, when homoeologous synapsis is also possible. Secondly, they showed that in the absence of *Ph1*, it was possible to increase the number of MLH1 sites progressing to COs by altering environmental growing condition. They also showed that higher nutrient levels in the soil or lower temperatures increased the level of both homologous and homoeologous COs.

It has been proposed that the effect of *Ph1* on synapsis is connected to altered histone H1 *CDK2*-dependent phosphorylation. Altered phosphorylation in the absence of *Ph1* was shown to affect chromatin structure and delay premeiotic replication and consequently, homologue synapsis, thus allowing homoeologous synapsis to take place (Greer et al. 2012; Martín et al. 2017). *Arabidopsis* lines carrying mutations in the *CDK2*-like homologue also exhibited reduced synapsis under specific conditions, suggesting a role for these genes in efficient synapsis (Zheng et al. 2014). Greer et al. (2012) previously proposed that the effect of *CDK2*-like genes on chromatin structure not only affects synapsis but might also affect the resolution of double Holliday Junctions (marked by *MLH1*) as COs. Okadaic acid treatment affects chromatin structure and can induce homoeologous CO in wheat-wild relative hybrids (Knight et al. 2010). However, given that the locus contains multiple copies of the *CDK2*-like and methyltransferase genes, it would be complex and laborious to identify EMS-induced mutants within these genes.

In contrast to Griffiths et al. (2006) and Martín et al. (2017) proposed that the *ZIP4* gene, located distally to the heterochromatin in the inserted segment on 5BL, is a more appropriate candidate for *Ph1* effects than the *cdk*-like genes. This is largely based on the evidence that the Zip1 protein is a major building block of the synaptonemal complex (SC) in *Saccharomyces cerevisiae* (Mitra and Roeder 2007), and, in its absence, SC fails to form, cells arrest or delay in meiotic prophase, and crossover is reduced.

Further studies on the effect of *Zip4* (designated *TaZIP4-B2*) on synapsis and crossover were performed by Rey et al. (2017). Although, there are *ZIP4* homologues on group 3 chromosomes, *TaZIP4-A1* in 3A, *TaZIP4-B1* in 3B, and *TaZIP4-D1* in 3D, the *ZIP4* paralogue (*TaZIP4-B2*) within the *Ph1* locus on chromosome arm 5BL is a single copy. The *TaZIP4-B2* gene is expressed during meiosis, has a higher level of expression than the ZIP4 homologues present on group 3 chromosomes, and its expression is significantly reduced upon *Ph1* deletion (Rey et al. 2017). Moreover, *ZIP4* has been shown to have a major effect on homologous COs, but not on synapsis, in both *Arabidopsis* and rice (Chelysheva et al. 2007; Shen et al. 2012).

Rey et al. (2017) searched for EMS-induced mutations in the *TaZIP4-B2* gene to determine whether they show reduced homologous CO with some homoeologous pairing and CO but exhibit homoeologous COs in hybrids with wild-relatives. For the crossings with wild relatives, the group used two mutant lines in the TaZIP4-B2 gene, selected from the mutants obtained by Rakszegi et al. (2010) in *T. aestivum* cv. Cadenza. Hybrids between the *Tazip4-B2* mutant Cadenza lines and *Ae. peregrina*, exhibited similar chiasma frequency to that observed in the *ph1b* mutant x *Ae. peregrina* hybrids, suggesting that *TaZIP4-B2* within the *Ph1* locus is involved in the suppression of homoeologous COs.

Since no multivalents and no significant increase in the number of univalents were observed at first meiotic metaphase of the *Tazip4-B2* mutant lines, it seems that homoeologous synapsis may not be significantly affected by *TaZIP4-B2* (Rey et al. 2017). Hence, *TaZIP4-B2* affects only one trait out of the array of phenotypic traits controlled by *Ph1*, indicating that this is only one gene in the *Ph1* gene cluster.

The *ph1b* mutant accumulates inter-subgenomic translocations due to homoeologous recombination, which reduces fertility (Sears 1977; Sánchez-Morán et al. 2001). It would therefore be most useful to use the *TaZIP4-B2* mutant lines, with reduced homoeologous synapsis and CO at meiosis, but which do exhibit homoeologous COs in hybrids with wild relatives. The absence of multivalents at meiosis in the *Tazip4-B2* Cadenza mutants shows that these mutants do not cause translocations between homoeologues and can be used instead of the *ph1b* mutant.

Sidhu et al. (2008), performing a detailed analysis of the 91 putative genes present within the 450-kb region on the rice R9 chromosome, which is orthologous to wheat chromosome arm 5BL, identified 26 candidates for the *Ph1* gene, including genes involved in chromatin reorganization, microtubule attachment, acetyltransferases, methyltransferases, DNA binding, and meiosis/anther-specific proteins. Four of these genes share domains/motifs with the meiosis-specific genes *Cor1*, *Scp1*, *Zip1*, and *RAD50*. *Cor1* codes for a protein of the axial element of the synaptonemal complex and *Scp1* from mammals and *Zip1* from yeast code for the transverse filaments that synapse the axial elements into the synaptonemal complex. The *RAD50* gene is required for the induction and processing of double-strand breaks and therefore, in a null mutant of this gene, crossovers cannot be formed.

Bhullar et al. (2014) identified a candidate *Ph1* gene (designated *C-Ph1*) in the *Ph1* region, whose silencing resulted in a phenotype which was, to a certain degree, characteristic of the *ph1b* and *ph1c* mutants, namely, an increased number of univalents, multivalent pairing, and interlocking bivalents. However, it also disrupted chromosome alignment on the first meiotic metaphase plate. Despite a highly conserved DNA sequence, the *C-Ph1* gene homoeologues on 5AL and 5DL exhibit a different structure and expression pattern, further supporting the claim that *C-Ph1* is indeed the candidate for the *Ph1* gene. Yet, the fact that *C-Ph1* is mostly expressed during first meiotic metaphase, rather than during premeiotic interphase and first meiotic prophase, sheds some doubt on this claim.

The suggestion that *C-Ph1* is the *Ph1* gene was mainly based on the observation of bivalent clumping at first meiotic metaphase of virus-induced gene silencing (VIGS) mutants of *T. aestivum*. Rey et al. (2017) drew attention to previous studies that had already shown that bread wheat contains a gene, termed *Raftin1*, with a phenotype similar to that of *C-Ph1*. This gene was characterized as a tapetal cell gene, whose maximal expression occurs around the first meiotic metaphase, when the tapetum is fully formed (Wang et al. 2003). Disruption of this gene results in chromosome clumping at first meiotic metaphase and consequently, high male sterility. Therefore, Rey et al. (2017) suggested that *C-Ph1* is not a *Ph1* candidate.

Yet, *C-Ph1* is not the *Raftin1* tapetal cell gene described by Wang et al. (2003), as claimed by Rey et al. (2017). The deletion of *Raftin1*, causes chromosome clumping in the equatorial plate at first meiotic metaphase, but no evidence exist that it triggers an increased number of univalents, multivalent pairing, and interlocking bivalents, as does the silencing of *C-Ph1*. Moreover, whereas *Raftin1*, whose protein product, RAFTIN, is essential for the late phase of pollen development, is expressed only in the anther, but not in root, stem, leaf tissues, or emasculated inflorescence

(Wang et al. 2003), *C-Ph1* is expressed during meiosis (maximum expression of the gene was observed during MI), and also in vegetative tissues, such as roots and flag leaves (Bhullar et al. 2014). Therefore, *C-Ph1* and *Raftin1* are not the same gene, but because *C-Ph1* exhibits maximum activity at metaphase I and triggers chromosome clumping at this stage, it cannot be *Ph1*. Yet, if *Ph1* phenotype stems from a cluster of linked genes, each controlling an aspect of *Ph1* phenotype, then *C-Ph1* may represent one of them. In this respect, it is worth mentioning that Rawale et al. (2019) observed that silencing of *C-Ph1* resulted in 26% recombinant gametes between 1BS of wheat and 1RS of rye in hybrids between two bread wheat lines, one carrying intact chromosome 1B and the second containing a translocation in which chromosome arm IRS of rye replaced the 1BS of wheat. No recombination between these two arms took place in hybrids carrying *Ph1*.

Rey et al. (2018) claimed that recombination between chromosomes of wild relatives and those of allohexaploid wheat can be increased in F_1 hybrids with a *ph1b* deletion, by treating the plants with Hoagland solution. A search for the element in the solution that is responsible for this increase revealed that irrigation of plants with a 1 mM Mg2 + solution caused a significant increase in homoeologous CO frequency in all analyzed wheat x wild relatives hybrids. These observations suggest a role for magnesium supplementation in improving the frequency of recombination in wheat-interspecific hybrids.

12.2.3 Other Suppressors of Homoeologous Pairing in Wheat

In addition to *Ph1*, there are several other suppressors of homoeologous pairing in *T. aestivum*, one located on chromosome 3D (Upadhya and Swaminathan 1967; Mello-Sampayo 1971a, b; Driscoll 1972; Mello-Sampayo and Canas 1973) and the other on 3A (Driscoll 1972; Mello-Sampayo and Canas 1973). The 3D and 3A genes, designated *Ph2* and *Ph3*, respectively (Sears 1982, 1984), are less potent than *Ph1* (Mello-Sampayo 1971a, b; Mello-Sampayo and Canas 1973). *Ph2* is more effective than *Ph3*, but only about half as effective as *Ph1*. The effect of *Ph2* and *Ph3* on homoeologous pairing in allohexaploid wheat is negligible, but in wheat inter-generic hybrids, they have a somewhat more pronounced impact. *Ph2* is located on the distal region of chromosome arm 3DS of *T. aestivum* (Driscoll 1972a, 1973) and presumably derived from the pairing suppressor gene on 3DS of *Ae. tauschii*, the donor of the D subgenome (Attia et al. 1977, 1979). The suppressor of homoeologous pairing on 3A is also located on the short arm (Driscoll 1972, 1973). It is most likely that the suppressors on 3DS and 3AS are homoeoalleles. Currently,

there is no evidence showing that 3BL carries a homoeologous-pairing suppressor.

Aside from the minor suppressors on 3DS and 3AS, there is also a suppressor on chromosome 4D (Driscoll 1973) that appears to be almost as effective as the one on 3AS. Evidence was also obtained that the long arm of chromosome 2D (2DL) and possibly also of 2A and 2B, may carry a minor suppressor(s) of pairing (Ceoloni et al. 1986).

The EMS-induced 10/13 mutation (Wall et al.1971a), that was believed to be in the *Ph1* gene (Wall et al. 1971b), was shown to be in the *Ph2* locus (Sears 1982, 1984). Accordingly, following Sears (1982), the correct designation of the mutation 10/13 is *ph2a*, and not *ph1a*. Another deletion, aside from *ph1b,* was recovered by Sears (1977), after X-irradiation of pollen of *T. aestivum* cv. Chinese Spring. This deletion was substantially less potent than *ph1b* and resulted in an intermediate level of pairing in F_1 hybrids with *Ae. peregrina* (designated *Ae. kotschyi* var. *variabilis* by Sears), i.e., approximately five bivalents per PMC, compared with one bivalent in the control with *Ph1* and *Ph2*, and approximately 13 bivalents in the same hybrid carrying *ph1b* (Sears 1977). Consequently, Sears (1982) concluded that this deletion is not a mutation of *Ph1* or of any other gene on chromosome 5B. Because Mello-Sampayo (1971a) and Driscoll (1972) found that chromosome arms 3DS and 3AS carry minor suppressors of homoeologous pairing, the mutation was suspected to involve one of these genes. Indeed, further studies showed that the mutation is on the terminal segment of the short arm of chromosome 3D (Sears 1982). Accordingly, this mutation was designated *ph2b* (Sears 1982). There is little or no homoeologous pairing when the mutation *ph2b* is homozygous in cv. Chinese Spring of *T. aestivum* (Sears 1977, 1982). Male transmission of the mutation is approximately normal, and fertility, while somewhat reduced, is sufficient for easy maintenance of the homozygous line (Sears 1982). The effect of *ph2b* on chromosomal pairing in wheat hybrids is quite similar to that of *ph1a* (Sears 1982). Using synteny with rice, Sutton et al. (2003) narrowed down the *Ph2* locus, to a terminal 80 Mb of the short arm of chromosome 3D. More recently, however, Svačina et al. (2020) showed that the deletion induced by Sears (1977), the *ph2b* mutant, is actually larger than expected, comprising about 125 Mb of terminal end of the short arm of chromosome 3D.

Hybrids of allohexaploid wheat with rye, lacking both 3DS and 3AS, exhibit a level of pairing almost as high as that obtained in the absence of *Ph1* (Mello-Sampayo and Canas 1973). There is evidently an interactive effect of the two deficiencies, as the pairing level when both 3DS and 3AS are missing is about twice as high as that observed in the absence of the more active suppressor, 3DS. However, no interaction or even additive action of *Ph1* with *Ph2* has been demonstrated; lack of both 5BL and 3DS resulted in no more pairing than lack of 5BL alone. Perhaps the absence of chromosome arm 5BL permits the maximum possible amount of pairing, and deletion of additional suppressors will therefore have no added effect (Mello-Sampayo and Canas 1973). In contrast, Ceoloni and Donini (1993) studied the effect of the combined deficiencies of *ph1b* and chromosome arm 3DS on homoeologous pairing in Chinese Spring and in hybrids between this line and *Ae. pergrina* and *Secale cereale*. They found that in both Chinese Spring itself and its hybrid with *Ae. peregrina,* the combined deficiencies reinforced the *ph1b* effect in promoting homoeologous pairing. On the other hand, no such effect was noted in the hybrid involving *S. cereale*.

It seems that *Ph2* operates in a different way than does *Ph1* (Benavente et al. 1998; Martinez et al. 2001a; Prieto et al. 2005). Benavente et al. (1998), using genomic in situ hybridization, studied the effect of *Ph1* and *Ph2* on wheat-wheat and wheat -rye chromosome pairing in first meiotic metaphase and in anaphase in wheat x rye hybrids carrying either *Ph1*, *ph1b*, and *ph2b*. Their observation revealed distinct mechanisms involved in the control of wheat homoeologous pairing by the two *Ph* genes. In accordance, it was suggested (Martinez et al. 2001a; Prieto et al. 2005) that *Ph2* has a different function to that of *Ph1*, as it is not involved in recognition of homologous chromosomes but instead affects the progression of synapsis. Thus, Martinez et al. (2001b) concluded that the *Ph1* and *Ph2* loci bring about cytological diploidization of allohexaploid wheat via a different mechanism, whereby *Ph2* affects synaptic progression while *Ph1* affects the correction process of multivalents to bivalents at the transition from zygotene to pachytene.

Ceoloni and Feldman (1987) tested the two mutant lines of *Ph2*, *ph2a* and *ph2b*, for their mitotic-spindle sensitivity to colchicine. The data showed clearly that plants deficient for *Ph2* or carrying *ph2* alleles, were less sensitive to colchicine treatment than those carrying *Ph2*. Interestingly, the two genes that suppress homoeologous pairing in allohexaploid wheat, *Ph1* and *Ph2*, affect the sensitivity of the mitotic spindle to colchicine, but in opposite ways: *Ph1* decreases sensitivity, whereas *Ph2* increases it. So, it can be assumed that *Ph1* and *Ph2* also operate by affecting microtubular stability, possibly interfering with the dynamic equilibrium of assembled-disassembled tubulin subunits, but in different manner. The 3DS *Ph2* locus, was recently isolated (Serra et al. 2021). It encodes for the wheat homolog of DNA mismatch repair gene *MSH7*, *TaMSH7*. This is consistent with an anti-recombination effect when genetic divergence between the two recombination partners is too high as is the case with homoeologs. Indeed, the mismatch repair machinery detects mismatches in heteroduplexes formed at Holliday junctions at sites of crossover and recruits proteins that disengage the recombination partners.

Sears (1982) proposed three possibilities for making use of the *ph2b* mutation: (1) inducing the transfer of genes to wheat from closely related but not fully homologous alien chromosomes, (2) assessing the degree of relationship between alien chromosomes and their wheat homoeologues, and (3) improving amphiploids that have a somewhat reduced level of chromosome pairing.

12.2.4 Pairing Promoters in Wheat

There are several pairing-promoting genes that are also involved in the control of chromosome pairing in allohexaploid wheat. The long arms of chromosomes 5A and 5D carry such promoters (Feldman 1966b, 1968). Promoters also exist on the short arms of chromosomes of group 5, 5AS, 5BS, and 5DS (Feldman 1966b; Riley et al. 1966; Feldman and Mello-Sampayo 1967; Dvorak 1976), and on the long arms of chromosomes of group 3, 3AL, 3BL, and 3DL (Mello-Sampayo 1971a; Driscoll 1972, 1973; Mello-Sampayo and Canas 1973). Promoters were also found on the short arms of chromosomes of group 2, 2AS, 2BS, and 2DS (Ceoloni et al. 1986). Hence, the control of pairing in allohexaploid wheat stems from a balance between several suppressors and promoters.

The 5BS promoter induces pairing, both in wheat itself and in interspecific hybrids (Feldman 1966b; Feldman and Mello-Sampayo 1967; Riley and Chapman 1967). Its effect is substantially smaller than that of *Ph1*, as shown by the increased homoeologous pairing when both are missing, as in nulli-5B. The effect of the 5DL promoter is evidently greater than that of 5BS, for ditelo-5BL (lacking 5BS) has nearly normal synapsis, whereas nulli-5D (lacking 5DL and 5DS) tends to be asynaptic (Feldman 1966b), particularly at low (and presumably at high) temperatures (Riley et al. 1966). Since extra doses of chromosome 5A in (tetrasomic) largely suppress the effect of 5D nullisomy (Feldman 1966b; Riley et al. 1966), there must be pairing promoters on each arm of 5A. Yet, the effect of the 5A promoters is smaller than that of the 5D promoters since nulli-5A has normal synapsis, even at low temperatures (Riley et al. 1966).

The promoters on 5DL and 5AL could be homoeoallelic to *Ph1* and those on 5DS and 5AS to the 5BS promoter. Rey et al. (2017) claimed that *Ph1* has no homoeoalleles on 5AL and 5DL. If the promoters of 5AL and 5DL are not homoeoalleles of *Ph1*, then they should be located in a different chromosomal region that *Ph1*. There is some evidence (Mello-Sampayo 1972) that the promoter on 5DL is located distally to *Ph1* and is not homoeoallelic to it. 5BL may also carry a promoter, distally to *Ph1*, that is homoeoallelic to the promoters of 5AL and 5DL.

The existence of promoter on the long arm of chromosome 3D was deduced from the fact that pairing in hybrids with *Ae. peregrina* was increased more by deficiency of the entire chromosome 3D than by deficiency of only the short arm of 3D carrying the pairing suppressor *Ph2* (Driscoll 1972). Hybrid plants carrying only 3DS (deficient for 3DL) exhibited a greater increase in pairing, indicating the presence of a pairing promoter on this arm. The data of Mello-Sampayo (1971a) and Mello-Sampayo and Canas (1973) regarding the same and other hybrids, also support a slight promoting effect of 3DL.

The presence of the pairing gene on chromosome arm 3BL is necessary for normal synapsis (Li et al. 1945; Sears 1944, 1954; Kempanna and Riley 1962; Kato and Yamagata 1982, 1983). Sears (1944, 1954) allocated this gene to the right arm (=long arm; Sears and Sears 1979) of chromosome 3B and reported that absence of this gene led to failed crossover. Yet, this gene apparently does not interact with *Ph1*, since it controls different meiotic processes (Kempanna and Riley 1962). Consequently, it was concluded (Kempanna and Riley 1962) that chromosome arm 3BL carries a gene that is responsible for crossover. Hence, this gene may be different from the putative promoter on 3BL, which is homoeoallelic to the promoters on 3AL and 3DL.

Chromosome arm 2AS also evidently carries a gene essential for normal pairing (Sears 1954), although its effect is not as great as that of the gene on 3BL. Some investigations (Upadhya and Swaminathan 1967; Kempanna 1963) have failed to confirm its existence altogether. But, while an extra dose of chromosome 2A of Chinese Spring, previously reported to carry a pairing promoter on its short arm, did not increase pairing between homoeologous chromosomes in F_1 hybrids between Chinese Spring and *Ae. peregrina*, two doses of chromosome 2D or 2B caused a significant increase in homoeologous pairing in this hybrid (Ceoloni et al. 1986). Evidently, chromosomes 2D and 2B carry a pairing promoter(s). Studies of F_1 hybrids between aneuploids of CS, either lacking chromosome 2D or carrying an extra dose of it, and *Ae. peregrina*, *Ae. longissima*, and *S. cereale*, supported the finding that this chromosome carries a pairing promoter. Using ditelosomic lines, the promoter was found to be located on the short arm of 2D (2DS) (Ceoloni et al. 1986). It was deduced that the promoter on 2B is also located on the homoeologous short arm, i.e., 2BS. Thus, 2AS, 2DS, and 2BS carry a pairing promoter(s), with the promoter(s) on 2AS seemingly weaker than those on 2DS and 2BS.

Luo et al. (1992) reported that a Chinese landrace of allohexaploid wheat possesses a promoter of homoeologous pairing. The landrace has only bivalents at meiosis but hybrids between this race and *S. cereale* or *Ae. peregrina* exhibited an increase in homoeologous pairing (Luo et al. 1992; Liu et al. 1997b, 2003; Xiang *et al.* 2005). Monosomic analysis indicated that the promoter is located on

chromosome 6A and, thus, it is not homoeoallelic to any of the other pairing genes (Liu et al. 2003; Hao et al. 2011). Yet, Fan et al. (2019), using two mapping populations, identified a QTL locus on 3AL, which is possibly responsible for the promotion of homoeologous pairing in hybrids with this race, and which is probably homoeoallelic to the promoter on 3AL.

Effects on spindle sensitivity to antimicrotubule agents seems to be a common feature of a number of genes that impact homoeologous pairing in hexaploid wheat and in its hybrids. For example, in addition to *Ph1* and *Ph2*, chromosome arm 5BS, which carries a promoter of homoeologous pairing (Feldman 1966b, 1968; Riley et al. 1966; Riley and Chapman 1967; Feldman and Mello-Sampayo 1967), was also found to affect spindle sensitivity to colchicine (Avivi et al. 1970; Ceoloni et al. 1984). Similarly, *Ae. speltoides* and *Amblyopyrum muticum* carry genes that suppress *Ph1* activity, induce a high degree of homoeologous pairing in hybrids with allohexaploid wheat (Riley 1966), and affect spindle characteristics. *Ae. speltoides* exhibits a higher mitotic sensitivity to colchicine than other Sitopsis species, which do not induce homoeologous pairing (L. Avivi and M. Feldman, unpublished data). In line with these reports, an *A. muticum* chromosome added to allohexaploid wheat induced homoeologous pairing and also caused alterations in the premeiotic spindle, while genotypes of allohexaploid wheat carrying other chromosomes of *A. muticum* did not exhibit homoeologous pairing and showed normal spindle functioning (Dover and Riley 1973).

On the other hand, no correlation between asynapsis and mitotic spindle features, as measured by colchicine sensitivity, was found in 3BL-deficient wheat genotypes. Kato and Yamagata (1983) found that absence of chromosome arm 3BL, which carries a pairing gene (Sears 1944, 1954), did not affect the sensitivity of the mitotic spindle to colchicine. This finding is in accord with earlier observations (Kempanna and Riley 1962) from which an effect of the 3BL gene on chromosome pairing, different from that of *Ph1*, was inferred and led to the conclusion that both the timing and action of the 3BL gene are clearly distinct from those of the suppressors and promoters of homoeologous pairing.

The striking correlation between the effect on homoeologous pairing and alteration of spindle characteristics indicates that the suppressors and promoters of this type of pairing exert their effect on some features of the mitotic and premeiotic spindle. These genes impact the sensitivity of the mitotic spindle to colchicine, vinblastine, and griseofulvin (Avivi et al. 1970; Avivi and Feldman 1973b; Ceoloni et al. 1984; Gualandi et al. 1984), all of which are antimicrotubule agents that specifically bind tubulin, the main protein subunit of microtubules, thus shifting the equilibrium towards depolymerization.

12.2.5 Pairing Genes in Wild Relatives of Wheat

In many of the studied hybrids of *T. aestivum* with their wild relatives, there has been little or no evidence that the alien genome has any effect on pairing. The amount of pairing observed in these hybrids has been essentially the same as in haploids of *T. aestivum* (Riley and Law 1965). However, genotypes possessing genes that promote pairing have been found in several diploid and allotetraploid species, namely, *Ae. speltoides* (Riley et al. 1961), *A. muticum* (Riley et al. 1961; Riley and Law 1965), *Ae. longissima* (Mello-Sampayo 1971b), and *Ae. peregrina* (Farooq et al. 1990). These promoters of pairing have no effect in the species themselves, where complete pairing of homologues is the rule. But in intergeneric hybrids with allopolyploid wheat, they promote homoeologous pairing.

The level of pairing induced by the high-pairing lines of *Ae. speltoides* and *A. muticum* is almost as high as that observed in the same hybrids in the absence of chromosome 5B. An early work (Riley and Law 1965) showed approximately the same degree of pairing in hybrids of *aestivum* with high-pairing *speltoides* and *muticum* when 5B was absent as when it was present. In contrast, Feldman and Mello-Sampayo (1967) observed significantly more pairing in *aestivum* x *speltoides* lacking 5B. Similar super-high pairing was subsequently found by Dover and Riley (1972a) in *aestivum* x *muticum* hybrids lacking chromosome 5B.

The pairing genes of *A. muticum* and *Ae. speltoides* promote pairing between homoeologous chromosomes in hybrids involving allopolyploid wheat, by counteracting the effect of *ph1* of allopolyploid wheat (Riley 1960; Feldman and Mello-Sampayo 1967; Dover and Riley 1972a; Dvorak 1972). Two such genes were identified in each of these species [Dover and Riley (1972a) in *A. muticum* and Dvorak (1972) in *Ae. speltoides*]. In *Ae. speltoides*, they were assigned to chromosomes 3S (*Su1-Ph1*) and 7S (*Su2-Ph1*) (Dvorak et al. 2006a). Interestingly, they were not found to be homoeoallelic to the suppressor and promoters of group 5 of allopolyploid wheat, but the promoter on 3S was suggested to be homoeoallele to the suppressors and promoters of group 3. The *speltoides* genes did not affect the level of pairing in the inter-specific diploid *Ae. speltoides* x *Ae. tauschii* and *Ae. speltoides* x *Ae. caudata* hybrids, in which *Ph1* is not present (Chen and Dvorak 1984). In contrast, studies of meiotic chromosomal paring in hybrids between *A. muticum* and other diploid species of the wheat group showed relatively high paring, presumably due to the promotion of pairing by the *muticum* genes (Ohta 1990, 1991). Kihara and Lilienfeld (1935) observed seven bivalents at first meiotic metaphase of the F₁ hybrid between *Ae. comosa* and *A. muticum*. In F₁ of all crosses of the wild and domesticated forms of *T. monococcum*, *Ae. speltoides*, *Ae.*

longissima, and *Ae. caudata* with *A. muticum*, Riley (1966) found high chromosome pairing in meiosis. Jones and Majisu (1968) reported high pairing in the F₁ hybrid between *Ae. tauschii* and *A. muticum* and Ohta (1990, 1991) and M. Feldman (unpublished) observed almost complete pairing in the *A. muticum* x *T. monococcum* hybrids.

The simplest assumption concerning the mode of action of the *A. muticum* and *Ae. speltoides* pairing genes is that they are promoters of pairing, irrespective of the presence or absence of the homoeologous pairing suppressor genes. In this view, the high-pairing alleles are almost as strong as *Ph1* and consequently largely neutralize the effect of *Ph1* when it is present. In the absence of *Ph1*, they raise the pairing level, but only slightly, because it is already near the maximum (Feldman and Mello-Sampayo 1967). Alternatively, Riley (1960) proposed that the *muticum* and *speltoides* genes act by suppressing *Ph1* activity. However, even the high-pairing alleles are unable to suppress *Ph1* completely, and, consequently, pairing is never as high when *Ph1* is present as when it is absent. The suggestion that the high-pairing alleles act by suppressing *Ph1* gained credence with the finding that in the absence of *Ph1*, there is no difference between low- and high-pairing genotypes of *speltoides* or *muticum* in their effect on pairing in the hybrid with allopolyploid wheats (Dover and Riley 1972a; Rubenstein 1976). This strongly suggests that these alleles act only on *Ph1* and have no effect when *Ph1* is absent.

It is possible, however, that the existence of strong promoters in *A. muticum* and *Ae. speltoides* may be favored by the predominance of cross-pollination in these species. Riley and Law (1965) suggested that genes for high pairing may have a selective advantage in out-pollinators, being necessary to maintain full pairing between highly heterozygous homologues. This argument, however, revokes the proposal that the pairing promoters only act by suppressing *Ph1*.

It is not unlikely that other diploid relatives of wheat would have a similar effect if subjected to a sensitive test. In fact, partial suppression of *Ph1* evidently occurs in hybrids with autotetraploid *Ae. caudata* (R. Riley, unpublished). Likewise, Upadhya (1966) found that *Ae. caudata* induces homoeologous pairing in hybrids with *T. aestivum* carrying chromosome 5B. This finding corroborates the reports of Kihara and Lilienfeld (1935) on the high degree of chromosome pairing in the triploid F₁ hybrid between *Ae. caudata* and *T. turgidum* ssp. *durum*.

Similarly, chromosome 5U of *Ae. umbellulata* is reported to have a promoting effect similar to that of 5D (Riley et al. 1973). Mochizuki (1962) also found that the addition of a particular pair of diploid *Agropyron elongatum* (currently *Elymus elongatus*) chromosomes to durum wheat resulted in a substantial amount of homoeologous pairing. Dvorak (1987) investigated chromosome pairing in haploids of *T. aestivum* lines with added or substituted chromosomes of

Elytrigia elongata (currently *Elymus elongatus, genome* EeEe), and found that promoters of homologous or homoeologous pairing on chromosome arms 3EeS, 3EeL, 4EeS, 5EeS, and 6Ee of *E. elongata*. Genes that suppressed pairing of homoeologous chromosomes were found on chromosome arms 4EeL and 7EeS of *E. elongatus*. Consequently, he suggested that genes promoting or suppressing pairing of homoeologous chromosomes are ubiquitous among diploid Triticeae species.

A pairing promoter was also detected in the wild subspecies of *T. monococcum*, ssp. *boeoticum* (currently ssp. *aegilopoides* (Chapman et al. 1976). These researchers proposed that the higher trivalent frequencies seen in the hybrids between allohexaploid wheat and ssp. *aegilopoides* could be due to homoeologous pairing and that the genotype of *ssp. aegilopoides* can partly suppress the activity of *Ph1* of wheat.

The genome of *S. cereale*, which has no apparent effect on the level of homoeologous pairing in normal hybrids with *T. aestivum*, increases such pairing if two or three sets of rye chromosomes are added (Riley et al. 1973). Interestingly, the effect of the rye genome appears to depend on a balance between the long and short arms of chromosome 5R, with 5RL having a suppressive effect on pairing and 5RS an promoting and stronger effect. Gupta and Fedak (1986) studied chromosomal pairing in hybrids between *T. aestivum* and Petkus and prolific lines of rye and suggested the presence of genes both with major and minor effects on pairing in rye, with the genetic system promoting pairing in Petkus rye differing from that in prolific rye. The genetic variation in rye observed by Gupta and Fedak (1986) was considered by be similar to that of *Ae. speltoides*. In this respect, Dvorak (1977) observed that some of the rye genotypes promoted homoeologous chromosome pairing in hybrids with *T. aestivum*, and from the absence of distinct segregation classes among the hybrids, concluded that these pairing genes constitute a polygenic system.

Lelley (1976) tested the effect of single rye chromosomes on the pairing of homoeologous wheat chromosomes by crossing the seven wheat-rye addition lines of Imperial rye to Chinese Spring with *S. cereale* and *S.montanum* (currently *S. strictum*). In euhaploid hybrid plants (2n = 4x = 28; genome ABDR), no homoeologous pairing was induced, whereas in the 29-chromosome hybrids, while some wheat genome-driven suppression of chiasma formation between homologous rye chromosomes was noted, unequivocal evidence for homoeologous pairing of wheat chromosomes was found in several F$_1$ plants. Lelley suggested that *S. cereale* and *S. strictum* possess a genetic system that suppresses *Ph1* and that this system consists of several genes, located on different chromosomes, which may act additively. Such genes seem to be more frequent in the wild species, *S. strictum*, than in the domesticated *S. cereale*.

Halloran (1966) and Yu et al. (1998, 2001) observed promotion of homoeologous pairing in hybrids between bread wheat and *Dasypyrum villosum* by *D. villosum* gene (s), even in the absence of *Ph1* or in the presence of its mutant *ph1b*. These results indicate that *D. villosum* contains promoter(s) of homoeologous pairing, irrespective of *Ph1*. Another case of induced homoeologous pairing in the presence of *Ph1* and *Ph2* was reported by Liu et al. (2011) and Koo et al. (2017), who observed frequent recombination between 5M and 5D chromosomes in substitution lines of bread wheat containing 5M° from *Ae. geniculata*, indicating that the promoter on 5M° can counteract *Ph1*. Further cytogenetic analysis showed that that the promoter may be located in proximal regions of chromosome 5M°. Later, Koo et al. (2017) used two different 5M° chromosomes from different accessions of *Ae. geniculata* in *T. aestivum* background and observed differential levels of pairing between 5M° and 5D in both lines, with chiasmata between 5M° and 5D detected in 6.7% versus 21.7% of studied meiocytes. This might have been caused by the presence of different alleles on 5M° that repress *Ph1*. Additionally, homoeologous pairing occurred only between the 5M° and 5D chromosomes, as no multivalents were detected (Koo et al. 2017).

Several accessions of allotetraploid species of *Aegilops* also carry promoters of homoeologous pairing, i.e., *Ae. peregrina* (Farooq et al. 1990), *Ae. triuncialis*, *Ae. crassa* (Claesson et al. 1990), and *Ae. geniculata* (Lacadena and Azpiazu 1969). Only a number of genotypes of each of the above diploid and allopolyploids *Aegilops* species possess the promoter genes, indicating the occurrence of genetic variation in this trait. It is not unlikely that other diploid relatives of wheat would have a similar effect if subjected to a sensitive cytogenetic test.

Of note, all the promoters of chromosome groups 5, 3, and 2 in allohexaploid wheat are active, implying that no genetic diploidization (mutation leading to neofunctionalization, inactivation or elimination) took place in these genes. It is presumably crucial for the regularity of pairing processes in allohexaploid wheat to maintain all these genes in an active state, despite genetic redundancy. All the common wheat promoters and suppressors are part of a well-coordinated gene system that affects some of the critical processes that are involved in homology recognition, pairing, and crossing over in allohexaploid wheat itself and in hybrids with related species.

Promoters of homologous and homoeologous chromosome pairing are present in many Triticineae species (Table 5.2). Yet, there is variation in the presence and intensity of homoeologous pairing promoters and suppressors (Naranjo and Benavente 2015). Ozkan and Feldman (2001) reported variation in the effect of *Ph1* of tetraploid wheat, *Triticum turgidum* subsp. *dicoccoides*, on the level of homoeologous pairing in F_1 hybrids with *Aegilops pereg-rina*. Likewise, promoters in different lines of the diploid species *Ae. speltoides* and *A. muticum* induce different levels of pairing (high-, intermediate- and low-pairing) in hybrids with common wheat (Kimber and Athwal 1972; Dvorak 1972; Vardi and Dover 1972; Dover and Riley 1972a). Similarly, low-pairing and intermediate-pairing lines of *Ae. longissima* induce different levels of pairing in hybrids with *T. aestivum* (Mello-Sampayo 1971b). Pairing promoters were also found in several lines of *Agropyron cristatum* (Ahmad and Comeau 1991; Chen et al. 1989) but not in others (Limin and Fowler 1990). Evidence of variation in the effect of the promoters on homoeologous pairing among *Secale* taxa was obtained by Cuadrado and Romero (1984), Naranjo et al. (1979), and Naranjo and Palla (1982). Similar evidence was obtained by Gupta and Fedak (1986), who studied the effect of two *S. cereale* cultivars on chiasma frequency in common wheat x *S. cereale* hybrids.

Viegas et al. (1980) reported that chromosome arm 5DL of a mutant line of *T. aestivum* carries a potent pairing suppressor. Suppressors of homoeologous pairing were also found in the diploid subspecies of *Elymus elongatus* (Dvorak 1987; Charpentier et al. 1988). Chromosome arms 4EeL, 7EeL, and 6Ee of this subspecies carry genes that suppress homoeologous pairing in hybrids with species of other genera (Dvorak 1987; Charpentier et al. 1988).

The existence of suppressors of homoeologous pairing in diploid species of the wheat group was suggested by several researchers (Okamoto and Inomata, 1974; Waines 1976). Waines interpreted pairing data of F_1 Triticeae diploid hybrids and amphiploids published by Sears (1941a, b), to indicate that diploid species of this group have genetic systems (suppressors) controlling homoeologous pairing in the inter-specific hybrids, which result in an immediate cytological diploidization in the derived amphiploids. These systems, which, according to Waines (1976), function at as isolating mechanisms, are already present in diploid species.

In accord with Waines, Shang et al. (1989) reported on the occurrence of pairing-control genes in diploid wheat, *T. monococcum*. Maan (1977) also concluded from his cytogenetic studies of *T. turgidum* ssp. *durum* x *Ae. comosa* ssp. *heldreichii* hybrids, that a *Ph1*-like gene exist in the latter.

Several studies attributed the low pairing in allopolyploid wheats x Emarginata species (*Ae. bicornis*, *Ae. sharonensis*, *Ae. longissima*, and *Ae. searsii*) to weak homoeologous pairing suppressors that exist in these diploid species, which, together with *Ph1* of allopolyploid wheats, suppress the pairing of homoeologous chromosomes in these hybrids (Feldman 1978). The assumption that several lines of *Ae. longissima* carry a weak suppressor is supported by Avivi's (1976) finding that an autotetraploid derived from a low-pairing type of *Ae. longissima* exhibited fewer multivalents and a larger number of bivalents than the autotetraploid derived from an intermediate-pairing type of *Ae. longissima*.

A correlation between increase in bivalent pairing on the account of quadrivalent pairing and an effect on mitotic spindle sensitivity was also observed in the low-pairing line of *Ae. longissima* (L. Avivi, unpublished data). She found that the two induced autotetraploid lines of *Ae. longissima*, differing in the degree of multivalent pairing, also differed in their mitotic spindle sensitivity to colchicine; the bivalent-forming autotetraploid, assumed to possess a suppressor (Avivi 1976), was less sensitive than the multivalent-forming one. Similarly, it was found (A. Charpentier, unpublished data) that the natural autotetraploid *Elymus elongatus* (=*Agropyron elongatum*), which forms only bivalents at meiosis, was much less sensitive to colchicine than an induced autotetraploid of this species that forms several multivalents in every meiocyte. Crosses of the two autotetraploid lines indicated that bivalentization in *E. elongatus* results from the activity of a recessive gene (Charpentier et al. 1986).

All the allopolyploid species of *Aegilops* are characterized by diploid-like meiotic behavior and exhibit strictly bivalent pairing at first meiotic metaphase (Riley and Law 1965). In the absence of *Ph1*, no genotype of an allopolyploid *Aegilops* species reproduced the effect this gene has on homoeologous pairing (Riley and Law 1965). This is in spite of the fact that they may contain weak suppressors with inconspicuous effects at the allopolyploid level, that were presumably contributed by their diploid parents. Based on the above results, Riley and Law (1965) concluded that if there is a system in these polyploids suppressing homoeologous pairing, then it must be functionally distinct from that of *T. aestivum*.

Indeed, none of the allopolyploid *Aegilops* genotypes tested by McGuire and Dvorak (1982) fully compensated for the absence of chromosome 5B, but genotypes of *Ae. cylindrica*, *Ae. juvenalis*, *Ae. triuncialis*, *Ae. geniculata*, *Ae. columnaris*, *Ae neglecta*, and *Ae. recta*, did have some suppressive effects on homoeologous pairing when 5B was absent from the hybrids. This suggests that the diploid-like meiosis in these allopolyploid species is also a consequence of genetic suppression. This conclusion is in accord with the absence of homoeologous pairing in a haploid of *Ae. geniculata* (Matsumura 1940; Kihara 1937), despite the pairing (2–5 bivalents and 0–2 trivalents) in the diploid hybrid between *Ae. umbellulata* and *Ae. comosa*, the assumed two parents of *Ae. geniculata* (Kihara 1949). It is, therefore, clear that in this allopolyploid species, completely regular meiosis, with pairing restricted to homologous partners, developed at the allotetraploid level, either by causing some structural changes among the homoeologues via elimination of DNA sequences, as was shown by

Feldman et al. (1997) and Ozkan et al. (2001), or via development of a genetic system that suppresses pairing of homoeologues.

12.2.6 The Effect of B Chromosomes on Pairing in Wheat Hybrids

Two diploid species of the wheat group, *Amblyopyrum muticum* and *Ae. speltoides*, have B (accessory) chromosomes which bear a suppressive effect, similar to that of *Ph1*, on chromosome pairing in hybrids with allopolyploid wheats. In *T. aestivum* x *A. muticum* or x *Ae. speltoides* hybrids lacking chromosome 5B but possessing B chromosomes, the level of pairing was generally the same as when 5B was present, irrespective of the presence or absence of *muticum* or *speltoides* pairing alleles (Vardi and Dover 1972). Up to six B chromosomes had little, if any, additive effect as compared to just one, on pairing in hybrids of *Ae. speltoides* with *T. aestivum* (Vardi and Dover 1972). However, six B chromosomes caused some pairing failure in *Ae. speltoides* itself (Dover 1975). There seems to be little, if any, interaction between the B chromosomes and the chromosome carrying *Ph1*; the pairing level is about the same when either or both are present. No supplementation of the effect of *Ph1* by B chromosomes could be found at any level of pairing (Dover 1973; Vardi and Dover 1972). Similarly, studies of Ohta (1990, 1991) on pairing in hybrids between *A. muticum* and all the diploid species of the wheat group, with or without B chromosomes, showed that *A. muticum* B chromosomes suppress pairing in these hybrids. Based on these results, Ohta (1990, 1991) exploited the effect of B chromosomes of *A. muticum* on meiotic chromosomal pairing in F_1 hybrids with diploid species of the wheat group, to evaluate the degree of genomic divergence between these diploid species and *A. muticum*.

Roothaan and Sybenga (1976) analyzed hybrids between *T. aestivum* cv. Chinese Spring and rye, with and without chromosome 5B of wheat and B-chromosomes of rye and found that the absence of 5B resulted in an increase in homoeologous pairing, alongside a decrease in chiasmata in the rye B chromosomes themselves. Two rye B chromosomes were entirely ineffective in compensating for the absence of 5B.

Neijzing and Viegas (1979) analyzed *T. aestivum* x *S. cereale* hybrids, with and without chromosome 5B or 5D and with and without B chromosomes of rye. In the absence of chromosome 5B, there was no effect of rye B chromosomes on meiotic synchrony and on chiasma frequency. Absence of 5D appeared to decrease synchrony at 20 and 15 °C, but genetic variation between plants played an important role as well.

Romero and Lacadena (1980) and Cuadrado et al. (1991) analyzed the interaction between homoeologous pairing of rye B chromosomes carrying the wheat pairing suppressors *Ph3*, *Ph2*, and *Ph1*, and the promoters on 3BL, 5AL, and 5DL in wheat-rye hybrids. These authors found that when the pairing suppressors of wheat were absent, rye B chromosomes had a suppressive effect on pairing, but behaved as promotors when the pairing promoters were absent.

Several researchers transferred B chromosomes of rye into an allopolyploid wheat background. Lindström (1965) managed to obtain plants of bread wheat with two B chromosomes of rye but, unfortunately, his sudden death, prevented him from studying the effect of these B chromosomes on wheat chromosome behavior. Müntzing et al. (1969) studied meiosis in this line and reported a reduction in pairing of the B chromosomes, with almost no noticeable effect on the pairing of the wheat chromosomes. Likewise, Niwa et al. (1997) transferred B chromosomes from Korean rye into cv. Chinese Spring of *T. aestivum* and reported that the presence of the Bs did not disturb pairing between wheat chromosomes. Mamun-Hossain et al. (1992) compared meiotic chromosome behavior in allotetraploid wheat, *T. turgidum* ssp. *durum*, carrying B chromosomes from rye, with that of hexaploid wheat with the same B chromosomes (from the Lindström strain). The B chromosomes induced some asynapsis of the wheat chromosomes, i.e., increased the number of rod bivalents and univalents per cell, with effect of Bs in tetraploid wheat was more pronounced than in hexaploid wheat.

An attempt to identify B chromosomes segments that affect homoeologous pairing in *T. aestivum* x *Ae. peregrina* hybrids, was made by Kousaka and Endo (2012) using the B-9 and B-10 segments of rye B chromosomes. The B-9 and B-10 segments are derived from reciprocal translocations between a wheat and B chromosomes; B-9 has the B chromosome pericentromeric segment and B-10 has the B distal segment. B-9, like the B chromosome, suppressed homoeologous pairing when chromosome 5B was absent. On the other hand, the B-9 and B-10 segments promoted homoeologous pairing when 5B was present. These results suggest that the effect of the B chromosomes on homoeologous pairing is not confined to a specific region and that the intensity of the effect varies, depending on the presence or absence of 5B and on the segment and dose of the B chromosomes.

12.2.7 Origin of the *Ph* Genes

It was assumed that the *Ph1* gene evolved at the allotetraploid level, in parallel to, or soon after, the allopolyploidization process (Riley 1960; Sears 1976b). Dover and Riley (1972b) proposed three obvious possibilities for the origin of the *Ph1* gene: (a) it arose as a mutation following the formation of allotetraploid wheat; (b) it was transferred to chromosome 5B of a nascent allotetraploid from a B chromosome of *Ae. speltoides* or *A. muticum*; or (c) it was already present in the diploid species that donated the B subgenome to allotetraploid wheat. The pairing suppressor in the diploid progenitor of allotetraploid wheat has a weak suppressive effect that was gradually or rapidly increased in the allotetraploid.

In the absence of any clear-cut evidence that the diploid species that contributed the B subgenome to allotetraploid wheat, possessed a *Ph1*-like gene, an origin of *Ph1* by mutation in the nascent allotetraploid seems a likely possibility. The presence of promoters, homoeoalleles to *Ph1*, on 5AL and 5DL (and possibly also in all the diploids of the wheat group) support the theory of evolution of *Ph1* through a mutation to an antimorphic allele. Riley (1960), Riley and Law (1965), and McGuire and Dvorak (1982) assumed that *Ph1* resulted from a mutation in a pairing-promoting gene that existed in the diploid donor of the B subgenome. The mutation may have altered the activity of the mutant gene, subsequent to the formation of allotetraploid wheat. Mutation of a promoter gene to *Ph1* would be a type of mutation (antimorphic) that occurs spontaneously in allopolyploids (Sears 1972). Such a mutation from a promoter on 5DL to a suppressor of homoeologous pairing was described by Viegas et al. (1980). Whichever type of mutation occurred, it presumably benefited from immediate, strong, favorable selection pressure, since it likely considerably increased the fertility of the nascent allotetraploid.

Another possibility for the origin of *Ph1* in the nascent allotetraploid wheat was recently proposed by Rey et al. (2017), who identified the *Ph1* region on 5BL in a segment containing a block of heterochromatin, the *Zip4* gene and several other genes that were duplicated in chromosome arm 5BL from chromosome arm 3BL. This segment has no similar segments in the homoeologous chromosome arms 5AL and 5DL. Mutations in the *Zip4* gene affect crossing over between homoeologues, similar to the effect of *ph1b*.

The likelihood that *Ph1* was transferred to chromosome 5B from a B chromosome was suggested by Dover and Riley (1972b), Dover (1973), and Riley et al. (1973), based on the similarity between the effects of *Ph1* and B chromosomes of *Ae. speltoides* and *A. muticum*. Dover (1973) speculated that *Ph1* originated from an ancestral interchange between a B chromosome and an A chromosome of a putative B-genome donor. But there is no evidence that the effect of B chromosomes of these two diploid species is due to a single gene. Moreover, there are some differences between the effects of these B chromosomes and *Ph1*; an extra dose of B with two doses of *Ph1* did not induce homoeologous pairing and interlocking of bivalents the way an extra dose of *Ph1* does alone.

Dover (1973) concluded that B chromosomes are not solely responsible for the reduction in homoeologous pairing in *T. aestivum* x *A. muticum* hybrids carrying B chromosomes. The presence of B chromosomes did not introduce an additional factor regulating the degree of pairing. Moreover, Vardi and Dover (1972) found that B chromosomes cause disturbances in the mitotic and meiotic spindle and Dover (1973) attributed this effect to the presence of satellited DNA in pericentromeric regions as opposed to the effect of a single gene.

The third possibility assumes the preexistence of weak homoeologous pairing suppressor genes in diploid species. Dover and Riley (1972a) proposed that *Ph1* may has evolved directly as the result of incorporation of a low-pairing allele of *Ae. speltoides* or *A. muticum* into nascent allotetraploid wheat. The discovery of suppressors other than *Ph1* in hexaploid wheat, and particularly the finding (Riley et al. 1973) that one chromosome of rye, chromosome 5R, carries a suppressor as well as a promoter of pairing, reinforces the proposal that the diploid relatives of wheat all carry suppressors. Indeed, such preexistence of suppressors was suggested by Okamoto and Inomata (1974) who, based on the formation of several bivalents in haploids of diploid wheat, barley and rye, concluded that these genomes include several duplications that are located in different chromosomes. Yet, despite the existence of these duplications, multivalents do not form in the diploids. Okamoto and Inomata (1974) therefore suggested that a gene or genes which suppress quadrivalent formation at meiosis is present in these diploid species. These genes suppress ectopic exchanges between the duplicated regions observed within the diploid genomes.

This possibility is strongly defended by Waines (1976) who, interpreted pairing data of F_1 Triticeae diploid hybrids and amphiploids, published by Sears (1941a, b), to indicate that diploid species of this group have genetic systems controlling homoeologous pairing in inter-specific and inter-generic hybrids, resulting in immediate cytological diploidization in the derived amphiploids. These systems, which, according to Waines (1976), function at the diploid level as isolating mechanisms, are already present in diploid species and did not necessarily arise by mutation de novo after allopolyploid formation, but rather, underwent changes increasing their efficiency at the tetraploid level.

The existence of low-potency suppressors of homoeologous pairing in diploid species of the wheat group was suggested by several researchers. The probable existence of a pairing suppressor gene in lines of *Ae. longissima* and possibly also in *Ae. bicornis, Ae. sharonensis, Ae. searsii*, and *Ae. speltoides*, may explain the low amount of pairing typically observed in hybrids between allopolyploid wheats and accessions of these species. In this respect, it is assumed that the donor of the B subgenome to allotetraploid wheat was a low-pairing type containing a weak suppressor of homoeologous pairing. As was suggested by Waines (1976), such genes may be a constituent of an inter-specific genetic barrier enabling the divergence of developing taxa into species. The F_1 hybrids between the donor of the B subgenome and that of the A subgenome exhibited very little pairing, presumably due to presence of a suppressor, and were sterile. Interspecific or intergeneric F_1 hybrids having low chromosomal pairing at meiosis tend to undergo chromosome doubling, due to the formation of unreduced gametes, more readily than hybrids with high pairing, and produce stable and fertile allopolyploids.

Alternatively, it is not improbable that a suppressor was transferred from one of these Sitopsis species to the donor of the B subgenome, if such a gene was not already in the B subgenome donor before, and from there to the B subgenome of allotetraploid wheat. Bernhardt et al. (2020) presented evidence indicating introgression from Sitopsis species, especially *Ae. longissima* and *Ae. searsii*, into the donor of the B subgenome of allopolyploid wheats or directly with the B subgenome of allotetraploid wheat.

If *Ph1* in tetraploid wheat derived from a suppressor of the diploid species, which is a relatively less potent suppressor, it was not sufficiently effective soon after formation of allotetraploid wheat. It is assumed, therefore, that only later on, a mutation(s) improved its efficiency. Similarly, the *Ph1* of hexaploid wheat presumably became stronger than that of the tetraploid. The absence of evidence for intergenomic recombination in tetraploid and hexaploid wheat, may indicate that the original weak *Ph1* gene was superimposed on another system which prevented pairing of homoeologous chromosomes, as suggested by Feldman et al. (1997) and Ozkan et al. (2001).

Attia et al. (1977, 1979) demonstrated that the *Ph2* gene is present in *Ae. tauschii*, the donor of the D subgenome to allohexaploid wheat. Their results and those of Ekingen et al. (1977) clearly indicated that several diploid biotypes of *Ae. tauschii* caused a suppression of homoeologous pairing similar to the effect of chromosome arm 3DS of *T. aestivum*. This shows that the 3DS suppressor of hexaploid wheats derived from *Ae tauschii*. These results are in accord with the assumption of Azpiazu and Lacadena (1970) that the 3DS suppressor originated in *Ae. tauschii*. The alternative suggestion (Driscoll 1972) of origin after the completion of hexaploid wheat, e.g., by mutation (Mello-Sampayo and Lorente 1968), seems to be inconclusive.

The evolution of *Ph1* at the tetraploid level likely occurred in several steps rather than in one event. This is inferred from the occurrence of variation in *Ph1* in allotetraploid wheat, which include high-, intermediate- and low-pairing alleles of this gene. Ozkan and Feldman (2001) reported on the discovery of genotypic variation in the control of homoeologous pairing among tetraploid wheats.

When comparing the levels of homoeologous pairing in hybrids between CS and *Ae. peregrina*, significantly higher levels were obtained in hybrids between *Ae. peregrina* and CS substitution lines in which chromosome 5B of CS was replaced by either 5B of wild allotetraploid wheat, *T. turgidum* ssp. *dicoccoides* line 09 (TTD09), or 5G of *Triticum timopheevii* ssp. *timopheevii* line 01 (TIM01). Similarly, a higher level of homoeologous pairing was found in the hybrid between *Ae. peregrina* and a substitution line of CS in which chromosome arm 5BL of line TTD140 of ssp. *dicoccoides* substituted 5BL of CS. The observed effect on the level of pairing is seemingly exerted by chromosome arm 5BL of ssp. *dicoccoides*, most probably by an allele of *Ph1*. Searching for variation in the control of homoeologous pairing among lines of wild allotetraploid wheats, *T. turgidum* ssp. *dicoccoides* or *T. timopheevii* ssp. *armeniacum*, showed that hybrids between *Ae. peregrina* and lines of these two wild allotetraploid exhibited three different levels of homoeologous pairing: low, low-intermediate, and high-intermediate. The low-intermediate and high-intermediate genotypes may possess less-effective alleles of *Ph1*. The variation in the activity of *Ph1* in different accessions of *T. turgidum* ssp. *dicoccoides* and in *T. timopheevii* ssp. *armeniacum* suggests a gradual, rather than a single-step, evolution of the homoeologous pairing suppressor in wild allotetraploid wheats.

Variation in *Ph1* activity was also found in bread wheat, *T. aestivum* ssp. *aestivum*. Martinez et al. (2005) analyzed chromosome pairing in haploids of three different cultivars of bread wheat (Thatcher, Chris, and Chinese Spring) that were obtained from crosses with *Zea mays* and found differences in their meiotic behavior. Thatcher and Chris haploids had significantly higher levels of pairing at first meiotic metaphase than CS haploids. This pairing correlated with higher levels of synapsis in the Thatcher and Chris first prophase nuclei as compared to the Chinese Spring nuclei. The authors concluded that variation exists in the effectiveness of the diploidizing mechanism among cultivars of bread wheat.

Rawale et al. (2019) characterized the structure and expression of *C-Ph1*, that according to Bhullar et al. (2014) is the candidate *Ph1* gene, in the ssp. *dicoccoides* accessions found by Ozkan and Feldman (2001) to differ in their effect on homoeologous pairing in hybrids with *Ae. pergrina*. The *C-TdPh1-5B* of ssp. *dicoccoides* transcribed three splice variants as observed in the hexaploid wheat (Rawale et al. 2019). Further, single-nucleotide changes differentiating accessions varying in homoeologous-pairing control were identified. Quantitative expression analysis showed that the wild emmer accessions that induced high homoeologous pairing, had ∼ 10,000-fold higher transcript abundance of the *C-TdPh1-5B* during first meiotic prophase compared to accessions that exhibited low homoeologous pairing. Based on these results, Rawale et al. (2019) concluded that the homoeologous-pairing control is mediated by transcriptional regulation of this gene during meiosis. The presence of genetic variation in the genetic control of homoeologous pairing in different accessions of wild emmer questioned the validity of the proposed single-step evolution of the homoeologous-pairing suppressor.

Both wild and domesticated *T. timopheevii* contain a *Ph1*-like gene (Feldman 1966a; Ozkan and Feldman 2001). One possibility is that *Ph1* independently originated in both wild allotetraploid *Triticum* taxa, *T. turgidum* ssp. *dicoccoides* and *T. timopheevii* ssp. *armeniacum*. This could have resulted from an independent mutation in a pairing promoter gene. Alternatively, the line the *Ae. speltoides*-related species which donated the B subgenome to ssp. *dicoccoides,* and the more modern *speltoides* line that contributed the G subgenome to ssp. *armeniacum,* contained a weak pairing suppressor. On the other hand, it is reasonable to assume that *Ph1* arose in only one wild allotetraploid taxon, most probably in the older ssp. *dicoccoides*, and was transferred to the younger ssp. *armeniacum* through introgression.

12.2.8 Conclusion

To sum up, cytological diploidization in allopolyploid *Triticum* species arises through two independent, complementary systems. One is based on the physical divergence of chromosomes, and the second, on the genetic control of pairing. The *Ph*-gene system superimposes itself on and takes advantage of, and thereby reinforces, the system of the physical differentiation of homoeologous chromosomes. The stringent selection for fertility in the allopolyploid *Triticum* species might favor the development of these two systems, to ensure prevention of multivalent formation and promotion of bivalent pairing in nature.

The use of aneuploid lines of *T. aestivum* enabled the identification of structural changes, mainly elimination, of DNA sequences in the chromosomes of synthetic and natural allopolyploids. The existence of the same changes in nascent and natural allopolyploids indicates that they were induced during the course of allopolyploidization. Elimination of DNA sequences from the homoeologues of one subgenome in allotetraploids and from the additional subgenome in allohexaploids, converts the remaining sequences into homologous-specific, which can presumably serve in homology search and in initiation of pairing at meiosis.

Superimposed on this physical system in allopolyploids *Triticum* species, is a genetic system comprising of a battery of genes that suppress and promote chromosome pairing. Although much information has accumulated on the occurrence and activity of these genes, some issues remain unclear, especially as to the ways in which these genes

operate. What is striking is the complexity of the system, which involves not only several suppressors and several promoters in the allopolyploid wheat itself, but also genes in related diploids that largely inhibit the major suppressor, and B chromosomes in the diploids that have an effect similar to that of the suppressor.

Relatively little information is available on pairing genes in the wild relatives of wheat, but it is clear that these species also carry both suppressors and promoters affecting chromosome pairing. As was shown by Sears (1941b), in at least some hybrids between diploids, the amount of pairing depends upon the particular biotypes used to make the crosses. Rarely have enough different biotypes been tested to adequately assess the potential for pairing in the combination concerned. Also, it is not clear which genetic system, if any, exists in allopolyploids of *Aegilops*.

The rapid process of cytological diploidization in the newly-formed allopolyploid species of the wheat group has been critical for their successful establishment in nature. The restriction of pairing to completely homologous chromosomes ensures regular segregation of genetic material, high fertility, genetic stability, and disomic inheritance, which prevents the independent segregation of chromosomes of the different subgenomes. This mode of inheritance leads to permanent maintenance of favorable inter-subgenomic genetic interactions, and thereby fixes heterotic interaction between subgenomes. On the other hand, disomic inheritance sustains the asymmetry in the control of many traits by the different subgenomes (Feldman et al. 2012). In addition, since cytological diploidization facilitates genetic diploidization, genes existing in double and triple doses can be diverted to new functions through mutations, thereby giving preferentiality to the creation of favorable, new inter-subgenomic combinations.

12.3 Genetic Diploidization

How the two or three divergent subgenomes, present in a single nucleus of a nascent allopolyploid, were led to operate in a harmonious manner? On the one hand, an extra dose of some homoeoalleles can be of positive adaptive value. On the other hand, overexpression of the increased gene dosage may lead to redundancy, waste, or, in some cases, even to a deleterious effect. To prevent such negative consequences, the expression of the increased dose of homoeoalleles is reduced, a phenomenon described as "gene dosage compensation". An example of gene-dosage compensation was presented by Galili et al. (1986) in genes coding for the subunits of high-molecular-weight glutenins in *T. aestivum* ssp. *aestivum*. Furthermore, because of genetic divergence of the wheat subgenomes at the diploid level and because of mutations in the allopolyploid level, some homoeoalleles in allopolyploid wheat may be variable, representing a different form of the gene, and thus, may differ from one another in their expression profile. In this case, activity of all the duplicated genes may be advantageous, producing favorable inter-subgenomic interactions. Inter-subgenomic gene interactions may be, in some cases, expressed in novel traits that do not exist in their parental diploids. Some of these traits may be of great adaptive value in nature. Inter-subgenomic gene interactions are also of direct relevance to wheat cultivation. For example, the baking quality of bread wheat is due to the unique properties of its gluten—a product derived from the combined contribution of the three subgenomes of ssp. *aestivum* and thus, exists only at the hexaploid level.

Moreover, allopolyploidy enables to generate a new type of genetic variation through recombination between homoeologous chromosomes. Using cytological and whole-genome sequence analyses, Zhang et al. (2020) identified 37 homoeologous exchange (HE) events in the progeny of a nascent synthetic allotetraploid *T. urartu—Ae. tauschii* (2n = 4x = 28; genome AADD). HEs exhibit typical patterns of homologous recombination hotspots, being biased toward low-copy, sub-telomeric regions of chromosome arms and showing association with known recombination hotspot motifs. But strikingly, while homologous recombination preferentially takes place upstream and downstream of coding regions, HEs are highly enriched within gene bodies, giving rise to novel recombinant transcripts, which in turn are predicted to generate new protein fusion variants. However, the amphidiploid *T. Urartu—Ae. tauschii* has a genomic combination (AADD) that does not exist in nature and might be an atypical case of inter-genomic recombination. Nevertheless, while HE is less frequent in natural wheat, several inter-genomic translocations were encountered in lines of natural allopolyploids of the wheat group, (Maestra and Naranjo 1999, and reference therein). Moreover, Intragenic recombination and formation of chimeric genes was detected in HEs of allopolyploids *Brassica* and *Arabidopsis suecica*, and in autopolyploids rice, banana, and peanut, indicating that homoeologous exchange may be a broad phenomenon that occurs also in natural allopolyploids. HE thus provides a mechanism for evolutionary novelty in transcript and protein sequences in nascent allopolyploids.

Another challenge of interspecific or intergeneric hybrids is to suppress activity in cases of deleterious effect or homoeoallele redundancy. The processes that bring redundant or unbalanced gene systems in allopolyploids toward a diploid-like mode of expression are called genetic or functional diploidization. Genetic diploidization is a regulatory process that brings redundant or unbalanced gene systems in allopolyploids toward a diploid-like mode of expression (Ohno 1970). It results either from elimination, mutation or repression of genes that, in many cases, restrict the activity

of sets of genes to only one subgenome (Liu et al. 1998b; Wendel 2000; Shaked et al. 2001; Levy and Feldman 2002, 2004; Feldman and Levy 2005, 2009; Comai 2005; Chen 2007). These genomic changes may increase the fitness and competitiveness of the newly formed allopolyploid, leading to its successful establishment in nature. Hence, successful allopolyploidizations are those that successfully trigger an array of genetic and epigenetic changes that confer evolutionary advantages. If heightened fitness is not achieved rapidly enough, the nascent allopolyploid would be selected against as in the case of hybrid incompatibility.

Studies with newly-formed allopolyploid wheats as well as genome sequencing data of natural wheats have indicated that a broad range of DNA alterations occurred during or soon after allopolyploidization, leading to genetic diploidization (Levy and Feldman 2004; Feldman and Levy 2005). These alterations included deletion or silencing of genes, or alternatively, neofunctionalization, occurrence of an adaptive mutation enabling one of the homoeoalleles to develop a new function that was not present in the ancestral gene. Silencing of one or more of the homoeoalleles can be in the form of pseudogenization, a process through which one of the homoeoalleles loses its function as a result of disruption of its regulatory or coding sequence, or subfunctionalization, a neutral mutation process causing an homoeoallele to maintain only a subset of its original ancestral gene. In addition, transposon activation and deactivation, are extensive and occur relatively rapidly. Levy and Feldman (2002, 2004), Feldman and Levy (2005, 2009, 2011, 2012), and Feldman et al. (2013)) distinguished between revolutionary changes, occurring during or immediately after allopolyploidization and evolutionary changes that take place throughout the life of the allopolyploid (Table 12.2). Revolutionary changes include genetic and epigenetic alterations that lead to cytological and genetic diploidization, thereby improving the harmonious functioning of the divergent subgenomes, stabilizing the nascent allopolyploid and facilitating its establishment as a new competitive species in nature—all of which are species-specific. Evolutionary changes comprise mostly genetic changes that promote genetic diversity, flexibility and adaptability—all of which are biotype- or population-specific. In contrast to genes that code for enzymes, genetic diploidization in allopolyploid wheats mainly involves genes that code for structural or storage proteins, e.g., histones, tubulin subunits, glutenins and gliadins subunits, and ribosomal RNA (and possibly also tRNA). In such genes, expression of all homoeoalleles might be redundant and even deleterious, due to over-production and inefficiency. In this case, traits controlled by genes from only or mostly one subgenome may have a higher adaptive impact. It is therefore expected that such gene loci would have been targets for genetic diploidization. The *Hardness*

(*Ha*) locus, controlling grain hardness in *Triticum* and *Aegilops* species, represents a classical example of a trait whose variation arose from gene loss after allopolyploidization (Chantret et al. 2005). The previously reported loss of *Pina* and *Pinb* genes from the *Ha* locus of allopolyploid wheat species, was caused by a large genomic deletion that likely occurred independently in the A and B subgenomes. Moreover, the *Ha* locus in the D subgenome of ssp. *aestivum* is 29 kb smaller than in the D genome of its diploid progenitor *Ae. tauschii*, principally because of transposable element insertions and two large deletions caused by illegitimate recombination. The data of Chantret et al. (2005) suggest that illegitimate DNA recombination, leading to various genomic rearrangements, constitutes one of the major evolutionary mechanisms in wheat species.

Likewise, Shitsukawa et al. (2007) reported that both genetic and epigenetic alterations occurred in the homoeologs of an allohexaploid wheat class *E mads* box gene. Two class E genes have been identified in wheat, i.e., wheat *Sepallata (WSEP)* and wheat *Leafy Hull Sterile1 (WLHS1)*. The three wheat homoeologs of *WSEP* show similar genomic structures and expression profiles, whereas the three homoeologs of *WLHS1* show genetic and epigenetic alterations. *WLHS1* of the A subgenome (*WLHS1-A*) has a structural alteration that contains a large novel sequence in place of the K domain sequence. A yeast two-hybrid analysis and a transgenic experiment indicated that the *WLHS1-A* protein has no apparent function. The *WLHS1-B* and *WLHS1-D* of the B and D subgenomes, respectively, have an intact *MADS* box gene structure, but *WLHS1-B* is predominantly silenced by cytosine methylation. Consequently, of the three *WLHS1* homoeoalleles, only *WLHS1-D* functions in allohexaploid wheat.

Lloyd et al. (2014) suggested that duplicates of meiotic genes return to a single copy following allopolyploidization, more rapidly than the genome-wide average. Therefore, it has been assumed that stabilization upon allopolyploidization of wheat also involved rapid changes in the content and expression of meiotic homoeoalleles. Such a genetic diploidization process would facilitate the correct pairing and synapsis of homoeologs during meiosis. However, the results of Alabdullah et al. (2019) do not support neither extensive gene loss nor changes in homeolog expression of meiotic genes upon wheat allopolyploidization.

In addition, an analysis of the sequences of homoeologous group 1 chromosomes of allohexaploid wheat showed significant deviations from synteny, with many of the non-syntenic genes representing pseudogenes (Wicker et al. 2011). Likewise, the discovery of premature termination codons in 38% of genes expressed in 3A double ditelosomic lines in the genetic background of bread wheat was consistent with ongoing pseudogenization of the wheat genome (Akhunov et al. 2013). Ramírez-González et al. (2018), analyzing genome-wide gene expression patterns in allohexaploid wheat, found expression asymmetries along wheat chromosomes, with homoeoalleles showing the largest inter-tissue, inter-cultivar, and coding sequence variations, most often located in high-recombination distal ends of chromosomes. These transcriptionally dynamic genes potentially represent the first steps toward neo- or sub-functionalization of wheat homoeologs.

Epigenetic alterations of one of the homoeoalleles can be achieved either through epigenetic silencing of one of the homoeoalleles via cytosine methylation, or activation of silenced genes due to their demethylation. Such changes in gene expression were observed in newly formed wheat allopolyploids (Shaked et al. 2001; Kashkush et al. 2002). Epigenetic changes can also result from chromatin modifications or remodeling as well as from alteration in the activity of small RNA molecules. In fact, changes in microRNAs, such as miR168 which targets the *Argonaute1* gene, were shown to occur in newly-synthesized hexaploid wheat (Kenan-Eichler et al. 2011). In addition, a high proportion of microRNAs showed non-additive expression upon allopolyploidization, potentially reflecting differential expression of important target genes (Li et al. 2014). These observations may provide insights into dynamic small RNA–mediated homoeologous regulation mechanisms that possibly contribute to heterosis in nascent hexaploid wheat.

Table 12.2 Types and characteristics of genome changes in allopolyploid wheat

	Revolutionary changes (triggered by allopolyploidization)	Evolutionary changes (facilitated by allopolyploidy)
	Occur immediately after allopolyploidization	Occur during the life of the allopolyploid species
	Genetic and epigenetic changes	Mostly genetic changes
	Species specific	Population or biotype specific
	Lead to diploid-like meiotic behavior (cytological diploidization)	–
	Improve harmonic functioning of the divergent genomes	–
	Stabilize the nascent allopolyploid and facilitate its establishment as a new species in nature	–

New interactions between regulatory factors of the parents, i.e., between the transfactor from one species and the cis or transfactors of the other parental species, may account for the observed inter-genomic suppression in allohexaploid wheat (Galili and Feldman 1984). Genetic suppression or reduction of gene activity can also be caused by DNA elimination (Liu et al. 2008; Kashkush et al. 2002). Methylation and demethylation of retrotransposons, affecting their state of activity (Sabot et al. 2005), were also observed in wheat allopolyploids (Yaakov and Kashkush 2011a, b). Activation of retrotransposons, that constitute most of the DNA of allopolyploids of the wheat group and which are normally transcriptionally silent, may silence or activate neighboring genes (Kashkush et al. 2003). Following allopolyploidization of wheat, the steady-state level of expression of LTR retrotransposons was massively elevated (Kashkush et al. 2002, 2003). This activation may promote either read-in transcripts of the transposon itself, or read-out transcripts into flanking host sequences (Kashkush et al. 2002, 2003; Kashkush and Khasdan 2007). Indeed, in many cases, read-out transcripts were associated with altered expression of adjacent genes, knocking-down or knocking-out the gene product if the read-out transcript was in the antisense orientation relative to the orientation of the gene transcript (such as the *iojap-like* gene), or over-expressing the gene if the read-out transcript was in the sense orientation (such as the *puroindoline-b* gene) (Kashkush et al. 2003). A recent study tracking methylation changes around a LTR retrotransposon in the first four generations of a newly formed wheat allopolyploid, indicated that this read-out activity is restricted to the first generations of the nascent polyploid species (Kraitshtein et al. 2010).

Likewise, Yuan et al. (2020) studied genome-wide DNA methylation landscapes in extracted tetraploid wheat (genome BBAA), natural hexaploid wheat from which the tetraploid was extracted, resynthesized hexaploid wheat, natural tetraploid wheat, and the diploid donor of the D subgenome of hexaploid wheat. In the endosperm, these authors found that levels of DNA methylation, especially in CHG (H = A, T, or C) context, were dramatically decreased in the extracted tetraploid relative to its allohexaploid parent. Interestingly, those demethylated regions in extracted tetraploid were remethylated in the resynthesized hexaploid wheat after the addition of the D subgenome. In the extracted tetraploid, hypo-demethylated regions correlated with gene expression, and TEs, dispersed in genic regions of the subgenomes, were demethylated and activated, and thus, may regulate the expression of TE associated genes. The genes that became expressed and TEs that became active in the extracted tetraploid, turned out to be silenced again in the newly synthesized allohexaploid. These dynamic and reversible changes in chromatin and DNA methylation correlate with altered gene expression and TE activity when the D subgenome was added to tetraploid wheat.

Functional diversification of duplicated genes, i.e., differential or partitioned expression of homoeoalleles in different tissues and in different developmental stages, is also a form of genetic diploidization. Botley et al. (2006) reported that differential expression of homoeoalleles in different plant tissues is a common phenomenon in allohexaploid wheat. The activity of several silenced genes could be restored in aneuploid lines, suggesting that no mutation was involved but, rather, new cis–trans interactions or reversible epigenetic alterations took place. Mochida et al. (2006) also presented evidence for differential expression of homoeoalleles in hexaploid wheat and suggested that inactivation of homoeoalleles is a non-random effect.

Allopolyploid patterns of gene expression might be intermediate (between that of the two parents), dominant (similar to one of the parents) or over-dominant (greater than that of the parents). Overdominance can produce novel traits not found in the parents and can be caused by novel cis–trans interactions between regulatory elements of the different genomes that coincide in the same nucleus, as shown in yeast (Tirosh et al. 2009). Many studies on gene expression compared the expression level in the allopolyploid to those of its parents and or to the average of its parents, expressed as the mid-parental value. In hexaploid wheat, Pumphrey et al. (2009) found that approximately 16% of the 825 analyzed genes displayed non-additive expression in the first generation of synthetic allohexaploid wheat. Chague et al. (2010) analyzed 55,052 transcripts in two lines of synthetic allohexaploid wheat and found that 7% of the genes had non-additive expression, while Akhunova et al. (2010) found that about 19% of the studied synthetic allohexaploid wheat genes showed non-additive expression. Li et al. (2014) reported that non-additively expressed protein-coding genes are rare but relevant to growth vigor, and that a high proportion of protein-coding genes exhibit parental expression-level dominance. Similar studies by He et al. (2003) showed that the expression of a significant fraction of genes (7.7%) was altered in the synthetic allohexaploid wheat, *T. turgidum-Ae. tauschii*, and that *Ae. tauschii* genes were affected much more frequently than those of *T. turgidum*. Interestingly, silencing of the same genes was also found in natural *T. aestivum*, indicating the reversibility of the effect and that the regulation of gene expression is established immediately after allohexaploidization and maintained over generations (He et al. 2003; Chagué et al. 2010). In accord with these results, increased small interfering RNA density was observed for transposable element–associated D-subgenome homoeologs in the progeny of newly formed allohexaploid wheat, which may account for biased repression of D homoeologs (Li et al. 2014). On the other hand, several genes, that are silent in the parental species, became active in the newly formed allo-hexaploid (He et al. 2003). Similarly, cDNA-AFLP gels also revealed several cDNAs that were expressed only in the allopolyploids and not in the diploid progenitors (Shaked et al. 2001; Kashkush et al. 2002).

The genetic system of the *Triticum* and *Aegilops* allopolyploid species facilitates the accumulation of genetic and epigenetic variation, through mutations or gene silencing and through interspecific or intergeneric hybridizations that can lead to introgressions or to new species formation. Such variation is tolerated more readily in allopolyploid wheats than in diploid species of the wheat group (Mac key 1954; Sears 1972). This genetic potential may contribute to genetic variability and to creation of populations with archipelagoes of genotypes, thereby increasing their adaptability, fitness, competitiveness and capacity to colonize newly-opened ecological niches.

12.4   Subgenomic Asymmetry in the Control of Various Traits

Most duplicated genes in the allotetraploid *T. turgidum* and triplicated genes in the allohexaploid *T. aestivum* remain active, contributing either to a favorable effect of an extra dosage or to the buildup of positive inter-subgenomic interactions when genes or regulation factors on homoeologous chromosomes are divergent. However, in a small number of loci (about 10–15%), genes of only one subgenome are active, while the homoeoalleles on the other subgenome(s) are either eliminated or partially or fully suppressed by genetic or epigenetic means. For several traits, the retention of some homoeoalleles is not random, with one subgenome favored over the other(s), as observed in many cases [morphological, physiological, molecular and agronomical traits, rRNA genes, storage protein production, interaction with pathogens (reviewed by Feldman et al.

2012)]. The allopolyploids of *Triticum* and *Aegilops* were classified into three cytogenetic clusters, with each cluster containing allopolyploids that share a subgenome and differ in the second subgenome(s) (Zohary and Feldman 1962; Table 10.7). The allopolyploids morphologically resemble the diploid donors of the shared (pivotal) subgenomes, namely, the donors of the U, D, and A subgenomes, and differ in other traits, e.g., eco-geographical and tolerance to biotic and abiotic stresses, controlled by the differential genome(s) (Feldman et al. 2012). The contribution of the A and B subgenomes to various traits in wild allotetraploid wheat *T. turgidum* ssp. *dicoccoides* was studied by Peng et al. (2003a, b), who found that the A subgenome controls morphological traits, including inflorescence structure, grain shape, free caryopsis, glumes with keels, plant habitus, and growth habit (Table 12.3). This genome also controls the autogamy of allotetraploid wheat (assuming that the donor of the B subgenome is allogamous, i.e., closely related species to *Ae. speltoides*) and harbors many domestication genes, such as the genes for non-brittle spike on 3AS (Rong 1999; Nalam et al. 2006; Millet et al. 2014), and free-threshing genes on 5AL (Sears 1954). The B subgenome regulates ecological adaptation and tolerance to biotic and abiotic stresses (Peng et al. 2003a, b) and plays a leading role in population adaptation to environmental conditions (Fahima et al. 2006).

There is evidence for molecular manifestation of genomic asymmetry in the allopolyploid wheat sub-genomes. The level of genetic diversity differs between the two or three subgenomes of the allotetraploid and allohexaploid *Triticum* species, respectively. The B subgenome exhibits a higher marker polymorphism than the A subgenome in allohexaploid wheat (Chao et al. 1989; Liu and Tsunewaki 1991; Devos et al. 1992), and the wild and domesticated allotetraploid wheat (Liu and Tsunewaki 1991; Huang et al. 1999; Rong et al. 1999). Such differences were most pronounced for loci revealed by gDNA rather than by cDNA

Table 12.3 Genome asymmetry in the control of various traits in the wild allotetraploid wheat, *T. turgidum* ssp. *dicoccoides* (genome BBAA)[*]

Traits under control of	
Genome A	Genome B
Inflorescence morphology	Regulation of ecological adaptation
Free caryopsis	Double the number of disease-resistance genes
Glumes with keels	Contains more stress-related genes?
The shape of the edge of the glumes (beaked glumes)	Higher polymorphism of molecular markers
Hairs at the base of every spikelet	Higher polymorphism of HMW glutenin genes
Plant habitus	Larger amount of repetitive sequences
Growth habit	Activity on nucleolar organizers
Autogamy	Larger number of rRNA genes
Many domestication genes	

[*] Taken from Table 12.2 in Feldman et al. (2012)

probes (Huang et al. 1999; Rong et al. 1999). Similarly, higher polymorphism in the B as compared to A subgenome microsatellites, was seen (Röder et al. 1998). B-subgenome chromosomes are characterized by more C-banding than chromosomes of A and D subgenomes (Gill 1987), reflecting a higher quantity of constitutive repetitive DNA sequences. Similarly, more retrotransposons and variations within them have been observed in the B subgenome, when compared with the A and D subgenomes (K Kashkush, personal communication).

Shaked et al. (2001) reported that sequence elimination in the newly formed allotetraploids *Ae. longissima–Ae. umbellulata* and *Ae. sharonensis–T. monococcum*, mainly affect one of the parental genomes. Likewise, a much higher level of sequence elimination, occurring immediately after formation of Triticale, was observed in the rye subgenome as compared to the wheat subgenome (Ma and Gustafson 2008). Shaked et al. (2001) reported that cytosine methylation was also asymmetric; twice as many sequences were affected in *T. monococcum*, when compared with those of *Ae. sharonensis* in the nascent amphiploid *Ae. sharonensis–T. monococcum*. Wicker et al. (2011) identified five times more non-syntenic genes on chromosome arm 1BS of *T. aestivum* than syntenic genes on the homoeologous arms. They proposed that this accumulation of genic sequences is driven by TE activity, and that these findings indicate that homoeologous wheat chromosomes can exhibit different evolutionary dynamics.

In *T. aestivum*, the nucleolar organizers of the B subgenome suppress the nucleolar organizers of the A and D subgenomes (Crosby 1957; Crosby-Longwell and Svihla 1960; Darvey and Driscoll 1972; Flavell and O'Dell 1979). Similarly, the nucleolar organizers of the B subgenome suppress those of the A subgenome in allotetraploid *T. turgidum* (Lacadena et al. 1984; Frankel et al. 1987) and those of the R genome in allotetraploid and allohexaploid triticale, respectively (Darvey and Driscoll 1972; Lacadena et al. 1984; Cermeño et al. 1984a; Martini and Flavell 1985; Appels et al. 1986). Nucleolar dominance of one subgenome was observed in all allopolyploid species of *Aegilops* (Cermeño et al. 1984b). In these species, the U subgenome from *Ae. umbellulata* completely suppresses the NOR activity of the M subgenome of *Ae. geniculata*, the S subgenome of *Ae. peregrina*, the D subgenome of *Ae. juvenalis*, and the C subgenome of *Ae. triuncialis* and that of one pair of the nucleolar organizer chromosomes of the M subgenome of *Ae. columnaris*, *Ae. biuncialis*, *Ae. juvenalis*, and *Ae. recta*. The nucleolar activity of the D subgenome is completely suppressed by the U subgenome in *Ae. juvenalis*, and the C subgenome in *Ae. cylindrica*.

Nucleolus formation is considered evidence for rRNA gene expression and their lack thereof, indicates the absence of rRNA gene transcription (Flavell et al. 1986). Moreover,

the relative size of nucleoli within the same nucleus has been taken as a measure of the differential activity between one NOR and another (Flavell et al. 1986). Nucleolar dominance in the allopolyploid species of the wheat group is achieved either by elimination of rRNA-encoding genes, as is the case of rRNA genes on 5AS, or by suppression of their activity. Gustafson and Flavell (1996) and Houchins et al. (1997) noted a correlation between inactivation of rRNA-encoding genes and increased cytosine methylation at their CCGG sites. Further evidence suggesting that nucleolar suppression is triggered by cytosine methylation came from the fact that the suppression of the NORs of genome R was reversed in wheat x rye hybrids and triticale by treatment with the demethylating agent 5-aza-cytosine (Vieira et al. 1990; Neves et al. 1995; Amado et al. 1997).

Chromosomes 1A and 5D of the standard laboratory cultivar Chinese Spring of *T. aestivum* contain a very small proportion (10%) of the rRNA-encoding genes, while chromosomes 1B and 6B possess 30% and 60% of these genes, respectively (2700 and 5500 copies, respectively) (Mohan and Flavell 1974; Flavell and O'Dell 1976). As a result, chromosomes 1A and 5D are associated with very small nucleoli or none at all in Chinese Spring (Crosby 1957; Crosby-Longwell and Svihla 1960).

Newly synthesized allopolyploids exhibit genetic and epigenetic changes in their rRNA-encoding genes similar to those occurring in natural allopolyploids, indicating that these changes are generated during allopolyploid formation (Baum and Feldman 2010). Likewise, Shcherban et al. (2008) detected rapid elimination of the *Aegilops sharonensis* rRNA-encoding genes in the synthetic allopolyploid *Ae. sharonensis–Ae. umbellulata*, which stands in agreement with the pattern in the natural allotetraploids carrying similar genomic combination, i.e., *Ae. peregrina* and *Ae. kotschyi*.

Wheat 5S DNA also undergoes immediate changes in response to allopolyploidization, followed by the differential elimination of unit classes of 5S DNA (Baum and Feldman 2010). This elimination was reproducible, i.e., the same unit classes were eliminated in natural and synthetic allopolyploids carrying the same genomic combinations, indicating that no further elimination occurred in the unit classes of the 5S DNA during the life of the allopolyploids.

The high molecular weight (HMW) glutenin subunits are encoded by the *Glu-A1* and *Glu-B1* gene clusters in allotetraploid wheat, and by the *Glu-A1, Glu-B1, and Glu-D1* gene clusters in allohexaploid wheat, located on the long arm of homoeologous-group-1 chromosomes (Payne et al. 1982; Galili and Feldman 1983a, and reference therein). In allohexaploid wheat, each of these gene clusters is composed of two multi-allelic gene loci: *Glu-A1-1* and *Glu-A1-2* on chromosome 1A, *Glu-B1-1* and *Glu-B1-2* on chromosome 1B, and *Glu-D1-1* and *Glu-D1-2* on chromosome 1D. The products of *Glu-A1-1, Glu B1-1*, and *Glu-D1-1* comprise the slow-migrating subunits (x) while those of *Glu-A1-2*, *Glu-B1-2*, and *Glu-D1-2* comprise the fast-migrating subunits (y). The genetic control of HMW glutenin subunits is another example for subgenome asymmetry. Galili and Feldman (1983b) analyzed 109 different lines of allohexaploid wheat, representing a wide spectrum of genetic backgrounds, and found that 22 lines (20.2%) had no HMW glutenin subunits controlled by chromosome 1A, 44 lines (40.4%) had only one such band and 43 lines (39.4%) had two bands. Moreover, in all lines bearing one subunit controlled by 1A, only the fast-migrating subunit was absent, i.e., only *Glu-1A-1* was active, while *Glu-A1-2* coding for the fast-migrating band, which is generally active in diploid wheat, was inactive (Waines and Payne 1987).

Likewise, on studying the HMW glutenin subunits in 456 accessions of wild emmer, originating from 21 different populations in Israel, Levy et al. (1988a, b) found that in 82% of these accessions the fast-migrating subunit of the A subgenome was absent, and in 17% of the accessions, the slow-migrating subunit of this subgenome was also absent. Namely, only the genes of the B subgenome were active. In addition, the fast-migrating subunit of the A subgenome was absent in all of the 11 studied lines of the primitive domesticated allotetraploid wheat, *T. turgidum* subsp. *dicoccon*. *Glu-A1-1*, the gene determining the slow-migrating subunit, was also inactive in 16% of the accessions. Thus, in both allotetraploid and allohexaploid wheat, inactivation of HMW glutenin genes is non-random and occurs in the genes of the A subgenome (Galili and Feldman 1983a, b; Feldman et al.1987; Levy et al. 1988a, b, and reference therein). This tendency has also been found among HMW gliadin genes in hexaploid wheat (Galili and Feldman 1983a, b). The order of inactivation was also non-random, starting with the fast-migrating subunits and continuing with the slow-migrating ones.

Galili and Feldman (1984) showed that inactivation of endosperm protein genes is also brought about by inter-subgenomic suppression. Endosperm protein genes, located in the A or the B subgenomes, were repressed by gene(s) of the D subgenome, immediately following the formation of allohexaploid wheat, about 9000 years ago. When the D subgenome was removed, as in the extracted tetraploids (Kerber 1964), these genes became active, indicating that they retained their potential for activity throughout the 9000 years. Similarly, Kerber and Green (1980) described the inter-genomic suppression of a rust resistance gene, located in the D-subgenome, by gene(s) of the A or B subgenomes.

Subgenome asymmetry also occurs in the control of various agronomic traits and of disease and pest resistance in domesticated allopolyploid wheats (Table 12.4). The B and D subgenomes control the most important genes associated with reduced plant height (*Rht*) and gibberelic acid insensitivity (*Ga*), yielding dwarf and semi-dwarf wheat, the main types of modern wheat cultivars. Dwarf and semi-dwarf wheat varieties are characterized by an improved harvest index and, consequently, are high-yielding varieties. These varieties have replaced the traditional tall, low-yield varieties in many parts of the world during the 'green revolution' and thus, have increased global wheat production. The B and D subgenomes control grain protein content (Law et al. 1978; Joppa and Cantrell 1990) and grain hardness (Morris et al. 1999; Chantret et al. 2005). These subgenomes also control wax production (Tsunewaki and Ebana 1999), an important trait that affects drought tolerance. The B and D genomes are also responsible for tolerance to abiotic stresses, with the B subgenome carrying genes associated with boron tolerance (Paull et al. 1991), low cadmium uptake (Penner et al. 1995), and tolerance to iron deficiency (Maystrenko 1992), and the D subgenome containing gene(s) conferring aluminum tolerance (Riede and Anderson 1996) and response to salinity (Dubcovsky et al. 1996). Most genes for herbicide resistance are located in the B subgenome (Snape et al. 1987), and those responsive to photoperiod and most of those responsive to vernalization are located on the B and D subgenomes. The A subgenome controls plant and spike morphology and the main traits of the domestication syndrome, e.g., non-brittle rachis (Nalam et al. 2006) and free threshing (Sears 1954).

The B subgenome harbors double the number of disease-resistance genes and resistance-gene analogue (RGA) loci than the A and D subgenomes (Peng et al. 2003b; Fahima et al. 2006). Screening the GrainGenes website (http://wheat.pw.usda.gov/) found that among 184 mapped wheat disease resistance genes, 88 (48%) are located in the B subgenome. Moreover, most genes conferring resistance to stem rust, stripe rust, and leaf rust, the most common wheat diseases, that cause significant global yield loss each year, are located in the B subgenome (Table 12.4).

In allopolyploid species of *Triticum* and *Aegilops*, as discussed above, different gene types show a differential propensity for homoeologous change or retention. Genes encoding functional proteins (enzymes) constitute one category of genes that shows a high degree of retention of homoeoalleles (Mitra and Bhatia 1971; Hart 1983a, b, 1987). Such retention enables inter-subgenomic interactions at both the transcriptional level and between gene products, giving rise to functional 'hybrid' multimeric enzymes consisting of subunits encoded by different subgenomes. These new heteromeric proteins may have new and desirable properties. Similarly, protein complexes, such as gluten, may also be 'hybrid'. Moreover, the retention of genes corresponding to *trans*-acting factors, such as transcription factors, suppressors, and microRNAs, may enable the generation of novel *trans* interactions that may lead to new expression patterns absent in the diploid parents, as seen in yeast (Tirosh et al. 2009).

Table 12.4 Genome asymmetry in the control of agronomic traits in durum (genome BBAA) and bread wheat (genome BBAADD)[*]

Traits	Traits under control of		
	Genome A	Genome B	Genome D
Elongated glumes	EgP1 on 7AL	EgP2 on 7BL (?)	
Branched spikes	Bh on 2AS		
Non-brittle rachis	Br-A1 on 3AS Br-A2 on 2A	Br-B1 on 3BS	Br-D1 on 3DS Br-D2 on 3DL
Free threshing	Q on 5L	q on 5BL	q on 5DL
Non-tenacious glume (lax glume)	Tg-A1 on 2AS	Tg-B1 on 2BS	Tg-D1 on 2DS
Reduce plant height	Rht7 on 2A Rht12 on 5AL	Rht-B1 on 4BS Rh4 on 2B1 Rht5 on 3BS Rht9 on 7BS Rht13 on7BS	Rht-D1 on 4DS Rht8 on 2DL
Grain protein content		Gpc-B1 on 6BS	Pro1 on 5DL Pro2 on 5DS
Grain Hardness			Ha on 5DS
Puroindolines and grain softness protein			Pin-D1 on 5DS
Gibberellic acid response		Gai1, Gai3 on 4BS	Gai2 on 4DS
Waxiness		W1 on 2BS	
Epistatic inhibitors of waxiness		W1[i] on 2BS W3i on 1BL	W2[i] on 2DS
Male sterility	Ms3 on 5AS Ms5 on 3A	Ms1 on 4BS	Ms2 on 4DS Ms4 on 4DS
Paring homoeologous		Ph1 on 5BL	Ph2 on 3DS
Hybrid necrosis		Net1 on 5BL Net2 on 2B	
Hybrid chlorosis	Ch1 on 2A		Ch2 on 3DL
Aluminum tolerance			Alt2 on 4DL
Boron tolerance		Bo1 on 7BL	
Low cadmium uptake		Cdu1 on 5BL	
Iron deficiency		Fe2 on 7BS	Fe1 on 7DL
Difenzoquat insensitivity		Dfg1 on 2BL	
Chlortoluron insensitivity		Su1 on 6BS	
Imidazolinone resistance	Imi3 on 6AL	Imi2 on 6BL	Imi1 on 6DL
Response to photoperiod		Ppd-B1 on 2BS	Ppd-D1 on 2DS
Response to vernalization	Vrn-A1 on 5AL	Vrn-B1 on 5BL Vrn-B3 on 7BS	Vrn-D1 on 5DL Vern-D4 on 5DL Vern-D5 on 5DL
Response to salinity			Kna1 on 4DL
Frost resistance	Fr1 on 5AL		Fr2 on 5DL

[*] Taken from Feldman et al. (2012); Table 12.3

For other categories of genes, lack of retention of parental genes or expression patterns, is frequent and non-random. This includes the genes that encode for ribosomal RNA, structural proteins, such as histones and subunits of tubulins, and storage proteins, such as subunits of glutenins and gliadins. In these cases, expression of all homoeoalleles may be redundant, resulting in over-production and even deleterious dose effects. In addition, activity of all homoeoalleles may produce intermediate phenotypes in several traits that decrease the viability of the plants (e.g., hybrid incompatibility genes). Hence, for some traits, control by genes from only one genome (genome asymmetry) may be of higher adaptive value than additive expression, by preventing a genomic clash or avoiding deleterious dosage effects.

Genetic diploidization is not a random process, distinctly affecting specific gene categories and their corresponding traits and forming a clear-cut division of tasks between the constituent subgenomes of allopolyploid wheats. The A-subgenome preferentially controls morphological traits, while the B-subgenome in allotetraploid wheat and the B and D subgenomes in allohexaploid wheat preferentially control the reaction to biotic and abiotic factors. Genetic diploidization may occur during or immediately after allopolyploidization (revolutionary changes), e.g., in rRNA-encoding genes, or through the life history of the species (evolutionary changes), for example, in HMW glutenin genes.

Genome asymmetry may be brought about by either transcriptional dominance of one of the parental genomes (Wang et al. 2006; Flagel and Wendel 2009, 2010; Rapp et al. 2009) or inter-genomic suppression of gene activity (Galili and Feldman 1984), due to incompatibility of regulatory elements (He et al. 2003; Tirosh et al. 2009), chromatin modification (Wang et al. 2006) or suppression of genes adjacent to transposable elements (Kashkush et al. 2003). Differential elimination or inactivation of coding sequences from one of the subgenomes in allotetraploids and from two of the subgenomes in allohexaploids also contributes to the asymmetrical control of the constituent subgenomes (Tate et al. 2006; Feldman and Levy 2009; Buggs et al. 2009, 2010a, b; Koh et al. 2010). Some major transcriptional suppressors, or small non-coding RNAs, such as microRNAs (Ha et al. 2009; Kenan-Eichler et al. 2011), may also have genome-wide effects on asymmetry through the suppression of several targets that, in turn, can affect a cascade of genes, thus leading to asymmetry.

The ability of one subgenome to suppress the activity of genes of another subgenome and thus, fully control a set of traits in allopolyploids, may prevent conflicting gene expression that could potentially lead to defective organ shapes. This protective mechanism ensures the development of viable plants. Diploid species that lack this adaptive ability might fail to produce viable allopolyploids. There are two diploid wheat species, 10 diploid *Aegilops* species, and one *Amblyopyrum* species (Eig 1929a, b; van Slageren 1994), most of which have geographical contact with one another (Kimber and Feldman 1987; van Slageren 1994). Many more allopolyploid species, apart from the currently existing ones, may have been generated over the 1–4 million years of the existence of the diploid species (Middleton et al. 2014; Gornicki et al. 2014; Marcussen et al. 2014). Allopolyploids involving the AD, AC, AM, AN, AU, AT, UD, UT, DS, and DT genomic combinations can be produced under artificial conditions but have not been found in nature. It is speculated that inter-genomic incompatibilities

leading to reduced fitness, due to failure to produce appropriate genomic asymmetry, hampers their establishment in nature. Rapid progress of structural and functional Triticeae genomics will provide further insights into the mechanisms and functional importance of genomic asymmetry in wheat allopolyploids and other allopolyploids of this group.

To sum up, cytological and genetic diploidization in allopolyploid *Triticum* and *Aegilops* species led to the construction of two contrasting and highly important genetic systems that contribute to their evolutionary success: (i) retention of expression of all homoeoalleles of those duplicated or triplicated gene loci whose extra gene dosage has a positive effect by itself or may facilitate the build-up of positive inter-subgenomic interactions between divergent regulation factors, i.e., build up and maintenance of lasting inter-subgenomic favorable genetic combinations (inter-genomic heterosis), and (ii) elimination or suppression of genes from one subgenome in allotetraploids and from two subgenomes in allohexaploids, whose extra dosage or new inter-genomic interactions are deleterious, thus bringing about genome asymmetry for various traits. The latter process may be tissue-specific (Buggs et al. 2011).

Inter-genomic pairing would have led to disruption of the linkage of the homoeoalleles that contribute to positive inter-subgenomic interactions and would have led to segregation of genes that participate in the control of certain traits by a single subgenome. Inter-subgenomic recombination would therefore result in many intermediate phenotypes that may negatively affect the functionality, adaptability and stability of the allopolyploids.

12.5 Evolution During the Life of the Allopolyploids

Allopolyploid species also undergo structural genomic changes during the lifetime of the taxon (evolutionary changes; Table 12.2), generating a new variation that scarcely exists in the diploid species. Allopolyploids harboring two or more different subgenomes within each nucleus, may facilitate inter-genomic horizontal transfer of chromosomal segments, transposable elements or genes. Inter-subgenomic invasion of chromatin segments from the B subgenome into the A subgenome was demonstrated by FISH in *T. turgidum* ssp. *dicoccoides* (Belyayev et al. 2000). Cytogenetic studies have shown that several inter-subgenomic translocations occur in the allopolyploids of the wheat group (Maestra and Naranjo 1999, and reference therein). Moreover, in contrast to most diploids, which are genetically isolated from each other and have undergone divergent evolution, allopolyploids in the wheat group exhibit convergent evolution because they contain genetic material from two or more different diploid genomes and can exchange genes with each other via hybridization and introgression, resulting in the production of new subgenomic combinations (Zohary and Feldman 1962).

The presence of duplicated or triplicated genetic material in allotetraploids and allohexaploids, respectively, has relaxed constraints on the function of the multiple genes, enabling, in the long run, continued genetic diploidization, achieved by silencing of one of the duplicated or triplicated genes or divergence of one homoeologous locus to a new function. Thus, the accumulation of genetic variation through mutations is more readily tolerated in allopolyploid than in diploid species (Mac Key 1954a; Sears 1972; Dubcovsky and Dvorak 2007).

The evolutionary changes might also occur in an accelerated manner, thanks to the buffering of mutations in the polyploid background that leads to rapid neo- or sub-functionalization of genes and to a further process of diploidization and of divergence from the diploid progenitor genomes. Akhunov et al. (2013) uncovered a high level of alternative splicing pattern divergence between the duplicated homoeologous copies of genes in common wheat. Their observations are consistent with the accelerated accumulation of alternative splicing isoforms, nonsynonymous mutations, and gene structure rearrangements in the wheat lineage, likely due to genetic redundancy created by allopolyploidization (Akhunov et al. 2013). While these processes mostly contribute to the degeneration of a duplicated genome and its diploidization, they have the potential to facilitate new functional variations, which, upon selection in the evolutionary lineage, may play an important role in the development of novel traits (Akhunov et al. 2013).

According to Stebbins (1950), newly formed allopolyploids are often characterized by limited genetic variation, a phenomenon he referred to as the "polyploidy diversity bottleneck." This bottleneck arises because only a few diploid genotypes were involved in the allopolyploid formation events, because the newly formed allopolyploid is immediately reproductively isolated from its two parental species and because time was not sufficient for the accumulation of mutations. However, despite this diversity bottleneck and despite the fact that all *Aegilops* and *Triticum* allotetraploids were formed much later than their ancestral diploids, e.g., during 1.3 MYA (Middleton et al. 2014; Gornicki et al. 2014; Marcussen et al. 2014), and allohexaploids were formed even more recently, most of the allopolyploids of the wheat group display greater genetic variation than their diploid progenitors (Zohary and Feldman 1962). It is most likely that the allopolyploid species were recurrently formed from different genotypes of their parental diploids, thus increasing their intra-specific variation. As a matter of fact, at least in *Ae. cylindrica*, there is evidence of multiple allopolyploidization events since some forms contain the D cytoplasm while others contain the C cytoplasm

(Caldwell et al. 2004; Gandhi et al. 2005). Moreover, allopolyploidy enables genome plasticity that in turn allows for accelerated evolution to take place, as observed in allopolyploid wheats (Dubcovsky and Dvorak 2007) and as was recently shown in experimental evolution studies in yeast (Selmecki et al. 2015).

One way in which the gene pool of the allotetraploid species was greatly enlarged and their evolutionary potentiality correspondingly increased, was by hybridization and introgression between related allotetraploids (Zohary and Feldman 1962). On the basis of plant habitus, spike morphology, and cytogenetic data, these authors classified the allopolyploid species of *Aegilops* and *Triticum* into three natural clusters (Table 12.1). Genome analysis of the allopolyploids within each cluster showed that they share one unaltered subgenome (the pivotal subgenome) and a subgenome(s) that is/are modified [the differential subgenome(s)]. In laboratory hybridization studies of allopolyploids, it was found that the homologous chromosomes of the shared subgenome paired ensuring some seed fertility following pollination by one of the parents (Feldman 1965b), while, the F_1 hybrids having 9–10 bivalents (7 bivalents involving chromosomes of the shared subgenome and 2–3 between chromosomes of the differential subgenomes) at meiosis, show that the chromosomes of the differential subgenomes, brought together from different parents, may pair to some extent and exchange genes (Feldman 1965c). Consequently, the differential subgenomes of these allopolyploids are recombinant subgenomes containing chromosomal segments that originated from two or more diploid genomes. Such genomic constitution reveals different evolutionary rates for each of the two or three subgenomes of every allopolyploid.

Thus, all seven alltetraploids and the one allohexaploid of the U-subgenome cluster share a genome homologous to that of diploid *Aegilops umbellulata* (Table 12.1; Kihara 1954), a weedy annual in the center of the genus distribution area, which possesses an unusually efficient method of seed dispersal (umbrella-type) in the form of a small spike with large number of awns on its glumes and lemmas. Seven distinct allotetraploid species contain a subgenome derived from *Ae. umbellulata*, and a second subgenome that derived from ancestral diploids which have been variously modified during evolution at the allotetraploid level (the modified subgenomes). These allotetraploids are aggressive weeds, the most common of which have become widespread in the Mediterranean basin.

Similarly, all the three allotetraploids and three allohexaploids of the D-subgenome cluster share a subgenome homologous to that of diploid *Ae. tauschii* (Table 12.1; Kihara 1954; Kihara et al. 1959), that grows in the eastern region of the genus distribution area (with the exception of *Ae. ventricosa*, which grows in western Mediterranean regions), and has a barrel-type dispersal unit. The two allotetraploids and two allohexaploids of the A-subgenome cluster, including all the wild and domesticated forms, sharing a subgenome closely related to that of diploid *Triticum urartu* (Table 12.1; Dvorak 1976; Chapman et al. 1976; Li et al. 2022), which grows in the center of the genus distribution area.

The allopolyploids of each cluster exhibit the same basic morphology (stature, leaf shape, and spike and spikelet morphology) seed dispersal unit structure of the diploid donor of the shared genome. They differ in features of the differential subgenome(s) that are primarily responsible for the eco-geographical adaptation of the various allopolyploid species in the cluster. Thus, inter-specific hybridization has played a decisive role in the production of a wide range of genetic variation in these allopolyploid species and probably significantly contributed to their evolutionary success.

The wild allotetraploid forms of the A-subgenome cluster. *T. turgidum* ssp. *dicoccoides* and *T. timopheevii* ssp. *armeniacum,* differ from the other species of the two subgenome clusters, by exhibiting relatively low morphological variation and distribution in a comparatively small area. One possible explanation for these differences is the fact that ssp. *dicoccoides,* formed 700,000–800,000 years ago (Gornicki et al. 2014; Marcussen et al. 2014), was, for many years, the only allotetraploid taxon in the A-subgenome cluster. The second wild form, ssp. *armeniacum*, was formed 300,000–400,000 years after ssp. *dicoccoides* (Table 10.7; Gornicki et al. 2014). Thus, sporadic hybridization and introgression between the two wild taxa of the A-subgenome cluster could only have occurred after the formation of ssp. *armeniacum,* and obviously, also after the migration of wild emmer to the northern part of the Fertile Crescent. Indeed, evidence for introgression between these two wild allotetraploid wheats, that currently grow in mixed stands in many locations of the northern part of the Fertile Crescent, were presented by several researchers. Genotypes of wild *T. turgidum* and wild *T. timopheevii* carrying recombinant B or G subgenomes were found by Sachs (1953), Wagenaar (1961, 1966), Rao and Smith (1968), Rawal and Harlan (1975), and Tanaka and Kawahara (1976). Cytogenetically, the Turkish–Iraqi race of wild *T. turgidum* shows a range of chromosome pairing capabilities, from complete affinity to *T. turgidum* to high affinity to *T. timopheevii* (Sachs 1953; Rao and Smith 1968; Rawal and Harlan 1975; Tanaka and Kawahara 1976). Likewise, the study of Gornicki et al. (2014) provides molecular evidence that evolution of these allopolyploid wheats was punctuated by chloroplast introgression. They reported that some of the chloroplast haplotypes of the Turkish–Iraqi race of *T. turgidum* are similar to haplotypes of *T. timopheevii*, suggesting an introgression from the latter (Gornicki et al. 2014). On the other hand, an accession of *T. turgidum* ssp. *dicoccoides*, for example, which shows high chromosome pairing with both *T. turgidum* and *T. timopheevii* (Rawal and Harlan 1975), carries the *T. timopheevii* chloroplast haplotype as a result of a cross between wild emmer and wild *timopheevii*. Conversely, one wild *timopheevii* accession carries the wild emmer chloroplast haplotype. Consequently, the northern ssp. *dicoccoides* presumably differs from the southern one by chromosomal segments of the G subgenome, whereas the B subgenome of the southern form may have introgressed with diploid *Aegilops* species of subsection Emarginata, mainly, *Ae. searsii* and *Ae. longissima*, and by the allotetraploid *Ae. peregrina*, which are common in the southern Levant. These inter-lineage and inter-generic introgressions are consistent with both allopolyploid wheat lineages sharing a weak *Ph1* locus (Ozkan and Feldman 2001).

In sharp contrast to the rarity of inter-specific hybridization at the diploid level, hybridization between allotetraploid species, particularly between those sharing a common subgenome, is common (Zohary and Feldman 1962; Feldman 1965a). The shared subgenome both increases the compatibility between the parents and the ease with which viable, fertile derivatives can be obtained by introgression (Feldman 1965b). In addition, the diploid species that contributes the common subgenome may possess some particular combination of adaptive characteristics which it transmits to all of its hybrids. Such allotetraploid species tend to grow in mixed stands, and many F_1 hybrids as well as backcrossed progeny have been repeatedly found in many localities in Israel, Turkey and Greece (Zohary and Feldman 1962; Feldman 1965a). Progeny raised from such hybrids segregate widely and recover almost complete fertility within one or two generations. Because of self-pollination, which is predominant in these species, these fertile introgressed genotypes become genetically fixed with relative ease and, if they are of adaptive value, they may enlarge the ecological amplitude and increase the gene pool of the concerned species. Of note, self-fertilization, while predominant in these species, is never complete; therefore, the introgressed genotypes, if inter-fertile with other individuals of the species concerned, can be regarded as part of its gene pool (Stebbins 1971). Yet, because of cytoplasmic divergence that may cause male sterility, introgression goes in the direction of the original female species. The occurrence of populations comprising several allotetraploid species in numerous habitats, increases the chance of likelihood of introgression between these species.

Additional evidence for the existence of introgressed genomes in allotetraploid *Aegilops* was obtained from C-banding analysis (Badaeva et al. 2004). An introgression of a DNA sequence from allohexaploid bread wheat to the allotetraploid *Aegilops* species, *Ae. peregrina,* was recently described (Weissmann et al. 2005). Hence, hybridization between allotetraploids, particularly between those sharing one common subgenome, and, to a lesser extent, between other allotetraploids, facilitates the rapid buildup of genetic variability at the tetraploid level. The differential subgenomes that are genetically and geographically isolated from one another at the diploid level, where emphasis is on divergence and specialization, are brought together and allowed to recombine at the tetraploid level. The ability to exchange genetic material through spontaneous interspecific hybridization, further promotes the convergent evolution of these species. This reticulate pattern of evolution tends to produce a range of morphological intermediates between allotetraploids and blurs the morphological boundaries between them, rendering it difficult to define the species border. Natural hybrids also occur between wild species and domesticated allopolyploid wheats. Hybrids have been recurrently recorded in *T. turgidum*, e.g., between subsp. *durum* and its progenitor, wild emmer (subsp. *dicoccoides,* in Israel (Percival 1921; Jakubziner 1932; Huang et al. 1999; Feldman 2001; Dvorak et al. 2006b; Luo et al. 2007), between wild emmer and *Aegilops* species (Cook 1913; Percival 1921), and between common wheat and wild emmer (Zohary and Brick 1962). On the basis of genome sequence analysis, recent studies have confirmed that a high rate of introgression took place from wild emmer wheat into the background of domesticated emmer wheat or of bread wheat, leading to a higher variability in these genomes (He et al. 2019 and Zhou et al. 2020). For example, the variation in domesticated-emmer genome was estimated to include ∼ 73% that of wild emmer (Zhou et al. 2020; Sharma et al. 2021; Keilwagen et al. 2022).

Substantial evidence of spontaneous hybridization between allotetraploids and diploid species in mixed natural populations of tetraploids and diploids was also obtained (Vardi and Zohary 1967; Vardi 1973; Zohary and Feldman, unpublished). The occurrence of hybrid derivatives in such populations, particularly as a result of backcrossing to the allotetraploid parents, implies the possibility of gene flow from the diploid to the tetraploid level.

The diploid species underwent divergent evolution and consequently, have diverse genomes that are, more or less, genetically isolated from one another. Most exhibit relatively limited morphological, cytological, and molecular variation, specialization in their spike and dispersal unit structure, occupation of few well-defined ecological habitats, and distribution throughout relatively small geographical areas (Zhukovsky 1928; Eig 1929a; Zohary and Feldman 1962; Kimber and Feldman 1987; Van Slageren 1994).

In contrast, the allopolyploids, comprising two or three diverse subgenomes in one nucleus, underwent genomic convergence. Nevertheless, most exhibit wider morphological, cytological, and molecular variation than their diploid parents, occupy a greater diversity of ecological habitats, and are distributed over larger geographical area than their

diploid progenitors (Zhukovsky 1928; Eig 1929a; Zohary and Feldman 1962; Kimber and Feldman 1987; Van Slageren 1994). The distribution areas of most of the tetraploids overlap, completely or partly, and extend beyond those of their two diploid parents. They grow well in a very wide array of edaphic and climatic conditions and so do not show the marked ecological specificity of the diploids. Their weedy nature is reflected in their ability to rapidly and efficiently colonize a variety of newly disturbed and secondary habitats. Undoubtedly, the expansion of agriculture and the opening up of many new habitats played a key role in the massive distribution of these allopolyploid species throughout the range of the group (Zohary and Feldman 1962; Kimber and Feldman 1987).

Therefore, students of wheat evolution have been fascinated by this paradox of "polyploidy diversity bottleneck" on the one hand and great genetic diversity, on the other, and have dedicated their research efforts to unraveling the processes and mechanisms that contributed to the rapid buildup of genetic diversity in the allopolyploids and to their great evolutionary success in term of proliferation and adaptation to new habitats, including under domestication.

In contrast to the wide distribution of the allotetraploid species, the distribution area of the natural allohexaploid species is, in all cases, smaller than that of their tetraploid and diploid parents. Also, their ecological amplitudes are much more restricted than those of the related tetraploids and even of the diploid parents. They grow in a smaller range of habitats. The morphological variation of the allohexaploids is also relatively limited. All these indicate a relatively recent origin of the allohexaploid species. *T. aestivum* ssp. *aestivum,* which was formed under cultivation, exhibits tremendous variation, due both to spontaneous hybridization with wild relatives and to modern breeding, and is currently grown in large parts of the world, and is exceptional in this respect.

References

Ahmad F, Comeau A (1991) A new intergeneric hybrid between *Triticum aestivum* L. and *Agropyron fragile* (Roth) Candargy: variation in *A. fragile* for suppression of the wheat *Ph*-locus activity. Plant Breeding 106:275–283

Akhunov ED, Sehgal S, Liang H, Wang S, Akhunova AR, Kaur G, Li W, Forrest KL, See D, Šimková H, Ma Y, Hayden MJ, Luo M, Faris JD, Dolezel J, Gill BS (2013) Comparative analysis of syntenic genes in grass genomes reveals accelerated rates of gene structure and coding sequence evolution in polyploid wheat. Plant Physiol 161:252–265

Akhunova AR, Matniyazov RT, Liang H, Akhunov ED (2010) Homoeolog specific transcriptional bias in allopolyploid wheat. BMC Genomics 11:505

Alabdullah AK, Borrill P, Martin AC, Ramirez-Gonzalez RH, Hassani-Pak K, Uauy C, Shaw P, Moore G (2019) A co-expression network in hexaploid wheat reveals mostly balanced expression and lack of significant gene loss of homoeologous meiotic genes upon polyploidization. Front Plant Sci 10:1325

Al-Kaff N, Knight E, Bertin I, Foote T, Hart N, Griffiths S, Moore G (2008) Detailed dissection of the chromosomal region containing the *Ph1* locus in wheat *Triticum aestivum*: with deletion mutants and expression profiling. Ann Bot 101:863–872

Amado L, Abrabches R, Neves N, Viegas W (1997) Development-dependent inheritance of 5-azacytidine-induced epimutations in triticale: analysis of rDNA expression patterns. Chromosome Res 5:445–450

Appels R, Moran LB, Gustafson JP (1986) The structure of DNA from the rye (*Secale cereale*) NOR R1 locus and its behaviour in wheat backgrounds. Can J Genet Cytol 28:673–685

Aragón-Alcaide L, Reader S, Miller T, Moore G (1997a) Centromeric behaviour in wheat with high and low homoeologous chromosomal pairing. Chromosoma 106:327–333

Aragón-Alcaide L, Reader S, Beven A, Shaw P, Miller T, Moore G (1997b) Association of homologous chromosomes during floral development. Curr Biol 7:905–908

Attia T, Ekingen H, Robbelen G (1979) origin of 3D-suppressor of homoeologous pairing in hexaploid wheat. Z. Pflanzenzücht 83:131–126

Attia T, Ekingen H, Röbbelen G (1977) Preexistence in diploid *Aegilops squarrosa* of regulators for chromosome pairing in the alloploid wheat. In: Sanchez-Monge E, Garcia-Olmedo E (eds) proceedings of the 8th Eucarpia congress on INTERSPECIFIC hybridization in plant breeding, pp 145–149

Avivi L (1976) The effect of gene controlling different degrees of homoeologous pairing on quadrivalent frequency in induced autotetraploid lines of *Triticum longissimum*. Can J Genet Cytol 18:357–364

Avivi L, Feldman M (1980) Arrangement of chromosomes in the interphase nucleus of plants. Hum Genet 55:281–295

Avivi L, Feldman M, Bushuk W (1969) The mechanism of somatic association in common wheat, *Triticum aestivum* L. I. Suppression of somatic association by colchicine. Genetics 62:745–752

Avivi L, Feldman M, Bushuk W (1970) The mechanism of somatic association in common wheat, *Triticum aestivum* L. II. differential affinity for colchicine of spindle microtubules of plants having different doses of the somatic association suppressor. Genetics 65:585–592

Avivi L, Efron Y, Feldman M (1972) Effect of somatic chromosomal association on the zymogram of alcohol dehydrogenase (ADH) in common wheat. Genetics 71:s2-3

Avivi L, Feldman M, Brown M (1982a) An ordered arrangement of chromosomes in the somatic nucleus of common wheat, *Triticum aestivum* L. I. Spatial relationships between chromosomes of the same genome. Chromosoma 86:1–16

Avivi L, Feldman M, Brown M (1982b) An ordered arrangement of chromosomes in the somatic nucleus of common wheat, *Triticum aestivum* L. II. Spatial relationships between chromosomes of different genomes. Chromosoma 86:17–26

Avivi L, Feldman M (1973a) Mechanism of non-random chromosome placement in common wheat. In: Sears ER, Sears LMS (eds) Proceedings of the 4th international wheat genetics symptoms, Missouri Agr Exp Sta, Columbia, Missouri, pp 627–633

Avivi L, Feldman M (1973b) The mechanism of somatic association in common wheat, *Triticum aestivum* L. IV. Further evidence for modification of spindle tubulin through the somatic-association genes as measured by vinblastine binding. Genetics 73:379–385

Azpiazu A, Lacadena JR (1970) Introduction of alien variation into wheat by gene recombination. Genet Iberica 22:1–7

Badaeva ED, Badaev NS, Gill BS, Filatenko AA (1994) Intraspecific karyotype divergence in *Triticum Araraticum* (Poaceae). Pl Sci Evol 192:117–145

Badaeva ED, Amosova AV, Samatadze TE, Zoshchuk SA, Shostak NG, Chikida NN, Zelenin AV, Raupp WJ, Friebe B, Gill BS (2004) Genome differentiation in *Aegilops*. 4. Evolution of the U-genome cluster. Plant Syst Evol 246:45–76

Bariah I, Keidar-Friedman D, Kashkush K (2020) Where the wild things are: transposable elements as drivers of structural and functional variations in the wheat genome. Front Plant Sci 18:585515

Baum BR, Feldman M (2010) Elimination of 5S DNA unit classes in newly formed allopolyploids of the genera Aegilops and Triticum. Genome 53:430–438

Belyayev A, Raskina O, Korol A, Nevo E (2000) Coevolution of A and B genomes in allotetraploid *Triticum dicoccoides*. Genome 43:1021–1026

Benavente E, Orellana J, Fernández-Calvin B (1998) Comparative analysis of the meiotic effects of wheat *ph1b* and *ph2b* mutations in wheat x rye hybrids. Theor Appl Gent 96:1200–1204

Bennett ST, Bennett MD (1992) Spatial separation of ancestral genomes in the wild grass Milium montianum Parl. Ann Bot 70:111–118

Bennett MD, Dover GA, Riley R (1974) Meiotic duration in wheat genotypes with or without homoeologous meiotic chromosome pairing. Proc R Soc London B 187:191–207

Bento M, Gustafson JP, Viegas W, Silva M (2011) Size matters in Triticeae polyploids: larger genomes have higher remodeling. Genome 54:175–183

Bernhardt N, Brassac J, Dong X, Willing E-M, Poskar CH, Kilian B, Blattner FR (2020) Genome-wide sequence information reveals recurrent hybridization among diploid wheat wild relatives. Plant J 102:493–506

Bhat P, Honson D, Guttman M (2021) Nuclear compartmentalization as a mechanism of quantitative control of gene expression. Nat Rev Mol Cell Biol 22:653–671

Bhullar R, Nagarajan R, Bennypaul H, Sidhu GK, Sidhu G, Rustgi S, von Wettstein D, Gill KS (2014) Silencing of a metaphase I-specific gene results in a phenotype similar to that of the pairing homoeologous 1 (*Ph1*) gene mutations. Proc Natl Acad Sci, USA 111:14187–14192

Borisy GG, Taylor EW (1967) The mechanism of action of colchicine. Binding of colchicine-^{3}H to cellular protein. J Cell Biol 34:525–533

Bottley A, Xia GM, Koebner RMD (2006) Homoeologous gene silencing in hexaploid wheat. Plant J 47:897–906

Boyko EV, Badaev NS, Maximov NG, Zelenin AV (1984) Does DNA content change in the course of triticale breeding. Cereal Res Commun 12:99–100

Boyko EV, Badaev NS, Maximov NG, Zelenin AV (1988) Regularities of genome formation and organization in cereals. I. DNA quantitative changes in the process of allopolyploidization. Genetika 24:89–97

Brown WV, Stack SM (1968) Somatic pairing as a regular preliminary to meiosis. Bull Torrey Bot Club 95:369–378

Buggs R, Doust A, Tate J et al (2009) Gene loss and silencing in Tragopogon miscellus (*Asteraceae*): comparison of natural and synthetic allotetraploids. Heredity 103:73–81

Buggs RJA, Chamala S, Wu W, Gao L, May GLD et al (2010a) Characterization of duplicate gene evolution in the recent natural allopolyploid *Tragopogon miscellus* by next-generation sequencing and Sequenom iPLEX MassARRAY genotyping. Mol Ecol 19:132–146

Buggs RJA, Elliott NM, Zhang L, Koh J et al (2010b) Tissue-specific silencing of homoeologs in natural populations of the recent allopolyploid *Tragopogon mirus*. New Phytol 186:175–183

Buggs RJA, Zhang L, Miles N, Tate JA, Gao L et al (2011) Transcriptomic shock generates evolutionary novelty in a newly formed, natural allopolyploid plant. Curr Biol 21:551–556

Caldwell KS, Dvorak J, Lagudah ES, Akhunov E, Luo M-C, Wolters P, Powell W (2004) Sequence polymorphism in polyploid wheat and their D-genome diploid ancestor. Genetics 167:941–947

Ceoloni C, Donini P (1993) Combining mutations for the two homoeologous pairing suppressor genes Ph1 and Ph2 in common wheat and in hybrids with alien Triticeae. Genome 36:377–386

Ceoloni C, Feldman M (1987) Effect of Ph2 mutants promoting homoeologous pairing on spindle sensitivity to colchicine in common wheat. Can J Genet Cytol 29:658–663

Ceoloni C, Avivi L, Feldman M (1984) Spindle sensitivity to colchicine of the Ph1 mutant in common wheat. Can J Genet Cytol 26:111–118

Ceoloni C, Strauss I, Feldman M (1986) Effect of different doses of group-2 chromosomes on homoeologous pairing in intergeneric wheat hybrids. Can J Genet Cytol 28:240–246

Cermeño MC, Orellana J, Santos JL, Lacadena JR (1984a) Nucleolar organizer activity in wheat, rye and derivatives analyzed by a silver-staining procedure. Chromosoma 89:370–376

Cermeño MC, Orellana J, Santos JL, Lacadena JR (1984b) Nucleolar activity and competition (amphiplasty) in the genus Aegilops. Heredity 53:603–611

Chague V, Just J, Mestiri I, Balzergue S, Tanguy AM, Huneau C, Huteau V, Belcram H, Coriton O, Jahier J, Chalhoub B (2010) Genome-wide gene expression changes in genetically stable synthetic and natural wheat allohexaploids. New Phytol 187:1181–1194

Chantret N, Salse J, Sabot F, Rahman S, Bellec A, Laubin B, Dubois I, Dossat C, Sourdille P, Joudrier P, Gautier MF, Cattolico L, Beckert M, Aubourg S, Weissenbach J, Caboche M, Bernard M, Leroy P, Chalhoub B (2005) Molecular basis of evolutionary events that shaped the hardness locus in diploid and polyploid wheat species (Triticum and Aegilops). Plant Cell 17:1033–1045

Chao S, Sharp PJ, Worland AJ, Warham EJ, Koebner RMD, Gale MD (1989) RFLP-based genetic maps of wheat homoeologous group 7. Theor Appl Genet 78:495–504

Chapman V, Miller TE, Riley R (1976) Equivalence of the A genome of bread wheat and that of Triticum urartu. Genet Res 27:69–76

Charpentier A, Feldman M, Cauderon Y (1986) Genetic control of meiotic chromosome pairing in tetraploid Agropyron elongatum. I. Pattern of pairing in natural and induced tetraploids and in F_1 triploid hybrid. Can J Genet Cytol 28:783–788

Charpentier A, Cauderon Y, Feldman M (1988) The effect of different Agropyron elongatum chromosomes on pairing in Agropyron-common wheat hybrids. Genome 30:978–983

Chelysheva L, Gendrot G, Vezon D, Doutriaux MP, Mercier R, Grelon M (2007) Zip4/Spo22 is required for class I CO formation but not for synapsis completion in Arabidopsis thaliana. PLoS Genet 3:802–813

Chen ZF (2007) Genetic and epigenetic mechanisms for gene expression and phenotypic variation in plant polyploids. Ann Rev Plant Biol 58:377–406

Chen Q, Jahier J, Cauderon Y (1989) Production and cytogenetical studies of hybrids between Triticum aestivum L. Thell. and Agropyron cristatum (L.) Gaertn. CR Acad Sci Paris 308 (III):425–430

Chen KC, Dvorak J (1984) The inheritance of genetic variation in Triticum speltoides affecting heterogenetic chromosome pairing in hybrids with Triticum aestivum. Can J Genet Cytol 26:279–287

Cheng H, Liu J, Wen J, Nie X, Xu L, Chen N, Li Z, Qilin Wang Q et al (2019) Frequent intra- and inter-species introgression shapes the landscape of genetic variation in bread wheat. Genome Biol 20:136

Claesson L, Kotimak M, Bothmer R, von (1990) Crossability and chromosome pairing in some interspecific Triticum hybrids. Hereditas 112:49–55

Colas I, Shaw P, Prieto P, Wanous M, Spielmeyer W, Mago R, Moore G (2008) Effective chromosome pairing requires chromatin remodeling at the onset of meiosis. Proc Natl Acad Sci USA 105:6075–6080

Comai L (2005) The advantages and disadvantages of being polyploid. Nat Rev Genet 6:836–846

Concia L, Veluchamy A, Ramirez-Prado JS, Martin-Ramirez A, Huang Y et al (2020) Wheat chromatin architecture is organized in genome territories and transcription factories. Genome Biol 21:104

Cook OF (1913) Wild wheat in Palestine. Bull Plant Indust US Dept Agric Washington 274:1–58

Crosby AR (1957) Nucleolar activity of lagging chromosomes in wheat. Amer J Bot 44:813–822

Crosby-Longwell AR, Svihla G (1960) Specific chromosomal control of the nucleolus and the cytoplasm in wheat. Exp Cell Res 20:294–312

Cuadrado MC, Romero C (1984) Interaction between different genotypes of allogamous and autogamous rye and the homoeologous pairing control of wheat. Heredity 52:323–330

Cuadrado C, Romero C, Lacadena JR (1991) Meiotic pairing control in wheat–rye hybrids. II. Effect of rye genome and rye B-chromosomes and interaction with the wheat genetic system. Genome 34:76–80

Cuñado N (1992) Analysis of metaphase I chromosome association in species of the genus Aegilops. Theor Appl Genet 85:283–292

Darvey NL, Driscoll CJ (1971) Evidence against somatic association in wheat. Chromosoma 36:140–149

Darvey NL, Driscoll CJ (1972) Nucleolar behaviour in Triticum. Chromosoma 36:131–139

Darvey NL, Driscoll CJ, Kaltsikes PJ (1973) Evidence against somatic and premeiotic association in hexaploid wheat. Proc Int Congr Genet 74:557–558

de Bodt S, Maere S, Van de Peer Y (2005) Genome duplication and the origin of angiosperms. Trends Ecol Evol 20:591–597

Devos KM, Atkinson MD, Chinoy CN, Liu CJ, Gale MD (1992) RFLP-based genetic map of the homoeologous group 3 chromosomes of wheat and rye. Theor Appl Genet 83:931–939

Dhaliwal HS (1977) The Ph gene and the origin of tetraploid wheats. Genetica 47:177–182

Dover GA (1975) The heterogeneity of B-chromosome DNA: no evidence for a B-chromosome specific repetitive DNA correlated with B-chromosome effects on meiotic pairing in the Triticineae. Chromosoma 53:153–173

Dover GA, Riley R (1972a) Variation at two loci affecting homoeologous meiotic chromosome pairing in Triticum aestivum × Aegilops mutica hybrids. Nature New Biol 235:61–62

Dover GA, Riley R (1972b) Prevention of pairing of homoeologous meiotic chromosomes of wheat by an activity of supernumerary chromosomes of Aegilops. Nature 240:159–161

Dover GA, Riley R (1973) The effect of spindle inhibitors applied before meiosis om meiotic chromosome pairing. J Cell Sci 12:143–161

Dover GA (1973) The genetics and interactions of A and B chromosomes controlling meiotic chromosome pairing in the Triticineae. In: Sears ER, Sears LMS (eds) Proceedings 4th international wheat genetics symptoms, Columbia, Missouri, pp 653–667

Driscoll CJ (1972) Genetic suppression of homoeologous chromosome pairing in hexaploid wheat. Can J Genet Cytol 14:39–42

Driscoll CJ (1973) Minor genes affecting homoeologous pairing in hybrids between wheat and related genera. Genetics 74:s66

Driscoll CJ, Darvey NL (1970) Chromosome pairing: effect of colchicine on an isochromosome. Science 169:290–291

Driscoll CJ, Quinn CJ (1968) Wheat-alien hybrids involving a chromosome 5B translocation. Can J Genet Cytol 10:217–220

Driscoll CJ, Darvey NL, Barber HN (1967) Effect of colchicine on meiosis of hexaploid wheat. Nature 216:687–688

Driscoll CJ, Bielig LM, Darvey NL (1979) An analysis of frequencies of chromosome configurations in wheat and wheat hybrids. Genetics 91:755–767

Dubcovsky J, Dvorak J (2007) Genome plasticity a key factor in the success of polyploid wheat under domestication. Science 316:1862–1866

Dubcovsky J, Luo MC, Zhong GY, Bransteitter R, Desai A, Kilian A, Kleinhofs A, Dvorak J (1996) Genetic map of diploid wheat, Triticum monococcum L., and its comparison with maps of Hordeum vulgare L. Genetics 143:983–999

Dvorak J (1972) Genetic variability in Aegilops speltoides affecting homoeologous pairing in wheat. Can J Genet Cytol 14:371–380

Dvorak J (1976) The relationship between the genome of Triticum urartu and the A and B genomes of Triticum aestivum. Can J Genet Cytol 18:371–377

Dvorak J (1977) Effect of rye on homoeologous chromosome pairing in wheat x rye hybrids. Can J Genet Cytol 19:549–556

Dvorak J (1987) Chromosomal distribution of genes in diploid Elytrigia elongata that promote or suppress pairing of wheat homoeologous chromosomes. Genome 29:34–40

Dvorak J (1998) Genome analysis in the Triticum-Aegilops alliance. In: Slinkard AE (ed) Proceedings 9th international wheat genetics symptoms. University Extension Press, University of Saskatoon, Saskatoon, Saskatchewan, Canada, pp 8–11

Dvorak J (2009) Triticeae genome structure and evolution. In: Feuiller C, Muehlbauer GJ (eds) Genetics and genomics of the Triticeae, plant genetics and genomics: crops and models, vol 7. Springer, Berlin, pp 685–711

Dvorak J, Akhunov ED, Akhunov AR, Deal KR, Luo M-C (2006b) Molecular characterization of a diagnostic DNA marker for domesticated tetraploid wheat provides evidence for gene flow wild tetraploid wheat to hexaploid wheat. Mol Biol Evol 23:1386–1396

Dvorak J, Zhang H-B (1992a) Application of molecular tools for study of the phylogeny of diploid and polyploid taxa in Triticeae. Hereditas 116:37–42

Dvorak J, Zhang H-B (1992b) Reconstruction of the phylogeny of the genusTriticum from variation in repeated nucleotide sequences. Theor Appl Genet 84:419–429

Dvorak J, Chen K-C, Giorgi B (1984) The C-band pattern of a Ph mutant of durum wheat. Can J Genet Cytol 26:360–363

Dvorak J, Deal KR, Luo M-C (2006a) Discovery and mapping of wheat Ph1 suppressors. Genetics 174:17–27

Eig A (1929b) Amblyopyrum Eig. A new genus separated from the genus Aegilops. PZE Inst Agricult Nat History Agricult Res 2:199–204

Eig A (1929a) Monographisch-Kritische Ubersicht der Gattung Aegilops. Reprium Nov Spec Regni Veg 55:1–288

Eilam T, Anikster Y, Millet E, Manisterski J, Feldman M (2008) Nuclear DNA amount and genome downsizing in natural and synthetic allopolyploids of the genera Aegilops and Triticum. Genome 51(8):616–627

Eilam T, Anikster Y, Millet E, Manisterski J, Feldman M (2010) Genome size in diploids, allopolyploids, and autopolyploids of Mediterranean Triticeae. J Bot 210:341380. https://doi.org/10.1155/2010/341380

Eilam T, Anikster Y, Millet E, Manisterski J, Feldman M (2007) Genome size and genome evolution in diploid Triticeae species. Genome 50:1029–1037

Ekingen H, Attia T, Robbelen G (1977) Suppressor of homoeologous pairing in diploid Ae. squarrosa. Z. Pflanzensücht 79:72–73

Endo TR (1988) Induction of chromosomal structural changes by a chromosome of Aegilops cylindtica L. in common wheat. J Hered 79:366–370

Fahima T, Cheng JP, Peng JH, Nevo E, Korol A (2006) Asymmetry distribution of disease resistance genes and domestication synrome QTLs in tetraploid wheat genome. 8th Inter. Congress of Plant Molecular Biol, Adelaide, Australia

Fan C, Luo J, Zhang S, Liu M, Li Q, Li Y et al (2019) Genetic mapping of a major QTL promoting homoeologous chromosome pairing in a wheat landrace. Theor Appl Genet 132:2155–2166

Farooq S, Iqbal N, Shah TM (1990) Promotion of homoeologous chromosome pairing in hybrids of Triticum aestivum x Aegilops variabilis. Genome 33:825–828

Feldman M (1965a) Fertility of interspecific F₁ hybrids and hybrid derivatives involving tetraploid species of Aegilops section Pleionathera. Evolution 19:556–562

Feldman M (1965b) Chromosome pairing between differential genomes in hybrids of tetraploid Aegilops species. Evolution 19:563–568

Feldman M (1965c) Further evidence for natural hybridization between tetraploid species of Aegilops section Pleionathera. Evolution 19:162–174

Feldman M (1966a) The mechanism regulating pairing in Triticum timopheevii. Wheat Inf Serv 21:1–2

Feldman M (1966b) The effect of chromosomes 5B, 5D and 5A on chromosomal pairing in Triticum aestivum. Proc Natl Acad Sci USA 55:1447–1453

Feldman M (1968) Regulation of somatic association and meiotic pairing in common wheat. In: Finlay KW, Shepherd KW (eds) Proceedings 3rd international wheat genetics symptoms, Canberra, Australia, pp 169–178

Feldman M (1978) New evidence on the origin of the B genome of wheat. In: Ramanujam RS (ed) Proceedings 5th international wheat genetics symptoms, New Delhi, pp 120–132

Feldman M (1988) Cytogenetic and molecular approaches to alien gene transfer in wheat. In: Proceedings 7th international wheat genetics symptoms, vol 1. Cambridge, England pp 23–32

Feldman M (1993) Cytogenetic activity and mode of action of the pairing homoeologous (Ph1) gene of wheat. Crop Sci 33:894–897

Feldman M (2001) Origin of cultivated wheat. In: Bonjean AP, Angus WJ (eds) The world wheat book. Lavoisier, Paris, pp 3–56

Feldman M, Avivi L (1973) The pattern of chromosomal arrangement in nuclei of common wheat and its genetic control. In: Sears ER, Sears LMS (eds) Proceedings of the 4th international wheat genetics symptoms, Missouri Agr Exp Sta, Columbia, Missouri, pp 675–684

Feldman M, Avivi L (1984) Ordered arrangement of chromosomes in wheat. Chromosomes Today, Vol. 8:181–189

Feldman M, Avivi L (1988) Genetic control of bivalent pairing in common wheat: the mode of Ph1 action. In: Brandham PE (ed) The 3rd kew chromosome conference. Kew, London, pp 269–279

Feldman M, Levy AA (2005) Allopolyploidy—a shaping force in the evolution of wheat genomes. Cytogenet Genome Res 109:250–258

Feldman M, Levy AA (2009) Genome evolution in allopolyploid wheat —a revolutionary reprogramming followed by gradual changes. J Genet Genomics 36:511–518

Feldman M, Levy AA (2011) Instantaneous genetic and epigenetic alterations in the wheat genome caused by allopolyploidization. In: Gissis SB, Jablonka E (eds) Transformations of Lamarckism. From Subtle Fluids to Molecular Biology, The MIT press, Cambridge MA, USA, pp 261–270

Feldman M, Levy AA (2012) Genome evolution due to allopolyploidization in wheat. Genetics 192:763–774

Feldman M, Mello-Sampayo T (1967) Suppression of homeologous pairing in hybrids of polyploid wheats × *Triticum speltoides*. Can J Genet Cytol 9:307–313

Feldman M, Galili G, Levy AA (1987) Genetic and evolutionary aspects of allopolyploidy in wheat. In: Barigozzi C (ed) The Origin and Domestication of Cultivated Plants, Elsevier, Amsterdam, Oxford, New York, Tokyo, pp. 83–100

Feldman M, Liu B, Segal G, Abbo S, Levy AA, Vega JM (1997) Rapid elimination of low-copy DNA sequences in polyploid wheat: a possible mechanism for differentiation of homoeologous chromosomes. Genetics 147:1381–1387

Feldman M, Lupton FGH, Miller TE (1995) Wheats. In: Smart J, Simmonds NW (eds) Evolution of crop plants, 2nd edn. Longman Scientific and Technical, pp 184–192

Feldman M, Levy AA, Fahima T, Korol A (2012) Genome asymmetry in allopolyploid plants—wheat as a model. J Exp Bot 63:5045–5059

Feldman M, Levy AA, Chalhoub B, Kashkush K (2013) Genome plasticity in polyploid wheat, In; Soltis PS, Soltis D (eds) Polyploidy and genome evolution, Springer, Berlin, pp 109–135

Feldman M, Mello-Sampayo T, Sears ER (1966) Somatic association in *Triticum aestivum*. Proc Natl Acad Sci USA 56:1192–1199

Finch RA, Smith JB, Bennett MD (1981) *Hordeum* and *Secale* genomes lie apart in a hybrid. J Cell Sci 52:391–403

Flagel LE, Wendel JF (2009) Gene duplication and evolutionary novelty in plants. New Phytologist 183:557–564

Flagel LE, Wendel JF (2010) Evolutionary rate variation, genomic dominance and duplicate gene expression evolution during allotetraploid cotton speciation. New Phytol 186:184–193

Flavell RB, O'Dell M (1976) Ribosomal RNA genes on homoeologous chromosomes of groups 5 and 6 in hexaploid wheat. Heredity 37:377–385

Flavell RB, O'Dell M (1979) The genetic control of nucleolus formation in wheat. Chromosoma 71:135–152

Flavell RB, O'Dell M, Thompson WF, Vingentz M, Sardana R, Barker RF (1986) The differential expression of ribosomal RNA genes. Philos Trans R Soc Lond B 314:385–397

Foote T, Roberts M, Kurata N, Sasaki T, Moore G (1997) Detailed comparative mapping of cereal chromosome regions corresponding to the *Ph1* locus in wheat. Genetics 147:801–807

Frankel OH, Gerlach WL, Peacock WJ (1987) The ribosomal RNA genes in synthetic tetraploids of wheat. Theor Appl Genet 75:138–143

Furuta Y, Nishikawa K, Tanino T (1974) Stability in DNA content of AB genome component of common wheat during the past seven thousand years. Jpn J Genet 49:179–187

Fussell CP (1987) The Rabl orientation: a prelude to synapsis. In: Moens PB (ed) Cell biology: a series of monographs: meiosis. Academic Press, Orlando, FL, pp 275–299

Gale MD, Devos KM (1998) Comparative genetics in the grasses. Proc Natl Acad Sci USA 95:1971–1974

Gale MD, Atkinson MD, Chinoy CN, Harcourt RL, Jia J, Li QY, Devos KM (1995) Genetic maps of hexaploid wheat. In: Li ZS, Xin ZY (eds) Proceedings of the 8th international wheat genetic symptoms. China Agricultural Scientech Press, Beijing, pp 29–40

Galili G, Feldman M (1983a) Genetic control of endosperm proteins in wheat. 2. Variation in high molecular weight glutenin and gliadin subunits of *Triticum aestivum*. Theor Appl Genet 66:77–86

Galili G, Feldman M (1983b) Diploidization of endosperm protein genes in polyploid wheats. In: Sakamoto S (ed) Proceedings 6th international wheat genetics symptoms. Faculty of Agriculture, Kyoto University, Kyoto, Japan, pp 1119–1123

Galili G, Feldman M (1984) Inter-genomic suppression of endosperm-protein genes in common wheat. Can J Genet Cytol 26:651–656

Galili G, Levy AA, Feldman M (1986) Gene-dosage compensation of endosperm proteins in hexaploid wheat *Triticum aestivum*. Proc Natl Acad Sci USA 83:6524–6528

Gandhi HT, Vales MI, Watson CJW, Mallory -Smith CA, Mori N et al (2005) Chloroplast and nuclear micro-satellite analysis of *Aegilops cylindrica*. Theor Appl Genet 111:561–572

Gelfand VI, Bershadsky AD (1991) Microtubule dynamics: mechanism, regulation, and function. Annu Rev Cell Biol 7:93–116

Gill BS (1987) Chromosome banding methods, standard chromosome band nomenclature, and applications in cytogenetic analysis. In: Heyne EG (ed) Wheat and wheat improvement, 2nd edn. American Society of Agronomy Inc., Madison, Wisconsin, USA, pp 243–254

Gill KS, Gill BS (1991) A DNA fragment mapped within the submicroscopic deletion of *ph1*, a chromosome pairing regulator gene in polyploid wheat. Genetics 129:257–259

Gill KS, Gill BS (1996) A PCR-based screening assay of *Ph1*, the chromosome pairing regulator gene of wheat. Crop Sci 36:719–722

Gill KS, Gill BS, Endo TR (1993a) A chromosome region-specific reveals gene-rich telomeric ends in wheat. Chromosoma 102:374–381

Gill KS, Gill BS, Endo TR, Mukai Y (1993b) Fine physical mapping of *Ph1*, a chromosome pairing regulator gene in polyploid wheat. Genetics 134:1231–1236

Gill BS, Friebe B, Endo TR (1991) Standard karyotype and nomenclature system for description of chromosome bands and structural aberrations in wheat (*Triticum aestivum*). Genome 34:830–839

Gillies CB (1987) The effect of *Ph* gene alleles on synaptonemal complex formation in *Triticum aestivum* × *T. kotschyi* hybrids. Theor Appl Genet 74:430–438

Giorgi B (1978) A homoeologous pairing mutant isolated in *Triticum durum* cv. Cappelli. Mutat Breed Newsl 11:4–5

Giorgi B, Barbera F (1981b) Increase of homoeologous pairing in hybrids between a *ph* mutant of *T. turgidum* var. *durum* and two tetraploid species of *Aegilops*: *Aegilops kotschyi* and *Ae. cylindrica*. Cereal Res Commun 9:205–211

Giorgi B, Cuozzo L (1980) Homoeologous pairing in a *ph* mutant of tetraploid wheat crossed with rye (x). Cereal Res Comm 8:485–490

Giorgi B, Barbera F (1981a) Use of mutants that affect homoeologous pairing for introducing alien variation in both durum and common wheat. In: Induced mutations—a tool in plant breeding, IAEA-SM 25½, pp 37–47

Giorgi B (1983) Origin, behaviour and utilization of a phI mutant of durum wheat, *Triticum turgidum* (L.) var. *durum*. In: Sakamoto S (ed) Proceedings of the 6th international wheat genetic symptoms, Kyoto, Japan, pp 1033–1040

Gornicki P, Zhu H, Wang J, Challa GS, Zhang Z, Gill BS, Li W (2014) The chloroplast view of the evolution of polyploid wheat. New Phytol 204:704–714

Grant V (1971) Plant speciation. Columbian University Press, New York, London, pp 1–435

Greer E, Azahara C. Martín AC, Pendle A, Colas I, Alexandra M.E. Jones AME, Moore G, Shaw P (2012) The *Ph1* locus suppresses *Cdk2*-Type activity during pre-meiosis and meiosis in wheat. The Plant Cell 24:152–162

Griffiths S, Sharp R, Foote TN, Bertin I, Wanous M, Reader S, Colas I, Moore G (2006) Molecular characterization of *Ph1* as a major chromosome pairing locus in polyploid wheat. Nature 439:749–752

Gualandi G, Ceoloni C, Feldman M, Avivi L (1984) Spindle sensitivity to isopropyl-*N*-phenyl-carbamate and griseofulvin of common wheat plants carrying different doses of the *Phl* gene. Can J Genet Cytol 26:119–127

Guo X, Han F (2014) Asymmetric epigenetic modification and elimination of rDNA sequences by polyploidization in wheat. Plant Cell 26:4311–4327

Gupta PK, Fedak G (1986) The inheritance of genetic variation in rye (*Secale cereale*) affecting homoeologous chromosome pairing in hybrids with bread wheat (*Triticum aestivum*). Can J Genet Cytol 28:844–851

Gupta PK, Kulwal PL, Rustgi S (2005) Wheat cytogenetics in the genomics era and its relevance to breeding. Cytogenet Genome Res 109:315–327

Gustafson JP, Flavell RB (1996) Control of nucleolar expression in triticale. In: Guedes-Pinto H, Darvey N, Carnide VP (eds) Triticale: today and tomorrow. Kluwer Academic Publishers, The Netherlands, pp 119–125

Gyawali Y, Zhang W, Chao S et al (2019) Delimitation of wheat *ph1b* deletion and development of ph1b-specific DNA markers. Theor Appl Genet 132:195–204

Ha M, Lu J, Tian L, Ramachandran V, Kasschau KD et al (2009) Small RNAs serve as a genetic buffer against genomic shock in *Arabidopsis* interspecific hybrids and allopolyploids. Proc Natl Acad Sci USA 106:17835–17840

Hadlaczky GY, Went M, Ringertz NR (1986) Direct evidence for the non-random localization of mammalian chromosomes in the interphase nucleus. Exp Cell Res 167:1–15

Halloran GM (1966) Pairing between *Triticum aestivum* and *Haynaldia villosa* chromosomes. J Hered 57:233–235

Han FP, Fedak G, Ouellet T, Liu B (2003) Rapid genomic changes in interspecific and intergeneric hybrids and allopolyploids of Triticeae. Genome 46:716–723

Han FP, Fedak G, Guo WL, Liu B (2005) Rapid and repeatable elimination of a parental genome specific DNA repeat (pGcIR-1a) in newly synthesized wheat allopolyploids. Genetics 170:1239–1245

Han J, Zhou B, Shan W, Yu L, Wu W, Wang K (2015) A and D genomes spatial separation at somatic metaphase in tetraploid cotton: evidence for genomic disposition in a polyploid plant. Plant J 84:1167–1177

Hao M, Luo J, Yang M et al (2011) Comparison of homoeologous chromosome pairing between hybrids of wheat genotypes Chinese Spring *ph1b* and Kaixian-luohanmai with rye. Genome 54:959–964

Harlan JR, Zohary D (1966) Distribution of wild wheats and barley. Science 153:1074–1080

Hart GE (1983b) Genetics and evolution of multilocus isozymes in hexaploid wheat. Isozymes 10:365–380

Hart GE, McMillin E, Sears ER (1976) Determination of the chromosomal location of a glutamate oxaloacetate transaminase structural gene using *Triticum-Agropyron* translocations. Genetics 83:49–61

Hart GE (1983a) Hexaploid wheat (*Triticum aestivum* L. em Thell.). In: Tanksley SD, OrtonTJ (eds) Isozymes in plant genetics and breeding, Part B, Elsevier Publishers B.V., Amsterdam, pp 35–56

Hart GE (1987) Genetic and biochemical studies of enzymes. In: Heyne EG (ed) Wheat and wheat improvement, 2nd edn. Amer Soc Agronomy, Madison, Wisconsin, USA

He P, Friebe BR, Gill BS, Zhou JM (2003) Allopolyploidy alters gene expression in the highly stable hexaploid wheat. Plant Mol Biol 52 (401–4):14

He F, Pasam R, Shi F, Kant S, Keeble-Gagnere G et al (2019) Exome sequencing highlights the role of wild-relative introgression in shaping the adaptive landscape of the wheat genome. Nat Genet 51:896–904

Hilliker AJ, Appels R (1989) The arrangement of interphase chromosomes: structural and functional aspects. Exp Cell Res 185:297–318

Hobolth P (1981) Chromosome pairing in allohexaploid wheat var. Chinese Spring. Transformation of multivalents into bivalents, a mechanism for exclusive bivalent formation. Carlsberg Res Commun 46:129–173

Holm PB, Wang X, Wischmann B (1988) An ultrastructural analysis of the effect of chromosome 5B on chromosome pairing in allohexaploid wheat. In: Brandham PE (ed) Kew chromosome conference III, HMSO, royal botanic gardens, kew, 1–8 Sept 1987, Kew Richmond, Surrey, UK, pp 281–291

Holm PB, Wang X (1988) The effect of chromosome 5B on synapsis and chiasma formation in wheat, *Triticum aestivum* cv. Chin. spring. Carls Res Commun 53:191–208

Houben A, Guttenbach M, Kress W, Pich U, Schubert I, Schmid M (1995) Immunostaining and interphase arrangement of field bean kinetochores. Chromosoma 3:27–31

Houchins K, O'Dell M, Flavell R, Gustavson PJ (1997) Cytosine methylation and nucleolar dominance in cereal hybrids. Mol Gen Genet 255:294–301

Huang L, Millet E, Rong JK, Wendel JF, Anikster Y, Feldman M (1999) Restriction fragment length polymorphism in wild and cultivated tetraploid wheat. Isr J Pl Sci 47:213–224

Jakubziner MM (1932) The wheats of Syria, Palestine and Transjordania, cultivated and wild. Trudy prikl Bot Genet Selek, Suppl 53, *Triticum dicoccoides* (Russian, with English summary), pp 150–178

Jampates R, Dvorak J (1986) Location of the *Ph1* locus in the metaphase chromosome map and the linkage map of the 5Bq arm of wheat. Can J Genet Cytol 28:511–519

Jauhar PP, Almouslem AB, Peterson TS, Joppa LR (1999) Inter- and intragenomic chromosome pairing in haploids of durum wheat. J Hered 90:437–445

Jia J, Xie Y, Cheng J, Kong C, Wang M et al (2021) Homology-mediated inter-chromosomal interactions in hexaploid wheat leads to specific subgenome territories following polyploidization and introgression. Genome Biol 22:26

John B (1976) Myths and mechanisms of meiosis. Chromosoma 54:295–325

Jones JK, Majisu BN (1968) The homoeology of *Aegilops mutica* chromosomes. Can J Genet Cytol 10:620–626

Joppa LR, Cantrell RG (1990) Chromosomal location of genes for grain protein content of wild tetraploid wheat. Crop Sci 30:1059–1064

Kashkush K, Khasdan V (2007) Large-scale survey of cytosine methylation of retrotransposons and the impact of readout transcription from long terminal repeats on expression of adjacent rice genes. Genetics 177:1975–1985

Kashkush K, Feldman M, Levy AA (2002) Gene loss, silencing and activation in a newly synthesized wheat allotetraploid. Genetics 160:1651–1659

Kashkush K, Feldman M, Levy AA (2003) Transcriptional activation of retrotransposons alters the expression of adjacent genes in wheat. Nature Genet 33:102–106

Kato T, Yamagata H (1983) Analysis of the action of 3B chromosome on meiotic homologous chromosome pairing in common wheat. In: Sakamoto S (ed) Proceedings of the 6th international wheat genetics symptoms. Faculty of Agriculture, Kyoto, Japan, pp 321–325

Kato T, Yamagata H (1980) Reduction of meiotic homologous chromosome pairing due to high temperature in common wheat. Jpn J Genet 55:337–348

Kato T, Yamagata H (1982) Effect of 3B chromosome deficiency on the meiotic pairing between the arms of an isochromosome in common wheat. Jpn J Genet 57:403–406

Keilwagen J, Lehnert H, Berner T et al (2022) Detecting major introgressions in wheat and their putative origins using coverage analysis. Sci Rep 12:1908

Kempanna C, Riley R (1962) Relationships between the genetic effects of deficiencies for chromosomes III and V on meiotic pairing in *Triticum aestivum*. Nature 195:1270–1273

Kempanna C (1963) Investigations into the genetic regulation of meiotic chromosome behaviour in *Triticum aestivum*. PhD thesis. Cambridge Univ, England

Kenan-Eichler M, Leshkowitz D, Tal L, Noor E, Cathy Melamed-Bessudo C, Feldman M, Levy AA (2011) Wheat hybridization and polyploidization results in deregulation of small RNAs. Genetics 188:263–272

Kerber ER (1964) Wheat: reconstitution of the tetraploid component (AABB) of hexaploids. Science 143:253–255

Kerber ER, Green GJ (1980) Suppression of stem rust resistance in hexaploid wheat cv. Canthach by chromosome 7DL. Can J Bot 58:1347–1350

Khawaja S, Gundersen GG, Bulinski JC (1988) Enhanced stability of microtubules enriched in detyrosinated tubulin is not function of detyrosination level. J Cell Biol 106:141–149

Kihara H (1919) Uber cytologische Studien bei einigen Getreidiarten. I. Species-bastarde des weizens und weizenroggen-bastarde. Bot Mag Tokyo 32:17–38

Kihara H (1924) Cytologische und genetische studien bei wichtigen getreidearten mit besonderer rücksicht ouf das verhalten der chromosomen und die sterilitat in den bastarden. Kyoto Imp Univ Bl, Mem Cell Sci, pp 1–200

Kihara H (1937) Genomanalyse bei *Triticum* und *Aegilops*. VII. Kurze ~ bersicht iiber die Ergebnisse der Jahre 1934–36. Mem Coll Agric Kyoto Imp Univ 41:1–61

Kihara H (1949) Genomanalyse bei *Triticum* und *Aegilops*. IX. Systematischer Aufbau der Gattung Aegilops auf genomanalytischer Grundlage. Cytologia 14:135–144

Kihara H (1954) Considerations on the evolution and distribution of *Aegilops* species based on the analyzer-method. Cytologia 19:336–357

Kihara H, Lilienfeld FA (1935) Genomanalyse bei *Triticum* und *Aegilops*. VI. Weitere Untersuchungen an *Aegilops* x *Triticum* und *Aegilops* x *Aegilops*-Bastarden. Cytologia 6:195–216

Kihara H, Yamashita H, Tanaka M (1959) Genomes of 6x species of *Aegilops*. Wheat Inf Ser 8:3–5

Kimber G (1974) A reassessment of the origin of the polyploid wheats. Genetics 78:487–492

Kimber G, Athwal RS (1972) A reassessment of the course of evolution of wheat. Proc Nat Acad Sci USA 69:912–915

Kimber G, Sears ER (1987) Evolution in the genus *Triticum* and the origin of cultivated wheat. In: Heyne EG (ed) Wheat and wheat improvement, 2nd edn. American Society of Agronomy, Madison, Wisconsin, USA, pp 154–164

Kimber G, Feldman M (1987) Wild wheats: an introduction. Special Report 353. College of Agriculture, Columbia. Missouri, USA, pp. 1–142

Kleckner N (1996) Meiosis, how could it work? Proc Natl Acad Sci USA 93:8167–8174

Knight E, Greer E, Draeger T, Thole V, Reader S, Shaw P, Moore G (2010) Inducing chromosome pairing through premature condensation: analysis of wheat interspecific hybrids. Funct Integr Genomics 10:603–608

Koh J, Soltis PS, Soltis DE (2010) Homeolog loss and expression changes in natural populations of the recently and repeatedly formed allotetraploid *Tragopogon mirus* (Asteraceae). BMC Genomics 11:97

Koo D-H, Liu W, Friebe B, Gill BS (2017) Homoeologous recombination in the presence of *Ph1* gene in wheat. Chromosoma 126:531–540

Kousaka R, Endo ER (2012) Effect of a rye B chromosome and its segments on homoeologous pairing in hybrids between common wheat and *Aegilops variabilis*. Genes Genetic Systems 87:1–7

Kraitshtein Z, Yaakov B, Khasdan V, Kashkush K (2010) Genetic and epigenetic dynamics of a retrotransposon after allopolyploidization of wheat. Genetics 186:801–812

Kreis TE (1987) Microtubules containing detyrosinated tubulin are lesss dynamic. The EMBO J 6:2597–2606

Lacadena JR, Azpiazu A (1969) Introduction of alien variation into wheat by gene recombination. II. Action of the 5B genetic system on the meiotic behaviour of mono 5B *Triticum aestivum* L. x *Aegilops ovata* L. hybrids. Genet Iberica 21:1–10

Lacadena JR, Cermeño MC, Orellana J, Santos JL (1984) Analysis of nucleolar activity in *Agropyron elongatum*, its amphiploid with *Triticum aestivum* and the chromosome addition lines. Theor Appl Genet 68:75–80

Law CN, Young CF, Brown JWS, Snape JW, Worland AJ (1978) The study of grain protein control in wheat using whole chromosome substitution lines. In: Seed protein improvement by nuclear techniques. International Atomic Energy Agency, Vienna, Austria, pp 483–502

Lelley T (1976) Induction of homoeologous pairing in wheat by genes of rye suppressing chromosome 5B effect. Can J Genet Cytol 18:485–489

Levy AA, Zaccai M, Millet E, Feldman M (1988b) Utilization of wild emmer for the improvement of grain protein percentage of cultivated wheat. In: Miller TE, Koebner RMD (eds) Proceedings of the 7th international wheat genetics symptoms, vol 2, Cambridge, UK, pp 969–974

Levy AA, Galili G, Feldman M (1988a) Polymorphism and genetic control of high molecular weight glutenin subunit in wild tetraploid wheat *Triticum turgidum* var. *dicoccoides*. Heredity 61:63–72

Levy AA, Feldman M (2002) The impact of polyploidy on grass genome evolution. Plant Physiol 130:1587–1593

Levy AA, Feldman M (2004) Genetic and epigenetic reprogramming of the wheat genome upon allopolyploidization. Biol J Linn Soc 82:607–613

Li HW, Pao WK, Li CH (1945) Desynapsis in the common wheat. Amer J Bot 32:92–101

Li DY, Zhang XY, Yang J, Rao GY (2000) Genetic relationship and genomic in situ hybridization analysis of the three genomes in *Triticum aestivum*. Acta Bot Sin 42:957–964

Li A, Liu D, Wu J, Zhao X et al (2014) mRNA and small RNA transcriptomes reveal insights into dynamic homoeolog regulation of allopolyploid heterosis in nascent hexaploid wheat. Plant Cell 26:1878–1900

Li L-F, Zhang Z-B, Wang Z-H et al (2022) Genome sequences of the five *Sitopsis* species of *Aegilops* and the origin of polyploid wheat B subgenome. Mol Plant 15:488–503

Lilienfeld F (1951) H. Kihara: genome analysis in *Triticum* and *Aegilops* X. Concluding review. Cytologia 16:101–123

Limin AE, Fowler DB (1990) An interspecific hybrid and amphiploid produced from *Triticum aestivum* crosses with *Agropyron cristatum* and Agropyron desertorum. Genome 33:581–584

Linde-Laursen I, Jensen J (1991) Genome and chromosome disposition at somatic metaphase in a Hordeum × Psathyrostachys hybrid. Heredity 66:203–210

Lindström J (1965) Transfer to wheat of accessory chromosomes from rye. Hereditas 54:149–155

Liu YG, Tsunewaki K (1991) Restriction fragment length polymorphism (RFLP) analysis in wheat. II. Linkage maps of the RFLP sites in common wheat. Jpn J Genet 66:617–633

Liu B, Segal G, Vega JM, Feldman M, Abbo S (1997a) Isolation and characterization of chromosome-specific DNA sequences from a chromosome arm genomic library of common wheat. Plant J 11:959–965

Liu DC, Luo MC, Yang JL, Yan J, Lan XJ, Yang WY (1997b) Chromosome location of a new paring promoter in natural populations of common wheat. SW China J Agric Sci 10:10–15

Liu DC, Zheng YL, Yan ZH, Zhou YH, Wei YM, Lan XJ (2003) Combination of homoeologous pairing gene *phKL* and *Ph2*-deficiency in common wheat and its mitotic behaviors in hybrids with alien species. Acta Bot Sin 45:1121–1128

Liu W, Rouse M, Friebe B, Jin Y, Gill BS, Pumphrey MO (2011) Discovery and molecular mapping of a new gene conferring resistance to stem rust, *Sr53*, derived from *Aegilops geniculata* and characterization of spontaneous translocation stocks with reduced alien chromatin. Chromosome Res 19:669–682

Liu B, Vega JM, Segal G, Abbo S, Rodova H, Feldman M (1998a) Rapid genomic changes in newly synthesized amphiploids of *Triticum* and *Aegilops*. I. Changes in low-copy noncoding DNA sequences. Genome 41:272–277

Liu B, Vega JM, Feldman M (1998b) Rapid genomic changes in newly synthesized amphiploids of *Triticum* and *Aegilops*. II. Changes in low-copy coding DNA sequences. Genome 41:535–542

Liu ZP, Chen ZY, Pan J, Li XF, Su M, Wang LJ, Li HJ, Liu GS (2008) Phylogenetic relationships in *Leymus* (Poaceae: Triticeae) revealed by the nuclear ribosomal internal transcribed space and chloroplast trnL-F sequences. Molecul Phylogenet Evol 46:278–289

Lloyd AH, Ranoux M, Vautrin S, Glover N, Fourment J, Charif D et al (2014) Meiotic gene evolution: can you teach a new-dog new-trick? Mol Biol Evo 31:1724–1727

Loidl J (1990) The initiation of meiotic chromosome pairing: the cytological view. Genome 33:759–778

Luo M-C, Yang Z-L, Yen C, Yang JL (1992) The cytogenetic investigation on F_1 hybrid of Chinese wheat landrace. In: Ren ZL, Peng JH (eds) Exploration of crop breeding. Science and Technology Press, Sichuan, pp 169–174

Luo M-C, Dubcovsky J, Dvorak J (1996) Recognition of homoeology by the wheat *Phl* locus. Genetics 144:1195–1203

Luo M-C, Yang Z-L, You FM, Kawahara T, Waines JG, Dvorak J (2007) The structure of wild and domesticated emmer wheat populations, gene flow between them, and the site of emmer domestication. Theor Appl Genet 114:947–959

Ma X-F, Gustafson JP (2005) Genome evolution of allopolyploids: a process of cytological and genetic diploidization. Cytogenet Genome Res 109:236–249

Ma X-F, Gustafson JP (2006) Timing and rate of genome variation in triticale following allopolyploidization. Genome 49:950–958

Ma X-F, Gustafson JP (2008) Allopolyploidization-accommodated genomic sequence changes in triticale. Ann Bot 101:825–832

Ma X-F, Fang P, Gustafson JP (2004) Polyploidization-induced genome variation in triticale. Genome 47:839–848

Maan SS (1977) Cytoplasmic control of cross incompatibility and zygotic and sporophytic sterility in interspecific hybrids. In: Proceedings 8th congress Eucarpia, Madrid, Spain, pp 201–214

Mac Key J (1954) Neutron and X-ray experiments in wheat and a revision of the speltoid problem. Hereditas 40:65–180

Mac Key J (1966) Species relationship in *Triticum*. In: Proceedings of the 2nd international wheat genetics symposts, vol 2, Lund, Sweden, Hereditas, pp 237–276

Madlung A, Tyagi AP, Watson B, Jiang HM, Kagochi T, Doerge RW, Martienssen R, Comai L (2005) Genomic changes in synthetic Arabidopsis polyploids. Plant Journal 41(2):221–230

Maestra B, Naranjo T (1999) Structural chromosome differentiation between *Triticum timopheevii* and *T. turgidum* and *T. aestivum*. Theor Appl Genet 98:744–750

Maguire MP (1967) Evidence for homologous pairing of chromosomes prior to meiotic prophase in maize. Chromosoma 21:221–231

Makino T (1970) On the location of asynaptic gene in wheats. Wheat Inf Serv 30:12–13

Mamun-Hossain ABM, Kishikawa H, Takagi Y (1992) Cytological behavior and genetic effects of the B chromosomes of rye in tetraploid wheat, *Triticum durum* Desf. Jpn J Breed 42:523–534

Manton I (1950) Problems of cytology and evolution in the Pteridophyta. Cambridge Univ Press, Cambridge, UK

Marcussen T, Sandve SR, Heier L, Spannagl M, Pfeifer M, Jakobsen KS, Wulff BBH, Steuernagel B, Klaus FX, Mayer KFX, et al (2014) Ancient hybridizations among the ancestral genomes of bread wheat. Science 345:288–291

Martín A, Rey M-D, Shaw P, Moore G (2017) Dual effect of the wheat *Phl* locus on chromosome synapsis and crossover. Chromosoma 126:669–680

Martín A, Shaw P, Phillips D, Reader S, Moore G (2014) Licensing MLH1 sites for crossover during meiosis. Nat Commun 5:4580

Martínez M, Naranjo T, Cuadrado C, Romero C (1996) Synaptic behaviour of the tetraploid wheat *Triticum timopheevii*. Theor Appl Genet 93:1139–1144

Martinez M, Cuñado N, Carcelén N, Romero C (2001a) The *Phl* and *Ph2* loci play different roles in the synaptic behaviour of hexaploid wheat *Triticum aestivum*. Theor Appl Genet 103:398–405

Martinez M, Naranjo T, Cuadrado C, Romero C (2001b) The synaptic behaviour of Triticum turgidum with variable doses of the *Phl* locus. Theor Appl Genet 102:751–758

Martinez M, Cuadrado C, Laurie DA, Romero C (2005) Synaptic behavior of hexaploid wheat with different effectiveness of the diploidizing mechanism. Cytogenet Genome Res 109:210–214

Martinez-Perez E, Shaw P, Reader S, Aragon-Alcaide L, Miller T, Moore G (1999) Homologous chromosome pairing in wheat. J of Cell Sci 112:1761–1769

Martinez-Perez E, Shaw P, Moore G (2001) The *Phl* locus is needed to ensure specific somatic and meiotic centromere association. Nature 411:204–207

Martini G, Flavell RB (1985) The control of nucleolus volume in wheat, a genetic study at three developmental stages. Heredity 54:111–120

Masterson J (1994) Stomatal size in fossil plants: evidence for polyploidy in majority of angiosperms. Science 264:421–424

Matsumura S (1940) Induced haploidy and autotetraploidy in *Aegilops ovata* L. Bot Mag Tokyo 54:404–413

Maystrenko OI (1992) The use of cytogenetic methods in ontogenesis study of common wheat. In: Ontogenetics of higher plants, Kishinev 'Shtiintsa' (In Russian), pp 98–114

McGuire PE, Dvorak J (1982) Genetic regulation of heterogenetic chromosome pairing in polyploid species of the genus *Triticum* sensu lato. Can J Genet Cytol 24:57–82

McIntosh JR, Hering GE (1991) Spindle fiber action and chromosome movement. Annu Rev Cell Biol 7:404–426

Mello-Sampayo T (1971a) Genetic regulation of meiotic chromosome pairing by chromosome 3D of *Triticum aestivum*. Nat New Biol 230:22–23

Mello-Sampayo T (1971b) Promotion of homoeologous pairing in hybrids of *Triticum aestivum* x *Aegilops longissima*. Genet Iber 23:1–9

Mello-Sampayo T (1972) Compensated monosomic 5B-trisomic 5A plants in tetraploid wheat. Can J Genet Cytol 14:463–475

Mello-Sampayo T (1973) Somatic association of telocentric chromosomes carrying homologous centromeres in common wheat. Theor Appl Genet 43:174–181

Mello-Sampayo T, Canas AP (1973) Suppressors of meiotic chromosome pairing in common wheat. In: Sears ER, Sears LMS (eds) Proceedings of the 4th international wheat genetics symposts.

Agricultural ExperimentStation, College of Agriculture and University of Missouri, Columbia, Missouri, pp 709–713

Mello-Sampayo T, Lorente R (1968) The role of chromosome 3D in the regulation of meiotic pairing in hexaploid wheat. EWAC Newsletter 2:19–24

Middleton CP, Senerchia N, Stein N, Akhunov ED, Keller B, Wicker T, Kilian B (2014) Sequencing of chloroplast genomes from wheat, barley, rye and their relatives provides a detailed insight into the evolution of the Triticeae tribe. PLoS ONE 9: e85761

Mikhailova EI, Naranjo T, Shepherd K, Wennekes-van EJ, Heyting C, de Jong H (1998) The effect of the wheat *Ph1* locus on chromatin organisation and meiotic pairing analysed by genome painting. Chromosoma 107:339–350

Millet E, Rong JK, Qualset CO, McGuire PE, Bernard M, Sourdille P, Feldman M (2014) Grain yield and grain protein percentage of common wheat lines with wild emmer chromosome-arm substitutions. Euphytica 195:69–81

Mitra R, Bhatia C (1971) Isoenzymes and polyploidy. 1. Qualitative and quantitative isoenzyme studies in the Triticinae. Genet Res 18:57–69

Mitra N, Roeder GS (2007) A novel nonnull *ZIP1* allele triggers meiotic arrest with synapsed chromosomes in *Saccharomyces cerevisiae*. Genetics 176:773–787

Mochida K, Kawaura K, Shimosaka E et al (2006) Tissue expression map of a large number of expressed sequence tags and its application to in silico screening of stress response genes in common wheat. Mol Genet Genom 276:304–312

Mochizuki A (1962) *Agropyron* addition lines of *durum* wheat. Seiken Zihô 13:133–138

Mohan J, Flavell RB (1974) Ribosomal RNA cistron multiplicity and nucleolar organizers in hexaploid wheat. Genetics 76:33–44

Moore G (2002) Meiosis in allopolyploids—the importance of 'Teflon' chromosomes. Trends in Genet 18:456–463

Mori M, Liu Y-G, Tsunewaki K (1995) Wheat phylogeny determined by RFLP analysis of nuclear DNA. 2. Wild tetraploid wheats. Theor Appl Genet 90:129–134

Morris R, Sears ER (1967) The cytogenetics of wheat and its relatives. In: Quisenberry KS, Reitz LP (eds) Wheat and wheat improvement. Amer Soc Agron, Madison, Wisconsin, USA, pp 19–87

Morris CF, DeMacon VL, Giroux MJ (1999) Wheat grain hardness among chromosome 5D homozygous recombinant substitution lines using different methods of measurement. Cereal Chem 76:249–254

Müntzing A, Jaworska H, Carlbom C (1969) Studies of meiosis in the Lindström strain of wheat carrying accessory chromosomes of rye. Hereditas 61:179–207

Murphy DB, Vallee R, Borisy G (1977) Identity and polymerization with microtubules. Biochem 15:2598–2605

Nalam VJ, Vales MI, Watson CJW, Kianian SF, Riera-Lizarazu O (2006) Map-based analysis of genes affecting the brittle rachis character in tetraploid wheat (*Triticum turgidum* L.). Theor Appl Gen 112:373–381

Naranjo T, Benaveste E (2015) The mode and regulation of chromosome pairing in wheat-Alien hybrids (*ph* genes, an updated view). In: Molnár-Láng M, Ceoloni C, Doležel J (eds) Alien introgrssion in wheat. Springer, Cytogentics, Molecular Biology and Genomics, pp 133–162

Naranjo T, Palla O (1982) Genetic control of meiotic pairing in rye. Heredity 48:57–62

Naranjo T, Lacadena JR, Giraldez R (1979) Interaction between wheat and rye genomes on homologous and homoeologous pairing. Z Pflan 82:289–305

Neijzing MG, Viegas WS (1979) The effect of rye B-chromosomes on meiotic stability of rye-wheat hybrids in normal, nulli 5B and nulli 5D background. Genetica 51:21–26

Neves N, Heslop-Harrison JS, Viegas W (1995) rRNA gene activity and control of expression mediated by methylation and imprinting during embryo development in wheat x rye hybrids. Theor Appl Genet 91:529–533

Nishikawa K, Furuta Y (1969) DNA content per nucleus in relation to phylogeny of wheat and its relatives. Jpn J Genet 44:23–29

Niwa K, Horiuchi G, Hirai Y (1997) Production and characterization of common wheat with B chromosomes of rye from Korea. Hereditas 126:139–146

Ohno S (1970) Evolution by gene duplication. Springer, New York

Ohta S (1991) Phylogenetic relationship of *Aegilops mutica* Boiss with the diploid species of congeneric *Aegilops-Triticum* complex, based on the new method of genome analysis using its B-chromosomes. Mem Coll Agric Kyoto Univ 137:1–116

Ohta S (1990) Genome analysis of *Aegilops mutica* Boiss based on the chromosome pairing in interspecific and intergeneric hybrids. PhD Thesis, submitted to the University of Kyoto, pp 1–267

Okamoto M (1957) Asynaptic effect of chromosome V. Wheat Inf Serv 5:6

Okamoto M, Inomata N (1974) Possibility of 5B-like effect in diploid species. Wheat Inf Serv 38:15–16

Okamoto M, Sears ER (1962) Chromosomes involved in translocations obtained from haploids of common wheat. Gan J Genet Cytol 4:24–30

Olmsted JB (1986) Microtubule-associated proteins. Ann Rev Cell Biol 2:421–457

Ozkan H, Feldman M (2001) Genotypic variation in tetraploid wheat affecting homoeologous pairing in hybrids with *Aegilops peregrine*. Genome 44:1000–1006

Ozkan H, Levy AA, Feldman M (2001) Allopolyploidy—induced rapid genome evolution in the wheat (*Aegilops-Triticum*) group. Plant Cell 13:1735–1747

Ozkan H, Tuna M, Arumuganathan K (2003) Nonadditive changes in genome size during allopolyploidization in the wheat (*Aegilops-Triticum*) group. J Hered 94:260–264

Paull JG, Rathjen AJ, Cartwright B (1991) Major gene control of tolerance of bread wheat (*Triticum aestivum* L.) to high concentrations of soil boron. Euphytica 55:217–228

Parisod C, Senerchia N (2012) Responses of transposable elements to poly- ploidy. In: Grandbastien MA, Casacuberta JM, editors. Plant transpos- able elements. Topics in current genetics. Vol. 24. Berlin: Springer- Verlag. p. 147–168

Payne PI, Holt LM, Worland AJ, Law CN (1982) Structural and genetical studies on the high-molecular-weight subunits of wheat glutenin. Part 3. Telocentric mapping of the subunit genes on the long arms of the homoeologous group 1 chromosomes. Theor Appl Genet 63:129–138

Peng J, Ronin Y, Fahima T, Röder MS, Li Y, Nevo E, Korol A (2003a) Genomic distribution of domestication QTLs in wild emmer wheat, *Triticum dicoccoides*. In: Pogna NE, Mcintosh RA (eds) Proceedings 10th international wheat genetics symptoms. Paestum, Italy, pp 34–37

Peng J, Ronin Y, Fahima T, Röder MS, Nevo E, Korol A (2003b) Domestication quantitative trait loci in *Triticum dicoccoides*, the progenitor of wheat. Proc Natl Acad Sci USA 100:2489–2494

Penner GA, Clarke K, Bezte LJ, Leisle D (1995) Identification of RAPD markers linked to a gene governing cadmium uptake in durum wheat. Genome 38:543–547

Percival J (1921) The wheat plant: a monograph. Duckworth, London, pp 1–46

Percival J (1923) Chromosome numbers in *Aegilops*. Nature 111:810

Pernickova K, Linc G, Gaal E, Kopecký D, Šamajová O, Lukaszewski A (2019) Out-of-position telomeres in meiotic leptotene appear responsible for chiasmate pairing in an inversion heterozygote in wheat (*Triticum aestivum* L.). Chromosoma 128:31–39

Piperno G, LeDizet M, Chang XJ (1987) Microtubules containing acetylated alpha-tubulin in mammalian cells in culture. J Cell Biol 104:289–302

Prieto P, Shaw P, Moore G (2005) Homologue recognition during meiosis is associated with a change in chromatin conformation. Nature Cell Biol 6:906–908

Pumphrey M, Bai J, Laudencia-Chingcuanco D, Anderson O, Gill BS (2009) Nonadditive expression of homoeologous genes is established upon polyploidization in hexaploid wheat. Genetics 181:1147–1157

Rakszegi M, Kisgyörgy BN, Tearall K et al (2010) Diversity of agronomic and morphological traits in a mutant population of bread wheat studied in the Health grain program. Euphytica 174:409–421

Ramírez-González RH, Borrill P, Lang D, et al (2018) The transcriptional landscape of polyploid wheat. Science 361:eaar6089

Rao PS, Smith EL (1968) Studies with Israeli and Turkish accessions of Triticum turgidium L. emend. var. dicoccoides (Korn.) Bowden. Wheat Inf Serv 26:6–7

Rapp RA, Udall JA, Wendel JF (2009) Genomic expression dominance in allopolyploids. BMC Biol 7:18

Rasmussen S, Holm PB (1978) Human meiosis. II. Chromosome pairing and recombination nodules in human spermatocytes. Carls Res Commun 43:275–327

Rawal K, Harlan JR (1975) Cytogenetic analysis of wild emmer populations from Turkey and Israel. Euphytica 24:407–411

Rawale KS, Khan MA, Gill KS (2019) The novel function of the Ph1 gene to differentiate homologs from homoeologs evolved in Triticum turgidum ssp. dicoccoides via a dramatic meiosis-specific increase in the expression of the 5B copy of the C-Ph1 gene. Chromosoma 128:561–570

Rey M-D, Martín AC, Higgins J, Swarbreck D, Uauy C, Shaw P, Moore G (2017) Exploiting the ZIP4 homologue within the wheat Ph1 locus has identified two lines exhibiting homoeologous crossover in wheat-wild relative hybrids. Mol Breed 37:95

Rey M-D, Martín AC, Smedley M, Hayta S, Harwood W, Shaw P, Moore G (2018) Magnesium increases homoeologous crossover frequency during meiosis in ZIP4 (Ph1 gene) mutant wheat-wild relative hybrids. Front Plant Sci 9:509

Riede CR, Anderson JA (1996) Linkage of RFLP markers to an aluminum tolerance gene in wheat. Crop Sci 36:905–909

Riley R (1960) The diploidization of polyploid wheat. Heredity 15:407–429

Riley R, Chapman V (1958) Genetic control of the cytologically diploid behaviour of hexaploid wheat. Nature 182:713–715

Riley R, Chapman V (1967) Effect of 5BS in suppressing the expression of altered dosage of 5BL on meiotic chromosome pairing in Triticum aestivum. Nature 216:60–62

Riley R, Kempanna C (1963) The homoeologous nature of the non-homologous meiotic pairing in Triticum aestivum deficient for chromosome V (5B). Heredity I8:287–306

Riley R, Law CN (1965) Genetic variation in chromosome pairing. Adv Genet 13:57–114

Riley R, Kimber G, Chapman V (1961) Origin of genetic control of diploid-like behavior of polyploid wheat. Hered 52:22–25

Riley R, Chapman V, Young RM, Belfield AM (1966) Control of meiotic chromosome pairing by the chromosomes of homoeologous group 5 of Triticum aestivum. Nature 212:1475–1477

Riley R, Chapman V, Miller TE (1973) The determination of meiotic chromosome pairing. In: Sears ER, Sears LMS (eds) Proceedings 4th international wheat genetics symposts, Columbia, Missouri, pp 731–738

Riley R (1966) The genetic regulation of meiotic behavior in wheat and its relatives. In: Mac Key J (ed) Proceedings of the 2nd international wheat genetic symposts, vol 2. Genetic Institute, University of Lund, Sweden, August 1963, Hereditas, pp 395–406

Riley R (1968) The basic and applied genetics of chromosome pairing. In: Finlay KW, Shepherd KW (eds) Proceedings 3rd international wheat genetics symposts. Aust Acad Sci, Canberra, Australia, pp 185–195

Roberts MA, Reader SM, Dalgliesh C, Miller TE, Foote TN, Fish LJ, Snape JW, Moore G (1999) Induction and characterization of Ph1 wheat mutants. Genetics 153:1909–1918

Röder MS, Korzun V, Wendehake K, Plaschke J, Tixier MH, Leroy P, Ganal MW (1998) A microsatellite map of wheat. Genetics 149:2007–2023

Romero C, Lacadena JR (1980) Interaction between rye B chromosomes and wheat genetic systems controlling homoeologous pairing. Chromosoma 80:33–48

Rong JK, Millet E, Feldman M (1999) Unequal RFLP among genomes, homoeologous groups and chromosome regions in wheat. In: Proceedings 9th ITMI workshop, August 25–27, Viterbo, Italy

Rong JK (1999) Mapping and tagging by DNA markers of wild emmer alleles that affect useful traits in bread wheat. PhD Thesis, submitted to the Scientific Council of The Weizmann Institute of Science, Rehovot, Israel, April 1999

Roothaan M, Sybenga J (1976) No 5-B compensation by rye B-Chromosomes. Theor Appl Genet 48:63–66

Rubenstein J (1976) Cytogenetic investigations of evolutionary relationships and the genetic control of chromosome pairing in the genus Triticum. PhD Thesis. Univ Missouri Columbia, USA

Sabot F, Guyot R, Wicker T, Chantret N, Laubin B, Chalhoub B, Leroy P, Sourdille P, Bernard M (2005) Updating of transposable element annotations from large wheat genomic sequences reveals diverse activities and gene associations. Mol Genet Gen 274:119–130

Sakamura T (1918) Kurze mitteilung über die chromosomenzahlen und die verwandtschaftsverhältnisse der Triticum Arten. Bot Mag 32:151–154

Salina EA, Adonina IG, Vatolina t, Kurata N (2004) A comparative analysis of the composition and organization of two sub-telomeric repeat families in Aegilops speltoides Tausch and related species. Genetica 122:227–237

Sánchez-Morán E, Benavente E, Orellana J (2001) Analysis of karyotypic stability of homoeologous-pairing (ph) mutants in allopolyploid wheats. Chromosoma 110:371–377

Sachs L (1953) Chromosome behavior in species hybrids with Triticum timopheevii. Heredity 7:49–58

Sax K (1927) Chromosome behavior in Triticum hybrids, Verhandlungen des V Int. Kongresses Für Vererbungswissenchaft 2:1267–1284

Sax K, Sax HJ (1924) Chromosome behavior in a genus cross. Genetics 9:454–464

Sax K (1921a) Chromosome relationships in wheat. Science 54:413–415

Sax K (1921b) Sterility in wheat hybrids. I. Sterility relationships and endosperm development. Genetics 6:399–416

Sax K (1922) Sterility in wheat hybrids. II. Chromosome behavior in partially sterile hybrids. Genetics 7:513–552

Scherthan H, Weich S, Schwegler H, Heyting C, Härle M, Cremer T (1996) Centromere and telomere movements during early meiotic prophase of mouse and man are associated with the onset of chromosome pairing. J Cell Biol 134:1109–1125

Schiemann E (1929) Zytologische Beitrage zur Gattung Aegilops. Chromosomenzahlen und morphologie. III Mitteilung. Ber Deutsch Bot Ges 47:164–181

Schwarzacher T (1997) Three stages of meiotic homologous chromosome pairing in wheat: cognition, alignment and synapsis. Sex Plant Report 10:324–331

Schwarzacher T, Leitch AR, Bennett MD, Heslop-Harrison JS (1989) In situ localization of parental genomes in a wide hybrid. Ann Bot 64:315–324

Schwarzacher T, Heslop-Harrison JS, K. Anamthawat-Jónsson K, Finch RA, Bennett MD (1992) Parental genome separation in reconstructions of somatic and premeiotic metaphases of *Hordeum vulgare* x *H. bulbosum*. J Cell Sci 101:13–24

Sears ER (1941a) Amphidiploids in the seven-chromosome Triticinae. Univ Missouri Agric Exper Stn Res Bull 336:1–46

Sears ER (1941b) Chromosome pairing and fertility in hybrids and amphidiploids in the Triticinae. Univ Missouri Agric Exp Stn Res. Bull 337:1–20

Sears ER (1944) Cytogenetic studies with polyploid species of wheat. II. Additional chromosomal aberrations in *Triticum vulgare*. Genetics 29:232–246

Sears ER (1954) The aneuploids of common wheat. Missouri Agric Exp Stn Res Bull 572:1–58

Sears ER (1966) Nullisomic-tetrasomic combinations in hexaploid wheat. In: Riley R, Lewis KR (eds) Chromosome manipulations and plant genetics. Springer, Boston MA, USA, pp 29–45

Sears ER (1977) An induced mutant with homoeologous chromosome pairing. Can J Genet Cytol 19:585–593

Sears ER, Sears LMS (1979) The telocentric chromosomes of common wheat. In: Ramanuam S (ed) Proceedings of the 5th international wheat genetics symptoms, New Delhi, 23–28 Feb 1978, pp 389–407

Sears ER (1972) The nature of mutation in hexaploid wheat. Symp Biol Hung 12:73–82

Sears ER, Okamoto M (1958) Intergenomic chromosome relationships in hexaploid wheat. In: Proceedings 10th international congress genetics, Montreal, Quebec 2:258–259

Sears ER (1952a) Homoeologous chromosomes in *Triticum aestivum*. Genetics 37:624

Sears ER (1952b) Misdivision of univalents in common wheat. Chromosoma 4:535–550

Sears ER (1976a) A synthetic hexaploid wheat with fragile rachis. Wheat Inf Serv 41-42:31–3

Sears ER (1976b) Genetic control of chromosome pairing in wheat. Ann Rev Genet 10:31–51

Sears ER (1982) A wheat mutation conditioning an intermediate level of homoeologous chromosome pairing. Can J Genet Cytol 24:715–719

Sears ER (1984) Mutations in wheat that raise the level of meiotic chromosome pairing. In: Gustafson JP (ed) Gene manipulation in plant improvement. Stadler Genetics Symposia Series. Springer, Boston, MA, pp 295–300

Segal G, Liu B, Vega JM, Abbo S. Rodova M, Feldman M (1997) Identification of a chromosome-specific probe that maps within the *Ph1* deletions in common wheat. Theor Appl Genet 94:968–970

Selmecki AM, Maruvka YE, Richmond PA, Guillet M, Shoresh N, Sorenson AL, De S, Kishony R, Michor F, Dowell R, Pellman D (2015) Polyploidy can drive rapid adaptation in yeast. Nature 519:349–352

Serra H, Svačina R, Baumann U et al (2021) *Ph2* encodes the mismatch repair protein MSH7-3D that inhibits wheat homoeologous recombination. Nat Commun 12:803

Shaked H, Kashkush K, Ozkan H, Feldman M, Levy AA (2001) Sequence elimination and cytosine methylation are rapid and reproducible responses of the genome to wide hybridization and allopolyploidy in wheat. Plant Cell 13:1749–1759

Shands H, Kimber G (1973) Reallocation of the genomes of *Triticm timopheevii* Zhuk. In: Sears ER, Sears LMS (eds) Proc 4th Inter Wheat Genet Symp., Agricultural Experiment Station, College of Agriculture, University of Missouri, pp 101–108

Shang XM, Jackson RC, Nguyen HT, Huang JY (1989) Chromosome pairing in the Triticum monococcum complex: evidence for pairing control genes. Genome 32:216–226

Sharma S, Schulthess A, Bassi F, Badaeva E et al (2021) Introducing beneficial alleles from plant genetic resources into the wheat germplasm. Biology 10:982

Shcherban AB, Sergeeva EM, Badaeva ED, Salina EA (2008) Analysis of 5S rDNA changes in synthetic allopolyploids *Triticum* x *Aegilops*. Mol Biol 4:536–542

Shen Y, Tang D, Wang K, Wang M, Huang J, Luo W, Luo Q, Hong L, Li M, Cheng Z (2012) *ZIP4* in homologous chromosome synapsis and crossover formation in rice meiosis. J Cell Sci 125:2581–2591

Shitsukawa N, Tahira C, Ken-ichiro Kassai K, Hirabayashi C, Tomoaki Shimizu T et al (2007) Genetic and epigenetic alteration among three homoeologous genes of a class *E MADS* box gene in hexaploid wheat. Plant Cell 19:1723–1737

Sidhu GK, Rustgi S, Shafqat MN, von Wettstein D, Gill KS (2008) Fine structure mapping of a gene-rich region of wheat carrying *Ph1*, a suppressor of crossing over between homoeologous chromosomes. Proc Natl Acad Sci 105:5815–5820

Skalická K, Lim KY, Matyasek R, Matzke M, Leitch AR, Kovarik A (2005) Preferential elimination of repeated DNA sequences from the paternal, *Nicotiana tomentosiformis* genome donor of a synthetic, allotetraploid tobacco. New Phytol 166:291–303. Skovmand B, Fox PN, Villareal RL (1984) Triticale in commercial agriculture: progress and promise. Adv Agron 37:1–45

Smith SG (1942) Polarization and progression in pairing. II. Premeiotic orientation and the initiation of pairing. Can J Res 20:221–229

Snape JW, Angus WJ, Parker B, Leckie D (1987) The chromosomal locations in wheatof genes confirming differential response to the wild oat herbicide, Difenzoquat. J Agric Sci 108:543–548

Soltis DE, Soltis PS (1993) Molecular data facilitate a reevaluation of traditional tenets of polyploid evolution. Crit Rev Pl Sci 12:243–273

Soltis DE, Soltis PS (1995) The dynamic nature of polyploid genomes. Proc Natl Acad Sci USA 92:8089–8091

Soltis DE, Albert VA, Leebens-Mack J et al (2009) Polyploidy and angiosperm diversification. Amer J Bot 96:336–348

Song K, Lu P, Tang K, Osborn TC (1995) Rapid genome change in synthetic polyploids of Brassica and its implications for polyploid evolution. Proc Natl Acad Sci USA 92:7719–7723

Sorokina ON (1937) Contribution to the synthesis of *Aegilops* species. Bull Appl Bot Genet Pl Br 7:151–160

Stebbins GL (1950) Variation and evolution in plants. Columbia University Press, New York

Stebbins GL (1971) Chromosomal evolution in higher plants. Edward Arnold Ltd., London

Su H, Liu Y, Liu C, Shi Q, Huang Y, Han F (2019) Centromere satellite repeats have undergone rapid changes in polyploid wheat subgenomes. Plant Cell 31:2035–2051

Sutton T, Whitford R, Baumann U, Dong C, Able JA, Langridge P (2003) The *Ph2* pairing homoeologous locus of wheat (*Triticum aestivum*): identification of candidate meiotic genes using a comparative genetics approach. Plant J 36:443–456

Svačina R, Karafiátová M, Malurová M, Serra H, Vítek D, Endo TR et al (2020) Development of deletion lines for chromosome 3D of bread wheat. Front Plant Sci 10:1756

Tanaka M, Kawahara T (1976) Wild tetraploid wheat from northern Iraq cytogenetically closely related to each other. Wheat Inf Serv 43:3–4

Tang H, Wang X, Bowers JE, Ming R, Alam M, Paterson AH (2008) Unraveling ancient hexaploidy through multiply-aligned angiosperm gene maps. Genome Res 18:1944–1954

Tate JA, Ni Z, Scheen A-C, Koh J, Candace A, Gilbert CA et al (2006) Evolution and expression of homoeologous loci in *Tragopogon miscellus* (*Asteraceae*), a recent and reciprocally formed allopolyploid. Genetics 173:1599–1611

Tirosh I, Reikhav S, Levy AA, Barkai N (2009) A yeast hybrid provides insight into the evolution of gene expression regulation. Science 324:659–662

Tsunewaki K (2009) Plasmon analysis in the *Triticum-Aegilops* complex. Breed Sci 59:455–470

Tsunewaki K, Ebana K (1999) Production of near-isogenic lines of common wheat for glaucousness and genetic basis of this trait clarified by their use. Genes Genet Systems 74:33–41

Tsunewaki K, Takumi S, Mori N, Achiwa T, Liu YG (1991) Origin of polyploid wheats revealed by RFLP analysis. In: Kinoshita T (ed) Sasakuma T. Nuclear and organellar genomes of wheat species, Kihara Memo Found, Yokohama, pp 31–39

Upadhya MD (1966) Altered potency of chromosome 5B in wheat-*caudata* hybrids. Wheat Inf Serv 22:7–9

Upadhya MD, Swaminathan MS (1967) Mechanism regulating chromosome pairing in *Triticum*. Biol Zentral 86:239–255

van Slageren MW (1994) Wild wheats: a monograph of *Aegilops* L. and *Amblyopyrum* (Jaub. and Spach) Eig (Poaceae). Agricultural University, Wageningen, The Netherlands

Vardi A (1973) Introgression between different ploidy levels in the wheat group. In: Sears ER, Sears LMS (eds) Proceedings 4th internal wheat genetics symptoms. College of Agriculture, University of Missouri, Agricultural Experiment Station, pp 131–141

Vardi A, Dover GA (1972) The effect of B chromosomes on meiotic and pre-meiotic spindles and chromosome pairing in *Triticum/Aegilops* hybrids. Chromosoma (berl) 38:367–385

Vardi A, Zohary D (1967) Introgression in wheat via triploid hybrids. Heredity 22:541–560

Vega JM, Feldman M (1998a) Effect of the pairing gene *Ph1* and premeiotic colchicine treatment on intra- and inter-chromosome pairing of isochromosomes in common wheat. Genetics 150:1199–1208

Vega JM, Feldman M (1998b) Effect of the pairing gene *Ph1* on centromere misdivision in common wheat. Genetics 148:1285–1294

Viegas WS, Mello-Sampayo T, Feldman M, Avivi L (1980) Reduction of chromosome pairing by a spontaneous mutation on chromosomal arm 5DL of Triticum aestivum. Can J Genet Cytol 22:569–575

Vieira R, Queiroz A, Morais L, Barão A, Melo-Sampayo T, Viegas WS (1990) 1R chromosome nucleolus organizer region activation by 5-azacytidine in wheat x rye hybrids. Genome 33:707–712

Wagenaar EB (1961) Studies on the genome constitution of *Triticum timopheevii* Zhuk. I. Evidence for genetic control of meiotic irregularities in tetraploid hybrids. Can J Genet Cytol 3:47–60

Wagenaar EB (1966) Studies on the genome constitution of *Triticum timopheevi* Zhuk. II. The *T. timopheevi* complex and its origin. Evolution 20:150–164

Waines JG (1976) A model for the origin of diploidizing mechanisms in polyploid species. Amer Naturalist 110:415–430

Waines JG, Payne PI (1987) Electrophoretic analysis of the high-molecular-weight glutenin subunits of *Triticum monococcum*, *T. urartu*, and the A genome of bread wheat (*Triticum aestivum*). Theor Appl Genet 74:71–76

Wall AM, Riley R, Chapman V (1971a) Wheat mutation permitting homoeologous meiotic chromosome pairing. Genet Res Camb 18:311–328

Wall AM, Riley R, Gale MD (1971b) The position of a locus on chromosome 5B of *Triticum aestivum* affecting homoeologous meiotic pairing. Genet Res Camb 18:329–39

Walters MS (1970) Evidence on the time of chromosome pairing from the preleptotene spiral stage in *Lilium longiflorum* "Cro." Chromosoma 29:375–418

Walters MS (1972) Preleptotene chromosome contraction in *Lilium longiflorum* "Cro." Chromosoma 39:311–332

Wang RR-C (1990) Intergeneric hybrids between *Thinopyrum* and *Psathyrostachys* (Triticeae). Genome 33:845–849

Wang X, Holm PB (1988) Chromosome pairing and synaptonemal complex formation in wheat-rye hybrids. Carls Res Commun 53:167–190

Wang G-Z, Miyashita N, Tsunewaki K (1997) Plasmon analyses of *Triticum* (wheat) and *Aegilops*: PCR–single-strand conformational polymorphism (PCR-SSCP) analyses of organellar DNAs. Proc Natl Acad Sci USA 94:14570–14577

Wang A, Xia Q, Xie W, Datla R, Selvaraj G (2003) The classical Ubisch bodies carry a sporophytically produced structural protein (RAFTIN) that is essential for pollen development. Proc Natl Acad Sci USA 100:14487–14492

Wang J, Tian L, Lee HS et al (2006) Genome wide nonadditive gene regulation in *Arabidopsis* allotetraploids. Genetics 172:507–517

Webster DR, Borisy GG (1989) Microtubules are acetylated in domains that turn over slowly. J Cell Sci 92:57–65

Weiner BM, Kleckner N (1994) Chromosome pairing via multiple interstitial interactions before and during meiosis in yeast. Cell 77:977–991

Weissman S, Feldman M, Gressel J (2005) Sequence evidence for sporadic inter-generic DNA introgression from wheat into wild *Aegilops* species. Mol Biol Evol 22:2055–2062

Wendel JF (2000) Genome evolution in polyploids. Plant Mol Biol 42:225–249

Wicker T, Mayer KFX, Gundlach H, Martis M, Steuernagel B, Scholz U, Simkova H, Kubalakova M, Choulet F, Taudien S, Platzer M, Feuillet C, Fahima T, Budak H, Dolezel J, Keller B, Stein N (2011) Frequent gene movement and pseudogene evolution is common to the large and complex genomes of wheat, barley, and their relatives. Plant Cell 23:1706–1718

Xiang ZG, Liu DC, Zheng YL, Zhang LQ, Yan ZH (2005) The effect of *phKL* gene on homoeologous pairing of wheat-alien hybrids is situated between gene mutants of *Ph1* and *Ph2*. Hereditas (Beijing) 27:935–940 (in Chinese)

Yaakov B, Kashkush K (2011a) Massive alterations of the methylation patterns around DNA transposons in the first four generations of a newly formed wheat allohexaploid. Genome 54:42–49

Yaakov B, Kashkush K (2011b) Methylation, transcription, and rearrangements of transposable elements in synthetic allopolyploids. Int J Plant Genomics 569826

Yacobi YZ, Mello-Sampayo T, Feldman M (1982) Genetic induction of bivalent interlocking in common wheat. Chromosoma 87:165–175

Yacobi YZ, Levanony H, Feldman M (1985a) An ordered arrangement of bivalents at first meiotic metaphase of wheat. I. Hexaploid wheat. Chromosoma 91:347–354

Yacobi YZ, Levanony H, Feldman M (1985b) An ordered arrangement of bivalents at first meiotic metaphase of wheat. II. Tetraploid wheat. Chromosoma 91:355–358

Yousafzai FK, Al-Kaff N, Moore G (2010) Structural and functional relationship between the *Ph1* locus protein 5B2 in wheat and *CDK2* in mammals. Funct Integr Genomics 10:157–166

Yu Y, Yang WY, Hu XR (1998) The effectiveness of ph1b gene on chromosome association in the F₁ hybrid of *T. aestivum* x *H. villosa*. In: Slinkard AE (ed) Proceedings 9th international wheat genetics symptoms, vol 2. University Extension Press, University of Saskatchewan, Saskatoon, Canada, pp 125–126

Yu MQ, Deng GB, Zhang XP, Ma XR, Chen J (2001) Effect of the *ph1b* mutant on chromosome pairing in hybrids between *Dasypyrum villosum* and *Triticum aestivum*. Plant Breed 120:285–289

Yuan J, Wu Jiao W, Liu Y et al (2020) Dynamic and reversible DNA methylation changes induced by genome separation and merger of polyploid wheat. BMC Biol 18:171

Zhang Z, Gou X, Kun H, Bian Y, Ma X, Lij LM, Gong L, Feldman M, Liu B, Levy AA (2020) Homoeologous exchanges occur through intragenic recombination generating novel transcripts and proteins in wheat and other polyploids. Proc Natl Acad Sci USA 117:14561–14571

Zheng T, Nibau C, Phillips DW, Jenkins G, Armstrong SL, Doonan JH (2014) CDKG1 protein kinase and male meiosis at high ambient temperature in *Arabidopsis thaliana*. Proc Natl Acad Sci USA 111:2182–2187

Zhou Y, Zhao X, Li Y et al (2020) *Triticum* population sequencing provides insights into wheat adaptation. Nat Genet 52:1412–1422

Zhukovsky PM (1928) A critical systematical survey of the species of the genus *Aegilops* L. Bull Appl Bot Genet and Pl Breed 18:417–609

Zohary D, Brick Z (1962) *Triticum dicoccoides* in Israel: notes on its distribution, ecology and natural hybridization. Wheat Inf Serv 13:6–8

Zohary D, Feldman M (1962) Hybridization between amphidiploids and the evolution of polyploids in the wheat (*Aegilops-Triticum*) group. Evolution 16:44–61

13.1 Introduction

Based on the chronological system used in the east Mediterranean, the seven millennia between 13,000 and 6200 uncalibrated years BP are divided into 4 periods (Harris 1998; Table 1.1): the late Epipalaeolithic (the Natufian) (13,000–10,300 BP), the Pre-Pottery Neolithic A (PPNA) (10,300–9500 BP), the Pre-Pottery Neolithic B (PPNB) (9500–7500 BP), and the Ceramic or Pottery Neolithic (7500–6200 BP). The archaeological characterization of these periods was described by Harris (1998), and the events related to the wheat culture were summarized by Kislev (1984), Bar-Yosef (1998), and Harris (1998). In the Natufian period, the hunter-gatherers of the Levant collected grains of wild cereals (wheat, barley, oat, rye and *Aegilops*) as well as seeds, fruits and roots of other plants. No evidence of cultivation has been obtained from this period (Harris 1998). It is in the PPNA that one finds the first indications of cultivation of wild cereals in the western part of the Fertile Crescent, wild emmer and wild barley in the southern part and wild einkorn and, to some of extent, also wild barley and wild *timopheevii* in the northern part (Fig. 13.1). Domesticated forms of these wheats, characterized by non-brittle rachis, appeared at the beginning of the PPNB (Table 13.1). Somewhat later, wheat with free-threshing grains (most probably tetraploid) also appeared in the western part of the Fertile Crescent, and together with domesticated emmer, einkorn, and *timopheevii* spread to the eastern wing of the Fertile Crescent (eastern Turkey, northern Iraq, and south-western Iran), and neighboring regions, where they established contact with *Ae. tauschii* Coss. (=*Ae. squarrosa* L.) resulting in the formation of hexaploid wheat. In the Pottery Neolithic, the wheat culture spread to Europe, Asia, and Africa.

The beginning of cultivation of wheat (and several other plant species) in the western part of the Fertile Crescent, i.e., the Levant, around 10,300 years ago, marked a dramatic turn in the development and evolution of human civilization, as it enabled the transition from a hunter-gatherer and nomadic pastoral society to a more sedentary agrarian one (Eckardt 2010). This change from hunting and gathering to cultivation of plants represents one of the most remarkable events that human society experienced. It triggered the development of human civilization, boosting sedentism, urbanization, and population growth. As such, the Neolithic shift, from an economy based on hunting and gathering to a system based on food production through the domestication of plants and, later also animals, was one of mankind's most dramatic transformations. It was a major socio-economic and cultural change that affected human evolution on the one hand and facilitated the development of human civilization on the other hand.

What brought hunter-gathers in the Levant, to start cultivation of wild plants during the early Neolithic period and become farmers? This question fascinated archeologists, anthropologists, plant geneticists and evolutionary biologists, starting with Darwin (1868). Several hypotheses have been raised concerning the reasons for the shift of early Neolithic man from hunting and gathering in a nomadic or semi-nomadic way of life to agricultural activities in sedentary village dwellings. A long-standing assumption on the origin of agriculture has been population pressure: the increase in population size towards the end of the Natufian period forced people to intensify food production. Moreover, considering (1) the reduction in food sources because of climatic changes that occurred during the end of the 11th millennium BP (the Younger Dryas during the end of the Pleistocene, 13,500–11,800 BP), (2) the over-exploitation of the immediate environment by the increasing human population, and (3) the development of relatively large communities with complex social organization brought about by accelerated sedentism, there was a great increase in food stress, partly also due to reduction in the number of big game (Dembitzer et al. 2022), and, thereby, a pressure to enhance food production. Despite various hypotheses, such as pressure on food resources due to increase in human population, and of climate change, the "why" of wheat cultivation remains a mystery. Climatic changes towards the end of the

© The Author(s) 2023
M. Feldman and A. A. Levy, *Wheat Evolution and Domestication*,
https://doi.org/10.1007/978-3-031-30175-9_13

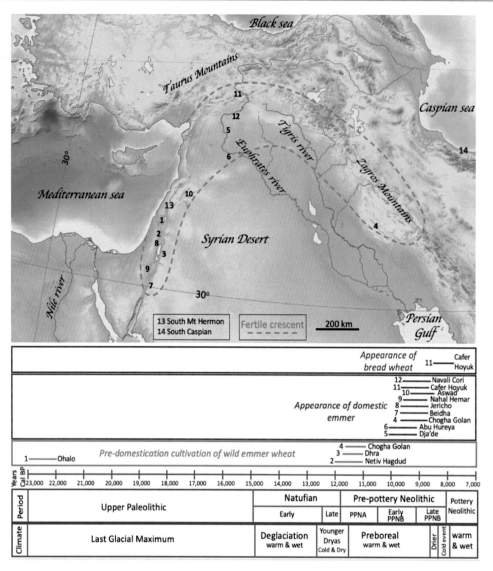

Fig. 13.1 Archaeological evidence for wheat cultivation and domestication in the near-east. The location of the fertile crescent is shown as dashed green lines. Its boundaries correspond to the distribution of wild progenitors of wheat, barley, and several legumes as well as to early domestication of these crops. The western part, called the Levant or levantine corridor (Bar-Yosef, 1998), goes south, around the Jordan valley between the dashed line on the side of the Syrian desert and the Mediterranean Sea. The south Levant is the region between Beidha (#6) and Aswad (#9) and the north Levant is north of Aswad (#10), for example, in Dja'de (#5) and Abu Hureya (#6). The northern area of the fertile crescent is also referred to as the upper Euphrates (e.g., Cafer Hoyuk, #11), and the east of the fertile crescent, in the Zagros mountain is represented by sites such as Chogha Golan (#4). The years on the blue horizontal axis correspond to "Calibrated years before present"

(Cal BP). The bottom boxes represent the climatic and the archeological periods when cultivation and domestication took place. Horizontal lines flanked by the location number (see map) and names indicate the relevant period when archeological evidence of cultivation or domestication was found. Numbers in red indicate two regions where evidence came from genomic data rather than archaeological data for the putative progenitors of domestic emmer (#13) and for the donor of the D subgenome of bread wheat (#14). Archaeological data were gathered from Nesbitt (2001); Willcox (2012); Zeder (2011); Riehl et al. (2013). A blank topographic map from Wikipedia (Middle East topographic map-blank 3000bc.svg, by Fulvio314, CCBY 3.0) served as the background on which text and data were added (Fig. 2 from Levy and Feldman (2022))

Pleistocene (Table 2.5) are regarded as important for their impact on the availability of the wild progenitors and human subsistence and are the favored component underpinning explanations for why cultivation began. Of particular importance is the Younger Dryas dry cold episode from

approx. 13,500–11,800 years BP (Bar-Yosef 1998, 2003; Harris 1998; Hillman et al. 2001).

A reduction in food sources, by itself, is not a sufficient cause to account for plant domestication (the process whereby wild plants have been evolved into crop through

Table 13.1 Uncalibrated and calibrated dates before present (BP) on remains of wild emmer, domesticated emmer and free-threshing tetraploid wheat from sites of the western flank of the Fertile Crescent (the Levantine Corridor) in pre-pottery Neolithic periods

Period uncalibrated (calibrated)[a] years BP	Wheat type	Levantine corridor region[b]	Site	Uncalibrated years BP	References
PPNA 10,300–9500 (12,000–10,800)	Wild emmer	Southern	Netiv Hagdud, Israel	10,000–9400	Kislev 1997
			Gilgal, Israel	10,000–9600	Weiss et al. 2006
			Iraq ed Dubb, Jordan	9950–	Colledge 2001
			Zad 2, Syria	9800–	Willcox 2005
			Aswad, Syria	9700–9300	Van Zeist and Bakker-Heeres 1982
			Jericho, Israel	9600–9200	Hopf 1983
Early PPNB 9500–9000 (10,800–10,200)	Mixture of wild and domesticated emmer	Southern	Jericho, Israel	9200–8650	Hopf 1983
			Nahal Heimar, Israel	9200–8100	Kislev 1988
			Ein Gazal, Jordan	9200–8600	Rollefson et al. 1985
			Beida, Jordan	9100–8550	Colledge 2001
			Aswad, Syria	8900–8500	Van Zeist and Bakker-Heeres 1982
		Northern	Dejade, Syria	9500–9000	Willcox 1996
			Nevali Çon, Turkey	9250–	Willcox 2005
			Cafer Höyük, Turkey	9200–8600	Moulins 1997
			Çayönü, Turkey	9200–8600	Van Zeist and de Roller 1995
			Abu Hureyra, Syria	9000–8300	Moulins 1997; Hillman 1978
			Jarmo, Iraq	8750-	Helbaek 1959, 1966
			Ali Kosh, Iran	8750-	Helbaek 1969
Middle and Late PPNB 9000–7500 (10,200–8300)	Mixture of wild and domesticated emmer and free-threshing tetraploid wheat	Southern	Aswad, Syria	8900–8600	Van Zeist and Bakker-Heeres 1982
			Ghoraifé, Syria	8800–8200	Van Zeist and Bakker-Heeres 1982
			Ramad, Syria	8300–7900	Van Zeist and Bakker-Heeres 1982
			Azraq, Jordan	8350–8300	Colledge 2001
			El Kowm, Syria	8200–8000	Moulins 1997
			Atlit-Yam, Israel	7500–6200	Kislev et al. 2004
		Northern	Cafer Höyuk, Turkey	9200–8600	Moulins 1997
			Halula, N. Syria	8700–7900	Willcox 1996
			Çayönü, Turkey	8600–8300	Van Zeist and de Roller 1994
			Sabi Abyad II, N. Syria	8500–8000	Van Zeist et al. 2000
			Ras Shamra, N. Syria	8500–8000	Van Zeist and Bakker-Heeres (1984)
			Bouqras, N. Syria	8400–7900	Van Zeist and Waterbolk-van Rooijen 1985
			Can Hasan, Turkey	8400–7700	Hillman 1972, 1978
			Abu Hureyra, N. Syria	8300–8000	Moulins 1997; Hillman 1978

[a]Calibrated dates for the start and end of each period were calculated using the calibration software OxCal v.4,2Bronk Ramsey ©2020 and the new dataset of the IntCal20 in Reimer et al. (2020)

[b]Levantine corridor regions—Southern Levantine corridor begins in the Damascus Basin in Syria and extending to the Jordan Rift Valley in Israel and Jordan; northern Levantine corridor begins in the Taurus foothills (Diyarbakir area) in southeastern Turkey, extending southward and incorporating the middle Euphrates

artificial selection). Climatic changes causing thinning out of food supply had occurred in the past and pre-Neolithic man reacted to them, most probably, by migration. Why had man not started plant domestication earlier? Presumably, at the end of the Natufian period, humans reached a cognitive stage that enabled them to comprehend the life cycle of plants and, thereby, to assume control over plant production. In addition to the development of tools for planting, harvesting, and food preparation as well as storage facilities and other agrotechnological skills, this required the ability to be engaged in more intensive socio-economic relationships. Apparently, the late Natufian or early PPNA humans became well familiar with the useful plants in their immediate environment through the gathering and harvesting of wild plant stands (Ladizinsky 1985). This know-how prepared them to start cultivation.

With the increase in agricultural activities, the economy of the early Neolithic communities became largely dependent on the products of cultivation. Hunting, fishing, and gathering of wild fruits, seeds, and plants supplemented the diet, which became, however, increasingly dependent on cultivated wheat, barley, and pulses (Bar-Yosef and Kislev 1989). The domesticated wheat, barley, rye, oat, and legumes provided mankind with a highly nutritive food which was low in water content and, therefore, easy to store, transport and process. This was a prerequisite for the development of human civilization.

Prior to this event, during the Epipalaeolithic phase, the climate was favorable, and dense populations of hunter-gatherers were settled in territories that included some possible year-round settlements (Hillman 1996). Due to a limited amount of available data, little is known of this pre-domestication period when human were gatherers and brought grains near their dwellings. The earliest data relevant to wheat comes from archeological records from the upper paleolithic Last Glacial Maximum period ∼23,000 years ago in the hunter-gatherer sedentary camp of Ohalo II on the shores of the lake of Galilee (Fig. 13.1), suggesting that there might have been a period of cultivation of wheat, that was not yet domesticated (Snir et al. 2015). The evidence is that extensive farming-related activity was detected at this site, as deduced from various flints, sickle blades and stone grinding tools, fauna remains, and large amounts of seeds (∼10,000 seeds from cereals) including wild emmer wheat, wild barley and wild oats, in what was a human-disturbed environment containing seeds from presumed weedy species. If this interpretation is correct, this would be the earliest known site of wild wheat cultivation. Alternatively, it might be a site where wild emmer wheat, harvested from nearby wheat stands, was brought, and processed. Ohalo II is a singular case as there is a gap of ∼10,000 years before other human sedentary settlements were found in the Late Natufian and early PPNA period

(Fig. 13.1). The Younger Dryas brought an end to this, with most sites being abandoned, and it is argued that a few groups may have resorted to cultivation during this period. Village populations reappeared throughout the area during the Pre-Pottery Neolithic A (PPNA) period (12,000–10,800 calibrated years ∼10,300–9500 uncalibrated years BP; Tables 1.1 and 13.1), and many of them appear to have been cultivators. Finds of domesticated plants are generally widespread in the subsequent Pre-Pottery Neolithic B (PPNB; 10,800–8300 calibrated years ∼ 9500–7500 uncalibrated years BP; Tables 1.1 and 13.1), and it was by this period that they began to spread beyond the domestication zone into central Turkey, Cyprus, Crete, and southern Greece (Fuller 2007). There is a consensus among archaeobotanists (not considering Ohalo II) that cultivation of wild cereals predated morphological domestication by >1000 years.

For many years it has been assumed that the transition to agricultural activities was a revolutionary process, and, accordingly, it was referred to as the "Neolithic or Agricultural Revolution". However, the accumulating archaeological data indicate that this shift was rather an evolutionary process (Kislev 1992; Hillman 1996; Harris 1998; Bar-Yosef 1998), a step-by-step domestication of different plants and animals. Each step, however, seems to have occurred during several hundred or even thousand years, and each crop may have been domesticated separately in time and place. Some PPNA villages were established in the absence of plant and animal domestication, which in turn occurred sometimes in the absence of village formation. Yet, on a prehistoric scale, sedentism and agriculture emerged virtually simultaneously. This is the concept of a diffuse beginning of agriculture. Only later, following the establishment of several domesticated crop plants, could these domesticated plants and animals have made such a powerful impact when spread from the western part of the Fertile Crescent (Kislev 1992). Although the recent accumulating archaeological data provide a clearer insight as to the origin and spread of wheat agriculture, still it is difficult to determine the exact date of the earliest agricultural activities, partly because only a few such operations were practiced in the early stages.

Several major events and phases are recognized in the development of wheat cultivation (Table 1.1): (i) harvesting from wild emmer and einkorn stands—a phase of agrotechnical development and preparation for cultivation; (ii) a pre-domestication cultivation period when wild emmer and einkorn were grown in small plots—the first phase of cultivation; (iii) appearance of non-brittle emmer and einkorn; cultivation of brittle and non-brittle types of emmer and einkorn wheats in mixture—the second phase of cultivation; (iv) appearance of free-threshing, naked tetraploid wheat; cultivation of wild and domesticated emmer and naked tetraploid wheat in mixture; expansion of wheat culture to all regions of the fertile crescent; appearance of non-brittle hexaploid wheat; significant increase in human population and site size—the third phase of cultivation; (v) spread of durum wheat culture to central Asia, southern Europe, and Egypt—expansion of durum agriculture; (vi) spread of bread wheat to many parts of the world and the accumulation of landraces—expansion of bread wheat agriculture; and (vii) modern breeding and the green revolution.

The agrotechnical phase was developed mainly during the Natufian period (15,000–12,000 calibrated years ~ 13,000–10,300 uncalibrated years BP; Table 1.1), although its beginning can be traced back several millennia earlier (Kislev 1984). Remnants found in Natufian sites in the Levant indicated that wheat as well as other plants were the important dietary constituents in this period. The unearthing of sickle blades, pestles, pounding stones, and querns as well as storage pits attest to the extensive collection of cereals by the Natufians. Several permanent or semi-permanent settlements that were based on hunting and gathering were formed during that period. One of these is the permanent settlement at Tell Abu Hureyra, in the Euphrates basin of northern Syria, first settled in 11,500 uncalibrated years BP. Remnants of many animals as well as of wild species of einkorn, barley, rye and legumes from late Natufian layers were identified at this site (Hillman 1975). It is therefore, assumed that the Natufian economy was based on hunting and on intensive collection and consumption of seeds. At that time, the flora and fauna on the border of the Mediterranean maquis were much richer than today's providing settlers with a large and reliable source of plant and animal food. Thus, hunters and gatherers could settle and increase in population, without engaging in agriculture. The invention of the sickle for harvesting and of other tools for grain-processing comprise major components of the agrotechnical revolution Kislev 1984. During this period the Natufians collected the large-grained wild grasses and legumes and became acquainted with their biology—a prerequisite for their cultivation. The initial attempts at plant cultivation, presumably mainly by women, might have taken place during this period.

The "domestication revolution" (12,000–8300 calibrated ~ 10,300–7500 uncalibrated years BP) can be divided into two sub-phases: cultivation of wild forms of cereals, and cultivation of domesticated forms Tables 1.1 and 13.1). Based on the archaeological evidence, the transition from the Natufian to the Neolithic culture occurred in the Jordan Valley, the Damascus basin, and the Middle Euphrates (Bar-Yosef and Kislev 1989; Bar-Yosef 1998; Harris 1998) and was rather rapid, although the early farmers were involved both in plant cultivation and in hunting. The earliest experiments with the cultivation of wild forms of

emmer and barley in the southern part of the Fertile Crescent and of einkorn, *wild timopheevii*, and barley in the northern part of this region, took place at early PPNA; wild einkorn in the north Levant as soon as 13,000 Cal-years BP (Abu Hureyat I and Mureybit I-III) and 11–12,000 Cal-Y BP, with wild emmer in the south Levant in Netiv Hagdud and Zaharat adh-Dhra (see reviews by Nesbitt 2001; Willcox 2012) and wild barley and to a lesser extent also of wild emmer, in the eastern part of the fertile crescent (Riehl et al. 2013). There was no clear morphological evidence for the presence of domesticated wheat. Domestication of animals (sheep and goats) occurred later (about 9000–8500 uncal years BP). Cereal farming itself may have originated in areas adjacent to rather than within the regions of greatest abundance of these wild cereals. Such prehistoric settlements peripherally to, rather than within, the current distribution areas of these species are Jericho, Gilgal, Netiv Hagdud, and Gesher in the Jordan Valley, Tell Aswad and Tell Ghoraife near Damascus, and Tell Abu Hureyra and Tell Mureybit in northern Syria and Chogha Golan in the Zagros mountains (Fig. 13.1).

Of the various species of cereals that grew in the Fertile Crescent and were harvested by the Natufians, only wild emmer and barley in the south, and barley and einkorn, and possibly also wild *timopheevii* in the north, were cultivated by the early PPNA people (van Zeist and Bakker-Heeres 1985; Kislev et al. 1986; Riehl et al. 2013). Assuming that the amount of grain collected per unit time was the most important criterion (Evans 1981), wild stands of wheat and barley were preferred over other cereals, because their large, heavy grains borne in spikes facilitated their harvesting. Indeed, Harlan (1967) in wild einkorn and Ladizinsky (1975) in wild emmer succeeded in gathering considerable amounts of grains per hour of harvesting. Wild emmer was a better candidate for domestication than barley, having larger grains and two grains in each dispersal unit compared to one in barley. Indeed, Ladizinsky (1975) harvested from wild stands twice as many grains of wild emmer as of wild barley. The taste of wheat, its high nutritional quality, and its large free grain might have also contributed to its preference over barley. In addition, judging from today's pattern of distribution, barley was probably very common and grew in abundance within a short distance from the early Neolithic settlements, while the dense stands of wild emmer were somewhat more distant. This might have created additional pressure to cultivate wheat, as sufficient quantities of barley could have been harvested from nearby wild stands.

The early Neolithic settlements in the Jordan Valley were located on the edges of alluvial fans or on alluvial terraces (Bar-Yosef and Kislev 1989). It is assumed that cultivation was based on sowing in such alluvial fans and terraces, as well as on the edges of freshwater ponds where the water table was always high, and the soil was periodically fertilized by floods of mud. The exploitation of wetted soils is also evidenced in other early agricultural sites such as Tell Aswad and Tell Ghoraife in the Damascus basin (van Zeist and Bakker-Heeres 1985).

The "invention" of cultivated fields by the early Neolithic people of the Levant and the development of agronomical practices associated with field preparation and sowing, as well as the selection of seeds to be sown, comprise the initial stages in agricultural technology. The conversion of alluvial terraces and, later on, also of grass-stands of annuals, into fields was undoubtedly a major change through which man affected not only cultivated plants but also plants that remained in the wild. The establishment of cultivated fields by early Neolithic man could have been inspired by the observation of natural herbaceous covers of annuals, such as wild emmer and barley, in open forest belts of deciduous oak. Man may have burnt off unwanted grass in the open forest and used the clear space for sowing wheat or barley and, later on, expanded this space to adjacent areas by clearing evergreen woods and shrubs. The transformation of the landscape from a high perennial evergreen vegetation to cultivated fields of annual grasses was perhaps the first and most important man-initiated change in the world of vegetation. The selection of grains for the next sowing season was the first event whereby man started to shape the genetic structure of the cultivated wheats. Whatever were the criteria for selection, these samples of early cultivated wheats became subjected to different selection pressures than those affecting their sibs that remained in the wild stands. The second sub-phase of the "domestication revolution", the cultivation of domesticated forms, presumably started several hundred years, or more, after the beginning of cereal cultivation. In the cultivated forms, the ripe ear no longer disarticulates into single spikelets as easily as in the wild forms, thus allowing the harvesting of intact spikes without the need to collect single spikelets.

We know little of the predomestication period, when human were gatherers and brought grains near their dwellings, due to a limited amount of available data. The earliest data relevant to wheat comes from archeological records from the upper paleolithic Last Glacial Maximum period approximately 23,000 years ago in the hunter–gatherer sedentary camp of Ohalo II on the shores of the lake of Galilee (Fig. 13.1), suggesting that there might have been a period of cultivation of wheat that was not yet domesticated (Snir et al. 2015). The evidence is that extensive farming-related activity was detected at this site, as deduced from various flints, sickle blades and stone grinding tools, fauna remains, and large amounts of seeds (approximately 10,000 seeds from cereals) including wild emmer wheat, wild barley, and wild oats, in what was a human-disturbed environment containing seeds from weedy species (presumably a field). If this interpretation is correct, this would be the earliest known site of wild wheat cultivation. Alternatively, it might be a site where wild emmer wheat, harvested from nearby wheat stands, was brought and processed (Piperno et al. 2004). Ohalo II is a singular case as there is a gap of approximately 10,000 years before other human sedentary settlements were found during the Younger Dryas in the Late Natufian early Pre-pottery Neolithic-A period throughout the fertile crescent (Map). Yet, it is generally accepted (Kislev 1992; Hillman 1996; Harris 1998; Bar-Yosef 1998) that domestication occurred about 1000 years, or more, after the initial cultivation of wild cereals, namely, at the beginning of the PPNB (ca. 9500 uncal years BP) when types with non-brittle spikes and some also with naked grains were clearly identified (Kislev 1992). On the other hand, for several legumes and flaxes the archaeological data indicate that these were already domesticated in the PPNA.

Regardless of the exact time of domestication, the most characteristic feature of domesticated cereals is the dependency on man for sowing: once the seed-dispersal mechanism was lost, the genetic donor to the next generation became solely determined by man. Man's selection and raising of domesticated wheat could have occurred only after ploughing and sowing had been practiced (Kislev 1984). It seems that at the end of the PPNA all essential agricultural practices had been already established. From this stage onward, the transformation of some of the wild cultivated forms into domesticated wheats proceeded very rapidly. This transformation involved not only the loss of the self-propagation mechanism, but also that of self-protection (stiff glumes), resulting in a free-threshing, naked grains. Increase in seed size and the loss of seed dormancy leading to a uniform and rapid germination was also achieved at this stage. Traits of strong negative adaptive value in the wild that had positive value under cultivation were favored by the early farmers. As such, the farmer's election imposed a new evolutionary pathway in the cultivated wheats, entirely distinct from that operating in wild wheats.

The replacement of the wild cultivated forms by domesticated ones was a slow process. According to Harris (1998), the establishment of domesticated populations of cereals was slowed down by constant gene flow from wild forms, which either grew in mixture or nearby the domesticated forms. Moreover, no severe selection in favor of domesticated forms was applied by the early farmers, possibly because of the techniques of harvest that were practiced at that time.

Hence, the establishment of agriculture in the Levant and neighboring regions was a very gradual process that took place over a period of 3000 years (from the beginning of the PPNA to the end of the PPNB). During these three millennia, agriculture became the main production system that supported man in southwest Asia.

The third phase of the agricultural revolution, the expansion of agriculture (in the 8th and 7th millennia BP; Kislev 1984; Harris 1998; Bar-Yosef 1998), was accompanied by a rapid and radical change in the economic organization of the Near East and the surrounding regions. Cereal culture spread from the western flank of the Fertile Crescent to other parts of this region (Civáň et al. 2013), and from there to central Asia through northern Iran, to Southeastern Europe through Transcaucasia, to Europe and North Africa through southwest Anatolia, and to Egypt through Israel and Jordan (Feldman 2001). According to Bar-Yosef (1998), expansion of agriculture was mainly due to the spread of farmers to new territories rather than the adoption of farming by the hunters–gatherers. Most wheats were the hulled domesticated emmer and einkorn; the more advanced naked tetraploid and hexaploid wheats were relatively rare.

Kislev (1994a) subdivided the "Agricultural Revolution" into three stages: the Agrotechnical Revolution, the Domestication Revolution, and the Expansion of Agriculture. Weiss et al. (2006) divided the process of domestication of plants into three stages: "gathering," in which people gathered annual plants from wild stands, "cultivation," in which wild plant genotypes were systematically sown in fields of choice, and "domestication," in which mutant plants with desirable characteristics were raised. Similar divisions of the process of domestication were suggested by Harris (1989) and Fuller (2007). Hillman and Davis (1990) defined domestication as a process causing populations of cultivated plants to lose features, particularly reproductive features, necessary for their survival in the wild habitats, i.e., a process which ultimately renders crop populations dependent on human intervention for their reproduction. Such a process involves genotypic changes in entire populations. Doebley et al. (2006) defined domestication of a plant species as the various genetic modifications in its wild progenitor made to meet human needs.

The most characteristic feature of domesticated wheats is their dependence on humans for gathering and planting. This development involved genetic changes in several crucial traits, defined by Hammer (1984) as the "domestication syndrome." In emmer wheat, this syndrome is characterized, sensu stricto, by suppression of seed dispersal, increased seed size, and free-threshing grains. While wheat domestication involves a limited number of chromosome regions (Table 13.2), though many relevant quantitative trait loci (QTLs) have also been detected (Peng et al. 2011).

One can distinguish between these sensu stricto syndrome traits and traits that were selected during a "process of optimization" for cultivated fields. These latter changes involved gradual genetic modifications, mainly due to non-conscious and sometimes conscious selection by early farmers, to increase crop adaptation to cultivated fields and to human consumption and needs. In tetraploid wheat, such optimization led to changes that resulted in rapid and uniform germination, erect plants, increased plant height, reduced tillering, and larger spikes and grains. Meyer and Purugganan (2013) made a distinction between domestication and diversification, with the former referring to the onset of evolutionary divergence from the wild ancestral species, and the latter referring to the subsequent evolution of new varieties.

Although many genes in various crops have been proposed to be domestication genes, only a few have been shown to have been targeted by selection (Olsen and Wendel 2013). When exposed to selection, both the favored genetic variant and other, neutral genes linked to it, increase in frequency, a phenomenon called a selective sweep (Wang et al. 1999; Olsen et al. 2006). Consequently, the related genetic region shows reduced genetic diversity.

13.2 Domestication of Tetraploid Wheat

13.2.1 History

The origin of domesticated plants and the development of agriculture have always stimulated man's curiosity and imagination. In the ancient world, where every phenomenon and every event were explained in a mythological manner, the philosophers and historians of many nations considered domesticated plants to be a generous gift from the gods. Thus, the ancient Egyptians were grateful to Isis and Osiris for bringing wild emmer wheat and wild barley from Mt. Tabor, Israel, to Egypt and teaching people how to cultivate them. Similarly, the ancient Greeks ascribed the gift of these important cereals to Demeter and the Romans to the goddess Ceres.

During the last few centuries, botanists and students of agricultural history have attempted to explain the origin of domesticated plants and their evolution under domestication on a scientific basis. Driven by the desire to find the historical and evolutionary truth, Link (1816), de La Malle (1826), and particularly de Candolle (1886), were the first to realize that historical, linguistic and folkloristic evidence is insufficient to trace the origin of domesticated plants. Botanical, geographical and archaeological studies are necessary to advance this field of science; the only definite

Table 13.2 Domestication syndrome genes

Trait	Genes (wild-type alleles)	Chromosomal-arm location	Wild-type phenotype	Species	References
Non-brittle rachis	Br1-3A	3AS	Brittle rachis (W-type)	T. turgidum ssp. dicoccoides	Watanebe et al. 2002, Watanebe et al. 2005 Nalam et al. 2006
	Br1-3B	3BS	Brittle rachis (W-type)		
	Br1-3D	3DS	Brittle rachis (W-type)	T. aestivum ssp. tibetanum	Chen et al. 1998
	Br2-3A?	3AL?	Null effect?		
	Br2-3B?	3BL?	Null effect?		
	Br2-3D	3DL	Brittle rachis (B-type)	Ae. tauschii, T. aestivum ssp. spelta	Watanabe et al. 2005 Li and Gill 2006
Free threshing	Q-5A	5AL	Tough glumes	T. turgidum ssp. dicoccoides and dicoccon; T. aestivum ssp. spelta	Sears 1954 Simonetti et al. 1999 Simons et al. 2006
	q-5B	5BL			
	q-5D	5DL			
	Tg -2A?	2AS?			
	Tg -2B	2BS	Tenacious glumes	T. turgidum. ssp. dicoccoides and dicoccon,	Simonetti et al. 1999
	Tg-2D	2DS	Tenacious glumes	Ae. tauschii, T. aestivum ssp. spelta, macha and vavilovii	Kerber and Rowland 1974

demonstration of the origin of a domesticated plant is the discovery and identification of its wild prototype. In cases where the wild prototype is unknown or extinct, the origin and full history of the domesticated crop can never be ascertained. In such a situation, it is very difficult, if not impossible, to identify the site(s) of domestication, the genetic changes that led to the formation of the primitive domesticated form, and the evolution of the domesticated crop under cultivation.

One of the first attempts to identify the progenitor of domesticated bread wheat, was in the middle of the nineteenth century, when several botanists regarded the natural hybrid derivatives of bread wheat x *Aegilops geniculata* (formerly *Ae. ovata*) as the ancestral forms of common wheat. The hybrid was first discovered by Requine, in 1821, in southern France, and later on, it was also collected in northern Italy and North Africa. Because of its resemblance to common wheat, Requine (see Fabre 1852) named it *Aegilops triticoides*. Due to the fact that there were also many intermediate forms between *Ae. triticoides* and common wheat, Fabre (1852, 1855) concluded that all domesticated wheats originated from *Ae. geniculata*. Since he found, in several cases, that grains from *Ae. geniculata* ears growing near wheat fields, yielded *Ae. triticoides* plants, he assumed that under cultivated conditions, *Ae. geniculata* gradually transformed into bread wheat. This hypothesis, which was accepted by several botanists, was disproved by Godron (1854, 1856, 1858a, b, 1869, 1876), who

demonstrated the hybrid nature of *Ae. triticoides* and all other intermediate forms between *Ae. geniculata* and bread wheat. He produced similar forms by crossing bread wheat with *Ae. geniculata* and backcrossing the hybrids to the two parental species.

These experiences with *Ae. triticoides* emphasize the need for a better definition of key features characterizing the wild prototype of the domesticated wheats. The wild prototype must be a valid species and, therefore, a self-propagating plant. It should contain a spike similar to that of domesticated wheats, but with a brittle rachis that disarticulates into single spikelets upon maturity. In addition, the wild prototype, like many other wild grasses, should have tightly closed glumes, resulting in a 'hulled grain' for protection against extreme climatic conditions and herbivores. Eco-geographically, the prototype should be characterized by a specific distribution and occupy well defined primary habitats.

In the second half of the nineteenth century, several botanists (e.g., Hausknecht) were of the opinion that all domesticated wheats derived from the single wild species, that was known to botanists at that time, *T. monococcum* ssp. *aegilopoides*. Yet, the majority of botanists interested in the origin of domesticated wheats, maintained the opinion that wheat origin is polyphyletic, and that at least two wild species serving as the progenitors.

The first evidence of wild wheat, presumably wild emmer, came from the Chaldean priest Berosus, who lived at about 2700 years before present (BP). He mentioned the occurrence of wild wheat in Mesopotamia (Syncellus, Frag. Hist. Graec., vol. 2, p. 416). In more recent times, Linnaeus (1753) cited Heintrelmann who found wild wheat in northwest Iran. Olivier (1807) found wheat, barley, and spelt in uncultivated areas northwest of Anah, on the right bank of the Euphrates, and mentioned that he had already seen such wild wheat several times in northern Mesopotamia. Since specimens of these cereals have not been preserved, it is not possible to identify the wheats to which Olivier referred, but it is highly probable that the spelt was wild emmer. In 1877, Andre Michoux saw spelt wheat growing wild north of Hamadan, western Iran, (see de Candolle 1886). Based on these reports, de Candolle (1886) assumed that the Euphrates basin was the distribution area of the two wild progenitors of the domesticated wheats.

Since wild tetraploid wheat was not known at the second half of the nineteenth century, most botanists and archaeologists agreed at that time with the theory of Solms-Laubach (1899), namely, that domesticated wheats, other than those derived from wild T. monococcum, originated in central Asia and that the wild progenitors of domesticated durum and bread wheats were lost as a result of drastic climatic changes. All previous evidence (mentioned above) were neglected. Moreover, no serious attention was given to the German botanist Friedrich August Körnicke, and several other botanists, who maintained, not only that a prototype of such a progenitor exists, but that they already had two spikelets of such wild wheat.

In 1855, the Austrian botanist Theodor Kotschy collected a plant of wild barley from Rashaya, the northwestern slopes of Mt. Hermon, Lebanon, and kept it in the herbarium of the National Museum of Vienna. In 1873, Körnicke, who analyzed barley plants in several European herbaria, found a segment of wheat spike among the culms of Kotschy's barley. Since this spike had a brittle rachis and two grains in each spikelet, Körnicke believed that it belonged to wild wheat, the progenitor of most domesticated wheats, and named it T. vulgare Vill. var. dicoccoides Körn. [This wild wheat is currently known as *T. turgidum* ssp. *dicoccoides* (Körn. ex Asch. and Graebn.) Thell.]. Körnicke believed that it grew in the Mt. Hermon area and, since he could not obtain support for a scientific expedition to that area, he asked Aaron Aaronsohn, an agronomist and amateur botanist who lived in Israel (then Palestine), to search for this wild wheat in the Mt. Hermon area. Schweinfurth (1906) reports that Aaronsohn found this wild wheat first in the settlement of Rosh Pina, northern Israel, and later the slopes of Mt. Hermon.

One year later, Aaronsohn found that this wild wheat grew abundantly in the southern Levant, namely, in northeastern Palestine, northwestern Jordan, southwestern Syria and southeastern Lebanon, having a wide range of morphological forms across the distribution area (Aaronsohn 1909, 1910). Later, it was also found in northern Syria, southeastern Turkey, northern Iraq and southwestern Iran (Harlan and Zohary 1966; Kimber and Feldman 1987). Specimens were also collected in 1910 by Theodor Strauss in the mountainous region of western Iran near Kerind. Thus, Körnicke's hypothesis that the wild progenitor of domesticated wheats, a two-grained wild wheat, still grows in the Near East, was fully confirmed. Aaronsohn's collections of wild wheat specimen from the southern Levant served as the basis for a series of botanical and genetic studies by Cook (1913), Schulz (1913a, b), von Tschermak (1914), Flaksberger (1915), Percival (1921), Vavilov 1932), and others.

Understanding the "where", "when", and "how" of wheat domestication was significantly advanced by the discovery of wild emmer wheat, *Triticum turgidum* (L.) Thell. ssp. *dicoccoides* (Körn. ex Asch. and Graebn.) Thell. in nature by Aaron Aaronsohn in 1906 (Aaronsohn and Schweinfurth 1906). Following this discovery, Schultz (1913b) was able to assemble the first natural classification of the wheat species in which he recognized three series: einkorn (one-grained wheat), emmer (two-grained wheat), and dinkel (Table 10.1).

Soon after, von Tschermak (1914) crossed wild emmer with the domesticated forms of tetraploid wheat, *T. turgidum* ssp. *dicoccon* (Schrank) Thell. and ssp. *durum* (Desf.) Husn., and concluded from the high fertility of the hybrids that the wild and the domesticated forms are genetically very close. Cook (1913) and Percival (1921) supported this notion by reporting on the occurrence of natural hybrids between these taxa in Israel (then Palestine) and Syria, as well as their spontaneous hybridization wherever they were brought into contact with each other in experimental fields. The discovery of the chromosome number of the wheats (Sakamura 1918) showed that the three series of Schulz represent a polyploid series in which the einkorn are diploids (2n = 14), the emmer are tetraploids (2n = 28), and the dinkel are hexaploids (2n = 42). Not surprisingly, analysis of first meiotic metaphase configurations in hybrids between wild emmer and domesticated tetraploid wheat showed that the wild chromosomes were fully homologous to those of all the domesticated subspecies of T. turgidum (Kihara 1924, 1937, 1940; Sax 1921a, b, 1922; Percival 1921). These studies supported the hypothesis of Körnicke (1889), Schweinfurth (1906, 1908), Aaronsohn and Schweinfurth (1906), Aaronsohn (1909, 1910), Schulz (1913b), von Tschermak (1914), and others, that domesticated emmer,

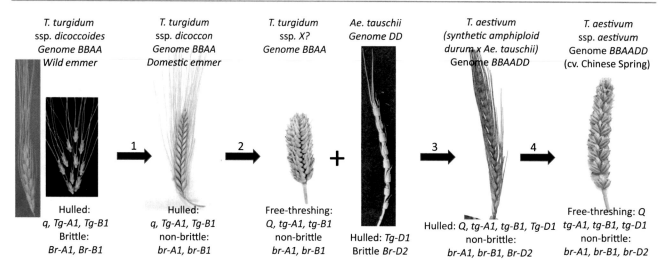

Fig. 13.2 Major mutations and morphological changes during wheat domestication: (1) The transition from ssp. *dicoccoides* to ssp. *dicoccon* involved mutations in the Brittle rachis loci. Some modern emmer wheat lines might also contain mutations in some but not all loci affecting free threshing. (2) Free-threshing tetraploid wheat, named ssp. *parvicoccum*, appears in the archeological record approximately 2000 years before ssp. *durum*. It is now extinct but might have resembled the tetraploid wheat (Genome BBAA) shown here as ssp. X that was extracted from hexaploid wheat and has a compact spike and small grains. Its genotype must have been similar to *durum*,

namely free threshing with soft glumes, with mutants *Q* and *tg-A1*, *tg-B1*. (3) The hybridization of this free-threshing tetraploid wheat with the DD subgenome donor, *Ae. tauschii*, gave rise to a primitive hulled hexaploid wheat, different from spelt wheat due to the *Q* factor, and absent from the archeological record. It likely resembled the picture shown from a synthetic hexaploid between ssp. *durum* and *Ae. tauschii* shown here. (4) Soon after its formation, hexaploid wheat became free threshing thanks to a mutation in *Tg-D1* and its rachis became thicker thanks to a mutation in *Br-D2* (Fig. 3 from Levy and Feldman (2022))

ssp. *dicoccon*, the hulled type of *T. turgidum* and therefore the primitive domesticated tetraploid wheat, derived from wild emmer, ssp. *dicoccoides*, by a series of mutations (Fig. 13.2). Genetic, cytogenetic, and more recently evidence from whole genome sequence (Avni et al. 2017) all indicate that ssp. *dicoccoides* is the direct progenitor of domesticated emmer, durum, and bread wheat.

Vavilov (1926) rejected the view that the domesticated forms of tetraploid wheat derived from wild emmer. His conclusion was based on sterility of hybrids between domesticated emmer or *durum* and what he thought was wild emmer. This latter species was actually another wild tetraploid wheat taxon, *T. timopheevii* (Zhuk.) Zhuk. ssp. *armeniacum* (Jakubz.) van Slageren, from Transcaucasia, which differs from wild emmer in its genomic composition (Lilienfeld and Kihara 1934). In Vavilov's view (1926), Aaronsohn's eco-geographical data were insufficient to regard wild emmer as the progenitor of domesticated tetraploid wheat. From the wealth of varieties and forms of domesticated emmer in Ethiopia, as compared with the relatively few variations in Israel and Syria, he considered Ethiopia, where wild emmer does not grow, to be the site of domestication of tetraploid wheat. Vavilov's concept, however, was rebutted by morphological, eco-geographical, cytogenetic, and molecular data.

13.2.2   Ecogeographical Characteristics of the Area of Wheat Domestication

Following the discovery of wild einkorn, *T. monococcum* ssp. *aegilopoides*, during the second half of the nineteenth century, and of wild emmer, *T. turgidum* ssp. *dicoccoides*, in the beginning of the twentieth century, the putative progenitors of the domesticated einkorn and emmer wheats, respectively, their distribution area was determined. It was then relatively easy to establish the geographical region that was the cradle of agriculture, and the specific regions from which these two-wheat subspecies were taken into cultivation.

The wild progenitors of cultivated wheats are natural constituents of some of the open oak-park belts and the herbaceous plant formations in southwest Asia. Their assumed center of origin and current center of distribution and diversity is in the "Fertile Crescent"—a hilly and mountainous region extending from the foothills of the Zagros mountains in south-western Iran, through the Tigris and Euphrates basins in northern Iraq and southeastern Turkey, continuing southwestward over Syria to the Mediterranean, and extending to central Israel and Jordan. Wild taxa belonging to 4 species of wheat and 17 species of the closely related genus *Aegilops* are endogenous to this region, which most likely was the arena of wheat domestication. This assumption is supported by ample archaeological evidence.

The Fertile Crescent is bounded by the Mediterranean in the west, by chains of large and high mountain ranges in the north and east (the Amanos in north-western Syria, the Taurus in southern Turkey, Ararat in north-eastern Turkey and the Zagros in western Iran), and in the south by the Syrio-Arabian desert with its western extension (Paran desert) in the Sinai Peninsula. Situated between the sea, the mountains, and the desert, the Fertile Crescent is under the influence of several different climates: on the one hand, it enjoys the temperate Mediterranean climate with a short, mild and rainy winter and long, hot and dry summer, yet, on the other hand, it is influenced by the more extreme steppical climate of the Iranian and Anatolian plateaus in the east and north, and by the desert climate in the south. Consequently, the Fertile Crescent encompasses two different phytogeographical regions, the Mediterranean in the south-western part and the Irano-Turanian in the north-eastern part, and is also affected by two other regions, the Saharo-Arabian in the south and the Euro-Siberian in the North. The Mediterranean part of the Fertile Crescent includes Israel, Jordan, Lebanon Syria and the western part of south-eastern Turkey and it centers around the Syrio-African rift (the Jordan rift valley in the south, the Beqa Valley of Lebanon and the Orentos valley in Syria). The Irano-Turanian part includes the eastern part of south-eastern Turkey, northern Iraq, and southwestern Iran and is influenced by the continental climate of the Iranian and Central Asiatic steppes.

No wonder, therefore, that this region is ecologically very diversified, comprising of a wide array of different habitats. Its versatile ecological conditions are manifested by its wide array of plant formations, ranging from well-developed Mediterranean forests and maquis, through open parks, shrubs and herbaceous formations, to small shrub and steppical plant formations. The open parks and the herbaceous formations, containing many annual grasses and legumes, occupy the open habitats in the edges and openings of the Mediterranean maquis, which presumably served in the past as the main pasture area for wild sheep, goats and gazelles—the game of pre-agricultural man.

It is generally accepted that Near Eastern agriculture originated within the distribution area of the wild progenitors in the Fertile Crescent region. Unfortunately, no signs of earlier farming communities were found there (with the possible exception of Ohalo) probably because of climate conditions, wild plants were not available in the same area but rather, further west and south (Hillman 1996; Bar-Yosef 1998). Indeed, new geological, climatic, and archaeological data from the east Mediterranean region indicate that from 13,500 to 11,500 years ago there was the Younger Dryas

climatic event, which was characterized by a cold and dry climate (for details see Hillman 1996; Bar-Yosef 1998). Hillman (1996), using palaeobotanical data, reconstructed the phytogeographical belts of this region during the Younger Dryas and concluded that the habitats of the annual cereals lie mainly in the open areas of the oak-park maquis in a relatively narrow strip of the east Mediterranean (Fig. 13.1). This narrow strip, called the "Levantine Corridor", begins in the Taurus foothills (Diyarbakir area) in south-eastern Turkey and extends along the Mediterranean southward, incorporating the middle Euphrates through the Damascus basin, the Lebanese mountains, the two sides of the Jordan Rift Valley into the Sinai Peninsula (Bar-Yosef 1998). The current distribution of wild wheats in this area is as follows: wild emmer is the dominant wild wheat in the central-southern part of the corridor (from north of Damascus to Jericho), on terra-rosa and basalt soils, while wild einkorn (*T. monococcum* ssp. aegilopoides), *T. urartu,* and wild *T. timopheevi* and also more sporadically wild emmer, are distributed in the northern part of the corridor. The distribution of the diploid species, ssp. *monococcum* and *T. urartu,* extended southwards up to Mt. Hermon, in the central part of the corridor.

The accumulating archaeological data indicate that agriculture originated in the Levantine Corridor (Table 13.1) This is clearly apparent from the distribution of the earliest Neolithic sites (10,300–9500 BP) (Fig. 13.1): Tell Mureybit and Tell Abu Hureyra in the Middle Euphrates in the northern part of the corridor, Tell Aswad and Tell Ghoraife in the Damascus basin, and Netiv Hagdud, Gilgal, and Jericho in the Jordan Valley, between the Lake of Galilee and the Dead Sea (Bar-Yosef and Kislev 1989; Hillman 1996; Bar-Yosef 1998). All these settlements were established along the ecotone between the relatively temperate Mediterranean phytogeographical region and the steppical Irano-Turanian region. The predominant plant formations in this transitional zone are open parks and herbaceous covers containing a large number of annual grasses and legumes.

13.2.3 Domestication of Emmer Wheat, ssp. *dicoccon*

13.2.3.1 Opening Remarks

Hillman (1996), using palaeobotanical data, reconstructed the phytogeographical belts of the east Mediterranean region during the Younger Dryas, 13,000–11,700 uncalibrated years BP. He concluded that the habitats of the annual cereals lay mainly in the open areas of the oak-park forest, in a relatively narrow strip of the east Mediterranean. Wild emmer was a natural constituent of this corridor but was more widespread in its central-southern part than in its northern part. This pattern of distribution of wild emmer wheat was also in existence about 2000 years later, when human in this region started to cultivate wild emmer.

Being native to the marginal Mediterranean habitats of the Fertile Crescent, wild emmer *T. turgidum* ssp. *dicoccoides* was preadapted for cultivation. It is an annual plant, which grows in mild winters and endures the dry, hot summer as seeds. Wild emmer is predominantly self-pollinated and has relatively large grains that assist the safe and rapid re-establishment of the stand. Their large seed size rendered them very attractive to the ancient gatherer. Its annual habit made it also amenable for dry farming, while its self-pollination system could have aided in the fixation of desirable mutants and recombinants resulting from rare outcrossing events. While wild emmer occupies poor, thin, rocky soils in its natural habitats, it responds well when transferred to richer habitats.

Cultivation imposed a new evolutionary direction on the wheats, whereby traits that had the greatest adaptive value in the cultivated field were preferred. Consequently, selection pressures exerted by farmers have operated in a different, and sometimes, even contradictory manner in cultivation and in wild. During the 10,000 years of wheat cultivation, the criteria for selection varied from time to time and from place to place, as suggested by Evans (1981) in regard to yield criteria. The first farmers selected plants with large grains and with more grains per spike. Later on, farmers selected for a higher ratio of grain harvested to grain sown, i.e., indirectly selecting for profuse tillering, many grains per spike, and strong grain retention. As the amount of arable land became a limiting factor and the crop monotypic, selection preferred higher number of grains per unit area. In this case, emphasis was given to lines which were weak competitors, yielded well in dense planting, and responding well to fertilizers and various agrochemicals. Thus, during the process of adaptation to the various cultivated environments and to the different demands of man, wheat has responded with a number of significant morphological and physiological changes. The main genetic changes reflect the process of adaptability to agriculture, particularly the loss of the ability of the plants to propagate and protect the seeds, which left wheat became fully dependent on the farmer for its survival.

The earliest data relevant to wheat comes from archeological records from the upper paleolithic Last Glacial Maximum (LGM) period ~23,000 years ago found in the hunter-gatherer's sedentary camp of Ohalo II on the shores of the lake of Galilee, suggesting that there might have been a period of cultivation of wheat, that was not yet domesticated (Snir et al. 2015). The evidence is that extensive farming-related activity was detected at this site, as deduced from various flints, sickle blades and stone grinding tools, fauna remains, and large amounts of seeds (~10,000 seeds from cereals) including wild emmer wheat, wild barley and

wild oats, in what was a human-disturbed environment containing seeds from weedy species (presumably a field). If this interpretation is correct, this would be the earliest known site of wild wheat cultivation. Alternatively, it might be a site where wild emmer wheat, harvested from nearby wheat stands, was brought and processed. Ohalo II (Snir et al. 2015) is a singular case as there is a gap of ~10,000 years before other human sedentary settlements were found in the Late Natufian and early Pre-pottery Neolithic-A (PPNA) (Table 1.1). There is a consensus among several archaeobotanists (not considering Ohalo II) that cultivation of wild cereals predated morphological domestication by >1000 years with wild einkorn in the north Levant as soon as 13,000 Cal-Y BP (Abu Hureyat I and Mureybit I-III) and 11–12,000 Cal-Y BP, with wild emmer in the south Levant (Tell Aswad, Netiv Hagdud, Zaharat adh-Dhra (see reviews by Nesbitt 2002; Willcox 2012). During these periods, there was no clear morphological evidence for the presence of domesticated wheat.

In recent decades, cytogenetic, genetic, and evolutionary studies supplemented by archaeological investigations, contributed to an improved understanding of events occurring during the transition from hunting/gathering to farming (Willcox 1998). In recent years, two important advances have occurred in archaeobotany: the recognition of pre-domestication cultivation and evidence for different sub-centers of crop domestication within the Fertile Crescent (Fuller 2007). Evidence for pre-domestication cultivation has been recognized through the statistical composition of wild seed assemblage (Colledge 1998, 2001, 2002; Harris 1998; Willcox 1999, 2002; Hillman 2000; Hillman et al. 2001). As is well known from later agricultural periods, archaeobotanical assemblages are made up predominately of crops and weeds, together with some gathered fruits and nuts. This pattern was already recognized in the PPNA and in some Late Epipalaeolithic sites, by samples dominated by wild cereals. It was suggested that the primary domesticated crops of the Neolithic Revolution, namely, einkorn wheat, emmer wheat, barley, lentil, pea, chickpea, and flax, appeared initially in a core area, from which they spread throughout the Middle East (Zohary and Hopf 2000; Salamini et al. 2002; Lev Yadun et al. 2000; Abbo et al. 2010). Recent archaeobotanical data, however, indicate that pre-domestication cultivation of the wild progenitors of these species was carried out autonomously in very early sites of the Near-Eastern PPNA (Pre-Pottery Neolithic A; Weiss et al. 2006). In accordance, Willcox (1998) concluded that the wild progenitors of Old-World cereals and legumes were exploited for several millennia, before the appearance of domestic counterparts and in each site, the local wild prototype were taken into cultivation.

At first, wild emmer was grown for several hundred years or more, until forms with a tough rachis and non-brittle spike gradually appeared, and for one millennium or more, were grown in a mixture with brittle forms and gradually replaced them throughout the Levantine Corridor (Kislev 1984; Tanno and Willcox 2006; Feldman and Kislev 2007). An archeological record shows that in ancient sites where agriculture was practiced, a mixture of spikelets with fragile and non-fragile rachises was found, and it took several millennia until the non-fragile spikes became prominent in farming of emmer wheat (Kislev 1984). Also, in agreement with this archaeological finding, genetic data showed that the loss of spike fragility in wild emmer was a gradual process (Nave et al. 2019) and was presumably spread very slowly from field to field by farmers. The gradual and prolonged replacement of wild emmer by the domesticated form was presumably caused by the fact that farmers used to collect the spikelets from the ground rather than harvest the spikes (Kislev 1984; Kislev et al. 2004), applying only a weak selection in favor of the latter. If wild cereals were harvested simply by passing through stands and shaking or beating spikes to knock spikelets into a basket, then the shattering, wild-type genotypes would be the ones to predominate in the next year's crop. Also, a good portion of wild spikelets that fell to the ground were not collected and germinated next season, contributing to the new generation of wild wheat.

Hole (1998) discussed the spread of agriculture from its apparent origin in the southern Levant, into the northeastern part of the Fertile Crescent. The first evidence of cultivation of wild emmer was found in PPNA sites in the southern Levant: Jericho, Netiv Hagdud, Gilgal, Gesher, all clustered within a 15 km radius on alluvial fans in the Jordan Valley, and sites in southern Syria (Kislev 1997; Colledge 2001; Willcox 2005; van Zeist and Bakker-Heeres 1982; Hopf 1983). All these sites are dated to approximately calibrated 10,300–9500 BP, just following the Younger Dryas cold interval which terminated ca. 11,000 BP or a little later (Becker et al. 1991). Mixtures of wild and domesticated emmer were found in the southern Levant in the early PPNB (9500–9000 BP) (Hopf 1983; Kislev 1988; Rollefson et al. 1985; Colledge 2001; van Zeist and Bakker-Heeres 1982). The first actual domesticates on the Euphrates appeared in Halula, north Syria, no earlier than 9000 BP, at least 500 years after their occurrence in the Jordan Valley. True agricultural villages appeared in the uplands of the Zagros at 8500 BP, about 1000 years later than in the Levant. Once agriculture began, it spread through diffusion to indigenous people (Harris 1996), or perhaps had been 'invented' repeatedly, or introduced by emigrant colonizers (Hole 1998).

Fuller (2007) suggested that the domestication syndrome in cereals usually meets the following six criteria: (1) Mutations forming a non-brittle rachis. This is often regarded as the single most important domestication trait, rendering a species dependent upon the farmer for survival; (2) Reduction in seed dispersal aids. Wild wheat and barley

have a range of structures that aid seed dispersal, including hairs, barbs, awns and even the general shape of the spikelet. Domesticated wheat spikelets are less hairy, have shorter or no awns and are plump, whereas in the wild, they are heavily haired, barbed, and aerodynamic in shape (Hillman and Davies, 1990). This can be considered to have come about by the removal of natural selection for effective dispersal, and once removed, metabolic 'expenditure' on these structures is reduced. These traits might evolve under initial cultivation and can be regarded as part of 'semi-domestication'; (3) Trends towards increasing grain size. This is likely to be selected for by open environments in which larger seedlings have advantages, surviving deeper burial within disturbed soils; thus, this trait is generally selected for by tillage and cultivation (Harlan et al., 1973). Larger seeds are strongly correlated with larger seedlings; (4) Loss of germination inhibition. Crops tend to germinate as soon as they are wet and planted: (5) Synchronous tillering and ripening. Planting at one time and harvesting at one time will favor plants that grow in synchronization; (6) More compact growth habit, erect versus prostrate plants.

13.2.3.2 Selection for Non-brittle Rachis

Several wild characters which had no pre-adaptive value for cultivation were selected against during domestication. The change from the wild emmer to the most primitive domesticated form domesticated emmer, *T. turgidum* ssp. *dicoccon*, involved the selection of phenotypic characters that suited the farmer rather than the wild environment. This group of traits represent the 'domestication syndrome' (Hammer 1984), and in wheat, mainly involved dramatic changes in seed dissemination, size and germination, and mode of seed protection (Zohary et al. 2012). Morphologically, domesticated emmer is similar to wild emmer, but differs from it by several of the domestication syndrome traits that affected the morphology and physiology of the evolving domesticated form.

Willcox (1998) stated that in the Late Natufian (11,000–10,300 BP), einkorn was dominant at Mureybit and Abu Hureyra, north Syria, while wild emmer and wild. barley was dominant at Ohalo II, Israel. In the Pre-Pottery Neolithic A (PPNA; 10,300–9500 BP), wild emmer was found in several sites of the southern Levantine Corridor in the western flank of the Fertile Crescent) (Table 13.1), namely, in Israel, Jordan, and southern Syria, while wild einkorn was found in the northern Levantine Corridor, e.g., Jerf el Ahmar and Nureybit, northern Syria (Willcox 1998). There was small-scale cultivation (Harris 1996), using of locally available wild cereals as seed stock. No evidence exists for domesticated cereals in this period. However, at early PPNB (9500–9000 BP), mixtures of wild and domesticated wheats were found in several sites (Kislev 1984; Tanno and Wilcox 2006).

Emmer domestication during early PPNB has been reported for several sites in the southern and northern parts of the Levantine corridor (Table 13.1). In many sites during this period, wild types remained at significant frequencies. The wheat crop consisted of a mixture of wild and domestic types. Naked wheat (free-threshing) was found from the middle to late PPNB (9000–7500 BP) at several sites of the Levantine corridor (Table 13.1). Emmer that was absent in the Euphrates Valley, northern Syria, during earlier periods, appears to have been introduced from elsewhere at Abu Hureyra and Halula, together with naked wheat (Willcox 1995, 1996). Taking the data from the PPNA and PPNB together, there is evidence for independent in situ cereal domestication at different sites, einkorn in the northern part of the Levantine corridor and emmer in the southern part (Willcox 1998).

Kislev (1989, 1992) and Nesbitt (2002) pointed out that it is extremely difficult to distinguish between brittle and non-brittle forms of emmer in archaeological material. This is because the ripe ear of both wild and domesticated forms of emmer disarticulates into single spikelets, spontaneously or after threshing, respectively, (Kislev et al. 2004). It is only after unearthing parts of spikes with a tough, non-fragile rachis that cultivation of non-brittle forms of emmer can be assumed. Thus, determining the exact time of origin of non-brittle forms of emmer is a central problem in the study of wheat domestication. Clear-cut evidence for domestication dates to the 9th millennium BP, with finds of the small-grain, free-threshing, naked, tetraploid wheat, ssp. *parvicoccum*, in Ramad, southwestern Syria, in 8300–8000 BP (van Zeist and Bakker-Heeres 1982), in Tell Aswad, southwestern Syria, in 8900–8500 BP (van Zeist and Bakker-Heeres 1982), and in Can Hassan, south Anatolia, in 8400–7700 BP (Hillman 1972, 1978; Kislev 1979/1980, 1992; see below). The free-threshing tetraploid wheat was presumably preceded by the more primitive, hulled, domesticated emmer wheat (Kislev 1979/1980).

Morphological and genetic evidence have indicated that the domesticated subspecies of *T. turgidum*, and particularly domesticated emmer, are closely related to wild emmer wheat *T. turgidum* ssp. *dicoccoides* (Feldman 2001; Feldman et al. 1995; Feldman and Kislev 2007). This closeness is apparent from the fact that hybrids between wild emmer and domesticated subspecies of *T. turgidum* exhibit regular chromosome pairing at meiosis, and are fully, or almost fully, fertile (von Tschermak 1914; Percival 1921). These close relationships are also apparent from the spontaneous hybridization and gene flow occurring occasionally when wild and domesticated *T. turgidum* grow side by side (Cook 1913; Percival 1921; Feldman and Kislev 2007; Luo et al. 2007). These data clearly imply that domesticated emmer derived from wild emmer and that all other domesticated subspecies of *T. turgidum* derived from domesticated emmer.

The most critical changes in the transition of a wild type to a domesticated form involves loss of wild-type seed dissemination method, seed size, and seed dormancy (Zohary et al. 2012). Among these changes, the most conspicuous trait differentiating between wild emmer and domesticated emmer, ssp. *dicoccon*, is rachis brittleness at maturity; in the wild subspecies, the rachis is brittle, while it is tough and non-brittle in the domesticated form. The brittle rachis in wild emmer is an essential trait, leading at maturity to disarticulation of the spike, from the top downwards, into individual arrowhead-shaped spikelets. Each spikelet, equipping with two long, strong, and straight awns above, a sharp rachis segment below, and with very stiff hairs that are bent backwards, is an effective seed-dispersal unit that is very valuable for seed dissemination and self-planting under wild conditions (Harlan et al. 1973; Zohary et al. 2012). This seed-dispersal unit facilitates self-burial in the soil through a lateral movement of the awns caused by daily changes in humidity (Elbaum et al. 2007). Buried spikelets are thus protected from birds, rodents and ants during the long, dry summer, and ensuring successful germination after the first rain. This must have proved a nuisance to the ancient farmer who had to collect most of the spikelets from the ground or cut the culms before the grains matured. No wonder, therefore, that types with brittle heads were selected against and, in contrast, plants with a tough rachis that did not disarticulate at maturity were favored. Consequently, in domesticated emmer the mature spike remains intact on the culm and breaks into individual spikelets only upon slight application of mechanical pressure at threshing (Dorofeev et al. 1980). This type of tough rachis was called a 'semi-tough' rachis, not to be confused with the fully tough rachis of the free-threshing tetraploid subspecies wheats which remains intact when threshed (Hillman and Davis 1990). This difference was caused by mutations for a semi-tough rachis, that may occasionally occur in the wild (Kamm 1974) and thus, prevents the dissemination of seeds. As such, in the wild, it has a negative adaptive value and it is soon selected against. The selective pressure is especially strong in grazed areas, where non-brittle mutants are presumably eliminated soon after appearance. If they survive the first summer, their grains may germinate on the culm after the first rain, and the seedlings then dry up. In contrast, under cultivation, it has a positive value, facilitating easy harvesting, and was therefore preferred by farmers. Thus, domestication transformed wheat to be human-dependent, that can survive only under cultivation in human agricultural niches to meet human needs (Peng et al. 2011). On the other hand, domestication of plants has prompted man to be dependent on food production through cultivation of plants rather than by gathering them from nature. Hence, the transition from seed dispersal through spike fragility to dispersal by farmers is a key event in the domestication of cereals (Konopatskaia et al. 2016), and presumably, was one of the first modifications during domestication.

Despite its strong negative adaptive value, mutants with non-brittle spikes were found repeatedly in Israeli populations of wild emmer (Kamm 1974), indicating that this mutation is not a rare event. von Tschermak (1914) showed that the main morphological difference between wild and domesticated emmer, i.e., the non-brittleness of the spike, is determined by a small number of genes. Indeed, spike non-brittleness in segregating F_2 populations between wild emmer and durum wheat was determined by two complementary recessive genes (Levy and Feldman, unpublished data). Studying a complete series of chromosome-arm substitution lines in which wild emmer chromosome arms substituted for their common wheat homologues, Rong (1999) and Millet et al. (2013) showed that spike non-brittleness is determined by two recessive genes, one on the domesticated chromosome arm 3AS and the second, with a somewhat stronger effect, on the domesticated arm 3BS. Nalam et al. (2006), using recombinant inbred line populations, obtained similar results, showing that the *Br* (brittle rachis) loci are located in wild emmer chromosome arms 3AS and 3BS, while the domesticated types possess the recessive *br* alleles. Genotypes homozygous for only one *br* allele (either in 3AS or in 3BS) exhibit semi-brittle spikes in which the upper part of the spike disarticulates at maturity and the lower part remains on the culm (Kamm 1974; Rong 1999; Millet et al. 2013).

The two loci, controlling the rachis character in tetraploid wheat, were first designated *Br2* and *Br3* (Watanabe and Ikebata 2000). However, in accordance with the rule for the symbolization of genes in homoeologous sets, Watanabe et al. (2002) proposed to designate the group 3 brittle rachis genes as follows: *Br-A1*, (formerly *Br2*), and *Br-B1* (-formerly *Br3*). The dominant alleles of these loci, located on the short arm of chromosome 3A and 3B, respectively, determine rachis brittleness in wild emmer wheat, which leads to spike disarticulation. Homozygosity for dominant alleles in one locus and for recessive alleles in the second locus, i.e., *Br-A1/Br-A1/br-b1/br-b1* or *br-A1/br-A1/Br-B1/Br-B1*, determines fragility only of the upper part of the spike (Rong 1999; Millet et al. 2013; Avni et al. 2017). Homozygosity for recessive alleles in both loci, *br-A1* and *br-B1*, determines a non-brittle rachis (Watanabe and Ikebata 2000; Watanabe et al. 2002; Nalam et al. 2006; Millet et al. 2013; Konopatskaia et al. 2016).

Wedge-type spikelets are formed when disarticulation occurs immediately below the rachis internode (above the rachis node) leading to a dispersal unit of a spikelet and a rachis segment below it, while barrel-type spikelets are formed when the breakage occurs above the rachis internode (below the rachis node), resulting in the formation of a dispersal unit of a spikelet and a rachis segment beside it

(Zohary and Hopf 2000). Sakuma et al. (2011) and Li and Gill (2006) provided an overview of the disarticulation systems and inflorescence characteristics, along with the genes underlying these traits, in the Triticeae tribe. Based on the observed phenotype of single chromosome additions into bread wheat, the orthologous brittle-rachis gene *Br1*, determining wedge-type disarticulation, has been located in various Triticeae species on the short arm of homoeologous group 3 chromosomes (Watanabe and Ikebata 2000; Li and Gill 2006), including *Dasypyrum villosum* chromosome 3 V (Urbano et al. 1988), *Thinopyrum bessarabicum* chromosome 3Eb (King et al. 1997), wild barley *Hordeum spontaneum* chromosome 3H (Takahashi and Hayashi 1964), *Ae. speltoides* on chromosome 3S (Friebe et al. 2000; Li and Gill 2006), *Ae. bicornis* chromosome $3S^b$ (Riley et al. 1966, cited in Urbano et al. 1988), and in *Ae. sharonensis* chromosome $3S^{sh}$ (reported by Miller and cited in Urbano et al. 1988). In *Ae. longissima*, the *Br1* gene determines a spike-type of disarticulation (umbrella-type dispersal unit) is also located on the short arm of chromosome $3S^l$ (Friebe et al. 1993; Urbano et al. 1988), but it causes fragility only in one site at the lower part of the spike. *Br1* genes determining umbrella-type disarticulation exist on the short arms of group 3 chromosomes in several other *Aegilops* species, namely, *Ae. searsii* chromosome $3S^s$ (Friebe et al. 1995), *Ae. peregrina* chromosome 3 Sv (Yang et al. 1996), on *Ae. geniculata* chromosome $3M^o$ (Friebe et al. 1999), *Ae. uniaristata* chromosome 3N (Miller et al. 1995; Iqbal et al. 2000b). Within *Triticum* itself, the above-rachis node disarticulation gene(s) have been located in wild *T. timopheevii* chromosome arms 3AS and 3GS (Li and Gill 2006; Nave et al. 2021), in wild emmer on chromosome arms 3AS and 3BS (Nalam et al. 2006; Millet et al. 2013, 2014), and in the feral hexaploid wheat, *T. aestivum* ssp. *tibetanum* on chromosome arm 3DS (Chen et al. 1998; Watanabe et al. 2002). All of these chromosomes carry a gene responsible for disarticulation below the rachis internode. Thus, there is ground for supposing that the disarticulation genes, as a whole form an orthologous set. Intriguingly, in *Ae. tauschii*, the above-rachis internode disarticulation trait leading to barrel-type disarticulation is controlled by a gene *(Br2)* mapping to the long arm of chromosome 3D (Li and Gill 2006) whereas the *Br1* gene is located ssp. *tibetanum* on 3DS (Chen et al. 1998; Watanabe et al. 2002), indicating the existence of this gene on chromosome 3DS of T. aestivum and obviously also on 3DS of *Ae. tauschii*. The *Br2* gene is paralogous to the orthologous *Br1* genes, probably formed in *Ae. tauschii* by a duplication followed by an intra-chromosomal transposition to the long arm. Such translocation of genes is a far from rare event during evolution (Tarchini et al. 2000; Li and Gill 2002; Pourkheirandish et al. 2007; Faris et al. 2008; Sakuma et al. 2010).

With the fully assembled the genome of wild emmer wheat, Avni et al. (2017), using a population derived from a cross between wild emmer (line Zavitan) and domesticated durum wheat (cv. Svevo), identified the mutations in the brittle rachis genes *Br-A1* and *Br-B1* that lead to non-fragile spikes in domesticated emmer. They revealed, in agreement with previous studies (Rong 1999; Watanabe et al. 2002, 2006; Nalam et al. 2006; Millet et al. 2013, 2014), genomic regions regulating the brittle rachis phenotype in wild emmer, on the short arm of chromosomes 3A and 3B (15.5 and 32.5 Mb, respectively). Yet, Avni et al. (2017) found out that each *Br1* gene is actually a compound locus, consists of duplicated genes, *TdBtr1* and *TdBtr2*, that exhibit homology to the *Btr1* and *Btr2* genes controlling brittle rachis in wild barley and wild *T. monococcum* (Pourkheirandish et al. 2015, 2018). Thus, Avni et al. (2017) identified the orthologous genes in tetraploid wheat, i.e., *TtBtr1-A* and *TtBtr2-A* on chromosome arm 3AS, and *TtBtr1-B* and *TtBtr2-B* on chromosome arm 3BS. The homoeology of *TtBtr1-A, TtBtr2-A, TtBtr1-B,* and *TtBtr2-B* from wild emmer wheat with *Btr1 and Btr2* on chromosome arm 3HS of wild barley (*Hordeum spontaneum*) and with *Btr1* and *Btr2* on chromosome arm 3AS of wild *T. monococcum*, suggests that the location of the genes for the brittle rachis trait in these species has been conserved.

In diploid wheat, *T. monococcum*, Pourkheirandish et al. (2018) reported that a single non-synonymous amino acid substitution at position 119 (alanine in wild form to threonine in domesticated form) of the protein product of *Btr1* is responsible for the loss of function mutation that leads to the non-brittle rachis trait. Substitution at position 10 in the protein product of *Btr2*, from an aspartic acid in wild to glutamic acid in domesticated einkorn, had no effect on rachis brittleness (Pourkheirandish et al. 2018). Their data supported the hypothesis that the substitution at position 119 of *Btr1* causes a functional change implying that the brittle/non-brittle rachis trait in diploid wheat is only controlled by the allelic status at *Btr1*.

Similar to the situation in diploid wheat, also in durum wheat (i.e., cv. Svevo) the mutant alleles *btr1-A* and *btr1-B* presumably are likely loss-of-function alleles (Avni et al. 2017). The causative mutation in *Btr1-A* is a 2 bp deletion in the coding sequence which causes a loss-of-function frame shift, and in *Btr1-B*, the loss of function is due to a 4 kbp insertion, 50 bp upstream of the stop codon (Avni et al. 2017). Diversity analysis of 113 wild emmer, 85 domesticated emmer, and 9 *durum* accessions showed that all domesticated accessions carry the loss-of-function alleles for both brittle rachis genes *btr-A1* and *btr-B1* (Avni et al. 2017). On the other hand, the fact that no polymorphisms were detected between the coding regions of wild (line Zavitan) and domesticated (cv. Svevo) *TtBtr2-A* or *TtBtr2-B* alleles, led Avni et al. (2017) to assume that the combination of the

mutations in the two TtBtr1 genes are complementary and sufficient to achieve the non-brittle rachis phenotype. These researchers developed a pair of near-isogenic lines (NILs), each carrying one functional allele (*TtBtr1-A* or *TtBtr1-B*) in the background of Svevo. Both NILs exhibited an intermediate brittle rachis phenotype, in which the upper part of the spike was brittle, and the lower part was non-brittle. Evidently, these two homozygous recessive mutations of the *TtBtr1* gene (but not in *TtBtr2*) appear to be required for transforming the brittle-rachis of wild emmer to non-brittle one in domesticated tetraploid wheat. Diversity analysis of 113 wild emmer, 85 domesticated emmer, and 9 durum accessions showed that all accessions of domesticated emmer and durum carry the loss-of-function alleles for both *btr1-A* and *btr1-B*. The requirement for two homozygous recessive mutations, suggests that selection for non-brittle rachis in tetraploid wheat may have been a gradual long process, first selection for plants showing only partial spike brittleness, due to a recessive mutation in one of the brittle rachis l genes, and later, selection for plants showing complete non-brittleness, arising from possession of two recessive mutations in both loci. Indeed, archaeological findings suggest that rachis non-brittleness took several hundred years to become established (Kislev 1984; Tanno and Willcox 2006; Feldman and Kislev 2007; Purugganan and Fuller 2011). The loss of fragility gave rise to the first known domesticated wheat, *T. turgidum* ssp. *dicoccon*, which is grown to this day, albeit on a small scale (de Vita et al. 2006).

Several studies have shown that some QTLs affect the brittle-rachis trait, namely, on chromosomes 2A (Peng et al. 2003; Peleg et al. 2011; Tzarfati et al. 2014), 3A (Watanabe et al. 2006), and 1B (Tzarfati et al. 2014). Thanh et al. (2013), analyzing F_2 plants derived from a cross between domesticated emmer and wild emmer, detected seventeen QTLs on chromosomes 1B, 2A, 2B, 3A, 3B, 4A, 5B, and 7B that affected plant and spike characteristics. Two regions on chromosomes 2A and 3B had a large effect on rachis fragility, and nine regions on chromosomes 2A and 5B affected traits related to seed production. Their results indicated that selection for these QTLs occurred during the domestication of emmer wheat, prior to the appearance of free-threshing forms. It was suggested that superimposition of the effect of all of the QTLs on that of the major genes *br-A1* and *br-B1* and the subsequent improvement or strengthening of the non-brittle phenotypes, might have evolved later, under domestication (Abbo et al. 2012, 2014).

13.2.3.3 Selection for Large Grain Size

An additional requirement for the newly domesticated wheat was increased grain size. It is well known that wild and domesticated cereal grains differ in size, and this has been used to infer the domesticated status of cereals, already in the earliest PPNB, including sites from the

Jordan Valley, the upper Euphrates in Syria, and the first settlements in Cyprus (Colledge 2001, 2004). Domesticated forms differ from the wild form in kernel morphology; in the domesticated forms, the grain tends to be wider, thicker and rounder in cross-section compared to the wild form (Willcox 1998, 2002, 2004). Current indications from available archaeobotanical evidence indicate that evolution of grain shape and size preceded the loss of wild-type seed dispersal mechanism (Willcox et al. 2008). It is possible that the early farmers tended to cultivate, and subsequently, to domesticate, large-seeded genotypes of wild emmer (Blumler 1992, 1994; Diamond 1997). In fact, Harlan and Zohary (1966) suggested that a large-seeded race of wild emmer, growing abundantly in the vicinity of the Upper Jordan Valley, is the likely progenitor of domesticated emmer.

Fuller (2007) explored the disjunction in wheat between seed size increase and loss of seed dispersal, and rates at which these features evolved were estimated from archaeobotanical data. He concluded that changes in grain size and shape evolved prior to non-brittleness of the rachis. Initial grain size increases may have evolved during the first centuries of cultivation of wild emmer, within perhaps 500–1000 years (Fuller 2007).

In accordance with the above, the growing morphometric database for wheat and barley from the Near East indicates that wheat and barley grains increased in size starting in the PPNA and earliest PPNB (Colledge 2001, 2004; Willcox 2004). This is before widespread evidence for the existence of tough, non-brittle rachises and loss of natural seed dispersal.

Willcox (1998) suggested that the evolution of large grains occurred over a few centuries. This explains why large domestic-type grains were already widespread and predominant on most Near Eastern sites by the start of the PPNB (approx. 9500 years BP). This evidence raises the question of how large-grained varieties of wild emmer evolved. One possibility is that methods of processing, such as use of sieves after threshing and winnowing, served to bias larger grains for stored cereals.

Gegas et al. (2010), analyzing morphometric and quantitative traits in several recombinant-doubled haploid populations of elite winter lines of common wheat and a comparison of grain material from primitive wheat species and modern elite varieties, concluded that grain shape and size are independent traits in both modern varieties and in primitive wheat species that are under the control of distinct genetic elements. They suggested that wheat domestication resulted in a switch from production of a relatively small grain with a long, thin shape to a more uniform, larger grain, with a short, wide shape. Their data illustrated the complex history of domesticated wheat evolution, suggesting that various traits arose independently at different stages. For example, these authors suggested that grain size increased early in domestication, through alterations both in grain width and length, followed, at later stages, by further modifications in grain shape, primarily, changes in grain length. In addition, the decrease in phenotypic diversity in grain morphology in modern commercial wheat is shown to be the result of a relatively recent and severe bottleneck that may have occurred either during the transition from hulled wheat to the free-threshing subspecies, or, more recently, through modern breeding programs.

13.2.3.4 Selection for Rapid and Synchronous Germination

Other requirements for the newly domesticated wheat were prevention of untimely germination, rapid and uniform germination after sowing, simultaneous ripening of grains, and erect rather than prostrate culms. Wild emmer wheat has two types of seed dormancy: a post-harvest type and a long-range type. The first type prevents premature and untimely germination—an important feature, pre-adapted to agriculture, particularly in view of the fact that seeds were often stored under unsuitable conditions. The second type of dormancy ensures a temporal distribution of germination in nature: it is invariably the larger grain of the second floret in each spikelet that germinates in the first autumn, while the smaller, darker grain of the first floret germinates the following year. Such temporal distribution of germination, referred to as differential dormancy (Nave et al. 2016), prevents, on the one hand, crowding of plants, and, on the other hand, ensures the occupation of sites for two years. Horovitz et al. (2013) reported that differential dormancy characterizing intact spikelets also exists in grains that were separated from the spikelet. This shows that differential dormancy is not the result of inhibitory factors extracted from the glumes and palea, but, rather, is caused by internal factors within the grains themselves.

Using phenotypic data from a wild emmer x durum wheat population and a high-density genetic map, Nave et al. (2016) exposed the genetic mechanism controlling differential grain dimensions and dormancy within wild tetraploid wheat spikelets. They showed that, in wild emmer, the lower grain within the spikelet is about 30% smaller and more dormant than the larger, upper grain. They also revealed a major locus on the long arm of chromosome 4B that explains >40% of the observed variation in grain dimensions and seed dormancy within spikelets. The domesticated variant of this locus, designated *QGD-4BL*, likely fixed during the domestication process, favors spikelets with seeds of uniform size and synchronous germination. In addition, a QTL for mean kernel weight was located on chromosome 5A (Tzarfati et al. 2014).

13.2.3.5 Monophyletic Versus Polyphyletic Origin of Domesticated Emmer

Harlan (1975) asserted that emmer wheat was domesticated many times in various parts of the Fertile Crescent. Yet, new evidence showed very clearly that in most regions where agriculture began, primary crops were domesticated only once or very few times. Concurrently, Blumler (1992, 1996) suggested that emmer was domesticated only once, and that diffusion of cultivation of emmer was far more important than independent invention. In agreement, the results of Haudry et al. (2007), who analyzed nucleotide diversity at 21 loci in wild emmer and domesticated wheats, are consistent with a monophyletic origin of all the domesticated lines studied, which is consistent with a single domestication event for emmer wheat (Zohary 1999). This is in agreement with recent archaeological data that support the view of emmer domestication as a geographically diffuse, gradual process, rather than an independent discovery and use of non-brittle types (Blumler 1996; Weiss et al. 2006

The genetic evidence reported by Salamini and coworkers (Salamini et al. 2002; Ozkan et al. 2002, 2005, 2011), identify the ancestral population of wild emmer from which domesticated emmer derived. Salamini and coworkers used amplified fragment length polymorphism (AFLP) fingerprinting of nuclear DNA to estimate the genetic similarity between wild and domesticated emmer populations (Ozkan et al. 2002; Salamini et al. 2002). They constructed neighbor-joining trees, from which they concluded that wild populations from the Karacadag Mountains in southeastern Turkey, are more similar to domesticated emmer than are other wild populations. This fact supports the hypothesis of a monophyletic origin of domesticated emmer (Zohary 1999). However, Allaby and Brown (2003, 2004) claim that a monophyletic origin might be erroneously inferred when populations are examined by AFLP genotyping and neighbor-joining analysis. Questioning the use of AFLP analysis for studies on the origin of crops, they argued, based on simulation data, that results resembling a monophyletic tree can be obtained when merging data from populations that were independently domesticated. In 172 out of 180 simulations, the domesticated hybrid formed a single clade in the resulting neighbor-joining tree (Allaby and Brown 2003). They argued that the combination of hybridization among populations, migration, and genetic drift may affect the shape of phylogenetic trees, which cannot be used to correctly reconstruct the history of the relationships between domesticated and wild populations (Allaby and Brown 2004).

All the domesticated wheats containing he BBAA or BBAADD genome have the same mutation, supporting a monophyletic origin. Later introgressions might have happened that change the phylogeny. But the seminal events are shared by all wheats. Yet, the study of Nave et al. (2019),

does not rule out a series of events occurring at different places and at different times. Willcox (2002) and Tanno and Willcox (2006) report that archaeological data support the view that domestication occurred independently in several places. Recent work on domesticated emmer wheat has identified two different lineages of a gluten genes which are so different that they are estimated to have evolved apart hundreds of thousands of years ago, i.e., long before domestication of wild emmer wheat. Such evidence implies two separate domestications of emmer (Allaby et al. 1999; Brown 1999; reviewed in Jones and Brown 2000).

Some domestication events may not be on the genetic record because the early cultivars have disappeared, and present-day populations represent only a fraction of those grown in the past. For instance, a large fraction of the ancient gene pool of domesticated tetraploid wheat was lost in the Near East about 2500 years ago, or earlier, when ssp. *durum* replaced domesticated emmer and the small-grained, free-threshing tetraploid wheat, ssp. *Parvicoccum* (Nesbitt 2002). Nonetheless, it may be reasonable to assume that the technology to cultivate wild plants originated in one of the sites of the Levantine Corridor and its spread to other Levantine sites motivated the local PPNA people to start cultivating wild plants that grew in their vicinity. Indeed, archaeological evidence indicates that different Near Eastern communities cultivated various local species during the PPNA (Weiss et al. 2006).

Archaeobotanical studies showed that acquisition of the full set of traits observed in domesticated emmer was a prolonged process, intermediate stages being seen at early farming sites throughout the Fertile Crescent (Brown et al. 2009). New genetic data are confirming the multiregional nature of cereal domestication, correcting a previous view that each crop was domesticated by a rapid, unique and geographically localized process. Brown et al. (2009) reviewed the evidence that has prompted this reevaluation of the origins of domesticated crops in the Fertile Crescent. Taken together, the archaeobotanical morphotypes and genetics suggest the occurrence of several domestications of wheat and barley in the Near Eastern Fertile Crescent region, and there is no reason to attribute them all to a single micro-region or a single process of agricultural origins, but to at least two or perhaps three (Willcox 2005).

Civáň et al. (2013) used datasets of nuclear gene sequences and novel markers detecting retrotransposon insertions in ribosomal DNA loci, to reassess the evolutionary relationships among subspecies of tetraploid wheat, *T. turgidum*. They concluded that domesticated emmer has a reticulated genetic ancestry, sharing phylogenetic signals with wild populations from all parts of the wild emmer range. Assuming that the extent of the genetic reticulation cannot be explained by post-domestication gene flow between domesticated emmer and wild emmer, they suggested that domesticated emmer originated from a hybridized population of different wild lineages. Consequently, they claimed that the phylogenetic relationships among tetraploid wheats are incompatible with simple linear descent of the domesticated emmer from a single wild emmer population and proposed an alternative model for the emergence of domesticated emmer. During a pre-domestication period, diverse wild populations were collected from a large area west of the Euphrates and were cultivated in mixed stands. Within these cultivated stands, hybridization gave rise to lineages displaying reticulated genealogical relationships with their ancestral populations. Domesticated emmer was derived from such reticulated lines of wild emmer. Then again, the mixtures of wild and domesticated emmer that existed for one millennium or more, enabled gene flow between the newly formed genotype bearing a non-brittle rachis and the different genotypes of wild emmer. Moreover, during the long period of cultivation, introgression of various wild emmer genes into domesticated emmer, from plants that grew close to domesticated emmer, was not a rare event (Blumler 1998; Luo et al. 2007).

Similar conclusion was reached by Oliveira et al. (2020). These researchers used genotyping-by-sequencing (GBS) to investigate the evolutionary history of domesticated tetraploid wheats and identified 1,172,469 single nucleotide polymorphisms (SNPs) in 189 wild and domesticated wheats. Principal component analyses separated wild *emmer* from e domesticated emmers and the naked wheats, showing that SNP typing by GBS is capable of providing robust information on the genetic relationships between subspecies of tetraploid wheat. Their data suggest that domesticated tetraploid wheats have closest affinity with wild emmers from the northern Fertile Crescent, consistent with the results of previous genetic studies on the origins of domesticated wheat. However, a more detailed examination of admixture and allele sharing between domesticates and different wild populations, along with genome-wide association studies (GWAS), showed that the domesticated tetraploid wheats have also received a substantial genetic input from wild emmer types from the southern Levant. Taking account of archaeological evidence that tetraploid wheats were first cultivated in the southern Levant, Oliveira et al. (2020) suggest that a pre-domesticated crop spread from the southern Levant to southeast Turkey and became mixed with a wild emmer population from the northern Fertile Crescent. Fixation of the domestication traits in this mixed population would account for the allele sharing and GWAS results that we report. Oliveira et al. (2020) also propose that feralization of the component of the pre-domesticated population that did not acquire domestication traits has resulted in the modern wild population from southeast Turkey displaying features of both the domesticated and wild emmer from the

southern Levant, and hence appearing to be the sole progenitor of domesticated tetraploids when the phylogenetic relationships are studied by methods that assume a treelike pattern of evolution.

13.2.3.6 Site(s) of Origin of Domesticated Emmer

Domestication may not have taken place where the wild cereals were most abundant (Harlan and Zohary 1966). Why should anyone cultivate a cereal where natural stands are as dense as a cultivated field? If wild cereal grasses can be harvested in almost unlimited quantities, why should anyone bother to till the soil and plant the seed? Harlan and Zohary (1966) suspected that harvesting of wild cereals lingered on long after some people had learned to farm, and that farming itself may have originated in areas adjacent to, rather than within, the regions of greatest abundance of wild cereals.

Already Harlan and Zohary (1966) assumed that the thin, sporadic stands of wild emmer in the Taurus-Zagros arc would hardly have been very attractive to the food-collecting cultures of the region. In contrast, the massive stands around the Lake of Galilee would surely have been more useful to a harvester of wild grass seeds. The archaeological and genetic evidence tends to point in the same direction (Table 13.1). While there is currently more knowhow on the genetic changes that led to the formation of domesticated emmer, the site(s) in which it was formed still require more clarification. Harlan and Zohary (1996) assumed that most of the domesticated subspecies of *T. turgidum* derived from the race of wild emmer now found in the upper Jordan watershed and thus, concluded that emmer was likely domesticated in that region.

Harlan and Zohary (1966) were not the first to propose the southern Levant as the place where wheat farming began. Vavilov, travelling the Levant in 1926, noticed a peculiar form of wild wheat that accompanied domesticated *durum* wheat in Israel (then Palestine) (Vavilov 1957), and concluded that it must be the wild progenitor of domesticated wheats because of its similarity to domesticated ssp. *durum*. This form was very distinct from other forms of wild emmer, which led Vavilov to classify it as a subspecies of wild emmer. Its spikes were large, with rough spikelets and large grains, undoubtedly representing the closest wild source of cultivated wheat, especially of durum wheat (Vavilov 1962). This type of wild emmer is currently considered as a derivative from hybridization of wild x domesticated durum wheat.

Archaeological data showed that the cultivation of wild emmer took place within or near the current geographical distribution area of wild emmer, i.e., in the Levantine Corridor (Hillman, 1996; Harris, 1998; Bar-Yosef, 1998; Bar-Yosef and Kislev, 1989; Bar-Yosef and Belfer-Cohen, 1992). The Levantine Corridor is the western part of the Fertile Crescent, encompassing a relatively narrow strip east of the Mediterranean Sea, from southeast Turkey in the north, to the Sinai Peninsula in the south (Bar-Yosef 1998). This corridor was divided into northern part including southeastern Turkey and northern Syria, and southern part including southwestern Syria, Southeastern Lebanon, eastern Israel and western Jordan., Bar-Yosef and Kislev (1989) and Kislev (1989, 1992) stated that wild emmer is the earliest wheat in archaeological material from early Neolithic sites in the southern Levantine Corridor (Table 13.1).

It is reasonable to assume that during the long cultivation of wild emmer (10,300–9500 BP) non-brittle types of emmer were derived from mutations of brittle types in various sites in the Levantine Corridor. Indeed, remnants of apparently non-brittle emmer wheat, mixed with remnants of brittle types, were found in several sites of the Near East, dating to the middle of the 9th millennium BP (Table 13.1).

Unfortunately, it is not known when man adopted the mutations that transformed wild emmer into domesticated emmer. In spite of being hulled and therefore, difficult to thresh, domesticated emmer was one of the most prominent crops for almost 6000 years, from the PPNB to the Iron Age, in farming villages throughout the Near East. In the 8th millennium BP, it was taken from the hilly and mountainous regions of the Fertile Crescent, in mixtures with naked tetraploid wheat, to the lowlands of Mesopotamia, from where it spread further to Central Asia and India, and westward to Anatolia, then spreading to the Mediterranean basin and Europe. During the 6th millennium BP, domesticated emmer was taken to Egypt, and later to Ethiopia, some 5000 years ago (Feldman 2001).

Domesticated emmer could have evolved from wild emmer by a monophyletic, diphyletic, or polyphyletic manner. Salamini and coworkers, using amplified fragment length polymorphism (AFLP) fingerprinting of nuclear DNA to estimate the genetic similarity between wild and domesticated emmer populations (Ozkan et al. 2002; Salamini et al. 2002), concluded that wild populations from the Karacadag Mountains in southeastern Turkey, are more similar to domesticated emmer than are other wild populations. On the other hand, Mori et al. (2003), using large-scale chloroplast DNA fingerprinting, analyzed larger sample of wild emmer from southeastern Turkey and Iraq. They concluded that two distinct maternal lineages were involved in the domestication of emmer, and that the domestication of emmer occurred independently in at least two locales; one site was in the Kartal Dagi Mountains, southern Turkey, about 280 km west of Karacadag, and the second could not be identified, but several closely related haplotypes of wild emmer were found in four geographically distant regions. After analyzing these same accessions by AFLP, Ozkan et al. (2005) concluded that wild emmer accessions from the Karacadag region and from the Sulaimanyia region in northern Iraq, are equally closely related to domesticated emmer. To revisit the

question of emmer domestication, Luo et al. (2007) pointed out that populations of wild emmer from the Kartal Dagi Mountains and Urfa plateau in southeastern Turkey, and from Iraq and Iran, were either inadequately sampled or not sampled at all by Ozkan et al. (2002), and consequently, the site at which emmer was domesticated remains inconclusive. They performed restriction fragment length polymorphism (RFLP) analysis at 131 loci and found that gene flow between wild and domesticated emmer occurred massively across the entire area of wild emmer distribution and concluded that emmer was likely domesticated in the Karacadag area (Diyarbakir region) in southeastern Turkey, which was followed by subsequent hybridization and introgression from wild to domesticated emmer in the southern Levant. Alternatively, although less likely from their point of view, emmer was domesticated independently in the Karacadag area and in the southern Levant (Luo et al. 2007). Thus, the data of Mori et al. (2003), Ozkan et al. (2002, 2005, 2011), and Luo et al. (2007) suggest a diphyletic or polyphyletic origin of domesticated emmer.

Nave et al. (2019) investigated, via haplotype analysis of a large collection of wild and domesticated emmer accessions, the geographical birthplace of the recessive mutations in the brittle rachis genes, *Br-A1* and *Br-B1* (=*Btr1-A* and *Btr1-B*), mutations that determine spike non-brittleness. The precursor of the domesticated haplotype of *br-A1* was detected in 32% of the wild accessions gathered throughout the Levant, from central Israel to eastern Turkey. In contrast, the precursor of the domesticated haplotype of *br-B1* was found in only 10% of the tested wild accessions, all from the southern Levant. Moreover, this haplotype is shared by all domesticated tetraploid and hexaploid wheats tested so far (Naveh et al. 2019). The phytogeographical results challenge the above thinking regarding the birthplace of domesticated emmer as well as regarding the concept of di- or poly-phyletic origins of domesticated emmer. Specifically, the precursor of *br-B1* presents a direct evolutionary link between domesticated wheat and wild emmer from the southern Levant, contrary to the widely held view that the northern Levant was the center of domestication of wild emmer. On the basis of the evidence presented by Nave et al. (2019), it is hypothesized that humans may have spread certain wild emmer genotypes from across the Fertile Crescent prior to domestication. Such a 'pre-domestication circulation of wild crop progenitors' theory could reconcile the wide distribution of the precursor of the domesticated *br–A1* haplotype and also aligns well with archaeological evidence of wild crop harvesting and utilization in the southern Levant prior to the Neolithic era. However, the identification of a single southern Levant wild emmer genotype that carried the progenitor haplotypes for the alleles *br-A1* and *br-B1* suggests that the first domesticated emmer appeared there. Later, domesticated emmer spread out to northern Levant, capturing local genetic diversity along the way, while maintaining the *br-A1 and br-B1* mutations.

Recently, Gornicki et al. (2014) sequenced whole chloroplast genomes isolated from twenty-five chloroplast genomes, and genotyped 1127 plant accessions representing 13 *Triticum* and *Aegilops* species. They detected a higher diversity of the chloroplast genome in the southern Levant populations of wild emmer than in the northern population and suggested that wild emmer originated there. They further suggested that the major chloroplast haplotype H1 of wild emmer was the founder of all domesticated species of the lineage, most of which also carry H1. This haplotype exists in the southern Levant as well as in southeastern Turkey. The geographic distribution of the chloroplast haplotypes of the wild tetraploid wheats, wild emmer and wild *timopheevii*, and those of *Ae. speltoides,* the assumed donor of the cytoplasm to the tetraploid wheats, demonstrates the possible geographic origin of the emmer lineage in the southern Levant and of the *timopheevii* lineage in northern Iraq.

13.2.4 Origin of Tetraploid Wheat with Free-Threshing Grains

13.2.4.1 Genetic Control of the Free-Threshing Trait

The loss of self-seed dissemination and the temporal control of germination was followed in the more advanced domesticated subspecies of *T. turgidum*, by the loss of self-seed protection. Both the wild form and the more primitive domesticated forms of *T. turgidum*, i.e., ssp. *dicoccon* and ssp. *paleocolchicum*, have hulled grains. Their grains are enclosed in the spikelet by tough glumes that do not break during threshing. In hulled wheats, the products of threshing are spikelets, rather than grains. Usually, the spikes of hulled domesticated wheats break during threshing at the same place at which their wild counterpart disarticulates spontaneously. In other words, threshing these still primitive wheats 'mimics' the shattering pattern of their wild progenitor; the individual spikelet with the internode segment at the base of the product of threshing. However, instead of the smooth abscission scars, which characterize the wild form, the surface of the breakage scars in the domestic hulled wheats is rough.

Domestication of tetraploid wheats went one step further, changing the hulled ssp. *dicoccon* to a free-threshing form. Hence, the free-threshing (or naked) tetraploid wheats emerged from already domesticated crops, i.e., domesticated emmer wheat, and are therefore termed secondary crops (Hillman and Davis 1990). All more advanced domesticated subspecies of *T. turgidum* are free threshing. Their glumes

are thinner and do not invest the grains tightly, and their rachis is fully tough. Consequently, threshing releases naked kernels. Because of this difference in threshing product, hulled wheats handling by the farmer is different from handling of the free threshing. In the hulled wheat, the grains have to be freed from the spikelets (usually by pounding) before they can be used. The utilization of naked wheat is simpler since threshing yields free grains. Because of the different appearances of the marketed products, hulled and free-threshing wheats were often regarded in antiquity as different cereals, and they were even called different names. For instance, in the bible (Exodus 9:32) free-threshing type was called 'wheat' (chitta in Hebrew) whereas hulled type was called 'rie' (=spelt; kosemeth in Hebrew).

The appearance of naked kernels was the second most important step in domestication of tetraploid wheat, after development of the non-brittle rachis. This trait could have been derived from a mutation that reduces the toughness of the glumes and increases the rigidity of the rachis (McFadden and Sears 1946; Morris and Sears 1967). The hulled, non-free-threshing emmer wheats contain the *q* gene, which determines a speltoid spike, characterized by an elongated spear-shaped spike, easily broken rachis and rigid, thick glumes. A clear distinction of glume shape has long been known between free-threshing ssp. *carthlicum*, which bears round glumes without keels, like free-threshing hexaploid wheats, and other free-threshing tetraploid wheats, which bear boat-shaped, keeled glumes (Muramatsu 1986). Watkins (1928, 1940) reported that, with the exception of ssp. *carthlicum*, all tetraploid wheats, including the free-threshing forms, had keeled glumes and therefore, possess the gene *K*, which was later determined to be the same as *q* (Mac Key 1954a). Mac Key (1954a, b, 1966, 1968, 1975) suggested that the free-threshing tetraploids contain a polygenic genetic system dictating this trait. However, Muramatsu (1978, 1979, 1985, 1986) showed that all the free-threshing forms of tetraploid wheat carry the dominant allele *Q*, but its expression may be modified, to some extent, by the genetic background (Muramatsu 1986). The indication is that the keeled glume in tetraploid wheat is due to genes of the A and B genomes and not to genes of the D genome (Fans et al. 2006). This is implied from the fact that glumes do not show a clear keel in *Ae. tauschii*, the D genome donor to the hexaploid wheats (Muramatsu 1986). Muramatsu assessed the effects of chromosome 5A of various tetraploids in the background of hexaploid wheat, cv. Chinese Spring, and found that not only ssp. *carthlicum* has the *QQ* genotype, but also ssp. *polonicum* and ssp. *durum*. There are some varieties of tetraploid wheat that have square-head spikes. Because squareheadedness is one of the pleiotropic effects of *Q*, it is unlikely in a plant with genotype *qq*. Indeed, Muramatsu (1979) demonstrated that ssp. *dicoccon* var. *liguliforme,* which has a semi-tough rachis with keeled glumes and square-headed spikes, contains Q. Based, on this, Muramatsu (1986) concluded that there is a wide phenotypic variation of characteristics in different *QQ* lines, and suggested that this range of variation is very narrow in the absence of *Q*, but when *Q* is present, they express an obvious phenotype.

Thus, all the extant, free-threshing tetraploids contain the dominant *Q* factor, determining the free-threshing trait (Muramatsu 1986; Simons et al. 2006). The *Q* factor, located on the long arm of chromosome 5A (Sears 1954), is one of the most significant domestication loci, as it controls the free-threshing characteristic and several other domestication-related traits. While the 5A homoeoallele has the most significant contribution, other homoeoalleles (on 5B in tetraploid and 5B and 5D in hexaploid wheat) were also shown to be involved in the domestication traits (Zhang et al. 2011). The *q* gene homoeoallele on chromosome 5B became a pseudogene after allotetraploidization. Expression analysis indicated that, whereas *Q* plays a major role in conferring domestication-related traits, the *q* homoeoallele on 5D contributes directly and *q* on 5B indirectly to suppression of the speltoid phenotype. Hence, according to Zhang et al. (2011), the evolution of the *Q/q* loci in polyploid wheat resulted in the hyper-functionalization of *Q* on 5A, pseudogenization of *q* on 5B, and sub-functionalization of *q* *on 5D*, all contributing to the domestication traits.

Q is thought to be a major regulatory gene for floral development (Muramatsu 1986). It encodes an AP2-like transcription factor that played an important role in the domestication of polyploid wheat (Simons et al. 2006; Zhang et al. 2011). Actually, Q has pleiotropic influences on many other domestication-related traits, such as spike density and length (square-head spike), fertility of the basal spikelets, glume shape and tenacity, glume keel formation, rachis fragility, plant height, spike emergence time, and chlorophyll pattern along nerves as well as grain size and shape, in a complicated interaction pattern (Mac Key 2005; Simons et al. 2006). This pleotropic effect stems from the fact that Q is a transcription factor that activates many different genes (Simons et al. 2006; Zhang et al. 2011). Actually, the Q gene has a high degree of similarity to members of the *AP2* family of transcription factors.

The mutation from q to Q occurred at the tetraploid level, from where it was transferred to hexaploid wheat (Simons et al. 2006). Muramatsu (1963) proposed that *Q* is a triplication of *q*, since five doses of *q* conferred the same phenotype as two doses of Q. However, the *Q* gene was recently isolated and characterized (Faris et al. 2003; Jantasuriyarat et al. 2004; Simons et al. 2006) and the data of the latter, involving Southern analysis and sequencing of a large bacterial artificial chromosomes (BACs) spanning *Q*, indicated that *Q* is not a duplication of q, but most likely arose through a gain-of-function mutation. The two alleles differ in a single

nucleotide (GAG in q and GCG in *Q*), leading to a change in one amino acid, namely, all q-containing forms have valine in position 329, whereas all *Q*-containing forms possess an isoleucine at this position (Simons et al. 2006). The mutation that gave rise to *Q* occurred only once leading to the world's cultivated wheats (Simons et al. 2006). Although Simons et al. (2006) considered a SNP, leading to the substitution of a valine by an isoleucine at position 329, as a possible cause for the *Q* mutation, they also noted a conserved SNP in the miRNA 172 binding site. Subsequent work by Debernardi et al. (2017) showed that in fact the SNP within the miRNA binding site is the causal polymorphism for the functional difference between the *Q* and *q* alleles. Moreover, *Q* is more abundantly transcribed than *q* (Simons et al. 2006), which lies in accord with the finding of Muramatsu (1963) who showed that extra doses (five or six) of q mimics the effect of *Q* in common wheat. Increased transcription of *Q* was most obviously associated with spike compactness and reduced plant height, as in plants tetrasomic for chromosome 5A (Sears 1954). The higher level of transcription of *Q* is also consistent with its dominant nature, as well as with the disruption of the miRNA 172 suppressive effect (Debernardi et al. 2017). Zhang et al. (2020) reported on new advances on *Q*'s mode of action using transcriptomics and phenotypic analyses. They show that modification of cell wall thickness and composition of glumes, e.g., lignin versus cellulose ratio, correlates with the expression of genes involved in secondary cell wall biosynthesis.

The discovery of a gene on the short arm of chromosome 2D of *Ae. tauschii*, designated *Tg*, that affects threshability by conferring tenacious glumes in synthetic $6 \times$ amphiploids *T. turgidum-Ae. tauschii* (Kerber and Dyck 1969; Kerber and Rowland 1974; see Sect. 13.3.2), promoted search for orthologous *Tg* genes in subgenomes A and B of *T. turgidum* and in its parental diploids. Consequently, Simonetti et al. (1999), analyzing a set of recombinant inbred lines, derived from a cross between the cv. Messapia of durum wheat and accession MG4343 of wild emmer, found that chromosome arm 2BS of wild emmer carries a gene that suppresses the free-threshing trait by determining tenacious glumes. Evidently, chromosome arm 2BS of wild emmer contains a *Tg* allele that determines tough glumes and 2BS of the free-threshing form ssp. *durum* contains the *tg* allele determining soft glumes (Simonetti et al. 1999). That the *Tg* allele exists on chromosome $2M^o$ of *Ae. geniculata* was shown by conferring tenacious glumes thru these chromosomes in addition and substation line in bread wheat (Friebe et al. 1999). Moreover, these authors reported that the spike morphology of disomic addition line $2M^o$ is similar to those of all of the homoeologous group 2 chromosomes of the Triticeae that have been added to CS wheat, in having tenacious glumes. Faris et al. (2014) and Sharma et al. (2019) reported that domesticated emmer, like wild emmer, also carries *Tg* alleles on chromosomes 2A and 2B. In accordance with the rule for the symbolization of genes in homoeologous sets, the Tg genes in emmer wheat should be designated *Tg-A1* and *Tg-B1*. The gene *Tg-B1* was found homoeoallelic to *Tg-D1* on chromosome arm 2DS of *Aegilops tauschii* (Faris et al. 2014).

Genetic analysis indicated that the effects of the three genes, *Q, tg-A1,* and *tg-B1,* are additive, with *Q* having the most profound effect on threshability, and that free-threshing alleles are necessary at all three loci to attain a complete free-threshing phenotype (Sharma et al. 2019). Thus, the free-threshing trait in tetraploid wheat is determined by the three complementary genes, *Q, tg-A1* and tg-B1, and therefore, three mutations were required to produce the free-threshing character in ssp. *parvicoccum,* the primitive free-threshing tetraploid wheat, changing the genotype *qqTgTgTgTg* to *QQtgtgtgtg* (Jantasuriyarat et al. 2004; Faris et al. 2014; Sharma et al. 2019). These mutations must have occurred within a relatively short period (Sharma et al. 2019). Domesticated emmer, ssp. *dicoccon*, first appeared in the Levant in the early PPNB period, i.e., from 9500 to 9000 uncalibrated years ago, and early free-threshing tetraploid wheat, presumably spp. *parvicoccum*, appeared in the middle to late PPNB, i.e., 9000–7500 uncalibrated years ago (Table 13.1). Muramatsu (1979) showed that a variety of domesticated emmer, ssp. *dicoccon* var. *liguliforme,* which has a semi-tough rachis with keeled glumes, contains the *Q* allele. This taxon is hulled presumably because it carries the *Tg* alleles on chromosome 2A and/or 2B. The presence of *Q* in some lines of domesticated emmer may indicate that the mutation from q to Q may have already occurred in this subspecies.

Another major gene affecting the free-threshing trait was identified and allocated to the short arm of chromosome 2A^m of diploid wheat, *T. monococcum* (Taenzler et al. 2002; Sood et al. 2009). A recessive mutation of this gene, called *sog* (soft glume), determines the *s*oft glume trait in *var. sinskajae* of domesticated *T. monococcum* [according to van Slageren (1994) this taxon should receive the rank of a variety rather than of a species]. Whereas Simonetti et al. (1999), Taenzler et al. (2002), and Jantasuriyarat et al. (2004) considered *tg* and *sog* to be orthologoues, Sood et al. (2009), comparing the map positions of *sog* and *Tg* using homoeologous group-2-specific RFLP markers, found these genes to be non-orthologous.

Several minor genes and modifiers also are involved in determining the threshability trait (Tzarfati et al. 2014). Similar to the finding of Jantasuriyarat et al. (2004) in hexaploid wheat, QTLs for threshing time and threshing efficiency were found on the long arm of chromosome 5A of tetraploid wheat, overlapping with the position of the *Q* gene, and on chromosomes 2A and 2B which, according to their locations, may correspond to the *sog* gene (Sood et al. 2009). Peleg et al. (2011) and Tzarfati et al. (2014) detected additional QTLs affecting threshability on chromosome 4B and chromosome 3A. QTLs affecting glume toughness were also found on chromosomes 4A, 6A and 7B (Simonetti et al. 1999; Peleg et al. 2011). Taken together, the threshability trait in tetraploid wheat seems to be under the control of several major and minor genes (Tzarfati et al. 2014).

Interestingly, the number of domestication-related QTLs mapped to the A subgenome was two-fold higher than those found on the B subgenome, i.e., 24 QTL effects for domestication and domestication-related traits in the A subgenome versus only 11 such QTLs in the B subgenome (Tzarfati et al. (2014). This is in accordance with the concept of 'genome asymmetry', implying that the A subgenome is dedicated to the control of morphological traits, house-keeping metabolic reactions and yield components (Peng et al. 2003; Feldman et al. 2012).

In addition to the above-mentioned classical domestication traits that were selected in the process of domestication of ssp. *dicoccon* and evolution of ssp. *parvicoccum, durum* and other free-threshing subspecies of *T. turgidum,* several other domestication traits that were advantageous to the farmer were selected over time. These include plant erectness versus the prostrate grassy types, simultaneous ripening of grains, increased number of seeds per spikelet, increased grain size, and reduced seed dormancy (Feldman 2001). Golan et al. (2015), compared grain weight, embryo weight, and the interaction between these two traits in ssp. *durum* and wild emmer wheat. They found that grain weight was increased under cultivation without any parallel change in embryo weight, resulting in a significantly reduced (30%) embryo weight/grain weight ratio in durum wheat. Using a population of recombinant inbred substitution lines, they found that a cluster of loci affecting grain weight and shape was located on the long arm of chromosome 2A, whereas a locus controlling embryo weight was mapped to the short arm of chromosome 2A. Their results suggest a differential selection of grain and embryo weight during the evolution of domesticated wheat.

At the end of the Pre-Pottery Neolithic B period in the Near East, about 7500 uncalibrated years BP, all the major evolutionary processes required to produce domesticated tetraploid wheat had already been completed (Kislev 1984). Notably, the origin and establishment of the main domesticated wheat, ssp. *dicoccon* and ssp. *parvicoccum*, occurred within the first millennium of cultivation, i.e., in the middle of the 9th millennium BP. The establishment of these new forms within such a short span of time must have been associated with an extremely high rate of evolution. This burst of evolutionary changes presumably resulted from the conditions prevailing in the cultivated field, which were entirely different from those in the wild. The selection pressures imposed by early farmers were understandably different from those operating in the wild. Characters with a negative advantage in nature were preferred under cultivation, thus establishing new evolutionary trajectories. This might explain the appearance of types with a non-brittle rachis, naked grains, and an erect stature, which are characterized by uniform, rapid germination. The spread of wheat culture to neighboring regions exposed the plant to new climatic, edaphic, and biotic conditions, and, hence, to new sets of selection pressures. The wheat genotypes might have reacted by increased mutability, possibly due to the temporal activation of various transposable elements, as a result of the new environmental stresses. In addition, the spread of domesticated tetraploid wheat to new regions facilitated contact between different domesticated related species, such as different types of tetraploid and hexaploid wheat, or even with wild species of related genera, with which they could exchange genes.

Several puzzling questions still remain regarding the evolution of the various domesticated tetraploid wheats: the time and site(s) of origin of domesticated emmer, the mode of origin of the first naked tetraploid wheat, ssp. *parvicoccum*, and the reasons for its extinction about 1900 years ago [Kislev (1986) reported this date as the latest cropping of ssp. *parvicoccum* in southern Judea, Israel], and the reasons for the late establishment of ssp. *durum* as a major crop in the Near East, despite its sporadic occurrence in older archaeological material from 7500 to 6500 years BP (Nesbitt and Samuel 1996).

13.2.4.2 Time of Complete Replacement of Cultivated Wild Emmer by Domesticated Tetraploid Wheats

At the dawn of cultivation, wild emmer wheat must have been sown from seeds gathered from wild stands, i.e., domestication occurred during the course of cultivation (Hillman and Davis 1990). Measured domestication rates in crops of wild cereals indicated that emmer and barley domestication would have occurred only if they were harvested in a partially ripe (or near-ripe) state, using specific harvesting methods (Hillman and Davis 1990; Zohary 1996). Hillman and Davies (1990, 1999) concluded that under certain conditions, namely, non-conscious selection of plants with non-brittle spikes, wild einkorn, emmer, and barley could become completely domesticated within 200 years, and perhaps even 20–30 years. The mutation rates in plants are commonly between 10^5 and 10^6 per base pair, per year (Rédei 1998). Assuming a mutation rate of 10^5 for genes that affect spike brittleness, and since non-brittleness in domesticated emmer is determined by two recessive homoeoalleles, the rate of spontaneous mutation in both genes is 10^{10}. Supposing a sowing of 200 grains per m^2, which is observed in traditional cropping systems (Gepts

2004), both of these mutations would be expected to appear in a total planting area of 5000 ha. Hillman and Davies (1999) estimated that areas sown in the PPNA for a family of five, ranged between approximately 0.5 and 2.8 ha. Assuming that the average family field was about 1.5 ha and that there were several families (five) in each of the approximately 30 PPNA sites, then the area of cultivated wild emmer at that period was about 225 ha. It would then take about 20–25 years for a plant containing the two mutant alleles to be formed. Due to the self-pollinating system that is predominant in emmer wheat, homozygous plants for the two mutant homoeoalleles would be fixed within a few generations. If people harvested wheat with a sickle and cut the entire spike, or plucked individual ears, or pulled plants up from the roots, this would tend to disperse shattering spikelets and retain all non-fragile mutants. In this manner, genotypes with a non-brittle rachis could be replanted the following year and over time, would dominate the population at the expense of wild, brittle types (Hillman and Davies 1990).

Yet, Fuller (2007) claimed a much slower replacement under cultivation of wild emmer plants by those with non-brittle rachis, i.e., 1000–2000 years later. The PPNA period lasted 800 years, during which several plants with non-fragile spikes could have independently formed in different fields of the Levant. Moreover, during the first millennium of the PPNB, wild and domesticated emmer were grown in mixed stands (Kislev 1984), and several additional mutants with non-brittle spikes could have independently formed. Mutation rates may, therefore, have not been a limiting factor in the multiple formations of domesticated emmer (Gepts 2004). Moreover, the mixture of a few genotypes of domesticated emmer with many genotypes of wild emmer in many fields, facilitated inter-genotypic hybridization, resulting in countless transfers of alleles for non-brittle spikes to other genotypes of wild emmer.

Archaeobotanical evidence clearly shows that the process of domestication was very slow (Willcox 1995) and finds indicate that domestic and wild cereals occurred as mixtures on several early Neolithic sites over a period of at least one millennium (Willcox 1998). Remains of wild emmer are identified by the smooth scar on the rachis segment, indicating normal abscission, whereas, in domesticated emmer, the scar is rough because the rachis had been broken apart by threshing. Archaeobotanical finds clearly show that late Epipalaeolithic and early Neolithic distributions of wild cereals were much more extensive than in earlier periods (Hillman 1996), and that the collected cereals differed on the various sites. But once cultivation of several preferred crops had been chosen, they became widespread on the account of others. For example, emmer, became more widespread at the expense of einkorn in southern part of the Fertile Crescent.

Gross and Olsen (2010) pointed out that while artificial domestication experiments conducted in cereals showed that classical domestication traits, such as the loss of rachis fragility and seed dormancy, can arise and increase in frequency over a short time period when subjected to strong selection (Hillman and Davis 1990), archaeological data indicate that the appearance of plants with non-brittle rachis was gradual, at least in wheat and barley (Kislev 1984; Tanno and Willcox 2006; Fuller 2007). In these crops, the non-brittle phenotype appeared only after an increase in grain size, a trait that, itself, reflects selection for germination under active cultivation conditions. Thus, although the loss of rachis fragility would be expected to greatly facilitate the harvesting of grains in planted fields, the phenotype did not actually appear in the initial stages of active cultivation and selection (Fuller 2007; Gross and Olsen 2010).

The archaeological evidence does not support the view that harvesting with a sickle was the selective force that led to rapid domestication of plants with a non-brittle rachis. Preserved sickles, or lithic sickle blades, are known from the Natufian period (13,000–10,300 years BP), in a period for which there is no evidence for domesticated wheats. Namely, in the Near East, sickles were in use prior to agriculture and were only applied to harvest wheat relatively late, after domestication (Fuller 2007). It is possible that sickles were used for harvesting of sedges (Cyperaceae) and reeds (Phragmites), as materials for basketry or thatching (Kislev 1984; Fuller 2007). As indicated by the archaeobotanical evidence (Fuller 2007), the rate of evolution of tough rachis einkorn and barley was far too slow to be accounted for by a model of conscious strong selective pressure that would be expected if sickling was used for cereal harvest, as modelled by Hillman and Davies (1990). Thus, it appears that early cultivators continued to employ the time-efficient harvesting methods associated with hunter-gatherers. Once cultivated populations had noticeably large proportions (majorities) of non-shattering types, then the transfer of the sickle technology to agriculture may have been seen as an obvious enhancement. As others have noted, the harvesting of cereals when green, i.e., immature, by plucking or beating, regardless of technique, will not select for domesticated types (Hillman and Davies, 1990; Willcox, 1999).

The gradual and prolonged replacement of wild by domesticated forms presumably was caused by the fact that farmers used to harvest immature spikes or collect the spikelets from the ground rather than harvested mature spikes (Kislev 1984; Kislev et al. 2004) applying only a weak selection in favor of the latter. If wild cereals were harvested simply by passing through stands and shaking or beating spikes to knock spikelets into a basket then the shattering, wild-type genotypes would be the ones to predominate in the next year's crop. Also, a good portion of wild spikelets that

fell to the ground were not collected and germinated next season, contributing to the new generation of wild wheat. All these practices delayed the selection for domesticated emmer.

The complete replacement of wild emmer by genotypes with non-brittle rachis was a gradual process (Fuller 2007). Actually, quantitative assessment of diploid wheat rachis remains from several sites suggested a gradual increase in the proportion of the domesticated-type spikes over the course of the PPNB (Tanno and Willcox 2006). Similar estimates were made for emmer wheat (Kislev 1984). In general, there is contrast between early sites, which largely or entirely contain wild-type chaff remains, while later sites are dominated by domesticated-type remains, with some intermediate proportions for sites chronologically in the middle. The rates of evolution do not come anywhere close to the 20–100 years estimated by Hillman and Davies (1990), who assumed sickle harvesting of morphologically wild near-mature plants or uprooting of whole plants. This vast difference in domestication rates raises questions about how to explain the absence of selection for domesticated non-shattering genotypes. From their data, Tanno and Willcox (2006) suggested that domestication occurred somewhat more quickly in wheat, perhaps around 1500 years, as opposed to barley, which shifted over a period of 2000 years or slightly more. In accord, Fuller (2007) drew attention to the fact that the shift to non-brittle rachis (full domestication) appears to have started about 500–100 years after large grain size had already evolved (semi-domestication), and this might therefore suggest a minimum estimate of 2000 years for the evolution of both aspects of the domestication syndrome.

The first free-threshing tetraploid wheat, ssp. *parvicoccum*, appeared several hundred years after the appearance of domesticated emmer, ssp. *dicoccon* (Kislev 1979/1980). These two tetraploid forms were the most important crops in the Mediterranean basin and Near East until the Hellenistic period, (ca. 2300 years ago), when they were gradually replaced by the more advanced, free threshing, with large grains, durum wheat, *T. turgidum* ssp. *durum* (Nesbitt 2002). This subspecies was already found by Hillman (1978) in layers of Pottery Neolithic Can Hassan III (7500–6200 years BP). It is perplexing that in spite of its early origin, ssp. *durum* was established as a major crop in the Mediterranean basin and the Near East only during the Hellenistic period. Perhaps the early types were less adapted to the conditions in these regions than were ssp. *parvicoccum* and ssp. *dicoccon*. Adapted genotypes of ssp. *durum* only evolved at a later period by a series of mutations, or through introgression of genes from the other two tetraploid subspecies. The cultivation of ssp. *durum* as an admixture with ssp. *parvicoccum* in the Near East may explain the morphology of some peculiar taxa of ssp. *durum*, such as

Horan wheat in the Levant, characterized by a short compact spike and plump grains, reminiscent the spike of ssp. *parvicoccum* (Kislev 1979/1980). Crosses between ssp. *durum* and ssp. *parvicoccum* might have resulted in such intermediate forms.

Today, ssp. *durum* is the principal tetraploid wheat. Its yield is relatively high under moderately dry conditions, and it grows as a major crop in the Mediterranean Basin, the Near East, India, the USSR, and in low rainfall areas of the great plains of the United States and Canada. Its large, hard-textured grains yield low-gluten flour suitable for macaroni and semolina products. Most other free-threshing subspecies of tetraploid wheats (*turgidum, polonicum, turanicum,* and *carthlicum*) are probably of a relatively recent origin and deviate from ssp. *durum* in only a few characters (Mac Key 1966; Morris and Sears 1967).

Maccaferri et al. (2019) genotyped the Global Tetraploid Wheat Collection consisting of 1856 accessions that represent the four main germplasm groups historically involved in tetraploid wheat domestication and breeding: wild emmer, domesticated emmer, durum landraces, and durum modern cultivars. The results show that two domesticated emmer populations from southern Levant displayed the closest relationship to all durum landrace populations, while the modern durum cultivars germplasm was mostly related to the two durum landrace populations from North Africa and Transcaucasia.

13.2.5 Domestication of Timopheevii Wheat

Compared with the available information concerning the domestication of *T. turgidum*, little is known about the domestication of the second allotetraploid wheat, *T. timopheevii*. Domesticated *timopheevii* wheat is hulled with stiff glumes which its current cultivation is restricted to a few localities in Western Georgia, Transcaucasia (Zhukovsky 1928). It is generally accepted that domesticated *timopheevii*, i.e., ssp. *timopheevii*, derived from wild *timopheevii, ssp. armeniacum* (Jakubziner 1932; Dorofeev et al. 1980). Nesbitt and Samuel (1996) alleged that the restricted cultivation of ssp. *timopheevii* to western Georgia may indicate that this crop was a secondary domesticate: when emmer cultivation spread to Transcaucasia, local populations of ssp. *armeniacum* could have grown as a weed in emmer fields and eventually became domesticated. Badaeva et al. (1994) suggest that the high karyotypic stability of domesticated *timopheevii*, as compared to the high degree of polymorphism of wild *timopheevii*, may be the result of its recent domestication or to its domestication in restricted area. Also, Mori et al. (2009), thought that ssp. *timopheevii* was domesticated later than emmer. They evaluated molecular variation at 23 microsatellite loci in the chloroplast

genome and found no variations among the analyzed six accessions of domesticated *timopheevii*, suggesting a monophyletic origin of this crop. Moreover, none of the wild *timopheevii* plastotypes collected in Transcaucasia were closely related to the plastotype of domesticated *timopheevii*. On the other hand, the plastotypes found in northern Syria and southern Turkey showed closer relationships with domesticated *timopheevi* suggesting that the cultivation of wild form of *T. timopheevii* and subsequent its domestication might have occurred in southern Turkey and northern Syria. Interestingly, Vavilov (1935) already suggested that *T. timopheevii* of western Georgia was probably originally introduced from northeastern Turkey. Likewise, Menabde and Ericzjan (1942; cited by Dorofeev et al. 1980) assumed that domesticated *T. timopheevii* was in the ancient kingdom of Urartu, eastern Turkey, and northwestern Iran. Immigrants from Urartu introduced it into western Georgia. This scenario of introduction of domesticated *timopheevii* into Georgia from the south should not be rejected (Dorofeev et al. 1980).

Furthermore, judging from the present geographical distribution of its wild progenitor, ssp. *armeniacum,* (Table 9.2), the carbonized grains, spikelets, and clay impressions found at Jarmo, northeastern Iraq, (ca. 8750 noncal years BP) and Cayonu Tepesi, southeastern Turkey, (ca. 9000 noncal years BP), that could belong to this taxon rather than to wild emmer, evidence that may indicate that during the down of agriculture both wild ssp. *dicoccoides* and wild ssp. *armeniacum* could have been taken into cultivation in southeastern Turkey and northern Iraq. If such assumed cultivation of ssp. *armeniacum* lasted sufficient time mutations for non-brittle *timopheevii* could have been happened. Then, this raises the following question: if domesticated *timopheevii* were indeed produced in the northern Fertile Crescent why were they replaced by domesticated emmer?

New archaeobotanical data (Jones et al. 2000), show that at three Neolithic sites and one Bronze Age site in northern Greece, spikelet bases of a "new" type of glume wheat (NGW) have been recovered. These spikelet bases are morphologically distinct from those of the typical domesticated einkorn, emmer, and spelt wheats. NGW were also recorded from Neolithic and Bronze Age sites in Turkey, Hungary, Austria, and Germany (Jones et al. 2000). Ulaş and Fiorentino (2021) analyzed morphologically remains of NGW spikelet bases from two Turkish settlements. attestating to its large-scale presence in Anatolia. It seems likely that the NGWs are tetraploids and have morphological features in common with *T. timopheevi*. Such domesticated *timopheevii* could have formed independently from the Georgian one by a separate domestication process(s) of wild *timopheevii* in the northern Fertile Crescent and its cultivation can spread westwards to Europe as a mixture with other wheats or as a pure crop. According to Jones et al. (2000), it is difficult to establish if the new type was cultivated as a pure crop or as a part of a mixture with other domesticated wheats. In any case, its cultivation has ceased over large geographical areas since the Bronze Age.

Czajkowska et al. (2020), using PCR primers specific for the wheat B and G subgenomes, detected DNA sequences from the G subgenome in two NGW accessions, the first comprising grain from the mid 9th millennium BP at Çatalhöyük in Turkey, and the second made up of chaff from the later 7th millennium BP site of Miechowice 4 in Poland. The Miechowice chaff also yielded a B genome sequence, which they ascribe to an admixture of emmer and the NGW. Their result therefore, support the conclusion of Jones et al. (2000) that the NGW is a member of the *T. timopheevii* group. Hence, domesticated *timopheevii* can no longer be looked upon as a minor crop, restricted to western Georgia, but instead must be viewed as a significant component of prehistoric Eurasian agriculture. It is therefore an important question why the cultivation of domesticated *timopheevii* come to an end in West Asia and Europe.

Nave et al. (2021) hypothesized that *T. timopheevii*, like *T. turgidum*, was also domesticated through mutations in the *Bt1* genes, but should carry distinct, novel *bt-A1* and *bt-G1* mutated alleles. To examine this hypothesis, they analyzed the sequence variation associated with the brittle rachis trait in various wild and domesticated *T. timopheevii* and *T. turgidum* accessions. Their analysis revealed a novel, recessive, loss-of-function *br-A1* allele in domesticated *T. timopheevii*, affecting a partially brittle rachis phenotype. This allele exists in all the studied accessions of domesticated *T. timopheevii* and was also found in one wild *timopheevii* accession that exhibits partial rachis brittleness. The mutation in this *timopheevii* allele is different from the mutation *br-A1* of all studied accessions of domesticated *T. turgidum*. This mutation, found exclusively in the coding sequence of the *bt-A1* haplotype of *T. timopheevii*, is responsible for changing seven amino acids in the C-terminal end of the protein coded by this mutation. Such a modification is expected to alter the function of the protein, contributing to the nonbrittle rachis phenotype. Using *T. turgidum* primers, the promoter region for *Bt-B1* could not be amplified in any *T. timopheevii* accessions, exemplifying the gene-level distance between the two species (Nave et al. 2021). Their results support the concept of independent domestication processes for the two allotetraploid wheat species. The structure at the *bt-A1* locus in each lineage supports the diphyletic origin of the two allotetraploid wheats (Mori et al. 2009; Jiang and Gill 1994).

A free-threshing mutant, designated *T. militanae* Zhuk. and Migush, was selected from a single specimen of ssp. *timopheevii*. According to van Slageren (1994), species described only on the basis of a mutation but never released as a commercial cultivar, should not be regarded as new species but rather, should be made synonym under the cultivated subspecies from which they were isolated. The genetic basis of the free-threshing trait in this mutant was not studied. It would be interesting to see if it controlled by the same genes system (Q and tg) as the free-threshing subspecies of *T. turgidum*.

A unique non-brittle cytotype of the *timopheevii* lineage is *T. zhukovskyi*, an auto-allohexaploid carrying genome GGAAA^{m}A^m, that was discovered in 1957, in Western Georgia, by Menabde and Ericzjan (Jakubziner 1959). This species is isolated from the Zanduri wheat (admixture of ssp. *monococcum*, ssp. *timopheevii*, and *T. zhukovskyi*) (Jakubziner 1959; Dorofeev 1966). It therefore must have been derived in the cultivated fields from hybridization between domesticated *timopheevii* (genome GGAA) and domesticated *monococcum* (genome A^{m}A^m). *T. zhukovskyi* is currently grown in a limited area in Transcaucasia and has never been cultivated alone.

13.3 Hexaploid Wheat

13.3.1 Introduction

Since no wild prototype of the hexaploid group is known to exist, many theories have been proposed as to the time, place, and way of origin of the various subspecies of *T. aestivum*. The fact that ssp. *spelta* is hulled wheat that disarticulates into spikelets when a slight mechanical pressure is appliedled as in thrashing, led de Candolle (1886), Hackel (1890), Schulz (1913b), and Carleton (1916) to consider it more primitive than the non-brittle, free-threshing hexaploid forms, and thus, as the oldest form of *T. aestivum*. This conclusion is further supported by the fact that all crosses of either hulled or free-threshing tetraploid wheats with all used lines of *Ae. tauschii* yielded only hulled forms resembling ssp. *spelta*, indicating that this subspecies is the prototype of hexaploid wheat (McFadden and Sears 1946; Kerber and Rowland 1974), and therefore, the predecessor of the more advanced, free-threshing forms. Already Schroder (1931), based on anatomical evidence, proposed that ssp. *aestivum* arose from ssp. *spelta*. With the understanding that ssp. *spelta* is the most primitive subspecies of *T. aestivum*, it was assumed that the free-threshing forms of *T. aestivum* derived from it as a result of mutations (McFadden and Sears 1946). Indeed, the principal differences between the major hexaploid taxa are due to one or two genes that affect gross morphology (Mac Key 1954b) (Table 10.11).

Yet, the genetic data suggesting that the first hexaploid wheats were hulled, spelt-type, and more primitive than the free-threshing forms, do not agree with the archaeological chronology. While free-threshing forms of *T. aestivum*, i.e., ssp. *aestivum*, were found at the middle of the 9th millennium BP and were abundant in the pre-historic Near East from the 8th millennium onwards, thus far there is archaeological evidence for ssp. *spelta* only a thousand years later (Kislev 1984). Neolithic, Near Eastern ssp. *spelta* is very rare and earlier evidence for the existence of ssp. *spelta* is still missing. There is evidence of *spelta* grains from Yarim Tepe II, northern Iraq, dating back to the 7th millennium BP and probably also from Yarim Tepe I, about one thousand years earlier (Kislev 1984). These discrepancies between the genetic and archaeological data pose some difficulties in tracing the early history of the hexaploids. Indeed, several researchers (see Tsunewaki 1968) postulated that spelt wheat could not be the progenitor of bread wheat, but rather, its derivative. On the other hand, assuming that the first hexaploids were hulled, their absence from the prehistoric remains of the Near East may indicate their lack of advantage over domesticated emmer and free-threshing forms of *T. turgidum* in that area. Ssp. *spelta* is grown today in extreme environments of the Near East, such as the high plateau of west-central Iran, eastern Turkey, and Transcaucasia. This cultivation is possibly of an ancient origin. The free-threshing ssp. *aestivum* was preferred by the early farmers of the region and quickly replaced the hulled forms. As man migrated to new areas, cultivated wheats encountered new environments, to which they responded with bursts of variation, resulting in many endemic forms.

13.3.2 Genetic Control of Non-Brittle Rachis in Hexaploid Wheat

McFadden and Sears (1946) were first to synthetize an allohexaploid wheat by crossing domesticated emmer, *T. turgidum* ssp. *dicoccon* (genome BBAA), with the wild diploid species, *Ae. tauschii* (genome DD) and doubled the chromosome number of the F$_1$ hybrid by colchicine treatment. The allohexaploid thus produced resembled morphologically *T. aestivum* ssp. *spelta*, exhibited full chromosome pairing and had high pollen and seed fertility. Also, its hybrids with several natural subspecies of *T. aestivum* had high pairing. McFadden and Sears (1946) thus demonstrated the origin of ssp. *spelta*, the primitive form of *T. aestivum*. The parents and the synthetic allohexaploid had different types of rachis fragility. The spike of ssp. *dicoccon* does not disarticulate at maturity but threshing breaks it upon the rachis node into wedge-type spikelets, each has the rachis internode below it, those of *Ae. tauschii* disarticulate at maturity below the rachis node into barrel-type spikelets, each carries a rachis internode besides it, and the synthetic *spelta* similar to the natural ssp. *spelta*, breaks after threshing into spikelets with rachis internode below, besides, below and besides, and without any rachis internode at all (McFadden and Sears 1946), indicating that the fragility

gene(s) of the two parents are codominant. Codominance of the fragility genes was also reported by Kihara and Lilienfeld (1949) in synthetic hexaploid wheats, produced by crossing wild emmer ssp. *dicoccoides* with *Ae. tauschii*. They found that the spike is fragile; the wedge-type disarticulation is seen in the main part of the spikes and the barrel-type in the upper spikelets.

Similar codominant effect was noted by Sears (1941a, b) in the amphiploid wild einkorn *T. monococcum* ssp. *Aegilopoides—Ae. tauschii*. This synthetic allotetraploid had a relatively semi-tough rachis, that at maturity breaks in different places, above the rachis node as in ssp. *aegilopoides*, and below the rachis node as in *Ae. tauschii*. The codominance effect of these fragility genes was noted also in hybrid between wild emmer and *Ae. tauschii* (Matsumoto et al. 1963) and in an accession of the semi wild wheat, *T. aestivum* ssp. *tibetanum* (Tsunewaki et al. 1990).

The tetraploid parent ssp. *dicoccon* has non-fragile rachis that breaks after threshing into wedge-type spikelets, i.e., each spikelet with the rachis internode below it, while the diploid parent *Ae. tauschii*, being a wild species, disarticulates at maturity into barrel-type dispersal units, i.e., each spikelet with the rachis internode beside it. Li and Gill (2006) mapped rachis disarticulation genes in wheat and its wild relatives. The *Br1* gene for wedge-type disarticulation was mapped to a region delimited by the DNA markers Xpsr598 and Xpsr1196 on the short arm of chromosomes 3A in *T. timopheevii*. The barrel-type disarticulation gene, designated by Li and Gill (2006) as *Br2*, was mapped in *Ae. tauschii* to an interval of 4.4 cm between Xmwg2013 and Xpsr170 on the long arm of chromosome 3D. Avni et al. (2017) found that the non-fragile rachis of the tetraploid parent is controlled by two compound recessive genes, *br-A1* that contains the genes *btr1-A and btr2-A and br-B1* containing the genes *btr1-B and btr2-B* (Avni et al. 2017). The btr1-A and btr1-B genes are complementary in bringing about loss-of-function of the dominant genes resulting in non-fragile rachis so that at maturity the spike remains intact on the culm and breaks into individual spikelets only upon slight application of mechanical pressure at threshing (Dorofeev et al. 1979). This type of tough rachis was called a 'semi-tough' rachis, not to be confused with the fully tough rachis of the free-threshing tetraploid subspecies wheats which remains intact when threshed (Hillman and Davis 1990).

In wild barley, *Hordeum spontaneum*, the formation of wedge-type dispersal units is genetically determined by the genes *Btr1* and *Btr2*, a pair of dominant, complementary, linked genes mapping to the short arm of chromosome 3H. A 1 bp deletion in the *Btr1* coding sequence, and one of 11 bp in the *Btr2* coding sequence are sufficient to convert a brittle rachis to a non-brittle one (Pourkheirandish et al. 2015). In wild *T. monococcum*, i.e., ssp. *aegilopoides*, the substitution of a single residue in the *Btr1* product converts a brittle to a non-brittle rachis (Pourkheirandish et al. 2018; Zhao et al. 2019). In the tetraploid wheat, mutations at both the A and B genome copies of *Btr1* are required for the formation of a non-brittle rachis (Avni et al., 2017). Hence, *Btr1* orthologs are required for disarticulation above the rachis nodes, since the loss-of-function *btr1* mutant forms a non-brittle rachis in *Hordeum, and in diploid and tetraploid wheat* (Pourkheirandish et al. 2015, 2018; Avni et al. 2017). *Ae. tauschii* lacks an intact copy of *Btr1* and disarticulates below the rachis nodes; the inference is that *Btr1* is not required to effect disarticulation below the rachis nodes.

All the Triticeae species that disarticulate at maturity into wedge-type spikelets have the *Br1* compound locus that is located on the short arm of group 3 chromosomes (Watanabe et al. 2002; Li and Gill 2006). Hence, the *Br1* genes are orthologous. *Ae. tauschii* exceptionally produces barrel-type dispersal units (Kihara 1954). Rather than mapping to the short arm of chromosome 3D, the locus responsible for this trait in *Ae. tauschii* maps to the long arm of chromosome 3D (Li and Gill 2006; Katkout et al. 2015; Amagai et al. 2015). Consequently, it was designated *Br2* (Li and Gill 2006). The gene *Br-D2* is paralogous to the genes *Br1* on 3AS, 3BS, and 3DS. It presumably resulted from a duplication of *Br-D1*, followed by an intra-chromosomal transposition to the long arm of 3D (Li and Gill 2006).

Zeng et al. (2020b) found that the *Br-D2* locus of *Ae. tauschii* lacks an intact copy of *Btr1* and disarticulation in this species occurs below, rather than above, the rachis node, resulting in barrel-type spikelets. Thus, the product of *Btr1* appears to be required for disarticulation to occur above the rachis node resulting in wedge-type spikelets. Zeng et al. (2020a) reason that *Ae. tauschii* could be an evolutionary intermediate between the Poeae/Aveneae and the Triticeae tribes, since members of the former two tribes also lack *Btr1*. However, unlike members of the Poeae/Aveneae, *Ae. tauschii* does harbor an intact copy of *Btr2*. The above reasoning would require that *Btr2*, and later *Btr1*, were acquired independently. An alternative evolutionary pathway assumed that the truncated *Btr1* sequences present in *Ae. tauschii* have a common origin with other diploid *Aegilops* species of the D- lineage (Marcussen et al. 2014), but it diverged from the D-lineage 5.37 MYA (Li et al. 2022). After this divergence *Ae. tauschii* lost its intact copy of *Btr1*, but retained the truncated one (Zeng et al. 2020a). Disarticulation below the rachis nodes could have evolved in *Ae. tauschii* following the de novo recruitment (or perhaps neofunctionalization) of a co-operating gene(s). The latter may include orthologs of genes known to be responsible for shattering in rice (see list of genes in Zeng et al. 2020a) since orthologs of these genes are present in *Ae. tauschii* and are transcribed in the immature spike. Especially, the *sh4* and *OsCPL1* orthologs showed higher

expression than the other ones in the immature spikes of *Ae. tauschii*. However, the genetic basis of the barrel type dispersal unit is not wholly unmistakable and has yet to be determined (Zeng et al. 2020b).

The presence of a *Btr2* gene in each of the Triticeae species examined, suggest that its product is involved in the determination of the brittle rachis trait above the rachis node. In barley, the finding that *Btr2* expression occurs in a thin cell layer above the rachis node has been taken to imply that this gene contributes to the formation of the disarticulation zone (Pourkheirandish et al. 2015). Whether *Btr2* in *Ae. tauschii* is involved in the same way below the rachis node remains an open question. However, it is clear that *Btr2* transcript is generated in immature *Ae. tauschii* spikes, although at a rather low abundance (Zeng et al. 2020b).

The fragility trait relies on the development of a disarticulation layer, in most species above the rachis node, resulting in wedge type dispersal units, but in some species below the rachis node, resulting in barrel type dispersal units (Fig. 2.3). Zeng et al. (2020b) showed that in *Ae. tauschii* *Btr2* transcript is present in a region below the rachis node where the abscission zone forms. The implication is that in this species, the *Btr2* product is involved in the formation of barrel type.

There are two forms of ssp. *spelta* Iranian and European. The Iranian spelt has a wedge-type disarticulation whereas the European spelt has a barrel-type disarticulation. It is speculated that the Iranian spelt originated from a cross between a tetraploid line(s) of *T. turgidum* having tough rachis (genotype *br-A1, br-B1*) and *Ae. tauschii* having brittle rachis (genotype *br-D1, Br-D2*) while the European spelt originated from a cross between common wheat carrying a loss-of-function mutation (see below; genotype *br-A1, br-B1, br-D1, br-D2*) and domesticated emmer wheat (genotype *br-A1, br-B1*). Consequently, the Iranian spelt has the rachis fragility genotype *br-A1, br-B1, br-D1, Br-D2* exhibiting codominance of the rachis brittle genes, whereas the European spelt as the genotype *br-A1, br-B1, br-D1, br-D2* exhibiting tough rachis.

Kuckuck (1964) suggested that the hexaploid wheat with a fragile rachis found by Dekaprelevich (1961) in Georgia, Transcaucasia, may have originated as an amphiploid between wild emmer, ssp. *dicoccoides*, and *Ae. tauschii*, independently of the origin of domesticated hexaploids, which are believed to have involved free-threshing tetraploid wheat as their tetraploid parent. Sears (1976) crossed wild emmer, ssp. *dicoccoides*, with *Ae. tauschii* and obtained a F_1 hybrid with a brittle rachis. Therefore, B- and W-type disarticulations are governed by two different paralogous loci on group-3 chromosomes.

After its formation, ssp. *spelta*, was not cultivated in southwest Asia on a large scale, presumably it could not compete with the free-threshing forms of tetraploid *T. tur-gidum*. After a relatively short time, about several tens of years, ssp. *spelta* was replaced by the free threshing type of *T. aestivum,* ssp. *aestivum*. In addition to a mutation in the *Tg-D1* gene that suppressed the free-threshing trait (see below), additional mutation should have occurred in hexaploid wheat that converted the codominant *Br-D2* gene to a recessive loss-of function allele, *br-D2,* that led to the formation of hexaploid wheat with fully tough rachis that remains intact when threshed, as that of the free-threshing subspecies of *T. turgidum*. These mutations have been critical for the establishment of bread wheat, ssp. *aestivum* as the most important wheat form. Ssp. *spel*ta is currently grown in several locations in Iran and Transcaucasia. The transformation from the hulled form with semi-tough rachis ssp. spelta to a free threshing with fully tough rachis ssp. *aestivum* was relatively a fast process.

The feral hexaploid wheat ssp. *tibetanum* has fragile rachis and tenacious glumes. The rachis disarticulates at maturity to yield wedge-type spikelets. Study of the genetic control of the fragile rachis and glume tenacity in the feral hexaploid wheat ssp. *tibetanum,* using progenies of crosses and backcrosses of ssp. *tibetanum* with *T. aestivum* ssp. *aestivum* cv. Columbus, indicated that the fragile rachis and the hulled character of ssp. *tibetanum* were dominant over the tough rachis and free-threshing character of bread wheat (Cao et al. 1997). However, rachis fragility and glume tenacity of ssp. *tibetanum* were each controlled by a single gene. In the cross between ssp. *tibetanum* and spp. *spelta,* the F$_2$ and F$_3$ populations did not segregate by glume tenacity but did segregate by rachis fragility. Cao et al. (1997) concluded that ssp. *tibetanum* differs from ssp. *aestivum* in rachis fragility and glume tenacity and from the hulled subspecies of *T. aestivum* (ssp. *spelta,* and ssp. *macha*) in the pattern and degree of rachis fragility.

Chen et al. (1998), using monosomic and ditelosomic lines of bread wheat, found that the gene controlling the brittle rachis of ssp. *tibetanum* is dominant and located on the short arm of chromosome 3D. Consequently, it is orthologous to the *Br1* gene of many Triticeae species and consequently, was designated *Br-D1*. As was found earlier (Tsunewaki et al. 1990), *Br-D1* is different from the gene determining barrel-type brittle rachis in *Ae. tauschii* and ssp. *spelta*. This gene, designated *Br-D2,* is located on the long arm of chromosome 3D and it is paralogous to the *Br1* genes (Li and Gill 2006). In the progenies of the cross *spelta* x *tibetanum,* F$_1$ plants exhibited the wedge and barrel types of dis-articulation, indicating that wedge-type disarticulation in ssp. *tibetanum* is codominant with the barrel type in spelt wheat. Similarly, mono-telosomic analysis indicated the rachis-fragilit genes were located at approximately 20 cM

from the centromere on the short arms of chromosomes 3A and 3B of wild emmer and on 3D of ssp. *tibetanum,* representing the orthologous locus *Br1* (Watanabe et al. 2002).

13.3.3 Genetic Control of Free-Threshing Grains in Hexaploid Wheat

The synthetic hexaploid wheat, involving hulled *T. turgidum* ssp. *dicoccon-Ae. tauschii,* produced by McFadden and Sears (1946), was a hulled form, resembling *T. aestivum* ssp. *spelta*. Yet, Kerber and Rowland (1974) reported that all 15 hexaploid wheats synthesized from various combinations of nine tetraploid wheats and seven forms of *Ae. tauschii,* were non-free threshing, regardless of the presence or absence of the *Q* allele in the tetraploid parent. Obviously, the genome of *Ae. tauschii,* genome D, carries a gene that suppresses in synthetic hexaploids the activity of the free-threshing gene *Q* that derived from the tetraploid parent. Indeed, Kerber and Dyck (1969) identified a single gene on the D genome that affected threshability in synthetic hexaploid wheat formed from hybridization of free threshing tetraploid wheat and *Ae. tauschii*. Monosomic and mono-telosomic analysis of this hexaploid line revealed the presence of a partially dominant gene for tenacious glumes, designated *Tg,* on the short arm of chromosome 2D (Kerber and Rowland 1974). This gene, should be designated *Tg-D1,* derived from the *Ae. tauschii* parent, inhibited the expression of *Q* in hexaploid wheat. These researchers concluded that primitive hexaploid wheat that carried the *Tg* allele on subgenome D, were hulled types. This finding reinforced the concept that the hulled forms of hexaploid wheat, ssp. *spelta* and ssp. macha, having the genotype QQTgTg, are the primitive forms of *T. aestivum*. A mutation from *Tg-D1* to *tg-D1* on chromosome arm 2DS is presumed to have occurred at the hexaploid level, resulting in the production of the free-threshing types.

As in tetraploid wheat, also in ssp. *aestivum* it is reasonable to assume that the effects of all the genes affecting free-threshing are additive, with *Q* having the most profound effect on threshability, and that the recessive alleles *tg-A1, tg-B1 and tg-D1* are necessary at all three loci to attain a completely free-threshing phenotype. The mutations from q to Q, from Tg-A1 to tg-A1, and from *Tg-B1* to *tg-B1,* occurred at the tetraploid level within a relatively short period (Sharma et al. 2019). The mutation to *Tg-D1* to *tg-D1* occurred at the hexaploid level.

While the *Q* gene on chromosome arm 5AL, has the most significant effct on free-threshing, its homoeoalleles on chromosome arms 5BL and 5DL, were also shown to be involved in this domestication trait (Zhang et al. 2011).

The *q* homoeoallele on chromosome 5B became a pseudo-gene after allotetraploidization. Expression analysis indicated that, whereas *Q* plays a major role in conferring domestication-related traits, the *q* homoeoallele on 5D contributes directly and *q* on 5B indirectly to suppression of the speltoid phenotype. Hence, according to Zhang et al. (2011), the evolution of the *Q/q* loci in polyploid wheat resulted in the hyperfunctionalization of *Q* on 5A, pseudogenization of *q* on 5B, and subfunctionalization of *q on 5D*, all contributing to the domestication traits.

The Asiatic ssp. *spelta* has the genotype *QQtg-A1tg-A1tg-B1tg-B1Tg-D1Tg-D1*, whereas the European ssp. *spelta,* being derived in Europe from spontaneous hybridizations of hexaploid wheat, ssp. *aestivum* (genotype *QQtg-A1tg-A1tg-B1tg-B1tg-D1tg-D1*) and *T. turgidum* ssp. *dicoccon* (genotype *qqtg-A1tg-A1tg-B1tg-B1*) is hulled *qqtg-A1tg-A1tg-B1tg-B1tg-D1tg-D1* having the genotype *qqtg-A1tg-A1tg-B1tg-B1tg-D1tg-D1* or *qqTg-A1Tg-A1Tg-B1Tg-B1tg-D1tg-D1*.

Jantasuriyarat et al. (2004), using recombinant inbred lines of the International Triticeae Mapping Initiative (ITMI) mapping population, derived from a cross between a common wheat cultivar and the synthetic hexaploid wheat ssp. *Durum-Ae. tauschii*, revealed two QTLs, one located on the short arm of chromosome 2D and the second on the long arm of chromosome 5A, that consistently affected threshability-associated traits. The QTL on 2DS, presumably representing the effect of the *Tg* allele that derived from *Ae. tauschii*, explained 44% of the variation in threshability, 17% of the variation in glume tenacity, and 42% of the variation in rachis fragility, whereas the QTL on 5AL, believed to represent the effect of *Q*, explained 21 and 10% of the variation in glume tenacity and rachis fragility, respectively. Overall, Jantasuriyarat et al. (2004) found that free-threshing-related characteristics were predominantly affected by *Tg* and, to a lesser extent, by *Q*.

Other QTLs, on chromosomes 2A, 2B, 6A, 6D, and 7B, were significantly associated with threshability-related traits. Four other QTLs on chromosomes 1B, 4A, 6A, and 7A consistently affected spike characteristics (Jantasuriyarat et al. 2004). The QTL on the short arm of chromosome 1B explained 18% and 7% of the variation in spike length and spike compactness, respectively. The QTL on the long arm of 4A explained 11%, 14%, and 12% of the variation in spike length, spike compactness, and spikelet number, respectively. The QTL on the short arm of 6A explained 27% of the phenotypic variance for spike compactness, while the QTL on the long arm of 7A explained 18% of the variation in spikelet number. QTLs on chromosomes 1B and 6A appeared to affect spike dimensions by modulating rachis internode length, while QTLs on chromosomes 4A and 7A did so by affecting the formation of spikelets. Other QTLs that were significantly associated with spike morphology-related traits, in at least one environment, were localized on chromosomes 2B, 3A, 3D, 4D, and 5A (Jantasuriyarat et al. 2004).

13.3.4 Selection for Yield in Hexaploid Wheat

Mac Key (1986) assumed that in stands of wild wheat, characters that contribute to the successful annual re-establishment of the stand are of the highest selective value. There is no selective advantage for the overproduction of seeds in wild dense stands, particularly when the seed dispersal mechanism is such that most seeds fall near the mother plant. On the other hand, the ability to compete well immediately upon germination, driven by large, energy- and protein-rich seeds, ensures the re-establishment of the stand. Hence, in wild wheat, efficiency per seed is more important than prolificacy, i.e., the ability to produce more viable seeds. Yet, from the dawn of agriculture, prolificacy became more important than seed efficiency, as successful seedling establishment in the cultivated field was achieved by providing the seedlings with optimal growth conditions.

According to Evans (1981), the improvement in grain productivity was achieved through increased leaf size and flag-leaf area, increased size of the vascular system in the spike, advanced flowering time, delayed flag-leaf senescence, increased duration of grain-filling period, increased rate and duration of assimilates translocation to the grains, increased grain size and grain number per spikelet, and increased spike number per plant or per unit area (Table 13.3). Early farmers tended to select for a higher ratio of grain harvested to grain sown, i.e., indirectly for profuse tillering, many grains per spike, and strong grain retention. As the amount of arable land became a limiting factor and the crop became monotypic, selection for higher number of grains per unit area was the preferred criterion. Emphasis was placed on lines that were weak intra-line competitors, yielded well in dense planting, and responded well to fertilizers and various agrochemicals. One of the most important changes during cultivation was an increase in the proportion of the dry matter allocated to harvested grains (harvest index). Continuous selection for increased grain production and size enhanced synthesis and translocation of carbohydrates to the developing grains, with no corresponding improvement in the translocation of amino acids. This obviously resulted in increased grain size, as well as in grain number, and in low grain-protein percentage (Feldman et al. 1990)—a trait common to all cultivated cereals compared to their wild progenitors. On the other hand, there were few, if any, significant changes during wheat cultivation in photosynthetic capacity per unit leaf area, growth rate, and accumulated dry crop weight (biomass) (Table 13.3).

Table 13.3 Modified (A) and conserved (B) yield-related traits in cultivated wheat (after Evans 1981)

A. modified traits		Plant height (increased and then decreased)
		Leaf size (increased)
		Flag-leaf area (increased)
		Flag-leaf senescence (delayed)
		Size of the vascular system in the spike (increased)
		Flowering time (advanced)
		Rate and duration of the grain-filling period (increased)
		Rate and duration of assimilates translocation to the grains (increased)
		Grain size (increased)
		Number of spikelets per spike (increased)
		Number of grains per spikelet (increased)
		Harvest index (increased
B. conserved traits		Plant biomass
		Rate of photosynthesis
		Ratio between photosynthesis and photorespiration

The increased yield in domesticated wheat, stems, to a large extent, from delayed flag-leaf senescence, which prolongs the duration of post-anthesis photosynthesis, resulting in higher production of carbohydrates for the developing grains. This trait can also be achieved by accelerating spike development, which advances anthesis by a few days. In semi-arid regions, where the available amount of water is often a limiting factor during grain development, senescence is delayed for only a short period, while in mesic regions, senescence and, consequently, maturity in general, are delayed for longer periods.

Compared to domesticated wheat, wild emmer, is characterized by later anthesis and earlier grain ripening, which avoids the heat of the late spring. This shorter grain-filling period is apparently the consequence of relatively early and rapid flag-leaf senescence occurring about 3 weeks after anthesis. The degradation of most leaf proteins at this stage reduces and eventually fully eliminates carbohydrate assimilation capacity in the leaves (mainly the flag leaf), resulting in a higher N/C ratio in the assimilates translocated to the grains. Indeed, all wild emmer lines analyzed had high grain protein percentage (GPP) (Avivi 1977, 1979a, 1979b; Feldman et al. 1990), that may contribute to seedling vigor (Millet and Zaccai 1991). Under the semi-arid conditions that prevail in the natural habitats of wild emmer stands, a small number of medium-sized grains with high grain protein content suffice to ensure rapid germination and successful establishment of the next stand.

Grain protein content (GPC) is much lower in domesticated durum and bread wheat than in wild emmer (Avivi 1977, 1979a, 1979b; Feldman et al. 1990). Joppa and colleagues (Joppa and Cantrell 1990; Joppa et al. 1991; Joppa 1993), using substitution lines of wild emmer chromosomes in the background of *T. turgidum* ssp. *durum* cultivar

Langdon (LDN), found that chromosomes 2A, 5B, 6A, and 6B of wild emmer increased GPC in the durum cultivar. Genotypes with wild emmer chromosomes 6B had the highest GPC (Cantrell and Joppa 1991). Using a population of recombinant inbred lines derived from cultivar LDN and the chromosome substitution line LDN (DIC6B), Joppa et al. (1997) mapped the higher GPC gene to chromosome arm 6BS. This gene also increased grain protein in several studied lines of tetraploid and hexaploid wheats (Joppa et al. 1997; Mesfin et al. 1999; Chee et al. 2001). Olmos et al. (2003) found that this factor for high-grain protein behaves as a simple Mendelian gene and designated it *Gpc-B1*, and mapped it proximal to the *Nor2* locus.

Recombinant substitution lines (RSLs) carrying the *Gpc-B1* allele of wild emmer exhibited on average, 12% higher concentration of Zn, 18% higher concentration of Fe, 29% higher concentration of Mn and 38% higher concentration of protein in their grains as compared to RSLs carrying the allele from durum wheat (Distelfeld et al. 2007). These authors also confirmed the effect of the wild emmer *Gpc-B1* allele on earlier senescence of the flag leaveand suggested that the *Gpc-B1* locus is involved in more efficient remobilization of protein, zinc, iron and manganese from the leaves to the grains, in addition to its effect on earlier senescence of the green tissues.

Brevis and Dubcovsky (2010) studied the grain yield, grain weight, protein yield (grain yield multiplied by grain GPC), and N harvest index (NHI) of bread and durum wheats using BC_6F_3 near-isogenic lines with and without the wild emmer *Gpc-B1* allele. All the studied lines in three California locations during the years 2005–2007, showed higher GPC and grain protein yield when the wild *Gpc-B1* allele was present. Bread and durum wheat lines having the wild emmer *Gpc-B1* allele showed a significant decrease in grain weight whereas the decrease in grain weight in the durum lines was not significant.

Brevis et al. (2010), evaluated the effects of *Gpc-B1* on bread-making and pasta quality in isogenic lines for the *Gpc-B1* introgression in six hexaploid and two tetraploid wheat genotypes. In bread wheat, the wild emmer *Gpc-B1* introgression was associated with significantly higher GPC, water absorption, mixing time and loaf volume, whereas in durum wheat, the introgression resulted in significant increases in GPC, wet gluten, mixing time, and spaghetti firmness, as well as a decrease in cooking loss. On the negative side, the wild emmer *Gpc-B1* introgression was associated, in some varieties, with a significant reduction in grain weight, test weight and flour yield, and significant increases in ash concentration.

Uauy et al. (2006a) reported that flag leaf chlorophyll degradation, and change in peduncle color, and spike water content were fully linked to the wild emmer *Gpc-B1* allele. The high GPC allele conferred a shorter duration of grain fill due to earlier flag leaf senescence and increased GPC in all four studied genetic backgrounds. The effect on grain size was more variable, depending on the genotype–environment combinations. These results are consistent with a model in which the wild-type *Gpc-B1* allele accelerated senescence in flag leaves, leading to pleiotropic effects, such as on nitrogen remobilization, total GPC, and grain size. At this point, the possibility of multiple genes that are present within the region governing different aspects of these responses cannot be ruled out. However, the authors assumed that it is more likely that these multiple traits are pleiotropic effects of a single gene rather than the result of multiple independent genes.

The wild-type *Gpc-B1* allele was isolated and found active in all studied wild emmer accessions, whereas in domesticated durum and bread wheat, it was mutated to a non-functional allele (Uauy et al. 2006a). In the free-threshing allotetraploid and allohexaploid wheats, the grain-filling period is longer than in wild emmer (about 3 weeks or more) and consequently, a greater amount of carbohydrates is assimilated and transported to the developing grains, thereby diluting the percentage of proteins and minerals in the grains (Uauy et al. 2006a). Since the wild type allele exists in several analyzed domesticated emmer (ssp. *dicoccon*) lines, it is assumed that the mutation to a loss-of-function allele occurred either in domesticated emmer or in durum wheat (Uauy et al. 2006a).

In a positional cloning of *Gpc-B1*, Uauy et al. (2006a) located the gene in a 7.4 kb region of chromosome arm 6BS, with complete linkage with the different phenotypes affected by *Gpc-B1*, verifying the assumption that *Gpc-B1* is indeed a single gene with multiple pleiotropic effects. The annotation of this 7.4 kb region identified a single gene encoding a NAC protein, characteristic of the plant-specific family of *NAC* (*NAM, ATAF* and *CUC*) transcription factors (Ooka et al. 2003). NAC proteins play important roles in developmental processes, auxin signaling, defense and abiotic stress responses, and leaf senescence (Olsen et al. 2005). Phylogenetic analyses revealed that the closest plant NAC proteins were the NAC transcription factor *ONAC01* in rice, and a clade of three *Arabidopsis* proteins, including No Apical Meristem (NAM) protein. On the basis of these similarities, Uauy et al. (2006b) designated the gene in *T. turgidum* and *T. aestivum* TtNAM-B1 and TaNAM-B1, respectively.

The NAC transcription factor coded by the wild-type *NAM-B1* allele accelerates senescence and increases nutrient remobilization from leaves to the developing grains, whereas free-threshing wheat cultivars carry a non-functional *NAM-B1* allele. Reduction in RNA levels of the multiple NAM homologs by RNA interference delayed senescence by more than 3 weeks and reduced wheat grain protein, zinc, and iron content by more than 30% (Uauy et al. 2006b).

Comparison of the wild-type and the LDN *TtNAM-B1* sequences revealed a 1-bp substitution within the first intron and a thymine residue insertion at position 11, generating a frame-shift mutation in the durum LDN allele (Uauy et al. 2006b). Since the wild type *TtNAM-B1* allele was found in all the 42 wild emmer accessions and in 17 of the 19 domesticated emmer accessions, while 57 durum studied lines lacked the functional allele, it can be assumed that the 1-bp frame-shift insertion was fixed during the domestication of durum wheat. The wild-type *TaNAM-B1* allele was also absent from a collection of 34 varieties of hexaploid wheat representing different cultivars of geographic locations. Twenty-nine of these showed no polymerase chain reaction (PCR) amplification products of the *TaNAM-B1* gene, which suggests that it was deleted, whereas the remaining five lines have the same 1-bp insertion observed in the durum lines.

In addition to the mutant *TtNAM-B1* on chromosome arm 6BS, the durum wheat genome includes an orthologous copy (*TtNAM-A1*) on chromosome arm 6AS and a paralogous one (*TtNAM-B2*), 91% identical at the DNA level to *TtNAM-B1*, on chromosome arm 2BS. These two copies have no apparent mutations. The studies of Uauy et al. (2006b) suggested that the reduced grain protein, Zn, and Fe concentrations were the result of reduced translocation from leaves, rather than a dilution effect caused by larger grains. This hypothesis was confirmed by analyzing the residual nitrogen, Zn, and Fe content in flag leaves.

There are three known alleles in the *NAM-B1* gene: the wild type (WT allele), a null allele, consisting of a deletion that covers 100 kb (deletion allele), and another null allele, bearing a 1 bp frame-shift insertion (insertion allele) (Uauy et al. 2006b). Lundström et al. (2017) studied the distribution of these alleles in wild and domesticated forms of *T. turgidum*. Out of 19 wild emmer accessions studied, 18 carried the wild-type allele and one carried the insertion allele. Among the 16 durum cultivars studied, 15 had the insertion allele and one carried the deletion allele. Among the 61 domesticated emmer accessions studied, 25 carried the deletion, 17 the insertion, and 19 the WT allele. Four landraces of hexaploid wheat were found to carry the wild-type allele (Asplund et al. 2013). The presence of the wild-type allele was also confirmed in several landraces of spring hexaploid wheat from Scandinavia (Hagenblad et al. 2012). Neither of the *NAM-B1* alleles appeared to be limited to specific geographic areas. These findings show that both null alleles exist in domesticated emmer and durum wheats. Moreover, based on the finding that the insertion allele exists in several accessions of wild emmer, Lundström et al. (2017) suggested that this allele arose in wild emmer.

Seed size is influenced by the *NAM-B1* gene, which has consequently been suggested to be a domestication gene (Dubcovsky and Dvorak 2007). Indeed, the *NAM-B1* gene fulfils some criteria for being a domestication gene, by encoding traits of domestication relevance (seed size and increased carbohydrate production due to delay in leaf senescence) and, as such, has been under positive selection (Dubcovsky and Dvorak 2007; Hu et al. 2012, 2013; Hebelstrup 2017; Lundström et al. 2017).

The flag leaf is the main source of protein and micronutrients in the wheat grain (Pearce et al. 2014). During senescence of the flag leaf, photosynthesis ceases, and the enzymes involved, mainly ribulose-1,5-bisphosphate carboxylase/oxygenase (rubisco), are degraded into amino acids that are transported, together with other nutrients and minerals, to the developing grains (Barneix 2007). The delay in the onset of senescence facilitates the production of more carbohydrates that are transferred to the grains. Since the uptake of nitrogen and minerals from the soil ceases at anthesis, and assimilation of carbohydrates continue for several more weeks, the levels of protein, Zn, and Fe are diluted in the mature grain. The constant selection for increased yield in domesticated durum and bread wheat, presumably unintentionally led to preference of mutants with a delayed flag-leaf senescence, which prolongs the duration of post-anthesis photosynthesis.

Coupling delayed flag-leaf senescence with post-anthesis nitrogen absorption might improve both grain yield as well as GPC (Feldman et al. 1990). In this respect, some variation in the duration of post-anthesis nitrogen uptake, up to several days after anthesis, was observed among accessions of wild emmer (Zaccai 1992). This trait should be exploited to increase GPC without affecting grain yield.

13.3.5 Summary of the Main Events in Wheat Cultivation and Domestication

Wild diploid wheat, *T. monococcum* ssp. *aegilopoides,* was apparently first taken into cultivation in the northern Levantine Corridor, i.e., in the Karacadag area, southern Turkey (Heun et al. 1997), while wild emmer, *T. turgidum* ssp. *dicoccoides,* was first cultivated in the watershed of the Jordan River in the southern Levantine Corridor (Bar-Yosef 1998). Based on Kislev (1981), Bar-Yosef and Kislev (1989), Bar-Yosef (1998) and Harris (1998), the domestication and evolution of the various wheat species can be summarized as follows: (a) The earliest evidence of cultivation of wild emmer is from the first half of the 10th millennium BP in the Jordan Valley and the Damascus basin (van Zeist and Bakker-Heeres 1985; Bar-Yosef and Kislev 1989). These sites are very close to the natural area of distribution of wild emmer today and 10 millennia ago. Later evidence of cultivation of wild emmer comes from the second half of the 10th millennium at Cayonu, east Anatolia, and from Ali Kosh, southwestern Iran. Around the middle of the 9th millennium BP, a non-brittle type of emmer, ssp. *dicoccon,* was derived from a brittle type by mutations. This domesticated wheat was grown for millennium or more in mixture with brittle forms and gradually replaced it throughout the Fertile Crescent. (b) Around the middle of the 9th millennium BP, small grains and rachis fragments of a dense ear of naked wheat were found in Tell Aswad, near Damascus, Syria (van Zeist and Bakker-Heeres 1985). These are assumed to be remnants of tetraploid naked wheat, *T. turgidum* ssp. *parvicoccum,* a taxon that presumably evolved from ssp. *dicoccon* by a series of mutations (Kislev 1979/1980). Because all existing naked tetraploid wheats apparently possess on chromosome-arm 5AL the *Q* factor, which confers a tough rachis, soft glumes, and free-threshing grains (Muramatsu 1986), it is assumed that the extinct ssp. *parvicoccum* also possessed this gene. Hence, mutation of the *q* gene to *Q* led to the main character change distinguishing *parvicoccum* from *dicoccon*. (c) In the ancient fields of ssp. *parvicoccum* in northwestern Iran, where *Ae. tauschii* grew as a weed, a hexaploid hulled wheat, containing *Q* from ssp. *parvicoccum* and *Tg* from *Ae. tauschii*, viz. *TgQ spelta,* could have arisen by hybridization. Neolithic Near Eastern ssp. *spelta* is very rare, but there is evidence of *spelta* grains from Yarim Tepe II, northern Iraq, back to the 7th millennium BP, and probably also from Yarim Tepe I, about one thousand years earlier (Kislev 1984). Earlier evidence for the existence of ssp. *spelta* is still missing. (d) The next evolutionary step was apparently the formation of the free-threshing bread wheat from *TgQ spelta* by mutation of the *Tg* gene to its recessive allele *tg* (Kerber and Rowland 1974). The earliest remnants of bread wheat, ssp. *aestivum,* are from Can Hassan III, southern Anatolia, from about the middle of the 9th millennium BP (Hillman 1996). These remnants indicate that free-threshing hexaploid wheat was formed a short time after the appearance of naked tetraploid wheat and its spread to northwestern Iran. (e) The naked tetraploid wheat ssp. *durum* might have arisen from ssp. *parvicoccum.* Ssp. *durum* appears somewhat later (Feldman and Kislev 2007). It is also possible that ssp. *durum* was derived directly from ssp. *dicoccon* or, more likely from hybridization between *dicoccon* and *parvicoccum.* Other free-threshing subspecies of tetraploid wheat, *turgidum, polonicum. turanicum,* and *carthlicum,* presumably derived from ssp. *durum.* (f) After bread wheat and domesticated emmer had been established in Europe, a second ssp. *spelta* appeared, mostly north of the Alps. It is therefore assumed that the European *spelta,* having genotype *tgtgqq,* arose from hybridization of bread wheat (*tgtgQQ*) with ssp. *dicoccon.* (*qq*). Alternatively, reverse mutation of *Q* back to *q* restored the character of hulledness to hexaploid wheat. (g) As a completely independent evolutionary development, cultivated einkorn, ssp. *monococcum,* arose from ssp. *aegilopoides.* This involved a series of events similar to those which occurred in the emmer group. The first records of domesticated einkorn are from ca. 9000 BP in Syria and Anatolia (review in Harris 1998; Bar-Yosef 1998). (h) Another side branch is a group of hulled, eastern wheat species, comprising of the wild ssp. *armeniacum* and its domesticated descendants ssp. *timopheevii* and *T. zhukovskyi.*

Vavilov (1926) coined the terms "center of origin" and "center of variation". The first is the site at which a given taxon evolved and the second is the site at which it responded to the environment with a burst of variation. In most cases the center of variation overlaps the center of origin. Accordingly, domesticated diploid wheat presumably originated in the northern Levantine Corridor, whereas domesticated tetraploid wheat originated at the watershed of the Jordan River in the southern Levantine Corridor. Naked tetraploid wheat presumably also evolved in this region. Hexaploid wheat originated southwest of the Caspian Sea. With their migration into new areas, cultivated wheats encountered new environments, towards some of which they responded with increased variation and formation of many endemic forms. Several such secondary centers of variation were described by Vavilov (1926), e.g., the Ethiopian plateau and the Mediterranean basin for tetraploids, and Afganistan (Hindu-Kush area) for hexaploids. Transcaucasia is a secondary center for tetraploid as well as for hexaploid types. Such secondary centers of diversity are valuable to wheat breeders as additional gene pools to those existing in the primary centers of variation.

There are still several puzzling questions concerning the evolution of the various domesticated wheats: the absence of non-brittle forms in the second diploid wheat, *T. urartu* (even if the mutations for non-brittleness occurred only in ssp. *aegilopoides,* the mutant alleles could have been transferred to *T. urartu* through hybridization); the mode of origin of the first naked tetraploid wheat and the reasons for its extinction about 1900 years ago (grains of the latest cropping of ssp. *parvicoccum* were found by Kislev in southern Judea, Israel, from the Bar- Kokhba period, 135 AD); the reasons for the late establishment of ssp. *durum* as a major crop in the Near East, in spite of its sporadic occurrence in archaeological material already from 7500 to 6200 BP; the reason for the relatively small contribution of *T. timopheevii* ssp. *armeniacum* to wheat cultivation, either as a tetraploid crop or as the donor of the AG genomes for an hexaploid crop; and the appearance of ssp. *aestivum* in archaeological excavations about 1000–2000 years before its presumed predecessor, ssp. *spelta.*

13.4   Molecular Changes in Domesticated Wheat

The last two decades witnessed remarkable progress in the development of a new arsenal of genomic tools, which have facilitated more in-depth study of various aspects of wheat genetics, genomics and evolution. One of the most significant tools was whole-genome sequencing of the A genome of diploid wheat, *T. urartu* (Ling et al. 2013), the D genome of *Ae. tauschii* (Jia et al. 2013; Luo et al. 2017), the genomes of wild emmer (Avni et al. 2017), ssp. *durum* (Maccaferri et al. 2019), and bread wheat (The International Wheat Genome Sequencing Consortium 2018). Single-nucleotide polymorphism (SNP) mapping of a broad collection of bread wheat landraces and modern varieties has indicated the genomic regions that underwent selective sweep (also known as genetic hitchhiking), which refers to a process that reduces variation in regions of neutral sites linked to a recently fixed beneficial mutation as it increases in frequency in the population (Nielsen et al. 2005; Cavanagh et al. 2013). A major recent advance in durum transcriptome analysis was the development of tools for the discrimination of A and B genome homeologues from expression sequence data such as RNA-Seq (Krasileva et al. 2013), and the first step towards a wheat pan-genome sequence has been made (Walkowiak et al. 2020). Small RNAs datasets are also becoming available (Kenan-Eichler et al. 2011; Yao and Sun 2012). These recent genomic tool developments have facilitated the identification of additional loci that control domestication-related traits in wheat.

The assembly of the genome of the modern durum cultivar Svevo (Maccaferri et al. 2019) and the availability of the assembly of the wild-emmer genome (line Zavitan; Avni et al. 2017) enabled Maccaferri et al. (2019) to analyze genome-wide genetic diversity, and to reveal changes between the domesticated and the wild genomes that were imposed by domestication processes and by thousands of years of selection and breeding (=diversification according to Meyer and Purugganan 2013). Processes leading to modern cultivars of ssp. *durum* were revealed by the four main germplasm groups of the Global Tetraploid Wheat Collection, namely, wild emmer, domesticated emmer, durum landraces, and durum cultivars. Combined genetic diversity and selection signature analyses yielded a dynamic description of the modifications imposed on the genome by domestication and diversification. More specifically, the comparison revealed strong overall synteny, with high similarity in total high-confidence (HC) gene number (durum 66,559; wild emmer 67,182), chromosome structure and transposable element composition. Yet, a number of HC genes displayed differences between Svevo and Zavitan. At least two-thirds of the varied genes displayed variation in intact gene number, whereas the complete gene loss, caused by large structural variations, was responsible for asymmetric gene distribution in only one-third of the cases. In Svevo, variation leading to intact gene number variation included 4811 genes, which represent 7.2% of all HC genes, a value similar to the 5% found after the comparison between two cultivars in a recent pangenome study of hexaploid wheat (Montenegro et al. 2017). When the Svevo-specific genes were mapped to the Zavitan genome, 1493 genes (31%) were not found on the Zavitan sequence, 1225 (26%) corresponded to shorter counterparts of annotated Zavitan HC genes and 1095 (23%) were annotated as low-confidence (LC) genes or pseudogenes. The remaining 965 genes (20%, that is, 1.4% of all Svevo HC genes) mapped to unannotated regions. The presumed HC gene losses were predominantly located in the more distal chromosomal regions. The distal highly recombinogenic regions of chromosomes were enriched in gene displaying variation of intact gene number. contain most of the known QTLs and the HC genes displayed a reduced expression breadth (that is, average expression value) across all tissue/treatment conditions.

Ayal et al. (2006), using cDNA microarrays, studied the alterations in gene expression that occurred during wheat domestication, and focused on the following aspects: (i) the extent of variation in gene expression that can be attributed to domestication, (ii) the specific genes whose expression was altered (up or down regulated) during domestication and (iii) the range of variation in the wheat transcriptome of wild (*T. turgidum* ssp. *dicoccoides*) versus domestic tetraploid wheats (*T. turgidum* ssp. *dicoccon* and ssp. *durum*). The group reported that the expression levels of 63 genes, which represent 2.53% of the total number (2490) of tested genes, were up- or downregulated; 24 genes (0.96%) were downregulated (Table 13.4A) and 38 genes (1.53%) were upregulated (Table 13.4B) in the wild compared to the domesticated lines. Assuming that the genome of tetraploid wheat contains approximately 67,000 HC genes (Maccaferri et al. 2019), this means that approximately 1700 genes are differentially expressed between wild and domestic wheat and, out of these, the expression of approximately 650 genes was reduced (i.e., lost, mutated, or silenced) by the domestication process. A high proportion of the genes that were upregulated in the domesticated wheats were related to carbon metabolism, such as the *Rubisco* large and small sub-unit and the *sucrose-synthase* (*SuSy*).

The intra-specific variation in gene expression within ssp. *dicoccoides* was about four times higher than within the domesticated species. This reduction in variation during domestication was in the same range found at the genomic level, for nucleotide polymorphism. Nevertheless, Ayal et al. (2006) found that certain genes were more variable in the domestic varieties than in the wild, suggesting that selection under domestication led to the fixation of new mutations that did not previously exist in the wild.

Ben-Abu et al. (2014) focused on genomic changes that correlated with the process of domestication and evolution of modern durum by comparing gene expression and copy number variation of genes and transposons in four genetic groups: wild emmer, domestic emmer, durum landraces and modern durum varieties. Genes were clustered based on their pattern of change in expression during durum evolution, e.g. gradual increase, or decrease, or increase at the onset of domestication and plateauing later on. Few genes changed > twofold in copy number. However, interestingly, the copy number of transposons increased with domestication, possibly reflecting the genomic plasticity that was required for adaptation under cultivation. Extensive changes in gene expression were seen in developing grains. For example, there was an enrichment for certain functions, e.g., genes involved in vesicle trafficking in the endosperm showed a gradual increase in expression during durum evolution and genes related to germination and germination inhibition increased in expression in the embryo in the more recent stages of durum evolution (Ben-Abu et al. 2014).

Yuan et al. (2015) described differential expression of the *CENH3* genes in wild versus domesticated tetraploid wheat. *T. timopheevii* ssp. *armeniacum*, the wild progenitor of domesticated *T. timopheevii*, had a higher transcript level of *cxCENH3* while the domesticated forms had a lower expression of *cxCENH3* and increased expression of *PCENH3*. Similar changes in the *CENH3* expression model were found in wild and in domesticated types of *T. turgidum*; the wild subspecies of *T. turgidum* exhibited a higher expression level of *cxCENH3* whereas in the domesticated forms, the differences in expression between *cxCENH3* and *PCENH3* were not so obvious. In contrast to the markedly higher expression level of *cxCENH3* in wild tetraploids,

Table 13.4 Changes in expression level of selected transcripts in domesticated wheat (After Ayal et al. 2006)

A. Higher expression in young wild spikes compared to domestic spikes

Clone	D/W[a]	Putative annotation[b]	P
WCC-2-23g	0.617	*3-isopropylmalate dehydratase*-like protein	0.0175
SSH2-u3b	0.628	*Ribosomal* protein	0.00011
WCC-2-23d	0.602	*30S ribosomal* protein S3	0.0147
SSH2-l5a	0.621	40S ribosomal protein S24	0.0136
WCC-2-111g	0.521	60S ribosomal protein L22	0.00966
SSH1-23e	0.54	Alpha/beta-gliadin precursor	0.00868
SSH1-23g	0.505	Alpha-amylase inhibitor	0.00782
SSH1-21d	0.517	ESTs AU032852	0.00914
SSH1-211a	0.592	Gamma-gliadin	0.0242
SSH2-a9h	0.608	Histone H2A	0.0127
WCC-2-28a	0.534	OsNAC6 protein—rice	0.00976
WCC-2-14d	0.595	Permease 1	0.0159
WCC-2-18g	0.603	Glyceraldehyde 3-phosphate dehydrogenase	0.0156
SSH2-t6d	0.608	Unspecific monooxygenase	0.0125

B. Higher expression in young domestic spikes compared to wild spikes

Clone	D/W[a]	Putative annotation[b]	P
SSH2-i2a	2.34	At1g19940/F6F9_1	1.19E-05
SSH2-q3a	1.872	Caffeic acid O-methyltransferase	4.76E-05
SSH2-s10g	2.286	*Rubisco* small subunit	6.49E-05
SSH2-s12g	3.314	Phenylalanine ammonia lyase	9.55E-07
SSH2-d11b	2.525	Probable peroxidase	0.00335
SSH2-j9g	1.801	Putative bifunctional nuclease	1.12E-08
SSH2-i8h	2.171	*Rubisco* chain precursor	0.000165
SSH2-K10d	1.91	*Rubisco* activase A	1.52E-05
WCC-2-18e	1.786	*Rubisco* large subunit	0.000895
SSH2-d2d	2.708	*Rubisco* small subunit	0.000118
SSH2-K2b	2.167	Absentia-like protein	0.00016
SSH2-u3g	1.799	*Sucrose synthase*	0.00118

[a]The average induction level in domesticated spikes compared to wild spikes
[b]Annotations are based on the TIGR database

expression of *PCENH3* was enhanced to a level near that of *cxCENH3* in the domesticated tetraploids (Yuan et al. 2015). These genome-wide analyses provide insights into the molecular basis of plant domestication, in particular for the non-obvious cellular functions that were selected during evolution under cultivation.

13.5 Founder Effect and Processes Contributing to Increased Genetic Variability of Domesticated Wheats

During their evolutionary history, wheats underwent several genetic bottlenecks: allotetraploidization, allohexaploidization, cultivation of wild emmer, domestication of emmer, selection of free-threshing tetraploids and hexaploids and replacement of the polymorphic fields of land races by high yielding cultivars (Table 13.5). It is reasonable to assume that the number of mutants that founded domesticated einkorn and emmer was not large and that the number of hexaploid plants produced by independent hybridization events between tetraploid wheat and *Ae. tauschii* was small. As pointed out by Ladizinsky (1985), one of the consequences of this "founder effect", namely, the establishment of a new taxon by a few individuals that necessarily represent only a small fraction of the genetic basis of the parental taxon, are narrow and very restricted genetic variability, a phenomenon described as a genetic bottleneck.

At the diploid level, Dhaliwal (1977), Zeven (1980) noted that domesticated einkorn is quite uniform while its wild progenitor, ssp. *aegilopoides,* exhibits a relatively wider variation. He ascribed this limited variability of

Table 13.5 Genetic bottlenecks during the evolution of domesticated wheat

Period	Event	Time (In years) before present)	Lost or reduced genepools
Pre-domestication	Formation of wild emmer wheat (ssp. *dicoccoides*) via allotetraploidization	800,000	From several individuals of B-Genome and A- genome donors
	Cultivation of wild emmer (ssp. *dicoccoides*)	10,300	Cultivation of genotypes representing only a fraction of the wild gene pool
Domestication	Selection of non-brittle, hulled emmer wheat (ssp. *dicoccon*)	9500	From several individuals of wild wheat
Post-domestication	Selection of free-threshing tetraploid wheat (ssp. *parvicoccum*)	9000	From several individuals of hulled tetraploid wheat
	Formation of hulled hexaploid wheat (ssp. *spelta*) via allohexaploidization	9000	From several individuals of domesticated tetraploid wheat and *Aegilops tauschii*
	Selection of free-threshing hexaploid wheat (ssp. aestivum)	8500	From several individuals of hulled hexaploid wheat
	Establishment of the large-grained, free-threshing tetraploid wheat (ssp. *durum*) as the main allotetraploid crop	2300	Loss of the gene pool of ssp. *parvicoccum* and a big part of that of ssp. *dicoccon*
	The green revolution	70	Loss of many hexaploid landraces

domesticated einkorn to the presumed small sample of wild einkorn genotypes that had been domesticated. Moreover, within a short period after its selection by man, domesticated einkorn was spread westward from the area of wild einkorn, thereby largely preventing the introgression of genes from its wild counterparts. Concurring with Dhaliwal, Zeven (1980) assumed that further erosion in the gene pool of cultivated einkorn resulted from intergenotypic competition in the polymorphic fields of ancient farmers. Moreover, because of higher yield, the more extensive cultivation of domesticated emmer and free-threshing hexaploid wheat in Europe reduced very much the cultivation of einkorn.

The pattern of diversity for tetraploid each germplasm group was recently assessed by Maccaferri et al. (2019). They used the whole genome of ssp. *durum* cv. Svevo as a reference to carry out a phylogenetic analysis of a collection of 1856 accessions, including wild and domestic emmer, durum landraces and modern durum cultivars. Data analysis could clearly distinguish between all four genetic groups. Wild emmer exhibits the highest diversity, while there was a strong reduction in diversity through domestication, thus providing a valuable reference for assessing the reduction of diversity associated with domestication. Compared to wild emmer, each of the subsequently domesticated germplasm groups showed several strong diversity decrements that arose independently and were progressively consolidated through domestication and breeding. With few exceptions,

the diversity depletions losses that occurred in the early transition of wild emmer to domesticated emmer, or domesticated emmer to free-threshing durum landraces, are confirmed. Domestic emmer showed a broad genetic variation compared to modern durum cultivars which was rather limited. Landraces of durum had also a rather broad genetic basis compared to modern durum cultivars, which were genetically related to a small group of landraces.

The genome of durum cultivars was characterized by a high number of regions that showed a strong reduction in diversity with near fixation of allelic diversity (genetic swift) presumably due to selection. This included 104 pericentric regions and 350 non-pericentric regions that overlapped significantly with genes known for their role in domestication, such as genes controlling spike fragility and threshability. This also included disease resistance loci, yellow pigment and some unexpected loci, such as a cadmium transporter, absent in wild emmer but widespread in durum cultivars.

Most of the strongest pericentromeric diversity depletions (chromosomes 2A, 4A, 4B, 5A, 5B, 6A and 6B) occurred during emmer domestication. Furthermore, one of the two brittle rachis regions marking the early domestication process (on 3BS) showed a sharp, localized reduction in diversity. The same region then underwent an extreme diversity reduction in the domesticated emmer-to-durum landrace transition. An additional 14 pericentromeric and 90 non-pericentromeric diversity depletions, including one harboring the major tough glume QTL governing threshability (Tg-2B), occurred during the domesticated emmer-to-durum landrace transition. Finally, several reductions in diversity were specifically associated with breeding of modern durum cultivars, including some associated with disease resistance and grain yellow pigment content loci.

Introgression from wild emmer into domesticated wheat is partly restricted by the mode of self-pollination. Moreover, the genes for rachis brittleness are dominant over those controlling non-brittleness, and therefore, the F_1 spikes disarticulate at maturity and only few F_2 spikes bear a non-brittle rachis. Accordingly, only a few of the segregants might fit the selection criteria of man and enrich the gene pool of the domesticated forms.

Although domesticated tetraploid wheat, *T. turgidum*, started with very restricted genetic variability, the original non-brittle emmer plants grew for 2–3 millennia, in mixed populations with wild forms and occasionally hybridized and exchanged genes with them. Moreover, even after the complete replacement of wild emmer by domesticated emmer, the latter and other subspecies of allotetraploid wheat continued to introgress with wild genotypes that grew in their vicinity and thereby, broadened their genetic basis (Huang et al. 1999; Dvorak et al. 2006; Luo et al. 2007). The F_1 hybrids between wild emmer and domesticated tetraploid wheat are fully fertile, and since the chromosomes of wild emmer are homologous to those of allotetraploid wheat many genes can be transferred from the wild into the domesticated chromosomes through crossover. As a result, over the 10,000 years of cultivation, there was an almost continuous flow of wild genes into domesticated tetraploid background, diversifying, to some extent, the domesticated tetraploid gene pool.

In contrast to allotetraploid wheat, allohexaploid wheat, *T. aestivum,* is partially isolated from its allotetraploid progenitor and wholly isolated from its diploid progenitor. Therefore, its primary genetic basis was more restricted. However, despite the partial sterility of the F_1 pentaploid hybrids between allohexaploid and allotetraploid wheats, few hybrid swarms resulting from spontaneous hybridization between hexaploid wheat and wild emmer were found in Israel and descried by Zohary and Brick (1962). Moreover, He et al. (2019), who performed targeted re-sequencing of 890 diverse accessions of hexaploid and tetraploid wheat to identify introgression from wild relatives, found that historic gene flow from wild relatives led to genome-wide increases in diversity and made a substantial contribution to the adaptive diversity of modern cultivars of hexaploid wheat, the principal domesticated subspecies of *T. aestivum*.

For several millennia, massive intra- and inter-specific gene flow has been facilitated by farming of mixtures of different genotypes and even different cytotypes, including representatives of two or even three different species of domesticated wheat, namely, *T. monococcum*, *T. turgidum* (emmer and free threshing) and *T. aestivum* (spelt and bread wheat), each represented by numerous different genotypes (Zeven 1980). These endless hybridizations during the long period of cultivation also enriched the domesticated genetic basis of both species. Walkowiak et al. (2020) used ten chromosome pseudomolecules and five scaffold assemblies of allohexaploid wheat to explore the genomic diversity among allohexaploid wheat lines from global breeding programs. Comparative analysis revealed extensive differences in gene content resulting from complex breeding histories aimed at improving adaptation to diverse environments, grain yield and quality, and resistance to stresses.

Spontaneous hybridization between domesticated wheat and the wild form of *T. timopheevii*, ssp. *armeniacum*, which shares with domesticated wheat the A subgenome but differs in the other subgenome(s), may have also occurred in the northeastern part of the Fertile Crescent and in Transcaucasia. The hybrids between these species are highly sterile, although a few seeds are produced upon backcrossing. Introgression of genes from various *Aegilops* species also contributed to the wide variation of domesticated allopolyploid wheats. Although intergeneric F_1 hybrids between domesticated wheat and more distant diploid and tetraploid species of *Aegilops*, as well as species of *Secale, Agropyron, Haynaldia* and other related genera, which grew within or near Mediterranean and Near Eastern wheat fields, are highly sterile, the few seeds produced upon backcrossing yield more-fertile plants. Such intergeneric hybridizations may often result in successful introgression, maintaining a weak but constant flow of genes into the domesticated background. Such introgression was recently well evidenced by Zhou et al. (2020), who identified composite introgression from wild populations contributing to a substantial portion (4–32%) of the bread wheat genome, which undoubtedly increased the genetic diversity of bread wheat and enabled its adaptation to new territories. Similarly, Walkowiak et al. (2020) presented evidence for introgressions from wild relatives to allohexaploid wheat.

Keilwagen et al. (2022) pointed out that intentional introgressions from crop wild relatives have been used to introduce valuable traits into domesticated plants. They demonstrated the utility of single nucleotide polymorphism-based methods to detect introgressions and predict the putative donor species. Analyzing ten publicly available wheat genome sequences with these methods they identified nine major introgressions from wild diploid and allotetraploid *Triticum* species, from diploid and allotetraploid *Aegilops* species and from *Elymus elongatus* ssp. *ponticum*. Keilwagen et al. (2022) traced introgressions to early wheat cultivars and show that natural introgressions, mainly those that harbour resistance genes, were utilized in early breeding history and still influence elite lines today.

Mutations have also played an important role in increasing genetic variability of domesticated wheats. The genetic structure of the allopolyploid species of wheat, i.e., four (in tetraploids) or six (in hexaploid s) doses of gene loci, reinforced by a diploid-like cytological behavior and predominantly self-pollination, has proven a very successful genetic system for facilitating rapid buildup of genetic diversity through genetic changes. In these wheat species, the accumulation of genetic variation through mutations or hybridizations is tolerated more readily than in diploid species. Moreover, polyploidy facilitates genetic diploidization, the process whereby existing genes in multiple doses can be diverted to new functions. Thus, tetraploid and hexaploid wheat can accumulate a significant amount of genetic variability through gene alterations. Mutations exerting a lethal or semi-lethal effect at the diploid level, such as Q, s, (the sphaerococcum gene) C, (the compactum gene) and $Ph1$, are viable at the polyploid level. Induction of mutations may have been accelerated by the activity of transposable elements (TEs) (e.g., Fedoroff 2000, 2012) and genome-restructuring genes (Feldman and Strauss 1983). Indications for the activity of such genes have been reported in wheat and in several of its wild relatives. The activation of TEs, mainly retrotransposons by various environmental and climatic stresses has an important evolutionary significance (e.g., Schrader and Schmitz 2019). In addition to the generation of genetic variability due to epigenetic changes, e.g., DNA methylation, chromatin acetylation and activity of various small RNA molecules, genome restructuring by TEs may lead to the formation of new linkage groups. Activity of genome -restructuring genes as well as TEs during wheat cultivation may explain the wide occurrence of chromosomal rearrangements among domesticated wheat taxa. Thus, domesticated allopolyploid wheat can accumulate a significant amount of genetic variation also through mutations.

The genetic diversity of the D subgenome of bread wheat is relatively limited, being 2–5 times lower than for the A and B subgenomes (Caldwell et al. 2004). Recently, there has been renewed interest in exploring the genetic diversity within the *Ae. tauschii* genome for the purpose of breeding. A study of a collection of 101 synthetic hexaploid wheat lines, made from crosses between tetraploid wheat and *Ae. tauschii* showed a high diversity, similar to that of the A and B subgenomes of hexaploid wheat (Bhatta et al. 2018). Whole genome sequencing was performed for 278 accessions of *Ae. tauschii*, including de novo sequencing for four accessions (Zhou et al. 2021). A core collection of 85 accessions, representing 99% of the species variation was defined and crossed with elite bread wheat cultivars to generate a new collection of Ae. tauschi—wheat synthetic octoploids (Zhou et al. 2021). This collection showed a

promising phenotypic variation of potential for wheat breeding for trait such as grain weight and pre-harvest sprouting. Nyine et al. (2021) tested introgressions of 21 *Ae. tauschii* accessions into hard winter wheat and, also found that some introgressions were positively associated with yield traits.

13.6 The Three Phases of Selection Under Cultivation

During the 10,000 years of wheat cultivation, the criteria for selection varied from time to time and from place to place, as suggested by Evans (1981) in regard to yield criteria. Selection pressures exerted by farmers in different locales have operated in different, and sometimes even in contradictory, manners in different cultivated fields. Thus, during the process of adaptation to the various cultivated environments and to the different demands of man, wheat responded with a number of significant genetic changes. The main changes reflect the process of adaptability to agriculture, particularly of the plants to propagate and protect its seeds, through which wheat became completely dependent on the farmer for its survival.

The modern gene pool of the domesticated wheats has developed through three main phases of selection: occasional and sometimes non-intentional selection, exerted by the earliest farmers, simply by the processes of harvesting and planting; alongside more deliberate selection by traditional farmers in polymorphic fields; and selection as part of scientifically planned modern breeding (Feldman 1976, 2001; Feldman et al. 1995).

During the first phase, a very significant sequence of changes occurred in the transition from the wild to the domesticated forms (Table 13.6). Grain size, an important domestication trait that has been associated with successful germination and seedling growth under cultivated conditions, was among the first traits selected. It resulted in a change from production of small grains with a long, thin shape in wild wheat to more uniform, larger and wider grains in domesticated wheat. Grain size increased early in domestication, whereas changes in grain shape occurred at somewhat later stages (Gegas et al. 2010). These authors found that grain size and shape are primarily independently controlled traits, by a limited number of QTLs in both wild and domesticated wheat.

Domestication of emmer, ssp. *dicoccon*, through the loss of rachis fragility was probably a gradual process, as suggested from both the archeological and genetic evidence. The archeological record (Kislev 1984; Tanno and Willcox 2006) showed a mixture of fragile and nonfragile types in ancient sites where agriculture was practiced, with nonfragile rachis becoming prominent in farming units only after 2–3 thousand years [Kislev (1984) in emmer wheat; Tanno and Willcox (2006) in einkorn]. Genetically, since the rachis-brittleness trait is controlled in wild emmer by two major dominant genes (Levy and Feldman 1989a, b; Watanabe and Ikebata 2000; Watanabe et al. 2002, 2005; Nalam et al. 2006; Millet et al. 2013), it probably took some time to obtain a homozygous recessive combination of these two genes.

According to Kislev (1984), the non-brittle types were probably selected by women, who were usually in charge of threshing. The non-brittle spikes were harder to thresh and were left aside as seeds for the next spring. These non-brittle types had been established in the cultivated field and spread gradually from farmer to farmer and from village to village throughout the Levant. The loss of rachis fragility gave rise to domesticated emmer one of the earliest forms of domesticated wheat, which is still grown today, albeit on a small scale (de Vita et al. 2006).

An additional requirement for the newly domesticated wheat was uniform and rapid germination. Wild wheat exhibits two types of dormancy: a post-harvest type and a next-year type. The first type prevents premature and untimely germination—an important feature, pre-adapted to agriculture, particularly in view of the fact that seeds were often stored under unsuitable storage conditions. The second type of dormancy ensures in wild wheat temporal distribution of germination; it is invariably the larger grain of the second floret in each spikelet that germinates in the first year, while the smaller grain of the first floret germinates in the second year. Induction of germination of all grains in the first year was of great advantage.

The loss of self-propagation and of the temporal control of germination was followed by the loss of self-protection of the grains. The wild forms have tightly closed tenacious glumes, resulting in a "hulled" post-threshing grain. Several primitive forms of domesticated wheat species, e.g., *T. monococcum* subsp. *monococcum*, *T. turgidum* subsp. *dicoccon* and *paleocolchicum*, *T. timopheevii* ssp. *timopheevii*, *T. aestivum* subsp. *spelta* and *macha*, retain this feature. The appearance of the free-threshing trait (naked kernels) in *T. turgidum* and *T. aestivum,* was the second-most important step in domestication of wheat after rachis non-brittleness. *T. turgidum* ssp. *parvicoccum*, a "fossil" subspecies of allotetraploid wheat was the first free-threshing form (Kislev 1979/1980). Ssp. *durum* may thus derive from hybridization between ssp. *parvicoccum* and ssp. *dicoccon*, receiving the free-threshing trait from *parvicoccum* and the large grains from *dicoccon*. The large grain of *durum* was probably preferred over the small grains of *parvicoccum*, leading to the prominence of *durum* as an allotetraploid wheat and to the extinction of *parvicoccum*. Similarly, it can be assumed that bread wheat, *T. aestivum* subsp. *aestivum*, received these mutations from tetraploid wheat but required

additional mutation in the *Tg* gene of genome D to become free threshing (Kerber and Rowland 1974).

In addition to the above-mentioned classical domestication traits, several other domestication traits were selected, including increased number of seeds per spikelet, simultaneous ripening of grains, and plant erectness versus the prostrate grassy types (Table 13.6).

During the second phase of selection under cultivation, wheat culture spread into new areas, an event that required the adaptation of domesticated wheat to new climatic, edaphic, and biotic conditions that were not encountered by wheat before. This created environmental stress that induced a variety of genetic and epigenetic changes. The wheat genotypes might have reacted by increased mutability, possibly due to the temporal activation of various transposable elements. During the spread to Europe, for example, photoperiodic and thermoperiodic responses were modified to achieve an optimum balance between vegetative and reproductive phases: the vegetative period was extended to benefit from the longer summer days and rainy season, while the need for high temperatures for maturation gradually disappeared. In addition, the grain-filling period became longer, and flag-leaf senescence was delayed. These latter adaptations, allowing for larger amounts of carbohydrates to be assimilated and translocated to the developing grains, greatly contributed to higher yields through further increase in grain size and spike counts.

The presence of diverse genotypes within a single polymorphic field, largely prevented the outbreak of epidemics and severe damage by ecological hazards. Hence, growing various genotypes, and even species, as a mixture in a single field heightens yield stability—an economic consideration ranked much more important than occasional high yield (Zeven 1980). The traditional farmers preferred a "safe", average yield each year, rather than a high yield for several years which might have been followed by crop failure. The lack of suitable means for long-term storage and for large-scale wheat import, translated to drastic consequences of crop failure in any given year. Hence, yield stability was of utmost importance.

The second phase of evolution under cultivation involved long and continuous selection for various agronomic and technological characters in the polymorphic fields of the traditional farmers. The main evolutionary advantage in such fields was the possibility for occasional hybridization between genotypes and even species. Because of the self-pollinating system, every hybridization resulted in a significant number of homozygous recombinants, thereby constantly providing the farmer with new genotypes for selection. Yet, in such fields, as numerous genotypes were grown in mixtures, the unit of selection was a combination of genotypes rather than a single one.

Table 13.6 Modifications that occurred in wheat during its three phases of cultivation (After Feldman 2001)

I. During the transition from the wild habitat to the cultivated field	Non-brittle spikes	
	Increased grain size	
	Free-threshing (naked grains)	
	Non-dormant seeds	
	Uniform and rapid germination	
	Erect plants	
	Increased spikelet number per spike (?)	
II. During 10,000 years of cultivation in polymorphic fields	Adaptation to new, sometimes extreme, regional environments	
	Increased tillering	
	Increased plant height	
	Development of canopy with wide horizontal leaves	
	Increased competitiveness with other wheat genotypes and weeds	
	Modifications in processes that control the timing of various growth stages	
	Increased grain number per spikelet	
	Improved seed retention (non-shattering)	
	Improved technological properties of grains	
III. During cultivation in monomorphic fields due to modern breeding procedures in the last century	Increased yield in densely planted fields; reduced intra-genotypic competition	
	Canopy with erect leaves	
	Reduced height	
	Enhanced response to fertilizers and agrichemicals	
	Increased resistance to grain shattering	
	Increased resistance to diseases and pests	
	Lodging resistance	
	Improved harvest index	
	Improved baking and bread-making quality	

In such polymorphic fields, inter-genotypic competition played a decisive role. Since the contribution of every genotype to the next year's seeds largely depended on the productivity of single plants, high tillering and vigorous vegetative growth were traits of high adaptive value. Plants with horizontal leaves, which shade weeds and competitors, had an advantage over genotypes with erect leaves.

The population in polymorphic fields had a genetic structure similar to that found in populations of wild wheats: a mixture of many genotypes partially isolated by the predominant self-pollination. According to Sewall Wright's model (Wright 1931), occasional gene exchange between partially isolated genotypes is one of the most effective processes in evolution. The genetic variation thus achieved was maintained in the traditional farming system; because the low yield of the landraces required the use of a greater proportion of the harvested seeds for yearly sowing, a large

proportion of the genetic variation was transferred to the next generation (Zeven 1980).

During the second phase of evolution under cultivation, selection efforts resulted in increased plant height, increased tillering, development of canopy with wide horizontal leaves, larger seed size, increased grain number per spikelet, better flour quality, improved seed retention (non-shattering), increased competitiveness with other wheat genotypes and weeds, and better adaptation to a wider range or climatic and farming regimes.

In the third phase of selection under cultivation, driven by scientifically planned modern breeding, starting at the second half of the nineteenth century, the wheat field became genetically uniform and no longer conducive to spontaneous intra-genotypic gene exchange. On the other hand, large-scale gene migration was promoted by worldwide-introduction services. Massive scientific screening aided in revealing desirable genes, and modern methods for manipulating and transferring these genes from one genetic background to another became available. Hybridizations became confined mainly to intra-specific crosses. Lately, however, some inter-specific and inter-generic crosses have also been performed. Individual genotypes, rather than mixtures, became the unit of selection in experimental-station fields. Selection was made mostly for traits that improved wheat performance in dense stands, such as minimum intra-genotypic competition, upright leaves to improve light penetration and prevent shadowing of neighboring plants, low tillering, a higher number of seminal roots whose development is independent of tillering. High-yielding cultivars were produced that owe their performance to an increase in the number of fertile florets per spikelet and, sometimes, to the length or density of the spike, reduction in shattering, and resistance to fungal, bacterial, and viral diseases as well as to pests.

One of the great achievements of this phase is the the Green Revolution that took place in the 1960s. During the course of this revolution, use of genes for reduced height (*Rht*), originally from the Japanese cultivar Norin 10, facilitated the production of semi-dwarf (90–120 cm) or dwarf (60–90 cm) high-yielding cultivars that replaced the conventional tall (120–140 cm) ones. These cultivars introduced to India, Pakistan, Iran, the Mediterranean basin, and other wheat-producing regions, replacing the numerous landraces in every locale. The Green Revolution succeed to improve yield, increase resistance to various diseases and tolerance to agrochemicals. These high-yielding cultivars respond well to new agrotechnical practices particularly to high fertilizer application rates, without lodging, and improved harvest index. Currently, the dwarfing genes have been widely incorporated into most existing cultivars, and are responsible for a very significant increase in wheat yield.

The main achievements in breeding for grain quality have been improvements in milling and baking characteristics. Certain modern cultivars are easily milled because the pericarp and seed coat are only loosely attached to the endosperm. Flour yield is particularly high in cultivars with short, almost spherical grains. To date, less progress has been made in improving the nutritional value of the grain, and further efforts are needed, particularly toward increasing protein content and remedying deficiencies in amino acid composition and some minerals. Among the traits affected by domestication are those affecting the storage proteins, in particular the high-molecular-weight glutenins whose subunit number is lower in domesticated allotetraploid wheat than in wild emmer (Levy and Feldman 1988, 1989b; Laido et al. 2013). Selection for product taste and quality have led to reduced numbers of subunits of high-molecular-weight glutenins in the A and D subgenomes of domesticated allotetraploid and allohexaploid wheats (Feldman et al. 1987).

Today's breeding techniques may achieve the objectives of the primitive and traditional farmers with greater prediction accuracy. The general goal was to achieve highest possible yields per area. Yet, the main limiting factor (globally) in achieving this goal was water availability, and plants have also been subsequently selected to address this goal and constraint.

13.7 Man-Made Allohexaploids

13.7.1 Synthetic *T. aestivum* Lines

Bread wheat, *T. aestivum* ssp. *aestivum*, has passed through two cycles of genetic bottlenecks, allohexaploidization that led to the formation of ssp. *spelta* from a relatively small number of individuals of tetraploid wheat and *Ae. tauschii*, and the evolvement of bread wheat, ssp. *aestivum*, by mutations from ssp. *spelta* (Table 13.5). However, during 9000 years of cultivation this narrow genetic basis became wider due to mutations and introgressions from domesticated and wild relatives. Yet, an additional reduction in genetic diversity of bread wheat occurred during the Green Revolution. This revolution achieved success in yield increase, better resistance to diseases and improved tolerance to various agrochemicals in many wheat growing countries. However, at the same time, this success was bought at the cost of an overall reduction in genetic diversity in bread wheat (Warburton et al. 2006). The loss in genetic diversity was mainly due to replacement of numerous land races by a relatively small number of elite cultivars and the lack of awareness to maintain the replaced land races in gene banks. The abandon of a large number of land races led to a sizeable genetic erosion in the genepool of bread wheat. The dangers of a narrow genetic base of domesticated wheats, mainly of bread wheat, have become a great concern in recent decades, and, consequently, increased the need to widen the genetic basis of bread wheat. In this endeavor, attempts have been made to transfer desirable genes from wild and other domesticated wheats and from various Triticeae species to bread wheat, and during the years, several successful gene transfers, mainly of disease resistant genes, were accomplished from various related species (Wulff and Moscou 2014). Yet, there are several obstacles in this endeavor that curtail transfer of desirable genetic material. As already pointed out by McFadden and Sears (1947), *T. turgidum* (genome BBAA), that has two homologous subgenomes to the A and B subgenomes of *T. aestivum*, its pentaploid hybrids with hexaploid wheat are only partially fertile. Moreover, the tetraploid F_1 hybrids between bread wheat and *Ae. tauschii* is extremely sterile.

To overcome the difficulties of partial sterility in crosses of *T. turgidum* x *T. aestivum* and the high sterility in *T.*

aestivum x *Ae. tauschii* hybrids, McFadden and Sears (1947) suggested the construction and use of synthetic *T. aestivum* lines that combines the BBAA genome of *T. turgidum* with the D genomes of *Ae. tauschii*. Hybrids between the synthetic lines and bread wheat are fully fertile, chromosome pairing is complete, and high rate of recombination between the synthetic and the domesticated chromosomes consistently occurs.

The pioneering production by McFadden and Sears (1946) of synthetic *T. aestivum* (STA), that resembled ssp. *spelta,* by crossing domesticated emmer (ssp. *dicoccon;* genome BBAA) with *Ae. tauschii* (formerly *Ae. squarrosa;* genome DD) and doubling the chromosome number of the F$_1$ hybrid with colchicine, indisputably indicated the two parental species of *T. aestivum* and the mode of its origin. As a direct consequence of this important discovery, McFadden and Sears (1947) proposed to produce synthetic *T. aestivum* lines to overcome the partial sterility in hybrids between tetraploid and hexaploid wheat, and the high sterility in *T. aestivum* x *Ae. tauschii* hybrids, and thus, to be able to transfer with ease desirable genes from the synthetic to the domesticated hexaploid wheat. Soon after, Kihara and Lilienfeld (1949) produced amphidiploids from a cross of ssp. *dicoccoides* var. *spontnneo-nigrum* with *Ae. tauschii* through the union of two unreduced gametes. Like the STA produced by McFadden and Sears (1946), also these amphiploids were fertile, exhibited regular chromosome pairing, and resembled hulled form of hexaploid wheat. The hybrids between the synthetic hexaploids with ssp. *Spelta* and ssp. *aestivum* were fertile and exhibited full or almost full pairing (McFadden and Sears 1946; Kihara and Lilienfeld 1949). These studies paved the way to produce a large number of STAs using elite cultivars of ssp. *durum* and diverse genotypes of *Ae. tauschii* (Fig. 13.3). Such genetic resource will facilitate transfer of desirable genes from the A and B subgenomes of ssp. *durum* to the corresponding subgenomes of bread wheat and from the D genome of *Ae. tauschii* the D subgenome of bread wheat (Mujeeb-Kazi et al. 1996; Trethowan and Mujeeb-Kazi 2008).

The use of ssp. *durum* as the maternal parent in the production of STA lines guarantees success in the cross with *Ae. tauschii* and also that the resultant STA will have the cytoplasm of *T. turgidum* that is identical to that of *T. aestivum*. Chromosome doubling of the triploid F$_1$ hybrids occurs either naturally due to production of unreduced meiotic cells or after colchicine treatment. The STAs thus produced comprise an important genetic resource for transferring novel genetic variation to bread wheat including desirable genes from wild and domesticated subspecies of *T. turgidum* and from *Ae. tauschii*. With no reproduction barrier and almost complete chromosomal homology, crossing these STAs with modern elite bread wheat cultivars facilitates the transfer economically important genes to the

domesticated background that may increase yield, widen resistance to fungal, bacterial and pests, improve tolerance to various abiotic stresses, and enhance nutritional and backing quality. No wonder, therefore, that STA has become an important component in wheat breeding (Mujeeb-Kazi et al. 2013; Ogbonnaya et al. 2013; Li et al. 2018; Rosyara et al. 2019; Aberkane et al. 2020).

New molecular technologies revealed a lower genetic diversity in the D subgenome than in the A and B subgenomes of bread wheat (e.g., Walkowiak et al. 2020; Gaurav et al. 2021). Nucleotide diversity in the A, B, and D subgenomes of bread wheat is considerably a lesser amount than that of the genetic variation in these genomes in the parental species (Zhao et al. 2020). It is important therefore, to introduce novel variation for desirable traits from the parental species of bread wheat. STAs can boost the genetic diversity in all three subgenomes of bread wheat. *Ae. tauschii* harbors substantial variation for many biotic and abiotic stress tolerance traits that are relevant in wheat breeding (Dudnikov and Kawahara 2006). In STA lines, the presence of the D subgenome, that derived from many liferent genotypes of *Ae. tauschii,* harbor unparalleled genetic diversity for addressing global wheat production constraints through genetic improvement (Mujeeb-Kazi et al. 2008).

From hundreds *Ae. tauschii* accessions hybridization efforts produced more than thousands STA combinations resulting from chromosome doubling of the F$_1$ hybrids with elite ssp. *durum* cultivars. During the last 50 years, more than several thousand different lines of STA were developed in the International Maize and Wheat Improvement Center (CIMMYT) (Warburton et al. 2006; Mujeeb-Kazi et al. 2008; Ogbonnaya et al. 2013; Das et al. 2016; Rosyara et al. 2019; Aberkane et al. 2020), in China (Li et al. 2018), and several other countries. This extensive production of STA represents a valuable resource of user-friendly genetic diversity (Warburton et al. 2006). Indeed, analysis of these STA lines provided encouraging diversity data for key abiotic constraints such as drought, salinity, and heat, ae well as for several biotic stresses (Mujeeb-Kazi et al. 2008). Likewise, Rosyara et al. (2019) estimated the contribution of the D subgenome of STA lines to derivative lines resulted from crosses with elite cultivars of bread wheat. Their results underline the importance of STA lines in maintaining and enhancing genetic diversity and genetic gain over years. STAs are a good source for novel resistance genes to fungal diseases and for pests most of which derived from *Ae. tauschii* (see review by Li et al. 2018).

This is an example of success utilizing wild relatives in mainstream breeding at large scale worldwide (Rosyara et al. 2019). However, STAs are hulled forms and cannot be used as cultivars because of the presence undesirable characters that derived from the wild parent *Ae. tauschii*, e.g., tenacious glumes, barrel-type rachis disarticulation, and other wild

Fig. 13.3 Spikes of synthetic *T. aestivum*; formed via hybridization of *T. turgidum* ssp. *durum* cv. Cappelli (lineTTR19) with *Ae. tauschii* ssp. *strangulata* (line TQ27)

traits. Thus, it is required to remove these wild undesirable characters through crossing and backcrossing to *T. aestivum* elite cultivars and selection in the segregating progeny. Several derivative cultivars having higher concentrations of both micronutrients and macronutrients and higher yield than their parental bread wheat cultivars, have been developed at CIMMYT (Guzman et al. 2014). By now, about several tens of lines that derived from such crosses have been registered as cultivars around the world, particularly in China (Yang et al. 2009; Li et al. 2018; Hao et al. 2019).

13.7.2 Triticale

13.7.2.1 General Description

Triticale [*Triticosecale* (Wittm. ex Camus) MK], is a man-made cereal crop, resembles wheat, whose name derived from the scientific name of its two parents, *Triticum* and *Secale*. The name Triticale Tsch. (von Tschermak 1937) was given to wheat-rye amphiploids, and since then has been used more and more as a common name (Villareal et al. 1990). Triticale obtained from hybridization of wheat as female, either tetraploid ssp. *durum* or hexaploid

ssp. *aestivum*, with rye, *Secale cereale*, as male, and double the chromosomes of the sterile F_1 hybrid by colchicine treatment to produce a fertile amphiploid. The idea was to combine the high yield potential, good grain quality, and disease resistance of wheat with the vigor and hardiness of rye and its resistance to various diseases. Both 4 × and 6 × triticales are produced with wheat as well as with rye cytoplasmic background, named *Triticosecale* and *Secalotriticum*, respectively.

Triticales have been synthesized at four different ploidy levels (4x, 6x, 8x, and 10x). Decaploid triticale (2n = 10x = 70; genome BBAADDRRRR) was obtained by Müntzing (1955) via crossing 8 × triticale with 2 × rye, or 6 × wheat with 4 × rye. It had cytological instability as well as poor seed set and could not be maintained and are thus lost (Müntzing 1955). Triticales at the other three ploidy levels are being used for a variety of studies. At early stages, octoploid triticale was the choice of geneticists and plant breeders but, with time, hexaploid triticale had greater potential to become a successful crop. Tetraploid triticale shows little promise to become a crop. Thus, most of the currently available triticales are hexaploids due to their superior vigor and reproductive stability compared to the octoploid type (Mergoum et al. 2009). Hexaploid triticales are being used for extensive cultivation in several countries, while octoploid triticales have been used for cultivation in a more limited scale in China (for review see Müntzing 1979; Gupta and Priyadarshan 1982). Tetraploid triticales, are produced by crossing 6 × triticale with 2 × or 4 × rye and backcrossed to rye (Krolow 1973). Some 4 × triticales that have mixogenomes A/B show considerable stability. Tetraploid triticales are still in different stages of development and may not be used for cultivation in the foreseeable future (Mergoum et al. 2009).

Detailed description of the history of triticale is given by Villareal et al. (1990) and Oettler (2005). Towards the end of the nineteenth century, breeders begun to cross wheat with rye, and the first triticale, which was octoploid, was produced in Germany by Rimpau in 1888, from crosses of *ssp. aestivum* and rye, followed by spontaneous chromosome doubling (Rimpau 1891). It was not until the 1960s that the first commercial releases became available for producers (Mergoum et al. 2009).

Triticale is a facultative autogamous allopolyploid species. As such, most lines are homozygous. The cytoplasm of triticale derived from the wheat maternal parent and is similar to that of allopolyploid wheats, T. *turgidum* and T. *aestivum*. Consequently, many *Secale* genes are not expressed in the background of wheat cytoplasm and the nuclear *Triticum* genome is predominant. Moreover, there is a sizeable reduction of the DNA amount of the rye subgenome in triticale, either 6x or 8x (Boiko et al. 1988; Ma and Gustafson 2005, 2006; Ma et al. 2004).

Taxonomical status of triticale

Chapter H of the international code of botanical nomenclature 2000 (ICBN 2000), dealing with names of hybrids, concluded in section H.6.2. that the nothogeneric name of a bigeneric hybrid is a condensed formula in which the names adopted for the parental genera are combined into a single word, using the first part or the whole of one and the last part or the whole of the other (but not the whole of both) and, optionally, a connecting vowel.

The problems concerning the taxonomy and nomenclature of triticale were discussed by Baum (1971), Gupta and Priyadarshan (1982), Gupta and Baum (1986) and Gupta (1986). In contrast to the current recommended usage of the nothogeneric name x *Triticosecale,* Gupta and Baum (1986) and Gupta (1986) advocated that triticale be not treated any longer as a nothogenus x *Triticosecale,* but rather, as a monotypic genus instead, with a new name, proper circumscription, and designation of a type specimen. Stace (1987) regarded that the main reason for not treating triticale as a genus stem from the inability to find characters that would distinguish the new genus from *Triticum* and the fact that the distinction between the two ``is becoming increasingly blurred' (Stace 1987), presumably due to continued introgression from hexaploid wheats used in triticale improvement programs. Stace (1987) proposed that the correct nothogeneric name for plants derived from *Triticum* x *Secale* crosses is x *Triticosecale* Wittmack ex A. Camus. No correct name at species level is available for the commonest crop triticales. These triticales are, however, still described under the nothogenus x *Triticosecale* Wittmack and its continued usage has recently been recommended by Stace (1987). However, a detailed study of inflorescence, glume, lemma, and lodicule characters, conducted by Baum and Gupta (1990) in 108 accessions of hexaploid and octoploid triticales, in 102 herbarium specimens representing 21 species of *Triticum* and *Aegilops,* and in 30 herbarium specimens representing 12 species of *Secale,* justify in their opinion a generic status for triticales. These authors provided, a key for distinguishing the genera *Secale, Triticum,* Aegilops, and the nothogenus x *Triticosecale.*

Mac Key (2005) disagree with Baum and Gupta (1990) that triticale deserves a separate genus rank and instead, incorporated the three different triticale species in the genus *Triticum.* He argued that triticale, like all *Triticum* species, has the typical growth habit and inflorescence of diploid wheat, and are all to be considered as hybrids carrying the pivotal AA genome as an essential part and thus, designed of it. From this aspect, they can be defined as forming a natural hybrid genus. The inclusion of triticale into *Triticum* is based both on commercial and scientific considerations (Mac Key (2005). According to him, the inclusion of triticale into *Triticum* is based both on commercial and scientific considerations. Mac Key (2005) knows that "such a delimitation of the genus Triticum is not strictly following the International Code of Botanical Nomenclature 2000 (ICBN 2000) but has at least the advantage of preserving the basic frame set by the old traditional taxonomy".

The basic principle of including triticale as a section in the genus *Triticum* has consistently (Mac Key 1954b, 1966, 1975, 1981, 1988) been to combine this genealogical aspect with trying to foresee a steadily ongoing, dynamic evolution. This idea appears now to have been supported also by the International Code of Nomenclature for Cultivated Plants 1995 (ICNCP). Mac Key (2005) included triticale as a separate section, *Triticosecale* (Wittm. ex Camas} Mac Key, *sectio nov. comb.* into the genus *Triticum.* Triticale is grouped into three different species, representing tetra- hexa- , and octoploid constitution, respectively. The 4 × triticale was not taxonomically treated before and is proposed to be named *Triticum semisecale* Mac Key, *comb. nov.* According to ICBN 2000, Art. 51, *Triticum neoblaringhemii* (Wittm. ex Camas} Mac Key, *comb. Nov.*for 6 × and *Triticum rimpaui* (Wittm.) Mac Key, *comb. nov.* for 8 × must be used.

Mac key's (2005) inclusion of triticale as a section in *Triticum* and its three species are as follows: Section Triticosecale (Wittm. Ex Camus) Mac Key, Sectio. nov. 2n = 4x/6x/8x = 28/42/56. *Triticum semisecale* Mac Key, comb. nov. 2n = 4x = 28 (AARR or A/BRR) (Subtriticale). *Triticum. neoblaringhemii* (Wittm. Ex Camus) Mac key, comb. nov. 2n = 6x = 42 (BBAARR) (Triticale). *Triticum rimpaui* (Wittm.) Mac Key, comb. nov. 2n = 8x = 56 (BBAADDRR) (Eutriticale).

The above treatment follows the taxonomic treatment of the genus *Triticum* by Mac Key (2005) using the biological species concept based on genome composition. The titicales resemble morphologically wheat more than rye (Baum and Gupta 1990).

T. semisecale is a less-stable species whereas *T. neoblaringhemii* is stable and very successful in cultivation. On the other hand, *T. rimpaui* is not completely stable and mainly has historical importance. It has glumes almost as small as in rye, i.e., much smaller than in wheat and in the 6 × and 8 × triticales, but the spike shape is wheat. The glumes nerves converge at the tip like in rye. This species has larger lodicules than on wheat and even on rye (Baum and Gupta 1990).

Triticum neoblaringhemii has a characteristic ear feature with more elongated spikelets than in wheat and capacity of a larger seed size than in rye. Glumes are clearly different from rye and have, like wheat, nerves not converging at the tip and with the midrib placed asymmetrically. This species has lemmas more in texture and appearance like wheat and different from rye. Larger lodicules than on wheat and even on rye (Baum and Gupta 1990).

Triticum rimpaui has more wheat-like glumes, which are less elongate, slender, and more tough than on *Triticum semisecale* and *T. neoblaringhemii* but still separable from wheat. Lemmas are in texture and appearance much like in wheat but the nerves converge in contrast to wheat at the tip. Since *T. rimpaui* is more consistently produced through *T. aestivum,* ssp. *aestivum,* as wheat parent, its basal part of glume nerves may be less sharply marked. Hairiness below ear is proof of rye dependence but less reliable (Baum and Gupta 1990). Shriveled seeds are more common in *T. rimpaui* than in *T. neoblaringhemii.*

13.7.2.2 Cultivation

In difference from the other cereal crops, triticale is a young crop whose evolution happened only during the last 140 years and its most dramatic evolutionary events were almost all directed by humans (Mergoum et al. 2009). Most triticale cultivars are hexaploids (Villareal et al. 1990). The first hexaploid triticales synthesized from tetraploid wheat, mainly from ssp. *durum* (genome BBAA) and rye, *Secale cereale* (genome RR), are called primary hexaploids, while hexaploid triticales synthesized from crosses between primary hexaploid triticales and/or between primary hexaploid triticale and hexaploid wheats or octoploid triticale are called secondary hexaploid triticales (Lukaszewski and Gustafson 1987). One advantage of the secondary hexaploid triticale is the increased genomic diversity, including the insertion of portions of the D subgenome from hexaploid wheats /into the R subgenome of triticale.

Triticale has either winter or spring growth habit, vary significantly in plant height, tend to tiller less, and generally have larger inflorescence in comparison to wheat. The majority of triticale cultivars have prominent awns, but a limited number of current both spring and winter types exhibit awnless traits (less than 5 mm). These types have increased potential for use as a hay forage for livestock (Villareal et al. 1990).

The first release of a commercial triticale cultivar occurred in Europe, whereas `Rosner' a Canadian release was the first triticale cultivar developed in North America. Europe is the major triticale producing region; Poland, Germany, Belarus, and France are the major producers of triticale in the European region (Table 13.7).

Triticale is gaining popularity among the livestock growers across the globe as it can be used in the animal feed. Triticale has the digestibility and water-soluble sugars similar to oats and cereal rye. Additionally, triticale can also thrive in drought, low-fertile soil which attracts more forage growers and livestock growers. This increasing adoption of triticale in animal feed is one of the major driving factors for the global triticale market.

13.7.2.3 Breeding
The history of triticale breeding for cultivar development has been an agronomic success story (Villareal et al. 1990). The first triticale cultivars were characterized by low yields, tall and weak straw, shrunken and shriveled kernels, high susceptibility to ergot [*Claviceps purpurea* (Fr.) Tul.], high protein, and high levels of the amino acid lysine, and preharvest sprouting (Oettler 2005). The advantage of high protein and high lysine in livestock food was nullified by the poor yield performance and the high incidence of ergot. Modern triticale has overcome most of these problems, some of which caused by cytological differences of the wheat and rye genomes such as different length of meiotic duration, difficulties in chromosome segregation and aneuploidy, and different in amount of constitutive telomeric heterochromatin that affects meiotic pairing and grain shriveling (Lukaszewski and Gustafson 1987; Gupta and Reddy 1991). After decades of additional breeding and gene transfer from wheat and rye, the more advanced triticale cultivars, released in the 70th–80th of the previous century, have improved agronomic traits including high yields, resistance to lodging and ergot, plump kernels, but at the expense of protein content, which is now comparable to wheat (Skovmand et al. 1984; Oettler 2005).

The first hexaploid triticale was reported in 1938, and breeding efforts soon after concentrated on the production of hexaploids from various cultivars of tetraploid wheat, mainly ssp. *durum,* and *Secale cereale,* as well as intercrossing of both octoploid and primary triticales (Villareal et al. 1990). More intensive breeding programs with the explicit objective of developing triticale into a commercial crop were initiated during the 1950s in Spain (Sanchez-Monge 1974), Canada (Shebeski 1974), and Hungary (Kiss 1974). In 1953, the University of Manitoba, Winnipeg, Canada, began the first North American triticale breeding program working mostly with durum wheat–rye crosses. Since Canada's program, other public and private programs have initiated both durum wheat–rye and common wheat–rye crosses. These early breeding efforts concentrated on developing a high yielding and drought-tolerant human food crop species suitable for marginal wheat-producing areas. Both winter and spring types were developed, with emphasis on spring types. The major triticale development program in North America is now at CIMMYT in Mexico, with some private companies continuing triticale breeding. The CIMMYT Triticale Improvement Program started in 1964 under the leadership of N.E. Borlaug, followed by F.J. Zillinsky in 1968 (Zillinsky and Borlaug 1971). This program, in cooperation with the University of Manitoba, led to the release of the first cultivars, triticale numbers 57 and 64 in Hungary in 1968, followed by "Cachirulo" in Spain, and "Rosner" in Canada, both in 1969. Towards the 20th of the current century, triticale production was concentrated in Europe with nearly 90% of the world production; more than 3 million hectares were planted, and more than 14 million ton harvested in 2020 (Table 13.7). Production trends do show steady growth

Table 13.7 World total and top triticale producing countries in 2020 (From FAOSTAT 2022)

Country	Hectare	Production in tones	Productivity (tons /ha)
Poland	1,306,025	5,246,647	4.02
Germany	418,200	2,972,200	7.11
Belarus	523,413	2,076,376	3.97
France	387,604	2,023,275	5.22
Russian Federation	247,553	654,136	2.64
Hungary	123,160	486,450	2.30
Spain	195,682	449,674	1.82
Lithuania	120,100	395,200	3.95
China	212,000	385,000	1.82
Australia	79,879	125,641	3.29
World total	3,812,724	15,361,341	4.029

over the last 40 years. The leading producers of triticale worldwide are Poland, Germany, Belarus, and France.

A first major breakthrough came by chance when a triticale plant resulting from a natural outcrossed to unknown Mexican semi-dwarf bread wheat was selected in 1967. The selected line designated "Armadillo," made a major contribution to triticale improvement worldwide since it was the first triticale identified to carry a chromosome substitution wherein a D-subgenome chromosome was substituted for the respective R-subgenome homeologue. Because of this drastic improvement in triticale germplasm, numerous cultivars were released, and the crop was promoted to farmers as a "miracle crop." However, by the late 1980s, data from international yield trials revealed that complete hexaploid triticale (AABBRR) was agronomically much superior to some D-chromosomal substitutions, particularly under marginal growing conditions. Thereafter, triticale germplasm at CIMMYT was gradually shifted towards complete R-genome types to better serve these marginal environments.

The last several decades of research on triticale initiated by CIMMYT in association with National Agricultural Research Systems around the world, have resulted in significant improvements of triticale crop. Triticale today is an international crop grown in more than 41 countries with the number of countries and the acreage under triticale production increasing. Accordingly, the Food and Agriculture Organization, 15.5 million tons of triticale were harvested in 2018 in 41 countries across the world (FAOSTAT 2018).

Triticale cultivars are classified into three basic types: spring, winter, and intermediate (facultative) (Villareal et al. 1990). Spring types are generally insensitive to photoperiod and have limited tillering. Yet, most of the world triticale acreage is under winter types (Mergoum et al. 2009). Winter types have prostrate type of growth in the early stages of development, and they require vernalization to initiate heading. In general, winter types yield more forage than

spring types mainly due to their long growth period. Intermediate (facultative) types are in-between spring and winter types (Mergoum et al. 2004; Salmon et al., 2004).

Current triticale breeding programs center on the improvement of grain yield, nutritional quality, plant height, biomass, and early maturity. These efforts make triticale a potential candidate for increasing global food production, particularly, for marginal and stress-prone growing conditions. Modern cultivars of triticale can be used for ethanol production (McKenzie et al. 2014).

Triticale breeding programs worldwide have emphasized improving the product quality and developing triticale cultivars for specific end-uses such as milling and baking purposes (Villareal et al. 1990). Emphasis has been also given to developing triticale for dual purpose (forage and feed grain), and grazing types (Villareal et al. 1990). Variability present in the triticale germplasm for preharvest sprouting and gluten quality has been exploited by breeders to develop cultivars with enhanced quality and sprouting resistance which has improved the bread-making qualities of triticale grain.

Lodging resistance in triticale has been successfully improved using the dwarfing genes from both *Triticum* and *Secale* species. This has resulted in a decrease of up to 20 cm in plant height and increasing yield as the semi-dwarf cultivars are high yielding and more responsive to inputs. Modern triticale cultivars are resistant to a wide range of biotic and abiotic stresses, resulted in increasing the acreage under triticale worldwide. Under marginal land conditions, where abiotic stresses related to environment (drought or temperature extremes) and soil conditions (extreme pH levels, salinity, toxicity, or deficiency of elements) are the limiting factors grain production, modern triticale cultivars have consistently shown its advantages and has outperformed the existing cultivated cereal crops (Mergoum et al. 2004; Estrada-Campuzano et al. 2012; Ayalew et al. 2018). Early maturity, a typical characteristic of modern triticale, allows escape from terminal developmental stresses, such as heat or frost, in highly productive environments, such as the irrigated subtropics and Mediterranean climates, which has contributed to triticale acceptance by farmers. Many triticale cultivars show tolerance to periods of drought. Substantial progress has continued to improve grain weight (Mergoum et al. 2004). Research reveals an increase in the adaptation and successful production of triticale to stressed environments, particularly to water stress (Barary et al., 2002). Both successful breeding and management have resulted in acceptance of triticale as a major alternate crop to traditional cereal crops.

In general, winter triticale produces higher forage biomass than spring types. Therefore, their use for forage (grazing), cut forage, silage, and grain or hay has been improved through the release of several forage-specific cultivars. In addition, in many countries cereal straw is a major feed source for animals and in some years can have greater value than grain. Under arid and semiarid conditions, triticale has been shown consistently to produce higher straw yields than wheat and barley (Mergoum et al. 1992).

In comparison with wheat, triticale appears to have good resistance to several common wheat diseases and pests including rusts (*Puccinia* sp.), Septoria complex, smuts (*Ustilago* and *Urocystis* sp.), bunts (*Tilletia* sp.), powdery mildew (*Blumeria graminis*), cereal cyst nematode (*Heterodera avenae*), and Hessian fly (*Mayetiola destructor*) (Skovmand et al. 1984). It also resists virus diseases, such as barley yellow dwarf, wheat-streak mosaic, barley-stripe mosaic, and brome mosaic (Skovmand et al. 1984). On the other hand, triticale has relatively greater susceptibility than wheat to diseases such as spot blotch (*Bipolaris sorokiniana*), scab (*Fusarium* sp.), and ergot (*Claviceps purpurea*) and bacterial diseases caused by *Xanthomonas* sp. and *Pseudomonas* sp. (Skovmand et al. 1984). In the past, susceptibility to ergot was a major limitation to triticales expansion since it was linked to floret sterility, but ergot is not seen as a major problem in current varieties.

In triticale, genetic diversity is increased through direct interspecific (bread wheat x triticale) and intraspecific (winter triticale x spring triticale) as well as octoploid x hexaploid triticale crosses. Such crosses have led to the formation of mixogenomes where some chromosomes from the B, A, R subgenomes of triticale have been replaced by some from the D subgenome. Hence, many modern triticale lines developed from such crosses carry D(A), D(B), and D (R) whole chromosome substitutions or chromosome translocations which add valuable traits to triticale.

The use of hybrid triticales as a strategy for enhancing yield in favorable as well as marginal environments has proven successful over time. Yield improvements of up to 20% have been observed (Oettler et al. 2001, 2003; Oettler 2005). Hybrid btriticale makes the optimum exploitation of heterosis possible and, with the aid of molecular markers, triticale germplasm is presently being investigated to establish genetically diverse heterotic groups (Oettler 2005).

13.7.2.4 Uses

Triticale is used for human food, animal feed, grazed or stored forage and fodder, silage, green-feed, and hay (Oelke et al. 1989; Mergoum et al. (2009). Quality evaluations of triticale grain for milling and baking show that it is inferior to bread-making wheat and to durum wheat for macaroni, but it is often considered superior to rye. Although, triticale contains gluten, it may play a role in the rising healthy food market due to its health benefits with its good essential amino acid balance, minerals and vitamins (Zhu 2018). Triticale is tested for possible use in breakfast cereals and for distilling or brewing (Oelke et al. 1989). During the last decades triticale has received attention as a potential energy crop, and research is currently also including the use of this crop biomass in bio-energy production (Villareal et al. 1990; McKenzie et al. 2014). As animal feed triticale is a good source of protein, amino acids, and vitamin B. The protein content of several triticale lines is somewhat higher than that of wheat and the amino acid composition of the protein is similar to wheat but may be slightly higher in lysine (Mergoum et al. 2009). Triticale has been and is increasingly grown for livestock grazing, cut forage (green chop), whole- plant silage, hay, and forage/grain dual purpose (Myer and Lozano del Rio, 2004). Straw is an important by-product of triticale grain production and is often overlooked (Myer and Lozano del Rio 2004). Triticale produces more straw than other small-grain cereals. Straw is frequently the only source of livestock feed in developing countries (Mergoum et al. 2004).

References

Aaronsohn A, Schweinfurth G (1906) Die Auffindung des wilden Emmers (Triticum dicoccum) in Nordpalästina. Altneuland Monatsschrift Wirtsch. Erschliessung Palästinas 5:213–220

Aaronsohn A (1909) Contribution à l'histoire des céréales, le blé, l'orge et le seigle à l'état sauvage. Bull Soc Bot France 56:196–203, 237–245, 251–258

Aaronsohn A (1910) Agricultural and botanical explorations in Palestine. Bull. Plant Industry, U.S. Dept. of Agriculture, Washington, No.180:1–63

Abbo S, Lev-Yadun S, Gopher A (2010) Agricultural origins: centers and non-centers; A near Eastern reappraisal. Critical Rev Plant Sci 29:317–328

Abbo S, Lev-Yadun S, Gopher A (2012) Plant domestication and crop evolution in the near East: on events and processes. Critical Rev Pl Sci 31:241–257

Abbo S, van Oss RP, Gopher A, Saranga Y, Ofner I, Peleg Z (2014) Plant domestication versus crop evolution: a conceptual framework for cereals and grain legumes. Trends Plant Science 19:351–360

Aberkane H, Payne T, Kishi M, Smale M et al (2020) Transferring diversity of goat grass to farmers' fields through the development of synthetic hexaploid wheat. Food Security 12:1017–1033

Allaby RG, Brown TA (2003) AFLP data and the origins of domesticated crops. Genome 46:448–453

Allaby RG, Brown TA (2004) Reply to the comment by Salamini et al. on AFLP data and the origins of domesticated crops. Genome 47:621–622

Allaby RG, Banerjee M, Brown TA (1999) Evolution of the high molecular weight glutenin loci of the A, B, D and G genomes of wheat. Genome 42:296–307

Amagai Y, Watanabe N, Kuboyama T (2015) Genetic mapping and development of near-isogenic lines with genes conferring mutant phenotypes in *Aegilops tauschii* and synthetic hexaploid wheat. Euphytica 205:859–868

Asplund L, Bergkvist G, Leino MW, Westerbergh A, Weih M (2013) Swedish spring wheat varieties with the rare high grain protein allele of NAM-B1 differ in leaf senescence and grain mineral content. PLoS ONE 8:e59704

Avivi L (1977) High grain protein content in wild wheat. Can J Gent Cytol 19:569–570

Avivi L (1979a) Utilization of *Triticum dicoccoides* for the improvement of grain protein quantity and quality in cultivated wheats. Monogr Genet Agr 4:27–38

Avivi L (1979b) High grain protein content in wild tetraploid wheat *Triticum dicoccoides* Korn. Proc 5th Int Wheat Genet Symp New Delhi, India 1:372–380

Avni R, Nave M, Barad O, Sven BK, O, Twardziok SO, et al (2017) Wild emmer genome architecture and diversity elucidate wheat evolution and domestication. Science 357:93–97

Ayal S, Ophir R, Levy AA (2006) Genomics of tetraploid wheat domestication. In: Tsunewaki K (ed) Frontiers of wheat bioscience, Wheat Inf Serv 100:185–203

Ayalew AH, Kumssa T, Butler TJ, Ma XF (2018) Triticale improvement for forage and cover crop uses in the Southern great plains of the United States. Front Plant Sci 9:1130

Badaeva ED, Badaev NS, Gill BS, Filatenko AA (1994) Intraspecific karyotype divergence in *Triticum Araraticum* (Poaceae). Pl Sci Evol 192:117–145

Barary M, Warwick NWM, Jessop RS, Taji AM (2002) Osmotic adjustment and drought tolerance in Australian triticales. In: Arseniuk E (ed) Proc 5th Int Triticale Symp June 30–July 5, 2002. Radzikow, Poland, pp 135–141

Barneix AJ (2007) Physiology and biochemistry of source-regulated protein accumulation in the wheat grain. J Plant Physiol 164:581–590

Bar-Yosef O (1998) On the nature of transitions: the middle to upper Palaeolithic and the Neolithic revolution. Cambridge Archaeol J 8:141–163

Bar-Yosef O (2003) Early colonizations and cultural continuities in the Lower Palaeolithic of Western Asia. In: Korisettar R, Petraglia MD (eds) Early human behaviour in global context. The Rise and Diversity of the Lower Palaeolithic Record, Routledge, London and New Yorh, pp 221–279

Bar-Yosef O, Belfer-Cohen A (1992) From foraging to farming in the Mediterranean Levant. In: Gebauer AB, Price TD (eds) Transitions to agriculture prehistory. Prehistory Press, Madison, Wisconsin, pp 21–48

Bar-Yosef O, Kislev ME (1989) Early farming communities in the Jordan Valley. In: Harris DR, Hillman GC (eds) Foraging and farming: the evolution of plant exploitation. Unwin-Hyman Ltd., London, pp 632–642

Baum BR (1971) The taxonomic and cytogenetic implications of the problem of naming amphiploids of *Triticum* and *Secale*. Euphytica 20:302–306

Baum BR, Gupta PK (1990) Taxonomic examination of Triticale (x Triticosecale). Can J Bot 68:1889–1893

Becker B, Kromer B, Trimborn P (1991) A stable-isotope tree-ring timescale of the Late Glacial/Holocene boundary. Nature 353:647–649

Ben-Abu Y, Tzfadia 0, Maoz Y, Kachanovsky D, Melamed-Bessudo C. Feldman M, Levy AA (2014) Durum wheat evolution—a genomic analysis. In: Porceddu E (ed) Proc Symp Genet Breed of durum. Rome, pp 29–44

Bhatta M, Morgounov A, Belamkar V et al (2018) Unlocking the novel genetic diversity and population structure of synthetic Hexaploid wheat. BMC Genomics 19:591

Blumler MA (1994) Evolutionary trends in the wheat group in relation to environment, quaternary climatic change and human impacts. In: Millington AC, Pye K (eds) Environmental change in dryland: biogeographical and geomorphological perspectives. John Wiley and Sons, Chichester, pp 253–269

Blumler MA (1996) Ecology, evolutionary theory and agricultural origins. In: Harris DR (ed) The origins and spread of agriculture and pastoralism in Eurasia. UCL Press, London, pp 25–50

Blumler MA (1992) Seed weight and environment in Mediterranean-type grasslands in California and Israel. UMI Dissertation Services, UC Berkeley, Ann Arbor

Blumler MA (1998) Introgression of durum into wild emmer and the agricultural origin question. In: Damania AB, Valkoun J, Willcox G, Qualset CO (eds) The origin of agriculture and crop domestication, Proceedings of the harlan symposium, 10–14 May,1997, ICARDA, IPGRI, FAO and UC/GRCP, pp 252–268

Boiko EV, Badaev NS, Maksimov NG, Zelenin AV (1988) Cereal genome formation and organization patterns. I. Change in DNA level during allopolyploidization. Genetica USSR 24:89–97

Brevis JC, Dubcovsky J (2010) Effects of the chromosome region including the *Gpc-B1* locus on wheat grain and protein yield. Crop Sci 50:93–104

Brevis JC, Morris CF, Manthey F, Dubcovsky J (2010) Effect of the grain protein content locus Gpc-B1 on bread and Lange pasta quality. J Cereal Sci 51:357–365

Brown TA (1999) How ancient DNA may help in understanding the origin and spread of agriculture. Phil Trans R Soc London B 354:89–98

Brown TA, Jones MK, Powell W, Allaby RG (2009) The complex origins of domesticated crops in the Fertile Crescent. Trends Ecol Evol 24:103–109

Caldwell KS, Dvorak J, Lagudah ES, Akhunov E, Luo M-C, Wolters P, Powell W (2004) Sequence polymorphism in polyploid wheat and their D-genome diploid ancestor. Genetics 167:941–947

Cantrell RG, Joppa LR (1991) Genetic analysis of quantitative traits in wild emmer (*Triticum turgidum* L. var. *dicoccoides*). Crop Sci 31:645-649

Cao W, Scoles GJ, Hucl P (1997) The genetics of rachis fragility and glume tenacity in semi-wild wheat. Euphytica 94:119–124

Carleton MA (1916) The small grains. The Macmillian Co, New York

Cavanagh CR, Chao S, Wang S, Huang BE, Stephen S, Kiani S, Forrest K, Saintenac C, Brown-Guedira GL, Akhunova A, See D, Bai G, Pumphrey M, Tomar L, Wong D, Kong S, Reynolds M, da Silva ML, Bockelman H, Talbert L, Anderson JA, Dreisigacker S, Baenziger S, Carter A, Korzun V, Morrell PL, Dubcovsky J, Morell MK, Sorrells ME, Hayden MJ, Akhunov E (2013) Genome-wide comparative diversity uncovers multiple targets of selection for improvement in hexaploid wheat landraces and cultivars. Proc Natl Acad Sci USA 110:8057–8062

Chee PW, Elias EM, Anderson JA, Kianian SF (2001) Evaluation of a high grain protein QTL from *Triticum turgidum* L. var. *dicoccoides* in an adapted durum wheat background. Crop Sci 41:295–301

Chen Q-F, Yen C, Yang J-L (1998) Chromosome location of the gene for brittle rachis in the Tibetan weed race of common wheat. Genet Resour Crop Evol 45:407–410

Civáň P, Ivaničová Z, Brown TA (2013) Reticulated origin of domesticated emmer wheat supports a dynamic model for the emergence of agriculture in the Fertile Crescent. PLoS ONE 8: e81955

Colledge S (2002) Identifying pre-domestication cultivation in the archaeobotanical record using multivariate analysis: Presenting the case for quantification. In: Cappers RTJ, Bottema S (eds) The dawn of farming in the Near East. Ex Oriente, Berlin, pp 141–152

Colledge S (2004) Reappraisal of the archaeobotanical evidence for the emergence and dispersal of the 'founder crops. In: Peltenburg E, Wasse A (eds) Neolithic revolution: new perspectives on South-West Asia in the light of recent discoveries in Cyprus. Oxbow, Oxford, pp 49–60

Colledge S (2001) Plant exploitation on epipalaeolithic and early neolithic sites in the Levant. BAR, International Series 986, Oxford

College S (1998) Identifying pre-domestication cultivation using multivariate analysis. In: Damania AB, Valkoun J, Willcox G, Qualset CO (eds) The origins of agriculture and crop domestication, Aleppo, ICARDA, pp 121–31

Cook OF (1913) Wild wheat in Palestine. Bull Plant Industry, US Dept Agric, Washington, no. 274: 1–58

Czajkowska BI, Bogaard A, Charles M et al (2020) Ancient DNA typing indicates that the "new" glume wheat of early Eurasian agriculture is a cultivated member of the *Triticum timopheevii* group. J Arch Sci 123:105258

Darwin CR (1868) The variation of animals and plants under domestication. John Murray, London

Das MK, Bai GH, Mujeeb-Kazi A, Rajaram S (2016) Genetic diversity among synthetic hexaploid wheat accessions (*Triticum aestivum*) with resistance to several fungal diseases Genet Resour Crop Evol 63:1285–1296

de Candolle A (1886) Origin of Cultivated Plants. London: Kegan, Paul, Trench and CO Second Edition du ble et de l'orge. Ann Sci Nat Ser I (Paris), IX:61–82

de La Malle D (1826) Histoire· ancienne, origine, et patrie des cereales et nommement du ble et de l'orge. Ann Sci Nat Ser I (Paris), IX:61–82

de Vita P, Riefolo C, Codianni P, Cattivelli L, Fares C (2006) Agronomic and qualitative traits of *T. turgidum* ssp. *dicoccum* genotypes cultivated in Italy. Euphytica 150:195–205

Debernardi JM, Lin H, Chuck G, Faris JD, Dubcovsky J (2017) MicroRNA172 plays a crucial role in wheat spike morphogenesis and grain threshability. Development 144:1966–1975

Dekaprelecich L (1961) Die Art *Triticum macha* Dek. et Men.im Lichte neuester Untersuchungen über die Herkuft der hexaploidenWeizen. Z Fur Pflanzenzüchtung 45:17–30

Dembitzer J, Barkai R, Ben-Dor M, Meiri S (2022) Levantine overkill: 1.5 million years of hunting down the body size distribution. Quaternary Sci Rev 276:107316

Dhaliwal HS (1977) Origin of *Triticum monococcum*. Wheat Inf Serv 44:14–17

Diamond J (1997) Location, location, location: the first farmers. Science 278(5341):1243–1244

Distelfeld A, Cakmak I, Peleg Z, Ozturk L, Yazici AM, Budak H, Saranga Y, Fahima T (2007) Multiple QTL-effects of wheat Gpc-B1 locus on grain protein and micronutrient concentrations. Physiol Plant 129:635–643

Doebley JF, Gaut BS, Smith BD (2006) The molecular genetics of crop domestication. Cell 127:1309–1321

Dorofeev VF (1966) Geographical localization and gene center of hexaploid wheat in Transcaucasia. (In Russian) Genetika 3:16–33

Dorofeev VF, Filatenko AA, Miguschova EF (1980) Wheat. In: Dorofeev VF, Korovina ON (eds) Flora of Cultivated Plants in theUSSR. "Kolos", Leningrad, Russia, Vol 1 (In Russian)

Dubcovsky J, Dvorak J (2007) Genome plasticity a key factor in the success of polyploid wheat under domestication. Science 316:1862–1866

Dudnikov AJ, Kawahara T (2006) *Aegilops tauschii*: Genetic variation in Iran. Genet Resour Crop Evol 53:579–586

Dvorak J, Akhunov ED, Akhunov AR, Deal KR, Luo M-C (2006) Molecular characterization of a diagnostic DNA marker for domesticated tetraploid wheat provides evidence for gene flow wild tetraploid wheat to hexaploid wheat. Mol Biol Evol 23:1386–1396

Eckardt NA (2010) Evolution of domesticated bread wheat. Plant Cell 22:993

Elbaum R, Zaltzman L, Burgert I, Fratzl P (2007) The role of wheat awns in the seed dispersal unit. Science 316:884–886

Estrada-Campuzano G, Slafer GA, Miralles DJ (2012) Differences in yield, biomass and their components between triticale and wheat grown under contrasting water and nitrogen environments. Field Crops Res 128:167–179

Evans LT (1981) Yield improvement in wheat: empirical or analytical? In: L.T. Evans LT, Peacock WJ (eds), Wheat science—today and tomorrow. Cambridge University Press, Cambridge, pp 203–222

Fabre E (1852) Des *Aegilops* du Midi de la France et de leur transformation [en *Triticum* (blé cultivé)]. Bonplandia 2:208–213

Fabre E (1855) On the species of *Aegilops* of the south of France and their translation into cultivated wheat. J Roy Agric Soc England 15:160–180

Fans JD, Simons KJ, Zhang Z, Gill BS (2006) The wheat super domestication gene Q. Wheat Inf Serv 100:120–148

FAOSTAT (2018) Food and agriculture organization of the United Nation (FAO), Statistical database, statistical division. Rome, http://faostat3.fao.org/browse/Q/QC/

Faris JD, Fellers JP, Brooks SA, Gill BS (2003) A bacterial artificial chromosome contig spanning the major domestication locus *Q* in wheat and identification of a candidate gene. Genetics 164:311–321

Faris JD, Zhang Z, Fellers JP, Gill BS (2008) Micro-colinearity between rice, *Brachypodium*, and *Triticum monococcum* at the wheat domestication locus Q. Funct Integr Genomics 8:149–164

Faris JD, Zhang Q, Chao S, Zhang Z, Xu SS (2014) Analysis of agronomic and domestication traits in a durum x cultivated emmer wheat population using a high-density single nucleotide polymorphism-based linkage map. Theor Appl Genet 127:2333–2348

Fedoroff NV (2000) Transposons and genome evolution in plants. Proc Natl Acad Sci, USA 97:7002–7007

Fedoroff NV (2012) Transposable elements, epigenetics, and genome evolution. Science 338:758–767

Feldman M (1976) Wheats *Triticum* spp. (Gramineae-Triticineae). In: Simmonds NW (ed) Evolution of crop plants. Longman Group Ltd., London, pp 120–128

Feldman M (2001) Origin of cultivated wheat. In: Bonjean AP, Angus WJ (eds) The world wheat book. Lavoisier, Paris, pp 3–56

Feldman M, Kislev M (2007) Domestication of emmer wheat and evolution of free-threshing tetraploid wheat. Isr J Pl Sci 55:207–221

Feldman M, Galili G, Levy AA (1987) Genetic and evolutionary aspects of allopolyploidy in wheat. In: Barigozzi C (ed) The Origin and Domestication of Cultivated Plants. Elsevier, Amsterdam, Oxford, New York, Tokyo, pp 83–100

Feldman M, Avivi L, Levy AA, Zaccai M, Avivi Y, Millet E (1990) High protein wheat. In: Bajaj YPS (ed) Biotechnology in agriculture and forestry, vol 6. Crops II. Springer-Verlag, Berlin, Heidelberg, pp 593–614

Feldman M, Lupton FGH, Miller TE (1995) Wheats. In: Smart J, Simmonds NW (eds) Evolution of Crop Plants. Second Edition, Longman Scientific & Technical, pp 184–192

Feldman M, Levy AA, Fahima T, Korol A (2012) Genome asymmetry in allopolyploid plants—wheat as a model. J Exp Bot 63:5045–5059

Feldman M, Strauss I (1983) A genome-restructuring gene in *Aegilops longissima*. In: Sakamoto S (ed) Proc 6th Intern Wheat Genet Symp, Kyoto, Japan, pp 309–314

Flaksberger CA (1915) Determination of wheats. Trudy prikl Bot Genet Selek VIII (1–2). *Triticum dicoccoides*: pp. 14–15, 26–27, 48–52. (Russian, with English summary)

Friebe B, Tuleen N, Jiang J, Gill BS (1993) Standard karyotype of *Triticum longissimum* and its cytogenetic relationship with *T. aestivum*. Genome 36:731–742

Friebe B, Tuleen NA, Gill BS (1995) Standard karyotype of *Triticum searsii* and its relationship with other S-genome species and common wheat. Theor Appl Genet 91:248–254

Friebe B, Tuleen NA, Gill BS (1999) Development and identification of a complete set of *Triticum aestivum—Aegilops geniculata* chromosome addition lines. Genome 42:374–380

Friebe B, Qi L, Nasuda S et al (2000) Development of a complete set of *Triticum aestivum-Aegilops speltoides* chromosome addition lines. Theor Appl Genet 101:51–58

Fuller DQ (2007) Contrasting patterns in crop domestication and domestication rates: recent archaeobotanical insights from the old world. Ann of Bot 100:903–924

Gaurav K, Arora S, Silva P, Sanchez-Martin J, et al (2021) Population genomic analysis of *Aegilops tauschii* identifies targets for bread wheat improvement. Nat Biotechnol. https://doi.org/10.1038/s41587-021-01058-4

Gegas VC, Nazari A, Griffiths S, Simmonds J, Fish L, Orford S, Sayers L, Doonan JH, Snape JH (2010) A genetic framework for grain size and shape variation in wheat. Plant Cell 22:1046–1056

Gepts P (2004) Crop domestication as a long-term selection experiment. Plant Breed Rev 24:1–44

Godron DA (1856) De l'*Aegilops triticoides* et de ses différentes forms. Ann Sci Nat Bot Ser 4(5):74–89

Godron DA (1858a) On the natural and artificial fertilization of *Aegilops* by *Triticum*. J Roy Agric Soc England 19:105–109

Godron DA (1858b) On *Aegilops triticoides* and its different forms. J Roy Agric Soc England 19:110–112

Godron DA (1869) Histoire des Aegilops hybrids. Mém De L'acad De Stanislas (nancy) 2:167–222

Godron DA (1876) Un Nouveau chapitre ajouté a l'histoire des Aegilops hybrids. Mém De L'acad De L'acad De Stanislas (nancy) 9:250–281

Godron DA (1854) Flore de Montpellier. Besançon, Imprimerie D'Outhenin-Chalandre Fils

Golan G, Oksenberg A, Peleg Z (2015) Genetic evidence for differential selection of grain and embryo weight during wheat evolution under domestication. J Exp Bot 66:5703–5711

Gornicki P, Zhu H, Wang J, Challa GS, Zhang Z, Gill BS, Li W (2014) The chloroplast view of the evolution of polyploid wheat. New Phytol 204:704–714

Gross BL, Olsen KM (2010) Genetic perspectives on crop domestication. Trends Plant Sci 15:529–537

Gupta PK (1986) Nomenclature, taxonomy and classification of triticales. In: Proc Intern Triticale Symposium, Sydney, Feb. 2-6, 1986. Australian Institute of Agricultural Science, Occas Publ, No. 24, pp. 22–30

Gupta Pk, Reddy VRK (1991) Cytogenetics of Triticale—A Man-Made Cereal. In: Gupta PK, Tsuchiya T (eds) Developments in Plant Genetics and Breeding Vol 2, Part A, pp. 335–359

Gupta PK, Baum BR (1986) Nomenclature and related taxonomic issues in wheats, triticales and some of their wild relatives. Taxon 35:144–149

Gupta PK, Priyadarshan PM (1982) Triticale-present status and future prospects. Adv Genet 21:255–334

Guzman C, Medina-Larque AS, Velu G, Gonzalez- Santoyo H et al (2014) Use of wheat genetic resources to develop biofortified wheat with enhanced grain zinc and iron concentrations and desirable processing quality. J Cereal Sci 60:617–622

Hackel E (1890) The true grasses, Illus. New York, p 228

Hagenblad J, Asplund L, Balfourier F, Ravel C, Leino MW (2012) Strong presence of the high grain protein content allele of NAM-B1 in Fennoscandian wheat. Theor Appl Genet 125:1677–1686

Hammer K (1984) Das Domestikationssyndrom. Genet Resourc Crop Evol 32:11–34

Hao M, Zhang L, Zhao L et al (2019) A breeding strategy targeting the secondary gene pool of bread wheat: introgression from a synthetic hexaploid wheat. Theor Appl Genet 13:2285–2294

Harlan JR (1967) A wild wheat harvest in Turkey. Archaeology 20:197–201

Harlan JR (1975) Our vanishing genetic resources. Science 188:618–622

Harlan JR, Zohary D (1966) Distribution of wild wheats and barley. Science 153:1074–1080

Harlan JR, de Wet JMJ, Price EG (1973) Comparative evolution of cereals. Evolution 27:311–325

Harris DR (1989) An evolutionary continuum of people-plant interaction. In: Harris DR, Hillman GC (eds) Foraging and Farming: The Evolution of Plant Exploitation. Unwin and Hyman, London, pp 11–26

Harris DR (1996) The origins and spread of agriculture and pastoralism in Eurasia. University College London Press, London

Harris DR (1998) The origins of agriculture in southwest Asia. Rev Archaeol 19:5–11

Haudry A, Cenci A, Ravel C, Bataillon T et al (2007) Grinding up wheat: a massive loss of nucleotide diversity since domestication. Mol Biol Evol 24:1506–1517

He F, Pasam R, Shi F, Kant S, Keeble-Gagnere G et al (2019) Exome sequencing highlights the role of wild-relative introgression in shaping the adaptive landscape of the wheat genome. Nat Genet 51:896–904

Hebelstrup KH (2017) Differences in nutritional quality between wild and domesticated forms of barley and emmer wheat. Plant Sci 256:1–4

Helbaek H (1959) Domestication of food plants in the old world. Science 130:365–372

Helbaek H (1966) Commentary on the phylogenesis of Triticum and Hordeum. Econ Bot 20:350–360

Helbaek H (1969) Plant collecting, dry-farming, and irrigation agriculture in prehistoric Deh Luran. In: Hole F, Flannery KV, Neely JA (eds) Prehistory and human ecology of the Deh Luran plain. An early village sequence from Khuzistan, Iran. Ann Arbor, MI: Museum of Anthropology, University of Michigan, Memoirs1, pp 383–426, plates 40–41

Heun M, Schäfer-Pregl R, Klawan D, Castagna R et al (1997) Site of einkorn wheat domestication identified by DNA fingerprinting. Science 278:1312–1214

Hillman GC (1972) Plant remains in French DH. Excavations at Can Hassan III 1969–1970. In: Higgs ES (ed) Papers in economic prehistory. Cambridge University Press, Cambridge, pp 182–190

Hillman GC (1975) The plant remains from Tell Abu Huretra: a preliminary report. Proc Prehistory Society 41:70–73

Hillman GC (1978) On the origins of domestic rye-*Secale cereale*: the finds from Aceramic Can Hasan III in Turkey. Anatolian Stud 28:157–174

Hillman CG (1996) Late Pleistocene changes in wild plant foods available to hunter-gatherers of the northern Fertile Crescent: possible preludes to cereal cultivation. In: Harris D (ed) The origin and spread of agriculture and pastoralism in Eurasia. UCL Press, London, pp 159–203

Hillman G (2000) Plant food economy of Abu Hureyra. In: Moore A, Hillman G, Legge T (eds) Village on the Euphrates, from foraging to farming at Abu Hureyra. Oxford Univ Press, New York, pp 372–392

Hillman GC, Davies MS (1990) Measured domestication rates in wild wheats and barley under primitive cultivation, and their archaeological implications. J World Prehist 4:157–222

Hillman CG, Davies MS (1999) Domestication rate in wild wheats and barley under primitive cultivation. In: Anderson P (ed) Prehistory of agriculture: new experimental and ethnographic approaches, vol Monograph. 40, Inst Archaeology. Univ California, Los Angeles, pp 70–102

Hillman GC, Hedges R, Moore A, Colledge S, Petitt P (2001) New evidence of Late Glacial cereal cultivation at Abu Hureyra on the Euphrates. The Holocene 11:383–393

Hole F (1998) The spread of agriculture to the eastern arc of the Fertile Crescent: food for the herders. In: Damania AB, Valkoun J, Willcox G, Qualset CO (eds) The Origin of Agriculture and Crop Domestication, Proc of the Harlan Symposium, 10–14 May 1997, ICARDA, IPGRI, FAO and UC/GRCP, pp 83–92

Hopf M (1983) Jericho plant remains. In: Kenyon KM, Holland TA (eds) Excavations at Jericho, vol 5. British School of archaeology in Jerusalem, London, pp 576–621

Horovitz A, Ezrati S, Anikster Y (2013) Are soil seed banks relevant for agriculture in our day? Crop Wild Relat 9:27–29

Hu X-G, Wu B-H, Liu D-C, Wei Y-M, Gao S-B, Zheng Y-L (2013) Variation and their relationship of *NAM-G1* gene and grain protein content in *Triticum Timopheevii* Zhuk. J Plant Physiol 170: 330–337

Hu X-G, Wu B-H, Yan Z-H, Dai S-F, Zhang L-Q, Liu D-C, Zheng Y-L (2012) Characteristics and polymorphism of NAM gene from *Aegilops* section Sitopsis species. African J Agric Res 7:5252–5258

Huang L, Millet E, Rong JK, Wendel JF, Anikster Y, Feldman M (1999) Restriction fragment length polymorphism in wild and cultivated tetraploid wheat. Isr J Pl Sci 47:213–224

ICBN (2000) Greuter W, Chairman, McNeill J, Vice-Chairman, Barrie FR, Burdet HM et al (2000) International code of botanical nomenclature, (Saint Louis Code), adopted by the Sixteenth International Botanical Congress, St Louis, Missouri, July–August 1999

Iqbal N, Reader SM, Caligari PDS et al (2000b) Characterization of *Aegilops uniaristata* chromosomes by Comparative DNA marker analysis and repetitive DNA sequence in situ hybridization. Theor Appl Genet 101:1173–1179

Jakubziner MM (1959) New wheat species. In: Jenkins BC (ed) Proc 1st intern wheat genetic symp. Public Press Ltd., Winnipeg, Manitoba, Canada, pp 207–220

Jakubziner MM (1932) A contribution to the knowledge of the wild wheat of Transcaucasia Trudy prikl Bot Genet Selek, Ser 5, 1:147–198. (Russian, with English summary)

Jantasuriyarat C, Vales MI, Watson CJW, Riera-Lizarazu O (2004) Identification and mapping of genetic loci affecting the free-threshing habit and spike compactness in wheat (*Triticum aestivum* L.). Theor Appl Genet 108:261–273

Jia J, Zhao S, Kong X, Li Y, Zhao G et al (2013) *Aegilops tauschii* draft genome sequence reveals a gene repertoire for wheat adaptation. Nature 496:91–95

Jiang J, Gill BS (1994) Different species-specific chromosome translocations in *Triticum timopheevii* and *T. turgidum* support the diphyletic origin of polyploid wheats. Chromosome Res 2:59–64

Jones M, Brown B (2000) Agricultural origins: the evidence of modern and ancient DNA. The Holocene 10:769–776

Jones G, Valamoti S, Charles M (2000) Early crop diversity: a "new" glume wheat from Northern Greece. Veg Hist Archaeobotany 9:133–146

Joppa LR (1993) Chromosome engineering in tetraploid wheat. Crop Sci 33:908–913

Joppa LR, Cantrell RG (1990) Chromosomal location of genes for grain protein content of wild tetraploid wheat. Crop Sci 30:1059–1064

Joppa LR, Hareland GA, Cantrell RG (1991) Quality characteristics of the Langdon *durum-dicoccoides* chromosome substitution lines. Crop Sci 31:1513–1517

Joppa LR, Du C, Hart GE, Hareland GA (1997) Mapping gene(s) for grain protein in tetraploid wheat (*Triticum turgidum* L.) using a population of recombinant inbred chromosome lines. Crop Sci 37:1586–1589

Kamm A (1974) Non-brittle types in a wild population of *Triticum dicoccoides* Körn in Israel. Isr J Bot 23:43–58

Katkout M, Sakuma S, Kawaura K, Ogihara Y (2015) *TaqSH1-D*, wheat ortholog of rice seed shattering gene *qSH1*, maps to the interval of a rachis fragility QTLon chromosome 3DL of common wheat (*Triticum aestivum*). Genetic Resour Crop Evol 62:979–984

Keilwagen J, Lehnert H, Berner T et al (2022) Detecting major introgressions in wheat and their putative origins using coverage analysis. Sci Rep 12:1908

Kenan-Eichler M, Leshkowitz D, Tal L, Noor E, Cathy Melamed-Bessudo C, Feldman M, Levy AA (2011) Wheat hybridization and polyploidization results in deregulation of small RNAs. Genetics 188:263–272

Kerber ER, Dyck PL (1969) Inheritance in hexaploid wheat of leaf rust resistance and other characters derived from *Aegilops squarrosa*. Can J Genet Cytol 11:639–647

Kerber ER, Rowland GG (1974) Origin of the free threshing character in hexaploid wheat. Can J Genet Cytol 16:145–154

Kihara H (1924) Cytologische und genetische studien bei wichtigcn getreidearten mit besonderer rücksicht ouf das verhalten der chromosomen und die sterilitat in den bastarden. Kyoto Imp Univ Bl, Mem Cell Sci, pp 1–200

Kihara H (1937) Genomanalyse bei Triticum und Aegilops. VII. Kurze ~ bersicht iiber die Ergebnisse der Jahre 1934–36. Mem Coll Agric Kyoto Imp Univ 41:1–61

Kihara H (1940) Anwendung der Genomanalyse fur die Systematik von *Triticurn* und *Aegilops*. Jpn J Genet 16:309–320

Kihara H (1954) Considerations on the evolution and distribution of *Aegilops* species based on the analyzer-method. Cytologia 19:336–357

Kihara H, Lilienfeld FA (1949) A new synthesized 6x-wheat. In: Larsson GBAR (ed) Proceedings of eighth international congress of genetics, Stockholm, Sweden, 1949. Hereditas (Suppl), pp 307–319

Kimber G, Feldman M (1987) Wild wheats: an introduction. Special Report 353. College of Agriculture, Columbia. Missouri, USA, pp 1–142

King IP, Forster BP, Law CC, Cant KA et al (1997) Introgression of salt-tolerance genes from *Thinopyrum bessarabicum* into wheat. New Phytol 137:75–81

Kislev ME (1981) The history of evolution of naked wheats. Z Archaeol 15:57–64

Kislev ME (1984) Emergence of wheat agriculture. Palaeorient 10:61–70

Kislev ME (1986) A barley store of the Bar-Kochba rebels (Roman period). Isr J Bot 35:183–196

Kislev ME (1988) Desicated plant remains: an interim report. Atiqot 18:76–81

Kislev ME, Bar-Yosef O, Gopher A (1986) Early neolithic domesticated and wild barley from the Netiv Hagdud region in the Jordan
Valley. Isr J Bot 35:197–201

Kislev ME, Weiss E, Hartman A (2004) Impetus for sowing and the
beginning of agriculture: ground collecting of wild cereals. Proc
Natl Acad Sci USA 101:2692–2695

Kislev ME (1979/1980) Triticum parvicoccum sp. nov, the oldest
naked wheat. Isr J Bot 28:95–107

Kislev ME (1989) Pre-domesticated cereals in the pre-pottery
Neolithic A period. In: Hershkovitz I (ed) People and culture
change. BAR Inter. Series 508I, Oxford, pp 147–151

Kislev ME (1992) Agricultural situation in the Near East in the VIIth
millennium BP. In: Anderson PC (ed) Prehistoire de l'agriculture:
nouvelles approches experimentales et ethnographiques. Édition du
CNRS, Monographie du Centre Recherches Archéologiques No. 6,
Paris, pp 87–93

Kislev ME (1997) The archaeology of Netiv Hagdud. In: Bar-Yosef O,
Gopher A (eds) An early neolithic village in the Jordan Valley,
Part I, Peabody Museum of archaeology and ethnology, Harvard
Univ., Cambridge, A, pp 209–236

Kiss A (1974) Triticale breeding experiments in Eastern Europe. In:
MacIntyre R, Campbell M (eds). Triticale: Proc Int Symp, 1973,
International Development Research Centre, El Batan, Mexico;
Ottawa, Ont, Canada, pp 41–50

Konopatskaia I, Vavilova V, Blinov A, Goncharov NP (2016) Spike
morphology genes in wheat Species (Triticum L.). Proc Latvian
Acad Sci B 70:345–355

Körnicke F (1889) Über Triticum vulgare var. dicoccoides. Bericht
über den Zustand und die Tätigkeit der Niederrheinischen Gesselschaft für Natur und Heilkunde (in Bonn) während des Jahres
1888 (1889), p. 21 (Sitzung von 11.3.1889)

Krasileva KV, Buffalo V, Bailey P, Pearce S, Ayling S, Tabbita F,
Soria M, Wang S, Consortium I, Akhunov E, Uauy C, Dubcovsky J
(2013) Separating homeologs by phasing in the tetraploid wheat
transcriptome. Genome Biol 14: R66

Krolow KD (1973) 4x Triticale, production and use in Triticale
breeding. In: Sears ER, Sears LMS (eds) Proc 4th Int Wheat Genet
Symp, Columbia, Mo, USA, pp 237–243

Kuckuck H (1964) Experimentelle Unersuchumgen zur Entstehungder
Kulturweizen. I. Die Variation des Iranischen Speltzweizns und
seine genetischen Beziehungen zu Triticum aestivum ssp. vulgare
(Vill-, HOSt) Mac Key, ssp. spelta (L.) Thell. und ssp. macha (Dek
et Men.) Mac Key mit einem Beitrag zur Genetik den Spelta-
Komplexes. Z Pflanzenzücht 51:97–140

Ladizinsky G (1975) Collection of wild cereals in the upper Jordan
Valley. Econ Bot 29:264–267

Ladizinsky G (1985) Founder effect in crop-plant evolution. Econ Bot
39:191–199

Laido G, Mangini G, Taranto F, Gadaleta A, Blanco A, Cattivelli L,
Marone D, Mastrangelo AM, Papa R, De Vita P (2013) Genetic
diversity and population structure of tetraploid wheats (Triticum
turgidum L.) estimated by SSR, DArT and Pedigree Data. Plos One
8

Levy AA, Feldman M (1988) Ecogeographical distribution of HMW
glutenin alleles in populations of the wild tetraploid wheat Triticum
turgidum var. dicoccoides. Theor Appl Genet 75:651–658

Levy AA, Feldman M (1989a) Genetics of morphological traits in wild
wheat, Triticum Turgidum Var. Dicoccoides. Euphytica 40:275–
281

Levy AA, Feldman M (1989b) Location of genes for high grain protein
percentage and other quantitative traits in wild wheat Triticum
turgidum var. dicoccoides. Euphytica 41:113–122

Levy AA, Feldman M (2022) Evolution and origin of bread wheat.
Plant Cell 34:2549–2567

Lev-Yadun S, Gopher A, Abbo S (2000) The cradle of agriculture.
Science 288:1602–1603

Li W, Gill B (2006) Multiple genetic pathways for seed shattering in
the grasses. Funct Integr Genomics 6:300–309

Li A, Liu D, Yang W, Kishii M, Mao L (2018) Synthetic hexaploid
wheat: yesterday, today, and tomorrow. Engineering 4:552–558

Li L-F, Zhang Z-B, Wang Z-H et al (2022) Genome sequences of the
five Sitopsis species of Aegilops and the origin of polyploid wheat B
subgenome. Mol Plant 15:488–503

Li W, Gill BS (2002) The colinearity of the Sh2/A1 orthologous region
in rice, sorghum and maize is interrupted and accompanied by
genome expansion in the Triticeae. Genetics 160:1153–1162

Lilienfeld F, Kihara H (1934) Genomanalyse bei Triticum and
Aegilops. V. Triticum timopheevii Zhuk. Cytologia 6:87–122

Ling H-Q, Zhao S, Liu D, Wang J, Sun H et al (2013) Draft genome of
the wheat A-genome progenitor Triticum urartu. Nature 496:87–90

Link HF (1816) Uber die. Altere Geschichte der Getreidearten.
Abhand. Akad. Wiss., Berlin Jahren 1816–1817, 17:122

Linnaeus C (1753) Species plantarum, Tomus I, May 1753: i-xii, 1–
560; Tomus II, Aug 1753: 561–1200, plus indexes and addenda,
1201–1231

Lukaszewski AJ, Gustafson JP (1987) Cytogenetics of Triticale. Plant
Breed Rev 5:41–93

Lundström M, Leino MW, Hagenblad J (2017) Evolutionary history of
the NAM-B1 gene in wild and domesticated tetraploid wheat. BMC
Genet 18:118

Luo M-C, Yang Z-L, You FM, Kawahara T, Waines JG, Dvorak J
(2007) The structure of wild and domesticated emmer wheat
populations, gene flow between them, and the site of emmer
domestication. Theor Appl Genet 114:947–959

Luo M-C, Gu YQ, Puiu D, Wang H, Twardziok SO, Deal KR et al
(2017) Genome sequence of the progenitor of the wheat D genome
Aegilops tauschii. Nature 551:498–502

Ma X-F, Gustafson JP (2005) Genome evolution of allopolyploids: a
process of cytological and genetic diploidization. Cytogenet
Genome Res 109:236–249

Ma X-F, Gustafson JP (2006) Timing and rate of genome variation in
triticale following allopolyploidization. Genome 49:950–958

Ma X-F, Fang P, Gustafson JP (2004) Polyploidization- induced
genome variation in triticale. Genome 47:839–848

Mac Key J (1954a) Neutron and X-ray experiments in wheat and a
revision of the speltoid problem. Hereditas 40:65–180

Mac Key J (1954b) The taxonomy of hexaploid wheat. Svensk Bot
Tidskr 48:579–590

Mac Key J (1966) Species relationship in Triticum. Proc 2nd Int Wheat
Genet Symp Lund, Sweden Hereditas Suppl 2:237–276

Mac Key J (1975) The boundaries and subdivision of the genus
Triticum. Proc 12th Inter Bot Congress, Leningrad: Abstract 2:509
(complete manuscript available at Vavilov Institute of Plant
Industry, Leningrad, pp. 1–23

Mac Key J (1981) Comments on the basic principles of crop taxonomy.
Kulturpflanze 29(S):199–207

Mac Key J (1988) A plant breeder's perspective on taxonomy of
cultivated plants. Biol Zentralblatt 107:369–379

Mac Key J (1968) Relationships in the Triticinae. In: Finlay KW,
Shepherd KW (eds) Proc 3rd Int Wheat Genet Symp, Canberra,
Aust Acad Sci, Canberra, pp. 39–50

Mac Key J (1986) Shoot, root and shoot: root interrelations in cereals
and the ideotype concept. In: Siddiqui KA, Faruqui AM (eds) New
genetical approaches to crop improvement. Atomic Energy Agricultural Research Center, Tandojam, Pakistan, pp 811–833

Mac Key J (2005) Wheat: its concept, evolution and taxonomy. In:
Royo C et al. (eds) Durum wheat, current approaches, future
strategies, CRC Press, Boca Raton, vol 1, pp 3–61

Maccaferri M, Harris NS, Twardziok SO, Pasam RK, Gundlach H et al (2019) Durum wheat genome highlights past domestication signatures and future improvement targets. Nature Genet 51:885–895

Marcussen T, Sandve SR, Heier L, Spannagl M, Pfeifer M. The International Wheat Genome Sequencing Consortium, Jakobsen KS, Wulff BBH, Steuernagel B, Klaus FX, Mayer KFX, Olsen OA (2014) Ancient hybridizations among the ancestral genomes of bread wheat. Science 345(6194): 288–291

Matsumoto K, Teramura T, Tabushi J (1963) Development analysis of the rachis disarticulation in *Triticum*. Wheat Inf Serv 15–16:23–25

McFadden ES, Sears ER (1947) The Genome approach in radical wheat breeding. Agron J 39:1011–1026

McFadden ES, Sears ER (1946) The origin of *Triticum spelta* and its free-threshing hexaploid relatives. J Hered 37(81–89):107–116

McKenzie RH, Bremer E, Middleton AB, Beres B et al (2014) Agronomic practices for bioethanol production from spring triticale in Alberta. Can J Plant Sci 94:15–22

Mergoum M, Ryan J, Shroyer JP (1992) Triticale in Morocco: potential for adoption in the semi-arid cereal zone. J Nat Res Life Sci Edu 21:137–141

Mergoum M, Singh PK, Peña RJ, Lozano-del Río AJ et al (2009) Triticale: a "new" crop with old challenges. In: Carena NJ (ed) Handbook of plant breeding, vol 3. Springer, Cereals, pp 267–287

Mergoum M, Pfeiffer WH, Peña RJ, Ammar K, Rajaram S (2004). Triticale crop improvement: the CIMMYT programme. In: Mergoum M, Gómez-Macpherson H (eds) Triticale improvement and production. FAO plant production and protection paper no. 179. Food and agriculture organization of United Nations, Rome, pp 11–26

Mesfin A, Frohberg RC, Anderson JA (1999) RFLP markers associated with high grain protein from *Triticum turgidum* L. var. *dicoccoides* introgressed into hard red spring wheat. Crop Sci 39:508–513; Mettin D, Blüthner WD, Schlegel G (1973) Additional evidence on spontaneous 1B/1R wheat-rye substitutions and translocations. In: Sears ER, Sears LMS (eds) Proc 4th Intern Wheat Genetic Symp, Columbia, Missouri, pp 179–184

Meyer RS, Purugganan MD (2013) Evolution of crop species: genetics of domestication and diversification. Nature Review Genet 14:840–852

Miller TE, Reader SM, Mahmood A, Purdie KA, King IP (1995) Chromosome 3N of *Aegilops uniaristata* - a source of tolerance to high levels of aluminium for wheat. In: Li ZS, Xin ZY (eds) Proc 8th Intern Wheat Genet Symp. Beijing, China Agricultural Scientech Press, Beijing, China, pp 1037–1042

Millet E, Zaccai M (1991) Effects of genotypically- and environmentally-induced differencesin seed protein content on seedling vigor in wheat. J Genet Breed 45:45–50

Millet E, Rong JK, Qualset CO, McGuire PE, Bernard M, Sourdille P, Feldman M (2013) Production of chromosome-arm substitution lines of wild emmer in common wheat. Euphytica 190:1–17

Millet E, Rong JK, Qualset CO, McGuire PE, Bernard M, Sourdille P, Feldman M (2014) Grain yield and grain protein percentage of common wheat lines with wild emmer chromosome-arm substitutions. Euphytica 195:69–81

Montenegro JD, Golicz AA, Mayer PE et al (2017) The pangenome of hexaploid bread wheat. Plant J 90:1007–1013

Mori N, Kondo Y, Ishii T, Kawahara T, Valkoun J, Nakamura C (2009) Genetic diversity and origin of timopheevi wheat inferred by chloroplast DNA fingerprinting. Breed Sci 59:571–578

Mori N, Ishi T, Ishido T, Hirosawa S, Watatani H, et al (2003) Origins of domesticated emmer and common wheat inferred from chloroplast DNA fingerprinting. In: Pogna NE, M Romano M, Pogna EA, G Galterio G (eds) Proceedings of 10th Intern Eheat Genetics

Symposium, Paestum, Italy. Rome, Instituto Sperimentale per la Cerealicoltura, pp 25–28

Morris R, Sears ER (1967) The cytogenetics of wheat and its relatives. In: Quisenberry KS, Reitz LP (eds) Wheat and wheat improvement. Amer Soc Agron, Madison, Wisconsin, USA, pp 19–87

Moulins de D (1997) Agricultural changes at Euphrates and syeps sites in the mid-8th to the 6th millennium B.C. Oxford Bar, International Series 683

Mujeeb-Kazi A, Rosas V, Roldan S (1996) Conservation of the genetic variation of *Triticum tauschii* (Coss.) Schmalh. (*Aegilops squarrosa* auct. non L.) in synthetic hexaploid wheats (*T. turgidum* L. × *T. tauschii*; 2n = 6x =42, AABBDD) and its potential utilization for wheat improvement. Genet Resourc Crop Evol 43:129–134

Mujeeb-Kazi A, Gul A, Farooq M, Rizwan S et al (2008) Rebirth of synthetic hexaploids with global implications for wheat improvement. Austral J Agric Res 59:391–398

Mujeeb-Kazi A, Kazi AG, Dundas I et al (2013) Genetic diversity for wheat improvement as a conduit to food security. Adv Agron 122:179–257

Müntzing A (1955) Mode of production and properties of a Triticale-Strain with 70 chromosomes. Wheat Inf Serv 2:1–12

Müntzing A (1979) Triticale: results and problems. Z Pflanzenziicht, Suppl 10:1–103

Muramatsu M (1963) Dosage effect of the spelta gene *q* of hexaploid wheat. Genetics 48:469–482

Muramatsu M (1985) Spike type in two cultivars of *Triticum dicoccum* with the *spelta* gene *q* compared with the *Q*-bearing variety *liguliforme*. Jpn J Breed 35:255–267

Muramatsu M (1986) The *vulgare* super gene, *Q*: its universality in *durum* wheat and its phenotypic effects in tetraploid and hexaploid wheats. Can J Genet Cytol 28:30–41

Muramatsu M (1978) Phenotypic expression of the *vulgare* gene *Q* in tetraploid wheat. In: Ramanujan S (ed), Proc 5th Intern Wheat Genet Symp, New Delhi 1:92–102

Muramatsu M (1979) Presence of the *vulgare* gene, *Q*, in a dense-spike variety of *Triticum dicoccum* Schubl. Report of the Plant Germ-Plasm Institute, Kyoto University, No. 4, pp 39–41

Myer R, Lozano del Rio AJ (2004) Triticale as animal feed. In: Mergoum M, Gómez-Macpherson H (eds). Triticale improvement and production. FAO plant production and protection paper no. 179. Food and Agriculture Organization of United Nations, Rome, pp 49–58

Nalam VJ, Vales MI, Watson CJW, Kianian SF, Riera-Lizarazu O (2006) Map-based analysis of genes affecting the brittle rachis character in tetraploid wheat (*Triticum turgidum* L.). Theor Appl Gen 112:373–381

Nave M, Avni R, Ben-Zvi B, Hale I, Distelfeld A (2016) QTLs for uniform grain dimensions and germination selected during wheat domestication are co-located on chromosome 4B. Theor Appl Genet 129:1303–1315

Nave M, Avni R, Çakır E, Portnoy V, Sela H, Pourkheirandish M, Ozkan H, Hale I, Takao Komatsuda T, Dvorak J, Distelfeld A (2019) Wheat domestication in light of haplotype analyses of the Brittle rachis 1 genes (BTR1-A and BTR1-B). Plant Sci 285:193–199

Nave M, Tas M, Raupp J, Tiwari VK et al (2021) The independent domestication of timopheev's wheat: insights from haplotype analysis of the *Brittle rachis 1* (*BTR1-A*) Gene. Genes 12:338

Nesbitt M (2001) Wheat evolution: integrating archaeological and biological evidence. In: Caligari PDS, Brandham PE (eds) Wheat taxonomy: the legacy of John Percival. The Linn Soc London, Burlington House, London, pp 37–60

Nesbitt M, Samuel D (1996) From staple crop to extinction? The archaeology and history of hulled wheat. In: Padulosi S, Hammer K, Heller J (eds) First International workshop on hulled wheats.

Castelvecchio Pascoli, Tuscany. Italy, International Plant Genetic Resources Institute, Rome, Italy, pp 41–100.

Nesbitt M (2002) When and where did domesticated cereals first occur in southwest Asia? In: Cappers RTJ, Bottema S (eds) The dawn of farming in the Near East. Studies in Near Eastern Production, Subsistence and Environment, 6, 1999. Ex Oriente, Berlin, pp 113–132

Nielsen R, Williamson S, Kim Y, Hubisz MJ, Clark AG, Bustamante C (2005) Genomic scans for selective sweeps using SNP data. Genome Res 15:1566–1575

Nyine M, Adhikari E, Clinesmith M et al (2021) The aplotype-based analysis of *Aegilops tauschii* introgression into hard red winter wheat and its impact on productivity traits. Front Plant Sci 12:716955

Oelke EA, Oplinger ES, Brinkman MA (1989) Triticale. Alternative field crops manual, Univ Wisconsin, CES, Madison, WI, Univ Minnesota CES, St. Paul

Oettler G (2005) The fortune of a botanical curiosity—Triticale: past, present and future. J Agric Sci 143:329–346

Oettler G, Becker HC, Hoppe G (2001) Heterosis for yield and other agronomic traits of winter triticale F_1 and F_2 hybrids. Plant Breed 120:351–353

Oettler G, Burger H, Melchiner AE (2003) Heterosis and combining ability for grain yield and other agronomic traits in winter triticale. Plant Breed 122:318–321

Ogbonnaya FC, Abdalla O, Mujeeb-Kazi A, Kazi AG, Xu SS et al (2013) Synthetic hexaploids: harnessing species of the primary gene pool for wheat improvement. Pl Breed Reviews 37:35–122

Oliveira HR, Jacocks L, Czajkowska BI, Kennedy SL, Brown TA (2020) Multiregional origins of the domesticated tetraploid wheats. PLoS ONE 15:e0227148

Olivier (1807) Voyage dans l'Empire Ottoman 3:46

Olmos S, Distelfeld A, Chicaiza O, Schlatter AR, Fahima T, Echenique V, Dubcovsky J (2003) Precise mapping of a locus affecting grain protein content in durum wheat. Theor Appl Gent 107:1243–1251

Olsen KM, Wendel JF (2013) A bountiful harvest: genomic insights into crop domestication phenotypes. Annu Rev Plant Biol 64:47–70

Olsen AN, Ernst HA, LoLeggio L, Skriver K (2005) NAC transcription factors: structurally distinct, functionally diverse. Trends Pl Sci 10:79–87

Olsen KM, Caicedo AL, Polato N, McClung A, McCouch S, Purugganan MD (2006) Selection under domestication: Evidence for a sweep in the rice waxy genomic region. Genetics 173:975–983

Ooka H, Satoh K, Doi K, Nagata T, Otomo Y et al (2003) Comprehensive analysis of *NAC* family genes in *Oryza sativa* and *Arabidopsis thaliana*. DNA Res 10:239–247

Ozkan H, Brandolini A, Schäfer-Pregl R, Salamini F (2002) AFLP analysis of a collection of tetraploid wheats indicates the origin of emmer and hard wheat domestication in southeast Turkey. Mol Biol Evol 19:1797–1801

Ozkan H, Brandolini A, Pozzi C, Effgen S, Wunder J, Salamini F (2005) A reconsideration of the domestication geography of tetraploid wheats. Theor Appl Genet 110:1052–1060

Ozkan H, Willcox G, Graner A, Salamini F, Kilian B (2011) Geographic distribution and domestication of wild emmer wheat (*Triticum dicoccoides*). Genet Resour Crop Evol 58:11–53

Pearce S, Tabbita F, Cantu D et al (2014) Regulation of Zn and Fe transporters by the *GPC1* gene during early wheat monocarpic senescence. BMC Plant Biol 14:368

Peleg Z, Fahima T, Korol AB, Abbo S, Saranga Y (2011) Genetic analysis of wheat domestication and evolution under domestication. J Exp Bot 62:5051–5061

Peng J, Ronin Y, Fahima T, Röder MS, Nevo E, Korol A (2003) Domestication quantitative trait loci in *Triticum dicoccoides*, the progenitor of wheat. Proc Natl Acad Sci USA 100:2489–2494

Peng JH, Sun D, Nevo E (2011) Domestication evolution, genetics and genomics in wheat. Mol Breed 28:281–301

Percival J (1921) The wheat plant: a monograph. Duckworth, London, pp 1–46

Piperno DR, Weiss E, Holst I, Nadel D (2004) Processing of wild cereal grains in the upper Palaeolithic revealed by starch grain analysis. Nature 430:670–673

Pourkheirandish M, Wicker T, Stein N et al (2007) Analysis of the barley chromosome 2 region containing the six-rowed spike gene *vrs1* reveals a breakdown of rice–barley micro collinearity by a transposition. Theor Appl Genet 114:1357–1365

Pourkheirandish M, Hensel G, Kilian B, Senthil N, Chen G et al (2015) Evolution of the grain dispersal system in barley. Cell 162:527–539

Pourkheirandish M, Dai F, Sakuma S, Kanamori H, Distelfeld A et al (2018) On the origin of the non-brittle rachis trait of domesticated einkorn wheat. Front Plant Sci 8:2031

Purugganan MD, Fuller DQ (2011) archaeological data reveal slow rates of evolution during plant domestication. Evolution 65:171–183

Rédei GP (1998) Genetic manual: Current theory, concepts, terms. World Scientific Publishing Co, pp 1–1113

Reimer PJ, Austin WEN, Bard E, Bayliss A et al (2020) The IntCal20 northern hemisphere radiocarbon age calibration curve (0–55 CAL kBP). Radiocarbon 62:725–757

Riehl S, Zeidi M, Conard NJ (2013) Emergence of agriculture in the foothills of the Zagros Mountains of Iran. Science 341:65–67

Riley R, Chapman V, Belfield AM (1966) Induced mutation affecting the control of meiotic chromosome pairing in *Triticum aestivum*. Nature 211:368–369

Rimpau W (1891) Kreuzungsprodukte Landwirtschaftlicher Kulturpflanzen. Land- Wirtschaftl Jahrb 20:335–371

Rollefson GO, Simmons AH, Donaldson ML, Gillespie W, Kafafi Z, Kohler-Rollefson IU, Mcadam E, Ralston SL, Tubb K (1985) Excavation at the Pre-Pottery Neolithic B village of 'Ain Ghazal (Jordan), 1983. Mitteilugen Der Deutschen Orient-Gesellschaft Zu Berlin 117:69–116

Rong JK (1999) Mapping and tagging by DNA markers of wild emmer alleles that affect useful traits in bread wheat. Ph.D. Thesis, submitted to the scientific council of the Weizmann institute of science, Rehovot, Israel, April 1999

Rosyara U, Kishii M, Payne T, Sansaloni CP, et al. (2019) Genetic contribution of synthetic hexaploid wheat to CIMMYT's spring bread wheat breeding germplasm. Scientific Reports 9: Article number: 12355

Sakamura T (1918) Kurze mitteilung über die chromosomenzahelen und die verwandtschaftsverhältnisse der Triticum Arten. Bot Mag 32:151–154

Sakuma S, Pourkheirandish M, Matsumoto T, Koba T, Komatsuda T (2010) Duplication of a well-conserved homeodomain-leucine zipper transcription factor gene in barley generates a copy with more specific functions. Funct Integr Genomics 10:123–133

Sakuma S, Salomon B, Komatsuda T (2011) The domestication syndrome genes responsible for the major changes in plant form in the Triticeae crops. Pl Cell Physiol 52:738–749

Salamini F, Ozkan H, Brandolini A, Schäfer-Pregl R, Martin W (2002) Genetics and geography of wild cereal domestication in the near East. Nat Rev Genet 3:429–441

Salmon DF, Mergoum M, Gómez Macpherson H (2004) In: Mergoum M, Gómez-Macpherson H (eds) Triticale production and management. FAO Plant production and protection paper no179, pp 27–36

Sanchez-Monge E (1974) Development of triticales in western Europe. In: MacIntyre R, Campbell M (eds)Triticale: Proc Int Symp,1973, International Development Research Centre, El Batan, Mexico, Ottawa, Ont, Canada, pp 31–39

Sax K (1921a) Chromosome relationships in wheat. Science 54:413–415

Sax K (1921b) Sterility in wheat hybrids. I Sterility relationships and endosperm development. Genetics 6:399–416

Sax K (1922) Sterility in wheat hybrids. II. Chromosome behavior in partially sterile hybrids. Genetics 7:513–552

Schrader L, Schmitz J (2019) The impact of transposable elements in adaptive evolution. Mol Ecol 28:1537–1549

Schroder E (1931) Anatomische Untersuchungen an den Spindeln der *Triticum*- und *Aegilops*-Arten zur Gewinnung neuer Gesichtspunkte für die Abstammung und Systematik der *Triticum*-Arten. Beih Bot Zentralbl 48:333–403

Schulz A (1913a) Über eine neue spontane Eutriticumform: *Triticum dicoccoides* Kcke. forma Straussiana, Berichte der Deutschen Bot. Gesselschaft 31:226–230

Schulz A (1913b) Die Geschichte der kultivierten Getreide, Nebert, Halle

Schweinfurth G (1906) Die Entdeckung des wilden Urweizens in Palästina. Vossische Zeitung 21(9):1906

Schweinfurth G (1908) Über die von A. Aaronsohn ausgeführten Nachforschungen nach dem wilden Emmer (*Triticum dicoccoides* Kcke). Ber. Deutsch. Bot. Ges. 26:309–324

Sears ER (1941a) Amphidiploids in the seven-chromosome Triticinae. Univ Missouri Agric Exper Stn Res Bull 336:1–46

Sears ER (1941b) Chromosome pairing and fertility in hybrids and amphidiploids in the Triticinae. Univ Missouri Agric Exp Stn Res Bull 337:1–20

Sears ER (1954) The aneuploids of common wheat. Missouri Agric Exp Stn Res Bull 572:1–58

Sears ER (1976) A synthetic hexaploid wheat with fragile rachis. Wheat Inf Serv 41–42:31–33

Sharma JS, Running KLD, Xu SS, Zhang Q et al (2019) Genetic analysis of threshability and other spike traits in the evolution of cultivated emmer to fully domesticated durum wheat. Mol Genet Genomics 294:757–772

Shebeski LH (1974) Future role of triticales in agriculture. In: MacIntyre R, Campbell M (eds) Triticale: Proc Int Symp, 1973, International Development Research Centre, El Batan, Mexico; Ottawa, Ont, Canada, pp 247–250

Simonetti MC, Bellomo MP, Laghetti G, Perrino P, Simeone R, Blanco A (1999) Quantitative trait loci influencing free-threshing habit in tetraploid wheats. Genet Res Crop Evol 46:267–271

Simons KJ, Fellers JP, Trick HN, Zhang Z, Tai YS, Gill BS, Faris JD (2006) Molecular characterization of the major wheat domestication gene Q. Genetics 172:547–555

Skovmand B, Fox PN, Villareal RL (1984) Triticale in commercial agriculture: progress and promise. Adv Agron 37:1–45

Snir A, Nadel D, Groman-Yaroslavski I, Melamed Y, Sternberg M, Bar-Yosef O, Weiss E (2015) The origin of cultivation and proto-weeds, long before Neolithic farming. PLoS ONE 10: e0131422

Solms-Laubach (1899) Weizen und Tulpe und deren Geschichte. Leipzig

Sood S, Kuraparthy V, Bai G, Gill BS (2009) The major threshability genes soft glume (*sog*) and tenacious glume (*Tg*), of diploid and polyploid wheat, trace their origin to independent mutations at non-orthologous loci. Theor Appl Genet 119:341–351

Stace CA (1987) Triticale: a case of nomenclatural mistreatment. Taxon 36:445–454

Taenzler B, Esposti RF, Vaccino P, Brandolini A, Evgen S, Heun M, Schafer-Pregl R, Borghi B, Salamini F, (2002) Molecular linkage

map of einkorn wheat: Mapping of storage-protein and soft-glume genes and bread-making quality QTLs. Genet Res, Camb 80:131–143

Takahashi R, Hayashi J (1964) Linkage study of two complementary genes for brittle rachis in barley. Ber Ohara Inst Landw Biol, Okayama Univ 12:99–105

Tanno K-I, Willcox G (2006) How fast was wild wheat domesticated? Science 311:1886

Tarchini R, Biddle P, Wineland R, Tingey S, Rafalski A (2000) The complete sequence of 340 kb of DNA around the rice *Adh1–Adh2* region reveals interrupted colinearity with Mmaize chromosome 4. Plant Cell 12:381–391

Thanh PT, Vladutu CI, Kianian SF, Thanh PT, Ishii T, Nitta M, Nasuda S, Naoki Mori N (2013) Molecular Genetic Analysis of domestication traits in emmer wheat. I: Map construction and QTL analysis using an F_2 pupulation. Biotechnol Biotechnol Equipment 27:3627–3637

Trethowan RM, Mujeeb-Kazi A (2008) Novel germplasm resources for improving environmental stress tolerances of hexaploid wheat. Crop Sci 48:1255–1265

Tsunewaki K, Yamada S, Mori N (1990) Genetical studies on a Tibetan semi-wild wheat *Triticum aestivum* ssp. *tibetanum*. Jpn J Genet 65:353–365

Tsunewaki K (1968) Origin and phylogenetic differentiation of bread wheat revealed by comparative gene analysis. In: Finley KW, Shepherd KW (eds) Proc 3rd Intern Wheat Genet Symp, Canberra, Australia, pp 71–85

Tzarfati RV, Barak V, Krugman T, Fahima T, Abbo A, Saranga Y, Korol AB (2014) Novel quantitative trait loci underlying major domestication traits in tetraploid wheat. Mol Breed 34:1613–1628

Uauy C, Brevis JC, Dubcovsky J (2006a) The high grain protein content gene *Gpc-B1* accelerates senescence and has pleiotropic effects on protein content in wheat. J Exp Bot 57:2785–2794

Uauy C, Distelfeld A, Fahima T, Blechl A, Dubcovsky J (2006b) A *NAC* gene regulating senescence improves grain protein, zinc, and iron content in wheat. Science 314:1298–1301

Ulaş B, Fiorentino G (2021) Recent attestations of "new" glume wheat in Turkey: a reassessment of its role in the reconstruction of Neolithic agriculture. Veget Hist Archaeobot 30:685–701

Urbano M, Resta P, Benedettelli S, Blanco A (1988) A *Dasypyrum villosum* (L.) Candargy chromosome related to homoeologous group 3 of wheat. In: Miller TE, Koebner RMD (eds) Proceedings of 7th Int Wheat Genet Symp, IPSR, Cambrifge Lab, Cambridge, UK, pp 169–173

Van Slageren MW (1994) Wild wheats: a monograph of *Aegilops* L. and *Amblyopyrum* (Jaub. and Spach) Eig (Poaceae). Agricultural University, Wageningen, The Netherlands

Van Zeist W, Bakker-Heeres JAH (1982) Archaeobotanical studies in the Levant. 1. Neolithic sites in the Damascus Basin: Aswad, Ghoraifé Ramad. Palaeohistoria 24:165–256

Van Zeist W, Bakker-Heeres JAH (1984) Archaeobotanical studies in the Levant. 2. Neolithic and Halaf levels at Ras Shamra. Palaeohistoria 26:151–170

Van Zeist W, Bakker-Heeres JAH (1985) Archaeobotanical studies in the Levant 1. Neolithic sites in the Damascus basin: Aswad, Choraife Ramad. Palaeohistoria 24:165–256

Van Zeist W, Waterbolk-van Rooijen W (1985) The palaeobotany of Tell Bouqras, Eastern Syria. Paléorient 11:131–147

Van Zeist W, de Roller GJ (1994) The plant husbandry of Aceramic Cayo ꞌnu ꞉. E. Turkey. Palaeohistoria 33:65–96

Van Zeist W, de Roller GJ (1995) Plant remains from Asikli Höyük, a pre-pottery Neolithic site in central Anatolia. Veg Hist Archaeobotany 4:179–185

Van Zeist W, de Roller GJ, Bottema S (2000) The plant remains. In: Verhoeven M, Akkermans PMMG (eds) Tell Sabi Abyad II: the prepottery Neolithic B settlement. Istanbul, Nederlands Historisch-Archaeolo- gisch Instituut, pp 137–147

Vavilov NI (1926) Studies on the origin of cultivated plants. Bull Appl Bot Genet Pl Breed 16:1–248 (In Russian and English summary)

Vavilov NI (1932) The process of evolution in cultivated plants. Proc 6th Intern Congr Genet 1:331–342

Vavilov NI (1935) Botaniko-geograficheskie osnovy selektsii [Botanical-geographical foundations of breeding]. Teoreticheskie osnovy selektsii rastenii [Theoretical foundations of plant breed ing], Moscow, Leningrad, pp 17–75 (in Russian)

Vavilov NI (1957) World resources of cereals, grains leguminous crops and flax and their utilization in plant breeding. General part: agroecological survey of the principal field crops. Moskva/Leningrad, p 462. (also transi, by M. Paenson and Z.S. Cole, Jerusalem, p 442)

Vavilov NI (1962) (Five continents) Moscow, State geograph Publ House (in Russian)

Villareal RL, Varughese G, Abdalla OS (1990) Advances in spring triticale breeding. Plant Breed Rev 8:43–90

von Tschermak E (1914) Die Verwertung der Bastardierung für phylogenetische Fragen in der Getreidegruppe. Zeitschr. Pflanzenzucht 2:291–312

von Tschermak E (1937) Wirkliche, abgeleitete unfrdaglicheWeizen-roggenbastarde (Triticale-Form). Cytologia, Fujii Jub 2:1003–1011

Walkowiak S, Gao L, Monat C et al (2020) Multiple wheat genomes reveal global variation in modern breeding. Nature 588:277–283

Wang RL, Stec A, Hey J, Lukens L, Doebley J (1999) The limits of selection during maize domestication. Nature 398:236–239

Warburton ML, Crossa J, Franco J et al (2006) Bringing wild relatives back into the family: recovering genetic diversity in CIMMYT improved wheat germplasm. Euphytica 149:289–301

Watanabe N, Ikebata N (2000) The effects of homoeologous group 3 chromosomes on grain colour dependent seed dormancy and brittle rachis in tetraploid wheat. Euphytica 115:215–220

Watanabe N, Sugiyama K, Yamagishi Y, Sakata Y (2002) Comparative telosomic mapping of homoeologous genes for brittle rachis in tetraploid and hexaploid wheats. Hereditas 137:180–185

Watanabe N, Takesada N, Fujii Y, Martinek P (2005) Comparative mapping of Genes for Brittle Rachis in Triticum and Aegilops. Czech J Genet Plant Breed 41:39–44

Watanabe N, Fujii Y, Kato N, Ban T, Martinrk P (2006) Microsatellite mapping of the genes for brittle rachis on homoeologous group 3 chromosomes in tetraploid and hexaploid wheats. J Appl Genet 47:93–98

Watkins AE (1928) The genetics of wheat species crosses. I. J Genet 20:1–27

Watkins AE (1940) The inheritance of glume shape in Triticum. J Genet 39:240–264

Weiss E, Kislev ME, Hartmann A (2006) Autonomous cultivation before domestication. Science 312:1608–1610

Willcox G (1995) Wild and domesticated cereal cultivation: new evidence from early Neolithic sites in the Northern LEVANT and South-Eastern Anatolia. ARX World J Prehist Ancient Stud 1:9–16

Willcox G (1996) Evidence for plant exploitation and vegetation history from three Early Neolithic pre-pottery sites on the Euphrates (Syria). Veget Hist Archaeobot 5:143–152

Willcox G (1999) Agrarian change and the beginnings of cultivation in the Near East: evidence from wild progenitors, experimental cultivation and archaeobotanical data. In: Hather J, Gosden C (eds) The prehistory of food. Routledge, London, pp 479–500

Willcox G (2004) Measuring grain size and identifying Near Eastern cereal domestication: evidence from the Euphrates valley. J Archaeol Sci 31:145–150

Willcox G (2005) The distribution, natural habitats and availability of wild cereals in relation to their domestication in the Near East:

multiple events, multiple centres. Veget Hist Archaeobot 14:534–541

Willcox G (2012) Pre-domestic cultivation during the late Pleistocene and early Holocene in the northern Levant. In: Gepts P, Famula TR, Bettinger RL, Brush SB, Damania AB, McGuire PE, Qualset CO (eds) Biodiversity in agriculture: domestication, evolution, and sustainability. Cambridge University Press, Cmbridge, pp 92–109

Willcox G, Fornite S, Herveux L (2008) Early Holocene cultivation before domestication in northern Syria. Vegetation History and Archaeology 17:313–332

Willcox G (1998) Archaeobotanical evidence for the beginnings of agriculture in southwest Asia. In: Damania AB, Valkoun J, Willcox G, Qualset CO (eds)The origin of agriculture and crop domestication, proceedings of the Harlan symposium, 10–14 May 1997, ICARDA, IPGRI, FAO and UC/GRCP, pp 25–38

Willcox G (2002) Geographical variation in major cereal components and evidence for independent domestication events in the Western Asia. In: Cappers RT, Bottema S (eds) The dawn of farming in the Near East. Studies in Near Eastern Production, Subsistence and Environment, 6, 1999. Ex Oriente, Berlin, pp 133–140

Wright S (1931) Evolution in Mendelian populations. Genetics 16:97–159

Wulff BBH, Moscou MJ (2014) Strategies for transferring resistance into wheat: from wide crosses to GM cassettes. Front Plant Sci 5: article 692

Yang YC, Tuleen NA, Hart GE (1996) Isolation and identification of Triticum aestivum L. em. Thell. cv. Chinese Spring–T. peregrinum Hackel disomic addition lines. Theor Appl Genet 92:591–598

Yang W, Liu D, Li J, Zhang L et al (2009) Synthetic hexaploid wheat and its utilization for wheat genetic improvement in China. J Genet Genomics 9:539–546

Yao Y, Sun Q (2012) Exploration of small non coding RNAs in wheat (Triticum aestivum L.). Plant Mol Biol 80:67–73

Yuan J, Guo X, Hu J, Lv Z, Han F (2015) Characterization of two CENH3 genes and their roles in wheat evolution. New Phytol 206:839–851

Zaccai M (1992) Physiological and genetic aspects of nitrogen economy in high-protein wheat. Ph.D. thesis, submitted to the Scientific Council of the Weizmann Institute of Science, Rehovot

Zeder MA (2011) The origins of agriculture in the Near East. Curr Anthropol 52:S221–S235

Zeng X, Tagiri A, Kikuchi S, Sassa H, Komatsuda T (2020b) The Ectopic Expression of Btr2 in Aegilops tauschii Switches the Disarticulation Layer From Above to Below the Rachis Node. Front Plant Sci 11: article 582622

Zeng X, Mishina K, Jia J, Distelfeld A, et al (2020a) The Brittle Rachis Trait in Species Belonging to the Triticeae and its controlling genes Btr1 and Btr2. Front Plant Sci 11: article 1000

Zeven AC (1980) Polyploidy and domestication: the origin and survival of polyploids in cytotype mixtures. In: Lewis WH (ed) Polyploidy —biological relevance. Plenum Press, New york, pp 385–407

Zhang ZC, Belcram H, Gornicki P, Charles M, Just J, Huneau C, Magdelenat G, Couloux A, Samain S, Gill BS, Rasmussen JB, Barbe V, Faris JD, Chalhoub B (2011) Duplication and partitioning in evolution and function of homoeologous Q loci governing domestication characters in polyploid wheat. Proc Natl Acad Sci, USA 108:18737–18742

Zhang Z, Li A, Song G, Geng S, Gill BS, Faris JD, Mao L (2020) Comprehensive analysis of Q gene near-isogenic lines reveals key molecular pathways for wheat domestication and improvement. Plant J 102:299–310

Zhao Y, Dong L, Jiang C et al (2020) Distinct nucleotide patterns among three subgenomes of bread wheat and their potential origins during domestication after allopolyploidization. BMC Biol 18:188

Zhao Y, Peng X, Guan P, Wang Y et al (2019) Btr1 -A induces grain shattering and affects spike morphology and yield-related traits in wheat. Plant Cell Physiol. 2019 Jun1; 60(6):1342-1353

Zhou Y, Zhao X, Li Y et al (2020) Nat Genet 52:1412–1422

Zhou Y, Bai S, Li H et al (2021) Introgressing the *Aegilops tauschii* genome into wheat as a basis for cereal improvement. Nat. Plants 7:774–786

Zhu F (2018) Triticale: nutritional composition and food uses. Food Chem 241:468–479

Zhukovsky PM (1928) A new species of wheat. Bull Appl Bot Genet PL Breed 19:59–66

Zillinsky FJ, Borlaug N (1971) Progress in developing triticale as an economic crop. CIMMYT Mexico DF, CIMMYT Res Bull 17:1–27

Zohary D (1996) The mode of domestication of the founder crops of southwest Asian agriculture. In: Harris DR (ed) The origin and spread of agriculture and pastoralism in Eurasia. Routledge, London and New York, pp 142–158

Zohary D (1999) Monophyletic vs. polyphyletic origin of the crops on which agriculture was founded in the Near East. Genet Resour Crop Evol 46:133–142

Zohary D, Brick Z (1962) *Triticum dicoccoides* in Israel: Notes on its distribution, ecology and natural hybridization. Wheat Inf Serv 13:6–8

Zohary D, Hopf M (2000) Domestication of plants in the Old World, 3rd edn. Oxford Science Publications, Oxford

Zohary D, Hopf M, Weiss E (2012) Domestication of Plants in the Old World-The origin and spread of domesticated plants in south-west Asia, Europe and the Mediterranean Basin, 4th edn. Oxford University Press, pp 1–243

Future Prospects

14

14.1 Introduction

Malthus prediction, that population growth would not be matched by growth in food production, causing major famines, did not materialize so far thanks to a series of advances in genetics and agrotechnology. Genetics breakthroughs were achieved, such as the introduction of the lodging-preventing dwarfing genes in breeding programs world-wide. Another mutation with a global impact was selected in genes controlling day length sensitivity that enabled the expansion of agriculture in regions that were not previously cultivated. Advances in engineering and chemistry enabled to implement mechanical harvest, fertilizers, herbicides and pesticides.

In this chapter we examine the question if and how we can beat Malthus one more time? We argue that wheat biodiversity will play an increasingly important role in wheat breeding as it will become necessary to enlarge the gene pool of wheat to adapt to climate change, new more extreme environment (e.g., salty, dry, hot, and cold), and new diseases. Unlike in the past, there is a very limited amount of suitable arable land left that can be used to expand wheat cultivation. There is even a reduction in arable land due to urbanization, roads, and construction of industrial areas. Wheat production should not be at the cost of destruction of biodiversity in natural prairies or through massive deforestation. Moreover, global warming and extreme events have caused desertification in many parts of the world making it unprofitable to grow wheat (e.g., parts of North Africa and the middle east). Increasing the amounts of fertilizers, herbicides, or pesticides is not an option due to their high cost, their environment-damaging effect with high greenhouse gas emission footprint and undesirable runoff in the aquifer or in the air. The formidable challenge for the coming generations will be "how to produce more with less input, less land, under climate change and in a sustainable manner"?

We present a carefully optimistic view, asserting that, like in the past, the impressive advances in technologies and genetics will enable one more time to improve wheat to feed the planet. Genetic advances will enable to fully exploit the potential of wheat and its wild relatives (Triticeae species) biodiversity. On the technological side, the on-going/next green revolution will make a wise use of robotics, drones, satellites with sensors, climate prediction, geo-engineering, etc. The computational tools and infrastructures, like internet, big data and artificial intelligence will promote both technological and genetic platforms. On the genetic side, on which we elaborate below, a plethora of new tools is available, such as whole genome sequences and dense SNP mapping for thousands of wheat lines, a broad range of mapping populations, transcriptomics, proteomics, epigenetic data on DNA methylation and chromatin modifications, non-coding RNA data, and integration of these data using systems biology tools, will boost gene discovery and functional identification of genes and gene networks from within the wide Triticeae gene pool, for wheat breeding. The Alpha-fold revolution for computational prediction of protein structures, together with precise modification of genomes, e.g., through gene editing or homologous or homoeologous recombination, and improved transgenesis technologies will undoubtedly accelerate wheat grain production and quality. In order to leverage all the recent scientific discoveries into new and improved wheat varieties, education in general and education of breeders in particular will be essential.

14.2 Wheat in the Post-omics Era

Advances in next-generation sequencing platforms and associated bioinformatics tools have already revolutionized wheat genomics (Adamski et al. 2019; Babu et al. 2020; Varshney et al. 2021) providing us with a wide range of whole genome sequences. However, the high cost of genome sequencing in wheat, due to a large size, a repetitive nature and polyploidy makes it still prohibitive to most wheat researchers and breeders to carry routine large scale genetic analyses, such as genome wide association study (GWAS),

M. Feldman and A. A. Levy, *Wheat Evolution and Domestication*,
https://doi.org/10.1007/978-3-031-30175-9_14

or analysis of segregating populations or characterization of wheat biodiversity. We envision a nearby future where the cost of sequencing would be reduced by one or two orders of magnitude, using high-quality long reads, making it possible to sequence and assemble the genomes of wheat lines, domesticated and wild, in a routine manner. While dense genetic maps are very useful and often sufficient for certain needs, there is a clear added value to a whole genome sequence, for example to look for rare alleles in exotic germplasm, to carry in depth functional analyses of genetic variants and phenotypes, to search for redundancy in the genome. As discussed in Chap. 3, gene copy number variation is an important and overlooked type of rearrangement, affecting gene dosage, sub- and neo-functionalization that can have profound impact on phenotypes; these can be detected only by whole genome sequences. Whole genome sequences together with transcriptome analysis can help understand how genetic diversity in the Triticeae translates into different phenotypes. Predicting how routine whole genome sequencing will impact wheat genetics and breeding is not too risky as we can look at the power of such data in advancing the genetic understanding of organisms with a small genome such as *Arabidopsis* or even budding yeast.

Transcriptomics became an accessible technology and there are good published "atlases" of the transcriptome in various wheat tissues (Ramírez-González et al. 2018), however there is still a need for improving methods so that full length, or differentially spliced transcripts, or non-coding RNAs can be analyzed on a large scale, in many diverse lines. In particular, the wheat community needs tools to better distinguish the activity of different homoeoalleles or duplicated genes. Single-cell transcriptomics captures the transcription of genes in specific cell types. This is critical to better understand differentiation processes in meristems or to study developmental processes such as meiosis. The cost of single-cell analysis is currently prohibitive and has not been used widely in wheat but the expected reduction in sequencing costs may give a boost to this approach and improve a knowledge-based amelioration of wheat.

Proteomics is a field where rapid advances are occurring and that had limited impact on wheat research so far. Current Mass Spectrometry technologies enable to fingerprint a wide range of proteins and even to determine the modifications of some of these (e.g., phosphorylation, acetylation, etc.). Likewise, computational structural biology, the Alpha-fold technology developed by Google, enables now to predict with high accuracy the structure of any protein. New tools are also being developed to predict how variants can affect protein structure and enzymatic activity. These methods will become powerful tools to study wheat biology. Of particular interest might be modelling of storage proteins structures to predict the gluten complex network formation and baking quality.

Lastly, metabolomics enables to profile a very broad range of small molecules, secondary and primary metabolites. It has also received limited attention so far in wheat research. This will help understand wheat metabolism, to dissect the genetic and enzymatic pathways involved in the biosynthesis of various compounds, as well as metabolism involvement in growth and response to biotic and abiotic stress. Altogether, such studies will be useful to improve both wheat yield and nutritional value.

All the above technologies will enable to explore and characterize the great diversity found in domesticated and wild wheat relatives.

14.3 Conservation of Biodiversity of Wheat and Its Wild Relatives

14.3.1 Opening Remarks

The wild relatives of wheat, namely, species of the grass tribe Triticeae, contain a large reservoir of genes that are lacking in domesticated wheats, which, upon transfer to domesticated background, may increase yield, improve nutritional quality, resistance and tolerance to biotic and abiotic stresses, and increase adaptablity to diverse environments and to a changing climate. The existence of a large reservoir of useful genes in the various Triticeae species is apparent from their adaptation to a wide variety of ecological niches and ecosystems. They thrive in a wide range of climatic regions, from cool, humid mountains and arctic-temperate regions, to hot, dry valleys of the Mediterranean and central Asiatic regions, as well as from areas with an annual rainfall of 1000 mm or more to arid regions with as little as 100 mm. Wild wheat relatives also grow on many different types of soils, in some places even on salty ones, e.g., *Elymus elongatus* and *E. farctus*. The evolution of wild wheat also points to their resilience as they survived many events of climatic changes, from periods warmer than nowadays to periods of glaciation.

These wild relatives of wheat will undoubtedly remain one of the main sources for new genes that will enable to achieve resilience to climatic changes and to continue to increase production. This means that we must conserve this diversity, characterize it and develop better methods for transferring valuable genes from the wild gene pool to modern wheats.

In the decades following the green revolution (60–80s), genetic erosion of domesticated wheats was observed due to the replacement of a genetically diverse pool of landraces by a limited number of dwarf or semi-dwarf elite varieties. This fairly rapid erosion of the gene pools not only reduced the possibility of further improvements in productivity but also rendered the world wheat crop increasingly vulnerable to new diseases and to adverse climatic changes. In recent decades, this trend is being reversed and breeders have made a broader use of wild wheat and its relatives, in particular in the search for disease-resistant genes. Conserving the existing genetic diversity is thus essential for ensuring future utilization. The growing pressure of expanding urbanization and intensified agricultural practices threatens the survival of these natural reservoirs of genetic reservoirs and their maintenance is an urgent necessity (Harlan 1975).

14.3.2 Ex-Situ Conservation

Ex-situ conservation in seed banks is the main source of wheat germplasm conservation. Globally, there are tens wheat collections with hundreds of thousands of accessions of domesticated and wild wheats stored in seed banks. The Consultative Group on International Agricultural Research (CGIAR) is a global partnership of several major seed banks dedicated at conserving, maintaining and distributing crops germplasm. The main wheat collections (see https://www.genebanks.org/resources/crops/wheat/) are in CIMMYT, Mexico; USDA; IPK, Germany; ICARDA; VIR, Russia; NPGR, India; IBBR, Italy; CGN, Netherland and in the Svalbard Global Seed Vault, in Norway. There is a high redundancy among seed banks, but this is something desirable as seed banks can be damaged as a result of natural or human catastrophic events. Seed banks are a last backup in case of loss of lines, e.g., landraces or of wild accessions, or extinction of populations in the wild or extinction of species. Anthropogenic activities, including overgrazing, urbanization, pollution, and greenhouse gas emission leading to global warming will put at risk the survival of many wild wheat populations in their natural habitat. It is thus essential to further support the long-term activities of the wheat-rich seed banks, as well as the activities of smaller seed banks or of collections in the laboratories of wheat geneticists, which might contain accessions that are not available in the large collections. Genotyping, or even sequencing, the collections in seed banks would also be a great contribution to the wheat community. This would enable to assess redundancy among collections and this would facilitate GWAS-types of analyses. Assuming that the cost of sequencing will further drop, this will be a possible goal. Note that sequencing data by itself is a form of conservation as extinct genes can be synthesized and reintroduced in a gene pool.

14.3.3 In Situ Conservation

This type of conservation is the on-site maintenance of natural populations of wild wheats and related species in their natural habitats, where they can evolve continuously in response to biotic and abiotic transformations of the ecosystem. It also involves the maintenance of landraces and cultivars of domesticated wheats via field gene banking and botanical gardening. Field gene banking and botanical gardens maintain a number of lines of domesticated wheats under cultivation in admixture populations. Yet, the collections in field gene banks and botanical gardens are relatively limited in number and are susceptible to loss due to artificial selection, genetic drift, and diseases. In situ conservation of landrace takes place also in farmer's fields, for niche markets (Negri 2003). Kamut is a successful example of a wheat landrace that is well conserved, through extensive cultivation and commercialization. Unfortunately, many other landraces could be lost and should be conserved. In recent years, several countries have invested intensive efforts to characterize their wheat landraces due to their historical, cultural and breeding value. Indeed, landraces are attractive because they show local adaptation, and are part of an ethnic cuisine, moreover they have full homology to the wheat subgenomes. For example, a thorough analysis of Turkish landraces was published as a book (Zencirci N, Baloch FS, Habyarimana E, Chung G (eds) Wheat Landraces, Springer Cham, https://doi.org/10.1007/978-3-030-77388-5). Conservation of landraces is mostly taking place in seed banks, with notable efforts to characterize them genetically (Kilian et al. 2021; Cavanagh et al. 2013; Dempewolf et al. 2017a, b; Frankin et al. 2020).

For wild wheats, in situ conservation in their natural habitat enables the evolution of a new variation that is not found in seed banks. It is conceivable that new mutations conferring a selective advantage, for example to warmer temperatures and/or to new diseases, could emerge in the wild and would not be found in seed banks. Unfortunately, this mode of conservation is quite limited. A recent long-term study of in situ conservation of wild emmer wheat showed that different genotypes were clustered in their specific micro-habitat for 36 years, and very few changes were observed during this period, pointing to the resilience of the population despite the increase of 1.5 °C in temperature and of 70 ppm in CO_2 concentration (Dahan-Meir et al. 2022). Insights from this study are that sampling a natural population should be done with consideration of the ecological niches and at relatively distant periods of time (one per 10 or 20 years). Moreover, it is essential that in situ conservation should be done in a protected area, with minimal grazing. Another insight is that a large proportion of the genetic variation in the population can be preserved on a relatively small area (e.g., one hectare). Natural reserves for in situ conservation of wild wheats can therefore be relatively small, requiring simple fencing and making it a feasible endeavor. Establishing natural reserves for in situ conservation of wild wheat species should become a high priority, in relevant regions, to maintain an evolving biodiversity that can be used for wheat breeding and to ensure food security.

14.4 Wheat Biodiversity Gene Pools

Many species of the grass tribe Triticeae, particularly those of the subtribe Triticineae, yield viable hybrids with domesticated wheats, and their genetic variability is thus readily exploitable. Since these species show different degrees of cytogenetic affinity and phylogenetic relatedness to the domesticated wheats, selection of the most advantageous gene transfer procedure for any given wild species should depend upon the degree of homology between the chromosome carrying the desirable gene(s) and the recipient wheat chromosome. Harlan and de Wet (1971) introduced the concept of gene pools, which is based on the genetic distance between the crop and the genetic resource. According to them, three types of gene pools exist: primary, secondary, and tertiary. The primary gene pool of wheat is composed of plants whose genome(s) is closely related (homologous) to one or more subgenomes of bread wheat, and includes wheat landraces, early domesticates, wild emmer, diploid wheat, and *Ae. tauschii*. Thus, in meiosis of hybrids between these species and bread wheat, and more so between synthetic hexaploid wheat wild emmer-*Ae. tauschii* (BBAADD genome) and bread wheat, or between addition or substitution lines carrying homologous chromosomes, most of the homologous chromosome's pair with the wheat chromosomes and genes can be transferred with comparatively little difficulty to the chromosomes of wheat by crossovers (Sears 1982; Feldman 1988). While up to now, only a small proportion of the existing genetic diversity of the primary gene pool has been utilized for wheat improvement. Species from the primary gene pool, such as wild emmer (Huang et al. 2016; Klymiuk et al. 2018; Faris et al. 2020; Li et al. 2020) and *Ae. tauschii* (Bhatta et al. 2018; Gaurav et al. 2021; Zhou et al. 2021) have been already shown to be an invaluable resource for breeders in particular when it comes to biotic and abiotic resistant genes not found in the domestic wheat gene pool. Likewise, the A genome of *T. urartu* is homologous to the A subgenome of bread wheat and can be used as a source of useful genes (Zeibig et al. 2021). Further exploitation of this gene pool should thus be expanded.

The secondary gene pool of wheat contains allopolyploid species that share at least one homologous subgenome with bread wheat, including *T. timopheevii* (GGAA), *T. zhukovskyi* (GGAAA^{m}A^m), the S-genome diploids of *Aegilops* section Sitopsis (close to the B subgenome), species of the allopolyploid *Aegilops* of section *Aegilops* containing the S

subgenome, and the *Aegilops* species of sections Vertebrata and Cylindropyrum that contain the D subgenome (Feldman 1988; Curtis et al. 2002; Qi et al. 2007; Badaeva et al. 2021a, b). Transferring genetic material from this genepool to bread wheat is comparatively more complex, with typical problems including hybrid seed death, sterility of F_1 hybrids and reduced recombination (Ogbonnaya et al. 2013). In addition, embryo rescue is often required to obtain F_1 hybrids. Recombination between the chromosomes of the secondary gene pool and those of bread wheat can be induced by promoters of homoeologous chromosomal pairing or by deficiency for *Ph1* or *Ph2* genes (Sears 1982, 1984; Feldman 1988).

The tertiary gene pool is composed of more distantly related diploids and polyploids with genomes that are distantly related to the subgenomes of bread wheat. In this case, transfer of genetic material to bread wheat is highly complex (Sears 1982; Feldman 1988; Feuillet et al. 2008). Usually, special techniques, such as irradiation, mutants for *Ph1* or use of genes that promote homoeologous pairing, as well as gametocidal chromosomes, are needed for gene transfer and embryo rescue is necessary (Jiang et al. 1994; Mujeeb-Kazi and Hettel 1995). Many of the germplasm in this group are perennials (Mujeeb-Kazi 2003). Although tertiary gene pool resources are highly complex to utilize, they have the potential of becoming an important means to develop new and diverse wheat germplasms (Mujeeb-Kazi 2006).

The restoration and enrichment of the gene pool of the domesticated wheats can be accomplished by far-reaching exploitation of the vast genetic resources found in the wild relatives of wheats (Feldman and Sears 1981; Sharma and Gill 1983; Feldman 1988; Jiang et al. 1994: Feuillet et al. 2008; Mujeeb-Kazi et al. 2013; Wulff and Moscou 2014; Zhang et al. 2017; Ceoloni et al. 2017; Rasheed et al. 2018) using new methods described below.

14.5 Gene Transfer from Exotic Germplasm to Elite Varieties

Wheat geneticists and breeders have transferred genes from at least 52 species, from gene pool 2 and 3, whose chromosomes do not pair regularly with domesticated wheat chromosomes (Walkowiak et al. 2020; Keilwagen et al. 2022). A major drawback of these wide transfers is the large size of the transferred fragments that may contain deleterious genes in addition to the desired ones (see review by Levy and Feldman 2022). Homologous recombination with the homologous pool, in particular in crosses with wild relatives, can also lead, to a lesser extent than homoeologous recombination, to the hitchhiking of undesirable genes linked to the selected one. We discuss below gene transfer methods and how they can be improved to reduce the size of introgressed fragments.

The transfer of desirable genetic material from wild relatives into domesticated wheat has been achieved so far by two main methodologies: cytogenetic manipulations (chromosomal engineering) and molecular techniques (genetic engineering) (Feldman 1988). We envision that future genome manipulation might occur through a third way, namely, precisely, by combining chromosomal engineering with molecular tools.

Homologous recombination (transfer from gene pool 1), can be enhanced by manipulating the recombination machinery, using QTLs (Gardiner et al. 2019); or mutants (Desjardin et al. 2022) or using viral vectors (Raz et al. 2021). QTLs and mutants are not convenient to use in breeding programs as they are not in the desired genetic background. However, viral vectors were shown to be used to silence genes that suppress homologous recombination and to increase crossover in various regions of the genome (Raz et al. 2021). Such manipulation is simple as it requires only the infection of premeiotic F_1 plants and the effect is obtained in the gametes of the next generation while the virus is not transmitted germinally. Another avenue that would enable to transfer genes and chromosome segments from one chromosome to its homolog in a precise manner has been developed in tomato and in *Arabidopsis* (Filler-Hayut et al. 2017, 2021; Ben shlush et al. 2021). It is based on DNA double-strand break induction at the target site in somatic tissues, followed by homologous repair and transmission to the germline. Targeted recombination is considered of high potential in plant breeding and could be easily applied to wheat (Taagen et al. 2020a, b). Chromosome engineering can also be achieved through CRISPR-induced cleavage and repair by non-homologous end-joining, enabling to obtain translocations, deletions or inversions or reciprocal exchange between homologous chromosomes at specific cleavage sites (Ronspies et al. 2021).

Gene transfer between homoeologous chromosomes (in gene pool 2) remains challenging. It is mostly based on induction of homoeologous pairing in the absence of *Ph1* (Sears 1981) or *Ph2* (Sears 1984). A targeted transfer, of smaller chromosomal segments, could be achieved as described above for homologous exchange, using CRISPR-mediated cleavage and repair via homoeologous recombination (in the *ph1* background) or via non-homologous end joining. In pool 3, transfer might be attempted as described above for gene pool 2. However, due to the difficulty, the transgenic option might be preferred (see below).

14.6 New Breeding Technologies

Transgenesis is not so new but has not been applied extensively in wheat, not commercially and not in research, due to various reasons: One is the regulation that makes it hard at best and in many countries impossible to grow commercially transgenic wheat. Another reason is that only few labs so far have succeeded to transform wheat. This is changing thanks to a recent breakthrough whereby proteins that promote regeneration are co-transformed and expressed transiently, together with the gene of interest (Debernardi et al. 2020), enhancing the rates for transformation and enabling to achieve it in a more routine manner. One type of transformation that has a huge potential in breeding is based on cis-genesis, namely the direct transfer of genes from the wheat gene pool into domesticated wheat. In particular, this approach is powerful for conferring broad spectrum resistance to multiple diseases/races in one single transformation (Luo et al. 2021). From the breeding point of view, all the introduced genes cloned on the same T-DNA, behave as a single Mendelian factor and become very easy to transfer. It is not clear when regulators, and public will be ready for such useful technology, however, what is important is that we have the tools to face future pandemics and to fully exploit the various wheat gene pools.

The new gene editing revolution has not yet been applied to wheat breeding however, the technology has been shown to work well in wheat. The singular contribution of the laboratory of Caixia Gao is worth mentioning: She was the first to show that gene editing can be achieved in wheat, knocking out all six homoeoalleles of the *Mlox* locus leading to resistance to powdery mildew (Wang et al. 2014; Li et al. 2022). In addition, she optimized the CRISPR technologies to achieve gene replacement, base editing and prime editing in wheat (see review Gao 2021). A step-by-step protocol for the implementation of CRISPR-mediated wheat gene editing and trait improvement is also provided by Bhowmik and Islam (2020). The applications of gene editing are immense and are less problematic from a regulation point of view. The future will probably see many wheat products derived from genome editing: Disease resistant plants, as well as wheat with modified nutrition value, including a targeted modification of the antigenic gluten domains. Genome editing tools will also be used for precise gene transfer as described above. An interesting application of editing, in relation to wide gene transfer is that new *ph1* alleles can be done, as shown by Rey et al. (2018) in any variety, so that wide transfer does not have to happen through the genetic background of Chinese Spring.

14.7 Hybrid Wheat

One of the promising possibilities to increase wheat yield is via the production of hybrid wheat. Hybrid wheat has not been implemented so far because limited heterosis (Gupta et al. 2019) has not justified the high cost of hybrid seeds production. Heterosis in wheat, might be increased by broadening the gene pool of wheat. Moreover, methods like gene editing might facilitate the production of male sterile and restorer lines. Wheat hybrids may utilize, in addition to the hybrid vigor occurring in bread wheat between subgenomes, positive intra-subgenomic interactions. Hybrid wheat lines may yield higher than pure, true breeding lines, and exhibit increased yield stability, improved quality and greater tolerance to environmental and biotic stresses (Wilson and Driscoll 1983; Pickett 1993; Bruns and Peterson 1998; Jordaan et al. 1999). Currently, there is no exploitation of heterosis between different alleles, as allohexaploid wheat is a self-pollinating plant and homozygous in all loci. During the 10,000 years of cultivation, selection for higher yield in this self-pollinating plant was carried out in homozygous combinations, whereas selection experiments in inter-varietal hybrids heterozygous for a large number of genes, were barely studied. Because wheat is a predominantly self-pollinator and each flower contains male and female organs, it is not practical to produce hybrid seeds on a commercial scale as is done with several other crops. Commercial hybrid wheat seed has been produced using chemical hybridizing agents (CHAs), plant-growth regulators that selectively interfere with pollen development, or by cytoplasmic male sterility system (CMS). By now, hybrid wheat has been a limited commercial success in Europe (particularly France), South Africa, and Australia. Male-sterility in bread wheat may be brought about also by a genic male sterility (Wilson and Driscoll 1983).

Several CHAs have been successfully utilized in recent years for inducing male sterility in the female parents to be used for commercial hybrid wheat production (Gupta et al. 2019). Some wheat hybrids derived using CHAs have been reported to give 10–20% yield advantage over the best available pure lines (Boeven et al. 2016; Longin 2016). Yet, utilization of a CHA to male-sterilize wheat plants is expensive, requires the treatment in a special short growth period, not always fully efficient, and pollutant.

CMS is exploited for hybrid seed production in alloplasmic lines, caused by the incompatible interaction of an alien cytoplasm with the bread wheat nuclear genome. Many attempts have been directed to producing hybrid seeds in bread wheat on the basis of CMS, mainly using the cytoplasm of *T. timopheevii*. Yet, the use of an alien cytoplasm as a sterilizing factor in bread wheat has a major drawback since various important traits including grain yield are negatively affected by the interaction between the wheat nuclear genome and the alien cytoplasm. In addition, it has been difficult to find stable fertility restoration genes for the alloplasmic male-sterile lines, which are highly effective in a wide range of wheat genotypes. Moreover, the system requires breeding of the male parent too [e.g., introduction of genes to the male parent line that can restore male-fertility to the alien cytoplasm (a *Rf* gene)], thus rendering hybrid seed production more expensive and limiting the number of male parents that can be tested for combining ability (contribution to a significant heterosis). Also, the CMS-based hybrid seed suffers from the problems of kernel shriveling/low test weight and low germination (Geyer et al. 2018). Despite the above limitations, there are several scattered reports of commercial cultivation of CMS-based hybrid wheats. A notable example is the production of hybrid wheats in China using *T. timopheevii* or *Ae. kotschyi-* based cytoplasm, albeit on a very small scale (Longin 2016; Tsunewaki 2015).

The following conditions are required for the production of hybrid wheat seeds by genetic means: (1) complete and stable male-sterility of the female parent; (2) complete and stable fertility restoration by the male parent; and (3) easy propagation of the male-sterile female parent by a male-fertile maintainer line. Genic male sterility is expressed in a normal wheat cytoplasm and does not involve deleterious effects on plant performance. It is operational in euplasmic lines, caused by a recessive mutation or deletion of a nuclear gene(s) that is essential for male-fertility in bread wheat. Using a female parent homozygous for a recessive male-sterility allele, any wheat cultivar which is by its nature homozygous for the dominant allele conferring male-fertility, can be used as a male parent that will restore complete fertility to the F$_1$ hybrids. There is no need to breed for male lines and no limitation exists for the number of males which can be crossed with the male-sterile females and evaluated for their combining ability.

Several chromosome arms have been described in bread wheat which carry genes affecting male-fertility. These arms are the long arm of chromosome 4A (4AL), the short arm of chromosome 4B (4BS) and the short arm of chromosome 4D (4DS), which carry the normal male-fertility *Ms-A1*, *Ms-B1* and *Ms-D1* genes, respectively. Also, the long arms of the group 5 chromosomes, 5AL, 5BL, and 5DL, carry the *Ms-A2*, *Ms-B2* and *Ms-D2* genes, respectively. However, until now, only in the *Ms-B1* locus, on the distal region of chromosome arm 4BS, three recessive alleles that cause male sterility were found or induced. These alleles, namely, *ms-B1-a*, *ms-B1-b* and *ms-B1-c* (often also called *ms1a*, *ms1b* and *ms1c*, respectively), were reported not to cause any effect, beyond male-sterility, on plant performance (reviewed by Wilson and Driscoll 1983). Maintenance of the male-sterile female lines remains the major obstacle for a successful hybrid production system based on genic male-sterility. Several such maintainer lines were suggested (e.g., Driscoll 1972, 1985).

The possibilities for future change in the wheat crop will then be limited only by the imagination of the researcher. Yet, it will also be essential to continue traditional field trials in research stations to demonstrate the advantage of newly released cultivars, so that they can be accepted by the farmers. Eventually, the farmer and the customer will remain the ultimate arbiters of any progress made.

References

Adamski NM, Borrill P, Brinton J et al (2019) A roadmap for gene functional characterisation in wheat. Peer J Preprints 7:e26877

Babu P, Baranwal DK, Pal HD et al (2020) Application of genomics tools in wheat breeding to attain durable rust resistance. Front Plant Sci 11:567147

Badaeva ED, Chikida NN, Belousova MK et al (2021a) A new insight on the evolution of polyploid *Aegilops* species from the complex Crassa: molecular-cytogenetic analysis. Plant Syst Evol 307:3. https://doi.org/10.1007/s00606-020-01731-2

Badaeva ED, Chikida NN, Fisenko AN, Surzhikov SA, Belousova MK, Özkan H, Dragovich AY, Kochieva EZ (2021b) Chromosome and molecular analyses reveal significant karyotype diversity and provide new evidence on the origin of *Aegilops columnaris*. Plants 10:956. https://doi.org/10.3390/plants10050956

Ben Shlush I, Samach A, Melamed-Bessudo C, Ben-Tov D, Dahan-Meir T, Shdema Filler-Hayut S, Levy AA (2021) CRISPR/Cas9 induced somatic recombination at the *CRTISO* Locus in Tomato. Genes 12(1). https://doi.org/10.3390/genes12010059

Bhatta M, Morgounov A, Belamkar V et al (2018) Unlocking the novel genetic diversity and population structure of synthetic Hexaploid wheat. BMC Genomics 19:591

Bhowmik PK, Islam MT (2020) CRISPR-Cas9-mediated gene editing in wheat: a step-by-step protocol. In: Islam MT, Bhowmik PK, Molla KA (eds) CRISPR-Cas methods, springer protocols handbooks. Springer Science + Business Media, LLC, part of Springer Nature, pp 203–222

Boeven PHG, Longin CF, Würschum T (2016) A unified framework for hybrid breeding and the establishment of heterotic groups in wheat. Theor Appl Genet 129:1231–1245

Bruns R, Peterson CJ (1998) Yield and stability factors associated with hybrid wheat. Euphytica 100:1–5

Cavanagh CR, Chao S, Wang S, Huang BE, Stephen S, Kiani S, Forrest K, Saintenac C, Brown-Guedira GL, Akhunova A, See D, Bai G, Pumphrey M, Tomar L, Wong D, Kong S, Reynolds M, da Silva ML, Bockelman H, Talbert L, Anderson JA, Dreisigacker S, Baenziger S, Carter A, Korzun V, Morrell PL, Dubcovsky J, Morell MK, Sorrells ME, Hayden MJ, Akhunov E (2013) Genome-wide comparative diversity uncovers multiple targets of

selection for improvement in hexaploid wheat landraces and cultivars. Proc Natl Acad Sci USA 110:8057–8062

Ceoloni C, Kuzmanovic L, Ruggeri R et al (2017) Harnessing genetic diversity of wild gene pools to enhance wheat crop production and sustainability: challenges and opportunities. Diversity 9:55

Curtis BC, Rajaraman S, MacPherson HG (eds) (2002) Bread wheat, improvement and production. Food and Agriculture Organization of the United Nations, Rome

Dahan-Meir T, Ellis TJ, Sela H, Mafessoni F et al (2022) The genetic structure of a wild wheat population has remained associated with microhabitats over 36 years. https://doi.org/10.1101/2022.01.10.475641

Debernardi JM, Tricoli DM, Ercoli MF, Hayta S, Ronald P, Palatnik JF, Dubcovsky J (2020) A GRF-GIF chimeric protein improves the regeneration efficiency of transgenic plants. Nat Biotechnol 38:1274–1279

Dempewolf H, Baute G, Anderson J, Kilian B, Smith C, Guarino L (2017a) Past and future use of wild relatives in crop breeding. Crop Sci 57:1070–1082

Dempewolf H, Baute G, Anderson J, Kilian B, Smith C, Guarino L (2017b) Past and future use of wild relatives in crop breeding. Crop Sci 57:1070–1082

Desjardins SD, Simmonds J, Guterman I et al (2022) FANCM promotes class I interfering crossovers and suppresses class II non-interfering crossovers in wheat meiosis. Nat Commun 13:3644

Driscoll CJ (1972) XYZ system of producing hybrid wheat. Crop Sci 12:516–517

Driscoll CJ (1985) Modified XYZ system of producing hybrid wheat. Crop Sci 25:1115–1116

Faris JD, Overlander ME, Kariyawasam GK, Carter A et al (2020) Identification of a major dominant gene for race-nonspecific tan spot resistance in wild emmer wheat. Theor Appl Genet 133:829–841

Feldman M (1988) Cytogenetic and molecular approaches to alien gene transfer in wheat. In: Proceedings of the 7th international wheat genetics symptoms, vol 1, Cambridge, England, pp 23–32

Feldman M, Sears ER (1981) The wild gene resources of wheat. Sci Am 244:102–112

Feuillet C, Langridge P, Waugh R (2008) Cereal breeding takes a walk on the wild side. Trend Genet 24:24–32

Filler Hayut S, Melamed Bessudo C, Levy AA (2017) Targeted recombination between homologous chromosomes for precise breeding in tomato. Nat Commun 8:15605

Filler-Hayut S, Kniazev K, Melamed-Bessudo C, Levy AA (2021) Targeted inter-homologs recombination in *Arabidopsis* euchromatin and heterochromatin. Int J Mol Sci 22:12096

Frankin S, Kunta S, Abbo S, Sela H, Goldberg BZ, Bonfil DJ, Levy AA et al (2020) The Israeli-Palestinian wheat landraces collection: restoration and characterization of lost genetic diversity. J Sci Food Agric 100:4083–4092

Gao C (2021) Genome engineering for crop improvement and future agriculture. Cell 184:1621–1635

Gardiner LJ, Wingen LU, Bailey P et al (2019) Analysis of the recombination landscape of hexaploid bread wheat reveals genes controlling recombination and gene conversion frequency. Genome Biol 20:69

Gaurav K, Arora S, Silva P, Sanchez-Martin J et al (2021) Population genomic analysis of *Aegilops tauschii* identifies targets for bread wheat improvement. Nat Biotechnol. https://doi.org/10.1038/s41587-021-01058-4

Geyer M, Albrecht T, Hartl L et al (2018) Exploring the genetics of fertility restoration controlled by *Rf1* in common wheat (*Triticum aestivum* L.) using high-density linkage maps. Mol Genet Genomics 293:451–462

Gupta PK, Balyan HS, Gahlaut V et al (2019) Hybrid wheat: past, present and future. Theor Appl Genet 132:2463–2483

Harlan JR (1975) Our vanishing genetic resources. Science 188:618–622

Harlan JR, de Wet JMJ (1971) Towards a rational classification of cultivated plants. Taxon 20:509–517

Huang L, Raats D, Sela H, Klymiuk V et al (2016) Evolution and adaptation of wild emmer wheat populations to biotic and abiotic stresses. Ann Rev Phytopathol 54:279–301

Jiang J, Friebe B, Gill BS (1994) Recent advances in alien gene transfer in wheat. Euphytica 73:199–212

Jordaan JP, Engelbrecht SA, Malan J, Knobel HA (1999) The genetics and exploitation of heterosis in crops. Madison, WI, USA, pp 411–421

Keilwagen J, Lehnert H, Berner T et al (2022) Detecting major introgressions in wheat and their putative origins using coverage analysis. Sci Rep 12:1908

Kilian B, Dempewolf H, Guarino L, Werner P, Coyne C, Warburton ML (2021) Crop science special issue: adapting agriculture to climate change: a walk on the wild side. Crop Sci 61:32–36

Klymiuk V, Yaniv E, Huang L, Raats et al (2018) Cloning of the wheat Yr15 resistance gene sheds light on the plant tandem kinase-pseudokinase family. Nature Commun 9:3735

Levy AA, Feldman M (2022) Evolution and origin of bread wheat. Plant Cell 34:2549–2567

Li S, Lin D, Zhang Y, Deng M, Chen Y, Lv B, Li B, Lei Y, Wang Y, Zhao L, Liang Y, Liu J, Chen K, Liu Z, Xiao J, Qiu JL, Gao C (2022) Genome-edited powdery mildew resistance in wheat without growth penalties. Nature 602(7897):455–460

Li M, Dong L, Li B, Wang Z et al (2020) A CNL protein in wild emmer wheat confers powdery mildew resistance. New Phytol 228:1027–1037

Longin CFH (2016) Future of wheat breeding is driven by hybrid wheat and efficient strategies for pre-breeding on quantitative traits. J Bot Sci 2308–2347

Luo M, Xie L, Chakraborty S et al (2021) A five-transgene cassette confers broad-spectrum resistance to a fungal rust pathogen in wheat. Nat Biotechnol 39:561–566

Mujeeb-Kazi A (2003) Wheat improvement facilitated by novel genetic diversity and In vitro technology. Plant Tissue Cult 13:179–210

Mujeeb-Kazi A (2006) Utilization of Genetic resources for bread wheat improvement, In: Singh RJ, Jauhar PP (eds) Genetic resources, chromosome engineering, and crop improvement, cereals, vol 2. CRC, Taylor & Francis, Boca Raton, Florida, London, New York, pp 61–97

Mujeeb-Kazi A, Hettel G (1995) Utilizing wild grass biodiversity in wheat improvement: 15 years of wide cross research at CIMMYT, pp 1–140

Mujeeb-Kazi A, Kazi AG, Dundas I et al (2013) Genetic diversity for wheat improvement as a conduit to food security. Adv Agron 122:179–257

Negri V (2003) Landraces in central Italy: where and why they are conserved and perspectives for their on-farm conservation. Genet Resourc Crop Evol 50:871–885

Ogbonnaya FC, Abdalla O, Mujeeb-Kazi A, Kazi AG, Xu SS et al (2013) Synthetic hexaploids: Harnessing species of the primary gene pool for wheat improvement. Pl Breed Reviews 37:35–122

Pickett AA (1993) Hybrid wheat—results and problems. Paul Parey Scientific Publication, Berlin

Qi LL, Friebe B, Zhang P, Gill BS (2007) Homoeologous recombination, chromosome engineering and crop improvement. Chromosome Res 15:3–19

Ramírez-González RH, Borrill P, Lang D, et al (2018) The transcriptional landscape of polyploid wheat. Science 361:eaar6089

Rasheed A, Mujeeb-Kazi A, Ogbonnaya FC (2018) Wheat genetic resources in the post-genomics era: promise and challenges. Ann Bot 121:603–616

Raz A, Dahan-Meir T, Melamed-Bessudo C, Leshkowitz D, Levy AA (2021) Redistribution of meiotic crossovers along wheat chromosomes by virus-induced gene silencing. Front Plant Sci 11:635139

Rey M-D, Martín AC, Smedley M, Hayta S, Harwood W, Shaw P, Moore G (2018) Magnesium increases homoeologous crossover frequency during meiosis in ZIP4 (Ph1 Gene) mutant wheat-wild relative hybrids. Front Plant Sci 9:509

Rönspies M, Dorn A, Schindele P, Puchta H (2021) CRISPR-Cas-mediated chromosome engineering for crop improvement and synthetic biology. Nat Plants 7:566–573

Sears ER (1981) Transfer of alien genetic material to wheat. In: Evans LT, Peacock WJ (eds) Wheat science—today and tomorrow. Cambridge University Press, pp 75–89

Sears ER (1982) A wheat mutation conditioning an intermediate level of homoeologous chromosome pairing. Can J Genet Cytol 24:715–719

Sears ER (1984) Mutations in wheat that raise the level of meiotic chromosome pairing. In: Gustafson JP (ed) Gene manipulation in plant improvement. Stadler Genetics Symposia Series. Springer, Boston, MA, pp 295–300

Sharma HC, Gill BS (1983) Current status of wide hybridization in wheat. Euphytica 32:17–31

Taagen E, Bogdanove AJ, Sorrells ME (2020a) Achieving controlled recombination with targeted cleavage and epigenetic modifiers. Trends Plant Sci 25:513–514

Taagen E, Bogdanove AJ, Sorrells ME (2020b) Counting on crossovers: controlled recombination for plant breeding. Trends Plant Sci 25:455–465

Tsunewaki K (2015) Fine mapping of the first multi-fertility-restoring gene, Rf^{multi}, of wheat for three Aegilops plasmons, using 1BS-1RS recombinant lines. Theor Appl Genet 128:723–732

Varshney RK, Bohra A, Yu J, Graner A, Zhang Q, Sorrells ME (2021) Designing future crops: genomics-assisted breeding comes of age. Trends Plant Sci 26:631–649

Walkowiak S, Gao L, Monat C et al (2020) Multiple wheat genomes reveal global variation in modern breeding. Nature 588:277–283

Wang S, Wong D, Forrest K, Allen A, Chao S et al (2014) Characterization of polyploid wheat genomic diversity using a high-density 90,000 single nucleotide polymorphism array. Plant Biotechnol J 12:787–796

Wilson P, Driscoll CJ (1983) Hybrid wheat. In: Frenkel R (ed) Heterosis, monographs on theoretical and applied genetics, vol 6. Springer-Verlag, pp 94–123

Wulff BBH, Moscou MJ (2014) Strategies for transferring resistance into wheat: from wide crosses to GM cassettes. Front Plant Sci 5:692

Zeibig F, Kilian B, Frei M (2021) The grain quality of wheat wild relatives in the evolutionary context. Theor Appl Genet. https://doi.org/10.1007/s00122-021-04013-8

Zhang W, Zhang M, Zhu X, Cao Y, Sun Q, Ma G, Chao S, Yan C, Xu SS, Cai X (2017) Genome-wide homology analysis reveals new insights into the origin of the wheat B genome. https://doi.org/10.1101/197640

Zhou Y, Bai S, Li H et al (2021) Introgressing the Aegilops tauschii genome into wheat as a basis for cereal improvement. Nat. Plants 7:774–786

Printed in the United States
by Baker & Taylor Publisher Services